COUNTY FINDER MAP OF MIS

Adair 17
Andrew 11
Atchison 1
Audrain 42
Barry 102
Barton 76
Bates 59
Benton 61
Bollinger 93
Boone 41
Buchanan 20
Butler 110
Caldwell 22
Callaway 43
Camden 68
Cape Girardeau 94
Carroll 31
Carter 108
Cass 49
Cedar 77
Chariton 32
Christian 99
Clark 9
Clay 29
Clinton 21
Cole 54
Cooper 52
Crawford 71
Dade 82
Dallas 79
Daviess 14
DeKalb 13
Dent 87
Douglas 100
Dunklin 113
Franklin 57
Gasconade 56
Gentry 12
Greene 83
Grundy 15
Harrison 4
Henry 60
Hickory 67
Holt 10
Howard 40
Howell 106
Iron 90
Jackson 37
Jasper 81
Jefferson 58
Johnson 50
Knox 18
Laclede 80
Lafayette 38
Lawrence 98
Lewis 19
Lincoln 46
Linn 24
Livingston 23
Macon 25
Madison 92
Maries 64
Marion 27
McDonald 101
Mercer 5
Miller 63
Mississippi 96
Moniteau 53
Monroe 34
Montgomery 44
Morgan 62
New Madrid 112
Newton 97
Nodaway 2
Oregon 107
Osage 55
Ozark 105
Pemiscot 114
Perry 75
Pettis 51
Phelps 70
Pike 36
Platte 28
Polk 78
Pulaski 69
Putnam 6
Ralls 35
Randolph 33
Ray 30
Reynolds 89
Ripley 109
St. Charles 47
St. Clair 66
St. Francois 73
St. Louis 48
Ste. Genevieve 74
Saline 39
Schuyler 7
Scotland 8
Scott 95
Shannon 88
Shelby 26
Stoddard 111
Stone 103
Sullivan 16
Taney 104
Texas 86
Vernon 65
Warren 45
Washington 72
Wayne 91
Webster 84
Worth 3
Wright 85

STEYERMARK'S

Flora of Missouri

STEYERMARK'S

Flora of Missouri

Volume 3

Revised edition

George Yatskievych

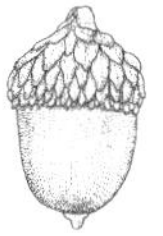

The Missouri Botanical Garden Press

St. Louis, Missouri

in cooperation with

The Missouri Department of Conservation

Jefferson City, Missouri

PRINTING HISTORY

Steyermark's Flora of Missouri, revised edition

Volume 1, published by the Missouri Department of Conservation, Jefferson City, Missouri, in cooperation with the Missouri Botanical Garden Press, St. Louis, Missouri.

Published February 1999

Volume 2, published by the Missouri Botanical Garden Press in cooperation with the Missouri Department of Conservation.

Published June 2006

Volume 3, published by the Missouri Botanical Garden Press in cooperation with the Missouri Department of Conservation.

Published August 2013

ISBN: 978-0-915279-13-5

Library of Congress Control Number: 2013942604

Technical editing and page composition by Kay Yatskievych

Missouri Botanical Garden Press
P.O. Box 299
St. Louis, MO 63166-0299

Missouri Department of Conservation
P.O. Box 180
Jefferson City, MO 65102-0108

CONTENTS

PREFACE

The present volume is the culmination of more than 25 years of effort. That may seem like a long time to complete a floristic encyclopedia, but it should be remembered that the late Julian Steyermark needed more than 30 years to complete the first edition. From its humble beginnings as a six-year project for a modest updating of Steyermark's *Flora of Missouri* the present project has evolved into an ongoing effort to collect information on Missouri's wonderful and diverse plant life that has not only expanded the content matter for the printed products, but has embraced the computer age and the world wide web. I have been blessed with a patient and supportive audience, which has allowed the gradual expansion of the project over the years to encompass aspects of floristic research far beyond the imaginings of the forward-thinking individuals who first conceived of such a work.

As outlined in the preface to Volume 2, a number of profound changes in our knowledge of Missouri plant life have occurred during the course of the Project, and this has complicated the accurate dissemination of knowledge in the books. Taxonomically, the widespread acceptance of the familial classification system of the Angiosperm Phylogeny Group (2003) has rendered some of the familial assignments in Volumes 1 and 2 obsolete. From the standpoint of vegetation, biologists have moved beyond a primary focus on natural divisions of Missouri as a way to subdivide the state biologically and physiographically and moved toward both a nationwide ecoregional classification and a finer set of land type associations within the state. In addition the classification of plant communities has seen a major update since the publication of the introductory chapter on vegetation in Volume 1. Such changes are inevitable as science marches onward. They fuel the continuing need for updated compilations of information on the flora of Missouri and set the stage for future activities.

Some of the alterations in familial classification from the system of Cronquist that have affected the contents of the present volume include: 1) a broad concept of the legumes as a single family Fabaceae with three subfamilies, instead of the recognition of three related families (Caesalpiniaceae, Fabaceae, and Mimosaceae); 2) a broad concept of the Ericaceae to include the achlorophyllous herbs previously segregated into the Monotropaceae; 3) the breakup of the traditional Boraginaceae into three families in Missouri: Boraginaceae, Ehretiaceae, and Heliotropiaceae; 4) the break-up of the traditional Scrophulariaceae and the transfer of many of its species into a series of other families (Linderniaceae, Orobanchaceae, Phrymaceae, and Plantaginaceae); 5) the submersion of the Buddlejaceae into the new version of the Scrophulariaceae and of the Callitrichaceae into the morphologically diverse Plantaginaceae; 6) the inclusion of the Aceraceae and Hippocastanaceae in the Sapindaceae; 7) the exclusion of *Polypremum* from both the Loganiaceae and Buddlejaceae and its inclusion in the small family Tetrachondraceae; 8) the inclusion of Sterculiaceae and Tiliaceae in the Malvaceae, and the submersion of Leitneriaceae in the seemingly dissimilar Simaroubaceae. Discussions justifying these changes are included in the treatments of the appropriate families. Table 1 provides a taxonomic summary of the families, genera, species, and infraspecific taxa included in Volume 3. This volume contains slightly more species than do either of the preceding ones.

The next logical step for the Flora Project will be an effort to update and condense the information in the three-volume encyclopedia into a one-volume manual. This manual will not only be a more portable resource for Missouri botanists, but will also provide a way to deal with the numerous new state records that continue to accumulate at an average rate of about nine species per year. It will also allow the presentation of a revised familial classification more in line with the one being adopted in other major state and national floristic manuals, as well as a way to account for changes in the taxonomy and nomenclature of Missouri plants that have been published over the years. A second future goal of the Project will be to expand its electronic offerings. One way to accomplish this will be to make the printed materials available as e-books in various formats. Another will be to expand the current offerings on the internet. Readers are invited to visit the Project's web pages to see what is available so far, at the following web site: http://www.tropicos.org/Project/MO. Here, will be found searchable text entries for families, genera, and species from the first two volumes (the third volume will take about a year to upload), as well as links to specimens, images, and a variety of types of maps. The Flora of Missouri database, which is mirrored in Tropicos, continues to grow steadily and, as of this writing, includes nearly 185,000 specimen records. The completeness and quality of the information served through the database and website can only be maintained if the Project continues to benefit from specimens documenting new distributional records, and users of the Flora are urged to document such new records when they are discovered. Similarly, readers are encouraged to submit reports of errors in the web site or the printed volumes when they are encountered, so that the Project may continue to improve the accuracy of the information that it shares with the world.

George Yatskievych
Missouri Botanical Garden
P.O. Box 299
St. Louis, MO 63166-0299
e-mail: George.Yatskievych@mobot.org

Table 1. Summary of the Missouri flora treated in Volume 3. The order of the 84 dicot families in the following table is the same as in the text. Native species are those considered to be native in at least part of their ranges in Missouri (they may be native in one part of the state and introduced in another). Those species discussed in the text for which the distributional status (native vs. non-native) in the state is uncertain are treated as native for purposes of this summary. Introduced species include all of the non-native species treated in the text, both those that are naturalized widely, as well as those that are known from one or few collections or from populations that may not have persisted to the present. Infraspecific taxa are those varieties and subspecies (excluding forms) in addition to the one that is understood to exist for each species accepted as part of the flora. The present summary includes only those species treated formally in the text and not those that are included only in discussions as potential additions to the flora.

Family	Genera	Species			Infraspecific Taxa		Hybrids	Total Taxa
		Native	Introduced	Total	Native	Introduced		
Fabaceae (Faboideae)	43	81	49	130	1	1	15	147
Fagaceae	3	23	3	26	1	0	42	69
Fumariaceae	3	6	1	7	2	0	0	9
Gentianaceae	7	13	0	13	1	0	2	16
Geraniaceae	2	2	6	8	0	0	0	8
Grossulariaceae	1	4	0	4	0	0	0	4
Haloragaceae	2	4	2	6	2	0	0	8
Hamamelidaceae	2	3	0	3	0	0	0	3
Heliotropiaceae	1	2	3	5	0	0	0	5
Hydrangeaceae	2	2	1	3	1	0	0	4
Hydroleaceae	1	2	0	2	0	0	0	2
Hydrophyllaceae	3	10	0	10	0	0	0	10
Iteaceae	1	1	0	1	0	0	0	1
Juglandaceae	2	11	0	11	0	0	4	15
Lamiaceae	31	50	25	75	10	0	4	89
Lardizabalaceae	1	0	1	1	0	0	0	1
Lauraceae	2	3	0	3	0	0	0	3
Lentibulariaceae	1	4	0	4	0	0	0	4
Limnanthaceae	1	1	0	1	0	0	0	1
Linaceae	1	5	2	7	0	0	0	7
Linderniaceae	1	1	0	1	0	0	0	1
Loasaceae	1	1	2	3	0	0	0	3
Loganiaceae	2	2	0	2	0	0	0	2
Lythraceae	6	7	2	9	0	0	0	9
Magnoliaceae	2	3	0	3	0	0	0	3

Family	Genera	Species			Infraspecific Taxa		Hybrids	Total Taxa
		Native	Introduced	Total	Native	Introduced		
Malvaceae	13	10	16	26	1	0	1	28
Martyniaceae	1	1	0	1	0	0	0	1
Melastomataceae	1	2	0	2	1	0	0	3
Menispermaceae	3	3	0	3	0	0	0	3
Menyanthaceae	2	1	1	2	0	0	0	2
Molluginaceae	2	0	2	2	0	0	0	2
Moraceae	4	1	4	5	0	0	0	5
Nelumbonaceae	1	1	1	2	0	0	0	2
Nyctaginaceae	2	3	1	4	0	0	0	4
Nymphaeaceae	2	2	0	2	1	0	0	3
Oleaceae	8	8	8	16	0	0	0	16
Onagraceae	4	32	4	36	1	0	1	38
Orobanchaceae	8	23	0	23	1	0	0	24
Oxalidaceae	1	4	1	5	0	0	0	5
Papaveraceae	6	2	7	9	0	1	0	10
Parnassiaceae	1	1	0	1	0	0	0	1
Passifloraceae	1	2	0	2	0	0	0	2
Paulowniaceae	1	0	1	1	0	0	0	1
Pedaliaceae	1	0	1	1	0	0	0	1
Penthoraceae	1	1	0	1	0	0	0	1
Phrymaceae	3	4	1	5	0	0	1	6
Phytolaccaceae	1	1	0	1	0	0	0	1
Plantaginaceae	17	33	23	56	0	0	0	56
Platanaceae	1	1	0	1	0	0	0	1
Plumbaginaceae	1	0	1	1	0	0	0	1
Polemoniaceae	4	8	4	12	2	0	0	14
Polygalaceae	1	4	0	4	1	0	0	5
Polygonaceae	7	28	17	45	1	2	2	50
Portulacaceae	4	5	2	7	0	0	0	7
Primulaceae	6	14	4	18	0	1	0	19
Ranunculaceae	16	43	13	56	3	0	3	63
Resedaceae	1	0	2	2	0	0	0	2

Family	Genera	Species			Infraspecific Taxa		Hybrids	Total Taxa
		Native	Introduced	Total	Native	Introduced		
Rhamnaceae	3	5	3	8	0	0	0	8
Rosaceae	26	94	38	132	23	0	28	183
Rubiaceae	8	25	5	30	0	0	0	30
Rutaceae	5	2	3	5	0	0	0	5
Salicaceae	2	14	4	18	2	0	11	31
Santalaceae	1	1	0	1	0	0	0	1
Sapindaceae	5	7	3	10	6	0	2	18
Sapotaceae	1	2	0	2	0	0	0	2
Saururaceae	1	1	0	1	0	0	0	1
Saxifragaceae	4	9	0	9	2	0	0	11
Scrophulariaceae	3	2	5	7	0	0	0	7
Simaroubaceae	2	1	1	2	0	0	0	2
Solanaceae	7	14	23	37	2	0	2	41
Sphenocleaceae	1	0	1	1	0	0	0	1
Staphyleaceae	1	1	0	1	0	0	0	1
Styracaceae	2	2	2	4	0	0	0	4
Tamaricaceae	1	0	1	1	0	0	0	1
Tetrachondraceae	1	1	0	1	0	0	0	1
Thymelaeaceae	1	2	0	2	0	0	0	2
Ulmaceae	3	9	2	11	0	0	1	12
Urticaceae	5	6	2	8	0	0	0	8
Valerianaceae	1	2	1	3	0	0	0	3
Verbenaceae	3	8	3	11	0	0	6	17
Violaceae	2	16	2	18	0	0	8	24
Viscaceae	1	1	0	1	0	0	0	1
Vitaceae	4	12	3	15	0	0	1	16
Zygophyllaceae	2	0	2	2	0	0	0	2
Total Vol. 3	**343**	**716**	**315**	**1031**	**65**	**5**	**134**	**1235**

ACKNOWLEDGMENTS

Over the decades, the efforts of countless individuals helped lead to the completion of the present work. As compiler of the information in these pages, however, responsibility for the content rests solely with the author, and any weaknesses in the treatments may be attributed to me. Because the work has extended over so many years, this list necessarily is incomplete, and I apologize for any names that I may have inadvertently omitted.

First and foremost, gratitude must be showered upon the Missouri Botanical Garden, where I have been headquartered since 1987. Peter Raven, President Emeritus of the Garden, provided the impetus for a revision of Missouri's flora and has been generous with his support through the completion of the three volumes. Peter and his wife, Pat, are an inspiration, and I feel fortunate to call them friends. Peter's vision of a Flora of Missouri Project has been supported well by his successor, Peter Wyse Jackson. My supervisor at the Garden, Bob Magill, and our administrative curator, Jim Solomon, have also been instrumental in keeping the Project on-track and going. I am also indebted to several of the curators (past and present) in the Garden's Science & Conservation Division who have lent support and expertise to the research, including Richard Abbott, Ihsan Al-Shehbaz, Matthew Albrecht, Fred Barrie, David Bogler, Tom Croat, Gerrit Davidse, Roy Gereau, Shirley Graham, Peter Hoch, Peter Jorgensen, Richard Keating, Gordon McPherson, Jim Miller, Olga Martha Montiel, John Pruski, Doug Stevens, Peter Stevens, Charlotte Taylor, Nick Turland, Carmen Ulloa, and Jim Zarucchi. Trisha Consiglio developed the custom scripts that allowed the generation of county dot maps for the printed volumes electronically directly from the Flora of Missouri Database.

A number of other present and former students and staff at the Garden also helped with the research in important ways, including Angela Brinker, Eric Feltz, Ron Liesner, Sandy Lopez, Mary McNamara, Anna Spencer, Andrea Voyer, and Laurel Zimmer. Computer support for the project has been provided by several members of the Garden's IT staff, including Bill Behrns, Zubin Chandra, Joe Ditto, Donna Miller, Jay Paige, and Mike Westmoreland. Chris Freeland, the former head of the Center for Biodiversity Informatics was a great sounding board for ideas and helped to sponsor the creation of the Project's web pages, to which Mike Blomberg, Stephanie Keil, Fred Keusenkothen, Randy Smith, and Wendy Westmoreland-Woods also contributed.

The raw materials upon which this revision is based originated in the libraries, the herbaria, and the landscape of Missouri. I am greatly indebted to the staff of the Missouri Botanical Garden's library, whose knowledge of their substantial accessions is truly impressive. Rosemary Armbruster, Andrew Colligan, Julie Crawford, Doug Holland, Victoria McMichael, Linda Oestry, Mary Stiffler, and Zoltan Tomory were especially helpful in helping to locate references. I also owe a great debt to the curators of the herbaria consulted both within Missouri and elsewhere in the United States for their courtesy in allowing me to examine and borrow specimens that were studied during the preparation of Volume 1.

My project assistants, interns, and volunteers have worked thousands of hours at jobs ranging from routine to very unusual and challenging, and all have performed above and beyond the call of duty. They have included: Alan Brant, Mary Broida, Karen Bynum, Germán Carnevali, Charles Caverly, Paige Cherry, Linda S. Ellis, Barbara Fry, Brad Guhr, Jack Harris, Claire Hemingway, Rex Hill, Emily Horton, Ann Larson, Nancy Parker, Bill Summers, Baadi Tadych, Charlotte Taylor, Joanna Turner, George Van Brunt, Matthew Wagner, Pat Walker, Andoni Westerhaus, Mark Williams, Jane Whitehill, and Kathleen Wood. Special mention must be given to Rex Hill, who not only designed and managed the Flora of Missouri Database, but did extensive proof-reading of the family treatments and has been an excellent sounding board for ideas. In addition, a number of specialists contributed treatments to the volume, as did several talented artists. These individuals are listed separately below.

Sadly, the Missouri Department of Conservation's role in the Project has waned in recent years. John Wylie was its first champion at the Department, and his efforts were followed by several able administrators there, including Jim H. Wilson, Rick Thom, Don Kurz, David Urich, Al Brohn, Ken Babcock, Eric Kurzejeski, and Vicki Heidy. Publication of the volumes would not have been possible without the unstinting support of the Department's Outreach and Education Division, where past and present staff, including Mike Huffman, Regina Knauer, Kathy Love, Bernadette Dryden, Libby Block, Tracy Ritter, Dan Witter, and Eric Kurzejeski, promoted the publication of the Flora of Missouri Project over the years. A number of present and former biologists with the Missouri Department of Conservation have been particularly helpful with plant records and other data. These include Craig Anderson, Mike Arduser, Cindy Becker, Nina Bicknese, Jeff Briggler, David Bruns, Steve Buback, Rick Clawson, Dan Drees, Randy Evans, Susan Farrington, John George, Bob Gillespie, Jenny Grabner, Greg Gremaud, Ethel Hickey, Sherry Holmes, Brad Jacobs, Randy Jensen, Emily Kathol, Karen Kramer, Sherry Leis, Hillary Loring, Norman Murray, Tom Nagel, Doug Newman, Tim Nigh, Krista Noel, Rhonda Rimer, Larry Rizzo, Deborah Rowan, Joe Ryan, C. D. Scott, Mike Skinner, Tim Smith, and Malissa Underwood.

Similarly, a number of past and present naturalists and biologists at the Missouri Department of Natural Resources have provided new records, including Chris Crabtree, Michael Currier, Wanda Doolen, Ken McCarty, Ron Mullikin, Bruce Schuette, Allison Vaughn, and Tim Vogt. Paul McKenzie of the U.S. Fish and Wildlife Service's Columbia Field Office continues to contribute important information and specimens to the Project and is also a wonderful source of information on plant rarity (and morale support!). Brian Davidson, David Moore, and Paul Nelson of the Mark Twain National Forest also have provided data and support. Doug Ladd and his associates at the Missouri Chapter of The Nature Conservancy have been very helpful both in providing new records and for insightful discussions on the vegetation and flora.

Since the start of the project, in 1987, more than 850 total contributors have collected Missouri specimens that have become accessioned in herbaria. Cumulatively, the Missouri Botanical Garden herbarium has averaged new accessions of about 5,000 new Missouri specimens annually. When one adds to this the new specimens at other herbaria around the state, it becomes apparent that a large number of amateur and professional botanists have generated valuable new information on the state's flora, including numerous distributional records. This lively interest, along with my interactions with these botanists, provided a strong impetus to see the work through to completion. In particular, the questions, discussions, and continual prodding by members of the Missouri Native Plant Society and the Botany Group of the Webster Groves Nature Study Society, provided a kind of positive support that kept me going at times when the project seemed a bit overwhelming. Among the legion of amateur and professional botanists too numerous to list, some individuals of special note have included Allan Bornstein, Michelle Bowe, Alan Brant, Art Christ, Carl and Dolly Darigo, Patrick Delozier, Edgar Denison, Karen Haller, Pat and Jack Harris, Nels Holmberg, Lisa Hooper, Stanton Hudson, Lance Jessee, Robin Kennedy, Rebekah Mohn, John Molyneaux, Jay Raveill, Paul Redfearn, Al Scoggin, James Sullivan, Justin Thomas, Steve Turner, George Van Brunt, Ann Wakeman, Jane Walker, Randy Walker, and Wally Weber.

The Missouri Department of Conservation provided the major financial support for the Project during the preparation of Volume 3. This funding was in the form of a five-year grant administered through the Department's Resource Science Division followed by a publication subsidy that has kept the retail price of the book down to a more affordable level than it might have been. When the Department created a funding shortfall for the project toward the end of 2005, a new source of major funding was secured through generous donations to the Missouri Botanical Garden from the Edward K. Love Foundation. without these funds, the project could not have reached this important milestone in its history.

No list of thank-yous could be complete without an acknowledgment of the late Julian Steyermark's role in the project. It was with his blessing that this revision became possible, and his support at the beginning was invaluable in getting me started. Julian Steyermark is a model of what a botanist should be, and his high standards of scholarship are something I continue to aspire to, but fear I will never reach. Would that I had known him while he was still living the questions that I would have liked to ask him now.

First and last, my wife Kay is responsible for this volume reaching completion. In addition to her prodigious knowledge of Midwestern flora, her love and support carried me through all phases of the work. For the third and final volume of the series, Kay went above and beyond in both acting as editor for the draft and in composing the final page layouts.

CONTRIBUTORS OF TREATMENTS

A number of talented botanists contributed their time and efforts as authors or co-authors of various treatments in the present volume. Their work is appreciated. However, as editor of the overall volume, I take full responsibility for any errors that readers may note in these treatments.

John D. Archer
Missouri Botanical Garden
P.O. Box 299
St. Louis, Missouri 63166

George W. Argus
Curator Emeritus
Canadian Museum of Nature
240 McLeod Street
Ottawa, Ontario K2P 2R1
Canada

Harvey E. Ballard, Jr.
Department of Environmental & Plant Biology
Ohio University
Porter Hall, Richland Avenue
Athens, Ohio 45701

David J. Bogler
Missouri Botanical Garden
P.O. Box 299
St. Louis, Missouri 63166

Alan E. Brant
Missouri Botanical Garden
P.O. Box 299
St. Louis, Missouri 63166

Carolyn J. Ferguson
Division of Biology
Ackert Hall
Kansas State University
Manhattan, Kansas 66506

Peter C. Hoch
Missouri Botanical Garden
P.O. Box 299
St. Louis, Missouri 63166

Robin C. Kennedy
Dunn-Palmer Herbarium
Museum Support Center
University of Missouri
Rock Quarry Road at Hinkson Creek
Columbia, Missouri 65211

James B. Phipps
Curator Emeritus
Herbarium, Department of Biology
University of Western Ontario
London, Ontario N6A 5B7
Canada

Jay A. Raveill
Department of Biology
University of Central Missouri
Warrensburg, Missouri 64093

Timothy E. Smith
Missouri Department of Conservation
P.O. Box 180
Jefferson City, Missouri 65102

Charlotte M. Taylor
Missouri Botanical Garden
P.O. Box 299
St. Louis, Missouri 63166

Michael A. Vincent
Dept of Botany
Miami University
316 Pearson Hall, 700 E High St
Oxford, Ohio 45056

Warren L. Wagner
Department of Botany, MRC-166
Smithsonian Institution
P.O. Box 37012
Washington, DC 20013

Alan Whittemore
U.S. National Arboretum
3501 New York Ave NE
Washington, DC 20002

Mark P. Widrlechner
Ada Hayden Herbarium, Botany Department
341A Bessey Hall
Iowa State University
Ames, Iowa 50011

ILLUSTRATION CREDITS

The revised edition features entirely new plates, and includes considerably more species illustrations than were in Steyermark's first edition. These drawings are a major contribution to the manual, and are an invaluable resource for both learning about various species and verifying specimen determinations. The 582 new plates of line drawings that appear in the three volumes of *Steyermark's Flora of Missouri* would not have been possible without generous contributions from the following funding sources:

The Academy of Science of St. Louis
Timon Primm Memorial Fund (at the Missouri Botanical Garden)
Mrs. John S. Lehmann
The John Allan Love Foundation
The Missouri Department of Conservation
The Wednesday Club of St. Louis

The following artists applied their considerable talents to the completion of the 195 plates that appear in Volume 3. Their dedication and patience with the author's often detailed instructions are greatly appreciated.

Phyllis Bick	Plates 388–418, 423, 426–429, 446–456, 459, 462–477, 479, 492–510, 521, 522, 552–558, 561–578
Linda S. Ellis	Plates 457, 458, 460, 461, 478, 480–491, 559, 560
Carole Fischer	Plates 523–544
Sheila Flinchpaugh	Plates 579–582
Kate Johnson	Plates 424, 545–551
Ellen Lissant	Plates 422, 445
John Myers	Plates 419–421, 425
Paul Nelson	Plates 511–520
Maryanne Rouff	Plates 430, 431
Rebecca Samson-Wagstaff	Plates 432–444

The key to dicot families
will appear in a supplementary publication.

FABACEAE (LEGUMINOSAE) (Bean Family)
3. subfamily Faboideae
(Leguminosae subfamily Papilionoideae L. ex DC.)

Contributed by David J. Bogler and George Yatskievych (except as noted)

Plants annual or perennial herbs, shrubs, vines, or trees, usually unarmed (sometimes with spines, thorns, or prickles elsewhere), sometimes with tendrils. Root nodules present. Leaves alternate (rarely appearing opposite on flowering branches) or occasionally all basal, the petiole (and leaflet stalks) commonly with a swollen base (pulvinus) but lacking large glands (sometimes dotted with scattered, minute, impressed glands). Stipules usually present but sometimes shed during leaf development, of various size, shape, and texture, sometimes modified into tendrils or spines. Leaf blades usually 1 time pinnately or palmately compound with 3 to numerous leaflets, rarely simple. Inflorescences various, axillary and/or terminal, commonly showy racemes, spikes, or headlike to umbellate clusters, sometimes of solitary flowers, usually with small bracts and sometimes also bractlets, these often shed early. Flowers perfect, perigynous. Calyces of (4)5-fused sepals, actinomorphic or often slightly to strongly zygomorphic, the lobes various, most commonly broadly to narrowly triangular, occasionally 2-lipped. Corollas usually well-developed, usually strongly zygomorphic, of 5 petals (1 in *Amorpha*), these often tapered to stalklike bases, differentiated (except in a few genera) into an upper banner petal, this usually larger than the others and external in the bud, a pair of lateral wing petals, these sometimes inconspicuous, and a pair of lower, innermost keel petals that are more or less fused along the lower margin, appearing as a single longitudinally keeled or folded structure enclosing the stamens and pistil (except in *Amorpha* and *Dalea*), this often appearing curved or rarely coiled. Stamens 10 (5 in *Dalea*), the filaments often fused most of the way, sometimes only 9 filaments fused and 1 filament free (diadelphous), less often all 10 filaments fused (monadelphous), rarely the filaments all free. Hypanthium short, shallow. Pistil sometimes stalked, the ovules 1 to many. Fruits usually legumes, less commonly loments or achenelike, variously shaped, dehiscent or indehiscent, usually dry. Seeds flattened or not, variously narrowly to broadly bean-shaped or sometimes nearly oblong or circular in outline, the seed coat usually hard, lacking a pleurogram (distinctive line or scar). About 440 genera, more than 12,000 species, worldwide.

The Faboideae are a very large and diverse subfamily, the dominant legume subfamily of the temperate zone, with many economically important members (see examples in the introduction to the Fabaceae in Volume 2 of the present work and in G. Lewis et al., 2005). Nearly all members (more than 95% of the genera), including all of the Missouri species, possess root nodules that contain bacteria capable of converting atmospheric nitrogen into nitrate compounds, compared to 85% in the Mimosoideae and 40% in the Caesalpinioideae genera (Allen and Allen, 1981). This ability to generate their own supply of nitrogen fertilizer has undoubtedly played a major role in the diversity, distribution, and abundance of the Faboideae, and made the group one of the most important for green manures and other soil enrichment practices. In addition, the abundant supply of nitrogen contributes to protein richness, and taxa of Faboideae are among the world's most important food and forage crop plants. Soybeans are the most important cash crop in Missouri, and alfalfa, clover, and other forages support the livestock industry in the state. The term *pulse* refers to food plants in the family Fabaceae that are cultivated for their dried seeds.

The subfamily Faboideae is believed to be derived from tropical woody ancestors resembling certain present-day Caesalpinioideae (Polhill et al., 1981). Primitive members of the Faboideae are woody and often have open, nearly actinomorphic flowers, free keel petals, and free stamens, as in Caesalpinioideae (G. Lewis et al., 2005). Indeed, this close similarity makes it difficult to divide the two groups, and is one of the main reasons why they are here recognized as subfamilies rather than distinct families. *Cladrastis* is one of these apparently archaic Faboideae present in the Missouri flora. The subfamily Faboideae has been divided into between 28 and 32 tribes (Polhill, 1981; G. Lewis et al., 2005). Relationships among the tribes and genera continue to receive close scrutiny using molecular and other approaches (Doyle et al., 1997; Kajita et al., 2001; G. Lewis et al., 2005). Most Missouri members of the subfamily are annual or perennial herbs, shrubs, or vines, with papilionaceous flowers (the corolla differentiated into banner, wings, and keel) and stamens united into a tube. Molecular data have shown that there was a broad radiation of these mostly herbaceous taxa of Faboideae in the temperate zone, including 45 genera in six tribes, all marked by the loss of a large inverted repeat portion of the chloroplast genome that is present in nearly all other land plants (Lavin et al., 1990; Wojciechowski et al., 1999).

Leaf characters have been used historically to delineate tribes (Polhill, 1981) and form the framework for the key to genera below, and it is important for readers to be familiar with leaf morphology in the group. Leaves of Faboideae are quite variable. Stipules usually are present and are persistent or are shed early. In some cases, the stipules are large and difficult to distinguish from true leaflets (*Lotus, Baptisia, Pisum*). In *Robinia*, the stipules are modified into sharp spines. The petiole is short to elongate, and usually has a swollen section known as a pulvinus. The leaf blades in the Faboideae most commonly are palmately or pinnately compound with threeto many leaflets, and they are never two times pinnately compound. The majority of taxa has pinnately compound leaves. Trifoliate (which technically should be called trifoliolate) leaves are often pinnately compound, with the terminal leaflet attached to a short to elongate stalklike extension of the rachis. In a few cases this stalklike extension may be very short, and the leaf mistakenly identified as palmately compound. Compound leaves may be odd-pinnate with a single terminal leaflet, or even-pinnate and lacking a terminal leaflet. If the leaflets are small it may be necessary to examine the leaf tip carefully. In a few genera, the terminal leaflet or leaflet pair is replaced by tendrils (*Lathyrus, Vicia, Pisum*). Some taxa are further distinguished by the presence of small glandular dots on the leaves, and/or on the stems and inflorescences. Stipels (small stipulelike flaps of tissue) may be present at the base of the leaflets. The stipels may be small (1–2 mm) and linear, so are easily overlooked.

Most Faboideae are recognizable by the strongly zygomorphic flowers that are termed papilionaceous (Latin for resembling butterflies), with the banner petal external to a pair of lateral wing petals and the two keel petals more or less fused together into a boat-shaped structure surrounding the stamens and pistil. Many Faboideae are pollinated by bees of one sort or another. Bees are attracted by the large banner petal, which is often blue, purple, yellow, or white. Nectar accumulates at the base of the ovary. The wing petals act as a landing platform. Pressure exerted by the weight of the bee depresses the keel and exposes the stamens or stigma, which come into contact with the underside of the insect. Self-pollination mechanisms have evolved in a number of species, with simplified flowers that bypass insect pollination altogether. A few species produce cleistogamous flowers that are self-fertile and never fully open.

Like all Fabaceae, members of the subfamily Faboideae produce legumes or fruits derived from legumes. Most commonly, the legume is elongate, dry at maturity, and dehisces by two longitudinal valves. Legumes vary greatly in size, wall thickness, presence or absence of a stalklike base or beak, hairiness, number of seeds, and degree of compression between the seeds. Less common are legume-derived fruits that break apart transversely into indehiscent single-seeded segments. Such fruits are called loments and have evolved in a number of different genera. Several tribes are characterized by small, usually indehiscent, single-seeded legumes that resemble achenes. The seeds of Faboideae vary from quite large to very small, and come in a variety of shapes. They are most commonly kidney-shaped, but also can be ovoid, globose, lens-shaped, or cylindric. The seeds never have the distinctive line or scar (pleurogram) that is a typical feature of the Mimosoideae and some Caesalpinioideae.

Three legume genera excluded from the flora deserve mention. A single plant of an unusual legume species with exclusively simple leaves was discovered in early bud along the Mississippi River floodplain in September 1992 during an inventory of the Riverlands National Environmental Demonstration Area (St. Charles County) by the late Guanghua Zhu and his colleagues. A small specimen consisting of a portion of one branch is accessioned in the herbarium of the Missouri Botanical Garden. The late Duane Isely, of Iowa State University, tentatively determined this specimen as some member of the genus *Alysicarpus* Neck. ex Desv. (alyce clover) Although the specimen was shown to several specialists and traveled as far as the herbarium at Kew Gardens in England, no further progress was made on its identity. All vegetation in the area where the plant grew was greatly impacted by the great flood of 1993 and similar plants have not been located there since that time, in spite of diligent searches. *Alysicarpus* comprises about 25 species native to the Old World tropics, and some species have been grown as minor forage crops in the southeastern United States. Because of the simple leaves, plants would key imperfectly to the genus *Crotalaria* in the generic key below. *Alysicarpus* differs from *Crotalaria* in a number of features, including its white to pink or reddish purple (vs. yellow) corollas, stamens with one free and nine fused (vs. all fused) filaments and fruits that are flattened loments (vs. inflated legumes). Migratory waterfowl may have been responsible for its introduction into Missouri and may bring plants northward again. See Isely (1998) for a discussion of *Alysicarpus* in the United States.

Steyermark (1963) included a treatment for *Anthyllis vulneraria* L. (kidney vetch) based on the mention of Missouri in the range of the taxon by Fernald (1950). However, Steyermark was unable to locate a specimen collected from the wild in Missouri and instead mapped a cultivated historical specimen collected on the grounds of the Missouri Botanical Garden. Yatskievych and Turner (1990) noted that there had not been any subsequent collections or reports of this species from the state and excluded it from the flora. *Anthyllis vulneraria* is a distinctive plant with rosettes of entire and pinnately lobed or compound leaves, and palmately lobed bracts subtending the umbellate clusters of flowers. It is a native of Europe that has become established sporadically in the United States, having been introduced as a seed contaminant of other imported legume seeds (Isely, 1998).

Steyermark (1963) reported the genus *Onobrychis* Mill. for Missouri based on a single collection from 1882 of *O. viciifolia* Scop. (sanfoin) purported to have originated from St. Louis County. *Onobrychis viciifolia* (Pl. 403 f–h) is a European species that has only rarely been grown in the United States for its forage value and has escaped sporadically in the northern states (Isely, 1999). The voucher specimen, which is accessioned at the herbarium of the Field Museum, was relocated during the present study. The collector, Joachim H. Schuette, never worked in Missouri, but was active in Michigan, Minnesota, and Wisconsin botany during the period of 1878–1902 (Vegter, 1986). The locality listed on the label, Fort Howard, was determined to be in present-day Brown County, Wisconsin. Thus, this species and genus are excluded from the flora in the present work.

The key to genera regrettably includes a mixture of fruiting and flowering characters. In most species, the sequence of flower maturation is such that fruits and flowers usually are both present on a given sample. Specimens collected toward the end of the season usually have at least a few persistent remains of flowers (often present at the base of the fruit). However, plants at the beginning of the flowering period may be difficult to determine using the key below.

1. Plants trees, shrubs, or lianas, the stems strongly woody well above the base
    2. Plants lianas

3. Leaf blades with 3 leaflets, the lateral leaflets usually conspicuously lobed; flowers appearing after the leaves .... 43. PUERARIA
3. Leaf blades with 7–19 leaflets, the lateral leaflets unlobed; flowers appearing before the leaves or as the leaves develop .... 54. WISTERIA
2. Plants trees or shrubs
4. Branches armed with paired spines or numerous stiff reddish spreading bristles .... 45. ROBINIA
4. Branches unarmed and lacking conspicuous bristles
5. Plants medium- to large-sized trees to 20 m tall........ 22. CLADRASTIS
5. Plants shrubs to 3(–4) m tall
6. Leaves with 9–40 leaflets; inflorescences dense spikelike racemes; corollas reduced to a solitary banner petal..... 14. AMORPHA
6. Leaves with 3 leaflets; inflorescences open clusters, racemes, or panicles; corollas not reduced, consisting of a banner petal, 2 wings, and a folded keel .... 34. LESPEDEZA
1. Plants annual or perennial herbs, sometimes with a woody base or from a woody rootstock
7. Plants vines, producing tendrils; leaf blades with an even number of leaflets, at least those of the upper leaves producing branched tendrils from an extension of the rachis tip
8. Style slender and hairlike, not appearing flattened, glabrous except for a dense ring of short hairs at the tip immediately below the stigma (entirely glabrous in *V. hirsuta* and the tuft of hairs mainly on one side in *V. sativa*); stems not winged .... 52. VICIA
8. Style relatively stout, appearing somewhat flattened or folded, with a dense longitudinal band of hairs along the inner side; stems longitudinally winged or not
9. Stipules noticeably smaller than the leaflets; style appearing somewhat flattened, not longitudinally folded; stems angled and/or winged .... 33. LATHYRUS
9. Stipules as large as or more commonly larger than the leaflets; style appearing longitudinally folded; stems round in cross-section, not winged .... 42. PISUM
7. Plants not true vines, if climbing or twining then not producing tendrils; leaf blades with an odd or even number of leaflets
10. Leaf blades all appearing simple (unifoliate).............. 24. CROTALARIA
10. Leaf blades all or mostly appearing compound, with 2 or more leaflets
11. Leaf blades with an even number of leaflets, lacking a terminal leaflet and usually with a short tapered to hairlike extension of the rachis past the uppermost pair of leaflets (occasional leaves may appear odd-compound because of damage during development, so more than 1 leaf should be examined)
12. Leaf blades with 4 leaflets .... 17. ARACHIS
12. Leaf blades with 16 to numerous leaflets
13. Stems and petioles pubescent with spreading to loosely ascending, bulbous-based glandular hairs (when fresh usually with a small globule of exudate at the tip); stipules more or less lanceolate, appearing attached along the lateral margin, with a long, free, downward-pointed lobe; calyces appearing strongly 2-lipped; fruit a loment breaking into 5–12, 1-seeded segments .... 13. AESCHYNOMENE

13. Stems glabrous or pubescent with appressed nonglandular hairs lacking a bulbous base; stipules narrowly triangular to awl-shaped, attached at the base; calyces broadly bell-shaped, actinomorphic or nearly so; fruit a short or elongate legume, dehiscing longitudinally by 2 valves, (1)2 or (15–)20–40-seeded . . . . 47. SESBANIA

11. Leaf blades with an odd number of leaflets, with a conspicuous terminal leaflet (occasional leaves may have lost their terminal leaflet because of damage with age, so more than 1 leaf should be examined)

14. Leaf blades with all or mostly 5 to numerous leaflets (occasional smaller leaves may have only 3 leaflets)

15. Leaf blades with all or mostly 5 or 7 leaflets

16. Leaf blades appearing palmately compound, all of the leaflets sessile . . . . 40. PEDIOMELUM

16. Leaf blades appearing pinnately compound, with a noticeable rachis

17. Calyces and usually also the stems and leaves minutely dotted with scattered glands; corollas only slightly zygomorphic (not papilionaceous), the lateral and lower petals similar in length and attached along the fused portion of the filaments; stamens 5 . . . . 25. DALEA

17. Plants not gland-dotted; corollas strongly zygomorphic (papilionaceous), the wing petals shorter than the keel and all of the petals free from the fused filaments; stamens 10

18. Stems 100–500 cm long, scrambling or climbing and twining; leaves with the petioles 1.5–7.0 cm long, the leaflets 2–9 cm long; corollas a mixture of brownish red, pale purple, and white . . . . 16. APIOS

18. Stems 15–60 cm long, spreading to loosely ascending, not twining; leaves appearing sessile, the leaflets 0.4–1.6 cm long corollas yellow to orangish yellow . . . . 35. LOTUS

15. Leaf blades with all or mostly 9 to numerous leaflets

19. Calyces, bracts, and usually also the leaflets with scattered to dense, minute, yellowish brown to black, glandular dots

20. Stems lacking glandular dots or with scattered yellowish brown to black glandular dots toward the tip; ovaries and fruits glabrous or hairy and sometimes also glandular but lacking prickles, the fruit not or only slightly exserted from the calyx at maturity . . . . 25. DALEA

20. Stems with dense, dark brown to black glandular dots toward the tip; ovaries and fruits with dense hooked prickles, the fruit long-exserted from the calyx at maturity . . . . 29. GLYCYRRHIZA

19. Plants not gland-dotted

21. Inflorescences axillary umbellate clusters; fruits slender loments, these circular in cross-section (not flattened), appearing transversely jointed and breaking into 2–10, cylindric, 1-seeded segments . . . . 46. SECURIGERA

21. Inflorescences axillary and/or terminal racemes, these sometimes spikelike or relatively short and dense; fruits legumes of various shapes, flattened or not, indehiscent or dehiscent longitudinally by 2 valves, not appearing jointed or breaking apart transversely into segments

22. Stems and usually also the leaves densely hairy with curved or spreading hairs (the leaves sometimes becoming nearly glabrous at maturity), the hairs unbranched and attached at their bases
    23. Corollas reduced to a solitary banner petal, this 4.5–6.0 mm long, bright bluish purple to purple; calyces 3–5 mm long . . . . 14. AMORPHA
    23. Corollas not reduced, 15–20 mm long, consisting of a banner petal, 2 wings, and a folded keel, the banner pale yellow to cream-colored, the other petals light pink or cream-colored to white but pinkish-tinged or striped; calyces (5–)7–8 mm long . . . . 50. TEPHROSIA
22. Stems and leaves glabrous or pubescent with mostly appressed hairs, the hairs technically with 2 branches spreading in opposite directions, but appearing as straight unbranched hairs attached at their midpoints, these sometimes mixed with basally attached unbranched loosely appressed hairs
    24. Leaves basal and also well-developed along the stems . . . . 18. ASTRAGALUS
    24. Leaves all basal or at most 1 stem leaf produced near the stem base
        25. Corollas 8–11 mm long, yellow; inflorescence stalk 2–10 cm long, shorter than to about as long as the basal leaves . . . . 18. ASTRAGALUS
        25. Corollas 15–20 mm long, purple to bluish purple; inflorescence stalk (10–)15–25 cm long, distinctly longer than the basal leaves . . . . 39. OXYTROPIS

14. Leaf blades with 3 leaflets (sometimes the lowermost or uppermost leaves may have only 1 or 2 leaflets)
    26. Leaflets finely toothed along the margins, at least above the midpoint, the teeth sometimes inconspicuous in *Trifolium*
        27. Inflorescences elongate racemes 5–15 cm long, the flowers well-spaced along the axis . . . . 37. MELILOTUS
        27. Inflorescences umbellate or headlike clusters or short dense racemes less than 2 cm long
            28. Stems somewhat 4-angled toward the tip; corollas not persistent at fruiting; fruits spirally curved or coiled, sometimes with tubercles or prickles, exserted from the calyx . . . . 36. MEDICAGO
            28. Stems circular or somewhat flattened toward the tip; corollas more or less persistent at fruiting, becoming brown and papery; fruits relatively straight, lacking tubercles and prickles, usually not exserted from the calyx . . . . 51. TRIFOLIUM
    26. Leaflets entire along the margins but occasionally with 1 or 2 coarse rounded lobes
        29. Leaf blades palmately trifoliate, the leaflets all sessile or with short stalks of equal or nearly equal length (note that the terminal leaflet may be slightly more tapered at the base than the lateral ones, but not noticeably longer-stalked than the lateral leaflets)
            30. Leaflets with both surfaces or at least the undersurface moderately to densely gland-dotted, the glands sometimes partially obscured by hairs . . . . 40. PEDIOMELUM

30. Leaflets lacking glandular dots
31. Corollas 20–30 mm long; fruits with the body 15–60 mm long, not flattened, appearing inflated, exserted from the calyx, appearing noticeably stalked . . . . 19. BAPTISIA
31. Corollas 3–18 mm long; fruits with the body 1–5 mm long, more or less flattened, not appearing inflated, not exserted from the calyx or if somewhat exserted then appearing sessile or nearly so
32. Inflorescences appearing as small axillary clusters of 2–5 flowers or of solitary flowers; fruits longer than and exserted from the calyx, indehiscent . . . . 31. KUMMEROWIA
32. Inflorescences appearing as axillary or terminal, dense, headlike clusters or spikelike racemes of 10 to numerous flowers, occasionally appearing as clusters of (3–)5–10 flowers; fruits not appearing exserted from the calyx or if long-exserted then dehiscing longitudinally with spirally twisting valves
33. Corollas white, pink, red, or pale cream-colored, persistent at fruiting, becoming papery and brown; fruits not appearing exserted from the calyx, indehiscent or, if dehiscing longitudinally, then the valves not becoming twisted . . . . 51. TRIFOLIUM
33. Corollas bright yellow to orangish yellow (often becoming marked with brownish red with age), not persistent at fruiting; fruits long-exserted from the calyx, dehiscing longitudinally with spirally twisting valves . . . . 35. LOTUS
29. Leaf blades pinnately trifoliate, the central leaflet short- to long-stalked, the lateral leaflets sessile or with stalks shorter than that of the central leaflet (when in doubt, check several leaves, especially the largest leaves on the plant; note also that in a few genera, the first few leaves produced by seedlings are palmately trifoliate)
34. Main stems erect or ascending, sometimes from a spreading base, not trailing, twining, or vinelike
35. Inflorescence axis and branches, flower stalks, and fruits (and sometimes also the stems and leaves) with sparse to moderate, short, hooked hairs (clingy to the touch), the vegetative portions often also with sparse to dense nonhooked hairs; fruits loments, appearing jointed and breaking transversely into 1-seeded segments
36. Stipels (small stipulelike structures at the leaflet bases) persistent at leaflet maturity; calyx lobes longer than the tube; stalklike bases of the fruits 1.0–6.5 mm long; connections between the fruit segments nearly symmetrical or the lower margin somewhat more deeply lobed between the segments than the upper margin . . . . 26. DESMODIUM
36. Stipels (small stipulelike structures at the leaflet bases) absent or shed before the subtending leaves fully mature; calyx lobes much shorter than the tube; stalklike bases of the fruits 4–18 mm long; connections between the fruit segments highly asymmetric, the lower margin very deeply lobed between the segments, the upper margin not lobed, at most slightly undulate . . . . 30. HYLODESMUM
35. Plants lacking hooked hairs; fruits legumes, not appearing jointed, dehiscing longitudinally by 2 valves or indehiscent
37. Corollas 45–55 mm long, the banner petal conspicuously (more than 2 times) longer than the wings and keel . . . . 23. CLITORIA

37. Corollas 3–20 mm long; banner petal less than 2 times as long as the wings and keel
38. Stipules conspicuous, 8–15 mm long, broadly lanceolate to ovate, persistent; inflorescences more or less jointed at the nodes; fruits 10–20 cm long . . . . 53. VIGNA
38. Stipules relatively inconspicuous, less than 5 mm long or if up to 10 mm then linear or hairlike and/or shed early; inflorescences not appearing jointed at the nodes; fruits 0.3–5.0 cm long
39. Inflorescences entirely of solitary axillary flowers; stipules glandular, a pair of dots or short lines . . . . 12. ACMISPON
39. Inflorescences axillary and sometimes also terminal spikelike racemes or clusters, if in axillary clusters then occasional nodes with only a solitary flower (users should examine several nodes); stipules scalelike or hairlike (sometimes shed early)
40. Leaflets with small (1–2 mm long) stipulelike flaps of tissue (stipels) at the base; plants annual, with a taproot . . . . 28. GLYCINE
40. Leaflets lacking stipulelike flaps of tissue at the base; plants perennial, with a woody rootstock, tuberous-thickened taproot, or rhizome
41. Calyces closely subtended by a pair of small bracts (these in addition to bracts present at the inflorescence node or the base of the flower stalk); fruits with more or less papery valves, these not wrinkled but often with a fine raised network of nerves, often hairy at maturity . . . . 34. LESPEDEZA
41. Calyces not subtended by bracts (however inflorescence bracts present at the nodes, these usually shed as the flowers mature); fruits with hardened valves, these cross-ribbed, the ribs sometimes connected to form a network, not hairy at maturity . . . . 38. ORBEXILUM
34. Main stems spreading to trailing or climbing, sometimes twining and vinelike
42. Corollas 25–55 mm long, the banner petal conspicuously (more than 2 times) longer than the wings and keel
43. Stems twining; flowers subtended by a pair of broadly lanceolate to ovate bractlets 8–12 mm long; calyces 10–16 mm long, the lobes as long as or longer than the tube; banner petal 25–30 mm long, with an inconspicuous spur at the base (note that cleistogamous flowers are not produced); fruits with the body (above the sessile or minutely stalklike base but including the beak) 7–12 cm long, 3–4 mm wide . . . . 21. CENTROSEMA
43. Stems trailing but mostly not twining; flowers subtended by a pair of lanceolate bractlets 3–7 mm long; calyces 14–21 mm long, the lobes shorter than the tube; banner petal 45–55 mm long, not spurred at the base (note that plants sometimes produce at least a few cleistogamous flowers that have a somewhat shorter calyx and lack a well-formed corolla); fruits with the body (above the pronounced stalklike base but including the beak) 3–6 cm long, 5–7 mm wide . . . . 23. CLITORIA
42. Corollas less than 20 mm long; banner petal less than 2 times as long as the wings and keel

44. Inflorescences and fruits with sparse to moderate, short, hooked hairs (clingy to the touch); fruits loments breaking transversely into 1-seeded segments at maturity . . . . 26. DESMODIUM
44. Plants lacking hooked hairs or if very minute hooked hairs are present on the stems and leaves (observable only under high magnification in *Phaseolus*), then the plants not clingy to the touch and the fruits legumes dehiscing longitudinally
    45. Leaflets with the undersurface moderately hairy but also with numerous minute more or less spherical yellow to orangish yellow glands; corollas lemon yellow to orangish yellow . . . . 44. RHYNCHOSIA
    45. Leaflets with the undersurface variously glabrous or hairy but lacking spherical glands, corollas variously colored
        46. Leaflets 1–7 cm long
            47. Corollas yellow to orangish yellow; stipules fused into a persistent tube around the stem with short triangular lobes . . . 49. STYLOSANTHES
            47. Corollas white, pink, lavender to purple, or some combination of these colors; stipules not fused, sometimes inconspicuous and/or shed early
                48. Inflorescences small headlike or umbellate axillary clusters at the tips of long stalks, these all or mostly much longer than the leaves . . . . 48. STROPHOSTYLES
                48. Inflorescences racemes, or if occasionally reduced to a loose cluster or solitary flower then the stalks shorter than the leaves
                    49. Flowers of 3 kinds, the open-flowering ones with well-developed corollas and in axillary racemes, the cleistogamous ones lacking a well-developed corolla and either subterranean or along trailing branches from near the stem base; calyces of open flowers without bractlets subtending them . . . . 15. AMPHICARPAEA
                    49. Flowers all similar, with well-developed corollas and in axillary racemes; calyces closely subtended by a pair of small bractlets
                        50. Calyces 4–7 mm long, appearing 4-lobed, the lobes similar in length but the uppermost lobe (consisting of 2 fused sepals) broader than the others; banner petal 9–10 mm long; fruits 2–4 cm long; plants moderately pubescent but none of the hairs hooked; lateral pair of leaflets symmetrical or nearly so . . . . 27. GALACTIA
                        50. Calyces 2.5–4.0 mm long, 5-lobed but often appearing somewhat 2-lipped, the 2-lobed lower lip longer than the shallowly 3-lobed upper lip; banner petal 5–7 mm long; fruits 3–8 cm long; plants moderately to densely pubescent, the hairs mostly short and with minutely hooked tips (visible only under strong magnification); lateral pair of leaflets asymmetrical . . . . 41. PHASEOLUS
        46. Leaflets 7–15 cm long
            51. Stipules 0.5–2.0 mm long, lanceolate to narrowly triangular, shed early (rarely observed); leaflets with minute, stipulelike flaps of tissue at the base, but these shed early and not observable at leaf maturity . . . . 20. CANAVALIA

51. Stipules 1.5–15.0 mm long, lanceolate to ovate or triangular, persistent; leaflets with small (2–10 mm long) stipulelike flaps of tissue (stipels) at the base
   52. Stems and petioles moderately to densely pubescent with relatively long (2–4 mm), spreading, light orange to reddish orange, minutely bulbous-based hairs; calyces 8–15 mm long . . . . 43. PUERARIA
   52. Stems and petioles glabrous or if pubescent then the hairs minute to short, appressed to spreading, white or off-white, slender-based; calyces 4–7 mm long
      53. Stipules 8–15 mm long, more or less persistent, appearing somewhat peltate or with a long downward-angled basal lobe; inflorescences very short racemes or clusters of 2–6 flowers at the tip of a long stalk . . . 53. VIGNA
      53. Stipules 1.5–7.0 mm long, persistent (in *Phaseolus*) or shed early (in *Lablab*), attached basally, sometimes somewhat asymmetrical at the base, but lacking a downward-angled lobe; inflorescences elongate racemes, with 8 to numerous flowers
         54. Stems glabrous or sparsely pubescent, the hairs more or less downward-curved or angled, not hooked at the tip; calyces 4–7 mm long; fruits 16–20 mm wide, the valves usually reddish- or purplish-tinged, especially before full maturity (sometimes becoming straw-colored with age), appearing finely warty along the upper suture . . . . . 32. LABLAB
         54. Stems moderately to densely pubescent, the hairs mostly with minutely hooked tips (visible only under strong magnification); calyces 2.5–4.0 mm long; fruits 7–12 cm long, the valves green to tan, light brown, or straw-colored, not reddish- or purplish-tinged, appearing smooth along the upper suture . . . . 41. PHASEOLUS

12. **Acmispon** Raf.

Eight or about 33 species, North America, South America.

The species of *Acmispon* formerly were included in the genus *Lotus*. Most botanists have treated the genus *Lotus* in a broad sense, with various numbers of infrageneric groups differing in stipule, floral, and fruit morphologies (Ottley, 1944; Callen, 1959; Isely, 1981). More recently, morphological and molecular evidence began to accumulate that the Old World species of *Lotus* (including the type species, *L. corniculatus*) are distinct from those in the New World. However, there has been disagreement on how many genera should be recognized among the New World species, with various authors accepting two to four or more genera (Allan and Porter, 2000; Sokoloff, 2000; Degtjareva et al., 2003; Arambarri et al., 2005; Brouillet, 2008). In part this has been because of poor resolution of some portions of the molecular phylogenies and because of disagreements about how many times some morphological character states have become derived. The solution of Brouillet (2008) to accept only the two best-resolved lineages of North American species as genera is adopted here. This results in a circumscription of *Acmispon* to include about 33 species. In the more restricted sense advocated by Sokoloff (2000) and some other authors, *Acmispon* comprises only eight species. Regardless of its circumscription, the genus is native to North America except for one Chilean endemic.

1. Acmispon americanus (Nutt.) Rydb. **var. americanus** (prairie trefoil)
Hosackia americana (Nutt.) Piper
H. purshiana Benth., an illegitimate name
H. unifoliolata Hook.
Lotus americanus (Nutt.) Bisch., an illegitimate name
L. purshianus (Benth.) F. Clements and E. Clements, an invalid combination
L. unifoliolatus (Hook.) Benth.

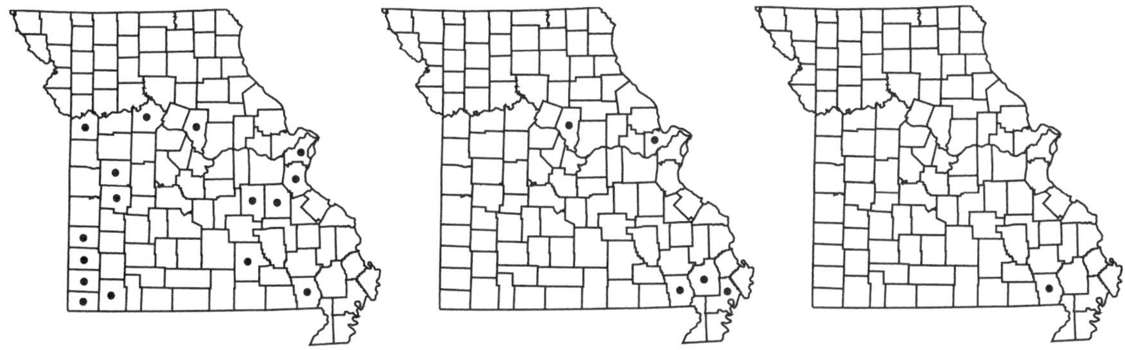

1705. Acmispon americanus 1706. Aeschynomene indica 1707. Aeschynomene rudis

Trigonella americana Nutt.

Pl. 401 e–g; Map 1705

Plants annual, with a long, thin taproot. Stems 15–45(–70) cm long, erect or strongly ascending, unbranched or with few to many branches and then appearing bushy, unarmed, moderately to densely pubescent with fine, more or less spreading to ascending hairs. Leaves alternate, pinnately trifoliate or the upper leaves simple (the blade then similar to the terminal leaflet of compound leaves), short-petiolate, the petiole 0.6–2.0 mm long. Stipules minute, represented by a pair of inconspicuous, sessile, circular to bean-shaped glands. Leaflets 6–20 mm long, 2.5–5.0(–8.0) mm wide, elliptic to lanceolate, often narrowly so, angled to a short-stalked base, angled or tapered to a sharply pointed tip, the margins entire, finely hairy, the surfaces moderately to densely and finely hairy, the venation pinnate but usually only the midvein readily visible. Terminal leaflet with the stalk 1–4 mm long, symmetric at the base; lateral leaflets with the stalk 0.2–1.5 mm long, often asymmetric at the base. Inflorescences axillary, of solitary flowers in most or only the upper axils, the stalks 5–20 mm long, hairy, with a simple, leaflike bract and a pair of minute glands at the tip, the flower with a stalk 0.2–1.0 mm long above the bract; bractlets absent. Calyces finely hairy, the tube 1.5–2.0 mm long, conic, the lobes 4–6 mm long, all similar, narrowly lanceolate. Corollas papilionaceous, white or pale cream-colored to pale pink, often with a dense pattern of darker pink nerves, the keel often yellowish toward the tip, the banner 5–7 mm long, 3.5–5.0 mm wide, the wings 4–6 mm long, 2–3 mm wide, obovate, the keel 4–6 mm long, 1.5–2.0 mm wide, boat-shaped, fused nearly to the narrowed tip. Stamens 10, all of similar length, 9 of the filaments fused and 1 free, the fused portion 3.5–4.0 mm long, the free portion 1.0–1.5 mm long, flattened and broadened at the tip, the anthers small, attached near the base. Ovary 2–3 mm long, the style 1–2 mm long, slender, slightly curved, the stigma small and terminal. Fruits legumes, 2–4 cm long, 2–3 mm wide, narrowly oblong to linear, beaked, more or less circular in cross-section, the sutures thickened and brown to reddish brown at maturity, dehiscing by 2 valves, these green at maturity, twisting spirally after dehiscence, mostly 5- or 6-seeded. Seeds 2.5–3.2 mm long, 1.5–2.0 mm wide, more or less kidney-shaped to nearly oblong in outline, the surface olive green to yellowish brown or reddish brown, mottled with darker brown or black, smooth, somewhat shiny. $2n=14$. May–October.

Uncommon south of the Missouri River; absent from much of the Ozark Division and the Mississippi Lowlands (U.S.; Canada, Mexico). Upland prairies, glades, tops of bluffs, and occasionally margins of ponds; also railroads, roadsides, and open disturbed areas.

Acmispon americanus is recognized by the trifoliate, somewhat hairy leaves with reduced, glandular stipules and the solitary pinkish flowers in the axils of simple bracteal leaves. The plants are inconspicuous and likely are more common in southern Missouri than has been documented by the herbarium record. This is the most widespread species of *Acmispon* in the United States, with major subranges in the western and Pacific states, the central states, and the southeastern states (Isely, 1990). Populations in the Southeast are glabrous or nearly so and have longer, narrower leaflets; they have been segregated as var. *helleri* (Britton) Brouillet (*Lotus americanus* var. *helleri* (Britton) Isely).

The nomenclature of this species is complicated. For many years, it was known as *Lotus purshianus* in most of the botanical literature (Ottley, 1944; Steyermark, 1963). However, Dorn (1988) and Kartesz and Gandhi (1991a) suggested that the name *Hosackia unifoliolata* represented an older original basionym for the same taxon within *Lotus*. The difficulty with using this name, as pointed out by Isely (1996), is that there was no type specimen designated for it and the name is of somewhat uncertain application. Dorn (1988), Kartesz and Gandhi (1991a), Sokoloff (2000) and Brouillet (2008) also noted that the name *Hosackia purshiana* represented an illegitimate renaming of *Trigonella americana*. The generic name *Trigonella* refers to an Old World genus more closely related to the sweet clovers (*Melilotus*), but the oldest validly published epithet, *T. americana*, must still be used for any subsequent renamings of the species. Within *Lotus*, this was not possible, because of the existence of an earlier publication of *L. americanus* Vell. for a different species. However, within *Acmispon*, this name is available for use.

13. **Aeschynomene** L.

Plants annual (perennial herbs or shrubby elsewhere), robust. Stems erect or strongly ascending (prostrate elsewhere), branched, unarmed, glabrous or more commonly hairy, the hairs simple or glandular with a swollen pustular base, the lower portion forming adventitious roots if submerged. Leaves sensitive to touch and light, alternate, even-pinnately compound (in our species), short-petiolate, the leaflets numerous (fewer elsewhere), mostly subopposite. Stipules conspicuous on young growth but usually shed with age, herbaceous, lanceolate, peltate, with a small lobelike appendage below the point of attachment, tapered to a sharply pointed tip, the margins often toothed and/or hairy; stipels lacking. Leaflets oblong to elliptic-oblong, oblique at the base, abruptly short-tapered to a minute sharp point at the tip, the margins usually entire, the surfaces glabrous, but the undersurface often minutely gland-dotted, 1-veined. Inflorescences few-flowered racemes, sometimes appearing as loose stalked clusters, spreading to somewhat pendant, sometimes with 1 to several small pinnately compound leaves toward the base of the stalk, the stalk often glandular-hairy and sticky, the bracts subtending the flowers similar to the stipules but often smaller. Calyces appearing 2-lipped, the tube short, the upper lip entire or 2-lobed, the lower lip entire or 3-lobed. Corollas papilionaceous, glabrous, yellow to orange, sometimes striped or tinged with red or purple, the petals tapered to a short stalklike base; the banner broadly ovate to nearly circular; the wings narrowly obovate; the keel oblanceolate and curved in outline, boat-shaped. Stamens 10, the filaments all fused, the tube split on 1 side or more commonly deeply on 2 sides into 2 groups of 5 stamens each, the anthers small, attached near the midpoint, all similar in size. Ovary linear, short-stalked, often hairy, the style glabrous, curved, the stigma small and terminal. Fruits loments, elongate, flattened, stalked at the base, the terminal segment short-tapered to a sharply pointed tip, the upper margin straight or slightly curved, the lower margin scalloped, the surfaces sparsely glandular-pustular hairy, often longitudinally wrinkled and/or with medial rows of small pustular-based warty projections, breaking into 5–12 indehiscent, 1-seeded segments. Seeds kidney-shaped, the surfaces smooth, somewhat shiny. About 130–160 species, widespread in tropical and subtropical regions of both hemispheres, with few native and several introduced species in temperate regions.

Aeschynomene species often grow in seasonally flooded areas such as rice paddies and marshes, and often dominate these communities. Neither of the two species present in Missouri was known to Steyermark (1963). They are very similar and can be difficult to distinguish. Useful characters for identification include the size of the fruits and segments, distribution of pubescence on the stems, size of the leaves and leaflets, floral bract morphology, and overall size of the flowers (Carulli et al., 1988; Perry et al., 1998). However, there is significant overlap in the size ranges of these characters and consequently much room for misidentification. Both *A. indica* and *A. rudis* are members of a wide-ranging, pantropical, weedy complex of weakly differentiated species (Rudd, 1955). However, allozyme data (Carulli and Fairbrothers, 1988) support the distinctness of *A. indica* and *A. rudis*, at least in the eastern coastal United

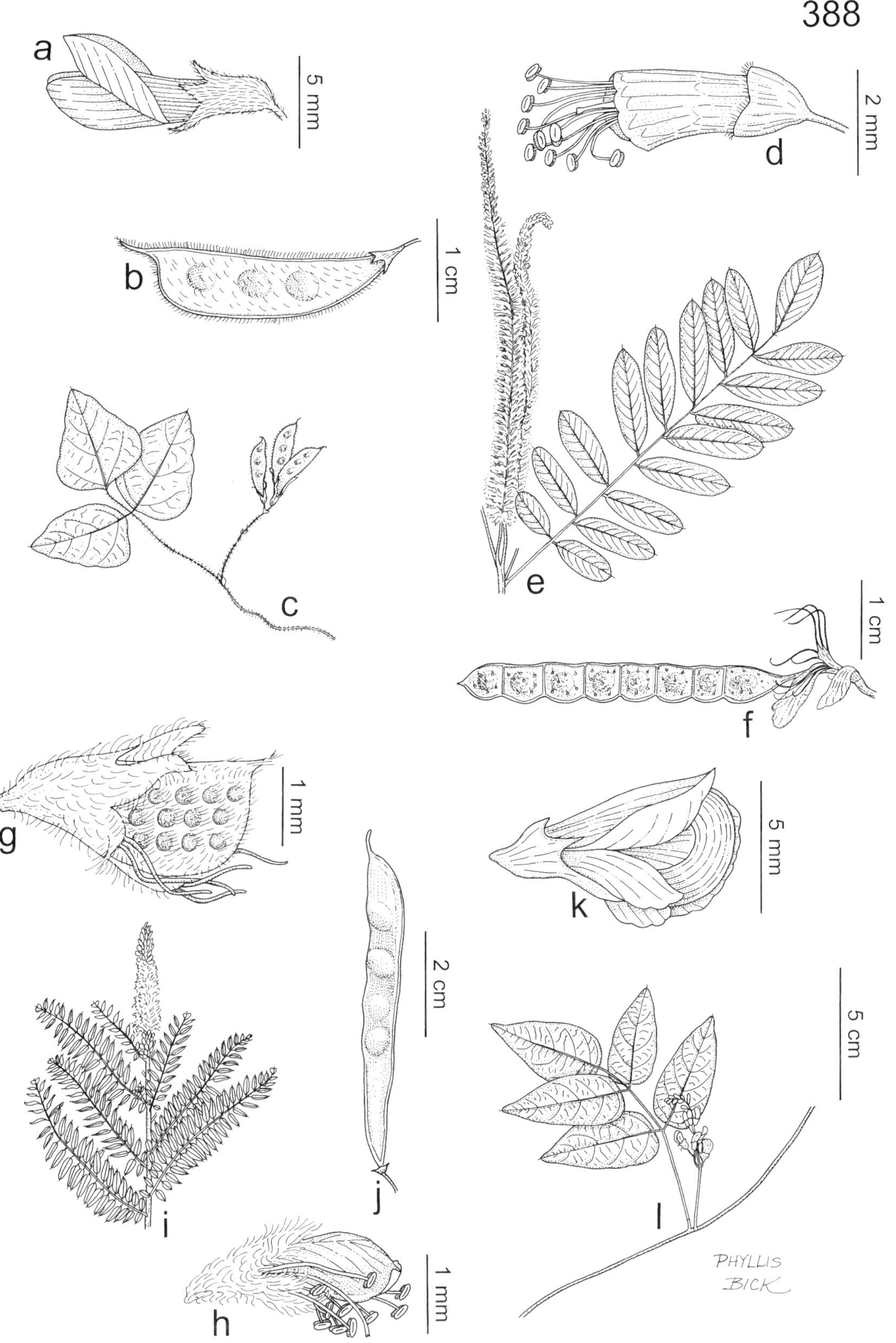

Plate 388. Fabaceae. *Amphicarpaea bracteata*, **a)** open (not cleistogamous) flower, **b)** fruit, **c)** fertile stem. *Amorpha fruticosa*, **d)** flower, **e)** fertile stem. *Aeschynomene indica*, **f)** fruit. *Amorpha canescens*, **g)** fruit, **h)** flower, **i)** fertile stem. *Apios americana*, **j)** fruit, **k)** flower, **l)** fertile stem.

States. Plants apparently are self-fertile, and crosses between the two species produced no offspring or infertile offspring. Both *A. indica* and *A. rudis* have the potential to become weeds in rice fields. This habitat is ideal for them ecologically and the seeds are about the size and weight of rice grains, making their removal from harvested rice difficult and expensive.

1. Upper portions of the stem glabrous or sparsely hairy; leaflets 5–10 mm long; hairs of petiole and inflorescence stalk mostly gland-tipped, yellowish; floral bracts entire or indistinctly toothed; mature fruits smooth or slightly wrinkled, sometimes with minute warty projections to 0.2 mm long, the segments 4–5 mm long, 4–5 mm wide . . . . 1. A. INDICA
1. Upper portions of the stem moderately pustular-hairy; leaflets 8–16 mm long; hairs of petiole and inflorescence stalk not gland-tipped, whitish; floral bracts distinctly toothed, the teeth tipped with pustular hairs; mature fruits usually with prominent warty projections to 1 mm long, the segments 5–6 mm long, 5–6 mm wide . . . . 2. A. RUDIS

1. Aeschynomene indica L. (Indian jointvetch)
Pl. 388 f; Map 1706

Stems 1–2 m tall, erect, often stout toward the base and pithy, the upper portion glabrous or sparsely and finely hairy. Leaves with 40–66 leaflets, the petiole 3–8 mm long, the rachis 4–9 cm long, the petiole and rachis pubescent with pustular-based gland-tipped, yellowish hairs. Stipules 10–14 mm long, 2–3 mm wide. Leaflets 5–10 mm long, 1.5–2.5 mm wide. Inflorescences usually on reduced lateral branches with few to several reduced pinnately compound leaves, appearing as branched, lax, racemose clusters 3–5 cm long, the inflorescence stalk 1–2 cm long, densely pubescent with pustular-based, gland-tipped, yellowish hairs, the bract subtending each flower 2–4 mm long, lanceolate, entire to minutely and indistinctly toothed, shed early, the flower stalk 3–4 mm long. Calyces with the tube 0.8–1.0 mm long; the upper lip 4–5 mm long, 2-toothed apically; the lower lip 4–5 mm long, shallowly 3-toothed apically. Corollas yellow, often reddish-tinged along nerves; the banner 6–8 mm long, 4.5–5.5 mm wide; the wings 6–7 mm long, 2.5–3.0 mm wide; the keel 7–8 mm long. Filaments 6–8 mm long, the tube split to about the midpoint. Ovary 4–5 mm long, the style 1.5–2.0 mm long. Fruits 3.5–4.5 cm long, 4–5 mm wide, the stalk 3–6 mm long, the surfaces at maturity sparsely pustular-hairy, smooth or occasionally with longitudinal wrinkles at maturity, with medial lines of minute, pustular-based, warty projections to 0.2 mm long, dehiscent into into 5–9 segments, these 4–5 mm long, 4–5 mm wide. Seeds 3–4 mm long, 2–3 mm wide, olive brown. $2n=40$. July–September.

Introduced, scattered in the Mississippi Lowlands Division (native of the New World Tropics; introduced in the southeastern U.S. north to Missouri and Virginia; also Asia, Africa, Australia). Margins of sloughs and oxbows; also ditches, rice fields, and roadsides.

Aeschynomene indica was first reported for Missouri by Dunn and Knauer (1975) from dispersed segments of loments that were collected in 1973 in the Stoddard County portion of Mingo National Wildlife Refuge and subsequently grown to maturity in a greenhouse at the University of Missouri.

2. Aeschynomene rudis Benth. (rough jointvetch)
Map 1707

Stems 1–2 m tall, erect, often expanded toward the base and pithy, the upper portion moderately pubescent with pustular-based spreading hairs to 2 mm long. Leaves with 40–50 leaflets, the petiole 10–12 mm long, the rachis 9–12 cm long, the petiole and rachis pubescent with pustular-based nonglandular, whitish hairs. Stipules 8–10 mm long, 4–5 mm wide. Leaflets 8–16 mm long, 2–3 mm wide. Inflorescences usually on reduced lateral branches with few to several reduced pinnately compound leaves, appearing as branched, lax, racemose clusters 3–5 cm long (rarely reduced to a solitary flower), the inflorescence stalk 0.5–1.0 cm long, densely pubescent with pustular-based, nonglandular, whitish hairs, the bract subtending each flower 2–4 mm long, lanceolate, coarsely toothed, usually shed early, the flower stalk 3–6 mm long. Calyces with the tube 0.8–1.0 mm long; the upper lip 4–5 mm long, shallowly 2-lobed apically; the lower lip 4–5 mm long, shallowly 3-lobed apically. Corollas yellow to orange, sometimes reddish- or purplish-striped or -tinged; the banner 8–9 mm long, 5–6 mm wide; the wings 7–8 mm long, 2–4 mm wide;

the keel 7–8 mm long. Filaments 7–9 mm long, the tube split to slightly below the midpoint. Ovary 5–6 mm long, the style 3–5 mm long. Fruits 4–6 cm long, 4–6 mm wide, the stalk 4–6 mm long, the surfaces at maturity sparsely pustular-hairy, usually with pronounced medial lines of stout pustular-based, warty projections to 1 mm long, dehiscent into 7–12 segments, these 5–6 mm long, 5–6 mm wide. Seeds 4–5 mm long, 2–3 mm wide, reddish brown to olive brown. July–September.

Introduced, uncommon in Butler County, to be expected elsewhere in the Mississippi Lowlands Division (native of South America; introduced in the U.S. north to California, Missouri, and Pennsylvania). Rice fields and ditches.

Aeschynomene rudis was first reported for Missouri by S. Hudson (1994), who noted that it had been reported anecdotally under the name northern jointvetch as early as 1987 by the University of Missouri Agricultural Extension Service. Carulli et al. (1988) documented its distribution in the southeastern United States and contrasted it morphologically with *A. indica* and *A. virginica* (L.) Britton, Sterns & Poggenb. (with which some collections of *A. rudis* had been confused by earlier botanists).

14. **Amorpha** L.
(Wilbur, 1975)

Plants shrubs or subshrubs (rarely perennial herbs elsewhere), with thick roots and rhizomes. Stems erect or ascending, unbranched or branched, unarmed, glabrous to densely hairy. Leaves alternate, odd-pinnately compound, subsessile to short-petiolate, aromatic when bruised, the leaflets 9 to numerous, variously alternate to opposite. Stipules inconspicuous but often persistent, 1–2 mm long, linear to hairlike, thin-textured and brown, attached at the base; stipels present, similar to the stipules, shed early. Leaflets ovate to oblong, rounded or angled to a short-stalked base, bluntly to sharply pointed at the tip, often with a minute sharp point (this sometimes gland-tipped) at the very tip, the margins entire, the surfaces variously hairy and gland-dotted, with a midvein and inconspicuous or conspicuous secondary veins. Inflorescences axillary and terminal, ascending to spreading, dense, spikelike racemes, often clustered, short- to moderately stalked, the bracts subtending flowers linear to hairlike and inconspicuous, shed early. Calyces 5-lobed, the lobes much shorter than to about as long as the tube, similar or the lowermost lobe slightly longer than the others. Corollas not papilionaceous, reduced to a solitary banner petal, this tapered to a short stalklike base, obovate to heart-shaped, folded around the stamens and gynoecium, the wing and keel absent. Stamens 10, the filaments all fused near the base, the anthers long-exserted, attached near the midpoint, all similar in size. Ovary ellipsoid to ovoid, sessile, the style slender, glabrous or hairy, the stigma small and terminal. Fruits modified legumes, slightly elongate, slightly flattened, sessile, more or less oblong and usually somewhat curved in outline, often conspicuously gland-dotted, indehiscent, 1-seeded. Seeds oblong-elliptic in outline, usually with a small notch at the attachment end, somewhat flattened, smooth, shiny. About 15 species, North America, most diverse in the U.S.

Amorpha is related to *Dalea* (Polhill, 1981). The species tend to be variable in degree of pubescence, leaf texture, density of glands, and number and length of racemes. Numerous additional species, varieties, and forms have been recognized in the past, but the general consensus is that most of these do not deserve recognition.

1. Plants subshrubs or shrubs, 30–70(–90) cm tall; leaves subsessile, the petiole 0.5–1.0 mm long, shorter than the width of the lowest leaflet, leaflets 9–17 mm long, often densely hairy, obscurely gland-dotted; calyx lobes 1.5–2.0 mm long, subequal; fruits 3–4 mm long . . . . 1. A. CANESCENS

1. Plants large shrubs, 100–200(–400) cm tall; leaves distinctly petiolate, the petiole 10–30 mm long, longer than the width of the lowest leaflet, leaflets 20–40 mm long, glabrous or sparsely hairy on the undersurface, noticeably gland-dotted; calyx lobes 0.2–1.0 mm long, 4 very short and 1 somewhat longer; fruits 4–7 mm long . . . . 2. A. FRUTICOSA

1. Amorpha canescens Pursh (lead plant)
A. canescens f. *glabrata* (A. Gray) Fassett
A. brachycarpa E.J. Palmer
Pl. 388 g–i; Map 1708

Plants small shrubs or subshrubs, from a knotty rootstock with a deep thick taproot. Stems 30–70(–90) cm tall, unbranched or sparsely branched, usually densely pubescent with short whitish to gray woolly hairs, rarely nearly glabrous, also sparsely and obscurely glandular. Leaves with 29–41 leaflets, subsessile, the petiole 0.5–1.0 mm long, much shorter than the width of the lowest leaflet, the rachis 8–13 cm long, the petiole and rachis densely pubescent with short whitish to gray woolly hairs. Leaflets 9–17 mm long, 4–8 mm wide, ovate-oblong to elliptic, rounded to cordate at the base and with a stalk 0.5–1.0 mm long, rounded or abruptly short-tapered to a minute sharp point at the tip, the margins entire, both surfaces sparsely to more commonly densely pubescent with short whitish to gray woolly hairs, also inconspicuously gland-dotted. Inflorescences 8–17 cm long, mostly in the upper leaf axils, commonly in clusters of 7–9, the flower stalks 0.8–1.2 mm long. Calyces with the tube 1.5–2.0 mm long, the lobes 1.5–2.0 mm long, all similar in size and shape. Corollas with the banner 4–5 mm long, 2.0–2.5 mm wide, obovate, folded around stamens and pistil, bluish purple to purple. Stamens with the free portion of the filaments 4–5 mm long, the anthers 0.3–0.5 mm long, yellow. Ovary 1.0–1.5 mm long, densely hairy, the style 2–3 mm long, glabrous or densely hairy. Fruits 3–4 mm long, 1.2–1.5 mm wide, exserted beyond the persistent calyx tube, hairy and gland-dotted. Seeds 2.0–2.4 mm long, 1.0–1.4 mm wide, olive to reddish brown. $2n=20$. May–August.

Scattered nearly throughout the state (Montana to New Mexico east to Michigan and Louisiana; Canada). Upland prairies, loess hill prairies, glades, tops of bluffs, savannas, and openings of dry upland forests; also pastures, railroads, and roadsides.

This showy species is a characteristic element of high-quality prairies in Missouri. In the typical form, the whitish gray plants with their contrasting purple inflorescences are striking and easily recognized. However, occasional plants with much sparser pubescence may be encountered nearly throughout the range. One such nearly glabrous variant in southwestern Missouri that also had slightly broader fruits and somewhat broader leaflets was described as a separate species, *A. brachycarpa* (Palmer, 1931), but too many intermediates with typical *A. canescens* exist to allow formal taxonomic recognition of this form (Wilbur, 1964, 1975; Isely, 1998). For a discussion of a putative hybrid with *A. fruticosa*, see the treatment of that species.

Amorpha canescens was once considered an indicator of lead ore in Wisconsin and Illinois. The flowers are protogynous, that is, the style matures in the flowers before the stamens. This beautiful species is well-adapted to sunny sites in the garden and is available commercially through many wildflower nurseries.

2. Amorpha fruticosa L. (false indigo)
A. fruticosa var. *angustifolia* Pursh
A. fruticosa var. *croceolanata* (P.W. Watson) Mouill.
A. fruticosa var. *emarginata* Pursh
A. fruticosa var. *oblongifolia* E.J. Palmer
A. fruticosa var. *tennesseensis* (Shuttlew. ex Kunze) E.J. Palmer
Pl. 388 d, e; Map 1709

Plants shrubs. Stems 1–2(–4) m tall, several-branched, glabrous or sparsely to moderately pubescent with minute curved and/or short straight hairs, not glandular or sparsely and inconspicuously gland-dotted. Leaves with 9–25 (often 17–19) leaflets, the petiole 10–30 mm long, longer than the width of the lowermost leaflet, the rachis 9–19 cm long, the petiole and rachis moderately short-hairy, often becoming nearly glabrous with age, occasionally sparsely pustular gland-dotted. Leaflets 20–45 mm long, 5–17 mm wide, oblong to elliptic, rounded or angled at the base and with a stalk 1–2 mm long, rounded to minutely notched at the tip but usually with a minute sharp point, the upper surface glabrous or nearly so, the undersurface sparsely to moderately short-hairy and pustular gland-dotted. Inflorescences 8–18 (–25) cm long, mostly in the upper leaf axils, solitary or in clusters of 3–9, the flower stalks 1–2 mm long. Calyces with the tube 2.0–2.4 mm long,

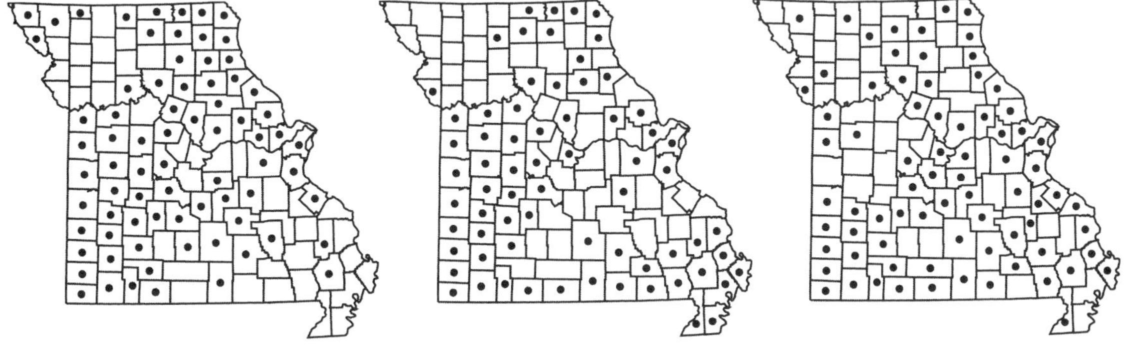

1708. Amorpha canescens 1709. Amorpha fruticosa 1710. Amphicarpaea bracteata

the lobes 0.2–1.0 mm long, the lowermost lobe conspicuously longer than the other 4 lobes. Corollas with the banner 4–5 mm long, 2.5–3.0 mm wide, obovate, arched, folded around the stamens and pistil, dark bluish purple to dark purplish blue. Stamens with the free portion of the filaments 3–4 mm long, the anthers yellowish orange to orange. Ovary 1–2 mm long, usually glabrous, the style 4–5 mm long, glabrous or more commonly with ascending hairs. Fruits 4–7 mm long, 1.6–2.0 mm wide, strongly exserted beyond the persistent calyx, glabrous but prominently pustular gland-dotted. Seeds 3.5–4.0 mm long, 1.4–1.6 mm wide, tan to reddish brown. $2n$=40. May–June.

Scattered nearly throughout the state (nearly throughout the U.S.; Canada, Mexico). Banks of streams and rivers, margins of ponds, lakes, and sloughs, bottomland prairies, bases and ledges of bluffs, and openings of swamps and bottomland forests; also ditches, railroads, roadsides, and moist disturbed areas.

In Missouri, *A. fruticosa* does well in gardens and is available commercially through wildflower nurseries. However, although widespread in the United States, this species is not native to New England and the Pacific Northwest, where it is considered an invasive exotic (Glad and Halse, 1993). In some northeastern states, *A. fruticosa* has escaped from cultivation to become a pest in sensitive riparian areas (Lapin, 1995). The plants propagate clonally by stems and roots washed up during floods, and the seeds have delayed germination, making them difficult to control.

This wide-ranging species is quite variable in the number and size of the leaflets, shape and size of fruits, and pubescence. Numerous varieties and forms have been described (Palmer, 1931; Steyermark, 1963; Wilbur, 1975). However, patterns of variation are not correlated with ecology, geography, or other morphological characters. The infraspecific taxa intergrade freely and are often difficult to distinguish in natural populations. For these reasons none of the infraspecific taxa is recognized in the present treatment.

Several closely related species in the *A. fruticosa* complex are recognized in states adjacent to Missouri (Wilbur, 1975; Isely, 1998), and the question arises as to their potential presence here. The taxonomy of the group requires more intensive study, and most of the segregates may not warrant recognition at the species level, if at all. Among them, *A. nitens* F.E. Boynton (shining false indigo) is the most likely to be discovered in the state and should be searched for in southern Missouri. This southeastern taxon is known from adjacent counties in Illinois, Kentucky, Tennessee, and Arkansas. It is said to differ from *A. fruticosa* in its generally fewer (mostly 9–15), usually slightly larger leaflets that are often somewhat shiny when fresh and more rounded at the tip. It is also unusual within the genus in that pressed samples tend to blacken upon drying.

A putative hybrid between *A. fruticosa* and *A. canescens* was reported by Palmer (1953), which he named *Amorpha* ×*notha* E.J. Palmer. This report was based on a locality near Webb City (Jasper County). Both parental species occurred in the same area, and the hybrid was growing between an upland prairie and an alluvial creek valley. The hybrid resembles *A. canescens* in its dense whitish gray pubescence, but is taller, has shorter calyx lobes, and a strongly curved fruit. Steyermark (1963) discussed additional morphological features in support of the hybrid status of the plant. Co-occurrence of the two parents is very uncommon, thus hybridization between the two species should be extremely rare.

15. **Amphicarpaea** Elliott ex Nutt.

Three species, North America, Europe, Asia, Africa.

The generic name was spelled *Amphicarpa* in much of the older botanical literature, but the spelling *Amphicarpaea* has been officially conserved.

1. Amphicarpaea bracteata (L.) Fernald (hog peanut)

A. bracteata var. *comosa* (L.) Fernald
A. monoica Nutt.
Glycine bracteata L.

Pl. 388 a–c; Map 1710

Plants annual, with a shallow taproot, frequently climbing on other vegetation. Stems (0.5–) 1.0–2.0 m long, twining, loosely spreading or climbing, unarmed, glabrous to densely pubescent with pale yellow to brownish yellow downward-pointed to spreading hairs. Leaves alternate, pinnately trifoliate, the petiole 2–5 cm long, hairy. Stipules 3–5 mm long, membranous or sometimes papery with age, persistent, broadly lanceolate to ovate, angled or tapered to a sharply pointed tip, the venation prominent; stipels 1–2 mm long, inconspicuous, often shed early. Leaflets (1.5–)3.0–7.0(–10.0) cm long, (1–)2–6(–7) cm wide, ovate to rhombic, broadly angled or rounded at the base, angled to a bluntly or less commonly sharply pointed tip, the margins entire, the surfaces sparsely to moderately hairy, more or less pinnately veined (with a pair of well-developed lateral veins produced at the base, especially on lateral leaflets). Terminal leaflet with the stalk 5–20 mm long, symmetric at the base; lateral leaflets with the stalk 1–2 mm long, usually asymmetric at the base. Inflorescences of 2 or 3 kinds, with 3 types of flowers and fruits (aerial and open-flowering, aerial and cleistogamous, and subterranean and cleistogamous). Open flowers in short dense racemes from the axils of upper leaves, sometimes appearing clustered, the inflorescence stalk 1–2 cm long, hairy, the bracts 2–3 mm long, 2–3 mm wide, fan-shaped, strongly veined, broadly angled at the tip, the flower stalks 1–2 mm long. Cleistogamous flowers in short dense racemes from the median stem nodes or solitary or few in open racemelike inflorescences from the lower stem nodes (or sometimes from subterranean nodes of the rootstock), the stalk of such inflorescences 10–30 cm or more long, stoloniferous in appearance, the flowers developing at the soil surface or underground, subtended by 2 small bracts. Calyces of open flowers 5-lobed, the lobes much shorter than the tube, similar in length but the upper 2 lobes fused nearly to the tip, the tube 4–6 mm long, somewhat asymmetric (pouched) at base, sparsely to densely hairy, the lobes 1–3 mm long, triangular, tapered to a sharply pointed tip, usually sparsely hairy; calyces of cleistogamous flowers reduced, the lobes about as long as the tube, 1–2 mm long, not fully opening at maturity. Corollas papilionaceous (absent or highly reduced in cleistogamous flowers), glabrous, white, tinged with pale pink or lavender, the petals with a stalklike base; the banner 11–14 mm long, 5–6 mm wide, obovate, abruptly bent outward above the stalklike basal portion; the wings 10–13 mm long, 2–3 mm wide, oblong, the blade portion with a pair of small auricles at the base; the keel 10–12 mm long, 2.5–3.0 mm wide, oblanceolate and straight in outline, boat-shaped. Stamens 10 (only 2–5 short stamens in cleistogamous flowers), 9 of the filaments fused and 1 free (often all free in cleistogamous flowers), the fused portion 8–10 mm long, the free portion 1–2 mm long, curved, the anthers small, attached near the midpoint, all similar in size. Ovary 4–5 mm long, linear-elliptic, short-stalked, hairy, the style 6–7 mm long (short and strongly recurved in cleistogamous flowers), slender, curved, glabrous, the stigma small and terminal. Fruits of 3 kinds: those maturing from open flowers 20–40 mm long, 6–8 mm wide, short-stalked above the persistent calyx, pointed at both ends, strongly flattened, glabrous or sparsely to moderately hairy (sometimes only along the margins), papery or somewhat leathery in texture, with (1)2–4 seeds, explosively dehiscent, the valves coiling after dehiscence; those of the aerial cleistogamous flowers similar but shorter and 1-seeded; those of the subterranean cleistogamous flowers 5–13 mm long, ovoid to ellipsoid or nearly globose, not flattened, spongy in texture, indehiscent. Seeds 3–5 mm long, 2–3 mm wide, kidney-shaped, shallowly notched, the surfaces smooth, glabrous, reddish brown or mottled lighter and darker brown. $2n=20$. August–September.

Scattered nearly throughout the state (eastern U.S. west to Montana and Texas; Canada; disjunct in southern Mexico). Banks of streams and rivers, fens, bottomland forests mesic upland forests, and bases of bluffs; also roadsides and shaded disturbed areas.

Amphicarpaea bracteata is variable in leaf size, stem robustness, and degree of hairiness, which has given rise to several named variants. Steyermark (1963) accepted two of these for Missouri. Plants growing in sunnier locations tend to be robust, with large leaves and dense, tawny hairs, and are sometimes recognized as var. *comosa*, whereas those in either sunny or shaded sites with smaller leaves and sparse pubescence have been called var. *bracteata*. These varieties were not accepted by B. L. Turner and Fearing (1964) because they often occur side by side, freely intergrade, and have no geographical correlation. However, preliminary comparative data from an allozyme study (Parker, 1996) indicated a genetic component to this morphological variation and morphometric analysis coupled with greenhouse and garden studies (Callahan, 1997) further suggested that the taxonomic distinctness of the two varieties should be studied in more detail. For the present, no attempt to distinguish varieties has been made during the present research.

Hog peanut has an interesting and somewhat complicated reproductive system. Individuals potentially produce more or less three kinds of inflorescences, although not always on the same plant. Open flowers with papilionaceous corollas are produced in short racemes from the upper nodes. These are pollinated primarily by bumblebees (Schnee and Waller, 1986). Reduced cleistogamous flowers are sometimes produced in short aerial racemes from the median nodes of some plants (a sort of intergradation into the next type). Both of these flower types produce flat, elongate, dehiscent legumes with two to four seeds. The seeds are dispersed ballistically when the valves suddenly dehisce and coil. Because of the height advantage, the seeds from open (chasmogamous) flowers are dispersed the greatest distance, possibly promoting cross fertilization (Trapp, 1988). Cleistogamous flowers also are sometimes produced from long thin runners that originate from the lowermost aboveground nodes of the plant. Thin shoots originating from these runners grow along the ground and into the soil to produce subterranean, cleistogamous flowers. Entirely subterranean cleistogamous flowers also can be produced from the underground nodes of the root system (Schnee and Waller, 1986). The fruits that develop from the subterranean flowers are large and fleshy with a relatively thin outer wall and a single fleshy seed. According to Steyermark (1963), these subterranean fruits are an important food source for mice and voles and were also gathered by Native Americans. He noted that when cooked, seasoned, and buttered they have a flavor similar to that of green beans. Production of the different flower and fruit types is correlated with the size of the plant and the relative amount of sun and shade.

Along some forest openings and margins large dense colonies of this species can develop in response to disturbance that appear to remain vegetative throughout the growing season. These thickets might be mistaken for the nonnative perennial *Pueraria* (kudzu), but *Amphicarpaea* is an annual (note the lack of dried stems from the previous season) that is much less noticeably hairy and has smaller leaves. Its population density usually declines within a few seasons. Steyermark (1963) also noted that vegetative specimens of *Amphicarpaea* might be confused with *Strophostyles umbellata*, but aside from the differences in flower and inflorescence morphology, the leaves of that species have a much shorter stalk at the base of the terminal leaflet and the stipules have fewer main veins (1–7 vs. 10–12).

16. Apios Fabr.

About 7 species, eastern North America, eastern Asia.

1. Apios americana Medik. (groundnut, American potato bean)

A. americana f. *pilosa* Steyerm.

A. americana var. *turrigera* Fernald

A. apios (L.) MacMill., an illegitimate name

A. tuberosa Moench

Pl. 388 j–l; Map 1711

Plants perennial herbs, frequently climbing on other vegetation, with white latex (the sap thus appearing milky), with long-creeping, subterranean rhizomes bearing few to many fleshy tubers, these 2–5 cm long, 1–2 cm wide, frequently appearing in chains. Stems 100–300 cm long or rarely longer, twining, forming loose mats or climbing, unarmed, glabrous to sparsely or occasionally densely pubescent with mostly short, fine hairs (but rarely with longer, more matted hairs). Leaves alternate, odd-pinnately compound with (3)5 or 7 leaflets, the petiole 1.5–7.0 cm long, glabrous or hairy, the hairs sometimes minute. Stipules 4–7 mm long, hairlike, shed early; stipels 1–2 mm long, inconspicuous, shed early. Leaflets 2–10 cm long,

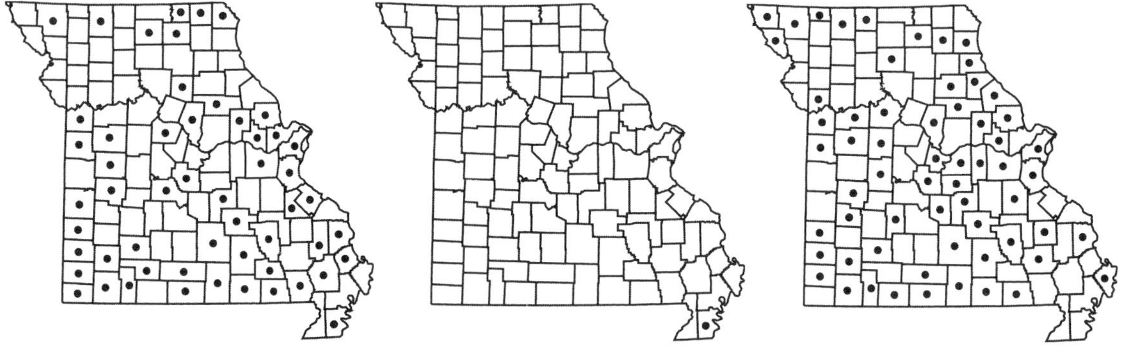

1711. Apios americana 1712. Arachis hypogaea 1713. Astragalus canadensis

2–6 cm wide, lanceolate to ovate, rounded at the base, angled or somewhat tapered to a sharply pointed tip, the margins entire, the surfaces glabrous or sparsely to moderately short-hairy, pinnately veined. Terminal leaflet with the stalk 15–30 mm long, symmetric at the base; lateral leaflets with the stalk 2–3 mm long, not or only very slightly asymmetric at the base. Inflorescences axillary racemes, these often relatively short and dense, the inflorescence stalk 2–5 cm long, hairy, the bracts 2–3 mm long, linear to lanceolate, shed early; bractlets 2–3 mm long, linear to narrowly lanceolate, closely subtending the flowers, shed early. Calyces irregularly 5-lobed, the tube 2.0–3.5 mm long, broadly bell-shaped to more or less hemispheric, glabrous or sparsely hairy, the lowest lobe up to 1/2 as long as the tube, broadly triangular, the other lobes narrower, very short or occasionally absent. Corollas papilionaceous, purple to brownish red, the banner usually pale purple to nearly white on the outer surface, 9.0–12.5 mm long, 10–14(–16) mm wide, the expanded portion broadly obovate, rounded to broadly notched at the tip, strongly concave and somewhat thickened apically (forming a small hoodlike structure), the wings 9.0–10.5 mm long, 4–5 mm wide, the expanded portion oblanceolate to obovate, downward angled and somewhat spreading, the keel 12–14 mm long, 2–4 mm wide, narrowly oblanceolate in outline and somewhat sickle-shaped, strongly curved upward, fused nearly to the rounded tip. Stamens 10, all of similar length, 9 of the filaments fused and 1 free, strongly curved, the fused portion 8–10 mm long, the free portion 1–3 mm long, the anthers small, attached near the base. Ovary 5–6 mm long, the style 1–3 mm long, relatively stout, strongly curved or coiled, the stigma capitate and terminal, covered by a roughened membrane (this ruptures and folds back during pollination). Fruits legumes, 4–6 cm long (to 10 cm elsewhere), 5–6 mm wide, linear, tapered (often asymmetrically so) at the tip, usually to a short beak, somewhat flattened, dehiscing by 2 valves, these green at maturity, twisting spirally after dehiscence, mostly 4- or 6-seeded. Seeds 4–5 mm long, 3.5–4.0 mm wide, broadly oblong-elliptic to oblong or occasionally nearly circular in outline, the surface dark brown, bluntly few-wrinkled, somewhat shiny. $2n$=22 (33 elsewhere). July–September.

Scattered nearly throughout the state, but less common north of the Missouri River (eastern U.S. west to North Dakota, Colorado, and Texas; Canada). Bottomland forests, bottomland prairies, banks of streams, rivers, and spring branches, margins of lakes and sinkhole ponds, fens, and marshes; also old fields, gardens, ditches, fencerows, roadsides, and moist, open disturbed areas.

Apios americana has an extensive rhizome system with numerous tubers, spaced evenly along the rhizome like beads on a string. The tubers are close to the surface and each is capable of sprouting to form a new stem when severed. They were an important food source for the Native Americans and settlers (Moerman, 1998). Tubers could be gathered at any season, but were most commonly collected in the autumn, often from rodent caches. They are sweet and are best when boiled or roasted like potatoes. Eating raw tubers is not recommended because they may contain toxic substances, which seem to be destroyed by cooking. Attempts have been made to domesticate and improve groundnuts (Blackmon and Reynolds, 1986). Although more nutritious than potatoes, groundnuts take two or three years to mature a crop of tubers. The beans can be eaten like peas and contain up to 18% protein by dry weight, although seed set in nature is relatively low. The plants also are eaten by livestock.

The flowers of *Apios* have an interesting pollination mechanism (Bruneau and Anderson, 1988). The stamens, pistil, and keel are strongly curved and held under pressure by a pocket in the basal fold of the banner. When an insect visitor trips the keel from its notch the stamens and pistil are forced out of the keel, depositing pollen on the insect and exposing the sticky stigma. Each flower can be tripped only one time. Abundant nectar is produced at the base of the pistil, and the flowers produce a sweet scent. Bees are the only insects that have been observed and recorded visiting and tripping the flowers (Bruneau and Anderson, 1994), but Westerkamp and Paul (1993) suggested that the maroon flower color and scent might be attractive to flies. Fruit set is everywhere rather low, and practically nonexistent in the northern part of the range (Seabrook and Dionne, 1976). Two chromosomal races have been found in *A. americana*. Populations north of Connecticut are more or less sterile triploids, set very few fruits, reproduce primarily by vegetative means, and may have been spread into this area by humans. Populations from the southern part of the range are mostly diploid and set more fruit (Bruneau and Anderson, 1988).

Harvey (2009) reported the presence of extrafloral nectaries in *A. americana* in populations in Georgia. These nectar-producing glands are located at the base of each flower cluster in the inflorescence and serve to attract ants. Harvey noted a correlation between the absence of ants on some plants and an increase in herbivory of fruits by insects. He suggested that elimination or reduction in density of ants by insecticides and other agricultural practices might have a negative effect on the ability of *A. americana* to reproduce by seeds.

Botanists in southeastern Missouri should search for the only other North American member of the genus, Price's groundnut, *A. priceana* B.L. Rob. This Federally Threatened species is found sporadically in portions of Kentucky, Tennessee, Mississippi, and Alabama (and historically in southern Illinois). It has a solitary tuber and no rhizomes, leaves with longer petioles (6.5–7.5 cm), large (2.0–2.5 cm long), greenish white to purplish green flowers, the banner with a fleshy apical knob, and longer fruits (12–18 cm) (M. Woods, 2005).

17. **Arachis** L.
(Valls and Simpson, 2005; Krapovickas and Gregory, 2007)

About 69 species, South America; cultivated widely.

1. Arachis hypogaea L. (peanut, groundnut)
Pl. 389 e; Map 1712

Plants annual, with stout taproots. Stems 10–35 cm long, erect to loosely ascending or trailing with runners, sometimes branched at the lower nodes, unarmed, sparsely to densely pubescent with fine, yellow to tan, mostly spreading hairs 1–2 mm long. Leaves alternate, 8–12 cm long, even-pinnately compound, the petiole 3–6 cm long, the rachis 4–9 cm long. Stipules 20–30 mm long, 2–3 mm wide, narrowly lanceolate, fused to the petiole toward the base, long-tapered at the tip, persistent and conspicuous, the venation prominent; stipels absent. Leaflets 4 in 2 pairs, 3–5 cm long, 2–3 cm wide, obovate, bluntly angled at the base, rounded to bluntly pointed at the tip, sometimes shallowly notched, the margins entire, the upper surface glabrous, the undersurface glabrous to sparsely hairy, pinnately veined. Inflorescences axillary, solitary or more commonly few-flowered spikelike clusters, but these often appearing as solitary flowers because only 1 flower develops at a time, each with a pair of conspicuous basal bracts, these 10–12 mm long, lanceolate, sheathing, hairy toward the base; bractlets absent. Flowers sessile (becoming long-stalked as the fruits develop), but appearing stalked because of a slender hypanthium below the perianth, this 2–3 cm long, hairy. Calyces deeply 2-lipped, the upper lip 5–6 mm long, with 4 short toothlike lobes, the lower lip 5–7 mm long, unlobed. Corollas papilionaceous, orange or yellow, the banner 10–12 mm long, 8–9 mm wide, the expanded portion broadly ovate, rounded and usually shallowly notched at the tip, the wings 7–8 mm long, 3–4 mm wide, obovate, the keel 5–6 mm long, 1.0–1.5 mm wide, boat-shaped, tapered at the tip, sharply curved upward. Stamens 10 but usually 2 of these nonfunctional and with small capitate tips, the filaments all fused, the fused portion attached at the tip of the hypanthium, 3–4 mm long, the free portion 3–4 mm long (shorter in nonfunctional stamens), the anthers of fertile stamens 0.7–1.0 mm long, oblong, attached near the midpoint. Ovary 3–4 mm long, more or less ovoid, the style 23–35 mm long, hairy, the stigma minute, terminal. Fruits developing underground following development of a long stalk at the ovary base after flowering, this burying the developing fruit, loments, 2–4 cm long, 1.0–1.5 cm wide, oblong, circular in cross-section, constricted between the seeds, the surface light brown, with a coarse, fibrous network, (1)2(3)-seeded, indehiscent, not breaking up into segments. Seeds 8–12 mm long, 4–6 mm wide, ovoid with a thin, papery, reddish brown seed coat. $2n=40$. July–September.

Introduced, uncommon in the Mississippi Lowlands Division (cultigen thought to have originated in Brazil, cultivated nearly worldwide; introduced sporadically in the southern U.S.). Roadsides and open disturbed areas.

Arachis hypogaea is distinguished from most other legumes (and most other plants) by the development of underground fruits, known as geocarpy. The flowers are self-fertile and sometimes cleistogamous. After fertilization occurs, the base of the ovary elongates forming a peg that bears the ovary underground. This requires hot summers and well-drained, friable, loamy or sandy soils. The seeds are very nutritious, with a protein content of 25–30% and an oil content of 45–50% (Sauer, 1993). The root nodules produce so much nitrogen that fertilization with inorganic nitrogen is seldom necessary. The cultivated peanut is unknown in the wild. It is an allotetraploid formed by the crossing of two wild diploid species, with subsequent chromosome doubling (Gregory et al., 1980; Smartt, 1990). Brazil was the primary center of peanut evolution and domestication. Peanuts were cultivated in South America as early as 2000 B.C., and later brought to the West Indies by the Arawaks. The Spanish introduced peanuts to Mexico and Africa, where they are an important crop with a large number of varieties. Peanuts are a major crop in Texas, Oklahoma, and several southeastern states, but are rarely planted in Missouri. The great American botanist, chemist, and educator, George Washington Carver (1864–1943), who was born in southwestern Missouri, was instrumental in promoting the uses of the peanut (Carver, 1916) and developed more than 300 commercial uses for the plant, its fruits, and its oil, revolutionizing agriculture in the southeastern United States as a result.

18. **Astragalus** L. (milk vetch, locoweed) (Barneby, 1964; Isely, 1983a, 1984, 1985, 1986a; Welsh, 2007)

Plants perennial herbs (annuals or shrubs elsewhere) sometimes woody at the base. Stems erect to prostrate, unarmed, glabrous or pubescent with unbranched or branched, hairs (these positioned flat along the stem and/or leaves and with 2 opposite branches, thus appearing as a straight line attached near the midpoint). Leaves alternate, odd-pinnately compound with 9 to numerous leaflets (sometimes trifoliate or simple elsewhere), short-petiolate. Stipules well-developed, free or fused to the petiole toward the base; stipels absent. Leaflets of various sizes and shapes, pinnately veined, but the lateral veins sometimes inconspicuous. Inflorescences axillary racemes, elongate and spikelike to short and dense (appearing as clusters), stalked, the bracts small and shed early; bractlets short and inconspicuous or absent. Flowers stalked, in some species appearing pendant or drooping. Calyces 5-lobed, the tube cylindric to bell-shaped, actinomorphic or zygomorphic, sometimes somewhat pouched basally on 1 side, the lobes usually shorter than tube, triangular. Corollas papilionaceous, purple, yellow, or white, the banner oblanceolate, longer than the wings, often curved or bent backward, rounded but often shallowly notched at the tip, the wings oblong, the keel shorter than wings, boat-shaped, fused, usually rounded or blunt at the tip, usually somewhat curved upward. Stamens 10, in an alternating set of 5 slightly shorter and 5 slightly longer stamens, 9 of the filaments fused and 1 free, the anthers attached near the base, mostly yellow to orange. Ovary sometimes short-stalked, the style straight or curved, glabrous, persistent at fruiting, the stigma minute, terminal. Fruits legumes, variable in shape from globose to elongate, often inflated, sometimes flattened, straight to arched or curved, with 1 or 2 locules (separated by a membranous partition), variously dehiscent or not. Seeds 1 to more commonly several to numerous, variously shaped, but most commonly shallowly notched at the attachment point. Perhaps 2300–2500 species (or more), nearly worldwide, except Australia; most diverse in arid and seasonally dry regions of the northern hemisphere.

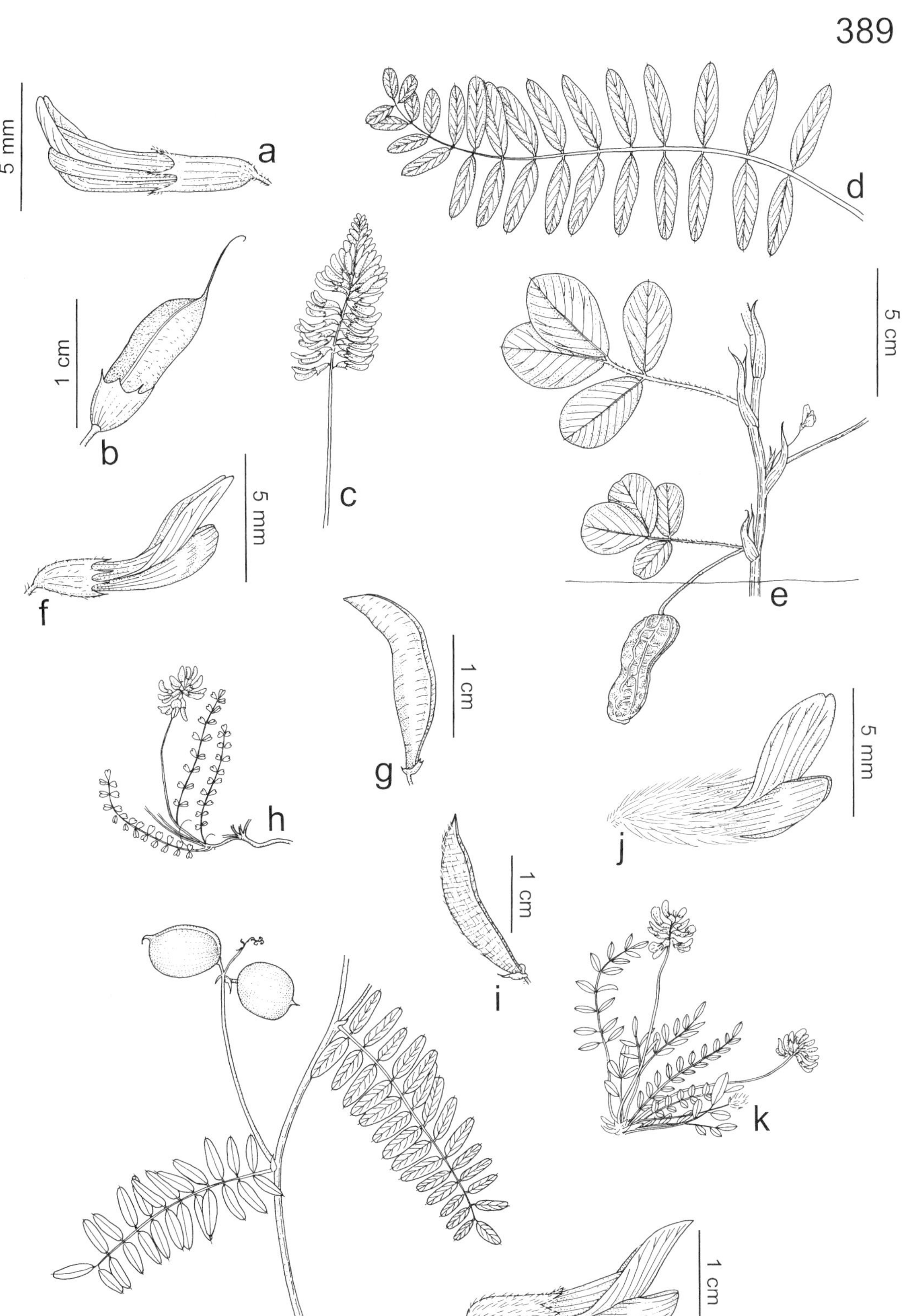

Plate 389. Fabaceae. *Astragalus canadensis*, **a)** flower, **b)** fruit, **c)** inflorescence, **d)** leaf. *Arachis hypogaea*, **e)** habit with subterranean fruit. *Astragalus distortus*, **f)** flower, **g)** fruit, **h)** habit. *Astragalus lotiflorus*, **i)** fruit, **j)** flower, **k)** habit. *Astragalus crassicarpus*, **l)** flower, **m)** fertile stem.

Astragalus is the largest genus of Fabaceae and one of the largest genera of flowering plants (Sanderson and Wojciechowski, 1996). The genus is very diverse in habit, pubescence, leaf morphology, and fruit characters. The majority of the approximately 375 species in the United States occurs as narrow endemics in arid, physiographically and geologically diverse regions in the western states, primarily in the Great Basin and Colorado Plateau. *Astragalus canadensis* is widespread across North America and closely related to some Old World species. Species such as *A. lotiflorus* and *A. crassicarpus* are found mostly in the central Great Plains, their ranges barely extending into western Missouri.

The large number of species in *Astragalus* makes it difficult to work with. It has been subdivided into smaller genera in the past (Rydberg, 1929), but most botanists currently recognize a single, very large genus (Barneby, 1964; Isely, 1998; Welsh, 2007). Most of the New World species are aneuploids with chromosome numbers of x=11–15. The Old World species are mostly euploids with a base chromosome number of x=8. Sequence data indicate that most of the New World species comprise a natural lineage (Wojciechowski et al., 1999). *Oxytropis* is very similar to *Astragalus*, but differs mainly in having the keel pointed at the tip and a different base chromosome number.

Many species of *Astragalus* and *Oxytropis* are toxic to humans and livestock. Toxic species are particularly problematic in western states, where ranching is prevalent and *Astragalus* is common. *Astragalus canadensis*, *A. crassicarpus*, and *A. lotiflorus* in Missouri are considered toxic. Consuming large amounts of plant material causes neurological damage and pronounced behavioral changes. Sick animals commonly become dazed, uncoordinated, frantic, or violent, and are referred to as loco, hence the common name locoweed. Animals often lose weight and may starve, or die of accidents, predation, or heart and respiratory failure. Fetuses may be aborted or born with deformities. There is no cure and damage is irreversible. The toxicity is caused primarily by several indolizidine alkaloids, especially swainsonine (Burrows and Tyrl, 2001). This alkaloid occurs throughout the plant, but levels are highest in the seeds and fruits. Even the pollen is toxic. Additional toxicity problems are associated with the production of 3-mitropropionic acid by some species, causing a condition known as cracker-heels for the sound made by the pelvic bones of afflicted animals. Some western species accumulate high levels of selenium from the soil, which is also toxic. Not all species of *Astragalus* are toxic, and some Asian species are considered valuable as forage plants. The Middle Eastern species *A. gummifer* Labill. and a few other Old World species are the source of gum tragacanth, a collection of water soluble polysaccharides that originates from the dried sap (Gentry, 1957) and that has a wide variety of uses, including as a burnishing compound in leatherworking, a textile stiffener, a binder in incense sticks and artists' pastels, and in the tobacco industry (as an adhesive for the paper or other outer layer in rolled cigars and cigarettes), as well as an emulsifier and thickener in foods and medicinally for coughs and burns.

Steyermark (1963) mentioned several additional species of *Astragalus* that had been reported from Missouri in the older botanical literature, but which he excluded because they could not be confirmed as growing in the state. The list included *A. gracilis* Nutt., *A. missouriensis* Nutt., *A. racemosus* Pursh, and *A. tennesseensis* A. Gray ex Chapm. As no evidence has been uncovered since Steyermark's treatment to support the addition of these species to the Missouri flora, they remain excluded for the present.

1. Stems mostly 60–150 cm long; stipules united at the base and sheathing the stems; inflorescences 12–30 cm long, elongate, spikelike racemes, with 30–70 flowers . . . . 1. A. CANADENSIS
1. Stems 5–60 cm long; stipules free, not sheathing the stem; inflorescences 1–8 (–12) cm long, subcapitate to somewhat elongated, appearing as clusters or short, spikelike racemes with 3–25 flowers

2. Stems 8–40(–60) cm long; calyces with the tube 3–9 mm long; fruits oblong-obovoid to nearly globose, fleshy when immature, becoming leathery with age, 2-locular . . . . 2. A. CRASSICARPUS
2. Stems 5–25 cm long; calyces with the tube 2–3 mm long; fruits elongate, tapered to sharply pointed ends, not fleshy, leathery or hardened, 1-locular
   3. Stems 8–25 cm long; leaflets with the undersurface sparsely pubescent with unbranched hairs, these mostly appressed, straight, attached at their bases (best observed carefully with magnification), occasionally glabrous; flowers all of one type, open, with well-developed petals, the corollas purple (rarely white); ovary glabrous; fruits arched or curved . . . . 3. A. DISTORTUS
   3. Stems 1–8(–12) cm long; leaflets pubescent with branched hairs, these appressed and with 2 opposite branches, thus appearing as a straight line attached toward the midpoint (best observed carefully with magnification); flowers of 2 types, the open ones with well-developed, greenish yellow petals, the cleistogamous ones (produced in short racemes from the stem base, either on separate plants or late in the growing season) with poorly developed corollas; ovary hairy; fruits more or less straight . . . . 4. A. LOTIFLORUS

1. Astragalus canadensis L. **var. canadensis** (rattleweed)

A. carolinianus L.

Pl. 389 a–d; Map 1713

Plants perennial herbs, with woody rhizomes, often colonial. Stems (30–)60–150 cm long, erect or strongly to loosely ascending, solitary or few per plant, often branched, reddish-tinged, sparsely pubescent with branched hairs, these appressed and with 2 opposite branches, thus appearing as a straight line attached near the midpoint (best observed carefully with magnification). Leaves with the petiole 10–30 mm long, with 15–31 leaflets, the rachis 14–24 cm long. Stipules 4–18 mm long, fused to the petiole toward the base and sheathing the stem, lanceolate to narrowly ovate, tapered to a sharply pointed tip, membranous when young, papery at maturity. Leaflets 10–50 mm long, 5–15 mm wide, lanceolate to oblong, elliptic, or (the smaller ones) ovate, rounded to less commonly bluntly pointed at the tip, the upper surface glabrous to sparsely hairy, the undersurface sparsely to densely branched-hairy. Inflorescences elongate, spikelike racemes, 12–30 cm long, the stalk 4–14 cm long, with 30–70 flowers, the bracts 3–5 mm long, hairy. Flowers slightly drooping, the stalks 1.0–2.5 mm long, hairy; cleistogamous flowers absent. Calyces with the tube 4–7 mm long, broadly cylindric, somewhat pouched, sparsely hairy, the lobes 1–4 mm long, linear. Corollas greenish yellow to cream-colored or white, sometimes tinged with purple or spotted with red, the banner 11–16 mm long, 5–6 mm wide, usually bent or curved upward, the wings 10–14 mm long, the keel 9–13 mm long. Stamens with the fused portion 7–8 mm long, the free portion 2–4 mm long. Styles 7–8 mm long. Fruits (10–)12–18 mm long, 5–7 mm wide, ovoid to oblong-ovoid, not curved, tapered to a beak at the tip, inflated, ascending to spreading, 2-locular, usually glabrous, yellowish brown, dehiscent tardily by apical valves. Seeds 1.5–2.0 mm long, the surface olive or reddish brown, smooth or nearly so. $2n=16$. June–August.

Scattered nearly throughout the state (nearly throughout the U.S. except for most southwestern states; Canada). Upland prairies, loess hill prairies, bases and ledges of bluffs, bottomland forests, and banks of streams; also pastures, railroads, and roadsides.

This was the first *Astragalus* species to be described from North America. It is one of the most widespread members of the genus and is related to a group of Old World taxa. It is recognized by the combination of relatively large size, pubescence of branched hairs (high magnification necessary), large leaflets, dense racemes of greenish yellow flowers, and bilocular fruits. Barneby (1964) recognized three varieties within the species. The widespread var. *canadensis*, which is the only one of the varieties to occur east of the Great Plains, has glabrous to hairy fruits lacking a longitudinal groove. Western plants with tall, slender stems, dorsally grooved, sparsely hairy fruits, and relatively short calyx teeth have been called var. *brevidens* (Gand.) Barneby. Short, stout plants with grooved, moderately hairy fruits and relatively prominent calyx teeth that grow in portions of Washington, Oregon, Idaho, and Montana have been called var. *mortonii* (Nutt.) S. Watson.

Astragalus canadensis apparently is palatable to livestock and deer, but has been suspected to cause locoism (Burrows and Tyrl, 2001).

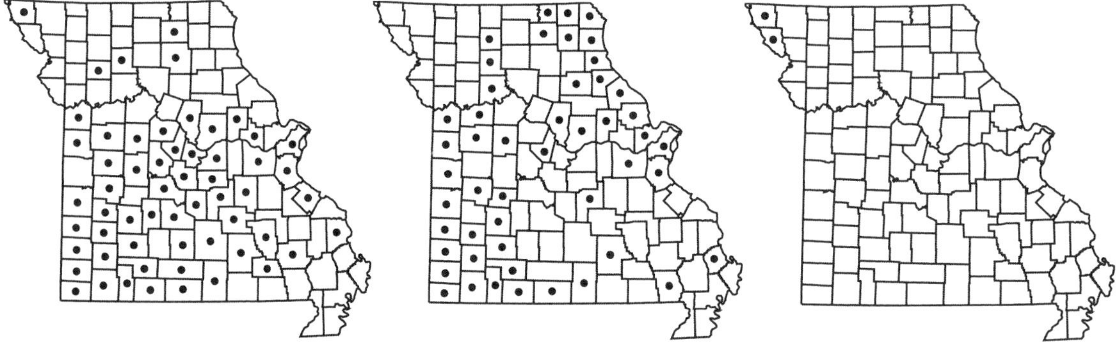

1714. Astragalus crassicarpus 1715. Astragalus distortus 1716. Astragalus lotiflorus

2. Astragalus crassicarpus Nutt. (ground plum)

A. caryocarpus Ker Gawl.

Pl. 389 l, m; Map 1714

Plants perennial herbs, with a thick woody taproot below a branched caudex. Stems 8–40(–60) cm long, erect or strongly to loosely ascending, sometimes from a spreading base, usually several per plant, unbranched or few-branched toward the tip, tan to grayish tan, sparsely to moderately pubescent with unbranched hairs, these appressed or strongly ascending, more or less straight to curved and woolly, attached at the base (best observed carefully with magnification). Leaves with the petiole 3–10 mm long, with 15–27(–33) leaflets, the rachis 5–12 cm long. Stipules 3–9 mm long, free from the petiole and not sheathing the stem, lanceolate to ovate, tapered to a sharply pointed tip, membranous when young, papery at maturity. Leaflets 3–17 mm long, 3–7 mm wide, oblanceolate to oblong, oblong-elliptic, or (the smaller ones) obovate to nearly circular, rounded to more commonly bluntly to sharply pointed at the tip, the upper surface glabrous to sparsely hairy, the undersurface moderately to densely branched-hairy. Inflorescences relatively short, spikelike racemes, 4–8(–12) cm long, the stalk 2–9 cm long, with 5–25 flowers, the bracts 2–5 mm long, usually hairy. Flowers ascending to spreading, the stalks 1–5 mm long (sometimes longer at fruiting), hairy; cleistogamous flowers absent. Calyces with the tube 3–9 mm long, cylindric to narrowly bell-shaped, not or only slightly pouched, moderately to densely hairy, the lobes 1–3 mm long, triangular. Corollas purple to bluish purple or pinkish purple, or greenish yellow to white, the keel and sometimes also the wings usually darker purple or purplish-tinged than the banner, the banner 14–24 mm long, 4–6 mm wide, somewhat curved upward to nearly straight, the wings 13–17 mm long, the keel 11–14 mm long. Stamens with the fused portion 9–12 mm long, the free portion 1–2 mm long. Styles 6–8 mm long. Fruits 15–28 mm long, 12–20 mm wide, oblong-obovoid to nearly globose, not curved, tapered abruptly to a slender beak at the tip, inflated, fleshy when immature, becoming leathery with age, spreading to drooping, 2-locular, glabrous, often reddish- to purplish-tinged, dehiscent tardily or indehiscent (dehiscing through decay after being shed). Seeds 2.0–2.2 mm long, the surface dark brown to black, smooth or nearly so. $2n=22$. April–May.

Scattered widely in the state, but apparently uncommon in portions of the Glaciated Plains Division and absent from the Mississippi Lowlands (Wisconsin to Montana south to Louisiana and Arizona; Canada).

Astragalus crassicarpus is recognized by the globose, bilocular, cherry-like fruits with a pronounced stylar beak. The young fruits are succulent and sweet, and can be consumed raw or boiled. They were eaten by both Native Americans and settlers, but given the potential problem of locoism, eating this plant in large quantities is not advisable. The fruits are also gathered, cached, and eaten by rodents.

Barneby (1964) treated *A. crassicarpus* as a morphologically variable species comprising five varieties, only two of which occur in Missouri. The other three varieties occur to the west of Missouri and differ variously in growth form, corolla color, and fruit shape. The two varieties recognized here have at times been treated as distinct species (Steyermark, 1963). *Astragalus crassicarpus* var. *crassicarpus* appears to be smaller and more prostrate than *A. crassicarpus* var. *trichocarpus*. The fruits in both varieties are almost identical, and there is some overlap in the corolla colors. The hairs on the calyx are distinctively different.

1. Plants relatively low-growing, the stems mostly 8–20 cm long, mostly loosely ascending, usually from a spreading base; inflorescence stalks 2–5 cm long; calyces sparsely to moderately pubescent with straight more or less appressed hairs, these mostly black or sometimes a mixture of black and white hairs; corollas usually purple to bluish purple . . . . 2A. VAR. CRASSICARPUS
1. Plants forming taller clumps, the stems 20–40(–60) cm long, mostly erect or strongly ascending (sometimes becoming more loosely ascending at fruiting); inflorescence stalks 5–9 cm long; calyces densely pubescent with fine, woolly hairs, these white to creamy white; corollas usually greenish yellow to creamy white (the tips sometimes purplish-tinged) . . 2B. VAR. TRICHOCALYX

2a. var. crassicarpus

Plants relatively low-growing, the stems mostly 8–20 cm long (often becoming greatly elongated after flowering), mostly loosely ascending, usually from a spreading base. Leaves with the rachis 5–8 cm long, with mostly 15–21 leaflets. Inflorescences mostly relatively short and dense, the stalk 2–5 cm long, densely hairy. Calyces with the tube 3–7 mm long, sparsely to moderately pubescent with straight, more or less appressed hairs, these mostly black or sometimes a mixture of black and white hairs. Corollas usually purple to bluish purple, rarely greenish white with purplish-tinged tips, the banner 14–24 mm long. Fruits mostly 2.0–2.8 cm long. $2n=22$. April–May.

Scattered in the Unglaciated Plains Division and the western half of the Glaciated Plains, uncommon or absent elsewhere in the state (Wisconsin to Montana south to Arkansas and New Mexico). Upland prairies, loess hill prairies, and glades; also roadsides.

2b. var. trichocalyx (Nutt. ex Torr. & A. Gray) Barneby ex Gleason

A. trichocalyx Nutt. ex Torr. & A. Gray

A. caryocarpus Ker Gawl. var. *trichocalyx* (Nutt. ex Torr. & A. Gray) Fernald

A. mexicanus A. DC. var. *trichocalyx* (Nutt. ex Torr. & A. Gray) Fernald

Plants forming bushy clumps, the stems mostly 20–40 cm long (becoming elongated to 30–60 cm at fruiting), erect to moderately ascending (sometimes reclining as the fruits mature). Leaves with rachis 8–12 cm long, with mostly (17–)21–27(–33) leaflets. Inflorescences somewhat more elongate than in var. *crassicarpus*, the stalk 5–9 cm, sparsely hairy. Calyces with the tube 6–9 mm long, densely pubescent with fine, woolly hairs, these white to creamy white. Corollas usually greenish yellow to creamy white (the tips sometimes purplish-tinged), the banner usually 14–18 mm long. Fruits mostly 1.5–2.5 cm long. April–May.

Scattered in the Ozark, Ozark Border, and Unglaciated Plains Divisions (Illinois to Kansas south to Louisiana and Texas). Glades, ledges and tops of bluffs, upland prairies, and openings of mesic to dry upland forests; often on calcareous substrates.

3. Astragalus distortus Torr. & A. Gray **var. distortus** (Ozark milk vetch, bent milk vetch)

Pl. 389 f–h; Map 1715

Plants perennial herbs, with a stout taproot below a short, branched caudex. Stems 8–25 cm long, prostrate to loosely ascending from a spreading base (sometimes forming mats), several per plant, several-branched, pale green to green, glabrous toward the base, sparsely pubescent with unbranched hairs, these appressed straight, attached at the base (best observed carefully with magnification). Leaves with the petiole 5–20 mm long, with 11–23(–27) leaflets, the rachis 6–10 cm long. Stipules 2–6 mm long, free from the petiole and not sheathing the stem, lanceolate to ovate, tapered to a sharply pointed tip, membranous when young, papery at maturity. Leaflets 3–11 mm long, 2–4 mm wide, obovate to elliptic-oblanceolate, rounded to truncate to shallowly notched at the tip, the upper surface glabrous, the undersurface sparsely pubescent with appressed, unbranched hairs or occasionally glabrous. Inflorescences short, spikelike racemes or appearing as capitate clusters, 2–10 cm long, the stalk 3–10(–15) cm long, with 5–15(–20) flowers, the bracts 2–3 mm long, usually hairy. Flowers spreading to somewhat drooping, the stalks 1–2 mm long (sometimes longer at fruiting), hairy; cleistogamous flowers absent. Calyces with the tube 2–3 mm long, narrowly bell-shaped, slightly pouched, pubescent with black and/or white, unbranched, mostly appressed hairs, the lobes 1.0–1.5 mm long, narrowly triangular. Corollas light purple to purple, often whitish toward the base, occasionally all white, the banner 10–14 mm long, 4–5 mm wide, somewhat curved or bent upward, the wings 7–11 mm long, the keel 6–9 mm long. Stamens with the fused portion 5–6 mm long, the free portion 1–2 mm long. Styles 6–8 mm long. Fruits 15–28 mm long, 4–5 mm wide, cylindric, short-tapered to a short beak at the tip, strongly arched or curved, not or only slightly inflated, leathery at maturity, ascending to spreading, 1-locular, glabrous, often reddish- to purplish-tinged when young, dehiscent tardily or more commonly indehiscent (dehiscing through decay after being shed). Seeds 1.5–2.5 mm long, the surface brown to reddish brown, smooth or with faint, coarse wrinkles. April–June.

Scattered widely in Missouri, but apparently absent from the northwestern portion of the Glaciated Plains Division and uncommon in the eastern half of the Ozarks and Mississippi Lowlands (Kansas to Illinois south to Texas and Mississippi, disjunct from West Virginia to Delaware). Upland prairies, sand prairies, glades, ledges and tops of bluffs, and openings of mesic to dry upland forests; also pastures, quarries, railroads and roadsides.

Astragalus distortus is recognized by the short calyx tube and arched or bent fruits. The var. *distortus* is found nearly throughout the species range. Plants in the southern portion of the species range, from Texas to Louisiana and adjacent Arkansas, with corollas only 9–12 mm long and fruits 12–18 mm long and somewhat fewer seeds have been treated as var. *engelmannii* (E. Sheld.) M.E. Jones (Barneby, 1964). Plants with white-flowered corollas have been called f. *albiflorus* McGregor.

4. Astragalus lotiflorus Hook. (low milk vetch, lotus milk vetch)

Pl. 389 i–k; Map 1716

Plants perennial herbs (but sometimes flowering the first year), with a taproot, sometimes below a branched caudex. Stems 1–8(–12) cm long, sometimes apparently absent and the nodes short and densely clustered, mostly prostrate, but the inner stems often loosely ascending to erect, several per plant, unbranched or more commonly few-branched, green or purplish-tinged, glabrous toward the base, moderately to more commonly densely pubescent with branched hairs, these appressed and with 2 opposite branches, thus appearing as a straight line attached near the midpoint (best observed carefully with magnification). Leaves with the petiole 10–30 mm long, with 9–15(–21) leaflets, the rachis 4–9 cm long. Stipules 2–8 mm long, free from the petiole and not sheathing the stem, narrowly lanceolate to ovate, tapered to a sharply pointed tip, membranous when young, papery at maturity. Leaflets 4–15 mm long, 2–6 mm wide, narrowly oblanceolate to lanceolate, elliptic, or obovate, rounded to bluntly pointed at the tip, the upper surface sparsely branched-hairy, the undersurface moderately to densely branched-hairy (the foliage leaves similar to those of the stem). Inflorescence of 2 types (produced on separate plants or those with cleistogamous flowers produced late in the season only from basal nodes), those with open flowers short, appearing as capitate clusters or short, dense, often more or less ovoid, spikelike racemes, 3–5 cm long, the stalk 2–6(–9) cm long, with 5–11 flowers; those with cleistogamous flowers appearing as capitate clusters or very short, spikelike racemes, 1–3 cm long, the stalk absent or to 2 cm long, with 2–4 flowers. Bracts 2–3 mm long, hairy. Flowers mostly spreading, the stalks 0.5–2.5 mm long, hairy. Calyces with the tube 2–3 mm long, narrowly bell-shaped, usually slightly pouched, densely hairy, the lobes 2–3 mm long, narrowly triangular. Corollas of open flowers well-developed, light greenish yellow, the banner 8–11 mm long, 4–5 mm wide, somewhat curved or bent upward, the wings 7–9 mm long, the keel 6–7 mm long; those of cleistogamous flowers reduced, not fully opening, white to pale cream-colored, the banner 4–7 mm long, the wings and keel 4–6 mm long. Stamens with the fused portion 4–5 mm long, the free portion 1.0–1.5 mm long. Styles 4–6 mm long. Fruits 16–25 mm long, 6–7 mm wide, cylindric, short-tapered to a short beak at the tip, straight or nearly so, not or only slightly inflated, leathery at maturity, ascending, 1-locular, hairy, sometimes reddish- to purplish-tinged when young, dehiscent from the tip, sometimes tardily or after being shed. Seeds 1.8–2.5 mm long, the surface brown to reddish grown, coarsely and bluntly wrinkled. $2n=26$. April–May.

Uncommon, known only from Holt and Atchison Counties (Montana to Minnesota south to New Mexico, Texas, and Arkansas; Canada). Loess hill prairies.

19. **Baptisia** Vent.
(Larisey, 1940; B. L. Turner, 2006)

Plants perennial herbs, often blackening upon drying, deeply rooted with thick, woody rhizomes and a woody crown. Stems erect or ascending to trailing or spreading, unarmed, mostly unbranched below the midpoint, glabrous or pubescent with short, fine hairs, sometimes glaucous. Leaves alternate, palmately trifoliate (the uppermost sometimes reduced to 1 or 2 leaflets, those toward the stem base usually reduced and scalelike), short- to long-petiolate, the lowermost and sometimes also the uppermost sessile or nearly so). Stipules variously poorly developed, inconspicuous, and shed early, or well-developed, similar to the leaflets, and persistent; stipels absent. Leaflets mostly oblanceolate to obovate, glabrous or hairy, sometimes glaucous, pinnately veined, the lateral veins sometimes inconspicuous. Inflorescences terminal or axillary racemes, stalked, the bracts herbaceous, small, lanceolate, and shed early, or larger, similar to the leaflets, and persistent; bractlets usually absent. Flowers short- to long-stalked. Calyces more or less 2-lipped, but usually appearing 4-lobed, the tube bell-shaped, slightly zygomorphic, the lobes usually shorter than the tube, the upper lobe broadly triangular to oblong-triangular, usually bluntly or broadly pointed at the tip, occasionally shallowly notched or shallowly 2-lobed, the other 3 lobes similar, triangular to broadly triangular, sharply pointed at the tips. Corollas papilionaceous, white, yellow, or purplish blue, the banner broadly obovate, to nearly circular or kidney-shaped, shorter than to about as long as the wings, somewhat curved upward, the sides curved or curled back, rounded or shallowly notched at the tip, the wings oblong, straight, the keel about as long as the wings, boat-shaped, fused toward their tips, usually rounded at the tip, straight or slightly curved upward. Stamens 10, all similar in length, the filaments not fused, the anthers attached near the base, mostly yellow. Ovary short-stalked, the style curved, glabrous, at least toward the tip, at least the basal portion persistent at fruiting, the stigma minute, terminal. Fruits legumes globose to oblong-ellipsoid or oblong-ovoid, sometimes asymmetrically so, strongly inflated, not flattened, straight, with 1 locule), dehiscent tardily, often from the tip, with few to more commonly numerous seeds. Seeds ellipsoid to irregularly kidney-shaped, usually shallowly notched at the attachment point, the surface more or less smooth, but with few to numerous pustular resinous dots. Fifteen to 17 species, temperate North America, most diverse in the eastern and central United States.

Baptisia is generally recognized by its trifoliate leaves that often blackening upon drying, racemes of large flowers, separate stamens, and greatly inflated fruits with a prominent beak. The species in Missouri are relatively easily distinguished. However, the taxonomy has been complicated by a series of nomenclatural changes at the level of both species and variety.

Baptisia has been the subject of important studies on interspecific hybridization. Larisey (1940) recognized eight hybrid taxa and noted that they are especially common in areas where the boundaries of widespread species overlap. In a groundbreaking chemosystematic study, Alston and Turner (1963) used flavonoid chromatography to study hybrids between four species of *Baptisia* in southeast Texas. They showed that the hybrids possessed a combination of chemical compounds found separately in the parental species.

Baptisia is closely related to the western genus *Thermopsis* R. Br., which differs mainly in having linear, uninflated fruits (B. L. Turner, 1981). The separate filaments in *Baptisia* are thought to be a primitive trait relating *Baptisia* with the relatively primitive woody genus *Sophora*, a relationship supported by some molecular evidence (Doyle et al., 1997). Chemically, *Baptisia* is similar to *Lupinus* and its relatives. *Baptisia* contains the quinolizidine lupine alkaloids cytisine and N-methylcytisine, as well as the pyridone alkaloid anagyrine in large quantities (Cranmer and Mabry, 1966). The use of *Baptisia* as a forage plant is limited by the presence of these bitter compounds. It is avoided by horses and cattle, although deer are said to eat the flowers.

Baptisia species are good indicators of former prairies, glades, and savannas. In wooded areas, they persist in openings and along edges for many years, but decrease as the canopy covers them. Several species are cultivated as ornamentals. Their deep taproots enable them to withstand drought and neglect. The generic name *Baptisia* originates from the Greek word *bapto*, meaning to dip or dye, in reference to the former use of these plants as a substitute for the blue dye plant indigo (*Indigofera* L.), hence the common name false or wild indigo.

1. Stems, leaves, and fruits persistently sparsely to densely and finely hairy; inflorescences spreading laterally, positioned below the level of the leaves; bracts 1–3 cm long, persistent; corollas cream-colored to pale yellow . . . . 3. B. BRACTEATA
1. Stems, leaves, and fruits glabrous or sparsely and inconspicuously hairy at maturity (the young growth short-hairy in *B. sphaerocarpa*); inflorescences erect or strongly ascending, positioned above the leaves; bracts 0.6–1.5 cm long, shed early; corollas white, blue to bluish purple, or bright yellow
   2. Stipules persistent; corollas blue to bluish purple; fruits with the beak tapered gradually, not hairlike, except toward the tip . . . . 2. B. AUSTRALIS
   2. Stipules mostly shed early; petals white or yellow; fruits with the beak tapered abruptly, hairlike nearly throughout
      3. Young (and mature) growth glabrous, somewhat glaucous; corollas white, the wings 18–23 mm long; mature fruits with the body 2.5–4.0 cm long, cylindric to ellipsoid . . . . 1. B. ALBA
      3. Young growth short-hairy, becoming glabrous or nearly so at maturity, not glaucous; corollas yellow, the wings 14–15 mm long; mature fruits with the body 1.2–2.0 cm long, subglobose . . . . 4. B. SPHAEROCARPA

1. Baptisia alba (L.) Vent. **var. macrophylla** (Larisey) Isely (white wild indigo)
B. lactea (Raf.) Thieret
B. leucantha Torr. & A. Gray
Pl. 390 e–g; Map 1717

Plants blackening upon drying, the herbage consistently glabrous. Stems 60–200 cm long, stout, finely ribbed, with ascending branches, somewhat glaucous. Leaves trifoliate, the petiole 4–25 mm long. Stipules (3–)5–25 mm long, lanceolate to narrowly ovate, mostly shed early. Leaflets 2–6(–8) cm long, oblanceolate to obovate or elliptic, rounded to bluntly or broadly pointed at the tips. Inflorescences 15–60 cm long, erect, positioned above the foliage, with many flowers, the stalk 4–12 cm long. Bracts 6–14 mm long, lanceolate to narrowly ovate, mostly shed early. Flowers with the stalk 3–14 mm long. Calyces with the tube 5–7 mm long, the lobes 2–3 mm long, the margins and inner surface finely white-hairy. Corollas white (sometimes mottled or tinged with purple), the banner 12–18 mm long, 10–16 mm wide, the wings 18–23 mm long, 4–7 mm wide, the keel 18–23 mm long, 5–8 mm wide. Stamens with the filaments 13–21 mm long, the anthers 1.2–1.4 mm long. Ovary 10–12 mm long, glabrous, the style 5–7 mm long. Fruits 2.0–3.5 cm long (excluding the stalklike base and beak), 10–15 mm wide, cylindric to ellipsoid, tapered to a stalklike base 10–12 mm long, tapered abruptly to the beak, this 5–7 mm long, hairlike nearly throughout, the walls relatively thin and brittle, glabrous and black at maturity. Seeds 4–5 mm long, 2.6–3.0 mm wide, greenish brown to yellowish brown with warty, resinous dots. $2n=18$. May–July.

Scattered nearly throughout the state, but apparently absent from most of the Mississippi Lowlands Division (New York to Minnesota south to Alabama and Texas; Canada). Bottomland prairies, upland prairies, savannas, glades, openings and edges of mesic to dry upland forests, tops of bluffs, and banks of streams; also pastures, railroads, and roadsides.

The name for this species has unfortunately changed several times (Isely, 1986b). For many years, it was known as *B. leucantha* (Steyermark,1963), with *B. alba* used for a different species from the southeastern United States that is distinguished by its wide fruits (Larisey, 1940). Thieret (1969a) argued that *Dolichos lacteus* Raf. was an older valid species name for B. *leucantha* and published the new combination, *B. lactea*. This name was used in several subsequent floristic works (McGregor, 1986; Gleason and Cronquist, 1991). Another change resulted when B. L. Turner (1982) viewed the type material in the British Museum upon which the name *B. alba* was based. Although the specimen lacked fruits, B. L. Turner determined that it was the same taxon as *B. lactea* (formerly *B. leucantha*). Because the name *B. alba* is the oldest validly published name in the group, it takes priority over both *B. lactea*

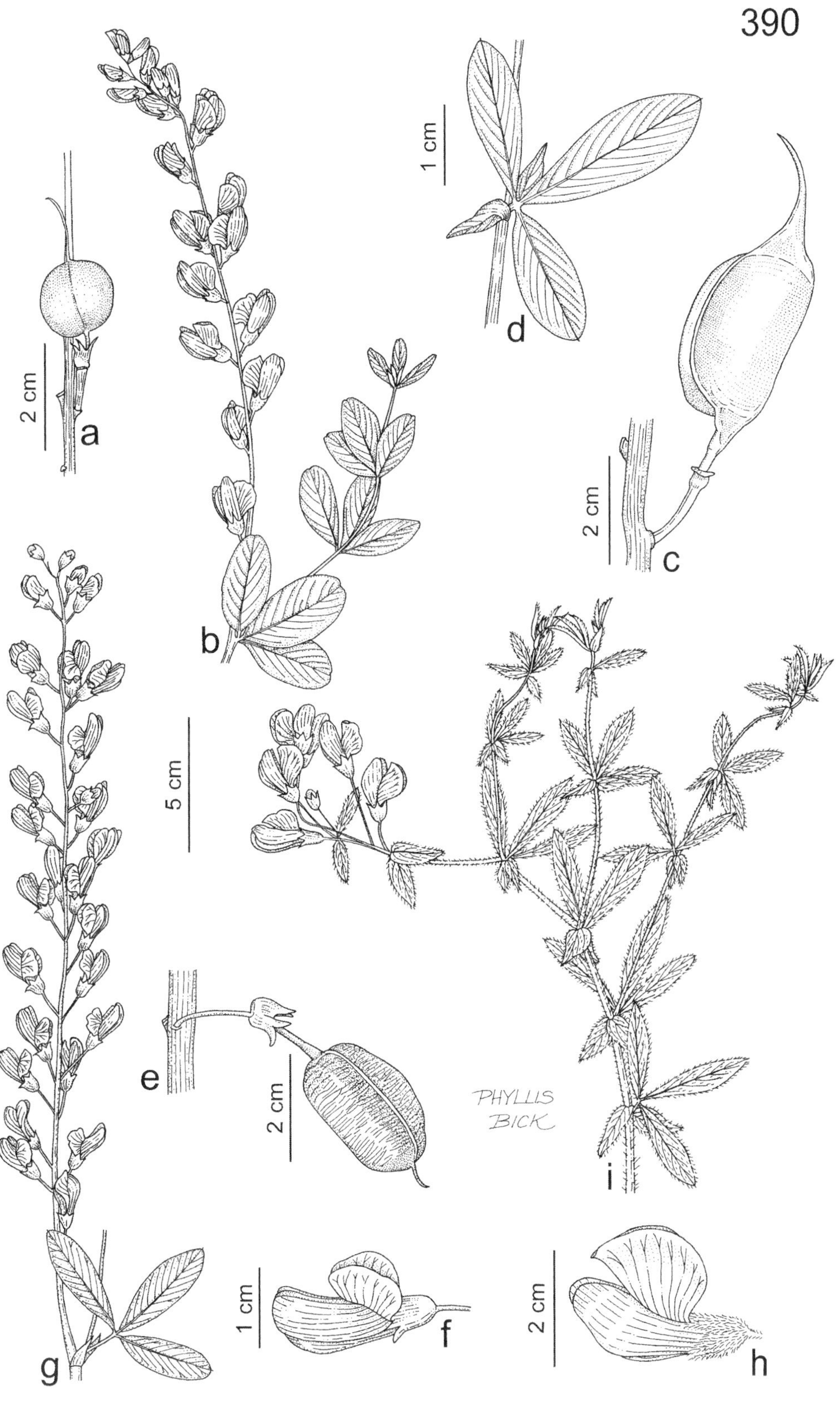

Plate 390. Fabaceae. *Baptisia sphaerocarpa*, **a)** fruit, **b)** branch with leaves and inflorescence. *Baptisia australis*, **c)** fruit, **d)** node with leaf and stipules. *Baptisia alba*, **e)** fruit, **f)** flower, **g)** branch with leaf and inflorescence. *Baptisia bracteata*, **h)** flower, **i)** branches with leaves and inflorescence.

1717. Baptisia alba

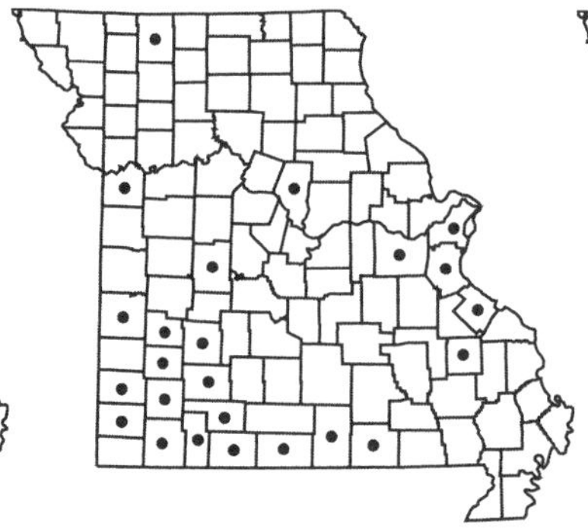

1718. Baptisia australis

1719. Baptisia bracteata

and *B. leucantha* when these are treated as a single species. The taxon formerly known as *B. alba* in the Southeast then became *B. albescens* Small (Isely, 1986b, 1990). Most recently, B. L. Turner (2006) has attempted to reapply the names involved, such that the name *B. albescens* becomes a synonym of *B. alba* and the name *B. lactea* becomes the correct name for the species that Isely and others have called *B. alba*. This reinterpretation of the type materials is complex and not at all clear. For now, the present treatment continues to follow that of Isely (1986b, 1990, 1998), in which the name *B. alba* corresponds to a widespread species of the eastern United States with broadly ellipsoid fruits, and the name *B. albescens* refers to a relatively uncommon, narrow-fruited species endemic to portions of the southeastern United States.

Two varieties of *B. alba* have been recognized by many botanists: var. *macrophylla* in Missouri and the central states; and var. *alba* along the Atlantic and Gulf Coastal Plains. In the northern portions of their respective ranges the varieties are easily distinguished, with var. *alba* differing mainly in its stouter (1.5–3.0 vs. 1.0–1.5 cm in diameter), thinner-walled fruits. However, where the distributions of the varieties overlap in the southern portion of their ranges, they exhibit some morphological overlap (Isely, 1990, 1998).

The pollination biology of *B. alba* and *B. bracteata* was studied and described by Haddock and Chaplin (1982). *Baptisia alba* flowers mostly in June and early July, somewhat later in the season than *B. bracteata*, when the surrounding vegetation is well-developed. There is about a week between flowering times when neither species is in flower. Individual flowers last three or four days, with most of the anthers dehiscing during the first two days, followed by development of the pistil. The major pollinators of both species are bumblebees (*Bombus* spp.). In *B. alba*, the bees visit the nectar-rich pistillate-phase flowers at the base of the upright racemes and work their way up to the nectar-poor staminate-phase flowers last, thereby promoting out-crossing. Both species of *Baptisia* suffer greatly from a variety of seed predators (Haddock and Chaplin, 1982).

For a discussion of putative hybridization between *B. alba* and *B. australis*, see the treatment of *B. australis*.

2. Baptisia australis (L.) R. Br. **var. minor** (Lehm.) Fernald (blue false indigo)

B. minor Lehm.

B. vespertina J. Small

Pl. 390 c, d; Map 1718

Plants blackening upon drying, the herbage consistently glabrous. Stems 40–60(–100) cm long, erect to loosely ascending or occasionally spreading, finely to moderately ribbed, with spreading branches, somewhat glaucous. Leaves trifoliate, the petiole 1–18 mm long. Stipules 5–20 mm long, lanceolate or rarely ovate, most of them persistent. Leaflets 2–4 cm long, oblanceolate to obovate or occasionally elliptic, rounded to bluntly or broadly pointed at the tips. Inflorescences 10–50 cm long, erect, positioned above the foliage, with relatively few to many flowers, the stalk 3–10 cm long. Bracts 6–9(–15) mm long, lanceolate to narrowly ovate, mostly shed early. Flowers with the stalk 3–16(–28) mm long. Calyces with the tube 4–7 mm long, the lobes 2–4 mm long, the margins and inner surface finely white-hairy. Corollas blue to bluish purple (the wings and keel sometimes yellowish-tinged), the banner 14–20 mm long, 12–20 mm wide, the wings 25–30 mm long, 7–9 mm wide, the keel 22–30 mm long, 7–10 mm wide. Stamens with the filaments 13–25 mm long, the anthers 1.2–1.4 mm long. Ovary 5–7 mm long, glabrous, the style 10–12 mm long. Fruits 3–5 cm long (excluding the stalklike base and beak), 18–25 mm wide, broadly cylindric to ellipsoid and appearing inflated, tapered to a stalklike base 4–10 mm long, tapered gradually to the beak, this 15–18 mm long, hairlike only toward the tip, the walls relatively thin and brittle, glabrous and black at maturity. Seeds 4–5 mm long, 2–3 mm wide, greenish brown to brown with warty, resinous dots. $2n=18$. April–June.

Scattered, mostly in the Ozark, Ozark Border, and Unglaciated Plains Divisions (Nebraska to Texas east to Indiana, Kentucky, and Arkansas; Canada). Bottomland prairies, upland prairies, glades, tops of bluffs, and banks of streams and rivers; also pastures and roadsides.

Baptisia australis is easily identified because it is the only species in the genus with blue corollas. Two varieties are sometimes recognized: var. *minor* is relatively short and spreading, with smaller leaves, and is found mostly in the central states; var. *australis* is relatively tall and erect, with large leaves, and occurs farther east in the species range. They have been treated as separate species (Larisey, 1940), but there is much intergradation between them.

Specimens with lighter flower color are sometimes encountered, and some of these have been named (Larisey, 1940; Kosnik et al., 1996). A hybrid between *B. australis* and *B. bracteata* has been found occasionally in Missouri and other states. This putative hybrid has been named *B.* ×*bicolor* Greenm. & Larisey and is usually found growing with both parental species. It is morphologically intermediate, with a blue banner, creamy yellow wings and keel, light pubescence, and inflorescence held at a 45-degree angle. Similarly, a putative hybrid of intermediate morphology between *B. australis* and *B. alba* has been collected in Polk County. Hybrids between *B. australis* and *B. sphaerocarpa* have been observed in Texas, with brick-red or multicolored flowers (Kosnik et al., 1996).

3. Baptisia bracteata Muhl. ex Elliott **var. leucophaea** (Nutt.) Kartesz & Gandhi (cream white indigo, long-bracted wild indigo, plains wild indigo)

B. bracteata Muhl. ex Elliott var. *glabrescens* (Larisey) Isely

B. leucophaea Nutt.

B. leucophaea var. *bracteata* (Muhl. ex Elliott) Isely

B. leucophaea var. *glabrescens* (Larisey) Isely

Pl. 390 h, i; Map 1719

Plants blackening upon drying, the herbage sparsely to densely pubescent with persistent, fine hairs. Stems 20–50 cm long, loosely ascending to spreading, finely to moderately ribbed, with mostly spreading branches, not glaucous. Leaves trifoliate, the uppermost rarely reduced to 1 or 2 leaflets, the petiole 1–9 mm long (but sometimes somewhat indistinct, because of winglike extensions of the lateral leaflets). Stipules 10–40 mm long, lanceolate to narrowly ovate, persistent (conspicuous, leaflet-like, usually with a noticeable network of veins). Leaflets 3–9 cm long, oblanceolate to narrowly obovate or occasionally elliptic, rounded to bluntly or sharply pointed at the tips. Inflorescences 10–30 cm long, spreading laterally, positioned below the foliage, with relatively few to many flowers, the stalk 3–6 cm long. Bracts 10–30 mm long, lanceolate to narrowly heart-shaped, persistent. Flowers with the stalk 10–40 mm long. Calyces with the tube 4–6 mm long, the lobes 3–5 mm long, the margins and outer surface finely hairy. Corollas cream-colored to pale yellow, the banner 16–22 mm long, 14–20 mm wide, the wings 22–26 mm long, 7–8 mm wide, the keel 20–26 mm long, 7–9 mm wide. Stamens with the filaments 18–20 mm long, the anthers 1.4–2.0 mm long. Ovary 8–12 mm long, hairy, the style 10–14 mm long. Fruits 3.5–5.0 cm long (excluding the stalklike base and beak), 14–25 mm wide, narrowly ellipsoid, tapered to a stalklike base 7–10 mm long, tapered gradually to the beak, this 15–20 mm long, hairlike toward the tip, the walls relatively thick and leathery to hardened, hairy when young, but sometimes nearly glabrous at maturity. Seeds 3–4 mm long, 2–3 mm wide, greenish brown to brown with warty, resinous dots. $2n=18$. April–June.

Scattered nearly throughout the state, but apparently absent from the Mississippi Lowlands Division (South Dakota to Texas east to Michigan and Mississippi). Upland prairies, glades, openings of mesic to dry upland forests, and banks of streams; also pastures, railroads, and roadsides.

Baptisia bracteata var. *leucophaea* is most easily identified by the large, persistent floral bracts and leaflet-like stipules. It also is distinguished by the hairiness of the leaves, stems, calyces, and/or fruits, the lateral inflorescences, long flower stalks, and large fruits with tapering tips. It tends to grow in low bushy clumps with the inflorescences hanging laterally or spreading along the ground. It matures in the late spring and flowers while the surrounding vegetation is still short.

For many years this taxon was known as *B. leucophaea* (Larisey, 1940, Steyermark, 1963). *Baptisia bracteata* was considered a closely related species of the Atlantic and Gulf Coastal Plains, distinguished by small differences in leaf shape and pubescence. Isely (1978) determined that *B. leucophaea* and *B. bracteata*, along with *B. leucophaea* var. *laevicaulis* A. Gray ex Canby of

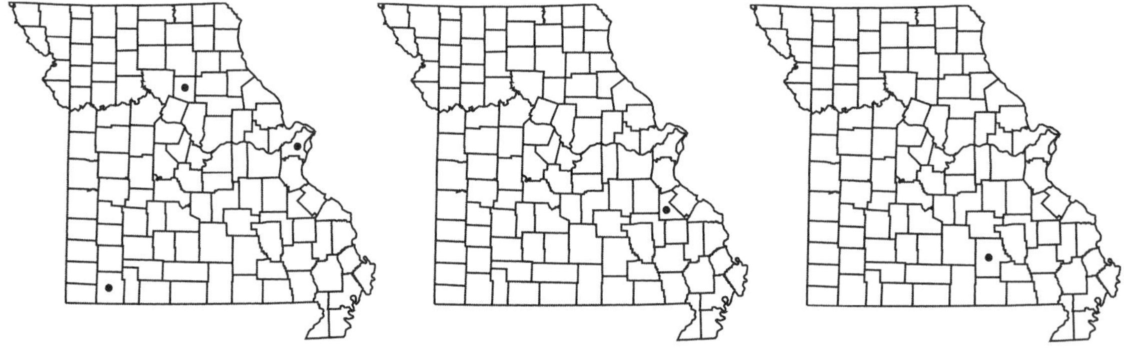

1720. Baptisia sphaerocarpa 1721. Canavalia ensiformis 1722. Centrosema virginianum

Texas and Louisiana, are all elements of a single species, and that *B. bracteata* is the oldest available specific epithet for the complex. The taxa were treated as three varieties of *B. bracteata* by Isely (1990, 1998), who used the name *B. bracteata* var. *glabrescens* for what was formerly *B. leucophaea*, and this name was taken up in several regional treatments (McGregor, 1986; Gleason and Cronquist, 1991). However, Kartesz and Gandhi (1991b) determined that the type specimen of *B. leucophaea* var. *glabrescens* actually fell within the morphological variation of var. *leucophaea*, a name automatically created when Larisey (1940) described *B. leucophaea* var. *glabrescens*. With the transfer of *B. leucophaea* to *B. bracteata* at the varietal level, the name *B. bracteata* var. *leucophaea* thus appears to be the correct name. B. L. Turner (2006) has continued to treat the two taxa as separate, closely related species.

The pollination biology of *B. bracteata* was studied by Haddock and Chaplin (1982). Individual flowers last about three to four days, with the stamens maturing before the pistillate parts. The flowers of *B. bracteata* produced less nectar than in *B. alba*. The major pollinators were queen bumblebees (*Bombus* spp.), which are actively engaged in nest-building when *B. bracteata* is flowering. *Baptisia alba* flowers later and is pollinated by worker bumblebees that are the offspring of these queens. The bees tended to land on the flowers with mature staminate parts and then move to the flowers with maturing pistillate parts, in contrast to the vertical movement of the bees on *B. alba*. *Baptisia bracteata* also suffers greatly from seed predation, but less so than in *B. alba*, perhaps because the maturing fruits are less visible (Haddock and Chaplin, 1982).

For a discussion of putative hybridization between *B. bracteata* and *B. australis*, see the treatment of *B. australis*.

4. Baptisia sphaerocarpa Nutt. (round-fruited yellow wild indigo)

B. viridis Larisey

Pl. 390 a, b; Map 1720

Plants turning olive brown upon drying, the young herbage short-hairy, becoming glabrous or nearly so at maturity. Stems 40–80 cm long, stout, finely ribbed, with loosely ascending to spreading branches, not glaucous, often reddish-tinged. Leaves trifoliate, the uppermost sometimes reduced to 1 or 2 leaflets, the petiole 2–4 mm long. Stipules 3–7 mm long, lanceolate to narrowly ovate, shed early. Leaflets 2–8 cm long, oblanceolate to obovate or elliptic, rounded to bluntly or broadly pointed at the tips. Inflorescences 10–30 cm long, erect or ascending, positioned above the foliage, with 5 to many flowers, the stalk 3–6 cm long. Bracts 4–7 mm long, lanceolate to narrowly ovate, shed early. Flowers with the stalk 20–60 mm long. Calyces with the tube 3–5 mm long, the lobes 3–4 mm long, the margins and inner surface finely white-hairy. Corollas bright yellow, the banner 15–18 mm long, 12–16 mm wide, the wings 14–15 mm long, 4–6 mm wide, the keel 11–18 mm long, 6–8 mm wide. Stamens with the filaments 14–16 mm long, the anthers 1.0–1.4 mm long. Ovary 4–6 mm long, minutely hairy, the style 10–12 mm long. Fruits 1.2–2.0 cm long (excluding the stalklike base and beak), 10–12 mm wide, subglobose, tapered to a stalklike base 7–8 mm long, tapered abruptly to the beak, this 10–13 mm long, hairlike nearly throughout, the walls thick, more or less fleshy when young, becoming woody or hardened at maturity, glabrous, brown or black. Seeds 4–5 mm long, 2.6–3.0 mm wide, yellowish green to yellowish brown with fine, warty, resinous dots. $2n$=18. May–June.

Uncommon, known thus far from Barry and Randolph Counties; introduced in St. Louis City and County (Texas to Mississippi, north to Oklahoma and Missouri). Upland prairies; also railroads and roadsides.

Baptisia sphaerocarpa is recognized by its erect racemes of bright yellow flowers and globose, woody fruits. There is some question as to its status in Missouri, given the relatively few specimens that mostly were collected at disturbed sites. Nevertheless, at least a few of the populations, including the one in Randolph County, appear to have been growing in natural plant communities. Steyermark's (1963) reports of this species in Greene and Ralls Counties could not be verified during the present study.

20. **Canavalia** Adans.
(Sauer, 1964)

About 60 species, tropical regions of the Old and New Worlds.

1. Canavalia ensiformis (L.) DC. (Jack-bean, wonder bean)

Dolichos ensiformis L.

Pl. 391 g–i; Map 1721

Plants annual or perennial herbs, bushy or more commonly reclining or climbing on other vegetation, the rootstock becoming potentially deep and spreading. Stems 80–200 cm long, variously erect to trailing, often twining, unarmed, glabrous. Leaves alternate, pinnately trifoliate, the petiole 6–10 cm long, glabrous or minutely hairy. Stipules 0.5–2.0 mm long, lanceolate to triangular, shed early; stipels 0.5–1.5 mm long, inconspicuous, shed early. Leaflets 9–15 cm long, 5–10 cm wide, ovate to elliptic, broadly angled or rounded at the base, rounded or angled to a bluntly pointed tip, sometimes with a minute, sharp extension of the midvein at the very tip, the margins entire, the surfaces glabrous or nearly so at maturity, but especially the undersurface often minutely hairy when young, pinnately veined. Terminal leaflet with the stalk 20–35 mm long, symmetric at the base; lateral leaflets with the stalk 5–7 mm long, often slightly asymmetric at the base. Inflorescences axillary clusters of mostly 4–8 flowers, these usually appearing racemose with swollen nodes, the inflorescence stalk 6–10 cm long, hairy, the bracts 1–2 mm long, lanceolate to ovate, shed early; bractlets 1–2 mm long, bluntly ovate, closely subtending the flowers, shed early. Calyces 2-lipped, the tube 6–7 mm long, asymmetrically bell-shaped, sparsely and minutely hairy, the upper lip 4–6 mm long, broadly rounded and usually shallowly notched at the tip, the lower lobe 5–7 mm long, shallowly 3-lobed, the lobes triangular. Corollas papilionaceous, pinkish or reddish purple, at least toward the tip, often paler or white toward the base, the banner 18–20 mm long, 16–18 mm wide, the expanded portion broadly obovate to nearly circular, rounded to shallowly notched at the tip, the margins relatively flat to curled backward, the wings 13–14 mm long, 4–5 mm wide, the expanded portion oblanceolate to obovate, positioned more or less parallel to the keel, the keel 14–15 mm long, 4–5 mm wide, oblanceolate in outline, somewhat curved upward, fused nearly to the rounded or bluntly pointed tip. Stamens 10, all of similar length, the filaments all fused, the fused portion 13–15 mm long, the free portion 4–5 mm long, the anthers small, attached near the base. Ovary 13–14 mm long, short-hairy, the style 8–10 mm long, glabrous, the stigma capitate and terminal. Fruits legumes, 10–30 cm long, 30–35 mm wide, narrowly oblong, straight to slightly arched, short-tapered (usually asymmetrically so) at the tip, usually to a short, blunt beak, flattened, the margins thickened, leathery to woody, more or less glabrous at maturity, dehiscing by 2 valves, these tan at maturity, twisting spirally after dehiscence, mostly 10–25-seeded. Seeds 20–22 mm long, 14–15 mm wide, oblong-elliptic to oblong, the surface pale yellow to nearly white with a brown line around the attachment point, smooth, usually not shiny. $2n=22$. June–September.

Introduced, known from a single historical specimen from St. Francois County (native of South America, cultivated worldwide; introduced sporadically in the southeastern U.S.). Habitat unknown but probably a disturbed area.

Sauer (1964) cited several specimens from the United States, however most of these populations probably have not persisted. Jack-bean is recognized by its robust bushy or viney habit, trifoliate leaves, and large legumes with large white seeds. Although not a traditional crop, Jack-bean is highly productive and commonly cultivated in tropical Asia and Japan. It is used for forage and green manure in crop rotations, grown separately or with other crops. The plants can be grown over wide ranges of temperature, rainfall, and elevation. They are deep-rooted and drought-resistant. The seeds sometimes are eaten by humans, but only with the proper cooking techniques (including several changes of water). They can cause severe and persistent diarrhea if eaten in large quantities. The seeds contain the toxic protein concanavalin, which binds to the mucous cells of the intestines, and the toxic, nonprotein amino acid canavanine. They also contain large amounts of the enzyme urease, which hydrolyzes urea to ammonia (Burrows and Tyrl, 2001).

21. Centrosema (DC.) Benth.

About 36 species, U.S. to South America, Caribbean Islands, introduced in the Old World.

1. Centrosema virginianum (L.) Benth.
(spurred butterfly pea)
Bradburya virginiana (L.) Kuntze

Map 1722

Plants perennial herbs, often reclining or climbing on other vegetation, with a branched rootstock. Stems 50–150 cm long, ascending to trailing, twining, unarmed, glabrous or sparsely pubescent with short, fine, straight hairs. Leaves alternate, pinnately trifoliate, the petiole 2–3 cm long, glabrous or short-hairy. Stipules 2–4 mm long, lanceolate, with prominent parallel venation; stipels 2–4 mm long, linear to hairlike, persistent. Leaflets 2.0–6.5 cm long, 10–25 cm wide, oblong-ovate to narrowly ovate, narrowly elliptic, or lanceolate, broadly angled or rounded at the base, angled to a bluntly or occasionally sharply pointed tip, the margins entire, the upper surface glabrous, the undersurface glabrous or finely pubescent with short, somewhat hooked hairs, pinnately veined, the undersurface with a noticeable network of veins. Terminal leaflet with the stalk 10–20 mm long, symmetric at the base; lateral leaflets essentially sessile above the pulvinus, often slightly asymmetric at the base. Inflorescences axillary, of solitary flowers or small clusters of 2(–4) flowers, the inflorescence stalk 0.8–1.0 cm long, glabrous or finely hairy, the bracts 4–6 mm long, ovate, shed early; bractlets 8–12 mm long, broadly lanceolate to ovate, closely subtending the flowers and often partially obscuring the calyx, persistent. Flowers all opening (cleistogamous flowers not produced), twisted at the base during development so that the top of the flower (banner) is oriented toward the bottom at maturity (resupinate). Calyces 10–16 mm long, 4-lobed, the tube more or less bell-shaped, slightly pouched on 1 side at the base, glabrous or minutely hairy, the lobes as long as or longer than the tube, somewhat unequal (the uppermost slightly larger than the others and sometimes with 2 narrow lobes apically), narrowly triangular, tapered to sharply pointed tips. Corollas papilionaceous, pinkish purple to lavender or pale to light purple (the banner with a white region toward the center, this often with irregular, darker purple lines), the banner 25–30 mm long, 25–35 mm wide, the expanded portion broadly obovate to nearly circular, with an inconspicuous spur at the base, shallowly notched at the tip, often with a shallow longitudinal keel, the margins flat to slightly curled backward, the wings 14–16 mm long, 4–5 mm wide, the expanded portion oblanceolate to oblong, somewhat cupped around the keel and fused to it toward the base, the keel 14–17 mm long, 7–8 mm wide, strongly curved downward, fused to the rounded or bluntly pointed tip. Stamens 10, all of similar length, 9 of the filaments fused and 1 free, the fused portion 14–16 mm long, the free portion 2–3 mm long, the anthers small, attached near the midpoint. Ovary 8–10 mm long, short-hairy, the style 6–10 mm long, finely hairy, the stigma appearing expanded and fringed. Fruits legumes, the portion above the sessile or minutely stalklike base but including the beak 7–13 cm long, 3–4 mm wide, linear, straight, tapered (often slightly asymmetrically so) at the tip to an elongate (20–40 mm), mostly straight beak, flattened, with a longitudinal ridge near each margin, leathery, glabrous at maturity, dehiscing by 2 valves, these green to tan at maturity, twisting spirally after dehiscence, with numerous seeds. Seeds 2.0–2.5 mm long, 1.5–2.0 mm wide, oblong to bluntly rectangular in outline, flattened, the surface dark brown, sometimes somewhat mottled, smooth, somewhat shiny, not sticky. $2n=18$. July–August.

Uncommon, known thus far only from a single historical collection from Shannon County (southeastern U.S. west to Oklahoma and Texas; Mexico, Central America, South America). Habitat unknown.

The habit and flowers of *Centrosema virginianum* and *Clitoria mariana* are superficially very similar, but there are several subtle differences. *Centrosema virginianum* has thinner, more twining stems, leaflets with a prominent network of veins, a pair of large bractlets closely subtending each flower, a shorter calyx tube with relatively long, narrow lobes, a spurred banner

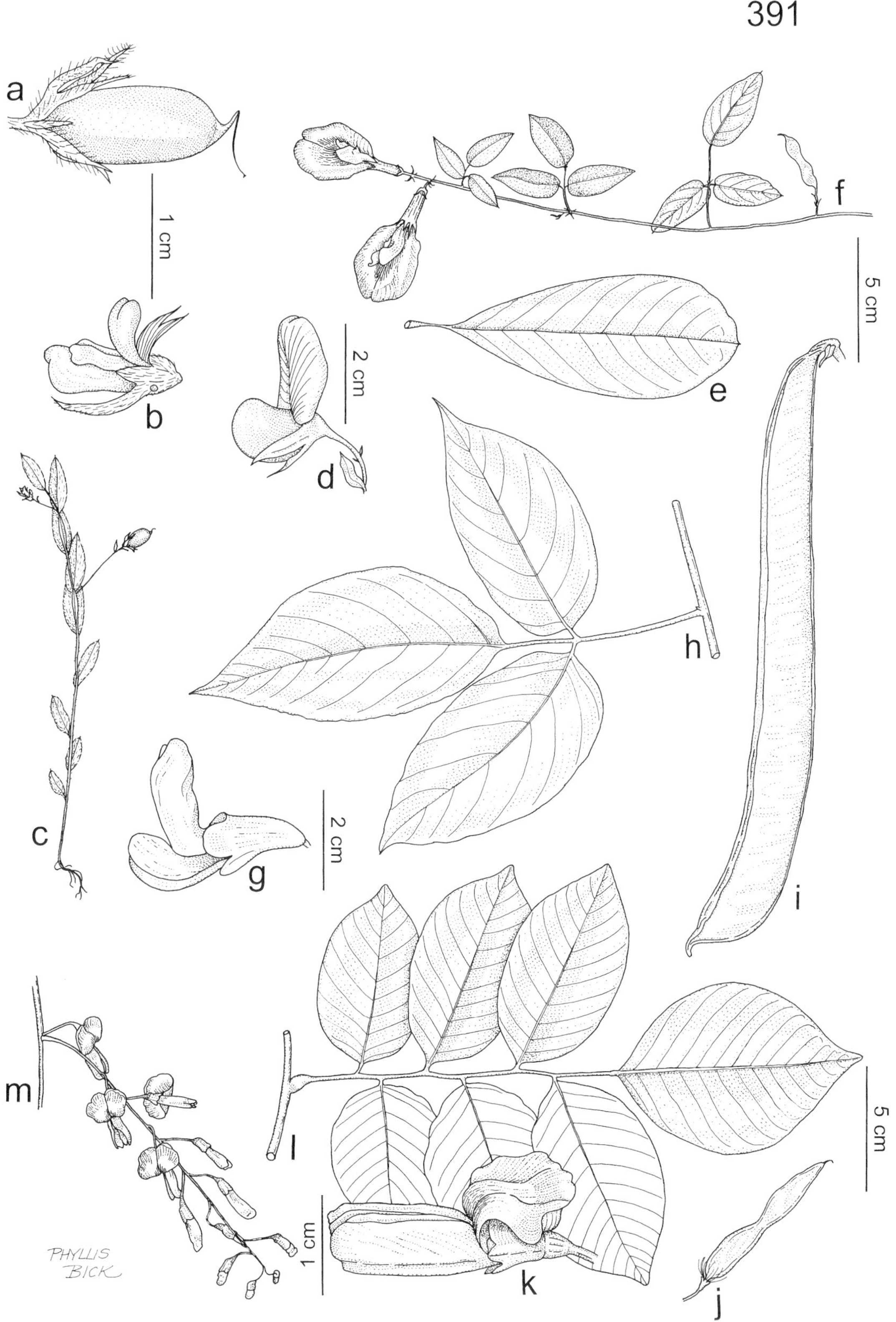

Plate 391. Fabaceae. *Crotalaria sagittalis*, **a)** fruit, **b)** flower, **c)** habit. *Crotalaria spectabilis*, **d)** flower, **e)** unifoliate leaf. *Clitoria mariana*, **f)** fertile stem. *Canavalia ensiformis*, **g)** flower, **h)** leaf, **i)** fruit. *Cladrastis kentukea*, **j)** fruit, **k)** flower, **l)** leaf, **m)** inflorescence.

1723. Cladrastis kentukea

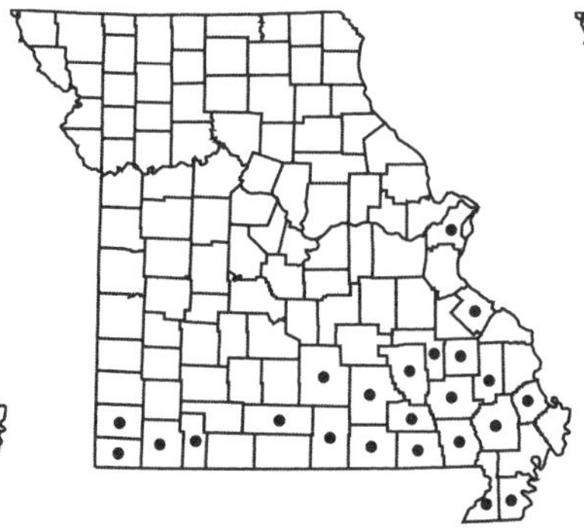

1724. Clitoria mariana

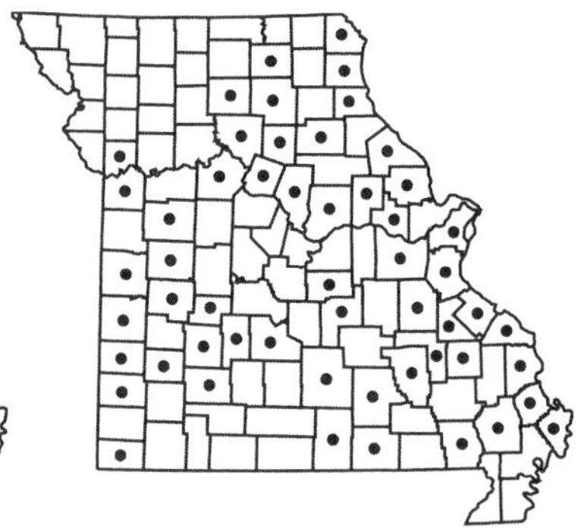

1725. Crotalaria sagittalis

petal, wings slightly shorter than to about equal in length to the keel, and a linear, prominently beaked legume. *Centrosema virginianum* is somewhat variable in leaflet shape and size, with some populations in other states having very narrow leaflets (Isely, 1998). The sole Missouri specimen originally was misdetermined as *Clitoria mariana*. It was redetermined by Paul Fantz of North Carolina State University and reported as part of a distributional summary of the group in temperate North America (Fantz, 2002a).

22. **Cladrastis** Raf.
(Duley and Vincent, 2003)

Six species, 1 in the U.S., the other 5 in Asia.

1. Cladrastis kentukea (Dum. Cours.) Rudd (yellowwood, American yellowwood)
C. lutea (Michx) K. Koch
C. lutea f. *tomentosa* Steyerm.
Sophora kentukea Dum. Cours.
Virgilia lutea Michx.
Pl. 391 j–m; Map 1723

Plants medium to large trees to 20 m tall, the main trunk commonly forking. Bark smooth, gray, often somewhat mottled with lighter gray, the branches more or less 2-ranked, somewhat zigzag, with alternate branching, unarmed. Buds small, enclosed in the hollow base of the subtending leaf's petiole (exposed after the leaves are shed), irregularly ovoid to obovoid, lacking scales, densely woolly. Twigs slender, initially green and sparsely to moderately short-hairy, becoming reddish brown and glabrous or nearly so at maturity, with scattered, pale, more or less circular lenticels. Leaves alternate, the petiole (10–)20–50 mm long, expanded and hollow at the base. Stipules absent; stipels absent. Leaf blades appearing odd-pinnately compound (but sometimes with an even number of leaflets), the rachis 5–18 cm long, glabrous, with 5–9(–11) leaflets, these alternate to less commonly subopposite on the rachis. Leaflets 3–14 cm long, 2–9 cm wide, broadly ovate or oval to elliptic or obovate, the terminal leaflet largest and broadly elliptic to broadly oval, rounded, angled, or tapered at the base, angled or tapered to a bluntly or sharply pointed tip, the margins entire, the upper surfaces glabrous, the undersurface glabrous or sparsely to moderately pubescent along the main veins with short, curved hairs, pinnately veined. Inflorescences terminal, open panicles 10–40 cm long, pendant, the stalk (2–)4–8 cm long. Bracts minute, shed early; bractlets absent. Flowers fragrant, with stalks 1–2 cm long, these sparsely to densely short-hairy. Calyces 5–8 mm long, cylindric to narrowly bell-shaped above a short hypanthium (this conic, usually darker colored, and somewhat asymmetric), densely pubescent with short, woolly hairs, the lobes much shorter than the tube, more or less similar (the lowermost slightly shorter than the others, the upper pair sometimes slightly fused basally), semicircular to broadly oblong, rounded at the tips. Corollas white, the banner 12–20 mm long, the expanded portion broadly obovate to nearly circular above the short, stalklike base, reflexed, usually shallowly notched at the tip, the inner surface with a yellow area below the midpoint, this usually with fine red spots, wings 12–18 mm long, 5–6 mm wide, oblong-lanceolate above the short, stalklike base, straight, the keel 11–18 mm long, 5–6 mm wide, oblong-lanceolate above the short, stalklike base, straight, folded and overlapping, but not fused. Stamens 10, the filaments fused only at the very base, 14–16 mm long, curved, graded in size, the anthers all similar, small, attached near the midpoint, yellow. Ovary 4–7 mm long, linear, densely short-hairy, the style 4–5 mm long, curved, glabrous, the stigma minute, terminal. Fruits legumes, (4–)7–10 cm long, 7–12 mm wide, linear to narrowly oblong, flattened, straight, sometimes shallowly indented between the widely spaced seeds, sometimes becoming loosely spirally twisted, glabrous at maturity, tapered to a short, stalklike base, tapered to a sharply pointed, sometimes short-beaked tip, the valves thin, papery, usually indehiscent (occasionally dehiscing tardily from the tip), with (1–)4–7 seeds. Seeds 7–8 mm long, 3–4 mm wide, oblong to somewhat asymmetrically kidney-shaped, flattened, the surface smooth, olive to dark brown, usually with a darker line around the circular attachment point. $2n=28$. May–June.

Uncommon in the southwestern portion of the Ozark Division, disjunct in Wayne County; introduced in Franklin County (Indiana and possibly Ohio south to South Carolina, west to Missouri and Oklahoma). Mesic upland forests, talus slopes, and bottoms and ledges of bluffs; usually on calcareous substrates.

Although most common in Tennessee, *C. kentukea* is not abundant anywhere. Yellowwood is identified by its graceful trunks with smooth bark (similar to that of the beech, *Fagus grandifolia* Ehrh., Fagaceae), odd-pinnately compound leaves with petioles that cover the buds, loose panicles of white flowers with separate petals and stamens, and relatively thin, few-seeded legumes.

The scientific name of yellowwood has undergone a change in recent decades (K. R. Robertson, 1977; Sand, 1992; Spongberg and Ma, 1997). For many years, it was called *C. lutea*, until Rudd (1971) discovered an overlooked earlier publication in which the taxon was named *Sophora kentukea*. Thus, the oldest correct name in the genus *Cladrastis* became *C. kentukea*.

Yellowwood is valued as an ornamental tree, although it is not common in cultivation because it is relatively slow-growing and does not flower until it is 10–20 years old. It is easy to propagate by cuttings or seed. A yellow dye was obtained from the wood and used for coloring homespun fabrics. It is seldom used for this purpose today, especially since this species is considered rare or imperilled in much of its range.

23. Clitoria L.
(Fantz, 1977)

About 62 species, tropical and warm-temperate regions of the Old and New Worlds, but most diverse in the New World tropics.

1. Clitoria mariana L. **var. mariana** (butterfly pea)

Ternatea mariana (L.) Kuntze

Vexillaria mariana (L.) Raf.

Pl. 391 f; Map 1724

Plants perennial herbs, sometimes reclining or climbing on other vegetation, with a branched rootstock. Stems 30–120 cm long, erect or ascending to trailing, mostly not twining, unarmed, glabrous or sparsely pubescent with short, fine, curved hairs. Leaves alternate, pinnately trifoliate, the petiole 1.0–4.5 cm long (the uppermost leaves occasionally nearly sessile), glabrous or sparsely short-hairy. Stipules 2–5 mm long, narrowly lanceolate to lanceolate, with prominent parallel venation; stipels 2–4 mm long, linear to hairlike, persistent. Leaflets 2–7 cm long, 1–4 cm wide, lanceolate to ovate or occasionally obovate, rounded at the base, angled to a bluntly or more commonly sharply pointed tip, the margins entire, the upper surface glabrous, the undersurface glabrous or finely pubescent with short, curved hairs along the main veins, pinnately veined, the undersurface lacking a noticeable network of veins. Terminal leaflet with the stalk 10–20 mm long, symmetric at the base; lateral leaflets essentially sessile above the pulvinus, often slightly asymmetric at the base. Inflorescences axillary, of solitary flowers or small clusters of 2(3) flowers, the inflorescence stalk 0.5–3.0 cm long, glabrous or sparsely and finely hairy, the bracts 1–3 mm long, linear to narrowly lanceolate, mostly shed early; bractlets 3–7 mm long, lanceolate, closely subtending the flowers, but not obscuring the calyx, persistent. Flowers mostly opening (inflorescences of cleistogamous flowers occasionally produced on the same or different plants), twisted at the base during development so that the top of the flower (banner) is oriented toward the bottom at maturity (resupinate). Calyces 14–21 mm long (6–9 mm in cleistogamous flowers), 5-lobed, the tube cylindric to narrowly bell-shaped, slightly pouched on 1 side at the base, glabrous, the lobes about 1/2 as long as the tube, more or less equal, lanceolate to ovate, tapered to sharply pointed tips. Corollas of open flowers papilionaceous, pinkish purple to lavender or pale to light purple (the banner, with a white or pale region toward the center, this often with irregular, darker purple lines), the banner 45–50 mm long, 25–40 mm wide, the expanded portion broadly obovate to nearly circular, lacking a spur at the base, shallowly notched at the tip, often with a shallow longitudinal keel, the margins flat to slightly curled backward, the wings 32–37 mm long, 9–11 mm wide, the expanded portion oblanceolate to oblong, more or less straight to somewhat spreading from the keel, but fused to it toward the base, the keel 25–28 mm long, 5–6 mm wide, curved downward, fused to the usually bluntly pointed tip. Corollas of cleistogamous flowers essentially lacking. Stamens 10, all of similar length, 9 of the filaments fused and 1 free, the fused portion 22–24 mm long (essentially absent in cleistogamous flowers), the free portion 2–3 mm long, the anthers small, attached near the midpoint. Ovary in open flowers 14–17 mm long (4–6 mm in cleistogamous flowers), densely pubescent with minute, somewhat hooked hairs, the style 14–16 mm long (4–6 mm in cleistogamous flowers), minutely hairy, the stigma appearing expanded and fringed. Fruits legumes, the portion above the noticeable stalklike base but including the beak 2–6 cm long, 5–8 mm wide, narrowly oblong, straight, tapered at the tip to an elongate (15–18 mm), curved beak, flattened and slightly indented between the seeds, lacking a longitudinal ridge near each margin, leathery, glabrous at maturity, dehiscing by 2 valves, these green to tan at maturity, twisting spirally after dehiscence, with 3–10 seeds (often only 1–3 in cleistogamous flowers). Seeds 4.0–4.5 mm long, 3–4 mm wide, oblong to bluntly rectangular in outline, flattened, the surface brown to black, smooth, shiny, and sticky. June–September.

Scattered in the southern portion of the Ozark and the Mississippi Lowlands, north historically to St. Louis County (eastern [mostly southeastern] U.S. west to Nebraska and Arizona; Mexico). Mesic to dry upland forests, savannas, banks of streams and rivers, ledges and tops of bluffs, and sand prairies; also roadsides.

Clitoria is closely related to *Centrosema*, but is distinguished by the narrower bractlets, glabrous calyx with a more cylindric tube and broadly lanceolate to ovate lobes that are much shorter than the tube, wings longer than the keel, broader legumes with a curved beak, and sticky seeds. Linnaeus adopted the generic name *Clitoria*, which refers to the fanciful similarity of the keel petals to human female sexual organs, and irreverently named the species for the Virgin Mary of the Christian religion. Attempts to change the name for the sake of modesty were unsuccessful (Fantz, 1977, 2000).

The distribution of *C. mariana* in temperate North America was detailed by Fantz (2002b). Fantz (1977, 1995) and Fantz and Predeep (1992) have treated *C. mariana* as comprising three varieties, with only the nominate variety occurring in most of the North American range (including Missouri). Plants in central and eastern Florida with moderately to densely hairy herbage have been called var. *pubescentias* Fantz, whereas the populations that are native to Asia, which differ in a suite of subtle quantitative characters and do not produce cleistogamous flowers, have been called var. *orientalis* Fantz. Plants of *C. mariana* found in the Ozarks are usually trailing and single- or few-stemmed, but, in sandy soils in the northern portion of the Mississippi Lowlands, plants can appear bushy and more upright, with several stems.

Roots and seeds of *Clitoria* species have been used medicinally as a laxative, vermifuges, and aphrodisiac, the last perhaps following the ancient but unscientific doctrine of signatures, the philosophy that plant parts resemble the human organs for which they are meant to be used (Fantz, 1991). This species should be investigated for its ornamental potential in Missouri gardens.

24. Crotalaria L.

Plants annuals (perennial herbs and shrubs elsewhere), with taproots. Stems erect or strongly ascending, unbranched or branched, usually 2- to several-angled or -ridged, unarmed, glabrous or hairy, often becoming reddish purple or purple with age. Leaves alternate, all appearing simple (unifoliate; palmately trifoliate elsewhere), short-petiolate. Stipules present (sometimes absent at the lower nodes), conspicuous or inconspicuous, free from the stem or long-decurrent below the nodes as wings of green tissue; stipels absent. Leaflets variously shaped, the margins entire, the surfaces glabrous or hairy, pinnately veined. Inflorescences relatively open racemes or sometimes reduced to a solitary flower, terminal or attached opposite the upper leaves, the bracts inconspicuous or conspicuous, linear to lanceolate or elliptic-lanceolate, usually persistent, the bractlets mostly inconspicuous, variously shaped, persistent. Flowers with short to relatively long, slender stalks. Calyces 2-lipped, the tube obliquely bell-shaped to conic, shorter than the lips, glabrous or hairy, the upper lip with 2 lobes, these usually slightly broader than the 3 lobes of the lower lip, which is concave and somewhat cupped around the wings and keel. Corollas papilionaceous, yellow (in our species), sometimes lined or tinged with red, sometimes fading to white, the banner larger than the wings and keel, the expanded portion obovate to nearly circular, bent upward or backward, short-tapered to a short, stalklike base, rounded or shallowly notched at the tip, with a shallow longitudinal groove, the wings positioned around the keel, the expanded portion asymmetrically and broadly oblong to rhombic-oblong, the stalklike base attached along the lower margin, the keel shorter than the wings (longer elsewhere), fused their entire length, usually finely hairy along the suture, strongly curved upward (curved downward elsewhere), twisted together at tip. Stamens 10, monadelphous, the filament tube split along the upper side, dimorphic, 5 stamens with longer filaments and larger, linear anthers attached near the base alternating with 5 on shorter filaments and smaller, broadly ovate anthers attached near the midpoint. Ovary short-stalked to nearly sessile, the style strongly bent at the base and strongly curved, hairy, the stigma terminal, minute, with a dense tuft of short, bristly hairs. Fruits legumes, oblong-cylindric or occasionally oblong-ellipsoid, tapered to a short-stalked base, tapered abruptly to a short, hairlike beak, inflated, the valves dry and papery to leathery and becoming black at maturity, glabrous, dehiscent longitudinally along the sutures, with 7 to numerous seeds. Seeds obliquely heart-shaped to asymmetrically broadly kidney-shaped, somewhat flattened, the surface brown to black, smooth, shiny. About 510 species, nearly worldwide, most diverse in tropical regions.

The Missouri species of *Crotalaria* are recognized by their unifoliate leaves, yellow papilionaceous flowers, and greatly inflated legumes that become black at maturity. The vernacular name rattlebox refers to the tendency of the seeds to become detached while still in the fruit prior to its dehiscence, resulting in a rattling sound when the fruits are disturbed or shaken. Species of *Crotalaria* are used for green manure and as shade or cover for other crops. Some Old World species were introduced into the United States in the 1930s for soil cover, and are now firmly established and sometimes weedy in other states (Senn, 1939; Windler, 1974). *Crotalaria* is not used as forage because it contains hepatotoxic pyrrolizidine alkaloids similar to those found in some members of the Asteraceae tribe Senecioneae (Burrows and Tyrl, 2001). All parts of the plant are toxic, and care must be taken to avoid contamination of animal feed with the leaves or seeds.

1. Stems and branches moderately to densely pubescent with long (to 2.5 mm), fine hairs mixed with shorter ones; leaflets linear to lanceolate or elliptic, 0.5–1.5 cm wide; stipules conspicuously decurrent as tapering wings of tissue below the nodes (often absent from the lower nodes); calyces hairy; banner 5–11 mm long; fruits 12–30 mm long, 5–12 mm wide . . . . 1. C. SAGITTALIS
1. Stems and branches glabrous or with inconspicuous, short (mostly less than 1 mm), stiff hairs; leaflets obovate to narrowly obovate, 2.5–4.0 cm wide; stipules inconspicuously short-decurrent below the nodes; calyces glabrous; banner 17–25 mm long; fruits 30–50 mm long, 10–20 mm wide . . . . 2. C. SPECTABILIS

1726. Crotalaria spectabilis

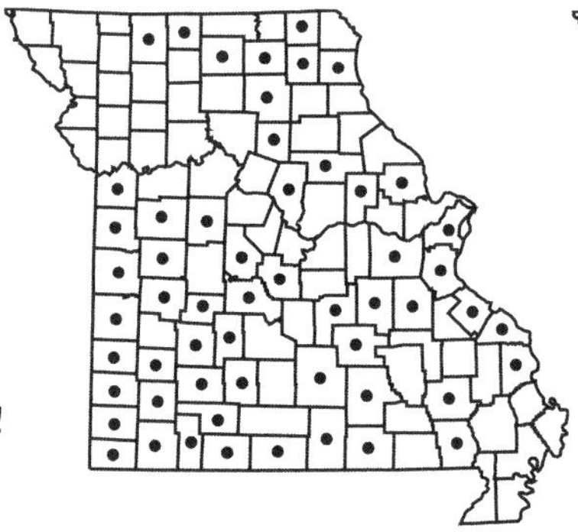

1727. Dalea candida

1728. Dalea enneandra

1. Crotalaria sagittalis L. (rattlebox)
C. sagittalis var. *fruticosa* (Mill.) Fawcett & Rendle

Pl. 391 a–c; Map 1725

Stems 10–40 cm long, 2–3 mm wide at base, unbranched or branched, moderately to densely pubescent with long (to 2.5 mm), spreading to loosely ascending, fine hairs mixed with shorter ones. Petioles absent or to 3 mm long. Stipules absent at the lower nodes, otherwise conspicuous, the free portion 2–7 mm long, narrowly ovate-triangular, tapered to a sharply pointed tip, decurrent basally below the node as a tapered wing of tissue extending to the next-lower node, sparsely to moderately hairy. Leaflets 1.5–5.5(–8.0) cm long, 0.5–1.5 cm wide, linear to more commonly narrowly lanceolate to narrowly elliptic or (the smaller ones) lanceolate to elliptic, rounded or angled at the base, angled or tapered to a bluntly or sharply pointed tip, sometimes with a minute, sharp extension of the midvein at the tip, the upper surface sparsely to moderately pubescent with long, fine hairs, the undersurface moderately hairy. Inflorescences opposite the upper and median leaves, usually attached slightly below the leaf node, of solitary flowers or short, clusterlike racemes of 2–4 flowers, the stalk 0.5–3.0 cm long. Bracts and bractlets 4–5 mm long, linear to narrowly lanceolate. Flowers with the stalk 3–8 mm long. Calyces moderately to densely shaggy-hairy, the tube 1–2 mm long, the lips 5–7 mm long, the lobes linear to lanceolate. Corollas with the banner 5–11 mm long, obovate, the wings 4–6 mm long, the keel 4–5 mm long, relatively sparsely hairy along the suture, sharply bent, the tips twisted, forming a short beak. Stamens with the fused portion of the filaments 2–3 mm long, the free portion 3–4 mm long. Fruits 12–30 mm long, 5–12 mm wide, glabrous at maturity. Seeds 2.3–3.0 mm in longest dimension, the surface greenish brown to brown, smooth and shiny. $2n=32$. June–September.

Scattered nearly throughout the state, but apparently absent from most of the western half of the Glaciated Plains Division (eastern U.S. west to South Dakota, Texas, and disjunctly Arizona; Mexico, Central America). Upland prairies, sand prairies, glades, openings of mesic to dry upland forests, savannas, talus slopes, and banks of streams; also pastures, old fields, mines, quarries, railroads, roadsides, and open disturbed areas.

Crotalaria sagittalis is somewhat polymorphic in leaflet shape, varying from broad to narrow depending on early vs. late in the growing season, the position on the plant, and other factors (Windler, 1973, 1974). It is an annual in the northern part of its range, but can become a short-lived perennial farther south.

2. Crotalaria spectabilis Roth (showy rattlebox)
C. retzii Hitchc.

Pl. 391 d, e; Map 1726

Stems 50–100 cm long, 5–8 mm wide at base, erect, branched, often strongly purplish-tinged, glabrous and glaucous or sparsely pubescent with short (mostly less than 1 mm), straight, loosely appressed-ascending to more or less spreading, stiff hairs. Petioles absent or to 6 mm long. Stipules present at all or most nodes, relatively inconspicuous, the free portion 4–7 mm long, ovate-triangular, tapered to a sharply pointed tip, only short-decurrent basally below the node, not tapered basally or appearing as a wing of tissue, glabrous or sparsely hairy. Leaflets 5–12(–15) cm long, 2.5–4.0 cm wide, obovate to narrowly obovate, angled or tapered at the base, rounded or broadly angled to a bluntly pointed tip, sometimes with a minute, sharp extension of the midvein at the tip, the upper surface glabrous, often somewhat glaucous, the undersurface moderately to densely appressed-hairy. Inflorescences terminal and occasionally also from the upper leaf axils (rarely attached between the leaf nodes), of elongate racemes of usually numerous flowers, the stalk 1–6 cm long. Bracts 6–12 mm long, narrowly heart-shaped, clasping the axis; bractlets 1–2 mm long, lanceolate to narrowly lanceolate. Flowers with the stalk 9–20 mm long. Calyces glabrous, glaucous, the tube 4–5 mm long, the lips 5–9 mm long, the lobes ovate to ovate-triangular. Corollas with the banner 17–25 mm long, broadly obovate to nearly circular, the wings 10–14 mm long, the keel 8–10 mm long, densely hairy (appearing short-fringed) along the suture, sharply bent, the tips twisted forming a beak. Stamens with the fused portion of the filaments 5–8 mm long, the free portion 4–6 mm long. Fruits 30–50 mm long, 10–20 mm wide, glabrous at maturity. Seeds 3.5–4.5 mm in longest dimension, the surface brown to more commonly black, smooth and shiny. $2n$=16, 32. September–October.

Introduced, uncommon in the Mississippi Lowlands Division and in a few counties bordering the Missouri River (native of Asia; introduced in the southeastern U.S. west to Oklahoma and Texas). Fallow fields, roadsides, and open disturbed areas.

25. **Dalea** L.
(Barneby, 1977)

Plants annual or perennial herbs (shrubs elsewhere). Stems prostrate to erect, branched, unarmed, glabrous or hairy, the hairs simple, sometimes also gland-dotted. Leaves alternate, odd-pinnately compound or rarely trifoliate, short-petiolate or sessile, the petiole and rachis often gland-dotted, the rachis usually appearing somewhat flattened or narrowly winged, the leaflets (3)5–35, mostly opposite. Stipules linear to narrowly triangular, mostly inconspicuous, sometimes shed early. Leaflets variously shaped, symmetric at the sessile or minutely stalked base, variously shallowly notched to truncate, rounded, or pointed at the tip, the margins entire, the surfaces glabrous or hairy, usually noticeably gland-dotted, 1-veined. Inflorescences variously open to dense, elongate or headlike spikes, in some species sometimes drooping with age, terminal on the stems and branches, the flowers subtended by often conspicuous bracts, these variously shaped, often gland-dotted, sometimes shed early; a pair of minute, inconspicuous bractlets also present in some species. Calyces not 2-lipped, 5-lobed, somewhat zygomorphic (the uppermost lobe often longer than the others), the tube shorter than to more commonly longer than the lobes, 10-ribbed, gland-dotted, the lobes variously shaped, but angled or tapered to sharply pointed tips, the margins often conspicuously hairy. Corollas papilionaceous (but sometimes very weakly so, with the wings and keel petals similar), variously colored; the banner with a long, stalklike base, the expanded portion variously heart-shaped or kidney-shaped to broadly oblong-ovate or nearly circular; the wings attached to the side or tip of the filament tube, elliptic to more or less strap-shaped; the 2 keel petals either similar to the wings or strongly overlapping to more or less fused, not boat-shaped. Stamens 5–10, the filaments all fused, the tube shorter than to longer than the free portions, sometimes split more deeply on 1 side, the anthers small, orange to yellow or blue, attached near the midpoint, all similar in size, the connective between the sacs sometimes with a gland. Ovary asymmetrically ovoid, sessile, glabrous or hairy, the style usually curved, the stigma small and terminal. Fruits short legumes more or less enclosed in the persistent calyx, asymmetrically obovoid, plump, sessile, broadly rounded to more or less truncate at the tip and usually with a minute tooth (the persistent style base) along the margin, the surfaces glabrous or hairy, sometimes gland-dotted, often thinner and more or less translucent toward the base, 1-seeded, indehiscent. Seeds 1.5–3.0 mm long, oblong-ovoid to asymmetrically kidney-shaped, somewhat flattened, the surfaces yellow to dark brown, smooth or in some species some of them wrinkled, shiny. About 160 species, North America to South America.

Dalea has a center of diversity in the southwestern United States and portions of Mexico. Many species have a thick root system which undoubtedly helps them survive in their often dry habitats. The genus is of little economic importance, although several species are important forage plants and a number are cultivated as ornamentals. McGregor (1986) noted that prairie clovers (in particular *D. candida* and *D. purpurea*), are selectively grazed by livestock and tend to disappear quickly when lands are overgrazed.

Barneby's (1977) circumscription of *Dalea* includes some groups that formerly were treated as separate genera, such as *Kuhnistera* Lam., *Parosela* Cav., *Petalostemon* Michx., and *Thornbera* Rydb. These were distinguished by details of petal shape and attachment on the stamen tube and the number of functional stamens. Shared characters that separate *Dalea* (as circumscribed here) from its relatives include the petals arising from the anther column, two ovules positioned side by side in the ovary (one of these abortive), and a base chromosome number of x=7. By this definition, some of the mostly shrubby, non-Missouri groups of species traditionally included in *Dalea* were segregated into the genera *Marina* Liebm. and *Psorothamnus* Rydb. Missouri species formerly included in *Petalostemon*, in which four of the petals arise from the rim of the stamen tube alternating with five anther-bearing filaments, include *D. candida*, *D. gattingeri*, *D. multiflora*, *D. purpurea*, and *D. villosa*. The wing and keel petals of these species are sometimes interpreted as staminodes, with the free banner petal all that remains of the true corolla (Wemple and Lersten, 1966). Alternatively, these may be true petals that have migrated to the rim of the staminal tube, with the loss of one whorl of stamens. In either case, such flowers are not truly papilionaceous in morphology because the wing and keel petals are not strongly differentiated and the keel petals are not fused. The other species of *Dalea* in Missouri have nine or ten stamens and petals attached to the sides of the stamen tube.

1. Blades of the main stem leaves (excluding any smaller leaves in axillary fascicles) with (9–)13–35 leaflets; roots yellowish orange to reddish orange
    2. Stems (excluding the inflorescence axes) and leaves glabrous; plants annual, taprooted; stamens 9 or 10; petals white or rarely pale bluish-tinged, the wing and keel petals attached along the side of the stamen tube . . . . . . 4. D. LEPORINA
    2. Stems and leaves noticeably hairy; plants perennial with a branched rootstock; stamens 5; petals pink to reddish purple, the wing and keel petals attached at the tip of the stamen tube . . . . 7. D. VILLOSA
1. Blades of the main stem leaves (excluding any smaller leaves in axillary fascicles) with (3)5–13 leaflets; roots dark brown to black (yellow in *D. enneandra*)
    3. Inflorescences loosely flowered, the axis readily visible between the flowers; stamens 9; wing and keel petals attached along the side of the stamen tube, the stamen tube noticeably longer than the free portion of the filaments; bracts broadly ovate to broadly obovate, strongly cupping the flowers, the margin with a conspicuous, thin, pale, narrow band; roots yellow . . . . 2. D. ENNEANDRA
    3. Inflorescences densely flowered, the axis not easily visible between the flowers except sometimes at fruiting (occasionally more or less visible in pressed specimens); stamens 5; wing and keel petals attached at the tip of the stamen tube, the stamen tube about as long as the free portion of the filaments; bracts variously linear to triangular or ovate, not strongly cupping the flowers the margin lacking a pale band; roots dark brown to black
        4. Petals white; calyces glabrous; axis of the inflorescence glabrous or occasionally minutely and inconspicuously hairy (viewing the axis requires removal of flowers or fruits)
            5. Plants with 1–4 stems; leaf blades with rachis 10–30 mm long, leaflets 5–30 mm long, the lowermost leaves often with larger leaflets than those of the upper leaves; inflorescences 1–7 per stem, 2–5 cm long, ovoid or cylindric; bracts subtending the flowers 5–6 mm long, longer than the calyces of the associated buds (usually shed by the time the flowers open) . . . . 1. D. CANDIDA

5. Plants in bushy clumps of 5 or more stems; leaf blades with the rachis 5–14 mm long, leaflets 3–7(–13) mm long, mostly uniform over the length of the stem; inflorescences more than 7 per stem, usually numerous, subglobose, 5–15 mm long; bracts subtending the flowers 1.4–2.4 mm long, shorter than the calyces of the associated buds (usually shed by the time the flowers open) . . . . . 5. D. MULTIFLORA

4. Petals purple; calyces densely hairy; axis of inflorescence densely pubescent with short, bristly hairs (viewing the axis requires removal of flowers or fruits)

6. Stems 10–40 cm long, low, loosely ascending from a spreading base or prostrate throughout; inflorescences 4–8 cm long, becoming lax or sinuous; bracts evenly hairy above the papery base . . . . 3. D. GATTINGERI

6. Stems 50–80 cm long, erect to strongly ascending or slightly arched; inflorescences 2–5 cm long, cylindric, dense, remaining straight or nearly so; bracts hairy in a narrow band above the base and below the tip . . . . 6. D. PURPUREA

1. Dalea candida Michx. ex Willd. (white prairie clover)

Petalostemon candidum Michx. ex Willd.

Pl. 392 c, d; Map 1727

Plants perennial herbs, with a knotty, dark brown to black caudex above a stout taproot. Stems 1–4, (20–)40–70 cm long, erect or slightly arched, unbranched or sparingly branched above the midpoint, prominently ribbed, glabrous, sparsely gland-dotted. Leaves 1.5–6.0 cm long, sometimes appearing in fascicles, with 2 smaller trifoliate leaves at the base of the main leaf. Leaf blades with the rachis 10–30 mm long, gland-dotted, those of the main leaves with (3)5–11 leaflets. Stipules 1–5 mm long. Leaflets 5–30 mm long, 2–7 mm wide, those of the lower leaves often larger than those of the upper leaves, linear to narrowly oblong or elliptic-oblanceolate, usually tapered at the base and tip, occasionally blunt but with a minute, sharp extension of the midvein at the tip, the upper surface glabrous, the undersurface gland-dotted but otherwise glabrous. Inflorescences dense spikes (the axis not visible between the flowers), 2–6 cm long, ovoid to cylindric, the stalk 5–11 cm long, the axis glabrous or occasionally minutely and inconspicuously hairy (viewing the axis requires removal of flowers or fruits), with scattered bracts and a whorl closely subtending the spike, these persistent, 4–10 mm long, 0.5–1.2 mm wide, linear to long-tapered above a slightly expanded base, green, gland-dotted, more or less pale-margined, grading into the bracts subtending flowers, these prominent in bud but shed early, 5–6 mm long (longer than the calyces of the associated buds), 1.5–2.0 mm wide, long-tapered above a narrowly elliptic-oblanceolate to obovate base, gland-dotted, the margins thin and pale. Calyces with the tube 1.9–2.5 mm long, glabrous, the ribs often reddish-tinged, gland-dotted around the rim, the lobes 1.0–1.8 mm long, linear to narrowly triangular, often reddish-tinged, the margins minutely hairy. Petals white, the banner with the expanded portion 1.5–2.0 mm long, the wing and keel petals similar, attached along the rim of the stamen tube, the expanded portion 1.8–2.0 mm long. Stamens 5, the filament tube 2–3 mm long, the free filaments 2–3 mm long, the anthers yellow. Fruits included in the persistent calyx or more commonly somewhat exserted, 2.5–4.0 mm long, firm above the membranous basal portion, with a transverse ring of glands separating the two regions. $2n=14$. June–September.

Scattered nearly throughout the state, but uncommon in the western portion of the Glaciated Plains Division and the Mississippi Lowlands (North Dakota to Texas east to Wisconsin and South Carolina; Canada). Upland prairies, glades, tops of bluffs, openings of mesic to dry upland forests, and savannas; also pastures, cemeteries, railroads, and roadsides.

Dalea candida is sympatric with *D. purpureum* throughout most of its range. This complex and widely dispersed species is very similar to *D. multiflora*, which has more numerous, shorter spikes. *Dalea occidentalis* (Rydb.) L. Riley}, a closely related taxon that occurs mostly to the west of the range of *D. candida* (Montana to Arizona east to North Dakota and Texas; Canada) has sometimes been treated as a western variety, *D. candida* var. *oligophylla* (Torr.) Shinners. It differs most noticeably in its spikes that become somewhat less dense with age and its more sharply ribbed, hairier calyces.

2. Dalea enneandra Nutt. (nine-anthered prairie clover)

D. enneandra var. *pumila* (Shinners) B.L. Turner

1729. Dalea gattingeri 1730. Dalea leporina 1731. Dalea multiflora

D. laxiflora Pursh
Parosela enneandra (Nutt.) Britton
Pl. 392 i, j; Map 1728

Plants perennial herbs, with a tough, somewhat woody, yellowish orange rootstock. Stems 1–3, 50–120 cm long, erect or somewhat arched, finely branched above the midpoint, not or only slightly angled, glabrous, sparsely gland-dotted toward the tip. Leaves 1.3–2.6 cm long, not appearing in fascicles. Leaf blades with the rachis 2–10 mm long, gland-dotted, those of the main leaves with 5–7(–11) leaflets. Stipules 0.5–1.3 mm long. Leaflets 3–11 mm long, 1–2 mm wide, those of the lower leaves often larger than those of the upper leaves, linear to narrowly oblong or narrowly elliptic-oblanceolate, often somewhat folded longitudinally, tapered at the base, rounded at the tip, but sometimes minutely notched or with a minute, sharp extension of the midvein at the tip, the upper surface glabrous, the undersurface gland-dotted but otherwise glabrous. Inflorescences slender, open spikes (the axis readily visible between the flowers), 4–8(–12) cm long, loosely cylindric, the stalk 1.0–2.5 cm long, the axis glabrous, with bracts subtending the flowers (additional sterile basal bracts absent), these persistent (shed with the fruit), 3–4 mm long, 3.5–4.5 mm wide, broadly ovate to broadly obovate, with a short, sharp extension of the midrib at the tip, strongly cupping the flower, green to more commonly dark purplish-tinged, strongly gland-dotted, the margin with a conspicuous, thin, pale, narrow band. Calyces with the tube 3–4 mm long, densely pubescent with relatively long, silky hairs, also gland-dotted, the ribs light brown to somewhat reddish-tinged, the lobes 3–4 mm long, linear to narrowly triangular, densely silky-hairy (appearing feathery). Petals white (the banner with greenish mottling toward the base on the inner surface), the banner attached at the base of the stamen tube, the expanded portion 2.5–3.0 mm long, the wings attached along the side of the stamen tube, the expanded portion 2–3 mm long, the keel petals attached along the side of the stamen tube, the expanded portion 4–5 mm long. Stamens 9, the filament tube 5–6 mm long, the free filaments 2–3 mm long, the anthers yellow. Fruits included in the persistent calyx, 3.0–3.7 mm long, papery above the membranous basal portion, glandless, silky-hairy toward the tip. $2n$=14. June–September.

Uncommon, currently known only from Atchison and Holt Counties, introduced historically in Jackson County (Montana to New Mexico east to Iowa and Missouri; Mexico). Loess hill prairies; also open, disturbed areas.

Dalea enneandra is easily distinguished from other Missouri species in the genus by its inflorescences of loosely flowered spikes. The calyx lobes are long-tapered with a feathery appearance because of the long-silky hairs, and perhaps function in wind dispersal of the fruits. The stem leaves wither and eventually are shed as the inflorescences and flowers mature.

3. Dalea gattingeri (A. Heller) Barneby
(Gattinger's prairie clover)
Kuhnistera gattingeri A. Heller
Petalostemon gattingeri (A. Heller) A. Heller
Map 1729

Plants perennial herbs, with a thick, black rootstock and sometimes stout rhizomes. Stems numerous, 10–40 cm long, loosely ascending from a spreading base or prostrate throughout (forming mats), usually many-branched, not or only slightly angled, glabrous or sparsely pubescent toward the tip with inconspicuous, fine, curved hairs, sparsely gland-dotted. Leaves 1.0–3.5 cm long, sometimes appearing in fascicles, with 2–4 smaller 3(5)-foliate leaves at the base of the main leaf. Leaf blades

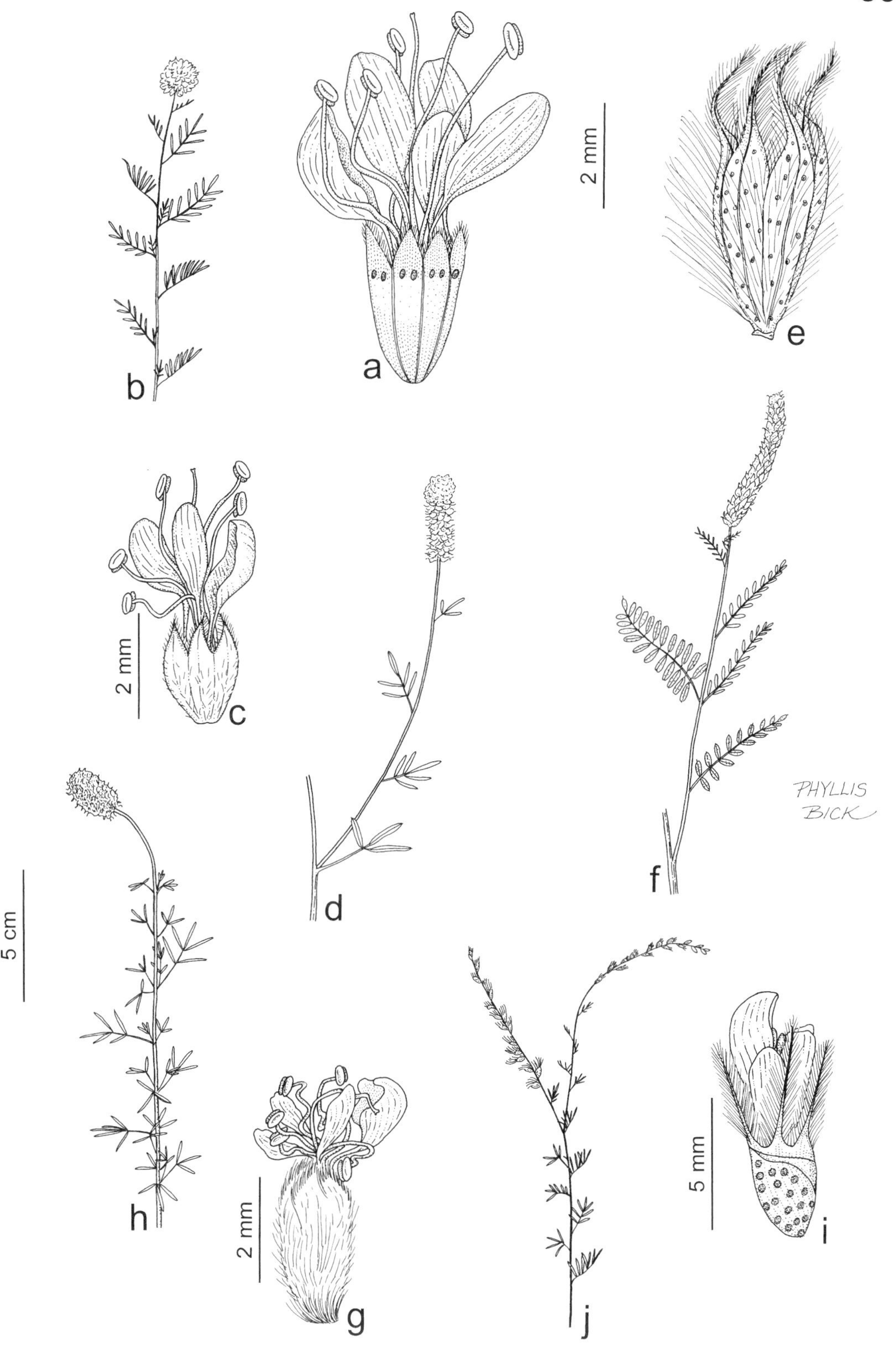

Plate 392. Fabaceae. *Dalea multiflora*, **a)** flower, **b)** fertile stem. *Dalea candida*, **c)** flower, **d)** fertile stem. *Dalea leporina*, **e)** fruiting calyx, **f)** fertile stem. *Dalea purpurea*, **g)** flower, **h)** fertile stem. *Dalea enneandra*, **i)** flower, **j)** fertile stem.

with the rachis 5–25 mm long, gland-dotted, those of the main leaves with 5–9 leaflets. Stipules 1.5–7.0 mm long. Leaflets 5–17 mm long, 0.5–1.5 mm wide, those of the lower leaves often slightly wider than those of the upper leaves, linear to narrowly oblong, narrowly lanceolate, or narrowly oblanceolate, more or less rounded at the base, usually somewhat tapered at the tip, the margins usually inrolled, the upper surface glabrous, the undersurface gland-dotted and usually also sparsely and inconspicuously hairy. Inflorescences dense spikes (the axis not visible between the flowers, but sometimes becoming slightly more visible as the fruits mature), 4–8 cm long, cylindric (becoming somewhat sinuous as the fruits mature), the stalk 1–2 cm long, the axis densely pubescent with short, bristly hairs (viewing the axis requires removal of flowers or fruits), with numerous bracts, the larger of these more or less persistent until the fruits are shed, others withering and shed early, 3–5 mm long, 1–2 mm wide, lanceolate-tapered, green above a membranous base to entirely straw-colored or grayish tan, evenly hairy above the base and sparsely gland-dotted, the margins also hairy. Calyces with the tube 2.0–2.5 mm long, densely hairy, the ribs often more or less obscured by the pubescence, not gland-dotted but often with minute reddish purple spots, the lobes 1.7–2.5 mm long, lanceolate to ovate, sometimes with minute reddish purple spots, the margins hairy and occasionally with scattered glands. Petals pale to deep pinkish purple, the banner with the expanded portion 2–3 mm long, the wing and keel petals similar, attached along the rim of the stamen tube, the expanded portion 2.2–3.0 mm long. Stamens 5, the filament tube 2–3 mm long, the free filaments 2–3 mm long, the anthers orange. Fruits more or less included in the persistent calyx, 2.0–2.4 mm long, firm and densely hairy above the membranous basal portion, inconspicuously and minutely gland-dotted. $2n=14$. May–June.

Uncommon, known thus far only from Howell County (Tennessee to Alabama and Georgia; disjunct in Arkansas and Missouri). Dolomite glades.

Dalea gattingeri is locally abundant in the cedar glades of central Tennessee and surrounding areas. It was discovered relatively recently in dolomite glades in Howell County (Summers et al., 1995), and subsequently in northern Arkansas. It and *Solidago gattingeri* (Asteraceae) are among the few glade endemics shared between the Missouri and Tennessee regions. *Dalea gattingeri* has numerous, short, spreading stems radiating from a stout rootstalk. The flowers resemble those of *D. purpurea*, but the spikes are often more elongate and become somewhat sinuous with age. Flowering in *D. gattingeri* occurs earlier and for a shorter period of time than in *D. purpurea*. There is some evidence that *D. gattingeri* produces an allelopathic substance that inhibits the seed germination and growth of other surrounding species of plants (B. H. Turner and Quarterman, 1975).

4. Dalea leporina (Aiton) Bullock (foxtail dalea, hare's foot dalea)
D. alopecuroides Willd.
D. lagopus (Cav.) Willd.
Parosela alopecuroides (Willd.) Rydb.
Pl. 392 e, f; Map 1730

Plants annuals, with a small, slender, yellowish orange taproot. Stems mostly 1–3, 30–70(150) cm long, erect or strongly ascending, many-branched above the midpoint, finely angled, glabrous, usually gland-dotted toward the tip. Leaves 2–10 cm long, not appearing in fascicles. Leaf blades with the rachis 20–65 mm long, lacking glands, those of the main leaves with 13–35 leaflets (sometimes as few as 9 in basal leaves). Stipules 1–3 mm long. Leaflets 3–12 mm long, 1.5–4.0 mm wide, those of the lower leaves often larger than those of the upper leaves, oblanceolate to oblong-lanceolate or obovate, often somewhat folded longitudinally, tapered at the base, mostly slightly notched at the tip, but sometimes with a minute, sharp extension of the midvein in the notch, the upper surface glabrous, the undersurface gland-dotted but otherwise glabrous. Inflorescences dense spikes (the axis not visible between the flowers), 2–5 cm long, ovoid to cylindric, the stalk 1–3 cm long, the axis densely short-hairy, with numerous bracts (additional sterile basal bracts absent), all but the lowermost shed early, 2.5–4.0(–7.0) mm long, 1.8–2.0 mm wide, abruptly long-tapered above a lanceolate to ovate base, green or dark purplish-tinged, at least the lowermost gland-dotted, glabrous or finely hairy, the margins lacking a conspicuous, pale band but sometimes hairy. Calyces with the tube 2.2–3.0 mm long, moderately to densely pubescent with relatively long, silky hairs, also gland-dotted, the ribs light brown to blackish purple, the lobes 1.0–2.5 mm long, lanceolate to ovate-triangular, sometimes hairy and also glandular along the margins (appearing feathery). Petals white or rarely pale bluish-tinged, the banner attached at the base of the stamen tube, the expanded portion 2.5–3.0 mm long, the wings attached along the side of the stamen tube, the expanded portion 2.5–3.0 mm long, the keel petals attached along the side of the stamen tube, the expanded portion 2–3 mm long. Stamens 9 or 10, the filament tube 5–6 mm long, the free filaments 0.6–1.0 mm long, the anthers pale blue to dark blue. Fruits more or less included in the persistent calyx, 2.5–3.0 mm long, thin and papery for most of its length, this region with an irregular transverse band of gland dots near the tip, the apical portion thicker, hairy. $2n=14$. July–August.

Scattered, mostly in counties along the Missouri and Mississippi Rivers (eastern [mostly northeastern] U.S. west to North Dakota and Arizona; Mexico, Central America, South America). Banks of rivers, sloughs, marshes, and bottomland prairies; also levees, borrow pits, ditches, margins of crop fields, roadsides, and open disturbed areas; often in sandy soil.

5. Dalea multiflora (Nutt.) Shinners (round-headed prairie clover)

D. candida Michx. ex Willd. var. *multiflora* (Nutt.) Rydb.

Kuhnistera multiflora (Nutt.) A. Heller

Petalostemon multiflorum Nutt.

Pl. 392 a, b; Map 1731

Plants perennial herbs, with a dark brown to black caudex above a thick, branched root system. Stems 5 to numerous, 30–60(–80) cm tall, erect or ascending, well-branched often mostly toward the tip, rounded to finely ridged, glabrous, sparsely gland-dotted. Leaves 1.5–4.0 cm long, usually not appearing in fascicles. Leaf blades with the rachis 5–14 mm long, gland-dotted, with 5–11 leaflets. Stipules 0.5–2.0 mm long. Leaflets 3–7(–13) mm long, 1–2 mm wide, relatively uniform over the length of the stem, narrowly oblong to oblong-oblanceolate or oblanceolate, tapered at the base, mostly rounded to broadly angled at the tip, but often with a minute, sharp extension of the midvein at the tip, the upper surface glabrous, the undersurface gland-dotted but otherwise glabrous. Inflorescences dense spikes (the axis not visible between the flowers), 7 to more commonly numerous per stem, 0.5–1.5 cm long, headlike, broadly oblong to nearly globose, the stalk 2–6 cm long, the axis glabrous (viewing the axis requires removal of flowers or fruits), with a few inconspicuous bracts and a whorl closely subtending the spike, these persistent, 1–2 mm long, 0.8–1.2 mm wide, tapered to short-tapered above an oblong to obovate base, usually reddish-tinged, usually not gland-dotted, grading into the bracts subtending flowers, these inconspicuous in bud and shed early, 1.5–2.5 mm long (shorter than the calyces of the associated buds), 0.8–1.5 mm wide, short-tapered above an oblong-obovate to obovate base, not gland-dotted, usually strongly reddish-tinged. Calyces with the tube 2.0–2.4 mm long, glabrous, the ribs green, gland-dotted around the rim, the lobes 1.0–1.4 mm long, triangular, with a gland at the tip, the margins minutely hairy. Petals white, the banner with the expanded portion 2.0–2.5 mm long, the wing and keel petals similar, attached along the rim of the stamen tube, the expanded portion 2–3 mm long. Stamens 5, the filament tube 2.0–2.4 mm long, the free filaments 2–3 mm long, the anthers yellow. Fruits included in the persistent calyx or more commonly somewhat exserted, 2–3 mm long, firm and gland-dotted above the membranous basal portion. $2n=14$. June–August.

Possibly introduced, uncommon, known thus far from Jackson and Polk Counties and the city of St. Louis (Colorado and Nebraska south to Texas; Mexico). Upland prairies; also railroads.

Dalea multiflora is distinguished from *D. candida* by the many branches, small globose spikes, smaller leaves and leaflets, and smaller bracts. The two species have overlapping ranges and possibly form hybrids (Wemple, 1970). Their flowers are very similar.

Round-headed prairie clover is quite rare in Missouri, known mostly from along railroads. The single Polk County specimen (*Timme 540* at the Missouri State University Herbarium) originated from La Petite Gemme Prairie Natural Area, where it has not been rediscovered since the original collection in 1980 despite numerous searches by several botanists. The plant may have been a chance, nonpersisting migrant into this high-quality upland prairie from a railroad track that crosses the western portion of the property.

6. Dalea purpurea Vent. **var. purpurea** (purple prairie clover)

Petalostemon molle Rydb.

P. purpureum (Vent.) Rydb.

Pl. 392 g, h; Map 1732

Plants perennial herbs, with a thick, dark brown to black, woody rootstock, sometimes with a thick, knotty caudex. Stems 1–5, 50–80 cm long, erect to strongly ascending or slightly arched, relatively few-branched toward the tip, occasionally unbranched, finely angled above the midpoint, the angles usually stronger and riblike below the nodes, glabrous or sparsely pubescent with inconspicuous, fine, curved hairs, rarely densely woolly, mostly lacking gland-dots. Leaves 1.5–4.0 cm long, sometimes appearing in fascicles, with 2–5 smaller trifoliate leaves at the base of the main leaf. Leaf blades with the rachis 10–20 mm long, gland-dotted, those of the main leaves with (3)5(7) leaflets. Stipules 2–4(–6) mm long. Leaflets 10–24 mm long, 1.0–1.4 mm wide, linear to narrowly oblanceolate,

1732. Dalea purpurea 1733. Dalea villosa 1734. Desmodium canadense

angled to somewhat tapered at the base and tip, folded longitudinally, the upper surface glabrous, the undersurface gland-dotted and sometimes also sparsely and inconspicuously hairy, rarely densely hairy. Inflorescences dense spikes (the axis not visible between the flowers), 2–5 cm long, cylindric, the stalk 1–6(–9) cm long, the axis densely pubescent with short, bristly hairs (viewing the axis requires removal of flowers or fruits), with numerous bracts, persistent until the fruits are shed, 2.5–5.0(–7.0) mm long, 1–2 mm wide, lanceolate-tapered to strongly tapered above an oblanceolate to ovate base, green or more commonly reddish-tinged above a membranous base, the expanded lower portion glabrous to sparsely hairy and sometimes gland-dotted, the tapered terminal portion densely hairy, with a pale thin band along the margin, this also hairy. Calyces with the tube 2.0–2.6 mm long, densely hairy, angled but not noticeably ribbed, not gland-dotted but often with minute reddish to brownish purple spots, the lobes 1.0–1.9 mm long, lanceolate to ovate, sometimes with minute reddish-purple spots, densely hairy. Petals pale to deep pinkish purple or reddish purple (rarely white elsewhere), the banner with the expanded portion 1.4–2.6 mm long, the wing and keel petals similar, attached along the rim of the stamen tube, the expanded portion 2.5–3.5 mm long. Stamens 5, the filament tube 2–3 mm long, the free filaments 2–3 mm long, the anthers orangish yellow. Fruits more or less included in the persistent calyx, 2.1–2.6 mm long, firm and densely hairy above the membranous basal portion, gland-dotted. $2n=14$. May–July.

Scattered nearly throughout the state, but uncommon in the Mississippi Lowlands Division (Montana to New Mexico east to Ohio and Alabama; Canada). Upland prairies, glades, tops of bluffs, openings of mesic to dry upland forests, and savannas; also pastures, cemeteries, railroads, and roadsides.

Dalea purpurea is a widespread and familiar wildflower of upland prairies, where it frequently co-occurs with *D. candida*. It exhibits great variation in degree of pubescence. Steyermark (1963) segregated densely hairy plants as f. *pubescens* (A. Gray) McGregor. A second variety with slender, long-stalked spikes that is mostly glabrous and occurs from Colorado and Nebraska south to New Mexico and Texas is known as var. *arenicola* (Wemple) Barneby. Rare plants with white corollas have been called f. *albiflorum* (Horr & McGregor) McGregor, but have only been recorded from Kansas thus far.

7. Dalea villosa (Nutt.) Spreng. **var. villosa** (silky prairie clover)

Map 1733

Plants perennial herbs, with ropelike, reddish orange roots below a branched caudex. Stems mostly 1 to several, 20–50 cm long, erect to loosely ascending, usually many-branched toward the tip, relatively strongly ribbed, densely pubescent with somewhat shaggy to woolly hairs, the gland-dots very sparse or absent. Leaves 2.0–4.5 cm long, appearing in fascicles, with 2–5 smaller 3- to 5-foliate leaves at the base of the main leaf. Leaf blades with the rachis 15–40 mm long, gland-dotted, those of the main leaves with 11–19 leaflets. Stipules 2–7 mm long. Leaflets 4–11 mm long, 1.5–3.5 mm wide, those of the lower leaves often larger than those of the upper leaves, elliptic to elliptic-oblanceolate, somewhat folded longitudinally, tapered at the base, angled to a bluntly or sharply pointed tip, usually also with a minute, sharp extension of the midvein at the tip, the surfaces moderately to densely pubescent with loosely appressed, silky to shaggy hairs, the upper surface glandless, the undersurface gland-dotted. Inflorescences dense spikes (the axis not visible between the flowers), 3–12 cm long (sometimes 10–14 cm at fruiting), cylindric (often becoming somewhat sinuous as the fruits mature), sessile or the stalk to 1 cm long, the axis densely short-hairy, with numerous bracts (additional sterile bracts often scattered along the inflorescence stalk), mostly shed early, 2.0–5.5 mm long, 1.0–1.4 mm wide, linear to narrowly lanceolate, abruptly tapered to a slender tip, the body green (this sometimes obscured by pubescence), the slender apical portion reddish purple, at least the lowermost gland-dotted, finely hairy, the margins lacking a conspicuous, pale, band. Calyces with the tube 2.0–2.7 mm long, densely pubescent with relatively long, silky hairs, not gland-dotted, but usually with minute reddish purple spots, the ribs light brown to purple (but mostly obscured by pubescence), the lobes 0.8–1.4 mm long, narrowly lanceolate to ovate-triangular, hairy and also glandular along the margins. Petals pink to reddish purple, the banner attached at the base of the stamen tube, the expanded portion 2.3–3.0 mm long, the wing and keel petals similar, attached along the rim of the stamen tube, the expanded portion 2.1–3.8 mm long. Stamens 5, the filament tube 2–3 mm long, the free filaments 2.5–4.5 mm long, the anthers yellow to orangish yellow. Fruits more or less included in the persistent calyx or occasionally slightly exserted, 2.5–3.2 mm long, thin and papery only near the base, somewhat thicker and papery, densely hairy, usually lacking gland-dots. $2n=14$. July–August.

Possibly introduced, uncommon, known thus far from a single historical specimen from Franklin County (Montana to New Mexico east to Wisconsin and Missouri; Canada). Habitat unknown, but probably in sandy soil.

The status of this species as a member of the Missouri flora is unclear. The sole voucher specimen was collected near Pacific in 1896 by George W. Letterman and is accessioned at the U.S. National Herbarium. The specimen label does not cite a habitat or more specific locality. It was overlooked by Steyermark (1963) and Yatskievych and Turner (1990), but the presence of a Missouri specimen was noted by Wemple (1970) and Barneby (1977). Wemple accepted Missouri as part of the taxon's distribution, but Barneby thought that its valid presence in Missouri was extremely doubtful. He also expressed doubts as to its disjunct presence in eastern Michigan, an opinion with which Voss (1985) concurred. Letterman botanized extensively in this portion of Missouri (see also the introductory chapter on the history of floristic botany in Missouri in Volume 1 of the present work [Yatskievych, 1999]) and collected several other taxa that have not been observed to grow in the state since his collections. The closest localities to the Missouri site are in eastern Kansas and eastern Iowa, more than 500 km to the west and north respectively. It is possible that the specimen was mislabeled or perhaps it originated from a nonnative plant growing as a waif along a railroad, but it cannot be excluded definitively from the state's flora from the presently available information. The native habitats of the taxon include dunes, prairies, banks of streams and rivers, and woodlands, always in sandy soils (McGregor, 1986).

Barneby (1977) treated *D. villosa* as comprising two varieties. The var. *grisea* (Torr. & A. Gray) Barneby is endemic to portions of eastern Texas and adjoining Louisiana and differs from var. *villosa* in its relatively tall, reddish, thinly hairy stems and somewhat larger, more strongly grayish-hairy leaflets.

26. **Desmodium** Desv. (beggar's lice, tick trefoil, stick-tights)
Contributed by Jay A. Raveill

Plants perennial herbs, with a short, often woody rootstock, often also taprooted, at least when young. Stems 1 to several, erect or ascending to less commonly spreading (not twining), often unbranched below the inflorescence, unarmed, glabrous or variously pubescent with glandular or nonglandular hairs, the nonglandular hairs often multicellular or with hooked tips. Leaves alternate, pinnately trifoliate or the lowermost (or rarely nearly all) leaves occasionally simple, short- to long-petiolate, with pubescence similar to that of the stem. Stipules shed as the leaves mature or persistent to various degrees, linear to broadly ovate, green and herbaceous when young, becoming brown or less commonly dark red and papery at maturity, the margins entire, often minutely hairy along the margins and on the outer surfaces, the venation prominent; stipels present, minute, usually persistent. Leaflets linear to broadly ovate or nearly circular, the lateral leaflets often slightly smaller and more rounded than the terminal one, rounded or angled to a usually well-developed (rarely nearly absent) stalk at the base, rounded to sharply pointed at the tip, sometimes with an abrupt, minute sharp point at the very tip, the margins entire, both surfaces glabrous or more commonly minutely to prominently hairy, pinnately veined. Inflorescences erect to ascending or arched (lateral inflorescences sometimes spreading), racemes (often grouped into panicles), rarely reduced to loose clusters, terminal and/or axillary, the axis with hooked hairs and sometimes also with straight or glandular hairs, the flowers paired or less commonly appearing as small clusters at each node, the bract(s) subtending each pair or cluster of flowers small, with prominent veins, shed as the flowers mature, the bractlet subtending each flower minute, linear, inconspicuous. Calyces 5-lobed but appearing 4-lobed, the short tube slightly inflated, 2-lipped, the upper 2 lobes fused except for a small notch at the tip, the lowermost lobe narrower and slightly longer than the others, hairy, sometimes inconspicuously so, usually inconspicuously nerved except for the midrib of each sepal, persistent at fruiting but not becoming enlarged. Corollas papilionaceous, glabrous, pale to dark pink, rarely white or cream-colored, often with a pair of contrasting purple, green, and/or yellow markings near the base of the banner petal (darker-colored corollas turning dull greenish blue or purplish with age, pale flowers turning cream-colored with age), the petals tapered or abruptly rounded (occasionally cordate) to a short, stalklike base, the banner obovate to oblong-obovate, the wings oblong, about as long as and usually somewhat fused to the keel basally, the keel oblanceolate and curved in outline, boat-shaped. Stamens 10, 9 of the filaments fused and 1 free nearly to the base, the anthers small, attached near the midpoint, all similar in size. Ovary linear to narrowly oblong-ellipsoid, sessile or short-stalked, the style slender, incurved, usually glabrous, more or less persistent at fruiting, the stigma small and terminal. Fruits loments, flattened, tapered abruptly to a short, often inconspicuous stalklike base 1.0–6.5 mm long, mostly 2- to several-seeded, indehiscent, divided into 1-seeded segments that separate at maturity, the segments with well-defined margins and a conspicuous network of nerves on each face, the connections between the segments nearly symmetrical or the lower margin somewhat more deeply lobed between the segments than the upper margin, the faces and sometimes also margins with hooked hairs. Seeds slightly kidney-shaped to nearly circular in outline, somewhat flattened, the surface smooth, green or pale yellow to dark brown, often turning black when dried. About 275 species, nearly worldwide but absent from portions of western North America and Europe.

The hooked hairs that are characteristic of the genus aid in dispersal of the fruit segments by adhesion to the fur of various passing mammals. A walk through an area with a dense population of *Desmodium* in the autumn is a testament to the efficacy of this dispersal mechanism, with socks and trousers becoming densely coated. The segments of the loments can be difficult to remove from fur or clothing when fresh; the process becomes easier once the segments have been allowed to dry. Steyermark (1963) noted that the fruits of some species are eaten by birds such as turkeys and quail and that the herbage is sometimes browsed by deer. He also noted that livestock graze on some of the species. Some of the Old World species are cultivated as green manure and fodder (Ohashi, 2005).

Identifications of species within this genus can present significant challenges. The taxon-specific differences can be subtle and/or subjective with a very small percentage of specimens displaying aberrant morphologies. Numerous varieties and forms have been named to account for extreme morphologies, but in the present work such infraspecific taxa are not considered sufficiently distinct to warrant formal recognition. The most obvious and frequently encountered variants are mentioned. Positive identification in *Desmodium* often requires the simultaneous examination of multiple characters including fruit, flower, and vegetative traits. The most significant difficulties in determinations involve species with similar fruit shape but contrasting pubescence and/or leaflet shape. Occasionally collectors have distributed several species under the same collection number, so caution should be used when making identifications based on duplicate collections (and in gathering samples from multiple plants in the field).

Some brief comments may aid in identifications. *Desmodium* flowers open in the morning and usually wilt by mid-afternoon, and often several consecutive nodes along the branches bloom on the same day. Although the flowers generally appear paired at each node, there often are one or more additional flower buds visible between the first pair; these buds only occasionally develop and flower later in the season. The number of segments in the fruit is dependent on the number of seeds that mature. Because in many fruits at least some ovules abort, the maximum number of segments should be considered the true number for identification purposes. The stalklike base of the fruit (stipe), can be difficult to measure due to the constriction of the lower part of the fruit, especially when the lowermost ovule is abortive. A change in texture and pubescence is used as the point of demarcation.

Herbarium specimens of other genera of legumes lacking reproductive structures are often misidentified as *Desmodium*. One distinguishing feature is that most *Desmodium* species have a pair of small stipels (stipulelike outgrowth at the base of each leaflet), which are not present in most other Missouri legume genera. Another helpful feature is that none of the Missouri *Desmodium* species are vines, so any evidence of climbing excludes this genus.

Descriptions of species are based on herbarium specimens collected from Missouri when possible. Some features that are often not well preserved in herbarium specimens, especially traits of the calyx, corolla, and bracts, also include data recorded from living plants observed in the field or grown in the greenhouse of the University of Central Missouri (Warrensburg). Corolla length, a critical trait in identification, is given for dried specimens, as the flowers usually shrink slightly upon drying.

Three of the Missouri species traditionally included in *Desmodium* (*D. glutinosum, D. nudiflorum*, and *D. pauciflorum*) are segregated into the genus *Hylodesmum* in the present work. For further discussion, see the treatment of that genus.

1. Stems prostrate or mostly spreading
    2. Stipules lanceolate to elongate-triangular, the base asymmetrically rounded but not clasping, semipersistent or shed early . . . . *D. ×humifusum* (see *D. rotundifolium*)
    2. Stipules ovate, with a short-tapered tip, the base slightly clasping, persistent
        3. Central leaflet ovate, the undersurface with a conspicuous network of raised veins; corollas white or cream-colored, drying pale yellow; fruits often twisted between the segments, with hooked hairs confined mostly to the margins of the segments . . . . 11. D. OCHROLEUCUM
        3. Central leaflet circular, broadly ovate, broadly obovate, or broadly rhombic, the undersurface with an inconspicuous to evident network of raised veins; corollas pink (rarely white), drying greenish blue to bluish purple; fruits flat, not twisted between the segments, with hooked hairs on both the margins and faces of the segments . . . . . 14. D. ROTUNDIFOLIUM
1. Stems erect or ascending to slightly arched
    4. Stipules large and conspicuous, mostly 7–20 mm long, broadly ovate to triangular-lanceolate, persistent or semipersistent
        5. Leaflets with the undersurface having a conspicuous network of raised veins, with hooked hairs along the veins throughout
            6. Inflorescences terminal as well as from the upper leaf axils, much-branched, the axis with dense, hooked and multicellular glandular hairs, as well as straight, spreading, nonglandular hairs; corollas 8–11 mm long; fruit segments 5–14 mm long, rhombic and angular with deeper indentation below than above . . . . 2. D. CANESCENS

6. Inflorescences terminal, not or few-branched, the axis with dense, hooked and multicellular glandular hairs, but lacking straight, nonglandular hairs; corollas 6–8 mm long; fruit segments 4–6 mm long, rounded, nearly equally indented above and below . . . . 6. D. ILLINOENSE

5. Leaflets with the undersurface having a faint to more evident network of at most slightly raised veins, lacking hooked hairs or these sparse and confined to larger veins near the leaflet base

7. Leaflets angled or slightly tapered to a bluntly or sharply pointed tip; stipules (5–)7–10 mm long; stipels 1.5–4.0 mm long; petioles 1–3 cm long; fruit segments 3–7 mm long, 3.5–5.0 mm wide . . . . 1. D. CANADENSE

7. Leaflets tapered to a sharply pointed tip; stipules 10–20 mm long; stipels 5–9 mm long; petioles 4–10 cm long; fruit segments 7–10 mm long, 4–6 wide . . . . 4. D. CUSPIDATUM

4. Stipules small, mostly 2–7 mm long, narrowly lanceolate to narrowly lanceolate-triangular, shed early or semipersistent

8. Fruit segments bluntly angled on the lower margin

9. Leaflets densely velvety-hairy on the undersurface, the central leaflet ovate-triangular, ovate, or rhombic; stipules semipersistent, often dark red and reflexed at maturity . . . . 17. D. VIRIDIFLORUM

9. Leaflets glabrous to densely hairy but not velvety to the touch, the central leaflet linear, oblong, lanceolate, ovate, or rhombic; stipules shed early or semipersistent, generally brown and appressed or slightly spreading at maturity

10. Stems and leaves glabrous or nearly so, glaucous (this sometimes not apparent in dried specimens); flower stalks (8–)10–20 mm long; fruits with the stalklike base 4.5–6.5 mm long; corollas 8–10 mm long; stipules shed early . . . . 7. D. LAEVIGATUM

10. Stems and leaves more or less hairy, not or only slightly glaucous; flower stalks 3–10(–17) mm long; fruits with the stalklike base 2–5 mm long; corollas 6–9 mm long; stipules generally semipersistent with at least some present in fruiting specimens

11. Stems and leaves with scattered appressed and straight and/or spreading and hooked hairs; at least the central leaflet lanceolate or narrowly oblong, generally 3–8 times as long as wide . . . . 12. D. PANICULATUM

11. Stems and leaves with moderate to dense, spreading, straight and/or hooked hairs; all leaflets ovate to broadly lanceolate or nearly rhombic, 1.5–4.0 times as long as wide (species difficult to separate)

12. Median portion of the stem with primarily hooked hairs, sometimes with a few straight hairs, all or mostly at the nodes; leaflets broadest at or slightly below the midpoint, the tip rounded or bluntly pointed, sometimes inconspicuously notched, the undersurface with a more or less obvious net work of raised veins . . . . 5. D. GLABELLUM

12. Median portion of the stem with relatively long, spreading, straight hairs and often with shorter, hooked hairs; leaflets widest well below the midpoint, the tip bluntly to sharply pointed, the undersurface usually with a faint network of raised veins . . . . 13. D. PERPLEXUM

8. Fruit segments rounded on the lower margin
  13. Fruits with the segments 4 or 5 (rarely fewer due to abortion), positioned in a more or less straight line; corollas 5–13 mm long
    14. Petioles of upper leaves 1–8 mm long, lower petioles to 30 mm long; flowers 8–13 mm long; primary bracts conspicuous before flowering, 5–10 mm long; stalklike base 1–2 mm long and hidden by the persistent calyx; stipules 5–10 mm long . . . . 1. D. CANADENSE
    14. Petioles of upper leaves 10–50 mm long; flowers 5–8 mm long; primary bracts inconspicuous before flowering, 1.5–3.0 mm long; stalklike base 2–5 mm long and extending above the persistent calyx; stipules 2–6 mm long . . . . 12. D. PANICULATUM
  13. Fruits with the segments 1–3 (sometimes 4 in *D. nuttallii*), appearing curved or arched downward in outline; corollas 3–5 mm long (to 7 mm in *D. nuttallii*)
    15. Leaflets relatively narrow, 4–10 times as long as wide, linear, narrowly lanceolate, or narrowly oblong
      16. Petioles nearly absent, 1–3 mm long; upper margin of each fruit segment convex (broadly rounded) . . . . 15. D. SESSILIFOLIUM
      16. Petioles 5–20 mm long; upper margin of each fruit segment straight to slightly concave; leaflets appearing strongly folded longitudinally . . . . 16. D. STRICTUM
    15. Leaflets relatively broad, 1–4 times as long as wide, oblong, lanceolate, ovate, deltoid, rhombic, or nearly circular
      17. Leaflets densely velvety-hairy on the undersurface; corollas 4–7 mm long; fruit segments 2–4; stipules often dark red and reflexed at maturity . . . . 9. D. NUTTALLII
      17. Leaflets glabrous to densely hairy but not velvety to the touch; corollas 3–5 mm long; fruit segments 1–3; stipules brown at maturity, appressed-ascending or slightly spreading
        18. Stems (below the inflorescence) and leaves glabrous or nearly so; leaflets with a barely visible network of raised veins; petioles of median leaves 5–30 mm long . . . . 8. D. MARILANDICUM
        18. Stems and leaves obviously hairy; leaflets with a conspicuous network of raised veins; petioles of median leaves 1–20 mm long
          19. Leaves with the central leaflet 0.5–3.0 cm long, usually less than twice as long as wide; petioles 1–10 mm long . . . . 3. D. CILIARE
          19. Leaves with the central leaflet 3–7 cm long, at least twice as long as wide; petioles of the median leaves 5–20 mm long . . . . 10. D. OBTUSUM

1. Desmodium canadense (L.) DC. (showy tick trefoil, Canada tick clover, giant tick clover)

Pl. 395 h, i; Map 1734

Stems 30–100 cm long, erect or ascending, rarely branched, the median portion with a mixture of sparse to dense, hooked and straight, spreading hairs. Petioles of the median leaves 1–3 cm long, progressively shorter toward the stem tip, those of the upper leaves only 0.1–0.8 cm long. Stipules 5–10 mm long, 0.8–1.0 mm wide, lanceolate, angled or slightly tapered to a bluntly or sharply pointed tip, at maturity brown and appressed-ascending or spreading, semipersistent. Leaflets flat or only slightly angled longitudinally, the undersurface with appressed to spreading, straight hairs and often also hooked hairs on the primary veins near the base, the network of raised veins inconspicuous to evident. Central leaflet (3–)5–8(–10) cm long, 1.0–3.5 cm wide, ovate to lanceolate (those of upper leaves narrower), the tip bluntly pointed to rounded. Lateral leaflets 2–7 cm long, 0.5–3.0 cm wide. Stipels 1.5–4.0 mm long. Inflorescences terminal and from upper axils, generally branched, the axis with hooked, multicellular glandular, and sometimes also spreading, straight hairs. Primary bracts 5–10 mm long, narrowly lanceolate, tapered to a sharply pointed tip.

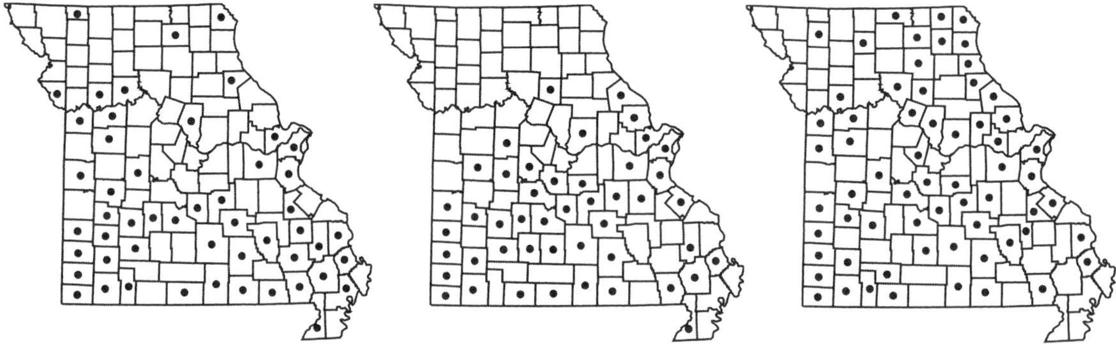

1735. Desmodium canescens 1736. Desmodium ciliare 1737. Desmodium cuspidatum

Secondary bracts 2.5–3.5 mm long wide, linear. Flower stalks 4–9 mm long. Calyces green or nearly white, sometimes with irregular purple markings, with sparse to dense, straight, spreading hairs and multicellular glandular hairs, as well as dense, very short hairs, the tube 1.5–3.0 mm long, the lobes 3.5–6.0 mm long. Corollas 8–13 mm long, pink or rarely white, the nectar guides pale yellow and white outlined in purple. Fruits straight in outline, the stalklike base 1–2 mm long and concealed by the persistent calyx, consisting of 1–5 segments, each 3–7 mm long and 3.5–5.0 mm wide, rounded above and below with deeper indentations below, uniformly covered with hooked hairs on the margins and faces. $2n=22$. June–September.

Scattered to uncommon throughout the state, more frequent north of the Missouri River, apparently absent from the Mississippi Lowlands Division (Northeastern U.S. west to North Dakota and Texas; Canada). Fens, banks of spring branches, and upland prairies; also old fields and roadsides.

This is the showiest species in the genus and is sometimes cultivated in perennial borders, especially in northern parts of Europe and North America, where it is one of the most cold-hardy species.

2. Desmodium canescens (L.) DC. (hoary tick trefoil, hoary tick clover)

Pl. 393 c, d; Map 1735

Stems erect, 50–200 cm long, erect or ascending, typically branched above, the median portion with moderate to dense, straight spreading and shorter hooked hairs. Petioles of median leaves 3–10 cm long, often only slightly shorter toward the stem tip (but the bracteal leaves often much reduced). Stipules (5–)7–15 mm long, 3.0–4.5 mm wide, broadly ovate with the base asymmetrically lobed and partially clasping the stem, angled or tapered to a sharply pointed tip, at maturity brown and reflexed, persistent. Leaflets flat or nearly so, the undersurface with many hooked and spreading hairs especially along the veins, the network of raised veins conspicuous. Central leaflet 4–12 (–14) cm long, 2.5–7.0 cm wide, ovate to ovate-lanceolate; tip sharply or bluntly pointed. Lateral leaflets 3–9 cm long, 1–5 cm wide. Stipels 2–4 mm long. Inflorescences terminal and from the upper leaf axils, branched, the axis with hooked, spreading and multicellular glandular hairs. Primary bracts 4–8 mm long, narrowly ovate, tapered to a sharply pointed tip. Secondary bracts 1.5–2.5 mm long, linear. Flower stalks 7–14 mm long. Calyces green, moderately long-hairy and densely very short-hairy, the margins conspicuously hairy, the tube 1–2 mm long, the lobes 3–5 mm long. Corollas 8–11 mm long, pink or rarely white, the nectar guides green and white outlined in purple. Fruit straight or slightly curved upward, the stalklike base 2–3 mm long, consisting of (1–)3–6 segments, each 5–14 mm long and 4–7 mm wide, rhombic with deeper indentations below, the hooked hairs confined mostly to the segment margins, the faces nearly glabrous. $2n=22$. July–September.

Scattered nearly throughout the state (Massachusetts, Wisconsin, and Nebraska south to Florida and Texas; Canada). Bottomland forests, mesic upland forests and forest margins, upland prairies, banks of stream and rivers, and ledges of bluffs; also fencerows, old fields, pastures, levees, railroads, and roadsides.

3. Desmodium ciliare (Muhl. ex Willd.) DC. (hairy small-leaf tick trefoil, slender tick clover)

Pl. 393 i, j; Map 1736

Stems erect, 40–100 cm long, often unbranched above the base, the median portion with sparse to dense, hooked hairs and frequently also straight, spreading hairs. Petioles 0.1–1.0 cm long. Stipules 3–6 mm long, 0.2–0.5 mm wide, linear, tapered to

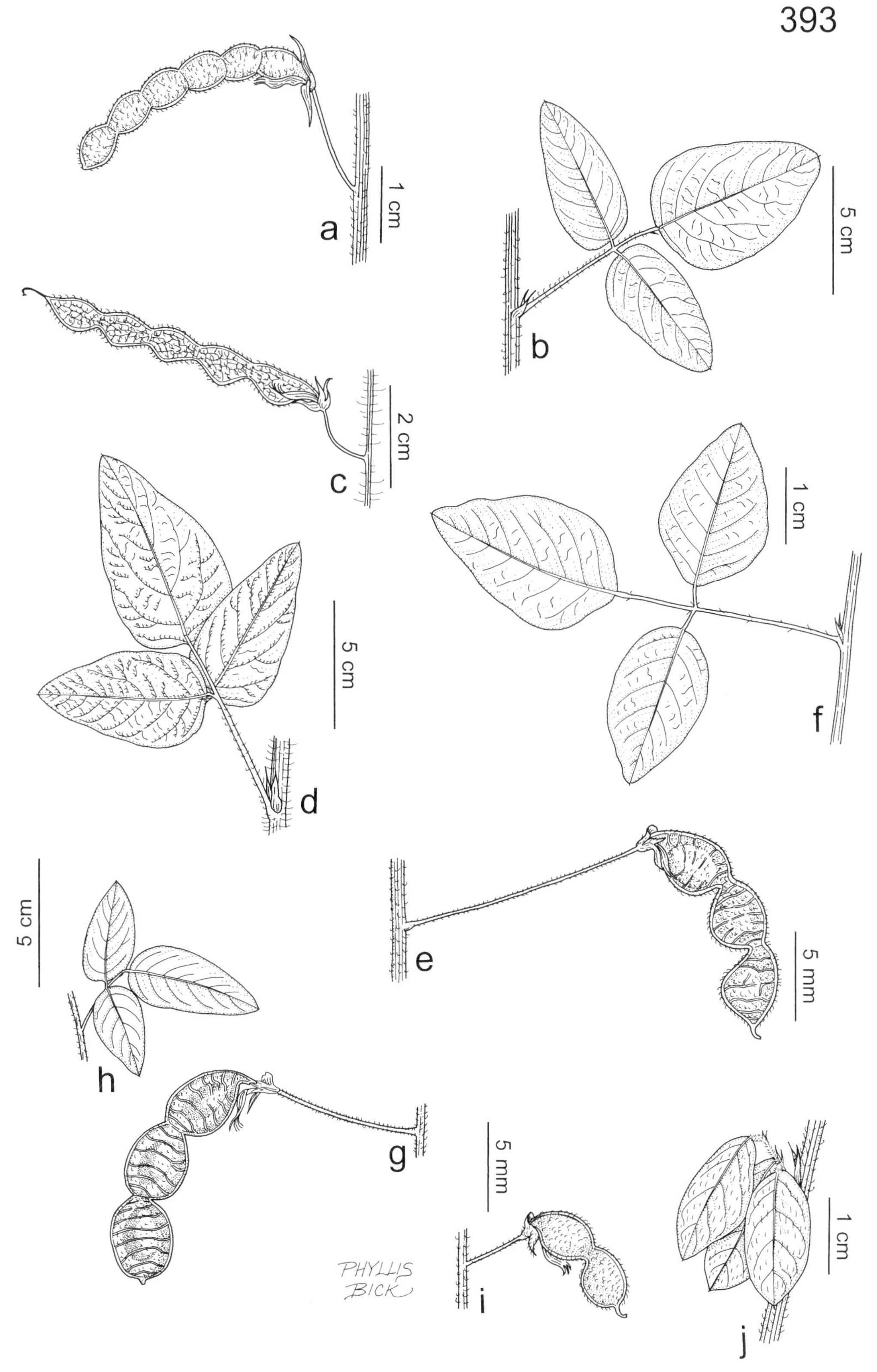

Plate 393. Fabaceae. *Desmodium illinoense*, **a)** fruit, **b)** leaf attached to stem. *Desmodium canescens*, **c)** fruit, **d)** leaf attached to stem. *Desmodium marilandicum*, **e)** fruit, **f)** leaf attached to stem. *Desmodium obtusum*, **g)** fruit, **h)** leaf attached to stem. *Desmodium ciliare*, **i)** fruit, **j)** leaf attached to stem.

a sharply pointed tip, at maturity brown and appressed-ascending, semipersistent or shed before the fruits mature. Leaflets flat or nearly so, the undersurface with sparse, straight, spreading hairs or nearly glabrous, the network of raised veins conspicuous. Central leaflet 0.5–3.0 cm long, 0.5–1.7 cm wide (mostly 1–2 times as long as wide), elliptic to rhombic or ovate, the tip bluntly pointed to rounded. Lateral leaflets 0.8–2.5 cm long, 0.3–1.5 cm wide. Stipels 0.5–1.5 mm long. Inflorescences all or mostly terminal, unbranched or few-branched, the axis with hooked and sometimes also straight spreading hairs. Primary bracts 1–2 mm long, ovate, tapered to a sharply pointed tip. Secondary bracts 1.0–1.5 mm long, linear to very narrowly lanceolate. Flower stalks 3–8 mm long. Calyces green, sometimes with irregular reddish patches, with dense, very short hairs and scattered, straight, spreading hairs, especially on the lower lobes, the tube 0.5–1.5 mm long, the lobes 1.0–1.8 mm long. Corollas 3–5 mm long, pink, the nectar guides green and yellow outlined in purple. Fruits arched downward in outline, the stalklike base 1–2 mm long, consisting of 1 or 2(3) segments, each 3–5 mm long and 2–3 mm wide, rounded above and below with deeper indentations below, uniformly covered with hooked hairs on the margins and faces. $2n=22$. August–September.

Scattered in the Mississippi Lowlands, Ozark, Ozark Border, and Unglaciated Plains Divisions (Massachusetts, and Nebraska south to Florida and Texas; Canada, Mexico, Cuba). Upland prairies, sand prairies, glades, openings of mesic to dry upland forests, and savannas; also pastures, old fields, fencerows, and railroads.

Desmodium ciliare and *D. marilandicum* frequently grow in close proximity in dry, acid soils in the Ozarks. Some specimens tend to combine traits of the two species making determinations difficult. Hybridization may be responsible for the morphological continuum (Isely, 1990), but documentation to support this is lacking. Isely (1990) also suggested the occurrence of putative hybrids between *D. ciliare* and the closely related *D. obtusum*.

4. Desmodium cuspidatum (Muhl. ex Willd.) Loudon (largebract tick trefoil, longleaf tick clover)

D. cuspidatum var. *longifolium* (Torr. & A. Gray) B.G. Schub.

Pl. 395 c–e; Map 1737

Stems 50–150 cm long, erect to somewhat arched, unbranched or branched above the midpoint, the median portion glabrous or sparsely (rarely densely) covered with straight, spreading hairs and short, hooked hairs. Petioles of the median leaves 4–10 cm long, those of the upper leaves often only slightly shorter. Stipules 10–20 mm long, 3–4 wide, lanceolate to ovate, tapered to a sharply pointed tip, at maturity brown and appressed to spreading, semipersistent, occasionally shed early. Leaflets flat or nearly so, the undersurface nearly glabrous or sparsely (densely) covered with straight spreading and/or hooked hairs, often glaucous, the network of raised veins inconspicuous to evident. Central leaflet 5–12 (–14) cm long, (2.0–)4.0–6.5 cm wide, ovate, long-tapered (rarely short-tapered) to a sharply pointed tip. Lateral leaflets 4–11 cm long, 1.5–4.5 cm wide. Stipels 5–9 mm long. Inflorescences terminal and from the upper leaf axils, unbranched or with a few branches, the axis with hooked hairs. Primary bracts 5–11 mm long, ovate, tapered to a sharply pointed tip. Secondary bracts 2.0–2.5 mm long, lanceolate to narrowly lanceolate. Flower stalks 2–8 mm long. Calyces green to nearly white, sparsely pubescent with glandular hairs, the tube 1.0–1.5 mm long, the lobes 2–4 mm long. Corollas 7–12 mm long, pink, the nectar guides white and yellow outlined in purple. Fruits straight in outline, the stalklike base 2–5 mm long, consisting of (1–)3–7 segments, each 7–10 mm long and 4–6 wide, rhombic, with nearly equal indentations on the upper and lower sides, with dense hooked hairs on the margins and scattered hooked hairs on the faces, the margins often purple. $2n=22$. July–September.

Scattered nearly throughout the state (eastern U.S. west to Minnesota and Texas; Canada). Mesic to dry upland forests, glades, tops of bluffs, land prairies, savannas, banks of streams and rivers, and margins of fens and calcareous seeps; also pastures, old fields, railroads, and roadsides.

Desmodium cuspidatum generally is usually distinguished by its long-tapered leaflet tips, more or less glabrous appearance, and rhombic fruit segments. More pubescent specimens have been designated as var. *longifolium*, but extensive intergradation prevents formal recognition of this variant. In addition to increased pubescence, some collections referred to this variety have leaflets with tips that are only slightly pointed and much shorter stipules and stipels than otherwise is typical of the species. Some of the more pubescent specimens can be difficult to distinguish from either *D. perplexum* or *D. canescens*. The possibility of hybridization between *D. cuspidatum* and other members of the genus should to be investigated. Specimens currently passing as var. *longifolium* may represent more than one hybrid combination.

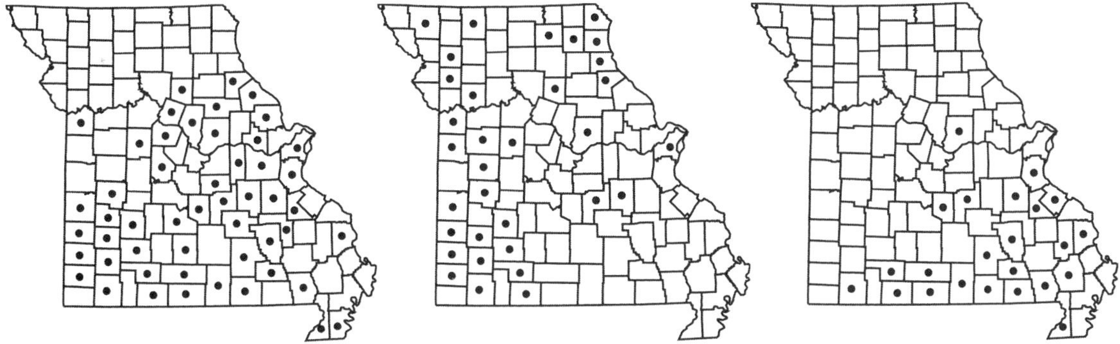
1738. Desmodium glabellum 1739. Desmodium illinoense 1740. Desmodium laevigatum

5. Desmodium glabellum (Michx.) DC. (tall tick clover, Dillenius' tick trefoil)

D. paniculatum (L.) DC. var. *dillenii* (Darl.) Isely (in part)

Pl. 394 g; Map 1738

Stems 50–150 cm long, erect to somewhat arched, branched or unbranched, the median portion with mostly hooked hairs, sometimes also with sparse, straight, spreading hairs, mainly at the nodes. Petioles 0.8–3.0 cm long, somewhat shorter toward the stem tip. Stipules 2–4 mm long, narrowly lanceolate to narrowly triangular, tapered to a sharply pointed tip, brown at maturity and appressed, falling before the fruits mature or less commonly some of them more persistent. Leaflets flat or nearly so, the undersurface with short appressed or spreading hairs, sometimes also with hooked hairs along the midvein, the network of raised veins on the undersurface evident to conspicuous. Central leaflet 2–8(–10) cm long, 1.5–5.0(–6.0) cm wide (generally 1.5–4.0 times as long as wide), rhombic or rarely ovate to lanceolate, broadest at or slightly below the midpoint, the tip rounded to bluntly pointed, sometimes inconspicuously notched. Lateral leaflets 2–8 cm long, 1.0–4.5 cm wide. Stipels 1–3 mm long. Inflorescences terminal and from the upper leaf axils, branched, the axis covered with hooked hairs. Primary bracts 1–3 mm long, ovate to narrowly ovate, tapered to a sharply pointed tip, shed early. Secondary bracts 1.0–1.2 mm long, linear, shed early. Flower stalks 3–10 mm long. Calyces green, with sparse, fine, long, straight hairs, the margins also hairy, the tube 1.0–1.8 mm long, the lobes 1–2 mm long. Corollas 6–8 mm long, pink, the nectar guides green and yellow outlined in purple. Fruits straight in outline, the stalklike base 3–5 mm long, consisting of (1–)3–5 segments, each 5–8 mm long and 3–5 mm wide, the upper margin shallowly and broadly rounded, the lower margin bluntly angled, with deeper indentations below, uniformly covered with hooked hairs on the margins and faces. $2n=22$. July–September.

Scattered, mostly south of the Missouri River, north locally to Pike County (eastern U.S. west to Iowa, Kansas, and Texas). Mesic to dry upland forests, tops of bluffs, glades, savannas, and banks of streams and rivers; also railroads and roadsides.

Desmodium glabellum is closely allied with *D. paniculatum* and *D. perplexum* and can be especially difficult to distinguish from the latter taxon. Further discussion on the complex is provided under *D. paniculatum*. Stem pubescence is the most frequently used trait to distinguish this species from *D. perplexum*, as indicated in the key, but this trait is not always diagnostic. Most populations contain at least a few individuals with straight, spreading hairs at the nodes and sometimes also in two lines below the node or even more generally distributed. The most consistent traits for distinguishing *D. glabellum* from *D. perplexum* are subtle, including the slightly thicker leaflets with a more obvious network of veins and rounded or even minutely notched leaflet tips. This species is generally well-defined in Missouri in habitats such as open woods and woodland edges, even when growing with *D. perplexum* and *D. paniculatum*. However morphologically ambiguous individuals can be abundant in habitats such as roadsides and gravel bars along streams, and some herbarium specimens cannot be assigned readily to any of the three species in the complex. The recognition of *D. glabellum, D. paniculatum*, and *D. perplexum* as distinct species seems the most prudent action until further research determines the degree of genetic distinctness among these three morphologically complex taxa. For a discussion of hybridization between members of the *D. paniculatum* complex and *D. rotundifolium*, see the treatment of *D. rotundifolium*.

6. Desmodium illinoense A. Gray (Illinois tick trefoil, Illinois tick clover)

Pl. 393 a, b; Map 1739

Stems 50–150 cm long, erect to slightly arched, generally unbranched, the median portion with hooked hairs and sometimes also with straight and spreading and/or multicellular glandular hairs. Petioles of median leaves 2.5–4.0 cm long, progressively shorter toward the stem tip (and base). Stipules 10–15 mm long, 3–6 mm wide, ovate to narrowly ovate (usually asymmetric at the base), tapered to a sharply pointed tip, at maturity brown and appressed to somewhat spreading but not reflexed, persistent. Leaflets flat or nearly so, the undersurface with relatively dense hooked hairs, especially on the veins, the network of raised veins conspicuous. Central leaflet (2–)4–10 cm long, (0.7–)2.0–5.5 cm wide, lanceolate to ovate; the tip usually broadly and bluntly pointed, sometimes with a minute sharp extension of the midvein at the very tip. Lateral leaflets (2.0–)4.0–6.5 cm long, 1.0–4.5 cm wide. Stipels 3–6 mm long. Inflorescences terminal, unbranched or less commonly few-branched, the axis with dense hooked hairs and muticellular, glandular hairs. Primary bracts 5.5–9.0 mm long, triangular to narrowly triangular, tapered to a sharply pointed tip, mostly shed early. Secondary bracts 2.5–4.0 mm wide, linear, shed early. Flower stalks 7–20(–25) mm long. Calyces green, with dense glandular hair and sparse long straight hairs, as well as usually dense minute, nonglandular hairs, the tube 1–2 mm long, the lobes 1.5–3.0 mm long. Corollas 6–8 mm long, pink or white, the nectar guides yellow and white outlined in purple. Fruits straight in outline, the stalklike base about 1 mm long and concealed by the persistent calyx, consisting of (2–)4–7 segments, each 4–6 mm long and 3.5–4.0 mm wide, rounded above and below, with nearly equal indentations on the upper and lower sides, uniformly covered with hooked hairs on the margins and face. $2n$=22. June–September.

Uncommon in the northern and western halves of the state; absent from most of the Ozark Division and the Mississippi Lowlands (South Dakota to Oklahoma east to Michigan, Ohio, and Arkansas; Canada). Upland prairies, banks of streams, and tops of bluffs; also pastures, railroads, and roadsides.

7. Desmodium laevigatum (Nutt.) DC. (smooth tick trefoil, smooth tick clover)

Pl. 394 c, d; Map 1740

Stems 40–120 cm long, erect or ascending to strongly arched, branched or unbranched, the median portion glabrous or occasionally with very sparse hooked hairs, slightly to more noticeably glaucous. Petioles 1–6 cm long. Stipules 5–7(–8) mm long, 2–3 mm wide, narrowly ovate (somewhat asymmetric at the base), tapered to a sharply pointed tip, green or purple, turning brown, appressed or spreading, shed early. Leaflets flat or nearly so, the undersurface glabrous or occasionally with sparse hooked hairs on the veins, usually conspicuously glaucous, the raised network of veins scarcely evident. Central leaflet 2–8 cm long, 2.5–5.5 cm wide, ovate; the tip rounded or broadly and bluntly pointed, sometimes with a minute sharp extension of the midvein at the very tip. Lateral leaflets (1.3–)2.0–8.0(–10.0) cm long, (0.9–)1.5–4.5(–5.5) cm wide. Stipels 1–4 mm long. Inflorescences terminal and from the upper leaf axils, branched, the axis with hooked hairs. Primary bracts 2–3 mm long, narrowly lanceolate, tapered to a sharply pointed tip, mostly shed early. Secondary bracts 1.0–1.5 mm long, linear to hairlike. Flower stalks (7–)10–20 mm long. Calyces green, with sparse short glandular and straight appressed nonglandular hair as well as dense fine minute hairs, the tube 1.5–2.2 mm long, the lobes 2–3 mm long. Corollas 8–10 mm long, dark pink to light purple, the nectar guides green, yellow, and white outlined in purple. Fruits straight in outline, the stalklike base 4.5–6.5 mm long, consisting of (1–)3–5 segments, each 4–8 mm long and 3.5–5.0 mm wide, the upper margin slightly rounded or broadly angled to nearly straight, the lower margin bluntly angled, with deeper indentations below, uniformly covered with hooked hairs on the margins and faces. $2n$=22. July–October.

Scattered in the Ozark, Ozark Border, and Mississippi Lowlands Divisions (eastern U.S. west to Missouri and Texas). Upland forests, sand savannas, and banks of streams and spring branches; also railroads and roadsides.

The fruit shape in this species is similar to that of *D. paniculatum* and related species. It generally is recognized easily by its smooth, glaucous appearance and slightly thicker and leathery leaves, stipules that are shed early, and fruits that are both longer-stalked and have a longer stalklike base. The nodes may be purple, but this trait is neither universal nor diagnostic, as several other species occasionally have similar purple nodes. An artificial interspecific cross between a purple-noded plant of this species and a plant of *D. perplexum* that lacked purple nodes indicates that this is a single-gene dominant trait (Raveill, 1995).

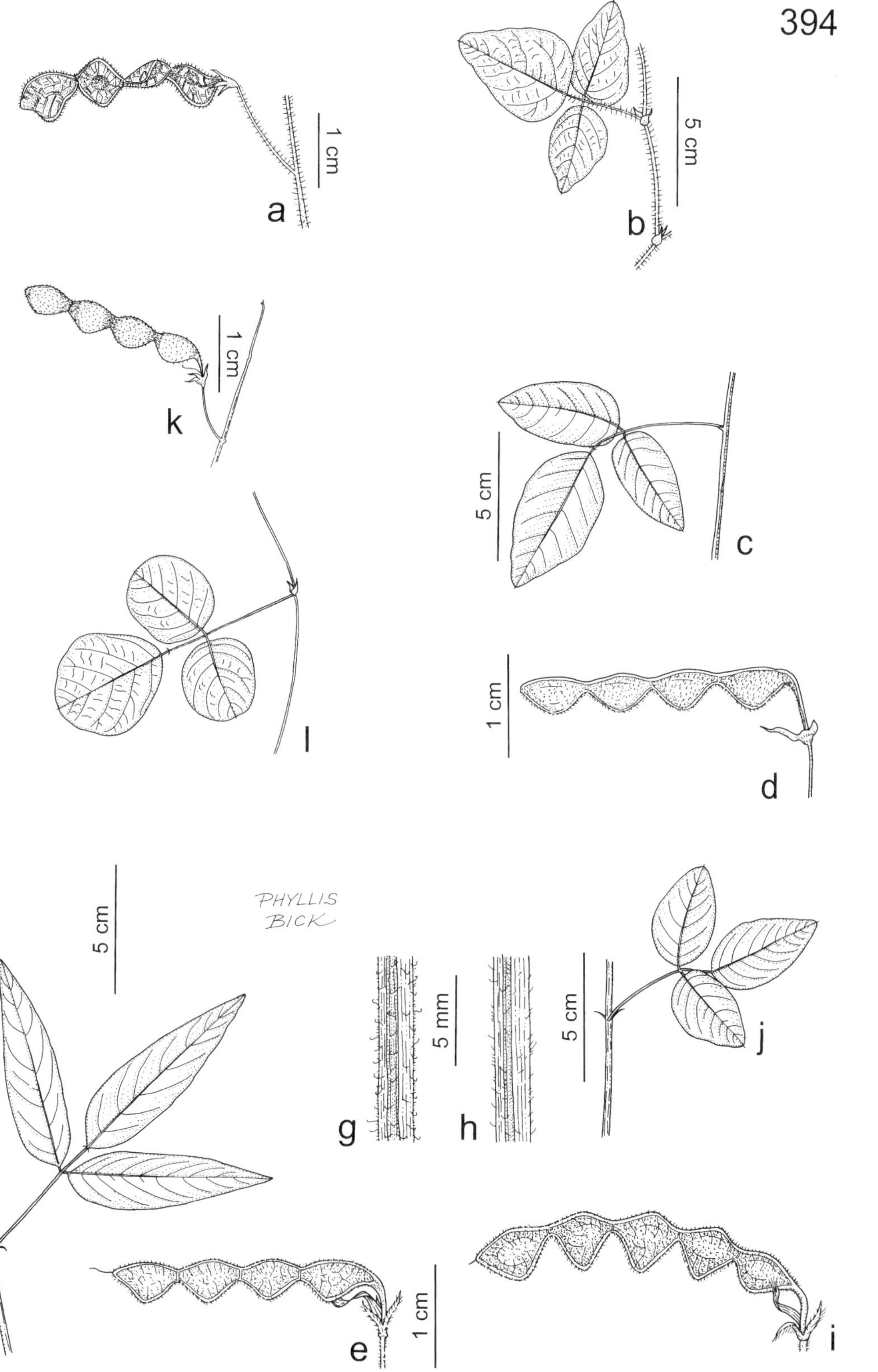

Plate 394. Fabaceae. *Desmodium ochroleucum*, **a)** fruit, **b)** leaf attached to stem. *Desmodium laevigatum*, **c)** leaf attached to stem, **d)** fruit. *Desmodium paniculatum*, **e)** fruit, **f)** leaf attached to stem. *Desmodium glabellum*, **g)** median portion of stem showing pubescence. *Desmodium perplexum*, **h)** median portion of stem showing pubescence, **i)** fruit, **j)** leaf attached to stem. *Desmodium rotundifolium*, **k)** fruit, **l)** leaf attached to stem.

1741. Desmodium marilandicum 1742. Desmodium nuttallii 1743. Desmodium obtusum

8. Desmodium marilandicum (L.) DC. (smooth small-leaf tick trefoil, Maryland tick clover)

Pl. 393 e, f; Map 1741

Stems 30–100 cm long, erect or ascending, often unbranched, the median portion glabrous or with sparse hooked hairs. Petioles of the median leaves 1–3 cm long, progressively shorter toward the stem tip. Stipules 2–5 mm long, 0.3–0.7 mm wide, narrowly triangular, tapered to a sharply pointed tip, at maturity brown and appressed or spreading, mostly shed before the fruits mature. Leaflets flat or nearly so, the undersurface glabrous or with sparse hooked hairs, the network of raised veins scarcely evident. Central leaflet 1–4 cm long, 0.7–2.5 cm wide, ovate to nearly circular (1–3 times as long as wide), the tip rounded or nearly so, sometimes with a minute sharp extension of the midvein at the very tip. Lateral leaflets 0.6–2.5 cm long, 0.4–1.8 cm wide. Stipels 1–3 mm long. Inflorescences terminal and sometimes from the upper leaf axils, often branched, the axis nearly glabrous or with sparse to dense hooked hairs. Primary bracts 1.0–1.5 mm long, lanceolate, tapered to a sharply pointed tip, mostly shed early. Secondary bracts 1.0–1.2 mm long, linear. Flower stalks 5–20 mm long. Calyces green, with sparse straight spreading nonglandular hairs and multicellular glandular hairs, as well as dense short hairs, the tube 1–2 mm long, the lobes 1–2 mm long. Corollas 3–5 mm long, dark pink, the nectar guides usually green with white near the base and outlined in purple. Fruits arched downward in outline, the stalklike base 1.0–2.5 mm long, consisting of (1)2 or 3 segments, each 3.5–5.0 mm long and 3–4 mm wide, the upper and lower margins rounded, with deeper indentation below, uniformly covered with hooked hairs on the margins and faces. $2n=22$. July–September.

Scattered mostly south of the Missouri River, north locally to Adair and Pike Counties (eastern U.S. west to Kansas and Texas; Canada). Mesic to dry upland forests, upland prairies, and tops of bluffs; also old fields and roadsides.

9. Desmodium nuttallii (Schindl.) B.G. Schub. (Nuttall's tick trefoil)

Pl. 396 c, d; Map 1742

Stems 40–150 m long, erect or ascending, sometimes somewhat arched, often unbranched, the median portion with sparse to dense hooked hairs and scattered straight spreading hairs, sometimes nearly glabrous. Petioles of the median leaves 1–3 cm long, only slightly reduced toward the stem tip. Stipules 4–7 mm long, 1.0–1.5 mm wide, lanceolate, tapered to a sharply pointed tip, at maturity brown or pale to dark red and reflexed or spreading, occasionally appressed, more or less persistent. Leaflets flat or nearly so (sometimes appearing slightly concave), the undersurface velvety-hairy, the network of raised veins evident, more or less visible through the pubescence. Central leaflet 2.5–9.0(–10.5) cm long, 2.5–5.5 cm wide (mostly 1–4 times as long as wide), lanceolate to more commonly ovate, rhombic, oblong-ovate, or triangular-ovate, the tip bluntly to sharply pointed. Lateral leaflets (1.0–)2.5–8.0 cm long, (0.8–)1.5–4.5 cm wide. Stipels 1–4 mm long. Inflorescences usually terminal and branched, the axis with dense hooked hairs and straight, spreading hairs. Primary bracts 2–4 mm long, ovate to narrowly ovate, tapered to a sharply pointed tip, more or less persistent. Secondary bracts 1–2 mm long, linear to very narrowly lanceolate. Flower stalks 4–8 mm long. Calyces green, with sparse to dense straight spreading hairs and appressed hairs, as well as dense very short hairs, the tube 1.0–1.5 mm long, the lobes 1–2 mm long. Corollas 4–7 mm long, pink, the nectar guides green and white outlined in purple. Fruits arched downward in outline, the stalklike base 2–4 mm long, consisting of (1)2–4 segments, each 3–5 mm long and 4.0–5.5 mm wide,

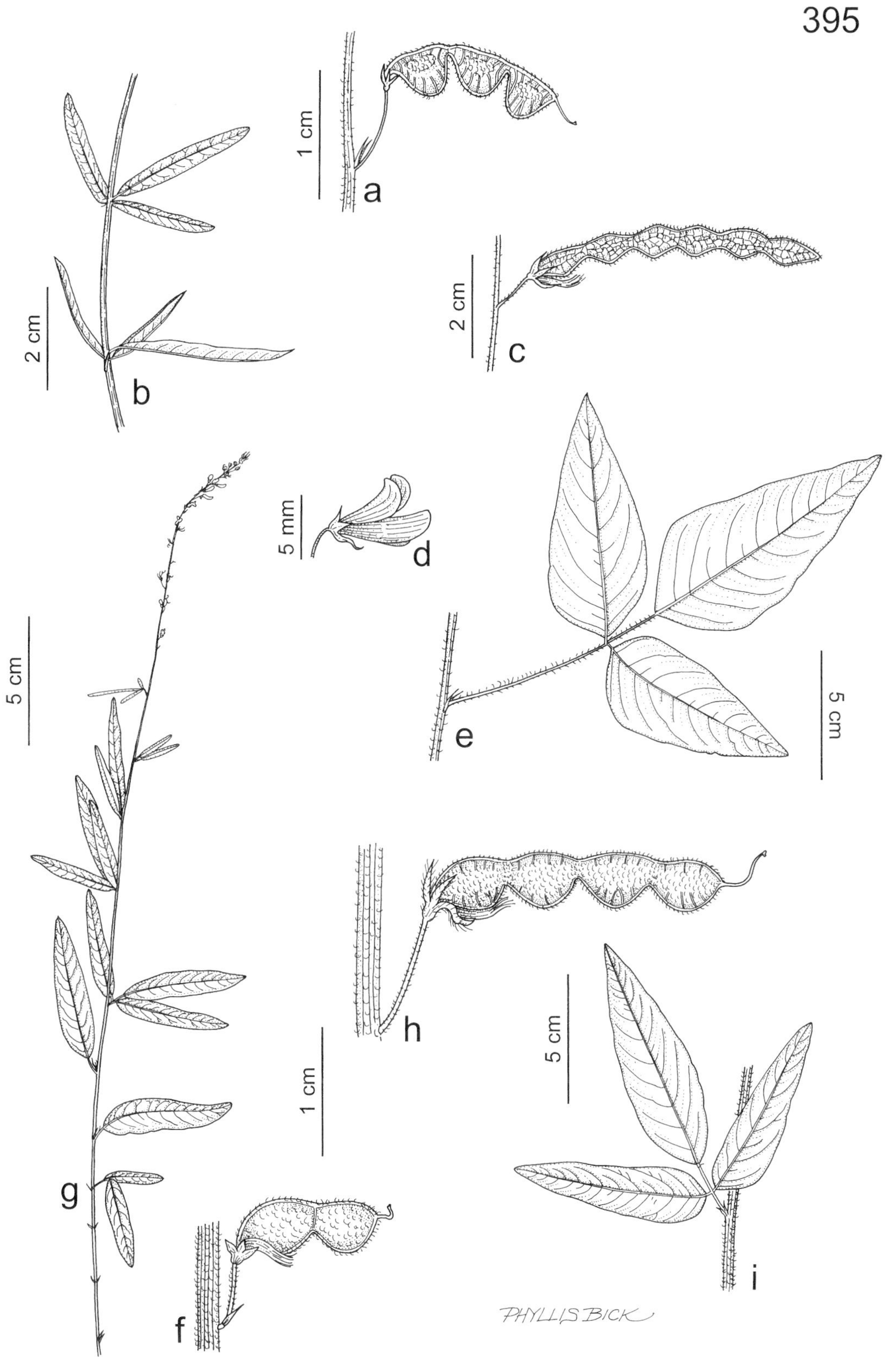

Plate 395. Fabaceae. *Desmodium strictum*, **a)** fruit, **b)** leaves attached to stem. *Desmodium cuspidatum*, **c)** fruit, **d)** flower, **e)** leaf attached to stem. *Desmodium sessilifolium*, **f)** fruit, **g)** habit. *Desmodium canadense*, **h)** fruit, **i)** leaf attached to stem.

rounded on the upper and lower margins, with deeper indentations below, uniformly covered with hooked hairs on the margins and faces. $2n=22$. August–September.

Uncommon in the Ozark and Ozark Border Divisions, as well as the northern portion of the Mississippi Lowlands (eastern U.S. west to Missouri and Texas). Mesic to dry upland forests, upland prairies, savannas, and banks of streams; also pastures and roadsides.

A close relationship exists between this species and *D. viridiflorum*, and mature fruit is usually needed for definitive determinations. Generally the pubescence on the leaflet undersurface is denser in *D. viridiflorum* than in *D. nuttallii*, obscuring the raised venation, but this character can be difficult to diagnose. Geographic location is somewhat useful in distinguishing these species in Missouri. *Desmodium nuttallii* is widespread in the southern and eastern portions of the Ozark Division, whereas the less common *D. viridiflorum* occurs in the Mississippi Lowlands. Unfortunately, both species occur in the Crowley's Ridge portion of Stoddard County, as well as the Sikeston Ridge in Scott County. Some specimens of *D. nuttallii* also resemble the more pubescent extremes of *D. perplexum*, but fruit shape is diagnostic.

10. Desmodium obtusum (Muhl. ex Willd.) DC. (stiff tick trefoil)

D. rigidum (Elliott) DC.

Pl. 393 g, h; Map 1743

Stems 30–100 cm long, erect or ascending, often unbranched, the median portion with dense hooked hairs. Petioles of the median leaves 0.5–2.0 cm long, progressively shorter toward the stem tip, the upper leaves nearly sessile. Stipules 4–6 mm long, 2.0–2.5 mm wide, lanceolate (asymmetric at the base), tapered to a sharply pointed tip, at maturity brown and appressed, often shed before the fruits mature. Leaflets flat or nearly so, the undersurface moderately pubescent with hooked and straight spreading hairs, especially along veins, the network of raised veins conspicuous. Central leaflet (0.5–)3.0–7.0 cm long, 0.5–3.5 cm wide, oblong, elliptic, or ovate-oblong (mostly 2–4 times as long as wide), the tip rounded to broadly and bluntly pointed, rarely sharply pointed. Lateral leaflets 2–4 cm long, 1–2 cm wide. Stipels 1–2 mm long. Inflorescences terminal and sometimes from the upper leaf axils, often sparingly branched, the axis with moderate to dense hooked hairs. Primary bracts 2.5–4.5 mm long, narrowly ovate to lanceolate, mostly persistent. Secondary bracts 1.0–1.5 mm long, lanceolate to narrowly ovate. Flower stalks 5–18 mm long. Calyces green, with dense, appressed, straight hairs and dense very short hairs, the margins also with spreading hairs, the tube 1.0–1.5 mm long, the lobes 1.5–2.5 mm long. Corollas 3–5 mm long, pink or white, the nectar guides green, yellow, and white outlined in purple. Fruits arched downward in outline, the stalklike base 1–3 mm long, consisting of 1–3 segments, each 4–5 mm long and 3–4 mm wide, rounded on the upper and lower margins with deeper indentations below, uniformly covered with hooked hairs on the margins and faces. $2n=22$. July–September.

Scattered nearly throughout the state, but absent from most of the western half of the Glaciated Plains Division (eastern U.S. west to Colorado and Texas). Mesic to dry upland forests, glades, upland prairies, sand prairies, and banks of streams; also pastures and roadsides.

Isely (1990) indicated the presence of plants in the southeastern United States with leaflet shapes and pubescence intermediate between *D. obtusum* and the closely related *D. ciliare*). He suggested that these arose through hybridization between the two species. Putative hybrids with this parentage have yet to be recorded from Missouri, but are likely to be found in the future.

11. Desmodium ochroleucum M. Curtis ex Canby (cream tick trefoil)

Pl. 394 a, b; Map 1744

Stems 40–100 cm long, trailing with ascending flowering branches, sometimes also loosely ascending toward the tip, branched, the median portion with dense, hooked hairs and moderate to dense, straight, spreading hairs. Petioles of the median leaves 1.5–5.0 cm long, only slightly reduced toward the stem tip. Stipules 6–10 mm long, 3.5–7.0 mm wide, ovate to ovate-lanceolate, broadened at the base and partially clasping the stem, at maturity brown and reflexed, with prominent, parallel, raised veins, persistent. Leaflets flat or nearly so, the undersurface sparsely to densely pubescent with hooked and straight spreading hairs, especially along the veins, the network of raised veins conspicuous. Central leaflet (1.5–)3.0–5.5 cm long, 1.5–5.0 cm wide, ovate to rhombic, the tip rounded to sharp pointed. Lateral leaflets (0.6–)2.0–4.5 cm long, (0.6–)1.5–2.5 cm wide, occasionally absent and the terminal leaflet then broadly ovate-triangular. Stipels 1–3 mm long. Inflorescences terminal and from the leaf axils, branched, the axis with dense hooked hairs and sparse to moderate, straight, spreading hairs, as well as glandular hairs. Primary bracts 3–5 mm long, ovate, tapered to a sharply pointed tip, with prominent, parallel, raised veins, persistent. Sec-

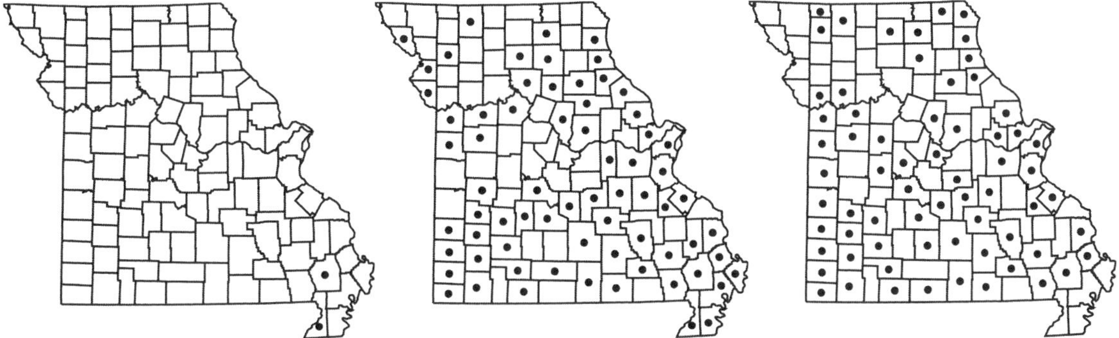

1744. Desmodium ochroleucum 1745. Desmodium paniculatum 1746. Desmodium perplexum

ondary bracts 2.5–3.0 mm long, loosely subtending the flowers, linear, more or less persistent. Flower stalks 8–20 mm long. Calyces green, with sparse to moderate, straight, glandular hairs and hooked hairs, as well as dense, minute hairs. Corollas 6–8 mm long, white or cream-colored, drying a pale yellow, lacking nectar guides. Fruits straight in outline, the stalklike base 2–3 mm long, consisting of 2–4 segments, each 5–8(–10) mm long and 4.5–6.0 mm wide, rounded to angular on the upper and lower margins, nearly equally indented above and below, often twisted at the joints, with dense hooked hairs on the margins and scattered hooked hairs on the faces, the faces sometimes glabrous or nearly so. August–September.

Uncommon, known only from historical collections from Butler and Dunklin Counties (New Jersey, West Virginia, Tennessee, and Missouri, south to Florida and Mississippi). Mesic upland forests; also roadsides.

The few Missouri collections are all more than 80 years old. Most historical specimens have only a general location on their labels, which has impeded searches for these historical populations. In other states where it is extant, it appears to grow best in pine savannas, a community type that was eliminated from Missouri during the era of widespread clear-cutting from the 1890s to the 1920s (for more information, see the chapter on *Analysis of the Flora* in Volume 1 of the present work [Yatskievych, 1999]). *Desmodium ochroleucum* is rare and sporadic throughout its highly fragmented range, which suggests that it possibly is the product of past hybridization, but to date no data exist to support this theory. It closely resembles *D. canescens* in general aspect, including traits such as the conspicuous network of veins on the lower leaflet surface and hooked hairs along the veins, as well as the large, persistent stipules that are reflexed at maturity. The two species are distinguished by habit and fruiting characters.

12. Desmodium paniculatum (L.) DC. (tall tick trefoil)

Pl. 394 e, f; Map 1745

Stems 50–150 cm long, erect or ascending, sometimes arched, rarely spreading with an ascending tip, often branched, the median portion sparsely pubescent with appressed hairs or with scattered straight, spreading and/or hooked hairs, rarely completely glabrous or densely covered with straight, spreading and/or hooked hairs. Petioles of the median leaves 1–5 cm long, only slightly reduced toward the stem tip, those of the upper leaves 1–5 cm long. Stipules 2–6 mm long, 0.3–1.0 mm wide, narrowly triangular, tapered to a sharply pointed tip, at maturity brown and appressed, shed early or less commonly semipersistent. Leaflets flat or nearly so, the undersurface sparsely pubescent with straight, appressed or rarely spreading hairs, the network of raised veins relatively inconspicuous. Central leaflet (1.5–)3.5–8.0(–10.0) cm long, (0.2–)0.5–2.5 cm wide (generally 3–8 times as long as wide), narrowly oblong, or lanceolate, the tip bluntly to sharply pointed. Lateral leaflets (1.5–)2.5–7.0(–8.0) cm long, (0.2–)0.5–2.0 cm wide. Stipels 1–4 mm long. Inflorescences terminal and from the upper leaf axils, branched, the axis with hooked and sometimes straight, spreading hairs. Primary bracts, 1.5–3.0 mm long, lanceolate, tapered to a sharply pointed tip, mostly shed early. Secondary bracts 1.0–1.5 mm long, linear or hairlike, shed early or semipersistent. Flower stalks 4–11 mm long. Calyces green, with scattered appressed hairs and dense very short hairs, the tube 1.0–1.5 mm long, the lobes 2–3 mm long. Corollas 5–8 mm long, pink, often fading to bluish purple or blue, the nectar guides green and white outlined in purple. Fruits straight in outline, the stalklike base 2–5 mm long (extending past the persistent calyx), consisting of (1–)4 or 5 segments, each 5–7 mm long and 3–5 mm wide, bluntly angled to more or less rounded on the upper margin, bluntly angled on the lower margin, with deeper indentation below (rarely rounded below and with nearly equal indentations above and below), the margins and faces uniformly covered with hooked hairs. $2n=22$. July–September.

Scattered to common nearly throughout the state (eastern U.S. west to Nebraska and Texas; Canada). Bottomland forests, openings and edges of mesic to dry upland forests, tops of bluffs, glades, savannas, upland prairies, banks of streams and rivers, edges of marshes and fens, and margins of ponds and lakes; also old fields, pastures, levees, railroads, and roadsides.

This species, along with the broader-leafleted *D. perplexum* and *D. glabellum*, presents one of the most difficult complexes in the Missouri flora. Although many specimens can be readily classified as one of the three species, a substantial proportion seems to bridge the morphological gaps between them (Isely, 1983b). Isely (1990) suggested that plants of intermediate morphology might represent hybrids between *D. paniculatum* and one or both of these other two species. Some other authors have chosen to combine all three taxa into *D. paniculatum* and to recognize only two varieties: the nominate variety with narrow leaflets and less conspicuous pubescence; and the other variety with broader leaflets and conspicuous pubescence (Steyermark, 1963). The latter has been called var. *dillenii* (Darl.) Isely and encompasses both *D. glabellum* and *D. perplexum*. The present recognition of three species is tentative and further research is needed. Adding to the complexity of the group is a tendency for some collections, especially from northeastern Missouri, to have larger flowers, longer stipules, and rounded fruit segments, somewhat similar to those of *D. canadense*. Isely (1990) noted the existence in the Southeast of occasional plants resembling *D. paniculatum*, but with very short petioles. He suggested that such plants might represent hybrids between this species and *D. ciliare*. Such plants eventually may be found in Missouri, as the two putative parents frequently grow in proximity. Evidence for hybridization with *D. rotundifolium* is discussed under that species.

13. Desmodium perplexum B.G. Schub. (tall tick clover, perplexing tick trefoil)

D. paniculatum (L.) DC. var. *dillenii* (Darl.) Isely (in part)

Pl. 394 h–j; Map 1746

Stems 50–150 cm long, erect or ascending to somewhat arched, branched, the median portion with sparse to moderate or occasionally relatively dense, straight, spreading hairs and often also shorter, hooked hairs. Petioles of the median leaves 2–5 cm long, usually only slightly reduced toward the stem tip. Stipules 2–6 mm long, 0.5–1.0 mm wide, linear to narrowly lanceolate, at maturity brown and appressed, shed early or semipersistent. Leaflets flat or nearly so, the undersurface sparsely to moderately pubescent with straight, spreading hairs, the network of raised veins relatively inconspicuous. Central leaflet 3–7 cm long, 2–4 cm wide (generally 1.5–4.0 times as long as wide), ovate to elliptic-ovate or rarely lanceolate (widest well below the midpoint), the tip bluntly to sharply pointed. Lateral leaflets 2–7 mm long, 1.2–4.0 mm wide. Stipels 1–2 mm long. Inflorescences terminal and from the upper leaf axils, branched, the axis with dense, hooked hairs and sparse to dense, straight, spreading hairs. Primary bracts 1.5–2.0 mm long, ovate, tapered to a sharply pointed tip, mostly shed early. Secondary bracts 0.6–0.8 mm long, linear to hairlike, often shed early. Flower stalks 4–8 mm long. Calyces green, sometimes with irregular red patches, with sparse, straight, spreading hairs and dense, very short hairs, the tube 1.0–1.2 mm long, the lobes 1.2–2.0 mm long. Corollas 6–9 mm long, pink, the nectar guides green and white outlined in purple. Fruits straight in outline, the stalklike base 3–5 mm long, consisting of (1–)4 or 5 segments, each 5–7 mm long and 3–5 mm wide, bluntly angled to more or less rounded on the upper margin, bluntly angled on the lower margin, with deeper indentations below, uniformly covered with hooked hairs on the margins and faces. July–September.

Scattered to common nearly throughout the state (eastern U.S. west to Nebraska and Texas; Canada). Bottomland forests, openings and edges of mesic to dry upland forests, tops of bluffs, glades, savannas, upland prairies, banks of streams and rivers, and margins of ponds; also old fields, pastures, railroads, and roadsides.

Desmodium perplexum and *D. glabellum* belong to the *D. paniculatum* complex, of which *D. perplexum* is the most widespread and abundant member in Missouri. These species can be difficult to distinguish morphologically. See the treatments of *D. glabellum* and *D. paniculatum* for further discussion. All of the Midwestern *Desmodium* species are eliminated by frequent mowing or intense grazing, but *D. perplexum* seems to be the most tolerant of disturbance and is found in lightly grazed pastures and hay meadows. For a discussion of hybridization between members of the *D. paniculatum* complex and *D. rotundifolium*, see the treatment of *D. rotundifolium*.

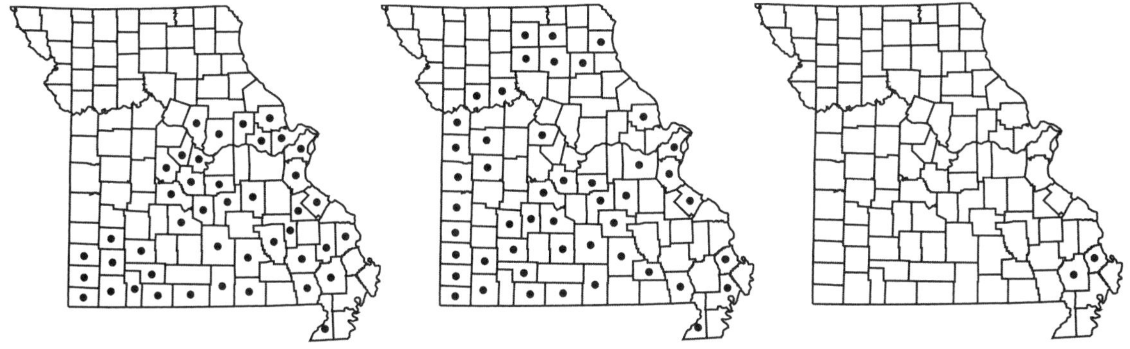

1747. Desmodium rotundifolium 1748. Desmodium sessilifolium 1749. Desmodium strictum

14. Desmodium rotundifolium (Michx.) DC. (prostrate tick trefoil, dollarleaf, low tick trefoil)

Pl. 394 k, l; Map 1747

Stems 50–150 cm long, prostrate, branched, the median portion moderately pubescent with straight, spreading hairs, sometimes also with sparse to moderate, hooked hairs. Petioles of the median leaves 1–5 cm long, somewhat reduced toward the stem tip. Stipules 5–12 mm long, 3–5 mm wide, ovate (the base partially clasping the stem), tapered to a sharply pointed tip, at maturity brown and spreading to more commonly reflexed, persistent. Leaflets flat or nearly so, the undersurface with straight, spreading hairs, the network of raised veins inconspicuous to evident. Central leaflet 2–7 cm long, 2–6 cm wide, often slightly larger than the lateral pair, more or less circular to broadly ovate, broadly obovate, or somewhat rhombic, the tip mostly rounded. Lateral leaflets 2–4 cm long 2–4 cm wide. Stipels 1–3 mm long. Inflorescences terminal and from the upper leaf axils, often unbranched, the axis with straight, spreading and hooked hairs. Primary bracts 4–6 mm long, ovate, tapered to a sharply pointed tip, mostly persistent. Secondary bracts 1.0–1.2 mm long, linear to hairlike, sometimes shed early. Flower stalks 6–22 mm long. Calyces green, densely pubescent with minute hairs and with sparse, long, straight, spreading hairs, especially on the lower lip, the tube 1.7–2.0 mm long, the lobes 2–3 mm long. Corollas 7–12 mm long, pink or rarely white, the nectar guides yellow and white outlined in purple. Fruits straight in outline, the stalklike base 3–5 mm long, consisting of 3–6 segments, each 5–7 mm long and 3–5 mm wide, angled or more or less rounded on the upper margin, angled on the lower margin, with deeper indentations below, less commonly rounded and nearly equally indented above and below, with dense hooked hairs on the margins and scattered hooked hairs on the faces. $2n=22$. July–September.

Scattered in the Ozark, Ozark Border, and Mississippi Lowlands Divisions (eastern U.S. west to Kansas and Texas; Canada). Mesic upland forests, sand savannas, margins of glades, and ledges and tops of bluffs; also railroads and roadsides.

Possible hybridization between *D. rotundifolium* and members of the *D. paniculatum* complex occurs in Missouri, with the hybrids having been called *D.* ×*humifusum* (Muhl. ex Bigelow) Beck. Steyermark (1963) assigned a few specimens from southern Missouri to this taxon, but treated it as a species rather than a hybrid. Yatskievych and Turner (1990) and Isely (1998) suggested that the Missouri materials were all misdetermined, poor specimens of some member of the *D. paniculatum* complex, although this has been shown to be incorrect in subsequent research. More recently, plants with hybrid morphology also have been collected in Indiana, Tennessee, and the District of Columbia. The hybrids tend to be robust plants that resemble *D. glabellum* and *D. perplexum* in their leaflet morphology, but generally are sprawling plants with persistent stipules that are longer and wider than in those two taxa. At the three New England locations for *D.* ×*humifusum* known in 1991, it occurred with *D. paniculatum* and *D. rotundifolium*, and allozyme evidence supports the hybrid status of these populations, with apparent first-generation and later-generation hybrids present (Raveill, 2002). A similar hybrid has now been collected in Missouri in Carter, Howell, Jasper, Pulaski, and Shannon Counties. It occurs along roadsides, on bluff tops, and in upland forests where hot fires have cleared most understory growth and thinned the canopy. Although hybrid plants from New England and Missouri produce viable seed (Raveill, 1995; personal observation), there is no evidence that a stable, new, genetically distinct species has arisen. Missouri plants of *D.* ×*humifusum* are being monitored to assess their long-term persistence. Exceptionally pubescent collections of *D.* ×*humifusum* made by the author in Tennessee occurred at a location where *D. perplexum* was the only other potential parent present, and Missouri collectors should be alert to the potential of *D. rotundifolium* hybridizing with other members of the *D. paniculatum* complex. Hybrids between *D. rotundifolium* and either *D. glabellum* or *D. perplexum* apparently have not been described with a binomial and although they would be morphologically nearly indistinguishable from true *D.* ×*humifusum*, technically they should not be called by this name if the three potential parental taxa in the complex are treated as separate species.

15. Desmodium sessilifolium (M. Curtis) Torr. & A. Gray (sessileleaf tick trefoil)

Pl. 395 f, g; Map 1748

Stems 50–120 cm long, usually stiffly erect, occasionally slightly arched, often unbranched, the median portion with dense, hooked hairs. Petioles all 1–3 mm long. Stipules 4–7 mm long, 0.5–0.8 mm wide, lanceolate to linear-lanceolate, tapered to a sharply pointed tip, at maturity brown and appressed to spreading, shed early or more commonly semipersistent. Leaflets flat to somewhat folded longitudinally, the undersurface with hooked and straight, spreading hairs, the network of raised veins relatively conspicuous. Central leaflet 3–6 cm long, 0.4–1.5 cm wide (4–10 times as long as wide), narrowly oblong to narrowly lanceolate or narrowly elliptic, the tip mostly bluntly pointed, sometimes with a minute sharp extension of the midvein at the very tip. Lateral leaflets 2–5 cm long, 0.3–1.1 cm wide. Stipels 1–2 mm long. Inflorescences terminal, branched or unbranched, the axis with dense, hooked hairs. Primary bracts 2–5 mm long, ovate, tapered to a sharply pointed tip, semipersistent or shed early. Secondary bracts 1.0–1.5 mm long, linear to hairlike, shed early. Flower stalks 1–4 mm long. Calyces green, with moderate to dense, short, spreading hairs and dense, minute hairs, the tube 1.0–1.5 mm long, the lobes 1.5–3.0 mm long. Corollas 4–6 mm long, pale pink or white, fading to a pale yellowish color, the nectar guides inconspicuous, pale green narrowly outlined in pink. Fruits arched downward in outline, the stalklike base 1–2 mm long, consisting of 1–3(4) segments, each 4–6 mm long and 3–4 mm wide, broadly rounded on the upper margin, rounded on the lower margin, with deeper indentations below, uniformly covered with hooked hairs on the margins and faces. $2n$=22. June–September.

Scattered nearly throughout the state, but apparently absent from most of the western half of the Glaciated Plains Division (eastern U.S. west to Nebraska and Texas; Canada). Upland prairies, sand prairies, glades, tops of bluffs, savannas, and sand savannas; also old fields, fencerows, railroads, and roadsides.

16. Desmodium strictum (Pursh) DC. (pinebarren tick trefoil, narrow-leaved tick clover)

Pl. 395 a, b; Map 1749

Stems 40–150 cm long, erect or strongly ascending, sometimes slightly arched, often unbranched, the median portion nearly glabrous to densely covered with hooked hairs. Petioles of median leaves 0.5–2.0 cm long, only slightly reduced toward the stem tip. Stipules 2–6 mm long, 0.3–0.5 mm wide, linear, tapered to a sharply pointed tip, at maturity brown and appressed or spreading, often shed early. Leaflets moderately to strongly folded longitudinally, the undersurface glabrous except for straight, appressed hairs along the margins and midrib, the network of raised veins conspicuous. Central leaflet 3–6 cm long and 0.3–0.8 cm wide (4–10 times as long as wide), linear, the tip short-pointed. Lateral leaflets (1.0–)2.0–4.5 cm long and 0.3–0.6 cm wide. Stipels 1–2 mm long. Inflorescences terminal, branched, the axis with dense, hooked hairs. Primary bracts 1–2 mm long, ovate, sharply or less commonly bluntly pointed. Secondary bracts 0.8–1.0 mm long, linear. Flower stalks 4–17 mm long, the flowers sometimes twisted sideways. Calyces green, often tinged with purple, with sparse, short, straight and hooked hairs and usually dense, minute hairs, the tube 0.8–1.0 mm long, the lobes 1.0–1.5 mm long. Corollas 3–5 mm long, bright pink, the nectar guides green and white outlined (sometimes irregularly) in purple. Fruits arched downward in outline, the stalklike base 1–2 mm long, consisting of 1–3 segments, each 3–4 mm long and 3–4 mm wide, straight to slightly concave on the upper margin, rounded on the lower margin, with deeper indentations below, uniformly covered with hooked hairs on the margins and faces. $2n$=22. July–August.

Uncommon, known thus far only from Scott and Stoddard Counties (New Jersey to Florida, west to Texas, mostly along the Atlantic and Gulf Coastal Plains; disjunct in Missouri). Sand prairies; also sand quarries.

This distinctive species was first reported from the state by Dunn (1982) and is restricted to open sandy sites in the southeastern portion of the state. The Missouri populations represent a substantial range disjunction from the nearest populations in Mississippi.

Vegetative specimens of *D. strictum* might be confused with the narrow-leafleted phase of *D. paniculatum* that is common in similar sandy habitats but these two species are easily distinguished. The leaflets of *D. strictum* are folded along the midrib, whereas those of *D. paniculatum* are flat. The venation of the leaflets is also diagnostic. The primary veins of *D. paniculatum* diverge from the midrib at a gradual angle before becoming less prominent, and no network of joined veins is apparent at the margins of the leaflets. The primary veins of *D. strictum* are nearly perpendicular as they diverge from midrib with some bending toward the leaflet base and some toward the tip before rejoining into a conspicuous network of veins along the margins. *D. strictum* also might be confused with *D. sessilifolium*, but that species is characterized by very short petioles.

17. Desmodium viridiflorum (L.) DC. (velvetleaf tick trefoil)

Pl. 396 a, b; Map 1750

Stems 50–150 cm long, erect, often unbranched, the median portion with dense, hooked hairs and sparse, straight, spreading hairs. Petioles of median leaves 1–4 cm long, only slightly reduced toward the stem tip. Stipules 3–5 mm long, 2.0–2.5 mm wide, ovate, tapered to a sharply pointed tip, at maturity dark red and reflexed, semipersistent. Leaflets flat or nearly so (sometimes appearing slightly concave) the undersurface velvety-hairy. the network of raised veins evident but usually obscured by the pubescence. Central leaflet (2–)4–9(–11) cm long, (0.8–)3.0–8.0 cm wide, triangular-ovate to broadly ovate or rhombic, the tip bluntly to sharply but often broadly pointed, occasionally rounded. Lateral leaflets (1.5–)3.0–5.0 cm long, (0.8–)2.0–4.0 cm wide. Stipels 1.5–2.5 mm long. Inflorescences terminal and from the upper leaf axils, branched, the axis with dense, hooked hairs. Primary bracts 2–4 mm long, 1.0–1.5 mm wide, ovate, tapered to a sharply pointed tip, more or less persistent. Secondary bracts 1–2 mm long, linear. Flower stalks 3–7 mm long. Calyces green, densely pubescent with straight, spreading hairs and scattered glandular hairs, the tube 1.0–1.5 mm long, the lobes 0.5–2.0 mm long. Corollas 6–9 mm long, pink, the nectar guides green and white, outlined in purple. Fruits straight or slightly curved upward in outline, the stalklike base 3–5 mm long, consisting of (3)4 or 5 segments, each 5–7 mm long and 3–5 mm wide, rounded on the upper margin, bluntly angled on the lower margin, with deeper indentations below, uniformly covered with hooked hairs on the margins and faces. $2n=22$. July–September.

Uncommon, known thus far only from Dunklin, Scott, and Stoddard Counties (New Jersey, Ohio, and Missouri, south to Florida and Texas; Mexico). Mesic upland forests; also roadsides and disturbed areas; often in sandy soils.

For further discussion of this species, see the treatment of the morphologically similar *D. nuttallii*.

27. Galactia P. Browne

Fifty to 60 species, nearly worldwide, mostly in warm tropical regions of the New World.

1. Galactia regularis (L.) Britton, Stearns & Poggenb. (downy milk pea)

G. mississippiensis Vail

G. volubilis (L.) Britton, misapplied

G. volubilis var. *mississippiensis* (Vail) Rydb.

Dolichos regularis L.

Pl. 396 e, g; Map 1751

Plants perennial herbs, scrambling or climbing on other low vegetation, with a slender, branched caudex above a taproot. Stems 30–90 cm long, prostrate or trailing to loosely ascending, frequently twining, unarmed, sparsely to densely pubescent with short, fine, spreading to downward-curved hairs. Leaves alternate, pinnately trifoliate, the petiole 0.5–2.0 cm long, hairy. Stipules 1–3 mm long, narrowly lanceolate to lanceolate, mostly shed early; stipels minute (less than 0.3 mm long), linear, more or less persistent. Leaflets 1.5–4.0 cm long, 0.5–2.5 cm wide, elliptic to oblong or oblong-ovate, rounded to broadly angled at the base, bluntly pointed to rounded or occasionally slightly notched at the tip, the midvein extended as a minute sharp point at the very tip, the margins entire, the upper surface moderately hairy to glabrous or nearly so, the undersurface moderately to densely and finely hairy, pinnately veined, the undersurface lacking a noticeable network of veins or the fine network occasionally slightly evident. Terminal leaflet with the stalk 3–8 mm long, symmetric at the base; lateral leaf-

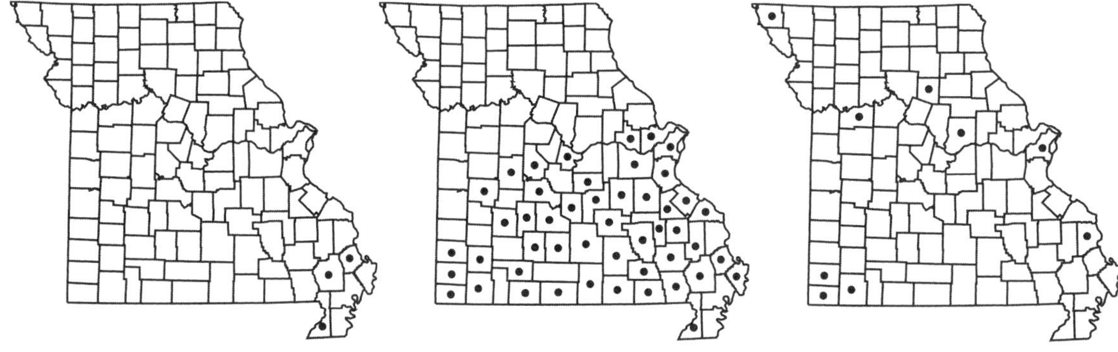

1750. Desmodium viridiflorum 1751. Galactia regularis 1752. Glycine max

lets essentially sessile above the pulvinus, mostly symmetric at the base. Inflorescences axillary, of clusters or racemes of 2–10 flowers, sometimes paired, the inflorescence stalk 0.5–3.0 cm long, hairy, the bracts 1.0–1.5 mm long, lanceolate, shed early; bractlets 0.6–1.0 mm long, ovate, loosely subtending the flowers, persistent. Cleistogamous flowers absent. Calyces 4–6 mm long, 5-lobed, but often appearing 4-lobed, because of fusion of the upper 2 lobes all or most of the way to the tip, the tube narrowly bell-shaped, hairy, the lobes about as long as to somewhat longer than the tube, the lowermost lobe longer than the others, ovate to triangular-lanceolate, angled or tapered to sharply pointed tips. Corollas papilionaceous, pink to pinkish purple (the banner, with a white or pale region toward the center, this often with an irregular, darker purple margin), fading to pale pink, the banner 6–9 mm long, 4–5 mm wide, the expanded portion broadly obovate, shallowly notched at the tip, often with a shallow longitudinal keel, bent backward, the wings 6–7 mm long, 1.5–2.0 mm wide, the expanded portion oblong, straight to somewhat incurved around the keel, the keel 6–8 mm long, 2–4 mm wide, straight to slightly curved upward, fused nearly to the usually bluntly pointed tip. Stamens 10, all of similar length, 9 of the filaments fused and 1 free, the fused portion 4–5 mm long, the free portion 1–2 mm long, the anthers small, attached near the midpoint. Ovary 3–4 mm long, finely hairy, the style 2–3 mm long, glabrous, the stigma minute, terminal. Fruits legumes, 2–5 cm long, 4–5 mm wide, narrowly oblong, straight, tapered at the tip to a short beak, flattened, herbaceous to somewhat papery, sparsely hairy at maturity, dehiscing by 2 valves, these green to tan at maturity, twisting spirally after dehiscence, with 3–8 seeds. Seeds 2.5–3.5 mm long, 1.5–2.5 mm wide, more or less kidney-shaped, flattened, the surface yellow to dark brown, sometimes mottled with purple, smooth, somewhat shiny. $2n$=20. July–August.

Scattered in the Ozark, Ozark Border, and Mississippi Lowlands Divisions (eastern [mostly southeastern] U.S. west to Kansas and Texas). Glades, savannas, dry upland forests, tops of bluffs, and sand prairies.

This taxon is sometimes confused with several other species, in particular, *G. volubilis* (Gleason and Cronquist, 1991). These species are members of a variable complex of species that Isely (1990, 1998) referred to as the *G. volubilis* group. They vary in pubescence, length of the inflorescence, and flower size, but they appear to hybridize and are difficult to define and characterize. W. H. Duncan (1979) noted that the type specimen of *G. regularis* is identical to specimens that some authors had been calling *G. volubilis* or *G. volubilis* var. *mississippiensis*, and the name *G. regularis* is thus used here for the Missouri materials. True *G. volubilis* has longer racemes and larger flowers, and is mostly confined to the Atlantic and Gulf Coastal Plains from Virginia to Louisiana (W. H. Duncan, 1979).

28. **Glycine** Willd.
(Hermann, 1962)

About 9 species, Asia to Australia, 1 taxon cultivated worldwide.

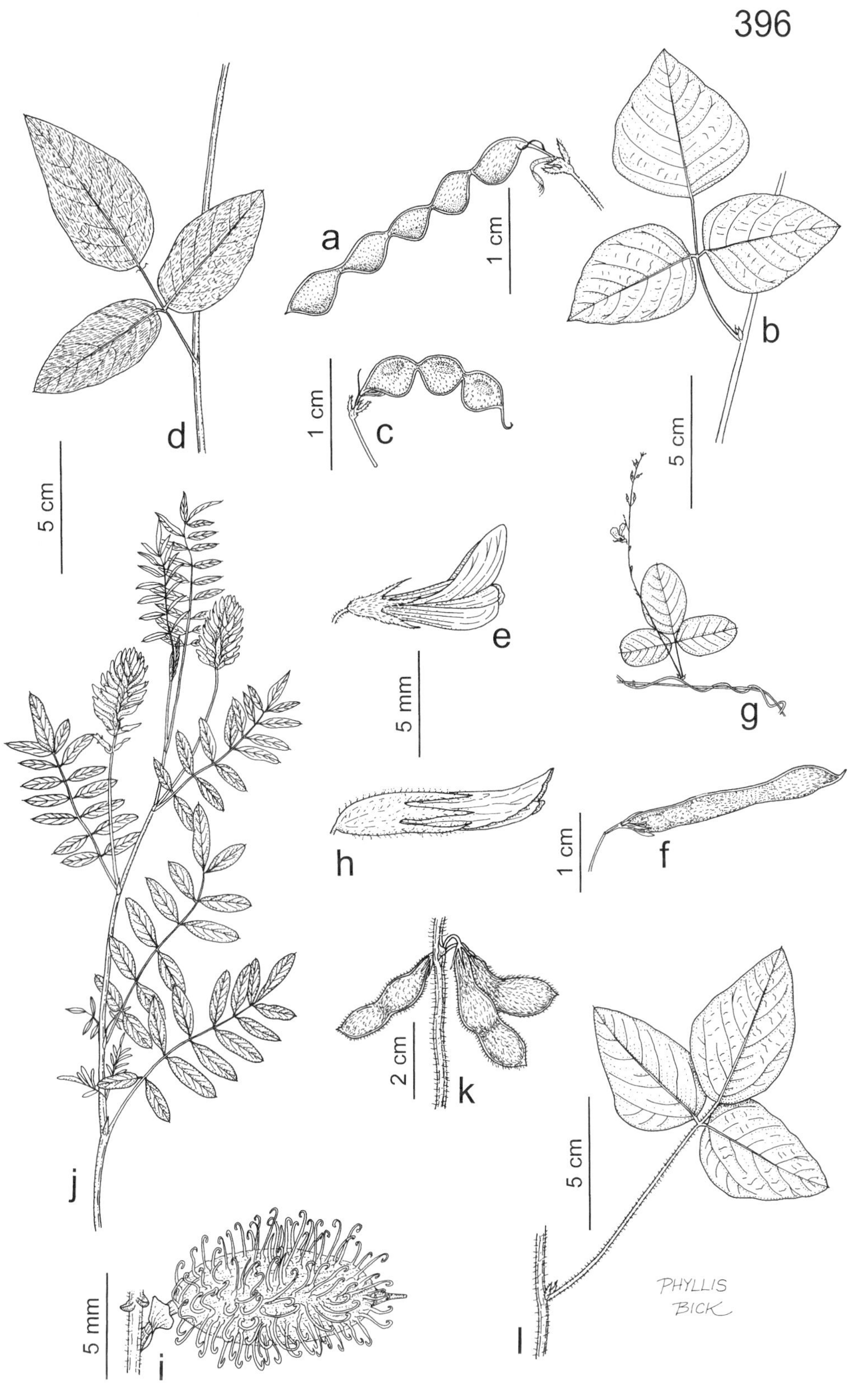

Plate 396. Fabaceae. *Desmodium viridiflorum*, **a)** fruit, **b)** leaf attached to stem. *Desmodium nuttallii*, **c)** fruit, **d)** leaf attached to stem. *Galactia regularis*, **e)** flower, **f)** fruit, **g)** habit. *Glycyrrhiza lepidota*, **h)** flower, **i)** fruit, **j)** habit. *Glycine max*, **k)** fruit, **l)** leaf attached to stem.

1. Glycine max (L.) Merr. (soybean)
Phaseolus max L.

Pl. 396 k, l; Map 1752

Plants annual, with taproots. Stems 20–60 cm long, stout, erect or ascending, sometimes reclining with age, branched, appearing bushy, unarmed, densely pubescent with fine, more or less spreading to somewhat matted, yellowish brown hairs. Leaves alternate, pinnately trifoliate, the petiole 5–15 cm long. Stipules 3–7 mm long, lanceolate, tapered abruptly from a broad base to a slender, elongate, sharply pointed tip; stipels 1–3 mm long, linear. Leaflets 4–10 cm long, 2–8 cm wide, broadly ovate to ovate or elliptic, the lateral leaflets sometimes broadly lanceolate, rounded to nearly truncate at the base, angled or slightly tapered to a bluntly or sharply pointed tip, the margins entire, finely hairy, the upper surface sparsely to moderately, finely hairy, the undersurface moderately to densely, finely hairy, the venation pinnate. Terminal leaflet with the stalk 4–19 mm long, symmetric at the base; lateral leaflets with the stalk 1.5–4.0 mm long, usually somewhat asymmetric at the base. Inflorescences axillary, short racemes or loose clusters of (1–)3–5(–8) flowers, the stalk 1–4 mm long, often obscured by dense hairs, the bracts 2–4 mm long, lanceolate, sometimes shed early, each flower with a stalk 2–4 mm long, the bractlets 2–3 mm long, narrowly lanceolate to hairlike. Calyces densely hairy, the tube 1.5–2.5 mm long, conic to bell-shaped, the lobes 1.5–4.5 mm long, lanceolate to narrowly lanceolate-triangular, sharply pointed at their tips, the 2 upper lobes fused to above the midpoint, the lowermost lobe somewhat longer than the others. Corollas papilionaceous, white to lavender or pale purple (rarely pale pink or bluish-tinged), sometimes with darker purple nerves, the banner 5–6 mm long, 3.0–3.5 mm wide, the expanded portion curved or bent backward, broadly obovate, shallowly notched at the tip, the wings 4–5 mm long, 1–2 mm wide, narrowly oblong, the keel 3–4 mm long, 1.0–1.5 mm wide, narrowly boat-shaped, fused to above the midpoint, curved upward toward the bluntly pointed tip. Stamens 10, all of similar length, 9 of the filaments fused and 1 free, the fused portion 2.5–3.0 mm long, the free portion 0.5–0.8 mm long, the anthers small, attached near the base, yellow. Ovary 2–3 mm long, densely hairy, the style 1.0–1.5 mm long, somewhat curved toward the tip, glabrous, the stigma small and terminal. Fruits legumes, 3–7 cm long, 7–13 mm wide, oblong, not beaked, turgid, but slightly flattened, usually somewhat constricted between the seeds, drooping, but usually somewhat arched upward, often dehiscing tardily by 2 valves from the tip, the valves becoming somewhat spirally twisted, brownish green to yellowish brown at maturity, bristly-hairy, mostly 2- or 3-seeded. Seeds 6–9 mm long, 5–7 mm wide, ovoid to nearly globose, the surface pale tan to olive green, sometimes tinged, mottled, or streaked with purple, gray, or black, smooth, somewhat shiny. $2n=40$. July–August.

Introduced, uncommon and widely scattered (cultigen of Asian origin; introduced widely but sporadically in the eastern U.S. west to Nebraska and Texas, Canada). Edges of bottomland prairies and banks of streams; also fallow fields, margins of crop fields, roadsides, and open disturbed areas.

The soybean is believed to have been domesticated in China, where a wild viney relative (*G. soja* Siebold & Zucc.) is still found. Soybeans were cultivated widely by 100 A.D. and became one of the major food sources supporting the large populations of Asia (Hymowitz, 1970). They are mostly self-pollinated and hundreds of genetically pure land-races have been developed. Soybeans are now probably the most important legume in the world, and have a multitude of uses. In Asian cuisines, soybeans are an important source of proteins and are consumed in the form of soymilk, tofu, miso, soy sauce, and tempeh. The raw seeds are poisonous and the trypsine inhibitors they contain must be neutralized by boiling water. Soybeans were a minor crop in the United States until World War II, when the shortage of butter created a demand for soybean oil for margarine. Production of soybeans in the U.S. has now surpassed that of Asia, and they are the most important cash crop in the country. The major use of soybeans in the United States is the oil, which is used for salad oil, cooking oil, and industrial applications. The residual cake is fed to livestock. Miscellaneous applications include soaps, cosmetics, resins, plastics, packaging materials, inks, crayons, solvents, and clothing. Soybeans also are an important source of biofuels.

29. **Glycyrrhiza** L.

Fifteen to 20 species, widespread in the temperate and subtropical regions of the world, most diverse in Europe, Asia.

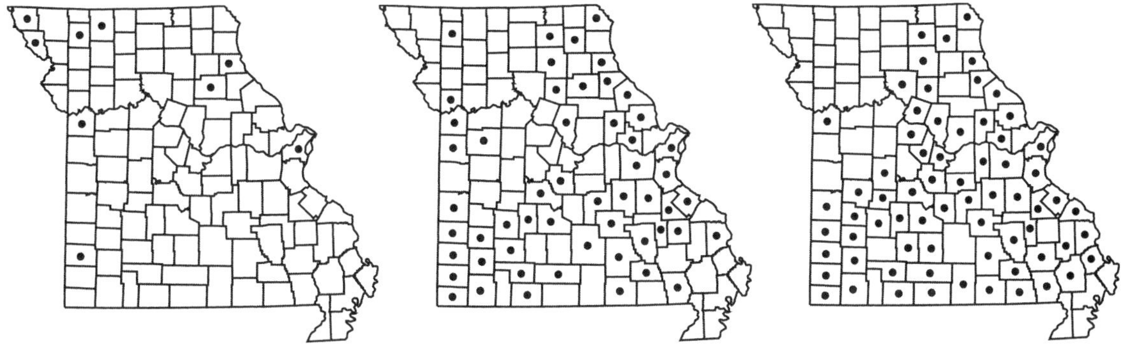

1753. Glycyrrhiza lepidota 1754. Hylodesmum glutinosum 1755. Hylodesmum nudiflorum

1. Glycyrrhiza lepidota (Nutt.) Pursh **var. lepidota** (wild licorice)

Pl. 396 h–j; Map 1753

Plants perennial herbs, colonial from deep, elongate, somewhat woody rhizomes. Stems 30–60(–100) cm long, erect or strongly ascending, often unbranched, unarmed, glabrous or sparsely to moderately pubescent with minute hairs, usually also gland-dotted, at least when young. Leaves alternate, odd-pinnately compound with 11–19 leaflets, the petiole 0.5–3.0 cm long, the rachis 4–12 cm long. Stipules 3–7 mm long, linear to narrowly lanceolate, shed early; stipels absent. Leaflets 1–4 cm long, 0.4–1.0 cm wide, lanceolate to oblong-lanceolate or oblong, angled to somewhat rounded at the base, angled to a narrowly rounded or bluntly to sharply pointed tip, the midvein usually extended into a short, sharp point at the very tip, the margins entire, the surfaces with sparse to moderate, minute scales when young, appearing gland-dotted after the scales are shed, sticky, the undersurface often also sparsely hairy, pinnately veined, but the secondary veins often relatively faint. Inflorescences axillary, spikelike racemes of numerous flowers, strongly ascending, the stalk 2–7 cm long, the bracts 5–7 mm long, lanceolate, more or less persistent, each flower with an inconspicuous stalk 0.5–1.0 mm long; bractlets absent. Calyces densely pubescent with short, spreading hairs, most or all of these gland-tipped, the tube 2–3 mm long, narrowly bell-shaped, more or less 2-lipped, the lobes 2.5–4.5 mm long, the upper 2 lobes shorter and fused to above the midpoint, the lower 3 lobes longer, narrowly lanceolate-triangular, sharply pointed at their tips. Corollas papilionaceous, yellowish white to greenish white, the banner 9–12 mm long, 3–4 mm wide, the expanded portion somewhat ascending, ovate to elliptic-ovate, angled to a mostly sharply pointed tip, the wings 8–10 mm long, 1–2 mm wide, narrowly oblong, the keel 7–9 mm long, 1.5–2.0 mm wide, narrowly boat-shaped, fused below the midpoint, more or less straight below the sharply pointed tip. Stamens 10, unequal in length, 9 of the filaments fused and 1 free, the fused portion 3–4 mm long, the free portion 2–3 mm long, the anthers small, attached above the base, orange. Ovary 2–3 mm long, with raised glandular areas, becoming bristly, the style 1–3 mm long, somewhat curved toward the tip, glabrous, the stigma small and terminal. Fruits legumes, burlike, the body 1.0–1.5 cm long, 4–6 mm wide, oblong-ellipsoid to narrowly ellipsoid, not beaked, somewhat flattened, straight, indehiscent, brown to dark brown, with dense, coarse, hooked prickles, mostly 2–4-seeded. Seeds 3–4 mm long, 2–3 mm wide, more or less kidney-shaped to somewhat trapezoid, the surface reddish brown to greenish brown, smooth, dull to slightly shiny. $2n=16$. May–June.

Uncommon, mostly in the western portion of the state, mostly introduced farther east (western U.S. east to Indiana and Texas, sporadically eastward to Maine and Virginia; Canada, Mexico). Upland prairies and margins of lakes; also pastures, railroads, and open disturbed areas.

In Missouri this species is known mostly from historical collections. It is easily recognized by the prickly fruits that resemble small cockleburs (*Xanthium*, Asteraceae), which are dispersed in the fur of mammals. In the western states, there is a second variety with conspicuous glandular hairs on the inflorescence, var. *glutinosa* (Nutt.) S. Watson, which grades into the widespread var. *lepidota* (Isely, 1998).

The roots of *G. lepidota* are very sweet because of the presence of the triterpenoid, glycyrrhizin, which is said to be 50 times sweeter than sucrose. Wild licorice was used medicinally as a laxative, expectorant, demulcent, and for toothaches. Commercial licorice extract is obtained from the European species, *G. glabra* L. Licorice extracts have been used to treat coughs, bronchial ailments, and ulcers, to stimulate glands of the adrenal cortex, and as a flavoring for various foods, confections, pills, cough syrup, and tobacco. In ancient Egypt, licorice was considered so valuable that the pharoahs were buried with a supply.

30. **Hylodesmum** H. Ohashi & R.R. Mill (beggar's lice, tick trefoil, stick-tights)

Contributed by Jay A. Raveill

Plants perennial herbs, with a short, sometimes woody rootstock, often taprooted, at least when young. Stems 1 or 2 to several, erect or strongly ascending to slightly arched or loosely ascending from a spreading base (in *H. pauciflorum*), unarmed, pubescent with short hooked or straight hairs. Leaves alternate or clustered at the tip of the stem in an apparent whorl, pinnately trifoliolate, petiolate. Stipules shed early or semipersistent, linear to narrowly lanceolate, the margins entire, olive green at maturity. Stipels absent or minute and shed very early. Leaflets ovate, obovate, rhombic, or broadly elliptic to nearly circular, stalked, angled or short- to long-tapered to a sharply pointed tip, the margins entire, hairy, the upper surface glabrous or sparsely pubescent with straight to curved (but not hooked) hairs, the undersurface sparsely pubescent with straight to curved (but not hooked) hairs, more densely so along the main veins, pinnately veined, the network of veins not strongly raised. Inflorescences terminal (from apparently separate, usually leafless flowering stems in *H. nudiflorum*) or both terminal and axillary from upper nodes (in *H. pauciflorum*), unbranched or less commonly few-branched, the axis sparsely to densely pubescent, often with a mixture of short and minute hairs, at least some of the hairs with hooked tips, the flowers paired or less commonly appearing as small clusters at each node, the bract(s) subtending each pair or cluster of flowers small, slender, shed as the flowers mature or more or less persistent; bractlets absent or minute and shed very early. Calyces 2-lipped, appearing shallowly 4-lobed (by fusion of the upper 2 lobes), the upper lip sometimes minutely notched, the lower lip with the 3 lobes usually broadly triangular, usually inconspicuously nerved, persistent at fruiting, but not becoming enlarged. Corollas papilionaceous, glabrous, pale to dark pink or white, sometimes with a pair of darker purple markings near the base of the banner, the petals tapered or abruptly rounded (occasionally cordate) to a short, stalklike base, the banner obovate to oblong-obovate, the wings not fused to the keel, the 2 keel petals free or fused above the stalklike bases, oblanceolate and curved in outline, angled apart when free, boat-shaped when fused. Stamens 10, the filaments all fused nearly to the tip (only to about the midpoint in *H. pauciflorum*), the anthers small, attached near the midpoint, all similar in size. Ovary linear to narrowly oblong-ellipsoid, short-stalked, the style slender, incurved, glabrous, more or less persistent at fruiting, the stigma small and terminal. Fruits loments, flattened, tapered abruptly to a prominent stalklike base 4–18 mm long, 1–4-seeded, indehiscent, flattened, divided into 1-seeded segments that separate at maturity, the segments with well-defined margins and a conspicuous network of nerves on each face, the connections between the segments highly asymmetric, the lower margin very deeply lobed between the segments, the upper margin not lobed between the segments, at most slightly undulate, the margins glabrous or with sparse, hooked hairs, the faces with dense, hooked hairs. Seeds slightly kidney-shaped in outline, flattened, the surface smooth, pale yellow to dark brown. Fourteen species, North America, Asia.

The species in this genus traditionally have been treated in *Desmodium* (Steyermark, 1963; Gleason and Cronquist, 1991), but have long been recognized as distinctive within that genus and classified into a separate section or subgenus (*Podocarpium*). They are distinguished from the rest of *Desmodium* by the prominent stalklike base of the fruit, the highly asymmetrical constrictions between the segments of the fruit, stamens with the filaments fused into a single group, calyx lobes shorter than the tube and often reduced to little more than teeth, wing petals not fused to the keel petals, and seed morphology (lack of a minute aril around the attachment scar) (Ohashi and Mill, 2000). Ecologically, the three Missouri species of *Hylodesmum* also differ from those remaining in *Desmodium* in that they tend to inhabit forests with relatively closed canopies, whereas true *Desmodium* species mostly are found in habitats with full sun or only partial shade. Although relationships among genera within the tribe Desmodieae Hutch. remain incompletely resolved, there seems ample evidence that *Hylodesmum* is a natural group that should be treated as a genus separate from *Desmodium*.

The hooked hairs that are characteristic of the genus aid in dispersal of the fruit segments by adhesion to the fur of various passing mammals. The segments also adhere strongly to socks and trousers. They can be difficult to remove from fur or clothing when fresh; the process becomes easier once the segments have been allowed to dry. Steyermark (1963) noted that the fruits of some species are eaten by birds, such as turkeys and quail, and that the herbage is sometimes browsed by deer.

1. Leaves alternate and relatively well-spaced; inflorescences terminal and also axillary from the upper nodes; corollas white, the keel petals not fused together . . . . 3. H. PAUCIFLORUM
1. Leaves mostly clustered or whorled at the tips of the stems (rare plants within a population may have the leaves more widely spaced along the stem); corollas pink (rarely white), the keel petals fused except for the stalklike bases; inflorescences terminal
    2. Inflorescence terminating a leafy stem (only 1 kind of aerial stem produced); stalklike bases of the fruits 4–8 mm long; stipules 9–12 mm long, semipersistent; flower stalks 3–6 mm . . . . 1. H. GLUTINOSUM
    2. Inflorescence terminating a usually leafless branch diverging from near or below ground level (usually appearing as though on a separate, leafless, fertile stem); stalklike bases of the fruits 8–15 mm long; stipules 2–4 mm long, shed early; flower stalks 8–20 mm long . . . . 2. H. NUDIFLORUM

1. Hylodesmum glutinosum (Muhl. ex Willd.) H. Ohashi & R.R. Mill (cluster tick trefoil, large-flowered tick clover, sticky tick clover)

Desmodium glutinosum (Muhl. ex Willd.) Alph. Wood

Pl. 397 e–g; Map 1754

Stems solitary, 30–100 cm long (below the inflorescence), erect, unbranched, the median portion moderately pubescent with minute, mostly hooked hairs and sparse, longer, straight, spreading hairs, both types sparser toward the sometimes nearly glabrous stem base and denser toward the tip. Leaves in a cluster at the base of inflorescence, usually appearing nearly whorled, rarely more widely spaced and alternate, the petiole (1–)4–15 cm long. Stipules 9–12 mm long, 0.3–0.8 mm wide, at maturity olive green and appressed, semipersistent. Leaflets with the undersurface somewhat lighter than the upper surface but not glaucous; the central leaflet (3.5–)6.0–12.0 cm long and 6–10 cm wide, ovate to nearly circular, short-tapered to a sharply pointed tip; lateral leaflets 3–10(–12) cm long and 2–7 cm wide, somewhat asymmetrically ovate (the basal side broader than the side adjacent to the central leaflet). Inflorescences terminal, erect or slightly arched, unbranched or less commonly with few, spreading branches, the axis pubescent with dense, minute, hooked hairs and scattered, straight, spreading hairs. Bracts 2–6 mm long and 0.3–0.5 mm wide, linear, mostly shed early; bractlets sometimes also present, these minute (to 0.2 mm long), shed early. Flower stalks 3–6 mm long. Calyces green, minutely hairy toward the base and with a few, longer, straight hairs above the midpoint, especially on the lower lip, the tube 1.5–2.0 mm long, the lobes nearly absent to at most 0.3 mm long. Corollas 4–7 mm long, pink or rarely white, the petals sometimes darker along the margins, the banner white near the clawed base, but no contrasting nectar guides present, the keel petals fused together. Fruits nearly straight or arched downward, tapered abruptly to a stalklike base 4–8 mm long, consisting of 1–4 segments, each 7–10 mm long and 4–6 mm wide, the upper side of each segment often slightly concave. $2n$=22. June–August.

Scattered to common nearly throughout the state (eastern U.S. west to North Dakota and Texas; Canada, Mexico). Bottomland forests, mesic upland forests, and bases and ledges of bluffs.

Occasional plants of this species are found with leaves relatively widely spaced along the stem and have been called *Desmodium glutinosum* f. *chandonetii* (Lunell) B.G. Schub. This form superficially resembles *H. pauciflorum* but can be readily distinguished by the persistent stipules.

2. Hylodesmum nudiflorum (L.) H. Ohashi & R.R. Mill (naked tick trefoil, naked-stemmed tick clover)

Desmodium nudiflorum (L.) DC.

Pl. 397 c, d; Map 1755

Stems solitary but usually appearing paired (rarely appearing as small clusters), 6–50 cm long, erect or ascending, in fertile plants with 1(–4) flowering branch(es) from just above or just below ground level, otherwise unbranched, the fertile branch(es) ascending to somewhat arched (often from a spreading base), usually overtopping the vegetative portion of the stem, usually leafless, the median portion of both types moderately to densely pubescent with minute, mostly hooked hairs and sparse to moderate, longer, straight, spreading hairs, both hair types usually sparser toward the stem base and denser toward the tip, the fertile branch occasionally nearly glabrous throughout. Leaves of vegetative portion in a cluster at the stem tip, usually appearing nearly whorled, often also 1 or few more widely spaced, alternate leaves present below the stem tip, the petiole 2–10 cm long. Stipules 2–4 mm long, 0.1–0.2 mm wide, at maturity olive green and appressed, mostly shed before flowering. Leaflets with the undersurface lighter than the upper surface and often somewhat glaucous; the central leaflet 3.0–8.5 cm long and 2–6 cm wide, rhombic, broadly elliptic, ovate, or obovate, angled or tapered (sometimes short-tapered) to a sharply pointed tip; lateral leaflets 2.0–6.5 cm long and 1.5–4.0(–5.0) cm wide, somewhat asymmetrically ovate (the basal side broader than the side adjacent to the central leaflet). Inflorescences solitary, terminal on the fertile branch, erect or arched, unbranched or less commonly with few, spreading branches, the axis variously glabrous or sparsely to moderately pubescent with minute, hooked hairs and occasionally also with very sparse, straight, spreading hairs. Bracts 3–5 mm long, 0.5–0.8 mm wide, linear to narrowly lanceolate-triangular, mostly shed early; bractlets often also present, these very short (to 1 mm long), shed early. Flower stalks 8–20(–25) mm long. Calyces green, nearly glabrous except for glandular hairs near the base and a few straight hairs on the lobes, the tube 1.5–2.0 mm long, the lobes 0.5–1.0 mm long. Corollas 5–9 mm long, pink or rarely white, the banner white near the clawed base, with a pair of slightly confluent, dark purple spots (nectar guides) above the base, the keel petals fused together. Fruits nearly straight or arched downward, tapered abruptly to a stalklike base 8–18 mm long, consisting of 1–4 segments, each 7–9 mm long and 4–5 mm wide, the upper side of each segment sometimes very slightly concave. $2n=22$. June–August.

Scattered nearly throughout the state but apparently absent from most of the Mississippi Lowlands Division and the western half of Glaciated Plains (eastern U.S. west to Minnesota and Texas; Canada). Mesic to dry upland forests, ledges of bluffs, edges of glades, and margins of sinkhole ponds.

Occasional plants are found with one or more leaves on the fertile shoot. These have been called *Desmodium nudiflorum* f. *foliolatum* (Farw.) Fassett. Rare plants with an inflorescence at the tip of the leafy shoot have been called f. *personatum* Fassett. The latter form is distinguished from *H. glutinosum* by the shape of the leaflets and fruit, as well as the absence of stipules at flowering time.

3. Hylodesmum pauciflorum (Nutt.) H. Ohashi & R.R. Mill (few-flowered tick trefoil, few-flowered tick clover, small-flowered tick clover)

Desmodium pauciflorum (Nutt.) DC.

Pl. 397 a, b; Map 1756

Stems solitary or few in loose clusters (then from a branched rootstock), (5–)10–50 cm long (below the inflorescence), loosely to strongly ascending, sometimes from a spreading base, unbranched or less commonly few-branched from below the midpoint, the median and basal portions sparsely pubescent with minute, mostly hooked hairs and longer, straight, downward-curved hairs, usually also sparsely hairy toward the tip. Leaves alternate and relatively well-spaced, not appearing whorled, the petiole 2–7 cm long. Stipules 2–5 mm long, 0.2–0.5 mm wide, at maturity olive green and appressed, shed early. Leaflets with the undersurface lighter than the upper surface but not glaucous; the central leaflet 2–8 cm long, 2–6 cm wide, obovate to rhombic or ovate, angled or tapered (often short-tapered) to a sharply pointed tip; lateral leaflets 3–7 cm long, 2.0–4.5 cm wide, somewhat asymmetrically ovate (the basal side broader than the side adjacent to the central leaf-

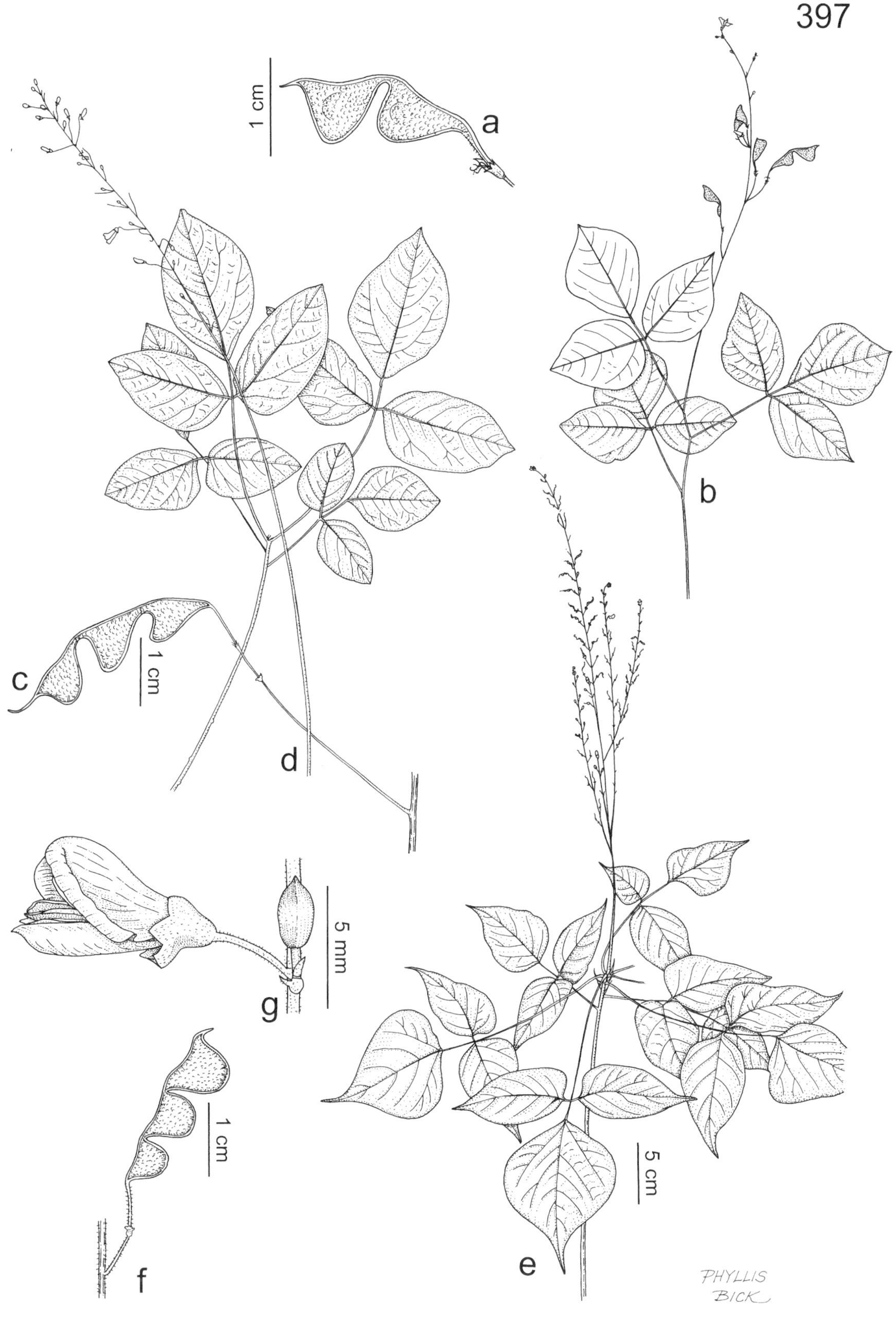

Plate 397. Fabaceae. *Hylodesmum pauciflorum*, **a)** fruit, **b)** habit. *Hylodesmum nudiflorum*, **c)** fruit, **d)** fertile and vegetative stems. *Hylodesmum glutinosum*, **e)** habit, **f)** fruit, **g)** flower.

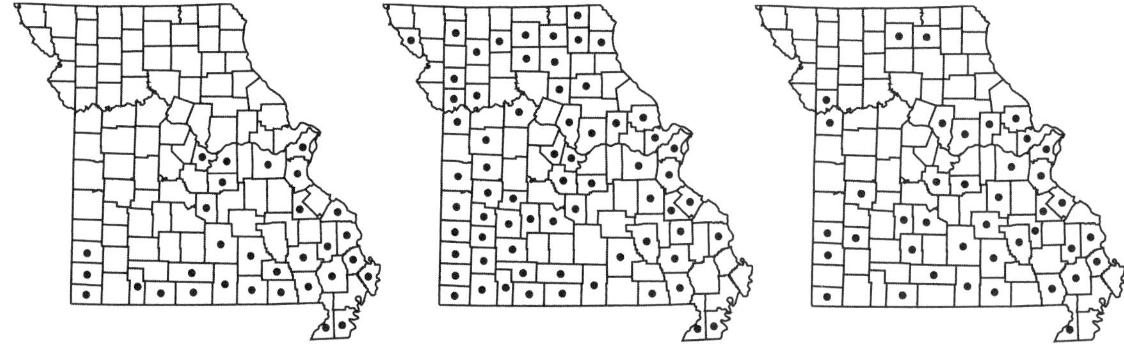

1756. Hylodesmum pauciflorum 1757. Kummerowia stipulacea 1758. Kummerowia striata

let). Inflorescences terminal and also axillary from the upper nodes, ascending, unbranched or with few, loosely ascending branches, the axis pubescent with sparse to moderate, minute, hooked hairs and often also widely scattered, straight, spreading hairs. Bracts 1–4 mm long, 0.2–0.3 mm wide, linear, sometimes shed early; bractlets sometimes also present, these somewhat shorter than the bracts (to 2.5 mm long), shed early. Flower stalks 2–11 mm long. Calyces olive green, minutely hairy toward the base and with sparse to moderate, longer, straight hairs throughout, the tube 1.5–2.0 mm long, the lobes 0.2–0.5 mm long. Corollas 5–7 mm long, white, the banner lacking contrasting nectar guides, the keel petals not fused together. Fruits nearly straight or nearly so, tapered abruptly to a stalklike base 3–7 mm long, consisting of 1 or 2(3) segments, each 9–12 mm long and 3.5–8.0 mm wide, the upper side of each segment usually not concave. $2n$=22. June–September.

Scattered in the Ozark, Ozark Border, and Mississippi Lowlands Divisions (eastern U.S. west to Kansas and Texas). Bottomland forests, mesic upland forests, bases of bluffs, banks of streams, rivers, and spring branches, and margins of sinkhole ponds.

This low, often strongly colonial species produces relatively few-flowered inflorescences of relatively small flowers, and sometimes is observed as colonies of all or mostly vegetative stems.

31. **Kummerowia** Schindl.
Contributed by Jay A. Raveill

Plants annual. Stems spreading to ascending or less commonly erect, with many well-developed branches, sparsely to densely hairy. Leaves alternate, short- to long-petiolate. Stipules persistent, ovate to narrowly ovate, herbaceous and green when young, becoming papery and straw-colored at maturity, the margins entire, inconspicuously and minutely appressed-hairy, the surfaces glabrous or nearly so, the venation prominent. Leaf blades appearing palmately trifoliate (the middle leaflet sometimes slightly longer than the lateral pair), the leaflets without stipels. Leaflets sessile or nearly so, obovate, narrowly elliptic, or narrowly oblong, usually rounded to shallowly notched at the tip, often with a minute hairlike extension of the midvein, the margins entire, hairy, the upper surface glabrous, the undersurface sparsely to moderately pubescent with relatively stiff hairs along the midvein, pinnately veined. Inflorescences axillary at most upper and some lower nodes, appearing as solitary flowers or small clusters of 2–5 flowers, each subtended by 3 or 4 minute papery bracts. Calyces 5-lobed but appearing 4-lobed because of the nearly complete fusion of the uppermost 2 sepals, the lobes similar in length, glabrous or hairy, usually prominently nerved, persistent and papery at fruiting. Corollas papilionaceous, pink to purplish pink (drying bluish purple) with large white patches at the base of the banner, the keel longer than the wings. Stamens 10, 9 of these with the filaments fused together most of their length, the remaining filament free, the anthers all of similar size. Fruits 1-seeded, indehiscent, flattened, the rim angled and somewhat thickened, the surface moderately pubescent with minute appressed hairs. Seeds 1.5–2.0 mm long, somewhat flattened, the surface smooth, brown to black, shiny. Two species, native to eastern Asia, widely introduced in North America.

The acceptance of *Kummerowia* as a genus distinct from *Lespedeza* is not universal. *Kummerowia* differs from *Lespedeza* primarily in its annual habit (vs. perennial or shrubby), conspicuous papery stipules, and details of the highly reduced inflorescence (Akiyama and Ohba, 1985; Nemoto and Ohashi, 1993). As in some *Lespedeza*, both species of *Kummerowia* can produce cleistogamous flowers.

Both species in the genus have been planted widely in the eastern United States for forage and soil improvement (Isely, 1948), and less commonly as an annual hay crop. They establish readily and thrive even in areas of thin and poor soils, but can out-compete native vegetation in such areas as sandstone glades. *Kummerowia striata* was introduced to the United States from Japan in 1846, but the later-introduced *K. stipulacea* (in 1919) has replaced it in importance in Missouri, both as an agricultural plant and as an escape (Steyermark, 1963).

1. Stems pubescent with forward-pointing hairs; leaves with conspicuous petioles, those of lower and median leaves 4–10 mm long; leaflets pubescent with conspicuous spreading hairs; calyx at maturity covering 1/3–1/2 of the fruit . . . . 1. K. STIPULACEA
1. Stem pubescent with recurved or backward-pointing hairs; leaves nearly sessile, the petioles of lower and median leaves 1–2(–4) mm long; leaflets glabrous or pubescent with inconspicuous appressed hairs; calyx at maturity covering 1/2–4/5 of the fruit . . . . 2. K. STRIATA

1. Kummerowia stipulacea (Maxim.) Makino (Korean clover, Korean lespedeza)
Lespedeza stipulacea Maxim.
Pl. 399 i, j; Map 1757

Stems 10–60 cm long, pubescent with stiff, forward-pointing, often appressed hairs. Leaves conspicuously petiolate, the lower and median leaves with petioles 4–10 mm long, these sparsely to moderately pubescent with forward-pointing hairs. Stipules of lower and median leaves 4–8 mm long. Leaflets 5–25 mm long, obovate to spatulate, tapered or narrowed at the base, rounded to shallowly notched at the tip, the upper surface glabrous, the undersurface glabrous or more commonly pubescent along the midvein, the margins with conspicuous, spreading, somewhat curved hairs. Calyces 1.6–2.2 mm long (slightly shorter in cleistogamous flowers), the tube about as long as the lobed, glabrous or sparsely hairy along the margins, at maturity covering 1/3–1/2 of the fruit. Corollas 6–7 mm long (shorter in cleistogamous flowers). Fruits 2.5–3.2 mm long, obovate to elliptic in outline, rounded to bluntly or broadly pointed at the tip, the surfaces with a conspicuous network of nerves. $2n$=20, 22. July–October.

Introduced, common nearly throughout the state (native of eastern Asia; introduced widely in the eastern U.S. west to Nebraska and Texas). Glades, rocky areas in upland prairies, and openings of dry upland forests; also pastures, old fields, roadsides, railroads, and disturbed open places.

2. Kummerowia striata (Thunb.) Schindl. (Japanese clover, Japanese lespedeza, common lespedeza)
Lespedeza striata (Thunb.) Hook. & Arn.
Pl. 399 k, l; Map 1758

Stems 10–50 cm long, pubescent with stiff, recurved or backward-pointing, usually appressed hairs. Leaves nearly sessile, the lower and median leaves with petioles 1–2(–4) mm long, these sparsely to moderately pubescent with recurved or backward-pointing hairs. Stipules of lower and median leaves 3–6 mm long. Leaflets 5–20 mm long, obovate to narrowly elliptic or narrowly oblong, narrowed or sometimes rounded at the base, rounded to shallowly notched at the tip, less commonly abruptly tapered to a short sharp point, the upper surface glabrous, the undersurface glabrous or sparsely to moderately pubescent with inconspicuous appressed hairs, the margins with inconspicuous appressed hairs. Calyces 1.6–2.2 mm long (slightly shorter in cleistogamous flowers), the tube about as long as the lobes, glabrous or sparsely hairy along the margins, at maturity covering 1/2–4/5 of the fruit. Corollas 4.5–6.0 mm long (shorter in cleistogamous flowers). Fruits 2.5–4.0 mm long, obovate to elliptic in outline, short-

1759. Lablab purpureus 1760. Lathyrus hirsutus 1761. Lathyrus latifolius

tapered to a beaklike tip, the surfaces with an inconspicuous network of nerves. $2n$=20, 22. July–October.

Introduced, scattered mostly south of the Missouri River (native of eastern Asia; introduced widely in the eastern U.S. west to Iowa, Kansas, and New Mexico). Glades, rocky areas in upland prairies, and openings of dry upland forests; also pastures, old fields, roadsides, railroads, and disturbed open places.

32. **Lablab** Adans.

One species, Old World, cultivated widely in warm climates.

1. Lablab purpureus (L.) Sweet (hyacinth bean, bonavist)
Dolichos lablab L.

Map 1759

Plants annual (perennial herbs farther south). Stems 200–600 cm long, trailing or climbing and twining (erect and bushy in some cultivated forms), branched, unarmed, usually purplish-tinged, glabrous or sparsely pubescent with more or less downward-curved or -angled hairs, these not hooked at the tip. Leaves alternate, pinnately trifoliate, the petiole 7–15 cm long. Stipules 3–7 mm long, lanceolate-triangular with an asymmetric base, tapered to a sharply pointed tip, shed early; stipels 3–4 mm long, narrowly oblong-elliptic, persistent. Leaflets 4–10 cm long, 3–9 cm wide, broadly ovate, the terminal leaflet sometimes somewhat rhombic, broadly angled to nearly truncate at the base, short-tapered to a sharply pointed tip, the margins entire, the surfaces glabrous or sparsely and inconspicuosuly hairy, with 3 main veins from the base, but otherwise more or less pinnately veined. Terminal leaflet with the stalk 25–35 mm long, symmetric at the base; lateral leaflets with the stalk 3–5 mm long, asymmetric at the base. Inflorescences axillary, narrow racemes 10–40 cm long, with 8 to numerous flowers in clusters of 2–5 per node, the stalk 50–70 mm long, the bracts 4–5 mm long, narrowly ovate, shed early, each flower with a stalk 3–4 mm long, the bractlets 3–4 mm long, ovate-elliptic, densely hairy along the margins, more or less persistent. Calyces short-hairy along the margins, usually strongly purplish-tinged or dark purple, the tube 3–4 mm long, broadly bell-shaped, 2-lipped, the lobes 2–3 mm long, unequal, the upper 2 lobes fused into a broadly ovate, slightly hooded structure, the lower 3 lobes shorter, similar, narrowly ovate-triangular, sharply pointed at their tips. Corollas papilionaceous, purple or white, the banner 12–16 mm long, 12–14 mm wide, the expanded portion curved or bent backward, broadly oblong-ovate to nearly circular, notched at the tip, with a pair of thickened areas toward the base, the wings 14–16 mm long, 6–8 mm wide, obliquely oblong-obovate, curled over the keel, the keel 12–14 mm long, 3–4 mm wide, boat-shaped, fused to above the midpoint, bent or curved upward abruptly near the midpoint, the tip bluntly to sharply pointed. Stamens 10, in 2 alternating, slightly shorter and longer series, 9 of the filaments fused and 1 free, the fused portion 11–13 mm long, bent upward at its tip, the free portion 5–7 mm long, the anthers small, attached at the base, yellow. Ovary 7–9 mm long, hairy, the style 10–12 mm long, bent or curved upward toward the tip, somewhat flattened, hairy on the inner side, the stigma small, terminal, oblique. Fruits legumes, 6–9 cm long, 16–20 mm wide, oblong, short-tapered asymmetrically to a beak, flattened, sometimes slightly constricted between the seeds, straight or more commonly curved upward, dehiscing tardily by 2 valves, the upper margin beaded with small, warty outgrowths, purple to brownish purple at maturity (more or less green in some cultivated races), glabrous or nearly so, often somewhat shiny, mostly 2–6-seeded. Seeds 10–12 mm long, 6–8 mm wide, oblong in outline, the surface white to dark purple, brown, or black, smooth, somewhat shiny, with a white aril. $2n=22$. July–September.

Introduced, uncommon, sporadic in southern Missouri (native of Africa, Asia, now cultivated and introduced throughout the tropics; introduced sporadically in the eastern U.S.). Alleys, roadsides, and open disturbed areas.

Lablab purpureus was first collected as an escape in Missouri by Paul Redfearn in 2001 from Greene County and by Bill Summers in 2002 from Howell County. It probably does not persist long outside of gardens. Hyacinth bean has been cultivated for thousands of years as a food for people and livestock, especially in India, China, and Japan (Tindall, 1983). The plants are hardy and do well in warm, dry climates. The young leaves can be picked and eaten like spinach. Some forms produce an edible tuber. The beans can be eaten, but only after being cooked in several changes of water because of the presence of toxic compounds. The species also is used as a cover crop and green manure. In the United States, it is planted mainly as an ornamental.

The more erect, bushy forms are sometimes called var. *lignosus* (Prain) Kumari, but these are better treated as cultivars until genetic relationships among the wild and cultivated races can be studied in more detail. Verdcourt (1970a), in his studies of *Lablab* for the *Flora of Tropical East Africa* Project, divided the species into three subspecies. He treated the native populations as ssp. *uncinatus* Verdc. and segregated an unusual African cultivated race with slender fruits as ssp. *benghalensis* (Jacq.) Verdc. The remaining polymorphic group of cultivated forms were included in ssp. *purpureus*.

33. **Lathyrus** L.
(Hitchcock, 1952)

Plants annual or perennial herbs, with taproots or more commonly rhizomes. Stems erect to spreading or climbing, angled or winged, unbranched or branched, unarmed, glabrous or pubescent with nonglandular hairs. Leaves alternate, even-pinnately compound with 2–10 leaflets (rarely with an extra lateral leaflet along 1 side), the petiole often expanded or winged, the rachis extended into a conspicuous, unbranched or branched tendril (this poorly developed elsewhere). Stipules leaflike, with a basal outgrowth or lobe of tissue on 1 side, this rounded or more commonly triangular, descending or clasping the stem, the margins otherwise entire or less commonly toothed, the venation mostly inconspicuous, persistent; stipels absent. Leaflets oblong to elliptic or linear, the margins entire, the surfaces glabrous or hairy, pinnately veined or only the midvein visible (often with 3 main veins in *L. sylvestris*). Inflorescences axillary, racemes or clusters, occasionally reduced to solitary flowers, the bracts 1–3 mm long, shed early, bractlets absent. Calyces 5-lobed, the tube bell-shaped, usually oblique, often somewhat pouched on 1 side at the base, more or less 2-lipped, the lobes subequal or the 3 lower lobes longer than the upper 2, variously shaped, sharply pointed at their tips, glabrous or hairy. Corollas papilionaceous, pink, red, purple, or white (yellow elsewhere), lacking conspicuous, contrasting markings near the base of the banner, the banner with a short, sometimes broad, stalklike base, the expanded portion obovate to nearly circular, notched at the tip, sharply curved or bent backward, the wings narrowly to broadly obovate, shorter than to slightly longer than the banner, usually curved over or around but not fused to the keel, the keel shorter than the wings, boat-shaped, curved upward, mostly tapered to a bluntly or sharply pointed tip. Stamens 10, 9 of the filaments fused and 1 free nearly to the base, all similar in length or in 2 alternating, slightly shorter and longer series, the free portions of the filaments slender, but usually broadened toward their tips, the anthers small, attached at the base, all similar in size, yellow or occasionally orange. Ovary sessile or very short-stalked, glabrous or occasionally hairy, the style abruptly curved or bent upward toward the base, somewhat flattened, hairy on the inner surface toward the tip, more or less persistent at fruiting, the stigma terminal, short or somewhat elongate. Fruits legumes, oblong to linear, tapered asymmetrically to a sharply pointed or more commonly beaked tip, flattened or in a few species turgid, not or only slightly constricted between the seeds, straight or slightly curved upward, 2- to numerous-seeded, dehiscing by 2 valves, these green to brown at maturity, usually twisting spirally after dehiscence. Seeds oblong to kidney-shaped, nearly circular or somewhat angular in outline, flattened or not, the surface smooth or wrinkled, olive brown to dark brown, sometimes mottled. About 160 species, nearly worldwide, most diverse in temperate regions.

Lathyrus is a member of the tribe Fabeae, which is one of the most economically important groups in the family and also includes the related genera *Lens* Mill. (lentil), *Pisum* L. (garden pea), and *Vicia* L. (vetch). Species of *Lathyrus* are widely planted for soil cover, green manure, and occasionally for human food. The seeds of *L. sativus* L. (grass pea) are high in protein and it has been cultivated for at least 8,000 years (Smartt, 1990). They are highly drought resistant and produce a crop when other crops fail completely. The use of the species as food for humans and fodder for livestock is limited by the presence of toxic nonprotein amino acids that cause a syndrome known as lathyrism, one of the oldest neurotoxic diseases known (Burrows and Tyrl, 2001). Symptoms include spasms, weakness, paralysis of the legs, and ultimately muscular atrophy. Lathyrism remains a problem in impoverished countries where people are forced to subsist on *Lathyrus* peas for lengthy periods in times of famine. Curiously, the evolution and divergence of *Lathyrus* species has been accompanied by a threefold increase in chromosome size and fourfold increase in the amount of DNA (Narayan, 1982; Nandini et al., 1997).

1. Leaflets 4–10
    2. Stipules 4–10(–18) mm wide, ovate to broadly ovate; leaflets mostly 6, some leaves occasionally with 4, 5, or 8 leaflets; inflorescences open racemes with 2–6(–9) flowers; calyces glabrous . . . . 3. L. PALUSTRIS
    2. Stipules 1–4 mm wide, narrowly lanceolate to lanceolate; leaflets (6)8–14; inflorescences dense racemes with 8–20 flowers; calyces hairy . . . . 7. L. VENOSUS
1. Leaflets 2
    3. Ovaries and fruits conspicuously hairy . . . . 1. L. HIRSUTUS
    3. Ovaries and fruits glabrous
        4. Stems unwinged; small tubers produced along the rhizomes . . . . 6. L. TUBEROSUS
        4. Stems winged; roots not producing tubers
            5. Stems narrowly winged, the wings at most 1 mm wide; corollas 5–9(–12) mm long; fruits 2.5–4.0 cm long; plants annual . . . . . . 4. L. PUSILLUS
            5. Stems broadly winged, the wings 1–4 mm wide, corollas 14–25 mm long; fruits 5–10 cm long; plants perennial
                6. Stipules 20–38 mm long, 5–12 mm wide, broadly lanceolate to ovate; corollas 18–25 mm long; fruits 6–10 cm long . . 2. L. LATIFOLIUS
                6. Stipules 8–14(–23) mm long, 1–3 mm wide, linear to narrowly lanceolate; corollas 13–20 mm long; fruits 4–7 cm long . . . . 5. L. SYLVESTRIS

1. Lathyrus hirsutus L. (Caley pea, singletary pea)

Pl. 398 d–f; Map 1760

Plants annual, with shallow taproot, the roots not producing tubers. Stems 15–100 cm long, trailing, ascending, or climbing, often branched at the base, glabrous or sparsely and inconspicuously hairy, angled or narrowly winged, the wings to 2 mm wide. Leaves with 2 leaflets, the petiole 2–4 cm long, broadly winged, the tendrils branched.

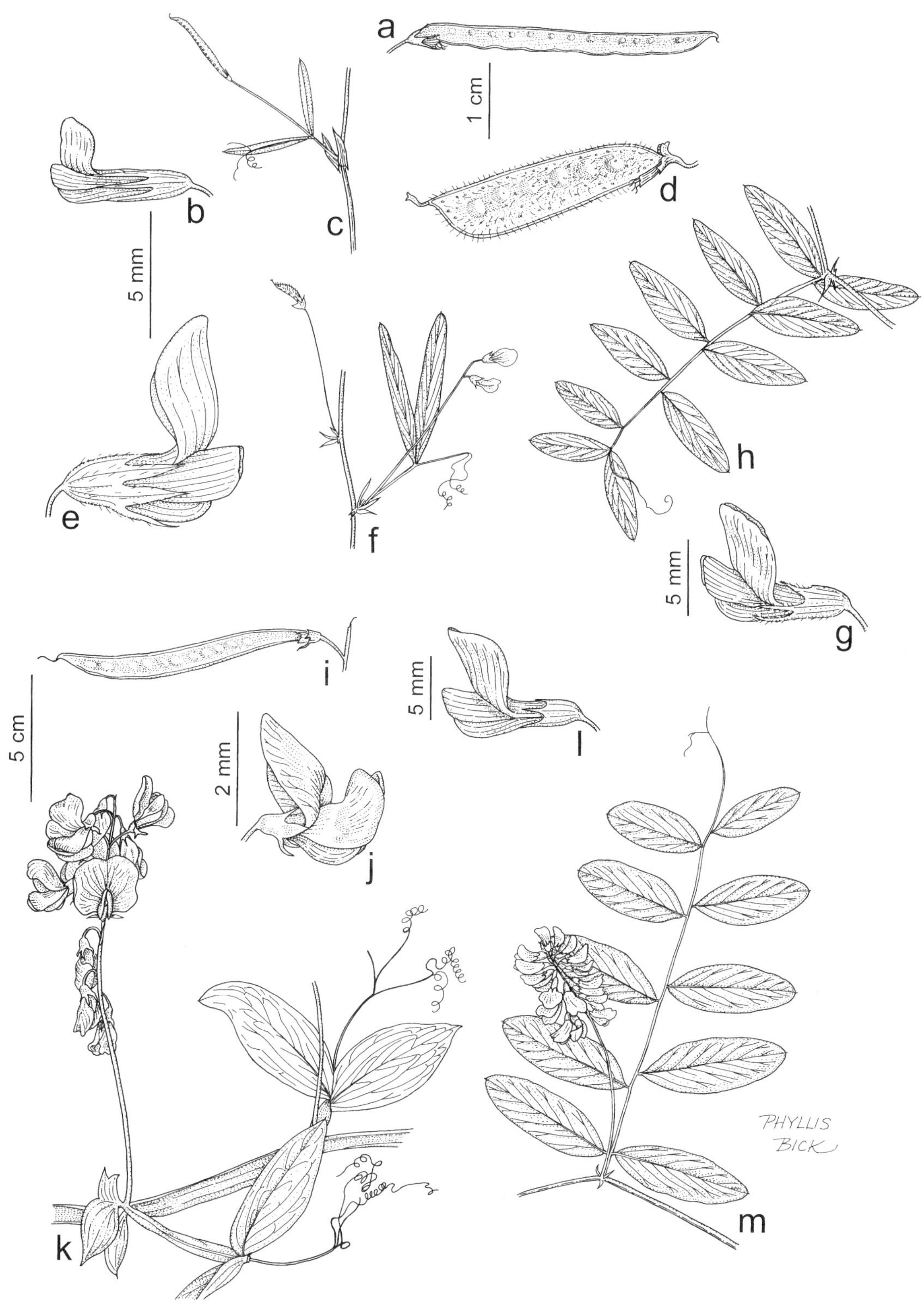

Plate 398. Fabaceae. *Lathyrus pusillus*, **a)** fruit, **b)** flower, **c)** fertile stem. *Lathyrus hirsutus*, **d)** fruit, **e)** flower, **f)** fertile stem. *Lathyrus palustris*, **g)** flower, **h)** leaves. *Lathyrus latifolius*, **i)** fruit, **j)** flower, **k)** fertile stem. *Lathyrus venosus*, **l)** flower, **m)** fertile stem.

Stipules 7–20 mm long, 1–3 mm wide, narrowly lanceolate, the basal lobe 3–10 mm long, narrowly triangular. Leaflets 2–9 cm long, 4–18 mm wide, narrowly elliptic to linear or oblong-lanceolate, angled at the base, angled to a bluntly or sharply pointed tip, the midvein sometimes extended into a minute sharp point at the very tip, glabrous. Inflorescences of solitary flowers or clusters of 2(3) flowers, the stalk 3–10(–20) cm long, usually extending past the attachment of the flowers as a short bristle, the flower stalks 4–6 mm long. Calyces with the tube 2–3 mm long, glabrous or sparsely appressed-hairy, the lobes 2–4 mm long, subequal, narrowly ovate to ovate. Corollas 9–14 mm long, pink to bluish purple, sometimes appearing bicolorous. Filaments with the fused portion 5–6 mm long, the free portion 2–3 mm long. Ovary densely hairy, the hairs developing pustular bases. Fruits 2–4 cm long, 6–8 mm wide, narrowly oblong, strongly flattened, conspicuously hairy, the hairs with pustular bases, 5–10-seeded. Seeds 3–4 mm long, globose to slightly angular, not flattened, the surface finely wrinkled, brown. $2n=14$, 28. May–June.

Introduced, uncommon, mostly in the southern portion of the state (native of Europe; introduced widely in the southern U.S. north locally to Oregon and Michigan). Edges of bottomland and mesic upland forests; also margins of crop fields, old fields, pastures, ditches, railroads, roadsides, and open disturbed areas.

Lathyrus hirsutus is easily recognized by the distinctive, pustular-based hairs on the ovary and fruits, along with the narrowly winged stems and leaves with two leaflets. It is widely cultivated in the United States for soil improvement and pasturage. This species was first collected in Missouri in Jasper County by Ernest J. Palmer in 1949, and Steyermark (1963) knew it from only one additional specimen collected by Viktor Mühlenbach in the city of St. Louis. Since that time it has become much more abundant in southern Missouri, especially in the Mississippi Lowlands Division.

2. Lathyrus latifolius L. (everlasting pea, perennial sweet pea)

Pl. 398 i, k; Map 1761

Plants perennial, with rhizomes, the roots not producing tubers. Stems 80–200 cm long, trailing or climbing, usually branched, glabrous, somewhat glaucous, conspicuously winged, the wings 2–4 mm wide. Leaves with 2 leaflets, the petiole 2–6 cm long, broadly winged, the tendrils branched. Stipules 20–38 mm long, 5–12 mm wide, broadly lanceolate to ovate, the basal lobe 8–22 mm long, elliptic-lanceolate to lanceolate or triangular-lanceolate. Leaflets 4–11 cm long, 10–50 mm wide, elliptic to broadly elliptic or oblong-elliptic, angled at the base, angled to a bluntly or more commonly sharply pointed tip, the midvein sometimes extended into a minute sharp point at the very tip, glabrous, usually somewhat glaucous. Inflorescences racemes of 4–16 flowers, the stalk 8–20 cm long, the flower stalks 10–25 mm long. Calyces with the tube 4–6 mm long, glabrous, the lobes 2.0–7.5 mm long, the upper 2 short and triangular, the lowermost lobe about twice as long as the upper 2, narrowly lanceolate-triangular, the lateral lobes of the lower lip intermediate in size and shape. Corollas 18–25 mm long, purple, red, pink, or white. Filaments with the fused portion 11–12 mm long, the free portion 5–6 mm long. Ovary glabrous. Fruits 6–10 cm long, 8–10 mm wide, linear to narrowly oblong, flattened, glabrous, 10–15-seeded. Seeds 5–6 mm long, oblong in outline (more or less truncate at each end), somewhat flattened, the surface finely wrinkled, dark brown. $2n=14$. June–October.

Introduced, scattered sporadically nearly throughout the state (native of Europe; introduced nearly throughout temperate North America) Edges of bottomland and mesic upland forests and banks of streams and rivers; also pastures, cemeteries, fencerows, levees, lawns, gardens, railroads, roadsides, and disturbed areas.

Lathyrus latifolius is recognized by its broadly winged stems, leaves with two, broad leaflets, very large stipules, and relatively large flowers and fruits. It is most similar to *L. sylvestris*, which has narrower stipules and smaller flowers and fruits. The flower color is variable, even within a population.

3. Lathyrus palustris L. (vetchling, marsh pea)

L. palustris var. *myrtifolius* (Muhl. ex Willd.) A. Gray

Pl. 398 g, h; Map 1762

Plants perennial, with rhizomes, the roots not producing tubers. Stems 30–100 cm long, trailing or climbing, unbranched or few-branched from the base, glabrous, angled or narrowly winged, the wings to 1 mm wide. Leaves with mostly 6 leaflets, some of them occasionally with 4, 5, or 8 leaflets, the petiole 0.8–3.0 cm long, unwinged, the tendrils branched or some of them unbranched. Stipules 6–18(–30) mm long, 4–10(–18) mm wide, ovate to broadly ovate, the basal lobe 4–8(–20) mm long, triangular to rounded-oblong and sometimes toothed. Leaflets 2–8 cm long, 3–20 mm wide, el-

1762. Lathyrus palustris 1763. Lathyrus pusillus 1764. Lathyrus sylvestris

liptic to narrowly elliptic, narrowly lanceolate, or linear, angled at the base, angled to a bluntly or sharply pointed tip, the midvein sometimes extended into a minute sharp point at the very tip, glabrous. Inflorescences open racemes of 2–6 (–9) flowers, the stalk 3–6 cm long, the flower stalks 2–6 mm long. Calyces with the tube 3–4 mm long, glabrous or hairy, the lobes 1–6 mm long, the upper 2 short and triangular, the lowermost lobe 2–3 times as long as the upper 2, narrowly lanceolate-triangular, the lateral lobes of the lower lip intermediate in size and shape. Corollas 12–20 mm long, reddish purple, to purple, lavender, or pink (rarely white elsewhere). Filaments with the fused portion 9–10 mm long, the free portion 3–4 mm long. Ovary glabrous. Fruits 4–6 cm long, 4–5 mm wide, linear to narrowly oblong, flattened, glabrous or with sparse, minute glandular hairs, 5–10-seeded. Seeds 3.0–3.5 mm long, more or less globose to slightly angular, not or only slightly flattened, the surface smooth, reddish brown to greenish brown or nearly black, sometimes faintly mottled. $2n=14$. May–June.

Uncommon in the eastern half of the state (eastern U.S. west to North Dakota and Texas, also Alaska to California; Canada, Europe, Asia). Fens, acid seeps, and bottomland forests; rarely also roadsides.

Lathyrus palustris is variable in leaf size and shape, stem size, and degree of pubescence, and several varieties have been described (Fernald, 1911; Hitchcock, 1952; Steyermark, 1963). Plants with slender, wingless stems and leaves with four or six broad, obtuse leaflets have been called var. *myrtifolius*. Plants with narrowly winged stems and six or eight narrow leaflets are var. *palustris*. However, Isely (1998) concluded that these varieties have no ecological or geographic basis and intergrade completely.

4. Lathyrus pusillus Elliott (singletary vetchling, tiny pea)

Pl. 398 a–c; Map 1763

Plants annual, with slender taproots, the roots not producing tubers. Stems 20–60 cm long, trailing or climbing, unbranched or few-branched from near the base, glabrous or sparsely and inconspicuously hairy, narrowly winged, the wings to 1 mm wide. Leaves with 2 leaflets, the petiole 1–2 cm long, ridged or narrowly winged, the tendrils mostly branched. Stipules 10–26 mm long, 2–6 mm wide, lanceolate to narrowly ovate, the basal lobe 7–10 mm long, lanceolate-triangular. Leaflets 2–5 cm long, 4–6 mm wide, narrowly lanceolate to narrowly elliptic or linear, angled at the base, angled or slightly tapered to a sharply pointed tip, the midvein sometimes extended into a minute sharp point at the very tip, glabrous or very sparsely hairy. Inflorescences of solitary or more commonly paired flowers, the main stalk 6–8 cm long, the flower stalks 1–4 mm long. Calyces with the tube 2.0–2.5 mm long, glabrous, the lobes 3–5 mm long, the upper 2 lobes slightly shorter than the lower 3, all lanceolate-triangular. Corollas 5–9(–12) mm long, pale lavender blue. Filaments with the fused portion 4–5 mm long, the free portion 1–2 mm long. Ovary glabrous. Fruits 2.5–4.0 cm long, 3–5 mm wide, linear to narrowly oblong, flattened, glabrous, 10–20-seeded. Seeds 1.5–2.5 mm long, more or less globose, not flattened, the surface finely wrinkled, brownish olive to dark brown. $2n=22$. May–June.

Uncommon, known only from historical collections from Jasper and Newton Counties (southeastern U.S. west to Kansas and Texas; Mexico; introduced in Oregon). Chert glades; also railroads.

Lathyrus pusillus is characterized by its winged stems, small flowers, and long calyx lobes. The fruits often are borne in pairs. The tendrils are weakly developed and often unbranched.

1765. Lathyrus tuberosus 1766. Lathyrus venosus 1767. Lespedeza bicolor

5. Lathyrus sylvestris L. (narrow-leaved vetchling, narrow-leaved everlasting pea) Map 1764

Plants perennial, with rhizomes, the roots not producing tubers. Stems 50–200 cm long, trailing or climbing, usually branched, glabrous, conspicuously winged, the wings 1–3 mm wide. Leaves with 2 leaflets, the petiole 1.5–3.0 cm long, narrowly to relatively broadly winged winged, the tendrils branched. Stipules 8–14(–23) mm long, 1–3 mm wide, linear to narrowly lanceolate, the basal lobe 3–6 mm long, linear to narrowly oblong-triangular. Leaflets 4–11 cm long, 5–8 mm wide, narrowly lanceolate to narrowly elliptic, angled at the base, angled to a usually sharply pointed tip, the midvein sometimes extended into a minute sharp point at the very tip, glabrous, sometimes slightly glaucous. Inflorescences racemes of 4–9(–12) flowers, the stalk 9–22 cm long, the flower stalks 8–20 mm long. Calyces with the tube 3–4 mm long, glabrous, the lobes 1–4 mm long, the upper 2 short and triangular, the lowermost lobe about twice as long as the upper 2, narrowly triangular, the lateral lobes of the lower lip intermediate in size and shape. Corollas 13–20 mm long, pink to pinkish or reddish purple. Filaments with the fused portion 9–11 mm long, the free portion 3–4 mm long. Ovary glabrous. Fruits 4–7 cm long, 8–9 mm wide, narrowly oblong, flattened, glabrous, 10–20-seeded. Seeds 3.5–4.0 mm long, more or less globose to slightly oblong in outline, sometimes slightly flattened, the surface finely wrinkled, dark brown. $2n=14$. June–August.

Introduced, uncommon, known thus far only from Franklin County (native of Europe; introduced and widely scattered in the U.S., Canada). Upland prairies.

Lathyrus sylvestris is similar to *L. latifolius*, but is distinguished mainly by its narrower stipules and smaller flowers. It was first collected in Missouri in 1995 by Doug Ladd as a weed in a constructed prairie at the Shaw Arboretum (now Shaw Nature Reserve).

6. Lathyrus tuberosus L. (tuberous vetchling) Map 1765

Plants perennial, with rhizomes, these producing small tubers. Stems 30–100 cm long, trailing or climbing, usually branched, glabrous, 2-angled but not winged. Leaves with 2 leaflets, the petiole 0.3–1.0 cm long, unwinged, the tendrils unbranched or branched. Stipules 3–12 mm long, 1–3 mm wide, lanceolate, the basal lobe 1–2 mm long, linear to narrowly oblong-triangular. Leaflets 1.5–4.0 cm long, 5–13 mm wide, lanceolate, narrowly lanceolate, narrowly elliptic, narrowly oblanceolate, or narrowly obovate, angled at the base, angled or slightly tapered to a bluntly or sharply pointed tip, the midvein sometimes extended into a minute sharp point at the very tip, glabrous. Inflorescences dense racemes of 2–10 flowers, the stalk 5–8 cm long, the flower stalks 3–5 mm long. Calyces with the tube 3–4 mm long, glabrous, the lobes 3–5 mm long, the upper 2 lobes slightly shorter than the lower 3, all triangular to broadly lanceolate-triangular. Corollas 10–16 mm long, reddish purple. Filaments with the fused portion 6–7 mm long, the free portion 4–5 mm long. Ovary glabrous. Fruits 1.5–4.0 cm long, 3–5 mm wide, narrowly oblong, flattened, glabrous, 1–5-seeded. Seeds 4–5 mm long, ovoid or oblong-ellipsoid to more or less globose or broadly oblong in outline, not flattened, the surface finely wrinkled, reddish brown to grayish brown or dark brown. $2n=14$. June–July.

Introduced, uncommon, known thus far only from Dallas County (native of Europe; introduced widely in the northern U.S., Canada). Fence rows and open, disturbed areas.

Lathyrus tuberosus is recognized by its wingless stems and the production of small tubers along the slender rhizomes. The tubers can be cooked and eaten. It was first reported for Missouri by Ladd (1994).

7. Lathyrus venosus Muhl. ex Willd. (bushy vetchling)

L. venosus var. *intonsus* Butters & H. St. John

Pl. 398 l, m; Map 1766

Plants perennial, with rhizomes, the roots not producing tubers. Stems 40–120 cm long, trailing or climbing, usually branched, sparsely to densely and finely hairy, angled but not winged. Leaves with (6)8–14 leaflets, the petiole 0.5–1.0 cm long, unwinged, the tendrils branched or unbranched. Stipules 4–10(–20) mm long, 1–4 mm wide, narrowly lanceolate to lanceolate, the basal lobe 1–6(–14) mm long, narrowly lanceolate to oblong-lanceolate. Leaflets 3–6 cm long, 10–30 mm wide, narrowly to broadly elliptic or oblong-lanceolate, rounded to angled or broadly angled at the base, angled to a usually bluntly pointed tip, the midvein sometimes extended into a minute sharp point at the very tip, finely hairy, at least on the pale undersurface. Inflorescences dense racemes of 8–20 flowers, the stalk 6–12 cm long, the flower stalks 3–5 mm long. Calyces with the tube 3.5–5.0 mm long, finely hairy, the lobes 1.0–4.5 mm long, the upper 2 short and triangular to broadly triangular, the lowermost lobe 2–3 times as long as the upper 2, narrowly triangular, the lateral lobes of the lower lip intermediate in size and shape. Corollas 12–20 mm long, pinkish purple, to lavender, occasionally somewhat bicolorous (the wings and keel lighter than the banner). Filaments with the fused portion 9–10 mm long, the free portion 4–5 mm long. Ovary glabrous. Fruits 4–6 cm long, 5–8 mm wide, linear to narrowly oblong, flattened, glabrous, 5–10-seeded. Seeds 4–5 mm long, more or less oblong in outline, slightly flattened, the surface smooth, reddish brown to dark brown. $2n$=28. May–June.

Uncommon in the Ozark Division; also known from a single historical collection from the city of St. Louis (eastern U.S. west to North Dakota and New Mexico, Alaska; Canada). Mesic to dry upland forests, margins of lakes, and ledges and tops of bluffs; also roadsides.

Lathyrus venosus is recognized by its bushy habit, leaves with 6–10 relatively large and usually finely hairy leaflets, relatively weakly developed tendrils, and hairy calyces. It tends to grow in drier habitats than does the morphologically similar *L. palustris*. Several infraspecific taxa have been recognized based on differences in pubescence (Hitchcock, 1952; Steyermark, 1963). The var. *venosus*, which is glabrous or nearly so, occurs to the east of Missouri. Missouri collections are at least sparsely hairy on the stems and foliage and correspond to var. *intonsus*. However, Isely (1998) concluded that the varieties are difficult to separate in many areas and thus abandoned their recognition.

Steyermark (1963) noted that this species sometimes occurs as relatively extensive colonies with few or no flowering stems. Thus, it is possible that the taxon has been undercollected in the state.

34. Lespedeza Michx.

(Clewell, 1966b)

Contributed by Jay A. Raveill

Plants herbaceous perennials or shrubs, often with multiple stems from a woody caudex. Stems erect to spreading, unbranched or branched, unarmed, sparsely to densely pubescent with appressed to spreading, unbranched hairs. Leaves pinnately trifoliate, short- to long-petiolate. Stipules hairlike to linear-triangular or less commonly lanceolate, herbaceous or more commonly papery, attached at the base, the margins entire, pubescent on the outer face, with 1 or rarely few prominent, unbranched vein(s), persistent; stipels lacking. Leaflets narrowly oblong to broadly ovate, the lateral leaflets often slightly shorter than the terminal one, rounded or angled to a stalked base, mostly rounded at the tip, sometimes shallowly notched or with an abrupt, minute, sharp point at the very tip, the margins entire, the surfaces glabrous or more commonly hairy, with a prominent midvein and conspicuous pinnate secondary veins. Inflorescences ascending to spreading, racemes or spikes (sometimes appearing grouped into panicles), or reduced to few-flowered clusters, well separated to tightly clustered and obscuring the inflorescence axis, the 3 bracts subtending each pair of flowers and 2 bractlets subtending each flower all minute, linear, inconspicuous. Cleistogamous flowers often present, usually mixed with open flowers in the same inflorescence. Calyces 5-lobed, the lobes as long as or longer than the tube, nearly equal in length, but the upper 2 lobes fused nearly to the tip, minutely hairy, persistent but not enlarging at fruiting. Corollas papilionaceous (highly reduced in cleistogamous flowers), glabrous, pinkish purple, yellowish, or cream-colored, often with purple near the base of the banner petal, the petals tapered to a short, stalklike base, the banner broadly obovate to oblong-obovate, abruptly curved-ascending above the midpoint, the wings oblong, about as long as and sometimes slightly fused to the keel toward the tip, the keel oblanceolate and curved in outline, boat-shaped. Stamens 10, 9 of the filaments fused and 1 free nearly to the base, the anthers small, attached near the midpoint, all similar in size. Ovary ellipsoid to ovoid, sessile or short-stalked, the style slender, usually glabrous, straight in chasmogamous flowers and recurved in cleistogamous flowers, persistent at fruiting, the stigma small and terminal. Fruits modified legumes, flattened, sessile or very short-stalked, those ripening from open flowers mostly elliptic and slightly longer, those from cleistogamous flowers broadly obovate to nearly circular and slightly shorter, often with a raised network of nerves, indehiscent, 1-seeded. Seeds slightly kidney-shaped to nearly circular in outline, sometimes with a shallow notch at the attachment point, somewhat flattened, the surface smooth, yellow or tan to nearly black, sometimes mottled. About 40 species, North America, Asia.

Lespedeza presents considerable challenges because the species display significant morphological variation. A number of additional species and infraspecific taxa have been proposed to accommodate this variation, but in most cases these do not seem discrete enough morphologically to warrant formal recognition. Much of the complexity apparently arises due to hybridization (Clewell, 1964, 1966a, c). The eight native species in Missouri could produce 27 possible biparental hybrids, and Clewell (1966b) has hypothesized each of these crosses based on progeny arrays reared in a garden from putative hybrid plants, field observations of co-occurring species, and examination of herbarium specimens. Offspring from crosses involving morphologically disparate species, for example from yellow-flowered taxa crossed with purple-flowered ones, can be conspicuous. However, hybridization between similar species may go undetected and could account for the seeming morphological continuum among similar species, for example the problems in distinguishing *L. virginica*, *L. stuevei*, and *L. violacea*.

Species frequently occur in mixed populations, but the number of individuals showing hybrid morphology is generally extremely low (less than 1%). Whenever aberrant plants are encountered, collectors are encouraged to record associated *Lespedeza* species, as field observations are among the greatest aids in determining the possible parental species of hybrids. The use of hypothesized hybrid parental formulas is encouraged in place of the many binomials that predate the appreciation of hybridization as a causal agent of morphological diversity (they originally were named as species rather than hybrids). The hybrid binomials, many of which were based on Missouri specimens (Mackenzie and Bush, 1902) may still be used, but because different crosses can result in similar hybrid morphology, some hybrid names may include multiple parental combinations.

Because of the morphological complexity of this genus, two keys are provided. One uses primarily vegetative features and the other uses largely reproductive traits. The dual keys provide different groupings, so important distinguishing morphology can be emphasized. Character states and measurements represent the morphological variation encountered within Missouri, and the following keys do not account for all of the variation found elsewhere in the ranges of the species.

Key based primarily on vegetative characters

1. Plants shrubs, 1–3 m tall
    2. Leaves oval or broadly elliptic, 1.5–2.0 times longer than broad . . . 1. L. BICOLOR
    2. Leaves elliptic to broadly oblong or rarely obovate, 2–3 times longer than broad . . . 9. L. THUNBERGII
1. Plants perennial herbs, generally up to 1.5 m tall
    3. Petioles 1–6 mm long, stout, 0.7–1.0 mm wide, densely pubescent. . . . 2. L. CAPITATA
    3. Petioles of at least the lower leaves 5–20 mm long (upper leaves may be nearly sessile), slender to stout, 0.2–0.7 mm wide, nearly glabrous to densely pubescent
        4. Stem pubescence conspicuous and mainly spreading
            5. Stems prostrate (usually with ascending flowering branches) . . . 6. L. PROCUMBENS
            5. Stems erect or ascending
                6. Leaflets broad in outline, 1.3–1.8 times longer than broad, mostly elliptic to broadly oblong or oval, rarely broadly ovate or broadly obovate, corollas cream-colored to pale yellow with purple markings on the banner petal . . . 5. L. HIRTA
                6. Leaflets narrow in outline, 1.5–3.5 times longer than broad, mostly oblong or elliptic, corollas pinkish purple . . . 8. L. STUEVEI
        4. Stem pubescence inconspicuous, mainly appressed, sometimes also with spreading hairs, or the stems nearly glabrous
            7. Stems mostly loosely ascending or spreading with ascending tips
                8. Stems mostly loosely ascending; small axillary leaves generally present; stipules 3–6 mm long . . . 4. L. FRUTESCENS
                8. Stems spreading with ascending tips, but young stems initially may be erect; small axillary leaves generally absent; stipules 1.5–3.0 mm long . . . 7. L. REPENS
            7. Stems erect or strongly ascending
                9. Leaflets mostly elliptic, ovate, obovate, or broadly oblong, less than 3 times longer than broad
                    10. Small axillary leaves numerous at most nodes; stems often branching below the midpoint; persistent calyx of cleistogamous flowers about 1/5 as long the fruit; some flowering branches much longer than the associated leaves . . . 4. L. FRUTESCENS
                    10. Small axillary leaves absent or few in number; stems branching only above the midpoint, persistent calyx of cleistogamous flowers 1/4–1/3 as long as the fruit; flowering branches shorter than to slightly longer than the associated leaves . . . 10. L. VIOLACEA
                9. Leaflets mostly narrowly oblong, narrowly oblanceolate, or linear, mostly more than 2.5 times as long as wide
                    11. Leaflets narrowly angled or tapered at the base, truncate or notched at the tip; stems thick, 2–5 mm wide near the base, green with conspicuous, vertical, white, longitudinal ridges, the pubescence mostly confined to the ridges; corollas creamy white with a purple spot on the banner petal; flowers in clusters of 1–4 from most of the median and upper leaf axils . . . 3. L. CUNEATA

11. Leaflets rounded or angled, but not tapered at the base, rounded apically or with an abrupt minute, sharp point at the very tip; stems slender, 1.5–3.0(–4.0) mm wide near the base, uniformly reddish brown below and green above or with inconspicuous, slightly lighter, longitudinal ridges, the pubescence evenly distributed or only slightly denser on the ridges; corollas pinkish purplish; flowers in clusters or short, dense racemes of 4–10 from most of the median and upper leaf axils . . . . 11. L. VIRGINICA

Key based mostly on reproductive characters (flower and fruit characters refer to open-flowered examples unless noted otherwise)

1. Corollas 8–15 mm long
    2. Corollas 8–12 mm long; calyx lobes slightly longer than or nearly equal to the tube in length . . . . 1. L. BICOLOR
    2. Corollas 12–15 mm long; calyx lobes conspicuously longer than the tube . . . . 9. L. THUNBERGII
1. Corollas 4–10 mm long
    3. Corollas pale yellow or cream-colored, but purplish-tinged or with purple markings on the banner petal
        4. Inflorescences axillary clusters of 1–4; calyces 3.5–5.0 mm long . . . . 3. L. CUNEATA
        4. Inflorescences loose to dense clusters or racemes of 10–40; calyces 5–14 mm long
            5. Flowers 16–40, in dense clusters, the inflorescence axis generally hidden by the flowers . . . . 2. L. CAPITATA
            5. Flowers 10–20, in loose clusters or racemes, the inflorescence axis generally visible between the flowers . . . . 5. L. HIRTA
    3. Corollas pinkish purple, sometimes with white streaks or markings
        6. Flowers in axillary and apparently terminal inflorescences that extend far beyond the associated leaves; stems prostrate to loosely ascending, often with ascending branches, the main stem sometimes more erect when young, often branching below the midpoint
            7. Flowers 6–10 per inflorescence; stems densely spreading pubescent . . . . 6. L. PROCUMBENS
            7. Flowers 2–8 per inflorescence; stems inconspicuously appressed-pubescent or nearly glabrous
                8. Keel noticeably longer than the wings; axillary clusters of leaves generally well-developed; stipules 3–6 mm long; persistent calyx covering lower 1/5 of fruits in cleistogamous flowers . . . . 4. L. FRUTESCENS
                8. Keel and wings of nearly equal length or the keel slightly longer; axillary clusters of leaves generally absent; stipules 1.5–3 mm long; persistent calyx covering at least 1/4 of the fruits in cleistogamous flowers . . . . 7. L. REPENS
        6. Flowers clustered in the axils of the upper leaves, these shorter than to slightly longer than the associated leaves; stems erect or ascending, often unbranched or branched only above the midpoint
            9. Leaflets linear to very narrowly oblong, 4–7 times longer than broad, except occasionally the lowermost leaves somewhat broader . . . . 11. L. VIRGINICA

9. Leaflets oblong, broadly elliptic, or obovate, 1.3–3.5 times longer than broad
   10. Stems and leaves conspicuously spreading-hairy .... 8. L. STUEVEI
   10. Stems and leaves inconspicuously appressed- to loosely appressed-hairy or nearly glabrous .... 10. L. VIOLACEA

1. Lespedeza bicolor Turcz. (shrubby lespedeza)

Pl. 399 f–h; Map 1767

Plants shrubs. Stems 100–300 cm long, 5–20 mm in diameter near the base, erect or arching, extensively branched, densely pubescent with appressed or somewhat spreading hairs, the hairs not confined to longitudinal ridges, nearly glabrous toward the base. Primary leaves passing gradually into bracteal leaves, the petiole (2–)10–90 mm long, becoming reduced above (bracteal leaves nearly sessile), 0.3–1.0 mm wide, glabrous or sparsely hairy. Stipules 2–7(–12) mm long, linear-triangular to nearly hairlike. Leaflets 2–4 cm long, 0.8–2.0 cm wide (1.5–2.0 times as long as wide), broadly elliptic or nearly oval, rounded at the base, rounded at the tip, occasionally with an abrupt, minute sharp point and/or a shallow notch at the very tip, the surfaces sparsely to occasionally densely pubescent with appressed hairs, the upper surface sometimes nearly glabrous. Axillary clusters of leaves absent, but side-branching extensive. Inflorescences axillary from the upper leaves, much longer than the subtending leaves, often leafy and branched (forming panicles of racemes). Flowers 12–28 per raceme or more in branched inflorescences, the axis visible between flowers. Calyces with the tube 1.5–2.0 mm long, the lobes 2–3 mm long. Corollas 8–12 mm long, dark rose-colored with nearly black areas near the base of the banner, rarely white, the keel much longer than the wings. Fruits from open flowers 6–10 mm long, the calyx covering the lower 1/4–1/2, fruits from cleistogamous flowers 6–7 mm long, the calyx covering about 1/4. Seeds 3–4 mm long, dark brown to purplish black. $2n$=22. July–October.

Introduced, scattered widely, mostly in the southeastern quarter of the state (native of Asia; introduced in the eastern U.S. west to Iowa and Texas, Canada). Banks of streams and rivers, sand prairies, sand savannas, and openings of mesic upland forests; also old fields, roadsides, and open disturbed areas.

The maintenance of *L. bicolor* as distinct from *L. thunbergii* in Missouri is controversial. Most Missouri specimens have morphological features that fall very close to the presumed boundary between these species and perhaps most plantings represent cultivated strains of hybrid origins (Clewell, 1966c; Isely, 1998). Both species are used in erosion control and increasingly as ornamentals (Yinger, 1992; T. Schwartz, 1995). Cold winters can kill the upper parts of plants, sometimes to ground level, but the roots survive and regrow.

2. Lespedeza capitata Michx. (round-headed bush clover, roundhead lespedeza)

L. capitata f. *argentea* Fernald
L. capitata var. *stenophylla* Bissell & Fernald
L. capitata var. *vulgaris* Torr. & A. Gray

Pl. 400 i–k; Map 1768

Plants perennial herbs. Stems 50–200 cm long, 3–10 mm in diameter near the base, erect or strongly ascending, generally unbranched below the midpoint, densely pubescent with spreading or appressed hairs, often with a silvery or tawny appearance, the hairs not confined to longitudinal ridges, becoming sparsely pubescent or even glabrous toward the base. Primary median leaves with the petiole short and stout, 1–6 mm long, 0.7–1.0 mm wide, densely spreading- or appressed-hairy. Stipules 2–6 mm long, linear-triangular to nearly hairlike. Leaflets 20–45 mm long, 6–10 mm wide, the shape more variable than in other species, elliptic to oblong, rounded or rarely slightly angled at the base, rounded or rarely somewhat pointed at the tip, often with a silvery or tawny sheen, the upper surface silvery-hairy to nearly glabrous, the undersurface moderately to densely appressed-hairy. Axillary leaves not well-developed. Inflorescences primarily from the axils of upper leaves, unbranched clusters or short, dense racemes, generally equaling or shorter than the subtending leaves but sometimes exceeding them. Flowers 16–40 per raceme, the axis usually hidden by the flowers. Calyces with the tube 0.7–1.0 mm long, the lobes 6–13 mm long. Corollas 8–10 mm long, cream-colored with purple markings on the banner, the keel nearly equal to or longer than the wings. Fruits from open and cleistogamous flowers 4–5 mm long, the calyx longer than the fruit. Seeds 2.5–3.0 mm long, olive green to brown or nearly black. $2n$=20. July–October.

Scattered nearly throughout the state (eastern U.S. west to South Dakota and Texas; Canada). Upland prairies, loess hill prairies, glades, savannas, sand savannas, tops of bluffs, openings of mesic to dry upland forests, banks of streams; also old fields, railroads, and roadsides.

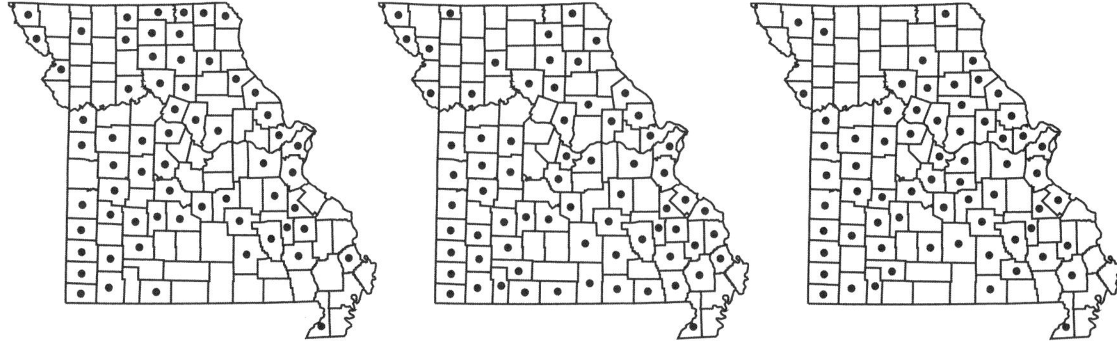

1768. Lespedeza capitata 1769. Lespedeza cuneata 1770. Lespedeza frutescens

The presumed hybrid between this species and *L. hirta* (*L.* ×*longifolia* DC.) has the congested inflorescences of *L. capitata*, but with the longer petioles and inflorescence stalks of *L. hirta* and leaves of intermediate shape. This hybrid was noted by Clewell (1966b) as "abundant." The presumed hybrid between *L. capitata* and *L. frutescens* (*L.* ×*manniana* Mack. & Bush) often has the calyx and corolla lengths similar to *L. capitata*, but with purple corollas and inflorescences that are slightly longer than the leaves. The putative cross between *L. capitata* and *L. virginica* (*L.* ×*simulata* Mack. & Bush) often resembles an extremely narrow-leafleted form of *L. capitata* vegetatively except for longer and more slender petioles. Another form more closely resembles *L. virginica* vegetatively but has calyx lobes more similar to those of *L. capitata*. The corollas often are purple, but may be yellow, and often are borne in relatively diffuse racemes at the tips of axillary branches that far exceed the subtending leaves, unlike either of the presumed parents.

3. Lespedeza cuneata (Dum. Cours.) G. Don (sericea lespedeza, Chinese lespedeza)

Pl. 401 c, d; Map 1769

Plants perennial herbs. Stem 50–150 cm long, 2–5 mm in diameter near the base, unbranched or branched above the midpoint, densely pubescent with appressed to somewhat spreading hairs mostly confined to conspicuous, white, longitudinal ridges running the length of the stem. Primary leaves with the petiole 2–12 mm long (shorter toward the stem tip), 0.3–0.5 mm wide, densely pubescent with appressed or slightly spreading hairs. Stipules 2–8 mm long, hairlike. Leaflets 1–3 cm long, 2–6 mm wide (mostly more than 3 times as long as wide), those of the uppermost leaves usually smaller, narrowly oblanceolate to narrowly oblong, narrowly angled or tapered at the base, truncate at the tip, but sometimes shallowly notched and usually with an abrupt, minute, sharp point at the very tip, often with a grayish appearance, both surfaces densely appressed-hairy or the upper surface sometimes becoming glabrous. Axillary clusters of leaves sometimes developed. Inflorescences axillary clusters of 1–4 flowers from the median and upper leaves, shorter than or nearly equal to the associated leaves. Calyces with the tube 1.0–1.5 mm long, the lobes 2.5–4.0 mm long. Corollas 5–9 mm long, creamy white with purple markings or markings on the banner, the keel about as long as the wings. Fruits from open flowers 3.0–4.5 mm long, the calyx nearly equal in length; fruits from cleistogamous flowers 2.0–3.5 mm long, the calyx slightly shorter than to somewhat longer than the fruit. Seeds 1.5–2.0 mm long, olive green to brown. $2n$=18, 20, 22. August–October.

Introduced, scattered to common nearly throughout the state (native of Asia; introduced in the eastern U.S. west to Nebraska and Texas, Canada). Upland prairies, loess hill prairies, glades, banks of streams and rivers, and margins of ponds and lakes; also old fields, mine spoils, quarries, railroads, roadsides, and open disturbed areas.

Although originally planted for forage, erosion control, soil enrichment, and wildlife food, *L. cuneata* escapes and readily invades native plant communities. The states of Kansas and Colorado have designated this species a noxious weed and it is considered an invasive exotic nearly everywhere that it grows in North America. The species is difficult to control, especially in grassland communities, as it responds well to the same environmental cues to which warm-season grasses and summer-flowering forbs respond. It spreads easily along roadsides, and also as a contaminant on road-grading equipment and farm combines. Until the recent past, sericea lespedeza was still planted widely following highway improvement

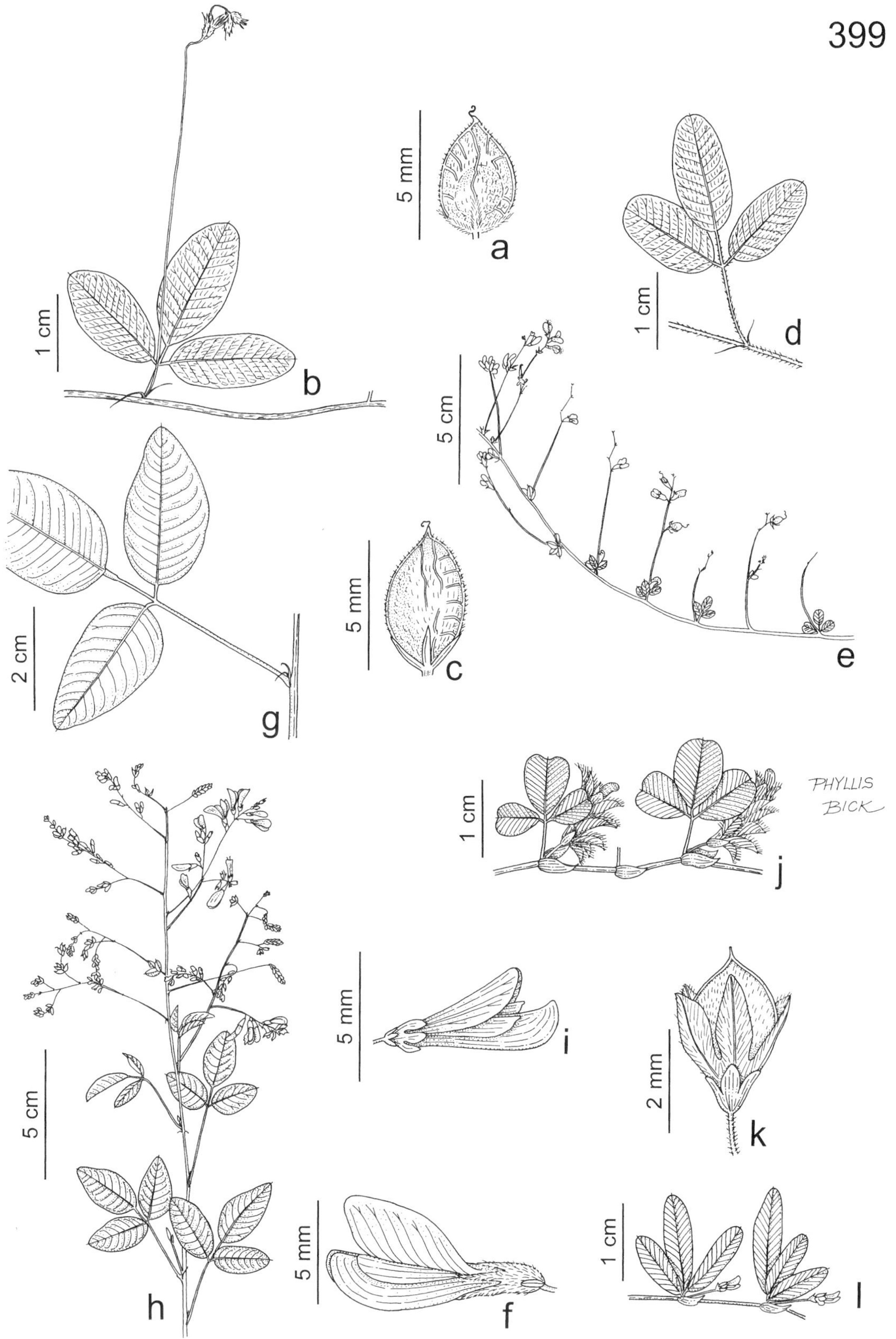

Plate 399. Fabaceae. *Lespedeza repens*, **a)** fruit, **b)** fertile stem. *Lespedeza procumbens*, **c)** fruit, **d)** leaf attached to stem, **e)** habit. *Lespedeza bicolor*, **f)** flower, **g)** leaf attached to stem, **h)** fertile stem. *Kummerowia stipulacea*, **i)** flower, **j)** fertile stem. *Kummerowia striata*, **k)** fruit, **l)** fertile stem.

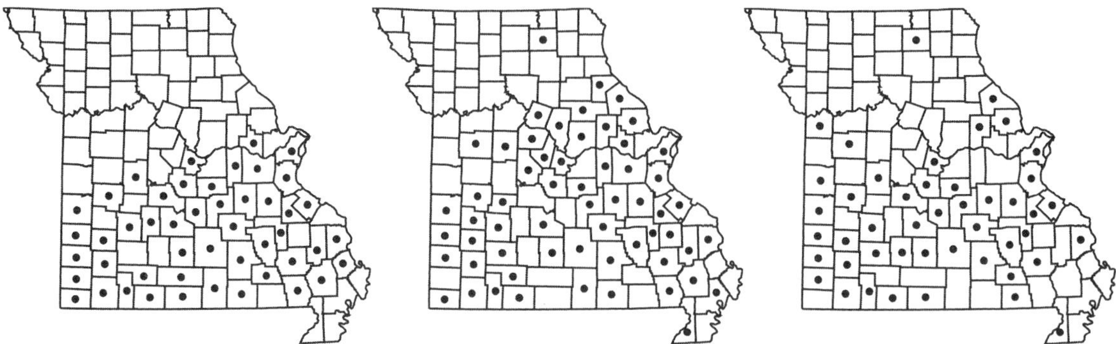

1771. Lespedeza hirta 1772. Lespedeza procumbens 1773. Lespedeza repens

projects, often in large monocultures. *Lespedeza cuneata* is not known to hybridize with any of the native North American species.

4. Lespedeza frutescens (L.) Hornem. (prairie lespedeza, prairie bush clover)

L. violacea (L.) Pers., misapplied

Pl. 400 a–c; Map 1770

Plants perennial herbs. Stems 20–60 cm long, 0.8–2.2 mm in diameter near the base, mostly loosely ascending, sometimes from a spreading base, often branched below the midpoint, sparsely pubescent with appressed or somewhat spreading hairs, often nearly glabrous at maturity. Primary median leaves with the petiole relatively long and slender, 8–22 mm long, 0.4–0.6 mm wide, sparsely appressed- to somewhat spreading-hairy. Stipules 3–6 mm long, linear-triangular to hairlike. Leaflets 20–50 mm long, 8–30 mm wide (mostly less than 3 times as long as wide), elliptic to less commonly oblong-obovate or oblong-ovate, rounded at the base, broadly rounded or minutely notched at the tip, the midvein usually extended as a minute, sharp point at the very tip, the upper surface with minute, appressed hairs along the midvein, otherwise usually glabrous, the undersurface sparsely to moderately appressed- to slightly spreading-hairy. Axillary leaves well-developed at most nodes. Inflorescences from the axils of the upper leaves and appearing terminal, often branched (grouped into panicles) and appearing leafy, at least some of them much-exceeding the leaves. Flowers mostly 4–8 per raceme or branch, the axis not hidden by the flowers. Calyces with the tube 1.5–2.0 mm long, the lobes 2–3 mm long. Corollas 6–9 mm long, pinkish purple to bluish purple, usually with a contrasting lighter splotch on the banner, the petals sometimes lighter colored toward the margins, the keel longer than the wings. Fruits from open flowers 6–9 mm long (including the stalk), the calyx covering about the lower 1/3; those from cleistogamous flowers 4–6(–7) mm long, the calyx covering the lower (1/6–)1/5(–1/4). Seeds 2.5–3.0 mm long, olive green to purplish brown. $2n=20$. July–October.

Scattered nearly throughout the state, but uncommon in the Mississippi Lowlands Division (eastern U.S. west to Nebraska and Texas; Canada). Bottomland forests, mesic to dry upland forests, edges of upland prairies, glades, savannas, tops of bluffs, and banks of streams; also old fields and roadsides.

Reexamination of type specimens by Reveal and Barrie (1992) showed that the name *L. frutescens* must be adopted for this taxon, even though it had long been known as *L. violacea*. The type of the name *L. violacea* was shown to represent a specimen of the taxon traditionally known as *L. intermedia*, and because the publication in which *L. violacea* was first described predates the publication of the name *L. intermedia*, it replaces that name.

Hybrids involving this species and *L. virginica* and *L. capitata* are reported as common, but all other hybrid combinations are considered rare (Clewell, 1966b). The very similar morphology makes putative hybrids with *L. repens* and *L. violacea* difficult to detect.

5. Lespedeza hirta (L.) Hornem. **ssp. hirta** (hairy lespedeza, hairy bush clover)

Pl. 400 f–h; Map 1771

Plants perennial herbs. Stems 30–150 cm long, 1.5–5.0 mm in diameter near the base, erect or strongly ascending, generally unbranched below the midpoint, densely pubescent with mostly spreading hairs, often with a silvery or tawny appearance, the hairs not confined to longitudinal ridges, sometimes becoming sparsely pubescent toward the base. Primary median leaves with the petiole mostly relatively long and slender, 5–20 mm long, 0.2–0.5 mm wide, densely spreading-

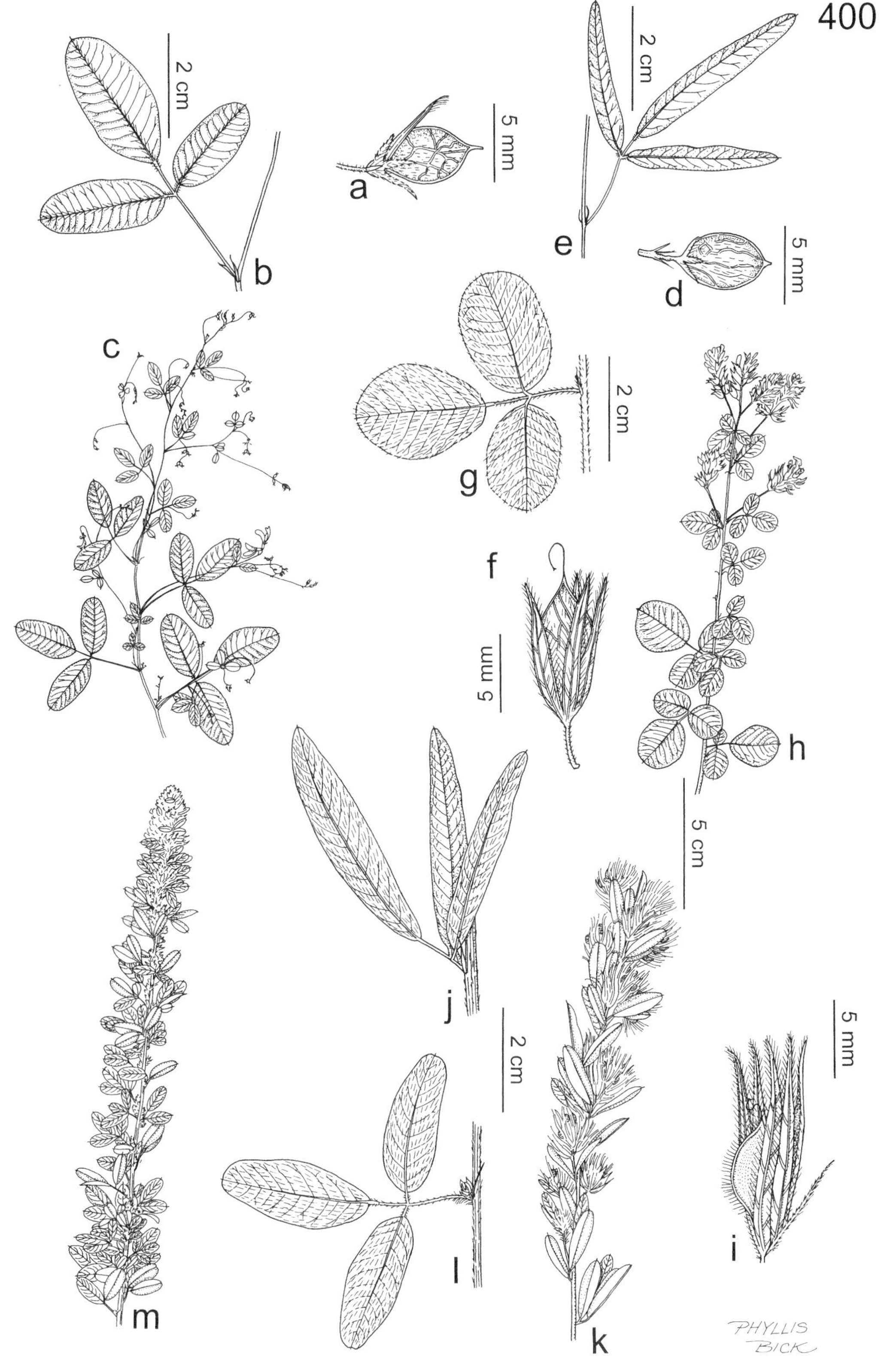

Plate 400. Fabaceae. *Lespedeza frutescens*, **a)** fruit, **b)** leaf attached to stem, **c)** fertile stem. *Lespedeza virginica*, **d)** fruit, **e)** leaf attached to stem. *Lespedeza hirta*, **f)** fruit, **g)** leaf attached to stem, **h)** fertile stem. *Lespedeza capitata*, **i)** fruit, **j)** leaf attached to stem, **k)** fertile stem. *Lespedeza violacea*, **l)** leaf attached to stem, **m)** fertile stem.

hairy. Stipules 3–8 mm long, narrowly triangular toward the stem base, grading to hairlike toward the tip. Leaflets (8–)15–40 mm long, (5–)10–30 mm wide (1.3–1.8 times as long as wide), elliptic to broadly oblong or oval, rarely broadly ovate or broadly obovate, those of the uppermost leaves sometimes nearly circular, rounded at the base, rounded or sometimes minutely notched at the tip, sometimes appearing grayish-tinged, but lacking a silvery or tawny sheen, the upper surface sparsely to moderately pubescent with mostly appressed hairs, the undersurface moderately to densely pubescent with appressed and usually also more or less spreading hairs. Axillary leaves usually absent. Inflorescences axillary or appearing terminal, unbranched clusters or racemes, relatively dense and ascending, longer than the subtending leaves. Flowers mostly 10–20 per cluster or raceme, the axis usually visible for most of the length, often hidden by the flowers near the tip. Calyces with the tube 1.0–1.5 mm long, the lobes 4–9 mm long. Corollas 6–8 mm long, cream-colored or pale yellow with purple markings on the banner, the keel shorter than the wings. Fruits from open flowers 5–8 mm long, the calyx about as long as or slightly longer than the fruit; fruits from cleistogamous flowers 4–7 mm long, the calyx about as long as or slightly longer than the fruit. Seeds 2.5–3.0 mm long, brown or purplish black. $2n=20$. July–October.

Scattered to common in the Ozark, Ozark Border, and Unglaciated Plains Divisions, but absent from most of the Glaciated Plains and uncommon in the Mississippi Lowlands (eastern U.S. west to Kansas and Texas; Canada). Upland prairies, sand prairies, glades, openings of mesic to dry upland forests, savannas, sand savannas, and tops of bluffs; also old fields and roadsides.

The ssp. *curtissii* Clewell (var. *curtissii* (Clewell) Isely) occurs mostly in sandy soils on the Atlantic and Gulf Coastal Plains. It differs from ssp. *hirta* in its leaflets with dense, silky, appressed hairs and its often appressed-hairy stems.

Lespedeza ×*nuttallii* Darl. is thought to represent crosses between *L. hirta* and *L. violacea*, as determined by examination of progeny arrays (Clewell, 1964). The potential exists that hybrids involving other purple-flowered species may also be confounded under this binomial (Clewell, 1966b; Isely, 1998). The hybrid can closely resemble *L. hirta* except for purple flowers and short calyx teeth. The inflorescences can be branched and either longer or shorter than the subtending leaves. Another form passing as this hybrid species can more closely resemble *L. stuevei* except for the inflorescences that well exceed the subtending leaves. For a discussion of putative hybridization with *L. capitata*, see the treatment of that species.

6. Lespedeza procumbens Michx. (trailing lespedeza, trailing bush clover)

Pl. 399 c–e; Map 1772

Plants perennial herbs. Stems 30–150 cm long, 0.8–1.5 mm in diameter near the base, prostrate with loosely ascending tips and branches, usually mat-forming, sometimes branched below the midpoint (more often short-branched toward the tip), moderately to densely pubescent with short, spreading hairs. Primary median leaves with the petiole mostly relatively short and slender, 4–10 (–22) mm long, 0.2–0.7 mm wide, sparsely to densely spreading-hairy. Stipules 2–4 mm long, linear to hairlike. Leaflets 10–20(–30) mm long, 5–15(–20) mm wide, elliptic to broadly oblong, rounded to broadly rounded at the base, broadly rounded or minutely notched at the tip, the midvein usually extended as a minute, sharp point at the very tip, the upper surface sparsely to moderately or rarely densely appressed-hairy, the undersurface moderately to densely pubescent with curved to loosely appressed hairs. Axillary leaves absent or poorly developed. Inflorescences from the axils of the upper and often also median leaves, also appearing terminal, unbranched or branched (grouped into panicles), sometimes appearing leafy, much-exceeding the leaves. Flowers mostly (2–)6–10(–14) per raceme or branch, the axis not hidden by the flowers. Calyces with the tube 1.0–1.5 mm long, the lobes 1–2 mm long. Corollas 5.5–7.0 mm long, pinkish purple, the banner darker purple toward the base, the wings and keel usually lighter colored below the tips, the keel about as long as or longer than the wings. Fruits from open flowers 5–6 mm long (including the stalk), the calyx covering about the lower 1/2; those from cleistogamous flowers 3.5–5.5 mm long, the calyx covering about the lower 1/3. Seeds 2–3 mm long, olive green to light brown. $2n=20$. July–October.

Scattered to locally common in the southern half of the state north locally to Adair County (eastern U.S. west to Kansas and Texas; Canada). Bottomland forests, openings of mesic upland forests, glades, upland prairies, and banks of streams and rivers; also old fields, ditches, railroads, and roadsides.

The putative hybrid between this species and *L. virginica* is reported to be common and has been named *L.* ×*brittonii* E.P. Bicknell (Clewell, 1966b). Such plants have few-flowered inflorescences extending beyond the leaves and are variable in habit and pubescence. Clewell (1966b) considered other

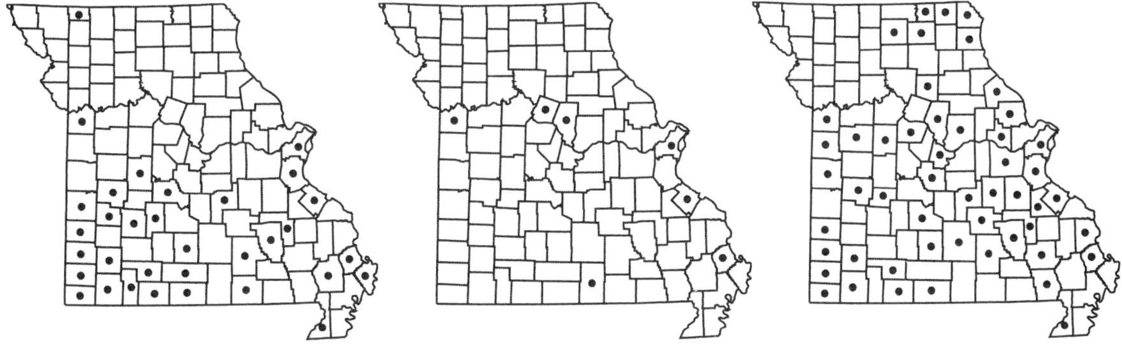
1774. Lespedeza stuevei
1775. Lespedeza thunbergii
1776. Lespedeza violacea

potential hybrid combinations to be "rare" with the exception of putative crosses with *L. capitata*, which he considered "abundant," and with *L. hirta*, which he stated to be "common." As pointed out by Isely (1998), the trailing habit of these proposed hybrids could have been contributed just as easily by *L. repens*.

7. Lespedeza repens (L.) W.P.C. Barton (creeping lespedeza, creeping bush clover)
Pl. 399 a, b; Map 1773

Plants perennial herbs. Stems 10–100 cm long, 0.7–2.0 mm in diameter near the base, prostrate or trailing with ascending tips and branches (sometimes more strongly ascending when young), usually mat-forming, branched above and below the midpoint, sparsely to moderately pubescent with appressed-ascending hairs. Primary median leaves with the petiole mostly relatively short and slender, 3–8(–10) mm long, 0.2–0.5 mm wide, sparsely appressed-hairy. Stipules 1.5–3.0(–4.0) mm long, linear to hairlike. Leaflets 8–18(–24) mm long, 5–12(–18) mm wide, obovate, oblong, or elliptic, rounded or occasionally broadly angled at the base, rounded or rarely slightly pointed at the tip, the midvein usually extended as a minute, sharp point at the very tip, the upper surface glabrous or sparsely appressed-hairy, the undersurface sparsely to moderately appressed-hairy. Axillary leaves absent or rarely present but then relatively poorly developed. Inflorescences from the axils of the upper and often also median leaves, unbranched or branched (grouped into panicles), appearing leafy, much-exceeding the leaves. Flowers mostly 2–8 per raceme or branch, the axis not hidden by the flowers. Calyces with the tube 1.5–2.0 mm long, the lobes 2–3 mm long. Corollas 5–7 mm long, pink to pinkish purple or rarely purple, the banner usually darker purple toward the base, the wings and keel usually darker colored at the tips, the keel about as long as or slightly longer than the wings. Fruits from open flowers 5–6(–7) mm long (including the stalk), the calyx covering about the lower 1/3–1/2; those from cleistogamous flowers (2–)3–4(–6) mm long, the calyx covering the lower 1/4–1/2. Seeds 2–3 mm long, olive green to light brown. $2n=20$. July–October.

Scattered in the southern half of the state north locally to Adair and Pike Counties (eastern U.S. west to Iowa, Kansas, and Texas; Canada). Openings of mesic to dry upland forests, glades, bottomland prairies, upland prairies, banks of streams and rivers, and margins of lakes; also old fields, railroads, and roadsides.

Lespedeza repens and *L. procumbens* are the only Missouri members of the genus to be consistently prostrate and mat-forming, although the stems can be upright when they first develop. At times *L. violacea* can be sprawling but the stems are generally at least partially ascending.

Aberrant plants with nearly glabrous stems that are trailing may well involve this species in combination with one of the purple-flowered species (excluding *L. stuevei* and *L. procumbens*, which are more strongly pubescent). Clewell (1966b) listed the hybrid between this species and *L. hirta* as "common."

8. Lespedeza stuevei Nutt. (tall lespedeza, tall bush clover)
Pl. 401 a, b; Map 1774

Plants perennial herbs. Stems 30–150 cm long, 2–4 mm in diameter near the base, erect or strongly ascending, unbranched or relatively few-branched only above the midpoint, densely pubescent with spreading and sometimes also appressed hairs, the hairs not confined to longitudinal ridges, sometimes becoming sparsely pubescent or nearly glabrous toward the base. Primary median leaves with the petiole mostly relatively long and slender, 5–20 mm long, 0.5–0.8 mm wide, densely spreading-hairy. Stipules 3–5 mm long, linear to hairlike. Leaflets 8–20(–50) mm long, 4–15(–20) mm wide (1.3–3.5 times as long as wide), elliptic to oblong, oblong-obovate, or narrowly obovate, those of the uppermost leaves sometimes oblanceolate, rounded, angled, or tapered at the base, rounded to truncate at the tip, occasionally broadly pointed or minutely notched, sometimes appearing grayish-tinged (especially the undersurface), but lacking a silvery or tawny sheen, the upper surface sparsely to moderately pubescent with mostly appressed hairs, sometimes nearly glabrous, the undersurface sparsely to densely appressed-hairy. Axillary leaves usually well-developed. Inflorescences axillary from the upper leaves, sometimes appearing as a leafy terminal cluster, unbranched, relatively dense, shorter than to slightly longer than the subtending leaves. Flowers mostly 5–15 per raceme, relatively dense, but the axis usually visible between the flowers. Calyces with the tube 1–2 mm long, the lobes 2–4 mm long. Corollas 6–7 mm long, pinkish purple, the banner darker purple toward the base, the wings often pale with purple tips, the keel also darker purple at the tip, shorter than the wings. Fruits from open flowers 5–7(–10) mm long, the calyx covering the lower 1/3–1/2; fruits from cleistogamous flowers 4–6 mm long, the calyx covering the lower 1/4–1/3. Seeds 2.5–3.0 mm long, brown or purplish black. $2n=20$. July–September.

Scattered south of the Missouri River (eastern U.S. west to Kansas and Texas). Upland prairies, sand prairies, glades, dry upland forests, savannas, and banks of streams and rivers; also old fields and roadsides.

Lespedeza stuevei is superficially similar to *L. capitata* and *L. hirta*, but differs in its pinkish purple corollas. The most common hybrids involving this species include those with *L. virginica* (*L.×neglecta* (Britton) Mack. & Bush) and *L. hirta*. The latter hybrid has been confirmed by progeny arrays (Clewell, 1966b).

9. Lespedeza thunbergii (DC.) Nakai (Thunberg's lespedeza)

Pl. 401 j, k; Map 1775

Plants shrubs. Stems 100–300 cm long, 5–30 mm in diameter near the base, erect or strongly ascending to slightly arching, extensively branched, densely pubescent with appressed or somewhat spreading hairs, the hairs not confined to longitudinal ridges, nearly glabrous toward the base. Primary leaves passing gradually into bracteal leaves, the petiole (2–)10–80 mm long, becoming reduced above (bracteal leaves nearly sessile), 0.2–1.0 mm wide, sparsely to moderately appressed-hairy. Stipules 2–7(–12) mm long, linear-triangular to nearly hairlike. Leaflets 2–5(–10) cm long, 0.8–2.5(–5.5) cm wide (2–3 times as long as wide), elliptic to oblong-elliptic, rarely obovate or ovate, rounded to somewhat angled at the base, rounded to broadly pointed at the tip, occasionally with an abrupt, minute sharp point and/or a shallow notch at the very tip, the surfaces sparsely to moderately pubescent with appressed hairs, the upper surface sometimes nearly glabrous. Axillary clusters of leaves absent, but side-branching extensive. Inflorescences terminal and axillary from the upper leaves, much longer than the subtending leaves, often leafy and branched (forming panicles of racemes). Flowers 12–28 per raceme or more in branched inflorescences, the axis visible between flowers. Calyces with the tube 1.5–2.5 mm long, the lobes 3.0–4.5 mm long. Corollas 12–15 mm long, dark rose-colored with nearly black areas near the base of the banner, rarely white, the keel much longer than the wings. Fruits from open and cleistogamous flowers 7–9 mm long, the calyx covering the lower 1/3–1/2. Seeds 3–4 mm long, dark brown to purplish black. $2n=22$. August–October.

Introduced, uncommon and widely scattered in the eastern half of the state (native of Asia; introduced in the eastern U.S. west to Wisconsin and Oklahoma, Canada). Bottomland forests; also mine spoils, roadsides, and open disturbed area.

Lespedeza thunbergii has been known in cultivation in Japan since the 1600s, but its native range has been the subject of considerable controversy (Ohashi, 1981). Some authors consider it derived from historical hybridization between unknown parents (Akiyama, 1988). For additional discussion, see the treatment of *L. bicolor*.

10. Lespedeza violacea (L.) Pers. (violet lespedeza)

L. intermedia (S. Watson ex A. Gray) Britton

Pl. 400 l, m; Map 1776

Plants perennial herbs. Stems 40–100 cm long, 1–4 mm in diameter near the base, erect to strongly ascending, unbranched or branched only above the midpoint, sparsely to occasionally densely pubescent with relatively inconspicuous, appressed to loosely appressed hairs, rarely nearly glabrous at maturity. Primary median leaves with the petiole most commonly relatively long and slender, 4–25 (–30) mm long, 0.3–0.5 mm wide (1.5–3.0 times as long as wide), sparsely to densely appressed- to loosely appressed-hairy. Stipules 2–4 mm long, linear to hairlike. Leaflets 8–30(–40) mm long, 3–17 mm wide (1.5–3.0 times as long as wide), broadly oblong to elliptic or obovate, those of the

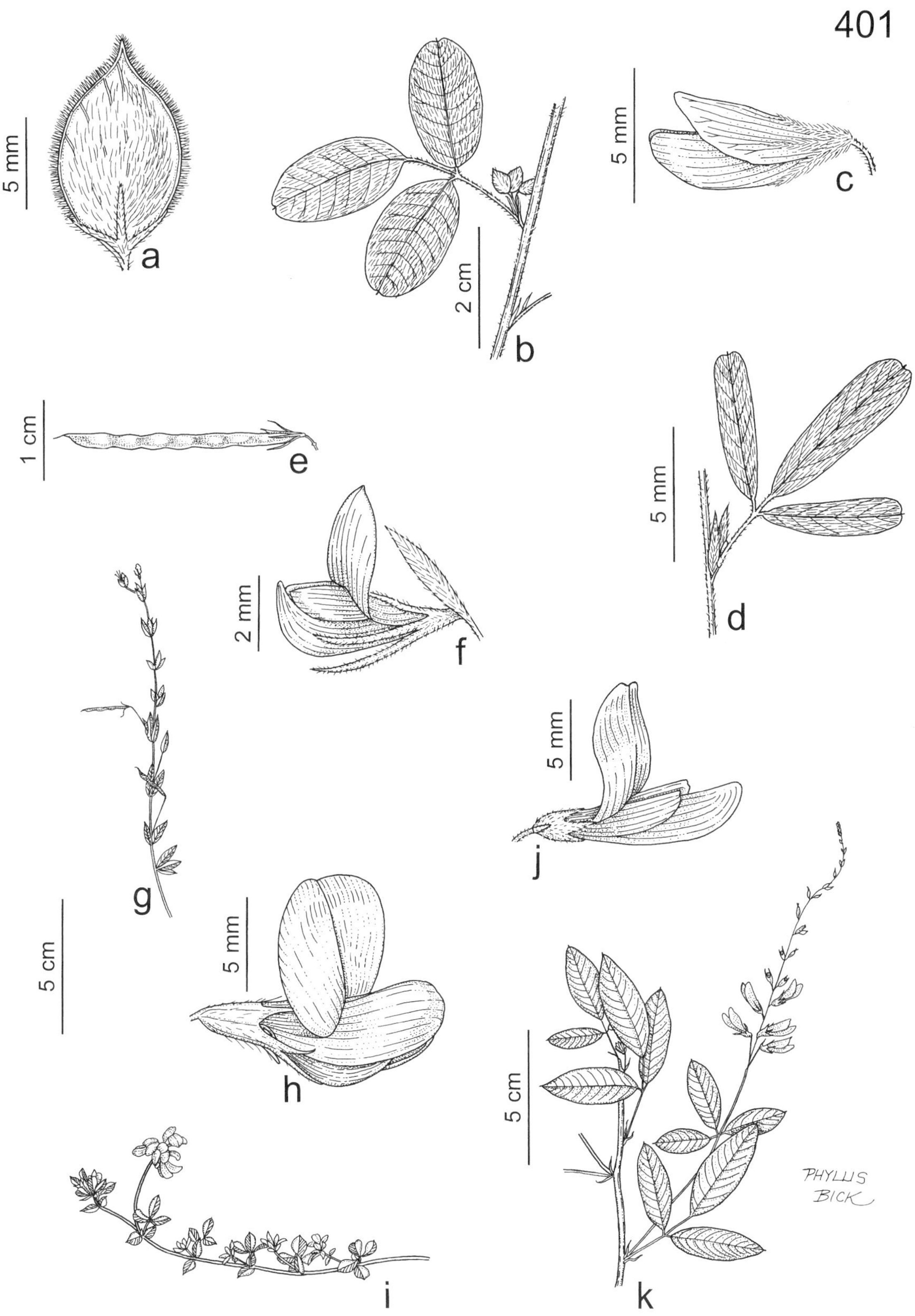

Plate 401. Fabaceae. *Lespedeza stuevei*, **a)** fruit, **b)** fertile stem. *Lespedeza cuneata*, **c)** flower, **d)** leaf attached to stem. *Acmispon americanus*, **e)** fruit, **f)** flower, **g)** habit. *Lotus corniculatus*, **h)** flower, **i)** habit. *Lespedeza thunbergii*, **j)** flower, **k)** fertile stem.

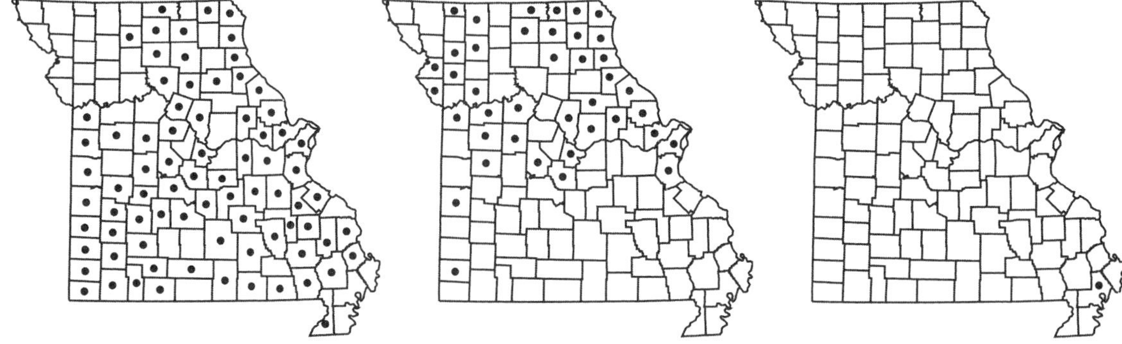

1777. Lespedeza virginica 1778. Lotus corniculatus 1779. Medicago arabica

uppermost leaves sometimes oblanceolate, rounded to angled at the base, rounded to nearly truncate at the tip, the midvein usually extended as a minute, sharp point at the very tip, the upper surface glabrous or sparsely appressed-hairy near the midvein, the undersurface sparsely to moderately or occasionally densely appressed-hairy. Axillary leaves absent or poorly developed at some nodes. Inflorescences from the axils of the upper leaves, but often appearing axillary and terminal on short branches toward the main stem tip (the cleistogamous flowers generally appearing in short, dense, axillary clusters), usually unbranched, sometimes appearing leafy, shorter than or only slightly exceeding the leaves. Flowers mostly 4–10 per raceme, the axis not hidden by the flowers (except sometimes in cleistogamous clusters). Calyces with the tube 1.5–2.0 mm long, the lobes 2–3 mm long. Corollas 5–7 mm long, pinkish purple, the banner usually darker purple toward the base, the keel sometimes darker purple at the tip, distinctly shorter than the wings. Fruits from open flowers 5–7 mm long (including the stalk), the calyx covering about the lower 1/4–1/3; those from cleistogamous flowers 4–5(–7) mm long, the calyx covering the lower 1/4–1/3. Seeds 2.5–3.0 mm long, olive green to brown. $2n=20$. June–October.

Scattered south of the Missouri River north locally to Sullivan and Clark Counties (eastern U.S. west to Kansas and Texas; Canada). Openings of mesic to dry upland forests, glades, savannas, tops of bluffs, upland prairies; also old fields and roadsides.

Reveal and Barrie (1992) examined type specimens in this complex and concluded that the types of *L. intermedia* and *L. violacea* actually represent the same taxon. Because the publication in which *L. violacea* was first described predates the publication of the name *L. intermedia*, it replaces that name as the oldest validly published epithet for the species. The plants formerly called *L. violacea* are a different species now called *L. frutescens*. The result has been pervasive confusion as some botanists have used the newer nomenclature in determining specimens but others have keyed plants in references still using the older nomenclature.

Potential hybridization with *L. capitata* and *L. hirta* are discussed under those species. Clewell (1966b) indicated rare hybridization between this species and the remaining purple-flowered native taxa, with some of these combinations noted as "identifications very tentative."

11. Lespedeza virginica (L.) Britton (slender lespedeza, slender bush clover)

Pl. 400 d, e; Map 1777

Plants perennial herbs. Stems 30–80 cm long, 1–3(–4) mm in diameter near the base, erect to strongly ascending, usually branched only above the midpoint, uniformly reddish brown below and green above or with inconspicuous, slightly lighter, longitudinal ridges, sparsely to moderately or less commonly densely pubescent with loosely appressed to more or less spreading hairs (sometimes slightly more densely so along the ridges), sometimes glabrous or nearly so toward the base. Primary median leaves with the petiole most commonly relatively long and slender, 5–13(–20) mm long, 0.2–0.5 mm wide, sparsely to moderately appressed-hairy. Stipules 2–4 mm long, linear to hairlike. Leaflets 9–30(–45) mm long, 3–9 mm wide, linear to narrowly oblong, those of the lower leaves rarely narrowly elliptic, rounded to angled at the base, rounded to nearly truncate or bluntly pointed at the tip, the midvein usually extended as a minute, sharp point at the very tip, the upper surface glabrous or sparsely appressed-hairy near the midvein, the undersurface moderately to densely appressed-hairy. Axillary leaves often well-developed at most nodes. Inflorescences from the axils of the upper and sometimes also median leaves, appearing as short, dense racemes or axillary clusters, unbranched, shorter than or only slightly exceeding the leaves. Flowers mostly 4–10 per raceme or cluster, the axis not hidden by the flowers. Calyces with the tube 1.0–1.5 mm long, the lobes 2.0–2.5 mm long. Corollas 5–7 mm long, pinkish purple to light pinkish purple, the banner usually darker purple toward the base, the keel sometimes darker purple at the tip, distinctly shorter than the wings. Fruits from open flowers 5–6 mm long (including the stalk), the calyx covering about the lower 1/3–1/2; those from cleistogamous flowers 4–6 mm long, the calyx covering the lower 1/5–1/4. Seeds 2.5–3.0 mm long, olive green to brown. $2n=20$. May–September.

Scattered to common nearly throughout the state, but apparently absent from most of the western portion of the Glaciated Plains Division (eastern U.S. west to Minnesota, Kansas, and Texas; Canada). Upland prairies, sand prairies, glades, mesic to dry upland forests, tops of bluffs, and banks of streams; also fencerows, pastures, old fields, quarries, railroads, and roadsides. $2n=20$. August–September.

Potential hybrid plants with very narrow leaflets often involve this species. One of the common examples has been called *L. ×acuticarpa* Mack. & Bush, which has very narrow and short leaflets and leafy branches from the upper axils with flowers near the tips of these side branches. The other parental species of this hybrid may be *L. frutescens* (Clewell, 1966b) or *L. repens* (Isely, 1998). The original description (Mackenzie and Bush, 1902) indicated that the keel is longer than the wings, making the former combination likely, at least for the type specimen.

Clewell (1996b) considered hybridization with all other native *Lespedeza* species to be "common" or "abundant" except for *L. violacea*, where hybrids were considered "rare," but the hybrid with *L. violacea* was indicated to be "relatively common" by Isely (1998). Plants with the aspect of *L. virginica* but with weak stems may be hypothesized as hybrids with *L. repens* if the stem has limited pubescence, or with *L. procumbens* if the stem pubescence is more pronounced. The latter has been named *L. ×brittonii* E.P. Bicknell. Some specimens have leaflet dimensions of this species or slightly broader, but with prominent pubescence, and may represent hybrids with *L. stuevei*; these have been called *L.×neglecta* (Britton) Mack. & Bush.

The prairie bush clover, *L. leptostachya* Engelm. ex A. Gray is a federally threatened species endemic to portions of Wisconsin, Minnesota, Illinois, and Iowa. The Iowa distribution extends southward to several south-central counties (Kartesz and Meacham, 1999); thus it is plausible that this species might be discovered in the future in northern Missouri. *Lespedeza leptostachya* inhabits upland prairies. Because of the widespread destruction of such prairies in the Glaciated Plains Division, it is possible that this species might have occurred in northern Missouri prior to European colonization of the region, but that it since has been extirpated from the state. The taxon has slender leaflets similar to those of *L. virginica*, but the stems are taller and more wandlike with the leaves more widely spaced, and the dense pubescence tends to give the plants a grayish appearance. *Lespedeza leptostachya* also differs from *L. virginica* in its stalked inflorescences of less-crowded flowers with the corollas cream-colored to pale pink.

35. **Lotus** L. (trefoil)

About 125 species, Europe, Asia, Africa, introduced nearly worldwide.

Most botanists have treated the genus *Lotus* in a broad sense, with various numbers of infrageneric groups differing in stipule, flower, fruit, and pollen morphologies (Ottley, 1944; Callen, 1959; Isely, 1981). More recently, evidence has begun to accumulate that the Old World species of *Lotus* (including the type species, *L. corniculatus*) are distinct from those in the New World. The species included by Steyermark (1963) and other authors as *Lotus purshianus* is treated in the segregate genus *Acmispon* in the present work. For further discussion of that species, see the treatment of *Acmispon*.

Lotus corniculatus is economically important in many countries as a forage and cover crop, and other species also are planted for erosion control.

1. Lotus corniculatus L. (bird's foot trefoil)
Pl. 401 h, i; Map 1778

Plants perennial herbs, from a woody taproot, sometimes with rhizomes. Stems 15–60 cm long, spreading to ascending, branched, often forming loose mounds, angled, unarmed, glabrous or sparsely to moderately pubescent with appressed-ascending hairs, especially toward the tip. Leaves alternate, odd-pinnately compound with 5 leaflets, sessile (sometimes misinterpreted as palmately trifoliate with leaflet-like stipules below a short petiole). Stipules inconspicuous, minute glandlike dots. Leaflets 4–20 mm long, 2.5–5.0(–8.0) mm wide, the basal pair lanceolate or more commonly asymmetrically ovate to broadly ovate (the upper margin angled at the base, the lower margin rounded to somewhat auriculate and clasping the stem), the upper lateral pair and terminal leaflet symmetrically oblong-oblanceolate to obovate, the margins entire, glabrous or sparsely hairy, the surfaces glabrous or sparsely hairy, angled or slightly tapered to a sessile base, angled or tapered (often abruptly short-tapered) to a sharply pointed tip, rarely rounded. Inflorescences axillary, umbellate to loosely headlike clusters of 3–8 flowers, the stalk 2–10 cm long, with fine appressed-ascending hairs or nearly glabrous, with a simple, leaflike bract and a pair of minute glands at the tip, the flower with a stalk 0.7–1.5 mm long above the bract; bractlets absent. Calyces sparsely hairy, the tube 2–3 mm long, conic, the lobes 2–3 mm long, more or less similar, linear to narrowly lanceolate. Corollas papilionaceous, bright yellow, sometimes with the main nerves red toward the base, often turning orange or becoming marked with brownish red with age, the banner 11–16 mm long, 6–8 mm wide, broadly ovate to nearly circular, the wings 10–14 mm long, 3–4 mm wide, oblong, the keel 11–14 mm long, 3–4 mm wide, boat-shaped, fused to above the midpoint, abruptly curved upward near the midpoint, narrowed toward the tip. Stamens 10, of 2 lengths (alternating longer and shorter filaments), 9 of the filaments fused and 1 usually shorter filament free, the fused portion 6–7 mm long, the free portion 2–4 mm long, those of at least the longer stamens flattened and broadened at the tip, the anthers relatively small, attached near the base. Ovary 5–6 mm long, the style 4–6 mm long, slender, curved upward abruptly at the base, hairy at the tip, the stigma small and terminal. Fruits legumes, 1.5–3.0 cm long, 1.5–2.5 mm wide, narrowly oblong to linear, beaked, not circular in cross-section, the sutures not thickened, dehiscing by 2 valves, these green to reddish brown at maturity, twisting spirally after dehiscence, mostly 5–14-seeded. Seeds 1.2–1.5 mm long, 0.9–1.2 mm wide, broadly kidney-shaped in outline, the surface olive green to reddish brown, often mottled with darker brown or black, smooth, somewhat shiny. $2n=24$. June–October.

Introduced, scattered to common in the northern half of the state, absent or uncommon farther south (native of Europe, Asia; introduced widely in North America). Upland prairies and banks of streams and rivers; also pastures, old fields, roadsides, and open, disturbed areas.

The bright yellow flowers in umbellate clusters are distinctive for this species. It also is recognized by the peculiar leaves, which sometimes have been interpreted as pseudo 5-foliate leaves (palmately trifoliate but with enlarged leaflet-like stipules) (Allen and Allen, 1981). However, other botanists have interpreted the leaves as pinnately compound with the lowermost pair of leaflets having migrated down the petiole to the stipular position and the true stipules represented by a pair of minute, glandlike processes at the leaf base (Isely, 1981, 1990; Sokoloff, 2000). Support for this latter hypothesis comes (among other sources) from the study of Chrtková-Zertová (1973), who noted that on rhizomes (which she called underground stems) of some forms the leaves are highly reduced, but the glandlike stipules are well-developed.

Steyermark (1963) knew *L. corniculatus* from only three counties, but it has spread widely, particularly in the northern half of the state, as a result of its use for soil stabilization following highway improvement projects. It is an important forage crop in Missouri, and when planted densely contributes significant amounts of soil nitrogen. Numerous varieties have been described within the native range of the species (Chrtková-Zertová, 1973) and there are also a number of cultivars. Because the material in the United States originated from various cultivars of uncertain origin, the present treatment does not attempt to recognize infraspecific taxa for Missouri materials.

36. **Medicago** L. (medick)
(E. Small and Jomphe, 1989)

Plants annual, biennial, or perennial herbs, with taproots. Stems erect or ascending to prostrate or trailing, sometimes forming mats, rounded to somewhat 4-angled, branched, unarmed, glabrous or variously hairy. Leaves alternate, pinnately trifoliate, short- to long-petiolate, the uppermost leaves sometimes sessile or nearly so, the terminal leaflet short-stalked.

Stipules conspicuous, lanceolate to ovate, mostly fused to the petiole to above or below their midpoints, the margins entire, toothed, or with slender, jagged lobes, persistent; stipels absent. Leaflets oblanceolate to obovate or nearly circular, the lateral leaflets usually similar to the terminal one, symmetric, angled or slightly tapered at the base, rounded to notched or occasionally nearly truncate at the tip, the margins sharply toothed above the midpoint, the surfaces glabrous or hairy, the venation pinnate, the secondary veins ending in the teeth. Inflorescences axillary, dense racemes, appearing headlike or short-cylindric, with 10 to numerous flowers (appearing as clusters of 3–8 flowers in *M. minima*), the stalks shorter than to longer than the subtending leaves, the bracts 0.5–2.0 mm long, linear to hairlike, inconspicuous, mostly persistent, the flower stalks 0.5–2.0 mm long; bractlets absent. Calyces glabrous or hairy, the tube shorter than the lobes, bell-shaped, 5-lobed, the lobes more or less similar, triangular to narrowly triangular or narrowly oblong-triangular. Corollas papilionaceous, yellow (often blue or purple in *M. sativa*), the banner obovate to oblong-obovate, narrowed to the base (lacking a stalklike basal portion), shallowly notched at the tip, curved or bent upward from about the midpoint, the wings in our species slightly shorter than the banner, with a stout, stalklike base, the expanded portion asymmetrically oblong-obovate, with a prominent auricle on one side, the keel shorter than to slightly longer than the wings, boat-shaped, only slightly curved upward, rounded at the tip. Stamens 10, all of similar lengths, 9 of the filaments fused and 1 free to the base or nearly so, the fused portion usually much longer than the free portion, oblique at the tip, the anthers relatively small, attached toward the base, yellow or white. Ovary short, sessile or nearly so, the style short, the stigma small and terminal. Fruits modified legumes, curved or more commonly spirally coiled 1–6 turns, exserted from the calyx, more or less flattened, rounded to thickened or with hooked prickles along the outer margin (the fruit then a bur), usually indehiscent, the surfaces papery to leathery, sometimes wrinkled, straw-colored to light brown or black, 1–8-seeded. Seeds 1.5–2.5 mm long, kidney-shaped in outline, somewhat flattened, the surface light yellow to reddish brown (sometimes olive green to nearly black in *M. lupulina*), smooth, usually shiny. About 83 species, Europe, Asia, Africa; several species widely cultivated and introduced nearly worldwide.

Medicago is closely related to *Melilotus* and *Trigonella* L. (which contains fenugreek, *T. foenum-graecum* L.) in the tribe Trifolieae (Bronn) Endl. and differs mainly in its coiled, often spinescent fruits. The flowers of *Medicago* have a characteristic tripping mechanism that promotes outcrossing (Lesins and Lesins, 1979; E. Small et al., 1987). During development, the staminal column is held under tension by the back of the keel, which is held in place by adhesion tissue on the edges of the keel petals and by spur-like outgrowths on the wings that fit into invaginations of the keel. The mechanism is tripped by insects trying to reach nectar produced by glands at the base of the stamens. When the wings are spread by an appropriately sized insect, the pressure releases the spurs and adhesion tissue and the stamen column flicks up and brushes the anthers and stigma against the insect's body. Although present in all species of *Medicago*, the tripping mechanism is best developed in the perennial species, such as *M. sativa*. The annual species are mostly self-fertile and have ways of self-tripping the mechanism, although outcrossing occasionally takes place. Many of the annual species have tightly coiled, burlike legumes with spiny projections. The hooked tips probably help anchor the legume to the ground, or possibly to become tangled in the fur of sheep and other animals. Some species have both prickly and smooth forms. Crossing studies indicate the spinescent character is controlled by a single gene, with the prickly genotype as the dominant allele (Lesins and Lesins, 1979).

Medicago contains many useful forage species, including alfalfa, which is considered to be the world's most important forage crop. The species are generally nutritious, adapted to a wide range of soil and climatic conditions, mostly drought resistant, thrive in open habitats, and bacteria in their root nodules fix large amounts of nitrogen that is returned to the soil and used by other plants.

1. Plants long-lived perennials, with deep, stout taproots; corollas 6–11 mm long, blue to purple, rarely yellow . . . . . 5. M. SATIVA
1. Plants annuals or biennials, with shallow, slender taproots; corollas 2–6 mm long, yellow
   2. Inflorescences with 10–30 flowers; fruits curved but not spirally coiled, unarmed, the surface strongly nerved (sometimes appearing shallowly wrinkled), black at maturity, 1-seeded . . . . . 2. M. LUPULINA
   2. Inflorescences with 2–5 flowers; fruits spirally coiled, usually armed with prickles, straw-colored to light brown at maturity, several-seeded
      3. Plants conspicuously hairy throughout; stipules entire or slightly toothed . . . . . 3. M. MINIMA
      3. Plants glabrous or nearly so; stipules deeply toothed or with slender, irregular lobes
         4. Stipules divided at most halfway to the midvein; leaflets with a central reddish purple to dark purple spot, about as long as wide, often conspicuously notched at the tip . . . . . 1. M. ARABICA
         4. Stipules divided more than halfway to the midvein; leaflets lacking a dark central spot, usually longer than wide, rounded or minutely notched at the tip . . . . . 4. M. POLYMORPHA

1. Medicago arabica (L.) Huds. (spotted medick, spotted bur clover)
M. maculata Willd.
Pl. 402 e–g; Map 1779

Plants annuals, with shallow, slender taproots. Stems 30–60 cm long, prostrate to loosely ascending from a spreading base, often mat-forming or low mound-forming, glabrous or occasionally with a few fine, nonglandular hairs. Petioles 15–60 mm long. Stipules 5–8 mm long, the margins sharply and deeply toothed or lobed at most about halfway to the midvein, fused to the petiole in about the lower 1/4–1/3. Leaflets 10–25 mm long, 8–25 mm wide (about as long as wide), broadly obovate to obtriangular, occasionally nearly inverse heart-shaped, broadly rounded to truncate or relatively broadly and deeply notched at the tip, usually with a small, narrowly triangular, sharply pointed tooth at the very tip, the upper surface glabrous, usually with a reddish purple to dark purple spot situated centrally (occasionally extending to the leaflet base or nearly so), the undersurface glabrous or sparsely appressed-hairy. Inflorescences appearing as small, headlike clusters, more or less globose, with 2–5 flowers, mostly shorter than the subtending leaves, the stalk 8–25 mm long. Calyces with the tube 1.0–1.5 mm long, the lobes 1.5–2.5 mm long. Corollas 4–5 mm long, yellow. Filaments with the fused portion 2.5–3.0 mm long, the free portion 0.5–1.0 mm long. Fruits with the body 3–6 mm long, more or less globose to slightly elongate, spirally coiled for mostly 3–7 turns, the outer margin with 3 inconspicuous, longitudinal ridges along the suture and usually 2 rows of prickles (occasionally smooth-margined elsewhere), these 1.5–3.0 mm long, with hooked tips, otherwise glabrous, straw-colored to light brown at maturity, 4–8-seeded. $2n=16$. April–May.

Introduced, uncommon, known thus far only from New Madrid County (native of Europe, Asia, Africa; introduced sporadically in the eastern [mostly southeastern] and western U.S. and elsewhere in the New World). Pastures, roadsides, and open, disturbed areas.

Medicago arabica is distinguished by the broadly obcordate leaflets, each with a purple spot. The presence and size of the purple patch appears to be related to light conditions and age of the plant, and elsewhere they are occasionally absent (Lesins and Lesins, 1979). The fruits are usually armed with prickles, but these are sometimes reduced or absent. There are three longitudinal, anastomosing nerves or ridges along the suture of the mature fruits, a character said to be unique in *Medicago* (Lesins and Lesins, 1979), but these usually are not easily observed.

2. Medicago lupulina L. (black medick)
M. lupulina var. *glandulosa* Neilr.
Pl. 402 k–n; Map 1780

Plants annuals or biennials (sometimes short-lived perennials elsewhere), with shallow, slender taproots. Stems 10–60 cm long, prostrate to loosely ascending (sometimes from a spreading base), often mat-forming or low mound-forming, glabrous or more commonly sparsely to moderately pubescent with fine, upward-curved to loosely appressed, nonglandular hairs and often also shorter, spread-

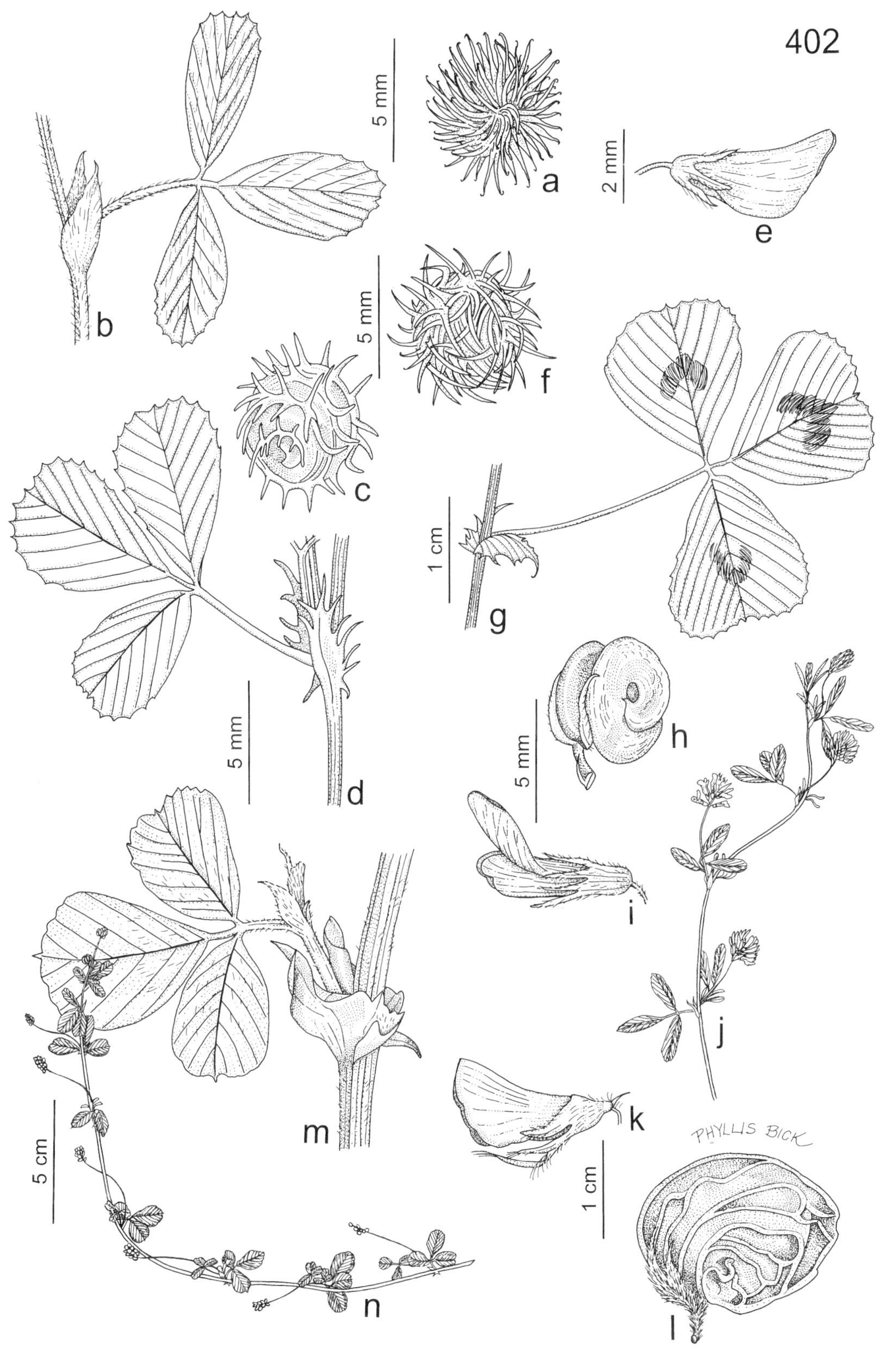

Plate 402. Fabaceae. *Medicago minima*, **a)** fruit, **b)** leaf attached to stem. *Medicago polymorpha*, **c)** fruit, **d)** leaf attached to stem. *Medicago arabica*, **e)** flower, **f)** fruit, **g)** leaf attached to stem. *Medicago sativa*, **h)** fruit, **i)** flower, **j)** fertile stem. *Medicago lupulina*, **k)** flower, **l)** fruit, **m)** leaf attached to stem, **n)** habit.

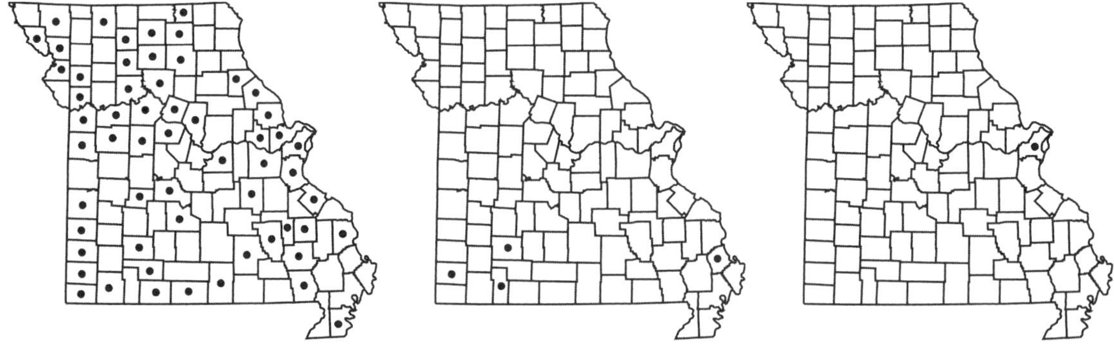

1780. Medicago lupulina 1781. Medicago minima 1782. Medicago polymorpha

ing, gland-tipped hairs. Petioles absent or to 4 mm long. Stipules 5–8 mm long, the margins entire or more commonly shallowly toothed, fused to the petiole at or below the midpoint. Leaflets 5–18 mm long, 2.5–15.0 mm wide, obovate to broadly obovate, oblong-obovate, elliptic, or nearly circular, more or less rounded at the tip, usually with a noticeable, triangular, sharply pointed tooth at the very tip, the upper surface glabrous or nearly so, lacking a red or purple spot, the undersurface sparsely to moderately pubescent with loosely appressed hairs and sometimes also shorter, spreading, gland-tipped hairs. Inflorescences dense racemes, appearing as headlike clusters, more or less globose to ovoid at flowering, becoming somewhat elongated and short-cylindric at fruiting, with 10–30 flowers, extending beyond the subtending leaves, the stalk 20–45 mm long. Calyces with the tube 0.4–0.6 mm long, the lobes 0.6–1.0 mm long. Corollas 2–3 mm long, yellow. Filaments with the fused portion 0.5–0.7 mm long, the free portion 0.1–0.3 mm long. Fruits 2.5–3.0 mm long, kidney-shaped, curved but not spirally coiled, longitudinally wrinkled, not prickly, strongly nerved (sometimes appearing shallowly wrinkled), glabrous or occasionally with scattered, gland-tipped hairs, black at maturity, 1-seeded. $2n=16$. May–November.

Introduced, scattered to common nearly throughout the state (native of Europe, Asia; introduced nearly throughout North America and elsewhere in the world). Banks of streams and rivers, openings of bottomland forests, upland prairies, glades, and ledges, and tops of bluffs; also pastures, levees, lawns, quarries, mine tailings, farmyards, edges of crop fields, railroads, roadsides, and open, disturbed area.

The foliage of black medick is palatable to deer and livestock, and the seeds are eaten by many birds. *Medicago lupulina* is distinguished by its low, trailing habit, pubescence (the hairs often gland-tipped), small yellow flowers, and unarmed, black, 1-seeded fruits. Not all individuals possess gland-tipped hairs, but there does not appear to be sufficient correlation with other characters or geographic distribution to support the recognition of varieties based on this trait (Shinners, 1956; Steyermark, 1963). Plants of *M. lupulina* are easily confused with two superficially similar, yellow-flowered hop-clovers, *Trifolium campestre* and *T. dubium*. Those species differ in their more or less entire stipules, persistent corollas that become brown and papery after flowering, and ovoid to oblong-ovoid, straight, dull, brown, stalked fruits.

3. Medicago minima (L.) Bartal. (small bur clover, downy bur clover, prickly medick)
M. polymorpha L. var. *minima* L.
Pl. 402 a, b; Map 1781

Plants annuals, with shallow, slender taproots. Stems 10–40 cm long, prostrate to loosely ascending from a spreading base, often mat-forming or low mound-forming, densely pubescent with fine, spreading to upward-curved, nonglandular hairs and often also shorter, spreading, gland-tipped hairs. Petioles 2–20 mm long. Stipules 4–8 mm long, the margins entire or shallowly toothed, fused to the petiole to below the midpoint. Leaflets 5–14 mm long, 3–10 mm wide, obovate to broadly obovate, broadly elliptic, or nearly circular, more or less rounded at the tip, usually with a small, narrowly triangular, sharply pointed tooth at the very tip, the upper surface sparsely to densely pubescent, lacking a red or purple spot, the undersurface moderately to densely pubescent, the hairs fine, loosely appressed, nonglandular. Inflorescences appearing as small, headlike clusters, more or less globose, with 2–5 flowers, mostly extending beyond the subtending leaves, the stalk 5–25 mm long. Calyces with the tube 0.8–1.3 mm long, the lobes

1.5–2.0 mm long. Corollas 3–5 mm long, yellow. Filaments with the fused portion 1.5–2.0 mm long, the free portion 0.5–1.0 mm long. Fruits with the body 3–5 mm long, more or less globose, spirally coiled for mostly 3–5 turns, the outer margin with 2 rows of prickles (occasionally smooth-margined or tuberculate elsewhere), these 1.5–2.0 mm long, with hooked tips, otherwise glabrous or with scattered, nonglandular and/or gland-tipped hairs, light brown at maturity, 3–6-seeded. $2n=16$. April–May.

Introduced, uncommon and sporadic in southern Missouri (native of Europe, Asia, Africa; introduced sporadically in the eastern U.S. west to Kansas and Texas, also Washington to California, Arizona). Lawns, railroads, roadsides, and open disturbed areas, often in sandy soils.

Medicago minima is recognized by its low, trailing habit, hairy stems, short petioles, small clusters of yellow flowers, and spiny, coiled legumes with several seeds. The length of the spines is somewhat variable. It is distinguished from *M. polymorpha* and *M. arabica* by the nearly entire stipules and hairy stems. This species was first reported from Missouri by Yatskievych (1990).

4. Medicago polymorpha L. (California bur clover, smooth bur clover)

M. polymorpha var. *vulgaris* (Benth.) Shinners

M. hispida Gaertn.

M. nigra Krock.

Pl. 402 c, d; Map 1782

Plants annuals, with shallow, slender taproots. Stems 10–60 cm long, prostrate to loosely ascending from a spreading base, often mat-forming or low mound-forming, glabrous. Petioles 15–70 mm long. Stipules 3–15 mm long, the margins deeply few-lobed more than halfway to the midvein, fused to the petiole near the base. Leaflets 6–25 mm long, 4–20 mm wide (mostly longer than wide), obovate to obtriangular, rounded or minutely notched at the tip, usually with a small, narrowly triangular, sharply pointed tooth at the very tip, the surfaces glabrous (occasionally with a few hairs at the base), lacking a red or purple spot. Inflorescences appearing as small, headlike clusters, more or less globose, with 2–5 flowers, mostly shorter than the subtending leaves, the stalk 8–25 mm long. Calyces with the tube 0.8–1.2 mm long, the lobes 1.5–2.0 mm long. Corollas 3.5–4.5 mm long, yellow. Filaments with the fused portion 1.5–2.0 mm long, the free portion 0.5–1.0 mm long. Fruits with the body 4–6 mm long, more or less globose, spirally coiled for mostly 2–4 turns, the outer margin thickened but lacking parallel ridges along the suture, usually with 2 rows of prickles (occasionally tuberculate or smooth-margined elsewhere), these 1.5–3.0 mm long, with hooked tips, otherwise glabrous, straw-colored to light brown at maturity, 3–6-seeded. $2n=14$. April–May.

Introduced, uncommon, known thus far only from the city of St. Louis (native of Europe, Asia, Africa; introduced widely in the U.S. [including Alaska, Hawaii], Canada, and elsewhere in the world). Railroads.

Medicago polymorpha is recognized by the glabrous stems, deeply incised stipules, small clusters of yellow flowers, and usually prickly fruits. There is some variation in the spine length, and plants with tuberculate or smooth fruits are not uncommon. The number of coils is also variable. Shinners (1956) recognized a number of varieties and forms, based on such characters, but subsequent authors mostly have treated these as minor variants unworthy of formal taxonomic recognition.

Medicago orbicularis (L.) Bartal. is another annual species with deeply lacerate stipules and few-flowered clusters of yellow flowers. It has been recorded as an escape in a number of states, including Texas, Oklahoma, and Illinois, and may be found in Missouri in the future. It differs from *M. polymorpha* in its relatively flat, larger (10–15 mm in diameter), loosely coiled, unarmed fruits.

5. Medicago sativa L. (alfalfa)

Pl. 402 h–j; Map 1783

Plants long-lived perennials, with deep taproots below a knobby or sometimes short-branched, woody caudex. Stems 30–60(–90) cm long, erect to loosely ascending, sometimes from a spreading base, sometimes clump-forming, glabrous or finely hairy. Petioles mostly 10–50 mm long (those of the uppermost leaves sometimes shorter). Stipules 6–18 mm long, the margins entire or shallowly toothed toward the base, fused to the petiole toward the base. Leaflets 10–32 mm long, 2–12 mm wide, oblanceolate to elliptic or narrowly obovate, rounded to minutely notched at the tip, the midvein often extended as a short, narrowly triangular, sharply pointed tooth at the very tip, the surfaces glabrous or sparsely hairy, the upper surface lacking a red or purple spot. Inflorescences dense racemes, appearing as headlike clusters, more or less globose to oblong-ovoid or short-cylindric at flowering, with 10–30 flowers, often extending beyond the subtending leaves, the stalk 10–30 mm long. Calyces with the tube 1.5–2.0 mm long, the lobes 2.5–3.5 mm long. Corollas 6–11 mm

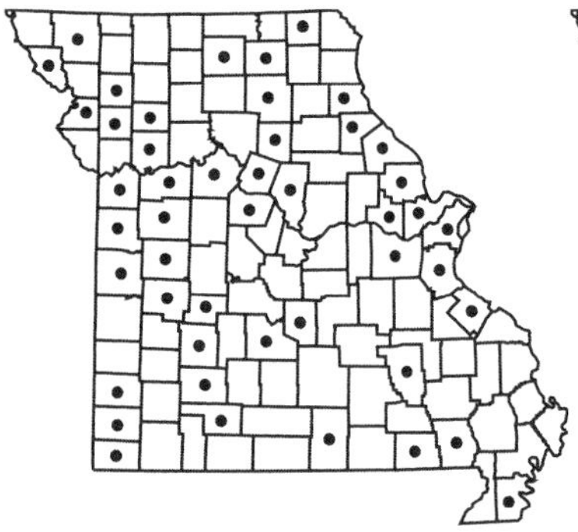

1783. Medicago sativa

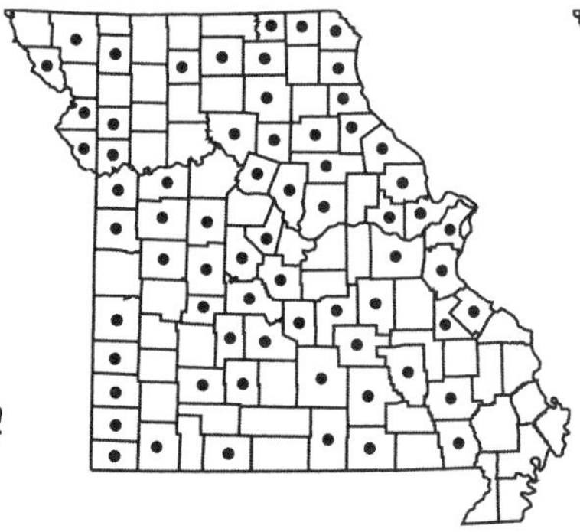

1784. Melilotus albus

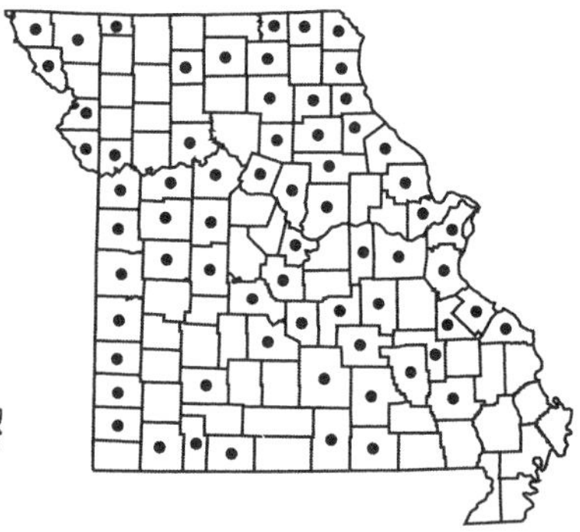

1785. Melilotus officinalis

long, blue to purple, rarely yellow (white, lavender, or variegated elsewhere). Filaments with the fused portion 4–5 mm long, the free portion 1–2 mm long. Fruits 4–6(–10) mm long, longer than wide, strongly curved to loosely coiled 1–3 turns, smooth, not prickly, glabrous or finely hairy, yellow to brown, 3–8-seeded. $2n=16$, 32. May–September.

Introduced, scattered nearly throughout the state (native of Europe, Asia; introduced nearly throughout the U.S. [including Alaska, Hawaii], Canada, and elsewhere in the world). Banks of streams and rivers, marshes, and tops of bluffs; also pastures, levees, railroads, roadsides, and open, disturbed areas.

The deep and strong taproots enable this species to withstand drought and to grow in dry, rocky habitats. Once established, it can persist for a long time. *Medicago sativa* is perhaps the oldest and most important cultivated forage crop (Quiros and Bauchan, 1988). Alfalfa was cultivated long before recorded history began in the area now known as Iran. Its development as a crop was tied to the increasing use of horses for transport and warfare. Early Roman writers reported that it was introduced into Greece as early as 490 B.C. by invading Mede and Persian armies. It was later taken to Italy and other European countries. Early attempts during the 1700s to grow alfalfa by colonists in the eastern portion of what is now the United States apparently were unsuccessful (Oakley and Westover, 1922; Barnes et al., 1988). However, the plant was successfully introduced into California from Chile by the Spanish during the Gold Rush era of the 1850s (Oakley and Westover, 1922; Barnes et al., 1988).

Yellow flowered specimens of *M. sativa* are occasionally collected in Missouri and have been difficult to name. The *M. sativa* polyploid complex consists of a series of diploid ($2n=16$) taxa and various polyploid derivatives. These have variously been treated as several closely related species (Lesins and Lesins, 1979; Isely, 1998) or mostly as a series of subspecies of *M. sativa* (E. Small and Jomphe, 1989). Among these, cultivated alfalfa (ssp. *sativa*) is thought to have arisen as an autotetraploid ($2n=32$) derivative from a wild, blue-flowered, somewhat smaller, diploid taxon known as ssp. *caerulea* (Less. ex Ledeb.) Schmalh., which is native to the Middle East and adjacent Europe. In contrast, yellow-flowered plants with sickle-shaped (rather than coiled) pods are known as ssp. *falcata* (L.) Arcang. (E. Small and Brooks, 1984; E. Small and Jomphe, 1989). These diploid plants apparently are native to northern portions of Europe and Asia, and have also been brought into cultivation. Molecular studies of the complex based on chloroplast and mitochondrial markers have been confounded by the unusual inheritance of the chloroplast genome, which is biparental with a strong paternal bias (rather than the much more widespread maternal inheritance), but Havananda et al. (2010) concluded that the blue- and yellow-flowered diploid taxa are likely better treated as subspecies than as species. Further complicating the issue of taxon recognition, the blue- and yellow-flowered taxa are known to hybridize readily, producing an intergrading swarm of offspring with mixtures of yellow, blue, violet, and variegated corolla colors, and the hybrid germplasm is now widely distributed. The hybrids have been called ssp. ×*varia* (Martyn) Arcang. (Rabeler, 1984). However the hybrids form a continuum, with frequent backcrossing, and are difficult to characterize morphologically (E. Small and Brooks, 1984). The Missouri specimens have legumes coiled about 1.5 times, which is consistent with some specimens of ssp. *sativa*, but have yellow corollas typical of ssp. *falcata*. Isely (1998) suggested that true *M. falcata* is probably rather rare in the U.S., and the Missouri plants probably fall somewhere within the broad continuum of independent character-sorting in cultivated lineages with contributions from both ssp. *sativa* and ssp. *falcata* that sometimes have been called ssp. ×*varia*. Because of the ambiguous nature of the situation, no infraspecific taxa of *M. sativa* are recognized formally in the current treatment.

37. **Melilotus** (L.) Mill.

Plants annual or biennial herbs (sometimes short-lived perennials farther south), with relatively stout taproots. Stems erect or strongly ascending, several-angled or ridged, branched, unarmed, glabrous or sparsely pubescent with short, curved hairs toward the tip. Leaves alternate, pinnately trifoliate, mostly relatively long-petiolate, the terminal leaflet short- to relatively long-stalked. Stipules linear above an expanded base (more or less awl-shaped), fused basally to the petiole base, the margins entire, often relatively thin and pale, persistent; stipels absent. Leaflets oblanceolate to obovate, oblong-elliptic, or elliptic, those of the uppermost leaves often narrowly so, the lateral leaflets usually slightly shorter than the terminal one, symmetric, angled or slightly tapered at the base, rounded at the tip, the midvein usually extended as a minute, sharp point at the very tip, the margins sharply and finely toothed, the basal portion often toothless, the upper surface glabrous, the undersurface moderately pubescent with fine, appressed hairs, the venation pinnate, the secondary veins ending in the teeth. Inflorescences axillary, elongate, spikelike racemes with numerous flowers, the stalks mostly longer than the subtending leaves, the bracts 0.8–1.5 mm long, linear to hairlike, inconspicuous, persistent, the flower stalks 0.5–2.0 mm long, downward-curved; bractlets absent. Calyces glabrous or finely hairy, 1.5–2.2 mm long, the tube about as long as to slightly longer than the lobes, bell-shaped, 5-lobed, the lobes more or less similar, triangular to narrowly triangular. Corollas papilionaceous, yellow or white, the banner with the expanded portion obovate to oblong-obovate, narrowed to a short, stout, stalklike basal portion (this sometimes nearly absent), shallowly notched at the tip, curved or bent upward from about the midpoint, the wings in our species slightly shorter than to about as long as the banner, with a stout, stalklike base, the expanded portion asymmetrically oblong to oblong-obovate, with a prominent auricle on 1 side, the keel about as long as the wings and more or less fused with them below the midpoint, boat-shaped, only slightly curved upward, rounded to bluntly pointed at the tip. Stamens 10, all of similar lengths, 9 of the filaments fused and 1 free to the base or nearly so, the fused portion usually much longer than the free portion, often oblique at the tip, the anthers relatively small, attached toward the base, yellow. Ovary short, sessile or very short-stalked, the style slender, usually longer than the ovary, glabrous, the stigma small and terminal. Fruits modified legumes, ovoid to oblong-ovoid, slightly flattened, sessile or tapered to a short, stout stalk at the base, the tip with a slender beak, indehiscent (occasionally dehiscing tardily and irregularly), the surfaces leathery, sometimes appearing wrinkled, with a pattern of raised nerves, light brown to dark brown, gray, or black, 1(2)-seeded. Seeds 1.5–2.5 mm long, ovoid (usually with a shallow, broad, asymmetric notch, somewhat flattened, the surface yellow to reddish brown, smooth, usually shiny. About 20 species, Europe, Asia, Africa, introduced nearly worldwide, mostly in temperate regions.

Melilotus is recognized by the upright habit with ascending branches, pinnately trifoliate leaves with toothed leaflets, elongate spikelike racemes of small, yellow or white flowers, and small, achenelike, 1-seeded legumes. *Melilotus* is closely related to *Medicago* and *Trigonella* L. (which contains fenugreek, *T. foenum-graecum* L.) in the tribe Trifolieae (Bronn) Endl. In fact, recent molecular phylogenetic research has suggested that the species of *Melilotus* represent merely a specialized group within *Trigonella* in which the fruits have become reduced, one-seeded, and indehiscent (Steele et al., 2010). If future studies support this conclusion, then the species of *Melilotus* will have to be transferred to *Trigonella*, as the latter is the older of the two generic epithets.

Cultivated sweet clovers are mostly biennials. During the first year's growth, the plants produce a central, much-branched stem. The root becomes fleshy as the plant stores food for the winter. In the second year, the plant produces vigorous, rapidly growing stems with dense foliage, flowers and fruits. Both *M. albus* and *M. officinalis* were introduced into North America by the 1600s (Turkington et al., 1978).

The drying herbage of sweet clovers often exudes a pronounced, sweet aroma. The species are winter hardy, drought tolerant, and highly valued as forage plants. Most species of *Melilotus* have a preference for alkaline soils. The foliage is bitter, but nutritious, and eaten by cattle and other livestock when they get used to it. The plants contain coumarin, which is responsible for the distinctive vanilla-like smell of the cut foliage. There is no problem with grazing the plants, but under conditions of high heat or spoilage coumarin is converted to dicoumarol, which interferes with vitamin K uptake and prevents blood from coagulating. Cattle fed large amounts of improperly cured or spoiled hay suffer uncontrollable bleeding with any cut or bruise. Low-coumarin cultivars have been developed, however they can interbreed with nearby high coumarin plants. *Melilotus* also is valuable for soil improvement and as a cover crop for eroded land. It can also become an unwanted weed in wheat fields and is considered an invasive exotic in prairie and glade habitats. The flowers of *Melilotus* species are attractive to bees, produce nectar over a long period, and are a major source of honey.

1. Corollas white, the banner 3.5–5.0 mm long, longer than wings and keel; ovaries narrowed at the base, but not stalked; mature fruits with a network of raised nerves . . . . 1. M. ALBUS
1. Corollas yellow, the banner 4–5 mm long, about as long as the wings and keel; ovaries noticeably stalked (but this often obscured by the calyx); mature fruits with a pattern of cross-nerves or with irregular cross-wrinkles . . . 2. M. OFFICINALIS

1. Melilotus albus Medik. (white sweet clover)
Trifolium officinalis L.

Pl. 403 a–c; Map 1784

Stems 30–200 cm long. Petioles 1–2 cm long, the terminal leaflet with the stalk 2–6 mm long. Stipules 5–9 mm long. Leaflets 10–30 mm long, 4–17 mm wide. Inflorescences (2–)4–14 cm long, becoming elongated at fruiting, the stalk 1–5 cm long. Corollas white, the banner 3.5–5.0 mm long, 2.0–2.2 mm wide, slightly longer than the wings and keel. Ovary narrowed at the base, but not stalked. Fruits 2.5–4.0 mm long, sessile or nearly so, the surfaces with a network of raised nerves. $2n=16$. June–December.

Introduced, scattered to common nearly throughout the state, but apparently absent from most of the Mississippi Lowlands Division (native of Europe, Asia; introduced nearly throughout the U.S., Canada). Upland prairies, glades, bases, ledges, and tops of bluffs, banks of streams and rivers, margins of ponds and lakes, and oxbows; also pastures, old fields, fallow fields, crop fields, levees, lawns, railroads, roadsides, and open, disturbed areas.

Melilotus albus is easily distinguished from *M. officinalis* by the white flowers and fruits with a network of raised nerves. It is very variable in height, thickness of the stems, degree of branching, period of flowering, shape and size of the leaves, and color of the seeds (Turkington et al., 1978). The flowers are self-fertile, but need an insect visitor to trip the pollination mechanism. The species name has been spelled *M. alba* in some of the botanical literature, but according to the International Code of Botanical Nomenclature *Melilotus* should be treated as a masculine word with the appropriate masculine spelling of the species epithet.

2. Melilotus officinalis (L.) Lam. (yellow sweet clover, yellow melilot)

Pl. 403 d, e; Map 1785

Stems 30–150 cm long. Petioles 0.5–2.0 cm long, the terminal leaflet with the stalk 2–7 mm long. Stipules 5–8 mm long. Leaflets 5–30 mm long, 2–12 mm wide. Inflorescences 5–16 cm long, becoming elongated at fruiting, the stalk 1–4 cm long. Corollas yellow, the banner 4–5 mm long, 2.2–

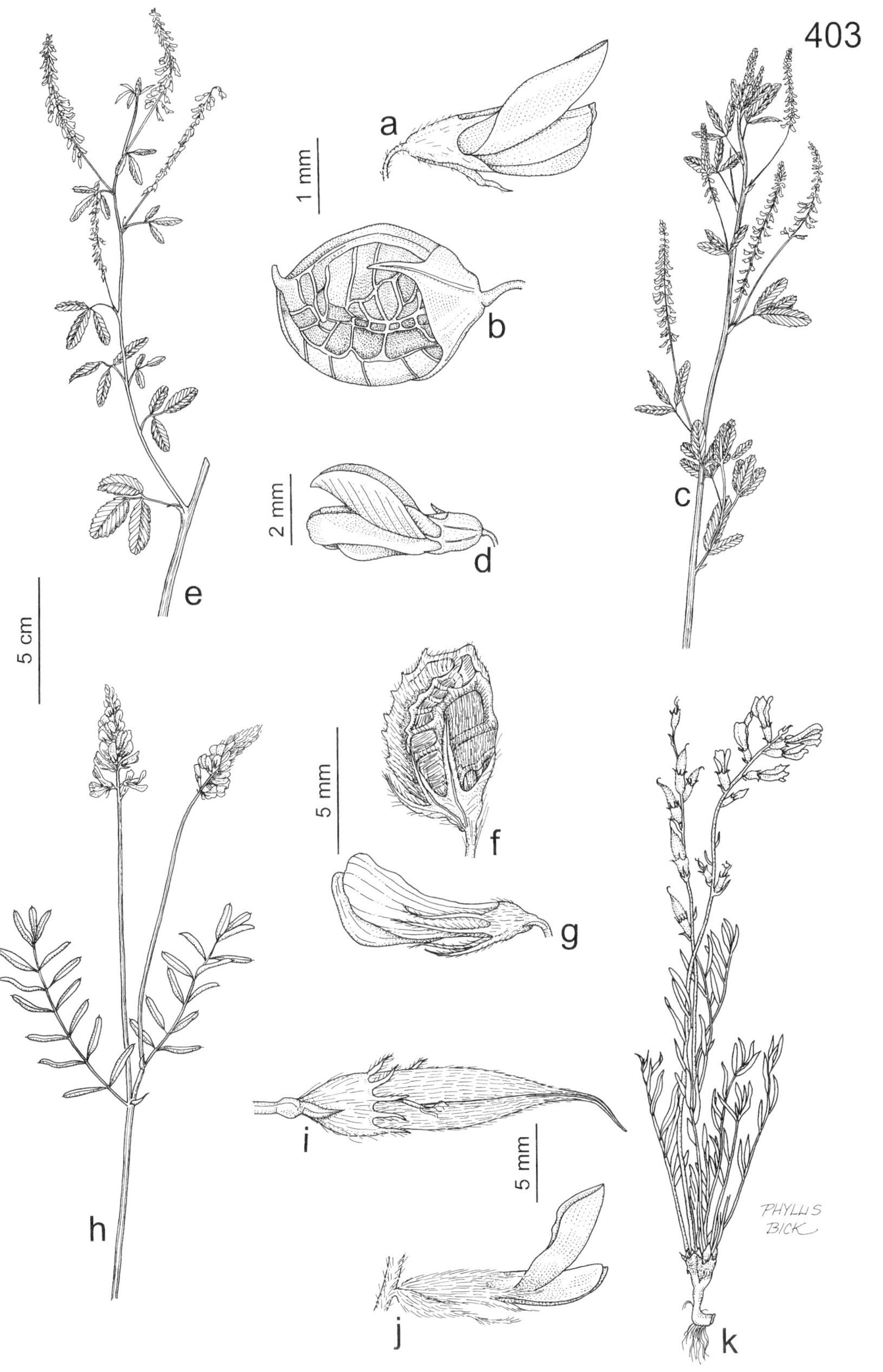

Plate 403. Fabaceae. *Melilotus albus*, **a)** flower, **b)** fruit, **c)** fertile stem. *Melilotus officinalis*, **d)** flower, **e)** fertile stem. *Onobrychis viciifolia*, **f)** fruit, **g)** flower, **h)** fertile stem. *Oxytropis lambertii*, **i)** fruit, **j)** flower, **k)** habit.

2.5 mm wide, about as long as the wings and keel. Ovary noticeably stalked at the base (but this often obscured by the calyx). Fruits 2.2–2.8 mm long, short-stalked, the surfaces with a pattern of cross-nerves or with irregular cross-wrinkles. $2n=16$. May–September.

Introduced, scattered to common nearly throughout the state, but apparently absent from most of the Mississippi Lowlands Division (native of Europe, Asia; introduced nearly throughout the U.S., Canada). Upland prairies, glades, openings of mesic upland forests, bases, ledges, and tops of bluffs, banks of streams and rivers, and margins of ponds and lakes; also pastures, old fields, fallow fields, crop fields, levees, ditches, strip mines, railroads, roadsides, and open, disturbed areas.

Melilotus officinalis is very similar to *M. albus* except for the yellow corollas and the mature legumes with a transverse banding pattern. Plants lacking flowers or mature fruits are often impossible to distinguish, but *M. officinalis* tends to be somewhat shorter and with thinner stems than *M. albus*; it starts to flower slightly earlier and ceases to produce flowers somewhat earlier, and it apparently can tolerate drier conditions than *M. albus*. The ovary of *M. officinalis* is distinctly stalked, a character best seen on the developing and mature fruits. The seeds of *M. officinalis* are said to be a little more rounded than those of *M. albus*, and may have purple specks (Turkington et al., 1978). There are reports (Geesink et al., 1999) of white flowered forms appearing within populations of *M. officinalis*, which apparently led Kartesz and Meacham (1999) to erroneously list *M. officinalis* and *M. albus* as a single species. Aside from the morphological differences in the flowers and fruits, Ha (1993) showed that the species could be distinguished based on seed protein profiles. Previously, Sano and Kita (1975) had shown that cytologically *M. albus* has a reciprocal translocation relative to the other species, and that artificial crosses between *M. albus* and *M. officinalis* (and the other species studied) resulted in sterile offspring with meiotic abnormalities. There are no reports of natural hybridization between the white and yellow sweet clovers.

Another yellow-flowered species that is similar to *M. officinale* is *M. indicus* (L.) All. (annual sweet clover). This Eurasian species has become established widely in the United States, mostly to the south of Missouri. It differs from *M. officinale* in its smaller flowers (1–3 mm long), globose fruits, and stipules of at least the lower leaves with more strongly flared, thin bases that partially encircle the stem. Although it has not yet been reported from regions particularly close to Missouri, it might be brought in as a contaminant of hay or grass seed in the future.

38. **Orbexilum** Raf.
(Grimes, 1990; B. L. Turner, 2008)

Plants perennial herbs, with deep, woody taproots or rhizomes. Stems erect or loosely ascending from a spreading base, usually branched (sometimes only at or near the base), unarmed, glabrous or hairy, sometimes minutely gland-dotted, the lowest nodes sometimes leafless or with small, scalelike outgrowths. Leaves alternate, pinnately trifoliate, the lower leaves occasionally appearing palmately trifoliate or simple, subsessile or short- to long-petiolate. Stipules mostly lanceolate-triangular to linear or the uppermost hairlike (those at the lowest, often leafless nodes sometimes broader and partly fused), attached at the base, mostly shed early; stipels absent. Leaflets elliptic to lanceolate, narrowly elliptic-lanceolate, or broadly ovate, the terminal leaflet usually somewhat longer than the lateral leaflets, all rounded at the base, bluntly to sharply pointed at the tip, the margins entire, short-hairy, the surfaces variously hairy, the undersurface sometimes also sparsely gland-dotted, the venation pinnate. Inflorescences axillary, dense spikelike racemes, variously elongate (at least at fruiting) to short and headlike, long-stalked, the bracts linear (sometimes above a short, expanded base) to lanceolate, hairy, shed early, the flower stalks 1–3 mm long; bractlets absent. Calyces 1.7–4.5 mm long, the tube bell-shaped, shorter than to longer than the lobes, 5-lobed, the lobes ascending at flowering, becoming slightly elongated and more spreading at fruiting, all similar or the lowermost lobe somewhat longer than the others. Corollas papilionaceous, usually blue to purple or lavender (the keel usually darker at the tip), rarely white, the banner with the expanded portion sometimes having white and yellowish green markings toward the base, ovate to obovate or nearly circular, tapered to a stalklike base, rounded to shallowly notched or broadly and bluntly pointed at the tip, the margins and tip curved or curled back, the wings slightly shorter than to as long as the banner, the expanded portion angled or curved over the keel, asymmetrically oblong to oblong-obovate, with a small auricle at the base, stalked, the keel shorter than the other petals, fused to the wings toward the base, short-stalked, boat-shaped, oblong to oblanceolate, only slightly curved upward, rounded at the tip. Stamens 10, the filaments all fused or 1 of the filaments more or less free above the midpoint, the free portion of alternating longer and shorter filaments, the anthers all similar and attached at the base (in *O. pedunculatum*) or in 2 alternating series (in *O. onobrychis*), those of the longer filaments attached at the base and those of the shorter series attached toward the midpoint, all yellow. Ovary short, asymmetrically ellipsoid to ovoid, very short-stalked, the style slender, often strongly curved, glabrous except for an inconspicuous ring of short hairs at the tip, the stigma small and terminal. Fruits modified legumes, asymmetrically ellipsoid or ovoid, slightly flattened, sessile or nearly so, short-tapered to a beaked tip, leathery or hardened, the surface glabrous, lacking gland-dots, strongly wrinkled, indehiscent (sometimes dehiscing irregularly with age after being shed), 1-seeded. Seeds oblong-elliptic to somewhat kidney-shaped in outline, somewhat flattened, smooth, somewhat shiny. Eight to 11 species, U.S., Mexico.

Traditionally, *Orbexilum*, *Pediomelum*, and *Psoralidium* were included in a broadly circumscribed *Psoralea* L. (Steyermark, 1963), but they are now recognized as distinct genera, mainly following Grimes (1990). *Psoralea* in the strict sense is a genus of about 50 species restricted to southern and eastern Africa. The most important morphological characters within the complex of genera are found in details of the fruits. *Orbexilum* is distinguished from these other genera by the pinnately trifoliate leaves and wrinkled fruits exserted beyond the calyces.

1. Plants with elongate rhizomes; stems usually solitary; leaflets 2–5 cm wide, terminal leaflet somewhat wider than the lateral leaflets; fruits obliquely ovate in outline, the surface wrinkled and distinctly warty . . . . 1. O. ONOBRYCHIS
1. Plants with stout taproots; stems usually several or solitary stems few- to several-branched from the base; leaflets 0.5–1.8 cm wide, the terminal leaflet similar in width but sometimes slightly longer than the lateral leaflets; fruits obliquely obovate in outline, the surface wrinkled, but not warty . . . . 2. O. PEDUNCULATUM

1. Orbexilum onobrychis (Nutt.) Rydb. (French grass)

Psoralea latifolia Torr.

P. onobrychis Nutt.

Pl. 404 g–i; Map 1786

Plants with elongate, branched rhizomes, strongly colonial, the stems mostly well-spaced. Stems 40–100 cm tall, erect, commonly reddish at the base, relatively few-branched, mostly above the midpoint, sparsely to moderately pubescent with short, strongly upward-curved to appressed hairs when young, becoming glabrous or nearly so with age. Leaves pinnately trifoliate or those below the midpoint sometimes appearing palmately trifoliate or simple (the lowermost nodes sometimes leafless and with the stipules fused into a single scale-like structure), the petioles of well-developed leaves 3–9 cm long, the terminal leaflet with the stalk 8–25 mm long. Stipules 2–5(–9) mm long, those subtending well-developed leaves downward-curved to reflexed. Leaflets 2–12 cm long, 2–5 cm wide, broadly lanceolate, the terminal leaflet somewhat wider and sometimes slightly longer than the lateral leaflets, angled or slightly tapered to a sharply pointed tip, the upper surface glabrous or sparsely short-hairy along the main veins, sometimes sparsely and inconspicuously gland-dotted, the undersurface pale, sparsely to moderately short-hairy, more densely so along the veins. Inflorescences (4–)9–24 cm long, elongating with age, the stalk 5–14 cm long, the bracts 2–3 mm long. Calyces short-hairy, the tube 1.2–1.6 mm long, the lobes 0.5–1.0 mm long, subequal, sharply pointed to more commonly bluntly pointed or rounded. Corollas with the banner 4–5 mm long, the wings 5–7 mm long, the keel 4–5 mm long. Filaments

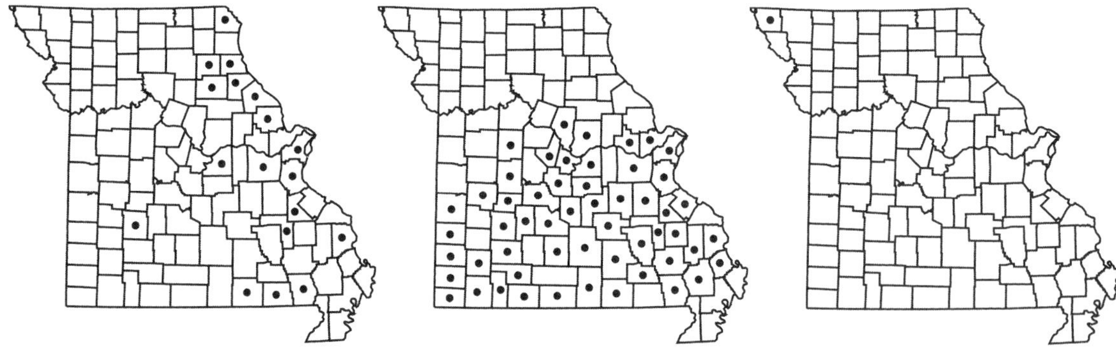

1786. Orbexilum onobrychis 1787. Orbexilum pedunculatum 1788. Oxytropis lambertii

with the fused portion 2–3 mm long, the free portion 0.4–0.6 mm long, the anthers in 2 alternating series, those of the longer filaments attached at the base and those of the shorter series attached toward the midpoint. Fruits with the body 6–8 mm long, 4–5 mm wide, obliquely ovate in outline, short-tapered to a stout beak, this 2–3 mm long, the surface with a network of cross-wrinkles, also distinctly warty, brown to black. Seeds 4–6 mm long. May–July.

Scattered, mostly in the eastern half of the state (Iowa to Arkansas east to Virginia and South Carolina). Banks of streams and rivers, margins of ponds and lakes, fens, bases and ledges of bluffs; also ditches, fallow fields, railroads, roadsides, and occasionally moist, open, disturbed areas.

Orbexilum onobrychis is recognized by its pinnately trifoliate leaves with large, mostly broadly lanceolate leaflets and strongly wrinkled, warty legumes. It is relatively infrequently encountered, but, where found, occurs in large clonal colonies from the widely creeping rhizomes.

2. Orbexilum pedunculatum (Mill.) Rydb. var. **pedunculatum** (Sampson's snakeroot)
Psoralea psoraloides (Walter) Cory var. *eglandulosa* (Elliott) Freeman

Pl. 404 a–c; Map 1787

Plants with woody taproots thickened toward the tip, not colonial, but the stems frequently 2 to several from the base, appearing clumped. Stems 25–70 cm tall, erect, green to straw-colored or light brown at the base, unbranched or less commonly few-branched above the base, moderately to densely pubescent with short, strongly upward-curved to appressed hairs toward the tip, often also inconspicuously gland-dotted, more sparsely hairy or sometimes nearly glabrous toward the base. Leaves pinnately trifoliate or those below the midpoint sometimes appearing palmately trifoliate or simple (the lowermost nodes sometimes leafless and with the stipules fused into a single scalelike structure), the petioles of well-developed leaves 0.5–5.0 cm long, the terminal leaflet with the stalk 6–10 mm long. Stipules 3–5(–8) mm long, linear to hairlike, those subtending well-developed leaves erect or ascending. Leaflets 2–7 cm long, 0.5–1.8 cm wide, narrowly elliptic to narrowly elliptic-lanceolate, those of the smaller leaves sometimes somewhat broader, the terminal leaflet similar in width but sometimes slightly longer than the lateral leaflets, rounded to bluntly pointed at the tip, the upper surface glabrous or more commonly sparsely to densely and loosely appressed-hairy, often also sparsely and inconspicuously gland-dotted, the undersurface not or only slightly lighter than the upper surface, moderately to densely hairy, not gland-dotted. Inflorescences 2–12(–20) cm long, elongating with age, the stalk 4–14 cm long, the bracts 4–7 mm long. Calyces short-hairy, the tube 1.0–1.5 mm long, the lobes 1–3 mm long, the lowermost lobe longer than the upper lobe, sharply pointed. Corollas with the banner 5–7 mm long, the wings 4.5–6.0 mm long, the keel 3–4 mm long. Filaments with the fused portion 2–3 mm long, the free portion 0.2–0.4 mm long, the anthers all similar and attached at the base. Fruits with the body 4–5 mm long, 3–4 mm wide, obliquely obovate in outline, short-tapered to a minute, strongly curved beak, the surface with a network of cross-wrinkles, but not warty, light brown to yellowish brown. Seeds 2.5–3.5 mm long. $2n$=22. May–July.

Scattered in the Ozark, Ozark Border, and Unglaciated Plains Divisions; also in the northern portion of the Mississippi Lowlands (eastern [mostly southeastern] U.S. west to Kansas and Texas). Bottomland and upland prairies, savannas, bottomland and upland forests, edges of fens and sinkhole ponds, banks of streams and rivers, and tops of bluffs; also railroads and roadsides.

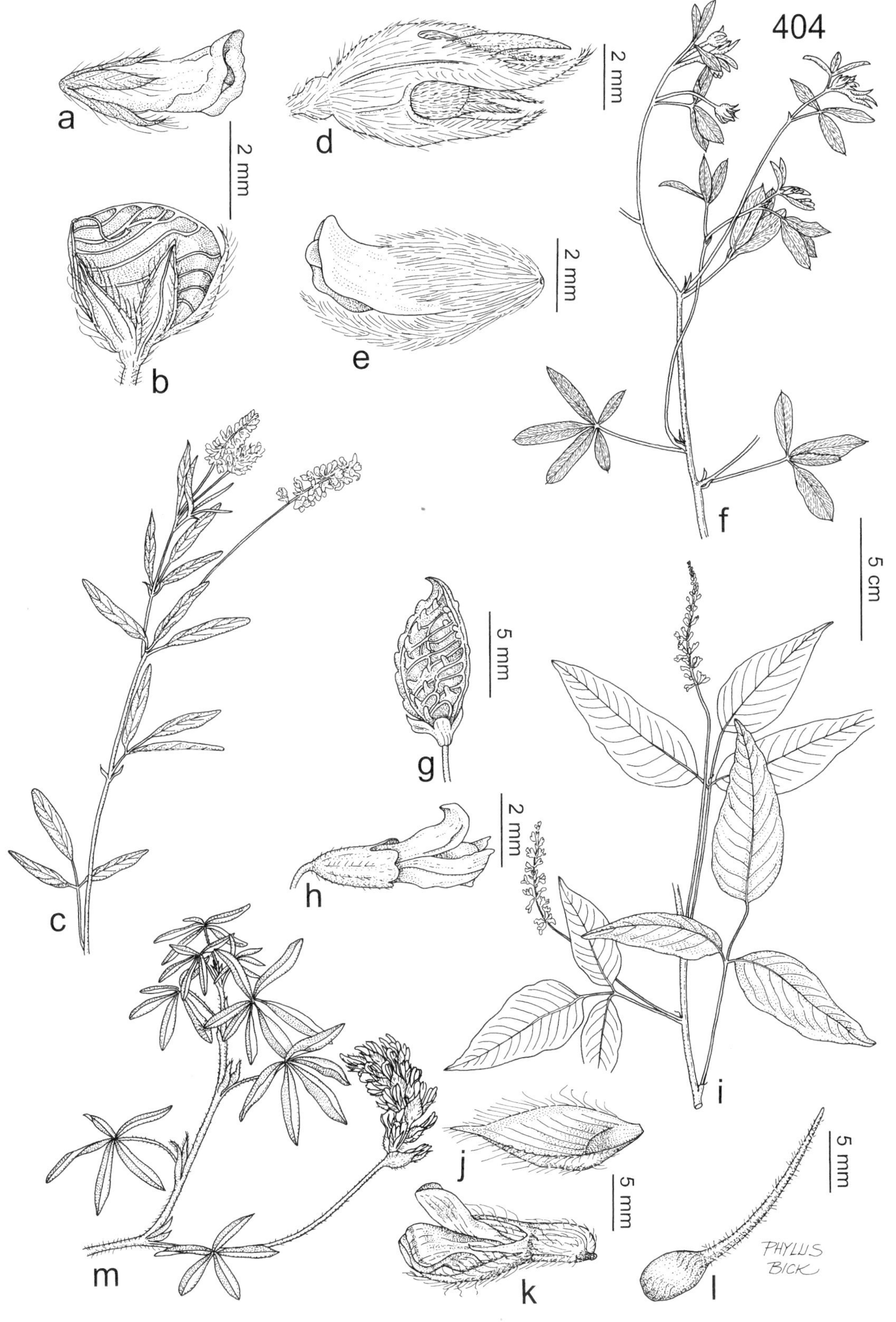

Plate 404. Fabaceae. *Orbexilum pedunculatum*, **a)** flower, **b)** fruit, **c)** fertile stem. *Pediomelum argophyllum*, **d)** fruit, **e)** flower, **f)** fertile stem. *Orbexilum onobrychis*, **g)** fruit, **h)** flower, **i)** fertile stem. *Pediomelum esculentum*, **j)** bract, **k)** flower, **l)** fruit, **m)** fertile stem.

This species is recognized by its pinnately trifoliate leaves with narrow leaflets and suborbicular, wrinkled fruits. The inflorescences appear long-stalked with the flowers bunched near the tips.

Orbexilum pedunculatum is commonly divided into two varieties based on the size of the bracts and whether the lower surface of the leaflets, calyces, and fruits are gland-dotted or not (Grimes, 1990; Isely, 1998). However, some authors consider these taxa to represent full species (B. L. Turner, 2008). Missouri specimens lack glands on these structures and have relatively large bracts, and thus conform with var. *pedunculatum*, which is widespread in the central United States. The other taxon, known variously as var. *gracile* (Torr. & A. Gray) J.W. Grimes or *O. gracile* (Torr. & A. Gray) Rydb., occurs mainly along the Atlantic Coastal Plain from Virginia to Florida. It is gland-dotted and has relatively small bracts.

39. Oxytropis DC.
(Barneby, 1952; Welsh, 2001)

Three hundred to 400 species, North America, Europe, Asia.

Oxytropis is very similar to *Astragalus*, but differs in having corollas with a strongly beaked keel and different base chromosome number. The chromosome number of *Oxytropis* is closer to that of some Old World species of *Astragalus*, suggesting that it is an independent lineage in the New World. Preliminary molecular evidence indicates *Oxytropis* is not nested within *Astragalus* (Wojciechowski et al., 1999).

1. Oxytropis lambertii Pursh var. lambertii
(purple locoweed)

Pl. 403 i–k; Map 1788

Plants perennial herbs, with a stout taproot below a usually branched, sometimes slightly woody caudex, sparsely to densely pubescent with 2-branched hairs (these appressed to somewhat spreading and with 2 opposite, unequal branches, thus appearing as a straight line attached near 1 end). Stems absent or nearly so. Leaves all basal or nearly so, odd-pinnately compound with 7–19 leaflets, leaves produced later in the season often noticeably smaller than those produced earlier in the year, the petiole 3–5 cm long, the rachis 5–12 cm long. Stipules 6–20 mm long, 3–4 mm wide, fused to the petiole toward the base, sheathing, usually sparsely hairy, becoming papery with age. Leaflets 5–40 mm long, 2–8 mm wide, linear to narrowly oblong or narrowly elliptic, angled at the base, angled or tapered to a sharply pointed tip, the margins entire. Inflorescences solitary to several, spikelike racemes with (6–)10–25 flowers, erect or strongly ascending, the stalk 5–18 cm long, the axis 5–12 cm long, the bracts 6–9 mm long, 1.0–1.5 mm wide, linear to narrowly lanceolate, hairy. Flowers sessile, ascending to spreading. Calyces 5-lobed, densely hairy, the tube 6–7 mm long, cylindric, with mostly whitish hairs, the lobes 2–3 mm long, narrowly triangular. Corollas papilionaceous, pinkish purple to violet, with an oblong pattern of dense, linear, white markings toward the banner base (rarely entirely white elsewhere), the banner 15–22 mm long, 5–11 mm wide, the expanded portion obovate to oblong-obovate, rounded to slightly notched at the tip, bent upward abruptly toward the base, the wings 12–18 mm long, 4–8 mm wide, oblanceolate, truncate to shallowly notched at the tips, the keel 13–18 mm long, 2–5 mm wide, boat-shaped, straight or somewhat curved upward, tapered to a pointed, beaklike tip 0.5–2.5 mm long. Stamens 10, all similar in length, 9 of the filaments fused and 1 free, the fused portion 10–12 mm long, the free portion 2–3 mm long, curved, the anthers 1.0–1.5 mm long, attached near the base, yellow. Ovary 4–5 mm long, hairy, the style 6–7 mm long, glabrous, curved, the stigma terminal, minute. Fruits legumes, ascending at maturity, the body 8–20 mm long, 4–6 mm wide, ovate to lanceolate in outline, tapered at the tip to a beak 3–7 mm long, the valves leathery to somewhat woody, hairy, with 2 locules, dehiscent from the tip. Seeds 1.8–2.0 mm long, broadly kidney-shaped, the surface brown, smooth, somewhat shiny. $2n=48$. May–June.

Uncommon, known thus far only from Atchison County (Montana to Wyoming east to Minnesota and Oklahoma; Canada). Loess hill prairies.

Steyermark (1963) also reported this species from Holt County, but did not cite a specimen. His report could not be confirmed during the present research.

Consumption of large amounts of *O. lambertii* causes loco poisoning in livestock, much as in *Astragalus* (Burrows and Tyrl, 2001). The sickness, which is often fatal, is caused by the presence of indolizidine alkaloids, especially swainsonine. Locoism is a serious problem in western states where *O. lambertii* and other locoweed species are more abundant. The flowers are attractive to butterflies.

Isely (1998) and Welsh (2001) followed Barneby (1952) in recognizing three varieties within *O. lambertii*. Missouri plants belong to var. *lambertii*, which is widespread in all but the southernmost portion of the species range and has a combination of features that include relatively small corollas, relatively long calyx teeth, mostly relatively narrow leaflets, and leathery to woody fruits that are much longer than the calyces at maturity. The var. *articulata* (Greene) Barneby consists of southern plants (Kansas to New Mexico, Texas, and Louisiana) with relatively narrow leaflets and woody fruits usually only slightly (if at all) longer than the calyces; var. *bigelovii* A. Gray refers to western and southern plants (Wyoming to Arizona, New Mexico, and Texas) with relatively broad leaflets and papery to leathery fruits that extend noticeably beyond the calyces. These differences are subtle and some specimens cannot be determined satisfactorily to variety.

40. **Pediomelum** Rydb.
(Grimes, 1990)

Plants perennial herbs, with deep, woody taproots, these usually thickened toward the tip, either strongly enlarged and turnip-shaped or forming a knobby and sometimes few-branched caudex. Stems solitary or few to several and clustered, erect or loosely ascending from a spreading base (occasionally nearly absent in *P. esculentum*), branched or unbranched, unarmed, variously hairy, sometimes also minutely gland-dotted, the lowest nodes sometimes leafless or with small, scalelike outgrowths. Leaves alternate (sometimes appearing in a basal cluster in *P. esculentum*), palmately 3–5(–7)-foliate, short- to long-petiolate, the petiole more or less jointed at the base. Stipules mostly lanceolate to narrowly elliptic or linear (those at the lowest, often leafless nodes usually longer, broader and partly fused), attached at the base, persistent or shed early; stipels absent. Leaflets elliptic to lanceolate, ovate, or obovate, rarely nearly linear, all those of a leaf more or less similar, mostly angled or slightly tapered at the base, variously shaped at the tip, the margins entire, hairy, the surfaces variously hairy, the upper surface sometimes also gland-dotted, the venation pinnate (but sometimes mostly obscured by the pubescence). Inflorescences axillary, dense spikes or spikelike racemes (sometimes reduced to dense clusters in *P. tenuiflorum*), relatively short, but usually becoming more elongate with age, mostly long-stalked, the bracts lanceolate to ovate, hairy, more or less persistent, the flower stalks absent or 1–2(–4) mm long; bractlets absent. Calyces 2.5–16.0 mm long at flowering, becoming somewhat enlarged at fruiting, the tube bell-shaped, sometimes asymmetrically so (appearing slightly pouched on 1 side), shorter than to longer than the lobes, 5-lobed (the upper pair sometimes fused toward the base), the lobes ascending at flowering, the lowermost lobe somewhat longer than the others. Corollas papilionaceous, usually blue to purple or lavender, rarely white (the keel often darker at the tip), usually fading to cream-colored or tan, the banner with the expanded portion sometimes having contrasting markings toward the base, oblanceolate to obovate, tapered to a stalklike base but with a pair of small basal auricles, rounded to broadly and bluntly pointed at the tip, occasionally shallowly notched, abruptly curved or bent upward at about the midpoint, the wings slightly shorter than to about as long as the banner, the expanded portion angled or curved over the keel, asymmetrically oblong to oblong-oblanceolate, with a small auricle at the base, stalked, the keel shorter than the other petals, fused to the wings toward the base, short-stalked, boat-shaped, oblong to oblanceolate, only slightly curved upward, rounded at the tip. Stamens 10, 9 of the filaments fused at flowering and 1 of the filaments free above the midpoint, the free portion of alternating longer and shorter filaments, the anthers in 2 alternating series, those of the longer filaments attached at the base and those of the shorter series attached toward the midpoint, all yellow. Ovary short, ellipsoid to ovoid, very short-stalked, hairy toward or at the tip, the style slender, often strongly curved, hairy toward the base, sometimes also with an inconspicuous ring of short hairs at the tip, the stigma small and terminal. Fruits modified legumes, elliptic to oblong or oblong-oblanceolate, flattened, sessile or nearly so, short-tapered to a beaked tip, the surfaces leathery or papery, glabrous or hairy, sometimes gland-dotted, more or less smooth (not wrinkled), indehiscent or dehiscing irregularly transversely below the midpoint, 1-seeded. Seeds oblong-elliptic to somewhat kidney-shaped in outline, somewhat flattened, olive green to reddish brown, sometimes with darker streaks or mottling, smooth, somewhat shiny. Twenty-three species, North America.

Pediomelum often has been included in *Psoralea* L., along with *Orbexilum* and *Psoralidium*, but Grimes (1990) recognized it as separate genus based on the palmately compound leaves, persistent bracts, pouched calyces, and the unusual dehiscence pattern of the fruits via a transverse rupture. The present treatment also includes *P. tenuiflorum*, a species placed by Grimes in another segregate genus, *Psoralidium,* because of its indehiscent fruits. For more discussion on this slightly expanded circumscription of *Pediomelum*, see the treatment of *P. tenuiflorum* below.

1. Leaf blades with the upper surface not gland-dotted; bracts 9–15 mm long; stems spreading-hairy; corollas 14–18 mm long; rootstock consisting of a taproot that is tuberous-thickened (turnip-shaped) toward the tip (often also with a short, few-branched caudex above the thickened portion) . . . . 2. P. ESCULENTUM
1. Leaf blades with the upper surface gland-dotted (but the glands sometimes obscured by dense hairs); stems appressed-hairy (sometimes also with scattered, longer, spreading hairs in *P. tenuiflorum*); bracts 2–9 mm long; corollas 5–11 mm long; rootstock consisting of a stout taproot below a thickened, often short-branched caudex
    2. Stems and undersurface of the leaflets densely pubescent with appressed, white to silvery, silky hairs; calyces 4–6 mm long at flowering, becoming enlarged to 7–10 mm at fruiting; fruits 5–8 mm long, not exserted from the calyx, woolly, lacking gland-dots, dehiscing irregularly transversely . . . . 1. P. ARGOPHYLLUM
    2. Stems and undersurface of the leaflets moderately to densely pubescent with appressed to upward-curved, white to off-white, dull hairs (sometimes sparsely hairy toward the base or with scattered, longer, spreading hairs); calyces 2.5–4.0 mm long at flowering, not becoming noticeably enlarged at fruiting; fruits glabrous, densely gland-dotted, indehiscent (shed along with the calyx and flower stalk) . . . . 3. P. TENUIFLORUM

1. Pediomelum argophyllum (Pursh) J.W. Grimes (silvery psoralea, silverleaf scurf-pea)

Psoralea argophylla Pursh

Psoralidium argophyllum (Pursh) Rydb.

Pl. 404 d–f; Map 1789

Rootstock consisting of a stout taproot below a thickened, often short-branched caudex, not tuberous-thickened. Stems 1–3(4), but sometimes appearing loosely colonial from root sprouts, 40–90 cm long, erect, well-branched above the midpoint, densely pubescent with appressed, white to silvery, silky hairs, also gland-dotted (but the glands somewhat obscured by the hairs), the lower few to several nodes often leafless (with the stipules fused). Leaves alternate, palmately 3–5(7)-foliate (3-foliate leaves mostly produced on branches), the petiole 0.5–5.0 cm long. Stipules of well-developed leaves 6–10 mm long, linear, densely hairy and gland-dotted. Leaflets 1.0–3.0(–4.5) cm long, narrowly elliptic to oblanceolate, obovate, or narrowly oblong-obovate, bluntly to sharply pointed at the tip, occasionally rounded or shallowly notched, the upper surface sparsely to densely appressed-hairy, also gland-dotted (but the glands often obscured by the hairs), the undersurface moderately to more commonly densely appressed-hairy. Inflorescences spikes with 2–5 usually well-separated clusters,

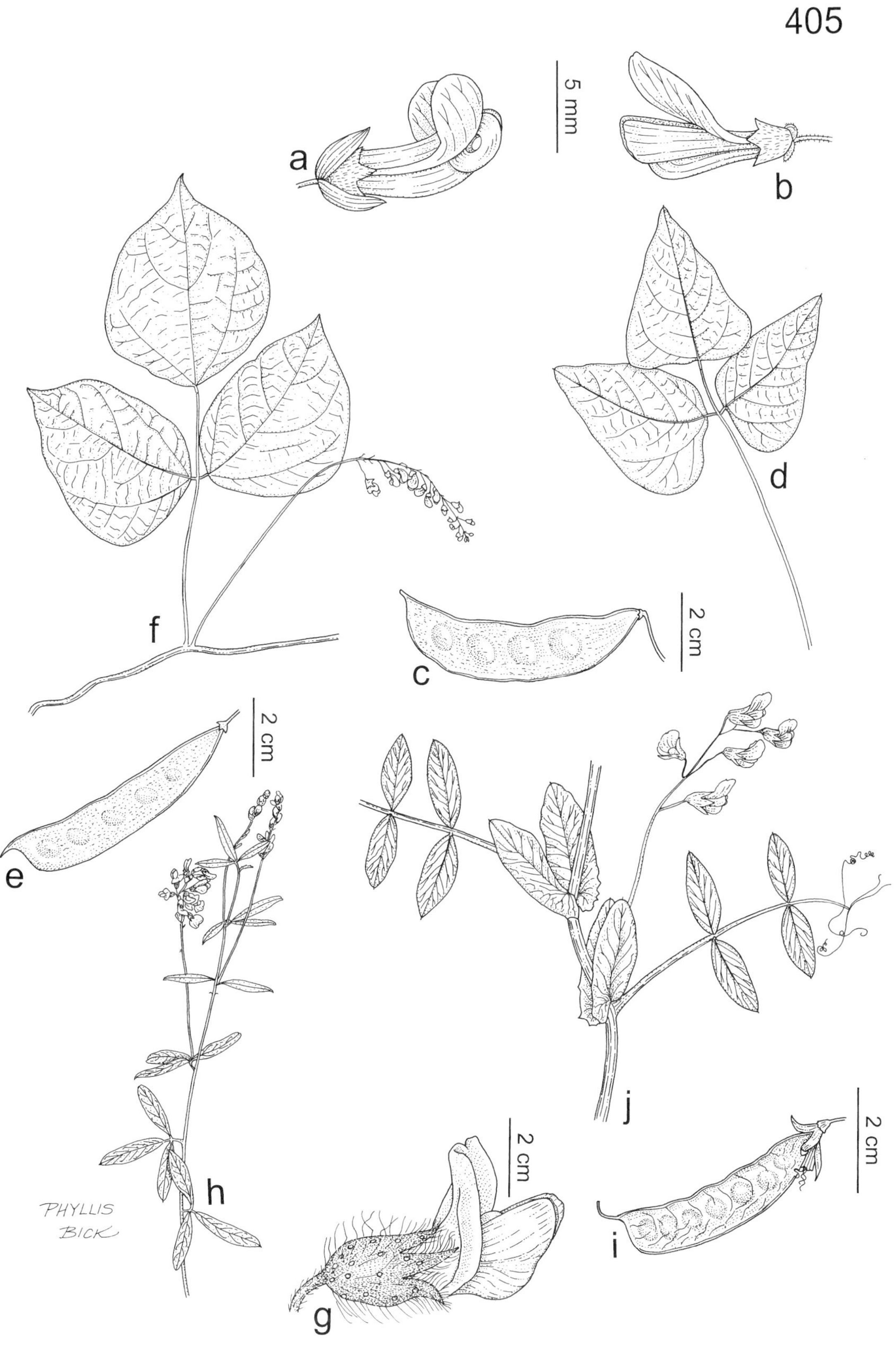

Plate 405. Fabaceae. *Phaseolus vulgaris*, **a)** flower. *Phaseolus lunatus*, **b)** flower, **c)** fruit, **d)** leaf. *Phaseolus polystachios*, **e)** fruit, **f)** fertile stem. *Pediomelum tenuiflorum*, **g)** flower, **h)** fertile stem. *Pisum sativum*, **i)** fruit, **j)** fertile stem.

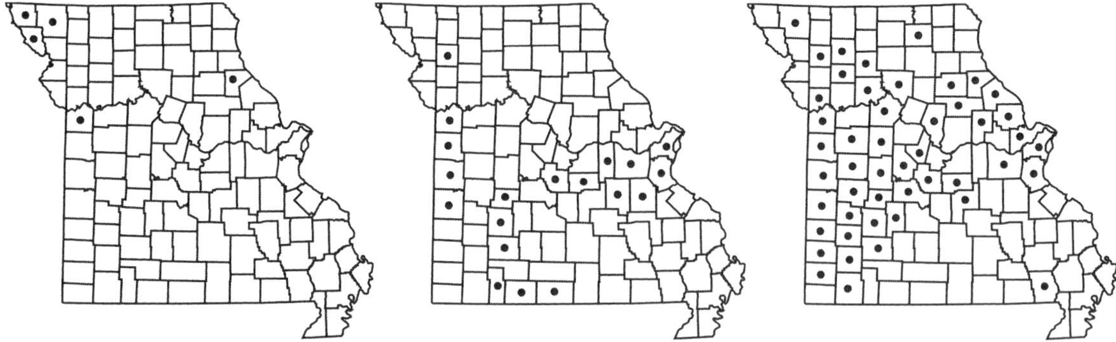

1789. Pediomelum argophyllum 1790. Pediomelum esculentum 1791. Pediomelum tenuiflorum

4–8 cm long and 1.0–1.5 cm wide at flowering, elongating slightly (if at all) with age, the stalk 3–7 cm long, the bracts 3–9 mm long, linear to broadly lanceolate, tapered at the tip, hairy, the flowers sessile or nearly so. Calyces uniformly woolly-hairy, 4–6 mm long at flowering (becoming enlarged to 7–10 mm at fruiting), the tube 2–3 mm long, somewhat pouched at the base, the upper lobes 1.0–1.5 mm long, the lowermost lobe 3–4 mm long. Corollas 7–11 mm long, usually deep blue to purplish blue, the banner sometimes with a white spot toward the base. Filaments with the fused portion 4–5 mm long, the free portion 0.5–1.0 mm long. Fruits with the body 5–8 mm long, 3–4 mm wide, oblong to oblanceolate in outline, woolly-hairy (sometimes less densely so toward the base), not gland-dotted, tapered abruptly to a short, straight, hairy beak 3–5 mm long, the surfaces somewhat leathery, dehiscing irregularly transversely. Seeds 4–5 mm long. $2n$=22. May–July.

Uncommon in northwesternmost Missouri and disjunct in Ralls County; introduced in Jackson County (Montana to New Mexico east to Wisconsin, Missouri, and Texas; Canada; introduced in Illinois). Loess hill prairies, bottomland prairies, and tops of bluffs; also railroads.

Pediomelum argophyllum is distinguished by the silvery-hairy, branched stems, palmately 3–5-foliate leaves, small bluish flowers, and densely hairy fruits. The species is variable in the density of the pubescence. McGregor (1986) noted that although the species has sometimes been described in the botanical literature as producing rhizomes, in his excavations new sprouts appeared exclusively from adventitious buds along the lateral roots.

2. Pediomelum esculentum (Pursh) Rydb. (prairie turnip, Indian breadroot, wild potato)

Psoralea esculenta Pursh

Pl. 404 j–m; Map 1790

Rootstock consisting of a taproot that is tuberous-thickened (turnip-shaped, 4–14 cm long, 2.5–8.0 cm wide) toward the tip, often with a short, branched caudex above the thickened portion. Stems 1–3, 5–30 cm long or rarely reduced and appearing nearly absent, when present erect, usually unbranched, densely pubescent with long (2.0–2.5 mm), fine, more or less spreading hairs, lacking gland-dots, the lower few to several nodes of well-developed stems often leafless (with the stipules fused). Leaves alternate, sometimes appearing in a basal cluster, mostly palmately 5-foliate, the upper leaves sometimes 3(4)-foliate, the petiole 2–15 cm long. Stipules 12–18 mm long, broadly lanceolate, the uppermost narrowly elliptic to narrowly lanceolate, sparsely hairy. Leaflets 2–5 cm long, narrowly ovate to elliptic, oblong-elliptic, or elliptic-oblanceolate, mostly sharply pointed at the tip, the upper surface glabrous except sometimes along the midvein, lacking gland-dots, the undersurface densely pubescent with long, loosely appressed hairs. Inflorescences dense spikelike racemes, 2–6 cm long and 0.8–2.5 cm wide at flowering, elongating slightly with age, the stalk (5–)7–11(–14) cm long, the bracts 9–15 mm long, broadly lanceolate to ovate, tapered at the tip, hairy, the flower stalks 1–3 mm long. Calyces uniformly long-hairy, the tube 4–5 mm long, somewhat pouched at the base, the upper lobes 4–7 mm long, the lowermost lobe 7–10 mm long. Corollas 14–18 mm long, lavender to purple or light bluish purple, the banner sometimes with a darker midnerve and the keel dark purple at the tip. Filaments with the fused portion 10–11 mm long, the free portion 1–3 mm long. Fruits slightly exserted from the calyces, the body 4–6 mm long, 2.5–4.0 mm wide, oblong in outline, glabrous below the tip, not gland-dotted, tapered abruptly to a slender, curved, hairy beak 10–15 mm long, the surfaces papery, dehiscing irregularly transversely. Seeds 4–5 mm long. $2n$=22. April–July.

Uncommon to scattered, mostly in the southwestern and east-central portions of the state (Montana to New Mexico east to Wisconsin and Louisiana; Canada; introduced in New York). Glades, upland prairies, and tops of bluffs; also rarely roadsides and disturbed, rocky areas.

Pediomelum esculentum is distinguished by its densely hairy stems, lack of gland-dots, large stipules, palmately compound leaves, large flowers with pouched calyces, and papery, beaked fruits. The substantial, tuberous, starchy root is highly nutritious and was an important food of the Native American tribes of the Great Plains. It was also popular with early French explorers, who called it *pomme de prairie* (prairie apple). Native Americans also used the roots ceremonially (Moerman, 1998).

3. Pediomelum tenuiflorum (Pursh) A.N. Egan (gray scurf pea)

Psoralea floribunda Nutt. ex Torr. & A. Gray

P. tenuiflora Pursh

P. tenuiflora var. *floribunda* (Nutt. ex Torr. & A. Gray) Rydb.

Psoralidium tenuiflorum (Pursh) Rydb.

Pl. 405 g, h; Map 1791

Rootstock consisting of a stout taproot below a thickened, often relatively long-branched caudex, not tuberous-thickened. Stems 1–5, 40–100 cm long, erect or loosely ascending, sometimes from a spreading base, well-branched above the base, moderately to densely pubescent with short, appressed to upward-curved, white to off-white, dull hairs, sometimes only sparsely hairy toward the base or with scattered, longer, spreading hairs, sometimes sparsely gland-dotted, the lower few nodes often leafless (with the stipules fused). Leaves alternate, palmately 3–5-foliate (the uppermost leaves occasionally appearing simple), the petiole 0.2–3.0 cm long, the uppermost leaves sometimes nearly sessile. Stipules of well-developed leaves 2–5 mm long, linear, hairy, and gland-dotted. Leaflets 1.5–4.0 cm long, elliptic to oblanceolate, rarely obovate or nearly linear, rounded to bluntly pointed or shallowly notched at the tip, occasionally a few of the leaflets sharply pointed, the upper surface sparsely and minutely appressed-hairy to nearly glabrous, also densely gland-dotted, the undersurface moderately appressed-hairy and sparsely to moderately gland-dotted. Inflorescences spikelike racemes with 2–4(–6) often relatively crowded clusters, sometimes reduced to a solitary cluster, 1.5–6.0 cm long and 0.8–1.2 cm wide at flowering, elongating to 3–10 cm with age, the stalk 2–9 cm long, the bracts 2–3 mm long, lanceolate to ovate, angled or tapered at the tip, hairy and usually also gland-dotted, the flower stalks 1.5–3.0(–6.0) mm long. Calyces uniformly woolly-hairy, 2.5–4.0 mm long at flowering (not becoming enlarged at fruiting), the tube 1–2 mm long, not pouched at the base, the upper lobes 0.5–1.0 mm long, the lowermost lobe 1.5–2.0 mm long. Corollas 5–7 mm long, usually bluish purple to purple, rarely white, the banner sometimes with a pair of white spots toward the base and lighter purple on the outer surface. Filaments with the fused portion 2.5–3.5 mm long, the free portion 0.5–1.0 mm long, the filaments somewhat broadened. Fruits with the body 5–7 mm long, 3–4 mm wide, elliptic in outline, glabrous, densely gland-dotted, tapered to a short, straight, triangular beak 1.5–2.0 mm long, the surfaces leathery, indehiscent (shed along with the calyx and flower stalk). Seeds 4–5 mm long. $2n$=22. April–September.

Scattered in the Glaciated Plains and Unglaciated Plains Divisions and uncommon in the Ozark Border; absent from nearly all of the Ozark Division and the Mississippi Lowlands (Montana to Arizona east to Indiana, Kentucky, and Mississippi; Mexico). Glades, upland prairies, banks of streams, rivers, and spring branches, and ledges and tops of bluffs; also railroads and roadsides.

This species traditionally was included in *Psoralea* (Steyermark, 1963), but it and two other species of the central and western United States were transferred to the resurrected genus *Psoralidium* Rydb. by Grimes (1990) in his dismemberment of *Psoralea* into smaller, more natural units. However, subsequent molecular and morphological research (Egan and Crandall, 2008; Egan and Reveal, 2009) revealed that *P. tenuiflorum* is more closely related to the species of *Pediomelum* than it is to the other two species that Grimes had placed into *Psoralidium*. Because *P. tenuiflorum* is the type species of *Psoralidium*, that generic name must be treated nomenclaturally as a synonym of *Pediomelum*. The new name, *Ladeania*, was published by Egan and Reveal (2009) to accommodate the other two former *Psoralidium* species. Species of *Ladeania* are characterized by calyces that do not become enlarged at fruiting, but instead tear longitudinally from the base to a lateral sinus and become broadly flared backward, and by the floral bracts that are shed as the calyces become recurved.

Pediomelum tenuiflorum has been planted for erosion control and in prairie gardens. It is showy when flowering and also is a good honey plant.

Steyermark (1963) noted the existence of a specimen in the Missouri Botanical Garden Herbarium (*Palmer 3367*) collected in 1911 in Jasper County that originally was determined as *Psoralea cuspidata* Pursh. He redetermined the immature plant material on this specimen as *P. tenuiflorum*. The taxon currently is known as *Pediomelum cuspidatum* (Pursh) Rydb. and is distributed to the west of Missouri, from Montana south to Texas. It differs from *P. tenuiflorum* in its dense, spikelike racemes and shorter, narrower stipules. In his taxonomic revision of the psoralioid legumes in the New World, Grimes (1990) cited this specimen as *Psoralidium lanceolatum* (Pursh) Rydb. in the catalog of collections at the end of the treatment, but did not include Missouri in the range that he presented for that taxon and also annotated the specimen as *P. tenuiflorum*. *Psoralidium lanceolatum* has since been renamed as *Ladeania lanceolata* (Pursh) A.N. Egan & Reveal (see the paragraph above). It occurs in the western United States and adjacent Canada east to North Dakota, Iowa, and Texas, and differs from *P. tenuiflorum* in its nearly globose fruits, corollas that are nearly white except for the purple keel tip, and calyces that become torn and reflexed at maturity. Although the main range of *L. lanceolata* is to the west of Missouri, it approaches the state border at a few sites in southwestern Iowa and eastern counties of Nebraska and Kansas. Missouri botanists may discover this species in northwestern counties in the future.

41. **Phaseolus** L.

Plants annual or perennial herbs, taprooted or the rootstock tuberous, sometimes climbing on other vegetation. Stems trailing or climbing, often twining, branched or unbranched, unarmed, glabrous or sparsely to densely pubescent with short, spreading to downward-curved or -angled hairs, some or all of the hairs usually minutely hooked at the tip. Leaves pinnately trifoliate, mostly long-petiolate, the petiole glabrous or hairy. Stipules 1.5–3.0 mm long, oblong-triangular to narrowly oblong-ovate, herbaceous, appearing basally attached, reflexed or spreading, persistent, strongly parallel-veined; stipels 0.8–2.5 mm long, persistent. Leaflets ovate to broadly ovate, the lateral ones usually asymmetrically so, the terminal leaflet sometimes somewhat rhombic, broadly rounded or broadly angled (obliquely so in lateral leaflets) at the base, tapered to a sharply pointed tip, unlobed (lobed elsewhere), the margins entire, the surfaces usually hairy. Inflorescences axillary, racemes or few-branched panicles, well-developed, with several to numerous flowers, and extending past the leaves or short, few-flowered, and appearing as clusters, the flower stalks with small bracts at the base, these 1–3 mm long, lanceolate to ovate or triangular, often with strong parallel venation, more or less persistent; each flower also closely subtended by 2 small bractlets, these 0.7–4.0 mm long, oblong to lanceolate or broadly ovate, persistent. Calyces with the tube broadly bell-shaped, more or less 2-lipped, the upper 2 lobes short and broad, fused with only a shallow notch at the tip, the lower 3 lobes triangular, unequal, the lowermost lobe longer than the others. Corollas papilionaceous, pink to purple, greenish white, or white (if colored, the petals often paler toward the base), the banner and wings cordate or abruptly tapered to an often relatively broad stalklike base, the banner with the expanded portion broadly ovate to nearly circular, usually shallowly notched at the tip, abruptly arched upward above the midpoint but generally appearing somewhat hooded over other petals, the wings oblong, usually longer than the banner and keel, the keel slender, spirally coiled. Stamens 10, 9 of the filaments fused to above the midpoint and 1 filament more or less free, the anthers small, attached near the base, all similar in size. Ovary sessile or nearly so, with a small nectary disc encircling the base, this sometimes somewhat lobed, the style coiled, often jointed toward the base, bearded on the upper side, the stigma more or less lateral or oblique, elongate. Fruits legumes, oblong or linear, flattened or not, straight or curved, at least along the lower margin, sessile or nearly so, asymmetrically short-tapered to a beak, the margins otherwise often parallel, glabrous or minutely hairy (at least when young), dehiscent by 2 valves, these papery to leathery in texture, becoming spirally twisted during dehiscence, with 2 to numerous seeds. Seeds kidney-shaped to broadly oblong in outline, flattened, the surface variously colored, often mottled, smooth. Sixty to 65 species, North America, Central America, South America, cultivated nearly worldwide.

Phaseolus once was thought to contain about 500 species, but it has been reorganized taxonomically, with many of the species transferred to other genera, such as *Vigna* (Lackey, 1981, 1983). The genus contains five main domesticated species of varying historical and modern economic importance as food plants: *P. acutifolius* A. Gray (tepary bean), *P. coccineus* L. (runner bean), *P. lunatus* L. (Lima bean), *P. polyanthus* Greenm. (year bean), and *P. vulgaris* L. (common bean). Gathering of wild beans for food has been documented as far back as 10,000 years ago and actual domestication may have occurred as long as 7,000 years ago (Kaplan and Kaplan, 1988). Bean domestication paralleled that of maize (*Zea mays* L., Poaceae). These two foods are complimentary in that individually their proteins are deficient in a few important amino acids, but in tandem the combination of amino acids in the two groups allows the human body to build complete proteins.

1. Stems twining and climbing; inflorescences slender racemes, often branched at the base (thus appearing paniculate); fruits 3–6 cm long; plants perennial, with tuberous roots . . . . 2. P. POLYSTACHIOS
1. Stems erect and bushy, at least toward base, the upper portions sometimes weakly twining; fruits 5–20 cm long; inflorescences stout racemes, well-developed or few-flowered and appearing as clusters, unbranched; plants annual, with taproots
    2. Inflorescences extending beyond the subtending leaves, with mostly 10–20 flowers; bractlets at base of calyx 1–2 mm long, inconspicuous; fruits (3–)5–9(–12) cm long, (10–)14–20 mm wide, oblong to broadly oblong, usually noticeably curved, flattened . . . . 1. P. LUNATUS
    2. Inflorescences shorter than the subtending leaves, with 2–4 flowers clustered toward the tip; bractlets at base of calyx 4–5 mm long, broadly elliptic to broadly ovate, conspicuous; fruits 5–20 cm long, 7–10 mm wide, linear to narrowly oblong, straight or only slightly curved, not or only slightly flattened . . . . 3. P. VULGARIS

1. Phaseolus lunatus L. (Lima bean, butter bean)

Pl. 405 b–d; Map 1792

Plants annual (perennial elsewhere), with a stout, sometimes branched taproot. Stems 20–60 cm long (to 300 cm or more elsewhere), erect to loosely ascending from an erect base, bushy, sometimes weakly twining toward the tip (more strongly twining and climbing elsewhere), glabrous or sparsely to moderately pubescent with short, spreading to downward-angled hairs, some or all of these sometimes hooked. Petioles 3–6(–15) cm long, glabrous or more commonly moderately to densely short-hairy, the terminal leaflet stalk 10–35 mm long. Leaflets 3–12 cm long, 2–9 cm wide, the upper surface glabrous or nearly so at maturity, the undersurface glabrous or sparsely short-hairy along the main veins. Inflorescences relatively stout, well-developed racemes, unbranched, extending beyond the subtending leaves, 5–20 (–35) cm long, with mostly 10–20 flowers, the flower stalks 4–10 mm long. Bractlets 1–2 mm long, narrowly oblong to oblong-lanceolate. Calyces usually moderately to densely and finely short-hairy on the outer surface, densely short-hairy on the inner surface and margins, the tube 1.5–2.5 mm long, the upper lip 0.5–0.8 mm long, the lobes rounded to broadly and bluntly pointed at the tips, the lowermost lobe 0.6–1.0 mm long, triangular to broadly triangular. Corollas white (pink to light purple elsewhere), the banner often greenish-tinged toward the tip on the inner surface and green on the outer surface, 5–7 mm long, 6–7 mm wide, the wings 10–12 mm long, 4–7 mm wide, the keel usually green toward the tip, tightly coiled. Fruits (3–)5–9(–12) cm long, 14–20 mm wide, flattened, oblong to broadly oblong, noticeably curved, sessile or nearly so, short-beaked at the tip, the

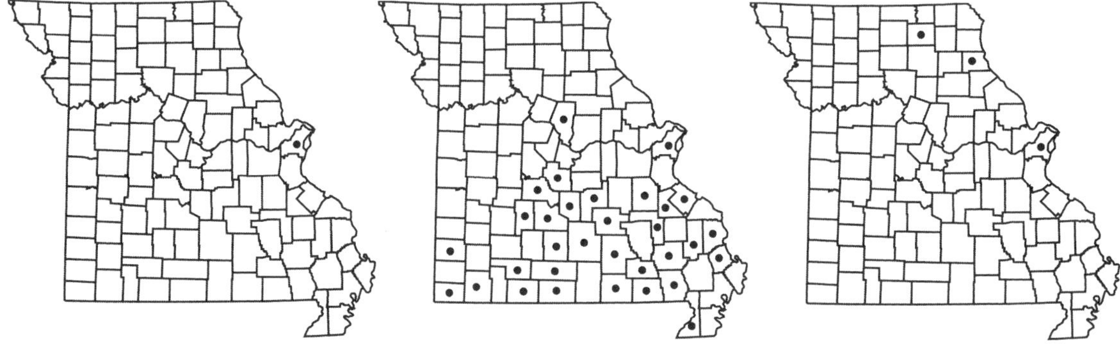

1792. Phaseolus lunatus 1793. Phaseolus polystachios 1794. Phaseolus vulgaris

valves relatively thick, pale cream-colored to pale greenish-tinged, densely and minutely hairy when young, but becoming glabrous at maturity, 2–4 (–6)-seeded. Seeds 8–11 mm long, 6–8 mm wide, oblong-elliptic to broadly kidney-shaped in outline, usually white in cultivated races (brown or with reddish streaks or mottling elsewhere). $2n=22$. June–September.

Introduced, uncommon, known thus far only from a single specimen from the city of St. Louis (native of Mexico, Central America, South America; introduced sporadically in the southeastern U.S. west to Missouri). Railroads.

Phaseolus lunatus is recognized by the usually bushy habit, 3-veined bractlets, short inflorescences, and crescent-shaped legumes, with white seeds. Lima beans apparently were domesticated independently in southern Mexico/Central America and Andean South America (Baudoin, 1988; Gutiérrez Salgado et al., 1995), where wild forms still exist. They spread from these areas and are now found throughout the humid tropics, where they are now a major legume crop. Many cultivars and forms have been selected. Pole types (with twining, climbing stems) are perennial, whereas bush types have a dwarfed habit and are grown as annuals. Steyermark (1963) noted that pole types are generally cultivated in tropical regions. The bush types are commonly grown in Missouri gardens. Steyermark (1963) called the Missouri specimen *P. lunatus* var. *lunonanus* L.H. Bailey, but that epithet refers to a cultivar rather than a variety (L. H. Bailey, 1923). The original species description was based on cultivated plants. Some authors have segregated the wild populations of the species into one or two additional varieties (Delgado-Salinas, 1985), but this treatment is still controversial (Freytag and Debouck, 2002). Lima beans probably continue to escape from Missouri gardens but do not persist for more than a season or two.

The raw beans and herbage contain varying amounts of linamarin, a cyanogenic glucoside whose breakdown results in the generation of toxic hydrocyanic acid, less so in most of the cultivated races. Cooking renders the beans and seeds edible by leaching away the toxin.

2. Phaseolus polystachios (L.) Britton, Sterns & Poggenb. **var. polystachios** (wild bean)
Dolichos polystachios L.

Pl. 405 e, f; Map 1793

Plants perennial, from a stout branched, tuberous rootstock. Stems 100–400(–600) cm long, twining, climbing or occasionally trailing, often purplish-tinged, usually moderately to densely pubescent with short, spreading hooked hairs, sometimes also with scattered, longer, spreading to downward-angled, straight hairs. Petioles 3–8 cm long, moderately to densely hairy, the terminal leaflet stalk 7–35 mm long. Leaflets 3–8(–10) cm long, 2–7 cm wide, the upper surface sparsely to moderately short-hairy along the main veins, sometimes nearly glabrous, the undersurface sparsely to moderately short-hairy, mostly along the veins. Inflorescences well-developed racemes, often branched at the base (thus appearing paniculate), extending beyond the subtending leaves, 5–20(–35) cm long, with 10 to numerous flowers, the flower stalks 4–8 mm long. Bractlets 0.7–1.2 mm long, oblong to oblong-lanceolate or narrowly oblong-ovate. Calyces sparsely to moderately and finely short-hairy on the outer surface, densely short-hairy on the inner surface and margins, the tube 1.6–2.2 mm long, the upper lip 0.3–0.8 mm long, the lobes rounded to broadly and bluntly pointed at the tips, the lowermost lobe 0.8–1.2 mm long, triangular. Corollas pink to light purple, the petals usually pale toward the base, rarely entirely white, the banner sometimes greenish-tinged on the outer surface, 6–9 mm long, 8–11 mm wide, the wings 10–13 mm long, 4–7 mm wide, the keel usually greenish yellow to greenish white toward the tip, tightly coiled. Fruits 3–6 cm long, 7–10 mm wide, flattened, linear to narrowly oblong, usually slightly curved, with a short stalk to 3 mm long, short-beaked at the tip, the valves relatively thin, tan, glabrous, 4–6-seeded. Seeds 6–10 mm long, 5–8 mm wide, oblong to broadly kidney-shaped in outline, dark reddish brown mottled with black or solid black. $2n=22$. July–September.

Scattered in the Ozark and Ozark Border Divisions, uncommon in the Mississippi Lowlands (eastern U.S. west to Missouri and Texas). Banks of streams and rivers, bases and ledges of bluffs, openings of bottomland and mesic upland forests; also roadsides.

Some authors have recognized one or two infraspecific taxa in addition to the widespread nominate one present in Missouri. Freytag and DeBouck (2002) accepted three subspecies, with the other two restricted to the southeastern states. Both ssp. *sinuatus* (Nutt. ex Torr. & A. Gray) Freytag and ssp. *smilacifolius* (Pollard) Freytag have lobed leaflets. Isely (1990) speculated that ssp. *smilacifolius*, which is known from a single specimen from northern Florida, may represent a hybrid between *P. polystachios* and a related species in the area, *P. sinuatus* Nutt. ex Torr. & A. Gray.

3. Phaseolus vulgaris L. **var. vulgaris** (bush bean, string bean)

Pl. 405 a; Map 1794

Plants annual (perennial elsewhere), with a taproot. Stems either 20–40 cm long, erect to loosely ascending from an erect base, bushy, sometimes weakly twining toward the tip, or 100–300 cm long or more, twining and climbing, both types sparsely to densely pubescent with short, spreading to downward-angled hairs, some or all of these sometimes hooked. Petioles 3–6(–12) cm long, moderately to densely short-hairy, the terminal leaflet stalk 10–25 mm long. Leaflets 3–5(–12) cm long, 2–5(–12) cm wide, the upper surface glabrous or nearly so at maturity, the undersurface glabrous or sparsely to moderately short-hairy along the main veins. Inflorescences relatively stout, poorly developed racemes, usually appearing as a cluster of 2–4 flowers (more elongate and with more flowers elsewhere), unbranched, shorter than the subtending leaves, 4–9(–20) cm long, the flower stalks 4–10 mm long. Bractlets 4–5 mm long, broadly elliptic to broadly ovate. Calyces sparsely to moderately or densely and finely short-hairy on the outer surface, densely short-hairy on the inner surface and usually also the margins, the tube 1.5–2.5 mm long, the upper lip 0.4–0.7 mm long, the lobes rounded to broadly and bluntly pointed at the tips, the lowermost lobe 0.6–0.9 mm long, triangular to broadly triangular. Corollas white to pink or light purple, the banner often greenish-tinged (especially toward the base) on the outer surface, 5–7 mm long, 6–7 mm wide, the wings 10–15 mm long, 4–8 mm wide, the keel usually green toward the tip, tightly coiled. Fruits 5–20 cm long, 7–10 mm wide, not or only slightly flattened, linear to narrowly oblong, straight or only slightly curved, sessile or nearly so, short-beaked at the tip, the valves relatively thick, pale cream-colored to pale greenish-tinged or tan, glabrous or sparsely to densely and minutely hairy, at least when young, (3)4–8(–12)-seeded. Seeds 6–10 mm long, 4–7 mm wide, oblong-ellipsoid to broadly kidney-shaped in outline, variously white to tan, yellowish brown, reddish brown, brown, or black, often with darker mottling. $2n=22$. June–October.

Introduced, uncommon, sporadic (native of Mexico, Central America, South America; introduced sporadically in the U.S.). Fallow fields, gardens, railroads, and roadsides.

Phaseolus vulgaris is one of the oldest and most important legume crops. It appears that the species was brought into domestication one or more times in Andean South America (Kaplan and Kaplan, 1988) and independently in Mesoamerica (Koenig and Gepts, 1989; Chacón S. et al., 2005), although some archaeologists still believe that the Mexican and Central American races originated through diffusion from the Andean crop plants. Hundreds of cultivars are known, including navy beans, kidney beans, pinto beans, black beans, and string beans. It is a polymorphic species with considerable variation in habit, size of plant, fruit and seeds, and flower color. The flowers of bush beans (the relatively nontwining races) are bunched together near the tip of the relatively stout racemes, but in pole beans (the twining races) the inflorescences are less congested. The corolla of the cultivated bush beans is usually white, and the flowers are self pollinating, but many races of pole beans have flowers with pink corollas. Unlike other *Phaseolus* species in Missouri, the legumes are nearly round in cross-section. Both climbing pole varieties and dwarf bush types are commonly cultivated in Missouri gardens, but only the bush type has been collected in waste areas. Steyermark (1963) called the low, nonclimbing, bushy kind of bean by the name var. *humilis* Alef., but Delgado-Salinas (1985) pointed out that many of the older varietal names were never validly published or described. He treated all cultivated and escaped forms as *P. vulgaris* var. *vulgaris*, and accepted only two wild varieties, one in Central America and Mexico and the other in South America.

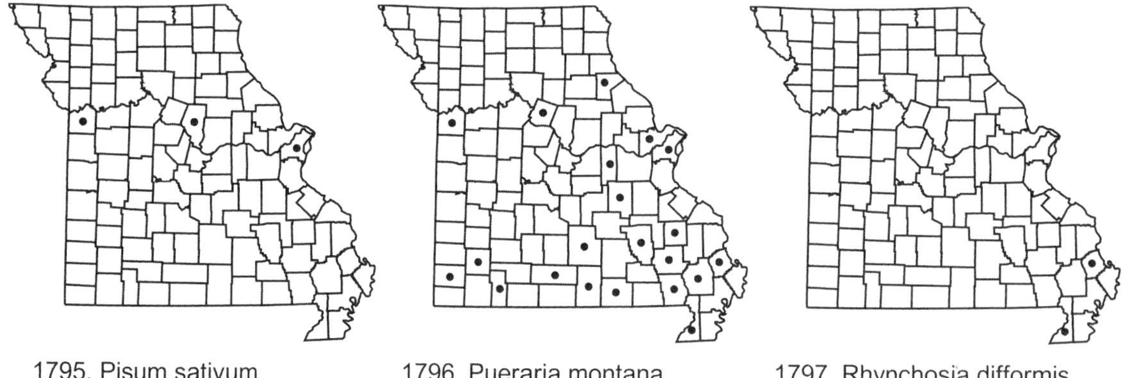

1795. Pisum sativum 1796. Pueraria montana 1797. Rhynchosia difformis

42. Pisum L.

About 5 species, Europe, Asia, Africa, introduced in the New World.

1. Pisum sativum L. (garden pea)
P. sativum var. *arvense* (L.) Poir.
P. arvense L.

Pl. 405 i, j; Map 1795

Plants annual, with taproots. Stems 15–60 (–150) cm long, spreading or climbing on other vegetation, usually branched, unarmed, glabrous, often somewhat glaucous. Leaves alternate, even-pinnately compound with (2)4 or 6 leaflets, the petiole 2–3 cm long, the rachis extended into a conspicuous, pinnately branched tendril. Stipules 28–45(–80) mm long, often larger than the leaflets, leaflike, asymmetrically ovate with a large basal auricle partially encircling the stem (those of adjacent stipules overlapping), the margins entire or toothed, persistent, the surfaces glabrous, often somewhat glaucous; stipels absent. Leaflets 2–4 cm long, 1–2 cm wide, elliptic to broadly elliptic, folded longitudinally during development, angled at the base, rounded or angled to a bluntly or broadly pointed tip, the midvein extended into a short, sharp point at the very tip, the margins entire or with small teeth, the surfaces glabrous, often somewhat glaucous, the venation pinnate. Inflorescences axillary, of solitary flowers or loose clusters or racemes of 2 or 3(4) flowers, ascending, the stalk 2–3 cm long, the bracts minute, shed early, each flower with a stalk 3–15 mm long; bractlets absent. Calyces glabrous, the tube 3–4 mm long, bell-shaped, the lobes 4–6 mm long, all similar, lanceolate, sharply pointed at their tips. Corollas papilionaceous, white or less commonly pink to purple (the wings and keel then often conspicuously darker than the banner), sometimes with green or purple nerves, the banner 12–18 mm long, 12–18 mm wide, the expanded portion abruptly curved or bent backward, broadly obovate to nearly circular, broadly notched at the tip, the wings 12–16 mm long, 6–9 mm wide, oblong, curved over the keel, the keel 10–13 mm long, 4–5 mm wide, boat-shaped, fused below the midpoint, curved upward abruptly and tapered to a slender, sharply pointed tip. Stamens 10, equal in length, 9 of the filaments fused and 1 free, the fused portion 6–7 mm long, the free portion 4–6 mm long, some or all of the filaments slightly broadened and densely hairy toward the tip, the anthers small, attached at the base, orange. Ovary 7–8 mm long, glabrous, the style 6–7 mm long, bent sharply at the base, curved toward the tip, flattened, with a fine, longitudinal groove, densely hairy on the inner side, the stigma small and subterminal. Fruits legumes, 5–10 cm long, 15–20 mm wide, obliquely oblong-ellipsoid, tapered to a short, stout beak, not or only slightly flattened, straight, indehiscent or dehiscent by 2 spirally twisting valves, the valves light green at maturity, 2–10-seeded. Seeds 4–7 mm long, 4–7 mm wide, more or less globose, the surface greenish yellow to green, sometimes mottled, smooth or wrinkled, shiny. $2n=14$. May–June.

Introduced, uncommon, sporadic (cultigen of apparently Asian origin; introduced sporadically in the U.S., Canada). Railroads and gardens.

Peas have been grown since prehistoric times in the cooler parts of southwestern Asia, and this was presumably the area of domestication. Remains have been recovered from some of the oldest known village sites. Peas are now a major crop throughout the cool temperate zone, especially in northern Europe and Russia. As with many cultivated species, there is a wide range of variation in *P. sativum* and a complex nomenclatural history (Makasheva, 1984; Smartt, 1990). The species has been divided and classified in several ways, none of which seem very satisfactory or stable. Eight or more varieties have been recognized, but these are better treated among the numerous cultivars. Two major kinds of peas are cultivated, the garden pea (mainly grown for human consumption of the seeds) and the field pea (grown mainly for livestock and soil improvement, but the seeds are also dried and used in soups). Garden pea tends to have relatively large pods and seeds, as well as white corollas, and is sometimes designated as var. *sativum*. Field pea has somewhat smaller pods and seeds, usually pink or purple, often bicolorous flowers, and is sometimes classified as var. *arvense*. Both varieties have been collected in Missouri, however it is unknown if they persist for more than a season or two. Other cultivars include the snow pea, sugar pea, and snap pea, which have relatively large fruits that are harvested prior to maturity and eaten intact. The young foliage also sometimes is used in Chinese cuisine as a component of stir-fried vegetables.

Peas are self-pollinating annuals. Studies of the inheritance of plant size, flower color, and seed wrinkles of peas by Gregor Mendel (1865) became the basis for the modern science of genetics. Today there are numerous pea cultivars developed for various types of garden cultivation, edible pods, mechanical harvesting, and freezing of the seeds and/or fruits.

43. **Pueraria** DC.
(Van der Maesen, 1985, 1988)

About 15 species, Asia to Australia, introduced widely.

1. Pueraria montana (Lour.) Merr. **var. lobata** (Willd.) Maesen & S.M. Almeida ex Sanjappa & Predeep (kudzu)
P. lobata (Willd.) Ohwi
P. thunbergiana (Sieb. & Zucc.) Benth.
Dolichos montana Lour.
Pl. 406 d–f; Map 1796

Plants perennial herbs, often woody at the base, with massive, tuberous roots. Stems to 30 m long, trailing, climbing, and twining, branched, unarmed, relatively stout, densely pubescent with bristly, usually yellowish brown hairs, these often with expanded, dark, bulbous bases. Leaves alternate, pinnately trifoliate, the petiole 8–23 cm long. Stipules 8–16 mm long, lanceolate to oblong-ovate, peltate, with an asymmetric base, angled or tapered to a bluntly or sharply pointed tip, sometimes toothed toward the tip, with conspicuous venation, hairy, mostly shed early; stipels 8–10 mm long, narrowly lanceolate to linear or hairlike, hairy, persistent. Leaflets 8–20 cm long, 5–19 cm wide, ovate to broadly ovate or rhombic, angled to broadly angled, broadly rounded, or nearly truncate at the base, mostly short-tapered to a sharply pointed tip, unlobed or shallowly to moderately 3-lobed, the lateral lobes rounded, the margins otherwise entire, the surfaces (especially the undersurface) appressed-hairy, with 3 main veins from the base, but otherwise more or less pinnately veined. Terminal leaflet with the stalk 20–70 mm long, symmetric at the base; lateral leaflets with the stalk 4–10 mm long, asymmetric at the base. Inflorescences axillary, dense racemes (rarely with a single basal branch), 9–30 cm long, with usually numerous flowers, the stalk 8–25 mm long (often appearing longer if the lowermost flowers fail to set fruit and are shed), the bracts 4–10 mm long, lanceolate to ovate, glabrous or hairy, shed early, each flower with a stalk 2–8 mm long, the bractlets 2–4 mm long, lanceolate to ovate, hairy (at least along the margins), persistent. Calyces appressed-hairy, the tube 3–5 mm long, bell-shaped, 2-lipped, the lobes 4–13 mm long, unequal, the upper 2 lobes fused into an ovate structure, this sometimes split at the tip, the lower lip with the 2 lateral lobes slightly shorter than the upper lip, lanceolate to narrowly ovate, the lowermost lobe much longer than the others, narrowly lanceolate, all of the lobes sharply pointed at their tips. Corollas papilionaceous, reddish purple to purple, sometimes appearing somewhat bicolorous with a lighter banner, the banner 12–20 mm long, 8–14 mm wide, the expanded portion curved upward but appearing somewhat concave (the margins curved inward), broadly ovate, notched at the tip, with a yellow spot toward the base, the wings 9–14 mm long, 3–6 mm wide, obliquely oblong-obovate, fused to the keel toward the base, the free portion cupped around it, the keel 12–17 mm long, 3–5 mm wide, boat-shaped, bent or curved upward, short-tapered to a sharply pointed tip. Stamens 10, in 2 alternating, slightly shorter and longer series, 9 of the filaments fused and 1 free to about the midpoint, the fused portion 10–12 mm long, the free portion 2–4 mm long, curved upward, the anthers small, attached near the base, yellow. Ovary 10–12 mm long, hairy, the style 3–5 mm long, bent upward toward the tip, slender, glabrous, the stigma small, terminal. Fruits legumes, 4–7 cm long, 5–9 mm wide, irregularly oblong (the shorter fruits narrowly oblong-elliptic), tapered to a beak, flattened, expanded over the seeds, straight or somewhat curved, dehiscing from the tip, the margins often irregularly wavy, with a dense fringe of long, reddish brown to yellowish brown hairs, the surfaces green at maturity, turning tan to reddish brown and papery with age, long-hairy, especially over the seeds, 1–7-seeded. Seeds 3–5 mm long, 2–3 mm wide, oblong to somewhat kidney-shaped in outline, the surface somewhat flattened, reddish brown with darker mottling, relatively dull. $2n=22$, 24. August–September.

Introduced, scattered in the southern half of the state, north locally to Jackson, Howard, and Ralls Counties (native from Asia to Australia; introduced widely in the eastern U.S. west to Kansas and Texas, also Oregon, Washington, Hawaii; also Mexico, Central America, South America, Europe). Edges of mesic upland forests and banks of rivers; also old fields, fencerows, cemeteries, quarries, railroads, roadsides, and open, disturbed areas.

Although originally described as a separate species, in recent years kudzu has been treated as a variety of *P. montana* (D. B. Ward, 1998). In his taxonomic revision of the genus, Van der Maesen (1985) accepted three subspecies within *P. montana* (as *P. lobata*), of which var. *lobata* is the only one to have become naturalized in temperate North America. The three taxa differ mainly in details of the inflorescences, flowers, and fruits.

Kudzu has been esteemed for centuries in Japan as a source of food, starch, medicine, fiber for baskets and cloth, paper, and as a soil cover (van der Maesen, 1985). The tubers, leaves, shoots, flowers and young pods are steamed or pickled and eaten as a vegetable (Shurtleff and Aoyagi, 1985). The starch from the roots is widely used as a thickening agent in sauces, soups, confections, and beverages. The roots, starch, and flower buds are used medicinally. The leaves are eaten by livestock. The flowers have a sweet fragrance.

Kudzu was introduced into the U.S. from Japan in 1876 at the Philadelphia Centennial Exposition. Its use was strongly promoted by the federal government, and by the 1930s the species was being planted widely throughout the southeastern states for erosion control and forage for livestock (R. J. Hill, 1985). Awareness of its invasive nature grew slowly, but by the 1950s kudzu was recognized as a significant or noxious weed. It has become a major nuisance in the southeastern United States, covering several million acres. It grows quickly in even the poorest soils, at a rate of up to a foot per day, smothering other vegetation, girdling stems, breaking branches, and uprooting entire trees. It covers fences, railroad tracks, power lines, and even whole buildings. Following the lead of several other states, the Missouri legislature declared kudzu a noxious weed in 2002. The strategy for controlling kudzu is to weaken and eventually kill the rootstalk by repeatedly cutting off all aboveground stems, followed by intense grazing for three to four years or application of large amounts of herbicides. Unfortunately, it is very difficult to completely eradicate kudzu once it becomes established without destroying all of the surrounding vegetation. The roots are extensive and can weigh up to 400 pounds, and are difficult to remove completely.

Harvey (2009) reported the presence of extrafloral nectaries in *P. montana* in populations in Georgia. These nectar-producing glands are located within the base of each flower stalk and are exposed when unpollinated flowers in a given cluster are shed. They serve to attract ants. Harvey noted that unlike the situation in *Apios americana*, in which there is a correlation between the presence of ants and a decrease in herbivory of fruits by insects, the ant species associated with kudzu apparently are not very efficient at protecting the fruits.

44. **Rhynchosia** Lour.
(Grear, 1978)

Plants perennial herbs (shrubs elsewhere), with thick, woody rootstalks. Stems prostrate or trailing to ascending, climbing, or rarely erect, usually twining, ridged or angled, unarmed, densely pubescent with mostly downward-angled or -curved hairs (variously hairy elsewhere).

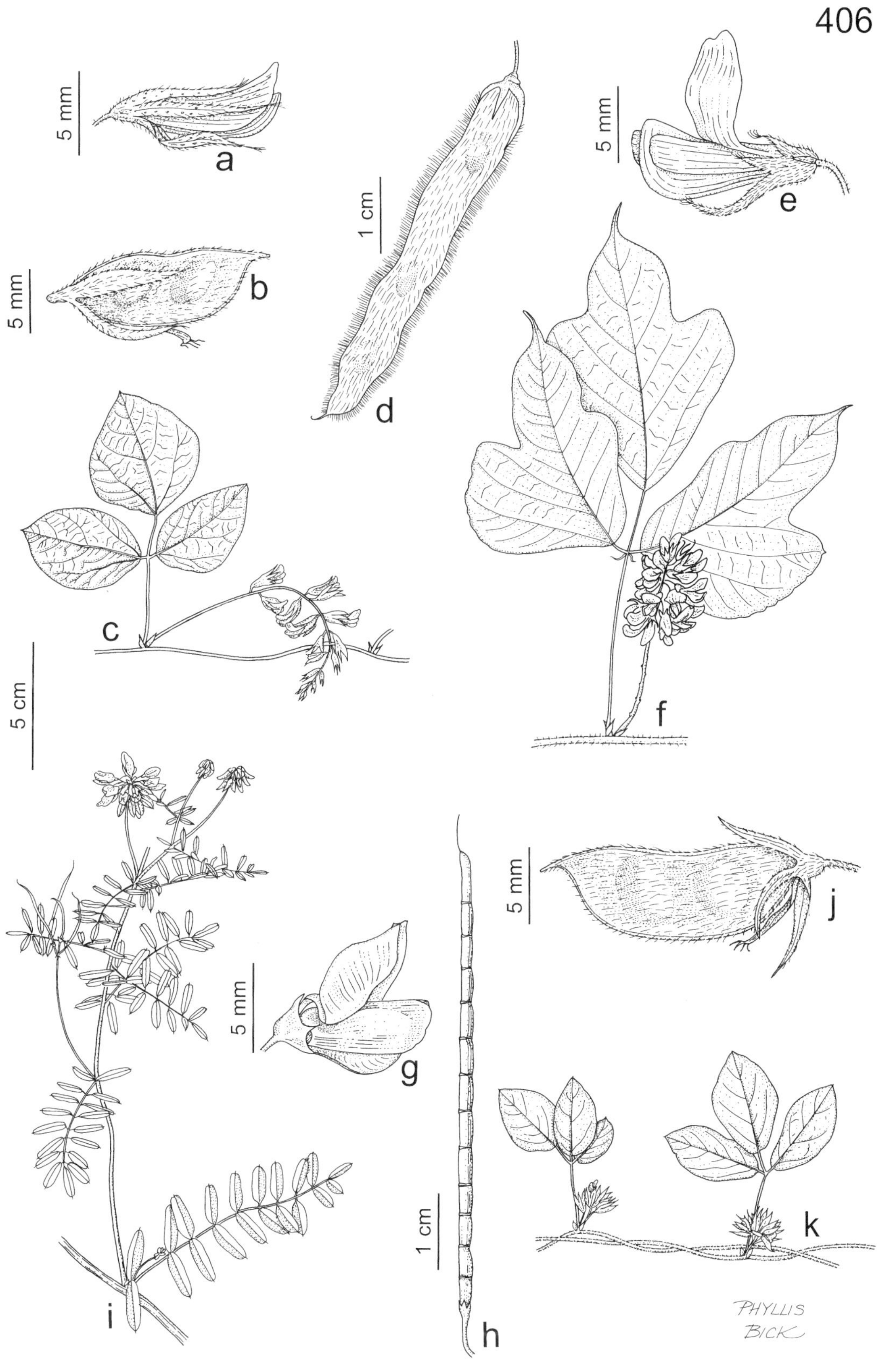

Plate 406. Fabaceae. *Rhynchosia latifolia*, **a)** flower, **b)** fruit, **c)** fertile stem. *Pueraria montana*, **d)** fruit, **e)** flower, **f)** fertile stem. *Securigera varia*, **g)** flower, **h)** fruit, **i)** habit. *Rhynchosia difformis*, **j)** fruit, **k)** habit.

Leaves pinnately trifoliate or the lowermost leaves sometimes lacking lateral leaflets and thus appearing simple, the petioles of the lowermost leaves long, progressively shorter toward the stem tip, the uppermost leaves short-petiolate, the petioles all densely hairy, the terminal leaflet stalk 5–20 mm long. Stipules narrowly lanceolate to ovate, sharply pointed at the tip, mostly shed early; stipels absent (present elsewhere). Leaflets broadly ovate to elliptic, somewhat rhombic, or nearly circular, the lateral leaflets sometimes somewhat asymmetric with an oblique base, otherwise rounded to broadly angled at the base, rounded or more commonly angled to a bluntly or sharply (but broadly) pointed tip, the margins entire, short-hairy, the upper surface moderately to densely pubescent with short, fine, spreading to curved hairs, the undersurface spreading-hairy, mostly along the veins, also dotted with minute, yellow to orange, more or less globose resin glands, the venation pinnate but with 3 main veins from the leaflet base, raised on the undersurface. Inflorescences axillary, short to elongate racemes or small clusters, the stalk mostly short, the bracts 3–7 mm long, linear to narrowly lanceolate, mostly shed early; bractlets absent. Calyces moderately to densely pubescent with fine, ascending hairs, sometimes mostly along the nerves and margins, also dotted with minute, yellow to orange, more or less globose resin glands, often becoming slightly enlarged at fruiting, the tube bell-shaped (cylindric elsewhere), shorter than the lobes, more or less 2-lipped, the upper 2 lobes fused to at or above the midpoint, the free portions narrowly triangular-ovate, the lower 3 lobes as long as or slightly shorter than the other lip, narrowly oblong-elliptic to lanceolate, all of the lobes angled or tapered to sharply pointed tips. Corollas papilionaceous, not much longer than the calyx, the petals lemon yellow to orangish yellow, the banner occasionally streaked with red or reddish-tinged on the outer surface, short-stalked, the expanded portion with a pair of small, incurved auricles at the base, obovate to nearly circular, rounded to minutely notched at the tip, shallowly keeled, strongly curved or bent backward from toward the base (but the flowers sometimes not fully opening), glabrous or short-hairy on the outer surface, the wings oblong-obovate with a minute auricle at the base, straight or nearly so, rounded at the tips, the keel oblong, somewhat curved upward, bluntly pointed at the tip. Stamens 10, all of similar lengths, 9 of the filaments fused and 1 free to about the midpoint, the fused portion longer than the free portion, curved upward toward the tip, the anthers small, attached below the midpoint, yellow, often darker-colored around the attachment. Ovary sessile, densely hairy and usually also glandular, the style curved upward, thickened toward the tip, glabrous, the stigma terminal, minute. Fruits legumes, oblong to elliptic-oblong or asymmetrically ovate in outline, flattened not or only slightly narrowed or indented between the seeds, tapered asymmetrically to a short beak at the tip, the 2 valves dark brown at maturity, densely hairy and dotted with minute, more or less globose glands, dehiscent, the valves becoming contorted or somewhat spirally twisted during dehiscence, (1)2-seeded. Seeds in our species 3–4 mm long, broadly oblong to more or less circular in outline, flattened, the surface brown, reddish brown, gray, or black, sometimes with darker mottling, smooth, shiny. About 230 species, nearly worldwide, most diverse in tropical regions.

Rhynchosia is recognized by the gland-dotted, trifoliate leaves, yellow corollas, and small, mostly 2-seeded legumes. Grear (1978) noted that in most of the species the flowers frequently do not open fully and that they do not produce significant quantities of nectar. During his extensive field work he never observed insects actually visiting *Rhynchosia* flowers. He concluded that the species are most commonly self-pollinated, which would limit opportunities for interspecific hybridization. Earlier bagging studies by Walraven (1967) indicated that all American species of *Rhynchosia* are self-fertile, with pollination occurring before the flowers open. There were no obvious meiotic irregularities noted in that study that might indicate that any hybridization had occurred.

In his revision of the genus, Grear (1978) cited a collection of *R. minima* (L.) DC. (least snout bean) made by Reverchon in 1903 at Sheldon, which he mistakenly mapped from Vernon

County, Missouri. Julien Reverchon (1834–1905) was an inveterate collector of the Texas flora, but although his personal herbarium was acquired by the Missouri Botanical Garden after his death there is no evidence that Reverchon ever botanized in the state of Missouri. The specimen cited by Grear (1978) actually originated from a Sheldon located near Houston, Texas, rather than the one in Missouri. *Rhynchosia minima* is an Old World species that is widespread as a presumed introduction in Latin America, the New World range extending northward sporadically into the southern United States, from Texas and southeastern Arkansas eastward to Georgia. Within the genus, it does not appear to be closely related to the two Missouri taxa, differing morphologically in a syndrome of floral features, especially its relatively short calyx lobes and smaller corollas, as well as its relatively long, curved fruits.

1. Stems prostrate to loosely ascending or rarely erect, not twining or only loosely twining; inflorescences 1–6 cm long, condensed, umbellate clusters or short, dense racemes, mostly shorter than the subtending leaves; calyces 8–11 mm long . . . . 1. R. DIFFORMIS
1. Stems trailing or climbing, twining; inflorescences 5–18 cm long, elongate, relatively open racemes, longer than the subtending leaves; calyces 10–14 mm long . . . . 2. R. LATIFOLIA

1. Rhynchosia difformis (Elliott) DC. (snout bean)

Arcyphyllum difforme Elliott

Pl. 406 j, k; Map 1797

Stems 20–90 cm long, prostrate to loosely ascending or rarely erect, not twining or more commonly loosely twining toward the tip, unbranched or branched. Leaves pinnately trifoliate or the lowermost sometimes appearing simple, the petiole 1–6 cm long. Stipules 3–8 mm long, lanceolate to ovate, angled or tapered at the tip, hairy. Leaflets 2–5 cm long, 1.5–6.0 cm wide. Inflorescences short umbellate clusters 1–3 cm long or short, dense racemes to 6 cm long, mostly shorter than the subtending leaves, with 3–8 flowers, the main stalk 1–2 mm long, the flower stalks 2–4 mm long. Calyces 8–11 mm long, the tube 1–2 mm long, the lobes 6–10 mm long. Corollas 8–10 mm long, lemon yellow to deep yellow, the banner glabrous or hairy on the outer surface. Filaments with the fused portion 5–6 mm long, the free portion 1–2 mm long. Fruits 14–20 mm long, 6–9 mm wide. $2n=22$. June–September.

Uncommon, known thus far only from Dunklin and Scott Counties (southeastern U.S. west to Missouri and Texas). Swamps, sand prairies, and openings of mesic upland forests on sandy substrates; also open, sandy disturbed areas.

The stems of *R. difformis* tend to be more slender and prostrate, and the leaflets smaller than in *R. latifolia*. However, there is much overlap in these characters. The species was first reported in Missouri by Steyermark (1963) from Dunklin County. Additional sites have been located in remnant sand prairies in Scott County. Isely (1998) mapped a specimen from Christian County, but this could not be located during the current study. The Arkansas distribution also is not well understood (E. B. Smith, 1988). There is some confusion about the identity of the Missouri specimens of *R. difformis*. Grear (1978) viewed the species as restricted mostly to states bordering the Atlantic and Gulf Coastal Plains and excluded Arkansas and Missouri from the range. He suggested that plants in southeastern Missouri and other interior sites represented material of *R. latifolia* whose inflorescences had not yet fully elongated at the time of collection. However, the situation in southeastern Missouri is more complicated, as documented by more recent collections. Some Missouri specimens have slightly more elongate racemes approaching those of *R. latifolia*. Plants with both umbellate and racemose inflorescences appear to be present within the same population at least at one Scott County site. This may perhaps indicate hybridization and introgression between the species. Casual field observations also suggest that plants can rebloom and that the inflorescences produced later in the season may be shorter than those produced earlier. Bagging studies by Walraven (1967) indicated that all American species of *Rhynchosia* are self-fertile, with pollination occurring before the flowers open. There were no obvious meiotic irregularities noted in that study that might indicate hybridization had been taking place. Grear (1978) also made a strong case that these plants tend to be self-pollinated, which limits the opportunity for hybridization The situ-

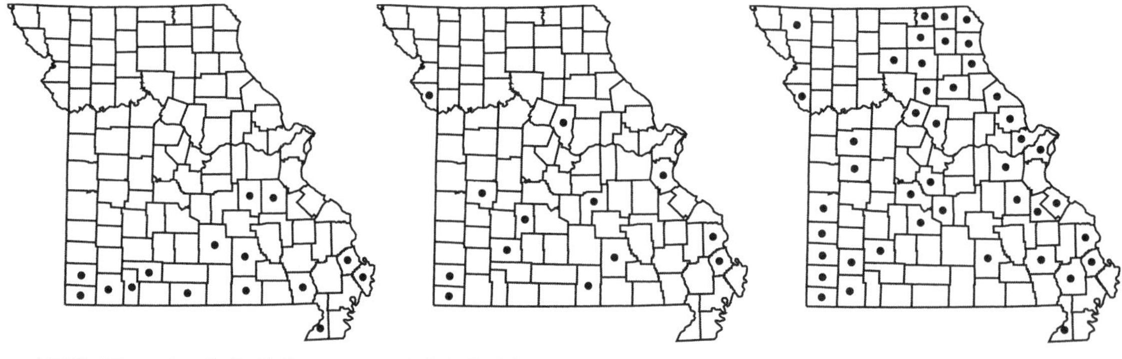

1798. Rhynchosia latifolia 1799. Robinia hispida 1800. Robinia pseudoacacia

ation requires more detailed population-level studies.

Erect specimens of *R. difformis* might also be confused with the polymorphic species, *R. tomentosa* (L.) Hook. & Arn. (Grear, 1978). However, that species is more densely hairy than is *R. difformis* and tends to have broader lateral calyx lobes. More careful analysis of these taxa and the characters used to identify them would help clarify these taxonomic issues.

2. Rhynchosia latifolia Nutt. ex Torr. & A. Gray (snout bean)

Pl. 406 a–c; Map 1798

Stems 30–150 cm long, trailing or climbing, twining, branched. Leaves pinnately trifoliate, the petiole 1.5–5.0 cm long. Stipules 2–5 mm long, lanceolate, tapered at the tip, hairy. Leaflets 2.0–7.5 cm long, 1.5–7.0 cm wide. Inflorescences elongate, relatively open racemes 5–18 cm long, longer than the subtending leaves, with mostly 10–20 flowers, the main stalk 5–20 mm long, the flower stalks 1–3 mm long. Calyces 10–14 mm long, the tube 1–3 mm long, the lobes 8–11 mm long. Corollas 8–13 mm long, yellow to orangish yellow, the banner hairy on the outer surface. Filaments with the fused portion 7–8 mm long, the free portion 2–3 mm long. Fruits 15–20 mm long, 7–9 mm wide. $2n=22$. June–September.

Scattered in the Ozark and Mississippi Lowlands divisions (Oklahoma and Texas east to Tennessee and Mississippi). Glades, savannas, dry upland forests, and sand prairies; also cemeteries, roadsides, and open, sandy, disturbed areas.

For a discussion of difficulties in distinguishing some plants of *R. difformis* from this species, see the treatment of *R. difformis*.

45. Robinia L.
(Isely and Peabody, 1984)

Plants trees or shrubs, often colonial by root suckers. Bark thick, furrowed and often with a network of slender ridges, gray to reddish brown or nearly black. Twigs slender, angled in cross-section, usually zigzag, sometimes armed with spines, variously hairy to glabrous, the winter buds sunken in the stem. Leaves often drooping and folding at night, odd-pinnately compound, short-petiolate, the leaflets 7–19, mostly opposite or subopposite. Stipules slender, papery, and shed with age or modified into spines and persistent; stipels present but shed early, hairlike or linear to very narrowly triangular. Leaflets ovate or oblong-ovate to oblong-elliptic or nearly circular, mostly short-stalked, rounded at the base, rounded at the tip but usually with a minute sharply pointed extension of the midvein, the margins entire, the surfaces hairy, at least when young. Inflorescences axillary racemes with (3)4–30 flowers, pendant, the stalk and axis variously hairy and sometimes also glandular, the flowers not subtended by bracts. Calyces appearing somewhat 2-lipped, the tube shorter than to longer than the lobes, cup-shaped to more or less bell-shaped, the lobes broadly to narrowly triangular, the upper 2 lobes partially fused. Corollas papilionaceous, glabrous or nearly so at maturity, white or pink to purple, the petals cordate or abruptly tapered to a stalklike base, the banner rounded or notched at the tip, abruptly arched upward toward the midpoint, the wings asymmetrically obovate, appearing curved, twisted above the stalk, the keel obovate, boat-shaped, strongly curved upward. Stamens 10, 9 of the filaments fused nearly to the tip and 1 filament free or fused only below the midpoint, the anthers small, attached at the base, all similar in size. Ovary oblong, nearly sessile, glabrous to glandular-hairy, the style bearded toward the tip, the stigma terminal and capitate, hairy. Fruits (when produced) legumes, narrowly oblong, slightly or strongly flattened, very short-stalked at the base, short-tapered to a bluntly or sharply pointed tip, sometimes short-beaked, the margins more or less parallel or irregularly indented between some of the seeds, glabrous or variously hairy and/or glandular, papery to leathery or slightly woody in texture, dehiscent, with 3–10 seeds (or the ovules aborting). Seeds ellipsoid to somewhat kidney-shaped, somewhat flattened, the surfaces brown to reddish brown or dark brown, often somewhat purplish-mottled, somewhat shiny. Four species, southeastern and western U.S. and adjacent Mexico; introduced farther north in North America and in the Old World.

The complex cytology and widespread cultivation of some members of *Robinia* has led to the naming of numerous additional species. Isely and Peabody (1984) and Lavin and Sousa S. (1995) accepted only the four basic biological elements within the genus as species.

1. Plants shrubs, 0.5–2.5 m tall; stems and inflorescence conspicuously pubescent with stiff, spreading, purplish hairs 2–5 mm long; corollas pink or purple; fruits usually not produced, densely bristly-hairy . . . . 1. R. HISPIDA
1. Plants trees, 5–15 m tall; stems and inflorescence glabrous or nearly so at maturity, not bristly; corollas white; fruits usually abundant, glabrous . . . . 2. R. PSEUDOACACIA

1. Robinia hispida L. **var. hispida** (bristly locust, rose acacia)

Pl. 407 k; Map 1799

Plants shrubs, 0.5–2.5 m tall, rhizomatous, root-suckering, and usually strongly colonial. Young stems conspicuously pubescent with stiff, spreading, purplish hairs 2–5 mm long, these gland-tipped when young, persistent for several seasons. Leaves with 9–13 leaflets, the petiole 1.5–2.0 cm long, the rachis 9–16 cm long, densely bristly-hairy. Stipules 4–8 mm long, linear, soft or occasionally spinescent. Leaflets 3.0–5.5 cm long, 1.5–3.0 cm wide, broadly elliptic to ovate, the upper surface glabrous, the undersurface sparsely short-hairy, at least when young. Inflorescences with (3)4–11(–15) flowers, the stalk and axis densely bristly-hairy and sometimes also short-hairy and/or glandular. Flowers with the stalk 5–8 mm long. Calyces with the tube 4–6 mm long, densely hairy, the lobes 4–8 mm long, tapered to a sharply pointed tip. Corollas pink or rosy purple (rarely white elsewhere), the banner 20–25(–30) mm long, often with a lighter spot toward the base, the wings 16–22 mm long, 7–9 mm wide, the keel 16–20 mm long, 8–10 mm wide. Filaments 10–18 mm long. Ovary 8–10 mm long, the style 10–12 mm long. Fruits usually not produced, when present 3–5 cm long, 3–5 mm wide, slightly flattened, densely pubescent with bristly purplish hairs 3–6 mm long. Seeds not produced in Missouri, the ovules aborting during seed development. $2n=30$. May–June.

Introduced, uncommon and widely scattered, mostly south of the Missouri River (native from Virginia to Georgia and Alabama; introduced farther north and west). Mesic upland forests and banks of streams; also old homesites, old fields, and roadsides.

Robinia hispida is easily recognized by its shrubby habit, densely bristly stems, and showy racemes of large pinkish flowers. The species was treated by Isely and Peabody (1984) and Isely (1998) as comprising five main groups that they treated as varieties for the sake of convenience. The var. *fertilis* (Ashe) R.T. Clausen (native to the mountains of North Carolina and Tennessee) and var. *kelseyi* (Cowell ex Hutch.) Isely (known with certainty only from cultivation) are fertile diploids ($2n=20$) that produce abundant fruits and seeds. The other varieties comprise a complex of diploid and triploid ($2n=30$) plants that are nearly entirely sterile (reproducing mostly vegetatively), producing few fruits and these with mostly abortive seeds. The var. *hispida*, which corresponds to triploid plants with long bristly hairs, is the most abundant and widespread member of the complex. Short-stemmed, few-branched diploid plants in the Southeastern Coastal Plain and adjacent Piedmont have been called var. *nana* (Elliott) DC., whereas taller sterile triploids mostly in the Appalachians are var. *rosea* Pursh. These last two sterile varieties both have shorter, softer, less spreading, and/or less persistent hairs than are found in var. *hispida*.

Steyermark (1963) knew *R. hispida* only from Howell and Platte Counties. It is now known from about a dozen counties, but continues to be relatively uncommon in Missouri, and shows little signs of moving from disturbed sites to invade more intact plant communities.

2. Robinia pseudoacacia L. (black locust)
Pl. 407 h–j; Map 1800

Plants trees, 5–15 m or more tall, root-suckering and often loosely colonial. Twigs sparsely and minutely hairy when young, not bristly, becoming glabrous or nearly so with age. Leaves with 11–19 leaflets, the petiole 2–3 cm long, the rachis 10–25 cm long, sparsely to moderately pubescent with minute, soft, curved hairs. Stipules initially 3–4 mm long, linear, and soft, but often becoming modified into slender to stout spines 5–10 mm long. Leaflets 3–7 cm long, 1.5–3.0 cm wide, ovate to elliptic, the surfaces glabrous or nearly so. Inflorescences with 10–35 flowers, the stalk and axis moderately to densely pubescent with minute curved and often velvety hairs, not bristly. Flowers with the stalk 5–10 mm long. Calyces with the tube 3–5 mm long, densely short-hairy, the lobes 1–3 mm long, angled or short-tapered to a usually sharply pointed tip. Corollas white, the banner 20–25(–30) mm long, with a yellow spot toward the base, the wings 15–22 mm long, 4–6 mm wide, the keel 15–22 mm long, 4–7 mm wide. Filaments 14–18 mm long. Ovary 9–12 mm long, the style 7–9 mm long. Fruits 5–11 cm long, 10–14 mm wide, strongly flattened, glabrous at maturity, 3–10-seeded. Seeds 5–6 mm long, 3.0–3.5 mm wide, kidney-shaped, the surface dark brown, often mottled with purplish brown, smooth. $2n=20$. May–June.

Scattered nearly throughout the state but apparently native only in the southern portion of the Ozark Division (eastern U.S. west to Missouri and Texas; introduced farther north and west, also in Europe). Bottomland forests, mesic to dry upland forests, banks of streams, spring branches, and rivers, margins of sinkhole ponds, bases and tops of bluffs, and edges of upland prairies and glades; also pastures, fencerows, mine spoils, roadsides, and disturbed areas.

Black locust is a common spring-flowering tree of the eastern United States. It was spread by Native Americans and early European settlers so extensively that the limits of its natural range are difficult to circumscribe. It is often planted as an ornamental shade tree and for its fragrant flowers, but Steyermark (1963) noted that its aggressive spread by root suckers makes it undesirable for many garden situations. Numerous horticultural variants have been described. Because of its ability to grow in poor soils, the species also has been used in mine revegetation projects. Although usually a small tree, it is capable of growing quite large and living for several centuries. The wood is very strong and resistant to decay, and in the past was commonly used for fence posts, railroad ties, ladders, tool handles, clubs, and handcrafts. The roots, bark, leaves, and young shoots are poisonous to livestock and humans, and even honey produced from the pollen is thought to be somewhat toxic (Kurz, 2003). However, Native Americans used the species medicinally, especially to induce vomiting and for treatment of rheumatism (Moerman, 1998).

46. Securigera DC.

About 13 species, Europe, Asia, introduced in North America.

Traditionally, the species of *Securigera* were included in *Coronilla* L., a related Old World genus which was thought to comprise about 22 species. On the basis of careful analysis of morphological characters, Lassen (1989) moved about 12 of these species to the formerly monotypic *Securigera* and 1 additional taxon to the larger genus *Hippocrepis* L. (about 34 species), leaving *Coronilla* in the restricted sense with only about nine species native to Europe, Asia, and Africa. *Securigera* differs from *Coronilla* in having angled stems, thinner leaves, free bracts with dark markings, and slender, glabrous styles. Preliminary molecular data also support the separation of *Hippocrepis* and *Securigera* from *Coronilla* (Allan and Porter, 2000; Degtjareva et al., 2003).

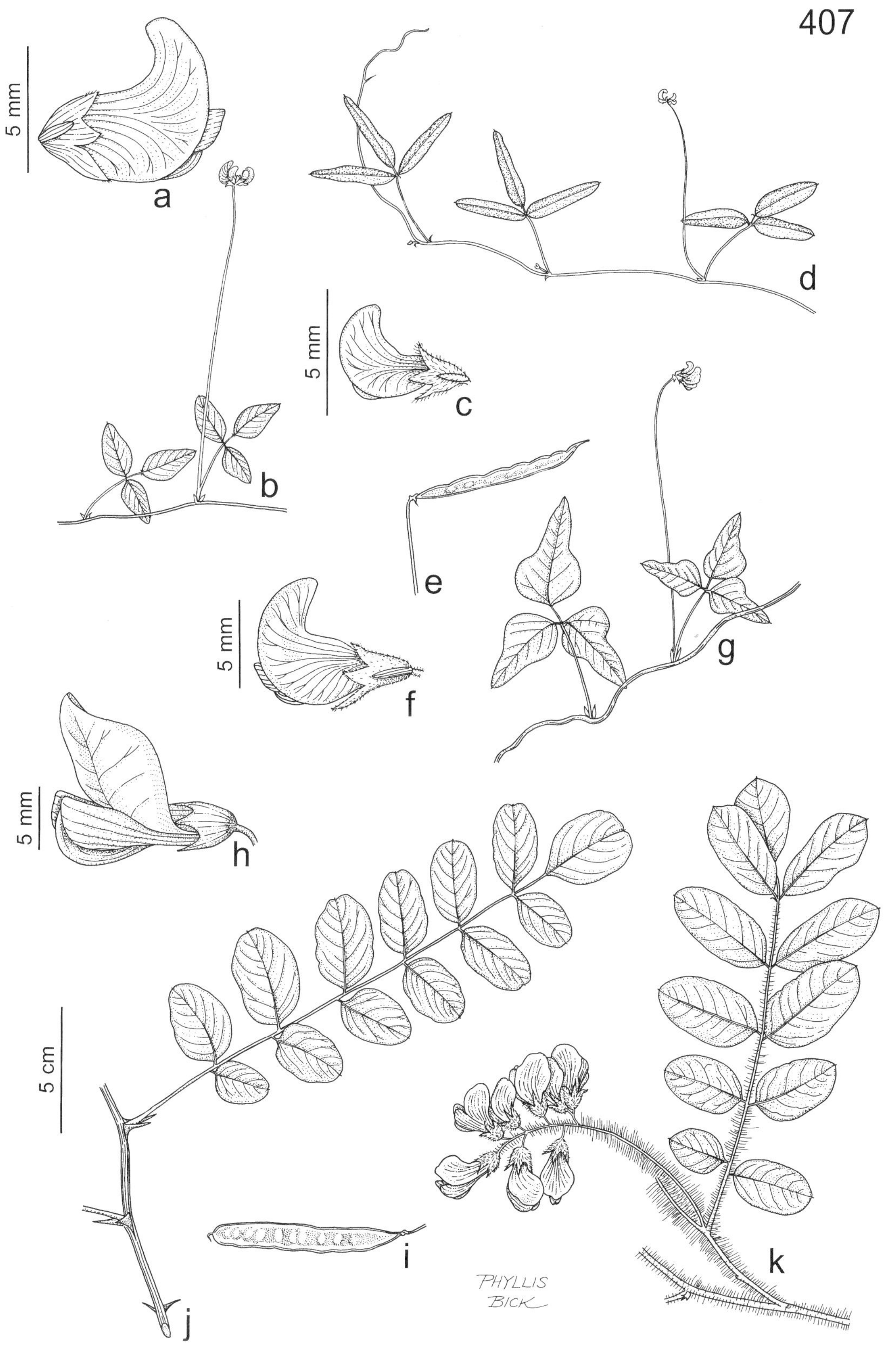

Plate 407. Fabaceae. *Strophostyles umbellata*, **a)** flower, **b)** fertile stem. *Strophostyles leiosperma*, **c)** flower, **d)** fertile stem. *Strophostyles helvola*, **e)** flower, **f)** fruit, **g)** fertile stem. *Robinia pseudoacacia*, **h)** flower, **i)** fruit, **j)** stem with leaf. *Robinia hispida*, **k)** fertile stem.

1801. Securigera varia 1802. Sesbania herbacea 1803. Sesbania vesicaria

1. Securigera varia (L.) Lassen (crown vetch)
Coronilla varia L.

Pl. 406 g–i; Map 1801

Plants perennial herbs, with a woody taproot below a short, branched caudex. Stems 30–120 cm long, spreading to loosely ascending or occasionally ascending, branched, often forming loose mats or mounds, angled or ridged, unarmed, glabrous or sparsely pubescent with spreading hairs. Leaves alternate, odd-pinnately compound with 9–25 leaflets, sessile or nearly so, the rachis 5–12 cm long. Stipules 1–4 mm long, narrowly lanceolate to narrowly oblong or oblong-triangular, usually reflexed, often with a green central stripe and pale margins, rounded to bluntly or sharply pointed at the sometimes darkened tip, the margins entire or with a few minute teeth (rarely with a single basal lobe); stipels absent. Leaflets 6–25 mm long, 2.5–9.0(–12.0) mm wide, the terminal leaflet usually narrowly oblanceolate to oblong-oblanceolate, the lateral leaflets oblong to oblong-oblanceolate, the shortest ones sometimes oblong-obovate, rounded or angled at the base, rounded to truncate or broadly and bluntly pointed at the tip, the midvein extended as a minute, sharp point at the very tip, the margins entire, the surfaces glabrous, the upper surface usually with microscopic, dark purple gland-dots (visible only under magnification), the undersurface often purplish-tinged or -mottled. Inflorescences axillary, umbellate to loosely headlike clusters of (5–)10–15(–20) flowers, the stalk 5–15 cm long, usually with sparse, short, spreading hairs, the tip with 1 bract per flower, these 0.5–1.0 mm long, linear to narrowly oblong or rarely forked apically, darkened at the tips, persistent, each flower with a stalk 3–7 mm long above the bract; bractlets absent. Calyces glabrous, the tube 1.0–1.5 mm long, broadly bell-shaped to hemispheric, more or less 2-lipped, the upper 2 lobes fused most of their length, 0.7–1.0 mm long, the unit broadly triangular with a forked tip, the lower 3 lobes 0.5–1.0 mm long, more or less similar, broadly triangular, the margins often thin and pale, glabrous or occasionally minutely and finely hairy, mostly in the sinuses between the lobes. Corollas papilionaceous, pink to purple, sometimes appearing bicolorous, the banner sometimes with a darker, small keel, the wings often lighter (sometimes nearly white), the keel dark purple toward the tip, the banner 9–13 mm long, 5–8 mm wide, the expanded portion broadly ovate, abruptly arched upward from below the midpoint, the wings 9–15 mm long, 4–5 mm wide, oblong, the keel 10–12 mm long, 2–4 mm wide, boat-shaped, fused to the tip, abruptly curved upward near the midpoint, narrowed or tapered to a bluntly pointed tip. Stamens 10, all of similar lengths, 9 of the filaments fused and 1 usually shorter filament free, the fused portion 6–7 mm long, the free portion 3–4 mm long, curved upward, the anthers small, attached at the base, yellow. Ovary 5–7 mm long, the style 5–6 mm long, slender, curved upward, glabrous, the stigma small and terminal. Fruits loments, 2–6 cm long, 1.5–2.2 mm wide, narrowly oblong to more commonly linear, short-stalked, tapered to a slender, beaked tip, bluntly 4(5)-angled, only very slightly constricted between the seeds, green to straw-colored or tan, glabrous, breaking apart at maturity into 3–12, 1-seeded segments, these indehiscent or sometimes dehiscing more or less longitudinally with age. Seeds 3–4 mm long, 0.9–1.5 mm wide, more or less cylindric (bluntly rectangular in outline), the surface reddish brown, smooth, dull. $2n=24$. May–August.

Introduced, scattered nearly throughout the state, though as yet uncommon or absent from portions of the Glaciated Plains and Mississippi Lowlands Divisions (native of Europe, Asia; introduced nearly throughout the U.S. [including Hawaii] and Canada). Banks of streams and rivers, margins of ponds and lakes, glades, and openings of mesic to dry upland forests; also pastures, old fields, old strip mines, railroads, roadsides, and open, disturbed areas.

Securigera varia is identified by the sprawling habit, pinnately compound leaves, umbellate inflorescence of pinkish flowers on a long stalk, and narrow, linear loments that break up into segments. Until recently, crown vetch was planted widely for erosion control, especially along highways following construction work. This practice has been discouraged, because although the plants form dense masses covering the ground, on steep slopes they frequently hide rather than eliminate soil erosion. Additionally, there have been concerns about the aggressive spread of the species into neighboring native plant communities, particularly along drainages. Invasive populations have been recorded in side-drainages of Ozark streams and rivers at points some 5 kilometers from the nearest road crossing.

Securigera varia provides cover for small mammals and birds. It is eaten by animals and has sometimes been cultivated for fodder. However, there are concerns about low-level toxicity to livestock from nitrotoxins (Burrows and Tyrl, 2001).

47. **Sesbania** Scop.

Plants annual (perennial herbs, shrubs, and trees elsewhere), with taproots. Stems erect, unbranched or less commonly branched from above the midpoint, unarmed, glabrous or with short appressed hairs lacking a bulbous base. Leaves often drooping and folding at night, even-pinnately compound, short-petiolate, the leaflets 16 to numerous, opposite. Stipules 4–10 mm long, linear, herbaceous to more commonly papery, appearing basally attached, persistent or shed early, the stipels minute, inconspicuous. Leaflets narrowly elliptic to oblong or occasionally linear, sessile or nearly so, angled or rounded at the base, rounded at the tip but usually with a minute sharply pointed extension of the midvein, the margins entire, the surfaces glabrous or less commonly with short appressed hairs, the undersurface sometimes purplish-tinged. Inflorescences axillary racemes or loose clusters with 2–7(–10) flowers, spreading to somewhat pendant, the stalk and axis glabrous or occasionally with sparse short appressed hairs lacking a bulbous base, the flower stalks subtended by linear bracts 3–6 mm long, these shed early, each flower also closely subtended by a pair of narrowly triangular bractlets 3–4 mm long, these also shed early. Calyces with the tube 3–4 mm long, broadly bell-shaped, glabrous, the lobes absent or shorter than the tube, triangular, and slightly unequal. Corollas papilionaceous, glabrous, yellow to greenish yellow, sometimes streaked or tinged with red or purple, the banner nearly circular above the stalklike base, shallowly notched at the tip, abruptly arched upward above the midpoint, the wings asymmetrically obovate, appearing somewhat curved, the keel boat-shaped and somewhat curved upward. Stamens 10, 9 of the filaments fused nearly to the tip and 1 filament fused below the midpoint, the anthers small, attached at the base, all similar in size. Ovary oblong to nearly linear, short-stalked, the style curved, glabrous, the stigma terminal and capitate. Fruits legumes, variously shaped, not or slightly flattened, stalked at the base, tapered to a prominent beak at the tip, the margins variously shaped, glabrous, papery to leathery in texture, dehiscent only with age, (1)2- or (15–)20–40-seeded. Seeds, broadly ellipsoid to oblong-ellipsoid, somewhat flattened, the surface brown, sometimes mottled with purplish brown, smooth. About 60 species, nearly worldwide, mostly in tropical and warm-temperate regions.

The two species of *Sesbania* recorded from Missouri are very similar in habit and vegetative features, although *S. herbacea* tends to have more leaflets than does *S. vesicaria*.

1. Corollas 11–15(–20) mm long; fruits 10–20 cm long, 3–4 mm wide, narrowly linear, appearing quadrangular in cross-section, parallel-sided, the faces slightly and narrowly indented between the seeds; seeds (15–)20–40 per fruit, 3–4 mm long, oblong-ellipsoid . . . . 1. S. HERBACEA

1. Corollas 8–10 mm long; fruits 1–8 cm long, 9–18 mm wide, narrowly oblong to elliptic, narrowed to acute winglike margins (narrowly biconvex in cross-section), convex or wavy-sided, the faces broadly flattened between the seeds; seeds (1)2 per fruit, 9–12 mm long, broadly ellipsoid .... 2. S. VESICARIA

1. Sesbania herbacea (Mill.) McVaugh (coffee weed, bequilla)
S. exaltata (Raf.) Rydb. ex A.W. Hill
S. macrocarpa Muhl.
Darwinia exaltata Raf.
Emerus herbacea Mill.

Pl. 408 c–e; Map 1802

Stems 70–200 cm long, the base often somewhat swollen and spongy. Leaves 8–30 cm long, with 30–70 leaflets. Leaflets 10–25 mm long, 2–6 mm wide, narrowly oblong to occasionally linear, the surfaces glabrous. Inflorescences 2–6 cm long, the flower stalks 5–9 mm long. Calyces with the tube 3–4 mm long, glabrous, sometimes finely purple-dotted, the lobes 1–2 mm long, slightly unequal. Corollas 11–15(–20) mm long, pale yellow to bright yellow, often streaked or tinged with purple. Fruits 10–20 cm long, 3–4 mm wide, narrowly linear, short-stalked at the base, appearing quadrangular in cross-section, parallel-sided, the faces slightly and narrowly indented between the seeds, (15–)20–40-seeded. Seeds 3–4 mm long, oblong-ellipsoid. $2n=12$. July–October.

Scattered in the Mississippi Lowlands division and disjunct in St. Charles County; introduced in the city of St. Louis (southern U.S. north to California, Kansas, Illinois, and Virginia; Mexico, Central America, South America; introduced north to Pennsylvania). Bottomland forests, banks of streams, and margins of oxbows and sloughs; also edges of crop fields, railroads, roadsides, and open disturbed areas.

Sesbania herbacea is recognized by its erect habit, evenly compound leaves with numerous leaflets, few-flowered racemes of yellow flowers, and long, linear, quadrangulate legumes. The taxonomy and nomenclature of this species are controversial (Wilbur, 1989; Isely, 1998). The name *S. macrocarpa* was used for this taxon for many years (McGregor, 1986). As reviewed by Wilbur (1989), some authors have maintained that Muhlenberg failed to validly publish this epithet (Merrill and Hu, 1949) and have substituted the name *S. exaltata* (Steyermark, 1963; Isely, 1990, 1998). Further confusing the issue has been the question of whether this taxon is a distinct species or merely an annual derivative of a tropical perennial species commonly known as *S. emerus* (Aubl.) Urb., distinguished by a larger, more branched habit, larger flowers, and flowering in the spring. Isely (1990; 1998) treated the two taxa as separate species, with *S. emerus* occurring within the United States only in Florida, but they have been combined by some authors under the name *S. emerus*. McVaugh (1987), proposed yet an older epithet for members of this complex in Mexico, *S. herbacea*. That name is tentatively accepted in the present work.

This species has been used as a groundcover for erosion control. The seeds contain saponins and other toxic compounds that cause severe diarrhea and internal hemorrhaging (Burrows and Tyrl, 2001).

2. Sesbania vesicaria (Jacq.) Elliott (bladderpod, bagpod)
Glottidium vesicarium (Jacq.) R.M. Harper
Robinia vesicaria Jacq.

Map 1803

Stems 70–200(–300) cm long, the base often somewhat swollen and hardened. Leaves 8–30 cm long, with 16–50 leaflets. Leaflets 10–40 mm long, 4–7 mm wide, narrowly elliptic to oblong-elliptic, the surfaces glabrous or less commonly with short appressed hairs. Inflorescences 6–14 cm long, the flower stalks 5–12 mm long. Calyces with the tube 2.5–4.0 mm long, glabrous, the margin more or less truncate, slightly asymmetrically wavy, the lobes essentially absent. Corollas 8–10 mm long, pale yellow to greenish yellow, often tinged with maroon or purple. Fruits 1–8 cm long, 9–18 mm wide, narrowly oblong to elliptic, relatively long-stalked at the base, narrowed to acute winglike margins (narrowly biconvex in cross-section), convex or wavy-sided, the faces broadly flattened between the seeds, (1)2-seeded. Seeds 9–12 mm long, broadly ellipsoid, $2n=12$. July–October.

Introduced, known thus far from a single collection from Taney County (southeastern U.S. west to Oklahoma and Texas). Gardens and disturbed areas.

Sesbania vesicaria is included in the Missouri flora based on a specimen collected in 2001 by Mary Heiss of Branson from plants that appeared spontaneously as weeds around a pile of soil and compost in her garden, apparently having been brought in as seed contaminants in the soil of bedding plants. These plants flowered, set seed, and grew again for at least one additional season. Although the presence of a species as a garden weed

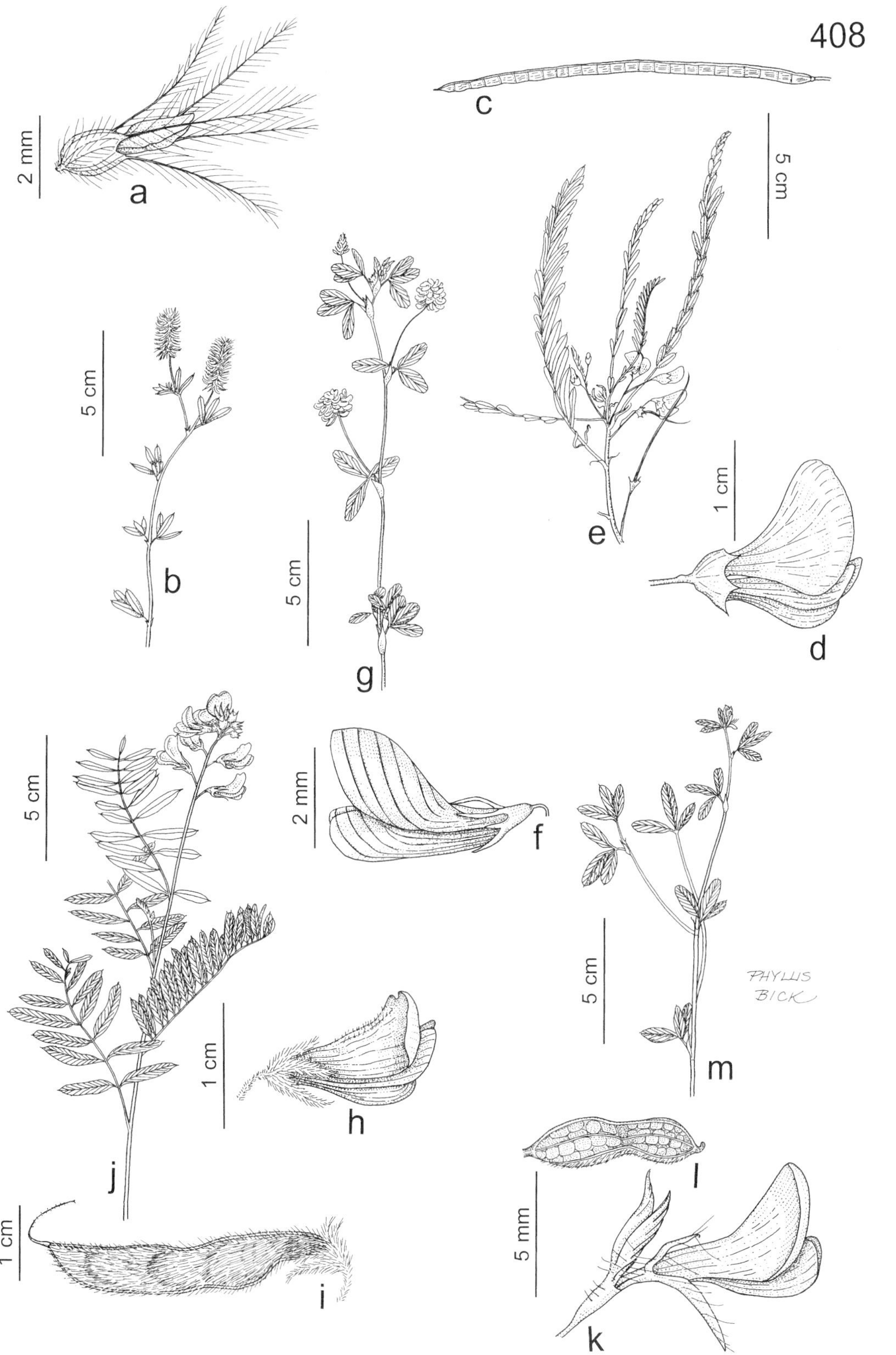

Plate 408. Fabaceae. *Trifolium arvense*, **a)** flower, **b)** fertile stem. *Sesbania herbacea*, **c)** fruit, **d)** flower, **e)** fertile stem. *Trifolium aureum*, **f)** flower, **g)** fertile stem. *Tephrosia virginiana*, **h)** flower, **i)** fruit, **j)** fertile stem. *Stylosanthes biflora*, **k)** flower, **l)** fruit, **m)** stem with leaves, flowers, and immature fruit.

at a single site is not usually grounds for its addition to the flora, the presence of *S. vesicaria* in the wild in northwestern Arkansas argues that it plausibly might be discovered elsewhere in southwestern Missouri in the future.

The taxonomy of *S. vesicaria* remains somewhat controversial. Most previous authors have treated this species as the sole member of the segregate genus *Glottidium* Desv. However, Isely (1998) noted that this segregation was problematic and required further research. Molecular phylogenetic studies by Lavin and Doyle (1991) and the morphologically based phylogenetic research of Lavin and Sousa S. (1995) provided evidence that *Glottidium* is merely a specialized element within *Sesbania*. That treatment is followed here.

48. **Strophostyles** Elliott
(Riley-Hulting et al., 2004)

Plants annual or perennial herbs, from short taproots, sometimes climbing on other vegetation. Stems trailing or climbing, twining, branched, sometimes rooting at the nodes, unarmed, moderately to densely pubescent with short, downward-curved to downward-angled and sometimes also more or less spreading hairs, these slender to somewhat tapered from the base, but lacking a bulbous base. Leaves sometimes drooping and folding at night, pinnately trifoliate, mostly long-petiolate, the petiole hairy. Stipules 2–4(–5) mm long, narrowly ovate-triangular, herbaceous to papery, appearing basally attached and somewhat spreading, mostly persistent, strongly parallel-veined; stipels 0.5–1.0(–1.5) mm long. Leaflets lanceolate to ovate, the lateral ones sometimes appearing asymmetrically rhombic or pear-shaped, entire or with a basal lobe. Inflorescences very short axillary racemes (rarely only a solitary flower), appearing as small headlike or umbellate clusters at the tip of a stalk that usually is much longer than the subtending leaf, the very short flower stalks with small bracts at the base, each flower also closely subtended by 2 small bractlets. Calyces with the tube bell-shaped, 2-lipped, the upper 2 lobes fused with only a shallow notch at the tip, the lower 3 lobes triangular, unequal, the lowermost lobe somewhat narrower and longer than the others, often as long as or longer than the tube. Corollas papilionaceous, pink to lavender (the keel sometimes darker purple), rarely white, often fading to green or yellow, the petals cordate or abruptly tapered to an often relatively broad stalklike base, the banner broadly ovate to nearly circular, rounded to shallowly notched at the tip, abruptly arched upward above the midpoint but generally appearing hooded over other petals, the wings oblong, shorter than the keel, the keel widest at the midpoint, abruptly constricted above the midpoint into a blunt, upward-arched beaklike tip that is slightly to strongly twisted to the side. Stamens 10, 9 of the filaments fused to above the midpoint and 1 filament more or less free, the anthers small, attached at or near the base, all similar in size. Ovary appearing arched, sessile, the style curved, jointed above the midpoint, bearded on upper side, the stigma terminal and capitate. Fruits legumes, linear, not or only slightly flattened, sessile, short-tapered on the lower side to a short slender beak, the margins otherwise parallel, glabrous or appressed-hairy, papery in texture, dehiscent, the valves becoming spirally twisted during dehiscence, 3–10-seeded. Seeds subcylindric, rectangular in outline, slightly flattened, the surface brown, often mottled with darker brown or purplish brown, smooth or minutely pubescent with white-woolly patches of hairs. Three species, North America, mostly in the U.S.

Strophostyles is very similar and probably closely related to *Phaseolus*, differing mainly in the twisted (vs. coiled) nearly vertical keel, lack of hooked hairs, short flower stalks, and somewhat swollen nodes of the inflorescence. Although palatable to deer and livestock, *Strophostyles* species generally are not abundant enough to be considered important forage plants. The genus sometimes is planted for erosion control.

1. At least some of the lateral leaflets (and often also the central leaflet) bluntly lobed at the base; bractlets subtending flowers as long or longer than the calyx tube . . . . 1. S. HELVOLA
1. Leaflets all unlobed; bractlets subtending the flowers shorter than the calyx tube
   2. Leaflets moderately to densely hairy on the upper surface; calyces densely hairy; corollas light pink to pale lavender, usually fading to grayish white, the banner 4.5–8.0 mm long, the wings 3.5–6.5 mm long; fruits hairy at maturity; seeds glabrous . . . . 2. S. LEIOSPERMA
   2. Leaflets glabrous or nearly so on the upper surface; calyces glabrous or nearly so; corollas pink to bright salmon pink, usually fading to greenish yellow, the banner 8–13 mm long, the wings 6–10 mm long; fruits glabrous or nearly so at maturity; seeds pubescent with patches of white-woolly hairs . . . . 3. S. UMBELLATA

1. Strophostyles helvola (L.) Elliott (wild bean, amberique bean)
S. helvola var. *missouriensis* (S. Watson) Britton
S. missouriensis (S. Watson) Small
Phaseolus helvolus L.

Pl. 407 e–g; Map 1804

Plants usually annual. Stems 70–300 cm long. Leaves with the petiole 2–5 cm long, hairy. Leaflets rounded or angled at the base, angled or tapered to a bluntly or sharply pointed tip, often with a minute extension of the midvein at the very tip, the upper surface glabrous or sparsely appressed-hairy, the undersurface sparsely to moderately appressed-hairy; the terminal leaflet with a stalk 6–15 mm long, the blade 1.8–6.5 cm long, 1–5 cm wide, ovate-triangular or narrowly ovate-triangular to oblong-triangular or pear-shaped, unlobed or with 1 or more commonly a pair of blunt basal lobes; the lateral leaflets sessile or with a stalk to 3 mm long, the blade 1.5–6.5 cm long, 0.6–4.0 cm wide, oblong-triangular or lanceolate to ovate or asymmetrically ovate, asymmetrically rhombic, or occasionally pear-shaped, unlobed or with 1 or less commonly a pair of blunt basal lobes. Inflorescences appearing as headlike clusters of 3–5 flowers, the inflorescence stalk 5–20 cm long, the flower stalks 1–2 mm long, the pair of bractlets closely subtending each flower 3.0–4.5 mm long, as long as or longer than the calyx tube. Calyces with the tube 1.5–3.0 mm long, sparsely hairy, the lobes 1.5–3.0 mm long. Corollas pinkish purple to pink, fading to greenish yellow, the banner 6.5–12.0 mm long, the wings 7–9 mm long, the keel 8–12 mm long, the beaklike apical portion relatively slender and usually sharply curved or twisted to the side. Fruits 3–9 cm long, 4–10 mm wide, sparsely hairy to nearly glabrous at maturity, 4–10 seeded. Seeds 5–8 mm long, the surface sometimes appearing waxy, brown, sometimes faintly mottled, pubescent with white-woolly patches. $2n=22$. June–October.

Scattered nearly throughout the state (eastern U.S. west to North Dakota and New Mexico; Canada). Bottomland forests, mesic upland forests, banks of streams, rivers, and spring branches, margins of ponds and lakes, bottomland and upland prairies, savannas, sand prairies, and edges of glades; also pastures, old fields, edges of crop fields, ditches, levees, railroads, roadsides, and open disturbed areas.

Strophostyles helvola is distinctive within the genus in its lobed leaflets, relatively large flowers, and brownish seeds with white-woolly patches. It should be noted that on some plants only a few of the leaves have lobed leaflets, so users should carefully examine specimens before scoring this character. Steyermark (1963) separated the var. *missouriensis*, which he characterized as a less common variant within the Missouri range of the species differing in its larger, mostly unlobed leaflets with blunter tips and its slightly larger fruits and seeds. Riley-Hulting et al. (2004) did not consider this variant worthy of formal taxonomic recognition.

The species epithet has been spelled *S. helvulus* in some of the botanical literature, in keeping with the spelling that Linnaeus originally used for his *Phaseolus helvulus*, the basis of the present species. However, many later botanists considered this a correctable spelling error (the derivation of the epithet is from the Latin *helvolus*, meaning pale brownish yellow), thus *S. helvolus* is also well represented in floristic works. In order to stabilize both the spelling and application of the name, Verdcourt (1997) formally proposed conservation

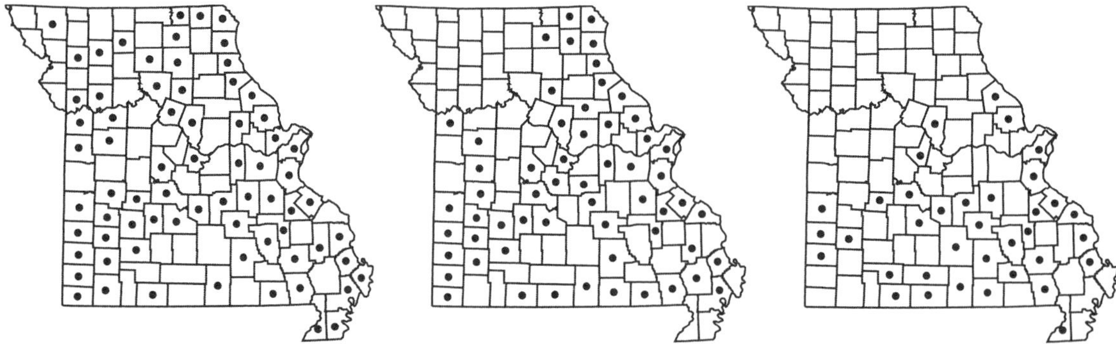

1804. Strophostyles helvola 1805. Strophostyles leiosperma 1806. Strophostyles umbellata

of the spelling *S. helvola*, and his proposal was approved at the subsequent International Botanical Congress.

2. Strophostyles leiosperma (Torr. & A. Gray) Piper (slickseed wild bean)
S. pauciflora (Benth.) S. Watson
Phaseolus leiospermus Torr. & A. Gray
Pl. 407 c, d; Map 1805

Plants usually perennial. Stems 40–100(–150) cm long. Leaves with the petiole 1.5–3.5 cm long, hairy. Leaflets rounded or angled at the base, angled or tapered to a rounded or bluntly pointed tip, often with a minute extension of the midvein at the very tip, the surfaces moderately to densely appressed- to loosely appressed-hairy; the terminal leaflet with a stalk 3–6 mm long, the blade 1.7–5.5 cm long, 0.5–1.5(–2.0) cm wide, triangular-lanceolate to lanceolate or narrowly lanceolate, unlobed; the lateral leaflets sessile or with a stalk to 1.5 mm long, the blade 1.5–5.5 cm long, 0.4–1.5 cm wide, triangular-lanceolate to lanceolate or narrowly lanceolate, unlobed. Inflorescences appearing as solitary flowers or more commonly headlike clusters of 2–5 flowers, the inflorescence stalk 4–12 cm long, the flower stalks 0.5–1.0 mm long, the pair of bractlets closely subtending each flower 1.0–1.5 mm long, shorter than the calyx tube. Calyces with the tube 1–2 mm long, densely hairy, the lobes 0.5–1.5 mm long. Corollas light pink to pale lavender, usually fading to grayish white, the banner 4.5–8.0 mm long, the wings 3.5–6.5 mm long, the keel 3.5–6.0 mm long, the beaklike apical portion relatively slender and only slightly curved or twisted to the side. Fruits 1.5–3.5 cm long, 2–4 mm wide, moderately to densely hairy at maturity, 3–8 seeded. Seeds 2–4 mm long, the surface smooth and not appearing waxy, olive brown, often mottled with darker brown or black, glabrous. $2n=22$. June–October.

Scattered in the southern half of the state and in the eastern portion of the northern half (North Dakota to New Mexico east to Kentucky and Alabama; Mexico). Openings of mesic upland forests, upland prairies, sand prairies, glades, savannas, and banks of streams; also old fields, fallow fields, edges of crop fields, railroads, roadsides, and open disturbed areas, often on acidic substrates.

Strophostyles leiosperma is identified by the narrowly lanceolate, unlobed leaflets, relatively small flowers with pale pink top pale lavender corollas, and smooth seeds. It has a more southern distribution than the other two species (Pelotto and Del Pero Martínez, 1998).

3. Strophostyles umbellata (Muhl. ex Willd.) Britton (wild bean)
Pl. 407 a, b; Map 1806

Plants usually perennial. Stems 50–150 cm long. Leaves with the petiole 1.5–3.0 cm long, hairy. Leaflets rounded or angled at the base, angled or tapered to a rounded or bluntly to less commonly sharply pointed tip, often with a minute extension of the midvein at the very tip, the upper surface glabrous or nearly so, the undersurface sparsely to moderately appressed-hairy; the terminal leaflet with a stalk 4–8 mm long, the blade 1.5–3.5 cm long, 0.8–2.0 cm wide, triangular-lanceolate to lanceolate or narrowly ovate, unlobed; the lateral leaflets sessile or with a stalk to 1.5 mm long, the blade 1.5–3.5 cm long, 0.8–2.0 cm wide, triangular-lanceolate to lanceolate or narrowly ovate, unlobed. Inflorescences appearing as solitary flowers or more commonly headlike clusters of 2–5 flowers, the inflorescence stalk 5–20 cm long, the flower stalks 0.5–2.0 mm long, the pair of bractlets closely subtending each flower 1–2 mm long, shorter than the calyx tube. Calyces with the tube 1.5–3.0 mm long, glabrous or nearly so, the lobes 1.0–2.5 mm long. Corollas pink to bright salmon pink, usually fading to greenish yellow, the banner 8–13 mm long, the wings 6–10 mm long, the keel 8–10 mm long,

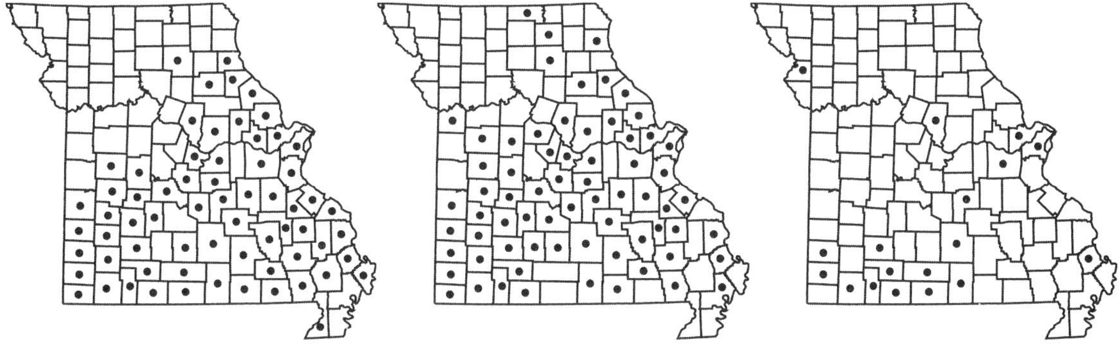

1807. Stylosanthes biflora 1808. Tephrosia virginiana 1809. Trifolium arvense

the beaklike apical portion relatively stout and only slightly curved or twisted to the side. Fruits 4–7 cm long, 2.5–5.5 mm wide, sparsely to moderately hairy at maturity, 5–10 seeded. Seeds 3.5–5.5 mm long, the surface sometimes appearing waxy, brown, sometimes faintly mottled, pubescent with white-woolly patches. $2n=22$. June–October.

Scattered in the southern half of the state, uncommon farther north in counties along the Mississippi River (eastern U.S. west to Iowa and Texas). Bottomland forests, mesic upland forests, banks of streams and rivers, bases of bluffs, sand prairies, and glades; also pastures, old fields, and roadsides, often on acidic substrates.

49. Stylosanthes Sw.
(Mohlenbrock, 1957, 1963)

Twenty-five to 50 species, nearly worldwide, most diverse in warm regions.

Generic limits of *Stylosanthes* remain poorly understood and there has been controversy as to its relationships within the family. There are only five species in the United States, which are relatively easily distinguished, but share several unusual characters, including stipules fused to the petioles and into a tube around the stem, short, dense inflorescences with bristly pubescence, yellow corollas, and dimorphic stamens. In tropical regions, some species are planted for fodder and soil stabilization.

1. Stylosanthes biflora (L.) Britton, Stearns & Poggenb. (pencil flower)

S. biflora var. *hispidissima* (Michx.) Pollard & C.R. Ball

S. riparia Kearney

Pl. 408 k–m; Map 1807

Plants perennial herbs, with a long, stout taproot below an often branched, woody caudex. Stems 10–40(–60) cm long, wiry, ascending to more commonly spreading, usually well-branched, sometimes forming loose mats, unarmed, glabrous or more commonly moderately to densely pubescent with fine, appressed-ascending hairs and often also stiff, bristly, spreading hairs. Leaves pinnately trifoliate, the petiole with the free portion 1–3 mm long, the basal half fused to the stipules. Stipules fused to the petiole and into a sheath around the stem, this 4–7 mm long, hairy, the free portions 3–6 mm long, linear, bristly-hairy along the margins; stipels absent. Leaflets 8–20(–40) mm long, 3–7(–18) mm wide, the terminal leaflet sometimes slightly larger than the lateral pair and with a stalk 1–2 mm long, narrowly elliptic to lanceolate, oblanceolate, or rarely narrowly ovate, angled at the base, angled or slightly tapered to a bluntly or sharply pointed tip, the midvein extended into a minute sharp point at the very tip, the margins entire, glabrous or sometimes sparsely bristly-hairy, the surfaces glabrous, the upward-angled lateral veins relatively prominent on the undersurface. Inflorescences axillary and terminal, of solitary flowers and/or short, dense spike-like clusters of 2(–4) flowers, the stalk 4–8 mm long, hidden in the stipular sheath (the inflorescences thus appearing sessile or nearly so), the bracts 3–5 mm long, 1–3 mm wide, linear to narrowly lanceolate, sometimes 3-lobed, usually bristly-hairy; bractlets 2–3 mm long, linear. Flowers sessile, but sometimes appearing short-stalked because of a hypanthium below the perianth, this 3–5 mm long, usually glabrous. Calyces (above the hypanthium) with the tube 2–3 mm long, bell-shaped, glabrous, 2-lipped, the upper 2 lobes 1.2–1.8 mm long, fused to about the midpoint, the free portions narrowly ovate-triangular, the lower lip with the lateral lobes similar in length to the upper ones, the lowermost lobe 2–3 mm long, narrowly oblong-triangular, its margins sometimes spreading-hairy. Corollas papilionaceous, orangish yellow to yellow, often fading to whitish or pinkish, the banner 5–9 mm long, 4–7 mm wide, the expanded portion broadly obovate to nearly circular, rounded to more commonly slightly and broadly notched at the tip, slightly keeled longitudinally, bent upward abruptly toward the base, the wings 3.5–5.0 mm long, 2–3 mm wide, oblong-oblanceolate, rounded at the tips, somewhat cupped around the keel, the keel 3–5 mm long, 2–3 mm wide, boat-shaped, curved upward, bluntly pointed at the tip. Stamens 10, the filaments all fused into a tube 5–6 mm long, the free portions 2–3 mm long, curved, the anthers of two kinds, 5 of them oblong and attached toward the midpoint, alternating with 5 that are nearly globose and attached near the base, yellow to orange. Ovary 1–2 mm long, usually minutely hairy, the style 6–8 mm long, glabrous or minutely hairy, curved or curled, the stigma terminal, minute. Fruits loments, 5–7 mm long, 2–3 mm wide, divided into 2 segments, the lower segment usually infertile, appearing stalklike, pale, and hairy, the upper segment 3.0–4.5 mm long, obliquely ovate in outline, flattened, tapered at the tip to a curved or curled beak 0.5–1.0 mm long, the outer wall papery, brownish yellow to brown, glabrous or minutely hairy, with a prominent network of rounded ridges (sometimes appearing wrinkled), indehiscent, shed as a unit, 1-seeded. Seeds 2.0–2.5 mm long, broadly oblong-circular in outline, flattened, the surface yellowish brown to brown, with irregular, blunt ridges, somewhat shiny. $2n=20$. June–September.

Scattered nearly throughout the state, but apparently absent from the northern half of the Unglaciated Plains Division and the western half of the Glaciated Plains (eastern U.S. west to Kansas, Texas, and possibly Arizona). Glades, upland prairies, sand prairies, savannas, dry upland forests, tops of bluffs, banks of streams and rivers, and rarely margins of ponds; also strip mines, old fields, and roadsides.

Stylosanthes biflora is recognized by the wiry stems with long, bristly hairs, orangish yellow flowers, and fruits that are 1-seeded loments. It is fairly uniform throughout its wide range, with only minor variation in leaf shape and habit (Mohlenbrock, 1957). Several species of *Stylosanthes* are important forage plants in the tropics.

50. **Tephrosia** Pers.
(Wood, 1949)

Three hundred and fifty to 400 species, nearly worldwide, most diverse in tropical and warm-temperate regions.

Many of the species of *Tephrosia* produce rotenone and related compounds in their roots and rhizomes, and have been used as fish poisons, insecticides, and herbal medicines in various parts of the world. A number of tropical species are planted for green manure and as cover and forage crops, and a few species are grown for their ornamental value.

1. Tephrosia virginiana (L.) Pers. (goat's rue, hoary pea, catgut)
T. virginiana var. *holosericea* (Nutt.) Torr. & A. Gray
Cracca virginiana L.
Pl. 408 h–j; Map 1808

Plants perennial herbs, with long, woody roots and/or rhizomes below a thick, knobby caudex. Stems 20–50 cm long, often clustered, erect or ascending, appearing unbranched above the base, unarmed, moderately to more commonly densely pubescent with fine, appressed and spreading hairs. Leaves odd-pinnately compound with (9–) 13–25(–31) leaflets, the petiole 1–2(–6) mm long, the rachis 4–12 cm long. Stipules 7–11 mm long, narrowly elliptic or lanceolate to linear, those of the upper leaves sometimes hairlike, often shed early; stipels absent. Leaflets 12–30 mm long, 2–10 mm wide, narrowly elliptic to oblanceolate, narrowly oblong, or linear, angled to rounded at the base, angled to a bluntly or sharply pointed tip, the terminal leaflet sometimes rounded or shallowly notched, the midvein often extended into a minute sharp point at the very tip, the margins entire, glabrous or more commonly ascending-hairy, the upper surfaces glabrous or moderately to densely pubescent with short, fine, curved hairs, the undersurface densely appressed-hairy and with noticeable lateral veins. Inflorescences terminal (occasionally at the tips of short upper branches and then appearing axillary), short, dense racemes of 12–40 flowers, the stalk 1–3 cm long, the bracts 5–12(–20) mm long, linear to hairlike, withering but usually persistent; bractlets absent, the flower stalks 4–20 mm long. Calyces with the tube 2–5 mm long, bell-shaped, densely hairy, 2-lipped, the upper 2 lobes 3–6 mm long, fused toward the base, the free portions long-tapered above a sometimes short, triangular base, the lower lip with the lateral lobes similar in length to the upper ones, the lowermost lobe 4–7 mm long, narrowly oblong-triangular. Corollas papilionaceous, bicolorous, the banner 15–19 mm long, 12–18 mm wide, the expanded portion broadly nearly circular to broadly kidney-shaped, broadly pointed to rounded or slightly and broadly notched at the tip, keeled longitudinally, bent upward abruptly toward the base, the outer surface hairy, green when young, turning lemon yellow to cream-colored at maturity, the inner surface white to cream-colored with a pinkish-tinged region surrounding a greenish yellow patch near the base, the wings 12–20 mm long, 5–7 mm wide, oblong-obovate, rounded at the tips, somewhat angled over the keel, rosy pink, the keel 12–15 mm long, 5–7 mm wide, boat-shaped, curved upward, rounded to bluntly pointed at the tip, light pink to light yellow with pinkish tinging or lines. Stamens 10, 9 of the filaments fused and 1 free to below the midpoint, the fused portion 8–11 mm long, the free portion 2–4 mm long, curved upward and with the filaments often broadened toward their tips, the anthers small, attached at the base, all similar in size, yellow. Ovary 8–10 mm long, densely hairy, with a collarlike nectary at the base, the style 8–11 mm long, hairy on the inner side, curved, the stigma terminal, minute. Fruits legumes, 3–5 cm long, 3–6 mm wide, narrowly oblong, flattened, straight to slightly curved, tapered asymmetrically at the tip to a slender, often curved beak, the surfaces papery, brown, sparsely to densely hairy, dehiscing by 2 valves, these straw-colored to brown at maturity, usually twisting spirally after dehiscence, 6–11-seeded. Seeds 3.0–4.5 mm long, kidney-shaped to more or less oblong in outline, flattened, the surface greenish brown to brown, often mottled with black, smooth, somewhat shiny. $2n=22$. May–July.

Scattered in much of the state, but absent from most of the western half of the Glaciated Plains and Mississippi Lowlands Divisions (eastern U.S. west to Minnesota, Kansas, and Texas; Canada). Upland prairies, sand prairies, savannas, glades, dry upland forests, openings of mesic upland forests, and tops of bluffs; also old strip mines, fallow fields, railroads, and roadsides; often on acidic substrates.

Tephrosia virginiana is recognized by the hairy stems, pinnately compound leaves, and the showy, bicolorous greenish yellow and pink flowers. This species is remarkably uniform in appearance, with only minor variation in the density of pubescence and leaflet size.

Native Americans extracted rotenone from the roots of this species for use as a fish poison, insecticide, and vermifuge (V. A. Little, 1931; Steyermark, 1963; Gard, 2010). At one time, it was under investigation for possible commercial cultivation as a source of rotenone, but this apparently did not prove fruitful, owing perhaps to differences in the concentration of the target compound among populations, between generations and seasons, and in response to different soil types (Sievers et al., 1938, 1940; Gard, 2010). In recent years, the species has instead become available horticulturally through some native plant nurseries. Steyermark (1963) noted that goat's rue favors acidic soils and that because of the long rootstock plants are very difficult to transplant.

51. **Trifolium** L. (clover)
(Zohary and Heller, 1984; J. M. Gillett and N. L. Taylor, 2001)
Contributed by Michael A. Vincent

Plants annual, biennial, or perennial herbs, with a taproot or fibrous roots. Stems 1 to several, often branched, unarmed. Leaves alternate (sometimes appearing opposite on flowering branches), palmately or less commonly pinnately compound, the leaflets 3 (rarely 4 or 5). Stipules usually conspicuous, persistent, variously shaped, usually herbaceous and often strongly veined, sometimes pale and membranous between the veins, partially fused to the petiole; stipels absent. Leaflets variously shaped, the margins toothed, sometimes very inconspicuously so. Inflorescences densely flowered, headlike or short spikes or racemes, occasionally appearing umbellate, especially at fruiting, axillary and/or terminal, long-stalked to nearly sessile, sometimes with an involucre of small to large, free or fused bracts, the individual flowers sometimes also subtended by small slender persistent bracts. Calyces 5-lobed, tubular to more or less bell-shaped, sometimes becoming somewhat inflated at fruiting, the lobes shorter to longer than the tube, nearly equal in length or the lower lobe longer than the others, the margins entire to toothed. Corollas papilionaceous, glabrous, white, pink, red, purple, or yellow, persistent and becoming brown and papery at fruiting, the petals (except the banner) tapered to a stalklike base and often fused basally with the filaments; the banner broadly oblong to obovate; the wings relatively narrow, usually longer than the keel; the 2-fused keel oblanceolate in outline, boat-shaped. Stamens 10, 9 with filaments fused nearly to the tip and 1 free (sometimes fused to the others at the base), some or all of the filaments dilated near the tip, the anthers small, attached at the base, all similar in size. Ovary linear, sessile or short-stalked, usually glabrous, the style glabrous, curved, the stigma more or less terminal, capitate to linear and curved. Fruits legumes, enclosed by the persistent calyx and corolla, oblong in outline to ovoid, dehiscent or indehiscent, 1–3(–9)-seeded. Seeds more or less spherical to kidney-shaped and somewhat flattened, smooth. About 250 species, widely distributed in temperate to subtropical areas, especially in the northern hemisphere; introduced nearly worldwide.

Most of the nonnative clovers were introduced into the United States for pasturage, forage, and hay, as well as for green manure and other soil improvement, ground covers, and for bees to produce clover honey. Some species are planted along highways for soil stabilization. For most of the commercially important taxa, a number of cultivars have been developed. These often differ morphologically in more pronounced ways than do the varieties and subspecies named for some of the species in nature. Thus, although infraspecific taxa are mentioned in the species treatments, they are not treated formally in the present work.

1. Plants perennial, the main stems prostrate, rooting at the nodes; inflorescences terminal on axillary, more or less erect branches or stalks; petals white to pinkish-tinged
    2. Inflorescence stalk solitary, arising directly from leaf axils of the main prostrate stem and lacking bract-like leaves; calyces 3–5 mm long, the teeth triangular-lanceolate, shorter than to about as long as the tube .... 10. T. REPENS
    2. Inflorescence stalk(s) 1 or 2, from a pair of bractlike leaves at the tip of an ascending branch arising from the leaf axils of the main prostrate stem; calyces 4–7 mm long, the teeth narrowly triangular and long-tapered, about 2 times as long as the tube .... 12. T. STOLONIFERUM
1. Plants annual, biennial, or perennial, erect to loosely ascending, sometimes from a spreading base, the stems not rooting at the nodes; inflorescences axillary and/or terminal; petals various colors
    3. Petals yellow; banner petal straight or incurved; fruit with an obvious stalk inside calyx; petioles mostly shorter than the leaflets
        4. Terminal leaflet sessile or nearly so; stipules nearly as long as to longer than the petiole; seeds ovoid .... 2. T. AUREUM
        4. Terminal leaflet stalked; stipules about half as long as the petiole; seeds ellipsoid
            5. Inflorescences 7–12 mm wide; corollas 3.5–7.0 mm long; stalk of the terminal leaflet 1–3 mm long .... 3. T. CAMPESTRE
            5. Inflorescences 5–7 mm wide; corollas 2.5–3.5 mm long; stalk of the terminal leaflet 0.8–1.5 mm long .... 5. T. DUBIUM

3. Petals white, pink, red, or purple; banner petal outcurved; fruit not or minutely stalked; petioles mostly longer than leaflets
    6. Flowers stalked, becoming sharply reflexed at fruiting, in umbels (in short dense headlike racemes in *T. hybridum*, but these appearing umbellate at fruiting)
        7. Plants annual or biennial, taprooted; corollas 8–14 mm long; inflorescences 2–4 cm wide; flower stalks 4–12 mm long; calyces 6–9 mm long, the teeth 2–3 times as long as the tube; leaflets 1.0–4.5 × 0.5–2.0 cm; stipules broadly ovate, leaflike . . . . 9. T. REFLEXUM
        7. Plants annual or perennial, the rootstock various; corollas 4–10 mm long, inflorescences 1.0–2.5 cm wide; flower stalks 1–6 mm long; calyces 3–4 mm long, the teeth 1–3 times as long as the tube; leaflets 0.6–3.5 × 0.3–2.0 cm; stipules narrowly obovate to lanceolate, not leaflike
            8. Plants annual; corollas 4–6 mm long, creamy white or pale pinkish lavender; calyces hairy, turning red at maturity, the teeth lanceolate to broadly triangular, 2–3 times as long as the tube . . 4. T. CAROLINIANUM
            8. Plants perennial; corollas 7–10 mm long, white, becoming rose pink at maturity; calyces glabrous or nearly so at maturity (sometimes slightly hairier in buds), with a few hairs at the mouth, remaining green at maturity, the teeth narrowly triangular to nearly linear, about as long as the tube . . . . 6. T. HYBRIDUM
    6. Flowers sessile or nearly so, erect to spreading in various directions at fruiting, in dense globose to elongate heads or headlike spikes
        9. Petals lavender to white; flowers resupinate, the banner appearing lowermost; calyces glabrous except for a band of hairs along the lower (dorsal) side, becoming inflated at fruiting, with 5–10 longitudinal nerves and numerous cross-nerves, forming a prominent network . . . . 11. T. RESUPINATUM
        9. Petals pink, red, or white; flowers not resupinate, the banner appearing uppermost; calyces glabrous or evenly hairy, not or only slightly inflated at fruiting (except in *T. vesiculosum*, which also has a network of nerves, but 20–30 longitudinal nerves)
            10. Corollas 5–7 mm long, slightly longer than to shorter than the calyces
                11. Corollas pale pink, the banner narrowly oblong, rounded to bluntly and broadly angled; leaflets linear-oblong to narrowly elliptic, sharply pointed at the tip or occasionally abruptly short-tapered to a minute sharp point . . . . 1. T. ARVENSE
                11. Corollas pink, the banner oblong, notched; leaflets obovate to oblong or elliptic, rounded to shallowly notched at the tip . . . . 13. T. STRIATUM
            10. Corollas 10–20 mm long, longer than the calyces
                12. Plants perennial, with a short stout rootstock; heads sessile or nearly so, globose to ovoid . . . . 8. T. PRATENSE
                12. Plants annual, taprooted; heads stalked, elongate-ovoid to cylindric
                    13. Leaflets broadly ovate to broadly obovate or nearly circular, broadly and bluntly pointed to rounded or shallowly notched at the tip, hairy; stipules broadly ovate to oblong-ovate; corollas red to dark red; calyces hairy, with 10 inconspicuous longitudinal veins (the venation not forming a network), not inflated at fruiting . . . . 7. T. INCARNATUM

13. Leaflets lanceolate to narrowly elliptic, angled or tapered to a sharply pointed tip, glabrous; stipules linear-lanceolate; corollas creamy white, pink, or purple; calyces glabrous, with 20–30 longitudinal nerves and numerous cross-nerves (forming a prominent network), becoming inflated at fruiting . . . . 4. T. VESICULOSUM

1. Trifolium arvense L. (rabbit's foot clover)
Pl. 408 a, b; Map 1809

Plants annual, taprooted. Stems 5–40 cm long, erect or ascending, not rooting at the nodes, often much-branched, with dense, appressed to spreading, silky hairs. Leaves petiolate toward the stem base to nearly sessile toward the tip, the longest petioles to 15 mm, shorter than the leaflets. Stipules longer than the associated petiole, narrowly lanceolate, the fused basal portion short and broad, the elongate free portions narrowly long-tapered at the tips, membranous between the veins, the margins usually entire. Leaflets 5–20 mm long, 2–4 mm wide, all sessile or nearly so, linear to narrowly lanceolate, angled at the base, angled to a sharply pointed tip, sometimes abruptly short-tapered to a minute sharp point, the margins with relatively few, minute, very inconspicuous, blunt teeth (sometimes visible only with magnification), the surfaces finely pubescent with long appressed to loosely appressed hairs. Inflorescences 8–30 mm long (usually elongating with age), 8–10 mm wide, ovoid to cylindric dense spikes, the stalk (1–)5–30 mm long. Flowers 10–150, sessile, spreading in various directions at fruiting. Calyces 4–7 mm long, the tube 1.5–2.0 mm long, long-hairy, often silvery to pinkish- or purplish-tinged, the teeth 1.5–2.0 times as long as the tube, equal or subequal, slender and long-tapered, plumose, lacking a prominent network of nerves and not becoming inflated at fruiting. Corollas 3–4 mm long, shorter than the calyx lobes, pale pink (white elsewhere), the banner outcurved, narrowly oblong, rounded to bluntly and broadly angled at the tip, faintly to moderately nerved, especially with age. Fruits 1.2–1.5 mm long, ovoid, sessile, the outer wall membranous to papery, 1-seeded. Seeds 0.9–1.3 mm long, more or less globose, pale yellow to pale yellowish green, shiny. $2n$=14, 16, 28. May–October.

Introduced, widely scattered, mostly in the Ozark and Ozark Border Divisions (native of Europe, Africa, Asia; introduced widely in the U.S. and Canada). Pastures, old fields, fallow fields, railroads, roadsides, and open disturbed areas.

Rabbit's foot clover is naturalized in many areas of the world and throughout much of North America. It sometimes is cultivated as a winter annual (Henson and Hollowell, 1960). It is adapted to infertile, dry, often sandy soil such as that found on roadsides, where is makes an attractive, silvery-pink display when in flower, and rose to buff when in fruit. It also has been called hare's foot, stone clover, old-field clover, and pussies.

Zohary and Heller (1984) tentatively recognized two varieties, which were said to differ somewhat in habit and pubescence density. Missouri specimens are all var. *arvense*. White-flowered plants, which have yet to be found in Missouri, have been called f. *albiflorum* Sylvén.

2. Trifolium aureum Pollich (yellow clover)
T. agrarium L., an officially rejected name
Pl. 408 f, g; Map 1810

Plants annual or biennial, taprooted. Stems 20–60 cm long, erect or ascending, not rooting at the nodes, often much-branched, with short appressed hairs. Leaves petiolate toward the stem base to short-petiolate above, the longest petioles to 12 mm, shorter than the leaflets. Stipules about as long as to somewhat longer than the associated petiole, oblong-lanceolate, fused to about the midpoint, the free portions narrowly long-tapered at the tip. Leaflets 15–25 mm long, 6–8 mm wide, all sessile or nearly so, oblanceolate to obovate or elliptic, angled at the base, broadly and bluntly pointed to rounded or shallowly notched at the tip, often with a minute sharp point at the very tip, the margins finely toothed above the midpoint, the surfaces usually glabrous. Inflorescences 10–25 mm long (elongating with age), 12–14 mm wide, ovoid to cylindric dense spikelike racemes, sometimes becoming flat-topped with age, the stalk 10–50 mm long. Flowers 10–40(–80), short-stalked, the stalk spreading or becoming reflexed at fruiting. Calyces 2–3 mm long, tube 0.8–1.0 mm long, glabrous, the (longest) teeth 2–3 times as long as the tube, unequal (the lower teeth 2–3 times the length of the upper ones), slender and moderately (shorter teeth) long-tapered (longer teeth), lacking a prominent network of nerves and not becoming inflated at fruiting. Corollas 5–8 mm long, longer than the calyx lobes, bright yellow, turning brown with age, the banner somewhat incurved, broadly obovate, broadly and shallowly notched at the tip, strongly parallel-nerved, especially with

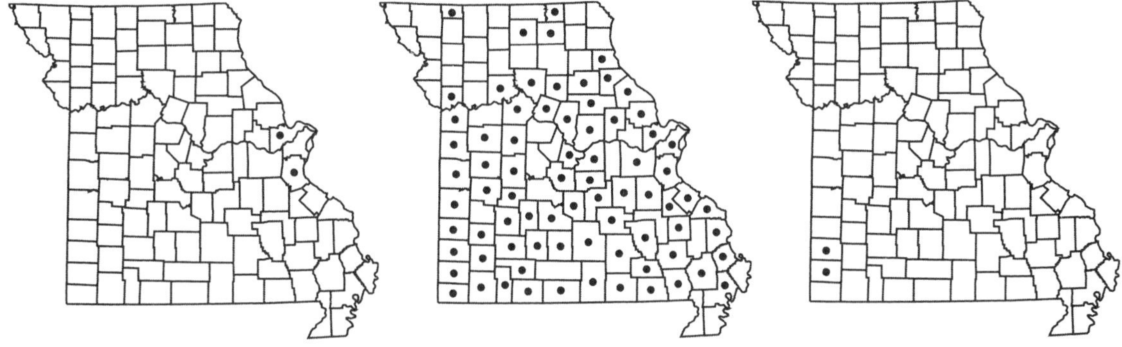

1810. Trifolium aureum

1811. Trifolium campestre

1812. Trifolium carolinianum

age. Fruits 3.0–3.5 mm long, oblong-ovoid, stalked, the outer wall membranous to papery, 1-seeded. Seeds 1.0–1.2 mm, ovoid, pale yellowish green to yellowish brown, shiny. $2n=14, 16$. June–September.

Introduced, uncommon, known thus far only from Jefferson and St. Charles Counties (native of Europe, Asia; introduced widely in the U.S. and Canada). Fallow fields and open disturbed areas.

Trifolium aureum also has been called large hop-clover, hop-clover, and palmate hop-clover. Yellow clover was an early introduction into North America. George Washington is known to have ordered seed of this species from Europe in 1786 (Pieters, 1920).

Zohary and Heller (1984) recognized two subspecies, which differ mainly in subtle details of the leaflet apices and style position. Missouri specimens are all *T. aureum* ssp. *aureum*. The ssp. *barbulatum* Freyn & Sint. ex Freyn is endemic to portions of southeastern Europe and adjacent Asia. Steyermark's (1963) report of a specimen from Christian County could not be confirmed during the present research.

3. Trifolium campestre Schreb. (large hop-clover, Bishop clover)

T. procumbens L., an officially rejected name

Pl. 409 h, i; Map 1811

Plants annual, taprooted. Stems 5–40 cm long, erect or ascending (rarely prostrate), not rooting at the nodes, few- to much-branched, glabrous or with loosely appressed hairs. Leaves long-petiolate toward the stem base to short-petiolate near the tip, the longest petioles to 25 mm, longer than the leaflets. Stipules shorter than the associated petiole, ovate, fused to about the midpoint, the free portions angled or somewhat tapered at the tips, often somewhat membranous in the basal half between the veins, the margins more or less entire. Leaflets 4–16 mm long, 4–8 mm wide, the terminal leaflet with a stalk 1–3 mm long, the lateral leaflets sessile or nearly so, oblong-obovate, angled at the base, rounded to more or less truncate or shallowly notched at the tip, usually with a short broad tooth at the very tip, the margins shallowly toothed above the midpoint, the surfaces glabrous or the undersurface finely hairy, sometimes only along the midvein. Inflorescence 7–15 mm long, 7–10 mm wide, globose to ovoid or cylindric dense spikelike racemes, the stalk 5–90 mm long. Flowers (10–)20–40(–50), short-stalked, the stalk spreading or becoming reflexed at fruiting. Calyces 1.2–2.5 mm long, the tube 0.5–1.0 mm long, glabrous or sparsely hairy, the (longest) teeth 2–3 times as long as the tube, unequal (the lower teeth 2–3 times the length of the upper ones), slender and moderately (shorter teeth) to long-tapered (longer teeth), each tooth often tipped with 1 or 2 stiff hairs, lacking a prominent network of nerves and not becoming inflated at fruiting. Corollas 3.5–6.0 mm, longer than the calyx lobes, pale to bright yellow, the banner incurved, obovate, usually shallowly notched at the tip and with a slightly toothed margin, strongly parallel-nerved, especially with age. Fruits 2.0–2.5 mm long, oblong-ovoid, stalked, the outer wall papery, 1-seeded. Seeds 1.0–1.5 mm long, oblong-ellipsoid, yellow, shiny. $2n=14$. April–September.

Introduced, common south of the Missouri River, widely scattered farther north (native of Europe; introduced widely elsewhere in the world, including most of the U.S., Canada). Banks of streams and rivers, margins of ponds, lakes, and oxbows, bases and tops of bluffs, disturbed portions of upland prairies, savannas, and glades; also fallow fields, old fields, pastures, lawns, levees, ditches, railroads, roadsides, and open disturbed areas.

Trifolium campestre is also called low hop-clover, pinnate hop-clover, Bishop clover, and small hop-clover. It and especially *T. dubium* often are confused with *Medicago lupulina*, which is encountered commonly in Missouri. That species differs from both of these clovers in its usually toothed stipules, nonpersistent corollas, and kidney-shaped, shiny, black fruits.

The Missouri collections are var. *campestre*. Two other varieties occurring within the native distribution have purplish corollas (Zohary and Heller, 1984).

4. Trifolium carolinianum Michx. (Carolina clover)

Pl. 409 f, g; Map 1812

Plants annual, taprooted. Stems 5–30 cm long, spreading to erect or ascending from a usually spreading base, not rooting at the nodes, often much-branched, glabrous or finely hairy. Leaves mostly long-petiolate, the longest petioles to 50 mm, longer than the leaflets. Stipules shorter than the associated petiole, lanceolate to ovate, tips acute to acuminate, fused in the lower $^{1}/_{4}$–$^{1}/_{3}$, the free portion tapered at the tips, all but the lowermost stipules remaining green between the veins. Leaflets 6–20 mm long, 3–13 mm wide, all sessile, obovate, angled at the base, bluntly pointed to rounded or shallowly notched at the tip, usually with a short broad tooth at the very tip, the margins shallowly toothed. Inflorescences 10–20 mm long and wide, umbels, the stalk 5–10 cm long. Flowers 5–40, short-stalked, the stalk becoming reflexed with age. Calyces 2–4 mm long, the tube 1.0–1.5 mm long, glabrous or more commonly finely hairy, unequal, the upper teeth 2–3 times as long as the tube, the shorter lower teeth 1–2 times as long as the tube, lanceolate to broadly triangular, lacking a prominent network of nerves and not becoming inflated at fruiting. Corollas 4–6 mm long, longer than the calyx teeth, creamy white to pinkish lavender, turning brown with age, the banner outcurved, broadly ovate, rounded to bluntly pointed at the sometimes slightly irregular tip, finely to moderately nerved, usually with age. Fruits 3.0–4.5 mm long, oblong in outline, very short-stalked, the outer wall papery to somewhat leathery and hairy toward the tip, 2–4-seeded. Seeds 1.0–1.5 mm long, more or less globose to slightly kidney-shaped, brown, dull and finely pebbled. $2n$=16. March–June.

Uncommon, known thus far only from historical collections from Boone and Jasper Counties and extant sites in Newton County (southeastern U.S. west to Kansas and Texas; introduced sporadically northward to Vermont). Upland prairies and glades; also pastures and open disturbed areas.

Carolina clover is uncommon throughout much of its native range in the southeastern United States (Isely 1998). The species had not been seen in Missouri since 1923 before being rediscovered in 2011 by members of the Missouri Native Plant Society in a system of chert glades in the Joplin area (Newton County). Steyermark's (1963) report of a collection by Viktor Mühlenbach from the St. Louis railyards was based on misdetermined plants of *T. campestre*.

5. Trifolium dubium Sibth. (little hop-clover)

Pl. 409 a–c; Map 1813

Plants annual, more or less taprooted. Stems 5–40 cm tall, erect or ascending from an often spreading base, sometimes rooting at the lower nodes, unbranched or branched, glabrous or with fine appressed hairs. Leaves short- or moderately petiolate toward the stem base to nearly sessile toward the tip, the longest petioles to 15 mm, mostly shorter than the leaflets. Stipules shorter than the associated petiole, ovate, the fused basal portion less than half the total length, the free portions tapered at the tip, herbaceous but paler toward the base. Leaflets 5–15 mm long, 4–7 mm wide, the terminal leaflet with a stalk 0.8–1.5 mm long, the lateral leaflets sessile or nearly so, obovate, angled at the base, rounded to shallowly notched at the tip, usually with a small tooth at the very tip, the margins shallowly and inconspicuously toothed, the surfaces glabrous or sparsely hairy along the midvein. Inflorescence 5–10 mm long, 6–8 mm wide, more or less ovoid to obovoid dense spikelike to headlike racemes, the stalk 12–30 mm long. Flowers 3–20, short-stalked, the stalk becoming sharply reflexed with age. Calyces 1.5–2.0 mm long, the tube 0.5–0.8 mm, glabrous, the teeth 1.5–2.0 times as long as the tube, unequal (the lower teeth about 2 times as long as the upper ones) slender and moderately (shorter teeth) to long-tapered (longer teeth), lacking a prominent network of nerves and not becoming inflated at fruiting. Corollas 3–4 mm long, longer than the calyx lobes, pale yellow, turning brown with age, the banner straight to more or less incurved, narrowly oblong-ovate, rounded at the tip, finely and relatively faintly nerved. Fruits 1.5–2.0 mm long, ovoid, stalked, the outer wall membranous, 1(2)-seeded. Seeds 1.0–1.5 mm long, ellipsoid, tan to dark brown, shiny. $2n$=16, 28. April–September.

Introduced, scattered in the southern half of the state, mostly in the Unglaciated Plains and Mississippi Lowlands Divisions (native of Europe; introduced nearly worldwide, including widely in the U.S., Canada). Banks of streams and rivers,

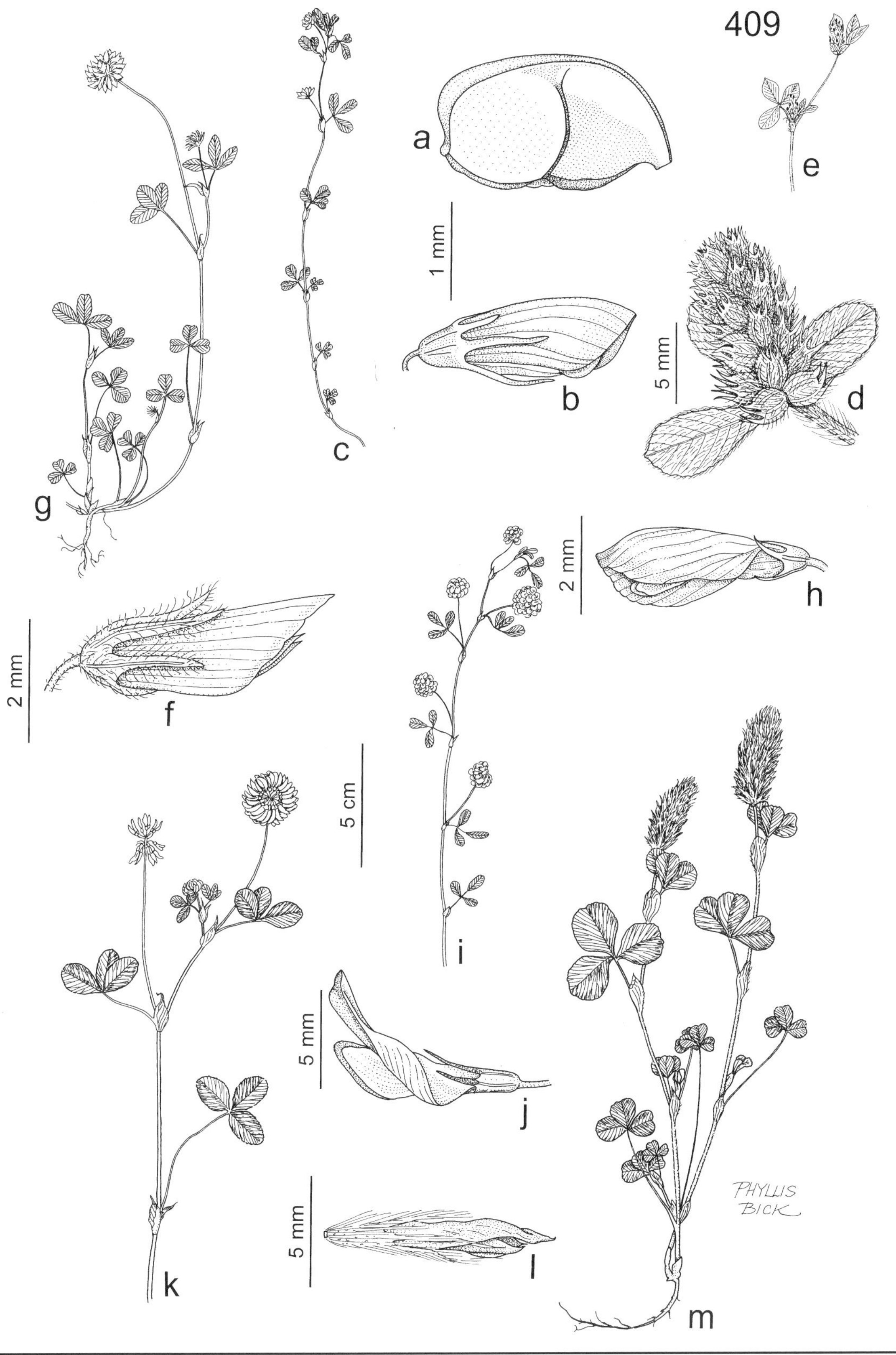

Plate 409. Fabaceae. *Trifolium dubium*, **a)** fruit, **b)** flower, **c)** habit. *Trifolium striatum*, **d)** inflorescence, **e)** fertile stem. *Trifolium carolinianum*, **f)** flower, **g)** habit. *Trifolium campestre*, **h)** flower, **i)** fertile stem. *Trifolium hybridum*, **j)** flower, **k)** fertile stem. *Trifolium incarnatum*, **l)** flower, **m)** habit.

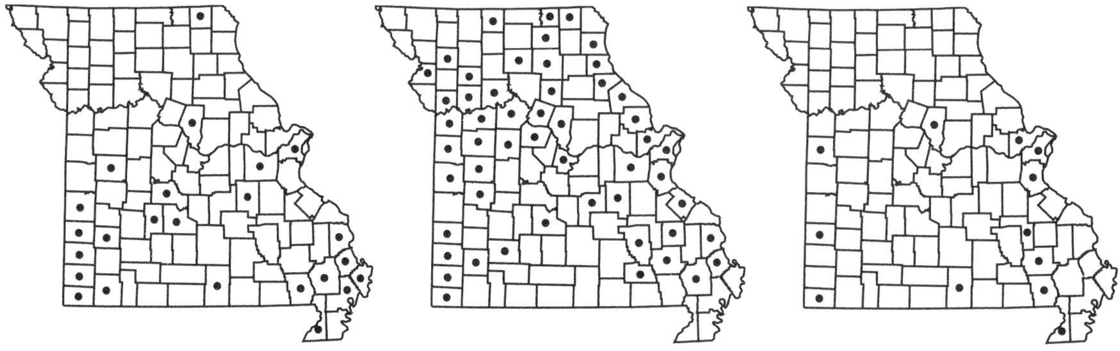

1813. Trifolium dubium 1814. Trifolium hybridum 1815. Trifolium incarnatum

margins of ponds and lakes, openings of bottomland forests, savannas, and glades; also lawns, levees, ditches, cemeteries, railroads, roadsides, and open disturbed areas.

Trifolium dubium is also called least hop-clover, small hop-clover, and shamrock. It is often confused with *T. campestre*, but can be distinguished from it by the smaller inflorescences with fewer flowers, as well a banner petal that is only finely and faintly nerved (vs. corrugated with strongly impressed nerves in *T. campestre*). It also commonly has been confused with *Medicago lupulina* (black medic), but can be distinguished as described in the treatment of *T. campestre*.

Trifolium dubium is thought by some to be the shamrock of Irish folklore, but others claim that the shamrock may be one of several species of *Trifolium*, *Medicago*, or *Oxalis* (Colgan, 1896; Everett, 1971; E. C. Nelson, 1991). Today, the houseplants most frequently sold in the United States as shamrocks are members of the genus *Oxalis* (Oxalidaceae). For further discussion, see the treatment of that genus in the present volume.

6. Trifolium hybridum L. (Alsike clover)
Pl. 409 j, k; Map 1814

Plants perennial, rhizomatous, Stems 15–60 (–80) cm long, loosely to strongly ascending, sometimes from a spreading base, not rooting at the lower nodes, glabrous or nearly so, sometimes somewhat fleshy. Leaves mostly long-petiolate, gradually reduced toward the stem tip, the longest petioles to 80(–100) mm, long, longer than the leaflets. Stipules shorter than the associated petiole, ovate to lanceolate, the fused basal portion about 1/3 of the total length, the free portions long-tapered at the tips, more or less herbaceous or sometimes becoming membranous between the veins. Leaflets 10–35 mm long, 10–20 mm wide, all sessile or nearly so, ovate to elliptic or rhombic, angled to broadly angled at the base, rounded to shallowly notched at the tip, usually with a minute triangular tooth at the very tip, the margins finely toothed, the surfaces glabrous or the undersurface sparsely hairy along the veins. Inflorescences 10–25 mm long and wide, globose headlike short dense racemes, appearing umbellate at fruiting, the stalk 20–80 mm long. Flowers 20–80, short- to moderately stalked, the stalk becoming reflexed with age. Calyces 2–4 mm long, the tube 1–2 mm long, glabrous, the teeth narrowly triangular, about as long as the tube, equal or subequal, long-tapered, inconspicuously hairy in the U-shaped sinuses, lacking a prominent network of nerves and not becoming inflated at fruiting. Corollas 6–11 mm long, longer than the calyx lobes, white or pinkish-tinged (commonly pink elsewhere), the banner outcurved, oblong-ovate, rounded or less commonly very shallowly notched at the tip, finely but sometimes relatively strongly parallel-nerved, usually with age. Fruits 3–4 mm long, oblong in outline, minutely stalked, 2–4-seeded. Seeds 1.0–1.3 mm long, ovoid to slightly kidney-shaped, mottled yellow and brown, red and brown, to nearly black, dull. $2n=16$. May–October.

Introduced, scattered nearly throughout the state, but absent or uncommon in portions of the Ozark Division (native of Europe and possibly Asia, introduced in temperate regions nearly worldwide). Bottomland and upland prairies, banks of streams and rivers, and edges and openings of bottomland forests; also pastures, old fields, railroads, roadsides, and open disturbed areas.

Alsike clover also has been called Alsatian clover and Swedish clover. Apparently, it was first cultivated in Sweden, and cultivated in England by about 1832 (N. L. Taylor, 1975). It was first brought to the United States around 1839 (N. L. Taylor, 1975). The plants may cause dermatitis in some persons (Steyermark, 1963; Hardin and

Arena, 1974). *Trifolium hybridum* also is said to cause photosensitivity and biliary fibrosis in some livestock (Fisher, 1995), but the connection between these diseases and the clover is not conclusive (Nation, 1989).

Zohary and Heller (1984) accepted three varieties within *Trifolium hybridum* differing mainly in growth form and branching patterns. Missouri collections seem referable mostly to var. *elegans* (Savi) Boiss. (var. *pratense* Rabenh.).

Trifolium nigrescens Viv. (ball clover, a Mediterranean species) is found with increasing frequency in southeastern United States (Isely 1990, 1998), and has been documented from numerous sites in Tennessee. It is possible that this species will be encountered in the southeastern counties. Ball clover is an annual, prostrate to ascending, glabrous to nearly glabrous species, which can be distinguished from Alsike clover by its habit, as well as by V-shaped sinuses between the calyx lobes (U-shaped in *T. hybridum*), white to cream or yellowish white (rarely pale pinkish) corollas (generally pinkish in *T. hybridum*), and stipules with sharply recurved, black to dark maroon, slender tips (straight, green tips in *T. hybridum*).

7. Trifolium incarnatum L. (crimson clover)
Pl. 409 l, m; Map 1815

Plants annual, taprooted. Stems 20–50(–90) cm tall, erect or ascending, not rooting at the lower nodes, unbranched or few-branched toward the base, densely pubescent with appressed to spreading tawny hairs. Leaves long-petiolate toward the stem base to nearly sessile toward the tip, the longest petioles to 60 mm, much longer than the leaflets. Stipules shorter than the associated petiole, broadly ovate to oblong-ovate, fused to above the midpoint and sheathing the stem, the free portions angled at the tips, white to pale green with dark green to reddish purple veins toward the base, the margins toothed toward the tip and rimmed with dark reddish purple or green. Leaflets 10–30(–40) mm long, 10–20(–30) mm wide, all sessile or nearly so, broadly ovate to broadly obovate or nearly circular, broadly angled at the base, broadly and bluntly pointed to rounded or shallowly notched at the tip, the margins irregular or shallowly toothed, the surfaces with relatively long tawny hairs. Inflorescence 20–60 mm long (elongating with age), 10–20 mm wide, dense narrowly ovoid to more or less cylindric spikes, the stalk 10–60 mm long. Flowers numerous (usually more than 150), sessile or nearly so, ascending at fruiting. Calyces 5–10 mm long, the tube 3–5 mm long, moderately to densely long-hairy, the teeth narrowly triangular to nearly linear, 1–2 times as long as the tube, equal or nearly so, long-tapered, plumose, inconspicuously 10-nerved (the venation not forming an obvious network) and not becoming inflated at fruiting. Corollas 10–17 mm long, longer than the calyx lobes, red to dark red (rarely white or pink elsewhere), the banner outcurved, linear-oblong to narrowly elliptic, usually sharply pointed at the tip, finely and relatively faintly nerved. Fruits 3–4 mm long, oblong-ovoid, sessile, the outer wall papery, 1-seeded. Seeds 1.9–2.3 mm long, ovoid to elliptic-ovoid, tan to brown. $2n=14$. April–July.

Introduced, widely scattered in the southern half of the state (native of Europe; introduced in temperate regions nearly worldwide, in North America most commonly in the southern U.S.). Pastures, roadsides, and open disturbed areas.

Crimson clover also has been called Italian clover and many other common names. The species has been cultivated since the 1700s in Europe and was introduced into the United States in 1818 (Knight, 1985). Crimson clover is used extensively as a ground cover in crop rotations, for green manure, as a nitrogen-fixing plant in fields, and as an annual hay crop.

Plants of *T. incarnatum* in Missouri all have red corollas and are referable to var. *incarnatum*. Zohary and Heller (1984) recognized slender-stemmed plants from southern Europe with cream-colored to pink corollas as var. *molineri* (Balb. ex Hornem.) Ser.

8. Trifolium pratense L. (red clover)
Pl. 410 i, j; Map 1816

Plants perennial, with a short stout rootstock. Stems 20–60(100) cm tall, erect or ascending, much-branched, glabrous or with appressed to spreading hairs. Leaves long-petiolate toward the stem base to nearly sessile toward the tip, the longest petioles to 80 mm, 3–4 times the length of the leaflets. Stipules much shorter than to about as long as the associated petiole, ovate to lanceolate, fused more than 2/3 of the way to the tips, the free portions short-tapered to long slender tips, pale with dark green to red veins, the margins usually entire. Leaflets 10–30(–50) mm long, 7–15(–25) mm wide, all sessile or nearly so, ovate to elliptic or obovate, broadly angled at the base, rounded or rarely minutely notched at the tip, usually with a minute broad tooth at the very tip, the margins minutely irregular or inconspicuously and broadly scalloped or toothed, often only near the tip, the surfaces glabrous or sparsely to moderately appressed-hairy. Inflorescences 10–30 mm long and

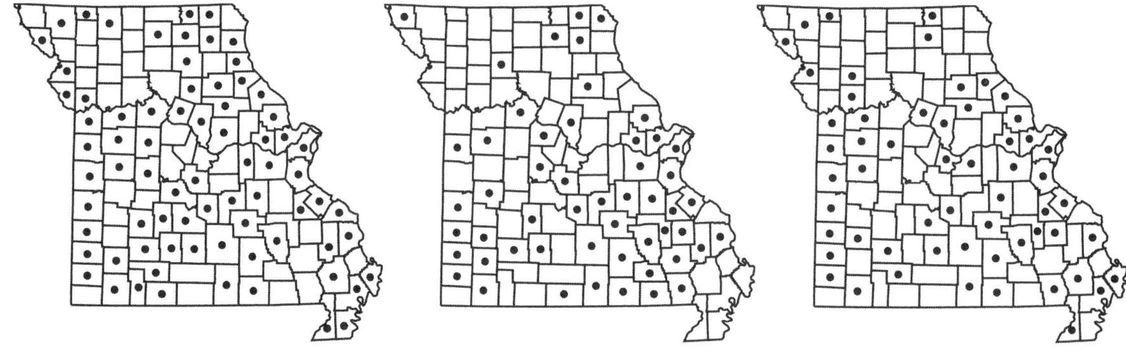

1816. Trifolium pratense 1817. Trifolium reflexum 1818. Trifolium repens

wide, dense globose to ovoid headlike spikes, sessile or the stalk 1–4 mm (closely subtended by a pair of bractlike leaves). Flowers 40–150, sessile, ascending to spreading at fruiting. Calyces 5–8 mm long, the tube 2.5–4.0 mm long, sparsely to moderately hairy, the teeth narrowly triangular to nearly linear, unequal, the lowest tooth about as long as the tube, the others nearly equal and much shorter, moderately hairy, lacking a prominent network of nerves and not becoming inflated at fruiting. Corollas 11–18 mm long, longer than the calyx lobes, reddish purple or rarely white, cream-colored, or pale pink, the banner outcurved, oblong-oblanceolate, shallowly notched at the tip, inconspicuously nerved. Fruits 2–3 mm long, oblong-obovoid, sessile, the outer wall membranous below a well-defined somewhat hardened shiny apical region, 1(2)-seeded. Seeds 1.5–2.0 mm long, ovoid to slightly kidney-shaped, tan to brown, dull. $2n$=14, 28, 56. April–October.

Introduced, common nearly throughout the state (native of Europe, Asia; introduced nearly worldwide). Upland prairies, glades, banks of streams and rivers, and margins of ponds, lakes, marshes, sloughs, and oxbows; also pastures, old fields, fallow fields, quarries, lawns, levees, ditches, roadsides, railroads, and open disturbed areas.

Red clover (also called purple clover) is grown in more areas of the world than any other species of *Trifolium* (N. L. Taylor, 1975; R. R. Smith et al., 1985), and it has been in cultivation since the third and fourth centuries, probably beginning in Spain, from where it spread to Holland and Lombardy, then to Germany. The species was introduced into England about 1645, from where it was brought to the New World by 1663 (N. L. Taylor and Quesenberry, 1996). It is a very important forage crop, but may also cause bloating in animals that overindulge in its young growth; a diet high in red clover also may cause infertility in sheep (N. L. Taylor and Quesenberry, 1996). Red clover has been used in a tea and to flavor vinegar (Coon, 1980), as well as medicinally as an ingredient in herbal cough syrup (Gibbons, 1962; Coon, 1980) and as a salve to treat eye and skin diseases (Reader's Digest Association, 1984). There are even claims that red clover can be used in cancer treatments (Duke, 1985; Ritchason, 1995). The young growth can be cooked as a vegetable (Coon, 1980). Dried flower heads have been ground and used in breads during times of famine (Millspaugh, 1974).

Trifolium pratense is morphologically very variable, and many infraspecific names have been published. Zohary and Heller (1984) recognized six varieties that differ in a confusing array of characters of the habit, stipules, calyces, and pubescence patterns. Missouri collections are most similar to var. *pratense* and var. *sativum* Schreb., but there is much intergradation in North American populations of the species. Rare white-flowered plants have been called f. *leucochraceum* Asch. & Prantl.

Trifolium hirtum All. (rose clover), an annual species resembling red clover, has been reported from nearby states and may be found in Missouri, especially as it is an occasional component of roadside seed mixtures. It differs from both *T. incarnatum* and *T. pratense* in its stipules, which have a long-tapered, slender, free apical portion.

9. Trifolium reflexum L. (buffalo clover)
Pl. 410 a, b; Map 1817

Plants annual or more commonly biennial, taprooted. Stems 20–50 cm long, loosely to strongly ascending, not rooting at the lower nodes, unbranched or few branched toward the base, glabrous to densely short-hairy. Leaves long-petiolate toward the stem base, somewhat reduced toward the tip, the longest petioles to 65 mm long, 3–4 times as long as the leaflets. Stipules shorter than the associated petiole, broadly ovate, fused nearly

Plate 410. Fabaceae. *Trifolium reflexum*, **a)** flower, **b)** habit. *Trifolium resupinatum*, **c)** fruiting calyx, **d)** habit. *Trifolium stoloniferum*, **e)** flower, **f)** habit. *Trifolium repens*, **g)** flower, **h)** habit. *Trifolium pratense*, **i)** flower, **j)** fertile stem.

to the midpoint and somewhat sheathing the stem, the free portions angled or tapered at the tips, herbaceous and leaflike, the margins entire or toothed. Leaflets 10–30(–45) mm long, 6–20(–25) mm wide, all sessile or nearly so, ovate to obovate, angled at the base, broadly rounded to sharply pointed at the tip, the margins sharply toothed. Inflorescences 20–35(–40) mm long and wide, dense globose umbels, the stalk 20–60(–80) mm long. Flowers 10–40, the stalk 4–8 mm long at flowering, elongating to 7–12(–15) mm and becoming sharply reflexed at fruiting. Calyces 4–9 mm long, the tube 1.0–1.5 mm long, glabrous to densely hairy, the teeth narrowly triangular to linear, 3–5 times as long as the tube, nearly equal, lacking a prominent network of nerves and not becoming inflated at fruiting. Corollas 8–14 mm long, longer than the calyx lobes, white to cream-colored or occasionally deep pink, the banner outcurved, oblong to elliptic, bluntly and broadly pointed to rounded or shallowly notched at the tip and with an entire or more commonly somewhat irregular margin, finely but relatively strongly parallel-nerved with age. Fruits 3–5 mm, ovoid to oblong-obovoid, minutely stalked, (1)2–4-seeded. Seeds 1.0–1.5 mm long, more or less globose, pale yellow, sometimes with darker mottling, dull and finely pebbled. $2n$=16. May–August.

Scattered widely in the state, but absent from most of the Mississippi Lowlands Division and most of the western portion of the Glaciated Plains (eastern U.S. west to Nebraska and Texas; Canada). Upland prairies, glades, openings and edges of bottomland to mesic and dry upland forests, savannas, banks of streams, and margins of ponds, lakes, and oxbows; also pastures, ditches, and roadsides.

Populations of this species often appear at sites after a burn, heavy logging, or some equally severe disturbance, and they also occur along logging roads. Without continued disturbance, plants usually disappear after a few years, but remain in the soil seed bank. According to N. L. Taylor et al. (1994), although plants are routinely visited by bees and other insects, buffalo clover is mostly self-pollinated.

Steyermark (1963) and some other authors have recognized two varieties (the hairy var. *reflexum*, and the more glabrous var. *glabrum* Lojac.), but the pubescence character is variable, with the more glabrous forms most common in the northeastern part of the range. Both types of plants, as well as apparent intermediates, are present in Missouri, and glabrous and pubescent plants may grow intermingled in some populations.

10. Trifolium repens L. (white clover)

Pl. 410 g, h; Map 1818

Plants perennial with fibrous roots and sometimes short rhizomes. Stems 10–40 cm long, prostrate and rooting at the nodes, glabrous or occasionally sparsely hairy. Leaves long-petiolate, the longest petioles to 200 mm, much longer than the leaflets. Stipules shorter than the associated petiole, ovate to lanceolate, fused most of the way into a sheathing tube (this rupturing as the inflorescence develops), the free portions short-tapered to the slender tips, membranous, whitish- to brownish-tinged, often with darker reddish to green veins, the margins entire. Leaflets 6–30 mm long, 10–25 mm wide, all sessile or nearly so, broadly elliptic to ovate, broadly angled at the base, bluntly pointed to more commonly rounded or shallowly to more deeply notched at the tip, the margins sharply and finely toothed, the surfaces glabrous or the undersurface sparsely hairy along the veins. Inflorescences 15–35 mm long and wide, dense globose (or nearly so) umbels or very short racemes, the stalk (which are mostly erect and develop singly from the leaf axil) 50–200 cm long. Flowers 20–50(–100), the stalk 1–2 mm long at flowering, elongating to 4–6 mm and becoming sharply reflexed at fruiting. Calyces 3–5 mm long, the tube 1.8–3.0 mm long, glabrous, the teeth triangular-lanceolate, shorter than (upper teeth) to about as long as (lower teeth) as the tube, unequal, often dark purple around the V-shaped sinuses, lacking a prominent network of nerves and not becoming inflated at fruiting. Corollas 7–12 mm long, longer than the calyx lobes, white to pinkish-tinged, the banner outcurved, elliptic-obovate, mostly rounded at the tip, the margins sometimes minutely irregular, finely and usually inconspicuously nerved. Fruits 3–5 mm long, narrowly oblong in outline, 3- or 4-seeded. Seeds 0.9–1.5 mm long, nearly globose to slightly kidney-shaped, yellowish tan to brown, somewhat shiny. $2n$=16, 28, 32, 48, 64. March–November.

Introduced, scattered to common throughout the state (native of Europe, Asia; introduced in temperate regions nearly worldwide). Banks of streams and rivers, margins of ponds, lakes, sinkhole ponds, and fens, and edges of bottomland forests; also old fields, fallow fields, pastures, lawns, gardens, roadsides, and open disturbed areas.

White clover (also called Dutch clover and Ladino clover) may be the most important temperate pasture plant (Piper, 1924; M. Baker and Williams, 1987). It was introduced so early and was so widely grown in North America that it was known to Native Americans as “white man’s foot grass” (Strickland, 1801). Its cultivation may have

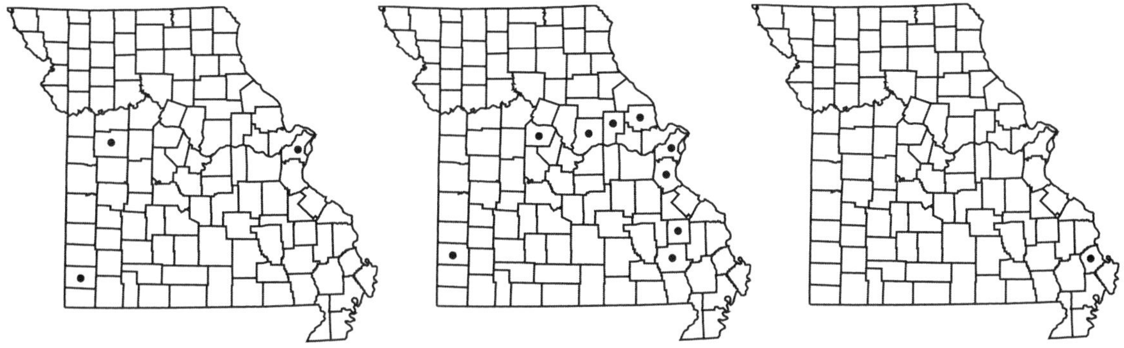

1819. Trifolium resupinatum 1820. Trifolium stoloniferum 1821. Trifolium striatum

begun in the early 1700s, and it was widespread by the middle of that century (Isely, 1998).

White clover may have some medicinal uses, although, according to Millspaugh (1974), human ingestion of powdered fresh flower heads resulted in "a sensation of fullness and congestion of the salivary glands with pain, and mumps-like pain followed by copious flow of saliva." Four-leaf and 5-leaf variants of this species are rare developmental abnormalities affecting individual leaves of plants otherwise possessing three leaflets per leaf. Four-leaf clovers have a very long history as symbols of good luck. The four-leaf clovers sometimes grown horticulturally mostly are an aquatic fern, *Marsilea quadrifolia* L. (water clover), in which the leaves all have four leaflets. *Oxalis deppei* Lodd., a Mexican species having leaves with mostly four leaflets, also occasionally is sold as a kind of four-leaf clover.

Trifolium repens is extremely plastic morphologically, and varies greatly in size of both leaves and flowers depending upon environmental conditions (J. M. Gillett and Cochrane, 1973), in light of which taxonomic recognition of forms or varieties probably is not warranted. Zohary and Heller (1984) recognized nine intergrading varieties. Most Missouri specimens appear to be var. *repens*, but plants with larger leaves and inflorescences seem to correspond to var. *giganteum* Lagr.-Fossar (these populations are often referred to as Ladino clover). A recent monograph on the species covers in great detail many aspects of its taxonomy, morphology, and cultivation (M. Baker and Williams, 1987).

11. Trifolium resupinatum L. (Persian clover)
Pl. 410 c, d; Map 1819

Plants annual, taprooted. Stems 10–60 cm long, variously erect or loosely to strongly ascending, few- to much-branched, often from near the base, glabrous or nearly so. Leaves long-petiolate toward the stem base to nearly sessile toward the tip, the longest petioles to 10 cm long, much longer than the leaflets. Stipules shorter than to longer than (near the stem tip) the associated petiole, lanceolate to narrowly ovate, fused to above the midpoint into a sheathing tube (this rupturing as the branches and inflorescences develop), the free portions tapered at the tips, membranous. Leaflets 5–20(–30) mm long, 2–4 mm wide, all sessile or nearly so, broadly obovate to elliptic, or occasionally lanceolate to rhombic, angled at the base, rounded to less commonly bluntly or sharply pointed at the tip, the margins finely toothed, sometimes only below the midpoint, the surfaces glabrous. Inflorescences 8–15 mm long and broad, dense hemispherical to globose heads, the stalk 20–50 mm. Flowers 6–20, sessile or nearly so, resupinate (the banner appearing lowermost), ascending to spreading in all directions at fruiting. Calyces 2–3 mm long at flowering (becoming enlarged at fruiting), the tube 1.5–2.0 mm at flowering, glabrous except for a band of hairs along the lower (dorsal) side, the teeth narrowly triangular with a hairlike tip, shorter than the tube, unequal, the apparent upper teeth noticeably longer than the lower ones, becoming enlarged and inflated at fruiting, the 5–10 longitudinal nerves with numerous cross-nerves, forming a prominent network. Corollas 4–9 mm long, longer than the calyx teeth, lavender to pink or rarely white, the banner outcurved, oblong, shallowly notched at the tip, inconspicuously nerved. Fruits 1.7–2.3 mm long, oblong to ovoid, sessile or nearly so, the outer wall membranous, 1(2)-seeded. Seeds 1.2–2.0 mm long, more or less ovoid, yellow to tan or purplish brown, dull. $2n$=14, 16, 32. May–July.

Introduced, uncommon, known thus far only from Newton and St. Louis Counties (native of Europe, Asia; introduced widely, in the U.S. most commonly in southeastern states). Lawns and open disturbed areas.

The resupinate corolla (inverted, with the banner below and the keel above) and inflated fruiting calyx makes *T. resupinatum* easy to distinguish from other clovers. The species also has been called reversed clover, shaftal clover, and birdseye clover.

12. Trifolium stoloniferum Muhl. ex Eaton (running buffalo clover)

Pl. 410 e, f; Map 1820

Plants perennial, fibrous-rooted but eventually developing a taproot, occasionally with short rhizomes. Stems of 2 kinds, the main stems 10–50 (–100) cm long, prostrate and stoloniferous, rooting at the nodes, with few prostrate branches (sometimes persisting as vegetative clones) but usually with few to several erect to ascending fertile branches (with 1 or occasionally 2 terminal inflorescences), glabrous or nearly so; fertile branches 10–40 cm long, unbranched. Leaves of the prostrate stems mostly well-spaced, long-petiolate, the longest petioles to 200 mm long; those of the fertile branches 2(3) and appearing nearly opposite (or whorled), the petiole shorter than to about as long as the leaflets. Stipules shorter than the associated petiole, those of the prostrate stems broadly lanceolate to somewhat rhombic, fused to about the midpoint or slightly more, the free portions tapered at the tip, more or less membranous, the margins entire or few-toothed; those of the upright stems ovate to oblong-ovate or somewhat rhombic, fused to below the midpoint, the free portions broadly triangular and tapered at the tips, the margins few- to several-toothed. Leaflets 10–40 mm long, 8–35 mm wide, all sessile or nearly so, obovate to broadly obovate, broadly angled at the base, rounded to broadly notched at the tip, the margins finely toothed, the surfaces glabrous or nearly so. Inflorescences 15–30 mm long and wide, dense short globose racemes, appearing umbellate, the stalk 10–80 mm long. Flowers 25–45; the stalk 1–4 mm long at flowering, elongating to 5–10 mm and becoming sharply reflexed at fruiting. Calyces 4–7 mm long, the tube 1.5–2.5 mm long, glabrous or sparsely and very finely hairy, the teeth about 2 times as long as the tube, equal or nearly so, narrowly triangular and long-tapered, lacking a prominent network of nerves and not becoming inflated at fruiting. Corollas 8–14 mm long, longer than the calyx lobes, white to pale cream-colored, sometimes pinkish-tinged with age, the banner outcurved, obovate to oblong-elliptic, rounded to slightly notched at the tip, often with a slender minute point at the very tip but the margins entire, inconspicuously nerved. Fruits 2.5–3.0 mm long, oblong in outline, stalked, the outer wall membranous to papery. 1- or 2-seeded. Seeds 1.3–2.0 mm long, more or less kidney-shaped, yellow to brown, somewhat shiny. $2n$=16, 32. April–July.

Uncommon, known historically from sporadic counties in the southern half of the state and from only a few extant sites in Callaway, Lincoln, Madison, and Montgomery Counties (Kansas to Arkansas east to Ohio and West Virginia). Banks of spring branches, streams and rivers; also semi-shaded disturbed areas.

Running buffalo clover was once far more common across its range than it is today (R. E. Brooks, 1983). Historically, it was most-collected in montane West Virginia and in the Ohio River drainage from Ohio south to central Kentucky. It formerly was found in great stands in Kentucky (J. J. N. Campbell et al., 1988) and Ohio (Cusick, 1989). Causes for its decline are not fully understood. This species was thought to be possibly extinct by the early 1980s, but was rediscovered in West Virginia in 1983 by Rodney Bartgis (1985) of The Nature Conservancy. In 1987, the species was listed as Endangered by the U.S. Fish and Wildlife Service under the Federal Endangered Species Act. However, as of this writing there are indications that the status of the species will soon be revised to Threatened because good progress has been made toward its recovery. Extant populations presently are known from Indiana, Kentucky, Missouri, Ohio, and West Virginia.

In Missouri, running buffalo clover originally was collected in 1830 by L. C. Beck in St. Louis County. Several historical collections were made subsequently, mostly in the 1880s, but the species was not observed after 1907 until plants appeared in 1990 in the yard of George and Kay Yatskievych in the city of St. Louis and later also in the yard of Nathan Pate in St. Louis County. These plants had sprouted from seeds lying long-dormant in soil along the Meramec River in Jefferson County, downstream from historical localities. The soil was excavated and sold for topsoil. Unfortunately, searches around the source site were fruitless. However, an extant population subsequently was discovered in Madison County by Deborah Rowan (1994, 1995), then with the Missouri Department of Conservation. More recently, viable colonies also have been discovered in Lincoln and Madison Counties by Bruce Schuette of the Missouri Department of Natural Resources. Two other small populations were found in the late 1990s in Maries and Phelps Counties in disturbed areas at public river accesses, but these plants did not persist.

Additionally, between 1990 and 1994, the Missouri Botanical Garden cooperated with the Missouri Department of Conservation and the Mark Twain National Forest to attempt the reintroduction of the species in experimental plantings at 38 sites on public lands in a number of counties (T. E. Smith, 1998a). Most of these plantings failed to thrive, but several produced fruits during one or more growing seasons.

N. L. Taylor et al. (1994) stated that running buffalo clover is a mostly outcrossing species that sets fewer seeds if self-pollinating, but that seed set in self-pollinated plants was still high enough to maintain the species in the wild. They also suggested that habitat loss and competition may contribute more to the decline of the species than inbreeding. Hickey et al. (1991) found that genetic diversity was low among many populations of the species, based on allozyme banding patterns, but Crawford et al. (1998), using RAPD molecular markers, found that most populations were not single clones and that even the smallest populations contained unique genetic information.

Trifolium stoloniferum sometimes has been confused with *T. repens* by collectors, from which it differs in the bract-like pair of leaves below the inflorescence on the upright stems, and by the larger size of the plants and the flowers.

13. Trifolium striatum L. (knotted clover)
Pl. 409 d, e; Map 1821

Plants annual, taprooted. Stems (5–)10–50 cm long, erect or ascending, sometimes from a spreading base, unbranched to much-branched, with long, spreading, shiny hairs. Leaves long-petiolate toward the stem base to nearly sessile toward the tip, the longest petioles to 10 cm, mostly longer then the leaflets. Stipules longer than to often shorter than (near the stem tip) the associated petiole, ovate to narrowly ovate, fused to about the midpoint, the free portions abruptly long-tapered toward the tip, mostly membranous with green or red veins, the margins mostly entire. Leaflets 5–20 mm long, 3–17 mm wide, narrowly obovate, oblong, or oblong-lanceolate, angled at the base, bluntly pointed to more or less truncate or shallowly notched, usually with a small broad tooth at the very tip, the margins finely toothed, the surfaces silky-hairy. Inflorescences 8–20 mm long, 6–15 mm wide, ovoid or oblong-ovoid, becoming elongate and more or less cylindric with age, the stalk absent or inconspicuous (1 or 2 heads closely subtended by a bractlike leaf). Flowers 50–120, sessile or nearly so, ascending at fruiting. Calyces 4–7 mm long, tube 2–4 mm, appressed-hairy, the teeth shorter than to about as long as the tube, unequal (the lowermost tooth usually noticeably longer than the others), narrowly triangular to nearly linear, tapered, with 10 prominent longitudinal nerves but lacking noticeable coss-nerves, becoming inflated (globose to urn-shaped) at fruiting. Corollas 4–7 mm, shorter than to about as long as the calyx lobes, pink, the banner outcurved, oblong, rounded to shallowly notched at the tip, inconspicuously nerved. Fruits 3.0–4.5 mm long, obovoid, sessile or nearly so, the outer wall membranous to papery, 1-seeded. Seeds 1.2–1.5 mm long, oblong-ovoid to nearly globose, reddish brown, shiny. $2n$=14. May–August.

Introduced, uncommon, known thus far only from Scott County (native of Europe, Asia; introduced in North America mostly along the eastern and western seaboards). Lawns and open disturbed areas.

Knotted clover is known from relatively few collections in North America. The first Missouri record was collected by Bill Summers in 1992. In nature, *T. striatum* is a morphologically variable species, and numerous varieties have been described. However, Zohary and Heller (1984) concluded that none of these is worthy of taxonomic recognition.

14. Trifolium vesiculosum Savi (arrowleaf clover)
Map 1822

Plants annual, taprooted. Stems 15–60 cm long, loosely ascending to erect, sometimes from a spreading base, not rooting at the lower nodes, unbranched to much-branched, glabrous or nearly so. Leaves long-petiolate toward the stem base, short-petiolate toward the tip, the longest petioles to 9 cm, mostly longer than the leaflets. Stipules longer than to sometimes shorter than (near the stem tip) the associated petiole, linear-lanceolate, fused for up to 2/3 of the way to the tip, the free portions long-tapered to the tip, membranous and whitish to brown, the margins entire. Leaflets 0.5–4.0(–5.0) mm long, 0.3–1.5 cm wide, obovate or more commonly oblanceolate to elliptic, angled or occasionally somewhat tapered at the base, rounded to sharply pointed at the tip, the margins finely and sharply toothed, the surfaces glabrous. Inflorescences 20–30 mm long (elongating to 30–60 mm with age), 15–30 mm wide, globose to ovoid or oblong-ovoid dense headlike spikes, the stalk 10–50 mm long (subtended by a pair of bractlike leaves). Flowers numerous (often more than 150), sessile, spreading in all directions at fruiting. Calyces 6–8(–10) mm long, the tube 3–5 mm long,

1822. Trifolium vesiculosum 1823. Vicia americana 1824. Vicia caroliniana

glabrous, the teeth about as long as the tube, equal or nearly so, slender, stiff, and sharply pointed, becoming enlarged and inflated at fruiting, the 20–30 longitudinal nerves with numerous cross-nerves, forming a prominent network. Corollas 12–18 mm long, longer than the calyx lobes, creamy white, pink, or purple, the banner outcurved, oblanceolate, sharply pointed at the tip and with an entire or slightly irregular margin, finely but strongly parallel-nerved with age. Fruits 1.5–2.0 mm long, obovoid to nearly globose, sessile, the outer wall membranous, 1 or 2(3)-seeded. Seeds 1.0–1.3 mm long, oblong-ovoid to slightly kidney-shaped, brown, dull and finely roughened. $2n$=16. April–July.

Introduced, uncommon, known thus far only from Howell County (native of Europe, Asia; introduced in the southeastern U.S. west to Oklahoma and Texas, also Washington to California). Open disturbed areas.

Arrowleaf clover is sometimes cultivated in the United States, but was more recently introduced than most of the other nonnative clovers (J. M. Gillett and N. L. Taylor, 2001). It occasionally is grown as a forage crop. In Missouri, the species was first discovered in 1996 by Larry Houf, now retired from the Missouri Department of Conservation.

Zohary and Heller (1984) discussed two varieties, differing in the degree of calyx inflation and cross-venation at fruiting, with var. *rumelicum* Griseb. having the calyces less-inflated and weaker-nerved. These are not treated here, but if they were separated then Missouri plants conform most closely to the more inflated and strongly nerved var. *vesiculosum*.

52. **Vicia** L.
(Hermann, 1960; Gunn, 1979)

Plants annual, biennial, or perennial herbs, with taproots or rhizomes. Stems erect to more commonly spreading or climbing, sometimes mat-forming, angled or ridged but not winged, unbranched or more commonly branched, unarmed, glabrous or pubescent with nonglandular hairs. Leaves alternate, even-pinnately compound with 2 to numerous leaflets, the petiole short or absent, unwinged, the rachis extended into a conspicuous, unbranched or branched tendril (this absent or poorly developed and bristlelike elsewhere). Stipules more or less leaflike, in some species with a basal outgrowth or lobe of tissue on 1 side, this rounded or more commonly triangular, descending or clasping the stem, the margins otherwise entire or toothed, the venation mostly inconspicuous, persistent; stipels absent. Leaflets variously shaped (sometimes even on the same plant), the margins entire, the surfaces glabrous or hairy, pinnately veined or only the midvein visible. Inflorescences axillary, spikelike racemes or clusters, sometimes reduced to solitary flowers, the bracts 1–3 mm long, shed early, bractlets absent. Calyces 5-lobed, the tube cylindric to bell-shaped, often at least slightly oblique, sometimes somewhat pouched on 1 side at the base, more or less 2-lipped, the lobes subequal or the 3 lower lobes longer than the upper 2, variously shaped, sharply pointed at their tips, glabrous or hairy. Corollas papilionaceous, blue, purple, red, yellow, or white, sometimes with darker nerves or appearing bicolorous), lacking conspicuous, contrasting markings near the base of the banner, the banner with a short, broad, stalklike base overlapping the wings, the expanded portion obovate to nearly circular, slightly to deeply notched at the tip, curved or bent upward or backward, the wings oblong to obovate, shorter than to slightly longer than the banner, usually curved over or around and fused to the keel below the midpoint, the keel shorter than the wings, boat-shaped, slightly to strongly curved upward, rounded or tapered to a bluntly or sharply pointed tip. Stamens 10, 9 of the filaments fused and 1 free nearly to the base, all similar in length, the free portions of the filaments slender, not broadened toward their tips the anthers small, attached at the base, all similar in size, yellow or occasionally orange. Ovary sessile or short-stalked, glabrous, the style abruptly curved or bent upward toward the base, not flattened, usually hairy toward the tip, the hairs encircling the style or in a tuft on the outer side, more or less persistent at fruiting, the stigma terminal, short. Fruits legumes, mostly oblong to elliptic, tapered asymmetrically to a sharply pointed or more commonly beaked tip (rounded elsewhere), flattened, slightly constricted between the seeds or not, straight or slightly curved upward, 2- to numerous-seeded, dehiscing by 2 valves, these green to brown at maturity, usually twisting spirally after dehiscence. Seeds oblong to oblong-elliptic, broadly ovate, or circular in outline, flattened or not, the surface smooth, reddish brown to dark brown, sometimes mottled, often dull. About 160 species, North America to South America, Europe, Asia, Africa, Pacific Islands, most diverse in temperate regions.

Vicia is a member of the tribe Fabeae, which is one of the most economically important groups in the family and also includes the related genera *Lathyrus* L. (sweet pea, vetchling), *Lens* Mill. (lentil), and *Pisum* L. (garden pea). *Vicia* is very similar to *Lathyrus*, but differs in having unwinged stems, generally smaller leaflets and flowers, filament tubes ending obliquely, and unflattened styles with hairs usually encircling the tip. Preliminary findings from molecular phylogenetic studies by Wojciechowski et al. (2004) suggest that, as currently circumscribed, the genus *Vicia* may not be natural, comprising two or more groups, some of which may be more closely related to other genera in the tribe than to the rest of *Vicia*. Determinations of taxa within *Vicia* can be a bit confusing, requiring multiple characters to separate species (Gunn, 1968, 1979; Isely, 1998). Leaf characters are generally relatively similar, with only minor variation in leaflet number, size, pubescence, and tendril branching. Self-pollinating species are particularly variable in plant size, number of leaflets, pubescence, flower color, and seed size and color. Seed characters are sometimes useful, but are overlapping (Lassetter, 1978). Some Old World species that were introduced long ago for cover crops have been collected a few times as escapes in Missouri, and might still persist in waste areas.

Vicia is widely cultivated as a cover crop and green manure for soil improvement and erosion control. The leaves of all species are palatable and nourishing to livestock, with few toxicity problems. However, the seeds of *V. sativa* and *V. villosa* are associated with neurotoxicity and skin problems when eaten by livestock (Burrows and Tyrl, 2001). Seeds of *V. faba* L. (fava bean, broad bean) are commonly eaten by people in Europe, the Mediterranean region, and China; however, unless cooked thoroughly they cause a kind of haemolytic anemia known as favism, particularly among a genetically predisposed segment of the population (Sokolov, 1984).

1. Inflorescences sessile or nearly so, of solitary or paired (rarely 3) flowers
    2. Stipules sometimes with an inconspicuous, translucent glandular spot; calyx tube 6–8 mm long; the lobes shorter than the tube, subequal; corollas 25–30 mm long, pale or light yellow, sometimes streaked or tinged with lavender . . . . 4. V. GRANDIFLORA
    2. Stipules sometimes with a prominent, purplish brown, glandular spot; calyx tube 4–5 mm long, the lobes about as long as the tube; corollas 12–18 mm long, pinkish purple to purplish blue, rarely white . . . . 8. V. SATIVA

1. Inflorescences conspicuously stalked, elongate racemes or small clusters (sometimes reduced to solitary flowers in *V. minutiflora*)
   3. Inflorescences with 1–10 flowers
      4. Inflorescences with 1 or 2 flowers; stipules with the margins entire or with a few shallow teeth . . . . 7. V. MINUTIFLORA
      4. Inflorescences with 2–10 flowers; stipules with the margins strongly toothed and/or lobed
         5. Corollas 12–25 mm long; calyces noticeably oblique at the base, slightly pouched on 1 side, the lobes unequal (the upper pair noticeably shorter than the lowermost); plants perennial, with rhizomes . . . . 1. V. AMERICANA
         5. Corollas 2.5–8.0 mm long; calyces not or only slightly oblique at the base, not pouched, the lobes subequal (the upper pair at most slightly shorter than the lowermost); plants annual, with short or slender taproots
            6. Corollas 2.5–4.5 mm long; style inconspicuously and sparsely short-hairy at the tip; fruits 6–10 mm long, finely hairy, (1)2(3)-seeded . . . . 5. V. HIRSUTA
            6. Corollas 4.5–8.0 mm long; style with a conspicuous band of dense, long hairs toward the tip; fruits 16–25 mm long, glabrous, 4–7-seeded . . . . 6. V. LUDOVICIANA
   3. Inflorescences with (8–)10–40(–60) flowers
      7. Calyx tube with the base strongly pouched on 1 side, the attachment strongly oblique, appearing lateral; corollas (10–)12–18 mm long; plants annual or biennial, with taproots . . . . 9. V. VILLOSA
      7. Calyx tube not or only slightly pouched, the attachment appearing basal, but sometimes somewhat off-center; corollas 8–13(–18) mm long; plants perennial, with rhizomes
         8. Tendrils mostly unbranched; inflorescences usually relatively open, with (8–)10–20 flowers; calyx lobes subequal in length, triangular to broadly triangular; corollas white to pale lavender-tinged, the keel strongly bluish-tinged toward the tip; fruits 4–6 mm wide . . . . 2. V. CAROLINIANA
         8. Tendrils mostly branched; inflorescences relatively dense, with (10–)15–30 flowers; calyx lobes unequal in length, the lowermost lobe narrowly triangular to lanceolate-triangular, 2–3 times as long as the broadly triangular upper lobes; corollas blue to bluish purple, rarely lilac or white; fruits 6–8 mm wide . . . . 3. V. CRACCA

1. Vicia americana Muhl. ex Willd. **var. americana** (American vetch)

V. americana var. *truncata* (Nutt. ex Torr. & A. Gray) Brewer ex S. Watson

Pl. 411 j, k; Map 1823

Plants perennial, with rhizomes. Stems 30–100 cm long, ascending or trailing to more commonly climbing, glabrous or sparsely and finely hairy. Leaves with 8–14 leaflets, the petiole absent or to 8 mm long, the tendrils unbranched or more commonly branched. Stipules 8–10 mm long, lacking a glandular spot, sharply several-lobed and/or toothed. Leaflets 5–35 mm long, (3–)4–14 mm wide, those of the lower leaves usually not reduced, the smallest leaflets toward the leaf tips, oblong-ovate to elliptic, narrowly elliptic or occasionally nearly linear, rounded to more commonly angled or tapered at the base, variously rounded, bluntly pointed, or truncate at the tip, rarely sharply pointed or slightly notched, the midvein often extended as a minute, sharp point at the very tip, the surfaces glabrous or occasionally sparsely and finely hairy, especially when young. Inflorescences racemes, the stalk 3–5 cm long, the flowers 3–10, each with a stalk 1–2 mm long. Calyces glabrous or sparsely and finely hairy, often bluish- to pur-

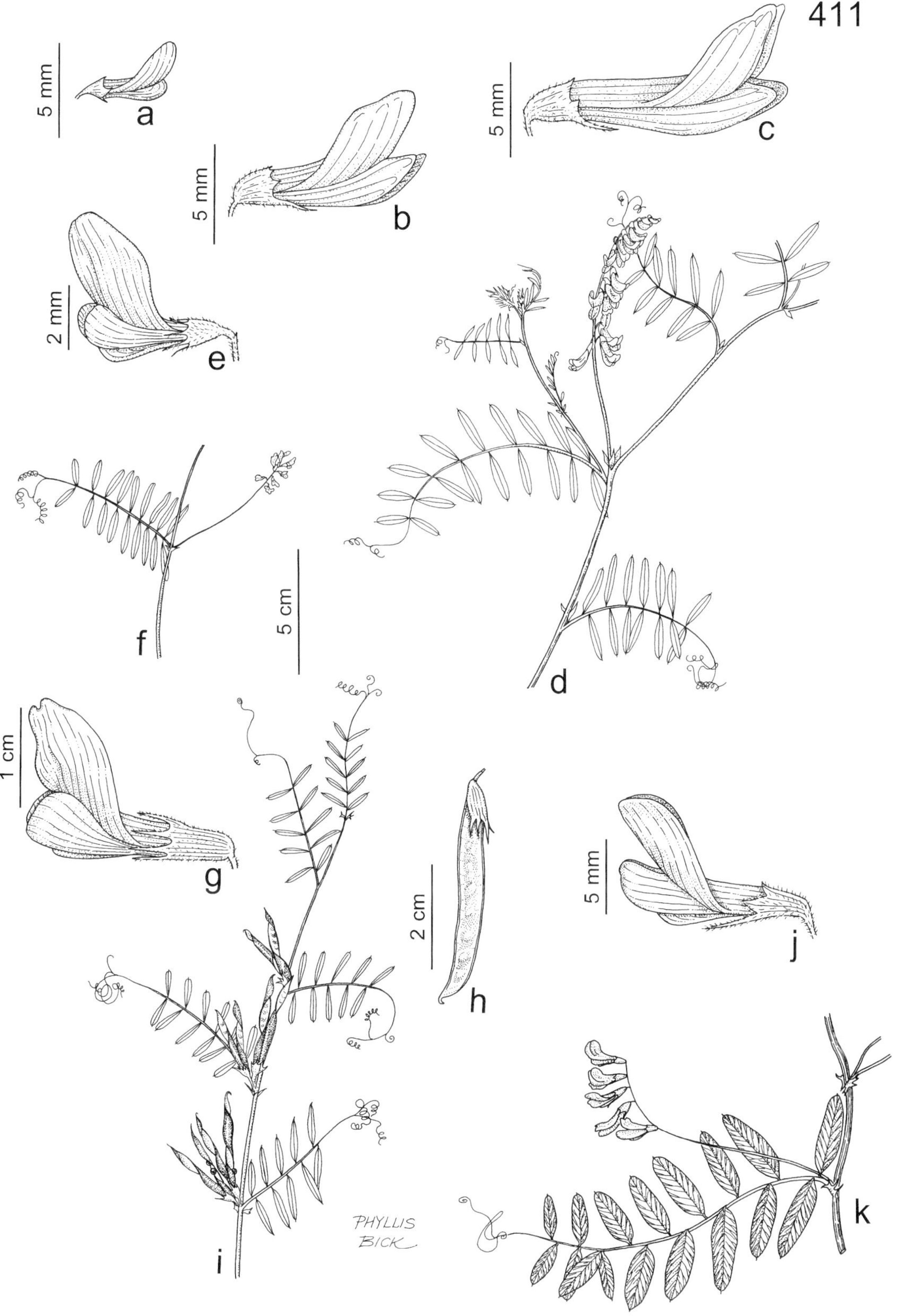

Plate 411. Fabaceae. *Vicia caroliniana*, **a)** flower. *Vicia cracca*, **b)** flower. *Vicia villosa*, **c)** flower, **d)** fertile stem. *Vicia ludoviciana*, **e)** flower, **f)** fertile stem. *Vicia grandiflora*, **g)** flower, **h)** fruit, **i)** fertile stem. *Vicia americana*, **j)** flower, **k)** fertile stem.

plish-tinged toward the tip, the tube 3.5–5.0 mm long, the base noticeably oblique and slightly pouched on 1 side, the attachment appearing basal but off-center, the lobes 0.5–3.0 mm long, unequal, the uppermost short and broadly triangular, the lowermost noticeably longer than the upper pair, triangular to narrowly triangular. Corollas 12–25 mm long, purplish blue to purple or rarely white, the keel and wings sometimes lighter or darker than the banner, the banner bent or curved upward, moderately to strongly curved around the wings and keel. Stamens with the fused portion 10–12 mm long, the free portion 2–3 mm long. Style encircled by a conspicuous band of dense hairs toward the tip. Fruits 25–35 mm long, 5–7 mm wide, short-stalked, glabrous, straw-colored to light brown at maturity, sometimes purplish-tinged, (4–)7–16-seeded. Seeds 2.5–3.5 mm long, reddish brown to purplish brown, often with darker mottling, more or less circular to broadly oblong in outline, not or only slightly flattened, the attachment scar not raised, dark brown to straw-colored, extending up to about 1/3 the circumference of the seed. $2n$=14, 28. April–June.

Uncommon in northwestern Missouri, disjunct in Bates and St. Louis Counties (northern U.S. [including Alaska] south to California, Texas, Missouri, and Virginia; Canada, Mexico). Bottomland prairies and edges of mesic upland forests; also open, disturbed areas.

Vicia americana is somewhat polymorphic, especially in the western states. Several weakly distinguished varieties or subspecies have been described based on variation in leaf shape and thickness, tendril branching, and number of flowers per inflorescence (Hermann, 1960; Gunn, 1968; Isely, 1998). Missouri materials all fall into var. *americana*. Gunn (1968, 1979) and Isely (1998) accepted only one other infraspecific taxon in the United States, var. *minor* Hook. (ssp. *minor* (Hook.) Piper & Beattie), with a shorter, more erect habit, fewer flowers, unbranched short tendrils, and narrower leaflets. It occurs from Montana to North Dakota south the New Mexico and Oklahoma. A third infraspecific taxon, ssp. *mexicana* C.R. Gunn, is endemic to northeastern Mexico and differs in its longer, subequal calyx lobes.

Vicia americana is relatively rarely encountered in Missouri. Steyermark (1963) reported an additional occurrence in Jefferson County without citation of a voucher specimen, but his could not be confirmed during the present study. The species is an important food plant for a variety of birds and mammals in portions of its range where it is more abundant. Gunn (1968) noted that it is relatively sensitive to grazing pressure and is among the first plants to disappear with heavy grazing by livestock.

2. Vicia caroliniana Walter (wood vetch, pale vetch)

Pl. 411 a; Map 1824

Plants perennial, with rhizomes. Stems 20–100 cm long, trailing or climbing, sparsely to moderately pubescent with short, fine, loosely appressed to more or less spreading hairs, at least on young growth, sometimes glabrous or nearly so at maturity. Leaves with 8–14(–20) leaflets, the petiole 1–6 mm long, the tendrils mostly unbranched. Stipules 2–4 mm long, lacking a glandular spot, unlobed or with a narrowly triangular basal lobe, the margins otherwise entire. Leaflets 8–20(–25) mm long, 2–8 mm wide, those of the lower leaves not strongly reduced, narrowly oblong to narrowly elliptic or oblong-lanceolate, rounded or angled at the base, variously rounded to bluntly or sharply pointed or occasionally nearly truncate at the tip, the midvein often extended as a minute, sharp point at the very tip, the surfaces finely hairy. Inflorescences racemes, usually relatively open, the stalk 2–4 cm long, the flowers 8–20, each with a stalk 0.5–1.5 mm long. Calyces finely hairy or occasionally glabrous, the tube 1.5–2.0 mm long, the base somewhat oblique and sometimes slightly pouched on 1 side, the attachment appearing basal but off-center, the lobes 0.5–1.0 mm long, subequal, the upper pair only slightly shorter than the lowermost, triangular to broadly triangular. Corollas 8–12 mm long, white, sometimes tinged with pale lavender, the keel usually strongly bluish-tinged toward the tip, the banner bent or curved upward at or above the midpoint, strongly curved around the wings and keel. Stamens with the fused portion 5–6 mm long, the free portion 1–2 mm long. Style encircled by a dense ring of short hairs toward the tip. Fruits 15–30 mm long, 4–6 mm wide, short-stalked, glabrous, straw-colored to light brown at maturity, 4–9-seeded. Seeds 2.5–4.0 mm long, reddish brown to purplish brown, sometimes nearly black, sometimes with darker mottling, more or less circular in outline, slightly to strongly flattened, the attachment scar slightly raised, yellowish brown to straw-colored, extending about 3/4 the circumference of the seed. $2n$=14. April–May.

Scattered in the Ozark and Ozark Border Divisions (eastern U.S. west to Minnesota and Texas; Canada). Mesic to dry upland forests, glades, ledges and tops of bluffs, margins of sinkhole ponds, and banks of streams and rivers; also fencerows and roadsides.

Vicia caroliniana is a native Missouri wildflower and should not be confused with the super-

1825. Vicia cracca 1826. Vicia grandiflora 1827. Vicia hirsuta

ficially similar, nonnative *V. villosa*. In addition to the somewhat smaller flowers with paler corollas and the other morphological characters outlined in the key to species above, plants of *V. caroliniana* do not form extensive tangles or mats and can grow in more shaded conditions than does *V. varia*.

3. Vicia cracca L. (tufted vetch, Canada pea, bird vetch)

Pl. 411 b; Map 1825

Plants perennial, with rhizomes. Stems 40–150 cm long, trailing or climbing, sparsely or more commonly moderately to densely pubescent with short, fine, curved hairs. Leaves with 14–22 leaflets, the petiole absent or to 3 mm long, the tendrils branched or rarely a few of them unbranched. Stipules 2–5 mm long, lacking a glandular spot, with a narrowly triangular basal lobe (rarely unlobed), the margins otherwise entire. Leaflets 10–25 mm long, 2–5 mm wide, those of the lower leaves not strongly reduced, linear to narrowly oblong, narrowly elliptic, or the smallest ones sometimes elliptic, rounded or angled at the base, variously rounded to bluntly or sharply pointed or occasionally nearly truncate at the tip, the midvein often extended as a minute, sharp point at the very tip, the surfaces finely hairy. Inflorescences racemes, usually relatively dense, the stalk 2–5 cm long, the flowers (10–)15–30, each with a stalk 1–3 mm long. Calyces finely hairy, often somewhat purplish-tinged, the tube 1.8–2.8 mm long, the base somewhat oblique and sometimes slightly pouched on 1 side, the attachment appearing basal but off-center, the lobes 0.5–2.5 mm long, unequal, the upper pair broadly triangular, the lowermost narrowly triangular to lanceolate-triangular, 2–3 times as long as the upper lobes. Corollas 9–13 (–18) mm long, blue to bluish purple, rarely lilac or white, the wings and keel occasionally lighter than the banner, the banner bent or curved upward at or above the midpoint, strongly curved around the wings and keel. Stamens with the fused portion 5–6 mm long, the free portion 1–2 mm long. Style encircled by a dense ring of short hairs toward the tip. Fruits 18–30 mm long, 6–8 mm wide, short-stalked, glabrous, light brown to grayish brown at maturity, 4–9-seeded. Seeds 2.5–3.0 mm long, yellowish brown to reddish brown or dark grayish brown, occasionally nearly black, often with darker mottling, more or less circular in outline, not or only slightly flattened, the attachment scar not raised, yellowish brown to brown with a lighter border, extending up to 1/3 the circumference of the seed. $2n$=12, 14, 21, 22, 28. July–August.

Introduced, uncommon, known thus far only from the city of St. Louis (native of Europe, Asia; introduced widely in the U.S. [including Alaska, Hawaii], Canada, Greenland, Mexico). Railroads.

Steyermark (1963) excluded this species from the Missouri flora, stating that specimens attributed to this species examined by him were misdetermined collections of *V. villosa*. However, Mühlenbach (1979) confirmed the presence of the species from the St. Louis railyards.

4. Vicia grandiflora Scop. (bigflower vetch)

V. grandiflora var. *kitaibeliana* W. Koch

Pl. 411 g–i; Map 1826

Plants annual, with taproots. Stems 30–60 cm long, erect or ascending, sometimes slightly climbing, finely hairy. Leaves with 6–12(14) leaflets, the petiole 4–6 mm long, the tendrils branched. Stipules 3–5 mm long, sometimes with an inconspicuous translucent, glandular spot, those of the lower leaves sharply several-lobed and/or toothed, those of the upper leaves usually with only the slender basal lobe, the margins otherwise entire or with a few faint teeth. Leaflets 6–20 mm long, 2–6 mm wide, those of the lower leaves often much shorter and broader than those of the upper leaves, narrowly to broadly oblanceolate to obovate, the broadest ones sometimes appearing heart-shaped (with the attachment opposite the notch), tapered at the base, notched to truncate at the tip, the midvein extended as a minute, sharp point at the very tip, the surfaces finely hairy. Inflorescences sessile, the flowers (1)2(3), each with a stalk 1–3 mm long. Calyces sparsely hairy, the tube 6–8 mm long, the base oblique and pouched, the attachment appearing basal but strongly off-center, the lobes 3–4 mm long, subequal, triangular. Corollas 25–30 mm long, pale or light yellow, sometimes streaked or tinged with lavender, the banner bent or curved upward relatively strongly, the keel usually tipped with purple. Stamens with the fused portion 14–15 mm long, the free portion 2–3 mm long. Style with a conspicuous tuft of dense hairs on the lower side. Fruits 25–45 mm long, 6–9 mm wide, sessile, minutely hairy when young, becoming glabrous or nearly so at maturity, dark brown to black at maturity, 6–8-seeded. Seeds 2.8–3.0 mm long, reddish brown, mottled, circular in outline, somewhat flattened, the attachment scar raised, white, and nearly encircling the seed. $2n=14$. March–June.

Introduced, uncommon in the Mississippi Lowlands Division (native of Europe, Asia; introduced in the eastern [mostly southeastern] U.S. west to Missouri and Texas). Levees, roadsides, and open, disturbed areas.

Vicia grandiflora is easily recognized by its large, oblanceolate, truncate leaflets and large yellow flowers borne in pairs in the upper axils. Steyermark (1963) knew it from a single specimen collected in 1954 in Cape Girardeau County, but several other populations have been documented since that time.

5. Vicia hirsuta (L.) Gray (tiny vetch, hairy vetch)

Endiusa hirsuta (L.) Alefield

Ervum hirsutum L.

Map 1827

Plants annual, with short taproots. Stems 20–70 cm long, loosely ascending to spreading or climbing, glabrous or sparsely and finely hairy. Leaves with (8)10–18 leaflets, the petiole absent or to 2 mm long, the tendrils mostly branched. Stipules 2–5 mm long, lacking a glandular spot, deeply 2–4-lobed, the margins otherwise entire or sharply few-toothed. Leaflets 5–14(–20) mm long, 1–3(–4) mm wide, those of the lower leaves often somewhat shorter than those of the upper leaves, linear to narrowly elliptic or narrowly oblong, angled or tapered at the base, truncate or slightly to broadly notched at the tip, the midvein often extended as a minute, sharp point at the very tip, the surfaces glabrous or sparsely and finely hairy. Inflorescences short racemes, the stalk 1.0–2.5 cm long, the flowers (2–)3–5(–8), each with a stalk 1–2 mm long. Calyces finely short-hairy, the tube 1.0–1.5 mm long, the base not or only slightly oblique, not pouched, the attachment appearing basal, the lobes 1.5–2.0 mm long, subequal (the lowermost only slightly longer than the other lobes), narrowly triangular. Corollas 2.5–4.5 mm long, pale blue to lavender or white, the keel sometimes slightly darker toward the tip, the banner somewhat curved upward, moderately to strongly curved around the wings and keel. Stamens with the fused portion 1.5–2.5 mm long, the free portion 0.5–1.0 mm long. Style with a few short hairs on the lower side at the tip. Fruits 6–10 mm long, 3–5 mm wide, sessile, finely hairy, brown to black at maturity, (1)2(3)-seeded. Seeds 1.5–2.5 mm long, brownish yellow to reddish brown and with strong, darker mottling (occasionally appearing nearly solid dark purplish brown), more or less circular in outline, not flattened or somewhat flattened, the attachment scar not raised, dark brown, extending less than 1/4 the circumference of the seed, obscured by the persistent, brown stalk. $2n=14$. May–June.

Introduced, uncommon in the Mississippi Lowlands Division (native of Europe; introduced widely in the U. S., Canada). Sand prairies; also fallow fields, roadsides, and open disturbed areas.

Steyermark (1963) reported this species from railroads in the St. Louis area. However, Mühlenbach (1979) noted that these collections were misdetermined and instead represented specimens of *V. villosa* ssp. *varia* (as *V. dasycarpa*). The presence of *V. hirsuta* in Missouri subsequently was first-confirmed by a collection made by Jay Raveill in 1985 in Dunklin County.

Vicia hirsuta is recognized by the leaves with several pairs of narrow leaflets, sharply lobed stipules, small flowers, nearly glabrous styles, hairy, mostly 2-seeded fruits, and seeds with a persistent attachment stalk. Because of these distinctive characters it has sometimes been treated historically as a separate genus (Gunn, 1979). The petals are small and shed early, and the fruit begins to develop before the flower is fully open, features indicative of self-fertilization.

6. Vicia ludoviciana Nutt. **ssp. leavenworthii** (Nutt. ex Torr. & A. Gray) Lassetter & Gunn (deer pea vetch)

V. leavenworthii Nutt. ex Torr. & A. Gray

Pl. 411 e, f; Map 1828

Plants annual, with short taproots. Stems 10–100 cm long, loosely ascending to trailing or climbing, glabrous or minutely hairy, especially on the

1828. Vicia ludoviciana

1829. Vicia minutiflora

1830. Vicia sativa

angles. Leaves with (6)8–14(–18) leaflets, the petiole absent or to 4 mm long, the tendrils branched. Stipules 3–4 mm long, lacking a glandular spot, unlobed or occasionally with a narrowly triangular basal lobe, the margins otherwise entire. Leaflets 5–12(–15) mm long, 1–4 mm wide, those of the lower leaves often somewhat shorter than those of the upper leaves, narrowly elliptic or narrowly oblong to narrowly oblanceolate, angled or tapered at the base, variously bluntly to sharply pointed or broadly notched at the tip, the midvein often extended as a minute, sharp point at the very tip, the surfaces sometimes sparsely and finely hairy when young, glabrous at maturity. Inflorescences short racemes or sometimes appearing as clusters, the stalk 3–6 cm long (sometimes slightly longer at fruiting), the flowers 2–5, each with a stalk 1–2 mm long. Calyces finely short-hairy, the tube 1.0–1.5 mm long, the base not or only slightly oblique, not pouched, the attachment appearing basal, the lobes 1–2 mm long, subequal (the lowermost only slightly longer than the other lobes), narrowly triangular. Corollas 4.5–8.0 mm long, lavender, pale pink, or white (bluish purple to purple elsewhere), the keel sometimes slightly darker toward the tip, the banner somewhat curved upward, moderately to strongly curved around the wings and keel. Stamens with the fused portion 2–3 mm long, the free portion 1.0–1.5 mm long. Style encircled by a conspicuous band of dense, long hairs toward the tip. Fruits 16–25 mm long, 4–7 mm wide, short-stalked, glabrous, straw-colored to brown at maturity, 4–7-seeded. Seeds 1.5–2.5 mm long, brownish yellow to dark brown or reddish brown, often with strong, darker mottling, more or less circular in outline, not or only slightly flattened, the attachment scar not raised, dark brown with a pale border, extending about 1/4 the circumference of the seed. $2n$=14. April–May.

Uncommon, known thus far from a single historical collection from Greene County (Missouri to Mississippi west to Kansas and New Mexico). Habitat unknown, but perhaps upland forests (Steyermark, 1963).

The *Vicia ludoviciana* complex consists of two subspecies, which have further been divided into a confusing array of weakly distinguished varieties or races, centered in Texas (Lassetter, 1984; Isely, 1998). The taxa intergrade and are difficult to separate. According to Lassetter (1984) the Missouri material falls into ssp. *leavenworthii*, which occupies a north-central portion of the species range and is characterized by 2–4, usually pinkish white flowers that open before the inflorescence stalks elongate, as well as relatively short styles, and leaflet numbers toward the upper end of the species range. The ssp. *ludoviciana* grows in the southern United States north to California, Arkansas, and Alabama, and also in Mexico. It differs in the early elongation of the inflorescence stalks, fruits not beginning to develop until after the flowers open, longer styles, and often slightly fewer leaflets.

Vicia ludoviciana is self-fertile (Lassetter, 1984). The pollen is released before the flower opens, with the young fruit already present and enlarging when the flower first opens. Pollinators rarely visit the flowers, and bagging experiments did not reduce the number of fruits.

7. Vicia minutiflora D. Dietr. (pygmy-flowered vetch)

V. micrantha Nutt., an illegitimate name

V. reverchonii S. Watson

Pl. 412 e, f; Map 1829

Plants annual, with taproots. Stems 20–80 cm long, erect or ascending to spreading or climbing, glabrous or sparsely and finely hairy. Leaves with (2)4–6 leaflets, the petiole 1–3 mm long, the tendrils well-developed and branched on the upper leaves, unbranched or poorly developed on the lower leaves. Stipules 2–4 mm long, lacking a glandular spot, with a prominent, triangular, basal lobe, the margins otherwise entire or less commonly shallowly few-toothed. Leaflets 5–24 mm long, 1–3 mm wide, those of the lower leaves often shorter and broader than those of the upper leaves, linear to narrowly elliptic or narrowly oblong, those of the lower leaves sometimes obovate or elliptic, angled or tapered at the base, variously rounded to truncate or bluntly pointed at the tip, the midvein often extended as a minute, sharp point at the very tip, the surfaces glabrous. Inflorescences with a stalk 1.0–3.5 cm long, the flowers 1 or 2 at the tip, each with a stalk 2–3 mm long. Calyces finely hairy, but sometimes becoming glabrous or nearly so at maturity, the tube 1.5–2.0 mm long, the base not or only slightly oblique, not pouched, the attachment appearing basal, the lobes 0.5–1.5 mm long, subequal (the lowermost only slightly longer than the other lobes), triangular. Corollas 4–6 mm long, pale blue to lavender, the banner sometimes with darker streaks, somewhat curved upward, strongly curved around the wings and keel. Stamens with the fused portion 2.5–3.5 mm long, the free portion 1–2 mm long. Style encircled by a conspicuous band of dense hairs toward the tip. Fruits 15–30 mm long, 4–5 mm wide, sessile, glabrous, brown to black at maturity, 3–6-seeded. Seeds 2.0–2.5 mm long, blackish purple or brown and with faint, darker mottling, circular in outline, not flattened, the attachment scar slightly raised, white, extending slightly more than 1/4 the circumference of the seed. $2n$=14. April–May.

Uncommon in southwesternmost Missouri; introduced in the city of St. Louis (southeastern U.S. west to Oklahoma and Texas). Chert and limestone glades; also railroads.

Vicia minutiflora is distinguished by the leaves with relatively few and narrow leaflets, slender inflorescences with 1 or 2 small, blue flowers, the usually solitary, slightly arched, glabrous fruits, and seeds with a raised attachment scar. *Vicia minutiflora* is vegetatively similar to the nonnative *V. hirsuta* and *V. tetrasperma*, but these are readily distinguished by fruit characters. Steyermark (1963) reported a specimen of *V. minutiflora* collected by B. F. Bush in 1891 as an adventive along a railroad in Courtney (Jackson County), but this specimen could not be located during the present study.

Steyermark (1963) reported *V. tetrasperma* (L.) Schreb. (sparrow vetch, slender vetch) from Jackson and Lawrence Counties but did not cite any specimens. Despite extensive searches in herbaria at institutions within and outside of Missouri, no records to document the occurrence of this species have been located to date, and it is thus excluded from the flora for the present. Superficially, this species is similar morphologically to *V. minutiflora*, but may be distinguished by its upper leaves with unbranched tendrils and its fruits, which are 10–13 mm long, rounded at the tip, nearly symmetrical, and 4-seeded. *Vicia tetrasperma* (Pl. 412 c, d) is further characterized by its tangled mats of fine stems, narrow leaflets, and inflorescences of 1 or 2 small flowers. It is native to Europe, but has become widely introduced in Asia; in the New World it is established in most eastern states and several western states, as well as in Canada.

8. Vicia sativa L. **ssp. nigra** (L.) Ehrh. (narrow-leaved common vetch, spring vetch)
V. angustifolia L.
V. angustifolia var. *segetalis* (Thuill.) W.D.J. Koch

Pl. 412 g, h; Map 1830

Plants annual, with taproots. Stems 25–100 cm long, erect or ascending to climbing, sparsely and finely hairy to glabrous or nearly so. Leaves with 8–16 leaflets, the petiole 1–3 mm long, the tendrils branched. Stipules 4–6 mm long, sometimes with a prominent, purplish brown, glandular spot on the undersurface, mostly sharply several-lobed and/or toothed. Leaflets 15–28 mm long, 2–7 mm wide, those of the lower leaves often shorter and broader than those of the upper leaves, linear to oblanceolate, those of the lower leaves sometimes obovate, tapered at the base, rounded to truncate or notched at the tip, the midvein extended as a minute, sharp point at the very tip, the surfaces glabrous or especially the undersurface sparsely and finely hairy. Inflorescences sessile, the flowers 1 or 2(3), each with a stalk 1–2 mm long. Calyces glabrous or finely hairy along the nerves and lobe margins, the tube 4–5 mm long, the base oblique but not or only slightly pouched, the attachment appearing basal but somewhat off-center, the lobes 3–5 mm long, subequal (the lowermost only slightly longer than the other lobes), narrowly triangular. Corollas 12–18 mm long, pinkish purple to purplish blue or rarely white, the banner only slightly curved upward, strongly curved around the wings and keel. Stamens with the fused portion 7–8 mm long, the free portion 1–2 mm long. Style encircled by a conspicuous band of dense hairs toward the tip. Fruits 25–50 mm long, 3–6 mm wide, sessile, minutely hairy when young, becoming sparsely hairy or glabrous at maturity, dark brown to black at maturity, 4–12-seeded. Seeds 2.5–3.0 mm long, dark brown and mottled

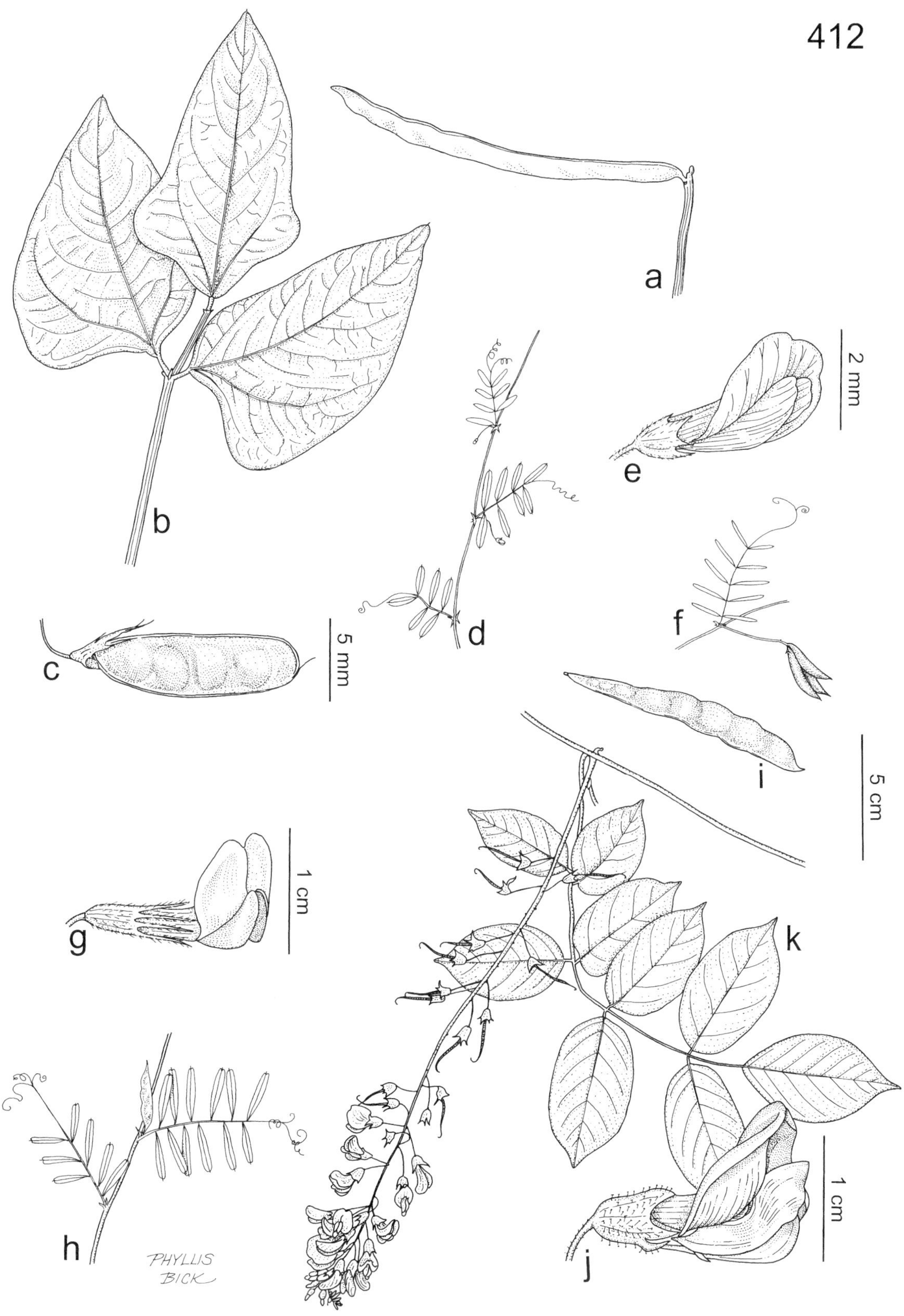

Plate 412. Fabaceae. *Vigna unguiculata*, **a)** fruit, **b)** leaf. *Vicia tetrasperma*, **c)** fruit, **d)** fertile stem. *Vicia minutiflora*, **e)** flower, **f)** fertile stem. *Vicia sativa*, **g)** flower, **h)** fertile stem. *Wisteria fructescens*, **i)** fruit, **j)** flower, **k)** fertile stem.

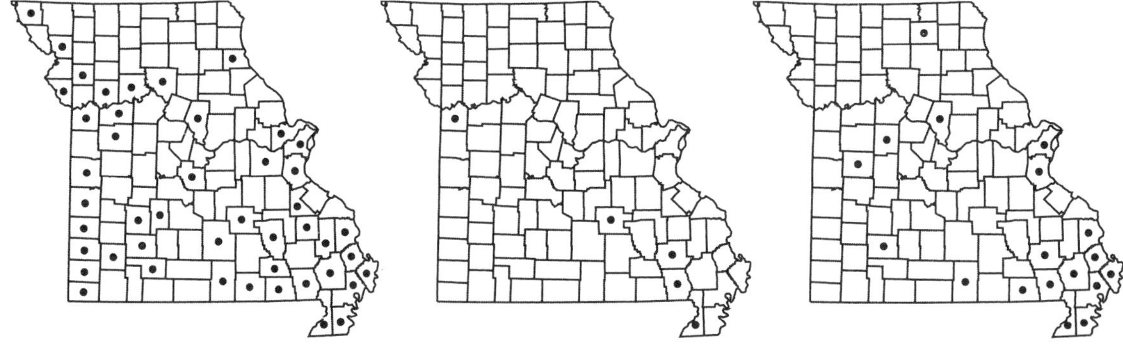

1831. Vicia villosa

1832. Vigna unguiculata

1833. Wisteria frutescens

with black to solid black, circular to bluntly oblong in outline, not flattened, the attachment scar not raised, white, extending only about 1/4 the circumference of the seed. $2n=12$. April–July.

Introduced, scattered, mostly south of the Missouri River (native of Europe; introduced nearly throughout the U.S. [including Alaska], Canada, Mexico). Edges and openings of mesic upland forests, banks of streams and spring branches, and tops of bluffs; also pastures, fencerows, ditches, gardens, railroads, roadsides and open disturbed areas.

Vicia sativa is recognized by the lobed, somewhat star-shaped stipules with a diagnostic dark, glandular spot (this sometimes absent in ssp. *nigra*) and the usually pinkish purple flowers paired in the upper axils. The ssp. *nigra* is the weedier phase of the species. The ssp. *sativa* (common vetch), with larger flowers (calyces 10–15 mm long, corollas 18–25 mm long) and generally fewer (mostly 2–6) leaflets, has escaped from cultivation sporadically but widely in the United States, but was excluded from the Missouri flora by Steyermark (1963). It has been recorded at least once from every state surrounding Missouri except Kansas and Nebraska and thus may be discovered in Missouri in the future.

The glandular spots on the stipules in this and some other *Vicia* species function as extrafloral nectaries. These were first studied by the German botanist and mycologist, Karl Wilhelm Gottlieb Leopold Fuckel (1846), whose father had first noticed bees visiting the stipules of this species. Subsequently, a number of nineteenth century botanists published research on the stipular nectaries of *Vicia*. Koptur (1978) discussed the composition of the nectar produced by glands of *V. sativa* in California and showed that a mutualistic relationship exists between this vetch and a nonnative ant, which harvests the nectar and defends the plants against herbivorous insects.

9. Vicia villosa Roth (woolly vetch, winter vetch)

Pl. 411 c, d; Map 1831

Plants annual or biennial, with taproots. Stems 40–100 cm long, erect to loosely ascending, trailing, or climbing, glabrous to sparsely pubescent with short, incurved hairs or moderately to densely pubescent with conspicuous, spreading hairs. Leaves with (10)12–18(20) leaflets, the petiole absent or to 3 mm long, the tendrils branched. Stipules 5–8 mm long, lacking a glandular spot, with a narrowly triangular basal lobe, the margins otherwise entire. Leaflets 10–30 mm long, 2–6 mm wide, those of the lower leaves not strongly reduced, narrowly oblong-lanceolate to narrowly elliptic or linear, rounded to more commonly angled or tapered at the base, variously rounded to bluntly or sharply pointed at the tip, the midvein often extended as a minute, sharp point at the very tip, the surfaces glabrous or finely hairy. Inflorescences racemes, the stalk 3–9 cm long, the flowers 10–40(–60), each with a stalk 1–2 mm long. Calyces glabrous or finely hairy, often bluish- or purplish-tinged, the tube 2–3 mm long, the base strongly oblique and conspicuously pouched on 1 side, the attachment appearing lateral, the lobes 0.5–4.0 mm long, strongly unequal, the upper pair short and triangular to broadly triangular, the lowermost much longer than the upper pair, narrowly triangular to lanceolate-triangular, or more or less hairlike above a short, triangular base. Corollas (10–)12–18 mm long, blue to purple or rarely pink or white, the keel and wings sometimes lighter than the banner, the banner bent or curved upward toward the tip, strongly curved around the wings and keel. Stamens with the fused portion 9–10 mm long, the free portion 1.5–3.0 mm long. Style with a dense ring of short hairs at the base and a patch of dense short hairs on the lower side near the tip. Fruits 25–30 mm long, 7–10 mm wide, short-stalked, sometimes finely hairy when young, glabrous at maturity, straw-colored to light brown at maturity, 3–7-seeded. Seeds 3.5–5.0 mm long, dark reddish brown to greenish brown or nearly black, often with darker mottling, more or less circular in outline, not or only slightly flattened, the attachment scar not raised, dark brown, inconspicuous, extending less than 1/4 the circumference of the seed. $2n$=14, 28. May–July.

Introduced, scattered south of the Missouri River, extending farther north mostly in counties along the Missouri and Mississippi Rivers (native of Europe, Asia; introduced widely in the U.S. [including Alaska], Canada, Mexico, Central America, South America, Africa). Banks of streams and rivers, edges of mesic upland forests, and rarely margins of upland prairies; also levees, fencerows, old fields, margins of crop fields, fallow fields, ditches, railroads, roadsides, and open disturbed areas.

This species is commonly planted along highways for erosion control and can become established along gravel bars when rains wash seeds or pieces of plants into drainages that cross the roads. Plants of *V. villosa* can form such dense, tangled mats that they smother other plant species. The species sometimes also has been planted as a green manure. It is reported to be outcrossing (Gunn, 1979).

Vicia villosa is distinguished by the densely flowered racemes of slender, mostly violet or blue flowers, the distinctly oblique and gibbous base of the calyx, narrow calyx lobes, and relatively broad fruits. Most authors divide *V. villosa* into two varieties or subspecies, which have sometimes been treated as separate species (Hermann, 1960; Steyermark, 1963). The two infraspecific taxa are about equally abundant in Missouri, although north of the Missouri River ssp. *villosa* is the variant more likely to be encountered.

1. Stems glabrous or sparsely pubescent with incurved to loosely appressed hairs 0.5–1.0 mm long; leaflets 2–4 mm wide; calyces with the lowermost lobe 2–3 mm long, sparsely hairy . . . . 9A. SSP. VARIA
1. Stems moderately to densely pubescent with spreading hairs mostly 1–2 mm long; leaflets 3–6 mm wide; calyces with the lowermost lobe 3–4 mm long, moderately to densely hairy . . . . 9B. SSP. VILLOSA

9a. ssp. varia (Host) Corb.

V. villosa var. *glabrescens* W.D.J. Koch

V. dasycarpa Ten.

Stems glabrous or sparsely pubescent with incurved to loosely appressed hairs 0.5–1.0 mm long. Leaves with mostly 12–18(20) leaflets, these mostly 10–25 mm long, 2–4 mm wide, the surfaces glabrous or sparsely appressed-hairy. Inflorescences mostly with 10–25 flowers. Calyces with the lobes 0.5–3.0 mm long, the lowermost lobe 2–3 mm long, narrowly triangular to lanceolate-triangular, glabrous or sparsely hairy. Corollas usually (10–)12–16 mm long, mostly violet or purple. $2n$=14, 28. May–July.

Introduced, scattered, mostly south of the Missouri River (native of Europe, Asia; introduced widely in the U.S., also occasionally in south America). Banks of streams and rivers, edges of mesic upland forests, and rarely margins of upland prairies; also levees, fencerows, old fields, margins of crop fields, fallow fields, ditches, railroads, roadsides, and open disturbed areas.

9b. ssp. villosa

Stems moderately to densely pubescent with spreading hairs mostly 1–2 mm long. Leaves with mostly 12–16 leaflets, these mostly 15–30 mm long, 3–6 mm wide, the surfaces moderately to densely pubescent with loosely appressed to spreading hairs. Inflorescences mostly with 20–40(–60) flowers. Calyces with the lobes 0.5–4.0 mm long, the lowermost lobe 2–4 mm long, often nearly hair-like above a short, triangular base, moderately to densely hairy. Corollas usually 14–18 mm long, blue to purple or rarely pink or white. $2n$=14, 28. May–July.

Introduced, scattered south of the Missouri River, extending farther north mostly in counties along the Missouri and Mississippi Rivers (native of Europe, Asia; introduced widely in the U.S. [including Alaska], Canada, Mexico, Central America, South America, Africa). Banks of streams and rivers, edges of mesic upland forests, and rarely margins of upland prairies; also levees, fencerows, old fields, margins of crop fields, fallow fields, ditches, railroads, roadsides, and open disturbed areas.

53. **Vigna** Savi

Eighty-five to 104 species, widespread in tropical and subtropical regions of the Old World; several species widely cultivated and occasionally escaped.

The generic circumscription of *Vigna* remains somewhat controversial. In the strict sense, there are probably 85–90 species in *Vigna* (Schrire, 2005). Phylogenetic study of selected members of the relatively large tribe Phaseoleae (Bronn) DC. using molecular data has suggested that a group of about 17 neotropical species traditionally included in *Vigna* is more closely related to some other New World genera than to the remainder of *Vigna* (Delgado-Salinas et al., 1993).

1. Vigna unguiculata (L.) Walp. (cowpea, black-eyed pea)

V. sinensis (L.) Endl. ex Hassk.

Pl. 412 a, b; Map 1832

Plants annual (perennial herbs farther south), usually with stout taproots. Stems 50–300 cm long, variously trailing, climbing, and/or twining to erect and bushy, branched, unarmed, rounded or somewhat ridged, sometimes purplish-tinged, glabrous or with patches of short, stout, stiff hairs at the nodes. Leaves alternate, pinnately trifoliate, the petiole 6–11 cm long. Stipules 8–15 mm long, ovate, peltate, with a long, ovate to triangular lobe oriented downward below the attachment point, angled or tapered to a sharply pointed tip, strongly veined, more or less persistent; stipels 2–3 mm long, oblong-elliptic, persistent. Leaflets 4–12 cm long, 3–7 cm wide, ovate to broadly ovate, the terminal leaflet sometimes rhombic with a pair of shallow, blunt, spreading lobes toward the base, broadly angled or (if lobed) broadly tapered at the base, angled or slightly tapered to a sharply pointed tip, the lateral leaflets strongly asymmetric, broadly angled to truncate at the base, angled or slightly tapered to a sharply pointed tip, the margins entire, the surfaces glabrous, with 3 main veins from the base, but otherwise more or less pinnately veined. Terminal leaflet with the stalk 15–30 mm long, symmetric at the base; lateral leaflets with the stalk 2–4 mm long, oblique at the base. Inflorescences axillary, very short racemes or clusters, with 2–6 flowers, the stalk 50–200 mm long, the bracts 4–10 mm long, lanceolate, shed early, each flower with a stalk 1–2 mm long, the bractlets 3–5 mm long, oblong-obovate, glabrous, more or less persistent. Calyces glabrous (sometimes minutely hairy along the margins when young), the tube 3–5 mm long, bell-shaped, somewhat 2-lipped, the lobes 2.5–5.0 mm long, unequal, the upper 2 lobes fused to about their midpoints, the free portions triangular to broadly triangular, sharply pointed at their tips, the lower 3 lobes slightly shorter, similar, triangular, sharply pointed at their tips. Corollas papilionaceous, pink or purple (the tips of the wings and keel sometimes darker than the other portions), rarely greenish white (yellow elsewhere), the banner 14–18 mm long, 13–22 mm wide, the expanded portion curved or bent backward, broadly oblong-ovate to nearly circular, notched at the tip, with a slender longitudinal keel, lacking thickened areas toward the base, the wings 15–17 mm long, 7–8 mm wide, obliquely ovate to obovate, somewhat curved around the keel, the keel 15–17 mm long, 9–10 mm wide, boat-shaped, fused to about the tip, bent or curved upward abruptly near the midpoint, the tip bluntly to sharply pointed. Stamens 10, in 2 alternating, slightly shorter and longer series, 9 of the filaments fused and 1 free, the fused portion 11–12 mm long, bent upward at or near its tip, the free portion 6–7 mm long, the anthers small, attached above the base, yellow. Ovary 9–12 mm long, glabrous, the style 10–12 mm long, bent or curved upward abruptly toward the tip, somewhat flattened, densely short-hairy on the inner side, the stigma small, terminal. Fruits legumes, 10–20(–30) cm long, 5–10 mm wide, linear to narrowly oblong, short-tapered somewhat asymmetrically to a beak, slightly flattened, somewhat constricted between the seeds, straight or slightly curved, indehiscent or dehiscing tardily by 2 valves, green to straw-colored or reddish brown at maturity, sometimes with darker mottling, glabrous, with 8 to numerous seeds. Seeds 8–12 mm long, 6–8 mm wide, oblong to kidney-shaped in outline, somewhat flattened, the surface variously brown, green, cream-colored, or white, sometimes with darker mottling, smooth or finely wrinkled, dull or somewhat shiny, with a white, slightly raised aril often framed with black at maturity. $2n=22$. August–September.

Introduced, uncommon and sporadic, mostly in the southeastern quarter of the state (probably a native of Africa, widely distributed in tropical and warm-temperate regions nearly worldwide, introduced in the eastern [mostly southeastern] U.S. west to Kansas and Texas). Banks of streams and rivers; also ditches, roadsides, and sandy, open, disturbed areas.

Vigna unguiculata is recognized by the coarse, bushy or twining habit, glabrous stems and leaves, large stipules with a basal lobe, flowers produced in small clusters at the end of an elongate stalk, and long, smooth legumes. It is one of the world's important species of legume crops. It probably originated in sub-Saharan Africa and was disseminated to India and southern Europe very early (Smartt, 1990). Cowpeas were introduced into the United States in the 1700s (Wight, 1907) and became a major legume crop in the southern states. They thrive in hot, wet weather. When the peas are harvested very young, they are called cream peas. When harvested near maturity, after the dark patch around the hilum has developed, they are called black-eyed peas.

Verdcourt (1970b) treated *V. unguiculata* as a complex of five subspecies (three cultivated and two wild) differing in their degree of twining and quantitative details of the calyces and pods. In his classification, plants cultivated in the New World are all members of the widespread ssp. *unguiculata*.

54. **Wisteria** Nutt. (wisteria)

Plants lianas (rarely shrubs elsewhere), often with rhizomes. Stems to 15 m or more long, high-climbing, strongly woody, stout, twining, unarmed, glabrous or nearly so, the bark relatively thin, initially smooth but developing a fine network of lines, furrows, or wrinkles, tan to brown, with conspicuous elongate, transverse lenticels, these becoming obscured with age. Leaves odd-pinnately compound, petiolate, with 7–13 leaflets. Stipules 1–3 mm long, linear, shed early; stipels similar to the stipules, mostly shed early. Leaflets ovate to elliptic, rounded to broadly angled at the base, tapered to a usually sharply pointed tip, the margins entire, the surfaces variously glabrous or hairy, the venation pinnate, the terminal leaflet with the stalk 10–25 mm long, the lateral leaflets very short-stalked. Inflorescences terminal racemes, showy and with usually numerous flowers, drooping or pendant, appearing before the leaves or as the leaves develop, the stalk usually short (less than 15 mm long), but often appearing longer if the lower flowers abort and are shed, the bracts 4–14 mm long, lanceolate, shed early; bractlets absent. Calyces 2-lipped, hairy, sometimes also with club-shaped, stalked glands, the tube shorter than to longer than the lobes, more or less bell-shaped, the upper lip slightly shorter than the lower lip, the 2 lobes fused to the tip or nearly so (often notched at the tip), broadly triangular, the lower lip with the 3 lobes triangular, the lowermost lobe the longest. Corollas papilionaceous, glabrous, blue to purple, lilac, or pink (white in some cultivars), the banner usually with a white and/or yellow central portion, the wings and keel often darker, at least toward the tip, the banner with the expanded portion broadly obovate to nearly circular, with a pair of thickened areas or small hornlike appendages at the base, rounded to very broadly angled at the tip and often finely longitudinally keeled, abruptly arched upward toward the base or midpoint, the wings about as long as the banner, asymmetrically oblong to oblong-obovate, appearing curved, not fused to the keel, the keel about as long as the banner, obovate, boat-shaped, strongly curved upward, rounded to bluntly pointed at the tip. Stamens 10, all of similar lengths, 9 of the filaments fused and 1 filament free, the fused portion 10–12 mm long, curved upward abruptly near the tip, the free portion 3–5 mm long, the anthers small, attached at or near the base, all similar in size. Ovary 10–12 mm long, linear, short-stalked, encircled by a ringed nectar gland below the midpoint, glabrous or hairy, the style glabrous, the stigma terminal, small. Fruits legumes, narrowly oblong (few-seeded fruits sometimes appearing narrowly oblanceolate), flattened, stalked at the base, tapered to a beaked tip, the margins appearing somewhat wavy, irregularly indented between the seeds, the surfaces also indented between the seeds, glabrous or hairy, hardened or woody in texture, dehiscent (sometimes tardily so) by 2 valves, these becoming loosely spirally twisted, with (2–)4–12 seeds. Seeds circular to ellipsoid, somewhat kidney-shaped, or occasionally more or less oblong in outline, flattened, the surfaces reddish brown to black, smooth or with a few faint wrinkles, shiny. About 6 species, temperate North America, Asia.

Stritch (1984, 1985) segregated the Asian species into the genus *Rehsonia* Stritch. There are a number of minor morphological differences that might justify this action, summarized in the key below. However, the new genus was not supported with a full taxonomic treatment of the group and the name has not been widely adopted by other botanists . Those who have studied the group have uniformly concluded that the Asian and American taxa of *Wisteria* in the broad sense accepted here are each other's closest relatives (Lavin and Sousa S., 1995; J.-m. Hu et al., 2002), but there has not yet been a detailed phylogenetic study of relationships among these taxa.

Several species of *Wisteria* are cultivated as ornamentals on trellises and arbors or trained as shrubs or trees. Care must be taken to provide a sturdy substrate for climbing plants, as the mass of stout stems can grow to be very heavy. The beautiful flowers are strongly scented and the volatile oils are used as a fragrance component in perfumes, soaps, incense, and other items. Strips of bark have been used in the manufacture of cloth, string, and handcrafts, especially basketry. The flowers are sometimes eaten, either fresh in salads or battered and fried (Stritch, 1985). However, in particular, the seeds apparently are moderately toxic, causing a variety of symptoms, including headache, dizziness, sweating, and severe distress of the digestive system (Burrows and Tyrl, 2001).

1. Calyces and flower stalks with club-shaped, stalked glands (nonglandular hairs also present); banner bent backward near the midpoint; ovaries glabrous; fruits (4–)7–10(–12) cm long, glabrous . . . . 1. W. FRUTESCENS
1. Calyces and flower stalks hairy, but lacking club-shaped, stalked glands; banner bent backward near the base; ovaries hairy; fruits 10–18 cm long, velvety-hairy . . . . 2. W. SINENSIS

1. Wisteria frutescens (L.) Poir. (American wisteria)

W. frutescens var. *macrostachya* Torr. & A. Gray

W. macrostachya Nutt. ex Torr. & A. Gray, an invalid name

Pl. 412 i–k; Map 1833

Stems twining in a counterclockwise direction. Leaves with the petiole 3–8 cm long, the rachis 15–25 cm long. Leaflets (2–)4–8 cm long, (1.5–)2.0–5.0 cm wide, upper surface glabrous or nearly so, the undersurface sparsely to moderately hairy, mostly along the main veins, the margins hairy. Inflorescences 7–25 cm long, the flowers maturing progressively from the base to the tip, the flower stalks 6–14 mm long, densely pubescent with short, stiff, mostly loosely ascending hairs and also scattered club-shaped, stalked glands (these absent elsewhere). Calyces 6–9(–11) mm long, usually white to pale green, often reddish- to purplish-tinged, densely pubescent with short, stiff, mostly loosely ascending hairs and also scattered club-shaped, stalked glands (these absent elsewhere). Corollas 15–20 mm long, bluish purple, or lilac, the banner with the expanded portion bent backward near midpoint, often with a white and/or light yellow region toward the base, the wings and keel sometimes darker at their tips. Fruits (4–)7–10 (–12) cm long, glabrous, with (2–)4–8. Seeds 11–14 mm long, 7–8 mm wide, kidney-shaped or broadly elliptic, sometimes more or less oblong in outline, only slightly flattened. $2n$=16. April–May.

Scattered in the southeastern quarter of the state, sporadic and mostly introduced farther north and west (eastern U.S. west to Iowa and Texas). Bottomland forests, banks of streams and rivers, bases of bluffs; also fencerows and roadsides.

Wisteria frutescens is variable in the number and size of the leaflets, size of the inflorescence, and the presence or absence of club-shaped glands. Some botanists (Steyermark, 1963; Gleason and Cronquist, 1991; Valder, 1995) have separated a second taxon as *W. macrostachya* or *W. frutescens* var. *macrostachya*. It is said to differ mainly in its longer inflorescences with more flowers and the presence of the club-shaped glands on the flower stalks and calyces. Steyermark (1963) noted that although the Missouri plants all possess the glands

1834. Wisteria sinensis
1835. Castanea dentata
1836. Castanea mollissima

characteristic of var. *macrostachya*, the inflorescences often are at the short end of the spectrum. In general, rangewide the characters separating the two extremes break down in so many populations that there is a growing consensus that only one variable taxon should be recognized (Gillis, 1980; Stritch, 1985; Isely, 1998).

The seeds are buoyant, which aids in their dispersal by water.

2. Wisteria sinensis (Sims) Sweet (Chinese wisteria)

Rehsonia sinensis (Sims) Stritch

Map 1834

Stems twining in a counterclockwise direction. Leaves with the petiole 3–7 cm long, the rachis 15–25 cm long. Leaflets (3–)5–8(–10) cm long, (1.5–) 2.0–4.0 cm wide, the surfaces moderately to densely appressed-hairy when young, becoming glabrous or nearly so at maturity. Inflorescences 10–20 cm long, the flowers maturing more or less at the same time, the flower stalks 8–20 mm long, densely pubescent with short, fine, curved to spreading hairs, lacking club-shaped, stalked glands. Calyces 4.0–7.5 mm long, light green, but often strongly purplish-tinged, moderately to densely pubescent with short, fine, curved to spreading hairs, lacking club-shaped, stalked glands. Corollas 15–22 mm long, bluish purple to lilac, or rarely pink (white in some cultivars), the banner with the expanded portion bent backward near the base, often with a white and/or light yellow to light green region toward the base, the wings and keel sometimes darker at their tips. Fruits 10–18 cm long, densely velvety-hairy, with 2–12 seeds. Seeds 11–14 mm long, 10–14 mm wide, more or less circular in outline, strongly flattened. $2n$=16. April–May.

Introduced, uncommon, mostly in the southeastern portion of the state (native of China, introduced in the eastern U.S. west to Missouri and Texas, also Hawaii). Mesic upland forests near former home sites; also fencerows and roadsides.

Missouri botanists should watch for escaped populations of *W. floribunda* (Willd.) DC, the widely cultivated Japanese wisteria. It is similar to *W. sinensis*, but has stems that twine in a clockwise (vs. counterclockwise) direction, leaves with 13–19 (vs. 9–15) leaflets, longer inflorescences (20–50 vs. 15–20 cm), and flowers that open sequentially from the inflorescence base to the tip (vs. more or less all at the same time). Interestingly, stems of the hybrid between these two taxa, *W. ×formosa* Rehder (which has not been reported as an escape), twines variably either clockwise or counterclockwise (Stritch, 1985).

FAGACEAE (Beech Family)

Contributed by Alan Whittemore

Plants shrubs or more commonly small to large trees, monoecious. Leaves alternate, mostly short-petiolate. Stipules present, membranous to papery or scalelike, often shed early. Leaf blades simple, unlobed or pinnately to rarely nearly palmately lobed, pinnately to rarely nearly palmately veined, the margins entire or toothed. Staminate and pistillate inflorescences separate or with pistillate flowers near the base and staminate flowers above them, with small bracts subtending at least the staminate flowers, consisting of heads or spikelike catkins or the pistillate flowers sometimes solitary, paired, or in small clusters. Flowers actinomorphic, imperfect, the pistillate ones epigynous. Staminate flowers with the calyx of (3)4–6(–8) lobes or sepals; corolla absent; the stamens (2–)6–18 or rarely more, free, with a slender filament and the anther attached at its base (but often in a deep basal notch); a rudimentary, nonfunctional pistil often present. Pistillate flowers with the calyx absent or more commonly of 4–6 lobes or sepals; corolla absent; stamens and staminodes absent; the ovary inferior (but sometimes appearing naked in the absence of a perianth), usually 3- or 6-locular, but occasionally appearing 1-locular toward the tip, with 2 ovules per locule, the placentation axile. Styles usually 3, flattened and often expanded laterally toward the tip, spreading, each with a stigmatic region along the upper side, sometimes only near the tip. Fruits nuts, these solitary or in clusters of 2–4, partly or completely enclosed in a leathery to woody cupule with a spiny or scaly outer surface. Ten or 11 genera, about 800 species, throughout the northern hemisphere temperate zone and in mountains in the northern hemisphere tropics.

Members of the Fagaceae are dominant trees in most Missouri forests. Many species produce valuable timber. The nuts are large and rich in oils. They are edible and nutritious, and are among our most important foods for wildlife, such as mammals, birds, and a large diversity of insects and other invertebrates.

1. Cupules scaly, never spiny, unlobed, usually only partially covering the solitary nut; leaves in several irregular ranks, the leaves and buds more or less crowded toward the stem apex; leaf blades entire, toothed, or lobed . . . . 3. QUERCUS
1. Cupules spiny, deeply lobed, completely enclosing 1–4 nuts; leaves in 2 regular ranks, the leaves and buds not crowded toward the stem apex (except sometimes in leading [not lateral] stems of *Castanea dentata*); leaf blades toothed
    2. Staminate inflorescences narrow, stiffly ascending or erect catkins; spines of the cupule 5–20 mm long, branched; leaf margins with the teeth bristle-tipped, 1–9 mm long; winter buds 2–5 mm long, the 2 basal scales opposite, the others alternate . . . . 1. CASTANEA
    2. Staminate inflorescences dense globose heads ca. 1 cm wide on drooping to pendant stalks 2–6 cm long; spines of cupule 1–3 mm long, unbranched; leaf margins with the teeth sharply pointed but not bristle-tipped, 0.5–1.0 mm long; winter buds 10–18 mm long, all of the scales alternate . . . . 2. FAGUS

1. Castanea Mill. (chestnut)
(G. P. Johnson, 1988)

Bark split into parallel ridges or sometimes smooth. Leaves and buds in 2 regular ranks, not crowded at the stem apex (except sometimes in leading [not lateral] stems of *C. dentata*). Buds 2–5 mm long, ovoid, rounded or bluntly pointed at the tip, with several overlapping scales, the 2 basal scales opposite, the others alternate, the terminal bud absent (the bud in the axil of the youngest leaf usually appears terminal). Stipules relatively large and prominent in spring, often shed early, herbaceous to papery, narrowly lanceolate to ovate. Leaf blades coarsely toothed, the secondary veins reaching the margin. Inflorescences staminate and pistillate or with pistillate flowers near the base and staminate flowers above them, the staminate catkins stiffly ascending or erect, elongate spikelike, the pistillate flowers sessile. Fruits ripening the first autumn after flowering. Cupules splitting into 2 or 4 valves, completely enclosing the nuts, spiny, the spines long and branched, crowded and hiding the surface of the cupule. Nuts 1–3 per cupule, more or less ovoid, sharply pointed at the tip and with a large attachment scar at the base, circular in cross-section or flattened on 1 or 2 sides, brown. About 10 species, eastern North America, Europe, Asia.

American chestnuts have been seriously impacted by chestnut blight, caused by an Asiatic fungus, *Cryphonectria parasitica* (Murrill) Barr. This disease was introduced into the United States during the late 1800s through nursery stock from Asia and was first reported to be killing trees in New York around 1904 (Burnham, 1988). All of the native chestnuts are very susceptible to the blight, and it spread rapidly through the eastern United States. Infected trees are typically killed down to ground level. The root system remains unaffected, and blighted trees may repeatedly resprout from the roots, the stems growing until they are reinfected and dying back again (Paillet, 1993). However, over successive diebacks and resproutings, the progressively weakened plants eventually perish. Burnham (1988) summarized efforts by the American Chestnut Society and other scientists to restore *C. dentata* by hybridizing plants with various Asian chestnut species and then repeatedly backcrossing the progeny to the American chestnut parent over several generations to produce a blight resistant tree with mostly a *C. dentata* genome. Other studies have focused on locating a less virulent form of the fungus to act as a biological control for chestnut blight. More recently, similar studies have begun through the Ozark Chinquapin Foundation and others with the intent of producing a disease resistant form of *C. pumila*.

1. Cupule 2-valved; nut 1 per cupule, more or less circular in cross-section .... 3. C. PUMILA
1. Cupule 4-valved; nuts 2 or 3 per cupule, flattened on 1 or 2 sides
    2. Twigs glabrous (except for minute sessile glands); buds with the outer scales glabrous or sparsely and minutely hairy; leaf blades narrowly elliptic, the undersurface glabrous or with a few hairs scattered on the largest veins, also finely glandular .... 1. C. DENTATA
    2. Twigs hairy; buds with the outer scales densely hairy; leaf blades oblong to narrowly obovate or rarely elliptic, the undersurface sparsely to densely hairy, not glandular .... 2. C. MOLLISSIMA

1. Castanea dentata (Marshall) Borkh. (American chestnut)

Pl. 413 h–k; Map 1835

Plants trees to 30 m tall (mostly much shorter now). Bark gray, deeply furrowed. Twigs very dark brown, with sessile glands, otherwise glabrous. Buds with the outer pair of scales brown or purplish, glabrous or sparsely and minutely hairy. Leaves with the petiole 13–22 mm long, glabrous or with a few long spreading hairs, also with very inconspicuous, sessile glands. Stipules lanceolate, shed early. Leaf blades 12–23 cm long, 4.5–7.5 cm wide, narrowly elliptic, rounded or broadly angled at the base, long-tapered at the tip, the marginal teeth 2–4 mm long, slenderly tapered, often hooked, the secondary veins 16–20 on each side of the midvein, both surfaces with the main veins glabrous or with a few long spreading hairs, also with very inconspicuous, sessile glands. Cupules 1–4 per spike, 3–4 cm wide at fruiting (excluding the spines), splitting into 4 valves, the spines 15–20 mm long. Nuts 2 or 3 per cupule, 18–25 mm long, flattened on 1 or 2 sides. $2n=24$. May–July.

Introduced, uncommon in eastern Missouri west locally to Howell County (eastern U.S. west to Wisconsin and Louisiana; introduced farther west). Mesic upland forests near old homesites; also margins of pastures and power line corridors.

American chestnut was once a dominant tree in mesic forests in the Appalachians. Its wood is strong and easily worked, and it was one of the most important timber trees in the eastern United States. It is highly susceptible to the chestnut blight, and trees have been killed to the ground in virtually all parts of its native range. Sprouts from surviving root systems are still commonly found in areas where the trees once grew, but these sprouts seldom survive long enough to flower and set seed, so the species is gradually dying out. Missouri is outside the native range of *C. dentata*, and it is known in the state only from a few escapes.

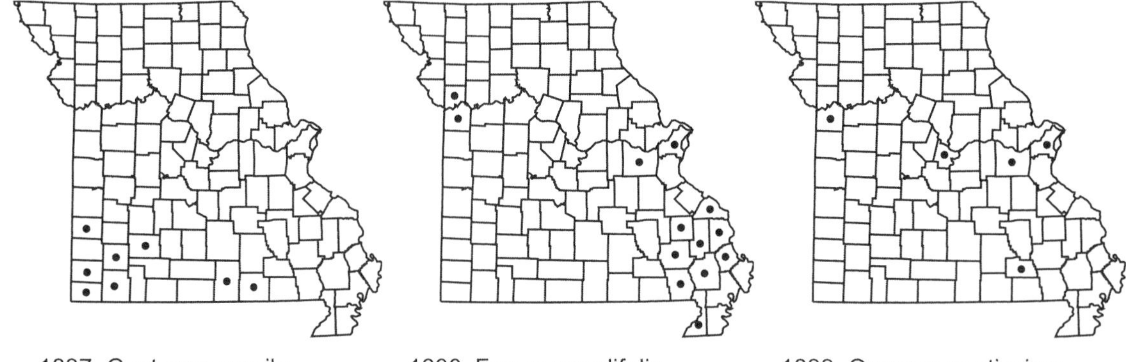

1837. Castanea pumila 1838. Fagus grandifolia 1839. Quercus acutissima

2. Castanea mollissima Blume (Chinese chestnut)

Map 1836

Plants trees to 20 m tall. Bark light gray, moderately to deeply furrowed. Twigs light gray to dark brown, with sparse longer hairs and rarely also dense minute hairs, usually not glandular. Buds with the outer pair of scales brown or purplish, densely hairy. Leaves with the petiole 5–20 mm long, moderately pubescent with long spreading hairs, sometimes also with dense shorter hairs, these sometimes all or partially gland-tipped. Stipules ovate, shed tardily. Leaf blades 12–19 cm long, 5.5–8.0 cm wide, oblong to narrowly obovate or rarely elliptic, broadly rounded to shallowly cordate at the base, abruptly short-tapered at the tip, the marginal teeth 1–2(–4) mm long, slenderly tapered, straight or curved, the secondary veins 14–19 on each side of the midvein, both surfaces with the midvein moderately pubescent with long spreading hairs, sometimes also with dense shorter hairs, these sometimes all or partially gland-tipped, the undersurface sometimes also with the tissue between the veins hairy. Cupules 1 or 2 per spike, 2.5–4.0 cm wide at fruiting (excluding the spines), splitting into 4 valves, the spines 14–19 mm long. Nuts 2 or 3 per cupule, 16–25 mm long, flattened on 1 or 2 sides. $2n=24$. June–July.

Introduced, known thus far only from Christian and Ozark Counties (native of eastern Asia; introduced in the eastern U.S. west to Illinois, Missouri, and Mississippi). Mesic upland forests, usually near streams and rivers.

Castanea mollissima is not cultivated widely in Missouri. It was first reported as an escape by Padgett and Parker (1998). A second Asian species, *C. crenata* Siebold & Zucc. (Japanese chestnut), also is sometimes cultivated, but is not known to escape in Missouri. *Castanea crenata* is a shrub or small tree, rarely over 10 m tall, with small glandular scales visible under high magnification on the stems, petioles, and the undersurface of the leaf blades.

3. Castanea pumila (L.) Mill. (chinquapin)

Pl. 413 d–g; Map 1837

Plants shrubs or trees to 20 m tall (mostly much shorter now). Bark gray to brown, smooth to deeply furrowed. Twigs dark brown or grayish, with inconspicuous, sessile glands and/or short gland-tipped hairs. Buds with the outer pair of scales dark purple, glabrous or hairy. Leaves with the petiole 2–19 mm long, with inconspicuous glands or short glandular hairs, sometimes also variously longer-hairy. Stipules lanceolate, shed early. Leaf blades 7.5–21.0 cm long, 3–9 cm wide, narrowly elliptic to oblanceolate or narrowly obovate, rounded, broadly angled, or rarely narrowly angled at the base, broadly to sharply angled or short-tapered at the tip, the marginal teeth 1–9 mm long, broadly to slenderly tapered or short-tapered above a triangular base, straight or curved, the secondary veins 11–23 on each side of the midvein, both surfaces with the main veins with inconspicuous glands or short glandular hairs, the undersurface also sparsely to densely hairy between the veins. Cupules 2–13 per spike, 1–2 cm wide at fruiting (excluding the spines), splitting into 2 valves, the spines 5–13 mm long. Nut 1 per cupule, 7–20 mm long, more or less circular in cross-section. $2n=24$. May–June.

Uncommon in southernmost Missouri (eastern U.S. west to Missouri and Texas). Mesic to dry upland forests; fencerows and roadsides; often on acidic substrates.

This species is less susceptible to chestnut blight than is *C. dentata*. The blight develops more slowly, and infected stems live much longer before

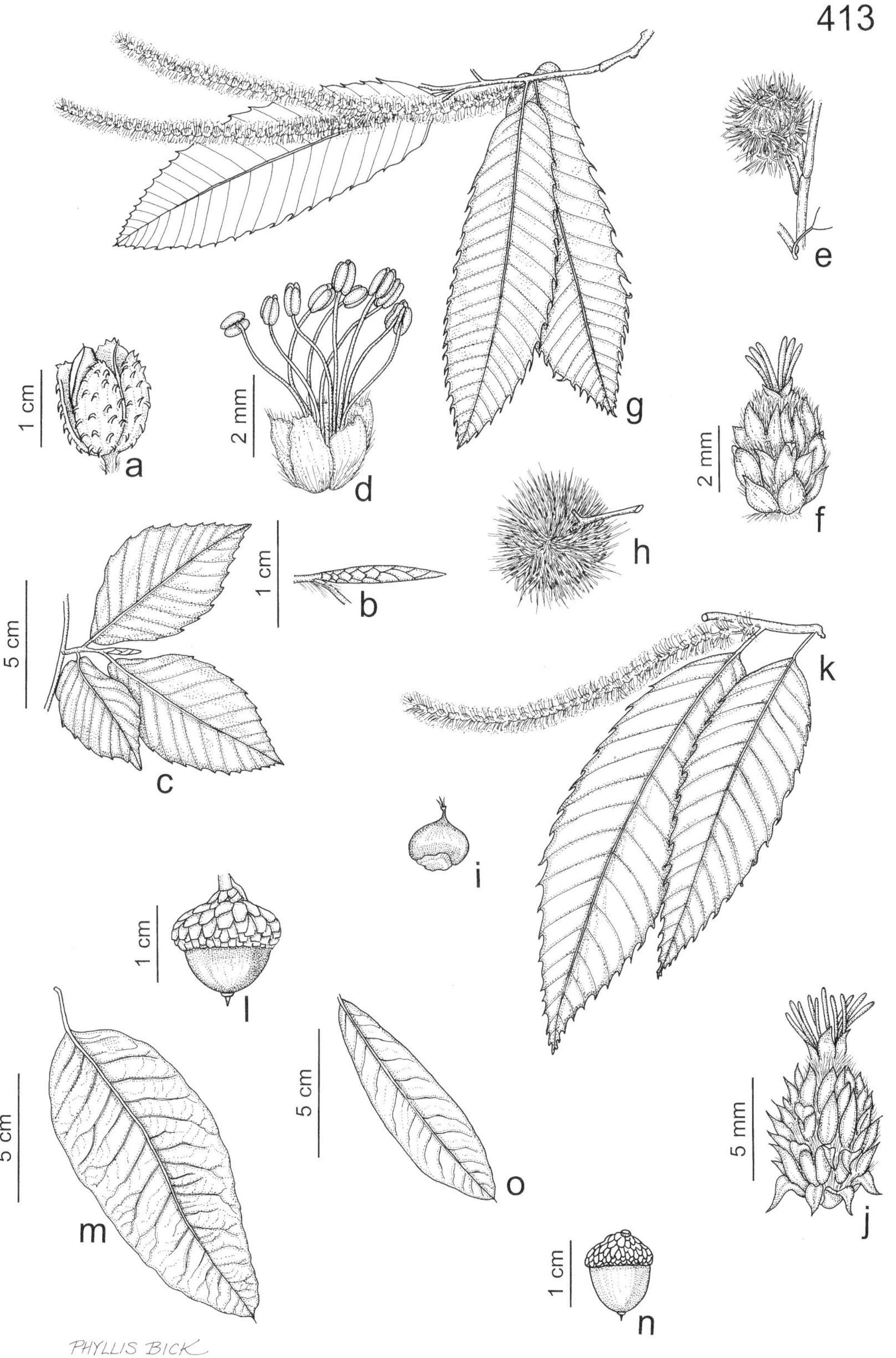

Plate 413. Fagaceae. *Fagus grandifolia*, **a)** fruit, **b)** winter bud, **c)** leaves. *Castanea pumila* var. *ozarkensis*, **d)** staminate flower, **e)** fruit, **f)** pistillate flower, **g)** branch with leaves and inflorescences. *Castanea dentata*, **h)** fruit, **i)** seed, **j)** pistillate flower, **k)** branch with leaves and inflorescence. *Quercus imbricaria*, **l)** acorn, **m)** leaf. *Quercus phellos*, **n)** acorn, **o)** leaf.

dying, so they are able to set seed. The characteristic growth form of the species has changed, however; due to the repeated death and replacement of the trunks. The tall, single-trunked trees present before the blight have been replaced by low, multitrunked small trees or shrubs. Large chestnut logs can still be found associated with living colonies of smaller trees in some parts of the Ozarks.

Castanea pumila is a puzzling and problematic species that has been split into six or more species by some authors (G. P. Johnson, 1988). In the Ozarks, there are two taxa that are generally well defined, which are treated as varieties in the present work. This is primarily because other races elsewhere in the range of the species intergrade with the Ozark race of var. *pumila*, and some of these combine characters of the Ozark race of var. *pumila* with characters of var. *ozarkensis* (larger, more tapered leaves, larger cupules, and sparser pubescence). As noted under var. *pumila*, the key and descriptions below hold only for Ozark material. In the literature, our two varieties are said to differ in twig indumentum (pubescent in var. *pumila*, glabrous in var. *ozarkensis*), but this character does not hold up in Missouri and Arkansas populations.

1. Leaf blades 14–21 cm long, 5.5–9.0 cm wide, short-tapered at the tip, the marginal teeth (2–)3–9 mm long; petiole (6–)9–19 mm long; spines of the cupule 9–13 mm long; bark brownish, moderately to deeply fissured . . . . 3A. VAR. OZARKENSIS
1. Leaf blades 7.5–16.0 cm long, 3–6 cm wide, broadly angled or weakly tapered at the tip, the marginal teeth 1–4 mm long; petiole 2–10 mm long; spines of the cupule 5–10 mm long; bark gray or brown, smooth or shallowly fissured . . . . 3B. VAR. PUMILA

3a. var. ozarkensis (Ashe) G.E. Tucker (Ozark chinquapin, Ozark chestnut)
C. ozarkensis Ashe

Pl. 413 d–g

Bark brownish, moderately to deeply furrowed. Twigs with inconspicuous glands or short gland-tipped hairs, sometimes also a few long spreading hairs. Petioles (6–)9–19 mm long, moderately to densely pubescent with inconspicuous, sessile glands or short gland-tipped hairs, sometimes also a few long spreading hairs. Leaf blades 14–21 cm long, 5.5–9.0 cm wide, narrowly elliptic to narrowly obovate, short-tapered at the tip, the marginal teeth (2–)3–9 mm long, straight or curved, the secondary veins 15–23 on each side of the midvein, the midvein with inconspicuous, sessile glands or short glandular hairs, sometimes also a few long spreading hairs, the undersurface also sparsely to densely hairy between the veins. Cupules 4–13 per spike, 1.5–2.0 cm wide at fruiting (excluding spines), the spines 9–13 mm long. $2n=24$. May–June.

Uncommon in southernmost Missouri (Missouri and Oklahoma south to Louisiana and Alabama, mostly extirpated outside the Ozarks). Mesic to dry upland forests; fencerows and roadsides; often on acidic substrates.

Leaves on young stump sprouts may be very sparsely pubescent, and they are sometimes mistaken for *C. dentata*. Juvenile var. *ozarkensis* may be distinguished by the hairy leaf margins and the presence of hairs on the minor veins (hairs are very sparse and confined to the midrib and secondary veins in sprout leaves of *C. dentata*).

3b. var. pumila (chinquapin, Allegheny chinquapin)

Bark gray to brown, smooth or slightly fissured. Twigs sometimes short-hairy, more often merely with sessile glands, sometimes also with longer, spreading hairs. Petioles 2–10 mm long. Leaf blade 7.5–16.0 cm long, 3–6 cm wide, narrowly elliptic to oblanceolate, broadly to sharply angled or somewhat short-tapered at the tip, the marginal teeth 1–4 mm long, straight or nearly so, the secondary veins 11–18 on each side of the midvein, the midvein with inconspicuous, sessile glands and scattered, long, spreading hairs, sometimes also short-hairy, the undersurface also densely hairy between the veins. Cupules 2–7 per spike, 1.0–1.2 cm wide at fruiting (excluding the spines); spines 5–10 mm long. $2n=24$. May–June.

Uncommon, known thus far only from a single historical collection from Howell County (eastern U.S. west to Missouri and Texas). Mesic upland forests in ravines.

As noted above, the varietal description here applies only to populations in the Ozarks, where this variety seems to be very constant in its morphology and quite distinct from var. *ozarkensis*. Elsewhere in its range, var. *pumila* forms a bewildering array of local races. Some of them are very dissimilar in their extreme forms, but they intergrade with one another very extensively.

2. **Fagus** L. (beech)

About 10 species, eastern North America, Europe, Asia.

1. Fagus grandifolia Ehrh. (American beech)
F. grandifolia var. *caroliniana* (Loudon) Fernald & Rehder
F. grandifolia f. *mollis* Fernald & Rehder
Pl. 413 a–c; Map 1838

Plants trees to 30 m tall. Bark smooth, light gray. Leaves and buds in 2 regular ranks, not crowded at stem apex. Terminal buds 10–18 mm long, very narrowly ellipsoid, sharply pointed at the tip, with several overlapping scales, these all alternate, glabrous except near the tips, brown, a terminal bud present. Twigs 1.5–2.0 mm thick, dark brown or dark purplish brown, at first densely pubescent with short, silky hairs, becoming glabrous or nearly so at maturity. Leaves with the petiole 3–10 mm long. Stipules 2–4 mm long, linear-lanceolate, the bases nearly encircling the twig, relatively prominent in the spring, shed as the leaves mature. Leaf blades 6–14 cm long, 4.0–7.5 cm wide, elliptic to ovate-elliptic or obovate-elliptic, broadly angled to rounded or occasionally slightly cordate at the often slightly asymmetric base, tapered to a sharply pointed tip, the secondary veins 11–17 on each side of the midrib, straight, parallel, each reaching the margin and ending in a small tooth 0.5–1.0 mm long (this sharply pointed but not bristle-tipped), at least the undersurface sparsely pubescent with short, silky hairs, these sometimes denser in the vein axils. Inflorescences of staminate and pistillate flowers separate. Staminate inflorescences dense, globose heads ca. 1 cm wide on drooping to pendant stalks 2–6 cm long. Pistillate inflorescences with an ascending stalk 8–12 mm long, consisting of a pair of flowers completely enclosed in a cupule. Fruits ripening the first autumn after flowering. Cupules 13–20 mm long, splitting into 4 valves, almost completely enclosing the nuts, spiny, the spines 1–3 mm long, unbranched, not crowded, not hiding the cupule surface. Nuts 2 per cupule, 15–20 mm long, more or less ovoid, sharply pointed at the tip, triangular in cross-section, the angles sometimes slightly winged, brown. $2n=24$. April–May.

Scattered in the Mississippi Lowlands Division and adjacent portions of the Ozarks and Ozark Border; introduced sporadically farther north (eastern U.S. west to Wisconsin, Missouri, and Texas; Canada, Mexico; introduced rarely farther west in the U.S.). Mesic upland forests, banks of streams and rivers, and tops of bluffs; also occasionally railroads and roadsides.

American beech is a dominant species in late-successional mesic upland forests in parts of the eastern deciduous forest of temperate North America. It also has been widely cultivated to the east of Missouri as an ornamental, for its smooth bark and its graceful foliage that turns yellow in the autumn, although plants can be relatively intolerant of transplantation. The bark of living trees has long been a favorite substrate on which vandals, lovers, and other extroverts have felt compelled to carve initials. The wood has been harvested for handcrafts, crates, barrels, clothespins, tool handles, fuel, and other uses, but is not durable enough to have major importance as lumber (Steyermark, 1963). Native Americans used various parts of the plant medicinally to treat worms, pulmonary problems, burns and other skin problems, venereal diseases, and blood deficiencies, as well as to induce abortions (Moerman, 1998). Beech nuts are valuable food for wildlife and also have been roasted and ground for use as a coffee substitute, as well as crushed and boiled to create a beverage. Beech oil, which is extracted from the nuts of the closely related European beech (*F. sylvatica* L.), is used as a table oil and food additive.

Morphological variation within the species has caused some botanists to suggest the recognition of various varieties and forms. Although on a local scale the subtle pubescence and fruit differences between some populations have been shown to have statistical validity (Cooper and Mercer, 1977), across the range of the species there is too much morphological overlap and too little correlation between characters to support formal taxonomic recognition for any of the named variants.

Throughout its range, *Fagus grandifolia* co-occurs with beechdrops, a harmless root parasite, *Epifagus virginiana* (L.) W.P.C. Barton, which parasitizes only this species. For further discussion of this relationship, see the treatment of *Epifagus* (Orobanchaceae).

3. **Quercus** L. (oak)

Bark split into parallel ridges, loose plates, or strips. Leaves and buds in several irregular ranks, more or less crowded at the stem apex, Buds 2–12 mm long, ovoid to ellipsoid, all of the scales alternate, a terminal bud present (sometimes difficult to distinguish in large apical bud clusters). Stipules small and inconspicuous, shed early in leaf development. Leaf blades entire, toothed, scalloped, or lobed, the secondary veins reaching the margin or turning aside before reaching it. Inflorescences of staminate and pistillate flowers separate; the staminate catkins drooping or pendant, elongate, spikelike; the pistillate flowers sessile or nearly so (sometimes a stalk developing as the fruit matures), solitary, paired, or in small clusters. Fruits ripening the first or second autumn after flowering. Cupules (usually called cups in *Quercus*) unlobed and not splitting at maturity, usually enclosing only the lower portion of the nut (except in *Q. lyrata*), scaly, the scales short and appressed or long and spreading, overlapping and covering the surface of the cupule. Nut 1 per cupule, variously narrowly ellipsoid to more or less globose, often with a prominent attachment scar at the base, in many species with a minute to small, blunt to sharply pointed, extension at the tip, circular in cross-section, brown. About 450 species, throughout the northern hemisphere, most diverse in temperate regions and montane portions of the tropics.

Oaks are the most important group of trees in Missouri. They dominate most of the forests, woodlands, and savannas in the state. Acorns are a very important food for many wild mammals, birds, and insects, as well as for domestic hogs. In prehistoric times acorns were a staple food for humans in many parts of the northern hemisphere. Acorns are quite nutritious, but they have high concentrations of tannins, which must be removed by leaching before they are palatable. Tannin concentration varies greatly from tree to tree; generally, black oaks have higher tannin concentrations than white oaks, and are more nutritious. Oaks produce high-quality hardwood lumber, which has been important for general construction, cabinet making and furniture making, veneer, pallets, and fenceposts. Other, specialized uses of some white oaks are discussed under section *Quercus*. Oaks also are a major source of high-quality firewood and charcoal.

Oaks are important as shade trees and ornamentals in places where native trees have been left standing after building. However, they are extremely vulnerable to construction injury. Every effort should be made to avoid severing roots, compacting the soil in the root zone, or smothering roots by earth fill. Many species are difficult to transplant or to start in containers, so they are somewhat underutilized in the nursery trade. The young seedling puts most of its growth into a deep taproot. If this develops abnormally (as in a pot) or is cut off (for transplantation), the plant is sometimes unable to compensate and dies within a few years. Oaks are best planted in the place where they are to grow, but care must be taken to protect the cotyledons of the seedling from squirrels, which frequently dig up the remains of the old nut and eat it.

In many oaks, particularly in members of the black oak group (see the classification below) the leaves are not totally deciduous. Instead, the plants retain their dried, brown leaves through the winter, especially on lower branches. This phenomenon, which also occurs in scattered other species of woody plants, is known as marcescence.

Oaks are very variable morphologically, and the species can be difficult to separate. Significant variation is seen between populations, among trees in the same population, and even among different parts of the same tree (Baranski, 1975; Braham, 1977; R. J. Jensen, 1989). The juvenile leaves of many species of oak are quite different in form from the adult leaves, and vigorous sprouts from the roots or trunk often revert to juvenile leaf morphology. Leaves growing in the shady inner part of the crown may be quite different from those in the sunny outer part of the crown, the shaded inner leaves being larger, thinner, more shallowly lobed, and more sparsely pubescent. Unfortunately, seedling leaves, sprouts, and shade leaves are sometimes the only leaves that are easily accessible from ground level. Specimens that consist only of such leaves are difficult to identify accurately.

Pubescence characters are very useful in oaks, which have a variety of conspicuous branched hairs on most of their organs. In stellate hairs, found only in white oaks, the branches of the hair (known as rays) spread in one plane, more or less parallel to the leaf or stem and often appressed to it. Other hair types include fasciculate hairs, multiradiate hairs, and dendroid hairs, all of which branch in three dimensions. These types are difficult to distinguish except under high magnification and are all referreed to in the present work as "branched spreading hairs." Many oaks also have soft, unbranched hairs appressed to the leaf surface. These are usually inconspicuous unless they are pigmented (common in *Q. marilandica* and *Q. stellata*, not in other species), so they are easily overlooked unless the leaf is examined carefully at high power. Hair structure can be difficult to observe except with high magnification and strong illumination, so these characters have been avoided in the keys as far as possible, but they are included in the descriptions because they are important for critical work, especially the accurate recognition of hybrids.

Pubescence descriptions here apply only to mature, fully expanded leaves and associated twigs. In many species whose leaves and twigs are glabrous at maturity, including *Quercus alba* and many red oaks, expanding leaves and twigs in the spring are densely hairy; these hairs are shed as the leaves mature. Other organs that provide important taxonomic characters develop later in the season. The terminal buds develop over the summer, usually attaining full size around the end of August. Acorns ripen in September and October.

Natural hybridization in North American oaks was reviewed by Palmer (1948). Several hybrid combinations not known to Palmer have been described since 1948, but his general comments on hybridization are still valid. Hybrids may be expected between any two species in the same section, but they are very uncommon. Palmer estimated that fewer than one tree in every 100,000, perhaps fewer than one in a million, is a first-generation hybrid. Hybrids are usually found in areas where one of the parental species is abundant and the other is uncommon and growing in an ecologically marginal habitat. Once a first-generation hybrid reaches reproductive maturity, it may backcross with one or both of its parents. Such backcross hybrids are much more variable and difficult to distinguish from the parental species than the first-generation hybrids. Over several successive generations, this process can lead to trees that are morphologically indistinguishable from one of the original parental species, but nevertheless have some of the other parent's genes incorporated into their genomes (Hardin, 1975; Whittemore and Schaal, 2001). Because of the large number of different hybrids documented from Missouri, it is more expedient to collect the data on these in a separate table than to disperse the information within the discussion of each species. It should be noted that in some cases one or more synonymous epithets have been published in reference to certain oak hybrids, but because of space considerations and the limited application of these names to Missouri plants, no attempt has been made here to list synonyny for these binomials. Thomson (1977) documented hybrids between the morphologically similar *Q. muehlenbergii* and *Q. prinoides* from a site in Lafayette County, but no voucher specimens to support this conclusion were discovered during the present research. Also, the putative hybrid between *Q. falcata* and *Q. imbricaria*, which Steyermark (1963) reported under the illegitimate name *Q.* ×*anceps* E.J. Palmer (a later homonym of *Q. anceps* Korth. [a Bornean taxon now usually treated as part of *Q. spicata* Sm.] that subsequently was legitimized as *Q.* ×*palmeriana* A. Camus), still requires documentation from Missouri. The specimen from Butler County cited by Steyermark has been redetermined as *Q. falcata*.

Putative parentage	Hybrid binomial	Documented counties
Q. alba × *Q. bicolor*	*Q.* ×*jackiana* C.K. Schneid.	Lafayette
Q. alba × *Q. macrocarpa*	*Q.* ×*bebbiana* C.K. Schneid.	Cole, Jackson, Pike, Putnam, St. Louis
Q. alba × *Q. michauxii*	*Q.* ×*beadlei* Trel. ex E.J. Palmer	Dunklin
Q. alba × *Q. muehlenbergii*		Clinton, Jackson, St. Louis
Q. alba × *Q. stellata*	*Q.* ×*fernowii* Trel.	Cole, Dade, Jackson, Lawrence, Macon, Marion, Polk, St. Clair, St. Louis City
Q. bicolor × *Q. lyrata*	*Q.* ×*humidicola* E.J. Palmer	Dunklin
Q. bicolor × *Q. macrocarpa*	*Q.* ×*schuettei* Trel.	Jackson, Johnson, St. Clair
Q. bicolor × [*Q. muehlenbergii* × *Q. prinoides*]	*Q.* ×*introgressa* P.M. Thomson	Lafayette
Q. bicolor × *Q. prinoides*	*Q.* ×*wagneri* Gaynor	Clinton, Jackson
Q. coccinea × *Q. velutina*		Carter, Oregon
Q. ellipsoidalis × *Q. velutina*		Harrison
Q. falcata × *Q. marilandica*	*Q.* ×*incomita* E.J. Palmer	Howell
Q. falcata × *Q. nigra*	*Q.* ×*garlandensis* E.J. Palmer	Butler, Dunklin
Q. falcata × *Q. phellos*	*Q.* ×*subfalcata* Trel.	Dunklin
Q. falcata × *Q. velutina*	*Q.* ×*willdenowii* Zabel	Howell
Q. imbricaria × *Q. marilandica*	*Q.* ×*tridentata* Engelm. ex A. DC.	Jackson, St. Louis, St. Louis City, Texas, Wayne
Q. imbricaria × *Q. palustris*	*Q.* ×*exacta* Trel.	Boone, St. Louis
Q. imbricaria × *Q. rubra*	*Q.* ×*runcinata* (A. DC.) Engelm.	Franklin, Jackson
Q. imbricaria × *Q. shumardii*	*Q.* ×*egglestonii* Trel.	Dade, Jackson, Laclede
Q. imbricaria × *Q. velutina*	*Q.* ×*leana* Nutt.	Adair, Boone, Cape Girardeau, Carter, Cole, Dade, DeKalb, Greene, Iron, Jackson, Jefferson, Johnson, Knox, Madison, St. Louis, Sullivan
Q. lyrata × *Q. michauxii*	*Q.* ×*tottenii* Melvin	Ripley
Q. lyrata × *Q. stellata*	*Q.* ×*sterrettii* Trel.	Carter
Q. macrocarpa × *Q. michauxii*	*Q.* ×*byarsii* Sudw. ex Trel.	Mississippi
Q. macrocarpa × *Q. muehlenbergii*	*Q.* ×*deamii* Trel.	Butler, Clinton, Greene, Holt, Jackson, Jasper, Johnson
Q. macrocarpa × *Q. prinoides*	*Q.* ×*beckyae* Gaynor	Grundy
Q. macrocarpa × *Q. stellata*	*Q.* ×*guadelupensis* Sarg.	Jackson, Jasper
Q. marilandica × *Q. velutina*	*Q.* ×*bushii* Sarg.	Adair, Barton, Boone, Cass, Christian, Dent, Franklin, Greene, Howell, Jackson, Jasper, Johnson, Knox, Lewis, McDonald, Monroe, Montgomery, Morgan, Newton, Putnam, Ralls, Randolph, St. Francois, St. Louis, St. Louis City, Washington
Q. nigra × *Q. phellos*	*Q.* ×*capesii* W. Wolf	Dunklin
Q. pagoda × *Q. velutina*		New Madrid, Scott

Q. palustris × *Q. phellos*	*Q.* ×*schochiana* Dieck	Stoddard, Wayne
Q. palustris × *Q. rubra*		Benton
Q. palustris × *Q. shumardii*	*Q.* ×*mutabilis* E.J. Palmer & Steyerm.	Bates
Q. palustris × *Q. texana*		Ripley
Q. palustris × *Q. velutina*	*Q.* ×*vaga* E.J. Palmer & Steyerm.	Nodaway
Q. phellos × *Q. rubra*	*Q.* ×*heterophylla* F. Michx.	Dunklin, Wayne
Q. phellos × *Q. shumardii*	*Q.* ×*moultonensis* Ashe	Ripley, Stoddard
Q. phellos × *Q. velutina*	*Q.* ×*filialis* Little	Butler
Q. prinoides × *Q. stellata*	*Q.* ×*stelloides* E.J. Palmer	Jackson, Lawrence, Sullivan
Q. rubra × *Q. shumardii*		Jackson, St. Louis
Q. rubra × *Q. velutina*	*Q.* ×*hawkinsiae* Sudw.	Clark, Dunklin, Jackson

Species of *Quercus* are divided among several subgenera or sections, but there is still some disagreement on the best classification to use (Oh and Manos, 2008). There are only two groups native to eastern North America, the red and black oaks (section *Lobatae* Loudon), which are found only in the New World, and the white oaks (section *Quercus*), the only subgroup that is native to both hemispheres. A third group of North American oaks, the intermediate oaks (section *Protobalanus* (Trel.) A. Camus), is found only in the southwestern United States and northern Mexico. Groupings in the Old World oaks are less clear. A morphologically distinctive group of species in tropical and subtropical Asia have the scales of the acorn cup fused as concentric rings or dense whorls. This group is subgenus *Cyclobalanopsis* (Oerst.) C.K. Schneid. Some of the remaining Old World species definitely belong to section *Quercus*, but others form one or more distinct sections. A species in one of these, section *Cerris* Loudon, is introduced in Missouri. See the key and descriptions below for morphological distinctions among the groups found in Missouri.

1. Acorns maturing during the first autumn after flowering (thus only 1 age group of acorns on the tree in summer and immature acorns not present on the tree during the winter); tips of veins at leaf margin never projecting as bristles; axils of major veins on leaf undersurface without conspicuous tufts of hairs; bark light to medium gray, splitting into loose or more or less persistent ridges, plates, blocks or strips (white oaks.) . . . . 3. QUERCUS
1. Acorns maturing in the second autumn after flowering (thus with 2 age groups of acorns on single twigs in summer and immature acorns present during the winter); tips of veins at leaf margin almost always projecting as 0.5–7.0 mm long bristles; axils of major veins on leaf undersurface usually with tufts of stalked 4–15-rayed hairs 0.3–0.5 mm high; bark medium to dark gray, splitting into persistent ridges or blocks
   2. Scales of acorn cup triangular, appressed, 1.5–5.0 mm long; leaves unlobed, with 10–18 bristles on each side . . . . 1. CERRIS
   2. Scales of acorn cup lanceolate or strap-shaped, strongly recurved, 8–10 mm long; leaf not as above; either deeply lobed or with 0–3 bristles on each side (black oaks) . . . . 2. LOBATAE

The following key to species emphasizes characters of the leaves and twigs wherever possible, so it is more difficult to use, but it will be more useful in spring and summer, when mature acorns and winter buds are not available.

1. Secondary veins of leaf all turning aside before reaching the margin, margin entire and unlobed (with 1or 2 shallow lobes on each side in *Q. nigra*)
   2. Petiole 5–21 mm long; leaf blades elliptic, the undersurface uniformly pubescent with spreading, branched hairs, the pubescence not different in the vein axils; acorn cups 6–8 mm long, the inner surface glabrous . . . . 5. Q. IMBRICARIA
   2. Petiole 1–10 mm long; leaf blades linear to narrowly elliptic or spatulate, the undersurface with prominent tufts of branched hairs in the axils of the major veins or sometimes all along the basal portion of the midvein, otherwise apparently glabrous (inconspicuous, unbranched appressed hairs sometimes visible with high magnification); acorn cup 4–6 mm long, its inner surface pubescent (sometimes glabrous near its margin)
      3. Leaf blades usually narrowly elliptic, sometimes lanceolate or oblanceolate, unlobed; buds glabrous except for the minutely hairy margins . . . . 10. Q. PHELLOS
      3. Leaf blades oblanceolate to narrowly spatulate, with 1 or 2 lobes on each side (these sometimes shallow and obscure); buds hairy on the surface (and often also the margins or the inner scales pubescent and the outer scales glabrous . . . . 7. Q. NIGRA
1. At least 1 secondary vein on each side of the leaf reaching the margin; leaf blades scalloped or with 1–5 strong lobes on each side (sometimes only 1[2] shallow lobes on each side in *Q. marilandica*)
   4. Secondary veins projecting beyond the leaf margins as bristles; acorns ripening the second autumn after flowering
      5. Leaf blades with no more than 9 marginal bristles in total
         6. Twigs glabrous; leaf blades 1.6–6.0 cm wide; acorn cups 4–6 mm high, covering 20–40% of the nut . . . . 7. Q. NIGRA
         6. Twigs pubescent; leaf blade 6.5–18.0 cm wide; acorn cups 6–10 mm high, covering 30–60% of the nut
            7. Leaf blades obovate to spatulate and indistinctly lobed or with 1(2) usually shallow lobes on each side of the leaf, the largest lobes usually near the leaf tip, usually a mere convexity on the leaf margin or, if distinct, then oblong to occasionally ovate, the lobe apex truncate or broadly rounded; leaf undersurface usually with only unbranched appressed hairs (these inconspicuous except with high magnification), sometimes with branched spreading hairs, smooth or slightly felty to the touch; petioles 3–15 mm long. . . . . 6. Q. MARILANDICA
            7. Leaf blades with 1 or 2 large lobes and 0–2 smaller lobes or teeth on each side, most of the leaves with a long, unlobed, narrowly (rarely broadly) rectangular apical portion, the lobes usually narrowly lanceolate and long-tapered, rarely ovate and obtuse; leaf undersurface densely and uniformly pubescent with branched hairs, felty to the touch; petioles 12–77 mm long . . . . 4. Q. FALCATA
      5. Leaf blades with 10–60 marginal bristles in total
         8. Leaves unlobed (but sharply and coarsely toothed); acorn cup scales 8–10 mm long, spreading or reflexed . . . . 1. Q. ACUTISSIMA
         8. Leaves lobed for 25–95% of their width; acorn cup scales 2–6 mm long, loosely to tightly appressed
            9. Leaf blades with the undersurface hairy, at least near the base and along the major veins

10. Leaf blades with the undersurface hairy in patches, especially near the leaf base and the major veins, green; major lobes of leaf oblong or rhombic, usually with 1 strong secondary lobe on the basalmost side; acorn cups 9–13 mm long, the inner surface hairy except near the rim, the scale tips more or less spreading . . . . 14. Q. VELUTINA
10. Leaf blades with the undersurface hairy throughout, white, tawny, or occasionally green; major lobes of the leaf narrowly triangular or lanceolate, rarely nearly oblong, always tapered toward the tip, undivided or with a strong tooth (rarely a secondary lobe) on the basalmost side; acorn cups 6–8 mm long, the inner surface pubescent throughout, the scale tips tightly appressed
  11. Each side of leaf with 1 or 2 large lobes and sometimes also 1 or 2 smaller lobes or teeth, these unevenly spaced, most of the leaves with a long, unlobed, narrowly (rarely broadly) rectangular apical portion; leaf base rounded to more or less broadly angled, rarely truncate; basalmost lobes 3–11 cm above the leaf base in most or all leaves . . . . 4. Q. FALCATA
  11. Each side of leaf with (2)3 or 4 lobes, these evenly spaced, without a long unlobed apical portion; leaf base broadly angled or broadly rounded to almost truncate; basalmost lobes 1.0–2.5 cm above the leaf base in most or all leaves . . . . 8. Q. PAGODA

9. Leaf blades with the undersurface glabrous, except for tufts of hairs in the axils of the major veins
  12. Buds hairy, at least in the upper half; inner bark yellow or orange; acorn cups covering 30–70% of the nut
    13. Twigs usually densely hairy (becoming more or less glabrous late in the season); buds hairy throughout; tufts of hairs in the axils of the major leaf veins large and conspicuous; acorn cups with the inner surface densely hairy except near the rim, the outer surface with the scales loosely appressed with weakly spreading tips, thin and flat throughout . . . . 14. Q. VELUTINA
    13. Twigs glabrous; buds with the lower scales glabrous (except along the margins) and only the upper scales hairy; tufts of hairs in the axils of the major leaf veins small or absent; acorn cups with the inner surface glabrous or sparsely hairy, the outer surface with the scales tightly appressed, thin or distinctly convex-thickened near the base
      14. Twigs usually dark purplish brown, sometimes dark brown or grayish; buds dark brown or reddish brown; leaves lobed for 25–90% of their width, the lobes oblong or lanceolate, rarely broadened outward; acorn cups 5–13 mm long, covering 20–40% of the nut; nuts 19–26 mm long . . . . 11. Q. RUBRA
      14. Twigs usually dark brown, sometimes dark reddish brown. Buds brown. Leaves lobed for 70–90% of their width, the lobes usually broadened outward, sometimes oblong; acorn cups 12–17 mm long, covering 40–70% of the nut; nuts 14–17 mm long
        15. Nuts ellipsoid, 10–11 mm wide, smooth at the tip; acorn cups 14–17 mm wide, enclosing about 40% of the nut . . . . . 3. Q. ELLIPSOIDALIS

15. Nuts ovoid or almost cylindrical, 15–16 mm wide, often with concentric grooves around tip; acorn cups 19–30 mm wide, enclosing 40–70% of the nut . . . . 2. Q. COCCINEA

12. Buds glabrous (except for the often hairy scale margins); acorn cups covering 10–50% of the nut

16. Leaf blades 25–60 mm wide, angled or tapered to a sharply pointed tip, rarely rounded or bluntly pointed, the lobes obovate and broadened outward (usually strongly so), rarely ovate, with 5–12 bristles per lobe, the sinuses between the lobes narrowly rounded, usually narrower than the lobes . . . . 12. Q. SHUMARDII

16. Leaf blades 10–40 mm wide, variously bluntly pointed or angled to narrowly tapered to a sharply pointed tip, the lobes oblong to lanceolate or triangular, seldom much broadened outward, with 1–9(–11) bristles per lobe, the sinuses between the lobes oblong or rounded, as wide as or wider than the lobes

17. Acorn cups thick, the inner surface smooth, the outer surface with the scales convex near the base; leaf blades lobed for 25–90% of the width, with 3–5 lobes on each side; tufts of hairs in the undersurface vein axils absent or small and inconspicuous, each hair with 5–9 rays . . . . 11. Q. RUBRA

17. Acorn cups thin, the inner surface dimpled all over, the outer surface with the scales plane; leaf blade lobed for 70–90% of the width, with 2 or 3(4) lobes on each side; tufts of hairs in the undersurface vein axils large and conspicuous, each hair with 9–19 rays

18. Nuts 9–14 mm long; acorn cup 3–6 mm long, saucer-shaped; twigs dark reddish brown . . . . 9. Q. PALUSTRIS

18. Nuts 14–23 mm long; acorn cup 8–10 mm long, bowl-shaped; twigs dark brown . . . . 13. Q. TEXANA

4. Secondary veins not projecting beyond the leaf margins, the margins without bristles

19. Leaf blades with the axils of the secondary veins on the undersurface having prominent tufts of hairs; acorns ripening the second autumn after flowering

20. Twigs glabrous; leaf blades 1.6–6.0 cm wide; acorn cups 4–6 mm high, covering 20–40% of the nut, the scales tightly appressed . . . . 7. Q. NIGRA

20. Twigs pubescent; leaf blades 6.5–14.0 cm wide; acorn cups 6–10 mm high, covering 30–60% of the nut, the scales loosely appressed (giving the cup a shaggy appearance) . . . . 6. Q. MARILANDICA

19. Leaf blades with the axils of the secondary veins on the undersurface without differentiated tufts of hairs (axil pubescence is similar to that of the rest of the undersurface); acorns ripening the first autumn after flowering

21. Leaf blades without visible hairs at maturity (except at very high magnification), more or less glaucous on both surfaces (immature leaves densely covered by stellate hairs with contorted, tangled rays that are shed in cobwebby masses as the leaves expand) . . . . 15. Q. ALBA

21. Leaf blades more or less hairy at maturity, at least on the undersurface, not glaucous but sometimes pale on the undersurface

22. Leaf blades unlobed or shallowly lobed (to 30% of their width), the margins bluntly toothed or scalloped, sometimes appearing wavy

23. Leaf blades with 3–8 scallops or shallow lobes per side; some secondary veins extending toward the sinuses and not reaching leaf margin; bark light to medium gray, split into more or less persistent ridges

24. Acorn cups with the marginal scales narrower than the other scales but not longer, never forming a conspicuous fringe; leaf blades usually without deep sinuses, generally merely coarsely scalloped, divided 20–40(–60)% of their width, the upper surface with a few simple or 2-branched, erect hairs; twigs usually glabrous, sometimes with scattered, long, spreading hairs; buds 2–3 mm long, glabrous except for the hairy margins of the scales .... 16. Q. BICOLOR

24. Acorn cups with the marginal scales elongate, linear, forming a fringe that projects 6–15 mm beyond the rim of the cup; leaf blades almost always with at least one pair of deep sinuses (60–90% of the leaf width) at or below the midpoint, the upper surface glabrous or with scattered stellate hairs; twigs usually pubescent with small stellate hairs, occasionally with spreading hairs or glabrous; buds (3–)4–6 mm long, hairy .... 18. Q. MACROCARPA

23. Leaf blades with 6–14 scallops or teeth per side; secondary veins all reaching leaf margin in tips of marginal teeth or scallops (except sometimes in the leaf base); bark light ashy gray, split into loose plates or strips or more or less persistent ridges on old trunks

25. Acorn cups 13–19 mm long, 24–33 mm wide, the scales 3–5 mm long, free to their bases; nuts 20–30 mm long, 15–24 mm wide, the fruit stalk 10–23 mm long; leaf blades with the undersurface green or pale green, sparsely to densely covered with spreading hairs but without appressed-stellate hairs, usually very felty to the touch; buds 4–7 mm long .... 19. Q. MICHAUXII

25. Acorn cups 5–12 mm long, 9–19 mm wide, the scales 1.5–3.0 mm long, fused to the cup toward their bases; nuts 10–19 mm long, 9–13 mm wide, the fruit stalk 0–8 mm long; leaf blades with the undersurface white or sometimes pale green, with appressed stellate hairs and sometimes also spreading hairs, sometimes felty to the touch; buds 3–5 mm long

26. Plants trees to 35 m tall (sometimes shrubby in exposed, rocky habitats); leaves 12–22 cm long, with (8–)10–14 scallops or shallow, rounded lobes on each side .... 20. Q. MUEHLENBERGII

26. Plants shrubs, 0.5–3.0(–5.0) m tall; leaves 7–13 cm long, with 6–9 scallops or shallow, rounded lobes on each side .... 21. Q. PRINOIDES

22. Leaf blades lobed for 60–90% of their width

27. Leaf blades thick and rigid, with 2 or 3 lobes on each side, the largest lateral lobes broadened toward their tips, truncate or slightly notched; leaf pubescence more or less roughened, sometimes even sandpapery to the touch .... 22. Q. STELLATA

27. Leaf blades thin and flexible, with 3–8 lobes on each side, the lobes acute to rounded (sometimes truncate or notched in *Q. macrocarpa*); leaf pubescence soft to the touch

28. Acorn cups completely enclosing the nut; leaf blades with 3–5(6) lobes per side, these acutely angled to broadly rounded; bark light ashy gray, splitting into loose plates.................................. 17. Q. LYRATA

28. Acorn cups enclosing 30–70% of the nut; lobes leaf blades with 3–8 lobes per side, these rounded or rounded-obtuse; bark light to medium gray, split into more or less persistent ridges
    29. Acorn cups with the marginal scales long, linear, forming a fringe that projects 6–15 mm beyond the rim of the cup; leaf blades almost always with at least 1 pair of deep sinuses (60–90% of leaf width) at midleaf, the upper surface glabrous or with scattered stellate hairs; twigs usually pubescent with small stellate hairs, occasionally with spreading hairs or glabrous; buds (3–)4–6 mm long, pubescent .... 18. Q. MACROCARPA
    29. Acorn cups with the marginal scales narrower than the median scales but not longer, never forming a conspicuous fringe, not or only slightly projecting beyond the rim of the cup; leaves without deep sinuses, usually merely scalloped, divided 20–40(–50)% of their width, the upper surface with a few simple or 2-branched erect hairs. Twigs usually glabrous, sometimes with scattered, long, spreading hairs; buds 2–3 mm long, glabrous except for the hairy margins of the scales .... 16. Q. BICOLOR

1. Section **Cerris** Loudon

About 50 species. Europe, Asia, Africa; introduced in North America.

Section *Cerris* includes several species that are cultivated as ornamentals or specimen trees in the United States, including *Q. cerris* L. (European turkey oak) and *Q. suber* L. (cork oak). However, only *Q. acutissima* appears to escape into natural landscapes, in part because it has also been planted for wildlife (Goelz and Carlson, 1997).

1. Quercus acutissima Carruth. (sawtooth oak)

Pl. 414 l, m; Map 1839

Plants trees to 30 m tall. Bark medium to dark gray, divided into narrow, persistent ridges. Twigs 2–3 mm thick, dark brown, minutely pubescent with 1–5-rayed, appressed (occasionally spreading) hairs, sometimes nearly glabrous. Buds 6–8 mm long, pubescent (at least on the apical half), brown, the scales glabrous on the surface, hairy along the margins. Leaves with the petiole 15–40 mm long. Leaf blades 14–21 cm long, 3.5–6.0 cm wide, oblong-lanceolate to lanceolate, rounded to truncate at the base, tapered to a sharply pointed tip, the margins sharply toothed, but not lobed, the teeth 10–18 per side, broadly angled to tapered, each ending in 1 bristle 2–5 mm long, the secondary veins (except the basalmost) reaching the margin at the tip of a tooth and extended into the bristle; the upper surface shiny, with scattered, inconspicuous, simple (rarely 2–4-rayed) hairs, the undersurface green, with inconspicuous, unbranched, appressed hairs between the veins, the veins with unbranched spreading hairs, the vein axils with small tufts of more or less 4-rayed stalked hairs. Styles linear or narrowly spatulate, the tip not or only weakly broadened. Pistillate flowers with the sepals fused to the ovary. Fruits ripening the second autumn after flowering, sessile or the stalk to 2 mm long. Acorn cups 14–15 mm long, 18–25 mm wide, hemispheric, covering 30–50% of the nut, the inner surface smooth, hairy, the outer surface with the scales 8–10 mm long, narrowly lanceolate or strap-shaped from a short, triangular, weakly convex base, the tips spreading to strongly recurved, those at the margin extending past the rim of the cup, longer than the other scales but not noticeably differentiated. Nuts 15–20 mm long, 13–17 mm wide, ovoid to ovoid-cylindrical, smooth or occasionally with weak concentric grooves around the tip, the inner surface of the shell free from the seedcoat (except basally), densely pubescent, with abortive ovules near the base. $2n$=24. April.

Introduced, uncommon and widely scattered (native of Asia; introduced sporadically in the eastern U.S.). Mesic upland forests; also parks, roadsides, and disturbed areas.

Quercus acutissima is becoming more common in horticulture. It was first reported to escape in Missouri by Yatskievych and Summers (1993). In addition to its use as an ornamental, the species

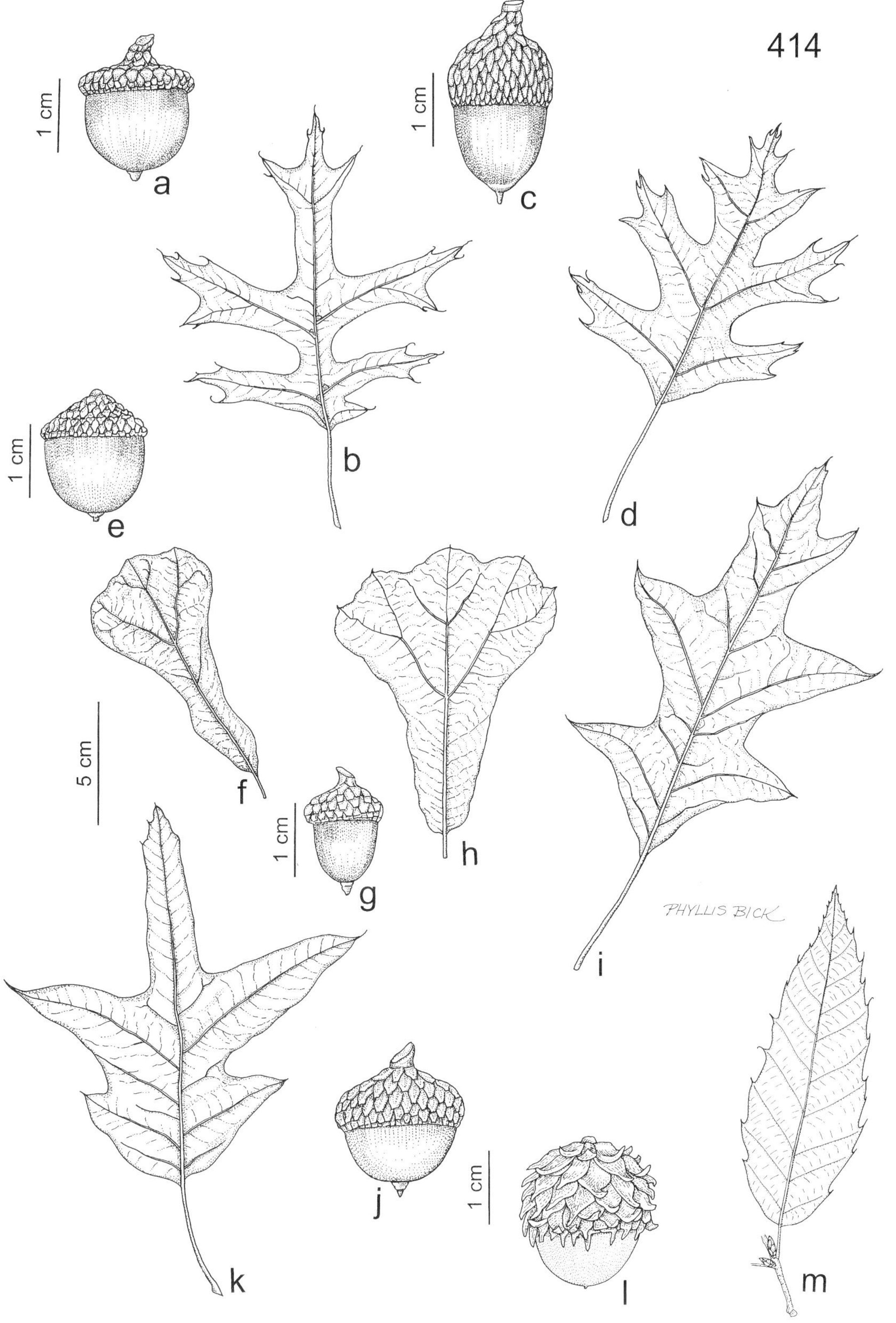

Plate 414. Fagaceae. *Quercus palustris*, **a)** acorn, **b)** leaf. *Quercus ellipsoidalis*, **c)** acorn, **d)** leaf. *Quercus nigra*, **e)** acorn, **f)** leaf. *Quercus marilandica*, **g)** acorn, **h)** leaf. *Quercus pagoda*, **i)** leaf. *Quercus falcata*, **j)** acorn, **k)** leaf. *Quercus acutissima*, **l)** acorn, **m)** leaf.

1840. Quercus coccinea 1841. Quercus ellipsoidalis 1842. Quercus falcata

also sometimes is grown in plantations to attract wildlife (especially turkeys), because some cultivars are relatively fast-growing and produce copious quantities of acorns at a relatively early age. Whittemore (2004) discussed the establishment of this potentially invasive exotic taxon since it was first recognized as spreading from cultivated plants in the early 1990s.

2. **Section Lobatae** Loudon (red oaks, black oaks)

Bark medium to dark gray, divided into persistent ridges or blocks. Leaf blades ovate to obovate, lanceolate to oblanceolate or spatulate, sometimes very narrow, entire to shallowly or deeply lobed, the lobes tapered to rounded or truncate, the major veins usually projecting beyond the leaf margin as bristles 0.5–7.0 mm long (sometimes only the midrib reaches the margin), the undersurface vein axils with tufts of 4–15-rayed stalked hairs in most species. Anthers usually notched at the tip, rarely short-tapered to a sharp point. Styles linear or narrowly spatulate, the tip not or weakly broadened. Pistillate flowers with the sepals free, forming a skirt or flange around the ovary. Fruits ripening the second autumn after flowering, sessile or the stalk to 7 mm long. Acorn cups saucer-shaped, bowl-shaped, or hemispheric, enclosing 20–70% of the nut, sometimes prolonged into a stalklike base, the inner surface smooth or dimpled, hairy or glabrous, the outer surface with the scales 2–6 mm long, triangular to ovate or lanceolate, plane or the base somewhat convex-thickened, free for most of their length, appressed, those at the margin not extending past the cup, not differentiated. Nuts occasionally with concentric grooves around the tip, the inner surface of the shell free from the seed, densely pubescent, with abortive ovules near the apex (lateral in a few species elsewhere). About 90 species, North America to northern South America.

Many of the upland species of the red oak group, particularly in the Ozark portion of Missouri (as well as Arkansas) began to experience dramatic die-offs beginning in the late 1990s. This phenomenon, called *oak decline,* is a complex situation with several contributing causes (Starkey et al., 2004). The situation probably began developing as a result of longterm fire suppression following European colonization of the region, which led to an increase in oak density in many places. Extended droughts during portions of the 1950s, 1990s, and 2000s weakened trees and made them more susceptible to disease and injury. Also contributing to the situation are increases in the populations of two beetles, the red oak borer (*Enaphalodes rufulus* (Haldeman)) and two-lined chestnut borer (*Agrilus bilineatus* Weber), whose larvae bore through the trunks of living trees, weakening the wood and further inviting disease and injury into the tree. Ultimately, many of the trees are dying as a result of infection by fungi, such as Armillaria root rot (*Armillaria* spp.), Hypoxylon canker (*Hypoxylon atropunctatum* (Schwein.) Cooke)), and oak wilt (*Ceratocystis fagacearum* (Bretz) Hunt)).

1. Leaf blades with the secondary veins all turning aside before reaching the margin, the margins entire and unlobed or (in *Q. nigra*) with 1 or 2 shallow lobes on each side
   2. Petioles 5–21 mm long; leaf blades elliptic, the undersurface uniformly pubescent with spreading, branched hairs, the pubescence not different in the vein axils; acorn cups 6–8 mm long, the inner surface glabrous .... 5. Q. IMBRICARIA
   2. Petioles 1–10 mm long; leaf blades linear to narrowly elliptic or spatulate, the undersurface with prominent tufts of branched hairs in the axils of the major veins or sometimes all along the lower portion of the midvein, otherwise apparently glabrous (inconspicuous, unbranched, appressed hairs sometimes visible with high magnification); acorn cups 4–6 mm long, the inner surface pubescent (sometimes glabrous near the margin)
      3. Leaf blades oblanceolate to narrowly spatulate, with 1(2) lobes on each side (these are sometimes shallow and obscure); buds hairy, or the inner scales hairy and the outer scales glabrous .... 7. Q. NIGRA
      3. Leaf blades usually narrowly elliptic, sometimes lanceolate or oblanceolate, unlobed; buds glabrous (except for the hairy scale margins) .... 10. Q. PHELLOS
1. Leaf blades with at least 1 secondary vein on each side reaching the margin and usually projecting as a bristle, the blade with 1–5 strong lobes on each side (sometimes only 1 or 2 shallow lobes in *Q. marilandica*)
   4. Acorn cups with the inner surface regularly dimpled over the whole surface
      5. Nuts 9–14 mm long, acorn cups saucer-shaped, 3–6 mm long .... 9. Q. PALUSTRIS
      5. Nuts 14–23 mm long, acorn cups bowl-shaped, 8–10 mm long .... 13. Q. TEXANA
   4. Acorn cups with the inner surface smooth
      6. Leaf blades at maturity with the undersurface uniformly covered with moderate to dense hairs, white, tawny, or occasionally green, usually very felty to the touch
         7. Leaf blades with 1 or 2 large lobes and 0–2 smaller lobes or teeth per side, mostly with a long, unlobed, narrowly (rarely broadly) rectangular apical portion, the base rounded-obtuse to rounded, rarely truncate, the basalmost lobes 3–11 cm above the blade base in most or all of the leaves .... 4. Q. FALCATA
         7. Leaf blades with with (2)3 or 4 lobes and no additional smaller lobes per side, the lobes evenly spaced, without a long, unlobed apical portion, the base broadly obtuse or rounded-obtuse to almost truncate, the basalmost lobes 1.0–2.5 cm above the blade base in most or all of the leaves .... 8. Q. PAGODA
      6. Leaf blades at maturity with the undersurface glabrous or with patches of hairs, especially near the blade base and the major veins, green, smooth or slightly felty to the touch (inconspicuous unbranched appressed hairs may be visible at high magnification, and expanding leaves in the spring are uniformly pubescent on both sides)
         9. Leaf blades with 1(2) broadly rounded lobes per side, the largest lobes above the midpoint, often near the blade tip, the lobes usually mere convexities on the leaf margin, if distinct then oblong or ovate, or sometimes the leaf merely spatulate and broad-tipped; petioles 3–15 mm long

10. Twigs pubescent; leaf blades 6.5–14.0 cm wide; acorn cups 6–10 mm high, covering 30–60% of the nut .... 6. Q. MARILANDICA
10. Twigs glabrous; leaf blades 1.5–6.0 cm wide; acorn cups 4–6 mm high, covering 20–40% of the nut .... 7. Q. NIGRA
9. Leaves with 3–5 lobes on each side, the lobes usually broadly angled to slenderly tapered, rarely rounded; petioles 18–72 mm long
11. Acorn cups with the inner surface densely pubescent except near the margin, the outer surface with the scales loosely overlapping, the tips slightly spreading, often giving the cup a shaggy appearance; leaf blades often (but not always) with the undersurface having patches of branched spreading hairs, especially near the blade base and the major veins; twigs usually densely pubescent (becoming glabrous or nearly so late in the season)............ 14. Q. VELUTINA
11. Acorn cups with the inner surface glabrous or sparsely pubescent, usually only near the nut scar, the outer surface with the scales tightly overlapping, the tips not spreading; leaf blades with the undersurface glabrous except for tufts in the vein axils (inconspicuous unbranched appressed hairs may also be visible at high magnification); twigs glabrous at maturity (sometimes hairy when very young)
12. Scales of the acorn cup thin and plane; nuts ellipsoid... 3. Q. ELLIPSOIDALIS
12. Many scales of the acorn cup distinctly convex-thickened near the base; nuts ovoid to cylindric
13. Leaf blades lobed for 25–90% of their width, the lobes (3)4 or 5 on each side, oblong or lanceolate, seldom at all broadened outward, the upper surface dull; buds dark brown or reddish brown; twigs usually dark purplish brown, sometimes dark brown or grayish .... 11. Q. RUBRA
13. Leaf blades lobed for (50–)60–90% of their width, the lobes 3 or 4 on each side, usually weakly to strongly broadened outward, the upper surface shiny; buds brown to grayish brown or sometimes dark brown; twigs usually yellowish brown, grayish brown, or dark brown, sometimes dark purplish brown or gray
14. Buds with at least the upper scales hairy; leaf blade undersurface with small tufts of hairs in the vein axils, these with 5–10 rays; acorn cups 12–17 mm long, covering 40–70% of the nut; nuts often with concentric grooves around the tip .... 2. Q. COCCINEA
14. Buds glabrous; leaf blade undersurface with prominent tufts of hairs in the vein axils, these with 7–15 rays; acorn cups 5–14 mm long, covering 20–50% of the nut; nuts smooth at the tip .... 12. Q. SHUMARDII

2. Quercus coccinea Muenchh. (scarlet oak)
Q. coccinea var. *tuberculata* Sarg.
Pl. 415 e, f; Map 1840

Plants trees to 30 m tall. Bark medium to dark gray, divided into persistent ridges, the inner bark pinkish orange. Twigs 2–3 mm wide, usually dark brown, sometimes dark purplish brown or grayish, glabrous. Buds 5–7 mm long, brown, the lower scales glabrous except along the margins, the upper scales pubescent. Petioles 30–70 mm long. Leaf blades 11–20 cm long, 9.0–17.5 cm wide, truncate or very broadly obtuse at the base, divided 70–90% of the width, the lobes 3 or 4 per side, the largest lobes usually the middle pair (of 3) or third pair from the base (of 4); well-developed lobes 22–45(–60) mm wide, usually somewhat broadened outward, sometimes oblong, tapered or sometimes narrowly angled apically, with a few teeth and often 1 secondary lobe on each margin, each with 4–11 bristles 2–6 mm long (the whole blade with 25–54 marginal bristles), the strongest secondary veins reaching the margin at the tips of the lobes and ending in bristles, others reaching toward sinuses and turning aside before reaching the margin; the upper surface rather shiny, glabrous, the undersurface green, glabrous, smooth to the touch,

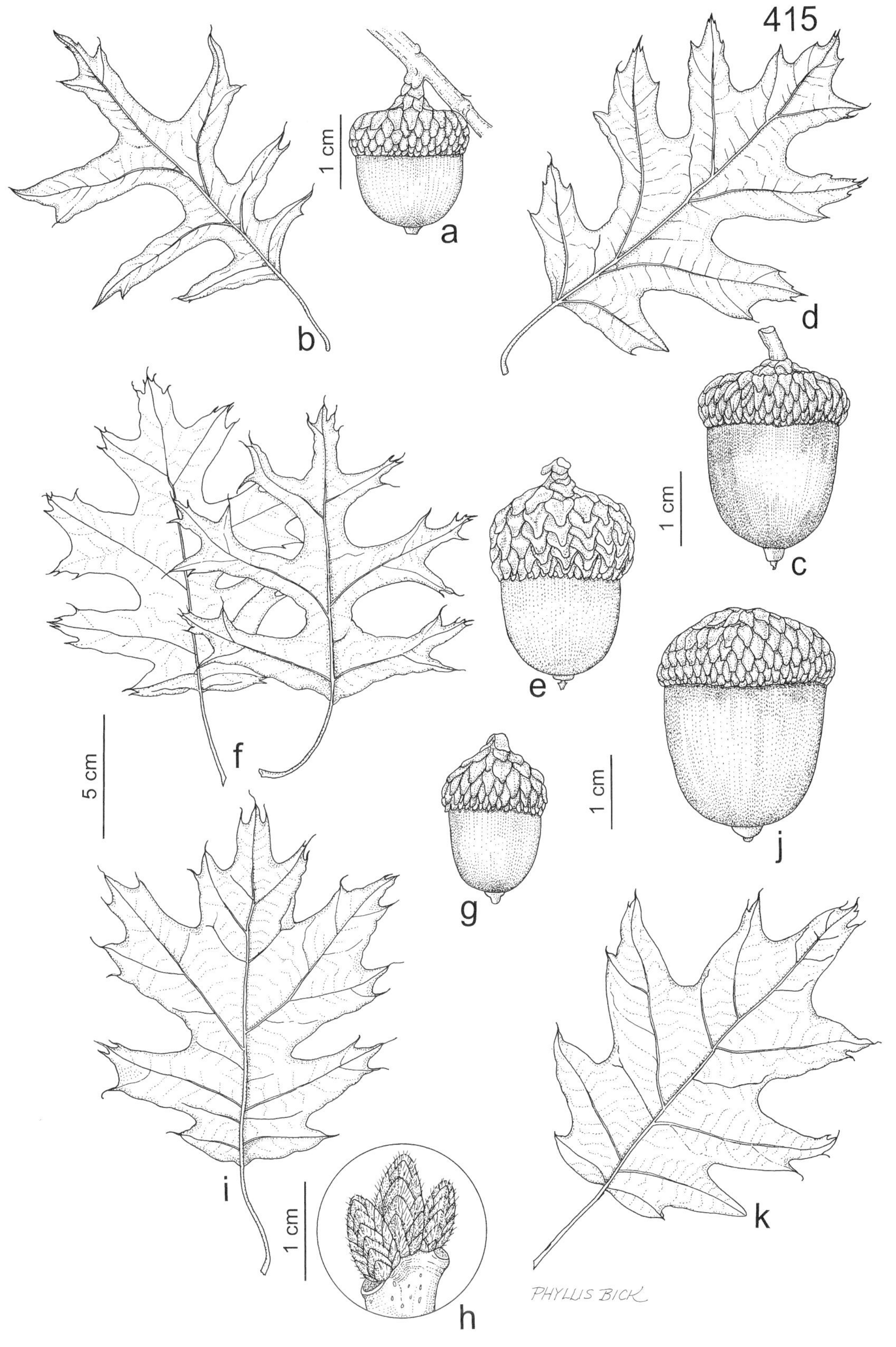

Plate 415. Fagaceae. *Quercus texana*, **a)** acorn, **b)** leaf. *Quercus shumardii*, **c)** acorn, **d)** leaf. *Quercus coccinea*, **e)** acorn, **f)** leaf. *Quercus velutina*, **g)** acorn, **h)** winter buds, **i)** leaf. *Quercus rubra*, **j)** acorn, **k)** leaf.

the vein axils with small tufts of 5–10-rayed, stalked hairs. Acorn cups 12–17 mm long, 19–30 mm wide, covering 40–70% of the nut, bowl-shaped, the base prolonged, the inner surface smooth, glabrous or with a few hairs near the nut scar, the outer surface with the scales mostly distinctly convex-thickened at the base, pubescent. Nuts 15–17 mm long, 15–16 mm wide, ovoid to nearly cylindric, often with concentric grooves around the tip. $2n=24$. April–May.

Scattered in the southeastern quarter of the state (eastern U.S. west to Wisconsin and Louisiana). Mesic to dry upland forests; often on acidic substrates.

Quercus coccinea is notable for the brilliant red color of its leaves in autumn. R. J. Jensen (1997) stated that the acorn cup encloses only 33–50% of the nut, but in many Missouri specimens the nut is more deeply (often much more deeply) enclosed. Plants in which the convex thickening of the acorn cup scale bases is particularly strong have been called var. *tuberculata*. In Missouri, only the hybrid between *C. coccinea* and *C. velutina* has been recorded; this may reflect the difficulty in distinguishing scarlet oak hybrids from other parents with deeply lobed leaves.

3. Quercus ellipsoidalis E.J. Hill (northern pin oak, Hill's oak, jack oak)

Pl. 414 c, d; Map 1841

Plants trees to 25 m tall. Bark medium to dark gray, divided into persistent ridges, the inner bark orangish. Twigs 2.0–2.5 mm wide, dark brown to reddish brown, glabrous. Buds 3.5–4.5 mm long, brown, the lower scales glabrous or nearly so, the upper scales pubescent. Petioles 30–45 mm long. Leaf blades 11–15 cm long, 8.5–12.0 cm wide, obtuse or broadly obtuse at the base, divided 80–90% of the width, the lobes 3 or 4 per side, evenly spaced, with well-developed lobes 32–45 mm wide, broadened outward, tapered apically, with 1 strong secondary lobe on the lower margin and several teeth, each with 5–9 bristles 4–5 mm long (the whole blade with 25–35 marginal bristles), the strongest secondary veins reaching the margin at the tips of the lobes and ending in bristles, others reaching toward sinuses and turning aside before reaching the margin; the upper surface rather shiny, glabrous or with a few hairs scattered along the main veins; the undersurface green, glabrous, smooth to the touch, the vein axils with small tufts of 6–9-rayed, stalked hairs, sometimes a few hairs also scattered along the sides of the midrib. Acorn cups 13–14 mm long, 14–17 mm wide, covering 40% of the nut, bowl-shaped, the base prolonged, the inner surface smooth, sparsely pubescent near the nut scar, the outer surface with the scales thin and plane, sparsely pubescent. Nuts 14–17 mm long, 10–11 mm wide, ellipsoid, without concentric grooves around the tip. April.

Uncommon, known only from Harrison County (north-central U.S. and adjacent Canada, from Ontario and Ohio west to Manitoba and Missouri). Mesic upland forests and savannas; also pastures.

Most Missouri specimens previously determined as *Q. ellipsoidalis* are actually the northern race of *Q. velutina*, which has a similar growth form (see discussion under that species). Additionally, at least one specimen from Harrison County (*Hershey* s.n. on 12 Sept. 1980, at the Missouri Botanical Garden Herbarium) is close to *Q. ellipsoidalis*, but appears to represent the result of past hybridization between that species and *Q. velutina*. *Quercus ellipsoidalis* is morphologically variable and confusing throughout its range (Overlease, 1977; R. J. Jensen, 1977, 1986; Hipp and Weber, 2008). R. J. Jensen (1977) showed that puzzling specimens intermediate between *Q. velutina* and *Q. ellipsoidalis* are found across the Great Lakes region. In their genetic study of *Q. ellipsoidalis* and its relatives, Hipp and Weber (2008) discussed evidence documenting past hybridization between this taxon and *Q. velutina*. Until more ample material is collected from Missouri, it is impossible to evaluate the status of this apparently very rare species in the state.

4. Quercus falcata Michx. (Spanish oak, southern red oak)

Pl. 414 j, k; Map 1842

Plants trees to 30 m tall. Bark medium to dark gray, divided into narrow persistent blocks or ridges, the inner bark orange. Twigs 2–3 mm wide, dark brown or tawny, with scattered or crowded, branched, spreading hairs. Buds 4–7 mm long, brown or dark brown, the lower scales pubescent or glabrous, except along the margins, the upper scales pubescent. Petioles 12–77 mm long. Leaf blades 9–21 cm long, 7–18 cm wide, rounded-obtuse to rounded at the base, rarely truncate, divided 60–90% of the width, with 1 or 2 large lobes and 0–2 smaller lobes or teeth per side, these usually unevenly spaced (most leaves with a long, unlobed, narrowly or rarely broadly rectangular apical portion), the basalmost lobes all or mostly 3–11 cm above the blade base; well-developed lobes 13–30 mm wide, narrowly lanceolate, rarely ovate, long-tapered or rarely obtuse apically, undivided or with a tooth on the lower margin, each with 1 or 2(3) bristles 3–4 mm long (the whole blade with 6–12 marginal bristles), with 2–4 secondary veins per side

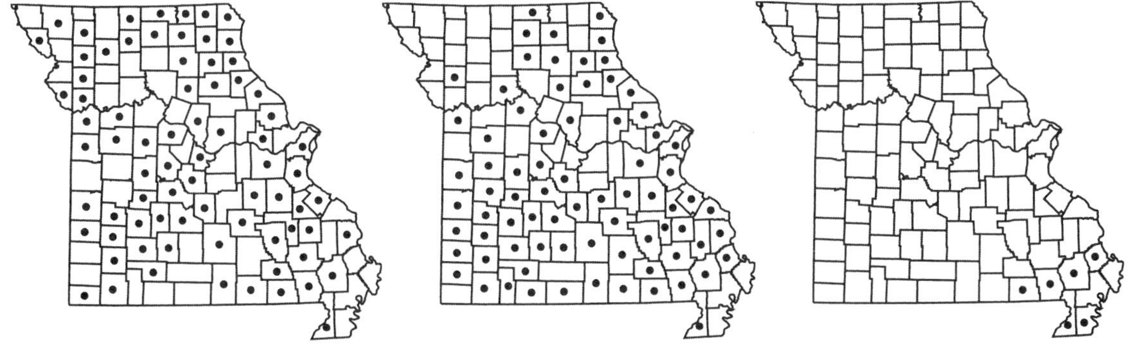

1843. Quercus imbricaria 1844. Quercus marilandica 1845. Quercus nigra

reaching the margin at the tips of lobes or teeth and ending in bristles, others reaching toward sinuses and turning aside before reaching the margin; the upper surface dark green and shiny, almost glabrous (with small branched spreading hairs mostly near the major veins), the undersurface tawny, with moderately to densely crowded 7–11-rayed, spreading hairs and often also inconspicuous unbranched appressed hairs over the whole surface, very felty to the touch, the vein axils with small tufts of 6–12-rayed, often stalked hairs. Acorn cups 6–8 mm long, 12–16 mm wide, covering 30–50% of the nut, bowl-shaped, the inner surface smooth, densely hairy, the outer surface with the scales thin and plane or weakly convex-thickened, pubescent. Nuts 10–13 mm long, 9–12 mm wide, ellipsoid to globose, without distinct concentric grooves around the tip. April–May.

Scattered mostly in southernmost Missouri, north locally to Camden and Perry Counties (southeastern U.S. west to Missouri and Texas). Bottomland forests, mesic upland forests, savannas, sand savannas, sand prairies, margins of glades, banks of stream and rivers, oxbows, sloughs, and swamps; also roadsides.

Morphological variation in *Q. falcata* was analyzed statistically by R. J. Jensen (1989). Typical leaves have 1 or 2 pairs of large lateral lobes on the lower part of the blade; the apical part of the leaf is longer than the lateral lobes. Leaf blades with large lateral lobes above the midpoint and an unlobed tip that is no larger than the lateral lobes are sometimes seen, especially leaves from deeply shaded branches on the lower side of the crown. Specimens with this leaf form have been called f. *triloba* (Michx.) E.J. Palmer & Steyerm. Plants having relatively narrow leaves with short lateral lobes have been called f. *angustior* E.J. Palmer & Steyerm. Hybrids have been recorded with four other species.

5. Quercus imbricaria Michx. (shingle oak)
Pl. 413 l, m; Map 1843

Plants trees to 30 m tall. Bark medium gray, divided into persistent ridges, the inner bark pinkish. Twigs 2–3 mm thick, brown or grayish brown, glabrous or occasionally pubescent with branched spreading hairs. Buds 3–5 mm long, brown, glabrous, the scales glabrous but often hairy along the margins. Petioles 5–21 mm long. Leaf blades 8.5–18.0 cm long, 3–8 cm wide, rounded or obtuse at the base, unlobed, entire, with only 1 apical bristle 0.5–3.0 mm long, the secondary veins all turning aside before reaching the margin; the upper surface usually rather shiny, glabrous, the undersurface green, with 3–6-rayed, stalked, branched spreading hairs and inconspicuous unbranched appressed hairs, usually rather felty to the touch, the hairs not different in the vein axils. Acorn cups 6–8 mm long, 14–18 mm wide, covering 20–40% of the nut, shallowly bowl-shaped to saucer-shaped, the base sometimes prolonged, the inner surface smooth or dimpled, glabrous, the outer surface with the scales plane or the base weakly convex-thickened, pubescent. Nuts 9–12 mm long, 10–12 mm wide, depressed-ovoid to short-cylindric or globose, with weak concentric grooves around the tip. $2n$=24. April–May.

Scattered to common nearly throughout the state (eastern U.S. west to Iowa, Kansas, and Texas). Bottomland forests, mesic to dry upland forests, banks of streams, rivers, and spring branches, and margins of upland prairies; also pastures, fencerows, and roadsides.

Although the wood of shingle oak is considered somewhat inferior to that of other red oak species, it has been used to make roofing shingles in some portions of its range (Steyermark, 1963), hence the common name shingle oak. There is some evidence, based on the paucity of older herbarium specimens relative to more recent collections, that shingle oak has been increasing in the state since about the

1930s. As it is rarely a dominant tree in the plant communities in which it occurs, the status of *Q. imbricaria* in Missouri prior to European settlement is still not well understood.

Hybrids between *Q. imbricaria* and other oaks are collected more frequently than those involving other parental species. This is likely because most hybrids involving shingle oak are relatively easily distinguished, as they constitute hybridization between entire-leaved and lobed-leaved parents. The hybrids thus produce leaves with relatively few, irregular, and often shallow lobes. Hybrids involving *Q. imbricaria* and five other species have been documented from Missouri.

6. Quercus marilandica Münchh. **var. marilandica** (blackjack oak)

Pl. 414 g, h; Map 1844

Plants trees to 15 m tall. Bark medium to dark gray, divided into narrow persistent ridges or blocks, the inner bark orangish. Twigs 2–4 mm wide, brown to dark brown, more or less pubescent with branched spreading hairs. Buds 4–9 mm long, tan or brown, pubescent or occasionally the lowermost scales almost glabrous. Petioles 3–15 mm long. Leaf blades 8–16 cm long, 6.5–14.0 cm wide, rounded to truncate or sometimes rounded-obtuse at the base; divided shallowly or up to 80% of the width, the lobes 1(2) per side, the largest lobes above the midpoint, usually near the blade tip, even well-developed lobes usually a mere convexity on the leaf margin, if distinct then 20–50 mm wide, usually oblong, sometimes ovate, truncate or broadly rounded, undivided or with a distinct bulge or shallow secondary lobe on the lower margin, each with 0–3 bristles 1–3 mm long (the whole blade with 0–7 marginal bristles), with only 1 pair of secondary veins reaching the margin at the tips of the lobes and ending in bristles, the others turning aside before reaching the margin; the upper surface rather dull or somewhat shiny, with scattered small branched spreading hairs, at least near the base, the undersurface green, with inconspicuous unbranched appressed hairs between the veins, sometimes also with 6–11-rayed, spreading hairs scattered in patches (especially near the base and along the main veins), smooth or slightly felty to the touch, the vein axils with prominent tufts of 6–10-rayed, stalked hairs. Acorn cups 6–10 mm long, 12–19 mm wide, covering 30–60% of the nut, bowl-shaped, the inner surface smooth, hairy, the outer surface with the scales thin and plane or weakly convex-thickened, pubescent. Nuts 9–16 mm long, 9–14 mm wide, short-ovoid, without concentric grooves around the tip. $2n$ probably = 24 (reported as $2x$=12 by Friesner [1930]). April–May.

Scattered to common nearly throughout the state, but uncommon to absent in the western portion of the Glaciated Plains Division and the southern portion of the Mississippi Lowlands (eastern U.S. west to Nebraska and Texas). Dry upland forests, savannas, glades, and occasionally banks of streams; also pastures, railroads, and roadsides; often on acidic substrates.

Small-leaved plants (with leaves only 4–7 cm long) found in dry habitats in Texas and Oklahoma have been called var. *ashei* Sudw. This variety was reported from Crawford County by Hunt (1990), but the specimen (*Hunt MO16*, at the University of Georgia Herbarium), falls into the size range of typical *Q. marilandica*. Hybrids involving *Q. marilandica* have been recorded in Missouri involving three other species.

7. Quercus nigra L. (water oak)

Pl. 414 e, f; Map 1845

Plants trees to 30 m tall. Bark medium gray, divided into low persistent ridges, the inner bark pinkish. Twigs 1–2 mm wide, dark brown, glabrous. Buds 3–5 mm long, dark brown or grayish, pubescent or the lower scales glabrous. Petioles 2–10 mm long. Leaf blade 5–11 cm long, 1.6–6.0 cm wide, narrowly angled to long-tapered at the base, divided shallowly or up to 70% of the width, the lobes 1(2) per side, the largest lobes above midpoint, often near the blade tip, even well-developed lobes usually a mere convexity on the leaf margin, if distinct then 10–20 mm wide, oblong or ovate, broadly rounded, undivided or with a distinct bulge on the lower margin, each with 0(–2) bristles 0.5–1.0 mm long (the whole blade with 0–7 marginal bristles), the secondary veins turning aside before reaching the margin or with 1 pair reaching the margin at the tips of the lobes and ending in bristles; the upper surface rather dull, glabrous, the undersurface green, glabrous or with inconspicuous unbranched appressed hairs, smooth to the touch, the vein axils with prominent tufts of 6–12-rayed, stalked hairs. Acorn cups 4–6 mm long, 13–14 mm wide, covering 20–40% of the nut, saucer-shaped to shallowly bowl-shaped, the inner surface smooth, hairy except near the rim, the outer surface with the scales tending to be distinctly convex-thickened at the base, pubescent. Nuts 11–13 mm long, 13–14 mm wide, depressed-ovoid, without concentric grooves around the tip. $2n$=24. April–May.

Scattered in the Mississippi Lowlands Division (southeastern U.S. west to Missouri, Oklahoma, and Texas). Swamps, bottomland forests, banks of streams and rivers, and margins of lakes, oxbows, and sloughs; also ditches and fencerows.

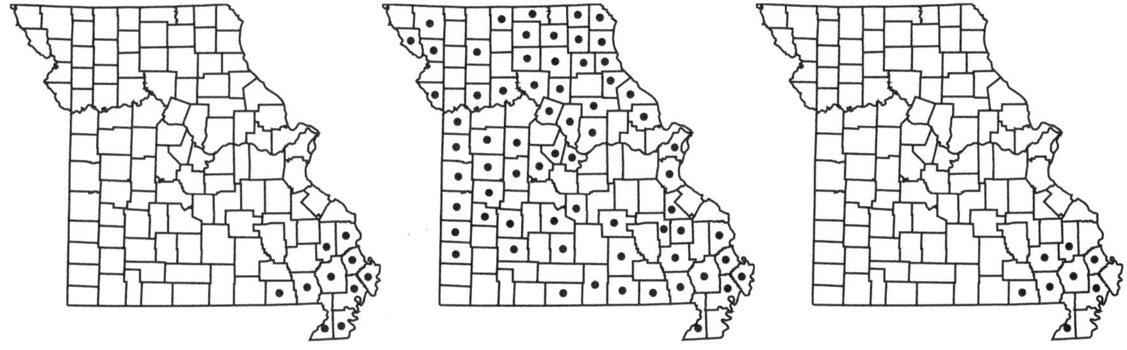

1846. Quercus pagoda 1847. Quercus palustris 1848. Quercus phellos

Leaves of *Q. nigra* are quite variable morphologically and a number of forms have been described from material in other states to account for extremes in variation. Juvenile leaves or leaves from vigorous sprouts may be deeply lobed. A specimen from Dunklin County (*Holmes 981,* at the Missouri Botanical Garden Herbarium) appears to represent a hybrid with either *Q. palustris* or *Q. texana*, but the parentage cannot be confirmed. Other hybrids in Missouri appear to be quite uncommon, only *Q. falcata* and *Q. phellos* have been recorded to hybridize with *Q. nigra* in the state.

8. Quercus pagoda Raf. (cherrybark oak)
Q. falcata Michx. var. *pagodifolia* Elliott
Pl. 414 i; Map 1846

Plants trees to 40 m tall. Bark medium to dark gray, divided into square blocks or plates, the inner bark orange. Twigs 2–4 mm wide, dark brown, densely pubescent with branched, spreading hairs when young, sometimes becoming more or less glabrous with age. Buds 5–7 mm long, brown, pubescent or the lower scales glabrous except along the margins. Petioles 27–45 mm long. Leaf blade 12–19 cm long, 10–15 cm wide, broadly obtuse or rounded-obtuse to nearly truncate at the base, divided 70–80% of the width, the lobes (2)3 or 4 per side, these evenly spaced, the median lobes usually the largest, the basalmost lobes 1.0–2.5 cm above the blade base in most or all leaves; well-developed lobes 15–30 mm wide, narrowly triangular, rarely almost oblong, usually tapered apically, occasionally rounded-obtuse or acute, undivided or with a strong tooth (rarely a secondary lobe) on the lower margin, each with 1–4 bristles 3–4 mm long (the whole blade with 10–20 marginal bristles), the strongest secondary veins reaching the margin at the tips of the lobes and ending in bristles, others reaching toward sinuses and turning aside before reaching the margin; the upper surface dull or somewhat shiny, with small branched spreading hairs when young, by midsummer usually with only scattered hairs near the midvein, the undersurface green or white, with moderately to densely crowded 7–11-rayed, spreading hairs and often also inconspicuous, unbranched, appressed hairs over the whole surface, usually felty to the touch, the vein axils with tufts of 5–15-rayed, often stalked hairs. Acorn cups 6–8 mm long, 17–18 mm wide, covering 30–60% of the nut, bowl-shaped, the inner surface smooth, densely hairy, the outer surface with the scales thin and plane or weakly convex-thickened, pubescent. Nuts 10–12 mm long, 11–12 mm wide, depressed-ovoid, without distinct concentric grooves around the tip. April–May.

Scattered in the Mississippi Lowlands Division and adjacent portions of the Ozarks and Ozark Border (southeastern U.S. west to Missouri, Oklahoma, and Texas). Bottomland forests, mesic upland forests, sand savannas, sand prairies, banks of stream and rivers, oxbows, sloughs, and swamps; also roadsides.

Morphological variation in *Q. pagoda* was analyzed statistically by R. J. Jensen (1989). It is very closely related to *Q. falcata*, which sometimes occurs in slightly drier habitats in the same region, and the two have sometimes been considered varieties of a single species (Steyermark, 1963). Hybrids involving *Q. pagoda* are rare in Missouri, with only the hybrid involving *Q. velutina* documented thus far.

9. Quercus palustris Münchh. (pin oak)
Pl. 414 a, b; Map 1847

Plants trees to 25 m tall. Bark medium to dark gray, divided into low persistent ridges, the inner bark pinkish. Twigs 1.5–3.5 mm wide, dark reddish brown, glabrous (rarely with scattered hairs). Buds 2.0–6.5 mm long, tan or brown, glabrous except along the margins. Petioles 15–52 mm long. Leaf blades 8.5–20.0 cm long, 6–23 cm wide, trun-

1849. Quercus rubra 1850. Quercus shumardii 1851. Quercus texana

cate to obtuse or sometimes tapered at the base, divided 70–90% of the width, the lobes 2 or 3(4) per side, evenly spaced or the lowest closer together, the second pair from the base the largest; well-developed lobes 10–33(–40) mm wide, oblong or tapered, seldom much broadened outward, narrowly tapered or sometimes merely acute apically, with a few teeth and sometimes a secondary lobe on each margin, each with 3–9 bristles 3–9 mm long (the whole blade with 17–38 marginal bristles); the strongest secondary veins reaching the margin at the tips of the lobes and ending in bristles, others reaching toward sinuses and turning aside before reaching the margin; the upper surface usually rather shiny, glabrous or with scattered hairs on the major veins, the undersurface green, glabrous, smooth to the touch, the vein axils with prominent tufts of 11–19-rayed, stalked hairs. Acorn cups 3–6 mm long, 12–19 mm wide, covering 10–30% of the nut, saucer-shaped, the inner surface regularly dimpled, the central portion hairy, the outer surface with the scales thin and plane, pubescent. Nuts 9–15 mm long, 10–15 mm wide, short-ovoid, without concentric grooves around the tip. $2n$=24. April–May.

Common nearly throughout the state (eastern [mostly northeastern] U.S. west to Wisconsin, Nebraska, and Oklahoma; Canada). Bottomland forests, swamps, banks of streams, rivers, and spring branches, margins of ponds, lakes, sinkhole ponds, oxbows, and sloughs, and occasionally mesic upland forests; also fencerows, edges of pastures, and roadsides.

Hybrids between *Q. palustris* and six other oaks have been documented from Missouri.

Pin oak is a common street tree in some cities, as it tends to grow somewhat more quickly than most other red oak species. However, in areas with soils having high pH the leaves can yellow, so care must be taken periodically to acidify such soils.

10. Quercus phellos L. (willow oak)
Pl. 413 n, o; Map 1848

Plants trees to 30 m tall. Bark medium gray, divided into persistent ridges, the inner bark light orange. Twigs 1–2 mm wide, dark brown or gray, glabrous or rarely with scattered, branched, spreading hairs. Buds 2–4 mm long, dark brown, glabrous, the scales hairy along the margins. Petioles 1–6 mm long. Leaf blade 6–12 cm long, 1–3 cm wide, rounded or acute at the base, unlobed, entire, with only 1 apical bristle 0.5–2.0 mm long, the secondary veins all turning aside before reaching the margin; the upper surface usually rather shiny, glabrous, the undersurface green, glabrous, or rarely with inconspicuous, unbranched, appressed hairs and inconspicuous hairs on the midrib, smooth to the touch, sometimes with small tufts of 5–15-rayed, often stalked hairs in the axils of the major veins or sometimes all along the lower midrib. Acorn cups 4–6 mm long, 13–16 mm wide, covering 20–40% of the nut, saucer-shaped to shallowly bowl-shaped, the inner surface smooth, densely hairy, the outer surface with the scales tending to be distinctly convex-thickened at the base, pubescent. Nuts 8–11 mm long, 10–13 mm wide, depressed-globose or broadly ellipsoid, without concentric grooves around the tip. April–May.

Scattered in the Mississippi Lowlands Division and adjacent portions of the Ozarks and Ozark Border (eastern [mostly southeastern] U.S. west to Missouri and Texas). Bottomland forests, swamps, banks of streams and rivers, and margins of oxbows and sloughs; also fencerows, ditches, and roadsides.

Juvenile leaves of *Q. phellos* or leaves from vigorous sprouts may be lobed, and they resemble those of the hybrid, *Q. nigra* × *Q. phellos*. In addition to this hybrid, *Q. phellos* has been documented to hybridize with five other oak species in Missouri. Because such hybrids involve a cross between a species having narrow, entire leaves, and a second parent with broader, lobed leaves, they may be more easily distinguishable than are hybrids in which both parental taxa have similar patterns of leaf division.

11. Quercus rubra L. (northern red oak)
Q. rubra var. *ambigua* (F. Michx.) Fernald
Q. borealis F. Michx.
Q. borealis var. *maxima* (Marsh.) Ashe
Pl. 415 j, k; Map 1849

Plants trees to 30 m tall. Bark medium gray, divided into persistent ridges, the inner bark pinkish. Twigs 1.5–4.0 mm wide, usually dark purplish brown, sometimes dark brown or grayish, glabrous. Buds 3–9 mm long, dark brown or reddish brown, glabrous or the upper scales pubescent. Petioles 18–58 mm long. Leaf blade 11.5–20.0 cm long, 9–16 cm wide, truncate or broadly obtuse at the base, divided 25–90% of the width, the lobes (3)4 or 5 per side, evenly spaced, the largest lobes usually at or above the midpoint; well-developed lobes oblong or lanceolate, seldom slightly broadened outward, 12–40 mm wide, obtuse to slenderly tapered apically, usually toothed, often with a secondary lobe on the lower (seldom also the upper) margin, each with 1–7(–11) bristles 2–6 mm long (the whole blade with 16–53 marginal bristles), the strongest secondary veins reaching the margin at the tips of the lobes and ending in bristles, others reaching toward sinuses and turning aside before reaching the margin; the upper surface dull, glabrous, the undersurface green, glabrous, smooth to the touch, the vein axils with small tufts of 5–9-rayed, stalked hairs, or glabrous. Acorn cups 5–13 mm long, 15–30 mm wide, covering 10–30% of the nut, saucer-shaped, the inner surface smooth, usually sparsely pubescent, at least near the nut scar, the outer surface with the scales mostly distinctly convex-thickened at the base, sparsely pubescent. Nuts 19–26 mm long, 13–23 mm wide, ovoid to cylindrical, without concentric grooves around the tip. $2n=24$. April–May.

Common throughout the state (eastern U.S. west to Minnesota, Nebraska, and Oklahoma; Canada). Mesic to dry upland forests, bases and tops of bluffs, banks of streams and rivers, margins of sinkhole ponds, and edges of glades, upland prairies, and loess hill prairies; also pastures and roadsides.

Steyermark's (1963) key to oak species stated that leaves of *Q. rubra* are never lobed for much more than half of their width. This is not accurate, even on specimens named by Steyermark himself, and has led to many subsequent misdeterminations. Hybrids involving *Q. rubra* can be difficult to distinguish from those involving other parents with similar leaves. Nevertheless, hybrids between *Q. rubra* and five other oak species have been collected in Missouri.

12. Quercus shumardii Buckley (Shumard oak)
Q. shumardii var. *schneckii* (Britton) Sarg.
Q. shumardii var. *stenocarpa* Laughlin
Pl. 415 c, d; Map 1850

Plants trees to 35 m tall. Bark medium to dark gray, divided into narrow persistent ridges, the inner bark pinkish. Twigs 2.0–3.5 mm wide, yellowish brown, grayish brown, or dark brown, glabrous. Buds 4–8 mm long, brown to grayish brown or sometimes dark brown, glabrous. Petioles 25–72 mm long. Leaf blades 11–17 cm long, 8–18 cm wide, truncate or broadly obtuse at the base, divided (50–)60–90% of the width, the lobes 3 or 4 per side, the largest lobes at or above the midpoint; well-developed lobes 25–60 mm wide, obovate and broadened outward (usually strongly so), rarely ovate, acutely angled or tapered apically, rarely rounded-obtuse, with several teeth, 1 or 2 secondary lobes on the lower margin and 0 or 1 on the upper margin, each with 5–12 bristles 2–5 mm long (the whole blade with 25–60 marginal bristles); the strongest secondary veins reaching the margin at the tips of the lobes and ending in bristles, others reaching toward sinuses and turning aside before reaching the margin; the upper surface glossy, glabrous, the undersurface green, glabrous or rarely with inconspicuous, unbranched, appressed hairs, smooth to the touch, the vein axils with prominent tufts of 7–15-rayed, stalked hairs. Acorn cups 5–14 mm long, 12–24 mm wide, covering 20–50% of the nut, saucer- or bowl-shaped, the inner surface smooth, glabrous or with a few hairs near the nut scar, the outer surface with the scales mostly distinctly convex-thickened at the base, sparsely pubescent. Nuts 14–27 mm long, 11–19 mm wide, ovoid or almost cylindric, without concentric grooves around the tip. April–May.

Scattered in the southern 2/3 of the state (eastern U.S. west to Nebraska and Texas; Canada). mesic to dry upland forests, savannas, edges of glades, tops of bluffs, banks of streams and rivers, and rarely margins of fens.

Forms with the acorn cup deeply bowl-shaped have been called var. *schneckii*. Steyermark (1963) noted that in Missouri such plants tend to occupy sites at the dry end of the ecological spectrum of the species. However, Hess and Stoynoff (1998)

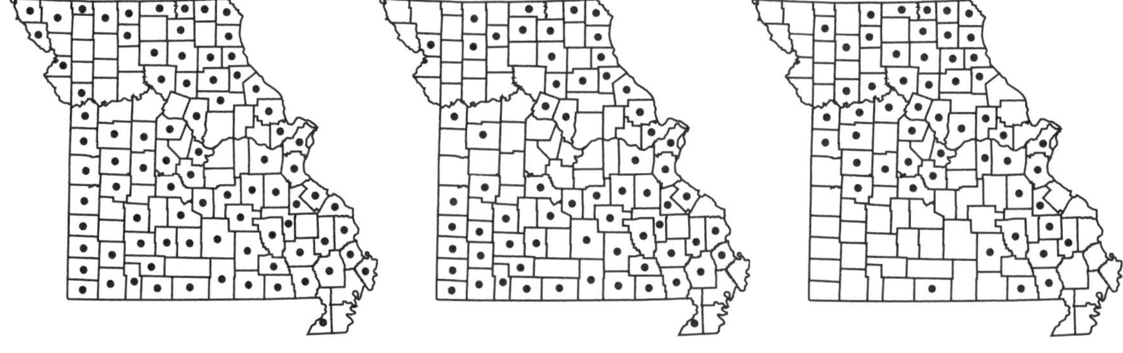
1852. Quercus velutina 1853. Quercus alba 1854. Quercus bicolor

noted that the type of var. *schneckii* originated from a bottomland site in southern Illinois, and they could not distinguish trees at the type locality of this segregate from typical var. *shumardii*. Similarly, Laughlin (1969) segregated var. *stenocarpa* to distinguish large trees in Missouri and Illinois having ellipsoid acorns with very shallow cups. Another variant that most botanists currently treat as a separate species is *Quercus acerifolia* (E.J. Palmer) Stoynoff & W.J. Hess (*Q. shumardii* var. *acerifolia* E.J. Palmer), maple-leaved oak, which differs in its nearly palmately compound leaves that are mostly wider than long. Although this taxon currently is thought to be endemic to west-central Arkansas, some specimens of *Q. shumardii* from southwestern Missouri come close to it in their leaf lobing pattern and dimensions. Morphological variation within this species complex requires further study before such variants can be recognized with confidence. It seems likely that in the future new data from molecular or other sources will result in a refinement of the taxonomy of *Q. shumardii* and its variants. Potentially complicating the issue is the fact that hybrids with four other oak species have been documented from Missouri.

13. Quercus texana Buckley (Nuttall's oak)
Q. nuttallii E.J. Palmer

Pl. 415 a, b; Map 1851

Plants trees to 25 m tall. Bark medium to dark gray, divided into persistent ridges, the inner bark pinkish. Twigs 1.5–3.0 mm wide, dark brown, glabrous. Buds 3.5–5.0 mm long, brown or grayish brown, the scales glabrous or minutely pubescent, also hairy along the margins. Petioles 20–48 mm long. Leaf blade 9–19 cm long, 7–18 cm wide, obtuse or tapered at the base, divided 70–90% of the width, the lobes 2 or 3(4) per side, evenly spaced or the lowest closer together, the second pair from the base the largest; well-developed lobes 11–26 (–34) mm wide, oblong or triangular, seldom much broadened outward, obtuse to slenderly tapered apically, often with a few teeth on each margin and sometimes a secondary lobe on the lower margin, each with 1–5 bristles 2–6 mm long (the whole blade with 13–30 marginal bristles), the strongest secondary veins reaching the margin at the tips of the lobes and ending in bristles, others reaching toward sinuses and turning aside before reaching the margin; the upper surface dull or rather shiny, glabrous, the undersurface green, glabrous, smooth to the touch, the vein axils with prominent tufts of 9–19-rayed, stalked hairs. Acorn cups 8–10 mm long, 16–20 mm wide, covering 20–50% of the nut, bowl-shaped from a prolonged base, the inner surface regularly dimpled, the central portion hairy, the outer surface with the scales thin and plane, pubescent. Nuts 14–23 mm long, 13–17 mm wide, ovoid, without concentric grooves around the tip. April–May.

Uncommon in the Mississippi Lowlands Division (Kentucky to Missouri south to Alabama and Texas). Bottomland forests and swamps.

This species is very similar to *Q. palustris*, and they are almost impossible to tell apart without acorns. *Quercus texana* has a generally more southerly distribution than does *Q. palustris*, but the ranges of the two species overlap in southeastern Missouri and adjoining portions of adjacent states. Acorn size can be very variable within some populations.

14. Quercus velutina Lam. (black oak, yellowbark oak)

Pl. 415 g–i; Map 1852

Plants trees to 25 m tall. Bark dark gray, divided into narrow, more or less persistent ridges, the inner bark yellow or orange. Twigs 2–4 mm wide, brown or dark brown, usually densely pubescent with branched spreading hairs when young, becoming glabrous with age. Buds 4–12 mm long, tan to light or rarely dark brown, pubescent. Petioles 25–60 mm long. Leaf blades 13–21 cm long, 11.0–16.5 cm wide, obtuse or truncate at the base, divided 40–90% of the width, the lobes 3 or 4(5) per side, the largest pair at or above midpoint; well-developed lobes 20–55 mm wide, oblong or rhombic, rarely much broadened outward, tapered to rounded apically, usually with 1 strong secondary lobe on the lower margin and a few teeth or small lobes on each margin, each with 2–5 bristles 2–7 mm long (the whole blade with 14–43 marginal bristles); the strongest secondary veins reaching the margin at the tips of the lobes and ending in bristles, others reaching toward sinuses and turning aside before reaching the margin; the upper surface usually rather shiny, with small, branched, spreading hairs scattered along the midvein and sometimes between the veins near the blade base, the undersurface green, glabrous or with 5–12-rayed, spreading hairs and inconspicuous, unbranched, appressed hairs along the main veins and scattered in patches on the blade (especially near the leaf base and the major veins), smooth or harsh-felty to the touch, the vein axils with conspicuous tufts of 6–8-rayed, stalked hairs. Acorn cups 9–13 mm long, 14–25 mm wide, covering 30–60% of the nut, hemispheric to bowl-shaped, the base prolonged, the inner surface smooth, hairy except near the rim, the outer surface with the scales thin and plane, pubescent. Nuts 11–16 mm long, 9–16 mm wide, ellipsoid to nearly cylindric, without concentric grooves around the tip. $2n=24$. April–May.

Common nearly throughout the state (eastern U.S. west to Minnesota and Texas; Canada). Mesic to dry upland forests, savannas, sand savannas, margins of upland prairies, sand prairies, loess hill prairies, glades, tops of bluffs, and less commonly banks of streams and rivers; also pastures, fencerows, and roadsides.

The common form of *Q. velutina,* with patches of stellate hairs scattered on the undersurface of the blade (especially near the leaf base and the major veins), has been called f. *missouriensis* (Sarg.) Trel. The inner bark of this species yields quercitrin, a yellow compound that was once widely used as a dye. Hybrids with nine other oak species have been documented from the state, which is the largest diversity of oak hybrids in Missouri.

Most Missouri populations of *Q. velutina* have relatively open, sparsely branched crowns. A few populations from the northernmost tier of counties in Missouri have denser, more closely branched crowns. Such populations are found elsewhere the northern part of the range of *Q. velutina*, especially in the Great Lakes region. This closely branched northern race of *Q. velutina* has sometimes been sometimes confused with *Q. ellipsoidalis* (Overlease, 1975), a species that is closely related to *Q. velutina* (Hipp and Weber, 2008).

3. Section Quercus (white oaks)

Bark light to medium gray, divided into loose or more or less persistent ridges, plates, blocks, or strips. Leaf blades obovate or elliptic, lobed, scalloped, or bluntly toothed, the lobes rounded or bluntly and broadly pointed, truncate or minutely notched, the major veins not projecting beyond the leaf margin, the undersurface vein axils without tufts of hairs (glabrous or with the same pubescence as the rest of the leaf). Anthers short-tapered to a sharply pointed tip, less commonly notched. Styles usually strongly broadened at the tip, rarely linear. Pistillate flowers with the sepals fused to the ovary. Fruits ripening the first autumn after flowering, sessile or the stalk to 100 mm long. Acorn cups bowl-shaped, hemispheric, or almost globose, enclosing 20% to all of the nut, not prolonged into a stalklike base, the inner surface smooth, densely hairy, the outer surface with the cup scales 1.5–5.0 mm long, triangular, weakly to very strongly convex-thickened or keeled, often fused to the cup for much of their length, appressed, those at the margin usually not differentiated (long, linear, and forming a fringe around the edge of the cup in *Q. macrocarpa*). Nuts without concentric grooves around the tip, the inner surface of the shell more or less fused to the seed coat, thus the sparsely hairy or glabrous surface not easily observed, the abortive ovules near the base. About 130 species, North America to Central America, Europe, Asia.

The acorns of most white oaks germinate immediately on falling from the tree, and require no cold treatment. *Quercus macrocarpa* apparently is an exception to this.

The wood of white oaks swells well when soaked, so in addition to its use in general construction and cabinet work, white oaks (especially *Q. alba*) were also of great importance for building the hulls and decking of wooden ships and in the construction of tight cooperage (barrels for holding liquids, used in the manufacture of wine and whiskey). Ships and casks are now usually made of steel, mostly as a result of the clearing of oak forests and the rise in wages for skilled labor.

1. Acorn cups broadly urn-shaped to almost globose, enclosing most or all of the nut, contracted to the mouth so the nut cannot be removed without breaking the cup . . . . 17. Q. LYRATA
1. Acorn cups hemispherical or bowl-shaped, enclosing 25–70% of the nut, the mouth broad so the nut can be removed without breaking the cup
    2. Acorn cups with the marginal scales long, linear, forming a fringe that projects 6–15 mm beyond the rim of the cup . . . . 18. Q. MACROCARPA
    2. Acorn cups with the marginal scales not elongated or linear, not forming a fringe and not or only slightly projecting beyond the rim of the cup
        3. Acorns moderately to long-stalked, the stalk 10–100 mm long
            4. Leaf blades with 11–14 coarse but relatively shallow scallops on each side, divided 0.1–0.2 of their width; acorn cup 13–19 mm long, 24–33 mm wide, the scales 3–5 mm long, free to their bases . . 19. Q. MICHAUXII
            4. Leaf blades with (2–)3–8 moderate to deep scallops or lobes on each side, divided 20–90% of their width; acorn cup 7–14 mm long, 11–27 mm wide, the scales 1.5–3.0 mm long, fused to the cup basally
                5. Leaf blades usually lobed, rarely deeply scalloped, divided (30–) 60–90% of their width, the undersurface lacking spreading hairs, glaucous, smooth to the touch; acorn cups with the scales toward the rim not noticeably narrower than the other scales . . . . . 15. Q. ALBA
                5. Leaf blades usually coarsely scalloped, rarely shallowly lobed, divided 20–40(–60)% of their width, the undersurface with spreading hairs, usually white but not glaucous, felty to the touch; acorn cups with the scales toward the rim much narrower than the other scales . . . . 16. Q. BICOLOR
        3. Acorns sessile or relatively short-stalked, the stalk 0–8 mm long
            6. Leaf blades lobed (rarely merely deeply scalloped), divided (30–)60–90% of their width
                7. Twigs glabrous, reddish brown to grayish brown, sometimes glaucous but never whitish; buds 2–5 mm long, glabrous or sparsely short-hairy; leaf blades with the lobes rounded or broadly angled at the tip; acorn cups enclosing 20–40% of the nut . . . . 15. Q. ALBA
                7. Twigs hairy, whitish tan or pale gray; buds 4–6 mm long, hairy. Leaf blades with the largest lobes truncate at the tip; acorn cups enclosing 50–70% of the nut . . . . 22. Q. STELLATA
            6. Leaf blades unlobed, merely coarsely scalloped or bluntly toothed, divided 10–30% of their width
                8. Plants trees to 35 m tall (sometimes shrubby in exposed, rocky habitats); leaves 12–22 cm long, with (8–)10–14 scallops on each side . . . . 20. Q. MUEHLENBERGII
                8. Plants shrubs, 0.5–3.0(–5.0) m tall; leaves 7–13 cm long, with 6–9 scallops on each side . . . . 21. Q. PRINOIDES

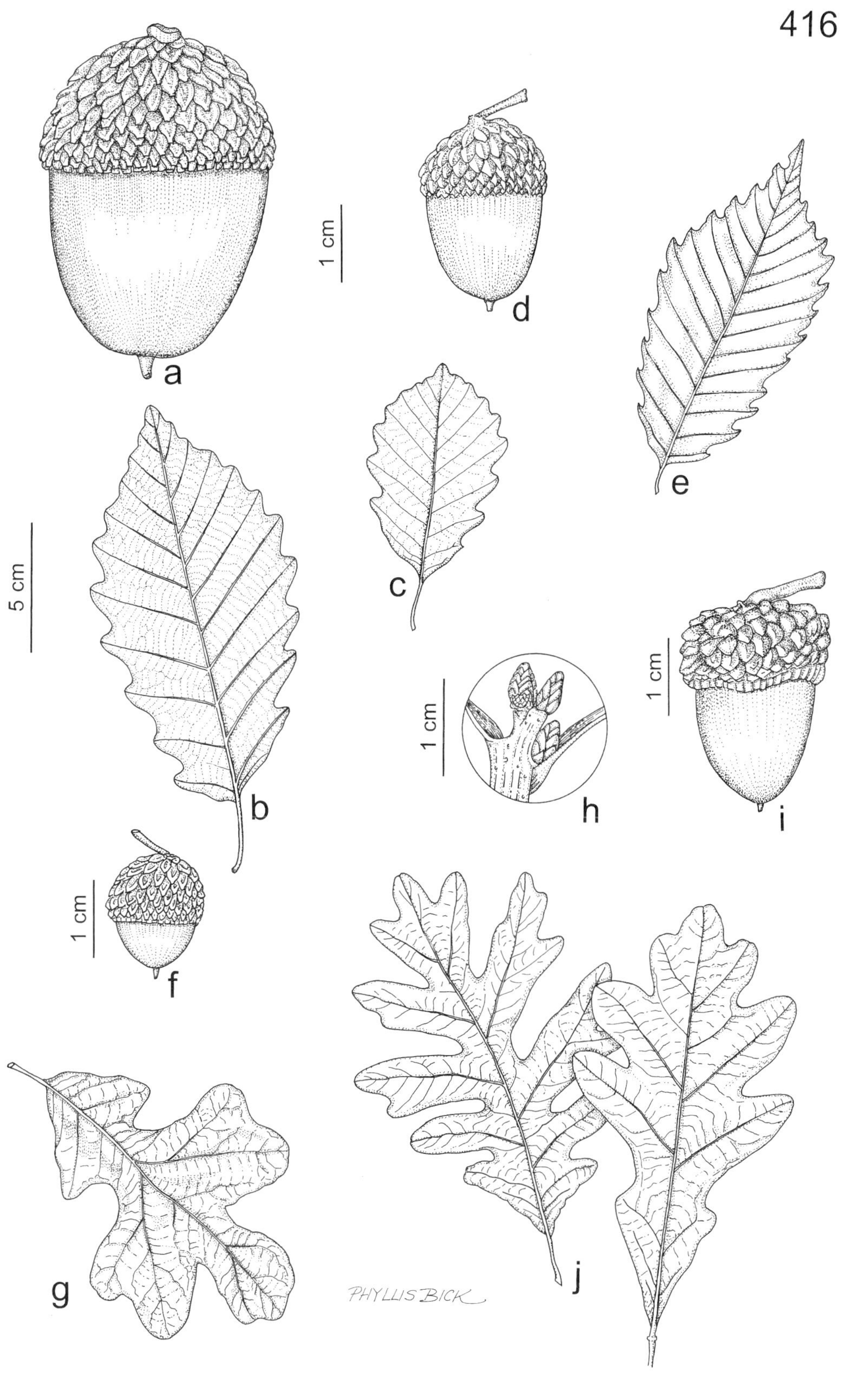

Plate 416. Fagaceae. *Quercus michauxii*, **a)** acorn, **b)** leaf. *Quercus prinoides*, **c)** leaf. *Quercus muehlenbergii*, **d)** acorn, **e)** leaf. *Quercus stellata*, **f)** acorn, **g)** leaf. *Quercus alba*, **h)** winter buds, **i)** acorn, **j)** leaves.

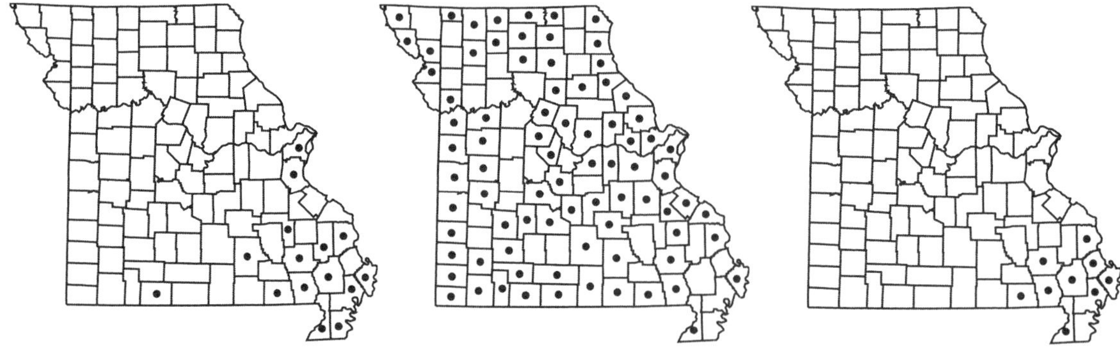
1855. Quercus lyrata 1856. Quercus macrocarpa 1857. Quercus michauxii

15. Quercus alba L. (white oak)

Pl. 416 h–j; Map 1853

Plants trees to 35 m tall. Bark light ashy gray, divided into loose plates or strips, or more or less persistent ridges on old trunks. Twigs 1.5–3.0 (–4.0) mm thick, reddish brown to grayish brown, sometimes glaucous, glabrous. Buds 2–5 mm long, reddish brown, the scales glabrous or sparsely pubescent, usually also hairy along the margins. Petioles 6–25 mm long. Leaf blades 12.5–21.0 cm long, 9.5–13.0 cm wide, relatively thin and flexible, obtuse to somewhat tapered at the base, divided (30–)60–90% of the width, the lobes (2)3–6 per side, deeper above the leaf base; well-developed lobes 13–36 mm wide, elliptic to narrowly oblong, rounded or rounded-obtuse apically, the largest lobes often with 1 or 2 small secondary lobes on the lower margin and sometimes 1 on the upper margin; secondary veins (3)4–6 per side, some reaching the margin at the tips of the lobes, usually others reaching toward sinuses and turning aside before reaching the margin; the upper surface dull, glabrous, the undersurface more or less glaucous, appearing glabrous (few to many appressed unbranched hairs visible at high magnification), smooth to the touch. Acorn stalks 2–41 mm long, the cups 7–12 mm long, 11–23 mm wide, covering 20–40% of the nut, bowl-shaped, the outer surface with the scales 1.5–3.0 mm long, those near the cup margin not differentiated. Nuts 16–23 mm long, 14–17 mm wide, barrel-shaped to bluntly ovoid. $2n=24$. April–May.

Common nearly throughout the state (eastern U.S. west to Minnesota, Nebraska, and Texas; Canada). Bottomland forests, mesic to dry upland forests, bases and tops of bluffs, banks of streams and rivers, and margins of sinkhole ponds; also pastures, railroads, and roadsides.

The leaves in *Quercus alba* vary a great deal in depth of lobing and width of the lobes (Baranski, 1975). Forms with leaves that are shallowly divided into broad lobes have been called f. *latiloba* (Sarg.) E.J. Palmer & Steyerm.

Hardin (1975) reviewed natural hybridization in this species. Hybrids with 11 other species of white oaks are known from throughout its range. Five of these have been documented from Missouri.

16. Quercus bicolor Willd. (swamp white oak)

Pl. 417 f–j; Map 1854

Plants trees to 30 m tall. Bark light to medium gray, divided into more or less persistent ridges. Twigs 2–3 mm wide, brown or gray, usually glabrous, sometimes with scattered, long, simple or branched, spreading hairs. Buds 2–3 mm long, tan to reddish brown, the scales glabrous, but hairy along the margins. Petioles 6–19 mm long. Leaf blades 12–22 cm long, 7–14 cm wide, relatively thin and flexible, rounded-obtuse to acute at the base, divided 20–40(–60)% of the width, the shallow lobes and/or coarse scallops 4–8 per side, smaller toward the blade tip; well-developed shallow lobes ovate or oblong, rounded or rounded-obtuse apically, sometimes with 1 or 2 teeth or secondary lobes on the lower margin; secondary veins (7)8–11(–15) per side, some reaching the margin at the tips of the lobes, usually others reaching toward sinuses and turning aside before reaching the margin; the upper surface dull or somewhat shiny, with a few simple or 2-branched, erect hairs, the undersurface usually white, rarely green, with erect, 2–7-branched hairs projecting from a mat of crowded, appressed, stellate hairs (rarely the stellate hairs absent), felty to the touch. Acorn stalks (15–)37–100 mm long, the cup 7–14 mm long, 17–27 mm wide, covering 20–40% of the nut, bowl-shaped (often deeply so), the outer surface with the scales 2–3 mm long, those near the cup margin much narrower. Nuts 18–28 mm long, 14–19 mm wide, ovoid. $2n=24$. April–May.

Scattered to common, particularly in the northern half of the state, but apparently ab-

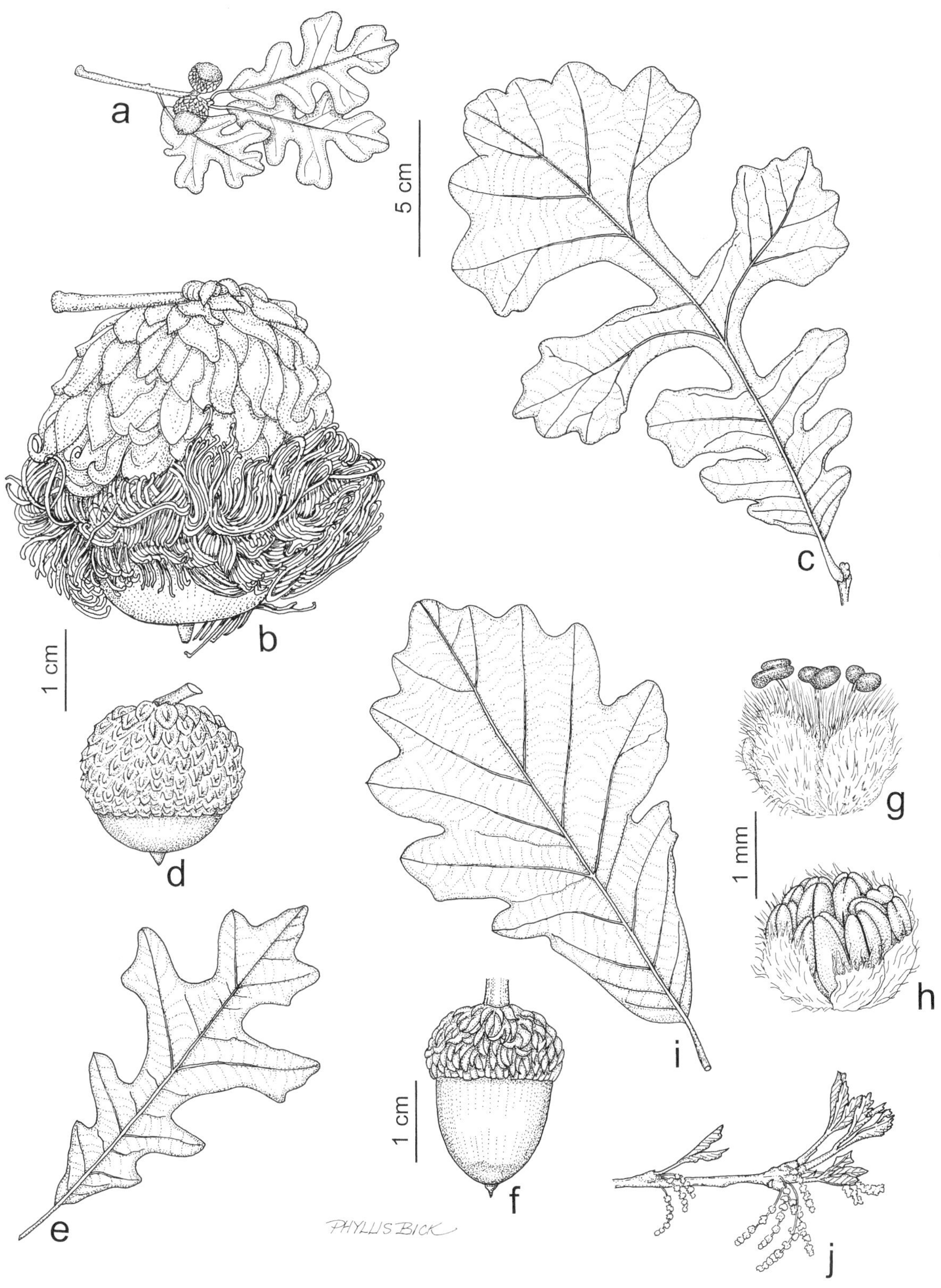

Plate 417. Fagaceae. *Quercus margarettiae*, **a)** branchlet with acorns and leaves. *Quercus macrocarpa*, **b)** acorn, **c)** leaf. *Quercus lyrata*, **d)** acorn, **e)** leaf. *Quercus bicolor*, **f)** acorn, **g)** mature staminate flower, **h)** immature staminate flower, **i)** leaf, **j)** branch tip with staminate inflorescences.

sent or uncommon in most of the Ozark, Mississippi Lowlands, and Unglaciated Plains Divisions (eastern U.S. west to Minnesota and Missouri; Canada). Bottomland forests, mesic upland forests in ravines, swamps, banks of streams and lakes, bases and ledges of bluffs, and margins of marshes and sloughs; also pastures, railroads, and roadsides.

In Missouri, hybrids have been collected between *Q. bicolor* and five other species.

17. Quercus lyrata Walter (overcup oak)
Pl. 417 d, e; Map 1855

Plants trees to 20 m tall. Bark light ashy gray, divided into loose plates. Twigs 2.0–3.5 mm wide, brown or gray, glabrous or with scattered spreading hairs. Buds grayish brown, 2–4 mm long, finely and minutely pubescent. Petioles 3–27 mm long. Leaf blades 11–24 cm long, 8.0–12.5 cm wide, relatively thin and flexible, usually acutely angled or tapered at the base, sometimes rounded-obtuse, divided 60–80% of the width, the lobes 3–5(6) per side, the sinuses often much shallower in the apical half of leaf; well-developed lobes 11–42 mm wide, ovate or oblong, sometimes broadened outward, acute to broadly rounded apically, sometimes with 1 or 2 teeth or secondary lobes on each margin; secondary veins 3–7 per side, some reaching the margin at the tips of the lobes, usually others reaching toward sinuses and turning aside before reaching the margin; the upper surface dull or somewhat shiny, glabrous or with scattered, single or paired hairs, the undersurface green or white, usually with scattered, 1–4-rayed, spreading hairs, sometimes with appressed-stellate hairs and with appressed unbranched hairs, usually felty to the touch. Acorn stalks 5–33 mm long, the cup (17–)21–31 mm long, (9–)18–21 mm wide, broadly urn-shaped to nearly globose, covering most or all of the nut, the outer surface with the scales 3–5 mm long, those near the cup margin much narrower. Nuts 9–20 mm long, 14–28 mm wide, broadly top-shaped. April–May.

Scattered in the southeastern quarter of the state, west locally to Taney County (southeastern U.S. west to Missouri, Oklahoma, and Texas). Bottomland forests, swamps, sloughs, banks of streams, and rarely sinkholes.

When the undersurface of the leaf is densely covered with stellate hairs, it appears white. If the stellate hairs are sparse or absent, the leaf is green on both sides. The green form has been called f. *viridis* Walter. Hybrids have been recorded with three other oak species in Missouri.

18. Quercus macrocarpa Michx. (bur oak, burr oak, mossycup oak)
Pl. 417 b, c; Map 1856

Plants trees to 35 m tall. Bark medium gray, divided into more or less persistent ridges. Twigs 3–4 mm wide, tan, brown or sometimes grayish, usually pubescent with small, stellate hairs, occasionally with branched spreading hairs or glabrous. Buds (3–)4–6 mm long, brown, pubescent. Petioles 8–30 mm long. Leaf blade 12–22 cm long, 9–13 cm wide, relatively thin and flexible, rounded to obtuse or occasionally acute at the base, divided (40–)60–90% of the width, the lobes 3–5(–7) per side, the median sinuses the deepest, the lower sinuses shallower, the upper sinuses shallow or the blade merely coarsely scalloped above the midpoint; well-developed lobes 20–45 mm wide, oblong to obovate, usually broadened outward, rounded or rounded-obtuse to truncate or shallowly notched apically, often scalloped, sometimes forked (2-lobed); secondary veins 5–11 per side, some reaching the margin at the tips of the lobes, usually others reaching toward sinuses and turning aside before reaching the margin; the upper surface dull, glabrous or with scattered, stellate hairs, the undersurface usually white, sometimes green, usually with 7–14-rayed, stellate hairs, sometimes with 2–7-rayed, spreading hairs, appressed linear hairs also present but usually concealed beneath the stellate hairs, usually felty to the touch,. Acorn stalks 5–33 mm long, the cup 19–25 mm long, 24–46 mm wide, covering 50–70% of the nut, hemispheric, the scales 2–5 mm long, those near the margin long, linear, forming a fringe that projects 6–15 mm beyond the rim of the cup. Nuts 24–32 mm long, 17–33 mm wide, ovoid. $2n=24$. April–May.

Scattered to common nearly throughout the state (eastern U.S. west to Montana and Texas; Canada). Bottomland forests, mesic upland forests, banks of streams and rivers, and bottoms and tops of bluffs; also pastures and roadsides.

Quercus macrocarpa is a widely distributed species that is quite variable in plant size and form, acorn size and degree of elongation of the marginal cups scales, and leaf size and division. In Missouri, the middle and/or lower parts of the blade are always lobed on well-developed leaves, but the upper part may be merely scalloped. Specimens that are deeply lobed to the tip have been called f. *oliviformis* (F. Michx.) Trel. A somewhat smaller, shrubbier variant with relatively small acorns having less-fringed cups that occurs in the northwestern portion of the overall species range has been called var. *depressa* (Nutt.) Engelm. The acorns are the largest produced by any oak native

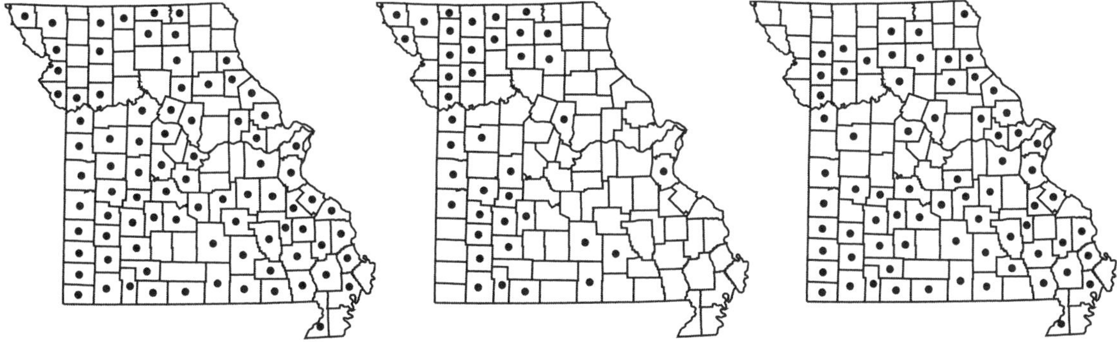

1858. Quercus muehlenbergii 1859. Quercus prinoides 1860. Quercus stellata

to the United States, but even within Missouri there is considerable size variation, with some populations from relatively upland sites in northern Missouri producing acorns that are significantly smaller than the enormous ones produced by trees growing in bottomlands in the southern half of the state.

In Missouri, hybrids have been recorded involving *Q. macrocarpa* and six other oak species.

19. Quercus michauxii Nutt. (swamp chestnut oak, basket oak, cow oak)

Pl. 416 a, b; Map 1857

Plants trees to 35 m tall. Bark light ashy gray, divided into loose plates or strips, or ± persistent ridges on old trunks. Twigs (1.0–)2.0–3.5 mm wide, grayish brown to brown, glabrous, finely hairy, or occasionally with a few branched, spreading hairs. Buds 4–7 mm long, reddish brown, finely short-hairy. Petioles 6–33 mm long. Leaf blades 13–21 cm long, 6.5–13.0 cm wide, relatively thin and flexible, obtuse, rounded, or truncate at the base; divided 10–20% of the width, the coarse scallops 11–14 per side, smaller toward the blade tip; well-developed scallops rounded or rounded-obtuse, undivided; secondary veins 12–17 per side, each (except the basalmost) reaching the margin at the tip of a scallop; the upper surface dull or somewhat shiny, glabrous or with scattered, 1–7-rayed, spreading hairs, the undersurface green or pale, sparsely to densely covered with 1–7-rayed, spreading hairs, never with appressed-stellate or unbranched appressed hairs, usually very felty to the touch. Acorn stalks 10–23 mm long, the cup 13–19 mm long, 24–33 mm wide, covering 30–50% of the nut, hemispheric, the outer surface with the scales 3–5 mm long, those near the cup margin not differentiated. Nuts 20–30 mm long, 15–24 mm wide, ovoid or ellipsoid. $2n$ probably = 24 (reported as $2x$=12 by Friesner [1930]). April–May.

Scattered in the Mississippi Lowlands Division and adjacent portions of the Ozarks and Ozark Border (eastern [mostly southeastern] U.S. west to Missouri, Oklahoma, and Texas). Bottomland forests, swamps, and banks of streams and rivers.

The name *Q. prinus* L. has been used at times for either for *Q. michauxii* or *Q. montana* Willd. (Hardin, 1979). In view of its persistent use for more than one species, Whittemore and Nixon (2005) proposed that it be formally rejected from further use.

The wood of *Q. michauxii* is of high quality, and it has the sweetest acorns (with the lowest tannin concentration) of any Missouri oak. Because of this, cows relish the acorns, giving rise to the common name cow oak. Because the wood tends to split naturally into long narrow strips, it was used in basketry in some of the southeastern states (Steyermark, 1963), hence the common name basket oak. In Missouri, hybrids have been recorded involving *Q. michauxii* and three other species.

20. Quercus muehlenbergii Engelm. (chinkapin oak)

Q. prinoides Willd. var. *acuminata* (Michx.) Gleason

Pl. 416 d, e; Map 1858

Plants trees (sometimes shrubby in exposed, rocky habitats) to 35 m tall. Bark light ashy gray, divided into loose plates, blocks or strips. Twigs 1.5–3.0 mm wide, brown or occasionally grayish brown, glabrous or with scattered, spreading hairs when young. Buds 3–5 mm long, reddish brown or grayish, the scales minutely pubescent or nearly glabrous, short-hairy along the margins. Petioles 14–33 mm long. Leaf blade 12–22 cm long, 5.0–14.5 cm wide, relatively thin and flexible, broadly obtuse, rounded, or truncate at the base; divided 10–30% of the width, the coarse scallops (8–)10–14 per side, equal or slightly shallower above midpoint; well-developed scallops rounded to acute, undivided; secondary veins 10–15 per side, each (except the basalmost) reaching the margin at the tip of a scallop; the upper surface shiny to rather dull, glabrous or nearly so (with a few inconspicuous, 3–6-rayed hairs), the undersurface white or sometimes pale green, with mostly 5–12-rayed, appressed, stellate hairs and sometimes also 2–6-rayed, spreading hairs, sometimes felty to the touch. Acorns sessile or the stalk to 8 mm long, the cup 5–12 mm long, 9–18 mm wide, covering 40–50% of the nut, bowl-shaped, the outer surface with the scales 2.0–2.5 mm long, those near the cup margin not differentiated. Nuts 11–19 mm long, 11–13 mm wide, ovoid or ellipsoidal. $2n=24$. April–May.

Common nearly throughout the state (eastern U.S. west to Wisconsin, Nebraska, and New Mexico; Canada, Mexico). Mesic to dry upland forests, savannas, sand savannas, upland prairies, loess hill prairies, glades, banks of streams, margins of lakes; also pastures, railroads and roadsides, often on calcareous substrates.

This species is very similar to *Q. prinoides*, and plants of *Q. muehlenbergii* that are immature or dwarfed by growing in dry sites with shallow soil can resemble that species very closely. *Quercus prinoides* spreads by underground runners to form large, clonal colonies, and the shrubs may flower and fruit when they are only 3–5 years old and less than 0.5 m tall (Nixon and Muller, 1997), but *Q. muehlenbergii* usually is slower to flower. The two species hybridize when they come into contact with one another, and extensive intermediate populations have been reported from other states, but it is hard to assess the degree of hybridization accurately because the parental species are so similar morphologically. The unusual secondary hybrid between *Q. bicolor* and the *Q. muehlenbergii*/*prinoides* hybrid (*Q.* ×*introgressa*) was described from a small colony of trees in a pasture in Lafayette County using morphometric data, and the presence of three individuals of *Q. muehlenbergii* × *Q. prinoides* was mentioned, but apparently voucher specimens of these were not preserved (Thomson, 1977). *Quercus muehlenbergii* has a much wider range than does *Q. prinoides*, both in Missouri and outside the state. In the northeastern United States the two species apparently are well separated ecologically, with *Q. prinoides* on acidic bedrock and *Q. muehlenbergii* on limestone. The ecological separation seems less sharp in Missouri (perhaps because so much dolomitic limestone with a high chert content is found in the state). Hybridization seems to be commoner in areas with intermediate geology, as with many other oaks (Muller, 1952). The only other hybrids involving *Q. muehlenbergii* that have been documented from Missouri are those with *Q. alba* and *Q. macrocarpa*.

Plants having unusually broad leaves have been called f. *alexanderi* (Britton) Trel. (*Q. prinoides* f. *alexanderi* (Britton) Steyerm.).

M. Smith and Parker (2005) reported *Q. montana* Willd. (rock chestnut oak) from four sites in Wayne County. This species, which is very similar to *Q. muehlenbergii* morphologically, is widespread in the eastern U.S., west to the bluffs and upland woods immediately east of the Mississippi River from Union County, Illinois south to northern Mississippi. It will key to *Q. muehlenbergii* in the keys above, but differs in its bark, which is deeply divided into thick, persistent ridges; the stellate hairs of the leaf undersurface, which are always sparse and have only 2–4 rays (or sometimes more on hairs close to the midrib); and its large acorns with the usually hemispherical cup 9–15 mm deep and the ellipsoid nut 15–30 mm long. Regrettably, repeated subsequent searches for these plants uncovered only juvenile plants of *Q. muehlenbergii* and the incomplete specimens documenting the supposed Missouri populations cannot be matched to *Q. montana* unequivocally. Thus, although this species may be confirmed as a member of the Missouri flora in the future, for the present it has been excluded from the flora.

21. Quercus prinoides Willd. (dwarf chestnut oak)

Pl. 416 c; Map 1859

Plants shrubs, 0.5–3.0(–5.0) m tall. Bark light ashy gray, divided into loose plates, blocks, or strips. Twigs 1.0–2.5 mm wide, orangish brown to reddish brown or grayish, glabrous or with branched spreading hairs. Buds 2–4 mm long, reddish brown or grayish, the scales glabrous or sparsely pubescent, usually hairy along the margins. Petioles 5–21 mm long. Leaf blades 7–13 cm long, 2.7–6.5(–9.0) cm wide, relatively thin and flexible, obtuse or rounded at the base, divided 10–25% of the width, the coarse scallops 6–9 per side, equal or slightly shallower above the midpoint; well-developed scallops rounded or rounded-obtuse, undivided; secondary veins 6–9 per side, each (except the basalmost) reaching the margin at the tip of a scallop; the upper surface rather dull, glabrous or nearly so (with a few inconspicuous, 4–8-rayed, spreading or appressed hairs), the undersurface white or sometimes pale green, covered (usually very densely so) with mostly 7–15-

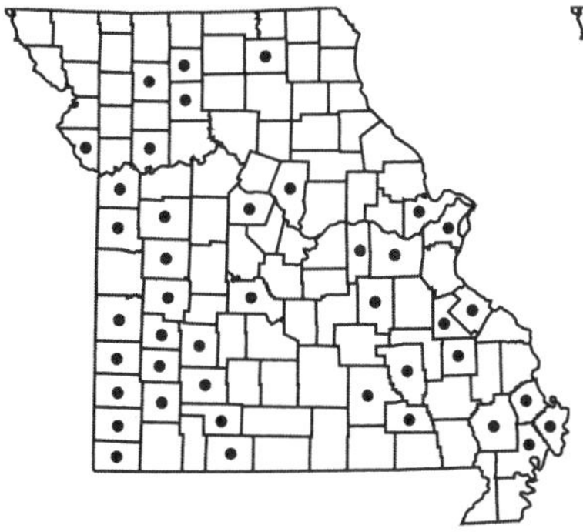

1861. Corydalis aurea

1862. Corydalis crystallina

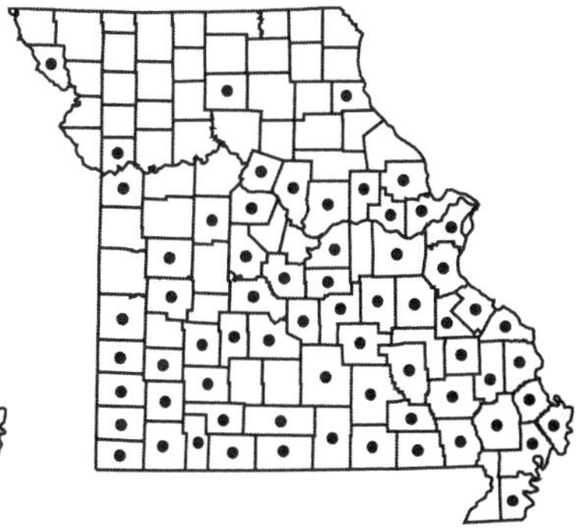

1863. Corydalis flavula

rayed, stellate hairs and sometimes also 5–7-rayed, spreading hairs (with rays 0.2–0.4 mm long), usually more or less felty to the touch. Acorn stalks 1–7 mm long, the cup 6–11 mm long, 12–19 mm wide, covering 40–50% of the nut, hemispheric or bowl-shaped, the outer surface with the scales 1.5–3.0 mm long, those near the cup margin not differentiated. Nuts 10–18 mm long, 9–12 mm wide, cylindric to ovoid or ellipsoid. April–May.

Scattered mostly in the western half of the state (eastern U.S. west to Nebraska and Oklahoma; Canada). Upland prairies, loess hill prairies, glades, savannas, openings of dry upland forests, and tops of bluffs; also roadsides.

For a discussion of the problems in separating *Q. prinoides* from the closely related *Q. muehlenbergii*, see the treatment of that species. Hybridization between the two is also discussed in the treatment of *Q. muehlenbergii*. In Missouri, hybrids involving *Q. prinoides* and three other oak species have been documented.

22. Quercus stellata Wangenh. (post oak)
Pl. 416 f, g; Map 1860

Plants trees, 3–20 m tall. Bark light gray, divided into loose or more or less persistent ridges or blocks. Twigs (2.5–)3.0–4.0 mm wide, whitish tan or gray, densely felted with small, stellate and branched, spreading hairs. Buds reddish brown, 4–6 mm long, the scales pubescent (the inner scales only sparsely so), hairy along the margins. Petioles 10–20 mm long. Leaf blades 12.5–20.0 cm long, 9–15 cm wide, relatively thick and rigid, obtuse to truncate at the base, divided 70–90% of the width, the lobes 2 or 3 per side, the largest lobes usually near the tip; well-developed lobes (29–)37–72 mm wide, with 1 secondary lobe as large as the main lobe on the lower margin, truncate or notched to 2-lobed apically; secondary veins 2–6 per side, some reaching the margin at the tips of the lobes, others reaching toward sinuses and turning aside before reaching the margin; the upper surface more or less shiny, with scattered, 4–10-rayed, spreading hairs, the undersurface green, with scattered, erect, 4–11-rayed hairs and few to many, white or orange, appressed unbranched hairs, harsh-felty or almost sandpapery to the touch. Acorn stalks absent or to 8 mm long, the cup 6–11 mm long, 12–16 mm wide, covering 50–70% of the nut, hemispheric, the outer surface with the scales 1.5–2.5 mm long, those near the cup margin not differentiated. Nuts 12–13 mm long, 8–12 mm wide, ellipsoid to barrel-shaped. April–May.

Scattered to common nearly throughout the state (eastern U.S. west to Iowa, Kansas, and Texas). Mesic to dry upland forests, glades, upland prairies, tops of bluffs, margins of sinkhole ponds, and banks of streams; also pastures and roadsides.

Steyermark (1963) reported the sand post oak, *Q. margarettiae* (Ashe) Small (*Q. stellata* var. *margarettiae* (Ashe) Sarg.; Pl. 417 a), from southwestern Missouri based on a few historical collections from Jasper and Lawrence Counties. Most of these represent merely depauperate plants of *Q. stellata*. However, the one Lawrence County specimen for which he cited collection data (*Palmer 51154*, at the University of Missouri Herbarium) appears to represent the hybrid, *Q. prinoides* × *Q. stellata*, or possibly a backcross between this hybrid and *Q. stellata*. *Quercus margarettiae* (often spelled *Q. margaretta* in the botanical literature) is a small tree mostly confined to loose, sandy soil. It resembles *Q. stellata*, but differs in lacking stellate hairs on the twigs (its twigs are usually glabrous but sometimes have scattered large, branched spreading hairs), and in having the hairs of the leaves stalked. It has not been documented to occur north of central Arkansas. To date, *Q. stellata* has been documented to hybridize with four other oak species in Missouri.

In 2012, late in the preparation of the present treatment, Justin Thomas collected a specimen of an unusual post oak from Otter Slough Conservation Area (Stoddard County) and subsequently located a few older herbarium records of similar trees from adjacent portions of the Mississippi Lowlands Division. These are morphologically identifiable with specimens from the southeastern United States that have been called Delta post oak or swamp post oak (*Q. similis* Ashe). *Quercus similis* differs most notably from typical *Q. stellata* in that it occurs in bottomland forests and swamps, where it forms canopy trees to 25 m tall. Its leaves also differ somewhat from those of typical post oak in their sometimes narrower outline and more irregular pattern of lobing (often with some of the blades more shallowly lobed or merely 3-lobed at the tip), as well as the somewhat narrower angle of branching of the secondary veins going into the largest lobes. However, the taxonomy of *Q. similis* presently is not well-understood and further research needs to be conducted to determine whether it represents a valid species, a hybrid between *Q. stellata* and some other oak, or an ecological variant of *Q. stellata*. The status and characteristics of this entity in southeastern Missouri also need to be assessed through more detailed field observations.

FUMARIACEAE (Fumitory Family)

Plants annual, biennial, or perennial herbs, glabrous (except for the fruits in *Corydalis crystallina*). Leaves alternate or basal, lacking stipules, pinnately or pinnately then ternately 2–4 (or more) times compound (often appearing highly dissected with relatively narrow ultimate segments). Inflorescences terminal (sometimes also appearing axillary in *Fumaria*), short to elongate racemes, sometimes reduced to clusters, the flowers subtended by small bracts. Flowers irregularly zygomorphic (bilaterally symmetrical in only 1 longitudinal plane or in 2 perpendicular planes), hypogynous, perfect. Calyces of 2 free sepals, these similar in size and shape, inconspicuous, often shed as the flower opens. Corollas of 4 petals, these free or fused at the base, dissimilar in size and shape, positioned in 2 whorls, the inner whorl of 2 shorter petals more or less apically hooded or fused (enclosing the anthers and stigma), the outer whorl of longer petals with 1 or both having a basal pouch or spur (except in cleistogamous flowers). Stamens 6, positioned in 2 clusters of 3, each cluster opposite an outer petal, the filaments within a cluster fused, at least toward the base (and sometimes also to the petals), often with a basal, spurlike, nectar-producing outgrowth, the anthers small, attached at or near the base, often more or less fused around the stigma, only the middle anther of each trio with 2 pollen sacs, the flanking pair with 1 pollen sac. pistil 1 per flower, of 2 fused carpels. Ovary superior, 1-locular, the placentation parietal. Style 1 per flower, unbranched, persistent at fruiting (except in *Fumaria*), the stigma usually flattened, 2–8-lobed, the lobes sometimes appearing as horns or wings. Ovules 2 to numerous. Fruits indehiscent and nutlike or capsules that are longitudinally dehiscent by 2 valves (sometimes tardily so), 1- to numerous-seeded. Seeds (except in *Fumaria*) usually somewhat flattened, nearly circular to more commonly finely notched or somewhat kidney-shaped, the embryo curved, the surface dark brown to black, usually with a pale elaiosome (oil-bearing aril-like appendage attractive to ants). About 19 genera, about 450 species, North America, Europe, Asia, Africa.

The Fumariaceae are closely related to the Papaveraceae and have been included within that family (usually as a subfamily) by a number of authors (Lidén, 1986; Judd et al., 2002). Relationships within the lineage that includes both of these groups is still somewhat controversial and it is unclear whether the best approach is to recognize one morphologically diverse family or split the lineage into two or more families. Within the group, Fumariaceae are distinguished morphologically by their asymmetrical or zygomorphic corollas, clear sap, and unusual stamen and stigma morphology. Like the Papaveraceae, members of the Fumariaceae produce a large variety of isoquinoline alkaloids. This accounts for their various medicinal uses (Moerman, 1998) as well as their toxicity to livestock and humans (Burrows and Tyrl, 2001). A number of species are cultivated as ornamentals.

1. Leaves all basal; flowers with both of the outer petals pouched or spurred at the base; plants perennial, the rootstock with clusters of small bulblets .... 2. DICENTRA
1. Stem leaves well-developed (basal leaves sometimes also present); flowers with only 1 of the outer petals swollen or spurred at the base; plants annual, with taproots, not producing swollen underground parts
   2. Corollas yellow, sometimes becoming reddish-tinged with age; fruits elongate, beaked, dehiscent (sometimes tardily so), 3- to numerous-seeded .... 1. CORYDALIS
   2. Corollas white to pinkish-tinged, grading to a dark reddish purple tip; fruits more or less globose, not beaked, indehiscent, 1-seeded .... 3. FUMARIA

1. **Corydalis** DC. (corydalis)

Plants annual or occasionally biennial (perennial elsewhere), with taproots. Stems loosely to strongly ascending. Leaves alternate, sometimes also basal, long-petiolate toward the stem base, grading into short-petiolate or sessile toward the stem tip. Leaf blades 2 or 3 times compound and lobed, oblong lanceolate, lanceolate, ovate, or ovate-triangular in outline (often on the same plant), the ultimate segments linear to narrowly oblong or narrowly lanceolate, occasionally elliptic to obovate, mostly sharply pointed at the tip, green or pale and glaucous. Inflorescences 1–15 cm long, usually relatively short-stalked and relatively densely (1–)4–30-flowered. Flowers bilaterally symmetrical in only 1 longitudinal plane, the stalks 1–15 mm long, ascending at flowering but sometimes spreading or pendant at fruiting, without a pair of bractlets. Sepals 1–2(–3) mm long, ovate to ovate-triangular or occasionally somewhat heart-shaped, attached basally, rounded to nearly truncate or occasionally somewhat cordate at the base, the margins entire or less commonly with several coarse, jagged teeth, membranous and white to pale yellow. Corollas pale to bright yellow, sometimes becoming reddish-tinged with age. Inner petals linear or nearly so toward the base, oblanceolate to spatulate above the midpoint with an inconspicuous keel, rounded at the concave tip. Outer petals dissimilar; the lower petal narrowly oblong to oblong-obovate, keeled or with a low irregular median wing (crest), the margins winged toward the usually somewhat spreading (except in cleistogamous flowers) tip; the upper petal with a short to long spur (this often absent or nearly so in cleistogamous flowers), slightly incurved, keeled or with an irregular median wing (crest), winged toward the usually abruptly spreading (except in cleistogamous flowers) tip. Style persistent, slender to relatively stout, the stigma 2-lobed, flattened and somewhat fan-shaped, with 4 or more commonly 8 winglike papillae. Fruits capsules, dehiscent (sometimes tardily so), 10–30 mm long, narrowly oblong-ellipsoid (tapered at each end, beaked at the tip), straight or curved, the surface not swollen over the seeds, smooth or appearing mealy (in *C. crystallina*), 3- to numerous-seeded. Seeds 1.4–2.3 mm long, somewhat flattened, more or less kidney-shaped in outline, rounded or bluntly to sharply angled along the rim, sometimes with a minute marginal ridge, the surface smooth or appearing finely pebbled, black, shiny, the elaiosome an irregular, somewhat conic, white mass attached in the notch, sometimes poorly developed. About 300 species, North America, Europe, Asia, Africa.

As in other members of the Fumariaceae, the inner petals of *Corydalis* species are cupped around the anthers and stigmas, and pollen grains are shed onto the stigma. However, most species of *Corydalis* are visited by a variety of insects (especially bumblebees) and thus are likely to be mostly or at least partly cross-pollinated (see the treatment of *Dicentra* for a similar pollination mechanism). This situation has not been studied in detail in most of the North American species. Macior (1978a, b) noted visitation of *C. flavula* flowers by queens of two species of bumblebees, but also determined that flowers bagged to exclude all insects still produced fruits. Thus, self pollination may be the main mode of reproduction in *C. flavula*. Additionally, plants of some species of *Corydalis* (in Missouri, principally *C. flavula* and *C. micrantha*) can produce inflorescences with some or all of the flowers cleistogamous. Such inflorescences tend to be short (often reduced to clusters rather than racemose). The tips of the outer petals do not spread and the spur is very short or absent.

Although specimens of *Corydalis* are most easily observed in the field when in flower, some species are best distinguished by their fruits and seeds. The best time to collect specimens is when both flowers and mature fruits are present. The size ranges for corolla lengths in the key to species include the spur in the measurement.

In addition to the species treated below, *C. curvisiliqua* (A. Gray) Engelm. ex A. Gray ssp. *grandibracteata* (Fedde) G.B. Ownbey should be searched for in northeastern and southeastern Missouri. This taxon is known from sand prairies in southeastern Iowa and western Illinois (Tyson and Ebinger, 1999) and also is found in sandy habitats from Kansas to Texas. It is most similar to *C. aurea*, but differs in having seeds that are roughened rather than pebbled (observed with magnification as having minute, blunt, tubercles rather than polygonal facets) and relatively conspicuous inflorescence bracts (the lowermost bracts 10–15 mm vs. 4–10 mm long).

1. Corollas 7–9 mm long, the spur 1.5–2.0 mm long, incurved; flower stalks (6–)9–22 mm long . . . . 3. C. FLAVULA
1. Corollas 10–22 mm long (except in cleistogamous flowers), the spur 4–8 mm long, straight or nearly so; flower stalks 1–6 mm long
    2. Ovaries and fruits moderately to densely pubescent with short, white, bladderlike, inflated hairs, appearing mealy . . . . 2. C. CRYSTALLINA
    2. Ovaries and fruits glabrous, not appearing mealy
        3. Corollas 14–20 mm long; seeds 1.8–2.1 mm in diameter, the surface smooth or nearly so, the rim bluntly to sharply angled or with a very minute raised ring around the rim . . . . 1. C. AUREA
        3. Corollas 11–14(–15) mm long (shorter in cleistogamous flowers); seeds 1.2–1.6 mm in diameter, the surface finely pebbled (best observed with magnification), the rim rounded . . . . 4. C. MICRANTHA

1. Corydalis aurea Willd. (golden corydalis)
Pl. 418 f; Map 1861

Plants green or more commonly gray and glaucous. Stems 10–35(–50) cm long, loosely to strongly ascending, often from a spreading base. Basal and lower stem leaves with the petiole 3–6 cm long, the upper leaves sessile or very short-petiolate. Leaf blades 1.5–10.0 cm long, with mostly 7–11 pinnae, these again 1 or 2 times deeply several-lobed, the ultimate segments linear or narrowly to occasionally broadly oblong-elliptic or lanceolate. Inflorescences extending past the foliage or not, all with open flowers, 5–30-flowered racemes. Flower stalks 2–4 mm long, ascending at flowering, ascending or pendant at fruiting. Corollas pale to bright yellow, the upper outer petal 13–18 mm long, the spur 4–9 mm long, straight or nearly so, the concave apical portion with a low, irregular crest or more often merely keeled. Fruits 15–30 mm long, straight or curved, glabrous, not appearing mealy. Seeds 1.8–2.1 mm long, the surface smooth or nearly so, the bluntly to sharply angled rim sometimes with a minute, raised ridge. $2n=16$. March–June.

Uncommon, widely scattered, mostly south of the Missouri River (western U.S. east to North Dakota and Texas and eastward to Illinois, Pennsylvania, and New Hampshire; Canada, Mexico).

The last monographer of North American *Corydalis*, Gerald B. Ownbey (1947) did not include Missouri in the range of this species. He treated *C. aurea* as comprising two intergrading subspe-

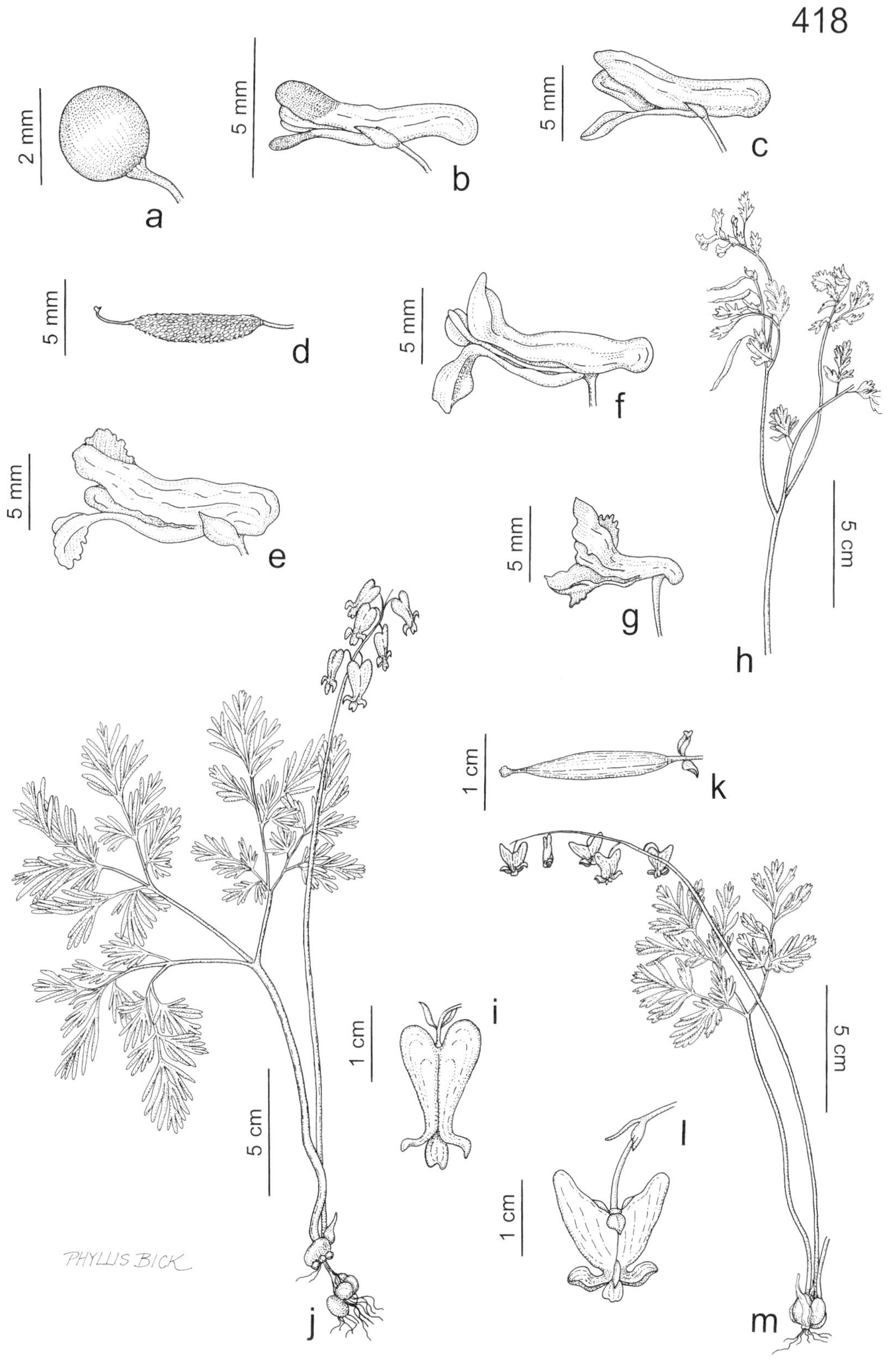

Plate 418. Fumariaceae. *Fumaria officinalis*, **a)** fruit, **b)** flower. *Corydalis micrantha*, **c)** flower. *Corydalis crystallina*, **d)** fruit, **e)** flower. *Corydalis aurea*, **f)** flower. *Corydalis flavula*, **g)** flower, **h)** flowering branch. *Dicentra canadensis*, **i)** fruit, **j)** habit. *Dicentra cucullaria*, **k)** fruit, **l)** flower, **m)** habit.

cies, with ssp. *aurea* occupying the western and northern portions of the overall range and ssp. *occidentalis* mainly in the southeastern portion of the species distribution. Steyermark (1963) apparently misinterpreted the characters separating *C. aurea* and its component subspecies from other Midwestern taxa of *Corydalis*, for he mapped abundant counties for each of them and also considered the taxa in the *C. aurea* complex to represent two separate species. Many of the specimens that Steyermark determined as *C. aurea* and *C. montana* have been annotated as *C. micrantha* by G. B. Ownbey and/or K. R. Stern. In his treatment for the Flora of North America Project, Kingsley R. Stern (1997a) adhered closely to the taxonomic scheme established by Ownbey (1947) but mapped a single disjunct Missouri locality for each of the two subspecies. For further discussion of present knowledge of the distributions of the two subspecies in Missouri, see the pertinent treatments below.

1. Fruits 18–25(–30) mm long, spreading to pendant at maturity; seeds lacking a marginal ring; inflorescences usually not extending past the foliage .... 1A. SSP. AUREA
1. Fruits 15–18(–20) mm long, ascending at maturity; seeds with a minute marginal ridge; inflorescences usually extending past the foliage .... 1B. SSP. OCCIDENTALIS

1a. ssp. aurea

Inflorescences with 10–30 flowers, usually not extending past the foliage. Corollas with the upper outer petal 13–16 mm long, the spur 4–5 mm long, the concave apical portion often lacking a crest, merely keeled. Fruits 18–30 mm long, straight or nearly so, spreading to pendant at maturity. Seeds bluntly angled along the rim, lacking a marginal ridge. $2n=16$. March–June.

Uncommon known thus far only from single historical collections from Jackson County and the city of St. Louis (western U.S. east to Texas and North Dakota and eastward to Illinois, Pennsylvania, and New Hampshire; disjunct in Missouri; Canada). Tops of bluffs.

As noted above, although Steyermark (1963) mapped this taxon from about 40 counties, virtually all of the specimens were misdetermined. Stern (1997a) mapped ssp. *aurea* from the St. Louis area. This apparently was based on two specimens accessioned at the New York Botanical Garden herbarium that were collected by Henry Eggert in 1887 in the stockyards of East St. Louis (Illinois). However, Stern annotated another specimen at the New York Botanical Garden as ssp. *aurea* that was collected by Kenneth K. Mackenzie in 1897 near Dodson (Jackson County).

1b. ssp. occidentalis (Engelm. ex A. Gray) G.B. Ownbey

C. curvisiliqua Engelm. ssp. *occidentalis* (Engelm.) W.A. Weber

C. montana Engelm.

Inflorescences with 5–12(–20) flowers, usually extending past the foliage. Corollas with the upper outer petal 14–18 mm long, the spur 5–9 mm long, the concave apical portion usually with a low irregular crest. Fruits 15–18(–20) mm long, usually curved, ascending at maturity. Seeds sharply angled along the rim, with a minute, marginal ridge. March–June.

Scattered widely, mostly south of the Missouri River (Nevada to Arizona east to South Dakota, Illinois, and Texas; Mexico). Glades, upland prairies, openings of dry upland forests, and banks of streams; also railroads and roadsides.

Corydalis aurea ssp. occidentalis is most often found growing on acidic substrates. Although it is less widely distributed than Steyermark (1963) thought it to be, it is much more abundant in the state than the single site in the Kansas City region that was mapped by Stern (1997a).

2. Corydalis crystallina (Torr. & A. Gray) Engelm. ex A. Gray (mealy corydalis)

Pl. 418 d, e; Map 1862

Plants green or more commonly gray and glaucous. Stems 10–40 cm long, loosely to strongly ascending, sometimes from a spreading base. Basal and lower stem leaves with the petiole 3–6 cm long, the upper leaves sessile or very short-petiolate. Leaf blades 1.5–8.0 cm long, with mostly 7 or 9 pinnae, these again 1 or 2 times deeply several-lobed, the ultimate segments linear to narrowly oblong or lanceolate. Inflorescences mostly extending past the foliage, all with open flowers, (4–)8–20-flowered racemes. Flower stalks 1–2 mm long, ascending at flowering and fruiting. Corollas bright yellow, the upper outer petal 15–22 mm long, the spur 6–8 mm long, straight or nearly so, the concave apical portion with an irregularly wavy or toothed crest. Fruits 10–20 mm long, straight or slightly curved, moderately to densely pubescent with short, white, bladderlike, inflated hairs, appearing mealy. Seeds 2.0–2.3 mm long, the surface finely pebbled (best observed with magnification), the bluntly to sharply angled rim lacking a marginal ridge. $2n=16$. April–June.

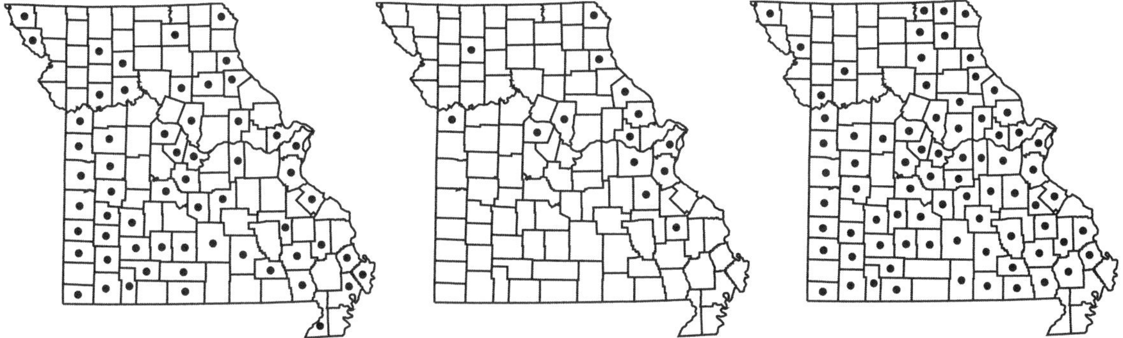

1864. Corydalis micrantha 1865. Dicentra canadensis 1866. Dicentra cucullaria

Scattered in western Missouri, mostly south of the Missouri River, east locally to Clark County and the city of St. Louis (Missouri and Kansas south to Oklahoma, Texas, and Arkansas). Glades, upland prairies; also railroads, roadsides, and open disturbed areas.

This distinctive species is most often found growing on acidic substrates.

3. Corydalis flavula (Raf.) DC. (pale corydalis, yellow fumewort, yellow harlequin)

Pl. 418 g, h; Map 1863

Plants green or sometimes dull gray and more or less glaucous. Stems 5–30 cm long, mostly loosely ascending from a spreading base. Basal and lower stem leaves with the petiole 4–8 cm long, the upper leaves usually sessile. Leaf blades 1–4 cm long, with 5 or 7 pinnae, these deeply several-lobed, the ultimate segments linear to narrowly oblong or less commonly broadly elliptic. Inflorescences mostly not extending past the foliage, those with open flowers (3–)6–12-flowered racemes, those with cleistogamous flowers 1–5-flowered clusters. Flower stalks (except sometimes in cleistogamous flowers) (6–)9–22 mm long, ascending at flowering, often pendant at fruiting. Corollas pale yellow or less commonly brighter yellow, the upper outer petal 7–9 mm long, the spur (except in cleistogamous flowers) 1.5–2.0 mm long, incurved, the concave apical portion usually with an irregular, wavy or toothed crest. Fruits 14–25 mm long, relatively straight, glabrous, not appearing mealy. Seeds 1.9–2.1 mm long, the surface finely pebbled, the sharply angled rim with a minute marginal ridge. $2n=16$. March–May.

Scattered, mostly south of the Missouri River (eastern [mostly northeastern] U.S. west to Nebraska and Oklahoma; Canada). Bottomland forests, mesic upland forests in ravines, bases of bluffs, banks of streams and rivers; also shaded roadsides.

This is the most abundant species of *Corydalis* in the state. It is a relatively inconspicuous wildflower. Some or all of the flowers on a plant often are cleistogamous.

4. Corydalis micrantha (Engelm. ex A. Gray) A. Gray (small-flowered corydalis, slender fumewort)

Pl. 418 c; Map 1864

Plants green or more commonly gray and glaucous. Stems 10–35 cm long, loosely to strongly ascending, often from a spreading base. Basal and lower stem leaves with the petiole 2–6 cm long, the upper leaves mostly sessile. Leaf blades 1–7 cm long, with mostly 5 or 7 pinnae, these deeply several-lobed, the ultimate segments linear or narrowly to broadly oblong-elliptic, lanceolate, or occasionally obovate. Inflorescences often extending past the foliage (except in those with mostly cleistogamous flowers), those with open flowers 5–20-flowered racemes, those with cleistogamous flowers 1–6-flowered clusters or short racemes. Flower stalks 2–6 mm long, ascending at flowering and fruiting. Corollas pale yellow to yellow, the upper outer petal 11–15 mm long, the spur 4–6 mm long, straight or nearly so, the concave apical portion with a low, irregular crest or occasionally merely keeled. Fruits 10–25 mm long, straight or curved, glabrous, not appearing mealy. Seeds 1.4–1.6 mm long, the surface minutely pebbled (best observed with magnification), the rounded rim lacking a marginal ridge. $2n=16$. April–June.

Ownbey (1947) and Stern (1997a) treated *C. micrantha* as consisting of three intergrading subspecies, two of which occur in Missouri. The two Missouri subspecies can be difficult to distinguish early in the flowering season, but appear to have fairly discrete distributions within the state. The third subspecies, ssp. *texensis* G.B. Ownbey, is endemic to Texas and consists of relatively robust plants with stout stems, elongate inflorescences, and slender fruits that are mostly 21–35 mm long.

1. Inflorescences (except those with all cleistogamous flowers) usually extending well past the foliage; spur of the upper outer petal blunt and not expanded at the tip; fruits 15–30 mm long, relatively slender . . . . 1A. SSP. AUSTRALIS
1. Inflorescences extending only slightly past the foliage; spur of the upper outer petal slightly expanded and rounded (appearing more or less globose) at the tip; fruits 10–15(–18) mm long, usually relatively stout . . . . . 1B. SSP. MICRANTHA

4a. ssp. australis (Chapman) G.B. Ownbey

C. halei (Small) Fernald & B.G. Schub.

Inflorescences (except those with all cleistogamous flowers) usually extending well past the foliage. Corollas with the upper outer petal 11–14 mm long, the spur blunt and not expanded at the tip. Fruits 15–25 mm long, relatively slender. April–June.

Scattered in the Mississippi Lowlands Division (Atlantic and Gulf Coastal Plains, also Texas and Louisiana north to Kansas and Missouri). Sand Prairies and edges of dry upland forests; also ditches, crop fields, railroads, roadsides, and disturbed sandy areas.

Subspecies *australis* is almost always found growing in sandy habitats. Steyermark (1963) knew it from only a few sites in Scott and Stoddard Counties, but more recent collections have indicated that it is found at most sites with sandy substrate in the Mississippi Lowlands Division.

4b. ssp. micrantha

Inflorescences extending only slightly past the foliage. Corollas with the upper outer petal 11–15 mm long, the spur slightly expanded and rounded (appearing more or less globose). Fruits 10–15 (–18) mm long, usually relatively stout. $2n=16$. April–June.

Scattered nearly throughout the state, but apparently absent from the Mississippi Lowlands Division (Wisconsin to South Dakota south to Illinois, Louisiana, and Texas; also locally in Tennessee, North Carolina, and South Carolina). Glades, upland prairies, savannas, openings of mesic to dry upland forests, banks of streams and rivers, and ledges of bluffs; also margins of crop fields, railroads, roadsides, and open disturbed areas.

Aside from *C. flavula*, this is the most commonly encountered taxon of *Corydalis* in Missouri.

2. Dicentra Bernh. (bleeding heart)

Plants perennial, the rootstock with small clusters of tuberlike bulblets. Stems absent (slender rhizomes sometimes present). Leaves all basal, long-petiolate, the petiole (5–)8–24 cm long. Leaf blades usually 4 times compound and lobed, 4.5–9.0 cm long, ovate to broadly triangular in outline, the ultimate segments linear to narrowly oblong-elliptic or narrowly oblanceolate, rounded (but often with a minute, sharply pointed projection) to sharply pointed at the tip, green or the undersurface glaucous. Inflorescences 10–35 cm long, long-stalked and loosely to densely 3–15-flowered. Flowers bilaterally symmetrical in 2 perpendicular longitudinal planes, the stalks 2–14 mm long, pendant at flowering and fruiting, with a pair of small bractlets positioned noticeably below the flower. Sepals 2–5 mm long, triangular to broadly ovate, attached basally, rounded to more or less truncate base, the margins entire, membranous and white. Corollas white, sometimes pinkish-tinged, with a yellow to orangish yellow tip. Inner petals linear or nearly so toward the base, expanded above the midpoint with a broadly winged margin and a well-developed, entire to slightly undulate crest, rounded at the concave tip. Outer petals similar, variously shaped, short- to long-spurred, with a differentiated, ascending to spreading, concave tip, the body more or less rounded, not keeled or crested. Style persistent, relatively slender, the stigma 4-lobed, the lobes appearing as horns and/or wings. Fruits capsules, dehiscent (sometimes tardily so), 5–15 mm long, narrowly more or less ellipsoid (tapered at each end), straight, the surface often slightly swollen over the seeds, otherwise smooth, 3- to numerous-seeded. Seeds 1.5–2.5 mm long, somewhat flattened, more or less kidney-shaped in outline, rounded along the rim, the surface smooth or nearly so, black, shiny, the elaiosome an irregular, somewhat conic, white mass attached in the notch. About 19 species, temperate portions of North America, Asia.

The genus *Dicentra* is morphologically diverse, including both annuals and perennials, some of which have tall leafy stems and more or less paniculate inflorescences, and others that have climbing stems or solitary flowers (Stern, 1961). The generic description above applies only to the species found in the state. The tuberlike underground bulblets of the two Missouri species of *Dicentra* are derived from swollen petiole bases rather than from thickened roots or stems. Stern (1961) discussed the variety of swollen storage structures produced in the genus and his terminology is followed in the present work.

Although the morphology of the flowers, with the stamens massed around the stigma and enclosed in the corolla tip, suggests that inbreeding should be the main reproductive strategy, species of *Dicentra* are outcrossers. Pollination is mostly accomplished by long-tongued bees, principally queen bumblebees foraging for nectar after they emerge from hibernation and begin searching for a nesting site (Macior, 1970a, 1978a, b). These open the flower tips in search of nectar present in the spur and brush past the stigma, picking up and depositing pollen. Honeybees also are effective pollinators. Interestingly, some bumblebees rob the flowers of nectar by chewing through the spur instead of opening the flower from the tip.

Stern (1997b) noted that an extract from the bulbs of *D. canadensis* and *D. cucullaria* has been used to treat chronic skin diseases and syphilis, as well as a tonic and diuretic. One of the alkaloids in *D. canadensis*, bulbocapnine, has been used in the treatment of Ménière's disease and muscle tremors, and as a pre-anesthetic. Some other species were used by Native Americans in the northwestern United States for worms. The species otherwise generally are considered highly poisonous.

Several species of *Dicentra* are cultivated as garden ornamentals, including the western North American *D. eximia* (Ker Gawl.) Torr. and *D. formosa* (Haw.) Walp., as well as the Asian *D. spectabilis* (L.) Lem. (all known as bleeding hearts for their pink, heart-shaped flowers). Steyermark (1963) excluded the last of these from the Missouri flora, stating that the sole historical collection from Linn County (*C.A. Benson s.n.* on 4 June 1930, in the University of Missouri herbarium) probably originated from a cultivated plant. A more recently collected specimen of *D. spectabilis* from St. Francois County (*J. Sheets 41* on 12 April 1987) in the herbarium of Southeast Missouri State University likely also represents cultivated material For *D. formosa* ssp. *formosa*, a single specimen in the Missouri Botanical Garden Herbarium (*W. H. Emig 330* on 23 April 1913) from the vicinity of Cliff Cave (Jefferson County) is part of a mixed collection that originally was mounted on the same sheet with *D. cucullaria*. In his monograph of the genus, Stern (1961) did not mention this specimen, but his annotation on the sheet indicates that he believed it to represent either a cultivated plant or more probably mislabeled material accidentally mounted on the wrong sheet. Because *D. formosa* occurs natively only from British Columbia to California and does not appear to escape from cultivation farther east, it is excluded from the Missouri flora in the present work.

1. Leaves moderately to strongly glaucous on the undersurface, the ultimate segments more or less rounded but often with a minute, sharply pointed projection at the tip; spurs 0.5–2.0 mm long, rounded (pouchlike), more or less parallel to slightly incurved; bulblets more or less globose, yellow . . . . 1. D. CANADENSIS
1. Leaves not or only slightly glaucous on the undersurface, the ultimate segments angled or tapered to a sharply pointed tip; spurs 3–10 mm long, triangular in outline, angled away from the flower stalk; bulblets teardrop-shaped, pink or less commonly white . . . . 2. D. CUCULLARIA

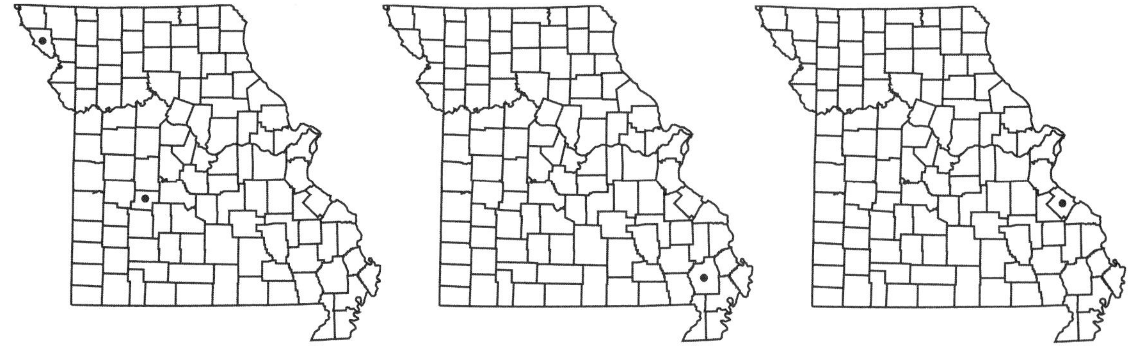

1867. Fumaria officinalis 1868. Bartonia paniculata 1869. Bartonia virginica

1. Dicentra canadensis (Goldie) Walp. (squirrel corn)

Pl. 418 i, j; Map 1865

Plants often producing slender rhizomes, the bulblets more or less globose, yellow. Leaf blades dark green to bluish green on the upper surface, pale and moderately to strongly glaucous on the undersurface, the ultimate segments more or less rounded but often with a minute, sharply pointed projection at the tip. Sepals 2–4 mm long, triangular to ovate. Outer petals 12–20 mm long, the concave apical portion 3–5 mm long, abruptly spreading, yellow to greenish yellow (occasionally pinkish-tinged), the remaining portion white, the short spurs 0.5–2.0 mm long, rounded (pouchlike), more or less parallel to slightly incurved. $2n=64$. March–May.

Uncommon in east-central Missouri but occurring sporadically westward to Jackson County (eastern [mostly northeastern] U.S. west to Minnesota and Missouri; Canada). Bottomland forests, mesic upland forests in ravines, and bases and lower ledges of mostly north-facing bluffs.

Steyermark noted that the flowers of squirrel corn produce a strong, sweet fragrance similar to that of hyacinths, but that in Missouri, in many cases, large populations produced relatively few inflorescences.

2. Dicentra cucullaria (L.) Bernh. (Dutchman's breeches)

Pl. 418 k–m; Map 1866

Plants occasionally producing slender to relatively stout rhizomes, the bulblets teardrop-shaped (asymmetrically ovoid and tapered to a sharply pointed tip), pink or less commonly white. Leaf blades bright green on the upper surface, somewhat paler (not or only slightly glaucous) and sometimes somewhat yellowish green on the undersurface, the ultimate segments angled or tapered to a sharply pointed tip. Sepals 2–5 mm long, broadly ovate. Outer petals 10–16(–20) mm long, the concave apical portion 2–5 mm long, abruptly spreading to reflexed, yellow to orangish yellow, the remaining portion white or pinkish-tinged, the spurs 3–10 mm long, triangular in outline, angled away from the flower stalk. $2n=32$. March–May.

Scattered to common nearly throughout the state (eastern [mostly northeastern] U.S. west to North Dakota and Oklahoma, disjunct in Idaho, Oregon, and Washington; Canada). Bottomland forests, mesic upland forests in ravines, and bases and ledges of bluffs.

This is a common spring wildflower of rich soils in mesic forested areas. Unlike those of *D. canadensis*, the flowers of *D. cucullaria* do not produce a fragrance.

3. **Fumaria** L. (fumitory)

About 50 species, Europe, Asia, Africa, Atlantic Islands.

1. Fumaria officinalis L. (fumitory, fumewort, earthsmoke)

Pl. 418 a, b; Map 1867

Plants annual, with taproots. Stems 20–50 (–80) cm long, loosely to strongly ascending. Leaves alternate, rarely also basal, short- to long-petiolate. Leaf blades 3 or 4 times compound and lobed, 1–5 cm long, lanceolate to ovate in outline, the ultimate segments linear to narrowly oblong or narrowly lanceolate, bluntly to sharply pointed at the tip, green or sometimes both surfaces glaucous. Inflorescences 2–6 cm long, relatively short-stalked and densely 10–

40-flowered. Flowers bilaterally symmetrical in only 1 longitudinal plane, the stalks 2–4 mm long, ascending at flowering and fruiting, without a pair of bractlets. Sepals 1.5–2.0 mm long, ovate, attached slightly above to rounded to more or less truncate base, the margins with several coarse, jagged teeth, membranous and white, often with a purplish-tinged tip. Corollas white to pinkish-tinged, grading to a dark reddish purple tip. Inner petals 5.0–5.5 mm long, slender above a slightly expanded base and more or less rounded at the spoon-shaped tip. Outer petals dissimilar; the lower petal linear but slightly expanded at each end, concave at the tip; the upper petal 7–9 mm long, the spur 2.0–2.5 mm long, slightly incurved, the body keeled but not or only shallowly and irregularly crested, the apical margins slightly to moderately irregular and winged. Style not persistent, slender, the stigma inconspicuously 2-lobed. Fruits indehiscent and nutlike (the seed not easily separable from the fruit wall), 1.5–2.0 mm long, globose to slightly depressed-globose, the surface pebbled or finely warty, 1-seeded. $2n$=32, 48. May–July.

Introduced, uncommon, known thus far from Hickory and Holt Counties (native of Europe, Africa; introduced widely but sporadically in the U.S., Canada). Gardens and roadsides.

This species was first reported for Missouri by Henderson (1980) from a highway embankment in loess soil. It has a long history of cultivation for herbal medicine in Europe. An infusion was used to treat a variety of ailments, including eye infections, eczema, liver and gallbladder problems, and (primarily in mixtures) constipation.

Lidén (1986) recognized three subspecies within *F. officinalis* differing in small details of petal shape and inflorescence development. The Missouri material would key to ssp. *officinalis*. Boufford (1997a) studied these in the context of North American specimens and concluded that these morphological variants did not differ significantly. He also discussed the occasional presence of weak-stemmed plants with mostly smaller, white, possibly cleistogamous flowers, but attributed this variant to a shade form.

GENTIANACEAE (Gentian Family)

Plants annual or perennial herbs (shrubs or small trees elsewhere), sometimes mycoheterotrophic (parasitic on mycorhizal fungi). Leaves simple, mostly opposite or whorled, sessile. Stipules absent, but the bases of leaves at each node usually joined by a transverse line around the stem. Leaf blades sometimes reduced to scales, the margins entire. Inflorescences terminal and/or axillary, of solitary flowers, clusters, racemes, or panicles, the flowers usually subtended by bracts. Flowers actinomorphic (zygomorphic elsewhere), perfect, hypogynous. Calyces with 2, 4, or 5(–14) lobes, sometimes deeply divided to form separate sepals. Corollas with 4 or 5(–14) lobes, frequently bearing nectar glands or scales on the inner (upper) surface, usually spirally twisted in bud. Stamens 4 or 5(–14), attached to corolla tube, alternating with the corolla lobes, the anthers attached near their midpoints or near their bases, distinct or sometimes fused, dehiscing longitudinally. Pistil 1 per flower, of 2 fused carpels, the ovary superior, 1-locular, the placentation parietal. Style 1 per flower (sometimes appearing 2-branched at the tip), the stigma usually 2-lobed. Ovules numerous. Fruits capsules (rarely fleshy), dehiscent longitudinally by 2 valves. Eighty genera, 1,100 species, nearly worldwide, especially in montane areas.

A number of genera contain species that are cultivated as oranmentals and several genera also contain species that have been used medicinally.

1. Leaves mostly in whorls of 4 or 5 per node; corolla lobes with large, fringed nectaries on upper surface . . . . 2. FRASERA
1. Leaves opposite or rarely alternate; nectaries absent or inconspicuous, fringeless, and borne at base of corolla.
    2. Calyces of 2 free, foliaceous sepals; petiole bases strongly decurrent along stem between nodes . . . . 5. OBOLARIA

3. Leaves not all scalelike (sometimes reduced to scales at base of stem), more than 5 mm long; corolla more than 7 mm long
    4. Styles short and stout, or absent; ovary stalked; corollas bluish purple to white or greenish; anthers not spirally coiled or twisted
        5. Corollas with pleated petaloid appendages between the lobes . . . 3. GENTIANA
        5. Corollas without folds, pleats, or petaloid appendages between the lobes . . . . 4. GENTIANELLA
    4. Styles filiform; ovary sessile; anthers spirally twisted or coiled; corollas pink, rarely white.
        6. Corollas saucer-shaped, the lobes much longer than the short tube . . . . 6. SABATIA
        6. Corollas trumpet-shaped, the lobes shorter than to about as long as the tube . . . . 7. ZELTNERA

1. **Bartonia** Muhl. ex Willd.
(J. M. Gillett, 1959; Mathews, 2009)

Plants apparently annual (but this not known with certainty), partially mycotrophic (receiving nutrients and water from associations with soilborne fungi) herbs, glabrous, with chlorophyll, but lacking root hairs. Stems erect or sometimes lax and twining. Leaves reduced to subulate scales, alternate or mostly opposite. Inflorescences paniculate or reduced to racemes. Inflorescences terminal, open racemes or slender panicles. Flowers with parts in whorls of 4. Calyces deeply lobed, the lobes nearly separate, subulate. Corollas narrowly funnelform to more or less bell-shaped, deeply lobed, white, yellowish white, or purplish-tinged. Fruits relatively thin-walled, ellipsoid to ovoid. Three species, eastern U.S., Canada.

In *Bartonia* the ephemeral stems are sometimes purplish-tinged at the base. Although said to be annual by most botanists, little is known of the life history of this genus. By analyzing the profiles of radioactive isotopes of carbon and nitrogen in plants of *B. virginica* relative to their abundance in the surrounding environment, Cameron and Bolin (2010) made a case for the preferential uptake of nitrogen from associated soilborne fungi and thus provided evidence that plants of *Bartonia* are able to parasitize these fungi. This phenomenon of parasitism of mycorhizal fungi by vascular plants is known as mycoheterotrophy or mycotrophy.

1. Leaves alternate; corolla lobes lanceolate, tapered to a sharp point at the tip . . . . 1. B. PANICULATA
1. Leaves all or mostly opposite; corolla lobes oblong, rounded at the tip, sometimes with a minute sharp point or somewhat irregularly toothed . . . . 2. B. VIRGINICA

1. Bartonia paniculata (Michx.) Muhl. **ssp. paniculata** (screwstem)

Pl. 420 a, b; Map 1868

Stems 10–40 cm long. Leaves 0.5–3.0 mm long, mostly alternate. Calyx lobes 2–3 mm long. Corolla lobes 2–5(–7) mm long, lanceolate, tapered to a sharp point at the tip. $2n=52$. August–October.

Uncommon, known only from Stoddard County (eastern U.S. west to Oklahoma). Acid seeps.

This inconspicuous plant is found on moist, mossy hummocks and banks in the sandy ravines at the base of Crowley's Ridge.

A second subspecies, ssp. *iodandra* (B.L. Rob.) J.M. Gillett, differs in its thicker stems and larger stamens, and occurs in northeastern North America. A third subspecies that is endemic to portions of Texas and Louisiana, ssp. *texana* (Correll) K.G. Mathews et al., historically was treated as a separate species, *B. texana* Correll. It differs morphologically from ssp. *paniculata* in its stiffly erect stems and shorter, broader corolla lobes. Mathews et al. (2009) provided evidence from a molecular study showing little genetic divergence between the three subspecies that they treated within *B. paniculata*.

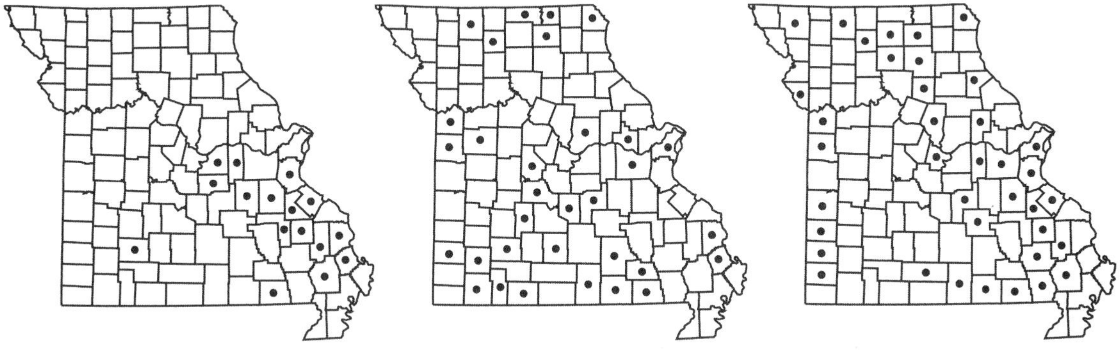

1870. Frasera caroliniensis 1871. Gentiana alba 1872. Gentiana andrewsii

2. Bartonia virginica (L.) Britton, Sterns & Poggenb. (Virginia bartonia, Virginia screwstem)

Map 1869

Stems 5–45 cm long. Leaves 0.9–4.5 mm long, mostly opposite. Calyx lobes 0.5–1.5 mm long. Corolla lobes 2.5–5.0 mm long, oblong, rounded at the tip, sometimes with a minute sharp point or somewhat irregularly toothed. $2n=52$. July–September.

Uncommon, known thus far only from Ste. Genevieve County (eastern U.S. and adjacent Canada west to Wisconsin and Louisiana). Mossy sandstone ledges.

Bartonia virginica is an inconspicuous species at its western geographic limit in Missouri. It was first collected in the state in 1993 by Bill Summers, and may eventually prove to be more common in the sandstone drainages of eastern Missouri.

2. **Frasera** Walter (columbo, green gentian) (St. John, 1941)

About 15 species. North America.

Segregation of the generic name *Frasera* for the New World species in this group is of long-standing controversy, and *F. caroliniensis* is included in the genus *Swertia* L. by some authors. In the broad sense, *Swertia* refers to a group of about 65 species occurring in North America, as well as portions of Europe, Asia, and Africa. Pringle (1990) presented a summary of the morphological evidence in support of a broad circumscription of *Swertia*. However, more recent molecular phylogenetic studies of the group (Chassot et al., 2001; von Hagen and Kadereit, 2002) have indicated that *Swertia* in the broad sense comprises several distantly related lineages and that the recognition of natural genera in this group is a considerably more complex problem than earlier workers had suspected.

1. Frasera caroliniensis Walter (American columbo)

Swertia caroliniensis (Walter) Kuntze

Pl. 421 i–k; Map 1870

Plants perennial, monocarpic herbs, producing an aerial stem only at the onset of floral development. Flowering stem greenish to purple, erect, thick, hollow, to 2.5 m long. Leaves of basal rosette oblanceolate, usually obtuse at tip, 20–40 cm long. Leaves of stems in whorls of (3)4 or 5, the lower much like the rosette leaves, 10–20 cm long, becoming progressively smaller, more lanceolate and pointed at tip, grading into linear bracts 1–4 cm long at the inflorescence. Inflorescences often appearing as a much-branched terminal panicle, but consisting of small, terminal and mostly axillary panicles. Flowers with parts in multiples of 4. Calyces lobed nearly to base, the lobes linear-lanceolate, 6–10 mm long. Corollas greenish yellow, purple-dotted, saucer-shaped, deeply lobed, the lobes lanceolate-ovate, 10–14 mm long, 3–5 times as long as the tube, each lobe bearing a large, fringed, purple nectary gland near the middle of the upper surface. Filaments of stamens elongate, fused to each other near attachment point to corolla. Ovaries ovoid, the style elongate, persistent in fruit, the stigma capitate, 2-lobed. Capsules ovoid, flattened, 15–25 mm long. $2n=78$. May to July.

Scattered in the eastern portion of the Ozark and Ozark Border Divisions and the northern portion of the Mississippi Lowlands, disjunct in Greene County (eastern U.S. west to Illinois and Oklahoma; Canada). Bottomland forests, mesic upland forests, glades, margins of sinkhole ponds, bases of bluffs, and banks of streams; also roadsides.

Frasera caroliniensis is a long-lived perennial often seen as colonies of plants with only sterile rosettes of leaves. Threadgill et al. (1981) studied populations in Kentucky and determined that the species is monocarpic, with rosettes resprouting annually for a number of years. Production of a flowering stem is followed by death of the individual after the fruits have matured. Steyermark (1963) noted that an individual that he transplanted into his garden persisted but did not flower for more than fifteen years.

In the spring, the rosettes of *F. caroliniensis* may be confused with those of *Mertensia virginica* (L.) Pers. (bluebells, Boraginaceae), but basal leaves of the latter taxon are usually a darker green and are more spatulate in shape, narrowing fairly abruptly to a petiole (as opposed to gradually tapering and essentially sessile). The roots of *F. caroliniensis* have been used medicinally as a tonic, emetic, and cathartic (Moerman, 1998).

3. **Gentiana** L. (gentian)
(Pringle, 1967)

Plants perennial herbs (annual elsewhere), usually several-stemmed from the base, glabrous or minutely hairy. Leaves opposite, with bases usually more or less clasping. Inflorescences terminal clusters, often also solitary flowers or small clusters in the upper leaf axils. Flowers with parts in multiples of 5. Calyces tubular, with ovate to linear lobes. Corollas tubular to funnelform, with toothed or fringed, pleated, petaloid appendages between the lobes. Anthers of stamens often fused to each other. Ovaries short-stalked, with inconspicuous nectary glands at bases, the style short or absent, the stigmas 2, persistent.

About 300 species, nearly worldwide (except Africa).

1. Corollas blue, with free portion of lobes conspicuous, 6–10(–15) mm long at maturity, spreading; anthers free; stems minutely hairy ...... 3. G. PUBERULENTA
1. Corollas blue, whitish, or greenish, with free portion of lobes inconspicuous, 0.5–6.0 mm long, erect or incurved; anthers partially fused together; stems glabrous
   2. Corollas blue to violet blue, with lobes shorter than the appendages, appearing closed at maturity; leaves with minutely ciliate margins .... 1. G. ANDREWSII
   2. Corollas white to greenish- or yellowish-tinged, with lobes longer than the appendages, appearing open at maturity; leaves glabrous .... 2. G. ALBA

1. Gentiana alba Muhl. ex Nutt. (pale gentian, white gentian, yellowish gentian)
G. flavida A. Gray
Pl. 419 d; Map 1871

Stems 30–90 cm long, erect or nearly so, glabrous. Leaves glabrous, ovate to lanceolate-ovate, oblong in reduced lower leaves, widest near the bases, 3–10 cm long; leaf margins sometimes minutely denticulate, but not ciliate. Calyces 11–21 mm long, the lobes ovate to lanceolate-ovate, as long as the tube. Corollas white to greenish or yellowish tinged, rarely pale lilac in upper half, open at maturity, narrowly funnelform, 3.5–4.7 cm long. Free portion of the corolla lobes broadly ovate, erect to somewhat incurved, longer than the appendages, 3–6 mm long. Anthers more or less fused to one another. $2n=26$. August–October.

Scattered nearly throughout the state (eastern and midwestern U.S.; Ontario). Upland prairies, glades, and openings of mesic upland forests.

A detailed nomenclatural history of this species was presented by Wilbur (1988), but subsequent bibiographic research by James Pringle and others has revealed the existence of an earlier valid combination for the name *G. alba* than was known to Wilbur.

Gentiana ×pallidocyanea J.F. Pringle, the rare hybrid between this species and *G. andrewsii*, has been found in several counties, mostly in northeastern Missouri. Its corollas are variable, but

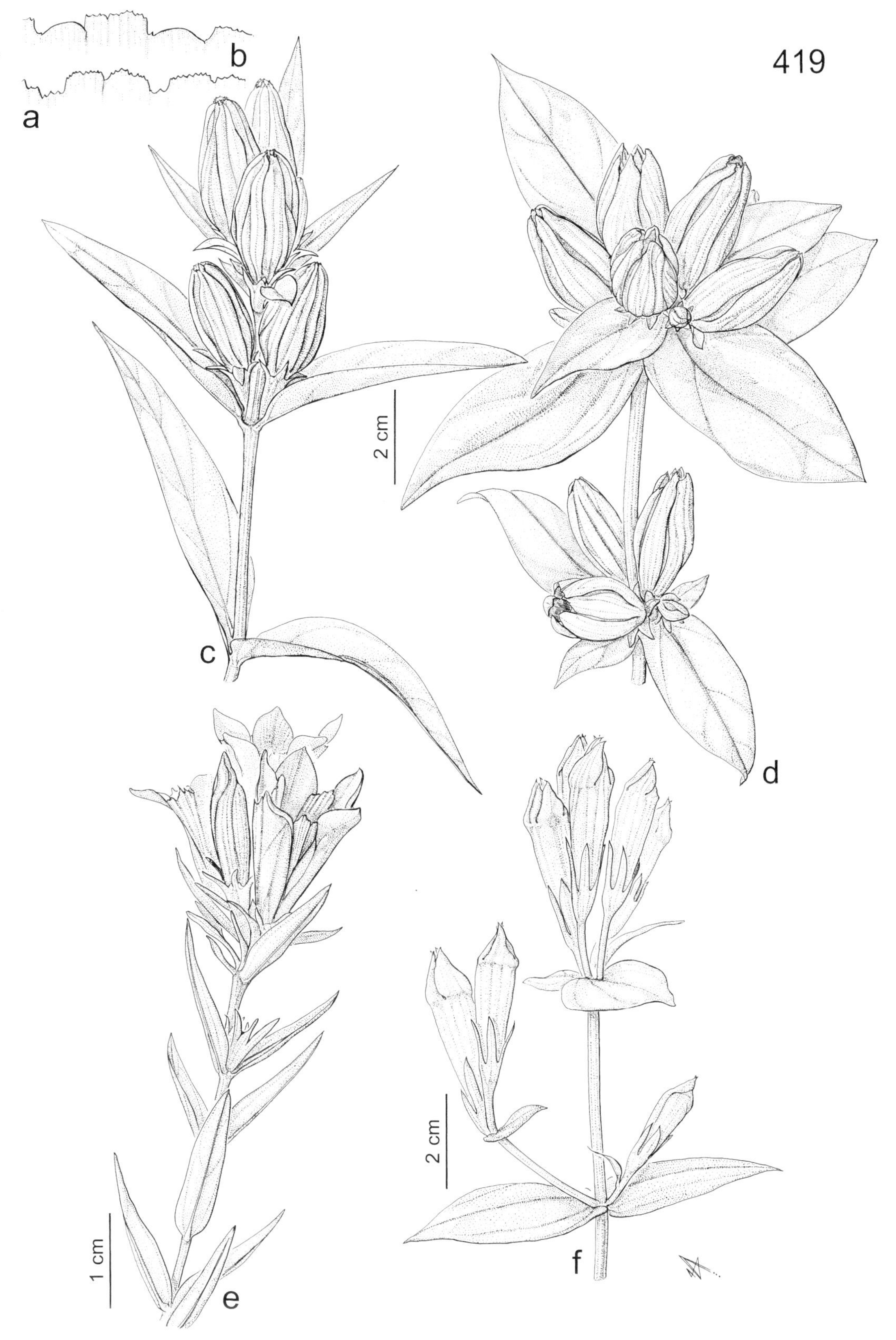

Plate 419. Gentianaceae. *Gentiana andrewsii* var. *andrewsii*, **a)** corolla margin. *Gentiana andrewsii* var. *dakotica*, **b)** corolla margin, **c)** habit. *Gentiana alba*, **d)** habit. *Gentiana puberulenta*, **e)** habit. *Gentianella quinquefolia*, **f)** habit.

1873. Gentiana puberulenta 1874. Gentianella quinquefolia 1875. Obolaria virginica

generally intermediate in color and relative appendage length between those of its parents. Another hybrid, *G. ×curtisii* J.F. Pringle (*G. alba* × *G. puberulenta*), was documented recently from an upland prairie in Adair county, and also occurs spontaneously at a site in Franklin County where both parental species have become naturalized from seedings a number of years ago.

2. Gentiana andrewsii Griseb. (closed gentian, bottle gentian)

G. clausa Raf., misapplied

Pl. 419 a–c; Map 1872

Stems (20–)30–70 cm long, erect or nearly so, glabrous. Leaves with minutely ciliate margins, lanceolate to lanceolate-ovate, 2.0–8.5 cm long, sometimes reduced to scales near stem bases. Calyces 14–25(–29) mm long, the lobes lanceolate to lanceolate-ovate, usually shorter than the tube. Corollas blue to bluish violet, closed at maturity, tubular, 2.8–4.0 cm long. Free portion of the corolla lobes shallowly triangular or reduced to a minute point, erect to somewhat incurved, 0.5–3.0 mm long, shorter than the finely toothed or fringed appendages. Anthers partially fused together. $2n$=26. August–October.

Scattered nearly throughout the state (northeastern U.S. west to North Dakota, Nebraska, and Missouri; Canada). Upland prairies, glades, and openings of mesic upland forests.

See the treatment of *G. alba* for a discussion of the hybrid with that species. Two varieties of *G. andrewsii* have been accepted (Pringle, 1967); both are present in Missouri.

1. Corolla lobes reduced to an inconspicuous point 0.5–1 mm long . . . . 2A. VAR. ANDREWSII
1. Corolla lobes broadly triangular, 1–3 mm long . . . . 2B. VAR. DAKOTICA

2a. var. andrewsii

Pl. 419 a

Corolla lobes reduced to an inconspicuous point 0.5–1.0 mm long. $2n$=26. August–October.

Uncommon and widely scattered in eastern Missouri and disjunctly in Platte County (northeastern U.S. west to North Dakota and Missouri; Canada). Ledges of bluffs and upland prairies.

2b. var. dakotica A. Nelson

Pl. 419 b, c

Corolla lobes broadly triangular, 1–3 mm long. $2n$=26. August–October.

Scattered nearly throughout the state (North Dakota to Nebraska east to Illinois and Missouri; Canada). Upland prairies, glades, and openings of mesic upland forests.

This is by far the most common variety in Missouri.

3. Gentiana puberulenta J.F. Pringle (prairie gentian, downy gentian)

G. puberula Michx.

Pl. 419 e; Map 1873

Stems 15–50 cm long, erect or strongly to loosely ascending, minutely hairy, usually in longitudinal lines. Leaves minutely hairy along basal portion of margins and on principle veins of lower surface, lanceolate, 1–6 cm long, sometimes reduced to scales near stem bases. Calyces 11–36 mm long, the linear to lanceolate lobes shorter than the tube. Corollas blue to bluish violet, sometimes darkening with age, open, funnelform, 3–4 cm long. Free portion of the corolla lobes 6–10(–15) mm long, spreading at maturity, longer than the appendages. Anthers free. $2n$=26. September–November.

Scattered nearly throughout the state, except in the Mississippi Lowlands Division (North Dakota to Oklahoma east to New York, Maryland,

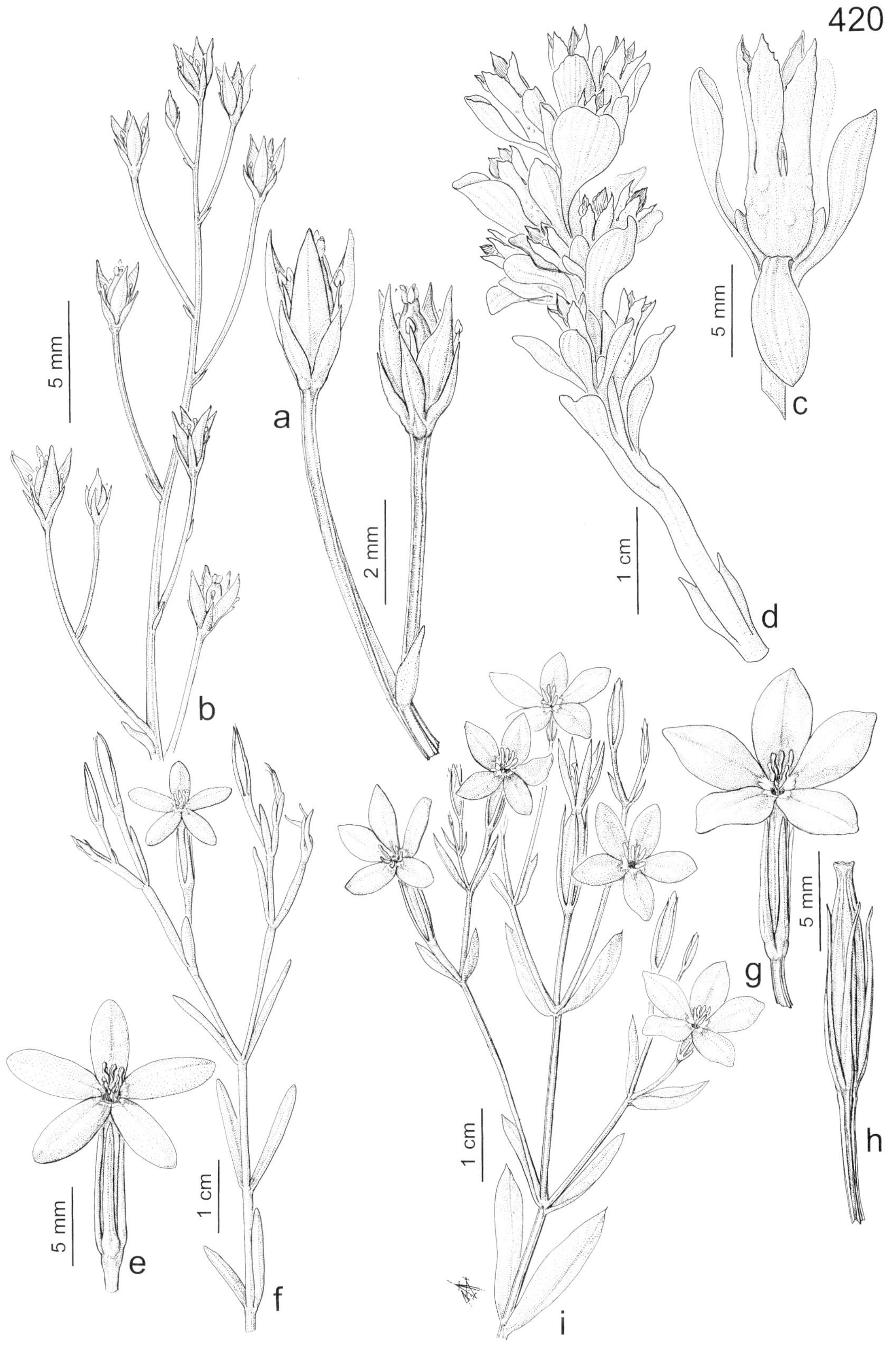

Plate 420. Gentianaceae. *Bartonia paniculata*, **a)** flowers, **b)** habit. *Obolaria virginica*, **c)** flower, **d)** inflorescence. *Zeltnera texensis*, **e)** flower, **f)** habit. *Zeltnera calycosa*, **g)** flower, **h)** fruit, **i)** habit.

Tennessee, and Louisiana; Canada). Upland prairies, savannas, and glades; usually on calcareous substrates.

Refer to the treatments of *G. alba* and *G. andrewsii* for discussions of hybrids with those species.

4. **Gentianella** Moench
(J. M. Gillett, 1957)

About 125 species, nearly worldwide (except Africa).

1. Gentianella quinquefolia (L.) Small **ssp. occidentalis** (A. Gray) J.M. Gillett (stiff gentian, ague weed)

Gentiana quinquefolia L. var. *occidentalis* A. Gray

Pl. 419 f; Map 1874

Plants annual or less commonly biennial herbs. Stems 20–80 cm long, often branched above, ridged or narrowly winged, square in cross-section, the ridges and wings minutely toothed. Leaves opposite, broadly ovate, 10–60 mm long, with clasping bases and minutely denticulate margins. Inflorescences terminal and axillary, clusters or small panicles, the lower ones sometimes reduced to a solitary flower. Flowers with parts in whorls of (4)5. Calyces 6–15 mm long, tubular, the elliptic-lanceolate to oblanceolate lobes slightly longer than the tube. Corollas 15–25 mm long, blue to whitish, commonly lavender blue, narrowly funnelform, the lobes erect or somewhat spreading, ovate to triangular and strongly tapered, about half as long as the tube; inconspicuous nectar glands present at base of corolla; stamens free, included; ovary elongate, short-stalked; style short or absent; stigmas 2, persistent; capsule fusiform, protected by the persistent calyx and dried corolla. $2n=36$. August–November.

Widely scattered in the eastern half of the state (Minnesota, Kansas, and Arkansas east to Ohio and Virginia; Canada). Ledges of bluffs, banks of streams, and bottomland forest; often on calcareous substrates.

5. **Obolaria** L.
(J. M. Gillett, 1959)

One species, eastern U.S.

1. Obolaria virginica L. (pennywort)

Pl. 420 c, d; Map 1875

Plants perennial mycoheterotrophic herbs, glabrous. Stems 4–17(–25) cm long, erect, somewhat fleshy, pale green to white. Leaves green, sometimes purplish-tinged, opposite, spathulate to obovate, 4–15 mm long, reduced to scales below. Leaf bases decurrent along stem between nodes. Inflorescences appearing spikelike, axillary and terminal clusters of 1–3 flowers. Calyces of 2, free, foliaceous sepals 6–9 mm long. Corollas 6–12(–15) mm long, white or greenish- to purplish-tinged, narrowly funnelform, the 4 ascending to slightly spreading lobes about as long as the tube. Stamens 4, not extending beyond corolla lobes, subtended by inconspicuous fringed scales near the corolla base. Ovaries sessile, the style short, the stigmas 2. Capsules covered by the persistent calyx and dried corolla. $2n=56$. March–May.

Uncommon, known only from Cape Girardeau, Stoddard, and possibly St. Francois Counties (eastern [mostly southeastern] U.S. west to Missouri and Texas). Bottomland forests and mesic upland forests.

This inconspicuous plant is probably more widely distributed in southeastern Missouri than present records indicate.

By analyzing the profiles of radioactive isotopes of carbon and nitrogen in plants of *O. virginica* relative to their abundance in the surrounding environment, Cameron and Bolin (2010) made a case for the preferential uptake of carbon from associated soilborne fungi and thus provided evidence that plants of *Bartonia* are able to parasitize these fungi. This phenomenon of parasitism of mycorhizal fungi by vascular plants is known as mycoheterotrophy.

1876. Sabatia angularis　　1877. Sabatia brachiata　　1878. Sabatia campestris

6. **Sabatia** Adans. (marsh pink) (Wilbur, 1955)

Plants annual or biennial herbs, glabrous. Stems erect, usually branched above; leaves opposite, often also in a basal rosette. Inflorescences open panicles, sometimes reduced to loose clusters. Flower with parts in multiples of (4)5. Corollas saucer-shaped, the lobes longer than the tubes (sometimes shorter elsewhere). Anthers spirally coiled after dehiscence. Ovaries sessile, the style slender, elongate, the stigma deeply 2-lobed. Capsules ovoid to cylindrical. Seventeen species, North America.

1. Branches of the inflorescence mostly alternate; calyx tube prominently ribbed or winged; stem square in cross-section . . . . 3. S. CAMPESTRIS
1. Branches of the inflorescence mostly opposite; calyx tube not or inconspicuously ribbed or winged
    2. Stem square in cross-section (at least in lower half), winged; leaves on stems widest near the clasping base . . . . 1. S. ANGULARIS
    2. Lower half of stem circular in cross-section (angled above), not winged; leaves on stems widest near the middle, the bases not clasping . . . . 2. S. BRACHIATA

1. Sabatia angularis (L.) Pursh (rose-pink)
Pl. 421 a–d; Map 1876

Stems (10–)30–80 cm long, nearly square in cross-section and winged, at least in the lower half, the branches mostly opposite. Leaves of stems ovate, widest near the rounded to cordate, clasping bases, 3–7-nerved, 15–40 mm long. Calyces deeply lobed, not or inconspicuously ribbed, 8–20 mm long, the lobes linear or rarely somewhat oblong, 9–18 mm long. Corollas pink to white, with a yellow spot at the base of each lobe, the lobes spathulate to elliptic, 10–22 mm long, 3–5 times as long as the short tube. $2n=38$. June–September.

Scattered in the southern half and northeastern quarter of the state (New York to Florida west to Kansas and Texas). Glades, openings of mesic to dry upland forests, banks of streams and rivers, bottomland prairies, moist depressions of upland prairies, and fens; also pastures, old fields, fallow fields, quarries, cemeteries, ditches, railroads, and roadsides.

Uncommonly encountered plants with white corollas have been called f. *albiflora* Raf. ex House. Steyermark (1963) remarked that the flowers of this species have a pleasant fragrance similar to that of *Prenanthes aspera* Michx. (Asteraceae).

2. Sabatia brachiata Elliott
Pl. 421 e, f; Map 1877

Stems 15–50(–60) cm long, the lower half round in cross-section, without wings (angled above), the branches mostly opposite. Leaves lanceolate to narrowly elliptic, with rounded to tapering bases, 1–3-nerved, 15–40 mm long. Calyces deeply lobed, not or inconspicuously ribbed, 6–11 mm long, the

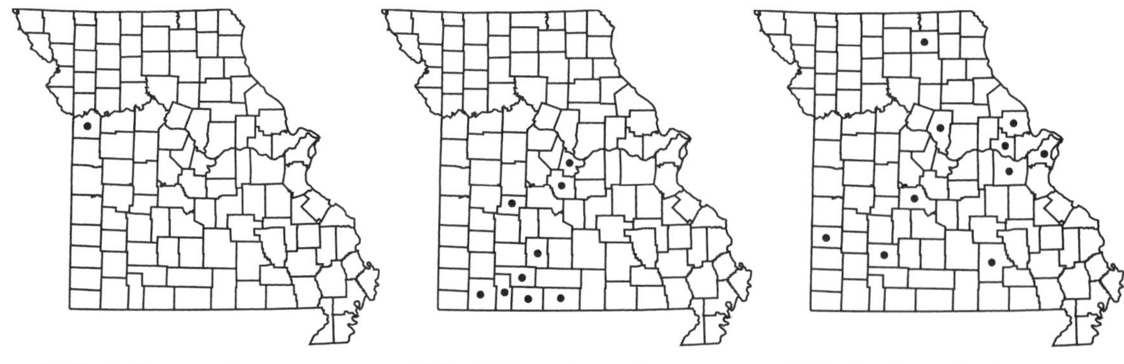

1879. Zeltnera calycosa

1880. Zeltnera texensis

1881. Erodium cicutarium

lobes linear, 5–10 mm long. Corollas pink, rarely white, with a yellowish green spot at the base of each lobe, the lobes oblong to spathulate, 7–14 mm long, 3–5 times as long as the short tube. $2n$=28, 32. June–August.

Uncommon, known thus far only from Butler County (Missouri to Virginia south to Georgia and Louisiana). Openings of upland forests; also roadsides and moist, open, disturbed areas.

3. Sabatia campestris Nutt. (prairie rose-gentian)

Pl. 421 g, h; Map 1878

Stems 15–35(–40) cm long, square in cross-section, often slightly winged, the branches mostly alternate. Leaves ovate to lanceolate-ovate, with rounded to cordate, clasping bases, widest near the base, 8–25(–30) mm long. Calyces deeply lobed, strongly 5-ribbed or winged, (11–)14–30 mm long, the lobes linear to lanceolate, (6–)10–22 mm long. Corollas pink to white, with a greenish yellow spot at the base of each lobe, the lobes broadly obovate to spathulate, 10–21 mm long, 2.0–2.5 times as long as the tube. $2n$=26. July–September.

Scattered in the Unglaciated Plains Division, uncommon elsewhere, mostly in counties along the Missouri River and the eastern portion of the Ozarks (Kansas to Texas east to Illinois and Mississippi; introduced in New England). Glades, upland prairies, sand prairies, mesic upland forests, and banks of streams and rivers; also pastures, ditches, railroads, roadsides, and open, disturbed areas.

7. **Zeltnera** G. Mans.
(Mansion, 2004)

Plants annual or biennial herbs, glabrous. Stems erect, branching, at least above. Leaves opposite, sometimes also forming a basal rosette. Inflorescences open to spikelike panicles, sometimes reduced to clusters. Flowers with parts in whorls of 4 or 5, lacking obvious nectary glands. Calyces deeply lobed, appressed to the corolla tube, the lobes keeled. Corollas trumpet-shaped, the tube narrow, longer than the lobes. Stamens exserted, anthers spirally twisted after dehiscence. Ovaries 1 per flower, elongate, sessile, the style 1, slender, elongate, the stigma capitate, 2-lobed. Fruits thin-walled, oblong to fusiform, protected by the persistent calyces and dried corollas. About 25 species, North America to South America.

Zeltnera formerly was included in a broadly circumscribed *Centaurium* Hill, but the molecular phylogenetic studies of Mansion (2004) and Mansion and Struwe (2004) have shown that some elements within the traditonal *Centaurium* are instead more closely related to *Sabatia*. The species of the New World *Zeltnera* differ from the mostly Old World taxa remaining in *Centaurium* in their undivided styles, with fan-shaped to rhombic stigma lobes that are not fleshy-thickened and in their broader capsules.

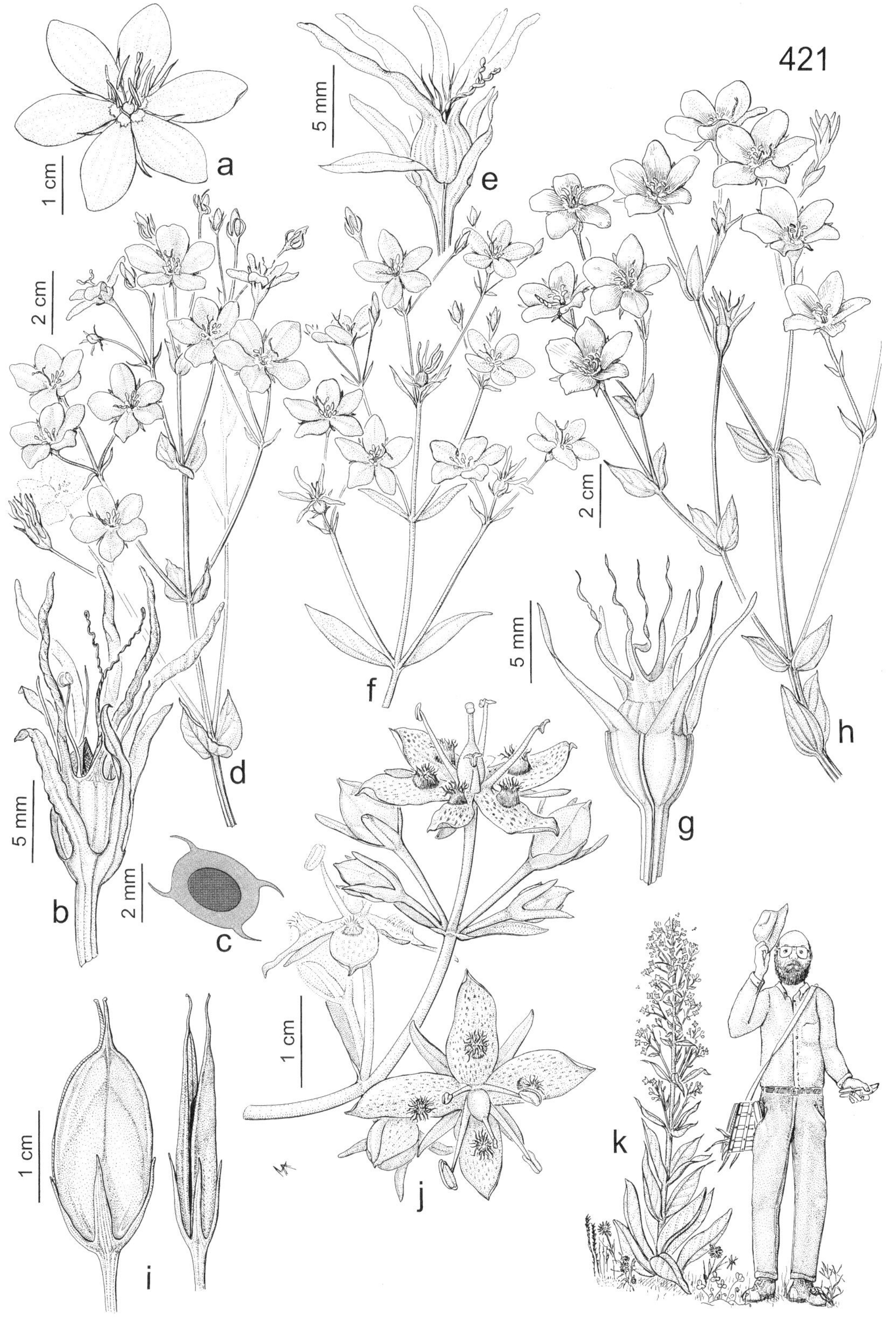

Plate 421. Gentianaceae. *Sabatia angularis*, **a)** flower, **b)** fruit, **c)** diagrammatic cross-section of stem, **d)** upper portion of stem with leaves and inflorescence. *Sabatia brachiata*, **e)** fruit, **f)** upper portion of stem with leaves and inflorescence. *Sabatia campestris*, **g)** fruit, **h)** upper portion of stem with leaves and inflorescence. *Frasera caroliniensis*, **i)** two views of fruit, **j)** portion of inflorescence, **k)** habit, with George Yatskievych (180 cm tall) for scale.

1. Stem leaves narrowly 4–15 mm wide, oblong to lanceolate; corollas 14–23 mm long. the lobes 7–10 mm long . . . . 1. Z. CALYCOSA
1. Stem leaves 1–8 mm wide, linear to narrowly lanceolate; corollas 12–18 mm long, the lobes 4–7 mm long . . . . 2. Z. TEXENSIS

1. Zeltnera calycosa (Buckley) G. Mans.
Centaurium calycosum (Buckl.) Fernald
Pl. 420 g–i; Map 1879

Stems (15–)20–60 cm long. Leaves of stems narrowly oblong to lanceolate, 4–15 mm wide. Calyx lobes 6–10 mm long. Corollas 14–23 mm long. pink or pinkish-tinged, rarely white or pale yellowish, the tube 7–13 mm long and lobes 7–10 mm long. July.

Introduced, uncommon, known thus far only from historical collections from Jackson County (southwestern U.S. and adjacent Mexico). Railroads.

Broome (1973) recognized three varieties within *Centaurium calycosum*, with only var. *calycosum* escaping in Missouri. However, B. L. Turner (1993a) suggested that the distinctions between var. *calycosa* and the dwarf var. *nana* were unworthy of formal taxonomic recognition and represented merely variation between various populations. Mansion (2004) presented evidence that Broome's (1973) var. *arizonicum* is better treated as a separate species, *Z. arizonica* (A. Gray) G. Mans.

2. Zeltnera texensis (Griseb.) G. Mans. ex J.S. Pringle
Centaurium texense (Griseb.) Fernald
Pl. 420 e, f; Map 1880

Stems 10–30 cm long. Leaves of stems linear to narrowly lanceolate, 1–8 mm wide. Calyx lobes 6–10 mm long. Corollas 12–18 mm long, pink, rarely white, the tube 8–11 mm long and lobes 4–7 mm long. $2n$=42. June–September.

Scattered in the southwestern portion of the state north to Miller County (Missouri to Texas). Calcareous glades.

GERANIACEAE (Geranium Family)

Plants annual or perennial herbs (shrubs elsewhere), often with glandular pubescence. Stems often with somewhat swollen nodes. Leaves basal and alternate or opposite, long-petiolate to nearly sessile. Stipules herbaceous, sometimes appearing scalelike, lanceolate to ovate or ovate-triangular. Leaf blades simple or compound, pinnately or more commonly palmately veined and divided or lobed, the margins usually toothed. Inflorescences axillary and often also terminal, small panicles or loose clusters, these sometimes umbellate, sometimes reduced to solitary flowers. Flowers actinomorphic (somewhat zygomorphic elsewhere), perfect, hypogynous. Calyces of 5 free sepals. Corollas of 5 free petals, usually alternating with minute nectaries at the base. Stamens 5 or 10, the filaments free or fused at the base, narrowly winged or expanded toward the base, the anthers attached toward their midpoints, bluish purple (appearing yellow as the pollen is released). Staminodes present in 5-stamened species, appearing as narrow scales similar to the filaments of fertile stamens but often somewhat shorter. Ovary 1 per flower, superior, of 5 fused carpels fused to a slender central column (this persistent, becoming elongated as the fruits mature), 5-lobed toward the tip, with 5 locules. Styles 5, but fused to the beaklike central column for most of their length, persistent and becoming elongated at fruiting, the stigmas 5, club-shaped or linear. Ovules 2 per locule. Fruits schizocarps, 5-lobed toward the base, splitting from the column into 5 mericarps at maturity, each with 1 seed, the stylar beak persistent on each mericarp. About 11 genera, about 700 species, nearly worldwide but most diverse in temperate or montane tropical regions.

Members of the genus *Pelargonium* L'Hér. are widely cultivated for their attractive flowers and foliage, mostly as indoor plants in homes and conservatories. They are the familiar geraniums of horticulture and belong to a genus of about 280 species native mostly to Africa.

Pelargonium differs from the closely related *Geranium* in its zygomorphic flowers and contains species with a bewildering variety of different morphologies, including shrubs, succulents, and tuberous or spiny plants.

1. Leaf blades pinnately divided or veined (sometimes appearing 3-veined in *E. texanum*); mericarps at maturity with the stylar beaks separating from the column and curling and/or twisting outward, the basal portion narrowly ellipsoid (tapered to a sharply pointed base), usually indehiscent . . . . 1. ERODIUM
1. Leaf blades palmately divided or veined; mericarps at maturity with the stylar beaks remaining attached to the apical portion of the column, the basal portion ovoid (rounded or broadly angled at the base), curling upward, dehiscent . . . . 2. GERANIUM

1. **Erodium** L'Hér. ex Aiton (stork's bill)

Plants annual (herbaceous or woody perennials elsewhere). Stems prostrate to loosely ascending at maturity, usually reddish- or purplish-tinged. Leaves basal and opposite. Leaf blades simple or pinnately compound, the main veins pinnate (sometimes appearing 3-veined in *E. texanum*), the veins and margins sometimes reddish- or purplish-tinged. Stipules mostly 3 at each node (by fusion of adjacent stipules on 1 side but not the other). Inflorescences axillary clusters of mostly 2–5 flowers, usually appearing umbellate. Sepals abruptly narrowed or tapered to a short awnlike extension at the tip. Stamens 5, the filaments free, gradually or abruptly broadened toward the base. Staminodes 5, scalelike, shorter than the filaments of fertile stamens. Mericarps at maturity with the stylar beaks separating from the column and curling and/or twisting outward, the basal portion narrowly ellipsoid (tapered to a sharply pointed base), usually indehiscent. Seeds ellipsoid, the surface smooth, brown. About 80 species, widespread in temperate and warm-temperate regions, especially in the Mediterranean region.

Seed dispersal in *Erodium* is accomplished by the shedding of intact mericarps, which split from the central column downward from the tip. The long slender stylar beak of each mericarp is hygroscopic, that is, it becomes spirally coiled as it dries and tends to uncoil when conditions become wetter or more humid. This action can effectively drill the spindle-shaped basal portion of the mericarp into the soil. The awns have been used as crude hygrometers to measure changes in atmospheric humidity.

1. Leaf blades pinnately compound with 4–8 pairs of deeply lobed leaflets; pubescence of gland-tipped hairs; petals 3–6 mm long, pink to lilac, lavender or pale purple . . . . 1. E. CICUTARIUM
1. Leaf blades simple, 3-lobed, the middle lobe often with additional shallow pinnate lobes; pubescence of nonglandular hairs; petals 6–12 mm long, reddish purple . . . . 2. E. TEXANUM

1. Erodium cicutarium (L.) L'Hér. ex Aiton (filaree, alfilaria, pink needle, pin clover)

Pl. 422 a–d; Map 1881

Plants moderately to densely pubescent with more or less spreading, minutely gland-tipped hairs. Stems 2–35 cm long, sometimes beginning to produce flowers when very short, but elongating and somewhat mat-forming later in the season. Leaves short- to long-petiolate. Leaf blades 2–10 cm long, narrowly oblong-oblanceolate in outline, pinnately compound with 4–8 pairs of opposite leaflets, these 4–13 mm long, mostly ovate to oblong-elliptic in outline, deeply pinnately lobed, the lobes toothed or shallowly lobed along the margins. Sepals 2.5–4.0 mm long at flowering, persistent and becoming elongated to 6.5 mm at

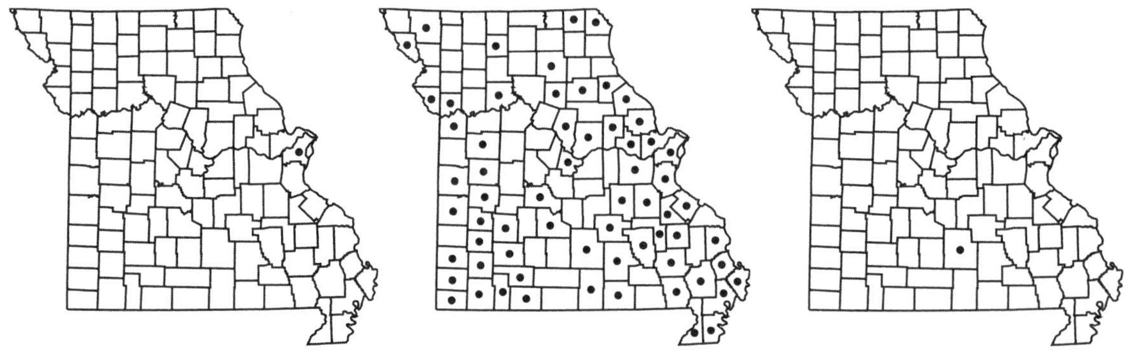
1882. Erodium texanum 1883. Geranium carolinianum 1884. Geranium columbinum

fruiting, lanceolate to oblong-ovate. Petals 3–6 mm long, obovate, narrowed to a stalklike base, pink to lilac, lavender or pale purple. Stamens with the filaments gradually broadened toward the base. Mericarps 25–45 mm long at maturity, the seed-containing basal portion 3.5–4.5 mm long, moderately to densely pubescent with short stiff ascending hairs, the stylar beak with inconspicuous short appressed hairs. Seeds 2–3 mm long. $2n=40$ ($2n=20$, 36, 48, 54 in some European populations). March–November.

Introduced, uncommon and sporadic, mostly south of the Missouri River (native of Europe, Asia, Africa; widely introduced in most temperate regions of the world, including the U.S. [including Alaska, Hawaii], Canada, Greenland, Mexico, etc.). Glades, also lawns, roadsides, and open disturbed areas.

Filaree is a widespread but mostly innocuous weed, although in the southwestern United States, it can interfere with seedling establishment of crop plants in cultivated fields. The petals are shed the same day that the flower opens.

2. Erodium texanum A. Gray (large-flowered stork's bill)

Pl. 422 e, f; Map 1882

Plants moderately to densely pubescent with more or less appressed, nonglandular hairs. Stems 2–30 cm long, sometimes beginning to produce flowers when very short, but elongating later in the season. Leaves mostly long-petiolate. Leaf blades 1.0–3.5 cm long, ovate, simple, truncate to shallowly cordate at the base, with 3 main, shallow to deep lobes, these rounded at the tip, the middle lobe longer than the lateral pair and often with 1 or more pairs of shallow pinnate lobes, the margins otherwise finely scalloped. Sepals 5–7 mm long at flowering, persistent and becoming elongated to 10 mm at fruiting, broadly lanceolate to oblong-oblanceolate or oblong-ovate. Petals 6–12 mm long, broadly obovate, narrowed but not stalklike at the base, reddish purple. Stamens with the filaments abruptly broadened toward the base. Mericarps 40–70 mm long at maturity, the seed-containing basal portion 7–9 mm long, moderately to densely pubescent with short stiff ascending hairs, these often with pustular bases, the stylar beak with relatively long, silky, ascending hairs along the inner side. Seeds 4–5 mm long. $2n=20$. March–May.

Introduced, known thus far from a single specimen from St. Louis City (native of the southwestern U.S. and adjacent Mexico east to Oklahoma and Texas; introduced in Missouri and South Carolina). Railroads.

2. Geranium L. (crane's bill)
(Aedo, 2000, 2001)

Plants annual, biennial, or perennial herbs, sometimes with rhizomes. Stems prostrate to erect or ascending at maturity, sometimes reddish- or purplish-tinged. Leaves basal and alternate or opposite. Leaf blades simple, shallowly to deeply palmately lobed with mostly 5–9 main lobes, the main veins palmate, the veins and margins sometimes reddish- or purplish-tinged. Stipules mostly 3 at each node (by fusion of adjacent stipules on 1 side but not the other). Inflorescences axillary clusters of mostly 2–5 flowers, usually appearing umbellate.

Plate 422. Geraniaceae. *Erodium cicutarium*, **a)** fruits, **b)** flower, **c)** habit, **d)** fertile stem. *Erodium texanum*, **e)** flower, **f)** leaf. *Geranium maculatum*, **g)** fruit, **h)** flower, **i)** habit. *Geranium dissectum*, **j)** fruit, **k)** fruit with separated mericarps, **l)** seed, **m)** leaf. *Geranium pusillum*, **n)** fruit. *Geranium molle*, **o)** fruit. *Geranium carolinianum*, **p)** flower, **q)** fruit, **r)** fertile branch.

Sepals narrowed or tapered (sometimes abruptly so) to a minute sharp point or short awnlike extension at the tip. Stamens 10 (5 in *G. pusillum*), the filaments free or fused at the base, gradually broadened toward the base. Staminodes absent (5 in *G. pusillum*, these scalelike, shorter than the filaments of fertile stamens). Mericarps at maturity with the stylar beaks remaining attached to the apical portion of the column, the basal portion ovoid (rounded or broadly angled at the base), curling upward, dehiscent. Seeds oblong-ellipsoid, the surface smooth or with a fine network of ridges and pits, brown. Three hundred to 430 species, nearly worldwide, but most diverse in temperate regions.

Seed dispersal in most *Geranium* species is explosive. As the mature fruit dries, the beak of each mericarp acts as a spring, placing outward tension on the basal portion. At dehiscence, the apical portion of the beak remains fused to the central column and the basal, seed-containing portion is suddenly released, curling upward, catapulting the seed up to several hundred cm away (K. R. Robertson, 1972).

In some species, the stylar beak is differentiated into 2 regions, a longer columnar portion above the seed-containing base, which is tapered abruptly into a slender beaklike extension above the column and below the stigmas, but in other species this extension is absent and the columnar portion narrows directly into the stigmatic region. This distinction is perhaps confusing when rendered into words, but is an important character in the determination of fruiting specimens.

1. Petals 12–23 mm long; plants perennial, with stout rhizomes .... 4. G. MACULATUM
1. Petals 3–7 mm long; plants annual or biennial, usually taprooted
    2. Sepals tapered or narrowed to an inconspicuous, minute, sharp point at the tip, this less than 0.2 mm long
        3. Mericarps with the seed-containing portion glabrous (except for a few hairs at the very base), the lateral surfaces cross-wrinkled, lacking a narrow riblike dorsal wing, the stylar portion with a slender beaklike extension 2–5 mm long above the column and below the stigmas; stamens 10; staminodes absent .... 5. G. MOLLE
        3. Mericarps with the seed-containing portion with the lateral surfaces moderately to densely and minutely hairy, smooth, with a narrow riblike dorsal wing (or this extending all the way around the mericarp), the stylar portion without a slender beaklike extension above the column, narrowing directly to the stigmas; stamens 5, alternating with 5 staminodes .... 6. G. PUSILLUM
    2. Sepals tapered or narrowed to a conspicuous, short, awnlike extension at the tip, this 0.7–3.0 mm long
        4. Individual flower stalks 3.5–10.0 times as long as the sepals, pubescent with downward-pointing, nonglandular hairs .... 2. G. COLUMBINUM
        4. Individual flower stalks 0.5–2.0 times as long as the sepals, pubescent with mostly spreading, gland-tipped hairs
            5. Petals 5.5–7.0 mm long, pale pink or pale lavender to nearly white; mericarps 19–25 mm long, the seed-containing portion 3.0–4.5 mm long, pubescent with relatively long (0.7–1.5) upward-pointing hairs .... 1. G. CAROLINIANUM
            5. Petals 5.0–5.5 mm long, reddish purple; mericarps 12–17 mm long, the seed-containing portion 2.0–2.5 mm long and pubescent with relatively short (0.2–0.5 mm) spreading hairs.......... 3. G. DISSECTUM

1. Geranium carolinianum L. (Carolina crane's bill)

G. carolinianum var. *confertiflorum* Fernald

Pl. 422 p–r; Map 1883

Plants annual, usually taprooted. Aerial stems 10–65 cm long, erect to loosely ascending, moderately pubescent with short to long (0.2–1.0 mm), spreading to somewhat downward-pointing,

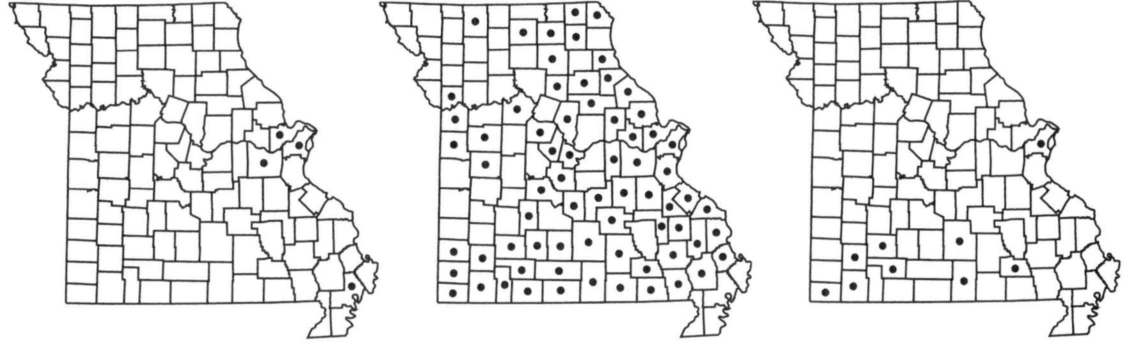

1885. Geranium dissectum 1886. Geranium maculatum 1887. Geranium molle

mostly nonglandular hairs (usually with at least some glandular hairs toward the tip). Leaves basal and opposite, the basal ones long-petiolate, those of the stems with progressively shorter petioles. Leaf blades 1–7 cm long, wider than long to about as long as wide, kidney-shaped to nearly circular in outline, deeply 5–9-lobed, the lobes more or less obovate, mostly deeply and sharply 3- or 5-lobed, sometimes with additional lobes and/or teeth along the margin, the surfaces sparsely to densely pubescent with more or less appressed nonglandular hairs. Inflorescences appearing axillary and often also terminal, short- to long-stalked, consisting of pairs of flowers, these sometimes condensed into small clusters. Individual flower stalks 3–11 mm long, 0.5–1.1 times as long as the sepals, pubescent with spreading, mostly gland-tipped hairs. Sepals 5–6.5 mm long, becoming enlarged to 11 mm at fruiting, ovate, tapered or narrowed to a conspicuous, short, awnlike extension 1.5–2.0 mm long at the tip, pubescent with short to long, loosely ascending, glandular and nonglandular hairs. Petals 5.5–7.0 mm long, obtriangular, truncate to shallowly notched at the tip, pale pink or pale lavender to nearly white. Stamens 10. Staminodes absent. Mericarps 19–25 mm long at maturity, the seed-containing basal portion 3.0–4.5 mm long, the lateral surfaces smooth, pubescent with relatively long (0.7–1.5 mm) upward-pointing hairs (these mostly or all nonglandular), lacking a dorsal ridge or wing, the stylar beak with spreading to loosely ascending, all or mostly nonglandular hairs, the slender extension between the columnar portion and the stigmas 1–2 mm long. Seeds 2.0–2.2 mm long, the surface appearing smooth or with a faint fine network of ridges and pits. $2n=52$. May–July.

Scattered nearly throughout the state (U.S., Canada, Mexico; introduced in South America, Caribbean Islands, and Asia). Glades, upland prairies, sand prairies, tops and ledges of bluffs, openings of mesic upland forests, and banks of streams and rivers; also pastures, fallow fields, roadsides, railroads, ditches, levees, and open disturbed areas.

Some earlier authors (Steyermark, 1963) recognized two or more varieties within *G. carolinianum*, based on subtle differences in pubescence and other floral characteristics. Aedo (2000) documented that pubescence, sepal nervation, and seed shape, among other characters varied on the same individual in some populations, thus precluding the formal recognition of such varieties. *Geranium carolinianum* is the only Missouri species of the genus with consistently pale-colored flowers.

2. Geranium columbinum L. (long-stalked crane's bill)

Map 1884

Plants annual, usually taprooted. Aerial stems 9–50 cm long, spreading to loosely ascending, sparsely to moderately pubescent with short (0.3–0.6 mm), downward-pointing, appressed, nonglandular hairs. Leaves basal and opposite, the basal ones long-petiolate, those of the stems with progressively shorter petioles. Leaf blades 1.5–5.5 cm long, wider than long to about as long as wide, kidney-shaped to nearly circular in outline, deeply 5- or 7-lobed, the lobes elliptic to more or less obovate, deeply and sharply 3–7-lobed, sometimes with additional lobes and/or teeth along the margin, the surfaces sparsely to densely pubescent with appressed nonglandular hairs. Inflorescences appearing axillary and often also terminal, long-stalked, consisting of pairs of flowers. Individual flower stalks 20–60 mm long, 3.5–10.0 times as long as the sepals, pubescent with downward-pointing, appressed, nonglandular hairs. Sepals 5–8 mm long, becoming enlarged to 11 mm at fruiting, ovate, tapered or narrowed to a conspicuous, short, awnlike extension 1.2–3.0 mm long at the tip, pubescent with appressed, nonglandular hairs.

Petals 8–10 mm long, obtriangular, notched at the tip, reddish purple. Stamens 10. Staminodes absent. Mericarps 20–25 mm long at maturity, the seed-containing basal portion 2.2–2.8 mm long, the lateral surfaces smooth, glabrous or sparsely pubescent with short (0.2–0.3 mm), spreading, nonglandular hairs, lacking a dorsal ridge or wing, the stylar beak with spreading to loosely ascending, nonglandular hairs, the slender extension between the columnar portion and the stigmas 3–5 mm long. Seeds 2.2–2.4 mm long, the surface finely pitted. $2n=18$. April–July.

Introduced, known thus far from a single specimen from Texas County (native of Europe, sporadically introduced in the U.S. and Canada). Old fields and open disturbed areas.

This species was first discovered growing in Missouri by Bill Summers in 1993. It is fairly easily recognized in the field by its long-stalked inflorescences, which overtop the foliage.

3. Geranium dissectum L. (cutleaf crane's bill)
Pl. 422 j–m; Map 1885

Plants annual, usually taprooted. Aerial stems 7–60 cm long, spreading to loosely ascending, moderately pubescent with short to long (0.2–0.8 mm), spreading to downward-pointing, nonglandular hairs. Leaves basal and opposite, the basal ones long-petiolate, those of the stems with progressively shorter petioles. Leaf blades 1.5–6.0 cm long, wider than long to about as long as wide, kidney-shaped to nearly circular in outline, shallowly to mostly deeply 3–7-lobed, the lobes more or less obovate, shallowly to mostly deeply and sharply 3–9-lobed, sometimes with additional lobes and/or teeth along the margin, the surfaces sparsely to densely pubescent with spreading to loosely appressed nonglandular hairs. Inflorescences appearing axillary and often also terminal, short- to long-stalked, consisting of pairs of flowers, these sometimes condensed into small clusters. Individual flower stalks 5–13 mm long, 1–2 times as long as the sepals, pubescent with spreading, glandular and nonglandular hairs. Sepals 3–5 mm long, becoming enlarged to 8 mm at fruiting, elliptic-ovate, tapered or narrowed to a conspicuous, short, awnlike extension 1.2–2.0 mm long at the tip, pubescent with short, more or less spreading, glandular and nonglandular hairs. Petals 5.0–5.5 mm long, obtriangular, notched at the tip, reddish purple. Stamens 10. Staminodes absent. Mericarps 12–17 mm long at maturity, the seed-containing basal portion 2.0–2.5 mm long, the lateral surfaces smooth, pubescent with relatively short (0.2–0.5 mm) spreading hairs (these glandular and nonglandular), lacking a dorsal ridge or wing, the stylar beak with spreading, glandular and nonglandular hairs, the slender extension between the columnar portion and the stigmas 1–2 mm long. Seeds 1.9–2.1 mm long, the surface with a conspicuous network of ridges and pits. $2n=22$. April–August.

Introduced, uncommon and sporadic in the eastern portion of the state (native of Europe, widely introduced in the eastern [mostly southeastern] U.S., western U.S., Canada, and Hawaii). Lawns and pen disturbed areas.

Steyermark (1963) knew this species only from a single site in St. Louis County, where he collected it in 1952, growing in a lawn along with *G. molle*.

4. Geranium maculatum L. (wild geranium, spotted crane's bill)
Pl. 422 g–i; Map 1886

Plants perennial, with stout rhizomes. Aerial stems 20–70 cm long, erect or ascending, moderately pubescent with mostly downward-pointing nonglandular hairs. Leaves basal and 1 or few opposite pairs on the stems, the basal ones long-petiolate, those of the stems short-petiolate. Leaf blades 4–15 cm long, wider than long, depressed-ovate to bean-shaped in outline, deeply 5-lobed, the lobes more or less obovate, mostly sharply pointed at the tip, the margins more shallowly lobed and sharply and coarsely toothed, the surfaces moderately pubescent with nonglandular hairs, those of the upper surface sometimes becoming sparse with age. Inflorescences appearing axillary or less commonly terminal, long-stalked, paired, loose clusters of 2–9 flowers. Individual flower stalks 10–30 mm long, 1.5–4.5 times as long as the sepals, pubescent with spreading to downward-pointing nonglandular hairs. Sepals 6–9 mm long, not becoming enlarged at fruiting, elliptic-ovate, tapered to a short awnlike extension 1.5–3.0 mm long at the tip, pubescent with a mixture of spreading to somewhat downward-pointing shorter nonglandular and longer glandular hairs. Petals 12–23 mm long, broadly obovate, rounded to very slightly notched at the tip, pink or rarely white. Stamens 10. Staminodes absent. Mericarps 25–40 mm long at maturity, the seed-containing basal portion 3.5–4.0 mm long, the lateral surfaces smooth, moderately to densely pubescent with spreading nonglandular hairs 0.2–0.6 mm long, lacking a dorsal ridge or wing, the stylar beak with spreading to loosely ascending nonglandular hairs, the slender extension between the columnar portion and the stigmas 5–8 mm long. Seeds 2.5–3.0 mm long, the surface with a fine network of ridges and pits. $2n=52$. April–June.

1888. Geranium pusillum 1889. Ribes americanum 1890. Ribes cynosbati

Scattered to common nearly throughout the state (eastern U.S. west to North Dakota and Oklahoma). Bottomland forests, mesic upland forests, banks of streams and rivers, and shaded ledges of bluffs).

This is a conspicuous and common spring wildflower of bottomlands. It is sometimes cultivated as a garden ornamental. Rare white-flowered plants have been called f. *albiflorum* (Raf.) House.

5. Geranium molle L. (dovesfoot crane's bill)
Pl. 422 o; Map 1887

Plants annual or biennial, usually taprooted. Aerial stems 20–50 cm long, spreading to ascending, moderately pubescent with relatively long (1.0–1.7 mm) loosely upward- to downward-pointing nonglandular hairs, often also with minute glandular hairs. Leaves basal and opposite, the uppermost leaves usually alternate, the basal ones long-petiolate, those of the stems with progressively shorter petioles. Leaf blades 0.8–4.0 cm long, wider than long to about as long as wide, kidney-shaped to nearly circular in outline, shallowly to deeply 5–9-lobed, the lobes more or less obtriangular, mostly shallowly but sharply 3-lobed or toothed at the tip, the surfaces moderately pubescent with longer nonglandular and shorter glandular hairs. Inflorescences appearing axillary, mostly long-stalked, consisting of pairs of flowers. Individual flower stalks 5–15 mm long, 2–4 times as long as the sepals, pubescent with spreading, longer nonglandular and shorter glandular hairs. Sepals 3–5 mm long, not becoming enlarged at fruiting, elliptic-ovate, tapered or narrowed to an inconspicuous, minute (0.1–0.2 mm), sharp point at the tip, pubescent with longer, more or less spreading, nonglandular and sparse shorter glandular hairs. Petals 4–7 mm long, obtriangular, notched at the tip, bright pinkish purple. Stamens 10. Staminodes absent. Mericarps 8–14 mm long at maturity, the seed-containing basal portion 1.8–2.1 mm long, the lateral surfaces obliquely cross-wrinkled, glabrous (except for a few hairs at the very base), lacking a dorsal ridge or wing, the stylar beak with spreading to loosely ascending nonglandular hairs, the slender extension between the columnar portion and the stigmas 2–5 mm long. Seeds 1.4–1.8 mm long, the surface appearing smooth or with a faint fine network of ridges and pits. $2n$=26. April–August.

Introduced, uncommon in the Ozark and Ozark Border Divisions (native of Europe, Asia, widely introduced in the U.S. and Canada but apparently still absent from the Great Plains; also Hawaii, South America, Africa, Australia, New Zealand). Roadsides, lawns, pastures, and open disturbed areas.

Steyermark (1963) knew this species only from a single site in St. Louis County, where he collected it in 1952, growing in a lawn along with *G. dissectum*.

6. Geranium pusillum L. (small crane's bill)
Pl. 422 n; Map 1888

Plants annual or biennial, usually taprooted. Aerial stems 10–50 cm long, spreading to ascending, moderately pubescent with short (0.1–0.3 mm) spreading glandular and nonglandular hairs. Leaves basal and opposite, the uppermost leaves often alternate, the basal ones long-petiolate, those of the stems with progressively shorter petioles. Leaf blades 1.5–4.0(–6.0) cm long, wider than long to about as long as wide, kidney-shaped to nearly circular in outline, shallowly to deeply 5–9-lobed, the lobes more or less obtriangular, mostly shallowly but sharply 3-lobed or toothed at the tip, the surfaces moderately pubescent with short nonglandular hairs, those of the upper surface sometimes becoming sparse with age. Inflorescences appearing axillary, short- to long-stalked, consisting of pairs of flowers. Individual flower stalks 5–15 mm long, 2–4 times as long as the sepals, pubescent with spreading, short, glandular and nonglandular hairs. Sepals 3.0–4.5 mm long, not becoming enlarged at fruiting, elliptic-ovate to ovate, tapered or narrowed to an inconspicuous, minute (to 0.1 mm), sharp point at the tip, pubescent with longer, more or less spreading, nonglandular and sparse shorter glandular hairs. Petals 2.5–5.0 mm long, obtriangular, notched at the tip, bright pinkish purple. Stamens 5(–7). Staminodes (3–)5. Mericarps 9–12 mm long at maturity, the seed-containing basal portion 1.7–1.9 mm long, the lateral surfaces smooth, moderately to densely and minutely hairy, with a narrow riblike dorsal wing (or this extending all the way around the mericarp), the stylar beak with minute ascending nonglandular hairs, lacking an extension, narrowing directly from the columnar portion to the stigmas. Seeds 1.6–1.8 mm long, the surface appearing smooth or with a faint fine network of ridges and pits. $2n=26$. April–August.

Introduced, scattered to uncommon, mostly in the Ozark and Ozark Border Divisions (native of Europe, Asia, widely introduced in the U.S. and Canada south to South Carolina, Arkansas, and California; also Hawaii, South America, Africa, New Zealand). Bottomland forests, openings of mesic upland forests, banks of streams and rivers, and glades; also old fields, lawns, pastures, roadsides, railroads, and moist open disturbed areas.

GROSSULARIACEAE (Currant Family)

Contributed by David Bogler

About 25 genera, 300–330 species, nearly worldwide.

Ribes traditionally was included in a broadly circumscribed Saxifragaceae (Steyermark, 1963). More recently, phylogenetic studies have indicated the woody members of the Saxifragaceae should be moved to several other families (Morgan and Soltis, 1993; Soltis and Soltis, 1997), with *Ribes* placed in the Grossulariaceae. The genus *Itea* has sometimes been included in the Grossulariaceae (Cronquist, 1991), but is treated in its own family, Iteaceae, in the present work (see that treatment for discussion).

1. **Ribes** L. (gooseberry, currant)

Plants shrubs. Stems erect to spreading, often armed with stiff internodal bristles (slender prickles) and slender straight to slightly curved nodal spines. Leaves alternate, often appearing fascicled at the tips of short shoots, the petiole often with a somewhat expanded base. Stipules absent. Leaf blades simple, 3- or 5-lobed, palmately veined, glabrous to densely pubescent. Inflorescences mostly axillary, small clusters (these sometimes appearing umbellate) or racemes, the short to long flower stalks with a minute glandular or herbaceous bract toward the midpoint. Flowers usually perfect, actinomorphic, epigynous, sometimes fragrant. Hypanthium bell-shaped to cylindrical. Sepals (4)5, shorter or longer than the hypanthium tube. Petals 5, shorter than the sepals and often shorter than the stamens. Stamens 5, short, alternating with the petals, attached to the hypanthium, the anthers attached toward their midpoint. Pistil 1 per flower, of 2 fused carpels. Ovary inferior, not grooved, glabrous to bristly, with 1 locule, with few to many ovules, the placentation parietal. Style 1, sometimes 2-lobed, elongating during flowering, the stigma(s) capitate. Fruits berries, globose, smooth or with bristly prickles, the veins of the outer layer sometimes appearing as stripes. Seeds few to numerous, small, angular, with a gelatinous aril and hard seed coat. About 150 species, North America, South America, Europe, Asia, Africa.

Ribes is sometimes divided into two genera, *Ribes* L. (the currants, with jointed flower stalks) and *Grossularia* Mill. (the gooseberries, with unjointed flower stalks), but this division has been shown to be oversimplified and unnatural (Spongberg, 1972; Sinnot, 1985). The fruits are eaten by a wide variety of animals. Seed dispersal is probably mostly by birds. The berries gathered from wild plants of some species were an important food source for Native Americans and early settlers. Numerous cultivated strains of currants and gooseberries have been developed by plant breeders. However, in the past many wild species of *Ribes* were the subject of eradication programs, particularly in the northeastern United States. This was in an attempt to control the spread of pine blister rust (*Cronartium ribicola* J.R. Fischer), an important disease of the commerically important white pine (*Pinus strobus* L.) and related 5-needle pines that utilizes species of *Ribes* as a host for part of its life cycle.

1. Twigs unarmed, lacking bristles or spines; inflorescences racemes with 3–15 flowers; flower stalks jointed toward the tip
   2. Leaves dotted with resinous glands on the undersurface; perianth greenish white; hypanthium narrowly bell-shaped, the portion above the ovary as long as or slightly shorter than the sepals . . . . 1. R. AMERICANUM
   2. Leaves not gland-dotted; perianth yellow, sometimes reddish-tinged; hypanthium narrowly cylindrical, the portion above the ovary much longer than the sepals . . . . 4. R. ODORATUM
1. Twigs armed with stiff bristles and large nodal spines; inflorescences clusters of 2–4 flowers (flowers occasionally solitary); flower stalks not jointed
   3. Petioles with many capitate-glandular hairs; hypanthium (above the ovary) as long as or slightly longer than the sepals; stamens not or only slightly exserted at flowering; ovary with gland-tipped hairs that become slender stiff prickles on the mature fruit . . . . 2. R. CYNOSBATI
   3. Petioles rarely with capitate-glandular hairs; hypanthium (above the ovary) much shorter than the sepals; stamens long exserted at flowering; ovary smooth, the fruit without prickles . . . . 3. R. MISSOURIENSE

1. Ribes americanum Mill. (American black currant, eastern black currant)

Pl. 423 e, f; Map 1889

Stems 1–2 m long, erect, ascending to spreading, unarmed. Bark relatively smooth, not peeling, reddish brown. Petioles 2–5 cm long, with minute, golden to brownish yellow, resinous glandular dots and minute hairs, but lacking capitate-glandular hairs. Leaf blades 1–8 cm long, broadly ovate to nearly circular in outline, the lobes sharply pointed at the tip, truncate to shallowly cordate at base, the margins sharply and doubly toothed, the upper surface sparsely hairy, the undersurface with minute, golden to brownish yellow, resinous, glandular dots and moderately pubescent with fine velvety hairs. Inflorescences drooping racemes with 6–15 flowers. Flower stalks 1–3 mm long, jointed toward the tip. Hypanthium narrowly bell-shaped, the more or less tubular portion above the ovary 3.5–4.5 mm long, minutely and sometimes sparsely hairy. Sepals 4.5–5.0 mm long, oblong-ovate, greenish white to yellowish brown, the outer surface minutely and sometimes sparsely hairy. Petals 2–3 mm long, creamy white. Stamens not exserted. Ovary glabrous. Fruits 6–10 mm long, ovoid, glabrous, without prickles, black. $2n=16$. April–June.

Uncommon, known thus far from a single site in Schuyler County and historical collections from St. Louis City and County (northern U.S. south to Virginia, Missouri, and New Mexico; Canada). Fens

Ribes americanum was rediscovered in Missouri in 1991 (although the existence of historical collections was not known at that time) and reported by Ladd et al. (1991) from a population in a deep muck fen along the Chariton River near the Iowa border. The habitats in which this species once occurred in the St. Louis area cannot be determined from the scanty label data on the specimens. The berries of *R. americanum* are not very palatable to humans.

2. Ribes cynosbati L. (prickly gooseberry, dogberry)

Grossularia cynosbati (L.) Mill.

Pl. 423 a–d; Map 1890

Stems 0.5–1.5 m long, erect to spreading, often armed with internodal bristles 2–6 mm long and nodal spines 5–8 mm long, these reddish

1891. Ribes missouriense 1892. Ribes odoratum 1893. Myriophyllum aquaticum

purple to reddish brown. Bark thin and papery, sometimes peeling somewhat with age, tan to brown. Petioles 1.0–3.5 cm long, with sparse nonglandular hairs and denser capitate-glandular hairs, especially on young leaves. Leaf blades 1.0–6.5 cm long, broadly ovate to nearly circular in outline, the lobes bluntly to sharply pointed at the tip, truncate to shallowly cordate at the base, the margins bluntly to sharply and doubly toothed, the upper leaf surfaces glabrous to finely hairy, the undersurface lacking glands, moderately pubescent with fine velvety hairs. Inflorescences small umbellate clusters of 2–4 flowers or occasionally the flowers solitary. Flower stalks 5–12 mm long, not jointed. Hypanthium cylindrical, the tubular portion above the ovary 3.0–4.5 mm long, greenish white, glabrous to minutely and sparsely nonglandular-hairy. Sepals 2.2–4.0 mm long, broadly oblong, green to greenish white, the outer surface glabrous to minutely and sparsely nonglandular-hairy. Petals 1–2 mm long, obovate, white, glabrous. Stamens not or only slightly exserted. Ovary with gland-tipped hairs that gradually become spinescent after flowering. Fruits 6–12 mm long, globose to somewhat ovoid, covered with long slender spines, greenish to pale red. $2n=16$. April–June.

Uncommon to scattered, mostly in the eastern half of the state (eastern U.S. west to North Dakota and Oklahoma; Canada). Ledges and crevices of shaded north-facing bluffs; usually on calcareous substrates.

A widely distributed species, *R. cynosbati* is somewhat variable in degree of pubescence and spines. Most of the Missouri material has dense bristles (slender prickles) along the stems.

3. Ribes missouriense Nutt. (Missouri gooseberry)

R. missouriense var. *ozarkanum* Fassett

R. gracile Michx.

Grossularia missouriense (Nutt.) Coville & Britton

Pl. 423 j–l; Map 1891

Stems 1–2 m long, erect and arching to sprawling, rarely armed with internodal bristles, more commonly with nodal spines 7–18 mm long, these reddish brown to brown. Bark smooth or somewhat scaly, peeling or shredding in narrow papery strips, dark gray to reddish brown. Petioles 0.6–2.0 cm long, often with dense nonglandular hairs but rarely with capitate-glandular hairs. Leaf blades 0.5–3.0 cm long, broadly obovate to circular or broadly ovate in outline, broadly narrowed to rounded or sometimes subcordate at the base, the lobes mostly rounded to bluntly pointed at the tip, the margins bluntly to sharply, sometimes irregularly, but mostly simply toothed, the upper surface sparsely to moderately minutely nonglandular-hairy, becoming glabrous or nearly so and somewhat shiny with age, the undersurface sparsely and minutely nonglandular-hairy, especially along the veins, becoming glabrous or nearly so with age. Inflorescences small umbellate clusters of 2–4 pendant flowers or occasionally the flowers solitary. Flower stalks 4–13 mm long, not jointed. Hypanthium shortly cylindric, the tubular portion above the ovary 1.5–2.5 mm long, greenish white to cream-colored, sometimes purplish-tinged, glabrous or sometimes with sparse minute nonglandular hairs. Sepals 5–8 mm long, oblong-linear, white to cream-colored. Petals 2.0–3.5 mm long, narrowly oblong-obovate, white, sometimes pinkish-tinged with age. Stamens long-exserted. Ovary glabrous. Berries 6–12 mm long, globose, glabrous, without prickles, green and translucent when young, becoming red to purple at maturity. $2n=16$. April–May.

Scattered nearly throughout the state (northeastern U.S. west to North Dakota and Oklahoma; Canada). Mesic to dry upland forests, banks of streams and rivers, and less commonly bottomland forests; also roadsides.

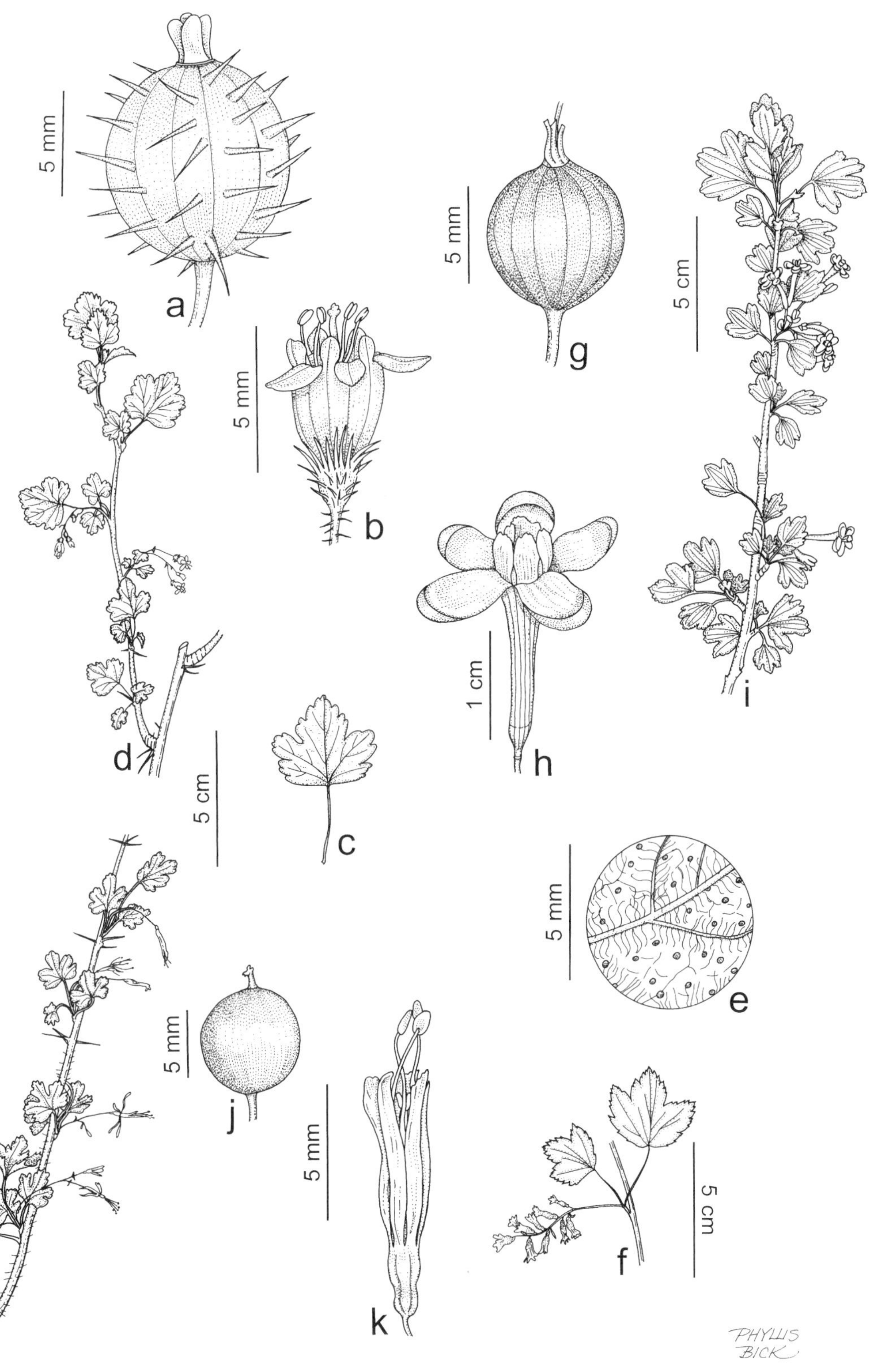

Plate 423. Grossulariaceae. *Ribes cynosbati*, **a)** fruit, **b)** flower, **c)** leaf, **d)** fertile branch. *Ribes americanum*, **e)** detail of leaf undersurface showing veins, hairs, and glands, **f)** node with leaves and inflorescence. *Ribes odoratum*, **g)** fruit, **h)** flower, **i)** fertile branch. *Ribes missouriense*, **j)** fruit, **k)** flower, **l)** fertile branch.

Ribes missouriense is the most common species of gooseberry in Missouri, and dense populations are an indication that a site has a history of grazing by cattle. The berries of this species can be eaten raw or used in cooking, however the taste is somewhat insipid. The var. *ozarkanum* refers to less pubescent plants that appear unworthy of formal taxonomic recognition.

4. Ribes odoratum H.L. Wendl. (golden currant, Missouri currant, buffalo currant)
R. aureum Pursh var. *villosum* DC.
R. fragrans Pall.
Chrysobotrya odorata (H.L. Wendl.) Rydb.
Pl. 423 g–i; Map 1892

Stems 1–2 m long, erect to arching, unarmed. Bark smooth to somewhat scaly, often peeling or shredding in narrow papery strips, gray to reddish brown. Petioles 1.5–5.0 cm long, glabrous or with sparse minute nonglandular hairs and/or longer capitate-glandular hairs. Leaf blades 1–5 cm long, broadly triangular-ovate to nearly circular in outline, the lobes mostly sharply pointed at the tip, broadly narrowed to truncate at the base, the margins entire to coarsely and mostly simply toothed, the upper surface glabrous, the undersurface glabrous or occasionally sparsely and minutely nonglandular-hairy. Inflorescences racemes with 3–10 flowers. Flower stalks 3–8 mm long, jointed toward the tip. Hypanthium narrowly cylindrical, the tubular portion above the ovary 12–16 mm long, bright yellow. Sepals 3–6 mm long, oblong-ovate, bright yellow, glabrous. Petals 2.0–3.5 mm long, yellow to bright red. Ovary glabrous. Fruits 7–9 mm long, globose, glabrous, without prickles, yellow to greenish yellow, becoming black at maturity. $2n=16$. April–May.

Uncommon in the western portion of the Ozark Division; introduced in Barton and Jefferson Counties and the city of St. Louis (Vermont to Montana south to Tennessee, Texas, and New Mexico; Canada). Ledges and tops of bluffs; also roadsides and railroads.

The species epithet refers to the fact that the flowers have a strong fragrance, described by Steyermark (1963) as resembling a combination of the spiciness of cloves and the sweet fragrance of carnations, especially in the afternoon. Golden currant is sometimes cultivated as a garden ornamental.

HALORAGACEAE (Water-milfoil Family)

Contributed by Alan E. Brant and George Yatskievych

Plants perennial herbs (woody elsewhere), sometimes monoecious or dioecious, often with rhizomes (these seldom collected in submerged aquatics), glabrous (in some species, young growth sometimes sparsely short-hairy). Stems unbranched or branched, sometimes rooting at the nodes. Leaves alternate, opposite, and/or whorled, sessile or short-petiolate. Stipules absent. Leaf blades simple and toothed to pinnately dissected (often both extremes present on the same plant). Flowers in terminal spikes or axillary and solitary or in small clusters, sessile or nearly so, perfect or imperfect (the lowermost flowers then usually pistillate, the others staminate), the pistillate and perfect ones epigynous, actinomorphic, each subtended by a pair of minute bracts. Calyces of 3 or 4 sepals, these free, sometimes shed as the flower opens, small, triangular. Corollas absent or of 4 petals, these free, shed early, membranous, small. Stamens 3–8, free, the filament short, the anthers attached at the base. Pistils 1 per flower, of 3 or 4 fused carpels. Ovary inferior, with 3 or 4 locules, each with 1 ovule, the placentation apical. Styles absent or short, the stigmas 4, appearing feathery and curved outward. Fruits nutlike, indehiscent or eventually splitting from the tip into 4 nutlets. Seeds 3 or 4, not easily separable from the fruit wall. Nine genera, 145 species, nearly worldwide, most diverse in the southern hemisphere.

Species of Haloragaceae are submerged or emergent aquatics, sometimes on muddy margins of wetlands. The submerged taxa provide valuable habitat for small fish and aquatic invertebrates and also generate oxygen for the waters in which they grow. However, the more aggressive species can choke out other aquatic plants and can contribute to eutrophication of ponds and lakes.

1. Leaves mostly whorled, occasionally on a given plant some leaves alternate or opposite; flowers mostly imperfect (a few occasionally perfect), most commonly in emergent spikes or in the axils of leaves on emergent branches, with 4 sepals, these often shed early; plants entirely or mostly submerged aquatics, often with only the flowering portions above water, occasionally stranded in mud along shores . . . . 1. MYRIOPHYLLUM
1. Leaves all alternate; flowers perfect, in the axils of emergent leaves, with 3 sepals, these persistent; plants of shallow water or wet shores, rarely if ever entirely submerged . . . . 2. PROSERPINACA

1. **Myriophyllum** L. (water-milfoil) (Aiken, 1981)

Plants mostly monoecious (dioecious in *M. aquaticum*), all or mostly submerged aquatics (occasionally stranded on muddy shores), often only the flowering portion emergent. Stem length dependent on depth of water, variously 5 cm to more than 3 m. Leaves whorled, or sometimes a few of them on a given plant opposite or alternate, the submerged ones pinnately dissected into threadlike segments; grading abruptly into reduced, emergent, bracteal leaves subtending the flowers (except in *M. aquaticum*, with flowers in the axils of unreduced leaves). Inflorescences emergent spikes or appearing as axillary clusters. Flowers in a given inflorescence often of 3 types, the lowermost pistillate, a few median ones perfect, and the upper ones staminate. Sepals 4, often shed early, minute, triangular. Petals 4 (lacking in pistillate flowers), inconspicuous but longer than the sepals, membranous and shed early, translucent, white, or pinkish-, reddish-, or purplish-tinged. Stamens 4 or 8. Ovary with 4 locules. Stigmas 4. Fruits 4-lobed, variously sculptured, eventually splitting into four nutlets.

Certain species of water-milfoil have been used in native American medicine, alone and in combination with other plants as a blood medicine (to improve circulation), and as an emetic. Also, the rhizomes, eaten raw, fried, or roasted are said to be sweet and crunchy and were a much relished food (Moerman, 1998).

Apparent instances of hybridization between species of *Myriophyllum*, notably *M. heterophyllum* × *M. pinnatum* and *M. sibiricum* × *M. spicatum*, have been detected from portions of New England and the upper Midwest (Moody and Les, 2002). These morphologically intermediate plants, if found in Missouri, might confound species determinations using the key below.

1. Emergent leaves similar to the submerged leaves, both types mostly 2–3 cm long, the overall shape linear-oblong (feather-like), with uniform segments, each 3–6 mm long . . . . 1. M. AQUATICUM
1. Emergent leaves shorter than the submerged ones, all less than 2 cm long, entire or more commonly toothed (the teeth less than 3 mm long); submerged leaves with the median divisions more than 6 mm long, the overall leaf shape narrowly to broadly ovate
    2. At least some of the leaves of the stem alternate or appearing subopposite; plants often terrestrial on mud or wet gravel; anthers 0.5–1.2 mm long . . . . 3. M. PINNATUM
    2. Leaves all whorled; plants commonly submerged aquatics, rarely stranded on mud; anthers 1.2–1.8 mm long
        3. Median bracts of the inflorescence toothed, 2–3(–6) times as long as the adjacent pistillate flowers; median internode length 3–8(–15) mm . . . . 2. M. HETEROPHYLLUM

3. Median bracts of the inflorescence entire, equal to or shorter than the adjacent pistillate flowers; median internode length 8–16 mm
    4. Median submerged leaves with (6–)10–22 segments; stem not thickened below the inflorescence, similar in width throughout .... 4. M. SIBIRICUM
    4. Median submerged leaves with 24–50 segments; stem slightly to strongly thickened below the inflorescence, up to twice the diameter of the lower portion of the stem .... 5. M. SPICATUM

1. Myriophyllum aquaticum (Vell.) Verdc.
(parrot's feather, water feather)
M. brasiliense Cambess.
M. proserpinacoides Gillies ex Hook. & Arn.
Pl. 424 a, b; Map 1893

Plants dioecious. Stems branched, relatively stout but uniform in diameter throughout, often with the apical vegetative portions extending several cm above the surface of the water, the median internodes 18–30 mm long, green to grayish tan or occasionally reddish-tinged toward the tip. Leaves all in whorls of 3–6, grayish green. Emergent leaves similar (occasionally slightly smaller) to the submerged leaves, both types whorled, mostly 2–3 cm long, the petiole 4–7 mm long, the bracteal leaves much longer than the associated flowers. Leaf blades all pinnately divided, linear-oblong (feather-like) in outline, the more or less uniform segments 20–36, each 3–6 mm long. Inflorescences of solitary flowers or small clusters in the axils of unreduced emergent leaves, not appearing spikelike. Flowers all pistillate (staminate flowers unknown in North America), appearing white, the bractlets dissected into 2 or 3 threadlike segments. Corollas absent. Fruits not formed. April–September.

Introduced, widely scattered in the southern half of the state (native of S. America, introduced widely in North America and Central America). Ponds, streams, and spring branches; also ditches; usually emergent aquatics.

This species is grown horticulturally for use in lily pools and fish ponds, and to a lesser extent as an aquarium plant. E.N. Nelson and Couch (1986) studied the historical establishment of plants in nature in the United States, noting that the oldest herbarium specimens of wild-collected material date back to the 1890s from New Jersey and surrounding states. However, they also noted an early collection from the Joplin area (Jasper County) dating to 1897. This specimen, housed at the Missouri Botanical Garden Herbarium, was collected by William Trelease along a local stream. By 1940, the species had become established in about 20 states and has continued to spread since then. However, in most states, parrot's feather remains sporadic in its distribution, reflecting the persistence of local escapes rather than active spread from original sites of escape. The fact that North American plants have only been recorded as pistillate clones that do not produce fruits probably has helped to inhibit the invasiveness of the species.

2. Myriophyllum heterophyllum Michx.
(twoleaf water-milfoil, coontail)
Pl. 424 c–e; Map 1894

Plants mostly monoecious (with occasional perfect flowers). Stems branched, relatively stout but uniform in diameter throughout, usually with only the inflorescence extending above the surface of the water, the median internodes 3–8(–15) mm long, green or tan toward the tip. Leaves all in whorls of 4–6, sessile or nearly so, green to dark green, seasonally often reddish- or brownish-tinged. Emergent leaves smaller than the submerged leaves, 3–15 mm long, the blade linear to narrowly oblong or narrowly lanceolate, the margins toothed, sometimes only toward the tip, the bracteal leaves 2–6 times as long as the associated flowers. Submerged leaves 6–25(–30) mm long, the blade pinnately divided, narrowly ovate to ovate in outline, the segments 12–20, variously 1–12 mm long (the median divisions more than 6 mm long). Inflorescences appearing as spikes. Flowers appearing green to tan, the bractlets triangular, coarsely toothed. Petals of staminate and perfect flowers translucent. Stamens 4, the anthers 1.5–1.7 mm long. Fruits 1.0–1.5 mm long, subglobose, the surfaces of the nutlets roughened and often with a few low tubercles. May–September.

Scattered in the southern half of the state, mostly in the Ozark and Ozark Border Divisions, disjunct in Scotland County (eastern U.S. west to North Dakota and Texas; Canada). Ponds, lakes, streams, sloughs and spring branches; also ditches; submerged aquatics, occasionally terrestrial on mud.

Myriophyllum heterophyllum is a common member of the floras of spring branches and other cool, running waters in the Ozarks. According to

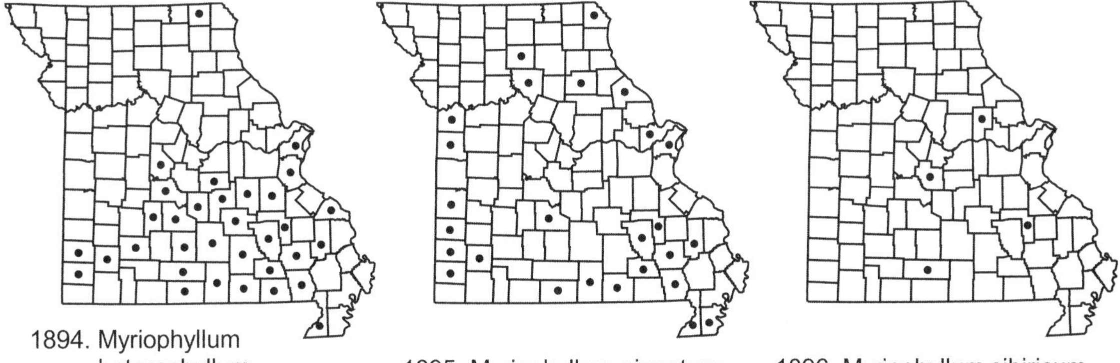
1894. Myriophyllum heterophyllum

1895. Myriophyllum pinnatum

1896. Myriophyllum sibiricum

Crow and Helquist (2000), terrestrial plants of this species can easily be confused with terrestrial plants of *M. spicatum*. However, *M. spicatum* is very uncommon in Missouri. A greater problem in Missouri is confusion with the more commonly encountered *M. pinnatum* (Steyermark, 1963), which is distinguished when flowering by having the uppermost leaves alternate or subopposite and the anthers 0.5–1.2 mm long.

3. Myriophyllum pinnatum (Walter) Britton, Sterns, & Poggenb. (rough water-milfoil, green parrot's feather)

Pl. 424 f–i; Map 1895

Plants mostly monoecious (with occasional perfect flowers). Stems branched or unbranched, variously slender or stout, uniform in diameter throughout or tapered toward the tip, sometimes with only the inflorescence extending above the surface of the water, the median internodes 2–15 mm long, usually green toward the tip. Leaves at least partially alternate to subopposite, the submerged leaves mostly in whorls of 3 or 4, sessile or the petiole to 3 mm long, green to dark green, occasionally reddish-tinged. Emergent leaves smaller than the submerged leaves, 5–18 mm long, the blade linear to narrowly oblong or lanceolate, the margins shallowly to deeply toothed, sometimes with slender lobes, the bracteal leaves 5–9 times as long as the associated flowers. Submerged leaves sometimes not produced, 10–20(–25) mm long, the blade pinnately divided, narrowly to broadly ovate in outline, the segments 10–20, variously 1–12 mm long (the median divisions more than 6 mm long). Inflorescences appearing as spikes or (in emergent forms) as axillary clusters. Flowers appearing white to pale green or pinkish- to purplish-tinged, the bractlets triangular, coarsely toothed. Petals of staminate and perfect flowers translucent. Stamens 4, the anthers 0.5–1.2 mm long. Fruits 1.0–1.5 mm long, subglobose, the surfaces of the nutlets with tuberculate ridges. May–October.

Scattered widely in the state but apparently absent from most of the western half of the Glaciated Plains Division and the northern portion of the Ozarks (eastern U.S. west to North Dakota and New Mexico). Ponds, lakes, sinkhole ponds, oxbows, sloughs, and margins of streams, spring branches, and rivers; also ditches; submerged or more commonly emergent aquatics, often terrestrial on mud or wet gravel.

As noted by Steyermark, this species frequently occurs as prostrate terrestrial plants on mud and is more likely to grow in warm standing water than in the cool waters of springfed streams. For a discussion of distinctions between *M. pinnatum* and *M. heterophyllum*, see the treatment of the latter.

4. Myriophyllum sibiricum Kom. (common water-milfoil)

M. exalbescens Fernald

M. spicatum var. *exalbescens* (Fernald) Jeps.

Map 1896

Plants mostly monoecious (with occasional perfect flowers). Stems branched, variously slender or stout, uniform in diameter throughout, with only the inflorescence extending above the surface of the water, the median internodes 8–30 mm long, white to tan or rarely reddish-tinged toward the tip. Leaves all in whorls of (3–)4–6, sessile or nearly so, grayish green, seasonally sometimes brownish-tinged. Emergent leaves (bracts) much smaller than the submerged leaves, 1.5–2.0 mm long, the blade elliptic to narrowly obovate, the margins entire or the lower ones inconspicuously toothed, the bracteal leaves equal to or shorter than the associated flowers. Submerged leaves 10–50 mm long, the blade pinnately divided, narrowly to broadly ovate in outline, the segments (6–)10–22,

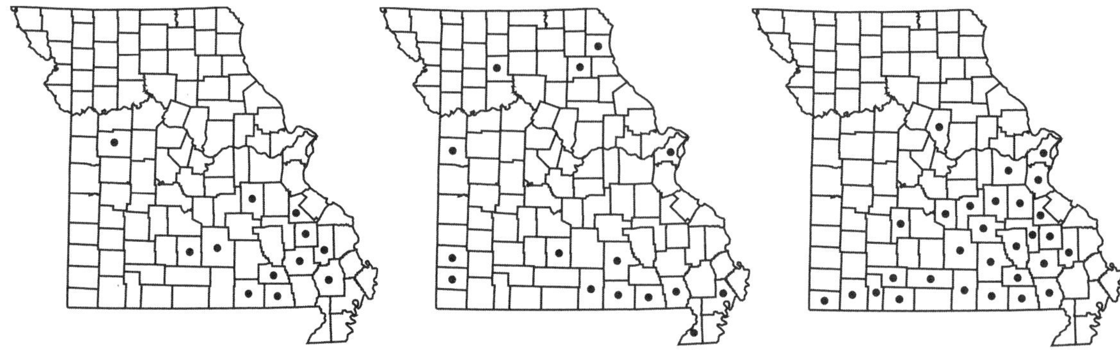

1897. Myriophyllum spicatum 1898. Proserpinaca palustris 1899. Hamamelis vernalis

variously 1–12 mm long (the median divisions more than 6 mm long). Inflorescences appearing as spikes. Flowers appearing white or pinkish- to reddish-tinged, the bractlets ovate to triangular, entire toothed. Petals of staminate and perfect flowers not translucent. Stamens 8, the anthers 1.2–1.8 mm long. Fruits 2–3 mm long, ovoid to subglobose, the surfaces of the nutlets smooth or roughened with minute tuberculate ridges. $2n=42$. June–September.

Uncommon, widely scattered in the southern half of the state (northern U.S. [including Alaska] south to Virginia, Oklahoma, and California; Greenland, Europe). Ponds and lakes.

Paradoxically, *M. sibiricum* is less common in Missouri than the noxious *M. spicatum*. See the treatment of that species for further discussion on distinctions between the two taxa. The status of *M. sibiricum* in Missouri is open to interpretation. It was first collected in the state by Grant Pyrah in 1979 from a lake in Douglas County and thus may be considered a recent introduction in Missouri. However, as a taxon native in adjacent states that presumably expanded its range into Missouri by natural vectors (waterfowl), it probably is more reasonable to consider it a native addition to the flora.

5. Myriophyllum spicatum L. (Eurasian milfoil, Eurasian water-milfoil)

Pl. 424 m, n; Map 1897

Plants mostly monoecious (with occasional perfect flowers). Stems branched, variously slender or stout, slightly to strongly thickened below the inflorescence (up to twice the diameter of the lower portion of the stem), with only the inflorescence extending above the surface of the water, the median internodes 10–30 mm long, white to tan or more commonly reddish-tinged toward the tip. Leaves all in whorls of (3)4–6, sessile or nearly so, grayish green, seasonally sometimes brownish-tinged. Emergent leaves (bracts) much smaller than the submerged leaves, 1.5–2.0 mm long, the blade elliptic to narrowly obovate, the margins entire or the lower ones inconspicuously toothed, the bracteal leaves equal to or shorter than the associated flowers. Submerged leaves 10–50 mm long, the blade pinnately divided, narrowly to broadly ovate in outline, the segments 24–50, variously 3–12 mm long (the median divisions more than 6 mm long). Inflorescences appearing as spikes. Flowers appearing white or pinkish- to reddish-tinged, the bractlets ovate to triangular, entire or toothed. Petals of staminate and perfect flowers not translucent. Stamens 8, the anthers 1.2–1.8 mm long. Fruits 2–3 mm long, ovoid to subglobose, the surfaces of the nutlets smooth or roughened with minute tuberculate ridges. $2n=42$. June–September.

Introduced, scattered in the southern half of the state (native of Europe, Asia, Africa; introduced in North America). Ponds, lakes, and less commonly streams, spring branches, and rivers.

This species closely resembles *M. sibiricum*, to which it is closely related. In addition to the key characters, it is also differs from that species in its production of specialized, small, dark green clusters of short leaves (turions) in the leaf axils and at the tips of sumerged branches toward the end of the growing season. These turions become detached and overwinter underwater. Molecular studies (Moody and Les, 2010) have confirmed the phylogenetic distinctions between the two species.

C. F. Reed (1977) and Couch and Nelson (1986) discussed the historical spread of this noxious weed in temperate North America following its introduction in both the eastern and western United States in the late 1930s or early 1940s. More recently, Padgett (2001) summarized information on the introduction and status of this species in Missouri, where it can become a problem plant, pri-

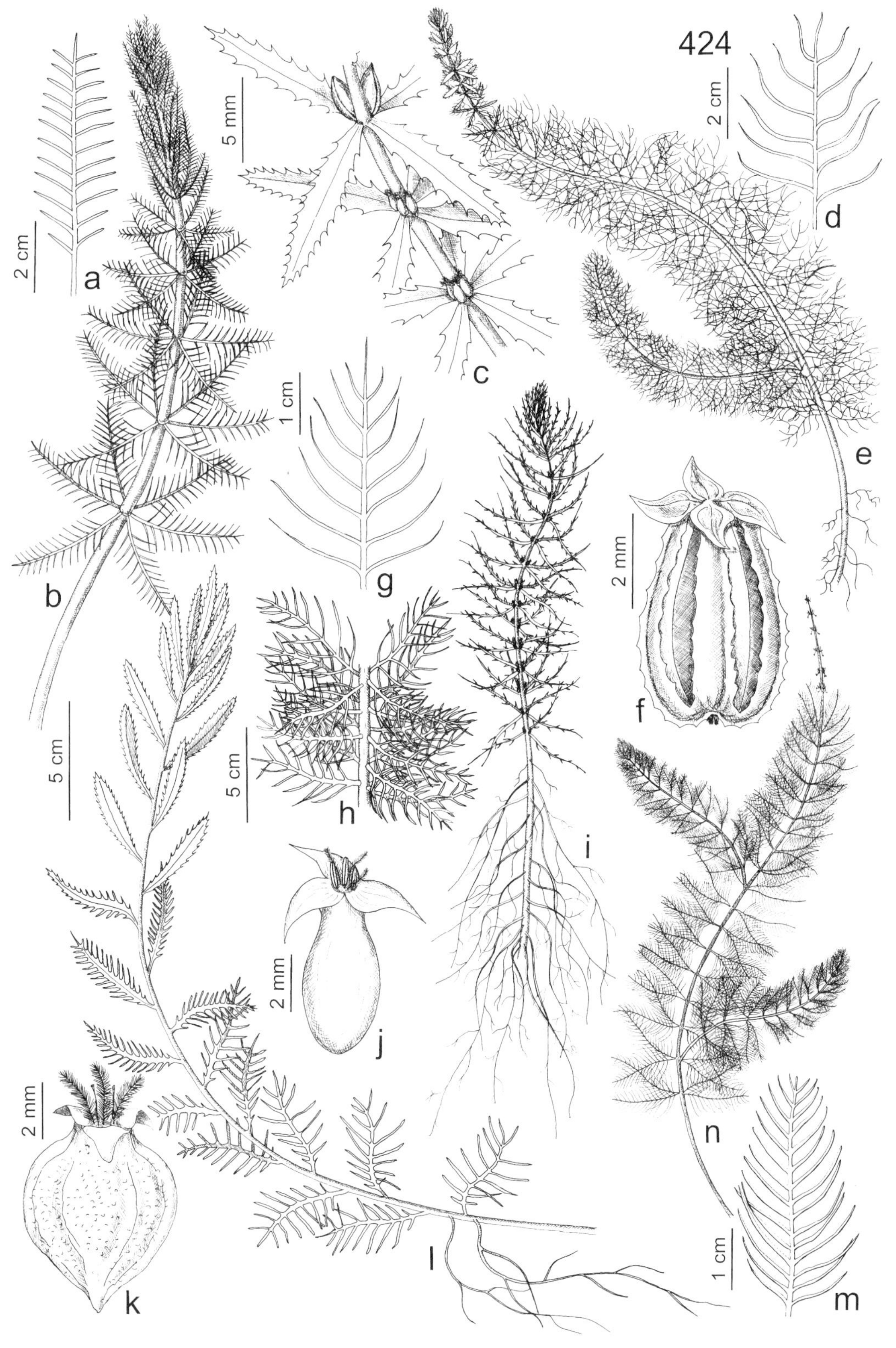

Plate 424. Haloragaceae. *Myriophyllum aquaticum*, **a)** leaf, **b)** habit. *Myriophyllum heterophyllum*, **c)** portion of inflorescence, **d)** leaf, **e)** habit. *Myriophyllum pinnatum*, **f)** fruit, **g)** leaf, **h)** portion of submerged phase with dissected leaves, **i)** terrestrial phase, habit. *Proserpinaca palustris* var. *palustris*, **j)** flower, **k)** fruit, **l)** habit. *Myriophyllum spicatum*, **m)** leaf, **n)** habit.

marily in ponds and lakes. The species was first collected in Missouri by Paul Redfearn in 1962 from a spring in Camden County. It probably is more widespread than the specimen record indicates, as few botanists collect submerged aquatics. Thus far, Missouri apparently does not have as severe an infestation of Eurasian milfoil as do some other states.

2. Proserpinaca L.

Two species (and their putative hybrid), North America.

1. Proserpinaca palustris L. (mermaid weed)
Pl. 424 j–l; Map 1898

Plants usually emergent aquatics, often long-rhizomatous and colonial. Aerial stems 10–40 cm long, erect or ascending, sometimes from a spreading base. Leaves 1–9 cm long, sessile or short-petiolate, often of 2 types, the blades of submerged leaves (these sometimes absent) deeply pinnately dissected into 16–28, threadlike or linear segments; grading fairly abruptly into those of the emergent leaves, which are simple, pinnately veined, the margins coarsely to finely and sharply toothed. Inforescences of solitary flowers or small clusters in the axils of unreduced emergent leaves. Flowers perfect. Calyces of 3 sepals, these persistent at fruiting. Petals absent. Stamens 3. Ovary 3-locular, the stigmas 3. Fruits 2–5 mm long, ovoid to ovoid-pyramidal, rhomboidal, or less commonly nearly spherical, 3-angled or narrowly 3-winged, not separating into nutlets. $2n=14$. May–October.

Scattered widely in the state, most abundantly in the eastern portion of the Ozark Division and the Mississippi Lowlands (eastern U.S. west to Wisconsin and Texas; Canada). Swamps, bottomland forests, marshes, sloughs, oxbows, and margins of ponds, lakes, and sinkhole ponds; rarely also margins of spring branches; also ditches and railroads.

Steyermark (1963) noted that ducks and other waterfowl relish the fruits. Although infraspecific taxa have been questioned or ignored by some authors in the recent past, Catling (1998) maintained three varieties based on examination of a large suite of specimens. However, it should be noted that vegetative plants cannot be separated into varieties, thus collectors must be careful to collect fertile specimens.

1. Faces of the fruit convex, bluntly angled (the angles sometimes nearly absent) . . . . 1A. VAR. AMBLYOGONA
1. Faces of the fruit flat or concave, sharply angled or with narrow wings
   2. Fruits 2.0–3.5 mm wide, as broad as long or at most 1.5 times as broad as long, tapered or rounded at the summit but lacking a distinct neck, with more or less flat faces and sharp angled but not winged . . . . 1B. VAR. CREBRA
   2. Fruits 3.0–4.5 mm wide, 1.3–2.0 times as broad as long, narrowed to a short neck at the tip, with concave faces and narrowly winged angles . . . . 1C. VAR. PALUSTRIS

1a. var. amblyogona Fernald

Fruits 2.5–3.5 mm wide, as broad as long or at most 1.5 times as broad as long, ovoid to nearly spherical, rounded or slightly tapered at the tip and lacking a distinct neck, bluntly angled (the angles sometimes nearly absent), the faces convex, usually appearing wrinkled or roughened. June–October.

Scattered in the eastern portion of the Ozark Division and the Mississippi Lowlands, with single disjunct occurences in Newton and Shelby Counties (Georgia to Texas, north to Ohio, Michigan, and Oklahoma; Canada). Swamps, marshes, oxbows, and margins of ponds, lakes, and sinkhole ponds; rarely also margins of spring branches.

1b. var. crebra Fernald & Griscom

Fruits 2.0–3.5 mm wide, as broad as long or at most 1.5 times as broad as long, ovoid to ovoid pyramidal, tapered or rounded at the summit but lacking a distinct neck, sharply angled but not winged, the sides more or less flat, usually appearing wrinkled or roughened. June–October.

Scattered widely in the state, most abundantly in the eastern portion of the Ozark Division and the Mississippi Lowlands (eastern U.S. west to Wisconsin and Texas; Canada). Swamps, bottomland forests, marshes, sloughs, oxbows, and margins of ponds, lakes, and sinkhole ponds; also ditches and railroads.

This is the most commonly encountered variety of *P. palustris* in Missouri.

1c. var. palustris

Pl. 424 j–l

Fruits 3.0–4.5 mm wide, 1.3–2.0 times as broad as long, ovoid to ovoid-pyramidal or rhomboidal, narrowed to a short neck at the tip, narrowly winged along the angles, the faces concave, usually appearing slightly wrinkled. June–October.

Uncommon in the Mississippi Lowlands Division and with a single, disjunct, historical collection from St. Louis County (eastern U.S. west to Minnesota and Texas; Canada). Swamps, bottomland forests, and sloughs, oxbows; also railroads.

Mississipi Lowlands and locally in St. Louis County historically (Florida to Texas, north to Missouri and Massachusetts; also West Indies and Colombia). Swamps and sloughs.

HAMAMELIDACEAE (Witch-Hazel Family)

Contributed by Alan Whittemore

Plants trees or shrubs, monoecious or with perfect flowers. Leaves alternate, simple, pinnately or palmately veined, the margins toothed or lobed. Stipules present. Inflorescences small clusters or dense headlike clusters, with individual inflorescences all pistillate or all staminate in monoecious species. Flowers actinomorphic. Calyces absent or of 4 sepals, these fused toward the base. Corollas absent or of 4 free petals. Stamens 4–8(–10), free. Staminodes sometimes present. Ovary partially inferior, 2-locular, the placentation apical or axile. Styles 2, often curled or reflexed, the stigmas small iregular areas at the style tip, sometimes extended somewhat along the side. Fruits 2-valved capsules, solitary or fused into groups of several to many. About 30 genera, about 100 species, widespread in warm-temperate and tropical areas.

1. Plants shrubs; leaves unlobed; flowers yellow, orange, or red; infructescences sessile or short-stalked . . . . 1. HAMAMELIS
1. Plants trees; leaves palmately 5- or 7-lobed; flowers green or dark purple; infructescences long-stalked . . . . 2. LIQUIDAMBAR

1. Hamamelis L. (witch-hazel)

Plants nonresinous shrubs or small trees with broad, rounded crowns and smooth or slightly roughened bark. Twigs stellate-hairy, never developing woody longitudinal wings. Buds short-stalked, naked or with 2 bud scales. Leaves pinnately veined, the margins coarsely toothed or with rounded shallow lobes. Inflorescences stalked clusters of mostly 3 flowers, in the leaf axils. Flowers perfect. Calyx 3–4 mm long, hemispherical, 4-lobed, brown, densely stellate-hairy, often persistent at fruiting. Petals 4, narrowly strap-shaped, yellow, orange, or red. Stamens 4, alternating with 4 inconspicuous scalelike staminodes, the filaments short and broad, the anther sacs separated by broad connecting tissue. Styles curved outward or reflexed, persistent on the fruit as short beaks. Ovules 1 per locule. Infructescence stalks 0.6–1.0 cm long. Capsules separate, 9–16 mm long, woody, brown, dehiscing explosively longitudinally along the midnerve of each locule, persistent long after the seeds are dispersed. Seeds 2 per fruit, 5–10 mm long, narrowly elliptic to oblong-elliptic in outline, black with a light-colored attachment scar toward the base, shiny, not winged. Four species, eastern North America and eastern Asia.

The common name witch-hazel comes from the frequent use of the twigs in water-divining (water witching). Ethnological sources in Missouri commonly mention the use of "hazel" or "witch-hazel" for this purpose. They generally do not distinguish the two species, but probably both have been used. Native Americans used *Hamamelis* as an analgesic for sore throats and other pains, as well as externally for treatment of sores and bruises (Moerman, 1998). Extracts of witch-hazel (which originate mostly from the bark and leaves of *H. virginiana*) currently are used commercially in some skin and shaving lotions, cosmetics, mouthwashes, eye ointments, soaps, and hemorrhoid creams.

The sudden splitting open of the fruits can cause the seeds to be thrown up to 10 m. The petals of witch-hazel also have an unusual adaptation in that the strap-shaped petals tend to coil inward from the tip during cold weather, unrolling again during warmer times. Because plants of *Hamamelis* bloom in the late autumn and winter, students often are curious about what types of insects are available to pollinate the flowers. De Steven (1983) studied the floral ecology of *H. virginiana*, noting that self-pollination is common, but that a wide variety of insects visited flowers during warmer times, particularly a few species of fungus gnats and small wasps. More recently, G. J. Anderson and Hill (2002) studied the reproductive biology of this species, noting that successful self-pollination apparently is quite rare and that fungus gnats, other small flies, wasps, and to a lesser extent beetles were the primary floral visitors and pollinators.

Our two species of *Hamamelis* may be difficult to distinguish if flowers are not available. Typically, *H. virginiana* grows on wooded slopes and flats (especially north-facing slopes) and *H. vernalis* on gravel bars along streams and rivers, but both species may be found in shaded, rocky places on streambanks and mesic lower slopes.

1. Petals 10–15 mm long when fresh (shrinking to 5–8 mm long in the herbarium), generally red or orange, rarely yellow; leaves turning brown in autumn but generally persisting through the winter; leaf blades with the base symmetrically obtuse or asymmetrical and rounded on only 1 side but narrowed on the other, the tip rounded or broadly and bluntly pointed; plants flowering January through April, rarely in December . . . . 1. H. VERNALIS
1. Petals 15–20 mm long when fresh (shrinking to 8–12 mm long in the herbarium); leaves generally falling promptly in autumn; leaf base usually strongly asymmetrical, rounded to cordate on one side, narrowed on the other, the tip broadly and bluntly tapered or narrowed; plants flowering November through December, rarely as early as September . . . . 2. H. VIRGINIANA

1. Hamamelis vernalis Sarg. (Ozark witch-hazel, vernal witch-hazel)

Pl. 425 d–f; Map 1899

Plants shrubs to 4 m tall, spreading by woody underground runners. Twigs moderately to densely stellate-hairy. Leaves turning yellow to brown in autumn but generally persisting through the winter. Petioles 3–15 mm long. Leaf blades 7–12 cm long, 5.5–8.0 cm wide, elliptic or obovate, the base obtuse to rounded, weakly to moderately asymmetrical, sometimes narrowly cordate on one side and narrowed on the other, the tip rounded or broadly and bluntly pointed, the undersurface often glaucous. Flowers strongly fragrant. Petals 10–15 mm long when fresh (shrinking to 5–8 mm long in the herbarium), generally red or orange, rarely yellow. Staminodes not or only slightly broadened toward the tip. $2n=24$. (December–)January–April.

Scattered in the southern portion of the state, in the Ozark and Ozark Border Divisons (Arkansas, Missouri, Oklahoma; endemic to the Ozark Plateau or nearly so). Gravelly and rocky places on banks of streams and rivers, rarely lower portions of wooded slopes.

Steyermark (1963) mapped this species from Stoddard County (presumably in Crowley's Ridge), but thus far no specimen of any *Hamamelis* species has been located to support this claim. Plants with uniformly dark red flowers have been called f. *carnea* Rehder, and plants with the undersurface of the leaf densely pubescent have been called f. *tomentella* Rehder.

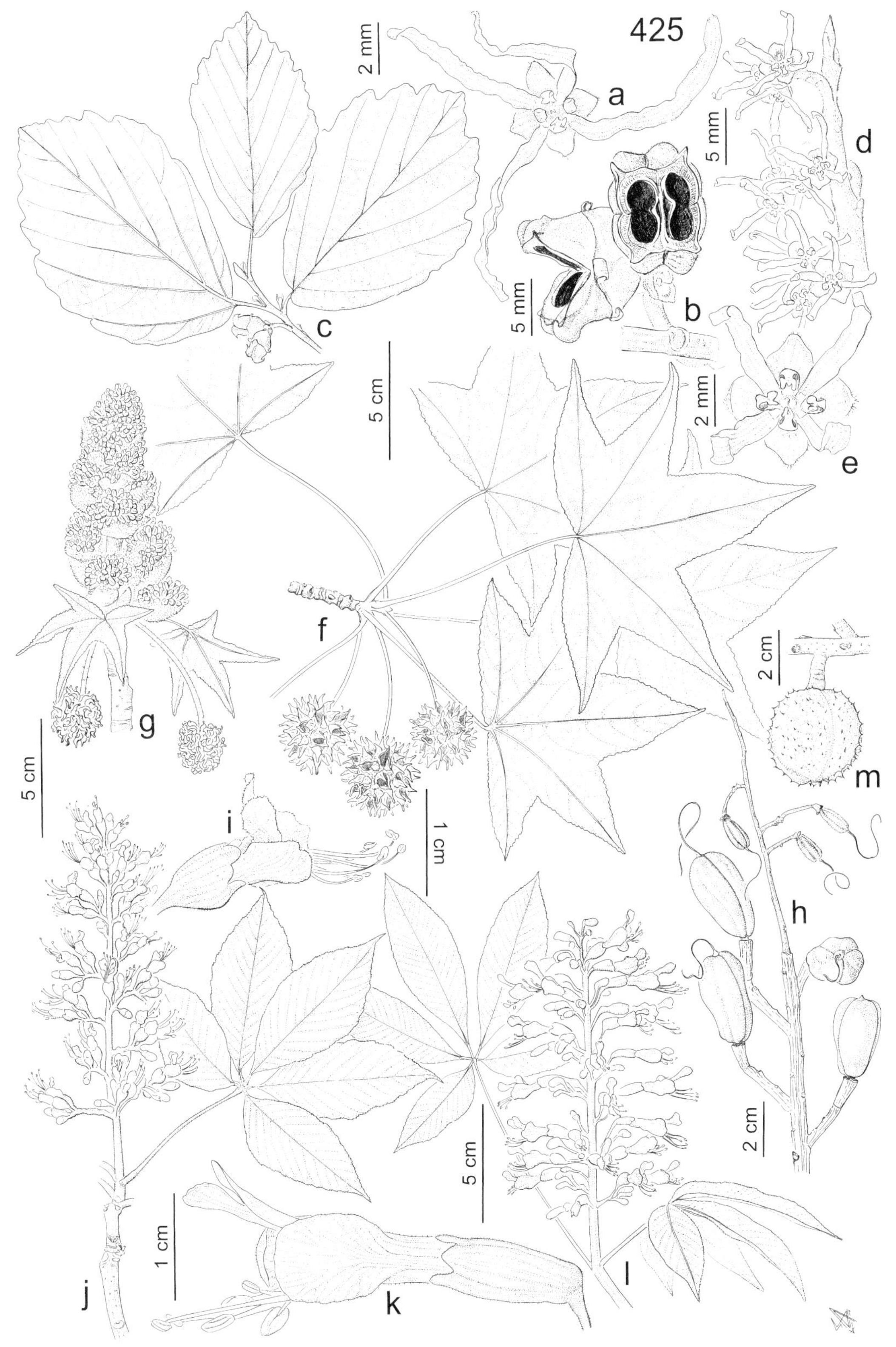

Plate 425. Hamamelidaceae. Sapindaceae. *Hamamelis virginiana*, **a)** flower, **b)** fruit, **c)** fruiting branchlet. *Hamamelis vernalis*, **d)** portion of branch with flowers, **e)** flower. *Liquidambar styraciflua*, **f)** fruiting branchlet, **g)** staminate and pistillate inflorescences. *Aesculus pavia*, **h)** developing infructescence with young fruits, **i)** flower, **j)** branch tip with leaf and inflorescence. *Aesculus glabra*, **k)** flower, **l)** branch tip with leaves and inflorescence, **m)** fruit.

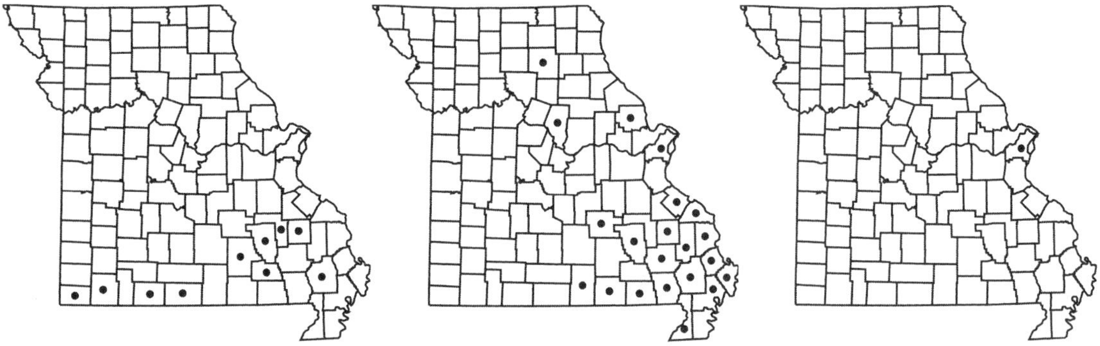

1900. Hamamelis virginiana

1901. Liquidambar styraciflua

1902. Heliotropium amplexicaule

2. Hamamelis virginiana L. (Eastern witch-hazel, American witch-hazel)

Pl. 425 a–c; Map 1900

Plants shrubs to 6 m tall, suckering but without underground runners. Twigs sparsely to moderately stellate-hairy. Leaves turning yellow to orangish yellow and generally falling promptly in the autumn. Petioles 7–18 mm long. Leaf blades 9–15 cm long, 5–10 cm wide, ovate to obovate, the base strongly asymmetrical, rounded to cordate on 1 side, narrowed on the other, the tip broadly and bluntly tapered or narrowed, the undersurface pale green but not glaucous. Flowers faintly fragrant. Petals 15–20 mm long when fresh (shrinking to 8–12 mm long in the herbarium), generally yellow, rarely reddish. Staminodes strongly broadened toward the tip. $2n=24$. (September–)November–December.

Scattered in the eastern portion of the Ozark Division, with a disjunct occurrence in Barry County (eastern U.S. west to Minnesota and Texas; Canada). Wooded slopes, flats, and creek banks, on limestone and granite.

Rebman and Weber (1988) discussed the disjunct occurrence of this species in southwestern Missouri and speculated that other populations may exist in the south-central Ozarks.

2. **Liquidambar** L. (sweet gum)

Four species, eastern North America, Mexico, Central America, and Asia.

Some botanists prefer to segregate this genus and two others into a separate family, Altingiaceae, based on their production of fragrant resins and unusual pistillate inflorescences and flowers (Judd et al., 2002).

1. Liquidambar styraciflua L. (sweet gum)

Pl. 425 f, g; Map 1901

Plants resinous trees, monoecious, with conic crowns and furrowed bark. Twigs with scattered simple hairs, older branchlets sometimes developing woody longitudinal wings. Buds sessile, dark brown, glossy, with about 8 bud-scales. Petioles 5–18 cm long. Leaf blades 7–16 cm long, 9–21 cm wide, palmately veined, 5- or 7-lobed, the sinuses extending 0.4–0.7 of the way to the leaf base, the base truncate to cordate, the lobes long-tapered to a sharp point, the margins finely toothed the undersurface with dense tufts of hairs at the base in the axils of the main veins, otherwise with sparse hairs, mainly along the veins. Inflorescences of many-flowered, dense, pistillate and staminate headlike clusters, the staminate ones short-stalked and clustered toward the ends of branches, the pistillate ones fewer and on long stalks immediately behind the staminate ones. Staminate flowers green; pistillate flowers dark purple, sometimes turning green after flowering. Calyx and corolla absent. Stamens 4–8(–10), the filaments relatively long and slender, the anther sacs attached at the base. Staminodes absent. Styles curled at flowering, persistent on the fruit as a beak, becoming enlarged to 5–7 mm long, straight, sharply pointed. Ovules several per locule. Infructesence stalks 4–8 cm long. Capsules fused together to form spherical multiple fruits 2.5–3.0 cm in diameter, woody, brown at maturity, dehiscing longitudinally between the locules but not explosively. Seeds 1 or 2 per fruit, 6–9 mm long, irregularly oblong-elliptic in outline, brown with a pattern of fine darker brown lines or mottling, the body thinning to a rounded wing at the tip. $2n$=32. April–May.

Scattered in southeastern Missouri in the Mississippi Lowlands, Ozark Border, Ozark, and Big Rivers Divisions, but escaping around planted trees north to the St. Louis area (eastern U.S. west to Illinois and Texas; Mexico, Central America). In woods, on bottomlands, river banks, lower slopes, and railroads.

This species is often grown as a street tree, prized for its deep red fall coloration. However, many homeowners have a passionate dislike for the long-persistent fruiting structures, which can clog drains, interfere with lawn mowing, and are painful if stepped on with bare feet. The lumber has been used for cabinetry, furniture, and interior finish. The inner bark produces a fragrant resin, known as American styrax or storax, which is used in cosmetics, soaps, perfumes, and tobacco, and as a fixative in lacquers and adhesives. Native Americans used an infusion of the bark medicinally for diarrhea and dysentary and also in poultices for cuts, sores, and bruises (Moerman, 1998).

HELIOTROPIACEAE (Heliotrope Family)

Contributed by David J. Bogler and George Yatskievych

Five to 8 genera, about 450 species, nearly worldwide, most diverse in tropical and warm-temperate regions.

The Heliotropiaceae are here treated as a family distinct from the Boraginaceae, in which many botanists traditionally placed them (Steyermark, 1963; Cronquist, 1981, 1991). Species of Heliotropiaceae, are distinguished from the Boraginaceae by a combination of morphological characters, including: a terminal style; fruits either fleshy to more commonly dry and drupelike or less commonly schizocarps, usually separating (sometimes tardily) into 2 or 4 nutlets or mericarps; and the frequent presence of a stigmatic appendage. See the treatment of Boraginaceae (in Volume 2) and Hydrophyllaceae for further discussion. Infrafamilial classification within the Heliotropiaceae is still controversial. Molecular studies (Diane et al., 2002; Hilger and Diane, 2003) have shown that most of the up to eight genera recognized by earlier workers (Förther, 1998) are not natural evolutionary lineages, but have not yet been able to resolve all of the species groups into a comprehensive new classification. The eventual number of genera recognized may be as few as four or as many as six, but most of these will not correspond to the traditionally circumscribed generic groups.

1. **Heliotropium** L. (heliotrope)

Plants annual or perennial herbs (shrubs elsewhere). Stems usually branched, often hairy, the hairs often lacking pustular bases, sometimes glandular. Leaves alternate or opposite, well-developed, sessile or short-petiolate. Stipules absent. Leaf blades simple, the margins entire or sometimes wavy, occasionally rolled under, the surfaces usually hairy, the hairs occasionally with persistent pustular bases (with calcified or silicified walls, known as cystoliths) and roughened to the touch, sometimes glandular. Inflorescences of solitary terminal flowers or more commonly terminal and sometimes also axillary spikes (sometimes appearing as dense clusters when young), these often appearing coiled (scorpioid) and uncoiling as the flowers develop, the flowers then all oriented toward the upper side of the axis, sometimes subtended by bracts. Flowers more or less actinomorphic, hypogynous, perfect; cleistogamous flowers absent. Calyces usually deeply 5-lobed, the lobes equal or unequal, persistent at fruiting. Corollas usually shallowly 5-lobed, saucer-shaped or funnel-shaped to trumpet-shaped, the inside of the throat often hairy but lacking appendages. Stamens 5, the filaments attached in the corolla tube, short, the anthers exserted, attached at their base, usually yellow. Pistil 1 per flower, of 2 fused carpels. Ovary not or only shallowly 4-lobed, 4-locular, with 1 ovule per locule, the placentation axile or sometimes appearing nearly basal. Style 1 or absent, situated at the tip of the ovary, usually not persistent at fruiting, the stigma usually with the receptive area in a band around the basal portion, crowned by a variously shaped sterile appendage. Fruits dry and drupelike (occasionally schizocarps elsewhere), unlobed or more commonly 2- or 4-lobed, usually separating (sometimes tardily) into 2 or 4 nutlets, these 1- or 2-seeded, glabrous or hairy, dark green to nearly black. Variously 260–420 species (depending on the generic circumscription), nearly worldwide, most diverse in tropical and warm-temperate regions.

Several species of heliotropes are cultivated widely as garden ornamentals. However. many species are noxious weeds and contain toxic pyrrolizidine alkaloids that can be a hazard to livestock (Burrows and Tyrl, 2001). Some species have been investigated for possible pharmaceutical value in the treatment of tumors (Al-Shehbaz, 1991).

1. Leaves 1–3 mm wide, linear; flowers solitary at the branch tips, noticeably stalked . . . . 5. H. TENELLUM
1. Leaves 3 mm wide or wider, not linear; flowers in scorpioid spikes (sometimes appearing as clusters when young), sessile
    2. Stems and leaves somewhat succulent, glabrous, usually glaucous . . . . 2. H. CURASSAVICUM
    2. Stems not succulent, hairy, not glaucous
        3. Stems and inflorescences with gland-tipped hairs; leaves sessile; corolla tube with the inner surface densely hairy near the tip; nutlets appearing finely wrinkled on the dorsal surface . . . . 1. H. AMPLEXICAULE
        3. Stems lacking gland-tipped hairs; at least some of the leaves petiolate; corolla tube with the inner surface glabrous; nutlets with various sculpturing
            4. Leaf blades broadly ovate to oblong-ovate or triangular-ovate, the margins often somewhat scalloped, wavy, or coarsely toothed; corollas blue to purplish blue; fruits mostly splitting into 2 pairs of nutlets, these with longitudinal ridges on the dorsal surface and glabrous . . . . 3. H. INDICUM
            4. Leaf blades elliptic to oblong-elliptic or oblanceolate, the margins usually entire; corollas white; fruits splitting into 4 nutlets, these smooth dorsally, but hairy . . . . 4. H. PROCUMBENS

1. Heliotropium amplexicaule Vahl (clasping heliotrope)

H. anchusifolium Poir.

Map 1902

Plants perennial herbs, with a deep, woody root. Stems 30–45 cm long, ascending, sometimes from a spreading base, finely hairy, with some or most of the hairs gland-tipped. Leaves alternate, sessile. Leaf blades 2–7 cm long, 5–20 mm wide, herbaceous (not succulentt), narrowly oblong-elliptic to oblanceolate, tapered at the base, angled or tapered to a bluntly or sharply pointed tip, the margins entire or more often somewhat wavy, the surfaces sparsely to moderately hairy. Inflorescences terminal clusters of 2–5 scorpioid spikes on a stalk 2–3 cm long, the spikes densely flowered, lacking bracts. Calyces 2–3 mm long, the lobes lanceolate, pubescent with both bristly hairs and smaller gland-tipped hairs, spreading after the fruit has dispersed. Corollas 6–7 mm long, 5–6 mm in diameter (measured across the spreading lobes), funnel-shaped to trumpet-shaped, blue, purple, or rarely white, hairy on the outer surface, the tube 3–4 mm long, hairy toward the tip on the inner surface, the lobes 1–2 mm long. Stamens attached at the base of the tube, the anthers 1.0–1.2 mm long. Stigma sessile, the sterile appendage broadly obconic, rounded to bluntly pointed at the tip, hairy. Fruits 2.5–3.0 mm long, 2–3 mm wide, depressed-globose to broadly ovoid, shallowly lobed at the tip, glabrous, eventually splitting into 2 nutlets (each 1- or 2-seeded). Nutlets with the dor-

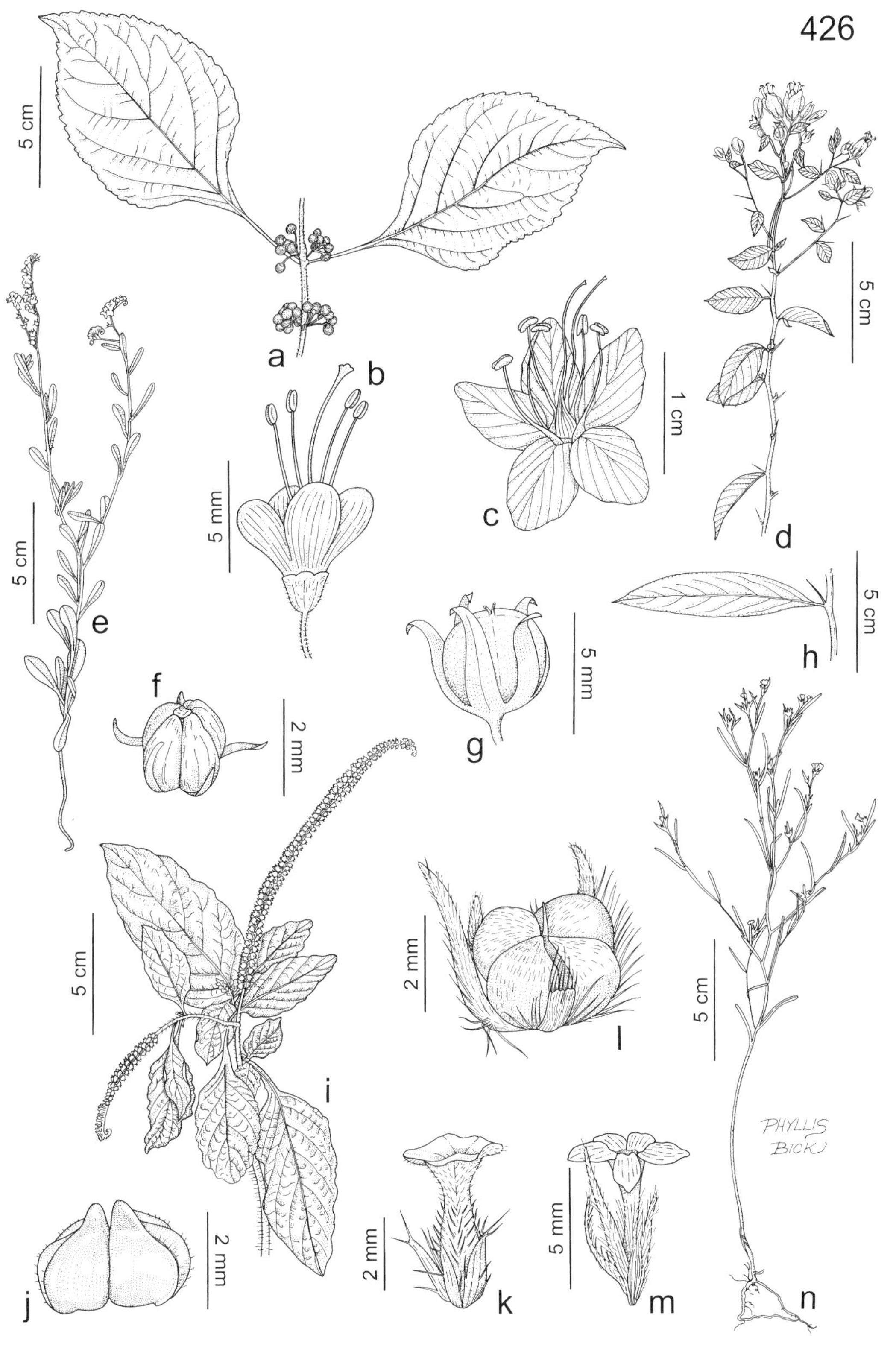

Plate 426. Heliotropiaceae. Hydroleaceae. Lamiaceae. *Callicarpa americana*, **a)** node with leaves and fruits, **b)** flower. *Hydrolea ovata*, **c)** flower, **d)** fertile stem. *Heliotropium curassavicum*, **e)** habit, **f)** fruit. *Hydrolea uniflora*, **g)** fruit, **h)** node with leaf and thorn. *Heliotropium indicum*, **i)** fertile stem, **j)** fruit, **k)** flower. *Heliotropium tenellum*, **l)** fruit, **m)** flower, **n)** habit.

1903. Heliotropium curassavicum

1904. Heliotropium indicum

1905. Heliotropium procumbens

sal surface roughened and finely wrinkled, brown. $2n$=26, 28. July–August.

Introduced, known thus far only from a single historical collection from the city of St. Louis (native of South America, introduced in the eastern [mostly southeastern] U.S. and California) Disturbed areas.

This species is commonly cultivated as an ornamental in gardens, mostly as an annual in Missouri's climate.

2. Heliotropium curassavicum L. **var. curassavicum** (seaside heliotrope)

Pl. 426 e, f; Map 1903

Plants perennial herbs (although usually functioning as annuals in Missouri), eventually with deepset rhizomes. Stems 10–40 cm long, prostrate or loosely ascending from a spreading base, somewhat succulent, glabrous, usually grayish-glaucous. Leaves opposite, subopposite, and/or alternate, sessile or with a short, winged petiole. Leaf blades 1–4 cm long, 3–12 mm wide, somewhat succulent, narrowly oblanceolate to narrowly elliptic, tapered at the base, rounded or more or less angled to a bluntly or occasionally sharply pointed tip, the margins entire, the surfaces glabrous, usually grayish-glaucous. Inflorescences terminal and sometimes also axillary scorpioid spikes, these solitary or paired, sessile or on a stalk 1–3 cm long, densely flowered, lacking bracts. Calyces 1.2–2.0 mm long (sometimes slightly longer at fruiting), the lobes lanceolate to narrowly oblong-lanceolate, glabrous, more or less spreading after the fruit has dispersed. Corollas 1.5–2.0 mm long, 2.5–3.5 mm in diameter (measured across the spreading lobes), funnel-shaped to trumpet-shaped, white or occasionally bluish-tinged, usually with a yellow spot at the base of the lobes, glabrous on the outer surface, the tube 0.7–1.0 mm long, glabrous on the inner surface, the lobes 0.5–1.0 mm long. Stamens attached toward the base of the tube, the anthers 0.6–0.8 mm long. Stigma sessile or nearly so, the sterile appendage very broadly obconic and rounded at the tip or appearing disc-shaped with a small knob in the center, glabrous. Fruits 1.5–2.0 mm long, 1.8–2.5 mm wide, globose to slightly depressed-globose, shallowly lobed at the tip, glabrous, eventually splitting into 4 nutlets (each 1-seeded). Nutlets with the dorsal surface smooth or faintly wrinkled, sometimes with 1 or 2 fine, blunt, longitudinal ribs, light brown to greenish brown. $2n$=26, 28. July–September.

Uncommon, known thus far only from Cooper County (southeastern U.S. west to Nebraska and New Mexico; Mexico, Central America, South America, Caribbean Islands; introduced sporadically farther north). Salt marshes and saline springs.

Heliotropium curassavicum is a widespread, variable species that has been divided into a number of infraspecific taxa, some of which have at times been considered separate species. Al-Shehbaz (1991) suggested that only the principal five variants should be accepted as varieties. These differ in a number of variable features, including degree of woodiness, corolla size, and stigma shape. Two of the varieties grow only in South America and two other North American varieties occur only from the western Great Plains westward. The var. *curassavicum* is the most widespread infraspecific taxon (see range above) and the only one to be found in Missouri thus far. Seaside heliotrope was first reported for Missouri by P. W. Nelson (1979a).

3. Heliotropium indicum L. (turnsole, Indian heliotrope)

Tiaridium indicum (L.) Lehm.

Pl. 426 i–k; Map 1904

Plants annual, with a slender taproot. Stems 20–60(–80) cm long, erect to moderately ascend-

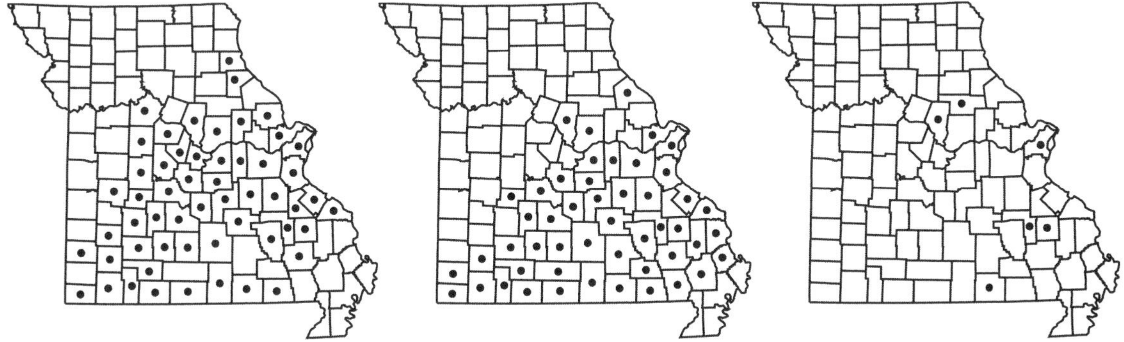

1906. Heliotropium tenellum 1907. Hydrangea arborescens 1908. Philadelphus coronarius

ing, sparsely to moderately pubescent with usually a mixture of longer, stiff, spreading, mostly pustular-based, nonglandular hairs and shorter, softer, spreading to loosely appressed, nonglandular hairs. Leaves usually alternate toward the stem base and opposite above, sometimes all mostly alternate or opposite, moderately to long-petiolate, the petiole usually winged toward the tip. Leaf blades 4–8(–12) cm long, 20–70 mm wide, herbaceous (not succulent), broadly ovate to oblong-ovate or triangular-ovate, abruptly tapered at the base, angled or tapered to a bluntly or sharply pointed or rarely rounded tip, the margins entire to somewhat scalloped, wavy, or with a few coarse, blunt teeth, the surfaces sparsely to moderately pubescent with fine, short hairs mostly along the veins, sometimes also with sparse longer, bristly hairs along the main veins toward the leaf base. Inflorescences terminal, solitary or rarely paired scorpioid spikes, these sessile or on a stalk to 2 cm long, densely flowered, lacking bracts. Calyces 1.5–3.0 mm long, the lobes linear to narrowly oblong-lanceolate, moderately to densely pubescent with stiff, loosely appressed, nonglandular hairs, more or less spreading after the fruit has dispersed. Corollas 2–4 mm long, 2–4 mm in diameter (measured across the spreading lobes), trumpet-shaped, blue to purplish blue, usually with a yellow to white spot at the base of the lobes, hairy on the outer surface, the tube 1–2 mm long, glabrous on the inner surface, the lobes 1.0–1.5 mm long. Stamens attached at the base of the tube, the anthers 0.5–0.8 mm long. Stigma on a short style, the sterile appendage short-cylindrical (about as long as wide) to slightly tapered, more or less truncate and minutely notched at the tip, glabrous. Fruits 2.5–4.0 mm long, 3–5 mm wide, appearing strongly 2-lipped above a more or less globose body, with 2, divergent, flattened lobes at the tip, glabrous, splitting into 2 pairs of nutlets, these sometimes breaking apart with age into 4, 1-seeded nutlets. Nutlets with the dorsal surface having 1 or 2(3) prominent longitudinal ridges, light brown ventrally with greenish brown to yellowish brown lobes and dorsal surface. $2n$=22, 24. August–October.

Scattered mostly south of the Missouri River, most abundant in the Mississippi Lowlands Division (native range uncertain but probably tropical South America; introduced nearly worldwide, mostly in tropical and warm-temperate regions; in the U.S. from New York to Florida west to Kansas and Texas). Banks of streams and rivers, margins of ponds, lakes, sinkhole ponds, and sloughs; also crop fields, fallow fields, ditches, farmyards, railroads, roadsides, and open disturbed areas.

This attractive species is distinguished by its thick, hollow stems, leaves that often appear somewhat wrinkled, elongate scorpioid spikes with dense, blue to purple flowers along the upper side, and oddly shaped fruits.

4. Heliotropium procumbens Mill. (fourspike heliotrope)

Euploca procumbens (Mill.) Diane & Hilger

Map 1905

Plants annual, with a slender taproot. Stems 5–35(–50) cm long, erect to loosely ascending, moderately to densely pubescent with stiffly appressed (to somewhat spreading toward the stem tip), mostly pustular-based, nonglandular hairs. Leaves alternate, sessile to long-petiolate, often on the same plant. Leaf blades 1–3(–4) cm long, 3–9(–20) mm wide, herbaceous (not succulent), oblanceolate or spatulate to narrowly elliptic or elliptic, tapered at the base, angled or tapered to a sharply pointed or less commonly rounded tip, the margins entire, the surfaces moderately to densely pubescent with appressed, pustular-based hairs. Inflorescences terminal (sometimes on a short branch positioned below a leafy node), solitary or paired scorpioid spikes, these sessile or on a stalk 1.0–1.5(–3.0) cm long, densely flowered, lacking bracts. Calyces 1.0–1.5 mm long (becoming enlarged to 2–3 mm at fruiting), the lobes linear to narrowly oblong-lanceolate, moderately to densely pubescent with stiff, mostly appressed, nonglandular hairs, spreading after the fruit has dispersed. Corollas 1.5–3.0 mm long, 2–3 mm in diameter (measured across the spreading lobes), trumpet-shaped, white, hairy on the outer surface, the tube 1–2 mm long, sparsely hairy toward the tip on the inner surface, the lobes 0.5–1.0 mm long. Stamens attached at the base of the tube, the anthers 0.3–0.5 mm long. Stigma sessile, the sterile appendage obconic, rounded to bluntly pointed at the tip, hairy. Fruits 1–2 mm long, 1–2 mm wide, globose to slightly depressed-globose, unlobed or shallowly lobed at the tip, short-hairy, eventually splitting into 4 nutlets (each 1-seeded). Nutlets with the dorsal surface smooth (lacking noticeable ridges, wrinkles, or tubercles), brown to dark brown. $2n$=14, 28. August–October.

Uncommon, known thus far from a single collection from Taney County (southern U.S. from California to Florida; Mexico, Central America, South America, Caribbean Islands, Pacific Islands; introduced in Maryland). Open disturbed areas.

This taxon was first-reported as new to Missouri by Raveill and Yatskievych (2008), but was misdetermined as *H. europaeum* L. (European heliotrope, potatoweed). *Heliotropium europaeum* is an Old World species that has become widely introduced in the New World, often as a crop weed. It differs from *H. procumbens* in its fruits with minute tubercles, more conic stigmas, generally broader leaves, and usually more spreading stem and leaf pubescence. The potential misdetermination of the Missouri material was first pointed out by Theo Witsell (Arkansas Natural Heritage Program) after a specimen was collected in 2010 on the Arkansas side of Bull Shoals Lake, across the reservoir from the Missouri station.

5. Heliotropium tenellum Torr. (pasture heliotrope)

Lithococca tennella Small

Lithospermum tenellum Nutt., an illegitimate name

Pl. 426 l–n; Map 1906

Plants annual, with a slender taproot. Stems 10–40 cm long, erect or strongly ascending, densely pubescent with straight, appressed, nonglandular, white hairs. Leaves alternate, sessile or with a short, poorly differentiated petiole. Leaf blades 0.8–4.0 cm long, 1–3 mm wide, herbaceous (not succulent), linear, tapered at the base, angled or tapered to a sharply pointed tip, the margins entire, sometimes rolled under, the surfaces moderately to densely pubescent with straight, appressed, nonglandular, white hairs. Inflorescences of solitary flowers at the tips of short branches (sometimes appearing as small clusters), each flower with a noticeable stalk 2–5 mm long, lacking bracts (but the larger calyx lobes appearing bractlike). Calyces deeply lobed, the lobes very unequal, 2–6 mm long, all linear and similar in appearance to the leaves, ascending after the fruit has dispersed. Corollas 4–5 mm long, 3–5 mm in diameter (measured across the spreading lobes), funnel-shaped to more or less trumpet-shaped, white, sometimes with a light yellow spot at the base of the lobes, hairy on the outer surface of the tube and with a hairy central band on the outer surface of each lobe, the tube 2–3 mm long, glabrous on the inner surface, the lobes 1.3–1.7 mm long. Stamens attached at the base of the tube, the anthers 0.7–1.2 mm long. Stigma on a short style, the sterile appendage narrowly conic, sharply pointed and often minutely notched at the tip, glabrous. Fruits 1.5–2.5 mm long, 2.5–3.0 mm wide, broadly ovoid to depressed-globose, finely hairy, splitting into 4 nutlets (each 1-seeded). Nutlets with the dorsal surface finely hairy, at least toward the base, otherwise smooth to shallowly pitted, yellowish brown to dark brown. July–October.

Scattered in the Ozark and Ozark Border Divisions north locally along the Mississippi River to Marion County (West Virginia to Georgia west to Iowa, Kansas, and Texas; Mexico). Glades, tops of bluffs, and openings of dry upland forests; also quarries; on limestone and dolomite substrates.

Heliotropium tenellum is somewhat similar in appearance to *Buglossoides arvensis* (Boraginaceae). That species differs in its oblanceolate to narrowly oblanceolate leaves (4–9 mm wide), inflorescences that eventually become elongate, leafy, spikelike racemes, and somewhat larger flowers with actinomorphic calyces. The leaves of *H. tenellum* are narrow and densely covered by white, reflective hairs, and have margins that become strongly curled under during times of drought. Baskin and Baskin (1978) showed that these features function to reduce leaf temperatures and thus limit water loss in the species, which tends to grow in relatively hot, dry places.

HYDRANGEACEAE (Hydrangea Family)

Contributed by David Bogler and George Yatskievych

Plants shrubs, sometimes suckering from the roots to form colonies. Stems erect to spreading, unarmed. Leaves opposite. Stipules lacking. Leaf blades simple, unlobed, the margins toothed, glabrous to more commonly pubescent. Inflorescences terminal and sometimes also axillary, short racemes or compound umbellate panicles, sometimes appearing as loose or dense clusters. Flowers more or less epigynous, perfect, actinomorphic, usually subtended by inconspicuous linear bracts, the marginal flowers of the inflorescence sometimes sterile and with enlarged petaloid sepals. Calyces of 4 or 5 distinct sepals. Corollas of 4 or 5 distinct petals. Stamens 8 to numerous, the anthers attached at their bases. Pistil of 3–5 fused carpels. Ovary inferior or partially inferior, with 2–5 locules, with numerous ovules, the placentation axile. Styles 1 or 2 per flower, if solitary then sometimes 4-lobed, the stigmas 1 or 4, if solitary then often 4-lobed, capitate or club-shaped to nearly linear. Fruits capsules (berries elsewhere). Seeds 1 to numerous, often winged. Sixteen or 17 genera, about 170 species, widespread in temperate and subtropical regions of the northern hemisphere.

The genera of Hydrangeaceae traditionally were included in a broadly circumscribed Saxifragaceae (Steyermark, 1963). More recently, phylogenetic studies have indicated the woody members of the Saxifragaceae should be moved to several other families (Morgan and Soltis, 1993; Hufford, 1997; Soltis and Soltis, 1997).

In addition to the two genera treated below, the genus *Deutzia* is frequently cultivated in Missouri. *Deutzia scabra* Thunb., the most commonly grown species, is a shrub to 2.5 m tall with ascending branches and the leaf blades roughened on both surfaces with minute stellate hairs. The flowers are all fertile and showy with usually white corollas, 3–5-locular ovaries, and 3–5 separate styles. Often the plants have flowers with a doubled perianth. Although this species has not yet been documented to escape in Missouri, it sometimes persists at old home sites and may eventually be recorded from a naturalized population in the state.

Uphof (1922) reported finding *Decumaria barbara* L. (climbing hydrangea) along a small stream in Carter County, but thus far no specimens have been located to verify any of the species found by this German ecologist during his vegetational studies in southeastern Missouri. This slender liana produces adventitious roots at some nodes and has ovate leaves and umbellate panicles of small flowers similar to those of some *Hydrangea* species. It differs in its climbing habit and in having all perfect flowers with 7–10 petals. As there are no reports from areas adjacent to the state, this species is presently excluded from the Missouri flora.

1. At least some of the marginal flowers of the inflorescence sterile or staminate and with enlarged showy sepals; petals 1.0–1.6 mm long; stamens (8)10; fruits strongly ribbed, dehiscing by a slit at the tip . . . . 1. HYDRANGEA
1. Flowers all fertile and similar; petals 12–18 mm long; stamens 20–45; fruits unribbed, dehiscing longitudinally . . . . 2. PHILADELPHUS

1. Hydrangea L. (hydrangea)
(McClintock, 1957)

About 25 species, North America to South America, Asia to Borneo.

Hydrangea occurs in temperate regions of eastern Asia and eastern North America, and extends southward into the tropics in both hemispheres. A group of species with deciduous leaves and shrubby habit has diversified in temperate eastern Asia and also includes the two species native to the United States (sect. *Hydrangea*). These species are considered relicts of the extensive Arcto-Tertiary forest that once extended continuously across the northern hemisphere. Another group of species with evergreen, leathery leaves and a climbing habit has diversified in subtropical montane regions of Central and South America (sect. *Cornidia* (Ruiz & Pav.) Engl.). The species in cultivation are mostly members of the deciduous group (McClintock, 1957). Recent molecular and morphological studies indicate that the species of *Hydrangea* do not form a monophyletic unit, and that some of the exotic species are more closely related to other genera of Hydrangeaceae (Soltis et al., 1995).

Many species and varieties of *Hydrangea* have been brought into cultivation. Some were cultivated in China and Japan long before their introduction into Europe. Among the most popular are cultivars in the hortensia group of *H. macrophylla* (Thunb.) Ser., which have large inflorescences in which all of the flowers have been replaced by showy white, pink, or blue sterile flowers and are sold under the names bigleaf hydrangea and blue snowball bush. Selections of *H. arborescens* with an inflorescence consisting of all sterile flowers are also popular. Many of these selections initially were given formal scientific names, leading to an abundance of nomenclature in the literature on the genus. *Hydrangea* is a pollination generalist and is visited by many different insects. The compound cyme serves as a stable platform for insects to move about. Bagging studies suggest that self-pollination is possible and probably common (Pilatowski, 1982).

1. Hydrangea arborescens L. (American hydrangea, smooth hydrangea, sevenbark)
Pl. 427 a, b; Map 1907

Plants sometimes colonial by rhizomes or stems that become prostrate and then root at the nodes. Stems 1–2 m long, erect to arching or spreading. Bark tan to grayish tan on older branches and peeling in long thin strips or sheets. Twigs light brown to brown, rarely reddish-tinged, sparsely to moderately hairy, the axillary buds small, with 3 or 4 scales visible, partially hidden by the U-shaped petiole bases. Leaves usually long-petiolate. Leaf blades 6–16 cm long, 3–12 cm wide, narrowly to more commonly broadly ovate or elliptic-ovate, tapered to a sharply pointed tip, mostly rounded to shallowly cordate (less commonly narrowed) at the base, the margins relatively coarsely toothed, the upper surface darker green, glabrous or sparsely hairy along the main veins, the undersurface light green, sparsely to densely hairy, sometimes only along the veins. Inflorescences terminal, much-branched, flat-topped to somewhat dome-shaped, umbellate panicles, with at least some of the peripheral flowers sterile or staminate and showier than the fertile flowers. Calyces of the fertile flowers of (4)5 sepals, these free, 0.3–0.6 mm long, triangular, glabrous, green; those of the sterile flowers of 3 or 4 sepals, these fused at the base, 6–12 mm long, broadly elliptic-ovate to angular-circular, glabrous, petaloid, white to greenish white. Corollas of 4 or 5 petals, these free, 1.0–1.6 mm long, oblong-elliptic, white, sometimes absent on sterile flowers. Stamens (8)10. Pistils 2 carpels, these united to the tip. Ovary completely inferior, 2-locular. Styles 2, the stigmas capitate to club-shaped. Fruits 2.0–2.5 mm long, broadly obconic to nearly globose, the surface with 8 or 9 prominent longitudinal ribs, dehiscing by a terminal slit developing between the styles. Seeds 0.6–0.8 mm long, oblong ellipsoid, more or less tapered at each end, the surface with several longitudinal ribs, brown, shiny. $2n=36$. May–June.

Scattered in the Ozark and Ozark Border Divisions and Crowley's Ridge (eastern U.S. west to Kansas and Oklahoma). Mesic upland forests, bases and shaded ledges of bluffs, and banks of streams, rivers, and spring branches.

Hydrangea arborescens was the first species described in the genus and was introduced into cultivation in England as early as 1736 (McClintock, 1957). Individuals within each subspecies with inflorescences consisting entirely of sterile flowers having enlarged petaloid sepals have been found rarely in various portions of the species range, including Missouri (f. *acarpa* H. St. John, f. *grandiflora* Rehder, f. *sterilis* (Torr. & A. Gray) H. St. John). Although they are of horticultural interest, these forms are not treated further here. The dried roots contain hydrangin, an alkaloid used medicinally as a diaphoretic and diuretic.

Hydrangea arborescens is quite variable in such characters as the relative presence or absence of sterile flowers, leaf size and shape, degree of pubescence, and hypanthium size, and several subspecies, varieties, and forms have been recognized

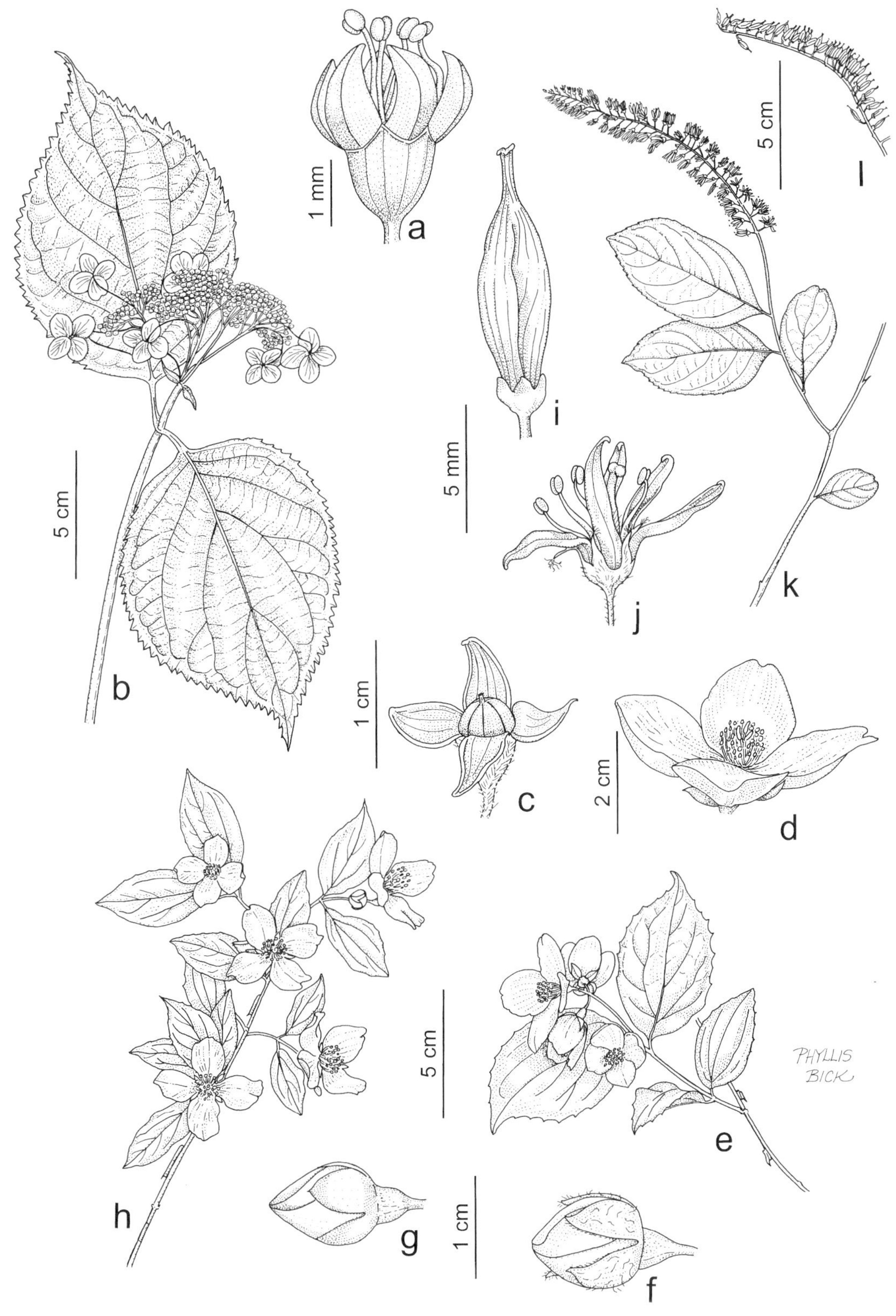

Plate 427. Hydrangeaceae. Iteaceae. *Hydrangea arborescens*, **a)** flower, **b)** flowering branch. *Philadelphus pubescens*, **c)** fruit, **d)** flower, **e)** flowering branch, **f)** bud. *Philadelphus coronarius*, **g)** bud, **h)** flowering branch. *Itea virginica*, **i)** fruit, **j)** flower, **k)** branch with leaves and inflorescence, **l)** infructescence.

(McClintock, 1957; Spongberg, 1972; Pilatowski, 1982). The present treatment follows that of McClintock (1957), who recognized three subspecies based upon differences in leaf pubescence patterns and trichome morphology, two of which grow in Missouri. The third taxon, ssp. *radiata* (Walter) E.M. McClint, is endemic to portions of the southeastern United States and differs in its leaf undersurfaces white or gray with dense matted hairs. Pilatowski (1982) noted that each of the three taxa has a distinctive flavonoid profile and maintains its morphological characteristics in the southern Appalachians where they grow together. He thus treated them as separate species. In our area, the distinctions between var. *arborescens* and var. *discolor* are not as clearcut, the two taxa occupy similar habitats in nearly the same geographic range, and intermediates are known.

1. Leaf blades with the undersurface pale or light green, but glabrous or nearly so, the hairs restricted to the main veins . . . . 1A. SSP. ARBORESCENS
1. Leaf blades with the undersurface appearing gray, moderately to densely pubescent, the hairs present both on and between the veins. . . . . 1B. SSP. DISCOLOR

1a. ssp. arborescens

H. arborescens var. *australis* Harb.
H. arborescens var. *oblonga* Torr. & A. Gray
H. vulgaris Michx., an invalid name

Leaf blades with the undersurface pale or light green, glabrous or nearly so, the hairs, when present, restricted to the main veins. $2n=36$. May–June.

Scattered in the Ozark and Ozark Border Divisions and Crowley's Ridge (eastern U.S. west to Kansas and Oklahoma). Mesic upland forests, bases and shaded ledges of bluffs, and banks of streams, rivers, and spring branches.

This is the commoner of the two subspecies in Missouri. Some specimens have been misdetermined as ssp. *discolor* because collectors occasionally misinterpret the color of the leaf undersurface. It is lighter than the upper surface in both subspecies. The key difference is in the pubescence density.

1b. ssp. discolor (Ser.) E.M. McClint.

H. arborescens var. *deamii* H. St. John
H. arborescens var. *discolor* Ser.
H. cinerea Small

Leaf blades with the undersurface appearing gray, moderately to densely pubescent, the hairs present both on and between the veins. May–June.

Scattered in the Ozark and Ozark Border Divisions and Crowley's Ridge (eastern U.S. west to Missouri and Oklahoma). Mesic upland forests, bases and shaded ledges of bluffs, and banks of streams, rivers, and spring branches.

This subspecies is less common in Missouri than ssp. *arborescens*. It tends to be encountered more frequently toward the northern and eastern portions of the species' distribution in the state, but occurs to some extent in nearly every county in which ssp. *arborescens* has been collected.

2. Philadelphus L. (mock orange)
(S. Y. Hu, 1954–1956)

Stems 1–3 m long, erect to more commonly arched-spreading, the branches often stiff, the bark gray to dark gray or reddish brown, often flaking or peeling in small plates or thin strips, leaving a tan to gray surface exposed. Twigs gray to reddish brown, glabrous or sparsely hairy, the axillary buds enclosed in the hollow petiole bases. Leaves short-petiolate. Leaf blades 1–9 cm long, 0.5–5.0 cm broad, lanceolate to elliptic or broadly ovate, narrowed or tapered to a usually sharply pointed tip, rounded or narrowed at the base, the margins with few (5–11 per side), widely spaced, fine or less commonly coarse, blunt or sharp teeth, sometimes entire or nearly so, the upper surface dark green, glabrous or sparsely hairy, the undersurface pale green, variously pubescent to nearly glabrous. Inflorescences small clusters or short racemes, each branch with (3–)5–9(–11) total flowers. Flowers showy, all similar and fertile. Sepals 4, ovate-triangular, tapered to a sharply pointed tip, the inner surface usually minutely hairy toward the tip, the outer surface variously glabrous or with longer incurved hairs, often densely and minutely hairy along the margins. Petals 4, 12–15(–20) mm long, elliptic-obovate, white. Stamens 20–45. Pistils usually of 4 carpels, these united nearly to the tip. Ovary inferior or nearly so, the hypanthium extending nearly to the broad tip of the ovary, 4-locular. Style 4-lobed to about the middle or only at the tip, the stigmas club-shaped to nearly linear. Fruits obconic, the surface smooth, not ribbed, dehiscing longitudinally. Seeds 2–3 mm long, narrowly oblong-cylindric, the body long-tapered to a slender wing, the other end truncate and with a small irregular crown, brown to dark brown. 50–70 species, North America, Europe, Asia.

Philadelphus is a difficult genus, with considerable taxonomic problems persisting in spite of a lengthy and detailed monograph (S. Y. Hu, 1954–1956). Species are wide-ranging and exhibit considerable morphological variation, and consequently can be difficult to identify. A number of species have long been cultivated and many of these originally were described from cultivated material. Two species are purported to occur in Missouri, the native *P. pubescens*, and a Eurasian taxon that has escaped from cultivation, *P. coronarius*. Both taxa belong to the sect. *Stenostigma* Koehne (S. Y. Hu, 1954–1956), and they can be thought of as New World and Old World analogs of the same species complex. In their native ranges, the "wild types" of the two apparently can be distinguished consistently by differences in pubescence of leaves and flowers, as well as bark color and flakiness. However, a number of cultivars, some possibly of hybrid origin, exist for both of the species, and these cultivars vary inconsistently for all of the features said to distinguish *P. coronarius* from *P. pubescens*. As far as can be determined from specimen labels, only the single population of *P. pubescens* from McDonald County reported by Steyermark (1963) is a native occurrence; all other specimens attributable to either species originated from the various cultivars that persisted at old home sites and eventually became naturalized in adjacent natural habitats. Accurate determination of these escapes has been equivocal. Because the characters of bark color and flakiness are difficult to interpret on herbarium specimens and do not seem to correlate well with other characters separating the species, the present treatment has arbitrarily used the characters in the key below to assign species determinations to the specimens vouchering the county distributions. Also, because the origin of various cultivars is poorly understood, no attempt has been made to categorize potential hybrids among the nonnative populations.

A wide variety of pollinators, including bees, flies, and butterflies, is attracted to the flowers, which secrete copious nectar and usually have a fragrance suggestive of orange blossoms. Although *Philadelphus* is cultivated as a hardy, disease-free ornamental, some gardeners object to the "scruffy" appearance of the vegetative shrubs.

1. Leaf blades with the undersurface moderately to densely hairy along the main veins, but glabrous or very sparsely hairy between the veins; hypanthium and calyx glabrous or very sparsely hairy, but the sepals often with minute dense hairs along the margins . . . . 1. P. CORONARIUS
1. Leaf blades with the undersurface moderately to densely hairy along and between the veins; hypanthium and calyx moderately to densely hairy . . . . 2. P. PUBESCENS

1. Philadelphus coronarius L. (sweet mock orange)

Pl. 427 g, h; Map 1908

Twigs with the bark often reddish brown, commonly flaking or peeling by the second growing season. Leaf blades with the undersurface moderately to densely hairy along the main veins, glabrous or very sparsely hairy between the veins. Inflorescence mostly racemes with 5–7 flowers, sometimes appearing as clusters. Hypanthium glabrous or very sparsely hairy. Sepals 4–8 mm long, glabrous or very sparsely hairy on the outer surface, glabrous or densely and minutely hairy along the margins. Style 4–7 mm long at flowering, becoming slightly elongated by fruiting, 4-lobed only toward the tip, the stigmas 2–3 mm long. Fruits 6–8 mm long, 5–6 mm wide. $2n=26$. May–June.

Introduced, uncommon and sporadic in Missouri (native of Europe, Asia; introduced in the eastern U.S. and adjacent Canada west to Minnesota and Missouri). Bottomland forests, mesic up-

1909. Philadelphus pubescens

1910. Hydrolea ovata

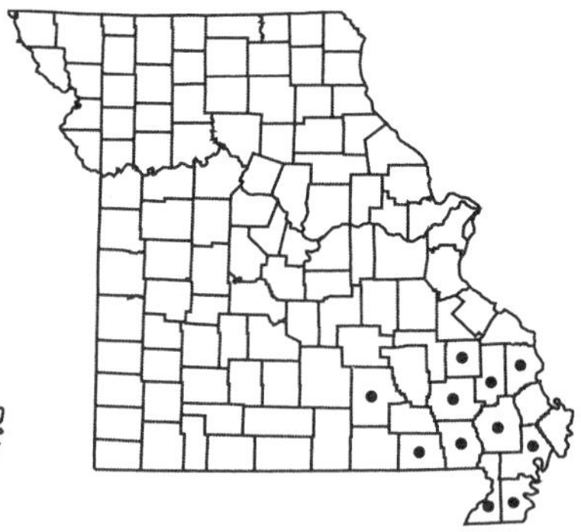

1911. Hydrolea uniflora

land forests, and banks of streams and rivers; also roadsides, railroads, and alleys.

Philadelphus coronarius has been widely cultivated in Europe and the United States, for so long in fact that the exact area of wild origin for the introduced plants is not known. Many varieties and forms have been named and numerous cultivars have been developed, including several dwarf and double-flowered forms (S. Y. Hu, 1954–1956). The species was first reported for Missouri by Mühlenbach (1979) from his collections in the St. Louis railyards.

Another species of *Philadelphus* has been noted to persist at old homesites, but apparently has not spread from original plantings and thus is not yet recognized formally as a member of the Missouri flora. *Philadelphus inodorus* L. (scentless mock orange) is a southeastern species that has long been cultivated and has escaped sporadically throughout the eastern half of the United States. Hu (1954–1956) noted specimens from plants cultivated in Hannibal (Marion County) and the city of St. Louis early in the 20th century, and a recent collection from a National Guard training site in Newton County confirms the long persistence of the taxon at homesites. This species most likely will be recorded as naturalized in Missouri in the future. It differs from *P. pubescens* in its flowers with a glabrous to sparsely hairy calyx and hypanthium, and from *P. coronarius* in its clusters of 1–3 flowers (vs. 5–7) per branchlet. It also differs from both taxa in its scentless (vs. fragrant) flowers. Hu (1954–1956) treated *P. inodorus* as comprising several varieties, which are now considered trivial variants unworthy of formal taxonomic recognition by most botanists. The Missouri materials mostly seem to correspond to *P. inodorus* in the strict sense (var. *inodorus* of Hu).

2. Philadelphus pubescens Loisel. **var. verrucosus** (Schrad. ex DC.) S.Y. Hu (hoary mock orange)

P. verrucosus Schrad. ex DC.

Pl. 427 c–f; Map 1909

Twigs with the bark often gray or reddish brown with a grayish tinge, usually not flaking or peeling until the third growing season. Leaf blades with the undersurface moderately to densely hairy along and between the veins. Inflorescence mostly racemes with 5–9 flowers, sometimes appearing as clusters. Hypanthium moderately to densely hairy. Sepals 5–9 mm long, moderately to densely hairy on the outer surface, densely and minutely hairy along the margins. Style 4–9 mm long at flowering, becoming slightly elongated by fruiting, 4-lobed to about the midpoint, the stigmas 3–5 mm long. Fruits 7–9 mm long, 5–6 mm wide. $2n=26$. April–June.

Uncommon, native in McDonald County, introduced sporadically farther north and east (Illinois, Kentucky, Tennessee, Alabama, Missouri, Arkansas, Oklahoma, and Texas; introduced farther north and west). Mesic upland forests and shaded bases of bluffs; also roadsides and railroads.

Although the native range of *P. pubescens* is relatively large, populations are very sporadic within each of the states listed. The single native population in Missouri is in the far southwestern corner of the state, within a mile of both Arkansas and Oklahoma. There has been some controversy over assignment of the name var. *verrucosus*. Rehder (1921), in discussing the occurrence of this taxon (as *P. verrucosus*) in southern Illinois, suggested that var. *verrucosus* referred to plants with exfoliating bark. However, in his monograph of the genus, Hu (1954–1956) disputed Rehder's observations, concluding that *P. pubescens* always had tardily exfoliating bark. He distinguished var. *verrucosus* primarily by its tendency toward narrowed (rather than rounded) leaf blade bases. As noted above, flaking and peeling of the bark is difficult to interpret from herbarium specimens and seems to be somewhat variable within the species. Some of the escapes in Missouri have the leaf blades more or less rounded at the base and may represent cultivars attributable to var. *pubescens*. The validity of these varieties requires further research.

HYDROLEACEAE (Hydrolea Family)

One genus, about 11 species, widespread in tropical and warm-temperate regions.

Traditionally, *Hydrolea* was included in the Hydrophyllaceae (Steyermark, 1963; Cronquist, 1981, 1991). However, because of its axile placentation, tendency to produce thorns, and nonscorpioid inflorescences (among other features), some botanists considered it a peripheral member of that family. Molecular data (D. M. Ferguson, 1999) have shown that *Hydrolea* is more closely related to the Convolvulaceae, Sphenocleaceae, and other members of the order Solanales than to the other genera of Hydrophyllaceae. See the treatments of Boraginaceae and Hydrophyllaceae for further discussion on the changing classification of these groups.

1. Hydrolea L. (hydrolea)
(Davenport, 1988)

Plants perennial herbs (shrubby or functionally annual elsewhere), with fibrous roots, sometimes with rhizomes. Stems branched, sometimes somewhat swollen toward the base, glabrous or minutely hairy, the hairs without pustular bases. Thorns often present, these solitary or rarely paired, positioned in the leaf axils, 4–15 mm long, straight, slender. Leaves alternate, well-developed. Stipules absent. Leaf blades simple, sessile or tapered to an often winged petiole, the surfaces glabrous or minutely, the hairs without pustular bases and not roughened to the touch, the margins entire or less commonly minutely toothed, sometimes also slightly wavy or minutely hairy. Inflorescences axillary or terminal, appearing as clusters or leafy panicles, not appearing coiled (scorpioid), seldom of solitary axillary flowers, some of the flowers appearing subtended by small leaflike bracts. Flowers actinomorphic, hypogynous, perfect; cleistogamous flowers not produced. Calyces very deeply 5-lobed, the lobes equal or nearly so, narrowly to broadly lanceolate and somewhat overlapping toward the base, persistent and becoming somewhat enlarged at fruiting. Corollas deeply 5-lobed, broadly bell-shaped to nearly saucer-shaped, blue to less commonly purplish blue (rarely white), sometimes greenish white near the base. Stamens 5, the filaments abruptly broadened at the base and attached near the base of the short corolla tube, the anthers more or less exserted, appearing 4-locular, attached near the dorsal midpoint, usually blue. Pistil 1 per flower, of 2 fused carpels. Ovary superior, unlobed, 2(–4)-locular and with numerous ovules per locule, the placentation axile. Styles 2(–4), situated at the tip of the ovary, elongate, entire, usually persistent at fruiting, the stigmas somewhat funnel-shaped or wedge-shaped. Fruits capsules, dehiscent longitudinally, either irregularly or more or less with 2 valves. Seeds numerous, 0.5–0.6 mm long, ovoid to nearly cylindrical, the surface with several longitudinal ridges, tan to brown, glabrous. About 11 species, eastern U.S. to South America, Caribbean Islands, Asia, Africa, most diverse in tropical or warm-temperate regions.

1. Inflorescences appearing as leafy panicles of flower clusters; corollas 11–17 mm long, minutely hairy on the outer surface; filaments and styles 10–15 mm long . . . . 1. H. OVATA
1. Inflorescences not appearing paniculate, of axillary and sometimes also terminal flower clusters; corollas 7–11 mm long, glabrous; filaments and styles 3.5–5.5 mm long . . . . 2. H. UNIFLORA

1. Hydrolea ovata Nutt. ex Choisy

Pl. 426 c, d; Map 1910

Plants herbacous or occasionally woody at the base, usually with rhizomes. Stems 30–60(–100) cm long, erect or ascending, usually much-branched above the midpoint, moderately to densely pubescent with fine, minute, straight, spreading, nonglandular hairs, rarely also with sparse longer gland-tipped hairs. Leaf blades 1–6 cm long, narrowly to broadly ovate, tapered to a sharply pointed tip, short-tapered at the base, the margins usually entire but sometimes slightly wavy and usually minutely hairy, the surfaces moderately to densely pubescent with fine, minute, straight, more or less spreading hairs. Inflorescences appearing as leafy panicles, the flowers in clusters at the branch tips and sometimes also in the upper leaf axils. Sepals 6–9 mm long, narrowly lanceolate, densely pubescent with minute gland-tipped hairs and longer, spreading, nonglandular hairs. Corollas 11–17 mm long, the lobes 5–9 mm wide, elliptic-ovate, rounded at the tip, sparsely to densely and minutely hairy on the outer surface. Filaments and styles 10–15 mm long. Fruits globose, 4–6 mm long, more or less glandular toward the tip. $2n=20$. June–September.

Uncommon in southeastern Missouri west to Howell County (southeastern U.S. west to Oklahoma and Texas). Margins of swamps, sloughs, and sinkhole ponds, and banks of streams; also ditches; sometimes emergent aquatics.

Hydrolea ovata is a beautiful wetland wildflower whose horticultural potential deserves investigation. Davenport (1988) noted the existence of occasional putative hybrids between *H. ovata* and *H. uniflora* in Louisiana and Texas. Given the rarity of these species in Missouri, such hybrids are unlikely to occur in Missouri.

2. Hydrolea uniflora Raf.

Pl. 426 g, h; Map 1911

Plants herbacous, sometimes with apparently rhizomatous lower stems. Stems 30–60(–100) cm long, the main stems prostrate with erect or ascending primary branches or ascending from a prostrate base, often rooting at the lower nodes, glabrous or sparsely to moderately pubescent with fine, minute, straight, spreading, nonglandular hairs, occasionally also with sparse, minute, sessile glands. Leaf blades 1.5–10.0 cm long, lanceolate to narrowly elliptic, tapered to a sharply pointed tip, angled or tapered at the base, the margins entire or minutely toothed, sometimes also slightly wavy and often minutely hairy, the surfaces glabrous or sparsely to moderately pubescent with fine, minute, straight, more or less spreading hairs. Inflorescences appearing as axillary and sometimes also terminal clusters. Sepals 5–8 mm long, broadly lanceolate to ovate, glabrous or sparsely to moderately pubescent with minute nonglandular hairs, occasionally also with sparse, minute, sessile glands. Corollas 7–11 mm long, the lobes 4–6 mm wide, broadly elliptic to broadly obovate, rounded to broadly and bluntly angled at the tip, glabrous. Filaments and styles 3.5–5.5 mm long. Fruits globose, 4.5–6.5 mm long, glabrous. $2n=20$. June–September.

Uncommon in southeastern Missouri, mostly in the Mississippi Lowlands Division (Indiana to Alabama west to Oklahoma and Texas). Bottomland forests, swamps, and margins of lakes and sloughs; also ditches and wet roadsides; sometimes emergent aquatics.

HYDROPHYLLACEAE (Waterleaf Family)

Plants annual, biennial, or perennial herbs (shrubs or trees elsewhere). Stems unbranched or more commonly branched, often hairy, sometimes appearing bristly, the hairs sometimes with persistent pustular bases. Leaves alternate and sometimes also basal (rarely the lowermost stem leaves opposite), well-developed. Stipules absent. Leaf blades pinnately compound

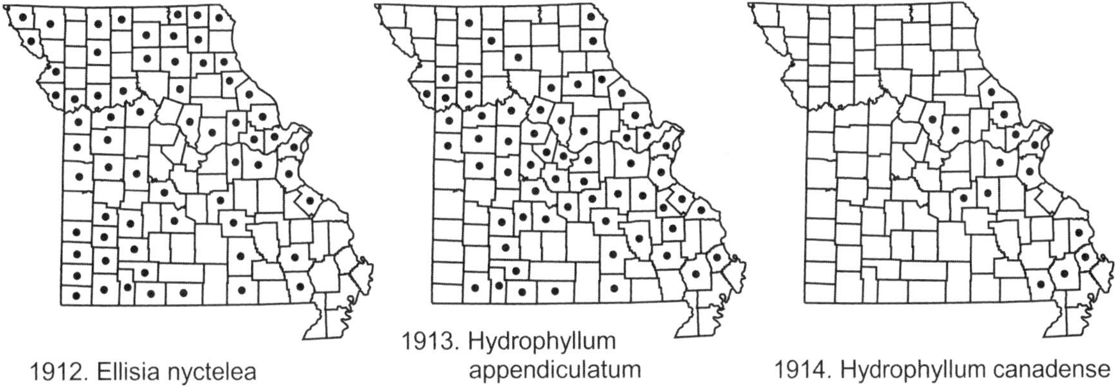

1912. Ellisia nyctelea

1913. Hydrophyllum appendiculatum

1914. Hydrophyllum canadense

or simple but pinnately or palmately lobed (unlobed elsewhere), sessile or petiolate, the surfaces usually hairy, the hairs sometimes stiff and with persistent pustular bases (with calcified or silicified walls, known as cystoliths) and then more or less roughened to the touch. Inflorescences terminal and/or axillary, of solitary flowers or more commonly racemes or clusters, these sometimes appearing paired (or paniculate elsewhere), often appearing coiled (scorpioid) and uncoiling as the flowers develop, the flowers usually not subtended by bracts. Flowers actinomorphic, hypogynous, perfect; cleistogamous flowers absent. Calyces deeply 5-lobed, sometimes with small triangular appendages in the sinuses, sometimes becoming enlarged at fruiting, the lobes equal, persistent. Corollas 5-lobed, narrowly to broadly bell-shaped. Stamens usually 5, the filaments attached in the corolla tube, sometimes subtended by 2 minute scales, short or long, the anthers exserted or not, appearing 2-locular, usually attached near the dorsal midpoint, white to pink, purple, or blue. Pistil 1 per flower, of 2 fused carpels. Ovary unlobed, 1-locular but sometimes appearing incompletely 2-locular by intrusion of the walls, often hairy, with 4–8 (numerous elsewhere) ovules, the placentation parietal (sometimes appearing nearly axile in species with an incompletely 2-locular ovary). Style 1, situated at the tip of the ovary, shallowly to deeply 2-branched, usually not persistent at fruiting, the stigmas capitate or discoid, sometimes minute. Fruits capsules, dehiscent longitudinally, either irregularly or with 2 valves, 1–8-seeded. About 15 genera, about 250 species, nearly worldwide, most diverse in the western U.S.

Classification of the Hydrophyllaceae remains controversial. Traditionally, most botanists thought that the Hydrophyllaceae should be grouped with the Convolvulaceae, Solanaceae, Polemoniaceae, and several other families in the order Solanales, but included the Boraginaceae along with the other "nutlet-producing families" Lamiaceae and Verbenaceae in the Lamiales (Cronquist, 1981, 1991). Molecular data (D. M. Ferguson, 1999) have shown that the Hydrophyllaceae, as traditionally treated, are not a cohesive natural group. Among the genera that occur in the midwestern United States, only *Hydrolea* continues to be a member of the Solanales. It is treated as the separate family Hydroleaceae in the present volume. The relationships of the remaining genera of Hydrophyllaceae require further study and were not classified into any existing order by the Angiosperm Phylogeny Group (2003). Some authors have treated them as more closely related to the Boraginaceae and have submerged them within that family (Judd et al., 2002; Angiosperm Phylogeny Group, 2009). Other authors have preferred to recognize several separate families within this group and have suggested that the Hydrophyllaceae are more closely related to the Ehretiaceae-Cordiaceae alliance than to Boraginaceae in the strict sense (Gottschling et al., 2001). See the treatment of Boraginaceae in Volume 2 for more details on the overall classification. Although the Hydrophyllaceae mostly share a distinctive scorpioid inflorescence with the Boraginaceae, they differ in having a capsular fruit (vs. nutlets) and parietal (vs. axile to nearly basal) placentation. The decision not to combine the Hydrophyllaceae with the Boraginaceae in the present work reflects the uncertainty of the present data in resolving relationships among all of the component taxa within the group and results in families that are relatively easily separable morphologically.

Nemophila Nutt. is a genus of 11 annual species that is mainly distributed in western North America. One of these, the attractive *N. menziesii* Hook. & Arn. (baby blue eyes), is cultivated as a bedding plant in gardens, but does not appear to escape into the wild in the eastern half of the United States. Palmer and Steyermark (1935) reported another species, *N. aphylla* (L.) Brummit (as *N. microcalyx* (Nutt.) Fisch. & C.A. Mey.) from Butler and Dunklin Counties. Constance (1949a) redetermined the Missouri specimens as *Phacelia ranunculacea* and Steyermark (1963) subsequently excluded the species from the state's flora. Gleason and Cronquist's (1991) report of the taxon (as *N. triloba* (Raf.) Thieret) from southeastern Missouri was a perpetuation of the initial misdeterminations. However, this southeastern species eventually may be confirmed as occurring in the state, as it is known to grow in closely adjacent portions of southern Illinois and western Kentucky (John Schwegman, personal communication), as well as in western Tennessee and northwestern Arkansas. It differs from the superficially similar *P. ranunculacea* in having the flowers solitary at the nodes and positioned opposite the leaves, vs. in small clusters or racemes at the branch tips.

1. Lowermost stem leaves often opposite; flowers solitary at the nodes, usually opposite the leaf . . . . 1. ELLISIA
1. Stem leaves all alternate; flowers in axillary or terminal clusters or racemes
   2. Inflorescences remaining compact, consisting of a pair of short dense clusters, these often appearing slightly scorpioid; ovaries 1-locular, the placenta-bearing portions of the wall not intruding into the locule . . . . 2. HYDROPHYLLUM
   2. Inflorescences continuing to elongate during the flowering season (except occasionally in *P. covillei* and *P. ranunculacea* and then not dense), consisting of solitary or less commonly paired open racemes, these often somewhat congested and scorpioid at the tip; ovaries incompletely (sometimes nearly completely) 2-locular, the placenta-bearing portions of the wall intruding into the locule as a partition . . . . 3. PHACELIA

1. **Ellisia** L.

One species, U.S., Canada.

1. Ellisia nyctelea (L.) L. (Aunt Lucy, waterpod, nyctelea)

Pl. 428 a–c; Map 1912

Plants annual, with slender taproots. Stems 5–40 cm long, spreading to loosely ascending or ascending from a spreading base, usually sparsely pubescent with stiff, straight, spreading to downward-angled, somewhat pustular-based hairs, occasionally also with sparse, fine curved, softer hairs. Basal leaves often withered at flowering, similar to the lower stem leaves. Stem leaves mostly alternate (often opposite toward the stem base), the petiole mostly shorter than the blade, stiffly hairy toward the base. Leaf blades 2–6 cm long, oblong-elliptic in outline, deeply pinnately lobed (rachis narrowly winged); the pinnae 7–13, mostly narrowly oblong, entire or more commonly with 1–3 pairs of coarse teeth or triangular lobes, angled or tapered to a sharply pointed tip, truncate and attached broadly at the base, the surfaces sparsely to moderately pubescent with stiff, straight, somewhat pustular-based hairs, sometimes only along the veins, the upper surface not appearing mottled. Inflorescences of solitary flowers at the nodes (occasionally also appearing paired at the branch tips), these usually positioned opposite the leaf, occasionally axillary, the flower stalks 5–10 mm long at flowering, becoming elongated to 2–6 cm at fruiting, stiffly hairy. Calyces 4–6 mm long at flowering, becoming enlarged to 7–10 mm at fruiting, lacking

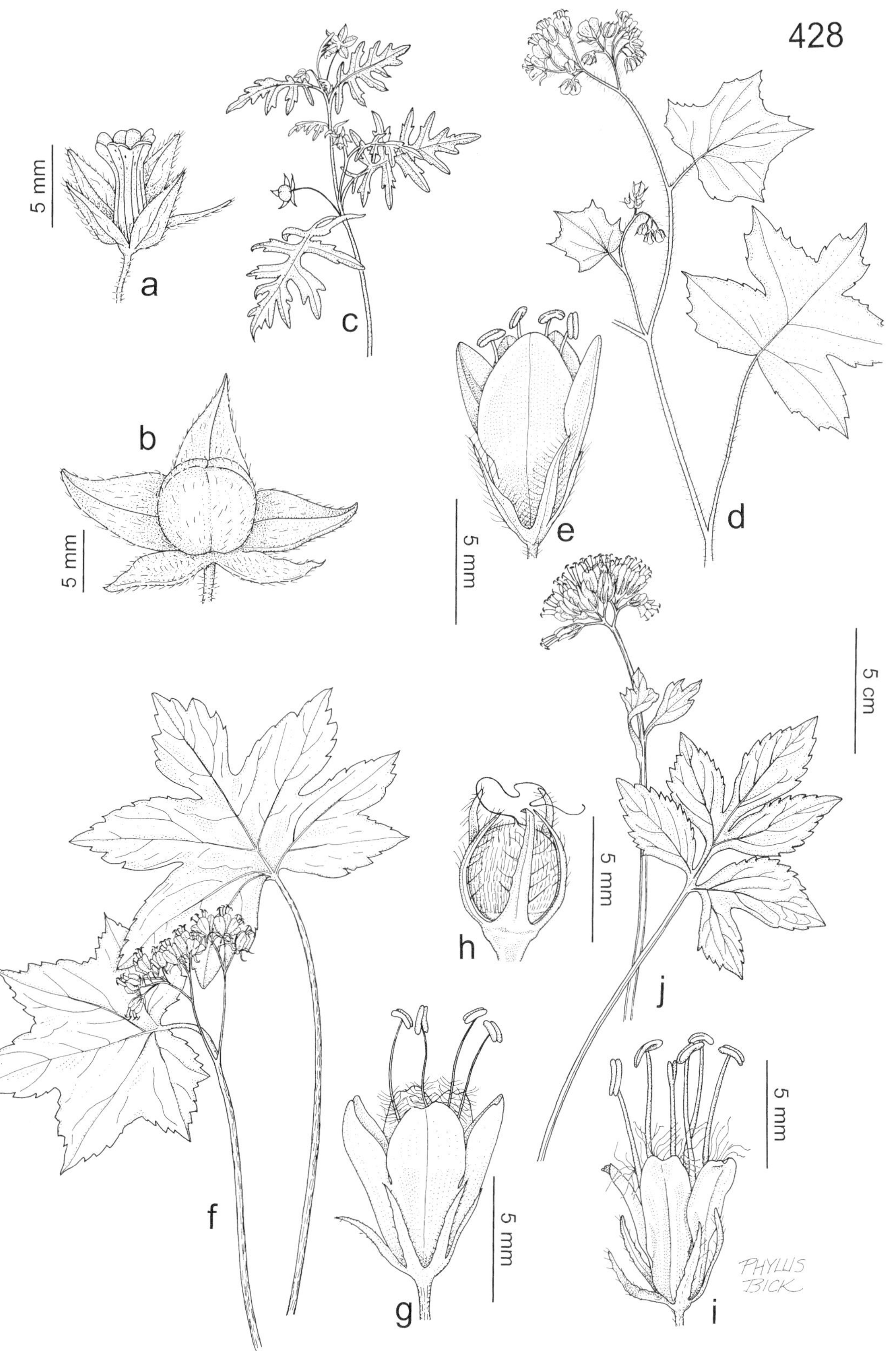

Plate 428. Hydrophyllaceae. *Ellisia nyctelea*, **a)** flower, **b)** fruit, **c)** fertile stem. *Hydrophyllum appendiculatum*, **d)** fertile stem, **e)** flower. *Hydrophyllum canadense*, **f)** flower, **g)** fertile stem and basal leaf. *Hydrophyllum virginianum*, **h)** fruit, **i)** flower, **j)** fertile stem and basal leaf.

appendages, the lobes narrowly triangular (becoming broader at fruiting), sharply pointed at the tip, the surfaces and margins sparsely to moderately and stiffly hairy. Corollas 5–8 mm long, narrowly bell-shaped to nearly funnel-shaped, white, often pale pinkish- or bluish-tinged, occasionally with blue spots. Stamens not exserted, the filaments attached at the base of the corolla tube, glabrous or nearly so, each with a pair of minute scales at the base. Ovary 1-locular, the placenta-bearing portions of the wall not intruding into the locule. Style not exserted, divided to about the midpoint. Fruits 5–6 mm long, globose, sparsely to moderately hairy, at least toward the tip, mostly 4-seeded. Seeds 2–3 mm long, globose, the surface with a fine network of ridges, dark brown. $2n=20$. April–June.

Scattered nearly throughout the state, but apparently absent from most of the Mississippi Lowlands Division (Massachusetts to Idaho south to Virginia, Texas, and Nevada; Canada). Bottomland forests, mesic upland forests in ravines, bases and ledges of bluffs, and banks of streams and rivers; also crop fields, gardens, roadsides, and moist, shaded, disturbed areas.

2. **Hydrophyllum** L. (waterleaf)

Plants biennial and taprooted or more commonly perennial and with short to long rhizomes, the roots usually somewhat fleshy. Stems loosely to strongly ascending or erect, variously pubescent, often at least in part with stiff, straight, spreading to downward-angled, somewhat pustular-based hairs, sometimes nearly glabrous with age. Basal leaves well-developed and usually persistent at flowering, relatively large (the blade usually more than 7 cm long), long-petiolate, the blade variously lobed or compound, the upper surface frequently with lighter mottling (as though water-stained). Stem leaves alternate, the petioles progressively shorter toward the stem tip, glabrous or more commonly hairy, the hairs more or less similar to those of the stem. Leaf blades variously shaped, deeply pinnately lobed or compound to shallowly palmately lobed; the pinnae or lobes mostly tapered to a sharply pointed tip, the margins coarsely toothed, the surfaces variously hairy, the upper surface sometimes with lighter mottling. Inflorescences remaining compact, consisting of a pair of short dense clusters, these irregularly dichotomously branched, but often appearing slightly scorpioid, the flower stalks 3–10 mm long at flowering, becoming only slightly elongated at fruiting, glabrous or hairy. Calyces becoming enlarged at fruiting, sometimes with small triangular to narrowly triangular, reflexed appendages in the sinuses, the lobes narrowly triangular to narrowly lanceolate (becoming slightly broader at fruiting), sharply pointed at the tip, the surfaces and margins glabrous or variously hairy. Corollas bell-shaped (narrowly bell-shaped in *H. virginianum*), greenish white or white to lavender or pale pinkish purple. Stamens exserted, the filaments attached at or near the base of the corolla tube, hairy below the midpoint, with a pair of usually well-developed, slender scales at the base, these fused to the corolla tube along adjoining margins. Ovary 1-locular, the placenta-bearing portions of the wall not intruding into the locule. Style more or less exserted, short-branched toward the tip. Fruits 3–4 mm long, globose, moderately to densely hairy, at least toward the tip, 1–3-seeded. Seeds 2–4 mm long, broadly ovoid to nearly globose, the surface with a fine network of ridges, light to dark brown or reddish brown. Nine species, eastern and western U.S. and adjacent Canada.

1. Stem leaves all or nearly all pinnately compound or very deeply lobed (the rachis slightly winged), the 5(7 or 9) lobes mostly narrowed toward the base; corollas narrowly bell-shaped . . . . 3. H. VIRGINIANUM
1. Stem leaves all shallowly to deeply palmately or pinnately lobed, the 5–9 lobes more or less truncate at the broad base; corollas bell-shaped to broadly bell-shaped

2. Plants biennial, with taproots; calyces with a well-developed appendage in each sinus; corollas narrowly bell-shaped; stamens only slightly exserted . . . . 1. H. APPENDICULATUM
2. Plants perennial, with rhizomes; calyces lacking appendages in the sinuses; stamens conspicuously exserted . . . . 2. H. CANADENSE

1. Hydrophyllum appendiculatum Michx. (woollen breeches, notchbract waterleaf)

Pl. 428 d, e; Map 1913

Plants biennial, with taproots. Stems 20–60 cm long, erect or ascending, pubescent with sparse to moderate longer and moderate to dense shorter, stiff spreading hairs. Basal leaves with the blade 5–17 cm long, oblong to broadly elliptic in outline, pinnately compound or deeply lobed with 5–9 leaflets or lobes, these truncate or slightly narrowed at the base, the margins coarsely toothed, the surfaces sparsely to moderately pubescent with longer spreading and shorter more or less appressed, stiff, straight hairs, the undersurface pale green. Stem leaves with the blade 3–14 cm long, broadly ovate to nearly circular in outline, short-pinnately to more or less palmately and moderately to deeply lobed, the 5 or 7 lobes tapered from the broad base, the pubescence similar to that of the basal leaves. Inflorescences usually extending above the leaves. Calyces 4–8 mm long, the lobes bristly-hairy along the margins and with shorter hairs on the outer surface near the base, each sinus with a well-developed narrowly lanceolate scale 1.0–1.5 mm long. Corollas 9–14 mm long, bell-shaped to broadly bell-shaped, lavender to purple. Stamens and style exserted only 1–3 mm beyond the corolla. $2n=18$. April–July.

Scattered nearly throughout the state (eastern [mostly northeastern] U.S. west to Minnesota, Kansas, and Arkansas; Canada). Bottomland forests, mesic upland forests, and bases of bluffs; rarely also quarries.

Steyermark (1963) discussed an occurrence of the rare white-flowered f. *album* Steyerm. in Johnson County.

2. Hydrophyllum canadense L. (broadleaf waterleaf)

Pl. 428 f, g; Map 1914

Plants perennial, with scaly, usually long rhizomes, often forming large colonies. Stems 15–50 cm long, ascending, nearly glabrous to sparsely pubescent with stiff, spreading to downward-curved hairs. Basal leaves with the blade 10–27 cm long, oblong to broadly elliptic in outline, pinnately deeply lobed with 7–11 lobes, these truncate or slightly narrowed at the base, the margins coarsely toothed, the surfaces sparsely pubescent with, stiff, straight hairs, the undersurface often nearly glabrous. Stem leaves with the blade 5–30 cm long, broadly ovate to nearly circular in outline, more or less palmately and shallowly to moderately lobed, the 5–9 lobes tapered from the broad base, the pubescence similar to that of the basal leaves. Inflorescences often overtopped by the stem leaves. Calyces 3–7 mm long, the lobes bristly-hairy along the margins and on the outer surface toward the base, the sinuses lacking appendages (uncommonly with minute appendages elsewhere). Corollas 7–11 mm long, bell-shaped to broadly bell-shaped, white to lavender or purplish tinged. Stamens and style exserted 3–5 mm beyond the corolla. $2n=18$. May–July.

Scattered in southeastern Missouri north to Boone and Pike Counties (eastern U.S. west to Iowa and Arkansas; Canada). Bottomland forests, mesic upland forests, and bases of bluffs.

Hydrophyllum canadense is much less common in Missouri than is *H. virginianum*. With supplemental watering, the basal leaves of this species last nearly throughout the growing season and it makes an attractive groundcover in the woodland garden. Steyermark (1963) noted that the young foliage can be cooked and eaten.

3. Hydrophyllum virginianum L. **var. virginianum** (waterleaf, Virginia waterleaf, John's cabbage)

Pl. 428 h–j; Map 1915

Plants perennial, with scaly, usually long rhizomes, often forming large colonies. Stems 25–60 (–80) cm long, ascending, nearly glabrous or sparsely to moderately pubescent with short, stiff, ascending hairs toward the tip, often also with sparse, slightly longer, spreading to downward-curved hairs. Basal and stem leaves similar, the blade 5–30 cm long, broadly ovate to broadly elliptic in outline, pinnately compound or very deeply lobed (the rachis slightly winged), the 5(7 or 9) lobes mostly narrowed toward the base, the margins coarsely toothed or less commonly also few-lobed, the surfaces sparsely to moderately pubescent with, short, stiff, straight, appressed hairs, the undersurface sometimes nearly glabrous and pale green. Inflorescences usually extending above

1915. Hydrophyllum virginianum 1916. Phacelia bipinnatifida 1917. Phacelia covillei

the leaves. Calyces 4–7 mm long, the lobes bristly-hairy along the margins and with shorter hairs on the outer surface, sometimes only near the base, the sinuses lacking appendages. Corollas 6–10 mm long, narrowly bell-shaped, white to purplish-tinged or purple. Stamens and style exserted 4–9 mm beyond the corolla. $2n=18$. April–July.

Scattered nearly throughout the state (eastern [mostly northeastern] U.S. west to North Dakota and Oklahoma; Canada). Bottomland forests, mesic upland forests, and bases of bluffs.

As in *H. canadense*, the basal leaves of this species can last nearly throughout the growing season with supplemental watering and *H. virginianum* makes an attractive groundcover in the woodland garden. Steyermark (1963) noted that the young foliage can be cooked and eaten. Missouri plants are all var. *virginianum*. From West Virginia to North Carolina, some populations have leaves with mostly 7 or 9 lobes or leaflets, dark purple corollas, and dark hairs in the flowers. These have been called var. *atranthum* (Alexander) Constance.

Another waterleaf species of the eastern United States is *H. macrophyllum* Nutt, which occurs as far west as southern Illinois and western Arkansas. It should be searched for in southern Missouri in habitats similar to those of the other *Hydrophyllum* taxa in the state. *Hydrophyllum macrophyllum* differs from *H. virginianum* in the long spreading hairs of the inflorescence and generally larger stem leaves with mostly 7–13 leaflets or lobes. It also tends to grow taller.

3. **Phacelia** Juss. (scorpionweed, phacelia)

Plants annual or biennial (perennial elsewhere), with slender taproots (sometimes stouter in *P. bipinnatifida*). Stems loosely to strongly ascending, sometimes from a spreading base, variously pubescent, at least in part with stiff, straight, spreading to downward-angled, somewhat pustular-based hairs, in some species nearly glabrous with age. Basal leaves usually present (in some species the oval, unlobed cotyledons also persistent) but often withered at flowering, usually relatively small (the blade usually less than 7 cm long, except in *P. bipinnatifida*), long-petiolate, the blade pinnately lobed or compound, the upper surface usually not mottled. Stem leaves alternate, long-petiolate to sessile, variously hairy. Leaf blades variously shaped, moderately to deeply pinnately lobed or compound (entire elsewhere); the pinnae or lobes narrowly rounded to more commonly bluntly or sharply pointed at the tip, the margins otherwise entire to deeply lobed, the surfaces variously hairy. Inflorescences initially sometimes compact, continuing to elongate during the flowering season (except occasionally in *P. covillei* and *P. ranunculacea*), consisting of solitary or less commonly paired open racemes with the flowers often mostly oriented along 1 side, often somewhat congested and scorpioid at the tip, the flower stalks 3–14 mm long at flowering, becoming elongated at fruiting, variously hairy. Calyces becoming somewhat enlarged at fruiting, lacking appendages, the lobes narrowly triangular to narrowly lanceolate (sometimes becoming slightly broader at fruiting), sharply pointed at the tip, the surfaces and margins variously hairy. Corollas narrowly to broadly bell-shaped (sometimes nearly cylindrical in *P. covilleii* and *P. ranunculacea*), white or lavender- or bluish-tinged. Stamens exserted or not, the filaments attached at or near the base of the corolla tube, glabrous or hairy, with a small linear gland between each pair of adjacent filament bases, this usually fused along all or most of its length with a minute pair of appendages or ridges. Ovary incompletely (sometimes nearly completely) 2-locular, the placenta-bearing portions of the wall intruding into the locule as a partition. Style exserted or not, shallowly or deeply branched. Fruits (2–)3–6 mm long, more or less globose, moderately to densely hairy toward the tip, at least when young, 2–8-seeded. Seeds 1.5–3.0 mm long, ovoid with a longitudinal line or ridge along the inner side to nearly globose, the surface with a network of ridges or occasionally appearing pitted, brown or black. One hundred fifty to 200 species, North America, Central America.

The Missouri species of this genus have all been classified as members of subgenus *Cosmanthus* (Nolte ex A. DC.) Constance, a group of about 15 species that is most diverse in the eastern half of the United States (Constance, 1949b). However, more taxonomic research is necessary to test whether the subdivision of the genus into three subgeneric groups based primarily on cytological patterns and annual vs. perennial habits reflects actual affinities (Sewell and Vincent, 2006).

Some species of *Phacelia* cause dermatitis in a few individuals when handled, but this may be due to the stiff sharply pointed hairs rather than because of reaction to some caustic chemical exudate. The genus contains a number of species considered good bee plants for the production of honey.

1. Plants relatively robust biennials, the basal and lowermost stem leaves with the blade often more than 7 cm long; flower stalks and calyces (also the inflorescence axis) densely glandular-hairy . . . . 1. P. BIPINNATIFIDA
1. Plants relatively slender annuals, the basal and lowermost stem leaves with the blade less than 7 cm long; flower stalks and calyces (usually also the inflorescence axis) moderately to densely nonglandular-hairy, occasionally also with a few, scattered, gland-tipped hairs
    2. Inflorescences with only 2–6 flowers; corollas 2–4(–5) mm long, narrowly bell-shaped to nearly cylindrical; stamens with the filaments 1.5–2.0 mm long, much shorter than the corolla; leaves all moderately to long-petiolate
        3. Uppermost stem leaf (immediately below the lowermost flower) with the terminal leaflet or lobe narrowly angled or tapered at the base; median portion of stem with mostly loosely ascending, nonglandular hairs, a few short gland-tipped hairs occasionally also present . . . . 2. P. COVILLEI
        3. Uppermost stem leaf (immediately below the lowermost flower) with the terminal leaflet or lobe broadly angled to rounded or nearly truncate at the base; median portion of stem with spreading to downward-angled, nonglandular hairs and moderate to dense, short, gland-tipped hairs . . . . 6. P. RANUNCULACEA
    2. Inflorescences with 8–30 flowers; corollas 6–10 mm long, bell-shaped to broadly bell-shaped; stamens with the filaments 4–7 mm long, about as long as or longer than the corolla; median and upper stem leaves short-petiolate or sessile
        4. Corolla lobes entire or minutely and inconspicuously toothed or scalloped . . . . 4. P. HIRSUTA
        4. Corolla lobes coarsely toothed to conspicuously fringed
            5. Corollas hairy on the outer surface; ovules 8; fruits mostly 6–8-seeded . . . . 3. P. GILIOIDES
            5. Corollas glabrous or very sparsely hairy on the outer surface; ovules 4; fruits mostly 2–4-seeded . . . . 5. P. PURSHII

1918. Phacelia gilioides 1919. Phacelia hirsuta 1920. Phacelia purshii

1. Phacelia bipinnatifida Michx. (forest phacelia)

Pl. 429 i, j; Map 1916

Plants relatively robust biennials. Stems 15–60 cm long, usually several-branched at the base and above, erect or strongly ascending from an often spreading base, moderately to densely pubescent with stiff, spreading to somewhat downward-angled, nonglandular hairs toward the base, these grading into dense, shorter, gland-tipped hairs above the midpoint. Basal (and lower stem) leaves often persistent at flowering, the blade (3–)7–12 cm long, broadly ovate to ovate-triangular in outline, pinnately compound with 3–7 pinnae, these mostly deeply 2- or 3-lobed and also coarsely several-toothed to moderately lobed, the teeth or lobes sharply pointed at the tips, the surfaces sparsely to moderately pubescent with loosely appressed to more or less spreading, stiff hairs, the undersurface sometimes nearly glabrous and usually pale green. Stem leaves all moderately to long-petiolate, the blade 1.5–9.0 cm long, progressively less divided toward the stem tip, the uppermost leaf blades often irregularly 3- or 5-lobed, the terminal lobe somewhat tapered at the base, the pubescence similar to that of the basal leaves. Inflorescences with 5–25 flowers, the axis and flower stalks densely pubescent with gland-tipped hairs, the stalks downward-arched or pendant at fruiting. Calyces 4–8 mm long, the lobes with the margins and surfaces pubescent with a mixture of nonglandular and gland-tipped hairs. Corollas 7–10 mm long, broadly bell-shaped, pale lavender to strongly bluish-tinged, the margins minutely irregular or scalloped, the outer surface finely hairy (this often inconspicuous upon drying). Stamens slightly exserted, the filaments usually 8–12 mm long, densely hairy for most of their length. Ovary with 4 ovules. Style shallowly to deeply branched. Fruits 4–6 mm long, the surface not appearing swollen, (3)4-seeded. Seeds 3–4 mm long, ovoid-angled, finely ridged and/or pitted, black. $2n=18$. April–June.

Uncommon in the eastern half of the Ozark Division north locally to St. Charles County (eastern U.S. west to Missouri and Arkansas). Banks of streams, bottomland forests, mesic upland forests in ravines, and bases and ledges of bluffs.

2. Phacelia covillei S. Watson

Map 1917

Plants relatively slender annuals. Stems 7–30 cm long, unbranched or more commonly few- to several-branched at the base and above, loosely to strongly ascending from a sometimes spreading base, moderately pubescent with stiff, loosely ascending to spreading, nonglandular hairs toward the base, these grading into dense, shorter, gland-tipped hairs toward the tip (median portion with mostly nonglandular hairs, a few gland-tipped hairs occasionally also present). Basal (and lower stem) leaves often persistent at flowering, the blade 1–3 cm long, broadly ovate to broadly elliptic or oblong-elliptic in outline, pinnately compound with 3–7 pinnae, these unlobed or (especially the terminal pinna) mostly moderately to deeply 2- or 3-lobed and sometimes also with a few additional teeth or smaller lobes, the teeth or lobes sharply to bluntly pointed or occasionally rounded at the tips, the surfaces sparsely to moderately pubescent with more or less appressed, stiff hairs. Stem leaves all moderately to long-petiolate, the blade 1–4 cm long, progressively less divided toward the stem tip, the uppermost leaf blades often irregularly 3-lobed, the leaflets or lobes sharply to bluntly pointed or occasionally rounded at the tips, the terminal lobe of at least the uppermost leaf (immediately below the lowermost flower) narrowly angled or tapered at the base, the pubescence similar to that of the basal leaves. Inflorescences with only 2–6 flowers, the axis and flower stalks densely glandular-hairy, sometimes also

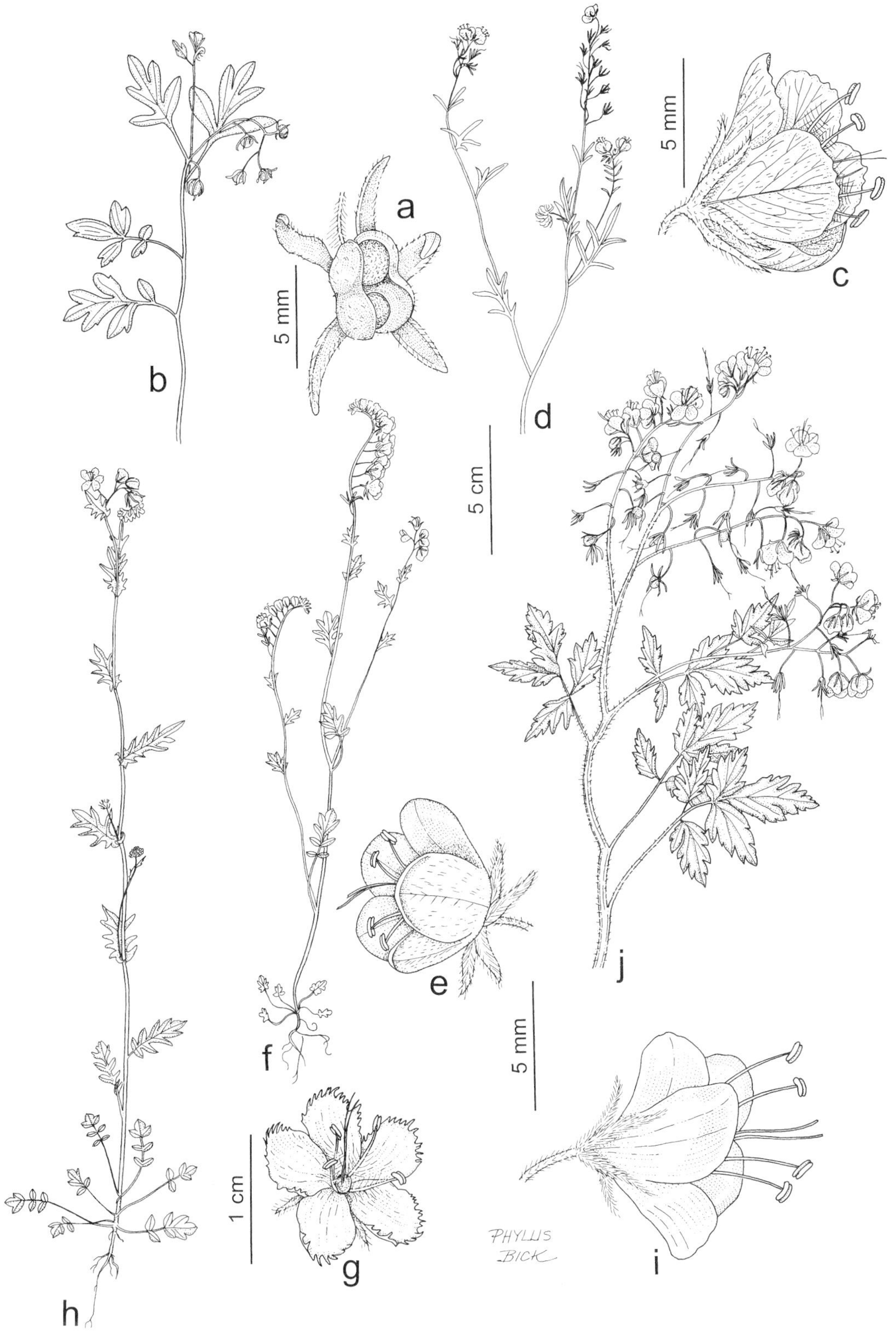

Plate 429. Hydrophyllaceae. *Phacelia ranunculacea*, **a)** flower, **b)** fertile stem. *Phacelia gilioides*, **c)** flower, **d)** fertile stem. *Phacelia hirsuta*, **e)** flower, **f)** habit. *Phacelia purshii*, **g)** flower, **h)** habit. *Phacelia bipinnatifida*, **i)** flower, **j)** fertile stem.

with scattered nonglandular hairs, the stalks spreading to downward-arched or pendant at fruiting. Calyces 3–7 mm long, the lobes with the margins and surfaces pubescent with mostly nonglandular hairs (glandular hairs more frequent at the calyx base). Corollas 2–4(–5) mm long, narrowly bell-shaped to nearly cylindrical, pale lavender to light purple, the margins entire, the outer surface glabrous. Stamens not exserted, the filaments 1.5–2.0 mm long, glabrous. Ovary with 4 ovules. Style branched to about the midpoint. Fruits 4–6 mm long, the surface usually irregularly swollen (distended by the seeds), 2–4-seeded. Seeds 2.0–2.5 mm long, broadly ovoid-angled to nearly globose, finely ridged, brown. $2n=28$. April–May.

Uncommon, known thus far only from single sites in Pulaski and Texas Counties (Maryland to North Carolina west disjunctly to Indiana, Illinois, and Missouri). Bottomland forests, mesic upland forests in ravines, and bases of bluffs.

Chuang and Constance (1977) were the first to document that plants called *P. ranunculacea* exhibit two distinct chromosome numbers in the eastern and western portions of the overall range, but they were not able to distinguish the two cytotypes morphologically. Sewell and Vincent (2006) showed that the eastern and western populations could be distinguished statistically based on a suite of subtle characters. By linking these relatively cryptic morphological differences to the different ploidy levels, they were able to use herbarium specimens to better circumscribe the ranges of the two cytotypes. Although the $n=14$ cytotype is mostly restricted to the eastern United States, there are scattered disjunct occurrences within the range of the $n=6$ cytotype in southern Indiana, southern Illinois, and southern Missouri. Because they could distinguish two entities morphologically and these entities have different base chromosome numbers, Sewell and Vincent (2006) chose to recognize them as distinct species. The underlying evolutionary relationship between the two is not understood and there are no data to address the conjecture that the high chromosome number in *P. covillei* is an indication of past hybridization between *P. ranunculacea* and some other species of *Phacelia*.

3. Phacelia gilioides Brand

Pl. 429 c, d; Map 1918

Plants relatively slender annuals. Stems 10–40 cm long, unbranched or few- to several-branched at the base and above, mostly strongly ascending from a sometimes spreading base, moderately pubescent with stiff, appressed or strongly ascending, nonglandular hairs toward the base, also with moderate to dense, fine, somewhat tangled, nonglandular hairs toward the tip. Basal leaves usually withered at flowering, the blade 1–3 cm long, elliptic to oblong in outline, pinnately compound or deeply lobed with 5–11 pinnae or lobes, these entire or occasionally with 1 or 2 lobes or coarse teeth (the terminal leaflet or lobe usually 3-lobed), the leaflets or lobes sharply to bluntly pointed or rounded at the tips, the surfaces sparsely to moderately pubescent with more or less appressed, stiff hairs. Stem leaves short-petiolate to sessile and usually clasping, the blade 0.8–5.0 cm long, progressively reduced toward the stem tip, the uppermost leaf blades usually 5 or 7(9)-lobed, the leaflets or lobes bluntly to more commonly sharply pointed at the tips, the terminal lobe of at least the uppermost leaf (immediately below the lowermost flower) truncate to slightly angled or tapered at the base, the pubescence similar to that of the basal leaves. Inflorescences with 8–25 flowers, the axis and flower stalks pubescent with dense, short, ascending to appressed hairs and sparse to moderate, longer, stiff, spreading hairs, the stalks ascending to spreading at fruiting. Calyces 4–8 mm long, the lobes with the margins and surfaces pubescent with longer and shorter hairs. Corollas 5–8 mm long, broadly bell-shaped, purple to bluish purple, the margins coarsely and irregularly toothed to fringed, the outer surface finely hairy, especially along the midvein and toward the tip. Stamens not or very slightly exserted, the filaments 4–7 mm long, densely hairy for most of their length. Ovary with mostly 8 ovules. Style branched to less than or about the midpoint. Fruits 3–4 mm long, the surface not appearing swollen, mostly 6–8-seeded. Seeds 1.5–2.0 mm long, ovoid-angled, finely ridged and/or pitted, dark brown. $2n=18$. April–June.

Scattered, mostly south of the Missouri River, most commonly in the Ozark Division (Illinois, Missouri, Arkansas, Kansas, and Oklahoma). Bottomland forests, mesic upland forests, banks of streams and rivers, margins of fens, bases and ledges of bluffs, thin-soil areas of upland prairies, sand prairies, and glades; also pastures, ditches, railroads, roadsides, and open disturbed areas.

Some specimens of *P. gilioides* can be difficult to distinguish from *P. hirsuta*. Constance (1949b) suggested that, based on distribution and morphology, *P. gilioides* might represent a species that evolved following past hybridization between *P. hirsuta* and *P. purshii*, but had no problems distinguishing all of these taxa. G. W. Gillett (1964, 1965) came to a somewhat different conclusion. His artificial hybridization program and greenhouse studies suggested that *P. purshii* is often

1921. Phacelia ranunculacea 1922. Itea virginica 1923. Carya aquatica

self-pollinated (other two taxa in the subgenus are mostly outcrossers), and that hybrids between it and either *P. gilioides* or *P. hirsuta* have reduced fertility. He found, however, that *P. gilioides* and *P. hirsuta* can produce relatively fertile hybrids and that some plants from central and southern Missouri were morphologically intermediate for three characters that he scored from herbarium specimens. Based on these data, he concluded that *P. gilioides* and *P. hirsuta* have formed hybrid swarms in the zone of geographic overlap of their ranges.

4. Phacelia hirsuta Nutt.

Pl. 429 e, f; Map 1919

Plants relatively slender annuals. Stems 10–45 cm long, unbranched or few- to several-branched at the base and above, mostly strongly ascending from a sometimes spreading base, densely pubescent with stiff, spreading, nonglandular hairs, also with moderate to dense, fine, somewhat tangled, nonglandular hairs toward the tip. Basal leaves usually withered at flowering, the blade 1–3 cm long, elliptic to oblong in outline, pinnately compound or deeply lobed with 5–9 pinnae or lobes, these entire or occasionally with 1 or 2 lobes or coarse teeth (the terminal leaflet or lobe usually 3-lobed), the leaflets or lobes sharply to bluntly pointed or rounded at the tips, the surfaces moderately to densely pubescent with more or less appressed, stiff hairs. Stem leaves short-petiolate to sessile and usually clasping, the blade 0.8–5.0 cm long, progressively reduced toward the stem tip, the uppermost leaf blades usually 5- or 7-lobed, the leaflets or lobes bluntly to sharply pointed at the tips, the terminal lobe of at least the uppermost leaf (immediately below the lowermost flower) truncate to somewhat angled or tapered at the base, the pubescence similar to that of the basal leaves. Inflorescences with 8–30 flowers, the axis and flower stalks pubescent with dense, short, ascending to appressed hairs and moderate to dense, longer, stiff, spreading hairs, the stalks ascending to spreading at fruiting. Calyces 4–9 mm long, the lobes with the margins and surfaces pubescent with longer and shorter hairs. Corollas 5–8 mm long, broadly bell-shaped, purple to bluish purple, often with a pale lower portion and sometimes also darker-spotted, the margins entire or minutely and inconspicuously toothed or scalloped, the outer surface finely hairy, especially along the midvein and toward the tip. Stamens not or very slightly exserted, the filaments 4–6 mm long, densely hairy for most of their length. Ovary with mostly 8 ovules. Style branched to less than or about the midpoint. Fruits 3–4 mm long, the surface not appearing swollen, mostly 6–8-seeded. Seeds 1.5–2.0 mm long, ovoid-angled, finely ridged and/or pitted, dark brown. $2n=18$. April–June.

Scattered in the Ozark Division north locally to Barton, Benton, Boone, and Macon Counties (Missouri and Kansas south to Louisiana and Texas). Bottomland forests, mesic upland forests, banks of streams and rivers, margins of fens, bases and ledges of bluffs, thin-soil areas of upland prairies, sand prairies, and glades; also pastures, ditches, railroads, roadsides, and open disturbed areas.

Rare white-flowered plants from southwestern Missouri have been called f. *albiflora* E.J. Palmer & Steyerm. For a discussion of possible relationships with *P. gilioides*, see the treatment of that species.

5. Phacelia purshii Buckley (Miami mist)

Pl. 429 g, h; Map 1920

Plants relatively slender annuals. Stems 10–40(–60) cm long, unbranched or few- to several-branched at the base and above, mostly strongly ascending from a sometimes spreading base, moderately pubescent with stiff, appressed or strongly ascending, nonglandular hairs toward the base, sometimes also with moderate to dense, shorter, nonglandular hairs toward the tip. Basal leaves usually withered at flowering, the blade 1–3 cm long, elliptic to oblong in outline, pinnately compound or deeply lobed with 3–7 pinnae or lobes, these entire or occasionally with 1 or 2 lobes or coarse teeth (the terminal leaflet or lobe usually 3-lobed), the leaflets or lobes sharply to bluntly pointed or occasionally rounded at the tips, the surfaces sparsely to moderately pubescent with more or less appressed, stiff hairs. Stem leaves short-petiolate to sessile and usually clasping, the blade 0.8–5.0 cm long, progressively reduced toward the stem tip, the uppermost leaf blades usually 5–9-lobed, the leaflets or lobes usually sharply pointed at the tips, the terminal lobe of at least the uppermost leaf (immediately below the lowermost flower) truncate to somewhat angled or tapered at the base, the pubescence similar to that of the basal leaves. Inflorescences with 8–30 flowers, the axis and flower stalks pubescent with dense, short, ascending to appressed hairs, sometimes also with sparse, longer, stiff, spreading hairs, the stalks ascending to spreading at fruiting. Calyces 3–7 mm long, the lobes with the margins and surfaces pubescent with longer and shorter hairs. Corollas 5–8 mm long, broadly bell-shaped, pale lavender to light purple or bluish-tinged, often with a pale center, the margins coarsely and irregularly toothed to fringed, the outer surface glabrous or very sparsely hairy. Stamens usually somewhat exserted, the filaments 4–7 mm long, densely hairy below the midpoint. Ovary with 4 ovules. Style shallowly to deeply branched. Fruits 3–5 mm long, the surface not appearing swollen, mostly 2–4-seeded. Seeds 2–3 mm long, ovoid-angled, finely ridged and/or pitted, dark brown. $2n=18$. April–June.

Scattered in eastern Missouri west locally to Boone and Pettis Counties and disjunct in McDonald County (eastern U.S. west to Illinois, Missouri, and Alabama; Canada). Bottomland forests, mesic upland forests, banks of streams and rivers, margins of fens, bases and ledges of bluffs, thin-soil areas of upland prairies, sand prairies, and glades; also pastures, ditches, railroads, roadsides, and open disturbed areas.

Plants of *P. purshi* tend to have the corolla lobes more deeply fringed than in *P. gilioides* and have slightly larger seeds. The stem leaves have lobes that often are slightly broader and more tapered than is typical for either *P. gilioides* or *P. hirsuta*. For further discussion of possible relationships among these taxa, see the treatment of *P. gilioides*. Steyermark (1963) indicated that all three of these species can be grown from seed in sunny gardens.

A single specimen of *P. fimbriata* Michx. collected by Samuel B. Buckley is present in the Missouri Botanical Garden Herbarium that carries no collection data other than the word Missouri. Buckley (1809–1884) was a naturalist who made extensive plant collections throughout the southeastern United States, including Texas. Although he collected briefly in Illinois in the 1830s, there is no evidence to suggest that he ever botanized in the state of Missouri. In 1947, the late Lincoln Constance, a specialist on the genus, annotated the sheet as probably instead having been collected in 1842 near Roan Mountain (Tennessee). Thus, the species is excluded from the Missouri flora. *Phacelia fimbriata* is endemic to the southern Appalachians from Virginia to Alabama and Georgia. It closely resembles *P. purshii*, but differs in its spreading (vs. appressed) stem hairs, white or rarely pale lavender (vs. pale lavender to light purple or bluish-tinged) corollas, and larger seeds (3.0–3.5 mm vs. 2–3 mm long).

6. Phacelia ranunculacea (Nutt.) Constance
Pl. 429 a, b; Map 1921

Plants relatively slender annuals. Stems 7–25 cm long, unbranched or more commonly few- to several-branched at the base and above, loosely to strongly ascending from a sometimes spreading base, moderately pubescent with stiff, loosely ascending to spreading, nonglandular hairs toward the base, these grading into dense, fine, gland-tipped hairs well below the midpoint (median portion with sparse to moderate nonglandular hairs and moderate to dense glandular hairs). Basal (and lower stem) leaves often persistent at flowering, the blade 1–3 cm long, broadly ovate to broadly elliptic or oblong-elliptic in outline, pinnately compound with 3–7 pinnae, these unlobed or (especially the terminal pinna) mostly moderately to deeply 2- or 3-lobed and sometimes also with a few additional teeth or smaller lobes, the teeth or lobes sharply to bluntly pointed or occasionally rounded at the tips, the surfaces sparsely to moderately pubescent with more or less appressed, stiff hairs. Stem leaves all moderately to long-petiolate, the blade 1–4 cm long, progressively less divided toward the stem tip, the uppermost leaf blades often irregularly 3-lobed, the leaflets or lobes sharply to bluntly pointed or occasionally rounded at the tips, the terminal lobe of at least the uppermost leaf (immediately below the lowermost flower) narrowly angled or tapered at the base, the pubescence similar to that of the basal leaves (but the petiole often with at least some glandular hairs). Inflorescences with only 2–6 flowers, the axis and flower stalks densely pubescent with gland-tipped hairs, rarely also with a few nonglandular hairs, the stalks spreading to downward-arched or pendant at fruiting. Calyces 3–7 mm long, the lobes with the margins and surfaces pubescent with mostly nonglandular hairs (glandular hairs more frequent toward the calyx base). Corollas 2–4(–5) mm long, narrowly bell-shaped to nearly cylindrical, white or pale lavender to bluish-tinged, the margins entire, the outer surface glabrous. Stamens not exserted, the filaments 1.5–2.0 mm long, glabrous. Ovary with 4 ovules. Style branched to about the midpoint. Fruits 4–6 mm long, the surface usually irregularly swollen (distended by the seeds), 2–4-seeded. Seeds 1.8–2.5 mm long, broadly ovoid-angled to nearly globose, finely ridged, light brown to brown. $2n=12$. April–May.

Uncommon in the Mississippi Lowlands Division west locally to Oregon County (Indiana to Tennessee and Mississippi west to Illinois, Missouri and Arkansas). Bottomland forests, mesic upland forests in ravines, and bases and ledges of bluffs; also gardens and shaded disturbed areas.

As noted above in the discussion of *Nemophila*, Palmer and Steyermark (1935) misdetermined the initial collections of this species in Missouri as a member of that genus. During his systematic studies of *Nemophila*, Constance (1949a, b) redetermined the Missouri specimens as *Phacelia ranunculacea* and Steyermark (1963) subsequently included the species in the state's flora under that name. For a discussion of the segregation of *P. covillei* from *P. ranunculacea*, see the treatment of that species.

ITEACEAE (Sweet-Spire Family)

Contributed by David Bogler

One genus, about 20 species, eastern North America, eastern Asia, and western Malaysia.

The genus *Itea* has been included in the Saxifragaceae by most authors (Steyermark, 1963) or sometimes in the Grossulariaceae (Cronquist, 1991) when that family has been segregated from the Saxifragaceae. However, recent research (Morgan and Soltis, 1993; Soltis and Soltis, 1997; Bohm et al., 1999) has shown that although it and the Grossulariaceae are related to the Saxifragaceae, these woody genera should be kept distinct from that family. The present treatment deviates from the standard practice in this manual of following the familial classification of Cronquist (1981, 1991) in treating *Itea* in its own family to reflect this new understanding of intergeneric relationships in the Saxifragales, as well as to make the families Grossulariaceae and Saxifragaceae more easily circumscribed and keyed.

1. Itea L.

About 20 species, eastern North America, eastern Asia, and western Malaysia.

1. Itea virginica L. (sweet-spire, Virginia willow, tassel-white)

Pl. 427 i–l; Map 1922

Plants shrubs, 1–3 m tall, the bark gray to brown, smooth or breaking into thin scales with age. Stems loosely ascending with spreading branches, unarmed. Twigs finely hairy, the pith broken into chambers at maturity. Leaves alternate, the petiole 2–6 mm long. Stipules minute, shed as the leaves develop. Leaf blades unlobed, 3–13 cm long, 1.5–4.5 cm wide, elliptic to oblong-lanceolate, narrowed or more commonly tapered to a sharply pointed tip, narrowed at the base, pinnately veined, the margins sharply and usually finely toothed, the upper surface glabrous, the undersurface finely hairy along the veins. Inflorescences terminal racemes 4–22 cm long, the axis hairy, the short flower stalks with inconspicuous small membranous bracts toward the base, these often shed during flower development. Flowers perfect, actinomorphic, perigynous (nearly hypogynous), fragrant. Hypanthium small, saucer-shaped. Sepals 5, 1.2–1.4 mm long, narrowly triangular from a broad base, the slender tips often broken off by fruiting. Petals 5, 4–6 mm long, somewhat longer than the stamens, linear, somewhat incurved, white. Stamens 5, short, alternating with the petals, attached to the hypanthium, the anthers attached toward their midpoint. Pistil 1 per flower, of 2 fused carpels. Ovary superior, deeply grooved, densely hairy, with 2 locules, with numerous ovules, the placentation axile. Style 1 at flowering but persistent and separating into 2 styles at fruiting, the stigma capitate. Fruits capsules, pendant at maturity, 7–10 mm long, 2–3 mm wide, the body narrowly elliptic in outline, tapered to the pair of short slender beaks, dehiscing along the groove between the carpels, with numerous seeds. Seeds 0.8–1.1 mm long, ovate to oblong-ovate in outline, somewhat flattened, the surface smooth or faintly and minutely pebbled, black, shiny. $2n=22$. May–June.

Uncommon in the Mississippi Lowlands Division, mainly along the base of Crowley's Ridge (southeastern U.S. west to Missouri, Oklahoma, and Texas). Swamps, bottomland forests, and acid seeps.

JUGLANDACEAE (Walnut Family)

Contributed by Alan Whittemore

Plants small to large trees, monoecious. Leaves alternate, short- to moderately long-petiolate. Leaf blades 1 time pinnately compound, the leaflets unlobed, pinnately veined, the margins toothed. Stipules absent. Staminate and pistillate inflorescences separate, with small bracts (and usually also minute bractlets) subtending and sometimes partially fused with each other and/or to the flowers, the staminate flowers in long, pendent, solitary or clustered catkins produced near the base of the current-year's growth, the pistillate flowers solitary or in very short spikes produced near the tip of the current-year's growth. Flowers actinomorphic, imperfect, the pistillate ones epigynous. Staminate flowers with the calyx absent or minutely 4-lobed; corolla absent; the stamens mostly 3–50, with a short filament and the anther attached at its base. Pistillate flowers with the calyx absent or minutely 4-lobed; corolla absent; stamens and staminodes absent; the ovary inferior (but sometimes appearing naked in the absence of a perianth), usually 2-locular toward the base but appearing 1-locular toward the tip, with 1 ovule, the placentation basal. Styles usually 2 (sometimes united basally), relatively stout and spreading, each with an expanded stigmatic region along the upper side. Fruits nuts, enclosed in a thick, fleshy to leathery or woody husk. About 8 genera, about 60 species, North America to South America, Europe, Asia, Pacific Islands.

The Missouri genera of this family all have large, oil-rich nuts that are an important food for wildlife, and have been important food for humans in the past. Various species are larval food plants for a variety of butterflies and moths, including the banded hairstreak butterfly (*Satyrium calanus* (Hübner)), walnut sphinx moth (*Amorpha juglandis* (J.E. Sm.)), luna moth (*Actias luna* (L.)), various underwing moths (*Catocala* spp.), dagger moths (*Acronicta* spp.), one-spotted variant moth (*Hypagyrtis unipunctata* (Haw.)), and giant regal moth (*Citheronia regalis* (Fabr.)) (Heitzman and Heitzman, 1987; Kurz, 2003). The last of these produces the largest caterpillar in the United States, the hickory horned devil, a fierce-looking green creature that can grow to nearly 15 cm in length. Members of the Juglandaceae are among the last trees to leaf out in the spring and among the first to lose their leaves in autumn.

1. Twigs with the pith solid and homogeneous; leaflets (3–)5–13 per leaf, the largest leaflets often near the tip of leaf, less commonly uniform in size or the median leaflets slightly larger; staminate catkins in clusters, stamens 3–10(–15) per flower; fruit husk splitting longitudinally for at least half of its length, more or less free from the smooth or shallowly wrinkled nut . . . . 1. CARYA

1. Twigs with the pith chambered (hollow between cross-partitions); leaflets (7–) 11–19(–23) per leaf, uniform in size or the median leaflets the largest; staminate catkins solitary, stamens 7–50 per flower; fruit husk not splitting, remaining more or less fused with the deeply grooved or ridged nut .... 2. JUGLANS

1. **Carya** Nutt. (hickory)

Twigs with the pith solid and homogeneous in texture. Leaflets (3–)5–13 per leaf, the largest leaflets often near the leaf tip, less commonly uniform in size or the median leaflets slightly larger than the others. Staminate catkins in clusters, sessile or stalked; each flower with 3–10(–15) stamens. Fruits with the husk splitting longitudinally for at least half of the fruit length, releasing the nut. Nuts often with 4 longitudinal ridges, the surface otherwise smooth or shallowly wrinkled. About 17 species, eastern U.S., Canada, Mexico, Asia.

Hickory nuts are composed of a large seed surrounded by several layers of tissue. The fleshy to leathery or somewhat woody husk of the hickory nut develops from the fused bractlets associated with each flower that become enlarged and thickened as the fruit matures. The husk splits and separates from the fruit. The fruit wall is hard (bony) and is known as the shell of the nut; it remains indehiscent and fused to the seed.

Species of *Carya* fall into two well-defined groups: the pecans (section *Apocarya* C. DC., including species 1, 2, and 4 below),with bud scales not overlapping, more numerous leaflets, and seeds that are often bitter; and the hickories (section *Carya*, including species 3 and 5–9) with strongly overlapping bud scales, fewer leaflets, and seeds that are always sweet-tasting.

The true hickories, not the pecans, were the staple plant food of many Missouri cultures for thousands of years (from about 6500 B.C. to 500 A.D.). The seeds are oily, and are rich sources of lipids, fiber, and starch. The protein content is moderate, and the amino-acid content of the protein is well balanced for human consumption. Unlike some other nuts, hickories do not need to be cooked or leached before eating. Hickories set seed in the fall; they are easy to gather and store, and are abundant in most years. Once the nut is free from the husk, it is impossible to remove it intact from the hard inner shell. Native Americans apparently smashed the nut, removed the large fragments of shell, and ground the nutmeat into a paste, so that smaller fragments of shell were finely ground and not troublesome. Contemporary Americans, with their preference for intact nuts, almost never eat hickories, and only the pecan, *Carya illinoinensis*, is much used now for human food.

Hickory nuts remain very important as an abundant and nutritious wildlife food. The wood of *Carya* spp. is strong and stands up well to impact and other shocks. It is widely used for tool handles, ladder rungs, baseball bats, agricultural implements, and other uses requiring shock resistance. It is also widely used for smoking meats.

Hickories are important street trees throughout the state. Their attractive form and often striking golden yellow fall color make them very desirable as street and park trees, but like oaks, the seedling forms a long taproot that makes transplantation difficult. Thus, although several of the species are available commercially, hickories are underutilized in the nursery trade.

1. Leaves with (5)7–13 leaflets (almost always at least some leaves with 9 or more leaflets), often appearing arched (asymmetrically tapered); terminal buds with the scales meeting at their margins and not or scarcely overlapping; fruits with four low wings along the sutures; shell of the nut no more than 1 mm thick (section *Apocarya*)
   2. Husk of fruit splitting for 0.5–0.8 of its length; terminal buds yellow or tan; bractlets surrounding the axillary buds separate (but often completely hidden by dense peltate scales) .... 2. C. CORDIFORMIS

2. Husk of fruit splitting to the base; terminal buds brown; bractlets surrounding axillary buds fused into a hood
   3. Bark exfoliating in long strips or large plates; leaflets with the underside of the midvein pubescent (occasionally nearly glabrous late in the season); staminate catkins more or less stalked; nuts flattened (oval on cross-section); seeds bitter . . . . 1. C. AQUATICA
   3. Bark remaining tight with persistent ridges or exfoliating in small plates; leaflets with the underside of the midvein glabrous or very sparsely pubescent; staminate catkins sessile or nearly so; nuts not flattened (more or less circular in cross-section), seeds sweet . . . . 4. C. ILLINOINENSIS

1. Leaflets (3)5 or 7(9) per leaf, more or less symmetrically tapered (not appearing arched); terminal buds with the scales strongly overlapping; fruits smooth or ridged along the sutures; shell of the nut 1.0–3.5 mm thick (section *Carya*)
   4. Leaflet margins pubescent with the hairs strongly tufted near the tips of the teeth; bark exfoliating from the trunk in long curly strips . . . . 6. C. OVATA
   4. Leaflet margins glabrous or pubescent with the hairs scattered evenly; bark shallowly to deeply furrowed to form ridges that remain firmly attached to the trunk or split off in small flakes (except in *C. laciniosa*)
      5. Leaflet undersurface evenly short-hairy; twigs 4–6 mm thick; terminal buds 9–24 mm long; leaflets mostly 7; husk of fruit 3–12 mm thick
         6. Bark exfoliating from the trunk in long curly strips; petiole and rachis usually sparsely pubescent except near the leaflet bases; fruits with the husk splitting to the base . . . . 5. C. LACINIOSA
         6. Bark splitting into ridges that remain firmly attached to the trunk; petiole and rachis usually densely (rarely sparsely) pubescent; husk splitting up to 80% of its length (sometimes one or two splits reaching base) . . . . 9. C. TOMENTOSA
      5. Leaflet undersurface glabrous or hairy only near the veins, rarely sparsely hairy throughout; twigs 2–4(–5) mm thick; terminal buds 4–10 mm long; leaflets usually 5–7; husk 1–3 mm thick
         7. Leaflet undersurface with many large round silvery gray scales; petiole and rachis sparsely to densely hairy (always densely so near leaflet bases) . . . . 7. C. PALLIDA
         7. Leaflet undersurface with the scales reddish and/or yellowish; petiole and rachis glabrous or sparsely hairy
            8. Winter buds dark brown, this often obscured by a dense covering of short, velvety, gray or tawny hairs; leaflet undersurface with many small red and yellow scales that are more or less circular and unlobed . . . . 3. C. GLABRA
            8. Winter buds dark brown but appearing golden-frosted due to a dense covering of small yellow scales, not hairy; leaflet undersurface with many large and small reddish scales that are irregularly circular and often more or less lobed . . . . 8. C. TEXANA

1. Carya aquatica (F. Michx.) Nutt. (water hickory)

Pl. 430 a–c; Map 1923

Plants trees to 35 m tall (to 45 m elsewhere). Bark light gray to brown, exfoliating from the trunk in long strips or large plates. Twigs 2–3 mm thick, brown to reddish brown or blackish, the terminal bud 8–10 mm long, dark brown, the bud scales meeting at their margins and not or scarcely overlapping, the bractlets around the axillary buds fused into a hood. Leaves with the petiole and rachis pubescent (sometimes becoming nearly glabrous late in the season), with 7–13 leaflets (always 9–13 leaflets in well-developed leaves). Leaf-

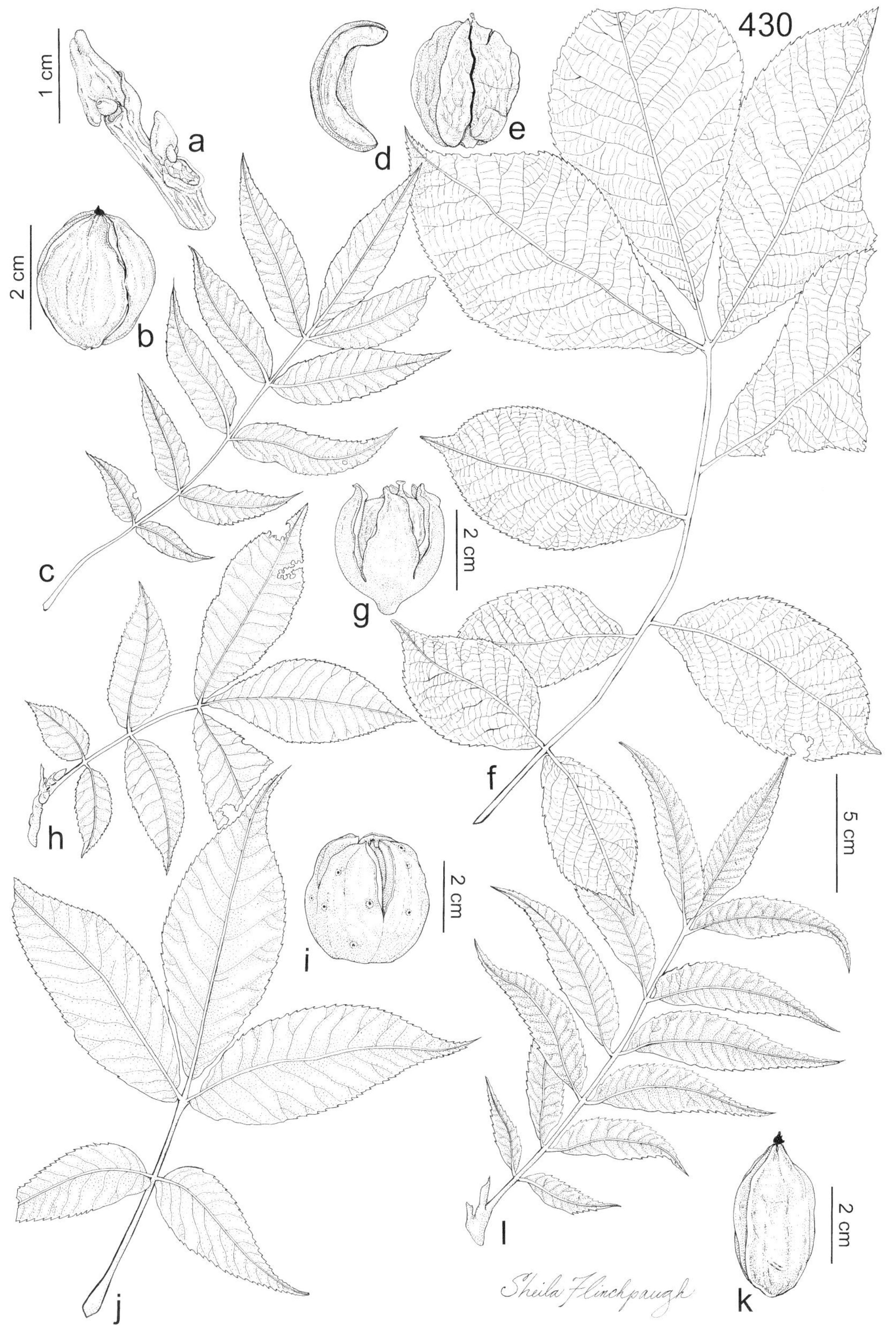

Plate 430. Juglandaceae. *Carya aquatica*, **a)** twig with winter buds and leaf scar, **b)** fruit, **c)** leaf. *Carya laciniosa*, **d)** section of husk of fruit, **e)** nut (fruit with husk removed), **f)** leaf. *Carya cordiformis*, **g)** fruit with splitting husk, **h)** twig with leaf. *Carya ovata*, **i)** fruit with splitting husk, **j)** leaf. *Carya illinoinensis*, **k)** fruit, **l)** twig with leaf.

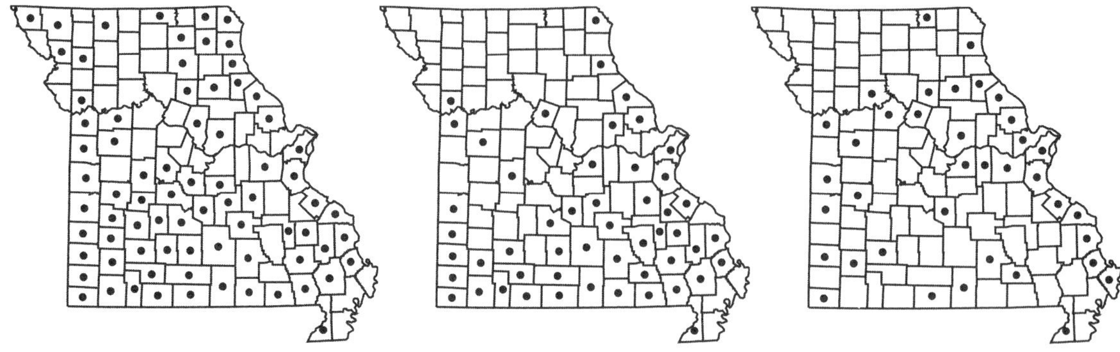

1924. Carya cordiformis 1925. Carya glabra 1926. Carya illinoinensis

lets 2–15 cm long, 1.0–4.5 cm wide, lanceolate to broadly lanceolate, weakly to strongly arched (appearing asymmetrically tapered), the margins entire to wavy or toothed, glabrous, the upper surface glabrous or with scattered hairs along the midvein near its base, sometimes also with scattered, small circular, pale yellow to reddish brown, peltate scales, the undersurface with scattered unbranched and branched (the branches appearing fasciculate) hairs along the midvein and usually also the secondary veins, also with scattered, small, circular, pale yellow to reddish brown, peltate scales. Staminate catkins more or less stalked. Fruits 2–3(–5) cm long, 2–3 cm wide, obovoid, noticeably flattened, with low wings along the 4 sutures, the husk 1–3 mm thick, splitting to the base, with small golden yellow scales that wear off with age. Nut flattened (oval in cross-section), the shell less than 1 mm thick. Seed bitter. April–May.

Uncommon in the Mississippi Lowlands Division (southeastern U.S. west to Missouri and Texas). Swamps, bottomland forests, and margins of oxbows and sloughs.

In Missouri, *C. aquatica* is known to hybridize occasionally with *C. illinoinensis* (*C. ×lecontei* Little).

2. Carya cordiformis (Wangenh.) K. Koch (bitternut hickory, pignut hickory)

Pl. 430 g, h; Map 1924

Plants trees to 35 m tall (to 50 m elsewhere). Bark gray to brownish gray, smooth or shallowly furrowed and exfoliating in small flakes. Twigs 3–4 mm thick, tan to light brown, the terminal bud 4–11 mm long, yellow or tan, the bud scales meeting at their margins and not or scarcely overlapping, the bractlets around the axillary buds free (but usually completely hidden by dense peltate scales). Leaves with the petiole and rachis pubescent (sometimes becoming nearly glabrous toward the petiole base), with (5)7 or 9 leaflets. Leaflets 4–18 cm long, 1–8 cm wide, lanceolate to oblanceolate or obovate, straight or weakly arched (appearing slightly asymmetrically tapered), the margins finely to coarsely toothed, glabrous or with scattered hairs, the upper surface glabrous or with scattered, small, circular, whitish to brownish, peltate scales, the undersurface with scattered to relatively dense, branched (the branches appearing fasciculate) hairs, sometimes mainly along the veins or becoming nearly glabrous with age, and also with scattered, small, circular, pale yellow to brownish, peltate scales. Staminate catkins noticeably stalked to nearly sessile. Fruits 2–3 cm long, 2–3 cm wide, globose to somewhat ellipsoid, not flattened, with low wings along the 4 sutures, the husk 1–2 mm thick, splitting 50–80% of its length, with small golden yellow scales that wear off with age. Nut not or only slightly flattened, the shell less than 1 mm thick. Seed bitter. $2n=32$. April–May.

Scattered to common nearly throughout the state (eastern U.S. west to Minnesota and Texas; Canada). Banks of streams and rivers, margins of ponds, lakes, sinkhole ponds, oxbows, and sloughs, bottomland forests, bases of bluffs, and less commonly mesic upland forests; also roadsides.

Plants with unusually broad leaflets (5 cm or broader) have been called f. *latifolia* (Sarg.) Steyerm.

3. Carya glabra (Mill.) Sweet (pignut hickory, broom hickory, black hickory)

C. glabra var. *odorata* (Marshall) Little
C. ovalis (Wangenh.) Sarg.
C. ovalis var. *odorata* (Marshall) Sarg.

Pl. 431 i–k; Map 1925

Plants trees to 35 m tall. Bark gray to light brownish gray, smooth or shallowly to deeply furrowed and sometimes exfoliating in small flakes. Twigs 3–4(–5) mm thick, reddish brown to brown,

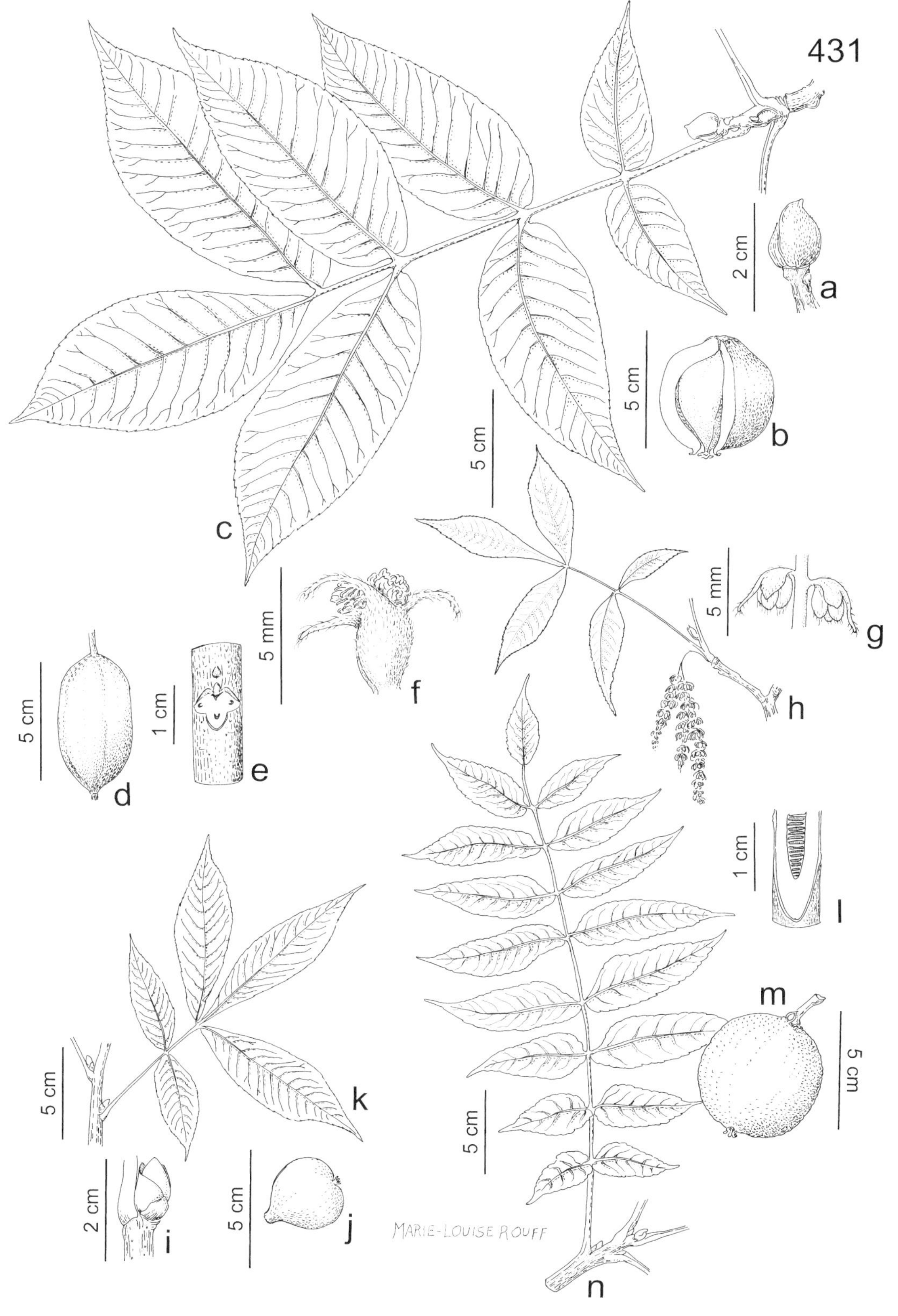

Plate 431. Juglandaceae. *Carya tomentosa*, **a)** winter bud, **b)** fruit with section of husk removed, **c)** twig with leaf. *Juglans cinerea*, **d)** fruit, **e)** section of twig with leaf scar, **f)** pistillate flower. *Carya texana*, **g)** staminate flowers, **h)** twig with leaf and staminate inflorescences. *Carya glabra*, **i)** winter bud, **j)** fruit, **k)** twig with leaf. *Juglans nigra*, **l)** cross-section of stem with chambered pith, **m)** fruit, **n)** twig with leaf.

the terminal bud 6–9 mm long, reddish brown to dark brown or tan (often appearing tawny because of the small yellow scales and/or velvety hairs), the bud scales strongly overlapping, the bractlets around the axillary buds fused for about half their length. Leaves with the petiole and rachis glabrous or inconspicuously short-hairy, with 5 or 7 leaflets. Leaflets 4–18 cm long, 2–8 cm wide, lanceolate to elliptic-lanceolate, straight (appearing symmetrically tapered), the margins finely to coarsely toothed, glabrous, the upper surface glabrous or with widely scattered, small, circular, yellow, peltate scales, the undersurface with scattered, branched (the branches appearing fasciculate) hairs mainly along the midvein or in the main vein axils (rarely more evenly pubescent), rarely glandular, also with scattered to dense, small, circular (unlobed), yellow, peltate scales. Staminate catkins noticeably stalked. Fruits 2.5–3.5 cm long, 2.5–3.0 cm wide, globose to ellipsoid, not or only slightly flattened, with low ridges or rarely low wings along the 4 sutures, the husk 2–3 mm thick, splitting to the base or rarely some of the sutures splitting only about 70% of their length, with small golden yellow scales that wear off with age. Nut usually slightly flattened, the shell 1.0–1.5 mm thick. Seed sweet. April–May.

Scattered from the Missouri River floodplain southward through the state and locally north along the Mississippi River to Clark County (eastern U.S. west to Iowa, Kansas, and Texas; Canada). Mesic to dry upland forests and less commonly banks of streams; also roadsides.

Most Missouri plants have scaly bark, leaves with mostly 7 leaflets, reddish petioles, and a husk that splits to the base; such plants have often been treated as a separate species, *Carya ovalis*. The other extreme of the species, with tight bark and large fruit with a husk that split for half its length, is well-developed in the southern states, especially on the Atlantic and Gulf Coastal Plains and lower Mississippi Embayment. Plants from the Mississippi Lowlands region in southeastern Missouri sometimes approach this extreme in having nonscaly bark, green petioles, and a large percentage of leaves with only 5 leaflets. These different forms intergrade so extensively that they now are seldom recognized taxonomically. A few populations in Madison and Iron Counties have the undersurface of the leaflets rather sticky and the husks winged along the sutures. This variant has been called var. *odorata* (red hickory).

A single specimen of *C. myristiciformis* (F. Michx.) Nutt.) exists in the Missouri Botanical Garden Herbarium that was collected in November, 1904, by R. B. Haugh. The specimen consists only of several nuts lacking hulls and bears a handwritten label that reads simply "Southern Missouri" without any further data. No additional information on the collector or occurrence could be located during the present research and it is not possible to determine whether the specimen originated from plants cultivated or growing wild in Missouri, or possibly from nuts collected elsewhere and submitted for identification by a resident of southern Missouri. Thus, the species is excluded from the Missouri flora for the present. *Carya myristiciformis* is a relatively uncommon species distributed sporadically throughout the southeastern United States and in northeastern Mexico. In Arkansas, it is considered uncommon and grows mainly in the southern half of the state. The species is somewhat similar to *C. glabra*, but differs in its fissured, peeling bark, leaves with mostly 7–9 leaflets that have relatively dense, peltate scales on the undersurfaces, and relatively small fruits with prominent ridges on the sutures.

4. Carya illinoinensis (Wangenh.) K. Koch (pecan)

Pl. 430 k, l; Map 1926

Plants trees to 35 m tall (to 45 m elsewhere). Bark light gray to brown, smooth or more commonly ridged, sometimes exfoliating from the trunk in small plates. Twigs 3–5 mm thick, tan to reddish brown, the terminal bud 6–10 mm long, yellowish brown, the bud scales meeting at their margins and not or scarcely overlapping, the bractlets around the axillary buds fused into a hood. Leaves with the petiole and rachis glabrous or sparsely pubescent, with 7–13 leaflets (always 9–13 leaflets in well-developed leaves). Leaflets 2–16 cm long, 1.0–4.5 cm wide, lanceolate to broadly lanceolate, moderately to strongly arched (appearing asymmetrically tapered), the margins finely to coarsely toothed, glabrous or with sparse, evenly scattered hairs, the surfaces rarely with scattered hairs along the midvein near its base, but usually with scattered, minute, gland-tipped hairs and scattered, small circular, pale yellow to reddish brown, peltate scales. Staminate catkins sessile or nearly so. Fruits 3–5 cm long, 1.5–2.5 cm wide, ellipsoid to subcylindrical, not flattened, with low wings along the 4 sutures, the husk 2–3 mm thick, splitting to the base, with small golden yellow scales that wear off with age. Nut not flattened (more or less circular in cross-section), the shell less than 1 mm thick. Seed sweet. $2n=32$. April–May.

Scattered nearly throughout the state, but apparently absent from much of the western portion of the Glaciated Plains Division (eastern [mostly

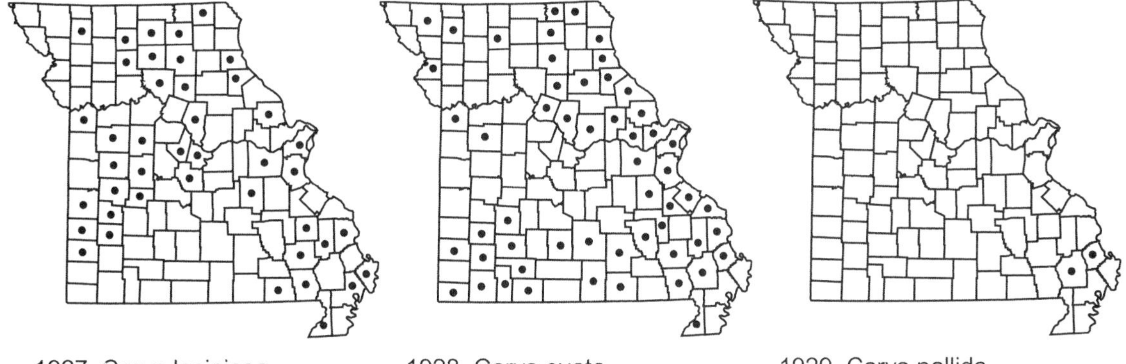

1927. Carya laciniosa 1928. Carya ovata 1929. Carya pallida

southeastern] U.S. west to Iowa and Texas; Mexico; introduced in Arizona, California). Bottomland forests, swamps, banks of streams and rivers, and margins of oxbows and sloughs, also pastures and roadsides.

Pecan is a nut crop of considerable economic importance. Large-scale commercial pecan production is mostly located in the southern states, but the tree is widely cultivated in Missouri.

In Missouri, *C illinoinensis* is known to hybridize occasionally with *C. aquatica* (*C.* ×*lecontei* Little), *C. laciniosa* (*C.* ×*nussbaumeri* Sarg.), and *C. tomentosa* (*C.* ×*schneckii* Sarg.). Note that Steyermark (1963) and many earlier botanists spelled the species epithet *C. illinoensis* rather than *C. illinoinensis* under the mistaken impression that this was a correctable orthographic error in the publication in which the taxon was first described.

5. Carya laciniosa (F. Michx.) Loudon (shellbark hickory, kingnut)

Pl. 430 d–f; Map 1927

Plants trees to 40 m tall. Bark light gray, exfoliating from the trunk in long curly strips. Twigs 4–6 mm thick, tan to dark brown, the terminal bud 15–24 mm long, tan or more commonly appearing grayish because of the dense hairs, the bud scales strongly overlapping, the bractlets around the axillary buds fused toward the base. Leaves with the petiole and rachis sparsely hairy becoming glabrous with age, with (5)7 or 9 leaflets. Leaflets 9–25 cm long, 4–11 cm wide, lanceolate to oblanceolate, elliptic, or obovate, straight (appearing symmetrically tapered), the margins coarsely toothed, glabrous or with scattered, evenly spaced hairs, the upper surface glabrous or with scattered unbranched and branched hairs, at least when young, also with occasional, small, circular, yellow, peltate scales, the undersurface with scattered unbranched and branched (the branches appearing fasciculate) hairs, also with scattered to dense, smaller, circular (unlobed), reddish brown, peltate scales and often also with scattered, larger yellow scales. Staminate catkins noticeably stalked. Fruits 4–6 cm long, 3.5–5.0 cm wide, globose to slightly ellipsoid, not or only slightly flattened, smooth along the 4 sutures, the husk 5–12 mm thick, splitting to the base, irregularly and minutely hairy. Nut flattened (oval in cross-section), the shell 3–6 mm thick. Seed sweet. $2n$=32. April–May.

Scattered widely in the state but absent from much of the Ozark Division (eastern U.S. west to Iowa, Kansas, and Texas; Canada). Bottomland forests, swamps, banks of streams and rivers, and margins of oxbows and sloughs.

In Missouri, *C. laciniosa* is known to hybridize occasionally with *C. illinoinensis* (*C.* ×*nussbaumeri* Sarg.). Kurz (2003) noted that this species has declined in abundance as the bottomlands of Missouri have been cleared for agriculture and other uses.

6. Carya ovata (Mill.) K. Koch **var. ovata** (shagbark hickory)

Pl. 430 i, j; Map 1928

Plants trees to 35 m tall (to 45 m elsewhere). Bark light gray, exfoliating from the trunk in long curly strips. Twigs 3–7 mm thick, tan to dark brown, hairy, the terminal bud (6–)8–14 mm long, tan to dark brown or more commonly appearing grayish to nearly black because of the dense hairs, the bud scales strongly overlapping, the bractlets around the axillary buds fused toward the base. Leaves with the petiole and rachis sparsely to moderately hairy, less commonly glabrous, with (3)5(7) leaflets. Leaflets 6–18 cm long, 3–8 cm wide, lanceolate to oblanceolate, elliptic, ovate, or obovate, straight (appearing symmetrically tapered), the margins finely to coarsely toothed, moderately pubescent with the hairs strongly tufted near the

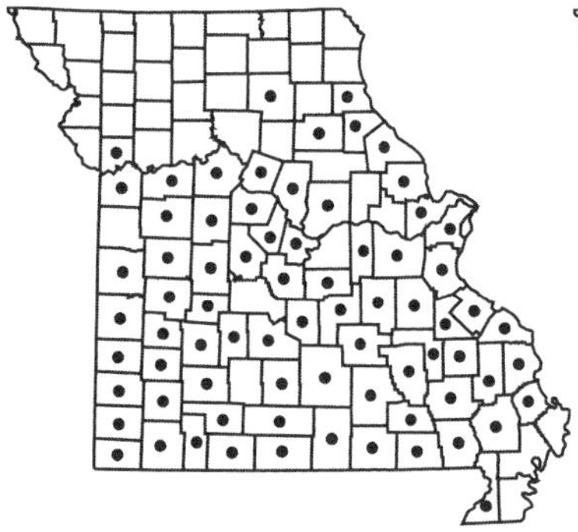

1930. Carya texana

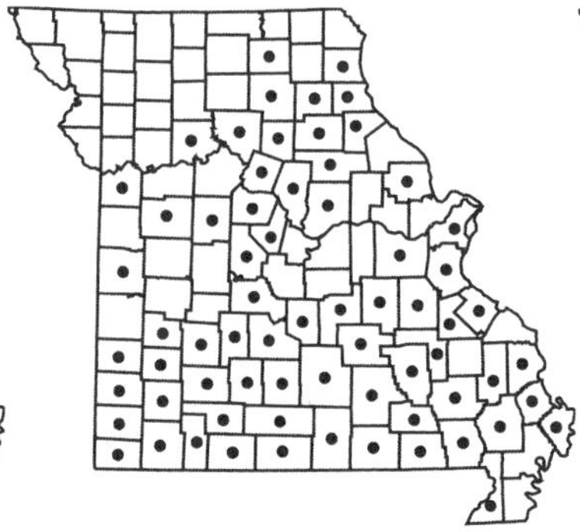

1931. Carya tomentosa

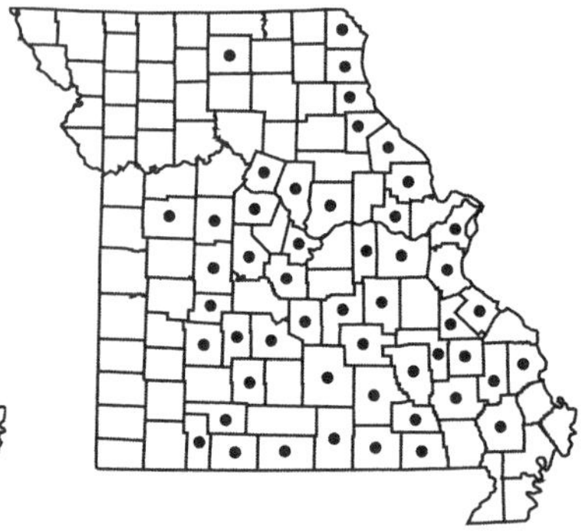

1932. Juglans cinerea

tips of the teeth, the upper surface glabrous or rarely with sparse hairs along the midvein, also usually with widely scattered, small, circular, reddish, peltate scales, the undersurface with scattered unbranched and branched (the branches appearing fasciculate) hairs, also with scattered to dense, smaller, circular (unlobed) to slightly irregular, reddish brown, peltate scales and often also with scattered, larger yellow scales. Staminate catkins noticeably stalked. Fruits 3–4 cm long, 3.0–4.5 cm wide, globose to slightly depressed-globose, not or only slightly flattened, smooth along the 4 sutures, the husk 4–12 mm thick, splitting to the base, glabrous or nearly so. Nut flattened (oval in cross-section), the shell 1.5–2.0 mm thick. Seed sweet. $2n=32$. April–May.

Common throughout the state (eastern U.S. west to North Dakota and Texas; Canada). Mesic to dry upland forests, edges of glades, banks of streams and rivers, margins of sinkhole ponds, and occasionally bottomland forests; also fencerows, pastures, and roadsides.

Names have been given to many trivial variants of the shagbark hickory. Plants with small leaves and fruits have been called f. *fraxinifolia* (Sarg.) Steyerm.; plants with elongate fruits have been called f. *ellipsoidalis* (Sarg.) Steyerm.; and plants with unusually small, strongly compressed fruit have been called f. *nuttallii* (Sarg.) Steyerm. A second variety, var. *australis* (Ashe) Little, occurs in the southeastern states and differs in its slender, glabrous twigs that tend to blacken upon drying, as well as its leaves with the leaflets glabrous or nearly so.

7. Carya pallida (Ashe) Engl. & Graebn. (sand hickory)

Map 1929

Plants trees to 30 m tall. Bark dark gray, deeply furrowed but remaining firmly attached (not exfoliating). Twigs 2–4 mm thick, reddish brown to brown, the terminal bud 5–10 mm long, dark brown to silvery gray or yellowish gray, the bud scales strongly overlapping, the bractlets around the axillary buds fused toward the base. Leaves with the petiole and rachis sparsely to densely hairy (densest at the leaflet bases), with (5)7 leaflets. Leaflets 3–14 cm long, 1–6 cm wide, lanceolate to elliptic or narrowly obovate, straight (appearing symmetrically tapered), the margins finely to coarsely toothed, glabrous or more commonly sparsely hairy (the hairs not noticeably clustered), the upper surface glabrous or occasionally with scattered hairs along the main veins (especially toward the leaflet base), also with scattered to dense, small, circular, yellow, peltate scales when young, the undersurface with scattered, branched (the branches appearing fasciculate) hairs mainly along the midvein or in the main vein axils, and also with a mixture of scattered to dense, larger, circular, silvery gray, peltate scales and smaller, circular, yellow, peltate scales. Staminate catkins noticeably stalked. Fruits 2.5–3.5 cm long, 2.5–3.0 cm wide, globose to somewhat pear-shaped, not or only slightly flattened, smooth or with low ridges along the 4 sutures, the husk 2–3 mm thick, splitting to the base or less commonly some of the sutures splitting 0.7–0.9 of their length, with small golden yellow scales that wear off with age. Nut usually slightly flattened, the shell 1–2 mm thick. Seed sweet. April–May.

Uncommon, known thus far only from Scott and Stoddard Counties (eastern [mostly southeastern] U.S. west to Missouri and Louisiana). Sand prairies, sand savannas, and less commonly mesic upland forests; usually in sandy soils.

This species was first reported for Missouri by T. E. Smith (1994), who noted that some trees in the state appear to be somewhat morphologically intermediate with *C. texana*.

8. Carya texana Buckley (black hickory, Ozark pignut hickory)

Pl. 431 g, h; Map 1930

Plants trees to 30 m tall (to 40 m elsewhere), but often much shorter. Bark dark gray to nearly black, smooth or shallowly to deeply furrowed or splitting into plates but remaining firmly attached (not exfoliating). Twigs 2–4(–5) mm thick, reddish brown, the terminal bud 4–8 mm long, reddish brown to dark brown (often appearing golden-frosted because of the abundant small yellow scales), the bud scales strongly overlapping, the bractlets around the axillary buds fused to above or below the midpoint. Leaves with the petiole and rachis glabrous or sparsely and inconspicuously hairy, with 5 or 7 leaflets. Leaflets 3–15 cm long, 1–6 cm wide, lanceolate to oblanceolate or narrowly obovate, straight (appearing symmetrically tapered), the margins finely to coarsely toothed, glabrous, the upper surface glabrous or with scattered to dense, small, irregularly circular, reddish, peltate scales, especially when young, the undersurface with scattered or rarely dense, branched (the branches appearing fasciculate) hairs along the midvein or in the main vein axils, and also with moderate to dense, larger and smaller, irregularly circular (more or less lobed), reddish, peltate scales. Staminate catkins noticeably stalked. Fruits 3–4 cm long, 2.5–3.0 cm wide, globose to more commonly obovoid or somewhat pear-shaped, not or only slightly flattened, smooth or with low ridges along the 4 sutures, the husk 1–3 mm thick, splitting to the base (sometimes only with age), with small, golden yellow scales that wear off with age. Nut not or more commonly slightly flattened, the shell 2–3 mm thick. Seed sweet. $2n$=64. April–May.

Scattered to common south of the Missouri River, uncommon farther north in the eastern portion of the Glaciated Plains Division (Indiana to Georgia west to Kansas and Texas). Mesic to dry upland forests, savannas, sand savannas, sand prairies, edges of glades, and margins of sinkhole ponds; also pastures and roadsides; usually on acidic substrates.

Carya texana is often difficult to separate from *C. glabra*. The scales on the undersurface of the leaf in *C. texana* are very heterogeneous in size and the large scales are usually crowded enough to give the leaf a rusty appearance, while the scales on the leaves of *C. glabra* are uniformly tiny, and the leaf undersurface merely looks very finely dark-speckled. The golden-yellow appearance of the winter buds is distinctive; the buds of *C. glabra* are gray or tawny, or merely dark brown. In the spring, when buds are absent, identifications made without a strong hand-lens or dissecting microscope should be considered tentative.

Plants with the petiole and rachis completely glabrous have been called *C. texana* f. *glabra* (E.J. Palmer & Steyerm.) Steyerm.

9. Carya tomentosa (Poir.) Nutt. (mockernut hickory, white hickory)

C. tomentosa var. *subcoriacea* (Sarg.) E.J. Palmer & Steyerm.

Pl. 431 a–c; Map 1931

Plants trees to 35 m tall. Bark dark gray, shallowly to deeply furrowed and splitting into plates but remaining firmly attached (not exfoliating). Twigs 3–5 mm thick, reddish brown, the terminal bud 9–14 mm long, grayish and densely hairy, the bud scales strongly overlapping, the bractlets around the axillary buds fused, at least toward the base. Leaves with the petiole and rachis usually densely hairy, with (3–)7(9) leaflets. Leaflets 5–21 cm long, 3–9 cm wide, oblanceolate to obovate, straight (appearing symmetrically tapered), the margins finely to coarsely toothed, glabrous or with evenly scattered hairs, the upper surface glabrous or with a few widely scattered hairs and/or small, circular, yellow, peltate scales, especially when young, the undersurface with sparse to moderate, unbranched and branched (the branches appearing fasciculate) hairs along and between the veins, and also with scattered to moderate, small, circular, whitish (translucent) and reddish, peltate scales. Staminate catkins noticeably stalked. Fruits 3.0–4.5 cm long, 3–4 cm wide, globose to somewhat ellipsoid or obovoid, not or only slightly flattened, smooth (rarely with faint ridges along 1 or more of the sutures), the husk 3–8 mm thick, splitting, splitting up to about 80% of its length (sometimes some of the splits reaching the base), with small golden yellow scales that wear off with age. Nut noticeably flattened (oval in cross-section), the shell 4–6 mm thick. Seed sweet. $2n$=64. April–May.

Scattered to common nearly throughout the state but apparently absent from the western portion of the Glaciated Plains Division (eastern U.S. west to Iowa, Kansas, and Texas). Mesic to dry upland forests, tops of bluffs, banks of streams and rivers, margins of ponds, lakes, and sinkhole ponds, and occasionally bottomland forests; usually on acidic substrates.

There has been controversy about the proper name for this species. Some authors have maintained that the correct name should be *C. alba* (L.) Nutt. The basis for this name, *Juglans alba*

L., was based on a mixture of plants here treated as *C. ovata* and *C. tomentosa*. Rehder (1945) noted this and suggested that the epithet *C. alba* should be rejected as a name of ambiguous application, and the recent proposal by Ward and Wiersema (2008) was approved at the 2012 International Botanical Congress. On the other hand, Wunderlin et al. (1985) concluded that the transfer of the epithet from *Juglans* into *Carya* established a priority for the name over *C. tomentosa*, based on their interpretation of the type element within the Linnaean name. However, this opinion has now been superseded by the official rejection of *Juglans alba* at the recent International Botanical Congress.

In Missouri, *C. tomentosa* is known to hybridize occasionally with *C. illinoinensis* (*C. ×schneckii* Sarg.). Names have been given to many variants of the mockernut. Plants with stalked fruits have been called f. *ficoidea* (Sarg.) E.J. Palmer & Steyerm.; plants with a tapering apex on the fruit have been called f. *ovoidea* (Sarg.) E.J. Palmer & Steyerm., and plants with thick leaflets and elongate nuts have been called var. *subcoriacea*. None of these trivial variants seems worthy of formal taxonomic recognition.

2. **Juglans** L. (walnut)

Twigs stout, brown to grayish brown, hairy, the pith chambered (hollow between cross-partitions). Leaflets (7–)11–19(–23) per leaf, relatively uniform in size or the median leaflets somewhat larger than the others. Staminate catkins solitary, sessile; each flower with 7–50 stamens. Fruits with the husk relatively thin and fleshy to papery, not splitting, remaining more or less fused with the nut. Nuts deeply grooved or ridged. About 20 species, North America to South America, Europe, Asia.

Two Old World walnut species with edible nuts are sometimes planted in Missouri but do not escape (Steyermark, 1963). The walnut of commerce is *J. regia* L., called Persian or English walnut; its leaves have 5–11 broad, entire leaflets. The Japanese walnut, *J. ailanthifolia* Carrière., has leaves with 9–17 toothed leaflets. Both differ from our native species in having nuts with thinner shells that are wrinkled, not grooved or ridged.

Both of our species provide high-quality lumber and edible nuts, as well as dyes (extracted from the bark and green fruits) that were once important in Missouri.

1. Fruits ellipsoid to ovoid or more or less cylindrical; surface of the nut with about 8 main, longitudinal ridges with narrower, interrupted, longitudinal ridges or folds between them; bark light gray or grayish brown, shallowly divided into smooth or scaly plates; terminal buds 12–18 mm; leaf scars with the upper edge straight, bordered by a well-defined velvety ridge . . . . 1. J. CINEREA
1. Fruits subglobose to globose or rarely ellipsoid; surface of the nut longitudinally irregularly many-grooved and coarsely warty between the grooves; bark medium to dark gray or brown, deeply split into narrow, rough ridges; terminal buds 8–10 mm long; leaf scars with the upper edge notched, glabrous or sometimes velvety, but never forming a prominent velvety ridge . . . . 2. J. NIGRA

1. **Juglans cinerea** L. (butternut, white walnut)

Pl. 431 d–f; Map 1932

Plants trees to 30 m tall. Bark light gray or grayish brown, shallowly divided into smooth or scaly plates. Pith dark brown. Terminal buds 12–18 mm long, conic, flattened. Leaf scars with the upper margin straight or nearly so, bordered by a well-defined, tan to gray, velvety ridge. Leaves 30–60 cm long, the petiole 3.5–12.0 cm long, glandular-hairy, with (7–)11–17 leaflets, including usually a large terminal leaflet. Leaflets (2.5–)5.0–11.0(–17.5) cm long, 1.5–6.5 cm wide, ovate to lanceolate or oblong-lanceolate, more or less symmetrical (not appearing asymmetrically tapered), mostly rounded at the base, tapered at the tip, the margins finely toothed, yellowish green, the upper surface with scattered fasciculate hairs or becoming nearly glabrous at maturity, the undersurface with abundant 4–8-branched hairs (the

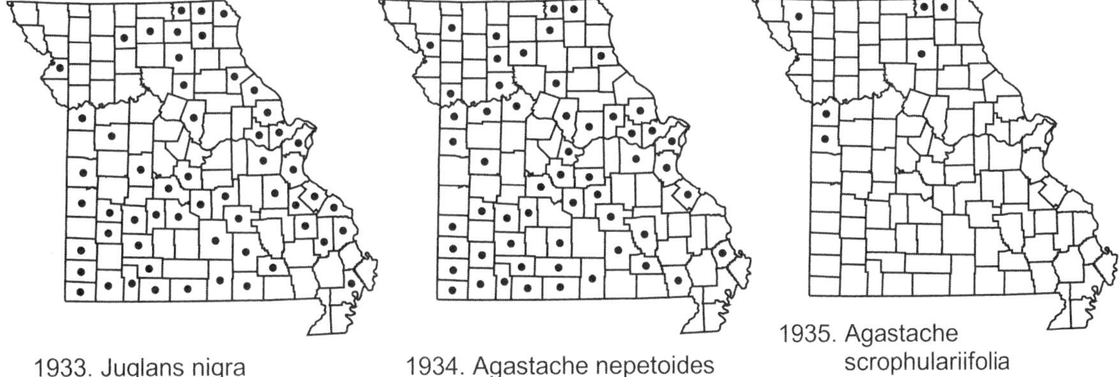

1933. Juglans nigra

1934. Agastache nepetoides

1935. Agastache scrophulariifolia

branches appearing fasciculate) and yellowish scales, sometimes also with gland-tipped hairs, the axils of the secondary veins with prominent tufts of fasciculate hairs. Staminate catkins 6–14 cm long, the staminate flowers with 7–15 stamens, the anthers 0.8–1.2 mm long. Fruits usually in clusters of 3–5, 4–8 cm long, ellipsoid to ovoid or more or less cylindrical, the husk smooth, densely covered with gland-tipped hairs, becoming slightly wrinkled and papery with age. Nuts 3–6 cm long, ellipsoid to ovoid or more or less cylindrical, with about 8 high, narrow, irregular, longitudinal ridges, the surface between the main ridges with narrower, interrupted, longitudinal ridges or folds. $2n=32$. April–May.

Formerly scattered nearly throughout the eastern 2/3 of Missouri, becoming increasingly uncommon (eastern U.S. west to Minnesota and Arkansas; Canada). Bottomland forests, mesic upland forests in ravines, bases of bluffs, and banks of streams and rivers.

Butternuts are difficult trees from a silvicultural perspective. They never form large stands but instead occur as scattered trees within their habitats. They are relatively intolerant of human disturbance, but they do not reproduce well in closed-canopy forests (Ostry et al., 1994). In recent decades, many trees have been killed by butternut canker, *Sirococcus clavigignenti-juglandacearum* Nair, Kostichka & Kuntz, a fungus that kills trees by girdling the limbs and trunk (Ostry et al., 1994). As a result, this imperiled species is becoming very uncommon throughout its range. Some of the butternuts in the horticultural trade apparently are hybrids between *J. cinerea* and the closely related Japanese walnut, *J. ailanthifolia* Carrière, which have improced resistence to the pathogen. These are not yet known to have escaped into the wild in Missouri, but are nearly impossible to distinguish from genetically pure butternut trees morphologically.

2. Juglans nigra L. (black walnut, eastern black walnut)

Pl. 431 l–n; Map 1933

Plants trees to 35 m tall (to 50 m elsewhere). Bark medium to dark gray or brown, deeply split into narrow, rough ridges. Pith light brown. Terminal buds 8–10 mm long, ovoid or subglobose, weakly flattened. Leaf scars with the upper margin notched, glabrous or sometimes velvety, but never forming a prominent velvety ridge. Leaves 20–60 cm long, the petiole 6.5–14.0 cm long, glandular-hairy, with (9–)15–19(–23) leaflets, sometimes including a small terminal leaflet. Leaflets (3–)6–15 cm long, 1.5–5.5 cm wide, lanceolate to narrowly ovate, more or less symmetrical or slightly arched (appearing slightly asymmetrically tapered), rounded to shallowly cordate (sometimes asymmetrically so) at the base, tapered at the tip, the margins finely toothed, yellowish green to green, the upper surface with scattered gland-tipped hairs and fasciculate hairs only along the midvein, the undersurface with moderate to abundant unbranched or 2-branched hairs (the branches appearing fasciculate) on and between the veins. Staminate catkins 5–10 cm long, the staminate flowers with 17–50 stamens, the anthers 0.8–0.9 mm long. Fruits usually solitary or paired, 4–8 cm long, subglobose to globose or rarely ellipsoid, the husk somewhat warty to nearly smooth, with scattered gland-tipped hairs and dense, minute scales, remaining fleshy or becoming somewhat leathery with age. Nuts 3–4 cm long, subglobose to globose or rarely ellipsoid, with numerous, irregular, rounded, longitudinal grooves and coarsely warty between the grooves. $2n=32$. April–May.

Scattered to common nearly throughout the state (eastern U.S. west to North Dakota and Texas; Canada; introduced farther west). Bottomland forests, mesic upland forests, bases of bluffs, and banks of streams and rivers; also margins of pastures and crop fields, railroads, and roadsides.

Black walnut is one of the premier hardwood timbers of North America. The nuts are a minor trade item; they surely would be used heavily if it were not so difficult to extract the nutmeat from the thick shell. In 1990, the Missouri legislature officially designated the eastern black walnut as the state nut, although at least one senator was of the opinion that this measure proved that Missouri had 197 state nuts sometimes also known as legislators (V. Young, 1990). The process of shelling the nuts (removing the husks) produces a strong unpleasant odor and stains the hands.

Walnut trees frequently have few plants growing close-by. A quinone compound known as juglone is produced mainly by the roots and fruit husks and has a strong allelopathic effect, inhibiting the establishment and growth of many other plant species (for a review, see Rice, 1984).

Plants with ellipsoidal fruits have been called f. *oblonga* (Marsh.) Fernald.

LAMIACEAE (LABIATAE) (Mint Family)

Plants annual or perennial herbs, shrubs, or small trees, rarely monoecious. Stems usually branched, usually strongly 4-angled (square) in cross-section, often hairy, sometimes bristly or roughened, or some or all of the hairs gland-tipped; apparently sessile spherical glands and/or flat to slightly impressed glandular dots sometimes also present. Leaves opposite or sometimes all or mostly basal, well-developed, sessile or petiolate. Stipules absent. Leaf blades simple (palmately compound in *Vitex*), sometimes deeply lobed, variously shaped, the margins entire or more commonly toothed, the surfaces often hairy, the hairs sometimes with pustular bases, sometimes stiff and calcified, some or all of the hairs sometimes gland-tipped; apparently sessile, spherical glands sometimes also present. Inflorescences terminal and/or axillary, variously solitary flowers, clusters, heads, racemes, spikes, or panicles, the flowers usually subtended by bracts. Flowers zygomorphic, sometimes weakly weakly so, hypogynous, usually perfect; cleistogamous flowers absent. Calyces mostly 5- or 10-lobed (toothed), occasionally only 2-lobed, persistent at fruiting, the lobes equal or nearly so to moderately unequal, sometimes spinescent. Corollas mostly 5-lobed, occasionally 4-lobed, often 2-lipped (sometimes only weakly irregular or the upper lip occasionally reduced or absent), variously shaped, the inside of the throat sometimes with a band of hairs, but not appendaged. Stamens 2 or 4, the filaments attached in the corolla tube (at its base in a few genera), short to elongate, the anthers exserted or not, attached at their base but the anther sacs sometimes spreading (attached toward the midpoint of a slender, asymmetrically elongated connective between the sacs in *Salvia*), or with 1 of the sacs abortive, yellow, purple, or blue, lacking a glandular appendage; staminodes absent or occasionally present (in some genera with 2 fertile stamens). Pistil 1 per flower, of 2 fused carpels, but usually appearing 4-carpellate. Ovary often with a basal nectar disc and/or a very short, stout stalk (gynophore), deeply 4-lobed from the tip (unlobed in *Callicarpa* and *Vitex*, shallowly lobed in *Ajuga, Teucrium*), 4-locular, with 1 ovule per locule, the placentation more or less basal. Style 1, usually not persistent at fruiting, situated in the deep apical pit between the ovary lobes and thus appearing attached laterally or nearly basally (at the tip of the ovary in *Ajuga, Callicarpa, Teucrium*, and *Vitex*), exserted or not, usually 2-lobed apically (1 of the lobes sometimes highly reduced), the lobes variously oriented, the stigmatic regions indistinct (not globose or swollen), linear along the lobe. Fruits usually schizocarps splitting into (1–)4 nutlets (drupelike with 4 stones in *Callicarpa* and *Vitex*), the nutlets 1-seeded, indehiscent, with a hardened often bony outer wall. About 250 genera, about 6,800 species, nearly worldwide.

In recent decades phylogenetic analyses based on both morphological (Cantino, 1992a, b; Judd et al., 1994, 2008) and molecular (Wagstaff and Olmstead, 1997) data sets has clarified the familial limits of the closely related families Lamiaceae and Verbenaceae. Interestingly, the revised classification is similar to one first suggested more than 75 years ago (Junell, 1934). Traditionally, the two families were distinguished morphologically mainly based on the position of the style relative to the ovary: toward the ovary base in a deep central depression in Lamiaceae vs. at the tip of the ovary in Verbenaceae. This character has since been reinterpreted as a specialization within a more broadly circumscribed Lamiaceae. Separation of the two families now rests on variation of a series of subtle characters, including determinate vs. indeterminate inflorescence axes, differences in the attachment patterns of the ovules, ultrastructure of the pollen grains, and the production of more or less indistinct receptive areas near the tips of the style branches in Lamiaceae vs. more or less discrete (often more or less globose), well-developed, stigmatic regions on 1 or both of the 2 stylar lobes in Verbenaceae.

The result of this reclassification has been an increase in the size of the Verbenaceae, from the traditional view of about 200 genera and 3200 species to about 250 genera and 6800 species. For Missouri, the effects of the changes have been relatively modest: the two woody genera, *Callicarpa* and *Vitex*, are treated in the Lamiaceae instead of the Verbenaceae in the present treatment.

Many species of Lamiaceae are useful economically. The principal genus of timber trees in the family is *Tectona* L. f. (teak), but a number of other genera are minor tropical hardwoods. Species in a large number of genera are cultivated as ornamentals (including some cut flowers and house plants), including *Ajuga* (bugle), *Callicarpa* (beautyberry), *Holmskioldia* Retz. (Chinese hat plant), *Horminum* L. (dragon mouth), *Lamium* (yellow archangel), *Lavandula* L. (lavender), *Leonotis* (Pers.) R. Br. (lion's ear), *Mentha* (mint), *Molucella* L. (shellflower), *Monarda* (beebalm), *Monardella* Benth. (mountain pennyroyal), *Nepeta* (catnip), *Perovskia* Kar. (Russian sage), *Physostegia* (obedient plant), *Plectranthus* L'Hér. (coleus), *Prostanthera* Labill. (mintbush), *Pycnanthemum* (mountain mint), *Rosmarinus* L. (rosemary), *Salvia* (sage), *Scutellaria* (skullcap), *Thymus* L. (thyme), and *Vitex* (monk's pepper tree).A number of different genera contain species with medicinal uses. Many of those members of the family cultivated as ornamentals also are grown for their essential oils, and are used in teas, sachets, potpourri, and most importantly as spices, seasonings, and flavorants in various cuisines. These include *Marrubium* (horehound), *Melissa* (lemon balm), *Ocimum* (basil), *Origanum* (marjoram, oregano), *Pogostemon* Desf. (patchouli), and *Satureja* L. (savory).

The key to genera unfortunately requires flowering material for successful determination of most genera. Fortunately, most mints have a relatively long flowering season; also in many cases the observant collector can find at least a few withered corollas persisting even after the main flowering period is over. Many of the individual species have distinctive aromas, but because scent descriptions are sometimes difficult to communicate accurately and because humans perceive and process odors so variably, this character type has been avoided in the key. In the present treatment, fruiting characters mostly have been deferred to the keys to species in some of the larger genera.

1. Plants shrubs or small trees, woody well above the base and not dying to the ground at the end of the growing season; ovaries unlobed; fruits, unlobed, fleshy, drupelike
    2. Plants shrubs with several arching stems 80–150(–250) cm long; leaves simple, unlobed; inflorescences axillary, dense clusters or small, dome-shaped, dichotomously branched panicles; fruits purplish red to pinkish purple, sometimes bluish-tinged . . . . 4. CALLICARPA
    2. Plants shrubs or small trees with 1 to few erect or ascending stems 200–500 cm long; leaves palmately compound with (3)5–9 leaflets; inflorescences terminal, dense, elongate, panicles; fruits olive green when young, becoming bluish black to black with age . . . . 31. VITEX

1. Plants annual or perennial herbs, sometimes woody at the very base, but dying back to the ground at the end of the growing season; ovaries shallowly to more commonly deeply 4-lobed; fruits lobed, dry, schizocarps splitting into 2 or 4 nutlets at maturity
   3. Calyces with a helmetlike or flaplike to caplike projection or extension along 1 side above the base
      4. Calyces 5-lobed, at least some of the lobes sharply pointed, the upper lobe enlarged, with the margins extending downward along the tube as a pair of crests, the whole structure appearing as a raised, more or less circular, flap or cap of tissue . . . . 20. OCIMUM
      4. Calyces 2-lobed, the lobes both rounded at the tip and similar in size and shape, the projection transverse, ridgelike or shieldlike, and extending from below the midpoint of the upper side of the calyx tube . . . . 27. SCUTELLARIA
   3. Calyces lacking extensions and projections (but sometimes asymmetric or somewhat pouched at the base)
      5. Corollas strongly zygomorphic, with the upper lip highly reduced and inconspicuous or apparently absent (the upper lobes then shifted so as to appear lateral)
         6. Plants mat-forming, producing spreading stolons that root at the nodes to form rosettes of leaves, the vertical flowering stems 7–30 cm long; corollas blue (rarely pink or white), the lower lip mostly 4-lobed . . . . 2. AJUGA
         6. Plants with the stems erect or strongly ascending, colonial from rhizomes, lacking basal rosettes at flowering, the vertical stems 35–120 cm long; corollas pinkish purple to pale lavender (rarely white), the lower lip 5-lobed . . . . 29. TEUCRIUM
      5. Corollas zygomorphic, but with the upper lip well-developed, entire or 2-lobed, or the corollas only slightly zygomorphic (and thus not appearing 2-lipped)
         7. Inflorescences all axillary (subtended by full-sized leaves), in loose to dense clusters or less commonly of solitary flowers
            8. Flowers and fruits all sessile
               9. Stems (at least toward the base) and leaves densely white-woolly, sometimes in patches; calyces 10-lobed, the lobes all spreading or hooked at the tip, stiff and and somewhat spinescent . . . . 15. MARRUBIUM
               9. Stems and leaves glabrous to densely hairy, but not appearing woolly; calyces 5-lobed (but may have 10 or more nerves), at least the upper lobes not spreading or hooked, relatively soft, not spinescent (except in *Galeopsis* and *Leonurus*)
                  10. Stamens 2
                      11. Corollas 2.5–5.0 mm long, appearing nearly actinomorphic, not 2-lipped, with 4 or 5 more or less similar lobes (if 4-lobed then the upper lobe slightly larger than the others and sometimes shallowly notched) . . . . 14. LYCOPUS
                      11. Corollas 15–50 mm long, appearing strongly zygomorphic, 2-lipped . . . . 18. MONARDA
                  10. Stamens 4

12. Calyces with the lobes tapered to slender, sharply pointed tips, but remaining soft (not spinescent) . . . . 12. LAMIUM
12. Calyces with the lobes tapered to slender, sharply pointed, stiff, spinescent tips
    13. Leaf blades unlobed; stems with relatively long, stiff, spreading hairs; corollas 15–22 mm long, with the lower lip producing a pair of small nipplelike projections on the upper surface toward the base; calyces with the spinescent lobes all more or less equally ascending . . . . . . 9. GALEOPSIS
    13. Leaf blades mostly 3-lobed or palmately 5–9-lobed; stems mostly with short, soft or stiff, curved or downward-angled hairs (occasionally glabrous); corollas 4–14 mm long, lacking nipplelike projections; calyces with the lower 3 lobes noticeably spreading (except sometimes in *L. japonicus*) . . . . 13. LEONURUS

8. Flowers and fruits stalked, the stalks at least 1 mm long
    14. Stems prostrate, rooting at the nodes and sometimes forming loose mats; leaf blades broadly kidney-shaped to nearly circular; flowers 2–6 per leaf node . . . . 10. GLECHOMA
    14. Stems erect or ascending, sometimes from spreading base, not rooting at the nodes or forming mats; leaf blades variously shaped, but not kidney-shaped or circular; flowers mostly 8 or more per node
        15. Calyces zygomorphic, 2-lipped or, if not 2-lipped, then the lobes differing in size and/or shape
            16. Stamens 2 . . . . 11. HEDEOMA
            16. Stamens 4
                17. At least the lower stem leaves long-petiolate, the blades of the largest leaves 4–7 cm long; calyces with the lobes of the upper lip broadly triangular . . . . 16. MELISSA
                17. Leaves sessile or short-petiolate, the blades 0.4–2.5 cm long; calyces with lobes of the upper lip narrowly triangular . . . . . . 5. CLINOPODIUM
        15. Calyces more or less actinomorphic, not 2-lipped, the lobes similar in size and shape or nearly so
            18. Stamens 2 . . . . 7. CUNILA
            18. Stamens 4
                19. Stamens not or only slightly exserted
                    20. Corollas 7–15 mm long, zygomorphic, 2-lipped, pale purple to lavender . . . . 5. CLINOPODIUM
                    20. Corollas 1.5–4.5 mm long, nearly actinomorphic, not 2-lipped, pink to purple or bluish purple . . . . 30. TRICHOSTEMA
                19. Stamens long-exserted
                    21. Corollas 2–7 mm long, nearly actinomorphic, not 2-lipped, white to pink; stamens with the filaments straight or slightly curved, the anther sacs parallel . . . . 17. MENTHA
                    21. Corollas 8–15 mm long, zygomorphic, 2-lipped, blue; stamens with the filaments strongly arched or curved, the anther sacs spreading . . . . 30. TRICHOSTEMA

7. Inflorescences terminal spikes, racemes, or panicles (these sometimes relatively short and/or with leaflike bracts that are significantly smaller or different in appearance from the foliage leaves), sometimes also axillary
 22. Stamens 2
  23. Stamens long-exserted; flowers and fruits stalked, the stalks at least 1 mm long
   24. Inflorescences panicles with open, racemose branches, the flowers 2 per node, each with a stalk 3–5 mm long; calyces 2.5–5.0 mm long at flowering, becoming enlarged to 6–8 mm at fruiting; corollas 12–15 mm long, mostly yellow to greenish yellowish brown, sometimes with dull purple to reddish brown lines, blotches, or streaks, the lower lip strongly fringed, often partially white . . . . 6. COLLINSONIA
   24. Inflorescences spikes or spikelike racemes, consisting of 1 to several dense clusters, these noticeably separate along the axis, each with 6 to numerous flowers, each sessile or with a short stalk to 2 mm long; calyces 1.5–11.0 mm long at flowering, not becoming enlarged at fruiting; corollas 6–13 mm long, white to pink, purple, or bluish purple, sometimes with darker spots on the lips, the lower lip 3-lobed but otherwise entire (not fringed)
    25. Corollas 8–14 mm long, strongly zygomorphic, the lips often spotted; calyces 4–11 mm long, 13-nerved, zygomorphic, 2-lipped, the upper lip noticeably longer than the lower lip, the lobes narrowly triangular to nearly hairlike . . . . 6. BLEPHILIA
    25. Corollas 6–8 mm long, weakly zygomorphic, not spotted; calyces 1.5–3.0 mm long, 10-nerved, actinomorphic, the 5 lobes all similar in size and shape, triangular . . . . 7. CUNILA
  23. Stamens not or only slightly exserted; flowers and fruits sessile or nearly so
   26. Calyces actinomorphic or nearly so, not 2-lipped, the lobes all similar . . . . 18. MONARDA
   26. Calyces zygomorphic, 2-lipped
    27. Inflorescences spikes consisting of 1–5(6) dense, headlike clusters, these noticeably separate along the axis, each with numerous flowers; stamens with the connective between the anther sacs very short and inconspicuous, the 2 sacs both fertile, appearing more or less sessile and spreading from the filament tip . . . . 3. BLEPHILIA
    27. Inflorescences spikes or racemes consisting of more than 5 clusters, these relatively open, more or less continuous along the axis, each with 1–12 flowers; stamens with the connective between the anther sacs slender and elongate, the upper sac fertile and fully formed, the lower sac either absent or reduced and nonfunctional, the connective appearing attached at or below its midpoint . . . . 25. SALVIA
 22. Stamens 4
  28. Corollas nearly actinomorphic, not 2-lipped, the lobes similar or nearly so
   29. Flowers in pairs on opposite sides of each node of the inflorescence axis (and its branches); calyces becoming enlarged and swollen or pouched basally on the lower side as the fruits develop . . . . . . 22. PERILLA
   29. Flowers all or mostly in clusters of 4 or more at each node of the inflorescence axis; calyces not becoming significantly enlarged, swollen, or pouched as the fruits develop

30. Leaf blades with the margins toothed (sometimes shallowly or bluntly so); stamens strongly exserted, the filaments straight or nearly so, the anther sacs parallel or nearly so .... 17. MENTHA

30. Leaf blades with the margins entire; stamens with the filaments not or only slightly exserted, or, if noticeably exserted, then strongly arched or curved, the anther sacs spreading ...... 30. TRICHOSTEMA

28. Corollas zygomorphic, 2-lipped

31. Inflorescences either loose, irregularly branched panicles, or dome-shaped panicles with the flowers in small, dense, ovoid spikelike racemes at the tips of the ultimate branches; flowers distinctly stalked (but sometimes short-stalked)

32. Inflorescences dome-shaped panicles with the flowers in small, dense, ovoid spikelike racemes at the tips of the ultimate branches; corollas pink to lavender or pale purple (occasionally white elsewhere); calyces actinomorphic .... 21. ORIGANUM

32. Inflorescences loose, irregularly branched panicles, the flowers solitary, paired, or in small loose clusters at the branch tips; corollas blue; calyces zygomorphic (2-lipped)............................. 30. TRICHOSTEMA

31. Flowers and fruits sessile; inflorescences spikes or panicles with spicate branches or dense, headlike clusters at the branch tips

33. Inflorescences composed of distinct clusters, these terminal on the stems and often appearing grouped into relatively flat-topped to broadly rounded panicles or widely to relatively densely spaced along an axis, but noticeably discrete and with the flowers not overlapping those of adjacent nodes

34. Inflorescences composed of dense, headlike (spherical or depressed-globose) clusters of numerous flowers at the branch tips, these solitary or appearing grouped into relatively flat-topped to broadly rounded panicles; stamens exserted.............. 25. PYCNANTHEMUM

34. Inflorescences composed of small clusters of 4 or 6(8) flowers per node (flowers 2 or 3[4] in the axil of each bract), these not headlike; stamens not exserted (hidden under the somewhat concave upper corolla lip)... 28. STACHYS

33. Inflorescences or their branches more or less continuous spikes, the few- to many-flowered individual clusters relatively dense along the axis (except sometimes in *Stachys*), sometimes the lowermost few nodes more widely spaced

35. Stamens long-exserted .... 1. AGASTACHE

35. Stamens not exserted (hidden under the somewhat concave upper corolla lip)

36. Calyces zygomorphic, 2-lipped

37. Bracts with the tip and sharply toothed margins spinescent; calyces with the central lobe of the upper lip noticeably longer and wider than the others; plants tap-rooted, annual (short-lived perennial elsewhere) .... 8. DRACOCEPHALUM

37. Bracts with the tip sharply pointed but not spinescent, the margins not toothed or spinescent; calyces with the 3 lobes of the upper lip all similar in size and shape; plants fibrous-rooted, perennial............................ 24. PRUNELLA

36. Calyces actinomorphic or nearly so, not 2-lipped, the lobes similar or sometimes differing slightly in size

38. Stems and the undersurface of the leaves densely pubescent with felted or somewhat woolly, sometimes matted, short, white to grayish-tinged hairs; calyces 15-nerved . . . . 19. NEPETA
38. Stems and/or leaves glabrous or variously pubescent, if so then not with felted or woolly, matted hairs; calyces 5- or 10-nerved
    39. Flowers 2 per node, solitary in the axil of each bract; stems and leaves glabrous . . . . 23. PHYSOSTEGIA
    39. Flowers 4 or 6 per node, in clusters of 2 or 3 in the axil of each bract; stems and/or leaves variously hairy, sometimes inconspicuously or sparsely so (sometimes totally glabrous in *Stachys hyssopifolia*). . . . . 28. STACHYS

1. **Agastache** J. Clayton ex Gronov. (giant hyssop) (Vogelmann, 1983)

Plants perennial herbs, with fibrous roots, often with short rhizomes. Stems erect or strongly ascending, usually stout, tall, and sharply 4-angled, unbranched or more commonly with ascending branches above the base, glabrous or hairy. Leaves mostly long-petiolate, the petiole unwinged. Leaf blades mostly narrowly to broadly ovate, less commonly especially those of the uppermost leaves lanceolate to elliptic-lanceolate, rounded to truncate or shallowly cordate at the base (angled in narrower leaves), tapered to a sharply pointed tip, the margins with relatively widely spaced, coarse, sharp teeth, the surfaces glabrous or more commonly short-hairy, also with inconspicuous, sessile glands. Inflorescences terminal, dense, more or less continuous spikes (sometimes a few of the lower nodes more widely spaced), cylindric (tapered at the tip), the flowers several to numerous at each node and strongly overlapping those of adjacent nodes. Calyces actinomorphic or nearly so (usually slightly oblique), lacking a lateral projection, more or less symmetric at the base, more or less cylindric, the tube 15-nerved, 5-lobed, the lobes shorter than the tube, similar in size and shape, variously shaped, not spinescent, becoming slightly enlarged and papery at fruiting. Corollas zygomorphic, greenish yellow to yellow or pink or purple, the outer surface sparsely to moderately short-hairy, the tube narrowly funnelform, relatively shallowly 2-lipped, the upper lip shallowly 2-lobed, slightly concave, the lower lip somewhat spreading, 3-lobed with a large central lobe and 2 small lateral lobes at its base. Stamens 4, strongly exserted, the lower 2 stamens with slightly shorter filaments that are slightly curved under the upper corolla lip, the upper 2 stamens with slightly longer, straight filaments (angled slightly downward), the anthers small, the connective very short, the pollen sacs 2, more or less parallel, white or pink. Ovary deeply lobed, the style appearing nearly basal from a deep apical notch. Style exserted, more or less equally 2-branched at the tip. Fruits dry schizocarps, separating into usually 4 nutlets, these 1.2–1.5 mm long, oblong-obovoid to oblong-ellipsoid, the surface yellowish brown, densely and minutely hairy at the tip, otherwise glabrous and slightly roughened. About 22 species, North America, Asia.

The two Missouri species belong to section *Agastache*, a group of seven species most diverse in the eastern United States that was studied by Vogelmann (1983). Several species of *Agastache* are cultivated as ornamentals, particularly as drought-resistant plants attractive to hummingbirds and other pollinators, but the two Missouri taxa tend to be too robust for most garden situations and also tend to be less tolerant than some of the taxa native farther west.

1. Corollas greenish yellow to light yellow; calyx lobes 1.0–1.5 mm long at flowering (to 2.5 mm at fruiting), ovate . . . . 1. A. NEPETOIDES

1. Corollas light pinkish purple to pale bluish purple or less commonly white; calyx lobes 2.0–2.5 long at flowering (to 3.2 mm at fruiting), narrowly triangular . . . . 2. A. SCROPHULARIIFOLIA

1. Agastache nepetoides (L.) Kuntze (yellow giant hyssop)

Pl. 432 c–e; Map 1934

Stems 80–150(–250) cm long, glabrous or more commonly with an inconspicuous band of short, mostly unicellular hairs at the nodes, occasionally sparsely hairy between nodes toward the tip. Petioles 0.8–7.0 cm long. Leaf blades 2–15 cm long, the upper surface glabrous or sparsely short-hairy along the main veins near the base, the undersurface moderately to densely pubescent with short, spreading hairs, more densely so along the veins. Inflorescences 4–20 cm long, 2.0–2.5 cm in diameter, the bracts elliptic-ovate to broadly ovate, shorter than to slightly longer than the calyces. Calyces 4.5–6.5 mm long, pale green to yellowish green, the lobes 2.0–2.5 long at flowering (to 3.2 mm at fruiting), narrowly triangular, tapered to sharply pointed tips. Corollas 5.5–8.0 mm long, greenish yellow to light yellow. $2n=18$. July–September.

Scattered nearly throughout the state (eastern U.S. west to South Dakota and Oklahoma; Canada). Bottomland forests, edges and openings of mesic upland forests, banks of streams and rivers, margins of lakes, and bases of bluffs; also pastures, roadsides, and moist disturbed areas.

2. Agastache scrophulariifolia (Willd.) Kuntze (purple giant hyssop, figwort giant hyssop)

Pl. 432 f, g; Map 1935

Stems 60–150(–200) cm long, moderately to densely pubescent with more or less spreading, multicellular hairs. Petioles 0.8–6.0 cm long. Leaf blades 2–15 cm long, the upper surface moderately pubescent with short, relatively stiff hairs along and between the veins, the undersurface moderately to densely pubescent with finer, somewhat crinkled, more or less spreading hairs. Inflorescences 3–17 cm long, 1.5–2.5 cm in diameter, the bracts elliptic-ovate to nearly circular, shorter than to more commonly slightly longer than the calyces. Calyces 4.0–6.5 mm long, green with the lobes often pale pinkish-tinged, the lobes 2.0–2.5 long at flowering (to 3.2 mm at fruiting), ovate, angled to bluntly to sharply pointed tips. Corollas 5–8 mm long, light pinkish purple to pale bluish purple or less commonly white. $2n=18$. July–September.

Uncommon in the Glaciated Plains Division, mostly north of the Missouri River (eastern [mostly northeastern] U.S. west to South Dakota and Kansas; Canada). Bottomland forests, openings of mesic upland forests, banks of streams, and swales in mesic upland prairies; also roadsides.

Steyermark (1963) called plants with slightly more hairy leaves f. *mollis* (var. *mollis* (Fernald) A. Heller), but that name apparently was never validly published.

2. Ajuga L. (bugle)

About 50 species, Europe, Africa, Asia, south to Australia; introduced nearly worldwide.

In addition to the commonly grown *A. reptans*, *A. genevensis* L. (standing bugle) and *A. chamaepitys* (L.) Schreb. (ground pine) also are cultivated ornamentals and ground covers that escape occasionally in the United States.

1. Ajuga reptans L. (carpet bugle)

Map 1936

Plants perennial herbs, with fibrous roots, usually strongly colonial and sometimes matforming. Main stems stoloniferous, to 60 cm long or longer, spreading, forming rosettes where rooting, bluntly to sharply 4-angled, the angles sometimes narrowly winged, leafy, glabrous or sparsely to densely pubescent with fine, straight to crinkly, multicellular hairs, sometimes mostly at the nodes. Flowering stems 1 to few per rosette, 7–30 cm long, erect or strongly ascending, bluntly to sharply 4-angled, usually unbranched, sparsely to moderately pubescent with fine, straight to crinkly, multicellular hairs, sometimes mostly at the nodes. Leaves in rosettes basal to the flowering stems and long-petiolate, also opposite on the stems and mostly short-petiolate to less commonly nearly sessile, the petiole winged, especially above the midpoint, often strongly purplish-tinged, less commonly copper-colored in the overwintering state. Leaf blades 0.8–6.0(–10.0) cm long, 0.4–4.0 cm

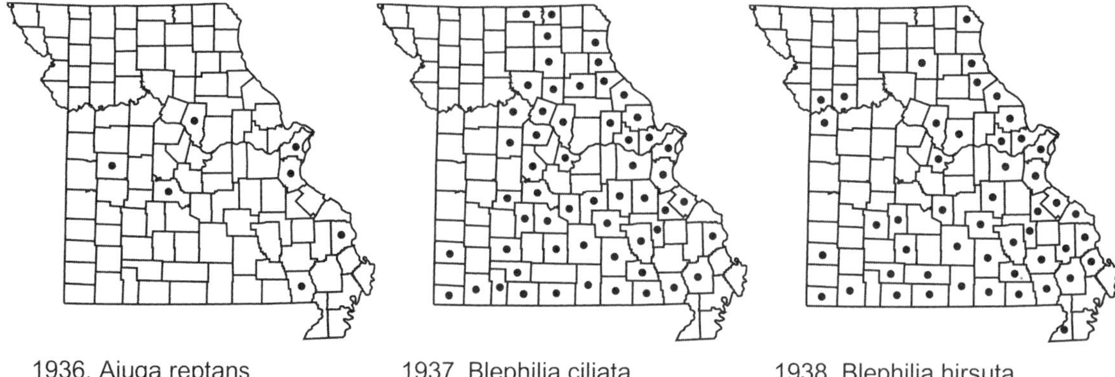

1936. Ajuga reptans 1937. Blephilia ciliata 1938. Blephilia hirsuta

wide (those of rosette leaves usually larger than those of flowering stems), variously narrowly to broadly elliptic, ovate, obovate, or nearly circular, sometimes appearing oblanceolate or spatulate, rounded or angled to tapered at the base, rounded or broadly angled to a bluntly pointed tip, the margins with fine to moderately coarse, rounded teeth or scallops, sometimes appearing wavy, the surfaces glabrous or sparsely to moderately pubescent with soft, more or less appressed, multicellular hairs, also with inconspicuous, sessile glands. Inflorescences appearing as terminal spikelike racemes, these appearing relatively dense but interrupted between some or most of the nodes, more or less cylindric, the flowers 6(–10) at each node and not or somewhat overlapping those of adjacent nodes, the stalks 0.5–1.0 mm long. Bracts leaflike, all but the uppermost noticeably longer than the flowers. Calyces actinomorphic or nearly so, 2.5–5.0 mm long, lacking a lateral projection, more or less symmetric at the base, bell-shaped, the tube inconspicuously 10-nerved, 5-lobed, the lobes as long as or somewhat longer than the tube, similar in size and shape, triangular to lanceolate-triangular, not spinescent, not becoming enlarged at fruiting. Corollas 12–20 mm long, zygomorphic, blue (rarely pink or white), the surfaces sparsely to moderately hairy, the tube funnelform, 2-lipped to about the midpoint, the upper lip reduced and inconspicuous, a small, relatively flat, unlobed flap of tissue, the lower lip prominent, spreading to somewhat recurved, mostly 4-lobed with 2 small lateral lobes and a large central lobe that in turn is notched (less commonly unlobed or shallowly 3-lobed apically). Stamens 4, slightly exserted (shorter than the lower lip), the lower 2 stamens with slightly shorter filaments, the anthers small, the connective very short, the pollen sacs 2, appearing angled to more or less spreading from their tips, dark blue. Ovary relatively shallowly lobed, the style appearing more or less lateral from a shallow apical notch. Style exserted, more or less equally 2-branched at the tip. Fruits dry schizocarps, often not separating into nutlets (or only tardily so) and often with only 1 or 2 of the ovules developing into fruits, these 1.5–2.0 mm long, oblong-obovoid, the surface brown, glabrous or sparsely short-hairy near the tip, with a prominent, honeycomblike network of ridges. $2n=32$. April–June.

Introduced, uncommon, widely scattered (native of Europe, Asia; introduced sporadically in the eastern and less abundantly western U.S., Canada). Bottomland forests, gardens, and old homesites.

Carpet bugle is commonly cultivated as a hardy ground cover, but can spread aggressively in some garden situations. The basal rosette leaves usually are singificantly larger than those of the flowering stems and often overwinter, turning dark purple or coppery-tinged.

3. **Blephilia** Raf.

Plants perennial herbs, usually with slender rhizomes. Stems erect or ascending, slender to relatively stout, sharply 4-angled, unbranched or branched, sometimes with stoloniferous, prostrate to loosely arched, basal branches, hairy. Leaves sessile or petiolate, the petiole unwinged or winged, with a faint or strong fragrance when crushed. Leaf blades lanceolate to elliptic, ovate, or oblong-lanceolate to oblong-ovate, the margins toothed, the surfaces hairy, sometimes only along the veins, also with sometimes conspicuous, sessile glands. Inflores-

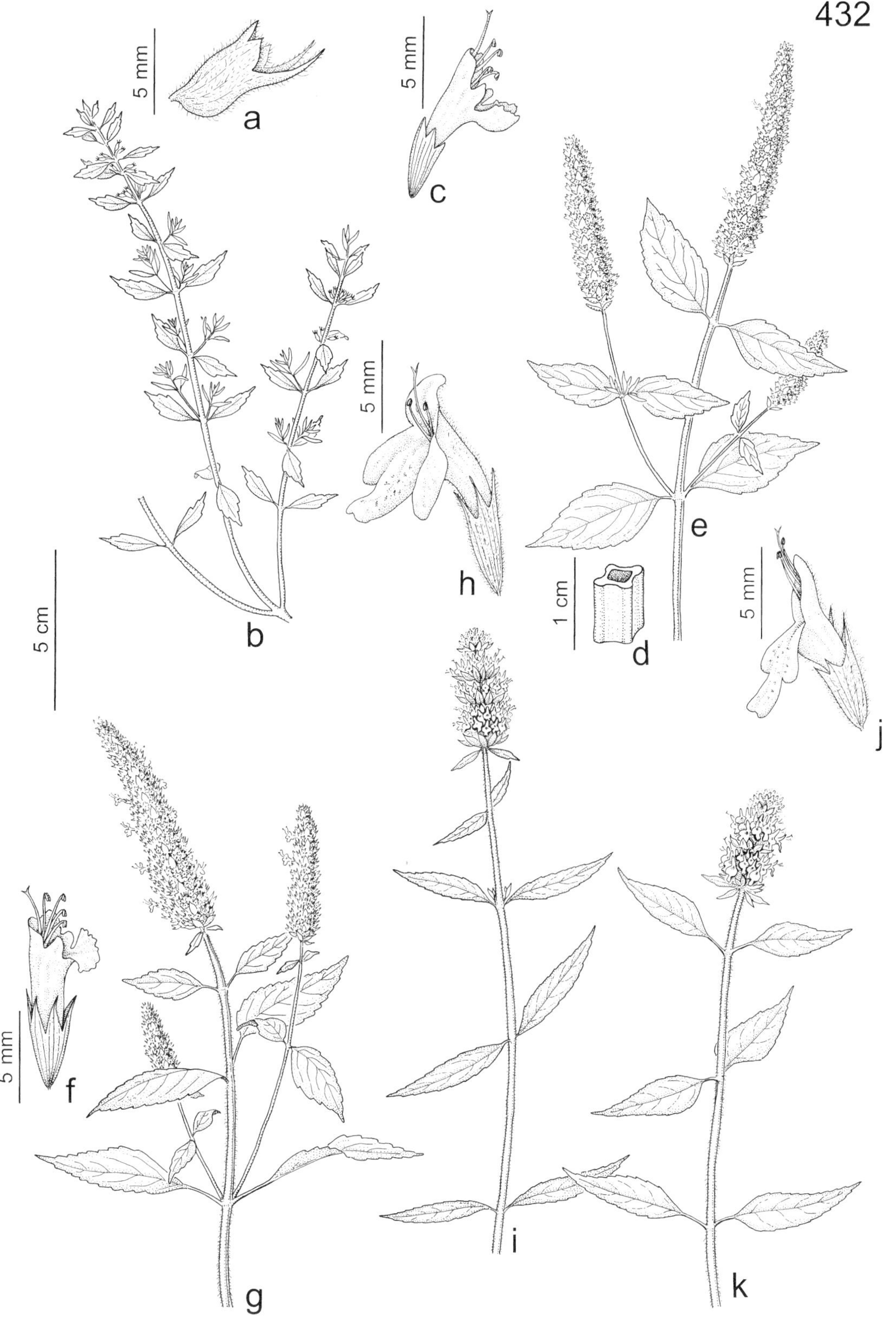

Plate 432. Lamiaceae. *Clinopodium acinos*, **a)** fruiting calyx, **b)** fertile stem. *Agastache nepetoides*, **c)** flower, **d)** section of stem, **e)** upper portion of stem with leaves and inflorescences. *Agastache scrophulariifolia*, **f)** flower, **g)** upper portion of stem with leaves and inflorescences. *Blephilia ciliata*, **h)** flower, **i)** fertile stem. *Blephilia hirsuta*, **j)** flower, **k)** fertile stem.

cences terminal spikes consisting of 1–5(6) dense, headlike clusters that are noticeably separate along the axis, each with numerous flowers, these sessile or nearly so, subtended by a pair of leaflike bracts and an involucre of conspicuous bractlets. Calyces zygomorphic, lacking a lateral projection, more or less symmetric at the base, more or less cylindric, the tube strongly 13-nerved (-ribbed), glabrous in the mouth, hairy externally, 2-lipped, the lobes shorter than to about as long as the tube, the upper lip 3-lobed, noticeably longer than the 2-lobed lower lip, the lobes narrowly triangular-tapered, not spinescent, not becoming enlarged and papery at fruiting. Corollas zygomorphic, white or pink to pale bluish purple with darker spots (rarely all white elsewhere), the outer surface moderately to densely pubescent with fine, spreading to somewhat tangled hairs, the tube funnelform, relatively shallowly 2-lipped, the upper lip shallowly 2-lobed, slightly concave to spreading to recurved, the lower lip more or less straight to slightly spreading, 3-lobed. Stamens 2, initially hidden under the upper corolla lip, but becoming exserted at maturity, the anthers small, the connective very short and inconspicuous, the pollen sacs 2, appearing more or less sessile and spreading from the filament tip, pink or purplish-tinged to white or pale yellow. Ovary deeply lobed, the style appearing lateral to nearly basal from a deep apical notch. Style exserted, unequally 2-branched at the tip. Fruits dry schizocarps, separating into usually 4 nutlets, these 0.9–1.2 mm long, oblong-obovoid (rounded at the tip, angled or tapered at the base), the surface tan to brown or black, glabrous, unevenly roughened or pebbled. Three species, eastern U.S., Canada.

The third species in the genus, *B. subnuda* Simmers & Kral, is endemic to limestone woodlands in northern Alabama. In describing the new species, Simmers and Kral (1992) provided a detailed morphological analysis of all three members of the genus.

1. Calyces with the lower lip extending beyond the sinuses between the lobes of the upper lip; stems usually unbranched, densely pubescent with more or less downward-curved hairs mostly 0.5–1.0 mm long (longer hairs sometimes also present at the nodes and along the angles); leaves (except the lower ones, which are usually withered by flowering) sessile or with the petiole to 12 mm long, the blade mostly tapered to narrowly angled at the base . . . . 1. B. CILIATA
1. Calyces with the lower lip not extending to the sinuses between the lobes of the upper lip; stems usually 2 to several-branched, moderately to densely pubescent with fine, more or less spreading hairs mostly 1–2 mm long (especially toward the tip); leaves all with the petiole mostly 10–30 mm long, the blade mostly broadly angled to rounded at the base . . . . 2. B. HIRSUTA

1. Blephilia ciliata (L.) Benth. (Ohio horse mint)

Pl. 432 h, i; Map 1937

Stems 25–60(–80) cm long, usually unbranched above the base, sometimes forming stoloniferous basal branches during the growing season, densely pubescent with tapered, downward-curved hairs mostly 0.5–1.0 mm long, sometimes also with sparse to moderate, longer, spreading hairs at the nodes and along the angles. Leaves sessile or with the petiole 1–12 mm long (to 25 mm on the lowermost leaves, which usually wither by flowering). Leaf blades 1–6(–8) cm long, lanceolate to oblong-elliptic or ovate, the lowermost, early-withering leaves often broadly ovate to nearly circular, mostly tapered to narrowly angled at the base (sometimes rounded or shallowly cordate in the early-withering lower leaves), rounded to bluntly or sharply pointed at the tip, the margins finely but sometimes sparsely toothed, the upper surface sparsely to moderately pubescent with short, stout, tapered hairs, the undersurface moderately to densely pubescent with finer, short hairs, often also with sparse long hairs along the main veins. Flower clusters with the bractlets 6–11 mm long, ovate to elliptic, mostly angled at the tip, the margins with long, spreading hairs toward their bases, usually strongly purplish-tinged. Calyces 7–11 mm long, the lower lip extending beyond the sinuses between the lobes of the upper lobe, the lobes narrowly triangular. Corollas 9–14 mm long, pale pink to pale lavender or pale bluish purple with darker reddish spots on the inner surface of the lower lip. Nutlets dark reddish brown to nearly black. May–August.

Scattered nearly throughout the state, but apparently absent from most of the Unglaciated

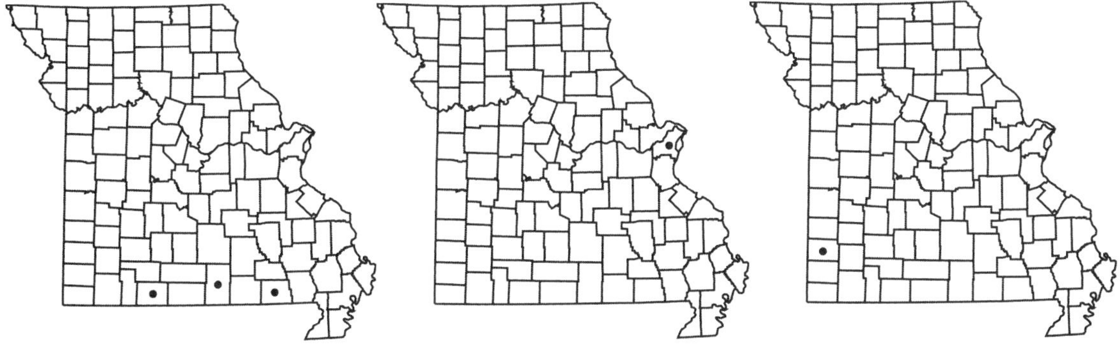

1939. Callicarpa americana 1940. Callicarpa dichotoma 1941. Clinopodium acinos

Plains Division and the western portion of the Glaciated Plains, as well as the southern portion of the Mississippi Lowlands (eastern U.S. west to Iowa and Oklahoma; Canada). Mesic to dry upland forests, ledges and tops of bluffs, margins of glades, banks of streams, and occasionally bottomland forests; also railroads and roadsides.

Steyermark (1963) noted that plants with entirely white corollas have been called f. *albiflora* House, but have not yet been reported from Missouri. He further noted the presence of a collection from Putnam County that might represent a rare hybrid between *B. ciliata* and *B. hirsuta*, based on its intermediate morphology.

According to Steyermark (1963), Ohio horse mint was eaten as a spring green by pioneers. He also noted that it makes an excellent ornamental in the shade garden. The crushed foliage tends to have a relatively strong, pleasant odor.

2. Blephilia hirsuta (Pursh) Benth. **var. hirsuta** (wood mint)

Pl. 432 j, k; Map 1938

Stems 40–120 cm long, usually 2- to several-branched above the base, sometimes forming ascending to spreading basal branches late in the growing season, moderately to densely pubescent with fine, more or less spreading hairs mostly 1–2 mm long (especially toward the tip) often also with spreading to downward-curved hairs 0.5–1.0 mm long (especially toward the base). Leaves with the petiole mostly 10–30 mm long. Leaf blades 1–8 cm long, narrowly to broadly ovate, mostly broadly angled to rounded at the base (occasionally a few of the blades shallowly cordate), angled or tapered to a sharply pointed tip, the margins finely toothed, the upper surface glabrous or sparsely pubescent with short, fine hairs (rarely with a few additional longer hairs), the undersurface sparsely to moderately pubescent with short, fine hairs, sometimes also with sparse longer hairs mostly along the main veins. Flower clusters with the bractlets 6–10 mm long, lanceolate to narrowly ovate, tapered at the tip, the margins and usually also the undersurface with spreading hairs, green or occasionally somewhat purplish-tinged. Calyces 4–7 mm long, the lower lip not extending to the sinuses between the lobes of the upper lobe, the lobes narrowly triangular to nearly linear. Corollas 8–12 mm long, white or slightly bluish-tinged with purple spots on the inner surface of the lower lip. Nutlets tan to reddish brown. May–September.

Scattered nearly throughout the state, but apparently absent from most of the Unglaciated Plains Division and uncommon in most of the Glaciated Plains (eastern U.S. west to Minnesota and Kansas; Canada). Bottomland forests, mesic upland forests, bases and ledges of bluffs, and banks of streams; also quarries, railroads, and roadsides.

As noted by Simmers and Kral (1992), the taxonomic status of the rare var. *glabrata* Fernald requires further study. It is known from only a few collections from Vermont and differs in its glabrous or nearly glabrous stems and leaves.

The crushed foliage of *B. hirsuta* tends to have a fainter, less pleasant odor than does that of *B. ciliata*. Steyermark (1963) corroborated earlier reports that plants of wood mint tend to be heavily foraged by insects, rendering the species less desirable as a garden ornamental.

4. **Callicarpa** L. (beautyberry)

Plants shrubs, incompletely monoecious (producing staminate and apparently perfect but functionally pistillate flowers in the same inflorescences). Stems several, loosely ascending to more commonly arched, branched, the bark light brown to grayish brown, relatively smooth, but with scattered vertically-oriented lenticels, on the oldest trunks eventually separating into small, thin plates. Twigs more or less circular to bluntly 4-angled, reddish to purplish brown, sparsely to densely pubescent with minute, scurfy to woolly, short-stalked, stellate (dendritic) hairs, these wearing away with age. Winter buds ovoid, angled to a sharply pointed tip, the usually 2 outer scales obscured by dense stellate hairs. Leaves opposite or in whorls of 3 at some nodes, shortly to moderately petiolate (1–35 mm), the petiole winged, at least above the midpoint, with a pungent, disagreeable odor when crushed. Leaf blades simple, unlobed, narrowly to broadly ovate, elliptic, oblong-elliptic, or occasionally lanceolate, tapered at the base, tapered to a sharply pointed tip, the margins with relatively coarse but sometimes shallow, blunt to sharp teeth, the surfaces moderately pubescent with mostly stellate hairs when young, sometimes mostly along the veins, also with inconspicuous, sessile glands, one or both surfaces becoming glabrous or nearly so at maturity, sometimes the stellate hairs shed but leaving the minute basal stalks. Inflorescences axillary, dense or open clusters or small, dome-shaped, dichotomously branched panicles, usually with numerous flowers, the flowers stalked, the stalk with a pair of bractlets at the base. Calyces actinomorphic, lacking a lateral projection, symmetric at the base, cup-shaped to broadly obconic, the tube more or less 4-angled, with obscure, branched nerves between the angles, very shallowly 4-lobed or sometimes entire, the lobes similar in size and shape, broadly triangular to broadly rounded (the margin appearing slightly undulate), not spinescent, not becoming enlarged at fruiting. Corollas actinomorphic, light pink to pale purple or blue (rarely reddish-tinged or white), the outer surface with sessile glands, the tube funnelform, the 4 lobes shorter than to longer than the tube, rounded to bluntly pointed at the tips and often slightly irregular along the margins. Stamens 4, present (but not necessarily functional) in all flowers, exserted, all similar in size, the filaments attached at the base of the corolla tube, the anthers small, the connective very short, the pollen sacs 2, parallel, yellow. Ovary present only in functionally pistillate flowers, unlobed, the style appearing terminal. Style exserted, unbranched at the tip, the stigma more or less capitate to peltate-flattened. Fruits fleshy, globose drupes 3–6 mm in diameter, the outer surface purplish red to pinkish purple, sometimes bluish-tinged, shiny, the stone eventually separating into usually 4 nutlets About 140 species, North America, Central America, South America, Caribbean Islands, Madagascar, Asia south to Australia.

As noted above, *Callicarpa* formerly was classified in the Verbenaceae, but has been transferred to the Lamiaceae, based on molecular data and a reassessment of the morphological features separating the family.

1. Inflorescences dense clusters, the inflorescence stalk 1–5 mm long; petioles 5–35 mm long, longer than the inflorescence stalks; leaf blades (at least on the undersurface) persistently hairy along and between the veins . . . . . 1. C. AMERICANA
1. Inflorescences more open, small, dome-shaped panicles, the inflorescence stalk 10–18 mm long; petioles 2–8 mm long, much shorter than the inflorescence stalks; leaf blades at maturity glabrous or sparsely to moderately hairy only along the main veins (often the hairs with the branched apical portion shed, leaving the minute stalks) . . . . C. DICHOTOMA

1. Callicarpa americana L. (American beautyberry, beautyberry, French mulberry)

Pl. 426 a, b; Map 1939

Stems 80–150(–250) cm long, usually relatively stout. Twigs with the surface partially or completely hidden by a dense, scurfy to woolly covering of short-stalked, stellate (dendritic) hairs, this wearing away with age. Leaves opposite or rarely in whorls of 3 at a few nodes, shortly to moderately petiolate, the petiole 5–35 mm (longer than the inflorescence stalks), usually densely stellate-hairy. Leaf blades (2–)5–23 cm long, (1–)2–13 cm wide, narrowly to broadly ovate or elliptic, tapered at the base, tapered to a sharply pointed tip, the margins with relatively coarse, blunt to sharp teeth, the surfaces moderately pubescent with mostly stellate hairs, the undersurface also with inconspicuous, sessile glands, the upper surface becoming glabrous or nearly so at maturity. Inflorescences appearing as small, dense, axillary, paniculate clusters of numerous flowers, the stalk 1–5 mm long. Calyces 1.5–1.8 mm long, very shallowly 4-lobed, the lobes broadly triangular. Corollas 4.0–4.5 mm long, light pink to pale purple or blue (rarely reddish-tinged or white), the 4 lobes shorter than the tube. Fruits 3–6 mm in diameter, the outer surface purplish red to pinkish purple, sometimes bluish-tinged. $2n=36$. June–August.

Uncommon in the southern portion of the state (southeastern U.S. west to Missouri, Oklahoma, and Texas). Ledges and tops of bluffs and openings of mesic upland forests on adjacent slopes; also bottomland forests.

This species was a favorite of Julian Steyermark, who remarked on the unusual color of its attractive, long-persistent fruits and its suitability for cultivation as an ornamental. Steyermark was adamantly opposed to the damming of Ozark watersheds by the U.S. Army Corps of Engineers for the creation of reservoirs. He wrote passionately decrying the destruction of scenic natural habitats and rare plants in the early 1950s when Bull Shoals Lake was created in the White River watershed in southwestern Missouri and adjacent Arkansas. *Callicarpa americana*, which he had discovered in 1949 growing on bluffs overlooking the river, was singled out as a prime example of a wonderful plant lost from the Ozarks as the lake waters rose (Steyermark, 1952). Although Steyermark believed that the species had become extirpated from the region, small populations still persist on the bluffs there above the high water line, and a few additional populations also have been discovered elsewhere in the state. Missouri is at the northwestern extreme of the overall range of the species.

2. Callicarpa dichotoma (Lour.) K. Koch (Chinese beautyberry)

Map 1940

Stems 50–150 cm long, usually relatively slender. Twigs with the surface partially hidden by a moderate to dense, scurfy covering of short-stalked, stellate (dendritic) hairs, this wearing away with age. Leaves opposite or commonly in whorls of 3 at some nodes, short-petiolate, the petiole 2–8 mm (much shorter than the inflorescence stalks), moderately stellate-hairy. Leaf blades (2–)5–10 cm long, 1–3(–4) cm wide, elliptic, oblong-elliptic, oblong-obovate, or occasionally lanceolate, tapered at the base, tapered to a sharply pointed tip, the margins with relatively coarse but often shallow, blunt to sharp teeth, mostly above the midpoint, the surfaces sparsely to moderately pubescent with stellate hairs, also with inconspicuous, sessile glands, becoming glabrous or nearly so at maturity except along the main veins (sometimes the branched apical portion of the hairs shed, leaving the minute stalks). Inflorescences appearing as small, axillary, dichotomously-branched, dome-shaped panicles of numerous flowers, the stalk 10–18 mm long. Calyces 0.9–1.4 mm long, very shallowly 4-lobed or sometimes entire, the lobes broadly triangular to broadly rounded (the margin appearing slightly undulate). Corollas 2.5–3.5 mm long, light pink to pale purplish pink, the 4 lobes about as long as to longer than the tube. Fruits 3–4 mm in diameter, the outer surface purplish red. $2n=36$. July–August.

Introduced, uncommon, known thus far only from St. Louis County (native of Asia, introduced very sporadically in the eastern U.S.). Disturbed bottomland forests.

This species was first collected by the author in 2010 near a foot path at Emmenegger Nature Park in Kirkwood. It and two other Asian species, *C. bodinieri* H. Lév. (Bodinier beautyberry) and *C. japonica* Thunb. (Japanese beautyberry), currently are cultivated as ornamentals in Missouri. As with *C. americana*, plants of all three Asian species sometimes die back during harsh winters, especially in the northern half of the state. They rarely, if ever, become established outside of cultivation, other than persisting from original plantings, and the other two have not been reported yet as escapes in Missouri.

5. Clinopodium L.

Plants annual or perennial herbs, usually with fibrous roots. Stems erect or strongly ascending, sometimes from a spreading base, slender, sharply 4-angled, branched throughout, sometimes with stoloniferous, prostrate, basal branches, glabrous or soft-hairy. Leaves sessile or petiolate, the petiole unwinged (except sometimes near the tip), with a faint or strong fragrance when crushed. Leaf blades variously shaped, sometimes dimorphic, the margins entire or toothed, the surfaces glabrous or hairy, also with sometimes conspicuous, sessile glands. Inflorescences axillary, small clusters of 2–8 per node, the flowers stalked, the stalk with a pair of bractlets at the base. Calyces zygomorphic, lacking a lateral projection, more or less symmetric to pouched along 1 side at the base, more or less cylindric (appearing somewhat flask-shaped at fruiting), the tube strongly 10–13-nerved (-ribbed), with a fringe of short, bristly hairs in the mouth, 2-lipped, the lobes shorter than to about as long as the tube, the upper lip shallowly to deeply 3-lobed, the lower lip deeply 2-lobed, the lobes narrowly to broadly triangular, not spinescent, not becoming enlarged and papery at fruiting. Corollas zygomorphic, pink to pale purple, rarely white or nearly so, the outer surface sparsely to moderately short-hairy and with inconspicuous sessile glands, the tube funnelform, relatively shallowly 2-lipped, the upper lip shallowly 2-lobed, slightly concave to spreading to recurved, the lower lip more or less straight to slightly spreading, 3-lobed. Stamens 4, not exserted, the upper 2 stamens with slightly shorter filaments than the lower pair, the anthers small, the connective about as broad as the pollen sacs and trapezoidal, the pollen sacs 2, broadly angled from the separated tips, pink or pinkish-tinged. Ovary deeply lobed, the style appearing lateral to nearly basal from a deep apical notch. Style usually slightly exserted, unequally 2-branched at the tip. Fruits dry schizocarps, separating into usually 4 nutlets, these 1.0–1.4 mm long, oblong-obovoid to oblong-ellipsoid, the surface brown to dark brown, glabrous, but sometimes with a mealy deposit. About 100 species, nearly worldwide.

The taxonomy of the so-called *Satureja* L. complex remains incompletely understood. It has been treated variously as a single large genus (Epling and Játiva, 1966) or as many as 17 small genera (Doroszenko, 1985). Based on a preliminary molecular study (Wagstaff et al., 1995), followed by an examination of morphology from herbarium specimens of a larger group of species, Cantino and Wagstaff (1998) concluded that the European *Satureja officinalis* L. (savory) represents a genus distinct from the approximately 225 other species placed into that genus by earlier botanists, but that the majority of the remaining species form a single group for which the oldest valid name is *Clinopodium*, rather than *Satureja*. However, generic limits remain unclear, as the molecular data failed to resolve a large lineage that includes such diverse genera as *Blephilia*, *Hedeoma*, and *Monarda*, and also failed to address the separation of *Acinos* and *Calamintha*, which some authors have maintained as distinct (DeWolf, 1954). The conclusions of Cantino and Wagstaff (1998) are followed here mainly for the pragmatic reason that our species fall into two complexes that are difficult to distinguish as genera.

1. Stems finely hairy throughout; main stem leaves elliptic to somewhat rhombic or less commonly slightly obovate, hairy, at least along the margins and veins; calyces pouched along 1 side at the base, hairy along the ribs .... 1. C. ACINOS
1. Stems glabrous, except for inconspicous, short hairs at some nodes; flowering stem leaves (except sometimes at the stem base) linear to narrowly oblanceolate, glabrous; calyces more or less symmetric at the base, not pouched, not hairy .... 2. C. ARKANSANUM

1. Clinopodium acinos (L.) Kuntze (mother of thyme)
Acinos arvensis (Lam.) Dandy
Satureja acinos (L.) Scheele
Pl. 432 a, b; Map 1941

Plants annual (short-lived perennials elsewhere), forming clumps. Stems not dimorphic, 10–40 cm long, loosely to moderately ascending from spreading bases, rooting at the lower nodes, moderately to densely pubescent throughout (sometimes more densely so on 2 opposing sides) with fine, spreading to slightly curved, short hairs. Leaves mostly short-petiolate, with a faint fragrance when crushed. Leaf blades 6–12 mm long, 4–8 mm wide, elliptic to somewhat rhombic or less commonly slightly obovate, those of the uppermost leaves often oblanceolate to narrowly oblanceolate, angled or tapered at the base, tapered to a sharply pointed tip, the margins entire or with a few shallow teeth, hairy, the surfaces moderately pubescent with short, curved hairs, at least along the veins, the sessile glands usually inconspicuous.

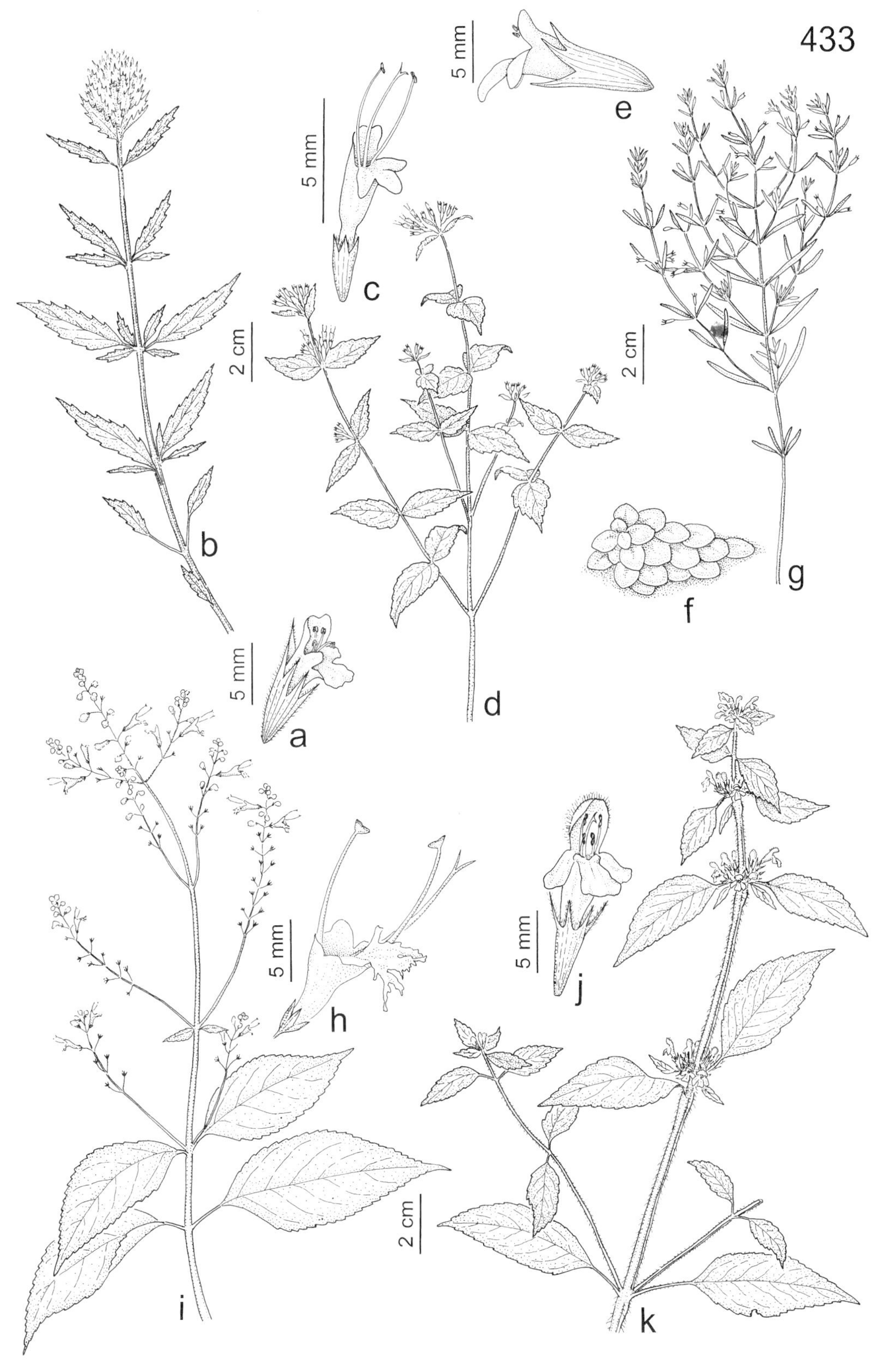

Plate 433. Lamiaceae. *Dracocephalum parviflorum*, **a)** flower, **b)** fertile stem. *Cunila origanoides*, **c)** flower, **d)** fertile stem. *Clinopodium arkansanum*, **e)** flower, **f)** overwintering branches with leaves, **g)** fertile stem. *Collinsonia canadensis*, **h)** flower, **i)** fertile stem. *Galeopsis tetrahit*, **j)** flower, **k)** fertile stem.

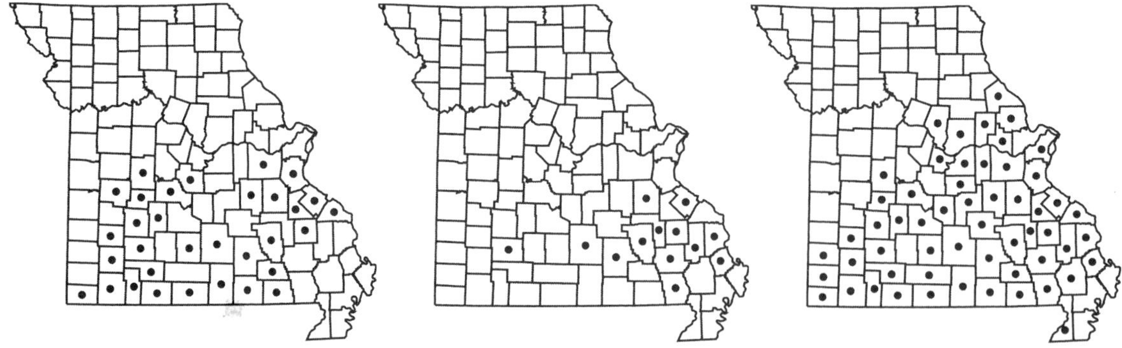
1942. Clinopodium arkansanum 1943. Collinsonia canadensis 1944. Cunila origanoides

Flowers in clusters of 2–6 per node, the stalks 2–4 mm long, the bracts 0.5–1.5 mm long, triangular to narrowly triangular. Calyces 5–6 mm long, pouched along 1 side at the base, hairy externally along the ribs, with sparse, inconspicuous, sessile glands between the ribs, the lobes narrowly triangular, often slightly inward-angled at fruiting. Corollas 7–10 mm long, the lips about half as long as the tube. Nutlets brown, the surface lacking a mealy coating, smooth. $2n=18$. May–September.

Introduced, known thus far only from a historical collection from Jasper County (native of Europe; introduced sporadically in the northern half of the U.S., Canada). Disturbed areas.

This species is cultivated occasionally as a groundcover and has a long history of various medicinal uses.

2. Clinopodium arkansanum (Nutt.) House (Arkansas calamint, satureja, low calamint)

Calamintha arkansana (Nutt.) Shinners
Satureja arkansana (Nutt.) Briq.
S. glabella (Michx.) Briq. var. *angustifolia* (Torr.) Svenson

Pl. 433 e–g; Map 1942

Plants perennial, with slender rhizomes (usually developing from stolons of previous years), usually forming loose colonies. Stems dimorphic, glabrous except for sparse, minute hairs at some nodes, the basal branches formed after flowering, 1–10 cm long, stoloniferous, prostrate, rooting at the tip and some nodes, unbranched or few-branched, the flowering stems 10–40 cm long, erect or strongly ascending, usually much-branched, mostly above the lower 1/4. Leaves more or less dimorphic, the margins entire, glabrous, sometimes reddish- or purplish-tinged (especially along the stolons), the sessile glands relatively conspicuous, with a strong, pleasant fragrance when crushed, those of the basal branches and sometimes also the basal portion of the flowering stems 3–10 mm long, short- to long-petiolate (the petiole shorter than to as long as the blade), the blade elliptic to ovate or nearly circular, relatively thick, rounded to angled at the base and tip, grading abruptly into the main leaves of the flowering stems, these 4–22 mm long, sessile or very short-petiolate, linear to narrowly oblanceolate, relatively thin, tapered at the base, angled or tapered to the rounded or bluntly to less commonly sharply pointed tip. Flowers in clusters of 2–8 per node, the stalks 3–10 mm long, the bracts 1–4 mm long, linear to threadlike. Calyces 4–6 mm long, more or less symmetric at the base, not pouched, glabrous externally except for the relatively conspicuous sessile glands between the ribs, the lobes narrowly triangular to triangular, straight or sometimes somewhat spreading at fruiting. Corollas 8–15 mm long, the lips about half as long as the tube. Nutlets dark brown, the surface often partially obscured by a granular or mealy, pale coating, otherwise appearing minutely pebbled or roughened. $2n=22$. May–September.

Scattered in the Ozark and Ozark Border Divisions (northeastern U.S. west to Minnesota and Arkansas, also Kansas, Oklahoma, Texas, and New Mexico; Canada). Glades, ledges and tops of bluffs, fens, and banks of streams, rivers, and spring branches; also roadsides; on calcareous substrates.

Epling and Játiva (1966) included Arkansas and Missouri in the range of the glade savory, *C. glabellum* (Michx.) Kuntze (as *Satureja glabella*), a close relative of *C. arkansanum* that some other authors have treated as only varietally distinct (Svenson, 1940; Gleason and Cronquist, 1991). Purportedly, *C. glabellum* differs in a more robust habit, lacking stoloniferous basal branches, but having minutely bearded nodes, leaves up to 5 cm long, and at least the lower leaves oblong-lanceolate to elliptic and often few-toothed. Steyermark (1963) had excluded this taxon from the

Missouri flora. Examination of Missouri materials during the present study revealed that most specimens have at least a few minute hairs at some nodes, but that none have densely bearded nodes. The basal branches tend to be produced during the latter half of the growing season and thus are not represented on some specimens. None of the Missouri material has toothed leaves, although occasional specimens from the southern portion of the Ozarks are relatively robust and have somewhat broader lower leaves than is typical of *C. arkansanum*. Neither E. B. Smith (1988) nor the Arkansas Vascular Flora Committee (2006) listed *C. glabellum* for Arkansas, and Kartesz and Meacham (1999) mapped it only from Kentucky, Tennessee, and historically Virginia. Taxonomic relationships between these two entities should be studied in greater detail. For the present, the two species are accepted as distinct, but *C. glabellum* is excluded from the Missouri flora. Additionally, Gleason and Cronquist (1991) noted that there is considerable morphological heterogeneity within the range of this taxon. Plants from the northeastern states tend to occur in sandy lakeshore habitats and to have shorter, more stiffly erect flowering stems with less branching. Rare, white-flowered plants were called *Satureja arkansana* f. *alba* by Steyermark, but that name apparently has not been validly published.

Walking through a population of this calamint is an olfactory treat. Steyermark (1963) commented on the strong pleasant fragrance of menthol produced by the plants, which he indicated had been used by early settlers in a tea and externally as an insect repellent. Kurz (1999) mentioned anecdotal evidence that the plants can cause dermatitis in some individuals and that the essential oil can be poisonous if ingested in sufficient quantity.

6. **Collinsonia** L. (horse balm)
(Peirson et al., 2006)

Four species, eastern tempterate North America.

1. Collinsonia canadensis L. (richweed, citronella horse balm)

Pl. 433 h, i; Map 1943

Plants perennial herbs, with short, thick, woody rhizomes. Stems 40–80(–120) cm long, erect or strongly ascending, bluntly to sharply 4-angled, unbranched or few-branched, glabrous or sparsely pubescent with inconspicuous, fine, crinkly, mostly multicellular hairs, sometimes only in lines or patches at the nodes. Leaves opposite, mostly long-petiolate (the uppermost leaves sometimes sessile), the petiole mostly unwinged. Leaf blades 3–18 (–25) cm long, 2–17 cm wide, elliptic to oblong-ovate or ovate, rounded or more commonly broadly short-tapered at the base, tapered to a sharply pointed tip, the margins coarsely toothed or occasionally scalloped, the upper surface glabrous or sparsely pubescent with short, more or less spreading, multicellular hairs, often also with moderate to dense, minute hairs along the main veins, especially toward the blade base, the undersurface sparsely to moderately short-hairy, also with relatively conspicuous sessile glands. Inflorescences terminal panicles with open, racemose branches, usually more or less pyramidal in overall outline, the flowers 2 per node and usually somewhat overlapping those of adjacent nodes, the stalks 3–5 mm long. Bracts inconspicuous (1–3 mm long), lanceolate to elliptic, shorter than the flowers. Calyces 2.5–5.0 mm long at flowering, somewhat zygomorphic to nearly actinomorphic, lacking a lateral projection, more or less symmetric at the base, bell-shaped, the tube faintly to strongly 15-nerved, glabrous in the mouth, usually slightly 2-lipped, the lobes variously shorter than to longer than the tube, narrowly triangular-tapered, not spinescent, minutely hairy and glandular on both surfaces, becoming enlarged to 6–8 mm and papery at fruiting. Corollas 12–15 mm long, zygomorphic, mostly yellow to greenish yellowish brown, sometimes with dull purple to reddish brown lines, blotches, or streaks, the surfaces sparsely to moderately and minutely hairy, also glandular, the tube narrowly funnelform, 2-lipped to at or slightly above the midpoint, the upper lip shorter, the 4 lobes spreading to loosely ascending, more or less ovate, rounded, with entire margins, the lower lip prominent, spreading to somewhat arched, 1-lobed, this strongly fringed, often partially white. Stamens 2 (2 minute staminodes sometimes also present), strongly exserted and somewhat spreading, the anthers small, the connective short, the pollen sacs 2, appearing more or less sessile and spreading from the filament tip, purple to bluish-tinged. Ovary deeply lobed, the style appearing nearly basal from a deep apical notch. Style exserted, with 2 slender branches at the tip. Fruits dry schizocarps, usually consisting of a solitary nutlet, this

1945. Dracocephalum parviflorum
1946. Galeopsis tetrahit
1947. Glechoma hederacea

2.0–2.5 mm long, globose, the surface dark brown, glabrous, smooth and somewhat shiny. $2n$=50, 52. July–September.

Scattered in the eastern half of the Ozark Division and the adjacent Ozark Border (eastern U.S. west to Wisconsin and Louisiana; Canada). Bottomland forests, mesic upland forests in ravines, banks of streams and rivers, and bases and ledges of bluffs.

This is the most broadly distributed species in the genus; the other three species are endemic mainly to localized areas in the southeastern United States and the southern Appalachians. Steyermark (1963) noted that the flowers of *C. canadensis* have a pleasant lemony fragrance. The crushed foliage also has a lemony odor. Steyermark further reported that the rootstock was used medicinally for kidney ailments. Moerman (1998) reported multiple other medicinal uses for the species by Native Americans ranging from pain relief to antidiarrheal, antirheumatic, and dermatologial properties, as well as for a stimulant and tonic.

7. **Cunila** D. Royen ex L. (dittany, stone mint)

About 15 species, North America to South America.

1. Cunila origanoides (L.) Britton (dittany)
Pl. 433 c, d; Map 1944

Plants perennial herbs, with an often somewhat woody rootstock and short, slender rhizomes. Stems 20–40(–60) cm long, erect or strongly ascending, bluntly (toward the stem base) to relatively sharply 4-angled, several- to numerous-branched, glabrous or sparsely pubescent with short, spreading and/or, crinkly hairs, sometimes mostly along the angles or along 2 opposing, slightly concave sides. Leaves opposite, sessile or nearly so. Leaf blades 1.5–4.0 cm long, 0.7–2.0 cm wide, narrowly to broadly ovate, rounded or occasionally shallowly cordate at the base, angled or tapered to a sharply pointed tip, the margins finely toothed or rarely entire, the surfaces glabrous or sparsely to moderately pubescent along the midvein or main veins with stiff, spreading, multicellular hairs (also sometimes with a few such hairs in the axils of the marginal teeth), also with relatively conspicuous sessile glands. Inflorescences terminal and axillary, dense clusters of 6 to numerous flowers, these short-stalked to nearly sessile (0.5–2.0 mm). Bracts inconspicuous (to 1 mm long), linear to narrowly lanceolate, shorter than the flowers. Calyces 1.5–3.0 mm long at flowering, actinomorphic, lacking a lateral projection, symmetric at the base, funnelform to narrowly bell-shaped, the tube strongly 10-nerved (-ribbed), with a fringe of short, bristly hairs in the mouth, the lobes variously shorter than the tube, narrowly triangular to triangular, not spinescent, glandular on the outer surface, not becoming enlarged or papery at fruiting. Corollas 6–8 mm long, weakly zygomorphic, pinkish purple to lavender or sometimes white, not spotted, the surfaces moderately to densely and minutely hairy, also glandular, the tube funnelform, slightly 2-lipped, the lips up to half as long as the tube, the upper lip notched at the broadly rounded tip, straight or slightly arched, the lower lip 3-lobed, arched to spreading. Stamens 2, strongly exserted, the anthers small, the connective short, the pollen sacs 2, parallel or nearly so, yellowish purple to nearly white. Ovary deeply lobed, the style appearing nearly basal from a deep apical notch. Style not or only slightly exserted, with 2 slender branches at the tip. Fruits dry schizocarps, separating into 2–4 nutlets (rarely the nutlet solitary), this 0.8–1.0 mm long, ellipsoid to ovoid, the surface yellowish brown to brown, glabrous, smooth or very finely pebbled. July–November.

Scattered throughout the Ozark and Ozark Border Divisions, also in portions of the Mississippi Lowlands and the southeasternmost part of the Glaciated Plains (eastern U.S. west to Kansas and Texas). Mesic to dry upland forests, savannas, upland prairies, sand prairies, ledges and tops of bluffs, and banks of streams and rivers; also old fields and roadsides; often on acidic substrates.

The crushed foliage of *C. origanoides* has a pleasant minty odor and the dried foliage has been used in teas, sachets, and potpourri. Acording to Moerman (1998), Native Americans used the plant as an antiseptic, for pain relief, to lower fevers, and as a tonic. The essential oils sometimes are extracted for medicinal use or for a flavorant in cooking. However, a number of Old World species are extracted commercially more commonly to produce the so-called oil of dittany, including the oregano relative, *Origanum dictamnus* L., and some members of the genus *Dictamnus* L. (Rutaceae).

Steyermark (1963) stated that *C. origanoides* is a desirable ornamental in wildflower gardens and rock gardens. He also noted that the species is noted as a frequent producer of spectacular *frost flowers* at the end of the growing season. These are formed when sudden overnight freezing temperatures cause the stems to burst, with the resultant release of quantities of sap that are pumped up from the still metabolically active roots. The exuded liquid freezes into intricate layered petal-like or ribbony shapes of ice sometimes more than 5 cm long. This phenomenon was described in fanciful detail by L. F. Ward (1893), who called the plant by the superfluous name *Cunila mariana* L. and referred to the exudate as frost freaks.

8. **Dracocephalum** L. (dragonhead)

About 45 species, North America, Europe Asia.

Some Old World species of *Dracocephalum* are cultivated as ornamental annuals in portions of the United States, notably *D. moldavica* L. (*Moldavica punctata* Moench) and *D. thymiflorum* L., but none of these has been reported to escape in Missouri thus far. Note that in some of the older botanical literature, species of the related genus *Physostegia* have mistakenly been called *Dracocephalum*.

1. Dracocephalum parviflorum Nutt. (American dragonhead, dragonhead)

Pl. 433 a, b; Map 1945

Plants annual (biennial or short-lived perennial elsewhere), with taproots. Stems 20–80 cm long, erect or ascending, relatively sharply 4-angled, usually branched, sparsely pubescent with minute, curved hairs, sometimes mostly along the angles, more densely hairy around the nodes. Leaves opposite, short- (toward the stem tip) to long- (toward the stem base) petiolate. Leaf blades 2–8 cm long, 1.0–2.5 cm wide, narrowly lanceolate or narrowly elliptic to ovate or broadly lanceolate-triangular, broadly angled or tapered (rarely rounded) at the base, narrowly angled or tapered to a sharply pointed tip, the margins coarsely and sharply toothed, often also hairy, especially toward the base, the upper surface glabrous or sparsely and minutely hairy along the main veins, the undersurface moderately to densely and minutely hairy, also with usually inconspicuous sessile glands. Inflorescences terminal spikelike racemes, more or less continuous (the lowermost few nodes sometimes appearing somewhat distinct), with numerous flowers, these short-stalked (0.5–4.0 mm), each node with a pair of leaflike bracts. Bractlets conspicuous, leaflike, 7–20(–30) mm long, elliptic to narrowly lanceolate, with spinescent teeth along the margins, shorter than to slightly longer than the adjacent calyces. Calyces 8–12 mm long at flowering, 2-lipped (the upper lip with the central lobe longer and broader than the 2 lateral lobes and the 2 lobes of the lower lip), lacking a lateral projection, symmetric at the base, more or less cylindric to narrowly bell-shaped, the tube strongly 15-nerved (-ribbed), glabrous in the mouth, the lobes mostly shorter than the tube, triangular, tapered to spinescent tips, glabrous on the outer surface, becoming slightly enlarged (to 16 mm) and papery at fruiting. Corollas 9–14 mm long, zygomorphic, pink to lavender or light purple, the lower lip often with darker nerves or central spot, the outer surface densely and minutely hairy, the tube funnelform, 2-lipped, the lips shorter than the tube, the upper lip notched at the broadly rounded tip, straight or slightly arched, the lower lip 3-lobed with the central lobe longer than the 2 lateral lobes, arched to spreading. Stamens 4, not exserted (curved under the upper lip), the anthers small, the connective short, the pollen sacs 2, spreading, sometimes becoming more or less parallel after the pollen has been shed), dark purple to bluish purple. Ovary deeply lobed, the style appearing nearly basal from a deep apical notch. Style not exserted, with 2 slender branches at the tip. Fruits dry schizocarps, separating into usually 4 nutlets, these 2.0–2.7 mm long, ellipsoid to narrowly ovoid, the surface black, glabrous, smooth or very finely pebbled. $2n=14$. May–August.

Introduced, uncommon, sporadic (northern U.S. [including Alaska] south to New York, Illinois, Nebraska, and Arizona; Canada; introduced sporadically farther south). Margins of crop fields and railroads.

9. **Galeopsis** L. (hemp nettle)

Ten species, Europe, Asia; introduced in the New World.

The stamens in the genus are unusual in their morphology and dehiscence. The two anther sacs are attached to the expanded connective in a vertical orientation and spreading from one another. Each anther sac dehisces by two valves in somewhat of a clamshell pattern. The larger valve of each pair, which is glabrous, dehisces away from the other anther sac of the stamen and becomes oriented parallel to the filament. The smaller valve of each pair, which is densely bearded with minute, bristly hairs, dehisces toward the other anther sac (the two smaller valves of the pair of anther sacs become oriented parallel to one another and are positioned more or less erectly after dehiscence). Although the anthers themselves usually are white or purplish-tinged, the pollen is yellow.

1. Galeopsis tetrahit L. (common hemp nettle)
Pl. 433 j, k; Map 1946

Plants annual, with taproots. Stems 20–80 cm long, erect or ascending, relatively sharply 4-angled, unbranched or more commonly branched, sparsely to moderately pubescent with relatively long, stiff, more or less spreading, multicellular hairs, mostly along the angles, also with shorter, finer, gland-tipped hairs, especially around the somewhat swollen nodes. Leaves opposite, short- to long-petiolate. Leaf blades 2–10 cm long, 0.8–5.0 cm wide, lanceolate or narrowly elliptic-lanceolate to rhombic-elliptic or ovate, angled or rounded at the base, angled or tapered to a sharply pointed tip, unlobed, the margins coarsely and bluntly toothed, also hairy, the upper surface moderately pubescent with stiff, multicellular hairs, the undersurface moderately pubescent with similar, but usually shorter, finer, mostly unicellular hairs, both surfaces also with usually inconspicuous sessile glands. Inflorescences axillary, dense clusters of 6–12 flowers per node, these sessile. Bractlets more or less leaflike, 3–12 mm long, narrowly oblong-elliptic or sometimes 3-lobed, spinescent at the tips, shorter than to longer than the adjacent calyces. Calyces 7–12 mm long at flowering, actinomorphic, lacking a lateral projection, symmetric at the base, more or less cylindric to narrowly bell-shaped, the tube strongly 10-nerved (-ribbed) and with finer nerves between the ribs, finely hairy in the mouth, the lobes about as long as the tube and more or less equally ascending, narrowly triangular to triangular, tapered to spinescent tips, moderately pubescent with fine, straight, mostly multicellular hairs on the outer surface and the inner surface of the lobes, becoming slightly enlarged (to 16 mm) and papery at fruiting. Corollas 15–22 mm long, zygomorphic, variously white, pink, or light purple (at least the lower lip usually partially white), the lower lip usually with darker nerves, also a pair of small nipplelike projections (often associated with yellow spots) on the upper surface toward the base, the outer surface moderately to densely pubescent with straight, mostly multicellular, spreading hairs, the tube narrowly funnelform, 2-lipped, the lips shorter than the tube, the upper lip occasionally shallowly notched at the broadly rounded tip, slightly hooded, the lower lip 3-lobed with the central lobe spreading and somewhat arched, longer than the 2, spreading, lateral lobes. Stamens 4, not exserted (curved under the upper lip), the filaments of 2 lengths, the anthers small, the connective somewhat expanded, the pollen sacs 2, oriented vertically and spreading, white or purplish-tinged. Ovary deeply lobed, the style appearing nearly basal from a deep apical notch. Style not exserted, with 2 slender branches at the tip. Fruits dry schizocarps, separating into 1–4 nutlets, these 2.5–3.5 mm long, ovoid to broadly ovoid, the surface brown with irregular, lighter, tan to yellow mottling, glabrous, smooth or the mottled areas appearing slightly raised. $2n$=32. June–September.

Introduced, uncommon, known thus far only from the city of St. Louis (native of Europe; introduced widely in the northern U.S., Canada). Railroads.

Galeopsis tetrahit is an allotetraploid species that arose in nature as a hybrid between two diploid progenitors, *G. pubescens* Besser and *G. speciosa* Mill. with a subsequent doubling of the chromosome number. The ancestry and cytological behavior of this taxon and its progenitors were studied by Müntzing (1930, 1932), who also successfully synthesized the hybrid through a breeding program. According to Mabberley (1997), it was the first naturally occurring allotetraploid to be artificially recreated by crossing of the parents under controlled conditions.

Burrows and Tyrl (2001) noted that when nutlets of this species contaminating animal feed are ingested in even fairly low doses, they can result in a variety of symptoms and occasional fatalities.

The Eurasian *G. bifida* Boenn., which has smaller corollas (to 15 mm) with the middle lobe of the lower lip notched or more deeply divided, sometimes has been treated as *G. tetrahit* var. *bifida* (Boenn.) Kudô. This taxon escapes sproadically in the northern United States, but has not been reported from Missouri. Several other infraspecific taxa have been named within the native range of *G. tetrahit*, but require further taxonomic study.

10. **Glechoma** L. (ground ivy)

About 10 species, Europe, Asia; introduced in the New World.

The generic name sometimes has been spelled *Glecoma*, but the orthography has been conserved as *Glechoma* (Adolphi, 1982).

1. Glechoma hederacea L. (ground ivy, gill-over-the-ground)

G. hederacea var. *micrantha* (Boenn. ex Rchb.) Nyman

Pl. 434 a, b; Map 1947

Plants perennial, with fibrous roots, sometimes forming loose mats. Stems 10–200 cm long, prostrate (sometimes with short, ascending branches), rooting at the nodes, mostly sharply 4-angled, branched, glabrous or minutely roughened along the angles, also usually with a line of longer bristly hairs at the nodes. Leaves opposite, mostly long-petiolate. Leaf blades 1.0–2.5(–4.0) cm long, 1.0–2.5(–5.0) cm wide, broadly kidney-shaped to nearly circular, shallowly to deeply cordate or occasionally broadly rounded to truncate at the base, rounded or broadly angled to a bluntly pointed tip, unlobed, the margins coarsely scalloped or bluntly toothed, sometimes also minutely hairy, the surfaces glabrous or less commonly sparsely short-hairy, the undersurface also with usually inconspicuous sessile glands. Inflorescences axillary, clusters of 2–6 flowers per node, these with stalks 1–3 mm long. Bractlets scalelike or hairlike, 1.0–1.5 mm long, linear to narrowly triangular, mostly shorter than the associated flower stalk. Calyces 4–7 mm long, slightly zygomorphic (the upper lobes slightly longer than the lower ones), lacking a lateral projection, symmetric at the base, more or less cylindric to narrowly bell-shaped, the tube 15-nerved, glabrous in the mouth, the lobes shorter than the tube and more or less equally loosely ascending, oblong-elliptic, tapered to short-spinescent tips, sparsely to moderately pubescent with minute crinkly hairs on the outer surface, not becoming enlarged at fruiting. Corollas 10–18(–22) mm long, zygomorphic, purplish blue to purple (rarely white), the lower lip usually with white and darker mottling or spots, the outer surface moderately to densely pubescent with short, straight, more or less spreading hairs, the tube funnelform, 2-lipped, the lips shorter than the tube, the upper lip narrowly obcordate, notched at the broadly rounded tip, straight and not noticeably hooded, the lower lip with 3 spreading lobes, the larger central lobe shallowly notched at the tip, longer and much broader than the 2 lateral lobes, with a beard of fine spreading hairs internally near its base. Stamens 4, not or only slightly exserted (ascending under the upper lip), the filaments of 2 lengths, the anthers small, the connective short, the pollen sacs 2, spreading, white or purplish- to bluish-tinged. Ovary deeply lobed, the style appearing nearly basal from a deep apiçal notch. Style usually slightly exserted, with 2 slender branches at the tip. Fruits dry schizo-

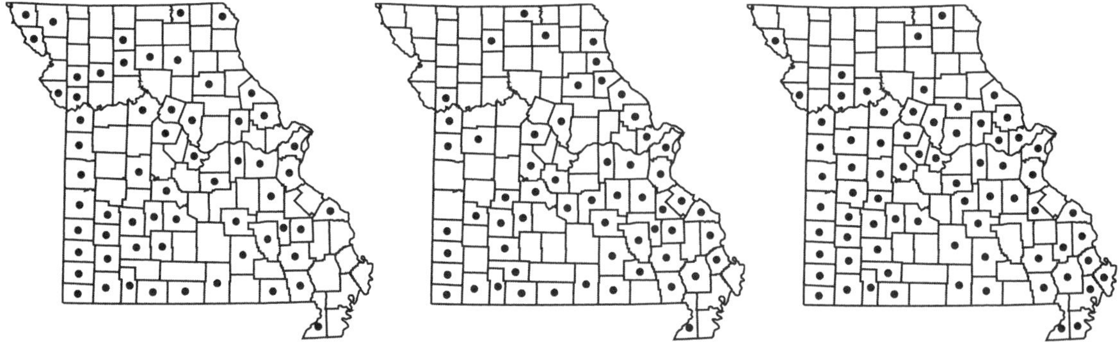

1948. Hedeoma hispida

1949. Hedeoma pulegioides

1950. Lamium amplexicaule

carps, separating into 2–4 nutlets (sometimes these failing to mature), these 1.5–2.0 mm long, narrowly ovoid to oblong-ovoid or oblong-ellipsoid, the surface light brown to yellowish brown, glabrous, smooth or finely pebbled. $2n$=18, 36 ($2n$=24, 45, 54 elsewhere). March–July.

Introduced, scattered nearly throughout the state, but mostly absent form the western portion of the Glaciated Plains Division (native of Europe, Asia; introduced nearly throughout the U.S., Canada, and in portions of the southern hemisphere). Bottomland forests, mesic upland forests, banks of streams, rivers, and spring branches, and bases of bluffs; also lawns, gardens, railroads, roadsides, and shaded, disturbed areas.

Böllmann and Scholler (2004) documented the distribution and historical spread of ground ivy in North America from its initial report from the eastern United States in 1814 to its present nearly cosompolitan distribution in temperate North America. According to them, the earliest specimen from Missouri was collected in 1868 along the Meramec River (St. Louis County). Böllmann and Scholler also noted that although the plant reproduces principally vegetatively by stem fragments, it can spread by nutlets that produce an adhesive mucilage when moistened. Today, the plant is considered a nuisance in lawns and gardens (and in disturbed floodplain habitats), capable of growing up to 2 m in a single year and continuing to elongate under the leaf litter during the winter. The foliage does produces a mildly unpleasant odor when crushed. Steyermark (1963) and Burrows and Tyrl (2001) noted livestock poisoning after animals ingested large quantities of the species fresh or in hay (but the causal toxins remain unknown).

The infraspecific taxonomy of *G. hederacea* is not well understood and several varieties and subspecies have been described. Polyploidy also plays a role in the morphological variability of the species (Iwatsubo et al., 2004). Currently, most Asian and European botanists do not divide *G. hederacea* into infraspecific taxa in its native range, but further study to correlate the cytological and morphological variation may result in the resurrection of one or more of the subspecies. Some North American botanists have recognized plants with smaller flowers as var. *micrantha* (Steyermark, 1963), but such individuals intergrade completely with larger-flowered plants. Both extremes have been recorded from Missouri.

11. **Hedeoma** Pers.(mock pennyroyal) (Irving, 1980)

Plants annual, with slender taproots (perennial herbs or low shrubs, sometimes with rhizomes or stolons, elsewhere). Stems erect or strongly ascending, slender, sharply 4-angled, unbranched or branched mostly below the midpoint, hairy. Leaves sessile or short-petiolate, the petiole winged toward the tip, with a slight to moderate minty fragrance when crushed. Leaf blades variously shaped, those uppermost on the stems usually somewhat reduced, the margins entire or few-toothed, the surfaces glabrous or more commonly hairy, also with sometimes conspicuous, sessile glands. Inflorescences axillary, small clusters of (2–)8–12(–20) at nearly every node, the flowers short-stalked, the stalk with a pair of short, linear to narrowly

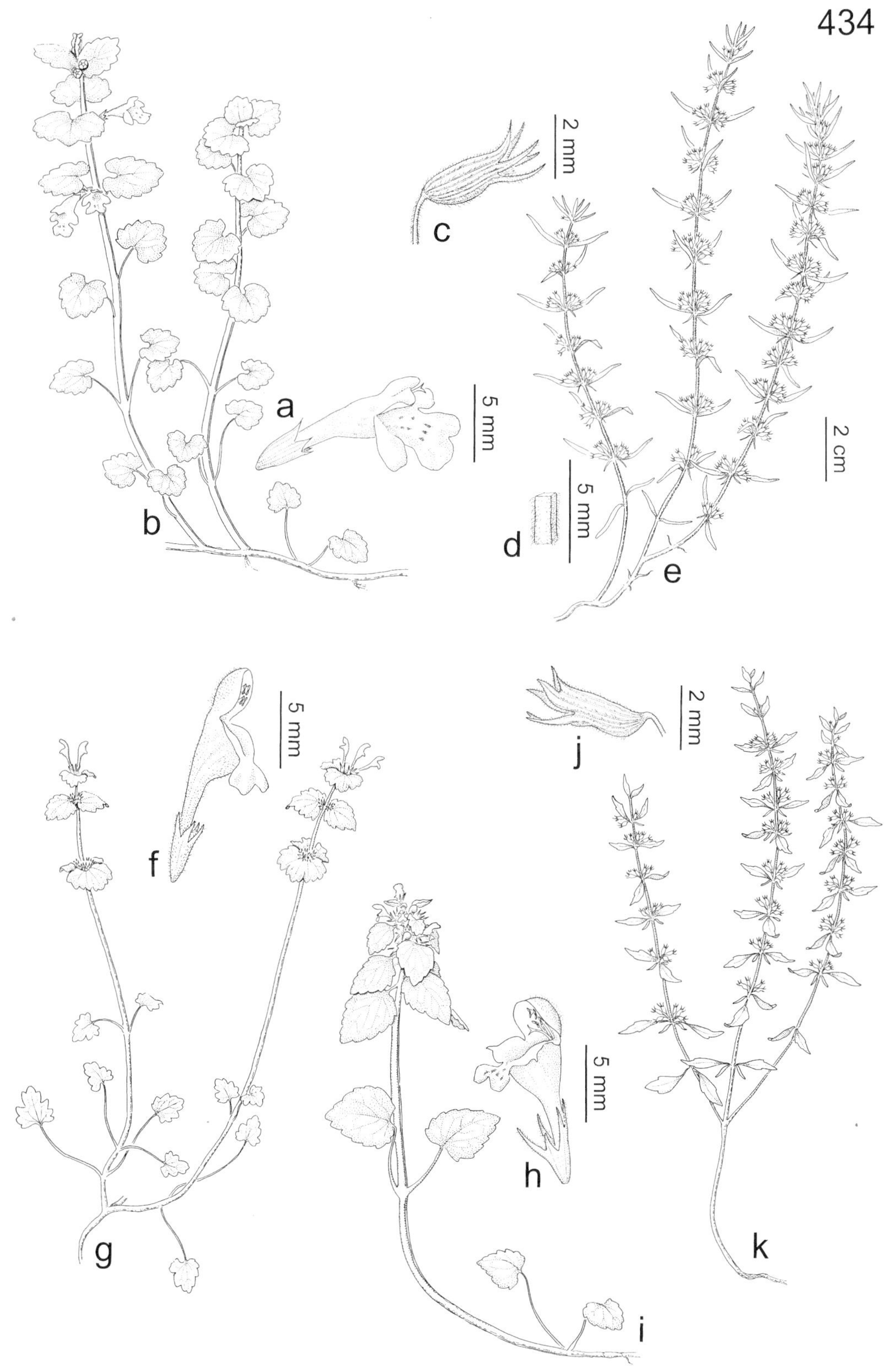

Plate 434. Lamiaceae. *Glechoma hederacea*, **a)** flower, **b)** habit. *Hedeoma hispida*, **c)** fruiting calyx, **d)** section of stem, **e)** habit. *Lamium amplexicaule*, **f)** flower, **g)** habit. *Lamium purpureum*, **h)** flower, **i)** habit. *Hedeoma pulegioides*, **j)** fruiting calyx, **k)** habit.

lanceolate bractlets at the base. Cleistogamous flowers sometimes produced, these with calyces similar to those of open flowers, but corollas that are only about as long as the calyx lobes and never fully open. Calyces zygomorphic, lacking a lateral projection, somewhat pouched along 1 side at the base, more or less cylindric, the tube strongly 13-nerved (-ribbed), with a fringe of short, bristly hairs in the mouth, 2-lipped, the lobes shorter than to about as long as the tube, the upper lip deeply 3-lobed and sometimes spreading upward, the lower lip deeply 2-lobed and straight to slightly arched upward, the lobes narrowly triangular to nearly linear, not spinescent, not becoming enlarged and papery at fruiting. Corollas (except in cleistogamous flowers) zygomorphic, lavender to pale bluish purple or nearly white, the lower lip sometimes with lighter or darker spots or mottling toward the base, the outer surface moderately short-hairy, the tube narrowly funnelform to nearly cylindric, relatively shallowly 2-lipped, the upper lip unlobed or shallowly notched, straight to slightly concave, the lower lip slightly concave (slightly scoop-shaped) to somewhat spreading, 3-lobed, the central lobe sometimes shallowly notched. Stamens 2 (a pair of small staminodes often also present), not exserted (ascending under the upper lip; somewhat exserted elsewhere), the anthers small, the connective short, the pollen sacs 2, spreading (parallel elsewhere), white or pinkish-tinged. Ovary deeply lobed, the style appearing nearly basal from a deep apical notch. Style not exserted, unequally 2-branched at the tip. Fruits dry schizocarps, separating into usually 4 nutlets, these 0.7–1.5 mm long, oblong-obovoid or oblong-ellipsoid to nearly spherical, the surface yellowish brown to dark brown or black, smooth or finely pebbled, glabrous, sometimes somewhat glaucous, otherwise somewhat shiny. Thirty-eight species, North America, South America.

1. Leaf blades linear to narrowly elliptic or narrowly oblong, the margins entire; upper lip of the calyx with the lobes linear, bristly-hairy along the margins; nutlets 1.0–1.5 mm long, oblong-obovoid to oblong-ellipsoid . . . . 1. H. HISPIDA
1. Leaf blades lanceolate to elliptic or ovate, the margins with at least a few irregular teeth; upper lip of the calyx with the lobes triangular, glabrous along the margins; nutlets 0.7–0.9 mm long, broadly ovoid to nearly spherical . . . . 2. H. PULEGIOIDES

1. Hedeoma hispida Pursh (mock pennyroyal, rough false pennyroyal)

Pl. 434 c–e; Map 1948

Stems 7–35 cm long, unbranched or few-branched below the midpoint, moderately to densely pubescent with mostly downward-curled hairs, sometimes with shorter, spreading hairs near the base. Leaves 5–20 mm long, sessile or nearly so, the blade linear to narrowly elliptic or narrowly oblong, bluntly to sharply pointed at the tip, the margins entire, both surfaces usually with sessile glands, sometimes very faintly so, the upper surface otherwise glabrous, the undersurface glabrous or finely hairy. Calyces 4.5–6.0 mm long, bristly-hairy along nerves, the upper lip with the 3 lobes 1.0–1.6 mm long, linear, bristly-hairy along the margins, the lower lip with the lobes 1.5–2.3 mm long. Corollas (except in cleistogamous flowers) 5.5–7.0 mm long. Nutlets 1.0–1.5 mm long, oblong-obovoid to oblong-ellipsoid, yellowish brown to dark brown, sometimes glaucous. $2n=34$. May–August.

Scattered nearly throughout the state (central U.S. west to Montana, Colorado, and Texas, east to Vermont, Connecticut, Ohio, and Alabama; Canada). Glades, upland prairies, savannas, openings of dry upland forests, tops of bluffs, and banks of streams and rivers; also pastures, old fields, fallow fields, railroads, roadsides, and open, disturbed areas.

According to Steyermark (1963), this species produces only a faint minty odor when the foliage is crushed.

2. Hedeoma pulegioides (L.) Pers. (pennyroyal, American false pennyroyal)

Pl. 434 j, k; Map 1949

Stems 10–35 cm long, unbranched or more commonly few- to several-branched below the midpoint, moderately to densely pubescent with minute spreading hairs. Leaves 7–25 mm long, the petiole 1–8 mm long, the blade lanceolate to elliptic or ovate, bluntly to sharply pointed at the tip, the margins with at least a few irregular teeth, the upper surface glabrous, the undersurface sparsely and finely hairy, also with usually conspicuous sessile glands. Calyces 3.5–5.0 mm long, glabrous or sparsely and finely hairy along the nerves, the upper lip with the 3 lobes 0.6–0.9 mm long, triangular, glabrous along the margins, the lower lip with the lobes 1–2 mm long. Corollas (except in cleistogamous flowers) 5–6 mm long. Nutlets 0.7–0.9 mm long, broadly ovoid to nearly spherical, mostly dark brown to black, not glaucous. $2n=36$. July–September.

Scattered nearly throughout the state, but apparently uncommon in the western half of the Glaciated Plains Division (eastern [mostly northeastern] U.S. west to Iowa, Kansas, and Arkansas; Canada). Glades, upland prairies, loess hill prairies, savannas, openings of mesic to dry upland forests, tops of bluffs, and banks of streams and rivers; also pastures, old fields, railroads, roadsides, and open, disturbed areas.

This species is a source of pennyroyal (which mainly is produced commercially from the European pennyroyal, *Mentha pulegium* L., thus the American extract also has been called oil of hedeoma), a volatile oil that has been used in perfumes and insect repellants. It also has been suspected of irritating the digestive systems of livestock when plants are ingested in quantity (Burrows and Tyrl, 2001).

12. **Lamium** L. (dead nettle)

Plants annual (perennial herbs elsewhere), with taproots. Stems erect or ascending, sometimes from a spreading base, sharply 4-angled, unbranched or branched, glabrous or hairy. Lowermost leaves moderately to long-petiolate, grading into the sessile upper leaves, the petiole unwinged or narrowly winged (in short-petiolate leaves), with a relatively weak, unpleasant fragrance when crushed. Leaf blades ovate to nearly circular or kidney-shaped, the margins toothed or scalloped and sometimes also lobed, the surfaces hairy, the undersurface usually also with inconspicuous, sessile glands. Inflorescences axillary, small clusters of mostly 4–12 flowers per node, the flowers sessile or nearly so, a pair of bractlets absent or inconspicuous, short, and slender. Cleistogamous flowers often produced, these with calyces similar to those of open flowers, but sometimes slightly shorter, and corollas that are only about as long as the calyx lobes and never fully open. Calyces actinomorphic or nearly so (the uppermost lobe often slightly longer than the lower lobes), lacking a lateral projection, symmetric at the base, cylindric to narrowly bell-shaped, the tube 5-nerved, glabrous or sparsely and finely hairy in the mouth, the lobes shorter than to longer than the tube at flowering, triangular to narrowly triangular, tapered to slender, sharply pointed tips, but remaining soft (not spinescent), not or only slightly becoming enlarged but often becoming papery at fruiting. Corollas zygomorphic, pale pink to lavender or pinkish purple, rarely entirely white, the lower lip sometimes lighter or white with purple spots or mottling, the outer surface sparsely to densely hairy, the tube funnelform, relatively shallowly 2-lipped, the upper lip entire or shallowly 2-lobed, concave (appearing hooded), the lower lip spreading, with a conspicuous central lobe that is narrowed at the base and entire to notched at the tip, the lateral lobes short and toothlike, sometimes reduced to small convexities along the lip margin. Stamens 4, not exserted (ascending under the hooded upper corolla lip), the upper 2 stamens with slightly shorter filaments than the lower pair, the anthers small, the connective short, the pollen sacs 2, angled to spreading, purple, reddish purple, or white, densely hairy. Ovary deeply lobed, the style appearing nearly basal from a deep apical notch. Style not exserted, equally 2-branched at the tip. Fruits dry schizocarps, separating into 2–4 nutlets, these 1.5–2.5 mm long, narrowly obovoid to obovoid, more or less truncate at the tip, often with thickened angles, with 2 flat sides and a rounded dorsal face, the surface light brown to grayish brown or olive brown, often with lighter mottling, finely pebbled, glabrous. About 40 species, Europe, Asia; introduced widely in the New World.

The genus *Lamium* was monographed by Mennema (1989), based primarily on the morphological study of herbarium specimens. Mennema accepted only 16 total species instead of

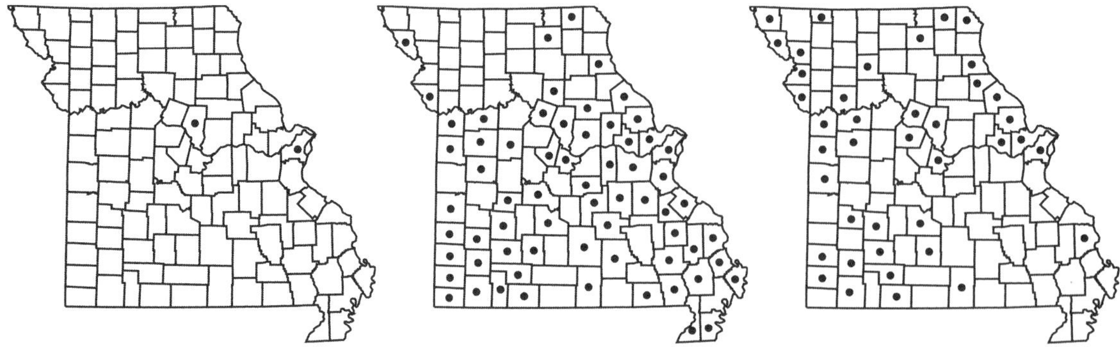

1951. Lamium hybridum 1952. Lamium purpureum 1953. Leonurus cardiaca

the approximately 40 species accepted my most earlier botanists, treating most of the complexes as single species, each divided into a series of subspecies or varieties. Because his taxonomic conclusions did not take into account the considerable body of biosystematic data on the genus, including the meticulous cytological analyses of artificially produced hybrids performed by Bernström (1955), Mennema's classification has not been fully embraced by most subsequent botanists.

A few of the rhizomatous perennial species are cultivated as ornamentals and groundcovers, notably *L. album* L. (white dead nettle), *L. galeobdolon* (L.) Crantz (*Lamiastrum galeobdolon* (L.) Ehrend. & Polatschek; yellow archangel) and *L. maculatum* L. (spotted dead nettle). The two species found in Missouri are disturbance-adapted annuals that often produce conspicuous displays of color in the spring, particularly in crop fields that have not yet been plowed.

1. Leaves subtending the flower clusters sessile or nearly so, sometimes more or less clasping the stem, usually wider than long; calyces densely hairy . . . . 1. L. AMPLEXICAULE
1. Leaves subtending the flower clusters distinctly petiolate, not clasping the stem, usually longer than wide; calyces sparsely hairy
   2. At least some of the leaves with the petiole winged above the midpoint and the blade irregularly lobed in addition to the toothed margins (sometimes appearing merely coarsely and doubly toothed) . . . . 2. L. HYBRIDUM
   2. None of the leaves with the petiole winged, the blade appearing singly toothed, not lobed or doubly toothed . . . . 3. L. PURPUREUM

1. Lamium amplexicaule L. (henbit)
Pl. 434 f, g; Map 1950

Stems 4–35 cm long, erect or ascending, often from a spreading base, unbranched or few-branched from at or near the base, glabrous or more commonly sparsely to moderately pubescent with short, downward-bent hairs. Leaves short- to long-petiolate (lower and often also median leaves), the shorter petioles often narrowly winged, the longer petioles usually unwinged, grading into sessile leaves (mostly those subtending flower clusters) that sometimes appear more or less clasping the stem. Leaf blades 5–15 mm long, unlobed or 3-lobed, the margins also with relatively coarse, blunt teeth or scallops, rounded to shallowly cordate at the base, rounded to bluntly pointed at the tip, the surfaces variously densely soft-hairy to nearly glabrous, those of the lower leaves ovate-triangular to broadly ovate or nearly circular, grading into those of the upper leaves, which are nearly semicircular to very broadly ovate or kidney-shaped, usually wider than long. Inflorescences with most of the nodes well-separated, the bractlets absent. Calyces 5–7 mm long, the tube and lobes densely bristly-hairy, the lobes shorter than to slightly longer than the tube. Corollas (except in cleistogamous flowers) 10–20 mm long, the outer surface sparsely to more commonly densely hairy, with a very dense patch of usually longer, darker hairs on the upper lip, the upper lip 3–5 mm long, rounded to shallowly notched at the tip, pinkish purple, rarely entirely white, the lower lip 1.5–3.0 mm long, usually lighter pink or white, with purple spots or mottling, rarely entirely white, the lateral lobes reduced to small convexities along the lip margin, the central lobe obcordate. 2*n*=18. February–May, rarely November–January.

Scattered to common nearly throughout the state, but apparently less common in most of the Glaciated Plains Division (native of Europe, Asia, Africa; introduced nearly throughout temperate North America). Edges of marshes, and openings of mesic upland forests; also crop fields, fallow fields, pastures, fencerows, lawns, sidewalks, railroads, roadsides, and open, disturbed areas.

Mennema (1989) treated *L. amplexicaule* as a complex of five varieties. In his classification, the introduced Midwestern material would correspond to var. *amplexicaule*. However, some of these varieties probably are better treated as separate species. More recently, based on a molecular-phylogenetic analysis of the genus, Bendiksby et al. (2011) determined that some of these non-Missouri taxa previously included within *L. amplexicaule* represent allopolyploid derivatives resulting from past hybridization between that species and other members of the genus. The situation requires further study. Rare plants with white corollas have been called f. *albiflorum* Dw. Moore (var. *album* Pickens and M.C.W. Pickens) and plants with all of the flowers cleistogamous have been called var. *clandestinum* Rchb.

Burrows and Tyrl (2001) reported that in Australian pastures livestock sometimes contract neurological problems (staggers) following ingestion of henbit, but that similar cases of toxicity are unknown from the native range of the species or from North America.

2. Lamium hybridum Vill.

Map 1951

Stems 7–35 cm long, erect or ascending, usually from a spreading base, unbranched or more commonly few- to several-branched from below the midpoint, glabrous or more commonly sparsely to moderately pubescent with short, downward-angled hairs (sometimes mainly on the angles). Leaves long-petiolate (lower and often also median leaves), at least some of the petioles winged above the midpoint, grading into short-petiolate leaves (mostly those subtending flower clusters) that are not clasping the stem. Leaf blades 5–35 mm long, those of some or all of the leaves irregularly lobed and also with fine to relatively coarse, blunt to sharp teeth (sometimes appearing merely coarsely and doubly toothed), broadly angled to truncate or shallowly cordate at the base, bluntly to sharply pointed or less commonly rounded at the tip, the surfaces moderately to densely pubescent with fine, loosely appressed hairs, those of the lower leaves ovate-triangular to ovate, slightly heart-shaped, or occasionally nearly circular, grading into those of the upper leaves, which are ovate-triangular to broadly ovate or slightly heart-shaped, usually longer than wide. Inflorescences with all or most of the nodes congested, the bractlets present, but inconspicuous, slender. Calyces 5–10 mm long, the tube and lobes sparsely bristly-hairy, sometimes also with sessile glands, the lobes usually longer than the tube. Corollas (except in cleistogamous flowers) 12–20 mm long, the outer surface sparsely to densely short-hairy, lacking a differentiated patch of longer, darker hairs on the upper lip, the upper lip 3–5 mm long, rounded to less commonly very shallowly notched at the tip, pale pink to lavender or pinkish purple, the lower lip 1.5–3.0 mm long, usually lighter pink or white, with purple spots or mottling, the lateral lobes short (to 0.5 mm), broadly triangular to more commonly narrowly triangular and toothlike, the central lobe obcordate and sometimes slightly uneven along the margins. 2*n*=36. March–May.

Introduced, uncommon, known thus far from the cities of Columbia (Boone County) and St. Louis (native of Europe, introduced sporadically in temperate North America). Railroads, roadsides, and disturbed areas.

Depauperate individuals of *L. purpureum* with cleistogamous flowers and unusually deeply toothed leaves might be confused with *L. hybridum* in the field. Mennema (1989) treated plants attributed by other botanists to *L. hybridum* as a mixture of two of his five varieties of *L. purpureum*: var. *hybridum* (Vill.) Vill. and var. *incisum* (Willd.) Pers. Earlier, Bernström (1955) had shown through cytological analysis of controlled crosses that *L. hybridum* is an allotetraploid species derived through past hybridization between *L. purpureum* and an unknown second progenitor. Working with nonnative populations in the state of Washington, R. J. Taylor (1990) completed a numerical analysis of morphological variation to reach a similar conclusion. Molecular data generated by Bendiksby et al. (2011) also confirm this hypothesis, but suggest that plants attributed to the name *L. hybridum* may represent the stabilized results of past hybridization between *L. purpureum* and more than one other parental taxon.

3. Lamium purpureum L. (dead nettle)
Pl. 434 h, i; Map 1952

Stems 7–30(–40) cm long, erect or ascending, often from a spreading base, unbranched or more commonly few- to several-branched from below the midpoint, glabrous or more commonly sparsely to moderately pubescent with short, downward-angled hairs (sometimes mainly on the angles), sometimes also with sparse longer, spreading hairs (especially near the nodes). Leaves long-petiolate (lower and often also median leaves), the petioles unwinged, grading into short-petiolate leaves (mostly those subtending flower clusters) that are not clasping the stem. Leaf blades 5–35 mm long, unlobed, the margins with fine to relatively coarse, blunt to sharp teeth, broadly angled to truncate or shallowly cordate at the base, bluntly to sharply pointed or less commonly rounded at the tip, the surfaces moderately to densely pubescent with fine, loosely appressed hairs, those of the lower leaves ovate-triangular to ovate, slightly heart-shaped, or occasionally nearly circular, grading into those of the upper leaves, which are ovate-triangular to broadly ovate or slightly heart-shaped, usually longer than wide. Inflorescences with the upper nodes often congested, the bractlets present, but inconspicuous, slender. Calyces 5–8 mm long, the tube and lobes sparsely bristly-hairy, sometimes also with sessile glands, the lobes usually longer than the tube. Corollas (except in cleistogamous flowers) 10–20 mm long, the outer surface sparsely to densely short-hairy, lacking a differentiated patch of longer, darker hairs on the upper lip, the upper lip 3–6 mm long, rounded to less commonly very shallowly notched at the tip, pale pink to lavender or pinkish purple, rarely entirely white, the lower lip 1.5–3.0 mm long, usually lighter pink or white, with purple spots or mottling, rarely entirely white, the lateral lobes short (to 0.5 mm), narrowly triangular and tooth-like, the central lobe obcordate and sometimes slightly uneven along the margins. $2n=18$. March–May, rarely November–January.

Scattered to common nearly throughout the state, but apparently uncommon in the western portion of the Glaciated Plains Division (native of Europe, Asia; introduced nearly throughout temperate North America). Bottomland prairies, bottomland forests, mesic upland forests, ledges of bluffs, and banks of streams; also crop fields, fallow fields, old fields, pastures, barnyards, fencerows, cemeteries, lawns, railroads, roadsides, and open, disturbed areas.

Mennema (1989) treated *L. purpureum* as a complex of four varieties. In this classification, the introduced Midwestern material would correspond to var. *purpureum*. However, these varieties probably are better treated as separate species (see also the treatment of *L. hybridum* for further discussion).

Comparing the current distribution to that mapped by Steyermark (1963), it is apparent that *L. purpureum* has expanded its distribution north of the Missouri River during the past few decades. An unnamed form with white corollas is encountered rarely within populations of pink-flowered plants.

13. Leonurus L.

Plants biennial or perennial herbs (sometimes functionally annual), with taproots or short rhizomes. Stems erect or ascending, sharply 4-angled, unbranched or more commonly branched, mostly with short, soft or stiff, curved or downward-angled hairs (occasionally glabrous). Leaves opposite, mostly short- to long-petiolate (the uppermost leaves sometimes sessile or nearly so), the petioles usually unwinged. Leaf blades variously shaped, unlobed or more commonly 3-lobed or palmately 5–9-lobed, the margins otherwise entire or sharply few-toothed, the surfaces glabrous or variously hairy, the undersurface also with usually conspicuous sessile glands. Inflorescences axillary, dense clusters of numerous flowers per node, these sessile or nearly so. Bractlets shorter than to about as long as the calyces, linear, often spinescent at the tips. Calyces slightly zygomorphic, lacking a lateral projection, symmetric at the base, more or less cylindric below the lobes, the tube 5- or 10-nerved or -ribbed (usually also strongly nerved on the inner surface), glabrous in the mouth (but note that the nutlets are densely bristly-hairy at the tip), the lobes shorter than to about as long as the tube, the upper 2 lobes more or less ascending, the lower 3 lobes noticeably spreading (except sometimes in *L. japonicus*), narrowly triangular to triangular, tapered to spinescent tips, glabrous or finely short-hairy on the outer surface, becoming slightly enlarged and papery or leathery at fruiting. Corollas 4–14 mm long, zygomorphic, variously white, pale pink, or lavender, the lower lip with darker purple spots, but lacking small nipplelike projections, the outer surface moderately to densely hairy, the tube narrowly funnelform, 2-lipped, the lips about as long as to slightly longer than the tube, the upper lip entire at the broadly rounded tip, somewhat hooded, the lower lip 3-lobed, spreading, with the central lobe longer than the 2, lateral lobes. Stamens 4, not exserted (the upper pair curved under the upper lip, the lower pair with the anthers positioned just above the corolla throat), the filaments of 2 lengths, the anthers small, the connective somewhat expanded, the pollen sacs 2, parallel or spreading (in *L. marrubiastrum*), dark purple. Ovary relatively shallowly lobed, the style appearing more or less lateral from a shallow apical notch. Style not exserted, with 2, short, equal or unequal branches at the tip. Fruits dry schizocarps, separating into usually 4 nutlets, these 1.5–2.5 mm long, more or less wedge-shaped to obovate in outline, obliquely truncate to broadly rounded at the tip, unequally triangular in cross-section, the surface yellowish brown to reddish brown or brown, sometimes with irregular, lighter, tan to yellow mottling, usually finely and minutely hairy except for the densely hairy tip, otherwise relatively smooth. Twenty-three to 27 species, Europe, Asia, Africa; introduced in the New World.

A case can be made for splitting *Leonurus* into three genera, segregating two Eurasian species that have not been introduced into the United States into *Panzerina* Soják and the widespread *L. marrubiastrum* into *Chaiturus* Willd. (Krestovskaja, 1992; H.-w. Li and Hedge, 1994). The morphological and cytological features that characterize these small segregates need to be evaluated in a phylogenetic context, and many recent authors have continued to treat the genus in the broad sense, a practice followed for the time being in the present treatment.

1. Calyces strongly 5-nerved, the nerves thickened into ribs; largest leaves usually with the blade palmately 5–9-lobed (and coarsely toothed); upper corolla lip densely pubescent with relatively long, spreading hairs; stamens with the anther sacs parallel . . . . 1. L. CARDIACA
1. Calyces finely 10-nerved, the nerves not thickened or riblike; leaf blades unlobed or deeply 3-lobed (the lobes coarsely toothed and/or narrowly lobed); upper corolla lip densely pubescent with short, loosely ascending to spreading, sometimes matted hairs; stamens with the anther sacs parallel
    2. At least the largest leaves with the blade deeply 3-lobed (the lobes coarsely toothed and/or narrowly lobed); corollas 10–14 mm long, distinctly longer than the calyces; stamens with the anther sacs parallel . . . . 2. L. JAPONICUS
    2. Leaves with the blade unlobed (but coarsely toothed); corollas 5–7 mm long, slightly shorter than to about as long as the calyces; stamens with the anther sacs usually spreading . . . . 3. L. MARRUBIASTRUM

1. Leonurus cardiaca L. (motherwort)
Pl. 435 a, b; Map 1953

Plants perennial, with fibrous roots. Stems 40–150(–200) cm long, glabrous or sparsely pubescent with short, fine, downward-angled hairs, often mostly on the angles. Leaf blades 1–12 cm long, those of the lower leaves palmately (3–)5–9 lobed and broadly ovate to nearly circular in outline, grading into those of the median and upper leaves, which are unlobed or 3-lobed and ovate to elliptic in outline, the margins otherwise with few, coarse, sharp teeth, variously rounded to narrowly angled or tapered at the base, angled or tapered to a sharply pointed tip (or lobe tips), the upper surface glabrous or sparsely to moderately pubescent with short, stiff, loosely appressed hairs near the margins, the undersurface sparsely to moderately pubescent with fine, loosely appressed hairs, mostly along the veins. Calyces 3.5–8.0 mm long, strongly 5-nerved, the nerves thickened into ribs, the teeth about as long as the tube, glabrous or sparsely and minutely hairy, also usually with numerous minute, club-shaped glands. Corollas 8–12 mm long, distinctly longer than the calyces, white to pale pink (the lower lip with reddish purple spots or mottling), moderately short-hairy

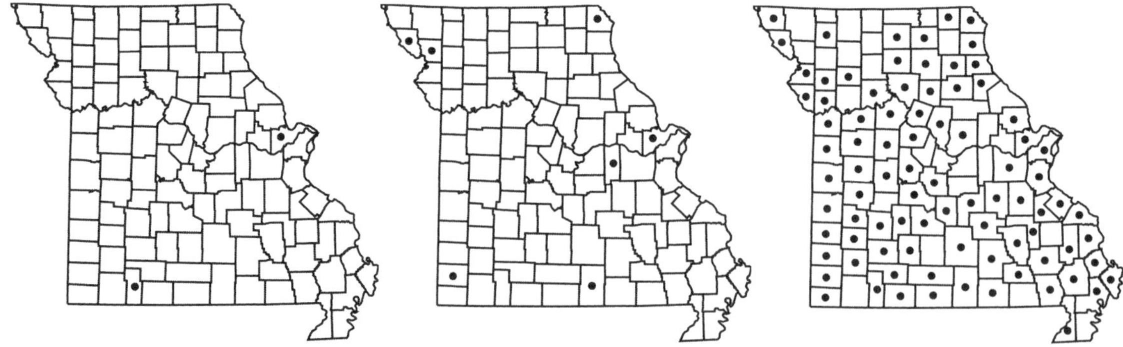

1954. Leonurus japonicus 1955. Leonurus marrubiastrum 1956. Lycopus americanus

externally, but the upper lip densely pubescent with longer, spreading hairs. Stamens with the anther sacs parallel. Nutlets 1.7–2.3 mm long, truncate at the tip. $2n$=18. May–August.

Introduced, scattered nearly throughout the state but apparently more abundant in the western half and absent from the Mississippi Lowlands Division (native of Europe, Asia; introduced widely in the New World). Bottomland forests, openings and edges of mesic upland forests, banks of rivers, marshes, ledges and tops of bluffs; also pastures, margins of crop fields, ditches, gardens, railroads, roadsides, and open disturbed areas.

Steyermark (1963) mentioned that the young foliage can be cooked or (in limited quantities) eaten fresh, and that the dried plants were sometimes used medicinally in tonics and as a menstrual aid. He noted that exposure to the plants caused dermatitis in some individuals and that the spinescent calyces sometimes caused mechanical injuries to the soft mouthparts of livestock. The species also is the source of a green dye.

2. Leonurus japonicus Houtt. (honeyweed)
L. heterophyllus Sweet
L. sibiricus L., misapplied
Pl. 435 d, e; Map 1954

Plants annual or biennial, with taproots. Stems 20–220 cm long, moderately to densely pubescent with short, fine, downward-angled hairs. Leaf blades 2–15 cm long, those of the lower leaves ovate to nearly circular in outline and deeply 3-lobed (the lobes coarsely toothed and/or divided into slender lobes), grading into those of the median and upper leaves, which are linear in outline and lacking lobes and teeth to elliptic or ovate in outline and deeply divided from at or near the base into 3 slender lobes, the margins then entire or with a few coarse, sharp teeth, variously angled or short-tapered (lower leaves) or narrowly tapered (upper leaves) at the base, angled or tapered to a sharply pointed tip (or lobe tips), the upper surface sparsely to moderately pubescent with short, fine, appressed hairs, the undersurface densely pubescent with fine, loosely appressed hairs. Calyces 5–8 mm long, finely 10-nerved, the nerves not thickened or riblike, the teeth shorter than the tube, densely and minutely hairy, also with numerous sessile glands. Corollas 10–14 mm long, distinctly longer than the calyces, pale pink to lavender (the lower lip usually with reddish purple spots or mottling), densely pubescent with short, fine, somewhat matted hairs externally except for the more or less glabrous lower lip. Stamens with the anther sacs parallel. Nutlets 1.5–1.8 mm long, truncate at the tip. $2n$=20. July–August.

Introduced, uncommon, known thus far only from St. Charles and Stone Counties (native of Europe, Asia; introduced widely in tropical and warm-temperate regions, sporadically in the eastern U.S. west to Minnesota, Nebraska, and Texas; Canada). Barnyards, roadsides, and disturbed areas.

Krestovskaja (1987), Holub (1993), and Harley and Paton (2001) discussed the taxonomy and nomenclature of this complex, noting that the name *L. sibiricus* has been applied to two different taxa, with the type corresponding to a species restricted to Siberia and adjacent portions of northern China. The taxon that is a widespread weed in the New World has sometimes been called *L. heterophyllus*, but that name is synonym of *L. japonicus*.

Steyermark (1963) noted that the species sometimes was used medicinally for pain relief and as a menstrual aid.

3. Leonurus marrubiastrum L. (biennial motherwort, lion's tail, horehound motherwort)
Cardiaca marrubiastrum (L.) Schreb.
Chaiturus marrubiastrum (L.) Spenn.
Pl. 435 c; Map 1955

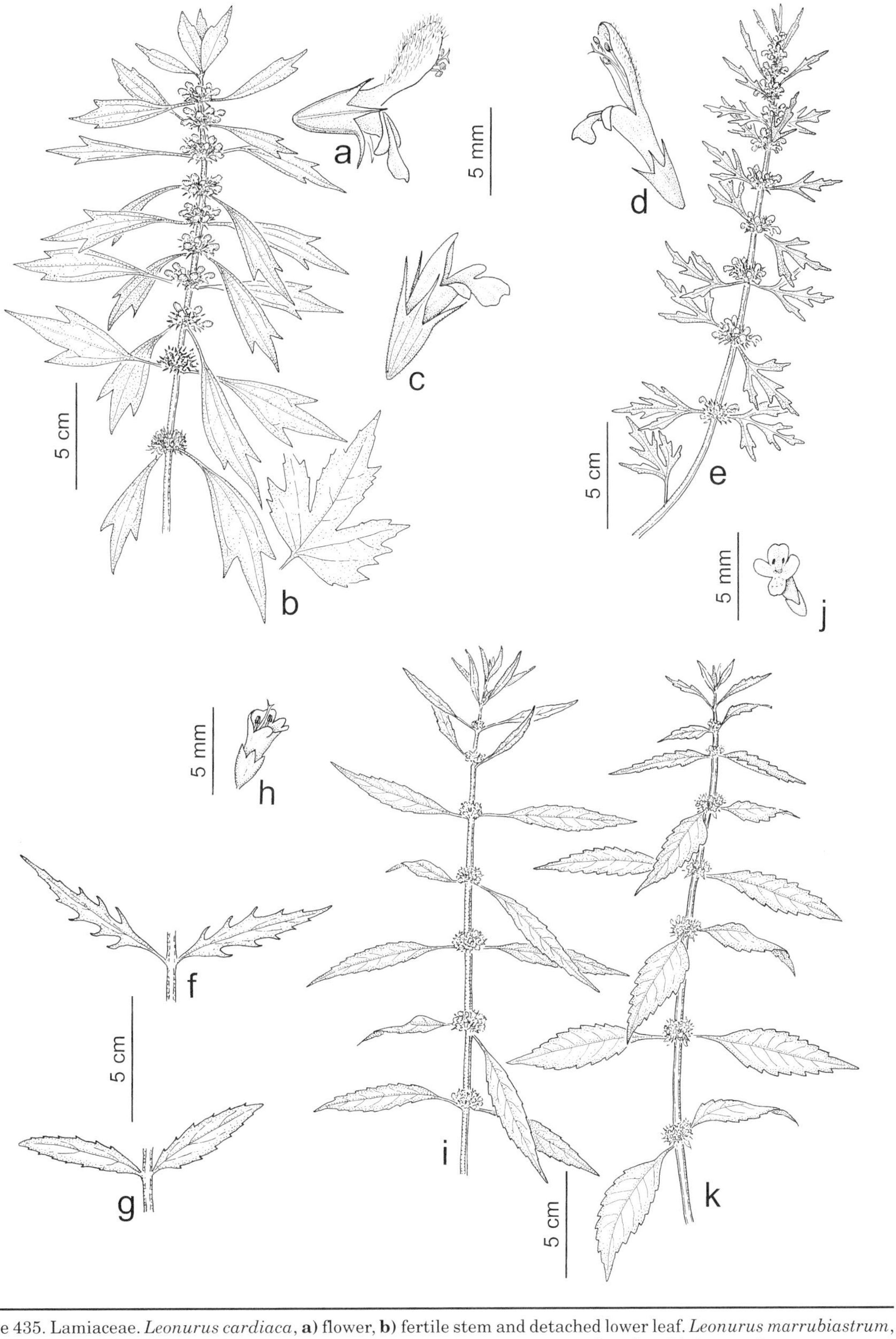

Plate 435. Lamiaceae. *Leonurus cardiaca*, **a)** flower, **b)** fertile stem and detached lower leaf. *Leonurus marrubiastrum*, **c)** flower. *Leonurus japonicus*, **d)** flower, **e)** ferile stem. *Lycopus americanus*, **f)** node with leaves. *Lycopus asper*, **g)** node with leaves. *Lycopus rubellus*, **h)** flower, **i)** fertile stem. *Lycopus virginicus*, **j)** flower, **k)** fertile stem.

Plants biennial, with taproots. Stems 20–100 (–150) cm long, moderately to densely pubescent with short, fine, downward-angled hairs. Leaf blades 1–9 cm long, all unlobed, those of the lower leaves ovate to broadly ovate in outline, grading into those of the median and upper leaves, which are lanceolate to elliptic or narrowly elliptic in outline, the margins otherwise with coarse, sharp teeth, variously broadly rounded to broadly angled (lower leaves) or narrowly angled to tapered (upper leaves) at the base, angled or tapered to a bluntly or sharply pointed tip, the upper surface sparsely to moderately pubescent with short, fine, appressed hairs, the undersurface densely pubescent with fine, loosely appressed hairs, sometimes also with sparse longer hairs. Calyces 4–7 mm long, finely 10-nerved, the nerves not thickened or riblike, the teeth about as long as the tube, densely and minutely hairy, also with numerous sessile glands. Corollas 4–7 mm long, slightly shorter than to about as long as the calyces, white, pale pink, or lavender (the lower lip usually with reddish purple spots or mottling), sparsely to moderately short-hairy externally, the upper lip densely pubescent with similar, fine, loosely appressed hairs. Stamens with the anther sacs usually spreading. Nutlets 1.5–2.5 mm long, broadly rounded at the tip. $2n=24$. July–October.

Introduced, uncommon and sporadic in the state (native of Europe, Asia, Africa; introduced widely in the New World). Bottomland forests and banks of streams and rivers; also barnyards, roadsides, and disturbed areas.

This species was first reported for Missouri by Castaner and Priesendorf (1989), based on a specimen collected in Holt County.

14. **Lycopus** L. (bugleweed) (Henderson, 1962)

Plants perennial herbs, with rhizomes, often with stolons, sometimes producing small tubers. Stems erect or ascending, occasionally reclining on surrounding vegetation, bluntly or sharply 4-angled (except sometimes in swollen, spongy, submerged-aquatic stems), unbranched or branched, variously glabrous to densely hairy. Leaves opposite, sessile or short-petiolate (the petioles occasionally relatively long in *L. rubellus*), the petioles then winged most of their length, with a pungent, disagreeable odor when bruised or crushed. Leaf blades variously shaped, unlobed to pinnately deeply several-lobed, the margins otherwise entire or sharply toothed, the surfaces glabrous or hairy, also with usually conspicuous sessile glands. Inflorescences axillary, dense clusters of numerous flowers per node, these sessile or nearly so. Bractlets shorter than to about as long as the calyces, linear to lanceolate, tapered to sharply pointed but not spinescent tips. Calyces not or only slightly zygomorphic, lacking a lateral projection, symmetric at the base, more or less cylindric to broadly bell-shaped, the tube with 4 or 5 fine main nerves (1 per lobe), in some species with a pair of fainter lateral nerves, glabrous in the mouth, the lobes much shorter than to about as long as the tube, ascending or spreading, broadly to narrowly triangular, rounded to more commonly angled or tapered to bluntly or sharply pointed but not spinescent tips, glabrous or finely short-hairy on the outer surface and along the lobe margins, sometimes also glandular, not becoming enlarged or papery at fruiting. Corollas 2.5–5.0 mm long, actinomorphic or nearly so (not 2-lipped), white, sometimes with faint to prominent, darker purple spots on some of the lobes, the outer surface glabrous or more commonly moderately to densely glandular, the tube funnelform, 4 or 5 lobes more or less similar (if 4-lobed then the upper lobe slightly larger than the others), shorter than the tube, the throat closed with a dense beard of multicellular hairs. Stamens 2, not or only slightly exserted, the anthers minute, the connective short, the pollen sacs 2, spreading, white, yellow, or dark purple. Ovary deeply lobed, the style appearing more or less basal from a deep apical notch. Style not or only slightly exserted, with 2, short, equal branches at the tip. Fruits dry schizocarps, separating into usually 4 nutlets, these 1.0–2.3 mm long, more or less tetrahedral, broadly rounded to truncate at the tip, unequally triangular in cross-section, with a corky winglike structure oriented vertically along the lateral margins and across the apex, the surface yellowish brown to reddish brown or brown (the corky band often somewhat lighter), relatively smooth (sometimes minutely roughened toward the tip), usually glandular. Fourteen to 16 species, North America, Europe, Asia; disjunct in Australia, Tasmania.

Steyermark (1963) noted that the tubers produced by most species of *Lycopus* are eaten by muskrats.

Species of *Lycopus* superficially resemble *Spermacoce glabra* Michx. (Buttonweed) in the Rubiaceae, which also has dense axillary clusters of small white flowers. *Spermacoce glabra* differs in its ovary position (inferior vs. superior); fruit type (unlobed and achenelike vs. nutlets), deeply lobed, unnerved calyces (vs. shallowly lobed or to about the midpoint and strongly nerved), and 4 (vs. 2) stamens, among other characters.

1. Calyces 1–2 mm long, the lobes triangular or broadly triangular, rounded or bluntly pointed at the tips (sometimes sharply pointed in *L. uniflorus*); nutlets extending beyond the calyx tube at maturity
   2. Calyces and corollas 5-lobed; corolla lobes spreading; stamens slightly exserted; nutlets with the tip not or only slightly oblique, the rim of the apex somewhat undulate or irregular . . . . 4. L. UNIFLORUS
   2. Calyces and corollas 4-lobed; corolla lobes ascending; stamens not exserted; nutlets with the tip noticeably oblique, the rim of the apex appearing tuberculate or toothed . . . . 5. L. VIRGINICUS
1. Calyces 1.8–4.5 mm long, the lobes narrowly triangular, tapered to sharply pointed tips; nutlets shorter than the calyx tube at maturity
   3. Corollas 5-lobed; nutlets with the corky band having 4–6 blunt teeth or tubercles; leaf blades finely toothed, not lobed, tapered concavely to a usually winged petiolar base . . . . 3. L. RUBELLUS
   3. Corollas 4-lobed; nutlets with the corky band entire or slightly undulate, lacking teeth or tubercles; leaf blades finely to coarsely toothed or lobed, sessile or petiolate
      4. At least the lower leaves with the blade deeply pinnately lobed (those of the upper and sometimes also median leaves often coarsely few-toothed), tapered concavely to a usually winged petiolar base; nutlets 1.0–1.4 mm long . . . . 1. L. AMERICANUS
      4. Leaves all with the blade unlobed (but finely to coarsely and sharply toothed), angled to a sessile or subsessile base; nutlets 1.8–2.3 mm long . . . . 2. L. ASPER

1. Lycopus americanus Muhl. ex W.P.C. Barton (American bugleweed)

L. americanus var. *longii* Benner

L. americanus var. *scabrifolius* Fernald

Pl. 435 f; Map 1956

Plants with elongate rhizomes, but lacking tubers and usually not producing stolons. Stems 30–90 cm long, bluntly or sharply 4-angled, the faces flat or more commonly shallowly concave to grooved, glabrous or sparsely to densely pubescent with short, appressed and/or longer, spreading hairs. Leaf blades 1.5–15.0 cm long, very variable, narrowly ovate to narrowly lanceolate, narrowly elliptic, or nearly linear in outline, those of at least the lower leaves deeply pinnately lobed (the lobes entire, few toothed, or occasionally pinnately few-lobed), those of the upper and sometimes also median leaves often merely coarsely few-toothed, tapered concavely at the base, usually to a winged, petiolar base, tapered to a sharply pointed tip (or lobe tips), the upper surface glabrous, the undersurface glabrous or sparsely short-hairy along the veins. Bractlets 1–3 mm long, narrowly lanceolate to elliptic-lanceolate. Calyces 2.0–3.2 mm long, 5-lobed to about the midpoint, the lobes more or less spreading, narrowly triangular, tapered to a sharply pointed tip. Corollas 2.5–3.5 mm long, 4-lobed, the upper lobe slightly broader than the others and shallowly notched, the lateral and lower lobes spreading. Stamens slightly exserted. Nutlets 1.0–1.4 mm long, shorter than the calyx tube at maturity, more or less oblique at the tip, the corky band entire or slightly undulate, lacking teeth or tubercles. $2n=22$. June–October.

Scattered throughout the state (throughout the U.S.; Canada, possibly also Asia; introduced rarely

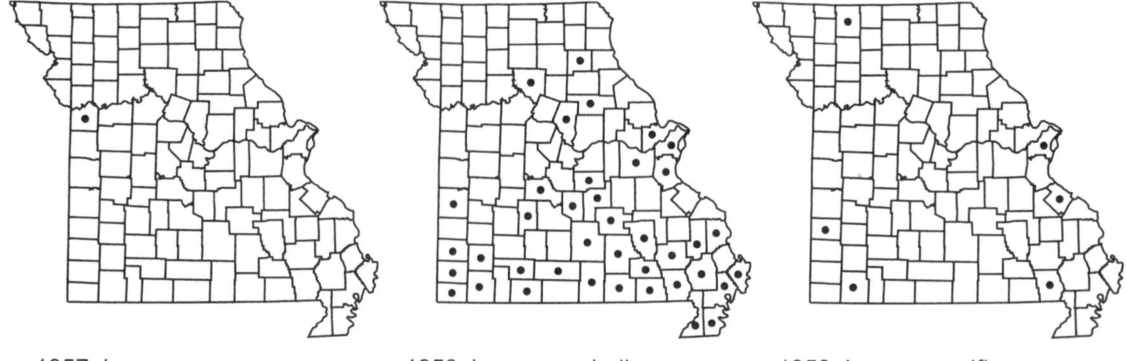

1957. Lycopus asper 1958. Lycopus rubellus 1959. Lycopus uniflorus

in South America). Bottomland forests, sloughs, oxbows, banks of streams and rivers, margins of ponds, lakes, and sinkhole ponds, fens, bottomland prairies, moist swales of upland prairies, and bases and ledges of bluffs; also ditches, fallow fields, railroads, roadsides, and moist disturbed areas.

This is the most common and widespread species of *Lycopus* in the state. Steyermark (1963) segregated somewhat hairier plants as var. *longii* and mentioned plants with somewhat roughened leaves as possibly representing var. *scabrifolius*, but Henderson (1962) concluded that the species is best treated as a single variable taxon.

2. Lycopus asper Greene (bugleweed, rough bugleweed)

Pl. 435 g; Map 1957

Plants with elongate rhizomes, these becoming tuberous-thickened at the tips, also producing slender stolons, these with small tubers at the tips. Stems 35–80(–130) cm long, bluntly 4-angled, the faces shallowly concave to grooved, sparsely to moderately pubescent with spreading, multicellular hairs, especially along the angles and at the nodes. Leaf blades 2–10 cm long, lanceolate to elliptic or narrowly, unlobed, the margins finely to coarsely and sharply toothed, angled to a sessile or subsessile base, tapered to a sharply pointed tip, the surfaces more or less roughened with sparse to moderate, minute, stout hairs, especially along the veins. Bractlets 2.5–3.5 mm long, narrowly lanceolate to elliptic-lanceolate. Calyces 2.5–4.5 mm long, 5-lobed to about the midpoint, the lobes more or less spreading, narrowly triangular, tapered to a sharply pointed tip. Corollas 3.5–5.0 mm long, 4-lobed, the upper lobe slightly broader than the others and usually shallowly notched, the lateral and lower lobes more or less spreading. Stamens slightly exserted. Nutlets 1.8–2.3 mm long, shorter than the calyx tube at maturity, oblique at the tip, the corky band entire or somewhat undulate, lacking teeth or tubercles. $2n=22$. July–August.

Uncommon, known thus far from a single historical collection from Jackson County (western U.S. east to Massachusetts, Ohio, and Texas; Canada). Banks of rivers.

3. Lycopus rubellus Moench (water horehound)

L. rubellus var. *arkansanus* (Fresen.) Benner

L. rubellus var. *lanceolatus* Benner

Pl. 435 h, i; Map 1958

Plants with elongate rhizomes, also producing slender stolons, these with small tubers at the tips. Stems 35–120 cm long, bluntly 4-angled, the faces shallowly concave to grooved, moderately to densely pubescent with minute, spreading hairs. Leaf blades 2.5–12.0 cm long, lanceolate to elliptic or narrowly ovate, unlobed, the margins finely and usually sharply toothed, tapered concavely to a usually winged petiolar base, angled or tapered to a sharply pointed tip, the upper surface glabrous or nearly so, the undersurface glabrous or sparsely to moderately short-hairy, sometimes only along the veins. Bractlets 1–2 mm long, linear. Calyces 1.8–2.5 mm long, 5-lobed to about the midpoint, the lobes more or less spreading, narrowly triangular, tapered to a sharply pointed tip. Corollas 2.5–5.0 mm long, 5-lobed, the lower lobe slightly longer than the others, the lateral and lower lobes spreading. Stamens slightly exserted. Nutlets 1.0–1.5 mm long, shorter than the calyx tube at maturity, not oblique at the tip, the corky band with 4–6 blunt teeth or tubercles along the nutlet apex. $2n=22$. July–October.

Scattered in the Ozark, Ozark Border, and Mississippi Lowlands Divisions, uncommon and sporadic elsewhere in the state (eastern U.S. west to Illinois, Kansas, and Texas; Canada). Bottomland forests, swamps, sloughs, oxbows, banks of streams and rivers, margins of ponds, lakes, and

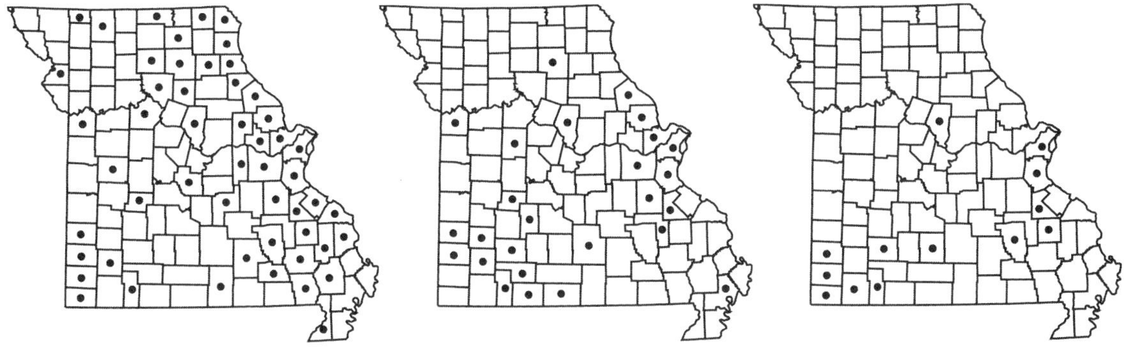

1960. Lycopus virginicus

1961. Marrubium vulgare

1962. Melissa officinalis

sinkhole ponds, bottomland prairies, and bases and ledges of bluffs; also ditches, edges of crop fields, and moist disturbed areas.

Steyermark (1963) treated plants with relatively hairy stems as var. *arkansanus* and somewhat less common plants with the stems glabrous or nearly so as var. *rubellus*. However, Henderson (1962) concluded that *L. rubellus* is best treated as a single variable taxon.

4. Lycopus uniflorus Michx. (northern bugleweed, one-flowered horehound)

Map 1959

Plants with elongate rhizomes, also producing slender stolons, these with small tubers at the tips. Stems 30–70 cm long, bluntly or sharply 4-angled, the faces shallowly concave to grooved, glabrous or sparsely to moderately pubescent with short, appressed-ascending hairs. Leaf blades 2–6(–10) cm long, narrowly to broadly lanceolate or elliptic, unlobed, the margins finely to coarsely and usually sharply toothed, angled or tapered to a sessile or short, winged petiolar base, angled or tapered to a sharply pointed tip, the upper surface glabrous or nearly so, the undersurface sparsely to moderately short-hairy only along the veins. Bractlets 0.5–1.0 mm long, linear. Calyces 1.3–1.8 mm long, 5-lobed from less than to about the midpoint, the lobes more or less spreading, triangular, angled to a bluntly or sharply pointed tip. Corollas 2.5–3.5 mm long, 5-lobed, the lobes all similar in size and shape, spreading. Stamens slightly exserted. Nutlets 1.0–1.2 mm long, extending beyond the calyx tube at maturity, not or only slightly oblique at the tip, the corky band somewhat undulate or irregular along the nutlet apex. $2n=22$. August–October.

Uncommon, known thus far from historical collections from the southern half of the state and recent collections from Harrison County (northern U.S. [including Alaska] south to South Carolina, Arkansas, and California; Canada, Asia). Bottomland forests, sloughs, banks of streams and rivers, and moist swales of upland prairies.

This species was first reported for Missouri by Henderson (1962). Recent botanical explorations by Justin Thomas and Doug Ladd at The Nature Conservancy's Dunn Ranch Preserve (Harrison County) have yielded new collections of *L. uniflorus* from Missouri.

5. Lycopus virginicus L. (bugleweed, Virginia bugleweed)

Pl. 435 j, k; Map 1960

Plants with elongate rhizomes, also producing slender stolons, these rarely with small tubers at the tips. Stems 30–90 cm long, bluntly 4-angled, the faces flat or very shallowly concave to grooved, moderately pubescent with short, appressed-ascending hairs, sometimes also with sparse longer, multicellular hairs. Leaf blades 1–10 cm long, lanceolate to ovate or elliptic, unlobed, the margins coarsely and sharply toothed, tapered concavely to a sometimes short, winged petiolar base, tapered to a sharply pointed tip, the upper surface glabrous or sparsely short-hairy along the midvein, the undersurface glabrous or more commonly sparsely to moderately short-hairy only along the veins. Bractlets 0.5–1.0 mm long, linear. Calyces 1–2 mm long, 4-lobed from less than to less commonly about the midpoint, the lobes ascending to somewhat spreading, triangular to broadly triangular, rounded or more commonly angled to a bluntly pointed tip. Corollas 1.8–2.2 mm long, 4-lobed, the upper lobe often shallowly notched at the tip, all of the lobes ascending. Stamens not exserted. Nutlets 1.6–2.0 mm long, extending beyond the calyx tube at maturity, noticeably oblique at the tip, the corky band with 4–6 blunt teeth or tubercles along the nutlet apex. $2n=22$. August–October.

Scattered nearly throughout the state (eastern U.S. west to Minnesota and Texas; Canada). Bottomland forests, swamps, sloughs, oxbows, banks of streams, rivers, and spring branches, margins of ponds, lakes, and sinkhole ponds, bottomland prairies, and bases and ledges of bluffs; also ditches, railroads, and moist disturbed areas.

An uncommon hybrid between *L. virginicus* and *L. uniflorus* (*L.* ×*sherardii* E.S. Steele) occurs where these species grow in proximity. It has been reported from the eastern Great Plains (including Kansas and Iowa) eastward and thus might eventually be discovered in Missouri. Hybrids tend to have the appearance of *L. virginicus* but are intermediate between the putative parents for corolla and nutlet characters.

15. **Marrubium** L. (horehound)

About 30 species, Europe, Asia, Africa, Macaronesia; introduced widely.

1. Marrubium vulgare L. (common horehound)
Pl. 436 a, b; Map 1961

Plants perennial herbs, with short, stout rhizomes, sometimes taprooted and flowering the first year. Stems 30–70(–100) cm long, erect to loosely ascending, often from a spreading base, usually bluntly 4-angled, unbranched or more commonly branched, densely pubescent with white, woolly hairs, sometimes in patches, the upper portion often somewhat less densely hairy. Leaves opposite, short- to long-petiolate, the petioles unwinged or winged toward the tip, with a pungent, odor usually described as pleasant when bruised or crushed. Leaf blades 1–6 cm long, broadly elliptic to broadly ovate or nearly circular, unlobed, the margins scalloped or bluntly to less commonly sharply toothed, the upper surface with strongly impressed veins, densely woolly when young, often becoming glabrous (sometimes only in patches) at maturity, the undersurface with strongly raised veins, usually persistently woolly, some of the hairs often stellate, also with usually conspicuous sessile glands, these often partially obscured by the hairs. Inflorescences axillary, dense clusters of numerous flowers per node, these sessile or nearly so. Bractlets 1.5–6.0 mm long, shorter than to about as long as the calyces, linear, sharply pointed, somewhat spinescent, and sometimes hooked at the tips. Calyces 5–8 mm long, not or only slightly zygomorphic, lacking a lateral projection, symmetric at the base, more or less cylindric to bell-shaped, the tube with 10 fine or faint nerves, with dense, relatively long, fine hairs in the mouth, the 10 lobes shorter than to nearly as long as the tube (frequently 5 longer lobes alternating with 5 somewhat shorter lobes), more or less spreading, triangular to narrowly triangular, tapered to sharply pointed, usually hooked, more or less spinescent tips, densely woolly on the outer surface, becoming slightly enlarged and leathery or hardened at fruiting. Corollas 4–6 mm long, zygomorphic, white to pale cream-colored, lacking darker purple spots, the outer surface densely hairy (except on the lowermost lobe), the tube narrowly funnelform, the lobes somewhat shorter than the tube, the upper lip straight or arching upward, divided into 2 slender lobes, the lower lip with the central lobe obovate and broader than the 2 lateral lobes, often shallowly notched at the tip. Stamens 4, not exserted, the anthers small, the connective short, the pollen sacs 2, spreading, usually yellow. Ovary deeply lobed, the style appearing more or less basal from a deep apical notch. Style not exserted, with 2, short, unequal branches at the tip. Fruits dry schizocarps, separating into usually 4 nutlets, these 1.8–2.5 mm long, obovoid to oblong-ellipsoid, broadly rounded at the tip, the surface tan to yellowish brown with darker mottling, relatively smooth or finely pebbled, glabrous. $2n$=34, 36. May–September.

Introduced, widely scattered, known mostly from historical collections (native of Europe, Asia, Africa, Macaronesia; introduced nearly worldwide, including throughout the U.S.). Glades and ledges and tops of bluffs; also pastures, fencerows, barnyards, railroads, roadsides, and open disturbed areas.

Steyermark (1963) noted that horehound has a long history of medicinal use as a stimulant, tonic, laxative, cold remedy, and an aid for rheumatism and dyspepsia. It formerly was a popular ingredient in some cough drops and a flavorant in some candies. Diggs et al. (1999) noted that the plant is most often observed in areas frequented by livestock and that the nutlets often become caught in animal fur and wool. It is thus a troublesome weed for wool producers.

The species is treated here in the traditional broad sense in which it has been included in the North American floristic literature. R. E. Brooks (1986) pointed out that most of the North American material attributed to *M. vulgare* does not key well in European floristic treatments, differing in

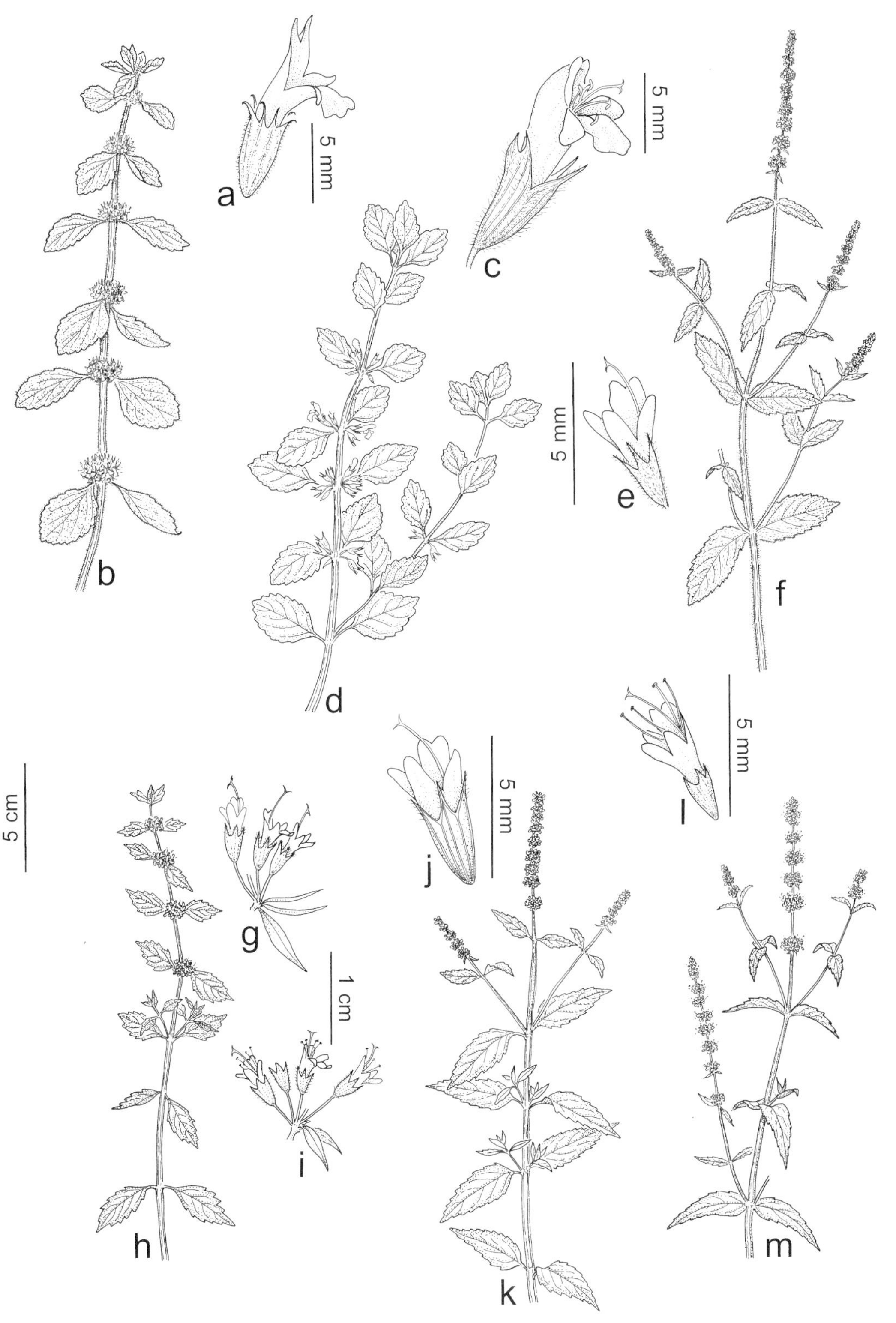

Plate 436. Lamiaceae. *Marrubium vulgare*, **a)** flower, **b)** fertile stem. *Melissa officinalis*, **c)** flower, **d)** fertile stem. *Mentha longifolia*, **e)** flower, **f)** fertile stem. *Mentha ×gracilis*, **g)** flower, **h)** fertile stem. *Mentha canadensis*, **i)** node with flowers. *Mentha ×piperita*, **j)** flower, **k)** fertile stem. *Mentha spicata*, **l)** flower, **m)** fertile stem.

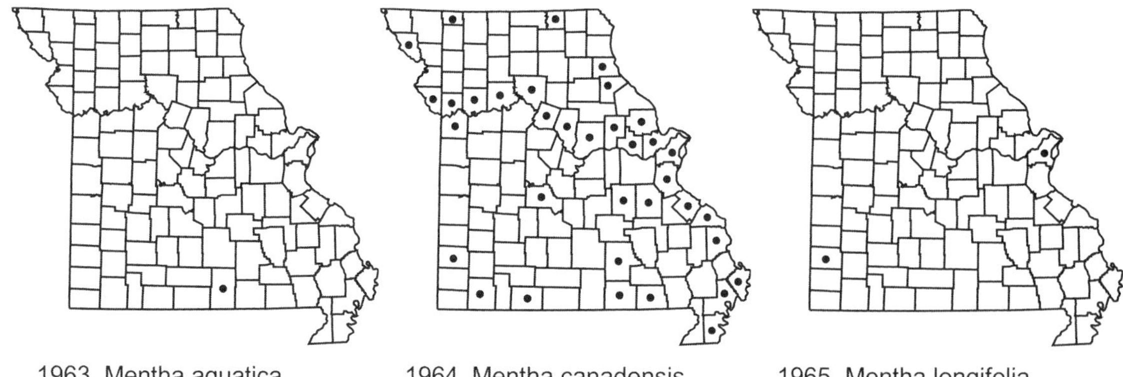

1963. Mentha aquatica 1964. Mentha canadensis 1965. Mentha longifolia

details of pubescence and calyx morphology. In the absence of a modern monograph of the genus, it is not possible to assess the introduced American plants in the context of variation among native populations of *M. vulgare* or other species. More studies are needed.

16. Melissa L. (balm)

Three species, Europe, Asia.

1. Melissa officinalis L. **var. officinalis** (lemon balm, beebalm)

Pl. 436 c, d; Map 1962

Plants perennial herbs, with rhizomes. Stems 30–100 cm long, erect to loosely ascending, often from a spreading base, sharply 4-angled, several-branched, densely pubescent with short, spreading, gland-tipped hairs and sparser, longer, nonglandular hairs (rarely nearly glabrous). Leaves opposite, the lower and often also median stem leaves long-petiolate, those of the inflorescence short-petiolate, the petioles unwinged, with a pleasant, lemony odor when bruised or crushed. Leaf blades 1–7 cm long, the transition from vegetative stems to inflorescences fairly abrupt, those in the inflorescences noticably smaller (1–3 cm) than the vegetative stem leaves (the largest 4–7 cm), heart-shaped or more commonly broadly ovate to triangular-ovate or elliptic, unlobed, the margins bluntly to sharply toothed, cordate to more commonly truncate or broadly angled at the base, angled to a bluntly or sharply pointed tip, the upper surface sparsely pubescent with relatively long, mostly multicellular hairs, the undersurface usually densely pubescent with minute, spreading, mostly gland-tipped hairs, sometimes also sparsely to moderately pubescent with long, fine, spreading hairs, also with usually conspicuous sessile glands. Inflorescences axillary, dense clusters of 4–14 flowers per node, these short-stalked. Bractlets 2–3 mm long, shorter than the calyces, linear to ovate, sharply pointed but not spinescent. Calyces 7–9 mm long, strongly zygomorphic, lacking a lateral projection, symmetric at the base, more or less bell-shaped, the tube strongly 13-nerved, glabrous in the mouth, the lips about 65% as long as the tube, the upper lip arched upward, with 3 shallow, broadly triangular lobes (teeth), the lower lip slightly arched upward, with 2 deeper, narrowly triangular lobes, the lobes all tapered to sharply pointed, slightly spinescent tips, sparsely to densely pubescent with minute, spreading, mostly gland-tipped hairs on the outer surface, also with sparse to moderate, long, spreading, nonglandular hairs, becoming somewhat enlarged and papery at fruiting. Corollas 8–15 mm long, zygomorphic, pale yellow, lacking darker purple spots, fading to white or pinkish-tinged, the outer surface moderately pubescent, with short, fine, spreading hairs, the tube funnelform and somewhat arched, densely glandular-hairy in the throat, the lobes shorter than the tube, the upper lip slightly hooded but spreading at the tip, shallowly to moderately 2-lobed, the lower lip with the central lobe broader than the 2 lateral lobes. Stamens 4, not exserted (ascending under the upper lip), the lower pair somewhat longer than the upper pair, the anthers small, the connective short, the pollen sacs 2, spreading, white or yellow. Ovary deeply lobed, the style appearing more or less basal from a deep apical notch. Style not exserted, with 2 slightly unequal branches at the tip. Fruits dry schizocarps, separating into usually 4 nutlets, these 1.5–2.0 mm long, oblong-obovoid to ellipsoid, rounded at the tip, the surface yellowish brown to greenish brown or dark brown, finely pebbled or finely pitted, glabrous. $2n=32$. June–September.

Introduced, uncommon, widely scattered, mostly in the southern half of the state (native of Europe; introduced sporadically in the U.S., Canada). Openings of mesic upland forests and banks of streams; also fencerows, old homesites, gardens, railroads, roadsides, and disturbed areas.

This species is cultivated for its pleasantly scented foliage, which is used in fragrances, potpourri, massage oils, and in liquors. It has also been used medicinally as a mild sedative and for hypertension. More recently there has been interest in its possible memory-enhancing properties as an adjunct treatment for the early stages of Alzheimer's Disease (S. Schwartz, 2005). Beekeepers sometimes have used *M. officinalis* to attract swarms of bees to new hives, as terpenoids in the essential oil are the same compounds as the natural pheromones produced in the Nasonov glands of honeybees (Mabberley, 1997).

The other subspecies, ssp. *altissima* (Sm.) Arcang., is endemic to portions of southern Europe. It is not cultivated and has not been introduced into North America. This tetraploid ($2n=64$) differs from ssp. *officinalis* in being densely white- to gray-woolly and has some of the upper leaves truncate (vs. angled) at the base, as well as differing slightly in the shape of the calyx lobes.

17. **Mentha** L. (mint)
(Tucker and Naczi, 2007)

Plants perennial herbs, with rhizomes. Stems erect or ascending, sometimes from a spreading base, sharply 4-angled, unbranched or branched, glabrous or hairy. Leaves sessile to long-petiolate, the petiole winged or unwinged, with various, strong, pleasant fragrances when crushed. Leaf blades variously shaped, unlobed, the margins toothed, the surfaces glabrous or hairy, also with usually conspicuous, sessile glands. Inflorescences ranging from small axillary clusters of mostly 8 to numerous flowers to terminal spikes or spikelike racemes, the individual flowers sessile (often in *M. spicata*) or short-stalked, with a pair of short, slender bractlets at the stalk base. Calyces actinomorphic or nearly so (sometimes slightly 2-lipped), lacking a lateral projection, symmetric at the base, cylindric to narrowly bell-shaped, the tube 10–13-nerved, glabrous in the mouth, the 5 lobes shorter than to longer than the tube, broadly to narrowly triangular, tapered to slender, sharply pointed tips, not spinescent, not becoming enlarged at fruiting, variously glabrous to hairy, usually also glandular. Corollas 2–7 mm long, nearly actinomorphic, not 2-lipped, the 4(5) lobes longer than the short, funnelform tube, the upper lobe (except in rare 5-lobed corollas) broader than the other 4 and often shallowly and broadly notched at the tip, white to lavender, pink, purple, or bluish purple, sometimes with darker spots or lines, the outer surface glabrous or hairy, sometimes also glandular. Stamens 4, long-exserted (sometimes nonfunctional and not exserted in hybrids or otherwise sterile plants), all similar in length, the filaments straight or slightly curved, the anthers small, the connective short, the pollen sacs 2, parallel, white or pink to dark bluish purple. Ovary deeply lobed, the style appearing nearly basal from a deep apical notch. Style long-exserted, equally 2-branched at the tip. Fruits dry schizocarps (rarely or not produced in hybrids or otherwise sterile plants), separating into 2–4 nutlets, these 0.5–1.3 mm long, broadly obovoid to broadly oblong-ellipsoid, more or less rounded at the tip, slightly and bluntly 3-sided to nearly circular in cross-section, the surface yellowish brown to reddish brown, dark brown, or black, with a network of fine, anastomosing ridges and pits, glabrous. About 18 species plus several widespread hybrids, nearly worldwide.

The genus *Mentha* has a long history of cultivation and a number of taxa are economically important as ornamentals and for their essential oils, which are used medicinally, as a fragrance, and as a flavorant. Hybridization and polyploidy are rampant (Harley and Brighton, 1977) and have been important in the formation of many of the commercially cultivated plants, most of which are propagated vegetatively by rhizome fragments. Taxa that are cultivated in the United States, but appear not to escape in Missouri, include *M. pulegium* L. (pennyroyal; see also the treatment of *Hedeoma*) and *M. suaveolens* Ehrh. (apple mint, pineapple mint). Because two of the hybrids are relatively frequently encountered in nature in Missouri, they are included in the key below. They are not given full species treatments, but are discussed under one of the parents.

1. Inflorescences of axillary clusters, the subtending leaves similar in size to other foliage leaves
   2. Calyces glabrous (but often with sessile glands) . . . . 2. M. CANADENSIS
   2. Calyces hairy above the midpoint externally . . . . . . *M. ×gracilis* (see *M. canadensis*)
1. Inflorescences of terminal spikes or spikelike racemes (axillary inflorescences sometimes also present), the bracteal leaves noticeably smaller than the foliage leaves
   3. Leaf blades with the undersurface moderately to densely hairy; calyces uniformly hairy externally . . . . 3. M. LONGIFOLIA
   3. Leaf blades with the undersurface glabrous or very sparsely hairy; calyces glabrous externally or hairy only above the midpoint
      4. Calyces glabrous externally (except for the margins of the lobes); inflorescences globose to ovoid, with 1–3 nodes . . . . 1. M. AQUATICA
      4. Calyces hairy above the midpoint externally (except sometimes in *M. ×piperita*); inflorescences elongate, with 5 to numerous nodes
         5. Bracteal leaves conspicuous, 2–3 times as long as the flower clusters, only slightly smaller than the foliage leaves . . . . *M. ×gracilis* (see *M. canadensis*)
         5. Bracteal leaves inconspicuous, shorter than to only slightly longer than the flower clusters, much smaller than the foliage leaves
            6. Calyces 3–4 mm long; foliage leaves with the petiole to 4–15 mm long . . . . *M. ×piperita* (see *M. spicata*)
            6. Calyces 1–3 mm long; foliage leaves sessile or the petiole to 3 mm long . . . . 4. M. SPICATA

1. Mentha aquatica L. **var. citrata** (Ehrh.) Fresen. (lemon mint, orange mint, bergamot mint)
M. citrata Ehrh.

Map 1963

Stems 20–80 cm long, glabrous or nearly so. Leaves sessile or very short-petiolate, the petiole (when present) usually unwinged. Leaf blades 2–7 cm long, broadly ovate to elliptic, rounded to shallowly cordate at the base, mostly angled to a bluntly pointed tip, the margins sharply toothed, sometimes also short-hairy, the surfaces glabrous. Inflorescences mainly terminal, short, dense, spikelike racemes with 1–3 closely spaced nodes confluent into a globose to ovoid mass, the bracteal leaves inconspicuous, shorter than to only slightly longer than the flower clusters, often also with a few dense headlike clusters on short stalks from the axils of the upper foliage leaves. Flowers usually functionally pistillate (the stamens not exserted, nonfunctional and with minute anthers). Calyces 2.5–4.0 mm long, hairy along the margins of the lobes, the outer surface often also with sessile glands. Corollas 3.5–5.0 mm long, lavender to light pink. Nutlets not produced. $2n=96$. July–October.

Introduced, known thus far only from a single historical collection from Howell County (native presumably of Europe, but the native distribution of this cultigen is poorly known; introduced widely but sporadically in the U.S., Canada, Asia). Bottomland prairies.

Most of the earlier botanical literature has interpreted the name *M. citrata* to apply to a series of hybrids between *M. aquatica* L. and *M. spicata* (Steyermark, 1963; Gleason and Cronquist, 1991), but Murray and Lincoln (1970) determined that the plants actually are a male-sterile variant of *M. aquatica* in which the biochemical pathway leading to the production of menthofuran (the major component of the essential oil) is disrupted, resulting in the production of the acyclic compounds linalyl acetate and linalool. Thus, the crushed foliage of var. *aquatica* has an unpleasant musty aroma, whereas that of var. *citrata* has been described as similar to lemon (similar to lavender or bergamot elsewhere). Morphologically, the principal difference between the two varieties is that var. *aquatica* is usually glandular on the leaves, flower stalks, and calyces, whereas var. *citrata* tends to be glabrous or sparsely pubescent with nonglandular hairs along the margins of the calyx lobes. Also, var. *aquatica* sometimes produces fruits. For further discussion of *M.* ×*piperita*, the hybrid between *M. aquatica* L. and *M. spicata*, see the treatment of *M. spicata*.

2. Mentha canadensis L. (corn mint)

M. arvensis L., misapplied
M. arvensis ssp. *canadensis* (L.) H. Hara
M. arvensis f. *glabra* (Benth.) S.R. Stewart
M. arvensis f. *glabrata* (Benth.) S.R. Stewart
M. arvensis ssp. *piperascens* (Malinv. ex Holmes) H. Hara
M. arvensis var. *villosa* (Benth.) S.R. Stewart

Pl. 436 i; Map 1964

Stems 15–80 cm long, glabrous or sparsely to densely hairy with shorter and/or longer, downward curved to spreading hairs, sometimes mostly along the angles. Leaves short-petiolate, the petiole often narrowly winged. Leaf blades 1.5–8 (–12.0) cm long, lanceolate to ovate, elliptic or somewhat rhombic, broadly angled to narrowly tapered at the base, tapered to a sharply pointed tip, the margins sharply toothed, sometimes also short-hairy, the surfaces glabrous or sparsely to moderately pubescent with short, loosely appressed hairs. Inflorescences of axillary clusters, the subtending leaves similar in size to other foliage leaves. Flowers almost always perfect. Calyces 2.5–3.2 mm long, glabrous, but often with sessile glands (hairy elsewhere). Corollas 4–7 mm long, white to lavender or pale pink. Nutlets 0.7–1.3 mm long, usually yellowish brown. $2n=96$. July–September.

Scattered nearly throughout the state (most commonly in counties along the Missouri and Mississippi Rivers), but apparently absent for the Unglaciated Plains Division and uncommon in the western half of the Ozarks (U.S. [including Alaska]; Canada, Mexico, Asia; introduced widely). Bottomland forests, banks of streams, rivers, and spring branches, banks of ponds, marshes, fens, marshes, and bases of bluffs; also pastures, railroads, roadsides, and moist, disturbed areas; sometimes emergent aquatics.

Mentha canadensis is the natural source of menthol, an important ingredient in cough drops, candies, chewing gums, tooth pastes, shaving products, perfumes, and mentholated cigarettes, among other uses. The species also is used medicinally. Currently, India and China are the main producers of menthol, but synthetic production also accounts for a significant share of the market (Hopp and Lawrence, 2007).

Mentha arvensis and its allies comprise a circumboreal polyploid complex within which a number of taxa have been recognized as species, subspecies, forms, and hybrids (Harley and Brighton, 1977; Tucker and Chambers, 2002; Tucker and Naczi, 2007). Tucker and Chambers (2002) performed artificial crosses between selected taxa and analyzed the cytology, phytochemistry, and morphology of the resulting hybrids. They determined that North American and Asian populations previously attributed to *M. arvensis* most likely represent an ancient, stabilized hybrid between the Old World native diploid ($2n=24$) species, *M. longifolia*, and a hexaploid ($2n=76$) population of the European *M. arvensis* in the strict sense. They resurrected the Linnaean name *M. canadensis* for this octoploid species. The American/Asian plants are the source of menthol, whereas the related compound, pulegone, is the main ingredient in the essential oils in most of the nine chemotypes of the European taxon (Tucker and Chambers, 2002), which tends to emit a lavender-like fragrance from the crushed foliage. Morphologically, true *M. arvensis* differs from *M. canadensis* relatively subtly. According to Gleason and Cronquist (1991) and Tucker and Naczi (2007), *M. arvensis* has less robust, less strongly ascending stems and ovate to nearly circular leaf blades relatively uniform in size along the stem (vs. progressively smaller toward the stem tip), those subtending flower clusters tending to have more broadly rounded bases. However, Tucker and Chambers (2002) noted that there was some overlap between the two taxa for the leaf characters that they studied. Gleason and Cronquist (1991) further noted that the European taxon is introduced in some northeastern states and provinces (but not Missouri), with intermediates fairly common in the region in which both taxa grow.

Mentha ×*gracilis* Sole ($2n=54$, 60, 72, 84, 96, 108, 120) (Pl. 436 g, h) represents a morphologically variable series of sterile hybrids between *M. arvensis* and *M. spicata*. In most of the earlier botanical literature, these have been called *Mentha* ×*gentilis* L. (which was based on a male-sterile clone of *M. arvensis* rather than a hybrid) or *M.* ×*cardiaca* J. Gerard ex Baker (a later name for *M.* ×*gracilis*) (Tucker and Fairbrothers, 1990; Tucker et al., 1991; Tucker and Naczi, 2007). Two independently derived clones within this hybrid complex known as American spearmint and Scotch spearmint are commercial sources of spearmint

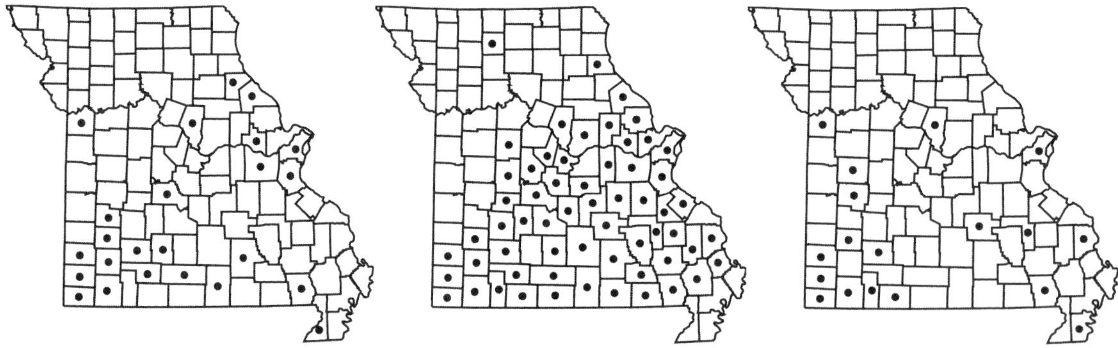

1966. Mentha spicata 1967. Monarda bradburiana 1968. Monarda citriodora

(Tucker and Fairbrothers, 1990; Tucker et al., 1991), although the parental taxon, *M. spicata*, also is an important source of this compound. See the treatment of that species for further discussion of spearmint. *Mentha ×gracilis* currently is known from six counties widely scattered in the state. It is spread vegetatively and commonly found in disturbed habitats such as ditches, railroads, and roadsides, but it also has been reported from banks of streams and fens.

3. Mentha longifolia (L.) Huds. **ssp. longifolia** (horse mint)
M. longifolia ssp. *mollissima* (Schübl. & G. Martens) Domin
Pl. 436 e, f; Map 1965

Stems 30–90 cm long, usually densely woolly. Leaves sessile or nearly so. Leaf blades 2–9 cm long, lanceolate to elliptic or oblong-ovate, those of the lowermost leaves sometimes nearly circular, rounded to shallowly cordate or occasionally broadly angled at the base, angled or tapered to a sharply pointed tip, those of the lowermost leaves sometimes rounded to bluntly pointed, the margins sharply toothed, the surfaces moderately to densely woolly. Inflorescences terminal, elongate, dense, spikelike racemes with usually numerous closely spaced nodes, the bracteal leaves inconspicuous, shorter than to only slightly longer than the flower clusters. Flowers occasionally functionally pistillate (the stamens not exserted, nonfunctional and with minute anthers). Calyces 1.5–2.5 mm long, uniformly hairy on the outer surface. Corollas 3.5–5.0 mm long, white to lavender. Nutlets sometimes not produced, when present, then 0.5–0.8 mm, yellowish brown to reddish brown or brown. $2n$=24. June–September.

Introduced, known thus far only from Jasper and St. Louis Counties (native of Europe, Asia; introduced sporadically in the U.S., Canada). Roadsides and moist, disturbed areas.

Infraspecific variation in this species is poorly understood and requires further study (Harley, 1972); Tucker and Naczi (2007) noted that the species has the largest native range of any *Mentha* species, with an overall distribution from western Europe to India, south to South Africa. They listed 22 principal variants, which they classified as subspecies. The few specimens collected in Missouri appear to be referable to ssp. *longifolia*. Specimens of *M. longifolia* can be confused with hairy variants of *M. spicata*. For a discussion of plants that Steyermark (1963) attributed to *M. longifolia* var. *undulata* (Willd.) Fiori & Paol., see the treatment of *M. spicata*.

4. Mentha spicata L. (spearmint)
M. longifolia var. *undulata* (Willd.) Briq.
Pl. 436 l, m; Map 1966

Stems 30–90 cm long, usually glabrous or nearly so (densely woolly in the rare ssp. *condensata* (Briq.) Greuter & Burdet), occasionally with scattered, minute, stalked glands. Leaves sessile or short-petiolate (to 3 mm long), the petiole (when present) unwinged. Leaf blades 1–7 cm long, narrowly ovate to elliptic, rarely broadly ovate, angled or rounded at the base, rarely shallowly cordate, angled or tapered to a sharply pointed tip, rarely angled to a bluntly pointed tip, the margins sharply toothed, sometimes somewhat curled under, the surfaces glabrous or the undersurface sparsely short-hairy along the main veins (surfaces densely woolly in the rare ssp. *condensata*). Inflorescences mainly terminal, elongate, dense to open spikelike racemes with 5 to numerous closely to more openly spaced nodes, the bracteal leaves inconspicuous, shorter than to only slightly longer than the flower clusters, occasionally also with a few headlike clusters or short spikes on short stalks from the axils of the upper foliage leaves. Flowers occasionally functionally pistillate (the stamens not exserted, nonfunctional and with minute anthers). Calyces 1–3 mm long, hairy above the midpoint (and along the margins of the lobes), the outer surface often also with sessile glands. Corollas 1.7–4.0 mm long, white to lavender or light pink. Nutlets often not produced, when present, then 0.7–1.0 mm, yellowish brown or more commonly dark brown to black. $2n=48$. June–October.

Introduced, scattered, sporadic, mostly south of the Missouri River (native of Europe; introduced nearly worldwide). Banks of streams, rivers, and spring branches, margins of ponds and lakes, fens, and bottomland forests; also roadsides and moist, disturbed areas; sometimes emergent aquatics.

Mentha spicata is grown commercially as the source of spearmint, which is used medicinally, as a fragrance, and as a flavorant. Harley and Brighton (1977) and Gobert et al. (2002) provided evidence that *M. spicata* developed long ago through hybridization between *M. longifolia* and *M. suaveolens*. The species is treated here in the broad sense. Tucker and Naczi (2007), who noted that infraspecific variation is poorly understood, treated *M. spicata* provisionally as comprising two subspecies (ssp. *spicata* and ssp. *condensata*), each divided into two varieties. Most of the Missouri materials are referable to var. *spicata*.

The other variant known to escape in Missouri was treated by Steyermark (1963) with some reservation as *M. longifolia* var. *undulata*, based on a single historical collection by Kenneth Mackenzie from Jackson County. Steyermark discussed that the original determination of this specimen was *M.* ×*alopecuroides* Hull (a name currently associated with the hybrid between *M. suaveolens* and *M. spicata*, and more properly called *M.* ×*villosa* Huds.), which he thought was plausible because of the prominent veins of the leaf undersurfaces, the open toothing, and the relatively rounded leaf tips. Subsequently, a morphologically similar specimen collected during Viktor Mühlenbach's botanical inventory of the St. Louis railyards was determined by Ray Harley of Kew Gardens (a specialist on *Mentha*) as "*M. spicata* complex." Tucker and Naczi (2007) treated *M. longifolia* var. *undulata* as a synonym of *M. spicata* var. *undulata* (Willd.) Lebeau. The plants in question are remarkable within *M. spicata* in their leaves, which are cordate or nearly so at the base, tend to be relatively broad and bluntly pointed at the tip, and have the undersurface densely woolly with unbranched hairs. Hairiness itself, which sometimes has been used as a character to separate the usually glabrous *M. spicata* from the usually hairy *M. longifolia* and *M. suaveolens*, was noted by Harley (1965) to be controlled by a single gene in *M. spicata* and thus does not alone provide a basis for taxonomic distinctions within the complex. Unfortunately, the distinctive differences in the odors of the crushed foliage cannot be evaluated from older, dried specimens; thus fragrance not useful for determination of the two specimens in question. For the present, these aberrant collections continue to be treated as unusual members of *M. spicata*, rather than as *M. longifolia, M. suaveolens*, or some sort of hybrids involving these species.

Mentha ×*piperita* L. ($2n=72$, 84, 108) (Pl. 436 j, k) represents a morphologically variable series of sterile hybrids between *M. aquatica* and *M. spicata* (Murray et al., 1972; Gobert et al., 2002). This hybrid is grown commercially as the source of peppermint, which is used medicinally, as a fragrance, and as a flavorant. In Missouri, this taxon is scattered, mostly south of the Missouri River. It occurs on banks of streams, rivers, and spring branches, along the margins of lakes, and in fens, sometimes as an emergent aquatic. It also grows in ditches, along railroads and roadsides, and in moist, disturbed areas.

18. **Monarda** L. (horse mint, wild bergamot)

Plants annual or more commonly perennial herbs, with slender taproots or rhizomes. Stems erect or ascending, bluntly to sharply 4-angled, unbranched or branched, sparsely to densely hairy or rarely glabrous. Leaves sessile or petiolate, the petiole unwinged or winged, usually with a strong fragrance when crushed. Leaf blades variously linear to ovate or ovate-triangular, the margins sharply toothed, the surfaces glabrous or hairy, also with conspicuous, sessile glands. Inflorescences terminal (some of the lower clusters sometimes associated with relatively large bracteal leaves and thus appearing axillary), consisting of 1 dense flower cluster or 2–7 clusters that are noticeably separate along the axis forming a spike, each cluster subtended by a conspicuous involucre of numerous bracts (the outer bracts more or less leaflike, but often whitened or strongly pinkish-, reddish- or purplish-tinged), the inner bracts shorter, narrower, and more scalelike or hairlike), each with numerous flowers, these sessile or nearly so, sometimes with 1 slender bractlet smilar to the innermost bracts at the base. Calyces actinomorphic or nearly so, lacking a lateral projection, more or less symmetric at the base, more or less cylindric, the tube strongly 13–15-nerved (-ribbed), glabrous or hairy in the mouth, hairy externally, not 2-lipped, the lobes all similar, much shorter than the tube (longer elsewhere), triangular to narrowly triangular-tapered, not spinescent, not becoming enlarged at fruiting, but usually becoming papery. Corollas 5–50 mm long, strongly zygomorphic, white to cream-colored, pale yellow, pink, lavender, purple, or red, the lower lip usually with darker spots or blotches, the outer surface glabrous or more commonly moderately to densely pubescent with fine, spreading, glandular and/or nonglandular hairs, in some species also with sessile glands, the tube narrowly funnelform with a tubular lower portion and a usually elongate, expanded throat, 2-lipped, the upper lip entire or shallowly notched at the tip, relatively elongate and slender, straight or arched downward, often somewhat hooded at the tip, the lower lip spreading or strongly arched downward, 3-lobed or appearing more or less entire, sometimes with short, terminal, toothlike extension(s). Stamens 2, ascending under the upper lip, not or only slightly exserted, the anthers small, the connective very short and inconspicuous, the pollen sacs 2, spreading, white, pink, purple, or yellow to yellowish brown. Ovary deeply lobed, the style appearing basal or nearly so from a deep apical notch. Style exserted, unequally 2-branched at the tip. Fruits dry schizocarps, separating into usually 4 nutlets, these 1.2–2.0 mm long, oblong-obovoid (rounded at the tip, angled or tapered at the base), the surface yellowish brown to brown or black, glabrous, smooth or finely pebbled. Fifteen to 20 species, U.S., Canada, Mexico.

The genus *Monarda* has been monographed taxonomically twice (McClintock and Epling, 1942; Scora, 1967), yet some workers have suggested that more research needs to be done on the group (Gill, 1977; B. L. Turner, 1994). Hypotheses of species limits and relationships based on morphological features and crossability should be tested using molecular markers.

A number of species are cultivated as ornamentals for their showy flowers. Native Americans used most of the species ceremonially and in a number of medicinal treatments, including as a cold and sore throat remedy, for sores, burns, and other skin problems, and for aches and fevers. A few of the species, notably *M. didyma*, are used in herbal teas. A few of the species are among several members of the Lamiaceae that have been investigated as sources of natural rubbers (Buchanan et al., 1978). Dried material of some species once had limited use as a fragrant stuffing for pillows, and more recently has been used in sachets, potpourri, perfumes, and aromatherapy. The genus also has been used as a spice and flavorant. However, the bergamot oil used to flavor Earl Grey tea is extracted from a variant of a widely cultivated tree native to southeastern Asia, the sour orange, *Citrus ×aurantium* L. ssp. *bergamia* (Risso & Poit.) Wight & Arn. ex Engl. (Rutaceae).

An aberrant specimen was collected by Jean-Baptiste Duerinck in St. Louis, possibly in the late 1830s (see the chapter on botanical history in the introductory section of Volume 1 of the present work [Yatskievych, 1999]). As discussed in detail by Dorr (1986), one duplicate of this collection in the National Botanical Garden of Belgium Herbarium was the basis for the description of *M. villosa* M. Martens, which today is considered a synonym of *M. bradburiana*. However, during his studies of the genus *Monarda*, Carl Epling annotated a duplicate accessioned in the Missouri Botanical Garden Herbarium as not corresponding to *M. bradburiana*, but instead perhaps representing *M. media* Willd. *Monarda media* was treated as a species by McClintock and Epling (1942), but Scora (1967) and Gill (1977) provided evidence that it represents a variable series of hybrids between *M. didyma* and either *M. clinopodia* or *M. fistulosa*. Such hybrids, if they exist in Missouri, almost certainly represent cultivated material rather than wild plants, given the rarity of *M. clinopodia* and *M. didyma* in the state. Plants attributed to this name otherwise are encountered sporadically in the northeastern United States, no closer to Missouri than northern Illinois and central Kentucky. This hybrid is thus excluded from the Missouri flora for the present.

1. Inflorescences all or mostly consisting of 2–7 flower clusters forming an interrupted terminal spike (some of the lower clusters sometimes associated with relatively large bracteal leaves and thus appearing axillary); corollas dotted with conspicuous sessile glands in addition to the hairs, the upper lip strongly arched downward; stamens mostly not exserted, hidden under the upper corolla lip
   2. Corollas cream-colored to pale yellow or rarely deep yellow, sometimes pinkish-tinged, the lips with prominent purplish brown to maroon or brownish purple spots or mottling; calyces with the lobes 1.0–1.5 mm long, angled to sharply pointed tips but lacking a bristlelike extension of the midnerve . . . . 7. M. PUNCTATA
   2. Corollas white to pink or pinkish purple, the lips lacking spots and mottling or with reddish purple spots; calyces with the lobes 2–7 mm long, angled or tapered to a prominent, bristlelike extension of the midnerve
      3. Bracts oblong-elliptic to oblanceolate, relatively abruptly narrowed or short-tapered to a bristlelike extension of the midvein, usually strongly pinkish- or purplish-tinged, the upper surface densely pubescent with minute, sometimes purplish hairs . . . . 2. M. CITRIODORA
      3. Bracts lanceolate to elliptic or narrowly ovate, gradually narrowed or tapered to a bristlelike tip, green or mostly greenish, the upper surface glabrous or nearly so . . . . 6. M. PECTINATA
1. Inflorescences consisting of only 1 terminal flower cluster (rarely 2 on robust plants); corollas with glandular and/or nonglandular hairs, but lacking or with inconspicuous sessile glands, the upper lip straight or only slightly arched; stamens conspicuously exserted from the corolla
   4. Leaves all sessile or the larger leaves sometimes with petioles to 5 mm long; upper corolla lip slightly arched, about as long as the tube; calyces with the lobes 2–4 mm long . . . . 1. M. BRADBURIANA
   4. All but the uppermost leaves with petioles 10–40 mm long; upper corolla lip straight or nearly so, somewhat shorter than the tube; calyces with the lobes 1–2 mm long
      5. Corollas pale lavender to pink, light purple, or pinkish purple, densely hairy on the outer surface (the lips conspicuously long-hairy toward the tips) . . . . 5. M. FISTULOSA
      5. Corollas white to pale cream-colored (often with purple spots or mottling, rarely pale pinkish-tinged) or bright red to purplish red, glabrous or sparsely to moderately hairy (sometimes mainly on the upper lip)
         6. Corollas 14–30 mm long, white to pale cream-colored (often with purple spots or mottling, rarely pale pinkish-tinged) . . . 3. M. CLINOPODIA
         6. Corollas 30–45 mm long, bright red to purplish red . . . . 4. M. DIDYMA

1. Monarda bradburiana L.C. Beck (beebalm, Bradbury beebalm)

M. russeliana Nutt. ex Sims, misapplied

Pl. 437 a, b; Map 1967

Plants perennial. Stems 25–50 cm long, unbranched or less commonly few-branched, glabrous or sparsely pubescent with fine, spreading hairs, more densely so around the nodes. Leaves all sessile or the larger leaves sometimes with petioles to 5 mm long, the median leaves the largest. Leaf blades 2–9 cm long, ovate to lanceolate or ovate-triangular, rounded to shallowly cordate at the base, angled or tapered to a sharply pointed tip, the margins with widely spaced, fine teeth, usually also hairy, the surfaces glabrous or sparsely to moderately pubescent with fine, spreading to loosely appressed hairs. Inflorescences consisting of only 1 terminal flower cluster (rarely 2 on robust plants), the outer bracts 15–20 mm long, narrowly ovate to linear-lanceolate, the margins entire but hairy, the surfaces usually glabrous, green to pale green or pinkish- to purplish-tinged. Innermost bracts 5–10 mm long, mostly linear, the margins hairy. Calyces 9–14 mm long,

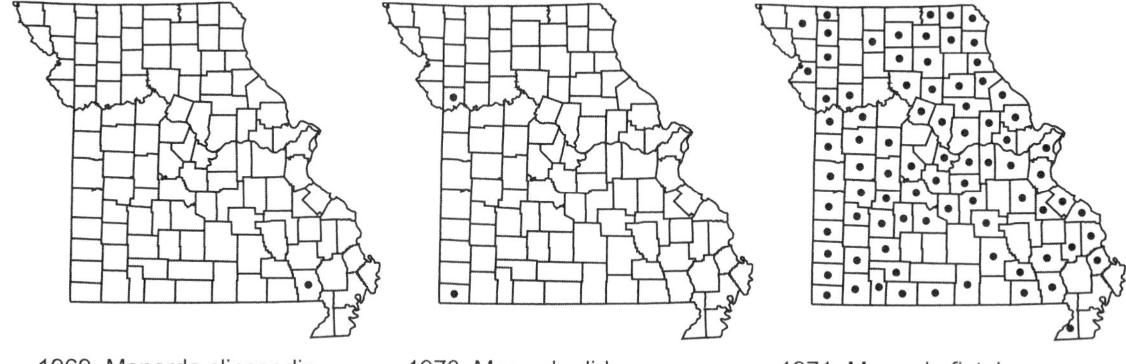

1969. Monarda clinopodia 1970. Monarda didyma 1971. Monarda fistulosa

the outer surface moderately to densely pubescent with minute, spreading hairs, sometimes with sparse longer hairs or minute, gland-tipped hairs toward the tip (and along the margins of the lobes), densely bristly-hairy in the throat, the lobes 2–4 mm long, long-tapered above a very short, triangular base, but lacking with a bristlelike extension. Corollas 24–38 mm long, white or pale pinkish- to lavender-tinged, usually with purple spots or mottling on the lower lip, sparsely to moderately pubescent with fine nonglandular hairs externally, lacking sessile glands, the lips with longer, denser hairs toward the tips, hairy in the throat, the upper lip about as long as the tube, slightly arched, the lower lip somewhat shorter than the tube, 3-lobed with a pair of short lateral lobes and a flangelike 2-toothed extension at the tip of the central lobe. Stamens conspicuously exserted from the corolla. Nutlets 1.5–2.0 mm long, yellowish brown to brown. April–June.

Scattered in the Ozark and Ozark Border Divisions, north locally along the Mississippi River to Marion County (Indiana to Iowa and Kansas south to Alabama, Louisiana, and Oklahoma). Edges of upland prairies and glades; mesic to dry upland forests, savannas, and less commonly banks of streams and rivers and margins of sinkholes; also old fields, railroads, and roadsides; often on acidic substrates.

Monarda bradburiana is gaining attention in the native plant nursery trade as an ornamental that has a shorter stature and earlier flowering period than does the more commonly grown *M. fistulosa*.

Steyermark (1963) called this species *M. russeliana*, following Fernald (1944), who reluctantly advocated the application of that name to the species present in Missouri and surrounding states, based on his interpretation of the type materials of *M. bradburiana* and *M. russeliana*. Scora (1965, 1967) presented strong evidence for a return to McClintock and Epling's (1942) interpretation of the nomenclature. His conclusions have been followed by some (R. E. Brooks, 1986) but not all (Gleason and Cronquist, 1991) subsequent authors and in the present work. True *M. russeliana* is restricted to southeastern Oklahoma and adjacent portions of Arkansas and Texas; it has mostly more than 1 flower cluster in the inflorescence and bears prominent stalked glands on the calyces.

2. Monarda citriodora Cerv. ex Lag. **var. citriodora** (lemon mint)

Pl. 437 c, d; Map 1968

Plants annual. Stems 20–80 cm long, usually branched, moderately to densely pubescent with minute, downward-curved hairs. Leaves usually short-petiolate, the petioles 2–20 mm long, the median and upper foliage leaves the largest. Leaf blades 2–6 cm long, lanceolate to elliptic-lanceolate or oblong-lanceolate, angled or tapered at the base, tapered to a sharply pointed tip (often with a bristlelike extension of the midvein), the margins with moderately to widely spaced, fine teeth, also minutely hairy (sometimes with longer, bristly hairs toward the base), the upper surface glabrous or sparsely pubescent with minute, curved hairs, the undersurface glabrous or more commonly sparsely to moderately pubescent with minute, curled hairs. Inflorescences consisting of (1)2–6 flower clusters forming an interrupted terminal spike (some of the lower clusters sometimes associated with relatively large bracteal leaves and thus appearing axillary), the bracts 15–35 mm long, oblong-elliptic to oblanceolate, relatively abruptly narrowed or short-tapered to a bristlelike extension of the midvein at the tip, the margins entire or occasionally finely toothed toward the tip, hairy, the upper surface densely pubescent with minute, sometimes purplish hairs, usually strongly pinkish- or purplish-tinged, the undersurface

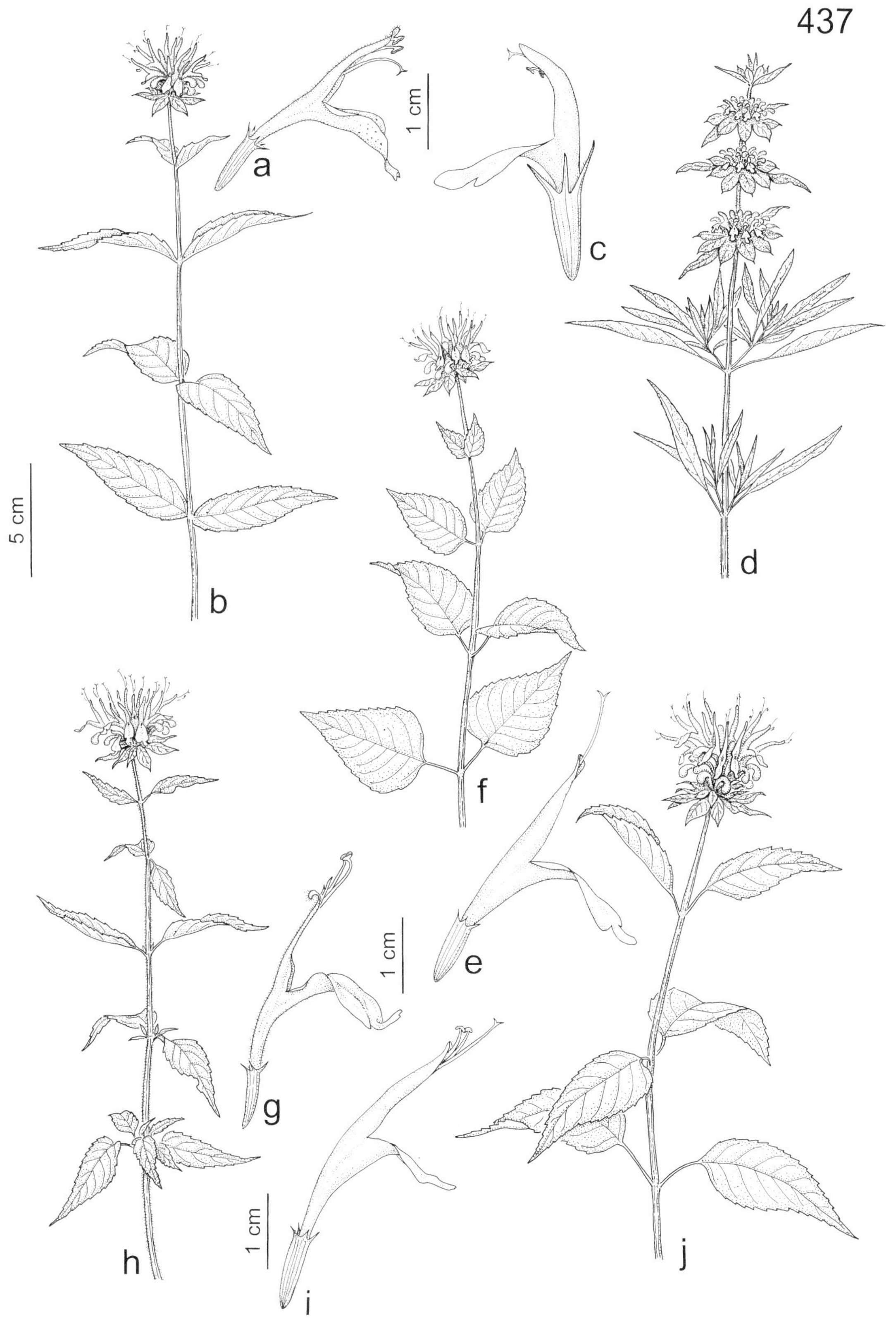

Plate 437. Lamiaceae. *Monarda bradburiana*, **a)** flower, **b)** fertile stem. *Monarda citriodora*, **c)** flower, **d)** fertile stem. *Monarda clinopodia*, **e)** flower, **f)** fertile stem. *Monarda fistulosa*, **g)** flower, **h)** fertile stem. *Monarda didyma*, **i)** flower, **j)** fertile stem.

sparsely to moderately and minutely hairy. Innermost bracts 6–10 mm long, linear, the margins hairy. Calyces 9–18 mm long, the outer surface moderately to densely pubescent with minute, curled hairs, densely bristly-hairy in the throat, the lobes 4–7 mm long, narrowly triangular, angled or tapered to a prominent, bristlelike extension of the midnerve. Corollas 15–28 mm long, white to pink or pinkish purple, the lower lip often with purple spots or lines, moderately pubescent with fine nonglandular hairs externally, also dotted with sessile glands, the lips lacking longer, denser hairs, hairy in the throat, slightly shorter than the tube, the upper lip strongly arched downward, the lower lip 3-lobed with a pair of short lateral lobes, the central lobe sometimes with a few fine teeth along the margin, lack a flangelike extension at the tip. Stamens not exserted from the corolla, hidden under the upper corolla lip. Nutlets 1.5–2.0 mm long, yellowish brown. May–August.

Scattered, mostly south of the Missouri River, those records from the western portion of the Ozark Division native, those farther north and/or east introduced (Missouri to Louisiana west to Arizona; Mexico). Upland prairies, glades, ledges and tops of bluffs, savannas, openings of mesic to dry upland forests, and banks of rivers; also fencerows, roadsides, and open, disturbed areas.

Scora (1967) and B. L. Turner (1994) treated *M. citriodora* as consisting of three infraspecific taxa. B. L. Turner's classification of these as varieties is followed in the present work. The other two varieties grow to the south and west of Missouri and differ in calyx length and corolla pubescence, as well as bract shape and position.

3. Monarda clinopodia L. (basil beebalm, beebalm)

Pl. 437 e, f; Map 1969

Plants perennial. Stems 30–120 cm long, unbranched or branched, glabrous or sparsely pubescent with fine, spreading or downward-curled hairs, more densely so around the nodes. Leaves (all but the uppermost) with petioles 10–40 mm long, the median leaves more or less the largest. Leaf blades 3–13 cm long, ovate to ovate-triangular, those of the uppermost leaves sometimes lanceolate to oblong-lanceolate, broadly angled to more commonly rounded to nearly truncate at the base, tapered to a sharply pointed tip, the margins with relatively closely spaced, usually fine teeth, also hairy, the upper surface sparsely pubescent with short, fine, loosely appressed hairs or nearly glabrous, the undersurface sparsely pubescent with longer, spreading to somewhat curved hairs, especially along the veins. Inflorescences consisting of only 1 terminal flower cluster, the bracts 8–25 mm long, lanceolate, the margins entire but hairy, the surfaces glabrous to densely pubescent with short, curved hairs, green, but often pale, whitened, or pinkish-tinged toward the base or along the midvein. Innermost bracts 5–11 mm long, mostly linear, the margins hairy. Calyces 6–10 mm long, the outer surface glabrous or sparsely to moderately pubescent with minute, gland-tipped hairs, glabrous or more commonly sparsely to moderately bristly-hairy in the throat, the lobes 1–2 mm long, long-tapered above a very short, triangular base, but lacking a bristlelike extension. Corollas 14–30 mm long, white to pale cream-colored, rarely pale pinkish-tinged, often with purple spots or mottling on the lower lip, glabrous or more commonly sparsely to moderately pubescent with minute, curled (occasionally cobwebby), nonglandular hairs on the outer surface, sometimes mainly on the upper lip, also usually with inconspicuous sessile glands, glabrous or sparsely hairy in the throat, the lips somewhat shorter than the tube, the upper lip straight or nearly so, the lower lip more or less entire, abruptly tapered to a short, terminal, toothlike extension. Stamens conspicuously exserted from the corolla. Nutlets 1.2–1.5 mm long, yellowish brown to brown. June–July.

Uncommon, known thus far only from two collections from Butler County (eastern U.S. west to Missouri and Alabama). Bottomland forests.

The flowers of *M. clinopodia* are fragrant. It is sometimes cultivated as an ornamental.

4. Monarda didyma L. (beebalm, Oswego tea)

Pl. 437 i, j; Map 1970

Plants perennial. Stems 60–150 cm long, unbranched or branched, glabrous or sparsely pubescent with fine, mostly spreading hairs, more densely so around the nodes. Leaves (all but the uppermost) with petioles 10–40 mm long, the largest leaves usually above the stem midpoint. Leaf blades 4–15 cm long, ovate to ovate-triangular, those of the uppermost leaves sometimes lanceolate, broadly angled to more commonly rounded at the base, tapered to a sharply pointed tip, the margins with closely to widely spaced, fine to relatively coarse teeth, also hairy, the upper surface sparsely pubescent with fine, more or less spreading hairs, the undersurface sparsely to moderately pubescent with spreading hairs, especially along the veins. Inflorescences consisting of only 1 terminal flower cluster (rarely 2 on robust plants), the bracts 12–25 mm long, lanceolate to narrowly

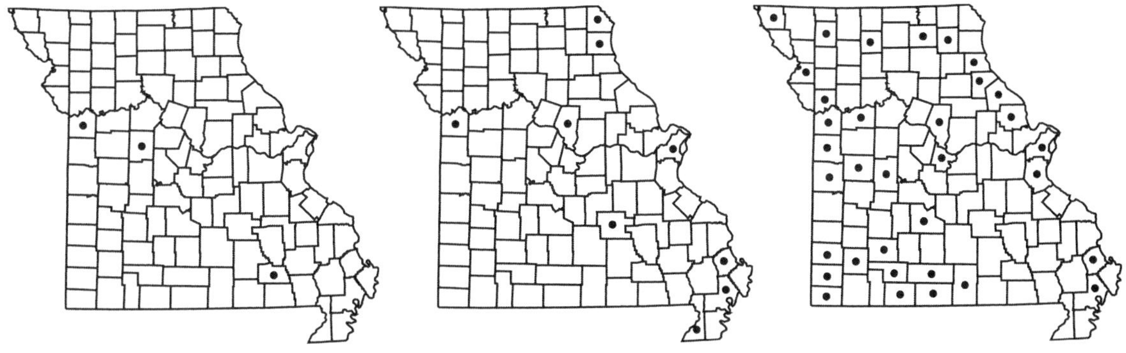

1972. Monarda pectinata 1973. Monarda punctata 1974. Nepeta cataria

ovate, the margins entire but hairy, the upper surface glabrous, usually strongly reddish- to purplish-tinged, the undersurface moderately pubescent with short, curved hairs. Innermost bracts 6–14 mm long, mostly linear, the margins hairy. Calyces 10–14 mm long, the outer surface glabrous or sparsely to moderately pubescent with minute, gland-tipped hairs, glabrous or nearly so in the throat, the lobes 1–2 mm long, long-tapered above a very short, triangular base, but lacking a bristlelike extension. Corollas 30–45 mm long, bright red or rarely purplish red, moderately to densely pubescent with minute, curled to spreading, nonglandular hairs on the outer surface, those on the lips often red, also with inconspicuous sessile glands, glabrous or sparsely hairy in the throat, the lips shorter than the tube, the upper lip straight or nearly so, the lower lip more or less entire, with a slender, flangelike 2-toothed extension at the tip. Stamens mostly conspicuously exserted from the corolla. Nutlets 1.6–2.0 mm long, yellowish brown. $2n$=36. June–September.

Introduced, known thus far only from a single historical collection from Clay County (Maine to Ohio and Minnesota south to Georgia and Tennessee; Canada; introduced farther west to Iowa and Missouri, also Washington, Oregon). Banks of streams and bottomland forests.

Monarda didyma is cultivated widely for its bright, hummingbird-pollinated flowers and also is popular for use in herbal teas. Steyermark (1963) noted that white- and purple-flowered variants are known in the horticultural trade, but the purple-flowered plants apparently often are the result of cross-breeding with *M. fistulosa*. Steyermark also reported the species without documentation from Pike and Ralls Counties, but specimens to support these reports could not be located during the present study.

5. Monarda fistulosa L. **ssp. fistulosa** (wild bergamot, horse mint)

M. fistulosa var. *mollis* (L.) Benth.

Pl. 437 g, h; Map 1971

Plants perennial. Stems 30–120 cm long, unbranched or branched, glabrous toward the base, sparsely to moderately pubescent toward the tip with fine, downward-curled hairs, sometimes also with scattered, longer, spreading hairs, the more glabrous variants often somewhat glaucous. Leaves with petioles 2–25 mm long (those of all but the uppermost leaves 10–25 mm), the median leaves more or less the largest. Leaf blades 2–10 cm long, ovate to broadly ovate or ovate-triangular, those of the uppermost leaves often lanceolate, broadly angled to rounded, broadly rounded, or nearly truncate at the base, angled or tapered to a sharply pointed tip, the margins with widely to relatively closely spaced, usually fine teeth, also hairy, the upper surface glabrous or sparsely pubescent with short, fine, loosely appressed hairs, the undersurface glabrous or sparsely to densely pubescent with short, fine, loosely appressed hairs and/or longer, spreading to somewhat curved hairs, especially along the veins. Inflorescences consisting of only 1 terminal flower cluster (rarely 2 on robust plants), the bracts 10–25 mm long, lanceolate to ovate, the margins entire but hairy, the surfaces glabrous or the undersurface sparsely to moderately pubescent with short, curved hairs, green, but often pale or pinkish- to purplish-tinged. Innermost bracts 6–10 mm long, mostly linear, the margins hairy. Calyces 7–10 mm long, the outer surface moderately pubescent with minute hairs, some or all of these gland-tipped, densely bristly-hairy in the throat, the lobes 1–2 mm long, long-tapered above a very short, triangular base, but lacking a bristlelike extension. Corollas 20–35 mm long, pale lavender to pink, light purple, or pinkish purple, lacking spots or mottling on the lower lip, densely pubescent with minute, curled, nonglandular hairs on the outer surface, the lips, especially the upper lip, with dense, longer, spreading hairs toward the tips, also usually with inconspicuous sessile glands, glabrous in the throat, the lips somewhat shorter than the tube, the upper lip straight or nearly so, the lower lip more or less entire, with a slender, flangelike, usually 2-toothed extension at the tip. Stamens conspicuously exserted from the corolla. Nutlets 1.5–2.0 mm long, brown to black. $2n$=32, 34, 36. May–August.

Scattered to common nearly throughout the state (eastern U.S. west to North Dakota and Texas; Canada, possibly also Mexico). Upland prairies, openings and edges of mesic to dry upland forests, savannas, glades, and banks of streams and rivers; also fallow fields, old fields, fencerows, railroads, and roadsides.

Monarda fistulosa is commonly cultivated for its showy flowers, which are sometimes hummingbird-pollinated. Horticulturally derived hybrids with *M. didyma* are known, as are naturally occurring hybrids, where the ranges of the two species overlap to the east of Missouri. Scora (1967) applied the name *M.* ×*media* Willd. to naturally occurring hybrids between *M. didyma* and *M. fistulosa*. In 1956, Julian Steyermark collected a fruiting specimen (*Steyermark 83306*, at the Missouri Botanical Garden Herbarium) in Butler County that he determined as this taxon; however, this specimen was redetermined as *M. fistulosa* during the present research.

Scora (1967) treated *M. fistulosa* as a complex of five varieties, two of which had been reported from Missouri by Steyermark (1963). The var. *mollis* was said to differ from var. *fistulosa* in its firmer leaves with the undersurface densely pubescent with minute, appressed hairs (vs. glabrous or sparsely to moderately pubescent with longer, less-appressed hairs). However, Gill (1977) noted that, at least in Canada, plants with the morphology of var. *fistulosa* are present within populations of var. *mollis* and var. *menthifolia*. He thus proposed a classification involving a mostly eastern ssp. *fistulosa* (including the name var. *mollis*) and a western ssp. *menthifolia* (Graham) L.S. Gill. The latter subspecies differs from ssp. *fistulosa* in its usually unbranched stems and sessile to short-petiolate leaves. Gill's classification is followed here provisionally, but the distinctness and relationships of these two taxa should be examined in more detail.

The closely related *M. stipitatoglandulosa* Waterf. was described from southeastern Oklahoma (Waterfall, 1970), but also occurs in the Ouachita Mountains region of west-central Arkansas (E. B. Smith, 1988). It tends to be a shorter plant (mostly 20–40 cm tall) than *M. fistulosa* and also differs in its white corollas and prominent stalked glands on the calyces and often also the bracts. Its relationship to *M. russeliana* should be investigated Although the closest known populations are some distance from Missouri, this taxon plausibly might be discovered in the state in the future

6. Monarda pectinata Nutt. (spotted beebalm, plains beebalm)

Pl. 438 a, b; Map 1972

Plants annual. Stems 15–40 cm long, usually branched, moderately to densely pubescent with minute, downward-curved hairs. Leaves usually short-petiolate, the petioles 2–15 mm long, the median leaves the largest. Leaf blades 1–5 cm long, lanceolate to oblong-lanceolate or elliptic, angled or tapered at the base, tapered to a bluntly or sharply pointed tip (this lacking a bristlelike extension of the midvein), the margins with moderately to widely spaced, fine teeth or nearly entire, also minutely hairy (sometimes with longer, bristly hairs toward the base), the upper surface glabrous or sparsely pubescent with minute, curved hairs, the undersurface glabrous or more commonly sparsely to moderately pubescent with minute, curled hairs. Inflorescences consisting of (1)2–6 flower clusters forming an interrupted terminal spike (some of the lower clusters sometimes associated with relatively large bracteal leaves and thus appearing axillary), the bracts 8–15 mm long, lanceolate to elliptic or narrowly ovate, gradually narrowed or tapered to a bristlelike extension of the midvein at the tip, the margins entire or finely toothed toward the tip, hairy, the upper surface glabrous or nearly so, green or mostly greenish (rarely pinkish-tinged), the undersurface sparsely to moderately and minutely hairy. Innermost bracts 3–8 mm long, linear, the margins hairy. Calyces 8–12 mm long, the outer surface moderately to densely pubescent with minute, curled hairs, densely bristly-hairy in the throat, the lobes 2.0–3.5 mm long, narrowly triangular, angled or tapered to a prominent, bristlelike extension of the midnerve. Corollas 12–20 mm long, white or more commonly pinkish-tinged to pink or purplish pink, the lower lip often with purple spots or lines, sparsely pubescent with fine nonglandular hairs externally, also dotted with sessile glands, the lips lacking longer, denser hairs, hairy in the throat, slightly shorter than the tube, the upper lip strongly arched downward, the lower lip 3-lobed with a pair of short, often somewhat pointed lateral lobes, the central lobe usually entire (rarely

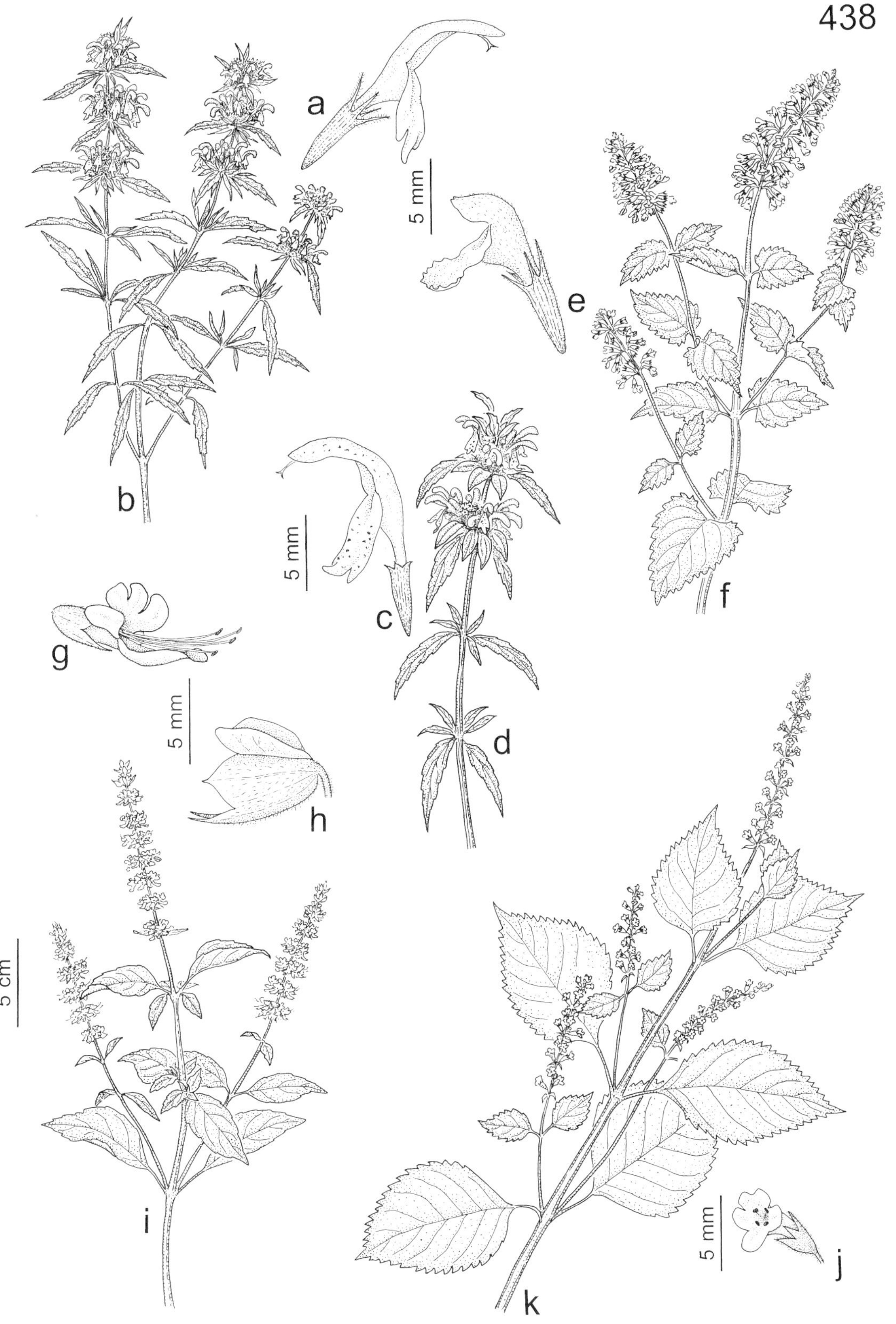

Plate 438. Lamiaceae. *Monarda pectinata*, **a)** flower, **b)** fertile stem. *Monarda punctata*, **c)** flower, **d)** fertile stem. *Nepeta cataria*, **e)** flower, **f)** fertile stem. *Ocimum basilicum*, **g)** flower, **h)** fruiting calyx, **i)** fertile stem. *Perilla frutescens*, **j)** flower, **k)** fertile stem.

minutely notched at the tip), lacking a flangelike extension at the tip. Stamens not exserted from the corolla, hidden under the upper corolla lip. Nutlets 1.2–1.5 mm long, yellowish brown to brown. June–October.

Introduced, uncommon, sporadic (South Dakota to Texas west to California; possibly also Mexico). Roadsides and open, disturbed areas.

This species is sometimes a component in wildflower mixes sown for roadside beautification projects by misguided highway departments.

7. Monarda punctata L. (dotted beebalm, horse mint)

Pl. 438 c, d; Map 1973

Plants annual or perennial. Stems 15–50 cm long, unbranched or branched, moderately to densely pubescent with minute, downward-curved hairs, rarely with sparse to dense, longer, bristly spreading hairs (sometimes only at the nodes). Leaves sessile or the petioles to 15 mm long, the median leaves the largest. Leaf blades 2–9 cm long, lanceolate to oblong-lanceolate or narrowly elliptic, rounded to more commonly angled or tapered at the base, angled or tapered to a sharply pointed tip, the margins with moderately to widely spaced, fine teeth, usually also hairy, the upper surface sparsely to moderately pubescent with minute, curved hairs, the undersurface sparsely to densely pubescent with minute, curled hairs, occasionally also with longer, somewhat woolly hairs along the main veins. Inflorescences consisting of (1)2–7 flower clusters forming an interrupted terminal spike (some of the lower clusters sometimes associated with relatively large bracteal leaves and thus appearing axillary), the bracts 15–40 mm long, oblong, elliptic, ovate, or obovate, the margins entire but hairy (at least near the base), the surfaces usually minutely hairy, whitish to pinkish- or purplish-tinged. Innermost bracts 3–8 mm long, linear, the margins hairy. Calyces 5–8 mm long, the outer surface sparsely to densely pubescent with minute, curled hairs, sometimes with sparse, longer hairs toward the tip, densely bristly-hairy in the throat, the lobes 1.0–1.5 mm long, triangular to narrowly triangular, sharply pointed but lacking a bristlelike extension. Corollas 15–28 mm long, cream-colored to pale yellow, sometimes pinkish-tinged, the lips with prominent purplish brown to maroon or brownish purple spots or mottling, moderately to densely pubescent with fine nonglandular hairs externally, also dotted with sessile glands, the lips with longer, denser hairs toward the tips, hairy in the throat, slightly shorter than the tube, the upper lip strongly arched downward, the lower lip 3-lobed with a pair of short lateral lobes and a flangelike, oblong (sometimes shallowly notched) extension at the tip of the central lobe. Stamens not exserted from the corolla, hidden under the upper corolla lip. Nutlets 1.2–1.5 mm long, usually brown. $2n=22, 24$. June–October.

Uncommon, mostly in counties along the Mississippi and Missouri Rivers (eastern U.S. west to Minnesota and New Mexico; Canada, Mexico). Sand prairies; also margins of crop fields, railroads, roadsides, and open disturbed areas; usually in sandy soil.

Monarda punctata consists of a messy and complicated series of morphologically overlapping infraspecific taxa. McClintock and Epling (1942) accepted eight subspecies, but reported only ssp. *villicaulis* Pennell from Missouri. Steyermark (1963) treated two varieties, var. *villicaulis* and var. *occidentalis*. Scora (1967) accepted eleven varieties and reported the same two taxa for Missouri as had Steyermark. In his study of the genus in Texas and Mexico, B. L. Turner (1994) raised four of the taxa treated earlier as subspecies or varieties to species level and also named a twelth variety endemic to southern Texas and adjacent Mexico. Thus, there is reason to believe that the dozen total taxa accepted in the *M. punctata* complex over the last few decades include at least six well defined entities (the five species accepted by B. L. Turner and his new variety), but the other six taxa include some varieties that exhibit a great deal of morphological overlap. Future taxonomic studies may show that most of these are better treated as a single polymorphic species, including the taxa reported from Missouri and accepted tentatively in the present work.

1. Stems densely pubescent with a mixture of minute (0.2–0.4 mm), downward-curled hairs and longer (0.8–1.5 mm), bristly hairs; leaf blades pubescent along the margins with minute, curled hairs and also with sparse to moderate longer, bristly hairs toward the base .... 7A. VAR. ARKANSANA
1. Stems moderately to densely pubescent with only minute (0.2–0.5 mm), downward-curled hairs, the longer (0.8–1.5 mm), bristly hairs absent or only in transverse rings at some of the nodes; leaf blades pubescent along the margins with usually only minute, curled hairs, lacking longer, bristly hairs (except occasionally in var. *villicaulis*)

2. Calyces with the lobes triangular, slightly longer than to about as long as wide, the margins densely hairy; leaf blades with the undersurface sparsely to densely pubescent with only minute, curled hairs, lacking longer, woolly hairs . . . . 7B. VAR. OCCIDENTALIS
2. Calyces with the lobes narrowly triangular, 1.5–2.0 times as long as wide, the margins glabrous or sparsely hairy (do not confuse the dense, bristly hairs in the throat with those of the calyx lobes); leaf blades with the undersurface densely pubescent with minute, curled hairs and, at least along the main veins, longer, somewhat woolly hairs . . . . 7C. VAR. VILLICAULIS

7a. var. arkansana (E.M. McClint. & Epling) Shinners (Arkansas beebalm)

M. punctata ssp. *arkansana* E.M. McClint. & Epling

Plants usually perennial. Stems densely pubescent with a mixture of minute (0.2–0.4 mm), downward-curled hairs and longer (0.8–1.5 mm), bristly hairs. Petioles also with a mixture of minute, curled and longer, bristly hairs. Leaf blades 2–9 cm long, the margins pubescent with minute, curled hairs and also with sparse to moderate longer, bristly hairs toward the base, the upper surface sparsely pubescent with short, fine hairs, the undersurface moderately pubescent with longer, somewhat woolly hairs, usually densely so along the main veins. Calyces with the lobes narrowly triangular, 1.5–2.0 times as long as wide, the margins glabrous or sparsely hairy (do not confuse the dense, bristly hairs in the throat with those of the calyx lobes). Corollas deep yellow. July–October.

Possibly introduced, known thus far from a single historical collection from St. Louis County (Arkansas, Georgia, Texas, and Missouri). Habitat unknown.

The sole Missouri record was collected by John Kellogg in 1884 at Allenton (St. Louis County) and is in the Missouri Botanical Garden Herbarium. It was annotated as ssp. *typica* (= var. *punctata*) by Epling and as var. *arkansana* by Scora, but was not cited in the published works of either author (and apparently was not examined by Steyermark). As with so many of the nineteenth century collections from the state, the specimen label includes no habitat data. A second historical specimen at the Missouri Botanical Garden Herbarium collected by Henry Eggert in 1878, apparently in the city of St. Louis, also was annotated by Epling as ssp. *typica*. Neither Scora nor Steyermark apparently studied the sheet. It appears, however, to represent var. *villicaulis*.

Scora (1967) reported var. *arkansana* from nearly throughout Arkansas and also from a few sites in Georgia and Texas. This variety is very similar to var. *punctata*, which is distributed from Pennsylvania to Florida west to southeastern Texas. The two differ from the other varieties attributed to Missouri thus far in their deep yellow corollas that are heavily marked with dark maroon spots or mottling, and in their stems with relatively long, spreading, bristly hairs. The var. *arkansana* differs from var. *punctata* only in its more densely bristly stems and somewhat larger leaves. It is possible that future studies will reveal that they should be combined. In that case, the older of the two names is var. *punctata*.

7b. var. occidentalis (Epling) E.J. Palmer & Steyerm.

M. punctata ssp. *occidentalis* Epling

Plants annual. Stems densely pubescent with only minute (0.2–0.5 mm), downward-curled hairs, the longer (0.8–1.5 mm), bristly hairs absent. Petioles also with only minute, curled hairs. Leaf blades 2–5 cm long, the margins pubescent with minute, curled hairs, the upper surface glabrous or sparsely pubescent with short, fine hairs, the undersurface sparsely to densely pubescent with short, fine hairs. Calyces with the lobes triangular, slightly longer than to about as long as wide, the margins densely bristly-hairy. Corollas cream-colored to pale yellow, sometimes pinkish-tinged. $2n=24$. June–October.

Introduced, uncommon, sporadic (Illinois to Colorado south to Texas; Mexico). Railroads, roadsides, and open, disturbed areas.

Missouri specimens referable to this variety tend to have relatively short leaves and to dry yellowish green. This variety is sometimes a component in wildflower mixes sown for roadside beautification projects by misguided highway departments.

7b. var. villicaulis (Pennell) Shinners

Plants annual or perennial. Stems densely pubescent with only minute (0.2–0.5 mm), downward-curled hairs, the longer (0.8–1.5 mm), bristly hairs absent or only in transverse rings at some of the nodes. Petioles with minute, curled hairs and sometimes also with scattered, longer, bristly hairs. Leaf blades 2–8 cm long, the margins pubescent with minute, curled hairs and sometimes also scattered, longer, bristly hairs toward the

1975. Ocimum basilicum 1976. Origanum vulgare 1977. Perilla frutescens

base, the upper surface sparsely to moderately pubescent with short, fine hairs, the undersurface moderately to densely pubescent with short, fine hairs, also with dense, longer, somewhat woolly hairs, at least along the main veins. Calyces with the lobes the lobes narrowly triangular, 1.5–2.0 times as long as wide, the margins glabrous or sparsely hairy (do not confuse the dense, bristly hairs in the throat with those of the calyx lobes). Corollas cream-colored to pale yellow, sometimes pinkish-tinged. $2n=22$. June–October.

Uncommon, in counties along the Mississippi and Missouri Rivers (northeastern U.S. west to Minnesota and Missouri; Canada). Sand prairies; also margins of crop fields, railroads, roadsides, and open disturbed areas; usually in sandy soil.

19. Nepeta L.

About 250 species, Europe, Asia; introduced nearly worldwide.

1. Nepeta cataria L. (catnip, catmint)
Pl. 438 e, f; Map 1974

Plants perennial, with taproots or occasionally branched rootstocks. Stems 30–100 cm long, erect or ascending, bluntly to sharply 4-angled, branched, densely pubescent with felted or somewhat woolly, sometimes matted, short, white to grayish-tinged hairs. Leaves opposite, mostly short- to more commonly long-petiolate, the petioles unwinged or winged only at the very tip, with a pungent odor when bruised or crushed. Leaf blades 2–8 cm long, ovate to ovate-triangular, unlobed, the margins mostly coarsely toothed (sometimes deeply so on larger leaves), truncate to shallowly cordate at the base, angled to a bluntly or sharply pointed tip, the upper surface green and sparsely to moderately pubescent with short, curved hairs, the undersurface usually pale, densely pubescent with felted or somewhat woolly, sometimes somewhat matted, short, white to grayish-tinged hairs, also with conspicuous sessile glands, these sometimes obscured by the hairs. Inflorescences terminal, often appearing as short, dense, headlike clusters when young, usually elongating into dense, but often interrupted, spikelike racemes, these sometimes with short branches (then paniculate), the flowers numerous per node, some sessile, others short-stalked. Bracts 6–30 mm long, often absent from the upper inflorescence nodes, extending past the flowers or not, narrowly ovate to lanceolate, sharply pointed but not spinescent; bractlets inconspicuous, minute, mostly linear. Calyces 5–7 mm long at flowering, zygomorphic, lacking a lateral projection, usually slightly pouched on 1 side at the base, more or less tubular (often slightly curved), the tube strongly 15-nerved (-ribbed), densely pubescent with short, fine, mostly spreading hairs on the outer surface, also with relatively dense, sessile glands, densely hairy in the mouth, the lips shorter than the tube, loosely ascending, the upper lip 3-lobed, longer than the 2-lobed lower lip, the lobes all similar, narrowly triangular, tapered to sharply pointed, but not spinescent tips, becoming papery but not enlarged at fruiting. Corollas 7–12 mm long, zygomorphic, 2-lipped, dull white to light cream-colored, the lower lip with pink to purple spots or mottling, the outer surface densely short-hairy, the tube funnelform, hairy in the throat, the lips shorter than to about as long as the tube, the upper lip shorter than the lower lip, entire or shallowly notched, slightly concave to more or less hooded, the lower lip spreading to arched, 3-lobed with a large central lobe and 2 small lateral lobes, the central lobe somewhat scoop-shaped, notched or more commonly with several small teeth at the broadly rounded to truncate tip. Stamens 4, short-exserted (more or less ascending under the upper corolla lip), the upper pair slightly longer than the lower pair, the anthers small, the connective short, the pollen sacs 2, spreading, purple. Ovary deeply lobed, the style appearing more or less basal from a deep apical notch. Style short-exserted, with 2 equal branches at the tip. Fruits dry schizocarps, separating into usually 4 nutlets, these 1.3–2.0 mm long, mostly oblong-obovoid, rounded at the tip, the surface reddish brown to dark brown, smooth or with few to several faint, longitudinal lines or grooves, glabrous. $2n$=32, 34, 36. June–September.

Introduced, scattered, mostly in the northern and western halves of the state (native of Asia; introduced widely, including nearly throughout the U.S. [including Alsaka], Canada). Mesic upland forests, banks of streams, and bases of bluffs; also fallow fields, old fields, pastures, fencerows, old homesites, railroads, roadsides, and disturbed areas.

Nepeta cataria has been used medicinally in teas for its mild sedative properties. However, it is best known for evoking the well-known *catnip response* in many different species of felines, which was reviewed in detail by Tucker and Tucker (1988). This psychosexual response (intoxication) apparently is under the control of a single dominent gene (thus some cats are not susceptible) and does not become active until the cat is about three months old. It is caused by nepetalactone, a cyclopentanoid monoterpene, present in the plant's essential oil. More recently, nepetalactone from catnip has shown promise as an insect repellent (C. J. Peterson et al., 2002; C. J. Peterson and Ems-Wilson, 2003), particularly of cockroaches, mosquitos, and termites.

20. **Ocimum** L.

About 150 species, widespread in tropical and warm-temperate regions, most diverse in South America and Africa.

1. Ocimum basilicum L. (basil, sweet basil, lemon basil)

Pl. 438 g–i; Map 1975

Plants annual, with taproots. Stems 20–70 cm long, erect or ascending, bluntly to sharply 4-angled, several- to many-branched, moderately pubescent with short, downward-curved, hairs, sometimes nearly glabrous toward the base. Leaves opposite, short- to long-petiolate, the petioles usually winged, at least toward the tip, with a characteristic, pungent odor when bruised or crushed. Leaf blades 1–7 cm long, ovate to narrowly ovate or elliptic, unlobed, the margins entire or with relatively sparse, fine, blunt teeth, angled to more commonly tapered at the base, angled or tapered to a bluntly or sharply pointed tip, the surfaces glabrous or sparsely short-hairy along the main veins, the undersurface also with conspicuous sessile glands. Inflorescences terminal, of slender, elongate spikelike racemes, the flowers 6 per node, these short-stalked. Bracts 5–8 mm long, shorter than to about as long as the flowers, ovate to lanceolate or elliptic, sharply pointed but not spinescent; bractlets absent. Calyces 3.5–4.0 mm long at flowering, strongly zygomorphic, slightly asymmetric at the base, more or less bell-shaped, the tube more or less 15-nerved (the upper lip with a dense network of veins, hairy in the mouth, 5-lobed, the lobes mostly shorter than the tube, the upper lobe enlarged, with the margins extending downward along the tube as a pair of crests, the whole structure depressed-ovate to circular, appearing as a raised flap or circular cap of tissue, the 2 lateral lobes broader and slightly shorter than the 2 lower lobes, these all narrowly ovate to ovate or triangular, tapered to sharply pointed, but not spinescent tips, sparsely short-hairy and usually also with sessile glands, becoming enlarged to 7–9 mm and papery at fruiting. Corollas 6–12 mm long, zygomorphic, 2-lipped, white to pink, sometimes purplish-tinged toward the throat or toward the middle of the lower lip, the outer surface sparsely to moderately hairy, more or less bell-shaped, the tube usually shorter than the lips, hairy in the throat, the upper lip 4-lobed, the lobes arched upward, somewhat irregular along the margins, the lower lip longer than the upper lip, unlobed, obovate, somewhat finely corrugated longitudinally, spreading. Stamens 4, exserted (arched along the lower lip), the lower pair slightly longer than the upper pair, the anthers small, the connective short, the pollen sacs 2, slightly angled to nearly parallel, usually white. Ovary deeply lobed, the style appearing more or less basal from a deep apical notch. Style exserted, with 2 more or less equal branches at the tip. Fruits dry schizocarps, separating into usually 4 nutlets, these 2.0–2.5 mm long, obovoid to oblong-obovoid, rounded at the tip, the surface dark brown to black, finely wrinkled, glabrous, but often with a lacquerlike irregular covering (becoming sticky when moistened). $2n$=48. August–September.

Introduced, known thus far only from the city of St. Louis (native of Asia, Africa; introduced sporadically in the eastern U.S. west to Missouri and Louisiana, also Arizona). Railroads.

Basil is a common herb used as a flavorant in various foods. The species also yields an essential oil used in perfumes and beverages (Steyermark, 1963). A number of cultivars are commonly cultivated. In its native range, *O. basilicum* sometimes is divided into two or more varieties. The introduced plants appear to correspond to var. *basilicum*, but the infraspecific taxonomy requires further study.

21. **Origanum** L. (marjoram)

About 38 species, Europe, Asia.

The genus *Origanum* includes both shrubby species and perennial herbs. Some species are cultivated as ornamentals or for other uses. A number of these produce essential oils that have been used medicinally, as scents in potpourris, and as flavorants for foods. Some of the herbaceous species also have been cultivated as potherbs in the Old World. The European *O. dictamnus* L. (dittany) has been used both medicinally as an antiseptic and as the main flavorant in vermouth. *Origanum marjorana* L. (sweet marjoram) is used to flavor meats, especially sausages and canned meats.

1. Origanum vulgare L. (oregano, wild marjoram)

Map 1976

Plants perennial herbs, with well-developed, somewhat woody rhizomes. Stems 40–90 cm long, erect or ascending, bluntly to sharply 4-angled, branched (often with short branches in most leaf axils, these with a pair of small leaves), moderately to densely pubescent with scattered, spreading, multicellular hairs toward the stem base and denser, fine, curled to somewhat matted hairs throughout, sometimes purplish-tinged. Leaves opposite, short-petiolate, the petioles unwinged, with a distinctive odor (of oregano) when bruised or crushed. Blades of main leaves 1.0–2.5 cm long (the axillary leaves mostly 0.5–1.2 cm long), ovate, unlobed, the margins entire or very broadly and shallowly scalloped, also hairy, broadly rounded at the base, rounded or angled to a bluntly pointed tip, the upper surface glabrous or with a few hairs at the base, the undersurface moderately pubescent with curved hairs and conspicuous, small gland-dots. Inflorescences terminal, relatively dense, dome-shaped panicles, the branching mostly in whorls of 3, the ultimate branches with short, dense, ovoid, spikelike, racemes (these often 4-angled), the flowers 2 per node (1 on each side of the axis), these short-stalked. Bracts 3–5 mm long, about as long as to nearly twice as long as the calyces, ovate to elliptic-obovate, bluntly pointed, strongly reddish- or purplish-tinged; bractlets absent. Calyces 2.5–3.5 mm long at flowering, actinomorphic or nearly so, lacking a lateral projection, symmetric at the base, more or less bell-shaped, short-hairy and with sessile glands on the outer surface, the tube 10–13-nerved, densely bristly-hairy in the mouth, the 5 teeth much shorter than the tube, oblong-ovate, angled to bluntly or sharply pointed, not spinescent tips, usually strongly reddish- or purplish-tinged, not or only slightly enlarged at fruiting. Corollas 5–8 mm long, strongly zygomorphic, 2-lipped, pink to lavender or pale purple (occasionally white elsewhere), lacking darker purple spots, the outer surface moderately to densely hairy, the tube glabrous or sparsely and inconspicuously hairy in the throat, the upper lip shallowly 2-lobed (notched), straight, more or less flat, entire or slightly irregular along the margins, the lower lip somewhat longer than the upper lip, deeply 3-lobed, more or less fan-shaped, spreading. Stamens 4, the lower pair exserted, the upper pair with shorter filaments and not or only slightly exserted, the anthers small, the connective short, the pollen sacs 2, spreading, pink to purple. Ovary deeply lobed, the style appearing lateral from a deep apical notch. Style long-exserted, with 2 equal branches at the tip. Fruits dry schizocarps, separating into usually 4 nutlets, these 0.5–0.7 mm long, ovoid, rounded at the tip, the surface brown, smooth, glabrous. $2n$=30, 32. August–September.

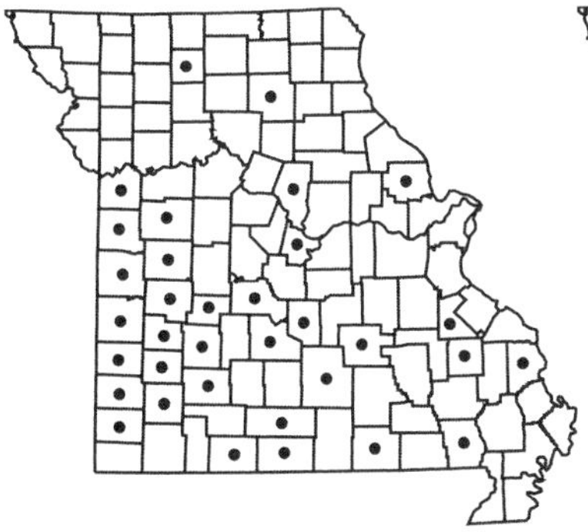

1978. Physostegia angustifolia  1979. Physostegia intermedia 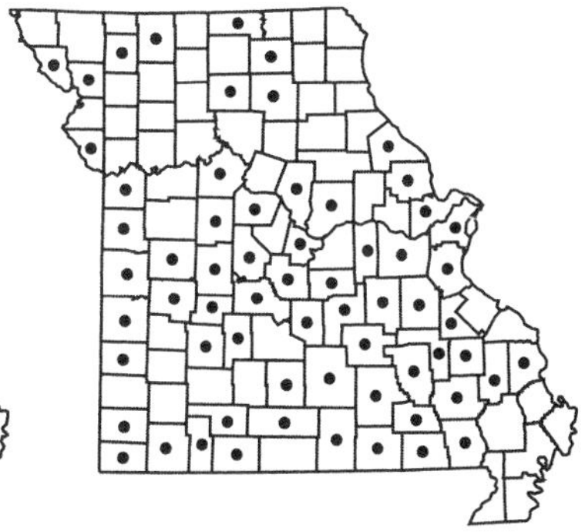 1980. Physostegia virginiana

Introduced, uncommon, known thus far only from a single specimen from Putnam County (native of Europe, Asia; introduced sporadically in the eastern U.S., Canada). Barnyards and gardens.

This species was first documented outside of cultivation in Missouri by Greg Gremaud, who discovered it in August 2008 naturalized near an old homesite. The dried leaves are crumbled and used as a seasoning, especially in Mediterranean cuisines. It is an important flavorant in pizzas and other Italian dishes involving tomato-based sauces. The inflorescences also yield a red dye.

22. Perilla L.

One or 4 species, Asia; introduced widely.

Perilla generally has been treated as comprising four taxa, which variously have been regarded as separate species or as components of *P. frutescens*. In the strict sense, *P. frutescens* is thought to represent an allotetraploid derived through past hybridization between as yet undetermined parents from among the other three diploid taxa in the genus (Nitta et al., 2005). However, there is no modern taxonomic revision of the genus available.

1. Perilla frutescens (L.) Britton (beefsteak plant, wild basil, summer coleus, rattlesnake weed)

Pl. 438 j, k; Map 1977

Plants annual, with taproots. Stems 20–80 (–100) cm long, erect or ascending, bluntly to sharply 4-angled, branched, sparsely to moderately pubescent with loosely downward-curved, multicellular hairs, more densely so toward the tip and around the nodes, often nearly glabrous toward the base, sometimes strongly purplish-tinged. Leaves opposite, mostly long-petiolate, the petioles unwinged, with an unpleasant, pungent odor when bruised or crushed. Leaf blades 3–15 cm long, sometimes entirely or mostly dark brownish purple, oblong-ovate to broadly ovate or occasionally broadly elliptic, unlobed, the margins coarsely toothed, rarely with slender irregular lobes (incised), broadly angled to angled or short-tapered at the base, angled or tapered to a sharply pointed tip, the surfaces glabrous or more commonly sparsely short-hairy, mostly along the veins, the undersurface also with conspicuous sessile glands. Inflorescences terminal and axillary, of relatively dense, slender, usually elongate spikelike racemes, the flowers 2 per node (1 on each side of the axis), these short-stalked. Bracts 3–6 mm long, not extending past the calyces, elliptic to ovate, sharply pointed but not spinescent; bractlets absent. Calyces 2–3 mm long at flowering, slightly zygomorphic, lacking a lateral projection, symmetric at the base at flowering, more or less bell-shaped, the tube 10-nerved, the nerves obscured by usually dense, spreading, multicellular hairs on the outer surface, sparsely hairy in the mouth, the lips slightly shorter than to slightly longer than the tube, the upper lip curved slightly upward, with 3 triangular lobes, the lower lip slightly curved upward, with 2 deeper, narrowly triangular lobes, the lobes all tapered to sharply pointed, but not spinescent tips, usually with relatively dense, sessile glands and sometimes with a few spreading, nonglandular hairs toward the base on the outer surface, becoming enlarged to 8–12 mm, papery, and swollen or pouched basally on the lower side at fruiting. Corollas 2.5–4.0 mm long, nearly actinomorphic, not 2-lipped, white to lavender or light purple, lacking darker purple spots, the outer surface glabrous or more commonly sparsely hairy, more or less bell-shaped, the tube hairy in the throat, the 5 lobes very short, all similar or nearly so (the lowermost lobe sometimes slightly larger and/or slightly more spreading), broadly rounded to bluntly pointed, somewhat spreading. Stamens 4, short-exserted, the 2 pairs equal or nearly so, the anthers small, the connective short, the pollen sacs 2, spreading, dark purple. Ovary deeply lobed, the style appearing more or less basal from a deep apical notch. Style short-exserted, with 2 unequal branches at the tip. Fruits dry schizocarps, separating into 2–4 nutlets, these 1.3–2.0 mm long, more or less globose, rounded at the tip, the surface orangish brown to reddish brown, with a network of fine nerves or slender, low ridges, glabrous. $2n=40$. June–September.

Scattered, mostly south of the Missouri River (native of Asia; introduced widely in the eastern U.S. west to Minnesota, Nebraska, Texas; also Canada, Europe). Bottomland forests, banks of streams and rivers, bases of bluffs, fens, and seeps; also fallow fields, pastures, cemeteries, railroads, roadsides, and moist, disturbed areas.

This species has a long history of use in Asian cuisine and traditional medicine, and as a source of a red dye. Steyermark (1963) noted that the seed oil has been important in the manufacture of lacquer, artificial leather, paper umbrellas, printer's ink, and waterproof clothing, and that in the United States it has been used as a substitute for linseed oil in the varnish and paint industries. It apparently was introduced into the United States in the late 1800s by Asian immigrants and was soon adopted by local residents in the Appalachians and Ozarks as a potherb and seasoning (Potts, 1996). However, Burrows and Tyrl (2001) reported that ingestion of the plant by livestock results in respiratory symptoms known as panting sickness and acute respiratory distress syndrome (ARDS), which can also lead to fatalities. Overconsumption in humans can cause similar poisoning. The plants contain a complex mixture of aromatic compounds (resulting in a pronounced disagreeable odor when the foliage is crushed), including cyanogenic glycosides, but apparently the toxicity is caused by several volatile furan ketones. As with many biochemically active plants, dosage is the key to distinguishing between a pleasant food flavorant and a toxic plant.

Over time, a complex series of infraspecific taxa has been described based on Asian materials, but many of these taxa appear to relate more to cultivated selections than to wild variants (Yu et al., 1997). Two main groups of variants generally are recognized. Plants called var. *crispa* (Benth.) W. Deane are characterized by dark brownish purple leaves with somewhat more finely toothed margins. Sometimes the leaf blades also are crisped along the margins. This variety is cultivated primarily as a food additive and for dye. Green plants with relatively coarsely toothed leaf margins are referable to var. *frutescens* and are cultivated mainly for their seed oils. Selections from both of these groups are used medicinally. However, in the United States, there appears to be a breakdown of these characters, in that plants with purplish leaves having relatively coarsely toothed margins are encountered and vice versa. In a genetic analysis of cultivated and weedy wild plants in Japan, Nitta and Ohnishi (1999) concluded that many of the wild plants may have arisen through past interbreeding of cultivated variants. For practical purposes, it does not appear reasonable to formally treat infraspecific taxa in the present work.

23. **Physostegia** Benth.
(Cantino, 1982)

Plants perennial herbs, with vertical rootstocks and sometimes slender rhizomes. Stems erect or ascending, bluntly to sharply 4-angled, usually with thickened, blunt, often pale angles, usually unbranched, glabrous or rarely minutely hairy at the very tip. Leaves progressively reduced from the stem base to its tip, sessile or the lowermost leaves short- to long-petiolate, the petiole winged, the petiolate leaves usually withered and shed by flowering time. Leaf blades variously linear to narrowly oblong, narrowly lanceolate, narrowly elliptic, or lanceolate, tapered to the base, sometimes narrowly truncate or with a pair of small, clasping auricles at the very base, tapered to a sharply pointed tip, the margins variously entire, somewhat wavy, or with bluntly or sharply pointed teeth, the surfaces glabrous, but usually with sparse, inconspicuous, sessile glands. Inflorescences terminal and occasionally also axillary, of elongate spikelike racemes, the terminal racemes sometimes in clusters of few to several from the inflorescence base, more or less continuous (except sometimes toward the base, the flowers 2 per node (solitary in the axil of each bract), sessile and short-stalked (to 2.5 mm) in the same inflorescence. Bracts 2–7 mm long, narrowly lanceolate to ovate, those at the inflorescence base occasionally to 20 mm long and linear; bractlets absent. Calyces actinomorphic or nearly so, often slightly oblique at the tip, lacking a lateral projection, more or less symmetric at the base, bell-shaped, the tube faintly to very faintly 10-nerved, 5-lobed, the lobes shorter than the tube, similar in size and shape, triangular to broadly lanceolate, not spinescent, becoming slightly enlarged and somewhat leathery at fruiting. Corollas zygomorphic, white, lavender, pale purple, pale pink, pink, or pinkish purple, the lower lip (and sometimes upper lip and throat) usually with reddish purple to purple spots or fine mottling, the outer surface glabrous or sparsely to moderately pubescent with minute hairs, the tube asymmetrically funnelform (somewhat pouched on the lower side above the base), 2-lipped, the lips shorter than the tube, the lips about the same length, the upper lip entire or shallowly notched, slightly concave, the lower lip spreading to arched, 3-lobed with a broad central lobe (this sometimes notched or slightly irregular along the margin) and 2 small lateral lobes. Stamens 4, not exserted, the lower pair with slightly longer filaments than the upper pair, all ascending under the upper corolla lip, the anthers small, the connective very short, the pollen sacs 2, parallel or nearly so, attached at their midpoints, dark purple. Ovary deeply lobed, the style appearing nearly basal from a deep apical notch. Style not exserted, more or less equally 2-branched at the tip. Fruits dry schizocarps, separating into usually 4 nutlets, these 2–4 mm long,ovoid to broadly ellipsoid, angled at the tip, 3-angled, the angles often ridged, the surface brown, glabrous, finely pebbled or smooth. Twelve species, North America.

The species of *Physostegia* are variable and can be very difficult to distinguish. Cantino's (1982) monograph of the genus included a key to species that requires 25 couplets to discriminate between only 12 species.

1. Leaf blades with the margins entire, somewhat wavy, or with irregular, very bluntly pointed teeth . . . . 2. P. INTERMEDIA
1. Leaf blades with the margins sharply toothed, the teeth coarse to relatively fine, sometimes relatively few and only toward the blade tip
    2. Inflorescence axis with 2 kinds of hairs: dense, minute hairs (visible even with 10× magnification only as minute nubs or slender tubercles to 0.05 mm long) and sparse to moderate, minute but distinctly longer (0.1–0.2 mm long); still requiring magnification to observe), slender hairs; blades of the largest leaves 3–9(–12) mm wide . . . . 1. P. ANGUSTIFOLIA
    2. Inflorescence axis uniformly pubescent with dense, minute hairs (visible even with 10× magnification only as minute nubs or slender tubercles); blades of the largest leaves (3–)10–40 mm wide (note that the largest leaves often are not collected and thus may be absent from herbarium specimens) . . . . 3. P. VIRGINIANA

1. Physostegia angustifolia Fernald (false dragonhead)

Pl. 439 a–c; Map 1978

Plants sometimes with rhizomes. Stems 40–170 cm long, with 9–18 nodes below the inflorescence. Leaves progressively shorter toward the stem tip, the foliage leaves grading into the inflorescence bracts, the inflorescences frequently often appearing elevated from the foliage, then with 2 to several pairs of short or elongate, usually widely spaced (10–50 mm apart), empty bracts. Blades of main foliage leaves 3–20 cm long, 3–9(–12) mm wide, relatively thick and stiff, the lowermost blades sometimes lanceolate to oblanceolate, those of the median and upper leaves mostly narrowly oblong-lanceolate or narrowly oblong-oblanceolate to linear, sometimes with small basal auricles that clasp the stem, more commonly angled to a truncate or abruptly rounded base as wide as or slightly wider than the stem node, the margins sharply but finely toothed, the teeth sometimes relatively few and/or only toward the blade tip. Axes of the inflorescences with 2 kinds of pubescence: dense, minute hairs (visible at 10× magnification only as minute nubs or slender tubercles to 0.05 mm long) and sparse to moderate, minute but distinctly longer (0.1–0.2 mm long), slender hairs. Bracts 4–7 mm long (except sometimes the empty basal bracts longer), mostly shorter than the calyces at flowering, lanceolate to ovate. Calyces mostly somewhat overlapping along the inflorescence axis, 6–8 mm long at flowering, becoming enlarged to 8–12 mm at fruiting, the outer surface densely pubescent with very minute hairs. Corollas (15–)20–30 mm long, white to pale lavender, occasionally pinkish-tinged or light pink. Nutlets 2.0–3.0(–3.5) mm long. $2n$=38. June–September.

Scattered, in the Unglaciated Plains, Ozark, and Ozark Border Divisions, north locally to Lincoln County (Kansas to Texas east to Illinois, Tennessee, and Georgia). Upland prairies, glades, tops of bluffs, mesic to dry upland forests, and occasionally banks of streams and bottomland forests; also ditches, roadsides, and disturbed areas.

On the whole, this is a less variable species morphologically than the closely related *P. virginiana*. Cantino (1982) noted that the pubescence of the inflorescence axis is the only stable character to separate this species from *P. virginiana* rangewide. Users of the key should not obsess over measuring the lengths of the hairs. The critical feature is that the hairs in *P. angustifolia* occur in two size classes not the absolute lengths of the two kinds of hairs. In Missouri, occasional plants of *P. virginiana* growing in fens are reduced and take on the appearance of *P. angustifolia*. Such individuals with narrow leaves and apparently well-elevated inflorescences still have the pubescence of the inflorescence axis characteristic of *P. virginiana*.

2. Physostegia intermedia (Nutt.) Engelm. & A. Gray (false dragonhead)

Pl. 439 d; Map 1979

Plants usually with rhizomes. Stems 40–120 cm long, with 9–20 nodes below the inflorescence, often somewhat swollen at the base, the basal portion also often with several closely spaced nodes (these leafless at flowering). Leaves progressively shorter toward the stem tip, the foliage leaves usually distinct from the inflorescence bracts, the inflorescences frequently often appearing elevated from the foliage, but the stalk usually lacking pairs of empty bracts. Blades of main foliage leaves 3–14 cm long, 3–14(–18) mm wide, relatively thin and flexible, narrowly lanceolate to narrowly oblong-elliptic or linear, occasionally narrowly oblanceolate, often shallowly cordate or with small basal auricles that clasp the stem, otherwise angled to a truncate or abruptly rounded base usually slightly wider than the stem node, the margins entire, somewhat wavy, or with irregular, very bluntly pointed teeth. Axes of the inflorescences with uniform, dense, very minute hairs. Bracts 2–5 mm long, mostly shorter than the calyces at flowering, lanceolate to occasionally ovate. Calyces mostly somewhat overlapping along the inflorescence axis, 3–6 mm long at flowering, becoming enlarged to 4–7 mm at fruiting, the outer surface densely pubescent with very minute hairs. Corollas 10–18 mm long (shorter or longer elsewhere), lavender to purplish pink or pinkish purple. Nutlets 2.0–3.0 mm long. $2n$=38. June–October.

Uncommon, restricted to the Mississippi Lowlands Division (Kentucky, Illinois, Missouri, Arkansas, Oklahoma, Louisiana, and Texas). Swamps and bottomland forests; also ditches, roadsides, railroads, and moist disturbed areas; sometimes emergent aquatics.

This is a less variable species morphologically than the closely related *P. virginiana*.

3. Physostegia virginiana (L.) Benth. (false dragonhead, obedient plant, Virginia lionsheart)

Pl. 439 e, f; Map 1980

Plants often with rhizomes. Stems 40–150 cm long, with 10–34 nodes below the inflorescence. Leaves progressively shorter toward the stem tip, the foliage leaves grading into the inflorescence bracts, the inflorescences appearing sessile or elevated from the foliage, then with 1 to several pairs of short or rarely elongate, closely to widely spaced (3–50 mm apart), empty bracts. Blades of main foliage leaves 2–18 cm long, (3–)10–40 mm wide, thin and flexible to thick and stiff, lanceolate to oblanceolate, oblong-lanceolate, often narrowly so, those of the lowermost leaves sometimes narrowly ovate, those of the median and upper leaves sometimes linear, sometimes with small basal auricles that clasp the stem, more commonly angled to a truncate or abruptly rounded base as wide as or slightly wider than the stem node, the margins sharply but sometimes finely toothed, sometimes mainly toward the blade tip. Axes of the inflorescences with uniform, dense, very minute hairs. Bracts 2–8 mm long (except sometimes the empty basal bracts longer), mostly shorter than the calyces at flowering, lanceolate to ovate. Calyces mostly somewhat overlapping along the inflores-

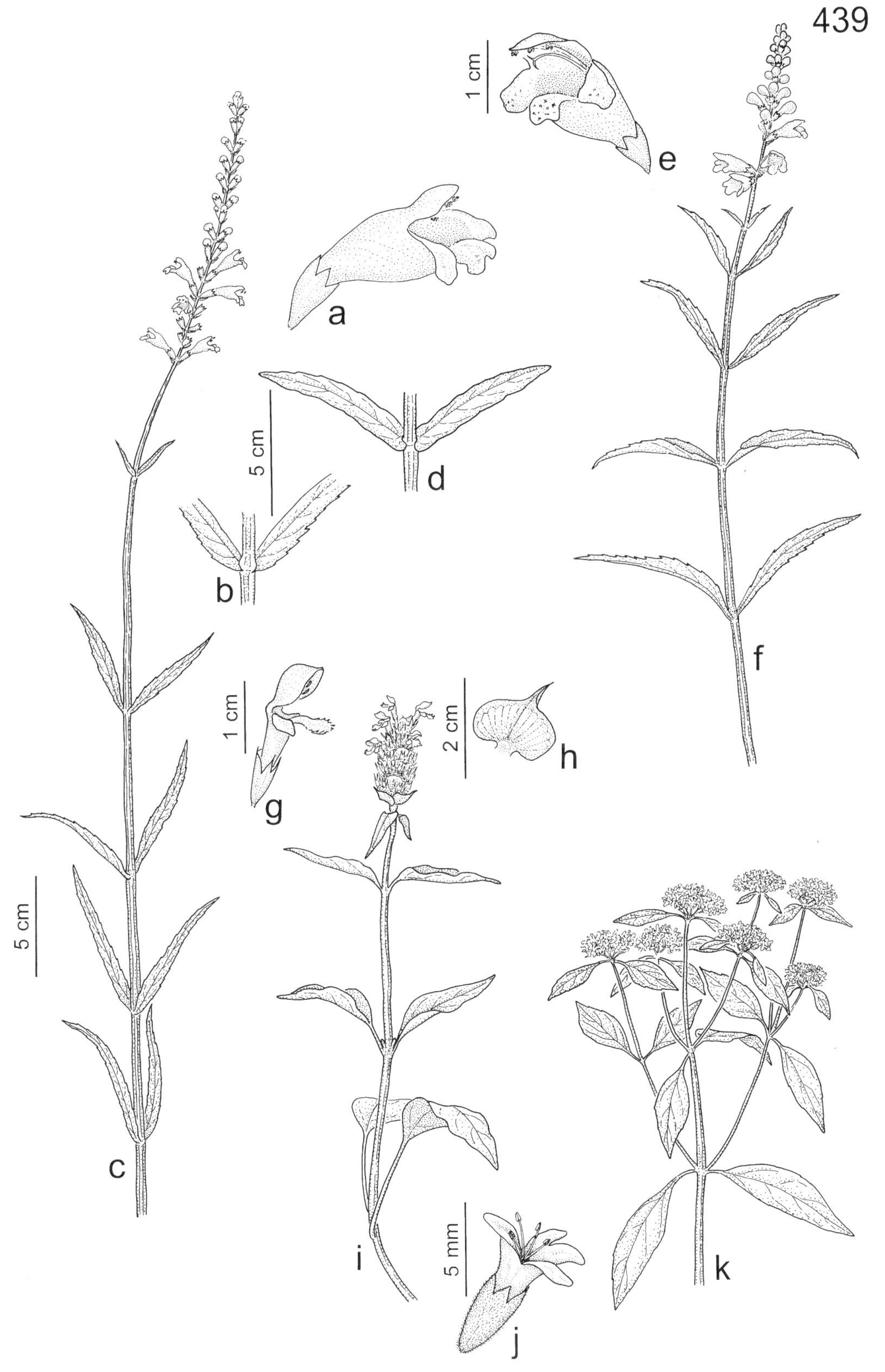

Plate 439. Lamiaceae. *Physostegia angustifolia*, **a)** flower, **b)** node with leaf bases, **c)** habit. *Physostegia intermedia*, **d)** node with leaves. *Physostegia virginiana*, **e)** flower, **f)** fertile stem. *Prunella vulgaris*, **g)** flower, **h)** bract, **i)** fertile stem. *Pycnanthemum albescens*, **j)** flower, **k)** fertile stem.

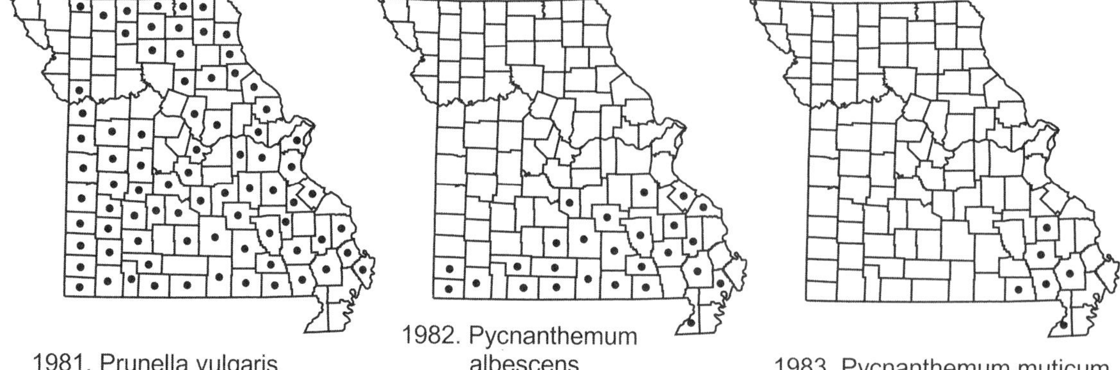

1981. Prunella vulgaris

1982. Pycnanthemum albescens

1983. Pycnanthemum muticum

cence axis, 3–9 mm long at flowering, becoming enlarged to 4–11 mm at fruiting, the outer surface densely pubescent with very minute hairs. Corollas 8–35 mm long, white to lavender, pink. or pinkish purple. Nutlets 2–4 mm long. $2n=38$. May–October.

Scattered nearly throughout the state, but uncommon in the Glaciated Plains and Mississippi Lowlands Divisions (eastern U.S. west to Montana, Utah, and New Mexico; Canada, Mexico). Glades, upland prairies, savannas, tops of bluffs, upland prairies, bottomland prairies, banks of streams, rivers, and spring branches, mesic upland forests, bottomland forests, swamps, acid seeps, and fens; also old fields, ditches, railroads, roadsides, and disturbed areas.

Physostegia virginiana is the most widespread and morphologically variable species in the genus. The Missouri plants (as keyed and described above) represent only a small portion of this variation. Cantino (1982) evaluated the several infraspecific taxa and segregate species accepted by some earlier authors and concluded that only two subspecies should be recognized. These equate roughly to southern and northern taxa, but there is a broad zone of geographic overlap (including Missouri). Steyermark (1963) segregated small-flowered plants as *P. formosior*, but Cantino (1982) indicated that these plants are part of a complex pattern of morphological variation within the species and that there exists no sharp discontinuities for corollas size within the species. Even restricting the infraspecific taxonomy to just a pair of subspecies, there are still a number of intermediate specimens. Complicating the issue is the existence of escapes from gardens. These often do not key well to subspecies. Occasional plants of *P. virginiana* growing in fens are reduced and take on the appearance of *P. angustifolia*. Such individuals with narrow leaves and apparently well-elevated inflorescences still have the pubescence of the inflorescence axis characteristic of *P. virginiana*.

1. Corollas 8–16 mm long . . . . 3B. SSP. . . . . VIRGINIANA
1. Corollas 16–35 mm long
   2. Inflorescences usually with 3 to numerous pairs of sterile bracts below the flowers, these usually at least in part closely spaced, often only 3–9 mm apart; plants often with the stems relatively densely clustered from a narrow, mostly vertical rootstock . . . . 3B. SSP. PRAEMORSA
   2. Inflorescences often lacking sterile bracts below the flowers, sometimes with 1–3 pairs of sterile bracts, these usually widely spaced, often 10–50 mm apart; plants with the stems often loosely colonial from long, branched, horizontal rhizomes . . . . 3B. SSP. VIRGINIANA

3a. ssp. praemorsa (Shinners) P.D. Cantino

P. virginiana var. *arenaria* Shimek

P. virginiana f. *candida* Benke

Plants often with the stems relatively densely clustered from a narrow, mostly vertical rootstock, the rhizomes, if produced, short, unbranched, and ascending. Blades of main foliage leaves (3–)10–25 mm wide Inflorescences occasionally lacking sterile bracts below the flower, more commonly with 3 to numerous pairs of sterile bracts, these usually at least in part closely spaced, often only 3–9 mm apart. Corollas 16–35 mm long. $2n=38$. May–October.

Scattered mostly in the Ozark and Ozark Border Divisions (New Mexico to Florida north to Nebraska, Iowa, Ohio, and Virginia; Mexico). Glades, upland prairies, savannas, tops of bluffs, banks of streams and rivers, mesic upland forests, bottomland forests, swamps, acid seeps, and fens; also roadsides and disturbed areas.

This subspecies tends to occur in drier habitats than does ssp. *virginiana*, but can also be found in wetter habitats.

3b. ssp. virginiana

P. virginiana var. *speciosa* (Sweet) A. Gray
P. formosior Lunell
P. speciosa Sweet

Plants often loosely colonial from long, branched, horizontal rhizomes. Blades of main foliage leaves (3–)10–40 mm wide. Inflorescences often lacking sterile bracts below the flowers, sometimes with 1–3 pairs of sterile bracts, these usually widely spaced, often 10–50 mm apart. Corollas 8–35 mm long. 2*n*=38. May–October.

Scattered nearly throughout the state, but uncommon in the Glaciated Plains and Mississippi Lowlands Divisions (Maine to Montana south to Kansas, Mississippi, and South Carolina; Canada). Banks of streams, rivers, and spring branches, bottomland forests, bottomland prairies, fens, and upland prairies; also old fields, ditches, railroads, roadsides, and disturbed areas.

This subspecies tends to occur in more mesic habitats than does ssp. *praemorsa*, but can also be found in drier habitats.

24. Prunella L.

Four to 7 species, nearly worldwide.

1. Prunella vulgaris L. (self-heal, heal-all)
Pl. 439 g–i; Map 1981

Plants perennial herbs, fibrous-rooted, occasionally with short rhizomes. Stems 10–50 cm long, erect or ascending to nearly prostrate, bluntly 4-angled (sometimes relatively weakly so), usually branched, glabrous or sparsely to moderately pubescent with short, ascending hairs or more or less spreading to somewhat curved, multicellular hairs, these sometimes only toward the stem tip or mostly along the angles and at the nodes. Leaves opposite and sometimes also basal, short- (toward the stem tip) to long- (toward the stem base) petiolate, the petioles often winged above the midpoint. Leaf blades 2–9 cm long, 0.7–4.0 cm wide, lanceolate to oblong-lanceolate, narrowly elliptic, ovate, or oblong-ovate, rounded, angled, or tapered at the base, angled or tapered to a bluntly or sharply pointed tip, the margins entire, shallowly wavy, or finely and sometimes irregularly toothed, the surfaces glabrous or sparsely short-hairy, the undersurface usually also with inconspicuous, sessile glands. Inflorescences terminal, dense spikes, more or less continuous, with (4)6 flowers per node, each node with a pair of leaflike bracts, these 5–15 mm long, 7–14 mm wide, depressed-ovate, abruptly short-tapered to a sharply pointed but not spinescent tip, the outer surface usually strongly purplish-tinged, the margins usually densely hairy, but not toothed or spinescent. Bractlets absent or minute, inconspicuous, and linear. Calyces zygomorphic, 6–10 mm long at flowering, 2-lipped, lacking a lateral projection, symmetric at the base, more or less cylindric to narrowly bell-shaped, the tube irregularly 10-nerved, hairy, often mainly along the nerves, glabrous in the mouth, the lips longer than the tube, the upper lip relatively broad and with 3, shallow, equal, triangular lobes at the tip, the lower lip narrower and deeply 2-lobed, the lobes narrowly triangular, the lobes all tapered to sharply pointed, soft or minutely spinescent tips, hairy along the margins, not or becoming only slightly enlarged and somewhat leathery at fruiting. Corollas 10–15(–19) mm long, zygomorphic, the lower lip with darker nerves or central spot, the outer surface glabrous or sparsely hairy, the tube funnelform, 2-lipped, the lips shorter than the tube, the upper lip entire or occasionally minutely notched, hooded, purplish blue to purple, lavender, or rarely white, often pale along the margin, the lower lip shorter than the upper lip, shallowly 3-lobed, spreading, white to pale lavender, usually with a darker line or markings on the inner surface, often somewhat irregular to shallowly fringed along the margins. Stamens 4, not or only slightly exserted (curved under the upper lip), the filaments (sometimes only the longer pair) more or less forked near the tip, only the lower forks bearing the anthers, the anthers small, the connective somewhat expanded, the pollen sacs 2, spreading, dark purple to bluish purple. Ovary deeply lobed, the style appearing nearly basal from a deep apical notch. Style not or only slightly exserted, with 2 slender branches at the tip. Fruits dry schizocarps, separating into usually 4 nutlets, these 1.6–2.0 mm long, ovoid to ellipsoid, the surface yellowish brown to brown with darker longitudinal lines that tend to converge at the ends, glabrous, smooth, shiny, the basal end with a small, pyramidal, white outgrowth (caruncle). $2n$=28, 32. May–September.

Scattered nearly throughout the state (nearly worldwide). Bottomland forests, mesic to dry upland forests, banks of streams and rivers, margins of ponds and sinkhole ponds, upland prairies, and tops of bluffs; also pastures, fencerows, old fields, ditches, railroads, roadsides, and disturbed areas.

As the vernacular name *heal* implies, this species has an extensive history of medicinal use for the treatment of most if not all ailments. *Prunella vulgaris* is a subcosmopolitan species within which numerous subspecies, varieties, and forms have been named. The plant has been transported between continents by humans, which has tended to confound attempts to create regionally based classifications to account for the morphological variation. The species deserves more detailed taxonomic studies. In North America, *P. vulgaris* can be divided into two principal varieties, one native and the other introduced from Europe. The taxonomic situation is complicated by the fact that both varieties apparently occur natively in portions of China (H. W. Li and Hedge, 1994). A. P. Nelson (1964), who studied plants from a mixed California population in the field and in common greenhouse culture concluded that the two varieties are genetically different, that they maintain themselves though high levels of self-pollination, but that limited hybridization gave rise to a small number of morphologically intermediate individuals.

1. Stem leaves lanceolate to oblong-lanceolate, 20–50% as wide as long, mostly angled or tapered at the base . . . . 1A. VAR. LANCEOLATA
1. Stem leaves ovate to oblong-ovate, 40–65% as wide as long, mostly rounded at the base . . . . 1B. VAR. VULGARIS

1a. var. lanceolata (W.P.C. Barton) Fernald
P. vulgaris var. *elongata* Benth.

Stem leaves lanceolate to oblong-lanceolate, 20–50% as wide as long, mostly angled or tapered at the base. $2n$=28, 32. May–September.

Scattered nearly throughout the state (throughout the U.S. [including Alaska]; Canada, possibly Asia). Bottomland forests, mesic to dry upland forests, banks of streams and rivers, margins of ponds and sinkhole ponds, upland prairies, and tops of bluffs; also pastures, old fields, ditches, railroads, roadsides, and disturbed areas.

White flowered plants have been called f. *candida* Fernald and those with strongly purplish-tinged calyces have been called f. *iodocalyx* Fernald.

1b. var. vulgaris

Stem leaves ovate to oblong-ovate, 40–65% as wide as long, mostly rounded at the base. $2n$=28, 32. May–September.

Scattered nearly throughout the state (native of Europe, Asia; introduced widely in the U.S., Canada). Mesic to dry upland forests and margins of sinkhole ponds; also pastures, fencerows, old fields, roadsides, and disturbed areas.

This variety is less common In Missouri than is var. *lanceolata* and tends to grow more in disturbed habitats.

25. Pycnanthemum Michx. (mountain mint)

Plants perennial herbs, with slender rhizomes. Stems erect or ascending, bluntly to sharply 4-angled, unbranched or branched mostly toward the tip, glabrous or more commonly hairy, sometimes only on the angles. Leaves sessile or short-petiolate, the petiole usually winged, at least toward the tip. Leaf blades variously linear to lanceolate, elliptic, or ovate, rounded to angled or tapered at the base (mostly rounded to cordate in *P. muticum*), angled or tapered to a usually sharply pointed tip, the margins entire or finely few-toothed, the surfaces glabrous or variously hairy, sometimes densely so, in a few species appearing whitened, the undersurface also with usually conspicuous, sessile glands. Inflorescences terminal, composed of dense, headlike (spherical or depressed-globose) clusters of numerous flowers at the stem or branch tips, these usually grouped into relatively flat-topped to broadly rounded, dense or more loosely branched panicles, the branches visible or mostly obscured. Bracts similar to the foliage leaves (except in *P. tenuifolium* with much smaller, broader bracts), but sometimes smaller. Bractlets usually 1 or 2 per flower cluster, much smaller than the bracts (except in *P. tenuifolium* with the bractlets not much smaller than the bracts), variously shaped. Calyces actinomorphic or zygomorphic and 2-lipped, lacking a lateral projection, more or less symmetric at the base, bell-shaped, the tube 10–13-nerved, 5-lobed, the lips and/or lobes shorter than the tube, similar or dissimilar in size and shape, triangular to narrowly triangular, not spinescent, not becoming enlarged or papery at fruiting. Corollas zygomorphic, white to pale pinkish-tinged or pale lavender, the lower lip with reddish purple to purple spots or mottling, the outer surface moderately to densely short-hairy, also with sessile glands, the tube funnelform, sparsely hairy in the throat, relatively deeply 2-lipped, the lobes slightly shorter than the tube, the upper lip shorter than to about as long as the lower lip, entire or shallowly notched, slightly concave, the lower lip spreading to arched, 3-lobed, the central lobe longer than the pair of lateral lobes. Stamens 4, sually exserted, the lower pair with slightly longer filaments than the upper pair, all relatively straight and not ascending under the upper corolla lip, the anthers small, the connective very short, the pollen sacs 2, parallel, usually light to dark purple. Ovary deeply lobed, the style appearing nearly basal from a deep apical notch. Style exserted, more or less equally 2-branched at the tip. Fruits dry schizocarps, separating into usually 4 nutlets, these 0.7–2.2 mm long, obovoid or oblong-ellipsoid, rounded or bluntly pointed at the tip, somewhat 3-angled, the surface dark brown to nearly black, glabrous or minutely hairy toward the tip, finely pebbled or finely pitted. Twenty species, North America.

1. Bracts subtending the inflorescences and sometimes the uppermost few pairs of foliage leaves strongly whitened on the upper surface, the lower and median leaves green or somewhat grayish on the upper surface
    2. Calyces zygomorphic, 2-lipped, the upper lip shallowly 3-lobed, the lower lip with 2 somewhat longer lobes; inflorescences relatively open, the branches mostly observable, at least at fruiting; leaf blades mostly angled to tapered at the base . . . . 1. P. ALBESCENS
    2. Calyces actinomorphic, the 5 lobes all similar in size and shape; inflorescences relatively dense, only the lowermost branches observable; leaf blades mostly rounded to cordate at the base . . . . 2. P. MUTICUM
1. Bracts subtending the inflorescences similar to the foliage leaves (except in *P. tenuifolium*, with short, relatively broad, grayish green bracts and slender, green leaves), green or grayish green on the upper surface
    3. Bracts subtending the inflorescences densely hairy on the upper surface; largest leaves with the blade 8–20 mm wide . . . . 3. P. PILOSUM
    3. Bracts subtending the inflorescences glabrous or nearly so on the upper surface (except in *P. tenuifolium*, with densely hairy bracts); largest leaves with the blade 1.5–11.0 mm wide
        4. Leaf blades glabrous, linear, those of the largest leaves 1.5–5.5 mm wide; stems glabrous or sparsely and minutely hairy on the angles . . . . 4. P. TENUIFOLIUM
        4. Leaf blades sparsely to moderately and minutely hairy along the veins on the undersurface, occasionally glabrous, lanceolate to linear-lanceolate, those of the largest leaves 5–11 mm wide; stems moderately to densely hairy on the angles, often with a mixture of straight and curled hairs
            5. Stems relatively evenly hairy on both the angles and sides; calyces with the teeth 1.0–1.5 mm long . . . . 5. P. TORREYI
            5. Stems hairy only on the angles or nearly so, the sides glabrous or rarely with a few hairs; calyces with the teeth 0.5–1.0 mm long . . . . 6. P. VIRGINIANUM

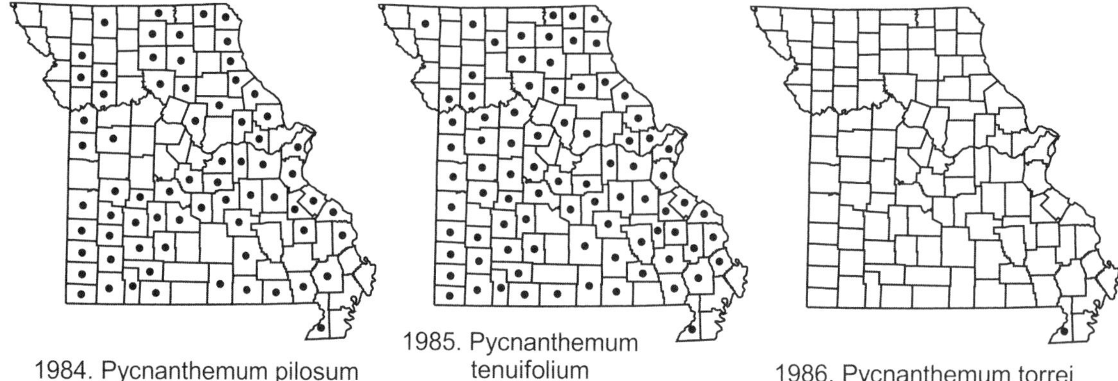
1984. Pycnanthemum pilosum

1985. Pycnanthemum tenuifolium

1986. Pycnanthemum torrei

1. Pycnanthemum albescens Torr. & A. Gray (white mountain mint)

Pl. 439 j, k; Map 1982

Stems 40–90(–150) cm long, pubescent on the angles and sides with a mixture of moderate to dense, short, curled hairs and sparser, longer, spreading hairs. Leaves sessile or short-petiolate, the largest leaves with the petioles 4–12 mm long. Leaf blades 2.5–7.0 cm long, 15–25 mm wide, ovate to broadly lanceolate or narrowly elliptic, angled or tapered at the base, the margins often finely toothed; the upper surface of the lower and median leaves sparsely to densely short-hairy and usually somewhat grayish, that of the uppermost leaves usually whitened with dense, short, curled hairs; the undersurface of all of the leaves pale or whitened with dense, short curled hairs and occasionally with a few longer, spreading hairs along the veins. Inflorescences relatively open, often appearing broadly rounded, the branches mostly observable, at least at fruiting. Bracts leaflike, whitened with dense, short, curled hairs on usually both surfaces. Bractlets 2–4 mm long, linear to narrowly lanceolate. Calyces 3.5–5.0 mm long, zygomorphic, 2-lipped, densely pubescent with minute, appressed hairs, lacking longer bristly hairs on the margins or tip, the upper lip shallowly 3-lobed, the lobes 0.4–0.6 mm long, triangular, bluntly pointed, the lower lip more deeply 2-lobed, the lobes 1.0–1.5 mm long, ovate to oblong-ovate, bluntly pointed. Corollas 5–8 mm long, white to pale lavender. Nutlets 1.0–1.4 mm long, hairy toward the tip. $2n$=38. July–September.

Scattered mostly in the Ozark and Mississippi Lowlands divisions (Kansas to Texas east to Kentucky and Florida). Mesic to dry upland forests, savannas, sand prairies, ledges of bluffs, and acid seeps; also pastures and old fields.

Where this species grows in proximity to *P. muticum* and *P. pilosum*, rare putative hybrids have been collected. E. Grant and Epling (1943) noted the existence of two races within *P. albescens*, with plants in the Gulf Coastal portion of the species range having stems with uniformly short, curled hairs, this pubescence type grading into plants with a mixture of such shorter hairs and longer, spreading ones. Missouri plants are typical of the latter race.

2. Pycnanthemum muticum (Michx.) Pers. (clustered mountain mint)

Pl. 440 a, b; Map 1983

Stems 40–110 cm long, moderately to densely pubescent on the angles and sides with mostly short, curled hairs, occasionally with a few longer, spreading hairs. Leaves sessile or nearly so. Leaf blades 3–8 cm long, 15–40 mm wide, ovate to broadly lanceolate or oblong-ovate, mostly rounded to cordate at the base, the margins usually finely toothed, the upper surface glabrous or sparsely short-hairy, mostly along the veins (rarely the uppermost pair whitened with dense, short, curled hairs), usually green, the undersurface usually green, glabrous or sparsely to moderately short-hairy, mostly along the veins. Inflorescences relatively dense, often appearing broadly rounded, often appearing sessile or nearly so, only the lowermost branches observable. Bracts leaflike, whitened with dense, short, curled hairs on the upper surface, glabrous or nearly so on the undersurface. Bractlets 1–3 mm long, linear to narrowly lanceolate. Calyces 3–5 mm long, actinomorphic, densely pubescent with minute, appressed hairs, mostly above the midpoint, lacking longer bristly hairs on the margins or tip, the lobes all similar in size and shape 0.5–1.5 mm long, narrowly triangular, sharply pointed. Corollas 4–7 mm long, white to pale pinkish-tinged or pale lavender. Nutlets 1.0–1.4 mm long, glabrous. $2n$=ca. 108. July–September.

Uncommon in the Mississippi Lowlands Division (eastern U.S. west to Missouri and Texas).

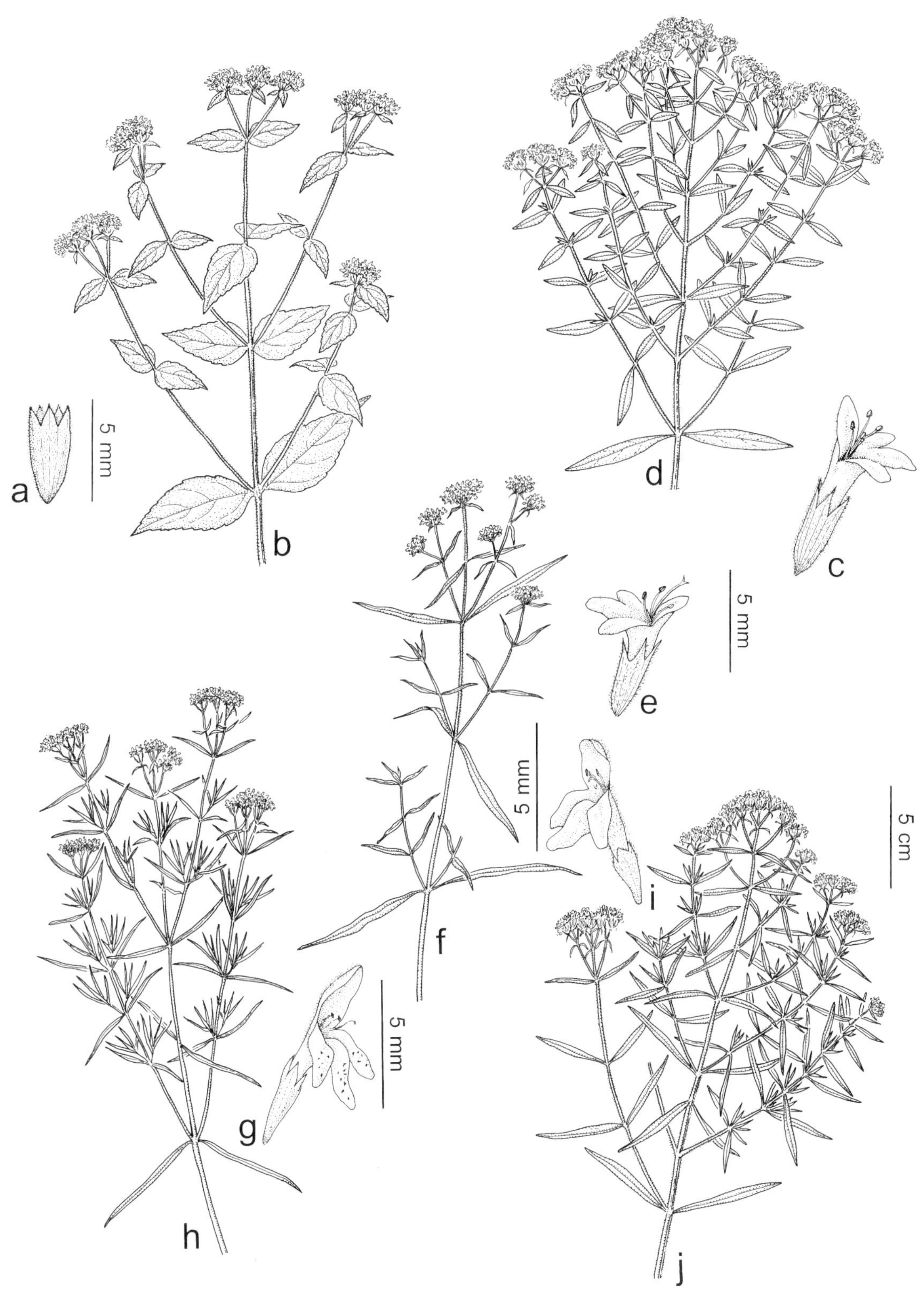

Plate 440. Lamiaceae. *Pycnanthemum muticum*, **a)** fruiting calyx, **b)** fertile stem. *Pycnanthemum pilosum*, **c)** flower, **d)** fertile stem. *Pycnanthemum torreyi*, **e)** flower, **f)** fertile stem. *Pycnanthemum tenuifolium*, **g)** flower, **h)** fertile stem. *Pycnanthemum virginianum*, **i)** flower, **j)** fertile stem.

Bottomland forests; also ditches, roadsides, and grassy disturbed areas.

Where this species grows in proximity to *P. albescens*, rare putative hybrids have been collected.

3. Pycnanthemum pilosum Nutt. (hairy mountain mint)

P. verticillatum (Michx.) Pers. var. *pilosum* (Nutt.) Cooperr.

Pl. 440 c, d; Map 1984

Stems 50–150 cm long, moderately to densely pubescent on the angles and sides with mostly short, spreading hairs. Leaves sessile or very short-petiolate, the largest leaves with the petioles 1–3 mm long. Leaf blades 2–8 cm long, 4–20 mm wide (those of the largest leaves 8–20 mm), lanceolate, elliptic-lanceolate, elliptic, or oblong-elliptic, mostly narrowly angled or tapered at the base, the margins usually entire, less commonly with a few shallow teeth, the upper surface glabrous or sparsely short-hairy, green, the undersurface grayish green, moderately to densely pubescent with relatively long, spreading hairs. Inflorescences relatively dense, often appearing broadly rounded, only the lowermost branches observable. Bracts similar to the foliage leaves, not whitened, but usually grayish green, densely pubescent with short, curled hairs on the upper surface, moderately to densely pubescent with relatively long, spreading hairs on the undersurface. Bractlets 3–5 mm long, linear to narrowly lanceolate. Calyces 3.5–4.5 mm long, actinomorphic or nearly so, densely pubescent with relatively long, woolly hairs, lacking longer bristly hairs on the margins or tip, the lobes all similar in size and shape 0.5–1.0 mm long, triangular, sharply pointed and usually with a minute, bristly extension of the midnerve. Corollas 5–8 mm long, white to pale lavender. Nutlets 1.0–1.3 mm long, sparsely hairy toward the tip. $2n$=ca. 76–78. July–September.

Scattered nearly throughout the state (eastern U.S. west to Nebraska and Texas; Canada). Mesic to dry upland forests, upland prairies, savannas, glades, ledges and tops of bluffs, banks of streams and rivers, margins of lakes, and occasionally bottomland forests; also old fields, cemeteries, railroads, roadsides, and open, disturbed areas.

Where this species grows in proximity to *P. albescens*, rare putative hybrids have been collected. Steyermark (1963) noted that hairy mountain mint is a favorite bee plant.

This species is very closely related to *P. verticillatum* of the eastern United States. Some authors have treated the two as varieties of that species (Gleason and Cronquist, 1991). According to E. Grant and Epling (1943), true *P. verticillatum* differs from *P. pilosum* in its stems that are hairy only on the angles and in its densely hairy bracts. The leaves of *P. verticillatum* also tend to be more sparsely hairy on the undersurface with the hairs confined mostly to the veins.

4. Pycnanthemum tenuifolium Schrad. (slender mountain mint)

Pl. 440 g, h; Map 1985

Stems 40–80(–100) cm long, glabrous or rarely sparsely and minutely hairy on the angles. Leaves sessile. Leaf blades 1.5–5.5 cm long, 1.0–5.5 mm wide (those of the largest leaves 1.5–5.5 mm), linear, angled or short-tapered at the base, the margins entire, the surfaces glabrous, green. Inflorescences relatively dense, often appearing relatively flat-topped, only the lowermost branches observable. Bracts dissimilar to the foliage leaves (2.5–4.0 mm long, lanceolate to narrowly ovate, with a thickened midvein and a nearly spinescent tip), not whitened, but grayish green, densely pubescent with short, curled hairs on the surfaces and longer, bristly hairs on the margins and tip. Bractlets 2–3 mm long, mostly lanceolate. Calyces 3.5–5.0 mm long, actinomorphic or nearly so, densely or occasionally sparsely pubescent with short, curled hairs below the lobes, lacking longer bristly hairs on the margins, the lobes all similar in size and shape 1.0–1.5 mm long, triangular, sharply pointed and sometimes with a minute, sharply pointed extension of the midnerve. Corollas 5–7 mm long, white or less commonly pale lavender. Nutlets 0.7–1.0 mm long, usually glabrous. $2n$=80. June–September.

Scattered nearly throughout the state (eastern U.S. west to Minnesota, Nebraska, and Texas; Canada). Mesic to dry upland forests, savannas, glades, bottomland prairies, upland prairies, banks of streams and rivers, fens, and ledges and tops of bluffs; also ditches, pastures, old fields, railroads, and roadsides.

Gleason and Cronquist (1991) noted that this species does not produce a strong odor when the foliage is crushed.

5. Pycnanthemum torreyi Benth. (Torrey's mountain mint)

Pl. 440 e, f; Map 1986

Stems 40–80(–120) cm long, moderately to densely and evenly pubescent on the angles and sides with a mixture of spreading and curled hairs. Leaves sessile or very short-petiolate, the largest leaves with the petioles 1–4 mm long. Leaf blades 3–9 cm long, 5–11 mm wide, lanceolate to linear-

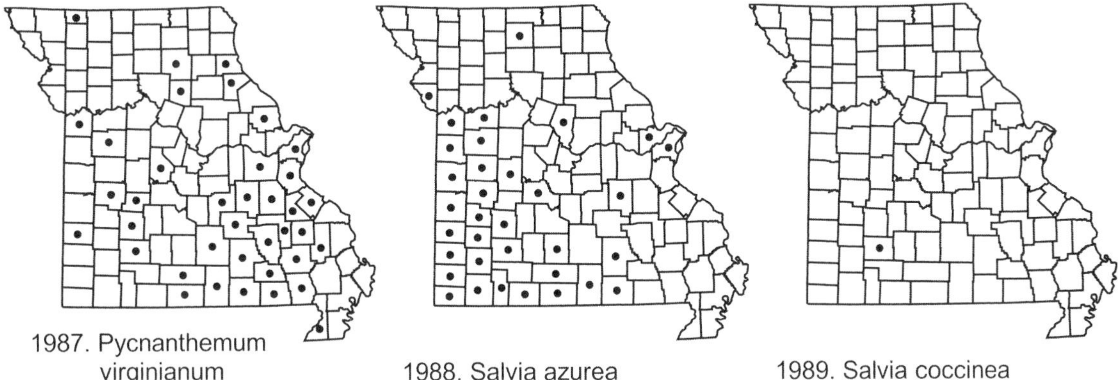

1987. Pycnanthemum virginianum

1988. Salvia azurea

1989. Salvia coccinea

lanceolate, angled or short-tapered at the base, the margins entire or occasionally with a few teeth, roughened with minute, angled hairs, the surfaces glabrous or the undersurface sparsely short-hairy along the midvein, green. Inflorescences relatively dense, often appearing relatively flat-topped, only the lowermost branches observable. Bracts similar to the foliage leaves, not whitened, green, densely pubescent with short, curled hairs on the upper surface, moderately to densely pubescent with relatively long, spreading hairs on the undersurface. Bractlets 3–5 mm long, lanceolate, with a thickened midvein, densely short-hairy along the margins and sometimes with a minute, somewhat spinescent tip. Calyces 4–5 mm long, nearly actinomorphic, densely pubescent, mostly above the midpoint, with short, spreading hairs below the lobes and spreading hairs along the lobe margins, the lobes more or less similar in size, but the 3 upper lobes slightly more fused at the base than the 2 lower ones, 1.0–1.5 mm long, triangular, sharply pointed and sometimes with a minute, sharply pointed extension of the midnerve. Corollas 4.0–6.5 mm long, white to pale pinkish-tinged or pale lavender. Nutlets 1.0–2.2 mm long, glabrous. $2n$=78–80. June–October.

Uncommon, known from a single historical collection from Dunklin County (New Hampshire to Illinois south to Georgia, Tennessee, and Missouri). Habitat unknown (dry, open woods, according to Steyermark [1963]).

E. Grant and Epling (1943) mapped this species mostly from a small region from Connecticut to West Virginia, but noted a small number of disjunct occurrences, based on a few specimens from South Carolina and adjacent Georgia and single specimens from Missouri and Kansas. Since then, additional records have accumulated from intervening states, but the species is considered extirpated or of conservation concern in most of the states in which it occurs. It should be noted that the name has been spelled *P. torrei* in most of the botanical literature, but more recently, this original spelling has been considered an orthographic error that is correctable to *P. torreyi* (the eipthet honors the famous nineteenth century North American botanist, John Torrey [1796–1873]).

Steyermark (1963) noted the existence of a second specimen that he referred to "*P. torreyi* var. *leptodon* (A. Gray) Boomhour" (a combination that has not been validly published). This sheet is accessioned at the Gray Herbarium of Harvard University and bears the label data: W. Missouri, December 1870, *Geyer s.n.* ex Herb. Engelmann). The so-called var. *leptodon* refers to rare plants in Virginia, West Virginia, and North Carolina with longer hairs on the stems and calyces than in var. *torreyi* and denser pubescence on the undersurface of the leaf blades. E. Grant and Epling (1943) included the basionym *P. leptodon* A. Gray in their list of uncertain names and putative hybrids, and suggested that it might represent a hybrid or polyploid derivative involving *P. virginianum*. Steyermark (1963) also chose to exclude this specimen from the Missouri flora, suggesting that it actually was not collected in the state of Missouri. Most likely, the information was written on the label in error after the specimen arrived at the Gray Herbarium, or the specimen was mixed up with some other sheet in a pile of *Pycnanthemum* prior to labeling. Adding to the potential confusion, E. Grant and Epling (1943) discussed occasional specimens from Missouri (and elsewhere) in which the stem pubescence was somewhat longer than typical for *P. virginianum* and distributed at least in part on the sides as well as the angles. They noted that such specimens are similar in appearance to *P. torreyi*, but more likely represent hybrids between *P. pilosum* and either *P. tenuifolium* (as *P. flexuosum*) or *P. virginianum*.

6. Pycnanthemum virginianum (L.) B.L. Rob. & Fernald (Virginia mountain mint)

Pl. 440 i, j; Map 1987

Stems 40–90(–120) cm long, moderately to densely and evenly pubescent on the angles or nearly so, the angles with mostly short, curled hairs, the sides glabrous or rarely with a few hairs. Leaves sessile. Leaf blades 2–6 cm long, 5–11 mm wide, lanceolate to linear-lanceolate, angled or short-tapered at the base, the margins entire or rarely with a few teeth, the surfaces glabrous or more commonly the undersurface sparsely short-hairy along the midvein, green. Inflorescences relatively dense, often appearing relatively flat-topped, only the lowermost branches observable. Bracts similar to the foliage leaves, not whitened, green, glabrous or sparsely short-hairy on the upper surface. Bractlets 2–4 mm long, lanceolate to narrowly ovate, lacking a thickened midvein, densely short-hairy toward the sharply pointed but non-spinescent tip. Calyces 3–5 mm long, actinomorphic or nearly so, densely pubescent, mostly above the midpoint, with short, curled hairs, the lobes all similar in size, 0.5–1.0 mm long, triangular, sharply pointed, lacking a sharply pointed extension of the midnerve. Corollas 4.0–6.5 mm long, white to pale pinkish-tinged or pale lavender. Nutlets 1.0–2.2 mm long, glabrous. $2n=80$. July–October.

Scattered, mostly in the Ozark and Ozark Border Divisions, uncommon in other parts of the state (eastern U.S. west to North Dakota, Colorado, and Oklahoma; Canada). Banks of streams, rivers, and spring branches, ledges of bluffs, fens, swamps, bottomland prairies, uplands prairies, and occasionally glades; also railroads; commonly on calcareous substrates.

E. Grant and Epling (1943) reported scattered specimens with foliage similar to that of *P. pilosum* but infloresacences more similar to those of *P. virginianum* that they interpreted as putative hybrids between the two species. For a discussion of uncommon specimens resembling *P. virginianum* but with stems having somewhat denser hairs on the sides, see the treatment of *P. torreyi*.

26. **Salvia** L. (sage, salvia)

Plants annual or perennial herbs (shrubs elsewhere), with taproots or somewhat thickened or woody rootstocks. Stems erect or ascending, bluntly to sharply 4-angled, unbranched or branched, glabrous or variously hairy. Leaves petiolate (sometimes the uppermost leaves sessile or nearly so), the petiole unwinged or winged, with a usually strong fragrance when crushed. Leaf blades variously shaped, unlobed to deeply pinnately lobed, the margins otherwise entire or more commonly toothed, the surfaces hairy, also with inconspicuous or sometimes conspicuous, sessile glands. Inflorescences terminal spikes or spikelike racemes, these sometimes grouped into loose clusters at their bases, the individual axes with more than 5 clusters, these relatively open, more or less continuous along the axis, each with 1–12 flowers, these sessile or short-stalked, subtended by a pair of short bracts, the individual flowers sometimes subtended by inconspicuous bractlets. Calyces zygomorphic, lacking a lateral projection, more or less symmetric at the base, bell-shaped, the tube strongly 10–15-nerved (-ribbed), glabrous in the mouth, hairy externally, 2-lipped, the lobes shorter than the tube, the upper lip unlobed or with 3 shallow, triangular lobes, the lower lip with 2 narrowly triangular lobes, the lobes sharply pointed and sometimes with a short, slender extension of the midnerve, but not spinescent, not or only slightly becoming enlarged (except in *S. reflexa*) or papery at fruiting. Corollas zygomorphic, white, red, lavender, or blue (pink or purple elsewhere), lacking markings on the lower lip or in some darker-flowered species with white mottling, the outer surface moderately to densely pubescent with minute, spreading hairs, especially on the upper lip, the tube funnelform to nearly cylindric, glabrous in the throat, shallowly to deeply 2-lipped, the upper lip usually shallowly notched at the tip (occasionally entire), either relatively flat (straight to slightly spreading) or hooded, the lower lip noticeably longer and broader than the upper lip, spreading or arched, 3-lobed, the middle lobe longer and broader than the lateral lobes, often shallowly notched at the tip. Stamens 2, not or only slightly exserted (more strongly exserted in *S. coccinea*), the anthers overall relatively large, the connective slender and elongate, the upper pollen sac fertile and fully formed, the lower sac either absent or reduced and nonfunctional (except sometimes in *S. lyrata*), the connective appearing attached at or below its midpoint, the fertile sac often yellow. Ovary deeply lobed, the style appearing nearly basal from a deep apical notch. Style not exserted (ascending under the upper lip), unequally 2-branched at the tip. Fruits dry schizocarps, separating into usually 4 nutlets, these 1.5–3.0 mm long, oblong-obovoid, rounded at the tip, the surface brown to dark brown, glabrous, smooth, minutely pebbled, or with dense, minute tubercles. About 900 species, nearly worldwide.

The genus *Salvia* is usually easily recognized by its characteristic, strongly 2-lipped calyces and often bright, showy corollas. Numerous species are cultivated as ornamentals in gardens (see Compton [1994] for a review of some of the Mexican species that are popular in gardens). The South American *S. splendens* Sellow ex Wied-Neuw. (scarlet sage) is among the most popular of annual bedding plants in Missouri. A number of salvias also have medicinal uses. *Salvia hispanica* L., *S. officinalis* L., and *S. fruticosa* Mill. are the main species grown as culinary herbs, the sages whose dried or fresh leaves are used to add flavor to cooked dishes. A few other species are used in beverages (mainly in teas. Some of the sages provide a laquerlike oil that has been used in the manufacture of oil paints. Sage is a popular ingredient in soaps, perfumes, toiletry products, sachets, and potpourris; the pleasantly scented *S. sclarea* L. is commonly used in these products and also is used as a flavorant in vermouth and some other liquors, as well as in an eye wash. *Salvia divinorum* Epling & Játiva (diviner's sage) has a long history of cultivation in Latin America as a hallucinogen; that use has spread to the United States in recent years and a number of states are in the process of making its possession illegal.

The unusual structure of the stamens and the associated pollination mechanism was first identified in 1793, but was reviewed by Claßen-Bockhoff et al. (2003). The two stamens develop before the stigmas. They are situated vertically in the throat of the corolla, with the fertile pollen sacs arching under the upper corolla lobe and the sterile arms of the connectives (these sometimes bearing a fertile pollen sac in *S. lyrata* and a few other species) positioned downward across the tube (and frequently becoming fused into a paddle-shaped structure). A pollinator (variously insects or hummingbirds) entering the flower moves the sterile arms backward, causing a leverlike action that pushes the fertile pollen sacs onto its back (or head), where a load of pollen is deposited. After the pollen has been shed, the stamens wither and the style elongates and arches downward, placing the stigmatic branches in a position to intercept the pollen on a pollinator that previously has visited a different flower during that flower's functionally staminate phase.

The unique staminal architecture has been used as a key character in defining the genus *Salvia* taxonomically. However, recent phylogenetic studies using molecular markers (Walker et al., 2004) have suggested that, as traditionally circumscribed, the genus is not a natural group, but instead consists of three independent species groups that are not particularly closely related within the tribe Mentheae. Under this scenario, the remarkable pollination mechanism described above developed independentally at least twice within this tribe of Lamiaceae, and the genus overall likely will be split into three genera in the future. The Missouri species would be classified in the genus *Calosphace* Raf., but new combinations will need to be published under that generic name.

Steyermark (1963) included *S. sclarea* L. (clary, clear-eye), based on a single historical collection made in 1910 by Earl Sherff in the city of St. Louis (with no further locality or habitat data). In the introduction to his *Flora of Missouri*, Steyermark discussed Sherrf's St. Louis collections, many of which were made from cultivated plants but not indicated as such, and he excluded a number of species from the Missouri flora that were documented only from Sherff's ambiguous collections. He included *S. sclarea* based on the anticipation that the species would be discovered by subsequent botanists elsewhere in the state. This optimistic prediction has not been realized. Thus, *S. sclarea* is here excluded from the Missouri flora. The species is a robust biennial with large ovate leaves that often are 6 cm or more wide and corollas often with yellowish markings in the throat (Pl. 441 g–i).

1. Stamens and stigmas strongly exserted; corollas red. . . . . 2. S. COCCINEA
1. Stamens and stigmas not or only slightly exserted; corollas white, lavender, or blue
   2. Leaves all or mostly basal, the stem leaves absent or only 1 or 2 pairs; leaf blades often pinnately lobed . . . . 3. S. LYRATA
   2. Leaves all or mostly on the stems, more than 2 pairs; leaf blade unlobed, the margins merely toothed
      3, Corollas 10–25 mm long, the tube much longer than the calyx; stems moderately to densely pubescent with short, curved hairs, at least toward the tip; plants perennial, with a thickened, often somewhat woody wootstock . . . . 1. S. AZUREA
      3. Corollas 6–10 mm long, the tube only slightly longer than the calyx; stems glabrous or sparsely pubescent wth short, curved hairs; plants annual, taprooted . . . . 4. S. REFLEXA

1. Salvia azurea Lam. **var. grandiflora** Benth. (blue sage)

S. azurea ssp. *pitcheri* (Torr. ex Benth.) Epling

Pl. 441 a, b; Map 1988

Plants perennial herbs, the rootstock thick and often somewhat woody. Stems 50–150 cm long, often wandlike, densely pubescent with short, curved hairs, sometimes more sparsely hairy toward the base, sometimes also with scattered, longer, spreading hairs. Basal and lower stem leaves usually absent at flowering, short- to long-petiolate, the blade 3–7 cm long, obovate to oblong-obovate or oblanceolate. Stem leaves 3 to several pairs, short-petiolate, the blade 2–8 cm long, linear-lanceolate to lanceolate or oblanceolate, tapered at the base, angled to a sharply pointed tip, the margins finely toothed to nearly entire, the upper surface sparsely pubescent with short, curved hairs or nearly glabrous, the undersurface densely pubescent with appressed or curved, somewhat felted, short hairs. Bracts usually persistent, 2–8 mm long, narrowly lanceolate to linear. Calyces 4–7 mm long at flowering, becoming elongated to 6–10 mm at fruiting, densely short-hairy, the upper lip entire, the lower lip shallowly 2-lobed. Corollas 10–25 mm long, the outer surface densely short-hairy, blue (rarely white, often with white or pale markings on the lower lip), the tube much longer than the calyx, the upper lip strongly hooded, entire or very shallowly notched at the tip, the lower lip with relatively well-developed lateral lobes and a broadly fan-shaped central lobe, this relatively deeply notched at the tip. Stamens and style not exserted. Nutlets 2.0–2.8 mm long, the surface brown, smooth or minutely pebbled, sometimes with sessile glands. $2n=20$. July–September.

Scattered, mostly in the Unglaciated Plains Division and the western half of the Ozarks (Utah and New Mexico east to Minnesota and Louisiana, also sporadically eastward to Georgia, disjunct in Illinois, Tennessee; introduced farther north to Indiana, Wisconsin, Michigan, and northeast to New York and Connecticut). Upland prairies, glades, ledges and tops of bluffs, savannas, openings of dry upland forests, and rarely banks of streams and rivers; also old fields, fencerows, railroads, roadsides and open, disturbed areas.

This attractive species is grown as an ornamental in wildflower gardens and especially in wildflower meadows.

Salvia azurea generally is considered to comprise two infraspecific taxa, recognized variously as varieties (Gandhi and Thomas, 1983) or subspecies (K. M. Peterson, 1978). The other variety, var. *azurea*, occurs in the southeastern United States and differs in its stems, which are glabrous or have sparse, asending to spreading hairs. K. M. Peterson (1978) noted that the native distribution of the species is difficult to circumscribe, as the plant has been cultivated as an ornamental for a long time and has escaped both within its native range and outside of it. Nonnative occurrences can be difficult to distinguish from native ones. White-flowered plants occur rarely within otherwise blue-flowered populations and have been called f. *albiflora* McGregor.

2. Salvia coccinea Buc'hoz ex Etl. (red sage, Texas sage)

Map 1989

Plants annual, with taproots. Stems 30–50 (–80) cm long, densely pubescent with short, curled hairs, also with sparse to moderate, longer, spreading, multicellular hairs, especially toward the tip. Basal leaves absent at flowering. Stem leaves 3 to several pairs, short- to long-petiolate, the blade 1–7 cm long, ovate to ovate-triangular, broadly rounded to truncate or cordate at the base, angled or tapered to a bluntly or sharply pointed tip, the

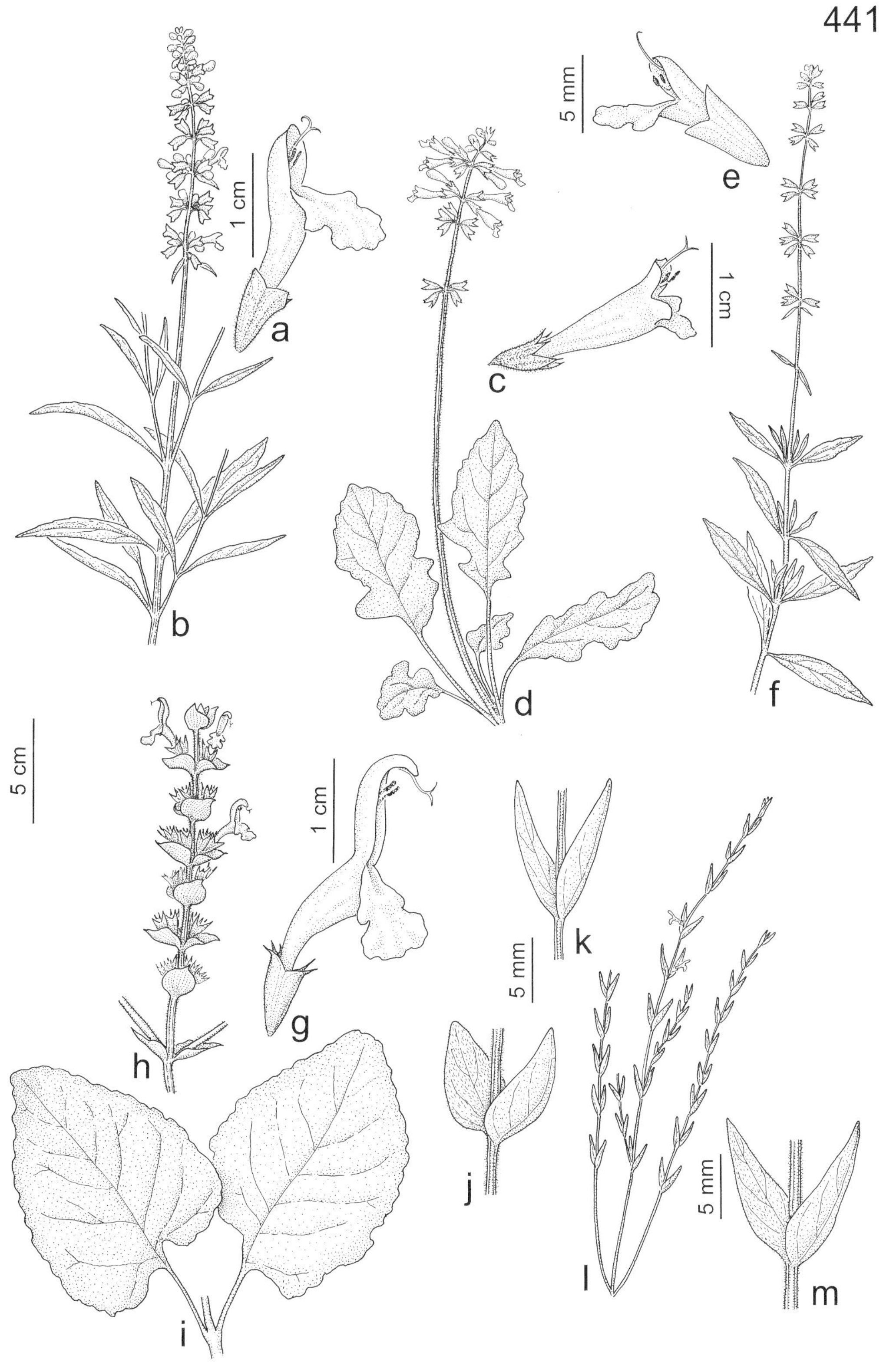

Plate 441. Lamiaceae. *Salvia azurea*, **a)** flower, **b)** fertile stem. *Salvia lyrata*, **c)** flower, **d)** habit. *Salvia reflexa*, **e)** flower, **f)** fertile stem. *Salvia sclarea*, **g)** flower, **h)** fertile stem, **i)** node with leaves. *Scutellaria parvula* var. *australis*, **j)** node with leaves. *Scutellaria parvula* var. *missouriensis*, **k)** node with leaves. *Scutellaria parvula* var. *parvula*, **l)** fertile stem, **m)** node with leaves.

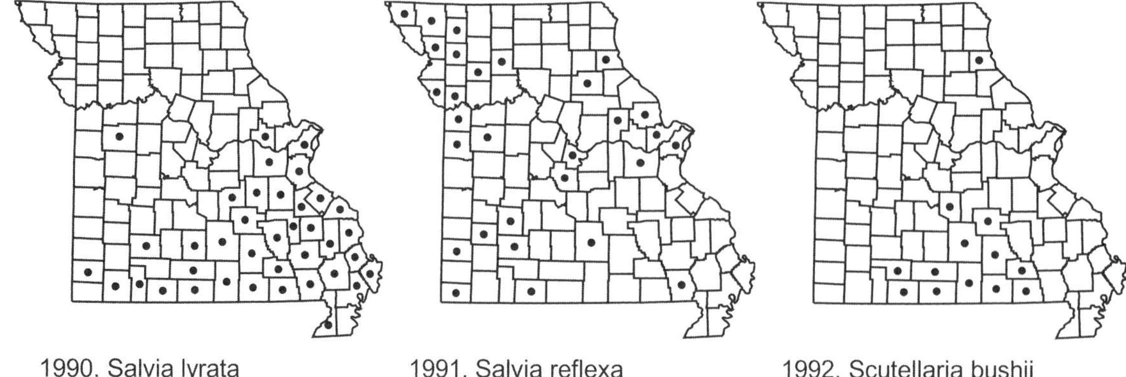

1990. Salvia lyrata 1991. Salvia reflexa 1992. Scutellaria bushii

margins finely to relatively coarsely, bluntly to sharply toothed, the upper surface green, moderately short-hairy, the undersurface grayish green, densely pubescent with short, woolly hairs. Bracts shed early, 6–10 mm long, lanceolate to ovate. Calyces 6–8 mm long at flowering, becoming elongated to 8–10 mm at fruiting, moderately to densely short-hairy, the upper lip entire, the lower lip shallowly 2-lobed. Corollas 20–25(–30) mm long, the outer surface densely pubescent with short, spreading, multicellular hairs, bright red (pink or white elsewhere), the tube much longer than the calyx, the upper lip slightly hooded, entire and truncate to broadly rounded at the tip, the lower lip with relatively well-developed lateral lobes and a depressed-oval central lobe, this usually relatively deeply notched at the tip. Stamens and style relatively strongly exserted. Nutlets 2.3–2.8 mm long, the surface dark brown, smooth or minutely pebbled. $2n=20$. May–October. $2n=24$. July–September.

Introduced, uncommon, known thus far only from Greene County (native of Latin America; introduced from Texas to South Carolina, north locally in Missouri and Ohio). Open, disturbed areas.

This garden ornamental was discovered in 2004 by Paul Redfearn growing as an escape from cultivation in the city of Springfield. Its native distribution is not well understood, as it became widespread in Latin America a long time ago. Epling (1938–1939) suggested Brazil as its country of origin, but Compton (1994) thought that Mexico was more likely. In the southeastern United States, it often grows along dry, open woodland margins where it appears native in spite of its nonnative status.

3. Salvia lyrata L. (lyre-leaved sage, cancerweed)

Pl. 441 c, d; Map 1990

Plants perennial, with a short, thick rootstock. Stems 30–50(–70) cm long, sparsely pubescent with short, more or less spreading hairs. Basal leaves several to many at flowering, the petiole 20–70 mm long, the blade 6–15 cm long, oblanceolate to elliptic, oblong-obovate, or oblong-elliptic in outline, the first set of leaves deeply pinnately lobed with a large terminal lobe and several pairs of smaller, spreading, rounded to broadly pointed, lateral lobes, the later leaves progressively less divided, sometimes unlobed and more or less spatulate, the margins otherwise entire or with a few, irregular, blunt teeth, the surfaces glabrous or sparsely to moderately short-hairy. Stem leaves absent or only 1 or 2 pairs, sessile or short-petiolate, the blade similar to those of the basal leaves, but only 2–7 cm long. Bracts usually persistent, 4–7 mm long, linear to lanceolate. Calyces 6–9 mm long at flowering, becoming elongated to 8–12 mm at fruiting, moderately short-hairy below the lobes, the upper lip shallowly 3-lobed, the lower lip more deeply 2-lobed. Corollas 18–30 mm long, the outer surface sparsely to moderately short-hairy mostly on the tube, pale lavender to pale blue or white, the tube much longer than the calyx, the upper lip usually folded longitudinally but relatively straight, truncate, and not hooded, entire, the lower lip with relatively small lateral lobes and a depressed-oval to depressed-obovate central lobe, this usually shallowly and broadly notched at the tip. Stamens and style not or only slightly exserted. Nutlets 1.7–2.2 mm long, the surface dark brown, with dense, minute tubercles, otherwise glabrous. $2n=36$. April–June.

Scattered, mostly in the Ozark, Ozark Border, and Mississippi Lolwands Divisions (eastern U.S. west to Kansas and Texas). Bottomland forests, mesic upland forests, banks of streams and rivers, margins of fens, bases, ledges, and tops of bluffs, and occasionally glades; also pastures, lawns, cemeteries, railroads, roadsides, and disturbed areas.

This attractive species sometimes is sold by wildflower nurseries, both for its interesting flowers and as a groundcover for its unusual basal leaves.

4. Salvia reflexa Hornem. (lance-leaved sage, Rocky Mountain sage)

Pl. 441 e, f; Map 1991

Plants annual, with taproots. Stems 20–70 cm long, glabrous or sparsely pubescent wth short, curved hairs. Basal leaves absent at flowering. Stem leaves 3 to several pairs, mostly short-petiolate, the blade 2–6 cm long, lanceolate to narrowly oblong or elliptic, occasionally ovate, angled or tapered at the base, rounded or angled to a bluntly pointed tip, the margins entire or sparsely and often bluntly toothed, the upper surface glabrous or sparsely pubescent with short, curved hairs along the main veins, the undersurface glabrous or sparsely to moderately pubescent with short, curved hairs. Bracts usually persistent, 2–6 mm long, lanceolate to narrowly ovate. Calyces 4–6 mm long at flowering, becoming elongated to 6–8 mm at fruiting, moderately short-hairy along the nerves, the upper lip entire, the lower lip shallowly 2-lobed. Corollas 6–10 mm long, the outer surface densely short-hairy, blue to pale blue or white, the tube only slightly longer than the calyx, the upper lip strongly hooded, entire or very shallowly notched at the tip, the lower lip with relatively well-developed lateral lobes and a depressed-oval central lobe, this not or only slightly notched at the tip. Stamens and style not exserted. Nutlets 2.0–2.5 mm long, the surface brown, smooth or minutely pebbled. $2n=20$. May–October.

Widely scattered in the state, most commonly in the western half (Ohio to Montana south to Texas and Arizona; Canada, Mexico; introduced farther west and east). Banks of streams and rivers, margins of ponds and lakes, marshes, bases and ledges of bluffs, bottomland prairies, openings of bottomland forests, and openings and edges of mesic upland forests; also pastures, old fields, edges of crop fields, railroads, roadsides, and open, disturbed areas.

Although this species has smaller flowers than does *S. azurea*, it is still an attractive plant for wildflower gardens and meadows.

27. Scutellaria L. (skullcap)

Plants perennial herbs (annuals or shrubs elsewhere), sometimes with slender rhizomes. Stems erect or ascending, sometimes from a spreading base, bluntly to sharply 4-angled, unbranched or branched, glabrous or variously hairy. Leaves petiolate or sessile, the petiole unwinged or narrowly winged, usually lacking a strong fragrance when crushed. Leaf blades variously shaped, unlobed, the margins entire or variously toothed, the surfaces glabrous or more commonly hairy, the undersurface or sometimes both surfaces often with inconspicuous sessile glands. Inflorescences axillary and/or terminal, the flowers 2 per node, either solitary in the axils of leaves or solitary in the axils of the bracts of slender, usually elongate racemes, these sometimes appearing in groups of 3 at the stem tip, the pair of flowers at each node short-stalked, subtended by a pair of short, leaflike bracts; bractlets absent or minute and inconspicuous. Calyces zygomorphic, with a prominent, transverse, ridgelike or shieldlike projection extending from below the midpoint on the upper side of the calyx tube, otherwise more or less symmetric at the base and bell-shaped, the tube not or faintly 5-nerved, glabrous in the mouth, often becoming closed apically after the corolla has been shed, variously hairy externally, 2-lipped, the lobes shallow and much shorter than the tube, similar in size and shape, both unlobed and broadly rounded, not spinescent, becoming slightly enlarged and leathery or papery at fruiting. Corollas strongly zygomorphic, light blue to blue, bluish purple, or purple (pink or other colors elsewhere), often with a white tube, rarely entirely white, the lower lip lacking markings or with irregular white mottling and sometimes darker bluish purple to blue spots, the outer surface moderately to densely pubescent with glandular and/or nonglandular hairs, the tube often somewhat S-shaped in outline (except in *S. lateriflora* with a nearly stright corolla tube), narrowly funnelform above the often abruptly upward-curved basal portion, glabrous or hairy in the throat (the apparent hairs in some species attached to the filaments rather than the corolla), shallowly to moderately 2-lipped, the upper lip entire or very shallowly notched at the tip, deeply hooded, the lower lip noticeably longer and broader than the upper lip, spreading or arched, 3-lobed, the lateral lobes usually relatively short, broad, and rounded, usually partially fused to the upper lip (and sometimes appearing more part of the upper lip than the lower one), the middle lobe often longer and/or broader than the lateral lobes, often somewhat folded or distorted at the base to more or less close the corolla throat, sometimes shallowly notched at the tip. Stamens 4, not exserted (ascending under/in the upper corolla lip), the anthers of the slightly shorter upper pair with 2 fertile pollen sacs, those of the lower pair with only a single fertile pollen sac, the anthers small, the connective short, the pollen sacs (of the upper stamens) slightly angled to nearly parallel. Ovary deeply lobed, the style appearing nearly basal from a deep apical notch. Style not or only slightly exserted (ascending under/in the upper lip), very unequally 2-branched at the tip (one of the branches sometimes absent or nearly so). Fruits dry schizocarps, separating into 1–4 nutlets, these 0.9–2.0 mm long, broadly depressed-obovoid to globose, rounded at the tip, the surface yellowish brown or more commonly dark brown to black, glabrous, finely warty or with prominent, blunt tubercles. About 350 species, nearly worldwide.

Some species of *Scutellaria* are cultivated as ornamentals and a small number of species have medicinal value. The genus is easily recognized by the combination of the calyces with a characteristic projection on one side near the base and the corollas with the upper lip strongly hooded (helmetlike). Steyermark (1963) is well-known for his description of the calyces of Missouri species of *Scutellaria* as helmetlike or resembling tractor seats, but regrettably in today's world few students or botanists have experience with tractors and thus do not know the appearance of a tractor seat (and most nonmilitary helmets today are shaped differently).

1. Inflorescences of axillary and/or terminal racemes
    2. Racemes all or mostly axillary; corollas with the tube nearly straight, not noticeably S-shaped in outline . . . . 5. S. LATERIFLORA
    2. Racemes all or mostly terminal; corollas with the tube noticeably S-shaped in outline, bent upward just above the calyx and strongly curved or oblique at or above the throat
        3. Largest leaves with the base of the blade broadly rounded to truncate or cordate, the petiole not winged at the tip; corollas lacking a ring of hairs in the throat (to observe this it is necessary to open the corolla) . . . 7. S. OVATA
        3. Largest leaves with the base of the blade angled or tapered at the base, or, if rounded to slightly cordate, then the petiole winged toward the tip; corollas with a ring of hairs in the throat (to observe this it is necessary to open the corolla)
            4. Calyces with the outer surface sparsely to densely pubescent with spreading, gland-tipped hairs, not becoming closed at fruiting . . . . 2. S. ELLIPTICA
            4. Calyces with the outer surface densely pubescent with appressed, nonglandular hairs, becoming closed at fruiting . . . . 4. S. INCANA
1. Inflorescences of axillary flowers, these 2 per node (solitary in the axils of full-sized foliage leaves)
    5. Leaf blades with the margins entire
        6. Leaf blades (1.5–)2.0–4.0 cm long, narrowly oblong to narrowly oblanceolate, tapered at the base; corollas 20–28 mm long . . . . 1. S. BUSHII
        6. Leaf blades 0.5–2.0 cm long, lanceolate to ovate, broadly ovate, or more or less triangular-ovate, rounded to truncate or shallowly cordate at the base, occasionally broadly angled; corollas 6–11 mm long . . . . . 8. S. PARVULA
    5. Blades of at least the larger leaves with the margins toothed
        7. Petioles 5–30 mm long; corollas with the tube nearly straight, not noticeably S-shaped in outline . . . . 5. S. LATERIFLORA

7. Petioles absent or to 4 mm long; corollas with the tube noticeably S-shaped in outline, bent upward just above the calyx and strongly curved or oblique at or above the throat
   8. Corollas 15–25 mm long; leaf blades mostly 2–4 times as long as wide . . . . 3. S. GALERICULATA
   8. Corollas 8–11 mm long; leaf blades mostly 1–2 times as long as wide . . . . 6. S. NERVOSA

1. Scutellaria bushii Britton (Bush's skullcap)

Pl. 442 a, b; Map 1992

Plants with knotty, woody rootstocks. Stems 20–50 cm long, erect or ascending, unbranched or few-branched, moderately to densely pubescent with short, upward-curved, nonglandular hairs. Leaves sessile or nearly so. Leaf blades (1.5–)2.0–4.0 cm long, narrowly oblong to narrowly oblanceolate, tapered at the base, rounded at the tip, the margins entire, the surfaces sparsely to moderately pubescent with short, curved hairs, also with dense sessile glands. Inflorescences of axillary flowers, these 2 per node, solitary in the axils of the upper foliage leaves. Calyces 4.0–5.5 mm long, becoming closed and enlarged to 6–7 mm at fruiting, the outer surface sparsely to moderately pubescent with short, curved hairs and sessile glands. Corollas 20–28 mm long, minutely glandular-hairy on the outer surface, deep blue to blue, the lower lip usually with a white patch or mottling toward the base, the tube S-shaped (bent upward just above the calyx and strongly curved or oblique at or above the throat), the lateral lobes relatively well-developed, ascending, the lower lip broadly fan-shaped, deeply notched at the tip. Nutlets 1 or 2(3) per calyx, 1.0–1.5 mm long, more or less globose, the surface dark brown to black, densely warty. May–June.

Scattered in the Ozark Division, with a single disjunct locality in Marion County (Arkansas, Missouri). Glades, savannas, and ledges and tops of bluffs; also roadsides; usually on calcareous substrates.

This attractive species is nearly endemic to the Ozarks. It is drought-resistent and grows in rocky situations. Bush's skullcap has only recently begun to gain attention in wildflower nurseries.

2. Scutellaria elliptica Spreng. (hairy skullcap)

S. ovalifolia Pers.

Pl. 442 c, d; Map 1993

Plants sometimes with short rhizomes. Stems 20–60(–80) cm long, erect or ascending, usually from a short, spreading base, unbranched or occasionally few-branched, moderately to densely pubescent with short, upward-curved or spreading, nonglandular or gland-tipped hairs, sometimes also with sessile glands. Leaves with the petioles 5–25 mm long, winged at the tip. Leaf blades 1.5–8.0 cm long, the lowermost heart-shaped to ovate-triangular, broadly rounded to truncate or cordate (but with a wedge of tissue in the notch extending down the petiole tip) at the base, rounded to more commonly bluntly pointed at the tip, the median and upper leaves ovate to rhombic-ovate or elliptic, angled or short-tapered at the base, rounded to bluntly or more commonly sharply pointed at the tip, all leaf blades with the margins finely to coarsely and bluntly toothed, the upper surface sparsely to moderately pubescent with short, more or less spreading, straight, sometimes multicellular, nonglandular or gland-tipped hairs, the undersurface sparsely to moderately pubescent with short, curved or straight, nonglandular or gland-tipped hairs, often mostly along the veins, both surfaces also with sessile glands. Inflorescences of slender racemes, these mostly terminal, sometimes in a cluster of 3 from the stem tip, the flowers 2 per node, solitary in the axils of bracts, the bracts 4–9(–12) mm long, narrowly lanceolate to narrowly elliptic or narrowly oblanceolate. Calyces 2–3 mm long, becoming enlarged to 6–8 mm at fruiting, but usually not becoming closed at maturity, the outer surface sparsely to densely pubescent with short, spreading, multicellular, gland-tipped hairs. Corollas 12–21 mm long, moderately to densely pubescent with minute, spreading, gland-tipped hairs on the outer surface, usually also with sessile glands, light blue or bluish purple above a usually white tube, the lower lip usually lighter-colored than the upper lip, with a white patch or mottling toward the base, the tube S-shaped (bent upward just above the calyx and strongly curved or oblique at or above the throat), with a ring of hairs in the throat, the lateral lobes not well-developed, ascending, the lower lip broadly fan-shaped, shallowly to moderately notched at the tip. Nutlets 1–4 per calyx, 1.5–2.0 mm long, more or less globose, the surface dark brown, with dense, low, rounded tubercles. May–July.

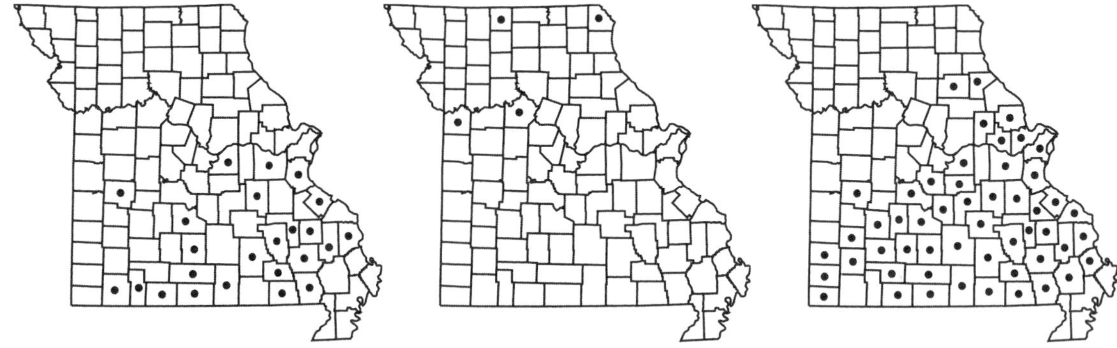

1993. Scutellaria elliptica 1994. Scutellaria galericulata 1995. Scutellaria incana

Scattered in the Ozark and Ozark Border Divisions (eastern U.S. west to Kansas and Texas). Mesic to dry upland forests, glades, ledges of bluffs, and banks of streams, rivers, and spring branches.

Steyermark (1963) included *S. montana* Chapm. as (*S. serrata* Andrews var. *montana* (Chapm.) Penland), based on a single historical specimen from Iron County. This species otherwise is endemic to portions of Alabama, Georgia, and Tennessee, and is distinguished from *S. elliptica* by its relatively large (2.4–3.3 cm) corollas and narrowly elliptic to narrowly oblanceolate bracts. Although the Missouri collection was not examined during his doctoral research, J. Leo Collins (Tennessee Valley Authority, pers. comm.) subsequently redetermined it as *S. elliptica*.

J. L. Collins (1976) treated *S. elliptica* as comprising three varieties, two of which occur in Missouri. However, his third variety, which segregates Ozarkian plants in Missouri, Arkansas, and Oklahoma having gland-dotted stems, was never published. It appears to represent a minor variant of var. *elliptica*.

1. Stems with the pubescence of short, curved, nonglandular hairs .... 2A. VAR. ELLIPTICA
1. Stems with the pubescence of relatively long, spreading, mostly gland-tipped hairs .... 2B. VAR. HIRSUTA

2a. var. elliptica

Stems moderately to densely pubescent with relatively short (mostly 0.5–1.0 mm), upward-curved hairs, sometimes also dotted with sessile glands. Leaf blades with sparse, nonglandular hairs. May–July.

Scattered in the Ozark and Ozark Border Divisions (eastern U.S. west to Kansas and Texas). Mesic to dry upland forests, ledges of bluffs, and banks of streams, rivers, and spring branches.

2b. var. hirsuta (Short) Fernald

Stems moderately to densely pubescent with relatively long (mostly 1–2 mm), spreading hairs. Leaf blades with sparse, spreading gland-tipped hairs, those of the undersurface mostly along the veins. May–July.

Uncommon, sporadic (eastern U.S. west to Missouri and Texas). Mesic to dry upland forests and glades.

3. Scutellaria galericulata L. (marsh skullcap)

S. epilobiifolia A. Ham.

Pl. 442 e, f; Map 1994

Plants with slender, inconspicuous rhizomes and/or stolons. Stems 20–80 cm long, erect or ascending, unbranched or few- to several-branched, glabrous or sparsely pubescent with short, downward-curved, nonglandular hairs. Leaves sessile or the lower leaves with petioles to 4 mm long. Leaf blades 1–4 cm long, mostly 2–4 times as long as wide, oblong-lanceolate to narrowly ovate, broadly rounded to truncate or shallowly cordate at the base, sharply pointed at the tip, the margins with shallow, usually rounded teeth, the upper surface glabrous, the undersurface moderately to densely pubescent with short, curved, nonglandular hairs, also with sessile glands. Inflorescences of axillary flowers, these 2 per node, solitary in the axils of the upper foliage leaves. Calyces 3–4 mm long, becoming closed and enlarged to 4–6 mm at fruiting, the outer surface densely pubescent with short, curved, nonglandular hairs. Corollas 15–25 mm long, densely pubescent with short, spreading, nonglandular hairs on the outer surface, blue or white, the lower lip with a white patch and purplish blue spots or mottlihng toward the base, the tube S-shaped (bent upward just above the calyx and strongly curved or oblique at or above the throat), with a ring of hairs in the throat, the lat-

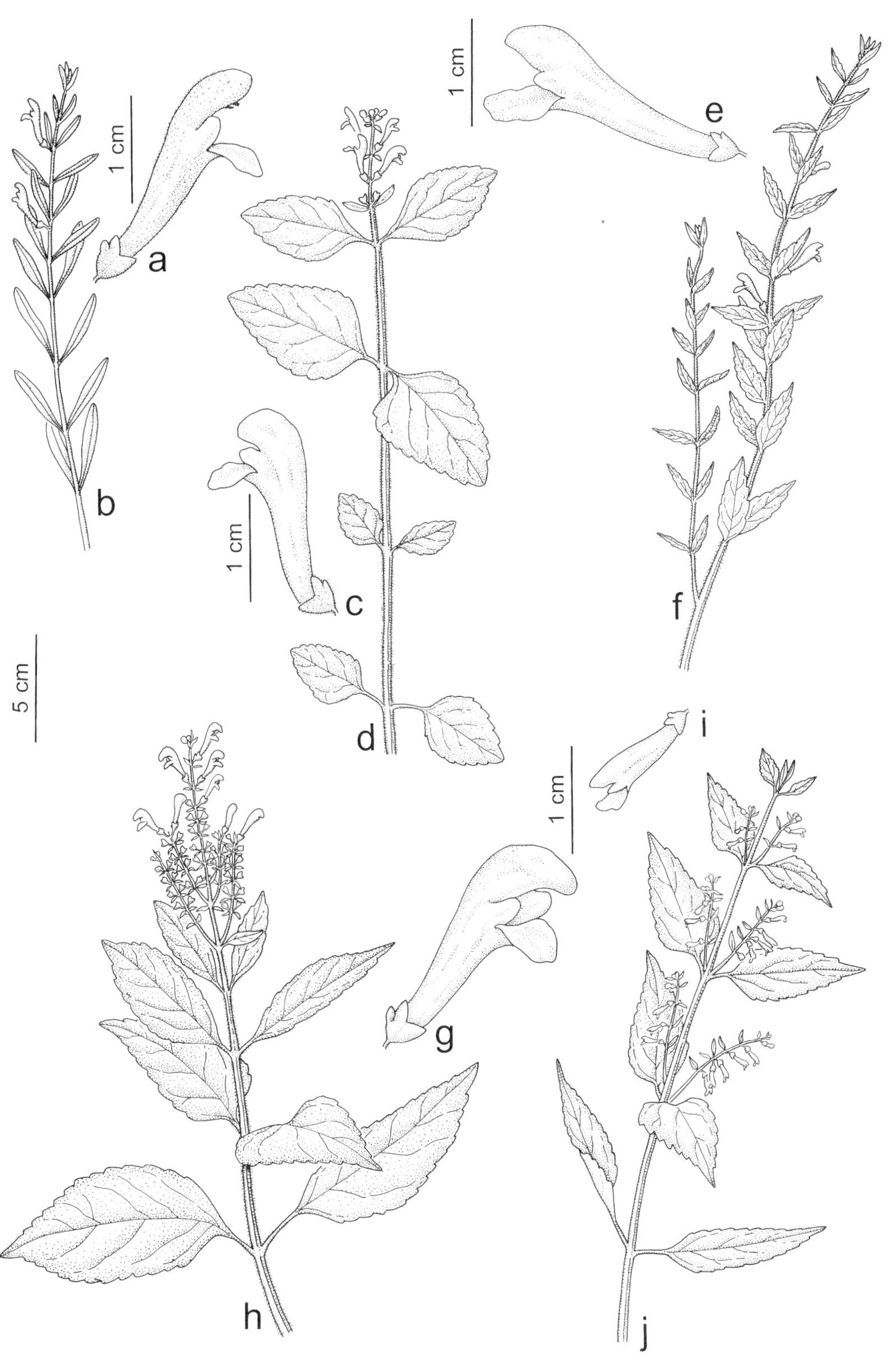

Plate 442. Lamiaceae. *Scutellaria bushii*, **a)** flower, **b)** fertile stem. *Scutellaria elliptica*, **c)** flower, **d)** fertile stem. *Scutellaria galericulata*, **e)** flower, **f)** fertile stem. *Scutellaria incana*, **g)** flower, **h)** fertile stem. *Scutellaria lateriflora*, **i)** flower, **j)** fertile stem.

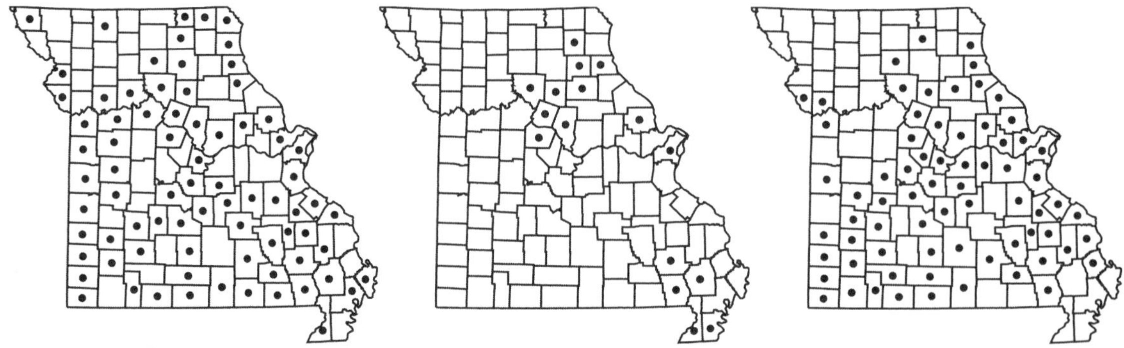

1996. Scutellaria lateriflora 1997. Scutellaria nervosa 1998. Scutellaria ovata

eral lobes not well-developed, ascending, the lower lip broadly fan-shaped to more or less semicircular, slightly irregular and sometimes slightly notched along the margin. Nutlets 1–4 per calyx, 1.2–1.6 mm long, more or less globose, the surface yellowish brown, densely covered with rounded tubercles, sometimes also with sessile glands. $2n=30$, 32. June–September.

Uncommon, sporadic in the northern half of the state (nearly throughout the U.S. [including Alaska but excluding some southewastern states]; Canada, Greenland, Europe, Asia). Marshes and fens.

The stems of this species often are relatively weak and supported by the surrounding vegetation. Although the flowers are showy, they tend to be produced sparsely and the plants thus are easily overlooked.

The name *S. epilobiifolia* was adopted by some botanists to distinguish North American plants from those in the Old World. However, Epling (1942) studied a large set of specimens and determined that there were no discrete morphological differences between populations in the Old and New Worlds. This situation may need to be revisited, as Old World plants appear to have a different chromosome number ($2n=30$) than those in North America ($2n=32$).

4. Scutellaria incana Biehler (downy skullcap, hoary skullcap)

S. incana var. *punctata* (Chapm.) C. Mohr

Pl. 442 g, h; Map 1995

Plants sometimes with short rhizomes. Stems 40–80(–120) cm long, erect or ascending, sometimes from a short, spreading base, unbranched or occasionally few-branched, densely pubescent with short, upward-curved, nonglandular hairs, often also with sessile glands. Leaves with the petioles 5–30 mm long, winged toward the tip. Leaf blades 4–11 cm long, ovate to elliptic (the uppermost narrowly so), angled or tapered at the base, those of the lowermost leaves occasionally asymmetrically rounded or shallowly cordate (but with a wedge of tissue in the notch extending down the petiole tip), angled or tapered to a sharply pointed tip, the margins finely to occasinally coarsely and more or less bluntly toothed, the upper surface green and sparsely to moderately pubescent with minute, curved, nonglandular hairs, mostly along the main veins or toward the blade base, the undersurface grayish green and moderately to densely pubescent with short, curved and more or less straight, nonglandular hairs, more densely so along the veins, both surfaces also with sessile glands. Inflorescences of slender racemes, these mostly terminal, sometimes in a cluster of 3 from the stem tip, the flowers 2 per node, solitary in the axils of bracts, the bracts 4–6 mm long, narrowly elliptic to lanceolate. Calyces 2–3 mm long, becoming closed and enlarged to 5–6 mm at fruiting, the outer surface densely pubescent with short, appressed to somewhat woolly, nonglandular hairs, sometimes also with sessile glands. Corollas 15–25 mm long, moderately to densely pubescent with minute, appressed to somewhat woolly, nonglandular hairs on the outer surface, often also with sessile glands, blue to bluish purple or purple above a usually white tube, the lower lip with a white patch toward the base, the tube S-shaped (bent upward just above the calyx and strongly curved or oblique at or above the throat), lacking a ring of hairs in the throat, the lateral lobes relatively well-developed, ascending, the lower lip broadly fan-shaped to nearly semicircular, usually shallowly notched at the tip. Nutlets 1–4 per calyx, 1.5–2.0 mm long, more or less globose, the surface dark brown, finely warty or with dense, low, rounded tubercles. $2n=30$. June–September.

Scattered, mostly in the Ozark and Ozark Border Divisions, north locally to Monroe and Pike Counties (eastern U.S. west to Iowa, Kansas, and Texas). Mesic to dry upland forests, banks of streams, rivers, and spring branches, fens, bases, ledges, and tops of bluffs; also ditches and roadsides.

Steyermark (1963) and some other authors have attempted to separate plants from throughout the species range with the leaf blades densely hairy on the undersurface (var. *incana*) form those in the southeastern United States with the pubescence sparser and mainly confined to the veins (var. *punctata*). Although striking in their extremes, these variants are not sharply distinct, and numerous intermediates exist.

5. Scutellaria lateriflora L. (mad dog skullcap)
Pl. 442 i, j; Map 1996

Plants with slender rhizomes. Stems 15–60 (–100) cm long, ascending, sometimes from a spreading base, usually branched, glabrous or sparsely pubescent on the angles with short, upward-curved, nonglandular hairs. Leaves with the petioles 5–30 mm long. Leaf blades 1–11 cm long, lanceolate to ovate, rounded to truncate or shallowly cordate at the base, sharply pointed at the tip, the margins finely to relatively coarsely toothed, the surfaces glabrous or the undersurface sparsely pubescent with short, appressed or curved, nonglandular hairs, the surfaces usually lacking sessile glands. Inflorescences of slender racemes, these mostly axillary, occasionally reduced to solitary axillary flowers, the flowers 2 per node, solitary in the axils of bracts or foliage leaves, the bracts 8–13 mm long and narrowly ovate toward the raceme base, progressively shorter and narrower toward the tip. Calyces 1.5–2.5 mm long, becoming closed and enlarged to 3–4 mm at fruiting, the outer surface moderately to densely pubescent with minute, curved, nonglandular hairs. Corollas 5–8 mm long, densely pubescent with minute, nonglandular hairs on the outer surface, pale blue or light bluish purple, rarely white, the lower lip usually lacking spots or mottling, the tube not S-shaped (nearly straight above the calyx, somewhat oblique at or above the throat), the lateral lobes not well-developed, ascending, the lower lip relatively short, oblong, usually very slightly notched at the tip. Nutlets 1–4 per calyx, 1.0–1.3 mm in diameter, depressed-globose or broadly obovoid, the surface yellowish brown, densely warty or with low, rounded tubercles. $2n$=88. June–October.

Scattered nearly throughout the state (nearly throughout the U.S. [including Alaska]; Canada). Banks of streams and rivers, margins of ponds, lakes, and sinkhole ponds, bottomland forests, bases of bluffs, sloughs, marshes, and swamps; also ditches; occasionally epiphytic on floating logs, hummocks of *Carex*, and the lower trunks of *Taxodium*.

Rare plants with white corollas have been called f. *albiflora* (Farw.) Fernald.

6. Scutellaria nervosa Pursh (veiny skullcap)
S. nervosa var. *calvifolia* Fernald
Pl. 443 a, b; Map 1997

Plants with slender, inconspicuous rhizomes. Stems 15–50(–60) cm long, erect or ascending, often from a spreading base, unbranched or few-branched, glabrous or sparsely pubescent on the angles and nodes with short, upward-curved, nonglandular hairs. Leaves sessile or the lower leaves with petioles to 4 mm long. Leaf blades 1–3 cm long, mostly 1–2 times as long as wide, narrowly to broadly ovate, rounded to truncate or shallowly cordate at the base or occasionally broadly angled, rounded or bluntly to sharply pointed at the tip, the margins of at least the larger leaves with broadly spaced, shallow, usually rounded teeth, the surfaces glabrous or sparsely pubescent with short, more or less spreading, nonglandular hairs, the undersurface usually also with sessile glands. Inflorescences of axillary flowers, these 2 per node, solitary in the axils of the upper foliage leaves. Calyces 2.5–3.5 mm long, becoming closed and enlarged to 5–7 mm at fruiting, the outer surface sparsely to moderately pubescent along the nerves with short, curved, nonglandular hairs. Corollas 8–11 mm long, minutely nonglandular-hairy on the outer surface, pale blue or white, the lower lip usually finely spotted with white or reddish purple toward the base, the tube not or only slightly S-shaped (relatively straight above the calyx, oblique at or above the throat), the lateral lobes not well-developed, ascending, the lower lip broadly fan-shaped, irregular or faintly scalloped along the margin. Nutlets 1–4 per calyx, 1.0–1.2 mm long, more or less globose, the surface dark brown to black, densely warty or with rounded tubercles, interrupted by a broad, smooth, transverse band. $2n$=20. May–July.

Scattered, mostly in the eastern half of the state (eastern U.S. west to Iowa and Louisiana; Canada). Banks of streams, rivers, and spring branches, margins of ponds and lakes, bottomland forests, bases of bluffs, sloughs, and swamps.

Steyermark (1963) and some other authors have attempted to separate this species into two varieties, based on whether the upper leaf surface is glabrous (var. *calvifolia*) or sparsely hairy (var. *nervosa*). There is no discontinuity for this character, and nearly glabrous-leaved plants are commonly encountered. Rare plants with white corollas have been called f. *alba* Steyerm.

7. Scutellaria ovata Hill (heart-leaved skullcap, egg-leaved skullcap)

Pl. 443 c; Map 1998

Plants with slender rhizomes. Stems (10–)25–80 cm long, erect or ascending, usually unbranched, densely pubescent with spreading, gland-tipped hairs. Leaves with the petioles 8–50 mm long, not winged at the tip. Leaf blades 1.5–7.0 cm long, heart-shaped to ovate, narrowly ovate, or triangular-ovate, broadly rounded to truncate or cordate at the base, bluntly or more commonly sharply pointed at the tip, the margins finely to relatively coarsely toothed, the surfaces densely pubescent with relatively long (and sometimes also shorter), spreading to somewhat curved, multicellular, mostly gland-tipped hairs, the undersurface sometimes also with sessile glands. Inflorescences of slender racemes, these mostly terminal, sometimes in a cluster of 3 from the stem tip, the flowers 2 per node, solitary in the axils of bracts, the bracts 3–9(–12) mm long, ovate to broadly ovate, sometimes finely few-toothed. Calyces 3–4 mm long, becoming closed and enlarged to 4–6 mm at fruiting, the outer surface densely pubescent with spreading, multicellular, mostly gland-tipped hairs. Corollas 17–25 mm long, densely pubescent with short, spreading, gland-tipped-hairs on the outer surface, pale blue to blue or bluish purple above a usually white tube, the lower lip variously white with bluish purple markings or blue to bluish purple with white and purple mottling and/or spots, the tube S-shaped (bent upward just above the calyx and strongly curved or oblique at or above the throat), lacking a ring of hairs in the throat, the lateral lobes not well-developed, ascending, the lower lip broadly fan-shaped, deeply notched at the tip. Nutlets 1–4 per calyx, 1.2–1.5 mm in diameter, depressed globose or broadly obovoid, the surface dark brown, densely warty or with low, rounded tubercles, these reddish brown to orangish brown. $2n=20$. May–October.

Scattered nearly throughout the state, but uncommon in the western portion of the Glaciated Plains Division and the Mississippi Lowlands (eastern U.S. west to Minnesota and Texas). Bottomland forests, mesic to dry upland forests, glades, banks of streams and rivers, and bases, ledges, and tops of bluffs; also old fields, railroads, and roadsides.

The infraspecific taxonomy of this species remains controversial. Epling (1942) recognized a confusing series of 12 subspecies differing only slightly morphologically and overlapping extensively geographically. At the other extreme, some botanists (Lane, 1986) have suggested that the species might best be treated as a single polymorphic taxon with no infraspecific entities. In his unpublished doctoral dissertation, Pittman (1988) performed a numerical analysis of morphological variation within *S. ovata* and concluded that three subspecies were supportable statistically. He further segregated four minor variants as varieties, but these seem scarcely worthy of attention. The subspecies circumscriptions proposed by Pittman are followed in the present work with some reservations, as many intermediate plants exist for each of the characters.

1. Bracts extending past the calyces of the flowers they subtend, occasionally even longer than the flowers .... 7A. SSP. BRACTEATA
1. Bracts not extending past the calyces of the flowers they subtend (except rarely in some late-season inflorescences)
   2. Corollas with the lower lip white with blue spots or fine mottling and sometimes a narrow blue band along the margins; leaf blades usually relatively smooth on the upper surface, the veins not strongly impressed; stems usually more than 40 cm long .... 7B. SSP. OVATA
   2. Corollas with the lower lip blue with a pair of longitudinal white bands along the midvein, these often with small blue spots or fine mottling; leaf blades usually with relatively strong topography on the upper surface (appearing puckered), the veins usually strongly impressed; stems usually less than 40 cm long .... 7C. SSP. RUGOSA

7a. ssp. bracteata (Benth.) Epling

S. ovata var. *bracteata* (Benth.) S.F. Blake

S. ovata ssp. *versicolor* (Nutt.) Epling, misapplied

Plants relatively robust, the stems mostly 30–80 cm long. Leaf blades mostly 3–7 cm long, not appearing strongly puckered, the veins slightly to moderately impressed on the upper surface, slightly to moderately raised on the undersurface. Inflorescences witrh the bracts extending past the calyces of the flowers they subtend, occasionally even longer than the flowers. Corollas with the lower lip blue to bluish purple, usually with white and purple mottling and/or spots. $2n=20$. May–October.

Scattered nearly throughout the state, but uncommon in the western portion of the Glaciated Plains Division and the Mississippi Lowlands

Plate 443. Lamiaceae. *Scutellaria nervosa*, **a)** flower, **b)** fertile stem. *Scutellaria ovata*, **c)** fertile stem. *Stachys aspera*, **d)** flower, **e)** fertile stem. *Stachys tenuifolia*, **f)** flower, **g)** node with leaves. *Stachys pilosa*, **h)** flower, **i)** fertile stem.

1999. Scutellaria parvula 2000. Stachys aspera 2001. Stachys hispida

(Kansas to Texas east to North Carolina and Florida). Mesic to dry upland forests, glades, banks of streams and rivers, and bases, ledges, and tops of bluffs; also old fields and roadsides.

Pittman (1988) studied typification in the group and concluded that the names *S. ovata* and *S. versicolor* Nutt. both represent the same taxon. The name *S. ovata* var. *versicolor* thus becomes a synonym of *S. ovata* and the next-oldest available epithet is *S. ovata* ssp. *bracteata*.

7b. ssp. ovata

S. ovata var. *versicolor* (Nutt.) Fernald

S. ovata ssp. *versicolor* (Nutt.) Epling

Plants relatively robust, the stems mostly 30–80 cm long. Leaf blades mostly 3–7 cm long, not appearing strongly puckered, the veins slightly to moderately impressed on the upper surface, slightly to moderately raised on the undersurface. Inflorescences with the bracts not extending past the calyces of the flowers they subtend (except rarely in some late-season inflorescences). Corollas with the lower lip white with blue spots or fine mottling and sometimes a narrow blue band along the margins. $2n=20$. May–October.

Scattered, mostly south of the Missouri River, north locally to Sullivan and Marion Counties (eastern U.S. west to Minnesota and Texas). Bottomland forests, mesic upland forests, glades, banks of streams and rivers, and bases, ledges, and tops of bluffs; also roadsides.

This subspecies tends to occur in somewhat more mesic habitats than the other two subspecies.

7c. ssp. rugosa (Alph. Wood) Epling

S. ovata var. *rugosa* (Alph. Wood) Fernald

S. ovata subsp. *rupestris* Epling, an invalid name

Plants relatively short, the stems mostly 10–40 cm long. Leaf blades mostly 1.5–5.0 cm long, appearing strongly puckered, the veins strongly impressed on the upper surface, strongly raised on the undersurface. Inflorescences with the bracts not extending past the calyces of the flowers they subtend (except rarely in some late-season inflorescences). Corollas with the lower lip blue, with a pair of longitudinal white bands along the midvein, these often with small blue spots or fine mottling. May–October.

Scattered, mostly in the Ozark and Ozark Border Divisions, north locally to Monroe and Marion Counties (Missouri to Arkansas east to West Virginia and North Carolina). Mesic to dry upland forests, glades, banks of streams and rivers, and bases, ledges, and tops of bluffs; also railroads and roadsides.

8. Scutellaria parvula Michx. (small skullcap)
Pl. 441 j–m; Map 1999

Plants with slender, inconspicuous rhizomes. Stems 8–20(–30) cm long, loosely to strongly ascending, often from a spreading base, unbranched or few-branched, glabrous or sparsely to densely pubescent with short, curved, nonglandular hairs and sometimes also longer, spreading gland-tipped hairs. Leaves sessile or nearly so (except on basal leaves, which are short-petiolate and usually absent at flowering). Leaf blades 0.5–2.0 cm long, lanceolate to ovate, broadly ovate, or more or less triangular-ovate, rounded to truncate or shallowly cordate at the base or occasionally broadly angled, rounded or bluntly pointed at the tip, the margins entire, the surfaces sparsely pubescent with short, curved hairs or more densely pubescent with short, gland-tipped hairs, sometimes also with dense sessile glands. Inflorescences of axillary flowers, these 2 per node, solitary in the axils of the upper foliage leaves. Calyces 2.0–3.5 mm long, becoming closed and enlarged to 4–5 mm at fruiting, the outer surface sparsely to moderately pubescent with short, curved, nonglandular or straight, gland-tipped hairs, sometimes also with sessile glands. Corollas 6–12 mm long, minutely nonglandular-hairy on the outer surface, bluish purple, the lower lip mottled and/or spotted with white and purple toward the base, the tube S-shaped (bent upward just above the calyx and strongly curved or oblique at or above the throat), the lateral lobes well-developed, spreading, the lower lip broadly depressed ovoid, deeply notched at the tip. Nutlets 1–4 per calyx, 0.9–1.2 mm long, more or less globose, the surface dark brown to black, densely warty or with rounded tubercles, interrupted by a broad, smooth, transverse band. $2n=20$. May–July.

Scattered nearly throughout the state (eastern U.S. west to North Dakota and Texas; Canada). Glades, bottomland prairies, upland prairies, savannas, ledges and tops of bluffs, bottomland forests, and mesic to dry upland forests; also pastures, railroads, and roadsides.

The *S. parvula* complex, including *S. nervosa*, presents a good topic for a future taxonomic and phylogenetic study. Most botanists admit that the four entities involved are closely related and that *S. nervosa* deserves separate species rank, given its longer stems with more frequent branching and larger leaves, but the treatment of the other three taxa remains controversial. Some authors have followed Epling (1942) in treating these as three separate species, *S. australis, S. leonardii*, and *S. parvula* in the strict sense. Others, including Steyermark (1963) and the present work have followed Fernald (1945) and treated them as three varieties of *S. parvula*. Gleason and Cronquist (1991) chose a compromise, treating two species, *S. leonardii* and *S. parvula*, but subsuming the third taxon as a variety of *S. parvula*. This last approach apparently is in response to Epling's (1942) comments that *S. leonardii* is morphologically similar in some ways to *S. nervosa* and that the two taxa sometimes are found growing in proximity. For the present, it seems prudent to follow Steyermark's (1963) approach, with a minor nomenclatural update for the former var. *leonardii*, following the historical research of Goodman and Lawson (1992).

1. Stems glabrous or more commonly sparsely to moderately pubescent with short, curved, nonglandular hairs; blades of the largest leaves with mostly 1 or 2 secondary veins on each side of the midvein; calyces pubescent with nonglandular hairs . . . . 8B. VAR. MISSOURIENSIS
1. Stems moderately to densely pubescent with a mixture of short, curved, nonglandular hairs and longer, spreading gland-tipped hairs; blades of the largest leaves with mostly 3–5 secondary veins on each side of the midvein; calyces pubescent with nonglandular and gland-tipped hairs
    2. Leaf blades not dotted with sessile glands, but with moderate to dense, spreading gland-tipped hairs on the undersurface, the lateral veins joined near the blade margins . . . . 8A. VAR. AUSTRALIS
    2. Leaf blades dotted with sessile glands and with moderate to dense, curved, nonglandular hairs on the undersurface, the lateral veins mostly not joined, free to their tips . . . . 8C. VAR. PARVULA

8a. var. australis Fassett (southern small skullcap)
S. australis (Fassett) Epling

Pl. 441 j

Stems moderately to densely pubescent with short, upward-curved, nonglandular hairs and sparse to moderate, spreading, gland-tipped hairs. Leaf blades ovate to broadly ovate, the margins usually flat, the undersurface with moderate to dense, spreading, gland-tipped hairs, not dotted with sessile glands, the blades of the largest leaves with mostly 3–5 secondary veins on each side of the midvein, the lateral veins joined near the blade margins. Calyces pubescent with a mixture of minute, nonglandular and somewhat longer, gland-tipped hairs. May–July.

Scattered, mostly south of the Missouri River (eastern [mostly southeastern] U.S. west to Kansas and Texas). Glades, upland prairies, savannas, ledges and tops of bluffs, bottomland forests, and openings of mesic to dry upland forests; also pastures.

8b. var. missouriensis (Torr.) Goodman & C.A. Lawson (Leonard's skullcap)
S. parvula var. *leonardii* (Epling) Fernald
S. leonardii Epling

Pl. 441 k

Stems glabrous or more commonly sparsely to moderately pubescent with short, upward-curved, nonglandular hairs. Leaf blades lanceolate to ovate or ovate-triangular, the margins curled under, the undersurface with sparse, curved, nonglandular hairs along the veins, sometimes nearly glabrous, the blades of the largest leaves with mostly 1 or 2 secondary veins on each side of the midvein, the lateral veins mostly not joined, free to their tips. Calyces pubescent with minute, nonglandular hairs. $2n=20$. May–July.

Scattered nearly throughout the state, but apparently absent from the Mississippi Lowlands Division (eastern U.S. west to North Dakota and Texas; Canada). Glades, bottomland prairies, upland prairies, ledges and tops of bluffs, and mesic to dry upland forests; also railroads and roadsides.

8c. var. parvula

Pl. 441 l, m

Stems moderately to densely pubescent with short, upward-curved, nonglandular hairs and sparse to moderate, spreading, gland-tipped hairs. Leaf blades ovate to narrowly ovate or narrowly oblong-ovate, the margins often somewhat curled under, the undersurface with moderate to dense, curved, nonglandular hairs, also dotted with sessile glands, the blades of the largest leaves with mostly 3–5 secondary veins on each side of the midvein, the lateral veins mostly not joined, free to their tips. Calyces pubescent with a mixture of minute, nonglandular and somewhat longer, gland-tipped hairs. $2n=20$. May–July.

Scattered, mostly south of the Missouri River (eastern U.S. west to Minnesota, Nebraska, and Texas; Canada). Glades, upland prairies, savannas, ledges and tops of bluffs, and openings of mesic to dry upland forests; also pastures and roadsides.

28. **Stachys** L. (hedge nettle) (Mulligan and Munro, 1989)

Plants perennial herbs (annual elsewhere), often with rhizomes, sometimes also with tubers. Stems erect or ascending, bluntly to sharply 4-angled, unbranched or branched, glabrous or hairy, the hairs not felted or woolly. Leaves sessile or short- to long-petiolate, the petiole unwinged. Leaf blades variously linear to narrowly oblong or narrowly lanceolate to broadly ovate, angled to rounded or occasionally shallowly cordate at the base, angled or tapered to a usually sharply pointed tip, the margins with relatively closely spaced, fine, sharp teeth, the surfaces glabrous, roughened, or sparsely to moderately short-hairy, also with inconspicuous, sessile glands. Inflorescences terminal, spikes, composed of distinct clusters, these widely or relatively densely spaced along the axis, but noticeably discrete and with the flowers not overlapping those of adjacent nodes, or less commonly dense, more or less continuous, the flowers in small clusters of 4 or 6(8) per node (flowers 2 or 3[4] in the axil of each bract), the clusters not headlike. Bracts similar to the foliage leaves but much smaller. Bractlets absent or very short, linear, and inconspicuous. Calyces actinomorphic or nearly so, lacking a lateral projection, more or less symmetric at the base, narrowly bell-shaped, the tube 5- or 10-nerved, 5-lobed, the lobes somewhat shorter than to about as long as the tube, similar in size and shape, triangular to narrowly triangular, not spinescent, not becoming enlarged or papery at fruiting. Corollas zygomorphic, white, light pink, or pale purple, the lower lip usually with reddish purple to dark purple or whitish (in darker corollas) spots or mottling, the outer surface sparsely to moderately pubescent with short, gland-tipped hairs, especially on the lobes, the tube funnelform, relatively deeply 2-lipped, the lobes slightly shorter than the tube, the upper lip usually slightly shorter than the lower lip, entire or shallowly notched, slightly concave to more or less hooded, the lower lip spreading to arched, 3-lobed with a large central lobe and 2 small lateral lobes. Stamens 4, not exserted, the lower pair with slightly longer filaments than the upper pair, all ascending under the upper corolla lip, the anthers small, the connective very short, the pollen sacs 2, spreading, usually dark purple. Ovary deeply lobed, the style appearing nearly basal from a deep apical notch. Style exserted, more or less equally 2-branched at the tip. Fruits dry schizocarps, separating into usually 4 nutlets, these 1.5–2.5 mm long, oblong-obovoid to broadly obovoid, rounded at the tip, somewhat 3-angled, the surface dark brown to nearly black, glabrous, finely pebbled or faintly irregular. About 300 species, nearly worldwide (but absent from southeastern Asia to Austalia), most diverse in temperate or mountainous regions of the northern hemisphere.

Some species of *Stachys* are cultivated as ornamentals for their foliage or flowers, including the Mediterranean *S. byzantina* C. Koch (lamb's ears), which has densely white-woolly leaves and dense spikes of pink flowers, and the bright red-flowered *S. coccinea* Ortega (Texas betony, scarlet hedge nettle), which is native to the southwestern United States and Mexico.

The Missouri species of *Stachys* are morphologically variable and difficult to distinguish. The present treatment has benefited greatly from an unpublished review of the genus prepared by Stacy Oglesbee, an undergraduate intern in 1991 at the Missouri Botanical Garden.

1. Stems (except sometimes toward the base) hairy on the sides in addition to the angles . . . . 3. S. PILOSA
1. Stems glabrous or hairy only along the angles (the sides glabrous)
   2. Calyces with the lobes hairy, sometimes only along the margins; leaves with the upper surface often noticeably and moderately hairy . . . . 2. S. HISPIDA
   2. Calyces with the lobes glabrous; leaves with the upper surface glabrous or inconspicuously and sparsely hairy
      3. Leaves sessile or short-petiolate, the petioles all less than 7 mm long . . . . 1. S. ASPERA
      3. Leaves short- to long-petiolate, the petioles 5–35 mm long, those of the largest leaves more than 8 mm long . . . . 4. S. TENUIFOLIA

1. Stachys aspera Michx. (hyssop hedge nettle, hedge nettle)

S. hyssopifolia Michx. var. *ambigua* A. Gray

Pl. 443 d, e; Map 2000

Stems 40–80 cm long, glabrous on the sides, roughened with sparse, short, downward-angled, pustular-based hairs along the angles, often with a transverse line of longer, slender, spreading hairs at the nodes, sometimes also with scattered, longer, spreading hairs. Leaves sessile or short-petiolate, the petioles all less than 7 mm long. Leaf blades 2–7 cm long, lanceolate to oblong, usually narrowly so, mostly angled at the base, angled or more commonly tapered to a sharply pointed tip, the surfaces glabrous or nearly so. Inflorescences interrupted spikes, the nodes mostly well-spaced. Calyces 5–7 mm long, the tube glabrous or with sparse, spreading, pustular-based hairs, the lobes 2.5–3.5 mm long, glabrous. Corollas 12–15 mm long. $2n$=68. June–August.

Uncommon, mostly in the northern half of the state (eastern U.S. west to Wisconsin, Missouri, and Alabama). Edges of bottomland forests, banks of streams, margins of ponds, lakes, and sinkhole ponds, and marshes; also pastures and moist, disturbed areas.

Steyermark (1963) knew this plant only from a few historical collections form the St. Louis area. The present distribution probably reflects the benefits of additional field and herbarium studies during the past few decades rather than an expansion of the taxon's range. Most of the Missouri records remain historical.

Mulligan and Munro (1989) reported Missouri in their distributional summary of the closely related *S. hyssopifolia* Michx. without documentation. Specimens referable to this species could not be located during the present study; it is thus excluded from the flora for the present. *Stachys hyssopifolia* occurs mostly in states along the Atlantric seaboard and also in states bordering Lake Michigan. It differs from *S. aspera* in its linear to narrowly oblong leaves (less than 7 mm wide) that are often entire or nearly so.

2. Stachys hispida Pursh (hairy hedge nettle)

S. tenuifolia Willd. var. *hispida* (Pursh) Fernald

Map 2001

Stems 40–100 cm long, glabrous on the sides, sparsely pubescent with relatively long, spreading to somewhat downward-angled, sometimes pustular-based hairs along the angles, usually glabrous at the nodes. Leaves sessile or short-petiolate, the petioles to 10 mm long. Leaf blades 3–12 cm long, lanceolate to oblong, often narrowly so, angled to rounded or rarely shallowly cordate at the base, angled or more commonly tapered to a sharply pointed tip, the upper surface moderately pubescent with loosely spreading hairs, the undersurface sparsely short-hairy. Inflorescences interrupted or more commonly loosely continuous spikes, the nodes well-spaced to somewhat crowded. Calyces 5–9 mm long, the tube sparsely to moderately pubescent with spreading, sometimes pustular-based hairs, sometimes also with minute, gland-tipped hairs, the lobes 3.0–4.5 mm

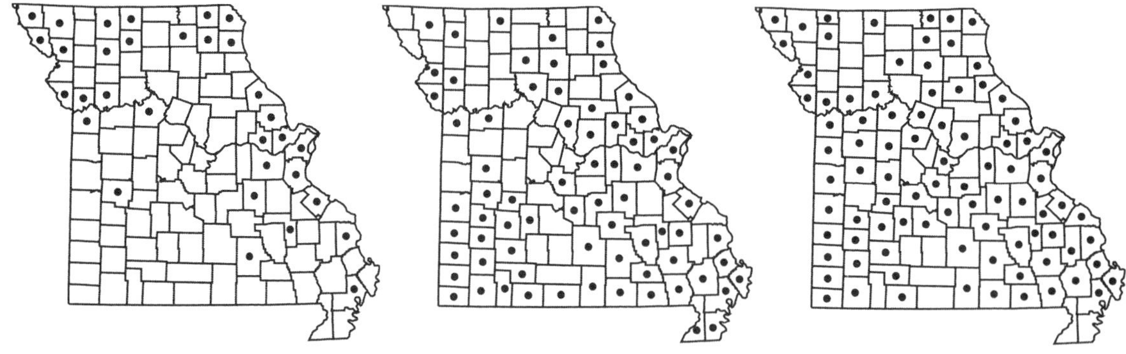
2002. Stachys pilosa 2003. Stachys tenuifolia 2004. Teucrium canadense

long, moderately pubescent with spreading, sometimes pustular-based hairs, sometimes only along the margins. Corollas 12–14 mm long. $2n=68$. June–September.

Scattered, mostly in the eastern half of the state, apparently absent from the Unglaciated Plains Division and most of the Ozarks (northeastern U.S. west to North Dakota and Missouri, south locally to Georgia; Canada). Bottomland forests, banks of streams, margins of ponds and lakes, bottomland prairies, and marshes; also ditches, railroads, roadsides, and moist, disturbed areas.

3. Stachys pilosa Nutt. (woundwort, marsh betony)

S. pilosa var. *arenicola* (Britton) G.A. Mulligan & D.B. Munro
S. arenicola Britton
S. palustris ssp. *pilosa* (Nutt.) Epling
S. palustris L. var. *arenicola* (Britton) Farwell
S. palustris var. *homotricha* Fernald
S. palustris var. *nipigonensis* Jenn.
S. palustris var. *phaneropoda* Weath. ex Fernald
S. palustris var. *pilosa* (Nutt.) Fernald

Pl. 443 h, i; Map 2002

Stems 30–100 cm long, pubescent on the sides and angles with moderate, shorter or longer, occasionally gland-tipped, downward-angled to loosely spreading hairs, some of these usually pustular-based, often with a denser transverse line of sometimes longer, finer hairs at the nodes. Leaves sessile or short-petiolate, the petioles all less than 7 mm long. Leaf blades 3–10 cm long, lanceolate, oblong-lanceolate, oblong-elliptic, oblong-oblanceolate, or narrowly ovate-triangular, angled to rounded or sometimes shallowly cordate at the base, angled or more commonly tapered to a sharply pointed tip, the upper surface moderately pubescent with short, loosely appressed, sometimes pustular-based hairs, the undersurface sparsely to moderately pubescent with short, fine hairs. Inflorescences interrupted or more commonly loosely continuous spikes, the nodes well-spaced to somewhat crowded. Calyces 6–10 mm long, the tube sparsely to moderately pubescent with fine, sometimes gland-tipped hairs and/or coarser, nonglandular, sometimes pustular-based hairs, the lobes 3.0–4.5 mm long, hairy similar to the tube. Corollas 12–15 mm long. $2n=68$. June–September.

Apparently absent from the Unglaciated Plains Division and most of the Ozarks, scattered elsewhere in the state (nearly throughout the U.S. [including Alaska], except some southeastern states; Canada). Bottomland forests, swamps, banks of streams, rivers, and sloughs, margins of ponds and lakes, bottomland prairies, moist swales in upland prairies and sand prairies, marshes, and fens; also ditches, railroads, roadsides, and moist, disturbed areas.

Stachys palustris and its allies form a polyploid complex that is morphologically very complex (Mulligan and Munro, 1983, 1989). True *S. palustris* is a hexaploid ($2n=102$) native to Europe that has become introduced sporadically in the northeastern United States and Canada. It differs from the native North American populations (*S. pilosa*) in a number of subtle features, including light purple (vs. white to light pink) corollas and calyx lobes tapered from near the base (vs. tapered from about the midpoint). Mulligan and Munro (1989) attempted to divide the North American populations into two groups, based primarily on differences in stem pubescence: the northeastern var. *arenicola* was characterized by stiff, downward-angled hairs; the western and northern var. *pilosa* was said to differ in softer mostly spreading hairs. The Missouri populations, which fall within the broad region of geographic overlap between these varieties, exhibit bewildering varia-

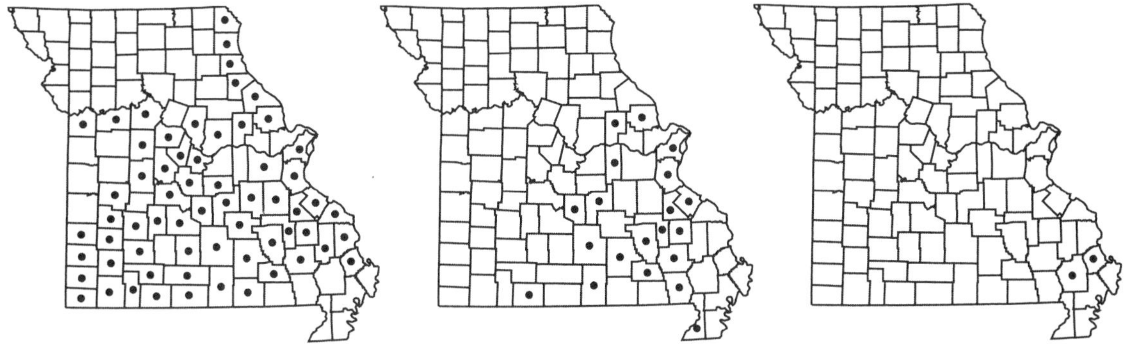

2005. Trichostema brachiatum 2006. Trichostema dichotomum 2007. Trichostema setaceum

tion in pubescence styles, from stiff, stout, pustular-based hairs to fine, nonpustular hairs, the latter sometimes gland-tipped. Many plants possess both types of hairs, with the density of each dependent upon position on the stems. Thus, it seems impractical to recognize varieties for this species.

4. Stachys tenuifolia Willd. (thinleaf betony, slenderleaf betony)

Pl. 443 f, g; Map 2003

Stems 30–100 cm long, glabrous or the angles sparsely pubescent with short, downward-angled, sometimes pustular-based hairs, often with a transverse line of longer, spreading hairs at the nodes. Leaves short- to long-petiolate, the petioles 5–35 mm long, those of the largest leaves more than 8 mm long. Leaf blades 2–12 cm long, lanceolate to oblong, oblong-elliptic, or occasionally ovate, angled or rounded at the base, mostly tapered to a sharply pointed tip, the surfaces glabrous or inconspicuously and sparsely pubescent with short, fine hairs. Inflorescences interrupted or very loosely continuous spikes, the nodes well-spaced to somewhat crowded. Calyces 4–7 mm long, the tube glabrous or with sparse, fine, spreading, often gland-tipped hairs, the lobes 1.5–3.5 mm long, glabrous or sparsely hairy along the margins. Corollas 10–13 mm long. $2n$=34, 68. June–September.

Scattered nearly throughout the state (eastern U.S. west to North Dakota and Texas; Canada). Bottomland forests, edges of mesic upland forests, swamps, banks of streams and rivers, bases of bluffs, margins of ponds and lakes, bottomland prairies, moist swales in upland prairies and sand prairies, marshes, and fens; also ditches, roadsides, and moist, disturbed areas.

29. Teucrium L. (germander)

About 100 species, nearly worldwide, most diverse in the Mediterranean region.

1. Teucrium canadense L. (wood sage, American germander)

Pl. 444 a, b; Map 2004

Plants perennial herbs, with rhizomes, sometimes also producing tubers, often colonial. Stems 35–120 cm long, bluntly to strongly 4-angled, erect or strongly ascending, unbranched or few-branched, moderately to densely pubescent with short, downward-angled hairs and sometimes also with longer, spreading hairs. Leaves opposite, sessile or short-petiolate, the petiole narrowly unwinged. Leaf blades 3–12(–16) cm long, 1–5 cm wide, lanceolate to ovate or oblong-lanceolate, angled at the base, angled or short-tapered to a sharply pointed tip, the margins finely to coarsely toothed, the upper surface glabrous or sparsely to moderately pubescent with short, loosely appressed hairs, the undersurface sparsely to densely pubescent with short, curved, loosely appressed hairs or relatively long, straight to woolly hairs, these tending to obscure the inconspicuous sessile glands. Inflorescences terminal, dense spikelike racemes, the flowers 2–6 per node and mostly overlapping those of adjacent nodes, the stalks 1–4 mm long. Bracts progressively shorter from the inflorescence base to tip, 4–25 mm long, leaflike, lanceolate to narrowly lanceolate or occasionally linear, usually longer than the calyces. Bractlets absent. Calyces 5–9 mm long at flowering, zygomorphic, lacking a lateral projection, more or less symmetric at the base, bell-shaped, the tube finely 10-nerved, with scattered longer, nonglandular hairs and sometimes also moderate, shorter, gland-tipped hairs in the mouth, the outer surface usually densely hairy, 2-lipped, the lobes shorter than the tube, the upper lip with 3 ovate to triangular lobes, straight or slightly arched upward, the lower lip with 2 narrowly triangular lobes, slightly arched upward, the lobes not spinescent, not becoming enlarged at fruiting, but becoming somewhat leathery. Corollas 10–18 mm long, strongly zygomorphic, corollas pinkish purple to pale lavender or rarely white, usually with darker purple spots or mottling on the lower lip, the surfaces usually moderately pubescent with short, gland-tipped hairs, sometimes also gland-dotted, the tube funnelform, divided to below the midpoint, the upper lip appearing absent, the lower lip prominent, 5-lobed, the 2 pairs of smaller lateral lobes ascending to arched upward, the longer central lobe obovate to somewhat rhombic, abruptly bent downward. Stamens 4, strongly exserted, the filaments somewhat arched downward, the anthers small, the connective not visible, the pollen sacs 2, parallel (back-to-back and dehiscing along opposite sides, reddish brown to orangish brown. Ovary shallowly lobed, the style appearing terminal or somewhat lateral from a shallow apical notch. Style strongly exserted, with 2 slender branches at the tip. Fruits dry schizocarps, consisting of usually 4 nutlets, these 1.5–2.5 mm long, broadly ovoid, the surface light brown to reddish brown, glabrous, with a network of shallow ridges or wrinkles. $2n$=32. June–September.

Scattered to common throughout the state (nearly throughout the U.S.; Canada, Mexico, Caribbean Islands). Bottomland forests, bottomland prairies, upland prairies, loess hill prairies, banks of streams and rivers, margins of sinkhole ponds, fens, and marshes; also pastures, old fields, margins of crop fields, fencerows, levees, railroads, and roadsides.

Teucrium canadense is sometimes cultivated as an ornamental, although it can spread aggressively in some garden situations. According to Steyermark (1963), plants were once used medicinally as a stimulant. The species is among several members of the Lamiaceae that have been investigated as sources of natural rubbers (Buchanan et al., 1978).

McClintock and Epling (1946), in their taxonomic revision of New World *Teucrium*, recognized three varieties of *T. canadense* based primarily upon differences in pubescence patterns, two of which occur in Missouri. Shinners (1963) combined two of these three varieties, but named a third, var. *nashii* (Kearney) Shinners, for plants of the southeastern United States similar to var. *canadense*, but with the leaves silvery-hairy on the undersurface.

1. Calyces with all of the hairs nonglandular, 0.5–0.8 mm long; leaf blades with the undersurface green, sparsely pubescent with short, curved, loosely appressed hairs .... 1A. VAR. CANADENSE
1. Calyces with all or some of the hairs gland-tipped, 1.0–1.5 mm long; leaf blades with the undersurface grayish green, densely pubescent with relatively long, straight to woolly hairs .... 1B. VAR. OCCIDENTALE

1a. var. canadense

T. canadense var. *virginicum* (L.) Eaton

Leaf blades with the undersurface green, sparsely pubescent with short, curved, loosely appressed hairs. Calyces with all of the hairs nonglandular, 0.5–0.8 mm long, more or less appressed and woolly. $2n$=32. June–September.

Scattered to common throughout the state (eastern U.S. west to Minnesota, South Dakota, and Arizona; Canada, Mexico, Caribbean Islands). Bottomland forests, bottomland prairies, upland prairies, loess hill prairies, banks of streams and rivers, margins of sinkhole ponds, fens, and marshes; also pastures, old fields, fencerows, levees, railroads, and roadsides.

1b. var. occidentale (A. Gray) E.M. McClint. & Epling

T. canadense var. *boreale* (E.P. Bicknell) Shinners

Leaf blades with the undersurface grayish green, densely pubescent with relatively long, straight to woolly hairs. Calyces with all or some of the hairs gland-tipped, 1.0–1.5 mm long, the nonglandular hairs usually curved or loosely appressed, the gland-tipped hairs more or less spreading. June–September.

Uncommon, mostly in counties bordering the Missouri River (northern U.S. south to Arizona, Texas, and Virginia; Canada). Banks of streams and rivers; also margins of crop fields.

McClintock and Epling (1946) used the name var. *occidentale* for this taxon. Shinners (1963) argued that the oldest available epithet at the varietal level is the one based on *T. boreale* A. Gray, but Kartesz and Gandhi (1989) noted that this was an error under the current rules for naming species.

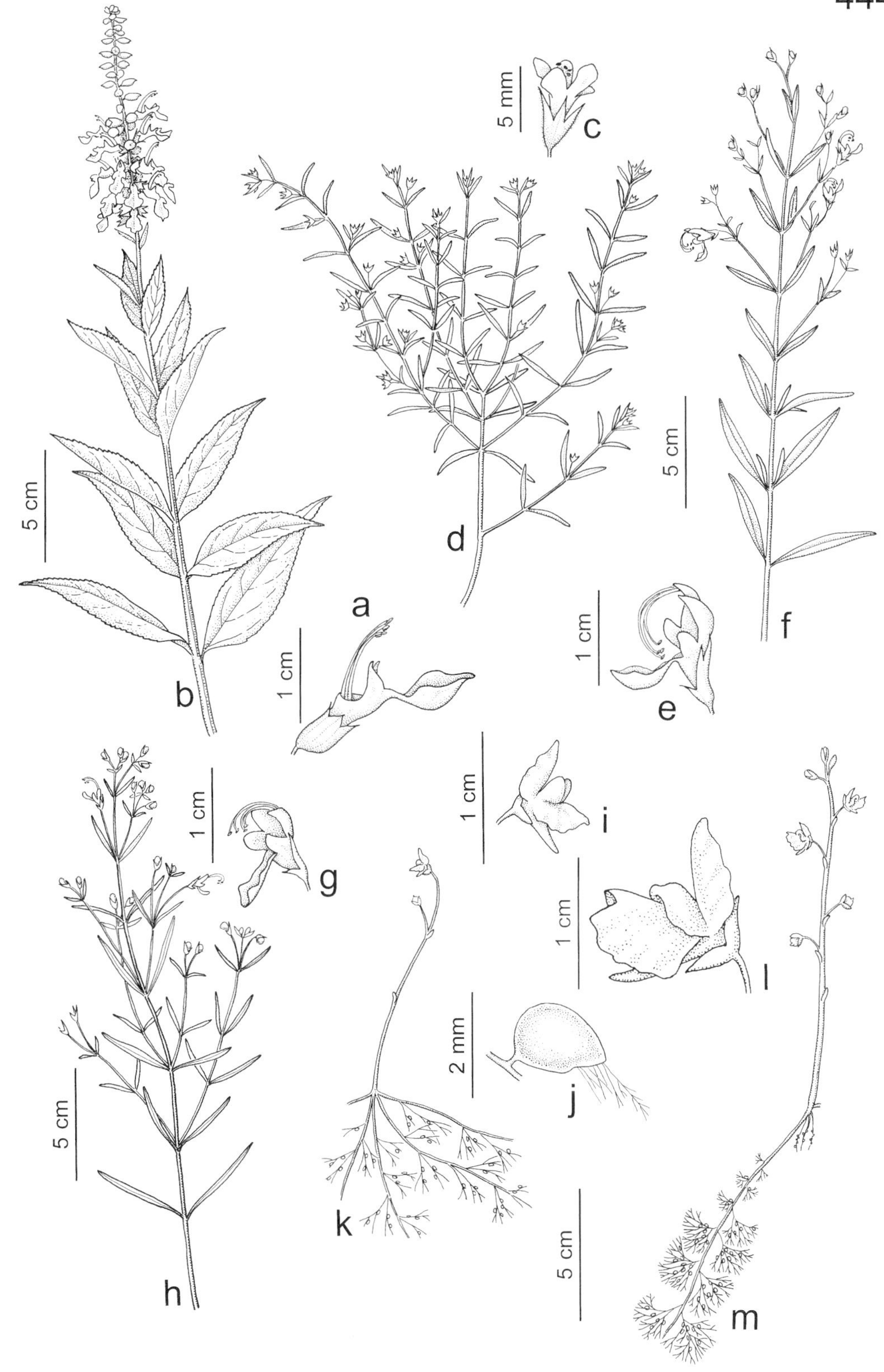

Plate 444. Lamiaceae. Lentibulariaceae. *Teucrium canadense*, **a)** flower, **b)** fertile stem. *Trichostema brachiatum*, **c)** flower, **d)** habit. *Trichostema dichotomum*, **e)** flower, **f)** fertile stem. *Trichostema setaceum*, **g)** flower, **h)** fertile stem. *Utricularia gibba*, **i)** flower, **j)** bladder, **k)** habit. *Utricularia vulgaris*, **l)** flower, **m)** habit.

30. **Trichostema** L. (blue curls)

Plants annual (perennial herbs or shrubs elsewhere), with slender taproots. Stems erect or ascending, slender, bluntly 4-angled (often obscurely so below the midpoint), unbranched or branched, moderately to densely pubescent with minute, gland-tipped hairs. Leaves sessile or petiolate, the petiole unwinged, with a usually strong fragrance when crushed. Leaf blades linear to lanceolate, oblong, or ovate, the margins entire (rarely minutely toothed in *T. brachiatum*), the surfaces hairy, some of the hairs often gland-tipped, also with sometimes conspicuous, sessile glands. Inflorescences axillary and/or apparently terminal, then appearing as open, irregularly branched panicles with leaflike bracts that are smaller than the foliage leaves, the nodes in both types with the flowers paired or with pairs of small, loose clusters or panicles, each with 3–7 flowers, the flowers stalked or some of them sessile, usually with a pair of inconspicuous, small, linear bractlets at the flower base. Calyces actinomorphic or zygomorphic and 2-lipped, lacking a lateral projection, more or less symmetric at the base, more or less bell-shaped, the tube finely 10-nerved, glabrous in the mouth, the lobes shorter than to longer than the tube, similar or dissimilar, narrowly to broadly triangular, not spinescent, becoming somewhat enlarged and papery at fruiting, in some species the flower becoming inverted as the fruits mature. Corollas 1.5–12.0 mm long, nearly actinomorphic or strongly zygomorphic, pink to lavender, purple, bluish purple, or blue, the outer surface sparsely to densely and minutely hairy, the tube narrowly funnelform to funnelform, slightly to strongly 2-lipped, the upper lip with 4, shorter, similar, spreading to slightly inward-arched lobes, the lower lip with 1, slightly to much longer, spreading lobe. Stamens 4, slightly to strongly exserted, the filaments straight to strongly downward-arched, the anthers small, the connective very short, the pollen sacs 2, spreading, dark purple or dark blue. Ovary deeply lobed, the style appearing lateral from a deep apical notch. Style slightly to strongly exserted, 2-branched at the tip. Fruits dry schizocarps, separating into usually 4 nutlets, these 1.5–3.0 mm long, obovoid, the surface olive green to greenish brown, yellowish brown, or dark brown, glabrous or inconspicuously and minutely hairy and with sessile glands near the rounded tip, with a coarse network of ridges or wrinkles. Eighteen species, North America.

1. Calyces actinomorphic or nearly so, the lobes all similar; corollas nearly actinomorphic, the lowermost lobe only slightly longer than the others; stamens slightly to moderately exserted, the filaments 2.5–4.5 mm long, straight or slightly arched . . . . 1. T. BRACHIATUM
1. Calyces zygomorphic, 2-lipped, the lobes not all similar; corollas strongly zygomorphic, 2-lipped, the lowermost lobe conspicuously longer than the others; stamens strongly exserted, the filaments 6–20 mm long, strongly downward-arched (curled)
    2. Leaf blades oblong to elliptic or ovate, mostly 5–25 mm wide, with more or less evident secondary veins branching from the midvein . . . 2. T. DICHOTOMUM
    2. Leaf blades linear to narrowly lanceolate, 1–6 mm wide, lacking evident secondary veins (only the midvein apparent) . . . . 3. T. SETACEUM

1. Trichostema brachiatum L. (false pennyroyal)

Isanthus brachiatus (L.) Britton, Sterns & Poggenb.

Pl. 444 c, d; Map 2005

Stems 15–30(–50) cm long, densely pubescent with minute, gland-tipped hairs, often also with longer, nonglandular hairs, these usually only on 2 opposing sides. Leaves sessile or short-petiolate. Leaf blades 1–5 cm long, 3–10(–16) mm wide, narrowly elliptic to elliptic or lanceolate, angled or tapered at the base and tip, the margins entire or rarely minutely toothed, minutely hairy, the surfaces moderately to densely pubescent with minute, gland-tipped hairs, also with inconspicuous sessile glands, usually also with longer, nonglandular hairs, the venation of a midvein and usually a pair of ascending, somewhat finer lateral veins from the blade base, the secondary venation obscure. Inflorescences axillary, of loose clusters of 2–6 flowers per node. Calyces 2.5–4.5 mm long at flowering, becoming enlarged to 3.5–7.0 mm at fruiting, actinomorphic or nearly so, slightly oblique, the lobes all similar, 1.5–2.5 mm long at flowering, triangular, the outer surface densely pubescent with a mixture of minute, gland-tipped and longer, nonglandular hairs, also with usually conspicuous sessile glands. Corollas 1.5–4.5 mm long, nearly actinomorphic, pink to lavender, purple, or occasionally bluish purple, lacking lighter or darker spots or mottling, the tube cylindric to narrowly funnelform, expanded abruptly in the throat, the lowermost lobe only slightly longer than the others, the lobes all apruptly spreading. Stamens slightly to moderately exserted, the filaments 2.5–4.5 mm long, straight or slightly arched, the anthers dark purple. Nutlets 2.3–3.0 mm long, the surface olive green to greenish brown or yellowish brown, inconspicuously and minutely hairy and with sessile glands toward the rounded tip. $2n=14$. July–October.

Scattered south of the Missouri River, but absent from most of the Mississippi Lowlands Division; uncommon farther north (eastern U.S. west to Minnesota and Arizona; Canada). glades, tops of bluffs, thin-soil areas of upland prairies, and banks of streams and rivers; also roadsides; often on calcareous substrates.

This species is sometimes segregated into its own genus, *Isanthus* Michx., based on its nearly actinomorphic calyces and short stamens. The present treatment follows that of H. Lewis (1945), who noted that some of the western species that have always been referred to as *Trichostema* have similar characteristics. This view recently received support from molecular data (Huang et al., 2008).

2. Trichostema dichotomum L. (blue curls, bastard pennyroyal)

T. dichotomum var. *puberulum* Fernald & Griscom

Pl. 444 e, f; Map 2006

Stems 10–50(–70) cm long, densely pubescent with gland-tipped and nonglandular hairs, these minute or more commonly somewhat longer. Leaves mostly short-petiolate, the uppermost leaves sometimes sessile or nearly so, the lowermost leaves occasionally long-petiolate. Leaf blades 1.5–6.0 cm long, mostly 5–25 mm wide, oblong to elliptic or ovate, angled or tapered at the base, narrowly rounded to angled or tapered to a bluntly or sharply pointed tip, the margins entire, minutely hairy, the surfaces moderately pubescent with minute, curved, nonglandular hairs, the undersurface also with inconspicuous sessile glands, the venation of a midvein and more or less evident, pinnately branching, secondary veins. Inflorescences appearing terminal and axillary, usually as open, irregularly branched panicles with leaflike bracts that are smaller than the foliage leaves, the flowers paired at the nodes or in pairs of small, loose clusters or panicles, each with 3–7 flowers. Flowers becoming inverted as the fruits mature. Calyces 3–5 mm long at flowering, becoming enlarged to 5–9 mm at fruiting, zygomorphic, 2-lipped, the upper lip (at flowering) with 3 triangular lobes, about 3 times as long as the lower lip, which has 2 narrowly triangular lobes, the outer surface densely pubescent with a mixture of minute, gland-tipped and longer, nonglandular hairs, also with usually conspicuous sessile glands. Corollas 4–12 mm long, strongly zygomorphic, the tube funnelform, expanded in the throat, the upper lip with 4, shorter, similar, spreading to slightly inward-arched lobes, these blue to purplish blue, the lower lip with 1, much longer, spreading lobe, this abruptly spreading, usually white below the blue to purplish blue tip, the white portion spotted or mottled with blue to purplish blue. Stamens strongly exserted, the filaments 6–16 mm long, strongly arched downward (curled), the anthers dark blue. Nutlets 1.5–3.0 mm long, yellowish brown to dark brown, glabrous. $2n=38$. August–October.

Scattered, mostly in the eastern half of the Ozark and Ozark Border Divisions (eastern U.S. west to Iowa and Texas; Canada). Glades, openings of mesic to dry upland forests, banks of streams and river, tops of bluffs, and sand prairies; also old fields; usually on acidic substrates.

Steyermark (1963) treated two varieties within *T. dichotomum* differing in whether the stem pubescence was of longer or shorter hairs. In his taxonomic study of the genus, H. Lewis (1945) documented four pubescence-based forms within the species overall distribution, but concluded that these were insufficiently distinct to be described formally as infraspecific taxa.

Trichostema dichotomum is an attractive species that is sometimes grown as an annual in gardens. In the Missouri Botanical Garden Herbarium, there are two specimens from Butler County that apparently represent volunteers that sprouted in a yard under a bird feeder.

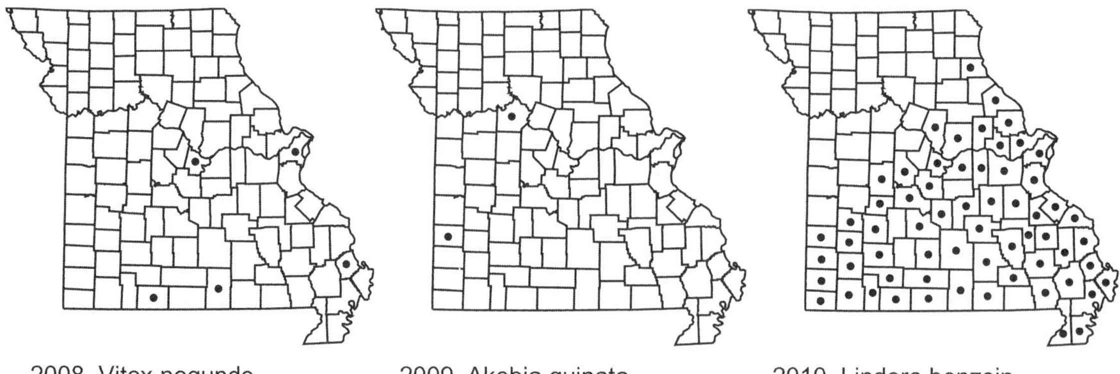

2008. Vitex negundo   2009. Akebia quinata   2010. Lindera benzoin

3. Trichostema setaceum Houtt.

Pl. 444 g, h; Map 2007

Stems 10–50 cm long, sparsely to densely pubescent with minute, downward-curved, nonglandular hairs. Leaves sessile or short-petiolate. Leaf blades 1–5 cm long, mostly 1–6 mm wide, linear to narrowly lanceolate, angled or tapered at the base, angled or tapered to a bluntly or sharply pointed tip, the margins entire, minutely hairy, the surfaces moderately pubescent with minute, curved, nonglandular hairs, the undersurface also with inconspicuous sessile glands, the venation of a solitary midvein, lacking evident secondary veins. Inflorescences appearing terminal and axillary, usually as open, irregularly branched panicles with leaflike bracts that are smaller than the foliage leaves, the flowers paired at the nodes or in pairs of small, loose clusters or panicles, each with 3–7 flowers. Flowers becoming inverted as the fruits mature. Calyces 3.5–5.5 mm long at flowering, becoming enlarged to 5.0–7.5 mm at fruiting, zygomorphic, 2-lipped, the upper lip (at flowering) with 3 triangular lobes, about 2–3 times as long as the lower lip, which has 2 narrowly triangular lobes, the outer surface densely pubescent with a mixture of minute, gland-tipped and nonglandular hairs, sometimes also with sessile glands. Corollas 5–12 mm long, strongly zygomorphic, the tube funnelform, expanded in the throat, the upper lip with 4, shorter, similar, spreading to slightly inward-arched lobes, these blue to purplish blue, the lower lip with 1, much longer, spreading lobe, this abruptly spreading, usually white below the blue to purplish blue tip, the white portion spotted or mottled with blue to purplish blue. Stamens strongly exserted, the filaments 8–20 mm long, strongly arched downward (curled), the anthers dark blue. Nutlets 1.5–2.1 mm long, yellowish brown to dark brown, glabrous. $2n$=38. September–October.

Uncommon, known thus far only from Scott and Stoddard Counties (eastern [mostly southeastern] U.S. west to Missouri and Texas). Sand prairies and sand savannas; also sandy disturbed areas.

Steyermark (1958a, 1963), who knew this species only from Stoddard County, considered it to be relictual in Missouri from a southeastern flora that was once more widespread in the state.

31. Vitex L. (monk's pepper tree)

About 250 species, nearly worldwide, most diverse in tropical regions.

1. Vitex negundo L. (vitex, chaste tree, hemp tree)

Map 2008

Plants shrubs or small trees. Main stems or trunk 1 to more commonly few to several, 100–500 cm long, erect or ascending, branched, the bark light brown to grayish brown, relatively smooth, but with scattered raised lenticels. Twigs mostly sharply 4-angled, reddish brown to dark brown with pale lenticels, but the surface more or less hidden by a dense, scurfy to felty covering of minute, spreading unbranched hairs, these often intermixed with sparse, longer hairs. Winter buds ovoid, angled to a sharply pointed tip, somewhat flattened, the 2–4 outer scales obscured by dense felty hairs. Leaves opposite, mostly long-petiolate (10–60 mm), the petiole unwinged. Leaf blades palmately compound with 3–9 leaflets, these 1–

15 cm long, 0.5–4.0 cm wide, mostly stalked (the basalmost leaflets usually markedly shorter than the others and often sessile or nearly so), lanceolate to elliptic or occasionally oblanceolate, tapered at the base, tapered to a sharply pointed tip, the margins variously entire to irregularly toothed or with slender pinnate lobes, the upper surface dark green, glabrous or more commonly sparsely to moderately short-hairy, sometimes only along the main veins, some of the hairs sometimes minutely gland-tipped, the undersurface appearing gray, densely pubescent with minute, spreading hairs, rarely glabrous, also with sparse, inconspicuous, sessile glands. Inflorescences terminal panicles, the short branches with clusters of several to numerous flowers mostly along the upper side. Bracts scalelike. Calyces actinomorphic, 2.5–3.5 mm long, lacking a lateral projection, symmetric at the base, bell-shaped, the tube more or less 5-nerved, the nerves obscured by dense, short hairs, shallowly 4-lobed, the lobes similar in size and shape, narrowly triangular, not spinescent, not becoming enlarged at fruiting. Corollas 6–8 mm long, zygomorphic, the tube funnelform, obliquely 2-lipped to about the midpoint, the upper lip shorter and 2-lobed, white to pale lavender (rarely darker), the lower lip longer, with a large central lobe and 2 short lateral lobes, pale lavender to more commonly purple or bluish purple, often with a yellow spot at the base (mouth), the outer surface (and mouth) densely grayish short-hairy. Stamens 4, exserted, all similar in size, the filaments attached in the corolla tube, the anthers small, the connective very short, the pollen sacs 2, appearing angled from their tips, dark purple. Ovary unlobed, the style appearing terminal. Style exserted, with a pair of slender, spreading branches at the tip. Fruits fleshy, more or less globose to very slightly obovoid (sometimes truncate to very slightly concave at the tip) drupes 3–5 mm in diameter, the outer surface dark brown to black, glabrous (sometimes with a few hairs toward the base and slightly glandular toward the tip), the stone 4-celled. $2n$=26, 32, 34. June–October.

Introduced, uncommon, widely scattered (native of Asia, Africa, Madagascar, Pacific Islands, introduced sporadically in the U.S.). Sand savannas and edges of mesic to dry upland forests; also old fields, fencerows, roadsides and disturbed areas.

This species was first reported for Missouri by Yatskievych and Summers (1993). It formerly was distributed by the Missouri Department of Conservation for use in wildlife plantings and also occasionally is cultivated as an ornamental in Missouri. Several varieties have been named, differing in leaflet shape and degree of toothing or lobing, as well as density and narrowness of the inflorescences (S.-l. Chen and Gilbert, 1994). The plants that escape from cultivation in temperate North America range from var. *negundo* (with leaflets entire or nearly so) to var. *heterophylla* (Franch.) Rehder (with leaflets having irregularly toothed and/or lobed margins. However, the infraspecific taxonomy of the species requires further study, as many intermediates are known between some of the named variants and the various characters said to separate the taxa sometimes do not appear well-correlated.

Vitex negundo is often confused with another commonly cultivated species, *V. agnus-castus* L., which is native to southern Europe and western Asia and escapes sporadically throughout the southern United States (farther north along both seaboards) and elsewhere in the world. As characterized by Wann and Akeroyd (2000), this species differs from *V. negundo* in: the stronger aromatic scent of the crushed foliage; usually slightly larger corollas (6–10 mm), more shallowly lobed calyces with broadly triangular lobes; and smaller fruits (2–3 mm in diameter). *Vitex agnus-castus* also has inflorescences that tend to be few branched from near the base, with the branches more strictly spicate than those of *V. negundo*, only the lowermost nodes having noticeably stalked flower clusters.

LARDIZABALACEAE (Lardizabala Family)

(Qin, 1997)

Nine genera, 35 species, South America, Asia.

The family is unusual in that members have floral parts mostly in multiples of three, a condition typical of monocots and quite uncommon among dicots.

1. **Akebia** Decne.

Four species, Asia.

1. Akebia quinata (Houtt.) Decne. (akebia)
Map 2009

Plants monoecious, lianas, glabrous. Stems to 15 m or more long, relatively slender, with purplish twigs, the bark of older stems thin, light gray or sometimes purplish tinged, with prominent lenticels (visible as small uneven bumps on the surface). Leaves alternate, lacking stipules, palmately compound with usually 5 leaflets, the petioles 3–14 cm long. Leaflets short-stalked, 2–8 cm long oblanceolate to obovate or more or less elliptic, tapered at the base, rounded and often with a small notch at the tip, the margins entire, dark green with a pale undersurface. Inflorescences pendant axillary racemes, sometimes appearing nearly umbellate. Flowers stalked, mostly with small slender bracts at the stalk base, the pistillate flowers usually fewer than the staminate ones. Sepals 3 (rarely 4 in pistillate flowers), elliptic to ovate, concave (cupped), bluntly pointed at the tip, spreading to somewhat reflexed, dark purple to reddish purple or rarely pale greenish pink, those of the staminate flowers 4–9 mm long, those of the pistillate flowers 8–16 mm long. Petals absent. Staminate flowers with 6 stamens incurved over 3–6 reduced nonfunctional pistils, the filaments very short, the anthers purplish black. Pistillate flowers with 6 minute staminodes and 3–9 separate pistils. Pistils dark bluish purple, consisting of 1 carpel, the superior, cylindrical ovary containing numerous ovules, tapered slightly at the tip to a capitate or donut-shaped stigma. Fruits generally not produced in Missouri material, berrylike follicles, 6–9 cm long, oblong in outline, the relatively thick wall purple to reddish purple with a white waxy coating, sometimes with small white spots, dehiscing along a longitudinal suture to expose the mass of seeds, these 4–6 mm long, ovate, somewhat flattened, reddish brown to black, shiny, embedded in a gelatinous pulp. $2n=32$. March–May.

Introduced, known thus far from single sites in Barton and Saline Counties (native to Asia, introduced sporadically in North America). Twining on and over other vegetation in open to somewhat shaded, disturbed areas.

Akebia quinata is sometimes cultivated as an ornamental on trellises and fences and for its unusual fragrant flowers. In Missouri, even cultivated plants rarely produce fruits, apparently because cross-pollination is necessary (Ernst, 1964). However, individuals can spread efficiently through vegetative growth, sometimes resulting in dense mats of intertwined stems. This species was first reported for Missouri by Timme (1997).

LAURACEAE (Laurel Family)

Plants shrubs and trees, sometimes colonial from rhizomes and/or root suckers, dioecious (with perfect flowers elsewhere), often incompletely so (pistillate inflorescences then with a few perfect flowers). Leaves alternate, simple, usually petiolate, the blades simple or lobed, the margins otherwise entire. Stipules absent. Inflorescences axillary clusters or terminal racemes produced before or as the leaves develop. Flowers all or mostly imperfect, more or less perigynous, actinomorphic, the buds lacking sheathing bracts, but some of the flowers sometimes with a pair of minute bracts, these shed early. Perianth not clearly differentiated into calyx and corolla, the tepals 6, in 2 whorls of 3, fused basally into a short perianth tube (this sometimes very short or essentially absent), this persistent at fruiting as a small disc or crown, the free portions shed after flowering (persistent elsewhere), relatively thin, usually yellow or greenish yellow (green or white elsewhere). Staminate flowers with 9 stamens (12 elsewhere) in 3 whorls and a highly reduced, nonfunctional ovary, the filaments somewhat flattened, attached to the perianth tube (when present), those of some stamens often with a pair of small nectar-producing basal appendages, the anthers attached at the base, yellow, dehiscing longitudinally by 2 or 4 small flaps. Pistillate flowers with 1 ovary and sometimes with varying numbers of staminodes sometimes also present, these varying from small filaments to seemingly well-formed (but nonfunctional) stamens, the ovary superior, apparently of 1 carpel and with 1 locule, the style and stigma 1. Ovule 1, the placentation more or less apical. Fruits drupes or berries, with 1 seed or stone. About 52 genera, about 2,850 species, nearly worldwide, most diverse in tropical regions.

The description above does not account for the unusual genus *Cassytha* L. (love vine), which is sometimes segregated into its own family, Cassythaceae. Plants in this genus are parasitic vines with orangish to yellowish, twining stems and small scalelike leaves, superficially similar in appearance to the dodders (*Cuscuta*, Convolvulaceae). The small, greenish white, perfect flowers are produced in small spikes or spikelike racemes, occasionally reduced to solitary flowers.

The Lauraceae contain many species responsible for several spices and flavorings (some of these mainly used in tropical countries), including cinnamon (*Cinnamomum verum* J. Presl and its relatives) and bay laurel (*Laurus nobilis* L.). A number of different tropical genera also are sources for aromatic oils. Camphor originally was extracted from the bark of *C. camphora* (L.) J. Presl. Some of the tropical tree species are harvested for timber. *Persea americana* Mill. is widely cultivated for its edible fruit, the avocado.

In 2003, conservationists and land managers in some southeastern states began noticing an alarming die-off of redbay, *Persea borbonia* (L.) Spreng. (and its segregates), another member of the Lauraceae that is native to the Atlantic Coastal Plain. Subsequently, it was discovered that the extensive mortality was being caused by an undescribed species in the pathogenic fungal genus *Raffaelea* Arx & Hennebert being spread by an invasive exotic ambrosia beetle, *Xyleborus glabratus* Eichhoff, an Asian insect with which the fungus has a symbiotic relationship (Fraedrich et al., 2008). Symptoms of the disease include the beetle entry holes in the stems, rapid wilting, branch dieback, black discolorations of the sapwood, and death of the affected plant. Currently the pathogen is still restricted to portions of South Carolina, Georgia, and Florida. However, research by plant pathologists has shown that other genera of Lauraceae also potentially are affected by this disease, which is known to infect members of the genera *Lindera* and *Sassafras*, as well as potentially cultivated avocado groves. There is concern that the insect vector eventually will disperse the pathogenic fungus throughout the eastern half of the United States and that there might be rangewide future declines of all of the species of Lauraceae that occur in Missouri.

1. Plants shrubs; leaves all unlobed; inflorescences dense axillary clusters, produced before the leaves develop; fruits red drupes, the stalk remaining slender or becoming slightly thickened at fruiting . . . . 1. LINDERA
1. Plants trees; leaves all unlobed or more commonly at least some of the leaves with 1 or 2 large lobes; inflorescences clusters of short terminal racemes, these occasionally appearing umbellate, produced as the leaves first begin to develop; fruits dark blue berries, the stalk becoming greatly thickened toward the tip at fruiting . . . . 2. SASSAFRAS

1. **Lindera** Thunb.

Plants shrubs (trees elsewhere), often colonial from rhizomes and/or root suckers. Twigs slender, reddish brown to dark brown, usually, with scattered, pale, slender, elongate lenticels, glabrous or sparsely to moderately pubescent with slender hairs, producing a spicy aroma when broken or bruised. Winter buds sessile, narrowly ovoid to narrowly ellipsoid, with a few overlapping scales, those producing inflorescences nearly globose, short-stalked, and with several scales. Leaves all unlobed, those produced first in the season often much smaller, more circular and more bluntly pointed (in *L. benzoin*) to more slender and more tapered (in *L. melissifolia*), than those produced afterward, the blades on expanding shoots ovate, elliptic, or obovate (somewhat more narrowly so in *L. melissifolia*), tapered to a sharply

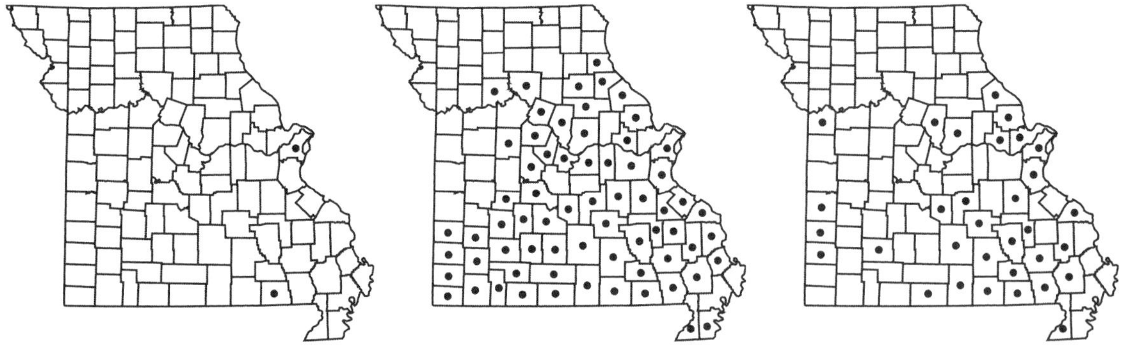

2011. Lindera melissifolia 2012. Sassafras albidum 2013. Utricularia gibba

pointed tip, rounded, angled, or tapered at the base, the surfaces variously glabrous or hairy, the venation with a single main vein and pinnate secondary veins, the network of tertiary veinlets between the secondary veins prominent or obscure. Inflorescences dense axillary clusters, produced before the leaves develop on portions of branches at least 1 year old. Flowers very short-stalked, the stalks elongating to 12 mm, remaining green, and slender or becoming slightly thickened as the fruit matures. Tepals 1–3 mm long (those of staminate flowers slightly longer than those of pistillate flowers), narrowly oblong-elliptic to oblong-ovate, bright yellow to lemon yellow or light yellow. Staminate flowers with 3 of the stamens having a pair of stalked, more or less 2-lipped, yellow nectaries at the base. Pistillate flowers with 6 to numerous staminodes, the ovary ovoid to ellipsoid, the style elongate, expanded apically into an asymmetrical, sometimes slightly 2-lobed stigma. Fruits red drupes, not appearing jointed to the tip of the not or only slightly expanded (nonbulbous), green to brownish green stalk, 6–12 mm long, ellipsoid, shiny, the seed 4–7 mm long, broadly ellipsoid to nearly globose, the surface dark brown or reddish brown, usually mottled with lighter or darker brown or with some portions covered with a pale deposit, with a faint pair of longitudinal ridges or these lacking, otherwise appearing smooth or slightly granular. About 100 species, North America, Asia, south to Australia.

1. Leaf blades angled or tapered at the base, the undersurface pale, the network of veinlets between the secondary veins inconspicuous on the undersurface; fruit stalks 3–5 mm long, not noticeably thickened toward the tip; fruits 6–10 mm long. . . . . 1. L. BENZOIN
1. Leaf blades rounded or broadly angled at the base, the undersurface not significantly lighter than the upper surface, the network of veinlets between the secondary veins conspicuous on the undersurface; fruit stalks 7–12 mm long, moderately thickened toward the tip; fruits 8–12 mm long . . . . . . 2. L. MELISSIFOLIA

1. Lindera benzoin (L.) Blume (spicebush)
Pl. 445 d–f; Map 2010

Plants shrubs, 1.2–3.0(–4.5) m tall. Leaves 1–15 cm long, oriented in ascending to spreading positions, the blade angled or tapered at the base, the upper surface glabrous or sparsely hairy along the midvein, dark green, the undersurface glabrous or sparsely to moderately hairy, mostly along the veins, pale and usually somewhat glaucous, the secondary veins all more or less parallel, the network of tertiary veinlets between the secondary veins inconspicuous. Fruit stalks 3–5 mm long, remaining slender throughout at maturity (1.0–1.5 mm in diameter at the tip), glabrous or sparsely hairy. Fruits 6–10 mm long. Seed usually broadly ellipsoid, the surface often partially covered with a pale deposit. $2n=24$. March–May.

Scattered to common in the Ozark, Ozark Border, Unglaciated Plains, and Mississippi Lowlands Divisions, but absent from all but the eastern portion of the Glaciated Plains (eastern U.S. west to Iowa and Texas; Canada). Bottomland forests, mesic upland forests, banks of streams, spring

Plate 445. Lauraceae. Limnanthaceae. Simaroubaceae. *Lindera melissifolia*, **a)** flowering branch, **b)** pistillate flower, **c)** fruiting branch. *Lindera benzoin*, **d)** flowering branch, **e)** staminate flower, **f)** fruiting branch. *Leitneria floridana*, **g)** staminate inflorescence, **h)** fruiting branch. *Floerkea proserpinacoides*, **i)** flower, **j)** fruit, **k)** habit. *Sassafras albidum*, **l)** flowering branch, **m)** pistillate flower, **n)** staminate flower, **o)** fruiting branch.

branches, and rivers, bases and ledges of bluffs, and occasionally margins of lakes; also shaded roadsides.

Lindera benzoin has sometimes been divided into two varieties, the mostly northern var. *benzoin*, with the leaves and twigs glabrous, and the more southerly var. *pubescens* (E.J. Palmer & Steyerm.) Rehder, with the leaves and twigs hairy (Steyermark, 1963). Missouri is in the broad region of geographic and morphological overlap between these morphotypes. Although the extremes are strikingly different, many Missouri specimens are only slightly hairy. Thus, it seems unreasonable to formally recognize infraspecific taxa, at least in our region.

Spicebush is an attractive ornamental in the shade garden that is early-flowering but produces attractive red fruits. Steyermark (1963) noted that staminate plants produce more showy inflorescences than do pistillate clones.

2. Lindera melissifolia (Walter) Blume (pondberry, southern spicebush)

Pl. 445 a–c; Map 2011

Plants shrubs, 0.6–2.0 m tall. Leaves 1–16 cm long, oriented in spreading to more commonly somewhat ascending positions, the blade rounded or broadly angled at the base, the upper surface sparsely to moderately hairy along the midvein and occasionally also the bases of the secondary veins, dark green, the undersurface not significantly lighter than the upper surface nor glaucous, the basal pair of secondary veins somewhat more strongly ascending than the other secondary veins, the network of tertiary veinlets between the secondary veins relatively conspicuous. Fruit stalks 8–12 mm long, becoming slightly thickened, especially toward the tip at maturity (2–3 mm in diameter at the tip), glabrous. Fruits 8–12 mm long. Seed usually nearly globose, the surface lacking a pale deposit. $2n$=24. March–April.

Uncommon, known thus far only from a single site in Ripley County (southeastern U.S. west to Missouri and [historically] Louisiana). Bottomland forests.

Steyermark (1949a) first discovered the Missouri population in 1948, where a large assemblage of mostly staminate plants was found growing in the swales between low sand dunes where water ponds during the winter and spring. *Lindera benzoin* occupies adjacent, higher, less swampy portions of the dune system. This population, which is among the largest in the world, extends southward across the Arkansas border. Steyermark brought notice to the discontinuous distribution of the species and its global rarity. About 40 extant populations currently are known from Missouri, Arkansas, Mississippi, Georgia, North Carolina, and South Carolina, but the species also was known historically from Louisiana, Alabama, and Florida. In 1986, pondberry was listed as Endangered under the federal Endangered Species Act, and it also is considered Endangered under the State Wildlife Code of Missouri. Much of the Missouri population has been brought into protection through land purchases by the Missouri Department of Conservation (Sand Ponds Natural Area and Conservation Area) and The Nature Conservancy (Nancy B. Altvater Pondberry Preserve).

Steyermark (1949a, 1963) catalogued a number of morphological differences between *L. melissifolia* and *L. benzoin*, in addition to those noted in the key to species above. Among these are the following: *L. melissifolia* tends to be a somewhat shorter, more strongly colonial plant with somewhat more drooping leaves; the lowermost pair of lateral veins branches from the midvein at an angle of about 35–45 (vs. 45–50) degrees, and this basal pair tends not to be fully parallel to the other secondary veins; the flower stalks of staminate flowers are glabrous (vs. sparsely hairy) and the stamens have the filaments of nearly equal width throughout (vs. basally broadened); and the perianth segments of pistillate flowers are longer (1.5–2.0 vs. 1.0–1.5 mm). In the older botanical literature, the species epithet sometimes was spelled *L. melissaefolium* (Steyermark, 1963), but when the generic name was conserved, the type species was listed as feminine (McNeill et al., 2006).

Steyermark (1949a, 1963) noted that the fruits of pondberry once were used locally by children as ammunition for so-called pop-guns, tubular devices constructed from the twigs of *Sambucus canadensis* L. (elderberry; Caprifoliaceae).

2. Sassafras J. Presl (sassafras)

Three to 5 species, eastern temperate North America, Asia.

1. Sassafras albidum (Nutt.) Nees (sassafras)

S. albidum var. *molle* (Raf.) Fernald

Pl. 445 l–o; Map 2012

Plants trees to 20(–40) m tall, often colonial from root suckers, the bark coarsely and deeply furrowed, reddish brown to gray, aromatic when broken or bruised. Twigs yellowish green to greenish brown (on older portions), glabrous or nearly so, producing a spicy aroma when broken or bruised. Winter buds sessile, ovoid or the lateral ones sometimes nearly globose, with a few overlapping scales. Leaves short- to more commonly long-petiolate, the petiole usually sparsely to moderately short-hairy. Leaf blades 3–12 cm long, variously all entire or more commonly with 1 or 2 large lobes, ovate to elliptic obovate in outline, long-tapered at the base, the blade or lobes angled or tapered to bluntly or sharply pointed tips; the exposed undersurface appearing strongly grayish and usually densely silky-hairy when young, the upper surface glabrous and often somewhat shiny at maturity, the undersurface pale and often somewhat glaucous at maturity, glabrous or sparsely to moderately short-hairy along the veins, the midvein with a pair of prominent, ascending branches from above the base (appearing 3-veined), the venation otherwise pinnate, with a network of fine veinlets between the secondary veins. Inflorescences clusters of short racemes, these occasionally appearing umbellate, appearing terminal (but below the season's new leaves), produced as the leaves first begin to develop. Flowers short-stalked, the stalks elongating, becoming red and strongly bulbous-thickened toward the tip as the fruit matures. Tepals 3–4 mm long, slender, greenish yellow. Staminate flowers with 3 of the stamens having a pair of bulbous, orange nectaries at the base. Pistillate flowers with 6 staminodes, the ovary ovoid, the style elongate, expanded apically into an asymmetrical, sometimes slightly 2-lobed stigma. Fruits dark blue berries, appearing somewhat jointed to the tip of the expanded, red stalk, 7–10 mm long, ovoid to broadly ellipsoid, shiny, the stone 4–7 mm long, broadly ellipsoid, with the surface uniformly light to dark brown, with a pair of angled longitudinal ridges, otherwise appearing somewhat roughened or granular. $2n=48$. April–May.

Scattered to common in the southern 2/3 of the state (eastern U.S. west to Wisconsin, Kansas, and Texas; Canada). Mesic to dry upland forests (mostly along the margins), bottomland forests, upland prairies, glades, savannas, tops of bluffs, slopes of sinkholes, and banks of streams and rivers; also fencerows, fallow fields, railroads, roadsides, and open to shaded disturbed areas.

Steyermark (1963) recognized two varieties within *S. albidum* differing in the leaf undersurface: the var. *albidum* (white sassafras), mostly of the northeastern United States, with the developing leaves relatively sparsely hairy and the mature leaves glabrous; and the widespread var. *molle* (red sassafras), with the developing leaves densely hairy and the mature leaves sparsely hairy along the veins. These two morphotypes are striking in their extremes but intergrade entirely, even within individual populations.

Sassafras has a long history of use medicinally and in foods. It was one of the earliest exports from the American colonies to England (Sokolov, 1981). The bark, leaves, and roots were used in teas and tonics for a variety of ills ranging from liver and stomach problems to fevers, vomiting, and venereal diseases. Colonists learned of the usefulness of sassafras form Native Americans, who used the plant to treat worms, diarrhea, rheumatism, halitosis, colds, bee stings, heart troubles, and numerous other conditions. (Moerman, 1998). Sassafras is also an important ingredient in gumbos and other dishes in Creole cuisine. The powdered leaves are called filé and have been used as a flavoring and thickener in place of (or in addition to) okra fruits (*Hibiscus esculentus* L.). However, the plant parts contain a substance known as American oil, which is mostly composed of safrole. Safrole is an irritant and in the 1950s was linked to liver lesions, necrosis, and cancer (Sokolov, 1981; Burrows and Tyrl, 2001). The use of sassafras extracts or plant parts containing safrole in teas, tonics, and extracts has been banned in many countries including the United States. Apparently filé remains legal, perhaps because it contains too little safrole to pose a problem (Sokolov, 1981).

Sassafras was also an ingredient in root beer, along with the bark of *Betula lenta* L. (sweet birch), the roots of *Smilax glauca* Walter (sarsparilla), and the stems and leaves of *Gaultheria procumbens* L. (wintergreen) (Sokolov, 1981). Aside from the toxicity and illegality of using sassafras in this pungent beverage, manufacturers eventually found it easier to make root beer with substitute colorants and flavorings, and with carbonation substituted for the fermentation process.

LENTIBULARIACEAE (Bladderwort Family)

Three genera, about 280 species, nearly worldwide.

The Lentibulariaceae are a family of carnivorous plants, trapping and digesting insects and other small invertebrates either with sticky glandular hairs on the leaf surface (*Pinguicula* L.), or in specialized pitchers (*Genlisea* A. St.-Hil.) or traps (*Utricularia*). The nutrients provided by decaying tissues and other organic matter facilitate the survival of the plants in otherwise nutrient-poor environments.

1. **Utricularia** L. (bladderwort)
(P. Taylor, 1989)

Plants annual or perennial, lacking true roots, with slender branched stems, some of these sometimes stoloniferous. Leaves (these sometimes interpreted as systems of flattened branches) absent or more commonly alternate (opposite or whorled elsewhere), sessile or short-petiolate, glabrous. Stipules absent. Leaf blades entire or dichotomously and/or pinnately dissected into few to numerous linear lobes. Traps few to numerous, small, globose to saclike, attached mostly laterally along the leaf divisions. Inflorescences racemes or sometimes reduced to a solitary flower, usually with a relatively short axis and a long stalk, the flowers subtended by small bracts, these also usually scattered along the inflorescence stalk. Flowers perfect, hypogynous. Calyces deeply divided into 2 lobes, the upper lobe often slightly broader than the lower one, persistent at fruiting. Corollas zygomorphic, 2-lipped with a very short tube, the lower lip with a raised, folded, and/or inflated, often slightly 2-lobed "palate" of tissue, which effectively closes the corolla throat, and a conic to narrowly cylindric spur at the base (this usually angled forward under the corolla). Stamens 2, attached at the tip of the corolla tube, the filaments short, the anthers attached at the midpoint. Pistil 1 per flower, of 2 carpels, the ovary superior, 1-locular, with free-central or basal placentation, the ovules numerous. Style 1, short, the stigma 2-lobed. Fruits capsules, with numerous tiny brown seeds. About 214 species, nearly worldwide.

Utricularia is the only genus of carnivorous plants to be reported thus far from Missouri. The traps generally operate by a suction mechanism, this triggered by disturbance to a pair of minute hairlike appendages located along the rim. Glands inside the trap secrete substances that digest the prey or other organic matter that may have entered the trap. Prey organisms are mostly minute aquatic or soilborne invertebrates and microorganisms.

1. Leaves absent or few, entire and linear; bracts peltate . . . . 3. U. SUBULATA
1. Leaves regularly present, dichotomously (occasionally palmately) divided or more or less pinnately dissected into linear segments; bracts attached basally (but the bases usually cordate or with basal auricles and usually also clasping)
    2. Leaf segments noticeably flattened (but often very slender and thus visible only with magnification); corolla spurs inconspicuous, 1–2 mm long, saclike or broadly and bluntly conic . . . . 2. U. MINOR
    2. Leaf segments not or only slightly flattened (more or less circular in cross-section, in *U. gibba*) or somewhat flattened (elliptic in cross-section, in *U. vulgaris*); corolla spurs relatively well-developed, 3–12 mm long, narrowly conic to more or less cylindric, bluntly or sharply pointed at the tip
        3. Inflorescences with 1–3(–4) flowers; corollas (including the spur) 5–12 mm long; leaves dichotomously divided or dissected into 2–4(–8) segments . . . . 1. U. GIBBA
        3. Inflorescences with (3–)6–20 flowers; corollas (including the spur) 12–20 mm long; leaves pinnately and dichotomously dissected into numerous segments . . . . 4. U. VULGARIS

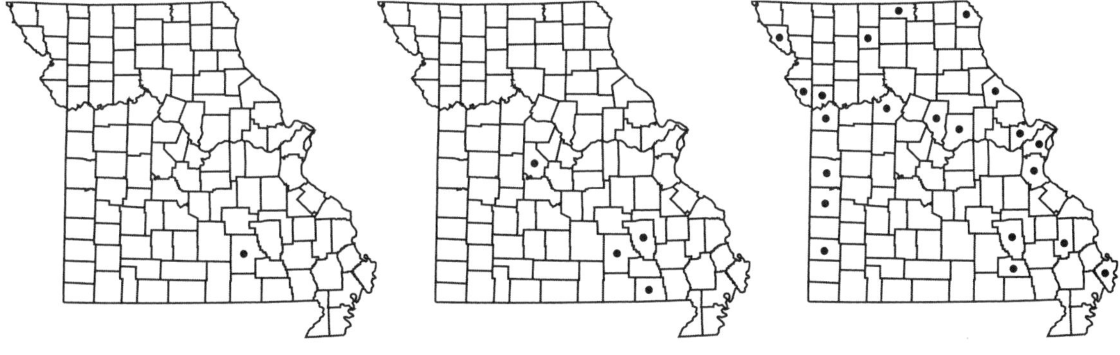

2014. Utricularia minor　　2015. Utricularia subulata　　2016. Utricularia vulgaris

1. Utricularia gibba L. (humped bladderwort, swollen-spurred bladderwort)

Pl. 444 i–k; Map 2013

Plants annual or more commonly perennial. Stems generally less than 20 cm long. Leaves regularly present, mostly 5–20 mm long, sessile, dichotomously divided into 2–4(–8) segments, the segments slender, not or only slightly flattened (more or less circular in cross-section), mostly tapered to minute, bristlelike tips. Inflorescences with 1–3(–4) flowers, the stalk relatively slender, glabrous. Bracts 0.8–1.2 mm long, oblong to broadly ovate, attached at the base, which partially encircles the axis, rounded to more or less truncate at the tip. Calyx lobes 1–3 mm long, broadly ovate to nearly circular, broadly rounded at the tip. Corollas (including the spur) 5–12 mm long, yellow, sometimes with reddish brown veins, the 2 lobes about the same size or the lower lip slightly smaller; the upper lip broadly ovate to nearly circular, entire or slightly 3-lobed; the lower lip broadly oblong to nearly circular, entire, the palate densely hairy, the spur 3–5(–7) mm long, shorter than to longer than the lower lip, conic to conic-cylindric, bluntly pointed (rarely minutely notched) at the tip. Fruits 2–3 mm long, globose. Seeds 0.8–1.0 mm in diameter, circular or irregularly angled in outline, strongly flattened, the rim broadly winged, the body smooth to finely warty. $2n$=28. May–September.

Scattered, mostly south of the Missouri River, most commonly in the Ozark Division (eastern U.S. west to Minnesota and Texas; also Washington to California; Canada, Mexico, Central America, South America, Caribbean Islands, Europe, Asia, Africa, New Zealand, Australia). Submerged aquatic (with emergent inflorescences) in ponds, sinkhole ponds, lakes, and sloughs; also ditches; rarely stranded on mud.

2. Utricularia minor L. (lesser bladderwort)

Map 2014

Plants perennial. Stems generally less than 20 cm long, but occasionally to 40 cm or more (difficult to observe as the brittle stems fragment easily when specimens are collected). Leaves regularly present, mostly 2–8 mm long, sessile, dichotomously or palmately and dichotomously divided into 7–16(–32) segments (those on branches below the soil surface often only 2–5-branched), the segments noticeably flattened (but often slender and thus only observed with magnification), angled or tapered to sharply pointed or bristlelike tips. Inflorescences with 2–6 flowers, the stalk relatively slender, glabrous. Bracts 1–2 mm long, broadly ovate to nearly kidney-shaped, attached at the base, but with a pair of well-developed, downward-angled auricles that partly encircle the axis, rounded to bluntly pointed at the tip. Calyx lobes 1.3–2.0 mm long, broadly ovate, broadly rounded at the tip (the upper lobe somewhat concave). Corollas (including the spur) 3.5–6.0(–8.0) mm long, yellow, lacking reddish veins, the 2 lobes about the same size or the lower lip slightly smaller; the upper lip broadly ovate to oblong-ovate, shallowly notched at the tip; the lower lip broadly obovate to nearly circular, entire or shallowly notched at the tip, the palate glabrous, the spur 1–2 mm long, inconspicuous, much shorter than the lower lip, saclike or broadly conic, rounded to broadly and bluntly pointed at the tip. Fruits 2–3 mm long, globose. Seeds 0.8–1.0 mm in diameter, circular or irregularly angled in outline, strongly flattened, the rim wingless or very narrowly winged, the body smooth to finely warty. $2n$=36, 40, 44. April–May (rarely to July).

Uncommon, known thus far from a single site in Shannon County (northern U.S. [including Alaska] south to Delaware, Arkansas, Arizona, and California; also North Carolina; Canada,

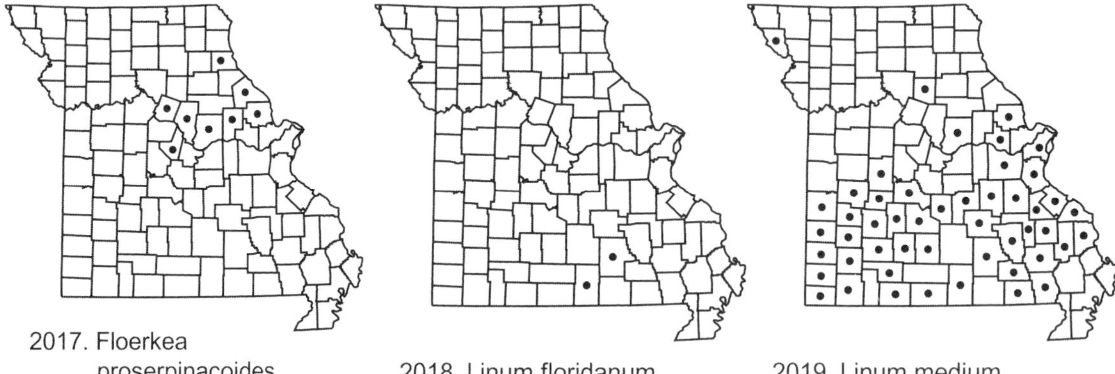
2017. Floerkea proserpinacoides
2018. Linum floridanum
2019. Linum medium

Greenland, Europe, Asia). Submerged aquatic (with emergent inflorescences) in very shallow water in fens.

This delicate species was discovered by Justin Thomas in 2010 in a fen that also supports *U. subulata* (Namestnik et al., 2012). *Utricularia minor* is visible as erect inflorescences above the water for only a short time, this mostly finished before *U. subulata* begins flowering. The plants form mats of intertwined stems in open areas that are permanently inundated with very shallow water. Stems that creep across the substrate surface generally are green, but other stoloniferous branches are subterranean. Bladders and leaves are produced on both stem types, but the leaves are more divided on stems positioned at the surface. Vegetatively the plants may be distinguished from *U. subulata* in their flattened leaves (visible with magnification).

3. Utricularia subulata L. (slender bladderwort, zigzag bladderwort)

Map 2015

Plants annual. Stems generally less than 10 cm long. Leaves absent or few, 3–12(–20) mm long, short-petiolate, entire, linear, slender to relatively stout, not flattened (circular in cross-section), mostly bluntly pointed at the tips. Inflorescences with 1–12 flowers, the stalk relatively slender, glabrous or sparsely appressed-hairy toward the base. Bracts 1–2 mm long, ovate to elliptic, peltate, more or less appressed to the axis, rounded to bluntly pointed at the ends. Calyx lobes 1.0–1.5 mm long, broadly ovate to nearly circular, broadly rounded at the tip. Corollas (including the spur) 8–14 mm long, yellow, the upper lip noticeably smaller than the lower one, broadly ovate, entire; the lower lip broadly ovate-triangular, 3-lobed, the palate densely hairy, the spur (which is often appressed to the lower lip and thus difficult to observe) 2.5–6.0 mm long, narrowly conic-tubular, relatively slender and bluntly to sharply pointed (rarely minutely notched). Fruits 1.0–1.5 mm long, globose. Seeds 0.2–0.3 mm long, obovate, not winged, the body with a reticulate pattern of low ridges. $2n=30$. May–June, occasionally again in September.

Uncommon, known thus far from Morgan, Reynolds, and Shannon Counties (U.S. [confined mostly to the Atlantic and Gulf Coastal Plains, north locally to Oklahoma and Missouri]; Canada, Mexico, Central America, South America, Caribbean Islands, Europe, Asia, Africa, Borneo, Australia). Open marly areas of fens.

This species is unusual among Missouri bladderworts in its more or less terrestrial habit (although because it occurs in water-saturated soils it is more correctly considered an emergent aquatic). The stems tend to be mostly subterranean and creeping, with only portions of the leaf blades and inflorescences above the ground. It was first reported for the state by J. M. Sullivan (1995) based on plants discovered that year by the Botany Group of the Webster Groves Nature Study Society during their annual spring field trip. The plants are quite attractive when in flower, dotting the ground with yellow, but can be impossible to locate a mere two weeks before or after flowering. P. Taylor (1989) stated that this is the most widespread species in the genus.

4. Utricularia vulgaris L. **ssp. macrorhiza** (Leconte) R.T. Clausen (common bladderwort, greater bladderwort)

U. vulgaris L. var. *americana* A. Gray

U. macrorhiza Leconte

Pl. 444 l, m; Map 2016

Plants perennial. Stems mostly 25–100 cm long. Leaves regularly present, mostly 15–60 mm long, sessile, usually dichotomously divided at the base, each half then more or less pinnately then dichotomously dissected into numerous segments, the segments somewhat flattened (elliptic in cross-section), tapered to minute, bristlelike tips. Inflorescences with (3–)6–20 flowers, the stalk relatively stout, glabrous. Bracts 3–7 mm long, broadly ovate, attached at the somewhat cordate base, bluntly to sharply pointed at the tip. Calyx lobes 4–6 mm long, broadly ovate, the upper lobe sharply pointed at the tip, the lower lobe rounded to shallowly notched at the tip. Corollas (including the spur) 12–20 mm long, yellow, the 2 lobes about the same size or the upper lip slightly smaller; the upper lip broadly ovate, entire or slightly 2-lobed; the lower lip broadly oblong-elliptic, usually slightly 2-lipped, the palate densely hairy and sometimes also glandular toward the base, often spotted or mottled red, the spur 7–12 mm long, shorter than the lower lip, conic, relatively stout and usually sharply pointed at the tip. Fruits 5–6 mm long, globose. Seeds about 0.4 mm long, square to depressed-rectangular in outline, strongly 4–7-angled, the angles and ends winged, the body smooth to finely roughened. $2n$=40, 44. May–September.

Scattered and sporadic, mostly south of the Missouri River (nearly throughout the U.S. [including Alaska]; Canada, Mexico, Asia). Submerged aquatic (with emergent inflorescences) in ponds, sinkhole ponds, lakes, sloughs, and slow-moving portions of streams; also ditches.

The taxonomy of this complex is still somewhat controversial. Steyermark (1963) treated *U. vulgaris* in the traditional sense as a morphologically variable, circumboreal species. P. Taylor (1989) accepted two species in the complex, the mostly American *U. macrorhiza* and the closely related Old World *U. vulgaris*, based on subtle differences in corolla spur shape, trap morphology, and seed ornamentation. Molecular data have suggested that the two taxa are closely related but apparently distinct genetically (Jobson et al., 2003). The present treatment takes a middle course, in line with the forthcoming treatment by Garrett Crow (University of New Hampshire, personal communication) in the Flora of North America series.

LIMNANTHACEAE (False Mermaid Family)

Two genera, 11 species, endemic to temperate North America.

1. **Floerkea** Willd. (false mermaid)

One species, endemic to temperate North America.

1. Floerkea proserpinacoides Willd.
Pl. 445 i–k; Map 2017

Plants annual, glabrous. Stems 5–30 cm long, erect or more commonly spreading to loosely ascending. Leaves alternate, short- to long-petiolate. Stipules absent. Leaf blades 2–7 cm long, pinnately deeply divided to once-compound with 3–7 leaflets, these 0.4–2.0 cm long, narrowly elliptic to oblanceolate, the margins entire. Inflorescences of solitary axillary flowers. Flowers perfect, hypogynous, long-stalked. Calyces of 3 free sepals, these 2.5–3.0 mm long at flowering, becoming enlarged to 6 mm at fruiting, lanceolate to narrowly ovate, narrowed to a sharply pointed tip. Corollas actinomorphic, of 3 free petals, these 1–2 mm long, oblanceolate, white. Stamens 3–6, free, the anthers yellow. Pistil of 2 or 3 carpels, these free nearly to the base. Ovaries superior, each with 1 locule containing 1 ovule. Style 1, attached toward the carpel base in the depression between the ovaries, the stigma 2-lobed. Fruits breaking apart into achene-like mericarps, these 2.2–2.6 mm long, broadly ovoid to nearly globose, somewhat fleshy, the surface green, finely warty. $2n$=10. April–May.

Uncommon in eastern Missouri, west locally to Boone, Callaway, and Howard Counties (northern U.S. south to Virginia, Louisiana, and California; Canada). Bottomland forests.

False mermaid is an easily overlooked species with inconspicuous flowers, often blending in with the foliage of other spring bottomland wildflowers. Steyermark (1958b) reported on his rediscovery of this species in Moniteau County in 1957 and of subsequent herbarium and literature review, which led to the discovery that Edwin James, second botanist on the Long Expedition (see the introductory chapter in Volume 1 on the history of floristic botany in Missouri), had collected the plant in Howard County in 1820 and had written about it in the expedition's report (James, 1823).

LINACEAE (Flax Family)

(C. M. Rogers, 1984)
Contributed by Robin C. Kennedy

Six to 8 genera, 220–250 species, nearly worldwide.

1. Linum L. (flax)

Plants annual or perennial herbs. Stems 1 to several from the base, erect or ascending, usually branched toward the tip. Leaves opposite or alternate, sessile, the stipules absent or present as small sessile glands (in *L. sulcatum*). Leaf blades linear to elliptic, the margins entire, the uppermost leaves rarely with small glandular teeth, the surfaces glabrous. Inflorescences panicles or loose paniculate clusters. Flowers perfect, actinomorphic, short- to long-stalked. Calyces of 5 free sepals, these in an outer whorl of 3 and an inner whorl of 2, often appearing somewhat overlapping, ascending, sometimes some or all with glandular teeth, usually persistent at fruiting. Corollas of 5 free petals, these often shed relatively quickly after the flower opens, curved outward to spreading, narrowed to a stalklike base, yellow or blue. Stamens 5, alternate with the petals, the filaments flattened, distinct or fused toward the base, often persistent at fruiting, the anthers attached above but near their base. Pistil 1 per flower, of mostly 5 fused carpels. Ovary superior, more or less 10-locular, each carpel with an incomplete “false septum,” the placentation axile or appearing apical. Styles mostly 5, distinct or slightly to nearly entirely fused, slender, the stigmas ovoid-capitate. Ovules 1 per locule (2 per carpel). Fruits capsules, ovoid or globose to depressed-globose, dehiscing along the false septa into 5 mericarps (each 2-seeded) or along all the septa into 10 mericarps (each 1-seeded), the mericarps wedge-shaped, flat or rounded across the dorsal surface. Seeds asymmetrically elliptic or obovate in outline, flattened, the surface smooth, often somewhat shiny. One hundred and fifty to 225 species, widely distributed in temperate and subtropical zones, especially in the northern hemisphere.

1. Petals blue; fruit stalks 20–38 mm long; sepals with the margins entire or with minute spreading hairs . . . . 6. L. USITATISSIMUM
1. Petals yellow; fruit stalks 5–10 mm long; some or all of the sepals with inconspicuous to prominent glandular-toothed margins
    2. Leaves with small black glandular stipules (visible as a pair of dots); styles fused only for about 1 mm at the base . . . . 5. L. SULCATUM
    2. Leaves lacking stipules; styles either fused almost to the tip or free to the base
        3. Leaves linear, 0.7–1.6 mm wide; styles 3–4 mm long, fused almost to the tip; all of the sepals with prominent glandular-toothed margins . . . . 3. L. RIGIDUM
        3. Leaves elliptic, more than 1.0 mm wide; styles 1–3 mm long, free to the base; outer sepals with the margins entire, the inner sepals inconspicuously to prominently glandular-toothed
            4. Styles 2–3 mm long; inner sepals with conspicuous glandular teeth; leaves 1.3–3.5 mm wide; fruits sometimes tardily dehiscent and remaining on the plant for some time after maturity
                5. Sepals narrowly lanceolate; fruits 2.3–3.0 mm long, ovoid to somewhat pear-shaped, longer than wide; leaf blades 1.3–2.0 mm wide . . . . 1. L. FLORIDANUM

5. Sepals ovate-lanceolate; fruits 1.5–2.5 mm long, depressed-globose, slightly wider than long; leaf blades 2.0–3.5 mm wide . . . . 2. L. MEDIUM

4. Styles 1–2 mm long; inner sepals entire or with sparse, minute, inconspicuous teeth above the midpoint; leaves 4–12 mm wide; fruits readily shattering, falling from the plant soon after maturity

6. Leaves 5–12 mm wide, elliptic; stem branches with fine longitudinal ridges or lines toward the tip, angled in cross-section, and with a small wing of tissue extending downward along the stem from the midrib of each leaf blade; mericarps rounded across the dorsal surface . . . . 4. L. STRIATUM

6. Leaves 4–7 mm wide, elliptic or occasionally elliptic-obovate; stem branches with inconspicuous longitudinal lines but not ridged, circular in cross-section, only occasionally with a small wing of tissue extending downward along the stem from the midrib of each leaf blade; mericarps flattened across the dorsal surface . . . . 7. L. VIRGINIANUM

1. Linum floridanum (Planch.) Trel. **var. floridanum** (Florida yellow flax)

L. virginianum L. var. *floridanum* Planch.

Map 2018

Plants usually perennial. Stems (20–)40–70 (–90) cm long, mainly solitary, glabrous, faintly longitudinally ridged (somewhat angled in cross-section), including an occasional ridge descending from the extension of a leaf midvein. Leaves alternate (occasionally the lowest few pairs opposite). Stipules absent. Leaf blades 1.2–2.0 cm long, 1.3–2.0 mm wide, narrowly elliptic to narrowly oblong-oblanceolate or nearly linear, narrowed or tapered to a sharp point at the tip, the margins entire. Sepals 2.0–3.5(–4.2) mm long, narrowly lanceolate, those of the outer whorl with entire margins, those of the inner whorl with conspicuous small glandular teeth along the margins. Petals 5.0–9.5 mm long, lemon yellow. Styles distinct, (2–)2.3–3.0 mm long. Fruits usually readily dehiscent, but remaining fused at the base and usually remaining on the plant for some time after maturity, 2.3–3.0 mm long, 1.8–2.8 mm in diameter, longer than wide, ovoid to somewhat pear-shaped, breaking into 10 mericarps, each 1-seeded, more or less rounded across the dorsal surface, the septa glabrous, the mature fruits strongly purplish-tinged. Seeds 1.6–2.1 mm long, reddish brown. $2n$=36. May–August.

Uncommon, known thus far only from Howell and Shannon Counties (southeastern U.S. west to Illinois, Missouri, and Texas; Caribbean Islands). Fens.

This species was first recognized as growing in Missouri in 2010 by Justin Thomas, although previously misdetermined specimens dating to 1990 were discovered during the course of his research. It is primarily a Coastal Plain species, but also has been reported from southern Illinois (Fernald, 1950; Mohlenbrock, 2002). C. M. Rogers (1963, 1984) discussed the strong morphological similarity of *L. floridanum* var. *floridanum* to *L. medium* var. *texanum*, which is the name under which earlier Missouri specimens had been masquerading. He noted that, in addition to the fruit dimensions, plants of *L. floridanum* tend to differ from those of *L. medium* in their slender, often taller stems with more numerous, narrower leaves, as well as in having smaller inflorescences with shorter branches. The fruits in *L. virginianum* dehisce readily, but the sections of fruit wall remain attached basally and tend to persist on the plants after dehiscence, whereas those of *L. medium* often dehisce tardily, but are usually shed soon after they break apart. C. M. Rogers also noted that pollen of the latter species regularly has three grooves, whereas in the former taxon, the pollen grains usually have ten grooves. Nevertheless, he went on to mention that some plants are morphologically intermediate between the two taxa and suggested that past interspecific hybridization probably has complicated the situation.

Linum floridanum is usually treated as comprising two varieities. The var. *chrysocarpum* C.M. Rogers tends to be restricted to the Atlantic and Gulf Coastal Plains and differs from var. *floridanum* in its more ovoid, slightly larger (3.0–3.2 mm long), thicker-walled, minutely pointed fruits that lack purple pigmentation, its slightly larger seeds (2.1–2.4 mm long), and its longer anthers (about 1.5 vs. 1.2 mm). C. M. Rogers (1963, 1984) noted that there is overlap between the varieties.

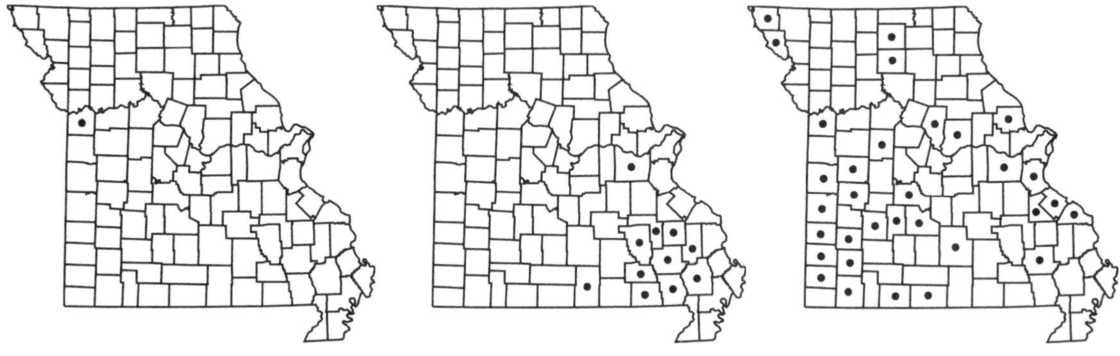
2020. Linum rigidum 2021. Linum striatum 2022. Linum sulcatum

2. Linum medium (Planch.) Britton **var. texanum** (Planch.) Fernald (stiff yellow flax, sucker flax)

L. virginianum L. var. *texanum* Planch.

Pl. 446 f, g; Map 2019

Plants usually perennial. Stems 20–70 cm long, mainly solitary, glabrous, faintly longitudinally ridged (somewhat angled in cross-section), including an occasional ridge descending from the extension of a leaf midvein. Leaves alternate (occasionally the lowest pair opposite). Stipules absent. Leaf blades 1.2–2.7 cm long, 2.0–3.5 mm wide, narrowly elliptic, narrowed or tapered to a sharp point at the tip, the margins entire. Sepals 2.0–3.5 mm long, lanceolate-ovate, those of the outer whorl with entire margins, those of the inner whorl with conspicuous small glandular teeth along the margins. Petals 4–8 mm long, lemon yellow. Styles distinct, 2–3 mm long. Fruits often tardily dehiscent, remaining on the plant for some time after maturity, 1.5–2.5 mm long, 2.0–2.5 mm in diameter, depressed-globose, breaking into 10 mericarps, each 1-seeded, more or less rounded across the dorsal surface, the septa glabrous, the mature fruits usually tinged with purplish stripes from the style bases downward along the less-prominent septa. Seeds 1.3–1.7 mm long, reddish brown. $2n=36$. May–September.

Scattered, mostly south of the Missouri River; apparently also absent from the Mississippi Lowlands Division (eastern U.S. west to Iowa and Texas; Canada; Caribbean Islands). Openings of mesic to dry upland forests, tops and ledges of bluffs, upland prairies, savannas, and glades, often on acidic substrates.

The var. *medium* is restricted to Ontario, Canada, and adjacent portions of New York and Pennsylvania. This tetraploid ($2n=72$) variety differs in its thicker blunter leaves and its sepals with less glandular-toothed margins. For discussion on the separation of this species from *L. floridanum*, see the treatment of that species.

3. Linum rigidum Pursh **var. compactum** (A. Nelson) C.M. Rogers (stiffstem yellow flax; Wyoming flax)

L. compactum A. Nelson

Pl. 446 h, i; Map 2020

Plants usually annual. Stems 8–25 cm long, several to numerous, glabrous or inconspicuously hairy toward the base, with prominent longitudinal ridges (angled in cross-section) descending from an extension of each leaf midvein. Leaves alternate. Stipules absent. Leaf blades 1.0–2.5 cm long, 1.0–1.5 mm wide, linear, narrowed or tapered to a sharp point at the tip, the margins entire or sparsely and shallowly toothed grading into glandular-toothed toward the tip. Sepals 5–9 mm long, linear-lanceolate (outer whorl) to lanceolate (inner whorl), those of both whorls with conspicuous glandular teeth along the margins and usually also with a noticeably ridged midvein. Petals 5–9 mm long; yellow. Styles fused almost to the tip, 3–4 mm long. Fruits tardily dehiscent, usually remaining on the plant for some time after maturity, 3.5–4.5 mm long, 2.6–3.4 mm in diameter, ovoid, breaking into five mericarps, each 2-seeded, more or less rounded across the dorsal surface, the septa glabrous, the mature fruits lacking purple stripes. Seeds 2.5–3.1 mm long, reddish brown. $2n=30$. May–July.

Introduced, known only from Jackson County (North Dakota to Montana south to Texas and New Mexico; Canada; introduced in Missouri and Illinois). Railroads and open disturbed areas.

Although in his later treatment for the North American Flora series, C. M. Rogers (1984) treated this taxon as a separate species, most authors have continued to follow the broader concept of *Linum rigidum* as comprising several intergrading varieties, as C. M. Rogers (1963, 1968) proposed in his earlier taxonomic revisions of the yellow-flowered flax species in North America and Central America. Steyermark (1963) and Gleason and Cronquist

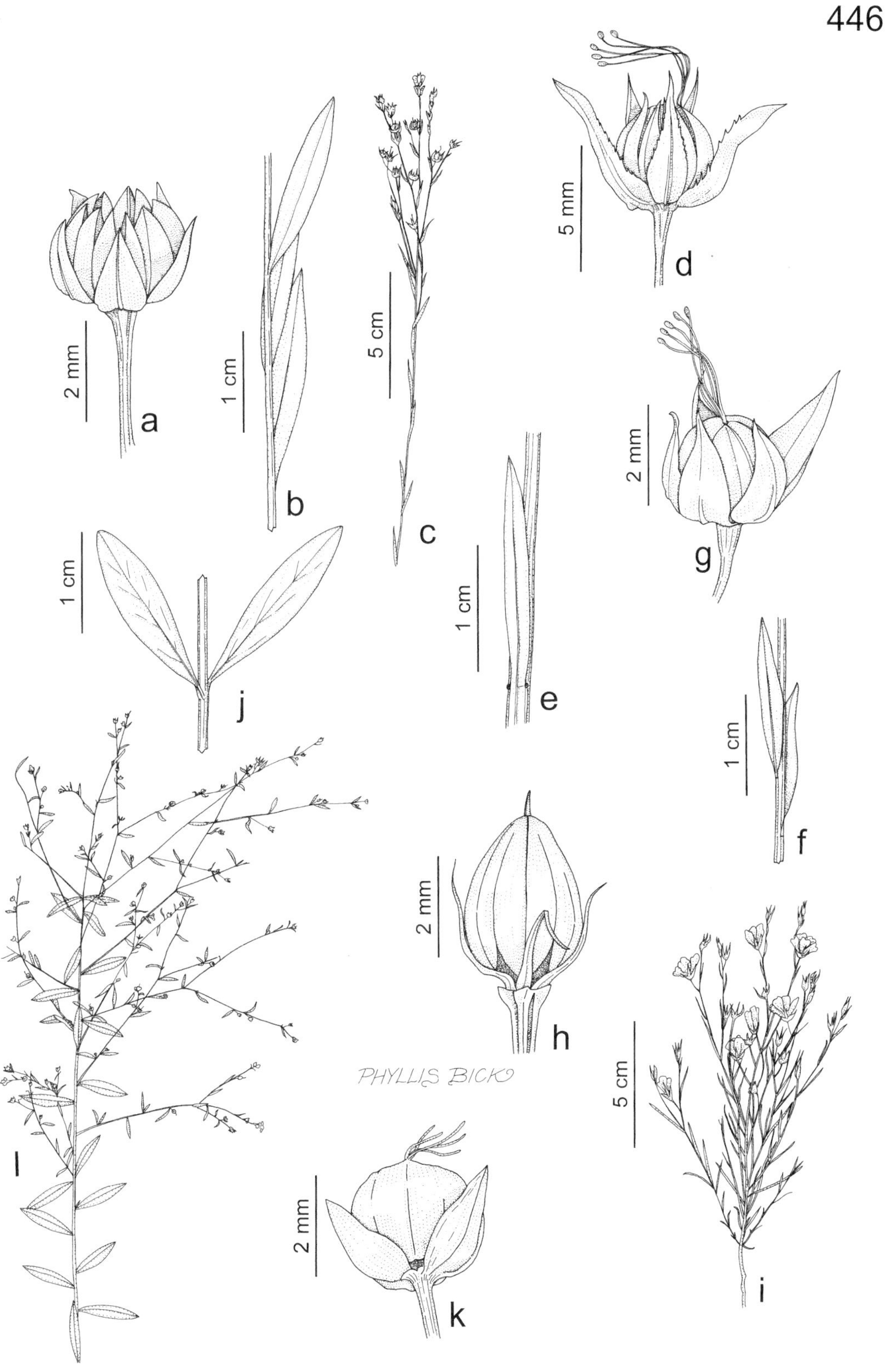

Plate 446. Linaceae. *Linum striatum*, **a)** fruit, **b)** portion of stem with leaves. *Linum sulcatum*, **c)** habit, **d)** fruit, **e)** node with leaf and glandular stipules. *Linum medium*, **f)** portion of stem with leaves, **g)** fruit. *Linum rigidum*, **h)** fruit, **i)** habit. *Linum virginianum*, **j)** portion of stem with leaves **k)** fruit, **l)** habit.

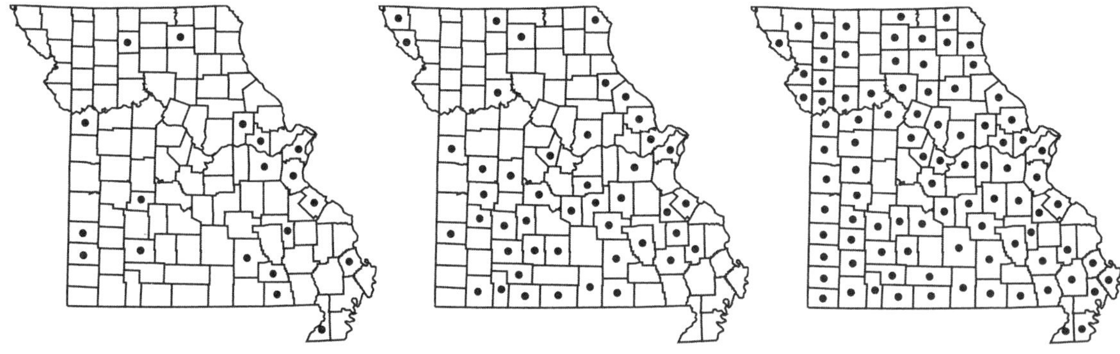
2023. Linum usitatissimum 2024. Linum virginianum 2025. Lindernia dubia

(1963, 1991) reported plants from Missouri as var. *rigidum,* which is widely distributed in the central United States and Canada (to the west of Missouri), but C. M. Rogers corrected the determination on the voucher specimens in various herbaria to var. *compactum*.

4. Linum striatum Walter

Pl. 446 a, b; Map 2021

Plants perennial. Stems 20–80 cm long, solitary or few, glabrous, finely longitudinally ridged or lined (angled in cross-section) above the midpoint, with small wings of tissue descending from an extension of each leaf midvein. Leaves mostly opposite below the stem midpoint, alternate above. Stipules absent. Leaf blades 1.5–3.5 cm long, (3–)5–12 mm wide, elliptic, narrowed to a sharply pointed tip, the margins entire. Sepals 1.5–3.5 mm long, elliptic, those of the outer whorl with entire margins, those of the inner whorl mostly entire but sometimes with inconspicuous minute teeth above the midpoint, these occasionally somewhat glandular. Petals 3–5 mm long, pale yellow to light yellow. Styles distinct, 1.2–2.0 mm long. Fruits readily shattering, falling from the plant soon after maturity, 1.3–1.9 mm long, 1.8–2.3 mm in diameter, depressed-globose, breaking into 10 mericarps, each 1-seeded, rounded across the dorsal surface, the septa glabrous, the mature fruits lacking purple stripes. Seeds 1.0–1.5 mm long, reddish brown. $2n$=36. May–September.

Scattered to uncommon in the southeastern quarter of the state (eastern U.S. west to Illinois, Oklahoma, and Texas; Canada). Fens, banks of streams and rivers, and bottomland forests.

5. Linum sulcatum Riddell **var. sulcatum**

(grooved yellow flax, ridged yellow flax)

Pl. 446 c–e; Map 2022

Plants annual. Stems 30–85 cm long, solitary, glabrous, usually with fine longitudinal lines but not prominently ridged (circular in cross-section). Leaves alternate. Stipules sessile glands (visible as a pair of small black dots at the leaf base). Leaf blades (1.0–)2.0–2.7 cm long, 2–3 mm wide, linear to linear-lanceolate, tapered to a sharply pointed tip, the margins entire (sometimes the bracteal leaves in the inflorescence with inconspicuous glandular teeth), the surfaces and margins glabrous. Sepals 3.5–7 mm long, lanceolate to narrowly lanceolate, those of both whorls with prominent glandular teeth along the margins. Petals 5–6(–10) mm long, pale yellow to light yellow. Styles fused for about 1 mm at the base, 3–5 mm long. Fruits readily shattering, falling from the plant soon after maturity, 2.5–3.5 mm long, 2–3 mm in diameter, globose to somewhat ovoid, breaking into 10 mericarps, each 1-seeded, more or less rounded across the dorsal surface, the septa moderately to densely hairy along the ventral margins, the mature fruits lacking purple stripes. Seeds 1.5–2.1 mm long, reddish brown. $2n$=30. May–October.

Scattered nearly throughout the state, but apparently absent from the Mississippi Lowlands Division and portions of the Glaciated Plains (eastern U.S. west to North Dakota and Texas; Canada). Glades, upland prairies, and less commonly openings of dry upland forests.

The var. *harperi* (Small) C.M. Rogers is endemic to portions of Florida and Georgia. It differs in its slightly shorter sepals and in its darker, purple-spotted stems (especially when specimens are dried).

6. Linum usitatissimum L. (common flax)

Pl. 447 g, h; Map 2023

Plants annual, Stems 25–80 cm long, solitary or more commonly several, glabrous, usually with fine longitudinal lines but not prominently ridged (circular in cross-section). Leaves alternate. Stipules absent. Leaf blades 2.5–3.0 cm long, 3–5 mm wide, linear-elliptic, narrowed or tapered to a

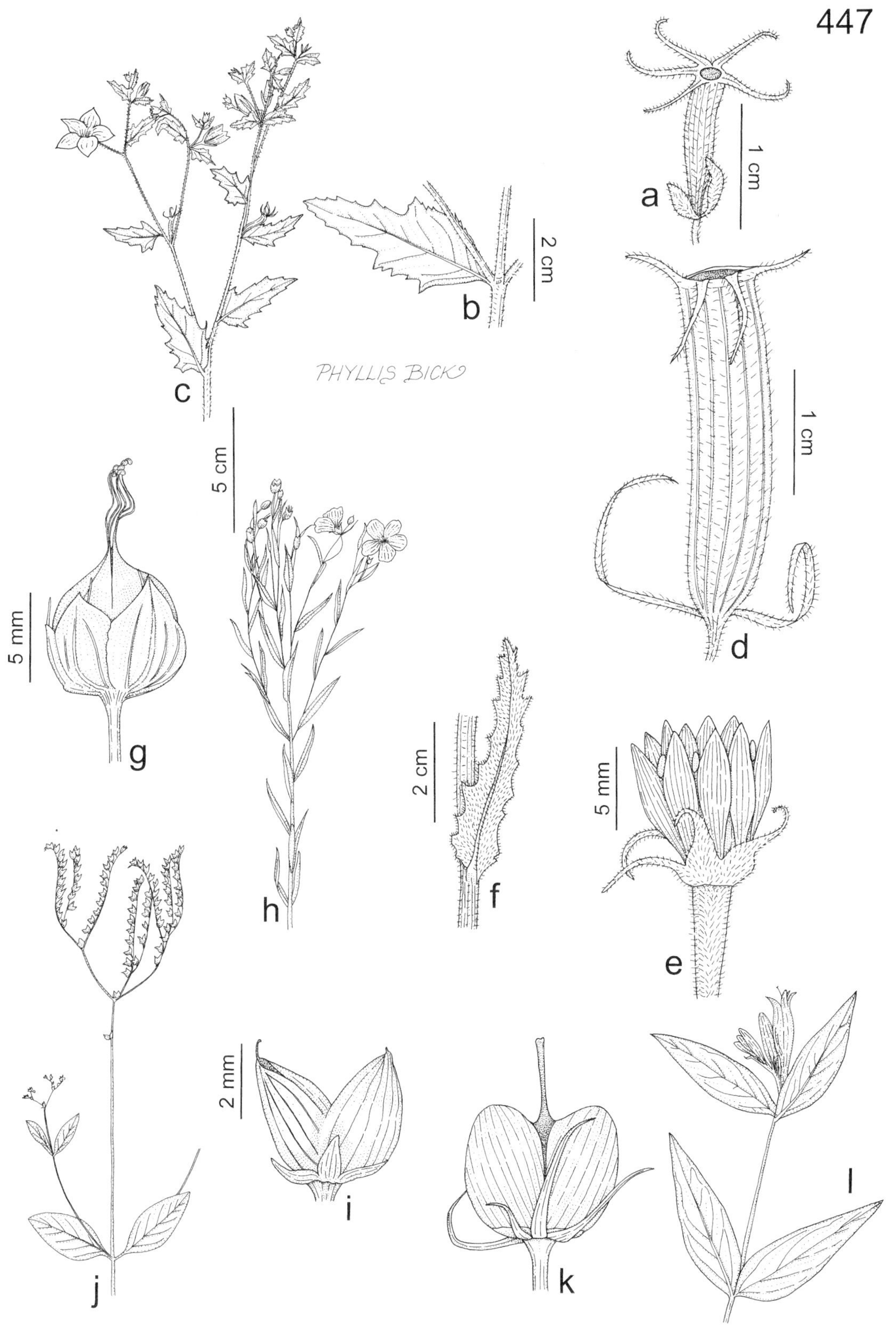

Plate 447. Linaceae. Loasaceae. Loganiaceae. *Mentzelia oligosperma*, **a)** fruit, **b)** node with leaf, **c)** fertile stem. *Mentzelia albescens*, **d)** fruit, **e)** flower, **f)** node with leaf. *Linum usitatissimum*, **g)** fruit, **h)** fertile stem. *Mitreola petiolata*, **i)** fruit, **j)** fertile stem. *Spigelia marilandica*, **k)** fruit, **l)** fertile stem.

sharply pointed tip, the margins entire. Sepals 7–9 mm long, ovate, those of both whorls with the margins thin and entire, those of the inner whorl often with minute spreading hairs. Petals 10–15 mm long, blue. Styles distinct, 2.5–5.0 mm long. Fruits usually tardily dehiscent, often remaining on the plant for some time after maturity, 7–10 mm long, 6–9 mm in diameter, ovoid to subglobose, breaking into 10 mericarps, each 1-seeded, rounded across the dorsal surface, the septa glabrous or occasionally sparsely hairy along the ventral margins, the mature fruits lacking purple stripes. Seeds 4–6 mm long, brown. $2n=30$. May–September.

Introduced, uncommon and sporadic, mostly in eastern and southwestern Missouri (an Old World cultigen unknown from natural populations, introduced nearly throughout North America). Roadsides, railroads, old fields, and open disturbed areas.

Common flax has an extremely long history of cultivation. It is not widely grown in Missouri, but is a common crop plant in the northern states from Minnesota to Montana. The seeds of flax are the source of linseed oil, and they are also a minor component of some animal feeds. They are also an ingredient in some bird seed mixes, which may be how some of the recently collected populations in Missouri became established. Linseed oil formerly was important as a drying agent in inks and paints, and in the manufacture of linoleum, varnishes, and soaps. Today it is no longer commonly used in human foods, but is still occasionally used in medicinal preparations. The stem fibers are used in linen cloth and sometimes also in canvas, thread, twine, paper (including cigarette rolling papers, fine writing papers, legal documents, and paper currency), and carpet material (Steyermark, 1963).

7. Linum virginianum L. (Virginia yellow flax) Pl. 446 j– l; Map 2024

Plants perennial. Stems 25–80 cm long, solitary or more commonly several, glabrous, with fine longitudinal lines but not ridged (circular in cross-section), occasionally with small wings of tissue descending from an extension of some leaf midveins. Leaves opposite toward the stem base, alternate toward the tip. Stipules absent. Leaf blades 2–3 cm long, 4.5–7.0 mm wide, elliptic or occasionally elliptic-obovate, narrowed to a sharply pointed tip, the margins entire. Sepals 1.5–3.5 mm long, lanceolate to narrowly ovate, those of the outer whorl with entire margins, those of the inner whorl with only a few minute sometimes glandular teeth toward the tip. Petals 3.5–5.5 mm long, yellow. Styles distinct, 1.0–1.6 mm long. Fruits readily shattering, falling from the plant soon after maturity, 1.3–1.8 mm long, 2.0–2.5 mm in diameter, depressed-globose, breaking into 10 mericarps, each 1-seeded, the mericarps flattened across the dorsal surface, the septa sparsely hairy along the ventral margins. Seeds 1.0–1.5 mm long, reddish brown. $2n=36$. May–September.

Uncommon to scattered in the Ozark and Ozark Border Divisions (eastern U.S. west to Iowa and Alabama; Canada). Glades, upland prairies, loess hill prairies, tops of bluffs, savannas, banks of streams and rivers, and rarely fens; also roadsides.

Steyermark (1963) overlooked the presence of *L. virginianum* in Missouri and misdetermined specimens as either *L. medium* or *L. striatum*. At about the same time, C. M. Rogers (1963) confirmed its presence in the state. The species is conspicuously uncommon in Illinois and the Missouri populations are somewhat disjunct from the eastern portion of the distributional range of the species.

LINDERNIACEAE (False Pimpernel Family)

Eight to 21 genera, about 300 species, nearly worldwide.

Molecular phylogenetic studies have revealed that *Lindernia* and a suite of related genera, which traditionally were classified as part of the *Gratiola* alliance (now in Plantaginaceae), represent a strongly supported lineage distinct from other families in the order Lamiales (Albach et al., 2005; Oxelman et al., 2005; Rahmanzadeh et al., 2005; Tank et al., 2006). However, the relationship of this lineage to other families in the order remains unclear and the number of genera to be included (and their circumscriptions) requires further research. The species examined to date appear to share characteristic stamens in which some or all of the filaments are strongly bent or kinked and often also with a small spur or protrusion at the bend.

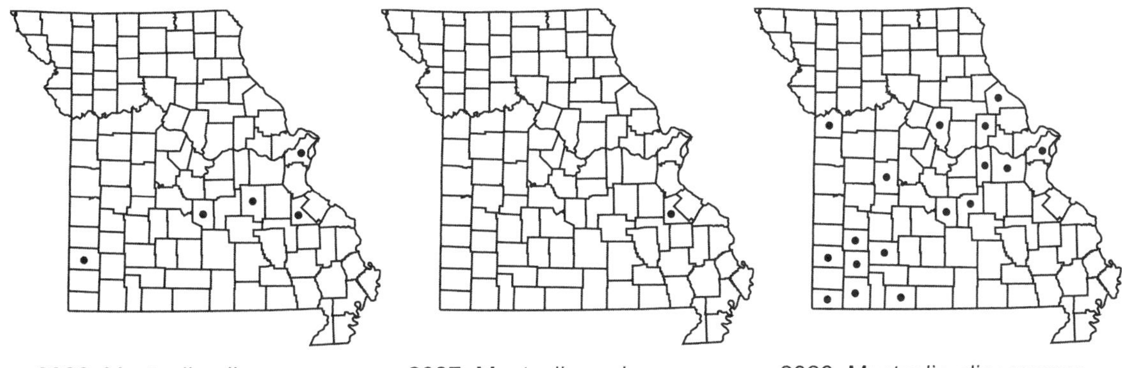

2026. Mentzelia albescens 2027. Mentzelia nuda 2028. Mentzelia oligosperma

Members of the Linderniaceae have little economic importance. Many of the species grow in wetlands and provide food for waterfowl and herbivorous wetland mammals. Some members of the pantropical genus *Torenia* L. (wishbone flower) are cultivated as ornamentals and aquatic members of a few other genera are sometimes grown as pond and aquarium plants. A few species have been used medicinally, mainly as salves for skin irritations.

1. **Lindernia** All. (false pimpernel) (D. Q. Lewis, 2000)

Perhaps 160 species, most diverse in the Old World tropics.

D. Q. Lewis (2000) accepted fourteen species as occurring in the New World, seven of these native to the western hemisphere and only five of them occurring in the United States.

1. Lindernia dubia (L.) Pennell

L. dubia var. *anagallidea* (Michx.) Cooperr.

L. dubia var. *riparia* (Raf.) Fernald

L. anagallidea (Michx.) Pennell

Pl. 457 f–h; Map 2025

Plants annuals, fibrous-rooted or sometimes with a slender taproot. Stems 1 to several from the base, 2–35 cm long, loosely to strongly ascending, sometimes lax when submerged, sometimes rooting at the lower nodes, unbranched or branched, 4-angled, glabrous or sparsely glandular-hairy, sometimes only near the base. Leaves opposite, sessile. Stipules absent. Leaf blades (2–)5–16(–30) mm long, elliptic to lanceolate, ovate, oblanceolate, or obovate, occasionally nearly circular, rounded to sharply pointed at the base and tip, sometimes slightly clasping the stem, the margins entire or sparsely scalloped or toothed, the surfaces glabrous, but finely and inconspicuously gland-dotted. Inflorescences axillary, of solitary flowers in the leaf axils (2 per node or sometimes 1 per node and alternating on opposite sides at successive nodes), these lacking additional bracts or bractlets, the stalks 0.5–30 mm long, glabrous or glandular-hairy. Cleistogamous flowers sometimes produced early and late in the growing season. Calyces 2.5–4.5 mm long at flowering, persistent and becoming slightly enlarged to 4.5–6.0 mm at fruiting, more or less actinomorphic, deeply 5-lobed, the lobes linear to narrowly lanceolate, slightly curved upward. Open flowers with the corolla 4–10 mm long, pale bluish-tinged or pale lavender to nearly white externally, internally white except sometimes for pale blue to light lavender or purple margins of the lobes and markings in the throat, zygomorphic, bilabiate, the lips 2–3 mm long, the upper lip shorter than the lower one, irregular or somewhat notched at the tip, more or less straight or slightly arched or hooded, the lower lip about equally 3-lobed, more or less straight (the lip somewhat scoop-shaped) to somewhat downward-curved, bearded basally with a pair of low, thickened ridges in the throat, the throat not appearing closed by the ridges. Stamens 2, the filaments attached in the corolla tube, not exserted (usually hidden under the upper corolla lip), the anthers attached more or less at their bases, loosely fused, light yellow. Staminodes 2, the filaments fused to the corolla tube, the free portion bent above its base, spurred and thus appearing forked, the spur tapered to a point, the other branch knob-shaped at the tip, light yellow, positioned along the ridges in the corolla throat. Pistil 1 per flower, 2-carpellate. Ovary 2-locular, the placentation axile, with numerous ovules. Style 1, slender, often persistent at fruiting, the stigma 2-lobed, appearing capitate, the lobes somewhat flattened. Fruits capsules, 3–6 mm long, ellipsoid to somewhat asymmetrically ovoid, sometimes narrowly so, tan at maturity, dehiscent along the 2 sutures. Seeds numerous, minute, 0.2–0.4 mm long, more or less oblong-ellipsoid to somewhat polygonal, the surface with a fine network of ridges (these visible only with magnification), yellow to reddish brown. $2n$=18, 32. June–October.

Scattered to common nearly throughout the state (nearly throughout the U.S., Canada, Mexico, Central America, South America, Caribbean Islands). Banks and shallow water of streams, rivers, and spring branches, margins of ponds, lakes, and sinkhole ponds, sloughs, openings of bottomland forests and swamps, bottomland prairies, moist depressions of upland prairies, marshes, fens; also ditches, fallow fields, railroads, roadsides, and moist, open, disturbed areas; sometimes emergent aquatics, rarely totally submerged.

There has been considerable controversy over the taxonomic limits of *L. dubia*. Traditionally, most botanists recognized two or three species in the complex, of which two were thought to grow widely in the United States (Pennell, 1935; Steyermark, 1963). Characters said to separate *L. anagallidea* and *L. dubia* included differences in leaf size and shape (more reduced toward the stem tip and more ovate in the former), flower stalk length (longer in the former), production of cleistogamous flowers (supposedly absent in the former), and seed color (darker in the former). Cooperrider and McCready (1975) studied the group in Ohio and enumerated the problems of distinguishing the two midwestern species consistently. They chose to treat the group in the Midwest as comprising two varieties of a single species. In her taxonomic revision of *Lindernia* in the New World, D. Q. Lewis (2000) treated the complex as one species with four varieties, two of which occur only from Mexico southward. However, she noted that var. *anagallidea* and var. *dubia* were essentially indistinguishable in some portions of their ranges and were not accepted as distinct by a number of botanists (Holmgren, 1986). This appears to be the case in Missouri, where numerous intermediate specimens have been collected. Thus, the present treatment accepts *L. dubia* in a broad sense, without formal recognition of varieties.

LOASACEAE (Loasa Family)

Contributed by David Bogler

About 14 genera, 200–260 species, North America to South America, Asia, Africa, mostly in temperate and tropical America.

Ovule morphology, the presence of iridoid compounds (Kooiman, 1974), and molecular data (Hempel et al., 1995) indicate that the Loasaceae are closely related to the Hydrangeaceae and other members of the Cornales. Members of the family are distinguished by their brittle stems and variety of specialized trichomes. Species in some genera, including *Mentzelia*, have remarkable barbed hairs that stick tightly to clothing and skin (even surviving trips to the laundromat), giving rise to such common names as stickleaf and velcro plant. Some other genera, like the closely related *Eucnide* Zucc., instead have stinging hairs, and the family also includes members with calcified hairs and various glandular trichomes (Ernst and Thompson, 1963).

1. **Mentzelia** L. (stick-leaf, blazing star)

Plants biennial or perennial herbs (annuals elsewhere). Stems branched, brittle, often whitish. Leaves usually alternate, simple, the petioles short and pubescent. Stipules absent. Leaf blades pinnately veined and lobed, with abundant minute barbed hairs. Inflorescences terminal, few-flowered clusters or solitary, these apparently arranged into small racemes or panicles. Flowers actinomorphic, perfect, epigynous. Sepals 5, sometimes persistent in fruit. Petals 5 or apparently 10 (including 5 petaloid staminodes), whitish yellow to deep orange. Stamens numerous, the filaments sometimes unequal, shorter toward flower center, fused together basally and to the petal bases, sometimes expanded and petaloid. Pistil of 3 fused carpels. Ovary inferior, with 1 locule, the placentation parietal, the ovules numerous. Styles 3, united most of their length, filiform, the stigmas represented by 3 furrows or tufts of hairs. Fruits capsules, more or less cylindrical, densely pubescent with minute barbed hairs, dehiscent by an apical valve. Seeds few to numerous, elongate and prismatic or flattened and winged. About 60 species, primarily in the southwestern U.S. and Mexico, a few species in Argentina and the Bahamas.

Sections within *Mentzelia* are distinguished by characters of the filaments, seeds, and placentae. In Missouri, the common native species *M. oligosperma* belongs to section *Mentzelia*, and the two introduced species belong to section *Bartonia* Torr. & A. Gray, which at one time was recognized as a separate genus (Darlington, 1934). Additional characters are found in the number of petals and petaloid staminodes, number of stamens, and seed morphology (Darlington, 1934; R. J. Hill, 1976; D. K. Brown and Kaul, 1981). Several species have edible seeds, which can be parched and ground into a flour, and a few species are cultivated for their showy flowers.

1. Petals 5, orange; sepals shed before fruiting; seeds 1–3 per fruit, not flattened or winged, the surface with many fine ridges; stamens 15–40 . . . . . 3. M. OLIGOSPERMA
1. Petals apparently 10, creamy white to lemon yellow; sepals persistent but withered at fruiting; seeds numerous per fruit, flattened and winged, the surface minutely pebbled; stamens 30 to numerous
    2. Flowers subtended by 1–3 entire bracts; sepals 3–6 mm long; petals 6–8 (–10) mm long, lemon yellow to pale yellow; stamens 30–40 . . . 1. M. ALBESCENS
    2. Flowers subtended by 3–5 leaflike, irregularly divided bracts; sepals 10–25 mm long; petals 15–50 mm long, white to creamy white; stamens 150–300 or more . . . . 2. M. NUDA

1. Mentzelia albescens (Gillies ex Arn.) Griseb. (wavyleaf blazing star)
M. wrightii A. Gray
Bartonia albescens Gillies ex Arn.
Nuttallia albescens (Gillies ex Arn.) Standl.
Pl. 447 d–f; Map 2026

Plants biennial or perennial herbs with a thickened taproot. Stems 25–60 cm long, minutely roughened or hairy, distinctly whitish green or grayish white. Leaf blades 3–15 cm long, 1.0–2.5 cm wide, linear to ovate-lanceolate, rounded to bluntly or sharply pointed at the tip, the margins coarsely and broadly toothed or lobed. Flowers subtended by 1–3 narrow entire bracts. Hypanthium 10–20 mm long. Sepals 3–6 mm long, linear, persistent but withered at fruiting. Petals apparently 10 (including 5 petaloid staminodes), 6–8(–10) mm long, 1–2 mm wide, lanceolate to spatulate, narrowed to a sharply pointed tip, lemon yellow to pale yellow. Stamens to 30–40, the outer 5 modified into petaloid staminodes (indistinguishable from the petals). Fruits 18–28 mm long, 6–7 mm in diameter. Seeds numerous, 2–3 mm in diameter, positioned horizontally, flattened, winged, the seed coat minutely pebbled, tan. $2n=22$. May–August.

Introduced, uncommon in the southwestern and eastern portions of the state (Oklahoma, Texas, Kansas; Mexico, South America. Railroads and open disturbed areas.

The flowers of this night-blooming species open around sunset. It was first discovered growing in Missouri by Palmer (1961) on old mine tailings in Jasper County. The record arrived too late for inclusion in the main body of Steyermark's (1963) Flora, but was mentioned in the short supplement at the end. The species has since been found along railroad right-of-ways in Crawford, Pulaski, and St. Louis Counties.

2. Mentzelia nuda (Pursh) Torr. & A. Gray (sand lily)
Bartonia nuda Pursh
Nuttallia nuda (Pursh) Greene
Torreya nuda (Pursh) Eaton
Map 2027

2029. Mitreola petiolata 2030. Spigelia marilandica 2031. Ammannia coccinea

Plants coarse biennial or perennial herbs, with a thickened taproot. Stems 35–100 cm long, roughened with minute barbed hairs, at least toward the tip, yellowish green to whitish gray. Leaf blades 4–10 cm long, 5–20 mm wide, oblanceolate to narrowly oblong-elliptic, bluntly to sharply pointed at the tip, the margins coarsely and broadly toothed or lobed. Flowers subtended by 3–5 leaflike, irregularly toothed or narrowly lobed bracts. Hypanthium 12–18 mm long. Sepals 10–25 mm long, narrowly triangular, tapered to a sharply pointed tip, persistent but withered at fruiting. Petals apparently 10 (including 5 petaloid staminodes), 15–50 mm long, 3–10 mm wide, oblanceolate to nearly spatulate, white to creamy white. Stamens 150–300 or more, the outer 5 modified into petaloid staminodes (indistinguishable from the petals). Fruits 20–30 mm long, 8–10 mm in diameter. Seeds numerous, flattened and broadly winged, 2.5–3.5 mm in diameter, the surface minutely pebbled, tan. $2n=20$. July–September.

Introduced, known thus far only from a single population in St. Francois County (Montana and South Dakota south to Arizona and Texas; introduced in Missouri, Illinois). Open sandy disturbed areas.

Mentzelia nuda was first reported for Missouri by Ladd (1994). Some authors divide the species into two varieties with substantially similar ranges, based upon the number and shape of the bracts subtending the flowers (Diggs et al., 1999). The typical variety is then restricted to plants with a single more or less entire bract per flower. Missouri plants correspond to var. *stricta* (Osterh.) H.D. Harr., with 3–5 toothed or lobed bracts. However, these morphotypes do not appear to be constant enough in the native range of the species to warrant formal taxonomic recognition.

This species is pollinated by a variety of bees and appears to be an obligate outcrosser, producing no seeds in the absence of pollinators (D. K. Brown and Kaul, 1981). The flowers open in late afternoon and close shortly after sunset. Nectar is produced while the flower is open and continues to be secreted during fruit development, which seems to attract ants for defense and enhances seed set (Keeling, 1981).

3. Mentzelia oligosperma Nutt. ex Sims (stick-leaf, chicken thief)

Pl. 447 a–c; Map 2028

Plants perennial with thick woody roots. Stems 20–80 cm long, the surface flaking when dry, roughened and densely pubescent with barbed hairs, whitish gray. Leaf blades 1–7 cm long, 0.5–3.5 cm wide, lanceolate to rhombic, bluntly to sharply pointed at the tip, the margins coarsely and irregularly toothed and often also few-lobed. Flowers subtended by 3–5 leaflike, irregularly toothed or narrowly lobed bracts. Hypanthium 4–6 mm long. Sepals 5–9 mm long, narrowly lanceolate, fused at the very base, shed as a unit before fruiting. Petals 5, 8–12 mm long, 5–6 mm wide, oblong-obovate, orange. Stamens usually about 25(15–40), none modified into staminodes. Fruits 7–14 mm long, 2–3 mm in diameter. Seeds 1–3 per fruit, 3–5 mm long, 1–2 mm wide, pendulous, not winged, narrowly oblong-elliptic in outline, 3-angled, the surface finely grooved with numerous wavy lines, brown. $2n=20, 22$. June–August.

Scattered in central and southwestern Missouri, mostly in the Ozark Border and Unglaciated Plains Natural Divisions (Illinois to Texas west to Wyoming and Arizona; Mexico). Ledges and tops of bluffs, openings of dry upland forests, and glades, usually on calcareous substrates; also railroads.

The species epithet refers to the relatively few seeds per capsule. Greenhouse experiments indicate that the species is self-compatible, with capsules full of seeds produced in the absence of pollinators (D. K. Brown and Kaul, 1981). The flowers open in mid-morning and generally close by late afternoon.

LOGANIACEAE (Logania Family)

Plants annual or perennial herbs (more commonly shrubs and trees elsewhere). Stems branched or unbranched, often angled or ridged, at least toward the base. Leaves opposite, sessile or nearly so, the bases of each pair connected by a small fused stipular membrane. Leaf blades simple. Inflorescences terminal or sometimes appearing axillary, dichotomously few-branched or unbranched, the main axis or branches consisting of 1-sided spikelike racemes, the axes straight or coiled (scorpioid) at the tip, sometimes short and the flowers then appearing clustered, with small linear bracts. Flowers actinomorphic, hypogynous, perfect, sometimes subtended by a minute linear bract. Calyces deeply 5-lobed or sometimes appearing as free sepals, glabrous, persistent at fruiting. Corollas 5-lobed. Stamens 5, alternating with the corolla lobes, the filaments attached in the corolla tube, the anthers not or barely exserted, attached at their midpoints, yellow. Pistil 1 per flower, of 2 fused carpels. Ovary superior, 2-locular, with usually numerous ovules, the placentation more or less axile. Style 1, the small stigma capitate, sometimes persistent at fruiting. Fruits capsules, obovate in outline, slightly or moderately flattened, deeply notched or 2-lobed (2-horned) at the tip, glabrous, smooth or less commonly with sparse minute papillae. Seeds mostly numerous, occasionally (in *Spigelia*) as few as 8. About 13 genera, about 350–380 species, nearly worldwide, most diverse in tropical and subtropical regions.

The Loganiaceae have been a controversial family in terms of genera contained and familial limits. The estimate of genera and species above is at the conservative end, but a more liberal interpretation might include up to 29 genera and 575 species (Leeuwenberg and Leenhouts, 1980). Most of the problem is not in the delimitation of genera, but in how many of the genera should be transferred to other families (Backlund et al., 2000). For example, Steyermark (1963) included the genus *Polypremum* in the Loganiaceae, but it is here treated in the Scrophulariaceae. See the treatment of that family for further discussion. Also, *Gelsemium* Juss., a small genus of shrubs and woody climbers with usually large yellow corollas, is often included in the Loganiaceae, although recent molecular phylogenetic studies (Oxelman et al., 1999; Backlund et al., 2000) suggest that it is better segregated as its own family, Gelsemiaceae. Of the three species of *Gelsemium*, which are collectively known as yellow jessamine, the southeastern native *G. sempervirens* (L.) J. St.-Hil. (Carolina jessamine) is the one most commonly cultivated as an ornamental. Uphof (1922) reported finding the species on mesic slopes near creeks in Butler and Ripley Counties. As no herbarium vouchers have been located to allow verification of any of Uphof's observations during his ecological and vegetational research in Missouri, *G. sempervirens* presently is not accepted as part of the Missouri flora. It is a liana with short-petiolate, lanceolate leaves; short, dense spicate clusters of flowers that open one or two at a time and have yellow trumpet-shaped corollas 2–3 cm long; and flattened, oblong capsules.

1. Leaves broadest at about the midpoint, the lower ones short-petiolate; corollas 1.5–3.0 mm long, white or slightly pinkish-tinged, the lobes erect or curved inward . . . . 1. MITREOLA
1. Leaves broadest below the midpoint, all sessile; corollas 30–60 mm long, red externally and yellow internally, the lobes spreading or curved outward . . . . . . 2. SPIGELIA

1. **Mitreola** L. (mitrewort)

Six to 8 species, North America to South America, Caribbean Islands, Asia, Madagascar, Borneo, Australia.

1. Mitreola petiolata (J.F. Gmel.) Torr. & A. Gray (lax hornpod)

Cynoctonum mitreola (L.) Britton

Pl. 447 i, j; Map 2029

Plants annual. Stems 8–40 cm long, erect or ascending, usually branched, glabrous. Lowermost leaves short-petiolate, the median and upper leaves sessile. Leaf blades 6–50 mm long, elliptic to very narrowly elliptic, broadest at about the midpoint, mostly narrowed or tapered to a sharply pointed tip (rarely blunt), narrowed or tapered at the base, glabrous. Inflorescences usually branched. Calyces 0.8–1.2 mm long, the lobes elliptic to ovate. Corollas 1.5–3.0 mm long, tubular to somewhat cup-shaped, white or slightly pinkish-tinged, the lobes erect or somewhat curved inward, the throat with a dense ring of short hairs. Stamens not exserted. Pistils initially united to the summit but soon becoming separated from the tip to below the midpoint and thus appearing 2-horned, the style initially 1, but often becoming split longitudinally and appearing as 2 units. Fruits 3–4 mm long, deeply 2-lobed, dehiscing longitudinally between the lobes. Seeds 0.2–0.4 mm in diameter, elliptic to nearly circular in outline, flattened, the attachment point in the middle of a concave face, the other face convex and rounded, brown, somewhat shiny. $2n=20$. June–November.

Uncommon, known from historical collections in Butler and Dunklin Counties and a single more recent station in Ripley county (southeastern U.S. west to Missouri and Texas; Mexico, Central America, South America, Caribbean Islands, Asia, Africa, Australia). Margins of ponds, swamps, and fens; also ditches.

Nigh and Ladd (1987) rediscovered this species in the open marly portion of a Ripley County fen more than 80 years after it was last seen in Missouri. J. B. Nelson (1980) summarized the nomenclatural history of the genus.

2. Spigelia L. (pinkroot)

About 50 species, North America to South America, Caribbean Islands.

1. Spigelia marilandica L. (woodland pinkroot, Indian pink, wormgrass)

Pl. 447 k, l; Map 2030

Plants perennial herbs. Stems 30–70 cm long, erect or ascending, unbranched or few-branched, minutely hairy at and around the nodes. Leaves sessile. Leaf blades mostly 3–12 cm long, the lowermost few pairs often highly reduced, sometimes less than 1 cm long, ovate to lanceolate, broadest toward the base, tapered to a sharply pointed tip (the reduced lowermost ones sometimes blunt or rounded), mostly rounded to truncate at the base, the upper surface glabrous, the undersurface (and the margins) minutely hairy along the main veins. Inflorescences unbranched or few-branched. Calyces 7–13 mm long, the lobes linear, gradually tapered to a sharply pointed tip. Corollas 30–60 mm long, more or less trumpet-shaped, bright red on the outer surface, yellow on the inner surface, the lobes spreading or curved outward, the throat glabrous. Stamens slightly exserted. Pistils notched or 2-lobed at the tip, the style usually shed before the fruit dehisces. Fruits 5–8 mm long, deeply 2-lobed, dehiscing longitudinally between the lobes and along the opposite margins, the valves eventually shed, leaving a shallow cuplike base. Seeds 2.0–2.8 mm in the longest dimension, irregularly rectangular to elliptic or nearly circular in outline, more or less rounded on the dorsal surface, the other sides angular, attached on the ventral side and forming ball-like masses, the surfaces roughened or with small flat irregular tubercles, dark brown, not shiny. $2n=46$. June–October.

Scattered in southeastern Missouri and disjunct locally north and west to Franklin and Cooper Counties (southeastern U.S. west to Missouri and Texas). Bottomland forests, mesic upland forests, and banks of streams, spring branches, and rivers; rarely in pastures.

Pinkroot is an attractive perennial that is gaining popularity in wildflower gardens. The beautiful flowers apparently are hummingbird-pollinated (G. K. Rogers, 1986). The species, along with a few others in the genus, also has a long history of medicinal use in the treatment of parasitic worms. G. K. Rogers (1986) warned, however, that poisonous compounds (presumably alkaloids) in the plants can cause delirium, vertigo, speech abnormalities, convulsions, vision problems, pain, spasms, and even death if an overdose is ingested. G. K. Rogers also noted that *S. marilandica* was disappearing from portions of its range because of overcollecting for medicinal and horticultural purposes. Fortunately, the species is currently being propagated at several native plant nurseries.

LYTHRACEAE (Toothcup Family)

Plants annual or perennial herbs, less commonly shrubs (small trees elsewhere). Stems usually branched, sometimes 4-angled. Leaves opposite or less commonly alternate or whorled, variously simple, sessile or less commonly short-petiolate, the leaf blade unlobed and with entire margins. Stipules absent. Inflorescences of solitary or clustered axillary flowers, sometimes appearing as elongate terminal spikes or racemes with leaflike bracts. Flowers perfect, strongly perigynous (the hypanthium appearing as a calyx tube), actinomorphic or less commonly zygomorphic, subtended by a pair of minute bractlets. Calyces of 4–7 small, ascending, triangular, toothlike sepals at the tip of the calyxlike hypanthium, these sometimes alternating with small appendages, the hypanthium and calyx usually persistent at fruiting. Corollas of 4–7 petals (absent in *Didiplis*), these sometimes appearing wrinkled in bud, attached on the inner surface of the hypanthium near its tip, alternating with the sepals, usually not persistent at fruiting. Stamens 4–14, as many as or twice as many as the sepals, the filaments sometimes in 2 series or of 2 or 3 different lengths in different flowers, attached to the inner surface of the hypanthium toward its base, the anthers small, attached near the midpoint, yellow or less commonly brown or purple. Pistil 1 per flower, composed of 2–4(–6) fused carpels, the superior ovary sometimes with a swollen nectar disc (this sometimes incomplete or absent) at the base, the style 1, slender, ranging from very short to relatively long and exserted, persistent at fruiting, sometimes of 2 or 3 different lengths in different flowers, the stigma usually capitate. Ovules several to numerous. Fruits capsules, dehiscing irregularly or longitudinally between the locules (indehiscent in *Didiplis*). Seeds 3 to numerous, small, sometimes winged. About 30 genera, about 600 species, nearly worldwide, most diverse in tropical and warm-temperate regions.

Some members of the Lythraceae (in Missouri principally the genera *Decodon* and *Lythrum*) have a condition known as tristyly, in which the relative staminal filament and style lengths differ in different flowers on the same plant. This phenomenon was first studied in *Lythrum salicaria* (and other unrelated species) by Charles Darwin (1877). Some flowers (so-called pin flowers) have a relatively long style and short filaments, such that the anthers are positioned well below the stigma. Other flowers (so-called thrum flowers) have the reverse situation, with a relatively short style and long filaments, such that the anthers are positioned well above the stigma. This mechanism promotes cross-pollination between flowers. A third flower type, intermediate between the other two with the stamens positioned near the stigma, is produced less frequently and is more often self-pollinated. The three flower types also differ in relative sizes of pollen grains, with those of thrum flowers the largest (Mulcahy and Caporello, 1970). In some members of the Lythraceae, the situation is complicated by the fact that individual flowers have stamens of two of the three possible lengths.

Some species of *Cuphea*, *Lagerstroemia* L. (crepe myrtle), and *Lythrum* are commonly cultivated as ornamentals. Paradoxically, the exotic *Lythrum salicaria* also becomes naturalized and invasive in North American wetlands (for further discussion, see the treatment of that species). Similarly, the Eurasian *Trapa natans* L. (water chestnut), a mostly submerged aquatic that is sometimes cultivated in ponds, has invaded North American ponds, lakes, streams, and rivers. The genus *Punica* L. (pomegranate) is often included in the family and is cultivated for its fruits that are popular fresh (the red pulp is mainly from the fleshy seed coats) or as juice. The juice sometimes also is used as a colorant in other beverages and foods. Some species of *Cuphea* are being investigated as an oilseed crop and for potential medical uses in weight reduction and control of cholesterol.

1. Stems with sticky, gland-tipped hairs; flowers zygomorphic, the hypanthium pouched or spurred basally on 1 side, oblique at the tip, the upper 2 petals longer than the other 4 petals . . . . 2. CUPHEA
1. Stems glabrous or, if hairy (in *Lythrum*), then the hairs not glandular or sticky; flowers actinomorphic, the hypanthium symmetric at the base (not pouched or spurred), not or only very slightly oblique at the tip, the petals all of similar lengths
    2. Plants submerged aquatics or less commonly stranded on mud, the stems then spreading to loosely ascending; leaves linear to narrowly elliptic, 1–3 mm wide; petals absent; calyces lacking small appendages between the sepals; fruits indehiscent . . . . 4. DIDIPLIS
    2. Plants terrestrial or strongly emergent aquatics, the main stems erect or strongly ascending (the lowermost branches sometimes spreading), rarely arched and rooting at the tip; leaves variously shaped, (2–)3–40 mm wide; petals present; calyces with small appendages between the sepals; fruits dehiscent
        3. Sepals and petals 4, the petals 1–3 mm long; hypanthium about as long as wide to slightly longer than wide; plants annual (but fibrous-rooted), the stems not woody, not wandlike, often branching from near the base, not producing offsets at the base
            4. Leaves all truncate to cordate at the base; appendages between the sepals more or less linear; fruits dehiscing irregularly, the outer wall smooth . . . . 1. AMMANNIA
            4. At least the upper leaves tapered at the base; appendages between the sepals triangular; fruits dehiscing longitudinally, the outer wall with fine transverse lines (best observed with magnification) . . . . . 6. ROTALA
        3. Sepals and petals (4)5–7, the petals 3–15 mm long; hypanthium about as long as wide or much longer than wide; plants perennial, often woody toward the base, usually wandlike, usually branched well above the base, producing offsets with age, sometimes arched and rooting at the tip
            5. Stems usually woody to well above the base (sometimes spongy at the base); leaves mostly in whorls of 3 or 4, narrowed or tapered at the base, often short-petiolate; hypanthium about as long as wide . . . . 3. DECODON
            5. Stems herbaceous, sometimes somewhat woody at the very base; leaves opposite or at least the upper leaves alternate (the lowermost leaves rarely in whorls of 3 in *L. alatum*), rounded to more commonly truncate or cordate at the base (narrowed to tapered in the rare *L. alatum* var. *lanceolatum*), sessile or very short-petiolate; hypanthium much longer than wide . . . . 5. LYTHRUM

1. **Ammannia** L. (toothcup)
(Graham, 1985)

Plants terrestrial or strongly emergent aquatics, annual, fibrous-rooted, sometimes rooting at the lower nodes, not producing offsets at the base, glabrous. Stems erect or strongly ascending, not wandlike, strongly 4-angled (square in cross-section), usually branched from near the base, the branches spreading to arched upward, not rooting at the tip. Leaves opposite, sessile. Leaf blades linear to narrowly lanceolate, narrowly oblong-triangular, or nar-

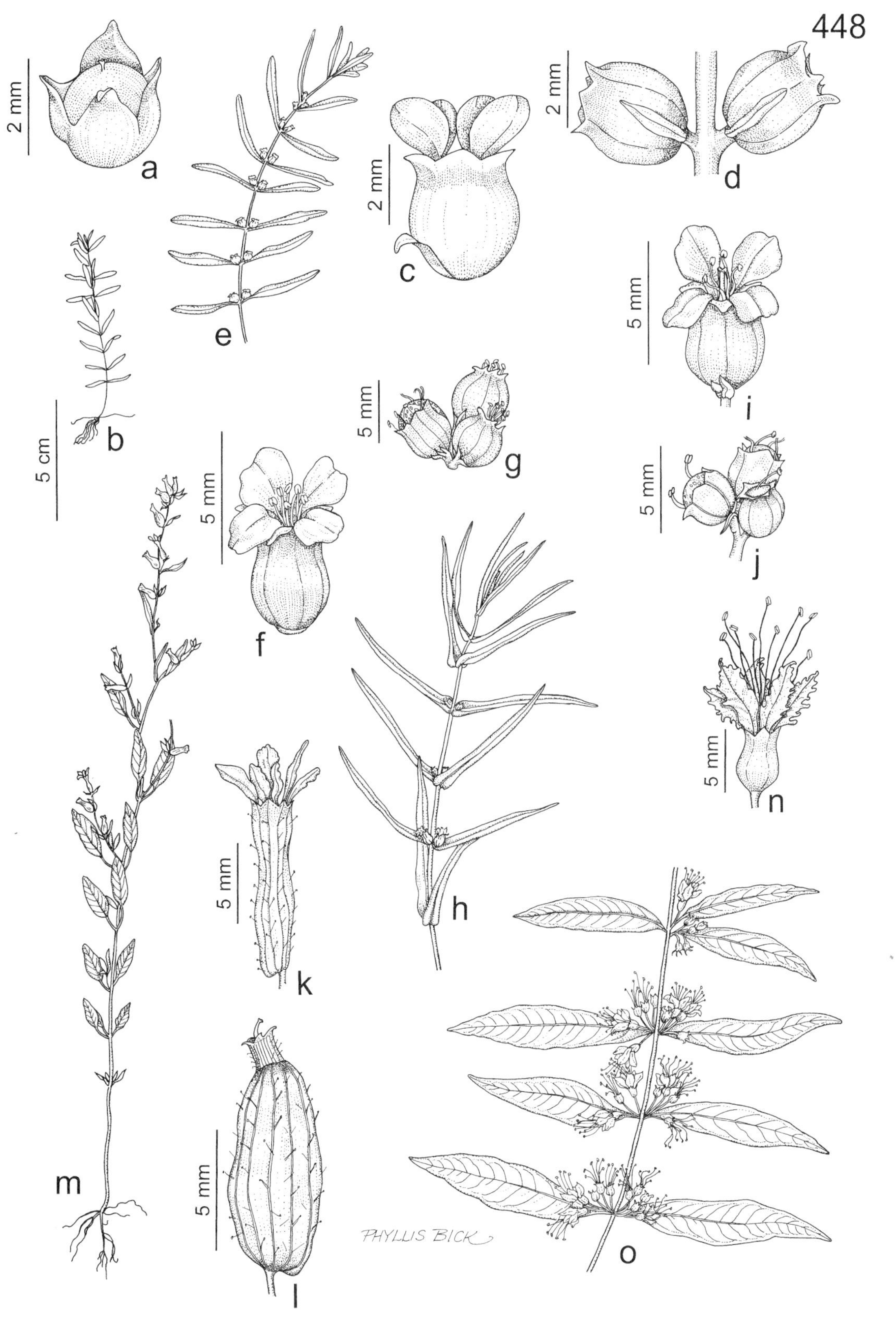

Plate 448. Lythraceae. *Didiplis diandra*, **a)** fruit, **b)** habit. *Rotala ramosior*, **c)** flower, **d)** node with fruits, **e)** fertile stem. *Ammannia robusta*, **f)** flower, **g)** cluster of fruits, **h)** fertile stem. *Ammannia coccinea*, **i)** flower, **j)** cluster of fruits. *Cuphea viscosissima*, **k)** flower, **l)** fruit, **m)** habit. *Decodon verticillatus*, **n)** flower, **o)** fertile stem.

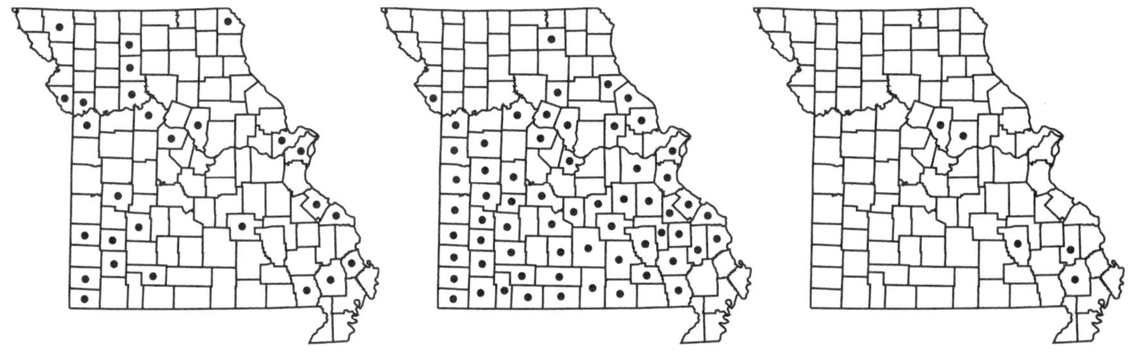

2032. Ammannia robusta 2033. Cuphea viscosissima 2034. Decodon verticillatus

rowly oblanceolate (in some lower leaves of *A. robusta*), those of the largest leaves always 3 mm or wider), narrowed or tapered to a usually sharply pointed tip, all truncate to more commonly auriculate-cordate at the base. Inflorescences of small axillary clusters of (1–)3–5 (–14) flowers, sessile or more commonly the cluster on a stalk to 4(–9) mm long. Flowers actinomorphic, the hypanthium about as long as wide, cup-shaped to urn-shaped (sometimes becoming nearly globose at fruiting), symmetric at the base (not pouched or spurred), not oblique at the tip, with 4 or 8 longitudinal ridges (these best developed at fruiting). Sepals 4, triangular, the appendages shorter than to about as long as the sepals, thickened, more or less linear. Petals 4, 1.5–3.0 mm long, pink to purple or pale lavender, not persistent at fruiting. Stamens 4(–8), those of different flowers with filaments the same length, the anthers exserted, yellow. Pistils lacking a nectary disc, the ovary incompletely 3- or 4-locular, the style relatively long and exserted (very short elsewhere). Fruits globose to oblong-globose (excluding the often persistent style) capsules, dehiscing irregularly with age, the outer wall smooth. Seeds numerous, 0.7–1.0 mm long, asymmetrically ovoid (1 side concave, the other convex), the surface with a network of fine, sometimes faint ridges, brown. About 25 species, nearly worldwide, most diverse in Africa.

1. Petals deep pink to purple; anthers deep yellow; flower clusters usually with an evident stalk 3–4(–9) mm long . . . . 1. A. COCCINEA
1. Petals pale lavender, sometimes with the midvein darker purple near the base; anthers light yellow; flower clusters sessile or rarely on an inconspicuous stalk to 2 mm long . . . . 2. A. ROBUSTA

1. Ammannia coccinea Rottb.

Pl. 448 i, j; Map 2031

Stems 10–40(–100) cm long. Leaves 1–8 cm long, 2–15 mm wide. Flower clusters usually with an evident stalk 3–4(–9) mm long, consisting of 3–5(–14) flowers, these with the stalk 0.5–2.0 mm long. Hypanthium plus sepals 2.5–4.0 mm long. Petals 1.5–2.3 mm long, deep pink to purple. Stamens 4(–7), the anthers deep yellow. Fruits 3.5–5.0 mm in diameter. $2n=66$. June–October.

Scattered nearly throughout the state (eastern U.S. west to South Dakota and Texas; disjunct in Arizona and California; Mexico to South America, Caribbean Islands). Banks of streams, rivers, and spring branches and margins of ponds, lakes, swamps, and sloughs; also ditches, fallow fields, edges of crop fields, gardens, and open disturbed areas.

Ammannia coccinea commonly grows in muddy soils, often colonizing receding shorelines. Occasionally, it is found as an emergent aquatic in very shallow water. Graham (1979, 1985) demonstrated that this polyploid species originated as a hybrid between *A. auriculata* Willd. ($2n=32$) and *A. robusta* ($2n=34$), followed by a doubling of the chromosome number (to $2n=66$) to restore fertility. *Ammannia auriculata* is a nearly cosmopolitan species with a distribution in the New World that extends northward through the central portion of the United States. It differs from both *A. robusta* and *A. coccinea* in being less robust, with more slender stems that usually are branched from well above the base; in having inflorescences with a slender stalk 3–9 mm long subtending a cluster of mostly 5–12 or more flowers; in having petals (similar in color to those of *A. coccinea*) 1.0–1.5 mm long; and in having fruits mostly less than 2.5 mm in diameter. Steyermark (1963) and many other botanists called plants *A. auriculata* that are now correctly determined as *A. robusta*.

2. Ammannia robusta Heer & Regel

Pl. 448 f–h; Map 2032

Stems 10–60(–100) cm long. Leaves 1–10 cm long, 3–15 mm wide. Flower clusters sessile or rarely on an inconspicuous stalk to 2 mm long, consisting of 1–3(–5) flowers, these sessile or with the stalk to 1 mm long. Hypanthium plus sepals 3–5 mm long. Petals 2–3 mm long, pale lavender, sometimes with the midvein darker purple near the base. Stamens 4(–8), the anthers light yellow. Fruits 4–6 mm in diameter. $2n$=34. July–October.

Scattered widely in the state, most commonly in counties in the southwestern quarter and along the Missouri and Mississippi Rivers (western U.S. east to Ohio, New Jersey, and Louisiana; Canada, Mexico to Central America, Caribbean Islands; introduced in Brazil). Banks of streams, rivers, and spring branches, margins of ponds, lakes, swamps, and sloughs, and edges of marshes and saline springs; also ditches, fallow fields, edges of crop fields, gardens, and open disturbed areas.

Ammannia robusta is less abundant in Missouri than *A. coccinea*. As noted in the treatment of the latter species, Steyermark (1963) and other earlier botanists confused *A. robusta* with *A. auriculata*. Steyermark reported that species from only three counties. Additionally, however, a number of specimens of *A. robusta* were misdetermined by Steyermark as *A. coccinea*. The two species occur in similar habitats and frequently grow together.

2. Cuphea P. Browne (cuphea)

About 260 species, North America to South America, Caribbean Islands, most diverse in the neotropics.

Some tropical species and hybrids of *Cuphea* are cultivated as garden annuals in Missouri, especially *C. hyssopifolia* Kunth (false heather), *C. ignea* A. DC. (cigar flower, firecracker plant), and *C. llavea* Lex. (bat-faced cuphea). The seed oils of many species in the genus are rich in short- and medium-chain fatty acids, which are important in the manufacture of detergents, lubricants, and other products (Graham et al., 1981; Graham and Kleiman, 1985). The sticky, glandular hairs in this genus cause livestock to avoid grazing on the plants and interfere with the ability of small insects to crawl into the flowers (Graham, 1964). The flowers of various species thus are pollinated mainly by hummingbirds and long-tongued insects.

1. Cuphea viscosissima Jacq. (clammy cuphea, blue waxweed)

C. petiolata (L.) Koehne, an illegitimate name

Pl. 448 k–m; Map 2033

Plants terrestrial, annual, fibrous-rooted, not rooting at the nodes, not producing offsets at the base. Stems 10–60 cm long, erect or strongly ascending, not wandlike, circular in cross-section, usually much-branched throughout, the branches spreading to ascending, not rooting at the tip, densely pubescent with longer, spreading, sticky, gland-tipped hairs and shorter, spreading, nonglandular hairs, purplish-tinged above the midpoint. Leaves opposite, petiolate, the petiole 3–15 mm long. Leaf blades 2.0–5.5 cm long, 5–20 mm wide, narrowly lanceolate to ovate, narrowed to a bluntly or more commonly sharply pointed tip, rounded or angled to short-tapered at the base, the surfaces sparsely to moderately pubescent with short, nonglandular hairs, sometimes somewhat roughened to the touch, more densely hairy along the main veins. Inflorescences of solitary, axillary flowers, these sometimes appearing grouped into open, leafy racemes, the flower stalks 1–5 mm long. Flowers zygomorphic, the hypanthium much longer than wide, cylindrical to narrowly urn-shaped, with a rounded pouch or spur at the base on 1 side, oblique at the tip, the hypanthium plus sepals 8–12 mm long, with 12 longitudinal ridges, dark purple toward the tip, yellowish, pinkish, or green toward the base, the outer surface densely pubescent with spreading, sticky, dark-colored, gland-tipped hairs, the inner surface with short, nonglandular hairs, splitting longitudinally along the upper side as the fruit dehisces. Sepals 6, the upper 1 larger than the other 5, triangular to broadly ovate-triangular, the appendages incon-

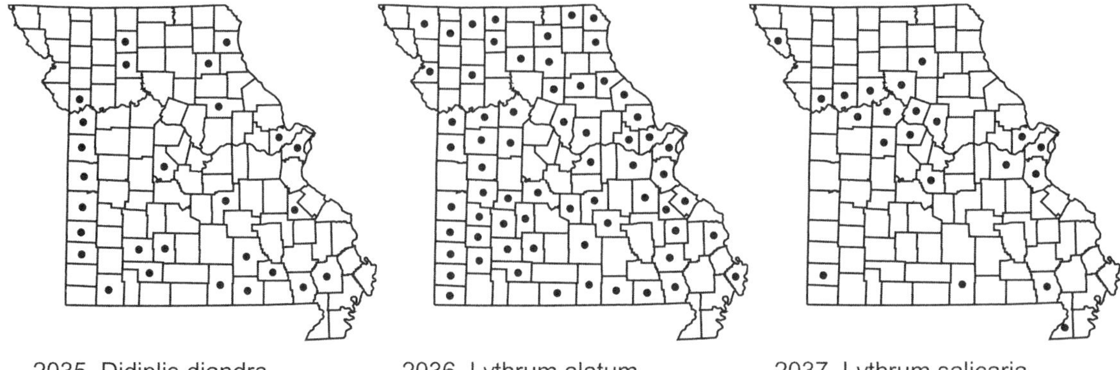

2035. Didiplis diandra 2036. Lythrum alatum 2037. Lythrum salicaria

spicuous, much shorter than the sepals, triangular. Petals 6, 3–6 mm long, the upper 2 longer than the other 4, purple, not persistent at fruiting. Stamens 11, those of different flowers with filaments of similar lengths, but the filaments alternately shorter and longer within each flower, the anthers slightly exserted, purple. Pistils with a curved nectary along the base of the upper side, the ovary incompletely 2-locular, usually appearing 1-locular at fruiting, the style 3–5 mm long, not or only slightly exserted. Fruits 5–8 mm in diameter, globose to ovoid capsules, dehiscing longitudinally along the upper side, the outer wall smooth, membranous to papery. Seeds 7–10, 2–3 mm in diameter, flattened, more or less circular in outline, the surface minutely tuberculate, dark brown. $2n=12$. July–October.

Scattered, mostly south of the Missouri River, but apparently absent from most of the Mississippi Lowlands Division (eastern U.S. west to Nebraska and Oklahoma; Canada). Glades, tops of bluffs, savannas, swales in sand prairies, openings of dry upland forests, and banks of streams and spring branches; also old fields, ditches, quarries, railroads, roadsides, and open disturbed areas.

Traditionally, this species was called *C. petiolata* (L.) Koehne, but that name is a homonym of the older *C. petiolata* Pohl ex Koehne, a different South American species (Graham, 1964).

3. **Decodon** J.F. Gmel.

One species, eastern U.S., Canada.

1. Decodon verticillatus (L.) Elliott (swamp loosestrife, water willow)
D. verticillatus var. *laevigatus* Torr. & A. Gray

Pl. 448 n, o; Map 2034

Plants strongly emergent aquatics, perennial (usually shrubs), fibrous-rooted, often rooting at the lower nodes and sometimes at or toward the tip, producing offsets at the base (at least with age), sometimes forming hummocks of stems. Stems 60–200(–250) cm long, wandlike and ascending, usually arched, usually woody well above the base, the submerged portion usually spongy-thickened, the emergent lower portion with the bark orangish brown and shed in narrow strips, the younger portions usually densely pubescent with short, irregularly twisted and pinnately branched (dendritic) hairs, becoming glabrous or nearly so with age, circular to somewhat 6-angled in cross-section, unbranched or few-branched well above the base when ascending, but producing numerous ascending branches if the main stem arches so strongly as to root at the tip. Leaves opposite or more commonly in whorls of 3 or 4 (rarely only subopposite near the stem tip), mostly short-petiolate, 3–20 cm long, 5–50 mm wide. Leaf blades lanceolate to elliptic, sometimes narrowly so, rarely ovate, tapered to a sharply pointed tip, angled or tapered at the base, the upper surface dark to bright green, glabrous except along the midvein and sometimes near the margins, the undersurface pale or light green, moderately to densely pubescent with short, irregularly twisted and pinnately branched (dendritic) hairs, especially along the veins. Inflorescences of 1–3 dense to loose clusters at each node, each cluster short-stalked (the stalk fused to the petiole base) and with few to several flowers, the flowers individually with stalks 3–11 mm long. Flowers actinomorphic, the hypanthium about as long as wide, cup-shaped to urn-shaped (often becoming nearly globose at fruiting), symmetric at the base (not pouched or spurred), not oblique at the tip, the hypanthium plus sepals 6–7 mm long, with (8)10(–14) longitudinal nerves (not ridged). Sepals (4)5(–7), triangular, ascending, the appendages up to twice as long as the sepals, thickened, narrowly conic-triangular, arched outward. Petals (4)5(–7), 9–15 mm long, reddish pink to pinkish purple, tending to appear ruffled or somewhat crinkled, not persistent at fruiting. Stamens 8–10, those of different flowers with filaments of potentially 3 different lengths (those of an individual flower usually of 2 lengths), the anthers exserted, yellow to yellowish brown. Pistils lacking a nectary disc, the ovary completely 3(–5)-locular, the styles of potentially 3 different lengths in different flowers, exserted or not. Fruits 4–6 mm in diameter, globose capsules, dehiscing longitudinally along 3(–5) partitions between locules, the outer wall with a network of minute ridges (observed with magnification). Seeds numerous (to about 30), 1.2–2.0 mm long, asymmetrically obpyramidal (1 side broader than the other 3), the surface with a network of minute ridges, olive green to yellowish brown or reddish brown (usually with a contrasting spot on 1 face. $2n$=32. July–September.

Uncommon, known thus far from Bollinger, Reynolds, and Stoddard Counties and introduced in Boone County (eastern U.S. west to Minnesota and Texas; Canada). Sinkhole ponds and margins of ponds and lakes.

Steyermark (1963) knew this species only from a pair of isolated, upland sinkhole ponds in Reynolds County, where these remarkable plants form dense hummocks and still grow today. Since that time, populations have been discovered at a few sites in the Mississippi Lowlands, where they often grow on logs floating along the margins of lakes derived from former swamps and sloughs. An unvouchered occurrence in Mississippi County is included in the Missouri Natural Heritage Program's database of species and communities of conservation concern.

Swamp loosestrife occasionally is cultivated as an ornamental in ponds and gardens in Missouri, but in such situations the individual plants tend to be relatively short-lived and to spread extensively by seed. Steyermark (1963) noted that the seeds and stems sometimes provide food for ducks and muskrats, respectively.

4. **Didiplis** Raf. (water purslane)

One species, eastern and central U.S.

Didiplis traditionally was included in the genus *Peplis* L. (Steyermark, 1963; Graham, 1964) along with seven other European species. These other species all have relatively reduced flowers lacking petals (or with small petals that are shed early) and have been determined to be closely related to *Lythrum* (D. A. Webb, 1967; Graham, 1975; Graham et al., 2005). The relationships of *Didiplis* to other genera requires further study, but may be closest to *Rotala* L. (Graham et al., 2005; J. A. Morris, 2007).

1. Didiplis diandra (Nutt. ex DC.) Alph. Wood (water purslane, water hedge)
Peplis diandra Nutt. ex DC.
P. diandra f. *aquatica* (Koehne) Fassett
P. diandra f. *terrestris* (Koehne) Fassett
Pl. 448 a, b; Map 2035

Plants submerged aquatics or less commonly stranded on mud, annual, fibrous-rooted, sometimes rooting at the nodes, glabrous. Stems 5–40 cm long, weak, usually ascending in aquatic plants, spreading to loosely ascending in terrestrial plants, not wandlike, 4-angled (more or less square in cross-section), unbranched or few- to several-branched, sometimes mostly from near the base, the branches not rooting at the tip, sometimes reddish-tinged toward the tip. Leaves opposite or the uppermost ones occasionally alternate, sessile, 0.3–3.0 cm long, 1–3 mm wide. Leaf blades linear to narrowly elliptic, sometimes reddish-tinged, narrowed or tapered to a usually sharply pointed tip, those in aquatic plants mostly truncate at the base, those in terrestrial plants often tapered at the base. Inflorescences of solitary flowers, these sessile. Flowers actinomorphic, the hypanthium about as long as wide, cup-shaped to urn-shaped (sometimes becoming nearly globose at fruiting), symmetric at the base (not pouched or spurred), not oblique at the tip, the hypanthium plus sepals 2–3 mm long, with 4 longitudinal lines or ridges (these often obscure, best developed at fruiting). Sepals 4, broadly triangular, the appendages absent. Petals absent. Stamens (2–)4, those of different flowers with filaments the same length, the anthers not exserted, light yellow to yellow. Pistils lacking a nectary disc, the ovary incompletely 2-locular, sometimes appearing 1-locular nearly throughout, the style minute and not exserted. Fruits 1–3 mm in diameter, globose, indehiscent, the outer wall smooth. Seeds 7–10 (rarely more), 0.4–0.8 mm long, flattened but somewhat curved above the base, wedge-shaped to spatulate, the surface finely pebbled, yellow. $2n=32$. May–October.

Scattered but relatively sporadic nearly throughout the state (eastern U.S. west to Minnesota and Texas; disjunct in Utah, Mexico). Pools in streams, ponds, lakes, and sinkhole ponds, swamps, and sloughs; also rice fields.

This nondescript species probably is overlooked frequently in the field, especially when growing as a submerged aquatic. However, it makes an attractive aquarium plant and sometimes is sold under the name "water hedge."

5. **Lythrum** L. (loosestrife)

Plants terrestrial or strongly emergent aquatics, perennial (annual elsewhere), fibrous-rooted, occasionally rooting at the lower nodes, with rhizomes and/or offsets, often forming colonies. Stems more or less wandlike, erect or ascending, not arched, herbaceous or sometimes woody at the very base, glabrous or hairy, more or less 4-angled in cross-section, usually few- to many-branched, mostly well above the base. Leaves opposite or at least the upper leaves alternate (the lowermost leaves rarely in whorls of 3 in *L. alatum*), sessile or very short-petiolate. Leaf blades linear or more commonly narrowly to broadly lanceolate or oblong-lanceolate or triangular-lanceolate, angled or tapered to a bluntly or sharply pointed tip, sometimes the lowermost leaves more or less rounded at the tip, rounded or more commonly truncate to shallowly cordate at the base, glabrous or hairy. Inflorescences either axillary or terminal, if terminal then appearing spicate, the flowers individually sessile or with stalks to 3 mm long. Flowers actinomorphic or nearly so, the hypanthium much longer than wide, cylindrical, symmetric at the base (not pouched or spurred), not or only slightly oblique at the tip, the hypanthium plus sepals 3–7 mm long, with 8–12 longitudinal ridges. Sepals (4–)6, triangular to broadly triangular, more or less ascending, the appendages about twice as long as the sepals, thickened, narrowly conic-triangular, arched outward. Petals (4–)6, 3–12 mm long, reddish pink to purple (rarely white elsewhere), sometimes somewhat contorted, but not appearing ruffled or crinkled, not persistent at fruiting. Stamens (4–)6 or 12, those of different flowers with filaments of potentially 3 different lengths (those of an individual flower of 1 or 2 lengths), the anthers exserted or not, yellow or dark purple. Pistils sometimes with a nectary disc, the ovary completely or more commonly incompletely 2-locular, the styles of potentially 3 different lengths in different flowers, exserted or not. Fruits 4–6 mm long, narrowly ovoid to oblong-ellipsoid capsules, dehiscing longitudinally (sometimes irregularly) along the partitions between the two locules, the outer wall with a network of minute ridges (observed with magnification). Seeds numerous (often 100 or more), 0.4–1.2 mm long, ovate to somewhat tetrahedral in outline, asymmetrically 4-sided (1 side broader than the other 3) with rounded angles, the surface minutely roughened or pebbled, shiny, tan to yellowish brown. About 30 species, North America, Europe, Asia, Africa, Australia.

1. At least some of the uppermost leaves alternate; plants glabrous; flowers solitary (rarely paired) in the axils of the upper and often also median leaves .... 1. L. ALATUM
1. All of the leaves opposite; plants hairy; flowers in pairs of small clusters in terminal inflorescences that appear as bracteose spikes or panicles with spicate branches .... 2. L. SALICARIA

1. Lythrum alatum Pursh (winged loosestrife)
Pl. 449 d, e; Map 2036

Plants glabrous, the stems and leaf undersurfaces sometimes somewhat glaucous, usually producing rhizomes. Stems 40–80(–150) cm long, ascending, wandlike, often narrowly winged, at least toward the tip, the wings interrupted and forming minute auricles at the nodes. Leaves mostly opposite, at least the uppermost leaves usually alternate, the lowermost leaves rarely in

Plate 449. Lythraceae. Magnoliaceae. *Lythrum salicaria*, **a)** flower, **b)** inflorescence, **c)** median portion of leafy stem. *Lythrum alatum*, **d)** flower, **e)** fertile stem. *Liriodendron tulipifera* , **f)** fruit, **g)** branch tip with leaves and flower. *Magnolia acuminata*, **h)** branch tip with leaves and flower.

whorls of 3, sessile. Leaf blades 0.5–4.0 cm long, narrowly lanceolate to oblong-lanceolate or oblong-triangular, the upper leaves sometimes linear or narrowly elliptic, rounded to truncate or very shallowly cordate (rarely narrowed to tapered) at the base, bluntly to sharply pointed (rarely those of the lowermost leaves nearly rounded) at the tip. Inflorescences of axillary flowers at the upper and often also median nodes, the flowers solitary or less commonly paired, in well-developed plants sometimes taking on the appearance of leafy terminal spikes, mostly short-stalked, each subtended by a pair of small bracts. Petals 3–7 mm long, lavender to reddish purple. Stamens (4–)6, the anthers usually dark purple. Pistils with a nectary disc. Seeds 0.4–0.6 mm long. $2n$=20. June–September.

Scattered nearly throughout the state (eastern [mostly northeastern] U.S. west to North Dakota, Wyoming, Colorado, and Oklahoma; Canada, Mexico, Caribbean Islands). Banks of streams, rivers, and spring branches, marshes, fens, bottomland prairies, and wet depressions of upland prairies and glades; also ditches, edges of crop fields, railroads, roadsides, and wet disturbed areas.

Graham (1975) discussed morphological variation across the range of *L. alatum*. Earlier botanists had recognized as many as four species in the complex. Graham concluded that only two reasonably discrete taxa could be separated, which she chose to recognize as varieties of *L. alatum*.

1. Leaf blades rounded to truncate or shallowly cordate at the base . . . . 1A. VAR. ALATUM
1. Leaf blades mostly angled or tapered at the base . . . . 1B. VAR. LANCEOLATUM

1a. var. alatum

Stems 40–80(–120) cm long. Leaf blades 0.5–4.0 cm long, narrowly lanceolate to more commonly oblong-lanceolate or oblong-triangular, the upper leaves sometimes linear or narrowly elliptic, rounded to truncate or very shallowly cordate at the base, bluntly to sharply pointed (rarely those of the lowermost leaves nearly rounded) at the tip. Flowers with the subtending bracts mostly below the midpoint of the short stalk. $2n$=20. June–September.

Scattered nearly throughout the state (eastern [mostly northeastern] U.S. west to North Dakota, Wyoming, Colorado, and Oklahoma; Canada). Banks of streams, rivers, and spring branches, marshes, fens, bottomland prairies, and wet depressions of upland prairies and glades; also ditches, edges of crop fields, railroads, roadsides, and wet disturbed areas.

1b. var. lanceolatum (Elliott) Torr. & A. Gray ex A. Gray

Stems 60–120(–150) cm long. Leaf blades 0.5–6.0 cm long, mostly lanceolate to narrowly lanceolate, the upper leaves sometimes linear or narrowly elliptic, narrowed to tapered at the base, usually sharply pointed at the tip. Flowers with the subtending bracts mostly above the midpoint of the short stalk. July–September.

Introduced, uncommon, known thus far only from a single collection from Butler County (southeastern U.S. west to Oklahoma and Texas; Mexico, Caribbean Islands). Ditches and margins of rice fields.

This variety has not previously been reported from Missouri. The Butler County specimen was collected by Stan Hudson in 1993.

2. Lythrum salicaria L. (purple loosestrife, spiked loosestrife)

Pl. 449 a–c; Map 2037

Plants moderately to densely pubescent with short, stiff, relatively broad-based, tapered, unbranched hairs, sometimes roughened to the touch, the lower portions of the stem sometimes becoming nearly glabrous with age, usually producing offsets. Stems 40–250(–300) cm long, erect with ascending branches, wandlike, usually more or less ridged, but not winged, lacking auricles at the nodes. Leaves all opposite, sessile or very short-petiolate. Leaf blades 1–10 cm long, narrowly to broadly lanceolate or oblong-lanceolate or triangular-lanceolate, the upper leaves sometimes nearly linear, truncate to shallowly cordate or rarely broadly rounded at the base, mostly sharply pointed at the tip. Inflorescences appearing as elongate, bracteose spikes or panicles with spicate branches, the leaves reduced fairly abruptly into the leaflike bracts, with a pair of small clusters of usually 3 flowers each at the nodes, the individual flowers sessile or with a minute stalk to 1 mm, usually subtended by a pair of inconspicuous, linear bractlets, these shed early. Petals 7–12 mm long, reddish pink to pinkish purple or reddish purple. Stamens usually 12, the anthers yellow. Pistils lacking a nectary disc. Seeds 0.6–1.2 mm long. $2n$=60 ($2n$=30, 50, 58 elsewhere). June–September.

Introduced, scattered widely in the state, most abundant in counties bordering the Missouri River (native of Europe, Asia; introduced nearly throughout the U.S.; also Canada, Africa, Australia). Banks of streams and rivers, margins of ponds and lakes, oxbows, sloughs, marshes, fens, and bottomland forests; also ditches, roadsides, and wet disturbed areas.

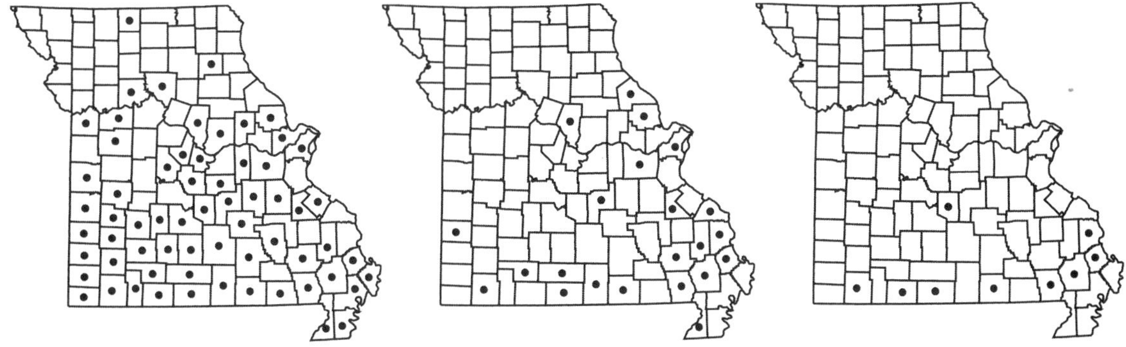

2038. Rotala ramosior 2039. Liriodendron tulipifera 2040. Magnolia acuminata

Purple loosestrife remains a controversial plant in Missouri. On the one hand, it has long been cultivated as an ornamental in gardens, prized for its showy inflorescences, long blooming period, and easy care. It also is a favorite plant among bee keepers, as the abundant pollen produces a superior honey. On the other hand, in 1989 the Missouri legislature officially declared purple loosestrife a noxious weed, and by law it is illegal (class A misdemeanor) to import, buy, sell, or plant seeds or plants of *Lythrum salicaria* or any of its hybrids and cultivars. In spite of this, it is still relatively easy to locate purple loosestrife for sale in plant nurseries in Missouri, in part because of a lack of effective means to enforce state noxious weed laws, inadequate public education on the environmental harm that the species can cause, and the sentimental attachment of many gardeners to heirloom plants. Attempts to develop sterile cultivars have failed thus far because such plants have proven to be only self-sterile and to produce copious seeds when crossed with other strains that may be growing nearby (Ottenbreit and Staniforth, 1994). Additionally, claims by some nurseries that certain cultivars represent the related, but supposedly less invasive Eurasian species, *L. virgatum* L., have merely confused the issue. *Lythrum virgatum*, which is easily distinguished from *L. salicaria* by its glabrous stems and foliage, as well as its less robust and less showy habit (Haining et al., 2007), apparently is rarely, if ever, sold in North America (Harper, 1986). *Lythrum virgatum* is represented as an escape only by a few historical collections from eastern Massachusetts.

Stuckey (1980) detailed the introduction and spread of *Lythrum salicaria* in North America. Originally brought into the United States as a garden ornamental from Europe, the plant was documented as an escape from cultivation in New England by the early 1800s. By the mid-1800s, it was invading wetlands in the northeastern States, and it has continued to spread and gain in population density since that time. As an invader of wetland communities, purple loosestrife crowds out other plant species, sometimes eventually developing into large stands that are nearly monocultures. The damage to these wetlands also includes secondary impacts to native pollinators that require particular native plant species and the reduction in food, nesting habitat, and cover for other animals. The effects of purple loosestrife on overall plant species diversity is still a controversial issue, with some researchers claiming that at least at the early stages of infestation, the presence of purple loosestrife in some wetlands can be correlated with a slight increase in the number of observed native plant species (Hager and Vinebrooke, 2004). Such short-term gains appear to be reversed dramatically as the infestation of *Lythrum* continues to gain in density and coverage.

Currently more than 30 states have legislated against *L. salicaria* in recognition of its detrimental effects on the environment. Attempts to control the species have relied most heavily on herbicides, but regrowth from surviving stem and root fragments, as well as from a long-lived seed bank in the soil, ensures that infestations must be controlled repeatedly over a span of many years. Various biological controls also have been developed, involving mainly an Asian root-mining beetle and a series of beetles that feed on leaves and flowers (Malecki et al., 1993), and these have proven effective in at least controlling populations in portions of the Great Lakes region and perhaps elsewhere. Concerns about the release of these biological agents into the environment center mainly on the potential for the insects also to attack native members of the Lythraceae, especially native American species of *Lythrum* and the genus *Decodon*.

In Missouri, the species originally was reported as an escape by L. P. Jensen (1932), based on a population that had become established at what is now the Shaw Nature Reserve (Franklin County). Palmer and Steyermark (1935) still recorded it only from that report. By the 1960s, Steyermark (1963) knew the species only from one additional population in Newton County. The extensive flooding that occurred in the floodplains of the big rivers in 1993 had a hand in increasing the number of populations, with the result that *L. salicaria* undoubtedly occurs in every county bordering the Missouri and Mississippi Rivers in Missouri. By 2002, the Missouri Department of Conservation's eradication program for this species was battling this species at more than 130 sites in 38 counties (personal communication from Preston Stogsdill, then coordinator of MDC's purple loosestrife control program).

At its original Missouri occurrence in Franklin County, the species first became established in 1930 from seeds contaminating a shipment of peat imported from New York (L. P. Jensen, 1932). Plants were first noted in 1930 beside a greenhouse where the peat had been dumped for storage. Following seasonal rains, seeds produced by these plants were washed down a small gully, resulting in its establishment along the shores of the Pinetum Lake. This population soon grew to form a continuous band along the lake margin, beautiful to behold when in flower but at the expense of most of the other emergent aquatics that had previously grown there. Efforts to rehabilitate the lake margin began in earnest in the mid-1980s with a combination of manual removal of *Lythrum* plants and annual herbicide applications. After more than three decades of effort, although the species has been controlled effectively, during most years a small number of individuals continue to appear that could result in a reinfestation if not dealt with.

From the preceding, it should be apparent that purple loosestrife represents a tangible and worsening environmental problem in the state. Gardeners and landowners are strongly urged to take steps to attempt to eradicate this species, both where it is cultivated and where it has escaped into nature.

6. **Rotala** L. (toothcup)

About 45 species, widespread in tropical and warm-temperate regions, most diverse in Asia, Africa.

1. Rotala ramosior (L.) Koehne

R. ramosior var. *interior* Fernald & Griscom

Pl. 448 c–e; Map 2038

Plants terrestrial or strongly emergent aquatics, annual, fibrous-rooted, sometimes rooting at the lower nodes, not producing offsets at the base, glabrous. Stems 5–40 cm long, erect or strongly ascending, not wandlike, strongly 4-angled (square in cross-section), often branched from near the base, the branches spreading to arched upward, not rooting at the tip. Leaves opposite, sessile, 1–5 cm long, 2–12 mm wide. Leaf blades linear to narrowly lanceolate, narrowly oblong-triangular, or oblanceolate, those of the largest leaves always 3 mm or wider), narrowed or tapered to a usually sharply pointed tip, the lower leaves truncate to more commonly auriculate-cordate at the base, the upper and often also median leaves tapered at the base. Inflorescences of solitary flowers, these sessile or nearly so. Flowers actinomorphic, the hypanthium about as long as wide, cup-shaped to urn-shaped (sometimes becoming nearly globose at fruiting), symmetric at the base (not pouched or spurred), not oblique at the tip, the hypanthium plus sepals 2–5 mm long, with 4 or 8 longitudinal ridges (these best developed at fruiting). Sepals 4, triangular to broadly triangular, the appendages about as long as the sepals, thickened, triangular. Petals 4, 1–2 mm long, white to light pink, not persistent at fruiting. Stamens 4(–6), those of different flowers with filaments the same length, the anthers not exserted, light yellow. Pistils lacking a nectary disc, the ovary incompletely 3- or 4-locular, the style relatively long and exserted (very short elsewhere). Fruits 3–5 mm in diameter, globose to oblong-globose capsules, dehiscing longitudinally along (2)3 or 4 of the partitions between locules, the outer wall with fine transverse lines (best observed with magnification). Seeds numerous, 0.4–0.8 mm long, asymmetrically ovoid (1 side concave, the other convex), the surface with a network of fine, sometimes faint ridges, yellow to reddish brown. $2n=32$. June–October.

Scattered widely in the state, most commonly south of the Missouri River (nearly throughout the U.S.; Canada, Mexico to South America, Caribbean Islands). Banks of streams, rivers, and spring branches, margins of ponds, lakes, swamps, and sloughs, and edges of marshes and saline springs; also ditches, fallow fields, crop fields, and open disturbed areas.

Inland plants are slightly more robust than those of coastal areas and have been called var. *interior*, but Graham (1975) noted that the quantitative differences between the varieties do not merit taxonomic recognition.

MAGNOLIACEAE (Magnolia Family)

Plants trees. Leaves alternate, simple, petiolate. Stipules large, membranous, enclosing the leaf buds, shed as the leaves develop and leaving a circular ring around the twig at the node, often partially fused to the petiole base. Flowers produced as the leaves develop (earlier elsewhere), solitary and terminal on branch tips, perfect, hypogynous, the buds enclosed in 1 or 2 sheathing bracts, these shed as the bud opens. Perianth sometimes not clearly differentiated into calyx and corolla, the tepals 9–15, in whorls of 3, usually shed after flowering and leaving noticeable scars, usually thickened and somewhat waxy or leathery. Stamens numerous, free, densely packed around the elongated receptacle, shed after dehiscence and leaving noticeable scars, the filaments relatively short, poorly differentiated from the anthers, the anthers dehiscing longitudinally. Pistils numerous, free, hairy, densely packed around the elongated receptacle, each with 1 carpel, the ovary superior and with 1 or 2 ovules, the style and stigma 1. Fruits conelike aggregates of samaras or follicles. Seeds (1)2 per samara or follicle. Six or 7 genera, 165–225 species, nearly worldwide.

1. Leaf blades about as wide as long, 4- or 6-lobed; fruits aggregates of samaras, indehiscent . . . . 1. LIRIODENDRON
1. Leaf blades longer than wide, unlobed; fruits conelike aggregates of drupes, the brightly colored seeds hanging by threads from the dehisced fruits . . . . . 2. MAGNOLIA

1. Liriodendron L.

Two species, eastern North America, eastern Asia.

1. Liriodendron tulipifera L. (tulip tree, yellow poplar)

Pl. 449 f, g; Map 2039

Plants trees to 40 m tall. Bark smooth on younger trees, becoming deeply furrowed with age, gray to dark gray. Twigs reddish brown to brown, glabrous, sometimes somewhat glaucous, the terminal buds somewhat flattened, rounded at the tip, with a characteristic "duck-bill" shape. Leaves long-petiolate. Leaf blades 6–15 cm long, about as wide as long, nearly square in outline, 4- or 6-lobed, the tip appearing broadly and shallowly notched, the base truncate to shallowly cordate, the margins otherwise entire or with a few additional points or small lobes on each side, the surfaces glabrous, the upper surface dark green to yellowish green, somewhat shiny, the undersurface pale green. Flowers with the perianth 3.5–6.0 cm long, consistingof 9 tepals, the outermost 3 sepal-like, spreading to reflexed at maturity, green, the inner 6 petal-like, ascending, greenish yellow with an orange base. Fruits 3.0–4.5 cm long, narrowly oblong-ovate in outline, dense aggregates of tightly overlapping indehiscent samaras, these eventually shed sequentially from base to tip and leaving noticeable scars on the persistent receptacle, the ascending, long, strap-shaped wing of each samara attached at an angle to the ripened ovary. $2n=38$. May–June.

Scattered in the Mississippi Lowlands Division and adjacent Big Rivers and Ozark Border, with possibly native stations in Barry and Ozark Counties, also rarely introduced in eastern Missouri (eastern U.S. and adjacent Canada west to Michigan, Illinois, Missouri, Arkansas, and Louisiana). Mesic upland forests, mostly in ravines, and bases of bluffs.

Rebman and Weber (1988) reported this species from Barry County and discussed the difficulty in determining whether these trees were a native population or escapes from an old homestead. Their assertion that this species may be native in southwestern Missouri appears to be borne out by the presence of another seemingly natural occurrence between the main distribution in southeastern Missouri and the Barry County station, this one in Ozark County along the North Fork River. Other specimens from outside the Mississippi Lowlands and adjacent Big Rivers Division clearly represent nonnative occurrences.

Tulip tree is an important timber tree of eastern deciduous forests and is thought to be the tallest hardwood species in temperate North America, with individuals reaching more than 65 m in height in some eastern states. In Missouri, it usually occurs in forests where *Fagus grandifolia* (Fagaceae) and *Liquidambar styraciflua* (Hamamelidaceae) grow, also trees are more common in the beech-maple forests that predominate farther to the east but are of restricted distribution in southeastern Missouri. *Liriodendron tulipifera* is frequently cultivated for its fast growth, tall narrow growth form, oddly-shaped leaves that turn yellow in the autumn, and beautiful faintly fragrant flowers (although these sometimes occur too high in the tree to be appreciated from the ground). The relatively light soft wood has a multitude of commercial uses in construction, canoes, furniture, flooring, musical instruments, boxes and crates, broom handles, and pulp for paper. The bitter inner bark of the roots has been used medicinally as a tonic and stimulant, and the bark contains alkaloids that have been used as a heart simulant (Steyermark, 1963).

2. **Magnolia** L. (magnolia)

Plants small to medium-sized trees. Terminal buds to 3 cm long, prominent. Leaves short-petiolate. Leaf blades longer than wide, unlobed, lacking teeth or sometimes with a pair of small points or spreading teeth above the middle, the tip narrowed or tapered to a point. Fruits conelike aggregates of follicles, dehiscing at maturity to release the seeds, which remain attached for some time, hanging by thin threads of tissue. Seeds relatively large, with a brightly colored aril covering the surface. About 120 species, North America south to Venezuela, Asia, Malaysia, south to Indonesia.

Magnolias generally are considered to be pollinated by beetles (Heiser, 1962), although a number of other types of insects visit the flowers.

Several species of *Magnolia* are cultivated in Missouri in addition to those treated here, but thus far none has become established outside cultivation. In addition to the two species included below, *M. macrophylla* Michx. (bigleaf magnolia, umbrella tree) should be searched for in southeastern Missouri. It occurs natively at a disjunct station on Crowley's Ridge in notheastern Clay County, Arkansas, less than ten miles from the Missouri border, where a small remnant population persists in the mesic upland forest of a drainage (Figlar, 1981). This species differs in its leaf blades with cordate bases and larger flowers (tepals 12–18 cm long).

1. Leaves scattered along the branches, the blades mostly (but not always) widest at or below the middle; tepals all greenish yellow, sometimes orange-tinged . . . . 1. M. ACUMINATA
1. Leaves mostly in whorl-like clusters at the branch tips, the blades widest at or more commonly above the middle; tepals white . . . . 2. M. TRIPETALA

1. Magnolia acuminata (L.) L. (cucumber tree, cucumber magnolia)

M. acuminata var. *ozarkensis* Ashe

Pl. 449 h; Map 2040

Plants trees to 25 m tall. Bark grayish brown to dark brown, at maturity with narrow flaky ridges and furrows. Twigs reddish brown, silvery-hairy when young, becoming glabrous at maturity, the terminal buds somewhat flattened, bluntly pointed at the somewhat asymmetric tip, silvery-hairy. Leaves scattered along the branches. Leaf blades 10–25 cm long, oblong-obovate to broadly elliptic-obovate, mostly widest at or below the middle (sometimes above the middle), broadly rounded to narrowed or truncate at the base, the upper surface glabrous or sparsely hairy when young, the undersurface pale green, hairy, sometimes becoming nearly glabrous with age. Flowers with the perianth 4–8 cm long, consistingof 6 tepals, all ascending, narrowly oblanceolate to oblong-spatulate, usually somewhat longitudinally cupped, especially the inner 3, greenish yellow, sometimes orange-tinged, often glaucous. Fruits 3–7 cm long, narrowly oblong and often somewhat asymmetric in outline, the follicles short-beaked, glabrous. Seeds 9–10 mm long, heart-shaped, often somehat flattened, the aril orangish red. $2n=76$. April–May.

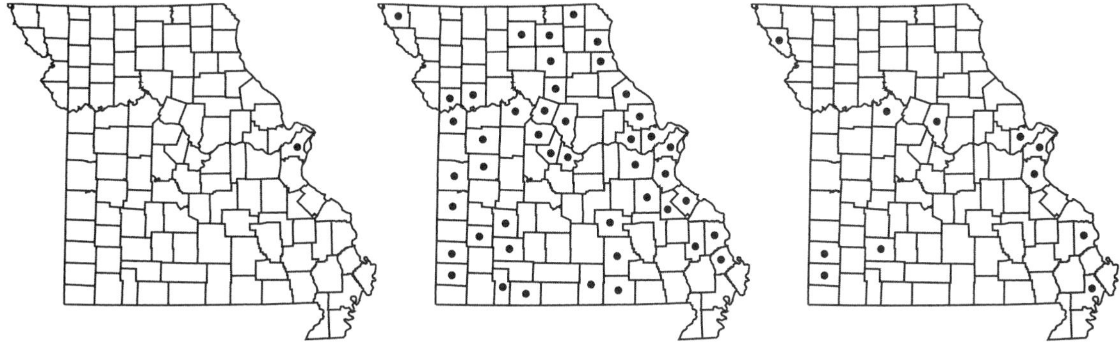

2041. Magnolia tripetala 2042. Abutilon theophrasti 2043. Alcea rosea

Uncommon in southern Missouri (eastern U.S. and adjacent Canada west to Illinois, Oklahoma, and Louisiana). Bottomland forests, mesic upland forests, usually in ravines, bases of bluffs, and banks of streams.

The wood of *Magnolia acuminata* has similar properties to that of *Liriodendron tulipifera* and is used similarly in the lumber industry (Settergren and McDermott, 1962). The common name "cucumber tree" refers to the cucumberlike appearance of the immature aggregations of fruits.

2. Magnolia tripetala L. (umbrella tree, umbrella magnolia)

Map 2041

Plants trees to 12 m tall. Bark light gray, thin, smooth, becoming finely warty with age. Twigs grayish brown, glabrous, the terminal buds somewhat flattened, bluntly pointed at the somewhat asymmetric tip, glabrous. Leaves mostly in whorl-like clusters at the branch tips. Leaf blades 10–20 cm long at flowering, to 60 cm long at maturity, oblong-elliptic to oblanceolate or narrowly obovate, widest at or more commonly above the middle, long-tapered to the base, the upper surface glabrous, the undersurface green, densely hairy, especially along the midvein. Flowers with the perianth 7–12 cm long, consistingof 6–9 tepals, the outermost usually the longest, all spreading to loosely ascending, oblanceolate to narrowly obovate, the outermost usually slightly cupped toward the tip, white. Fruits 6–12 cm long, narrowly oblong-ovate in outline, the follicles long-beaked, glabrous. Seeds 9–12 mm long, ovate, often somehat flattened, the aril pinkish red to red. $2n$=38. April–May.

Possibly introduced, known from a single site in St. Louis County (eastern U.S. west to Oklahoma and Louisiana). Mesic upland forests.

Steyermark (1963) excluded this species from the Missouri flora, stating that the only historical collection was mislabeled and originated from Tennessee. However, Harriman (1969) reported a single mature individual in a remnant forest stand in St. Louis County, which he suggested was a disjunct native occurrence, based on the associated species at the site and the lack of previous development in the immediate area. Flowers of this species have an unpleasant scent.

MALVACEAE (mallow family)

(Fryxell, 1997)

Plants annual or perennial herbs or less commonly shrubs or trees, often with stellate pubescence. Leaves alternate and sometimes also basal, simple, petiolate, sometimes shallowly to deeply lobed (occasionally appearing nearly compound), the main veins or lobes usually palmate. Stipules hairlike, scalelike, or somewhat leaflike, sometimes shed early. Inflorescences terminal and/or axillary racemes, panicles, clusters, or solitary flowers, sometimes with bracts at the branch points, these sometimes shed early. Flowers mostly perfect (pistillate flowers sometimes mixed with the perfect ones), hypogynous, actinomorphic. Sepals 5, fused into a cup or tube, at least toward the base (free in *Tilia*), sometimes closely subtended by 2 or more bractlets. Petals 5, often fused to the base of a staminal tube. Stamens numerous (5 in *Melochia*), the filaments united most of the way into a tubular column (except in *Tilia*), the anthers usually each with a single locule opening by a longitudinal slit. Staminodes absent (present and petaloid in *Tilia*). Pistil 1 per flower, composed of 5 to numerous fused carpels (in some genera these loosely fused and tending to separate in fruit), the superior ovary with 3 to numerous locules, the styles as many as the locules, fused into a tube toward the base (entirely fused in *Gossypium* and *Tilia*, free in *Melochia*), each branch with 1 globose to disc-shaped or club-shaped stigma or a linear stigmatic region toward the tip. Fruits longitudinally dehiscent, globose to ovoid capsules with 1 to several seeds per locule, or schizocarps with a flattened ring of as many mericarps as locules, each with 1 to several seeds, or nutlike drupes (in *Tilia*). About 204 genera, about 2,330 species, nearly worldwide, especially in tropical regions.

Pubescence types are an important feature in distinguishing some closely related species in the family. For purposes of the following treatment, three different hair types are mentioned. Simple hairs are unbranched individual hairs, which may be relatively stouter, longer, and more or less spreading, or may be fine and appressed. The bases may be unmodified or somewhat expanded and bulbous (pustular). Stellate hairs have three or more arms from the base and are mostly relatively fine (exceptions exist) and appressed. Fasciculate hairs are generally spreading, relatively stout and long, and usually have pustular bases. They presumably represent stellate hairs with spreading arms, as well as clusters of simple hairs with the bases fused. For the most part, these hair types are distinguishable only with magnification.

The Malvaceae are here defined in a broad sense, to include Missouri genera traditionally segregated in the Sterculiaceae and Tiliaceae (Cronquist, 1981, 1991). The family is recognized by the presence in many (but not all) members of a staminal tube, as well as palmate leaf venation, stellate pubescence, and the production of mucilaginous sap. Molecular studies (summarized by C. Bayer et al., 1999; Alverson et al., 1999) have suggested that the woody families Bombacaceae, Sterculiaceae, and Tiliaceae, among others, should be included in an expanded concept of Malvaceae, which has rendered what was once an easily recognizable group into one whose morphological characteristics are somewhat less easily defined. However, molecular evidence is fairly strong that such families are embedded within Malvaceae in the traditional sense, and that, if separated, none of these families would comprise a natural group. To avoid potential accusations of narrow-mindedness (Mabberley, 2008), the family is here grudgingly treated in the broad sense. Within the Malvaceae, there exist nine major lineages that have been treated as informal groups equivalent to subfamilies. The "Tilioideae" group comprises the former Tiliaceae, and members of the former Sterculiaceae are included in two groups, "Byttnerioideae" and "Sterculioideae" (Whitlock et al., 2001).

Numerous species of Malvaceae have showy flowers and are cultivated as ornamentals. *Tilia* is a popular shade tree that also is used for fiber, furniture, musical instruments, utensils, and crafts. Other economically important timber species include *Ochroma pyramidalis* (Cav.ex Lam.) Urb. (balsa, formerly Bombacaceae). Cotton, one of the world's most important natural fibers, is derived from the long hairs covering the seeds of several *Gossypium* species, and the genus also is the basis for cottonseed oil. Cultivated food plants in the family (in the broad sense) include okra (*Abelmoschus esculentus* (L.) Moench) and durian (*Durio zibethinus* Rumph. ex Murray). Species in several genera also have economic importance as crop weeds. Marshmallow, originally prepared from the mucilaginous root extract of the European *Althaea officinalis* L. (marsh mallow) whipped with sugar and egg whites, is now made instead from corn syrup, gelatine, sugar, and egg whites. Cocoa is derived from the seeds of the neotropical tree *Theobroma cacao* L. and kola nuts (responsible for the flavor of cola beverages) comes from the edible seeds of species of the African genus *Cola* Schott & Endl. (both of these formerly Sterculiaceae). A number of species in this group contain compounds that are cardiac stimulants and thus have been used medicinally and pharmaceutically.

1. Plants trees; inflorescences appearing to rise from the midvein of leaflike bracts; stamens with the filaments united at the base into usually 5 groups, but not fused into a tube . . . . 13. TILIA
1. Plants annual or perennial herbs, or less commonly shrubs; inflorescences not appearing fused to bracts or to arise from the midvein of a bract (but individual flowers sometimes closely subtended by sepaloid bracts); stamens with the anthers united into a noticeable tube
    2. Leaves not lobed, the margins usually toothed
        3. Calyx not subtended by bractlets
            4. Leaf blades heart-shaped, deeply cordate at the base, mostly more than 4 cm wide; stipules shed very early; mericarps 10–18 mm long, each with 1 conspicuously spreading beak, 3–9-seeded . . . . . 1. ABUTILON
            4. Leaf blades linear to ovate, angled or rounded to truncate or shallowly cordate at the base, all less than 3 cm wide; stipules persistent, linear; mericarps 3–5 mm long, each with a pair of short erect beaks, 1-seeded
        3. Calyx closely subtended by 2 or more bractlets, these sometimes relatively inconspicuous
            5. Bractlets 8–12(–15); petals showy, 15–120 mm long. . . . . 6. HIBISCUS
            5. Bractlets (2)3–6; petals 2.5–6.5 mm long
                6. Stamens numerous; calyces appearing winged; petals inconspicuous, 2.5–5.0 mm long, yellow; style 1, with 5(6) short branches near the tip . . . . 9. MALVASTRUM
                6. Stamens 5; calyces not appearing winged; petals conspicuous, 4.0–6.5 mm long, light pink to nearly white (sometimes turning reddish as the flowers fade); styles 5, not fused . . . . 10. MELOCHIA
    2. At least some of the leaves shallowly to deeply lobed, the margins often also toothed
        7. Calyces not subtended by bractlets
            8. Calyces saucer-shaped at fruiting, the lobes becoming flattened and spreading horizontally; mericarps with the dorsal beak spreading, the lateral walls thin, smooth to slightly roughened, and shattering irregularly at maturity . . . . 3. ANODA
            8. Calyces cup-shaped at fruiting (sometimes fully enclosing the developing fruits), the lobes ascending to somewhat spreading but not becoming flattened horizontally; mericarps with the dorsal beak absent or erect to ascending or incurved, the lateral walls having a reticulate pattern of thickenings (faintly so in *Sphaeralcea*), indehiscent or dehiscent longitudinally along the dorsal and ventral margins
                9. Petals with broadly rounded to truncate tips, these with an irregular to somewhat fringed margin; mericarps 1-seeded; stigma a linear area along the inner side toward the tip of each style branch . . . . 4. CALLIRHOE
                9. Petals with asymmetrically rounded tips, the margins entire or nearly so; mericarps 1 or 2(3)-seeded; stigma terminal on each style branch, globose to club-shaped
                    10. Calyces 5-angled, the sepals also somewhat keeled; petals yellow to yellowish orange; fruits with 5–12 mericarps; mericarps with a pair of conspicuous beaks at the tip, 1-seeded . . . . 11. SIDA

10. Calyces not angled, the sepals not keeled; petals bright pink, usually whitened at the base, drying reddish purple; fruits with 11–15 mericarps; mericarps with a single small, inconspicuous beak at the tip (this siometimes absent), 2(3)-seeded . . . . 12. SPHAERALCEA

7. Calyces closely subtended by 2 or more bractlets, these sometimes inconspicuous
 11. Bractlets subtending the calyx fused to above the middle . . . . 7. LAVATERA
 11. Bractlets subtending the calyx free to the base
  12. Calyces closely subtended by 6 or more conspicuous bractlets
   13. Bractlets 6–9, broadly triangular; ovary with 18–40 locules; fruit a schizocarp; stigma a linear area along the inner side toward the tip of each style branch . . . . 2. ALCEA
   13. Bractlets 8–12(–15), linear; ovary with 5 locules; fruit a capsule; stigma terminal on each style branch, globose or disc-shaped . . . . 6. HIBISCUS
  12. Calyces closely subtended by 2–6 bractlets, these conspicuous or inconspicuous
   14. Bractlets broadly ovate-cordate, the margins incised with deep narrow lobes, much exceeding the calyx; style unbranched, with 3–5 linear, longitudinal stigmatic areas near the tip; ovary with 3–5 locules; seeds densely pubescent with long fibrous hairs . . . 5. GOSSYPIUM
   14. Bractlets linear to ovate or spatulate, the margins entire, shorter than to slightly longer than the calyx, or if noticeably longer than the calyx then linear; style branched toward the tip, the stigmas terminal or lateral along the branches; ovary with 9–23 locules (5 in *Melochia*); seeds glabrous to densely short-hairy
    15. Petals with asymmetrically rounded tips; locules (mericarps) 2(3)-seeded; stigma terminal on each style branch, globose to club-shaped . . . . 11. SPHAERALCEA
    15. Petals with broadly rounded, truncate, or shallowly notched tips; locules 1-seeded (sometimes 2-seeded in *Melochia*); stigma a linear to narrowly club-shaped area along the inner side toward the tip of each style or style branch
     16. Petals with broadly rounded to truncate tips, these with an irregular to somewhat fringed margin; mericarps differentiated into a basal fertile portion and an apical sterile incurved beaklike portion (except sometimes in *C. triangulata*); roots usually tuberous-thickened . . . . 4. CALLIRHOE
     16. Petals with shallowly notched to truncate or occasionally broadly and bluntly pointed tips, the margins otherwise entire or nearly so; mericarps (or capsular locules) not differentiated into apical sterile and basal fertile portions, beakless; roots not tuberous-thickened
      17. Stamens numerous, the anthers usually white; style 1, 8–20-branched toward the tip; leaf blades shallowly to deeply, palmately 3–7-lobed . . . . 8. MALVA
      17. Stamens 5, the anthers yellow; styles 5, not fused; leaf blades mostly entire, the larger blades occasionally very shallowly 3-lobed . . . . 10. MELOCHIA

1. **Abutilon** Mill. (Indian mallow)

About 160 species, nearly worldwide.

1. Abutilon theophrasti Medik. (velvetleaf, butter-print)

Pl. 450 g–i; Map 2042

Plants annual, densely pubescent throughout with stellate hairs. Stems 20–200 cm long, ascending to erect, branched or unbranched. Leaves long-petiolate, the blades 2.5–20.0 cm long, heart-shaped, unlobed, the base strongly cordate, abruptly long-tapered at the tip, the margins entire or shallowly toothed. Stipules shed before leaf maturity, 3–8 mm long, linear. Flowers solitary in the leaf axils or in loose terminal and axillary clusters, the bractlets subtending the calyx absent. Calyces 5–12 mm long, cup-shaped or becoming reflexed at fruiting, the sepals free nearly to the base, the lobes ovate. Petals 6–15 mm long, the tips truncate or more commonly shallowly notched, the margin otherwise entire or somewhat irregular, yellow to orangish yellow. Stamens numerous, the staminal column circular in cross-section, without a low crown of teeth at the tip, the anthers yellow. Pistils with 9–15 locules, the carpels arranged in a loose apically flattened ring. Styles fused most of their length, each branch with a globose terminal stigma. Fruits schizocarps breaking into 9–15 mericarps. Mericarps 10–18 mm long, wedge-shaped, becoming blackened at maturity, with a prominent horizontally spreading beak toward the tip, the dorsal surface lacking a longitudinal groove, oblong to kidney-shaped in profile, the lateral walls thin, smooth to slightly roughened, dehiscing apically from the center to the beak at maturity (the fruit also eventually breaking apart into individual mericarps), 3–9-seeded. Seeds 3–4 mm long, kidney-shaped to nearly triangular, the surfaces minutely stellate-hairy (appearing granular under lower magnification), black. $2n=42$. June–October.

Introduced, scattered to common nearly throughout Missouri (native of Asia, widely introduced in the U.S.). Crop fields, roadsides, railroads, and open disturbed areas; rarely margins of lakes and banks of streams.

Velvetleaf originally was introduced into the United States early in the nineteenth century as a fiber plant for the production of twine, thread, and other cordage, but was soon abandoned as cheaper sources of fiber for this purpose became available. Presently, it is one of the most important broadleaf weeds of crop fields in the United States, particularly in corn and soybean fields. It is resistant to some herbicides and the seeds may persist in the soil for up to 50 years.

2. **Alcea** L.
(Zohary, 1963a, b)

About 60 species, Europe, Asia.

1. Alcea rosea L. (hollyhock)

Althaea rosea (L.) Cav.

Pl. 450 c, d; Map 2043

Plants biennial or perennial herbs, variably pubescent throughout with smaller stellate hairs and coarse simple hairs (these mostly in fascicles), somewhat roughened to the touch. Stems 100–300 cm long, mostly erect, usually unbranched. Leaves long-petiolate, the blades 2–10 cm long, broadly ovate to circular or kidney-shaped in outline, at least the lower leaves shallowly 5- or 7-lobed, the base rounded to shallowly cordate, pointed to more commonly rounded at the tip, the margins shallowly scalloped or toothed. Stipules shed before leaf maturity, 3–9 mm long, lanceolate. Flowers solitary or in small clusters in the leaf axils, often also in a terminal spikelike raceme, the bractlets subtending the calyx 6–9, conspicuous, 9–12 mm long, fused toward the base, the lobes broadly triangular. Calyces 16–24 mm long, cup-shaped at fruiting, the sepals fused in the basal $1/3$–$1/4$, the lobes triangular. Petals 35–50 mm long, the tips broadly rounded to truncate or shallowly notched, the margin otherwise entire or nearly so, white to dark purplish black, often pink. Stamens numerous, the staminal column 5-angled in cross-section, without a low crown of teeth at the tip, the anthers yellow. Pistils with 18–40 locules, the carpels arranged in a loose apically flattened ring. Styles fused most of their length, each branch with

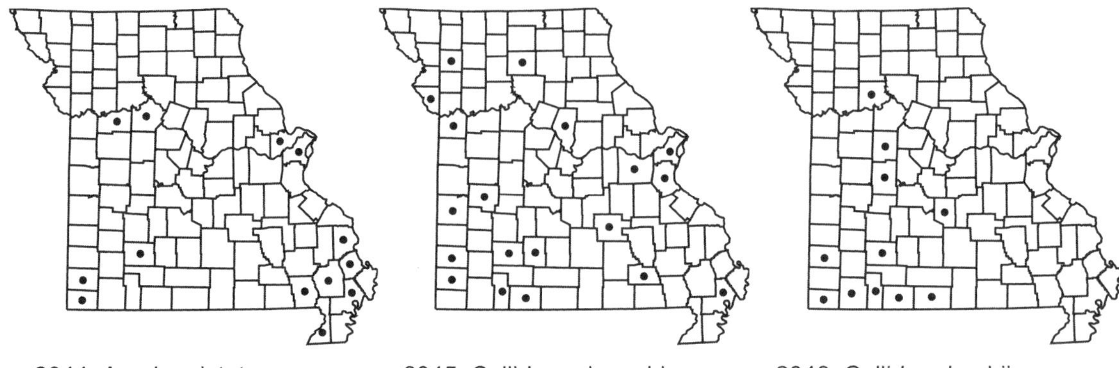

2044. Anoda cristata 2045. Callirhoe alcaeoides 2046. Callirhoe bushii

a single linear stigmatic area along the inner side toward the tip. Fruits schizocarps breaking into 18–40 mericarps. Mericarps 4–8 mm long, wedge-shaped, tan to brown, beakless, the dorsal surface with a longitudinal groove, kidney-shaped to nearly circular in profile, the lateral walls thin and smooth but with a reticulate pattern of thickenings toward the dorsal margin, which is angled into a narrow wing, the thin portion shattering irregularly at maturity (the fruit also eventually breaking apart into individual mericarps), 1-seeded. Seeds 3–4 mm long, kidney-shaped, the dorsal portion densely and minutely pubescent with mostly spreading fascicles of hairs, brown. $2n$=42. May–September.

Introduced, uncommon and widely scattered in the state (native range unknown but presumably southwestern Asia; widely and sporadically escaped from cultivation in the U.S.). Roadsides, railroads, and open disturbed areas.

Hollyhocks are widely cultivated as ornamentals in Missouri, but rarely become established outside gardens. A large number of cultivars exist varying in plant height, corolla colors, and leaf morphology, including some with doubled corollas, and the few specimens of escaped plants reflect this variability. Some authors continue to maintain *Alcea* as part of an expanded concept of the genus *Althaea*, but Zohary (1963a, b) and D. A. Webb (1968), among others, have separated the two groups based on differences in inflorescence and flower structure and size.

3. **Anoda** Cav.
(Fryxell, 1987)

Twenty-three species, United States to South America, Caribbean Islands.

1. Anoda cristata (L.) Schltdl. (spurred anoda)
Pl. 450 a, b; Map 2044

Plants annual. Stems 25–100 cm long, loosely ascending to erect, usually branched, sparsely pubescent with mostly simple, spreading to downward-pointing pustular-based hairs. Leaves long-petiolate, very variable, the blades 1.5–8.5 cm long, narrowly triangular-hastate to broadly ovate, sometimes with 1 or 2 pairs of palmate lobes, the margins otherwise entire to irregularly and shallowly toothed or scalloped, the surfaces sparsely to moderately pubescent with mostly simple appressed hairs, the upper surface sometimes with a purple area along the midrib. Stipules persistent, 6–10 mm long, linear. Flowers solitary in the leaf axils, the bractlets subtending the calyx absent. Calyces 5–10 mm long at flowering, expanding to 20 mm long at fruiting, becoming saucer-shaped with the lobes flattened and spreading horizontally, the sepals fused to about the middle, the fused portion whitened, strongly reticulate-veined at maturity, and sometimes with purple spots, the lobes narrowly lanceolate-triangular at flowering, becoming broadly triangular at fruiting, the outer surface pubescent with simple spreading pustular-based hairs. Petals 8–12(–20) mm long, the tips truncate or more commonly shallowly notched, the margin otherwise entire or somewhat irregular, light bluish purple, rarely white, often drying blue. Stamens numerous, the staminal column circular in cross-section, without a low crown of teeth at the tip, the anthers white or light bluish purple. Pistils with 8–18 locules, the carpels arranged in a loose, strongly flattened ring. Styles fused most of their length, each branch with a globose terminal stigma.

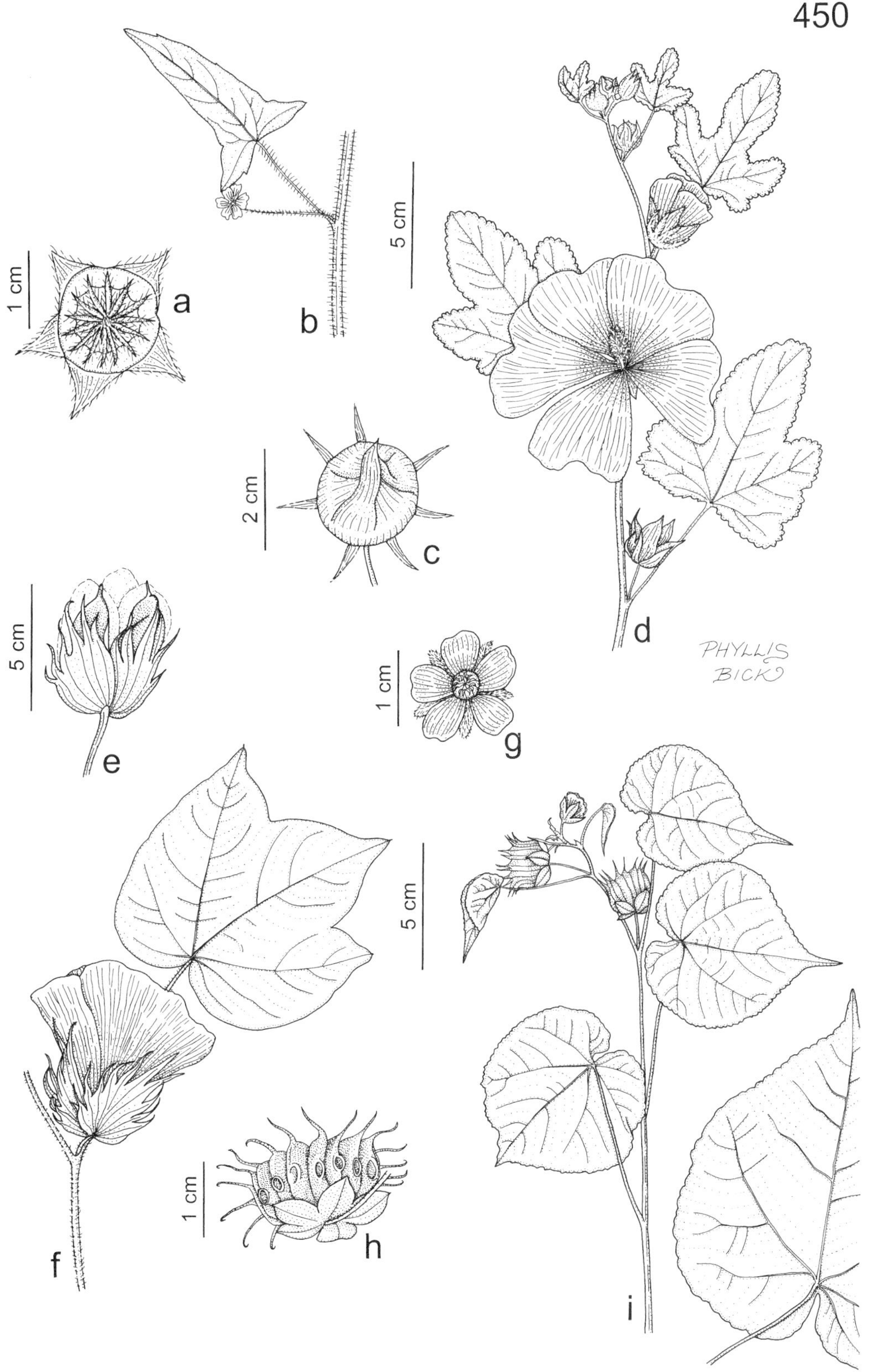

Plate 450. Malvaceae. *Anoda cristata*, **a)** fruit, **b)** node with leaf and flower. *Alcea rosea*, **c)** fruit, **d)** fertile stem. *Gossypium hirsutum*, **e)** fruit, **f)** node with leaf and flower. *Abutilon theophrasti*, **g)** flower, **h)** fruit, **i)** fertile stem and lower leaf.

Fruits schizocarps breaking into 8–18 mericarps. Mericarps 2–4 mm long, wedge-shaped, pubescent with spreading pustular-based simple hairs, the dorsal surface green, lacking a longitudinal groove, with a prominent horizontally spreading beak, oblong to kidney-shaped in profile, the lateral walls thin, whitened, smooth to slightly roughened, shattering irregularly at maturity, 1-seeded. Seeds 3–4 mm long, oblong-rectangular to kidney-shaped, the surfaces minutely pebbled or warty, black or dark brown. $2n$=30, 60, 90. June–September.

Introduced, uncommon and widely scattered in Missouri (southwestern U.S. east to Texas; Mexico, Central America, South America, Caribbean Islands; introduced in the U.S. east to Pennsylvania and also in the Old World). Roadsides, railroads, ditches, margins of crop fields, and open disturbed areas.

Steyermark (1963) reported this species from a single specimen from McDonald County. It appears to have become more common and widely dispersed in recent years.

4. **Callirhoe** Nutt. (poppy mallow, wine cup) (Dorr, 1990; M. Morris and Yatskievych, 2000)

Plants perennial herbs (annual or biennial elsewhere), pubescent with simple and/or stellate hairs, the roots usually thickened and tuberous. Stems prostrate to erect, usually unbranched below the inflorescence. Leaves petiolate, the blades entire to deeply palmately lobed, the margins entire to irregularly undulate, toothed, or lobed again. Stipules linear to broadly triangular or rhombic, often asymmetric at the base, shed early in some species. Inflorescences terminal and sometimes also axillary racemes, less commonly panicles or condensed and appearing as stalked clusters or umbellate. Flowers perfect or uncommonly only pistillate, in some species the calyx closely subtended by 3 bractlets. Calyces cup-shaped at fruiting, the lobes ascending to somewhat spreading but not becoming flattened horizontally, lobed 2/3–3/4 of their length, lanceolate to broadly triangular, the outer surface glabrous or variously hairy, the inner surface usually with a mat of stellate hairs, especially near the margins. Petals wine red to reddish purple or less commonly pink, pale lavender, or white, the broadly rounded to truncate tips with an irregular to somewhat fringed margin. Stamens numerous, the staminal column circular in cross-section, without a low crown of teeth at the tip, glabrous or hairy toward the base, the anthers white, red, or purple. Pistils with 9–23 locules, the carpels arranged in a loose flattened ring. Styles fused most of their length, each branch with a single linear stigmatic area along the inner side toward the tip. Fruits schizocarps breaking into 9–23 mericarps. Mericarps 3–6 mm long, indehiscent or dehiscent, wedge-shaped, the dorsal surface usually with a longitudinal groove and an inconspicuous inflexed beak (this often absent), oblong to kidney-shaped in profile, each differentiated into an incurved sterile upper cell (this rarely absent) and a lower cell containing 1 seed, the upper sterile cell smooth-walled to finely roughened and also usually with a shallow dorsal groove, the lower cell with a prominent reticulate pattern of thickenings on the sides (except in *C. triangulata*). Seeds 2–3 mm long, kidney-shaped, black or less commonly dark brown. Nine species, endemic to the central and southeastern United States and adjacent northeastern Mexico.

Species of *Callirhoe* have a long history of sporadic cultivation as garden ornamentals. Native Americans also appreciated the flowers of some species for their aesthetic appeal (Moerman, 1998) and used a decoction of the roots as an analgesic (Dorr, 1990; Moerman, 1998). The roots of all of the perennial species are both edible and palatable, and were eaten by Native Americans and early European travelers in the Great Plains and southern states.

1. Calyx without subtending bractlets; stipules shed before flowering or, if persistent, then 5–8 mm long
   2. Stems densely pubescent with stellate hairs; stipules persistent 5–8 mm long; Inflorescences racemes, sometimes condensed and appearing as stalked

clusters or umbels; outer surface of calyx densely hairy; petals nearly white to light pink or pale lavender . . . . 1. C. ALCAEOIDES

2. Stems glabrous or nearly so; stipules 6.0–7.5 mm long, shed early; inflorescences panicles; outer surface of calyx glabrous; petals wine red to bright red, rarely light red . . . . 3. C. DIGITATA

1. Calyx closely subtended by three bractlets; stipules persistent, 4–15 mm in length

3. Stem leaves triangular to ovate-cordate or ovate-hastate in outline, usually not deeply lobed, the margins entire, undulate, scalloped, bluntly toothed, or shallowly lobed; inflorescences panicles; bractlets spatulate to obovate . . . . 5. C. TRIANGULATA

3. Stem leaves broadly deltoid to broadly ovate-cordate or obovate in outline, the margins deeply lobed and often irregularly lobed again or deeply toothed; inflorescences racemes; bractlets linear to ovate

4. Stems ascending; buds with the sepal tips joined to form a beaklike projection 6–10 mm in length; basal leaves with mostly simple hairs . . . . 2. C. BUSHII

4. Stems spreading with ascending tips; buds with the sepal tips free and somewhat spreading, not joined into a beak; basal leaves with many stellate hairs in addition to simple hairs . . . . 4. C. INVOLUCRATA

1. Callirhoe alcaeoides (Michx.) A. Gray (pink poppy mallow, pale poppy mallow, plains poppy mallow)

Pl. 451 a, b; Map 2045

Stems 10–45 cm long, erect or ascending, densely pubescent with stellate hairs. Basal leaf petioles 7–19 cm long, pubescent with stellate hairs. Basal leaf blades 2.5–10.0 cm long, deltoid-cordate to ovate in outline, with 3–5 shallow to deeply palmate lobes (rarely unlobed), these often irregularly lobed again, the ultimate segments usually relatively broad, the margins entire to scalloped, the upper surface pubescent with simple hairs, the undersurface with stellate and sometimes also scattered, simple hairs. Leaves of the aerial stems with the blades 4–8 cm long, triangular-cordate to broadly obovate in outline, with 3–5 shallow to deeply palmate lobes, these often pinnately lobed again, the margins entire to scalloped, the pubescence as in the basal leaves. Stipules persistent, often partially fused to the petiole, 5–8 mm long, lanceolate to narrowly lanceolate. Inflorescences racemes with 4–14 flowers, sometimes condensed and appearing as stalked clusters or umbellate. Bractlets subtending the calyx absent. Buds ovate, the sepal tips valvate, joined to form a short beaklike projection 1.5–4.0 mm long. Calyces 7–10 mm long, the outer surface densely pubescent with simple hairs and often also a few 4-rayed stellate hairs, the lobes 5–8 mm long, lanceolate. Petals 8–20 mm long, nearly white to light pink or pale lavender. Fruits 6–9 mm in diameter, with 10–13 mericarps. Mericarps indehiscent, 4.0–5.5 mm long, the dorsal surface hairy, the sides of the fertile portion with a reticulate pattern of thickenings, separated from the prominent sterile portion by a well-developed collar. $2n=28$. May–August.

Scattered in the western portion of the Glaciated Plains Division and the Unglaciated Plains; also introduced at scattered sites in the rest of Missouri, but apparently still absent from the Mississippi Lowlands (Iowa to Louisiana, west to South Dakota and Texas, adventive eastward to Indiana, Tennessee, and possibly Alabama). Dry upland forests, upland prairies, and calcareous glades; also stream banks, roadsides, railroads, old quarries, and pastures.

2. Callirhoe bushii Fernald (Bush's poppy mallow)

C. involucrata (Torr. & A. Gray) A. Gray var. *bushii* (Fernald) R.F. Martin

C. papaver (Cav.) A. Gray var. *bushii* (Fernald) Waterf.

Pl. 451 c–e; Map 2046

Stems 35–50 cm long, ascending, densely pubescent with simple hairs and sometimes also, stellate hairs. Basal leaf petioles 10–23 cm long, pubescent with stellate hairs. Basal leaf blades 4–12 cm long, triangular-cordate to broadly ovate in outline, with 3–7 deep palmate lobes, these sometimes irregularly lobed again, relatively broad, the margins usually sparsely and coarsely toothed, the

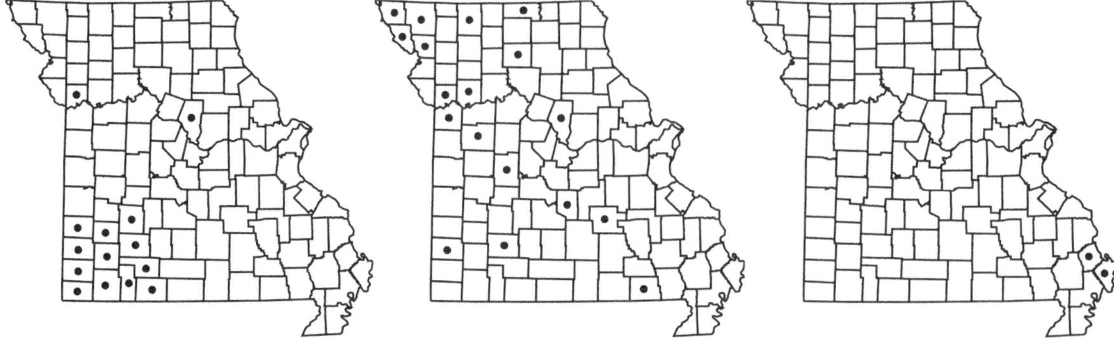
2047. Callirhoe digitata 2048. Callirhoe involucrata 2049. Callirhoe triangulata

upper and lower surfaces pubescent with simple and occasionally also a few stellate hairs. Leaves of the aerial stems with the blades 3–12 cm long, broadly triangular to ovate-cordate or obovate in outline, with 3(5) deep palmate lobes, these often pinnately lobed again, the margins entire to sparsely and coarsely toothed, the upper and lower surfaces pubescent with simple and usually also stellate hairs. Stipules persistent, 9–15 mm long, ovate, sometimes auriculate. Inflorescences racemes with 7–17 flowers. Bractlets subtending the calyx 8–14 mm long, linear to narrowly elliptic-lanceolate. Buds ovate to broadly ovate, the sepal tips valvate, forming a beaklike projection 6–10 mm long. Calyces 15–27 mm long, the outer surface densely pubescent with simple hairs and sometimes also with sparse 4-rayed stellate hairs near the tip, the lobes 10–17 mm long, lanceolate to narrowly lanceolate. Petals 17–32 mm long, pale rose to more commonly wine red or purplish red. Fruits 9–13 mm in diameter, with 16–23 mericarps. Mericarps indehiscent, 4.0–5.5 mm long, the dorsal surface glabrous, the sides of the fertile portion with a reticulate pattern of thickenings, separated from the inconspicuous sterile portion by a well-developed collar. $2n$=56. May–August.

Scattered in the southwestern quarter of the state north to Carroll County and east to Pulaski County (Iowa to Arkansas, west to adjacent Kansas and Oklahoma). Bottomland forests to dry upland forests, bottomland prairies, upland prairies, and calcareous glades; also roadsides, road cuts, railroads, pastures, and open disturbed areas.

This species is sometimes confused with *C. involucrata*, but the two species differ in a number of subtle characters. In addition to the habit and bud tip characters in the key to species, they also differ in details of the calyx. In *C. involucrata*, the calyx lobes are strongly nerved, the nerves raised and often whitened. The pubescence on the outer surface of the lobes consists of long, spreading simple hairs that have somewhat pustular bases and are somewhat fused into irregular fascicles, as well as shorter 4-rayed stellate hairs that are mostly loosely appressed. In *C. bushii*, the calyx lobes are less strongly veined and generally not raised into whitened ridges. The pubescence of the outer surface is mostly of simple spreading hairs, these not fasciculate, with shorter stellate hairs absent or sometimes sparse and near the tips.

Fernald (1950), Gleason (1952), Steyermark (1963), and some other authors included Missouri in the distribution of the closely related *C. papaver*. That species occurs only to the south and east of Missouri, and is distinguished from *C. bushii* by its narrower leaf lobes, narrower (or absent) floral bractlets, and less pubescent stems (Dorr, 1990). The character of appressed vs. spreading pubescence used by Steyermark (1963) to distinguish *C. papaver* from *C. bushii* is too variable to differentiate these species.

3. Callirhoe digitata Nutt. (fringed poppy mallow, finger poppy mallow, finger wine cup)

Pl. 451 h–j; Map 2047

Stems 30–120 cm long, erect or ascending, glabrous, glaucous, rarely sparsely pubescent with simple hairs near the nodes. Basal leaf petioles 7–30 cm long, glabrous or more commonly pubescent with simple hairs. Basal leaf blades 3–12 cm long, kidney-shaped to nearly circular in outline, with 3–9 deep palmate lobes, these often 1 or 2 times (2)3-lobed again, the ultimate segments usually linear, the margins entire or with few shallow lobes or teeth, the upper surface glabrous or pubescent with simple hairs, the undersurface usually pubescent with simple hairs. Leaves of the aerial stems with the blades 3–13 cm long, kidney-shaped to nearly circular or broadly ovate in outline, with

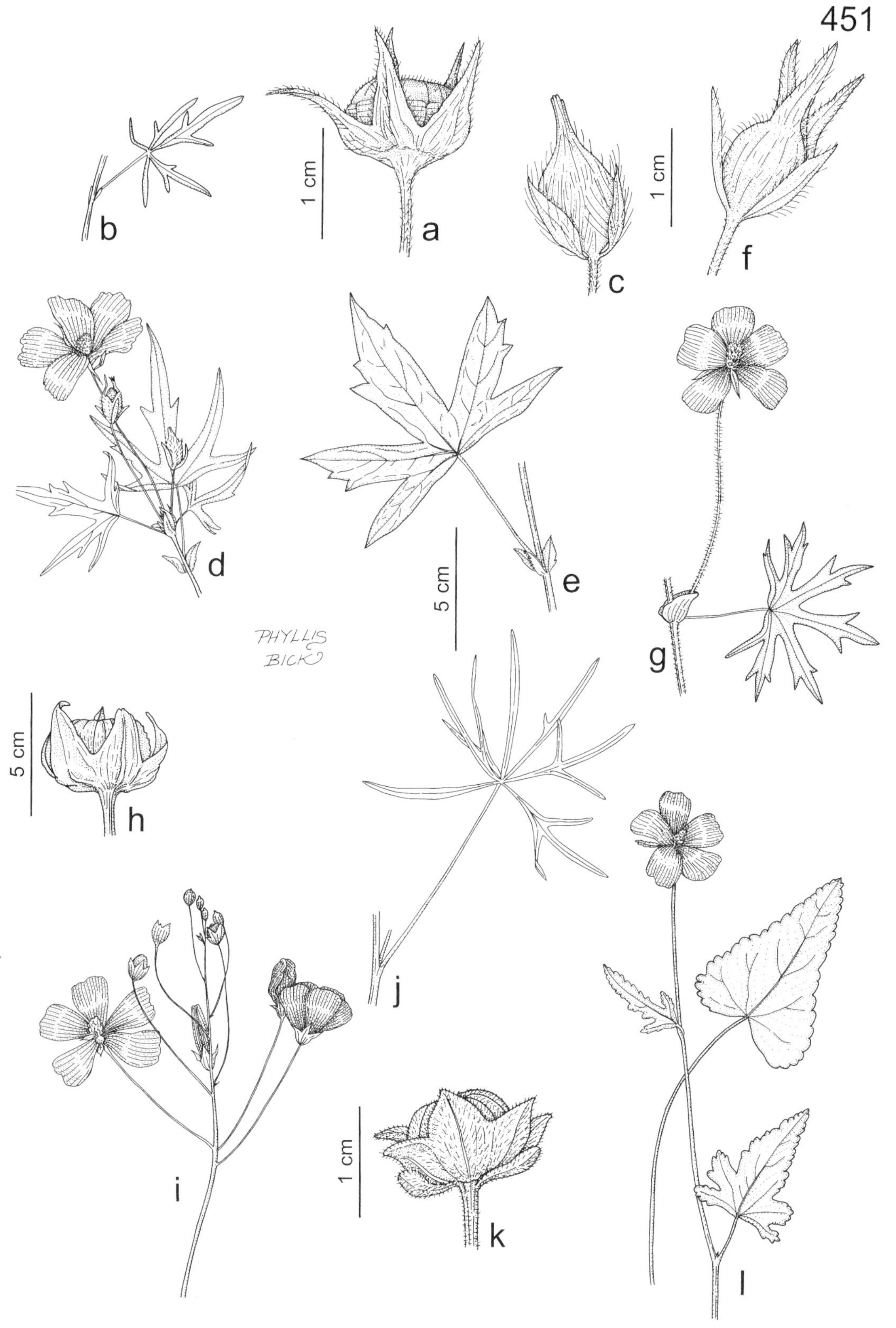

Plate 451. Malvaceae. *Callirhoe alcaeoides*, **a)** fruit, **b)** leaf. *Callirhoe bushii*, **c)** fruit, **d)** fertile branch, **e)** node with leaf. *Callirhoe involucrata*, **f)** fruit, **g)** node with leaf and flower. *Callirhoe digitata*, **h)** fruit, **i)** inflorescence, **j)** leaf. *Callirhoe triangulata*, **k)** fruit, **l)** fertile stem tip and basal leaf.

3–9 deep palmate lobes, these often 1 or 2 times (2)3-lobed again, the ultimate segments usually linear, the margins entire or with few shallow lobes or teeth, the pubescence as in the basal leaves. Stipules mostly shed before flowering, 6.0–7.5 mm long, linear to narrowly lanceolate. Inflorescences panicles with 6–20 flowers. Bractlets subtending the calyx absent. Buds ovate, the sepal tips valvate, forming a short beaklike projection 1.0–2.5 mm long. Calyces 7–10 mm long, the outer surface glabrous, the lobes 3.5–7.0 mm long, lanceolate to narrowly ovate or narrowly triangular. Petals 15–26 mm long, deep rose to wine red or purplish red, usually with a lighter area toward the base. Fruits 7–9 mm in diameter, with 12–16 mericarps. Mericarps indehiscent, 3.5–4.5 mm long, the surfaces glabrous, the sides of the fertile portion with a reticulate pattern of thickenings, separated from the inconspicuous sterile portion but without a differentiated collar. $2n=28$. May–September.

Scattered in the western portion of the Ozark Division and the Unglaciated Plains; introduced in Boone and Clay Counties (Missouri to Arkansas, west to adjacent Kansas and Texas; adventive in Illinois, Indiana, and possibly Louisiana). Dry upland forests, upland prairies, and calcareous glades; also roadsides and railroads.

4. Callirhoe involucrata (Torr. & A. Gray) A. Gray **var. involucrata** (purple poppy mallow)

Pl. 451 f, g; Map 2048

Stems 7–50 cm long, spreading with ascending tips, densely pubescent with mostly stellate hairs. Basal leaf petioles 5–9(–12) cm long, pubescent with mostly stellate and sparse simple hairs. Basal leaf blades 2–6 cm long, kidney-shaped to nearly circular in outline, with 3 or 5 deep palmate lobes, these sometimes irregularly lobed again, broad or narrow, the margins entire or sparsely and coarsely toothed, the upper and lower surfaces pubescent with dense 4-rayed stellate and sparse simple hairs. Leaves of the aerial stems with the blades 3–8 cm long, broadly ovate-cordate to obovate in outline, with 3 or 5 deep palmate lobes, these usually pinnately lobed again, the margins entire to sparsely and coarsely toothed, the upper surface pubescent with simple hairs, the lower surface pubescent with stellate and simple hairs. Stipules persistent, 7–13 mm long, narrowly to broadly ovate, sometimes somewhat auriculate. Inflorescences racemes with 4–10 flowers. Bractlets subtending the calyx 8–13 mm long, linear to narrowly ovate or oblanceolate. Buds ovate to broadly ovate, the terminal 3–10 mm of the sepals free and loosely ascending to spreading. Calyces 12–19 mm long, the outer surface densely pubescent with simple hairs (these often in irregular fascicles along the strongly raised nerves) and also with 4-rayed stellate hairs, the lobes 8–16 mm long, lanceolate to narrowly lanceolate or less commonly narrowly ovate. Petals 19–33 mm long, pale rose to more commonly wine red or purplish red, sometimes with a lighter area toward the base. Fruits 9–12 mm in diameter, with 14–20 mericarps. Mericarps indehiscent, 3.0–4.5 mm long, the dorsal surface glabrous or pubescent with mostly simple hairs, the sides of the fertile portion with a reticulate pattern of thickenings, separated from the inconspicuous beaklike sterile portion but without a well-developed collar. $2n=30$, 60. May–August.

Scattered in the western portion of the Glaciated Plains Division east to Putnam County; introduced nearly throughout Missouri, but apparently still absent from the Mississippi Lowlands and the eastern portion of the Glaciated Plains (Indiana to Arkansas, west to South Dakota, Wyoming, and New Mexico; adventive eastward to Pennsylvania and disjunctly in Oregon). Upland prairies; also roadsides, railroads, pastures, and open disturbed areas.

Dorr (1990) accepted three varieties of *C. involucrata* differing in details of leaf division and involucral bract characters. Of these, only the var. *involucrata* occurs in Missouri. The var. *involucrata* occupies most of the species' range, but is absent from Mexico and most of Texas, where var. *tenuissima* Baker f. and var. *lineariloba* (Torr. & A. Gray) A. Gray occur respectively. The natural range of var. *involucrata* has become somewhat blurred because it is cultivated as an ornamental and has become naturalized in a number of states to the east and south of its presumed native distribution. In Missouri, most of the specimens are from disturbed sites representing adventive populations.

See the treatment of *C. bushii* for a discussion of differences between that species and *C. involucrata*.

5. Callirhoe triangulata (Leavenw.) A. Gray (clustered poppy mallow)

Pl. 451 k, l; Map 2049

Stems 10–45 cm long, spreading to ascending, densely pubescent with stellate hairs. Basal leaf petioles 10–22 cm long, pubescent with stellate hairs. Basal leaf blades 3.5–10.5 cm long, triangular to ovate-cordate or ovate-hastate in outline, unlobed or less commonly with 3 shallow to deeply palmate lobes, these broad or narrow, the margins

2050. Gossypium hirsutum

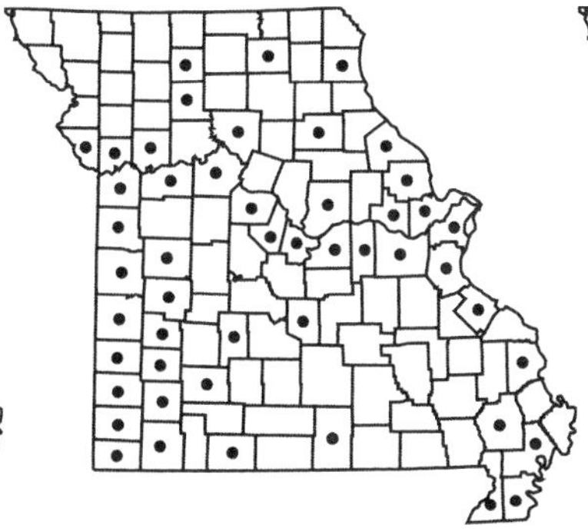

2051. Hibiscus laevis

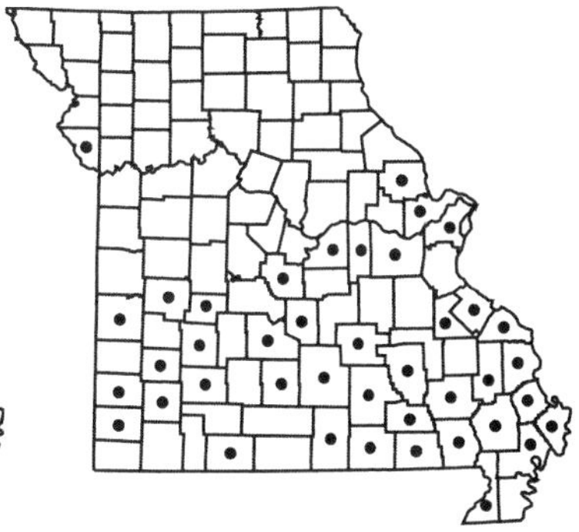

2052. Hibiscus lasiocarpos

entire or more commonly scalloped, the upper and lower surfaces pubescent with dense (3)4(–8)-rayed stellate hairs. Leaves of the aerial stems with the blades 3–12 cm long, triangular to ovate-cordate or ovate-hastate in outline, with 3 or 5 shallow or less commonly deeply palmate lobes, the margins entire or more commonly undulate, scalloped, scalloped-toothed, or shallowly lobed again, the pubescence as in the basal leaves. Stipules persistent, 3.5–7.2 mm long, elliptic-lanceolate to less commonly ovate. Inflorescences panicles with 2–8 flowers, sometimes condensed and appearing as stalked clusters or umbellate. Flower stalks 0.6–2.0 cm long. Bractlets subtending the calyx 4–7 mm long, narrowly obovate to spatulate. Buds ovate to broadly ovate, the sepal tips valvate, forming a short beaklike projection 1–2 mm long. Calyces 5–8 mm long, the outer surface densely pubescent with 4–6-rayed stellate hairs, the lobes 2–6 mm long, triangular-ovate, often with abruptly short-pointed tip. Petals 15–27 mm long, red to purplish red, sometimes with a lighter area toward the base. Fruits 6–9 mm in diameter, with 10–13 mericarps. Mericarps dehiscing longitudinally along the dorsal surface, 3.0–3.8 mm long, the dorsal surface pubescent with simple hairs and also with irregularly 2-branched hairs, the sides of the fertile portion thin and smooth to slightly granular, separated from the inconspicuous sterile portion (this rarely absent) but without a well-developed collar. $2n=30$. July–August.

Uncommon, known only from historical specimens from Mississippi and Scott Counties (southeastern U.S. from North Carolina to Mississippi (except Florida) and midwestern U.S. from Wisconsin and Iowa south to Indiana and Missouri). Sand prairies; also roadsides.

The specimens attributed to this species and mapped by Steyermark (1963) from Franklin and St. Louis Counties could not be located during the present study. *Callirhoe triangulata* apparently is no longer extant in Alabama, Iowa, and Missouri, and is of conservation concern in most of the rest of its range (except in Illinois).

5. **Gossypium** L. (cotton)

About 50 species, nearly worldwide in tropical and warm-temperate regions.

Several species of *Gossypium* have long been cultivated in different parts of the world as a source of cotton fiber, cottonseed oil, and seed meal. The cytogenetics and evolution of the cultivated taxa have been studied in detail. Those grown in the New World are tetraploid taxa that originated from past hybridization between diploid parents native to the Old World and New World tropics respectively, whereas those cultivated in the Old World are diploids originating from Old World diploid taxa. In both areas, human selection has resulted in a large number of races and cultivars. For a fascinating review of the natural history of this group, see Fryxell (1979).

1. Gossypium hirsutum L. (upland cotton)

Pl. 450 e, f; Map 2050

Plants annual (shrubs elsewhere). Stems 30–130 cm long, erect or ascending, usually branched, pubescent with spreading, simple or fasciculate hairs. Leaves mostly long-petiolate, the blades 3–10 cm long, broadly ovate to nearly circular in outline, 3- or 5-lobed, the base cordate, the lobes broadly triangular, tapered to pointed tips the margins entire, the surfaces pubescent with sparse to dense stellate hairs. Stipules tending to shrivel before flowering, 3–9 mm long, linear to elliptic-lanceolate. Flowers solitary or in small clusters in the leaf axils, the bractlets subtending the calyx 3, conspicuous and exceeding the calyx, 20–45 mm long, broadly ovate-cordate, overlapping, the margins incised with deep narrow lobes. Calyces 5–24 mm long, cup-shaped at fruiting, the sepals fused nearly to the tip, the lobes broadly triangular. Petals 25–55 mm long, the tips broadly rounded, the margin usually slightly irregular, white or more commonly pale yellow, sometimes pinkish- or purplish-tinged, often with minute dark glandular dots. Stamens numerous, the staminal column circular in cross-section, with a low crown of teeth at the tip, the anthers yellow. Pistils with 3–5 locules, the carpels closely fused. Style unbranched, with 3–5 linear longitudinal stigmatic areas toward the tip. Fruits capsules, broadly ovoid to nearly globose, dehiscing longitudinally from the tip, glabrous. Seeds 5–11 per locule, 7–10 mm long, narrowly obovoid, the body hidden by a dense covering of shorter white matted hairs and long fluffy white fibrous hairs. $2n=56$. July–September.

Introduced, uncommon in southern Missouri north to the St. Louis and Kansas City areas (native of Mexico and Caribbean islands; widely cultivated and escaping sporadically). Roadsides, railroads, and open disturbed areas.

Gossypium hirsutum is the principal species of cotton grown in North America and has a history of cultivation dating far back into pre-Columbian times. In Missouri, cotton production occurs in the Mississippi Lowlands Division and to a lesser extent in adjacent Ozark counties. Lower-growing cultivars adapted to sandy soils seem to have replaced most of the taller bushier plants represented in the older specimens of escapes from cultivation.

Steyermark (1963) excluded this species from the state's flora in spite of the existence of several older specimens, reasoning that it was unlikely to become established in Missouri for any great length of time. However, in the supplement to his flora, he readily accepted the inclusion of *G. herbaceum* L. (an Old World species), based on a single misdetermined collection of *G. hirsutum* made by Viktor Mühlenbach in the St. Louis railroad yards. Mühlenbach (1979) later reported additional specimens and corrected the report.

6. **Hibiscus** L. (hibiscus)

Plants annual or perennial herbs, shrubs, or small trees. Stems spreading or ascending to erect, branched or less commonly unbranched. Leaf blades 1.5–12.0 cm wide, narrowly to broadly ovate in outline, unlobed or shallowly to deeply lobed (sometimes appearing nearly compound in *H. trionum*) broadly rounded to truncate or shallowly cordate at the base, rounded or pointed at the tip, the margins entire to more commonly toothed, scalloped, or lobed. Stipules 1–8 mm long, linear, shed before leaf maturity or persistent. Flowers solitary in the leaf axils (rarely in small clusters in *H. syriacus*), sometimes also appearing terminal, the bractlets subtending the calyx 8–12(–15), conspicuous, linear. Calyces closely cup-shaped at fruiting, appearing rounded or sometimes somewhat 5-angled at fruiting, the sepals fused to above the middle, the lobes triangular. Petals showy, 15–120 mm long, rounded to broadly rounded at the tip, the margin otherwise somewhat undulate or irregular, sometimes slightly to strongly asymmetric, variously colored, usually with a dark reddish purple area at the base. Stamens numerous, the staminal column circular in cross-section, with a low crown of teeth at the tip. Pistils with 5 locules, the carpels closely fused. Styles fused most of their length, the tube usually curved or bent obliquely upward at the tip, each branch with a globose or disc-shaped terminal stigma. Fruits capsules, dehiscing longitudinally from the tip. About 200 species, nearly worldwide, most diverse in tropical and warm-temperate regions.

The elongate, mucilaginous fruits of *Hibiscus esculentus* L. are known as okra and are popular in particularly Creole cuisine. The fruits are sliced and included in gumbos, with the mucilage acting as a thickener.

1. Plants strongly woody shrubs or small trees . . . . 3. H. SYRIACUS
1. Plants annual or perennial herbs

2. Stems 25–55 cm long, mostly spreading to loosely ascending; leaf blades deeply 3-lobed (often appearing nearly compound), the main lobes often shallowly to deeply lobed again; petals 1.5–4.0 cm long . . . . 4. H. TRIONUM
2. Stems 80–200 cm long, erect or ascending; leaf blades appearing simple, unlobed or with a pair of lobes at the base, these shallow or moderately deep, not lobed again; petals 5–20 cm long
   3. Calyces and the undersurface of leaf blades glabrous or sparsely hairy along the main veins; fruits glabrous except along the inner margins of the carpel walls (seen after dehiscence); seeds hairy. . . . . 1. H. LAEVIS
   3. Calyces and the undersurface of leaf blades densely hairy; fruits hairy; seeds glabrous . . . . 2. H. LASIOCARPOS

1. Hibiscus laevis All. (rose mallow, halberd-leaved rose mallow)
H. militaris Cav.
Pl. 452 a–d; Map 2051

Plants perennial herbs, sometimes slightly woody at the base. Stems 80–200 cm long, erect or ascending, glabrous, often somewhat glaucous. Leaf blades 4–20 cm long, ovate or less commonly elliptic-lanceolate in outline, shallowly to deeply 3-lobed (hastate) at the base, less commonly unlobed, the margins relatively finely and sharply toothed, the upper surface glabrous, the undersurface glabrous or sparsely pubescent with stellate hairs along main veins. Stipules shed during leaf development. Bractlets subtending the calyx 8–12(–15), 12–25 mm long, glabrous or pubescent with unbranched or fasciculate hairs along the margins. Calyces 25–35 mm long at flowering, becoming slightly enlarged to 40 mm and slightly inflated at fruiting, glabrous or sparsely and finely pubescent with stellate hairs along the veins. Petals 6–8 cm long, white or more commonly pink, usually with dark reddish purple bases. Fruits 2.2–3.2 cm long, ovoid to ovoid-cylindric, glabrous except for dense simple and fasciculate hairs along the inner side of the margins of the valves (visible after dehiscence). Seeds 12–18 per locule, 3.2–3.8 mm long, broadly kidney-shaped to nearly globose, the surface densely covered with short stiff orangish brown hairs, dark brown. $2n=38$. July–October.

Scattered, mostly outside the Ozark Division (eastern U.S. west to Nebraska and Texas). Openings of bottomland forests, swamps, banks of streams and rivers, and margins of ponds, lakes, sinkhole ponds, and sloughs, sometimes in shallow water; also ditches and wet roadsides.

Hibiscus laevis frequently forms large colonies. Its seeds are eaten by wildlife, particularly birds. Although this species and *H. lasiocarpos* are sometimes found growing together, especially in disturbed habitats like roadside ditches, hybridization between them is rare and the hybrids (of intermediate general morphology) produce shrunken abortive seeds. O. J. Blanchard (1976), who examined more than 3,000 herbarium specimens from throughout the provenance of these species and conducted extensive field work, located only 8 specimens rangewide representing putative hybrid plants. From Missouri, he located only a single historical collection from waste ground along a railroad in Jasper County.

2. Hibiscus lasiocarpos Cav. (rose mallow, hairy rose mallow)
H. moscheutos L. ssp. *lasiocarpos* (Cav.) O.J. Blanch.
H. moscheutos var. *occidentalis* Torr.
Pl. 452 e–g; Map 2052

Plants perennial herbs, sometimes slightly woody at the base. Stems 80–200 cm long, erect or ascending, densely pubescent with mostly stellate hairs, less commonly glabrous or nearly so. Leaf blades 4–20 cm long, narrowly lanceolate to broadly ovate in outline, unlobed or less commonly shallowly 3-lobed at the base, the margins finely to coarsely scalloped or bluntly toothed, the upper surface moderately to densely pubescent with stellate hairs, sometimes becoming glabrous or nearly so with age, the undersurface densely pubescent with stellate hairs. Stipules shed during leaf development. Bractlets subtending the calyx 8–12 (–14), 18–35 mm long, densely pubescent with stellate hairs and usually also with simple or fasciculate hairs along the margin. Calyces 15–40 mm long at flowering, not becoming enlarged or inflated at fruiting, finely pubescent with stellate hairs, especially along the veins. Petals 6.5–12.0 cm long, white or less commonly pink, sometimes with dark reddish purple bases. Fruits 1.5–3.2 cm long, ovoid to ovoid-cylindric, densely pubescent with stellate hairs and also with longer simple or fasciculate hairs, especially along the margins of the valves. Seeds 18–35 per locule, 2.5–3.0 mm long, obovoid to nearly globose, the surface minutely pebbled or roughened and with a faint pat-

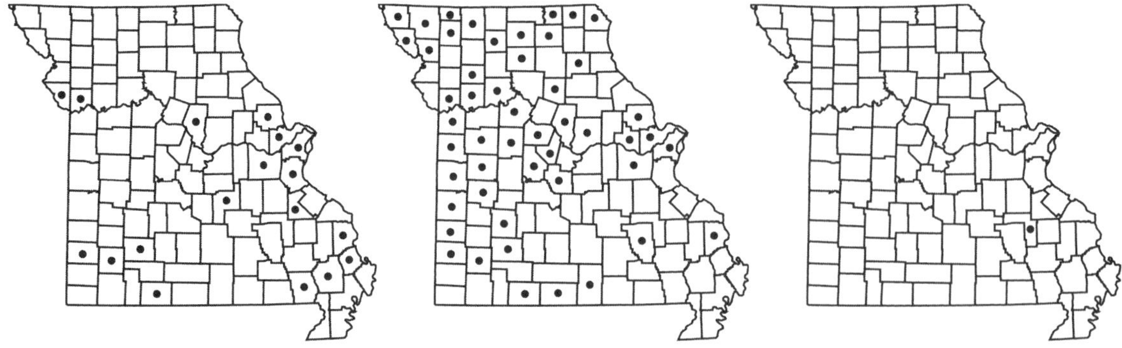
2053. Hibiscus syriacus
2054. Hibiscus trionum
2055. Lavatera trimestris

tern of fine parallel ridges, orangish brown to black, glabrous. $2n=38$. July–October.

Scattered, mostly south of the Missouri River (eastern U.S. west to Missouri and Texas; disjunct in New Mexico and California; Mexico). Openings of bottomland forests, swamps, banks of rivers, and margins of ponds, lakes, sinkhole ponds, and sloughs, sometimes in shallow water; also ditches and wet roadsides.

Some authors prefer to combine *H. lasiocarpos* with the closely related *H. moscheutos* L. or to treat it as var. *occidentalis* or ssp. *lasiocarpos* of that species. True *H. moscheutos* occurs only to the east of Missouri and differs from *H. lasiocarpos* in its glabrous fruits, bractlets without longer spreading hairs along the margins, and more glabrous upper leaf surfaces. For a discussion of hybrids with *H. laevis*, see the treatment of that species.

Hibiscus lasiocarpos frequently forms large colonies. Its seeds are eaten by wildlife, particularly birds.

3. Hibiscus syriacus L. (rose of Sharon)
Pl. 452 h; Map 2053

Plants shrubs or small trees. Stems 200–600 cm long, erect or ascending, sparsely to moderately hairy when young, becoming glabrous or nearly so with age, the twigs usually light brown. Leaf blades 3–10 cm long, ovate in outline, mostly deeply 3-lobed, the lobes sometimes shallowly lobed again, the margins coarsely scalloped or bluntly toothed, the upper surface glabrous, the undersurface glabrous or sparsely pubescent with stellate or simple hairs. Stipules more or less persistent. Bractlets subtending the calyx 8–10, 8–18 mm long, glabrous or finely pubescent with stellate hairs. Calyces 10–15 mm long at flowering, not becoming enlarged or inflated at fruiting, finely pubescent with stellate hairs. Petals 3–5 cm long, white or cream-colored to pink or purple, usually with dark reddish purple bases. Fruits 1.5–2.5 cm long, ovoid to ovoid-cylindric, noticeably beaked, hairy, yellow. Seeds 5–8 per locule, 4.0–4.5 mm long, broadly kidney-shaped to nearly, semicircular in outline, the surface minutely roughened or with a faint pattern of reticulate ridges, dark brown, glabrous, the margin densely pubescent with a line of spreading, simple or fasciculate, orangish tan hairs. $2n=80$, 90, 92. July–September.

Introduced, uncommon and widely scattered in the state (native of Asia, sporadically escaped from cultivation in North America). Banks of streams and rivers, bottomland forests, and mesic upland forests; also roadsides, railroads, and open disturbed areas.

Hibiscus syriacus is commonly cultivated as an ornamental shrub. Flower color is highly variable and some cultivars have doubled corollas. It often persists at old home sites, but only rarely becomes established in self-reproducing populations outside cultivation.

4. Hibiscus trionum L. (flower-of-an-hour, Venice mallow)
Pl. 453 c, d; Map 2054

Plants annual. Stems 25–55 cm long, mostly spreading to loosely ascending, moderately to densely hairy when young, becoming glabrous or nearly so with age. Leaf blades 1–6 cm long, broadly ovate in outline, deeply 3-lobed (often appearing nearly compound), the main lobes often shallowly to deeply lobed again, the margins coarsely scalloped or bluntly toothed, the upper surface sparsely pubescent with simple and/or fasciculate hairs along the veins, the undersurface moderately pubescent with stellate and fasciculate hairs. Stipules persistent. Bractlets subtending the calyx 10–12(–15), 4–10 mm long, bristly pubescent especially along the margins with simple or fasciculate hairs. Calyces 9–12 mm long at flowering, becoming greatly enlarged

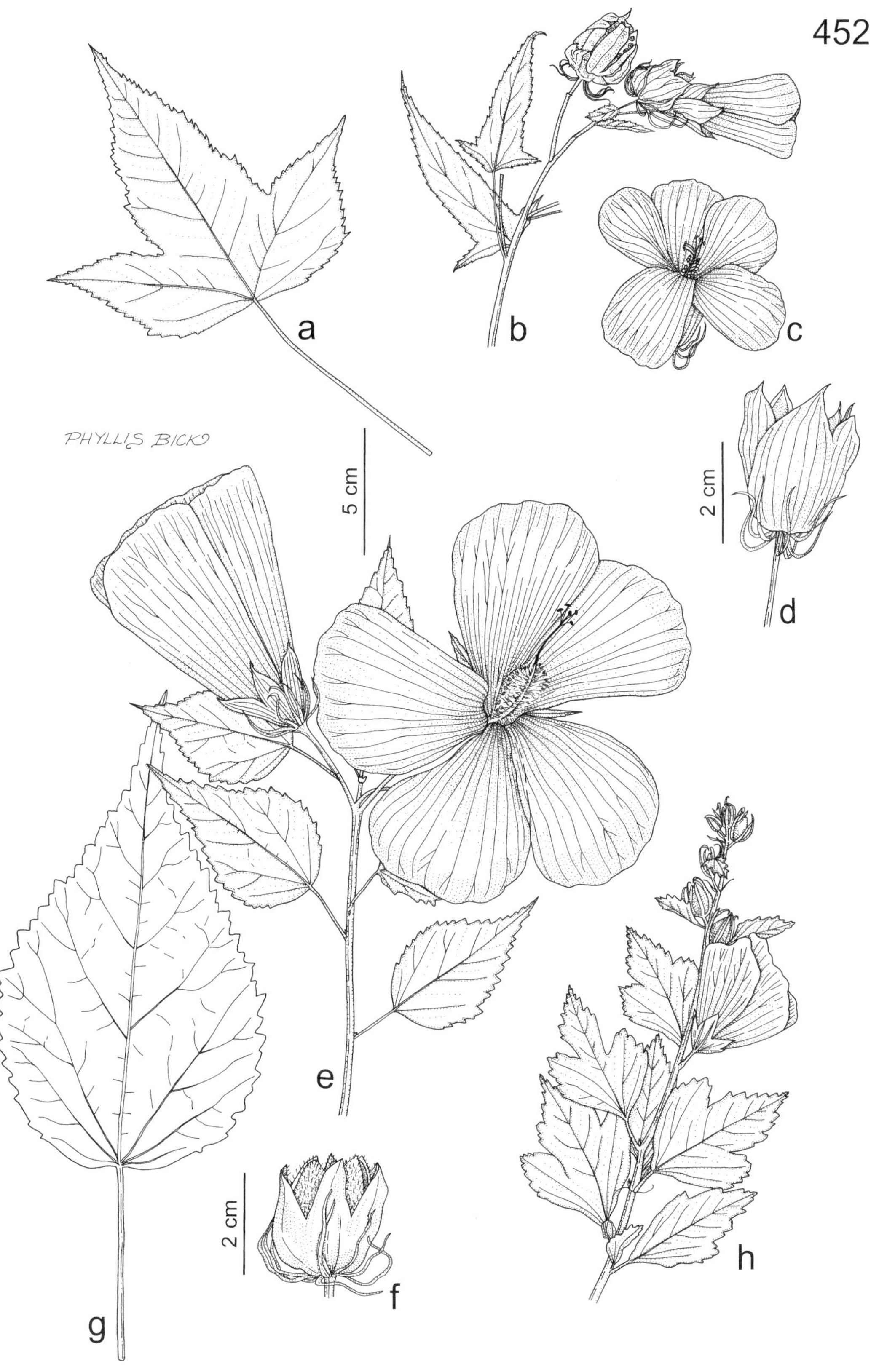

Plate 452. Malvaceae. *Hibiscus laevis*, **a)** median leaf, **b)** branch tip with leaves, flower, and fruit, **c)** flower, **d)** fruit. *Hibiscus lasiocarpos*, **e)** stem tip with flowers, **f)** fruit, **g)** median leaf. *Hibiscus syriacus*, **h)** fertile branch.

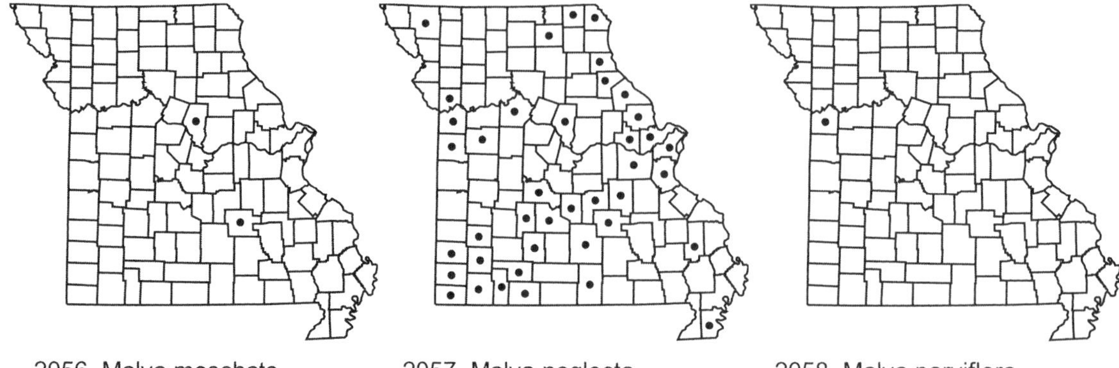

2056. Malva moschata

2057. Malva neglecta

2058. Malva parviflora

to 16–22 mm, papery, and inflated at fruiting, the main veins raised and ridgelike, dark green to dark purple, bristly pubescent with simple or fasciculate hairs. Petals 1.5–4.0 cm long, cream-colored to yellow with dark reddish purple bases. Fruits 0.9–1.6 cm long, ovoid-cylindric to nearly globose, hairy. Seeds 5–8 per locule, 2.0–2.5 mm long, broadly kidney-shaped to nearly, triangular in outline, the surface minutely warty and usually also with minute stellate hairs, dark brown or grayish black. $2n=28$, 56. June–September.

Introduced, scattered to common nearly throughout the state, although apparently still absent from the Mississippi Lowlands Division (native of Europe, widely introduced in North America). Margins of ponds and lakes and banks of streams and rivers; also crop fields, fallow fields, gardens, roadsides, railroads, and open disturbed areas.

This species originally was introduced into the United States as a garden ornamental, but has become a problem weed of horticultural crops. Several cultivars are still grown as annual bedding plants for their beautiful flowers.

7. **Lavatera** L. (tree mallow)

Thirteen to 40 species, North America, Europe, Africa, Asia, Australia.

The relationship between *Lavatera* and the closely related *Malva* is still not fully understood. Most authors have circumscribed *Lavatera* as a group of 25–40 species with an unusual character of united bractlets subtending the calyx. More recently, M. F. Ray (1995) suggested that the two genera should be reclassified using details of fruit structure as the basic separation (in addition to molecular characters). Under this scheme, *Lavatera* becomes restricted to about 13 species of the Mediterranean region, with the remaining taxa transferred to an expanded *Malva*. Further studies involving more species in the group are necessary to test this hypothesis.

1. Lavatera trimestris L.

Map 2055

Plants annual. Stems 30–120 cm tall, erect or ascending, branched or unbranched, pubescent with downward-pointing simple or fasciculate hairs, these usually with somewhat pustular bases. Leaves long-petiolate. Stipules 3–5 mm long, narrowly elliptic-lanceolate, shed early, hairy. Leaf blades 2–6 cm long, ovate to nearly circular in outline, at least the upper ones 3-angled or shallowly 3-lobed, the base broadly rounded or truncate to shallowly cordate, narrowed to a pointed tip, the margins finely scalloped or toothed, the surfaces glabrous or pubescent with appressed, simple or less commonly fasciculate hairs, usually with stellate hairs at the base. Flowers solitary in the leaf axils, the bractlets subtending the calyx fused into a cup with 3(6) broadly triangular lobes, shorter than the calyx, becoming enlarged and flattened horizontally at fruiting. Calyces 9–14 mm long at flowering, expanding to 25 mm long at fruiting, lobed 2/3–3/4 of their length, initially cup-shaped, but becoming enlarged and flattened horizontally at fruiting, the lobes lanceolate to broadly triangular, the outer surface pubescent with stellate hairs and simple or fasciculate hairs along the

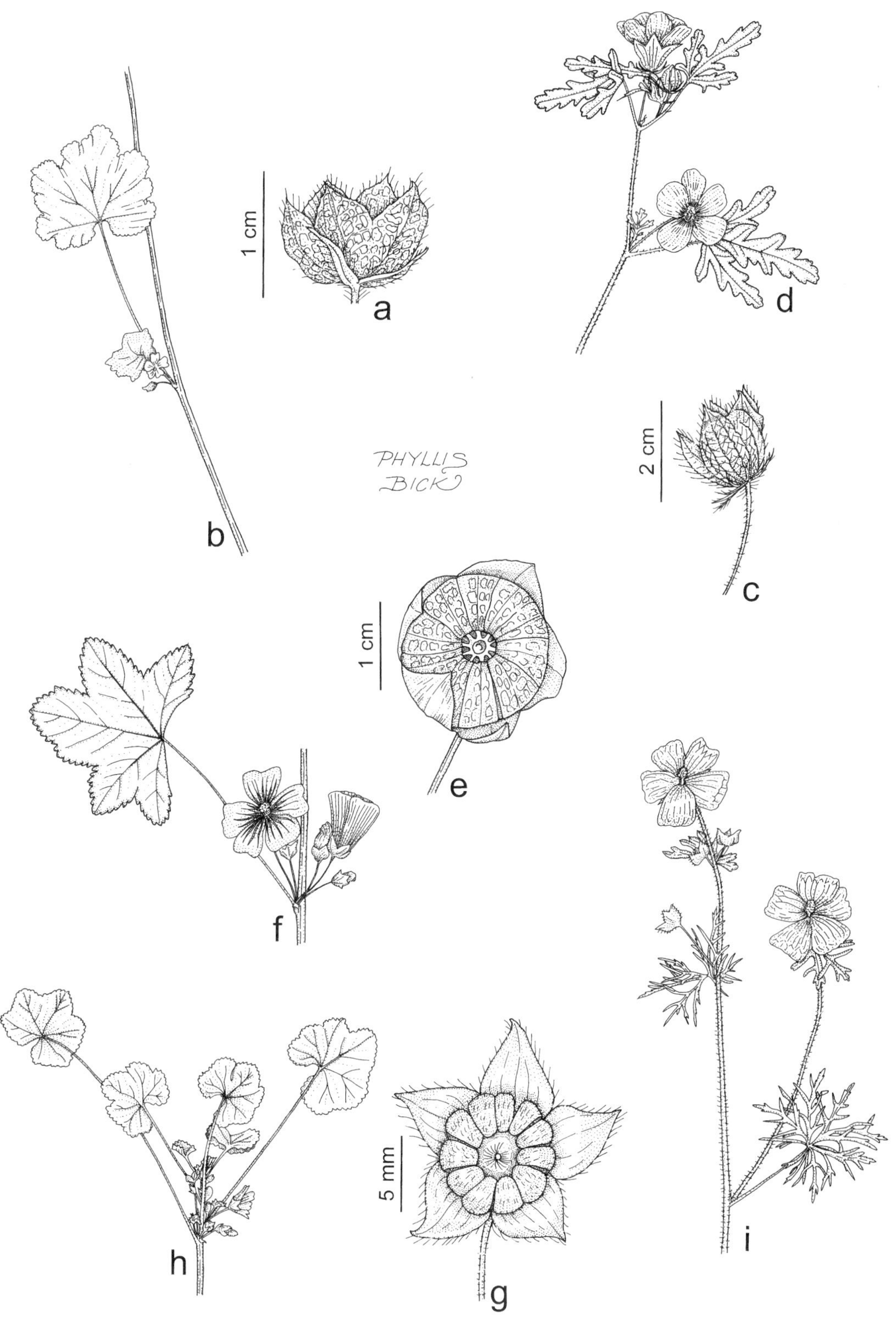

Plate 453. Malvaceae. *Malva verticillata*, **a)** fruit, **b)** node of fertile stem. *Hibiscus trionum*, **c)** fruit, **d)** fertile stem. *Malva sylvestris*, **e)** fruit, **f)** node of fertile stem. *Malva neglecta*, **g)** fruit, **h)** fertile stem. *Malva moschata*, **i)** fertile stem.

margins. Petals 2.5–4.5 cm long, reddish pink or less commonly pinkish purple, red, or white, the broadly rounded to truncate or less commonly shallowly notched tips with a more or less entire margin. Stamens numerous, the staminal column circular in cross-section, without a low crown of teeth at the tip, glabrous or hairy toward the base, the anthers white, red, or purple. Pistils with 5–12 locules, the carpels arranged in a flattened ring about an expanded central column, this becoming enlarged and expanded at fruiting into a broad concave disc covering and hiding the fruits. Styles fused most of their length, each branch with a single linear stigmatic area along the inner side toward the tip. Fruits schizocarps breaking into 5–12 mericarps. Mericarps 3.0–3.5 mm long, indehiscent or tardily dehiscent longitudinally, more or less wedge-shaped, oblong to kidney-shaped in profile, beakless, the dorsal and rounded lateral surfaces thin and papery, with a fine but noticeable reticulate pattern of thickenings, 1-seeded. Seeds 2.5–3.0 mm long, kidney-shaped to nearly circular in outline, black or less commonly dark brown. $2n=14$. August–September.

Introduced, known from a single specimen from Iron County (native of Europe, Asia; escaped sporadically in the southern U.S.). Open disturbed areas.

The single Missouri specimen of this species, which was originally determined as the superficially similar *Malva sylvestris*, was collected from a picnic ground in 1993. The species is sometimes cultivated as an annual bedding plant and is also a component of some wildflower seed mixes, but rarely becomes established outside cultivation in the United States.

8. **Malva** L. (mallow)

Plants annual or perennial herbs, variously hairy, the roots not tuberous-thickened. Stems prostrate to erect, branched or unbranched. Leaves long-petiolate, the blades shallowly to deeply palmately 3–7-lobed, cordate or less commonly truncate at the base. Stipules linear to broadly ovate-triangular, often asymmetric at the base, entire or less commonly lobed, persistent. Flowers solitary or in small clusters in the leaf axils, less commonly (in *M. moschata*) also in a short terminal raceme, the bractlets subtending the calyx (2)3, shorter than to about as long as the calyx, linear to narrowly obovate or oblong-ovate. Calyces cup-shaped at flowering, sometimes becoming somewhat enlarged and flattened horizontally at fruiting, mostly lobed to at or below the middle, the lobes ascending to somewhat spreading at flowering, broadly ovate-triangular, overlapping at flowering, the outer surface hairy. Petals white, light pink, pale lavender, or reddish purple, sometimes drying blue, the tips truncate or more commonly shallowly notched, otherwise entire or nearly so. Stamens numerous, the staminal column circular in cross-section, without a low crown of teeth at the tip, glabrous or hairy toward the base, the anthers usually white. Pistils with 8–20 locules, the carpels arranged in a flattened ring around a relatively broad central axis. Styles fused most of their length, each branch with a single linear stigmatic area along the inner side toward the tip. Fruits schizocarps breaking into 8–20 mericarps. Mericarps indehiscent, wedge-shaped, the dorsal surface rounded or broadly grooved, beakless, oblong to kidney-shaped in profile, not differentiated into sterile and fertile portions, 1-seeded, the sides thinner than the dorsal surface and usually papery, sometimes with a reticulate pattern of thickenings. Seeds kidney-shaped to nearly circular in outline, black or commonly dark brown, the surfaces usually finely roughened, glabrous. Twenty-five to 40 species, Europe, Asia, Africa, naturalized in the New World.

See the treatment of *Lavatera* for a discussion of uncertainties in the circumscriptions of these two genera.

1. At least the uppermost leaf blades deeply lobed (more than 1/2 way to the base); petals 2.0–3.5 cm long . . . . 1. M. MOSCHATA
1. Leaf blades unlobed or shallowly lobed (much less than 1/2 way to the base); petals 0.4–2.5 cm long

2. Bractlets subtending the calyx oblong-ovate to narrowly obovate; petals 1.5–2.5 cm long, reddish purple . . . . 5. M. SYLVESTRIS
2. Bractlets subtending the calyx linear to narrowly lanceolate; petals 0.4–1.2 cm long, white, light pink, or pale lavender
   3. Leaf blades strongly and irregularly crisped or curled; flowers all nearly sessile at flowering (the stalks elongating as the fruits mature) . . . . 6. M. VERTICILLATA
   3. Leaf blades flat or slightly crisped along the margins; flowers noticeably short- to long- stalked; species sometimes difficult to distinguish
      4. Petals 0.6–1.2 cm long, at flowering about twice as long as the calyx; fruits rounded on the dorsal surface, slightly roughened or with a faint reticulate pattern of thickenings, but not noticeably transversely wrinkled . . . . 2. M. NEGLECTA
      4. Petals 0.3–0.6 cm long, at flowering slightly shorter than to slightly longer than the calyx; fruits flattened on the dorsal surface, strongly transversely wrinkled and with a reticulate pattern of thickenings
         5. Calyx becoming pale, enlarged, and papery at fruiting, with a noticeable network of veins; pubescence of the calyx mostly of stellate hairs, the marginal hairs all less than 0.5 mm long . . . . 3. M. PARVIFLORA
         5. Calyx not or only slightly enlarged at fruiting, at least the lobes remaining green and herbaceous, without a noticeable network of veins; pubescence of the calyx mostly of simple and fasciculate hairs, the marginal hairs 0.6–1.2 mm long (mostly more than 1 mm long) . . . . 4. M. PUSILLA

1. Malva moschata L. (musk mallow)

M. moschata f. *laciniata* (Desr.) Hayek

Pl. 453 i; Map 2056

Plants perennial. Stems 40–100 cm long, erect or ascending, sparsely pubescent with mostly simple hairs toward the base, with mostly fasciculate or stellate hairs toward the tip. Stipules 3–8 mm long, linear to narrowly oblong-lanceolate. Leaf blades 2–6 cm long, flat or nearly so, circular to kidney-shaped in outline, at least the uppermost ones deeply 5- or 7-lobed (more than 1/2 way to the base), the lobes mostly deeply lobed again, the margins otherwise sparsely and bluntly toothed or lobed, the surfaces glabrous or sparsely pubescent with mostly simple or fasciculate hairs. Flowers in axillary clusters and also usually in a short dense terminal raceme, long-stalked, the bractlets subtending the calyx linear to narrowly lanceolate or oblanceolate, glabrous or nearly so except along the margins. Calyces 6–8 mm long at flowering, expanding to 15 mm long at fruiting, initially cup-shaped, but becoming enlarged and flattened horizontally at fruiting, with a distinct network of veins, the outer surface pubescent with simple, fasciculate, and stellate hairs, the marginal hairs 0.6–1.2 mm long. Petals 2.0–3.5 cm long, white to pink or pale purple. Fruits 1.5–2.0 mm long, rounded on the dorsal surface, densely hairy but lacking wrinkles or a pattern of thickenings, the sides thin and papery, without noticeable veins or thickenings. Seeds 1.2–1.5 mm long. $2n=42$. May–October.

Introduced, known thus far from historical collections from Boone and Dent Counties (native of Europe, escaped sporadically in the northern U.S. and Canada). Margins of crop fields, roadsides, and open disturbed areas.

Malva moschata is cultivated as an ornamental in gardens and occasionally escapes from cultivation.

2. Malva neglecta Wallr. (common mallow, cheeses)

Pl. 453 g, h; Map 2057

Plants annual or perennial. Stems 15–100 cm long, spreading to ascending, sparsely pubescent with simple and stellate hairs. Stipules 3–6 mm long, narrowly triangular to ovate-triangular. Leaf blades 1–5 cm long, flat or slightly crisped along the margins, circular to broadly kidney-shaped in outline, unlobed or broadly and very shallowly 5-lobed (much less than 1/2 way to the base), the margins finely scalloped or toothed, the surfaces glabrous or sparsely pubescent with mostly stellate hairs, especially at the base. Flowers in axillary clusters, long-stalked at flowering, the bractlets subtending the calyx linear to narrowly oblong-lanceolate, stellate-hairy on the under-

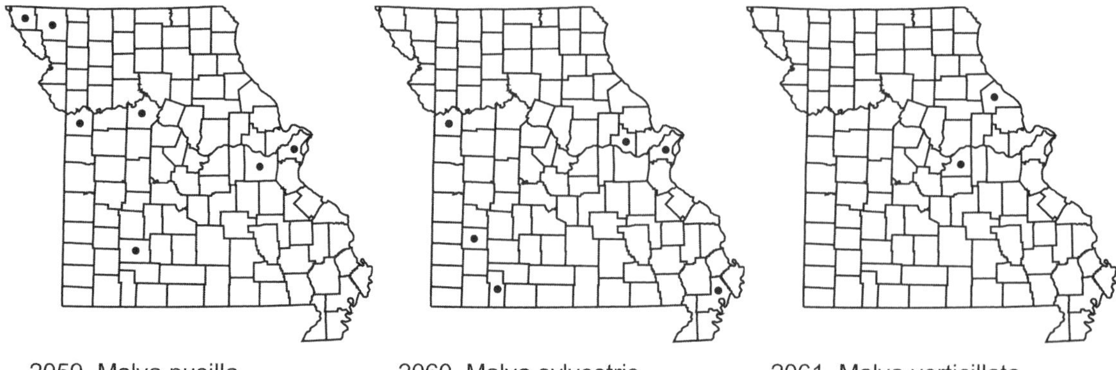
2059. Malva pusilla
2060. Malva sylvestris
2061. Malva verticillata

surface and with mostly simple or fasciculate hairs along the margins. Calyces 4–7 mm long at flowering, expanding to 8 mm long at fruiting, initially cup-shaped, not or only slightly enlarged and flattened horizontally at fruiting, at least the lobes remaining green, herbaceous, and without a distinct network of veins, the outer surface pubescent with mostly stellate hairs, the marginal hairs all less than 0.5 mm long. Petals 0.6–1.2 cm long, about twice as long as the calyx at flowering, white, light pink, or pale lavender. Fruits 1.5–2.0 mm long, the dorsal surface rounded, slightly roughened or with a faint reticulate pattern of thickenings, not transversely wrinkled, usually minutely stellate-hairy, the junction between the dorsal and lateral surfaces rounded or bluntly angled, the sides thin and papery, without visible veins. Seeds 1.0–1.5 mm long. $2n$=42. April–October.

Introduced, scattered nearly throughout the state (native of Europe, Asia, Africa, widely naturalized in the U.S. and Canada). Banks of rivers and margins of ponds and lakes; also roadsides, railroads, crop fields, pastures, lawns, farm yards, and open disturbed areas.

This is the commonest and most widely distributed species of *Malva* in Missouri. The depressed central axis to which the carpels are attached is unusually broad in this species, up to 1/3 the diameter of the ring of fruits (less than 1/5 the diameter in *M. pusilla* and most *M. parviflora*). See the treatment of *M. pusilla* for a discussion of the confusion surrounding the name *M. rotundifolia*, which was misapplied to plants of *M. neglecta* by many earlier authors.

3. Malva parviflora L. (small-flowered mallow)
Pl. 454 g–i; Map 2058

Plants annual. Stems 40–100 cm long, erect or ascending, less commonly spreading, glabrous or sparsely to moderately pubescent with mostly stellate hairs. Stipules 4–5 mm long, linear to narrowly triangular. Leaf blades 2–7 cm long, flat or slightly crisped along the margins, circular to broadly kidney-shaped in outline, unlobed, or broadly and shallowly 5- or 7-lobed (much less than 1/2 way to the base), the margins finely scalloped or toothed, the surfaces glabrous or sparsely to moderately pubescent with simple and/or stellate hairs, especially at the base and along the veins. Flowers in axillary clusters, short-stalked at flowering, the stalks elongating somewhat as the fruits mature, the bractlets subtending the calyx linear to narrowly lanceolate, glabrous or with mostly simple hairs along the margins. Calyces 3.0–4.5 mm long at flowering, expanding to 8 mm long at fruiting, initially cup-shaped, but becoming papery, enlarged, and flattened horizontally at fruiting, with a distinct network of veins, the outer surface pubescent with mostly stellate hairs, the marginal hairs all less than 0.5 mm long. Petals 0.3–0.6 cm long, mostly slightly longer than the calyx at flowering, white, light pink, or pale lavender. Fruits 2.0–2.5 mm long, the dorsal surface flat, glabrous or finely stellate-hairy, strongly transversely wrinkled, and with a reticulate pattern of thickenings, the junction between the dorsal and lateral surfaces with a narrow toothed or undulate wing (the ring of fruits thus appearing finely ribbed between the carpels at maturity), the sides thin and papery, with a radiating network of thickened veins. Seeds 1.5–2.0 mm long. $2n$=42. May–October.

Introduced, known thus far only from a historical collection from Jackson County (native of Europe, Asia, Africa, widely naturalized in the central and western U.S. south to South America, sporadic in the eastern U.S.). Roadsides and moist open disturbed areas.

The Jackson County specimen is somewhat atypical in having the calyx less enlarged at fruiting than is typical of the species farther west in its naturalized North American range.

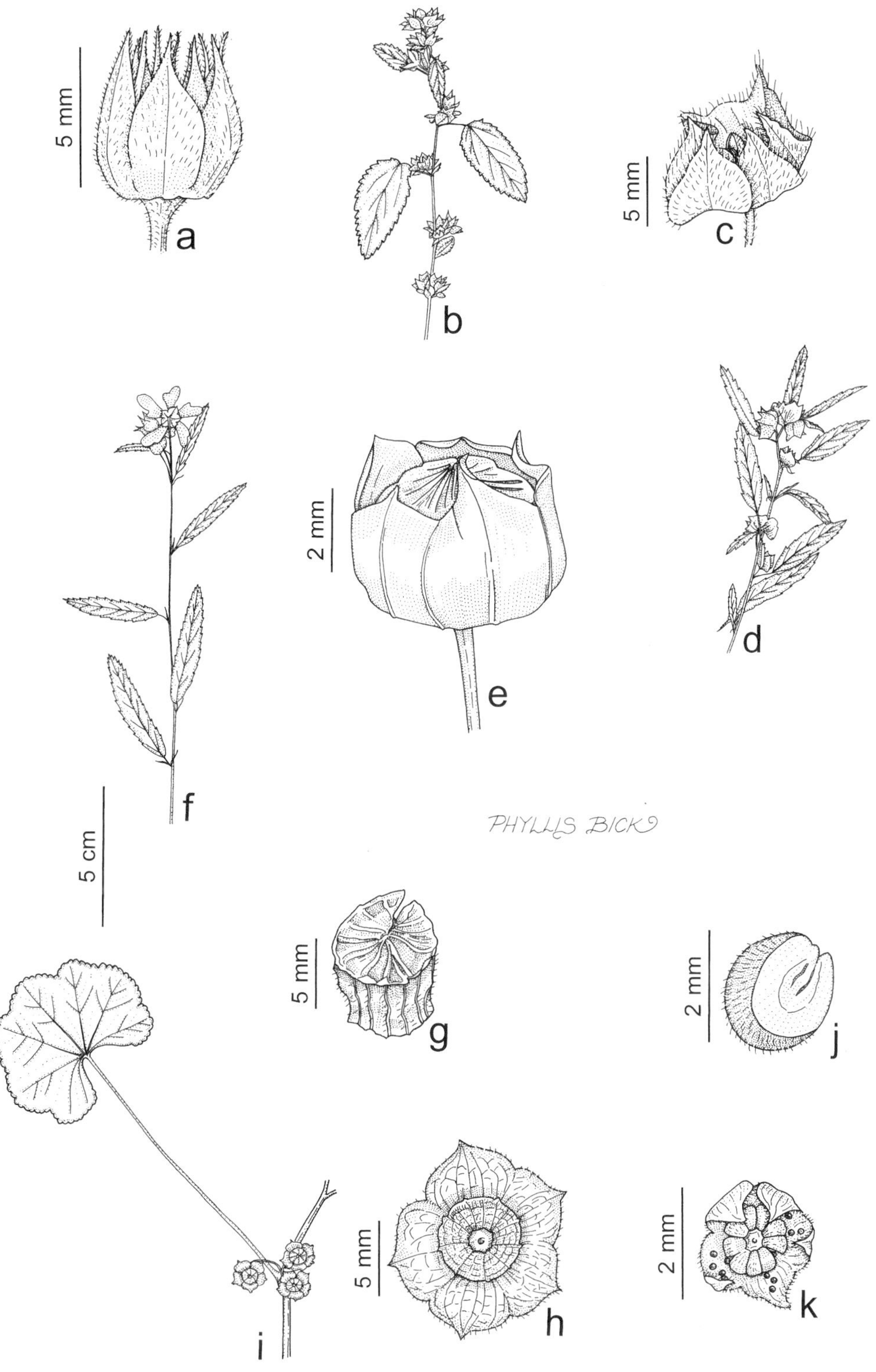

Plate 454. Malvaceae. *Sida spinosa*, **a)** fruit, **b)** fertile stem. *Malvastrum angustum*, **c)** fruit, **d)** fertile stem. *Sida elliottii*, **e)** fruit, **f)** habit. *Malva parviflora*, **g)** mericarp, **h)** fruit, **i)** node of fertile stem. *Malva pusilla*, **j)** mericarp, **k)** fruit.

4. Malva pusilla Sm. (dwarf mallow)

Pl. 454 j, k; Map 2059

Plants annual or perennial. Stems 15–100 cm long, spreading to ascending, sparsely pubescent with simple and stellate hairs. Stipules 3–5 mm long, narrowly triangular to ovate-triangular. Leaf blades 1–5 cm long, flat or slightly crisped along the margins, circular to broadly kidney-shaped in outline, unlobed or broadly and very shallowly 5-lobed (much less than 1/2 way to the base), the margins finely scalloped or toothed, the surfaces glabrous or sparsely pubescent with mostly fasciculate and stellate hairs, especially at the base. Flowers in axillary clusters, short- to long-stalked at flowering, the stalk elongating as the fruits develop, the bractlets subtending the calyx linear, with simple and fasciculate hairs along the margins. Calyces 3–5 mm long at flowering, initially cup-shaped, not or only slightly enlarged and flattened horizontally at fruiting, at least the lobes remaining green, herbaceous, and without a distinct network of veins, the outer surface pubescent with mostly simple and fasciculate hairs, the marginal hairs 0.6–1.2 mm long (mostly more than 1 mm long). Petals 0.3–0.5 cm long, slightly shorter than to slightly longer than the calyx, white, light pink, or pale lavender. Fruits 1.5–2.0 mm long, the dorsal surface flat, glabrous or finely stellate-hairy, strongly transversely wrinkled, and with a reticulate pattern of thickenings, the junction between the dorsal and lateral surfaces narrowly angled and sometimes appearing toothed (but generally not noticeably winged), the sides thin and papery, with a radiating network of thickened veins. Seeds 1.0–1.5 mm long. $2n$=42. May–October.

Introduced, uncommon and widely scattered in Missouri (native of Europe, sporadically naturalized in the U.S.).

At one time, the name *M. rotundifolia* L. was routinely applied to plants presently known as either *M. neglecta* or *M. pusilla* in the botanical literature (Voss, 1985; Turland, 1996a), resulting in confusion as to correct application of that name and many misdetermined herbarium specimens. This was the case in Missouri, where *M. pusilla* is less common than reported in the earlier literature (Palmer and Steyermark, 1935). Turland (1996a) proposed that the name *M. rotundifolia* should be rejected officially as an ambiguous name, and his proposal subsequently was approved at the 1999 International Botanical Congress.

The character of whether the margin between the dorsal and lateral surfaces of the mericarps is merely angled or narrowly winged, which has been used by some authors (Steyermark, 1963) to distinguish *M. pusilla* from *M. parviflora*, is very difficult to interpret. In spite of the seemingly easily observed key characters, some specimens can be difficult to discriminate among *M. neglecta, M. parviflora*, and *M. pusilla*.

5. Malva sylvestris L. (high mallow)

M. sylvestris var. *mauritiana* (L.) Boiss.

Pl. 453 e, f; Map 2060

Plants biennial (perennial in warmer climates). Stems 40–150 cm long, erect or ascending, glabrous or sparsely pubescent with stellate hairs. Stipules 3–8 mm long, linear to ovate-triangular. Leaf blades 2–8 cm long, flat or somewhat crisped or curled toward the margins, circular to broadly kidney-shaped in outline, unlobed to 3–7-angled or broadly and shallowly lobed (much less than 1/2 way to the base), the margins finely scalloped or toothed, the surfaces glabrous or sparsely pubescent with mostly simple or fasciculate hairs, also usually with stellate hairs at the base. Flowers in axillary clusters, long-stalked, the bractlets subtending the calyx oblong-ovate to narrowly obovate, glabrous or sparsely hairy, except along the margins. Calyces 5–6 mm long at flowering, expanding to 9 mm long at fruiting, initially cup-shaped, but becoming slightly enlarged and somewhat flattened horizontally at fruiting, with a distinct network of veins toward the base, the outer surface pubescent with mostly stellate hairs, the marginal hairs stellate and fasciculate, all less than 0.5 mm long. Petals 1.5–2.5 cm long, reddish purple. Fruits 2.0–2.5 mm long, flat on the dorsal surface, glabrous or somewhat hairy, moderately to strongly wrinkled or with a reticulate pattern of thickenings, angled at the junction between the dorsal and lateral surfaces, the sides thin and papery, with a radiating network of thickened veins. Seeds 1.5–2.2 mm long. $2n$=42. May–July.

Introduced, widely scattered in Missouri (native of Europe, Asia; escaped sporadically in North America). Roadsides, railroads, and open disturbed areas.

Malva sylvestris is cultivated as an ornamental in gardens and occasionally escapes from cultivation.

6. Malva verticillata L. **var. crispa** L. (curled mallow, clustered mallow)

Pl. 453 a, b; Map 2061

Plants annual. Stems 40–100 cm long, erect or strongly ascending, pubescent with simple to stellate hairs, sometimes nearly glabrous. Stipules 4–7 mm long, ovate-triangular. Leaf blades 3–8(–15) cm long, strongly and irregularly crisped or curled, circular to broadly kidney-shaped in outline,

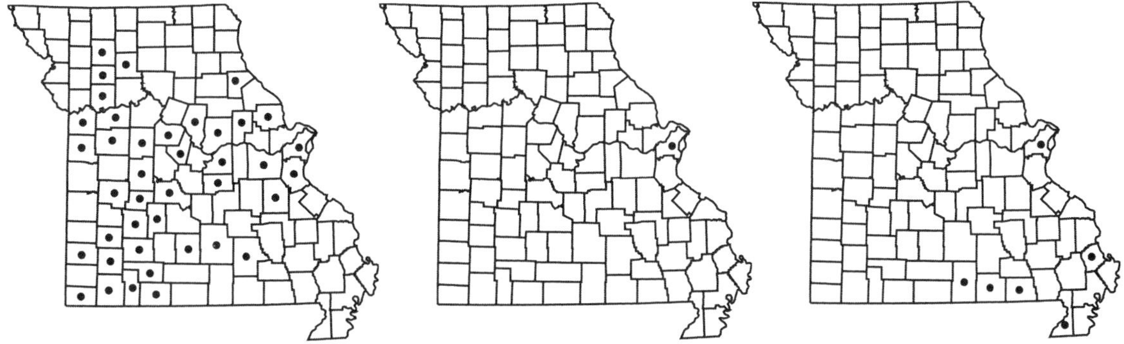
2062. Malvastrum angustum 2063. Melochia corchorifolia 2064. Sida elliottii

unlobed or broadly and shallowly 5- or 7-lobed (much less than 1/2 way to the base), the margins with numerous fine spreading teeth, the surfaces sparsely to moderately pubescent with fasciculate and stellate hairs, especially at the base and along the veins. Flowers in axillary clusters, nearly sessile at flowering, the stalks elongating as the fruits mature, the bractlets subtending the calyx linear to narrowly lanceolate, stellate-hairy. Calyces 4–6 mm long at flowering, expanding to 10 mm long at fruiting, initially cup-shaped, but becoming enlarged, papery, and flattened horizontally at fruiting, with a distinct network of veins, the outer surface pubescent with mostly stellate hairs, the marginal hairs fasciculate, all less than 0.5 mm long. Petals 0.6–0.8 cm long, white or pale lavender. Fruits 2.5–3.0 mm long, more or less rounded on the dorsal surface, glabrous or sparsely hairy, faintly to moderately wrinkled or with a reticulate pattern of thickenings, rounded or bluntly angled at the junction between the dorsal and lateral surfaces, the sides thin and papery, with a radiating network of thickened veins. Seeds 2.0–2.5 mm long. $2n$=84, 112. July–September.

Introduced, known thus far only from historical collections from Osage County (apparently native of Asia, widely cultivated in Europe and Asia, and sporadically in North America, escaped sporadically in the northern U.S.). Roadsides and open disturbed areas.

There is no consensus on the taxonomic level at which var. *crispa* should be recognized, and many workers recognize it as a distinct species, *M. crispa* (L.) L. Typical *M. verticillata*, which apparently is native to southeastern Asia and has a long history of cultivation there for use as a cooked vegetable, a salad plant, and in traditional medicine, differs in its flatter, more deeply lobed leaf blades and pink petals. The var. *crispa* is widely naturalized in Europe and Asia, but apparently is a cultigen derived from wild *M. verticillata* stock and is unknown from native populations.

9. **Malvastrum** A. Gray (false mallow) (S. R. Hill, 1982)

About 14 species, nearly worldwide, but most diverse in the American tropics.

1. Malvastrum angustum A. Gray (narrow-leaved false mallow, hispid false mallow)
M. hispidum (Pursh) Hochr., misapplied
Sphaeralcea angusta (A. Gray) Fernald
Pl. 454 c, d; Map 2062

Plants annual (perennial herbs and shrubs elsewhere), densely pubescent with stellate hairs throughout (also with simple hairs along the leaf margins). Stems 40–140 cm long, ascending to erect, often branched in the basal half. Leaves short-petiolate, the blades 1.8–5.5 cm long, 0.5–1.3 cm wide, linear to narrowly oblong-elliptic or lanceolate, rounded or narrowed at the base, mostly pointed at the tip, unlobed, the margins usually toothed. Stipules shriveling at maturity, 2–7 mm long, linear. Flowers solitary or in small clusters in the leaf axils, the bractlets subtending the calyx (2)3, 3–7 mm long, linear. Calyces 3.5–7.0 mm long at flowering, becoming enlarged and inflated to 9.0–12.0 mm at fruiting, the sepals fused below the midpoint, the lobes ascending, broadly ovate-cordate and overlapping, giving the calyx a strongly angular-winged appearance. Petals inconspicuous, 2.5–5.0 mm long, the tips shallowly and somewhat asymmetrically notched, the margin otherwise entire or nearly so, yellow. Stamens numerous, the staminal column circular in cross-section, without a low crown of teeth at the tip, the anthers yellow. Pistils with 5(6) locules, the carpels arranged in a ring. Styles fused most of their length, each branch with a globose terminal stigma. Fruits deeply lobed schizocarps breaking into 5(6) mericarps. Mericarps dehiscing longitudinally from the tip along the dorsal and ventral surfaces, 2.6–3.0 mm long wedge-shaped, tan to brown, the dorsal surface lacking a longitudinal groove, beakless, nearly circular to broadly kidney-shaped in profile, not differentiated into sterile and fertile cells, 1-seeded, the sides smooth-walled to finely roughened. Seeds 2.0–2.5 mm long, kidney-shaped, glabrous, dark reddish brown to black. $2n$=36. July–September.

Scattered nearly throughout the state, but apparently absent from the Mississippi Lowlands Division and northern portions of the Glaciated Plains (Illinois to Alabama west to Nebraska, Kansas, and Oklahoma). Upland prairies, glades, and tops of bluffs, frequently but not exclusively on calcareous substrates, less commonly stream banks; also roadsides and open moist to dry disturbed areas.

Malvastrum angustum is among the few species (including also *Plantago cordata*; Plantaginaceae) that are reasonably widespread in Missouri but quite uncommon elsewhere in their distributional ranges. Recent field work in Missouri indicates that, although some historical populations have disappeared, it is still well represented in the state's flora. This species has been classified in several genera over the years, including being lumped into *Sida* and *Sphaeralcea* or segregated from the rest of *Malvastrum* in the monotypic *Sidopsis* Rydb. Bates (1967) and S. R. Hill (1982) reviewed the generic taxonomy and relationships, both concluding that the species is best treated as a specialized annual species within *Malvastrum*. The proper species epithet for the plant also has been a controversial topic. The taxon has long been referred to as *M. hispidum* (based on *Sida hispida* Pursh), but S. R. Hill (2002) suggested that this name actually refers to a species of true *Sida* from the southeastern United States. He advocated adopting the next earliest name, *M. angustum*, for the taxon, but this has only recently begun to be adopted by other botanists.

Steyermark (1963) suggested that the corolla is sometimes absent, but this has not been confirmed by subsequent workers. S. R. Hill (1982) used garden transplants of Missouri plants to show that the flowers are open for a brief period in late morning and early afternoon, that they are effectively self-pollinated, and that some corollas remain closed and the flowers are then cleistogamous. He also noted that plants are often relatively short-lived, producing mature fruits about three months after the seeds germinate.

10. Melochia L.
(Goldberg, 1967)

About 60 species, nearly worldwide, most diverse in the New World tropics.

Traditionally, *Melochia* has been treated in the segregate family Sterculiaceae, a mostly tropical group of some 65 genera and more than 1,000 species, these mostly shrubs or trees. However, molecular studies have shown that the genera group into two main lineages, one of which appears more closely related to one of the two major groups of former Tiliaceae and the other of which is only distantly related within the Malvaceae in the broad sense and closer to the other portion of the former Tiliaceae (Whitlock et al., 2001).

It should be noted that the flowers of some species of *Melochia* do not possess an epicalyx of closely subtending bracts and such species cannot be correctly identified to genus in the key above. Also, members of *Melochia* tend to produce long-styled and short-styled flowers (heterostyly). The staminal tube can be relatively short in long-styled flowers.

1. Melochia corchorifolia L. (chocolate weed)
Pl. 455 d, e; Map 2063

Plants annual (perennial herbs and subshrubs elsewhere). Stems 10–60(–120) cm long, loosely ascending to more commonly erect, often branched from the base, sparsely pubescent with minute stellate hairs and sometimes also scattered simple or fasciculate hairs, especially near the nodes. Leaves short- to long-petiolate (those of axillary fascicles often very short-petiolate), the blades 0.8–6.0 cm long, 0.5–2.5 cm wide, oblong-lanceolate to ovate, rounded to shallowly cordate at the base,

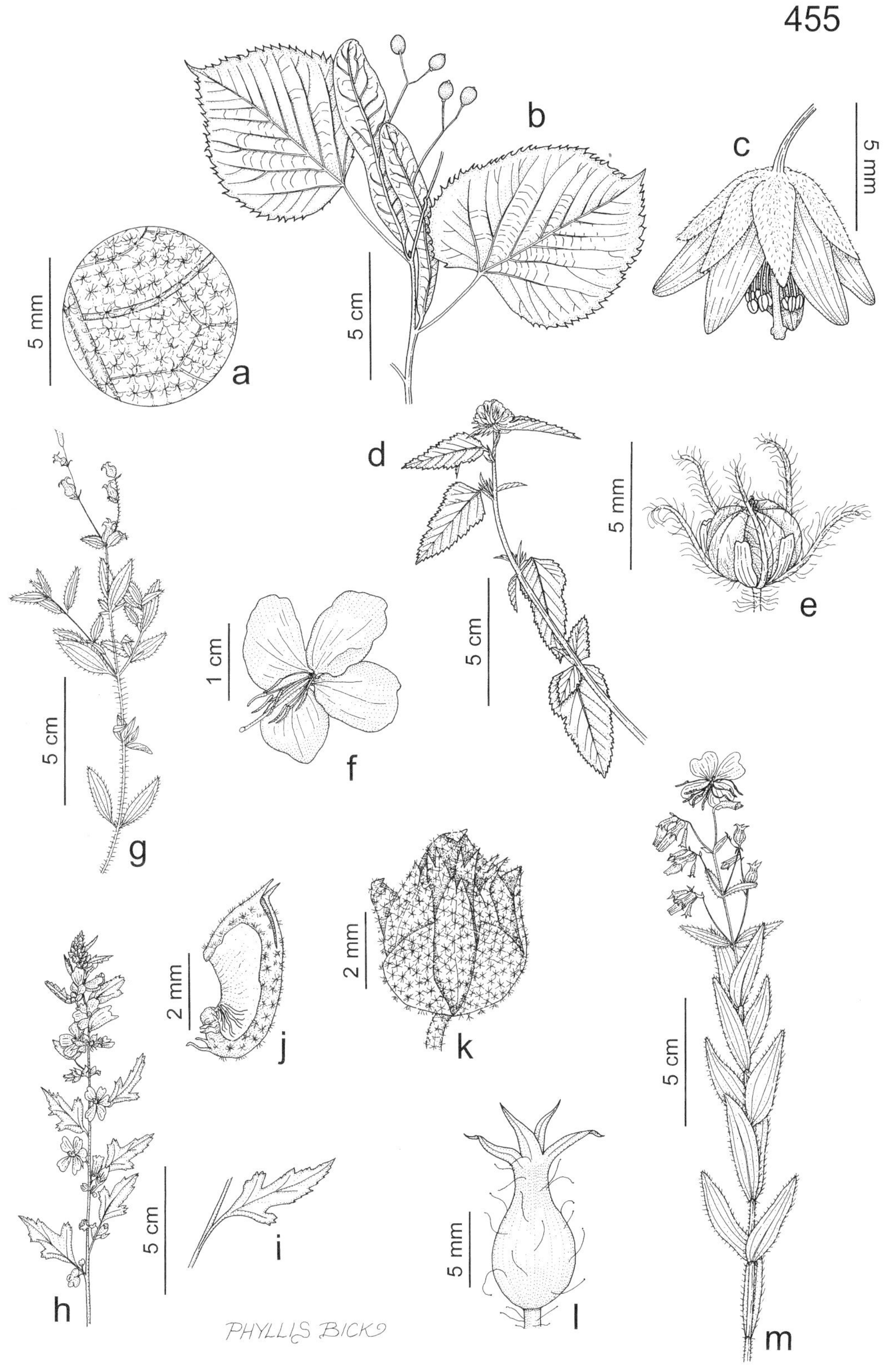

Plate 455. Malvaceae. Melastomaceae. *Tilia americana* var. *heterophylla*, **a)** detail of leaf undersurface. *Tilia americana*, **b)** branch tip with leaves, bracts, and fruits, **c)** flower. *Melochia corchorifolia*, **d)** fertile stem, **e)** fruit. *Rhexia mariana*, **f)** flower, **g)** fertile stem. *Sphaeralcea fendleri*, **h)** fertile stem, **i)** node with leaf, **j)** mericarp, **k)** fruit. *Rhexia virginica*, **l)** fruit, **m)** fertile stem.

rounded to more commonly bluntly or sharply pointed at the tip, unlobed or rarely very slightly 3-lobed, the margins toothed. Stipules mostly persistent at maturity, 2–6 mm long, linear. Flowers small, dense clusters, these axillary and/or terminal (then sometimes forming an irregular raceme, the bractlets subtending the calyx 4–6, 4–7 mm long, linear, bristly-hairy along the margins. Calyces 2.0–2.8 mm long at flowering, not becoming enlarged or inflated at fruiting, the sepals fused to above the midpoint, the lobes more or less ascending, broadly triangular with broad, U-shaped sinuses and sharply pointed tips, not overlapping, the calyx not appearing angled or winged. Petals relatively small but conspicuous, 4.0–6.5 mm long, the tips shallowly notched to truncate or broadly and very bluntly pointed, often slightly irregular, the margin otherwise entire or nearly so, light pink to nearly white (sometimes turning reddish as the flowers fade). Stamens 5, the staminal column circular in cross-section, without a low crown of teeth at the tip, the anthers yellow, positioned opposite the petals. Pistils with 5 locules. Styles 5, separate, each with a linear to slightly club-shaped stigmatic area toward the tip. Fruits bluntly 5-lobed capsules, these 3.5–5.0 mm long, depressed-globose, beakless, dehiscing longitudinally between the valves and usually shattering completely at maturity, each locule 1- or 2-seeded, the valves variously green to reddish brown or dark brown at maturity, smooth-walled, but hairy. Seeds 2–3 mm long, asymmetrically ovate-elliptic in profile, wedge-shaped in cross-section, dark brown, glabrous, but usually with a thin, papery, lighter brown covering, this eroding irregularly to expose the underlying seed coat. $2n=36$, 46. August–October.

Introduced, known thus far only from St. Louis City (native of Africa and Asia south to Australia, introduced sporadically in the eastern [mostly southeastern] U.S. and also tropical America). Railroads.

This species was reported by Mühlenbach (1979) from his botanical inventory of the St. Louis railyards. It has not been seen as an escape in Missouri since the original documentation.

11. **Sida** L. (sida)
(Fryxell, 1985, 1988)

Plants perennial herbs (annuals or shrubs elsewhere). Stems spreading to more commonly ascending or erect, branched. Leaf blades unlobed, ovate to linear, narrowed to truncate or shallowly cordate at the base, rounded or pointed at the tip, all less than 3 cm wide, the margins toothed and sometimes with a narrow band of purplish red color. Stipules persistent, linear. Flowers solitary in the leaf axils or in small terminal and axillary clusters, the bractlets subtending the calyx absent. Calyces cup-shaped when the fruits are mature (enclosing the developing fruits), strongly 5-angled between the sepals, the sepals also somewhat keeled, fused about half way or more, the lobes triangular, the margins (and sometimes also the keels) sometimes with a narrow band of purplish red color. Petals with the tips strongly asymmetric, broadly rounded with a shorter, shallowly concave, notched or single-toothed portion toward one side, the margin otherwise more or less entire, yellow to yellowish orange. Stamens numerous, the staminal column circular in cross-section, without a low crown of teeth at the tip, the anthers yellow. Pistils with 5–11 locules, the carpels arranged in a ring. Styles fused most of their length, each branch with a globose terminal stigma. Fruits schizocarps breaking into 5–12 mericarps. Mericarps 3–5 mm long, wedge-shaped, brown to dark brown, with a pair of short erect beaks toward the tip (beaks single elsewhere), the dorsal surface broadly rounded to somewhat flattened or with a shallow longitudinal groove, oblong to kidney-shaped in profile, minutely pubescent with stellate and fasciculate simple hairs toward the tip, differentiated into a sterile upper cell and a lower cell containing 1 seed, the upper sterile cell smooth-walled to finely roughened and the lower cell with a reticulate pattern of thickenings on the sides, dehiscent toward the tip between the spines. Seeds 2.0–2.5 mm long, ovate with a small notch at the tip, the surfaces glabrous, brown to dark brown. About 150 species, nearly worldwide, most diverse in tropical and warm-temperate regions.

In addition to the species treated below, *S. rhombifolia* L. (arrowleaf sida) may also be expected to be found in southern Missouri in the future. Like *S. spinosa*, this is a weedy

2065. Sida spinosa 2066. Sphaeralcea fendleri 2067. Tilia americana

pantropical species, and it has become naturalized in the southeastern United States as far north as northern Arkansas. It resembles *S. elliottii* in its more robust habit and greater number of mericarps and style branches (10–14), but has the broader leaf blades and shorter petals (4–9 mm long) more typical of *S. spinosa*. *Sida rhombifolia* differs from both of the species presently known from the state in having a single short beak toward the tip of each mericarp, which would cause problems in the key to genera of Missouri Malvaceae.

1. Ovary locules, style branches, and mericarps (8–)10(–12); flowers in the leaf axils solitary; petioles shorter than the accompanying axillary flower stalk . . . . 1. S. ELLIOTTII
1. Ovary locules, style branches, and mericarps 5; flowers in the leaf axils mostly in small clusters; petioles shorter than the accompanying axillary flower stalk . . . . 2. S. SPINOSA

1. Sida elliottii Torr. & A. Gray **var. elliottii** (Elliott's sida)

Pl. 454 e, f; dust jacket; Map 2064

Stems 35–100 cm long, sparsely to moderately pubescent with minute stellate hairs. Leaves with the petiole shorter than the axillary flower stalk. Leaf blades 1–7 cm long, linear to narrowly elliptic, rarely the largest blades oblong-ovate, the base angled to narrowly rounded, rarely truncate, glabrous or with sparse simple hairs on the upper surface, pubescent with stellate hairs on the undersurface. Flowers solitary in the leaf axils and usually also in small terminal clusters. Calyces 5–7 mm long, the lobes tapered to a sharply pointed tip, pubescent with minute stellate hairs, the margins and keels with longer simple or fasciculate hairs. Petals 10–16 mm long. Pistils with (8–)10 (–12) locules and style branches. Fruits with (8–) 10(–12) mericarps, these minutely pubescent with stellate and fasciculate simple hairs toward the tip. August–October.

Uncommon in the Crowley's Ridge Section of the Mississippi Lowlands Division and the southeastern portion of the Ozarks (southeastern U.S. west to Missouri, Arkansas, and Louisiana). Sand prairies and glades; also roadsides and railroads.

This species is more robust and usually more branched than *S. spinosa*. At two Oregon County populations, the flowers were observed to open at about sunset, which may account in part for the paucity of specimens of this species. The other variety, var. *parviflora* Chapm., occurs in Texas, Florida, Mexico, and Guatemala, and is a shorter plant with exclusively axillary flowers having calyces hairy only toward the base (Siedo, 1999).

2. Sida spinosa L. (prickly sida)

Pl. 454 a, b; Map 2065

Stems 10–60 cm long, moderately to densely pubescent with minute stellate hairs. Leaves with the petiole longer than the axillary flower stalk. Leaf blades 0.5–5.0 cm long, linear to ovate, the base rounded to truncate or shallowly cordate, glabrous or pubescent with stellate hairs on the upper surface, pubescent with stellate hairs on the undersurface. Flowers solitary or more commonly in small clusters in the leaf axils and usually also in small terminal clusters. Calyces 5–7 mm long, the lobes angled to a sharply pointed tip, pubescent with minute stellate hairs. Petals 4–6 mm long. Pistils with 5 locules and style branches. Fruits with 5 mericarps. $2n$=14, 28. June–October.

Introduced, common nearly throughout the state (native of the New and Old World tropics, naturalized widely in the eastern and central U.S.). Banks of streams and rivers and margins of ponds and lakes, rarely savannas; also roadsides, railroads, margins of crop fields, pastures, and open disturbed areas.

The species epithet and common name are in reference to a short spinelike projection at the base of the leaves in well-developed plants. The northern limits of the natural distribution of this species are unclear. In Missouri, it was present at least as early as 1833, but the oldest specimens originated from disturbed sites.

12. **Sphaeralcea** A. St.-Hil. (globe mallow)

About 40 species, North America, South America.

1. Sphaeralcea fendleri A. Gray **ssp. fendleri** Pl. 455 h–k; Map 2066

Plants perennial herbs, moderately to densely pubescent throughout with minute stellate hairs. Stems 40–140 cm long, loosely ascending to erect, unbranched or few-branched toward the base. Leaves short- to long-petiolate, the blades 1.8–5.5 cm long, narrowly oblong-ovate to broadly ovate, with 1 pair of usually deep lobes at the base and often 1–3 pairs of additional shallow pinnate lobes, the margins otherwise irregularly toothed, the undersurface noticeably lighter-colored than the upper surface. Stipules shed before leaf maturity, 3–5 mm long, linear. Flowers in small clusters in the leaf axils, the bractlets subtending the calyx 2 or 3 (sometimes absent), 1–3 mm long, linear, sometimes shed before flowering. Calyces 4–6 mm long, the sepals fused to about the middle, the lobes ascending to somewhat spreading but not becoming flattened horizontally, narrowly ovate to triangular at flowering. Petals 8–13 mm long, the tips asymmetrically rounded, the margin otherwise entire or nearly so, bright pink, usually whitened at the base, drying reddish purple. Stamens numerous, the staminal column circular in cross-section, without a low crown of teeth at the tip, the anthers yellow or purple. Pistils with 11–15 locules, the carpels arranged in a ring. Styles fused most of their length, each branch with a globose or club-shaped terminal stigma. Fruits schizocarps breaking into 11–15 mericarps. Mericarps dehiscing longitudinally from the tip along the dorsal and ventral surfaces, 4–5 mm long, wedge-shaped, tan to brown, the dorsal surface lacking a longitudinal groove, with an inconspicuous erect or ascending beak (this sometimes absent), oblong to narrowly kidney-shaped in profile, each weakly differentiated into a sterile upper cell and a lower cell containing 2(3) seeds, the sides smooth-walled to finely roughened, the lower cell usually with a faint reticulate pattern of thickenings toward the base. Seeds 1.3–1.5 mm long, kidney-shaped, the surfaces minutely pubescent with stellate hairs and with simple or fasciculate spreading hairs along the dorsal margin, dark brown. $2n$=10, 20, 30. June–September.

Introduced, known thus far from a single collection from Vernon County (Colorado to Arizona east to Kansas; Mexico). Roadsides.

The single collection documenting the existence of this species was made in 1965, but not determined as *S. fendleri* until 1992. The locality where it was collected apparently was destroyed during road construction for the widening of U.S. Highway 71 during the 1980s. Kearney (1935) divided the species into four subspecies based on minor differences in leaf shape and cutting, as well as flower color and density of pubescence. Most botanists presently believe these characters to be too variable to merit formal recognition of infraspecific taxa, but some of them may turn out to have merit following further study. For splitters the Missouri plants correspond to ssp. *fendleri*.

13. **Tilia** L. (basswood, linden) (Hardin, 1990; McCarthy, 1995)

About 23 species, North America, Europe, Asia, Africa.

As traditionally circumscribed, the family Tiliaceae comprised about 450 species grouped into about 50 genera. Molecular studies have shown that these genera are not a natural lineage, with various groups of genera more closely related to different components of the order

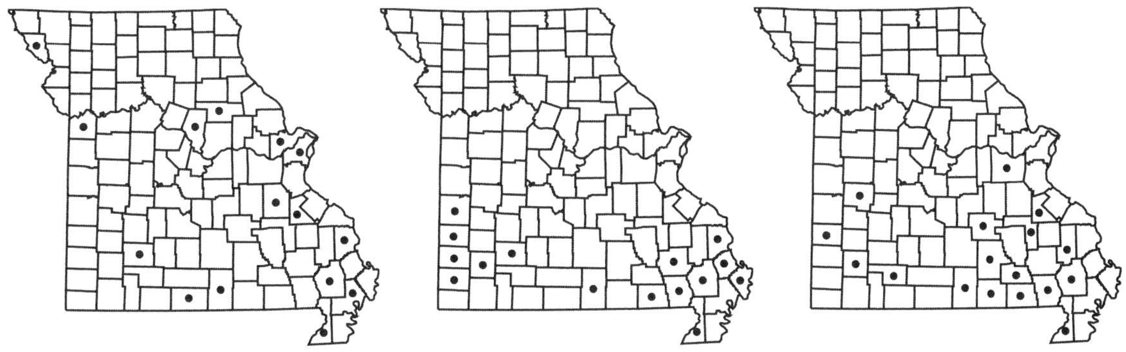

2068. Proboscidea louisiana 2069. Rhexia mariana 2070. Rhexia virginica

Malvales (Alverson et al., 1999; C. Bayer et al., 1999; Whitlock et al., 2001). There are only two genera of trees supported by the DNA sequence data as close relatives of *Tilia*: *Craigia* W.W. Sm. & W.E. Evans (two species of southern China and Vietnam) and *Mortoniodendron* Standl. & Steyerm. (twelve species from Mexico to northern South America).

Species of *Tilia* are often grown as shade trees. In addition to the native species, the most commonly encountered species in cultivation in the Midwest is *T. cordata* Mill. (small-leaved linden), which produces flowers lacking staminodes, has generally smaller leaves (usually less than 10 cm long), produces axillary tufts of short hairs in the leaf axils, and has nuts with a relatively thin shell. Several other species are occasionally cultivated as specimen plants.

The light wood of *Tilia* species has been used for furniture, veneers, plywood, crates, musical instruments, utensils, and crafts, and as pulp for paper. The fibrous inner bark was harvested by Native Americans to make rope and twine (Steyermark, 1963). The flowers often have a distinctive fragrance that has been described as sweet and peppery, and can be used to make an herbal tea. They are popular as a pollen source for honey-bees. The leaves and fruits provide food for wildlife.

G. J. Anderson (1976) studied the breeding system and pollination biology of *Tilia* species in North America. He concluded that the flowers are outcrossing, with the stamens shedding pollen before the stigmas become receptive. The biochemically complex fragrances of the flowers become stronger and more pungent as evening approaches. The flowers appear to lack pollinator specificity and can be pollinated by a variety of bees, wasps, butterflies, skippers, and flies during the day, as well as nocturnally by moths.

1. Tilia americana L. (American basswood)
Pl. 455 a–c; Map 2067

Plants medium to less commonly large trees to 25 m tall. Bark deeply ridged and grooved, light brown to gray. Twigs slender to moderately stout, often appearing somewhat zig-zag, green to yellowish or reddish brown, eventually becoming gray, smooth but with abundant small lenticels, glabrous or sparsely to moderately pubescent with minute, stellate hairs, often somewhat shiny. Winter buds axillary (true terminal bud absent, but the uppermost bud often appearing asymmetrically terminal), oblong-ovoid, with a pair of exposed outermost buds scales, these orangish brown to reddish brown or brownish purple, glabrous or more commonly at least sparsely stellate-hairy. Leaves alternate and appearing 2-ranked, long-petiolate, the fall foliage turning bright yellow. Leaf blades 5–16(–20) cm long, 4–13(–19) cm wide, broadly ovate to heart-shaped, truncate to more commonly cordate at the base, but often somewhat asymmetric and sometimes with 1 side truncate to broadly angled; short-tapered to broadly angled at the usually sharply pointed tip, unlobed, the margins sharply and often finely toothed, also minutely hairy, especially in the sinuses between teeth, the upper surface dark green, glabrous and somewhat shiny, the undersurface green or pale green, variously glabrous at maturity except for scattered, minute glands and/or small tufts of hairs in the axils of main veins (then sometimes inconspicuously pubescent with unbranched hairs during development), to persistently and moderately to densely woolly with mostly small, stellate hairs, the venation conspicuously palmate with mostly 5 veins from the blade base. Stipules not persistent at maturity, 6–10 mm long, narrowly strap-shaped to lanceolate, glabrous or hairy. Inflorescences axillary, flat-topped to dome-shaped panicles, these more or less pendant, stalked, the stalk partially fused to and appearing to originate from the midvein of a prominent, bract, this 6–11 cm long, strap-shaped to narrowly oblong-oblanceolate, initially pale green to white and membranous to somewhat leaflike, but soon turning tan to pale brown and papery. Flowers not subtended by bractlets. Calyces bell-shaped, not appearing angled or winged, the sepals free to the base, not overlapping in bud, 4–6 mm long, not persistent at fruiting, triangular-lanceolate, sharply pointed at the tip, relatively thick and leathery, the upper surface white and densely woolly, the undersurface pale green and moderately to densely pubescent with minute, stellate hairs. Petals 7–9 mm long, narrowly elliptic, the tips rounded to nearly truncate, sometimes slightly irregular, the margin otherwise entire or nearly so, white or pale cream-colored, occasionally faintly pinkish-tinged, not persistent at fruiting. Stamens numerous, the filaments not fused into a tube, united at the base into usually 5 groups, often forked near the tip so that the 2 anther sacs appear separate, these small, attached near their midpoints, yellow. Staminodes usually present, usually 1 per fused cluster of stamens, this innermost and positioned opposite the petals, petaloid, slightly shorter than the petals. Pistils with 5 locules, each with 2 ovules. Style 1, the stigma small, capitate, shallowly 5-lobed or 5-angled. Fruits nutlike drupes, 6–10 mm long, broadly ellipsoid to nearly globose, often with a short, slender beak at the tip, indehiscent, the outer shell relatively thick and hard, brown to grayish brown at maturity, the outer wall obscured by dense stellate hairs, 1-locular, with 1(–3) seeds. Seeds 3–7 mm long, subglobose (variously shaped in 2- and 3-seeded fruits), usually with a shallow longitudinal ridge along 1 side, brown to reddish brown, glabrous. $2n$=82. May–July.

Scattered to common nearly throughout the state (eastern U.S. west to North Dakota and Texas; Canada, Mexico). Bottomland forests, mesic upland forests, banks of streams and rivers, and bases and ledges of bluffs; also edges of pastures.

Steyermark (1963) and Jones (1968) treated the basswoods native to North America as comprising four closely related species, two of which occur in Missouri. In a detailed study of morphological variation across the range of the complex, Hardin (1990) documented considerable overlap for every key character and concluded that these taxa were more properly treated as varieties. The var. *caroliniana* (Mill.) Castigl. occurs to the south and east of Missouri, mainly on the Atlantic and Gulf Coastal Plains, and differs in its tendency to produce stalked-stellate hairs and/or fascicles of nonstellate hairs on the undersurface of the leaves. The var. *mexicana* (Schltdl.) Hardin is scattered but widespread in Mexico and is distinctive in its relatively densely and mostly persistently hairy twigs. Most recently, the morphological, molecular, and ecological studies of McCarthy (2012) resulted in the finding of little congruence between the data sets, as well as the lack of correlation among morphological characters across the distributional range of the complex, and a suggestion that much of the observed variation may be environmentally induced rather than genetically determined. Thus the varietal treatment accepted in the present work remains somewhat controversial.

1. Leaf blades with the undersurface green, glabrous at maturity except for scattered, minute, glands and/or small tufts of hairs in the axils of main veins (then sometimes inconspicuously pubescent with unbranched hairs during development), less commonly with widely scattered, small, sessile, stellate hairs; stalks of flowers and fruits usually glabrous .... 1A. VAR. AMERICANA
1. Leaf blades with the undersurface pale green, persistently and moderately to densely woolly with mostly small, sessile, stellate hairs; stalks of flowers and fruits usually persistently pubescent with minute, sessile, stellate hairs, occasionally nearly glabrous at fruiting .... 1B. VAR. HETEROPHYLLA

1a. var. americana (American basswood)

T. americana var. *neglecta* (Spach) Fosberg

Leaf blades with the undersurface green, glabrous at maturity except for scattered, minute glands and/or small tufts of hairs in the axils of main veins (then sometimes inconspicuously pubescent with unbranched hairs during development), less commonly with widely scattered, small, sessile, stellate hairs. Bracts glabrous or nearly so. Stalks of flowers and fruits usually glabrous. $2n$=82. May–July.

Scattered to common nearly throughout the state (northeastern U.S. west to North Dakota and Oklahoma). Bottomland forests, mesic upland forests, banks of streams and rivers, and bases and ledges of bluffs; also edges of pastures.

This variety occupies the northeastern portion of the overall species range. In Missouri, it is by far the most widespread and abundant variety. Rare plants with the leaves sparsely stellate-hairy on the undersurface were segregated by Steyermark (1963) as var. *neglecta*, but were considered to possibly represent the results of past intervarietal hybridization with var. *heterophylla* by Hardin (1990).

1b. var. heterophylla (Vent.) Loudon (white basswood, white linden, silver-leaved linn)
T. heterophylla Vent.

Pl. 455 a

Leaf blades with the undersurface pale green, persistently and moderately to densely woolly with mostly small, sessile, stellate hairs. Bracts often stellate-hairy, at least when young. Stalks of flowers and fruits usually persistently pubescent with minute, sessile, stellate hairs, occasionally nearly glabrous at fruiting. May–July.

Uncommon in the eastern and southwestern portions of the state and somewhat disjunct in Jackson County (eastern U.S. west to Missouri and Louisiana). Bottomland forests, mesic upland forests, banks of streams and rivers, and bases and ledges of bluffs; also edges of pastures.

MARTYNIACEAE (Unicorn Plant Family)

About 5 genera, 16 species, North America to South America.

The present treatment differs from that of Cronquist (1981, 1991) in treating the Martyniaceae as separate from the Pedaliaceae. In the traditional, broad sense, the Pedaliaceae have been divided into three morphologically distinct subfamilies that some authors treated as three distinct families (Thieret, 1977; Manning, 1991). Molecular studies have determined that each of the three groups is more closely related to different families within the order Lamiales than to each other. Pedaliaceae in the strict sense (subfamily Pedalioideae) contains about 13 genera with about 60 species (see the treatment of that family for further discussion), all native to the Old World, including *Sesamum* L. This group appears to be related to the Acanthaceae (Olmstead et al., 2001). The Martyniaceae (Pedaliaceae subfamily Martynioideae) are all native to the New World genera and differ from the Pedaliaceae in the strict sense primarily in their inflorescence structure, ovary construction and placentation, and pollen morphology. Molecular studies suggest an affinity of this group with the Verbenaceae (Gutierrez, 2008). The other segregate, Trapellaceae (Pedaliaceae subfamily Trapelloideae), contains only the Asian genus, *Trapella* Oliv., with 1 or 2 species of prostrate aquatics having modified flowers with 2 stamens and an inferior ovary. Its relationships within the Lamiales have not been fully resolved, but some authors have suggested affinities with the Plantaginaceae (see Ihlenfeldt [2004] and Mabberley [2008] for brief reviews).

In the decades around the turn of the twentieth century, plants of *Ibicella lutea* (Lindl.) Van Eselt. (*Proboscidea lutea* (Lindl.) Stapf) were cultivated at the Missouri Botanical Garden. A 1908 specimen accessioned there records the fact that the species once escaped onto a nearby compost pile. There is no evidence that it ever became truly established outside cultivation in Missouri, thus it is excluded from the flora. This South American native differs from *Proboscidea* species in its free sepals, yellow corollas, and spiny fruits.

1. **Proboscidea** Schmidel (unicorn plant, proboscis flower) (Bretting, 1981)

Six to 8 species, North America to South America.

Fibers from the ripe fruit walls of some species, particularly *P. parviflora* (Wooton) Wooton & Standl., are used by native Americans in the southwestern United States and Mexico in basketry, and the species has a history of domestication for this purpose (Bretting, 1981; Nabhan et al., 1981). The immature fruits of some species have been pickled or cooked for human consumption and the seeds are sometimes eaten raw. Some species are also cultivated as garden ornamentals and the unusual mature fruits used in dried flower arrangements. The unusual large "devil's claw" fruits in this genus are dispersed by becoming attached to large mammals. In some cases, these fruits have caused injury to the legs and mouthparts of livestock and there are reports of cattle starving to death when fruits became attached to their mouth parts and prevented feeding (Bretting, 1981). Fruits that become tangled in the fur of sheep can also interfere with shearing for wool.

1. Proboscidea louisiana (Mill.) Thell. **ssp. louisiana** (devil's claw, ram's horn)

Pl. 477 g, h; Map 2068

Plants annual, densely pubescent throughout with multicellular hairs, many of these with a bulbous glandular tip containing a gummy resin. Stems 10–200 cm long, erect or ascending and unbranched when young (sometimes flowering at this stage), developing numerous spreading to ascending branches later in the season. Leaves opposite or sometimes alternate toward the branch tips, petiolate. Stipules absent. Leaf blades 6–25 cm long, broadly ovate to nearly circular in outline, cordate at the base, rounded to bluntly pointed at the tip, unlobed or shallowly 3- or 5-lobed, the margins otherwise entire or to somewhat undulate. Inflorescences racemes, terminal or sometimes appearing axillary. Flowers perfect, hypogynous, subtended by small bracts. Calyces 14–20 mm long, somewhat zygomorphic, unequally 5-lobed, deeply split between the 2 lower lobes, the lobes broadly triangular to broadly ovate. Corollas 3.5–5.5 cm long, zygomorphic, the tube bell-shaped, the lobes shorter than the tube, appearing obliquely 2-lipped, white to yellowish white, sometimes pinkish- or purplish-tinged toward the base, with yellow and reddish purple spots in the throat and the base of the upper lobes. Stamens 4, the filaments fused to the corolla tube, the anthers more or less fused in 2 pairs, attached basally, the anther sacs spreading. Staminode 1, reduced and inconspicuous, fused to the corolla tube. Pistil 1 per flower, of 2 fused carpels. Ovary superior, 1-locular at flowering, becoming divided into 4 or 5 locules at fruiting. Style 1, the stigma deeply 2-lobed. Fruits drupelike capsules (initially with a fleshy outer coating, this deciduous, the inner woody layer capsular), 10–25 cm long at maturity, the body 6–12 cm long, asymmetrically elliptic in outline, more or less circular in cross-section, with an irregular fringelike longitudinal crest along the upper side, the remainder with a network of ridges and pits, tapered at the tip and splitting from the tip into a pair of long upwardly curled tusk-like beaks. Seeds numerous, 7–10 mm long, angled in cross-section (2 sides flattened, the other convex), broadly elliptic in outline, glabrous, the surfaces wrinkled and with irregular tubercles, dark gray to black. $2n=30$. June–September.

Scattered, mostly south of the Missouri River (eastern U.S. west to South Dakota, Colorado, and Texas). Banks of streams and rivers; also pastures, fallow fields, roadsides, railroads, and open sandy disturbed areas.

Most botanical works have spelled the species epithet as *P. louisianica* (Steyermark, 1963; Gleason and Cronquist, 1991). However, the original publication of the taxon epithet (the basionym) used the unlikely but nomenclaturally valid spelling *Martynia louisiana* Mill. An attempt was made by Thellung to correct this when he published the combination in *Proboscidea*, but according to the International Code of Botanical Nomenclature (McNeill et al., 2006) the original spelling must be preserved.

As circumscribed by Bretting (1981, 1983), *P. louisiana* consists of two well-marked subspecies with intermediate plants in the zone of contact between them. The ssp. *fragrans* (Lindl.) Bretting occurs from central Mexico north to southern Texas and differs in its more reddish purple corollas with longer lobes, white (vs. reddish purple) anthers, and more noticeably lobed leaves with toothed margins. The precise range of ssp. *louisiana* is difficult to determine, owing to its long history of association with mankind. The species apparently was dispersed from Texas through the southern Great Plains by bison (Bretting, 1981), but plants were also cultivated as far east as South Carolina as early as 1767, so were thus in a position to become naturalized eastward prior to most documentation of the North American flora. Native Americans may even have dispersed the plant prior to European colonization. In Missouri, many of the available specimens were collected in floodplains of streams and rivers, which are naturally disturbed habitats. In the absence of evidence to the contrary, *P. louisiana* is treated as native to Missouri in the present work. Interestingly, the species has become naturalized in California, Cuba, and Australia.

Plants of *P. louisiana*, including the flowers, have a rank, musty odor. The flowers are pollinated by various bees, especially bumblebees. The hairy spreading stigma lobes are sensitive to touch, closing together rapidly when placed in contact with a fingertip or pollinator, an apparent mechanism to minimize potential self-pollination (Thieret, 1977).

MELASTOMATACEAE (Melastome Family)

About 200 genera, 4,000–4,500 species, nearly worldwide, mostly in tropical and warm-temperate regions.

Species in this family mostly may be easily recognized by two unusual sets of features. The leaves have a characteristic pattern of venation, with usually 3 or 5 main veins (to 9 elsewhere), these palmately arranged and running parallel to the leaf margin (diverging above the base and arching so as to converge toward the tip), with the secondary veins parallel to one another and arranged in a ladderlike pattern between the main veins. In the flowers, the stamens are zygomorphic in that they become twisted downward (toward the lower side of the open flower) as the flower matures. The linear anthers are attached to the deflexed filaments toward their bases (with a short section extended past the attachment point), and a small sterile spur of tissue is usually also present at the filament tip. In most species, the anthers are curved upward and shed pollen through an apical pore (which is thus positioned at the lower edge of the open flower). These specialized stamens are an adaptation to pollination by insects (especially bees), with the hinged anther swinging up as the insect lands and shedding pollen on it.

1. **Rhexia** L. (meadow beauty)
(Kral and Bostick, 1969)

Plants perennial herbs, sometimes slightly woody at the base. Pubescence of relatively long, spreading, bristly hairs, these all or mostly with dark glandular tips. Stems solitary or few, erect or ascending, often 4-angled or -winged above the basal portion. Leaves opposite, sessile or short-petiolate. Stipules absent. Leaf blades simple, with 3(5) palmate main veins more or less parallel to the margins, narrowed or tapered to a sharply pointed tip, narrowed or rounded at the base, the margins sharply and finely toothed, the teeth often hair-tipped. Inflorescences usually appearing as small panicles, with small loose clusters subtended by small leaflike bracts at the nodes (these often shed by fruiting). Flowers with the perianth actinomorphic and the stamens and to some extent the pistil zygomorphic, perfect, epigynous (but the hypanthium often becoming somewhat free from the mature fruits and the ovary then appearing superior). Hypanthium tubular at flowering, persistent and becoming urn-shaped as the fruit matures, extended past the ovary as a short necklike tube. Calyces of 4 free lobelike sepals at the hypanthium tip, these 2–4 mm long, triangular, sharply pointed at the tip, persistent at fruiting. Corollas of 4 free petals, these spreading, somewhat asymmetrically oblong to broadly obovate, rounded or with an abrupt small narrow point at the tip, pink to rose-purple (rare white forms not yet found in Missouri). Stamens 8, strongly exserted at flowering, subequal in size, the filaments fused to the inner apical portion of the hypanthium, S-shaped and curved downward, the anthers 5–8 mm long, attached just above their bases, strongly curved outward, yellow, dehiscing by terminal pores. Pistil of 4 fused carpels, inferior (but often appearing superior at fruiting). Ovary 4-locular, the placentation axile. Style 1 per flower, somewhat curved downward, about as long as the stamens (including the anthers), the stigma depressed-capitate to somewhat disc-shaped, entire or nearly so. Ovules numerous. Fruits capsules, dehiscing irregularly and longitudinally between the sutures. Seeds numerous, 0.4–0.7 mm long, flattened and spiral-shaped (snail-shell-shaped), the surface brown, with several concentric, spiralled ridges along the sides and especially along the keel, these varying from smooth to warty or tubercled. Eleven species, U.S., Canada, Caribbean Islands, most diverse along the Atlantic and Gulf Coastal Plains.

Meadow beauties often occur in dense colonies and are very attractive when flowering. They can be cultivated successfully in the garden if planted in moist sandy or otherwise acidic soils. The stems of *Rhexia* species often persist through the winter, leafless but with noticeable, persistent, urn-shaped hypanthia. These are sometimes used in dried flower arrangements.

1. Stem angles not or only narrowly winged (up to about 0.25 mm); necklike free portion of the hypanthium usually longer than the body (the portion fused to the ovary) at fruiting . . . . 1. R. MARIANA
1. Stems angles noticeably winged (0.3–2.0 mm); necklike free portion of the hypanthium usually shorter than than the body (the portion fused to the ovary) at fruiting . . . . 2. R. VIRGINICA

1. Rhexia mariana L. (Maryland meadow beauty, dull meadow beauty)

Pl. 455 f, g; Map 2069

Rhizomes present, sometimes shallow and stolonlike, the roots usually lacking tubers, the stem bases not spongy-thickened. Stems 20–100 cm long, equally or unequally 4-angled, sometimes appearing nearly circular in cross-section, the angles not or inconspicuously winged, the wings less than 0.25 mm wide, sparsely to moderately glandular-hairy, especially at the nodes. Leaf blades (1–)2–8 cm long, elliptic or ovate to narrowly ovate, becoming elliptic lanceolate to narrowly lanceolate toward the stem tip, glabrous or more commonly sparsely to moderately glandular-hairy. Hypanthium 6–12 mm long at fruiting, glabrous or sparsely to moderately glandular-hairy, the necklike free portion usually longer than than the body (the portion fused to the ovary) at fruiting. Petals 12–18 mm long, glabrous or with sparse glandular hairs on the outer surface and margins. $2n=22$, 44. June–October.

Scattered in the southern third of the state (eastern [mostly southeastern] U.S. west to Kansas and Texas). Sandy banks of streams and rivers, wet depressions of upland prairies, sand prairies, and sandstone glades, open margins of ponds and sinkhole ponds, and less commonly openings in bottomland forest; also ditches, roadsides, railroads, and moist sandy open disturbed areas.

1. Angles of the stem more or less equal, the sides all flat or slightly concave, the angles relatively sharp above the stem midpoint and sometimes narrowly winged; petals often with sparse glandular hairs on the outer surface . . . . 1A. VAR. INTERIOR
1. Angles of the stem unequal, with 2 broader, slightly convex to rounded sides alternating with 2 narrower, concave sides, appearing nearly circular in cross-section or the angles blunt and not winged; petals glabrous, except sometimes for a minute hairlike point at the tip . . . . 1B. VAR. MARIANA

1a. var. interior (Pennell) Kral & Bostick

R. interior Pennell

Stems with the angles more or less equal, the sides all flat or slightly concave, the angles relatively sharp above the stem midpoint and sometimes narrowly winged (the wings less than 0.25 mm wide), glandular-hairy mostly at the nodes, the internodes often glabrous or nearly so. Petals often with sparse glandular hairs on the outer surface and margins. Hypanthium mostly 10–12 mm long at fruiting. $2n=44$. June–October.

Scattered to uncommon in southwestern Missouri, nearly entirely in the Unglaciated Plains Division (Indiana to Alabama west to Kansas and Texas). Sandy banks of streams and rivers and wet depressions of upland prairies and sandstone glades; also ditches, roadsides, railroads, and moist sandy open disturbed areas.

This taxon originally was described as a separate species based on collections made by B. F.

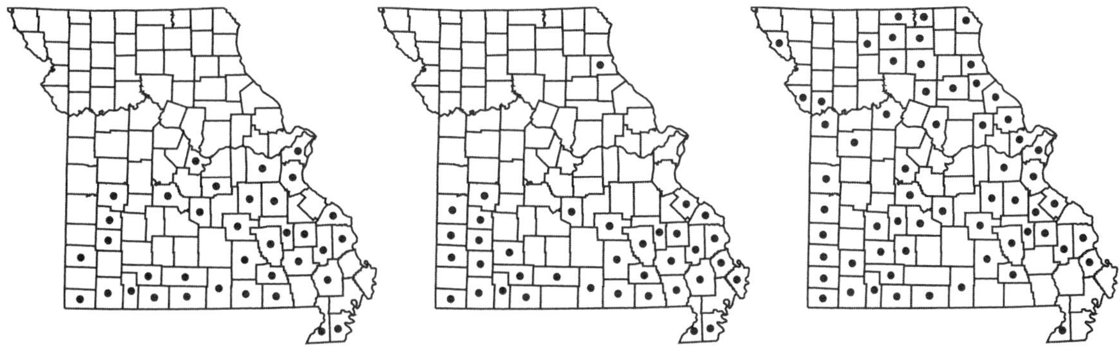

2071. Calycocarpum lyonii 2072. Cocculus carolinus 2073. Menispermum canadense

Bush in Jasper County. Within the context of Missouri, it may seem relatively distinct, but taken as a whole across the entire geographic range of the *R. mariana* complex, the differences seem less clear. Occasional intermediate specimens are encountered that are difficult to score for the key characters said to separate these taxa. Kral and Bostick (1969) treated *R. mariana* as a complex of three morphologically overlapping varieties, recognizing two varietal segregates within the distributional range of the widespread var. *mariana*. The var. *ventricosa* (Fernald & Griscom) Kral and Bostick is endemic to the eastern seaboard from New Jersey to South Carolina and differs from var. *interior* mainly in details of seed sculpturing, even though the two are widely separated geographically.

1b. var. mariana

Stems with the angles unequal, with 2 broader, slightly convex to rounded sides alternating with 2 narrower, concave sides, appearing nearly circular or the angles blunt above the stem midpoint, not winged, fairly evenly glandular-hairy at the nodes and internodes. Petals glabrous, except sometimes for a minute hairlike point at the tip. Hypanthium mostly 6–10 mm long at fruiting. $2n$=22, 44. June–October.

Scattered to uncommon in southeastern Missouri west locally to Howell County (eastern [mostly southeastern] U.S. west to Kansas and Texas). Sandy banks of streams, wet depressions of sand prairies and sandstone glades, open margins of ponds and sinkhole ponds, and less commonly openings in bottomland forest; also ditches, roadsides, railroads, and moist sandy open disturbed areas.

2. Rhexia virginica L. (wing-stemmed meadow beauty, handsome Harry)

Pl. 455 l, m; Map 2070

Rhizomes usually absent, but the roots often producing small tubers, the stem bases sometimes somewhat spongy-thickened. Stems 20–100 cm long, more or less equally 4-angled, the angles noticeably winged, the wings mostly 0.3–2.0 mm wide, sparsely to moderately glandular-hairy at the nodes, the internodes pubescent with similar hairs to nearly glabrous. Leaf blades (1–)2–7 cm long, elliptic or more commonly ovate, becoming narrowly ovate or less commonly lanceolate toward the stem tip, glabrous or more commonly sparsely to moderately glandular-hairy. Hypanthium 7–10 mm long at fruiting, glabrous or sparsely to moderately glandular-hairy, the necklike free portion usually shorter than the body (the portion fused to the ovary) at fruiting. Petals 15–20 mm long, often with sparse glandular hairs on the outer surface and margins. $2n$=22, 44. June–October.

Scattered in the southern portion of the state northward locally to Franklin and St. Clair Counties (eastern U.S. west to Wisconsin and Texas; Canada). Fens, acid seeps, sandy banks of streams and rivers, wet depressions of sand prairies and sandstone glades, and open margins of sinkhole ponds; also fallow fields, ditches, roadsides, railroads, and moist sandy open disturbed areas.

At a few sites in southern Missouri, this species co-occurs with *R. mariana*, but thus far no instances of hybridization have been reported. Kral and Bostick (1969), who studied both natural and artificially produced hybrids between these two species, found that interspecific hybrids are rare in nature and mostly sterile, with abortive seeds. They indicated that such hybrids could be recognized vegetatively by their unusually narrow leaves. In Missouri, *R. virginica* tends to flower slightly earlier than does *R. mariana*, but there is overlap in flowering times during most years.

MENISPERMACEAE (Moonseed Family)

Contributed by Alan Whittemore

Plants dioecious, perennial vines, often woody toward the base, lacking tendrils or spines. Leaves alternate, simple, petiolate. Stipules absent. Inflorescences axillary, racemes or panicles. Flowers actinomorphic, hypogynous; never with bractlets adjacent to the calyx. Calyces of usually 6 free sepals. Corollas absent or of usually 6 free petals, these relatively inconspicuous, white. Staminate flowers with the stamens 6–36, free and distinct, the anthers opening by longitudinal slits. Pistillate flowers with the separate pistils usually 3 or 6, each with 1 locule, the placentation marginal. Style 1 per pistil, short. Fruits 1-seeded drupes. About 80 genera and 500 species, nearly worldwide.

Fruits in the family frequently have the hard inner layer (endocarp) surrounding the seed with unusual shape or ornamentation, giving rise to such common names as cup seed, moon seed, and snail seed.

1. Petals absent or minute; anthers 2-locular; leaf blades with the lobes tapered to a usually slender sharp point; fruits ellipsoidal, the hard inner layer smooth, deeply bowl-shaped . . . . 1. CALYCOCARPUM
1. Petals well-developed; anthers 4-locular; leaf blades with the tips or lobes rounded or narrowed to a blunt point (occasionally broadly acute) and with an abrupt, minute, sharp point at the very tip; fruits subglobose, the hard inner layer transversely ridged, both sides concave
   2. Leaves not peltate, the petiole attached at the margin of the blade; fruits red . . . . 2. COCCULUS
   2. Leaf bases peltate, the petiole attached to the underside of the blade near the base; fruits bluish black . . . . 3. MENISPERMUM

1. **Calycocarpum** Nutt. ex Torr. & A. Gray

One species, endemic to the southeastern United States.

1. Calycocarpum lyonii (Pursh) A. Gray (cupseed)

Pl. 456 k–m; Map 2071

Plants relatively stout robust climbers. Stems to 20 m or more long. Leaves not peltate, the petiole attached at the margin of the blade. Leaf blades 7–20 cm long, 7–28 cm wide, broadly ovate-cordate in outline, shallowly to deeply 3–5(–7)-lobed, with 5 or 7 main veins from the base, glabrous or sparsely to less commonly moderately hairy, the base deeply cordate, tapered to a usually slender sharp point at the tips of the lobes. Inflorescences 2.5–24.0 cm long. Sepals 1.5–2.5 mm long. Petals absent or minute. Stamens 6–12, the anthers 2-locular. Pistils 3. Fruits 15–18 mm long, ellipsoidal, green (drying bluish black), the endocarp smooth, deeply bowl-shaped. June–July, occasionally also September–October.

Scattered in the southern half of Missouri, in the Mississippi Lowlands, Big Rivers, Ozark, and Ozark Border Divisions (southeastern U.S., from South Carolina and Georgia west to Missouri and Louisiana). Bottomland forests, mesic upland forests, and banks of streams and rivers; also pastures, old fields, margins of crop fields, roadsides, and disturbed areas, sometimes climbing on telephone poles.

2. **Cocculus** DC.

Eleven species, U.S., Mexico, Asia, Africa.

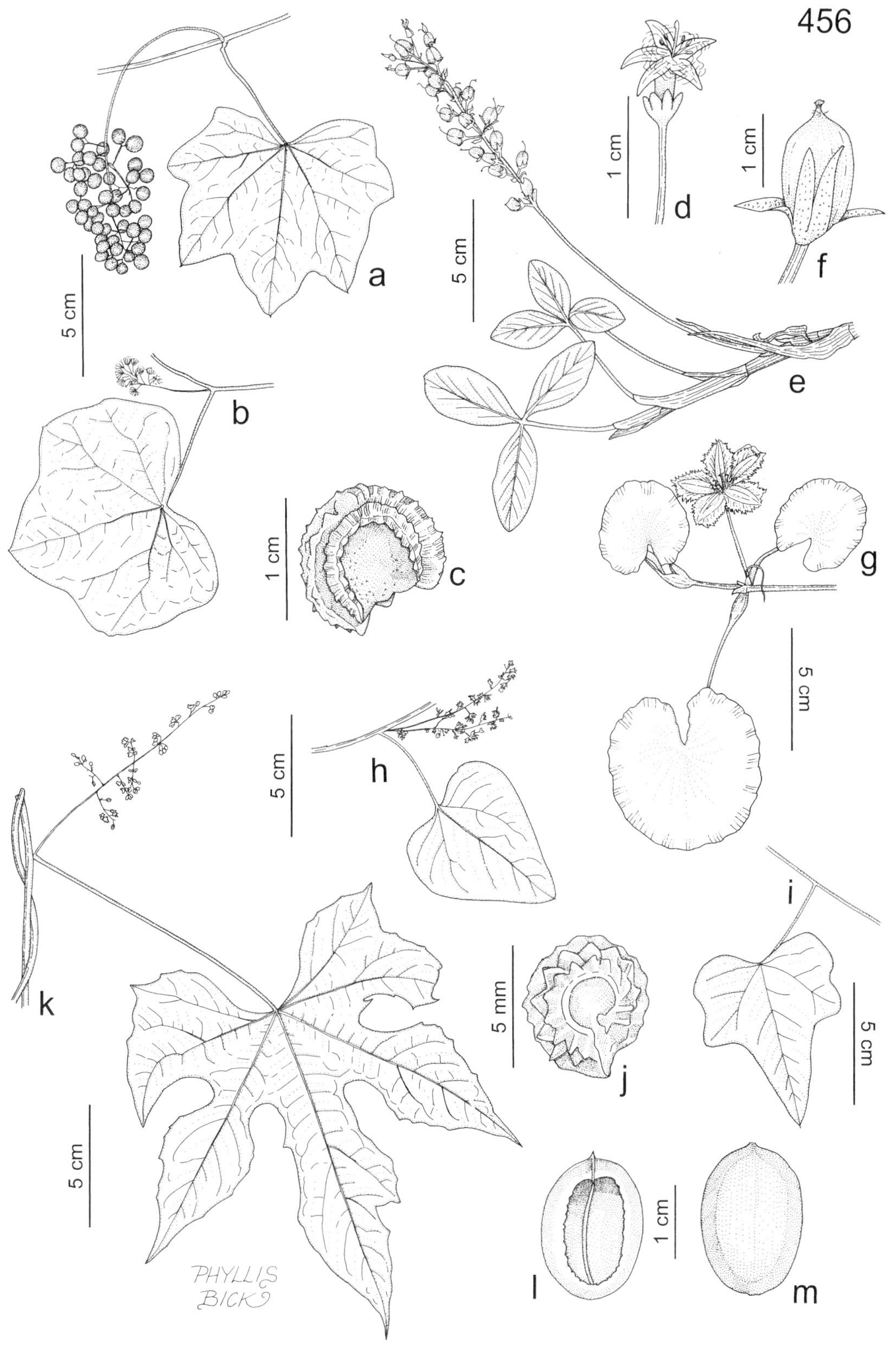

Plate 456. Menispermaceae. Menyanthaceae. *Menispermum canadense*, **a)** stem with leaf and fruits, **b)** stem with leaf and flowers, **c)** seed. *Menyanthes trifoliata*, **d)** flower, **e)** habit. *Nymphoides peltata*, **f)** fruit, **g)** habit. *Cocculus carolinus*, **h)** node with unlobed leaf and inflorescence, **i)** node with lobed leaf, **j)** seed. *Calycocarpum lyonii*, **k)** stems with leaf and flowers, **l)** seed, **m)** fruit.

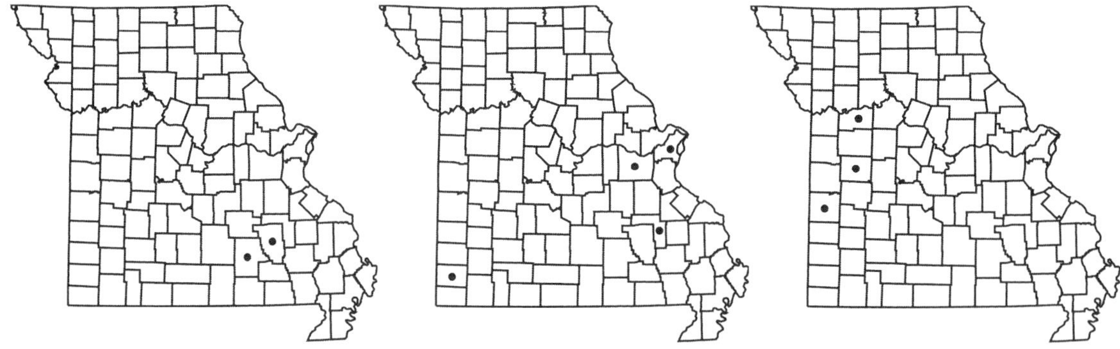

2074. Menyanthes trifoliata 2075. Nymphoides peltata 2076. Glinus lotoides

1. Cocculus carolinus (L.) DC. (Carolina moonseed, Carolina snailseed, fishberry)

Pl. 456 h–j; Map 2072

Plants relatively slender climbers. Stems to 5 m or more long. Leaves not peltate, the petiole attached at the margin of the blade. Leaf blades 4–15 cm long, 3.0–8.5(–14.0) cm wide, triangular or ovate-cordate in outline, undivided or shallowly 3- or 5-lobed, with 5 main veins from the base, the upper surface glabrous or sparsely to densely hairy, the undersurface sparsely to densely hairy, the base truncate to moderately cordate, rounded to bluntly pointed at the tip or the tips of the lobes, usually with an abrupt minute sharp point at the very tip. Inflorescences 2–12 cm long. Sepals 1.5–2.5 mm long. Petals 6. Stamens usually 6, the anthers 4-locular. Pistils 6. Fruits 5–8 mm long, more or less globose, red, the endocarp discoid, the rim and thickened margin transversely ridged, both sides concave. $2n=78$. July–August.

Scattered in southern and eastern Missouri, in the Mississippi Lowlands, Big Rivers, Ozark, and Ozark Border Divisions (southeastern U.S. west to Kansas, Texas, and disjunctly Arizona; Mexico). Bottomland forests, mesic upland forests, swamps, upland prairies, glades, tops, ledges, and bases of bluffs, and banks of streams; also old fields.

3. Menispermum L.

Three species, eastern North America, Mexico, Asia.

1. Menispermum canadense L. (moonseed, yellow parilla)

Pl. 456 a–c; Map 2073

Plants relatively slender climbers. Stems to 5 m long. Leaves narrowly peltate, the petiole attached to the underside of the blade near the base. Leaf blades 5–13 cm long, 6.0–17.5 cm wide, triangular to almost reniform, shallowly 5- or 7-lobed or sometimes unlobed, with 7 or 9 main veins from the attachment point of the petiole, the upper surface glabrous or sparsely to densely hairy, the undersurface sparsely to densely hairy, the base shallowly cordate, rounded to bluntly pointed at the tip or the tips of the lobes, usually with an abrupt minute sharp point at the very tip. Inflorescences 3–8 cm long. Sepals 1–3 mm long. Petals 4–12. Stamens 12–36, the anthers 4-locular. Pistils 2–4. Fruits 9–15 mm long, more or less globose, bluish black, the endocarp discoid, with a transversely ridged crest around the rim, both sides concave. $2n=52$. May–June.

Scattered to common nearly throughout Missouri (eastern U.S. west to North Dakota and Texas; Canada). Bottomland forests, mesic upland forests, bases and ledges of bluffs, banks of streams and rivers; also roadsides, railroads, and shaded disturbed areas.

MENYANTHACEAE (Buckbean Family)

(Wood, 1983)

Plants perennial herbs, with long-creeping, branched rhizomes. Leaves alternate, short- to long-petiolate, the petioles sheathing the stem. Leaf blades simple or trifoliate, glabrous. Stipules absent. Inflorescences racemes, umbels, or appearing as loose clusters. Flowers actinomorphic, perfect, slightly perigynous, often dimorphic (some flowers with exserted stamens and short styles, others with exserted styles and included stamens with short filaments). Calyces deeply 5-lobed, fused to the basal portion of the ovary. Corollas deeply 5-lobed, fused to the basal portion of the ovary. Stamens 5, attached to corolla tube, alternating with the lobes, the anthers attached at or near their bases, dehiscing longitudinally. Staminode-like structures sometimes present. Pistil 1 per flower, of 2 fused carpels. Ovary superior, with 1 locule, the placentation parietal. Style 1, sometimes very short, the stigma 2-lobed, persistent at fruiting. Ovules numerous. Fruits capsules, more or less 2-valved, dehiscing more or less irregularly. Seeds numerous. Five genera, about 45 species, nearly worldwide.

Many earlier authors, including Steyermark (1963), treated these genera as a tribe or subfamily of the Gentianaceae, but more recent morphological, biochemical, and molecular studies (M. H. G. Gustafsson and Bremer, 1995; M. H. G. Gustafsson et al., 1996) have indicated that the two groups are not very closely related. Surprisingly, the Menyanthaceae apparently instead are distant relatives of the lineage leading to the Asteraceae and the non-Missouri families in that group.

1. Leaves trifoliate, emergent from the water; inflorescences racemes; corollas white or pinkish-tinged . . . . 1. MENYANTHES
1. Leaves simple, the blades floating; inflorescences umbellate or appearing as loose clusters; corollas yellow . . . . 2. NYMPHOIDES

1. Menyanthes L.

One species, circumboreal.

1. Menyanthes trifoliata L. (buckbean, bogbean)

M. trifoliata var. *minor* Fernald

Pl. 456 d, e; Map 2074

Plants with stout rhizomes, the persistent leaf bases somewhat overlapping. Leaves emergent, produced only near the tips of the rhizome branches, mostly long-petiolate. Leaf blades palmately trifoliate, the leaflets 3–10 cm long, 1–5 cm wide, oblong-elliptic to ovate-elliptic or oblanceolate, rounded, bluntly pointed, or rarely shallowly notched at the tip, narrowed or tapered at the base, the margins entire or slightly wavy. Inflorescences racemes, terminal on the rhizome branches, dense at flowering time but elongating as the fruits mature, the flowers subtended by small bracts. Calyces 2–5 mm long, the lobes oblong-ovate, somewhat spreading or curved outward at flowering. Corollas 10–16 mm long, the margins entire or nearly so, white, the outer surface sometimes pinkish-tinged, especially in bud, the lobes narrowly triangular, arched outward to somewhat recurved, the inner surface appearing bearded or fringed with dense long white hairs. Staminodes absent. Fruits emergent, 6–10 mm long, globose, the wall somewhat corky, dehiscing irregularly more or less along the 2 sutures. Seeds 2.0–2.5 mm long, broadly ovate-elliptic in outline, somewhat flattened, the margins and surfaces smooth, light brown to yellowish brown, shiny. $2n$=54, 108. April–May.

Uncommon, known only from one historical and one extant site in Reynolds County (northern U.S. [including Alaska] south to North Carolina, Missouri, and California; Canada, Greenland, Europe, Asia). Emergent aquatics in fens and sinkhole ponds.

Steyermark (1963), who considered this species a relict from past glacial times when the climate in Missouri was cooler, discovered a population growing in a marly “boglike” fen near the town of Bunker. Over the course of the next two decades, he revisited the site repeatedly, collecting and distributing an excessive number of specimens, but he reported finding it in flower on only one occasion in 1952. By the time that Orzell (1982) attempted to relocate the plants, the population was no longer extant, and he considered the species to have become extirpated from the state's flora. During a Natural Features Inventory of Reynolds County conducted by the Missouri Department of Conservation, Joe Ryan (1994) discovered a new population in a sinkhole pond near Ellington. Here, flowering is also quite sporadic, but the plants have spread abundantly by vegetative means, intertwining densely with *Dulichium arundinaceum* (L.) Britton (Cyperaceae) and a species of *Sphagnum* (peat moss), forming a floating mat in the center of the pond sufficiently dense and thick enough to support the weight of a botanist.

Some authors, including Steyermark (1963) have chosen to segregate North American plants from those in the Old World as var. *minor*. The two varieties are said to be differentiated in subtle details of corolla size, color, and degree of fringing, but these characters do not appear to adequately separate plants into two groups, either singly or in combination.

2. **Nymphoides** Ség. (floating heart)

About 30 species, North America to South America, Europe, Africa, Madagascar, Asia, south to Australia.

1. Nymphoides peltata (S.G. Gmel.) Kuntze (yellow floating heart)
Limnanthemum peltatum S.G. Gmel.
Pl. 456 f, g; Map 2075

Plants with moderately slender rhizomes bearing widely spaced, long-petiolate leaves and also ascending, branched, petiolelike stems, each branch bearing 1 or more shorter-petiolate leaves at the tip and often also an inflorescence and a cluster of tuberous roots (these appearing spurlike or bananalike), the leaf bases not persistent on the stems. Leaf blades floating, simple, 4–15 cm long and wide, more or less circular, deeply cordate at the base, the undersurface often faintly to strongly purplish-tinged or mottled. Inflorescences umbels or appearing as loose umbellate clusters, terminal on the ascending stem branches, the flowers lacking bracts. Calyces 8–12 mm long, the lobes elliptic-lanceolate, loosely ascending at flowering. Corollas 20–28 mm long, the margins appearing somewhat wavy and irregularly cut or loosely fringed, bright yellow, the lobes appearing somewhat corrugated, oblong, spreading, the inner surface glabrous. Staminodes 5, attached to the corolla tube opposite the lobes, each appearing as a small tuft or fringe of hairs. Fruits maturing underwater, 12–25 mm long, ovoid, the wall somewhat fleshy, eventually dehiscing irregularly. Seeds 4–5 mm long, obovate in outline, strongly flattened, the margins with a dense fringe of long hairs, the surface finely pebbled, light brown to yellow, not shiny. $2n=54$. May–September.

Introduced, known thus far only from historical collections from Iron, Newton, and St. Louis Counties and an extant site in Franklin County (native of Europe and Asia; introduced sporadically in the United States). Floating-leaved aquatics in ponds.

The species epithet has been misspelled *"peltatum"* in some of the earlier literature (Steyermark, 1963), but the generic ending "oides" is to be considered feminine. The flowering cycle is structurally complex and difficult to describe. From a node along the rhizome develops an ascending petiole-like stem, which is often few-branched. Each fertile branch has at its tip a pair of leaves, these often unequal in size. The umbellate cluster of long-stalked flowers also arises from this branch tip, and usually the plant also produces a small cluster of spurlike or bananalike tuberous roots (in the aquarium trade, plants are often marketed as "underwater banana plants"). The flower buds mature at different times and each flower lasts but a single day, opening at the water surface around mid-morning and withering by late afternoon. Afterward, the flower stalk recurves and the fruit develops over a two- to three-month period underwater, eventually bursting irregularly and releasing seeds into the water. In spite of this complicated phenology, the species is often spread vegetatively by stem fragments.

Nymphoides peltata is widely cultivated as an ornamental in aquariums and ponds, although it can spread so aggresively as to interfere with oxygen exchange across the pond surface, penetration of light to reach submerged flora and fauna, and the activities of fishing enthusiasts. Stuckey (1974) summarized the North American distribution and spread of the species, from the first noncultivated specimens collected in New York City in the 1880s to its sporadic spread westward to Arizona and Washington (and more recently also California). He suggested that it had escaped from cultivation independentally several times in different regions and that individual populations often do not persist for very many years. The extant population in Franklin County apparently was not intentionally planted at the site and has existed there for well over a decade, resisting attempts at mechanical removal and the application of herbicides.

MOLLUGINACEAE (Carpetweed Family)

Plants annual (perennial elsewhere), not or only slightly succulent, with short taproots, glabrous or with stellate hairs. Stems usually repeatedly dichotomously branched, often slightly thickened at the nodes. Leaves alternate, opposite, or whorled, the leaves at each node usually somewhat unequal in size. Stipules absent. Leaf blades simple, the margins entire. Flowers in small clusters in the leaf axils, sessile or stalked, without bracts, actinomorphic, perfect, hypogynous. Calyx with 5 sepals, these sometimes slightly fused at the base, green or petaloid and white, persistent in fruit. Corolla absent. Stamens 3–10, free or the filaments fused at the very base. Pistil 1 per flower, the ovary superior, consisting of 3(–5) fused carpels, with 3(–5) locules, the placentation axile. Styles absent or very short and as many as the carpels, the stigmas as many as the carpels. Ovules numerous. Fruits capsules (achenes elsewhere), dehiscing longitudinally by 3(–5) valves, not winged at the tip. Seeds minute, somewhat flattened, somewhat kidney-shaped (the embryo appearing curved or coiled) to nearly circular in outline. Thirteen genera, 95–130 species, mostly in tropical and subtropical regions, but especially diverse in southern Africa.

The genera *Glinus* and *Mollugo* were treated in the family Aizoaceae by Steyermark (1963) and many earlier authors, but are now considered by nearly all botanists to belong to a group of genera best treated as a separate family, Molluginaceae (Bogle, 1970; Cronquist, 1981, 1991). Aside from differences in flower and fruit morphology (free vs. fused perianth, sepal number, capsule dehiscence), as well as in habit and pubescence types, the two families differ in details of stem anatomy and produce different classes of red pigments (unusual compounds known as betalains in Molluginaceae vs. the more widespread anthocyanins in Aizoaceae).

1. Plants with sparse to dense stellate hairs; leaves mostly opposite (less commonly alternate or appearing whorled), the blade narrowly to broadly obovate, sometimes nearly circular in outline; flowers sessile or with stout stalks 1–4 mm long . . . . 1. GLINUS
1. Plants glabrous; leaves whorled, the blade broadly to narrowly oblanceolate, rarely nearly linear; flowers with slender stalks 5–15 mm long . . . . 2. MOLLUGO

1. Glinus L.

Twelve or 13 species, tropics and subtropics of both the Old and New Worlds, introduced in temperate regions.

1. Glinus lotoides L. (glinus)

Pl. 457 a–c; Map 2076

Plants with sparse to dense stellate hairs on stems, leaves, and perianth. Stems 10–35 cm long, prostrate or less commonly ascending. Leaves mostly opposite (less commonly alternate or appearing whorled), the blade 5–25 mm long, narrowly to broadly obovate, sometimes nearly circular in outline, rounded or abruptly pointed at the tip, narrowed at the base to a petiole 3–20 mm long. Flow-

2077. Mollugo verticillata 2078. Broussonetia papyrifera 2079. Fatoua villosa

ers sessile or with stout stalks 1–4 mm long. Sepals 4–7 mm long, lanceolate, green with thin lighter margins. Stamens 5–10. Stigmas sessile on the ovaries. Fruits 3.0–4.5 mm long, ovoid to ellipsoid. Seeds partially embedded in a bladderlike caruncle, a linear appendage also wrapped around the seed, the seed coat smooth to pebbled or tuberculate, shiny, reddish brown to black. $2n=36$. July–September.

Introduced, uncommon in the western portion of the state (native to Europe and Africa; widely introduced in the New World, mostly in tropical to warm-temperate regions). Banks of rivers and margins of oxbow lakes.

Steyermark (1963) noted that this species superficially resembles *Tidestromia lanuginosa* (Nutt.) Standl. (Amaranthaceae). However, that species differs in having flowers subtended by three papery, hairy bracts and in its 1-locular, indehiscent fruits containing a single seed.

2. **Mollugo** L., carpetweed

About 35 species, tropics and subtropics of both the Old and New Worlds, introduced in temperate regions.

1. Mollugo verticillata L. (carpetweed, Indian chickweed)

Pl. 457 d, e; Map 2077

Plants glabrous. Stems 5–40 cm long, prostrate and mat-forming. Leaves in whorls of 3–8, the blade 8–40 mm long, broadly to narrowly oblanceolate, rarely nearly linear, rounded or pointed at the tip, narrowed gradually at the base, sessile or with a short, winged petiole 1–4 mm long. Flowers with slender stalks 5–15 mm long, these drooping in fruit. Sepals 2.0–2.5 mm long, elliptic or narrowly oblong, white or pale green. Stamens 3 or 4. Stigmas on very short, curved styles. Fruits 2.5–3.0 mm long, ovoid to ellipsoid. Seeds without a caruncle or linear appendage, but with a minute peglike protuberance at the attachment point, the seed coat with several arching ridges along the sides and back (these rarely absent), shiny, dark reddish brown. $2n=64$. April–December.

Introduced, common throughout Missouri (apparently native to tropical America, now naturalized nearly worldwide). Banks of rivers and streams, often in sandy soils; also crop fields, fallow fields, roadsides, railroads, and open disturbed areas.

Bogle (1970) noted that plants of *M. verticillata* can progress from seed germination to the flowering state within the space of a few days. The species is doubtlessly present in every county in the state.

MORACEAE (Mulberry Family)

Contributed by Alan Whittemore

Plants trees or less commonly annuals (shrubs or vines elsewhere), monoecious or dioecious, unarmed or (in *Maclura*) armed with axillary thorns, sometimes producing milky sap.

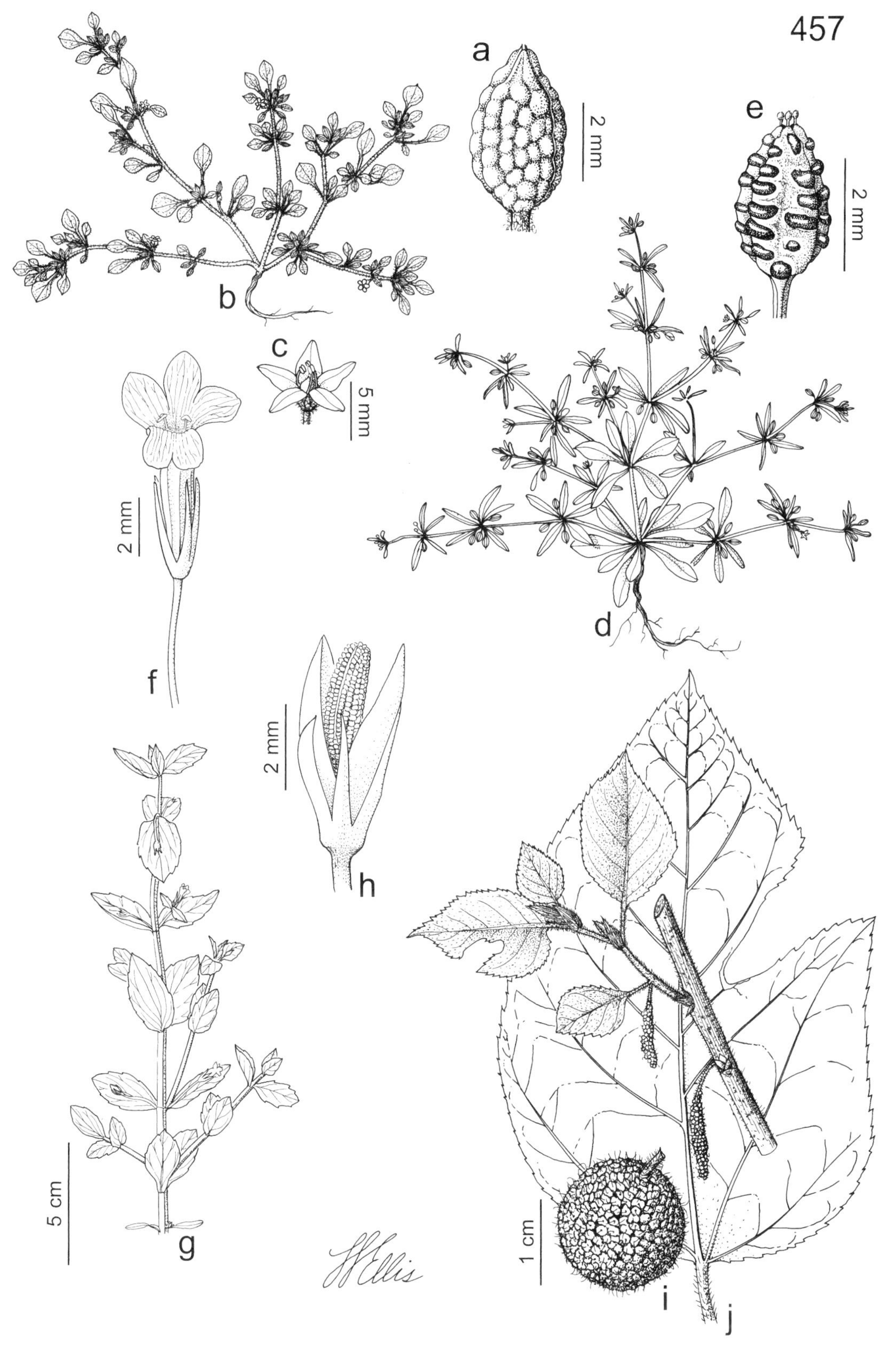

Plate 457. Linderniaceae. Molluginaceae. Moraceae. *Glinus lotoides*, **a)** fruit, **b)** habit, **c)** flower. *Mollugo verticillata*, **d)** habit, **e)** fruit. *Lindernia dubia*, **f)** flower, **g)** fruit, **h)** habit. *Broussonetia papyrifera*, **i)** fruit, **j)** inflorescence and leaf.

Leaves alternate, petiolate. Stipules present, small, scalelike or herbaceous, often shed early. Leaf blades simple, palmately or pinnately veined, the margins entire or toothed. Inflorescences variously short racemes, spikes, heads, or dense headlike clusters, sometimes the flowers on an expanded receptacle, this variously shaped. Flowers imperfect, actinomorphic, hypogynous, usually subtended by small narrowly lanceolate to narrowly elliptic bracts. Staminate flowers with calyces of 4(5) small, free or united sepals. Pistillate flowers with calyces of 4(5) free or fused sepals, persistent at fruiting and sometimes becoming enlarged and fleshy. Corollas absent. Stamens 4 (absent in pistillate flowers), free, the filaments in our species curved inward in the bud, the anthers attached basally, white to pale yellow, dehiscent by longitudinal slits. Pistil 1 per flower (absent in staminate flowers), of 2 fused carpels or 1 of these failing to develop, the ovary superior, 1- or 2-locular, with 1 ovule per locule, the placentation apical or nearly so. Styles 1 or 2, when only 1 then unbranched or 2-branched. Multiple fruits composed of drupelets or achenes, each subtended by or enclosed in the persistent calyx and either densely spaced or fused into groups of several to many. About 38 genera, about 1,150 species, nearly worldwide.

The Moraceae are a large, important family in the tropics, with few species in the north-temperate zone. They are a complex group and include many genera that do not resemble the temperate mulberries very closely.

The fruits of our species of Moraceae are complex in structure, with much of the apparent fruit being accessory tissue (developed from the calyces and bracts of the flowers). The best-known fruits are figs, in which the numerous minute fruits become enclosed within the expanded, fleshy receptacle.

Commercial silk fiber is produced by the pupae of silkworms, which are moths in the genus *Bombyx* L. (family Bombycidae Latreille), especially *B. mori* (L.) Silkworm larvae feed only on Moraceae, and persistent efforts through most of the seventeenth and eighteenth centuries to establish a silk industry in the southeastern colonies resulted in the introduction and widespread planting of food plants. The main species used in silkworm farming is *Morus alba*, the primary food plant for silkworms in the silk-producing regions of China, but some other species of *Morus* and *Maclura* are acceptable to the larvae and have been used at times.

1. Plants herbaceous; each inflorescence with staminate flowers on one side and pistillate flowers on the other; fruits dry . . . . 2. FATOUA
1. Plants trees or shrubs; each inflorescence with only staminate or pistillate flowers; fruits appearing fleshy or juicy
    2. Leaves with a single midvein (pinnately veined), the margins entire and unlobed; stems often with solitary, stout, straight thorns to 2 cm long in the leaf axils; multiple fruits (aggregates of fruits) 9–13 cm long, green . . . . 3. MACLURA
    2. Leaves 3- or 5-veined from the base, the margins toothed and sometimes lobed; stems unarmed; multiple fruits (aggregates of fruits) 1–4 cm long, white, orange, red, purple, or black
        3. Twigs and petioles pubescent with spreading hairs; leaves alternate and/or opposite, the upper surface grayish green and dull, the undersurface densely pubescent with grayish, woolly hairs; multiple fruits consisting of dense, globose masses of fleshy, orange to orangish red drupes, these long-exserted from the papery calyces, the aggregates not shed together . . . . 1. BROUSSONETIA
        3. Twigs and petioles minutely hairy or nearly glabrous; leaves always alternate, the upper surface bright green, dull or shiny, the undersurface sparsely to densely pubescent with short, nonwoolly hairs; multiple fruits consisting of crowded, ovoid to short-cylindric clusters of ovoid achenes, these covered by the fleshy, white, red, purple, or black calyces, each aggregate shed as a unit . . . . 4. MORUS

1. **Broussonetia** L'Hér. ex Vent.

About 7 species, Asia to Polynesia, introduced elsewhere in the world.

1. Broussonetia papyrifera (L.) L'Hér. ex Vent. (paper mulberry)

Pl. 457 i, j; Map 2078

Plants trees or shrubs to 15 m tall, unarmed, with milky sap. Bark smooth or somewhat grooved longitudinally, yellowish brown to grayish brown. Twigs relatively stout, often somewhat zigzag, grayish green with circular, orange lenticels, densely pubescent with spreading hairs, the winter buds bluntly ovoid, with several overlapping scales, densely hairy. Leaves alternate and/or opposite. Petioles 3–11 cm long, densely pubescent with long spreading hairs. Leaf blades 8–19 cm long, 5–16 cm wide, ovate, unlobed or shallowly to deeply 3-lobed and with 3 main veins from the base, abruptly short-tapered to a sharply pointed tip, the lateral lobes (when present) also abruptly short-tapered, broadly angled to rounded, truncate, or occasionally shallowly cordate at the base, the margins otherwise toothed, the upper surface dark grayish green, roughened, with the main veins more or less hairy, the undersurface pale green or off-white, usually densely hairy, felty to the touch. Inflorescences entirely staminate or pistillate. Staminate inflorescences solitary in the leaf axils, dense catkins 4–8 cm long, pendant, the calyces 1.5–2.5 mm long, deeply 4-lobed, hairy. Pistillate inflorescences solitary in the leaf axils, dense, globose clusters 1–2 cm in diameter, the calyces 1.5–2.5 mm long, deeply 4-lobed, the lobes oblanceolate, densely hairy at their tips, the stigma 1, unbranched, linear. Fruits fused into compound, fleshy spherical masses, these 1.5–3.0 cm in overall diameter at maturity, the individual drupes more or less club-shaped, long-exserted from the persistent calyces, orange to orangish red. $2n=26$. April–May.

Introduced, scattered in the southeastern quarter of the state, uncommon or absent elsewhere (native of Asia, introduced in the eastern United States west to Nebraska and Texas). Bottomland forests, mesic upland forests, and edges of sand prairies; also ditches, railroads, roadsides, and disturbed areas.

Paper mulberry persists at disturbed sites, most commonly in the Mississippi Lowlands Division, often forming large, dense clonal colonies from root sprouts. Pistillate inflorescences can be relatively showy, with the calyces bright red, but nearly all of the fertile specimens collected in Missouri thus far have been staminate.

Fiber from the inner bark is made into cloth and paper in the Old World and the sap has been used in the formulation of a glue. Steyermark (1963) noted that the pollen can cause hay fever.

2. **Fatoua** Gaudich.

Two species, Asia, Madagascar, Australia; introduced in North America.

1. Fatoua villosa (Thunb.) Nakai (mulberry weed)

Map 2079

Plants annuals, unarmed, taprooted, the sap not milky. Stems 8–45 cm long, erect or ascending, minutely hairy, sometimes becoming nearly glabrous with age. Leaves alternate, the petiole 0.8–6.0 cm long, minutely hairy. Leaf blades 1–10 cm long, 0.8–7.0 cm wide, triangular-ovate, unlobed, 3-veined from the base, angled or tapered to a sharply pointed tip, truncate to shallowly cordate at the base, the margins scalloped or bluntly toothed, upper surface dark green, often somewhat roughened and with scattered, inconspicuous, appressed hairs, the undersurface pale green, roughened, hairy along the veins, or sometimes somewhat felty to the touch. Inflorescences with both staminate and pistillate flowers, solitary in the leaf axils, consisting of dense, hemispheric clusters with staminate and pistillate flowers on opposite sides. Staminate flowers with the calyces 1.0–1.2 mm long, 4-lobed to below the midpoint, the lobes broadly elliptic-ovate and concave, glandular- and nonglandular-hairy. Pistillate flowers with the calyces 1–2 mm long, deeply 4-lobed, the lobes ovate to broadly elliptic-ovate and somewhat concave, glandular- and nonglandular-hairy, the style 1, attached laterally, unbranched (usually with a vestigial outgrowth representing an abortive second branch near its base), the stigma linear. Fruits crowded in dry, brown, dense hemispheric clusters 0.4–0.6 cm long and 0.6–1.0 cm wide, the individual achenes depressed-ovoid, bluntly 3-angled. August-October.

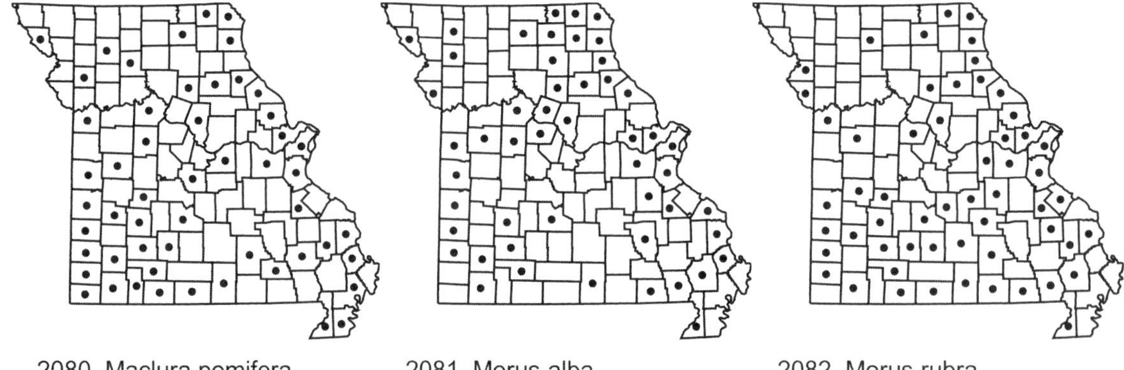
2080. Maclura pomifera 2081. Morus alba 2082. Morus rubra

Introduced, scattered, mostly in and around urban areas (native of Asia, introduced in the eastern U.S. west to Missouri and Texas, also California, Washington, and Caribbean Islands). Edges of mesic upland forests; also lawns, gardens, greenhouses, and open disturbed areas.

This species was first reported as a weed in North America by Thieret (1964). Since then, it has spread relatively rapidly to 26 or more states in the eastern United States, mainly as a weed or seed contaminant in the soil of bedding plants sold horticulturally (Vincent, 2004a). It was first reported from Missouri by Yatskievych and Raveill (2001). There is no evidence that *F. villosa* becomes naturalized along edges of upland forests, although plants occasionally spread there from adjacent highly disturbed habitats.

3. **Maclura** Nutt.

About 12 species, North America to South America, Africa, Asia to Australia.

1. Maclura pomifera (Raf.) C.K. Schneid. (bois d'arc, Osage-orange, hedge apple, bowwood)

Pl. 458 a, b; Map 2080

Plants trees or shrubs to 20 m tall, when young, the branches often armed with stout, straight thorns to 2 cm long in the leaf axils, with milky sap. Bark thick, developing deep furrows and coarse ridges on older trunks, the ridges with the surface often peeling in thin strips with age, brown to brownish orange. Twigs relatively stout, sometimes developing into short, very congested shoots, on elongate shoots often somewhat zigzag, greenish yellow to orangish brown with circular to oval, lighter lenticels, glabrous or minutely hairy, the winter buds often paired (then of unequal sizes), globose, with several overlapping scales, these minutely hairy along the margins. Leaves alternate, but appearing more or less whorled on short shoots. Petioles 1–4 cm long, short-hairy, sometimes also with scattered, longer, spreading hairs. Leaf blades 4–14 cm long, 2–7 cm wide, ovate to elliptic or narrowly ovate, unlobed and with 1 main vein (pinnately veined) from the base, abruptly short-tapered to a sharply pointed tip, rounded to truncate or broadly angled at the base, the margins otherwise entire, the upper surface green, smooth, with the main veins more or less hairy, the undersurface pale green, sparsely short-hairy, at least along the main veins. Inflorescences entirely staminate or pistillate. Staminate inflorescences clustered on short shoots, dense, more or less globose clusters (occasionally slightly elongate), 1.3–2.0 cm long, oriented in several directions, the calyces 1.0–1.5 mm long, deeply 4-lobed, hairy. Pistillate inflorescences solitary in the leaf axils, dense, globose clusters 1.0–1.5 cm in diameter (but appearing larger because of the elongate stigmas), the calyces 2–3 mm long, deeply 4-lobed, the lobes obovate and clasping the ovary, densely hairy at their tips, style 2-branched, the stigmas unbranched, linear. Fruits fused into massive, compound, fleshy spherical masses, these 9–14 cm in overall diameter at maturity, the individual achenes enclosed in the enlarged, thickened calyces, which become fused and sunken into the enlarged receptacle, the surface of the multiple fruit yellowish green to green, with a convoluted, irregular network of shallow grooves and short, rounded ridges, and sometimes also scattered, short remains of the stigmas. $2n=56$. May–June.

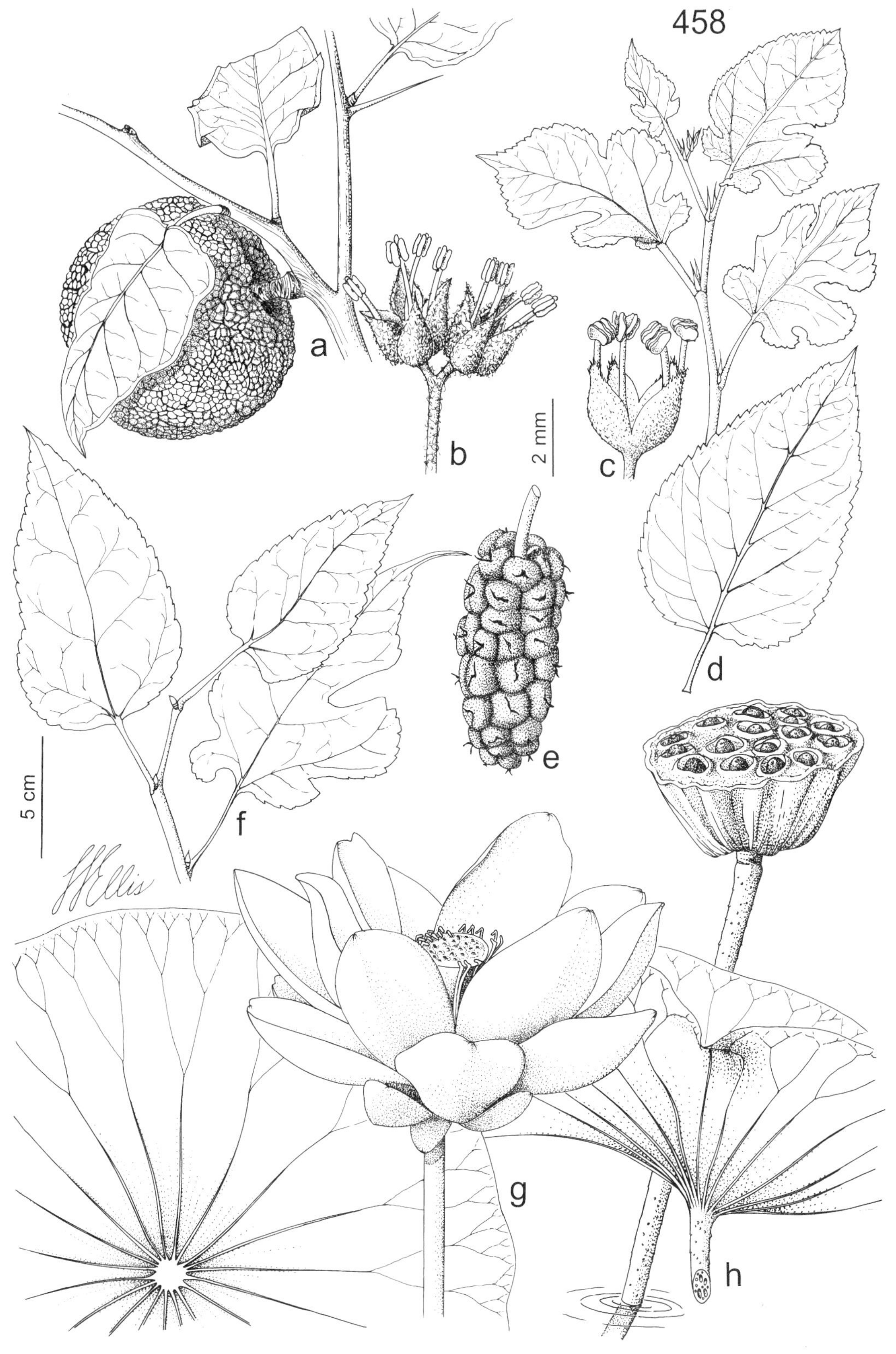

Plate 458. Moraceae. Nelumbonaceae. *Maclura pomifera*, **a)** branch with thorn, leaves, and multiple fruit, **b)** staminate flowers. *Morus alba*, **c)** staminate flower, **d)** branch tip with lobed leaves and separate unlobed leaf. *Morus rubra*, **e)** multiple fruit, **f)** branch tip with leaves. *Nelumbo lutea*, **g)** leaf (upper surface) and flower. *Nelumbo nucifera*, **h)** leaf (side view) and fruit.

Probably ntroduced, scattered to common nearly throughout the state (native of Texas, Oklahoma, Arkansas, and perhaps portions of some adjacent states, introduced nearly throughout the remaining U.S., Canada). Bottomland forests, mesic upland forests, bottomland and upland prairies, banks of streams and rivers, marshes, and margins of sinkhole ponds; also fencerows, pastures, railroads, and roadsides.

Bois d'arc has played a key role in Missouri economies over many centuries. The species may have been native to southern Missouri or more likely was introduced in the state before the arrival of European explorers. Seeds and living plants were transported widely in pre-Columbian times, primarily because of the value of the wood for making bows (Moerman, 1998). It is the premier bow wood of North America. Bois d'arc bows were a valuable trade item, and they were found over much of eastern and central North America, often many hundreds of miles from the tribes that grew the trees and carved the bows. The hard, strong wood was also used for war clubs and tomahawks, as well as for ceremonial staffs. The inner bark and sapwood yield a yellow dye and rough cordage was made from the fibrous inner bark.

Early European settlers were quick to appreciate the quality of the wood and dyes, and the species also was widely planted as a street and shade tree. Bois d'arc became even more important when settlement reached the prairie regions of the state in the mid-nineteenth century. The scarcity of wood in the prairie districts made fencing impractical, and settlers turned to thorn hedges to keep their ploughed land safe from their stock. Bois d'arc quickly became the favorite hedge plant. The dense growth form and strong thorns of vigorous young plants made a hedge "pig tight, horse high, and bull strong," (W. P. Webb [1931], quoting an anonymous writer of 1872) and it was widely planted in northern and western Missouri (and adjacent states) for this purpose. The hedges also protected land from prairie fires, which would often clear the vegetation from large areas of land in northern and western Missouri. As hedges matured, however, the large size of the plant and the extensive shallow root system took too much land away from the crops. Thorn hedges mostly were replaced by fences after the introduction of mass-produced barbed wire in the late 1870s, and the few hedges planted after this date were mostly roses (*Rosa*, Rosaceae), not bois d'arc. Bois d'arc continued to be planted as a shade tree and street tree, and to be used for a variety of purposes: the wood for railroad ties, fenceposts, wheel hubs and rims, and foundation blocks for buildings; the leaves were sometimes used for feeding silkworms; and the dye extracted from the bark continued to be important well into the twentieth century. Today, *M. pomifera* is still planted as a shade, street, and specimen tree. The species sometimes is considered to be invasive, forming thickets in upland prairies, especially in draws.

Lewis and Clark encountered the Osage-orange growing at the St. Louis home of Pierre Chouteau, Thomas Jefferson's local Agent of Indian Affairs (Earle and Reveal, 2003). Chouteau, a member of one of the city's most influential families, handled trade negotiations with the Osage and other tribes for the Chouteau business empire. He advised and helped to outfit the Corps of Discovery for its historic voyage across the country. Chouteau provided cuttings of *Maclura* that Meriwether Lewis shipped back to Monticello on 16 March 1804 (W. W. Phillips, 2003). Jefferson not only grew the trees himself, but provided material to the Philadelphia nurseryman Bernard M'Mahon. Although apparently there were a few specimen trees already present in the eastern United States before this time, it was the plant material from St. Louis that formed the basis for much of the commercial cultivation of the species in that part of the country.

4. **Morus** L. (mulberry)

Plants trees to 20 m tall (shrubs elsewhere), unarmed, with milky sap. Bark shallowly grooved longitudinally, light to dark brown, sometimes yellowish-, orangish-, or reddish-tinged. Twigs slender to stout, not or only slightly zigzag, reddish brown to greenish brown, usually with circular to elongate, light or dark brown lenticels, minutely hairy or nearly glabrous, the winter buds 3–7 mm long, ovoid, bluntly to sharply pointed at the tip, with several overlapping scales, glabrous or hairy, especially along the margins, sometimes shiny. Leaves alternate, the petiole 2–5 cm long, minutely hairy or nearly glabrous. Leaf blades ovate, unlobed or shallowly to deeply (2)3–5-lobed and with 3 or 5 main veins from the base, the margins otherwise toothed, the upper surface bright green, roughened or smooth and glabrous to sparsely hairy, the undersurface lighter green, sparsely to densely pubescent with short, nonwoolly hairs. Inflorescences entirely staminate or pistillate, dense catkins, the calyces, deeply 4-lobed (occasionally 5-lobed in *M. alba*), 1–2 mm long, the lobes ovate to broadly elliptic-ovate and somewhat concave, hairy, often reddish-tinged. Staminate inflorescences solitary in the leaf axils. Pistillate inflorescences solitary in the leaf axils (but sometimes appearing clustered on short shoots), the style 2-branched, the stigmas linear. Fruits consisting of crowded, ovoid to short-cylindric clusters of ovoid achenes, these covered by the fleshy, white, red, purple, or black calyces, each aggregate shed as a unit. About 10 species, North America, Europe, Asia.

The mulberry fruit superficially resembles a blackberry, but its structure is totally different. The mulberry consists of tightly clustered fruits of many adjacent flowers, and the fruitlets are not true drupes; the juicy part develops from the calyx and the true fruit is an achene, the dry, so-called stone inside it. In contrast, a blackberry consists of numerous ripened ovaries on an expanded receptacle within a single flower and the individual fruitlets are true drupes.

Mulberries are hardy, fast-growing small trees that produce sweet fruit in very high yield from a young age. The fruit is attractive to many species of birds and mammals. Both of our species are eaten by humans, but they are considered much inferior in flavor to those of the black mulberry (*Morus nigra* L.), an Old World species. Mulberries can become aggressive and weedy in gardens.

Quickly growing saplings or root sprouts sometimes produce relatively large leaves that are deeply and irregularly palmately 5-lobed, with the relatively slender lobes bluntly and irregularly lobed, scalloped, and/or bluntly toothed. These sometimes are mistaken by gardeners for ornamental figs (but *Ficus* L. species generally are not winter-hardy in Missouri) or young maples (but *Acer* species have opposite leaves). Lobed leaves are seen relatively commonly in *M. alba*, but are far less frequently encounted in *M. rubra*.

1. Leaf blades with the undersurface glabrous except for scattered hairs on the main veins, the upper surface smooth or nearly so, often shiny, glabrous or with a few hairs on the main veins; blades 3–9 cm wide, those of lobed leaves with the lateral lobes rounded or broadly angled (except sometimes in very deeply lobed blades) . . . . 1. M. ALBA
1. Leaf blades with the undersurface short-hairy on and between the veins, upper surface often moderately to strongly roughened (rarely smooth), dull, with scattered inconspicuous hairs; leaf blades 7–25 cm wide, those of lobed leaves with the lateral lobes short-tapered to slender tips . . . . 1. M. RUBRA

1. Morus alba L. (white mulberry, silkworm mulberry)

Pl. 458 c, d; Map 2081

Twigs orangish brown to dark green, sometimes reddish-tinged, minutely hairy or nearly glabrous, the lenticels usually darker brown. Petioles 0.7–3.5(–4) cm long. Leaf blades 4–12 cm long, 3–9 cm wide, unlobed or shallowly to deeply 3–5-lobed, abruptly tapered to a sharply pointed tip, the lateral lobes (when present) rounded or broadly angled (occasionally narrowly angled to short-tapered in very deeply lobed blades), truncate to cordate or occasionally broadly rounded at the base, the upper surface smooth or nearly so and often shiny, glabrous or with a few hairs along the main veins, the undersurface with scattered hairs along the main veins Staminate inflorescences 25–40 mm long, cylindric. Pistillate inflorescences 5–8 mm long, short-cylindric to subglobose. Multiple fruits 0.6–1.8 cm long, 0.5–0.7 cm wide, short-cylindric to subglobose, white, red, or purple, the achenes 2–3 m long. $2n$=28. April–May.

Introduced, scattered nearly throughout the state, but apparently uncommon in some counties of the Ozark Division (native of Asia, introduced nearly throughout North America, Europe). Bottomland forests, mesic upland forests, banks of streams and rivers, margins of ponds, lakes, sinkhole ponds, and marshes, sloughs, and edges of upland prairies; also ditches, fencerows, gardens, farm yards, railroads, roadsides, and disturbed areas.

2083. Nelumbo lutea
2084. Nelumbo nucifera
2085. Boerhavia erecta

White mulberry is one of the most important commercial food plants for silkworms in the silk-producing regions of China. It was introduced to North America before 1620 as a food plant for silkworms, and it is now thoroughly naturalized throughout the eastern United States, as well as in scattered areas across the West. The species became naturalized so long ago that Native Americans used it medicinally as a laxative and purgative, and for the treatment of dysentery (Moerman, 1998). White mulberry was once important on farms as food for pigs and poultry, but in modern homes with no livestock the heavy crops of soft, sweet fruit are a nuisance, especially when the trees overhang paths and driveways. It is difficult to clean up the stains from red- or purple-berried forms, and from the droppings of birds that have fed on them.

Morus alba and *M. rubra* hybridize where they occur together (Salah, 2006). Hybrids are intermediate betweent the parental species in leaf size and pubescence. A number of horticultural forms of *M. alba* have been developed. Rehder (1940) and other authors reported strains of this species in cultivation with very large unlobed leaves—reportedly to 22 cm long, far outside the normal size range for the species. These have been called cv. 'Macrophylla'; they likely are the result of past hybridization with another species of mulberry. Steyermark (1963) identified a single collection (*Palmer 50832*, from Lawrence County) as this form, but the specimen in Palmer's herbarium at the University of Missouri, Columbia, has leaves less than 10 cm long. Uncommon plants with all of the leaves deeply divided into relatively slender, narrowly pointed lobes have been called f. *skeletoniana* (C.K. Schneid.) Rehder.

2. Morus rubra L. (red mulberry)
M. murrayana Saar & Galla
M. rubra var. *murrayana* (Saar & Galla) Saar

Pl. 458 e, f; Map 2082

Twigs reddish brown to light greenish brown, minutely hairy or nearly glabrous, the lenticels usually pale. Petioles 1–4 cm long. Leaf blades 7–28 cm long, 7–25 cm wide, unlobed or shallowly to deeply (2)3–5-lobed, abruptly tapered to a sharply pointed tip, the lateral lobes (when present) short-tapered to slender tips, truncate to shallowly cordate at the base, the upper surface smooth to more commonly moderately to strongly roughened, dull, with scattered inconspicuous hairs, the undersurface sparsely to densely short-hairy, smooth or somewhat felty to the touch Staminate inflorescences 30–50 mm long, cylindric. Pistillate inflorescences 8–12 mm long, short-cylindric. Multiple fruits 1.0–1.9(–3.5) cm long, 0.6–0.8(–1.2) cm wide, short-cylindric, dark purple to black, the achenes 1.5–2.0 m long. $2n=28$. April–May.

Common nearly throughout the state, except in the far northwestern and southeastern corners (eastern U.S. west to Nebraska and Texas; Canada). Bottomland forests, mesic upland forests, banks of streams, rivers, and spring branches, margins of sinkhole ponds, ledges and tops of bluffs, edges of glades, and savannas; also fencerows, pastures, and roadsides.

Deciduous woods and edges of pastureland, often along streams, roadsides, and railroads.

The wood of red mulberry is light and durable, and it has been used for fence posts, barrels, boats, and tools of various kinds. The inner bark yields good-quality fiber that has sometimes been used in the manufacture of cordage and cloth. Native Americans used red mulberry medicinally as a laxative and purgative, and also to treat dysentery, urinary problems, and ringworm (Moerman, 1998). Red mulberry occasionally has been used to feed silkworms, but is considered much inferior to white mulberry for this purpose.

Unfortunately, *M. rubra* recently has been redescribed under a new name, *M. murrayana*. The descriptions and illustrations of the supposed new species given by Galla et al. (2009) are a perfect match for typical *M. rubra*. The descriptions and illustrations labeled *M. rubra* in their paper clearly are not that species. Instead, they appear to be represent a misdetermination of the hybrid between *M. alba* and *M. rubra*. The controversy was addressed recently by Nepal et al. (2012) using both morphological and molecular characters, and an attempt by Saar et al. (2012) to reinterpret their taxon as a variety of *M. rubra* was unconvincing. *Morus murrayana* is thus treated as a synonym of *M. rubra* in the present work.

NELUMBONACEAE (Lotus Family)

One genus, 2 species, northern hemisphere, but in the Old World also south to Australia.

Traditionally (Steyermark, 1963), many botanists considered *Nelumbo* to represent a subfamily of the water lily family, Nymphaeaceae. Details of the flower and pollen morphology and the anatomy of the plants, as well as some aspects of phytochemistry, suggested that the two groups are not closely related and had been linked mostly because of the shared "water lily" growth form (Cronquist, 1981). However, where best to place the Nelumbonaceae within a classification of angiosperms has not fully been understood until recently. The current hypothesis (summarized in Judd et al., 2008) links the Nelumbonaceae loosely with two very different, woody families, the Platanaceae (sycamores and plane trees) and the Proteaceae (banksias, silk trees, and macadamia nuts). The three currently are considered to represent a separate order, Protealaes, which, in turn, appears to be loosely related to the Ranunculales (Berberidaceae, Menispermaceae, Papaveraceae, Ranunculaceae) and a few other families (Buxaceae, Trochodendraceae) at the base of the so-called eudicots (see Judd et al. [2008] for a summary). The Nelumbonaceae have a very long fossil history. As noted by C. L. Anderson et al. (2005), fossils from both the New and Old Worlds with the characteristic leaves and expanded receptacle date back to the early Cretaceous period (110 million years ago).

1. Nelumbo Adans. (lotus)

Plants perennial, herbaceous aquatics, with extensive, branching, spongy rhizomes, rooting at the nodes and bearing large, banana-shaped tubers (produced in autumn), producing milky latex. Leaves spirally alternate, mostly long-petiolate, the petiole to 3 m or more, relatively stiff and stout, reaching the water surface or emergent to about 80 cm, sometimes prickly. Stipules apparently absent, but perhaps represented by small, sheathing scales at the petiole bases. Leaf blades 20–60(–80) cm in diameter, centrally peltate, more or less circular, the margin entire but slightly undulate, usually with a shallow sinus containing a small, broadly triangular, blunt to sharp point, usually also with a narrow, pale to yellowish or brownish differentiated band, the upper surface flat (in floating blades) to moderately concave (in emergent blades) about the central attachment point (visible as a pale, scarlike mark), glabrous, often somewhat glaucous, the undersurface pale but with small brown spots, the main veins numerous, palmate from the petiolar attachment, mostly branched dichotomously 2 or more times, raised on the blade undersurface, connected by a network of finer crossveins. Inflorescences of solitary flowers, these long-stalked directly from the nodes of the rhizome, emergent (usually slightly overtopping the emergent leaves), hypogynous, more or less perfect, actinomorphic. Perianth showy, of numerous, spirally arranged, overlapping tepals; the outermost 3–5 tepals 0.8–1.4 cm long, more or less ovate, somewhat concave, often green or greenish-tinged; grading into the median tepals, these 6.0–8.5 cm long, obovate; grading into the innermost ones, these somewhat shorter, lanceolate, grading into the stamens. Stamens numerous, arranged in a dense, overlapping spiral, the filament 6–12 mm long, slender, the anther 9–18 mm long, attached at the base, orangish yellow, with a terminal appendage, this 3–7 mm long, club-shaped, to narrowly club-shaped, usually somewhat arched or curved, pale yellow. Pistils several to numerous, simple, free from one another, embedded in deep pits along the upper surface of a large, flat-topped, conic, spongy, expanded receptacle, this continuing to expand as the fruits mature, becoming nodding, dark brown, and somewhat leathery to hardened at fruiting. Style 1 per ovary, very short, the stigma protruding through an apical pore in the receptacle pit, more or less disc-shaped, slightly concave and with a central pore. Ovule 1 per carpel, the placentation apical. Aggregate fruits consisting of nuts embedded in the long-persistent, greatly expanded receptacle, these 9–15 mm long, ovoid to nearly spherical, sometimes somewhat flattened, indehiscent, hard-walled, brown to nearly black. Two species, northern hemisphere, but in the Old World also south to Australia.

The flowers of *Nelumbo* are unusual in both their morphology and function. The conic, flat-topped, enlarged receptacle is thermogenic, that is, it has been shown to generate heat and to moderate the flower temperature to about 31 degrees Celsius independent of the ambient temperature. This apparently is related to two main functions, increased release of volatile, odoriferous compounds produced by the flower tissues and the breakdown of starch in the receptacle through an alternate respiratory pathway (see Vogel and Hadacek [2004] and Watling et al. [2006] for reviews). The stamens produce unusual, club-shaped appendages that also contain starch. Although it might be suspected that these appendages would function as a lure for pollinators, foraging of beetles or other insects upon these structures has not been observed in nature. Thus, these structures may have lost their former function in attracting insects. C. R. Robertson (1889a) recorded a series of more than 30 insect visitors to flowers of *N. lutea* in Illinois, mainly bees and flies, but later observers (Sohmer and Sefton, 1978; Schneider and Buchanan, 1980) also noted visitation by various beetles. Unlike in the true water lilies, in which pollination usually involves entrapment of beetles in flowers that close overnight, no such specialized pollination mechanism has been observed to function in *Nelumbo*.

Today, the principal use of *Nelumbo* in the United States is as an ornamental aquatic in ponds and lakes. Cultivation of lotus for various uses goes back to prehistoric times. However, the plants spread very aggressively, forming a dense, thick layer of intertwined rhizomes and tubers, and can become a nuisance, both because of excessive shading of the water and interference in canoe and other small boat traffic. Hall and Penfound (1944) estimated that a small patch in Tennessee enlarged radially by more than 45 feet during a single growing season. Steyermark (1963) noted that the rhizomes and tubers are eaten by wildlife, especially beavers, and that some waterfowl eat the fruits. He further noted that the plants form excellent cover and habitat for fish, waterfowl, and other animals. Native Americans made extensive use of *N. lutea* for food, boiling or baking the starchy tubers and cooking the young foliage. The immature fruits were eaten raw prior to hardening of the fruit wall; mature fruits were parched, dehulled, and the seeds cooked or ground into a flour (Steyermark, 1963; Moerman, 1998). Native foods enthusiasts continue to use the foliage, tubers, and young fruits in similar fashion (J. Phillips, 1998). Tubers of *N. nucifera* are cooked and sliced as a starchy vegetable in Asian cuisines. They also are dried and powdered for use in teas and herbal tonics.

The seeds of *Nelumbo* in their hard-walled fruits are extremely long-lived. Steyermark (1963) noted that in Missouri seeds of *N. lutea* had germinated in the bottoms of St. Louis County when areas that had been farmed for more than 40 years were allowed to go fallow. He also stated that other seeds that had been dormant for as long as 200 years retained the ability to germinate. However, since that time, other studies involving *N. nucifera* have greatly extended knowledge of seed longevity in the group. Shen-Miller et al. (2002) excavated seeds from deposits in a dry lakebed that were carbon-dated to be up to 466 years old and were able to successfully germinate them and to grow plants from them (although plants exhibited reduced vigor). Even more amazingly, Shen-Miller and her associates had earlier managed to germinate lotus seeds up to 1,300 years old. The seeds also are unusual in two other features: the embryo is green within the seed prior to germination, and the radicle portion of the embryo aborts prior to maturity, so the seedling never develops an initial taproot.

Although several additional taxa have been described, the prevailing view has been that *Nelumbo* contains only two morphologically similar species native to the New World and the Old World respectively. Borsch and Barthlott (1994) analyzed morphological variation in the genus and concluded that these two taxa should be considered subspecies of a single species. However, their careful analysis did disclose a pattern of subtle differences between the two taxa for a number of features, mostly overlapping quantitative differences in characters such as anther appendage length and perianth size. Additionally, the two taxa have been resolved in molecular studies involving both restriction fragment analysis of the mitochondrial genome (Kanazawa et al. (1998) and *rbcL* sequence data (C. L. Anderson et al., 2005). The situation requires more intensive study, but for now it seems prudent to continue to recognize two species (Wiersema, 1997).

1. Tepals cream-colored to pale yellow (sometimes with pinkish margins); flower stalks and petioles smooth . . . . 1. N. LUTEA
1. Tepals pink or strongly pinkish-tinged (sometimes fading to white); flower stalks and petioles roughened . . . . 2. N. NUCIFERA

1. Nelumbo lutea Willd. (American lotus, yanquapin, water chinquapin, lotus lily)
N. nucifera Gaertn. ssp. *lutea* (Willd.) Borsch & Barthlott

Pl. 458 g; Map 2083

Leaves with the petioles smooth. Flower stalks smooth. Tepals (except the outermost) cream-colored to pale yellow (sometimes with pinkish margins), the longest tepal 7–12 cm, the outermost tepals tending to persist at fruiting. Anthers with the sterile appendage 3–5 mm long. Receptacle with 8–32 embedded pistils, becoming enlarged to 10 cm at fruiting, tapering from slightly below the rim to the base, the sides usually shallowly fluted longitudinally, smooth. Fruits 9–16 mm long, 6–13 mm wide, mostly nearly circular in outline. $2n$=16. June–September.

Scattered nearly throughout the state (eastern U.S. west to Nebraska and Texas; Mexico, Central America, Caribbean Islands). Ponds, lakes, sloughs, oxbows, sinkhole ponds, and marshes.

Because of its large size and aquatic habit, *N. lutea* is undercollected and more widespread in the state than the distribution map for the species indicates. Some botanists have applied the older name, *N. pentapetala* Walter, to this taxon, but D. B. Ward (1977) presented a compelling argument that Walter's description is of uncertain application, having been based all or in part on material of *N. nucifera*.

2. Nelumbo nucifera Gaertn. (sacred lotus, East Indian lotus)

Pl. 458 h; Map 2084

Leaves with the petioles having scattered, slender prickles, these often breaking off at maturity, leaving a roughened surface. Flower stalks with scattered prickles similar to those of the petioles. Tepals (except the outermost) pink or strongly pinkish-tinged (sometimes fading to white), the longest tepal 7–14 cm, all of the tepals usually shed as the fruits mature. Anthers with the sterile appendage 3–7 mm long. Receptacle with 10–38 embedded pistils, becoming enlarged to 10 cm at fruiting, tapering from the upper rim to the base, the sides usually not or only very slightly fluted longitudinally, often finely pebbled or roughened. Fruits 10–20 mm long, 7–13 mm wide, mostly elliptic to ovate in outline. $2n$=16, 24. June–September.

Introduced, uncommon, known thus far from Iron County and the Kansas City region (Europe, Asia, south to Australia; introduced sporadically in the eastern [mostly southeastern] U.S. west to Missouri and Louisiana, also South America, Caribbean Islands). Ponds and lakes.

NYCTAGINACEAE (Four-O'clock Family)

Plants annual or perennial herbs, sometimes slightly woody at the base. Stems usually branched, with slightly to moderately swollen nodes. Leaves opposite, lacking stipules, 1 of each pair often slightly smaller than the other, petiolate. Leaf blades simple, with more or less entire margins. Inflorescences terminal and axillary panicles (sometimes reduced to solitary axillary clusters in *Mirabilis*) with small sessile or irregularly umbellate clusters of flowers at the ends of the branches, each of the clusters sometimes subtended by a calyxlike involucre. Flowers actinomorphic, perfect, hypogynous, but often appearing epigynous. Calyces usually brightly colored and corollalike, fused basally into a narrow tube with a constriction above the ovary, the upper portion ascending to spreading, usually 5 lobed. Petals absent. Stamens 1–5, the filaments free or sometimes fused into a ring at the base, the anthers attached basally. Pistil 1 per flower, the ovary superior but appearing inferior because of the closely enveloping perianth tube, consisting of 1 carpel, with 1 locule, the placentation basal. Style 1, the stigma globose or disclike. Ovule 1. Fruits achenes surrounded by the persistent, hardened or somewhat fleshy perianth tube (the rest of the perianth shed). Thirty to 34 genera, about 350 species, nearly worldwide, mostly in tropical and subtropical regions, particularly in the New World.

1. Flowers subtended by minute free bracts, these usually shed early; perianth 1.0–1.5 mm long . . . . 1. BOERHAVIA
1. Flower clusters subtended by a persistent calyxlike involucre of fused bracts; perianth 5–10 mm long. . . . . 2. MIRABILIS

1. **Boerhavia** L. (spiderling, wine flower)

Three to 30 species, widespread in tropical and warm-temperate regions around the world, especially the southwestern U.S. and adjacent arid and seasonally dry regions of Mexico.

Boerhavia is in need of a thorough taxonomic revision. The generic name is often incorrectly spelled *Boerhaavia* in the older literature (Bogle, 1974).

1. Boerhavia erecta L. (spiderling)
Pl. 459 a, b; Map 2085

Plants annual. Stems 20–100 cm long, erect or ascending, glabrous or minutely hairy toward the base, the nodes sometimes with a sticky band. Leaf blades 2–7 cm long, the lowermost broadly ovate, progressively reduced (ultimately linear) toward the stem tip, the tip rounded to pointed, the base rounded or narrowed to truncate, the margins entire or somewhat undulate, glabrous or sparsely and minutely hairy, the lower surface pale green. Flowers sessile or irregularly umbellate in small clusters of 2–6 at the ends of inflorescence branches, subtended by inconspicuous free bracts 0.6–1.2 mm long, these usually shed early. Perianth 1.0–1.5 mm long, white, sometimes tinged with pink, glabrous, the expanded portion bell-shaped at flowering. Fruits (including the leathery perianth tube) 3–4 mm long, club-shaped or narrowly obtriangular in outline, bluntly 5-angled, glabrous. July–October.

Introduced, known only from a single collection from the city of St. Louis (widespread in tropical and warm-temperate regions around the world, northward into the southwestern U.S.; introduced eastward to South Carolina). Railroads.

The limits of the native distribution of this weedy pantropical species are unclear as it inhabits disturbed habitats everywhere it grows. Steyermark (1963) noted that it is used as a potherb in the American tropics.

2. **Mirabilis** L. (umbrellawort, four-o'clock)

Plants perennial herbs, sometimes slightly woody at the base, the roots woody (somewhat tuberous in *M. nyctaginea*). Stems erect or ascending, glabrous or hairy, sometimes

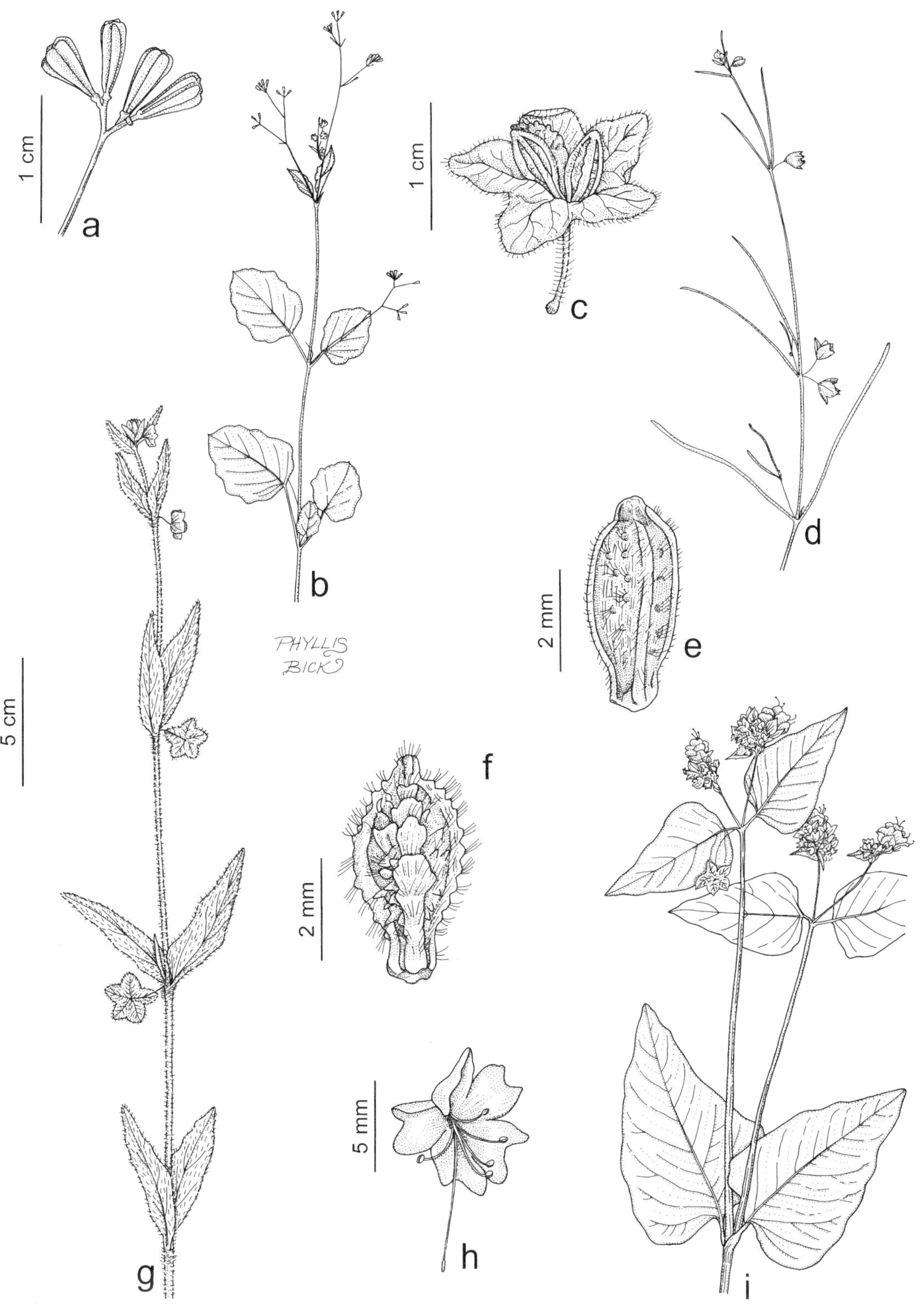

Plate 459. Nytaginaceae. *Boerhavia erecta*, **a)** cluster of fruits, **b)** fertile stem. *Mirabilis linearis*, **c)** cluster of fruits in involucre, **d)** fertile stem, **e)** fruit. *Mirabilis albida*, **f)** fruit, **g)** fertile stem, **h)** flower. *Mirabilis nyctaginea*, **i)** fertile stem.

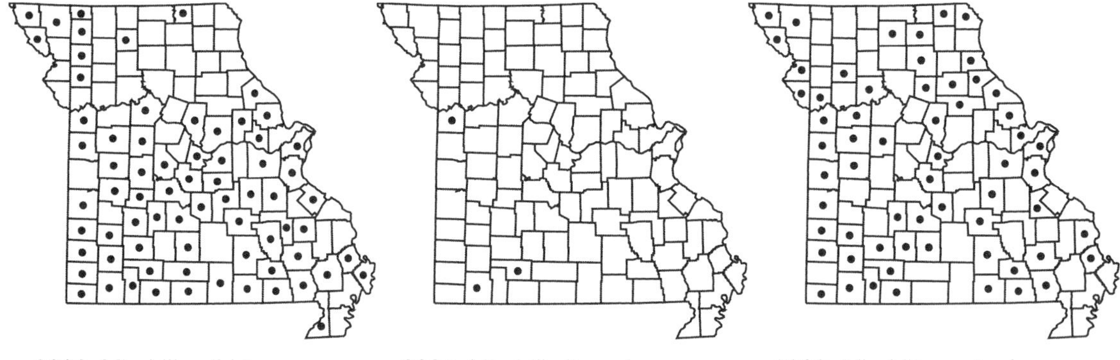

2086. Mirabilis albida 2087. Mirabilis linearis 2088. Mirabilis nyctaginea

glaucous. Leaf blades variously shaped, the margins entire or nearly so, glabrous or hairy. Flowers sessile or minutely stalked in small clusters at the ends of inflorescence branches, the overall inflorescences terminal and appearing as dense clusters, umbels, or less commonly small panicles, and/or sometimes as solitary, long-stalked flower clusters in the leaf axils. Each flower cluster with (1)2–6 flowers, subtended by a calyxlike involucre of fused bracts, this broadly bell-shaped at maturity, with 5 broad shallow lobes, persistent and becoming enlarged, somewhat flattened, and somewhat papery at fruiting. Perianth 5–10 mm long, white, pink, or reddish purple, the expanded portion bell-shaped to saucer-shaped at flowering, the lobes notched at the tip. Fruits (including the hardened perianth tube) narrowly obovoid to narrowly ellipsoid, bluntly 5-angled or ribbed, variously ornamented and hairy. Forty-five to 60 species, North America to South America, Asia.

Mirabilis jalapa L. (common four-o'clock) is a perennial neotropical species that is commonly cultivated in the United States as an annual bedding plant. It has 1-flowered, deeply 5-lobed involucres and flowers with the perianth 3–6 cm long. Because this species has tuberous roots and can seed itself, it may eventually be recorded as an escape from cultivation in Missouri.

The Missouri species are part of a confusing complex of about 25 species that is treated as *Oxybaphus* L'Hér. ex Willd. in some of the older literature, one of several genera sometimes segregated from *Mirabilis*. These species have broadly bell-shaped shallowly lobed involucres that tend to become enlarged at fruiting and subtend (1)2–6 flowers, flowers with the calyx tube not elongated, and fruits with the outer surface tending to turn gelatinous when wet. The common name "four-o'clock" refers to the flowers opening in late afternoon, and our species tend to begin flowering from late afternoon to early evening, with flowers withering early the following morning. Taxonomy of the group is based mostly on characters of the fruits.

1. Leaves (except the uppermost ones) with petioles 1–5 cm long, the blades oblong-ovate to triangular-ovate, broadly rounded to truncate or cordate at the base . . . . 3. M. NYCTAGINEA
1. Leaves sessile or with petioles less than 1 cm long, the blades linear to lanceolate or oblong-elliptic, rarely narrowly ovate, narrowed or tapered at the base
    2. Fruits noticeably warty on the angles and sides, each tubercle with a tuft of short hairs at the tip, otherwise minutely hairy . . . . 1. M. ALBIDA
    2. Fruits evenly and finely short-hairy, the angles lacking tubercles, the sides with fine cross-wrinkles . . . . 2. M. LINEARIS

1. Mirabilis albida (Walter) Heimerl (white four-o'clock, pale umbrellawort, hairy four-o'clock)

Oxybaphus albidus (Walter) Sweet

Pl. 459 f–h; Map 2086

Stems 20–100 cm long, glabrous to densely hairy, the hairs sometimes in 2 vertical lines, when hairy sometimes also glandular, when glabrous appearing light brown, tan or silvery-whitened. Leaves sessile or with petioles to 0.8 cm long. Leaf blades 3–10(–12) cm long, linear to more commonly lanceolate or oblong-elliptic, rarely narrowly ovate, narrowed or tapered at the base, narrowed or abruptly tapered to a rounded or bluntly pointed (rarely sharply pointed) tip, glabrous to densely hairy, sometimes also somewhat glandular, when glabrous the undersurface pale or glaucous. Inflorescences terminal and axillary or sometimes only axillary. Involucres 4–5 mm long at flowering, becoming enlarged to 8–12 mm long at fruiting, glabrous or sparsely to densely hairy on the surfaces, sometimes also glandular, with (1)2–5 flowers. Perianth 6–10 mm long, white to pink. Fruits (including the hardened perianth tube) 4–6 mm long, noticeably warty on the angles and sides, each tubercle of the angles with a tuft of short hairs at the tip, otherwise minutely hairy, olive green to dark brown. $2n$=58. May–October.

Scattered nearly throughout Missouri, but apparently absent from most of the northeastern quarter of the state (eastern [mostly southeastern] U.S. and adjacent Canada west to North Dakota, Nebraska, and Texas). Upland prairies, sand prairies, glades, tops and exposed ledges of bluffs, banks of streams, and less commonly savannas and openings of dry to mesic upland forests; also pastures, roadsides, railroads, and open disturbed areas.

Mirabilis albida is quite variable in most vegetative features, including hairiness. Steyermark (1963) and most other earlier authors accepted *M. hirsuta* (Pursh) MacMill. (hairy four-o'clock) as a species separate from *M. albida* (white four-o'clock, pale umbrellawort), based on its densely hairy leaves and stems (vs. glabrous or nearly so). B. L. Turner (1993b), who treated the genus in Texas, accepted this species with reservation and remarked that, "...I suspect that, as treated by most American workers, it is a hodge-podge of hirsute specimens belonging to several species, mainly *M. albida* and *M. nyctaginea*. For example, Steyermark (1963), in his *Flora of Missouri*, retained the species, but it seems clear from his key and distribution maps that it might be better treated as a leaf form of *M. albida*." In his studies of the genus for the Flora of North America Project, Spellenberg (2003) also noted that fruits of plants attributed to *M. hirsuta* always seem to resemble those of other species more common in the region, and placed *M. hispida* into synonymy under *M. albida*, based on examination of the type specimen.

2. Mirabilis linearis (Pursh) Heimerl **var. linearis** (narrow-leaved four-o'clock, narrow-leaved umbrellawort)

Oxybaphus linearis (Pursh) B.L. Rob.

Pl. 459 c–e; Map 2087

Stems 20–100 cm long, glabrous or less commonly sparsely hairy and glandular toward the tip, glaucous and/or silvery-whitened. Leaves sessile or with petioles to 0.4 cm long. Leaf blades 3–10 cm long, linear to linear-lanceolate, narrowed or tapered at the base, narrowed or tapered to a rounded to sharply pointed tip, glabrous or rarely glandular, usually glaucous. Inflorescences terminal and axillary or sometimes only axillary. Involucres 3.8–4.2 mm long at flowering, becoming enlarged to 6–10 mm long at fruiting, hairy and usually also glandular on the surfaces, with (1–)3(4) flowers. Perianth 8–10 mm long, pink to reddish purple, rarely white. Fruits (including the hardened perianth tube) 4–5(–7) mm long, evenly and finely short-hairy, the angles lacking tubercles, the sides with fine cross-wrinkles, olive green to dark brown. $2n$=52. May–October.

Uncommon in southwestern Missouri and introduced in Jackson County (Minnesota to Montana south to Texas and Arizona, rarely introduced farther east; Mexico). Glades, upland prairies, tops of bluffs, and openings of dry upland forests, often on calcareous substrates.

Gleason and Cronquist (1991) considered this species to be introduced in Missouri, but it is here interpreted as occurring natively in southwestern Missouri (although the status there requires confirmation). The few specimens available were collected more than 100 years ago and contain few habitat data. The habitats listed above were taken from Steyermark (1963), but some of the specimens upon which he circumscribed this species' range in Missouri were misdetermined collections of *M. albida*. Narrow-leaved specimens of *M. albida* routinely have been misdetermined as *M. linearis*, but this species may be differentiated reliably based on the fruit characters in the key. In addition, *M. linearis* has thicker, somewhat succulent leaves that tend to wrinkle upon drying.

Spellenberg (2002, 2003) divided *M. linearis* into three varieties, with only var. *linearis* reported from Missouri. The other two varieties occur to the southwest of Missouri : var. *decipiens* (Standl.) S.L. Welsh differs from var. *linearis* in its slightly broader leaves and deeper pink corollas; var.

2089. Nuphar advena

2090. Nymphaea odorata

2091. Chionanthus virginicus

subhispida (Heimerl) Spellenb. differs from the other two varieities in its more densely pubescent stems with somewhat longer hairs.

3. Mirabilis nyctaginea (Michx.) MacMill. (wild four-o'clock, heart-leaved umbrellawort)

Oxybaphus nyctagineus (Michx.) Sweet

Pl. 459 i; Map 2088

Stems 30–120 cm long, glabrous or sparsely hairy, sometimes slightly glaucous. Leaves (except the uppermost ones) with petioles 1–5 cm long. Leaf blades 3–15 cm long, oblong-ovate to triangular-ovate, broadly rounded to more commonly truncate or cordate at the base, narrowed or abruptly tapered to a bluntly or sharply pointed tip, glabrous or nearly so. Inflorescences terminal. Involucres 5–6 mm long at flowering, becoming enlarged to 10–15 mm long at fruiting, glabrous or sparsely hairy on the surfaces, densely hairy along the margins, with 3–6 flowers. Perianth 8–10 mm long, pink to reddish purple. Fruits (including the hardened perianth tube) 4–6 mm long, densely hairy and roughened or finely warty on the sides and ribs, grayish brown to less commonly nearly black. $2n$=58. May–October.

Scattered nearly throughout the state (eastern U.S. and adjacent Canada west to Montana, Colorado, and New Mexico; Mexico). Upland prairies and banks of streams and rivers; also pastures, roadsides, railroads, and disturbed areas.

NYMPHAEACEAE (Water Lily Family)

Plants perennial, herbaceous aquatics, with extensive, branching rhizomes, rooting at and between the usually closely spaced nodes and sometimes bearing tubers (produced in autumn), often with lacticifers (latex canals), but the sap generally clear or at least not milky. Leaves spirally alternate, long-petiolate (sometimes short-petiolate in submerged leaves or in plants stranded on mud), the petiole to 2 m or more, in some leaves reaching the water surface or slightly emergent, sometimes with reddish or brownish streaks or lines. Stipules sometimes produced, when present represented by often relatively conspicuous, sheathing scales, these sometimes fused into a single unit. Leaf blades attached at the base of a deep sinus (peltate elsewhere), variously shaped, the margin entire but sometimes slightly undulate, usually with a narrow, pale to yellowish or brownish differentiated band, the upper surface flat (the margins rarely curled upward slightly), usually with a noticeable scar at the petiolar attachment, glabrous, often somewhat shiny, rarely reddish-tinged, the undersurface green or slightly to strongly reddish- or purplish-tinged, also sometimes with small brown spots, glabrous or less commonly short-hairy, the venation usually with a single pronounced midvein, the secondary veins numerous, pinnate from the midvein or partially palmate from the petiolar attachment, branched dichotomously 2 or more times above the midpoint, connected by a network of finer crossveins. Inflorescences of solitary flowers, these usually long-stalked directly from the nodes of the rhizome, floating or short-emergent, hypogynous or perigynous (epigynous elsewhere), perfect, actinomorphic. Calyces of 4–8(–14) free sepals, some or all of these usually petaloid in *Nuphar*, variously shaped. Corollas of numerous free petals, these spirally arranged and overlapping, large and showy or (in *Nuphar*) inconspicuous and scale-like, variously colored, sometimes grading into the stamens. Stamens numerous, arranged in a dense, overlapping spiral, the innermost and/or outermost sometimes nonfunctional (staminodes), the filament usually flattened, strap-shaped, and 3-nerved, the anther often not sharply differentiated, mostly appearing more or less embedded along the sides or upper surface of the apical portion of the filament, yellow, lacking a differentiated terminal appendage. Pistil 1 per flower, of 5 to numerous fused carpels, the superior or half-inferior (fully inferior elsewhere) ovary with 5 to numerous locules, constricted slightly at the tip below a flattened, expanded, disc-shaped apex, this with a radial pattern of raised stigmatic regions or deep lobes. Placentation diffuse (laminar), the numerous ovules attached all over the partitions between locules. Fruits berrylike or sometimes more or less capsular, usually somewhat spongy or leathery at maturity, indehiscent, or irregularly dehiscent near the base with age. Seeds numerous, narrowly to broadly ovoid, variously colored, sometimes (in *Nymphaea*) with a membranous, saclike aril. Six genera, about 65 species, nearly worldwide.

The genera *Brasenia* and *Cabomba* form an easily distinguished group that is closely related to, but well-separated from the Nymphaeaceae. Some authors have continued to treat these two groups as a single family, Nymphaeaceae (Judd et al., 2008), but many others treat *Brasenia* and *Cabomba* in the segregate family Cabombaceae (Les et al., 1999), a practice followed in the present work. Steyermark (1963) and many earlier authors also included the genus *Nelumbo* in the traditional concept of Nymphaeaceae, but that genus has been shown to be more closely related to a small group of families that includes the Platanaceae (sycamores, plane trees). For more details see the treatment of the Nelumbonaceae and Judd et al. (2008).

1. Sepals 6(–9), ascending to somewhat spreading, remaining cupped around the pistil at flowering; petals inconspicuous, reduced to scalelike or stamenlike organs positioned below the stamens, shorter than the sepals; stigmatic disc not lobed (but with raised, linear to narrowly oblong stigmatic areas); leaf blade with the venation mostly pinnate . . . . 1. NUPHAR
1. Sepals 4, they and the petals spreading to reflexed, individually flat to somewhat concave but not cupped around the pistil at flowering; petals showy, slightly shorter to slightly longer than the sepals; stigmatic disc strongly radially lobed; leaf blade with the venation palmate but also pinnate along the midvein . . . . . 2. NYMPHAEA

1. **Nuphar** Sm. (cowlily, spatterdock, yellow pond lily, yellow water lily) (Padgett, 2007)

Seven species, U.S. (including Alaska), Canada, Mexico, Caribbean Islands, Europe, Asia, Africa.

Some species of *Nuphar* are cultivated as ornamentals in water gardens and ponds.

1. Nuphar advena (Aiton) W.T. Aiton

N. lutea (L.) Sm. ssp. *advena* (Aiton) Kartesz & Gandhi

Pl. 460 b, c; Map 2089

Plants mostly emergent aquatics (sometimes floating-leaved or submerged), sometimes stranded on mud. Rhizomes relatively stout, sometimes somewhat flattened, light yellow, not producing tubers. Leaves submerged or more commonly floating or emergent; the submerged (overwintering and juvenile) leaves membranous (the margins appearing moderately to strongly undulate or crisped), short-petiolate; the floating and emergent leaves leathery and mostly long-petiolate (shorter in plants stranded on mud). Leaf blades 6–40 cm long and wide, nearly circular to ovate or broadly elliptic-ovate, rounded at the tip, attached in the sinus of a V-shaped, deeply cordate base, the lobes rounded at the tip, mostly bright green (often olive green to yellowish green in submerged leaves), the upper surface (except in submerged leaves) usually shiny, the undersurface glabrous or occasionally sparsely short-hairy, the venation mostly pinnate. Flowers emergent or appearing floating, 1.8–4.5 cm in diameter when open, hypogynous, the stalk relatively stout, not coiling as the fruits mature, glabrous or occasionally sparsely hairy. Sepals 6(–9), 1.0–2.3 cm long, elliptic to oblong-elliptic, broadly ovate or nearly circular, rounded to truncate or very shallowly notched at the tip, strongly concave, ascending to somewhat spreading and remaining cupped around the pistil at flowering, the outer 3 or 4 sepals smaller, green or reddish-tinged on one or both surfaces, the inner 3–5 sepals larger, green (occasionally reddish-tinged) toward the base and on the outer surface, yellow or sometimes reddish-tinged above the often green basal portion on the inner surface, more or less persistent at fruiting. Petals numerous, inconspicuous, reduced to relatively thick, oblong, scale-like organs positioned below and similar in size and shape to the stamens, shorter than the sepals. Stamens 3–9 mm long, mostly truncate at the tip. Ovary superior, the perianth and stamens attached at its base. Stigmatic disc 6–28 mm in diameter, circular, unlobed, with 8–24, raised, linear to narrowly oblong or narrowly lanceolate, stigmatic regions. Fruits capsular, 1.5–5.5 cm long, globose to ovoid, somewhat ribbed longitudinally, green or reddish-tinged, irregularly dehiscent around the basal portion with age. Seeds obovoid, 3–6 mm long, the surface smooth, green to brown, lacking an aril. $2n=34$. May–October.

Scattered in the Ozark, Ozark Border, and Unglaciated Plains Divisions, north locally to Lincoln County and southeast locally to Dunklin County (eastern U.S. west to Wisconsin and Texas; Mexico, Caribbean Islands). Ponds, lakes, slow-moving to stagnant portions of streams, rivers, and spring branches, sloughs, and fens.

C. R. Robertson (1889a) observed Illinois plants of *N. advena* and suggested that pollination was mainly by bees and flies, with beetles consuming floral tissues but not effective pollinators. In contrast, Schneider and Moore (1977) studied Texas plants and concluded that beetles were the dominant pollinator, with flies and bees also contributing to pollination. Lippok et al. (2000) studied pollination and insect visitation in *N. advena*, including a population of ssp. *ozarkana* in Butler County. They documented visitation by a number of bees, flies, and beetles, all of which served as pollinators. These authors concluded that the relative abundance of local pollinators was a more important factor than the types of insect in determining which insects accounted for the most pollination. Flowers of *Nuphar* open only slightly to uncover the stigma during their first day and are then functionally pistillate. They close in the evening, often trapping beetles in the flower overnight. The second day, the flowers open more fully and are functionally staminate. Flowers continue to open for one to three subsequent days and presumably continue to shed pollen until the point when the ovary begins to swell and the perianth begins to decompose (Schneider and Moore, 1977).

Until recently, species limits in the *N. lutea* complex were controversial, with some authors preferring to consider the North American and Eurasian populations part of the same species under the name *N. lutea* (L.) Sm. (Beal, 1956). Padgett et al. (1999) and Padgett (2007) provided convincing evidence from morphological and molecular analyses that the New World plants are not as closely related to those in the Old World as previously thought and that they should be treated as two separate species. A number of infraspecific taxa also have been accepted over time; for example, Beal (1956) recognized seven subspecies as occurring in North America, some of which subsequently have been considered full species. Most recently, Padgett (2007) restricted *N. advena* to only four subspecies, the widespread ssp. *advena* and three regional endemics, one of which occurs in Missouri. The other two subspecies, ssp. *orbiculata* and ssp. *ulvacea*, are endemic to northern Florida and adjacent portions of Alabama and Georgia. The former has nearly circular, deeply notched leaf blades that are densely hairy beneath and relatively large flowers; the latter has more or less lanceolate, shallowly notched leaf blades that are glabrous beneath and relatively small flowers.

1. Sepals green to yellowish green; open flowers mostly 2.0–4.5 cm in diameter; fruits 1.9–5.5 cm long, mostly ovoid, usually green at maturity .... 1A. SSP. ADVENA
1. Sepals reddish- to purplish-tinged on the outer surface; open flowers mostly 1.8–3.0 cm in diameter; fruits 1.5–2.5 cm long, globose to broadly ovoid, usually strongly reddish-tinged at maturity .... 1B. SSP. OZARKANA

1a. ssp. advena

N. lutea ssp. *macrophyllum* (Small) E.O. Beal

Submerged (overwintering) leaves produced uncommonly. Flowers mostly 2.0–4.5 cm in diam-

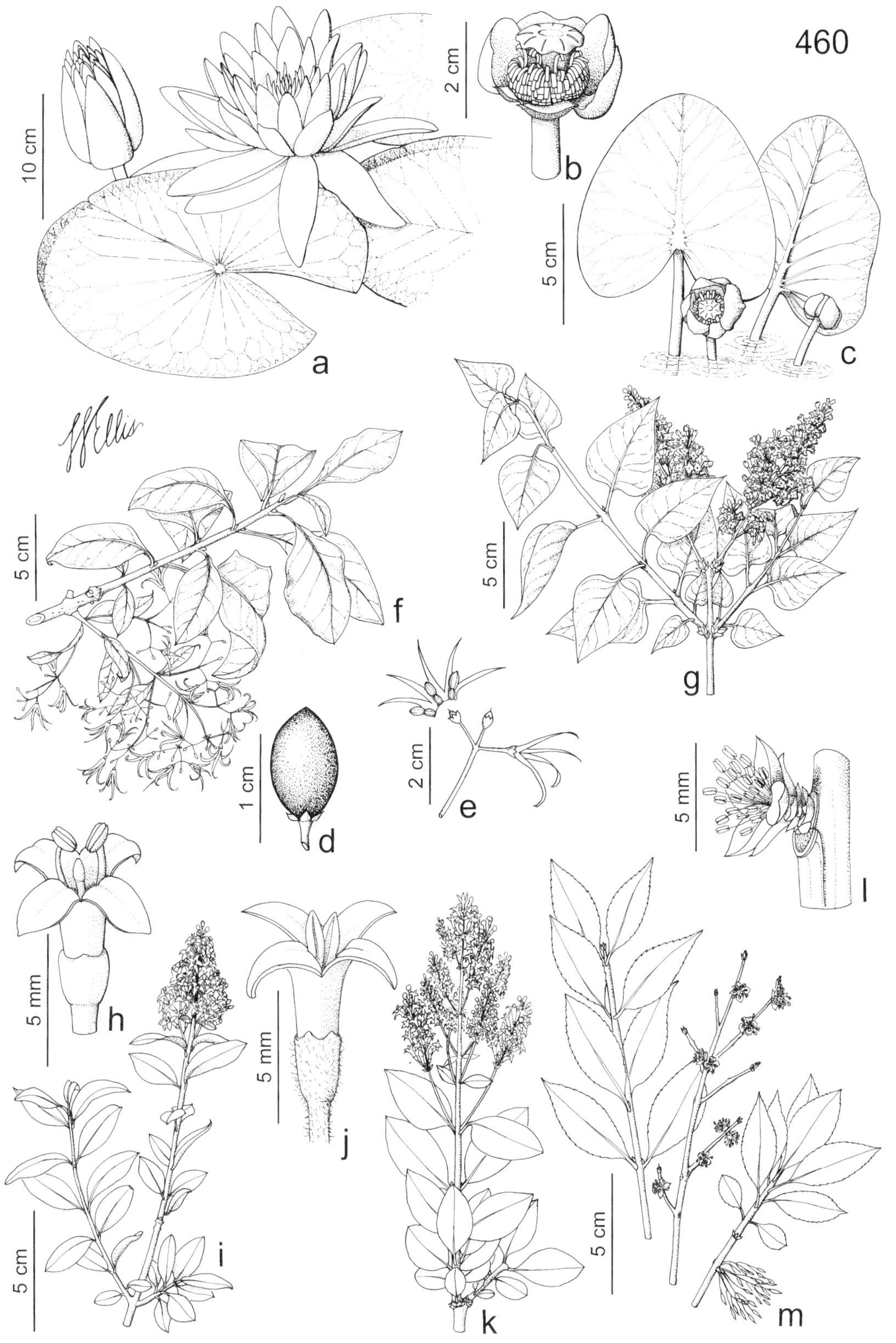

Plate 460. Nymphaeaceae. Oleaceae. *Nymphaea odorata*, **a)** habit. *Nuphar advena*, **b)** dissected flower, **c)** habit. *Chionanthus virginicus*, **d)** fruit, **e)** cluster of flowers (one of them dissected), **f)** fertile branch. *Syringa vulgaris*, **g)** fertile branch. *Ligustrum vulgare*, **h)** flower, **i)** habit. *Ligustrum obtusifolium*, **j)** flower. *Ligustrum ovalifolium*, **k)** habit. *Forestiera acuminata*, **l)** flower, **m)** vegetative, flowering, and fruiting branches.

eter when open. Sepals green to yellowish green on both surfaces, occasionally reddish-tinged on the inner surface. Stigmatic disc 18–50 mm in diameter, with 10–24 stigmatic regions, these linear to narrowly oblong or narrowly lanceolate, mostly terminating 1–3 mm from the disc margin. Fruits 1.9–5.5 cm long, mostly ovoid, usually green at maturity. May–October.

Scattered in the Ozark, Ozark Border, and Unglaciated Plains Divisions, north locally to Lincoln County and southeast locally to Dunklin County (eastern U.S. west to Wisconsin and Texas; Mexico, Caribbean Islands). Ponds, lakes, slow-moving to stagnant portions of streams and rivers, and sloughs.

1b. ssp. ozarkana (G.S. Mill. & Standl.) Padgett

N. lutea ssp. *ozarkana* (G.S. Mill. & Standl.) E.O. Beal

N. ozarkana (G.S. Mill. & Standl.) Standl.

Submerged (overwintering) leaves produced commonly. Flowers mostly 1.8–3.0 cm in diameter when open. Sepals green to yellowish green or sometimes reddish-tinged on the inner surface, reddish- to purplish-tinged on the outer surface. Stigmatic disc 9–12 mm in diameter, with 8–12 stigmatic regions, these linear to less commonly narrowly oblong, mostly terminating 1.0–1.5 mm from the disc margin. Fruits 1.5–2.5 cm long, globose to broadly ovoid, usually strongly reddish-tinged at maturity. May–October.

Scattered in the Ozark Division (Arkansas and Missouri). Ponds, lakes, slow-moving to stagnant portions of streams, rivers, and spring branches, sloughs, and fens.

Although ssp. *advena* is more widely distributed in Missouri and elsewhere in the United States, ssp. *ozarkana* is more commonly encountered in Missouri and is the characteristic species of spring-fed Ozark streams and rivers, as well as fens. The ssp. *ozarkana* is nearly endemic to the Ozarks, with a few occurrences farther south in Arkansas.

2. Nymphaea L. (water lily)

Thirty-five to 40 species, nearly worldwide.

Padgett (2001) reported a small population of the canary water lily, *N.* ×*marliacea* Wildsmith (reported as possibly cv. 'Chromatella'), in a small pond in Greene County. It is a popular plant in horticulture that arose through crosses between *N. alba* L. (European white water lily) and *N. mexicana* Zucc. (yellow water lily) and is propagated mainly by the slender, elongate tubers, which are produced in dense clusters from the tips of the rhizome branches in the autumn. This taxon differs from the water lilies native to Missouri in its yellow petals and purple mottling on the upper leaf surface. Numerous other water lilies are cultivated in ponds around the state, including cultivars with various flower colors ranging from light to bright pink and purple to blue, as well as various leaf shapes and colorings. However, to date, none of these has been documented as an escape in more natural environments.

1. Nymphaea odorata Aiton (fragrant water lily, white water lily)

N. odorata var. *gigantea* Tricker

N. odorata f. *rubra* Guillon

N. odorata ssp. *tuberosa* (Paine) Wiersema & Hellq.

N. tuberosa Paine

Pl. 460 a; Map 2090

Plants mostly floating-leaved aquatics (sometimes emergent or submerged, particularly in dense populations), rarely stranded on mud. Rhizomes relatively stout, not flattened, white to off-white (this often obscured by blackish hairs), sometimes producing tubers, these actually simple or compound branches to 8 cm that are strongly constricted at the base and easily dispersed. Leaves submerged, floating or occasionally emergent; the submerged leaves similar to the floating ones but usually somewhat more membranous; the floating and emergent leaves leathery and mostly long-petiolate. Leaf blades 6–20(–30) cm long and wide, circular or nearly so, rounded at the tip, attached in the sinus of a narrowly V-shaped, deeply cordate base, the lobes asymmetrically angled at the tip or sometimes with a small, slender, tapered extension at the very tip, both surfaces glabrous, the upper surface bright green to dark green, usually somewhat shiny, the undersurface often strongly tinged with red to purplish brown, the venation palmate but also pinnate along the midvein. Flowers floating or less commonly slightly emergent, mostly 6–19 cm in diameter when open, deeply perigynous, the stalk usually relatively slender, becoming coiled and submerged as the fruits mature, glabrous. Sepals 4, 2.5–8.0 cm long, ovate or narrowly ovate, rounded at the tip, flat to shallowly concave, spreading to reflexed and not cupped around the pistil at flowering, often reddish- to purplish-tinged on the outer surface, the margins thin and usually relatively pale, not persistent. Petals numerous, showy, spreading to reflexed, 3–9 cm long, slightly shorter to slightly longer than the sepals, ovate to narrowly elliptic or elliptic-lanceolate, flat to slightly concave, rounded to narrowly rounded at the tip, grading into the stamens (the innermost staminodial petals narrowly oblanceolate), white or rarely pink, not persistent. Stamens 10–40 mm long, rounded to bluntly pointed at the tip. Ovary nearly inferior, the perianth and stamens attached to a hypanthium that extends nearly to the ovary tip and is fused to the ovary sides. Stigmatic disc 5–22 mm in diameter, with 10–25 slender lobes, these stigmatic toward the tip. Fruits berrylike, 2.5–4.5 cm long, depressed-globose, shallowly and broadly concave at the tip, not ribbed, but the lateral walls with transverse scars from the shed perianth parts and stamens, green, indehiscent. Seeds ovoid, 1.5–4.5 mm long, the surface smooth, olive green to orangish brown, covered by a membranous, bag-shaped, white aril. $2n$=56, 84. May–September.

Scattered sporadically nearly throughout the state (nearly throughout the U.S.; Canada, Mexico, Central America, Caribbean Islands; introduced in South America). Ponds, lakes, sloughs, and slow-moving to stagnant portions of rivers; also ditches.

C. R. Robertson (1889a, b) observed visitation of *Nymphaea* flowers in Illinois by a variety of bees, flies, and beetles. Schneider and Chaney (1981) observed a similar range of insects visiting flowers in Texas, but concluded that the most important pollinator is a small halictid bee, *Lasioglossum versatum* Robertson. Flowers of the white water lily open in the morning for usually three successive days and close by mid-afternoon. During the first day, the inward-arched stigmatic lobes are receptive and more or less erect, guiding the insect to a small pool of liquid secreted onto the concave tip of the ovary. Insects frequently drown in this liquid as the flowers close the first afternoon. During subsequent days, the ovary tip is dry and the stigmatic lobes are curved inward strongly over the tip of the ovary and are no longer receptive, but the stamens shed pollen (Schneider and Chaney, 1981).

The taxonomy of this complex remains controversial. Field studies indicated that many, if not all, of the characters previously thought to separate taxa are at least partially environmentally controlled (G. R. Williams, 1970; Mitchell and Beal, 1979). However, Wiersema and Hellquist (1994) countered anecdotally that plants from populations showing morphological extremes within the complex retained at least some of their characteristics when grown in common culture. Two species have been recognized traditionally, with a number of infraspecific taxa as well. Steyermark (1963) reported the segregate species, *N. tuberosa*, from two historical specimens collected in Dunklin and St. Louis Counties, and Wiersema and Hellquist (1997) and K. Woods et al. (2005a) suggested an additional potential distribution in northernmost Missouri (as *N. odorata* ssp. *tuberosa*). In Steyermark's key, this taxon was separated from the more widespread *N. odorata* in the strict sense by its nonfragrant, larger flowers with relatively broad staminal filaments, as well as leaves with the undersurface green and the petioles green or brownish-striped toward the tip. None of these characters appears to hold up across the supposed ranges of the two taxa (G. R. Williams, 1970; Voss, 1985). The character of whether short tuberous rhizome branches that are constricted at the base are produced commonly (in *N. tuberosa*) or rarely (in *N. odorata* in the strict sense), may be more constant, but rhizomes are present on only a small proportion of the specimens in herbaria. The existence of cultivars and hybrids that have been planted in natural bodies of water or escaped into them has further confused the issue. Based on their examination of herbarium specimens, Wiersema and Hellquist (1994, 1997) attempted to justify a compromise approach by treating the two taxa as subspecies of *N. odorata*, because at least some plants can be assigned to one or the other readily. However, even these authors admitted that many specimens are intermediate for one or more features. Thus, it does not seem practical to separate the two water lilies taxonomically at any level. Most recently, two different molecular approaches and a morphometric study (K. Woods et al., 2005a, b) suggested that there was at least some genetic separation between the two morphological extremes within the *N. odorata* complex, but these studies also provided evidence of hybridization between the two and the existence of some populations that are intermediate or in which the morphological and molecular data are discordant. For those wishing to attempt to segregate two subspecies within *N. odorata*, the quantitative studies of morphological variation by K. Woods et al. (2005a) suggested a focus on the following character set to separate ssp. *odorata* and ssp. *tuberosa*, respectively: 1) leaf blade shape (very broadly oval, slightly longer than wide vs. circular or nearly so, about as long as wide); 2) petiole markings (generally lacking vs. green or brownish-striped); and the tips of the basal lobes of the leaf blades (rounded to bluntly pointed vs. sharply pointed).

OLEACEAE (Olive Family)

Contributed by Timothy E. Smith and George Yatskievych

Plants shrubs or trees (lianas elsewhere), often incompletely monoecious or dioecious (with at least a few perfect flowers among the pistillate and/or staminate ones. Branches sometimes angled or ridged, at least when young. Leaves opposite (rarely subopposite), sessile or petiolate. Stipules absent. Leaf blades simple to ternately or pinnately compound. Inflorescences terminal and/or axillary, variously dense to loose clusters, racemes, or panicles, rarely reduced to solitary flowers, the axes not coiled at the tip, the inflorescence branch points often with small, leaflike or scalelike bracts. Flowers actinomorphic, hypogynous, perfect or imperfect, lacking bracts (except in *Jasminum*). Calyces small or minute, less commonly absent, when present shallowly to more commonly deeply 4–6-lobed, sometimes merely toothed or unlobed and truncate, usually persistent at fruiting (shed early in some *Fraxinus* species). Corollas absent or more commonly present and shallowly to deeply 4–6-lobed. Stamens 2 (rarely 1 or 3 in *Forestiera* or 3–5 elsewhere), the filaments attached at or above the base of the corolla tube, the anthers not exserted or short-exserted, attached at their bases, dehiscing by a pair of opposite, longitudinal slits, sometimes with a small, toothlike, sterile, terminal extension, usually yellow. Pistil 1 per flower, of 2 fused carpels. Ovary superior, sometimes with a small nectar disc surrounding the base, 2-locular, with usually 2 ovules per locule (sometimes numerous in *Forsythia*), the placentation axile. Style 1 (rarely absent), either unbranched with a single, capitate stigma or more commonly forked at the tip with a pair of small stigmas, usually withered or absent at fruiting. Fruits drupes, berries, samaras, or longitudinally dehiscent capsules, variously shaped, usually glabrous. Seeds 1(–4) per fruit (usually 2 in *Syringa*; numerous in *Forsythia*). About 25 genera, about 600 species, nearly worldwide.

The family Oleaceae contains a number of economically important species. Several genera are cultivated as ornamentals, including at least some species in all of those treated in the present flora, as well as *Osmanthus* Lour. (osmanthus, fragrant olive) and *Noronhia* Stadman ex Thouars (Madagascar olive). *Fraxinus* contains commercially important timber trees whose wood is used in furniture, flooring, veneers, baseball bats, hockey sticks, canoe paddles, other implement handles, and handcrafts. Perhaps the most important species economically is the European olive (*Olea europaea* L.), which is cultivated as an ornamental in warmer climates, but is also the source of olives and olive oil. Some of the introduced species, principally of *Ligustrum*, are considered invasive exotics, especially in the southeastern states.

1. Leaves pinnately compound; fruits slender, elongate samaras . . . . 5. FRAXINUS
1. Leaves simple or ternately compound; fruits drupes, berries, capsules, or, if samaras (in *Fontanesia*), then less than 2 times as long as wide
    2. Leaves or leaflets with the margins toothed or, if entire, then the flowers lacking corollas
        3. Leaves simple, the margins entire or shallowly and bluntly toothed; corollas absent . . . . 3. FORESTIERA
        3. Leaves simple or ternately lobed and/or compound, the leaflet margins irregularly and sharply toothed; corollas present . . . . 4. FORSYTHIA
    2. Leaves or leaflets with the margins entire; flowers producing corollas
        4. Leaves mostly ternately compound (simple leaves often present at the bases of branchlets); corollas 5- or 6-lobed . . . . 6. JASMINUM
        4. Leaves all simple; corollas all or nearly all 4-lobed
            5. Corollas lobed nearly to the base, not forming a noticeable tube
                6. Corollas 12–30 mm long, the lobes linear or narrowly strap-shaped . . . . 1. CHIONANTHUS

6. Corollas 2–3 mm long, the lobes lanceolate to narrowly ovate . . . . 2. FONTANESIA
5. Corollas lobed to about the midpoint or less, forming a prominent tube
   7. Corollas white or occasionally cream-colored; fruits berrylike drupes; petioles of largest leaves to 1–16 mm long . . . . 7. LIGUSTRUM
   7. Corollas lilac, bluish-lavender, or rarely white; fruits capsules; petioles of largest leaves 19–30 mm long . . . . 8. SYRINGA

1. **Chionanthus** L.

About 90 species, eastern North America, eastern Asia, Africa.

1. Chionanthus virginicus L. (fringe tree, old man's beard, grancy gray beard)
Pl. 460 d–f; Map 2091

Plants shrubs or occasionally small trees, 2–5(–10) m tall, dioecious or with some perfect flowers mixed with the pistillate and/or staminate ones. Trunks usually several, mostly spreading, at least toward the base, loosely to strongly ascending toward the tip, the bark light brown to gray, thin and breaking into small plates, becoming thinly ridged or furrowed with age. Twigs stout, gray or tan, glabrous or short-hairy, more or less circular in cross-section, with raised leaf scars and raised, oval, darker brown lenticels. Terminal buds sometimes closely flanked by the uppermost, shorter pair of axillary buds, ovoid to ellipsoid, with several, overlapping, sharply pointed scales that are spreading at the tips, the axillary buds ovoid, with scales that are broadly to bluntly pointed at the tips. Leaves opposite or occasionally subopposite, sometimes appearing clustered toward the branch tips, short-petiolate. Leaf blades simple, 8–20 cm long, 2.5–12.0 cm wide, lanceolate-elliptic to ovate, elliptic, or somewhat obovate, broadly angled to short-tapered to the bluntly to sharply pointed tip, angled or tapered at the base (the upper portion of the petiole often winged), the margins entire but sometimes somewhat wavy, the upper surface dark green and essentially glabrous, the undersurface lighter green or pale and sparsely to densely short-hairy, sometimes mainly along the veins. Inflorescences axillary, many-flowered, usually drooping panicles 10–15 cm long, produced from 1-year-old branches, developing with or before the leaves, the lower branch points with small, leaflike bracts, the flowers with slender stalks 5–7 mm long, slightly fragrant. Calyces 4-lobed, 1–2 mm long, the lobes triangular. Corollas 4(–6)-lobed nearly to the base, 15–30 mm long, more or less bell-shaped, the lobes linear to narrowly strap-shaped, at first greenish white, turning white at maturity. Style 1–2 mm long, with a pair of ascending branches at the tip. Fruits drupes, 10–15 mm long, ellipsoid to ovoid, olive green, turning bluish black or dark purple, usually somewhat glaucous at maturity. $2n$=46. April–May.

Scattered in disjunct portions of southwestern and southeastern Missouri (eastern U.S. west to Missouri and Texas; Canada). Glades, tops of bluffs, banks of streams, and bottomland forests, often on calcareous substrates.

Fringe tree is planted ornamentally in temperate North America and in Europe and is becoming increasingly available at nurseries. In full flower, it rivals any native shrub or small tree in beauty, and it tolerates urban conditions well. Staminate specimens are considered more showy in flower. The fall foliage is deep yellow, often with greenish splotches. Native Americans used the bark in poultices for skin ailments and wounds, and as a fever remedy (Moerman, 1998).

2. **Fontanesia** Labill.
(K.-j. Kim, 1998)

Two species, Europe, Asia.

1. Fontanesia phillyreoides Labill. **ssp. fortunei** (Carrière) Yalt. (Persian lilac)
F. phillyreoides var. *fortunei* (Carrière) Koehne
F. fortunei Carrière
Map 2092

Plants shrubs or less commonly small trees, 1.5–8.0 m tall, dioecious or with some perfect flowers

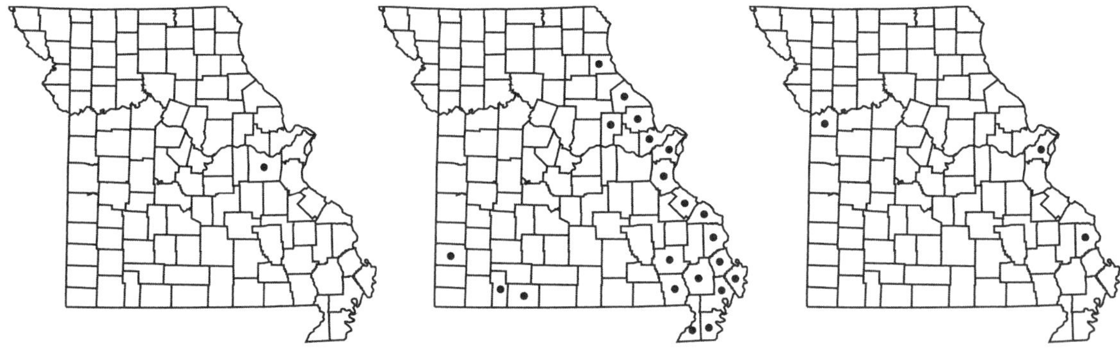
2092. Fontanesia phillyreoides 2093. Forestiera acuminata 2094. Forsythia suspensa

mixed with the pistillate and/or staminate ones. Trunks few several, mostly strongly ascending, the bark light brown to gray, thin and breaking into small plates, becoming thinly ridged or furrowed with age. Twigs relatively slender, purplish to light brown and somewhat shiny, glabrous, more or less 4-angled in cross-section, at least when young (usually with 4 slender ridges), with the leaf scars not or only slightly raised and the lenticels inconspicuous, small, and not raised. Terminal buds narrowly ovoid to ellipsoid, with several, overlapping, sharply pointed scales, the axillary buds ovoid, with scales that are broadly to bluntly pointed at the tips. Leaves opposite, short-petiolate. Leaf blades simple, 2–6(–12) cm long, 0.8–2.5 cm wide, lanceolate to lanceolate-elliptic or occasionally narrowly ovate, angled or more commonly tapered at the sharply pointed tip, mostly angled at the base, the margins entire, the upper surface dark green, glabrous, and often somewhat shiny, the undersurface lighter green and glabrous. Inflorescences terminal, several- to many-flowered, ascending panicles 2–6 cm long developing with or after the leaves, the lower branch points with small, leaflike or scalelike bracts, the flowers with slender stalks 1.5–2.0 mm long, not or slightly fragrant. Calyces deeply 4-lobed, 0.5–0.8 mm long, the lobes ovate. Corollas 4-lobed nearly to the base, 2–3 mm long, more or less bell-shaped, the lobes narrowly ovate to narrowly elliptic or strap-shaped, cream-colored or greenish white. Style 1.5–3.0 mm long, with a pair of ascending branches at the tip. Fruits samaras, 5–9 mm long, flattened and narrowly winged around the rim, elliptic to oblong-elliptic in outline, sometimes minutely notched at the tip, olive green, turning yellowish brown with age, glabrous. $2n=26$. May–June.

Introduced, known thus far from a single site in Franklin County (native of China; introduced uncommonly in the U.S. [Ohio]). Open disturbed areas.

This species is no longer planted extensively in North America because it is less showy than some related genera and not as easily maintained as a hedge. The Missouri plants were discovered in 2006 by Glenn Beffa at the Shaw Nature Reserve in Gray Summit, where a large, old, planted specimen had given rise to a population of escapes, these forming a locally reproducing population.

The two subspecies of *F. phillyreoides* formerly were treated as separate species, but now usually are considered components of a single species (Chang et al., 1996; K.-j. Kim, 1998). The ssp. *phillyreoides* is native to the Mediterranean region in eastern Europe and adjacent Asia, and differs primarily in its angled to only slightly tapered leaf tips. It is not cultivated extensively. The second species in the genus, the recently described *F. longicarpa* K.-J. Kim (Kim 1998), is endemic to a small region in southeastern China. It differs from *F. phillyreoides* in its fruits, which are 1.5–2.5 cm long, slender, more or less unwinged, and forked apically.

3. **Forestiera** Poir.
(C. J. Brooks, 1977)

Eleven to 15 species, North America, Central America, Caribbean Islands.

1. Forestiera acuminata (Michx.) Poir. (swamp privet)

Pl. 460 l, m; Map 2093

Plants shrubs or occasionally small trees, 2–8 m tall, dioecious or with some perfect flowers mixed with the pistillate and/or staminate ones. Trunks often only 1 or few but frequently branched from near the base, erect or irregularly ascending (the branches sometimes arched), the bark brown to dark brown, thin, becoming thinly ridged with age. Twigs relatively slender, gray or light brown to reddish brown, glabrous or short-hairy, more or less circular in cross-section, with raised leaf scars and oval, pale lenticels. Terminal buds often closely flanked by the uppermost, shorter pair of axillary buds, narrowly ovoid to narrowly ellipsoid, with several, overlapping, sharply pointed scales that often are slightly spreading at the tips, the axillary buds broadly ovoid, with scales that are broadly to bluntly pointed at the tips. Leaves opposite or occasionally subopposite, short- to moderately petiolate. Leaf blades simple, 2–9(–11) cm long, 1–3(–4) cm wide, elliptic to ovate, or somewhat rhombic, tapered to the sharply pointed tip, tapered at the base (the upper portion of the petiole often winged), the margins entire or shallowly and bluntly toothed, the upper surface medium green and glabrous or less commonly sparsely to moderately hairy along the midvein, the undersurface lighter green to yellowish green and glabrous. Inflorescences axillary, many-flowered, yellow, those with staminate flowers appearing as dense clusters to 1.5 cm in diameter, those with pistillate flowers appearing as loose clusters or small panicles to 2 cm long (to 3 cm long at fruiting), produced from 1-year-old branches, developing before the leaves or just as the leaves begin to expand, some of the branch points with small, scalelike bracts (these shed early), the flowers sessile or with slender stalks to 2 mm long, not fragrant. Calyces absent or rarely present in pistillate flowers, then 4-lobed or -toothed, 0.5–1.5 mm long, the lobes or teeth triangular. Corollas absent. Style 0.5–1.5 mm long, with a pair of ascending branches at the tip. Fruits drupes, 7–15 mm long, usually slightly flattened, narrowly and often somewhat asymmetrically ellipsoid (often slightly curved), green, turning dark purple to purplish black. $2n=46$. March–May.

Scattered in the eastern portion of the state, mostly in the Mississippi Lowlands Division and counties bordering the Mississippi River (north to Pike County), also uncommon and disjunct in the southwestern portion of the state (Kansas to Texas east to Kentucky, South Carolina, and Florida). Bottomland forests, swamps, and banks of rivers and sloughs; also ditches, railroads, and moist, open, disturbed areas.

The species was reported by Sargent (1922) as occasionally cultivated and hardy at the Arnold Arboretum in Boston. Steyermark (1963) encouraged that it be planted more extensively due to its early blooming. It apparently has not gained much in popularity and is not included in several modern references on cultivated plants. Swamp privet purportedly has close-grained, hard wood that is suitable for use in wood-turning projects.

4. **Forsythia** Vahl (forsythia, golden bell)

Seven to 11 species, Europe, Asia.

1. Forsythia suspensa (Thunb.) Vahl (weeping forsythia)

Map 2094

Plants shrubs, 1–3 m tall, with perfect flowers. Main stems numerous, variously erect and wandlike to strongly arched and sometimes rooting at the tips, the bark light brown to tan, thin, the outer layer usually peeling in thin strips with age. Twigs relatively stout, green to reddish brown, becoming yellowish brown, glabrous, more or less 4-angled in cross-section, at least when young, with raised leaf scars and warty, oval, pale lenticels. Terminal buds ovoid to ellipsoid, with several, overlapping, sharply pointed scales, the axillary buds broadly ovoid, with scales that are bluntly to sharply pointed at the tips. Leaves opposite, short- to moderately petiolate. Leaf blades simple and at least in a few leaves ternately lobed and/or compound, 2–10 cm long, 1.5–5.0 cm wide, the blade or leaflets broadly ovate (in lobed leaves) to ovate or oblong-ovate, angled or tapered to the usually sharply pointed tip, rounded or angled at the base, the margins irregularly and sharply toothed (sometimes sparsely so), the surfaces medium green to yellowish green and glabrous. Inflorescences axillary, of solitary flowers or more commonly small clusters of 2 or 3(–5) flowers, produced from near the branch tip to older portions of branches, developing before the leaves, but often persisting as the leaves begin to expand; bracts absent, the flow-

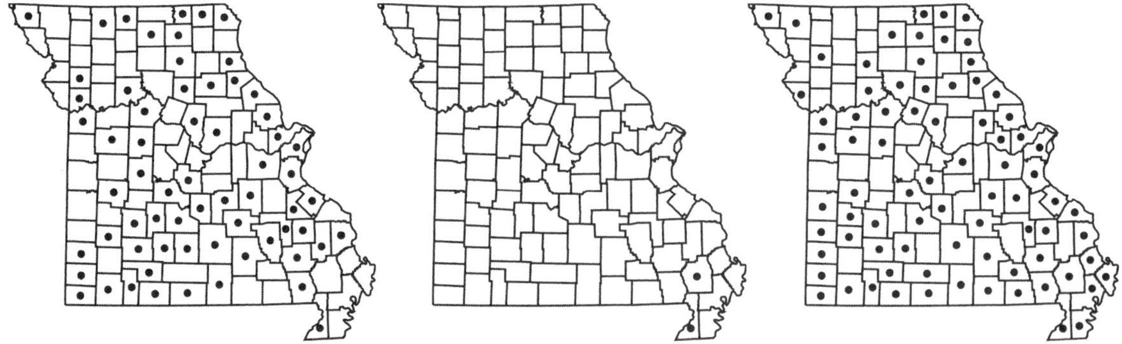

2095. Fraxinus americana 2096. Fraxinus biltmoreana 2097. Fraxinus pennsylvanica

ers with slender stalks 3–6 mm long, not or only slightly fragrant. Calyces deeply 4-lobed, 4–7 mm long, the lobes oblong. Corollas deeply 4-lobed, 15–25 mm long, bell-shaped to more or less saucer shaped, the lobes narrowly oblong to strap-shaped, yellow (the 4–5 mm long tube often with orange stripes along the inner surface). Style 1–2 mm long, with a pair of ascending branches at the tip. Fruits very rare in North American plants, 2-locular, longitudinally dehiscent capsules, 15–25 mm long, narrowly ellipsoid to narrowly ovoid, beaked, brown with scattered, pale lenticels (seeds numerous, minute, narrowly winged). $2n$=28. March–April.

Introduced, known thus far only from St. Louis County (native of Asia; introduced sporadically in mostly the eastern half of the U.S.). Mesic upland forests and banks of streams; also roadsides.

This taxon was first collected in 2004 by Rex Hill and the Botany Group of the Webster Groves Nature Study Society in a large wooded drainage where stem fragments probably washed down the slope from gardens in a subdivision along the adjacent ridge. Hardin (1974) and Chang et al., 1996) noted that each individual of *Forsythia* produces only one of two flower types, either with longer stamens and shorter styles or vice versa. This distylous behavior is correlated with an outcrossing breeding system. Hardin (1974) noted that unless individuals with both flower types are grown in proximity, they cannot effectively become pollinated, and that this might account for the rarity of fruits in North American cultivated and escaped plants.

Other commonly cultivated taxa in the genus include *F. viridissima* Lindl. and its hybrid with *F. suspensa*, *F. ×intermedia* Zabel. Numerous cultivars exist. Plantings may persist for years around old homesites, but escapes from cultivation appear to be rare, considering the popularity of these shrubs for landscaping. *Forsythia viridissima* differs from *F. suspensa* by having chambered pith throughout the younger branches (vs. hollow or with a few irregularly spaced cross-walls along the internodes and solid at the nodes), a somewhat more erect growth form, and in having only simple leaves. A specimen at William Jewell Collage collected in a forested area of Kansas City in 1971 may represent an escape, but the label data are ambiguous. Hybrid plants generally are intermediate, and tend to have chambered pith along the internodes and solid pith at the nodes. Pith characters can be confusing, with chambered, solid, and hollow types present in a single stem.

5. **Fraxinus** L. (ash)
(G. N. Miller, 1955)

Plants trees (rarely shrubs elsewhere), to 40 m tall, dioecious or with some perfect flowers mixed with the pistillate and/or staminate ones (usually with mostly perfect flowers in *F. quadrangulata*; sometimes monoecious elsewhere). Trunk usually 1 (more following damage or upon resprouting after logging), the branches spreading to ascending, the bark gray, relatively thick, developing a network of ridges and diamond-shaped furrows, sometimes becoming scaly with age. Twigs relatively stout, gray to brown, glabrous or hairy, more or less circular to sharply 4-angled in cross-section, with raised leaf scars and inconspicuous to conspicuous, pale lenticels. Terminal buds usually closely flanked by the uppermost, shorter pair of axillary buds, variously shaped, with several, overlapping, sharply pointed scales, the axillary buds also variously shaped. Leaves opposite, short- to moderately petiolate. Leaf blades odd-pinnately compound with mostly 5–11 leaflets, to 40 cm long and 25 cm wide, more or less oval in outline, the leaflets mostly short-stalked, variously shaped, angled or tapered to the sharply pointed tip, rounded or angled at the sometimes asymmetric base, the margins entire or shallowly toothed, the surfaces variously glabrous or hairy, the upper surface medium to dark green, the undersurface pale or lighter green. Inflorescences axillary, many-flowered, appearing yellowish green to green or occasionally purplish red (pistillate) or red to purplish red (staminate, shedding yellow pollen), consisting of fascicles, clusters, or short racemes, these usually grouped into small panicles, those with staminate flowers usually appearing denser and consisting of more capitate clusters than those with pistillate flowers, produced from 1-year-old branches, developing before the leaves or as the leaves expand, some of the branch points with small, scalelike bracts (these shed early), the flowers stalked, not fragrant. Calyces absent or, if present, then sometimes shed early, shallowly to deeply 4-lobed (sometimes cleft nearly to the base along 1 side), 0.5–1.5 mm long (sometimes to 4 mm in pistillate flowers of *F. profunda*), the lobes or teeth narrowly triangular to triangular. Corollas absent (present and with 2–6 petals or deep lobes elsewhere). Style 1–3 mm long, with a pair of ascending branches at the tip. Fruits slender, elongate samaras, usually pendant, 25–80 mm long (including the wing), the body flattened or more commonly turgid, with a conspicuous, slender, sometimes somewhat asymmetric wing at the tip, the wing extending narrowly and laterally along the body, pale green, turning straw-colored to grayish tan or brown with age. Sixty to 65 species, North America, Central America, Caribbean Islands, Europe, Africa, Asia, Malaysia; most diverse in temperate regions.

Ashes are common components of most upland and bottomland forests. The wood is of commercial importance for its great strength and shock resistance, and is used in furniture, flooring, veneers, baseball bats, hockey sticks, canoe paddles, other implement handles, and handcrafts. Moerman (1998) noted that Native Americans also used various parts of the plants medicinally, ceremoniously, and as a minor food source. The windborne pollen of ashes is considered a seasonally important cause of hayfever. White and green ash are both easily transplanted and commonly grown ornamentally, with green ash being used more commonly because of its faster growth. The foliage turns bright yellow in the autumn. Cultivars are available for both species featuring lack of fruit production and, in white ash selections, purple fall foliage. Both species, however are subject to a number of disease and insect pest problems. The Eurasian *Fraxinus ornus* L. (flowering ash, manna ash) occasionally is planted in Missouri for its large, terminal panicles of fragrant flowers with slender, white petals, but is not known to escape from cultivation.

All North American ashes have become seriously threatened in recent years by the spread of a destructive, invasive, wood-boring beetle from Asia called the emerald ash borer (*Agrilus planipennis* Fairmaire). This species was first detected in 2002 in southeastern Michigan and soon after in adjacent Ontario, Canada (Poland and McCullough, 2006). It presumably entered North America as a contaminant in wood packing material or pallets used in the transport of other imports from China, and it has since spread into portions of Illinois, Indiana, Ohio, West Virginia, Virginia, Maryland, and Pennsylvania. The first report for Missouri was in 2008 from the southeastern portion of the state (Wayne County), and by 2012 the pest had been documented from the adjacent counties of Madison and Reynolds, as well as from western Missouri in Platte County. According to the official emerald ash borer website (http://www.emeraldashborer.info), more than 20 million trees have been killed thus far. Adult females of *A. planipennis* deposit eggs in bark crevices of ash trees during the summer. The larvae form elaborate, meandering, S-shaped tunnels under the bark, feeding primarily on the phloem tissues. This results in various symptoms, including branch dieback, peeling and splitting of the bark, sparse foliage, anomalous branching, and eventually death (R. Lawrence, 2005). First- or second-year larvae pupate to form adults that emerge through the bark in the spring, leaving characteristic D-shaped exit holes. Efforts currently are underway to develop biological controls (mainly other insects and fungi that might parasitize emerald ash borer). However, this insect pathogen has demonstrated an ability for relatively rapid long-distance dispersal into new habitats, apparently by the transport of infested ash saplings and firewood by unsuspecting humans (Muirhead et al., 2006).

1. Young twigs 4-sided, sometimes narrowly winged; plants with all or mostly perfect flowers; calyces absent or shed early; samaras with the wing 6–12 mm wide . . . . 5. F. QUADRANGULATA
1. Twigs circular in cross-section, never winged; plants completely or incompletely dioecious; calyces present but often minute, persistent at fruiting; samaras with the wing 3–8 mm wide (7–13 mm in *F. profunda*)
   2. Samara wing extending more than 1/3 of the way along each side of the body, leaflets with the undersurface yellowish green to pale green but not whitened; leaf scars truncate to shallowly and broadly concave on the apical side
      3. Samaras with the wing 3–6 mm wide, extending mostly 1/3–1/2 way along each side of the usually slender body; young twigs and leaves glabrous or hairy; mature lateral leaflets with the mostly winged stalk 2–6 mm long; pistillate flowers with the calyx 0.5–1.5 mm long . . . . 3. F. PENNSYLVANICA
      3. Samaras with the wing 7–13 mm wide, extending more than 1/2 way along each side of the usually stout body; young twigs and leaves hairy; mature lateral leaflets with the mostly unwinged stalk 5–15 mm long; pistillate flowers with the calyx (1.0–)2.5–4.0 mm long . . . . . . 4. F. PROFUNDA
   2. Samara wing extending 1/4–1/3 of the way along each side of the body; leaflets with the undersurface usually somewhat whitened; leaf scars variously shallowly to deeply notched on the apical side
      4. Leaf scars U- or V-shaped to nearly semicircular, deeply notched on the apical side (this best observed on older twigs); winter buds dark brown to black; fruits (19–)25–32(–38) mm long, the body 5–11 mm long, the wing 3–6 mm wide; twigs glabrous . . . . 1. F. AMERICANA
      4. Leaf scars oblong-obovate to depressed-obovate, shallowly notched to nearly truncate on the apical side; winter buds brown; fruits 32–54 mm long, the body (7–)10–15 mm long, the wing 5–8 mm wide; twigs glabrous or hairy
         5. Twigs and petioles moderately to densely short-hairy . . . . 2. F. BILTMOREANA
         5. Twigs and petioles glabrous . . . . 6. F. SMALLII

1. Fraxinus americana L. (white ash)
Pl. 461 g–i; Map 2095

Plants trees to 30 m tall with an oval crown, dioecious or incompletely dioecious. Twigs circular in cross-section, unwinged, glabrous, sometimes somewhat glaucous, gray to brown with relatively inconspicuous, somewhat paler, circular to elongate lenticels, the leaf scars U- or V-shaped to nearly semicircular, deeply notched on the apical side, the associated lateral buds dark brown to more commonly black, relatively deeply sunken into the twig. Terminal buds 4–6 mm long, broadly ovoid to broadly conic, wider than long, rounded to bluntly pointed at the tip, dark reddish brown, usually covered with minute, peltate yellow to brownish yellow scales (the surface often appearing mealy), with 3 or 4 pairs of scales, the outermost pair relatively short and tightly appressed. Leaves (5–)9–30(–35) cm long, the petiole glabrous. Leaflets mostly (5)7(9), 2–15 cm long, 1–7 cm wide, variable in shape but mostly lanceolate to ovate or elliptic, rounded or angled to the sometimes slightly apically winged stalk (this mostly 8–12 mm long on the terminal leaflet and 2–9 mm long on lateral leaflets), relatively thin to slightly leathery, the upper surface glabrous, dull to slightly

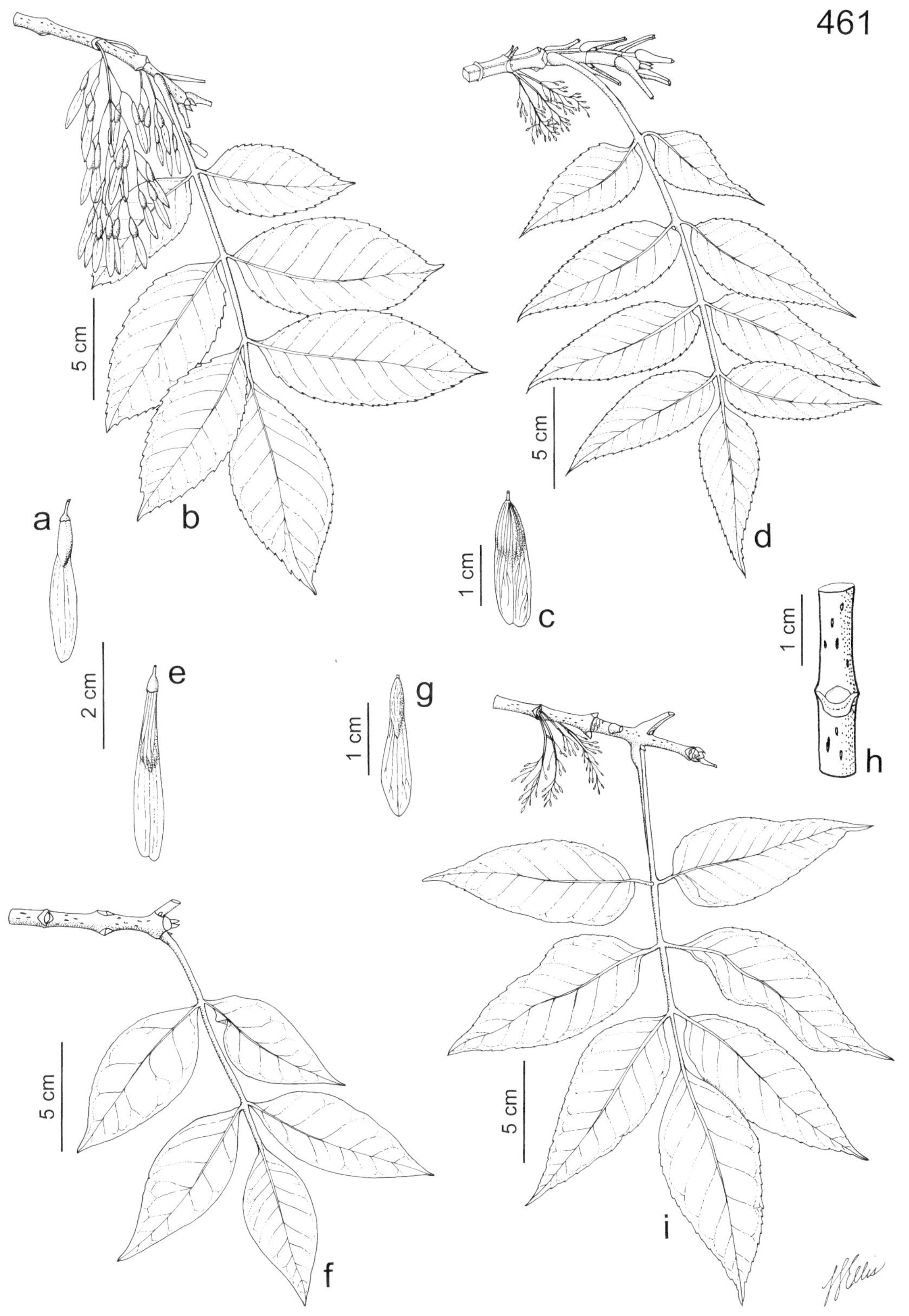

Plate 461. Oleaceae. *Fraxinus pennsylvanica*, **a)** fruit, **b)** fertile branch. *Fraxinus quadrangulata*, **c)** fruit, **d)** fertile branch. *Fraxinus profunda*, **e)** fruit, **f)** fertile branch. *Fraxinus americana*, **g)** fruit, **h)** twig with leaf scar and bud, **i)** fertile branch.

shiny, the undersurface glabrous or sparsely to moderately short-hairy when young, usually whitened, the margins entire or with blunt teeth. Calyces present, persistent at fruiting, 0.5–1.5 mm long. Fruits (19–)25–32(–38) mm long, the slender stalk 5–10 mm long, the body 5–11 mm long, 2–4 mm wide, slender, narrowly oblong in outline, not flattened, the wing 3–6 mm wide, narrowly oblanceolate to narrowly oblong-lanceolate, more or less rounded at the tip, less commonly with a small notch or bluntly pointed, extending less than 1/3 of the way along each side of the body. $2n=46$. April–May.

Common throughout the state (eastern U.S. west to Nebraska, Colorado, and Texas; Canada, Mexico). Mesic to less commonly dry upland forests, edges of glades, bottomland forests, banks of streams, rivers, and spring branches, margins of ponds, lakes, and sinkhole ponds, bases and ledges of bluffs, and loess hill prairies; also pastures, roadsides, and disturbed areas.

Black-Schaefer and Beckmann (1989) observed that individual trees of white ash in their study sometimes switched from staminate to pistillate, or vice versa, or to mixed flowers from one season to the next. It is not clear how commonly this occurs or whether other ash species are capable of changing gender.

Several authors (Wright, 1944; Hardin and Beckmann, 1982) have described the cuticular structure of the leaflet undersurface as a useful character in distinguishing even sterile specimens of some ashes. White ash can be distinguished from the other Missouri ashes by the presence of minute papillae, usually with interconnecting ridges that form an intricate cuticular network. However, observation of this character requires 30× or greater magnification

The present treatment follows that of Hardin (1974) in not recognizing infraspecific taxa within *F. americana*, but differs in recognizing *F. biltmoreana* and *F. smallii* as segregate species. See the treatments of these two taxa for further discussion. Other variants outside Missouri, such as the ones with somewhat smaller leaves and fruits (var. *microcarpa* A. Gray, var. *texensis* A. Gray), also seem unworthy of formal taxonomic recognition.

White ash wood is preferred for baseball bats and various tool and implement handles and is used to a lesser extent for furniture. The species has landscaping value but may be more suited to larger areas than typical residential lots because of its large size and susceptibility to various diseases and insect pests. In nature, it tends to occur at somewhat drier sites than is typical for *F. pennsylvanica*. Its fruits mature as early as June and are wind-dispersed in autumn, with some persisting on the tree into winter. It also sometimes is cultivated in South America and in the Old World.

2. Fraxinus biltmoreana Beadle (Biltmore ash)

F. americana L. ssp. *biltmoreana* (Beadle) A.E. Murray

F. americana var. *biltmoreana* (Beadle) J. Wright ex Fernald

Map 2096

Plants trees to 35 m tall with an oval crown, dioecious or incompletely dioecious. Twigs circular in cross-section, unwinged, moderately to densely pubescent with short, felty or somewhat matted hairs, not glaucous, gray to brown with relatively inconspicuous, paler, circular to elongate lenticels, the leaf scars oblong-obovate to depressed-obovate, shallowly notched to nearly truncate on the apical side, the associated axillary buds brown, not strongly sunken into the twig. Terminal buds 4–6 mm long, broadly ovoid to broadly conic, wider than long, rounded to bluntly pointed at the tip, dark reddish brown, hairy, often also with scattered, minute, peltate scales (sometimes occurring in patches), with 3 or 4 pairs of scales, the outermost pair relatively short and tightly appressed. Leaves (5–)9–40 cm long, the petiole moderately to densely short-hairy. Leaflets mostly (5)7–9, 2–20 cm long, 1–9 cm wide, variable in shape but mostly lanceolate to ovate or elliptic, rounded or angled to the sometimes slightly apically winged stalk (this mostly 8–18 mm long on the terminal leaflet and 3–15 mm long on lateral leaflets), relatively thin to slightly leathery, the upper surface glabrous or less commonly sparsely short-hairy, dull to slightly shiny, the undersurface glabrous or sparsely to moderately short-hairy, usually whitened, the margins entire or with blunt teeth. Calyces present, persistent at fruiting, 0.5–1.5 mm long. Fruits 35–54 mm long, the slender stalk 5–10 mm long, the body (7–)11–15 mm long, 2–4 mm wide, slender, narrowly oblong in outline, not flattened, the wing 6–8 mm wide, narrowly oblanceolate to narrowly oblong-lanceolate, more or less rounded at the tip, less commonly with a small notch or bluntly pointed, extending less than 1/3 of the way along each side of the body. $2n=138$. April–May.

Uncommon, known thus far only from Dunklin County (eastern U.S. west to Illinois, Missouri, Arkansas, and Louisiana). Swamps and bottomland forests.

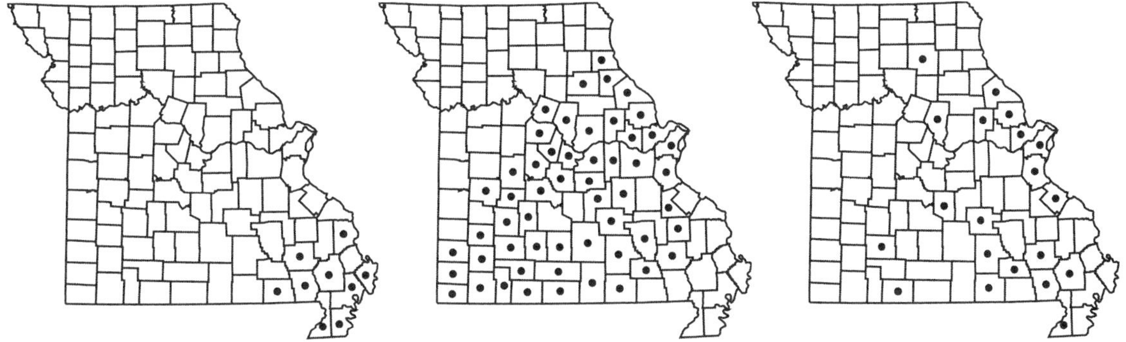

2098. Fraxinus profunda 2099. Fraxinus quadrangulata 2100. Fraxinus smallii

Fraxinus biltmoreana has been included in synonymy under *F. americana* by many botanists (Hardin and Beckmann, 1982; Gleason and Cronquist, 1991) or treated as a variety of that species (Fernald, 1950; Steyermark, 1963). G. N. Miller (1955) and Santamour (1962) discussed morphological distinctions between members of the *F. americana* polyploid complex. Nesom (2010a) expanded on these earlier studies and provided a more incisive morphological analysis. *Fraxinus biltmoreana* is a hexaploid ($2n$=138) member of the complex and, according to Nesom, is characterized by the pubescence on its twigs, petioles, and leaf rachises, and also by averaging the largest fruits in the complex. G. N. Miller (1955) suggested that the Biltmore ash originated through past hybridization between *F. americana* and *F. pennsylvanica* (an origin that also has been suggested for *F. profunda*). Black-Schaeffer and Beckmann's (1989) study of secondary biochemistry in the leaves of ash species showed the presence of flavonoid compounds in *F. biltmoreana* that were more or less additive between those found in *F. americana* and *F. pennsylvanica*. Wallander's (2008) molecular analysis of the genus also did not contradict this interpretation. If the parentage of *F. biltmoreana* should become confirmed as the result of a cross between a diploid *F. pennsylvanica* parent and a tetraploid member of the *F. americana* complex, then that tetraploid likely represents the taxon treated here as *F. smallii*.

3. Fraxinus pennsylvanica Marshall (green ash, red ash)

F. pennsylvanica var. *subintegerrima* (Vahl) Fernald

Pl. 461 a, b; Map 2097

Plants trees to 25 m tall with a somewhat spreading, oval crown, dioecious or incompletely dioecious. Twigs circular in cross-section, unwinged, glabrous or minutely hairy, not glaucous, gray to grayish green or brown with relatively conspicuous, pale, circular to elongate lenticels, the leaf scars truncate to shallowly concave on the apical side on both new and older twigs. Terminal buds 3–8 mm long, ovoid to conic, longer than wide, bluntly to sharply pointed at the tip, brown to rust-colored, glabrous or scurfy-hairy, with usually 3 pairs of scales, the outermost pair relatively short and tightly to loosely appressed. Leaves 5–30 cm long, the petiole glabrous or hairy. Leaflets mostly (5)7(9), 1.5–16.0 cm long, 0.5–7.5 cm wide, variable in shape but mostly lanceolate to narrowly ovate, narrowly elliptic, or elliptic (the basalmost pair sometimes broader), mostly angled or tapered to the often narrowly winged stalk (this mostly 5–12 mm long on the terminal leaflet and 2–6 mm long on lateral leaflets), relatively thin to slightly leathery, the upper surface glabrous or sparsely short-hairy, dull to slightly shiny, the undersurface glabrous or sparsely to densely short-hairy, sometimes only along the veins, yellowish green to pale green but not whitened, the margins entire or more commonly with blunt teeth. Calyces present, persistent at fruiting, 0.5–1.5 mm long. Fruits 30–60 mm long, the slender stalk 5–10 mm long, the body 10–25 mm long, 1.5–3.5 mm wide, slender, narrowly oblong in outline, not flattened, the wing 3–6 mm wide, narrowly oblanceolate to narrowly oblong-lanceolate, more or less rounded at the tip, less commonly with a small notch, extending mostly $\frac{1}{3}$–$\frac{1}{2}$ the way along each side of the body. $2n$=46. April–May.

Common throughout the state (eastern U.S. west to Montana and New Mexico; Canada, Mexico). Bottomland forests, swamps, banks of streams, rivers, and spring branches, margins of ponds, lakes, and sinkhole ponds, bases of bluffs, bottomland prairies, loess hill prairies, edges of marshes and fens; also ditches, fallow fields, pastures, railroads, roadsides, and moist disturbed areas.

Green ash leaves and twigs are variable in pubescence. In Missouri, most plants have glabrous leaves and twigs and have been segregated by some authors as var. *subintegerrima* (Steyermark, 1963). Pubescent individuals are less commonly encountered, but still widely distributed in the state, and are not considered worthy of formal taxonomic recognition (G. N. Miller, 1955; Hardin, 1974). In contrast, the two other subspecies of *F. pennsylvanica* accepted by G. N. Miller (1955) from the western United States, now generally are regarded as separate species.

Vegetative specimens of green ash can be difficult to identify. The cuticular structures of the lower leaflet surfaces of white ash (see discussion under that taxon) are lacking in green ash, a character that can be helpful in laboratory determinations.

Green ash wood is slightly inferior to white ash but is used for many of the same purposes and is faster growing. The species is grown commonly along streets and in other horticultural plantings, even at upland sites, because of its tolerance of a wide range of environmental conditions. The fruits mature as early as June and are wind-dispersed in autumn, with many persisting on the tree into winter. Seedlings can be weedy and are a nuisance to gardeners with pistillate trees in their neighborhoods. The species also is sometimes cultivated in Central America, South America, and the Old World.

4. Fraxinus profunda (Bush) Bush (pumpkin ash, red ash)

F. profunda var. *ashei* E.J. Palmer

F. americana var. *profunda* Bush

F. tomentosa F. Michx., an illegitimate name

Pl. 461 e, f; Map 2098

Plants trees to 40 m tall with a narrow, open crown and spreading branches, the trunk often swollen at the base, dioecious or incompletely dioecious. Twigs circular in cross-section, unwinged, minutely velvety-hairy (at least when young), not glaucous, gray to grayish brown with relatively conspicuous, pale, circular to elongate lenticels, the leaf scars truncate to shallowly concave on the apical side on both new and older twigs. Terminal buds 3–9 mm long, ovoid to conic, slightly longer than wide, mostly bluntly pointed at the tip, reddish brown, velvety-hairy, with usually 3 pairs of scales, the outermost pair relatively short and tightly to loosely appressed. Leaves 15–40 cm long, the petiole velvety-hairy. Leaflets 7 or 9, 5–25 cm long, 2.5–12.0 cm wide, variable in shape but mostly lanceolate to narrowly ovate, ovate, or elliptic, rounded, angled or short-tapered to the sometimes slightly apically winged stalk (this mostly 8–20 mm long on the terminal leaflet and 5–15 mm long on lateral leaflets), relatively thick and somewhat leathery, the upper surface glabrous, often somewhat shiny, the undersurface moderately to densely short-hairy, yellowish green to pale green but not whitened, the margins entire or rarely with sparse, blunt teeth. Calyces present, persistent at fruiting, 0.5–1.5 mm long in staminate flowers, (1.0–)2.5–4.0 mm long in pistillate flowers (sometimes becoming enlarged to 6 mm at fruiting). Fruits 45–80 mm long, the slender stalk 5–12 mm long, the body 10–30 mm long, 2–3 mm wide, relatively stout, narrowly oblong in outline, not flattened, the wing 7–13 mm wide, narrowly oblanceolate to narrowly oblong-lanceolate or somewhat spatulate, more or less rounded at the tip, less commonly with a small notch or an abrupt, minute, sharp point, extending narrowly more than 1/2 the way along each side of the body, sometimes nearly to the base. $2n=138$. April–May.

Uncommon, restricted to the Mississippi Lowlands Division (eastern U.S. west to Illinois, Missouri, and Louisiana; Canada). Swamps, bottomland forests, and uncommonly acid seeps; also wet roadsides.

Pumpkin ash, which originally was described based on specimens collected in the Missouri Bootheel (Nesom, 2010b), may have originated as an autopolyploid of *F. pennsylvanica* or may represent an allopolyploid derived from hybridization between *F. americana* and *F. pennsylvanica* (Hardin, 1974; G. N. Miller, 1955; Santamour, 1962; Wright, 1965; K. A. Wilson and Wood, 1959). Molecular data (Wallander, 2008) appear to support the hybrid hypothesis. The species has the largest leaves and fruits among Missouri ashes, but can be difficult to distinguish from *F. pennsylvanica*. It grows in association with other lowland trees such as *Taxodium distichum* (L.) Rich. (Cupressaceae), *Planera aquatica* (Ulmaceae), and *Nyssa aquatica* L. (Cornaceae). The common name derives from the appearance of the swollen base of the trunk at sites that are flooded for long periods of time. Its wood is of some commercial value for tool and implement handles. No horticultural uses are known. Ducks, other birds, and rodents reportedly eat the seeds.

5. Fraxinus quadrangulata Michx. (blue ash)

Pl. 461 c, d; Map 2099

Plants trees to 30 m tall (usually much shorter) with a narrow, rounded crown, with all or most of the flowers perfect. Twigs strongly 4-angled (square) in cross-section, the angles sometimes

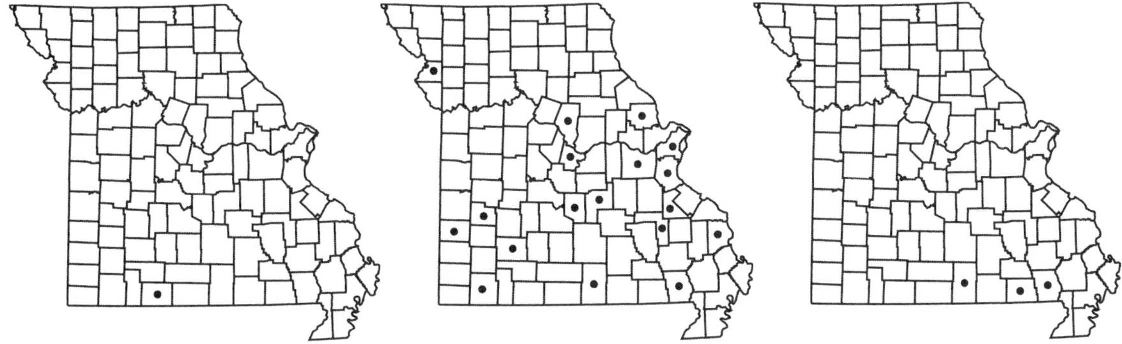
2101. Jasminum nudiflorum 2102. Ligustrum obtusifolium 2103. Ligustrum ovalifolium

with narrow, corky wings, glabrous or more commonly minutely velvety-hairy when young, not glaucous, gray to reddish brown, eventually becoming tan to gray with age (as they expand to become more circular), with relatively conspicuous, pale, oval lenticels, the leaf scars broadly concave on the apical side on both new and older twigs. Terminal buds 4–7 mm long, ovoid to conic, slightly longer than wide, mostly bluntly pointed at the tip, greenish gray to brownish gray or reddish brown, velvety-hairy, with usually 3 pairs of scales, the outermost pair relatively long and loosely appressed. Leaves 8–30 cm long, the petiole glabrous or minutely hairy. Leaflets (5–)7–11, 4–12 cm long, 1.5–6.5 cm wide, mostly lanceolate to narrowly ovate or ovate, rounded or broadly angled (often asymmetrically so) above the usually narrowly winged stalk (this mostly 1–6 mm long on the terminal and lateral leaflets), relatively thick and somewhat leathery, velvety-hairy when expanding, at maturity the upper surface glabrous or nearly so, somewhat shiny, the undersurface moderately short-hairy toward the base along the main veins, yellowish green to pale green but not whitened, the margins with numerous, fine, blunt teeth. Calyces absent or, if present, shed early, 0.5–1.5 mm long. Fruits 25–45(–60) mm long, the slender stalk 4–10 mm long, the body 10–20 mm long, about 2 mm wide, narrowly lanceolate, relatively poorly differentiated from the wing and somewhat flattened, the wing 6–12 mm wide, narrowly oblanceolate to narrowly oblong, usually shallowly notched at the otherwise rounded to truncate tip, extending to the base of the body or nearly so. $2n=46$. March–May.

Scattered to common in the Ozark and Ozark Border Divisions, northward locally in the eastern portion of the Glaciated Plains to Marion County; apparently absent from the remainder of the state (eastern U.S. west to Minnesota and Oklahoma; Canada). Mesic to dry upland forests, glades, savannas, and ledges and tops of bluffs; less commonly banks of streams and rivers, bases of bluffs, and bottomland forests; often on calcareous substrates.

The name blue ash refers to the fact that the inner bark or branches can be macerated in water to yield a blue dye, a practice of early pioneers (Steyermark, 1963). The wood is used commercially for purposes similar to those of white ash, but the species is less abundant and of less economic importance due to its smaller size. Blue ash has had limited use horticulturally as a species tolerant of drought and alkaline soils.

6. Fraxinus smallii Britton (Sullivan's ash)
F. americana L. var. *subcoriacea* Sarg.
Map 2100

Plants trees to 25 m tall with an oval crown, dioecious or incompletely dioecious. Twigs circular in cross-section, unwinged, glabrous, sometimes slightly glaucous, gray to brown with relatively inconspicuous, paler, circular to elongate lenticels, the leaf scars oblong-obovate to depressed-obovate, shallowly notched to nearly truncate on the apical side, the associated axillary buds brown, not strongly sunken into the twig. Terminal buds 4–6 mm long, broadly ovoid to broadly conic, wider than long, rounded to bluntly pointed at the tip, dark reddish brown, moderately to densely pubescent with minute, yellowish hairs or more commonly covered with minute, peltate yellow to brownish yellow scales (the surface often appearing mealy), with 3 or 4 pairs of scales, the outermost pair relatively short and tightly appressed. Leaves 8–35 cm long, the petiole glabrous. Leaflets mostly (5)7(9), 2–15 cm long, 1–7 cm wide, variable in shape but mostly lanceolate to ovate or elliptic, rounded or angled to the sometimes slightly apically winged stalk (this mostly 8–12 mm long on the terminal leaflet and 2–9 mm long on lateral leaflets), relatively thin to slightly leathery, the upper surface glabrous or less commonly sparsely short-hairy, dull to slightly shiny, the undersurface glabrous or sparsely to moderately short-hairy, usually whitened, the margins entire or with blunt teeth. Calyces present, persistent at fruiting, 0.5–1.5 mm long. Fruits 32–44 mm long, the slender stalk 5–10 mm long, the body 10–13 mm long, 2–4 mm wide, slender, narrowly oblong in outline, not flattened, the wing 5–7 mm wide, narrowly oblanceolate to narrowly oblong-lanceolate, more or less rounded at the tip, less commonly with a small notch or bluntly pointed, extending less than 1/3 of the way along each side of the body. $2n$=92. April–May.

Scattered in the southern half of the state (eastern U.S. west to Iowa, Kansas, and Texas). Mesic to less commonly dry upland forests, edges of glades, bottomland forests, banks of streams, rivers, and spring branches, margins of ponds, lakes, and sinkhole ponds, and bases and ledges of bluffs; also pastures, roadsides, and disturbed areas.

Fraxinus smallii has been included in synonymy under *F. americana* by most botanists (Steyermark, 1963; Hardin and Beckmann, 1982; Gleason and Cronquist, 1991). Santamour (1962) discussed morphological distinctions between members of the *F. americana* polyploid complex. Nesom (2010a) expanded on this brief, earlier study and concluded that the name *F. smallii* refers to tetraploid ($2n$=92) plants in the complex. According to Nesom, *F. smallii* differs consistently from *F. americana* in its more shallowly notched leaf scars and somewhat larger fruits. Interestingly, differences in winter buds and leaf scars were noted many years ago by the eminent St. Louis area amateur natural-historian, Fr. James Sullivan, who also observed that the buds of true *F. americana* are much darker in color (black or nearly so) and tend to be more sunken into the twigs than in trees now called *F. smallii* (Van Brunt, 2011). Members of the Webster Groves Nature Study Society's Botany Group were the first to suggest adoption of the vernacular name Sullivan's ash for this brown-budded member of the white ash complex.

Nesom (2010a) suggested that *F. smallii* may have arisen through past hybridization between *F. americana* and *F. pennsylvanica*, but was unable to rule out the possibility that the taxon developed following autopolyploidy (the doubling of the chromosome number within a single species) in *F. americana*. Autopolyploids are more frequently treated as infraspecific taxa within a given species. The taxon is treated here provisionally as a species, in line with Guy Nesom's forthcoming treatment of the genus in a future volume of the Flora of North America series.

6. **Jasminum** L. (jasmine, jessamine)

About 200 species, South America, Africa, Asia, south to Australia.

Burrows and Tyrl (2001) listed *Jasminum* as a genus of questionable toxicity, noting that the plants sequester extensive monoterpenoids and their glycosides, especially jasminin, jasminoside, jasmoside, and jasmolactones. These compounds, when ingested in quantity, can cause irritation to the digestive tract.

1. Jasminum nudiflorum Lindl. (winter jasmine)

Map 2101

Plants shrubs, 0.4–1.5 m tall (but the spreading stems up to 5.0 m long), with perfect flowers. Main stems numerous, intricately branched, mostly loosely ascending to spreading or arched, sometimes rooting at the tip, rarely climbing on other vegetation, the bark tending to remain dark green, but eventually turning purplish brown to dark brown or brown with age, smooth or with a few, irregular, longitudinal furrows. Twigs relatively slender, dark green or occasionally purplish- or brownish-tinged, glabrous, 4-angled in cross-section (usually with 4 slender ridges), with the leaf scars not or only slightly raised and the lenticels inconspicuous, small, and not raised. Terminal buds narrowly ovoid to narrowly ellipsoid, with several, overlapping, sharply pointed scales, the axillary buds similar to the terminal ones. Leaves opposite, short-petiolate to nearly sessile. Leaf blades mostly ternately compound (simple leaves often present at the bases of branchlets), 0.7–3.0 cm long, 0.4–1.5 cm wide, the blade or leaflets ovate to oblong-obovate or less commonly nearly circular, angled or short-tapered at the blunt tip (sometimes with an abrupt, minute, sharp point at the very tip), angled at the base, the margins entire but usually minutely hairy, the upper surface dark green, glabrous, and shiny, the undersurface sometimes slightly lighter green and glabrous. Inflorescences axillary, of solitary flowers, developing before the leaves, each usually with 1 or few, small, sepaloid bract(s), the flowers with slender stalks 2–6 mm long, not fragrant. Calyces deeply 5- or 6-lobed, 4–6 mm long, the lobes narrowly lanceolate to narrowly oblong-lanceolate, often strongly reddish-tinged. Corollas 5- or 6-lobed less than 1/2 way to the base, more or less trumpet-shaped (the tube often slightly curved), the lobes elliptic to oblong-elliptic or oblong-obovate, yellow (the outer surface and the tube frequently reddish-tinged). Style 2–5 mm long, either with a solitary, capitate stigma or more commonly with a pair of ascending branches at the tip. Fruits berries, 5–7 mm long, ellipsoid to ovoid, olive green, turning blackish with age, glabrous. $2n$=26, 52 (possibly also $2n$=29, 34, 48). February–March.

Introduced; known thus far from a single site in Taney County (native of Asia, introduced uncommonly and sporadically in the eastern half of the U.S.). Glades; also old homesites.

Jasminum nudiflorum is most often grown as a mounding ground cover, but sometimes is trained on trellises or staked into an upright form. The flowers appear in the late winter, when few other plants are flowering. It is also used in bonsai, where it is trained into a small treeform plant with weeping branches. In Missouri it is not cultivated as frequently as it is farther south, as it is only semihardy from USDA Hardiness Zone 5 northward (for a chart of hardiness zones, see the introductory section on climate of Missouri in Volume 1 of the present work [Yatskievych, 1999]). The first Missouri record was collected in 2004 by Michael Skinner of the Missouri Department of Conservation at a site near Branson where plants apparently had escaped from a planting at an old homesite and colonized an adjacent dolomite glade.

Some botanists segregate a dwarfed, prostrate variant of the species that occurs at high elevations in southwestern China as var. *pulvinatum* (W.W. Sm.) Kobuski.

7. **Ligustrum** L. (privet)
(Yatskievych and Summers, 1991)

Plants shrubs (small trees elsewhere), sometimes evergreen or semi-evergreen. Trunks few to several, ascending, sometimes arched, the bark grayish brown to gray, thin, relatively smooth, but with raised leaf scars and lenticels. Twigs relatively slender, variously yellowish green to gray, brown, or nearly black, glabrous or hairy, more or less circular in cross-section, the leaf scars raised and the lenticels conspicuous and raised. Terminal buds ovoid to conic-ovoid, with scales that are sharply pointed at the tips, the axillary buds similar to the terminal ones, but smaller. Leaves opposite or occasionally subopposite, mostly short-petiolate, the petioles of the largest leaves 1–16 mm long. Leaf blades simple, 1–8 cm long, 0.5–2.5 cm wide, oblong-elliptic to elliptic, oblong-ovate, or less commonly lanceolate to ovate, angled or tapered (rarely rounded or shallowly notched) to the bluntly to sharply pointed tip, rounded or angled to broadly angled at the base, the margins entire (sometimes minutely hairy), the upper surface green to dark green, glabrous, the undersurface slightly lighter green to yellowish green, glabrous or hairy, often also inconspicuously gland-dotted. Inflorescences terminal, few- to more commonly many-flowered, variously ascending to spreading or nodding panicles (occasionally reduced and nearly racemose) 1.5–10 cm long developing after the leaves, the lower branch points with small, leaflike or scalelike bracts, the flowers mostly with slender stalks 1–5 mm long (occasionally some of them sessile), moderately but unpleasantly fragrant. Calyces truncate at the tip (unlobed or sometimes appearing slightly wavy) or with 4 very shallow lobes 1.5–2.0 mm long, the lobes irregular, blunt or toothlike. Corollas 4-lobed to about the midpoint or less, 5–12 mm long, trumpet-shaped, the lobes oblong to oblong-ovate, white or occasionally cream-colored. Style 1–2 mm long, with a pair of ascending branches at the tip or more frequently the fused stigmas appearing as a club-shaped mass. Fruits berry-like drupes, 5–8 mm long, globose to broadly ellipsoid, green to olive green, turning bluish black or black, glabrous, sometimes slightly glaucous. About 45 species, Europe, Africa, Asia south to Australia.

Many species are grown in cultivation for their foliage in hedges, as screens or as specimen plants. The heavily scented flowers of *Ligustrum* were described by Mabberley (1997) as having an unpleasant, ammonia-like or fishy smell (caused by methylamine) that reportedly can taint the honey of honeybees that visit the flowers. Species of *Ligustrum* are considered toxic to livestock and humans, but there are conflicting reports on how poisonous the various species may be. Terpenoid glycosides, especially ligustrin derivatives of oleanolic acid, are suspected as the chemical basis of intoxication (Burrows and Tyrl, 2001). Symptoms are mostly gastrointestinal, including stomach and intestinal irritation, vomiting, and diarrhea, but also may progress to paralysis of the limbs, fluid in the lungs, and rarely death.

Privets are popular in horticulture as background shrubs, windbreaks, hedges, and specimen plants. A number of other *Ligustrum* species in addition to those treated below are cultivated in the United States, including *L. amurense* Carrière (Amur privet), *L. japonicum* Thunb. (Japanese privet), *L. lucidum* W.T. Aiton (glossy privet), and *L. quihoui* Carrière (waxy-leaved privet). Cultivars are available with variegated leaves of green and white or green and yellow.

The inclusion by Steyermark (1963) of only *L. ovalifolium* and *L. vulgare* (as an excluded species) for the state has led to the misidentification of some Missouri specimens. Additional collections of naturalized plants are needed to more accurately determine the species occurring in the state and their respective distributions. Hardin (1974) and K. A. Wilson and Wood (1959) pointed to a need for further taxonomic work on the genus. The following key is adapted from Hardin's work and requires flowering material for accurate determinations.

1. Twigs glabrous
    2. Corollas with the tube 2–3 times as long as the lobes . . . . 2. L. OVALIFOLIUM
    2. Corollas with the tube shorter than to about as long as the lobes . . 4. L. VULGARE
1. Twigs short-hairy (sometimes becoming glabrous or nearly so by second year)
    3. Corollas with the tube slightly longer than to about 3 times as long as the lobes; calyces short-hairy . . . . 1. L. OBTUSIFOLIUM
    3. Corollas with the tube shorter than to about as long as the lobes; calyces glabrous or nearly so
        4. Twigs pubescent with short, curved hairs; leaf blades with the undersurface hairy along the midvein . . . . 3. L. SINENSE
        4. Twigs pubescent with minute, spreading hairs; leaf blades glabrous . . . . 4. L. VULGARE

1. Ligustrum obtusifolium Siebold & Zucc. (border privet)

Pl. 460 j; Map 2102

Plants 1–2(–4) m tall and about as wide, the main stems usually numerous, arched, with spreading branches. Twigs moderately to densely pubescent with minute, spreading hairs, the new growth grayish green, becoming blackish gray with lighter colored lenticels. Winter buds with the scales tawny, short-hairy. Petioles 1–3 mm long, glabrous or short-hairy, narrowly winged. Leaf blades 2–6 cm long, 7–25 mm wide, relatively thin and herbaceous, elliptic to oblong-ovate, rounded to bluntly or broadly angled at the tip, the upper surface glabrous, not shiny, the undersurface short-hairy, at least along the midvein, also minutely but often faintly gland-dotted. Inflorescences spreading to nodding, relatively slender panicles, 1.5–4.0 cm long, with several to numerous flowers. Calyces moderately to densely short-hairy. Corollas 6–10(–12) mm long, the tube slightly longer than to about 3 times as long as the lobes, white. Stamens not exserted. $2n=46$. May–June.

Introduced, scattered, mostly south of the Missouri River (native of Asia, introduced in the northeastern U.S. west to Nebraska and Missouri). Bottomland forests, mesic to dry upland forests, glades, and banks of streams, spring branches, and rivers; also old fields, ditches, railroads, roadsides, and disturbed areas.

Yatskievych and Summers (1991) first reported this species for Missouri from plants in Franklin County that previously had been thought to represent *L. vulgare*. Other naturalized Missouri specimens of *L. obtusifolium* that initially had been misdetermined as *L. vulgare* were found in the review of specimens for the present treatment. Fruiting specimens may be difficult to separate from *L. sinense*, but that species has longer, more open, often drooping panicles.

A compact variant that has been called var. *regelianum* Rehder (Regel privet) is promoted horticulturally as a trouble-free shrub with lustrous foliage for border, hedge, or specimen plantings, but has not, so far, escaped from cultivation. Within its native range, a few additional subspecies are recognized (Chang et al., 1996), but

2104. Ligustrum sinense

2105. Ligustrum vulgare

2106. Syringa vulgaris

the application of this classification to the plants cultivated and escaped in North America is not well understood.

2. Ligustrum ovalifolium Hassk. (California privet)

Pl. 460 k; Map 2103

Plants 1–3(–5) m tall, often forming thickets, the main stems usually numerous, ascending to somewhat arched, with usually ascending branches. Twigs glabrous, the new growth grayish green, becoming gray or grayish brown with sparse, pale lenticels. Winter buds with the scales tawny to brown, glabrous. Petioles 1–5 mm long, glabrous, narrowly to broadly winged. Leaf blades 2–6 cm long, 7–25 mm wide, relatively thick and leathery, elliptic to ovate or oblong-ovate, angled or slightly tapered to a sharply but sometimes broadly pointed tip, the upper surface glabrous, shiny, the undersurface glabrous, usually faintly gland-dotted. Inflorescences stiff, erect or ascending, relatively broad panicles, 4–10 cm long, with numerous flowers. Calyces glabrous. Corollas 6–9 mm long, the tube 2–3 times as long as the lobes, white or cream-colored (especially while in bud). Stamens short-exserted. $2n=46$. May–July.

Introduced, known thus far only from single collections from Butler, Howell, and Ripley Counties (native of Asia, introduced sporadically in the southeastern U.S.). Edges of fens and mesic upland forests; also roadsides and banks of ditches.

California privet is used horticulturally for border, hedge, or screen planting, including cultivars featuring variegated leaves and compact growth form, but its winter hardiness is questionable in our region. Steyermark (1963) originally included *L. ovalifolium* in the Missouri flora based on a 1957 specimen that he collected in Texas County, but this specimen could not be located during the present research. The three specimens accepted in the present treatment were all collected subsequent to Steyermark's first report.

3. Ligustrum sinense Lour. (Chinese privet)

Map 2104

Plants 1–5 m tall, usually taller than wide, the main stems 1 or more commonly few to several, erect or ascending, with spreading to ascending branches. Twigs moderately to densely pubescent with short, curved hairs, the new growth yellowish green, becoming blackish gray with pale lenticels. Winter buds with the scales tawny, short-hairy. Petioles 1–6 mm long, short-hairy, sometimes narrowly winged. Leaf blades 2–7 cm long, 7–25 mm wide, relatively thick and often somewhat leathery, elliptic to oblong-elliptic, lanceolate, or occasionally ovate, angled or slightly tapered to a bluntly or less commonly sharply pointed tip (rarely rounded or shallowly notched), the upper surface glabrous, slightly to moderately shiny, the undersurface short-hairy along the midvein, occasionally faintly gland-dotted. Inflorescences spreading to nodding, relatively broad panicles, 4–10 cm long, with numerous flowers. Calyces glabrous or with sparse, short hairs. Corollas 4–7 mm long, the tube shorter than to about as long as the lobes, white. Stamens short-exserted. $2n=46$. May–June.

Introduced, scattered in the southern half of the state, mostly in the Mississippi Lowlands Division and adjacent portion of the Ozarks (native of Asia; introduced in the eastern [mostly southeastern] U.S. west to Kansas and Texas). Bottomland forests, mesic upland forests, bottomland prairies, margins of sand prairies, and banks of streams and rivers; also old fields, old homesites, ditches, railroads, roadsides, and disturbed areas.

Yatskievych and Summers (1991) first reported the species from Missouri based on a Dunklin County collection. As judged by its aggressive spread in the Gulf Coastal Plain, this privet is probably more abundant in southeastern Missouri than current records indicate. It is capable of forming dense stands in forest understories, undoubtedly due to seed dispersal by birds. For a comparison with the morphologically similar *L. obtusifolium*, see the treatment of that species.

This species is no longer used commonly for ornamental plantings because it spreads too aggressively by seeds, but a less-vigorous cultivar, cv. 'Variegatum,' is planted for its white-margined leaves, and other cultivars are available with specific growth forms. Within its native range, a complex classification of six or more varieties has been developed (Chang et al., 1996). The application of these varietal names to the North American escapes is uncertain and the feasibility of recognizing infraspecific taxa for plants in the wild requires further study.

4. Ligustrum vulgare L. (common privet; European privet)

Pl. 460 h, i; Map 2105

Plants 1–5 m tall and about as wide, the main stems usually numerous, erect or ascending, with spreading to broadly ascending branches. Twigs densely pubescent with minute, spreading hairs (becoming glabrous or nearly so by second year), the new growth green, becoming gray with light grayish brown lenticels. Winter buds with the scales tawny to brown, glabrous. Petioles 3–16 mm long, glabrous, sometimes narrowly winged. Leaf blades 2–8 cm long, 6–20 mm wide, relatively thick and somewhat leathery, narrowly ovate to narrowly elliptic or lanceolate, angled or tapered to a sharply or occasionally bluntly pointed tip, the upper surface glabrous, shiny, the undersurface glabrous, faintly gland-dotted. Inflorescences ascending to spreading or nodding, relatively broad panicles, 3–6 cm long, with numerous flowers. Calyces glabrous. Corollas 5–9 mm long, the tube shorter than to about as long as the lobes, white. Stamens not exserted or short-exserted. $2n=46$. May–June.

Introduced, uncommon, widely scattered (native of Europe; introduced widely but sporadically in the U.S., Canada). Banks of spring branches; also old homesites and disturbed areas.

This privet is the most widely planted *Ligustrum* species for hedges east of Missouri. Numerous cultivars have been developed for variegated leaves as well as various growth habits and fruit colors.

Steyermark (1963) discussed *L. vulgare* as an excluded species that he expected to eventually be discovered in Missouri outside of cultivation. The present study has confirmed this prediction, although two other species not included in Steyermark's treatment, *L. obtusifolium* and *L. sinense*, have proven to escape cultivation more commonly in our region.

8. Syringa L. (lilac)

About 23 species, Europe, Asia.

Several other *Syringa* taxa are cultivated commonly in the United States, including *S. emodi* Wall. ex Royle (Himalayan lilac), *S. josikaea* J. Jacq. (Hungarian lilac), *S. reticulata* (Blume) H. Hara. (Japanese tree lilac), and some horticulturally derived interspecific hybrids. These rarely, if ever, become established outside cultivation.

1. Syringa vulgaris L. (common lilac)

Pl. 460 g; Map 2106

Plants shrubs, 1.5–3.0(–7.0) m tall. Trunks few to several, ascending, the bark grayish brown to more commonly gray, thin, relatively smooth, but with raised leaf scars and lenticels. Twigs relatively stout, reddish brown to dark brown, with a pale, waxy coat, tending to peel in strips with age, glabrous, more or less 4-angled in cross-section (usually with 4 slender ridges), the leaf scars raised and the lenticels conspicuous and raised. Terminal buds usually absent, suppressed by an apical pair of relatively large axillary buds, these ovoid to broadly ovoid, with scales that are broadly but sharply pointed at the tips (those lower on the twigs similar, but smaller). Leaves opposite, moderately to long-petiolate, the petioles of the largest leaves 19–30 mm long. Leaf blades simple, 3–10 cm long, 1–6 cm wide, ovate, tapered to the sharply pointed tip, broadly rounded to truncate or shallowly cordate at the base, the margins entire, the upper surface green to dark green, glabrous, the undersurface lighter green, glabrous. Inflorescences terminal (often paired), many-flowered, ascending to spreading or drooping panicles 8–20 cm long developing with or after the leaves, the lower branch points with small, leaflike or scalelike bracts, the flowers with slender stalks 1–4 mm long, strongly fragrant. Calyces shallowly 4-lobed, 1.5–2.2 mm long, the lobes narrowly triangular to triangular, often toothlike. Corollas 4-lobed to slightly above the midpoint, 9–16 mm long, trumpet-shaped, the lobes oblong to oblong-elliptic, purple to bluish lavender, pale lavender, or

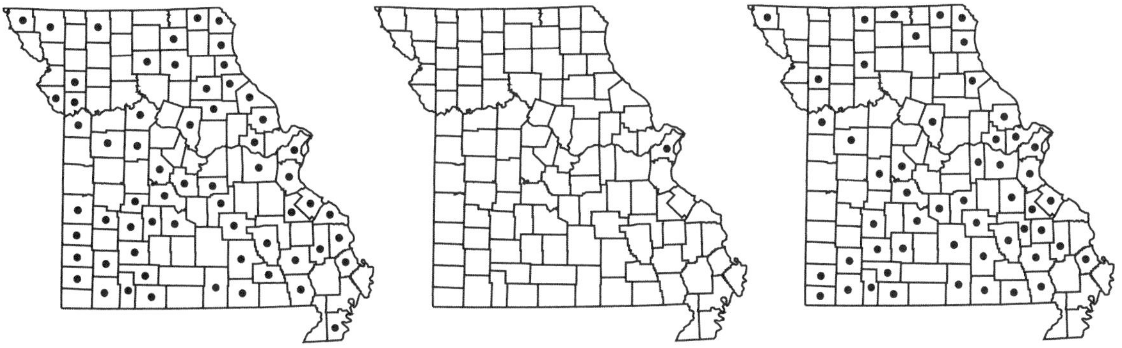

2107. Circaea canadensis

2108. Epilobium ciliatum

2109. Epilobium coloratum

rarely white. Style 1.5–2.5 mm long, with a pair of ascending branches at the tip. Fruits capsules, 10–18 mm long, slightly flattened, not winged, oblanceolate to narrowly obovate in outline, beaked at the tip, brown, glabrous, often somewhat shiny, dehiscing longitudinally. Seeds 10–14 mm long, flattened, narrowly winged toward the midpoint, tapered at each end, brown. $2n$=44–48. April–June.

Introduced, uncommon in the eastern half of the state (native of Europe; introduced sporadically nearly throughout the U.S., Canada). Edges of mesic upland forests; also old homesites, railroads, and roadsides.

This species frequently persists for long periods of time at abandoned farmsteads, but in Missouri it rarely reproduces itself. Lilac is cultivated for its showy flowers with their characteristic, strong, sweet fragrance. A very large number of cultivars exists, varying in flower color, flower structure (single and double flowers), growth form, and disease resistance. Lilac fragrance, which is composed of a mixture of furanoterpenoid derivatives, is sold as an essential oil and is used extensively in perfumes, soaps, bath products, scented candles, and potpourri.

ONAGRACEAE (Evening Primrose Family)

Contributed by Warren L. Wagner and Peter C. Hoch

Plants annual or perennial herbs, sometimes woody near the base (shrubs or trees elsewhere); stems branched or less commonly unbranched. Leaves alternate, basal, or opposite (rarely whorled elsewhere), simple, sessile or petiolate, the blade entire to pinnately lobed, the margins entire or toothed. Stipules absent or inconspicuous, then herbaceous, hair-like, or glandular, and often shed early. Inflorescences of solitary axillary flowers or terminal and/or axillary spikes, racemes, or panicles, the flowers then often subtended by leaflike bracts. Flowers perfect (uncommonly imperfect elsewhere), epigynous, actinomorphic or less commonly zygomorphic, not subtended by bractlets (except sometimes in *Ludwigia*). Hypanthium absent or more commonly well-developed, appearing as a floral tube, it and the perianth shed after flowering (except the calyx persistent in *Ludwigia*). Calyces of (2)4–5(–7) variously shaped sepals at the tip of the floral tube, some of these sometimes remaining irregularly and partially fused as the buds open (in some species of *Oenothera*). Corollas of (2)4–5(–7) petals or rarely absent, when present attached at the tip of the floral tube, alternating with the sepals. Stamens as many as or twice as many as the sepals (occasionally reduced to 1 elsewhere), the filaments sometimes in 2 series and often of 2 different lengths in the same flower, attached to the inner surface of the floral tube or less commonly to a nectar disc at the tip of the ovary, the anthers small or large, attached near the midpoint or less commonly (in small anthers) near the base of the dorsal side, usually yellow, the pollen often shed as small groups of grains held together loosely with cobwebby filaments (viscin threads). Pistil 1 per flower, composed of (2)4 or 5(–7) fused carpels, the inferior ovary sometimes with a small nectar disc (at the tip) at the base, the style 1, slender, usually relatively long, not persistent at fruiting, the stigma variously disc-shaped or more or less capitate to deeply 4-lobed. Ovules numerous or less commonly 1 to few. Fruits capsules (sometimes appearing nutlike or berrylike), indehiscent or dehiscing longitudinally between the locules, sometimes incompletely or only with age. Seeds 1 to numerous, small, in *Epilobium* often with a tuft of long, silky hairs at the tip. Twenty-two genera, about 660 species, nearly worldwide.

Many of the more conspicuous members of the Onagraceae are recognized easily by their four sepals and petals, eight stamens, inferior ovary, and well-developed floral tube. However, the family has a diversity of floral morphologies, and a number of the Missouri taxa deviate from this standard for one or more sets of floral organs. Particularly in *Oenothera*, the pollen grains are coherent into loose masses that are held together by strands of a substance called viscin. These very fine filaments may be observed with the naked eye in flowers that are actively shedding pollen or by touching the pollen from a dehiscing anther with a fingertip.

A number of the genera are cultivated as ornamentals for their flowers, including *Clarkia, Fuchsia, Ludwigia*, and *Oenothera*. The family also includes many species that have been used intensively in cytogenetic research on chromosomes pairing at meiosis. Many members of the family exhibit an unusual phenomenon in which portions of one chromosome become translocated to a different chromosome. When this behavior involves several chromosomes, the homologous portions of different chromosomes tend to pair at meiosis, creating a ring of chromosomes rather than the more usual separate pairs.

Beginning in the 1970s, Peter Raven and his students and colleagues began a long series of systematic studies (some of which are cited in the present generic treatments) that resulted in the Onagraceae being regarded by botanists as among the best-studied taxonomically in the plant kingdom. More recently, the application of molecular techniques to the study of phylogeny in the family has resulted in a substantial refinement of the generic classification (W. L. Wagner et al., 2007). Some genera, such as *Calylophus*, that had been segregated from *Oenothera* by the Raven group based on morphological evidence alone, are now regarded as representing merely specialized groups within *Oenothera*. More surprisingly, the genera *Gaura* and *Stenosiphon*, which had been thought to be distinct genera by nearly all botanists based on their relatively small, nutlike fruits and in *Gaura* unusual zygomorphic corollas, also have been shown to represent subgroups within *Oenothera*. The present treatment follows the revised classification presented in the recent comprehensive generic monograph of the family by W. L. Wagner et al. (2007).

Key to genera based mainly on vegetative features

1. Leaves alternate or all basal
    2. Floral tube well-developed, the perianth appearing attached to the tip of an elongate tube well above the ovary . . . . 4. OENOTHERA
    2. Floral tube absent or very short, the perianth appearing attached at the tip of the ovary or a short, crownlike tube . . . . 3. LUDWIGIA
1. Leaves opposite, at least along the main stem
    3. Leaf blades with the margins toothed
        4. Petioles well-developed, 10–50 mm long; leaf blades broadly ovate to ovate or oblong-ovate, 20–60 mm wide; sepals, petals, and stamens 2 per flower; fruits globose to broadly pear-shaped. . . . . 1. CIRCAEA
        4. Petioles absent or inconspicuous, to 5 mm long; leaf blades linear to narrowly lanceolate or oblong-lanceolate, 2–25 mm wide; sepals, petals, and stamens 4 per flower; fruits linear . . . . 2. EPILOBIUM
    3. Leaf blades with the margins entire

5. Leaf blades linear to very narrowly elliptic, the surfaces densely short-hairy; stems erect or ascending . . . . 2. EPILOBIUM
5. Leaf blades narrowly to more commonly broadly elliptic or obovate-elliptic, the surfaces glabrous; stems floating, creeping or loosely ascending . . . . 3. LUDWIGIA

Key to genera based mainly on flowers and fruits

1. Sepals persistent after flowering; floral tube absent . . . . 3. LUDWIGIA
1. Sepals shed (along with the other flower parts) after flowering; floral tube present, often elongate
   2. Stipules present, often shed early; fruits indehiscent, burlike, the surface with hooked hairs; flowers with 2 sepals and petals; leaves opposite, long-petiolate . . . . 1. CIRCAEA
   2. Stipules absent; fruits dehiscent capsules or, if indehiscent, then not burlike; flowers with (3)4 sepals and petals; leaves basal, alternate, or opposite, sessile or short-petiolate
      3. Leaves mostly opposite below the inflorescence; seeds with a dense tuft of hairs; sepals remaining erect at flowering . . . . 2. EPILOBIUM
      3. Leaves alternate or basal; seeds lacking tuft of hairs; sepals becoming reflexed as the flowers open . . . . 4. OENOTHERA

1. Circaea L.

Eight species, North America, Europe, Asia.

1. Circaea canadensis (L.) Hill (enchanter's nightshade)

C. lutetiana L. ssp. *canadensis* (L.) Aschers. & Magnus

C. quadrisulcata (Maxim.) Franch. & Sav. var. *canadensis* (L.) H. Hara

Pl. 462 i–k; Map 2107

Plants perennial herbs, with long basal stolons. Stems 20–90 cm long, erect or strongly ascending, simple or rarely branched, glabrous toward the base, sparsely pubescent with short, glandular hairs toward the tip. Leaves opposite, the petiole (1.3–)2.5–5.5 cm long. Stipules minute, glandular, usually shed early. Leaf blades 5–16 cm long, 2.5–8.5 cm wide, narrowly to broadly ovate or oblong-ovate, rounded to slightly cordate at the base, gradually tapered to a usually sharply pointed tip, the margins finely toothed, the surfaces minutely glandular-hairy to nearly glabrous. Inflorescences terminal racemes, these often grouped into open, few-branched panicles, the axis 2.5–30 cm long, glandular-hairy. Flowers zygomorphic, opening during the day; the stalk 2.5–6.5 mm long, spreading at flowering, reflexed or downward-curved at fruiting. Floral tube (0.4–)0.7–1.2 mm long, funnelform. Sepals 2, 1.9–3.8 mm long, 1.2–2.4 mm wide, very broadly elliptic, oblong, or oblong-ovate, green or purple, spreading to reflexed at flowering. Petals 2, (1.3–)1.6–2.9 mm long, (1.5–)2.2–4.0 mm wide, commonly white, rarely pink, broadly ovate-triangular to broadly obovate or heart-shaped; apical notch 1/3 to slightly more than 1/2 the length. Stamens 2; the filaments 1.2–2.8 mm long. Ovary 2-locular, the style 2.5–5.5 mm long, the stigma entire or shallowly 2-lobed, capitate. Fruits indehiscent, burlike, 2.8–4.5 mm long, 1.9–3.6 mm wide, pear-shaped to subglobose, rounded at the tip, tapered obliquely at the base, the surface longitudinally ribbed, with short, dense, hooked hairs. Seeds 2, adhering to inner ovary wall, lacking an apical tuft of hairs. Chromosome number: $2n=22$. June–August.

Scattered nearly throughout the state (eastern U.S. and Canada, west to Manitoba, North Dakota, Wyoming, Oklahoma, and Louisiana). Bottomland forests, mesic upland forests, and bases of bluffs.

Boufford (1982) treated this circumboreal species under the name *C. lutetiana* L. as comprising three subspecies, only one of which occurs in North America. Later, Boufford (2005) reevaluated the taxonomy of the group based on molecular and other data and concluded that the three subspecies should be reclassified as two species, segregating *C. lutetiana* (Asian in the strict sense) based on differences in flower and fruit morphology, and leaving *C. canadensis* with two subspecies: ssp. *canadensis* in the New World and ssp. *quadrisulcata* (Maxim.) Boufford in the Old World (northeastern Europe across Asia to Japan). The Old World populations are distinguished from *C. canadensis* in the strict sense only in having flower stalks with the bractlet absent or microscopic (to 0.2 mm) vs. minute (0.2–0.7 mm). A recent molecular phylogenetic analysis supported the further separation of the *C. canadensis* complex into two species (Xie et al., 2009) and provided evidence that these were not closely related within the group of *Circaea* species possessing 2-locular ovaries. Thus, despite the strong morphological similarities of taxa within the *C. canadensis* complex, the Asian component is best treated as a separate species, *C. quadrisulcata* (Maxim.) Franch. & Sav.

2. **Epilobium** L. (willow herb)

Plants perennial herbs (tap-rooted annuals elsewhere), rarely woody at the base, sometimes with short rhizomes or elongate stolons. Stems erect to ascending, occasionally from a spreading base, unbranched or branched, sparsely to densely pubescent with short, appressed to upward curved, nonglandular hairs, sometimes also glandular, occasionally glabrous or nearly so. Leaves mostly opposite below the inflorescence (occasionally appearing in fascicles in *E. leptophyllum*), sessile to short-petiolate. Stipules absent. Leaf blades linear or narrowly oblong to lanceolate, elliptic, or those of the lowermost leaves sometimes obovate. Inflorescences terminal, short to elongate spikes, racemes, or panicles, or of solitary axillary flowers. Flowers actinomorphic (rarely zygomorphic elsewhere), opening during the day, sessile or stalked, erect or ascending. Floral tube short (elongate elsewhere), usually with hairs, scales, or a ring of tissue within, shed (with the sepals, petals, and stamens) after flowering. Sepals 4, lanceolate to ovate, green or rarely reddish- to purplish-tinged, ascending at flowering. Petals 4, white or pinkish-tinged (pink to pinkish purple or orangish red elsewhere), notched at the tip. Stamens 8, in two unequal series. Ovary 4-locular, the stigma entire or nearly so, narrowly club-shaped to capitate. Fruits capsules, narrowly cylindric, circular to sharply 4-angled in cross-section, dehiscing longitudinally to the base, leaving an intact central column. Seeds usually numerous per locule, with a dense tuft of long, silky hairs at the tip (the coma, this sometimes lacking elsewhere). About 165 species, nearly worldwide.

1. Leaf blades linear to very narrowly elliptic, 1.5–7.0 mm wide, the margins entire or nearly so; lower portion of stems evenly and densely hairy; plants forming threadlike, nearly leafless stolons that terminate in compact fleshy turions .... 3. E. LEPTOPHYLLUM
1. Leaf blades narrowly lanceolate to lanceolate or narrowly ovate, 5–35 mm wide, the margins sharply, irregularly or finely toothed; lower portion of stems nearly glabrous or the hairs mostly in longitudinal lines; plants forming leafy basal rosettes, lacking elongate stolons
    2. Leaf blades less densely toothed, the teeth mostly 1–5 per cm; seeds with a short beak, the apical tuft of hairs 2–8 mm long, white to off-white .... 1. E. CILIATUM
    2. Leaf blades more densely toothed, the teeth mostly 4–9 per cm; seeds lacking a beak, the apical tuft of hairs 8–12 mm long, tan to reddish brown .... 2. E. COLORATUM

1. **Epilobium ciliatum** Raf. **ssp. ciliatum** (fringed willow herb)

Map 2108

Plants variable in stature, forming leafy basal rosettes, these sessile or at the tips of very short rhizomes, lacking elongate stolons. Stems (20–)40–85(–120) cm long, unbranched or branched above, sparsely to moderately hairy toward the base, the hairs in lines decurrent from the leaf bases, grading into dense, short, downward-curved hairs to-

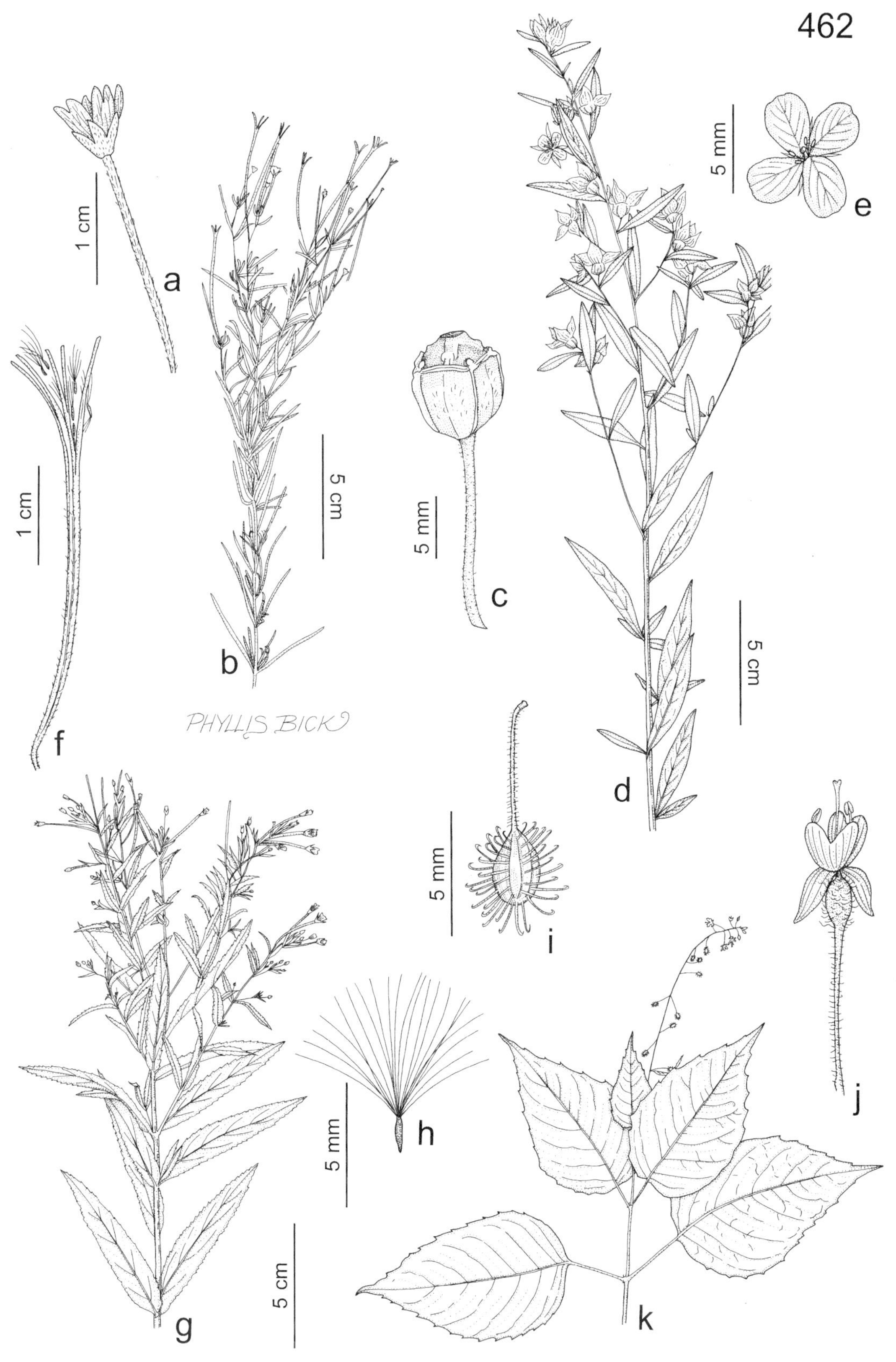

Plate 462. Onagraceae. *Epilobium leptophyllum*, **a)** flower, **b)** fertile stem. *Ludwigia alternifolia*, **c)** fruit, **d)** flower, **e)** fertile stem. *Epilobium coloratum*, **f)** dehiscing fruit, **g)** fertile stem, **h)** seed. *Circaea canadensis*, **i)** fruit, **j)** flower, **k)** fertile stem.

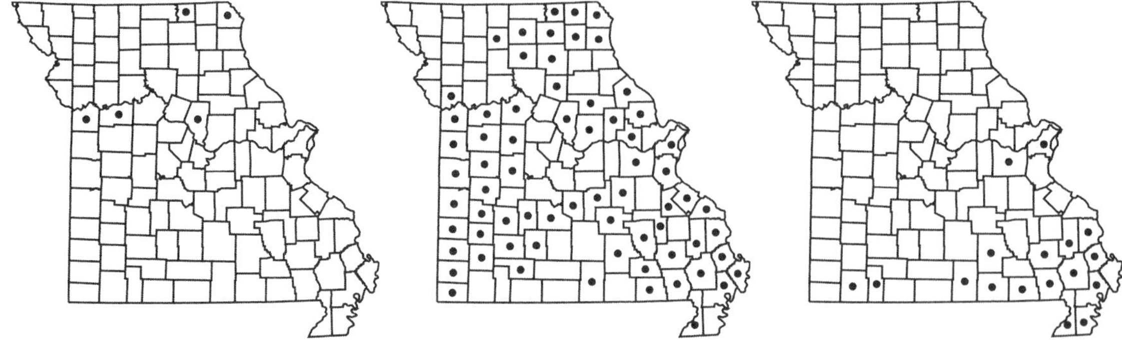

2110. Epilobium leptophyllum 2111. Ludwigia alternifolia 2112. Ludwigia decurrens

ward the tip, sometimes also with longer, gland-tipped hairs. Petioles 1–5 mm long, the upper leaves usually sessile or nearly so. Blades of stem leaves 3–12 cm long, 5–35 mm wide, narrowly lanceolate to narrowly ovate, rounded or angled at the base, angled or tapered to a sharply pointed tip, the margins with short, sharp, irregularly spaced teeth, these mostly 1–5 per cm, with short hairs on the veins and margins or nearly glabrous, the venation with prominent lateral veins, these often raised on the undersurface. Inflorescences short to elongate panicles, sometimes condensed and appearing as dense, large clusters, the bracts smaller than the stem leaves, the axis densely short-hairy, often also with longer, gland-tipped hairs, the flower stalks 3–17 mm long. Floral tube 0.5–2.6 mm long. Sepals 2–5 mm long, 0.7–2.5 mm wide, the tips free in bud. Petals 1.5–5.0 mm long, 1.5–3.0 mm wide, white or pinkish-tinged, the notch 0.7–1.5 mm deep. Stamens with the filaments white or purple, the anthers 0.4–1.8 mm long. Fruits 4–10 cm long, short-hairy. Seeds 0.8–1.6 mm long, 0.3–0.6 mm wide, narrowly oblong-obovoid, abruptly tapered to a short, broad beak at the tip, the surface brown to olive brown, with longitudinal ridges, the apical tuft of hairs 2–8 mm long, white to off-white. $2n$=36. June–August.

Introduced, known thus far only from St. Louis County (nearly throughout the U.S. [including Alaska] except for some southeastern states; Canada, eastern Asia; introduced in Europe, elsewhere in Asia, and Australia). Open, disturbed area.

This species was first collected by Kay Yatskievych in 2009 growing among low shrubs in a newly planted suburban shopping center parking lot median, where plants likely had originated from seed contaminants in mulch. The population appears not to have persisted, but the species has been recorded from most surrounding states and is thus expected to reappear again in the future.

Epilobium ciliatum is usually treated as comprising three subspecies. The ssp. *ciliatum* is widespread. The other two, ssp. *glandulosum* (Lehm.) Hoch & P.H. Raven and ssp. *watsonii* (Barbey) Hoch & P.H. Raven, which have American distributions to the north and/or west of Missouri, differ in their somewhat larger, more strongly pink to pinkish purple corollas, denser inflorescences, and generally broader leaves that are only slightly reduced in the inflorescence.

2. Epilobium coloratum Biehler (purple-leaved willow herb, cinnamon willow herb)
Pl. 462 f–h; Map 2109

Plants often robust and rank, forming leafy basal rosettes, these sessile or at the tips of very short rhizomes, lacking elongate stolons. Stems (20–)40–85(–120) cm long, freely branched above, sparsely hairy to nearly glabrous toward the base, grading into moderate to dense, short, incurved hairs toward the tip, the hairs sometimes in lines decurrent from the leaf bases. Petioles 4–10 mm long, the upper leaves often sessile or nearly so. Blades of stem leaves 3–10(–15) cm long, 5–20(–30) mm wide, narrowly lanceolate to lanceolate or narrowly elliptic, rounded or angled at the base, angled or tapered to a sharply pointed tip, the margins with short, sharp, irregularly spaced teeth, these mostly 4–9 per cm, with short hairs on the veins and margins or nearly glabrous, the venation with prominent lateral veins, these often raised on the undersurface. Inflorescences short to elongate panicles, sometimes condensed and appearing as dense, large clusters, the bracts smaller than the stem leaves, the axis densely short-hairy, the flower stalks 5–10 mm long. Floral tube 0.3–0.6 mm long. Sepals 1.3–3.2 mm long, 0.5–1.5 mm wide, fused to the tip in bud. Petals 2.5–5.5 mm long, 2.0–3.8 mm wide, white or pinkish-tinged, the notch 1.0–1.3 mm deep. Stamens with the filaments white, the anthers 0.3–0.45 mm long. Fruits 4.0–

6.5 cm long, short-hairy. Seeds 1.2–1.7 mm long, 0.3–0.5 mm wide, very narrowly obovoid, abruptly rounded at the tip, lacking a beak, the surface brown to olive brown, with evenly spaced minute papillae, the apical tuft of hairs 8–12 mm long, tan to orangish or reddish brown. $2n=36$. July–October.

Scattered nearly throughout the state (eastern U.S. west to North Dakota and Texas; eastern Canada). Banks of streams and spring branches, bottomland prairies, marshes, swamps, fens, and seeps; sometimes emergent aquatics in shallow water.

3. Epilobium leptophyllum Raf. (narrow-leaved willow herb, bog willow herb)

Pl. 462 a, b; Map 2110

Plants often robust and rank, forming threadlike, nearly leafless stolons that terminate in compact fleshy turions. Stems 15–95 cm long, unbranched to much-branched in larger plants, densely pubescent throughout with short, incurved hairs, sometimes also with gland-tipped hairs toward the tip, the hairs evenly distributed, not in lines. Petioles absent or nearly so. Blades of stem leaves 2.0–7.5 cm long, 1.5–7.0 mm wide, linear to very narrowly elliptic, angled or tapered at the base, angled to a bluntly or sharply pointed tip, the margins entire or nearly so, sometimes rolled under, the surfaces densely short-hairy, the venation with inconspicuous lateral veins, these not appearing raised on the undersurface. Inflorescence racemes (occasionally clustered into few-branched panicles or reduced to solitary flowers), the bracts not much smaller than the stem leaves, the axis densely short-hairy and sometimes also glandular-hairy, the flower stalks 10–15 mm long. Floral tube 0.8–1.5 mm long. Sepals 2.5–4.5 mm long, 0.9–1.3 mm wide, green, the tips free in bud. Petals 3.5–7 mm long, 1.6–4 mm wide, white to light pinkish-tinged, the notch 1.0–1.8 mm deep. Stamens with the filaments white or cream-colored, the anthers 0.6–0.9 mm long. Fruits 3.5–8.0 cm long, densely short-hairy. Seeds 1.5–2.2 mm long, 0.5–0.7 mm wide, narrowly obovoid, abruptly tapered to a short, broad beak at the tip, the surface brown to olive brown, with evenly spaced minute papillae, the apical tuft of hairs 6–8 mm long; off-white. $2n=36$. July–September.

Uncommon, known thus far only from two counties in northeastern Missouri and three counties along the Missouri River (northern U.S. [including Alaska], less common in the southeastern U.S., sporadically west to California, and Arizona; Canada). Fens and marshes.

3. Ludwigia L. (false loosestrife, seedbox)

Plants annual or more commonly perennial herbs (shrubs or rarely small trees elsewhere). Stems glabrous to densely hairy, erect to prostrate and then often rooting at the nodes, sometimes floating, the underwater parts sometimes swollen and spongy or with inflated, white, spongy, rootlike structures (pneumatophores). Leaves alternate (opposite in *L. palustris* and a few non-Missouri species), sessile or petiolate. Stipules minute, scalelike or succulent and reddish purple, often withering and often turning dark brown, usually shed early. Leaf blades linear to lanceolate, oblong, obovate, or rarely triangular-ovate, the margins entire but sometimes glandular, less commonly with scattered, minute glandular teeth. Inflorescences axillary, of solitary flowers (occasionally appearing terminal and racemose). Flowers actinomorphic, opening during the day, sessile or stalked, ascending to spreading, sometimes with a pair of bractlets at the base of the ovary. Floral tube absent. Sepals 4 or 5(–7), green, sometimes turning yellow at maturity, spreading to nearly erect at flowering, persistent at fruiting. Petals absent or more commonly 4 or 5(–7), yellow (white elsewhere), shed after flowering, rounded to truncate at the tip, occasionally minutely notched or with a minute sharp point at the very tip. Stamens 4 in 1 series or 8 or 10(12, 14) in 2 subequal series. Ovary 4- or 5(6)-locular (rarely with more locules elsewhere), the stigma entire or irregularly lobed, capitate or hemispheric. Fruits capsules, obconic, cylindric, or globose, circular or with 4 or more sharp angles in cross-section, dehiscing irregularly, by a terminal pore, or longitudinally from the tip along the angles. Seeds numerous, in 1 to several rows per locule, lacking an apical tuft of hairs. About 82 species, nearly worldwide, most diverse in tropical and warm-temperate regions.

Traditionally, many North American authors regarded species with stamens twice as many as the petals as a separate genus, *Jussiaea* L. (Steyermark, 1963). Hara (1953) summarized information, mostly based on Old World taxa, indicating that this separation was artificial and that the combined group should be called *Ludwigia*.

Aquatic species of *Ludwigia* typically form structures from the roots or submerged or floating portions of the stems known as pneumatophores. These usually white rootlike structures are involved in gas exchange. As they develop, they elongate to the water surface and the terminal portion is floating rather than emergent.

1. Stamens twice as many as the sepals
    2. Sepals 4; seeds free; stem sharply 4-angled and 4-winged ..... 2. L. DECURRENS
    2. Sepals 5(–7); seeds more or less embedded in the inner layer of the fruit at maturity, not free; stem circular in cross-section or nearly so, never winged
        3. Seeds loose in horseshoe-shaped pieces of inner fruit tissue; stems 30–250 cm long, erect or strongly ascending, the plants not producing rhizomes ..... 5. L. LEPTOCARPA
        3. Seeds firmly embedded in woody blocks of inner fruit tissue; plants producing rhizomes or creeping or floating stems 20–150(–300) cm long, the flowering stems (produced from the base of the plant in *L. grandiflora*) or branches sometimes erect or ascending
            4. Flowering stems usually erect; bractlets obovate; sepals 8–12(–15) mm long; petals mostly 15–18(–20) mm long .... 4. L. GRANDIFLORA
            4. Flowering stems floating or creeping, occasionally loosely ascending toward the tip; bractlets deltoid; sepals 8–12 mm long; petals mostly 9–15(–22) mm long ..... 9. L. PEPLOIDES
1. Stamens as many as sepals
    5. Fruits subglobose to nearly cubic, dehiscing by an apical pore (sometimes eventually breaking into 4 valves with age); flower stalks 2.2–7.0 mm long ..... 1. L. ALTERNIFOLIA
    5. Fruits cylindric to obconic or club-shaped, dehiscing irregularly by decay of the walls or by an irregular zone of weakness at the base of the sepals; flowers sessile or the stalks less than 1 mm long
        6. Leaves opposite; petals absent; stems prostrate to loosely ascending ..... 8. L. PALUSTRIS
        6. Leaves alternate; petals absent or 4; stems erect or ascending
            7. Fruits at least twice as long as wide
                8. Petals absent; leaf blades narrowly elliptic to narrowly lanceolate, 0.2–2.1 cm wide; fruits more or less cylindric (sometimes with 4 shallow, longitudinal grooves, but not wider toward the tip) ..... 3. L. GLANDULOSA
                8. Petals 4; leaves linear or nearly so, 0.1–0.4 cm wide; fruits narrowly club-chaped (wider toward the tip) .... 6. L. LINEARIS
            7. Fruits less than twice as long as wide
                9. Leaf blades 0.4–1.7 cm long, 2–10 mm wide, obovate-spatulate to oblanceolate; sepals 0.9–2.0 mm long .... 7. L. MICROCARPA
                9. Leaf blades 3.5–11.0 cm long, 4–10(–17) mm wide, narrowly elliptic to narrowly oblong-elliptic or narrowly oblanceolate; sepals 2.5–4.5 mm long ..... 10. L. POLYCARPA

1. Ludwigia alternifolia L. (bushy seedbox)

L. alternifolia var. *pubescens* E.J. Palmer & Steyerm.

Pl. 462 c–e; Map 2111

Plants perennial, with fleshy, somewhat thickened roots. Stems 60–150 cm long, erect or strongly

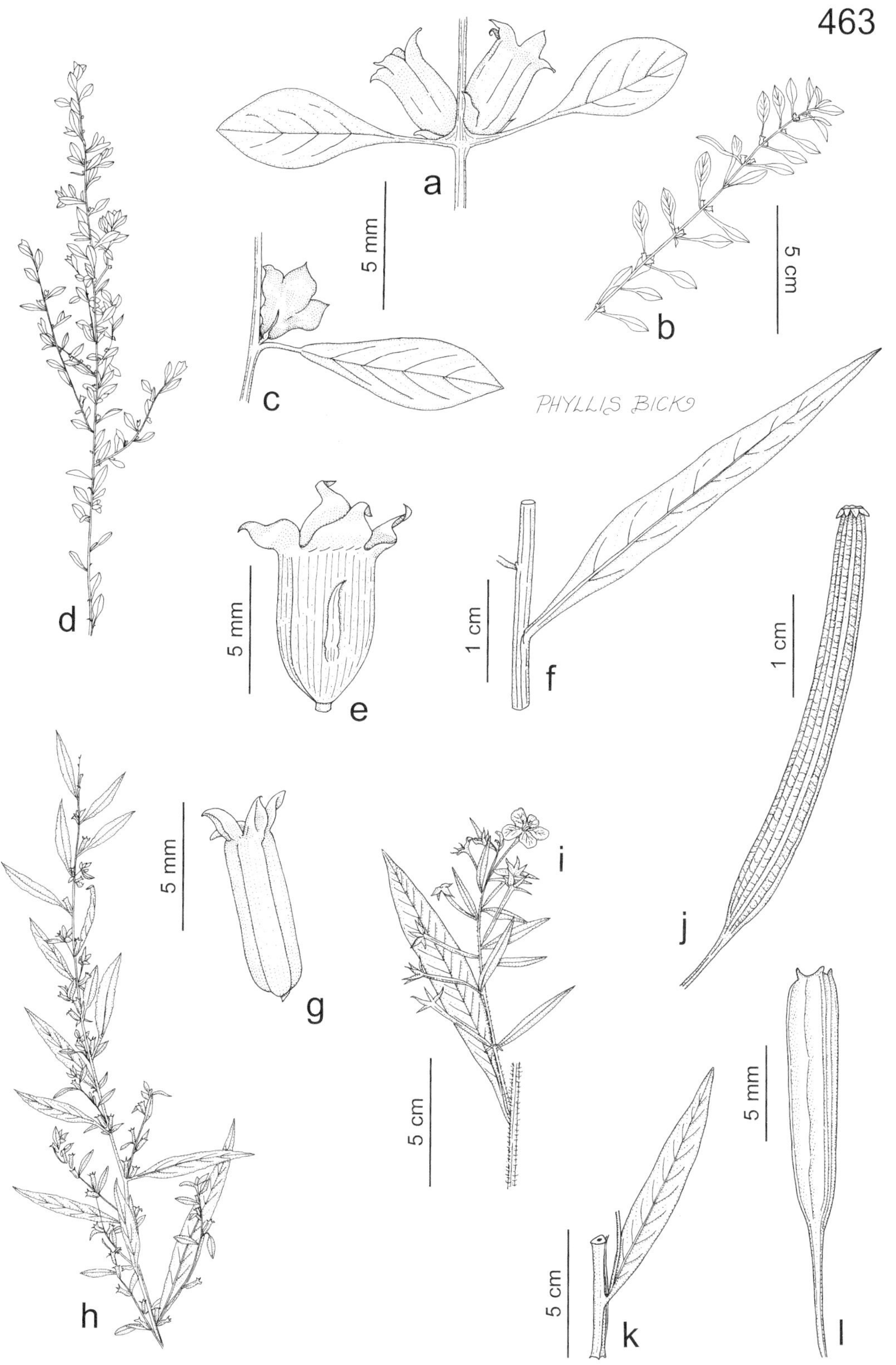

Plate 463. Onagraceae. *Ludwigia palustris*, **a)** node with pair of leaves and fruits, **b)** fertile stem. *Ludwigia microcarpa*, **c)** node with leaf and fruit, **d)** fertile stem. *Ludwigia polycarpa*, **e)** fruit, **f)** portion of stem with leaf. *Ludwigia glandulosa*, **g)** fruit, **h)** fertile stem. *Ludwigia leptocarpa*, **i)** fertile stem, **j)** fruit. *Ludwigia decurrens*, **k)** portion of stem with leaf, **l)** fruit.

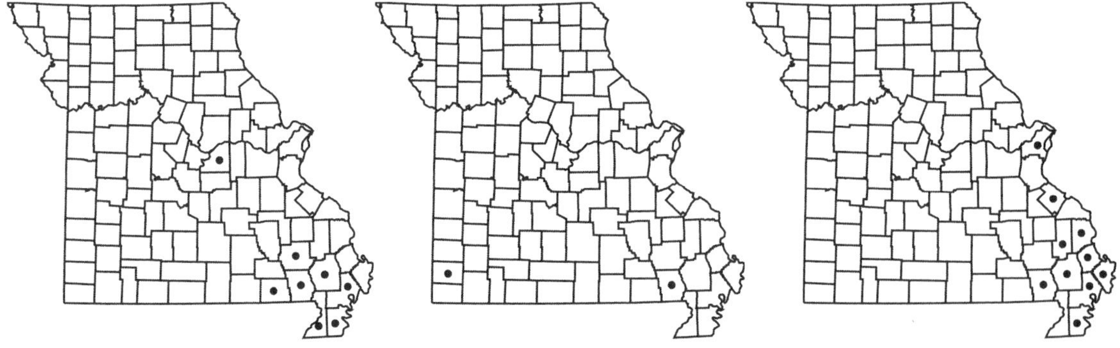

2113. Ludwigia glandulosa 2114. Ludwigia grandiflora 2115. Ludwigia leptocarpa

ascending, unbranched to much-branched toward the tip, not swollen or spongy, but the bark splitting near the base in older plants, glabrous or pubescent with short hairs in lines decurrent from the leaf bases. Leaves alternate, the petiole 1–3 mm long. Stipules 0.2–0.3 mm long, ovate-triangular to peg-like, dark reddish purple, succulent. Leaf blades (0.6–)2.0–12.0 cm long, (3–)10–12(–25) mm wide, lanceolate to narrowly lanceolate or lanceolate-elliptic, tapered at the base, angled to a sharply pointed tip, the margins entire, the surfaces variably short-hairy, especially along the veins, the venation pinnate, but the secondary veins more or less fused to form a submarginal vein or loops. Flower stalks 2–7 mm long at flowering and fruiting, the bractlets narrowly lanceolate, green, short-hairy. Sepals 4, (5.9–)6.5–9.6 mm long, 4.0–6.8 mm wide, green or reddish-tinged, ovate-triangular, bluntly or sharply pointed at the tip. Petals 4, 10–14 mm long, 8–12 mm wide, obovate to obtriangular or more or less heart-shaped, broadly rounded to truncate or with a shallow notch at the tip, yellow. Stamens 4, the filaments white, 1.0–3.0 mm long; the anthers 1.0–1.7 mm long. Fruits 4–7 mm long, 4–6 mm wide, subglobose to nearly cubic, 4-angled and often narrowly 4-winged, hard-walled, glabrous or short-hairy, dehiscing by a terminal pore (sometimes eventually breaking into 4 valves with age). Seeds 0.5–0.8 mm long, oblong-cylindric, sometimes slightly curved, free from the fruit tissue at maturity, the surface yellowish brown, shiny, with a fine network of ridges. $2n$=16. June–August.

Scattered nearly throughout the state, but apparently absent from most of the northwestern quarter (eastern U.S. west to Nebraska and Texas; Canada). Banks of streams, rivers, and spring branches, margins of ponds, lakes, and sinkhole ponds, marshes, fens, seeps, bottomland prairies, and swamps; also ditches, railroads, and roadsides.

This species is the only member of the endemic North American sect. *Ludwigia* in Missouri. Steyermark (1963) separated somewhat hairy plants as var. *pubescens*. However, Lutz (1986) indicated that there is too much intergradation between the glabrous and pubescent races of *L. alternifolia* to permit formal recognition of varieties. The species appears to be self-fertile and facultatively self-pollinated.

2. Ludwigia decurrens Walter (erect primrose willow)

Jussiaea decurrens (Walter) DC.

Pl. 463 k, l; Map 2112

Plants annual, the roots and lower stem sometimes inflated and spongy. Stems 30–200 cm long, erect or strongly ascending, unbranched to densely branched, sharply 4-angled and 4-winged (from the decurrent leaf bases), glabrous. Leaves alternate, sessile. Stipules 0.4–0.5 mm long, ovate-triangular. Leaf blades 2–20 cm long, 2–50 mm wide, lanceolate to narrowly lanceolate or sometimes broadly lanceolate to elliptic, rounded or angled at the base, angled or tapered to a sharply pointed tip, the margins entire, but often minutely roughened, the surfaces glabrous or the undersurface occasionally minutely hairy along the veins, the relatively prominent venation pinnate, but with the secondary veins fused to form a submarginal vein. Flower stalks 1–5 mm long, the bractlets obovate, usually glabrous, usually shed after flowering. Sepals 4, 7–12 mm long, 1.5–4.0 mm wide, ovate or less commonly lanceolate, angled or short-tapered to a sharply pointed tip. Petals 4, 10–20 mm long, 10–18 mm wide, broadly obovate, broadly rounded to truncate or with a shallow notch at the tip, yellow. Stamens 8, the filaments yellow, the anthers 1.3–1.6 mm long. Fruits 10–25 mm long, 3–5 mm wide, narrowly obconic, 4-angled and 4-winged, straight or rarely sharply curved, thin-walled, glabrous, dehiscing longitudinally. Seeds 0.5–0.6 mm long, broadly oblong-cylindric, sometimes slightly wider toward 1 end, free from the fruit tissue at maturity, the surface yellowish brown, nearly smooth, with a faint longitudinal patttern of fine lines. $2n=16$. June–October.

Scattered in the southeastern quarter of the state, north locally to Franklin and St. Louis Counties and west locally to Barry and Stone Counties (eastern [mostly southeastern] U.S. west to Missouri and Texas). Banks of streams, rivers, and sloughs, margins of ponds, lakes, and sinkhole ponds, marshes, fens, and swamps; also ditches, borrow pits, and crop fields.

This species, which occurs throughout the southeastern United States, Central America, and the Caribbean, to northeastern Argentina, is a member of sect. *Pterocaulon* Ramamoorthy, other members of which occur mainly in South America, one species also extending to Africa (Raven, 1963; Ramamoorthy and Zardini, 1987).

3. Ludwigia glandulosa Walter **ssp. glandulosa** (cylindric-fruited primrose willow)

Pl. 463 g, h; Map 2113

Plants perennial, forming leafy basal stolons up to 20 cm long, the roots fibrous, often spongy when submerged. Stems (20–)40–80(–100) cm long, erect or strongly ascending (sometimes spreading and rooting at the nodes in young plants), densely branched, glabrous or sparsely pubescent with short hairs in lines decurrent from the leaf bases, often reddish. Leaves alternate, sessile or the petiole to 15 mm long. Stipules 0.2–0.4 mm long, ovate-triangular, reddish purple, succulent. Leaf blades 1–12 cm long, 3–21 mm wide, elliptic to oblanceolate-elliptic or very narrowly elliptic, tapered at the base, angled to a usually sharply pointed tip, the margins appearing nearly entire, but hairy and with minute, glandular teeth, the venation with inconspicuous secondary veins. Flowers sessile or nearly so, the bractlets 0.3–1.0 mm long, linear to lanceolate, green. Sepals 4, 1.2–2.3 mm long, 1.1–1.7 mm wide, ovate-triangular, tapered to a sharply pointed tip. Petals absent. Stamens 4, the filaments green, the anthers 0.3–0.5 mm long. Fruits (4–)5–7(–9) mm long, 1.6–2.0(–3.0) mm wide, subcylindric, bluntly 4-angled to nearly circular in cross-section, stiff-walled, short-hairy to nearly glabrous, dehiscing irregularly by disintegration of the wall. Seeds 0.5–0.7 mm long, oblong-cylindric but curved (sometimes appearing more or less kidney-shaped), free from the fruit tissue at maturity, the surface light brown to yellowish brown, appearing faintly pitted, under magnification with oblong longitudinally elongate cells. $2n=16$. June–September.

Uncommon in the Mississippi Lowlands Division and adjacent portions of the Ozarks, with a single disjunct occurrence in Osage County (southeastern U.S. west to Colorado and Texas). Bottomland forests, swamps, fens, and margins of ponds; also ditches and cemeteries.

This taxon, a member of the North American sect. *Microcarpium* Munz, is common throughout the Atlantic and Gulf coastal plains from Texas to Virginia. Peng (1989) recognized *L. glandulosa* ssp. *brachycarpa* (Torr. & A. Gray) C.I. Peng for populations with smaller capsules on the western edge of this distribution.

Ludwigia glandulosa is sometimes cultivated as a submerged aquatic in aquaria, especially selections with abundant, reddish stem and leaf coloration, as well as a compact growth form.

4. Ludwigia grandiflora (Michx.) Greuter & Burdet (large-flowered primrose willow)

L. grandiflora (Michx.) Zardini, H. Gu, & P.H. Raven, an invalid name

L. uruguayensis (Cambess.) H. Hara

Jussiaea uruguayensis Cambess.

Pl. 464 c, d; Map 2114

Plants perennial, fibrous-rooted. Stems 20–120 cm long, sometimes woody at the base (but the bark not peeling), often of 2 types: vegetative stems floating and/or submerged in water or creeping on mud, rooting at the lower nodes, forming pneumatophores, sometimes transitioning into fertile stems or branches, these erect or strongly ascending, unbranched or more commonly well-branched above the midpoint, rounded or sometimes angled toward the tip, densely pubescent with short, spreading glandular hairs (sticky) or sometimes nearly glabrous toward the base (or throughout if submerged). Leaves alternate, sometimes appearing in fascicles, the petiole (1–)5–21 mm long. Stipules 0.6–2.0 mm long, lanceolate to ovate-triangular, dark reddish purple, succulent. Leaf blades (1.7–)3.0–8.0(–10.0) cm long, 8–28 mm wide, lanceolate to elliptic, narrowly elliptic, oblanceolate, or narrowly oblanceolate, rarely nearly linear or very narrowly elliptic, angled or tapered at the base, angled or slightly tapered to a bluntly or more or less sharply pointed tip, often with a minute glandular extension of the midvein at the very tip, the margins entire, the surfaces (except when submerged) sticky, densely hairy, the venation pinnate, but the relatively conspicuous secondary veins more or less fused to form a submar-

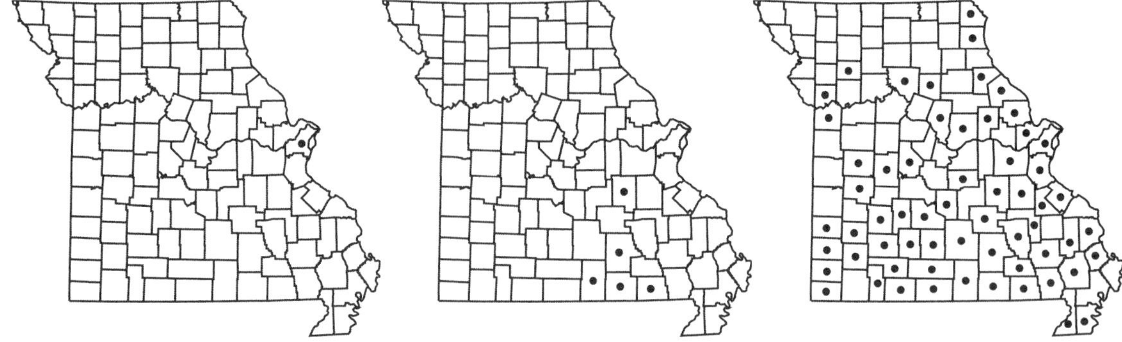

2116. Ludwigia linearis 2117. Ludwigia microcarpa 2118. Ludwigia palustris

ginal vein. Flower stalks 10–25 mm long (becoming elongated to 15–45 mm at fruiting), the bractlets 0.5–1.0 mm long, obovate, green or reddish-tinged. Sepals 5(6), 8–12(–15) mm long, 2–4(–5) mm wide, lanceolate, angled or slightly tapered to a sharply pointed tip. Petals 5(6), (12–) 15–18(–25) mm long, 11–21 mm wide, broadly obovate, broadly rounded to truncate or with a shallow notch at the tip, yellow. Stamens 10(12) in 2 unequal whorls, the filaments yellow, the anthers 1.0–2.5 mm long. Fruits (11–)14–25 mm long, 3–5 mm wide, cylindric, more or less circular in cross-section, truncate at the tip and narrowed abruptly at the base, thick-walled and rather woody, densely short-hairy, dehiscing tardily and irregularly. Seeds 1.0–1.3 mm long, oblong-cylindric, firmly embedded in wedge-shaped pieces of woody fruit tissue. $2n$=48. August–October.

Uncommon, known thus far only from historical collections from Newton County and a single more recent collection from Butler County (eastern U.S. west to Missouri and Texas, also disjunctly in California to Washington; Mexico, Central America, South America; introduced in Europe). Streams and ponds; usually emergent to floating aquatics (the basal portions rooted along the banks, the nonbasal portions of the stems often floating on the water).

Based on morphological and cytological data, Zardini et al. (1991) recognized two distinct, widely distributed species within the extremely diverse species formerly known as *L. uruguayensis*. The Missouri specimens correspond to a hexaploid ($2n$ = 48) taxon that is hairy and usually sticky when fresh, with mainly lanceolate leaves and smaller flowers. The oldest name for this taxon, which includes the type specimen of *L. uruguayensis*, was determined to be *L. grandiflora* (Zardini et al., 1991; Zardini and Raven, 1992). Populations of dodecaploid ($2n$ = 80) plants that are glabrous and have mainly oblanceolate leaves and somewhat larger flowers (sepals (8–)13–19 mm, petals (15–) 20–29 mm) have been called *L. hexapetala* (Hook. & Arn.) Zardini, H.Y. Gu & P.H. Raven or *L. grandiflora* ssp. *hexapetala* (Hook. & Arn.) G.L. Nesom & Kartesz. This larger-flowered variant, which is widespread in the southeastern United States (and from there to South America), has not yet been documented from Missouri, but eventually may be found in the state.

Both of these species are members of the cosmopolitan sect. *Oligospermum* (Micheli) H. Hara, which includes some of the most widespread and invasive taxa in the genus. Farther south in its range, *L. grandiflora* can form large, dense monocultures along the margins of ponds and other still to slow-moving bodies of water. It is thus considered a noxious weed in some southeastern states. Steyermark (1963) did not consider *L. grandiflora* to be native in the United States. However, there is no evidence from specimen labels that the collectors of the Missouri specimens thought that their vouchers represented nonnative occurrences, and subsequent authors have treated these populations as the northernmost native occurrences of the species.

5. Ludwigia leptocarpa (Nutt.) H. Hara (hairy primrose willow)

Jussiaea leptocarpa Nutt.

Pl. 463 i, j; Map 2115

Plants annual or perennial, sometimes woody toward the base, not producing rhizomes, when aquatic usually forming pneumatophores from the roots. Stems 30–250 cm long, erect or strongly ascending, usually well-branched, circular in cross-section or nearly so, unwinged, with longitudinal lines, densely pubescent with fine, mostly spreading hairs (glabrous elsewhere). Leaves alternate, the petiole 2–35 mm long. Stipules 0.2–0.5 mm long, ovate-triangular. Leaf blades 3–18 cm long, 10–40 mm wide, narrowly to broadly lanceolate,

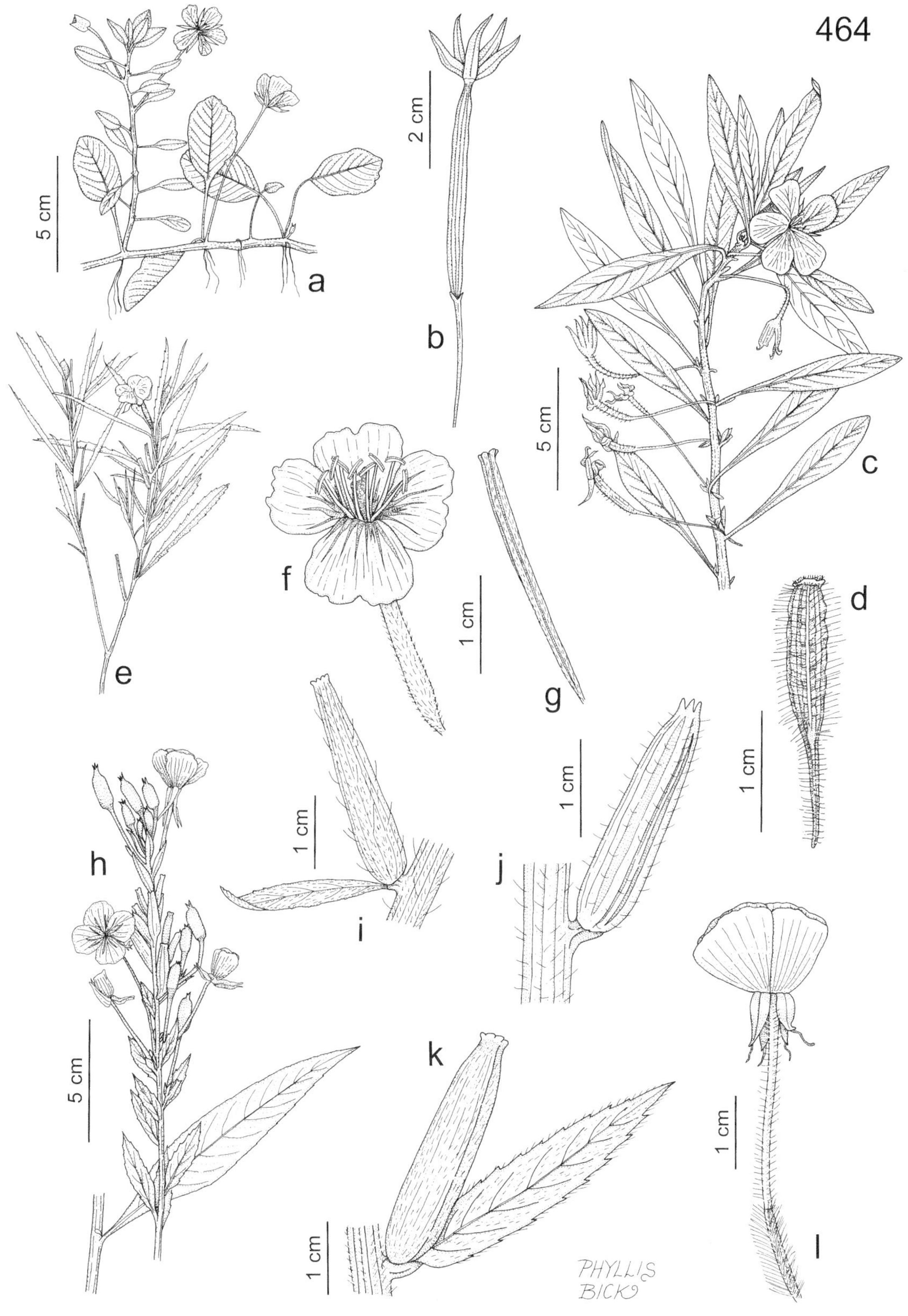

Plate 464. Onagraceae. *Ludwigia peploides*, **a)** habit, **b)** fruit. *Ludwigia grandiflora*, **c)** fertile stem, **d)** fruit. *Oenothera serrulata*, **e)** fertile stem, **f)** flower, **g)** fruit. *Oenothera biennis*, **h)** fertile stem tip and median node with leaf, **i)** node with bract and fruit. *Oenothera nutans*, **j)** node with fruit. *Oenothera villosa*, **k)** node with bract and fruit. *Oenothera parviflora*, **l)** flower.

narrowly angled or somewhat tapered at the base, angled or tapered to a bluntly or sharply pointed tip, the margins entire, the surfaces moderately to densely hairy when young, at maturity the pubescence mostly restricted to the main veins, but the tissue between the veins often appearing minutely pebbled (slightly raised where hairs formerly were attached), the relatively inconspicuous venation pinnate, but with the secondary veins fused to form a series of submarginal loops. Flower stalks 1–4 mm long, becoming elongated to 2–20 mm at fruiting, the bractlets usually absent (narrowly ovate-triangular, hairy, and usually shed after flowering elsewhere). Sepals 5(–7), 5.5–11.0 mm long, 1.5–3.0 mm wide, lanceolate, tapered to a sharply pointed tip. Petals 5(–7), 5–11 mm long, 4–8 mm wide, broadly obovate, broadly rounded or with a shallow notch at the tip, yellow to orangish yellow. Stamens (8)10(–14), the filaments yellow, the anthers 1.2–1.6 mm long. Fruits 15–20 mm long, 2.5–4.0 mm wide, more or less cylindric, rounded or slightly several-angled (with 10–14 prominent nerves), unwinged, straight or slightly curved, thin-walled, hairy, dehiscing tardily and irregularly. Seeds 1.0–1.2 mm long, obovoid, loosely embedded in horseshoe-shaped pieces of inner fruit tissue, the surface pale brown, finely pitted, shiny. $2n$=32, 48. August–October.

Uncommon in the Mississippi Lowlands Division north locally along the Mississippi River to Ste. Genevieve County; recently discovered in St. Louis City (eastern [mostly southeastern] U.S. west to Missouri and Texas; Mexico, Central America, south america, Caribbean Islands). Banks of rivers and sloughs, margins of ponds and lakes, and swamps; also ditches and fallow fields.

This species, which is widespread in warmer regions of the New World, is the only member of sect. *Seminuda* P.H. Raven to occur in North America; the remaining species are distributed in South America and Africa.

6. Ludwigia linearis Walter (narrow-leaved primrose willow)

Map 2116

Plants perennial, forming leafy basal stolons to 20 cm long, the roots fibrous, sometimes spongy when submerged. Stems (20–)30–60(–140) cm long, erect or strongly ascending, few- to many-branched, appearing glabrous or nearly so without strong magnification, but usually microscopically hairy, often reddish-tinged. Leaves alternate, essentially sessile. Stipules 0.2–0.3 mm long, linear to linear-elliptic, green or reddish-tinged, not succulent. Leaf blades 1.5–6.0(–8.5) cm long, 1–4(–6) mm wide, linear or nearly so, angled or short-tapered at the base, narrowly angled to a sharply pointed tip, the margins appearing entire, but with minute, glandular teeth, the venation with inconspicuous secondary veins that mostly are fused into submarginal veins (these more or less parallel to the midvein). Flowers sessile or nearly so, the bractlets 0.4–4.0(–7.5) mm long, linear, green or reddish-tinged. Sepals 4, 2.3–5.0(–6.0) mm long, 1.0–3.0(–3.5) mm wide, triangular-ovate to narrowly triangular, tapered (often abruptly so) to a sharply pointed tip. Petals 4, 3–6 mm long, 2–5 mm wide, obovate to nearly circular, broadly and bluntly angled to rounded at the tip, sometimes with a shallow notch, yellow. Stamens 4, the filaments yellowish green, the anthers 1–2 mm long. Fruits 5–10(–12) mm long, 2.0–5.5 mm wide (at least twice as long as wide), narrowly club-shaped (obpyramidal), bluntly 4-angled in cross-section and often with a shallow longitudinal groove on each face, stiff-walled, glabrous or minutely hairy, dehiscing by disintegration of a ring of cells at the tip between the sepal bases and style base. Seeds 0.4–0.6 mm long, oblong-ellipsoid but with slightly curved ends, free from the fruit tissue at maturity, the surface light brown to yellowish brown, appearing faintly pitted, under magnification with oblong longitudinally or transversely elongate cells. $2n$=16. July–September.

Uncommon, known thus far only from two historical collections from St. Louis County (southeastern U.S. west to Missouri and Texas). Habitat unknown, but presumably some type of wetland.

This taxon, a member of the North American sect. *Microcarpium*, is included in the Missouri flora based on two specimens in the Missouri Botanical Garden Herbarium collected in the Allenton area by George Letterman (one of them dated 10 August 1894), but lacking other data on the labels. Letterman collected a number of species in the Allenton area that have not been rediscovered in Missouri in modern times (see also the History of Missouri Botany section in Volume 1 of the present work [Yatskievych, 1999]). Steyermark apparently did not have the opportunity to examine these specimens during his research on the Missouri flora. They were discovered and determined in 2002 by Nancy Parker, a volunteer for the Flora of Missouri Project. The closest localities for this mostly Coastal Plain species are in central Arkansas. Peng (1989) documented four races within *L. linearis* differing in their pubescence patterns. He chose not to formally name these because of widespread intergradation and co-occurrence within some populations. The Missouri specimens correspond to his “completely glabrous morph.”

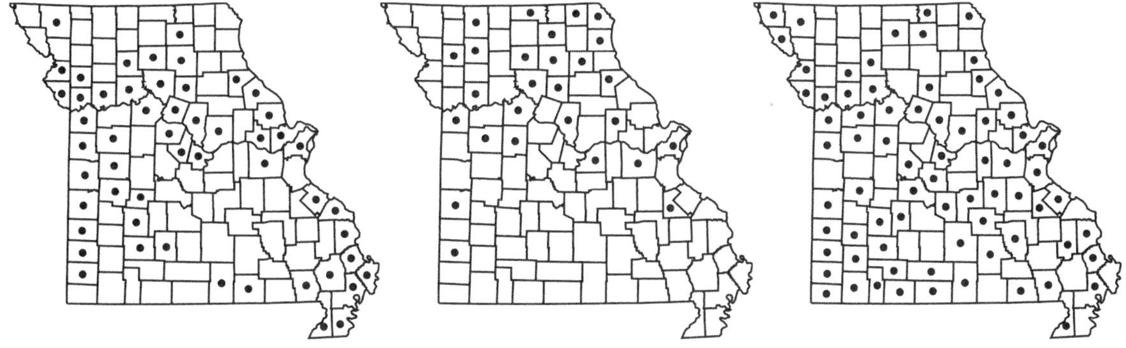

2119. Ludwigia peploides 2120. Ludwigia polycarpa 2121. Oenothera biennis

7. Ludwigia microcarpa Michx. (small false loosestrife)

Pl. 463 c, d; Map 2117

Plants perennial, forming leafy basal stolons to 15(–25) cm long, the roots fibrous, occasionally spongy when submerged. Stems 5–60 cm long, erect or strongly ascending, occasionally reclining with age, rarely prostrate, unbranched to well-branched, glabrous but with raised lines decurrent from the leaf bases, usually green. Leaves alternate, the petiole 0.5–5.0 mm long, narrowly winged. Stipules 0.1–0.2 mm long, lanceolate-triangular, reddish purple, succulent. Leaf blades 0.4–1.7 cm long, 2–10 mm wide, obovate-spatulate to oblanceolate, those of the largest leaves sometimes narrowly oblanceolate-elliptic, tapered at the base, angled or short-tapered to a usually sharply pointed tip, the margins appearing entire, but with minute, glandular teeth, the venation with inconspicuous secondary veins, these mostly free but occasionally the lowermost fused submarginally to the adjacent vein at their tips. Flowers sessile or nearly so, the bractlets 0.3–1.2 mm long, linear to narrowly oblong, usually reddish purple. Sepals 4, 0.9–2.0 mm long, 1.0–1.9 mm wide, triangular-ovate, short-tapered to a sharply pointed tip. Petals absent. Stamens 4, the filaments green and somewhat translucent, the anthers 0.1–0.2 mm long. Fruits 1.0–1.5 mm long, 1.4–1.9 mm wide (usually slightly wider than long), more or less obconic, sometimes slightly and bluntly 4-angled in cross-section but more often appearing irregularly distended over the seeds, thin-walled, appearing glabrous (microscopically hairy), dehiscing by disintegration of a ring of cells at the tip between the sepal bases and style base. Seeds 0.5–0.6 mm long, oblong-ovoid, free from the fruit tissue at maturity, the surface reddish brown, appearing glabrous but faintly pitted under magnification with oblong transversely elongate cells. $2n=16$. July–October.

Uncommon in the southeastern portion of the Ozark Division (southeastern U.S. west to Missouri and Texas). Fens.

The occurrence in Missouri of this self-pollinating species, a member of the North American sect. *Microcarpium* is quite disjunct, with its primary distribution to the southeast (Peng, 1989).

8. Ludwigia palustris (L.) Elliott (water purslane)

L. palustris var. *americana* (DC.) Fernald & Griscom

L. palustris var. *nana* Fernald & Griscom

Pl. 463 a, b; Map 2118

Plants perennial, fibrous-rooted,the stems often rooting at the nodes roots. Stems 10–50(–70) cm long, loosely ascending to more commonly prostrate, occasionally submerged or floating in shallow water, usually well-branched and forming mats, not swollen or spongy, glabrous or nearly so. Leaves opposite, the petiole 1.2–25 mm long, narrowly winged. Stipules 0.1–0.3 mm long, lanceolate to ovate-triangular, reddish purple, succulent. Leaf blades 0.5–4.5 cm long, 2.5–23.0 mm wide, narrowly to more commonly broadly elliptic or obovate-elliptic, tapered at the base, angled or slightly short-tapered to a bluntly or sharply pointed tip, the margins entire but short-hairy, the surfaces glabrous, the venation pinnate, the inconspicuous secondary veins more or less fused to form a submarginal vein or loops toward the blade tip. Flower stalks absent or to 0.5 mm long at flowering and fruiting, the bractlets minute, linear to linear-lanceolate, not succulent, glabrous. Sepals 4, 1.1–2.0 mm long, 1.0–2.1 mm wide, green, broadly lanceolate to broadly ovate-triangular, sharply pointed at the tip. Petals absent. Stamens 4, the filaments green, somewhat translucent; the anthers 0.2–0.4 mm long. Fruits 2–5 mm long, 1.4–3.0(–3.5) mm wide, oblong to slightly obconic with an abruptly constricted base, slightly and bluntly 4-angled, with a longitudinal, green, central band on each face, thin-walled, appearing glabrous (microscopically hairy), dehiscing irregularly and tardily by decay of the walls. Seeds 0.5–0.7 mm long, ellipsoid, free from the fruit tissue at maturity, the surface yellowish brown, shiny, with a fine network of ridges. $2n=16$. June–September.

Scattered to common south of the Missouri River, uncommon to absent farther north (U.S. [except some western states]; Canada, Mexico, Central America, South America, Caribbean Islands, Europe, Asia, Africa; introduced in Hawaii, New Zealand). Banks of streams, rivers, and spring branches and margins of ponds, lakes, and sinkhole ponds; also crop fields and ditches; often emergent aquatics in shallow water, occasionally submerged aquatics.

This species in the primarily North American sect. *Isnardia* (L.) W.L. Wagner & Hoch (formerly *Dantia* (DC.) Munz) is widespread in eastern North America into Central America and the Caribbean region, secondarily spreading to western North America and into Eurasia and New Zealand as an invasive. Steyermark (1963) followed the treatment of this species proposed by Fernald and Griscom (1935), which recognized four varieties based on supposed differences in leaf shape and fruit size, two of which were said to occur in Missouri (see synonymy above). After an exhaustive study of living and herbarium materials, Peng et al. (2005) concluded that there were no morphological discontinuities that would allow the recognition of such variants.

Steyermark (1963) reported *L. natans* Elliott based on an undocumented occurrence in Greene County. Raven (1965) subsequently determined that *L. repens* J.R. Forst. is the oldest correct name for this taxon. In their monograph of *Ludwigia* sect. *Dantia* (now sect. *Isnardia*), Peng et al. (2005) performed extensive herbarium searches, but failed to locate a voucher specimen to verify the Missouri occurrence. Thus, *L. repens* is excluded from the flora in the present treatment. *Ludwigia repens* occurs from Mexico and the Caribbean islands north across the southern United States. The closest populations to Missouri are in northeastern Texas and central Oklahoma (Peng et al., 2005). The species is vegetatively similar to *L. palustris*, but differs in producing flowers with small, yellow petals (1–3 mm) that are shed soon after the flowers open, as well as somewhat larger sepals (2–5 mm) and fruits (4–10 mm).

9. **Ludwigia peploides** (Kunth) P.H. Raven **ssp. glabrescens** (Kuntze) P.H. Raven (floating primrose willow, creeping primrose willow)

Jussiaea repens L. var. *glabrescens* Kuntze

Pl. 464 a, b; Map 2119

Plants perennial, fibrous-rooted. Stems 15–150(–300) cm long, floating on water or creeping on mud, occasionally loosely ascending toward the tip, rooting at the lower nodes, usually not forming pneumatophores, mostly rounded and often slightly succulent, glabrous or emergent stems occasionally with scattered hairs toward the tip. Leaves alternate, the petiole (15–)20–45(–60) long. Stipules 0.6–1.6 mm long, ovate-triangular to nearly semicircular or rarely 3-lobed, dark reddish purple, succulent. Leaf blades glabrous, often shiny, those of the floating leaves 2–3 cm long, 7–23 mm wide, broadly elliptic to broadly oblanceolate, broadly obovate, or nearly circular, short- to long-tapered at the base, rounded to bluntly pointed at the tip; those of emergent leaves 3–8(–10) cm long, 4–40 mm wide, narrowly oblong or narrowly elliptic to ovate, obovate, circular, or oblanceolate, angled or tapered at the base, angled or slightly tapered to a bluntly or more or less sharply pointed tip, often with a minute glandular extension of the midvein at the very tip, the margins entire, the surfaces not sticky, glabrous, the venation pinnate, but the inconspicuous secondary veins more or less fused to form a submarginal vein. Flower stalks 10–60 mm long (becoming elongated to 35–90 mm at fruiting), the bractlets 0.5–2.0 mm long, ovate-triangular, green or reddish-tinged. Sepals 5(6), 8–12 mm long, 1.5–4.0 mm wide, narrowly to broadly lanceolate, angled or slightly tapered to a sharply pointed tip. Petals 5(6), (7–)9–15(–24) mm long, 4–13 mm wide, obovate, broadly rounded to truncate or with a shallow notch at the tip, yellow. Stamens 10(12) in 2 unequal whorls, the filaments yellow, the anthers 0.8–1.4 mm long. Fruits 20–40 mm long, 3–5 mm wide, cylindric or nearly so, often somewhat curved, usually bluntly 5-angled in cross-section, truncate at the tip and narrowed abruptly at the base, thick-walled and often slightly woody, glabrous, dehiscing tardily and irregularly. Seeds 1.0–1.3 mm long, oblong-cylindric, firmly embedded in wedge-shaped pieces of woody fruit tissue. $2n=16$. May–October.

Common nearly throughout the state (eastern U.S. west to Nebraska and Texas; Mexico, Central America, South America). Slow-flowing portions of streams and rivers, ponds, lakes, sinkhole ponds, sloughs, oxbows, marshes, and swamps; also ditches; usually emergent to floating aquatics (the basal portions rooted along the banks, the nonbasal portions of the stems often floating on the water).

Like several other members of the cosmopolitan sect. *Oligospermum*, this taxon has the potential for invasiveness and has proliferated nearly statewide along the margins of ponds. The absence of dots from some of the Ozarkian counties on the present distribution map is an artifact of undercollection of the species.

This taxon formerly was called *Jussiaea repens* (Steyermark, 1963). According to Raven (1963), the type of that name represents an Asian species now referred to as *L. adscendens* (L.) H. Hara. The current treatment of *L. peploides* recognizes four subspecies, one of which (ssp. *stipulacea* (Ohwi) P.H. Raven) is restricted to eastern Asia, another (ssp. *montevidensis* (Spreng.) P.H. Raven) occurs primarily in southern South America but adventively in Australia and New Zealand, Europe, and the southern United States, especially in California; the latter is easily distinguished from other North American subspecies by its dense villous and viscid pubescence. In addition to subsp. *glabrescens* as treated here, the fourth subspecies (ssp. *peploides*) occurs widely from Argentina and Chile, through the Andean South American countries to Central America, Mexico, and the Greater Antilles, and sporadically from west Texas through California to Oregon. The latter taxon is distinguished from ssp. *glabrescens* mainly by having shorter petioles (5–25 mm), leaf blades (1–4 cm), flower stalks (10–30 mm), and capsules (10–25 mm); however, this is an extremely variable group on which more work is needed.

10. Ludwigia polycarpa Short & R. Peter
Pl. 463 e, f; Map 2120

Plants perennial, forming leafy basal stolons to 15(–20) cm long, the roots fibrous, usually spongy when submerged. Stems (10–)25–60 (–85) cm long, erect or strongly ascending, well-branched, sparsely pubescent with short, appressed hairs along raised lines decurrent from the leaf bases, usually green. Leaves alternate, the petiole 1–10 mm long, narrowly winged. Stipules 0.2–0.4 mm long, narrowly to broadly ovate, reddish purple, succulent. Leaf blades 3.5–11.0 cm long, 4–10(–17) mm wide, narrowly elliptic to narrowly oblong-elliptic or narrowly oblanceolate, narrowly angled or long-tapered at the base, narrowly angled or tapered to a sharply pointed tip, the margins appearing entire, but with minute, glandular teeth, the venation with inconspicuous secondary veins, these more or less fused to form a submarginal vein or loops. Flowers sessile or nearly so, the bractlets 3.5–6.5 mm long, linear-lanceolate, usually green. Sepals 4, 2.5–4.5 mm long, 1.5–3.2 mm wide, triangular-ovate, tapered to a slender, sharply pointed tip. Petals absent. Stamens 4, the filaments yellowish green, the anthers 0.5–0.8 mm long. Fruits 4–7 mm long, 2.5–5.0 mm wide (less than 2 times as long as wide), oblong-obovoid, slightly and bluntly 4-angled in cross-section, usually with a slender, shallow, longitudinal groove on each face, stiff-walled, appearing glabrous but sometimes microscopically hairy, dehiscing irregularly by deterioration of the walls. Seeds 0.5–0.6 mm long, oblong-ovoid, but slightly curved at 1 or both ends, free from the fruit tissue at maturity, the surface light brown to yellowish brown, appearing smooth, under magnification with slender, longitudinally elongate cells, glabrous. $2n=16$. June–September.

Scattered north of the Missouri River and in the Unglaciated Plains Division, but absent from most of the Ozarks and Mississippi Lowlands (eastern [mostly northeastern] U.S. west to Minnesota, Nebraska, Kansas, Arkansas, and Alabama; disjunct in Idaho; Canada). Banks of streams and rivers, margins of ponds, lakes, and sloughs, swamps, and bottomland prairies; also ditches, edges of crop fields, and railroads.

Like other species of the North American sect. *Microcarpium*, this species occurs in wet habitats but is not truly aquatic, spreading by leafy basal stolons.

4. Oenothera L. (evening primrose)

Plants annual, biennial, or perennial herbs. Stems absent or more commonly present, the outer wall sometimes shredding and peeling with age, otherwise glabrous to densely hairy, erect or ascending, occasionally loosely ascending from a creeping base and then often rooting at the nodes. Leaves basal (all basal in stemless species, the rosette sometimes withered or absent at flowering in stemmed species) and alternate, the basal leaves sessile or variously petiolate, the stem leaves sessile or short-petiolate. Stipules absent. Blades of basal leaves variously shaped, in stemmed species usually longer and wider than those of the stem leaves, those of the stem leaves linear to lanceolate or elliptic, the margins entire or more commonly toothed or shallowly to deeply pinnately lobed. Inflorescences of solitary axillary flowers and/or terminal spikes, racemes, or clusters, these often somewhat leafy-bracted. Flowers actinomorphic or zygomorphic (in sect. *Gaura*), opening during the day or at dusk, sessile or (in some members of sect. *Gaura*) stalked (do not confuse the floral tube with the stalk), the stalk sometimes better developed at fruiting, ascending (the buds sometimes nodding or reflexed or the inflorescence tip sometimes nodding); bractlets absent. Floral tube elongate-cylindric with a usually flared tip, glabrous or ocasionally with dense, woolly hairs within, shed (with the sepals, petals, and stamens) after flowering. Sepals (3)4, green or tinged, striped, mottled, or spotted with red or purple, reflexed individually, in pairs, or as a unit (then reflexed to the side) at flowering, not persistent at fruiting. Petals (3)4, white, yellow. or pink (red or purple elsewhere), often changing color with age, shed after flowering, rounded to truncate at the tip, sometimes broadly or shallowly notched at the tip. Stamens (6)8 in 1 or 2 subequal or unequal series. Ovary (3)4-locular (appearing 1-locular in sect. *Gaura*), the stigma with 4 linear lobes or (in sect. *Calylophus*) peltate-discoid and obscurely or shallowly 4-lobed. Fruits capsules, variously shaped, circular or with 4 angles or wings, dehiscing longitudinally (sometimes tardily so) or indehiscent and nutlike in sect. *Gaura*. Seeds usually numerous (1–4[–8] per capsule in sect. *Gaura*), in 1 or 2(3) rows per locule, lacking an apical tuft of hairs. About 145 species, North America to South America, especially diverse in western North America, a number of species widely naturalized worldwide.

Oenothera is a morphologically variable genus that is currently classified into fourteen sections (W. L. Wagner et al., 2007). Of these, eight occur in Missouri: sect. *Calylophus* (Spach) Torr. & A. Gray, sect. *Gaura* (L.) W.L. Wagner & Hoch, sect. *Hartmannia* (Spach) W.L. Wagner & Hoch, sect. *Kneiffia* (Spach) Endl., sect. *Lavauxia* (Spach) Endl., sect. *Megapterium* (Spach) Endl., sect. *Oenothera*, and sect. *Peniophyllum* (Pennell) Munz. As noted in the discussion of the family, some earlier treatments of Missouri species treated sect. *Calylophus* as the genus *Calylophus* Spach (Yatskievych and Turner, 1990) and treated members of sect. *Gaura* as the genera *Gaura* L. and *Stenosiphon* Spach (Steyermark, 1963; Yatskievych and Turner, 1990). The sectional affiliations of the Missouri species are noted in the key to species below.

1. Fruits indehiscent, 3–15 mm long; flowers zygomorphic, with all of the petals positioned in the upper half of the flower or, if nearly actinomorphic, then the petals white to pink and short (4–13 mm); petals tapered abruptly to a pronounced, stalklike base; seeds 1–4(–8) (sect. *Gaura*)
    2. Fruits with a slender stalklike base 2–8 mm long; plants aggressively rhizomatous, often forming extensive colonies . . . . 18. O. SINUOSA
    2. Fruits sessile or with a thick, cylindric base; plants taprooted or with a woody rootstock, this sometimes branched underground but not strongly rhizomatous, not colonial or at most forming small, open colonies
        3. Flowers nearly actinomorphic; sepals 2–6 mm long
            4. Plants biennial or perennial; stems glabrous below inflorescence, but glaucous, at least toward the base; fruits ovoid, somewhat flattened, 4-angled . . . . 6. O. GLAUCIFOLIA
            4. Plants annual; stems hairy, not glaucous; fruits narrowly obovoid, not flattened, weakly 4-angled in the apical 1/3, the angles becoming broad and rounded below . . . . 3. O. CURTIFLORA
        3. Flowers strongly zygomorphic; sepals 5–18 mm long
            5. Sepals 7–18 mm long; fruit ellipsoid, 4-angled throughout; plants annual or biennial, from a fleshy taproot, stems several from the base . . . . 4. O. FILIFORMIS
            5. Sepals 5–10 mm long; fruit pyramidal in upper half, constricted sharply to a stout, cylindric base; plants perennial, clumped from a thick taproot, often branching below ground or only at surface, less commonly unbranched. . . . . 20. O. SUFFRUTESCENS

1. Fruits longitudinally dehiscent, sometimes tardily so, 10–70(–115) mm long (shorter in *O. linifolia* and *O. perennis*); flowers actinomorphic, petals yellow, white, or pink (in *O. speciosa*), 10–68 mm long (usually only 3–5 mm in the yellow-petaled *O. linifolia*), lacking a stalklike base; seeds more than 8, typically numerous
   6. Fruits club-shaped, rarely ellipsoid, ellipsoid-rhomboid, or subglobose, the lower part sterile, narrowed and stalklike; seeds clustered in each locule, not in definite rows
      7. Petals white to pink; fruits with a prominent rib along the upper part of each face; petals 25–40 mm long (sect. *Hartmannia*) . . . . 9. O. SPECIOSA
      7. Petals pale to bright or dark yellow; fruits not prominently ribbed on the faces, but sometimes with an inconspicuous rib or thickened nerve
         8. Stem leaves linear to threadlike, 0.5–1.0 mm wide; petals 3–5(–7) mm long (sect. *Peniophyllum*) . . . . 10. O. LINIFOLIA
         8. Stem leaves lanceolate to ovate, elliptic, or oblanceolate to obovate, rarely linear, 2–20(–50) mm wide; petals 5–30 mm long (sect. *Kneiffia*)
            9. Petals 5–10 mm long; stigma surrounded by the anthers at flowering . . . . 14. O. PERENNIS
            9. Petals (8–)15–30 mm long; stigma well elevated above the anthers at flowering
               10. Sepals with the free tips 0.5–1.0(–6.0) mm long; plants with fibrous or occasionally somewhat fleshy roots, rarely producing rhizomes; fruit body club-shaped or oblong in outline . . . . 5. O. FRUTICOSA
               10. Sepals with the free tips 1–4 mm long; plants with a thickened rootstock and usually producing rhizomes; fruit body narrowly club-shaped to ellipsoid . . . . 15. O. PILOSELLA
   6. Fruit lanceoloid to ellipsoid, ovoid, or cylindric, the lower part fertile and not stalklike (tapered abuptly to a stalklike base 2–6(–12) mm long in *O. macrocarpa*); seeds in 1 or 2(3) definite rows per locule
      11. Fruit winged; floral tube (21–)40–140 mm long; seeds (2.1–)2.5–5.0 mm long
         12. Leaves entire or shallowly toothed; fruit winged throughout, the wings 15–30 mm wide; petals (40–)50–68 mm long; plants with well-developed, spreading to loosely ascending stems 15–60 cm long (sect. *Megapterium*) . . . . 11. O. MACROCARPA
         12. Leaves deeply and irregularly lobed, rarely subentire; fruit winged only in the upper 2/3, the wings 5–10 mm wide; petals (10–)12–30(–38) mm long; plants stemless or rarely with inconspicuous (mostly hidden by the dense basal rosette), ascending stems to 10(–20) cm long (sect. *Lavauxia*) . . . . 21. O. TRILOBA
      11. Fruit not winged; floral tube (2–)12–40(–50) mm long; seeds 0.6–1.8(–2.0) mm long
         13. Stigma peltate-discoid and obscurely or shallowly 4-lobed, 1–2 mm in diameter; sepals with the midrib keeled (sect. *Calylophus*) . . . . 17. O. SERRULATA
         13. Stigma divided into four linear lobes, the lobes 1.5–13 mm long; sepals not keeled (sect. *Oenothera*)
            14. Seeds irregularly prismatic, the surface with an irregular network of ridges and pits, angled; fruits usually lanceoloid

15. Sepal tips subapical, divergent in bud; inflorescences usually somewhat curved . . . . 13. O. PARVIFLORA
15. Sepal tips terminal, usually appressed-ascending in bud; inflorescences erect
    16. Bracts shed before the flowers open; inflorescences with minute gland-tipped hairs, occasionally also with short, straight, mostly pustular-based, nonglandular hairs; fruit often glabrous or nearly so at maturity . . . . 12. O. NUTANS
    16. Bracts persistent; inflorescences with short, straight, sometimes pustular-based, nonglandular hairs, sometimes also glandular-hairy, rarely only glandular; fruit hairy, rarely becoming nearly glabrous at maturity
        17. Inflorescence not appearing conspicuously bracteate, usually glandular-hairy and also pubescent with shorter and/or longer, nonglandular hairs; stem leaves with inconspicuous secondary veins; plants sparsely to moderately hairy, green . . . . 1. O. BIENNIS
        17. Inflorescence appearing conspicuously bracteate, usually lacking glandular pubescence; stem leaves with distinct veins; plants densely hairy, grayish green . . . . 22. O. VILLOSA

14. Seeds ellipsoid to suborbicular, the surface pitted, never angled; fruits cylindric to narrowly lanceoloid
    18. Flowers appearing solitary in the axils of the upper leaves; buds adjacent to the most mature one curved upward by the floral tube; petals truncate or notched at the tip
        19. Petals 25–40 mm long; stigma elevated above the anthers at flowering . . . . 7. O. GRANDIS
        19. Petals 5–20(–22) mm long; stigma surrounded by the anthers at flowering . . . . 9. O. LACINIATA
    18. Flowers in dense spikes, mature buds and the adjacent ones straight; petals rounded to pointed at the tip
        20. Petals 5–16 mm long; stigma surrounded by the anthers at flowering . . . . 2. O. CLELANDII
        20. Petals 15–35 mm long; stigma elevated above the anthers at flowering
            21. Sepals and floral tube often pubescent with long spreading hairs, these with red pustular bases, also glandular-hairy and with fine, short, nonglandular hairs, occasionally glabrous, rarely with only short nonglandular hairs; mature buds overtopping the tip of the inflorescence axis; sepal tips 2–6 mm long . . . . 8. O. HETEROPHYLLA
            21. Sepals and floral tube pubescent with short, nonglandular hairs, sometimes also glandular-hairy; mature buds not overtopping the tip of the axis; sepal tips in bud 0.5–2.0(–3.0) mm long . . . . 16. O. RHOMBIPETALA

1. Oenothera biennis L. (common evening primrose)

O. biennis var. *pycnocarpa* (G.F. Atk. & Bartlett) Wiegand

Pl. 464 h, i; Map 2121

Plants biennial, with taproots. Stems 1 to several, 30–200 cm long (including the inflorescence), erect or strongly ascending, unbranched or branched, sparsely to more commonly densely pubescent with short, upward-curved to more or less appressed, nonglandular hairs and longer, spreading to loosely appressed, mostly pustular-based, nonglandular hairs (but the plants generally appearing green), the inflorescence with short,

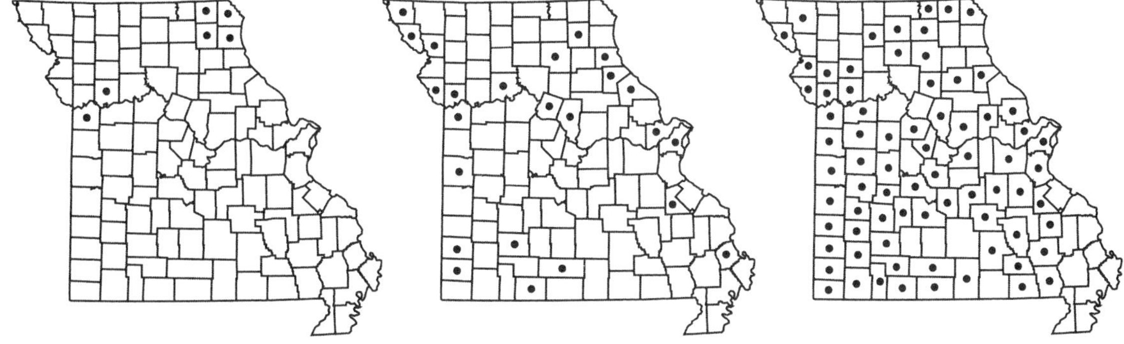

2122. Oenothera clelandii 2123. Oenothera curtiflora 2124. Oenothera filiformis

straight, sometimes pustular-based, nonglandular hairs, usually also with minute glandular hairs, rarely only glandular. Leaves basal and alternate, the rosette leaves 10–30 cm long, 20–50 mm wide, the blade narrowly oblanceolate to oblanceolate, tapered to the petiole, the margins sparsely to moderately and often somewhat irregularly toothed, sometimes few-lobed toward the base, the surfaces and margins sparsely to moderately pubescent with short, appressed to curved, nonglandular hairs; stem leaves 5–22 cm long, (10–)15–50(–60) mm wide, narrowly oblanceolate to oblanceolate or narrowly elliptic to elliptic, tapered to the sessile or short-petiolate base, the margins sparsely to moderately and irregularly toothed, those of the lower leaves sometimes with a few lobes, the surfaces and margins with pubescence similar to that of the basal leaves, the secondary veins relatively inconspicuous. Inflorescences short to elongate spikes, relatively dense, erect and relatively straight, sometimes grouped into panicles with ascending branches, not appearing conspicuously bracteate, the bracts persistent, 12–50 mm long, narrowly lanceolate to narrowly ovate or narrowly elliptic, the margins entire or finely and irregularly toothed, the surfaces sparsely to moderately pubescent with short, appressed to curved, nonglandular hairs, sometimes also with longer, spreading, nonglandular hairs and minute glandular hairs. Flowers actinomorphic, opening at dusk, the floral tube (20–)25–40 mm long, sparsely to densely pubescent with minute glandular hairs and with sparse, short, ascending, nonglandular hairs, sometimes also with short or long, more or less spreading hairs. Sepals 12–22(–28) mm long, the midribs not keeled, with pubescence similar to the floral tube, the free tips in bud 1.5–3.0 mm long, terminal, usually erect and appressed. Petals 12–25(–30) mm long, 14–27(–32) mm wide, broadly obovate to broadly heart-shaped (broadly but sometimes only slightly notched at the tip), lacking a stalklike base, yellow or occasionally light yellow, fading to pale yellow or orange. Stamens with the filaments 8–15(–20) mm long, glabrous at the base, the anthers 3–6(–9) mm long, yellow. Style 30–55 mm long, the stigma positioned at about the same level as the anthers, deeply 4-lobed, the lobes 3–6 mm long. Fruits 20–40 mm long, 4–6 mm wide, narrowly lanceoloid to lanceoloid (tapered toward the tip), straight, longitudinally dehiscent nearly the entire length, 4-locular, not flattened or winged, more or less circular in cross-section, the surface green to dull green and often with reddish stripes at maturity, not blackening upon drying, moderately to densely pubescent with various mixtures of minute glandular hairs, short, appressed to ascending, nonglandular hairs, and/or longer, more or less spreading, mostly pustular-based, nonglandular hairs, rarely becoming nearly glabrous at maturity, not tapered to a sterile, stalklike base. Seeds numerous in each locule, arranged in 2 rows, 1.1–2.0 mm long, 0.6–1.1 mm wide, irregularly prismatic, angled, the surface brown to dark brown or nearly black, with an irregular network of ridges and pits. Self-compatible. $2n=14$. June–October.

Common throughout the state (eastern U.S. west to North Dakota and Texas; Canada; introduced sporadically farther west in the U.S., Canada, also South America, Europe, Asia, Africa, Madagascar, Réunion, the Azores, Madeira, Canary Islands, New Zealand). Openings and margins of mesic to dry upland forests, margins of bottomland forests, glades, bluffs, upland prairies, bottomland prairies, marshes, sand prairies, and banks of streams and rivers; also ditches, pastures, old fields, edges of crop fields, fallow fields, mine spoils, gardens, railroads, roadsides, and open disturbed areas.

This is the most common and widespread evening primrose in the state. It is very variable morphologically.

2. Oenothera clelandii W. Dietr., P.H. Raven & W.L. Wagner (sand primrose)

Pl. 466 b–d; Map 2122

Plants biennial, with taproots. Stems 1 to several, 20–70(–100) cm long (including the inflorescence), erect or strongly ascending, the outer stems of a clump sometimes arched upward, unbranched or more commonly well-branched, densely pubescent with short, appressed, nonglandular hairs (the plants generally appearing green or grayish green). Leaves basal and alternate, the rosette leaves 5–16 cm long, 5–15 mm wide, the blade narrowly oblanceolate, long-tapered to the sometimes short or indistinct petiole, the margins more or less entire to deeply pinnately lobed with the terminal lobe larger than the lateral ones, the surfaces and margins sparsely to densely pubescent with short, appressed, nonglandular hairs; stem leaves 2–12 cm long, 5–20 mm wide, very narrowly elliptic to narrowly lanceolate, tapered to the sessile or short-petiolate base, the lower ones nearly entire to deeply pinnately lobed, grading to the more or less entire or sparsely toothed upper leaves, the pubescence similar to the basal leaves, the secondary veins relatively inconspicuous. Inflorescences dense spikes, usually unbranched, the mature buds not overtopping the tip of the inflorescence axis, straight, the subtending bracts persistent, 10–25 mm long, narrowly lanceolate to lanceolate, the margins entire to bluntly few-toothed, the pubescence similar to that of the basal leaves. Flowers actinomorphic, opening at dusk, the floral tube 15–40 mm long, sparsely to densely pubescent with short, more or less appressed, nonglandular hairs, sometimes also with minute glandular hairs. Sepals 6–13 mm long, the midribs not keeled, with pubescence similar to the floral tube, the free tips in bud 0.5–2.0 mm long, terminal, erect and appressed. Petals 5–16 mm long, 3–11 mm wide, broadly elliptic to rhombic-ovate (rounded to more commonly pointed at the tip), lacking a stalklike base, light yellow to yellow, fading to pale yellow. Stamens with the filaments 4–18 mm long, glabrous at the base, the anthers 2.0–3.5 mm long, yellow. Style 20–40 mm long, the stigma positioned at about the same level as the anthers, deeply 4-lobed, the lobes 1.5–4.0 mm long. Fruits 10–20 mm long, 2–3 mm wide, narrowly lanceoloid, straight to curved, longitudinally dehiscent nearly the entire length, 4-locular, not flattened or winged, more or less circular to very bluntly 4-angled in cross-section, the surface green, not blackening upon drying, moderately to densely pubescent with short, appressed, nonglandular hairs, not tapered to a sterile, stalklike base. Seeds numerous in each locule, arranged in 2 rows, 1.0–1.9 mm long, 0.4–0.8 mm wide, ellipsoid, not angled, the surface brown, often flecked with dark brownish red spots, pitted. Self-compatible. $2n$=14. June–October.

Uncommon, known from historical collections in the Kansas City area and extant occurrences in northeastern Missouri (northeastern U.S. west to Iowa and Arkansas). Sand prairies and banks of rivers; also cemeteries, railroads, roadsides, and open, sandy, disturbed areas.

Steyermark (1963) treated the Missouri occurrences of this species as part of a broadly circumscribed *O. rhombipetala*. Dietrich & Wagner (1988) summarized the reasons for recognizing two species in the complex. See the treatment of *O. rhombipetala* for further discussion.

3. Oenothera curtiflora W.L. Wagner & Hoch (velvety gaura)

Gaura parviflora Douglas ex Lehm.

G. parviflora f. *glabra* Munz

G. parviflora f. *lachnocarpa* Weath.

Pl. 465 c–f; Map 2123

Plants annual, with stout taproots (to 3 cm or more in diameter). Stems solitary or few, (20–)30–150(–300) cm long (including the inflorescence), erect or strongly ascending, sometimes reclining with age, unbranched or more commonly several-branched above the midpoint, densely pubescent with short, glandular hairs and scattered long, spreading, nonglandular hairs, not glaucous, sometimes somewhat woody toward the base and then with peeling or shredding bark. Leaves alternate and occasionally basal (the rosette and lower stem leaves usually withered or absent by flowering); rosette leaves 3–15 cm long, 10–30 mm wide, broadly oblanceolate, the margins entire to more commonly slightly wavy or with a few teeth, also hairy, the surfaces finely nonglandular-hairy and with a mixture of short, glandular and long, nonglandular hairs along the veins; stem leaves 1.5–12.5 cm long, 5–40 mm wide, narrowly elliptic to narrowly ovate, not clasping at the base, the margins entire or the largest slightly wavy or with a few teeth, also hairy, the surfaces finely nonglandular-hairy and with a mixture of short, glandular, and long, nonglandular hairs along the veins, the secondary veins inconspicuous. Inflorescences dense, elongate, often somewhat wandlike spikes, these sometimes grouped into panicles, the axes variously glabrous to densely pubescent with glandular and nonglandular hairs. Bracts relatively inconspicuous, 2–6 mm long, 0.5–2.0 mm wide, linear to lanceolate. Flowers slightly zygomorphic, the petals similar in size but grouped toward the upper half of the flower, opening at

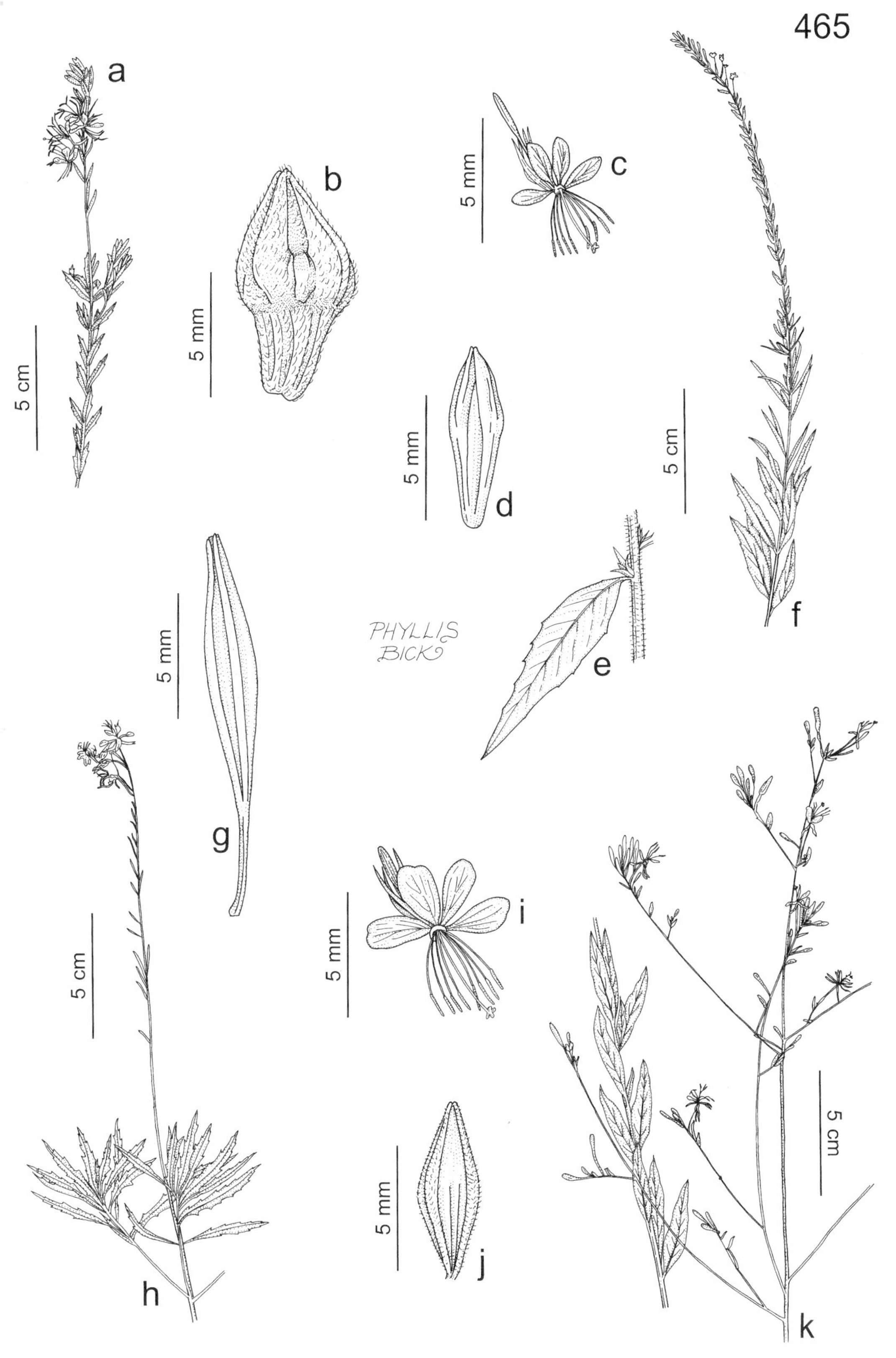

Plate 465. Onagraceae. *Oenothera suffrutescens*, **a)** fertile stem tip, **b)** fruit. *Oenothera curtiflora*, **c)** flower, **d)** fruit, **e)** median node with leaf, **f)** fertile stem tip. *Oenothera sinuosa*, **g)** fruit, **h)** fertile stem tip. *Oenothera filiformis*, **i)** flower, **j)** fruit, **k)** fertile stem tip and median portion of stem with leaves.

2125. Oenothera fruticosa

2126. Oenothera glaucifolia

2127. Oenothera grandis

dusk, the floral tube 1.5–5.0 mm long, glabrous or with short spreading hairs. Sepals 2.0–3.5 mm long, glabrous or with short, loosely appressed hairs, lacking free tips in bud. Petals 1.5–3.0 mm long, 1–2 mm wide, oblong-obovate to elliptic-oblanceolate, tapered abruptly to a pronounced stalklike base, white to light pink, fading pink or red. Stamens with the filaments 1.5–3.0 mm long, ascending, glabrous at the base, the anthers 0.5–1.0 mm long, yellow or reddish-tinged. Style 3–9 mm long, the stigma positioned well beyond the anthers (but the anthers maturing first and usually depositing pollen directly onto the elongating stigma), deeply 4-lobed, the lobes 0.5–0.8 mm long. Fruits indehiscent, appearing 1-locular, 5–11 mm long, 1.5–3.0 mm in diameter, the body narrowly obovoid, not flattened, weakly 4-angled in the apical 1/3, the angles becoming broad and rounded below, lacking a sterile, stalklike base, the surfaces glabrous or short-hairy. Seeds 3 or 4 per fruit, 2–3 mm long, 1.0–1.5 mm wide, oblong-ellipsoid to oblong-ovoid, the surface reddish brown or tan with reddish brown mottling or streaks, pebbled. Self-compatible. $2n=14$. June–October.

Scattered, mostly in the western half of the state and in counties bordering the Mississippi River (western U.S. east to Indiana and Louisiana; Mexico; introduced eastward to Massachusetts and Florida, also South America, Australia). Glades, bluffs, and banks of rivers; also old fields, railroads, roadsides, and open disturbed areas.

The native range of *Oenothera curtiflora* is not clear. It is a self-pollinated species that spreads easily in disturbed areas. Raven and Gregory (1972) suggested that it originally may have been native to the shortgrass prairies of the Great Plains region, but if that was the case, it was spread widely farther east and west relatively long ago.

This taxon has long been known as *Gaura parviflora*. Goodman and Lawson (1995) noted the existence of an earlier epithet for the species, *G. mollis* E. James. However, W. L. Wagner and Hoch (2000) presented arguments that nomenclatural stability would be served best if *G. mollis* were to be officially rejected in favor of the long-standing *G. parviflora*. This proposal was approved at the 2005 International Botanical Congress in Vienna. Unfortunately, this meant that when the genus *Gaura* was reclassified as a section within *Oenothera*, the epithet *G. mollis* was unavailable to use as the basis for a new combination in that genus. As there already existed a different species called *O. parviflora* L. (see treatment below), W. L. Wagner et al. (2007) were forced to coin a new name for the taxon within *Oenothera*.

4. Oenothera filiformis (Small) W.L. Wagner & Hoch (large-flowered gaura)

Gaura biennis L. var. *pitcheri* Torr. & A. Gray

G. longiflora Spach

Pl. 465 i–k; Map 2124

Plants annual or rarely biennial, with fleshy taproots. Stems 1 to several from the base, 50–200 cm long (including the inflorescence), erect or somewhat arched, usually many-branched well above the base, moderately to more commonly densely pubescent with a mixture of appressed to upward-curved, somewhat woolly nonglandular hairs and short glandular hairs, sometimes also with scattered longer, spreading to loosely upward-curved hairs. Leaves in a rosette and alternate (the stem leaves also often with small fascicles of axillary leaves); rosette leaves usually withered at flowering, 5–15(–40) cm long, 10–35 mm wide, often highly irregular in outline, the margins variously entire to wavy or with broadly spaced, irregular teeth or shallow lobes, the surfaces glabrous or with short, more or less appressed hairs, the undersurface often also with slightly longer hairs along the midvein; stem leaves 1.5–13 cm long, 2–28 mm wide, narrowly lanceolate to ellip-

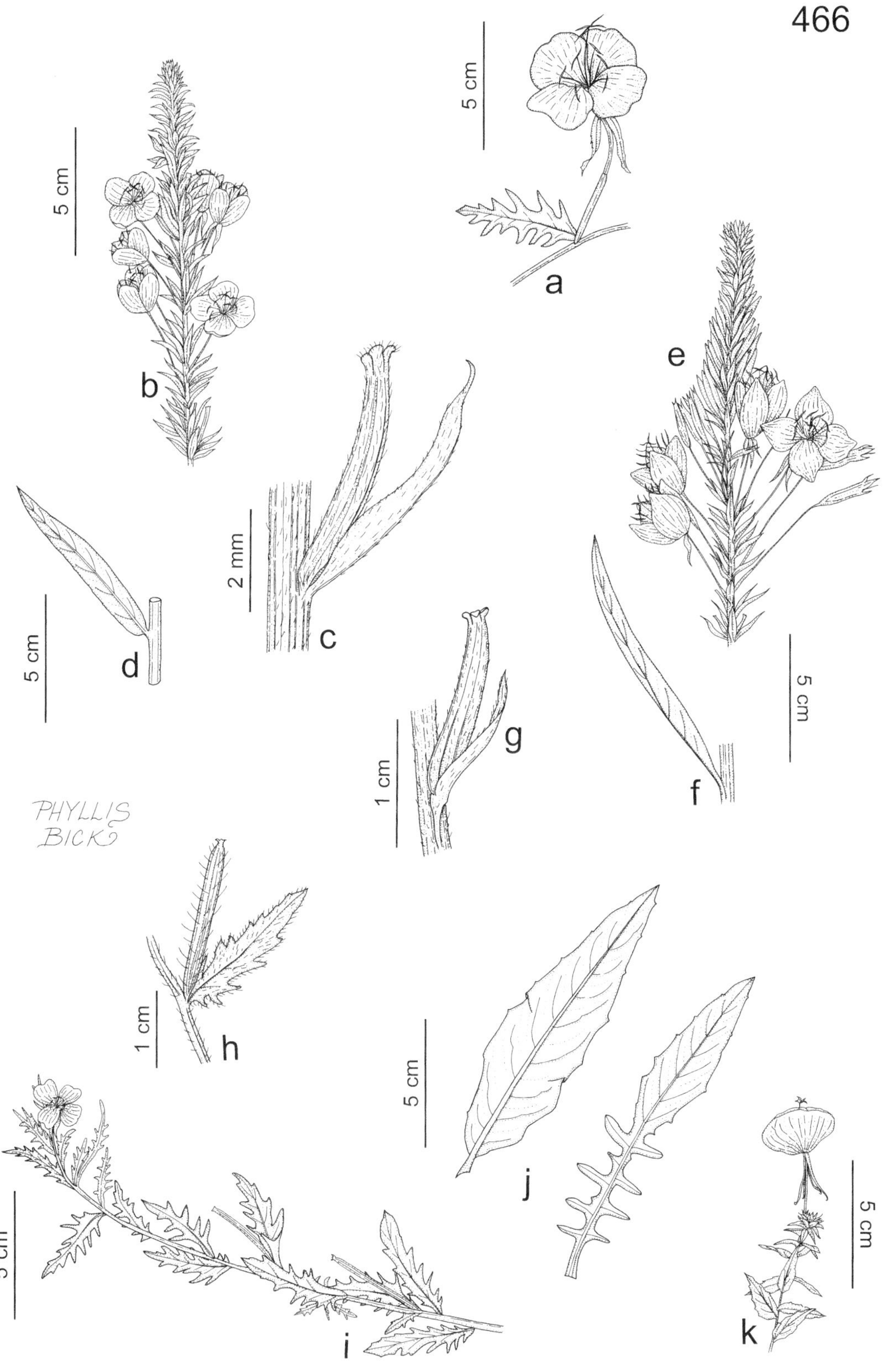

Plate 466. Onagraceae. *Oenothera grandis*, **a)** node with bract and flower. *Oenothera clelandii*, **b)** inflorescence, **c)** node with bract and fruit, **d)** median node with leaf. *Oenothera rhombipetala*, **e)** inflorescence, **f)** node with bract and fruit, **g)** median node with leaf. *Oenothera laciniata*, **h)** node with bract and fruit, **i)** habit. *Oenothera heterophylla*, **j)** less and more divided rosette leaves, **k)** inflorescence.

tic, the smaller leaves sometimes ovate, the margins entire to wavy or with broadly spaced, irregular teeth or shallow lobes, the surfaces glabrous or with short, more or less appressed to curved hairs, the undersurface often also with slightly longer hairs along the midvein, the secondary veins of at least the larger blades usually relatively conspicuous. Inflorescences well-branched panicles of open to moderately dense, often wandlike spikes (rarely reduced to a solitary raceme), often half or more of the total length of the plant, the axes usually densely and minutely glandular- and/or glandular-hairy. Bracts relatively inconspicuous, 1–6 mm long, 0.5–2.0 mm wide, narrowly lanceolate to ovate. Flowers strongly zygomorphic, with all of the petals positioned in the upper half of the flower, opening at dusk, the floral tube 4–13(–15) mm long, short-hairy. Sepals 7–18 mm long, short-hairy, lacking free tips in bud. Petals 6.5–15.0 mm long, 2–7 mm wide, elliptic-oblanceolate, tapered abruptly to a pronounced stalklike base, white, fading to pink or red. Stamens with the filaments 5–13 mm long, spreading to pendant, glabrous at the base, the anthers 1.5–5.0 mm long, brownish red. Style 12–34 mm long, the stigma positioned well beyond the anthers, deeply 4-lobed, the lobes 0.5–1.5 mm long. Fruits indehiscent, appearing mostly 1-locular, 4.5–7.0 mm long, 1.5–2.5 mm in diameter, the body ellipsoid, 4-angled throughout, not constricted to a sterile, stalklike base, but sometimes tapered to a short, thick, cylindric base. Seeds 2–4 per fruit, not in definite rows, 1.2–3.0 mm long, 1.5–2.5 mm wide, ovoid, sometimes irregularly flattened (by crowding in fruit), the surface yellowish brown to reddish brown, more or less smooth. Self-incompatible. $2n$=14. June–October.

Scattered nearly throughout the state, but apparently absent from the Mississippi Lowlands Division (Nebraska and Colorado to Texas, east to Michigan and Alabama; Canada; probably introduced very sporadically farther east to Massachusetts). Glades, upland prairies, bottomland prairies, sand prairies, openings and edges of mesic to dry upland forests, edges of bottomland forests, fens, and banks of streams and rivers; also ditches, fallow fields, margins of crop fields, old fields, railroads, roadsides, and open disturbed areas.

As discussed above in the introduction to the family Onagraceae, the genus *Gaura* is now considered a section of the genus *Oenothera*. The name *O. longiflora* L. already was in use for a South American species, *O. longiflora* L., thus the next-oldest species epithet was used when the species was transferred into *Oenothera*.

In their taxonomic study of the genus *Gaura*, Raven and Gregory (1972) separated *G. longiflora* and *G. biennis* based on a series of subtle morphological differences and the generally more western distribution of *G. longiflora*. They hypothesized that *G. biennis* was derived from *G. longiflora* following a series of complex cytogentic events. *Gaura longiflora* is self-incompatible and thus outcrossing, whereas *G. biennis* is self-compatible and in nature apparently is mostly inbred (Carr et al., 1986). True *G. biennis* (now better known as *Oenothera gaura* W.L. Wagner & Hoch) is widespread in the northeastern United States. Although mapped by Steyermark (1963) from a few, scattered, mostly southwestern counties, it has not yet been confirmed as occurring in the state (Steyermark's specimens were all subsequently redetermined as *O. filiformis*). However, eventually *O. gaura* may be discovered in Missouri (most likely in a northeastern county). It should be noted that the ranges of the two taxa overlap substantially in Iowa and Illinois and that within this region they can be very difficult to distinguish morphologically. Hybrids exhibiting a wide range of fertility also have been recorded (Raven and Gregory, 1972; Carr et al., 1986). Therefore, it is not surprising that some botanists have preferred to regard the two as varieties of a single, widespread species (Gleason and Cronquist, 1991). Collectors seeking to document *O. gaura* in the state should focus on plants with somewhat less wandlike stems than is usual for Missouri plants and with relatively long and shaggy stem pubescence, mostly glandular hairs in the inflorescences, and fruits containing 3–6 seeds.

5. Oenothera fruticosa L. (sundrops, narrowleaf evening primrose, shrubby sundrops)

Pl. 467 c–e; Map 2125

Plants perennial, with fibrous or occasionally somewhat thickened roots, rarely with rhizomes. Stems solitary or few, (10–)30–80 cm long (including the inflorescence), erect or ascending, sometimes from a more or less spreading base, unbranched or with few to several branches, glabrous or nearly so or sparsely to densely pubescent with short nonglandular and/or glandular hairs. Leaves basal and alternate, the basal leaves overwintering and often withered at flowering; rosette leaves 3–12 cm long, 5–30 mm wide, oblanceolate to obovate, short- or occasionally longer-petiolate, the margins entire to slightly irregular or wavy or the broader ones with a few coarse teeth, hairy, the surfaces glabrous to densely hairy; stem leaves 2–6(–11) cm long, 2–20(–50) mm wide, very narrowly elliptic to broadly ovate, mostly short-petiolate, the margins entire to slightly irregular or wavy or the

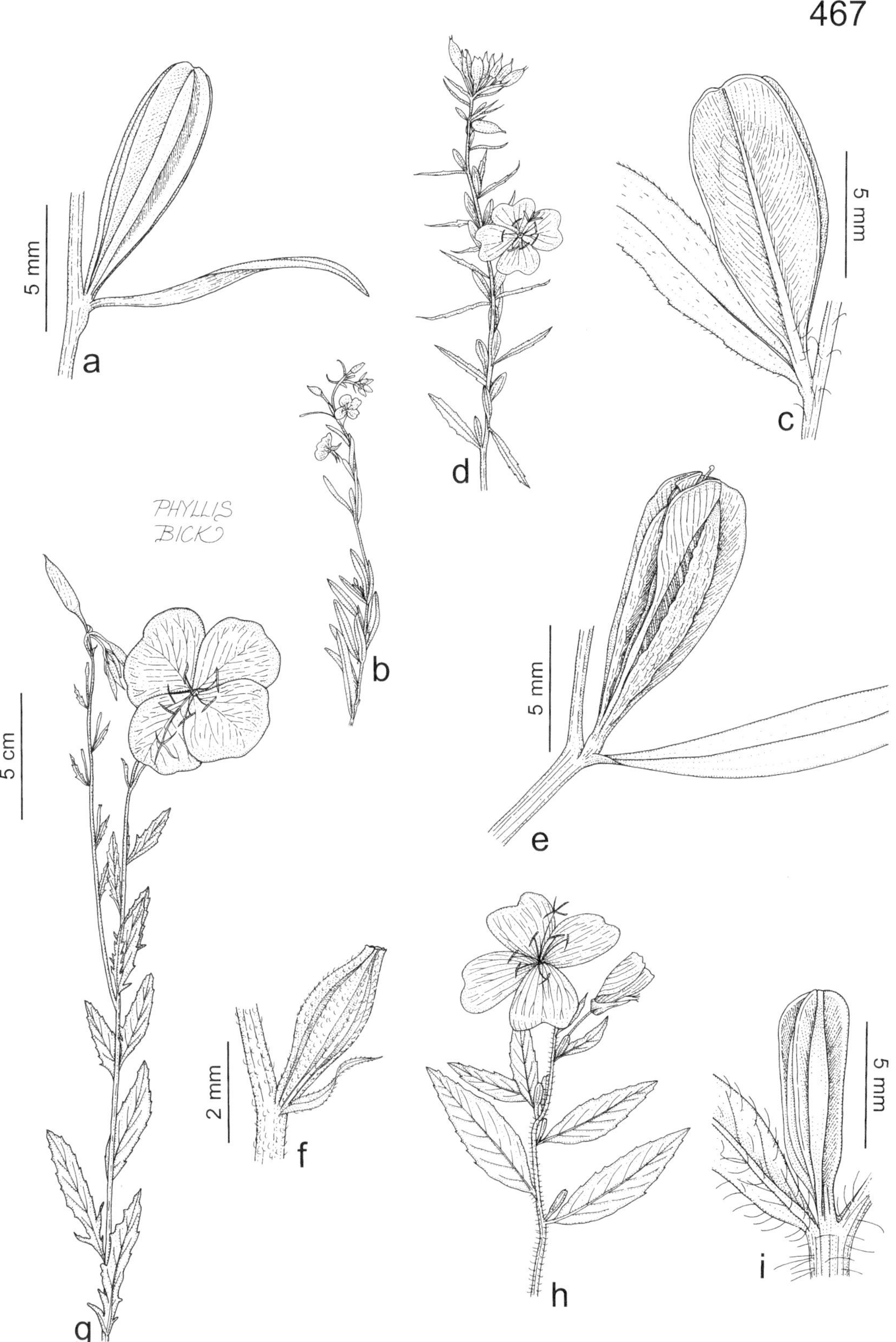

Plate 467. Onagraceae. *Oenothera perennis*, **a)** fruit, **b)** inflorescence and upper portion of stem. *Oenothera fruticosa* ssp. *glauca*, **c)** node with fruit, **d)** inflorescence. *Oenothera fruticosa* ssp. *fruticosa*, **e)** node with fruit. *Oenothera speciosa*, **f)** node with fruit, **g)** fertile stem. *Oenothera pilosella*, **h)** inflorescence, **i)** fruit.

broader ones with a few coarse teeth, hairy, the surfaces glabrous to densely nonglandular and/or glandular-hairy, the secondary veins inconspicuous or (on broader leaves) moderately conspicuous. Inflorescences open spikes, not grouped into panicles, usually relatively short and sometimes few-flowered, the portion in bud erect or rarely nodding, the axis glabrous to densely pubescent with short, nonglandular and/or glandular hairs. Bracts relatively conspicuous, 5–40 mm long, 1–10 mm wide, linear to lanceolate. Flowers actinomorphic, opening in the morning, the floral tube 5–20 mm long, short-hairy. Sepals 5–20(–22) mm long, with short, spreading, glandular or nonglandular hairs, the free tips in bud 0.5–1.0(–6) mm long, erect and appressed or somewhat spreading. Petals (8–)15–30 mm long, (6–)10–30 mm wide, broadly obovate to broadly oblong-obovate (truncate to shallowly and broadly notched and sometimes somewhat undulate at the tip), lacking a stalklike base, light yellow to deep yellow, fading to yellow or lavender. Stamens with the filaments 5–15 mm long, ascending, glabrous at the base, the anthers 4–7 mm long, yellow. Style 12–20 mm long, the stigma positioned well above the anthers, deeply 4-lobed, the lobes 3–5 mm long. Fruits (5–)12–19(–25) mm long, the main body longitudinally dehiscent toward the tip and eventually more or less the entire length, 4-locular, (5–)10–17(–20) mm long, (2–)3–4(–6) mm in diameter, club-shaped or oblong in outline, not flattened, strongly 4-angled or 4-winged, the surfaces glabrous or finely glandular- or nonglandular-hairy, gradually or abruptly tapered to a sterile, stalklike base 1–10 mm long, this also glabrous or hairy. Seeds numerous in each locule, clustered and not in definite rows, 0.8–1.0 mm long, 0.3–0.5 mm wide, irregularly rhombic-ellipsoid to prismatic-ovoid, the surface dark reddish brown, pebbled. Self-compatible. $2n$=28, 42. June–October.

Uncommon in the southern portion of the Ozark Division disjunctly north to Henry County (eastern U.S. west to Michigan, Oklahoma, and Louisiana; Canada). Openings of mesic to dry upland forests.

Oenothera fruticosa is grown as a garden ornamental and a number of cultivars exist. According to Straley (1977), *O. fruticosa* can be separated into two subspecies with strongly overlapping ranges, both of which have been recorded from Missouri.

1. Fruits club-shaped, widest above the midpoint, moderately to densely pubescent with mostly nonglandular hairs; leaves usually moderately to densely hairy .... 5A. SSP. FRUTICOSA
1. Fruits more or less oblong in outline, equally wide for most of the length or slightly wider near the midpoint, glabrous or sparsely to occasionally densely hairy with mostly gland-tipped hairs; leaves glabrous or sparsely hairy .... 5A. SSP. GLAUCA

5a. ssp. fruticosa

O. fruticosa var. *linearis* (Michx.) S. Watson

Pl. 467 e

Stems moderately to densely hairy with mostly spreading, nonglandular hairs, sometimes also with sparse glandular hairs toward the tip. Stem leaves narrowly elliptic to narrowly ovate, the margins mostly subentire, the surfaces usually moderately to densely hairy. Inflorescences erect or rarely nodding, the axis mostly densely pubescent with appressed to spreading, nonglandular hairs. Sepals 5–20 mm long, the free tips in bud usually erect and appressed. Petals (8–)15–25 mm long. Ovary club-shaped to narrowly oblong in outline, (6–)10–15(–18) mm long, 1–2 mm wide, sparsely to densely nonglandular-hairy or occasionally glabrous, rarely also with sparse glandular hairs. Fruits club-shaped, widest above the midpoint, strongly 4-angled, rarely narrowly 4-winged, gradually narrowed to a sterile stalklike base 3–10 mm long, the surface sparsely to densely nonglandular-hairy or occasionally glabrous, rarely also with sparse glandular hairs. $2n$=28, 42. June–October.

Uncommon in the southern portion of the Ozark Division disjunctly north to Henry County (eastern U.S. west to Indiana, Oklahoma, and Louisiana). Openings of mesic to dry upland forests.

5b. ssp. glauca (Michx.) Straley

O. tetragona Roth var. *hybrida* (Michx.) Fernald

Pl. 467 c, d

Stems, sparsely to densely pubescent with short, spreading, glandular hairs, occasionally also sparsely nonglandular-hairy, sometimes nearly glabrous. Stem leaves narrowly elliptic to broadly ovate, the margins subentire to wavy or coarsely few-toothed, the surfaces glabrous to sparsely hairy. Inflorescences erect, the axis glabrous or more commonly moderately to densely pubescent with mostly short, spreading, glandular hairs. Sepals 8–22 mm long, the free tips in bud erect and appressed to somewhat spreading. Petals (8–)15–20(–30) mm long. Ovary narrowly to broadly obovoid, (3–)4–8(–13) mm long, 1–3 mm in diameter, sparsely to densely glandular-hairy and sometimes also with spreading or appressed, nonglandular hairs. Fruits more or less oblong in outline, widest at the middle, strongly 4-angled to 4-winged, usually abruptly tapered to a sterile stalklike base 1–3(–7) mm long, the surface glabrous or sparsely to occasionally densely hairy with mostly gland-tipped hairs. $2n=28$. June–September.

Uncommon, known thus far only from historical specimens from Barry and Shannon Counties (eastern [mostly northeastern] U.S. west to Michigan and Missouri; Canada). Openings of mesic to dry upland forests.

Steyermark (1963) reported this taxon (as *O. tetragona*) from Henry County without citing a voucher. The specimen on which he based his report has been redetermined as ssp. *fruticosa*.

6. Oenothera glaucifolia W.L. Wagner & Hoch (false gaura)

Gaura linifolia Nutt. ex E. James

Stenosiphon linifolius (Nutt. ex E. James) Heynh.

S. virgatus Spach, an illegitimate name

Pl. 468 a–d; Map 2126

Plants biennial or perennial, with stout taproots having thickened lateral branches. Stems solitary or rarely 2 or 3, 30–200(–300) cm long (including the inflorescence), erect to slightly arched, unbranched or occasionally few-branched toward the tip, glabrous below inflorescence but glaucous, at least toward the base, sometimes somewhat woody toward the base and then with peeling or shredding bark. Leaves in a rosette and alternate (stem leaves often with axillary fascicles of small leaves); rosette leaves 3–7 cm long, 5–20 mm wide, oblong to oblong-lanceolate, the margins entire, the surfaces glabrous or occasionally minutely glandular-hairy; stem leaves 2–8(–10) cm long, 4–18 mm wide, oblong to lanceolate, often somewhat clasping at the base, the margins entire but usually short-hairy, the surfaces glabrous or minutely glandular-hairy, the secondary veins inconspicuous. Inflorescences moderately dense, elongate wandlike spikes, these sometimes grouped into panicles, the axes sparsely to densely pubescent with minute, glandular hairs. Bracts relatively inconspicuous, 4–8(–12) mm long, 0.6–2.0 mm wide, linear. Flowers slightly zygomorphic, 1 of the petals slightly longer than the others, opening in the morning, the floral tube 6–13 mm long, with short spreading hairs. Sepals 4–6 mm long, sparsely short-hairy, mainly along the margins, lacking free tips in bud. Petals 4–6 mm long, 1.5–4.0 mm wide, rhombic-ovate, tapered abruptly to a pronounced stalklike base, white, sometimes pink at the base of the stalklike portion, fading to off-white. Stamens with the filaments 5–8 mm long, ascending, glabrous at the base, the anthers 1.5–2.0 mm long, yellow. Style 2–12 mm long, the stigma positioned well beyond the anthers, deeply 4-lobed, the lobes 0.6–1.0 mm long. Fruits indehiscent, appearing 1-locular, 3–4 mm long, 1.5–2.3 mm in diameter, the body ovoid, slightly flattened, 4-angled with a heavier rib along each angle and a narrower rib on each face, these connected by irregular cross-ridges, lacking a sterile, stalklike base, the surfaces minutely hairy. Seed 1 per fruit, 2.4–2.6 mm long, 1.0–1.5 mm wide, oblanceoloid, the surface whitish yellow, with obscure, fine, parallel grooves. Self-incompatible. $2n=14$. July–October.

Uncommon, restricted to Christian, Ozark, and Taney Counties; also Caldwell County, where apparently introduced (Wyoming to New Mexico east to Missouri and Arkansas; introduced in Indiana, Ohio). Limestone and dolomite glades; also roadsides and open disturbed areas.

As discussed above in the introduction to the family Onagraceae, the monotypic genus *Stenosiphon*, which has been thought to be a close relative of *Gaura*, has now been submerged in a recircumscribed *Oenothera*, where it is treated as a subsection within sect. *Gaura*. The names *O. linifolia* Nutt. (see treatment below) and *O. virgata* (for a South American species) already were in use, so a new epithet was coined for this taxon when it was transferred to *Oenothera*.

Unvouchered reports of this species from Cole and Stone Counties exist in the Missouri Natural Heritage Program's database of species and communities of conservation concern. The Stone County report refers to a single plant observed on a glade in 1988 by Tim Smith of the Missouri Department of Conservation. The Cole County report was based on a small population in a rocky roadside area that was observed annually for several years beginning in 2000 by Dennis Figg, also of the Missouri Department of Conservation.

7. Oenothera grandis (Britton) Smyth (cut-leaved evening primrose)

O. laciniata Hill var. *grandiflora* (S. Watson) B.L. Rob.

Pl. 466 a; Map 2127

Plants annual, with taproots. Stems 1 to several, 50–200 cm long (including the inflorescence), erect or strongly ascending, unbranched or branched, densely pubescent with short, appressed, nonglandular hairs, sometimes also with scattered longer, spreading to loosely appressed, pustular-based, nonglandular hairs (the plants generally appearing grayish green to light

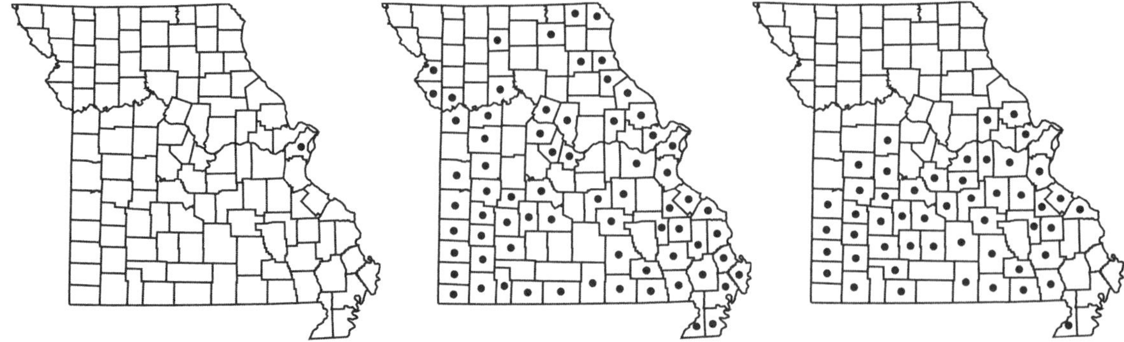

2128. Oenothera heterophylla 2129. Oenothera laciniata 2130. Oenothera linifolia

green), the inflorescence rarely also with minute glandular hairs. Leaves basal and alternate, the rosette leaves 10–30 cm long, 12–40(–50) mm wide, the blade narrowly oblanceolate to oblanceolate, tapered to the usually long petiole or occasionally rounded, the margins sparsely to moderately and often somewhat irregularly toothed, sometimes few-lobed toward the base, the surfaces and margins moderately to densely pubescent with short, appressed, nonglandular hairs, rarely the hairs loosely appressed to loosely ascending (shaggy in appearance); stem leaves 5–20 cm long, 10–25(–40) mm wide, the lower ones narrowly oblanceolate to oblanceolate, grading to narrowly lanceolate or lanceolate toward the stem tip, tapered to the sessile or short-petiolate base, the margins sparsely to moderately and irregularly toothed, those of the lower leaves sometimes with a few lobes, the surfaces and margins with pubescence similar to that of the basal leaves, the secondary veins relatively conspicuous and occasionally reddish-tinged. Inflorescences appearing as solitary, axillary flowers toward the stem tips, with buds adjacent to the most mature one curved upward by the floral tube, the subtending leaves (conspicuous leaflike bracts) persistent, 20–90 mm long, narrowly lanceolate to lanceolate or narrowly ovate to elliptic, the margins irregularly toothed or more commonly lobed, the surfaces with pubescence similar to that of the basal leaves. Flowers actinomorphic, opening at dusk, the floral tube 25–45 mm long, densely pubescent with short, more or less straight, spreading, nonglandular hairs and minute glandular hairs, sometimes also with more or less appressed, nonglandular hairs toward the base. Sepals 15–30 mm long, the midribs not keeled, with pubescence similar to the floral tube, the free tips in bud 1.5–5.0 mm long, terminal, stout (hornlike) and ascending. Petals 25–40 mm long, 30–55 mm wide, very broadly obovate (more or less truncate to shallowly and broadly notched at the tip), lacking a stalklike base, yellow, fading to orange or red. Stamens with the filaments 12–22 mm long, glabrous at the base, the anthers 4–11 mm long, yellow. Style 40–75 mm long, the stigma positioned well above the anthers, deeply 4-lobed, the lobes 5–13 mm long. Fruits 25–50 mm long, 2–3 mm wide, cylindric, straight to somewhat curved, longitudinally dehiscent nearly the entire length, 4-locular, not flattened or winged, more or less circular in cross-section, the surface dull green, not blackening upon drying, moderately to densely pubescent with short, appressed, nonglandular hairs and longer, more or less spreading, nonglandular hairs, not tapered to a sterile, stalklike base. Seeds numerous in each locule, arranged in 2 rows, 0.8–1.5 mm long, 0.5–0.9 mm wide, broadly ellipsoid to subglobose, not angled, the surface yellowish brown to brown, pitted. Self-incompatible. $2n=14$. May–October.

Uncommon and sporadic (Wyoming to Texas east to Illinois and Louisiana; Mexico; introduced sporadically west to California, east to Connecticut and Florida). Upland prairies; also pastures and railroads.

Steyermark (1963) and many other botanists treated *O. grandis* as a large-flowered variant of *O. laciniata*. Dietrich and Wagner (1988) noted other differences between the two taxa including chromosomal features and differences in the breeding system. *O. grandis* flowers have the stigmas elevated well above the anthers at flowering, the pollen is highly fertile, and the species is thus mostly outcrossing. In contrast, the flowers of *O. laciniata* have the stigmas surrounded by the anthers at flowering, about half of the pollen is abortive, and the species is self-pollinated.

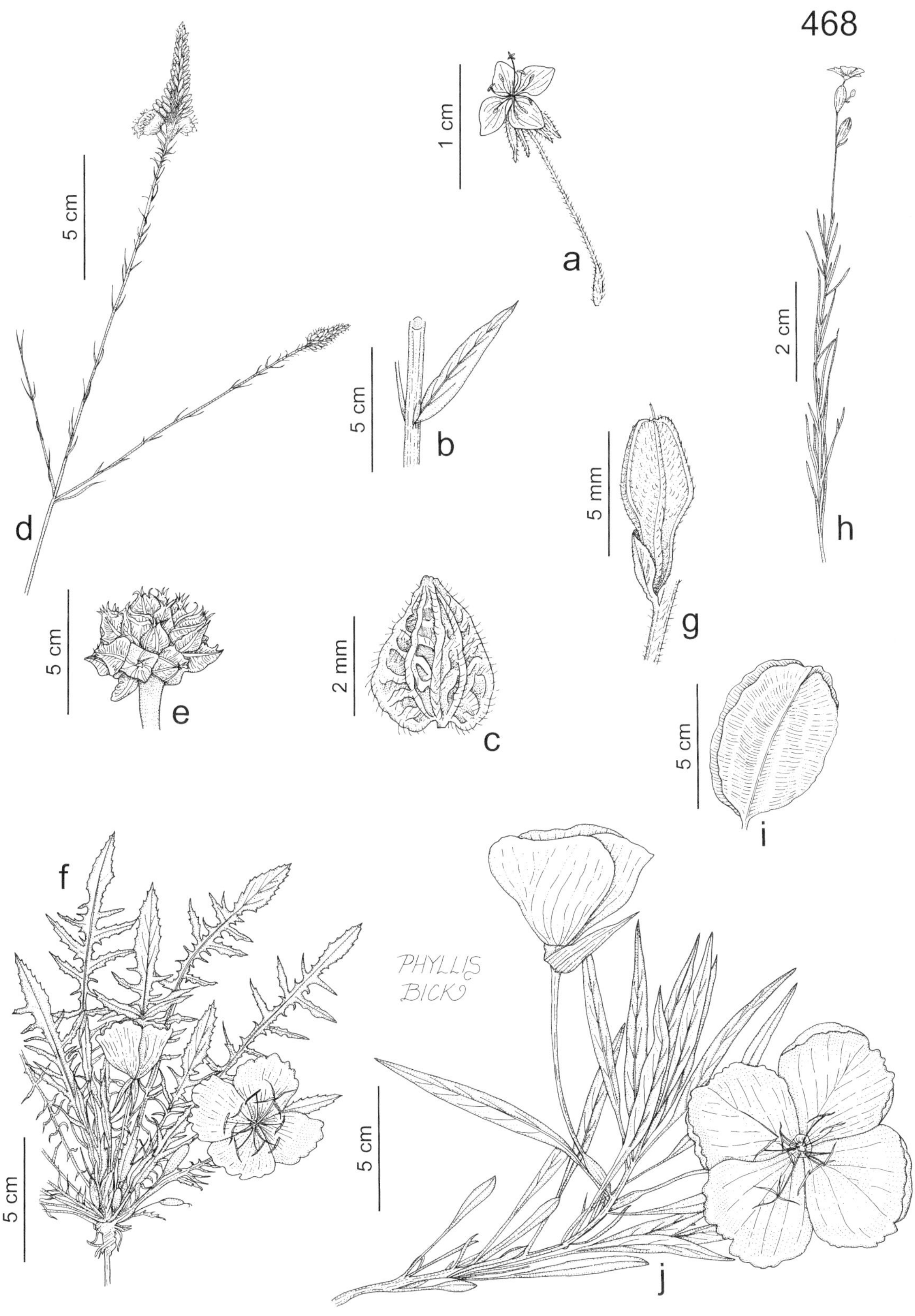

Plate 468. Onagraceae. *Oenothera glaucifolia*, **a)** flower, **b)** median node with leaf, **c)** fruit, **d)** inflorescence. *Oenothera triloba*, **e)** infructescence, **f)** habit. *Oenothera linifolia*, **g)** fruit, **h)** fertile stem. *Oenothera macrocarpa*, **i)** fruit, **j)** habit.

8. Oenothera heterophylla Spach **ssp. heterophylla**

Pl. 466 j, k; Map 2128

Plants annual or short-lived perennials, with taproots. Stems 1 to several, 25–70 cm long (including the inflorescence), erect or strongly ascending, the outer stems sometimes arched upward, unbranched or branched mainly above the midpoint, sparsely to densely pubescent with short, appressed, nonglandular hairs (the plants generally appearing green), the inflorescence axis sometimes nearly glabrous. Leaves basal and alternate, the rosette leaves sometimes absent or withered at flowering, 7–15 cm long, 10–25 mm wide, the blade narrowly oblanceolate to nearly linear, tapered to the sometimes short or indistinct petiole, the margins more or less entire or more commonly coarsely toothed to deeply pinnately lobed with the terminal lobe larger than the lateral ones, the surfaces and margins sparsely pubescent with short, appressed, nonglandular hairs, sometimes nearly glabrous; stem leaves 3–13 cm long, 4–23 mm wide, very narrowly lanceolate to lanceolate or very narrowly elliptic to elliptic, narrowly angled to rounded at the base, with pubescence similar to that of the basal leaves, the secondary veins relatively inconspicuous. Inflorescences dense spikes, sometimes clustered into panicles, the mature buds overtopping the tip of the inflorescence axis, straight, the subtending bracts persistent, 10–30 mm long, narrowly lanceolate to narrowly ovate or ovate, the margins entire to few-toothed, truncate to shallowly cordate at the base, the pubescence similar to that of the basal leaves. Flowers actinomorphic, opening at dusk, the floral tube 25–42 mm long, sparsely pubescent with minute glandular hairs, also with sparse to moderately dense, long, more or less spreading, nonglandular hairs having red pustular bases, often also irregularly flecked with red spots. Sepals 15–28 mm long, the midribs not keeled, often moderately pubescent with long spreading hairs, these with red pustular bases, also glandular-hairy and with fine, short, nonglandular hairs, occasionally glabrous, rarely with only short nonglandular hairs, often also irregularly flecked with red spots, the free tips in bud 2–6 mm long, terminal, usually more or less spreading. Petals 18–35 mm long, 20–30 mm wide, broadly elliptic to more or less rhombic (rounded to more commonly pointed at the tip), lacking a stalklike base, light yellow to yellow, fading yellow to pale orange. Stamens with the filaments 15–30 mm long, glabrous at the base, the anthers 3–8 mm long, light yellow to yellow. Style 25–40 mm long, the stigma usually well-elevated above the anthers, deeply 4-lobed, the lobes 2–5 mm long. Fruits 13–25 mm long, 2.5–4.0 mm wide, lanceoloid, usually curved, longitudinally dehiscent nearly the entire length, 4-locular, not flattened or winged, more or less circular to very bluntly 4-angled in cross-section, the surface green, not blackening upon drying, sparsely to densely pubescent with short, appressed, nonglandular hairs and minute glandular hairs, not tapered to a sterile, stalklike base. Seeds numerous in each locule, arranged in 2 rows, 1.1–1.8 mm long, 0.4–0.8 mm wide, ellipsoid to broadly ellipsoid, not angled, the surface brown, often flecked with darker brown spots, pitted. Self-incompatible. $2n=14$. July–October.

Introduced, known thus far only from two historical specimens from St. Louis City and County (Texas, Louisiana). Railroads and disturbed areas.

This species was first reported for Missouri by Mühlenbach (1983) based on his collection in 1956 from the St. Louis rail yards, although it was first collected in 1893 by George Letterman in the Allenton area. Dietrich and Wagner (1988) also treated a second subspecies, which is endemic to disjunct areas in western Alabama and southern Arkansas. The ssp. *orientalis* W. Dietr., P.H. Raven & W.L. Wagner lacks pustular-based hairs and red dots on the floral tube and sepals that characterize ssp. *heterophylla* and the flower buds have somewhat shorter free sepal tips. It also tends to have more basal leaves that are more deeply divided than those of ssp. *heterophylla*.

9. Oenothera laciniata Hill (cut-leaved evening primrose, ragged evening primrose)

O. sinuata L.

Pl. 466 h, i; Map 2129

Plants annual or short-lived perennials, with taproots. Stems 1 to several, 5–50 cm long (including the inflorescence), erect to spreading with ascending tips, unbranched or branched, sparsely to moderately pubescent with short, appressed, nonglandular hairs and usually also with longer, spreading nonglandular hairs (the plants generally appearing green to light green, but the stems often strongly reddish-tinged), sometimes also with minute glandular hairs toward the tip. Leaves basal and alternate, the rosette leaves sometimes absent or withered at flowering, 4–15 cm long, 10–30 mm wide, the blade narrowly oblanceolate to nearly linear or occasionally lanceolate, tapered to the usually long petiole, the margins more or less entire or more commonly coarsely toothed to deeply pinnately lobed, the surfaces and margins sparsely to densely pubescent with short, more or less appressed, nonglandular hairs and often also

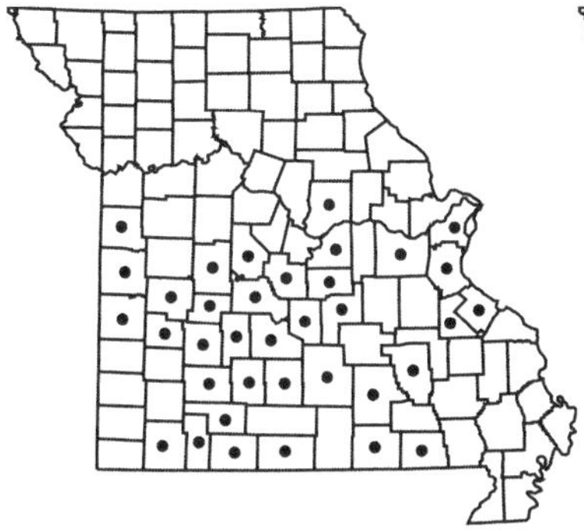

2131. Oenothera macrocarpa

2132. Oenothera nutans

2133. Oenothera parviflora

with longer, more or less spreading hairs, sometimes also with minute glandular hairs; stem leaves 2–10 cm long, 5–35 mm wide, narrowly oblanceolate to oblanceolate or narrowly oblong to narrowly elliptic, tapered to the sessile or short-petiolate base, otherwise similar to the basal leaves, the secondary veins relatively inconspicuous. Inflorescences appearing as solitary, axillary flowers toward the stem tips, with buds adjacent to the most mature one curved upward by the floral tube, the subtending leaves (conspicuous leaflike bracts) persistent, 20–70 mm long, narrowly oblong to narrowly ovate, otherwise similar to the basal leaves. Flowers actinomorphic, opening at dusk, the floral tube 12–35 mm long, sparsely to densely pubescent with short, more or less straight, spreading, nonglandular hairs and minute glandular hairs, sometimes also with more or less appressed, nonglandular hairs. Sepals 5–15 mm long, the midribs not keeled, with pubescence similar to the floral tube, the free tips in bud 0.3–3.0 mm long, terminal, usually more or less spreading. Petals 5–20(–22) mm long, 7–20 mm wide, very broadly obovate (more or less truncate to shallowly and broadly notched at the tip), lacking a stalklike base, pale yellow to yellow, fading pale orange to red. Stamens with the filaments 3–14 mm long, glabrous at the base, the anthers 2–6 mm long, yellow. Style 20–50 mm long, the stigma positioned at about the same level as the anthers, deeply 4-lobed, the lobes 2.5–5.0 mm long. Fruits 20–50 mm long, 2–4 mm wide, cylindric, straight to somewhat curved, longitudinally dehiscent nearly the entire length, 4-locular, not flattened or winged, more or less circular in cross-section, the surface dull green, not blackening upon drying, moderately to densely pubescent with short, appressed, nonglandular hairs and scattered, longer, more or less spreading, nonglandular hairs, not tapered to a sterile, stalklike base. Seeds numerous in each locule, arranged in 2 rows, 0.9–1.8 mm long, 0.4–0.9 mm wide, broadly ellipsoid to subglobose, not angled, the surface brown to dark brown, pitted. Self-compatible. $2n=14$. May–October.

Scattered nearly throughout the state (eastern U.S. west to North Dakota and New Mexico; Canada, Mexico; introduced in California and Hawaii, also Central America, South America, Europe, Asia, Africa, Australia). Glades, upland prairies, sand prairies, banks of streams and rivers, and margins of ponds; also pastures, fallow fields, margins of crop fields, gardens, railroads, roadsides, and open disturbed areas.

Oenothera laciniata is closely related to *O. grandis*. For further discussion, see the treatment of that species.

10. Oenothera linifolia Nutt. (sundrops, thread-leaved sundrops)

Pl. 468 g, h; Map 2130

Plants annual, with slender taproots. Stems solitary or few, 10–50 cm long (including the inflorescence), erect or nearly so, unbranched or with few to many ascending branches, minutely nonglandular-hairy when young, sometimes also minutely glandular-hairy toward the tip, sometimes becoming glabrous or nearly so at maturity. Leaves basal and alternate (the stem leaves also usually with small fascicles of axillary leaves), the basal leaves sometimes withered at flowering; rosette leaves 1–2(–4) cm long, 2–6 mm wide, ovate to narrowly elliptic or obovate, petiolate, the margins entire or with a few small teeth, also hairy, the surfaces glabrous or sparsely pubescent with short, appressed to curved, nonglandular hairs and/or minute, spreading, glandular hairs; stem leaves 1–4 cm long, 0.5–1.0 mm wide, linear to threadlike, sessile, the margins entire, the secondary veins not apparent. Inflorescences open spikes, not grouped into panicles, short to elongate, the axes sparsely to densely pubescent with short, appressed to upward-curved nonglandular hairs and/or minute, spreading glandular hairs. Bracts relatively inconspicuous, 0.5–2.0 mm long, 1–3 mm wide, ovate to ovate-triangular. Flowers actinomorphic, opening at dusk (but remaining open through the following day), the floral tube 1–2 mm long, sparsely hairy. Sepals 1.5–2.0 mm long, with short, spreading, glandular or nonglandular hairs, lacking free tips in bud. Petals 3–5(–7) mm long, 1–3(–4) mm wide, more or less heart-shaped to oblong-obovate (notched at the tip), lacking a stalklike base, yellow, fading pink. Stamens with the filaments 1–2 mm long, ascending or slightly incurved, glabrous at the base, the anthers 0.5–1.0 mm long, light yellow. Style 1–2 mm long, the stigma positioned at about the same level as the anthers, deeply 4-lobed, the lobes 0.5–1.0 mm long. Fruits 4–6(–10) mm long, the main body tardily longitudinally dehiscent toward the tip (the fruit sometimes shed before dehiscence), 4-locular, 4–6 mm long, 1.5–3.0 mm in diameter, elliptic-rhomboid, not flattened, 4-ridged, the surfaces finely glandular- and/or nonglandular-hairy, sessile or rarely with a sterile, stout, stalklike base to 4 mm long, this also glandular- and/or nonglandular-hairy. Seeds numerous in each locule, clustered and not in definite rows, 0.9–1.4 mm long, 0.4–0.6 mm wide, irregularly rhombic-ellipsoid, the surface pale reddish brown, smooth to somewhat pebbled or minutely warty. Self-compatible (often self-fertilizing in bud or during flowering). $2n$=14. May–July.

Scattered, mostly south of the Missouri River (southeastern U.S. west to Kansas and Texas). Glades, upland prairies, tops and ledges of bluffs, openings of dry upland forests, and savannas; also pastures, cemeteries, railroads, and roadsides; usually on acidic substrates.

11. Oenothera macrocarpa Nutt. **ssp. macrocarpa** (Missouri evening primrose, Missouri primrose)

O. missouriensis Sims

Pl. 468 i, j; Map 2131

Plants perennial, with a stout, woody, often somewhat branched, vertical rootstock, sometimes producing new shoots from lateral roots. Stems solitary or more commonly few to several, 15–60 cm long (sometimes longer in cultivated plants), spreading to loosely ascending, unbranched or with few to several branches, glabrous to densely pubescent with short, appressed, nonglandular hairs, strongly reddish- to dark purplish-tinged. Leaves alternate, 4–12(–15) cm long, 4–25 mm wide, lanceolate-elliptic to broadly elliptic or sometimes linear or lanceolate, short- to moderately petiolate, the margins entire to somewhat irregular or wavy or with short, broad, broadly spaced teeth, hairy, the surfaces pubescent with appressed, nonglandular hairs, densely so on young leaves, often less densely so at maturity (but often grayish-tinged, as well as with reddish splotches), the secondary veins usually relatively inconspicuous. Inflorescences of axillary flowers, the bracts not differentiated from foliage leaves. Flowers actinomorphic, opening at dusk (but often remaining open into the following morning), the floral tube (78–)95–115(–140) mm long, pubescent with short, appressed, nonglandular hairs and minute, glandular hairs. Sepals (45–)50–65(–75) mm long, glabrous or with dense, appressed, nonglandular hairs, the free tips in bud (4–)8–10(–12) mm long, erect and appressed. Petals (40–)55–65(–68) mm long, (43–)48–64(–69) mm wide, broadly obovate to very broadly obovate (more or less truncate to broadly and shallowly notched or somewhat irregular at the tip), lacking a stalklike base, yellow to bright yellow, not fading or fading to orange. Stamens with the filaments (25–)30–40(–44) mm long, ascending to somewhat S-shaped or curved toward the top of the flower, glabrous at the base, the anthers (15–)17–24 mm long, yellow. Style (45–)55–190 mm long, the stigma positioned well above the anthers, deeply 4-lobed, the lobes (3–)7–13(–19) mm long. Fruits 52–75(–120) mm long, overall broadly oblong-elliptic to oblong in outline, the main body longitudinally dehiscent about 1/4–1/3 the length, 4-locular, 50–70(–115) mm long, 6–8 mm in diameter, narrowly ellipsoid or lanceoloid, not flattened, strongly 4-winged, each wing (14–)18–28(–34) mm wide, flat to somewhat undulate, rounded or truncate at the tip, becoming tan and papery at maturity, tapered abruptly to a sterile, stalklike base 2–6(–12) mm long. Seeds numerous in each locule, arranged in a single row, 3–5 mm long, mostly 1.8–2.3 mm wide, obovoid, the surface grayish brown to dark brown, coarsely wrinkled and usually somewhat corky. Self-incompatible. $2n$=14. May–August.

Scattered, mostly south of the Missouri River, but apparently absent from the Mississippi Lowlands Division (Wyoming to Illinois south to Texas, Arkansas, and Tennessee). Glades, tops of bluffs, rocky portions of upland prairies, and banks of streams and rivers; also quarries and roadsides; on calcareous substrates.

Thomas Nuttall first collected fruits of this taxon during the winter of 1810, when he and John Bradbury were botanizing somewhere south of St. Louis while awaiting the departure of the fur-trading expedition they were to join the following spring (see also the History of Missouri Botany section in Volume 1 of the present work [Yatskievych, 1999]). The large, winged fruits that frequently become detached from the stems with age and tumble across the landscape are persistent and conspicuous. Plants later were grown in England from the Missouri seeds, and both Nuttall and John Sims of Kew Gardens independently described these as a species new to science. W. L. Wagner (1981) presented an interesting historical review of this situation. The species subsequently became a popular ornamental in European gardens. It was not until interest grew in the 1980s for the use of native perennials in midwestern gardens that *O. macrocarpa* became available through wildflower nurseries in the United States. Because of its drought tolerance, creeping habit, silvery foliage, exquisite flowers, and unusual fruits, the species most recently has become available widely in mainstream horticulture.

Oenothera macrocarpa belongs to a small section of four species that are characterized by relatively large winged fruits, and unique seed morphology. Fruits disarticulate from the plant at maturity, and when dry they open and blow around, dispersing seeds. *Oenothera macrocarpa* is a variable species that has differentiated extensively in the Great Plains region. Each of the five distinctive subspecies occupies a different geographical and ecological situation. The entities are, in general sharply distinct and each is characterized by a number of features. The differences are primarily those of pubescence, leaf features, flower and floral tube size, and size and morphology of the fruits and seeds. They are treated as subspecies primarily because of their complete interfertility and extensive intergradation in any area of marginal contact (W. L. Wagner, 1983; W. L. Wagner et al., 2007). The most similar to ssp. *macrocarpa*, is ssp. *mexicana* W.L. Wagner, which is endemic to a small area of northeastern Mexico. It is distinctive in possessing smaller flowers and especially narrow leaves with dense short, appressed pubescence, and has smaller fruits and narrower leaves than ssp . *macrocarpa*. The three other subspecies have mostly distinct ranges and are more divergent from ssp. *macrocarpa*. The ssp. *oklahomensis* (Norton) W.L. Wagner is a completely glabrous plant endemic to central Oklahoma additionally characterized by leaves that are usually undulate and also have conspicuous, short teeth, as well as divergent and often twisted free sepal tips. Another of the subspecies is ssp. *incana* (A. Gray) W.L. Wagner, which occurs in the Texas panhandle, western Oklahoma, and adjacent Kansas. It is characterized by short, silky, gray, dense, appressed pubescence (rarely glabrous) and leaves that are broadly elliptic to suborbicular, rarely oblanceolate or elliptic, and 20–43 mm wide. The most distinctive subspecies is ssp. *fremontii* (S. Watson) W.L. Wagner, which is characterized by a tufted habit, free sepal tips 1–2(–5) mm long, petals 1.7–3.3(–3.7) cm long, and floral tube (2.1–)3.5–6.5(–8.0) cm long. Its fruits often are twisted, 1.3–3.0(–6.5) cm long, the wings 2–5(–9) mm wide, and seeds with vestigial wing toward the tip. It occurs in western Kansas and Nebraska.

12. Oenothera nutans G.F. Atk. & Bartlett (nodding evening primrose)

O. biennis var. *austromontana* (Munz) Cronquist

O. biennis L. var. *nutans* (G.F. Atk. & Bartlett) Wiegand

Pl. 464 j; Map 2132

Plants biennial, with taproots. Stems 1 to more commonly several, 30–200 cm long (including the inflorescence), erect or strongly ascending, often branched, often appearing glabrous to the naked eye, but usually pubescent with either: a) short, appressed, nonglandular hairs toward the base and scattered longer, spreading, pustular-based, nonglandular hairs throughout; or b) dense, short, appressed, nonglandular hairs and scattered, longer, more or less appressed, nonglandular hairs toward the base and glabrous to densely glandular-hairy toward the tip; the inflorescence with minute gland-tipped hairs, occasionally also with short, straight, mostly pustular-based, nonglandular hairs. Leaves basal and alternate, the rosette leaves 10–32 cm long, 30–70 mm wide, the blade narrowly oblanceolate to narrowly obovate, tapered to the petiole, the margins sparsely to moderately and often somewhat irregularly toothed, sometimes few-lobed toward the base, the surfaces sparsely pubescent with short, appressed to curved, nonglandular hairs or glabrous except for the undersurface midvein; stem leaves 6–20 cm long, 20–80 mm wide, narrowly lanceolate or lanceolate to narrowly elliptic, or narrowly oblanceolate to oblanceolate, tapered to the sessile or (on lower leaves) short-petiolate base, the margins sparsely to moderately and irregularly toothed, those of the lower leaves sometimes with a few lobes, the surfaces with pubescence similar to that of the basal leaves, the secondary veins relatively conspicuous. Inflorescences short to elongate spikes, relatively dense, erect and relatively straight, sometimes grouped into panicles with ascending branches, the bracts shed before flowering, 10–25 mm long, narrowly lanceolate to narrowly ovate, the margins entire or finely and irregularly toothed, the surfaces with pubescence

2134. Oenothera perennis 2135. Oenothera pilosella 2136. Oenothera rhombipetala

similar to that of the basal leaves. Flowers actinomorphic, opening at dusk, the floral tube 30–43 mm long, glabrous to sparsely pubescent with minute glandular hairs or short, nonglandular hairs. Sepals 10–23 mm long, the midribs not keeled, glabrous to sparsely pubescent with minute glandular hairs or short, nonglandular hairs, the free tips in bud 1.5–6.0 mm long, terminal, usually erect and appressed. Petals 14–25(–30) mm long, 15–28 mm wide, broadly obovate to broadly heart-shaped (broadly notched at the tip), lacking a stalklike base, yellow, fading pale yellow to straw-colored and often somewhat translucent. Stamens with the filaments 10–25 mm long, glabrous at the base, the anthers 4–10 mm long, yellow. Style 35–63 mm long, the stigma positioned at about the same level as the anthers, deeply 4-lobed, the lobes 3–7 mm long. Fruits 12–36 mm long, 3–6 mm wide, lanceoloid (tapered toward the tip) to occasionally narrowly ovoid, straight, longitudinally dehiscent nearly the entire length, 4-locular, not flattened or winged, more or less circular in cross-section, the surface green to dull green at maturity, not blackening upon drying, sparsely to densely glandular-hairy and sometimes also with scattered, appressed or spreading, nonglandular hairs, the sterile, stalklike base absent or very short and poorly differentiated. Seeds numerous in each locule, arranged in 2 rows, 1.1–1.9 mm long, 0.6–0.9 mm wide, irregularly prismatic, angled, the surface brown to nearly black, with an irregular network of ridges and pits. Self-compatible. $2n=14$. June–October.

Uncommon and widely scattered (eastern U.S. west to Indiana, Missouri, and Mississippi; Canada). Banks of rivers; also open disturbed areas.

Dietrich et al. (1997) mapped *O. nutans* from three widely scattered counties based on historical specimens. They speculated that the somewhat disjunct Missouri occurrences might represent introductions. More recently, a collection of this species from a sandy area along the Missouri River in Holt County suggests that it is more likely a disturbance-adapted native that is overlooked by collectors because of its morphological similarity to the nearly ubiquitous *O. biennis.*

Steyermark (1963) reported *Oenothera grandiflora* L'Hér. ex Aiton from a single purportedly introduced record at Willard (Greene County). The apparent voucher for this report was located at the herbarium of Drury University and was collected in 1889 by Joseph W. Blankinship. Interestingly, although plants of *O. grandiflora* are morphologically most similar to *O. nutans* in the Missouri flora, the Blankinship specimen instead represents a misdetermined plant of *O. villosa*. It should be noted that during the times both when the specimen was collected and when Steyermark was completing his research on the Missouri flora, the taxonomy and species circumscriptions within the *O. biennis* complex were still poorly understood. *Oenothera grandiflora* is native to the southeastern United States, but has long been cultivated as an ornamental in gardens in the United States and Europe. It has escaped sporadically in the northeastern states, but never as far west as Missouri (Dietrich et al., 1997). This species is thus excluded from the Missouri flora for the present. It differs from *O. biennis* and *O. nutans* in its larger flowers (sepals 2.2–4.6 cm long, petals mostly 3.0–4.5 cm long) and in having its stigmas elevated well above the stamens as the flowers open (thus promoting cross-pollination rather than selfing). It also lacks the conspicuously dense, appressed hairs on the floral tubes and fruits that are characteristic of *O. villosa*.

13. Oenothera parviflora L. (northern evening primrose)

O. parviflora ssp. *angustissima* (R.R. Gates) Munz

Pl. 464 l; Map 2133

Plants biennial, with taproots. Stems 1 to several, 30–150 cm long (including the inflorescence), erect or strongly ascending, unbranched or branched, sparsely pubescent with a mixture of short, appressed to upward-curved, nonglandular hairs, minute glandular hairs, and longer, spreading, pustular-based hairs, sometimes mostly toward the base. Leaves basal and alternate, the rosette leaves 10–30 cm long, 10–40 mm wide, the blade narrowly oblanceolate to narrowly elliptic, tapered to the petiole, the margins sparsely to moderately and often somewhat irregularly toothed, sometimes few-lobed toward the base, the surfaces sparsely pubescent with short, more or less appressed, nonglandular hairs, the upper surface sometimes nearly glabrous; stem leaves 4–18 cm long, 10–30 mm wide, lanceolate to narrowly ovate, very narrowly to narrowly elliptic, or narrowly oblong, tapered to the sessile or short-petiolate base, the margins sparsely to moderately and irregularly toothed or those of the upper leaves occasionally nearly entire, the surfaces sparsely pubescent with short, appressed, nonglandular hairs, the upper surface sometimes nearly glabrous, the secondary veins relatively conspicuous. Inflorescences short to elongate spikes, relatively dense, usually somewhat curved, sometimes grouped into panicles with ascending branches, the bracts persistent, 20–80 mm long, narrowly lanceolate to narrowly ovate, the margins entire or slightly irregular or occasionally few-toothed, the surfaces minutely glandular hairy, sometimes also with short appressed, nonglandular hairs toward the tip and/or scattered, longer, spreading, nonglandular hairs with pustular bases. Flowers actinomorphic, opening at dusk, the floral tube 22–40 mm long, glabrous to densely pubescent with minute glandular hairs and sometimes also with sparse, short, nonglandular hairs. Sepals 7–17 mm long, the midribs not keeled, glabrous or appressed-hairy toward the base, the free tips in bud 0.5–5.0 mm long, usually subterminal, divergent. Petals 8–15(–20) mm long, 9–20 mm wide, broadly obovate to broadly heart-shaped (broadly notched at the tip), lacking a stalklike base, light yellow to yellow, fading yellow to orange. Stamens with the filaments 7–13 mm long, glabrous at the base, the anthers 3.5–6.0 mm long, yellow. Style 25–50 mm long, the stigma positioned at about the same level as the anthers, deeply 4-lobed, the lobes 2.5–6.0 mm long. Fruits 20–40 mm long, 3.5–5.0 mm wide, usually lanceoloid (tapered toward the tip), straight, longitudinally dehiscent nearly the entire length, 4-locular, not flattened or winged, more or less circular in cross-section, the surface dark green at maturity but often blackening upon drying, variously glabrous or pubescent with a mixture of glandular and nonglandular hairs, lacking a sterile, stalklike base. Seeds numerous in each locule, arranged in 2 rows, 1.1–1.8 mm long, 0.5–1.0 mm wide, irregularly prismatic, angled, the surface brown to dark brown, with an irregular network of ridges and pits. Self-compatible. $2n=14$. June–September.

Uncommon and widely scattered, known thus far only from single historical specimens from Boone, Clinton, and St. Louis Counties (northeastern U.S. west to Minnesota and Missouri; Canada; introduced in Europe, Asia, Africa, New Zealand). Openings of mesic upland forests; also roadsides.

Palmer and Steyermark (1935) included this species in the Missouri flora with a statement of “general and common.” Steyermark (1963) later excluded it with a note that all of the specimens had been redetermined as *O. biennis*. However, Dietrich et al. (1997) reported a single historical collection from St. Louis County to justify the continued inclusion of *O. parviflora* in the state's flora, and two others were located during the present research. Elsewhere in its range, the species occurs in disturbed habitats similar to those in which *O. biennis* can be found.

14. Oenothera perennis L. (sundrops, small sundrops)

Pl. 467 a, b; Map 2134

Plants perennial, with fibrous roots. Stems solitary or few, 15–40(–70) cm long (including the inflorescence), erect or ascending, unbranched or more commonly with few to several branches above the midpoint, moderately to densely pubescent with short, appressed to upward-curved, nonglandular hairs. Leaves basal and alternate, the basal leaves overwintering and often withered at flowering; rosette leaves 2–4 cm long, 2–12 mm wide, oblanceolate to obovate, short-petiolate, the margins entire or slightly irregular, hairy, the surfaces glabrous; stem leaves 3–7 cm long, 2–12 mm wide, narrowly oblanceolate to obovate, mostly short-petiolate, the margins entire, hairy, the surfaces sparsely short-hairy, the secondary veins inconspicuous. Inflorescences open spikes, not grouped into panicles, relatively short and few-flowered, the portion in bud nodding, the axes moderately to densely pubescent with minute, glandular hairs. Bracts relatively conspicuous, 8–18 mm long, 1–2 mm wide, linear to narrowly oblong-elliptic or narrowly oblanceolate. Flowers actinomorphic, opening in the morning, the floral tube 3–10 mm long, short-hairy. Sepals 2–4 mm long, with short, spreading, glandular or nonglandular hairs, the free tips in bud 0.4–0.9 mm long, erect and appressed. Petals 5–10 mm long, 4–10 mm wide, broadly obovate to broadly oblong-obovate (truncate to shallowly and broadly notched at the tip), lacking a stalklike base, bright yellow to deep yellow, fading to yellow or lavender. Stamens with the filaments 3–4 mm long, ascending, glabrous at the base, the anthers 1–2 mm long, yellow. Style 3–4 mm long, the stigma positioned at about the same level as the anthers, deeply 4-lobed, the lobes 0.9–1.4 mm long. Fruits 5–10 mm long, the main body longitudinally dehiscent toward the tip and eventually more or less the entire length, 4-locular, 4–9 mm long, 2–3 mm in diameter, club-shaped, not flattened, strongly 4-angled or narrowly 4-winged, the surfaces finely glandular- or nonglandular-hairy, tapered to a sterile, stalklike base 1–2 mm long, this also hairy. Seeds numerous in each locule, clustered and not in definite rows, 0.7–0.8 mm long, 0.2–0.3 mm wide, irregularly rhombic-ellipsoid to prismatic, the surface bright reddish brown, pebbled. Self-compatible. $2n$=14. June–August.

Uncommon and sporadic, known thus far from Dent, Johnson, and Shannon Counties (northeastern U.S. west to Minnesota and Missouri; Canada). Fens, margins of sinkhole ponds, and moist portions of upland prairies.

Botanists in southwestern Missouri eventually may locate the closely related *O. spachiana* in that portion of the state. This species occurs from eastern Texas north to northeastern Oklahoma and east to Alabama. It differs from *O. perennis* most notably in its axillary flowers and annual habit.

15. Oenothera pilosella Raf. (sundrops, prairie sundrops)
O. pilosella f. *laevigata* E.J. Palmer & Steyerm.

Pl. 467 h, i; Map 2135

Plants perennial, with a thickened rootstock, usually producing well-developed rhizomes. Stems solitary or few to occasionally several, 20–80 cm long (including the inflorescence), erect or ascending, unbranched or with few to several branches, sparsely to densely pubescent with spreading, nonglandular hairs 1–2 mm long, rarely glabrous. Leaves basal and alternate, the basal leaves overwintering and usually withered at flowering; rosette leaves 4–8 cm long, 2–5 mm wide, lanceolate to ovate, short- to moderately petiolate, the margins entire or the broader ones coarsely toothed, hairy, the surfaces nonglandular-hairy, rarely glabrous; stem leaves 2–10(–13) cm long, 10–20(–40) mm wide, mostly lanceolate, occasionally linear or ovate, mostly short-petiolate, the margins entire or slightly irregular to coarsely toothed, hairy, the surfaces nonglandular-hairy, rarely glabrous, the secondary veins usually relatively conspicuous (except on the narrowest leaves). Inflorescences open spikes, not grouped into panicles, usually relatively short and sometimes few-flowered, the portion in bud erect, the axis usually moderatelty to densely pubescent with relatively long, spreading, nonglandular hairs. Bracts relatively conspicuous, 5–45 mm long, 1–10 mm wide, linear to lanceolate. Flowers actinomorphic, opening in the morning, the floral tube 10–25 mm long, usually spreading-hairy. Sepals 10–20 mm long, with spreading to appressed, nonglandular hairs, the free tips in bud 1–4 mm long, usually somewhat spreading. Petals 15–30 mm long, 15–25 mm wide, broadly obovate to somewhat heart-shaped (broadly but sometimes shallowly notched at the tip), lacking a stalklike base, yellow to deep yellow, fading to yellow or lavender. Stamens with the filaments 7–15 mm long, ascending, glabrous at the base, the anthers 4–8 mm long, yellow. Style 10–20 mm long, the stigma positioned well above the anthers, deeply 4-lobed, the lobes 2–5 mm long. Fruits (5–)10–15(–28) mm long, the main body longitudinally dehiscent toward the tip and eventually more or less the entire length, 4-locular, (5–)10–14(–26) mm long, 2–4(–5) mm in diameter, narrowly club-shaped to ellipsoid, not flattened, strongly 4-angled or rarely slightly 4-winged, the surfaces glabrous to densely pubescent with appressed or spreading, nonglandular hairs, sessile or tapered to an indistinct, sterile, stalklike base to 2 mm long, this also glabrous or hairy. Seeds numerous in each locule, clustered and not in definite rows, 0.8–1.0 mm long, 0.3–0.5 mm wide, irregularly rhombic-ellipsoid to prismatic-ovoid, the surface dark reddish brown, pebbled. Self-incompatible. $2n$=56. May–July.

Scattered in the eastern half of the state (eastern [mostly northeastern] U.S. west to Iowa and Louisiana; Canada). Fens, marshes, swamps, and bottomland prairies; also ditches, old fields, railroads, and roadsides.

This attractive species is sometimes cultivated as an ornamental in gardens, and several cultivars are available. Straley (1977) treated *O. pilosella* as consisting of two subspecies. The ssp. *sessilis* (Pennell) Straley occupies the southwestern portion of the species range in Arkansas, Louisiana, and Texas. It differs from ssp. *pilosella* in having dense, minute, appressed hairs, as well as flowers with a shorter ovary (4.5–6.5 vs. 9–12 mm at flowering) and buds with the free sepals tips incurved to erect (vs. ascending-spreading). Straley (1977) discussed a problematic historical collection made near Corning, in Clay County, Ar-

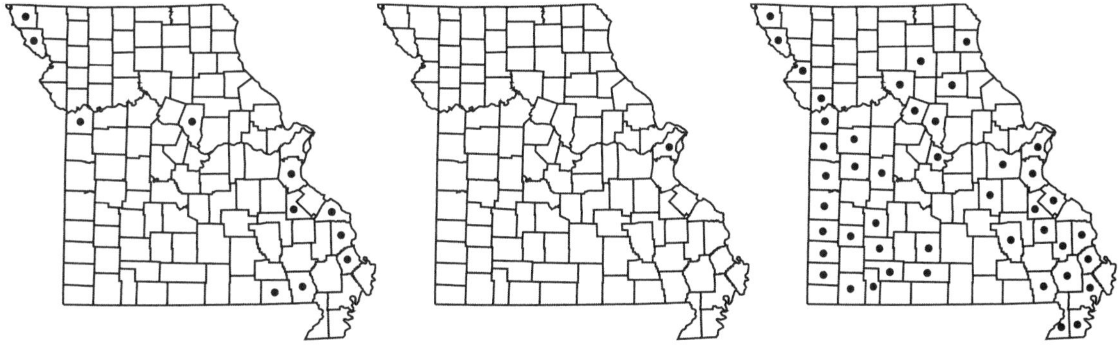
2137. Oenothera serrulata 2138. Oenothera sinuosa 2139. Oenothera speciosa

kansas, which is very close to southeastern Missouri. Thus, Missouri botanists should be on the lookout for ssp. *sessilis* in the Mississippi Lowlands Division in the future. Recent studies by Krakos (2011) have shown that plants attributable to ssp. *pilosella* are outcrossers pollinated during the day by bees, whereas plants of ssp. *sessilis* are self-pollinated. Further, her molecular phylogenetic studies of *Oenothera* have shown that ssp. *pilosella* is more closely related to *O. fruticosa* than to *O. sessilis*. Krakos therefore concluded that the two taxa should be treated as separate species, *O. pilosella* and *O. sessilis* (Pennell) Munz.

Straley (1977, as *O. pilosella* ssp. *sessilis*) discussed a problematic historical collection of *O. sessilis* made near Corning, in Clay County, Arkansas, which is very close to southeastern Missouri. Thus, Missouri botanists should be search for populations of *O. sessilis* in the Mississippi Lowlands Division in the future.

16. Oenothera rhombipetala Nutt. ex Torr. & A. Gray (sand primrose)

Pl. 466 e–g; Map 2136

Plants biennial, with taproots. Stems 1 to several, 30–100(–150) cm long (including the inflorescence), erect or ascending, sometimes arched upward, unbranched to well-branched, sparsely to densely pubescent with short, appressed, nonglandular hairs (the plants generally appearing green or grayish green), sometimes also with minute glandular hairs toward the tip. Leaves basal and alternate, the rosette leaves occasionally absent or withered at flowering, 6–20 cm long, 6–20 mm wide, the blade narrowly oblanceolate, long-tapered to the sometimes short or indistinct petiole, the margins bluntly toothed to deeply pinnately lobed with the terminal lobe larger than the lateral ones, the surfaces and margins usually densely pubescent with short, appressed, nonglandular hairs; stem leaves 3–15 cm long, 8–25 mm wide, narrowly elliptic to narrowly lanceolate or narrowly oblanceolate to ovate, narrowly angled to truncate at the base, the margins variously toothed to pinnately lobed or occasionally nearly entire, with pubescence similar to that of the basal leaves, the secondary veins relatively inconspicuous. Inflorescences dense spikes, usually unbranched, the mature buds not overtopping the tip of the inflorescence axis, straight, the subtending bracts persistent, 8–25 mm long, narrowly lanceolate to narrowly ovate, the margins entire to few-toothed, broadly angled or rounded at the base, the pubescence similar to that of the basal leaves. Flowers actinomorphic, opening at dusk, the floral tube 30–45 mm long, sparsely to densely pubescent with short, more or less appressed, nonglandular hairs (these lacking pustular bases) and sometimes also minute glandular hairs. Sepals 15–30 mm long, the midribs not keeled, sparsely to moderately pubescent with short, more or less appressed, nonglandular hairs, sometimes also minute glandular hairs, sometimes flecked with red spots, the free tips in bud 0.5–2.0(–3.0) mm long, terminal, erect and appressed. Petals 15–35 mm long, 12–30 mm wide, broadly elliptic to more or less rhombic (usually pointed at the tip), lacking a stalklike base, yellow, fading to yellow to pale orange. Stamens with the filaments 13–23 mm long, glabrous at the base, the anthers 3–8 mm long, light yellow to yellow. Style 25–50 mm long, the stigma elevated above the anthers, deeply 4-lobed, the lobes 2–5 mm long. Fruits 13–25 mm long, 2.5–3.0 mm wide, narrowly lanceoloid, usually curved, longitudinally dehiscent nearly the entire length, 4-locular, not flattened or winged, more or less circular to very bluntly 4-angled in cross-section, the surface green, not blackening upon drying, moderately to densely pubescent with short, appressed, nonglandular hairs, sometimes also with sparse, minute glandular hairs, not tapered to a sterile, stalklike base. Seeds numerous in each locule, arranged in 2 rows, 1.0–1.7 mm long, 0.4–0.7 mm wide, ellipsoid, not angled, the surface brown, sometimes flecked with dark red spots, pitted. Self-incompatible. 2*n*=14. June–October.

Uncommon, known only from a few historical collections from Jackson County and a more recent, introduced occurrence in Douglas County (South Dakota to New Mexico east to Missouri and Arkansas). Native habitat unknown, but possibly upland prairies or banks of rivers; also open disturbed areas.

The historical status of *O. rhombipetala* in Missouri is not known with certainty, as the few bonifide historical collections made by B. F. Bush in the Kansas City area lack habitat information or an indication whether the plants were native or not. Apparently, the species was uncommon in Missouri at the end of the 1800s when Bush collected it. The Douglas County occurrence was first noted in 2001 by Arlena Maggard, a landowner who discovered plants volunteering on a pile of sand brought to her property in 2000 from Oklahoma.

Steyermark's (1963) report of occurrences of *O. rhombipetala* in the St. Louis area was based on misdetermined specimens of *O. heterophylla* and his reports of the species from northeastern Missouri was based on specimens since redetermined to *O. clelandii*. Steyermark and most other earlier botanists did not discriminate *O. rhombipetala* from the closely related, but smaller-flowered *O. clelandii*. The two taxa differ in the constitutions of their chromosomes, resulting in different configurations at meiosis. Also, *O. rhombipetala* is a mainly cross-pollinated species with mostly fertile pollen and the stigma positioned well above the stamens at flowering. In contrast, *O. clelandii* is a self-pollinated species with about half of the pollen abortive and the stigma surrounded by the anthers at flowering. The distribution of *O. clelandii* is mostly to the east of that of *O. rhombipetala*, but the ranges of the two taxa overlap in portions of the Midwest, including Missouri. They are not known to grow together (Dietrich and Wagner, 1988).

17. Oenothera serrulata Nutt. (plains yellow primrose, toothed evening primrose)

Calylophus serrulatus (Nutt.) P.H. Raven

Pl. 464 e–g; Map 2137

Plants perennial, often woody at the base, with a stout, woody, often branched rootstock. Stems 1 to many, 10–60(–80) cm long (including the inflorescence), loosely to strongly ascending, unbranched or branched, glabrous to densely pubescent with short, appressed to upward-curved, nonglandular hairs, especially toward the tip. Leaves alternate (the main stem leaves also often with small fascicles of axillary leaves), the lower ones sometimes withered or shed by flowering, 1–10 cm long, 1–9 mm wide, linear to narrowly lanceolate or oblanceolate, often somewhat folded lengthwise, sessile or nearly so, the margins with few to several sharp teeth or occasionally nearly entire, the surfaces glabrous to densely pubescent with short, appressed, nonglandular hairs, the secondary veins inconspicuous. Inflorescences short, compact spikes, the bracts not markedly differentiated from foliage leaves, sometimes appearing as axillary flowers from the upper, leafy nodes. Flowers actinomorphic, opening at dusk, the floral tube (2–)12–16 mm long, glabrous to sparsely appressed-hairy. Sepals 1.5–9.0 mm long, the midribs keeled, glabrous or appressed-hairy toward the base, the free tips in bud absent or to 4 mm long, when present erect and appressed. Petals 5–12(–20) mm long, 5–15(–20) mm wide, obovate to nearly circular or somewhat heart-shaped (finely irregular and broadly rounded to truncate or shallowly notched at the tip, sometimes with a minute sharp point), often appearing somewhat corrugated lengthwise, lacking a stalklike base, yellow, fading to dark yellow to orange. Stamens with the filaments in 2 unequal series, those opposite the sepals 1–5(–7) mm long, those opposite the petals 0.5–3.0 mm long, ascending, glabrous at the base, the anthers 1.5–4.0(–7.0) mm long, yellow to dark yellow. Style 2–15(–20) mm long, the stigma positioned at about the same level as the anthers, peltate-discoid and obscurely or shallowly 4-lobed, 1–2 mm in diameter. Fruits 10–35 mm long, 1–3 mm wide, cylindric, straight or sometimes somewhat curved, tardily longitudinally dehiscent nearly the entire length, 4-locular, not flattened or winged, bluntly 4-angled (each valve rounded on the back), becoming somewhat woody at maturity, lacking a sterile, stalklike base. Seeds numerous in each locule, arranged in 2 rows, 1.0–1.8(–2.0) mm long, mostly 0.6–1.3 mm wide, asymmetrically obovoid, sharply angled or occasionally narrowly winged, the surface dark brown with the angles often lighter brown, finely roughened or with minute tubercles. Self-compatible. 2*n*=14. May–September.

Uncommon, mostly in counties bordering the Missouri and Mississippi Rivers (Montana to Arizona east to Michigan and Arkansas; Canada, Mexico; introduced in Indiana). Loess hill prairies, tops and ledges of bluffs, margins of dry upland forests; also railroads and roadsides.

Steyermark (1963) and earlier botanists treated this taxon in the genus *Oenothera*, but soon thereafter Raven (1964) resurrected the generic name *Calylophus* Spach in which to segregate a group of species having a combination of distinctive floral and other features that he believed indicated a closer relationship to *Gaura* and some other genera than to other species of *Oenothera*. Towner (1977) published a taxonomic revision of this segregate to include six total species, all native to North America. As noted above in the introduction to the family Onagraceae, both *Calylophus* and *Gaura* are now considered to represent sections within *Oenothera* (W. L. Wagner et al. 2007), so our species has come full circle nomenclaturally and taxonomically.

18. Oenothera sinuosa W.L. Wagner & Hoch (wavy-leaved gaura)

Gaura sinuata Nutt. ex Ser.

Pl. 465 g, h; Map 2138

Plants perennial, with taproots, spreading aggressively by rhizomes, often forming extensive colonies. Stems 1 to several from the base, 20–60 cm long, erect or ascending, unbranched or more commonly well-branched, glabrous or sparsely to moderately pubescent with short, appressed to upward-curved, nonglandular hairs, sometimes also with longer, spreading hairs, especially toward the base. Leaves in a rosette and alternate; rosette leaves 5–9 cm long, 10–20 mm wide, oblanceolate to oblong-lanceolate, the margins entire or with broadly spaced, irregular lobes and/or teeth, the surfaces glabrous or minutely hairy; stem leaves 1–10 cm long, 1–20 mm wide, linear to narrowly oblanceolate, the margins with sparse, broadly spaced, irregular lobes and/or teeth, rarely subentire, often wavy, the surfaces densely short-hairy to nearly glabrous, the secondary veins inconspicuous. Inflorescences moderately dense spikelike racemes 10–50(–100) cm long, simple or branched. Bracts inconspicuous, 1–5 mm long, 0.5–2.0 mm wide, lanceolate to narrowly ovate. Flowers strongly zygomorphic, with all of the petals positioned in the upper half of the flower, opening at dusk, the floral tube 2.5–3.0(–5.0) mm long, short-hairy. Sepals 7–14 mm long, short-hairy, lacking free tips in bud. Petals 7–15 mm long, 3–7 mm wide, elliptic-oblanceolate, tapered abruptly to a pronounced stalklike base, white, fading pink to red. Stamens with the filaments 5–11 mm long, spreading to pendant, densely hairy at the very base, the anthers 3–5 mm long, brownish red. Style 12–19 mm long, spreading to somewhat arched or pendant, positioned well beyond the anthers, deeply 4-lobed, the lobes 0.4–0.6 mm long. Fruits indehiscent, appearing mostly 1-locular, 8–12(–15) mm long, 1.5–3.5 mm in diameter, the body ovoid, narrowly winged, abruptly constricted to a sterile, stalklike base, this 2–8 mm long, not angled or winged. Seeds (1–)2–4 per fruit, not in definite rows, 2–3 mm long, 1–1.5 mm wide, ellipsoid, the surface light to reddish brown, finely pebbled. Self-incompatible. $2n=28$. June–October.

Introduced, uncommon, known thus far only from the city of St. Louis (Oklahoma, Texas, Arkansas, and Louisiana; introduced west in California and east to South Carolina and Florida, also Europe, Africa). Railroads.

This species was first reported for Missouri by Mühlenbach (1969), based on a specimen from his St. Louis rail yard collections. As discussed in the introduction to the family Onagraceae, the genus *Gaura* has been reclassified as a section within *Oenothera*. Because there already was a different validly published species in *Oenothera* to which the specific epithet was attached, *O. sinuata* L. (currently regarded as a taxonomic synonym of *O. laciniata*), it became necessary to coin a new name for this taxon in *Oenothera*.

19. Oenothera speciosa Nutt. (white evening primrose, showy evening primrose)

Pl. 467 f, g; Map 2139

Plants perennial, with a branched vertical rootstock, spreading by shoots arising from lateral root branches. Stems 1 to several, 10–50 cm long (including the inflorescence), erect or ascending, unbranched or more commonly well-branched, glabrous or more commonly sparsely to densely pubescent with short, appressed nonglandular hairs and sometimes also long, spreading, nonglandular hairs. Leaves basal and alternate (the stem leaves also often with small fascicles of axillary leaves), the rosette leaves longer-petiolate than the stem leaves; rosette leaves 2–9 cm long, 3–32 mm wide, oblanceolate to obovate, the margins pinnately lobed with few to several blunt lateral lobes below a larger terminal lobe, this with the margins wavy or bluntly and coarsely toothed, also hairy, the surfaces sparsely to moderately and finely nonglandular-hairy, the hairs often a mixture of short, loosely appressed and longer, spreading hairs, mostly along the veins; stem leaves 1–10 cm long, 3–35 mm wide, narrowly elliptic to ovate, the lowermost similar to the rosette leaves grading to the upper with 1 or few narrow basal lobes and the central lobe with few to several, short to longer, more sharply pointed teeth, sometimes unlobed and only with short teeth, the secondary veins relatively conspicuous, at least on the undersurface. Inflorescences open, usually relatively short, the apical portion usually strongly nodding while in bud, sometimes reduced and appearing as solitary

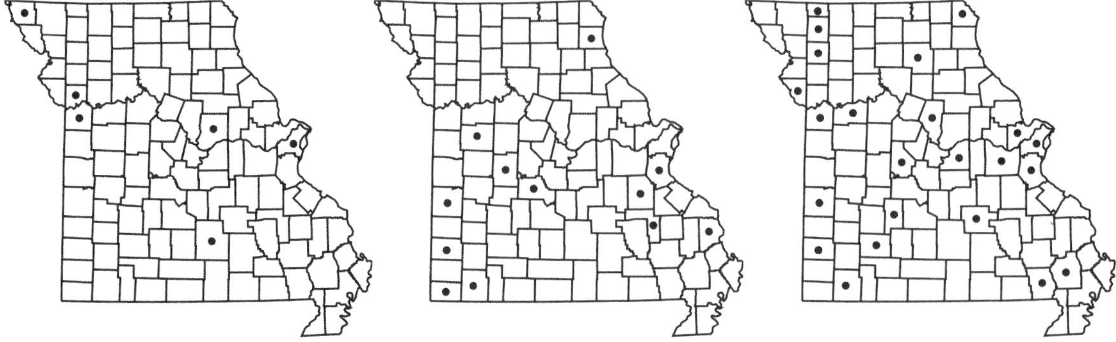

2140. Oenothera suffrutescens 2141. Oenothera triloba 2142. Oenothera villosa

flowers from the upper 1–3 leaf axils, the axes glabrous or sparsely to densely pubescent with short, appressed nonglandular hairs. Bracts relatively conspicuous, 10–20 mm long, 2–6 mm wide, linear-lanceolate to elliptic. Flowers actinomorphic, opening in the morning (plants with pink corollas) or at dusk (plants with white corollas), the floral tube 12–25 mm long, with short appressed hairs. Sepals 15–20 mm long, appressed-hairy, the free tips in bud 1–4(–5) mm long, erect and appressed. Petals 25–40 mm long, 15–40 mm wide, broadly obovate to oblong-circular (broadly rounded to broadly and shallowly cordate at the tip), lacking a stalklike base, white or pink, fading pink or purple. Stamens with the filaments 10–22 mm long, ascending, glabrous at the base, the anthers 10–16 mm long, light or pale yellow. Style 25–55 mm long, the stigma positioned well beyond the anthers, deeply 4-lobed, the lobes 5–14 mm long. Fruits 11–25 mm long, the main body longitudinally dehiscent, 4-locular, 8–15 mm long, 3–6 mm in diameter, narrowly obovoid to narrowly rhombic-ellipsoid, not flattened, 8-ribbed (ribs along the angles and along the midnerve of each face), the surfaces appressed-hairy, tapered to a sterile, slender, stalklike base 3–10 mm long, this glabrous or appressed-hairy. Seeds numerous in each locule, clustered and not in definite rows, 1.0–1.2 mm long, 0.5–0.6 mm wide, narrowly ellipsoid, the surface reddish brown, pebbled. Self-incompatible. $2n$=14, 28, 42. May–July.

Scattered in the state but absent or uncommon in much of the Glaciated Plains Divison and the central portion of the Ozarks (Nebraska and Iowa south to Texas and Louisiana; Mexico; introduced west to California and east to Connecticut and Florida). Upland prairies and glades; also pastures, railroads, roadsides, and open disturbed areas.

This species was recommended for use gardens by Steyermark (1963), but can spread aggressively. The seeds also have been included in various wildflower mixes, including some that have been planted along highways, and many of the Missouri occurrences may represent recent introductions by such means. In the northern half of its range, the flowers tend to open in the evening and have white corollas that fade to pink by the following morning. Especially in the southern portion of the range, populations exist with pink flowers that open at dawn. A number of Missouri populations have pink flowers that remain open during the day, lending support to the hypothesis that at least some of the Missouri plants represent nonnative germplasm.

The discussion of sect. *Hartmannia* by W. L. Wagner et al. (2007) suggests that further taxonomic studies are needed in the *O. speciosa* complex. Munz (1932) divided the species into two varieties, with plants growing from Oklahoma to New Mexico and Mexico characterized by smaller corollas, slightly narrower fruits, and often more loosely ascending stems as var. *berlandieri* (Spach) Munz. Many subsequent authors have treated this taxon as a separate species, *O. berlandieri* (Spach) Spach ex D. Dietr., which is also the name that it is usually listed under in the horticulatural trade. Nomenclatural confusion has arisen because of the existence of a different *O. berlandieri* (Spach) Steud., an illegitimate name that refers to a yellow-flowered taxon in sect. *Calylophus* that occurs in the south-central United States and adjacent Mexico.

Steyermark (1963) reported a single specimen of *Oenothera kunthiana* (Spach) Munz from the St. Louis rail yard collections of Viktor Mühlenbach, but this specimen subsequently was redetermined as *O. speciosa* by Warren L. Wagner.

20. Oenothera suffrutescens (Ser.) W.L. Wagner & Hoch (scarlet gaura)
Gaura coccinea Pursh

Pl. 465 a, b; Map 2140

Plants perennial, clumped from a thick taproot, often branching below ground or only at surface, less commonly unbranched, the underground stems often spreading more or less horizontallly and giving rise to new plants to form small open colonies. Stems usually several from the base, 20–50(–120) cm long, erect to loosely ascending, well-branched, variously from the base or mainly toward the tip, densely pubescent (often appearing grayish) with short, appressed to upward-curved, nonglandular hairs, also with long, spreading hairs toward the base. Leaves alternate (rosette leaves usually only present in seedlings), 0.7–6.5 cm long, 1–15 mm wide, linear to narrowly elliptic, the margins entire or with broadly spaced, coarse, irregular teeth, the surfaces moderately to more commonly densely short-hairy (often appearing grayish), the secondary veins inconspicuous. Inflorescences moderately dense, often somewhat wandlike spikes, 5–40 cm long, these sometimes grouped into panicles, the axes densely short-hairy. Bracts inconspicuous, 2–5 mm long, 0.4–1.1 mm wide, linear. Flowers strongly zygomorphic, with all of the petals positioned in the upper half of the flower, opening at dusk, the floral tube 4–11(–13) mm long, densely pubescent with appressed to upward-curved hairs. Sepals 5–10 mm long, more or less appressed-hairy, lacking free tips in bud. Petals 3–8 mm long, 2–4 mm wide, elliptic-oblanceolate, tapered abruptly to a pronounced stalklike base, white, fading to orange or dark maroon, very rarely pale cream-colored. Stamens with the filaments 3–7 mm long, spreading to pendant, glabrous at the base, the anthers 2.5–5.5 mm long, brownish red. Style 10–21 mm long, the stigma positioned well beyond the anthers, deeply 4-lobed, the lobes 0.6–2.0 mm long. Fruits indehiscent, appearing mostly 1-locular, 4–9 mm long, 1.5–3.0 mm in diameter, the body pyramidal, 4-angled, constricted sharply to a sterile, stalklike base, this stout, cylindric. Seeds (1–)3 or 4 per fruit, not in definite rows, 1.5–3.0 mm long, 1.0–1.5 mm wide, ovoid, sometimes irregularly flattened (by crowding in fruit), the surface light brown to reddish brown, smooth to finely pebbled. Self-incompatible. $2n$=14, 28, 42, 56. May–August.

Uncommon, known thus far only from Atchison County; introduced in the Kansas City and St. Louis areas (Minnesota to Louisiana west to Montana and California; Canada, Mexico; introduced farther east to New York and Indiana, also South America, Europe). Loess hill prairies; also railroads and open disturbed areas.

As discussed in the introduction to the family Onagraceae, the genus *Gaura* has been reclassified as a section within *Oenothera*. Because there already was a different species in *Oenothera* to which the specific epithet was attached, *O. coccinea* Britton (a South American species more correctly known as the earlier *O. campylocalyx* C. Koch & Bouché), another name had to be selected when the taxon was transferred into *Oenothera*.

Steyermark's (1963) report of this species from Holt County could not be confirmed during the present study, but its presence there is probable, at least historically when intact loess hill prairies were more widespread.

21. Oenothera triloba Nutt. (stemless evening primrose)

Pl. 468 e, f; Map 2141

Plants annual, with taproots. Stems usually absent, rarely 1 to several, short and inconspicuous (mostly hidden by the dense basal rosette), to 10(–20) cm long, ascending, unbranched, densely leafy, glabrous to moderately pubescent with short, appressed to upward-curved, nonglandular hairs. Leaves basal and rarely alternate (the stem leaves somewhat shorter than the basal ones), (2–)6–25 (–32) cm long, (6–)15–40(–50) mm wide, oblanceolate to elliptic or oblong-oblanceolate, short-petiolate or occasionally relatively long-petiolate, the blade with several deep, irregular pinnate lobes, rarely nearly entire, the terminal lobe much larger than the others, lanceolate to ovate, often toothed along the margins, the larger lateral lobes often interspersed with much smaller lobes, sharply pointed, glabrous or sparsely to moderately pubescent with short, more or less appressed, nonglandular hairs and often also minute, glandular hairs, especially on the veins and margins, occasionally with scattered, long, spreading, nonglandular hairs along the midvein, the secondary veins (midveins of the lobes) often relatively inconspicuous. Inflorescences of solitary axillary flowers. Flowers actinomorphic, opening at dusk (but often remaining open into the following morning), the floral tube (20–)28–95(–138) mm long, sparsely to densely pubescent with short, more or less appressed, nonglandular hairs and minute, glandular hairs, often also with scattered, long, spreading, nonglandular hairs having red pustular bases. Sepals (6–)10–30(–35) mm long, glabrous or with sparse to dense, more or less appressed, nonglandular hairs, sometimes also with minute glandular hairs, the free tips in bud 3–7 mm long, erect and appressed. Petals (10–)12–30(–38) mm long, (10–)12–30(–38) mm wide, broadly obovate to very broadly obovate (broadly rounded to truncate or occasionally minutely notched at the tip but then with a minute sharp point in the notch), lacking a stalklike base, light yellow to yellow, fad-

2143. Agalinis aspera 2144. Agalinis auriculata 2145. Agalinis fasciculata

ing to orange and sometimes eventually lavender. Stamens with the filaments (5–)8–15(–18) mm long, ascending to slightly S-shaped or outward-curved, glabrous at the base, the anthers 3.5–11.0 mm long, yellow. Style (34–)42–115(–163) mm long, the stigma positioned at the same level as the anthers or occasionally elevated above the anthers, deeply 4-lobed, the lobes (1.5–)3.0–5.0 mm long. Fruits (10–)15–25(–28) mm long 4–8 mm in diameter (excluding the wings), rhomboid or rhombic-obovoid to nearly obpyramidal (including the wings), constricted abruptly to a short beak, longitudinally dehiscent in the apical 1/8–1/3, becoming woody (the whole plant eventually forming an overwintering woody cluster of fruits), 4-locular, not flattened, lacking a sterile, stalklike base, strongly 4-winged only in the upper 2/3, each wing 5–10 mm wide, flat, broadly triangular, often with a conspicuous, narrow, sometimes hooked tooth at the widest point. Seeds numerous in each locule, arranged in 2(3) rows, (2.1–)2.5–3.0(–3.3) mm long,(1.0–)1.4–1.8 mm wide, asymmetrically wedge-shaped to irregularly rhomboid, slightly concave but irregularly winged on 1 side (also winged at the tip), the surface reddish brown to dark purplish brown or nearly black, pebbled. Self-incompatible. $2n$=14. April–May.

Uncommon, widely scattered in the state, mostly south of the Missouri River (Kansas to New Mexico east to Kentucky and Alabama; Mexico; apparently introduced from Illinois to Pennsylvania and Virginia). Glades and openings of dry upland forests; also roadsides and dry, open, disturbed areas; usually on calcareous substrates.

Steyermark (1963) noted that this species makes a fine addition to a rock garden or garden border. The dried fruiting heads are long-persistent and together with the dried taproot somewhat resemble a mace.

22. Oenothera villosa Thunb. **ssp. villosa** (common evening primrose)

O. biennis L. var. *canescens* Torr. & A. Gray

O. biennis var. *hirsutissima* A. Gray ex S. Watson

Pl. 464 k; Map 2142

Plants biennial, with taproots. Stems 1 to several, 50–200 cm long (including the inflorescence), erect or strongly ascending, unbranched or branched, densely pubescent with short, appressed to upward-curved, nonglandular hairs, sometimes also with scattered longer, spreading to loosely appressed, pustular-based, nonglandular hairs (the plants generally appearing gray or strongly grayish-tinged), the inflorescence rarely also with minute glandular hairs. Leaves basal and alternate, the rosette leaves 10–30 cm long, 12–40(–50) mm wide, the blade narrowly oblanceolate to oblanceolate, tapered to the petiole or occasionally rounded, the margins sparsely to moderately and often somewhat irregularly toothed, sometimes few-lobed toward the base, the surfaces and margins moderately to densely pubescent with short, appressed, nonglandular hairs, occasionally the hairs loosely appressed to loosely ascending (shaggy in appearance); stem leaves 5–20 cm long, 10–25(–40) mm wide, the lower ones narrowly oblanceolate to oblanceolate, grading to narrowly lanceolate or lanceolate toward the stem tip, tapered to the sessile or short-petiolate base, the margins sparsely to moderately and irregularly toothed, those of the lower leaves sometimes with a few lobes, the surfaces and margins with pubescence similar to that of the basal leaves, the secondary veins relatively conspicuous and occasionally reddish-tinged. Inflorescences short to elongate spikes, usually relatively dense, erect and straight, appearing conspicuously bracteate, the bracts persistent, 20–70 mm long, narrowly lanceolate to narrowly ovate or narrowly elliptic, the margins entire or finely and irregularly toothed, the surfaces densely pubescent with short, more or less appressed, nonglandular hairs. Flowers actinomorphic, opening at dusk, the floral tube 23–44 mm long, sparsely to densely pubescent with short, appressed, nonglandular hairs and often also sparse to dense, longer, loosely ascending, nonglandular hairs (some of these usually pustular-based), sometimes also with minute glandular hairs. Sepals 9–18 mm long, the midribs not keeled, with pubescence similar to the floral tube, the free tips in bud 0.5–3.0 mm long, terminal, erect and appressed. Petals 7–20 mm long, 8–21 mm wide, broadly obovate to broadly heart-shaped (broadly but sometimes only slightly notched at the tip), lacking a stalklike base, yellow to light yellow, fading to yellow or pale orange. Stamens with the filaments 7–15 mm long, glabrous at the base, the anthers 4–10 mm long, yellow. Style 30–55 mm long, the stigma positioned at about the same level as the anthers, deeply 4-lobed, the lobes 3–9 mm long. Fruits 20–43 mm long, 4–7 mm wide, lanceoloid (tapered toward the tip), straight, longitudinally dehiscent nearly the entire length, 4-locular, not flattened or winged, more or less circular in cross-section, the surface grayish green to dull red, not blackening upon drying, moderately to densely pubescent with short, appressed, nonglandular hairs, and often also longer, more or less spreading, nonglandular hairs, not tapered to a sterile, stalklike base. Seeds numerous in each locule, arranged in 2 rows, 1–2 mm long, 0.5–1.2 mm wide, irregularly prismatic, angled, the surface brown to nearly black, with an irregular network of ridges and pits. Self-compatible. $2n=14$. June–October.

Scattered widely in the state (eastern U.S. west to Montana and Wyoming; Canada; introduced in Europe, Asia, Africa, Canary Islands). Openings and margins of mesic to dry upland forests, upland prairies, sand prairies, and banks of rivers; also old fields, railroads, roadsides, and open disturbed areas.

Steyermark (1963) treated *O. villosa* as varietally distinct within his concept of *O. biennis*, but most later botanists have followed Raven et al. (1979) and Dietrich et al. (1997) in treating them as separate species. Dietrich et al. also noted that *O. villosa* hybridizes readily with *O. biennis*, including sporadic collections from Missouri, but that although such hybrids can persist locally they seldom become widely established.

Dietrich et al. (1997) treated *O. villosa* as comprising two subspecies. The other infraspecific taxon, ssp. *strigosa* (Rydb.) W. Dietr. & P.H. Raven (*O. strigosa* (Rydb.) Mack. & Bush), occurs to the north and west of Missouri. It differs from ssp. *villosa* in its green appearance, leaf venation inconspicuous, relatively open inflorescences with an obtuse apex, yellowish to reddish-tinged sepals, longer, narrower bracts, and production of at least some pustular-based and gland-tipped hairs on the stems and/or leaves. Plants of ssp. *villosa* are usually self-pollinated and the flowers are sometimes functionally cleistogamous, with the pollen shed onto the stigma while still in bud.

For a discussion of the superficially similar *O. grandiflora*, a species that is sometimes cultivated but has not yet become established in the wild in Missouri, see the treatment of *O. nutans*.

OROBANCHACEAE (Broomrape Family)

Plants parasitic on the roots of other plants, sometimes lacking chlorophyll, annual or perennial herbs, sometimes woody and/or tuberous at the base, the roots sometimes brittle, and appearing dense and coralloid, occasionally reduced or absent. Stems various, sometimes thickened or succulent, sometimes white or yellowish brown, sometimes strongly purplish- to blackish-tinged or blackening upon drying. Leaves alternate or opposite and sometimes also basal, sometimes appearing densely spiraled, then reduced to linear to ovate scalelike structures. Stipules absent. Leaf blades various. Inflorescences terminal and/or axillary, spikes, racemes, or of solitary flowers, at least the lower nodes subtended by bracts, inconspicuous bractlets sometimes also present below each flower. Flowers perfect, hypogynous, zygomorphic. Calyces 2–5-lobed, the lobes sometimes minute and toothlike, persistent at fruiting. Corollas often bilabiate, variously colored, 4- or 5-lobed, the tube usually well-developed, sometimes persistent at fruiting. Stamens 4, the filaments attached in the corolla tube, sometimes slightly unequal, the anthers usually not exserted (exserted elsewhere), attached near their midpoints, the anther sacs sometimes appearing spreading or asymmetrical (one of the sacs then reduced and nonfunctional), white or yellow. Staminodes absent. Pistil 1 per flower, of 2 fused carpels. Ovary usually 1-locular, with numerous ovules, the placentation parietal. Style 1, often persistent at fruiting, the stigma 1, variously shaped, entire or 2-lobed. Fruits capsules, dehiscent longitudinally from the tip. Seeds numerous, minute. About 96 genera, about 2,100 species, nearly worldwide.

With the few possible exceptions noted below, members of the Orobanchaceae (as currently delimited) are all root parasites. In some cases, the haustorial attachments to the host root are so fine that they are routinely lost when plants are excavated. For practical reasons, the species often are considered to belong to two broad physiological classes. The first, *hemiparasitic,* are plants that contain chlorophyll and thus merely supplement the sugars produced through photosynthesis with those that they receive from their hosts. The second, *holoparasitc,* are plants that contain no chlorophyll and thus have lost their ability to photosynthesize, siphoning all of their nutrition from the host. *holoparasitic*. Several studies have shown that for selected species of hemiparasites in the family it is possible to grow plants to reproductive maturity in pot culture in the absence of a host, especially when the plants are provided with extra fertilizer (Lackney, 1981; Mann and Musselman, 1981; Crank, 1990). However, in all cases, such plants underperform those grown with host plants and there is no evidence to suggest that such plants are successful in completing their life cycles in the wild.

Traditionally, the Orobanchaceae were treated in a much more restricted sense, comprising only about 17 genera and 230 species of holoparasites (Cronquist, 1981, 1991). However, a number of taxonomists had long questioned the separation of Orobanchaceae from the traditional, broadly circumscribed Scrophulariaceae, especially from the hemiparasitic genera in that family (Boeshore, 1920; Kuijt, 1969; Thieret, 1971; E. Fischer, 2004). These parasitic genera had been included in one or two tribes in the traditional classification of Scrophulariaceae. Characters such as axile vs. parietal placentation, the number of seeds per fruit, and other minor features were thought to vary too greatly to be of use in distinguishing families and some genera were ambiguously placed in either family. Even the transition from hemiparasitism to holoparasitism did not serve to separate the families, as the mainly African genus *Harveya* Hook. was noted to contain both chlorophyllous and achlorophyllous members. Molecular studies have resulted in the break-up of the former Scrophulariaceae (see the treatment of that family for more discussion) and have agreed on three things: 1) the parasitic lifestyle evolved only once within the overall group; 2) loss of chlorophyll evolved several times within the lineage of parasitic genera; 3) Orobanchaceae (in the broad sense to include all of the parasitic taxa) are more closely related to families such as Phrymaceae, Paulowniaceae, and even Acanthaceae and Bignoniaceae, than to the core group of genera remaining in Scrophulariaceae in the restricted modern sense (dePamphilis et al., 1997; Wolfe and dePamphilis, 1998; Nickrent et al., 1998; N. D. Young et al., 1999; Wolfe et al., 2005; Bennett and Mathews, 2006; Tank et al., 2006; Albach et al., 2009). These studies further suggested that the small paleotropical genus *Lindenbergia* Lehm. is the closest extant nonparasitic genus to the rest of the Orobanchaceae. Most recently, two other small autotrophic Asian genera have been implicated as related to the lineage that includes the parasitic genera: *Rehmannia* Libosch. ex Fisch. & C.A. Mey. and *Triaenophora* (Hook. f.) Soler. (A. R. Jensen et al., 2008; Albach et al., 2009; Xia et al., 2009).

On a global scale there remain many interesting problems of generic circumscription that require further study. The genus *Orobanche* itself has been shown to consist of five well-separated lineages, each of which might better be classified as a separate genus (J.-m.Park et al., 2008).

Some members of the Orobanchaceae, especially some Old World taxa of *Orobanche* and some species of *Striga* Lour. (witchweed) are important parasitic weeds of crop plants and are considered noxious weeds by the U.S. Department of Agriculture. However the North American species of *Orobanche* have not been implicated as crop pests. The negative effects of native hemiparasitic genera such as *Agalinis, Aureolaria*, and *Dasistoma*, upon the establishment and survival of tree seedlings in plantations and other plantings were documented by Musselman and Mann (1978), but their relative importance in Missouri forests and pine plantations is not known.

1. Plants lacking chlorophyll, the stems and leaves white to yellow, yellowish brown, or brown; leaves reduced and scalelike
    2. Flowers of 2 types, the lower ones cleistogamous and with the corollas not opening, the upper ones open-flowering and with 4-lobed corollas. . . . 6. EPIFAGUS
    2. Cleistogamous flowers not produced, the flowers all similar and with 5-lobed corollas . . . . 7. OROBANCHE
1. Plants with chlorophyll, the stems and/or leaves green, the green color sometimes partially obscured by purplish or blackish tinging; leaves not scalelike, with a well-developed blade and often also a petiole
    3. Stem leaves alternate
        4. Corollas with the upper lip straight, not or only slightly hooded; leaf blades entire or deeply pinnately divided or lobed into 3–7 divisions or lobes, these linear, entire along the margins, sharply pointed at the tips . . . . 4. CASTILLEJA
        4. Corollas with the upper lip appearing hooded or helmet-shaped; leaf blades moderately to deeply pinnately divided or lobed into 13–35 lobes or the uppermost merely coarsely toothed, the lobes oblong to ovate, toothed and/or lobed along the margins, mostly rounded at the tips . . . . 8. PEDICULARIS
    3. Stem leaves opposite
        5. Leaf blades linear to threadlike, all unlobed (do not confuse fascicles of simple leaves on very short, axillary branches with lobed or divided leaves), the margins entire . . . . 1. AGALINIS
        5. Leaf blades variously shaped but broader than linear, at least some of them pinnately lobed (in *Agalinis* with only a pair of basal lobes) or if all unlobed then the margins toothed
            6. Corollas with the upper lip fused into a hooded or helmet-shaped structure, this with a pair of small teeth near the otherwise truncate tip or unlobed and tapered to a beaklike tip; calyces appearing bilabiate, the lips variously short-lobed or entire . . . . 8. PEDICULARIS
            6. Corollas with the upper lip variously straight to spreading, if somewhat incurved or slightly arched, then strongly 2-lobed; calyces 5-lobed, not bilabiate, the lobes sometimes short and toothlike
                7. Corollas yellow, sometimes with reddish markings in the throat
                    8. Corollas 30–60 mm long, the tube sparsely to densely short-hairy on the inner surface, the hairs not blocking the throat; anthers hairy; fruits 9–23 mm long . . . . 2. AUREOLARIA
                    8. Corollas 14–16 mm long, the tube densely hairy on the inner surface, the beard tending to partially block the throat; anthers glabrous; fruits 6–11 mm long . . . . 5. DASISTOMA
                7. Corollas pink, purple, bluish purple, or rarely white, sometimes with yellowish markings in the throat
                    9. Leaf blades with 1 or 2 basal lobes, the margins otherwise entire; corollas more or less bell-shaped, pink to purplish pink; anthers hairy . . . . 1. AGALINIS
                    9. Leaf blades coarsely few-toothed; corollas trumpet-shaped, deep or dark purple to bluish purple; anthers glabrous . . . . 3. BUCHNERA

1. **Agalinis** Raf. (gerardia)

Plants annual herbs (perennial elsewhere), hemiparasitic, often lacking a well-developed taproot, yellowish green to green or dark green, sometimes purplish-tinged, sometimes blackening upon drying. Stems erect or ascending, branched, usually appearing 4-angled (with pairs of ridges decurrent from the leaf bases), glabrous or roughened with minute hairs, these often broadened basally, sometimes pustular-based. Leaves opposite, sessile. Leaf blades linear and entire-margined or (in *A. auriculata*) lanceolate to narrowly ovate with the margins entire or basally 1- or 2-lobed. Inflorescences open, terminal racemes with leafy bracts (these usually reduced progressively toward the axis tip), the flowers paired at the nodes, sometimes appearing as axillary flowers toward the stem tip or as solitary terminal flowers; variously very short-stalked to long-stalked, the stalks often somewhat thickened toward the tips, lacking bractlets. Cleistogamous flowers absent. Calyces 5-lobed, only slightly zygomorphic, the tube relatively slender at flowering, often 5-angled, the lobes sometimes small and toothlike, persistent, often becoming somewhat enlarged and prominently veiny at fruiting. Corollas 8–33 mm long, 5-lobed, more or less bell-shaped, pink to pinkish purple, rarely white, the tube glabrous or hairy on the inner surface (at least near the tip), the hairs not blocking the throat, the lobes usually slightly shorter than the tube, the lower 3 lobes spreading, the upper 2 lobes either spreading to bent or angled backward or straight to slightly incurved, glabrous or more commonly hairy, the throat usually pale internally with darker pinkish red spots or markings, usually also with a pair of light yellow, longitudinal lines. Stamens with the filaments of 2 lengths, hairy (at least toward the base), the anthers with 2 sacs, these more or less parallel, variously blunt to tapered at 1 end, light yellow, hairy. Style somewhat curved downward and often slightly exserted, the stigmatic portion elongate and somewhat flattened. Fruits 3–15 mm long, globose to subglobose or less commonly elliptic, glabrous. Seeds oblong-ellipsoid to more or less trapezoid, the surface with a network of ridges and pits, variously yellow to light brown, dark brown, or black. About 40 species, North America to South America, Caribbean Islands.

Flowers of *Agalinis* species are mostly bee-pollinated (C. R. Robertson, 1891). Members of the genus have an extremely broad host range, including a diverse array of herbaceous monocots and dicots, as well as many pines and woody flowering plants (Musselman and Mann, 1978; Musselman et al., 1978; Cunningham and Parr, 1990).

Steyermark (1963) and many other authors chose to treat this genus under the name *Gerardia* L. Thieret (1958) reviewed the nomenclatural history and pointed out that the type species of *Gerardia* corresponds with a different taxon more properly included in *Stenandrium* Nees, a mostly neotropical genus in the Acanthaceae, and the name has been officially rejected against the conserved *Stenandrium*. In order to stabilize the nomenclature of the genus, Thieret proposed that the name *Agalinis*, which had been used for the group in some older floras, be conserved against the older but obscure generic epithet *Chytra* C.F. Gaertn., which might instead have replaced the misapplied name *Gerardia*. The name *Agalinis* has since been officially conserved against *Chytra*.

Thieret (1958) also noted that the group has been treated variously in a broad sense or with some anomalous species groups segregated into the genera *Aureolaria* Raf., *Tomanthera* Raf., and *Virgularia* Ruiz & Pav. Further research has supported the separation of plants with large, yellow corollas as *Aureolaria* (Neel and Cummings, 2004; Bennett and Mathews, 2006), with a closer relationship to some of the other yellow-flowered genera in the group than to *Agalinis*. However, *Virgularia* (based on a South American species) is now considered part of *Agalinis* and the generic name *Agalinis* has been officially conserved against it (D'Arcy, 1978, 1979; Canne-Hilliker, 1988). Likewise, the small segregate *Tomanthera* is now thought to represent merely a broader-leaved condition within the range of variation of *Agalinis* (Pennell,

1928; Bentz and Cooperrider, 1978; Canne, 1981; Canne-Hilliker and Kampny, 1991; Neel and Cummings, 2004; Pettengill and Neel, 2008).

The present treatment follows closely the excellent summary of *Agalinis* in the Ozark Region by John Hays (1998). His insights into morphological variation in the Missouri species have been very helpful.

1. Leaf blades 5–25 mm wide, lanceolate to narrowly ovate, those of the upper leaves with 1 or 2 basal lobes; stems roughened with minute, downward-angled hairs (usually also pubescent with sparse, longer, spreading hairs) . . . . . . 2. A. AURICULATA
1. Leaf blades 0.5–6.0(–8.0) mm wide, threadlike to linear or less commonly lanceolate, all entire or those of the lower leaves sometimes with a pair of deep basal lobes or divisions (do not confuse these with fascicles of entire leaves on very short axillary branches); stems glabrous or roughened with minute, upward-angled hairs
    2. Flower stalks 1–6 mm long at flowering, shorter than to about as long as the calyces
        3. Calyces 5–10 mm long, bell-shaped to narrowly bell-shaped, longer than wide at flowering (becoming distended as the fruits mature)
            4. Leaf lades threadlike to linear, all 0.5–1.5 mm wide, all entire; calyx lobes 1–2(–3) mm long, much shorter than the tube . . . . 1. A. ASPERA
            4. Leaf lades linear-lanceolate to lanceolate, those of the largest leaves 2–6 mm wide, those of the lower leaves sometimes deeply 3-lobed or -parted; calyx lobes 3–6 mm long, slightly shorter than to noticeably longer than the tube . . . . 5. A. HETEROPHYLLA
        3. Calyces 3–5(–6) mm long, broadly bell-shaped to hemispheric, slightly longer than wide to about as long as wide at flowering (becoming even wider as the fruits mature)
            5. Stems strongly roughened with moderate to dense, minute hairs on the angles and faces; leaf axils mostly with well-developed fascicles of leaves, these often nearly as long as (or even longer than) the subtending leaves . . . . 3. A. FASCICULATA
            5. Stems glabrous or more commonly slightly roughened with sparse, minute hairs, mostly along the angles; leaf axils with fascicles of leaves absent or poorly developed, when present then noticeably shorter than the subtending leaves . . . . 6. A. PURPUREA
    2. Flower stalks mostly 5–25 mm long at flowering, noticeably longer than the calyces
        6. Calyces 5–10 mm long, bell-shaped to narrowly bell-shaped, longer than wide at flowering (becoming distended as the fruits mature), the lobes densely short-hairy on the inner surface and margins; fruits 7–12 mm long, ellipsoid to oblong-ellipsoid . . . . 1. A. ASPERA
        6. Calyces 3.0–5.5 mm long, broadly bell-shaped to hemispheric, slightly longer than wide to about as long as wide at flowering (becoming even wider as the fruits mature), the lobes glabrous or sparsely short-hairy on the inner surface and margins; fruits 3.5–7.0 mm long, globose to broadly ellipsoid (obovoid in *A. viridis*)
            7. Corollas pink to more commonly purplish pink to reddish purple (rarely white), glabrous in the throat adjacent to the upper lobes, the upper 2 lobes straight or somewhat incurved to slightly arched; stems and leaves green to dark green, sometimes purplish- or blackish-tinged, often blackening upon drying . . . . 8. A. TENUIFOLIA

7. Corollas pink to light pink, hairy in the throat adjacent to the upper lobes, the upper 2 lobes spreading to bent backward at full flowering (check several flowers to avoid misdiagnosing buds that are beginning to open); stems and leaves green to yellowish green, not blackening upon drying
   8. Plants broadly bushy, with numerous branches, green; leaves mostly soft and spreading to arched or curled downward; inflorescences relatively short, the flowers often appearing solitary or in small open clusters at the branch tips . . . . 4. A. GATTINGERI
   8. Plants relatively slender, with few to several short branches, yellowish green; leaves mostly stiff, ascending; inflorescences usually more or less elongate, appearing as racemes at the branch tips
      9. Fruits 3.5–5.0 mm long, globose or nearly so; stems minutely roughened along the angles; flower stalks mostly longer than the subtending bracts . . . . 7. A. SKINNERIANA
      9. Fruits 5–7 mm long, obovoid to broadly oblong-obovoid; stems smooth (glabrous); flower stalks shorter than to about as long as the subtending bracts . . . . 9. A. VIRIDIS

1. Agalinis aspera (Douglas ex Benth.) Britton (rough gerardia)

Gerardia aspera Douglas ex Benth.

Pl. 470 g, h; Map 2143

Plants relatively slender, often blackening upon drying, dark green, sometimes purplish- or blackish-tinged. Stems 20–60(–80) cm long, erect, with numerous, ascending branches, mostly above the midpoint, circular to bluntly 4-angled toward the base, more strongly angled and ridged above the lower branch points, sparsely to moderately roughened with minute, ascending hairs along the angles toward the base, the roughening progressively denser and also between the angles toward the tip. Primary leaves mostly with well-developed axillary fascicles of leaves, these shorter than the primary leaves. Leaf blades stiffly ascending, 15–40 mm long, 0.5–1.5 mm wide, threadlike to linear, entire, the upper surface and usually also the undersurface strongly roughened. Inflorescences narrow racemes, the flower stalks 5–11 mm long at flowering (shorter than to about as long as the calyces), elongating to 9–18 mm at fruiting, curved upward to more or less straight and strongly ascending. Calyces 5–10 mm long, bell-shaped to narrowly bell-shaped, longer than wide at flowering (becoming distended as the fruits mature), the lobes 1.0–2.0(–3.0) mm long, much shorter than the tube, relatively thick and triangular, densely short-hairy on the inner surface and margins, the sinuses between the lobes at flowering variously narrowly to broadly V-shaped or somewhat U-shaped, often narrowly so. Corollas 16–25 mm long, reddish purple to deep purple, the tube densely pubescent with fine, purple, multicellular hairs externally, the throat with darker spots, glabrous, the lobes glabrous, but fringed along the margins, the upper 2 lobes spreading to somewhat bent backward. Anthers 1.9–2.5 mm long. Fruits 7–12 mm long, ellipsoid to oblong-ellipsoid. Seeds 0.9–1.1 mm long, black to dark brown. $2n=28$. August–October.

Uncommon, mostly in the western half of the state, most abundantly in northwestern Missouri (Wisconsin to North Dakota south to Arkansas and Texas; Canada). Glades, upland prairies, and loess hill prairies.

2. Agalinis auriculata (Michx.) S.F. Blake (auriculate false foxglove)

Gerardia auriculata Michx.

Tomanthera auriculata (Michx.) Raf.

Pl. 469 a–c; Map 2144

Plants slender to moderately stout, usually not blackening upon drying, dull green to dark green, sometimes purplish-tinged. Stems 20–60(–80) cm long, loosely to strongly ascending, unbranched or with few to several, spreading to loosely ascending branches, mostly above the midpoint, 4-angled moderately to densely roughened with minute, downward-angled hairs, also with sparse to moderate, longer, softer, spreading hairs. Primary leaves sometimes with poorly developed axillary fascicles of leaves, these much shorter than the primary leaves. Leaf blades loosely ascending to spreading, 15–60 mm long, 5–25 mm wide, lanceolate to narrowly ovate, those of the upper leaves with 1 or 2, narrowly oblong basal lobes, the surfaces and margins moderately to densely pubescent (roughened) with short, stiff hairs. Inflorescences interrupted spikelike racemes with leaflike bracts, often appearing as solitary flowers (2 per

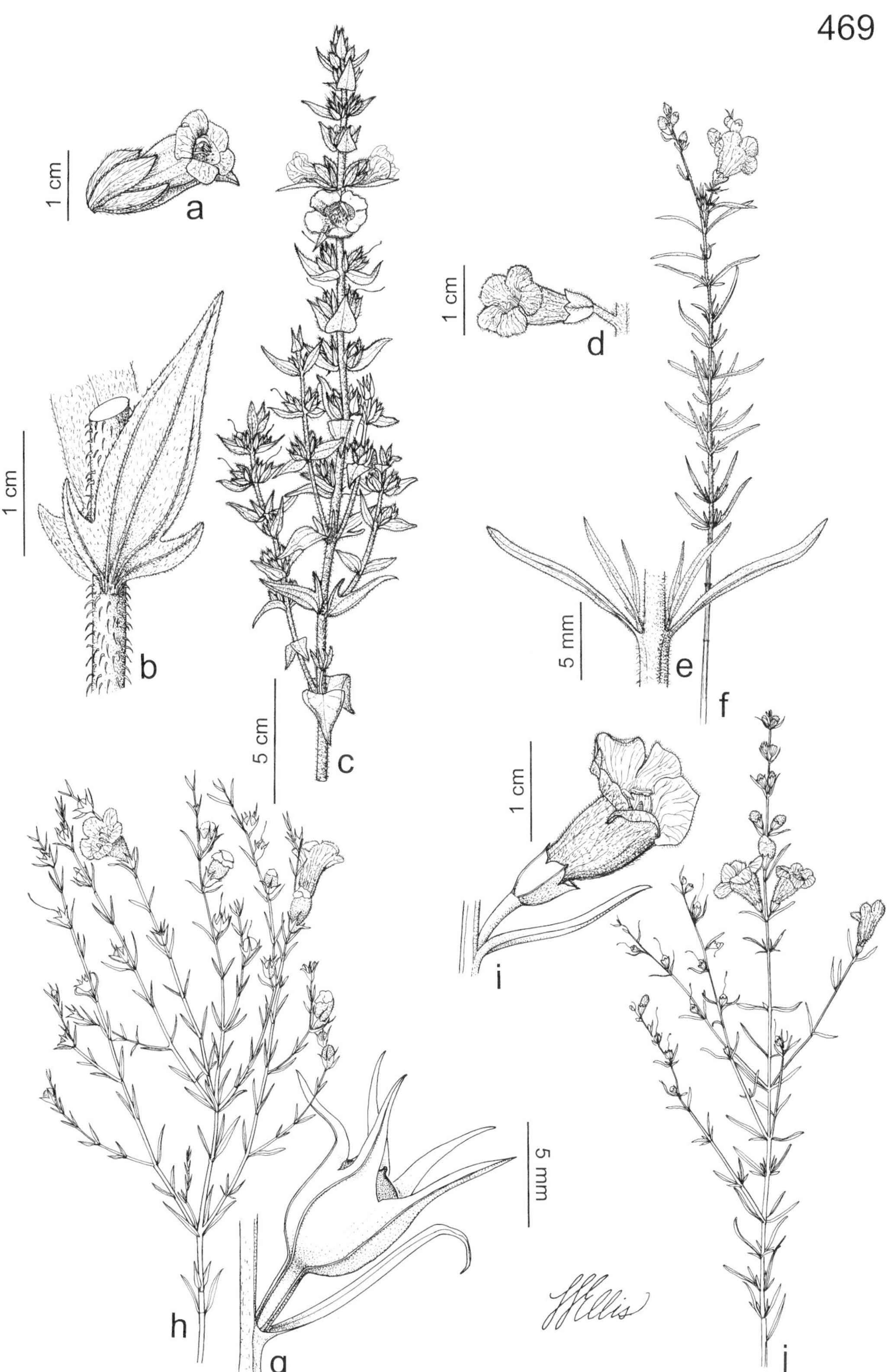

Plate 469. Orobanchaceae. *Agalinis auriculata*, **a)** flower, **b)** node with leaves, **c)** fertile stem. *Agalinis fasciculata*, **d)** flower, **e)** node with leaves, **f)** fertile stem. *Agalinis heterophylla*, **g)** node with fruit, **h)** fertile stem. *Agalinis purpurea*, **i)** flower, **j)** fertile stem.

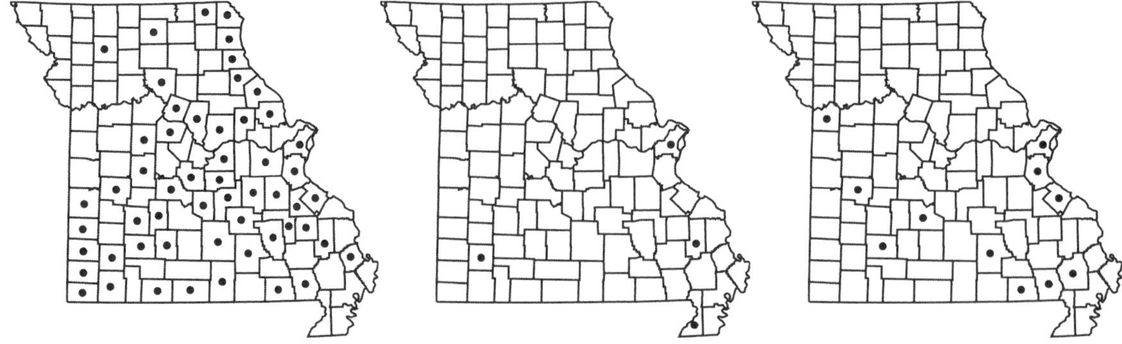

2146. Agalinis gattingeri 2147. Agalinis heterophylla 2148. Agalinis purpurea

node) in the axils of foliage leaves, the flower stalks 0.5–1.5 mm long (much shorter than the calyces), not noticeably elongating at fruiting, more or less straight and spreading to loosely ascending. Calyces 10–18 mm long, bell-shaped, somewhat longer than wide at flowering, the lobes 6–12 mm long, relatively strongly unequal, slightly longer than the tube, herbaceous and lanceolate to oblong-lanceolate, densely short-hairy on the surfaces and margins, the sinuses between the lobes at flowering narrowly V-shaped. Corollas 20–28 mm long, pink to pinkish purple, the tube densely and finely hairy externally, the throat pale to white with dark purple to brown spots, hairy, the lobes hairy on the outer surface and margins, the upper 2 lobes spreading to somewhat bent backward. Anthers 1.9–2.5 mm long. Fruits 9–15 mm long, broadly ovoid to ovoid. Seeds 1.2–1.6 mm long, light brown. $2n=26$. August–September.

Uncommon, widely scattered, mostly in the northern and western halves of the state (eastern [mostly northeastern] U.S. west to Minnesota and Texas). Upland prairies, savannas, glades, and edges of dry upland forests; also old fields, pastures, roadsides, and open disturbed areas.

This species has a relatively broad distribution, but apparently is not common anywhere. Populations appear to benefit from disturbances that create open ground, such as fire or even machines for brush removal, and plants can disappear from a site for many years as succession crowds them out, only to reappear from seeds in the soil following new disturbance.

3. Agalinis fasciculata (Elliott) Raf. (fasciculate false foxglove)

Gerardia fasciculata Elliott

Pl. 469 d–f; Map 2145

Plants relatively slender to less commonly broadly bushy, often blackening upon drying, dark green, sometimes purplish- or blackish-tinged. Stems 25–70(–120) cm long, erect, with numerous, ascending to occasionally spreading branches, mostly above the midpoint, circular to bluntly 4-angled toward the base, more strongly angled and ridged above the lower branch points, strongly roughened with moderate to dense, minute, ascending hairs along and between the angles. Primary leaves mostly with well-developed axillary fascicles of leaves, these often nearly as long as (or even longer than) the primary leaves. Leaf blades ascending or arched to curled outward, 10–35(–40) mm long, 1–2(–4) mm wide, linear, entire, the upper surface strongly roughened, the undersurface usually roughened along the midvein. Inflorescences narrow racemes (occasionally appearing as solitary flowers [2 per node] in the axils of foliage leaves), the flower stalks 2–5 mm long at flowering (shorter than to about as long as the calyces), not or only slightly elongating at fruiting, more or less straight and strongly ascending or curved outward. Calyces 3–5(–6) mm long, bell-shaped, slightly longer than wide to about as long as wide at flowering (becoming distended as the fruits mature), the lobes 0.5–1.5(–2.0) mm long, much shorter than the tube, relatively thick and triangular, glabrous, the sinuses between the lobes at flowering broadly U-shaped. Corollas 15–30 mm long, pink to purplish pink or light purple, rarely white, the tube moderately to densely and finely hairy externally, the throat with a pair of longitudinal, pale lines and darker, purple to reddish purple spots, finely pubescent with relatively long, pink to purple, multicellular hairs at the base of the upper lobes, the lobes minutely hairy on the outer surface, fringed along the margins, the upper 2 lobes spreading to somewhat bent backward. Anthers 2.5–3.5 mm long. Fruits 4.5–6.0(–7.0) mm long, globose to subglobose. Seeds 0.6–1.0 mm long, black to dark brown. $2n=28$. August–October.

Scattered, mostly in the southern half of the state, most abundantly in the Unglaciated Plains division (eastern [mostly southeastern] U.S. west to Iowa and Texas). Upland prairies (often in swales), savannas, openings and edges of mesic to dry upland forests, banks of streams, fens, and margins of sinkhole ponds; also old fields, ditches, roadsides, and moist, sandy, open disturbed areas.

Occasional white-flowered plants have been called *Gerardia fasciculata* f. *albiflora* E.J. Palmer, a name that has not been transferred to *Agalinis*.

4. Agalinis gattingeri (Small) Small ex Britton (rough-stemmed gerardia)
Gerardia gattingeri Small
Pl. 470 n, o; Map 2146

Plants usually relatively broadly bushy, not blackening upon drying, green to slightly yellowish green, occasionally somewhat purplish-tinged. Stems 10–60 cm long, erect, with numerous, loosely ascending to spreading branches, mostly above the midpoint, rounded to bluntly 4-angled toward the base, somewhat more strongly angled and sometimes ridged above the lower branch points, slightly to moderately roughened with sparse to moderate, ascending, minute hairs, mostly along the angles, sometimes nearly glabrous. Primary leaves usually lacking fascicles of leaves. Leaf blades mostly spreading or arched to curled outward, 15–35 mm long, 0.5–1.5 mm wide, threadlike to linear, entire, relatively soft, the upper surface slightly roughened, the undersurface glabrous or occasionally slightly roughened along the midvein. Inflorescences relatively short, the flowers often appearing solitary or in small open clusters at the branch tips, the flower stalks 5–25 mm long at flowering (noticeably longer than the calyces), elongating to 10–30 mm at fruiting, slender, more or less straight and ascending to spreading. Calyces 3–5 mm long, broadly bell-shaped to hemispheric, slightly longer than wide to about as long as wide at flowering (becoming distended as the fruits mature), the lobes 0.5–1.5 mm long, much shorter than the tube, relatively thick and triangular, glabrous or sparsely short-hairy on the inner surface and margin, the sinuses between the lobes at flowering broadly U-shaped. Corollas 10–16 mm long, pink to light pink, the tube sparsely to moderately hairy externally, the throat with darker, reddish purple spots, occasionally also with a pair of longitudinal, pale yellow lines, finely pubescent with relatively long, pink to purple, multicellular hairs at the base of the upper lobes, the lobes (especially the 3 lower ones) finely hairy on the outer surface, fringed along the margins, the upper 2 lobes spreading to bent backward at full flowering. Anthers 1.3–2.0 mm long. Fruits 4–5 mm long, globose to subglobose. Seeds 0.5–0.9 mm long, yellow to yellowish brown. $2n=26$. August–October.

Scattered nearly throughout the state, more abundantly south of the Missouri River (Ohio to Alabama west to Minnesota and Texas; Canada). Savannas, edges and openings of mesic to dry upland forests, upland prairies, glades, and rarely margins of ponds; also roadsides and open, disturbed areas; often on acidic substrates.

Holmgren (1986) treated *A. gattingeri* as a synonym of *A. skinneriana*, but Canne-Hilliker (1987), Canne-Hilliker and Kampny (1991), and Hays (1998) provided convincing anatomical and morphological data to support the continued recognition of two species.

5. Agalinis heterophylla (Nutt.) Small
Gerardia heterophylla Nutt.
Pl. 469 g, h; Map 2147

Plants relatively slender to broadly bushy, often blackening upon drying, dark green, sometimes purplish- or blackish-tinged. Stems 30–80(–100) cm long, erect, with numerous, spreading to ascending branches, mostly above the midpoint, 4-angled and ridged, glabrous or more commonly sparsely roughened with minute, ascending hairs, mostly along the angles, toward the tip. Primary leaves usually lacking axillary fascicles of leaves. Leaf blades ascending, 15–35 mm long, those of the largest leaves 2–6(–8) mm wide, linear-lanceolate to lanceolate, those of the lower leaves sometimes deeply 3-lobed or -parted (these often withered by flowering), the upper surface and margins moderately roughened, the undersurface roughened along the midvein. Inflorescences spikelike racemes, the flower stalks 1–3 mm long at flowering (shorter than the calyces), not or only slightly elongating at fruiting, usually curved outward. Calyces 5–10 mm long, bell-shaped, longer than wide at flowering (becoming distended as the fruits mature), the lobes 3–6 mm long, slightly shorter than to more commonly noticeably longer than the tube, stiff, narrowly lanceolate-triangular, glabrous, the sinuses between the lobes at flowering variously narrowly to broadly V-shaped or somewhat U-shaped, often narrowly so (often becoming somewhat distended at fruiting). Corollas 20–32 mm long, pink to pinkish purple, the tube densely and finely short-hairy externally, the throat pale or white with dark purple to brownish purple spots, glabrous, the lobes glabrous internally, but fringed along the margins, the upper 2 lobes spreading to somewhat bent backward. Anthers 2.7–3.5 mm long. Fruits 5–8 mm long, globose to subglobose. Seeds 0.7–1.1 mm long, black to dark brown. $2n=28$. August–September.

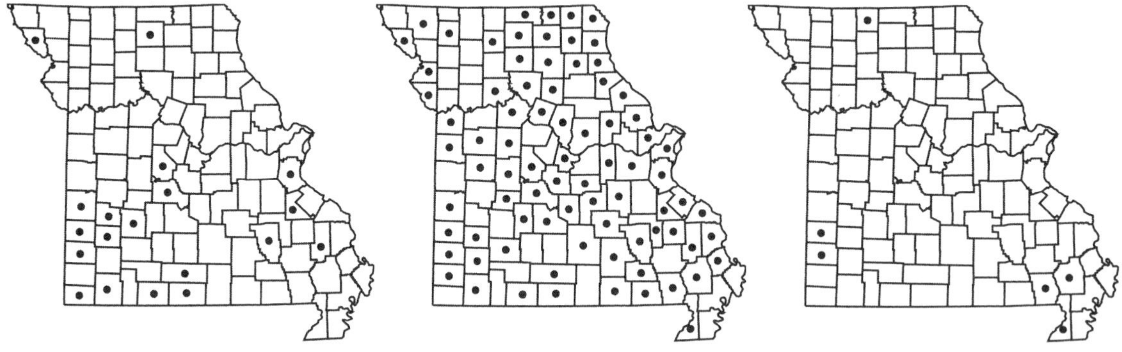

2149. Agalinis skinneriana 2150. Agalinis tenuifolia 2151. Agalinis viridis

Uncommon, sporadic, mostly in southern Missouri (Kansas and Missouri south to Texas and Florida). Upland prairies, sand prairies, savannas, and edges of dry upland forests.

Steyermark (1963) knew this species only from southeastern Missouri, but it is as abundant in the southwestern part of the state. He excluded a specimen collected by John Letterman in 1886 near Allenton (St. Louis County), but given that the species has been documented from adjacent St. Clair County, Illinois (Hays, 1998), the presence of the taxon historically in the St. Louis area seems plausible.

6. Agalinis purpurea (L.) Pennell (smooth gerardia)
Gerardia purpurea L.

Pl. 469 i, j; Map 2148

Plants usually relatively broadly bushy, often blackening upon drying, dark green, sometimes purplish-tinged. Stems (10–)30–100(–120) cm long, erect, with numerous, ascending to spreading branches, mostly above the midpoint, circular to bluntly 4-angled toward the base, more strongly angled and ridged above the lower branch points, glabrous or more commonly slightly roughened with sparse, ascending, minute hairs, mostly along the angles. Primary leaves lacking fascicles of leaves or these noticeably shorter than the primary leaves. Leaf blades loosely ascending to spreading or arched outward, 15–70 mm long, 0.5–4.0(–6.0) mm wide, linear, entire, the upper surface moderately roughened, the undersurface usually roughened along the midvein. Inflorescences narrow racemes, the flower stalks 2–5 mm long at flowering (shorter than to about as long as the calyces), not or only slightly elongating at fruiting, more or less straight and ascending or curved outward. Calyces 4–6 mm long, bell-shaped, slightly longer than wide to about as long as wide at flowering (becoming distended as the fruits mature), the lobes 0.8–2.2 mm long, much shorter than the tube, relatively thick and triangular, glabrous, the sinuses between the lobes at flowering broadly U-shaped or occasionally broadly V-shaped. Corollas 17–33 mm long, pink to purplish pink (rarely white elsewhere), the tube moderately to densely but inconspicuously and minutely hairy externally, the throat with a pair of longitudinal, pale yellow lines and darker, purple to reddish purple spots, finely pubescent with relatively long, pink to purple, multicellular hairs at the base of the upper lobes, the lobes glabrous or sparsely to densely and minutely hairy on the outer surface, densely fringed along the margins, the upper 2 lobes spreading to somewhat bent backward. Anthers 2.0–3.5 mm long. Fruits 4–6 mm long, globose to subglobose. Seeds 0.7–1.2 mm long, black to dark brown. $2n=28$. July–September.

Scattered in the southern half of the state (eastern U.S. west to Wisconsin, Nebraska, and Texas; Canada). Bottomland prairies, swales in upland prairies, edges and openings of bottomland and mesic upland forests, and rarely fens; also roadsides and moist, open disturbed areas.

Gleason and Cronquist (1991) treated *A. purpurea* as comprising three or four poorly marked varieties, of which only var. *purpurea* is found in Missouri. The infraspecific taxonomy requires further study. Steyermark (1963) noted that rare, white flowered plants, which are known as f. *albiflora* House, have not yet been recorded from Missouri.

7. Agalinis skinneriana (Alph. Wood) Britton (pale gerardia, midwestern gerardia)
Gerardia skinneriana Alph. Wood

Pl. 470 a–c; Map 2149

Plants relatively slender, not blackening upon drying, yellowish green, not purplish- or blackish-tinged. Stems 10–50 cm long, erect or ascending, with few to several, short, ascending branches,

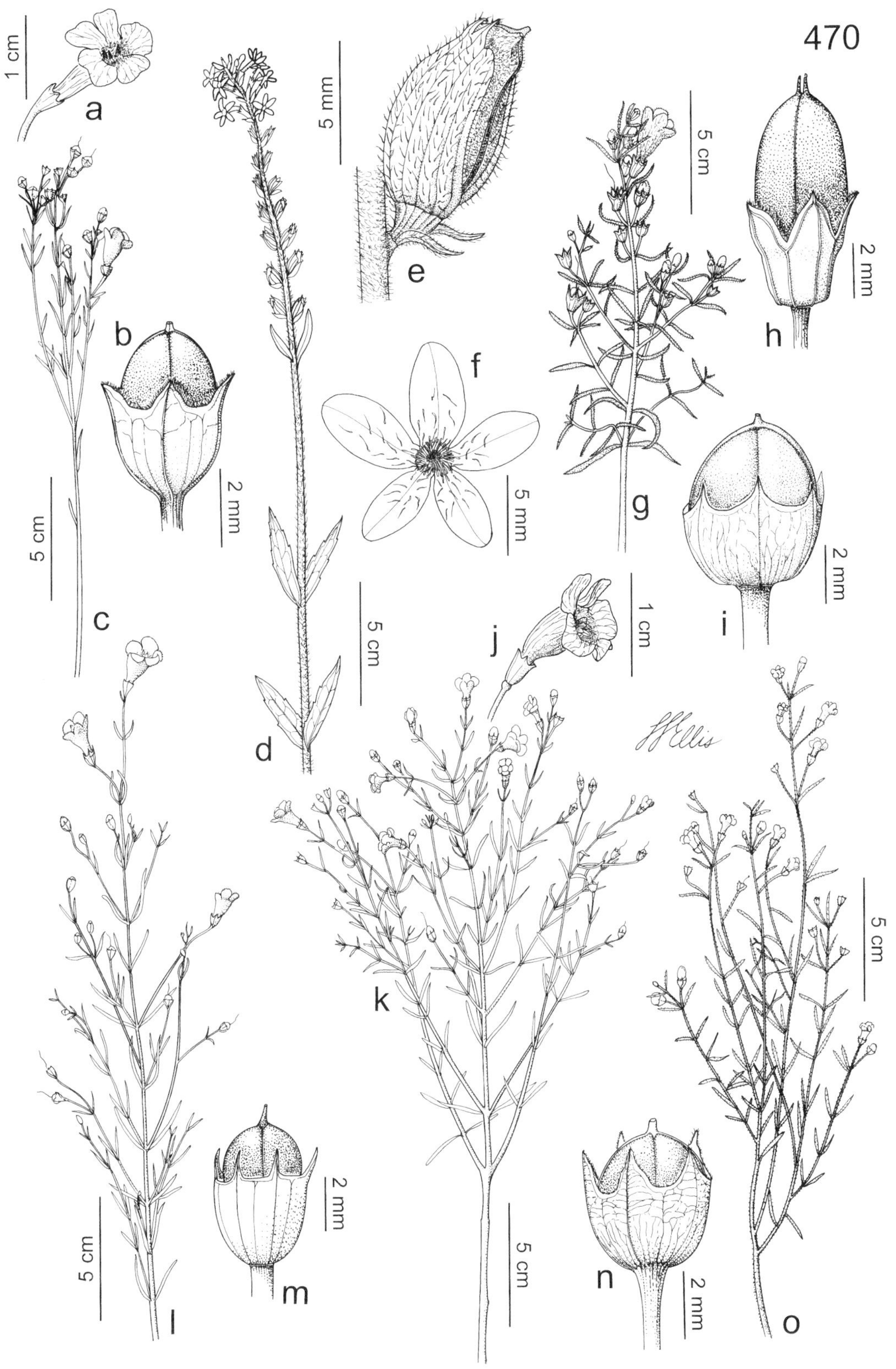

Plate 470. Orobanchaceae. *Agalinis skinneriana*, **a)** flower, **b)** fruit, **c)** fertile stem. *Buchnera americana*, **d)** flower, **e)** fertile stem, **f)** fruit. *Agalinis aspera*, **g)** fertile stem, **h)** fruit. *Agalinis tenuifolia*, **i)** fruit, **j)** flower, **k)** fertile stem. *Agalinis viridis*, **l)** fertile stem, **m)** fruit. *Agalinis gattingeri*, **n)** fruit, **o)** fertile stem.

mostly above the basal 1/3 of the stem, sharply 4-angled, often strongly ridged or narrowly winged above the lower branch points, moderately roughened with ascending, minute hairs along the angles. Primary leaves lacking fascicles of leaves. Leaf blades ascending to strongly ascending, straight, 10–25 mm long, 0.5–1.5 mm wide, linear, entire, relatively stiff, the upper surface moderately roughened, the undersurface roughened along the midvein. Inflorescences usually more or less elongate, appearing as racemes at the branch tips, the flower stalks 5–15 mm long at flowering (noticeably longer than the calyces), mostly longer than the subtending bracts, elongating to 10–25 mm at fruiting, more or less straight and ascending. Calyces 2.5–4.5 mm long, broadly bell-shaped to hemispheric, slightly longer than wide to about as long as wide at flowering (becoming distended as the fruits mature), the lobes 0.5–1.2 mm long, much shorter than the tube, relatively thick and triangular, glabrous or sparsely short-hairy on the inner surface and margin, the sinuses between the lobes at flowering broadly U-shaped. Corollas 8–16 mm long, pink to light pink, the tube glabrous or sparsely and inconspicuously short-hairy toward the tip externally, the throat with darker, reddish purple spots, sometimes also with a pair of longitudinal, pale yellow lines, finely pubescent with relatively long, pink to purple, multicellular hairs at the base of the upper lobes, the lobes (especially the 3 lower ones) glabrous on the outer surface, fringed along the margins, the upper 2 lobes spreading to bent backward at full flowering. Anthers 0.6–1.2 mm long. Fruits 3.5–5.0 mm long, globose to subglobose. Seeds 0.6–1.0 mm long, yellow to yellowish brown. $2n=26$. July–October.

Scattered, mostly south of the Missouri River (Ohio to Mississippi west to Nebraska, Oklahoma, and Louisiana; Canada). Glades, upland prairies, tops of bluffs, and occasionally edges of mesic to dry upland forests; also roadsides; frequently on calcareous substrates.

8. Agalinis tenuifolia (Vahl) Raf. (slender gerardia, common gerardia)

A. tenuifolia var. *macrophylla* (Benth.) S.F. Blake
A. tenuifolia var. *parviflora* (Nutt.) Pennell
A. besseyana (Britton) Britton
Gerardia tenuifolia Vahl
G. tenuifolia var. *macrophylla* Benth.
G. tenuifolia var. *parviflora* Nutt.

Pl. 470 i–k; Map 2150

Plants usually relatively broadly bushy, often blackening upon drying, green to dark green, often purplish- or blackish-tinged. Stems 20–50 (–80) cm long, erect, with numerous, ascending to spreading branches, mostly above the midpoint, bluntly 4-angled toward the base, more strongly angled and ridged above the lower branch points, glabrous or more commonly slightly roughened with sparse, ascending, minute hairs, mostly along the angles. Primary leaves mostly lacking fascicles of leaves or these poorly developed and noticeably shorter than the primary leaves. Leaf blades loosely ascending to spreading or arched to curled outward, 10–50(–70) mm long, 1–3(–5) mm wide, threadlike to linear, entire, the upper surface moderately roughened, the undersurface sometimes roughened along the midvein. Inflorescences narrow racemes, the flower stalks 6–25 mm long at flowering (noticeably longer than the calyces), elongating to 15–30 mm at fruiting, slender, more or less straight and ascending to spreading or sometimes curved upward toward the tip. Calyces 3.0–5.5 mm long, broadly bell-shaped to hemispheric, slightly longer than wide to about as long as wide at flowering (becoming distended as the fruits mature), the lobes 0.3–2.0 mm long, much shorter than the tube, relatively thick and triangular, glabrous or sparsely short-hairy on the inner surface and margin, the sinuses between the lobes at flowering broadly U-shaped. Corollas 10–16 mm long, pink to more commonly purplish pink to reddish purple (rarely white), the tube moderately to densely short-hairy externally, the throat with a pair of longitudinal, pale yellow lines and darker, reddish purple spots, glabrous, the lobes glabrous or sparsely to moderately and minutely hairy on the outer surface, short-hairy along the margins, the upper 2 lobes straight or somewhat incurved to slightly arched over the stamens. Anthers 1.0–2.3 mm long. Fruits 3–7 mm long, globose to subglobose. Seeds 0.7–1.0 mm long, tan to dark brown. $2n=28$. August–October.

Scattered nearly throughout the state (eastern U.S. west to North Dakota, Wyoming, and New Mexico; Canada). Openings and edges of bottomland and mesic to dry upland forests, upland prairies, sand prairies, savannas, glades, ledges and tops of bluffs, banks of streams and rivers, margins of ponds and lakes, and fens; also old fields, pastures, roadsides and open, sandy to gravelly, disturbed areas.

Agalinis tenuifolia is by far the most abundant member of the genus in Missouri. Morphological variation within *A. tenuifolia* is complex and requires further study. Pennell (1929, 1935) divided the species into a series of five subspecies, based mostly on quantitative differences among populations. Two of these were endemics with small ranges within the distributions of the other three,

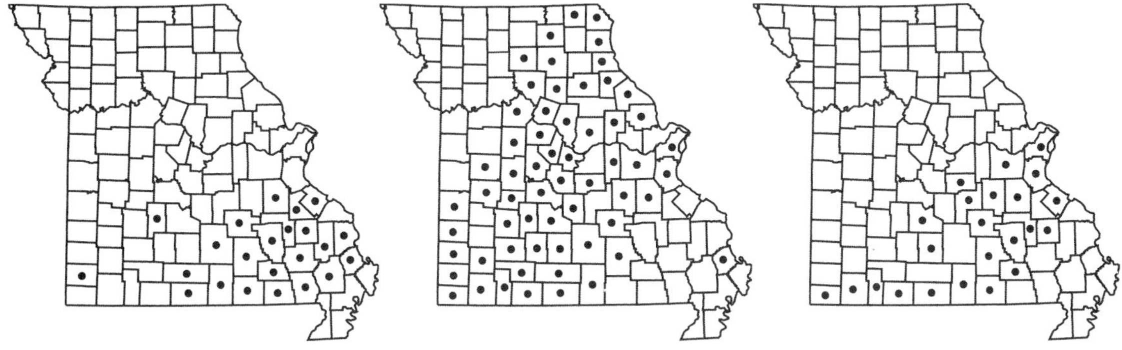

2152. Aureolaria flava 2153. Aureolaria grandiflora 2154. Aureolaria pectinata

widespread subspecies. Steyermark (1963) recognized these three as varieties within his concept of *Gerardia tenuifolia*. He referred small plants mostly lacking axillary leaf fascicles and with relatively small calyces and fruits to var. *tenuifolia*. Among the larger-flowered and -fruited populations, those with only moderately hairy anthers, medium-sized stems, well-developed axillary leaf fascicles, and more or less ascending branches were called var. *parviflora*, whereas large plants with densely hairy anthers, relatively large, stout stems, less well-developed axillary leaf fascicles, and relatively spreading branches were called var. *macrophylla*. In practice, it is possible to separate the more extreme populations, but many specimens appear intermediate for one or more of these characters. Rare white-flowered plants have been called f. *albiflora* House.

9. Agalinis viridis (Small) Pennell
Gerardia viridis Small
Pl. 470 l, m; Map 2151

Plants relatively slender to broadly bushy, not blackening upon drying, yellowish green, not purplish- or blackish-tinged. Stems 10–60 cm long, erect or ascending, with few to numerous short to elongate, ascending to spreading branches, mostly above the midpoint of the stem, sharply 4-angled, often strongly ridged or narrowly winged above the lower branch points, smooth (glabrous) along and between the angles. Primary leaves lacking fascicles of leaves. Leaf blades ascending, straight, 10–25(–30) mm long, 0.5–2.0(–3.0) mm wide, linear, entire, relatively stiff, the upper surface moderately roughened, the undersurface roughened along the midvein. Inflorescences usually more or less elongate, appearing as racemes at the branch tips, the flower stalks 5–15 mm long at flowering (noticeably longer than the calyces), shorter than to about as long as the subtending bracts, elongating to 8–25 mm at fruiting, more or less straight and loosely to strongly ascending. Calyces 3.0–4.5 mm long, broadly bell-shaped to hemispheric, slightly longer than wide to about as long as wide at flowering (becoming distended as the fruits mature), the lobes 0.8–2.0 mm long, shorter than the tube, relatively thick and triangular, glabrous on the inner surface and margin, the sinuses between the lobes at flowering broadly U-shaped. Corollas 8–12 mm long, pink to light pink, the tube glabrous or sparsely and inconspicuously short-hairy near the tip externally, the throat with darker, reddish purple spots, usually also with a pair of longitudinal, pale yellow to white lines, finely pubescent with relatively long, pink to purple, multicellular hairs at the base of the upper lobes, the lobes (especially the 3 lower ones) glabrous on the outer surface, fringed along the margins, the upper 2 lobes spreading to bent backward at full flowering. Anthers 0.8–1.3 mm long. Fruits 5–7 mm long, obovoid to broadly oblong-obovoid. Seeds 0.7–1.0 mm long, yellow to yellowish brown. August–October.

Uncommon in the southeastern and southwestern portions of the state (Missouri to Mississippi west to Oklahoma and Texas). Bottomland prairies, moist swales in upland prairies, and rarely edges of mesic upland forests; also roadsides.

2. Aureolaria Raf. (false foxglove)

Plants annual or perennial herbs, hemiparasitic, green or dark green, sometimes purplish-tinged, sometimes blackening upon drying. Stems erect or ascending, usually with several to many branches, rounded or bluntly 4-angled, glabrous (then often somewhat glaucous) or densely pubescent with glandular or nonglandular hairs, not roughened to the touch. Leaves opposite, sessile or short-petiolate, the petioles usually winged, sometimes nearly to the base. Leaf blades lanceolate to ovate in outline, more or less unlobed to deeply 1 or 2 times pinnately lobed or divided, the margins otherwise entire or toothed. Inflorescences open, terminal racemes with leafy bracts (these reduced progressively toward the axis tip), the flowers paired at the nodes, usually appearing as axillary flowers toward the stem tip, variously short-stalked to moderately long-stalked, the stalks somewhat thickened toward the tips, lacking bractlets. Cleistogamous flowers absent. Calyces 5-lobed, slightly to moderately zygomorphic, bell-shaped, the lobes slightly shorter than to longer than the tube, entire to toothed or lobed, persistent, becoming distended and slightly enlarged at fruiting. Corollas 30–60 mm long, 5-lobed, more or less bell-shaped, yellow, the tube somewhat curved or oblique, sparsely to densely short-hairy on the inner surface, the hairs not blocking the throat, the lobes shorter than the tube, spreading, their surfaces glabrous or glandular-hairy, hairy along the margins, the throat (also the outer surface of the tube) sometimes tinged with brownish markings. Stamens with the filaments of 2 lengths, hairy (at least toward the base), the anthers with 2 sacs, these more or less parallel, tapered to an awnlike base, light yellow to yellow, hairy. Style somewhat curved downward and often slightly exserted, the stigma club-shaped to more or less capitate, unlobed. Fruits 9–20 mm long, ovoid or ellipsoid, glabrous at maturity or glandular-hairy. Seeds 0.8–2.7 mm long, ellipsoid to oblong-ellipsoid or more or less trapezoid, usually slightly flattened, the surface with a fine to coarse network of ridges and pits, the ridges sometimes appearing winglike, brown to dark brown or black. Nine to 11 species, eastern U.S., Canada, Mexico.

The taxonomy of *Aureolaria* is in need of thorough revision. Pennell (1928) recognized eleven species, with most of the more widespread ones subdivided into a complex series of infraspecific taxa. Since then, there has been no comprehensive study of the systematics of the genus. However, some authors working on regional floras have eliminated some of the less widely distributed species or reduced these to infraspecific taxa under more widespread relatives. The treatment below is thus very preliminary, pending future studies.

Steyermark (1963) and many other authors also chose to treat this genus under the name *Gerardia* L. For further discussion, see the treatment of *Agalinis* above.

Species of *Aureolaria* have been documented to parasitize the roots of a number of woody species, including both pines and hardwoods, but have been documented most commonly on the roots of oaks (*Quercus*, Fagaceae) (Musselman and Mann, 1978; Werth and Riopel, 1979). The large, relatively open-throated flowers of *Aureolaria* are pollinated mainly by bumblebees (C. R. Robertson, 1891; Pennell, 1935). Ballard and Pippen (1991) documented a sterile putative hybrid between *A. pectinata* (as *A. pedicularia*) and *A. flava* in southern Michigan. Previously, Bell and Musselman (1982) had studied the breeding system and potential for hybridization among four *Aureolaria* species in Virginia. They found that interspecific hybrids among the perennial, obligately outcrossing species were generally fairly fertile. However, when pollen from the self-compatible annual or biennial *A. pectinata* (as *A. pedicularia*) was placed on the stigmas of perennial species, the crosses failed and when pollen from the perennial species served as pollen parents in crosses with *A. pectinata*, the resultant hybrid offspring had relatively low fertility. Although natural hybrids involving *A. pectinata* should be searched for in Missouri, they are likely to be very uncommon. Conversely, the degree of interspecific hybridization between *A. flava* and *A. grandiflora* has not been studied and may be relatively frequent. It may account for some of the patterns of pubescence and variation in leaf morphology exhibited by these two species. Members of the genus are mostly pollinated by bumblebees.

1. Plants annual; stems (and other parts) densely pubescent with gland-tipped hairs; calyx lobes longer than the tube, toothed or pinnately lobed . . 3. A. PECTINATA
1. Plants perennial; stems glabrous or moderately to densely pubescent with nonglandular hairs; calyx lobes shorter than to slightly longer than the tube, entire
   2. Stems, flower stalks, and calyces glabrous, the stems sometimes somewhat glaucous . . . . 1. A. FLAVA
   2. Stems, flower stalks, and calyces densely short-hairy . . . . 2. A. GRANDIFLORA

1. Aureolaria flava (L.) Farw. (smooth false foxglove)
A. flava var. *macrantha* Pennell
A. calycosa (Mack. & Bush) Pennell
Gerardia flava L.
G. flava var. *calycosa* (Mack. & Bush) Steyerm.

Pl. 471 b–d; Map 2152

Plants perennial. Stems 50–150 cm long, glabrous, usually somewhat glaucous. Leaf blades 6–15 cm long, deeply pinnately lobed, the lobes linear to narrowly elliptic or narrowly lanceolate, the margins otherwise entire or few-toothed, the upper surface moderately roughened with minute, stiff, broad-based, nonglandular hairs, the undersurface sparsely roughened to glabrous or nearly so. Bracts linear to narrowly lanceolate, entire or with a few teeth or slender lobes toward the base. Flower stalks 4–12 mm long at flowering, elongating to 5–16 mm at fruiting, relatively stout, at least toward the tip, straight or more commonly curved upward, glabrous. Calyces 9–16 mm long, glabrous, the lobes shorter than to slightly longer than the tube, entire. Corollas 35–60 mm long, the lobes glabrous, except along the margins. Fruits 12–20 mm long, glabrous. Seeds 1.7–2.7 mm long, with coarse, winglike ridges. $2n=24$. June–September.

Scattered in the southeastern portion of the state north and west to Ste. Genevieve, Washington, Douglas, and Ozark Counties; also a single specimen from Newton County (eastern U.S. west to Wisconsin and Texas; Canada). Mesic to dry upland forests and edges of glades, fens, and sinkhole ponds.

Pennell (1928) separated *A. flava* into three varieties, of which he recorded only var. *macrantha* from Missouri, based on single specimens from Madison, Scott, and Wayne Counties. However, Pennell also segregated populations from mostly the Ozark portions of Arkansas and Missouri as a separate species, *A. calycosa*. The relatively widespread var. *macrantha* was said to have somewhat larger flowers than in var. *flava*. *Aureolaria calycosa*, to which Pennell (1928) assigned the majority of the Missouri populations, was said to differ in a suite of mostly quantitative characters, including relatively deeply lobed leaves, relatively small (40–50 mm) corollas glabrous within, relatively long awnlike anther bases, and relatively small seeds. Based on his experience with Missouri material in the field and herbarium (as *Gerardia*), Steyermark (1963) concluded that there was no basis for the segregation of var. *macrantha* from the typical variety and that the other taxon should be treated as var. *calycosa* rather than as a species. However, even he admitted that there was considerable morphological overlap between his concepts of var. *calycosa* and var. *flava* and that the characters said to distinguish the taxa varied independently of one another.

2. Aureolaria grandiflora (Benth.) Pennell (big-flowered gerardia)
Gerardia grandiflora Benth.

Pl. 471 a; Map 2153

Plants perennial. Stems 40–150 cm long, moderately to densely pubescent with minute, grayish, curved hairs, not glaucous. Leaf blades 5–18 cm long, variously unlobed to shallowly few-lobed but sharply toothed or deeply pinnately lobed, the lobes linear to narrowly elliptic or narrowly lanceolate, otherwise entire or few-toothed, those of the largest leaves occasionally lobed again, the surfaces moderately roughened with minute, stiff, broad-based, nonglandular hairs. Bracts linear to lanceolate, entire, with a few teeth, or with slender lobes. Flower stalks 4–9 mm long at flowering, elongating to 8–14 mm at fruiting, relatively stout, straight or more commonly curved upward, densely and minutely nonglandular-hairy. Calyces 12–18 mm long, densely and minutely nonglandular-hairy, the lobes about as long as to slightly longer than the tube, entire (toothed elsewhere). Corollas 40–55 mm long, the lobes glabrous, except along the margins. Fruits 10–23 mm long, glabrous. Seeds 1.7–2.2 mm long, with coarse, winglike ridges. June–September.

Scattered in most of Missouri, but uncommon or apparently absent in the northwestern and southeastern portions of the state (Minnesota to Indiana south to Texas and Mississippi). Mesic to

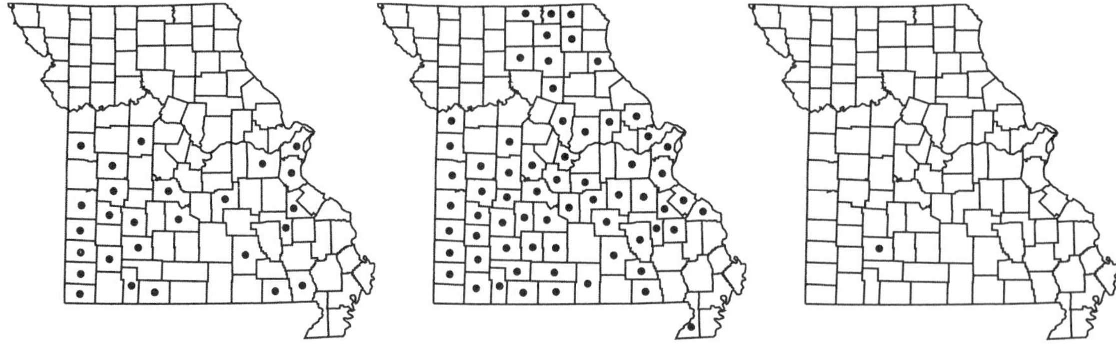

2155. Buchnera americana 2156. Castilleja coccinea 2157. Castilleja purpurea

dry upland forests, savannas, ledges and tops of bluffs, banks of streams, and edges of glades, fens, and sinkhole ponds; occasionally also old fields and roadsides.

This is the most commonly encountered member of the genus in Missouri. Pennell (1928) separated *A. grandiflora* into three subspecies, two of which occur in Missouri. The third, ssp. *grandiflora*, is restricted to Texas and differs in its broadly lanceolate calyx lobes with toothed margins and very short petioles. The taxonomic status and relationships of the infraspecific taxa within this species are poorly understood. There are a number of Missouri specimens that are intermediate between the two varieties, but the majority of collections are easily determined.

1. Blades of all the leaves and at least the lower bracts deeply lobed; fruits 15–23 mm long . . . . 2A. VAR. PULCHRA
1. Blades of at least the upper leaves and bracts with the margins entire, toothed, or with a few shallow lobes toward the base (those of the lower leaves sometimes more deeply lobed); fruits 10–17 mm long . . . . 2B. VAR. SERRATA

2a. var. pulchra Pennell
Gerardia grandiflora var. *pulchra* (Pennell) Fernald

Stems moderately to densely pubescent with minute, grayish, curved hairs. Blades of the lower leaves deeply lobed, the lobes entire, toothed, or occasionally lobed again. Blades of the upper leaves gradually smaller than those of the lower and median leaves, the margins moderately to deeply lobed. Bracts with the margins lobed. Fruits 15–23 mm long. June–September.

Scattered in most of Missouri, but uncommon or apparently absent in the northwestern and southeastern portions of the state (Minnesota to Missouri east to Wisconsin and Indiana). Mesic to dry upland forests, savannas, ledges and tops of bluffs, banks of streams, and edges of glades, fens, and sinkhole ponds; occasionally also old fields and roadsides.

This variety occupies the northern portion of the range of the species.

2b. var. serrata (Benth.) Pennell
Gerardia grandiflora var. *cinerea* (Pennell) Cory

Stems very densely pubescent with minute, grayish, curved hairs. Blades of the lower leaves unlobed or moderately to deeply lobed, the lobes entire or toothed. Blades of the upper leaves often relatively abruptly smaller than those of the lower and median leaves, the margins entire or more commonly sharply toothed, occasionally with a few shallow lobes toward the base. Bracts with the margins entire or toothed. Fruits 10–17 mm long. June–September.

Scattered in most of Missouri, but uncommon or apparently absent in the northwestern and southeastern portions of the state (Kansas and Missouri south to Texas and Mississippi). Mesic to dry upland forests, savannas, ledges and tops of bluffs, banks of streams, and edges of glades, fens, and sinkhole ponds; occasionally also old fields and roadsides.

This variety occurs mostly to the south of the range of var. *pulchra*, but the two have almost identical ranges in Missouri.

3. Aureolaria pectinata (Nutt.) Pennell
A. pectinata var. *ozarkensis* Pennell
Gerardia pedicularia L. var. *pectinata* Nutt.
Pl. 471 h, i; Map 2154

Plants annual. Stems 40–100 cm long, densely pubescent with a mixture of longer and shorter, slender, multicellular, gland-tipped hairs, not glaucous. Leaf blades 2–7 cm long,

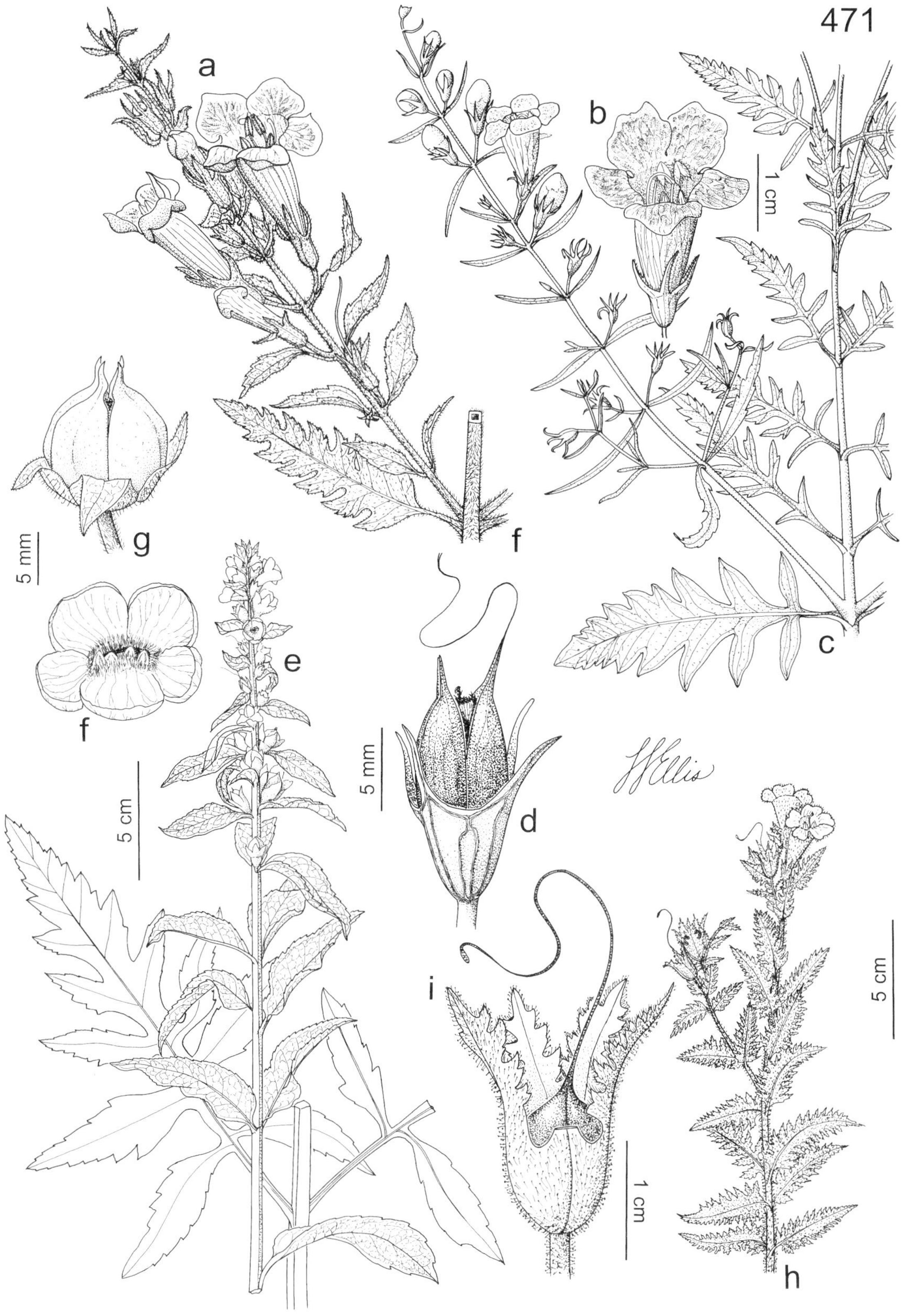

Plate 471. Orobanchaceae. *Aureolaria grandiflora*, **a)** fertile stem with cross-section. *Aureolaria flava*, **b)** flower, **c)** fertile stem, **d)** fruit. *Dasistoma macrophylla*, **e)** tip of fertile stem and lower node with leaves, **f)** flower, **g)** fruit. *Aureolaria pectinata*, **h)** fertile stem, **i)** fruit.

deeply pinnately lobed, the lobes linear to elliptic or broadly lanceolate, variously entire, toothed, or lobed again, the surfaces and margins densely pubescent with a mixture of longer and shorter, slender, multicellular, gland-tipped and nonglandular hairs. Bracts 1–4 cm long, otherwise similar to the leaves. Flower stalks 4–10 mm long at flowering, elongating 5–12(–20) mm at fruiting, relatively stout, at least toward the tip, straight or curved upward, densely glandular-hairy. Calyces 11–20 mm long, densely glandular-hairy, the lobes longer than the tube, toothed or pinnately lobed. Corollas 30–40 mm long, glandular-hairy on the outer surface, the upper lobes glandular-hairy on the inner surface near the base, the margins usually also hairy. Fruits 9–15 mm long, glandular-hairy. Seeds 0.6–1.2 mm long, with relatively short, fine ridges enclosing relatively large areoles. August–September.

Scattered in the Ozark and Ozark Border Divisions (southeastern U.S. west to Missouri and Texas). Mesic to dry upland forests, savannas, glades, tops of bluffs, and banks of streams.

Pennell (1928) treated *A. pedicularia* as comprising five confusingly similar varieties, none of which were reported to grow in Missouri. He segregated *A. pectinata* as a closely related species and divided this into four equally confusing varieties. Plants from the Ozarks were assigned to *A. pectinata* var. *ozarkensis*, the only member of the complex that he reported from Missouri. More recently, some botanists have taken a more conservative approach to the taxonomy of the group. Steyermark (1963, as *Gerardia*) and Gleason and Cronquist (1991) both treated the complex as a single species, *A. pedicularia*, with various numbers of varieties. However, although Cronquist (1991) treated our plants as var. *pectinata* (Nutt.) Gleason, that combination apparently has never been validly published within *A. pedicularia*. The main characters indicated to separate *A. pectinata* from *A. pedicularia* include the calyces hemispheric vs. conic, leaf blade margins sharply vs. bluntly toothed, ovoid vs. ellipsoid fruits, and denser and more spreading pubescence. The group of annual members of the genus requires a careful taxonomic revision.

3. **Buchnera** L. (blue hearts)
(Philcox, 1965)

About 100 species, North America to South America, Caribbean Islands, Africa, Asia south to Australia.

1. Buchnera americana L. (blue hearts)
Pl. 470 d–f; Map 2155

Plants annual, hemiparasitic, dark green, sometimes purplish-tinged, sometimes blackening upon drying. Stems 30–90 cm long, solitary or few, usually unbranched, erect or strongly ascending, circular to bluntly 4-angled, moderately to densely pubescent with short, stiff, upward-angled, nonglandular, pustular-based hairs, roughened to the touch. Leaves opposite, sessile. Leaf blades 2–6(–10) cm long, lanceolate to broadly ovate, unlobed, the margins coarsely few-toothed, the surfaces moderately to densely pubescent with short, stiff, upward-angled, nonglandular, pustular-based hairs, roughened to the touch. Inflorescences open to moderately dense, terminal spikes with small bracts shorter than the calyces, these lanceolate to narrowly ovate, unlobed, roughened-hairy, the flowers paired at the nodes or subopposite, stalkless; bractlets 2 per flower, shorter than the bracts, linear to narrowly lanceolate. Cleistogamous flowers absent. Calyces 6–9 mm long, tubular at flowering, becoming distended and somewhat urn-shaped at fruiting, moderately roughened-hairy on the outer surface, 5-lobed, actinomorphic, the lobes much shorter than the tube, triangular. Corollas 15–23 mm long, 5-lobed, weakly zygomorphic, trumpet-shaped, deep or dark purple to bluish purple, the tube hairy on the inner surface, roughened-hairy on the outer surface, the throat densely hairy, the lobes shorter than the tube, abruptly spreading, their outer surface hairy, especially toward the base, the inner surface glabrous or sparsely hairy, the margins glabrous. Stamens with the filaments of 2 lengths, short, not exserted, the anthers appearing to have only 1 sac, narrowly lanceolate, yellow, glabrous. Style relatively short and stout, not exserted, the stigma club-shaped, unlobed. Fruits 6–8 mm long, asymmetrically oblong-ovoid (appearing somewhat pouched on 1 side at the base), usually slightly flattened, glabrous. Seeds 0.6–0.8 mm long, angular, irregularly oblong-ellipsoid to oblong-obconic, the surface with a fine network of low ridges and narrow, elongate areolae, brown to dark brown. $2n$=42. June–September.

Scattered south of the Missouri River, but absent from the Mississippi Lowlands Division (eastern U.S. west to Illinois, Kansas, and Texas; Canada, Mexico). Upland prairies, glades, and savannas; occasionally also old fields and roadsides.

Pennell (1935) suggested that although the slender tube of *Buchnera* flowers appeared to be an adaptation for butterfly pollination, the interior pubescence and short staminal filaments likely resulted in significant amounts of self-pollination.

4. **Castilleja** Mutis ex L.f. (paintbrush, painted cup)

Plants annual, biennial, or perennial herbs, with fibrous roots, sometimes with woody bases, hemiparasitic, often lacking a well-developed taproot, light green to dark green, sometimes purplish-tinged, sometimes blackening upon drying. Stems solitary to several, often unbranched (branched elsewhere), erect or ascending, rounded or sometimes slightly ridged from the leaf bases, 2 of the sides sometimes slightly concave, moderately to densely grayish-pubescent with slender, weak, multicellular, nonglandular hairs, these often appearing woolly or cobwebby, sometimes wearing away in patches. Leaves alternate, sometimes also basal, sessile. Leaf blades variously shaped, entire or deeply pinnately divided or lobed into 3–7 divisions or lobes, these linear, entire along the margins, sharply pointed at the tips. Inflorescences dense, short to more commonly elongate, terminal spikes or spikelike racemes with at least the lowermost bracts more or less leafy, the median and upper bracts in some species becoming more incised and brightly colored, the flowers solitary in the axil of each bract; sessile or very short-stalked (to 2 mm in our species), lacking bractlets. Cleistogamous flowers absent. Calyces zygomorphic, tubular, strongly oblique at the tip, deeply divided into 2 primary lobes, each of these variously further lobed or divided, the lips often brightly colored (usually colored similarly to the bracts), persistent, becoming somewhat distended and papery at fruiting. Corollas 20–55 mm long, strongly bilabiate, green to yellow (red elsewhere), narrowly tubular, the upper lip fused into a slender, structure folded around the stamens, this not or only slightly hooded, tapered to an unlobed, beaklike tip, the lower lip variously well-developed or reduced, shorter than the upper one, ascending or spreading, usually 3-lobed or appearing 3-toothed, the tube straight to slightly arched, glabrous on the inner surface, the tube and lips minutely to shortly hairy on the outer surface (sometimes appearing somewhat mealy). Stamens hidden under the upper corolla lip, the filaments of 2 lengths, glabrous or hairy near the base, the anthers with 2 sacs, these unequal (1 attached near the midpoint, the other attached at its tip), blunt at the ends, yellow, glabrous. Style extended under the upper corolla lip, usually slightly exserted, the stigmatic portion capitate, unlobed or 2-lobed. Fruits (in our species) 8–17 mm long, somewhat obliquely oblong-ovoid, glabrous. Seeds variously shaped (even within a single capsule), asymmetrically oblong to oblong-ovoid or trapezoidal in outline, the outer wall loosely attached, with somewhat enlarged, translucent cells that typically lose their outer wall at maturity, the surface thus with a fine network of prominent, polygonal ridges and pits, tan to yellowish brown, orangish brown, or brown, sometimes appearing somewhat iridescent. About 190 species, North America, Central America, South America, Caribbean Islands, Europe, Asia.

At least the Missouri species of *Castilleja* typically have a very broad host range that includes everything from grasses to prairie forbs, and woody plants, and studies have shown that plants will even attempt to form haustoria on inert objects, such as pebbles and organic remains of dead plants (Malcolm, 1966).

1. Bracts green; corollas (35–)40–55 mm long, extending past the subtending bracts, the tube extending noticeably past the calyces . . . . 3. C. SESSILIFLORA

1. Bracts red, orange, yellow, white, or purple; corollas 21–35(–40) mm long, shorter than or occasionally extending only slightly past the subtending bracts, the tube shorter than to about as long as the calyces
   2. Plants annual or biennial, with a soft rootstock, usually single-stemmed; basal leaves usually present at flowering, but sometimes withered; calyces with the 2 primary lobes entire or occasionally minutely notched, rounded at the tips . . . . 1. C. COCCINEA
   2. Plants perennial, often with a woody rootstock, usually several-stemmed; basal leaves absent at flowering; calyces with the 2 primary lobes deeply parted, the ultimate lobes linear to lanceolate, sharply pointed at the tips . . . . 2. C. PURPUREA

1. Castilleja coccinea (L.) Spreng. (Indian paintbrush, painted cup, Indian blanket)
Pl. 472 e–h; dust jacket; Map 2156

Plants annual or more commonly biennial, the basal rosette usually persistent at flowering, but sometimes withered, with a soft rootstock. Stems usually solitary, 15–50(–70) cm long. Basal rosette usually present at flowering, but sometimes withered, the leaves oblanceolate to obovate, unlobed. Stem leaves with the blade 2–7(–8) cm long, the main body linear to narrowly lanceolate, unlobed or with 1 or 2 pairs of linear lateral lobes, the surfaces moderately to densely pubescent with short, multicellular hairs, some of these sometimes gland-tipped. Bracts with the main body lanceolate, with 1 or 2 pairs of lobes toward the tip, all or mostly red, orange, yellow, white, or rarely reddish purple. Calyces (13–)17–25 mm long, the 2 primary lobes 5–10 mm long, entire or occasionally minutely notched, rounded at the tips. Corollas 20–27 mm long, shorter than or occasionally extending only slightly past the subtending bracts, the tube shorter than the calyces, white to pale yellow or pale green, the upper lip 6–10 mm long, yellow with a green dorsal band to almost entirely green, the lower lip 2.0–3.5 mm long. Fruits 8–10 mm long. Seeds 0.8–1.4 mm long, tan to yellowish brown. $2n=48$. April–July.

Scattered in most of Missouri, but absent or uncommon in the northwestern and southeastern portions of the state (eastern U.S. west to Minnesota and Texas; Canada). Upland prairies, glades, savannas, fens, and banks of streams and rivers.

The brightly colored inflorescences of this species are pollinated mainly by hummingbirds (C. R. Robertson, 1891). Bract color varies greatly both within and between populations over time. The typical form has scarlet bracts, but various shades of red and orange have been documented. Plants lacking anthocyanins and thus having the bracts yellow have been called f. *lutescens* Farw. These frequently occur in mixed populations with reddish-bracted plants, sometimes along with a wide array of intermediates. Rarely, plants with darker, reddish purple bracts also are encountered.

2. Castilleja purpurea (Nutt.) G. Don **var. purpurea** (purple paintbrush)
Pl. 472 i–k; Map 2157

Plants perennial, the basal rosette absent at flowering, with a sometimes somewhat woody rootstock. Stems usually several, 15–30(–40) cm long. Basal rosette absent at flowering. Leaves with the blade 2–7(–9) cm long, the main body linear to narrowly oblong-lanceolate, unlobed or with 1 or 2(3) pairs of linear lateral lobes, the surfaces densely pubescent with short, multicellular hairs. Bracts with the main body linear, with 1 or 2 pairs of lobes toward the tip, all or mostly reddish purple (yellow, greenish yellow, yellowish orange, or red elsewhere). Calyces (20–)25–34 mm long, the 2 primary lobes (10–)13–22 mm long, deeply parted, the ultimate lobes linear to lanceolate, sharply pointed at the tips. Corollas 25–40 mm long, shorter than or occasionally extending only slightly past the subtending bracts, the tube shorter than the calyces, white to pale yellow or pale green, the upper lip 9–13 mm long, green or yellow with a broad, green dorsal band, the lower lip 1.5–3.0 mm long. Fruits 10–12 mm long. Seeds 1.0–1.9 mm long, brown to dark brown. April–May.

Uncommon, possibly extirpated, known only from a single, historical collection from Greene County (Missouri, Kansas, Oklahoma, and Texas; Mexico). Habitat unknown, but possibly a glade or blufftop overlooking a cave opening.

Shinners (1958) recognized two additional varieties, neither of which has been reported from Missouri. His var. *citrina* (Pennell) Shinners is characterized by a well-developed, flared lower corolla lip 3–7 mm long and yellow to greenish yellow bracts and calyces; the var. *lindheimeri* (A. Gray) Shinners is a minor variant with the bracts and calyces yellowish orange to red.

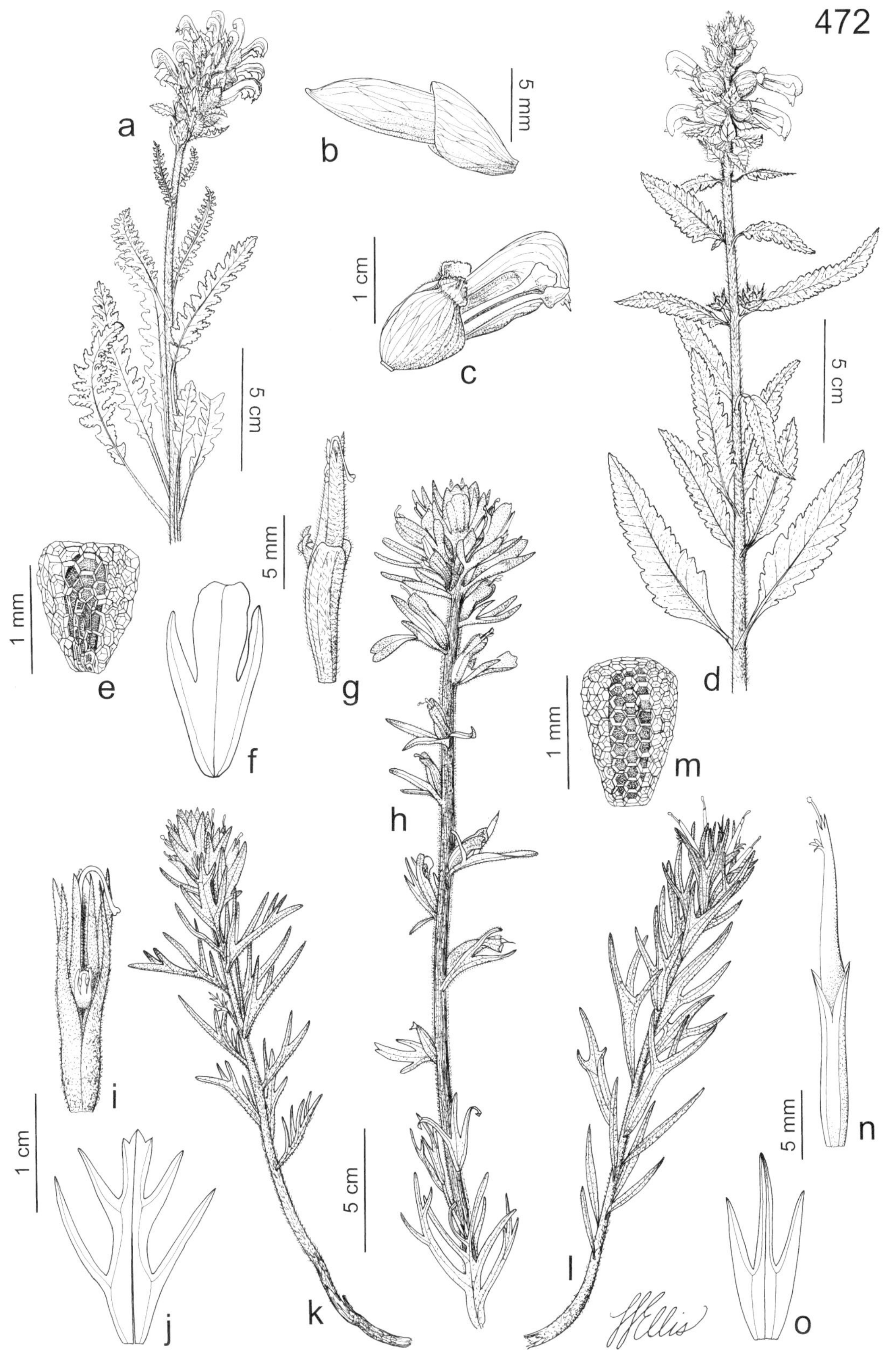

472. Orobanchaceae. *Pedicularis canadensis*, **a)** fertile stem, **b)** fruit. *Pedicularis lanceolata*, **c)** flower, **d)** fertile stem. *Castilleja coccinea*, **e)** seed, **f)** bract, **g)** flower, **h)** fertile stem. *Castilleja purpurea*, **i)** flower, **j)** bract, **k)** fertile stem. *Castilleja sessiliflora*, **l)** fertile stem, **m)** seed, **n)** flower, **o)** bract.

2158. Castilleja sessiliflora

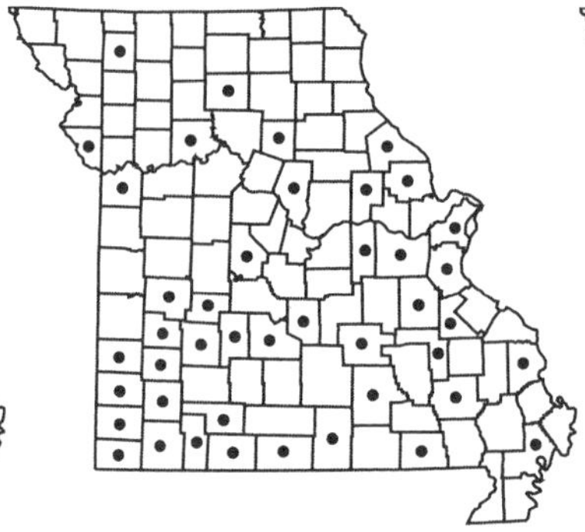

2159. Dasistoma macrophylla

2160. Epifagus virginiana

3. Castilleja sessiliflora Pursh (downy painted cup)

Pl. 472 l–o; Map 2158

Plants perennial, the basal rosette absent at flowering, with a sometimes somewhat woody rootstock. Stems usually several, 10–30(–40) cm long. Basal rosette absent at flowering. Leaves with the blade 3–6 cm long, the main body linear to narrowly oblong-oblanceolate, unlobed or the those of the upper leaves with 1 pair of linear lateral lobes, the surfaces densely pubescent with short, multicellular hairs. Bracts with the main body linear to narrowly oblong, with 1 pair of lobes toward the tip, green (sometimes pinkish- or pale purplish-tinged toward the tip elsewhere). Calyces 25–40 mm long, the 2 primary lobes 12–20 mm long, deeply parted, the ultimate lobes linear, sharply pointed at the tips. Corollas 35–55 mm long, extending noticeably past the subtending bracts, the tube longer than the calyces, cream-colored to pale yellow or nearly white (pinkish- to purplish tinged toward the tip elsewhere), the upper lip 9–12 mm long, usually light green to olive green (at least toward the tip), sometimes with pale yellow to cream-colored or nearly white margins, the lower lip 5–6 mm long, often slightly pinkish-tinged. Fruits 12–17 mm long. Seeds 1–2 mm long, yellowish brown to brown. $2n=24$. April–July.

Uncommon, known thus far only from Atchison and Holt Counties (Montana to Wisconsin south to Arizona and Texas; Canada, Mexico). Loess hill prairies.

Steyermark (1963) noted that the flowers of this species have a sweet fragrance reminiscent to that of rose gentian (*Sabatia angularis*, Gentianaceae).

5. Dasistoma Raf. (mullein foxglove)

One species, eastern U.S.

1. Dasistoma macrophyllum (Nutt.) Raf. (mullein foxglove, false foxglove)

Seymeria macrophylla Nutt.

Pl. 471 e–g; Map 2159

Plants annual, hemiparasitic, green or dark green, sometimes purplish-tinged, sometimes blackening upon drying. Stems 100–220 cm long, usually with many branches, erect or strongly ascending, mostly bluntly 4-angled, moderately to densely pubescent with fine, curved, nonglandular hairs, sometimes only on 2 opposing sides, not roughened to the touch. Leaves opposite, sessile or short-petiolate, the petioles winged, at least toward the tip. Leaf blades 10–35 cm long, lanceolate to broadly ovate in outline, those of the lower leaves moderately to deeply 1 or 2 times pinnately lobed or divided, the margins also toothed, those of the upper leaves often only shallowly lobed or toothed, the surfaces sparsely to moderately pubescent with short, slightly broad-based, nonglandular hairs, mostly along the veins. Inflorescences open, terminal, spikelike racemes with leafy bracts, these reduced progressively toward the axis tip, entire or few-lobed, the flowers paired at the nodes, usually appearing as axillary flowers toward the stem tip, the stalks 1–4 mm long, stout, thickened toward the tip, hairy, lacking bractlets. Cleistogamous flowers absent. Calyces 6–8 mm long, becoming distended and slightly elongated to 7–10 mm at fruiting, 5-lobed, slightly zygomorphic, bell-shaped, the tube moderately short-hairy on the outer surface, densely and mi-

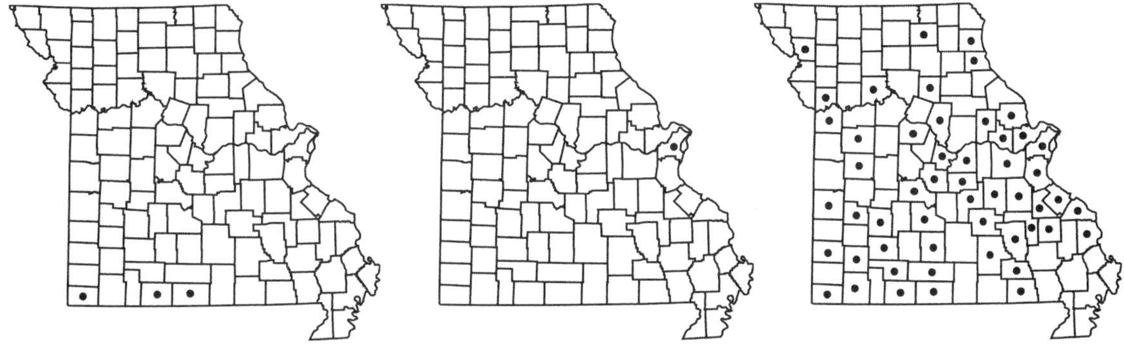

2161. Orobanche ludoviciana 2162. Orobanche riparia 2163. Orobanche uniflora

nutely hairy on the inner surface, the lobes about as long as to longer than the tube at flowering, longer than the tube at fruiting, entire or with 1 or 2 blunt teeth, the surfaces glabrous or sparsely short-hairy, the margins hairy in the sinuses between the lobes. Corollas 14–16 mm long, 5-lobed, weakly bilabiate, yellow, the tube densely hairy on the inner surface, the beard tending to partially block the throat, the lobes shorter than the tube, more or less spreading (1 of the lobes often arched upward), their surfaces glabrous, the margins glabrous or minutely hairy. Stamens with the filaments of 2 lengths, hairy, the anthers with 2 sacs, these more or less parallel, tapered to a minute, awnlike base, light yellow to yellow, glabrous. Style relatively short and stout, not exserted, the stigma more or less capitate, 2-lobed. Fruits 6–11 mm long, ovoid to nearly globose, usually slightly flattened, glabrous. Seeds 2.0–2.5 mm long, irregularly oblong-ellipsoid to more or less trapezoid, usually slightly flattened, the surface with a coarse network of often fine ridges and pits, the ridges sometimes appearing winglike, brown to dark brown or black. June–September.

Scattered nearly throughout the state (eastern U.S. west to Nebraska and Texas). Bottomland forests, mesic upland forests, banks of streams and rivers, and occasionally glades; also roadsides.

Dasistoma macrophylla parasitizes the roots of angiosperm tree species, including such diverse genera as *Acer* and *Aesculus* (Sapindaceae), and *Ulmus* (Ulmaceae) (Piehl, 1962). the flowers are pollinated mainly by larger-bodied and longer-tongued bees (C. R. Robertson, 1891).

6. **Epifagus** Nutt.

One species, eastern U.S. and adjacent Canada; Mexico.

1. Epifagus virginiana (L.) W.P.C. Barton (beechdrops)

Pl. 473 c, d; Map 2160

Plants annual, holoparasitic, lacking chlorophyll, minutely glandular-hairy, the base enlarged into an irregular, tuberous structure, usually with yellow to orange, coralloid roots. Stems 8–40 cm tall, tan to brown, usually striped or tinged with purple or yellowish brown. not succulent, usually with several ascending branches. Scale-leaves 2–4 mm long, triangular to ovate. Inflorescences spikelike racemes, the flower stalks 1–3 mm long, the flowers subtended by 2 small bracteoles, these 1–2 mm long, triangular-ovate. Flowers of 2 types, the lower ones cleistogamous, the upper ones open-flowering (but usually not producing fruits). Calyces 3–5 mm long, brown to yellowish brown or reddish brown, 5-lobed, the lobes much shorter than the tube. Corollas of cleistogamous flowers 4–6 mm long, caplike and not opening, brown to purple. Corollas of open flowers 9–12 mm long, 4-lobed, white, usually with a pair of purple or reddish brown stripes, the tube somewhat curved or arching, the lobes shorter than the tube. Stamens 4, the filaments sparsely to moderately hairy, not or only slightly exserted, the anthers with 2 sacs, these slightly divergent, tapered to a minute, awnlike base, sparsely hairy, yellow. Style usually slightly exserted, the stigma capitate, slightly 2-lobed. Fruits 4–6 mm long, globose to broadly and somewhat obliquely ovoid, somewhat flattened, glabrous, dehiscing into 2 halves from the tip. August–October.

Scattered in southeastern Missouri (eastern U.S. west to Wisconsin and Texas; Canada; Mexico). Mesic upland forests, usually on acidic substrates.

The range of this unusual parasite is almost entirely coincident with that of its only host, *Fagus grandifolia* (American beech, Fagaceae) (Thieret, 1969b).

7. **Orobanche** L. (broomrape)

Plants annual (perennial elsewhere), holoparasitic, lacking chlorophyll, variously glabrous to short-hairy and/or short glandular-hairy, the base sometimes somewhat thickened, but not enlarged into a tuberous structure (tuberous-based elsewhere), lacking coralloid roots or nearly so. Stems white to yellow, yellowish brown, or brown, sometimes tinged with purple. Scale-leaves relatively densely spirally alternate, appressed, lanceolate to ovate, broadly or obovate. Inflorescences spikes, racemes, or the flowers 1 per stem, each flower stalk subtended by a bract. Flowers with 2 small bractlets at the base of the calyx or lacking bractlets. Cleistogamous flowers absent. Calyces 5-lobed (2–4-lobed elsewhere), the lobes usually longer than the tube. Corollas variously colored, the tube somewhat curved or arching, the 5 lobes shorter than the tube, usually papery, brown, and persistent at fruiting. Stamens 4, usually included. Style included or slightly exserted, the stigma capitate to slightly concave, somewhat 2-lobed. Fruits ovoid, tapering abruptly to the persistent style, dehiscing longitudinally. About 150 species, nearly worldwide.

1. Stems not greatly thickened; flowers 1–3 per stem; flower stalks 40–160 mm long . . . . 3. O. UNIFLORA
1. Stems greatly thickened; flowers numerous, in spikes or racemes; flower stalks 0–6(–25) mm long
    2. Corolla lobes narrowly oblong to broadly ovate, rounded to bluntly pointed at the tips . . . . 1. O. LUDOVICIANA
    2. Corolla lobes triangular to lanceolate-triangular, sharply pointed at the tips . . . . 2. O. RIPARIA

1. Orobanche ludoviciana Nutt. (prairie broomrape)

Pl. 473 a, b; Map 2161

Stems 6–30 cm tall, 1–2 cm in diameter near base, greatly thickened and succulent, white to tan, sometimes tinged with purple, often with numerous, ascending branches. Inflorescences spikes or racemes, the flowers numerous on the branches. Flower stalks absent or to 3 mm long. Flowers with 2 small bracteoles at the base of the calyx. Calyx 10–17 mm long, the lobes linear-lanceolate, longer than the tube. Corollas 18–26 mm long, white or cream-colored with 2 yellow pleats in the throat and pink to purple lobes, the lobes narrowly oblong to broadly ovate, obtuse. Fruits 6–10 mm long. $2n$=24, 48, 72, 96. June–July.

Uncommon in the southwestern portion of the state, known thus far only from McDonald, Ozark, and Taney Counties (Minnesota to Texas, west to Colorado and Washington, southern Canada). Parasitic on the roots of *Grindelia lanceolata* and perhaps other Asteraceae, on dolomite glades.

The presence of this taxon in Missouri was first discussed by T. E. Smith (1990) under the name *O. multiflora* Nutt. Application of the name *O. ludoviciana* for the upland taxon present in southwestern Missouri was discussed by L. T. Collins et al. (2009). True *O. multiflora* (*O. ludoviciana* ssp. *multiflora* (Nutt.) L.T. Collins ex H.L. White & W.C. Holmes) remains a valid taxon, but occurs only to the southwest of Missouri. The Missouri populations of *O. ludoviciana* represent the easternmost localities for the species, which is primarily distributed in the Great Plains.

2. Orobanche riparia L.T. Collins (river broomrape)

Pl. 473 h–j; Map 2162

Stems 8–35 cm tall, 1–3 cm in diameter near base, greatly thickened and succulent, white to yellow or tan, sometimes tinged with purple, unbranched or more commonly few-branched (rarely with more numerous branches) from near the base. Inflorescences spikelike racemes, the flowers mostly

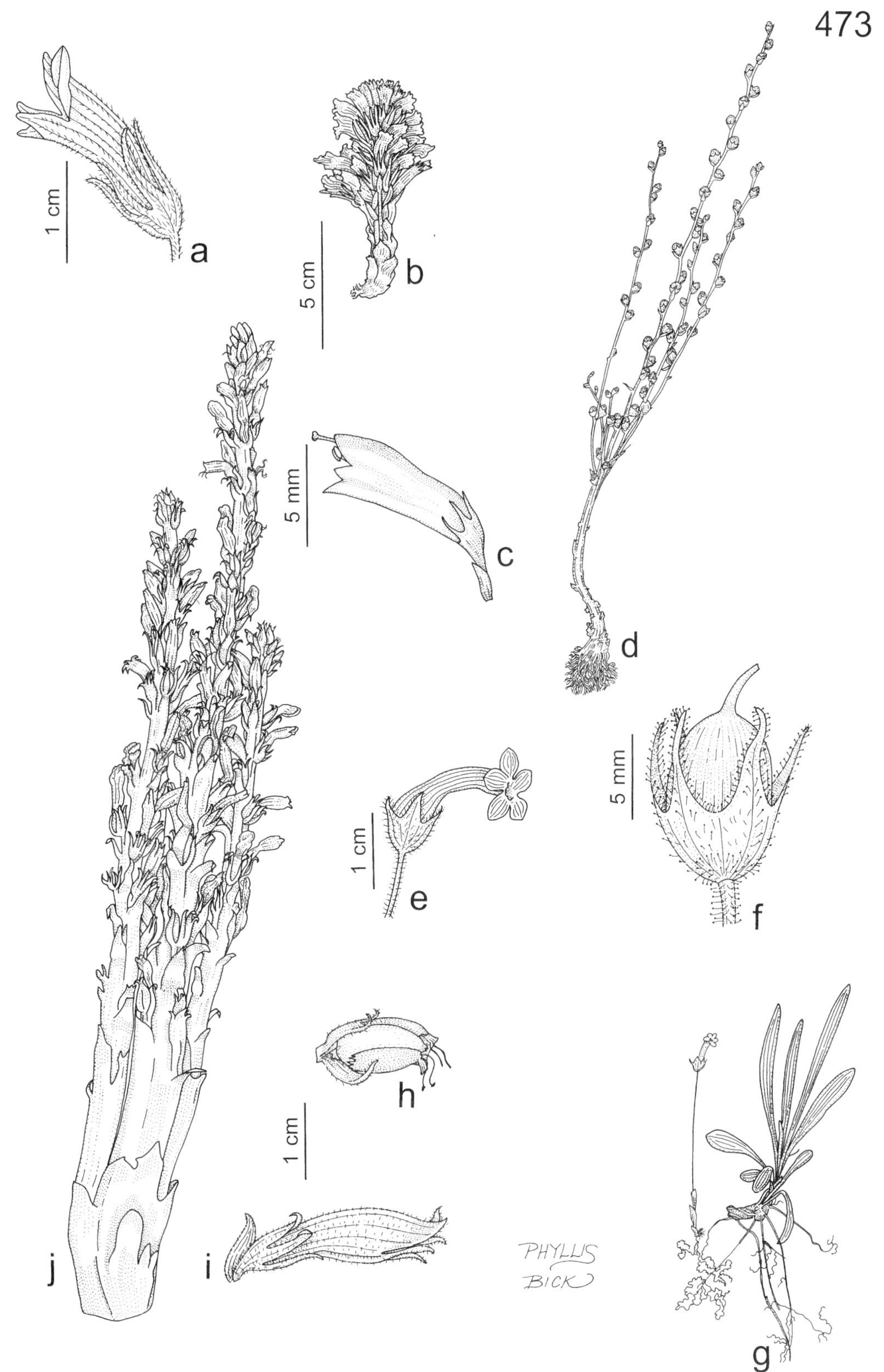

Plate 473. Orobanchaceae. *Orobanche ludoviciana*, **a)** flower, **b)** habit. *Epifagus virginiana*, **c)** open flower, **d)** habit. *Orobanche uniflora*, **e)** flower, **f)** fruit, **g)** habit, parasitic on roots of young *Rudbeckia missouriensis* (Asteraceae). *Orobanche riparia*, **h)** fruit, **i)** flower, **j)** habit.

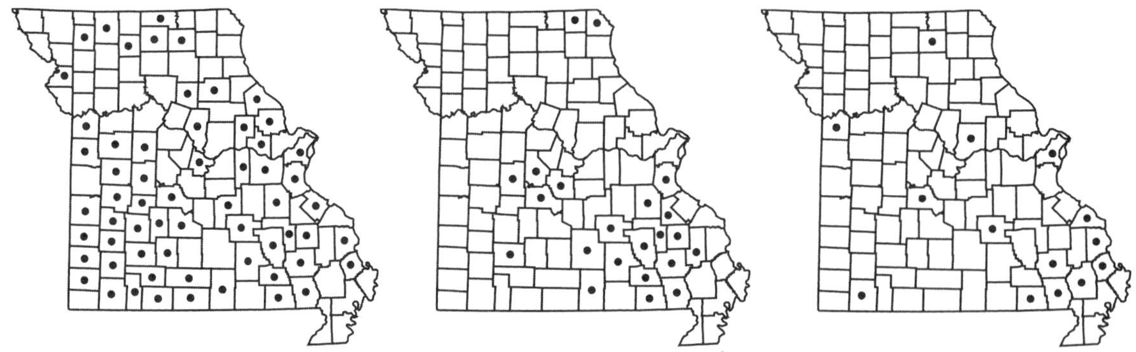
2164. Pedicularis canadensis 2165. Pedicularis lanceolata 2166. Oxalis corniculata

numerous on the branches. Flower stalks absent or to 6 mm long. Flowers with 2 small bracteoles at the base of the calyx. Calyces 7–11(–13) mm long, the lobes linear-lanceolate, longer than the tube. Corollas 13–22 mm long, white or cream-colored, sometimes purplish-tinged, with 2 yellow pleats in the throat and lavender to purple lobes, the lobes triangular to lanceolate, sharply pointed at the tips. Fruits 7–10 mm long. $2n$=48. August-September.

Uncommon, St. Louis County (Nebraska to Ohio). Parasitic on the roots of *Ambrosia trifida* L. and *Xanthium strumarium* L., on banks of Rivers. Elsewhere, this taxon rarely parasitizes cultivated tobacco (*Nicotiana tabacum* L.).

The taxonomy of the *Orobanche ludoviciana* complex has recently been somewhat clarified. L. T. Collins (1973) suggested that the primarily mid-western and eastern plants of riparian habitats should be segregated from *O. ludoviciana*, but the new name for this segregate, *O. riparia*, was not published until much later (L. T. Collins et al., 2009). Thus, what had been called *O. ludoviciana* in the earlier Missouri literature (Steyermark, 1963; T. E. Smith, 1990) is now known by the name *O. riparia*. For further discussion of the correct application of the name *O. ludoviciana*, see the treatment of that species.

This species currently is known only from a small stretch of the Meramec River in western St. Louis County. However, its hosts are extremely widespread in Missouri bottomlands and adjacent to crop fields. The overall distribution of the parasite extends both eastward and westward from Missouri. Thus, *O. riparia* should be searched for elsewhere in the state, especially in the floodplains of the Missouri and Mississippi Rivers. The species has been collected along the Mississippi River in Lauderdale County, Tennessee, less than 50 km south of the Missouri border (Piehl and Piehl, 1973).

3. Orobanche uniflora L. (one-flowered cancer-root, small cancer-root)

Pl. 473 e–g; Map 2163

Stems 1–6(–8) cm tall, 3–8 mm in diameter near base, not thickened, somewhat succulent, tan to brown, sometimes tinged with yellow, often with scattered branches. Inflorescences of 1–3 flowers per branch tip. Flower stalks 3–16 cm long, erect. Flowers lacking bracteoles at the base of the calyx. Calyx 5–10 mm long, the lobes triangular, acuminate, slightly longer than the tube. Corollas 14–24 mm long, white or cream-colored with 2 yellow pleats in the throat and violet to purple lobes, the lobes broadly obovate to oblong or nearly circular, rounded or broadly and bluntly pointed. Fruits 9–12 mm long. $2n$=36, 48, 72. April–June.

Scattered nearly throughout the state except in the Mississippi Lowlands Division (nearly throughout the U.S.; Canada). Parasitic on the roots of various hosts, rarely in mesic forests, prairies, and disturbed sandy soil, more commonly on dolomite and limestone glades and bluffs, and in dry, upland forests. Confirmed hosts in Missouri include members of several families, including various species of *Ostrya* Scop., *Potentilla, Quercus, Rudbeckia* L., *Solidago* L., and *Symphyotrichum* Nees, but host specificity appears to be poorly developed and other genera are to be expected.

8. Pedicularis L. (lousewort)

Plants perennial herbs, with fibrous roots (tuberous-rooted elsewhere), sometimes with rhizomes or stolons, hemiparasitic, often lacking a well-developed taproot, olive green to green or dark green, sometimes purplish-tinged, sometimes blackening upon drying. Stems solitary to several, unbranched or with short branches near the tip, erect or ascending, rounded or sometimes slightly ridged from the leaf bases, usually glabrous or nearly so toward the base, progressively more pubescent toward the tip with relatively stout, multicellular, nonglandular hairs, the upper portions and inflorescence axes often densely hairy or appearing woolly. Leaves alternate or opposite, sometimes also basal, lowermost stem leaves variously long-petiolate or short-petiolate to sessile, those of the upper nodes short-petiolate to sessile. Leaf blades narrowly oblong-elliptic to lanceolate or oblanceolate, moderately to deeply pinnately divided or lobed into 13–35 lobes or the uppermost merely coarsely toothed, the lobes oblong to ovate, toothed and/or lobed along the margins, mostly rounded at the tips. Inflorescences relatively dense, short to elongate, terminal (often also appearing axillary in *P. lanceolata*) spikes or spikelike racemes with at least the lowermost bracts more or less leafy (reduced progressively toward the axis tip), the flowers solitary in the axil of each bract; sessile or very short-stalked, lacking bractlets. Cleistogamous flowers absent. Calyces strongly bilabiate, oblique at the tip, bell-shaped to more commonly more or less tubular, the lips variously toothed or with 2 variously shaped lobes (4 or 5 teeth or lobes elsewhere), persistent, not or only slightly enlarged at fruiting. Corollas 15–27 mm long, strongly bilabiate, white or yellow, sometimes partially pinkish-, reddish-, or purplish-tinged or at least the upper lip dark purple, the upper lip fused into a hooded or helmet-shaped structure, this with a pair of small teeth near the otherwise truncate tip or unlobed and tapered to a beaklike tip, the lower lip slightly shorter than the upper one, ascending or spreading, 3-lobed, longitudinally grooved between the lobes, the tube in our species usually somewhat arched or curved, hairy on the inner surface (at least near the base), the hairs not blocking the throat, the lobes glabrous. Stamens hidden under the upper corolla lip, the filaments of 2 lengths, hairy near the base, the anthers with 2 sacs, these parallel, blunt at the ends, light yellow, glabrous. Style curved under the upper corolla lip, sometimes slightly exserted, the stigmatic portion small, capitate, unlobed. Fruits (in our species) 8–17 mm long, obliquely ovoid to more or less oblong-ovoid, glabrous. Seeds asymmetrically oblong to oblong-ovoid, sometimes winged, the surface with a fine to very fine network of sometimes faint ridges and pits, yellowish brown to brown or dark brown. About 500 species, North America, South America, Europe, Asia.

The Missouri species of *Pedicularis* have very large host ranges that include both herbaceous and woody species (Piehl, 1963). Lackney (1981) showed that plants of *P. lanceolata* could even be cultivated in pots successfully on the roots of wheat and red clover.

1. Upper corolla lip with a pair of small, slender teeth just below the otherwise more or less truncate tip; stem leaves mostly alternate, the lower ones long-petiolate; basal leaves well-developed at flowering .... 1. P. CANADENSIS
1. Upper corolla lip unlobed (rarely minutely notched), narrowed to a beaklike tip; stem leaves opposite or subopposite; lower stem leaves sessile or short-petiolate; basal leaves absent at flowering .... 2. P. LANCEOLATA

1. Pedicularis canadensis L. (common lousewort, wood betony)

P. canadensis var. *dobbsii* Fernald

Pl. 472 a, b; Map 2164

Plants often colonial, with short rhizomes or stolons. Stems 10–30(–40) cm long, sparsely hairy toward the base, densely hairy or appearing woolly toward the tip and inflorescence axis. Basal leaves well-developed at flowering, long-petiolate. Stem leaves mostly alternate. Lower stem leaves long-petiolate. Leaf blades 2–9(–15) cm long, lanceolate to narrowly oblong-elliptic or occasionally oblanceolate, moderately to deeply lobed, those of the larger leaves lobed more than halfway to the midvein, the lobes finely toothed or scalloped to lobed again, usually with a narrow, thickened pale band along the margins, the surfaces sparsely hairy to nearly glabrous. Inflorescences terminal spikes or spikelike racemes, the flower stalks absent or to 3 mm long, relatively short and dense when the first flowers open, continuing to elongate and appearing less dense as the fruits develop.

Calyces 7–9 mm long, strongly oblique, the tube with a pair of shallow, rounded teeth at the tip of the upper side (sometimes appearing merely shallowly notched), the lower side deeply notched or divided. Corollas 18–25 mm long, yellow or variously tinged with pink, red, or purple, rarely white, the upper lip 11–15 mm long, sometimes completely purple to dark purple, narrowed to a more or less truncate tip, but with a pair of short, slender, toothlike lobes just below the tip, the lower lip 7–10 mm long, the lobes spreading, the central lobe shorter and more sharply pointed than the lateral ones. Fruits 10–17 mm long. Seeds 1.6–2.2 mm long, more or less oblong-ellipsoid, not flattened, with a short, sharp point at the tip, truncate to concave at the base, unwinged, the surface with a fine but noticeable network of ridges, brown. $2n=16$. April–May.

Scattered nearly throughout the state, less abundantly north of the Missouri River (eastern U.S. west to North Dakota and Texas; Canada). Glades, upland prairies, ledges and tops of bluffs, savannas, openings of mesic to dry upland forests, and less commonly banks of streams and rivers and openings of bottomland forests; also old fields, pastures, and roadsides.

Viewed from above, the inflorescences of this species often appear similar to pinwheels, with the flowers strongly spiraled. Flower color can vary greatly within populations. Rare plants with the corollas white or mostly white have been called f. *albescens* Steyerm. Steyermark (1963) noted anecdotal reports that *P. canadensis* is toxic to sheep.

2. Pedicularis lanceolata Michx. (swamp lousewort, swamp wood betony)
Pl. 472 c, d; Map 2165

Plants not colonial, lacking rhizomes and stolons. Stems (15–)30–80 cm long, glabrous toward the base, moderately to densely hairy toward the tip and inflorescence axis. Basal leaves absent at flowering. Stem leaves opposite or subopposite. Lower stem leaves short-petiolate or sessile. Leaf blades 3–12 cm long, narrowly oblong-lanceolate to narrowly elliptic, lobed less than half way to the midvein, the lobes finely toothed or scalloped, unlobed, with a narrow, pale band along the margins, the surfaces sparsely to moderately hairy, sometimes mostly along the veins. Inflorescences spikes or spikelike racemes, these terminal or appearing lateral at the tips of short branches produced from the upper leaf axils, the flower stalks absent or to 2 mm long, relatively dense. Calyces 8–12 mm long, oblique, the tube with a pair of rounded, appendagelike, lateral lobes, the upper and lower margins truncate or broadly notched. Corollas 15–27 mm long, pale yellow to cream-colored or nearly white, the upper lip 8–13 mm long, narrowed to a beaklike, unlobed tip (rarely minutely notched), the lower lip 7–12 mm long, ascending to slightly spreading, somewhat incurved, the central lobe extending past the lateral ones. Fruits 8–11 mm long. Seeds 2.5–3.2 mm long, more or less semicircular to slightly kidney-shaped in outline, strongly flattened, the brown to dark brown, somewhat thickened body surrounded on 3 sides by a prominent, lighter brown wing, this rounded at each end, the surface with a very fine (visible only under magnification) and often faint network of ridges. August–October.

Scattered in the Ozark and Ozark Border Divisions, also disjunct in the Glaciated Plains in Clark and Scotland Counties (northeastern U.S. west to North Dakota, Nebraska, and Arkansas; Canada). Fens and calcareous seeps; also occasionally seepy roadsides.

OXALIDACEAE (Wood Sorrel Family)

Contributed by George Yatskievych and John D. Archer

Five genera, 500–800 species, nearlyworldwide.

Although represented in Missouri by a small number of herbaceous species, elsewhere this morphologically variable family includes shrubs and small trees, and even succulent, cushion-forming, and liana-forming species, as well as at least one floating-leaved aquatic. Some South American *Oxalis* species are cultivated as food plants for their starchy tubers (see below). The carambola (star fruit), which is usually sold as a fresh fruit in American markets but can also be pickled or prepared into jams and jellies, comes from a tree species, *Averrhoa carambola* L., which is native to the Old World tropics but is also grown widely in the New World. The juice from both it and the other species in the genus, *A. bilimbi* L. (bilimbing, cucumber tree), additionally has been used as a stain remover. A number of Oxalidaceae species also have horticultural value.

1. **Oxalis** L. (wood sorrel, lady's sorrel)

Plants annual or more commonly perennial herbs (occasionally woody elsewhere), sometimes with rhizomes, stolons, or bulbs (sometimes with tubers elsewhere). Aerial stems absent (in species with bulbs) or present and variously erect to prostrate. Leaves all basal or alternate on the aerial stems, then sometimes appearing fasciculate, mostly long-petiolate, the petiole jointed at the base. Stipules absent or, if present, then relatively small, scalelike or herbaceous, fused to the petiole for part or all of the length. Leaf blades palmately trifoliate (sometimes the first leaves simple; also with 4 to several palmate or pinnate leaflets elsewhere). Leaflets narrowly to broadly obtriangular to shallowly obcordate, angled at the base to a minute, thickened stalklike portion (pulvinus), broadly rounded to truncate or shallowly notched at the tip, often reddish- or purplish-tinged. Inflorescences basal (in stemless species, then arising directly from the bulb) or axillary, usually relatively long-stalked, consisting of a simple umbel or small panicles (these with a central flower flanked by a pair of branches bearing 2 or more flowers), occasionally reduced to a solitary flower. Flowers actinomorphic, hypogynous, perfect, the stalk often hinged and with a small, slender bract at the base. Calyces of 5 free sepals, these usually slightly overlapping laterally, persistent but not becoming enlarged at fruiting. Corollas bell-shaped to broadly funnelform, of 5 free petals, these tapered to short stalklike bases and attached to the base of the stamen tube, overlapping laterally, yellow or pink to purple, rarely white, sometimes with reddish or green markings toward the base on the upper side. Stamens 10 in 2 cycles of alternating longer and shorter ones, the filaments fused into a shallow ring toward the base, the anthers not exserted, attached toward their midpoints, yellow. Pistil 1 per flower, of 5 fused carpels. Ovary superior, 5-locular, somewhat 5-angled in cross-section, with several to numerous ovules in a single longitudinal series per locule, the placentation axile. Styles 5, sometimes persistent at fruiting, the stigmas 1 per style, often more or less 2-lobed. Fruits capsules, cylindrical (oblong-ellipsoid in *O. violacea*), usually tapered to a short beak, somewhat 5-angled in cross-section, dehiscent by a longitudinal slit on each valve. Seeds 2 to several per locule, surrounded by a translucent aril, ejected from the fruit when the aril abruptly turns inside out, 1–2 mm long, more or less elliptic with pointed ends, somewhat flattened, the surfaces with 5–13, broad, transverse (longitudinal elsewhere) ridges usually connected irregularly to form a raised network (often appearing somewhat wrinkled or only faintly ridged in *O. violacea*), reddish brown to dark brown, the tops of the ridges sometimes grayish or whitened. Five hundred or more species, nearly worldwide.

Species of *Oxalis* exhibit a high degree of morphological variation in most of their vegetative structures. The groups that do not form bulbs tend to vary in their stem thickness, orientation, degree of branching, and hairiness. The leaves can vary similarly in size, vestiture and degree of anthocyanin (purplish pigments) production. This creates difficulty in species determinations. The leaflets usually exhibit a *circadian rhythm*: at the end of each day they droop or fold downward from a small thickened basal portion (pulvinus), becoming oriented parallel to the petiole, only to spread again the following morning (Johnsson et al., 2006).

Species of *Oxalis* tend to accumulate oxalates (oxalic acid, potassium oxalate, calcium oxalate) in their tissues, resulting in a tart or sour flavor. For this reason, fresh leaves are sometimes used as an ingredient in salads and greens, similar to the use of *Rumex acetosella* (sheep sorrel, Polygonaceae), which has a similar flavor. The leaves of some species traditionally were used medicinally for scurvy. The South American species, *O. tuberosa* Molina (oca), is cultivated as a root crop for its starchy tubers, particularly in the Andean region. However, readers should note that *Oxalis* tissue that has not been treated to break down or leach out the oxalates can be toxic to humans and livestock when consumed in large quantities, as a build-up of oxalate crystals can lead to kidney damage and other symptoms (Burrows and Tyrl, 2001).

It is unclear whether the original shamrock of Irish folklore was a species of *Trifolium*, some other member of the Fabaceae, or a species of *Oxalis* (Colgan, 1896; Everett, 1971; E. C. Nelson, 1991). Whatever the case, the shamrocks currently in the horticultural trade mostly are cultivars of several *Oxalis* species, commonly the South American *O. regnellii* Miq. (with three, strongly and broadly obtriangular leaflets) and the Mexican *O. tetraphylla* Cav. (*O. deppei* Lodd.; with four leaflets). A specimen at the Missouri Botanical Garden Herbarium collected in 1962 by Frederick Comte (*#4633*) documents another cultivated species, *O. corymbosa* DC. (*O. martiana* Zucc.; *O. debilis* Kunth var. *corymbosa* (DC.) Lourteig), as a weed in a greenhouse in Kirkwood (St. Louis County). This species is known as lilac oxalis and pink wood sorrel for its relatively showy, light purple to reddish purple or pink corollas. It is a perennial, bulb-forming native of tropical South America, but is widely cultivated as a houseplant and has long been a widespread weed in tropical and warm-temperate portions of the Old World. Although this species occasionally becomes established as an escape in some southeastern states (K. R. Robertson, 1975), it is not cold-hardy in Missouri's climate and thus is not considered likely to become a member of the flora.

Various species of *Oxalis* act as hosts in the complex life cycles of the common rusts of maize, sorghum, and related grasses (*Puccinia sorghi* Schwein. and related fungi), some of which are commerically important crop pathogens. Several species also can be aggressive weeds of greenhouses, lawns, crop fields, and disturbed ground.

The taxonomy of the yellow-flowered wood sorrels with aerial stems has long been controversial. Differences in interpretation of type specimens (K. R. Robertson, 1975; D. B. Ward, 2004) and thus the application of various names to plants of differing morphologies by the two most recent monographers of the group (Eiten, 1963: Lourteig, 1979) have been complicated by the general morphological variability of the plants. Although the reduced fertility of artificially produced interspecific hybrids has been studied (Lovett Doust et al., 1981), the frequency of natural hybridization is poorly known. Many of the species also are heterostylous (Eiten, 1963), that is, two or more, commonly three, different kinds of flowers are produced, differing in the lengths of the styles relative to the stamens (long and ascending above the relatively short filaments, short and curved outward between the long filaments, and sometimes also intermediate). In spite of this phenomenon, potential inbreeding is high (Ornduff, 1972; Lovett Doust et al., 1981) and facultative apomixis is known to occur in at least *O. dillenii* (Lovett Doust et al., 1981). At the species level, the present treatment substantially follows that of D. B. Ward (2004), who studied the genus in Florida, and Nesom (2009), who studied the group for a forthcoming treatment in the Flora of North America series. Both of these works differ markedly from the Missouri treatments of Steyermark (1963) and Yatskievych and Turner (1990), and neither agrees entirely with either of the last two taxonomic revisions of the group (Eiten, 1963; Lourteig, 1979).

1. Plants lacking aerial stems, the leaves all basal, arising directly from a scaly bulb; corollas pink, violet, or purple, rarely white; sepals bright orange at the tips . . . . 5. O. VIOLACEA
1. Plants with aerial stems, at least some of the leaves attached to an erect to creeping stem; bulbs absent; corollas yellow, rarely with reddish markings toward the base; sepals green or translucent at the tips

2. Inflorescences paniculate (with a central flower flanked by a pair of branches bearing 2 or more flowers), if rarely reduced to 3 flowers then the central flower sessile or much shorter-stalked than the outer pair; total flowers (3–)5–8(–15) per inflorescence; stems usually with at least a few relatively long, more or less spreading, slender, multicellular hairs (these sometimes dense) in addition to sparse to moderate, short, upward-curved to loosely appressed, unicellular hairs . . . . 4. O. STRICTA
2. Inflorescences umbellate, occasionally reduced to a solitary flower; total flowers 1–5(–8) per inflorescence; stems lacking long, multicellular hairs, ranging from nearly glabrous to densely pubescent with appressed and/or upward-curved, short, unicellular hairs
    3. Seeds entirely brown, the ridges not whitened; stipules represented by a pair of oblong, rounded auricles at the petiole base; stems commonly 2 to several from the rootstock, creeping (sometimes with ascending branches), rooting at all or most of the nodes . . . . 1. O. CORNICULATA
    3. Seeds brown, but with the ridges whitened; stipules absent or, if present, then represented by slight thickenings or inconspicuous wings at the petiole base; stems 1 or less commonly 2 (sometimes several in *O. dillenii*), variously prostrate to erect, if prostrate, then not or rarely rooting at the nodes
        4. Fruits densely pubescent at maturity, the short, appressed to more or less spreading, unicellular and/or multicellular hairs often hiding a layer of microscopic hairs; stems moderately pubescent with appressed to upward-curved hairs, the pubescence often denser near the stem tips . . . . 2. O. DILLENII
        4. Fruits glabrous or sparsely pubescent with minute, curved hairs, sometimes mostly along the sutures; stems sparsely (often nearly glabrous) to moderately pubescent with fine, spreading to upward-curved hairs, the pubescence sometimes somewhat denser near the stem base. . . . 3. O. FLORIDA

1. Oxalis corniculata L. (creeping wood sorrel)
O. corniculata var. *atropurpurea* Planch.
O. repens Thunb.

Pl. 474 e-g; Map 2166

Plants perennial, but flowering the first year and sometimes appearing annual, with small taproots, lacking bulbs. Aerial stems commonly 2 to several from the rootstock, 4–10(–25) cm long, the main stems green or darkened and creeping (sometimes with ascending branches), stoloniferous, rooting at all or most of the nodes, sparsely pubescent with mostly appressed hairs (the ascending branches usually more densely hairy). Leaves basal (on young plants) and alternate, those on older stems often appearing fasciculate from the stem nodes, the petiole moderately to densely pubescent with appressed to strongly ascending hairs. Stipules represented by a pair of small (1.5–2.5 mm), oblong, rounded to truncate auricles at the petiole base, these usually brown. Leaflets 4–12 mm long, obcordate, the apical notch to 1/3 of the total length, the upper surface glabrous or sparsely pubescent with short, curved to loosely appressed hairs, the undersurface moderately to densely pubescent with mostly appressed hairs, green or the surfaces or margins purplish- to brownish-tinged. Inflorescences umbellate with 2 or 3(–6) flowers, occasionally reduced to a solitary flower. Sepals 2.5–5.0 mm long, oblong-lanceolate to narrowly ovate, green or translucent at the tip. Petals 4–8 mm long, yellow. Fruits 7–20 mm long, cylindrical at maturity, sparsely pubescent with short, curled or curved, unicellular hairs or sometimes glabrous. Seeds 1.0–1.8 mm long, brown, the ridges not whitened. $2n$=24, 36, 42, 46, 48. April–November.

Introduced, uncommon and widely scattered, mostly in and around urban areas (native range poorly known but possibly the Old World tropics, presently known nearly worldwide). Banks of streams and rivers, margins of ponds, lakes, and sinkhole ponds, and occasionally disturbed openings of bottomland and mesic upland forests; also pastures, fallow fields, greenouses, gardens, lawns, roadsides, and open disturbed areas.

Opinions have varied on the origin of this cosmopolitan weed. K. R. Robertson (1975) and some other authors considered it to be native to portions

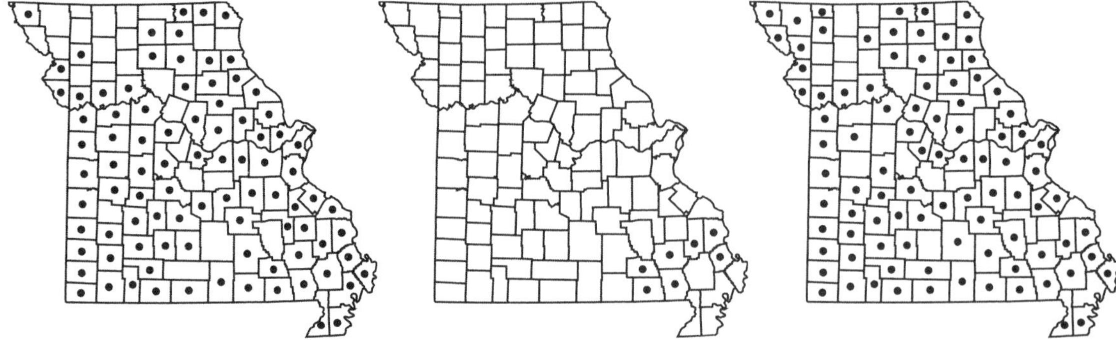

2167. Oxalis dillenii 2168. Oxalis florida 2169. Oxalis stricta

of the Old World, Eiten (1963) and Lourteig (1979) both suggested that the species is native in both the Old and New Worlds, including potentially at least the southeastern United States. However, authors of regional floristic treatments, even for states such as Florida (D. B. Ward, 2004) mostly have considered the species introduced in their regions, based on its discontinuous occurrences and restriction to disturbed habitats.

B. L. Turner (1994) applied the name *O. corniculata* var. *wrightii* (A. Gray) B.L. Turner to plants from the central portion of the United States (including Missouri), but that name applies to a western species that Eiten (1963) called *O. albicans* Kunth. Turner's concept of the taxon appears to be an amalgamation of plants here called *O. dillenii* and *O. stricta*.

2. Oxalis dillenii Jacq. (yellow wood sorrel, gray-green wood sorrel)
O. corniculata var. *dillenii* (Jacq.) Trel.
O. lyonii Pursh

Pl. 474 h–j; Map 2167

Plants perennial, but flowering the first year and sometimes appearing annual, initially with taproots, but usually developing blackish brown rhizomes, lacking bulbs. Aerial stems 1 to several, 8–25 cm long, variously erect or becoming prostrate, then not or rarely rooting at the nodes, moderately pubescent with short, unicellular, appressed to upward-curved hairs, the pubescence often denser near the stem tips. Leaves basal (on young plants) and alternate, those on older stems often appearing fasciculate from the stem nodes, the petiole moderately to densely pubescent with appressed to strongly ascending hairs. Stipules absent or, if present, then represented by slight thickenings or inconspicuous wings at the petiole base, these green. Leaflets 4–15(–20) mm long, obcordate, the apical notch to 1/3 of the total length, the upper surface usually glabrous, the undersurface sparsely to moderately pubescent with mostly appressed hairs, yellowish green to grayish green, usually lacking purplish to brownish markings. Inflorescences umbellate with 2–5(–8) flowers, occasionally reduced to a solitary flower. Sepals 3–6 mm long, oblong-lanceolate to narrowly oblong-elliptic, green or translucent at the tip. Petals 5–11 mm long, yellow. Fruits 12–20(–25) mm long, cylindrical at maturity, densely pubescent at maturity, the short, appressed to more or less spreading, unicellular and/or multicellular hairs often hiding a layer of microscopic hairs. Seeds 1–2 mm long, brown, the ridges grayish or whitened. $2n$=18, 20, 22, 23, 24. May–November.

Scattered to common nearly throughout the state, but apparently uncommon in northwestern Missouri (nearly throughout the U.S.; Canada, Mexico; introduced in the Caribbean Islands, Europe). Bottomland forests, mesic to dry upland forests, savannas, upland prairies, glades, and banks of streams and rivers; also pastures, fallow fields, gardens, railroads, roadsides, and disturbed areas.

Plants of *O. dillenii* with prostrate stems sometimes have been misdetermined as *O. corniculata*, but the noticeable stipules of that species are not present in *O. dillenii*. For a discussion of problems with the separation of *O. dillenii* and *O. florida*, see the treatment of the latter species.

Mühlenbach (1979) reported the closely related *O. texana* (Small) Fedde (as *O. priceae* Small ssp. *texana* (Small) Eiten) based on a specimen from the St. Louis railyards. Yatskievych and Turner (1990) noted that this report was incorrect, but mistakenly refered the specimen to *O. corniculata*. The specimen has since been redetermined as *O. dillenii*. *Oxalis texana* is endemic to portions of Arkansas, Louisiana, and Texas and differs from *O. dillenii* in its usually larger number of flowers per umbel (mostly 3–8) and larger corollas (12–16 mm long). Lourteig (1979) applied the name *O. lyonii* Pursh to this taxon (which Yatskievych and

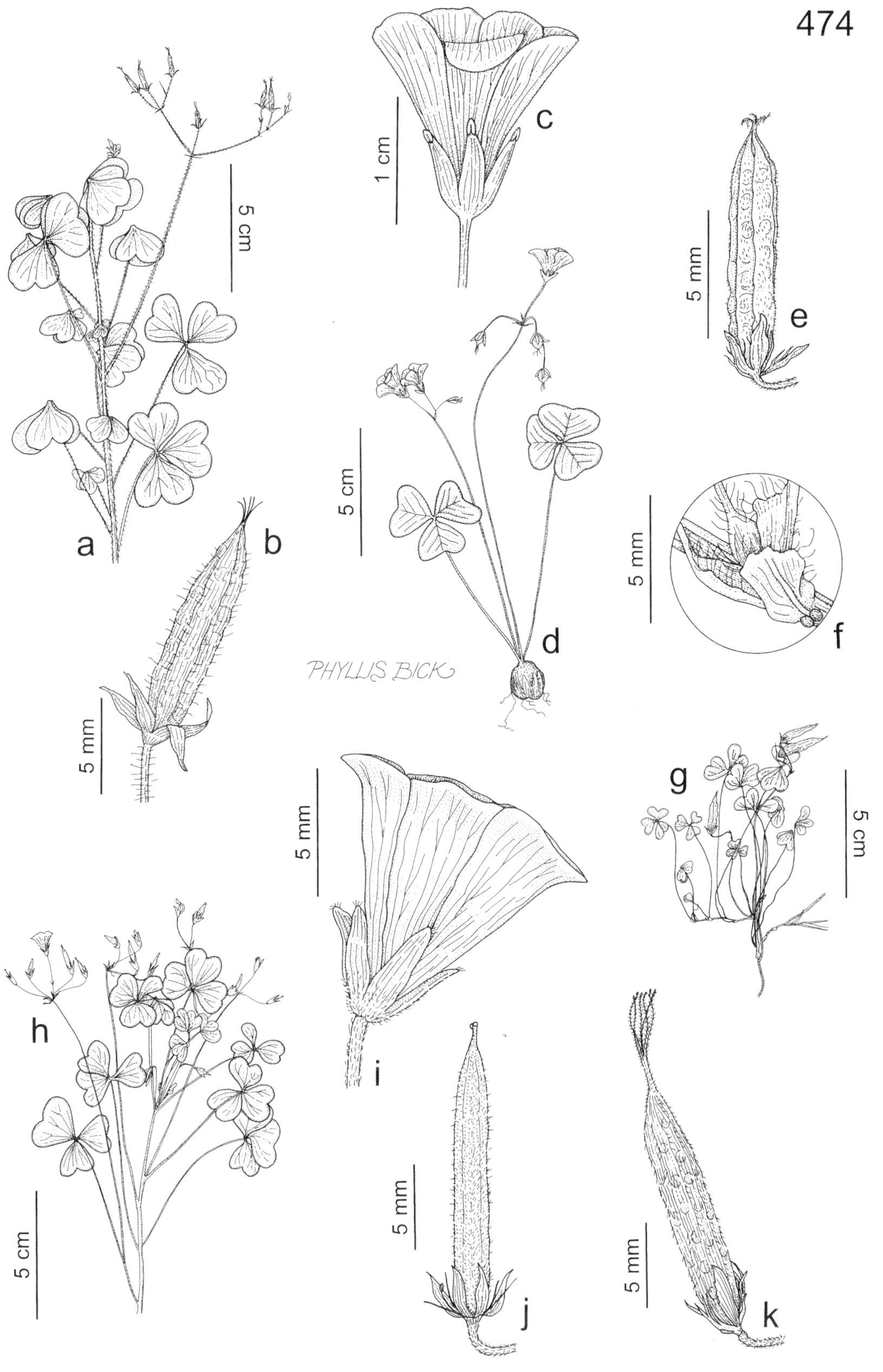

Plate 474. Oxalidaceae. *Oxalis stricta*, **a)** fertile stem, **b)** fruit. *Oxalis violacea*, **c)** flower, **d)** habit. *Oxalis corniculata*, **e)** fruit, **f)** node with stipules, **g)** habit. *Oxalis dillenii*, **h)** fertile stem, **i)** flower, **j)** fruit. *Oxalis florida*, **k)** fruit.

2170. Oxalis violacea   2171. Argemone albiflora   2172. Argemone mexicana

Turner [1990] followed), based on an invalid type designation; that name is better treated as a synonym of *O. dillenii* (D. B. Ward, 2004).

3. Oxalis florida Salisb. (yellow wood sorrel)
O. dillenii Jacq. ssp. *filipes* (Small) G. Eiten
O. filipes Small

Pl. 474 k; Map 2168

Plants perennial, but flowering the first year and often appearing annual, with taproots, usually lacking dark rhizomes, but sometimes developing stolons with age, lacking bulbs. Aerial stems 1(2), 5–20(–30) cm long, usually erect or strongly ascending, not rooting at the nodes, sparsely (often nearly glabrous) to moderately pubescent with fine, spreading to upward-curved, unicellular hairs, the pubescence sometimes somewhat denser near the stem base. Leaves basal and alternate, sometimes appearing fasciculate from the stem nodes, the petiole glabrous or sparsely pubescent with appressed to strongly ascending hairs. Stipules absent or, if present, then represented by slight thickenings or inconspicuous wings at the petiole base, these green. Leaflets 4–15 mm long, obcordate, the apical notch to 1/3 of the total length, the upper surface glabrous, the undersurface sparsely to moderately pubescent with mostly appressed hairs, yellowish green to grayish green, usually lacking purplish to brownish markings. Inflorescences umbellate with 2 or 3(–5) flowers, sometimes reduced to a solitary flower. Sepals 3–6 mm long, oblong-lanceolate to narrowly oblong-elliptic, green or translucent at the tip. Petals 5–11 mm long, yellow. Fruits 8–15 mm long, cylindrical at maturity, glabrous or sparsely pubescent with minute, curved hairs, sometimes mostly along the sutures. Seeds 1–2 mm long, brown, the ridges grayish or whitened. $2n=16$. April–September.

Uncommon in the southeastern portion of the Ozark Division; known thus far from relatively few specimens collected in Bollinger, Butler, Carter, Ripley, and Wayne Counties (eastern U.S. west to Missouri, Arkansas, and possibly Texas). Bottomland forests, mesic upland forests, and margins of ponds and lakes.

Eiten (1963) and D. B. Ward (2004) considered this taxon a subspecies of *O. dillenii*. The two are closely related and future taxonomic studies may support this view. *Oxalis dillenii* is most easily distinguished from *O. florida* when fruits are present. At least in Missouri, where populations are disjunct from the main center of distribution for the species, plants of *O. florida* also tend to be rather slender-stemmed, delicate plants that do not appear to form well-developed rhizomes.

4. Oxalis stricta L. (yellow wood sorrel)
O. stricta var. *bushii* (Small) Farw.
O. bushii Small
O. cymosa Small
O. europaea Jord.
O. europaea var. *bushii* (Small) Wiegand
O. europaea f. *cymosa* (Small) Wiegand
O. europaea f. *pilosella* Wiegand
O. europaea f. *subglabrata* Wiegand
O. europaea f. *vestita* Wiegand
O. europaea f. *villicaulis* Wiegand
O. fontana Bunge
O. fontana var. *bushii* (Small) H. Hara

Pl. 474 a, b; Map 2169

Plants annual or perennial, with taproots when young, but usually developing slender, white to pinkish-tinged rhizomes, these sometimes becoming thickened and blackish brown with age, lacking bulbs. Aerial stems 1(–3), 10–50(–90) cm long, variously erect or becoming prostrate, then not or rarely rooting at the nodes, sparsely to moderately pubescent with short, unicellular, appressed to upward-curved hairs and also at least a few (sometimes dense) relatively long, more or less spreading, slender, multicellular hairs (these often with reddish purple crosswalls), the pubescence sometimes denser near the stem tips, some of the hairs occasionally gland-tipped. Leaves basal (on young plants) and alternate, those on older stems often appearing fasciculate from the stem nodes, the petiole moderately to densely pubescent with appressed to strongly ascending hairs and often also sparse to dense, spreading, multicellular hairs. Stipules absent or, if present, then represented by slight thickenings or inconspicuous wings at the petiole base, these green. Leaflets (5–)8–30 mm long, obcordate, the apical notch to 1/3 of the total length, the upper surface usually glabrous, the surfaces sparsely to moderately pubescent with appressed to loosely appressed hairs, yellowish green to grayish green, usually lacking purplish to brownish markings. Inflorescences paniculate (with a central flower flanked by a pair of branches bearing 2 or more flowers), with 5–8(–15) total flowers, if rarely reduced to 3 flowers then the central flower sessile or much shorter-stalked than the outer pair. Sepals 2.5–6.0 mm long, oblong-lanceolate to narrowly oblong-elliptic, green or translucent at the tip. Petals 4–11 mm long, yellow. Fruits 8–20 mm long, cylindrical at maturity, variously glabrous to densely pubescent with short, appressed to more or less spreading, unicellular and/or multicellular hairs. Seeds 1.0–1.8 mm long, brown, the ridges often at least faintly grayish or whitened. $2n$=18, 24. May–October.

Scattered nearly throughout the state (nearly throughout the U.S.; Canada, Mexico; introduced widely in the Old World). Bottomland forests, mesic to dry upland forests, savannas, upland prairies, glades, and banks of streams and rivers; also pastures, fallow fields, gardens, railroads, roadsides, and disturbed areas.

This is the most common species of yellow wood sorrel in the state and also forms the longest stems. Specimens with poorly developed inflorescences can easily be confused with *O. dillenii*. In cases where only three flowers are produced in the inflorescences, plants of *O. stricta* usually have the central flower sessile or relatively short-stalked and sometimes have an additional pair of bracts along one or both lateral branches. Collectors should also check for the presence of slender stolons, which are easily lost when a specimen is collected. As in other species of *Oxalis*, plants of *O. stricta* are quite variable morphologically. Particularly distinctive are populations in which plants have relatively robust, erect, unbranched stems that are densely pubescent with spreading multicellular hairs. However, as shown in Steyermark's (1963) discussion, the stem pubescence character states do not correlate with variations in leaf, inflorescence, and sepal pubescence or degree of glandularity.

Oxalis illinoensis Schwegman is superficially similar to *O. stricta* in its relatively long, often erect stems, usual presence of multicellular hairs, relatively large leaves, and branched inflorescences. It differs from *O. stricta* in its larger flowers (corollas 12–18 mm long) and in the production of small white tubers along the rhizome. This species was described relatively recently (Schwegman, 1982) as a segregate of another large-flowered, but generally more eastern species, *O. grandis* Small, which lacks tubers, has more rounded, often purplish-tinged or -marked leaflets, and has less prominently red-marked petals. *Oxalis illinoensis* is endemic to a small region from southern Indiana to central Tennessee, west to southern Illinois and western Kentucky, where it occurs close to the Missouri border. Thus, Missouri botanists should search for populations of this large-flowered wood sorrel in the southeastern portion of the state.

5. Oxalis violacea L. (violet wood sorrel, purple wood sorrel)

Pl. 474 c, d; Map 2170

Plants perennial, with scaly, ovoid, brown or pinkish-tinged bulbs, occasionally with inconspicuous stolons. Aerial stems absent. Leaves all basal, few to numerous (occasionally absent at flowering), the petiole glabrous or sparsely to densely pubescent with short, more or less spreading, multicellular, gland-tipped hairs, especially toward the base. Stipules represented by inconspicuous, narrow, tapered (occasionally broader and truncate apically) wings at the petiole base, these translucent. Leaflets 5–20 mm long, broadly obcordate (usually wider than long), the apical notch to 1/3 of the total length and usually with an orange to orangish brown thickening (a concentration of oxalate crystals), the upper surface glabrous, green to dark green and often reddish- to purplish-tinged or with purplish red to reddish purple or dark purple cross-markings, the undersurface glabrous or sparsely pubescent with short, more or less spreading, multicellular, gland-tipped hairs, green to dark green, but usually mottled to strongly tinged with purplish red to reddish purple or dark purple, the leaflet base sometimes with scattered, stiff, unicellular, nonglandular hairs, usually lacking purplish to brownish markings. Inflorescences umbellate with (2–)4–16 flowers, rarely reduced to a solitary flower. Sepals 4–6 mm long, oblong-lanceolate to narrowly oblong-elliptic or occasionally narrowly ovate, with usually 2 bright orange thickenings (concentrations of oxalate crystals) at the tip. Petals 9–20 mm long, pink, violet, or purple, rarely white;. Fruits 4–6 mm long, oblong-ellipsoid, glabrous. Seeds 1.0–1.5 mm long, brown, the surface somewhat wrinkled or with a faint network of ridges, these not grayish or whitened. $2n=28$. April–July, sometimes also September–November.

Scattered nearly throughout the state, but less common in the Glaciated Plains Division (eastern U.S. west to South Dakota, Oklahoma, and Arizona). Openings and edges of mesic to dry upland forests, savannas, upland prairies, glades, tops and ledges of bluffs; also pastures, fallow fields, railroads, and roadsides.

Rare plants with white petals have been called f. *albida* Fassett, and those with gland-tipped petiolar hairs have been called f. *trichophora* Fassett. This species is very variable in leaflet size, petiole length, degree of red pigmentation, and corolla size and color. The inflorescences usually are held above the level of the leaf blades. Plants are sometimes encountered flowering while leafless, especially individuals that reflower from September to November.

PAPAVERACEAE (Poppy Family)

Plants annual, biennial, or perennial herbs (shrubs elsewhere), often with white or colored sap. Leaves alternate (often subopposite in *Stylophorum*; opposite elsewhere) or basal, sessile or short-petiolate. Stipules absent. Leaf blades simple or compound, the blade often 1–4 times pinnately or ternately lobed or dissected (the blade more or less palmately lobed in *Stylophorum*). Inflorescences terminal (sometimes also axillary in *Chelidonium*), of solitary flowers or loose clusters (these sometimes appearing as small panicles or umbels), the flowers usually subtended by small bracts. Flowers actinomorphic, hypogynous (the receptacle sometimes expanded and forming a thin, small cup- or saucer-shaped disc below the flower), perfect. Calyces of 2 or 3 sepals, these free or fused to well above the midpoint, similar in size and shape, mostly obovate, shed as the flower opens. Corollas of 4 petals (6 in *Argemone*, 7–12 in *Sanguinaria*, numerous in rare forms with doubled corollas), these free, similar in size and shape (somewhat unequal in *Sanguinaria*), narrowly oblanceolate to obovate, lacking a basal pouch or spur, variously colored. Stamens 12 to numerous, the filaments free, the anthers small, attached at or near the base, usually yellow. Pistil 1, composed of 2–18 carpels, the ovary superior, 1-locular or appearing 2- to multi-locular by intruding placentae, the placentation parietal. Style 1 per flower or absent, when present unbranched, the stigmas or stigma lobes as many as the carpels, variously shaped, sometimes persistent at fruiting. Ovules numerous. Fruits capsules that are dehiscent longitudinally by valves or by pores near the tip (circumscissile or breaking into indehiscent 1-seeded segments elsewhere), usually with numerous seeds. Seeds small, variously shaped, the embryo curved, the surface dark brown to black, sometimes with a pale aril or elaiosome (oil-bearing aril-like appendage attractive to ants). About 23 genera, about 230 species, nearly worldwide, most diverse in north-temperate regions.

The Papaveraceae are sometimes circumscribed broadly to include the genera here treated in the closely related Fumariaceae. For further discussion, see the treatment of that family.

Numerous members of the family are cultivated as ornamentals and many have a long history of medicinal use (see treatments of the genera).

1. Leaf 1 per plant (rarely 2), basal, palmately 3–7(9)-lobed (rarely unlobed); flower solitary, with 7–12 petals (except in rare doubled forms) . . . . 5. SANGUINARIA
1. Leaves several to numerous, basal and alternate, the blades 1–3 times pinnately and/or ternately compund and/or lobed; flowers solitary or in clusters, the petals 4 or 6 (except in rare doubled forms)

2. Leaf blades with the margins prickly, the surface often with lighter mottling along the main veins; fruits prickly . . . . 1. ARGEMONE
2. Leaf blades unarmed, the surface uniformly green; fruits unarmed (with bristly hairs in *Stylophorum*)
   3. Petals pink, red, or purple, occasionally white or pale lavender blue; stigma forming a stout, sessile, crownlike or disclike, 5–18-lobed structure, this especially evident at fruiting; fruits dehiscing by pores near the tip . . . 4. PAPAVER
   3. Petals yellow or yellowish orange; stigma shallowly to deeply 2- or 4-lobed, usually at least slightly elevated from the ovary and not forming a crownlike or disclike structure; fruits dehiscing longitudinally by valves from the tip or the base
      4. Sepals fused into a cap that is shed as a unit as the flower opens; receptacle expanded into a persistent, small, thin, cup-shaped to saucer-shaped disc under the flower; leaves pinnately and/or ternately 2–4 times dissected, the ultimate lobes slender; stigmas deeply 4-lobed, the lobes linear, erect or ascending . . . . 3. ESCHSCHOLZIA
      4. Sepals free, shed separately as the flower opens; receptacle not expanded into a noticeable disc under the flower; leaves pinnately deeply divided into 5 or 7 broad, rounded lobes, these variously scalloped, irregularly and bluntly toothed, or occasionally slightly lobed; stigmas shallowly 2- or 4-lobed, the lobes short, more or less spreading to downward-angled
         5. Sepals 8–12 mm long, glabrous; petals 8–14 mm long; fruits narrowly cylindric, glabrous, dehiscent from the base . . . . 2. CHELIDONIUM
         5. Sepals 12–15 mm long, hairy; petals 20–30 mm long; fruits ellipsoid, bristly-hairy, dehiscent from the tip . . . . 6. STYLOPHORUM

1. **Argemone** L. (prickly poppy)
(Ownbey, 1958)

Plants annual or biennial (perennial herbs elsewhere), with taproots; sap yellow (white to orange elsewhere). Aerial stems strongly ascending, usually sparsely to moderately prickly, also somewhat glaucous. Leaves basal and alternate on the stems, the stem leaves progressively shorter toward the stem or branch tips, all sessile, at least the uppermost leaves usually with a pair of rounded auricles at the base, clasping the stem. Leaf blades moderately to deeply pinnately lobed with relatively broad, often U-shaped sinuses, the lobes variously shaped, with irregularly toothed margins, these armed with staw-colored, slender prickles, the surfaces glabrous or armed with scattered slender prickles along the main veins, glaucous, often with lighter mottling along the main veins. Flowers terminal, loose clusters (sometimes appearing as small panicles) at the stem or branch tips, sometimes reduced to a solitary flower, the flowers short- to long-stalked, the stalk erect or ascending at flowering, mostly subtended by 1 or 2 bracts, these similar in appearance to leaves but smaller, the receptacle slightly expanded at the tip but not forming a cup or disc. Sepals 3, free, shed individually as the flower opens, the body 9–18 mm long, oblong elliptic, broadly pointed at the tip, sometimes armed with prickles, also with a prominent, ascending, conic, dorsal horn near the tip. Petals 4, broadly obovate, broadly rounded and often somewhat uneven or slightly ruffled at the tip, white to pale yellow or yellow. Stamens numerous. Ovary lacking a well-differentiated style at flowering, the stigma more or less capitate, with 4–6, shallow, spreading lobes. Fruits erect or ascending, oblong-elliptic to elliptic, variously nearly truncate to angled to a sharply pointed tip, the surface usually with numerous stout, ascending prickles, these often attached to shallow nipplelike structures, also longitudinally 4–6-ribbed, dehiscing to about 1/3 of the way from the tip by valves, leaving the persistent stigma attached to a network of vascular tissue exposed around the seeds. Seeds 1.5–2.0 mm long, more or less globose but abruptly tapered to a small point at the base, the short aril inconspicuous, the surface with a prominent network of ridges and pits, brown to reddish brown, not shiny. Thirty-two species, North America, South America, Hawaii; introduced in the Old World.

1. Petals white; stamens 100 or more . . . . 1. A. ALBIFLORA
1. Petals yellow or rarely pale yellow to cream-colored; stamens 20–50 . . . . 2. A. MEXICANA

1. Argemone albiflora Hornem. (white prickly poppy)

Pl. 475 c–e; Map 2171

Stems 40–100(–150) cm long, sparsely to moderately prickly. Leaf blades 4–25 cm long, the surfaces glabrous or more commonly sparsely prickly along the main veins, glaucous, lacking pale mottling along the veins. Flowers closely or distantly subtended by the bracts. Buds 12–18 mm long, broadly ellipsoid to nearly globose, the horns 3–6(–10) mm long, unarmed below the tip or sparsely prickly. Petals 20–40 mm long, white (occasionally pinkish- or lavender-tinged elsewhere). Stamens 100–250. Fruits 10–17 mm long, narrowly ellipsoid to oblong-ellipsoid or oblong in outline, the surface moderately prickly. $2n=28$. May–September.

Introduced, uncommon and widely scattered in the southern half of the state (native of the southeastern U.S., introduced in the midwestern and northeastern states). Railroads, roadsides, and open, often sandy, disturbed areas.

Ownbey (1958) recognized two subspecies within *A. albiflora* that can be difficult to distinguish.

1. Fruits oblong in outline to oblong-ellipsoid, the prickles more or less similar in size, all relatively large and relatively evenly distributed . . . . 1A. SSP. ALBIFLORA
1. Fruits usually narrowly ellipsoid, the prickles not uniform in size and distribution, the scattered larger ones interspersed with irregularly spaced, denser shorter ones of varying sizes . . . . 1B. SSP. TEXANA

1a. ssp. albiflora

Pl. 475 d

Stems sparsely prickly. Sepals with the horns 3–5 mm long, unarmed except at the tip. Fruits oblong in outline to oblong-ellipsoid, the surface with prickles more or less similar in size, all relatively large and relatively evenly distributed. $2n=28$. June–September

Introduced, uncommon and widely scattered in the southern half of the state (native of the southeastern U.S.; introduced north to Missouri, Illinois, and Connecticut). Railroads, roadsides, and open, disturbed areas.

1b. ssp. texana G.B. Ownbey

Stems sparsely to moderately prickly. Sepals with the horns 4–6(–10) mm long, unarmed below the tip or sparsely prickly. Fruits usually narrowly ellipsoid, the surface with prickles not uniform in size and distribution, the scattered larger ones interspersed with irregularly spaced, denser shorter ones of varying sizes. $2n=28$. May–August.

Introduced, uncommon in the southern portion of the state (native of Texas; introduced north to Missouri). Railroads, roadsides, and open, disturbed areas.

2. Argemone mexicana L. (devil's fig, Mexican prickly poppy)

Pl. 475 h–j; Map 2172

Stems 25–80 cm long, sparsely prickly or occasionally unarmed. Leaf blades 4–25 cm long, the surfaces glabrous or sparsely prickly along the main veins, glaucous, the upper surface with pale mottling along the main veins. Flowers usually closely subtended by the bracts. Buds 10–15 mm long, broadly ellipsoid to nearly globose, the horns 5–10 mm long, unarmed below the tip or sparsely prickly. Petals 15–35 mm long, yellow or rarely pale yellow to cream-colored. Stamens 20–50. Fruits 25–45 mm long, oblong in outline to oblong-ellipsoid or ellipsoid, the surface unarmed or sparsely to moderately prickly. $2n=28$. May–August.

Introduced, uncommon and sporadic, mostly in the southern half of the state (native of the southeastern U.S., Mexico, Central America, Caribbean Islands; introduced north to Nebraska and Massachusetts). Railroads and open, disturbed areas.

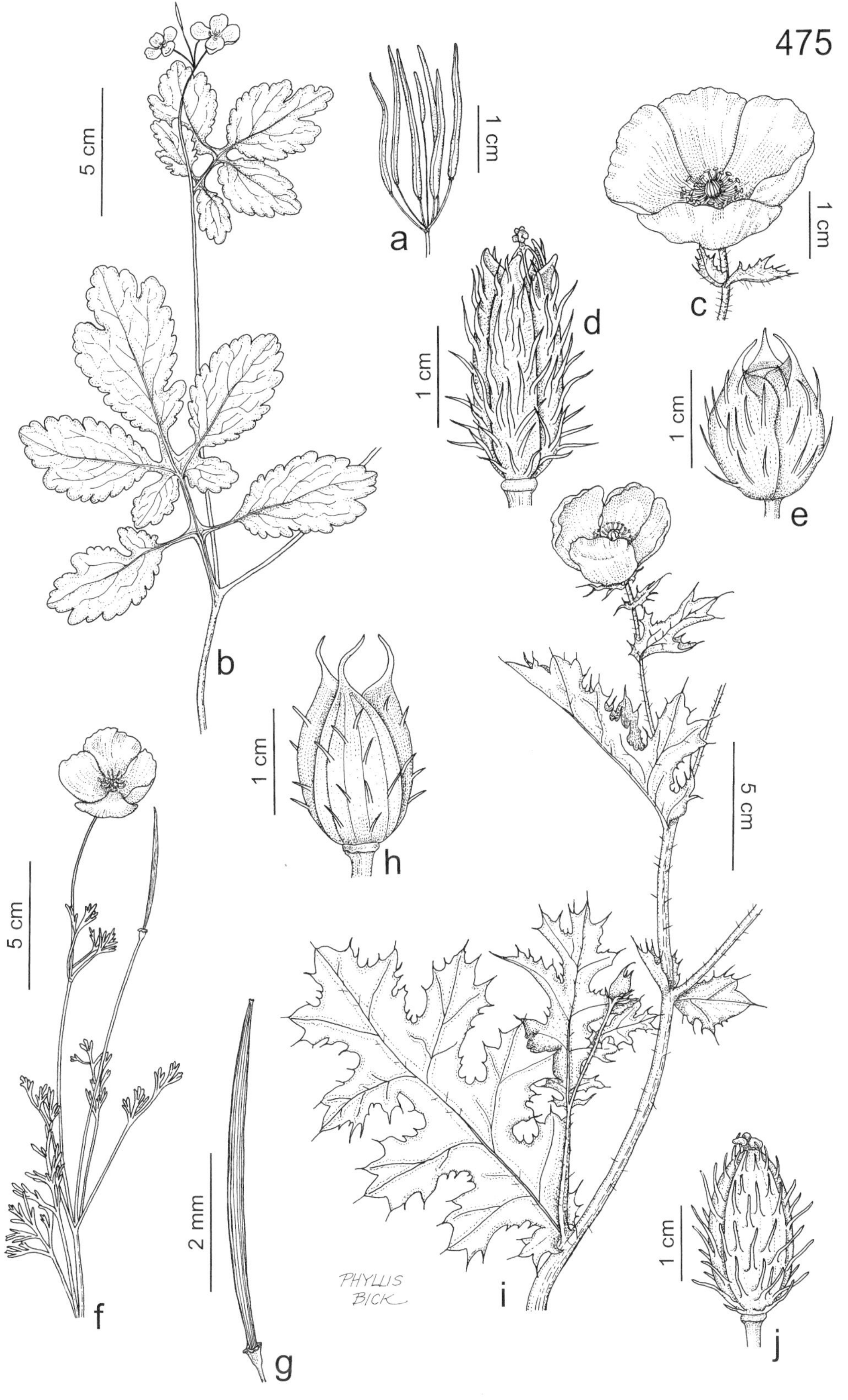

Plate 475. Papaveraceae. *Chelidonium majus*, **a)** fruits, **b)** fertile stem. *Argemone albiflora*, **c)** stem tip with flower, **d)** fruit, **e)** bud. *Eschscholzia californica*, **f)** habit, **g)** fruit. *Argemone mexicana*, **h)** bud, **i)** fertile stem, **j)** fruit.

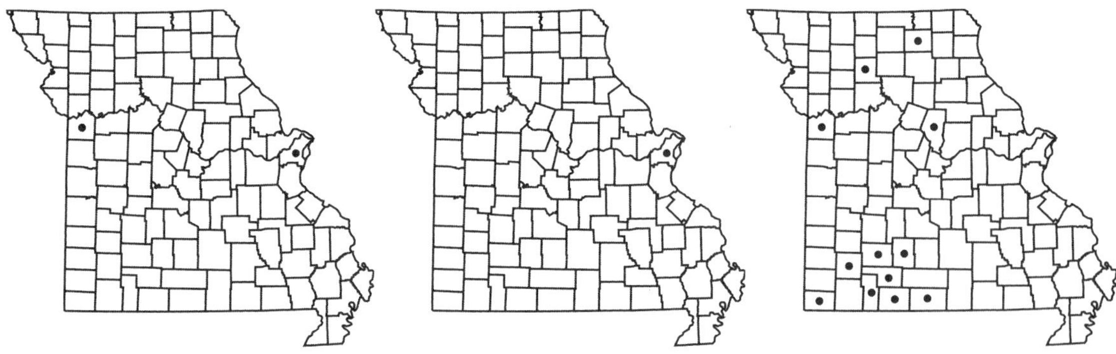

2173. Chelidonium majus 2174. Eschscholzia californica 2175. Papaver dubium

2. Chelidonium L.

One species, native of Europe, Asia, introduced in North America.

1. Chelidonium majus L. (celandine, greater celandine, tetterwort)

Pl. 475 a, b; Map 2173

Plants perennial herbs, with thick, sometimes branched rhizomes; sap yellow to yellowish orange. Aerial stems 30–80 cm long, sometimes branched, loosely to strongly ascending, sparsely pubescent with fine, more or less spreading, multicellular hairs. Leaves basal (several) and alternate (few to several) along the stems, 15–35 cm long, the basal leaves long-petiolate the stem leaves mostly short-petiolate to sessile. Leaf blades pinnately deeply lobed or compound into 3–7 lobes or leaflets, these oblong-obovate, rounded at the tip, the margins irregularly scalloped and/or bluntly toothed, usually also shallowly to deeply few-lobed, the upper surface green, glabrous, the undersurface pale and glaucous, glabrous or sparsely hairy. Inflorescences terminal and sometimes also axillary, umbellate, the umbels sessile or stalked, few-flowered, occasionally reduced to a solitary flower, the individual flower stalks 0.5–3.5 cm long, ascending, slightly expanded at the tip but not forming a cup or disc, each with a bract at the base, this 0.5–2.0 mm long, broadly oblong-ovate, glabrous. Sepals 2, free, shed individually as the flower opens, 8–12 mm long, broadly elliptic-ovate and deeply concave (cupped around the flower), broadly pointed at the tip. Petals 4, 8–14 mm long and wide, broadly obovate, rounded at the tip, yellow. Stamens 12 to numerous. Ovary tapered to a minute, persistent style 0.5–1.0 mm long, the stigma more or less capitate, with 2 shallow, ascending lobes (appearing notched). Fruits 2–5 cm long, erect or ascending, narrowly cylindric, glabrous, dehiscing longitudinally from the base by 2 valves. Seeds 1.2–1.8 mm long, ovoid, with a short, winglike aril along 1 side, the surface otherwise with a network of relatively coarse but shallow ridges and pits, reddish brown to brown, shiny. $2n=12$. April–August.

Sepals 8–12 mm long, glabrous; petals 8–14 mm long; fruits narrowly cylindric, glabrous, dehiscent from the base

Introduced, uncommon, known thus far only from Franklin, Jackson, and St. Louis Counties (native of Europe, Asia; introduced widely in the northeastern and northwestern U.S.; Canada). Gardens and disturbed areas.

The brightly colored sap is a skin irritant and has been used medicinally to remove warts. The species is sometimes cultivated as an ornamental in gardens, but tends to spread aggressively by seeds at suitable sites.

3. Eschscholzia Cham. (gold poppy)

Twelve species, western U.S.; Mexico.

Gold poppies are a conspicuous component of the spring bloom in the southwestern states, especially in the Sonoran Desert and southern California grasslands. The generic name has been spelled *Eschscholtzia* in some of the older botanical literature, but Chamisso's original spelling is to be retained, although the name honors Johann F. G. von Eschscholtz (note extra letter t). See Ernst (1962) for further discussion.

1. Eschscholzia californica Cham. **ssp. californica** (California poppy)
Pl. 475 f, g; Map 2174

Plants annual (perennial herbs elsewhere), with taproots; sap colorless or with an orangish tinge. Aerial stems 5–40(–60) cm long, usually unbranched, loosely to strongly ascending, glabrous, often glaucous. Leaves basal and sometimes also 1 or few alternate along the stems, 15–45(–60) cm long, the basal leaves short- to long-petiolate, the stem leaves mostly short-petiolate to sessile. Leaf blades pinnately and/or ternately 2–4 times dissected into numerous lobes, these narrowly oblong to linear, bluntly to sharply pointed at the tip, the margins of the ultimate segments entire or with 1 or 2 basal lobes, the surfaces glabrous, green or more commonly 1 or both surfaces pale and slightly to moderately glaucous. Inflorescences terminal, few-flowered, loose clusters or reduced to a solitary flower, the individual flower stalks 3–15 cm long, erect or ascending, usually lacking bracts, the receptacle expanded into a persistent, small, thin, cup-shaped to saucer-shaped disc under the flower. Sepals 2, fused into a cap that is shed as a unit as the flower opens, 10–40 mm long, the unit conic, tapered to a sharply pointed tip, often with a slender extension at the very tip. Petals 4, 20–60 mm long, 15–50 mm wide, broadly obovate-obtriangular to obtriangular, broadly rounded to more or less truncate at the tip, yellow to orangish yellow, usually orange toward the base. Stamens numerous. Ovary lacking a well-differentiated style at flowering, the stigma deeply 4-lobed, the lobes linear, erect or ascending. Fruits 30–90 mm long, erect or ascending, narrowly cylindric, slightly tapered into a stout, beaklike sterile tip, glabrous, longitudinally 10-ribbed, dehiscing longitudinally from the base by 2 valves (the seeds often dispersed explosively). Seeds 1.4–1.8 mm long, globose to broadly ovoid or broadly ellipsoid, lacking an aril, the surface with a network of coarse ridges and pits, greenish brown to black, not shiny. $2n=12$. April–July.

Introduced, uncommon, known thus far only from the city of St. Louis (native of the western U.S. from California to Washington; introduced widely but sporadically farther east and in Canada). Railroads.

This taxon is an ingredient in some wildflower seed mixes and thus might be found elsewhere in the state in the future. However, it probably does not persist long at any site in the region. The other subspecies, ssp. *mexicana* (Greene) C. Clark (*E. mexicana* Greene) occurs in the southwestern United States and adjacent Mexico, and differs in its usually annual habit, unlobed (vs. 2-lobed) cotyledons, and in having a less conspicuous expanded receptacle tip.

4. **Papaver** L. (poppy)
(Kiger, 1975, 1997)

Plants annual (perennial herbs elsewhere), with taproots; sap white, orange, or red. Aerial stems loosely to strongly ascending (absent elsewhere), glabrous or hairy, sometimes glaucous. Leaves basal and alternate on the stems, the basal leaves short- to long-petiolate (sessile in *P. somniferum*), the stem leaves mostly short-petiolate to sessile. Leaf blades pinnately lobed (often merely coarsely toothed in *P. somniferum*), the lobes toothed to dissected, variously shaped, the surfaces glabrous or hairy, sometimes glaucous. Flowers solitary, terminal on the stem or branches, long-stalked, the stalk erect or nodding in bud, erect or ascending at flowering, lacking bracts, the receptacle slightly expanded at the tip but not forming a cup or disc. Sepals 2, free, shed individually as the flower opens, 10–35 mm long, broadly elliptic-ovate and deeply concave (cupped around the flower), broadly pointed at the tip. Petals 4, broadly obovate, broadly rounded and often somewhat uneven or slightly ruffled at the tip, pink, red, or purple, occasionally white, pale lavender blue, or orange, often with a pronounced dark or light spot at the base. Stamens numerous. Ovary lacking a well-differentiated style at flowering, the stigma forming a stout, sessile, crownlike or disclike, 5–18-lobed structure, this especially evident at fruiting. Fruits erect or ascending, narrowly to broadly obovoid to nearly globose, more or less truncate to slightly convex at the tip, crowned by the persistent stigmatic stucture, glabrous or hairy, sometimes longitudinally ribbed, dehiscing by pores near the tip. Seeds 0.4–0.7 mm long, kidney-shaped, lacking an aril, the surface with a network of ridges and pits, dark brown to black, shiny. Seventy to 100 species, North America, Europe, Asia, Africa, Australia.

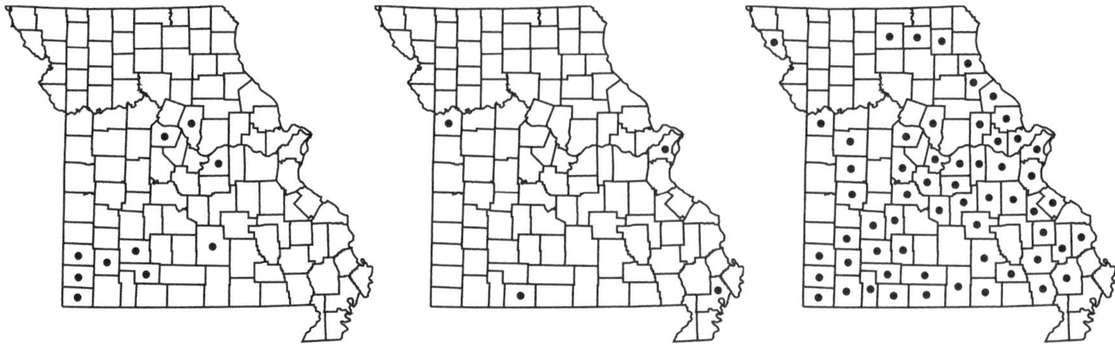

2176. Papaver rhoeas   2177. Papaver somniferum   2178. Sanguinaria canadensis

A number of poppy species are cultivated as bedding plants in gardens or as components of wildflower meadows or highway median plantings. In addition to the three species that escape from cultivation in Missouri that are treated below, three other species might in the future become established in the state. *Papaver atlanticum* (Ball) Coss. (Moroccan poppy) is similar to *P. dubium*, but differs in its perennial habit, yellow (vs. dark purple to black) anthers, and strongly ribbed fruits. *Papaver glaucum* Boiss. & Hausskn. (tulip poppy) is similar to *P. somniferum*, but differs in its smaller buds (1–2 vs. 2–4 cm long), and smaller fruits (1.5–2.0 cm long at maturity). *Papaver orientale* L. (Oriental poppy) is similar to *P. rhoeas*, but differs in its perennial habit, bigger buds (2.0–3.5 vs. 0.5–2.0 cm long), and bigger petals (4–8 cm long).

1. Stem leaves with clasping bases . . . . 3. P. SOMNIFERUM
1. Stem leaves not clasping at the base
   2. Fruits narrowly obovoid; flower stalks with appressed hairs toward the tip, these grading into spreading hairs toward the base; stigmatic crown with (4–)7–9 lobes . . . . 1. P. DUBIUM
   2. Fruits broadly obovoid to nearly globose; flower stalks with spreading hairs throughout; stigmatic crown with (5–)8–18 lobes . . . . 2. P. RHOEAS

1. Papaver dubium L. (blind eyes, longhead poppy)

Pl. 476 a–c; Map 2175

Sap usually white. Stems 15–60 cm long, moderately pubescent with relatively long, spreading, broad-based hairs. Basal leaves with the blade 3–8 cm long, 1 or 2 times pinnately deeply lobed (rarely fully compound toward the base), variously oblanceolate to elliptic or ovate in outline, the ultimate segments lanceolate to oblong-lanceolate, or narrowly elliptic, tapered to sharply pointed tips, the margins otherwise entire or with a few coarse teeth, the surfaces and margins sparsely to densely hairy with relatively coarse hairs. Stem leaves similar to the basal ones, sessile or short-petiolate, with shorter blades, the margins sometimes more densely toothed, not clasping the stems at the base. Flower stalks 12–20 cm long, with relatively long, spreading hairs toward the base, these grading into appressed hairs toward the tip. Sepals 10–18 mm long, with relatively coarse, more or less appressed hairs. Petals 15–25 mm long, red to pinkish red or orange, usually with a pronounced dark spot at the base. Anthers dark purple to black. Stigmatic crown with (4–)7–9 lobes. Fruits 12–22 mm long, narrowly obovoid, longitudinally lined or slightly ribbed, glabrous, occasionally slightly glaucous when young. $2n=42$. May–July.

Introduced, uncommon and sporadic, mostly in the southwestern portion of the state (native of Europe, Asia; introduced widely but sporadically in the eastern and northwestern U.S., Canada). Glades; also pastures, railroads, roadsides, and open, disturbed areas.

Kiger (1997) noted that in its native range, *P. dubium* is a polyploid complex that has been split into several morphologically intergrading subspecies. He stated that these have no geographic integrity among populations introduced in North America and thus are impossible to adequately

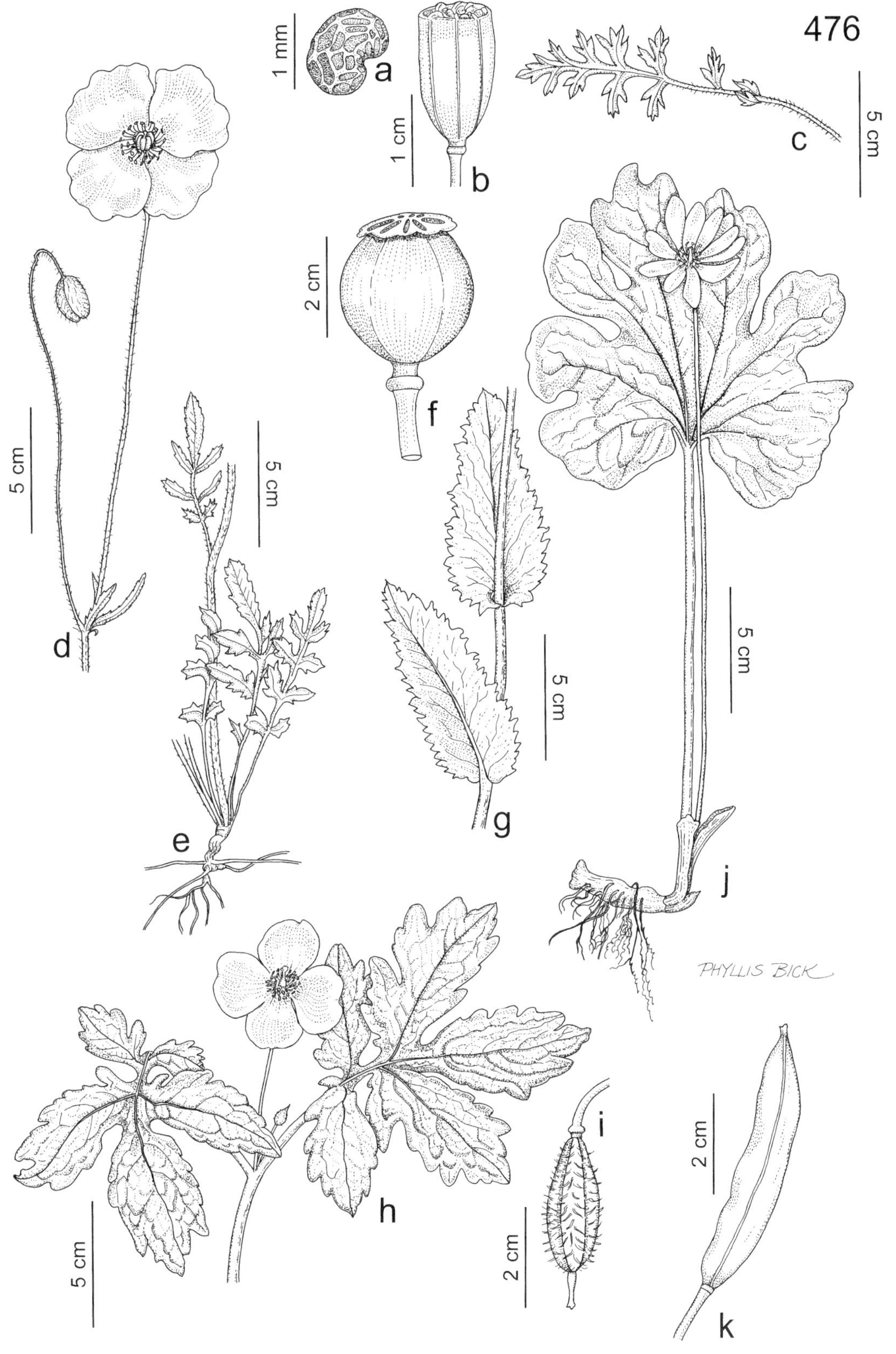

Plate 476. Papaveraceae. *Papaver dubium*, **a)** seed, **b)** fruit, **c)** leaf. *Papaver rhoeas*, **d)** stem tip with flower and bud, **e)** basal portion of plant with leaves. *Papaver somniferum*, **f)** fruit, **g)** median portion of stem with leaves. *Stylophorum diphyllum*, **h)** fertile stem, **i)** fruit. *Sanguinaria canadensis*, **j)** habit, **k)** fruit.

discriminate and thus unworthy of formal recognition.

2. Papaver rhoeas L. (corn poppy, field poppy, Shirley poppy)

Pl. 476 d, e; Map 2176

Sap white or pale orange. Stems 20–80 cm long, moderately pubescent with relatively long, spreading, broad-based hairs. Basal leaves with the blade 3–8 cm long, 1 or 2 times pinnately deeply lobed (rarely fully compound toward the base), variously oblanceolate to elliptic or ovate in outline, the ultimate segments lanceolate to oblong-lanceolate, or narrowly elliptic, tapered to sharply pointed tips, the margins otherwise entire or with a few coarse teeth, the surfaces and margins moderately to densely hairy with relatively coarse hairs. Stem leaves similar to the basal ones, sessile or short-petiolate, with shorter blades, the margins sometimes more densely toothed, not clasping the stems at the base. Flower stalks 12–25 cm long, with relatively long, spreading hairs throughout. Sepals 8–20 mm long, with relatively coarse, loosely ascending hairs. Petals 20–40 mm long, red to pink or purple, usually with a pronounced dark spot at the base, sometimes white or streaked with white, rarely orange. Anthers yellow, usually brownish-tinged. Stigmatic crown with (5–)8–18 lobes. Fruits 12–20 mm long, broadly obovoid to nearly globose, longitudinally lined or slightly ribbed, glabrous, occasionally slightly glaucous when young. $2n=14$. May–October.

Introduced, uncommon, mostly in the southwestern and central portions of the state (native of Europe, Asia, Africa; introduced widely but sporadically in the U.S., Canada). Glades and banks of streams and spring branches; also pastures, old mines, railroads, roadsides, and open, disturbed areas.

This red-flowered species grows abundantly in some European meadows (and cemeteries) and was the inspiration for the famous World War I era poem “In Flanders Fields” written in 1915 by the Canadian physician and officer, John McCrae (1919). It subsequently became an international symbol for the sacrifices of Armed Services war veterans.

3. Papaver somniferum L. (common poppy, opium poppy)

Pl. 476 f, g; Map 2177

Sap white. Stems 30–100 cm long, glabrous or nearly so, usually somewhat glaucous. Basal and stem leaves sessile, the blade 15–35 cm long (the stem leaves progressively shorter to the stem or branch tips), often appearing strongly crisped or corrugated, sometimes with a few relatively deep lobes toward the base, but more frequently with coarse, irregular, jagged teeth along the margins, variously narrowly obovate to oblong-obovate or ovate in outline, the stem leaves with rounded auricles at the base clasping the stem, the surfaces glaucous, glabrous or the undersurface occasionally with sparse to moderate bristly hairs along the midvein. Flower stalks 12–25 cm long, sparsely pubescent with relatively long, spreading, broad-based hairs, sometimes only toward the tip. Sepals 18–35 mm long, usually glabrous (rarely with a few spreading hairs near the base), glaucous. Petals 30–60 mm long, white, pink, red, or purple, often with a pronounced dark (less commonly light) spot at the base. Anthers yellow. Stigmatic crown with 8–18 lobes. Fruits 25–60 mm long, broadly obovoid to nearly globose, sometimes longitudinally faintly or finely lined but not ribbed, glabrous, glaucous. $2n=22$. May–September.

Introduced, (cultigen of probable origin in Europe or Asia; introduced widely in the northern hemisphere). Railroads, roadsides, and open, disturbed areas.

Papaver somniferum has a very long history of ethnobotanical use. Currently, it is grown as a garden annual in sunny garden beds (technically, this is illegal, because of the plant’s drug properties). It also is the source of poppy seeds, which are used widely as a flavorant in baked goods and other foods. Poppyseed oil is extracted from the seeds and used both in cooking and as a carrier in oil-based paints. Alkaloids in the sap of this species have strong narcotic properties and are the source of opium and its derivatives, including heroin, morphine, and codeine. Although morphine and codeine have legitimate pharmaceutical uses, the opiates also are the basis of an immense, illegal, international drug trade.

5. Sanguinaria L.

One species, eastern U.S. Canada.

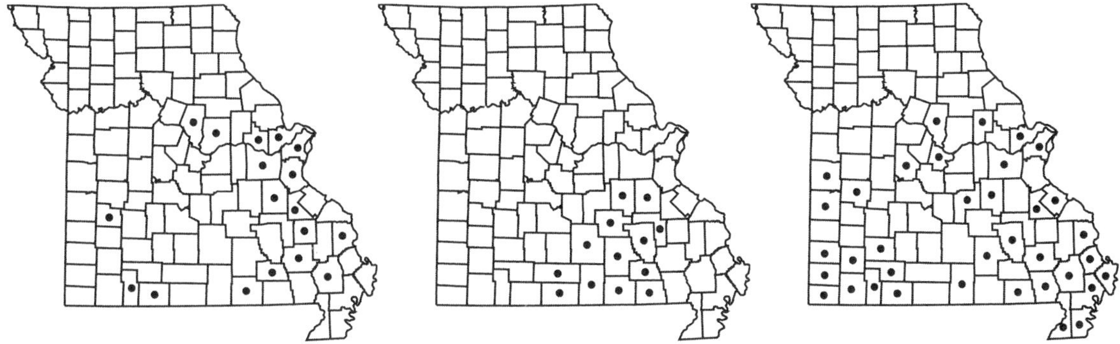

2179. Stylophorum diphyllum 2180. Parnassia grandifolia 2181. Passiflora incarnata

1. Sanguinaria canadensis L. (bloodroot, red puccoon)
S. canadensis var. *rotundifolia* (Greene) Fedde

Pl. 476 j, k; Map 2178

Plants perennial herbs, with thick, sometimes branched rhizomes; sap reddish orange to red. Aerial stems absent. Leaves 1(2) per year from the apical portion of the rhizome or rhizome branches, the petioles 5–15 cm long, erect. Leaf blades 6–17 cm long, 8–20 cm wide, curved around the scape, more or less circular to broadly kidney-shaped in outline, variously shallowly 3-lobed at the tip (rarely unlobed) or with 3–7(–9) palmate lobes, cordate to deeply cordate at the base, the lobes rounded to bluntly and broadly pointed, the margins otherwise shallowly scalloped or wavy, glabrous, the upper suface green, the undersurface pale, somewhat glaucous. Flowers solitary, long-stalked, the stalk 5–15 cm long at flowering, erect, lacking subtending bracts, glabrous, the receptacle slightly expanded at the tip but not forming a cup or disc. Sepals 2, free, shed individually as the flower opens, 8–12 mm long, elliptic-ovate, bluntly pointed at the tip. Petals 7–12 (rarely numerous), 12–30 mm long, 5–12 mm wide, narrowly oblanceolate to oblong-oblanceolate, rounded at the tip, white. Stamens numerous. Ovary tapered to a persistent style 1–3 mm long, the stigma more or less capitate, with 2, shallow, downward-angled lobes. Fruits 30–55 mm long, erect, narrowly ellipsoid, glabrous, somewhat glaucous, dehiscing longitudinally from the base by 2 valves. Seeds 3.0–3.8 mm long, ovoid to nearly globose, with a crestlike aril along 1 side, the surface otherwise smooth or faintly pebbled, reddish brown to nearly black, shiny. $2n=18$. March–April.

Scattered in the Ozark and Ozark Border Divisions, uncommon elsewhere in the state (eastern U.S. west to North Dakota and Texas; Canada). Bottomland forests, mesic upland forests, banks of streams and rivers, and bases and ledges of bluffs.

Plants with the leaves unlobed or shallowly 3-lobed at the tip have been called var. *rotundifolia*, but this trivial variant grades completely into plants with more divided leaves. Rare plants with doubled flowers (with numerous petals) have been called f. *multiplex* (E.H. Wilson) Weath. and a form with pink petals has been called f. *colbyorum* Benke.

Sanguinaria canadensis is an attractive wildflower when grown in shade and woodland gardens, but flowers early and for a relatively brief time, and dies back soon after the fruits mature. Native Americans and early European settlers used bloodroot internally for sore throats, externally as a reddish skin dye and antiseptic, and had a number of other uses for the plant. The sap can irritate the skin and has been used to treat warts and skin cancers. More recently, the alkaloid sanguinarine (first extracted and characterized from *Sanguinaria*) has been used commercially for its anti-plaque and antiseptic properties as an ingredient in some toothpastes and mouthwashes. This has not been without controversy, as some authorities have argued that such alkaloids may be carcinogenic when ingested. The uses of bloodroot extract and its potential hazards were discussed by Duke (1993), who also noted that excessive harvesting of plants from wild populations might eventually prove detrimental to the species.

6. **Stylophorum** Nutt. (wood poppy)

Three species, U.S.; Canada, Asia.

1. Stylophorum diphyllum (Michx.) Nutt. (celandine poppy, wood poppy)

Pl. 476 h, i; Map 2179

Plants perennial herbs, with thick, sometimes branched rhizomes; sap yellow to yellowish orange. Aerial stems 30–50 cm long, sometimes branched, loosely to strongly ascending, often purplish-tinged at the base, moderately to densely pubescent with fine, more or less spreading, multicellular hairs. Leaves with the upper surface green, glabrous, the undersurface pale and glaucous, sparsely hairy; basal leaves several, 30–50 cm long, long-petiolate (the petiole often purplish-tinged at the base), the blade pinnately deeply lobed or compound into 5 or 7 lobes or leaflets, these oblong-obovate, rounded at the tip, the margins irregularly scalloped and/or bluntly toothed, sometimes also shallowly few-lobed; stem leaves 2, subopposite, similar to the basal leaves, but somewhat smaller and usually short-petiolate. Inflorescences terminal, umbellate, the umbel sessile, few-flowered, sometimes reduced to a solitary flower, the individual flower stalks 2–5 cm long, ascending, hairy, slightly expanded at the tip but not forming a cup or disc, each with a bract at the base, this 3–5 mm long, narrowly oblong-elliptic, hairy along the entire margins and usually also sparsely so on the undersurface. Sepals 2, free, shed individually as the flower opens, 12–15 mm long, broadly elliptic-ovate and deeply concave (cupped around the flower), broadly pointed at the tip. Petals 4, 20–30 mm long and wide, broadly obovate, rounded at the tip, yellow. Stamens numerous. Ovary tapered to a persistent style 3–6 mm long, the stigma more or less capitate, with 3 or 4(5), shallow, spreading lobes. Fruits 20–30 mm long, nodding, ellipsoid, moderately to densely pubescent with relatively stout, bristly, multicellular hairs, dehiscing longitudinally from the tip by 4(5) valves. Seeds 1.6–2.1 mm long, ovoid, with a crestlike aril along 1 side, the surface otherwise with a network of fine ridges and pits, light brown to brown, not shiny. $2n=20$. April–June.

Scattered to common in the eastern half of the Ozark and Ozark Border Divisions, less common elsewhere in the Ozarks and the Big Rivers, apparently absent from most of the Glaciated Plains, Unglaciated Plains, and Mississippi Lowlands Divisions (Missouri to Michigan and Ohio south to Arkansas, Alabama, Georgia, and Virginia; Canada). Bottomland forests, mesic upland forests, banks of streams and rivers, and bases and ledges of bluffs.

This attractive wildflower has become popular in native shade gardens. However, it can spread aggressively by seed in suitable habitats. Vegetatively, it can be very difficult to distinguish from *Chelidonium majus*.

PARNASSIACEAE (Grass-of-Parnassus Family)

One genus (but see below), about 50 species, North America, Europe, Asia.

The genus *Parnassia* has been included in the Saxifragaceae by most authors (Steyermark, 1963; Cronquist, 1991). However, recent phylogenetic studies (Morgan and Soltis, 1993; Soltis and Soltis, 1997; Zhang and Simmons, 2006) have shown that it is more closely related to the Celastraceae. The present treatment deviates from the standard practice in this manual of following the familial classification of Cronquist (1981, 1991) in treating *Parnassia* in its own family to reflect this new understanding of intergeneric relationships in the Saxifragales, as well as to make the Saxifragaceae more easily circumscribed and keyed.

Lepuropetalon spathulatum Muhl. ex Elliott, is a tiny annual with a highly disjunct distribution in wet, sandy, disturbed areas in the southeastern United States, Mexico, and portions of South America. It is often placed in the Parnassiaceae, but sometimes segregated into its own family, the Lepuropetalaceae. The biology and systematics of this odd plant, which is easily overlooked in the field, require further study.

1. Parnassia L. (grass-of-Parnassus)

About 50 species, North America, Europe, Asia.

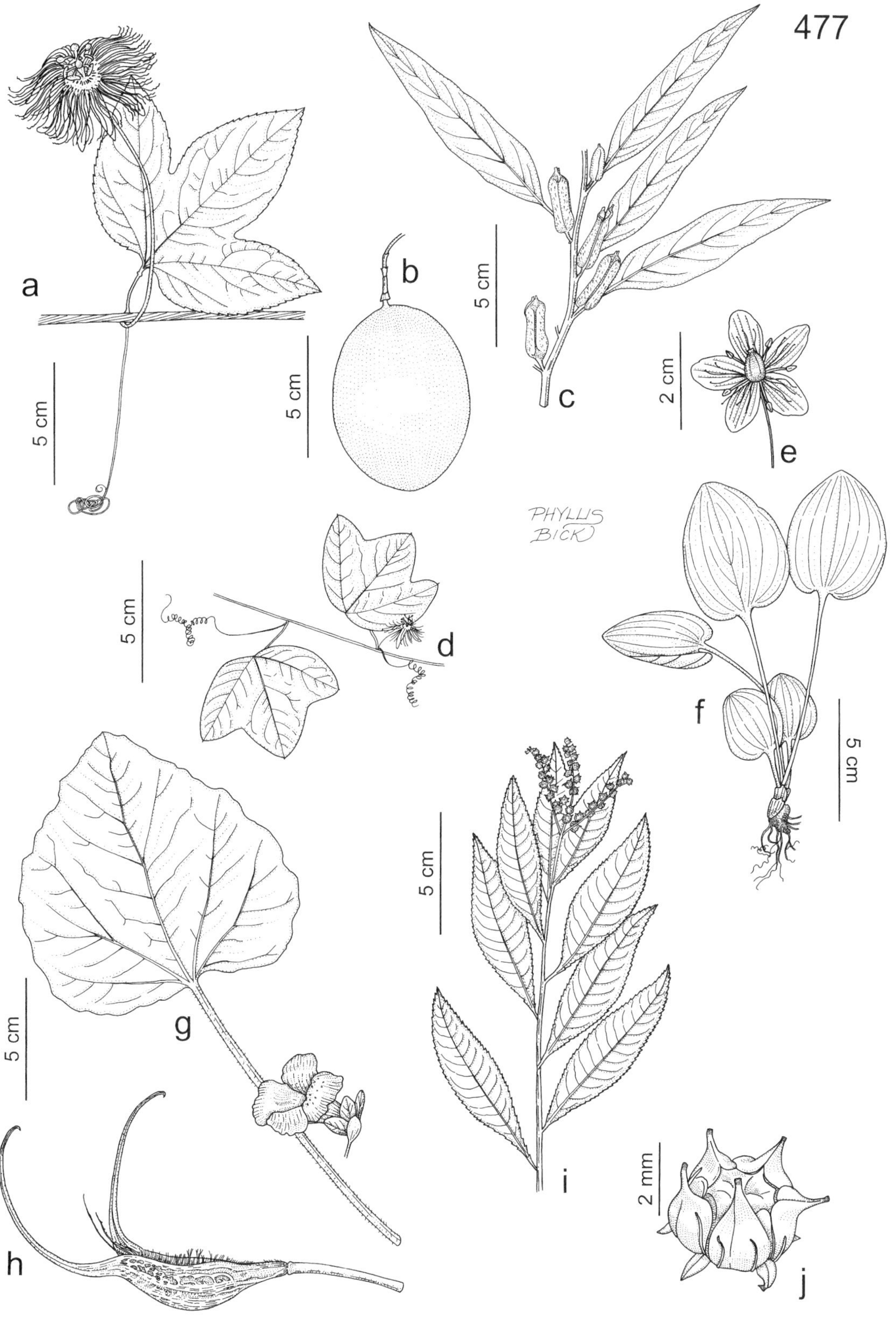

Plate 477. Martyniaceae. Parnassiaceae. Passifloraceae. Pedaliaceae. Penthoraceae. *Passiflora incarnata*, **a)** node with leaf, tendril, and flower, **b)** fruit. *Sesamum indicum*, **c)** fruiting branch. *Passiflora lutea*, **d)** nodes with leaves, tendrils, and flower. *Parnassia grandifolia*, **e)** flower, **f)** vegetative habit. *Proboscidea louisiana*, **g)** leaf and flower, **h)** fruit. *Penthorum sedoides*, **i)** flower, **j)** fertile stem.

1. Parnassia grandifolia DC.

Pl. 477 e, f; dust jacket; Map 2180

Plants perennial herbs, usually with short rhizomes, glabrous, but often with areas of minute rusty brown dots. Aerial stems 12–45 cm long, erect, unbranched. Leaves in a basal rosette, long-petiolate, the petiole broadened toward the tip, also a solitary bractlike leaf positioned at or below the midpoint of the flowering stem, this sessile or nearly so. Stipules absent. Leaf blades simple, relatively thick and leathery, those of the basal leaves 2.5–10 cm long, those of the stem leaves 0.6–4.0 cm long, both types oblong-ovate or broadly ovate to nearly circular, rounded to broadly and bluntly pointed at the tip, broadly rounded to truncate or shallowly cordate at the base, mostly with 7 or 9 palmate main veins, the margins entire. Inflorescences of solitary flowers naked at the aerial stem tips. Flowers perfect, actinomorphic, hypogynous or slightly perigynous. Hypanthium absent or nearly so. Sepals 5, fused at the very base, 3–4 mm long, elliptic-ovate, usually prominently veined, with broad, thin, white to transparent margins, ascending at flowering, persistent and becoming reflexed at fruiting. Petals 5, 12–22 mm long, spreading, oblong-elliptic to elliptic-ovate, white, with 7–11, prominent, green, mostly parallel veins (the outermost pair with several branches toward the margin). Stamens 5, alternating with the petals, the relatively large anthers attached toward the cordate base, pink. Staminodes 5, opposite the petals, somewhat longer than the fertile stamens, each divided at or toward the base into 3 filament-like stalks, each stalk with a glistening capitate tip. Pistil 1 per flower, of 4 fused carpels. Ovary superior, bright green, with 1 locule, with numerous ovules, the placentation parietal. Style absent or nearly so, the 4 stigmas club-shaped to more or less capitate, outwardly curved. Fruits capsules, 11–16 mm long, ovoid, dehiscing longitudinally, with numerous seeds. Seeds 0.7–1.4 mm long, irregularly ovate to tetrahedral in outline, somewhat flattened, the outer covering loose, somewhat wrinkled, and winglike, the surface, with a minute network of hexagonal pits, brown. $2n$=32. August–October.

Scattered in the south-central and eastern portions of the Ozark Division (southeastern U.S. west to Missouri and Texas). Fens and moist calcareous seeps along streams and bluffs.

Steyermark (1963) wrote that, "This species brightens the swampy meadows in autumn with the open white flowers..." The white petals with their prominent green veins make this a most attractive wildflower. The rosettes of thick leaves also are distinctive, even when the plants are not flowering. Regrettably, *P. grandifolia* is considered of conservation concern in nearly every state within its distributional range. In Missouri, where its status appears to be reasonably secure, it is considered an indicator of high-quality calcareous seepage communities, especially fens.

PASSIFLORACEAE (Passion Flower Family)

Contributed by David Bogler

About 18 genera, 575–660 species, widespread, mostly in tropical and warm temperate regions of the New and Old Worlds.

1. Passiflora L. (passion flower)

Plants herbaceous perennial vines, with axillary tendrils. Leaves alternate, simple, palmately lobed. Stipules scalelike or more commonly herbaceous, inconspicuous. Extrafloral nectar glands present on the petioles of some species. Inflorescences axillary, the flowers solitary or often in pairs. Flowers actinomorphic, perfect, perigynous, with a saucer-shaped to cup-shaped hypanthium. Sepals 5, free, spreading to somewhat reflexed, attached to the rim of the hypanthium, petaloid or green on the outer surface and colored on the inner surface. Petals 5, free spreading to somewhat reflexed, attached to the rim of the hypanthium, usually brightly colored. Conspicuous corona present between the petals and stamens, this structurally complex, of filamentous and/or membranous appendages, and often brightly colored. Stamens and pistil attached to the tip of a noticeable central stalk (known as the *androgynophore*, an extension of the receptacle). Stamens 5, inserted at base of ovary, the anthers attached toward the midpoint, pendant. Pistil 1 per flower, composed of 3 fused carpels, with 1 locule and numerous ovules, the placentation parietal. Styles 3, elongate, spreading to somewhat pendant, the stigmas capitate. Fruits berries, the seeds surrounded by a pulpy aril. About 500 species, mostly in tropical and warm-temperate regions of the New World.

The common name refers to the imaginative correlation of the flower parts with components of the Christian crucifixion rite, or Passion of Christ, in a period when the Doctrine of Signatures was an accepted philosophy (Vanderplank, 2000). The corona represents the crown of thorns, including a fringe of blood. The 10 perianth segments represent the 10 faithful apostles. The 3 stigmas represent the Trinity or nails, and the 5 stamens correspond to the 5 wounds. The tendrils might represent whips or ropes. Doctrine of Signatures refers to the medieval European philosophy of cosmology and herbalism in which the "essential virtues" or "purpose" of living things could be inferred through interpretation of selected morphological features in light of superficial similarities with medical, religious, or other themes.

Passiflora exhibits a wide range of flower morphologies (Brizicky, 1961; MacDougal, 1994). In bee-pollinated species, the corona acts as a landing platform and the concentric bands of purple guide the bee to the relatively concentrated (40% sugar) nectar, which is secreted by a glandular ring at the base of the androgynophore into the cup-shaped hypanthium. The corona forces the bee to probe around the flower to get the nectar. The anthers are in position to brush against the insect's body, and move away after pollen release. The styles then curve downward and position themselves where they will brush against a bee carrying pollen from another plant. The flowers open in the morning and last only one day. In the tropics, flowers pollinated by hummingbirds are characterized by having a red color, small corona, and elevated androgynophore. Pollination by bats or wasps also occurs in a few species.

In the tropics, the leaves of *Passiflora* are eaten by herbivorous insects, notably *Heliconius* butterflies, and the plants have developed a number of chemical and morphological defensive mechanisms in a kind of coevolutionary "arms race" (W. W. Benson et al., 1975; Gilbert, 1982). The extrafloral nectaries on the leaf petioles attract ants, which in turn protect the plant from herbivores (McLain, 1983).

Many species are used as ornamentals and several species are grown commercially for their edible fruits,which are eaten fresh or pressed for juice. Although once considered close to the Cucurbitaceae because of their tendrils and parietal placentation, recent molecular studies indicate that the Passifloraceae are instead related to Turneraceae, Violaceae, Flacourtiaceae, and Salicaceae.

1. Petioles with 2 conspicuous glands toward the tip; leaves deeply 3 lobed, the lobes pointed, finely toothed; flowers white and purple, large, 4–9 cm wide; glandular floral bracts present; sepals awned; fruit 3–6 cm in diameter . . . . 1. P. INCARNATA
1. Petioles lacking glands; leaves shallowly 3 lobed; the lobes rounded, entire; flowers greenish yellow, small, 1.5–2.5 cm across; glandular floral bracts absent; sepals not awned; fruits 1.0–1.2 cm diameter . . . . 2. P. LUTEA

1. Passiflora incarnata L. (May pops, apricot vine, passion flower)

Pl. 477 a, b; Map 2181

Stems 3–10 m long, glabrous to finely hairy. Petioles to 3–8 cm long, finely hairy, with two prominent glands toward the tip. Leaf blades deeply 3-lobed, 6–15 cm long along the midvein, about as long as wide, the lobes usually tapered to a sharp point at the tip, the margins finely glandular-toothed, the upper surface glabrous, dark green, the undersurface sparsely and minutely hairy, lighter green, often also somewhat glaucous. Stipules 2–3 mm long, linear to narrowly lanceolate, withering and sometimes shed as the leaves develop. Flower stalks 4–10 cm long, relatively stout, elongating somewhat at fruiting. Flowers 4–9 cm wide, subtended by 2–4 leaflike bracts, these 4–6 mm long, elliptic to oblanceolate, the margins with minute glandular teeth and a pair of larger glands toward the base. Sepals 25–

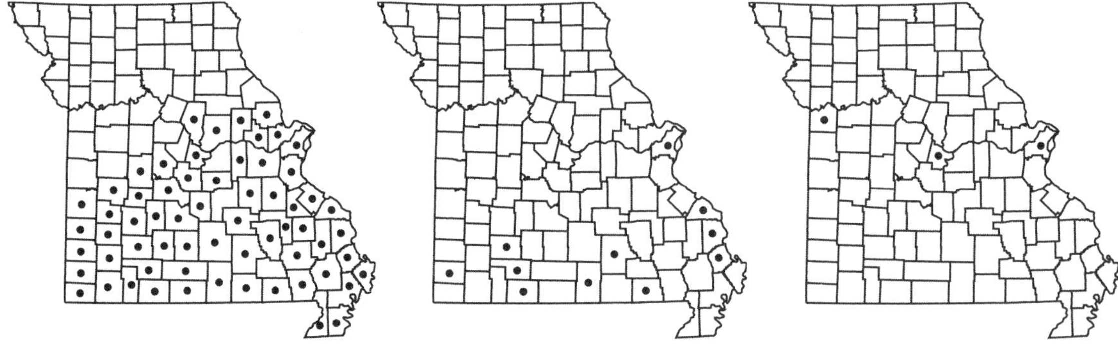
2182. Passiflora lutea 2183. Paulownia tomentosa 2184. Sesamum orientale

30 mm long, 6–8 mm wide narrowly oblong-elliptic, green on the outside, whitish lavender on the inside, each with the midrib having a prominent hornlike awn to 5 mm long toward the tip. Petals 22–25 mm long, 5–6 mm wide, narrowly oblong-elliptic, white or more commonly pale lavender to lavender. Corona filaments in several series, the outer 2 series similar, 20–30 mm long, pink or lavender to purple with white to whitish yellow crossbands toward the base, the innermost series about 2 mm long, white to whitish yellow with purple tips. Ovary densely velvety-hairy. Fruits 3–6 cm long, globose to oblong-ellipsoid, greenish yellow to yellowish orange when ripe. Seeds numerous, 4–6 mm long, 3–4 mm wide, obovate in outline, broadly rounded to more or less truncate at the tip, the surface coarsely pitted, white to light brown. $2n$=18, 36. June–September.

Scattered, mostly south of the Missouri River (eastern [mostly southeastern] U.S. west to Kansas and Texas). Banks of streams and rivers, margins of bottomland forests, and moist depressions of upland prairies; also fallow fields, fencerows, railroads, roadsides, and open disturbed areas, often in sandy soil.

Steyermark (1963) considered this species to occur natively only in southeastern and southwestern Missouri, stating that populations elsewhere in the southern half of the state were introduced. Although the native range of *P. incarnata* in Missouri is not completely understood, apparently native occurrences have been recorded at least as far north as Morgan County.

This species is probably the hardiest of all passion flowers. The large fruits are edible, either eaten fresh or used to make jams. Extrafloral nectaries occur on the petioles, floral bracts, and occasionally the leaf blades in this species. Nectar is secreted for the life of the leaf and is relatively high in amino acids (Durkee, 1983). The functional importance of extrafloral nectaries has been studied in this species (McLain, 1983). Plants in which the extrafloral nectaries were experimentally removed attracted fewer ants, experienced greater herbivory, and set fewer fruits than plants with intact extrafloral nectaries. The leaves have antispasmodic and sedative properties and are used in herbal preparations to treat nervous disorders. This species is frequently visited by large carpenter bees and bumble bees.

2. Passiflora lutea L. **var. glabriflora** Fernald (yellow passion flower)

Pl. 477 d; Map 2182

Stems 0.5–4.0 m or more long, glabrous or very sparsely hairy. Petioles 1–6 cm long, glabrous, without glands. Leaf blades shallowly 3 lobed, 1–12 cm along the central midvein, 1–16 cm wide, usually wider than long, the lobes rounded to bluntly pointed at the tip, sometimes with a minute, abrupt, sharp point, the upper surface glabrous, dark green, often mottled with irregular lighter areas, the undersurface glabrous, lighter green, not glaucous, the margins entire. Stipules 4–5 mm long, linear to narrowly lanceolate. Flower stalks 1.5–5.0 cm long, slender, elongating slightly at fruiting. Flowers 1.5–2.5 cm wide, without subtending bracts. Sepals 5–10 mm long, 2–3 mm wide, narrowly oblong, pale greenish yellow, not awned. Petals 3–7 mm long, 1–2 mm wide, linear-oblong, pale greenish yellow. Corona filaments in several series, the outer series 5–10 mm long, linear, pale yellow to yellow, the middle series 3–5 mm long, club-shaped, pale yellow to yellow, the inner membranous series 1.5–2.5 mm long, pale yellow to yellow, often purple-tinged toward the base. Ovary glabrous. Fruits 0.9–1.5 cm long, globose to ovoid, dark purple when ripe. Seeds 4–7, 3–5 mm long, 3.0–4.5 mm wide, the surface with 6 or 7 relatively coarse cross-ridges and grooves, the ridges with numerous finer longitudinal grooves, white to dark brown. $2n$=24, 84. May–August.

Scattered, mostly south of the Missouri River (eastern [mostly southeastern] U.S. west to Kansas and Texas). Bottomland forests, mesic upland forests, bases and ledges of bluffs, margins of glades, and banks of streams and rivers; also occasionally old fields, fencerows, and roadsides.

The weakly separable var. *glabriflora* has glabrous calyces, petioles, and usually also the young stems. The hairier var. *lutea* occurs to the east of Missouri. The flower is commonly visited by wasps, which may be an important pollinator (J. MacDougal, personal communication).

PAULOWNIACEAE (Empress Tree Family)

One genus, about 7 species, Asia.

The genus *Paulownia* has long vexed botanists, with some treating it as a peripheral taxon in the Scrophulariaceae (Steyermark, 1963; Armstrong, 1985) and others as an aberrant member of the Bignoniaceae (Cronquist, 1981, 1991). The presence of endosperm in the seeds, details of floral anatomy, and embryo structure are characteristic of the Scrophulariaceae, but the arborescent habit, leaf morphology, winged seeds, and chromosome number are more suggestive of the Bignoniaceae. Recent molecular research (Olmstead and Reeves, 1995) has not fully resolved the issue of familial classification, but has provided some support for treatment of the genus in its own family, a placement that has been accepted with some reluctance in the present work. *Paulownia* is most easily distinguished from the superficially similar *Catalpa* by its bluish purple corollas and ovoid fruits.

1. **Paulownia** Siebold & Zucc. (empress tree)

About 7 species, Asia.

1. Paulownia tomentosa (Thunb.) Steud. (royal paulownia, empress tree, princess tree)

Pl. 480 a, b; Map 2183

Plants trees to 15 m tall. Bark smooth when young, developing a network of irregular shallow furrows with age, brownish gray. Twigs stout, with prominent white lenticels, hairy and somewhat sticky when young, the leaf scars prominent. Leaves opposite or less commonly whorled, lacking stipules, simple, long-petiolate. Leaf blades 8–70 cm long, ovate, sometimes very shallowly 3(5)-lobed or angled toward the base, cordate at the base, tapered to the sharply pointed tip, the margins otherwise entire, the upper surface sparsely pubescent with mostly fascicled hairs, the undersurface densely pubescent with stellate and fascicled hairs. Inflorescences large terminal panicles. Flowers perfect, hypogynous, without subtending bracts. Calyces actinomorphic or very slightly zygomorphic, 5-lobed to about the middle, densely pubescent with orangish brown stellate hairs, the lobes broadly triangular-ovate, rounded to bluntly pointed at the tip. Corollas 5–7 cm long, zygomorphic, glandular on the outer surface, 5-lobed, the tube bell-shaped, the lobes shorter than the tube, appearing obliquely 2-lipped, light violet to blue with darker purple spots and longitudinal yellow stripes in the throat. Stamens 4, the filaments fused to the corolla tube, the anther sacs attached at the base, conspicuously spreading. Staminodes absent. Pistil 1 per flower, of 2 fused carpels. Ovary superior, with 2 locules, the placentation axile. Style 1 per flower, the stigma 1, entire or shallowly 2-lobed. Ovules numerous. Fruits capsules, 3–4 cm long, 2-valved, ovoid, tapered to a beak at the tip, longitudinally dehiscent, the valves glabrous, with a woody texture. Seeds 2.5–4.0 mm long, oblong-elliptic in outline, flattened, dark brown, with a thin irregular longitudinal wing around the middle. $2n=40$. April–May.

Introduced, uncommon and widely scattered, mostly in the southern half of the state (native of China, widely cultivated and escaped sporadically in the southern and eastern U.S.). Mesic upland forests; also roadsides, railroads, fencerows, and open disturbed areas.

Empress tree is commonly cultivated as a shade tree and ornamental for its beautiful, sweetly scented flowers. In the northern portion of its introduced range (including Missouri), the pithy branch tips are sometimes damaged by cold temperatures during harsh winters, especially on younger plants or stump sprouts. Flower buds are formed in the autumn and overwinter, and the plants flower in the spring before the leaves are fully expanded. After dehiscence, the large displays of woody fruits also persist during the winter months. In some southeastern states, *Paulownia* is also cultivated as an extremely fast-growing timber tree. The soft lightweight whitish wood is mostly exported. In some portions of southeastern Missouri, it is somewhat invasive in mesic upland forests.

PEDALIACEAE (Sesame Family)

About 13 genera, about 60 species, native to the Old World, most diverse in tropical regions; a few species introduced widely in the New World.

The Pedaliaceae traditionally have been divided into three distinctive subfamilies that are often treated as separate families (Thieret, 1977; Manning, 1991). Subfamily Pedalioideae is the largest and most morphologically diverse of the groups and includes trees, shrubs, and herbs. The genus *Proboscidea* and 3 or 4 other New World genera were included in the Martynioideae, differing from Pedalioideae primarily in their inflorescence structure, ovary construction and placentation, and pollen morphology. The third subfamily, Trapelloideae, contains only the Asian genus, *Trapella* Oliv., with 1 or 2 species of prostrate aquatics having modified flowers with 2 stamens and an inferior ovary. Recent molecular studies have provided support for the view that these three groups are each more closely related to other families in the order Lamiales than to each other. Pedaliaceae in the strict sense appear allied with the Acanthaceae (Olmstead et al., 2001). The Martyniaceae apparently are related to the Verbenaceae (Gutierrez, 2008). Relationships of *Trapella* within the Lamiales have not been fully resolved, but some authors have suggested affinities with the Plantaginaceae (see Ihlenfeldt [2004] and Mabberley [2008] for brief reviews).

1. **Sesamum** L. (sesame, bené)

About 20 species, Africa, Asia.

1. Sesamum indicum L.

S. orientale L.

Pl. 477 c; Map 2184

Plants annual, sparsely to more commonly densely pubescent throughout with multicellular hairs and also with sessile glands. Stems 40–150 cm long, erect, usually unbranched. Leaves mostly opposite, usually alternate toward the stem tip, petiolate. Stipules absent. Leaf blades 3–20 cm long, narrowly elliptic-lanceolate to broadly ovate, rounded to tapered at the base, tapered to a sharply pointed tip, the margins entire or with few coarse teeth toward the base, the lowermost leaves occasionally shallowly 3-lobed. Flowers solitary in the upper leaf axils, perfect, hypogynous, subtended by small bracts. Calyces 4.5–7.0 mm long, nearly actinomorphic, deeply lobed, the lobes narrowly triangular. Corollas 2.0–2.5 cm long, zygomorphic, the tube bell-shaped, the lobes shorter than the tube, appearing obliquely 2-lipped, white or yellowish white (pink to light purple elsewhere), without markings in the throat. Stamens 4, the filaments fused to the corolla tube, the anthers free, attached basally, the anther sacs more or less parallel. Staminode 1, reduced and inconspicuous, fused to the corolla tube. Pistil 1 per flower, of 2 fused carpels. Ovary superior, with axile placentation, 2-locular at flowering, becoming more or less divided into 4 locules at fruiting. Style 1, the stigma 2-lobed. Fruits capsules, 2–3 cm long at maturity, narrowly oblong in outline, bluntly 4-angled or 4-lobed in cross-section, tapered abruptly

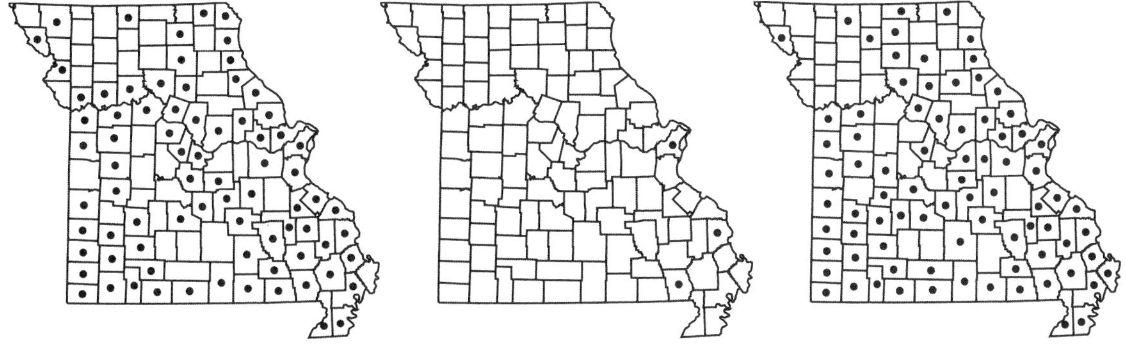

2185. Penthorum sedoides 2186. Mazus pumilus 2187. Mimulus alatus

to a single, short, straight beak at the tip, longitudinally dehiscent from the tip, the valves rigid but not woody. Seeds numerous, 2.5–3.0 mm long, flattened, ovate in outline, with a narrow ridge around the margin and a faint midrib along 1 surface and, the surfaces otherwise glabrous, with a faint reticulate pattern of veins, light brown to white (darker elsewhere). $2n$=26. July–September.

Introduced, widely scattered in and around urban areas (cultivated in tropical to warm-temperate regions around the world, escaped sporadically in North America north to New York). Railroads, yards, and sandy disturbed areas.

Sesame is an important crop plant in many of the warmer parts of the world. It is presumed to have been a native of India, but presently is unknown from natural populations (Manning, 1991). Its many uses have been summarized by Steyermark (1963), Manning (1991), and Moerman (1998), among others. The seeds are rich in oil and have the added advantage of having a relatively long storage life without becoming rancid. The whole seeds are used on baked goods and as a garnish, and also as feed for livestock and a component of some bird seed mixes (from which volunteers sometimes grow). They are a principal ingredient of halvah (a Middle Eastern confection) and tahini (sesami paste). The oil is used as a salad and cooking oil, as an ingredient in margarine and shortening, and also in the manufacture of some soaps and paints, as well as a lubricant. It is also used as a fixative in some perfumes and as a carrier for fat-soluble substances in some pharmaceuticals. The lignans sesamin, sesamol, and sesamolin, present in the oil, are sometimes added to pyrethrin-based insecticides to increase their effectiveness. Medicinally, the plants have been used as an antifungal agent, and some Native Americans used seeds and oil as a cathartic, for diarrhea and gynecological problems, and as a pediatric aid. Although the showy flowers are attractive to various bees, they are mostly self-pollinated.

This species has been called *S. orientale* in many older botanical works. Nicolson and Wiersema (2004) published a proposal to officially conserve the name *S. indicum* for this species, and this was adopted at the following International Botanical Congress.

PENTHORACEAE (Ditch Stonecrop Family)

One genus, 2 species, eastern North America, eastern Asia.

The genus *Penthorum* has been included in the Crassulaceae by some authors (Van Bruggen, 1986) and in the Saxifragaceae by others (Steyermark, 1963; Cronquist, 1991), depending upon which features have been emphasized. However, most authors have agreed that it is anomalous in either family, and recent research (Morgan and Soltis, 1993; Soltis and Soltis, 1997) has shown that it is probably more closely related to the aquatic family Haloragaceae. The present treatment deviates from the standard practice in this manual of following the familial classification of Cronquist (1981, 1991) in treating *Penthorum* in its own family to reflect this new understanding of intergeneric relationships in the Saxifragales, as well as to make the families Crassulaceae and Saxifragaceae more easily circumscribed and keyed.

1. Penthorum L.

Two species, eastern North America, eastern Asia.

1. Penthorum sedoides L. (ditch stonecrop)
Pl. 477 i, j; Map 2185

Plants perennial herbs, with rhizomes. Aerial stems 15–70 cm long, erect or ascending, unbranched or few-branched, often slightly succulent and reddish-tinged, somewhat angled from decurrent leaf bases, usually with scattered, toothlike stalked glands along the angles, otherwise glabrous. Leaves alternate, short-petiolate, the petiole glabrous or more commonly with scattered toothlike stalked glands. Stipules absent. Leaf blades simple, 3–15 cm long, lanceolate to narrowly elliptic, tapered to a sharply pointed tip, tapered to the petiole at the base, the margins sharply and finely toothed, the surfaces glabrous, usually somewhat shiny. Inflorescences panicles, mostly terminal, 2–4-branched toward the base, each branch a 1-sided raceme that is somewhat arched or curled (scorpioid) at the tip, the axes minutely and often densely glandular. Flowers perfect, actinomorphic, perigynous. Hypanthium relatively short, fused to the base of the pistil, the free portion shallow and saucer-shaped. Sepals 5(–7), attached along the hypanthium rim, 1.3–1.9 mm long, ovate to triangular-ovate, persistent at fruiting. Petals absent or less commonly 1–7, inconspicuous, attached along the hypanthium rim between the sepals, 0.7–1.5 mm long, linear to narrowly oblanceolate, green or greenish white. Stamens 10, the filaments attached along the inner margin of the hypanthium, not noticeably tapered, white, the anthers attached at their bases, orange. Staminodes absent. Pistil 1 per flower, of 5(–7) carpels, these fused to about or slightly below the midpoint, more or less erect at flowering, spreading obliquely at fruiting. Ovaries partially inferior, green to greenish white, each with 1 locule, with numerous ovules, the placentation more or less parietal. Styles 1 per carpel, the ovary tip abruptly tapered, persistent at fruiting, the stigmas capitate, outwardly curved. Fruits capsules, 4–7 mm in diameter, wider than long, radially 5(–7) angled and beaked, with circumscissile dehiscence (a cap including the beak and top of the ovary is shed from each carpel), yellowish green, often reddish-tinged, with numerous seeds. Seeds 0.6–0.8 mm long, narrowly ellipsoid to obovoid, the surface, with dense fine tubercles or blunt spines, pinkish tan. $2n=18$. July–October.

Scattered nearly throughout the state (eastern U.S. west to North Dakota and Texas; Canada; introduced in Oregon). Margins of sloughs, ponds, and lakes, swamps, marshes, fens, banks of streams and rivers, and bottomland forests; also ditches, wet fallow fields, and railroads; sometimes emergent aquatics.

PHRYMACEAE (Lopseed Family)

Plants annual or perennial herbs. Stems variously erect to spreading, mostly 4-angled, the angles sometimes narrowly winged. Leaves opposite and often also basal, usually alternate toward the stem tips in *Mazus*, sessile or petiolate. Stipules absent. Leaf blades simple, unlobed or pinnately lobed, the margins usually toothed or scalloped. Inflorescences terminal and/or axillary spikelike racemes (these often long-stalked and bracteate) or of solitary axillary flowers. Flowers perfect, zygomorphic, hypogynous, usually subtended by 1 or 2 bractlets. Calyces (3)4- or 5-lobed, actinomorphic or zygomorphic (the upper lobes usually somewhat longer than the lower ones), often ridged or winged along the midrib of each sepal, persistent at fruiting. Corollas 4- or 5-lobed, zygomorphic, bilabiate, the lower lip usually thickened or bluntly ridged on the upper surface and the throat thus all or partially closed, usually not persistent at fruiting. Stamens 4 (fewer elsewhere), the filaments attached in the corolla tube, those of the lower pair slightly longer than those of the upper pair, usually none of them exserted, the anthers attached at their midpoint or spreading and appearing attached between the sacs. Staminodes absent. Pistil 1 per flower, of 2 fused carpels (but appearing 1-carpellate in *Phryma*). Ovary superior, 1- or 2-locular, placentation axile (basal in *Phryma*; parietal or apical elsewhere). Style 1, slender, the stigmas 2, appearing flattened and broadly spatulate. Fruits relatively slender capsules, dehiscing longitudinally between the sutures, or achenes (in *Phryma*). Seeds 1 (in *Phryma*) or numerous, then small, narrowly ellipsoid to oblong-ellipsoid, sometimes somewhat flattened. About 18 genera, about 230 species, North America, South America, Africa, Asia south to Australia.

As currently circumscribed, the generic constitution of the family Phrymaceae, may seem startling to many North American botanists. Molecular data have provided evidence for the relatively close relationship between *Phryma* (formerly thought to be related to or even included within the Verbenaceae) and *Mimulus* (formerly classified in the Scrophulariaceae), along with several non-Missouri genera (Beardsley and Olmstead, 2002). The family Phrymaceae, as thus defined, is morphologically diverse and difficult to discriminate from some other families in the order Lamiales. A further complication has been that some molecular studies of the order have suggested that two mainly Asian genera, *Mazus* and *Lancea* Hook. f. & Thomson, may not be as closely related to the remaining genera of Phrymaceae as had been indicated by earlier, less inclusive research (Oxelman et al., 1999; Albach et al., 2009). In fact, the family name Mazaceae Reveal (2011) was recently validated to accommodate *Mazus* and *Lancea* for those botanists who felt uncomfortable including these genera in the Phrymaceae. Thus, not only is the family difficult to recognize morphologically, but its limits remain somewhat controversial. Characters that have been said to unite members of the Phrymaceae include a tubular calyx with relatively short, toothlike lobes, capsular fruits that dehisce longitudinally along the lines between the carpels, and stamens with flattened filaments that release pollen on the inner side; however none of these is constant across all of the genera (Tank et al., 2006). The authors of papers on the group published to date all agree that although the genera currently included in the Phrymaceae cannot be included in other families such as Verbenaceae or Scrophulariaceae, further research is needed to develop a stable classification. Thus the current treatment (which includes *Mazus* in the Phrymaceae) should be viewed as tentative.

A number of species of *Mimulus* (in the broad sense) are cultivated as ornamentals in the United States, mostly in the western states. Similarly, some of the nonnative genera, such as *Mazus*, are grown in gardens.

1. Calyces strongly reflexed and appressed to the inflorescence axis at fruiting, the 3 upper lobes hooked at their tips; ovaries appearing 1-carpellate (with a suture on only 1 side), containing 1 ovule; fruits achenes . . . . 3. PHRYMA
1. Calyces spreading to ascending at fruiting, none of the lobes hooked at their tips; ovaries appearing 2-carpellate (with a pair of sutures on opposing sides), containing numerous ovules; fruits capsules
    2. Calyces somewhat angled or thickened along the midrib of each sepal, but not winged, the lobes slightly shorter than to slightly longer than the tube; uppermost stem leaves usually alternate; inflorescences terminal racemes . . . . 1. MAZUS
    2. Calyces strongly angled and often narrowly winged along the midrib of each sepal, the lobes much shorter than the tube; uppermost stem leaves opposite; inflorescences of solitary, axillary flowers . . . . 2. MIMULUS

1. **Mazus** Lour.

About 30 species, Asia south to Australia.

1. Mazus pumilus (Burm. f.) Steenis (mazus)
M. japonicus (Thunb.) Kuntze
Pl. 478 e–g; Map 2186

Plants annual or possibly short-lived perennial herbs, usually with a slender taproot. Stems 1 to several from the base, 5–20 cm long (including the inflorescence), loosely ascending to ascending, individually usually unbranched or few-branched, bluntly 4-angled, the angles slightly ridged, not winged, sparsely pubescent with short, curved hairs, also glandular toward the tips (and inflorescence axes). Leaves in a basal rosette and opposite, usually becoming alternate toward the stem tip, sessile or the basal ones with an indistinct, winged petiole. Leaf blades 1–4 cm long, those of the basal leaves noticeably larger and spatulate to oblong-spatulate or occasionally obovate-spatulate, tapered at the base, rounded at the tip, those of the stem leaves shorter, oblong to elliptic or somewhat spatulate, variously tapered or truncate at the base, sometimes somewhat clasping the stem, rounded to bluntly pointed at the tip, the margins of both types irregularly wavy to bluntly toothed (sometimes the upper leaf blades nearly entire), the surfaces sparsely to moderately short-hairy. Inflorescences terminal, open racemes of 3–15(–20) alternate flowers, each subtended by an inconspicuous, linear bract 1–3 mm long, the stalks 3–10 mm long at flowering, elongating to 7–16 mm at fruiting. Calyces 3–5 mm long at flowering, becoming enlarged to 5–9 mm at fruiting, not becoming papery, actinomorphic, bell-shaped, the 5 lobes slightly shorter than to slightly longer than the tube, elliptic to ovate-elliptic, short-hairy, and somewhat angled or thickened along the midrib of each sepal, but not winged, somewhat spreading at flowering and fruiting, sharply pointed but not hooked at their tips. Corollas 7–10 mm long, lavender to purple and somewhat glandular on the outer surface, the upper lip somewhat hooded, shallowly notched at the tip, lavender to purple on the inner surface, the lower lip with the lateral lobes much larger than the middle lobe, white or very pale lavender on the inner surface, the base of the lip with a pair of prominent, rounded, longitudinal ridges, these with a series of large, yellow spots. Ovaries appearing 2-carpellate (with a pair of sutures on opposing sides), containing numerous ovules. Fruits capsules, 3.0–4.5 mm long, globose, glabrous. Seeds 0.3–0.4 mm long, oblong-ellipsoid to narrowly ellipsoid, the surface appearing pebbled or with a fine network of ridges, brown. $2n$=16–52.

Introduced, uncommon and sporadic in eastern Missouri (native of Asia, introduced widely in the U.S. except for most of the Rocky Mountains and Great Plains regions). Banks of streams; also lawns and open disturbed areas. May–October.

Mazus pumilus is sometimes cultivated as a ground cover or rock garden plant. It was long known as *M. japonicus* (H.-l. Li, 1954; Steyermark, 1963), but in his studies of the flora of New Guinea, van Steenis (1958) discovered that an older epithet was available. Botanists in China sometimes recognize several varieties, in which case the North American materials appear to correspond to var. *pumilus*. However, the distinctness of the varieties is not clear.

2. **Mimulus** L. (monkey flower)

Plants perennial herbs (variously annuals and shrubs elsewhere), with fibrous roots, sometimes with rhizomes or stolons. Stems 1 to several from the base, erect to strongly ascending or spreading to creeping (in *M. glabratus*), usually branched (unbranched elsewhere), sharply 4-angled, the angles sometimes narrowly winged, sometimes hollow between the nodes, glabrous or rarely minutely hairy (commonly glandular elsewhere). Leaves all opposite, sessile or petiolate. Leaf blades variously shaped, glabrous (minutely hairy and/or glandular elsewhere). Inflorescences axillary, of solitary flowers (racemes elsewhere), appearing 2 per node, lacking additional bracts and bractlets, short- to long-stalked at flowering, the stalks continuing to elongate as the fruits develop. Calyces becoming at least slightly enlarged at fruiting and often eventually papery, slightly to moderately zygomorphic, mostly funnelform, the 5 lobes much shorter than the tube, the upper lobe the longest, mostly narrowly to broadly

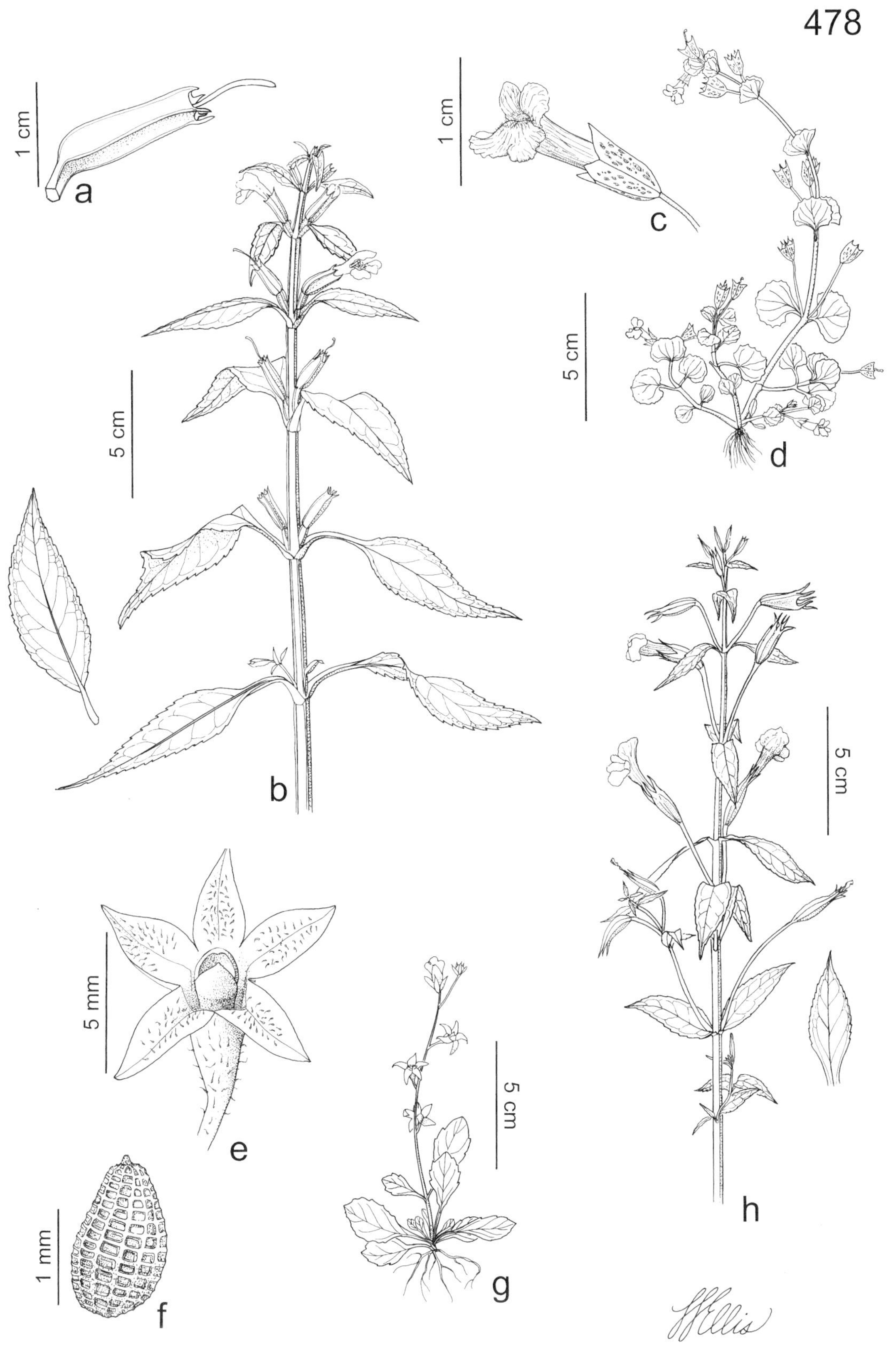

Plate 478. Phrymaceae. *Mimulus alatus*, **a)** fruit, **b)** fertile stem and detached leaf. *Mimulus glabratus*, **c)** flower, **d)** habit. *Mazus pumilus*, **e)** flower, **f)** seed, **g)** habit. *Mimulus ringens*, **h)** detached leaf and fertile stem.

triangular, strongly angled and often narrowly winged along the midrib of each sepal, erect or slightly incurved at flowering and fruiting, sharply pointed or even short-awned, but not hooked at their tips. Corollas yellow or light blue to light lavender blue or purplish blue, rarely pale pink or white (red elsewhere), the upper lip usually smaller and narrower than the lower one, angled or arched outward, 2-lobed or relatively deeply notched at the tip, the lobes often curved backward along the sides, the lower lip with the lateral lobes usually narrower than the middle lobe, the base of the lip with a pair of prominent, rounded, longitudinal ridges, often with an area of contrasting color (pale yellow to whitish in the bluish-flowered species) and sometimes with red spots. Ovaries appearing 2-carpellate (with a pair of sutures on opposing sides), containing numerous ovules. Fruits capsules, ellipsoid to ovoid, glabrous. Seeds 0.3–0.4 mm long, oblong-ellipsoid to ellipsoid-ovoid, the surface with a fine network of ridges, light brown to brown. About 120 species (in the broad sense), nearly worldwide, most diverse in the western United States.

In the broad sense, *Mimulus* is by far the largest genus in the Phrymaceae. However, its taxonomic limits remain less than fully understood. Historically, two subgenera with ten total sections were erected to accommodate the morphological diversity in the group (A. L. Grant, 1924). Some of these infrageneric taxa subsequently were segregated as genera, but there was no consensus on classification within the group. More recently, molecular studies have shown that two main lineages exist, but relationships and classification between and within these major groups is still being refined (Beardsley and Olmstead, 2002; Beardsley et al., 2004, Nie et al., 2006; Barker et al., 2012). The type of the genus is *M. ringens*; thus, if *Mimulus* becomes more narrowly circumscribed taxonomically in the future, the group that includes the eastern North American species with purplish blue corollas will retain the generic name, but nearly all of the yellow- to red-flowered species whose center of diversity is in the western United States will need to be segregated into one or more other genera.

Species of *Mimulus* have long been model organisms for the study of various genetic, ecological, and evolutionary processes (Beardsley and Olmstead, 2002; Beardsley et al., 2004). Some of the species are cultivated as ornamentals, particularly in the western United States. The name monkey flower is an allusion to the red pollinator markings on the corolla lips of some of the yellow-flowered species, which fancifully resemble a stylized simian face.

1. Corollas 8–12 mm long, yellow; stems spreading or creeping with ascending tips and branches; leaf blades circular to broadly oval or kidney-shaped, about as long as wide . . . . 2. M. GLABRATUS
1. Corollas 20–40 mm long, light blue to light lavender blue or purplish blue, rarely pale pink or white; stems erect or strongly ascending; leaf blades variously narrowly oblong to broadly lanceolate, oblong-obovate, ovate, or elliptic-ovate, 3–7 times as long as wide
    2. Leaves short-petiolate, the blade base angled or tapered into the petiole, not clasping the stem; flower stalks 2–10 mm long, sometimes elongating to 3–17 mm at fruiting; calyx lobes 0.8–2.5 mm long . . . . 1. M. ALATUS
    2. Leaves sessile, the blade base broadly angled to rounded and often somewhat clasping the stem; flower stalks 15–40 mm long, elongating to 20–60 mm at fruiting; calyx lobes (2.0–)2.5–8.0 mm long . . . . 3. M. RINGENS

1. Mimulus alatus Aiton (sharpwing monkey flower)

Pl. 478 a, b; Map 2187

Plants often with rhizomes or stolons. Stems 25–120 cm, long, erect or strongly ascending, glabrous, the angles mostly narrowly winged. Leaves short-petiolate, the petiole 1–2 cm long, sometimes narrowly winged, not clasping the stem. Leaf blades 4–13 cm long (the uppermost often shorter and bractlike), 3–5 times as long as wide, broadly lanceolate to ovate or elliptic-ovate, angled or tapered at the base, tapered to a sharply pointed tip, the margins relatively sharply and finely to coarsely toothed, the surfaces glabrous, the vena-

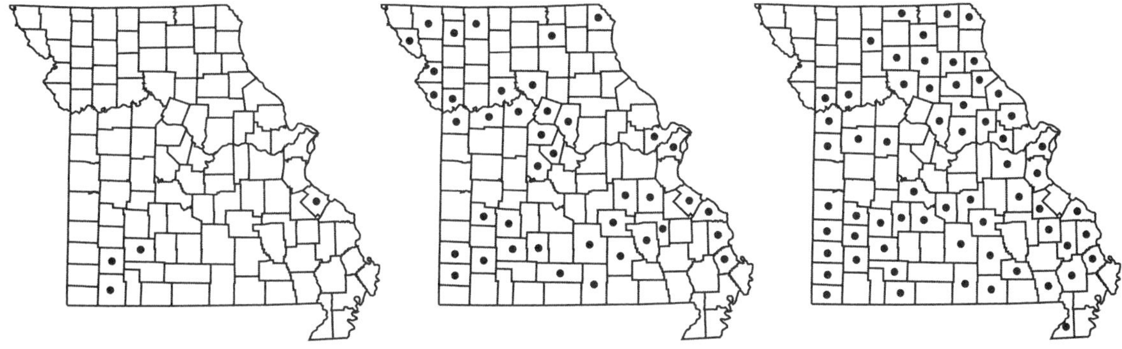

2188. Mimulus glabratus 2189. Mimulus ringens 2190. Phryma leptostachya

tion pinnate. Flower stalks 2–10 mm long, sometimes elongating to 3–17 mm at fruiting. Calyces 11–19 mm long, nearly actinomorphic, narrowly funnelform, not becoming inflated at fruiting, the lobes 0.8–2.5 mm long, broadly ovate-triangular to semicircular, the thickened midrib of each sepal often extending past the body of the lobe as a short, stiff, straight awn, glabrous externally. Corollas 20–40 mm long, light blue to light lavender blue or purplish blue (rarely pale pink or white), finely glandular-hairy, the upper lip usually strongly curved or bent outward, the lower lip with a pale yellow area toward the base, this usually at least faintly red-spotted, the throat usually appearing more or less closed by the prominent, blunt ridges of the lower lip. Fruits 8–12 mm long, ovoid to narrowly ellipsoid-ovoid. $2n$=22. June–September.

Scattered nearly throughout the state (eastern U.S. west to Nebraska and Texas; Canada). Banks of streams, rivers, and spring branches, margins of ponds, lakes, sloughs, and oxbows; fens, acid seeps, swamps, and bottomland forests; also ditches and wet roadsides.

Occasional white-flowered plants have been called f. *albiflorus* House. When these rarely cross with blue-flowered individuals, the offspring tend to have pale pink corollas. For a discussion of hybrids with the closely related *M. ringens*, see the treatment of that species.

2. Mimulus glabratus Kunth **var. jamesii** (Torr. & A. Gray ex Benth.) A. Gray (roundleaf monkey flower, yellow monkey flower)

M. glabratus var. *fremontii* (Benth.) A.L. Grant

M. glabratus ssp. *fremontii* (Benth.) A.L. Grant ex Pennell

Erythranthe geyeri (Torr.) G.L. Nesom

Pl. 478 c, d; Map 2188

Plants lacking rhizomes, occasionally with short stolons. Stems 10–40 cm, long, spreading or creeping with ascending tips and branches, sometimes rooting at the lower nodes, glabrous or finely short-hairy toward the tip, the angles sometimes narrowly winged. Leaves short- to long-petiolate toward the stem base, the petioles progressively shorter toward the stem tip, the uppermost leaves sessile, the petiole 0.1–0.6(–1.5) cm long, often narrowly winged toward the tip, the sessile leaves sometimes slightly clasping the stem. Leaf blades 0.8–3.0 cm long, about as long as wide, circular to broadly oval or kidney-shaped, rounded to truncate or cordate at the base, rounded to broadly and bluntly pointed at the tip, the margins variously entire or slightly wavy to finely scalloped, the surfaces glabrous or finely short-hairy, the venation palmate. Flower stalks 8–40 mm long, sometimes elongating slightly at fruiting. Calyces 5–11 mm long, zygomorphic, broadly funnelform to somewhat urn-shaped, becoming enlarged to 8–16 mm long and inflated at fruiting, the upper lobe 1.0–2.5 mm long, triangular-ovate, rounded to bluntly pointed at the tip (but usually appearing sharply pointed when viewed in profile), the thickened sepal midrib sometimes extending past the body of the lobe as a short, straight awn, the other 4 lobes minute or absent, glabrous or finely short-hairy externally. Corollas 8–12 mm long, yellow, glabrous except for the finely bearded lower lip, the upper lip usually strongly curved or bent outward, the lower lip lacking spots or areas of contrasting color, the throat usually appearing partially closed by the prominent, blunt ridges of the lower lip. Fruits 5–9 mm long, ovoid to ellipsoid. $2n$=30 (species includes $2n$=28–92, but many counts not assigned to variety). May–October.

Uncommon, known only from widely scattered historical collections, mostly from southwestern Missouri (Montana to Arizona east to Michigan, Missouri, and Texas; Canada, Mexico). Spring outlets, spring branches, and wet ledges, often emergent aquatics.

Additional Missouri records from Bates and Cass Counties mapped for this rare taxon by W. R. Weber et al. (2000) could not be verified during the present study. *Mimulus glabratus* is a widespread, morphologically variable species. A. L. Grant (1924) accepted three varieties, the widespread North American var. *glabratus*, the var. *fremontii* in the eastern portion of the species range in North American, and a little-known South American taxon, var. *parviflorus* (Lindl.) A.L. Grant. Pennell (1935) treated the species in North America as comprising four subspecies. Within *M. glabratus*, var. *jamesii* is separable by the following suite of characters (each of which overlaps with various other varieties to at least a small extent): stems spreading (vs. ascending or erect); leaf blades circular to broadly oval or kidney-shaped, more or less rounded at the tip and entire or finely scalloped along the margins; and corollas 8–12 mm long (vs. 12–22 mm), lacking reddish markings in the throat.

As noted in the discussion under the genus *Mimulus*, if the genus is circumscribed in the narrow sense, as suggested by molecular studies, then the yellow-flowered western groups require segregation under one or more segregate generic names. Under this scenario, the correct name for this taxon becomes *Erythranthe geyeri* (Barker et al., 2012).

Missouri botanists should be on the watch for another yellow-flowered species that may eventually be discovered in the southern portion of the state. *Mimulus floribundus* Douglas ex Lindl. (*Erythranthe floribunda* (Douglas ex Lindl.) G.L. Nesom) has been reported from several counties in northwestern Arkansas, although the Ozarkian occurrences of the species are disjunct by more than 1,000 km from the closest populations in Colorado and New Mexico (E. B. Smith, 1988). This widespread species of the western United States is similar to *M. glabratus* var. *jamesii* in its yellow corollas, but differs from that species in its annual habit, glandular-hairy (sticky) flower stalks and calyces, broadly ovate leaf blades that are more pointed at the tips and more sharply toothed along the margins, and corollas with reddish pollinator guides.

3. Mimulus ringens L. (Allegheny monkey flower)

Pl. 478 h; Map 2189

Plants often with rhizomes or stolons. Stems 20–130 cm, long, erect or strongly ascending, glabrous, the angles sometimes narrowly winged, the wings often poorly developed. Leaves sessile, often somewhat clasping the stem. Leaf blades 4–14 cm long (the uppermost often shorter and bractlike), 3–7 times as long as wide, narrowly oblong to oblong-lanceolate or narrowly oblong-obovate, broadly angled to rounded at the base, angled or tapered to a sharply pointed tip, the margins relatively bluntly and finely toothed or scalloped, the surfaces glabrous, the venation pinnate. Flower stalks 15–40 mm long, elongating to 20–60 mm at fruiting. Calyces 10–17 mm long, nearly actinomorphic, narrowly funnelform, not becoming inflated at fruiting, the lobes (2.0–)2.5–8.0 mm long, triangular to broadly triangular, the thickened midrib of each sepal often extending past the body of the lobe as a short, stiff, straight awn, glabrous externally. Corollas 20–40 mm long, light blue to light lavender blue or purplish blue (rarely pale pink or white), finely glandular-hairy, the upper lip usually strongly curved or bent outward, the lower lip with a pale yellow area toward the base, this red- or brownish red-spotted, the throat usually appearing more or less closed by the prominent, blunt ridges of the lower lip. Fruits 9–12 mm long, ovoid to narrowly ellipsoid-ovoid. $2n$=22, 24. June–September.

Scattered nearly throughout the state (eastern U.S. west to North Dakota and Oklahoma, west locally to Montana, Idaho, and Washington; Canada; possibly introduced in California). Banks of streams, rivers, and spring branches, margins of ponds, lakes, sloughs, and oxbows; fens, swamps, and bottomland forests; also ditches.

Occasional white-flowered plants have been called f. *peckii* House, but have yet to be documented from Missouri. Although *M. alatus* and *M. ringens* both occur nearly statewide, the former is more abundant in Missouri. Plants treated by Steyermark (1963) and some other botanists as *M. ringens* var. *minthodes* (Greene) A.L. Grant appear instead to represent interspecific hybrids between *M. ringens* and the closely related *M. alatus*. The name *M.* ×*minthodes* Greene has been applied to such plants, which appear to be at least partially fertile and are variable morphologically. Putative hybrids tend to have the leaf blades tapered or angled at the base, but lacking a true petiole. They also have calyx lobes intermediate in length between those of the two parents. Because the parents frequently grow in mixed populations, hybrid individuals probably are relatively common but undercollected.

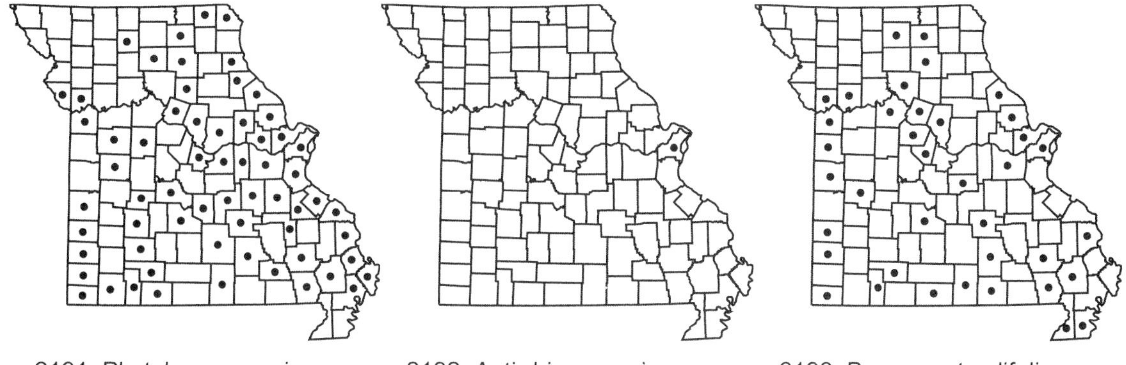

2191. Phytolacca americana 2192. Antirrhinum majus 2193. Bacopa rotundifolia

3. **Phryma** L. (lopseed)
(Thieret, 1972)

One species, eastern temperate North America, Asia.

The morphology of the genus *Phryma* is unusual in the context of Lamiales in that apparently during the evolution of the genus a two-carpellate ovary with several ovules became reduced to a single carpel with a single ovule and a capsular fruit became modified into an achene. The developing fruits are further modified by the enlarged, hardened, hooked, three upper teeth of the calyces. The buds are ascending, but begin to droop very quickly. The open flowers are horizontally spreading and the mature fruits are strongly reflexed and appressed to the inflorescence axis. This syndrome of features apparently represents an adaptation to dispersal in the fur of mammals (some species of which forage on the plants as well).

1. Phryma leptostachya L. **var. leptostachya**
Pl. 479 a–c; Map 2190

Plants perennial herbs, with fibrous roots. Stems usually solitary, 40–100 cm long, erect or strongly ascending, unbranched or few-branched toward the tip, 4-angled, the angles not winged, slightly swollen and often purplish-tinged above each node, moderately to densely pubescent with short hairs, sometimes glabrous or nearly so at maturity. Leaves opposite, the lower leaves long-petiolate, the petioles progressively shorter toward the stem tip, the uppermost leaves sometimes sessile. Leaf blades 3–16 cm long, ovate to elliptic-ovate or narrowly ovate, truncate or broadly angled to broadly tapered at the base, angled or tapered to a sharply or less commonly bluntly pointed tip, the margins scalloped or bluntly toothed (often coarsely or irregularly so), the surfaces sparsely to moderately short-hairy or glabrous. Inflorescences terminal (and sometimes also axillary from the upper leaves), interrupted, long-stalked spikelike racemes of numerous opposite flowers, each subtended by an inconspicuous, linear bract 2–3 mm long and a pair of very short, toothlike bractlets on the upper side, the stalks 0.5–0.8 mm long at flowering, not elongating at fruiting. Flowers ascending in bud, becoming horizontally spreading at flowering and strongly reflexed (appressed to the inflorescence axis) at fruiting. Calyces 5–8 mm long, not noticeably enlarged at fruiting, but becoming somewhat hardened, glabrous or minutely hairy, 5-lobed, bilabiate, cylindric, the 3 upper lobes 2–3 mm long, about as long as or slightly shorter than the tube, slender, the lower 2 lobes 0.3–0.5 mm long, triangular, thickened along the midrib of each sepal, but not winged, erect at flowering and fruiting, the upper 3 lobes outward-hooked at the tips and purplish-tinged. Corollas 6–10 mm long, the upper lip 1.5–2.0 mm long, arched outward, notched at the tip, lavender to pale purple or rarely reddish purple, the lower lip 2.5–3.5 mm long, with the lateral lobes about equal to or somewhat wider than the middle lobe, white, rarely reddish-tinged, the base of the lip with a pair of rounded, longitudinal ridges, these lacking contrasting lines or spots, bearded. Ovaries appearing 1-carpellate (with a solitary suture often asymmetrically along 1 side), containing 1 ovule. Fruits achenes, 3–4 mm long, narrowly ellipsoid, glabrous. $2n=28$. June–September.

Scattered nearly throughout the state but apparently uncommon in northwesternmost Missouri (eastern U.S. west to North Dakota and Texas; Canada; Asia). Bottomland forests, mesic upland forests, banks of streams and rivers, oxbows; bases and ledges of bluffs, and rarely savannas.

Many botanists divide *P. leptostachya* into two varieties. The eastern Asian plants correspond to var. *asiatica* H. Hara (ssp. *asiatica* (H. Hara) Kitam.). The two infraspecific taxa differ in a suite of minor quantitative characters, including leaf size, calyx tooth length, corolla size, and degree of corolla lobing, but there is at least some overlap for every morphological feature that has been studied (Thieret, 1972; Nie et al., 2006). Genetic data suggest that *Phryma* is a relatively old genus, but that the current disjunctional distribution of *P. leptostachya* is of relatively recent origin (Nie et al., 2006).

PHYTOLACCACEAE (Pokeweed Family)

Three to 20 genera, 30–120 species, nearly worldwide, most diverse in tropical and subtropical regions of the New World.

Traditionally, the Phytolaccaceae have been classified as consisting of five or six subfamilies (Nowicke, 1969) with 17–20 total genera. More recent taxonomic studies involving morphological and anatomical characters (G. K. Brown and Varadarajan, 1985) and gene sequence data (Manhart and Rettig, 1994), have suggested that most of these genera are not closely related to the three core members of the subfamily Phytolaccoideae (*Phytolacca* and two small South American genera, *Anisomeria* D. Don and *Ercilla* A. Juss.). With the family circumscription restricted to this small group, the remaining genera are relegated to up to five other families. However, these studies are considered somewhat preliminary and further research involving more intensive taxon sampling and more characters may result in another reevaluation of familial limits.

1. **Phytolacca** L.

About 25 species, North America to South America, Caribbean Islands, Europe, Asia, Africa, Madagascar.

1. Phytolacca americana L. **var. americana** (pokeweed, poke)

Pl. 479 d, e; Map 2191

Plants perennial herbs, with stout taproots. Stems 20–300 cm long, often branched, erect or ascending, glabrous, often reddish- or purplish-tinged. Leaves alternate, short- to long-petiolate. Stipules absent. Leaf blades 4–30 cm long, oblong-lanceolate to elliptic or ovate, narrowed or tapered to a sharp point at the tip, narrowed at the base, the margins entire or slightly undulate, glabrous, the midvein sometimes reddish-tinged. Inflorescences racemes, terminal or appearing axillary or opposite the leaves, 4–25 cm long, often long-stalked, drooping or arched, glabrous. Flowers actinomorphic, mostly perfect, hypogynous. Calyx of 5 sepals, these 2–3 mm long, ovate to nearly circular, white, sometimes pinkish-tinged, often darkening after flowering. Petals absent. Stamens (8–)10, the anthers attached toward their midpoints. Pistil 1 per flower, the ovary superior, consisting of (8–)10 carpels in a ring (visible as lobes), each with 1 locule, the placentation more or less basal. Styles (8–)10, each with a linear stigmatic area along the inner side toward the tip. Ovule 1 per carpel. Fruits berries, 5–10 mm in diameter, depressed-globose, sometimes slightly lobed at maturity, dark reddish purple to nearly black at maturity, usually somewhat shiny, the juice reddish purple. Seeds (8–)10, 2.5–3.5 mm long, ovoid, slightly flattened, the surface smooth, black. $2n=36$. May–October.

Common nearly throughout the state (eastern U.S. west to Minnesota, Nebraska, and Texas; Canada, Texas; introduced westward to California and Oregon and in the Old World). Banks of streams and rivers, margins of ponds, sinkhole ponds, and lakes, bases and ledges of bluffs, and

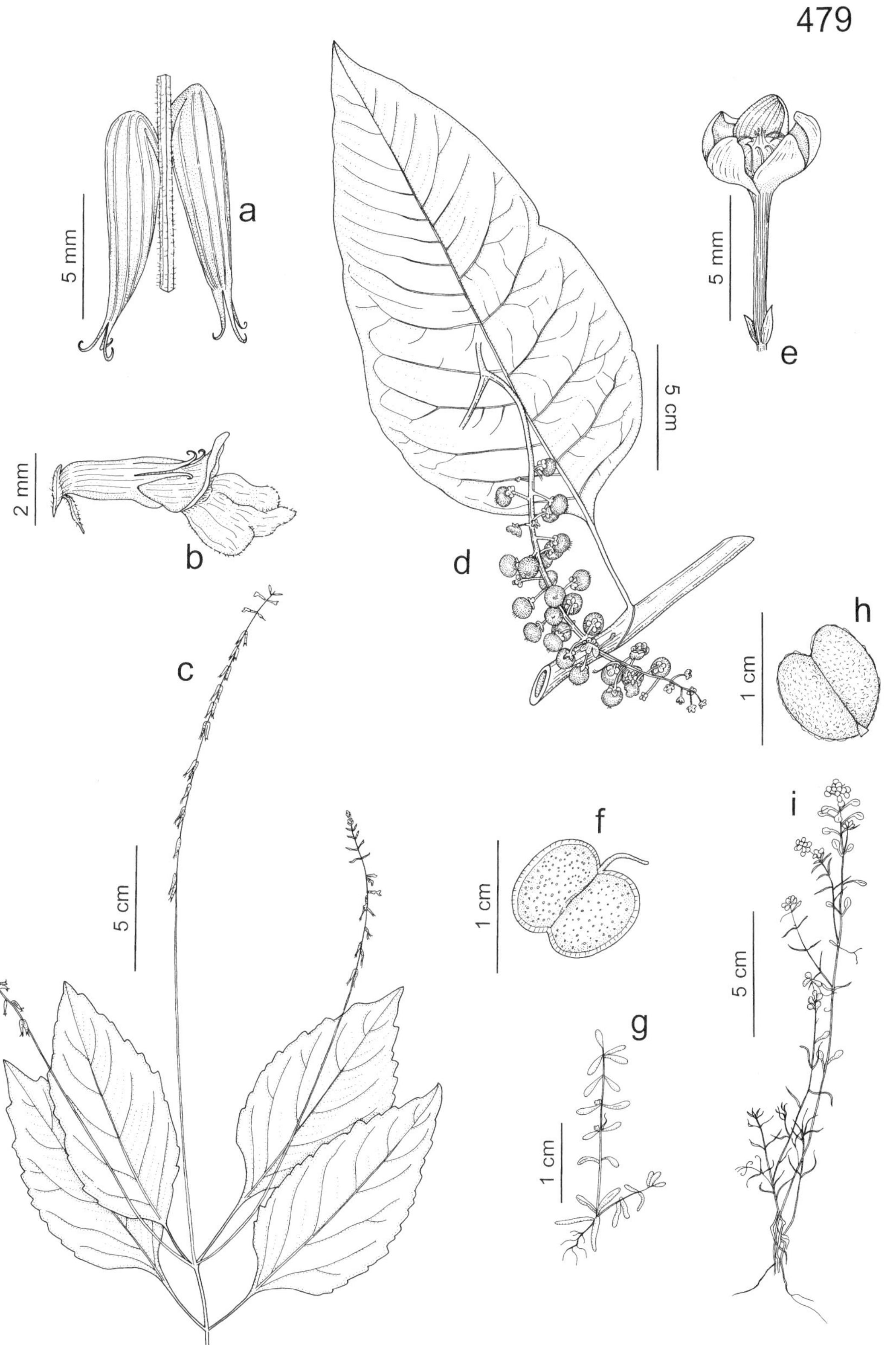

Plate 479. Phrymaceae. Phytolaccaceae. Plantaginaceae. *Phryma leptostachya*, **a)** node with fruits, **b)** flower, **c)** fertile stem. *Phytolacca americana*, **d)** flower, **e)** node with leaf and inflorescence with flowers and fruits. *Callitriche terrestris*, **f)** fruit, **g)** habit. *Callitriche heterophylla*, **h)** fruit, **i)** habit.

edges of bottomland and mesic upland forests; also fencerows, pastures, fallow fields, roadsides, railroads, and open disturbed areas.

The var. *rigida* (Small) Caulkins & R.E. Wyatt has often been treated as a separate species. It is endemic to the Atlantic and Gulf Coastal Plains and differs from var. *americana* in its ascending inflorescences and short fruit stalks. Caulkins and Wyatt (1990) studied the taxonomy of this complex and concluded that the two taxa are best treated as varieties of a single species.

Sauer (1950) and G. K. Rogers (1985) summarized the ethnobotany and biology of pokeweed. All parts of *Phytolacca* plants should be considered poisonous to both humans and livestock, including the roots, stems, foliage, and berries, although the effects of ingestion apparently vary with different people (G. K. Rogers, 1985). In spite of this, the young foliage of pokeweed was used by both Native Americans and pioneers as a vegetable, and is still gathered as a wild green in some parts of the United States. The fruits occasionally have been baked in pies. Boiling or baking apparently serves to deactivate the toxins (but caution is advised). There also are reports of dermatitis in some individuals from the handling of plants. The juice of the berries has been used as a water-soluble dye in handcrafts and particularly as a colorant in food and beverages. The array of toxic compounds in the plants, which is still not fully understood, accounts for their narcotic properties, as well as medicinal uses for skin ailments, as a purgative, and as an antitumor agent. The pharmaceutical industry has been investigating selected proteins in the species for their potential antiviral and antitumor properties.

Pokeweed is mostly self-pollinated, with pollen grains released from the anthers in the flower buds. The seeds, which may survive in the soil for 40 or more years, are dispersed mainly by birds (which are seemingly immune to the toxins in the fruits). The species thus is an aggressive and efficient colonizer of disturbed sites (Sauer, 1952; G. K. Rogers, 1985).

PLANTAGINACEAE (Plantain Family)

Plants annual, biennial, or perennial herbs (shrubs elsewhere). Leaves alternate, opposite, or less commonly whorled, sometimes also or only basal, short- to long-petiolate. Stipules absent (the bases of the pair of leaves at a node often connected by a narrow membrane in *Callitriche*). Leaf blades unlobed or less commonly lobed, variously shaped, the margins usually scalloped or toothed. Inflorescences terminal or axillary, clusters, racemes, spikes, or panicles, sometimes of solitary flowers; bracts and/or bractlets sometimes subtending the flowers, these sometimes inconspicuous. Flowers perfect or less commonly imperfect (monoecious in *Callitriche*, occasionally incompletely monoecious in *Plantago*), hypogynous. Cleistogamous flowers absent (except sometimes in *Plantago*). Calyces actinomorphic or somewhat zygomorphic (absent in *Callitriche*), (3)4- or 5-lobed, persistent at fruiting. Corollas actinomorphic (in some *Plantago*) or slightly to strongly zygomorphic and sometimes bilabiate (absent in *Callitriche*), variously colored, 4- or 5-lobed, the tube variously short to long, sometimes with a basal nectar spur. Stamens 2 or 4 (1 in *Callitriche*), the filaments attached in the corolla tube (sometimes near its base), sometimes of 2 lengths, the anthers exserted or not, attached near their midpoints or basally, sometimes appearing U- or V-shaped, variously colored. Staminodes absent or present, then 1, well differentiated from the fertile stamens. Pistil 1 per flower, of 2 fused carpels. Ovary 2-locular, with 1 or 2 to more commonly several to numerous ovules per locule, the placentation axile (occasionally appearing basal in *Plantago*). Style 1 (2 in *Callitriche*), often persistent at fruiting, the stigma 1 per style, club-shaped to capitate or occasionally linear, entire or 2-lobed. Fruits capsules, variously dehiscent (indehiscent and separating into 4 achenelike nutlets in *Callitriche*). Seeds numerous or occasionally reduced to 1 or 2 per locule, mostly minute. About 104 genera, 1,820–1,900 species, nearly worldwide.

Many of the genera now included in the Plantaginaceae formerly were classified as members of the Scrophulariaceae. Molecular studies showed that the traditional circumscription of Scrophulariaceae included many genera that had closer affinities to other plant families than to the relatively few genera related to Scrophulariaceae (as reviewed by Tank et al. [2006]). The family that experienced the biggest generic expansion in the order Lamiales as a result of these studies was the Plantaginaceae (Albach and Chase, 2004; Albach et al., 2005; Oxelman et al., 2005), which sometimes was called Veronicaceae in the molecular literature before the issue of nomenclatural priority was clarified. In addition to genera transferred from the Scrophulariaceae, molecular work resulted in the inclusion of two small families of aquatics in the Plantaginaceae, the Callitrichaceae and the non-Missouri family, Hippuridaceae. The resultant expanded circumscription of Plantaginaceae thus includes plants with a variety of pollination syndromes (wind, water, insects, birds) and a related suite of adaptations in floral morphology. For further discussion of the break-up of the former Scrophulariaceae, see the treatment of that family.

As currently circumscribed, Plantaginaceae contains a number of horticulturally important genera, including *Angelonia* Bonpl. (angelonia, summer snapdragon), *Antirrhinum* (snapdragon), *Chelone* (turtlehead), *Collinsia* (blue-eyed Mary), *Digitalis* L. (foxglove), *Linaria* (butter and eggs), *Penstemon* (beard-tongue), *Russelia* Jacq. (firecracker plant), *Veronica* (speedwell), and *Veronicastrum* (culver's root). Conversely, some of the genera, including *Kickxia, Plantago*, and *Veronica* contain species that are weeds of crop fields and pastures or invasive in rangelands and native plant communities. Several genera have been used medicinally. The Old World genus *Digitalis* is a rich source of cardiac glycosides, such as digoxin, which can be potent toxins but in proper dosage form the basis for important medications to control the heartbeat.

1. Leaves all basal . . . . 15. PLANTAGO
1. Stem leaves present (the stems sometimes relatively short)
    2. At least some of the stem leaves (ignore any inflorescence bracts or leaves subtending axillary flowers) alternate; corollas with a spur or saclike outgrowth at the base
        3. Stems more or less hairy
            4. Stems erect or strongly ascending, not twining; leaves sessile or nearly so; leaf blades linear to narrowly oblong-oblanceolate . . . . 4. CHAENORHINUM
            4. Stems prostrate, or loosely spreading, sometimes twining and thus climbing on surrounding vegetation; leaves noticeably petiolate; leaf blades nearly circular to broadly ovate, broadly triangular-ovate, more or less heart-shaped, or halberd-shaped . . . . 9. KICKXIA
        3. Stems glabrous below the inflorescence
            5. Stems prostrate or loosely spreading, sometimes twining and thus climbing on surrounding vegetation; leaf blades more or less kidney-shaped to nearly circular, palmately 5(–9)-lobed or coarsely toothed, the venation palmate . . . . 7. CYMBALARIA
            5. Stems erect or strongly ascending (short, spreading stems produced in addition to the longer, erect ones in *Nuttallanthus*); leaf blades linear to narrowly lanceolate, oblanceolate or spatulate, the margins entire, the venation pinnate or with only a midvein apparent (sometimes totally obscure in *Nuttallanthus*)
                6. Corolla tube with a short, broad, saclike outgrowth at the base; at least the larger stem leaves with the blade lanceolate, 6–13 mm wide . . . . 1. ANTIRRHINUM
                6. Corolla tube with an elongate, slender spur at the base; leaf blades linear, 0.5–6.0 mm wide

7. Leaf blades 2–7 cm long, 1.5–6.0 mm wide; corollas 20–30 mm long (including the spur), yellow and orange . . . . 11. LINARIA
7. Leaf blades 0.5–3.0 cm long, those of flowering stems 0.5–2.0 mm wide (occasionally slightly wider only on prostrate, vegetative stems); corollas 8–22 mm long (including the spur), pale blue to light purplish blue, rarely white 13. NUTTALLANTHUS

2. Leaves all opposite or whorled below the inflorescence; corollas not spurred (except in *Nuttallanthus*)
8. Flowers minute (ovary and fruit less than 1.5 mm long), imperfect, lacking a calyx and corolla; fruits indehiscent and separating into 4 achenelike nutlets; leaf bases mostly connected around the stem by a narrow herbaceous membrane . . . . 3. CALLITRICHE
8. Flowers larger (ovary or fruit more than 1.5 mm long), perfect, with both a calyx and corolla; fruits capsules, variously dehiscent, not separating into nutlets; leaf bases not connected around the stem
9. Inflorescences axillary, dense spikes of numerous small flowers, these short and headlike or more elongate; corollas deeply 4-lobed, papery . . . . 15. PLANTAGO
9. Inflorescences terminal or axillary, if axillary then various but not dense spikes of numerous flowers; corollas mostly 5-lobed, if 4-lobed then only shallowly so
10. Corollas with an elongate, slender spur at the base; stems of 2 types: flowering stems with alternate leaves above the base; and short, spreading, vegetative stems with the leaves opposite or in whorls of 3 . . . . 13. NUTTALLANTHUS
10. Corollas lacking a spur or saclike outgrowth at the base (corolla tube slightly swollen or pouched on the upper side above the base in *Collinsia*); stems uniform, the plants not producing modified vegetative stems different from the flowering ones, the leaves all opposite or whorled
11. Leaves all or mostly in whorls of 3–7 . . . . 17. VERONICASTRUM
11. Leaves all opposite
12. Leaf blades deeply lobed or parted . . . . 10. LEUCOSPORA
12. Leaf blades entire, toothed, or lobed less than halfway to the midvein
13. Corollas moderately to relatively deeply 4-lobed, not bilabiate; calyces often 4-lobed; fruits flattened, at least slightly notched at the tip . . . . 16. VERONICA
13. Corollas shallowly to relatively deeply 5-lobed, sometimes bilabiate (the upper lip 2-lobed, the lower lip 3-lobed); calyces 5-lobed; fruits not or only slightly flattened, variously tapered to rounded at the tip, but not concave or notched
14. Inflorescences terminal and sometimes also axillary, at least the terminal ones spikes, racemes, panicles, or clusters of more than 2 flowers
15. Calyces with a well-developed tube (but this shorter than the lobes; lower corolla lip with the middle lobe folded lengthwise around the stamens and partially obscured by the broad lateral lobes . . . . 6. COLLINSIA

15. Calyces divided nearly to the base, lacking a well-developed tube; lower corolla lip with the middle lobe similar to the lateral ones, not folded, obscured or hiding the stamens (corollas sometimes not strongly bilabiate in *Penstemon*)
    16. Flowers sessile or nearly so (inflorescences spikes), each inflorescence usually subtended by a pair of slightly reduced, leaflike bracts, each flower subtended by 2 or 3 prominent, sepaloid bractlets positioned immediately below the calyx . . . . 5. CHELONE
    16. Flowers short- to long-stalked (inflorescences panicles or occasionally racemes), each node a pair of leaflike or scalelike bracts, the individual flowers not subtended by bracts or bractlets (in racemose inflorescences, at most 1 bract positioned at the base of each flower stalk) . . . . 14. PENSTEMON

14. Inflorescences axillary, the flowers solitary or paired (rarely to 4 in *Bacopa*) at each node
    17. Fertile stamens 2 (2 staminodes often present, these appearing as antherless filaments); corollas bilabiate; flowers with a pair of small bractlets positioned immediately below the calyx at the tip of the flower stalk . . . . 8. GRATIOLA
    17. Fertile stamens 4; corollas not bilabiate, weakly zygomorphic; flowers lacking bractlets or these positioned at the base of the flower stalk
        18. Stems usually submerged or floating on water, when stranded on mud then prostrate; leaf blades broadly obovate to nearly circular; flowers lacking bractlets . . . . 2. BACOPA
        18. Stems terrestrial or emergent from shallow water, erect or ascending; leaf blades mostly oblanceolate; flowers with a pair of bractlets positioned at the base of the stalk . . . . 12. MECARDONIA

1. **Antirrhinum** L. (snapdragon) (Sutton, 1988)

About 20 species, Mediterranean region in Europe, Asia, Africa.

1. Antirrhinum majus L. (snapdragon, common snapdragon)

Pl. 480 c, d; Map 2192

Plants perennial herbs (frequently flowering the first year), with fibrous roots, terrestrial. Stems erect or ascending, glabrous or more commonly sparsely to moderately glandular-hairy toward the tip. Leaves opposite toward the stem base, alternate toward the stem tip, short-petiolate, the petiole bases somewhat expanded, those of the opposite leaves frequently connected around the stem by a narrow herbaceous membrane. Leaf blades 2–7 cm long, 4–13 mm wide, linear to narrowly lanceolate, angled or tapered to a sharply pointed tip, tapered at the base, unlobed, the margins entire, the venation pinnate or with only a midvein apparent, glabrous or the undersurface sparsely glandular-hairy along the midvein. Inflorescences terminal racemes, the flower stalks 3–9 mm long at flowering, becoming elongated to 6–15 mm at fruiting, each subtended by a short, broadly lanceolate to ovate bract; bractlets absent. Flowers perfect. Calyces 4.5–8.0 mm long, deeply 5-lobed nearly to the base, the lobes slightly unequal in length, broadly ovate to oblong-ovate or elliptic-ovate, rounded or bluntly pointed at the tip, moderately to densely glandular-hairy. Corollas 27–45 mm long, bilabiate, 5-lobed, often glandular-hairy on the outer surface, especially on the tube, the tube about as long as the lobes, usually multicolored, the lower lip and throat often yellow, the remainder variously white and pink, red, reddish purple, or bluish purple, the tube with a short, broad, saclike outgrowth at the base (this positioned between the lower 2 calyx lobes), the throat mostly closed by the strongly convex base of the lower lip, the upper lip more or less straight, often somewhat folded longitudinally between the 2 lobes, the lower lip with the lobes more or less recurved. Fertile stamens 4, the filaments of 2 lengths, not exserted, the anther sacs more or less spreading; staminodes absent. Style 1, not exserted, the stigma capitate, unlobed. Fruits capsules, 10–15 mm long, obliquely ovoid, usually glandular-hairy on the outer surface, the 2 locules unequal in size, dehiscent by pores near the tip, the larger upper locule with 2 pores, the smaller, lower locule with 1 pore. Seeds numerous, 0.8–1.4 mm long, bluntly triangular to more or less trapezoid in profile, not flattened, the surface reddish brown to brown, with a network of coarse, jagged, winglike ridges. $2n$=16, 32. June–November.

Introduced, uncommon, known thus far only from the city of St. Louis (native of Europe; introduced widely but sporadically). Railroads.

Snapdragon is widely cultivated as a bedding plant but only rarely escapes into the wild in the Midwest. Numerous color forms and cultivars exist, including those with doubled corollas. According to Sutton (1988), the origin of ssp. *majus* is not known with certainty, but the plant likely was brought into cultivation first in southwestern Europe. The fruits, when viewed from the top, have been described fancifully as resembling a face, with the two pores of the upper locule the eyes, the single pore of the lower locule the mouth, and the persistent style base the nose (Holmgren, 1986).

Sutton (1988) separated wild populations into 5 subspecies differing in a suite of stem, leaf, inflorescence, flower, and seed characters. Of these, the widespread ssp. *majus* is thought to have given rise to the cultivated plants.

2. **Bacopa** Aubl.

About 56 species, nearly worldwide, most diverse in tropical and warm-temperate regions.

For a discussion of the potential segregation of this genus and its relatives from the Plantaginaceae, see the treatment of *Gratiola*. Steyermark (1963) included a second species for Missouri, *B. acuminata* (Walter) B.L. Rob., but that species currently is treated in the genus *Mecardonia* by most botanists, following Pennell (1946). For further discussion, see the treatment of *Mecardonia*.

A few species of *Bacopa* are cultivated as aquarium and pond plants, notably *B monnieri* (L.) Wettst. In Asia, the seeds of some species are ingested, purportedly to enhance memory function.

1. Bacopa rotundifolia (Michx.) Wettst. (water hyssop)

Pl. 480 e, f; Map 2193

Plants annual or perennial herbs, with fibrous roots, aquatic or sometimes terrestrial when stranded on mud. Stems (5–)10–60 cm long, at least the lower portion usually submerged, the apical portion weak and floating on water (occasionally entire plant floating), when terrestrial then prostrate to loosely ascending, rooting at the lower nodes, not angled, moderately to densely and finely short- hairy, at least on newer growth. Leaves opposite, sessile. Leaf blades 1.0–3.5 cm long, (8–)10–28 mm wide, relatively thick (but drying very thin and fragile), broadly spatulate to more commonly broadly oval or circular, unlobed, rounded at the tip, shallowly cordate to rounded or broadly angled at the base, mostly at least slightly clasping the stem, the margins entire, but sometimes finely short-hairy, the venation palmate with 5–9 main veins, the surfaces glabrous or nearly so at maturity, the upper surface somewhat shiny. Inflorescences axillary, the flowers 1 or 2(–4) per leaf axil, the flower stalks 6–23 mm long; bractlets absent. Flowers perfect. Calyces 3.0–4.5 mm long at flowering, becoming enlarged to 4–6 mm at fruiting, deeply 5-lobed nearly to the base, glabrous, the outer 3 lobes leaflike, ovate to broadly ovate, the inner 2 inconspicuous, lanceolate. Corollas 4.5–9.0 mm long, not bilabiate, weakly zygomorphic, 5-lobed, glabrous, the tube slightly shorter than to about as long as the lobes, white or less commonly very pale pinkish-tinged with a yellow throat, spurless, the throat open, the lobes

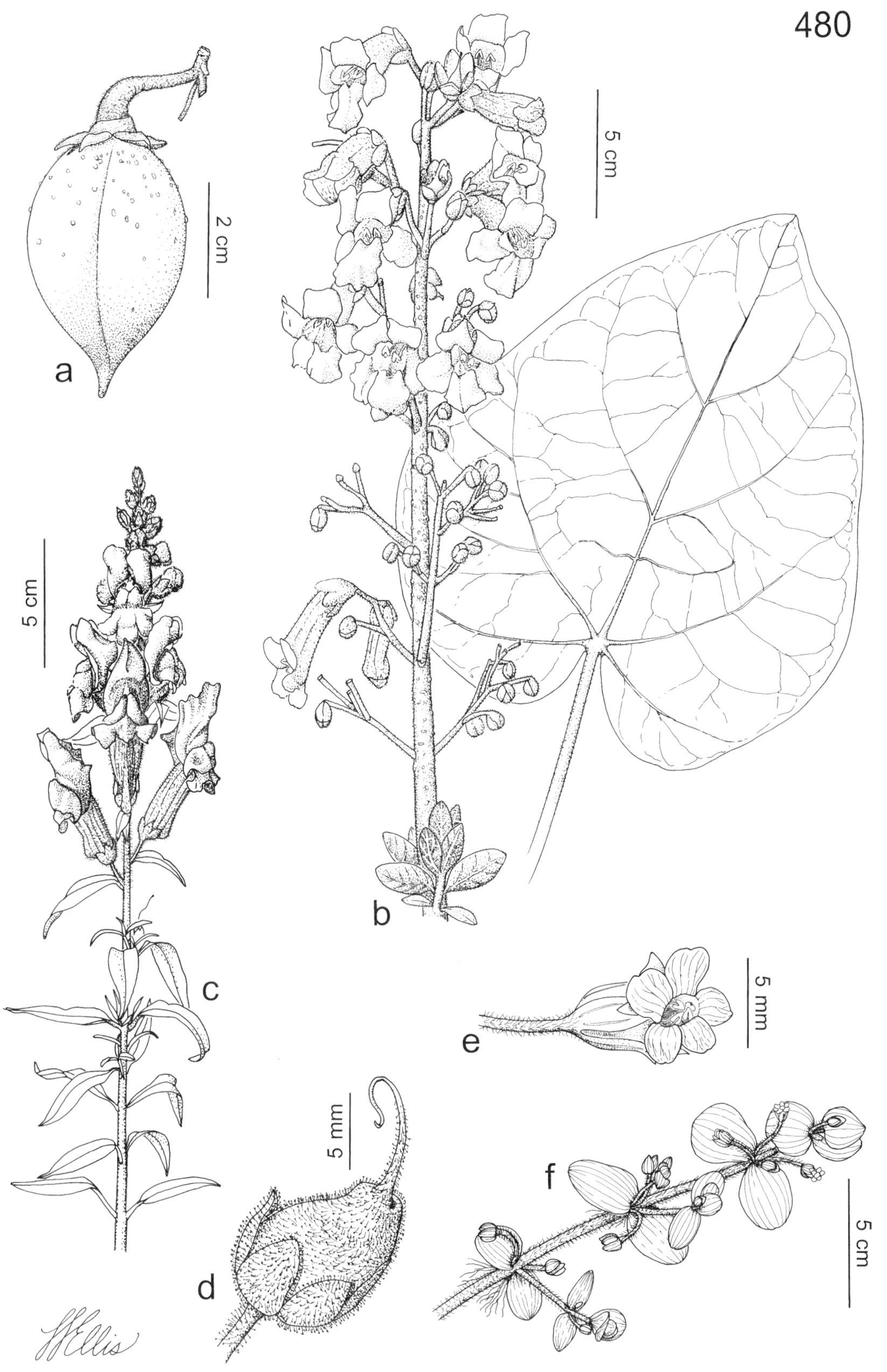

Plate 480. Paulowniaceae. Plantaginaceae. *Paulownia tomentosa*, **a)** fruit, **b)** leaf and inflorescence. *Antirrhinum majus*, **c)** fertile stem, **d)** fruit. *Bacopa rotundifolia*, **e)** flower, **f)** habit.

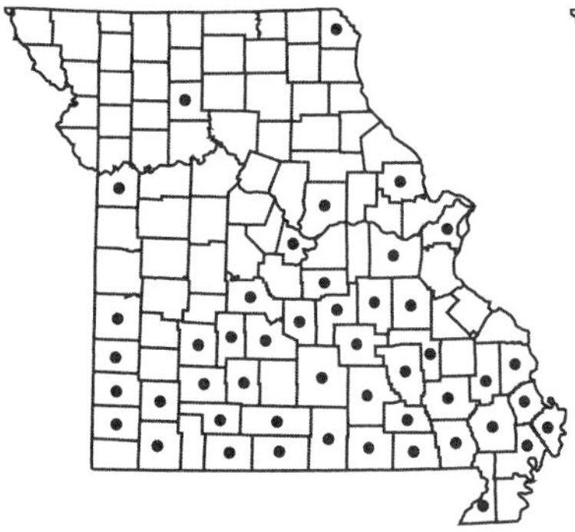

2194. Callitriche heterophylla    2195. Callitriche palustris 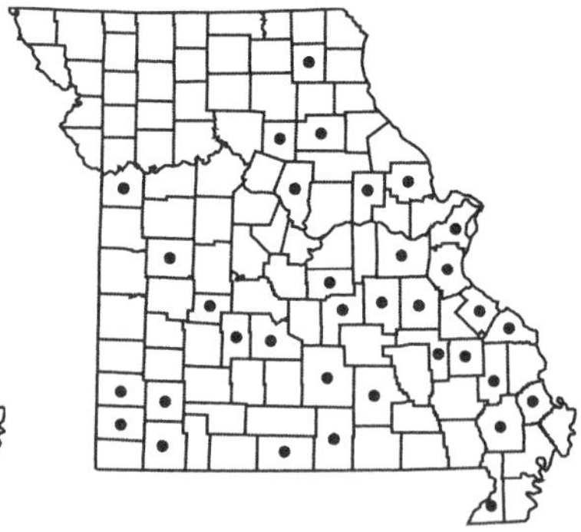   2196. Callitriche terrestris

collectively bell-shaped above the slender tube, individually oblong-obovate with more or less truncate but shallowly notched tips. Fertile stamens 4, the filaments of 2 lengths, not exserted, the anthers attached near their midpoints, the anther sacs parallel; staminodes absent. Style 1, not exserted, forked near the tip, each branch with an expanded, irregularly oval, cushionlike stigma at the tip. Fruits capsules, 3.5–5.5 mm long, globose to broadly ellipsoid, glabrous, the 2 locules equal in size, dehiscent longitudinally by 4 valves. Seeds numerous, 0.4–0.6 mm long, ellipsoid to cylindric with a minute tail-like appendage at each end, not flattened, the surface yellowish brown, with a network of fine ridges. $2n=36$. May–September.

Scattered south of the Missouri River, uncommon farther north (eastern U.S. [except some northeastern states] west to Idaho and Arizona; Canada; introduced in California and Asia). Floating or mostly submerged aquatics in shallow water of ponds, lakes, sinkhole ponds, swamps, and sloughs, less commonly in backwater areas of streams; also ditches and rice fields; sometimes stranded on wet mud.

3. **Callitriche** L. (water starwort)
(Philbrick, 1989; N. G. Miller, 2001)
Contributed by Alan E. Brant

Plants annual or less commonly perennial herbs, monoecious, glabrous but sometimes appearing minutely scaly or yellowish-glandular spotted at high magnification. Stems simple to much branched, often rooting at the nodes. Leaves opposite or sometimes appearing whorled at the stem tip, sessile or short- to long-petiolate. Leaf blades simple, entire, linear to spatulate, the bases mostly connected around the stem by a narrow herbaceous membrane. Stipules absent. Flowers imperfect, tiny, solitary or in small clusters of 2 or 3 per leaf axil, the staminate and pistillate flowers usually together in the same axil, in some species subtended by a pair of minute bracts. Perianth absent. Staminate flower(s) maturing before the adjacent pistillate flower(s), consisting of 1 stamen, the slender filament attached at the anther base. Pistillate flower(s) consisting of 1 pistil, the ovary superior, of 2 carpels, somewhat flattened and 4-lobed, each carpel with 2 ovules. Styles 2, elongate, often persistent at fruiting, the stigmatic region toward the slender tip. Fruits depressed-elliptic or more or less heart-shaped in outline, the ovary separating into 4 achenelike nutlets, these oblong-elliptic to kidney-shaped or obovate in outline, flattened and sometimes narrowly winged, the surface with a very fine network of ridges (observable only with magnification), light brown to brown. About 50 species, nearly worldwide.

Some species of water starwort are used horticulturally in outdoor pools and ponds. The herbage is eaten by waterfowl and some species of fish. The seeds are dispersed by water and by animals, and they reportedly survive passage through the digestive system of ducks. Studies have shown that species of *Callitriche* have developed an unusual breeding system where pollen germinates within intact, unopened anthers, and the pollen tubes grow downward inside the filament and then through the vegetative tissue and into adjacent pistillate flowers to effect pollination (Philbrick, 1984). This presumably helps to ensure abundant seed production in populations where extreme and rapid environmental fluctuations occur. Also, this helps to explain observations of seed production in plants whose stamens appear to be rudimentary and incapable of dehiscing to release pollen.

As with many groups of aquatic angiosperms having reduced floral structures, the relationships of *Callitriche* have been controversial. Traditionally, the most widely accepted alignment was as a separate family, Callitrichaceae, close to the Euphorbiaceae (Steyermark, 1963), presumably because of superficial similarities in fruit morphology. Studies in ovule and embryo development suggested a placement as a separate family close to the Lamiaceae and Verbenaceae, which has been followed more recently by a number of botanists (Cronquist, 1981, 1991). The most recent evidence from molecular studies has classified *Callitriche* into the expanded Plantaginaceae. For further discussion, see the paragraph under the family description above.

1. Plants usually terrestrial (sometimes submerged for short periods in temporary pools); fruits short-stalked, equally rounded at the tip and base; leaves monomorphic . . . . 3. C. TERRESTRIS
1. Plants usually aquatic (sometimes stranded on mud); fruits sessile, more rounded at the tip than the base; leaves not monomorphic, the floating ones shorter and broader than the submerged ones
    2. Fruits 0.5–0.9 mm long, about as long as wide, the margins thin but mostly unwinged, not appearing thickened toward the base . . . . . 1. C. HETEROPHYLLA
    2. Fruits 1.1–1.4 mm long, slightly longer than wide, the margins narrowly winged, at least above the midpoint, the basal, unwinged portion appearing thickened . . . . 2. C. PALUSTRIS

1. Callitriche heterophylla Pursh **var. heterophylla** (greater water starwort)

Pl. 479 h, i; Map 2194

Stems 5–35 cm long, prostrate to ascending. Leaves 3–20 mm long, polymorphic; the submerged leaves mostly linear, shallowly notched at the tip, 1-nerved, usually grading into the floating leaves; these obovate to spatulate, rounded at the tip, 3(5)-nerved; forming rosettes on the surface of the water (the rosette leaves progressively shorter and more rounded); leaves on stranded plants linear to obovate. Flowers subtended by two minute, pale, inflated bracts, these shed before the fruits mature. Fruits 0.5–0.9 mm long, 0.6–0.9 mm wide, sessile, more or less heart-shaped, with the tip more rounded than the base, the nutlets narrow-margined but often unwinged. $2n=20$. March–November.

Scattered, mostly south of the Missouri River; apparently absent from portions of the Unglaciated Plains and Glaciated Plains Divisions (New England west to southern British Columbia and south to northern Mexico). Mostly entirely submerged aquatics (stem tips usually with floating leaves) in spring branches, streams, rivers, sloughs, ponds, sinkhole ponds, and rarely small pools in depressions of sandstone glades; also ditches and seasonally flooded fallow fields; occasionally stranded on wet mud.

Callitriche heterophylla is a characteristic plant of spring branches in the Ozarks, where it tends to inhabit inlets or pools with slower-flowing water. However, it is also found in a variety of other aquatic habitats. Completely submerged plants might at first be confused with other narrow-leaved aquatics but can be distinguished readily by their unusual fruits and opposite, mostly narrowly perfoliate leaves. The other variety, *C. heterophylla* var. *bolanderi* (Hegelm.) Fassett, is restricted to the western United States, adjacent Canada, and Alaska, and it differs only in its slightly longer (0.9–1.1 mm) fruits.

2. Callitriche palustris L. (vernal water starwort)

C. palustris var. *verna* (L.) Fenley ex Jeps.
C. verna L.

Map 2195

Stems 10–30 cm long, prostrate to ascending. Leaves 3–20 mm long, polymorphic; the submerged

2197. Chaenorhinum minus

2198. Chelone glabra

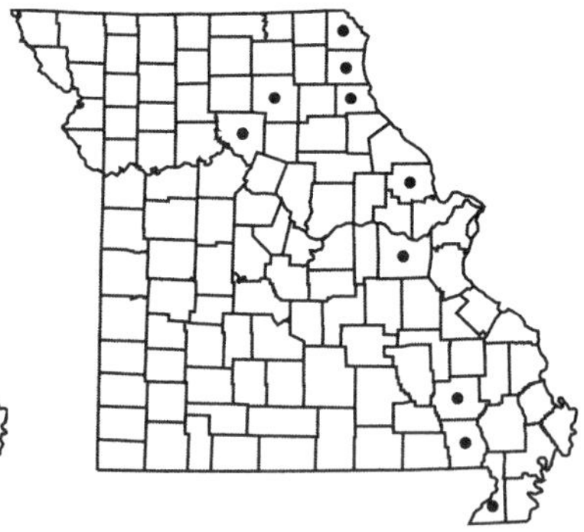

2199. Chelone obliqua

leaves mostly linear, shallowly notched at the tip, 1-nerved, usually grading into the floating leaves; these obovate to spatulate, rounded at the tip, 3(5)-nerved; forming rosettes on the surface of the water (the rosette leaves progressively shorter and more rounded); leaves on stranded plants linear to obovate. Flowers subtended by 2 minute, pale, inflated bracts, these shed before the fruits mature. Fruits 1.1–1.4 mm long, 0.9–1.2 mm wide, sessile, more or less heart-shaped, with the tip more rounded than the base, the nutlets narrowly winged, at least above the midpoint, the basal, unwinged portion appearing somewhat thickened. $2n=20$. March–September.

Known thus far from a single site in St. Louis County (northern U.S. south to Virginia, Texas, and California; Alaska, Canada, Greenland, South America, Europe, Asia). Mostly submerged aquatics (stem tips usually with floating leaves) in spring branches.

Missouri specimens of this species were first collected by Bill Summers in 1979, but these were first determined as the closely related *C. heterophylla* and went unnoticed until recently. Vegetatively, the two species are indistinguishable, although the Missouri specimens of *C. palustris* collected thus far tend to be relatively robust with a greater proportion of more broadly obovate leaves (this probably due to environmental influences). *Callitriche palustris* also grows in portions of Nebraska, Iowa, and Illinois, and thus should be searched for elsewhere in northern and eastern Missouri.

3. Callitriche terrestris Raf. **ssp. terrestris** (terrestrial starwort)

Pl. 479 f, g; Map 2196

Stems 1.5–5.0 cm long, prostrate. Leaves 1.5–4.0 mm long, monomorphic, oblanceolate to obovate or spatulate; faintly 3-nerved, not forming rosettes at the stem tips. Flowers without bracts. Fruits 0.5–0.9 mm long, 0.6–0.9 mm wide, with stalks 0.3–0.6 mm long, more or less heart-shaped, equally rounded at the tip and the base, the nutlets usually narrowly winged. $2n=10$. March–August.

Scattered in the southern ⅔ of the state (eastern U.S. west to Wisconsin, Kansas, and Texas). Terrestrial on banks of streams and rivers, margins of ponds, bottomland forests, and mesic upland forests; also fallow fields, crop fields, roadsides, paths, and disturbed areas; rarely submerged aquatics in temporarily flooded depressions or a weed in greenhouses; most commonly in acidic, bare, moist soils at shaded sites.

Plants of this species are small and easily overlooked or mistaken for a moss or liverwort. They probably are more common than collections indicate. Two other subspecies, ssp. *subsessilis* (Fassett) Bacig. and ssp. *turfosa* (Bertero ex Hegelm.) Bacig., occur in Central and South America respectively.

4. Chaenorhinum (DC.) Rchb.

About 21 species, Mediterranean region in Europe, Asia, Africa.

1. Chaenorhinum minus (L.) Lange **ssp. minus** (dwarf snapdragon)

Pl. 481, a, b; Map 2197

Plants annuals, with taproots, terrestrial. Stems erect or strongly ascending, moderately to densely glandular-hairy. Leaves alternate or the

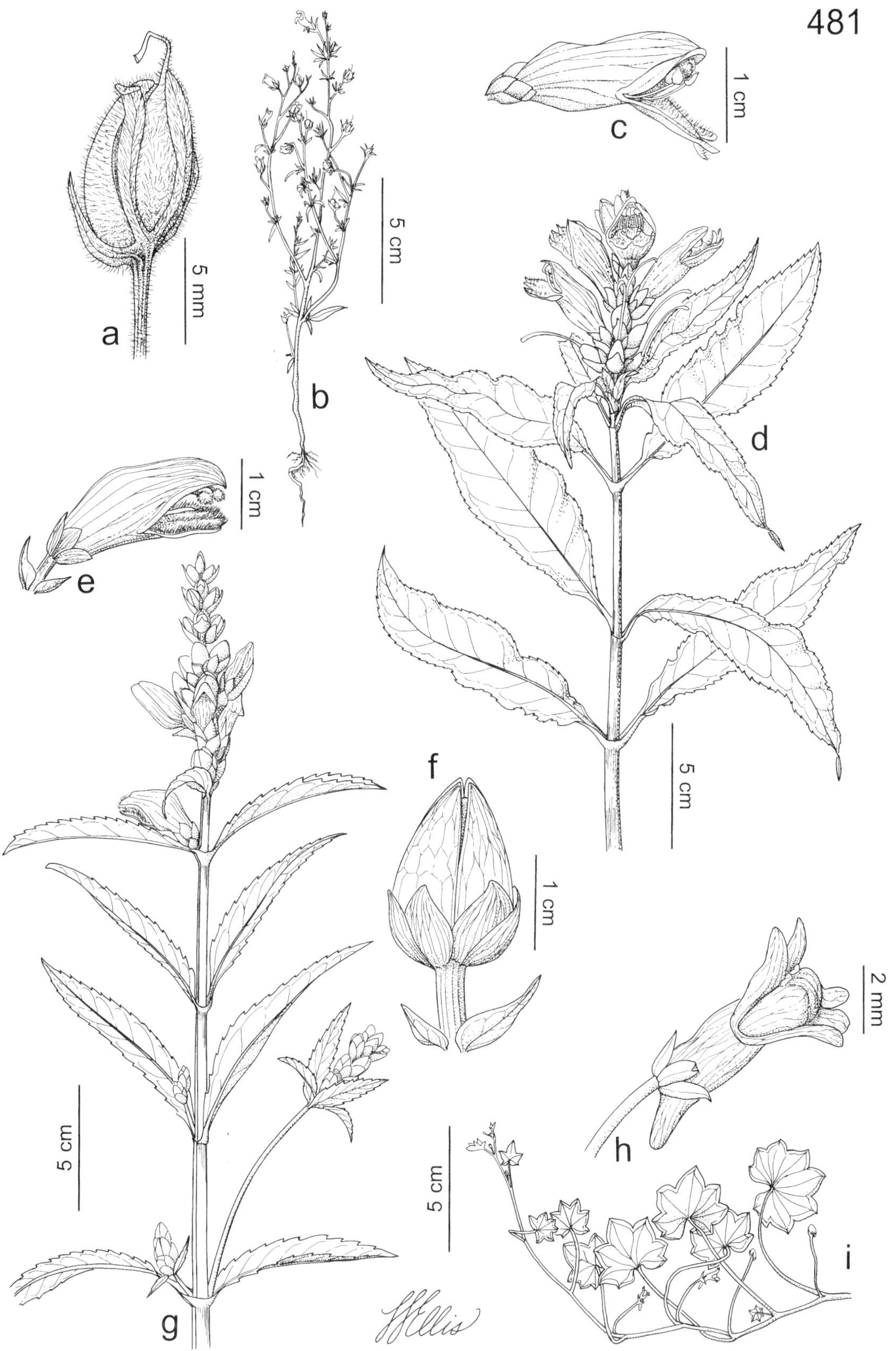

Plate 481. Plantaginaceae. *Chaenorhinum minus*, **a)** fruit, **b)** habit. *Chelone obliqua*, **c)** flower, **d)** fertile stem. *Chelone glabra*, **e)** flower, **f)** fruit, **g)** fertile stem. *Cymbalaria muralis*, **h)** flower, **i)** habit.

lowermost few opposite, sessile or very short-petiolate, not expanded at the base. Leaf blades 0.5–3.0 cm long, 1–3 mm wide, linear to narrowly oblong-oblanceolate, mostly rounded or angled to a bluntly pointed tip, tapered at the base, unlobed, the margins entire, mostly with only a midvein apparent, the surfaces and margins moderately glandular-hairy. Inflorescences terminal racemes (often appearing as solitary axillary flowers), the flower stalks 4–13 mm long at flowering, becoming elongated to 5–23 mm at fruiting, each subtended by a leaflike, linear bract; bractlets absent. Flowers perfect. Calyces 2.0–3.5 mm long at flowering, becoming enlarged to 3.5–4.5 mm at fruiting, deeply 5-lobed nearly to the base, the lobes slightly unequal in length, linear to narrowly oblanceolate, rounded or bluntly pointed at the tip, moderately to densely glandular-hairy. Corollas 4.5–8.0 mm long, bilabiate, 5-lobed, glandular-hairy on the outer surface, especially on the tube, the tube about as long as or slightly longer than the lobes, white to lavender or purple, the lower lip and throat white or with pale yellow markings, sometimes pale purplish- or pinkish-tinged, the remainder variously white, lavender, or light purple, the tube with a well-developed, narrowly conic spur 1.7–2.8 mm long at the base (this positioned between the lower 2 calyx lobes), the throat not closed by the noticeably convex base of the lower lip, the upper lip more or less straight, the lower lip with the lobes more or less spreading. Fertile stamens 4, the filaments of 2 lengths, not exserted, the anther sacs more or less spreading; staminodes absent. Style 1, not exserted, the stigma capitate, unlobed. Fruits capsules, 3–6 mm long, broadly oblong-ovoid to globose, usually glandular-hairy on the outer surface, the 2 locules unequal in size, dehiscent by irregular pores at the tip, each locule with 1 pore. Seeds numerous, 0.5–0.8 mm long, oblong to oblong-ovate in profile, not flattened, the surface dark brown, with several longitudinal ridges, also with minute papillae. $2n=14$, 28. May–September.

Introduced, widely scattered in the state (native of Europe, Asia, introduced widely in the U.S. [most abundantly in northeastern states]; Canada). Bottomland prairies and upland prairies; also railroads, roadsides, and disturbed areas.

Sutton (1988) treated *C. minus* in its native range as comprising three subspecies, with only ssp. *minus* naturalizing in the New World. He and some other authors also used an altered but incorrect spelling of the generic name as *Chaenorrhinum*. The other subspecies differ in details of calyx and corolla size, spur length, and seed size. Widrlechner (1983) summarized observations from literature and herbarium records to conclude that from its initial discovery in New Jersey in 1874 (where it probably arrived as seeds contaminating ship's ballast), *C. minus* has spread effectively and rapidly using railroad corridors as its main avenue of dispersal. The species was first collected in Missouri in 1946, along railroad tracks in Audrain and Pike Counties (Steyermark, 1949b).

5. **Chelone** L. (turtlehead)

Plants perennial herbs, with rhizomes, terrestrial or occasionally emergent aquatics. Stems erect or ascending (sometimes sprawling with age), unbranched or branched toward the tip, glabrous. Leaves opposite, sessile or short-petiolate, sometimes slightly expanded at the very base but not clasping the stem. Leaf blades simple, unlobed, linear to elliptic or ovate, the margins entire or toothed, the surfaces glabrous or the undersurface minutely hairy, mostly or exclusively along the main veins, the venation pinnate. Inflorescences terminal spikes (occasionally also at the tips of mostly leafless branches, then appearing more or less axillary), 3–9 cm long, dense (the axis not visible between the flowers), the inflorescence usually subtended by a pair of slightly reduced, leaflike bracts, the individual flowers sessile or the uppermost sometimes minutely stalked, each subtended by 2 or 3 prominent, sepaloid bractlets positioned immediately below the calyx. Flowers perfect. Calyces sometimes becoming slightly enlarged at fruiting, deeply 5-lobed, the lobes equal or slightly unequal in length, broadly elliptic to broadly ovate-elliptic, rounded to broadly and bluntly pointed at the tip, glabrous or the margins fringed with usually minute, nonglandular hairs. Corollas 25–35 mm long, bilabiate, 5-lobed but often appearing more or less 4-lobed, glabrous externally, the tube longer than the lobes, white to all or partially pinkish-tinged to pink or reddish purple, lacking colored nectar guides (these present elsewhere), the tube gradually enlarged from near the base, lacking a spur, the throat mostly closed, appearing somewhat flattened, the lower lip shallowly 3-lobed, convex, and bearded with woolly hairs, the upper lip slightly to moderately keeled and arched downward or slightly helmet-shaped, minutely notched to shallowly 2-lobed at the tip. Fertile stamens 4, the filaments of 2 lengths, not exserted, flattened and hairy, curved inward toward their tips, the anther sacs spreading, densely woolly; staminode 1, shorter than the stamens, linear to narrowly strap-shaped, positioned along the lower side of the corolla tube, glabrous (except occasionally for a few minute hairs at the tip), the tip straight and usually truncate. Style 1, not exserted, the stigma capitate, unlobed. Fruits capsules, broadly ovoid, broadly ovoid-elliptic, or nearly globose, tapered to a sharply pointed or minutely beaked tip, glabrous, the 2 locules equal in size, dehiscent longitudinally along the 2 sutures. Seeds numerous, 3–4 mm long, broadly oval to more or less circular in outline, strongly flattened, the body brown to dark brown, usually only on 1 of the surfaces, sometimes lighter-mottled or -streaked, encircled by a broad wing, this tan to light brown, often with a faint to more prominent, radiating pattern of somewhat translucent, fine lines or streaks, the surface otherwise glabrous, more or less smooth. Four species, eastern temperate North America.

The genus *Chelone* is related to *Penstemon*. Both possess relatively well-developed staminodes. For further discussion of similarities and differences between the two genera, see the treatment of *Penstemon*.

Pennell (1935) described the pollination mechanism in the genus, which is mainly pollinated by bumblebees. A bee landing on the flower grasps the lower lip, the weight pulling apart the otherwise closed throat. In entering the throat of the flower the bee causes the anthers to separate, which dust the thorax of the insect with pollen (in addition to pollen it collects intentionally).

The two Missouri turtlehead species are both sold as garden ornamentals, particularly *C. obliqua*. They can tolerate relatively dense shade, but require a moist site. *Chelone glabra* was a minor ceremonial and medicinal plant for Native Americans, who prepared an infusion for the treatment of worms and unwanted pregnancies, among other uses (Moerman, 1998).

1. Corollas white or pale cream-colored, sometimes greenish yellow to green toward the tip or pinkish- to rarely purplish-tinged toward the tip or in the throat; staminode green . . . . 1. C. GLABRA
1. Corollas pink to reddish purple throughout, sometimes somewhat lighter toward the tip or base; staminode white or very pale yellow . . . . 2. C. OBLIQUA

1. Chelone glabra L. (white turtlehead)
C. glabra var. *linifolia* N. Coleman
C. glabra f. *tomentosa* (Raf.) Pennell
Pl. 481 e–g; Map 2198

Stems 50–150(–200) cm long. Leaves sessile or short-petiolate, the petioles usually less than 5 mm long. Leaf blades (except those of the bracts immediately below the inflorescences) 2–20 cm long, 5–30(–40) mm wide, linear to lanceolate or narrowly elliptic-lanceolate, angled or tapered to a sharply pointed tip, angled at the base, the margins sharply toothed to nearly entire, the teeth ascending to spreading, sometimes small and/or widely spaced; the surfaces glabrous or the undersurface occasionally minutely hairy. Bracts 1.5–8.0 cm long, linear to narrowly lanceolate or narrowly oblong-lanceolate. Bractlets 4–10(–22) mm long, mostly ovate (the lowermost sometimes narrower), all but the uppermost pointed at the tip. Calyces 5–11 mm long, the lobe margins sometimes thin and translucent, glabrous or minutely hairy. Corollas 25–35 mm long, white or pale cream-colored, sometimes greenish yellow to green toward the tip or pinkish- to rarely purplish-tinged toward the tip or in the throat. Staminode green, often darker toward the tip and pale green toward the base. $2n$=28. July–October.

Uncommon in the southeastern quarter of the state west locally to Camden and Saline Counties (eastern U.S. west to Minnesota and Arkansas; Canada). Banks of streams and spring branches, bottomland forests, fens, bases and ledges of bluffs.

Pennell (1935) treated *C. glabra* as comprising a complex series of seven subspecies. Crosswhite (1965) studied variation within the species in Wisconsin and concluded that there was too much morphological overlap to permit formal recognition of infraspecific taxa. A broader popula-

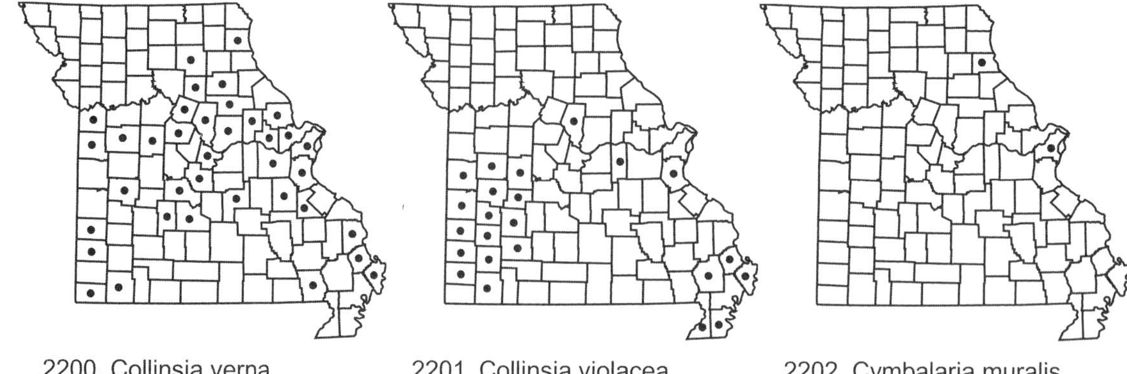

2200. Collinsia verna

2201. Collinsia violacea

2202. Cymbalaria muralis

tional sampling studied for morphological and allozyme variation led A. D. Nelson and Elisens (1999) to reach a similar conclusion.

2. Chelone obliqua L. (rose turtlehead, pink turtlehead)

C. obliqua var. *speciosa* Pennell & Wherry

Pl. 481 c, d; Map 2199

Stems 40–150 cm long. Leaves short-petiolate, the petioles mostly 5–15 mm long, the petioles often winged toward the tip or for most of their length. Leaf blades (except those of the bracts immediately below the inflorescences) 4–20 cm long, 12–60 mm wide, lanceolate to oblong-lanceolate, elliptic, or narrowly oblong-ovate, angled or tapered to a sharply pointed tip, angled or short-tapered at the base, the margins sharply toothed, the teeth ascending, usually relatively closely spaced; the surfaces glabrous (minutely hairy elsewhere). Bracts 1–10 cm long, oblong-lanceolate to elliptic or narrowly ovate. Bractlets 4–10(–17) mm long, mostly ovate (the lowermost sometimes narrower), usually pointed at the tip. Calyces 7–11 mm long, the lobe margins sometimes thin and translucent, minutely to short-hairy. Corollas 28–37 mm long, pink to reddish purple throughout, sometimes somewhat lighter toward the tip or base. Staminode white, sometimes very pale yellow toward the tip. $2n=84$ ($2n=56$ elsewhere). August–October.

Uncommon in the eastern half of the state, most abundant in the northeastern counties (eastern [mostly southeastern] U.S. west to Minnesota and Arkansas; Canada). Bottomland forests, sloughs, oxbows, fens, and banks of streams and spring branches; also levees, power line cuts, and roadsides.

Pennell (1935) treated *C. obliqua* as consisting of three subspecies. The Missouri plants generally have been treated as belonging to ssp. *speciosa* (Steyermark, 1963), which was said to be characterized by relatively large, bright corollas, calyx lobes strongly fringed with short hairs along the margins, and leaf blades relatively sharply toothed. However, even within the Missouri populations there is considerable variation for each of these characters. In their study of morphological and allozyme variation within the genus, A. D. Nelson and Elisens (1999) concluded that there was no basis for the formal recognition of infraspecific taxa in *C. obliqua*. Populations of *C. obliqua* in the Blue Ridge region are tetraploids ($2n=56$), whereas those that have been sampled from elsewhere in the species range are hexaploid ($2n=84$) (A. D. Nelson et al., 1998; A. D. Nelson and Elisens, 1999).

6. **Collinsia** Nutt.

Plants annual, with taproots, terrestrial. Stems erect or strongly ascending, sometimes from a spreading base, unbranched or branched, moderately to densely pubescent with minute, glandular hairs toward the tip, glabrous toward the base. Leaves opposite, the uppermost sessile and clasping the stem, grading into the lower, short- to long-petiolate ones. Leaf blades simple, unlobed, variously lanceolate to oblong, elliptic, or ovate (those of the lowermost leaves sometimes broadly ovate to nearly circular), rounded or bluntly to sharply pointed at the tip, the margins otherwise entire to shallowly scalloped or toothed, the surfaces glabrous or sparsely and minutely hairy toward the base of the midvein, the venation pinnate, but the lateral veins sometimes very faint. Inflorescences terminal racemes, but often appearing axillary because of the pair of leaflike bracts at each node, the flowers opposite or whorled at the nodes, long-stalked; bractlets absent. Flowers perfect. Calyces somewhat irregularly 5-lobed to below the midpoint, bell-shaped, nearly actinomorphic but oriented slightly obliquely, the lobes narrowly lanceolate to lanceolate, sharply pointed at the tip, glabrous except sometimes minutely hairy along the margins, often pinkish- or purplish-tinged. Corollas bilabiate, 5-lobed, glabrous or the tube sparsely hairy, the upper lip 2-lobed, angled upward to recurved, white (occasionally faintly bluish- or purplish-tinged), the lower lip 3-lobed, purple or blue to purplish blue (sometimes grading to white basally or with a white central line on each lateral lobe, rarely entirely white), the lateral lobes arched somewhat downward to nearly straight, the central lobe shorter and folded lengthwise (enclosing the stamens and style), partially obscured by the lateral lobes, the tube shorter than the lobes, white or pale purplish- or bluish-tinged, lacking a spur, but slightly swollen or pouched on the upper side above the base, the throat closed by the convex or ridged base of the upper lip, often with yellow to brownish yellow mottling or spots on the upper side. Fertile stamens 4, the filaments of 2 lengths, hidden in the folded lowermost lobe, the anther sacs spreading, confluent at their tips; staminodes absent or more commonly 1 present as a short knoblike or ribbonlike structure. Style 1, hidden in the folded lowermost lobe, unbranched, the stigma small, capitate, unlobed or shallowly 2-lobed. Fruits capsules, more or less globose, circular in cross-section, glabrous, the 2 locules equal in size, dehiscent longitudinally along and between the 2 sutures. Seeds 2 to numerous, somewhat asymmetrically broadly oblong-ellipsoid to ovoid, more strongly convex on 1 side than on the other, the surface dark brown, with a minute and sometimes faint network of fine ridges in longitudinal rows (this often visible only with strong magnification). About 20 species, North America.

The genus *Collinsia* is most diverse in the western states, especially California. Our species are both winter annuals with seeds that germinate during the autumn, producing flowering stems the following spring and dying soon after the fruits mature. Pennell noted the structural similarities of the corollas to the papilionaceous flowers of many members of the legume family Fabaceae. He reviewed the floral adaptions to bee pollination, in which the weight of the landing insect causes the boat-shaped central lobe of the lower lip to spread, making the pollen accessible and bringing the anthers and stigma into contact with the lower abdomen of the insect.

Some species of *Collinsia* are prized for their attractive 2-colored corollas and are cultivated in sunny garden beds and in rock gardens. In Missouri, the species most-often grown is *C. verna*, which usually seeds itself readily. Steyermark (1963) noted that during his time populations of *C. violacea* frequently formed dense, showy patches along some roadsides in southwestern Missouri, but this is not usually the case today. However, *C. verna* is a conspicuous wildflower in some rich ravines and valleys, forming colorful carpets of blue and white.

1. Corolla lobes shallowly notched at the tip (except the folded central lobe of the lower lip); upper corolla lip about as long as the lower lip, the lower lip blue to purplish blue; main leaves broadest just above the base; fruits with (2)4 seeds, these 2.5–3.2 mm long . . . . 1. C. VERNA
1. Corolla lobes noticeably notched at the tip (except the folded central lobe of the lower lip); upper corolla lip noticeably shorter than the lower lip, the lower lip purple (rarely white); main leaves widest well above the base, often near the midpoint; fruits with 6–12 seeds, these 1.2–1.6 mm long . . . . 2. C. VIOLACEA

1. Collinsia verna Nutt. (blue-eyed Mary, eastern blue-eyed Mary)
Pl. 482 c–e; Map 2200

Stems 10–40 cm long, sparsely to moderately and minutely glandular-hairy toward the tip, the hairs frequently in longitudinal lines. Leaf blades

1–3(–5) cm long, those of the lower petiolate leaves broadly ovate to nearly circular, grading into the sessile main stem leaves, which are ovate to narrowly ovate, triangular-ovate, or occasionally oblong-ovate, broadest just above the base. Flower stalks 9–25 mm long at flowering, becoming elongated to 15–40 mm at fruiting. Calyces 3.5–5.0 mm long. Corollas 10–14(–17) mm long, the lobes shallowly (0.5–1.0 mm) notched, the upper lip about as long as the lower one, white or occasionally faintly bluish-tinged, the lower lip blue to purplish blue, sometimes with a longitudinal, white stripe or line on each lateral lobe. Fruits 4.0–6.5 mm long. Seeds (2)4 per fruit, 2.5–3.2 mm long. $2n$=14. April–June.

Scattered in the state, but absent from much of the Ozark Division and most of the western half of the Glaciated Plains (northeastern U.S. west to Wisconsin and Oklahoma; Canada). Banks of streams, rivers, and spring branches, bottomland forests, mesic upland forests, and bases and ledges of bluffs; also pastures, railroads, and roadsides.

2. Collinsia violacea Nutt. (violet collinsia, Ozark blue-eyed Mary)

Pl. 482 a, b; Map 2201

Stems 10–30(–45) cm long, sparsely to moderately and minutely glandular-hairy toward the tip, the hairs relatively uniform, not in lines. Leaf blades 1.0–3.0(–4.5) cm long, those of the lower petiolate leaves ovate to broadly ovate, grading into the sessile main stem leaves, which are oblong-lanceolate to oblong-elliptic or elliptic, broadest well above the base, often near the midpoint. Flower stalks 6–15 mm long at flowering, becoming elongated to 10–25 mm at fruiting. Calyces 5–7 mm long. Corollas 9–13 mm long, the lobes noticeably (2–3 mm) notched, the upper lip noticeably shorter than the lower one, white or occasionally faintly purplish-tinged, the lower lip purple, rarely white. Fruits 4–5 mm long. Seeds 6–12 per fruit, 1.2–1.6 mm long. $2n$=14. April–June.

Scattered in the southwestern quarter of the state, mostly in the Unglaciated Plains division, and also in the Mississippi Lowlands; sporadic in other counties adjacent to the Mississippi and Missouri Rivers (Kansas to Texas east to Illinois and Louisiana). Glades, dry upland forests, savannas, sand prairies, thin-soil areas of upland prairies, and ledges and tops of bluffs; also railroads and roadsides; often on acidic substrates.

Unlike most other members of the Plantaginaceae, the calyces of *C. violacea* frequently are not persistent at fruiting. Uncommon plants with entirely white corollas have been called f. *pallida* E.J. Palmer.

7. Cymbalaria Hill

Nine species, Europe, Asia.

1. Cymbalaria muralis P. Gaertn., B. Mey. & Scherb. (kenilworth ivy)

Pl. 481 h, i; Map 2202

Plants annuals, with fibrous roots (young plants may also have small taproots), terrestrial. Stems 10–40 cm long, prostrate or loosely spreading, sometimes twining and thus climbing on surrounding vegetation, rooting at the nodes, glabrous. Leaves alternate, long-petiolate, the petiole bases not expanded. Leaf blades 1.0–2.5 cm long, 10–35 mm wide, kidney-shaped to nearly circular in outline, shallowly and palmately 5(–9)-lobed or coarsely toothed, the lobes or teeth mostly short-tapered to a bluntly or sharply pointed tip, truncate to more commonly cordate at the base, the margins otherwise entire, sometimes with a narrow reddish marginal line, the venation palmate with as many main veins as lobes, glabrous. Inflorescences axillary, of solitary flowers, the flower stalks 10–35 mm long at flowering, becoming elongated to 22–70 mm at fruiting; bractlets absent. Flowers perfect. Calyces 1.5–2.0 mm long at flowering, becoming enlarged to 2–3 mm at fruiting, deeply 5-lobed nearly to the base, the lobes slightly unequal in length, linear to narrowly lanceolate, mostly sharply pointed at the tip, glabrous. Corollas 8–11 mm long (including the spur), bilabiate, 5-lobed, glabrous, the tube shorter than to about as long as the lobes, the tube bluish purple to purplish blue, the lobes similar, but the throat with white and yellow markings, the tube with a short, blunt spur at the base (this positioned between the lower 2 calyx lobes), the throat mostly closed by the strongly convex base of the lower lip, the upper lip more or less straight to somewhat arched outward, the lower lip with the lobes spreading to strongly arched. Fertile stamens 4, the filaments of 2 lengths, not exserted, the anther sacs spreading; staminodes absent. Style 1, not exserted, the stigma capitate, unlobed. Fruits capsules, 2.5–4.0 mm long, globose, glabrous, the 2 locules equal in size, dehiscent by pores near the

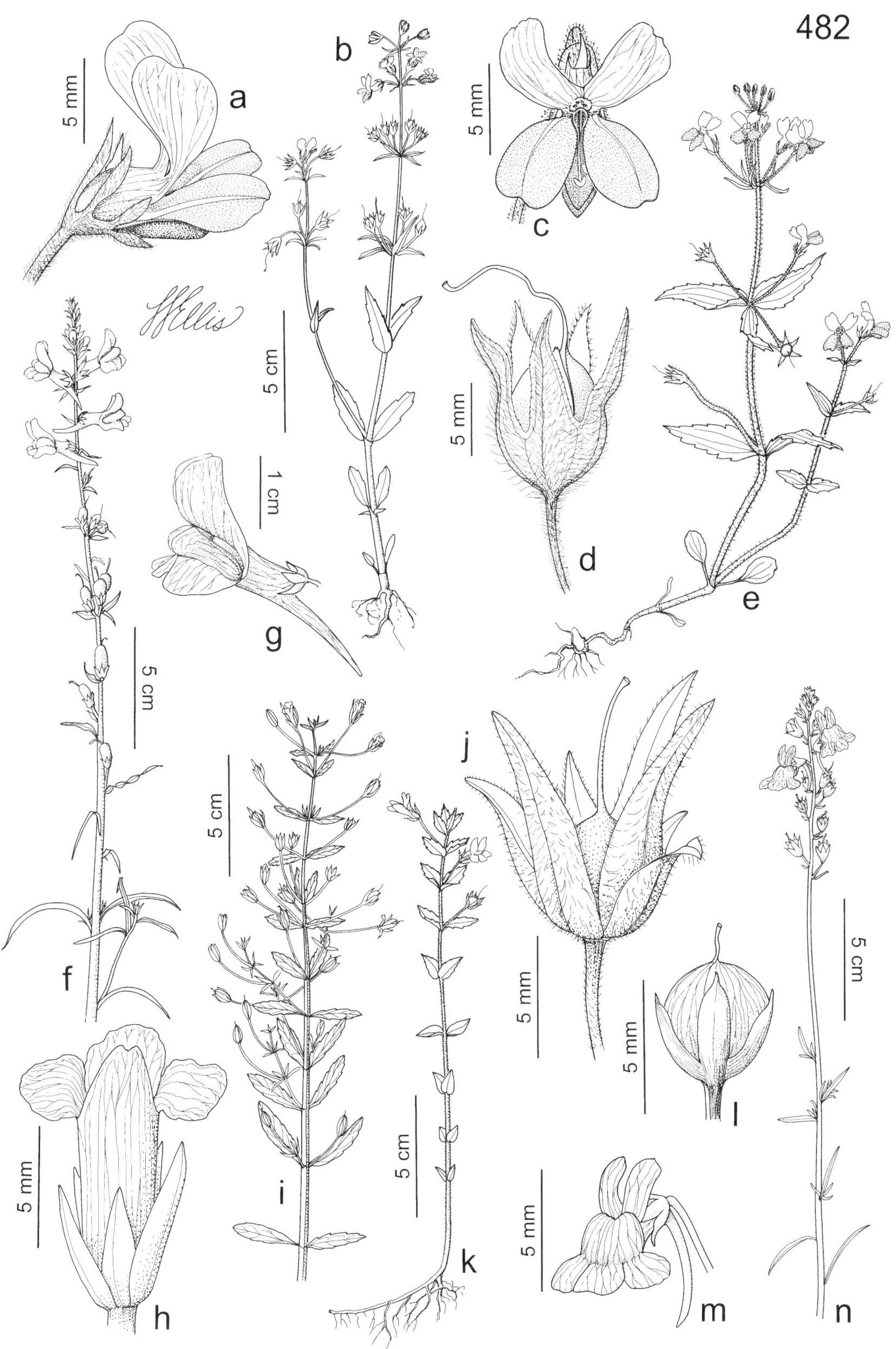

Plate 482. Plantaginaceae. *Collinsia violacea*, **a)** flower, **b)** habit. *Collinsia verna*, **c)** flower, **d)** fruit, **e)** habit. *Linaria vulgaris*, **f)** fertile stem, **g)** flower. *Mecardonia acuminata*, **h)** flower, **i)** fertile stem. *Gratiola viscidula*, **j)** fruit, **k)** habit. *Nuttallanthus texanus*, **l)** fruit, **m)** flower, **n)** fertile stem.

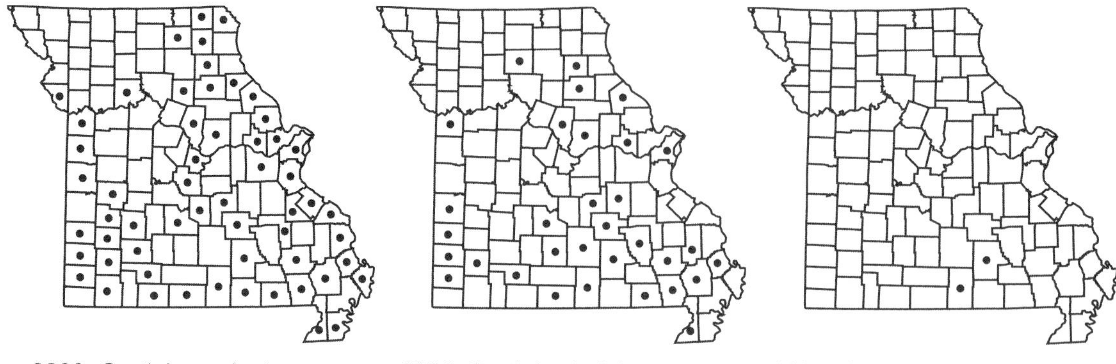

2203. Gratiola neglecta 2204. Gratiola virginiana 2205. Gratiola viscidula

tip, each locule with 1 pore, the valves eventually separating completely with age. Seeds numerous, 0.7–1.0 mm long, ellipsoid to nearly circular in profile, not flattened, the surface dark brown, with a network of usually mostly longitudinal, coarse ridges. $2n=14$. May–October.

Introduced, known thus far only from historical collections from Marion and St. Louis Counties (native of Europe, Asia; introduced widely but sporadically in the U.S., Canada). Banks of rivers; also roadsides and open, disturbed areas.

Kenilworth ivy is an attractive groundcover that is widely cultivated in gardens and rarely escapes into natural plant communities. Sutton (1988) recognized three subspecies within the native range based on differences in pubescence density and flower stalk lengths, but noted that the species is morphologically variable and occurs in a number of different habitats. Although the Missouri materials all appear to correspond to the glabrous ssp. *muralis*, it does not seem prudent to attempt to classify the North American plants below the species level.

8. **Gratiola** L. (hedge hyssop)

Plants annual or perennial herbs, with fibrous roots, terrestrial or emergent aquatics in shallow water, sometimes with rhizomes. Stems erect or ascending, less commonly spreading with ascending tips, then rooting, at least at the lower nodes, bluntly to sharply 4-angled near the tips, but rounded for most of their lengths, glabrous or glandular-hairy, sometimes only toward the tips, sometimes swollen and spongy toward the bases. Leaves opposite, sessile. Leaf blades not thickened or leathery, variously shaped, those of submerged leaves sometimes much narrower than the others, unlobed, mostly bluntly to sharply pointed at the tip, variously tapered to rounded at the base, sometimes clasping the stem, the margins entire or toothed, mostly above the midpoint, the venation palmate with 3 or 5 main veins, the surfaces glabrous or variously hairy, not shiny, the undersurface usually inconspicuously gland-dotted. Inflorescences axillary, the flowers 1 or 2 per node, the flower stalks short to elongate; bractlets 2, closely subtending the calyx, sepaloid. Flowers perfect. Calyces becoming somewhat enlarged at fruiting, deeply 5-lobed nearly to the base, glabrous or more commonly glandular- or nonglandular-hairy, the lobes subequal, variously shaped. Corollas bilabiate, 5-lobed, but sometimes appearing nearly 4-lobed, more or less tubular to somewhat trumpet-shaped, the tube longer than the lobes, usually slightly arched upward, white or cream-colored to pale yellow externally (occasionally pale lavender-tinged) with dark purple to dark brownish purple venation, white to pale pinkish-tinged or light yellow in the throat, spurless, the throat open, hairy on the inner surface of the upper side, the upper lip nearly straight to somewhat recurved very shallowly notched or 2-lobed, the lower lip straight to somewhat spreading or slightly arched upward, with 3 deeper lobes, these usually shallowly and broadly notched (the lowermost lobe sometimes with a pair of minute notches), the lobes usually white, often becoming purplish-tinged upon drying. Fertile stamens 2, the filaments not exserted, the anthers positioned on the expanded, flattened, membranous filament tip, the anther sacs parallel; staminodes usually present, appearing as a pair of antherless filaments. Style 1, not exserted, forked near the tip, the short branches flattened and broad (appearing similar to a pair of lips). Fruits capsules, ovoid to globose, glabrous, the 2 locules equal in size, dehiscent longitudinally along the 2 sutures and often also between them (then appearing 4-valved). Seeds numerous, 0.5–0.9 mm long, more or less cylindric to somewhat conic, not flattened, but usually appearing several-angled, the surface yellowish brown to brown, with a network of prominent ridges, these arranged into vertical ranks. About 25 species, nearly worldwide.

Rahmanzadeh et al. (2005) suggested that *Gratiola, Bacopa*, *Mecardonia*, and thirteen other non-Missouri genera of the Scrophulariaceae tribe Gratioleae should be segregated into a separate family, Gratiolaceae, based on a limited sampling of taxa for molecular analysis. However, they indicated that the group is closely allied to the Plantaginaceae lineage, and most recent authors have continued to interpret it as part of that family (Olmstead et al., 2001; Tank et al., 2006; Judd et al., 2008; Angiosperm Phylogeny Group, 2009). As noted by Tank et al. (2006), the systematics of this group are still not well understood.

1. Plants perennial, with rhizomes; main stem leaves broadest at or near the rounded or clasping base; fruits 3–5 mm long, noticeably shorter than the calyces . . . . 3. G. VISCIDULA
1. Plants annual, lacking rhizomes; main stem leaves (ignore submerged foliage) broadest near the midpoint, narrowed basally and not clasping the stems; fruits 3–7 mm long, slightly shorter than to about as long as the calyces
    2. Flower stalks slender (threadlike), 8–20 mm long, becoming elongated to 25 mm at fruiting; stems relatively slender, not inflated or spongy, glandular-hairy toward the tip . . . . 1. G. NEGLECTA
    2. Flower stalks relatively stout, 1–4 mm long, becoming elongated to 9(–14) mm long at fruiting; stems relatively stout, often somewhat inflated and spongy, glabrous or nearly so . . . . 2. G. VIRGINIANA

1. Gratiola neglecta Torr. (clammy hedge hyssop, hedge hyssop)

Pl. 483 c, d; Map 2203

Plants annuals, fibrous- to somewhat fleshy-rooted, lacking rhizomes. Stems 10–40 cm long, erect, usually several-branched, relatively slender (1–2 mm wide), not inflated or spongy, densely pubescent with short, gland-tipped hairs, at least toward the tip. Leaf blades 1–6 cm long, linear to narrowly elliptic, or oblanceolate, those of the uppermost leaves sometimes elliptic-obovate, those of the main leaves (ignore submerged foliage) broadest near the midpoint, narrowed basally and not clasping the stems, sharply pointed at the tip, the margins with few to several, shallow to moderately coarse, sharp teeth, the surfaces sparsely to moderately short-hairy, sometimes appearing glabrous or nearly so. Flower stalks 8–20 mm long, becoming elongated to 25 mm at fruiting, slender, glandular-hairy. Bractlets 3–6 mm long, linear to narrowly lanceolate-elliptic. Calyces 3–7 mm long, the lobes lanceolate to narrowly oblong-lanceolate, the margins entire. Corollas 7–10(–12) mm long, cream-colored to pale yellow externally (occasionally pale lavender-tinged) with dark purple to dark brownish purple venation, the throat and lobes white to pale pinkish-tinged. Fruits 3–7 mm long, about as long as the calyces, ovoid. $2n$=14, 16. May–October.

Scattered to common nearly throughout the state, but uncommon or absent from most of the western half of the Glaciated Plains Division (nearly throughout the U.S., Canada). Margins of ponds, lakes, and sinkhole ponds, sloughs, banks of streams and rivers, bottomland forests, mesic upland forests in ravines, fens, marshes, bottomland prairies, wet swales in sand prairies, and seepy depressions of glades; also ditches, fallow fields, low margins of crop fields, quarries, wet roadsides, and wet, disturbed areas; usually terrestrial, but occasionally emergent aquatics in shallow water.

2. Gratiola virginiana L. (round-fruited hedge hyssop, hedge hyssop)

Pl. 483 a, b; Map 2204

Plants annuals, fibrous- to somewhat fleshy-rooted, lacking rhizomes. Stems 10–40(–50) cm

2206. Kickxia elatine 2207. Kickxia spuria 2208. Leucospora multifida

long, erect or strongly ascending from a spreading base (stems sometimes becoming prostrate following inundation and developing ascending branches thereafter), unbranched or more commonly few- to several-branched, relatively stout ([1–]2–18 mm wide), often somewhat inflated and spongy, glabrous or nearly so. Leaf blades 1.5–5.0 cm long, narrowly lanceolate to narrowly oblong-elliptic, elliptic, or oblanceolate, those of the uppermost leaves sometimes ovate-elliptic to obovate, those of the lower (often submerged) leaves usually narrower and grading into the broader emergent leaves toward the stem tips, those of the main leaves (ignore submerged foliage) broadest at or above the midpoint, narrowed basally and not clasping the stems (those of the lower leaves often broadest below the midpoint and clasping the stems), rounded to more commonly bluntly or sharply pointed at the tip, the margins entire or with few to several, shallow, blunt to sharp teeth, the surfaces glabrous. Flower stalks 1–4 mm long, becoming elongated to 9(–14) mm long at fruiting, relatively stout, minutely glandular-hairy to nearly glabrous. Bractlets 2–6(–9) mm long, linear to very narrowly elliptic. Calyces 3–6 mm long (often becoming slightly enlarged at fruiting), the lobes lanceolate to narrowly oblong-lanceolate, the margins entire. Corollas 8–14 mm long, cream-colored to pale yellow externally with dark purple to dark brownish purple venation, the throat and lobes white to pale lavender-tinged. Fruits 3–6 mm long, slightly shorter than to about as long as the calyces, more or less globose. $2n=16$. April–October.

Scattered south of the Missouri River, uncommon and sporadic farther north (eastern [mostly southeastern] U.S. west to Kansas and Texas; introduced in Iowa). Ponds, lakes, sinkhole ponds, sloughs, swamps, marshes, bottomland forests, and wet swales in sand prairies; also ditches; usually emergent aquatics in shallow water, less commonly terrestrial along receding shorelines.

3. Gratiola viscidula Pennell (sticky hedge hyssop, hedge hyssop)

G. viscidula ssp. *shortii* Pennell

Pl. 482 j, k; Map 2205

Plants perennial herbs, with rhizomes. Stems 10–70 cm long, the rhizomatous portions prostrate to loosely ascending, the apical portions and branches usually more strongly ascending, the apical portions unbranched or few-branched, relatively stout (1.5–2.5 mm wide) but not inflated or spongy, moderately to densely short-hairy, especially toward the tip, the hairs often gland-tipped. Leaf blades 1.0–2.5 cm long (occasionally to 4.0 cm long on submerged stems), narrowly oblong-lanceolate to lanceolate or ovate, those of the main emergent leaves broadest at or near the rounded or clasping base, mostly bluntly pointed at the tip, the margins with few to several, shallow, sharp teeth, the surfaces sparsely to moderately short-hairy. Flower stalks 9–16 mm long, becoming only slightly elongated to 10–18 mm at fruiting, slender, glandular-hairy. Bractlets 3–6 mm long, lanceolate to narrowly lanceolate-elliptic. Calyces 4–7 mm long, the lobes lanceolate to narrowly oblong-lanceolate, the margins entire or more commonly minutely few-toothed toward the tip. Corollas 8–13 mm long, white or occasionally pale lavender-tinged externally. Fruits 3–5 mm long, noticeably shorter than the calyces, more or less globose. $2n=14$. June–September.

Uncommon, known thus far only from single extant populations in Howell and Shannon Counties (southeastern U.S. west to Kentucky and Alabama; Missouri). Margins of sinkhole ponds; terrestrial or occasionally emergent from shallow water.

Pennell (1928) segregated especially robust plants from Kentucky and southern Ohio as var. *shortii*, however Spooner (1984) studied morphological variation across the species range and concluded that populations were too variable to war-

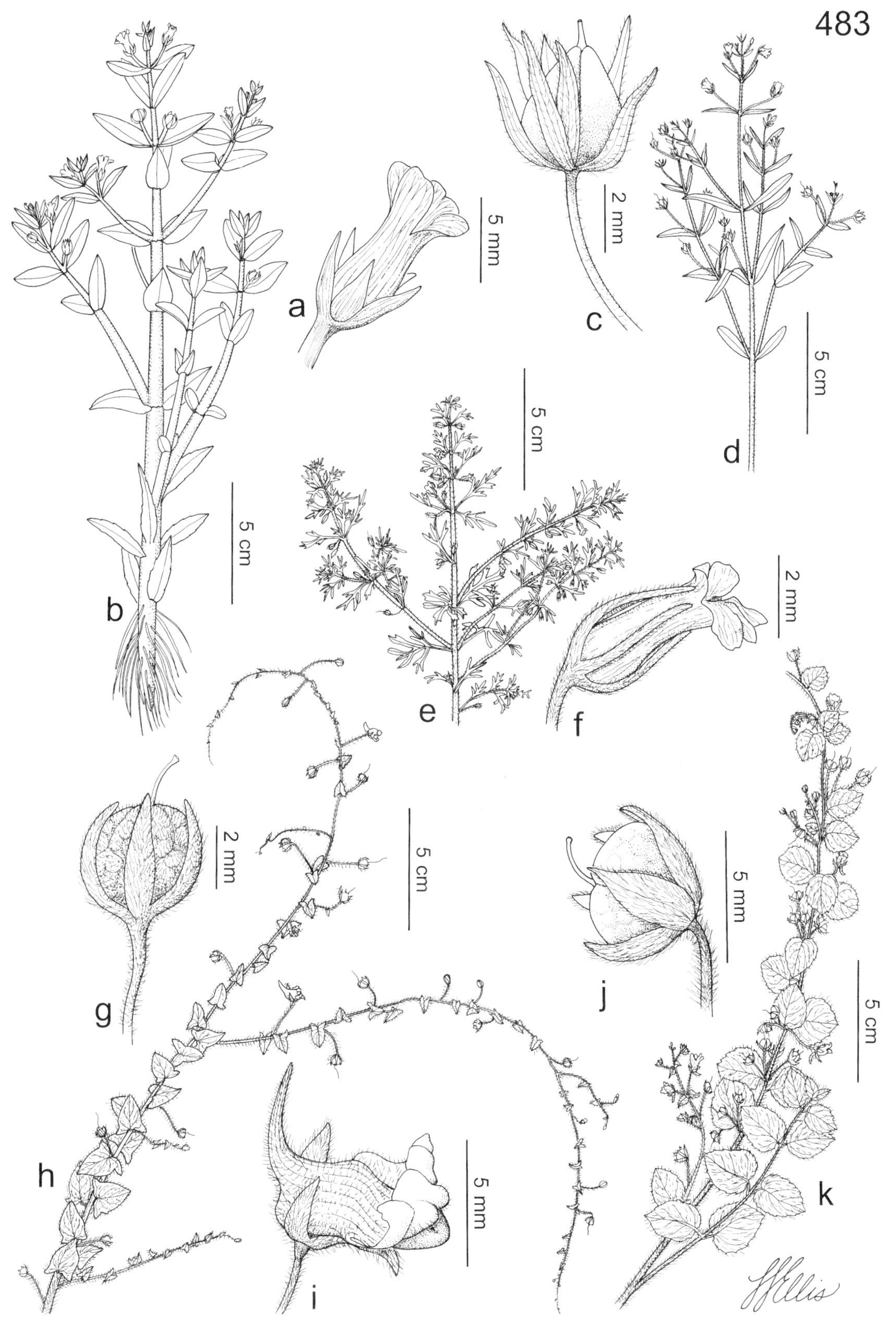

Plate 483. Plantaginaceae. *Gratiola virginiana*, **a)** flower, **b)** habit. *Gratiola neglecta*, **c)** fruit, **d)** habit. *Leucospora multifida*, **e)** habit, **f)** flower. *Kickxia elatine*, **g)** fruit, **h)** habit, **i)** flower. *Kickxia spuria*, **j)** fruit, **k)** habit.

rant the recognition of infraspecific taxa. The Missouri populations are disjunct from the closest populations in central Kentucky and Tennessee (Steyermark, 1963; Spooner, 1984).

9. **Kickxia** Dumort. (canker-root)

Plants annuals (perennial herbs or subshrubs elsewhere), with taproots, terrestrial. Stems prostrate, or loosely spreading, sometimes twining and thus climbing on surrounding vegetation, densely pubescent with relatively long, multicellular, gland-tipped hairs. Leaves alternate or the lowermost few opposite, noticeably short-petiolate, not or only slightly expanded at the base. Leaf blades nearly circular to broadly ovate, broadly triangular-ovate, more or less heart-shaped, or halberd-shaped, rounded or less commonly angled to a bluntly pointed tip, rounded to truncate or cordate at the base, unlobed or with a small pair of slender, spreading basal lobes, the margins otherwise entire or with relatively coarse, blunt teeth, pinnately veined or sometimes the smaller leaves appearing more or less palmately 3- or 5-veined from at or near the base, the surfaces and margins densely pubescent with fine, spreading, multicellular hairs, some of the hairs sometimes gland-tipped. Inflorescences axillary, of solitary flowers, the flower stalks 5–14 mm long at flowering becoming elongated to 8–25(–35) mm at fruiting; bractlets absent. Flowers perfect. Calyces deeply 5-lobed nearly to the base, the lobes slightly unequal in length, lanceolate to ovate, sharply pointed at the tip, moderately to densely pubescent with fine, spreading, multicellular hairs, the hairs usually nonglandular. Corollas 6.5–11.0 mm long (including the spur), bilabiate, 5-lobed, spreading-hairy on the outer surface, especially on the tube, the tube somewhat shorter than the lobes, light yellow, the upper lip purple or bluish-tinged, the tube with a well-developed, slender spur 3.5–5.5 mm long at the base (this positioned between the lower 2 calyx lobes), the throat closed by the noticeably convex base of the lower lip, the upper lip usually arched or bent upward, the lower lip with the lobes more or less spreading. Fertile stamens 4, the filaments of 2 lengths, not exserted, the anther sacs spreading, hairy along the margins; staminodes absent. Style 1, not exserted, the stigma capitate, unlobed. Fruits capsules, 3.0–4.5 mm long, globose or nearly so, minutely glandular-hairy on the outer surface, sometimes only toward the tip, the 2 locules equal in size, dehiscent circumscissilely above the midpoint (the upper half of each valve shed). Seeds numerous, 0.8–1.2 mm long, oblong to oblong-ovate, oblong-elliptic, or more or less rectangular in profile, not flattened, the surface dark brown, with a network of convoluted ridges (with tubercles elsewhere), these sometimes appearing winglike. Nine species, Europe, Asia, Africa.

Sutton (1988) and other earlier botanists treated *Kickxia* in a broad sense to include about 46 species in two well-marked sections. Ghebrehiwet (2001) used molecular markers, cytological, data, and an analysis of morphological features to show that these two sections were sufficiently distinct to represent two sister genera. He chose to segregate the majority of the species into the genus *Nanorrhinum* Betsche, which is distributed in Africa, Asia, and various oceanic islands. Within *Kickxia*, Sutton (1988) also chose to recognize various poorly differentiated subspecies within the more widely distributed taxa, but Ghebrehiwet (2001) chose not to treat these.

1. Leaf blades variously broadly ovate to halberd-shaped, at least some of them with a pair of spreading basal lobes, the margins often also with relatively coarse, blunt teeth; flower stalks usually hairy only toward the tip . . . . 1. K. ELATINE
1. Leaf blades ovate to heart-shaped or nearly circular, unlobed, the margins entire; flower stalks uniformly hairy . . . . 2. K. SPURIA

1. Kickxia elatine (L.) Dumort. (cancer-root, canker-root)

Pl. 483 g–i; Map 2206

Stems 10–50 cm long. Petioles 1–5 mm long. Leaf blades 1–3 cm long, variously broadly ovate to halberd-shaped, at least some of them with a

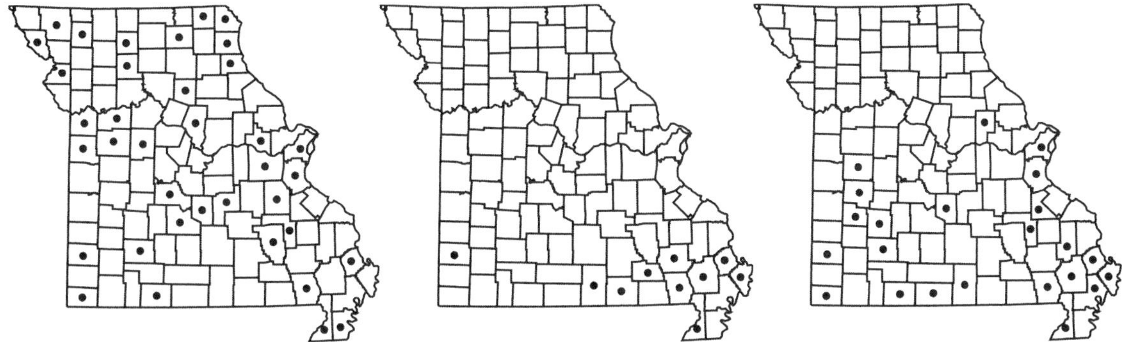

2209. Linaria vulgaris 2210. Mecardonia acuminata 2211. Nuttallanthus canadensis

pair of spreading basal lobes, the margins often also with relatively coarse, blunt teeth, broadly rounded to more commonly truncate or shallowly cordate at the base. Flower stalks usually hairy only toward the tip, occasionally also sparsely to moderately spreading-hairy below the midpoint. $2n=36$. May–October

Introduced, scattered in the Ozark, Ozark Border, and Big Rivers Divisions (native of Europe, Asia; introduced in the eastern U.S. west to Wisconsin and Texas; also Washington to California; Canada). Banks of streams and rivers, bottomland prairies, and margins of sinkhole ponds; also railroads and open, disturbed areas.

In some western states, this species is considered a noxious weed in crop fields. In Missouri, it is more commonly encountered than is the morphologically similar *K. spuria*. The flowers and fruits of the two species are similar in size and appearance.

2. Kickxia spuria (L.) Dumort. (female fluellin)
Pl. 483 j, k; Map 2207

Stems 10–60 cm long. Petioles 2–6 mm long. Leaf blades 0.8–4.0 cm long, ovate to heart-shaped or nearly circular, unlobed, the margins entire, broadly rounded to truncate or shallowly to moderately cordate at the base. Flower stalks uniformly and relatively densely pubescent with spreading hairs. $2n=18$. May–September

Introduced, widely scattered (native of Europe, Asia; introduced in the western U.S. from Idaho to California, also sporadically in the eastern states west to Wisconsin and Missouri; Canada). Banks of streams and rivers, sloughs; also open, disturbed areas.

10. Leucospora Nutt.

One species, eastern temperate North America.

1. Leucospora multifida (Michx.) Nutt.
(narrowleaf paleseed, Obi-Wan-Conobea)
Conobea multifida (Michx.) Benth.
Stemodia multifida (Michx.) Spreng.
Pl. 483 e, f; Map 2208

Plants annual, with slender taproots, terrestrial. Stems 5–20 cm long, erect or strongly ascending, variously unbranched to diffusely branched, densely pubescent with short, spreading, glandular hairs. Leaves opposite, mostly long-petiolate, the petiole winged. Leaf blades 1–3 cm long, 1 or 2 times more or less pinnately deeply lobed or parted, broadly triangular-ovate in outline, the ultimate segments linear to narrowly oblong or narrowly oblanceolate, rounded or bluntly to sharply pointed at the tips, the margins otherwise mostly entire, the surfaces moderately to densely pubescent with short, gland-tipped hairs, the venation overall pinnate, each lobe or segment with a single vein. Inflorescences axillary, of solitary or paired flowers, the flower stalks 3–7 mm long at flowering, not or only slightly elongated at fruiting; bractlets absent. Flowers perfect. Calyces 2.5–4.0 mm long, becoming elongated to 3.5–5.5 mm at fruiting, deeply 5-lobed nearly to the base, nearly actinomorphic, the lobes linear to narrowly lanceolate, rounded to bluntly pointed at the tip, glandular-hairy. Corollas 3–5 mm long, bilabiate, 5-lobed, the upper lip shallowly 2-lobed, more or less spreading to somewhat recurved, the lower lip more deeply 3-lobed, slightly spreading, glabrous, the tube longer than the lobes, white to pale cream-colored, but usually at least partially pale pinkish- to lavender-tinged, the tube lacking a spur, the throat open, with yellow markings and often small red spots. Fertile stamens 4, the filaments of 2 lengths, not exserted, the anther sacs more or less parallel; staminodes absent. Style 1, not exserted, 2-lobed, the stigmas wedge-shaped. Fruits capsules, 3.5–4.5 mm long, ovoid, more or less circular in cross-section, thin-walled, glabrous, the 2 locules equal in size, dehiscent longitudinally along the 2 sutures. Seeds numerous, 0.2–0.4 mm long, oblong-ovoid, more or less circular in cross-section, the surface pale greenish yellow, with a minute and sometimes faint network of fine ridges in longitudinal rows (this often visible only with strong magnification). May–October.

Scattered south of the Missouri River north sporadically to Harrison, Mercer, and Pike Counties (eastern U.S. west to Nebraska and Texas; Canada). Banks of streams, rivers, and spring branches, margins of ponds and lakes, oxbows, swamps, and rarely glades; also crop fields, fallow fields, gardens, railroads, roadsides, and open disturbed areas.

This species was first segregated from the small, neotropical genus *Conobea* Aubl. by Pennell (1935) based on a lengthy series of morphological differences in both the vegetative and reproductive portions of the plants. B. L. Turner and Cowan (1993) suggested that *Leucospora* might better be treated as a member of the widespread genus *Stemodia* L., but this hypothesis seems not to have gained traction with North American botanists. Relationships of the genus would benefit from further study.

Pennell (1935) noted that the floral morphology suggests bee-pollination, but that the small corollas may also be adapted for self-pollination.

11. **Linaria** Mill. (toadflax)

About 150 species, Europe, Asia, Africa.

1. Linaria vulgaris Mill. (butter and eggs)
Pl. 482 f, g; Map 2209

Plants perennial herbs (frequently flowering the first year), with fibrous roots, terrestrial. Stems 30–80 cm long, erect or ascending, glabrous below the inflorescence (this occasionally sparsely glandular-hairy). Leaves relatively dense, alternate or occasionally the lowermost leaves appearing opposite, sessile or very short-petiolate, the petiole bases not expanded. Leaf blades 2–7 cm long, 1.5–6.0 mm wide, linear, angled or slightly tapered to a sharply pointed tip, tapered at the base, unlobed, the margins entire, usually only a midvein apparent, glabrous. Inflorescences dense terminal spikelike racemes, the flower stalks 1–4 mm long at flowering, not becoming elongated at fruiting, each subtended by a linear to lanceolate bract (3–25 mm long); bractlets absent. Flowers perfect. Calyces 2.5–4.5 mm long, deeply 5-lobed nearly to the base, the lobes slightly unequal in length, lanceolate to narrowly ovate, sharply pointed at the tip, glabrous. Corollas 20–30 mm long (including the spur), bilabiate, 5-lobed, glabrous except for the bearded base of the lower lip, the tube shorter than to about as long as the lobes, light yellow with a darker yellow and orange throat, the tube with a straight or somewhat curved, slender spur 8–14 mm long at the base (this positioned between the lower 2 calyx lobes), the throat mostly closed by the strongly convex base of the lower lip, the upper lip more or less straight, strongly folded longitudinally between the 2 lobes, the lower lip with the lobes more or less recurved. Fertile stamens 4, the filaments of 2 lengths, not exserted, the anther sacs spreading; staminodes absent (but the lower lip bearded at the base). Style 1, not exserted, the stigma capitate, unlobed. Fruits capsules, 5–10 mm long, ovoid to more or less globose, glabrous, the 2 locules equal or unequal in size, each dehiscent from the tip by 3–5 teeth. Seeds numerous, 1.5–2.5 mm long, circular to broadly oblong-elliptic in profile, flattened and with a broad marginal wing, the surface black, the main body with scattered small warts or tubercles. $2n=12$. May–November.

Introduced, scattered nearly throughout the state (native of Europe, Asia; introduced throughout temperate North America). Ditches, fallow fields, pastures, fencerows, cemeteries, old homesites, gardens, railroads, roadsides, and open, disturbed areas.

Butter and eggs is an attractive ornamental, but is considered to be a significant invader of native plant communities in some states. In 1742, a young Swedish botanist brought to the attention of Carolus Linnaeus an abnormal form of *Linaria vulgaris* that he had discovered while botanizing near Uppsala. In this plant, the flowers were actinomorphic and 5-spurred rather than the usual condition of a zygomorphic corolla with a single spur. Although he recognized it as an unusual kind of *Linaria*, Linnaeus was unable to account for this developmental abnormality and subsequently named it as a new genus *Peloria* (derived from the Greek, for monster). The phenomenon of developmental monstrosities that affect floral symmetry has since become known as pelorism (Å. Gustafsson, 1979) and is known from several other species, including *Antirrhinum majus*. *Linaria vulgaris* has been an important research tool in the study of the developmental biology of flowers, mutations that affect flower morphology, and the inheritance of genes that affect floral development (Meyerowitz et al., 1989).

12. **Mecardonia** Ruiz & Pav.

Ten species, North America, South America, Caribbean Islands.

Steyermark (1963) and many earlier authors treated *Mecardonia* as part of the genus *Bacopa*. Pennell (1946) provided a lucid discussion of the problems associated with generic limits in the group and concluded that *Mecardonia* was not, in fact, very closely related to *Bacopa*. The two genera differ in a suite of floral characters, as well as in the kinds of secondary compounds produced.

1. Mecardonia acuminata (Walter) Small

Bacopa acuminata (Walter) B.L. Rob.

Pl. 482 h, i; Map 2210

Plants perennial herbs, with fibrous roots, terrestrial. Stems 20–55 cm long, erect or ascending, usually not rooting at the lower nodes, 4-angled, glabrous. Leaves opposite, sessile. Leaf blades 1–5 cm long, 2–11(–15) mm wide, somewhat thickened and leathery, linear to more commonly oblanceolate, spatulate, elliptic, or narrowly rhombic, unlobed, rounded or bluntly to sharply pointed at the tip, angled or tapered at the base, not clasping the stem, the margins sharply toothed, mostly above the midpoint, the venation pinnate but often with the lowermost pair of secondary veins branching well above the base and relatively prominent, the surfaces glabrous, not shiny, the undersurface inconspicuously gland-dotted. Inflorescences axillary, the flowers 1 per leaf axil, the flower stalks 7–35 mm long; bractlets 2, at the base of the flower stalk, 6–10 mm long, linear to narrowly oblanceolate. Flowers perfect. Calyces 6–10 mm long at flowering, not or only slightly enlarged at fruiting, deeply 5-lobed nearly to the base, glabrous but minutely gland-dotted, the lobes subequal (the outer 3 lobes slightly broader than the inner 2), linear to narrowly lanceolate. Corollas 6–10 mm long, somewhat bilabiate, 5-lobed, weakly zygomorphic, narrowly bell-shaped, the tube longer than the lobes, white or with pinkish or purplish lines or streaks and a yellow throat, spurless, the throat open, hairy on the inner surface of the upper side, the upper lip shallowly 2-lobed, the lower lip with 3 broader, deeper lobes. Fertile stamens 4, the filaments of 2 lengths, not exserted, the anthers attached near their midpoints, the anther sacs parallel; staminodes absent (perhaps represented by the bearded throat). Style 1, not exserted, forked near the tip, the short branches flattened and broad (appearing similar to a pair of lips). Fruits capsules, 5–8 mm long, narrowly ovoid to narrowly oblong-ovoid, glabrous, the 2 locules equal in size, dehiscent longitudinally along the 2 sutures. Seeds numerous, 0.5–0.6 mm long, more or less cylindric to somewhat conic, not flattened, but usually appearing several-angled, the surface dark brown, with a network of prominent ridges, these arranged into vertical ranks. $2n=42$. July–September.

Uncommon in the southwestern and southeastern portions of the Ozark Division and the Mississippi Lowlands (southeastern U.S. west to Kansas and Texas). Swamps, bottomland forests, margins of ponds, lakes, and sinkhole ponds, sloughs, fens, wet swales in sand prairies and glades, and occasionally banks of spring branches; also gardens and wet portions of old fields; terrestrial or emergent aquatics in shallow water.

Steyermark (1963) noted that specimens of this species tend to blacken as they dry.

13. **Nuttallanthus** D.A. Sutton

Plants annual, with small taproots, terrestrial. Stems of 2 kinds: a rosette of few to several vegetative stems (sometimes withered or absent at flowering), these 1–6 cm long (sometimes elongating to 10 cm with age), mostly prostrate (sometimes with ascending tips), usually unbranched; and 1 or few fertile stems, these 10–40(–60) cm long, erect or strongly ascending, unbranched or with ascending branches toward the tip; both types glabrous (sometimes sparsely glandular-hairy in the inflorescence). Leaves simple, unlobed and entire, glabrous, with at most a midvein apparent, of 2 kinds: those of the vegetative stems mostly opposite or in whorls of 3, sessile or occasionally short-petiolate, the blades 0.2–1.0 cm long, linear to oblanceolate or spatulate; those of the fertile stems mostly alternate (except sometimes the lowermost), sessile, 0.5–3.0 cm long, 0.5–2.0 mm wide, linear or threadlike. Inflorescences terminal, slender racemes, these sometimes relatively short and dense at the start of flowering but later usually becoming elongated, the flower stalks 1–6(–9) mm long at flowering, not becoming elongated at fruiting, each subtended by a linear to lanceolate bract; bractlets absent. Flowers perfect. Calyces deeply 5-lobed nearly to the base, the lobes slightly unequal in length, linear or narrowly lanceolate to lanceolate or elliptic-lanceolate, sharply pointed at the tip, glabrous. Corollas 8–22 mm long (including the spur), bilabiate, 5-lobed, glabrous or sparsely hairy toward the base of the lower lip, the tube shorter than to about as long as the lobes, pale blue to light purplish blue, sometimes with a paler or white throat (rarely entirely white elsewhere), the tube with a curved, slender spur 2–11 mm long at the base (this positioned between the lower 2 calyx lobes), the throat partially closed by the strongly convex base of the lower lip, the upper lip relatively short, more or less straight, flat, the lobes erect or angled slightly backward, the lower lip with the lobes spreading and somewhat curved. Fertile stamens 4, the filaments of 2 lengths, not exserted, somewhat curved inward toward their tips, the anther sacs spreading; staminodes absent. Style 1, not exserted, the stigma capitate, unlobed. Fruits capsules, oblong-ovoid to broadly oblong-ovoid, glabrous, the 2 locules equal in size, each dehiscent from the tip by 3–5 teeth. Seeds numerous, minute, ovate to rectangular, more or less trapezoidal, or triangular in profile, 4–7-angled in cross-section, the surface black to dark gray, variously ridged and/or with small tubercles, the tubercles sometimes poorly developed and the seed then nearly smooth (except under high magnification). Four species, North America, South America.

Sutton's (1988) segregation of the New World *Nuttallanthus* from the otherwise Old World genus *Linaria* has become accepted by most botanists. The flowers of the two genera differ in the shapes and relative lengths of the lips, as well as their spurs, and the seeds of the two genera are markedly different (prismatic vs. flattened in cross-section). The two Missouri species are closely related and have been treated as components of a single species by some botanists. They are occasionally difficult to distinguish and sometimes have been recorded as growing in mixed populations.

1. Corollas 8–11(–13) mm long, including the 2–6 mm spur; seeds with low ridges along the angles, the faces with minute tubercles between them, sometimes appearing nearly smooth . . . . 1. N. CANADENSIS
1. Corollas 13–22 mm long, including the 6–11 mm spur; seeds bluntly angled, lacking ridges, the faces with dense, small, relatively prominent tubercles . . . . 2. N. TEXANUS

1. Nuttallanthus canadensis (L.) D.A. Sutton (blue toadflax, old-field toadflax)

Linaria canadensis (L.) Dum. Cours.
L. canadensis f. *albina* Fernald
L. canadensis f. *cleistogama* Fernald

Pl. 484 g–i; Map 2211

Inflorescences with the flowers relatively widely spaced, even toward the start of flowering. Bracts 1–7 mm long, linear, bluntly pointed at the tip. Flower stalks 2–5 mm long, strongly ascending. Calyces 2–4 mm long, the lobes linear to narrowly lanceolate, green with a narrow, thin, pale or white border along the margins. Corollas 8–11(–13) mm long (including the spur), the upper lip 1.2–1.6 mm long, the lower lip 2–6 mm long, the spur 2–6 mm long. Fruits 2.0–3.5 mm long. Seeds 0.4–0.5 mm long, with low ridges along the angles, the faces with minute tubercles between them, sometimes appearing nearly smooth. $2n=12$. May–November.

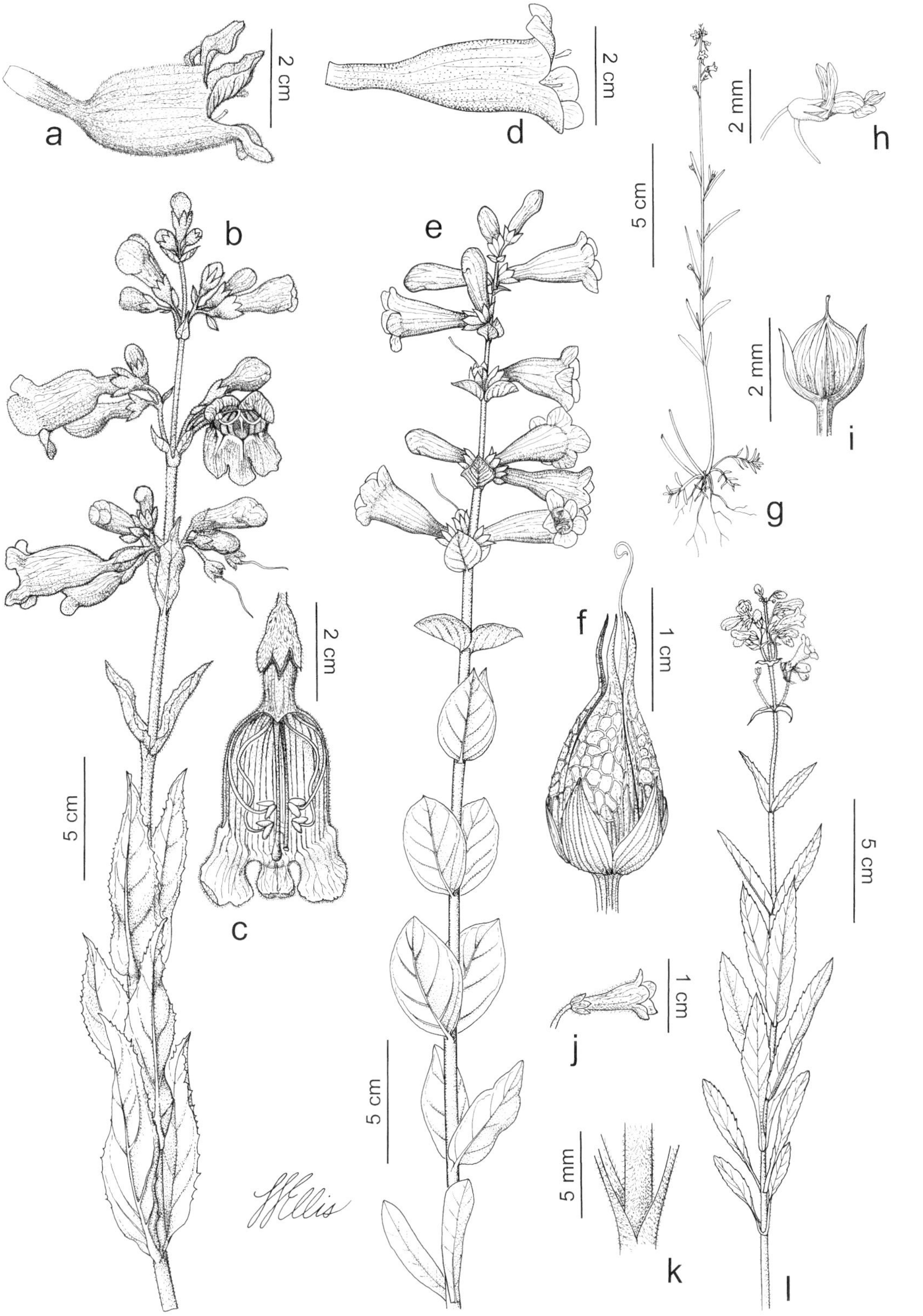

Plate 484. Plantaginaceae. *Penstemon cobaea*, **a)** flower, **b)** fertile stem, **c)** dissected flower. *Penstemon grandiflorus*, **d)** flower, **e)** fruit, **f)** habit. *Nuttallanthus canadensis*, **g)** habit, **h)** flower, **i)** fruit. *Penstemon arkansanus*, **j)** flower, **k)** node with leaf bases, **l)** fertile stem.

2212. Nuttallanthus texanus 2213. Penstemon arkansanus 2214. Penstemon cobaea

Scattered, mostly in the Ozark, Ozark Border, and Mississippi Lowlands Divisions (eastern U.S. west to Minnesota and Texas, also Washington to California; Canada). Glades, ledges and tops of bluffs, sand prairies, bottomland forests, openings of mesic to dry upland forests, savannas, banks of streams and rivers, margins of ponds, and fens; also ditches, fallow fields, pastures, railroads, roadsides, and open, disturbed areas; on calcareous and acidic substrates.

2. Nuttallanthus texanus (Scheele) D.A. Sutton (southern blue toadflax)
Linaria canadensis (L.) Dum. Cours. var. *texana* (Scheele) Pennell
Linaria texana Scheele

Pl. 482 l–n; Map 2212

Inflorescences with the flowers often relatively dense toward the start of flowering, usually becoming less densely flowered with age. Flower stalks 2–6(–9) mm long, strongly ascending. Bracts 2–3 mm long, linear, bluntly pointed at the tip. Calyces 2–4 mm long, the lobes lanceolate or elliptic-lanceolate, linear to narrowly lanceolate, green with a narrow, thin, pale or white border along the margins. Corollas 13–22 mm long (including the spur), the upper lip 3–5 mm long, the lower lip 6–11 mm long, the spur 6–11 mm long. Fruits 2.0–3.5 mm long. Seeds 0.4–0.5 mm long, bluntly angled, lacking ridges, the faces with dense, small, relatively prominent tubercles. $2n=12$. May–November.

Scattered, mostly south of the Missouri River, apparently absent from most of the central portion of the Ozark Division (western U.S. east to South Dakota and Louisiana, eastward sporadically to Tennessee and Florida; Canada, Mexico, South America; introduced sporadically northeast of the native range). Glades, ledges and tops of bluffs, thin-soil areas of upland prairies, sand prairies, dry upland forests, savannas, and margins of ponds; also pastures, railroads, roadsides, and open, disturbed areas; often on acidic substrates.

14. Penstemon Schmidel (beard-tongue)

Plants perennial or rarely biennial herbs, often taprooted, terrestrial. Stems erect or loosely to strongly ascending, usually unbranched below the inflorescence, variously glabrous or hairy, sometimes glandular-hairy or glaucous. Leaves opposite or rarely in whorls of 3 (in *P. digitalis*) (overwintering basal rosettes usually produced but often absent at flowering), sessile (except sometimes in basal leaves), sometimes clasping the stem. Leaf blades simple, unlobed, variously shaped, the margins entire or toothed, the surfaces variously glabrous or hairy, sometimes glandular-hairy or glaucous, the venation pinnate or occasionally only the midvein apparent. Inflorescences terminal panicles, with mostly whorls of branches along an elongate central axis, these terminating in clusters of flowers, the branches sometimes relatively short and ascending with relatively dense flower clusters (then appearing more or less racemose), each node with a pair of inconspicuous or conspicuous, leaflike bracts, the individual flower stalks sometimes short but always noticeable at flowering, not becoming elongated at fruiting; bractlets absent. Flowers perfect. Calyces sometimes becoming slightly enlarged at fruiting, deeply 5-lobed, the lobes equal or slightly unequal in length, lanceolate to ovate, usually sharply pointed at the tip, glabrous or glandular-hairy. Corollas 15–50 mm long, bilabiate or nearly actinomorphic, 5-lobed, glabrous or glandular-hairy externally, the tube longer than the lobes, white or lavender to pale bluish purple or purple, the throat often lined with pink to purple and yellow nectar guides, the tube variously shaped, lacking a spur, the throat more or less open (the lower lip sometimes thickened, convex, and/or bearded toward the base), the lobes variously angled outward, spreading, or somewhat reflexed (those of the lower lip in *P. pallidus* often not or only slightly spreading). Fertile stamens 4, the filaments of 2 lengths, not exserted (exserted elsewhere), somewhat curved inward toward their tips, the anther sacs spreading in our species; staminode 1, about as long as the stamens, linear to narrowly strap-shaped, sometimes very slightly thickened toward the tip, positioned along the lower side of the corolla tube, usually with a dense beard of yellow hairs, at least toward the tip. Style 1, not exserted, the stigma capitate, unlobed. Fruits capsules, ovoid, tapered to a beaked tip, glabrous (sometimes with minute papillae along the sutures), the 2 locules equal in size, dehiscent longitudinally along the 2 sutures. Seeds numerous, minute, ovate to rectangular, more or less trapezoidal, or triangular in profile, rounded or 3–6-angled in cross-section, the surface dark brown to black, often with lighter ridges, also usually with a fine network of minute ridges. About 270 species, North America, Central America.

Penstemon is a relatively large genus that is almost entirely distributed in North America (Straw, 1966). The greatest species diversity is in the Intermountain region of the western United States and many of the species are narrow endemics. The genus is relatively closely related to *Chelone* and both are characterized by the presence of a staminode in the flower. One morphological difference between the two genera is that in *Penstemon* nectar is produced by glandular hairs that occur on the staminal filaments, whereas in *Chelone* the nectar is produced by a glandular disc (nectary) at the base of the ovary. *Chelone* also differs in its closed (vs. usually open) corolla throat; unbranched, spicate inflorescences; increased production of bracts, flattened, winged seeds, and base chromosome number.

The eastern and midwestern species of *Penstemon* are in need of detailed taxonomic study. Species limits within the polyploid complex of white-flowered penstemons in the region are not well understood and the role of past hybridization in the formation of species also requires more research. Several segregate taxa, some of these relatively rare and of potential conservation concern, have been accepted by some authors but not by others. The present treatment should be viewed as preliminary.

Species of *Penstemon* exhibit a variety of floral morphologies that correspond to specialized pollination syndromes involving bees, moths, butterflies, wasps, beetles, and hummingbirds. The species native to the eastern and midwestern United States are mainly bee-pollinated (Clinebell and Bernhardt, 1998; Dieringer and Cabrera R., 2002), which is considered the primitive condition in the genus (Wolfe et al., 2006). However, exceptions occur. In *P. tubaeflorus*, with its slender corolla tube, the main pollinators appear to be certain butterflies, such as the spicebush swallowtail, *Pterourus troilus* (L.), although some bees also contribute to pollination (Clinebell and Bernhardt, 1998). In the remaining Missouri species (except *P. arkansanus*, which has not been studied yet), Clinebell and Bernhardt (1998) reported that bumblebee queens are major pollinators, but other pollinators include smaller bees, bee-flies, the *Penstemon* wasp (*Pseudomasaris occidentalis* (Cresson)), and the bumblebee flower beetle (*Euphoria sepulcralis* (Fab.)). These authors also showed that self-pollination can be successful in *P. digitalis*, but not in four other Missouri species that they studied. Dieringer and Cabrera R. (2002), who studied pollination in *P. digitalis*, documented a diversity of bees of different body sizes as pollinators and also showed the importance of the staminode in facilitating pollination by most bees.

Penstemon species are popular native wildflowers in gardens, particularly in the western states. In the Midwest, the main species that are grown include *P. cobaea, P. digitalis, P. grandiflorus*, and *P. tubaeflorus*. Several cultivars involving primarily variations in flower color and/or overall plant color have been developed in *P. digitalis* and *P. cobaea*. In general, beard-tongues are easily-grown, sun-loving plants whose main drawback is that as perennials they tend to be relatively short-lived and usually must reseed themselves to remain present in a garden for more than a few years.

Some species of *Penstemon* were used medicinally by Native Americans, mainly in the treatment of sexually transmitted diseases (Moerman, 1998; Nold, 1999).

1. Corollas 35–55 mm long; calyces 7–16 mm long at flowering
    2. Stems glabrous and strongly glaucous; leaves glabrous and strongly glaucous; corollas lavender to pale bluish purple, glabrous internally and externally . . . . 4. P. GRANDIFLORUS
    2. Stems hairy, the hairs sometimes minute and gland-tipped; leaves variously glabrous or hairy, but not glaucous; corollas either white or purple (then sometimes with a white ring at the base of the corolla lobes), rarely pink or light purple, minutely glandular-hairy externally and in the throat . . . 2. P. COBAEA
1. Corollas 15–30 mm long; calyces 2.5–8.0 mm long at flowering
    3. Stems glabrous, at least toward the base
        4. Corollas nearly actinomorphic, only slightly bilabiate, the tube slender, expanded evenly toward the tip, the throat densely and minutely glandular-hairy, white, lacking colored nectar guides . . . . 7. P. TUBAEFLORUS
        4. Corollas zygomorphic, strongly bilabiate, the tube broadened abruptly (often asymmetrically so) at or below the midpoint, the throat sparsely to densely pubescent with bristly, nonglandular hairs, white or occasionally pale pinkish- to purplish-tinged, usually lined with pale to bright pink or purple and/or yellow nectar guides
            5. Corollas 15–18 mm long, the throat appearing slightly flattened and relatively strongly 2-ridged; calyces 2–4 mm long at flowering . . . . 1. P. ARKANSANUS
            5. Corollas (18–)20–30 mm long, the throat not appearing flattened and only slightly 2-ridged; calyces (3–)4–8 mm long at flowering . . . . 3. P. DIGITALIS
    3. Stems hairy from tip to base, the hairs sometimes minute and/or glandular
        6. Corolla tube strongly broadened at or below the midpoint, the throat not appearing flattened and only slightly 2-ridged, the lower lip spreading to strongly arched downward, thus appearing to project about the same distance as the spreading to slightly recurved upper lip; calyces (3–)4–8 mm long at flowering
            7. Leaf blades with the margins entire or finely to relatively coarsely but unevenly or sparsely toothed; calyx lobes narrowly ovate to ovate; fruits 10–15 mm long. . . . . 3. P. DIGITALIS
            7. Leaf blades with the margins evenly but very finely many-toothed; calyx lobes lanceolate-triangular; fruits 4–7 mm long . . . . 6. P. TENUIS
        6. Corolla tube slightly but noticeably expanded at or below the midpoint, the throat appearing slightly flattened and relatively strongly 2-ridged, the lower lip more or less erect and straight, thus appearing to project beyond the spreading upper lip; calyces 2–5 mm long at flowering
            8. Leaf blades with the upper surface glabrous or nearly so; stems usually glabrous and strongly reddish-tinged; corolla tube expanded well below the midpoint; calyces 2–4 mm long at flowering; corollas 15–18 mm long . . . . 1. P. ARKANSANUS
            8. Leaf blades with the upper surface hairy; stems hairy and grayish green to green; corolla tube expanded at about the midpoint; calyces 3–5 mm long at flowering; corollas 17–23 mm long . . . . 5. P. PALLIDUS

1. Penstemon arkansanus Pennell (Arkansas beard-tongue)

Pl. 484 j–l; Map 2213

Stems 25–60 cm long, erect or ascending, densely pubescent with minute, nonglandular hairs, sometimes also with slightly longer glandular hairs toward the tip, occasionally glabrous or nearly so toward the base, olive green to grayish green or more commonly purplish-tinged, not glaucous. Basal leaves 2–6(–9) cm long, the blade oblanceolate to spatulate or obovate, rounded to bluntly pointed at the tip, tapered basally to a usually winged petiole, the margins bluntly to sharply and usually finely toothed, the surfaces glabrous to more commonly sparsely to moderately nonglandular-hairy, mostly along the margins and main veins, not glaucous. Stem leaves 2–8(–10) cm long, the lowermost with the blade oblanceolate to narrowly oblong-elliptic, grading into lanceolate or narrowly lanceolate at the stem tip, rounded to sharply pointed at the tip, sessile or nearly so, the base of the lower blades tapered, grading through rounded to those of the uppermost leaves sometimes shallowly cordate and clasping, the margins variously bluntly to sharply and sometimes minutely toothed, at least above the midpoint, the surfaces sparsely to densely and minutely nonglandular-hairy, sometimes only along the main veins, occasionally also dotted with widely spaced, sessile glands, not glaucous. Inflorescences narrow to occasionally more broadly pyramidal panicles, the central axis minutely glandular-hairy, green or purplish-tinged, not glaucous, with (2–)3–7(–9) nodes, each with a pair of relatively small, linear to narrowly lanceolate, somewhat clasping bracts, the branches ascending or arched upward, with a pair of few- to several-branched clusters per node, each main branch with 3–9(–13) flowers. Calyces 2–4 mm long at flowering, moderately to densely glandular-hairy, not glaucous, the lobes ovate. Corollas 15–18 mm long, the tube abruptly but relatively slightly enlarged below the midpoint (usually very slightly so on the lower side), strongly bilabiate, the upper lip spreading to somewhat recurved, the lower lip spreading to somewhat reflexed, white or occasionally pale pinkish- to purplish-tinged, the throat appearing slightly flattened, relatively strongly 2-ridged and usually lined with purple nectar guides on the lower side, minutely glandular-hairy externally and sparsely so in the throat. Staminode white, strongly flattened toward the tip, bearded on the upper surface, the hairs yellow or purple toward the base, slightly curved downward apically and bearded with yellow hairs. Fruits 5–7 mm long. Seeds 0.7–1.0 mm long, dark brown to black, the reddish brown ridges poorly developed. $2n=16$. April–June.

Uncommon in the southwestern portion of the Ozark Division eastward to Howell County; also disjunct in Cape Girardeau County (Texas, Oklahoma, Missouri, Arkansas, and Illinois). Glades, ledges and tops of bluffs, savannas, openings of dry upland forests, and banks of streams; also roadsides; usually on calcareous substrates.

Gleason and Cronquist (1991) and some other authors have treated *P. arkansanus* as a small-flowered form of *P. pallidus*. Steyermark (1963) and McWilliam (1967) presented strong morphological evidence for keeping these two species separate. Koelling (1964) did not fully treat the taxon, but mentioned that he considered it a distinct species. Both are apparently diploids and thus might possibly have been involved in the parentage of such polyploids as *P. digitalis*. Pending more detailed taxonomic studies, it has been accepted as distinct in the present treatment. Interestingly, specimens and photographs attributed to *P. arkansanus* are more frequently misdeterminations of *P. digitalis* than of *P. pallidus*.

2. Penstemon cobaea Nutt. (cobaea beard-tongue, cobaea penstemon)

P. cobaea var. *purpureus* Pennell

Pl. 484 a–c; Map 2214

Stems 20–80 cm long, erect or ascending, sometimes from a spreading base, densely pubescent with mostly glandular hairs, the hairs slightly longer toward the stem tip, green, not glaucous. Basal leaves 3–24 cm long, the blade oblanceolate to spatulate, rounded to sharply pointed at the tip, tapered basally to a winged petiole, the margins entire to more commonly sharply toothed, the surfaces glabrous to more commonly glandular-hairy (rarely pubescent with nonglandular hairs), not glaucous. Stem leaves 3–15 cm long, the lowermost with the blade oblanceolate to narrowly oblong-elliptic, grading into lanceolate to ovate at the stem tip, sharply pointed at the tip, sessile or nearly so, the base of the lower blades tapered, grading through rounded to those of the uppermost leaves cordate and clasping, the margins sharply toothed, at least above the midpoint, the surfaces glabrous to more commonly glandular-hairy (rarely pubescent with nonglandular hairs), not glaucous. Inflorescences narrow panicles, the central axis glabrous or more commonly minutely glandular-hairy, green to yellowish green, not glaucous, with 3–6(–8) nodes, each with a pair of leaf-like, ovate, clasping bracts, the branches ascend-

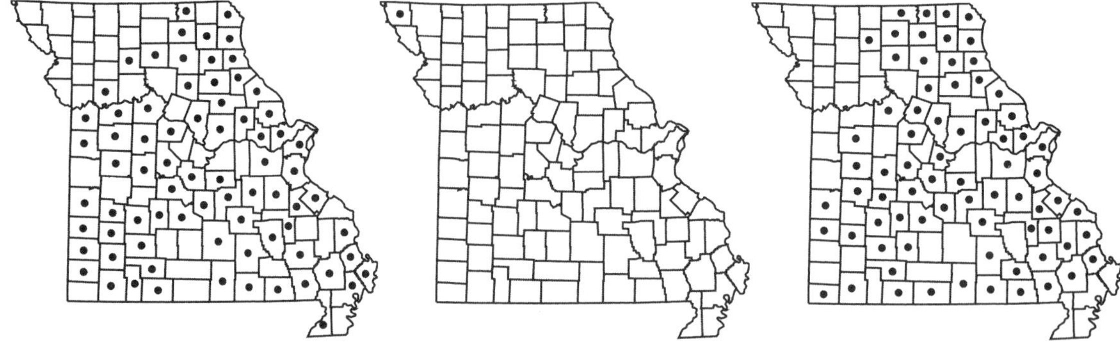

2215. Penstemon digitalis　　2216. Penstemon grandiflorus　　2217. Penstemon pallidus

ing, with a pair of mostly branched clusters per node, each with 2–6 flowers. Calyces (8–)10–16 mm long at flowering, densely glandular-hairy, not glaucous, the lobes lanceolate to narrowly ovate. Corollas 35–55 mm long, the tube abruptly enlarged well below the midpoint, strongly bilabiate, the upper lip spreading to somewhat arched forward, the lower lip spreading to somewhat reflexed, either white or purple (then sometimes with a white ring at the base of the corolla lobes), rarely pink or light purple, the throat lined with darker reddish purple nectar guides and noticeably ridged on the lower side, minutely glandular-hairy externally and in the throat. Staminode white, strongly flattened toward the tip, bearded on the upper surface, the hairs shorter and yellow to light brown toward the base, curled-under apically and bearded with white or pale yellow hairs. Fruits 13–18 mm long. Seeds 2.5–3.5 mm long, black, the black ridges well-developed. $2n$=16, 64. April–June.

Uncommon to scattered in the southern portion of the Ozark Division and disjunct in Bates County; introduced elsewhere in the state (Nebraska to Texas east to Iowa and Arkansas; introduced farther east and west). Limestone and dolomite glades and rock outcrops; also roadsides.

Missouri populations are on the eastern edge of the distribution of this Central United States endemic. Pennell (1935) and Steyermark (1963) treated plants from southwestern Missouri and adjacent Arkansas with purple corollas as a distinct variety, var. *purpureus*. Pennell stated that this variety was the showiest of all the beard-tongues. Escapes from gardens involving purple-flowered plants have expanded the range of this morph. However, no additional characters have been shown to distinguish these from white-flowered plants and they are better treated merely as trivial color forms.

3. Penstemon digitalis Nutt. ex Sims (smooth beard-tongue, tall white beard-tongue)

P. alluviorum Pennell

Pl. 485 f, g; Map 2215

Stems 25–90(–140) cm long, erect or ascending, usually glabrous, but sometimes moderately to densely pubescent toward the tip with minute, usually nonglandular hairs, these often in longitudinal lines, green to olive green or more commonly purplish-tinged, at least toward the base, often somewhat shiny, not glaucous. Basal leaves 3–20 cm long, the blade oblanceolate to spatulate or obovate, rounded to bluntly or sharply pointed at the tip, tapered basally to a usually winged petiole, the margins entire or nearly so or bluntly to sharply and sometimes finely toothed, the surfaces glabrous or occasionally minutely hairy along the main veins, the undersurface often dotted with minute sessile glands, not glaucous. Stem leaves 2.5–18.0 cm long, the lowermost with the blade oblanceolate to narrowly oblong-elliptic, grading into ovate, lanceolate, or narrowly lanceolate toward the stem tip, mostly sharply pointed at the tip, sessile or nearly so, the base of the lower blades tapered, grading through rounded to those of at least the uppermost leaves often shallowly cordate and clasping, the margins variously entire to bluntly or sharply, irregularly, and sometimes minutely toothed, the surfaces glabrous or occasionally minutely hairy along the main veins, the undersurface often dotted with minute sessile glands, not glaucous. Inflorescences narrow to more broadly pyramidal panicles, the central axis sparsely to moderately glandular-hairy, sometimes also nonglandular-hairy, green or purplish-tinged, not glaucous, with (2–)3–6 nodes, each with a pair of relatively small, linear to narrowly lanceolate, somewhat clasping bracts, the branches ascending or arched upward, with a pair of few- to several-branched clusters per node, each main branch

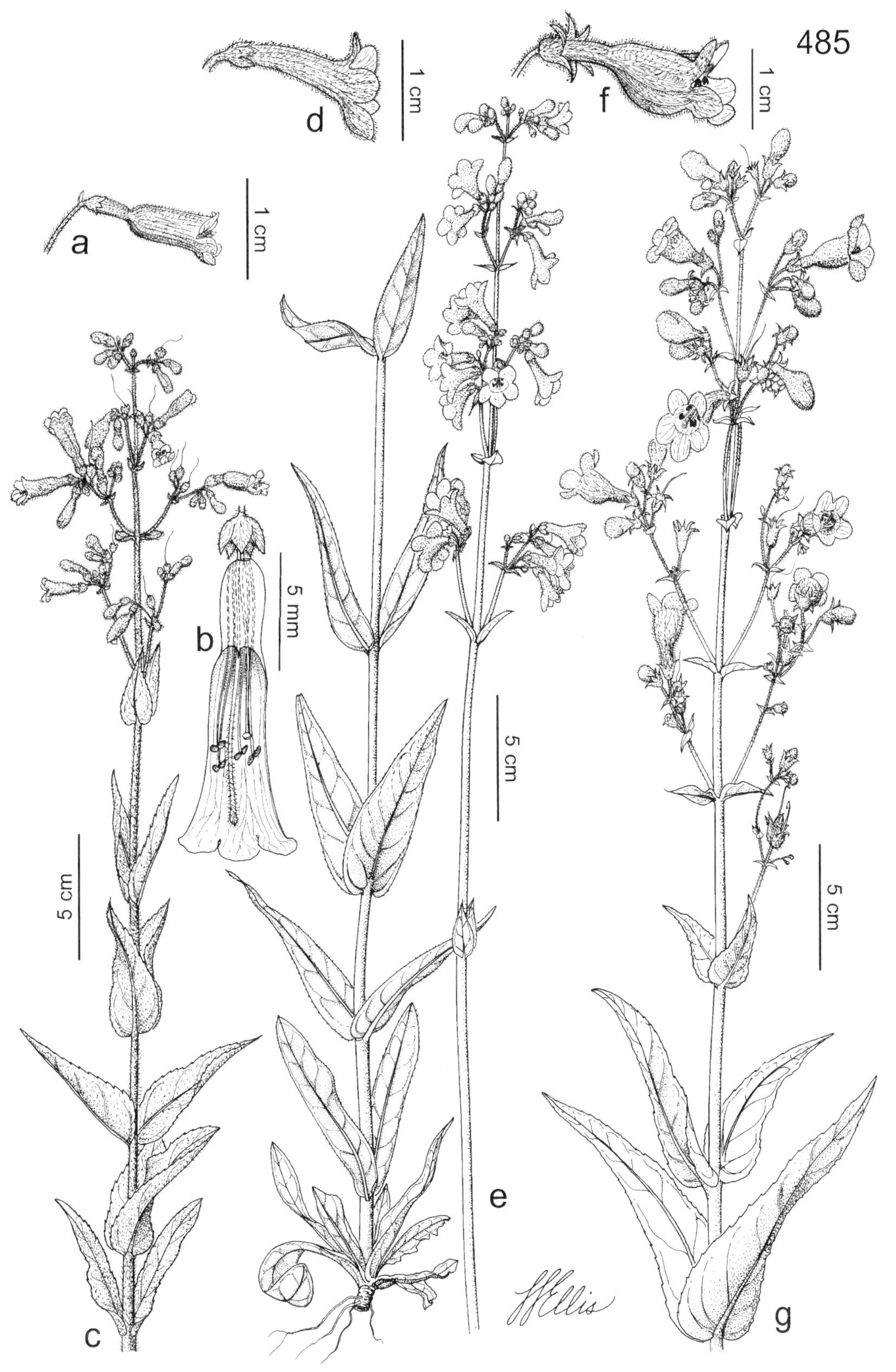

Plate 485. Plantaginaceae. *Penstemon pallidus*, **a)** flower, **b)** dissected flower, **c)** fertile stem. *Penstemon tubaeflorus*, **d)** flower, **e)** fertile stem. *Penstemon digitalis*, **f)** flower, **g)** fertile stem.

with 3–18 flowers. Calyces (3–)4–8 mm long at flowering, moderately to densely glandular-hairy, not glaucous, the lobes narrowly ovate to ovate, angled or only slightly tapered to the sharply pointed tips. Corollas (18–)20–30 mm long, the tube abruptly enlarged at or more commonly below the midpoint, strongly bilabiate, the upper lip spreading to slightly recurved, the lower lip spreading to strongly arched downward, not projecting significantly beyond the upper lip, white or occasionally pale pinkish- to purplish-tinged, the throat not appearing flattened, only slightly 2-ridged and usually lined with purple and/or pink and occasionally also yellow nectar guides on the lower side, minutely glandular-hairy externally and sparsely to moderately nonglandular-hairy in the throat. Staminode white, strongly flattened toward the tip, sometimes somewhat curved or curled downward apically, bearded on the upper surface above the midpoint, the hairs yellow or purple toward the base, yellow to yellowish brown toward the tip. Fruits 10–15 mm long. Seeds 0.8–1.3 mm long, tan to dark brown or black, the tan to reddish brown ridges sometimes well-developed. $2n$=96. April–June.

Scattered to common nearly throughout the state, but absent or uncommon in much of the western half of the Glaciated Plains Division (eastern U.S. west to South Dakota and Texas; Canada). Bottomland forest, mesic to dry upland forests, savannas, glades, ledges and tops of bluffs, upland prairies, sand prairies, bottomland prairies, sloughs, oxbows, banks of streams, marshes, and fens; also pastures, old fields, fallow fields, fencerows, ditches, railroads, and roadsides.

This is the most abundant and morphologically variable species of *Penstemon* in the state. Pennell (1935) and Steyermark (1963) maintained *P. alluviorum* as a separate species, but most recent authors (Koelling, 1964; Cooperrider, 1976; Gleason and Cronquist, 1991) have considered it to represent merely small-flowered plants of *P. digitalis*. The rare f. *baueri* Steyerm. was described from a single historical population in Osage County in which the leaves occur in whorls of three at most stem nodes. Pennell (1935) hypothesized that the original distribution of *P. digitalis* was confined largely to the western and southern portions of its present range and that its presence in much of the eastern United States is a relatively recent phenomenon. It is difficult to know with certainty what the extent of the species' distribution might have been prior to European colonization, but it apparently has been successful in using disturbance corridors such as railroads and roads to spread naturally into new areas.

4. Penstemon grandiflorus Nutt. (large beard-tongue)

Pl. 484 d–f; dust jacket; Map 2216

Stems 40–100 cm long, erect or strongly ascending, glabrous, grayish green to grayish brown or grayish purple, strongly glaucous. Basal leaves 3–16 cm long, the blade oblanceolate to spatulate or obovate, rounded to bluntly pointed at the tip (sometimes with an abrupt, minute, sharp point), mostly tapered basally to a winged petiole, the margins entire, the surfaces glabrous, strongly glaucous. Stem leaves 1.5–11 cm long, the lowermost with the blade oblanceolate to spatulate, grading into circular or depressed-ovate at the stem tip, rounded to bluntly pointed at the tip (sometimes with an abrupt, minute, sharp point), the base rounded to shallowly cordate and then clasping the stem, the margins entire, the surfaces glabrous, strongly glaucous. Inflorescences usually appearing racemose, the central axis glabrous, grayish brown or grayish purple, glaucous, with 3–9 nodes, each with a pair of leaflike, broadly ovate to circular or depressed-ovate, clasping bracts, the relatively short branches ascending, with 2 or 3 flowers per node. Calyces 7–11 mm long at flowering, glabrous, glaucous, the lobes lanceolate to narrowly ovate. Corollas 35–50 mm long, the tube abruptly enlarged well below the midpoint, strongly bilabiate, the upper lip spreading to somewhat recurved, the lower lip spreading, projecting slightly beyond the upper lip, lavender to pale bluish purple, the throat lined with darker reddish purple nectar guides but not noticeably ridged, glabrous internally and externally. Staminode white, strongly flattened toward the tip, curled-under and bearded with yellow hairs apically. Fruits 16–24 mm long. Seeds 2.5–4.0 mm long, brown to black, the brown to black ridges well-developed. $2n$=16. May–June.

Uncommon, known thus far only from Atchison County (Montana to New Mexico east to Michigan, Indiana, and Texas; introduced sporadically eastward). Loess hill prairies.

5. Penstemon pallidus Small (pale beard-tongue)

Pl. 485 a–c; Map 2217

Stems 25–60 cm long, erect or ascending, densely pubescent with minute, nonglandular hairs, also with moderate to dense, longer, glandular hairs, green to grayish green, not glaucous. Basal leaves 2–12(–18) cm long, the blade oblanceolate to spatulate, obovate or elliptic, rounded to bluntly pointed at the tip, tapered basally to a usually winged petiole, the margins entire or nearly so to bluntly or sharply and sometimes

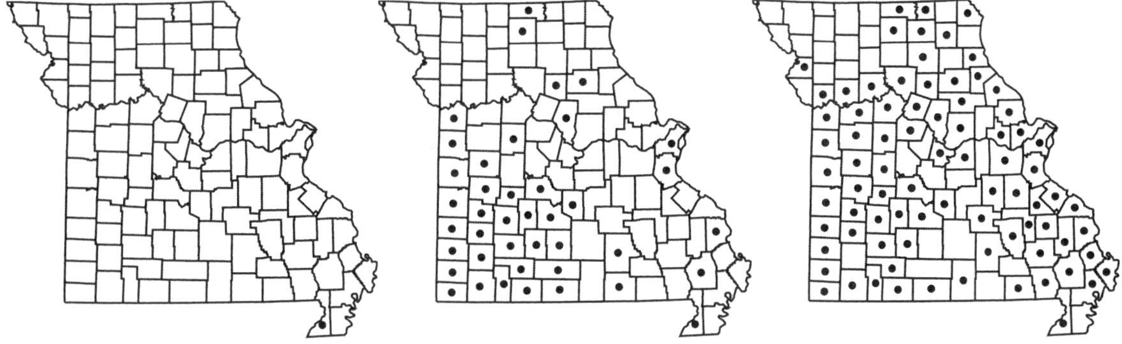

2218. Penstemon tenuis   2219. Penstemon tubaeflorus   2220. Plantago aristata

minutely toothed, the surfaces moderately to densely pubescent with shorter nonglandular and longer glandular hairs, sometimes mostly along the margins and main veins, not glaucous. Stem leaves 2–10 cm long, the lowermost with the blade oblanceolate to narrowly oblong-elliptic or narrowly ovate, grading into lanceolate or narrowly lanceolate at the stem tip, bluntly to sharply pointed at the tip, sessile or nearly so, the base of the lower blades tapered, grading through rounded to those of the uppermost leaves usually shallowly cordate and clasping, the margins entire or nearly so to bluntly or sharply and sometimes minutely toothed, at least above the midpoint, the surfaces pubescent with shorter nonglandular and longer glandular hairs, sometimes mostly along the margins and main veins, not glaucous. Inflorescences narrow to occasionally more broadly pyramidal panicles, the central axis densely glandular-hairy, green to grayish green, not glaucous, with 3–8 nodes, each with a pair of relatively small, linear to narrowly lanceolate, somewhat clasping bracts, the branches loosely to strongly ascending or arched upward, with a pair of few-to several-branched clusters per node, each main branch with 2–9(–16) flowers. Calyces 3–5 mm long at flowering, densely glandular-hairy, not glaucous, the lobes ovate. Corollas 17–23 mm long, the tube abruptly but relatively slightly enlarged at about the midpoint (usually very slightly so on the lower side), strongly bilabiate, the upper lip spreading to somewhat recurved, the lower lip slightly spreading to nearly straight, white or occasionally pale pinkish- to purplish-tinged, the throat appearing slightly flattened, relatively strongly 2-ridged and usually lined with purple nectar guides on the lower side, minutely glandular-hairy externally and sparsely to moderately nonglandular-hairy in the throat. Staminode white, strongly flattened toward the tip, white-hairy on the upper surface, slightly curved downward apically and bearded with yellow hairs. Fruits 5–7 mm long. Seeds 0.5–0.8 mm long, brown to dark brown, the tan to reddish brown ridges usually poorly developed. $2n=16$. April–July.

Scattered to common nearly throughout the state, but absent or uncommon in the southern portion of the Mississippi Lowlands Division and the western half of the Glaciated Plains (eastern U.S. west to Minnesota, Kansas, and Arkansas; Canada). Glades, ledges and tops of bluffs, upland prairies, savannas, and openings of dry upland forests; also pastures, old fields, quarries, railroads, and roadsides; on both acidic and calcareous substrates.

This species is morphologically similar to *P. arkansanus*. For further discussion, see the treatment of that species. Steyermark (1963) noted that this species usually appears hairier than other members of the genus in Missouri. Pennell (1935) hypothesized that the colonization of *P. pallidus* into much of the easternmost states occurred relatively recently.

6. Penstemon tenuis Small (Gulf Coast beard-tongue)

Map 2218

Stems 25–60(–80) cm long, erect or ascending, moderately to occasionally densely pubescent with minute, usually nonglandular hairs, especially toward the tip, the hairs relatively uniformly distributed, green to olive green or more commonly purplish-tinged, often somewhat shiny, not glaucous. Basal leaves 3–10 cm long, the blade oblanceolate to spatulate, obovate, or ovate-elliptic, rounded to bluntly or sharply pointed at the tip, tapered basally to a sometimes winged petiole, the margins entire or nearly so or bluntly and very finely toothed, the surfaces glabrous or nearly so, the undersurface sometimes sparsely dotted with minute sessile glands, not glaucous. Stem leaves 2–10 cm long, the lowermost with the blade oblanceolate to narrowly oblong-elliptic, grading into ovate-triangular or lanceolate-triangular toward the stem tip, the lower leaves angled to a bluntly pointed tip, the upper leaves tapered to a sharply pointed tip, sessile or nearly so, the base of the lower blades tapered or angled to rounded, grading to those of the median and upper leaves, which are shallowly cordate and clasping, the margins evenly but very finely, bluntly to sharply many-toothed, the surfaces glabrous, the undersurface sometimes sparsely dotted with minute sessile glands, not or only slightly glaucous. Inflorescences narrow to more broadly pyramidal panicles, the central axis glabrous or sparsely and minutely nonglandular-hairy toward the base, grading to sparsely pubescent with slightly longer, mostly glandular hairs toward the tip, usually purplish-tinged, not glaucous, with (2)3–6 nodes, each with a pair of relatively small, linear to narrowly lanceolate-triangular, clasping bracts, the branches ascending or arched upward, with a pair of few- to several-branched clusters per node, each main branch with 3–17 flowers. Calyces (3–)4–6 mm long at flowering, sparsely to moderately glandular-hairy, not glaucous, the lobes lanceolate-triangular. Corollas 14–18 mm long, the tube abruptly enlarged near the midpoint, strongly bilabiate, the upper lip spreading to slightly recurved, the lower lip spreading to strongly arched downward, not projecting significantly beyond the upper lip, white light pinkish purple to purple elsewhere), the throat not appearing flattened, only slightly 2-ridged and usually lined with purple and or pink and occasionally also yellow nectar guides on the lower side, minutely glandular-hairy externally and sparsely to moderately nonglandular-hairy in the throat. Staminode white, strongly flattened toward the tip, sometimes somewhat curved or curled downward apically, bearded on the upper surface above the midpoint, the hairs yellow to yellowish brown. Fruits 5–9 mm long. Seeds 0.7–1.0 mm long, tan to dark brown or black, the tan to reddish brown ridges sometimes well-developed. $2n=16$. April–May.

Uncommon, known thus far only from a single collection from Dunklin County (southeastern U.S. west to Missouri and Texas). Ditches and roadsides.

The specimen collected by Jay Raveill in 1986 in southeastern Dunklin County originally was determined as *P. alluviorum*, but recently was redetermined as *P. tenuis* by Dwayne Estes (Austin Peay State University, Clarksville, Tennessee). This bottomland species has a relatively restricted range from eastern Texas, Louisiana, and northwestern Mississippi north to northern Arkansas. It closely resembles the small-flowered forms of *P. digitalis* that have been called *P. alluviorum* by some botanists, but differ from *P. digitalis* in several features (Koelling, 1964; D. Estes, personal communication): 1) corollas usually pale pink to light purple (vs. usually white); 2) the larger leaves consistently with the margins finely and regularly toothed (vs. variably entire or, if toothed, then often unevenly so); 3) upper leaves and bracts tapered evenly from a somewhat expanded base; 4) anthers glabrous or with small papillae (vs. usually bearded); 5) calyx lobes lanceolate-triangular (vs. narrowly ovate to ovate); 6) fruits 5–9 (vs. 10–15) mm long; and 7) a diploid ($2n=16$) vs. hexaploid ($2n=96$) chromosome number. It should be noted that the Missouri specimen is a bit atypical in its white corollas. Future field work will be necessary to confirm the abundance and distribution of *P. tenuis* in Missouri.

7. Penstemon tubaeflorus Nutt. (trumpet beard-tongue)

Pl. 485 d, e; Map 2219

Stems 30–100 cm long, erect or strongly ascending, glabrous (occasionally with 2 longitudinal lines of minute, glandular hairs extending partway to the stem base elsewhere), green or rarely slightly purplish-tinged, not glaucous. Basal leaves 2–11 cm long, the blade obovate to spatulate, rounded to bluntly pointed at the tip, tapered basally to a winged petiole, the margins entire or occasionally with sparse, minute, teeth, the surfaces glabrous, not glaucous. Stem leaves 1.5–14.0 cm long, relatively few toward the stem tip, variously lanceolate to oblong-lanceolate or narrowly ovate, bluntly to more commonly sharply pointed at the tip, sessile and mostly somewhat clasping, variously angled to rounded or cordate, the margins entire or with sparse, minute teeth, the surfaces glabrous, not glaucous. Inflorescences narrow panicles, the central axis glabrous or sparsely glandular-hairy near the nodes, green to dark green or rarely somewhat purplish-tinged, not glaucous, with 4–8 nodes (dense and with the upper nodes not clearly separable in robust plants), each with a pair of relatively small, linear to lanceolate, somewhat clasping bracts, the branches strongly ascending, with a pair of branched clusters per node, each with 3–10 or more flowers. Calyces 2.5–5.0 mm long at flowering, moderately to densely glandular-hairy, not glaucous, the lobes broadly lanceolate to ovate. Corollas 15–25 mm long, the tube slender, expanded evenly only toward the tip, nearly actinomorphic, only slightly bilabiate, the lobes spreading, the 3 lower lobes usually projecting slightly farther forward than the upper 2, white or the tube sometimes pale purplish-tinged toward the base, the throat lacking colored nectar guides and ridges, minutely glandular-hairy externally and in the throat. Staminode white, slightly flattened toward the tip, curled-under apically and relatively sparsely bearded with yellow or brown hairs. Fruits 7–10 mm long. Seeds 1.0–1.5 mm long, tan to dark brown or black, the tan to reddish brown ridges usually poorly developed. $2n=32$. May–June.

Scattered to common in the southwestern quarter of the state, progressively less common to the north and east (Nebraska to Maine south to Texas and Mississippi; Canada). Upland prairies, glades, savannas, openings of mesic to dry upland forests, and rarely fens and openings of bottomland forests; also pastures, railroads, and roadsides.

Pennell (1935) hypothesized that the original distribution of *P. tubaeflorus* was confined largely to the Ozarks and surrounding areas, and that its present, broader distribution is the result of a relatively recent range expansion northward and eastward. This appears to be a natural spread along disturbance corridors such as roadsides. Plants in the northeastern states that are relatively tall and slender have sometimes been segregated as var. *achoreus* Fernald, but this distinction does not appear worthy of formal taxonomic segregation. Note that the species epithet sometimes has been erroneously spelled *tubiflorus* (Yatskievych and Turner, 1990; Gleason and Cronquist, 1991).

15. **Plantago** L. (plantain)

Plants annual, biennial, or perennial herbs (shrubs elsewhere), with taproots or fibrous roots (the roots somewhat fleshy in *P. cordata*). Aerial stems absent or very short (except in *P. indica*). Leaves basal and sometimes also alternate (opposite in *P. indica*), sessile or petiolate. Stipules absent, but the leaf bases sometimes somewhat expanded and sheathing. Leaf blades simple, the margins entire, wavy or toothed, sometimes with slender, rounded pinnate lobes, the venation pattern consisting of a more or less prominent midvein and often a series of only slightly less prominent, more or less parallel, lateral veins (these branching from near the blade base and rejoining the midvein near the blade tip). Inflorescences terminal (axillary in *P. indica*) spikes, usually elongate (sometimes shorter and somewhat headlike in *P. indica*), with numerous dense flowers, at least above the basal portion (more open in *P. cordata*), long-stalked at maturity. Flowers individually inconspicuous, actinomorphic or somewhat zygomorphic, hypogynous, perfect or occasionally incompletely monoecious (with functionally staminate and perfect flowers intermingled), each subtended by a scalelike bract. Cleistogamous flowers sometimes present (noticeable because the anthers and stigmas are not strongly exserted at flowering). Calyces deeply 4-lobed (appearing 3-lobed in *P. lanceolata*), persistent at fruiting, the 2 lobes adjacent to the bract sometimes fused entirely or to above the midpoint, the other lobes free nearly to the base. Corollas 4-lobed, papery, mostly persistent at fruiting, white to tan or slightly grayish-tinged, the short tube slender, the lobes (or 3 of them) abruptly spreading (except in cleistogamous flowers), overlapping in bud, sometimes slightly differing in shape. Stamens 2 or 4, alternating with the corolla lobes, the filaments attached in the corolla tube, the anthers exserted (except in cleistogamous flowers), often somewhat heart-shaped or horned, attached toward their midpoints (or at least above the base), yellow or occasionally dark purple. Pistil 1 per flower, of 2 fused carpels. Ovary superior, 2-locular, with usually 1 to several ovules per locule, the placentation axile or appearing more or less basal. Style 1, the stigma often 2-lobed, it or its lobes mostly linear to club-shaped. Fruits membranous capsules (achenes elsewhere), ovoid to narrowly ellipsoid, with circumscissile dehiscence. Seeds 1 to several per locule, often strongly mucilaginous when moistened. About 270 species, nearly worldwide.

A recent molecular study (Ishikawa et al., 2009) provided data that indicate that some of the polyploid taxa in the genus originated from hybridization events long ago, in some cases involving relatively distantly related species groups. The Missouri species of *Plantago* are wind-pollinated (except in cleistogamous flowers) and some of the more abundant, weedy species produce sufficient pollen to contribute to hayfever allergies. Steyermark (1963) noted that the young foliage of the broader-leaved species can be cooked as a vegetable.

1. Aerial stems well-developed; leaves opposite; inflorescences axillary, more or less elongate spikes or sometimes relatively short and headlike . . . . 5. P. INDICA
1. Aerial stems absent or very short and inconspicuous; leaves all basal and (when aerial stems are present) alternate; inflorescences at maturity elongate spikes (sometimes appearing relatively short when first developing)
   2. Bracts and often also calyces distinctly hairy, at least along the midrib
      3. At least the lowermost bracts extending past the flowers
         4. Leaves with the upper surface glabrous; lowermost bracts 10–30 mm long . . . . 1. P. ARISTATA
         4. Leaves with the upper surface woolly- or silky-hairy; lowermost bracts 5–12 mm long . . . . 8. P. PATAGONICA
      3. Bracts all shorter than to about as long as the flowers
         5. Main leaves linear to very narrowly oblanceolate (those of seedlings and overwintering rosettes sometimes narrowly obovate), 1–7 mm wide; corolla lobes remaining spread or becoming reflexed after flowering
            6. Leaves with the upper surface moderately to densely woolly- or silky-hairy; corolla lobes 1.2–2.2 mm long . . . . 8. P. PATAGONICA
            6. Leaves with the upper surface glabrous or occasionally sparsely short-hairy; corolla lobes 2.4–3.2 mm long. . . . . 12. P. WRIGHTIANA
         5. Leaves all elliptic to oblanceolate or obovate, 4–30 mm wide; corolla lobes erect or overlapping after flowering, appearing beaklike on the fruit
            7. Bracts 3.0–4.5 mm long; corolla lobes 2–3 mm long; seeds 2.2–2.8 mm long, red to reddish black, with lighter margins . . . . 9. P. RHODOSPERMA
            7. Bracts 1.0–2.5 mm long; corolla lobes 0.8–2.3 mm long; seeds 1.4–2.0 mm long, yellowish brown to black . . . . 11. P. VIRGINICA
   2. Bracts and sometimes also calyces glabrous or inconspicuously hairy only along the margins
      8. Main leaves linear to very narrowly oblanceolate (those of seedlings sometimes narrowly obovate); stamens 2; corolla lobes mostly erect or overlapping after flowering, appearing beaklike on the fruit
         9. Seeds mostly 4 per fruit, 1.0–1.8 mm long; bracts shorter than to only slightly longer than the calyces . . . . 3. P. ELONGATA
         9. Seeds mostly 10–25 per fruit, 0.5–0.8 mm long; bracts mostly noticeably longer than the calyces . . . . 4. P. HETEROPHYLLA
      8. Leaves all elliptic to oblanceolate, obovate, or heart-shaped; stamens 4; corolla lobes remaining spread or becoming reflexed after flowering
         10. Leaf blades with some of the lateral veins arising from the midvein well above the blade base; inflorescences usually becoming more or less trailing or prostrate after flowering, the axis hollow, the flowers relatively widely spaced for its entire length. . . . . 2. P. CORDATA
         10. Leaf blades with all of the lateral veins arising from the midvein at the blade base; inflorescences remaining erect or ascending after flowering, the axis not hollow, the flowers dense for its entire length or the lowermost flowers sometimes more widely spaced
            11. Calyces appearing 3-lobed, the 2 sepals adjacent to the bract fused into a single structure with 2 midnerves; bracts and calyx lobes not keeled or only slightly and inconspicuously angled along the midnerve; corolla lobes 2.0–2.5 mm long; seeds mostly 2 per fruit . . . . 6. P. LANCEOLATA

11. Calyces 4-lobed, the 2 sepals adjacent to the bract not fused into a single structure; bracts and calyx lobes with a prominent raised keel along the midnerve; corolla lobes 0.7–1.2 mm long; seeds 4–30 per fruit
    12. Fruits 2.5–4.0 mm long, rhombic-ovoid, the line of circumscissile dehiscence at about the midpoint; seeds 6–18(–30) per fruit, 0.7–1.0 mm long; bracts mostly broadly ovate; petioles usually green at the base . . . . 7. P. MAJOR
    12. Fruits 4–7 mm long, narrowly ellipsoid or narrowly ovoid, the line of circumscissile dehiscence well below the midpoint; seeds 4–10 per fruit, 1.5–2.5 mm long; bracts narrowly lanceolate-triangular; petioles usually reddish- or purplish-tinged at the base . . . . 10. P. RUGELII

1. Plantago aristata Michx. (bracted plantain, buckhorn)

Pl. 486 i, j; Map 2220

Plants annual, with taproots. Aerial stems absent or very short and inconspicuous (to 3 cm long with age), then unbranched or rarely with a basal branch. Leaves in a dense basal rosette (aerial stem leaves alternate but crowded), sessile or with a short, poorly differentiated petiole, strongly ascending. Leaf blades 3–17 cm long, 1–7(–10) mm wide, linear or narrowly oblanceolate, angled or tapered to a sharply pointed tip, long-tapered at the base, the margins entire or occasionally with a few inconspicuous teeth toward the base, hairy, the upper surface glabrous, usually appearing dark green, the undersurface moderately to densely pubescent with shaggy or woolly hairs, appearing uniformly gray, with 1 main vein. Inflorescences 1 to several per plant, terminal, elongate spikes, 3–15 cm long (3–)7–15 mm in diameter (excluding the bracts), densely flowered (the axis not visible between the flowers), the stalk 2–25 cm long, hairy, the axis solid. Lowermost bracts 10–30 mm long, the bracts progressively shorter toward the spike tip, all or most extending past the flowers, linear above a short, inconspicuous, translucent pair of basal wings, long-tapered to the loosely ascending to somewhat arched tip, hairy, more densely so on the undersurface. Cleistogamous flowers usually abundant. Calyces deeply 4-lobed, 2.0–2.5 mm long, slightly zygomorphic, the lobes narrowly oblong-obovate, rounded at the tip, the upper pair with somewhat broader, papery margins than the lower pair. Corollas more or less zygomorphic, the lobes 1.4–2.5 mm long, broadly ovate with a shallowly cordate base, rounded at the tip, the margins entire, each with an inconspicuous brown base, otherwise white to somewhat translucent, the upper lobe slightly shorter than the others and ascending at flowering, the other lobes spreading, spreading to reflexed after flowering. Stamens 4, the anthers somewhat heart-shaped. Fruits 2.8–3.5 mm long, ellipsoid to ovoid, circumscissile just below the midpoint. Seeds usually 2 per fruit, 2–3 mm long, oblong-elliptic, the surface deeply concave on 1 side, otherwise relatively smooth, reddish brown to brown, with a pair of lighter longitudinal stripes on either side of the concave portion. $2n=20$. May–November.

Scattered nearly throughout the state, but apparently absent from the northwestern portion of the Glaciated Plains Division (eastern U.S. west to South Dakota and Texas; Canada; introduced in the western U.S., Hawaii, Europe, Asia). Glades, savannas, upland prairies, openings of mesic to dry upland forests, ledges and tops of bluffs, banks of streams, and occasionally margins of marshes, fens, and sloughs; also old strip mines, pastures, old fields, fallow fields, banks of ditches, railroads, roadsides, and open disturbed areas.

This species is very variable in size, sometimes flowering when plants are only about 5 cm tall, at other times reaching a total height of more than 40 cm. It is distinctive in its elongate bracts, but these sometimes are not well-developed on young plants. The bracts and inflorescence axis continue to elongate as the flowering season progresses.

2. Plantago cordata Lam. (heartleaf plantain)

Pl. 487 e–g; Map 2221

Plants perennial, with several thick (sometimes to more than 1 cm), fleshy, main roots. Aerial stems absent or very short. Leaves in a dense basal rosette, with long, winged and grooved petioles, strongly wine-colored to purplish-tinged at the base, arched or spreading to loosely ascending. Leaf blades 12–25(–40) cm long, 7–20 cm wide (much shorter and only 1–3 mm wide in overwintering rosettes), ovate or broadly elliptic-ovate to broadly heart-shaped (those of overwintering rosettes rhombic-spatulate), angled to a bluntly pointed or rounded tip, rounded to shallowly cordate at the base, the margins entire or with irregular, blunt teeth, sometimes appearing undulate or scalloped, the surfaces glabrous, appearing green to dark green, with 1 midvein and several pairs of strong secondary veins, some of these arising from the midvein well above the blade base,

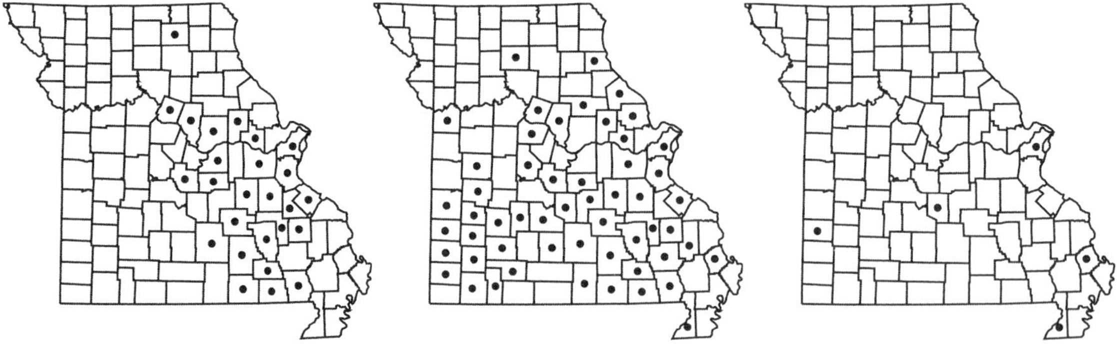

2221. Plantago cordata 2222. Plantago elongata 2223. Plantago heterophylla

the others arising from the blade base. Inflorescences 1 to few per plant, terminal, elongate spikes, 12–30 cm long, 8–14 mm in diameter, loosely flowered for its entire length (the axis visible between the flowers), the stalk 8–25 cm long, erect to loosely ascending at flowering, becoming trailing or prostrate at fruiting, glabrous, the axis becoming hollow as the flowers develop. Bracts 1.6–2.5 mm long, similar in length, shorter than the flowers (slightly shorter than the calyces), obovate, mostly with narrow brown to translucent, papery margins, not or only slightly keeled, often with a slender, slightly raised midnerve, truncate or broadly rounded at the tip, glabrous. Cleistogamous flowers absent. Calyces deeply 4-lobed, 2.0–2.8 mm long, actinomorphic, ovate to elliptic-obovate, rounded or bluntly pointed at the tip, the slender, poorly developed keel glabrous, the narrow, brown to translucent margins thin and papery. Corollas not noticeably zygomorphic, the lobes 1.5–1.8 mm long, ovate-triangular with a truncate base, sharply pointed at the tip, the margins entire and often slightly inrolled, translucent white to tan, all of the lobes spreading at flowering, spreading or more commonly becoming somewhat reflexed at fruiting. Stamens 4, the anthers not horned. Fruits 5–10 mm long, ovoid to broadly ovoid, circumscissile at the midpoint or nearly so. Seeds 2 or less commonly 3 per fruit, 3–4 mm long, oblong-elliptic, the surface with a flattened or slightly concave area on 1 side near 1 end, otherwise smooth, strongly mucilaginous, tan to brown, more or less shiny, shed as a unit along with the associated placental matter. $2n=24$. May–July.

Scattered in the eastern half of the Ozark and Ozark Border Divisions, also disjunct in Adair County (eastern U.S. west to Iowa and Arkansas; Canada). Banks of streams and spring branches (usually gravelly) and rarely bottomland forests; often emergent aquatics in shallow water.

This species is rare and known mostly from historical collections everywhere it occurs outside of Missouri. In contrast, in Missouri it is found in numerous headwater streams and spring branches, usually those in partial shade and flowing over dolomite substrates. The late Herb Wagner (University of Michigan, personal communication) suggested that this was mainly because in other states streams have become much more silted-in as a result of changes in hydrology, agricultural runoff, and other human-mediated perturbations, whereas in Missouri there were still abundant streams with gravelly bottoms and banks. The relatively stout, hollow inflorescences start out ascending to erect, but usually trail in the water at fruiting, a presumed adaptation to seed dispersal by water. Mymudes and Les (1993) reported that plants have pronounced leaf dimorphism, with those of overwintering rosettes smaller and having more rhombic-spatulate blades. Tessene (1968) noted that immature individuals sometimes do not have the characteristic leaf venation pattern of mature plants and that the inflorescence axis starts as a solid structure that eventually becomes hollow as the flowers mature.

Mymudes and Les (1993) studied genetic and morphological variation in *P. cordata*. They concluded that it likely is an allopolyploid, but there have been no suggestions of the possible diploid parents. These authors further found that although the flowers are protogynous, that is, the stigma of a given flower becomes receptive before the stamens of that flower are mature, there was evidence of frequent pollinations between different flowers on the same plant. Morphologically, the species is relatively uniform across its range.

Steyermark (1963) noted that this is the most ornamental of the plantain species native to Missouri. It is a long-lived perennial that is easily raised from seed, but it requires a moist environment in which to grow.

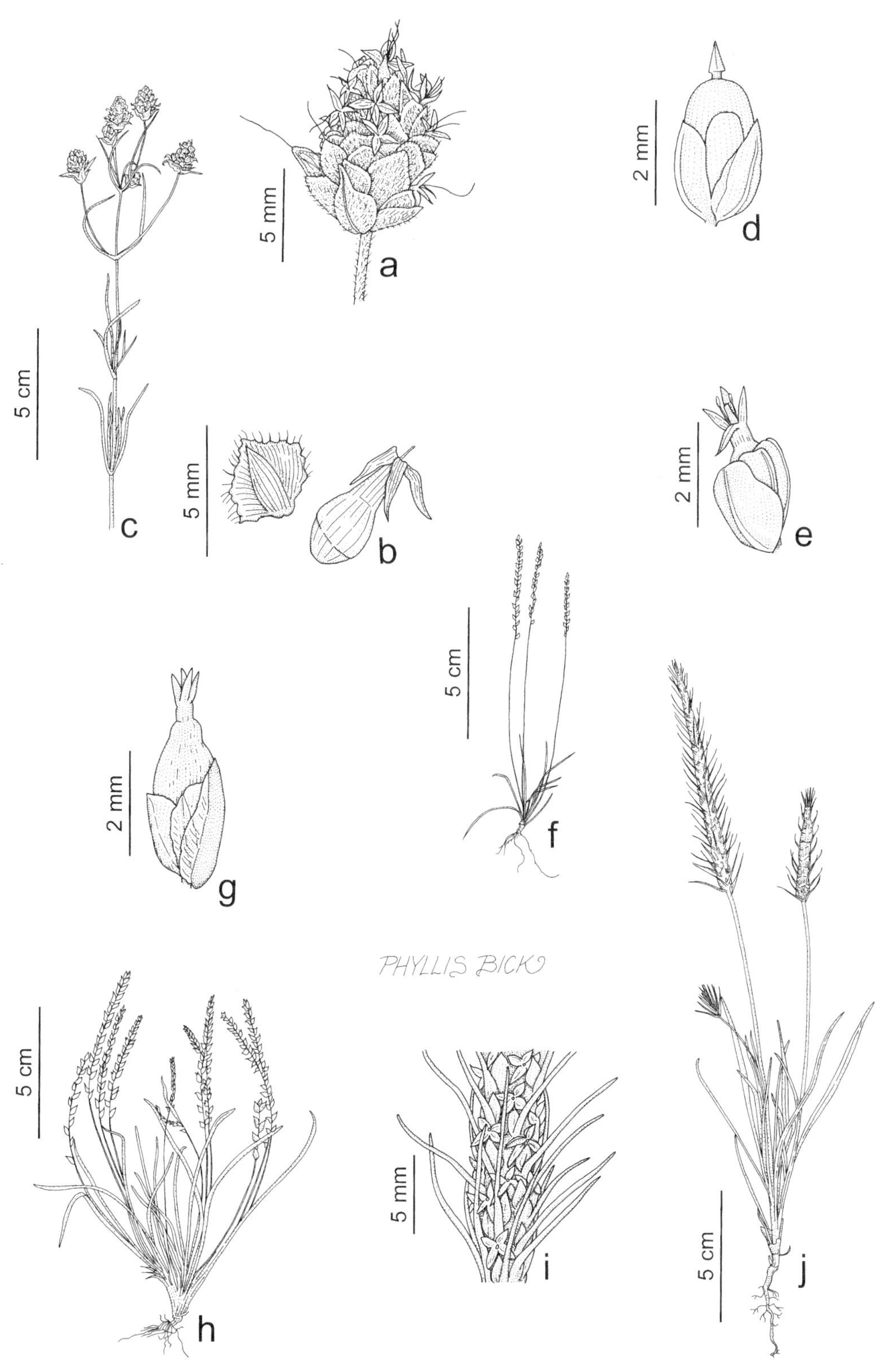

Plate 486. Plantaginaceae. *Plantago indica*, **a)** inflorescence, **b)** fruit and detached bract, **c)** fertile stem. *Plantago elongata*, **d)** fruit with bract, **e)** flower, **f)** habit. *Plantago heterophylla*, **g)** fruit with bract, **h)** habit. *Plantago aristata*, **i)** portion of inflorescence, **j)** habit.

3. Plantago elongata Pursh (small plantain, slender plantain)

P. pusilla Nutt.

P. pusilla var. *major* Engelm.

Pl. 486 d–f; Map 2222

Plants annual, usually with slender taproots (these sometimes disappearing quickly, leaving only a small patch of fibrous roots). Aerial stems absent or very short. Leaves in a dense basal rosette, mostly lacking discrete petioles (but the bases usually slightly expanded and sheathing), usually pale at the base, erect to ascending. Leaf blades 1–6(–9) cm long, 0.5–2.5 mm wide, linear to threadlike (those of seedlings sometimes broader), angled to a sharply pointed tip, tapered at the base, the margins entire or rarely inconspicuously few-toothed, the surfaces glabrous or sparsely to moderately pubescent with short, appressed hairs, appearing green to dark green, with 1 main vein. Inflorescences 1 to several per plant, terminal, elongate spikes, 1–7(–12) cm long, 2–4 mm in diameter, loosely flowered (the axis visible between the flowers), the stalk 2–8(–12) cm long, erect or strongly ascending, sparsely appressed-hairy, the axis solid. Bracts 1.0–2.5 mm long, similar in length, shorter than to about as long as the flowers (shorter than to only slightly longer than the calyces), lanceolate to ovate, with narrow to broad, translucent margins and a thickened, bluntly keeled, green midnerve, angled to a bluntly or sharply pointed tip, glabrous. Cleistogamous flowers usually ubiquitous (open flowers with spreading corollas rarely intermingled with cleistogamous ones). Calyces deeply 4-lobed, 1.5–2.5 mm long, slightly zygomorphic, oblong-ovate to obovate, rounded or bluntly pointed at the tip, the keel glabrous, the relatively broad margins thin and papery. Corollas not noticeably zygomorphic, the lobes 0.5–1.0 mm long, narrowly lanceolate to lanceolate with a truncate base, sharply pointed at the tip, the margins entire, tan, all of the lobes erect and overlapping at flowering, appearing beaklike at fruiting (rarely those of a few flowers spreading during flowering). Stamens 2, the anthers not horned. Fruits 1–3 mm long, narrowly ellipsoid, circumscissile below the midpoint. Seeds mostly 4 per fruit, 1.0–1.8 mm long, narrowly ovate to narrowly oblong-elliptic, the surface with a differentiated band on 1 side, but not appearing concave, otherwise finely pitted, tan to dark brown or black, shiny. $2n=12$, 36. April–June.

Scattered, mostly south of the Missouri River (nearly throughout the U.S., Canada). Glades, savannas, upland prairies, tops of bluffs, openings of mesic to dry upland forests, banks of streams and rivers, and rarely margins of sinkhole ponds; also old fields, fallow fields, pastures, and roadsides; on acidic substrates or in sandy soils.

The taxonomy and nomenclature of the *P. elongata* complex require further study. The nomenclature would be clarified by stable typification of various names whose applications currently remain somewhat ambiguous. Traditionally, some botanists have attempted to separate more western populations as *P. elongata* and eastern populations as *P. pusilla*, based on slightly larger seeds and spreading corolla lobes in the former, vs. smaller seeds and erect corollas lobes in the latter (Bassett, 1966; Gleason and Cronquist, 1991). Shinners (1967) was unable to locate type specimens for either name but based on circumstantial evidence concluded that the original materials for both names were collected while early botanists Nuttall and Bradbury were traveling together (see the History of Botany chapter in Volume 1 of the present work [Yatskievych, 1999]) and thus must have originated from the same geographic area in eastern Nebraska. Hoggard (1998), in his study of the genus in Oklahoma, noted that many populations produce corollas of both types, as well as intermediates, further casting doubt on the distinctness of the two taxa. There appears little merit in maintaining two species. Plants that have been called *P. pusilla* var. *major* represent merely the larger extreme in the range of size variation within the species (Hoggard, 1998). For further discussion of the complex, see the treatment of *P. heterophylla*.

4. Plantago heterophylla Nutt. (small plantain)

Pl. 486 g, h; Map 2223

Plants annual, with slender taproots (these sometimes disappearing quickly, leaving only a small patch of fibrous roots). Aerial stems absent or very short. Leaves in a dense basal rosette, mostly lacking discrete petioles (but the bases usually slightly expanded and sheathing), usually pale at the base, ascending or sometimes spreading with age. Leaf blades (1–)3–9 cm long, 0.5–2.5 mm wide, linear to threadlike (those of seedlings sometimes narrowly obovate), angled to a sharply pointed tip, tapered at the base, the margins entire or with sparse, spreading to loosely ascending, slender teeth, the surfaces glabrous or sparsely to moderately pubescent with short, appressed hairs, appearing green to dark green, with 1 main vein. Inflorescences 1 to several per plant, terminal, elongate spikes, 1–6(–8) cm long, 2–4 mm in diameter, loosely flowered (the axis visible between the flowers), the stalk 2–10 cm long, erect or strongly ascending, sparsely appressed-hairy,

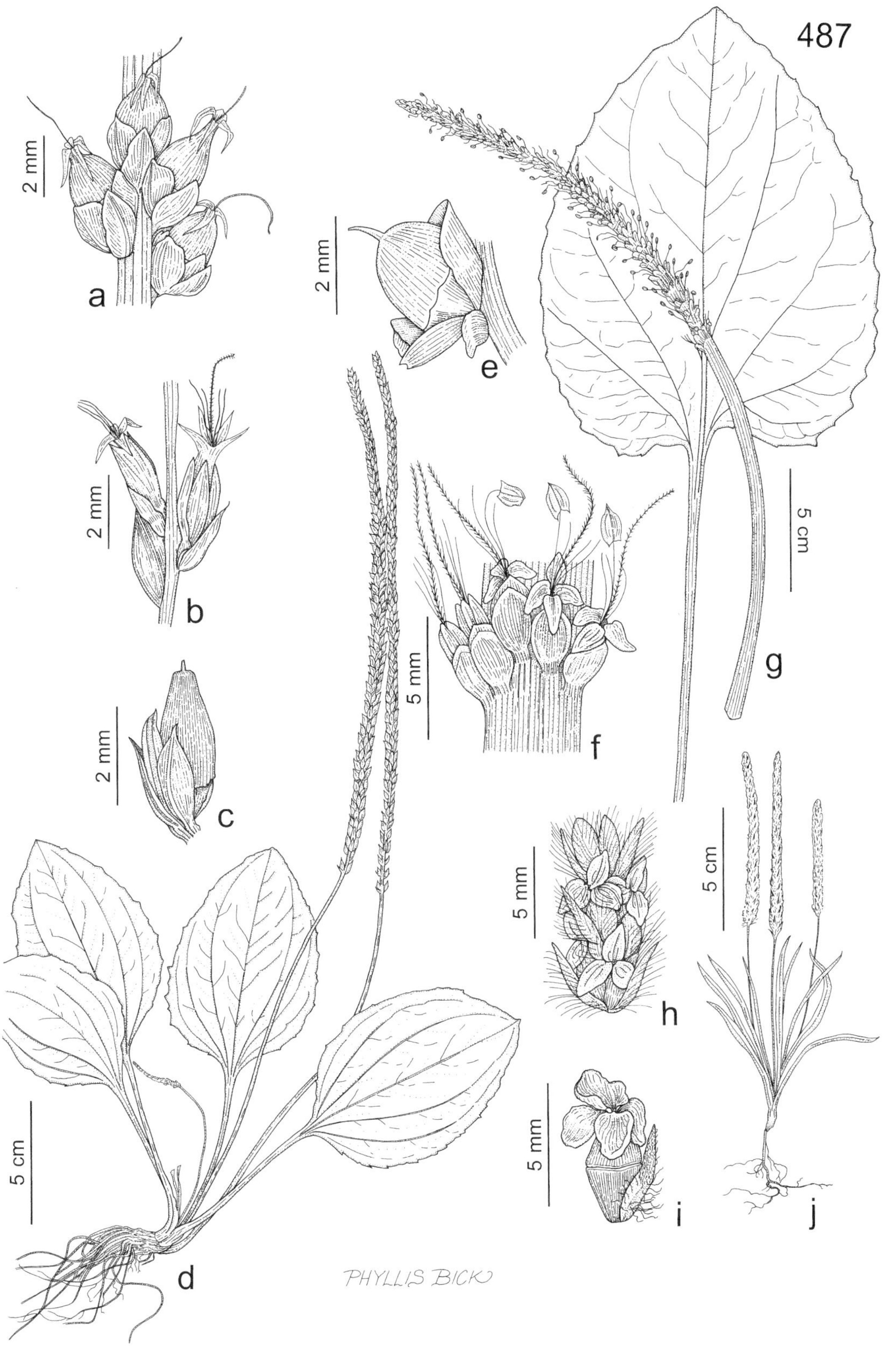

Plate 487. Plantaginaceae. *Plantago major*, **a)** portion of inflorescence with flowers and bracts. *Plantago rugelii*, **b)** portion of inflorescence with flowers and bracts, **c)** fruit with bract, **d)** habit. *Plantago cordata*, **e)** node with fruit and bract, **f)** portion of inflorescence with flowers and bracts, **g)** inflorescence and leaf. *Plantago patagonica*, **h)** portion of inflorescence with flowers and bracts, **i)** fruit with persistent corolla and bract, **j)** habit.

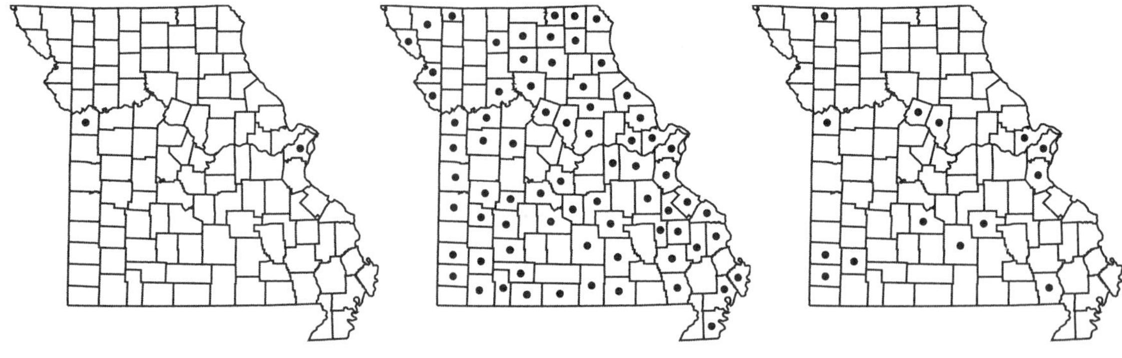
2224. Plantago indica 2225. Plantago lanceolata 2226. Plantago major

the axis solid. Bracts 1.5–3.0 mm long, similar in length, about as long as to more commonly somewhat longer than the flowers (mostly noticeably longer than the calyces), lanceolate to ovate, with broad, translucent margins and a thickened, bluntly keeled, green midnerve, rounded or angled to a bluntly pointed tip, glabrous. Cleistogamous flowers usually ubiquitous (open flowers with spreading corollas rarely intermingled with cleistogamous ones). Calyces deeply 4-lobed, 1.5–2.5 mm long, slightly zygomorphic, oblong-ovate to obovate, rounded at the tip, the keel glabrous, the relatively broad margins thin and papery. Corollas not noticeably zygomorphic, the lobes 0.4–0.8 mm long, narrowly lanceolate to lanceolate with a truncate base, sharply pointed at the tip, the margins entire, tan, all of the lobes erect and overlapping at flowering, appearing beaklike at fruiting (rarely those of a few flowers spreading during flowering). Stamens 2, the anthers not horned. Fruits 1–3 mm long, narrowly ellipsoid, circumscissile below the midpoint. Seeds mostly 10–25 per fruit, 0.5–0.8 mm long, rhombic-ovate to rhombic-elliptic, the surface with a differentiated band on 1 side, but usually not appearing concave, otherwise coarsely pitted, dark brown or black, shiny. $2n$=12, 22. April–May.

Uncommon, sporadic in the southern half of the state (eastern [mostly southeastern] U.S. west to Missouri and Texas; Mexico). Old fields and open, sandy disturbed areas.

The relationship of *P. heterophylla* to *P. elongata* requires further study. Some authors have suggested that the seed number per capsule does not work well to distinguish the two taxa (Shinners, 1967). However, in a morphometric study of the complex in Oklahoma, Hoggard (1998) concluded that a suite of quantitative vegetative and floral characters also differs statistically between the two. He suggested that eventually the two taxa might best be treated as varieties of a single species, but concluded that more detailed research was necessary before taxonomic changes could me made with confidence. Thus the present treatment continues to maintain them as separate, but closely related species.

It should be noted that the name *P. hybrida* W. Barton also requires further evaluation. Although most botanists have treated it as a synonym of *P. elongata* (including *P. pusilla*), Shinners (1967) took it to represent an older name for the plants usually called *P. heterophylla*. Rosatti (1984) correctly pointed out that the original description does not mention seed number, thus the question of whether this name refers to plants of *P. elongata* or *P. heterophylla* remains unresolved. If it should prove to refer to plants currently called *P. heterophylla*, then *P. hybrida* is the older name and has nomenclatural priority over *P. heterophylla*.

5. Plantago indica L. (Indian plantain, whorled plantain)
P. arenaria Waldst. & Kit.
P. psyllium L.

Pl. 486 a–c; Map 2224

Plants annual, with taproots. Aerial stems well-developed, 10–60 cm long, erect or ascending, with more or less ascending branches, moderately to densely pubescent with short, curved hairs, often also glandular toward the branch tips. Leaves opposite (basal rosette absent at flowering), well-spaced, sessile or nearly so, spreading to loosely ascending. Leaf blades 1–8 cm long, 1–3(–5) mm wide, linear or very narrowly oblanceolate, angled or tapered to a sharply pointed tip, angled or tapered at the base, the margins entire but hairy, the surfaces moderately to densely pubescent with short, curved hairs, appearing uniformly grayish green. Inflorescences usually several per plant, axillary, more or less elongate spikes or sometimes relatively short and headlike, 0.5–1.5 cm long 8–11 mm in diameter (including the flowers), densely flowered (the axis not visible between the flowers), the stalk 1–6 cm long, grayish-hairy, the axis solid. Bracts 3–6 mm long, the lowermost 2 bracts 6–10 mm long, extending past the flowers, the others progressively shorter, broadly ovate to nearly circular, rounded at the tip or with a short, bluntly pointed extension of the midnerve, the margins broad, thin, and papery or translucent, the midnerve green, strongly raised into a rounded ridge, the outer surface moderately to densely pubescent with short, spreading hairs, at least some of which are gland-tipped. Cleistogamous flowers absent. Calyces deeply 4-lobed, 3–4 mm long, zygomorphic, the upper pair of lobes slightly shorter, ovate, the lower pair slightly longer, narrowly ovate, otherwise similar to the bracts. Corollas more or less zygomorphic, the lobes 1.5–2.0 mm long, broadly to narrowly ovate-triangular, tapered to a sharply pointed tip, the margins often minutely scalloped, each with a brown base and midnerve, otherwise white to somewhat translucent, the upper lobe ascending at flowering, the others spreading, spreading to reflexed after flowering. Stamens 4, the anthers usually horned. Fruits 3–4 mm long, ellipsoid to broadly ellipsoid, circumscissile at about the midpoint. Seeds 1 or 2 per fruit, 2–3 mm long, oblong-elliptic, the surface smooth, reddish brown to nearly black. $2n=12$. July–September.

Introduced, known thus far only from historical collections from Jackson County and the city of St. Louis (native of Europe, Asia; introduced widely but sporadically in the eastern [mostly northeastern] and western U.S., Canada). Railroads and open, usually sandy, disturbed areas.

This species was first reported from Missouri by Mühlenbach (1983). Its nomenclature has been controversial. Many authors have treated the species under the name *P. psyllium*, but others have considered that epithet to be of uncertain application because of ambiguous and conflicting data in the publications of Linnaeus that included the taxon. Another Linnaean name for the taxon, *P. indica*, has been thought by some to be illegitimate, leaving *P. arenaria* as the next-oldest valid name for the species The history of this problem was discussed by Rosatti (1984). More recently, Applequist (2006) presented arguments against the use of the name *P. psyllium* and proposed its official rejection under the International Code of Botanical Nomenclature. Brummitt (2009) reported on the decision of the nomenclatural Committee for Vascular Plants to approve Applequist's proposal to reject *P. psyllium*, but to accept *P. indica* as validly published.

Plantago indica has a long history of medicinal use in the eastern Mediterranean region to which it is native, often for gastrointestinal problems. It should be noted that other Old World plantains have been used similarly, but that *P. psyllium* (along with *P. ovata* Forssk.) is one of the two main species in cultivation. Today, the species is commercially important as a dietary aid and medicinal supplement. The seed hulls, known as psyllium fiber, are rich in soluble fiber. On the one hand, they are effective as a mild laxative, but they also absorb water, and can thus be used to treat diarrhea. The mucilaginous layer of the seeds also is said to provide a coating for the inner intestinal walls. Supplements containing psyllium fiber sometimes are a component of the treatment for gastrointestinal diseases such as Crohn's disease and ulcerative colitis. There also is evidence that regular doses of psyllium fiber lower cholesterol levels. More recently, it has begun gaining popularity as an appetite inhibitor and weight loss supplement. Ingested before meals, the mucilaginous coating of the seeds absorbs water and swells, providing a feeling of fullness.

6. Plantago lanceolata L. (English plantain, buckhorn, rib grass)

P. lanceolata f. *eriophora* (Hoffmanns. & Link) Beck

P. lanceolata var. *sphaerostachya* Mert. & W.D.J. Koch

Pl. 488 c, d; Map 2225

Plants perennial (but often flowering the first year), with numerous, slender, fibrous roots and sometimes also a taproot, the rootstock occasionally branched at the tip (this often appearing woolly with tan hairs). Aerial stems absent or very short. Leaves in a dense basal rosette, sessile or with obscure, broadly winged petioles, usually wine-colored to purplish-tinged at the base, ascending to arched or spreading to loosely ascending. Leaf blades (5–)10–40 cm long, 7–35 mm wide, narrowly elliptic to narrowly lanceolate or elliptic-lanceolate (sometimes ovate in seedlings and overwintering rosettes), angled to a usually sharply pointed tip, tapered at the base, the margins entire or with widely spaced, short and broad or rarely long and slender teeth, the surfaces glabrous or sparsely to moderately pubescent with curved to spreading hairs (sometimes more densely so on the undersurface), appearing green to dark green, with 3 to several main veins, these all arising from the blade base and appearing more or less parallel. Inflorescences 1 to more commonly several to many per plant, terminal, elongate spikes, 1.5–8.0 cm long, 6–10 mm in diameter,

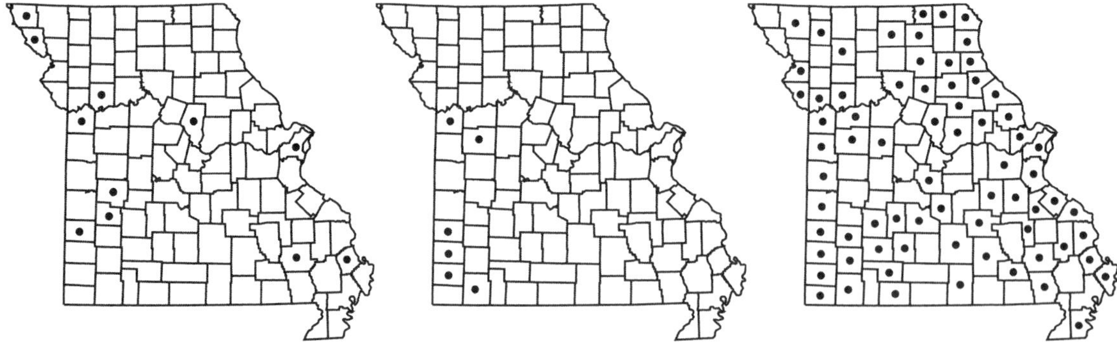

2227. Plantago patagonica 2228. Plantago rhodosperma 2229. Plantago rugelii

densely flowered for the entire length (the axis not visible between the flowers) or the lowermost flowers sometimes more widely spaced, the stalk 12–55 cm long, erect or strongly ascending at flowering and fruiting, hairy, the axis solid, somewhat 5-angled in cross-section. Bracts with the body 1.5–3.0 mm long, similar in length, shorter than to longer than the flowers (the body slightly shorter than the calyces), ovate, with broad, translucent margins and a slender, not or only slightly keeled, green midnerve, often tapered into a slender awnlike or hairlike tip up to as long as the body, glabrous. Cleistogamous flowers absent. Calyces appearing 3-lobed (derived from 4 lobes but with the 2 lobes adjacent to the bract fused into a usually apically notched structure with 2 midnerves), 1.8–3.0 mm long, zygomorphic, ovate (the fused pair broadly oblong-ovate), sharply pointed at the tip, the slender, not or only slightly keeled midnerve glabrous or more commonly hairy toward the tip, the broad, translucent margins thin and papery. Corollas not noticeably zygomorphic, the lobes 2.0–2.5 mm long, narrowly ovate with a shallowly cordate base, sharply pointed at the tip, the margins entire, translucent white to tan, all of the lobes spreading to reflexed at flowering, reflexed at fruiting. Stamens 4, the anthers not horned. Fruits 3–4 mm long, ovoid to obovoid, circumscissile near the base. Seeds mostly 2 per fruit, 1.7–2.2 mm long, oblong-elliptic to ovate, the surface with a flattened or slightly concave area on 1 side, otherwise finely pitted, yellowish brown to brown, shiny. $2n$=12, 13, 24, 96. April–October.

Introduced, common nearly throughout the state (native of Europe, Asia, introduced nearly worldwide). Banks of streams and rivers, margins of ponds and lakes, marshes, and openings of mesic upland forests; also old fields, fallow fields, pastures, edges of crop fields, lawns, gardens, fencerows, railroads, roadsides, and open disturbed areas.

Some botanists recognize a number of infraspecific taxa for *P. lanceolata* (Steyermark, 1963), based on slight differences in inflorescence shape and pubescence patterns. Tessene (1968) corroborated earlier research showing that these can be accounted for mainly by environmental variables, rather than genetically based variation.

7. Plantago major L. (common plantain, broad-leaved plantain, white man's foot)

P. major var. *paludosa* Bég.

P. major ssp. *pleiosperma* Pilger

Pl. 487 a; Map 2226

Plants perennial (but often flowering the first year), with numerous, slender, fibrous roots, the rootstock unbranched and erect at the glabrous tip. Aerial stems absent. Leaves in a dense basal rosette, with short to long, winged and shallowly grooved petioles, usually green at the base, arched or spreading to loosely ascending. Leaf blades 4–18(–25) cm long, 1.5–11.0(–14.0) cm wide, elliptic to ovate to broadly ovate, rounded or broadly angled to a bluntly pointed tip, rounded or occasionally shallowly cordate at the base, the margins entire or occasionally with widely spaced, small teeth, the surfaces glabrous or especially the undersurface sparsely and inconspicuously short-hairy, appearing green to dark green, with mostly 3 or 5 main veins, these all arising from the blade base and appearing arched or curved. Inflorescences 1 to more commonly several or many per plant, terminal, elongate spikes, (3–)5–25 cm long, 4–9 mm in diameter, densely flowered for the entire length (the axis not visible between the flowers) or the lowermost flowers sometimes more widely spaced, the stalk 5–30 cm long, erect or strongly ascending at flowering and fruiting, glabrous or inconspicuously short-hairy, the axis solid, circular in cross-section. Bracts 1.2–2.0 mm long, similar in length, shorter than the flowers (usually shorter than the calyces), mostly broadly

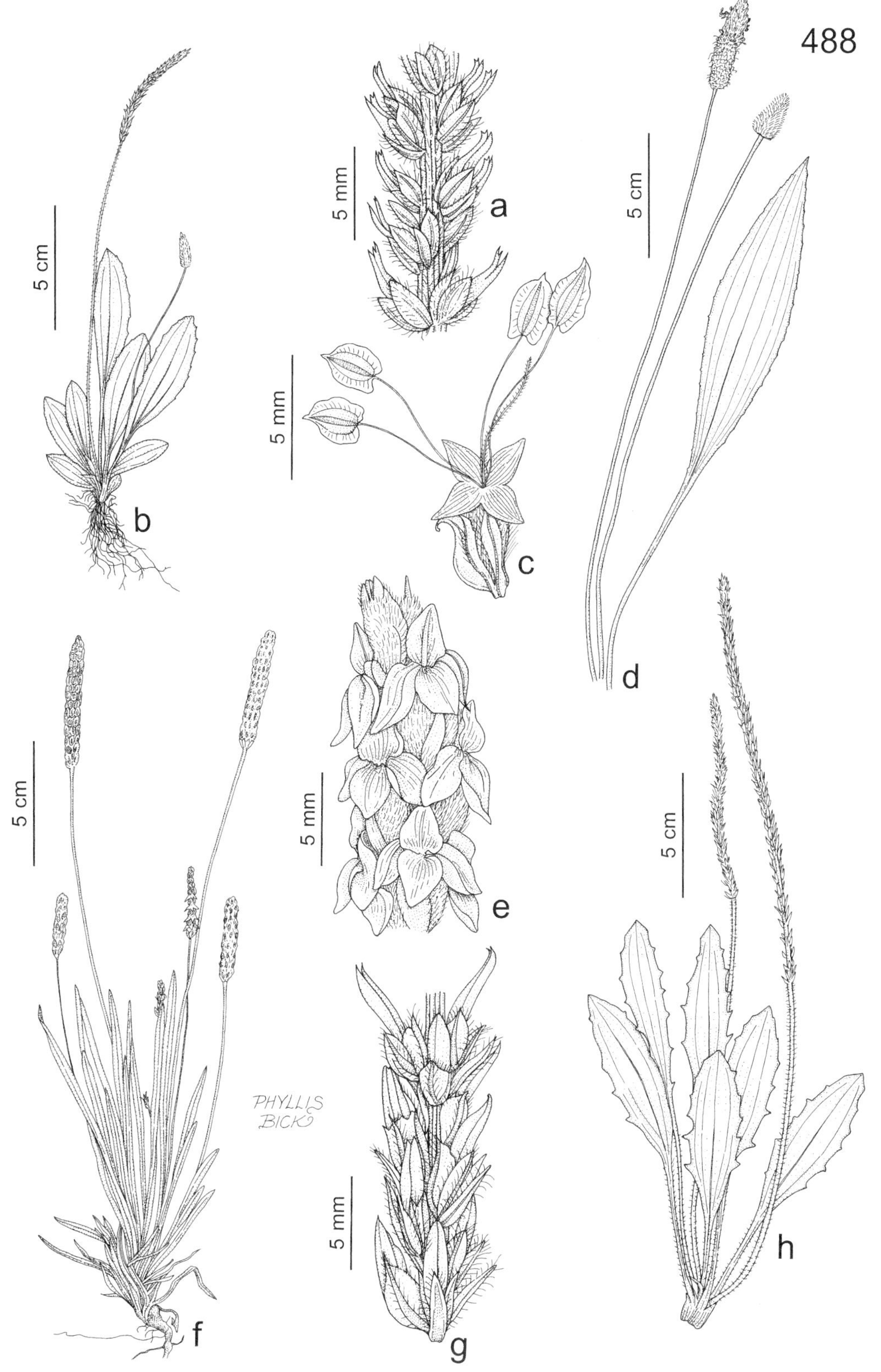

Plate 488. Plantaginaceae. *Plantago virginica*, **a)** portion of inflorescence with flowers and bracts, **b)** habit. *Plantago lanceolata*, **c)** flower, **d)** inflorescences and leaf. *Plantago wrightiana*, **e)** portion of inflorescence with flowers and bracts, **f)** habit. *Plantago rhodosperma*, **g)** portion of inflorescence with flowers and bracts, **h)** habit.

ovate, with broad, translucent margins and a prominently keeled, green midnerve, bluntly pointed or less commonly rounded at the tip, glabrous. Cleistogamous flowers absent. Calyces deeply 4-lobed (the 2 lobes adjacent to the bract not fused into a single structure), 1.5–2.5 mm long, actinomorphic, mostly broadly ovate, rounded or broadly angled to a bluntly pointed tip, the prominently keeled, green midnerve glabrous, the broad, translucent margins thin and papery. Corollas not noticeably zygomorphic, the lobes 0.7–1.2 mm long, lanceolate to narrowly ovate-triangular with a truncate base, sharply pointed at the tip, the margins entire, translucent white, all of the lobes spreading to reflexed at flowering, reflexed at fruiting. Stamens 4, the anthers horned. Fruits 2.0–3.5 mm long, rhombic-ovoid, circumscissile at about the midpoint. Seeds 6–18(–30) per fruit, 0.7–1.0 mm long, irregularly ovate to somewhat quadrate, angled, the surface lacking a well-defined flattened or concave area, otherwise finely ridged, dark brown to black, shiny. $2n$=12, 24. May–October.

Introduced, uncommon in southern Missouri and in counties bordering the Missouri and Mississippi Rivers (native of Europe, Asia, introduced nearly throughout New World). Banks of streams and rivers; also farmyards, lawns, sidewalks, railroads, roadsides, and open, disturbed areas.

Plantago major is superficially very similar to the native *P. rugelii*, especially when depauperate or juvenile plants are present. Tessene (1968), who studied the genus in Wisconsin, noted that juvenile or depauperate plants were difficult to distinguish when in a vegetative state, but that larger or fertile plants could be distinguished by some combination of the following characters (the states for *P. major* listed first): 1) Leaf blades ovate, broadly rounded to cordate at the base vs. elliptic to broadly elliptic, mostly tapered at the base; 2) Leaf blade margins usually entire vs. usually with 3–7 small teeth on each side; 3) petiole bases usually green vs. usually reddish- or purplish-tinged; 4) main veins mostly 3 or 5 vs. mostly 5 or 7; 5) calyx lobes and bracts broadly ovate, broadly angled at the tips vs. narrowly lanceolate-triangular, tapered at the tip; 6) capsules rhombic-ovoid, dehiscent near the midpoint vs. narrowly ellipsoid or narrowly ovoid, dehiscent near the base; 7) seeds 6–18(–30) per fruit, 0.7–1.0 mm long vs. 4–10 per fruit, 1.5–2.5 mm long.

Steyermark (1963) noted that a complex series of subspecies and varieties have been named to account for morphological variation within the native range of the species. However, most North American botanists (and many in Europe) instead treat the taxon as a widespread, morphologically variable species without segregating infraspecific taxa.

8. Plantago patagonica Jacq. (salt-and-pepper plant, Pursh plantain, Patagonian plantain)

P. patagonica var. *breviscapa* (Shinners) Shinners
P. patagonica var. *gnaphalioides* (Nutt.) A. Gray
P. patagonica var. *spinulosa* (Decne.) A. Gray
P. purshii Roem. & Schult.
P. spinulosa Decne.

Pl. 487 h–j; Map 2227

Plants annual or short-lived perennial, with taproots and later sometimes a short, sometimes branched rootstock. Aerial stems absent or very short and inconspicuous (to 4 cm long with age), then unbranched. Leaves in a dense basal rosette (aerial stem leaves alternate but crowded), sessile or with a short, poorly differentiated petiole, ascending. Leaf blades 2–15(–20) cm long, 1–7(–12) mm wide, linear or narrowly oblanceolate (those of seedlings and overwintering rosettes sometimes narrowly obovate), angled or tapered to a sharply pointed tip, long-tapered at the base, the margins entire, hairy, the surfaces moderately to more commonly densely woolly- and/or silky-hairy, appearing uniformly gray, with 1 main vein. Inflorescences 1 to several per plant, terminal, elongate spikes, 2–15 cm long, 7–11 mm in diameter (not including the bracts), densely flowered (the axis not visible between the flowers), the stalk (1–)5–20(–26) cm long, hairy, the axis solid. Lowermost bracts 4–10(–14) mm long, the other bracts 1–5 mm long, progressively shorter toward the spike tip, only the lowermost few to several sometimes extending past the flowers, variously lanceolate triangular with translucent marginal bands and a thickened midnerve to linear above a short, inconspicuous, translucent pair of basal wings, angled to long-tapered at the tip (this loosely ascending to somewhat arched in elongate bracts), densely hairy on both surfaces. Cleistogamous flowers usually abundant. Calyces deeply 4-lobed, 1.4–2.5 mm long, slightly zygomorphic, the lobes narrowly oblong-obovate to obovate, rounded at the tip, the upper pair with somewhat broader, papery margins than the lower pair. Corollas zygomorphic, the lobes 1.2–2.2 mm long, broadly ovate to nearly circular with a shallowly cordate base, rounded to very bluntly pointed at the tip, the margins entire, the surfaces often appearing finely wrinkled, each with an inconspicuous brown base, otherwise white to somewhat translucent, the upper lobe slightly shorter than the others and ascending at flowering, the other lobes spreading, spreading to reflexed after flowering. Stamens 4, the anthers horned. Fruits 3.0–3.5 mm long, ellipsoid to ovoid, circumscissile at or just below the midpoint. Seeds usually 2 per fruit, 2.0–2.8 mm long, oblong-elliptic, the surface deeply concave on 1 side, otherwise finely pitted, reddish brown, lighter in the concave portion. $2n=20$. May–August.

Uncommon in the southwestern and northwestern portions of the state, introduced sporadically farther east (western U.S. east to Wisconsin, Indiana, Missouri, and Texas; Canada, Mexico, South America; introduced farther east in the U.S.). Upland prairies, loess hill prairies, and glades; also old quarries, railroads, roadsides, and open, sandy, disturbed areas.

Some authors (Steyermark, 1963; McGregor and Brooks, 1986; Gleason and Cronquist, 1991) have recognized plants with longer bracts as var. *spinulosa* and dwarf plants have been called var. *breviscapa*. However, Rahn (1978) thought these differences to be unremarkable in the context of overall morphological variation across the very large geographic range of the species.

9. Plantago rhodosperma Decne. (red-seeded plantain)

Pl. 488 g, h; Map 2228

Plants annual or biennial, with taproots. Aerial stems absent. Leaves in a dense basal rosette, mostly with a short, relatively well-differentiated petiole (1–6 cm), reddish- to purplish-tinged at the base, spreading to loosely ascending. Leaf blades (1–)5–20 cm long, (4–)7–30 mm wide, narrowly to broadly oblanceolate or less commonly elliptic to elliptic-lanceolate, rounded or angled to a bluntly or less commonly sharply pointed tip, tapered at the base, the margins entire, sparsely toothed, or occasionally with several slender, rounded lobes, hairy, the surfaces moderately pubescent with more or less spreading to angled hairs, especially along the veins, appearing green, with 3–7 main veins. Inflorescences 1 to several per plant, terminal, elongate spikes, (1–)3–15 cm long, 5–9 mm in diameter, densely flowered (the axis not visible between the flowers), the stalk (1–)3–15 cm long, hairy, the axis solid. Bracts 3.0–4.5 mm long, similar in length, shorter than to about as long as the flowers, narrowly lanceolate, or lanceolate–triangular, with narrow to broad, translucent margins and a thickened, bluntly keeled, green midnerve, rounded or angled to a bluntly or abruptly sharply pointed tip, hairy along the midnerve. Cleistogamous flowers usually the only type produced. Calyces deeply 4-lobed, 2.5–3.2 mm long, slightly zygomorphic, the lobes broadly oblanceolate to obovate, rounded at the tip but with the hairy keel extended into a short sharp point, the margins papery, the upper pair broader than the lower pair. Corollas slightly zygomorphic, the lobes 2–3 mm long, narrowly to broadly lanceolate with a shallowly cordate base, sharply pointed at the tip, the margins entire, tan, lighter at the base, the upper lobe slightly shorter than the others all of the other lobes erect and overlapping (rarely spreading elsewhere) during and after flowering. Stamens 4, the anthers not horned. Fruits 2.5–3.8 mm long, narrowly ellipsoid, circumscissile below the midpoint. Seeds usually 2 per fruit, 2.2–2.8 mm long, ovate, the surface shallowly concave on 1 side, otherwise finely pitted or finely wrinkled to nearly smooth, red to reddish black, with lighter margins. $2n=24$. May–June.

Introduced, uncommon in the western portion of the state (native of the southwestern U.S. east to Oklahoma, Texas; Mexico; introduced farther north to Nebraska and Illinois and east to Kentucky, Tennessee, and possibly Georgia). Banks of rivers and openings of mesic upland forests; also open, sandy or rocky, disturbed areas.

Steyermark (1963) knew *P. rhodosperma* only from a single historical collection from Jackson County. This species is very similar morphologically to *P. virginica*, which is much more abundant in Missouri. It may be routinely overlooked by collectors and thus more abundant in the state than the meager herbarium record indicates.

10. Plantago rugelii Decne. (Rugel's Plantain, broad-leaved plantain)

P. rugelii var. *asperula* Farw.

Pl. 486 b–d; Map 2229

Plants perennial (but often flowering the first year), with numerous, slender, fibrous roots, the rootstock unbranched and erect at the glabrous tip. Aerial stems absent. Leaves in a dense basal rosette, with short to long, winged and shallowly grooved petioles, usually reddish- or purplish-tinged at the base, arched or spreading to loosely ascending. Leaf blades 4–20(–35) cm long, 1.5–12.0(–18.0) cm wide, elliptic to broadly elliptic, rounded or broadly angled to a bluntly pointed tip, mostly short-tapered at the base, less commonly rounded or narrowly cordate, the margins entire or more commonly with 3–7, widely spaced, small, blunt teeth per side, the surfaces glabrous or nearly so, rarely somewhat roughened with minute hairs, appearing green to dark green, with mostly 5 or 7 main veins, these all arising from the blade base and appearing arched or curved. Inflorescences 1 to more commonly several or many per plant, terminal, elongate spikes, (3–)5–35 cm long, 4–12 mm

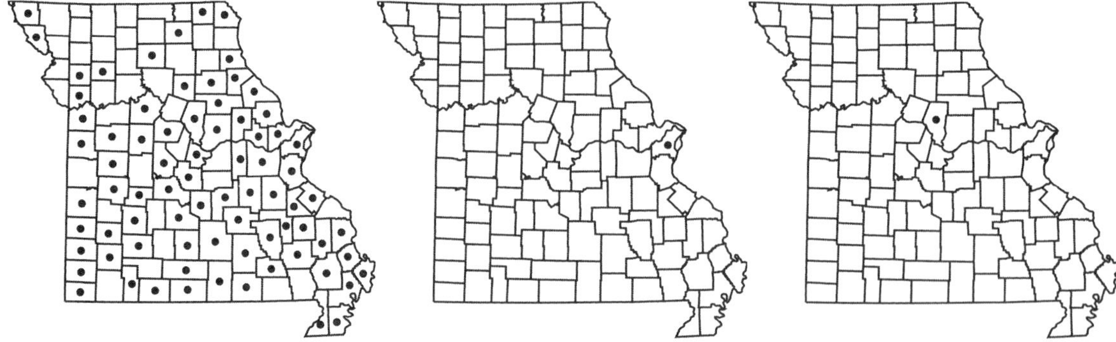

2230. Plantago virginica 2231. Plantago wrightiana 2232. Veronica americana

in diameter, densely flowered for the entire length (the axis not visible between the flowers) or the lowermost flowers sometimes more widely spaced, the stalk (3–)5–30 cm long, erect or strongly ascending at flowering and fruiting, glabrous or nearly so, rarely roughened with minute hairs, the axis solid, circular in cross-section. Bracts 1.0–2.5 mm long, similar in length, shorter than the flowers (usually shorter than the calyces), narrowly lanceolate-triangular, with narrow, translucent or dark purple margins and a prominently keeled, green midnerve, angled or more commonly tapered to a sharply pointed tip, glabrous. Cleistogamous flowers absent. Calyces deeply 4-lobed (the 2 lobes adjacent to the bract not fused into a single structure), 1.5–2.5 mm long, actinomorphic, narrowly lanceolate-triangular, angled or more commonly tapered to a sharply pointed tip, the prominently keeled, green midnerve glabrous, the narrow, translucent or dark purple margins thin and papery. Corollas not noticeably zygomorphic, the lobes 0.7–1.2 mm long, lanceolate to narrowly ovate-triangular with a truncate base, sharply pointed at the tip, the margins entire, translucent white, all of the lobes spreading to reflexed at flowering, reflexed at fruiting. Stamens 4, the anthers horned. Fruits 4–7 mm long, narrowly ellipsoid or narrowly ovoid, circumscissile well below the midpoint. Seeds 4–10 per fruit, 1.5–2.5 mm long, irregularly ovate to somewhat quadrate, angled, the surface lacking a well-defined flattened or concave area, otherwise very finely pitted or appearing pebbled, dark brown to black, shiny. $2n=24$. May–October.

Common throughout the state (eastern U.S. west to North Dakota and Texas; Canada). Banks of streams and rivers, margins of ponds, lakes, and sinkhole ponds, oxbows, marshes, and openings and edges of bottomland and mesic upland forests; also old fields, ditches, fencerows, lawns, sidewalks, railroads, roadsides, and open, disturbed areas.

Steyermark (1963) noted that this is the most abundant of the broad-leaved plantains in Missouri. Plants with the leaf blades and often also the inflorescence stalks more or less roughened-hairy have been called var. *asperula*. Such plants generally occur as individuals within populations of otherwise typical plants. Superficially, the species bears a strong resemblance to the nonnative *P. major*. See the treatment of that species for a discussion of distinctions between them.

11. Plantago virginica L. (hoary plantain, pale-seeded plantain)

P. virginica var. *viridescens* Fernald

Pl. 488 a, b; Map 2230

Plants annual, with taproots. Aerial stems absent or very short. Leaves in a dense basal rosette, sometimes the larger leaves with a more or less well-differentiated petiole (1–6 cm long), usually pale at the base, spreading to ascending. Leaf blades 1–10(–20) cm long, 4–35 mm wide, narrowly to broadly oblanceolate, obovate, or occasionally spatulate, rarely elliptic-oblanceolate, angled to a bluntly or sharply pointed tip, tapered at the base, the margins entire or sparsely and inconspicuously toothed (rarely with relatively large, widely spaced, spreading teeth), hairy, the surfaces moderately pubescent with more or less spreading hairs, appearing green to yellowish green or pale green, with 3–7 main veins. Inflorescences 1 to several per plant, terminal, elongate spikes, (1–)3–15(–25) cm long, 5–9 mm in diameter, densely flowered (the axis not visible between the flowers), the stalk (0.5–)2–15 cm long, spreading-hairy, the axis solid. Bracts 1.0–2.5 mm long, similar in length, mostly shorter than the flowers, broadly lanceolate to ovate, with narrow to broad, translucent margins and a thickened, bluntly keeled, green midnerve, rounded or angled to a bluntly pointed tip, hairy, at least along the midnerve. Cleistogamous flowers usually abundant. Calyces deeply 4-lobed, 1.5–

2.7 mm long, slightly zygomorphic, oblong-obovate, rounded at the tip, the keel hairy, the relatively broad margins thin and papery. Corollas slightly zygomorphic, the lobes 0.8–2.3 mm long, narrowly to broadly lanceolate with a shallowly cordate base, sharply pointed at the tip, the margins entire, tan, lighter at the base, the upper lobe slightly shorter than the others, all of the other lobes erect and overlapping or less commonly spreading during flowering, but becoming erect soon afterward. Stamens 4, the anthers not horned. Fruits 2.5–3.8 mm long, narrowly ellipsoid, circumscissile below the midpoint. Seeds usually 2 per fruit, 1.4–2.0 mm long, narrowly ovate, the surface relatively deeply concave on 1 side, otherwise finely pitted or finely wrinkled to nearly smooth, yellowish brown to black, dull. $2n$=12, 24. April–June.

Common throughout the state (eastern U.S. west to South Dakota, Arizona, California and Oregon; Canada, Mexico, Caribbean Islands; introduced in Europe, Asia). Glades, upland prairies, loess hill prairies, sand prairies, savannas, banks of streams and rivers, margins of sinkhole ponds, and ledges and tops of bluffs; also pastures, old fields, fallow fields, railroads, roadsides, and open, disturbed area.

This short-lived plant is among the most variable species in the genus in terms of overall size, leaf shape, and pubescence. In some dry or sandy habitats, tiny plants occur that nevertheless successfully flower and fruit.

12. Plantago wrightiana Decne. (Wright's plantain)

Pl. 488 e, f; Map 2231

Plants annual or short-lived perennial, with taproots and later sometimes a short, sometimes branched rootstock. Aerial stems absent or very short and inconspicuous (to 5 cm long with age), then unbranched. Leaves in a dense basal rosette (aerial stem leaves alternate but crowded), sessile or with a short, poorly differentiated petiole, strongly ascending. Leaf blades 3–12(–18) cm long, 1–7 mm wide, linear or narrowly oblanceolate (those of seedlings and overwintering rosettes sometimes narrowly obovate), angled or tapered to a bluntly or more commonly sharply pointed tip, long-tapered at the base, the margins entire, hairy, the upper surface glabrous or occasionally sparsely short-hairy, usually appearing dark green (usually darkening upon drying), the undersurface moderately to densely pubescent with shaggy or woolly hairs, appearing uniformly gray, with 1 main vein. Inflorescences 1 to several per plant, terminal, elongate spikes, 1–10 cm long 7–9 mm in diameter, densely flowered (the axis not visible between the flowers), the stalk 5–20 cm long, hairy, the axis solid. Bracts 2–3 mm long, similar in length, shorter than to about as long as the flowers, triangular-ovate, with relatively broad, translucent margins and a thick, green midnerve, angled to a bluntly or sharply pointed tip, glabrous or hairy along the midnerve. Cleistogamous flowers usually abundant. Calyces deeply 4-lobed, 2.7–3.5 mm long, slightly zygomorphic, narrowly oblong-obovate to obovate, rounded at the tip, the upper pair with somewhat broader, papery margins than the lower pair. Corollas zygomorphic, the lobes 2.4–3.2 mm long, broadly ovate with a shallowly cordate base (the lateral lobes asymmetric), rounded to very bluntly pointed at the tip, the margins entire, each with an inconspicuous brown base, otherwise white to somewhat translucent, the upper lobe slightly shorter than the others and ascending at flowering, the other lobes spreading, spreading to reflexed after flowering. Stamens 4, the anthers horned. Fruits 3.5–4.2 mm long, ellipsoid to ovoid, circumscissile below the midpoint. Seeds usually 2 per fruit, 2.5–3.2 mm long, oblong-elliptic with a shallow groove around the circumference just below the midpoint, the surface shallowly to deeply concave on 1 side, otherwise finely pitted, reddish brown, lighter in the concave portion. $2n$=20. May–August.

Introduced, known thus far only from the city of St. Louis (native of Arizona, New Mexico, Oklahoma, Texas; Mexico; introduced widely in the southeastern U.S. and Missouri). Railroads.

This species was discovered in Missouri by Viktor Mühlenbach during his inventories of the St. Louis railyards. It was first reported by Steyermark (1963) in the supplement of species documented too late for inclusion in the main text of his *Flora of Missouri*.

16. **Veronica** L. (speedwell)

Plants annual or perennial herbs, terrestrial or emergent aquatics. Stems variously erect to prostrate, unbranched or branched, glabrous or hairy, the hairs sometimes glandular. Leaves opposite (the leaflike inflorescence bracts usually alternate), sessile or short-petiolate. Leaf blades simple, unlobed or shallowly to moderately palmately 3- or 5-lobed, variously shaped, the margins otherwise entire to scalloped or toothed, the surfaces glabrous or hairy, the venation palmate or pinnate, but the lateral veins then sometimes very faint. Inflorescences terminal or axillary racemes, sometimes appearing spikelike, the flowers alternate along the axis (the lowermost sometimes subopposite), short- to long-stalked; each subtended by a leaflike or reduced bract, but bractlets absent. Flowers perfect. Calyces deeply 4- or less commonly 5-lobed nearly to the base, actinomorphic or only slightly zygomorphic (the lower pair of lobes then slightly longer than the upper pair), the lobes variously shaped, glabrous or hairy. Corollas not bilabiate, weakly zygomorphic (the upper lobe larger and the lower lobe smaller than the lateral lobes), moderately to relatively deeply 4-lobed, usually glabrous, the lobes spreading to loosely ascending, blue, purple, pink, or white, often with darker veins, lacking a spur, the throat open, sometimes abruptly much lighter than the lobes. Stamens 2, the filaments relatively attached at the tip of the tube and exserted from the corolla, the anther sacs more or less parallel; staminodes absent. Style 1, unbranched, the stigma small, capitate, unlobed or shallowly 2-lobed. Fruits capsules, heart-shaped to nearly circular or rarely broadly ovate in outline, slightly to strongly notched at the tip, moderately to strongly flattened, glabrous or hairy, the 2 locules equal in size, dehiscent longitudinally mostly along the 2 sutures (along the rim) and sometimes also along the locule wall (then splitting into 4 more or less equal valves). Seeds 2 to numerous, oblong to oblong-elliptic or oblong-ovate in outline (nearly circular in *V. heterophylla*), flattened or concave on 1 side, flattened to convex on the other side, the surface pale brown to yellowish brown, brown, or nearly black, smooth, pebbled with a network of minute ridges, or with few to several coarse cross-wrinkles. About 450 species, nearly worldwide.

Veronica has been considered a taxonomically difficult genus with a large number of infrageneric groups containing relatively cryptic species, these sometimes involving polyploid complexes. Even the taxonomic limits of the genus have remained controversial (Albach et al., 2004). Most of the Missouri species are Old World natives that either have escaped from cultivation or are weedy taxa that were introduced to the New World long ago. A number of species in the genus are popular in horticulture, thus new escapes are likely to be recorded in the future. Interestingly, six of the fourteen species found in the state were first reported since 1980 and a seventh state record is reported for the first time in the present work, thus the number of species documented for the state has doubled in the past few decades.

1. Inflorescences axillary racemes, these usually in opposite pairs at the stem nodes (1 per leaf), the bracts much smaller than the foliage leaves
    2. Stems moderately to densely pubescent with nonglandular hairs throughout (sometimes less densely so toward the base); fruits dehiscing along the sutures into 2 valves
        3. Main stems erect or strongly ascending . . . . 14. V. TEUCRIUM
        3. Main stems prostrate or loosely ascending, the flowering branches sometimes more strongly ascending
            4. Stems with the hairs spreading, mostly confined to a pair of longitudinal lines, at least between the lower nodes; leaf blades ovate to oblong-ovate or broadly ovate, the margins flat; calyx lobes 4; fruits hairy along the margins . . . . 6. V. CHAMAEDRYS
            4. Stems with the hairs curled upward, evenly distributed between the nodes (not in lines); leaf blades linear to narrowly oblong-elliptic or narrowly lanceolate, the margins curled or curved under; calyx lobes (4)5; fruits glabrous . . . . 12. V. SCHEERERI
    2. Stems glabrous or sparsely to moderately pubescent with gland-tipped hairs only toward the tip (and inflorescence axis); fruits usually dehiscing both along and between the sutures into 4 valves

5. Leaves all short-petiolate
 6. Leaf blades lanceolate to narrowly ovate or triangular-ovate (the lowermost occasionally elliptic to elliptic-ovate), mostly broadest toward the base; stems mostly strongly ascending to erect, only a short basal portion spreading (occasionally plants flattened by flooding may have a more spreading growth form) . . . . . . 1. V. AMERICANA
 6. Leaf blades elliptic to obovate or oblong-elliptic, mostly broadest at or above the midpoint; stems mostly prostrate to loosely ascending, more strongly ascending to erect only near the tips . . . . 4. V. BECCABUNGA
5. At least the upper and median leaves sessile (vegetative, late-season stems with mostly short-petiolate leaves are sometimes produced in *V. anagallis-aquatica* and might be misdetermined as *V. beccabunga*)
 7. Leaf blades broadly lanceolate to ovate (mostly 1.5–3.0 times as long as wide); stalks of fruits ascending or curved upward . . . . 2. V. ANAGALLIS-AQUATICA
 7. Leaf blades lanceolate to oblong-oblanceolate (mostly 3–5 times as long as wide); stalks of fruits spreading or occasionally some of them slightly ascending or upcurved . . . . 5. V. CATENATA

1. Inflorescences terminal, open or spikelike racemes, sometimes with additional, apparently axillary racemes (these usually terminating short branches) or with the flowers widely spaced and subtended by bracts that resemble foliage leaves but are alternate along the axis (the inflorescences then appearing as solitary axillary flowers)
 8. Blades of foliage leaves (not bracts) (3–)4–12(–15) cm long, lanceolate to broadly lanceolate, usually grading into narrowly lanceolate or narrowly elliptic-lanceolate toward the stem tip, the margins densely and sharply toothed; inflorescences dense elongate racemes, the axis mostly hidden by the flowers; corollas with the tube slender, longer than wide; style 4–10 mm long . . . . 8. V. LONGIFOLIA
 8. Blades of foliage leaves 0.8–3.5 cm long, elliptic to ovate, broadly ovate, or nearly circular (narrowly oblong to oblanceolate in *V. peregrina*), the margins variously entire to scalloped, bluntly few-toothed, or few-lobed; inflorescences more open (the axis readily visible, or if appearing denser at the start of flowering then not elongate; corollas with the tube appearing broader, wider than long; style 0.1–3.0 mm long
  9. Inflorescences with the bracts abruptly much smaller than the foliage leaves, thus appearing as well-differentiated racemes; plants perennial, with rhizomes, with a well-developed vegetative portion . . . . 13. V. SERPYLLIFOLIA
  9. Inflorescences with the bracts similar to the foliage leaves or gradually reduced toward the axis tip, the inflorescence thus frequently appearing as flowers solitary in the leaf axils; plants annuals, with fibrous roots or short taproots but lacking rhizomes, often producing flowers nearly to the stem base
   10. Leaf blades palmately 3–7(9)-lobed
    11. Stems prostrate to loosely ascending (often forming loose mats); bracts subtending open flowers and fruits not noticeably smaller than the foliage leaves, the flowers appearing solitary in the leaf axils; fruits 2.5–4.0 mm long, very shallowly notched; style 0.6–0.9 mm long . . . . 7. V. HEDERIFOLIA

11. Stems erect or ascending, sometimes from a spreading base; bracts subtending open flowers gradually reduced, those toward the inflorescence tip noticeably smaller than the foliage leaves, the flowers appearing in racemes (these short and condensed at the start of flowering); fruits 4.2–5.5 mm long, relatively deeply notched; style 1.2–1.6 mm long . . . . 15. V. TRIPHYLLOS

10. Leaf blades unlobed, the margins entire or scalloped to bluntly toothed
12. Flower and fruit stalks 0.5–2.0 mm long, shorter than the calyces; calyx lobes linear to lanceolate
13. Stems hairy, at least the hairs below the stem midpoint nonglandular; corollas blue to bluish purple; style 0.4–1.0 mm long; calyx lobes unequal, the lower pair noticeably longer than the upper pair . . . . 3. V. ARVENSIS
13. Stems glabrous or glandular-hairy; corollas white or pale yellow; style 0.1–0.4 mm long; calyx lobes subequal . . . . 9. V. PEREGRINA
12. Flower and fruit stalks 4–40 mm long, longer than the calyces; calyx lobes broadly lanceolate or oblanceolate to ovate
14. Flower and fruit stalks 15–40 mm long; corollas 8–11 mm wide; style 2–3 mm long . . . . 10. V. PERSICA
14. Flower and fruit stalks 4–10(–15) mm long; corollas 4–8 mm wide; style 0.8–1.5 mm long . . . . 11. V. POLITA

1. Veronica americana Schwein. ex Benth. (American brooklime)

V. beccabunga L. var. *americana* Raf.

Pl. 490 a, b; Map 2232

Plants perennial, with rhizomes. Stems 10–100 cm long, mostly strongly ascending to erect, only a short basal portion spreading (occasionally plants flattened by flooding may have a more spreading growth form), glabrous. Leaves all short-petiolate. Leaf blades (0.5–)1.5–5.0(–8.0) cm long, mostly 2–4 times as long as wide, lanceolate to narrowly ovate or triangular-ovate (the lowermost occasionally elliptic to elliptic-ovate), mostly broadest toward the base, bluntly to sharply pointed at the tip, rounded to truncate or shallowly cordate at the base, not clasping the stems, the margins unlobed, flat, finely and bluntly to more commonly sharply toothed (those of the lowermost leaves sometimes subentire), the surfaces glabrous. Inflorescences axillary racemes, these usually in opposite pairs at the stem nodes (1 per leaf), open at maturity, with 10–25 flowers, the bracts 2–6 mm long, much smaller than the foliage leaves, linear to narrowly lanceolate. Flower stalks 5–10 mm long at flowering (to 15 mm long at fruiting), ascending at flowering, but mostly spreading at a right angle to the axis at fruiting. Calyces 2.5–5.0 mm long, the lobes slightly unequal, the upper 2 lobes slightly shorter than the lower 2 lobes, deeply 4-lobed, the lobes lanceolate or oblong-oblanceolate to ovate or oblong-obovate, glabrous. Corollas 5–10 mm wide, blue to purplish blue with darker veins, the throat white or sometimes pale greenish-tinged at the base, the lobes spreading to loosely cupped upward. Style 2.0–3.5 mm long at fruiting. Fruits 2.5–3.5 mm long, mostly slightly wider than long, depressed-obovate to more or less circular in profile, somewhat turgid, the notch very shallow, the surfaces and margins glabrous, usually dehiscing both along and between the sutures into 4 valves. Seeds numerous, 0.4–0.7 mm long, strongly flattened on one side and somewhat convex on the other, the surfaces appearing smooth or slightly pebbled, light brown to brown. $2n=36$. June–August.

Uncommon, known thus far only from historical collections from Boone County (nearly throughout the U.S. [including Alaska]; Canada). Habitats unknown (limose places listed on specimen labels) but presumably banks of streams and spring branches.

2. Veronica anagallis-aquatica L. (water speedwell)

Pl. 490 j, k; Map 2233

Plants perennial, with rhizomes. Stems 10–60(–100) cm long, mostly strongly ascending to erect (occasionally plants flattened by flooding may have a more spreading growth form), glabrous or sparsely to moderately pubescent with minute, gland-tipped hairs toward the tip (and along the inflorescence axis). At least the upper and median leaves sessile, the lower leaves sometimes short-petiolate (vegetative, late-season stems with

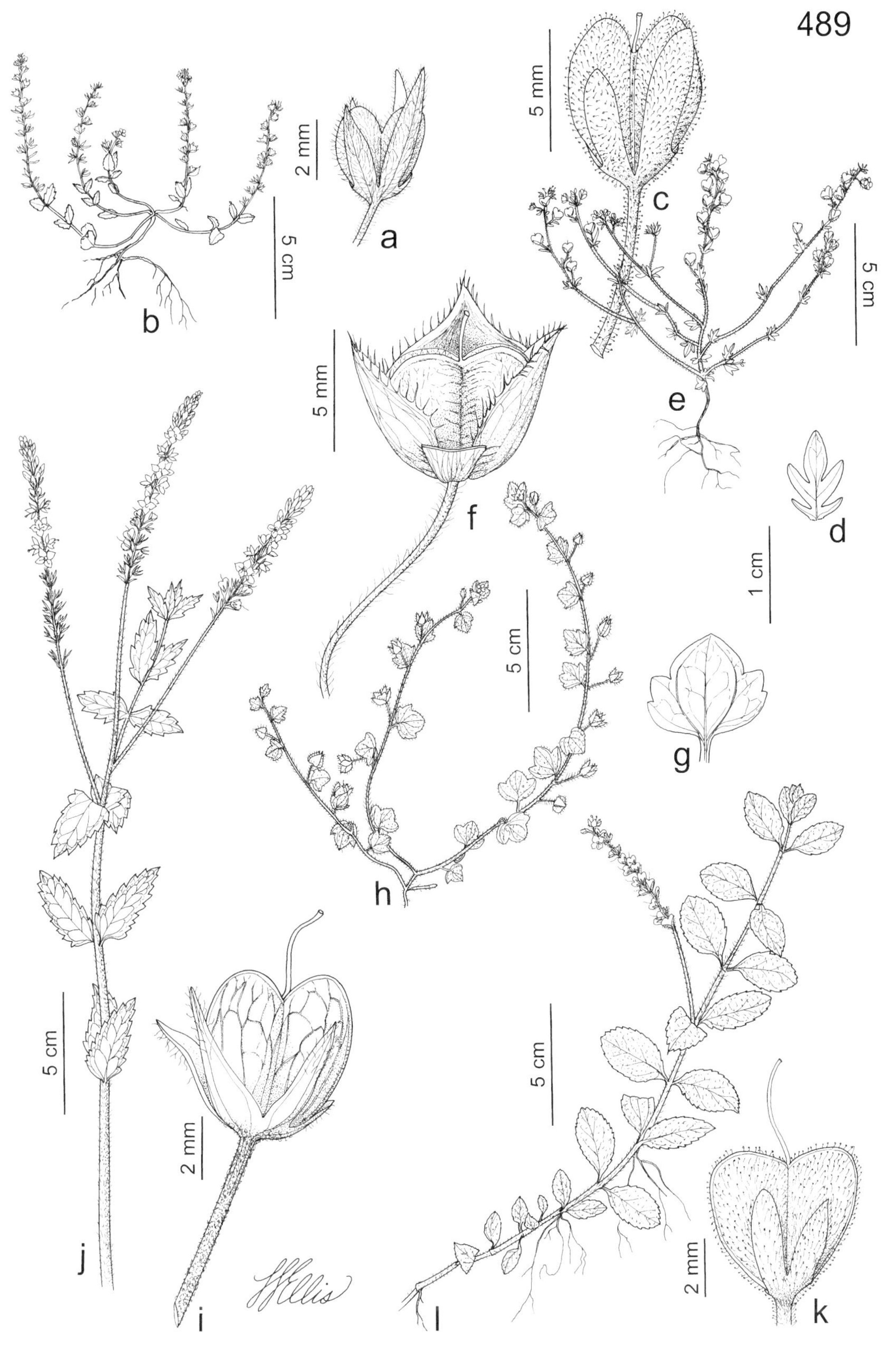

Plate 489. Plantaginaceae. *Veronica arvensis*, **a)** fruit, **b)** habit. *Veronica triphyllos*, **c)** fruit, **d)** leaf, **e)** habit. *Veronica hederifolia*, **f)** fruit, **g)** leaf, **h)** habit. *Veronica teucrium*, **i)** fruit, **j)** habit. *Veronica chamaedrys*, **k)** fruit, **l)** habit.

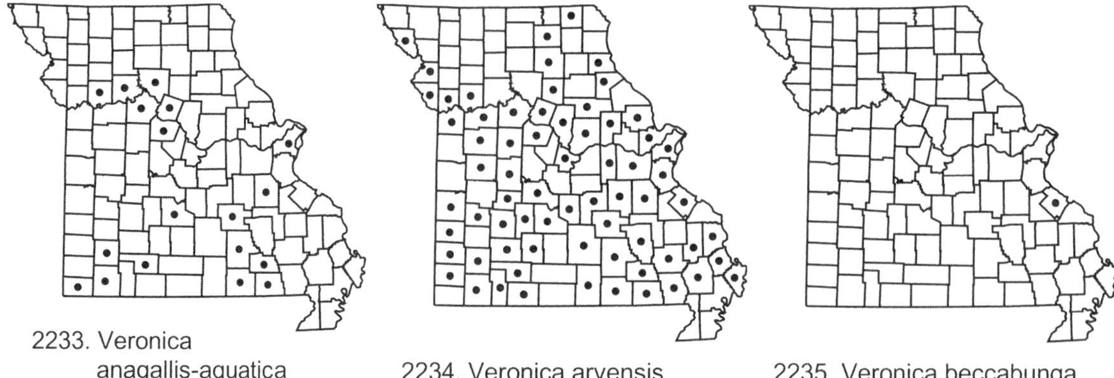
2233. Veronica anagallis-aquatica
2234. Veronica arvensis
2235. Veronica beccabunga

mostly short-petiolate leaves are sometimes produced). Leaf blades 1.5–8.0 cm long, mostly 1.5–3.0 times as long as wide, broadly lanceolate to ovate or occasionally oblong-ovate, broadest at or more commonly below the midpoint, sharply pointed at the tip (sometimes rounded or bluntly pointed on vegetative, late-season stems), rounded to truncate or shallowly cordate at the base, those of the sessile leaves more or less clasping the stems, the margins unlobed, flat, mostly subentire, usually with at least a few, widely spaced, minute teeth (occasionally the teeth more numerous), the surfaces glabrous or the undersurface inconspicuously glandular-hairy toward the base of the midvein. Inflorescences axillary racemes, these usually in opposite pairs at the stem nodes (1 per leaf), open at maturity, with (15–)25–60 flowers, the bracts 1.5–4.0 mm long, much smaller than the foliage leaves, linear to narrowly lanceolate. Flower stalks 2.5–5.0 mm long at flowering (to 8 mm long at fruiting), ascending or curved upward at flowering and fruiting. Calyces 3.0–5.5 mm long, the lobes unequal or slightly unequal, the upper 2 lobes then slightly shorter than the lower 2 lobes, deeply 4-lobed, the lobes lanceolate to broadly lanceolate, glabrous or minutely glandular-hairy toward the base. Corollas 5–10 mm wide, pale blue to light blue, pale purple, or light bluish purple with darker veins, the throat white or sometimes pale greenish-tinged at the base, the lobes loosely cupped upward. Style 1.5–3.0 mm long at fruiting. Fruits 2.5–4.0 mm long, mostly slightly longer than wide, broadly obovate to more or less circular in profile, somewhat turgid, the notch very shallow, the surfaces and margins glabrous or a few minute glandular hairs present along the margins, usually dehiscing both along and between the sutures into 4 valves. Seeds numerous, 0.3–0.5 mm long, strongly flattened on one side and somewhat convex on the other, the surfaces appearing smooth or slightly pebbled, light brown to brown. $2n$=36. April–September.

Introduced, uncommon to scattered in the Ozark Division, north locally to a few counties along the Missouri River (native of Europe, Asia; introduced nearly throughout the U.S., Canada). Banks of streams, rivers, and spring branches, margins of lakes and sloughs, moist bases of bluffs, and disturbed wetlands; often emergent aquatics.

This taxon was first reported from Missouri by Nightingale and Olson (1984), but previously misdetermined herbarium specimens exist dating back to the 1930s. R. E. Brooks (1976), Gleason and Cronquist (1991), and others have noted the existence of sterile putative hybrids between *V. anagallis-aquatica* and the native *V. catenata*, but although these are to be expected in Missouri they have not yet been documented in the state. The two species are closely related members of a circumboreal complex and are separable by characters in the key above.

3. Veronica arvensis L. (corn speedwell)
Pl. 489 a, b; Map 2234

Plants annual, with slender taproots. Stems 5–30 cm long, erect or ascending, sometimes from a spreading base (then often rooting at the lower nodes), moderately pubescent with longer, spreading, nonglandular hairs toward the base, grading into short, often glandular hairs toward the tip. Leaves mostly sessile, the lowermost leaves sometimes short-petiolate. Leaf blades 0.5–1.8 cm long, 1–2 times as long as wide, ovate to broadly ovate, broadly elliptic, or nearly circular, broadest at or below the midpoint, rounded or bluntly to sharply (but broadly) pointed at the tip, broadly angled to rounded, truncate, or shallowly cordate at the base, those of the sessile leaves sometimes slightly clasping the stems, the margins unlobed, more or less flat (the veins impressed on the upper surface), finely to relatively coarsely scalloped or bluntly toothed, the upper surface glabrous or sparsely pubescent with more or less spreading, nonglandular hairs, the undersurface and margins

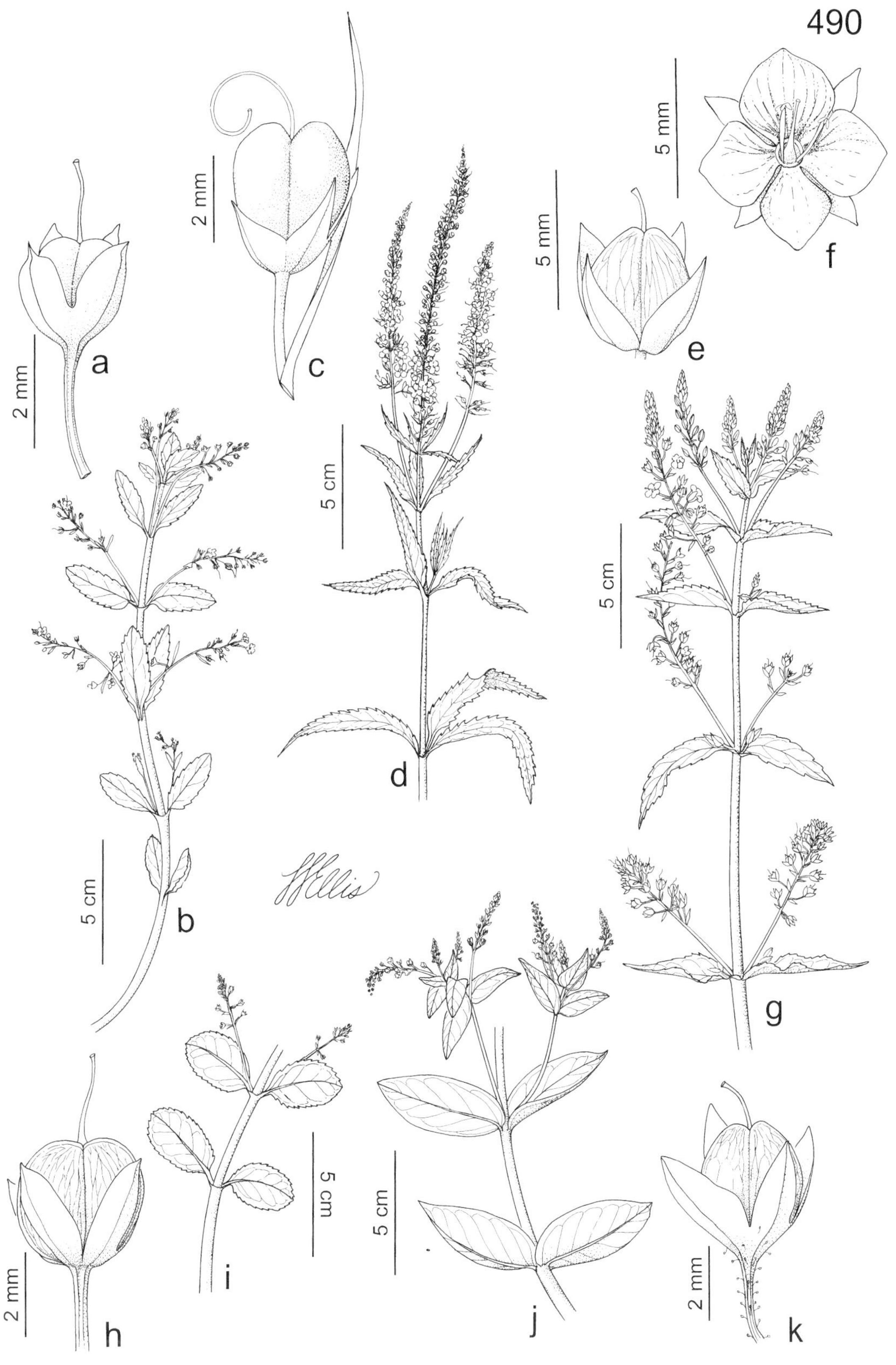

Plate 490. Plantaginaceae. *Veronica americana*, **a)** fruit, **b)** fertile stem. *Veronica longifolia*, **c)** fruit with bract, **d)** fertile stem. *Veronica catenata*, **e)** fruit, **f)** flower, **g)** fertile stem. *Veronica beccabunga*, **h)** fuit, **i)** fertile stem. *Veronica anagallis-aquatica*, **j)** fertile stem, **k)** fruit.

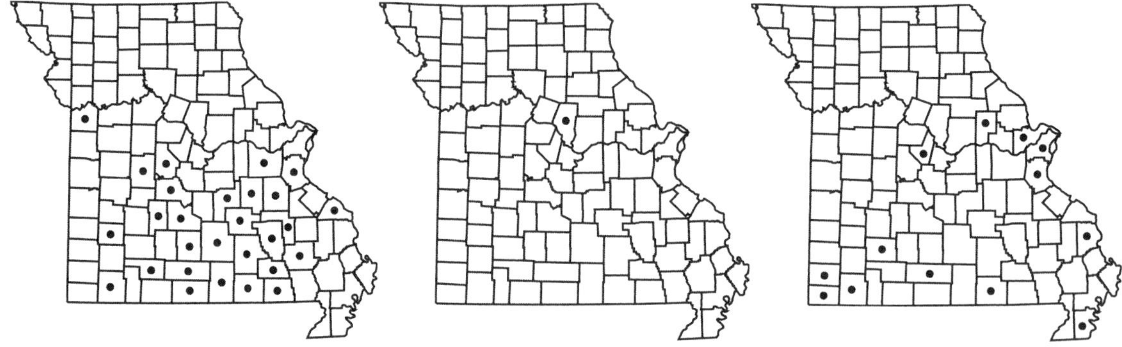

2236. Veronica catenata 2237. Veronica chamaedrys 2238. Veronica hederifolia

usually more densely hairy. Inflorescences terminal, elongate, moderately dense (when young) to open, spikelike racemes, with 6–15 flowers, the axis visible between the flowers (except at the start of flowering), the bracts 3–10 mm long, gradually reduced toward the axis tip, the lower ones similar to the adjacent foliage leaves, the upper bracts narrowly lanceolate to narrowly oblong-lanceolate. Flower stalks absent or to 1 mm long at flowering (to 2 mm at fruiting), shorter than the calyces, when present more or less ascending at flowering and fruiting. Calyces 3–6 mm long, the lobes unequal, the upper 2 lobes noticeably shorter than the lower 2 lobes, deeply 4-lobed, the lobes linear to more commonly lanceolate, hairy, the hairs often glandular. Corollas 2–3 mm wide (2.5–3.5 mm long), blue to bluish purple with darker veins, the throat white or greenish-tinged, the tube appearing relatively broad, wider than long, the lobes loosely ascending to broadly bell-shaped. Style 0.4–1.0 mm long at fruiting. Fruits 2.5–3.5 mm long, wider than long, more or less heart-shaped in profile, flattened, the notch relatively deep (0.4–0.8 mm), the margins with spreading, glandular hairs, dehiscing along the sutures into 2 valves. Seeds 8–12 per locule, 0.7–1.0 mm long, strongly flattened on both sides, the surfaces appearing smooth or nearly so (faintly and finely ridged and pitted), orangish brown to brown. $2n$=14, 16. April–August.

Introduced, scattered to common in the southern 2/3 of the state, uncommon or absent farther north (native of Europe, Asia; introduced nearly throughout the U.S. [including Alaska, Hawaii], Canada, Greenland). Bottomland forests, mesic upland forests, banks of streams and rivers, disturbed portions of upland prairies, sand prairies, savannas, glades, and tops of bluffs; also crop fields, fallow fields, old fields, lawns, railroads, roadsides, and open, disturbed areas.

4. Veronica beccabunga L. (European brooklime)

Pl. 490 h, i; Map 2235

Plants perennial, with rhizomes. Stems 10–60(–100) cm long, mostly prostrate to loosely ascending, more strongly ascending to erect only toward the tip, glabrous. Leaves all short-petiolate. Leaf blades (1–)2–4(–6) cm long, mostly 2–4 times as long as wide, elliptic to obovate or oblong-elliptic, mostly broadest at or above the midpoint, rounded or bluntly pointed at the tip, short-tapered to rounded at the base, not clasping the stems, the margins unlobed, flat, subentire or more commonly finely scalloped, the surfaces glabrous. Inflorescences axillary racemes, these usually in opposite pairs at the stem nodes (1 per leaf), open at maturity, with 10–20 flowers, the bracts 2–5 mm long, much smaller than the foliage leaves, linear. Flower stalks 3–6 mm long at flowering (to 8 mm long at fruiting), spreading to loosely ascending at flowering, but mostly spreading at a right angle to the axis at fruiting. Calyces 2.5–4.0 mm long, the lobes slightly unequal, the upper 2 lobes slightly shorter than the lower 2 lobes, deeply 4-lobed, the lobes lanceolate or oblong-oblanceolate, glabrous. Corollas 5–7 mm wide, pale blue to blue or purplish blue with darker veins (rarely pinkish-tinged), the throat white or sometimes pale greenish-tinged at the base, the lobes spreading to loosely cupped upward. Style 1.8–2.2 mm long at fruiting. Fruits 2.0–3.5 mm long, slightly wider than long, depressed obovate to more or less circular in profile, somewhat turgid, the notch very shallow, the surfaces and margins glabrous, usually dehiscing both along and between the sutures into 4 valves. Seeds numerous, 0.3–0.7 mm long, strongly flattened on one side and convex on the other, the surfaces appearing smooth or slightly pebbled, light brown to yellowish brown. $2n$=18, 36. June–September.

Introduced, uncommon, known thus far only from Ste. Genevieve County (native of Europe, Asia; introduced sporadically in the northeastern and western U.S.). Banks of streams and spring branches.

This taxon was first reported from Missouri by Yatskievych and Summers (1991). It is morphologically similar to *V. americana*, but differs markedly in its more spreading habit and in its relatively thick and almost succulent stems and leaves.

5. Veronica catenata Pennell (water speedwell)

V. catenata var. *glandulosa* (Farw.) Pennell

V. comosa Richt., misapplied

V. connata Raf., misapplied

Pl. 490 e–g; Map 2236

Plants perennial, with rhizomes. Stems 10–60(–100) cm long, mostly strongly ascending to erect (submerged plants or those flattened by flooding may have a more spreading growth form), sometimes from a spreading base, glabrous or sparsely to moderately pubescent with minute, gland-tipped hairs toward the tip (and along the inflorescence axis). At least the upper and median leaves sessile, the lower leaves sometimes short-petiolate. Leaf blades 2–8 cm long, (2.5–)3.0–5.0 (–8.0) times as long as wide, lanceolate to oblong-oblanceolate, mostly broadest below the midpoint, sharply to less commonly bluntly pointed at the tip (rounded on submerged stems), rounded to truncate or shallowly cordate at the base, those of the sessile leaves more or less clasping the stems, the margins unlobed, flat, subentire or with widely or irregularly spaced, minute teeth, often mainly toward the tip, the surfaces glabrous or the undersurface inconspicuously glandular-hairy toward the base of the midvein. Inflorescences axillary racemes, these usually in opposite pairs at the stem nodes (1 per leaf), open at maturity, with 15–25 flowers, the bracts 1.5–4.0 mm long, much smaller than the foliage leaves, linear to narrowly lanceolate. Flower stalks 2.5–5.0 mm long at flowering (to 7 mm long at fruiting), spreading to ascending at flowering, spreading at a right angle to the axis at fruiting or occasionally some of them slightly ascending or upcurved. Calyces 3.0–3.5 mm long, the lobes subequal or slightly unequal, the upper 2 lobes then slightly shorter than the lower 2 lobes, deeply 4-lobed, the lobes broadly lanceolate to ovate, glabrous or minutely glandular-hairy toward the base. Corollas 3–6 mm wide, white to pinkish-tinged or pale blue to pale purple, with darker, pinkish purple to blue veins, the throat greenish-tinged to yellowish-tinged, the lobes loosely cupped upward. Style 1.3–2.0 mm long at fruiting. Fruits 2.5–3.0 mm long, slightly wider than long, depressed-obovate to more or less heart-shaped in profile, somewhat turgid, the notch shallow (0.1–0.3 mm), the surfaces and margins glabrous or a few minute glandular hairs present along the margins, usually dehiscing both along and between the sutures into 4 valves. Seeds numerous, 0.3–0.7 mm long, strongly flattened on one side and somewhat convex on the other, the surfaces appearing smooth or appearing slightly pebbled, light brown to yellowish brown. $2n=36$. April–October.

Scattered in the Ozark and Ozark Border Divisions, and disjunct in Jackson County (nearly throughout the U.S. [except for some southeastern states, but including Alaska]; Canada, Europe, Asia). Banks of streams, rivers, and spring branches; often emergent aquatics, sometimes submerged.

Some earlier botanists called this taxon *V. connata* (Steyermark, 1963). Burnett (1950) studied descriptions and type specimens of several binomials in this nomenclaturally confusing complex. He concluded that although the description of *C. connata* was relatively poor and a type specimen was not discovered, it was fairly certain that the name applied to plants otherwise known by the earlier name *V. scutellata* L. (narrow-leaved speedwell), a circumboreal species not yet known from Missouri. Pennell (1921) had come to this conclusion earlier, but later became confused in speculating on the intent of the protologue describing the binomial *V. connata*. *Veronica scutellata* has a broad range in temperate North America and has been documented in Nebraska, Iowa, and Illinois, as well as in Tennessee and apparently Louisiana. It should be searched for in the state, particularly in fens and other wetlands in northern Missouri. It differs from *V. catenata* in its narrower (mostly 4–20 times as long as wide) leaf blades and in its fruits, which are more strongly flattened, conspicuously notched apically, and contain only 5–9, larger (1.5–1.8 mm) seeds per locule.

Similarly, Burnett (1950) also placed into synonymy under *V. scutellata* plants that Pennell (1935) had called *V. comosa* ssp. *glaberrima* Pennell. Burnett also studied the literature and type specimens associated with the name *V. comosa* and concluded that this name referred to a different species, probably *V. anagallis-aquatica*. Burnett also eliminated from consideration several additional names applied to the species in question in various parts of its range and concluded that the correct name must be *V. catenata*. The var. *glandulosa* was resurrected by Holmgren (1986) for plants with the upper stems glandular-hairy (vs. glabrous) and sometimes slightly narrower fruits. Such plants do not appear distinct enough to warrant formal taxonomic recognition.

Veronica catenata is the most common of the perennial speedwell species in Missouri and is commonly encountered along Ozark waterways. Plants are sometimes encountered, especially in spring branches, as vegetative colonies that are totally submerged. Such plants have lax, often much-branched stems and thin leaves that tend to be oblong to oblong-lanceolate and rounded at the tips. Presumably, plants of *V. anagallis-aquatica* also can occur as submerged aquatics, but such plants would be impossible to distinguish morphologically from submerged individuals of *V. catenata*.

6. Veronica chamaedrys L. (germander speedwell)

Pl. 489 k, l; Map 2237

Plants perennial, with rhizomes. Stems 10–35 cm long, the main stems prostrate or loosely ascending, the flowering branches sometimes more strongly ascending, moderately to densely pubescent with spreading nonglandular hairs, these mostly confined to a pair of longitudinal lines, at least between the lower nodes. Leaves sessile or nearly so. Leaf blades 1.0–3.5 cm long, 1–2 times as long as wide, ovate to oblong-ovate or broadly ovate, broadest near the base, rounded to bluntly pointed at the tip, broadly rounded to truncate or shallowly cordate at the base, those of the larger leaves usually clasping the stems, the margins unlobed, relatively coarsely and usually bluntly toothed, the surfaces and margins spreading-hairy, less densely so on the upper surface. Inflorescences axillary racemes, these mostly in opposite pairs at the stem nodes (1 per leaf) but often the lowermost alternate, open at maturity, with 10–20 flowers, the bracts 3–7 mm long, much smaller than the foliage leaves, linear to narrowly oblanceolate. Flower stalks 5–9 mm long at flowering (to 12 mm long at fruiting), loosely ascending at fruiting. Calyces 3–5 mm long, the lobes subequal or the lower pair slightly larger than the upper pair, deeply 4-lobed, the lobes linear to narrowly oblong-lanceolate, usually glandular-hairy. Corollas 8–12 mm wide, blue to bluish purple with darker veins, the throat white and often light green toward its base, the lobes spreading to loosely arched upward. Style 3–5 mm long at fruiting. Fruits rarely produced, 3.0–3.5 mm long, wider than long, broadly heart-shaped in profile, flattened, the notch very shallow, the margins with short, gland-tipped hairs, dehiscing along the sutures into 2 valves. Seeds about 6 per locule, 1.0–1.7 mm long, strongly flattened on both sides, the surfaces appearing smooth or nearly so (faintly and finely pebbled), tan to brown. $2n=16$, 32. April–June.

Introduced, known thus far only from Boone County (Native of Europe; introduced sporadically but widely in the northeastern and western U.S., Canada). Lawns and gardens.

This taxon was first reported from Missouri by Dunn (1982) as a casual escape in an urban yard.

A superficially similar species that eventually may be recorded from Missouri is *V. officinalis* L. (common speedwell, common gypsyweed). This native of Europe has escaped from gardens to become widespread in temperate North America, including several of the states surrounding Missouri. It differs from *V. chamaedrys* in its oval to elliptic or elliptic-obovate leaf blades that are widest at or slightly above the midpoint and angled to a petiolar base; stems with relatively uniform and dense, spreading hairs; flower stalks shorter than the subtending bracts; smaller (4–8 mm wide), lighter blue corollas; and slightly larger (3.5–4.5 mm long) fruits with glandular hairs on the surfaces as well as the margins.

7. Veronica hederifolia L. (ivy-leaved speedwell)

Pl. 489 f–h; Map 2238

Plants annual, with slender taproots, often forming loose mats. Stems 5–40 cm long, the branches and tips often loosely ascending, the main stems mostly more or less prostrate (occasionally rooting at the lower nodes), sparsely to moderately pubescent with spreading, nonglandular hairs. Leaves short-to long-petiolate. Leaf blades 0.6–1.5 cm long, 1.0–1.5 times as long as wide, ovate to broadly ovate, or depressed-ovate, mostly broadest below the midpoint, broadly rounded to shallowly cordate at the base, not clasping the stems, palmately 3- or 5-lobed, the lobes rounded or broadly and mostly bluntly pointed at their tips, the margins flat, otherwise entire or the lateral lobes occasionally with a small, blunt tooth along the outer margin, the surfaces sparsely to moderately pubescent with nonglandular hairs. Inflorescences terminal, elongate, open, spikelike racemes, but (because the bracts are unreduced and the inflorescence frequently extends nearly to the stem base) appearing as flowers solitary in the leaf axils, with 4–25 flowers, the axis visible between the flowers, the bracts 5–12 mm long, similar to the adjacent foliage leaves and somewhat reduced only toward the axis tip, ovate to broadly ovate. Flower stalks 5–12 mm long at flowering (to 18 mm at

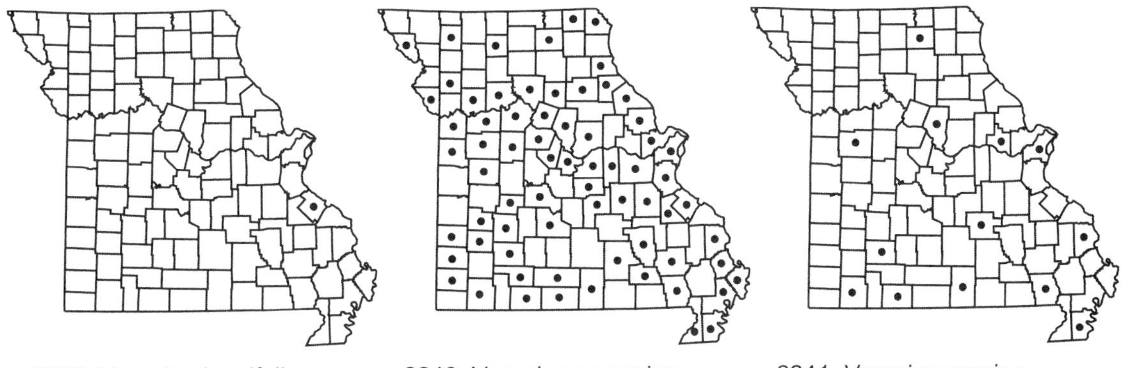

2239. Veronica longifolia 2240. Veronica peregrina 2241. Veronica persica

fruiting), much longer than the calyces and bracts, more or less spreading or loosely upcurved to loosely ascending at flowering and fruiting. Calyces 4–7 mm long, the lobes subequal, deeply 4-lobed, the lobes broadly ovate-triangular to slightly heart-shaped, pubescent with spreading, nonglandular hairs along the margins. Corollas 3–6 mm wide (1.5–4.0 mm long), pale blue to less commonly pale purplish blue with darker veins, the throat white, often light greenish at the center, the tube appearing relatively broad, wider than long, the lobes spreading to shallowly cupped. Style 0.6–0.9 mm long at fruiting. Fruits 2.5–4.0 mm long, slightly to noticeably wider than long, depressed-obovate to nearly circular in profile, turgid, the notch very shallow (0.2–0.3 mm), the surfaces glabrous, dehiscing along the sutures into 2 valves. Seeds 1 or 2 per locule, 2.3–3.0 mm long, cup-shaped (deeply concave on one side, convex on the other), the convex surface appearing cross-wrinkled, dark brown to black. $2n$=36. March–May.

Introduced, uncommon, sporadic, mostly in the southern half of the state (native of Europe, introduced widely in the eastern U.S. west to South Dakota and Oklahoma, also several western states, Canada). Bottomland forests and banks of streams and rivers; also lawns, cemeteries, railroads, roadsides, and shaded disturbed areas.

This taxon was first reported from Missouri by Key (1980) and Dunn (1982). The oldest Missouri specimen encountered during the present study was collected in 1941 in Jefferson County. The species has become locally common around some urban areas, sometimes forming large, loose mats in disturbed bottomland forests.

8. Veronica longifolia L. (garden speedwell)
Pseudolysimachion longifolium (L.) Opiz
Pl. 490 c, d; Map 2239

Plants perennial, forming dense clumps. Stems 40–120 cm long, erect or strongly ascending, glabrous or more commonly sparsely to densely pubescent with short, curly, nonglandular hairs, especially toward the tip. Leaves short-petiolate. Leaf blades (3–)4–12(–15) cm long, 3–11 times as long as wide, those of the lower and median leaves lanceolate to broadly lanceolate, grading into narrowly lanceolate or narrowly elliptic-lanceolate toward the stem tip, broadest below the midpoint, sharply pointed at the tip, angled to rounded, truncate, or shallowly cordate at the base, not clasping the stems, the margins unlobed, flat (however the leaves sometimes more or less folded lengthwise), sharply and finely to coarsely many-toothed, the surfaces sparsely to moderately hairy with short, curved or curly, nonglandular hairs, these sometimes only along the main veins, occasionally glabrous or nearly so. Inflorescences terminal, elongate, dense, spikelike racemes, sometimes with additional, apparently axillary racemes (these usually terminating short branches), with numerous flowers, the axis mostly hidden by the flowers, the bracts 1.5–6.0 mm long, much smaller than the foliage leaves, linear. Flower stalks 0.5–2.0 mm long at flowering (not elongated at fruiting), strongly ascending at flowering and fruiting. Calyces 2.0–3.5 mm long, the lobes subequal, deeply 4-lobed, the lobes narrowly oblong-lanceolate to ovate-triangular, glabrous or minutely nonglandular-hairy, sometimes only along the margins. Corollas 4–5 mm wide (5.5–8.0 mm long), lavender or more commonly purple to blue, lacking conspicuous, darker veins, the throat slightly lighter and densely hairy, the tube slender, longer than wide, the lobes ascending to broadly bell-shaped. Style 4–10 mm long at fruiting. Fruits 2.5–3.0 mm long, longer than wide, obovate to more or less heart-shaped in profile, somewhat turgid, the notch very shallow to moderate (0.1–0.4 mm), the surfaces and margins glabrous, dehiscing both along and between the sutures into 4 valves. Seeds numerous, 0.7–0.8 mm long, strongly flattened on both sides, the surfaces appearing smooth or slightly pebbled, tan to light brown. $2n=34$, 68. April–August.

Introduced, uncommon, known thus far only from a single historical collection from Ste. Genevieve County (native of Europe; introduced in the northern U.S. south to Oregon, Missouri, Maryland, and apparently also Florida; Canada). Roadsides.

This species is commonly cultivated in gardens and is an occasional escape. The specimen upon which Steyermark (1963) based his report of this taxon for Missouri could not be located during this study, but it seems very unlikely to have been misdetermined. The collector, Robert Mohlenbrock, believes that the specimen (*Mohlenbrock 5926*) he collected while visiting Chimney Rocks on 30 April 1955 was accessioned into the herbarium at Southern Illinois University–Carbondale, but likely was destroyed in error at a later date during a purging of the SIU herbarium by an irrational former staff person.

9. Veronica peregrina L. (neckweed, purslane speedwell)

V. peregrina var. *xalapensis* (Kunth) Pennell

V. peregrina ssp. *xalapensis* (Kunth) Pennell

Pl. 491 d–f; Map 2240

Plants annual, with fibrous roots or slender taproots. Stems 5–20(–30) cm long, erect to loosely ascending, sometimes from a spreading base (then usually not rooting at the lower nodes), glabrous to moderately pubescent with short, glandular hairs. Leaves mostly sessile, the lowermost leaves sometimes short-petiolate. Leaf blades 0.5–2.2 cm long, 1–4 times as long as wide, mostly relatively thick, narrowly oblong to oblong-oblanceolate, those of the petiolate leaves often lanceolate to ovate, broadest below to above the midpoint, mostly rounded or broadly and bluntly pointed at the tip, angled or short-tapered to rounded, or occasionally nearly truncate at the base, not clasping the stems, the margins unlobed, flat, entire or shallowly and irregularly scalloped or bluntly toothed, the surfaces glabrous or sparsely to moderately pubescent with short, glandular hairs. Inflorescences terminal, elongate, open, spikelike racemes, but (because the bracts are unreduced and the inflorescence frequently extends nearly to the stem base) appearing as flowers solitary in the leaf axils, with 6–20(–30) flowers, the axis visible between the flowers, the bracts 4–10 mm long, similar to the adjacent foliage leaves and only slightly reduced toward the axis tip, narrowly oblong to narrowly oblanceolate, the uppermost bracts sometimes linear. Flower stalks absent or to 1 mm long at flowering (to 1.5 mm at fruiting), shorter than the calyces, when present more or less ascending at flowering and fruiting. Calyces 3–6 mm long, the lobes subequal, deeply 4-lobed, the lobes narrowly elliptic to lanceolate, glabrous or glandular-hairy. Corollas 2.0–2.5 mm wide (2.5–3.0 mm long), white or occasionally pale yellow, lacking darker veins, the throat white or sometimes light greenish-tinged, the tube appearing relatively broad, wider than long, the lobes curved outward. Style 0.1–0.4 mm long at fruiting. Fruits 3–4 mm long, about as long as wide, heart-shaped in profile, flattened, the notch relatively broad and moderately deep (0.2–0.5 mm), the surfaces and/or margins glabrous or glandular-hairy, dehiscing along the sutures into 2 valves. Seeds numerous, 0.5–1.0 mm long, strongly flattened on both sides, but with a slender longitudinal ridge on one side, the surfaces appearing smooth or slightly pebbled, light brown to yellowish brown. $2n=52$. April–August.

Scattered nearly throughout the state (nearly throughout the U.S. [including Alaska]; Canada, Mexico, Central America, South America, Caribbean Islands; introduced in Hawaii, Europe, Asia). Bottomland forests, margins of sinkhole ponds, banks of rivers, bases of bluffs, and disturbed marshes and fens; also crop fields, fallow fields, old fields, lawns, pastures, ditches, railroads, roadsides, and open, disturbed areas.

Most botanists have divided this species into two weakly distinct varieties. Typically, plants of var. *peregrina* are glabrous, whereas those of var. *xalapensis* have at least the stems and fruits glandular-hairy. However, especially the populations east of the Rocky Mountains appear to exhibit a great deal of variation in the abundance and distribution of the pubescence. Plants are encountered that are totally glabrous or with varying densities of pubescence occurring on only the stems or also on various combinations of the stems, leaves, inflorescence axes, calyces, and fruits. Even Pennell (1935) noted that intergradation between the two morphotypes was extensive during his early studies. In Missouri, glabrous plants are currently more common and have a broader distribution, whereas plants that are variously glandular-hairy are concentrated in counties bordering the Missouri and upper Mississippi Rivers. Comparing the overall North American ranges circumscribed by Pennell (1935) and Gleason and Cronquist (1991) it appears that the mostly western and southern var. *xalapensis* has been expanding its range into the eastern United States. It seems likely that the two varieties have come into broader contact in recent decades and the distinc-

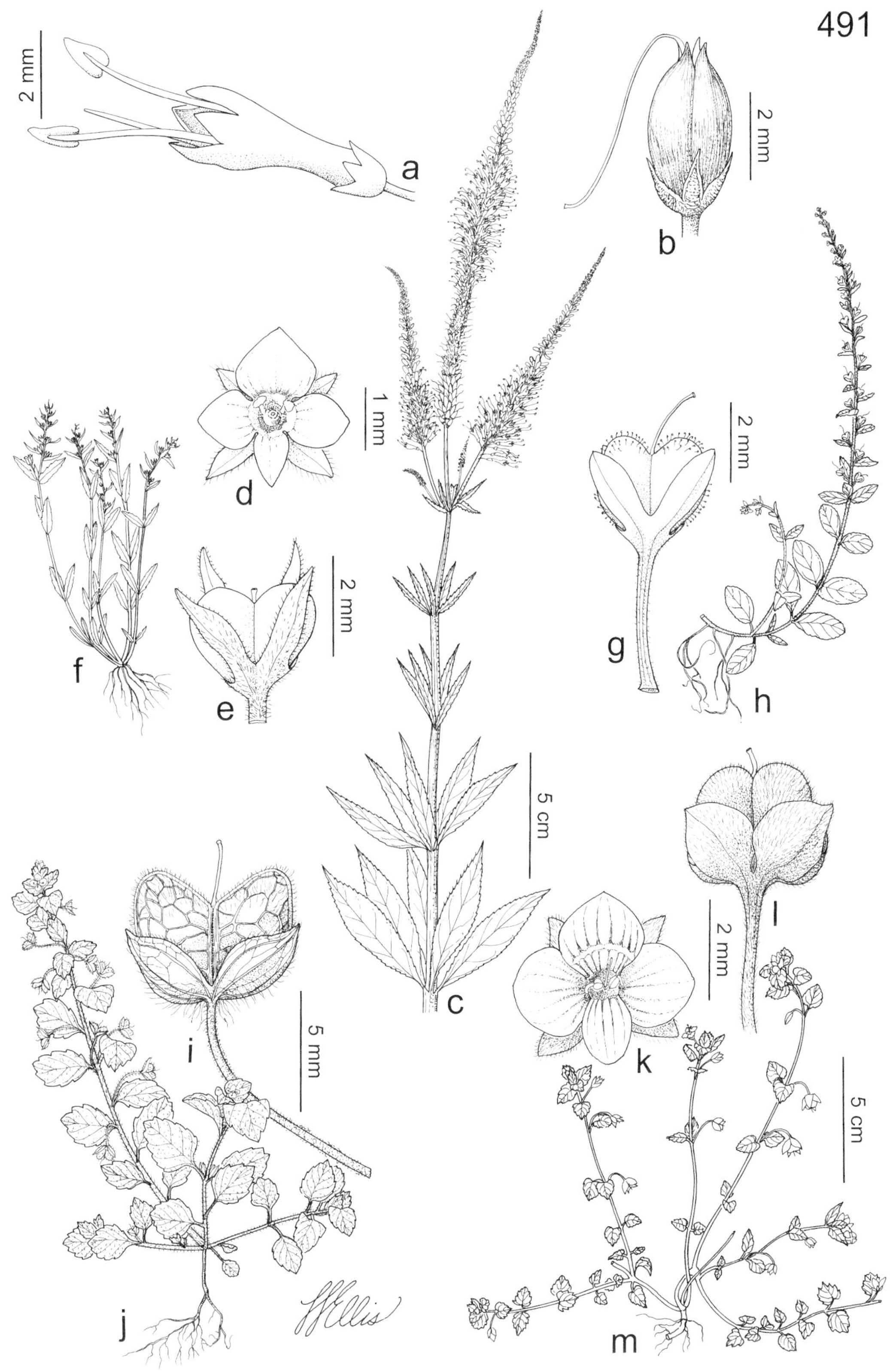

491. Plantaginaceae. *Veronicastrum virginicum*, **a)** flower, **b)** fruit, **c)** fertile stem. *Veronica peregrina*, **d)** flower, **e)** fruit, **f)** habit. *Veronica serpyllifolia*, **g)** fruit, **h)** habit. *Veronica persica*, **i)** fruit, **j)** habit. *Veronica polita*, **k)** flower, **l)** fruit, **m)** habit.

2242. Veronica polita 2243. Veronica scheereri 2244. Veronica serpyllifolia

tions between them have become even more blurred. Thus, in the present work no attempt has been made to segregate infraspecific taxa.

10. Veronica persica Poir. (bird's-eye speedwell)

Pl. 491 i, j; Map 2241

Plants annual, with slender taproots. Stems 8–40 cm long, the branches and tips loosely ascending, the main stems mostly more or less prostrate (often rooting at the lower nodes), sparsely to moderately pubescent with spreading, nonglandular hairs. Leaves mostly short-petiolate, the uppermost foliage leaves sometimes sessile or nearly so. Leaf blades 0.7–2.5 cm long, 1.0–1.5 times as long as wide, ovate to broadly ovate or nearly circular, mostly broadest below the midpoint, rounded or broadly and bluntly pointed at the tip, broadly rounded to truncate or shallowly cordate at the base, mostly not clasping the stems, the margins unlobed, flat, entire or more commonly coarsely scalloped or bluntly toothed, the surfaces usually sparsely to moderately pubescent with nonglandular hairs. Inflorescences terminal, elongate, open, spikelike racemes, but (because the bracts are unreduced and the inflorescence frequently extends nearly to the stem base) appearing as flowers solitary in the leaf axils, with 6–20 flowers, the axis visible between the flowers, the bracts 5–12 mm long, similar to the adjacent foliage leaves and somewhat reduced only toward the axis tip, ovate to broadly ovate. Flower stalks 15–25 mm long at flowering (to 40 mm at fruiting), much longer than the calyces and bracts, more or less spreading or loosely upcurved at flowering and fruiting. Calyces 4.5–8.0 mm long, the lobes subequal, deeply 4-lobed, the lobes broadly lanceolate, pubescent with nonglandular hairs along the margins. Corollas 8–11 mm wide (5–9 mm long), blue (rarely slightly purplish tinged) with darker veins, the lower lobe often paler or white, the throat white, often light greenish at the center, the tube appearing relatively broad, wider than long, the lobes spreading to shallowly cupped. Style 2–3 mm long at fruiting. Fruits 3.5–4.5 mm long, noticeably wider than long, broadly heart-shaped in profile, flattened, the notch relatively broad and deep (0.7–1.2 mm), the surfaces glabrous, dehiscing along the sutures into 2 valves. Seeds mostly 5–11 per locule, 1.2–2.0 mm long, cup-shaped (deeply concave on one side, convex on the other), the convex surface appearing cross-wrinkled, brown. $2n=28$. April–June.

Introduced, uncommon and widely scattered (native of Europe, Asia; introduced nearly worldwide). Lawns, ditches, roadsides, and open, disturbed areas.

This taxon was first reported from two sites in Missouri by Castaner (1982b) and Dunn (1982). Since that time, it has spread to at least eleven counties in Missouri, probably as a contaminant in grass seed or garden soil. M. A. Fischer (1987) studied the species in Europe and southwestern Asia and concluded that it was an allotetraploid that likely had arisen through past hybridization between *V. polita* and a related species that has not become established yet in the New World, *V. ceratocarpa* C.A. Mey. Superficially, plants of *V. persica* resemble those of *V. polita*, but the larger corollas and longer flower stalks serve to distinguish it from that species.

11. Veronica polita Fr. (wayside speedwell)

Pl. 491 k–m; Map 2242

Plants annual, with fibrous roots or slender taproots. Stems 5–20(–30) cm long, the branches and tips sometimes loosely ascending, the main stems mostly prostrate (often rooting at the lower nodes), sparsely to moderately pubescent with spreading or curved, nonglandular hairs. Leaves mostly short-petiolate, the uppermost foliage leaves sometimes sessile or nearly so. Leaf blades

0.4–1.0(–2.0) cm long, 0.8–1.5 times as long as wide, ovate or broadly ovate to depressed-ovate or nearly circular, mostly broadest below the midpoint, rounded or broadly and bluntly to occasionally sharply pointed at the tip, broadly rounded to truncate or shallowly cordate at the base, mostly not clasping the stems, the margins unlobed, flat, coarsely and bluntly to sharply toothed, the surfaces sparsely to moderately pubescent with mostly nonglandular hairs. Inflorescences terminal, elongate, open, spikelike racemes, but (because the bracts are unreduced and the inflorescence frequently extends nearly to the stem base) appearing as flowers solitary in the leaf axils, with 4–12(–20) flowers, the axis visible between the flowers, the bracts 3–10 mm long, similar to the adjacent foliage leaves and somewhat reduced only toward the axis tip, ovate to broadly ovate or nearly circular. Flower stalks 4–10 mm long at flowering (to 15 mm at fruiting), longer than the calyces, more or less spreading or loosely upcurved to loosely ascending at flowering and fruiting. Calyces 3–6 mm long, the lobes subequal, deeply 4-lobed, the lobes broadly lanceolate, to ovate or rarely broadly oblanceolate to obovate, pubescent with short, curved, nonglandular hairs. Corollas 4–8 mm wide (3–5 mm long), blue with darker veins, the lower lobe occasionally paler, the throat white, often light greenish at the center, the tube appearing relatively broad, wider than long, the lobes spreading to shallowly cupped. Style 0.8–1.5 mm long at fruiting. Fruits 2.5–4.0 mm long, noticeably wider than long, broadly heart-shaped in profile, flattened, the notch relatively broad and deep (0.8–1.3 mm), the surfaces moderately to densely pubescent with a mixture of longer, glandular and shorter, nonglandular hairs, dehiscing along the sutures into 2 valves. Seeds mostly 6–12 per locule, 1.3–1.7 mm long, cup-shaped (deeply concave on one side, convex on the other), the convex surface appearing cross-wrinkled, tan to yellowish brown. $2n=14$. March–June, rarely also November–December.

Introduced, scattered south of the Missouri River, uncommon farther north (native of Europe, Asia; introduced nearly worldwide). Bottomland forests, banks of streams and rivers; also lawns, cemeteries, gardens, barnyards, ditches, roadsides, and open, disturbed areas.

Missouri collectors should keep watch for the closely related *V. agrestis* L. (field speedwell), another weedy Eurasian species that has become introduced in temperate North America. It differs from *V. polita* in its larger (4–6 mm long), sparsely glandular-hairy fruits that are more narrowly and deeply notched at the tip, and in its white to partially pink or pale blue corollas. In addition, *V. agrestis* is a tetraploid ($2n=28$) taxon (Albach et al., 2008). The introduced North American range of this species has become confused because some floristic manuals have treated it as a synonym of the widespread *V. polita* (Holmgren, 1986). According to Gleason and Cronquist (1991), *V. agrestis* has become established in the northeastern United States and adjacent Canada west to Pennsylvania and Michigan.

12. Veronica scheereri (J.-P. Brandt) Holub (prostrate speedwell)

V. prostrata L. ssp. *scheereri* J.-P. Brandt

Map 2243

Plants perennial, with rhizomes, often forming loose mats. Stems 10–30 cm long, the main stems short and prostrate, the flowering branches more elongate and strongly ascending, moderately to densely pubescent with upward-curled, nonglandular hairs, these relatively uniformly distributed along the internodes. Leaves sessile or nearly so. Leaf blades 1.0–2.5 cm long, appearing 4–10 times as long as wide (sometimes wider, but the margins curved or curled-under), linear to narrowly oblong-elliptic or narrowly lanceolate, broadest at or below the midpoint, mostly sharply pointed at the tip, truncate to angled at the base, those of the larger leaves sometimes clasping the stems, the margins unlobed, relatively coarsely and usually bluntly toothed (but this sometimes partially concealed by the curled margins), the upper surface sparsely short-hairy, the undersurface sparsely to densely pubescent with short, curly hairs. Inflorescences axillary racemes, these mostly alternate, 3–6(–10) cm long, dense to open at maturity, with (4–)8–16(–25) flowers, the bracts 3–7 mm long, much smaller than the foliage leaves, linear to narrowly oblong-lanceolate. Flower stalks 3–6 mm long at flowering (to 10 mm long at fruiting), loosely ascending at fruiting. Calyces 2–4 mm long, the lobes unequal, the lower pair somewhat larger than the upper pair, deeply (4)5-lobed, the lobes linear to narrowly oblong, glabrous. Corollas 8–11 mm wide, blue to bluish purple with darker veins, the throat white and occasionally pale greenish-tinged toward its base, the lobes spreading to loosely arched upward. Style 3.5–5.2 mm long at fruiting. Fruits 3.5–4.5 mm long, as long as or slightly longer than wide, obovate to more or less heart-shaped or nearly circular in profile, flattened, the notch shallow, the margins and surfaces glabrous, dehiscing along the sutures into 2 valves. Seeds about 6 per locule, 0.9–1.2 mm long, strongly flattened, the sides both flat, the surfaces appearing smooth or nearly so (with an

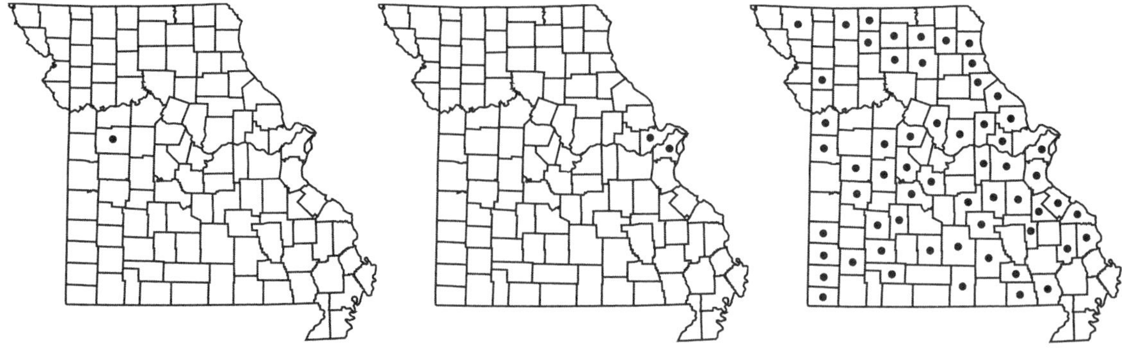

intricate, microscopic pattern of blunt, contorted ridges), tan to orangish brown or yellowish brown. $2n=32$. May–June.

Introduced, uncommon, known thus far only from a single historical collection from Jefferson County (native of Europe; introduced uncommonly in the midwestern U.S.) Habitat unknown, but presumably disturbed areas.

This taxon was long treated as a subspecies of the closely related *V. prostrata*, which, in the strict sense, differs from *V. scheereri* in its ovate to oblong-lanceolate leaf blades and smaller (mostly 6–8 mm vs. mostly 8–11 mm in diameter), pale blue corollas (Walters and Webb, 1972). Both taxa are native to Europe, but *V. scheereri* tends to be restricted to the western portion of the range of *V. prostrata*. Chromosomal studies have suggested that *V. prostrata* is a diploid ($2n=16$) and *V. scheereri* is a tetraploid ($2n=32$) (Albach et al., 2008), which, coupled with the differences in morphology has resulted in the treatment of the two taxa as separate species in some of the recent botanical literature. Both *V. prostrata* and *V. scheereri* are grown as perennial groundcovers under the name prostrate speedwell. However, they rarely escape from cultivation. The sole voucher specimen to document the occurrence of this taxon in the United States was collected by former Missouri Botanical Garden director George T. Moore near De Soto in 1941.

13. Veronica serpyllifolia L. var. **serpyllifolia** (thyme-leaved speedwell)

Pl. 491 g, h; Map 2244

Plants perennial, with rhizomes, sometimes forming mats. Stems 10–30 cm long, the tips or flowering branches ascending from prostrate main stems, moderately pubescent with short, curly or upcurved, nonglandular hairs. Leaves short-petiolate or the median and upper leaves more or less sessile. Leaf blades 0.8–2.5(–3.5) cm long, 1–2(–3) times as long as wide, elliptic to ovate, broadly ovate, or nearly circular, broadest at or below the midpoint, rounded to broadly and bluntly pointed at the tip, broadly angled to rounded or nearly truncate at the base, not clasping the stems, the margins unlobed, flat, entire or sparsely and very shallowly scalloped, the surfaces glabrous or sparsely hairy with short, curved or curly, nonglandular hairs, these sometimes only along the veins and/or margins. Inflorescences terminal, elongate, moderately dense (when young) to open, spikelike racemes, with 5–20(–40) flowers, the axis usually visible between the flowers, the bracts 2–6 mm long, noticeably smaller than the foliage leaves, linear to narrowly lanceolate, narrowly oblanceolate, or lanceolate. Flower stalks 2.5–5.0 mm long at flowering (to 8 mm at fruiting), strongly ascending at flowering and fruiting. Calyces 2–4 mm long, the lobes slightly unequal, the upper 2 lobes slightly shorter than the lower 2 lobes, deeply 4-lobed, the lobes narrowly ovate to elliptic or oblong-elliptic, minutely glandular-hairy, often mainly along the margins. Corollas 4–8 mm wide (2.5–4.0 mm long), white (light blue elsewhere) with purple veins, the throat densely hairy, the tube appearing relatively broad, wider than long, the lobes loosely ascending to broadly bell-shaped. Style 2.2–3.0 mm long at fruiting. Fruits 2.8–4.0 mm long, wider than long, more or less heart-shaped in profile, flattened, the notch shallow to moderate (0.3–0.8 mm), the surfaces and margins sparsely glandular-hairy, dehiscing along the sutures into 2 valves. Seeds numerous, 0.6–1.1 mm long, strongly flattened on both sides, the surfaces appearing smooth or nearly so (faintly and finely pitted), brown. $2n=14$, 28. April–August, rarely also November.

Introduced, scattered in the southeastern quarter of the state, uncommon and sporadic elsewhere (native of Europe, Greenland; introduced nearly throughout the U.S., Canada). Banks of streams, rivers, and spring branches, bottomland forests; also lawns, roadsides, and open, disturbed areas.

Steyermark (1963) knew this taxon only from the southeastern counties. The sporadic records from farther north and west represent a recent spread in Missouri. A second variety, var. *humifusa* (Dicks.) Vahl, is recognized by many North American botanists and tends to occur at higher elevations in montane meadows and streamsides. It appears to be circumboreal, with a broad distribution in Europe and northern North America. It differs from var. *serpyllifolia* in its smaller stature, more strongly spreading stems, fewer-flowered inflorescences, longer, sparsely glandular-hairy flower stalks, slightly larger, bright blue corollas, and shorter stamens (filaments 2–4 vs. 1.0–2.5 mm long). In their study of the *V. serpyllifolia* complex, Martínez-Ortega and Rico (2001) accepted a third infraspecific taxon within *V. serpyllifolia*. The ssp. *repens* (DC.) Hartl, which is endemic to Corsica, is similar to var. *humifusa* in its habit, but differs in its lower stems glabrous or nearly so and its slightly larger corollas that are white to pale blue or pink with darker veins.

14. Veronica teucrium L. (saw-leaved speedwell)

V. austriaca L. ssp. *teucrium* (L.) D.A. Webb

V. latifolia L., an officially rejected name

Pl. 489 i, j; Map 2245

Plants perennial, with rhizomes. Stems 30–100 cm long, erect or strongly ascending, moderately to densely pubescent with spreading to curly, nonglandular hairs throughout (sometimes less densely so toward the base). Leaves sessile or nearly so. Leaf blades 2–5(–7) cm long, 2–5 times as long as wide, lanceolate to ovate, broadest near the base, bluntly to sharply pointed at the tip, rounded to truncate or shallowly cordate at the base, those of the larger leaves usually clasping the stems, the margins unlobed, flat, relatively coarsely and bluntly to more commonly sharply toothed, the surfaces sparsely to moderately hairy. Inflorescences axillary racemes, these usually in opposite pairs at the stem nodes (1 per leaf), dense when young, becoming more open with age, mostly with 12–30 flowers, the bracts 2–4 mm long, much smaller than the foliage leaves, linear to narrowly lanceolate. Flower stalks 2–5 mm long at flowering (to 11 mm long at fruiting), loosely ascending at fruiting. Calyces 2.5–4.0 mm long, the lobes noticeably unequal, the upper 2 lobes shorter than the lower (2)3 lobes, deeply (4)5-lobed, the lobes linear to narrowly oblong-lanceolate, usually short-hairy. Corollas 10–13 mm wide, blue to purplish blue with darker veins, the throat white or pale yellow, the lobes spreading. Style 6–8 mm long at fruiting. Fruits 4–6 mm long, longer than wide, obovate in profile, flattened, the notch very shallow the surfaces and margins usually pubescent with short, nonglandular hairs, dehiscing along the sutures into 2 valves. Seeds mostly about 6 per locule, 1.5–2.0 mm long, strongly flattened, the sides both flat, the surfaces appearing smooth or nearly so (with a very fine network of ridges), tan to brown. $2n$=48, 64. May–June.

Introduced, uncommon, known thus far only from historical collections from Johnson County (native of Europe, Asia; introduced sporadically in the northeastern U.S. west to South Dakota and Missouri, Canada). Margins of lakes.

This species is cultivated in gardens as an ornamental, and several cultivars have been developed. It was first reported as an escape in Missouri by Castaner (1982b, as *V. latifolia*). Its nomenclature has been controversial. Pennell (1921, 1935) accepted the name *V. latifolia* for this taxon and discussed typification of that epithet, while noting that some authors had applied the name to a different taxon. However, Pennell's interpretation of the typification was later shown to be incorrect because it was based on two different specimens. D. A. Webb (1972) later correctly assigned a type to the name *V. latifolia*, but also discussed the desirability of its rejection as an ambiguous name. More recently, this name was officially rejected under the International Code of Botanical Nomenclature (Martínez-Ortega et al., 2006; McNeill et al., 2006). The next-oldest available specific epithet for the taxon then became *V. teucrium*. Some authors also have included *V. teucrium* as one of several infraspecific variants within a broadly defined *V. austriaca* (Walters and Webb, 1972; D. A. Webb, 1972).

15. Veronica triphyllos L.

Pl. 489 c–e; Map 2246

Plants annual, with slender taproots. Stems 5–17 cm long, erect or ascending, sometimes from a spreading base (not rooting at the lower nodes), moderately to densely pubescent with spreading, mostly glandular hairs. Leaves sessile or short-petiolate. Leaf blades 0.7–1.5 cm long, 1.0–1.2 times as long as wide, ovate to broadly ovate, mostly broadest below the midpoint, broadly angled to broadly rounded or truncate at the base, not clasping the stems, palmately 3–7(9)-lobed, the lobes rounded or broadly and bluntly to occasionally sharply pointed at their tips, the margins flat, otherwise entire or rarely with a small, blunt tooth along 1 or both margins, the surfaces moderately to densely pubescent with short, glandular hairs.

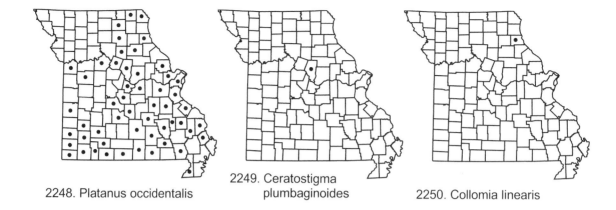

2248. Platanus occidentalis

2249. Ceratostigma plumbaginoides

2250. Collomia linearis

Inflorescences terminal, spikelike racemes, these short and condensed at the start of flowering, but elongating with age, often extending nearly to the stem base but not appearing as solitary axillary flowers, with 4–12 flowers, the axis visible between the flowers (except sometimes at the start of flowering), the bracts 3–10 mm long, the lowermost similar in size to the foliage leaves, gradually reduced, those toward the inflorescence tip noticeably smaller than the foliage leaves, more or less ovate in outline, deeply 3(5)-lobed, the lobes slender. Flower stalks 2–5 mm long at flowering (to 8[–12] mm at fruiting), mostly shorter than the calyces at flowering, upcurved to ascending at flowering and fruiting. Calyces 3–4 mm long (to 8 mm at fruiting), the lobes unequal, the upper 2 lobes shorter than the lower 2 lobes, deeply 4-lobed, the lobes broadly oblanceolate to oblong-oblanceolate, pubescent with spreading, glandular hairs. Corollas 3.0–4.5 mm wide (2.5–3.5 mm long), blue with darker veins, the throat white or more commonly yellowish-tinged or light green, the tube appearing relatively broad, wider than long, the lobes more or less bell-shaped. Style 1.2–1.6 mm long at fruiting. Fruits 4.2–5.5 mm long, about as long as wide or more commonly slightly longer, more or less heart-shaped in profile, flattened, the notch relatively deep (0.8–1.1 mm), the surfaces glandular-hairy, dehiscing along the sutures into 2 valves. Seeds mostly 6–12 per locule, 1.5–2.0 mm long, cup-shaped (deeply concave on one side, convex on the other), the convex surface appearing cross-wrinkled, brown. $2n=14$. April–June.

Introduced, uncommon, known thus far only from St. Charles and St. Louis Counties (native of Europe; introduced sporadically in the western, central and southeastern U.S.). Edges of crop fields, roadsides, and open, disturbed areas.

This species was first reported for Missouri by Yatskievych and Sullivan (1992).

17. **Veronicastrum** Heist. ex Fabr.

About 20 species, temperate North America, Asia.

The genus *Veronicastrum* traditionally was considered to comprise only two species (one in Asia, the other in North America), but Yamazaki (1957) expanded the generic concept to include about 18 additional Asian species previously classified into the genus *Botryopleuron* Hemsl. Molecular studies (Albach et al., 2004) and the unusual (for the tribe) chromosomal base number of $x=17$ (Albach et al., 2008) lend support for the expanded concept of the genus.

1. Veronicastrum virginicum (L.) Farw.
(culver's root)

Pl. 491 a–c; Map 2247

Plants perennial herbs, with rhizomes, terrestrial. Stems of only 1 kind, 60–150 cm long, erect or strongly ascending, unbranched or with ascending branches toward the tip, glabrous or less commonly sparsely to moderately pubescent with mostly short, nonglandular hairs. Leaves in whorls of 3–7 (rarely a few nodes with opposite leaves), sessile or short-petiolate. Leaf blades 4–14 cm long, simple, unlobed, lanceolate to elliptic-lanceolate, tapered to a sharply pointed tip, tapered at the base, the margins sharply toothed, the surfaces glabrous or sparsely to moderately short-hairy, rarely more densely pubescent with longer hairs, the venation pinnate. Inflorescences terminal, spikelike racemes, these solitary or in panicles consisting of whorls of 3 to several racemes at 1–3 nodes, elongated, the centralmost raceme longer than the others, each raceme stalked, each node with a whorl of small leaflike bracts, the flower stalks 0.3–1.5 mm long at flowering, not becoming elongated at fruiting, each subtended by a minute, lanceolate to triangular-ovate bractlet. Flowers perfect. Calyces 1.5–3.0 mm long, deeply (4)5-lobed nearly to the base, the lower 2 lobes noticeably longer than the (2)3 upper lobes, all narrowly oblong-lanceolate to lanceolate, bluntly to sharply pointed at the tip, glabrous. Corollas 4–6 mm long, nearly actinomorphic (the upper 2 lobes slightly wider than the lower 2), 4-lobed, glabrous, the tube 2–3 times as long as the lobes, white, cream-colored, or pale to light pink, the tube lacking a spur, the throat open, the lobes more or less spreading at maturity. Fertile stamens 2, the filaments exserted, straight, the anther sacs parallel, confluent at the tips; staminodes absent. Style 1, more or less exserted, the stigma minute, unlobed. Fruits capsules, 2.5–4.5 mm long, ovoid or ellipsoid, more or less circular in cross-section, glabrous, the 2 locules equal in size, dehiscent from the tip by 4 teeth. Seeds numerous, 0.5–0.7 mm long, ovoid to oblong-ovoid, more or less circular in cross-section, the surface light brown, appearing granular or with a minute network of fine ridges in longitudinal rows. $2n=34$. June–September.

Scattered nearly throughout the state, but apparently absent from most of the Mississippi Lowlands Division (eastern U.S. west to North Dakota and Texas; Canada). Bottomland prairies, upland prairies, openings and edges of mesic to dry upland forests, savannas, ledges and tops of bluffs, banks of streams and rivers, margins of ponds and lakes, and rarely fens; also pastures, railroads, and roadsides.

Culver's root has a long history of medicinal use as an emetic, cathartic, diuretic, tonic, gastrointestinal aid, and for treatment of various conditions ranging from liver problems to tuberculosis (Steyermark, 1963; Moerman, 1998). Today, however, its principal use is as a sun-loving ornamental in gardens. The flowers are pollinated mainly by bees, but are also visited by a number of other insects (Pennell, 1935). Typical specimens of this species have the leaf blades glabrous or inconspicuously pubescent with short hairs. Less commonly, plants scattered in the range of the species have leaf blades that are relatively densely hairy on the undersurface have been called f. *villosum* (Raf.) Pennell (Pennell, 1935; Steyermark, 1963).

PLATANACEAE (Plane Tree Family)

Contributed by Alan Whittemore

One genus, about 10 species; North America, Europe, Asia.

The family has an extremely long and extensive fossil record, with both leaves and reproductive materials very similar to extant species of *Platanus* dating back more than 100 million years to Lower Cretaceous times (Schwarzwalder, 1986).

1. **Platanus** L. (sycamore, plane tree)

About 10 species, North America, Europe, Asia.

1. Platanus occidentalis L. (sycamore, American plane tree)

P. occidentalis f. *attenuata* Sarg.

P. occidentalis var. *glabrata* (Fernald) Sarg.

Pl. 492 d, e; Map 2248

Plants large trees with widely spreading branches, monoecious. Trunks to 50 m tall, the bark flaking off as large scales or thin plates, otherwise smooth, white where recently exposed, greenish gray with age, the trunk thus a patchwork of white and gray areas. Twigs 2–4 mm thick, brown to dark brown or gray, finely hairy, becoming nearly glabrous with age. Leaves alternate, simple. Stipules to 5 cm wide, leaflike, coarsely toothed or lobed, shed during leaf development and leaving a scar visible as a narrow ring extending all the way around the twig. Petioles 1–7 cm long, their bases swollen, hollow, and completely enclosing the buds. Leaf blades 9–25 cm long, 12–33 cm wide, broadly triangular to nearly kidney-shaped, with 3–7 main veins from the base, palmately 3–7-lobed, each lobe broadly triangular and tapered toward the sharply pointed tip, the margins coarsely toothed or entire, the sinuses extending no more than halfway to the leaf base, sometimes mere shallow indentations, the leaf base rounded or broadly angled to deeply cordate, the surfaces pubescent when young with minute stellate hairs, at maturity becoming nearly glabrous or hairy only along the main veins. Inflorescences unisexual heads, these dense, globose, usually occurring singly from the tips of short lateral twigs, pendant; the staminate heads 0.8–1.2 cm in diameter, on stalks 1–2 cm long, greenish yellow; the pistillate heads 0.8–1.2 cm in diameter at flowering, 2.5–3.0 cm in diameter at fruiting, on stalks 3–10 cm long, reddish brown at flowering, tan to yellowish brown at fruiting, usually persisting through the winter. Flowers actinomorphic, hypogynous, the calyx of 3–6 sepals, these 0.3–0.5 mm long, scalelike, free or sometimes united toward the base, hairy, the corolla absent or (on some staminate flowers) of 3–6 extremely minute segments. Staminate flowers with 3–6 stamens and sometimes a reduced abortive pistil, the filaments absent or very short, the anthers attached at the base, with 2 locules, the connecting tissue prolonged into a small hairy peltate appendage at the tip. Pistillate flowers with 3 or 4 small staminodes and 3–9 pistils. Pistils with the ovary superior, 1-locular, the placentation apical. Styles 1 per pistil, elongate, linear, the stigmatic area lateral toward the tip. Fruits achenes, 7–8 mm long, club-shaped, light brown, each with a basal tuft of hairs almost as long as the fruit. $2n=42$. April–June.

Common throughout the state (eastern U.S. west to Wisconsin, Nebraska, and Texas; Canada). Banks of streams and rivers, margins of ponds and lakes, and bottomland forests; also roadsides, railroads, and moist disturbed areas.

In natural vegetation, this species is mostly confined to riparian and floodplain forests. However, its ability to colonize bare mineral soils at sunny sites has made it a successful invader of bulldozed habitats, such as roadsides and railroad grades. The achenes, which may remain on the trees in winter, are eaten by many species of birds. Plants in which the main lobes of the leaves are entire or nearly so have been called var. *glabrata*, a taxon of uncertain merit (Hsiao, 1973; Schwarzwalder, 1986) that has been collected from a few scattered counties in southern Missouri. Plants with obtuse leaf bases have been called f. *attenuata*, but this variant seems unworthy of formal taxonomic recognition. Older trees develop massive trunks (up to more than 14 m in circumference, in historical reports), and early descriptions of the forests often mentioned massive sycamores. The wood is very resistant to splitting and has been used to make things that must stand up to a great deal of stress, such as ox yokes and butcher's blocks. The fruits are sometimes inadvertently dispersed by children, who throw the mature, fruiting "itchy-bombs" at each other for entertainment.

The London plane tree, *P.* ×*acerifolia* (Aiton) Willd., is a common street tree that apparently does not escape from cultivation. It is probably descended from hybrids between *P. occidentalis* and *P. orientalis* L. of Asia Minor. London plane was very heavily used for urban plantings earlier in the last century because of its ability to grow in areas polluted by coal smoke, but it is not nearly as commonly planted as it once was. It differs from *P. occidentalis* in having somewhat more deeply lobed leaves and pistillate heads in pairs on each stalk.

PLUMBAGINACEAE (Leadwort Family)

Twelve to 29 genera, about 730 species, nearly worldwide.

Species in several genera of Plumbaginaceae are cultivated as garden ornamentals and used in the cut flower industry, including *Acantholimon* Boiss. (prickly thrift), *Armeria* Willd. (thrift, sea pink), *Ceratostigma*, *Limonium* Mill. (statice, everlasting, sea lavender), and *Plumbago* L. (plumbago, leadwort)

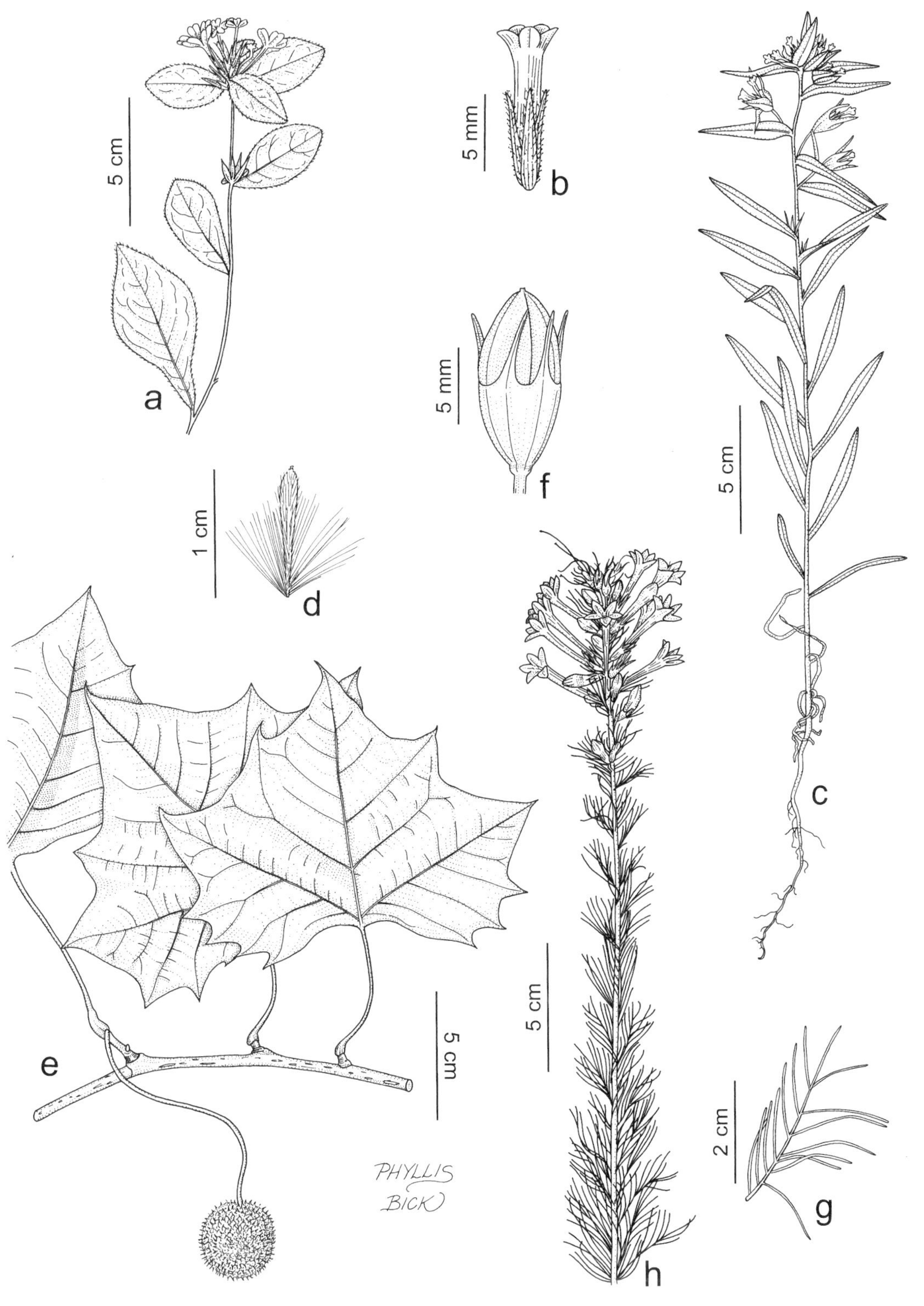

Plate 492. Platanaceae. Plumbaginaceae. Polemoniaceae. *Ceratostigma plumbaginoides*, **a)** fertile stem. *Collomia linearis*, **b)** flower, **c)** habit. *Platanus occidentalis*, **d)** fruit, **e)** portion of branch with head of fruits and leaves. *Ipomopsis rubra*, **f)** fruit, **g)** leaf, **h)** fertile stem.

1. **Ceratostigma** Bunge
(G. H. M. Lawrence, 1954)

Eight species, Asia, Africa.

1. Ceratostigma plumbaginoides Bunge (leadwort)

Pl. 492 a; Map 2249

Plants perennial herbs, the rootstock and lower stems often somewhat woody. Stems 15–50 cm long, unbranched or branched, spreading to loosely ascending, sometimes trailing on surrounding vegetation, angled (with blunt longitudinal ridges), glabrous or sparsely pubescent with short, bristly hairs, somewhat more densely so on young growth, sometimes also sparsely mealy, the winter buds naked. Leaves alternate, sessile or with a short, winged petiole, the base expanded and soemwhat sheathing the stem. Stipules absent. Leaf blades simple, 2–10 cm long, obovate to somewhat rhombic or less commonly elliptic, tapered at the base, rounded or angled to short-tapered to a usually bluntly pointed tip, the margins entire, pubescent with stiff ascending hairs, the surfaces glabrous or nearly so, but usually sparsely mealy, sometimes somewhat reddish-tinged (becoming bronze-colored to strongly reddish-tinged in the autumn). Inflorescences terminal and often also in the upper leaf axils, consisting of dense headlike clusters. Bracts present, surrounding the flowers and groups of flowers, variously 6–12 mm long, the outermost bracts shorter and oblong to elliptic, the inner bracts longer and lanceolate, mostly papery to hardened in texture, brown to brownish red, the margins with bristly, ascending hairs. Flowers actinomorphic, perfect, hypogynous. Calyces 12–14 mm long, of 5 sepals, these fused into a slender tube that is strongly 5-ribbed, the erect lobes narrowly triangular to nearly linear, angled or tapered to short-awned tips, uniformly papery to hardened and brown to brownish red except for narrow, membranous bands between the ribs and along the lobe margins, persistent at fruiting. Corollas 23–28 mm long, 5-lobed, spirally twisted in bud, the slender, pale reddish-tinged tube noticeably longer than the sepals, the lobes 8–10 mm long, abruptly spreading, obtriangular, shallowly concave at the tip with a short, slender extension of the midnerve, bright blue to deep blue (drying purplish). Stamens 5, opposite the petals, the filaments attached to the base of the corolla tube, the anthers usually slightly exserted from the corolla tube, linear, dark blue to dark purplish blue. Staminodes absent. Pistil 1 per flower, the ovary superior, of 5 fused carpels, with 1 locule, the placentation basal. Style 1, 5-branched toward the tip, the stigmas 1 per branch, linear. Fruits seldom observed, capsules, 5–6 mm long, enclosed in the persistent calyx, 1-seeded, dehiscing longitudinally from the base into 5 valves. Seeds 3–4 mm long, narrowly oblong-ellipsoid, brown, the embryo straight. $2n$=14. July–September.

Introduced, known thus far only from Boone County (native of Asia, introduced uncommonly in the U.S.). Fencerows and disturbed areas.

This species was first reported for Missouri by Dunn (1982). It has not been collected again since Dunn's original report. It is commonly cultivated in flower beds (where it generally forms loose mats) and in hanging baskets. According to G. H. M. Lawrence (1954), the species was brought into cultivation in England in about 1846. It rarely escapes from cultivation and appears to be propagated primarily vegetatively by divisions from existing plants.

POLEMONIACEAE (Phlox Family)

Contributed by Carolyn J. Ferguson

Plants annual, biennial, or perennial herbs (woody elsewhere). Stems unbranched or more commonly branched, glabrous or hairy and/or glandular, the hairs lacking persistent pustular bases. Leaves alternate or opposite and sometimes also basal, well-developed, sessile or petiolate. Stipules absent. Leaf blades simple or pinnately deeply divided to compound, the surfaces usually hairy (sometimes becoming glabrous or nearly so at maturity), the hairs lacking persistent pustular bases and not roughened to the touch. Inflorescences terminal and/or axillary, of solitary flowers or clusters, these sometimes grouped into panicles, not appearing paired or coiled, the individual flowers often not subtended by bracts (but the branch points sometimes with bracts). Flowers actinomorphic (somewhat zygomorphic elsewhere), hypogynous, perfect; cleistogamous flowers absent. Calyces usually 5-lobed, often with thicker green lobes extending the length of the tube and intervening thin, translucent areas (these delicate and often rupturing as the fruits mature), usually not becoming enlarged at fruiting, the lobes equal, more or less persistent at fruiting. Corollas usually 5-lobed, usually spirally twisted in bud, trumpet-shaped or bell-shaped. Stamens usually 5, the filaments attached in the corolla tube, not subtended by scales, usually relatively short, but occasionally somewhat unequal in length, the anthers not exserted or some or all of them somewhat exserted, appearing 2-locular, usually attached near the dorsal midpoint, white, red, dark purple, or blue. Pistil 1 per flower, of usually 3 fused carpels. Ovary unlobed to bluntly 3-angled or 3- or 6-ribbed, 3-locular, with 1–10 ovules per locule, the placentation axile. Style 1, attached at the tip of the ovary, usually 3-branched at the tip, persistent or not at fruiting, the branches stigmatic on the inner surface for most of their length, slender. Fruits capsules, dehiscent longitudinally (sometimes tardily or incompletely so), usually with 3 valves, with 1–12 seeds per locule. About 26 genera, about 375 species, most diverse in the New World, but also in Europe, Asia.

Economic importance of the Polemoniaceae is chiefly horticultural, with numerous cultivated ornamentals in the family (particularly within *Phlox*, *Ipomopsis*, *Polemonium*). The family has received much attention from evolutionary biologists over the decades, and is of particular interest due to diversification in varied ecological habitats of western North America, its center of diversity (e.g., V. Grant 1959; V. Grant and K. Grant, 1965). The family is taxonomically complex, and generic delimitation has been unstable through history. Recent, extensive phylogenetic work for the family and its constituent genera has advanced our understanding of diversity in the group (see L. A. Johnson et al., 2008), and a phylogenetic classification of three subfamilies and 26 genera has emerged (J. M. Porter and Johnson, 2000). Most of these genera, including all members of the Missouri flora, are grouped within a large temperate subfamily Polemonioideae, with the remaining genera in the tropical subfamily Cobaeoideae Arn. and the monotypic subfamily Acanthogilioideae J.M. Porter & L.A. Johnson (Baja California, Mexico). Continued study of the family hopefully will clarify relationships of the genera. The systematics of particular genera remains complicated, presenting challenges relating to hybridization, polyploidy, presence of cryptic species, etc. Ongoing work within genera can be expected to further highlight diversification and species boundaries within this interesting family. Wilken (1986) is noted as a reference for information on the non-*Phlox* taxa.

1. Leaf blades simple, not lobed, the margins entire or nearly so
    2. All of the main stem leaves of the stem alternate . . . . 1. COLLOMIA
    2. At least the lower stem leaves opposite (the upper ones alternate in *P. drummondii*) . . . . 3. PHLOX
1. Leaf blades pinnately dissected or compound
    3. Leaf divisions linear to threadlike, corollas red, sometimes with pink or white markings on the lobes, calyx lobes narrowly triangular . . . . 2. IPOMOPSIS
    3. Leaf divisions oval to lanceolate, corollas blue to lavender, calyx lobes triangular-ovate (sometimes narrowly so at fruiting) . . . . 4. POLEMONIUM

1. **Collomia** Nutt. (trumpet)

About 15 species, North America.

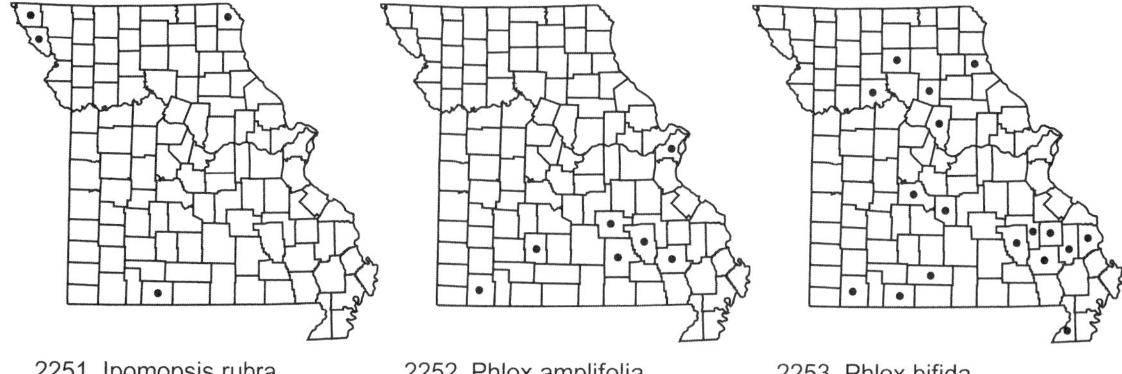

2251. Ipomopsis rubra 2252. Phlox amplifolia 2253. Phlox bifida

1. Collomia linearis Nutt.

Pl. 492 b, c; Map 2250

Plants annuals, taprooted. Stem 5–60 cm long, solitary, erect or strongly ascending, sometimes branched at the lower nodes, glabrous to short-hairy toward the base, moderately pubescent with slightly longer, fine, gland-tipped hairs toward the tip. Leaves alternate, sessile. Leaf blades 1–8 cm long, simple, those of the lower leaves linear to lanceolate, those of the upper ones lanceolate to ovate, angled or tapered to a sharply pointed tip, the bases of the larger leaves sometimes somewhat clasping, the margins entire or nearly so, the upper surface glabrous but often somewhat sticky, the undersurface with short, often glandular hairs, mostly along the midvein, the lateral veins obscure. Inflorescences terminal, solitary, headlike clusters surrounded by leaflike bracts. Calyces 5-lobed to above or near the midpoint, 4–7 mm long at flowering, becoming enlarged to 6–9 mm at fruiting, narrowly bell-shaped, the tube uniform in texture (becoming papery at fruiting), the lobes triangular to narrowly triangular, glandular-hairy. Corollas 5-lobed, white to pink, trumpet-shaped (but often appearing more or less tubular), the tube 8–15 mm long, the lobes 1–4 mm long. Stamens with the filaments attached unequally in the tube, the anthers included or 1 or 2 slightly exerted. Style not exserted. Seeds 3, 2.0–2.5 mm long, oblong-ovoid, slightly flattened, with a shallow longitudinal groove along 1 side and usually a minute winglike ridge at each end, the surface brown, smooth to faintly and finely wrinkled, becoming sticky when moistened. $2n=16$. May–August.

Introduced, uncommon, known thus far only from historical collections from Marion County (western U.S. east to North Dakota and Nebraska; Canada; introduced farther east). Pastures and open, disturbed areas.

Steyermark (1963) mentioned an occurrence in Christian County without further documentation. No specimens from this county could be located during the present study.

2. Ipomopsis Michx. (ipomopsis)

About 29 species, North America.

1. Ipomopsis rubra (L.) Wherry (standing cypress)

Gilia rubra (L.) A. Heller

Pl. 492 f–h; Map 2251

Plants biennials, with stout taproots. Stems 10–40(–90) cm long, solitary, erect or strongly ascending, unbranched below the inflorescence, sparsely pubescent with minute, nonglandular hairs. Leaves in a prominent basal rosette and alternate, sessile or nearly so. Leaf blades 4–8 cm long, oblong-elliptic to oblong-ovate in outline, pinnately dissected into 10–15 segments, the segments 5–20 mm long, linear or threadlike, short-tapered to sharply pointed tips, the midvein extended into a short, spinelike projection, the margins otherwise entire, the undersurface glabrous or sparsely and minutely hairy along the midvein toward the base, each segment appearing 1-veined. Inflorescences terminal, solitary, spikelike panicles with short, strongly ascending branches, each branch with (1–)3–5 flowers, these horizontally spreading. Calyces 5-lobed to below the midpoint, narrowly bell-shaped to nearly tubular, the tube 3–4 mm long, differentiated into 5 thicker, green bands (extending into the lobes), these sparsely hairy and somewhat sticky, separated by intervening thin, translucent areas (these delicate and often rupturing as the fruits mature), the lobes 4–6 mm long, narrowly triangular, tapered to sharply pointed tips. Corollas 5-lobed, bright red (the tube usually orangish yellow internally, the lobes sometimes with pink or white markings), trumpet-shaped, the tube 20–25 mm long, the lobes 9–11 mm long. Stamens with the filaments attached equally in the tube, but unequal in lengths, the anthers thus unequally exserted (2 stamens shorter than the other 3). Style exserted. Seeds numerous, 2–3 mm long, oblong-ovoid to narrowly ovoid, irregularly 3-angled, the angles ridged or narrowly winged, the surface tan to straw-colored, more or less smooth to finely pitted, not becoming sticky when moistened. $2n=14$. May–August.

Introduced, uncommon and widely scattered (native in the southern United States from Texas to southwestern Oklahoma, east to Florida, North Carolina). Open, disturbed areas.

This species is cultivated for its bright red flowers, which contrast with the lacy leaves, and are pollinated by hummingbirds (Estes and Hall, 1975).

3. **Phlox** L. (phlox)
(Wherry, 1955)

Plants perennial herbs (a few species annual). Stems erect to loosely ascending from a spreading base, with rhizomes (except in annual species, which are taprooted), sometimes forming dense mats or low mounds (to cushion-like elsewhere), unbranched or branched (at the base or toward the tip). Leaves opposite, the uppermost sometimes subopposite or rarely alternate, sessile to subsessile, simple, unlobed, the margins entire, the bases of the opposing leaves sometimes forming a narrow ridge or membrane across the node. Inflorescences terminal and often also axillary from the upper leaves, consisting of clusters or small panicles, sometimes appearing narrowly racemose or reduced to a solitary flower. Calyces 5-lobed to above or near the midpoint (in some species some of the lobes separating tardily after the flower opens), tubular to narrowly bell-shaped, differentiated into 5 thicker, green bands (extending into the lobes), these separated by intervening thin, translucent areas (these delicate and often rupturing as the fruits mature), glabrous to densely pubescent with short, glandular and/or nonglandular, spreading hairs. Corollas trumpet-shaped, white to more commonly pink, purple, or occasionally blue (sometimes red in *P. drummondii*; yellow elsewhere), often with lighter or darker markings at the mouth of the tube and lobe bases, the tube slender, somewhat expanded near the tip, the lobes abruptly spreading, rounded or angled to a broad, blunt tip, sometimes abruptly tapered to a minute, sharply pointed extension, in a few species coarsely notched apically. Stamens with the anthers positioned unequally along the tube, either all included or some of them positioned near or less commonly slightly exerted from the mouth of the tube. Style either short and not extending past the midpoint of the corolla tube or longer and extending to near the mouth of the corolla tube. Seeds 3 (rarely 1–12 elsewhere), 2–4 mm long, oblong-ovate to occasionally broadly ellipsoid, slightly flattened, the surface yellowish brown to dark brown, faintly to moderately wrinkled, not becoming sticky when moistened. About 60 species, North America, Asia (Siberia), one species (*P. drummondii*) introduced sporadically nearly worldwide.

Phlox is the largest genus of the Polemoniaceae, and exhibits interesting diversity relative to ecology and geography across much of North America. Approximately a third of the species occur in eastern North America, and the genus is well-represented in the Missouri flora. *Phlox* is known for hybridization and polyploidy, and these factors have likely contributed to a complicated phylogenetic history (C. J. Ferguson and Jansen 2002). Most of the species in our flora, however, are readily distinguished, and an encounter with a hybrid in the field would be very rare. A valuable recent work on the genus, including horticultural information, was provided by Locklear (2011).

1. Plants taprooted annuals; upper leaves alternate . . . . 4. P. DRUMMONDII
1. Plants perennial herbs; leaves opposite throughout (occasionally subopposite near the inflorescence)
    2. Stems spreading, 4–30 cm tall, forming mats; corolla lobes often noticeably notched at the tip
        3. Plants forming open mats, with the nodes of the prostrate stems uncongested at maturity; leaf lengths typically variable on the plant, the largest leaves 25–55 mm long; corolla lobes each with a conspicuous notch 1.5–5.0 mm deep . . . . 2. P. BIFIDA
        3. Plants forming dense mats, with the prostrate stems typically producing short axillary shoots and thus appearing congested; leaves not noticeably unequal in length on the plant, the largest leaves 8–20 mm long; corolla lobes each with a notch to 1.0–1.5(–3.0) mm deep, rarely rounded and lacking a notch . . . . 9. P. SUBULATA
    2. At least the flowering stems erect or strongly ascending, 25–120 cm tall; corolla lobes rounded or abruptly tapered to a sharp point (sometimes slightly notched in *P. paniculata*)
        4. Style short (0.4–1.5 mm), the stigma positioned within the lower half of the corolla tube and below all of the anthers; anthers all included well within the corolla tube
            5. Plants producing vegetative stems that are partially spreading and rooting at the nodes, differentiated from the flowering stems, with elliptic, in part evergreen, leaves; flowering stems with 4–6 nodes; corolla tube glabrous on the outer surface . . . . 3. P. DIVARICATA
            5. Plants with few or no vegetative stems, when produced, these erect or strongly ascending and morphologically similar to the flowering stems; flowering stems with 6–17 nodes, corolla tube usually hairy on the outer surface . . . . 8. P. PILOSA
        4. Style elongate (9–29 mm), the stigma located among the anthers in the upper portion of the corolla tube; uppermost anther(s) positioned near the mouth of the corolla tube
            6. Leaves narrow elliptic to oval or ovate, the largest 20–65 mm wide, with visible veins connecting toward the margins to form a distinctive looped pattern
                7. Stems with 20–35 nodes; leaves narrowly elliptic, usually glabrous, calyces glabrous or pubescent with nonglandular hairs; corolla tube usually hairy externally . . . . 7. P. PANICULATA
                7. Stems with 8–14 nodes, leaves elliptic toward the stem base becoming oval or ovate toward the stem tip, sparsely pubescent with soft or bristly hairs; calyces pubescent with glandular hairs; corolla tube glabrous externally . . . . 1. P. AMPLIFOLIA
            6. Leaves linear to lanceolate (the uppermost sometimes narrowly elliptic), the largest 4–20 mm wide, with inconspicuous veins forming a pinnate pattern without connecting loops, often only the midvein visible
                8. Inflorescences loosely clustered and about as wide as long, sometimes appearing as domed panicles; uppermost leaves narrowly lanceolate or narrowly elliptic (the bases not cordate), stems not noticeably streaked with red or purple . . . 5. P. GLABERRIMA
                8. Inflorescences cylindric (with all of the clusters short-stalked), much longer than wide, appearing narrowly racemose; uppermost leaves lanceolate (the bases sometimes cordate), stems reddish- or purplish-streaked . . . . 6. P. MACULATA

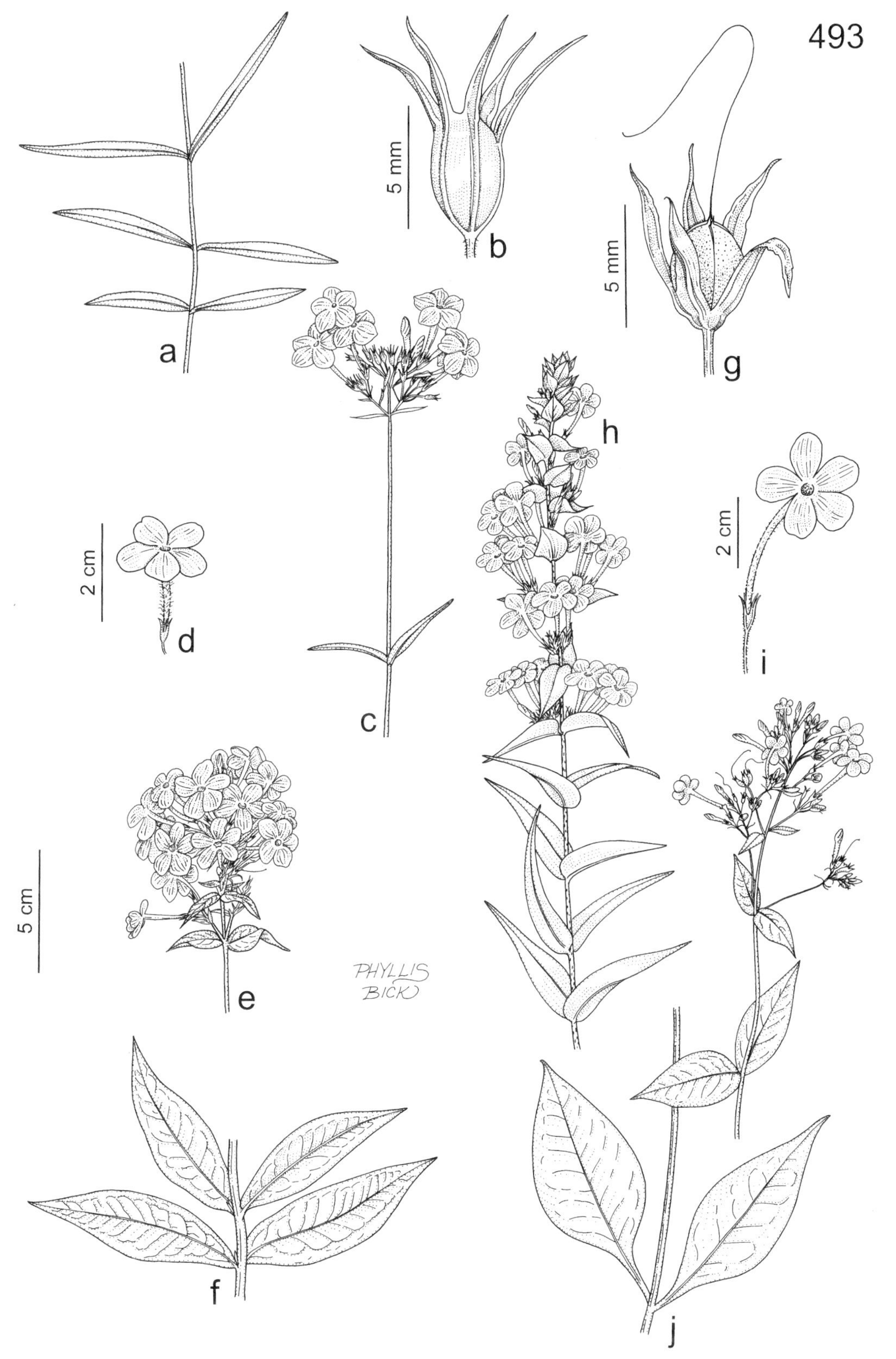

Plate 493. Polemoniaceae. *Phlox glaberrima*, **a)** portion of stem with leaves, **b)** fruit, **c)** fertile stem. *Phlox paniculata*, **d)** flower, **e)** inflorescence, **f)** median portion of stem with leaves. *Phlox maculata*, **g)** fruit, **h)** fertile stem. *Phlox amplifolia*, **i)** flower, **j)** stem tip with inflorescence and median node with leaves.

2254. Phlox divaricata 2255. Phlox drummondii 2256. Phlox glaberrima

1. Phlox amplifolia Britton (broadleaf phlox, largeleaf phlox)

Pl. 493 i, j; Map 2252

Plants perennial herbs with short, thick rhizomes. Vegetative stems not produced or, if present, then similar to the flowering ones. Stems typically 1–3, 45–100 cm tall, erect, with 8–14 nodes, glabrous toward the base, hairy above the midpoint, with the hairs becoming shorter and glandular toward the inflorescence, sometimes with red streaks. Leaves all opposite, the blade elliptic toward the stem base, grading to oval or ovate toward the stem tip, those of the largest leaves 9.0–11.5 cm long and 35–65 mm wide, angled or more commonly tapered to a sharply pointed tip, mostly tapered at the base, the surfaces sparsely pubescent with soft or bristly hairs, the margins short-hairy, the secondary veins forming closed loops. Inflorescences with 25–150 flowers, consisting of clusters, the aggregate of clusters often appearing as domed panicles. Flower stalks 2–8 mm. Calyces 6–9 mm long, the lobes slender, tapered evenly to sharply pointed tips, glandular-hairy. Corollas bright pink to pinkish purple, the tube 15–30 mm long, glabrous externally, lacking a basal constriction, the lobes 8–11 mm long and 5–8 mm wide, obovate, rounded at the tips. Stamens with the filaments 12–27 mm long, the anthers positioned from below to at or above the stigma near the mouth of the tube (not exserted). Style 13–23 mm long, the stigmas 0.7–1.0 mm long. $2n$=14. June–August.

Uncommon in the Ozark Natural Division north locally to St. Louis County (eastern U.S. west to Missouri and Arkansas). Mesic upland forests; rarely also roadsides.

2. Phlox bifida L.C. Beck (cleft phlox, sand phlox)

P. bifida ssp. *stellaria* (A. Gray) Wherry

P. bifida var. *stellaria* (A. Gray) Wherry

Pl. 494 a, b; Map 2253

Plants perennial herbs with woody, often branched rootstocks, forming open mats or low, loose mounds, with short primary rhizomes. Vegetative stems spreading to slightly ascending, sometimes initially forming tufts of long leaves at congested nodes (but the nodes becoming uncongested at maturity), these leaves otherwise morphologically similar to those of the flowering stems. Flowering stem branches numerous, produced from the vegetative stems, 4–30 cm tall, with 4–6 nodes, erect or ascending, usually densely pubescent with curved or crinkled, multicellular hairs, these often all or mostly gland-tipped, at least toward the stem tip (and inflorescence). Leaves all opposite, the blade typically variable on a given plant, those of the largest leaves 2.5–5.5 cm long and 1–4 mm wide, linear to narrowly lanceolate, the surfaces and margins of the uppermost leaves moderately to densely pubescent with curved or crinkled, multicellular, often gland-tipped hairs, progressively less hairy toward the stem base, those of the lowermost leaves often glabrous or nearly so. Inflorescences relatively few-flowered, short and appearing as clusters, with 3–25 flowers or occasionally reduced to solitary flowers. Flower stalks 3–30 mm long. Calyces 5.0–9.5 mm long, the lobes each tapered evenly to a short sharp extension of the midnerve, usually glandular-hairy. Corollas pale bluish purple or light lavender to white, the tube 8–14 mm long, glabrous externally, lacking a basal constriction, the lobes 5.0–12.5 mm long and 3–8 mm wide, obovate, with a conspicuous apical notch 1.5–5.0 mm deep. Stamens with the filaments 5–11 mm long, the anthers positioned from below to at or above the stigma at the mouth of the tube (sometimes 1 or 2 slightly exerted). Style 5–10 mm, the stigmas 0.5–1.0 mm long. $2n$=14. March–May.

Scattered in the Ozark Natural Division, northward locally to Linn and Marion Counties; also known from a single collection from Dunklin County (Iowa to Oklahoma and Arkansas east to Michigan and Tennessee). Glades, rocky slopes in mesic to dry upland forests, bases, ledges, and tops of bluffs, and banks of streams and rivers; also roadsides.

This species is sometimes cultivated as an ornamental and occasional occurrences in Missouri may represent naturalized rather than truly native plants. Two infraspecific taxa have been widely recognized within *P. bifida*: ssp. *bifida* and ssp. *stellaria*. The latter name has been applied to material with nonglandular inflorescence pubescence and somewhat less deeply notched petals. The vast majority of Missouri populations are assignable to ssp. *bifida*, but a minority may be considered ssp. *stellaria*. In our area, nonglandular plants tend to occur adjacent to populations with glandular pubescence (often in relatively thin-soiled areas) or in mixed populations. The author does not view infraspecific taxon recognition to be warranted, although *P. bifida* does exhibit interesting morphological and ecological variation across its range that requires more detailed study at the population level.

3. Phlox divaricata L. (blue phlox, wild sweet William)

P. divaricata ssp. *laphamii* (Alph. Wood) Wherry

P. divaricata var. *laphamii* Alph. Wood

Pl. 494 f; Map 2254

Plants perennial herbs with slender rhizomes. Vegetative stems 1 to several, morphologically distinct from the flowering ones, usually overwintering (the leaves at least partly evergreen), 2–18 cm long, spreading or loosely ascending from an often spreading base, rooting at least at the lower nodes, densely pubescent with short, appressed to curved hairs. Flowering stems typically 1–8, 25–45 cm tall, erect, with 4–6 nodes, moderately to densely pubescent with curved or crinkled hairs, these often gland-tipped toward the stem tip. Leaves all opposite, those of the vegetative stems with the blade narrowly to broadly elliptic, angled to a bluntly or sharply pointed tip, angled at the base, the largest 1.5–5.0 cm long and 10–20 mm wide, the margins usually hairy, the surfaces moderately hairy to nearly glabrous (with age); those of the flowering stems with the blade oblanceolate to elliptic or ovate toward the stem base and mostly angled or tapered at the base, grading to lanceolate toward the stem tip and broadly angled to rounded or truncate at the base, variously rounded to bluntly pointed at the tip, the largest 2.5–5.0 cm long, at least the uppermost hairy along the margins, the surfaces sparsely to densely pubescent with curved or crinkled hairs (sometimes nearly glabrous with age), the hairs often gland-tipped on the uppermost leaves, the secondary veins obscure, pinnate and not forming conspicuous loops. Inflorescences with 8–25 flowers, consisting of clusters, the aggregate of clusters sometimes appearing as small, domed panicles. Flower stalks 4–13 mm. Calyces 6–10 mm long, the lobes slender, each tapered evenly to a short sharp extension of the midnerve, glandular-hairy. Corollas lavender to light purple or pale blue, rarely white, the tube 11–17 mm long, glabrous externally, with a slight constriction 1.5–3.0 mm above the base, the lobes 10–15 mm long and 5–9 mm wide, obovate, rounded at the tips, sometimes with an abrupt, minute point at the very tip (shallowly notched elsewhere). Stamens with the filaments 5–14 mm long, the anthers positioned above the stigma within the tube (not exserted). Style 0.5–1.5 mm long, the stigmas 1.0–1.5 mm long. $2n=14$. April–June.

Scattered to common nearly throughout the state (eastern United States west to South Dakota and Texas; Canada). Bottomland forests, mesic upland forests, banks of streams and rivers, bases of bluffs, and rarely edges of upland prairies; also roadsides.

This species is sometimes cultivated as an ornamental. Plants in portions of the range to the east of Missouri commonly exhibit notched (to ca. 3 mm) corolla lobes and correspond to ssp. *divaricata*, whereas plants in Missouri and some other portions of the Midwest and southern states with notchless, apically rounded corolla lobes have been called ssp. *laphamii*. However, there appears to be too much intergradation between these extremes to allow the formal recognition of subspecies. Rare individuals within populations produce white corollas and have been called f. *candida* E.J. Palmer & Steyerm. *Phlox divaricata* hybridizes with *P. pilosa* on rare occasions when the two, which have different ecological affinities, come into close contact.

4. Phlox drummondii Hook. (annual phlox)

P. drummondii var. *peregrina* Shinners

Pl. 494 e; Map 2255

Plants annuals with taproots. Vegetative stems not produced. Stems mostly solitary (occasionally few- to many-branched from the base and then appearing multi-stemmed), 10–30 cm long, erect or ascending, with 8–14 nodes, moderately to densely pubescent with spreading to somewhat curved or crinkled, multicellular hairs, these often gland-tipped toward the stem tip. Leaves opposite toward the stem base, grading to alternate at or above the midpoint, the blade oblanceo -late to oblong, elliptic or lanceolate, angled to a usually sharply pointed tip, tapered to angled,

2257. Phlox maculata

2258. Phlox paniculata

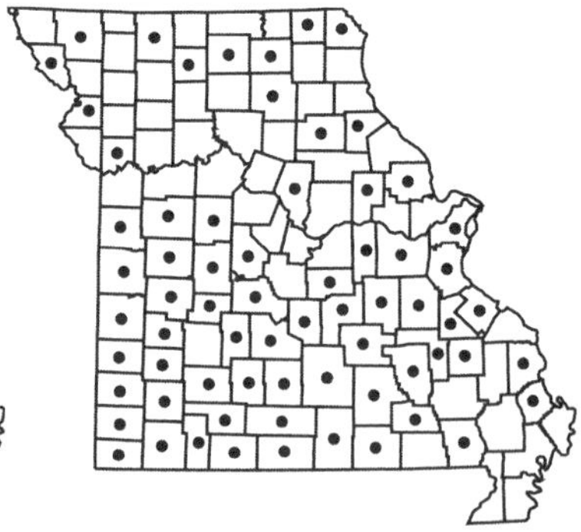

2259. Phlox pilosa

rounded, or truncate at the base, occasionally slightly cordate and clasping, the largest 3.5–7.5 cm long and 6–20 mm wide, the margins of at least the upper leaves hairy, the surfaces moderately to densely hairy or those of the lowermost leaves sometimes nearly glabrous, the hairs of the upper leaves usually gland-tipped. Inflorescences with 5–30 flowers, consisting of clusters, the aggregate of clusters sometimes appearing as flat-topped to domed panicles. Flower stalks 2–10 mm. Calyces 8–12 mm long, the lobes narrowly triangular, each tapered evenly from just above the base to a slender, sharply pointed tip, glandular-hairy. Corollas variously red to pink, purple, lavender, or white (often variable within populations), the tube 12–17 mm long, densely hairy externally, the hairs often gland-tipped, with a slight constriction 2–4 mm above the base, the lobes 9–15 mm long and 7–11 mm wide, obovate, tapered abruptly to a short, often slender, sharply pointed tip. Stamens with the filaments 4–15 mm long, the anthers positioned above the stigma within the tube (not exserted). Style 0.8–1.5 mm long, the stigmas 1.0–1.4 mm long. $2n=14$. June–September.

Introduced, uncommon and sporadic (native of Texas; Mexico; introduced sporadically nearly worldwide, including much of the eastern U.S.). Railroads and open, disturbed areas.

Phlox drummondii is a popular ornamental in gardens and also has been used as a component in wildflower seed mixes, including those used for "roadside beautification" projects along highways. Breeders have taken advantage of corolla color variation in the species to develop a large array of cultivars, and escaped and naturalized populations often exhibit striking color variations and patterns. *Phlox drummondii* var. *peregrina* was described by Shinners (1951) for plants of cultivated origin that had become established in the wild but whose morphology could not be accounted for in the native populations. These morphological variants are better treated as cultivars than as a variety. Five subspecies are recognized based on native populations (B. L. Turner, 1998b; Locklear, 2011).

5. Phlox glaberrima L. **ssp. interior** (Wherry) Wherry (smooth phlox)

P. glaberrima var. interior Wherry

P. carolina L.

P. carolina var. *angusta* (Wherry) Steyerm.

Pl. 493 a–c; Map 2256

Plants perennial herbs with short, thick rhizomes. Vegetative stems not produced. Stems typically solitary, 35–100 cm tall, erect, with 10–22 nodes, glabrous or rarely sparsely and inconspicuously hairy, not noticeably streaked with red or purple. Leaves all opposite, the lowermost with the blades mostly linear, grading to narrowly lanceolate or narrowly elliptic (rarely more broadly lanceolate) toward the tip, angled to a sharply pointed tip, tapered to angled or rounded at the base, those of the largest leaves 6–14 cm long and 4–20 mm wide, the surfaces glabrous (the upper surface shiny), the margins sometimes somewhat rolled under, appearing glabrous but with microscopic, forward-angled teeth (visible only with magnification), the secondary veins obscure, pinnate and not forming conspicuous loops. Inflorescences with 12–100 flowers, consisting of clusters, the aggregate of clusters often appearing as domed panicles (about as long as wide). Flower stalks 2–7 mm. Calyces 5.5–10.5 mm long, the lobes triangular to narrowly triangular, tapered evenly to sharply pointed tips, glabrous. Corollas light pink to pink or pinkish purple, the tube 14–22 mm long, glabrous externally, lacking a basal constriction, the lobes 7–10 mm long and 6–9 mm wide, obovate, rounded at the tips. Stamens with the filaments 9–21 mm long, the anthers positioned from below to at or above the stigma near the mouth of the tube (occasionally 1 or 2 slightly exserted). Style 9–20 mm long, the stigmas 0.6–1.3 mm long. $2n=14$. May–July.

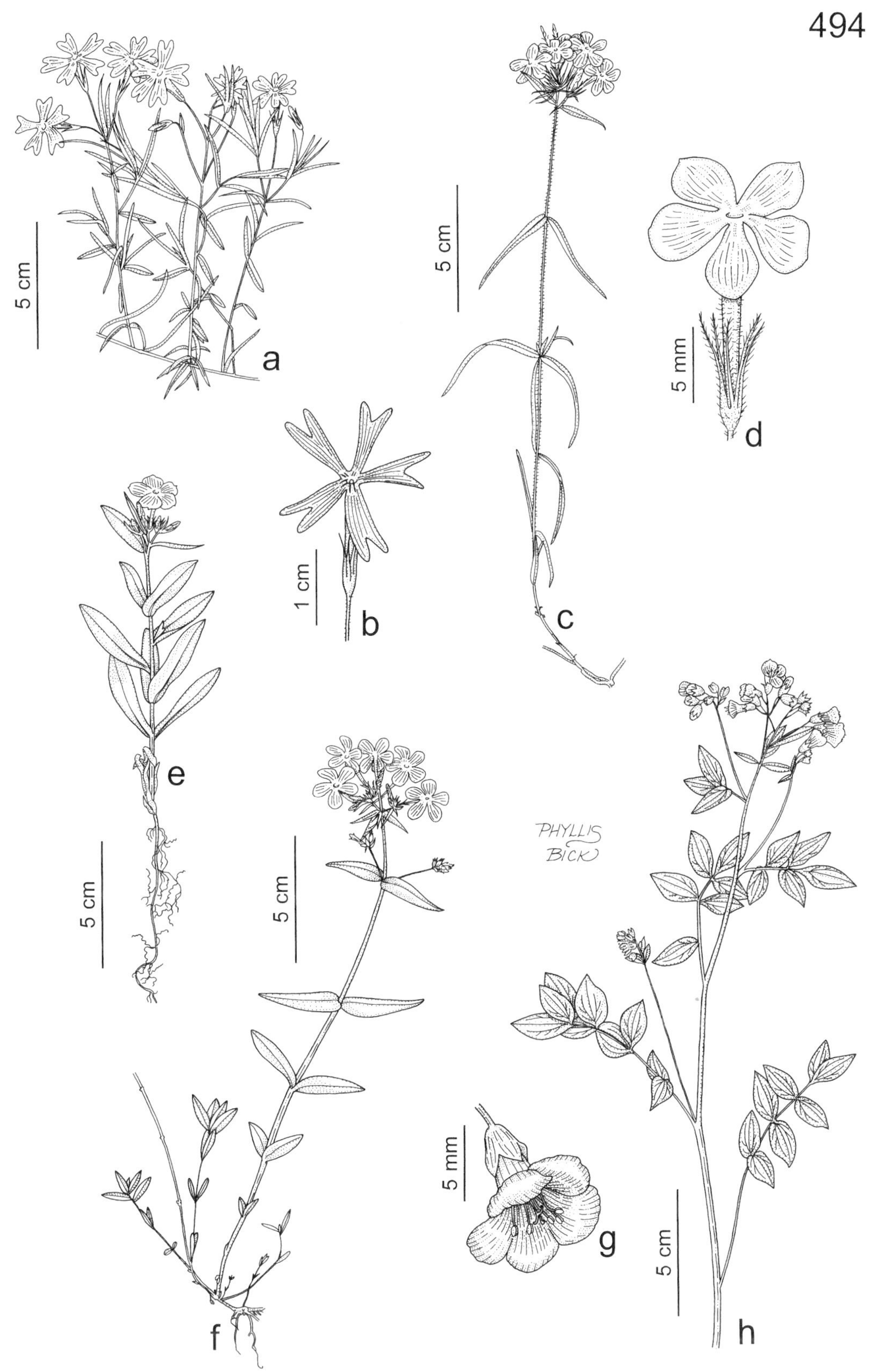

Plate 494. Polemoniaceae. *Phlox bifida*, **a)** habit, **b)** flower. *Phlox pilosa*, **c)** habit, **d)** flower. *Phlox drummondii*, **e)** habit. *Phlox divaricata*, **f)** habit. *Polemonium reptans*, **g)** flower, **h)** habit.

Scattered in the eastern half of the Ozark Division and adjacent Mississippi Lowlands west locally to Lawrence and Polk Counties (Indiana to Tennessee west to Missouri and Arkansas). Fens, acid seeps, banks of streams, rivers, and spring branches, bottomland prairies, bottomland forests, and swamps; rarely also railroads and roadsides.

Phlox glaberrima, in the broad sense, is a highly variable species distributed widely in the eastern United States. It has a complicated taxonomic history, with numerous epithets at the specific and infraspecific levels having been established and recombined historically by different workers (Wherry, 1955). Two species have been recognized widely in the complex, *P. glaberrima* and *P. carolina*, but morphological variation presents challenges with regard to species boundaries. Botanists have variously assigned material to each of these taxa (Wherry, 1955; Locklear, 2011) or recognized a single, variable *P. glaberrima* (Gleason and Cronquist, 1991; C. J. Ferguson, 1998). In his monograph of the genus, Wherry (1955) recognized seven infraspecific taxa among *P. glaberrima* (with three subspecies) and *P. carolina* (with four subspecies). This contrasted with his previous focused treatments of the group (Wherry, 1931, 1932, 1945) in terms of both alignment and recognition of infraspecific entities, and he noted the taxonomic challenges in the complex (Wherry 1955). Overall, morphological variation within the group across its range is noteworthy, and *P. glaberrima* warrants further taxonomic investigation. For the present, it does not seem prudent to accept two species.

Phlox glaberrima is purported to hybridize with *P. maculata* in some areas, but the species are highly distinct in Missouri. Published chromosome counts for *P. glaberrima* (including multiple counts for *P. glaberrima* ssp. *interior*) have established it as diploid (e.g., D. M. Smith and Levin, 1967; with the exception of some cultivated material), although flow cytometry data suggest there are some tetraploid populations in the Appalachians (C. J. Ferguson, unpublished).

Taxonomic challenges aside, the vast majority of material in Missouri corresponds to what has been called *P. glaberrima* ssp. *interior*. Steyermark (1963), following Wherry (1955), included *P. carolina* in his treatment based on limited material from southeastern Missouri. He recognized *P. carolina* var. *angusta* (Wherry) Steyerm. from Ripley County and *P. carolina* var. *carolina* from Carter County. *Phlox carolina* ssp. *angusta,* as circumscribed by Wherry (1955), can be difficult to distinguish from *P. glaberrima* ssp. *interior,* and it is not clear whether Ripley County populations represent a link with a more southerly entity. One specimen from Carter County with wide leaves and bracts indeed suggests a link with populations farther to the east that were considered by Wherry to be *P. carolina* ssp. *carolina*. A satisfying treatment of varieties within a broadly circumscribed *P. glaberrima* has not yet been proposed, and focused collection and study of material from areas showing interesting variation (such as southeastern Missouri) will be valuable to these efforts.

6. Phlox maculata L. **ssp. pyramidalis** (Sm.) Wherry (wild sweet William, meadow phlox)

P. maculata var. *pyramidalis* (Sm.) Wherry

Pl. 493 g, h; Map 2257

Plants perennial herbs with slender rhizomes. Vegetative stems not produced. Stems typically solitary, 65–100 cm tall, erect, with 18–35 nodes, glabrous or less commonly minutely and inconspicuously hairy toward the tip, with prominent dark reddish to reddish purple spots and/or streaks. Leaves all opposite, the lowermost with the blades linear to narrowly lanceolate, grading to lanceolate toward the tip, angled or tapered to a sharply pointed tip, tapered to angled, rounded, or shallowly cordate (those toward the stem tip) at the base, those of the largest leaves 7.5–12.5 cm long and 5–20 mm wide, the surfaces glabrous or nearly so (the upper surface shiny), the margins of the lower leaves sometimes slightly rolled under, glabrous, the secondary veins obscure, pinnate and not forming conspicuous loops. Inflorescences with 40–150 flowers, consisting of sessile to short-stalked clusters, the aggregate of clusters appearing cylindric and narrowly racemose (several times longer than wide). Flower stalks 1–6 mm. Calyces 5.5–9.5 mm long, the lobes triangular to narrowly triangular, tapered evenly to sharply pointed tips, the midnerve sometimes extended as a short, sharply pointed tip, glabrous. Corollas light pink to pink, light pinkish purple, or lavender (rarely all or partly white, mostly in cultivated forms), the tube 16–22 mm long, glabrous externally, lacking a basal constriction, the lobes 5–9 mm long and 5–9 mm wide, obovate, rounded at the tips. Stamens with the filaments 9–19 mm long, the anthers positioned from below to at or above the stigma near the mouth of the tube (occasionally 1 or 2 slightly exserted). Style 11–17 mm long, the stigmas 0.6–1.0 mm long. $2n$=14. July–September, rarely also May.

Uncommon, mostly in the eastern half of the Ozark Division, west locally to Greene, Morgan, and Pettis Counties (eastern [mostly southeastern] U.S. west to Missouri and Mississippi). Fens and less commonly banks of streams and spring branches.

This species is cultivated occasionally as an ornamental. Most botanists accept two infraspecific taxa within *P. maculata*, although there is considerable morphological and distributional overlap between them. The other subspecies, ssp. *maculata*, is more commonly encountered in the northern half of the overall range of the species (which extends to southern Canada). It differs in its less congested growth form with fewer nodes, slightly broader leaves with the upper leaves more strongly cordate, and somewhat earlier blooming time. Hybridization and intergradation between *P. maculata* (particularly ssp. *maculata*) and *P. glaberrima* have been noted in some areas, but such hybrids are not known to occur in the state. Some characters that are useful to distinguish *P. maculata* from *P. glaberrima* within Missouri (for example, reddish purple markings on the stems and inflorescence shape) are less consistent elsewhere.

7. Phlox paniculata L. (perennial phlox, summer phlox, garden phlox)
Pl. 493 d–f; Map 2258

Plants perennial herbs with short, thick rhizomes. Vegetative stems not produced or, if present, then similar to the flowering ones. Stems typically 3–7 (more in cultivated forms), 70–120 cm tall, erect, with 20–35 nodes, glabrous toward the base, grading to sparsely to densely pubescent above the midpoint with short, curved, usually nonglandular hairs, sometimes with red streaks. Leaves all opposite or sometimes the uppermost subopposite, the blade narrowly elliptic to oblong-lanceolate, angled or more commonly tapered to a sharply pointed tip, variously angled or tapered at the base, the uppermost sometimes ovate, those of the largest leaves 9–14 cm long and 20–40 mm wide, the upper surface usually glabrous, the undersurface glabrous or less commonly sparsely short-hairy, the margins short-hairy, the secondary veins forming closed loops. Inflorescences with mostly 35 to more than 200 flowers, consisting of clusters or small panicles, the aggregate of clusters usually appearing as broad, domed to hemispheric or pyramidal panicles, occasionally narrower and appearing somewhat racemelike. Flower stalks 2–8 mm. Calyces 6–10 mm long, the lobes slender, tapered evenly to sharply pointed tips, glabrous or pubescent with short, nonglandular hairs. Corollas bright pink to light pinkish purple, purple, or light purplish blue, rarely white, the tube 18–25 mm long, sparsely to densely hairy externally, lacking a basal constriction, the lobes 6–12 mm long and 4–11 mm wide, obovate to broadly obovate, rounded at the tips, occasionally with a very shallow apical notch. Stamens with the filaments 15–21 mm long, the anthers positioned from below to at or above the stigma near the mouth of the tube (occasionally 1–3 slightly exserted). Style 12–24 mm long, the stigmas 0.8–1.3 mm long. $2n=14$. June–September.

Scattered south of the Missouri River and in the northeastern quarter of the state (eastern U.S. west to Missouri and Louisiana; introduced farther north and west, also Canada). Bottomland forests, mesic upland forests, banks of streams and rivers, bases of bluffs, and fens; also roadsides and shaded, disturbed areas.

Phlox paniculata is the most commonly cultivated species in midwestern gardens. Some of the Missouri populations undoubtedly arose as escapes from gardens or compost. Many cultivars have been developed. Steyermark (1963) noted that although the wild species inhabits mostly shaded sites, the horticultural variants frequently tolerate full sunlight. Rare plants with white corollas have been called f. *alba*, but this epithet does not appear ever to have been officially published as a forma.

8. Phlox pilosa L. (downy phlox, prairie phlox)
Pl. 494 c, d; Map 2259

Plants perennial herbs with slender rhizomes. Vegetative stems not produced or, if 1 or a few present, then shorter than but otherwise similar to the flowering ones. Flowering stems typically 1–3, 25–60 cm tall, erect or ascending, with 6–17 nodes, moderately to densely pubescent with multicellular, spreading to curved or crinkled hairs, some or all of these sometimes gland-tipped. Leaves all opposite or the uppermost rarely subopposite, the lowermost (sometimes all of the leaves on vegetative stems) with the blades linear to narrowly elliptic, grading to narrowly lanceolate, lanceolate, or ovate toward the tip, angled or tapered to a sharply pointed tip, tapered to angled, rounded, or shallowly cordate (those toward the stem tip) at the base, those of the largest leaves 7.5–12.5 cm long and 5–20 mm wide, the surfaces glabrous or nearly so (the upper surface shiny), the margins of the lower leaves sometimes slightly rolled under, glabrous, the secondary veins obscure, pinnate and not forming conspicuous loops, the blade narrowly to broadly elliptic, angled or tapered to a sharply pointed tip, angled or tapered at the base (the uppermost sometimes rounded to nearly truncate), the largest 3.5–8.0 cm long and 2–20 mm wide, the margins usually short-hairy (except sometimes on the lowermost leaves), the surfaces sparsely (occasionally on the lowermost

2260. Phlox subulata

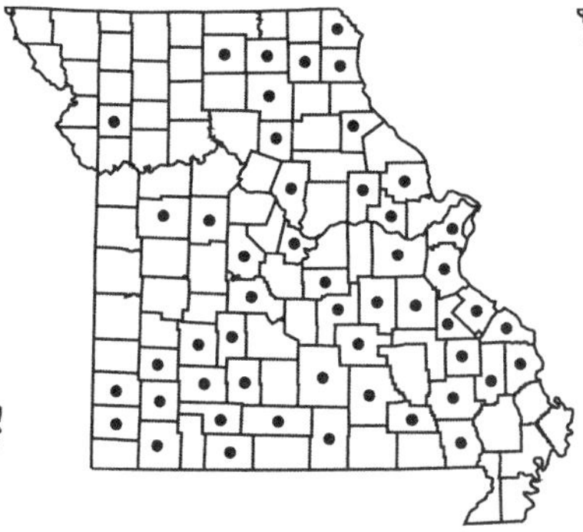

2261. Polemonium reptans

2262. Polygala incarnata

leaves) or moderately to densely hairy, the hairs nonglandular or some or all of them gland-tipped, the secondary veins obscure, pinnate and not forming conspicuous loops. Inflorescences with 15–60 flowers, consisting of sometimes spreading or somewhat drooping clusters, the aggregate of clusters sometimes appearing as small, domed panicles. Flower stalks 1–9 mm. Calyces 7–14 mm long, the lobes slender, tapered to sharply pointed tips, glandular- or nonglandular-hairy. Corollas light pink to bright pink, lavender, or light purple, rarely white, the tube 10–16 mm long, usually moderately to densely hairy externally, with a slight constriction 1.5–3.0 mm above the base, the lobes 6–14 mm long and 4–11 mm wide, obovate or less commonly oblanceolate, rounded at the tips, sometimes with an abrupt, short point at the very tip. Stamens with the filaments 4–15 mm long, the anthers positioned above the stigma within the tube (not exserted). Style 0.4–1.0 mm long, the stigmas 0.9–1.6 mm long. $2n$=14, 28. April–June.

Scattered to common nearly throughout the state (eastern U.S. west to North Dakota and Texas; Canada, Mexico). Glades, savannas, upland prairies, bottomland forests, mesic to dry upland forests, bases, ledges, and tops of bluffs, margins of sinkhole ponds, banks of streams and rivers, and fens; also pastures, old fields, ditches, railroads, and roadsides.

Phlox pilosa is cultivated occasionally as an ornamental (mainly ssp. *pilosa*). The species exhibits great morphological variation across its range and several infraspecific entities have been recognized, with three of these (ssp. *fulgida*, ssp. *ozarkana* and ssp. *pilosa*) reported from Missouri. Taxa not occurring in Missouri are narrow in range (two occurring in dry limestone soils of central Texas [one known from adjacent Coahuila], one in quartzitic hills in southwestern Oklahoma, one in a floodplain area in central Illinois, and one in open woods of northwestern Tennessee, western Kentucky and southern Indiana; see C. J. Ferguson [1998]). Variation in ploidy level occurs within *P. pilosa* (the central Texas taxa are tetraploid; and the existence of both diploid and tetraploid populations of *P. pilosa* ssp. *pilosa*, as currently circumscribed, is evidenced by some reports of chromosome counts [Arkansas and Texas; D. M. Smith and Levin, 1967; Levy and Levin, 1974] and by flow cytometry data [Oklahoma and Texas; C.J. Ferguson and L. Worcester, unpublished]), but the species appears to be diploid over most of its range, and polyploidy has not been documented for Missouri populations.

Rare plants of *P. pilosa* with white corollas have been called f. *albiflora* MacMill. Within Missouri, *P. pilosa* ssp. *fulgida* is well differentiated ecologically and morphologically, but ssp. *ozarkana* and ssp. *pilosa* intergrade extensively in the Ozark region. Both of these latter taxa occasionally co-occur with *P. divaricata* and putative hybrids can occur. Furthermore, some of the variation within *P. pilosa* ssp. *ozarkana* may be a result of past hybridization with *P. divaricata*. Population genetic studies are needed to help address this possibility.

1. Inflorescence and calyx hairs nonglandular (but appearing lustrous with magnification) .... 8A. SSP. FULGIDA
1. Inflorescence and calyx hairs gland-tipped
    2. Leaves narrowly elliptic toward the stem base, grading to lanceolate or ovate toward the stem tip, the maximum width 8–20 mm, the bases of the upper leaves often shallowly cordate; stems pubescent with gland-tipped hairs throughout (except occasionally near the stem base) .... 8B. SSP. OZARKANA

2. Leaves linear or narrowly elliptic toward the stem base, grading to narrowly lanceolate or lanceolate toward the stem tip, the maximum width 2–14 mm, the bases of the upper leaves angled or rounded, but not cordate; stems usually pubescent with nonglandular hairs below the midpoint, grading to gland-tipped hairs only toward the stem tip . . . . 8C. SSP. PILOSA

8a. ssp. fulgida (Wherry) Wherry

P. pilosa var. *fulgida* Wherry

Plants typically lacking vegetative stems. Stems with 10–17 nodes, sometimes with short axillary branches at many of the nodes, sparsely to densely pubescent with nonglandular hairs, these fine, white, and typically somewhat lustrous (when viewed with magnification). Leaves opposite or rarely subopposite toward the stem tip, linear toward the stem base, grading to narrowly lanceolate or lanceolate toward the stem tip, the bases of the uppermost leaves angled to rounded, not cordate, the lowermost hairy to nearly glabrous, the uppermost moderately to densely pubescent with nonglandular leaves, the largest leaves 4–8 cm long and 2–9 mm wide. Leaves subtending flower clusters linear lanceolate to lanceolate, the bases not cordate. Inflorescences pubescent with nonglandular hairs, these fine, white, and typically somewhat lustrous (when viewed with magnification). Calyces 7–12 mm long, nonglandular-hairy. Corollas with the tube nonglandular-hairy, the lobes 6–10 mm long and 4–8 mm wide. $2n=14$. May–June.

Scattered to common in the Unglaciated Plains Division, scattered to uncommon elsewhere in the state, absent from many counties in the Ozark and Mississippi Lowlands Divisions (North Dakota to Oklahoma east to Indiana; Canada). Glades, savannas, upland prairies, mesic to dry upland forests, and bases and ledges of bluffs; also pastures, railroads, and roadsides.

8b. ssp. ozarkana (Wherry) Wherry

P. pilosa var. *ozarkana* Wherry

Plants sometimes with 1 or a few vegetative stems. Stems with 6–12 nodes, sometimes with short axillary branches at many of the nodes, sparsely to densely pubescent with fine, gland-tipped hairs throughout (except occasionally near the base). Leaves opposite, narrowly elliptic toward the stem base, grading to lanceolate or ovate toward the stem tip, the bases of the upper leaves often shallowly cordate, the lowermost hairy to nearly glabrous, the uppermost moderately to densely pubescent with sometimes gland-tipped hairs, the largest leaves 3.5–7.5 cm long and 8–20 mm wide. Leaves subtending flower clusters linear-lanceolate to lanceolate or occasionally ovate, the bases sometimes cordate. Inflorescences pubescent with fine, gland-tipped hairs. Calyces 8–14 mm long, glandular-hairy. Corollas with the tube pubescent with at least some of the hairs gland-tipped, rarely glabrous or nearly so, the lobes 8–13 mm long and 4–11 mm wide. April–May.

Scattered to common in the southern half of the state (Oklahoma east to Missouri, Arkansas, and Louisiana). Glades, savannas, upland prairies, bottomland forests, mesic to dry upland forests, bases, ledges, and tops of bluffs, margins of sinkhole ponds, banks of streams and rivers, and fens; also pastures, old fields, ditches, railroads, and roadsides.

8c. ssp. pilosa

P. pilosa var. *amplexicaulis* (Raf.) Wherry

P. pilosa var. *virens* (Michx.) Wherry

P. argillacea Clute & Ferris

Plants sometimes with 1 or a few vegetative stems. Stems with 9–14 nodes, usually lacking axillary branches, sparsely to densely pubescent with nonglandular hairs below the midpoint, grading to gland-tipped hairs only toward the tip. Leaves opposite, linear to less commonly narrowly elliptic toward the stem base, grading to narrowly lanceolate or lanceolate toward the stem tip, the bases of the upper leaves not cordate, the lowermost hairy to nearly glabrous, the uppermost moderately to densely pubescent with sometimes gland-tipped hairs, the largest leaves 3.5–7.5 cm long and 2–14 mm wide. Leaves subtending flower clusters linear-lanceolate to lanceolate, the bases not cordate. Inflorescences pubescent with fine, gland-tipped hairs (rarely nonglandular-hairy). Calyces 7–12 mm long, glandular-hairy. Corollas with the tube pubescent with at least some of the hairs gland-tipped, rarely glabrous or nearly so, the lobes 7–14 mm long and 4–10 mm wide. $2n=14$. April–May.

Scattered to common in the southern half of the state, disjunct in Nodaway and Scotland Counties (eastern U.S. west to Iowa and Texas; Canada). Glades, savannas, upland prairies, bottomland forests, mesic to dry upland forests, bases, ledges, and tops of bluffs, margins of sinkhole ponds, banks of streams and rivers, and fens; also pastures, old fields, ditches, railroads, and roadsides.

Steyermark (1963) recognized rare plants of *P. pilosa* ssp. *pilosa* in Missouri with relatively coarse, nonglandular hairs as *P. pilosa* var. *amplexicaulis*, and listed *P. pulcherrima* (Lundell) Lundell in synonymy. The present author considers var. *amplexicaulis* a synonym of *P. pilosa* ssp. *pilosa*, but recognizes *P. pulcherrima* as a distinct species not occurring in Missouri (a narrowly endemic tetraploid taxon in eastern Texas).

9. Phlox subulata L. (moss phlox, moss pink, rock pink)

Map 2260

Plants perennial herbs with woody, often branched rootstocks, forming dense mats, with short primary rhizomes. Vegetative stems spreading to slightly ascending, usually producing many short, axillary shoots, these with tufts of leaves morphologically similar to those on flowering stems. Flowering stem branches numerous, produced from the vegetative stems, 4–10 cm tall, with 3–6 nodes, erect or ascending, usually densely pubescent with curved or crinkled, multicellular hairs, these sometimes all or mostly gland-tipped. Leaves all opposite, the blade typically relatively uniform on a given plant, those of the largest leaves 0.8–2.0 cm long and 1–2 mm wide, linear to narrowly elliptic, the surfaces mostly glabrous, the margins moderately nonglandular-hairy. Inflorescences relatively few-flowered, short and appearing as clusters, with 3–6 flowers or occasionally reduced to solitary flowers. Flower stalks 5–20 mm long. Calyces 5–9 mm long, the lobes narrowly elliptic-oblong, each tapered to a short, sharp extension of the midnerve, densely hairy, the hairs sometimes glandular. Corollas pink to pale bluish purple, lavender, or white, the tube 8–16 mm long, glabrous externally, lacking a basal constriction, the lobes 6–12 mm long and 3–8 mm wide, obovate, with a notch 1.0–1.5(–3.0) mm deep, rarely rounded and lacking a notch. Stamens with the filaments 6–10 mm long, the anthers positioned from below to at or above the stigma at the mouth of the tube (sometimes 1 or 2 slightly exserted). Style 7–11 mm, the stigmas 0.5–1.0 mm long. $2n=14$, 28. March–May.

Introduced, uncommon and sporadic, known thus far only from Jefferson and Pike Counties (northeastern U.S. west to Michigan and Tennessee; Canada). Cemeteries and roadsides.

Although *P. subulata* is not considered to be fully naturalized in the state, it is commonly cultivated and may persist from old plantings. It may be difficult to determine whether plants at a site are reproducing and spreading or merely persisting from plantings. Steyermark (1963) excluded it from the state's flora, based on two historical collections from Clay and Platte Counties that he was certain represented cultivated plants.

Cultivars resulting from hybridization between *P. subulata* and *P. bifida* are known (Locklear, 2011).

4. Polemonium L. (Jacob's ladder, Greek valerian)

About 28 species, North America, Europe, Asia.

1. Polemonium reptans L. (Jacob's ladder, Greek valerian)

Pl. 494 g, h; Map 2261

Plants perennial herbs, the rootstock woody. Stems 15–50 cm long, solitary or 2–5, erect or loosely ascending, often few- to several-branched above the midpoint, glabrous or sparsely pubescent with fine, spreading, nonglandular, multicellular hairs. Leaves basal and alternate, the basal leaves mostly long-petiolate, the stem leaves short-petiolate or the uppermost ones occasionally sessile. Leaf blades 3–15 cm long, narrowly oblong in outline or those of the smallest leaves broadly ovate-triangular in outline, pinnately compound with 7–19 leaflets (occasionally reduced to 3 leaflets in the uppermost leaves), the leaflets 7–50 mm long, oval to ovate or lanceolate, angled to broad or narrow, usually sharply pointed tips, the midvein extended into a short, thickened projection, the margins otherwise entire, the surfaces glabrous or sparsely and minutely hairy along the main veins, the venation pinnate or appearing palmate, with the relatively few secondary veins looping to rejoin the midvein at the tip. Inflorescences terminal and often also from the uppermost few leaves, stalked clusters or small, open panicles with mostly 5–17 flowers, these variously ascending to nodding. Calyces 5-lobed to above or near the midpoint, 4–6 mm long at flowering, becoming enlarged to 8–12 mm at fruiting), bell-shaped (becoming papery and somewhat inflated at fruiting), the tube uniform in texture, the lobes triangular-ovate (sometimes narrowly so at fruiting), angled to sharply pointed tips, glabrous or sparsely and minutely hairy. Corollas 5-lobed, blue to lavender, bell-shaped, the tube 5–7 mm long, the lobes

5–8 mm long. Stamens with the filaments attached equally in the tube, but unequal in lengths, the anthers thus unequally exserted (2 stamens slightly shorter than the other 3). Style exserted. Seeds 15–21, 2–3 mm long, asymmetrically narrowly ellipsoid to oblong-ellipsoid, with a fine longitudinal line or groove along 1 side and narrow wings at one or both ends, the surface dark brown, with short wrinkles, not becoming sticky when moistened. $2n=18$. April–June.

Scattered in most counties south of the Missouri River, less common farther north (eastern U.S. west to Minnesota and Oklahoma; Canada). Bottomland forests, mesic upland forests, bases of bluffs, and banks of streams, rivers, and spring branches.

Steyermark (1963) noted that the roots of this species have been used medicinally for kidney ailments and as a diuretic. This graceful perennial makes a nice addition to a woodland garden and is becoming available more frequently at wildflower nurseries.

Some authors have recognized two varieties in this species. Plants of var. *reptans* are widespread in the eastern United States and adjacent Canada. In portions of Kentucky and Ohio, plants with densely glandular-hairy inflorescences and often also stems, as well as slightly smaller corollas, have been called var. *villosum* E.L. Braun. The taxonomic status of these plants requires further investigation and formal infraspecific taxa are not recognized in the present treatment.

POLYGALACEAE (Milkwort Family)

Nineteen to 23 genera, about 925 species, nearly worldwide.

A few genera of Polygalaceae contain species that are cultivated as ornamentals or are used medicinally. The genus *Xanthophyllum* Roxb. (sometimes segregated into its own family), which occurs from southeastern Asia to Australia, contains some species used for dyes and seed oils.

1. **Polygala** L. (milkwort)

Plants annual or perennial herbs (shrubs elsewhere). Stems unbranched or branched, in our species glabrous or inconspicuously hairy. Leaves alternate or less commonly opposite or whorled (sometimes appearing leafless or nearly so at flowering in *P. incarnata*), sessile or very short-petiolate, the lowermost leaves sometimes reduced and scalelike. Stipules absent. Leaf blades simple, unlobed, variously shaped, the margins entire or minutely toothed, glabrous (variously hairy elsewhere). Inflorescences terminal, often spikelike racemes, the flowers usually short-stalked, subtended by small, inconspicuous, membranous to scalelike bracts. Flowers hypogynous, perfect. Calyces zygomorphic, of 5 sepals, these free, the upper sepal and the lower pair small and green or whitish green, the lateral pair (referred to as wings) larger and petaloid. Corollas zygomorphic, of 3 petals, these usually fused into a U-shaped tube open along 1 side, the upper (appearing lateral) 2 lobes similar, the lower (appearing central) lobe (referred to as a keel) boat-shaped, often fringed or crested on the outer surface (or mainly at the tip), greenish white to white, pink, or pinkish purple (other colors elsewhere). Stamens 8 (6 elsewhere), the filaments fused into a tube just inside the keel that is split longitudinally along the upper side, also fused to the corolla tube, the anthers attached at their bases, usually yellow, the pollen usually shed by apical pores or short slits. Pistil 1 per flower, of 2 fused carpels. Ovary superior, 2-locular, often flattened, the placentation axile (appearing more or less apical). Style 1, often curved toward the tip, usually unequally 2-lobed near the tip, the upper lobe bearing a more or less capitate stigma, the other lobe sterile, with a fringe of hairs. Ovule 1 per locule. Fruits capsules, variously shaped, dehiscing longitudinally. Seeds oblong-obovoid to obovoid, the surface dark brown to black, moderately to densely pubescent with loosely appressed to more or less spreading, fine, straight hairs, also with a white or yellowish aril at the attachment point. About 325 species, nearly worldwide.

The estimate of number of species in *Polygala* cited above is based on the traditional broad circumscription of the genus. However, molecular data have suggested that the genus is actually an artificial assemblage of several groups more closely related to other genera than to *Polygala* in the strict sense (Persson, 2001). Abbott (2011) has recently advocated for the recognition of four segregate genera in the New World, based on preliminary molecular data and morphological analysis. In Abbott's restricted sense, *Polygala* worldwide contains only about 200 species. However, all of the Missouri species remain within *Polygala*, even in the strict sense.

Some species of *Polygala* are cultivated as ornamentals. The Missouri species tend not to produce cleistogamous flowers, but elsewhere a number of species produce cleistogamous flowers, including some that produce them on subterranean branches.

1. Leaves in whorls of 3–7, at least at the lower stem nodes . . . . 4. P. VERTICILLATA
1. Leaves all alternate (plants sometimes appearing leafless or nearly so in *P. incarnata*)
    2. Plants perennial, several-stemmed from a thick, somewhat fleshy or woody rootstock; leaves (except for the highly reduced, scalelike lowermost ones) mostly 4–35 mm wide, narrowly lanceolate to narrowly ovate, the margins minutely toothed (visible with magnification) . . . . 3. P. SENEGA
    2. Plants annual, single-stemmed from a slender taproot; leaves 0.5–5.0 mm wide, linear or occasionally narrowly elliptic, the margins entire
        3. Stems appearing leafless or few-leaved at flowering, the leaves mostly shed early, glaucous, appearing bluish green to grayish green; flowers 7–10 mm long, the corolla longer than the calyx . . . . 1. P. INCARNATA
        3. Stems appearing relatively densely leafy at flowering, the leaves not shed early, not glaucous, green to olive green; flowers 4.5–6.0 mm long, the calyx wings longer than the corolla . . . . 2. P. SANGUINEA

1. Polygala incarnata L. (pink milkwort, slender milkwort)

Pl. 495 a, b; Map 2262

Plants annuals, single-stemmed, with slender taproots. Stems 10–60 cm long, erect or ascending, unbranched or few-branched toward the tip, glabrous, glaucous, appearing bluish green or grayish green. Leaves widely spaced, alternate, 4–12 mm long, narrowly linear, 0.5–1.0 mm wide, the margins entire, shed early, the plants appearing leafless or few-leaved at flowering. Inflorescences dense spikelike racemes, 1–4 cm long. Wings 3–4 mm long, narrowly oblong-oblanceolate, pale pinkish purple or rarely white with a pale green central band. Corollas 7–10 mm long, pale pinkish purple or rarely white, the slender fused portion 4–5 mm long. Fruits 2.5–3.0 mm long, ovoid-orbicular, not flattened. broadly rounded to nearly truncate and very shallowly notched at the tip. Seeds 1.8–2.2 mm long, the aril 0.6–1.0 mm long, membranous, more or less unlobed. May–November.

Uncommon, widely scattered in the Unglaciated Plains and Ozark Divisions, as well as a few counties in the southwestern portion of the Glaciated Plains (eastern U.S. west to Iowa and Texas; Canada). Glades and upland prairies; also pastures, old fields, and roadsides.

2. Polygala sanguinea L. (field milkwort, blood polygala)

Pl. 495 c–e; Map 2263

Plants annuals, single-stemmed, with slender taproots. Stems 10–40 cm long, erect or ascending, unbranched or few- to several-branched toward the tip, glabrous, not glaucous, green to olive green. Leaves relatively closely spaced, alternate, 5–40 mm long, linear to narrowly elliptic, 1–5 mm wide, the margins entire, not shed early, the plants appearing leafy at flowering and fruiting. Inflorescences dense headlike or spikelike racemes, 1.0–2.5 cm long. Wings 4.5–6.0 mm long, ovate to nearly oval, white, usually slightly to

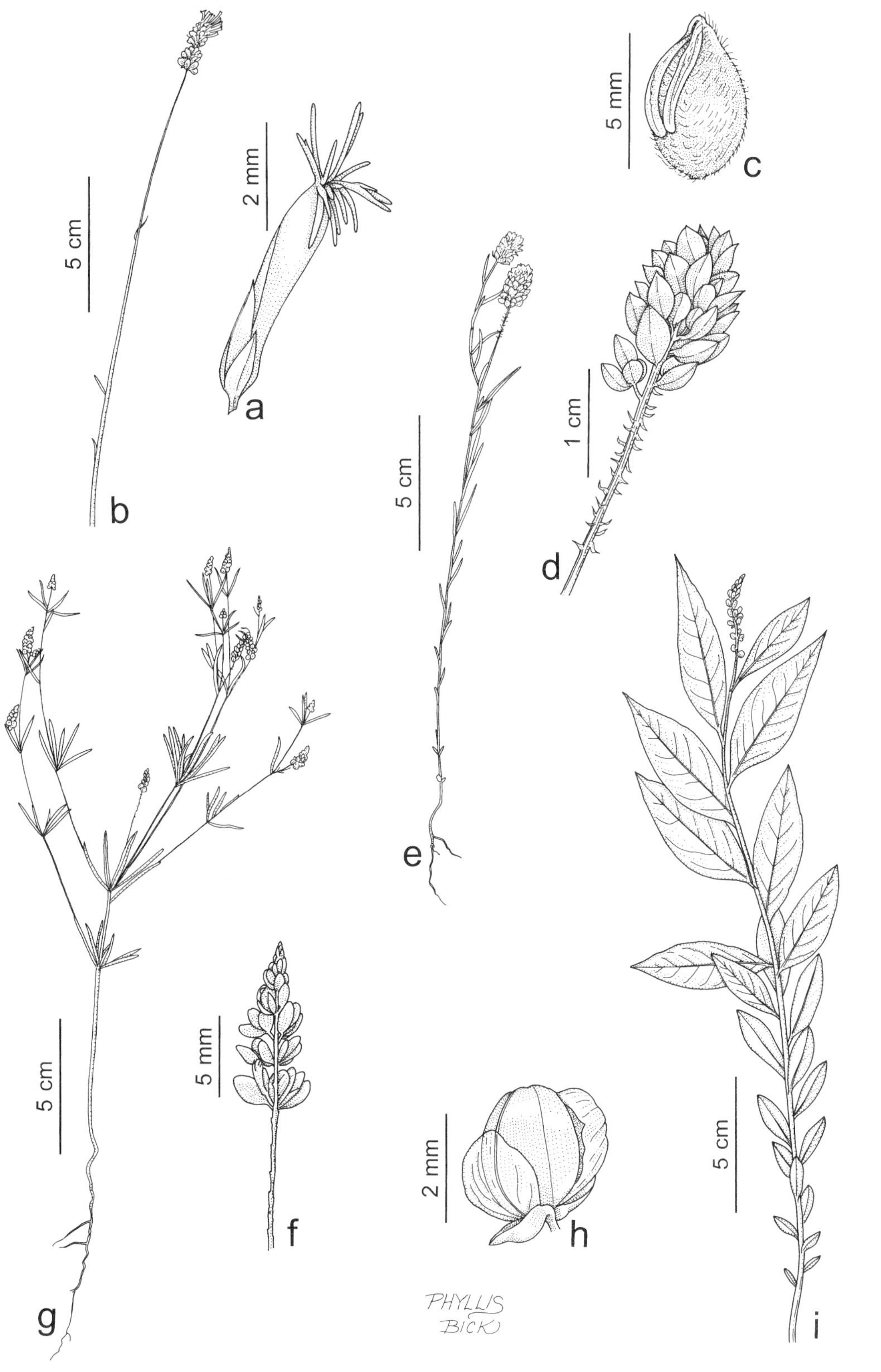

Plate 495. Polygalaceae. *Polygala incarnata*, **a)** flower, **b)** fertile stem. *Polygala sanguinea*, **c)** fruit, **d)** inflorescence, **e)** habit. *Polygala verticillata*, **f)** inflorescence, **g)** habit. *Polygala senega*, **h)** fruit, **i)** fertile stem.

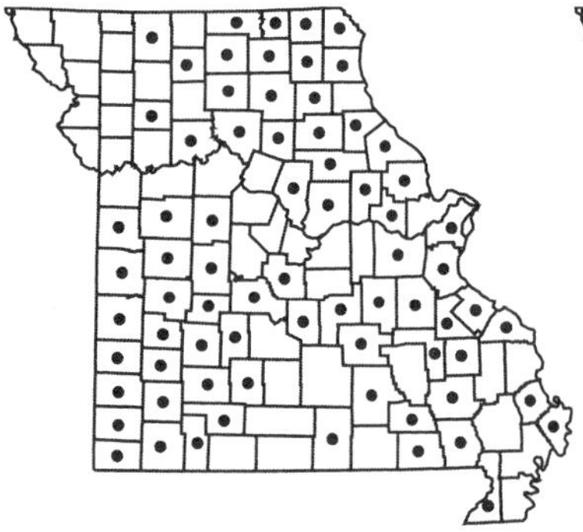

2263. Polygala sanguinea

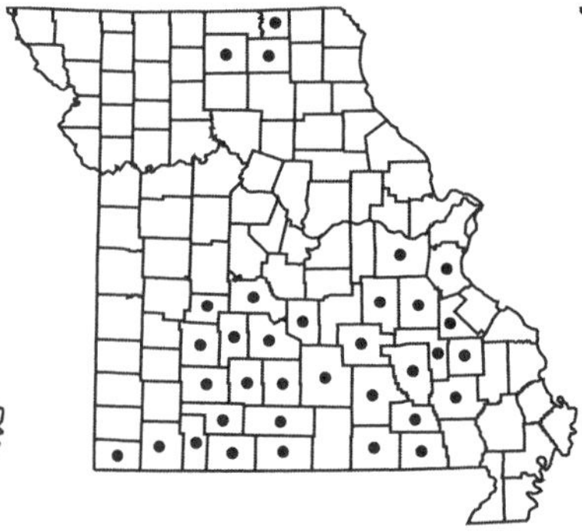

2264. Polygala senega

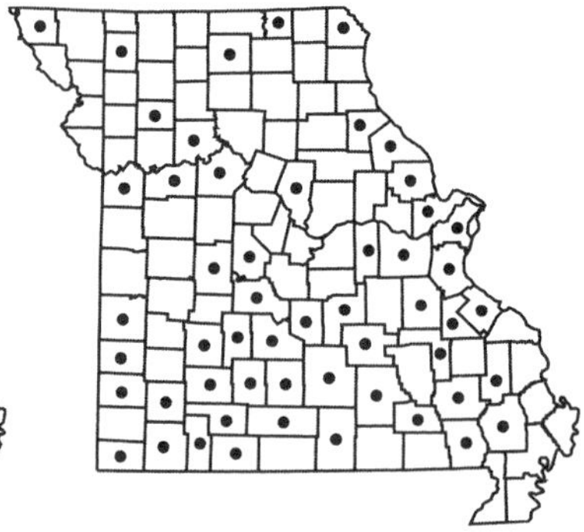

2265. Polygala verticillata

strongly pinkish-tinged toward the tip, rarely greenish white, often with an inconspicuous slender pale green central band. Corollas 2.5–2.7 mm long, white, usually slightly to strongly pinkish-tinged toward the tip, rarely greenish white, the fused portion 1.5–2.2 mm long. Fruits 2.0–2.5 mm long, obovoid-orbicular, not or only slightly flattened. shallowly notched at the tip. Seeds 1.5–1.7 mm long, the aril 1.0–1.2 mm long, papery, with 2 linear lobes. May–October.

Scattered nearly throughout the state, but apparently absent from the northwesternmost counties (eastern U.S. west to South Dakota and New Mexico; Canada). Glades, upland prairies, openings of mesic to dry upland forest, and occasionally banks of rivers, sloughs, and fens; also pastures, old fields, fallow fields, ditches, railroads, and roadsides.

Two trivial variants of *P. sanguinea* have been reported uncommonly in Missouri (Steyermark, 1963). White-flowered plants have been called f. *albiflora* Millsp. and those with greenish corollas have been called f. *viridescens* (L.) Farw.

The superficially similar *P. cruciata* L. (drum heads) occurs in the eastern United States and adjacent Canada west to Minnesota and Texas. It has been found in every state that adjoins Missouri except for Kansas and Nebraska and thus might be located in Missouri in the future. It may be distinguished from *P. sanguinea* most easily by its whorled leaves and sharply pointed (vs. broadly rounded, sometimes with a minute sharp point) wings that are only 3–4 mm long.

3. Polygala senega L. (Seneca snakeroot)
P. senega var. *latifolia* Torr. & A. Gray
Pl. 495 h, i; Map 2264

Plants perennial herbs several-stemmed from a thick, somewhat fleshy or woody rootstock. Stems 10–50 cm long, ascending, unbranched, moderately to densely pubescent with inconspicuous, minute, curved, sometimes slightly club-shaped hairs, not glaucous, yellowish green to whitish green. Leaves moderately spaced, alternate, the lowermost reduced and scalelike, grading into the main leaves, these 15–70 mm long, narrowly lanceolate to narrowly elliptic or narrowly ovate, 4–35 mm wide, the margins with closely spaced, minute teeth, not shed early, the plants appearing leafy at flowering and fruiting. Inflorescences dense spikelike racemes, 1.5–5.0 cm long. Wings 3.0–3.7 mm long, broadly elliptic to nearly circular, white to greenish white. Corollas 2.2–3.0 mm long, white to greenish white or pale cream-colored, the fused portion 1.5–2.2 mm long. Fruits 2.5–4.2 mm long, obovoid-orbicular, somewhat flattened. shallowly notched at the tip. Seeds 2.5–3.5 mm long, the aril 2.5–3.8 mm long, membranous, with 2 linear lobes (these usually positioned in parallel fashion and thus sometimes difficult to distinguish). May–July.

Scattered in the Ozark and Ozark Border Divisions, also in a few counties in the north-central portion of the state (eastern U.S. west to North Dakota, Wyoming, and Oklahoma; Canada). Mesic to dry upland forests, bottomland forests, tops, ledges, and bases of bluffs, banks of streams, rivers, and spring branches, upland prairies, and glades; also railroads.

The infraspecific taxonomy of this species remains controversial. Steyermark (1963) and some earlier authors segregated var. *latifolia*, based on its broader leaves and wing sepals, slightly larger fruits, absence from northern Missouri, and occurrence mainly in woodlands. However, in his work on the Indiana flora Deam (1940) found that he could not reliably distinguish plants in that state and noted that plants growing in more open areas tended to have broader leaves. Trauth-Nare and Naczi (1997, 1998) performed a morphometric analysis on the complex and were able to consistently distinguish two taxa based on a small set of quantitative and qualitative characters, as well as

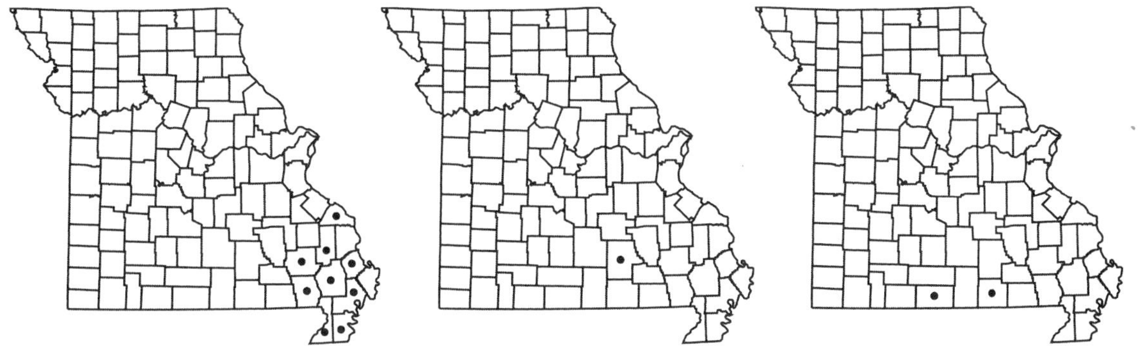

2266. Brunnichia ovata   2267. Eriogonum annuum   2268. Eriogonum longifolium

slight differences in blooming dates; they went so far as to suggest that there might be two morphologically cryptic species involved. In his study of Ohio *Polygala*, J. F. Burns (1986) reviewed the literature to that time and concluded that it was most prudent to treat *P. senega* as a single, morphologically variable species until more conclusive evidence was presented to support the segregation of a second taxon. That conclusion appears still to be valid.

Native Americans used the roots of this species medicinally for a wide variety of treatments (Moerman, 1998), including as a respiratory aid, abortifactant, antirheumatic, diuretic, anticonvulsive, blood and kidney aid, and cold remedy, among others.

4. Polygala verticillata L. (whorled milkwort)
Pl. 495 f, g; Map 2265

Plants annuals, single-stemmed, with slender taproots. Stems 5–25(–40) cm long, erect or ascending, unbranched or few- to several-branched toward the tip, glabrous, not glaucous, green. Leaves at relatively widely spaced nodes, those of at least the lower 1–3 (often of nearly all) in whorls of 3–7, those of at least the uppermost nodes grading through opposite to alternate, 8–30 mm long, linear to narrowly oblong, narrowly elliptic, or narrowly oblanceolate, 1–3 mm wide, the margins entire, not shed early, the plants appearing leafy at flowering and fruiting. Inflorescences dense spikelike racemes or sometimes with relatively widely spaced flowers, at least toward the axis base, 1.0–2.5(–4.0) cm long. Wings 1.5–2.2 mm long, obovate to nearly oval, white, occasionally purplish-tinged, sometimes with a slender to broad green central band. Corollas 1.2–1.5 mm long, white, the fused portion 0.4–0.6 mm long. Fruits 1.8–2.5 mm long, oblong-ovoid, somewhat flattened. shallowly notched at the tip. Seeds 1.5–2.2 mm long, the aril 0.5–1.0 mm long, papery, with 2 linear lobes. $2n=34$. May–October.

Scattered nearly throughout the state, but apparently absent from most of the Mississippi Lowlands Division (eastern U.S. west to Montana, Utah, and Texas; Canada; introduced in Asia). Glades, upland prairies, loess hill prairies, sand prairies, savannas, openings of mesic to dry upland forests, and banks of streams and spring branches; also pastures, old fields, and fallow fields.

Steyermark (1963) followed some earlier authors in accepting five varieties of this species in Missouri, but noted that there was extensive morphological intergradation between some of these taxa. J. F. Burns (1986) presented a detailed account of the convoluted nomenclatural and taxonomic history of the *P. verticillata* complex, and his sensible treatment is here followed for Missouri plants until future research can shed more light on variation within the complex.

1. Leaves mostly alternate, only the lowermost nodes with whorls of leaves; inflorescences with relatively widely spaced flowers, at least toward the axis base; wing sepals as long as or longer than the fruits .... 4A. VAR. AMBIGUA
1. Leaves all or mostly in whorls of 3–7 per node; inflorescences densely flowered throughout; wing sepals shorter than the fruits .... 4B. VAR. VERTICILLATA

4a. var. ambigua (Nutt.) Alph. Wood
P. verticillata var. *dolichoptera* Fernald
P. ambigua Nutt.

Stems unbranched or few-branched toward the tip. Leaves of the lower 1–3 nodes in whorls of 3–7, those of most nodes alternate. Inflorescences with relatively widely spaced flowers, at least toward the axis base. Wing sepals as long as or longer than the fruits. May–October.

Scattered in the Ozark and Ozark Border Divisions north locally to Pike County (eastern U.S.

west to Wisconsin and Texas; Canada). Glades, upland prairies, openings of dry upland forests, and occasionally banks of streams and spring branches.

4b. var. verticillata

P. verticillata var. *isocycla* Fernald

P. verticillata var. *sphenostachya* Pennell

Pl. 495 g

Stems mostly few- to several-branched toward the tip. Leaves of all or nearly all of the nodes in whorls of 3–7, those of the uppermost nodes sometimes grading through opposite to alternate. Inflorescences densely flowered throughout. Wing sepals shorter than the fruits. $2n=34$. May–October.

Scattered nearly throughout the state, but apparently absent from most of the Mississippi Lowlands Division (eastern U.S. west to Montana, Utah, and Texas; Canada; introduced in Asia). Glades, upland prairies, loess hill prairies, sand prairies, savannas, openings of mesic to dry upland forests, and banks of streams and spring branches; also pastures, old fields, and fallow fields.

POLYGONACEAE (Smartweed Family)

Contributed by George Yatskievych and Alan E. Brant

Plants annual or perennial herbs or shrubs (trees elsewhere), sometimes herbaceous vines or lianas, occasionally dioecious, sometimes tinged with pink to purple, glabrous or variously pubescent. Stems often with swollen nodes, sometimes with longitudinal angles, stripes, or ridges. Leaves alternate (occasionally opposite or whorled elsewhere), sessile or petiolate, the petiole sometimes jointed at the base; the blade simple, often unlobed, but sometimes with a pair of basal lobes, the margins variously entire or occasionally finely toothed or wavy to crisped. Stipules absent or more commonly present, then modified into a tubular, papery sheath called an ocrea. Inflorescences axillary and/or terminal, commonly consisting of fascicles or clusters subtended by a bract or an involucre (in *Eriogonum*), these sometimes arranged into racemes, umbels, or panicles, rarely reduced to solitary axillary flowers. Flowers individually subtended by a small, papery, sheathing bract called an ocreola, usually with a short, jointed stalk, occasionally also with a pair of small bractlets, perfect or less commonly imperfect, hypogynous or more commonly perigynous. Perianth more or less actinomorphic, of (2–)5 or 6 tepals, these sometimes differentiated slightly by size, in 1 or 2 whorl(s), all sepaloid or all petaloid, usually fused toward the base, the noticeable or inconspicuous cuplike portion commonly tapered into the stalk, usually persistent and sometimes becoming enlarged and leathery or papery at fruiting. Stamens (2–)5–9, absent or reduced to minute staminodes in pistillate flowers, the filaments free or sometimes fused toward the base, the fused portion occasionally with small, nectar-producing, toothlike glands between the filaments, the anthers attached basally or toward their midpoints, variously white, pink, red, purple, yellow, or green. Pistil 1 per flower (absent in staminate flowers), the ovary superior, consisting of (2)3 fused carpels, with 1 locule, the placentation basal. Styles (2)3, sometimes fused below the midpoint, the stigmas 1 per style, club-shaped or capitate, sometimes divided into 2 or more lobes or appearing fringed. Ovule 1 per flower. Fruits achenes or less commonly nutlike (enclosed in a fleshy hypanthium and drupelike elsewhere), usually at least partially enclosed by the persistent perianth, usually somewhat flattened or 3-angled, sometimes winged, often beaked, indehiscent. Seed 1, often somewhat flattened, variously shaped (the embryo straight or more commonly appearing curved, but usually not easily observed). About 43 genera, 1,100 species, worldwide.

The Polygonaceae generally are easily recognized variously by combinations of the following characters: the production of ocreae, small flowers with persistent perianth and grouped into fascicles or clusters, and angled, sometimes winged fruits. However, there is a lot of morphological variation within the family that has led to various classification schemes and generic circumscriptions. The characteristic ocrea of the Polygonaceae has been interpreted variously as an outgrowth of the sheathing base of the petiole, as the product of the fusion of either a row of adjacent stipules or two lateral stipules, or as an expanded axillary stipule (Graham and Wood, 1965). Traditionally, the genera have been grouped into two subfamilies, with *Eriogonum* and about 19 allied genera of subfamily Eriogonoideae Arn. lacking or producing very reduced ocreae and the remaining 23 or so genera of subfamily Polygonoideae producing mostly well-developed ocreae. However, recent molecular studies (Lamb Frye and Kron, 2003; Sanchez and Kron, 2008) have suggested that this classification represents an oversimplified system not reflecting true taxonomic relationships. In general, the present treatment follows that in the Flora of North America series (Freeman and Reveal, 2005, and included generic treatments), which contains excellent discussions on the taxonomy and classification of the temperate North American genera.

Numerous species provide food for a large array of mammals, birds, and other wildlife, and in particular the wetland species produce fruits that are a staple for waterfowl during their annual migration. Many species are popular with bees and their keepers as honey plants. In tropical regions, some of the woody genera are minor sources of lumber. Several genera of Polygonaceae contain species that are economically important for human food, including *Coccoloba* P. Browne (sea grape), *Fagopyrum* (buckwheat), and *Rheum* L. (rhubarb). A few genera contain medicinally important species, especially *Persicaria* (smartweed) and *Rumex* (sorrel). Other genera are cultivated as ornamentals, including some species of *Antigonon* Endl. (coral vine), *Coccoloba*, *Eriogonum* (wild buckwheat), *Persicaria*, *Polygonum*, and *Rumex*. A number of genera contain species that have negative economic and ecological impacts as crop weeds and/or exotic invaders of native plant communities, including *Fallopia* (climbing buckwheat, Mexican bamboo), *Polygonum* (knotweed), *Persicaria* (smartweed), and *Rumex* (dock). A variety of weedy species in the family have been documented to suppress germination and growth of other nearby plant species through production of allelopathic compounds by the roots and decaying aboveground portions of the plants (reviewed briefly by Hussain et al., 1997) Some species of *Rumex* produce abundant tannins in their rootstocks and were once important in the tanning industry.

It is surprisingly difficult to construct a key to the genera of Polygonaceae in Missouri that takes into account the exceptional character states present in various species in the larger genera, in spite of the fact that most of the species can be assigned readily to a genus in the field by the experienced botanist. Generally, fertile material is desirable for generic determinations and necessary to determine species within the larger genera. Fruiting specimens are more useful than material just beginning to flower, and the tepals frequently change shape, size, and orientation as the fruits develop.

1. Plants shrubs or lianas, the stems woody well above the base
    2. Plants lianas, climbing by tendrils; leaf blades lanceolate to ovate or somewhat heart-shaped, 10–80 mm wide; perianth green to yellowish green, leathery or hardened at fruiting, with an elongate, tapered, flattened, basal wing (this consisting of the winged stalk and the tapered perianth base) . . . . 1. BRUNNICHIA
    2. Plants shrubs, not climbing or producing tendrils; leaf blades linear to narrowly oblanceolate, 0.5–1.0 mm wide; perianth white to pink, papery at fruiting, not winged . . . . 6. POLYGONUM
1. Plants herbaceous, the stems climbing or not, but not producing tendrils; perianth at fruiting papery to slightly leathery, unwinged or the 3 outer tepals forming lateral wings

3. Perianth and leaf undersurface (also the young stems and basal portions of the mature stems) densely pubescent with white, matted, woolly or felty hairs; ocreae absent; inflorescences panicles with the flowers in small clusters subtended by cuplike involucres (additional inconspicuous bracts present at most inflorescence nodes) . . . . 2. ERIOGONUM
3. Perianth, leaves, and stems glabrous or variously pubescent, but not woolly- or felty-hairy; ocreae well-developed, sometimes becoming shredded with age; inflorescences various, but the flower not in clusters subtended by involucres (although fascicles of flowers may be subtended by an ocreola, this not cuplike)
    4. Stems with abundant, small, downward-curved (hooked) prickles . . . . 5. PERSICARIA
    4. Stems lacking prickles
        5. Perianth of 6 tepals in 2 series that are free, the inner (not outer) series of 3 tepals becoming enlarged and with a prominent network of nerves at fruiting (except in *R. acetosella*), the margins becoming winged or dissected into a series of teeth or slender lobes (except in *R. acetosella*); 1 or more of the enlarged inner tepals developing a prominent, differentiated, central swelling at fruiting (here called a tubercle), except in *R. acetosella* and *R. hastatulus* . . . . 7. RUMEX
        5. Perianth of (4)5 or 6 tepals in 1 or 2 series that are free or partially fused, these not becoming enlarged at fruiting or, if enlarged, then the outer (not inner) series of 3 tepals sometimes becoming somewhat enlarged and winged; none of the tepals developing differentiated, swollen areas centrally at fruiting
            6. Leaf blades tapered, angled, or rounded at the base, unlobed
                7. Inflorescences terminal and sometimes also axillary, consisting of elongate spikes or spikelike racemes, these sometimes grouped into panicles, usually lacking bracteal leaves (except sometimes reduced leaves present at the branch points of panicles); leaves not jointed at the base; ocreae unlobed or sometimes split along 1 side, occasionally with an asymmetrical green flange or appendage at the tip (frequently sometimes becoming shredded or disappearing with age); perianth white or pink to red, less commonly greenish white or green basally and white, cream-colored, or tan toward the tip . . . . 5. PERSICARIA
                7. Inflorescences axillary, the flowers solitary or in small clusters, the subtending leaves similar to or slightly smaller than the vegetative leaves (sometimes appearing as terminal spikelike racemes in *P. ramosissimum* and *P. tenue*); leaves jointed at the base; ocreae 2-lobed when young (frequently becoming shredded or disappearing with age); perianth green, sometimes white or pinkish-tinged at the tip . . . . 6. POLYGONUM
            6. Leaf blades cordate to truncate at the base, sometimes with a pair of rounded or triangular basal lobes
                8. Stems twining or climbing on other vegetation, occasionally forming loose mats. . . . . 4. FALLOPIA
                8. Stems erect or ascending, not climbing or twining, not forming mats
                    9. Perianth of tepals fused only toward the base, much shorter than the mature fruit; inflorescences mostly of clusters or short panicles with arched or spreading, racemose branches, at least the terminal ones usually appearing wider than long . . . . 3. FAGOPYRUM

9. Perianth of tepals that are fused below the midpoint, longer than to slightly shorter than the mature fruit; inflorescences of elongate, spikelike racemes, these usually grouped into panicles, appearing all or mostly longer than wide
   10. Outer series of 3 tepals enlarged and winged at fruiting; styles 3, the stigmas lobed, appearing fringed; fruits 3-angled . . . . 4. FALLOPIA
   10. None of the tepals enlarged or winged at fruiting; styles 2, the stigmas entire, capitate; fruits somewhat flattened (biconvex with rounded margins) to nearly circular in cross-section . . . . 5. PERSICARIA

1. **Brunnichia** Banks ex Gaertn. (buckwheat vine)

Four species, southeastern U.S., Africa.

Some botanists segregate the three African species into the genus *Afrobrunnichia* Hutch. & Dalziel, based on their 2-winged fruits.

1. Brunnichia ovata (Walter) Shinners (ladies' eardrops, American buckwheat vine, redvine)

B. cirrhosa Gaertn.

Rajania ovata Walter

Pl. 496 f, g; Map 2266

Plants lianas, with thin, fibrous roots, climbing by unbranched or more commonly branched tendrils produced at the tips of short, axillary branches. Stems to 15 m long, usually woody only toward the base, frequently spreading across soil or water and rooting at the nodes, but also climbing on other vegetation (or telephone poles), branched, lacking prickles, glabrous or finely hairy toward the tip. Leaves alternate, the petiole 7–25 mm long, somewhat expanded basally and partially sheathing the stem. Ocreae reduced to a short ring to 1 mm long, this usually with dense reddish brown hairs along the margin. Leaf blades 2–15 cm long, 1–8 cm wide, ovate to narrowly ovate or occasionally (in young plants) somewhat heart-shaped, those near the stem tips sometimes lanceolate, usually appearing somewhat folded longitudinally, tapered to a sharply pointed tip, truncate to shallowly cordate at the base, in the narrowest leaves sometimes angled at the base, the margins entire, glabrous or the undersurface sparsely and inconspicuously short-hairy. Inflorescences of axillary spikelike racemes and terminal panicles with spicate branches, the flowers usually 3–5 in small fascicles subtended by ocreolae (often only 1 flower per fascicle developing into a fruit), often pendant at maturity, sometimes appearing positioned on 1 side of the axis, subtended by inconspicuous, scalelike bracts 1–3 mm long. Flowers perfect, relatively long-stalked (the perianth also tapered at the base), the stalk jointed in the lower third, 3-winged above the joint, 1 of the wings becoming enlarged at fruiting creating the appearance of an elongate, flattened, basal wing. Tepals 5 in 2 whorls, 3–4 mm long at flowering, more or less similar in size and shape, fused toward the base, with a slender stalklike basal extension, greenish white and spreading as the flower opens, but quickly turning green to yellowish green, glabrous, at fruiting becoming slightly enlarged, incurved, and leathery or hardened, not developing lateral wings or a prominent network of nerves, not developing a central swelling (grain). Stamens 8, exserted at flowering, but later appearing included in the enlarged, incurved perianth, the anthers attached toward their midpoints. Styles 3, the stigmas capitate to somewhat disc-shaped. Fruits achenes enclosed in the persistent perianth, 8–10 mm long, 3-angled, unwinged, narrowly ovate in outline, short-beaked, brown, glabrous. $2n=48$. May–August.

Scattered to common in the Mississippi Lowlands Division (southeastern U.S. west to Missouri, Oklahoma, and Texas). Swamps, bottomland forests, and banks of sloughs and oxbows; also ditches, fallow fields, crop fields, roadsides, and moist, disturbed areas.

This species was long-known by the name *B. cirrhosa*. Shinners (1967) reviewed the nomenclatural history and determined that although both the epithets were published the same year, the name *Rajania ovata* appeared a few months earlier, requiring its use as the oldest validly published epithet for the species.

Brunnichia ovata sometimes can be a problem weed in crop fields, particularly in soybeans. It is relatively resistant to glyphosate herbicides and thus grows well in fields of transgenic, glyphosate resistant soybeans. However, because plants of

Brunnichia have relatively fine roots, deep-tilling of the soil prior to planting of the crop has been shown to reduce infestations markedly (Reddy, 2005).

2. **Eriogonum** Michx. (wild buckwheat)

Plants annual, biennial, or perennial herbs (shrubs elsewhere; occasionally dioecious elsewhere). Stems erect or ascending (other growth forms elsewhere), in our species unbranched or occasionally with a few, ascending branches below the inflorescence, lacking prickles, densely woolly- or felty-hairy, at least toward the base and/or when young. Leaves in our species basal (these sometimes absent or withered by flowering) and alternate, short- to long-petiolate. Ocreae absent. Leaf blades variously shaped, tapered at the base, the margins entire or nearly so, sometimes somewhat curled-under, at least the undersurface densely pubescent with white, matted, woolly or felty hairs. Inflorescences in our species open panicles with dichotomous and/or trichotomous, ascending to spreading branches, with usually 3 inconspicuous, scale-like, basally fused bracts 1–5 mm long at most nodes, the flowers clustered into small fascicles, each fascicle subtended by a small cuplike involucre with 3–6 short teeth. Flowers perfect, sessile or short-stalked (the perianth sometimes tapered to a stalklike base), usually jointed at the perianth base. Tepals 6 in 2 whorls, fused toward the base, sometimes with a slender stalklike basal extension, white to pink or yellow (other colors elsewhere), often hairy, at fruiting not becoming enlarged or winged, not developing a prominent network of nerves, and not developing a central swelling (grain). Stamens 9, exserted or not, the anthers attached toward their midpoints. Styles 3, the stigmas capitate. Fruits achenes, 3-angled (2-angled elsewhere), unwinged, narrowly ovate in outline, short-beaked, brown, glabrous or hairy. About 250 species, North America.

Eriogonum is one of the largest genera in the Flora of North America region, with only about 26 of the approximately 250 total species in the genus restricted to Mexico, which is outside the geographic scope of the Flora of North America region. The greatest species diversity is in the western United States, where several of the native species also are cultivated as ornamentals. The genus has a well-justified reputation as a taxonomically difficult one, and some of the widespread species have been subdivided into complex series of infraspecific taxa.

1. Plants annual or less commonly biennial, with a slender taproot, the basal rosette leaves absent or withered by flowering; perianth 1.0–2.5 mm long, not extending to a stalklike base, white or pinkish-tinged, the inner surface densely hairy, the outer surface glabrous; stamens not exserted; fruits 1–2 mm long, glabrous . . . . 1. E. ANNUUM
1. Plants perennial, with a stout, sometimes branched, ascending, woody caudex above a stout, woody taproot, the basal rosette leaves persistent at flowering; perianth 4.5–11.0 mm long, including the tapered, stalklike base (this 0.5–3.0 mm), yellow to yellowish green, the inner surface glabrous, the outer surface densely hairy; stamens exserted; fruits 4–6 mm long, hairy . . . . . 1. E. LONGIFOLIUM

1. Eriogonum annuum Nutt. (annual wild buckwheat)

Map 2267

Plants annual or less commonly biennial, with a slender taproot, the basal rosette leaves nonpersistent, absent or withered by flowering. Stems solitary or few, 10–50(–100) cm long below the inflorescence, often unbranched, densely pubescent with silvery grayish, woolly or felty, matted hairs throughout, the hairs persistent or sometimes wearing away in patches. Basal rosette leaves few, 2–5 cm long, the blade oblanceolate, tapered to a petiole 0.3–1.2 cm long, the surfaces densely pubescent with silvery grayish, woolly or felty, matted hairs. Stem leaves mostly positioned below the stem midpoint, short-petiolate (2–5 mm), the blade 1–5 cm long, oblanceolate to narrowly elliptic or narrowly oblong-oblanceolate, the surfaces ini-

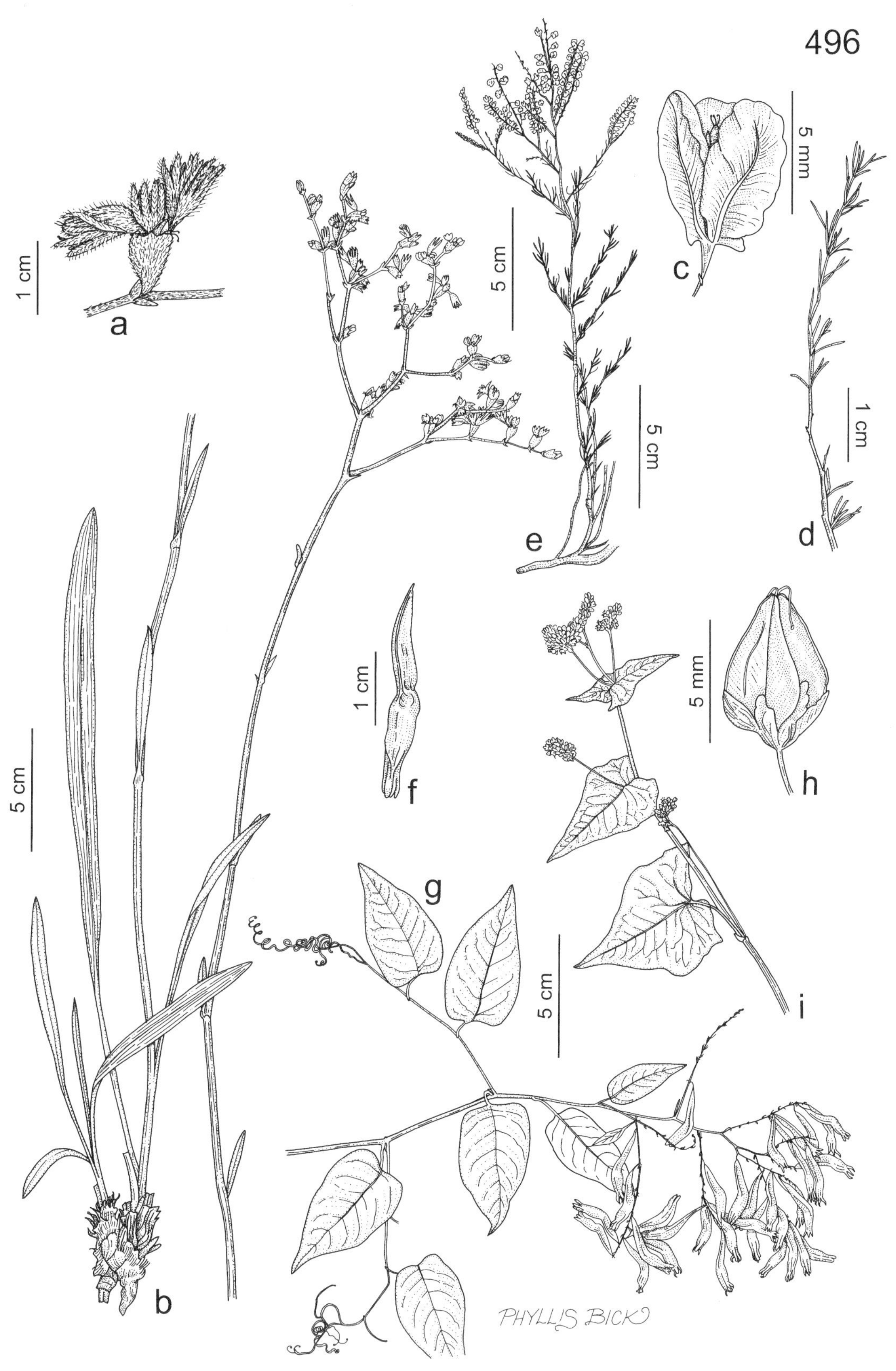

Plate 496. Polygonaceae. *Eriogonum longifolium*, **a)** fascicle of flowers, **b)** habit. *Polygonum americanum*, **c)** fruit, **d)** branch tip with leaves, **e)** habit. *Brunnichia ovata*, **f)** fruit, **g)** habit. *Fagopyrum esculentum*, **h)** flower, **i)** fertile stem.

2269. Fagopyrum esculentum

2270. Fallopia convolvulus

2271. Fallopia dumetorum

tially densely pubescent with silvery grayish, woolly or felty, matted hairs, those of the upper surface often wearing away (sometimes in patches) at maturity. Involucres 2.5–4.0 mm long, with 5 or 6 small teeth, densely woolly on the outer surface, glabrous on the inner surface. Perianth 1.0–2.5 mm long, not extending to a stalklike base, the outer series of 3 tepal lobes slightly broader than the inner series, white or occasionally pinkish-tinged, densely hairy on the inner surface, glabrous on the outer surface. Stamens not exserted. Fruits 1–2 mm long, glabrous. $2n=40$. September–October.

Introduced, uncommon, known thus far from a single specimen from Shannon County (Minnesota to Montana south to Arkansas, Texas, and New Mexico; Mexico; introduced sporadically eastward to New Jersey and in Arizona). Mesic upland forests and savannas.

Eriogonum annuum was discovered in 1996 by a group of Natural History Biologists of the Missouri Department of Conservation at an upland site being converted from hardwood forest to pine savanna. It is possible that the plants were dispersed into the site on equipment being used for habitat restoration. Although it is considered a recent introduction in Missouri, the species is native in portions of Arkansas, Kansas, and Oklahoma.

2. Eriogonum longifolium Nutt. **var. longifolium** (umbrella plant, long-leaved wild buckwheat)

Pl. 496 a, b; Map 2268

Plants perennial, with a stout, sometimes branched, ascending, woody caudex above a stout, woody taproot, the basal rosette leaves persistent and prominent at flowering and fruiting. Stems solitary or few, 50–170 cm long below the inflorescence, usually unbranched, initially densely pubescent with grayish, woolly or felty, matted hairs throughout, the hairs often wearing away (sometimes in patches) above the basal portion of the stem. Basal rosette leaves several to numerous, 1–20 cm long, the blade lanceolate to oblong or oblanceolate, tapered to a petiole 1–17 cm long. Stem leaves much shorter than the rosette leaves, gradually reduced toward the stem tip, mostly sessile or short-petiolate (to 50 mm), the blade 0.5–12.0 cm long, lanceolate to narrowly oblong or narrowly lanceolate, the uppermost sometimes somewhat scalelike, the upper surface sparsely to densely pubescent with grayish, woolly or felty hairs, the undersurface densely pubescent with grayish, woolly or felty, matted hairs. Involucres 4–6 mm long, with 3 or 5 small teeth, densely woolly on the outer surface, glabrous on the inner surface. Perianth 4.5–11.0 mm long, including the tapered, stalklike base (this 0.5–3.0 mm), the tepal lobes all similar, yellow to yellowish green, glabrous on the inner surface, densely hairy on the outer surface. Stamens exserted. Fruits 4–6 mm long, the brown surface more or less obscured by the dense, white, woolly hairs. July–October.

Uncommon, known only from the southern portion of the Ozark Division (Missouri and Kansas south to Louisiana, Texas, and New Mexico). Glades and tops of bluffs, on calcareous substrates.

Several of infraspecific taxa have been described within *E. longifolium*, but Reveal (2005) accepted only three varieties. The other two varieties are both regional rarities listed as Threatened under the federal Endangered Species Act. The var. *harperi* (Goodman) Reveal occurs in widely scattered populations in portions of Alabama, Tennessee, and historically Kentucky, and is unique in its relatively small involucres (3.0–3.5 mm) and small flowers (5–7 mm) with very short stalklike bases. On the other hand, var. *gnaphalifolium* Gand. is restricted to the central

Florida scrub and is characterized by relatively large involucres (6–7 mm) and large flowers (8–15 mm) with relatively long stalklike bases. Steyermark (1963) noted that the species is very uncommon in Missouri and that he encountered it mostly as rosettes that can persist in the vegetative state for many years, producing flowering stems relatively infrequently.

3. **Fagopyrum** Mill. (buckwheat)

Sixteen species, Europe, Asia, Africa; introduced nearly worldwide.

The taxonomy and phylogeny of this economically important genus have been studied by Prof. Ohmi Ohnishi of Kyoto University and his colleagues and students (among other papers: Ohnishi and Matsuoka, 1996; Yasui and Ohnishi, 1998; and Ohsako and Ohnishi, 2000).

1. Fagopyrum esculentum Moench (common buckwheat)

F. sagittatum Gilib., an illegitimage name

Pl. 496 h, i; Map 2269

Plants annual, with slender taproots, not climbing or producing tendrils, not forming mats. Stems 10–50(–90) cm long, erect or ascending, usually branched, sometimes with pinkish lines or stripes, glabrous or with a ring of inconspicuous hairs at some nodes. Leaves alternate, the lower ones long-petiolate, with progressively shorter petioles toward the stem tip, the uppermost leaves sessile or nearly so. Ocreae tan to white, loosely wrapped around the stem, somewhat obconic and oblique at the tip, usually split longitudinally along 1 side, glabrous or more commonly minutely hairy on the outer surface. Leaf blades 2–8 cm long and wide, variously heart-shaped or triangular with a pair or spreading or downward-pointed, rounded to pointed, triangular lobes, angled or tapered to a bluntly or sharply pointed tip, truncate to more commonly broadly cordate at the base, the margins otherwise somewhat irregular and usually minutely hairy, the surfaces glabrous or more commonly inconspicuously hairy along the veins, the venation palmate with a thick midvein and 6 or 8 slightly narrower main veins from the blade base, the finer veins forming a complex network. Inflorescences axillary and terminal, mostly of clusters or short panicles with arched or spreading, racemose branches, at least the terminal ones usually appearing wider than long, the flowers usually in fascicles of 2–6 at the nodes, subtended by ocreolae, spreading in all directions along the axis or sometimes appearing positioned mostly along the upper side of the branches, lacking other bracts. Flowers perfect (usually 2 kinds produced varying in the relative lengths of stamens and styles), short-stalked (the perianth somewhat tapered at the base), the stalk not jointed, unwinged. Tepals 5 in 2 whorls, fused only toward the base, subequal, 3–5 mm long, not becoming enlarged at fruiting, elliptic to obovate, more or less spreading at flowering, white, cream-colored, or pink, usually green toward the base, at fruiting becoming papery, not developing wings or a prominent network of nerves, not developing a central swelling (grain). Stamens 8, not or only slightly exserted (of 2 different lengths in different flowers), the anthers attached toward their midpoints. Styles 3, usually downward curved, the stigmas capitate. Fruits achenes, not enclosed in the persistent perianth, 4–7 mm long, 3-angled, unwinged, ovate in outline, short-beaked, light brown to dark brown, sometimes with darker streaks, glabrous, shiny. $2n$=16. May–September.

Introduced, scattered widely in the state (native of Asia; introduced nearly worldwide, including nearly throughout the U.S. [including Alaska], Canada). Edges of marshes; also pastures, fallow fields, roadsides, railroads, and moist, disturbed areas.

Common buckwheat is an important crop in various portions of Europe, Asia, and the New World, grown for its edible fruits, which commonly are ground into flour and/or used for animal feed. Archaeological studies have documented that the species was cultivated as long as 4,600 years ago in China, and that the wild ancestor most likely came from southwestern China (Ohnishi, 1998). Steyermark (1963) noted that in Missouri the species frequently is grown as a green manure rather than as a pseudocereal crop. It also is planted in wildlife food plots, particularly as a seed source for migratory waterfowl, and is a popular honey plant. Steyermark further noted that both the foliage and the cooked flour can cause dermatitis in susceptible persons and photosensitive dermatitis in some livestock. Buckwheat hulls also are an ingredient in some mulches and the dried hulls are used as fill for certain types of pillows that are considered ecologically friendly and/or to have health benefits.

4. **Fallopia** Adans. (false buckwheat)

Plants annual or perennial, sometimes with rhizomes, sometimes climbing and/or twining (then lacking tendrils). Stems branched, spreading to erect, glabrous or hairy. Leaves alternate, short-petiolate, jointed at the petiole base. Ocreae white to tan or greenish brown, papery, unlobed and more or less oblique at the tip when young (sometimes becoming shredded or disappearing with age), glabrous, roughened, or minutely hairy. Leaf blades ovate to triangular-ovate, halberd-shaped, or heart-shaped, not leathery, mostly tapered to a sharply pointed tip, angled to truncate or cordate at the base, the margins entire to somewhat wavy, sometimes more or less with a pair of rounded or pointed basal lobes, the surfaces glabrous, roughened, or minutely hairy along the veins (sometimes with a mealy deposit in *F. cilinodis* (Michx.) Holub). Inflorescences terminal and/or axillary, of racemes or panicles, mostly erect or ascending, the nodes with small clusters of flowers, these subtended by ocreolae, ascending or spreading along the axis, lacking other bracts. Flowers perfect or more commonly not, short-stalked (the perianth tapered at the base or not), jointed at or below the perianth base, winged toward the tip or unwinged. Tepals 5 in 2 whorls, fused mostly only near the base, with entire or less commonly wavy to corrugated margins, glabrous; ascending during and after flowering, the outer whorl of 3 tepals, with boat-shaped to hooded tips, ridged or winged along the midnerve, more or less enclosing the inner whorl of 2 shorter, relatively flat tepals, those of both whorls elliptic to obovate, white to greenish white (then sometimes pinkish-tinged along the margins and/or bases) or pink to pinkish red, at fruiting, becoming slightly to strongly enlarged, developing narrow to broad wings, but not a prominent network of nerves or a central swelling (grain). Stamens (in staminate or perfect flowers) (6–)8, not exserted, the anthers attached toward their midpoints, the filaments sometimes hairy toward their bases. Styles 3, sometimes fused toward the base or nearly to the tip, the stigmas capitate or (in *F. japonica*) divided into numerous, slender fringelike lobes. Fruits achenes, enclosed in the persistent perianth, 2–6 mm long, 3-angled, unwinged, elliptic to ovate in outline, short-beaked, dark brown to black, glabrous, smooth or minutely pebbled, usually shiny. About 12 species, North America, South America, Europe, Asia, Africa.

As noted in the discussion under the treatment of the genus *Polygonum*, *Fallopia* has become accepted by most modern botanists as a separate genus. However, the generic limits of *Fallopia* remain controversial. Two well-marked subgroups are represented in the Midwest and these in turn have a close relationship to the genus *Muehlenbeckia* Meisn. (23 mostly tropical species of herbaceous climbers). The climbing buckwheats are members of *Fallopia* in the strict sense. The Mexican bamboo and its relatives, which are robust, nonclimbing, rhizomatous perennials, have sometimes been segregated into the genus *Reynoutria* Houtt. or treated as *Fallopia* sect. *Reynoutria* (Houtt.) Ronse Decr. Freeman and Hinds (2005a) reviewed this situation in their treatment for the Flora of North America Project. Most recently, molecular data have suggested that the two groups should be treated as separate genera (Schuster et al., 2011), but the situation requires more detailed study. The present treatment with its broad circumscription of *Fallopia* should thus be viewed as provisional.

1. Stems stout, erect or ascending (rarely loosely ascending), not climbing or twining; plants strongly rhizomatous perennials with somewhat woody rootstocks, forming dense colonies; stigmas divided into numerous, slender fringelike lobes . . . . 3. F. JAPONICA
1. Stems slender, climbing on other vegetation or spreading (on open ground), twining; plants annuals or short-lived perennials, sometimes forming mats; stigmas entire along the margins

2. Fruits with the surface dull, appearing finely roughened or with dense, minute tubercles; outer tepals ridged dorsally at maturity, unwinged or rarely with narrow wings to 0.9 mm wide . . . . 1. F. CONVOLVULUS
2. Fruits with the surface shiny, smooth; outer tepals at maturity with the dorsal wings 1–3 mm wide
    3. Fruiting perianth usually truncate at the base, not or only slightly decurrent along the stalklike base (with a noticeable concave shoulder), the wings relatively flat; leaf blades cordate at the base, with the pair of basal lobes bluntly to sharply pointed and somewhat spreading . . . . 2. F. DUMETORUM
    3. Fruiting perianth noticeably tapered and decurrent along the stalklike base, the wings relatively flat or somewhat crisped, wavy, and/or undulate; leaf blades truncate or more commonly cordate at the base, with the pair of basal lobes rounded and parallel to slightly spreading . . . . 4. F. SCANDENS

1. Fallopia convolvulus (L.) Á. Löve (black bindweed, climbing buckwheat, wild buckwheat)

Polygonum convolvulus L.

Pl. 497 h–j; Map 2270

Plants annual, with slender taproots, forming mats in open areas or more commonly climbing on adjacent vegetation. Stems 20–200 cm long, slender, spreading or climbing, twining, sparsely to moderately pubescent with minute, downward-curved hairs, occasionally also somewhat mealy. Ocreae 2–4 mm long, oblique at the tip, the surface glabrous or minutely roughened, tan to greenish brown. Leaf blades 1–7 cm long, 0.8–6.0 cm wide, more or less heart-shaped to ovate-triangular, cordate at the base with the downward-pointing or occasionally spreading basal lobes rounded or pointed, tapered to a sharply pointed tip, the upper surface glabrous, the undersurface minutely roughened and often somewhat mealy, also gland-dotted. Inflorescences axillary, spikelike racemes, the flowers 3–6 per fascicle, perfect, the flower stalks 1–3 mm long, not winged, jointed above the midpoint. Tepals 3–4 mm long at flowering, sometimes becoming enlarged to 4–5 mm at fruiting, those of the inner whorl mostly elliptic, those of the outer whorl obovate, angled to very short-tapered and not decurrent along the stalklike base, glabrous or sparsely and minutely hairy, white, sometimes pinkish- or purplish-tinged toward the base, the outer 3 tepals ridged dorsally or rarely with inconspicuous, slender wings to 0.9 mm wide at fruiting. Styles very short, fused most of their length, the stigmas capitate, entire along the margins. Fruits 2.3–3.5 mm long, the surface dull, appearing finely roughened or with dense, minute tubercles, black. $2n=40$. May–November.

Introduced, scattered nearly throughout the state (native of Europe, Asia; introduced nearly worldwide in temperate regions). Bottomland forests, banks of streams and rivers, and ledges of bluffs; also crop fields, fallow fields, gardens, railroads, roadsides, and disturbed areas.

A related species that might be documented from Missouri in the future is *F. cilinodis* (Michx.) Holub (fringed black bindweed), which is native and widespread in the eastern United States and Canada, and has been documented from Indiana, Illinois, Kentucky, and Tennessee. It differs from *F. convolvulus* in having ocreae bases that are fringed with reflexed hairs and bristles (Freeman and Hinds, 2005a).

2. Fallopia dumetorum (L.) Holub (false buckwheat, crested buckwheat)

Bilderdykia dumetorum (L.) Dumort.

Polygonum dumetorum L.

P. scandens L. var. *dumetorum* (L.) Gleason

Reynoutria scandens var. *dumetorum* (L.) Shinners

Tiniaria dumetorum (L.) Opiz

Map 2271

Plants annual, with taproots, forming mats in open areas or more commonly climbing on adjacent vegetation. Stems 20–300 cm long, slender, spreading or climbing, twining, glabrous or sparsely to moderately roughened or pubescent with minute, usually downward-curved hairs. Ocreae 1.5–3.5 mm long, oblique at the tip, the surface glabrous or minutely roughened, tan to greenish brown. Leaf blades 2–8 cm long, 1–5 cm wide, ovate-triangular, cordate at the base, with the pair of basal lobes bluntly to sharply pointed and somewhat spreading, tapered to a sharply pointed tip, the surfaces glabrous or less commonly minutely roughened, the undersurface also gland-dotted. Inflorescences axillary, spikelike racemes, the flowers 2–6 per fascicle, perfect, the flower stalks 3–8 mm long, winged toward the tip, jointed

2272. Fallopia japonica 2273. Fallopia scandens 2274. Persicaria amphibia

above the midpoint. Tepals 3–5 mm long at flowering, becoming enlarged to 3.5–7.0 mm at fruiting, those of the inner whorl mostly elliptic, those of the outer whorl obovate, truncate or rarely very short-tapered to the stalklike base (with a noticeable concave shoulder), glabrous, white, sometimes pinkish-tinged, the outer 3 tepals with relatively flat or occasionally slightly undulate dorsal wings 1.5–2.0 mm wide at fruiting. Styles very short, fused most of their length, the stigmas capitate, entire along the margins. Fruits 2–4 mm long, the surface shiny, smooth, black. $2n=20$. July–November.

Introduced, uncommon, sporadic, mostly north of the Missouri River (native of Europe, Asia; introduced in the eastern U.S. west to Wisconsin and Texas; Canada). Openings and edges of bottomland forests, banks of streams and rivers, and tops of bluffs; also roadsides and open disturbed areas.

For a discussion of the taxonomy of the *F. scandens* complex, see the treatment of that species.

3. Fallopia japonica (Houtt.) Ronse Decr. (Japanese knotweed, Mexican bamboo, Japanese bamboo)

Polygonum cuspidatum Siebold & Zucc.

Reynoutria japonica Houtt.

Pl. 502 g, h; Map 2272

Plants perennial herbs, with somewhat woody rootstocks and widely creeping rhizomes, forming dense colonies, usually functionally dioecious (staminate plants occasionally with a few pistillate or perfect flowers). Stems 100–250 cm long, stout, erect or ascending (loosely ascending in rare plants attributed to var. *compacta* (Hook. f.) J.P. Bailey), not climbing or twining, glabrous, often somewhat glaucous. Ocreae 4–10 mm long, oblique at the tip, the surface glabrous or minutely hairy, tan. Leaf blades 4–15 cm long, 2–10 cm wide, ovate, broadly angled or broadly rounded to truncate at the base (rarely a few of the leaves very shallowly cordate), tapered or short-tapered to a sharply pointed tip, the upper surface glabrous, the undersurface roughened or sparsely pubescent along the veins with minute, unicellular, stout, knob-shaped to bluntly pointed hairs, also gland-dotted, sometimes also somewhat glaucous. Inflorescences terminal and axillary, panicles or rarely reduced to racemes, the flowers 3–10 per fascicle, mostly staminate or pistillate (rarely perfect), the flower stalks 3–5 mm long, winged to below the midpoint, jointed at or below the midpoint. Tepals 2–4 mm long at flowering, becoming enlarged to 4–10 mm at fruiting, those of the inner whorl mostly elliptic, those of the outer whorl obovate, tapered and decurrent along the stalklike base, glabrous, white, greenish white, or pink to reddish pink, the outer 3 tepals with dorsal wings 1.4–2.0 mm wide at fruiting. Styles fused toward the base, the stigmas divided into numerous, slender fringelike lobes. Fruits 2.3–3.5 mm long, the surface shiny, smooth, dark brown. $2n=44, 66, 88$. July–October.

Introduced, scattered, mostly in the eastern half of the state (native of Asia; introduced widely in the U.S.; Canada). Bottomland forests, edges of mesic upland forests, banks of streams and rivers, margins of saline seeps; also ditches, fencerows, old homesites, gardens, alleys, railroads, roadsides, and open disturbed areas.

Mexican bamboo spreads aggressively, forming dense thickets. Plants are too aggressive for most garden uses, and the species is considered an invasive exotic by land managers and conservationists in many of the states where it has become established. In fact, Mexican bamboo has been declared a noxious weed in several states. Murrell et al. (2011) noted that members of this group produce a number of bioactive (antimicrobial and/or antifungal) chemical compounds and found that in garden plots the related *F. ×bohemica* (Chrtek & Chrtková) J.P. Bailey (see below) acted allelopathically to suppress the germination and growth of other plant species. Steyermark (1963) docu-

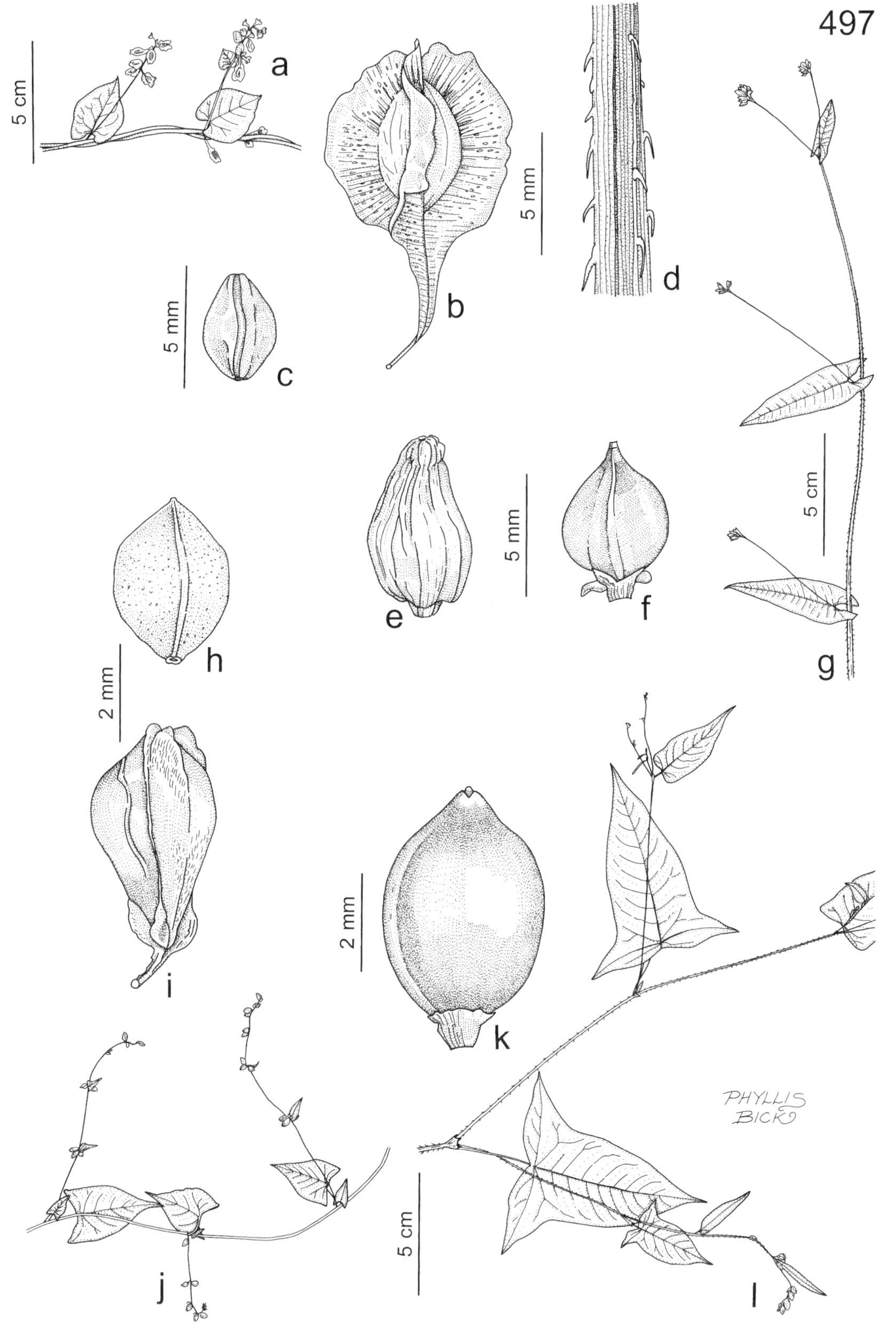

Plate 497. Polygonaceae. *Fallopia scandens*, **a)** habit, **b)** fruit in calyx, **c)** fruit without calyx. *Persicaria sagittata*, **d)** stem detail, **e)** flower, **f)** fruit, **g)** habit. *Fallopia convolvulus*, **h)** fruit, **i)** flower, **j)** habit. *Persicaria arifolia*, **k)** fruit, **l)** habit.

mented *F. japonica* (as *Polygonum cuspidatum*) from only 5 counties, but that range has since expanded to include nearly 20 counties. Fortunately, in Missouri it has thus far colonized mainly highly disturbed sites, but it has the potential to spread more widely into grassland and wetland habitats.

Some botanists have split this species into four or more varieties (J. P. Bailey and Stace, 1992), but the situation is still poorly understood. Most of our plants would be referable to var. *japonica*. However, Mühlenbach (1983) reported a collection of var. *compacta* (as *Polygonum cuspidatum* var. *compactum* (Hook. f.) L.H. Bailey) from the St. Louis railyards. This amounts to a dwarf selection, with shorter, somewhat more spreading stems and somewhat smaller leaves. It was brought into the horticultural trade from plants originally found native in eastern Asia.

Two other taxa in the *F. japonica* complex probably will be documented in Missouri in the future. The closely related *F. sachalinensis* (F. Schmidt) Ronse Decr. (*Polygonum sachalinense* F. Schmidt; giant knotweed) has escaped widely in the northeastern states (also in the Pacific Northwest) and has been reported from Illinois, Kentucky, and Tennessee (Freeman and Hinds, 2005a). This species is superficially similar to *F. japonica*, but differs in having leaf blades with multicellular hairs along the veins abaxially and with regularly cordate bases. A hybrid of horticultural origin between *F. japonica* and *F. sachalinensis* is widespread as an escape and has been reported from all of the states surrounding Missouri except Arkansas and Oklahoma (Freeman and Hinds, 2005a). *Fallopia ×bohemica* (Bohemian knotweed), as it is called, can be difficult to separate from *F. japonica*, but usually has leaf blades with minute, sharply pointed, relatively soft hairs along the veins abaxially (this requires strong magnification to observe, but is usually apparent to the touch). Often, it also has at least some of the leaf blades shallowly cordate at the base. See Zika and Jacobson (2003) for a detailed morphological analysis of the complex. Where the parents and/or the hybrid grow in proximity, hybrid swarms have been documented (Gammon et al., 2007), further confounding attempts to determine specimens.

4. Fallopia scandens (L.) Holub (false buckwheat, crested buckwheat)

F. cristata (Engelm. & A. Gray) Holub
Bilderdykia scandens (L.) Greene
Polygonum scandens L.
P. cristatum Engelm. & A. Gray
P. scandens var. *cristatum* (Engelm. & A. Gray) Gleason
Reynoutria scandens (L.) Shinners
R. scandens var. *cristata* (Engelm. & A. Gray) Shinners
Tiniaria scandens (L.) Small
T. cristata (Engelm. & A. Gray) Small

Pl. 497 a–c; Map 2273

Plants annual or perennial herbs, with fibrous roots, sometimes taprooted, forming mats in open areas or more commonly climbing on adjacent vegetation. Stems 20–500 cm long, slender, spreading or climbing, twining, sparsely to moderately pubescent with minute, downward-curved hairs, occasionally also somewhat mealy. Ocreae 1–6 mm long, oblique at the tip, the surface glabrous or minutely roughened, tan to greenish brown. Leaf blades 1–14 cm long, 0.8–7.0 cm wide, more or less heart-shaped to less commonly ovate-triangular, truncate or more commonly cordate at the base, with the parallel or slightly spreading basal lobes rounded at the tips, the upper surface glabrous, the undersurface minutely roughened and often somewhat mealy, also gland-dotted. Inflorescences axillary, spikelike racemes, the flowers 3–6 per fascicle, perfect, the flower stalks 3–8 mm long, winged or not, jointed above the midpoint. Tepals 3–5 mm long at flowering, sometimes becoming enlarged to 4–10 mm at fruiting, those of the inner whorl mostly elliptic, those of the outer whorl obovate, angled to tapered and decurrent or not along the stalklike base, glabrous or sparsely and minutely hairy, white, greenish white, or pinkish-tinged, the outer 3 tepals with dorsal wings to 1–3 mm wide at fruiting. Styles very short, fused most of their length, the stigmas capitate, entire along the margins. Fruits 2–6 mm long, the surface shiny, smooth, dark brown to black. $2n$=20. June–November.

Scattered to common nearly throughout the state (eastern U.S. west to North Dakota, Wyoming, and Texas; Canada). Openings and edges of bottomland forests and mesic upland forests, savannas, glades, banks of streams and rivers, margins of ponds, lakes, sloughs, and seeps, and ledges and tops of bluffs; also crop fields, fencerows, ditches, quarries, railroads, roadsides, and open disturbed areas.

The *F. scandens* complex consists of three closely related entities and is circumboreal, with both native and introduced populations present in the eastern half of the United States and Canada. It has been treated variously as three separate species, two species, with the native North American species comprising a single taxon or split into two varieties or subspecies, or as a single species, sometimes divided into two or three infraspecific taxa (Graham and Wood, 1965; Horton, 1972;

Mitchell and Dean, 1978; Gleason and Cronquist, 1991; S.-t. Kim et al., 2000; Freeman and Hinds, 2005a). The present treatment follows the taxonomic concepts of Freeman and Hinds (2005a) in treating populations native to the Old World as a distinct species, *F. dumetorum*, but not formally recognizing infraspecific taxa within the native North American populations. Readers should note that there is continuing disagreement as to the taxonomic treatment of this complex. For example, based on their numerical analysis of variation in 20 morphological characters, S.-t. Kim et al., 2000) proposed retaining the New and Old World taxa as separate species, but suggested that the New World populations should be treated as two varieties. Within *F. scandens*, as treated here, these two variants exhibit somewhat overlapping morphology, which caused Freeman and Hinds (2005a) to decide against segregating them as varieties. The extremes can be distinguished using fruiting characters as follows:

1. Fruiting perianth 4–7 mm long, the wings 1.0–1.7 mm wide, somewhat crisped, wavy, and/or undulate . . . . *F. cristata*
1. Fruiting perianth 7–10 mm long, the wings 1.5–3.0 mm wide, flat or nearly so . . . . *F. scandens*

Steyermark (1963) and others who have treated the complex in a broadly circumscribed *Polygonum*, accepted the two variants as varieties of *P. scandens*. Under *Fallopia, a* few sources, mainly on the World Wide Web, list the taxon with small, crisped fruiting sepals as "var. *cristata* (Engelm. & A. Gray) H.R. Hinds," but although the late Harold Hinds intended to publish this combination, apparently he never did so. In Missouri, plants corresponding to the *F. cristata* segregate are encountered uncommonly in scattered counties, mostly south of the Missouri River.

5. **Persicaria** (L.) Mill. (smartweed)

Plants annual or perennial herbs, with taproots or fibrous roots, sometimes with rhizomes. Stems unbranched or branched, loosely ascending to arched or erect, in a few species spreading or climbing (but not twining), usually noticeably swollen at the nodes, often angled or ribbed, glabrous or hairy, in a few species with small, downward-curved (hooked) prickles. Leaves alternate, sessile or petiolate, not jointed at the base. Ocreae usually white, tan, or brown, sometimes reddish-tinged or -streaked, mostly papery, fused basally to the petiole, unlobed, but sometimes with bristly hairs along the margin or split along 1 side (frequently becoming shredded or disappearing with age), occasionally with an asymmetrical green flange or appendage at the tip, glabrous, roughened, or hairy. Leaf blades lanceolate to elliptic or ovate, in a few species arrowhead-shaped to halberd-shaped with a pair of parallel or spreading basal lobes, herbaceous to slightly leathery, variously angled, tapered, rounded, truncate, or cordate at the base, angled or tapered to a sharply pointed tip, the margins otherwise entire, the surfaces glabrous or hairy, in a few species the undersurface with small prickles. Inflorescences terminal and sometimes also axillary, consisting of elongate spikes or spikelike racemes, these sometimes grouped into panicles or condensed into globose clusters (often in P. *sagittata*) usually lacking bracteal leaves (but reduced leaves sometimes present at the branch points of panicles), the flowers usually in small fascicles or clusters of 2–14 (rarely solitary), these subtended by ocreolae, ascending or spreading along the axis, lacking other bracts. Flowers perfect (in *P. amphibia* and *P. hydropiperoides*), sessile or short-stalked (the perianth not tapered at the base), the stalk unwinged, jointed at the perianth base, the joint sometimes swollen. Tepals 4 or more commonly 5, in 2 whorls, fused variously only in the basal 1/3 or to well above the midpoint, with entire margins, glabrous but sometimes gland-dotted, ascending or less commonly spreading at flowering, ascending at fruiting, the outer whorl of 2 or 3 tepals (except in *P. virginiana*) longer, with more or less boat-shaped to hooded tips, partially enclosing the inner whorl of 2 shorter, relatively flat tepals, those of both whorls variously shaped, white, greenish, pink, or red, not becoming enlarged at fruiting, unwinged, the nerves not forming a prominent network (with a prominent anchor-shaped pattern of nerves in *P. lapathifolia*) or a central swelling (grain). Stamens 5–8, exserted or more commonly not exserted, the anthers attached toward their midpoints, the filaments usually with dilated bases.

Styles 2 or 3, sometimes fused toward the base, the stigmas capitate. Fruits achenes, exserted or more commonly enclosed in the persistent perianth, 2.0–3.5 mm long, 3-angled or 2-angled (biconvex with rounded margins to nearly circular in cross-section), unwinged, elliptic or ovate in outline, not beaked (except in *P. virginiana*), brown to dark brown or black, glabrous, dull or shiny. About 100 species, nearly worldwide.

As noted in the discussion under the treatment of the genus *Polygonum*, *Persicaria* has become accepted by most modern botanists as a separate genus comprising three well-marked subgroups. The smartweeds are members of *Persicaria* in the strict sense. The tearthumbs and their relatives, which are annuals with climbing stems beset with small, downward-curved (hooked) prickles, usually are classified into the section *Echinocaulon* (Meisn.) H. Gross. *Persicaria virginiana* (Virginia knotweed) and two or three Asian relatives characterized by long, spikelike inflorescences with well-separated nodes, as well as flowers with the four tepals similar in size and relatively long, exserted styles, have been classified into the section *Tovara* (Adans.) H. Gross. Some botanists have treated this group as a separate genus, *Tovara* Adans. However, that name has been officially rejected as a nomenclatural homonym of the nearly identically spelled generic name *Tovaria* Ruiz & Pav. in the small neotropical family Tovariaceae. Freeman and Hinds (2005b) reviewed the infrageneric taxonomy of *Persicaria* in their treatment for the Flora of North America Project.

Smartweeds are among the most valuable marsh and aquatic wildlife foods in the United States. They are especially important to migratory waterfowl and are an important component of seed mixes planted by government fish and game departments in managed wetlands designed to promote populations of these birds. More than 65 mammal and bird species have been found to consume smartweed as part of their diets (Martin et al., 1951). The fruits of a few of the species also were harvested by Native Americans for food. Some of the species are excellent honey plants. A number of species were used medicinally by Native Americans for a variety of treatments, including colds, coughs, pain, and others (Moerman, 1998). A few of the species are cultivated as ornamentals, notably *P. virginiana*, but most of the species (especially the annuals) are considered aggressive weeds in gardens and some also are important weeds in crop fields.

In general, species of *Persicaria* should be collected when mature inflorescences are present (those with at least some mature fruits). Thomas (2005) created a key to the species of *Polygonum* (including *Fallopia* and *Persicaria*) in Missouri that can be very useful when mature inflorescences are absent. The present treatment borrows heavily from that of Hinds and Freeman (2005) in the Flora of North America series.

1. Stems spreading or climbing, with small, downward-curved (hooked) prickles
    2. Leaf blades triangular to halberd-shaped, usually with a pair of spreading, sharply pointed, basal lobes; fruits lenticular (biconvex in cross-section) . . . . 2. P. ARIFOLIA
    2. Leaf blades broadly oblong-lanceolate to arrowhead-shaped, cordate at the base or more commonly with a pair of parallel, rounded, basal lobes; fruits trigonous (triangular in cross-section) . . . . 14. P. SAGITTATA
1. Stems not climbing, erect or ascending, occasionally loosely ascending to spreading with ascending tips, unarmed
    3. Inflorescences open spikelike racemes, the nodes well-separated, the fascicles of flowers not overlapping those of the adjacent nodes; tepals 4; styles ex serted, persistent, the fruits with hooked beaks . . . . 16. P. VIRGINIANA
    3. Inflorescences relatively dense spikelike racemes, the nodes close together, the fascicles of flowers overlapping those of the adjacent nodes (except occasionally at the inflorescence base and in *P. punctata*); tepals (4)5 (usually 4 in *P. lapathifolia*); styles not or only slightly exserted, withering as the fruits develop, the fruits beakless

4. Some or all of the ocreae green and herbaceous toward the tip, often with a stipulelike flange of tissue at the tip; leaf blades broadly ovate, often rounded to truncate or cordate; inflorescences terminal; tepals bright reddish pink to bright pink
 5. Stems 20–120 cm long (to 300 cm in aquatic plants), often rooting at the lower nodes, glabrous or with short, appressed hairs; leaf blades glabrous or appressed-hairy on the undersurface; inflorescences solitary or paired at the stem tips, more or less straight . . . . 1. P. AMPHIBIA
 5. Stems (60–)120–300 cm long, not rooting at the nodes, with soft, spreading hairs; leaf blades soft-hairy; inflorescences usually several in a loose cluster at the stem tips, nodding . . . . 10. P. ORIENTALIS
4. Ocreae tan to white and papery throughout, lacking a flange of tissue at the tip; leaf blades linear to lanceolate, oblong-lanceolate, or narrowly ovate, angled or tapered at the base; inflorescences terminal and axillary (terminal in *P. amphibia*); tepals white, pale green, or pink (bright reddish pink in *P. longiseta*)
 6. Ocreae entire along the margin or nearly so, lacking bristles (occasionally with short, slender, extensions of the veins to 0.8 mm long)
  7. Tepals 4(5), the outer tepals with the nerves branched apically into an anchor-shaped pattern; inflorescences strongly arched or nodding . . . . 7. P. LAPATHIFOLIA
  7. Tepals (4)5, the outer tepals with the nerves irregularly few-branched (not anchor-shaped); inflorescences straight or slightly arched
   8. Ocreae truncate at the tip, not oblique, the outer surface dotted with scattered, minute glands (these sometimes relatively pale) . . . . . 4. P. GLABRA
   8. Ocreae oblique at the tip, the outer surface glabrous or hairy, not gland-dotted
    9. Plants perennial, with rhizomes; stems often rooting at the lower nodes; inflorescences terminal, solitary or paired; tepals bright pink to reddish pink . . . . 1. P. AMPHIBIA
    9. Plants annual, with taproots; stems usually not rooting at the lower nodes; inflorescences terminal and axillary from the upper leaves, usually several to numerous per main stem; tepals white or pinkish-tinged to light pink
     10. Plants of 2 kinds within a population, some with the stamens exserted and others with the styles exserted; fruits with the faces flat to slightly convex, 1 of the faces usually with a central hump . . . . 3. P. BICORNIS
     10. Plants all similar, usually with neither the stamens or the styles exserted (the fruits sometimes slightly exserted); fruits with the faces somewhat concave, none of the faces with a central hump . . . . 11. P. PENSYLVANICA
 6. Ocreae with erect bristles along the margins, these 1–12 mm long
  11. Tepals moderately to densely dotted with scattered, minute glands (these sometimes relatively pale)
   12. Fruits with the surface minutely roughened, dull; inflorescences usually nodding toward the tip . . . . 5. P. HYDROPIPER
   12. Fruits with the surface smooth, shiny; inflorescences straight or nearly so
    13. Inflorescences relatively open, the ocreolae mostly not overlapping, their margins with bristles 1–2 mm long; leaf blades 6–20 (–24) mm wide . . . . 12. P. PUNCTATA

13. Inflorescences relatively dense, the ocreolae mostly overlapping, their margins glabrous or those of the lower ocreolae sometimes with bristles to 1 mm long; leaf blades 2.0–4.5 cm wide . . . . 13. P. ROBUSTIOR

11. Tepals not gland-dotted (rarely with a few gland-dots toward the base in *P. hydropiperoides*)

14. Inflorescences terminal, solitary or paired at the stem tips; plants of 2 kinds within a population, some with the stamens exserted and others with the styles exserted; tepals bright pink to reddish pink . . . . 1. P. AMPHIBIA

14. Inflorescences terminal and axillary, usually several to numerous per main stem; plants all similar, usually with neither the stamens or the styles exserted (the fruits sometimes slightly exserted); tepals white, greenish, pinkish-tinged to light pink (bright reddish pink in *P. longiseta*)

15. Inflorescences 7–12 mm in diameter; tepals with the nerves forming a network toward the base; ocreolae glabrous or with bristles 0.2–1.0 mm long along the margins . . . . 9. P. MACULOSA

15. Inflorescences 2–8 mm in diameter; tepals with the nerves open-branched (not forming a network); ocreolae with bristles 1.0–3.5 mm long along the margins

16. Tepals bright reddish pink; plants annual, with taproots (but the stems sometimes rooting at the lower nodes); stems loosely ascending or spreading with ascending tips; inflorescences mostly 10–40 mm long; ocreolae with bristles mostly 2.0–3.5 mm long along the margins . . . . 8. P. LONGISETA

16. Tepals white, greenish, pinkish-tinged to light pink; plants perennial, usually with rhizomes (the stems often rooting at the lower nodes); stems erect or ascending (sometimes loosely ascending in *P. hydropiperoides*); inflorescences mostly 30–80 mm long; ocreolae with bristles mostly 1–2 mm long along the margins

17. Ocreae with the surface glabrous or more commonly appressed-hairy; petioles appressed-hairy . . . . 6. P. HYDROPIPEROIDES

17. Ocreae more or less spreading-hairy, at least toward the base; petioles spreading-hairy . . . . 15. P. SETACEA

1. Persicaria amphibia (L.) Gray (water smartweed, shoestring smartweed, swamp smartweed)

Polygonum amphibium L.

Pl. 498 a–f; Map 2274

Plants perennial, usually with rhizomes and/or stolons. Stems 20–120 cm long in terrestrial plants, to 300 cm in aquatic plants, prostrate to ascending or erect, glabrous or short-hairy, often rooting at the lower nodes. Ocreae persistent, usually tearing with age, 5–50 mm long, cylindric or flared toward the tip, tan to white and papery most of the length, green and herbaceous toward the tip, sometimes with a stipulelike flange of tissue at the otherwise oblique tip, the margin lacking bristles (entire or nearly so) or with erect bristles 0.5–4.5 mm long, the surface glabrous or short-hairy, not gland-dotted. Petioles absent or to 3(–7) cm long. Leaf blades 2–15(–23) cm long, 1–6(–8) cm wide, broadly lanceolate to elliptic, oblong-lanceolate, or broadly ovate, angled or tapered to rounded, truncate, or cordate at the base, rounded or more commonly angled or tapered to a bluntly or sharply pointed tip, the surfaces glabrous or the undersurface appressed-hairy, unarmed (but the margins roughened to the touch), lacking impressed glands, lacking a reddish or purplish area on the upper surface. Inflorescences terminal, solitary or occasionally paired, 1–15(–18) cm long, 8–20 mm wide, erect or ascending, dense, uninterrupted or interrupted toward the base, the stalk 1–5 cm long, glabrous or hairy, the hairs often gland-tipped. Ocreolae overlapping except sometimes toward the axis base, the margins with bristles 0.5–1.0 mm long, the surface usually glabrous, not gland-dotted. Flowers sometimes functionally pistillate or staminate, 1–3(4) per fascicle, those of some plants with the stamens exserted

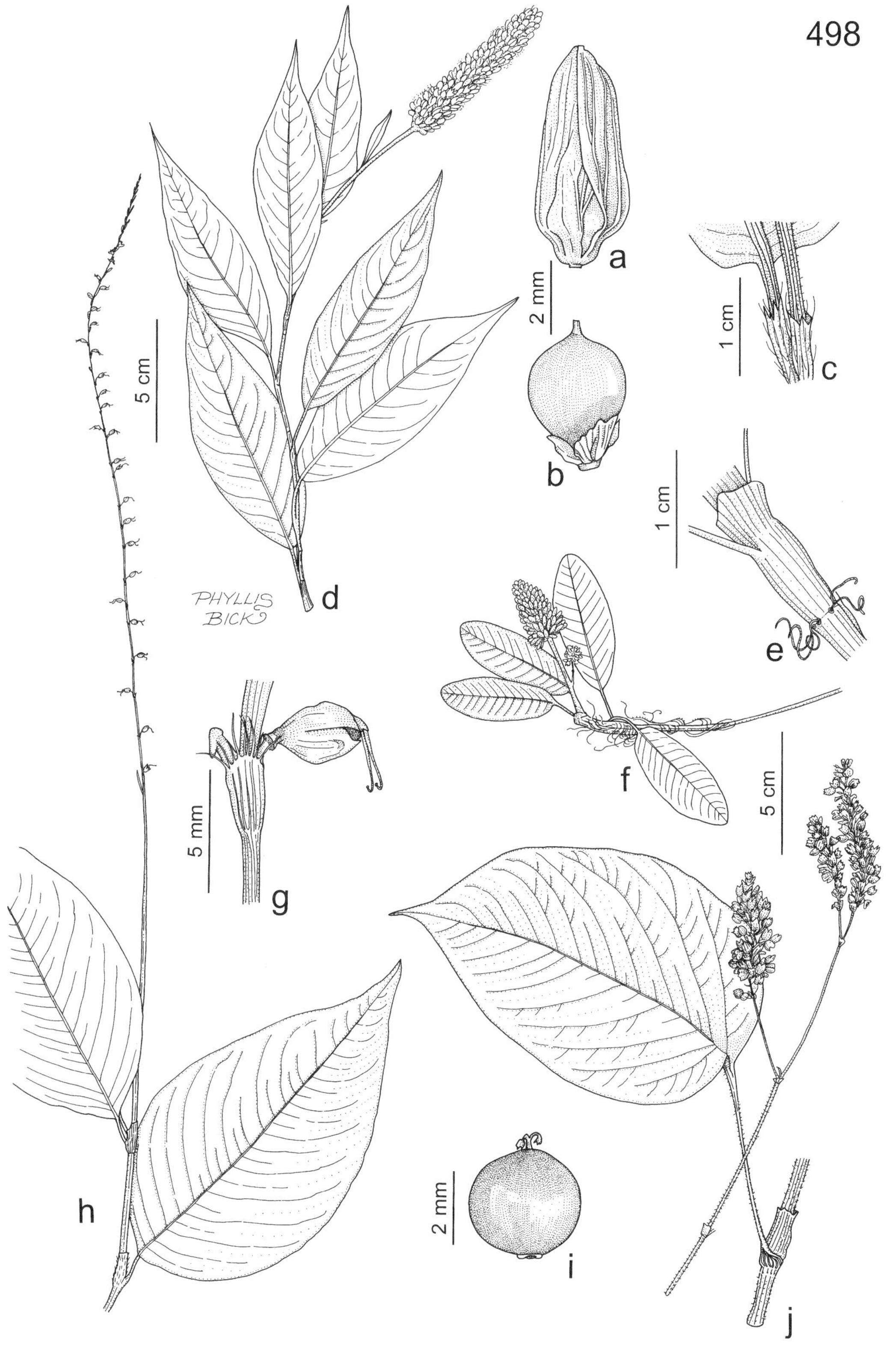

Plate 498. Polygonaceae. *Persicaria amphibia* var. *emersa*, **a)** flower, **b)** fruit, **c)** node with leaf base and ocrea, **d)** habit. *Persicaria amphibia* var. *stipulacea*, **e)** node with roots, leaf base, and ocrea, **f)** habit. *Persicaria virginiana*, **g)** detail of inflorescence with fruit and ocreola, **h)** habit. *Persicaria orientalis*, **i)** fruit, **j)** node with leaf and detached inflorescence.

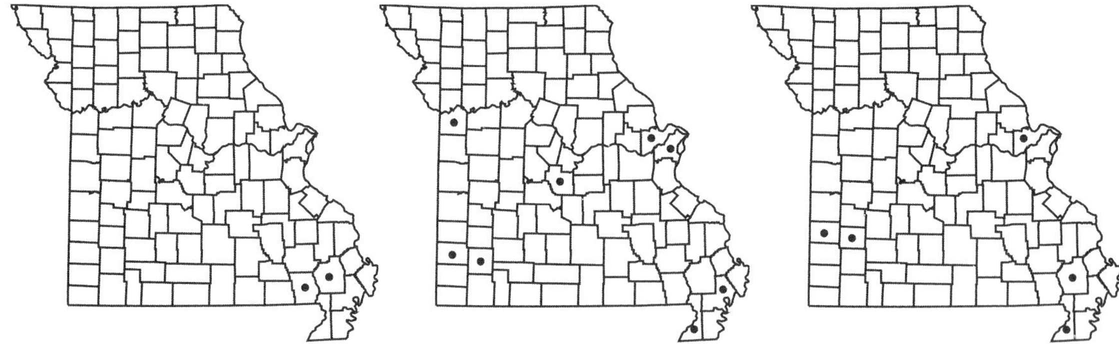

2275. Persicaria arifolia 2276. Persicaria bicornis 2277. Persicaria glabra

and those of others with the styles exserted. Perianth bright reddish pink to bright pink, bell-shaped, not gland-dotted, the tepals 5, 4–6 mm long, fused in the basal 1/3, the nerves noticeable, not anchor-shaped. Stamens 5, exserted or not, the anthers pink or red. Styles 2-branched above the midpoint, not or only slightly exserted, not persistent. Fruits 2–3 mm long, (1.5–)1.8–2.6 mm wide, not exserted, bluntly 2-angled, beakless, the faces convex, the surface smooth or minutely granular, dark brown, shiny or dull. $2n$=66, 132. June–October.

Scattered nearly throughout the state (nearly worldwide). Swamps, sloughs, oxbows, marshes, ponds, lakes, sinkhole ponds, bottomland prairies, streams, rivers, and edges of botttomland forests; also ditches, railroads, roadsides, and wet disturbed areas. Frequently emergent aquatics, less commonly floating-leaved.

In a series of ecological, morphological, and anatomical studies, Mitchell (1968, 1976) showed that the *P. amphibia* complex (as *Polygonum*) is best treated as a single polymorphic species, with most of the named variants describing plants with adaptations to different hydrological regimes. He chose to recognize three intergrading varieties, two North American taxa and the Old World var. *amphibia*, whose morphology is somewhat intermediate between the other subspecies. Hinds and Freeman (2005) and others have suggested that intergradation between these subspecies is too great to permit formal recognition of such variants. Even Mitchell (1978) mentioned that some of the differences between the aquatic and terrestrial extremes disappeared in common garden experiments. The genetic (vs. environmental) basis for the morphological variation deserves further study. For the present, the two North American varieties are accepted in the present treatment.

1. Plants emergent aquatics or terrestrial in wet soil; stems erect or ascending, usually stiff; leaves usually finely appressed-hairy; ocreae cylindric, papery; inflorescences 4–15(–18) cm long . . . . 1A. VAR. EMERSA
1. Plants floating-leaved aquatics or emergent to stranded at sites with rapidly receding shorelines; stems loosely ascending to spreading, often weak; leaves usually glabrous; ocreae usually flared, herbaceous when first formed (turning papery with age); inflorescences 1–4(–12) cm long . . . . 1B. VAR. STIPULACEA

1a. var. emersa (Michx.) J.C. Hickman
Persicaria coccinea (Muhl. ex Willd.) Greene
Polygonum amphibium var. *emersum* Michx.
Po. coccineum Muhl. ex Willd.
Po. coccineum var. *pratincola* (Greene) Stanford
Po. emersum (Michx.) Britton

Pl. 498 a–d

Stems erect or ascending, usually stiff, rarely spongy-thickened (in aquatic plants). Ocreae cylindric (not flared), tan to white and papery, even when first formed. Leaf blades lanceolate to elliptic, angled to tapered at the base, tapered to a usually sharply pointed tip, the surfaces usually pubescent with short, appressed hairs. Inflorescences 4–15(–18) cm long, short- to long-cylindric. $2n$=66. June–October.

Scattered nearly throughout the state (throughout the U.S. [including Alaska]; Canada, Mexico). Margins of swamps, sloughs, oxbows, marshes, ponds, lakes, and sinkhole ponds, banks of streams and rivers, bottomland prairies, and edges of botttomland forests; also ditches, railroads, roadsides, and wet disturbed areas. Frequently emergent aquatics, rarely in deeper water.

1b. var. stipulacea (N. Coleman) H. Hara
Polygonum amphibium f. *fluitans* (Eaton) Fernald
Po. amphibium var. *stipulaceum* N. Coleman
Po. coccineum f. *natans* (Wiegand) Stanford
Pl. 498 e, f

Stems loosely ascending to more commonly spreading, usually weak, sometimes spongy-thickened. Ocreae often flared, at first usually green and herbaceous, at least toward the tip. Leaf blades lanceolate to elliptic, angled to truncate or cordate at the base, rounded or angled to tapered to a bluntly or sharply pointed tip, the surfaces usually glabrous. Inflorescences 1–4(–12) cm long, mostly ovoid-conic to short-cylindric. $2n=132$. June–September.

Uncommon, widely scattered (throughout the U.S. [including Alaska]; Canada, Mexico). Swamps, ponds, lakes, sinkhole ponds, and pools in streams, rivers, and spring branches. Frequently emergent or floating-leaved aquatics, rarely stranded on mud.

2. Persicaria arifolia (L.) Haraldson (halberd-leaved tearthumb)
Polygonum arifolium L.
Po. arifolium var. *pubescens* (R. Keller) Fernald
Tracaulon arifolium (L.) Raf.
Truellum arifolium (L.) Soják
Pl. 497 k, l; Map 2275

Plants annual, taprooted (but sometimes rooting at the lower nodes). Stems 20–150 cm long, spreading or climbing, glabrous, but armed with small, downward-curved (hooked) prickles. Ocreae persistent or disintegrating with age, 8–15 mm long, cylindric, tan or brownish, papery, with minute prickles at the base, the margin with erect bristles 0.5–2.5 mm, the surface glabrous or appressed- to spreading-hairy. Petioles 1–7 cm long. Leaf blades (2.0–)6.5–20.0 cm long, (1–)6–16 cm wide, triangular to halberd-shaped, usually with a pair of spreading, sharply pointed, basal lobes, truncate to shallowly cordate at the base, tapered to a usually sharply pointed tip, the margins otherwise entire, but hairy and/or prickly, the upper surface appressed-hairy or rarely glabrous, the undersurface stellate-hairy or rarely glabrous, sometimes minutely prickly along the main veins, lacking impressed glands, lacking a reddish or purplish area on the upper surface. Inflorescences terminal and axillary, solitary or grouped into small panicles, 0.5–1.2 cm long, 3–8 mm wide, dense and headlike, uninterrupted, the stalk 1–8 cm long, stellate-hairy and reddish or pinkish glandular-hairy toward the tip, prickly toward the base. Ocreolae usually overlapping except sometimes toward the axis base, the margins smooth or with bristles to 0.5 mm long, not gland-dotted. Flowers 2–4 per fascicle, the perianth pink or red, often whitish green toward the base, bell-shaped, not gland-dotted, the tepals 4, 5–6 mm long, fused below the midpoint, the nerves not prominent, not anchor-shaped. Stamens (6–)8, not exserted, the anthers pink. Styles 2 (distinct to the base), not exserted, not persistent. Fruits 3.5–6.0 mm long, 3–4 mm wide, not exserted, bluntly 2-angled, beakless, the faces convex, the surface smooth, dark brown to black, shiny. July–October.

Uncommon, known thus far only from Butler and Stoddard Counties (eastern [mostly northeastern] U.S. west to Minnesota and Louisiana; Canada). Bottomland forests, banks of acidic seeps.

As noted in the discussions of the genera *Persicaria* and *Polygonum*, *Pe. arifolia* and *Pe. sagittata* usually are treated as section *Echinocaulon*, the tearthumbs, which are characterized by curved prickles on the stems and leaves. This group of 21 species was monographed (as *Po.* section *Echinocaulon*) by C.-w. Park (1988). It includes *Pe. perfoliata* (L.) H. Gross (mile-a-minute vine), an Asian species that was introduced into the United States horticulturally and that escaped to become a major invasive exotic problem plant in the northeastern United States. The species is expanding its range, having reached portions of Louisiana, and eventually may invade Missouri. It tends to form large mats that strangle other vegetation. The species is distinguished by its often somewhat peltate leaves with the blade triangular and truncate to broadly and shallowly concave at the base, the surfaces often glaucous. It also has at least some of the ocreae green and herbaceous.

3. Persicaria bicornis (Raf.) Nieuwl. (pink smartweed)
Pe. longistyla (Small) Small
Polygonum bicorne Raf.
Po. longistylum Small
Pl. 499 h, i; Map 2276

Plants annual, taprooted. Stems 20–180 cm long, ascending or erect, sometimes from a spreading base, glabrous or appressed to spreading-hairy, most of the hairs usually gland-tipped. Ocreae persistent, usually tearing with age, 6–20 mm long, somewhat inflated toward the base, tan to white

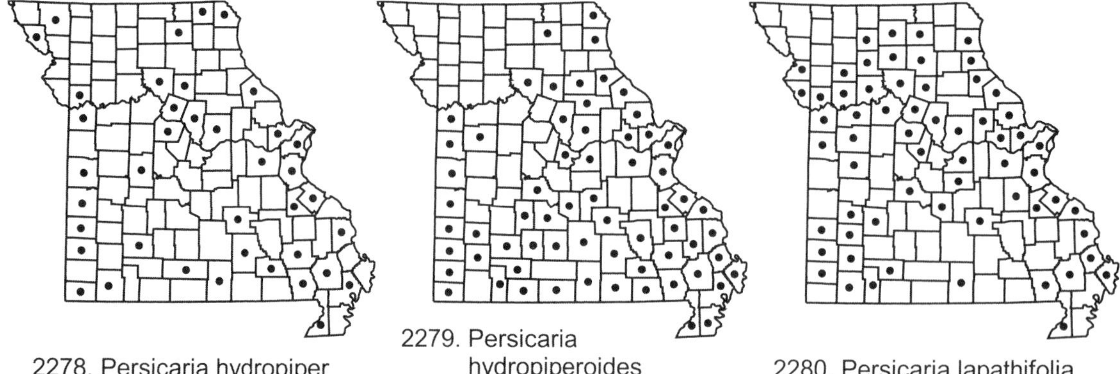
2278. Persicaria hydropiper
2279. Persicaria hydropiperoides
2280. Persicaria lapathifolia

and papery throughout, truncate and not oblique at the tip, the margin entire or nearly so, lacking bristles or occasionally with short, slender, extensions of the veins to 0.8 mm long, the surface glabrous or minutely roughened-hairy toward the base, not gland-dotted. Petioles 0.1–2.2 cm long. Leaf blades 2–13(–18) cm long, (0.4–)1.0–2.5 cm wide, linear-lanceolate to broadly lanceolate, angled or tapered at the base, angled or tapered to a sharply pointed tip, the surfaces glabrous or appressed-hairy along the main veins, unarmed, the undersurface with impressed glands, the upper surface sometimes with a reddish or purplish, chevron- to crescent-shaped or triangular area. Inflorescences terminal and axillary from the upper leaves, usually several to numerous per main stem, 0.8–6 cm long, 10–18 mm wide, erect or ascending, dense, uninterrupted, the stalk 0.8–7.0 cm long, glabrous or hairy, the hairs usually gland-tipped. Ocreolae overlapping, the margins entire or with bristles 0.1–0.8 mm long, the surface usually glabrous, not gland-dotted. Flowers sometimes functionally pistillate or staminate, 2–11 per fascicle, those of some plants with the stamens exserted and those of others with the styles exserted. Perianth light pink, bell-shaped, not gland-dotted, the tepals 5, 4–6 mm long, fused below the midpoint, the nerves relatively prominent but not anchor-shaped. Stamens 6–8, exserted or not, the anthers pink or red. Styles 2-branched from near the base, not or only slightly exserted, not persistent. Fruits 2–3 mm long, 2.0–2.8 mm wide, not or only slightly exserted, discoid or appearing slightly 3-angled, beakless, the faces flat to slightly convex, 1 of the faces usually with a central hump, the surface smooth, blackish brown to black, shiny. July–October.

Uncommon, widely scattered (Illinois to Louisiana west to Wyoming and New Mexico). Swamps and margins of ponds; also railroads and moist disturbed areas.

This species is morphologically similar to *P. pensylvanica*. According to Hinds and Freeman (2005), it differs from that species in having flowers variously with longer or shorter styles (heterostylous) and usually in having a pronounced convexity (hump) centrally on one of the faces. The Missouri records are mostly historical.

4. Persicaria glabra (Willd.) M. Gómez
Pe. densiflora (Meisn.) Moldenke
Polygonum densiflorum Meisn., an illegitimate name
Po. glabrum Willd.
Po. portoricense Bertero ex Small, an illegitimate name

Pl. 499 a–d; Map 2277

Plants perennial, with rhizomes, the stems also occasionally rooting at the lower nodes. Stems 30–170 cm long, ascending or erect, often from a spreading base, glabrous or rarely hairy toward the tip, sometimes also gland-dotted. Ocreae persistent, usually tearing with age, 12–25 mm long, somewhat inflated toward the base, tan and papery throughout, truncate, not oblique at the tip, lacking a flange of tissue at the tip, the margin entire or nearly so, lacking bristles, the surface glabrous, but dotted with scattered, minute glands. Petioles 0.2–2.0 cm long. Leaf blades (10–)15–35 cm long, 1.5–5.5 cm wide, lanceolate, tapered at the base, angled or tapered to a sharply pointed tip, the upper surface glabrous, the undersurface glabrous or minutely roughened along the midvein, unarmed, the undersurface with impressed glands, the upper surface lacking a reddish or purplish, chevron- to crescent-shaped or triangular area. Inflorescences terminal and sometimes also axillary from the upper leaves, usually several per main stem, 3–10 cm long, 5–9 mm wide, erect or ascending, straight or only slightly arched toward the tip, dense, uninterrupted, the stalk 1–5 cm long, glabrous or minutely roughened, also gland-

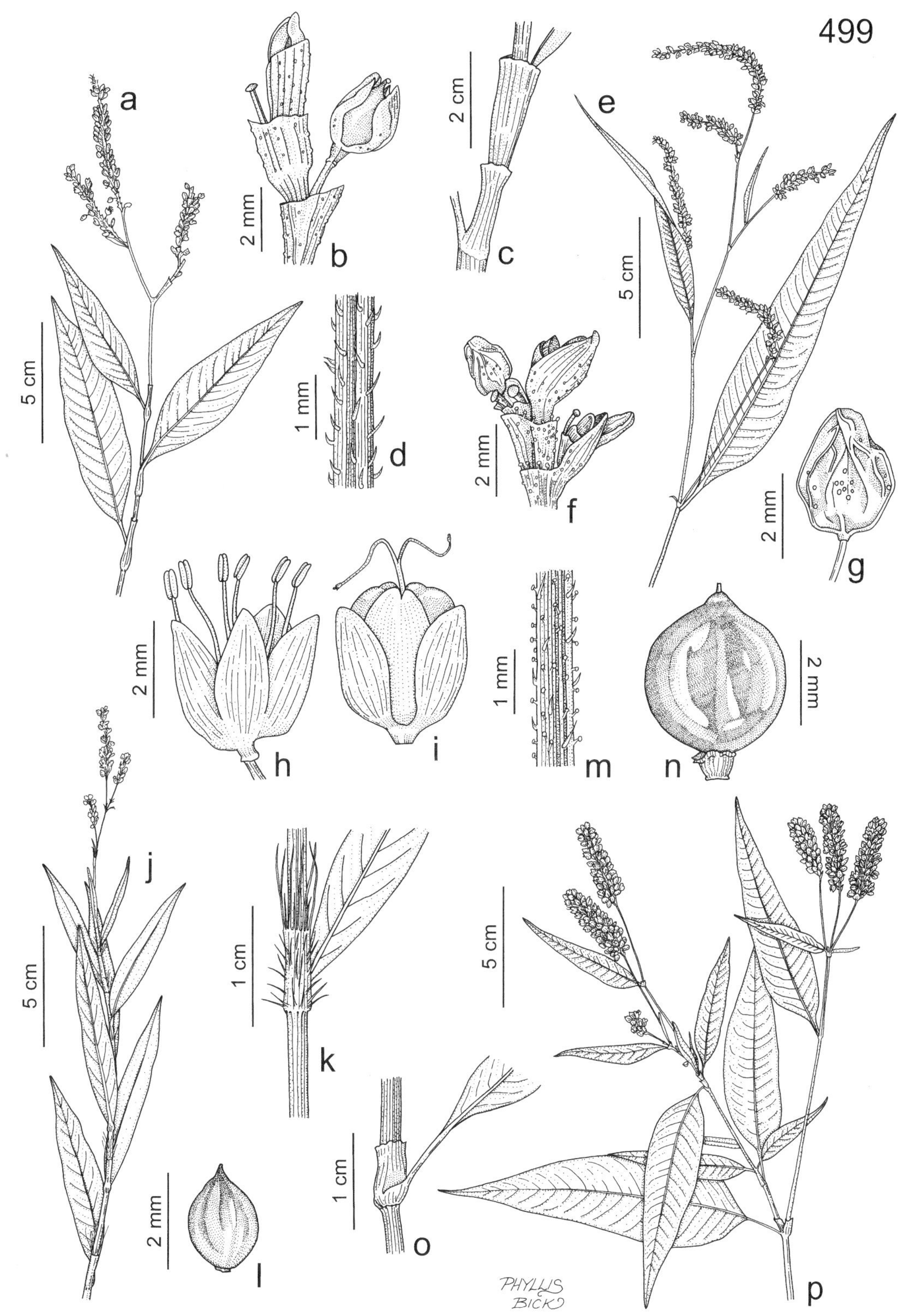

Plate 499. Polygonaceae. *Persicaria glabra*, **a)** fertile stem, **b)** detail of inflorescence with fruits and ocreola, **c)** nodes with leaf bases and ocrea, **d)** stem detail. *Persicaria lapathifolia*, **e)** fertile stem, **f)** detail of inflorescence with fruit and ocreola, **g)** flower. *Persicaria bicornis*, **h)** staminate flower, **i)** pistillate flower. *Persicaria setacea*, **j)** fertile stem, **k)** node with leaf and ocrea, **l)** fruit. *Persicaria pensylvanica*, **m)** stem detail, **n)** fruit, **o)** node with leaf base and ocrea, **p)** fertile stem.

dotted. Ocreolae usually overlapping, the margins entire, the surface glabrous, sometimes sparsely gland-dotted (the dots usually inconspicuous and pale). Flowers perfect, (1–)3–8 per fascicle, all similar. Perianth white to greenish white or somewhat pinkish-tinged, bell-shaped, sometimes gland-dotted, the tepals 5, 3.0–3.6 mm long, fused below the midpoint, the nerves usually relatively inconspicuous, irregularly few-branched (not anchor-shaped). Stamens 5–7, not exserted, the anthers pink or red. Styles 2-branched from near the base, not exserted, not persistent. Fruits 2.0–2.3 mm long, 1.3–1.6 mm wide, not exserted, discoid, beakless, the faces slightly convex, 1 of the faces usually with a central hump, the surface smooth, brown to nearly black, shiny. $2n$=66. August–October.

Uncommon, widely scattered (southeastern U.S. west to Missouri and Texas; Mexico, Central America, South America, Asia, Africa, Pacific Islands). Bottomland forests, swamps, marshes, sloughs, and banks of rivers; usually emergent aquatics.

Steyermark (1963) knew this taxon only from a few historical collections, mostly from the Mississippi Lowlands Division. Recently, it was rediscovered by McKenzie et al. (2006), who discussed its ecology and morphology. The species is robust and colonial, with minute, dark glands dotting the ocreae and slender, straight to arched inflorescences. The species has long been known as *P. densiflora*, a New World segregate of the *P. glabra* complex. K. L. Wilson (1990) concluded that the American plants (as *Polygonum*) are insufficiently distinct to warrant separate species status. However, S.-t. Kim and Donoghue (2008b) found that North American accessions of the complex appeared in a different lineage from those originating from South America and Asia and suggested that there might be merit in the continued segregation of the North American materials as *P. densiflora*. The molecular studies of these authors further suggested that this hexaploid species likely arose through past hybridization between the diploid, *P. lapathifolia* and some tetraploid member of the *P. hydropiperoides* complex. Further studies are neeeded.

5. Persicaria hydropiper (L.) Delarbre (water pepper, smartweed)
Polygonum hydropiper L.
Po. hydropiper var. *projectum* Stanford
Pl. 500 f–h; Map 2278

Plants annual, with taproots, the stems also occasionally rooting at the lower nodes. Stems 20–90 cm long, spreading to erect, glabrous, but gland-dotted. Ocreae persistent, usually tearing with age, (8–)10–15 mm long, somewhat inflated toward the base, tan to brown and papery throughout, truncate, sometimes slightly oblique at the tip, lacking a flange of tissue at the tip, the margin with erect bristles 1–4 mm long, the surface glabrous or minutely roughened, usually gland-dotted. Petioles 0.2–2.0 cm long. Leaf blades 2–10(–15) cm long, 0.4–2.5 cm wide, linear to more commonly lanceolate or narrowly rhombic, angled or tapered at the base, angled or tapered to a sharply pointed tip, the surfaces glabrous or the undersurface minutely roughened along the midvein, unarmed, the undersurface with impressed glands (these sometimes pale), the upper surface lacking a reddish or purplish, chevron- to crescent-shaped or triangular area. Inflorescences terminal and axillary from the upper leaves, usually several per main stem, 3–8(–18) cm long, 5–9 mm wide, erect or ascending, straight or more commonly nodding toward the tip, not dense, uninterrupted or sometimes somewhat interrupted, the stalk absent or to 5 cm long, glabrous, but gland-dotted. Ocreolae not overlapping or somewhat overlapping toward the inflorescence tip, the margins entire or with short bristles to 1 mm long, the surface gland-dotted (the dots sometimes pale). Flowers perfect, mostly 1–3 per fascicle, all similar. Perianth greenish-tinged toward the base, white or pinkish-tinged to pink toward the tip, bell-shaped, moderately to densely dotted with scattered, minute glands (these sometimes pale and inconspicuous), the tepals 4 or 5, 2.5–4.0 mm long, fused below the midpoint, the nerves conspicuous or inconspicuous, irregularly few-branched (not anchor-shaped). Stamens 6–8, not exserted, the anthers pink or red. Styles 2-branched or more commonly 3-branched from near the base, usually not exserted, not persistent. Fruits 2–3 mm long, 1.5–2.0 mm wide, not or only slightly exserted, 3-angled or occasionally discoid, beakless, the faces concave (slightly convex on discoid fruits), lacking a central hump, the surface minutely roughened, dark brown to nearly black, dull. May–November.

Scattered nearly throughout the state (nearly throughout the U.S. and adjacent Canada; Europe; introduced elsewhere nearly worldwide). Banks of streams and rivers, margins of ponds, lakes, and sinkhole ponds, swamps, sloughs, and openings of bottomland forests; also sloughs, crop fields, fallow fields, pastures, railroads, roadsides, and open disturbed areas.

This species can be confused with *P. punctata*. According to Thomas (2005), *P. hydropiper* typically has reddish-tinged stems (vs. usually green), leaf blades often elliptic-lanceolate (vs. lanceolate), and ocreae flared toward the tip (vs. tightly cylin-

Plate 500. Polygonaceae. *Persicaria longiseta*, **a)** fertile stem, **b)** flower. *Persicaria punctata*, **c)** flower, **d)** fertile stem, **e)** node with leaf and ocrea. *Persicaria hydropiper*, **f)** flower, **g)** fertile stem, **h)** fruit. *Persicaria hydropiperoides*, **i)** fruit, **j)** node with leaf base and ocrea, **k)** fertile stem. *Persicaria maculosa*, **l)** flower, **m)** node with leaf base and ocrea, **n)** fertile stem.

dric). The gland-dots on the perianth (and ocreae) can be relatively light in color and thus difficult to observe on fresh plants. They tend to darken somewhat upon drying.

6. Persicaria hydropiperoides (Michx.) Small (wild water pepper, mild water pepper)

Pe. opelousana (Riddell ex Small) Small
Polygonum hydropiperoides Michx.
Po. hydropiperoides var. *bushianum* Stanford
Po. hydropiperoides var. *opelousanum* (Riddell ex Small) W. Stone
Po. hydropiperoides f. *strigosum* (Small) Stanford
Po. opelousanum Riddell ex Small

Pl. 500 i–k; Map 2279

Plants perennial, often with rhizomes, the stems often also rooting at the lower nodes. Stems 15–100 cm long, erect to loosely ascending, sometimes ascending from a spreading base, glabrous or sparsely appressed-hairy toward the tip, not glandular. Ocreae persistent, usually tearing with age, 5–12 mm long, somewhat inflated toward the base, light brown to brown and papery throughout, truncate, not oblique at the tip, lacking a flange of tissue at the tip, the margin with bristles (2–)4–10 mm long, the surface glabrous or appressed-hairy, not gland-dotted. Petioles essentially absent or to 2 cm long, appressed-hairy. Leaf blades 5–22 cm long, 0.5–3.7 cm wide, linear or more commonly narrowly to broadly lanceolate, angled or tapered at the base, tapered to a sharply pointed tip, the surfaces glabrous or appressed-hairy, sometimes only along the main veins, unarmed, the undersurface sometimes with impressed glands, the upper surface lacking a reddish or purplish, chevron- to crescent-shaped or triangular area. Inflorescences terminal and axillary from the upper leaves, solitary or more commonly few to numerous per main stem, 3–8 cm long, 2–5 mm wide, erect or ascending, straight or somewhat curved or arched, relatively dense, uninterrupted or sometimes interrupted toward the base, the stalk 1–3 cm long, glabrous or minutely roughened to appressed-hairy, not glandular. Ocreolae mostly overlapping (except sometimes toward the inflorescence base), the margins with bristles mostly 1–2 mm long (mostly shorter than the ocreolae), the surface glabrous or more commonly appressed-hairy, not gland-dotted. Flowers perfect or often some of them staminate, 2–6 per fascicle. Perianth pinkish green to pink toward the base, white to more commonly green or pinkish-tinged toward the tip, bell-shaped, not gland-dotted or sometimes gland-dotted only on the inner 2 tepals, the tepals 5, 1.5–3.0 mm long, in perfect flowers becoming enlarged to 2.5–4.0 mm at fruiting, fused below the midpoint, the nerves inconspicuous, irregularly few-branched, open-branched, not forming a network, not anchor-shaped. Stamens 8, exserted only in staminate flowers, the anthers pink to red. Styles 3-branched from near the base, not exserted, not persistent. Fruits 1.5–3.0 mm long, 1.0–2.3 mm wide, not or only slightly exserted, 3-angled, beakless, the faces flat or slightly concave, lacking a central hump, the surface smooth, dark brown to black, shiny. $2n$=40. June–November.

Scattered to common nearly throughout the state, but uncommon in the western half of the Glaciated Plains Division (nearly throughout the U.S.; Canada, Mexico, Central America, South America). Banks of streams, rivers, and spring branches, margins of ponds, lakes and sinkhole ponds, bases of bluffs, swamps, marshes, fens, openings and edges of bottomland forests, and moist swales in upland prairies and sand prairies; also ditches and railroads; sometimes emergent aquatics.

Persicaria hydropiperoides is extremely variable morphologically and a number of infraspe cific taxa and segregate species have been named to account for this variation. McDonald (1980) reached a similar conclusion to that of Steyermark (1963) that *Po. opelousanum* cannot be maintained as distinct from *P. hydropiperoides*. Horton (1972), Hinds and Freeman (2005), and many other authors have further suggested that the named varieties intergrade too much to warrant recognition. The exception is var. *setaceum*, which is regarded as a distinct species (McDonald, 1980) and a possible diploid progenitor of tetraploid *P. hydropiperoides*.

7. Persicaria lapathifolia (L.) Delarbre (pale smartweed)

Polygonum lapathifolium L.
Po. nodosum Pers.
Po. scabrum Moench

Pl. 499 e–g; Map 2280

Plants annual, taprooted, the stems also occasionally rooting at the lower nodes. Stems (5–)10–120 cm long, ascending or erect, sometimes from a spreading base, glabrous or occasionally short-hairy toward the tip, most of the hairs sometimes gland-tipped. Ocreae persistent, usually tearing with age, 4–30 mm long, somewhat inflated toward the base, tan and papery throughout, truncate and slightly oblique at the tip, lacking a flange of tissue at the tip, the margin entire or nearly so, lacking bristles or occasionally with short, slender, extensions of the veins to 0.8 mm long, the sur-

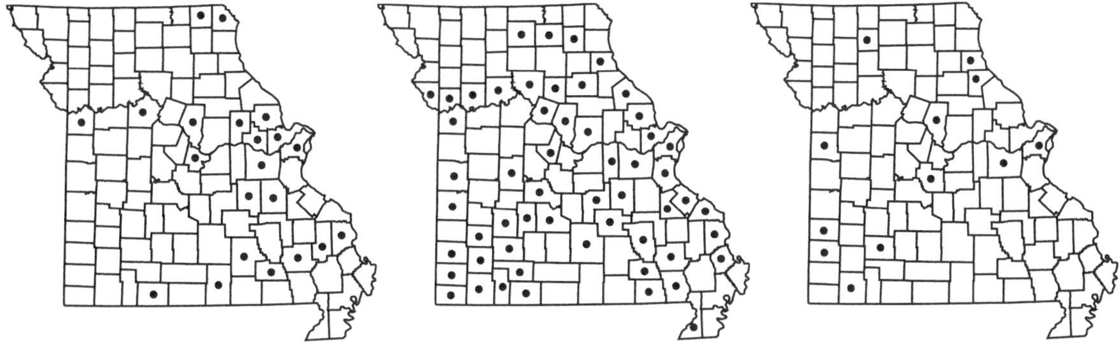

2281. Persicaria longiseta 2282. Persicaria maculosa 2283. Persicaria orientalis

face glabrous or rarely appressed-hairy, not gland-dotted. Petioles 0.1–1.8 cm long. Leaf blades 3–12(–22) cm long, 0.4–4.5(–6.0) cm wide, narrowly to broadly lanceolate, angled or tapered at the base, angled or tapered to a sharply pointed tip, the upper surface usually glabrous, the undersurface glabrous or appressed-hairy along the main veins, unarmed, the undersurface with impressed glands, the upper surface sometimes with a reddish or purplish, chevron- to crescent-shaped or triangular area. Inflorescences terminal and axillary from the upper leaves, usually several to numerous per main stem, 3–8 cm long, 5–9(–12) mm wide, ascending but strongly arched or nodding toward the tip, dense, uninterrupted, the stalk 0.2–2.5 cm long, glabrous or more commonly hairy, the hairs gland-tipped. Ocreolae usually overlapping, the margins entire or with bristles 0.1–0.4 mm long, the surface glabrous, not gland-dotted. Flowers perfect, 4–14 per fascicle, all similar. Perianth usually white to greenish white (sometimes appearing grayish-tinged), rarely pale pinkish-tinged, more or less urn-shaped, not gland-dotted, the tepals 4(5), 2–3(–4) mm long, fused below the midpoint, the 3 nerves relatively prominent, the midnerve branched apically into an anchor-shaped pattern. Stamens 5 or 6, not exserted, the anthers pink or red. Styles 2(3)-branched from near the base, not exserted, not persistent. Fruits 1.5–3.2 mm long, 1.5–2.8 mm wide, not or only slightly exserted, discoid or rarely 3-angled, beakless, the faces flat or slightly convex, 1 of the faces usually with a central hump, the surface smooth to finely pebbled, brown to black, shiny or dull. $2n=22$. June–November.

Scattered nearly throughout the state (nearly worldwide). Banks of streams and rivers, margins of ponds and lakes, sloughs, marshes, swamps, and edges of bottomland forests; also margins of crop fields, ditches, railroads, roadsides, and open disturbed areas; sometimes emergent aquatics.

Mühlenbach (1979) reported plants that he called *Po. scabrum* from the St. Louis railyards. This name corresponds to an Old World taxon occasionally introduced into North America. However, Consaul et al. (1991) found that it should not be segregated from *Pe. lapathifolia*, based on their study of allozyme variation in the complex. Similarly, most botanists have not segregated *Pe. nodosa* (Pers.) Opiz (*Po. nodosum* Pers.), a variant with relatively strongly purple-spotted stems, pink perianth, and relatively coarse pubescence. However, S.-t. Kim and Donoghue (2008b) have instead characterized it as a tetraploid taxon putatively derived through past hybridization between *Pe. lapathifolia* and the Asian *Pe. viscofera* (Makino) H. Gross ex Nakai. More studies are needed.

Steyermark (1963) reported a specimen (as *Polygonum*) from Moniteau County that he considered a hybrid between *P. lapathifolia* and *P. pensylvanica*, but this has been redetermined as *P. lapathifolia*.

The inflorescences of *P. lapathifolia* characteristically are slender and nodding. The petioles and undersurface midveins of the leaf blades typically are roughened with minute, conic hairs.

8. Persicaria longiseta (Bruijn) Kitag.

Pe. caespitosa (Blume) Nakai var. *longiseta* (Bruijn) C.F. Reed

Polygonum caespitosum Blume var. *longisetum* (Bruijn) Danser

Po. longisetum Bruijn

Pl. 500 a, b; Map 2281

Plants annual, taprooted, the stems often also rooting at the lower nodes. Stems 15–50(–80) cm long, loosely ascending or spreading with ascending tips, glabrous, not glandular. Ocreae persistent, usually tearing with age, 5–12 mm long, somewhat inflated toward the base, white to light brown and papery throughout, truncate, not oblique at the tip, lacking a flange of tissue at the tip, the margin with bristles 4–12 mm long, the surface glabrous or appressed-hairy, not gland-dotted. Petioles absent or to 0.6 cm long. Leaf blades 2–8 cm long, 1–3 cm wide, linear-lanceolate to broadly elliptic-lanceolate, angled to tapered or rarely rounded at the base, angled or tapered to a sharply pointed tip, the surfaces glabrous or the undersurface appressed-hairy along the main veins, unarmed, lacking impressed glands, the upper surface lacking a reddish or purplish, chevron- to crescent-shaped or triangular area. Inflorescences terminal and axillary from the upper leaves, 1 to several per main stem, 1–4(–6) cm long, 3–7 mm wide, erect or ascending, straight, relatively dense, uninterrupted or occasionally with 1 or a few less-crowded areas, the stalk 1–5 cm long, glabrous, not glandular. Ocreolae mostly overlapping, the margins with bristles mostly 2.5–3.5 mm long, the surface glabrous, not gland-dotted. Flowers perfect, 4–14 per fascicle, all similar. Perianth pinkish green toward the base, bright reddish pink toward the tip, bell-shaped, not gland-dotted, the tepals 5, 2.2–2.8 mm long, fused below the midpoint, the nerves inconspicuous, irregularly few-branched, not forming a network, not anchor-shaped. Stamens 5, not exserted, the anthers yellow. Styles 3-branched from near the base, not exserted, not persistent. Fruits 1.6–2.5 mm long, 1.1–1.6 mm wide, not exserted, 3-angled, beakless, the faces usually somewhat concave, lacking a central hump, the surface smooth, dark brown to black, shiny. $2n$=40. May–October.

Introduced, scattered, mostly in the eastern half of the state (native of Asia; introduced in the eastern U.S. west to Nebraska and Texas; Canada, Europe). Banks of streams and rivers and bottomland forests; also lawns, gardens, railroads, roadsides, and disturbed areas.

Steyermark (1963) knew this species only from a few collections from the city of St. Louis. It has become widely distributed in Missouri since that time. Paterson (2000) documented the spread of the taxon in the United States, beginning with the first specimens collected in 1910 near Philadelphia, with a subsequent southward and westward range expansion through the eastern half of the country. It often has sprawling stems and is characterized by its slender inflorescences of bright pink flowers. *Persicaria longiseta* can be an annoying weed in lawns and gardens.

Hinds and Freeman (2005b) followed recent Asian botanists (A.-j. Li et al., 2003) in separating *P. longiseta* at the species level from the closely related Asian taxon, *P. posumbu* (Buch.-Ham. ex D. Don) H. Gross (*P. caespitosa* (Blume) Nakai), which differs most notably in its narrower leaves with more strongly tapered tips. Some Asian botanists further separate a morphotype with the leaf bases rounded rather than angled as *Po. longisetum* var. *rotundatum* A.J. Li, but both leaf types are encountered occasionally within introduced populations in North America.

9. Persicaria maculosa Gray (lady's thumb, heart's ease, smartweed, heartweed)
Pe. vulgaris Webb & Moq.
Polygonum persicaria L.

Pl. 500 l–n; Map 2282

Plants annual, taprooted, the stems also occasionally rooting at the lower nodes. Stems (5–)10–70(–120) cm long, spreading to erect, glabrous or appressed-hairy, not glandular. Ocreae persistent, usually tearing with age, 4–30 mm long, somewhat inflated toward the base, tan to light brown and papery throughout, truncate, not oblique at the tip, lacking a flange of tissue at the tip, the margin with bristles 1.0–3.5(–5.0) mm long, the surface glabrous or appressed-hairy, not gland-dotted. Petioles absent or to 0.8 cm long. Leaf blades (1–)5–10(–18) cm long, (0.2–)1.0–2.5(–4.0) cm wide, linear-lanceolate to lanceolate or narrowly ovate, angled or tapered at the base, angled or tapered to a sharply pointed tip, the surfaces glabrous or the undersurface appressed-hairy along the main veins, unarmed, the undersurface sometimes with impressed glands, the upper surface often with a reddish or purplish, chevron- to crescent-shaped or triangular area. Inflorescences terminal and axillary from the upper leaves (sometimes also with small clusters of flowers at the median and lower nodes), solitary or more commonly several to numerous per main stem, 1.0–4.5 cm long, 7–12 mm wide, erect or ascending, straight, dense, uninterrupted, the stalk 1–5 cm long, glabrous (occasionally appressed-hairy elsewhere), not glandular. Ocreolae mostly overlapping, the margins glabrous or more commonly with bristles 0.2–1.0 mm long, the surface glabrous, not gland-dotted. Flowers perfect, 4–14 per fascicle, all similar. Perianth white to greenish white toward the base, pinkish-tinged to pink toward the tip, bell-shaped, not gland-dotted, the tepals 4 or 5, 2.0–3.5 mm long, fused below the midpoint, the nerves irregularly few-branched, forming a network toward the base (not anchor-shaped). Stamens 4–8, not exserted, the anthers yellow or pink. Styles 2- or 3-branched from near the base, not exserted, not persistent. Fruits 2–3 mm long, 1.5–2.5 mm wide, not or only slightly exserted, discoid or 3-angled, beakless, the faces flat or slightly concave, lacking a central hump, the surface smooth, dark brown to black, shiny. $2n$=44. May–October.

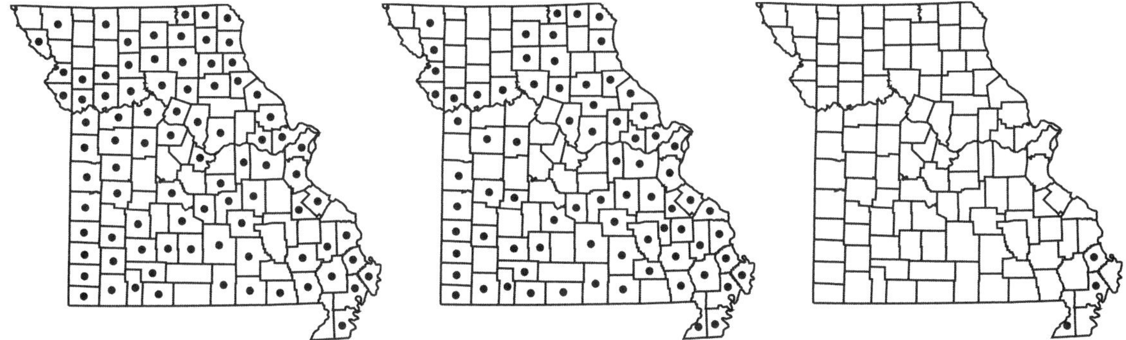

2284. Persicaria pensylvanica 2285. Persicaria punctata 2286. Persicaria robustior

Introduced, scattered nearly throughout the state (native of Europe, Asia, Africa; introduced nearly worldwide). Banks of streams, rivers, and spring branches, sloughs, fens, bottomland forests, openings of mesic upland forests, and moist swales in sand prairies; also crop fields, fallow fields, old fields, gardens, ditches, railroads, roadsides, and open disturbed areas; occasionally emergent aquatics.

Persicaria maculosa is an allotetraploid species that apparently arose through past hybridization between *Pe. lapathifolia* and a second, as yet undetermined, diploid parent (Consaul et al., 1991). This taxon has long been known by the name, *Po. persicaria*. However, when *Persicaria* is segregated from *Polygonum*, it requires a different name. There has been controversy surrounding the proper epithet in *Persicaria*, with some authors having used *Pe. vulgaris*, which however was published 25 years later than was *Pe. maculosa*.

According to Thomas (2005), this species frequently has a reddish to purplish spot on the upper leaf surface.

10. Persicaria orientalis (L.) Spach (prince's feather, kiss-me-over-the-garden-gate)
Polygonum orientale L.

Pl. 498 i, j; Map 2283

Plants annual, taprooted. Stems (60–)120–300 cm long, erect or ascending, with soft, spreading hairs (these sometimes sparse toward the stem base), not rooting at the lower nodes. Ocreae persistent, usually tearing with age, 10–20 mm long, inflated toward the tip, tan to white and papery most of the length, green and herbaceous toward the tip, sometimes with a stipulelike flange of tissue at the truncate (not oblique) tip, the margin with bristles 1–3 mm long, the surface appressed- or spreading-hairy, not gland-dotted. Petioles 1–9(–14) cm long. Leaf blades 6–30 cm long, 3–17 cm wide, broadly ovate, broadly rounded to truncate or cordate at the base, short-tapered to a pointed tip, the surfaces densely hairy, especially along the main veins of the undersurface, unarmed (but the margins often roughened to the touch), lacking impressed glands, lacking a reddish or purplish area on the upper surface. Inflorescences terminal, usually several in a loose cluster, 1–15 cm long, 8–18 mm wide, nodding, dense, uninterrupted, the stalk 2–10 cm long, hairy. Ocreolae overlapping, the margins with bristles 0.2–1.0 mm long, the surface glabrous or minutely appressed-hairy, not gland-dotted. Flowers (1–)2–5 per fascicle, all similar. Perianth bright reddish pink to bright pink, bell-shaped, not gland-dotted, the tepals 5, 3.0–4.5 mm long, fused in the basal 1/3, the nerves noticeable, not anchor-shaped. Stamens 6–8, exserted or not, the anthers pink or red. Styles 2-branched above the midpoint, not or only slightly exserted, not persistent. Fruits 2.5–3.5 mm long, 3.0–3.5 mm wide, not exserted, flattened, beakless, only slightly biconvex in cross-section, the surface smooth or minutely granular, dark brown to black, shiny or dull. June–October.

Introduced, uncommon, widely scattered (native of Asia; introduced widely in the eastern U.S. west to Nebraska and Texas, also Oregon and California; Canada). Fallow fields, railroads, and open disturbed areas.

This tall, large-leaved species is cultivated as an ornamental for its nodding spikes of bright pink flowers. It escapes occasionally, but does not appear to persist for many years outside of cultivation in Missouri. The plants characteristically have relatively densely hairy stems and leaves.

11. Persicaria pensylvanica (L.) M. Gómez (pinkweed, common smartweed, Pennsylvania smartweed)
Polygonum pensylvanicum L.
Po. pensylvanicum f. *albineum* Farw.

Po. pensylvanicum var. *eglandulosum* Myers
Po. pensylvanicum var. *laevigatum* Fernald
Pl. 499 m–p; Map 2284

Plants annual, with taproots, the stems usually not rooting at the lower nodes. Stems 10–200 cm long, ascending or erect, sometimes from a spreading base, glabrous or appressed-hairy toward the tip, not also gland-dotted. Ocreae persistent, usually tearing with age, 5–20 mm long, somewhat inflated toward the base, tan to light brown and papery throughout, truncate, oblique at the tip, lacking a flange of tissue at the tip, the margin entire or nearly so, lacking bristles or occasionally with short, slender, extensions of the veins to 0.5 mm long, the surface glabrous or appressed-hairy, not gland-dotted. Petioles 0.1–3.0 cm long. Leaf blades 4–18(–25) cm long, (0.5–)1.0–4.8 cm wide, narrowly to broadly lanceolate, angled or tapered at the base, angled or tapered to a sharply pointed tip, the surfaces glabrous or variously appressed- to spreading-hairy, the spreading hairs sometimes gland-tipped, unarmed, the undersurface sometimes with impressed glands, the upper surface sometimes with a reddish or purplish, chevron- to crescent-shaped or triangular area. Inflorescences terminal and usually also axillary from the upper leaves, usually several to numerous per main stem, 0.5–5.0 cm long, 5–15 mm wide, erect or ascending, straight or rarely somewhat nodding toward the tip, dense, uninterrupted, the stalk 1–7 cm long, glabrous or pubescent with conspicuous, dark, stalked glands. Ocreolae overlapping, the margins entire or with minute bristlelike teeth to 0.5 mm long, the surface glabrous, occasionally sparsely gland-dotted (the dots usually inconspicuous and pale). Flowers perfect, 2–14 per fascicle, all similar (elsewhere those of some plants with the stamens exserted and those of others with the styles exserted). Perianth greenish white, pinkish-tinged, or light pink, bell-shaped, not gland-dotted, the tepals 5, 3.0–3.6 mm long, fused below the midpoint, the nerves relatively prominent, irregularly few-branched (not anchor-shaped). Stamens 6–8, not exserted, the anthers yellow, pink, or red. Styles 2(3)-branched from near the base, not exserted, not persistent. Fruits 2.0–3.5 mm long, 1.8–3.0 mm wide, not or only slightly exserted, discoid or rarely 3-angled, beakless, the faces somewhat concave, lacking a central hump, the surface smooth, brown to black, shiny. $2n$=22, 44, 88. May–October.

Scattered to common throughout the state (nearly throughout the U.S. [including Alaska]; Canada; introduced in South America, Europe). Banks of streams and rivers, margins of ponds, lakes, and sinkhole ponds, swamps, sloughs, fens, marshes, bottomland prairies, and bottomland forests; also ditches, pastures, fencerows, railroads, roadsides, and disturbed areas; sometimes emergent aquatics.

Consaul et al. (1991) provided evidence from allozyme data that *P. pensylvanica* is an allopolyploid species that apparently arose through past hybridization involving *P. lapathifolia* and a second, as yet undetermined parent. They refuted an earlier diploid ($2n$=22) count for this species. Several varieties have been accepted in some of the botanical literature (Steyermark, 1963), but there do not appear to be clear discontinuities for the morphological characters said to separate these taxa. Thus, most recent authors (Taylor-Lehman, 1987; E. B. Smith, 1988; Gleason and Cronquist, 1991; Hinds and Freeman, 2005) have accepted the species as a polymorphic taxon with a complex pattern of morphological variation expressed in a large series of mostly inbreeding populations. Rare, white-flowered plants have been called *Polygonum pensylvanicum* f. *albineum*.

12. Persicaria punctata (Elliott) Small (water smartweed)
Polygonum punctatum Elliott
Po. punctatum var. *confertiflorum* (Meisn.) Fassett
Pl. 500 c–e; Map 2285

Plants annual or perennial, sometimes with rhizomes and/or stolons, the stems sometimes rooting at the lower nodes. Stems 15–90(–120) cm long, ascending to erect, often from a spreading base, glabrous, but gland-dotted. Ocreae persistent, usually tearing with age, (4–)9–18 mm long, somewhat inflated toward the base, brown and papery throughout, truncate, not oblique at the tip, lacking a flange of tissue at the tip, the margin with erect bristles 2–12 mm long, the surface glabrous or sparsely hairy, gland-dotted. Petioles absent or to 1 cm long. Leaf blades 4–10(–15) cm long, 0.6–2.0(–2.4) cm wide, lanceolate to narrowly ovate or somewhat rhombic, angled or tapered at the base, angled or tapered to a sharply pointed tip, the surfaces glabrous or the undersurface minutely roughened along the midvein, unarmed, the undersurface with impressed glands (these sometimes pale), the upper surface lacking a reddish or purplish, chevron- to crescent-shaped or triangular area. Inflorescences terminal and sometimes also axillary from the upper leaves, solitary or few per main stem, 5–20 cm long, 4–8 mm wide, erect or ascending, straight or sometimes slightly zigzag, mostly open with irregularly spaced flowers, usually interrupted, the stalk 3–6 cm long, glabrous, but gland-dotted. Ocreolae mostly not overlapping,

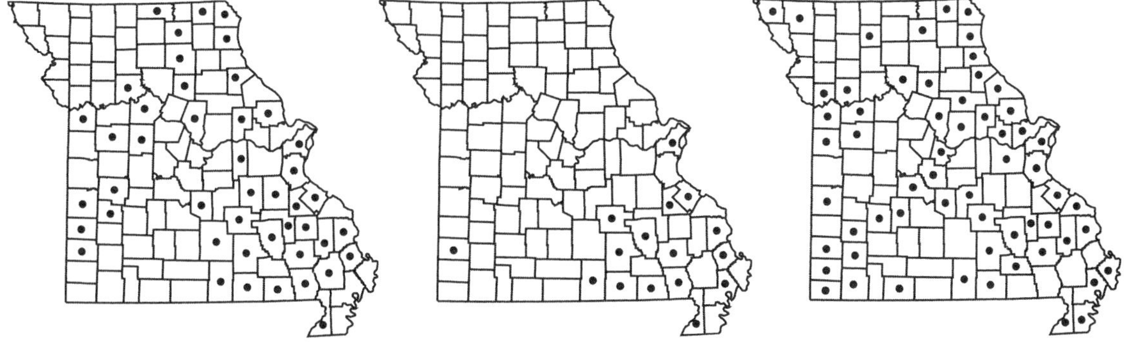

2287. Persicaria sagittata 2288. Persicaria setacea 2289. Persicaria virginiana

the margins usually with short bristles 1–2 mm long, the surface gland-dotted (the dots sometimes pale). Flowers perfect, mostly 2–6 per fascicle, all similar. Perianth green toward the base, white or rarely pinkish-tinged toward the tip, bell-shaped, moderately to densely dotted with scattered, minute glands (these sometimes pale and inconspicuous), the tepals 5, 3.0–3.5 mm long, fused below the midpoint, the nerves conspicuous or inconspicuous, irregularly few-branched (not anchor-shaped). Stamens 6–8, not exserted, the anthers pink or red. Styles 2- or 3-branched from near the base, not exserted, not persistent. Fruits 1.8–3.2 mm long, 1.5–2.2 mm wide, not or only slightly exserted, 3-angled or discoid, beakless, the faces concave (in 3-angled fruits) or slightly convex (in discoid fruits), lacking a central hump, the surface smooth, dark brown to nearly black, shiny. $2n=44$. July–November.

Scattered to common nearly throughout the state (nearly throughout the U.S.; Canada, Mexico, Central America, South America, Caribbean Islands). Banks of streams, rivers, and spring branches, margins of ponds, lakes, and sinkhole ponds, swamps, sloughs, fens, marshes, bottomland forests; also crop fields, fallow fields, ditches, and disturbed areas; sometimes emergent aquatics.

Fassett (1949) divided *P. punctata* (as *Polygonum*) into a confusing series of 12 varieties. Hinds and Freeman (2005) concluded that most of these were not sufficiently distinct to warrant taxonomic recognition. However, he treated one variant as a separate species, *P. robustior*, a conclusion that is followed in the present work.

This species can be confused with *P. hydropiper*. For further discussion, see the treatment of that species. Based on their molecular analyses, S.-t. Kim and Donoghue (2008b) suggested that *P. punctata* might have originated through past hybridization between *P. hydropiper* and some other diploid species, perhaps *P. setacea* or *P. hirsuta* (Walter) Small (of the southeastern United States). Thomas (2005) noted that this species can tolerate more shade than most other Missouri smartweeds. The gland-dots on the tepals (and ocreae) can be relatively light in color and thus difficult to observe on fresh plants. They tend to darken somewhat upon drying.

13. Persicaria robustior (Small) E.P. Bicknell

Pe. punctata var. *robustior* (Small) Small

Polygonum punctatum Elliott var. *majus* (Meisn.) Fassett

Po. punctatum var. *robustius* Small

Po. robustius (Small) Fernald

Map 2286

Plants perennial, with often woody rhizomes and sometimes also stolons, the stems sometimes rooting at the lower nodes. Stems 30–200 cm long, ascending to erect, often from a spreading base, glabrous, but gland-dotted. Ocreae persistent, usually tearing with age, 10–15 mm long, somewhat inflated toward the base, tan to light brown and papery throughout, truncate, not or slightly oblique at the tip, lacking a flange of tissue at the tip, the margin with erect bristles 3–12 mm long, the surface appressed-hairy, gland-dotted. Petioles 0.2–2.0 cm long. Leaf blades 4–20 cm long, 2.0–4.5 cm wide, lanceolate to elliptic-lanceolate, angled or more commonly tapered at the base, angled or tapered to a sharply pointed tip, the surfaces glabrous or the undersurface minutely roughened along the midvein, unarmed, the undersurface with impressed glands (these sometimes pale), the upper surface lacking a reddish or purplish, chevron- to crescent-shaped or triangular area. Inflorescences terminal and sometimes also axillary from the upper leaves, solitary or few to several per main stem, 2–8 cm long, 4–10 mm wide, erect or ascending, straight or sometimes slightly arched, dense, not interrupted, the stalk 0.5–4.0

cm long, glabrous, but gland-dotted. Ocreolae mostly overlapping, the margins lacking bristles or those of the lower ocreolae with short bristles to 1 mm long, the surface gland-dotted (the dots sometimes pale). Flowers perfect, mostly 2–4 per fascicle, all similar. Perianth green toward the base, white toward the tip, narrowly bell-shaped to slightly urn-shaped, moderately to densely dotted with scattered, minute glands (these sometimes pale and inconspicuous), the tepals 5, 3.2–4.2 mm long, fused below the midpoint, the nerves conspicuous or inconspicuous, irregularly few-branched (not anchor-shaped). Stamens 6–8, not exserted, the anthers pink or red. Styles 3-branched from near the base, not exserted, not persistent. Fruits 2.7–3.5 mm long, 2.0–2.5 mm wide, not exserted, 3-angled, beakless, the faces concave, lacking a central hump, the surface smooth, dark brown to nearly black, shiny. July–November.

Uncommon, restricted to the Mississippi Lowlands Division (northeastern U.S. west to Michigan and Missouri; disjunct in Florida and Texas; Canada, Mexico, Central America, South America, Caribbean Islands). Swamps.

This species often has been treated as a variety of *P. punctata* (Steyermark, 1963). It differs most notably in its larger leaves and flowers, mostly unbranched, robust stems, and denser inflorescences. The gland-dots on the tepals (and ocreae) can be relatively light in color and thus difficult to observe on fresh plants. They tend to darken somewhat upon drying. The Missouri specimens are all historical, although it is likely that the species still grows in the state but has been overlooked by collectors.

14. Persicaria sagittata (L.) H. Gross (arrow-leaved tearthumb)

Polygonum sagittatum L.
Tracaulon sagittatum (L.) Small
Truellum sagittatum (L.) Soják

Pl. 497 d–g; Map 2287

Plants annual, taprooted (but sometimes rooting at the lower nodes). Stems 30–200 cm long, spreading or climbing, glabrous, but armed with small, downward-curved (hooked) prickles. Ocreae mostly persistent, 5–13 mm long, cylindric, tan or brownish, papery, unarmed at the base, the margin entire or with erect bristles 0.2–1.0 mm, the surface glabrous. Petioles 0.5–4.0 cm long. Leaf blades 2.0–8.5 cm long, 1–3 cm wide, broadly oblong-lanceolate to arrowhead-shaped, cordate at the base or more commonly with a pair of parallel, rounded, basal lobes, tapered to a bluntly or sharply pointed tip, the margins otherwise entire, but sometimes hairy, the surfaces glabrous or moderately to densely appressed-hairy, the undersurface sometimes also minutely prickly along the main veins, lacking impressed glands, lacking a reddish or purplish area on the upper surface. Inflorescences terminal and axillary, solitary or grouped into small panicles, 0.5–1.5 cm long, 4–12 mm wide, dense and headlike, uninterrupted, the stalk 1–8 cm long, usually glabrous, occasionally prickly toward the base. Ocreolae usually overlapping, the margins smooth, not gland-dotted. Flowers 2 or 3 per fascicle, the perianth white or greenish white, sometimes pinkish- or reddish-tinged, bell-shaped, not gland-dotted, the tepals 5, 3–5 mm long, fused below the midpoint, the nerves not prominent, not anchor-shaped. Stamens 8, not exserted, the anthers pink. Styles 3-branched above the midpoint, not exserted, not persistent. Fruits 2.5–4.0 mm long, 1.8–2.5 mm wide, not exserted, 3-angled (triangular in cross-section), beakless, the faces flat or slightly concave, the surfaces smooth or minutely pitted, brown to black, shiny or dull. $2n=40$. June–October.

Uncommon in the western portions of the Glaciated Plains and Ozark Divisions, scattered elsewhere in the state (eastern U.S. west to North Dakota and Texas; Canada, Asia). Banks of streams and spring branches, margins of ponds, lakes, and sinkhole ponds, swamps, sloughs, fens, bases of bluffs, bottomland forests, bottomland prairies, and moist swales in upland prairies; also ditches and roadsides.

As noted in the discussions of the genera *Persicaria* and *Polygonum*, *Pe. arifolia* and *Pe. sagittata* usually are treated as section *Echinocaulon*, the tearthumbs, which are characterized by curved prickles on the stems and leaves. This group of 21 species was monographed (as *Polygonum* section *Echinocaulon*) by C.-w. Park (1988). Some botanists have segregated from *Pe. sagittata* closely related Asian plants, which differ in having leaves with glabrous margins and in their relatively dull, minutely pitted fruits, as *Po. sieboldii* Meisn. However C.-w. Park (1988) regarded these (as *Polygonum*) as representing merely a minor variant within a morphologically variable *Pe. sagittata*.

15. Persicaria setacea (Baldwin) Small (bog smartweed)

Polygonum setaceum Baldwin
Po. setaceum var. *interjectum* Fernald
Po. hydropiperoides var. *setaceum* (Baldwin) Gleason

Pl. 499 j–l; Map 2288

Plants perennial, with rhizomes and sometimes stolons (in emergent aquatic plants), the stems

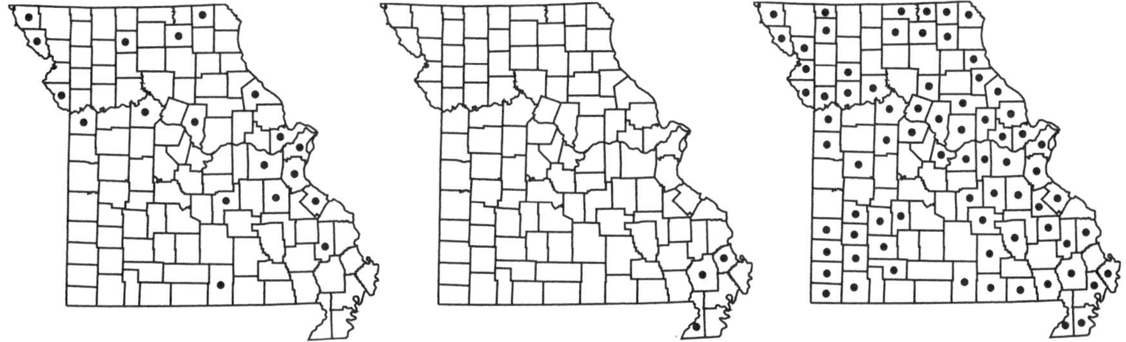

2290. Polygonum achoreum 2291. Polygonum americanum 2292. Polygonum aviculare

often also rooting at the lower nodes. Stems 50–150 cm long, erect or ascending, glabrous or sparsely to moderately pubescent with loosely ascending to more or less spreading hairs toward the tip, not glandular. Ocreae persistent, usually tearing with age, 10–20 mm long, somewhat inflated toward the base, brown and papery throughout, truncate, not or only slightly oblique at the tip, lacking a flange of tissue at the tip, the margin with bristles 6–12 mm long, the surface appressed-hairy and also more or less spreading-hairy, at least toward the base, not gland-dotted. Petioles essentially absent or to 0.5 cm long, spreading-hairy. Leaf blades (3–)6–15(–18) cm long, (1.5–)2.0–3.2(–5.0) cm wide, lanceolate, angled or tapered to broadly rounded or truncate at the base, angled or tapered to a sharply pointed tip, the surfaces sparsely to moderately pubescent with spreading to loosely appressed, fine hairs (those of submerged leaves sometimes glabrous), unarmed, the undersurface lacking impressed glands, the upper surface sometimes with a reddish or purplish, chevron- to crescent-shaped or triangular area. Inflorescences mostly terminal, occasionally also axillary from the upper leaves, solitary or more commonly few to several per main stem, (2–)3–8 cm long, 4–8 mm wide, erect or ascending, straight or somewhat curved or arched, relatively dense, usually uninterrupted, the stalk 1–7 cm long, appressed-hairy, not glandular. Ocreolae overlapping, the margins with bristles mostly 0.5–3.0(–5.0) mm long, the surface glabrous or appressed-hairy, not gland-dotted. Flowers perfect, (1)2–4(5) per fascicle. Perianth green toward the base, cream-colored to tan toward the tip, bell-shaped, not gland-dotted, the tepals 5, 2–3 mm long, fused toward the base, the nerves irregularly few-branched, open-branched, not forming a network, not anchor-shaped. Stamens 5, not exserted, the anthers pink to red. Styles 3-branched from near the base, not exserted, not persistent. Fruits (1.5–)2.0–2.5 mm long, 1.2–1.7 mm wide, not or only slightly exserted, 3-angled, beakless, the faces flat or slightly concave, lacking a central hump, the surface smooth, brown to black, shiny. $2n=20$. July–November.

Scattered in the southeastern quarter of the state, disjunct in Jasper County and in St. Louis County and City (eastern U.S. west to Indiana and Texas, disjunct in Washington). Banks of streams, margins of ponds, lakes, and sinkhole ponds, swamps, fens, and bottomland forests.

Steyermark (1963) and some other botanists have treated this taxon as a variety of the closely related *P. hydropiperoides*. Horton (1972) treated it as a separate species, based primarily on its relatively small fruits. McDonald (1980) also concluded that it should be treated as a species, based on his studies of cytology, morphology, and hybridization in the complex.

16. Persicaria virginiana (L.) Gaertn. (Virginia knotweed, jumpseed)

Antenoron virginianum (L.) Roberty & Vautier

Polygonum virginianum L.

Po. virginianum var. *glaberrimum* (Fernald) Steyerm.

Tovara virginiana (L.) Raf.

Pl. 498 g, h; Map 2289

Plants perennial, with rhizomes, the stems not rooting at the lower nodes. Stems 40–60(–90) cm long (below the inflorescence), erect or ascending, glabrous or densely appressed-hairy toward the tip toward the tip, not glandular. Ocreae persistent, often tearing with age, 10–20 mm long, loosely cylindric but not noticeably inflated, light brown to brown and papery throughout, more or less truncate but often noticeably oblique at the tip, lacking a flange of tissue at the tip, the margin with bristles 0.5–4.0 mm long, the surface densely appressed-hairy, the hairs sometimes appearing somewhat woolly, not gland-dotted. Petioles of the upper leaves absent or very short progressively longer on lower leaves to 1–3 cm on lowermost leaves, appressed-hairy, mostly on the upper surface. Leaf blades 5–18 cm long, 2–10 cm wide, ovate to elliptic or obovate, those of the uppermost leaves sometimes lanceolate, angled or rounded at the base, tapered to a sharply pointed tip, the upper surface glabrous or more commonly with scatted, stiff, bristlelike hairs (similar to those of the margins), the undersurface glabrous or sparsely to moderately pubescent with softer, short, more or less appressed hairs, mostly along the veins, unarmed, the undersurface often appearing finely bumpy or pebbled (especially in dried specimens) but lacking impressed glands, the upper surface sometimes with a reddish or purplish, chevron- to crescent-shaped or triangular area. Inflorescences terminal and axillary from the upper leaves, solitary or few per main stem, 40–130 cm long, 6–15 mm wide, erect or more commonly somewhat curved or arched, open, the nodes well-separated, the fascicles of flowers not overlapping those of the adjacent nodes, the stalk 2–40(–60) cm long, appressed-hairy, sometimes glabrous or nearly so toward the tip, not glandular. Ocreolae not overlapping, the margins with bristles 1–3 mm long, the surface appressed-hairy, not gland-dotted. Flowers perfect, solitary or 2 or 3 per fascicle. Perianth white, greenish white, or rarely pink, bell-shaped at flowering but becoming more or less urn-shaped soon after flowering, not gland-dotted, the tepals 4, 2.5–3.5 mm long, fused toward the base, the nerves inconspicuous, irregularly few-branched, open-branched, not forming a network, not anchor-shaped. Stamens 4, somewhat exserted, the anthers yellow or pink. Styles 2-branched from the base, exserted. Fruits 3.5–4.0 mm long, 2.0–2.8 mm wide, not or more commonly slightly exserted, bluntly 2-angled, tapered to a prominent, hooked beak, the faces convex, lacking a central hump, the surface smooth or finely wrinkled, brown to black, shiny or dull. $2n$=44. June–October.

Scattered nearly throughout the state (eastern U.S. west to Minnesota, Nebraska, and Texas; Canada, Mexico). Bottomland forests, mesic upland forests, bases of bluffs, marshes, and banks of streams and rivers; rarely also fallow fields and ditches.

As noted in the discussions of the genera *Persicaria* and *Polygonum*, this species and two Asian relatives sometimes have been segregated into the genus *Tovara* (an officially rejected name). This group has an interesting mechanism of seed dispersal, in which tension builds at the joint of the flower stalk as the fruit matures, which acts as a spring to ballistically disperse the fruit up to 4 m (Reed and Amoot, 1906). Often, passing animals trigger fruit dispersal. The hooked beak of the fruit then becomes tangled in the fur of the animal.

Persicaria virginiana sometimes is cultivated in woodland gardens. Cultivars with reddish inflorescences or variegated leaves have been developed. In nature, a rare minor variant with relatively thick leaves that are glabrous or nearly so at maturity have been called var. *glaberrima*. It occurs sporadically in the southern portion of the species range and was reported from Pemiscot County by Steyermark (1963).

6. **Polygonum** L. (knotweed)

Plants usually annual, with sometimes woody taproots, or shrubby (in *P. americanum*), not climbing or producing tendrils. Stems usually densely branched, prostrate to erect, glabrous (sometimes minutely roughened in *P. tenue*), sometimes somewhat glaucous (note that nonglaucous plants infected with powdery mildew [especially the common, widespread *Erysiphe polygoni* DC.], usually appear uniformly pale or grayish). Leaves alternate, sometimes heterophyllous, then relatively dense and sometimes noticeably smaller on the branches produced later in the growing season and sparser and larger on the main stems produced earlier in the season, sessile or short-petiolate, jointed at the base. Ocreae usually pale, white, or silvery above a green base, papery, usually fused to the petiole and 2-lobed when young (frequently becoming shredded or disappearing with age), glabrous. Leaf blades linear to ovate or obovate, usually somewhat leathery (somewhat thickened and succulent in *P. americanum*), variously rounded to sharply pointed at the tip, angled or tapered at the base, the margins entire (usually minutely toothed in *P. tenue*), the surfaces glabrous, sometimes somewhat glaucous (see note on powdery mildew above). Inflorescences axillary or terminal, of solitary flowers, small clusters of 2–6 flowers, or spikelike racemes of several to numerous flowers (these sometimes grouped into panicles in *P. americanum*), the individuals or clusters subtended by ocreolae, ascending or spreading along the axis, lacking other bracts. Flowers perfect, sessile or short-stalked (the perianth not tapered at the base), jointed at or just below the perianth base, unwinged. Tepals (4)5 in 2 whorls (in some species appearing more or less in 1 whorl), fused variously only near the base or to below or above the midpoint, with more or less entire margins, glabrous; ascending during and after flowering, the outer whorl of 2 or 3 tepals sometimes shorter or longer than the inner whorl, sometimes with boat-shaped to hooded tips, then more or less enclosing the inner whorl of relatively flat tepals (the outer whorl becoming reflexed in *P. americanum*), those of both whorls variously shaped, green with white or occasionally pinkish-tinged margins and/or tips (uniformly yellowish green in *P. achoreum*), not becoming enlarged at fruiting, remaining green, not developing wings or a prominent network of nerves, not developing a central swelling (grain). Stamens 3–8, not exserted (except in *P. americanum*), the anthers attached toward their midpoints, the filaments with dilated bases. Styles (2)3, the stigmas capitate. Fruits achenes, more or less enclosed in the persistent perianth (sometimes well-exserted in *P. ramosissimum*), 2–4 mm long (late-season fruits to 5.0 mm), 3-angled (late-season fruits sometimes biconvex and 2-angled), unwinged, ovate in outline, short-beaked, variously yellowish green to brown or black, glabrous, dull or shiny. About 75 species, nearly worldwide, most diverse in temperate portions of the northern hemisphere.

The segregation of more homogeneous groups from the formerly broadly circumscribed genus *Polygonum* has received a lot of attention in recent decades. Beginning with the groundbreaking work of Haraldson (1978), whose detailed and insightful study used a novel set of correlated morphological and anatomical characters to redefine generic and infrageneric limits in subfamily Polygonoideae, there has been increasing acceptance of the idea that the traditional view of *Polygonum* encompasses several well-circumscribed taxonomic lineages that should be recognized as separate genera. The dissection of *Polygonum* has since received support from independent research involving floral morphology (Ronse Decraene and Akeroyd, 1988; Hong et al., 1998), fruit morphology and anatomy (Ronse De Craene et al., 2000), and various molecular markers (Lamb Frye and Kron, 2003; S.-t. Kim and Donoghue, 2008a; Sanchez and Kron, 2008). Botanists have differed somewhat on how many segregates to recognize at generic rank, but the molecular work has clarified the hierarchy of lineages within the group and the relationships among the various segregates. In the modern classification, as discussed for the temperate North American species by Freeman (2005), three genera are represented in the Missouri flora that formerly were treated as several sections in the genus *Polygonum* by Steyermark (1963): *Polygonum* is restricted to the knotweeds; the genus *Persicaria* comprises three well-marked groups, the smartweeds, the tearthumbs, and Virginia knotweed; and *Fallopia* consists of two groups, the climbing buckwheats and Mexican bamboo. For further discussions of the generic circumscriptions of *Fallopia* and *Persicaria*, see the treatments of those genera.

However, recent studies have strongly suggested that *Polygonella* Michx., a group of eleven species most diverse in the southeastern United States (Horton, 1963), should be treated as a specialized group within *Polygonum*. The transfer of *Polygonella* to a section of *Polygonum* was first suggested primarily on the basis of similarities in fine structure of the flowers (Ronse De Craene et al., 2004). The molecular studies of Lamb Frye and Kron (2003) and Schuster et al. (2011) also provided evidence that generic limits needed to be reassessed. Thus, *Polygonella* is here treated within a slightly expanded concept of *Polygonum*.

1. Plants shrubs or lianas, the stems woody well above the base .... 2. P. AMERICANUM
1. Plants herbs (sometimes with woody taproots)
    2. Leaf blades with 2 longitudinal folds or grooves parallel to the midvein, the margins with minute, spinescent teeth .... 6. P. TENUE

2. Leaf blades without longitudinal folds or grooves parallel to the midvein (the secondary venation obscure), the margins entire
    3. Outer 3 perianth lobes about as long as the inner 2 lobes, flat at the tips or nearly so . . . . 3. P. AVICULARE
    3. Outer 3 perianth lobes noticeably longer than the inner 2 lobes, boat-shaped to hooded at the tips
        4. Plants heterophyllous, the leaves produced along mid- to late-season branches significantly smaller than those on the main stems
            5. Leaf blades lanceolate to oblanceolate or linear, mostly 4–12 times as long as wide; surfaces of fruits smooth to slightly and irregularly roughened along the angles . . . . 5. P. RAMOSISSIMUM
            5. Leaf blades ovate to obovate, mostly 2–4 times as long as wide; surfaces of fruits pebbled or with close-set, fine lines of minute papillae
                6. Tepals green with usually white, reddish, or occasionally brownish-tinged margins and/or tips . . . . 3. P. AVICULARE
                6. Tepals yellowish green, often with yellow margins . . . . . 4. P. ERECTUM
        4. Plants homophyllous or nearly so, the leaves with more or less continuous size variation during the growing season
            7. Leaf blades lanceolate to oblanceolate or linear, mostly 4–12 times as long as wide; fruits with the surface smooth to slightly and irregularly roughened along the angles . . . . 5. P. RAMOSISSIMUM
            7. Leaf blades ovate to obovate, mostly 2–4 times as long as wide; fruits with the surface appearing finely pitted or pebbled or with minute papillae, except in late-season fruits (which are smooth)
                8. Perianth lobes shorter than the fused portion; stems erect or ascending, sometimes loosely ascending to spreading with age, forming clumps . . . . 1. P. ACHOREUM
                8. Perianth lobes as long as or longer than the fused portion; stems prostrate, the branch tips sometimes ascending, forming mats or low mounds . . . . 3. P. AVICULARE

1. Polygonum achoreum S.F. Blake

Po. erectum L. ssp. *achoreum* (S.F. Blake) Á. Löve & D. Löve

Pl. 501 i, j; Map 2290

Plants herbaceous. Stems 15–50(–70) cm long, erect or ascending, sometimes loosely ascending to spreading with age, forming dense clumps, homophyllous or nearly so (with more or less continuous size variation during the growing season). Leaf blades 8–40 mm long, mostly 2–4 times as long as wide, ovate to obovate, bluish green to light green, lacking 2 longitudinal folds or grooves, the secondary venation obscure, the margins entire. Inflorescences axillary, the flowers solitary or in small clusters. Tepals 2.5–4.0 mm long, more or less uniformly yellowish green, the margins yellowish- or occasionally pinkish-tinged, shorter than the fused portion, usually incurved and somewhat narrowed above the fruit (except in late-season fruits), the outer 3 lobes longer than the 2 inner lobes and more or less enclosing them, boat-shaped to hooded at the tips, the inner 2 lobes shorter and flat. Fruits enclosed or sometimes exserted, the surface finely pitted or pebbled, or with minute papillae (except sometimes in late-season fruits), light brown to yellowish green, not shiny. $2n=40$, 60. May–September.

Scattered in the state, most commonly in counties bordering the Mississippi and Missouri Rivers (northern U.S. south to Nevada, Missouri, and West Virginia; Canada). Bottomland forests, bottomland prairies, marshes, and banks of streams and rivers; also pastures, crop fields, fallow fields, levees, farmyards, railroads, roadsides, and disturbed areas.

Steyermark (1963) included this species based upon a single collection from Jackson County, and thought it to be introduced in the state. It subsequently was reported from three additional counties: Adair (Conrad, 1984), Grundy (Dierker, 1989), and Platte (Castaner and LaPlante, 1992). Currently it is known from nearly 20 counties.

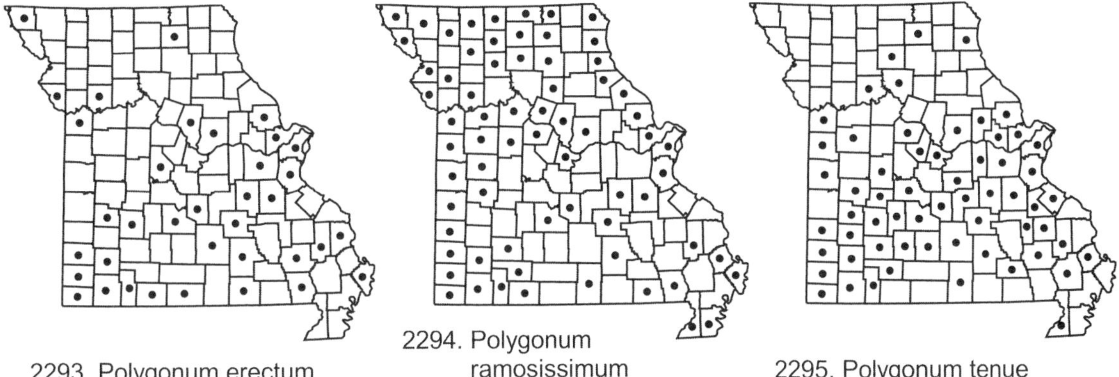
2293. Polygonum erectum

2294. Polygonum ramosissimum

2295. Polygonum tenue

2. Polygonum americanum (Fisch. & C.A. Mey.) T.M. Schust. & Reveal
Polygonella americana (Fisch. & C.A. Mey.) Small

Pl. 496 c–e; Map 2291

Plants shrubby, with woody taproots. Stems 40–100 cm long, densely branched, the main stems erect to more or less spreading, the ultimate branches usually relatively short, herbaceous, grayish green to yellowish green, sometimes slightly glaucous, homophyllous (leaves dense on the ultimate branches, sparser on the main stems, sometimes shed early). Leaf blades 0.4–1.2(–1.8) cm long, mostly 8–12 times as long as wide, linear to narrowly oblanceolate, somewhat thickened and succulent, yellowish green to green, angled to a bluntly or sharply pointed, thin and somewhat translucent tip, green, occasionally slightly glaucous, lacking 2 longitudinal folds or grooves, the secondary venation obscure, the margins entire. Inflorescences terminal spikelike racemes, these sometimes grouped into panicles, the flowers usually solitary at the nodes and spreading in all directions along the axis. Tepals (4)5, fused at the very base, with somewhat irregular margins, glabrous; the outer whorl of 2 tepals 1.2–2.0 mm long at flowering, becoming persistently reflexed as the flower opens, elliptic to ovate, white with a green midnerve, sometimes pinkish-tinged at fruiting; the inner whorl of 3 tepals 1.5–3.0 mm long and loosely ascending at flowering, becoming enlarged to 3.0–4.5 mm at fruiting, ovate to nearly circular, notched at the tip, usually white at flowering, but often becoming pinkish-tinged or pink as the fruit matures (pinkish brown with age). Fruits more or less enclosed, the surface smooth, reddish brown, glabrous, shiny. $2n=36$. July–October.

Uncommon, restricted to the Crowley's Ridge Section of the Mississippi Lowlands Division (southeastern U.S. west to Missouri, Oklahoma, and New Mexico). Sand prairies and sand savannas; also borrow pits.

Steyermark's (1963) undocumented report of a disjunct Iron County station for this species could not be verified during the present study. Steyermark also noted that the species does well in well-drained, acidic soils and is desirable as an ornamental in gardens with dry, sunny sites and appropriate soils. He observed that it is grown easily from seeds.

Species of the *Polygonella* group exhibit an unusual character in which the branch bases tend to be fused to the subordinate stems for a short distance above the actual branch point, creating the illusion that branches are produced between, rather than at, the nodes. For a discussion of the submersion of *Polygonella* in *Polygonum*, see the discussion above following the generic description.

3. Polygonum aviculare L. (common knotweed, knotgrass, dooryard weed)

Pl. 501 a–h; Map 2292

Plants herbaceous. Stems 10–120 cm long, prostrate to ascending, forming mats or loose clumps, homophyllous or heterophyllous. Leaf blades 4–60 mm long (in heterophyllous plants those of main stems 20–60 mm, those of secondary branches 4–20 mm), mostly 2–9 times as long as wide, linear to lanceolate, narrowly oblong, elliptic, or obovate, green to grayish green, lacking 2 longitudinal folds or grooves, the secondary venation obscure (sometimes more apparent in ssp. *depressum*), the margins entire. Inflorescences axillary, the flowers solitary or in small clusters. Tepals 1.8–5.0 mm long, green with usually white, reddish, or occasionally brownish-tinged margins and/or tips, longer than to about as long as the fused portion, incurved or spreading above the fruit, about as long as the inner 2 lobes (the lobes then all relatively flat) or (in ssp. *buxiforme*) the outer 3 lobes longer than the 2 inner lobes, more or less enclosing them, and boat-shaped to hooded at the tips (the inner 2 lobes then shorter and flat). Fruits enclosed or sometimes slightly exserted, the surface with small papillae (these sometimes faint in late-season fruits), light to more commonly dark brown, not shiny. $2n=40$, 60. May–October.

Common throughout the state (U.S., Canada, Europe, Asia; introduced nearly worldwide). Bottomland forests, marshes, swamps, banks of streams and rivers, margins of ponds and lakes, bottomland prairies, moist depressions of upland prairies, saline seeps, and ledges and tops of bluffs; also pastures, crop fields, levees, ditches, farmyards, gardens, lawns, sidewalks, railroads, roadsides, and disturbed areas.

Taxonomic treatments of the *P. aviculare* complex have varied greatly over time. Löve and Löve (1956) and Styles (1962) were the first to use cytological and morphological data to support the recognition of segregate species from *P. aviculare* in the broad sense. Mertens and Raven (1965) refined the classification of the North American representatives and McNeill (1981) and Wolf and McNeill (1986, 1987) further clarified the taxonomy of the group in the eastern United States and Canada. Regional accounts, such as those of Savage and Mertens (1968) for Indiana and Wisconsin, Mitchell and Dean (1978) for New York, and Yatskievych and Brant (1994) for Missouri, added additional details of species morphology and distributions. The preceding authors segregated a number of taxa at the species level from *P. aviculare* in the strict sense, based on ploidy levels and morphology of the various segregates as well as the native ranges of the proposed segregates. Yatskievych and Turner (1990) and Yatskievych and Brant (1994) included five of these species in the Missouri flora. More recently, based on additional chromosome counts and preliminary population-genetic data, some botanists have questioned whether some of these segregates are as distinct as earlier authors thought them to be. The present treatment follows that of Costea and Tardif (2003a) and Costea et al. (2005), who treated *Polygonum* sect. *Polygonum* in North America as comprising 13 native and nonnative species, and *P. aviculare* as consisting of 6 native and nonnative subspecies.

1. Perianth incurved above the fruit, the outer 3 lobes longer than the 2 inner lobes, more or less enclosing them, somewhat pouched at the perianth base and boat-shaped to hooded at their tips (the inner 2 lobes shorter and flat) . . . . 3B. SSP. BUXIFORME
1. Perianth ascending to somewhat spreading above the fruit, the lobes all similar in length and not or only slightly overlapping, not pouched at the perianth base and all relatively flat at their tips (the outer 3 lobes sometimes slightly hooded at the tips)
   2. Plants heterophyllous, the leaves produced along branches of mid- to late-growing season noticeably shorter than those of the main stems produced earlier in year . . . . 3A. SSP. AVICULARE
   2. Plants homophyllous or nearly so, the leaves with more or less continuous size variation during the growing season
      3. Leaf blades broadly oblanceolate to elliptic or obovate, mostly 2–4 times as long as wide; stems usually prostrate and forming mats . . . . 3C. SSP. DEPRESSUM
      3. Leaf blades linear to narrowly elliptic or narrowly oblong-oblanceolate, mostly 4–9 times as long as wide; stems often ascending toward the tips and forming loose mounds . . . . . . 3D. SSP. NEGLECTUM

3a. ssp. aviculare (northern knotweed)

Po. aviculare var. *littorale* (Link) W.D.J. Koch
Po. aviculare var. *vegetum* Ledeb.
Po. monspeliense Thiéb.-Bern. ex Pers.

Pl. 501 a, b

Stems prostrate to ascending, forming mats or more commonly loose clumps, heterophyllous. Leaves with the blade 4–60 mm long, those of main stems 20–60 mm long (with petioles 2–5 mm long), those of later season branches 4–20 mm long (sessile or nearly so), mostly 2–4 times as long as wide, broadly oblanceolate to elliptic or obovate. Perianth 2.8–5.0 mm long, divided $^{2}/_{3}$–$^{4}/_{5}$ of its length at fruiting time, the lobes ascending to somewhat spreading above the fruit, all similar in length and not or only slightly overlapping, not pouched at the perianth base and all relatively flat at the tips (the outer 3 lobes sometimes slightly hooded at the tips). Fruits 2.5–3.7 mm long, enclosed or only slightly exserted, dark brown. $2n=40$, 60. June–October.

Introduced, scattered in the state (native of Europe, Asia; introduced widely in the U.S., Canada, and elsewhere). Bottomland forests, marshes, banks of streams, and margins of ponds

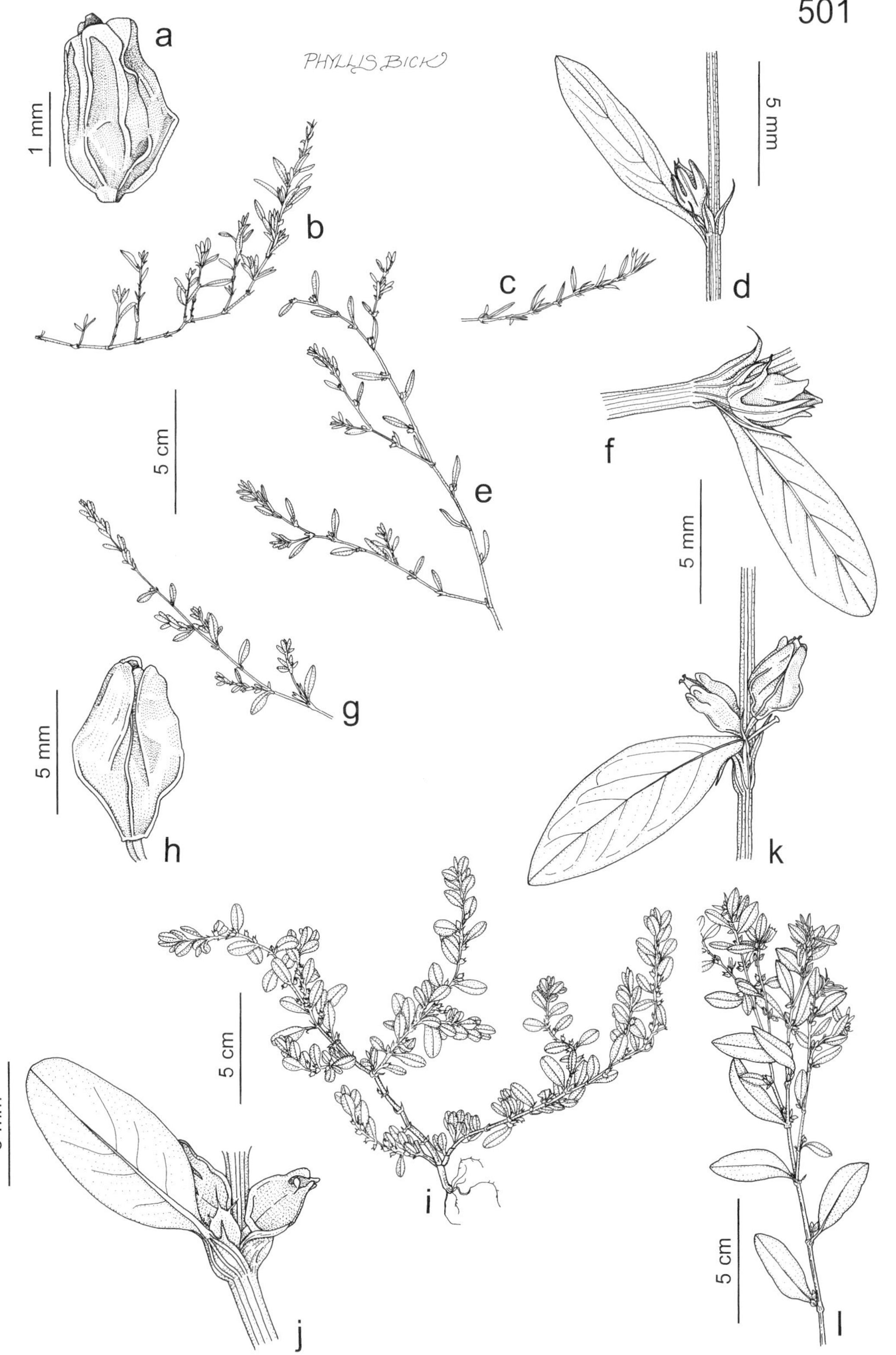

Plate 501. Polygonaceae. *Polygonum aviculare* ssp. *aviculare*, **a)** habit, **b)** flower. *Polygonum aviculare* ssp. *neglectum*, **c)** habit, **d)** node with flower, ocrea, and leaf. *Polygonum aviculare* ssp. *depressum*, **e)** node with flower, ocrea, and leaf, **f)** habit. *Polygonum aviculare* ssp. *buxiforme*, **g)** habit, **h)** flower. *Polygonum achoreum*, **i)** node with flower, ocrea, and leaf, **j)** habit. *Polygonum erectum*, **k)** node with flowers, ocrea, and leaf, **l)** fertile stem.

and lakes; also levees, ditches, gardens, lawns, sidewalks, roadsides, and disturbed areas.

Steyermark (1963) treated the name *P. aviculare* var. *angustissimum* Meisn. as a synonym of *P. aviculare* in the strict sense, but nomenclaturally it is instead a synonym of *P. aviculare* ssp. *rurivagum* (Jord. ex Boreau) Berher, as circumscribed by Costea et al. (2005), a taxon not yet reported from Missouri (Costea and Tardif, 2003b).

3b. ssp. buxiforme (Small) Costea & Tardif (American knotweed)

Po. buxiforme Small

Pl. 501 g, h

Stems prostrate, sometimes loosely ascending at the tips, forming mats or less commonly low mounds, homophyllous or nearly so. Leaves with the blade 5–37 mm long, mostly 2–4 times as long as wide, narrowly ovate to elliptic or obovate. Perianth 2.0–3.5 mm long, divided $^{2}/_{3}$–$^{4}/_{5}$ of its length at fruiting time, the lobes incurved above the fruit, the outer 3 lobes longer than the 2 inner lobes, more or less enclosing them, somewhat pouched at the perianth base and boat-shaped to hooded at the tips (the inner 2 lobes shorter and flat). Fruits 2–3 mm long, enclosed or only slightly exserted, light to dark brown. $2n$=60. May–October.

Scattered nearly throughout the state (U.S.; Canada). Bottomland forests, swamps, banks of streams and rivers, bottomland prairies, moist depressions of upland prairies, margins of lakes, saline seeps, and ledges and tops of bluffs; also pastures, crop fields, lawns, farmyards, levees, railroads, roadsides, and disturbed areas.

Steyermark (1963) placed this taxon under synonymy with *P. aviculare* var. *littorale*, but that name refers to plants with leaf and perianth features typical of *P. aviculare* ssp. *aviculare*, whereas *P. buxiforme* may be distinguished by characters presented in the key above. This subspecies is a native North American component of the mainly European *P. aviculare* complex. It has often been treated as a separate species. Costea and Tardif (2003a) summarized the overlapping morphological characters that caused them to treat the taxon as a subspecies of *P. aviculare*.

3c. ssp. depressum (Meisn.) Arcang. (common knotweed, oval-leaved knotweed)

Po. aviculare ssp. *calcatum* (Lindman) Thell.
Po. aviculare var. *depressum* Meisn.
Po. arenastrum Boreau

Pl. 501 e, f

Stems usually prostrate, forming mats, homophyllous or nearly so. Leaves with the blade 4–30 mm long, mostly 2–4 times as long as wide, broadly oblanceolate to elliptic or obovate. Perianth 1.8–3.5 mm long, divided $^{2}/_{5}$–$^{3}/_{5}$ of its length at fruiting time, the lobes ascending to somewhat spreading above the fruit, all similar in length and not or only slightly overlapping, not pouched at the perianth base and all relatively flat at the tips (the outer 3 lobes sometimes very slightly hooded at the tips). Fruits 1.8–3.0 mm long, sometimes slightly exserted, dark brown. $2n$=40, 60. June–October.

Introduced, scattered to common nearly throughout the state (native of Europe; introduced widely in the U.S., Canada, and elsewhere). Bottomland forests and banks of streams and rivers; also pastures, gardens, lawns, sidewalks, railroads, roadsides, and disturbed areas.

The ssp. *depressum* is more common and widespread in Missouri than is ssp. *aviculare*, with which it was combined by many earlier authors prior to the research of Styles (1962) and McNeill (1981), among others.

3d. ssp. neglectum (Besser) Arcang. (narrow-leaved knotweed)

Po. neglectum Besser

Pl. 501 c, d

Stems prostrate, often with the branch tips loosely ascending, forming mats or loose mounds, homophyllous or nearly so. Leaves with the blade 8–30 mm long, mostly 4–9 times as long as wide, linear to narrowly elliptic or narrowly oblong-oblanceolate. Perianth 2.0–3.5 mm long, divided $^{1}/_{2}$–$^{3}/_{4}$ of its length at fruiting time, the lobes usually somewhat spreading above the fruit, all similar in length and not or only slightly overlapping, not pouched at the perianth base and all relatively flat at the tips. Fruits 1.5–2.5 mm long, often slightly exserted, dark brown. $2n$=40, 60. May–October.

Introduced, uncommon, sporadic, most abundant in the Kansas City and St. Louis metropolitan areas (native of Europe; introduced widely in the U.S., Canada). Banks of rivers, saline seeps, and tops of bluffs; also pastures, levees, gardens, lawns, sidewalks, railroads, roadsides, and disturbed areas.

Steyermark (1963) included this taxon in synonymy under *P. aviculare* in the strict sense. It is more commonly confused with ssp. *arenastrum* by collectors, but differs from both of these taxa in its narrower leaves.

4. Polygonum erectum L. (erect knotweed)

Po. aviculare var. *erectum* (L.) Roth

Pl. 501 k, l; Map 2293

Plants herbaceous. Stems 15–45(–75) cm long, erect or ascending, sometimes loosely ascending

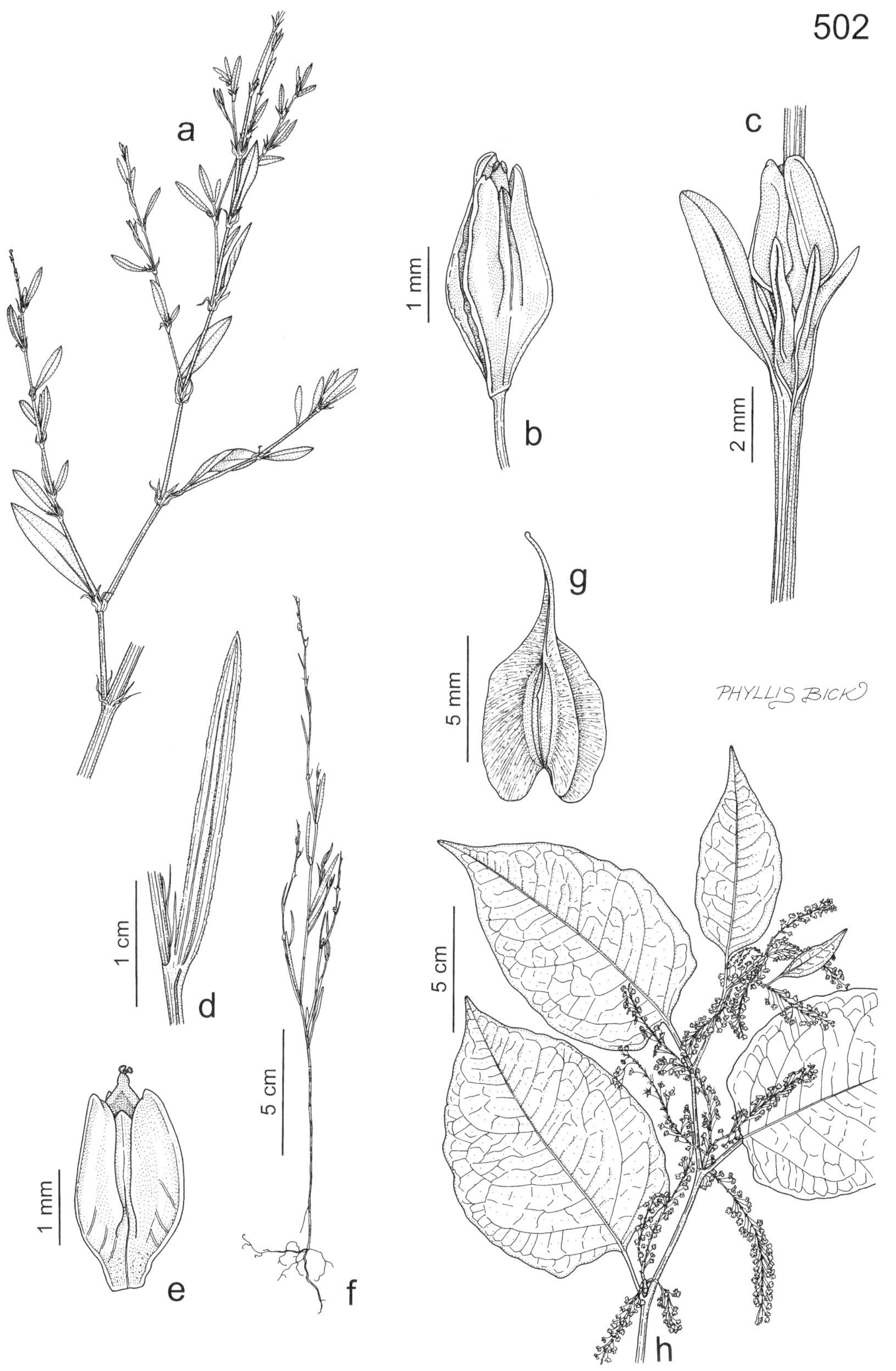

Plate 502. Polygonaceae. *Polygonum ramosissimum*, **a)** fertile stem, **b)** longer-stalked flower, **c)** node with leaf, ocrea, and short-stalked flower. *Polygonum tenue*, **d)** node with leaf and ocrea, **e)** flower, **f)** habit. *Fallopia japonica*, **g)** fruit in calyx, **h)** fertile stem.

to spreading with age, forming dense clumps, heterophyllous. Leaf blades 4–60 mm long (those of main stems 18–60 mm, those of secondary branches 4–10 mm), mostly 2–4 times as long as wide, ovate to elliptic or obovate, yellowish green to bright green, lacking 2 longitudinal folds or grooves, the secondary venation obscure, the margins entire. Inflorescences axillary, the flowers solitary or in small clusters. Tepals 2.8–4.0 mm long, yellowish green, often with yellow margins, longer than the fused portion, usually incurved above the fruit, the outer 3 lobes longer than the 2 inner lobes and more or less enclosing them, boat-shaped to hooded at the tips, the inner 2 lobes shorter and flat. Fruits enclosed, the surface with close-set, fine lines of minute papillae, light to dark brown, not shiny. $2n$=40. June–October.

Scattered in the Ozark and Ozark Border Divisions, uncommon elsewhere in the state (U.S.; Canada). Bottomland forests, and banks of streams, rivers, and spring branches; also pastures, lawns, farmyards, and open disturbed areas.

This species is not encountered commonly today, but apparently was more widespread historically in Missouri, based on herbarium records. It apparently is not very abundant throughout most of its range (Gleason and Cronquist, 1991). Vegetatively, it has been confused most frequently with *P. achoreum* by collectors. Aside from the characters presented in the key above, *P. erectum* may be distinguished readily by the color of the leaves, which are yellowish green to green, as opposed to the bluish green coloration of *P. achoreum*.

5. Polygonum ramosissimum Michx. **ssp. ramosissimum** (bushy knotweed, long-fruited knotweed)

Po. ramosissimum f. *atlanticum* B.L. Rob.
Po. exsertum Small
Po. triangulum E.P. Bicknell ex Small

Pl. 502 a–c; Map 2294

Plants herbaceous. Stems 20–170 cm long, erect or ascending, rarely loosely ascending to spreading, usually not forming clumps, homophyllous or heterophyllous. Leaf blades 4–70 mm long (in heterophyllous plants, those of main stems 15–70 mm, those of secondary branches 4–10 mm), mostly 4–12 as long as wide, lanceolate to oblanceolate or linear, rarely narrowly ovate, yellowish green to green or bluish green, rarely light green, often glaucous, lacking 2 longitudinal folds or grooves, the secondary venation obscure, the margins entire. Inflorescences axillary, the flowers solitary or in small clusters, but these sometimes appearing as terminal spikelike racemes. Tepals 2.2–4.0 mm long, yellowish green with sometimes white or pinkish-tinged tips or margins, longer than the fused portion, usually incurved above the fruit, the outer 3 lobes longer than the 2 inner lobes and more or less enclosing them, boat-shaped to hooded at the tips, the inner 2 lobes shorter and flat. Fruits enclosed or sometimes exserted, the surface smooth to slightly and irregularly roughened along the angles, light to dark brown, shiny or dull. $2n$=20, 60. May–October.

Scattered to common nearly throughout the state (U.S.; Canada). Bottomland prairies, moist depressions of sand prairies, marshes, saline seeps, banks of rivers, margins of lakes, and swamps; also levees and roadsides.

Steyermark (1963) and Yatskievych and Turner (1990) included ssp. *prolificum* (Small) Costea & Tardif (as *P. prolificum* (Small) B.L. Rob.) for Missouri, based on collections from disturbed areas in the Missouri River floodplain in Clay and Jackson Counties. These collections were redetermined as ssp. *ramosissimum* during the course of the present study. Likewise, vouchers for Henderson's (1980) report of this taxon from Lawrence County, Solecki's (1984) report from Vernon County, and Ohmart's (1987) report from Scott County were redetermined as ssp. *ramosissimum*. A specimen collected by the author in 1998 at a saline spring in Ralls County was determined as *P. prolificum* by Dr. O. V. Yurtseva (Moscow State University) in 2008, but appears to be a somewhat atypical plant of ssp. *ramosissimum*. The taxonomic distinction of ssp. *prolificum* from ssp. *ramosissimum* remains controversial. Although many botanists have considered it a separate species, others have treated it as a variety (var. *prolificum* Small) or subspecies of *P. ramosissimum* (Mitchell and Dean, 1978; Costea and Tardif, 2003a; Costea et al., 2005), and a few workers suggest that it might be unworthy of any taxonomic segregation (Kaul, 1986). Plants of ssp. *prolificum* are said to differ in their bluish green color and tendency to blacken upon drying, as well as having slightly shorter flower stalks and more or less continuous variation in leaf sizes during the growing season. However, these characters do not appear to correlate well. The situation requires further study rangewide.

Whatever the taxonomic outcome, true ssp. *prolificum* has not been confirmed to grow in Missouri. Costea et al. (2005) reported it from nearly the same overall range as ssp. *ramosissimum* and from every state surrounding Missouri. Two apparent ecotypes of ssp. *ramosissimum* in Missouri that display some characteristics of ssp. *prolificum*

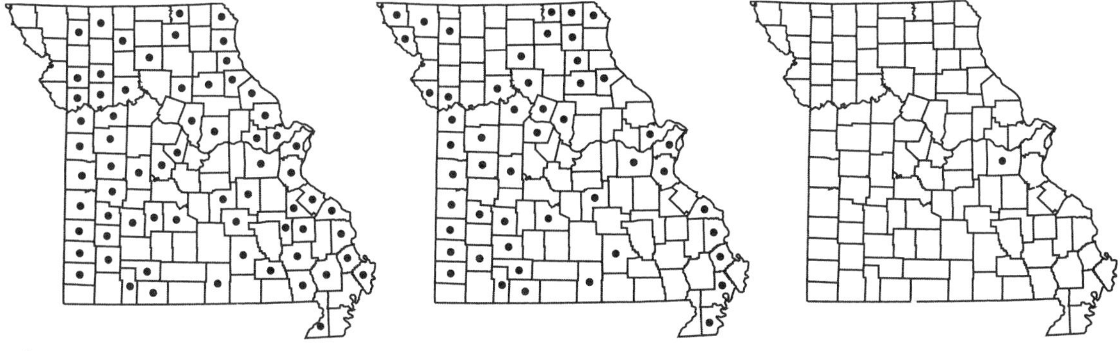

2296. Rumex acetosella 2297. Rumex altissimus 2298. Rumex conglomeratus

are worthy of note. Occasional collections from Jasper, Shannon, and Taney Counties represent a morphotype that has flowers with very short flower stalks (Pl. 502 c), as in the latter taxon, but have the narrowly angled to tapered leaves and a tendency toward a heterophyllous growth form more typical of the former subspecies. Field experience suggests that these represent short-stalked individuals within populations of otherwise normal plants of ssp. *ramosissimum*. A second unusual morphotype consists of highly colonial plants with relatively fleshy, light green leaves that tend to be obtuse at the tips, but that have flower stalks longer than is typical in ssp. *prolificum*. These plants are confined to saline seeps in Cooper, Howard, and Saline Counties, where colonies grow in hypersaline, shallow water supporting few other vascular plant species. Steyermark (1963) mistakenly treated these populations as a strain of *P. aviculare*. Although these plants appear somewhat different from other Missouri material, they seem to fit within the variation exhibited by *P. ramosissimum* ssp. *ramosissimum* elsewhere in its distributional range, particularly in brackish, coastal areas. It may be that these particular populations represent a separate introduction into Missouri of plants from elsewhere in the species range. Taxonomically, they appear unworthy of separation at the varietal or species levels.

6. Polygonum tenue Michx. (slender knotweed)

Pl. 502 d–f; Map 2295

Plants herbaceous. Stems 10–45 cm long, erect or strongly ascending, not forming clumps, homophyllous. Leaf blades 5–40 mm long, mostly 4–12 times as long as wide, linear to narrowly lanceolate, green to pale green or light green (turning orange in autumn), with 2 longitudinal folds or grooves parallel to the midvein, the margins with minute, spinescent teeth. Inflorescences axillary, the flowers solitary or in small clusters, but these sometimes appearing as terminal spikelike racemes. Tepals 2.5–4.2 mm long, green with white, pinkish-tinged, or brown margins, much longer than the fused portion, often somewhat incurved above the fruit, the outer 3 lobes longer than the 2 inner lobes and more or less enclosing them, boat-shaped to hooded at the tips, the inner 2 lobes shorter and flat. Fruits enclosed or exserted, the surface finely pebbled or with lines of minute papillae, dark brown to black, not shiny. $2n$=20, 30, 32. July–October.

Scattered in the Ozark, Ozark Border, and Unglaciated Plains Divisions, uncommon elsewhere in the state (eastern U.S. west to Wyoming and Texas; Canada). Glades, savannas, openings of dry upland forests, sand prairies, and banks of streams and rivers; also railroads and open disturbed areas; usually on acidic substrates.

7. **Rumex** L. (dock, sorrel)

Plants annual or perennial herbs, sometimes monoecious or dioecious, with slender to thickened taproots or rhizomes. Stems unbranched or branched, loosely ascending to erect, usually noticeably swollen at the nodes, glabrous or minutely hairy, unarmed. Leaves basal (except in a few species) and alternate, all but the upper stem leaves usually petiolate, not jointed at the base. Ocreae white, silvery, tan, or brown to reddish brown, mostly papery, fused basally to the petiole, entire or in a few species irregularly divided into numerous slender lobes, usually glabrous. Leaf blades variously linear to broadly ovate, unlobed or (in *R. acetosella* and *R. hastatulus*) with a pair of slender, spreading basal lobes, herbaceous to slightly fleshy or slightly leathery, variously angled, tapered, rounded, or cordate at the base, rounded or angled to tapered to a bluntly or sharply pointed tip, the margins otherwise entire or occasionally finely scalloped, sometimes wavy or crisped, the surfaces glabrous or the undersurface less commonly minutely hairy. Inflorescences terminal and sometimes also axillary, panicles with the branches of narrow racemes (axillary inflorescences sometimes reduced to solitary racemes), the bracteal leaves absent or relatively well-developed, the flowers in small fascicles or clusters of (1–)4 to numerous, these subtended by ocreolae, spreading or nodding along the axis, lacking other bracts. Flowers perfect or imperfect, short-to long-stalked (the perianth not tapered at the base), the stalk unwinged, jointed variously toward the base, midpoint, or tip, the joint occasionally obscure. Tepals 6, in 2 whorls, these free, glabrous, green to greenish white at flowering, becoming brown to reddish at fruiting, mostly spreading at flowering, the inner whorl of 3 tepals becoming ascending, enlarged and with a prominent network of nerves at fruiting (except in *R. acetosella*), ovate to triangular or somewhat heart-shaped in outline, the margins becoming winged or dissected into a series of teeth or slender lobes (except in *R. acetosella*); 1 or more of the enlarged inner tepals developing a prominent, differentiated, central swelling at fruiting (here called a tubercle) (except in *R. acetosella, R. hastatulus*), the outer whorl of 3 not becoming enlarged at fruiting, ascending or spreading to more commonly reflexed, inconspicuous, lanceolate to narrowly lanceolate or narrowly oblong. Stamens (in staminate or perfect flowers) 6, exserted, the anthers attached toward their bases, the filaments short, unwinged. Styles (in pistillate or perfect flowers) 3, free, the stigmas deeply divided into numerous slender lobes, appearing fringed. Fruits achenes, enclosed in the persistent perianth (often slightly exserted in *R. acetosella*), 0.9–4.5 mm long, 3-angled, unwinged or slightly winged, ovate to nearly triangular in outline, not beaked, tan to dark brown, glabrous, usually shiny. About 200 species, nearly worldwide, most diverse in temperate regions.

The young foliage of some species can be eaten raw (sorrel) or cooked (dock), but the mature leaves of most *Rumex* are rich in tannins and thus bitter. A few species of dock whose enlarged rootstocks accumulate tannins were once used in the tanning of hides for leather. Members of the genus are nearly entirely wind-pollinated; thus their pollen can contribute to the incidence of hayfever. Some of the species are aggressive weeds of gardens and crop fields. Several species have been used medicinally for various treatments. The fruits of *Rumex* are important wildlife foods.

The genus *Rumex* has been divided into four or more subgenera, some of these with several sections. In Missouri, three main groups are present. The majority of the species fall into subgenus *Rumex*, the docks. The two species of sorrel, *R. acetosella* and *R. hastatulus*, appear superficially similar, but generally are classified into separate subgenera, *Acetosella* (Meisn.) Rech. f. and *Acetosa* (Mill.) Rech. f., respectively, based on differences in the fruiting perianths (Mosyakin, 2005). The infrageneric classification of *Rumex* is in need of more detailed studies using molecular markers.

The present treatment borrows heavily from that of Mosyakin (2005) in the Flora of North America series. Generally, fruiting material is necessary to determine specimens of *Rumex* to species; thus vegetative or flowering material should be avoided when possible. For most species, it is also important to note or collect the basal leaves (if present) and to note whether the stems are branched below the inflorescences.

1. Flowers staminate or pistillate and produced on separate plants (plants dioecious, rarely incompletely so); leaves mostly with a pair of spreading basal lobes (unlobed or with additional lobes in occasional leaves); tepals lacking tubercles at fruiting

2. Inner whorl of tepals not becoming enlarged, winged or prominently nerved at fruiting; outer whorl of tepals ascending to somewhat spreading at fruiting . . . . 1. R. ACETOSELLA
2. Inner whorl of tepals becoming enlarged, winged, and with a network of prominent nerves at fruiting; outer whorl of tepals reflexed at fruiting . . . . 8. R. HASTATULUS

1. Flowers perfect or, if imperfect, then staminate and pistillate flowers present in the same inflorescence (plants monoecious, but the staminate flowers withered and usually not apparent at fruiting); leaves not lobed at the base; 1 or more of the inner tepals with a tubercle at fruiting
   3. Wings of the inner tepals at fruiting with conspicuous long, often slender, bristlelike or spinelike teeth along the margins
      4. Leaf blades 1.5–3.0(–4.0) cm wide, linear-lanceolate to lanceolate or occasionally oblong-lanceolate, 4 or more times as long as wide; teeth 1.5–2.5 times as long as the width of the undivided portion of each inner tepal; plants annual, with tapered taproots . . . . 7. R. FUEGINUS
      4. Blades of main leaves (2–)3–15 cm wide, narrowly to broadly ovate or oblong-ovate, mostly 2–4 times as long as wide; teeth shorter than to less commonly slightly longer than the width of the undivided portion of each inner tepal; plants perennial, with thickened, somewhat turnip-shaped, vertical rootstocks
         5. Stems 60–120(–150) cm long; blades of main leaves mostly 20–40 cm long, 10–15 cm wide; inflorescence branches ascending (diverging from the main axis at an angle of 30–45°); stalks conspicuously longer than the fruiting perianth; fruiting perianth with usually only 1 tubercle (2 of the inner tepals lacking tubercles), rarely with 1 larger and 2 smaller tubercles, all smooth . . . . 9. R. OBTUSIFOLIUS
         5. Stems 20–60(–70) cm long; blades of main leaves 4–10(–15) cm long, (2–)3–5 cm wide; inflorescence branches spreading to loosely ascending (diverging from the main axis at an angle of 60–90°, then arched upward); stalks shorter than to about as long as the fruiting perianth; fruiting perianth with (1–)3 tubercles, these similar in size or some times 1 larger than the other 2, all finely warty . . . . 11. R. PULCHER
   3. Wings of the inner tepals at fruiting entire, slightly irregular, or with triangular to narrowly triangular teeth that are not bristlelike or spinelike along the margins
      6. Inner tepals 2–3 mm long, 1–2 mm wide at fruiting, very narrowly winged, the tubercles covering nearly the entire tepals . . . . 3. R. CONGLOMERATUS
      6. Inner tepals 2.5–5.0 mm long, 2.0–4.5(–5.0) mm wide at fruiting, moderately to broadly winged, the tubercles covering up to 1/2 of the width of the tepals (including the wings) (note that in a few species only 1 or 2 of the tepals produce well-developed tubercles)
         7. Plants lacking a well-developed basal rosette at flowering; stems with few to several branches below the inflorescence
            8. Stalks 2–5 times as long as the fruiting perianth, mostly strongly arched downward or drooping at maturity . . . . 14. R. VERTICILLATUS
            8. Stalks shorter than to about 2 times as long as the fruiting perianth, straight to somewhat curved, but not uniformly arched downward or drooping at maturity

9. Inner tepals at fruiting broadly triangular to ovate-triangular, mostly bluntly pointed to narrowly rounded at the tip; fruiting perianth with 3 tubercles, these usually similar in size; fruits 2.5–3.5 mm long, 1.8–2.3 mm wide .... 2. R. ALTISSIMUS
9. Inner tepals at fruiting broadly triangular, mostly sharply pointed at the tip; fruiting perianth usually with 1 tubercle, rarely with 2 or 3 tubercles but then with only 1 fully formed tubercle, the other(s) much smaller; fruits 1.7–2.2 mm long, 1.0–1.5 mm wide .... 13. R. TRIANGULIVALVIS

7. Plants with a well-developed basal rosette (this occasionally withered by fruiting); stems unbranched below the inflorescence (but sometimes few to several from the rootstock)
10. Wings of the inner tepals at fruiting entire or slightly irregular along the margins, not toothed
11. Inner tepals broadly ovate to ovate-triangular, 3.5–6.0 mm long, 3–5 mm wide, truncate or very slightly cordate at the base; fruiting perianth with 3 tubercles, 1 of these usually somewhat larger than the other 2; leaf blades strongly undulate and appearing crisped along the margins; stems 40–100(–150) cm long .... 4. R. CRISPUS
11. Inner tepals 5.5–10.0 mm long, 5–9(–10) mm wide, broadly ovate to nearly circular, usually distinctly cordate at the base; fruiting perianth with 1 tubercle; leaf blades flat or slightly undulate along the margins; stems 80–200 cm long .... 10. R. PATIENTIA
10. Wings of the inner tepals at fruiting with triangular to narrowly triangular teeth along the margins
12. Inner tepals at fruiting 6–9 mm long; fruiting perianth with 3 tubercles, but 1 of these noticeably larger and plumper than the other 2; fruits 2.8–3.5 mm long, 2.0–2.5 mm wide .... 5. R. CRISTATUS
12. Inner tepals at fruiting 3.0–5.5 mm long; fruiting perianth with 3 tubercles, these all similar in size
13. Wings of the inner tepals at fruiting with narrowly triangular to spinelike teeth 1–3(–5) mm long; blades of main leaves 3–8(–12) cm long, mostly less than 4 times as long as wide, mostly truncate to shallowly cordate at the base .... 6. R. DENTATUS
13. Wings of the inner tepals at fruiting with triangular to narrowly triangular teeth 0.2–1.5 mm long; blades of main leaves 15–25(–30) cm long, mostly more than 4 times as long as wide, mostly angled to truncate at the base......................... 12. R. STENOPHYLLUS

1. Rumex acetosella L. (sheep sorrel, red sorrel, field sorrel, sour dock)
Acetosa acetosella (L.) Mill.
Pl. 503 b, c; Map 2296

Plants perennial (sometimes flowering the first year), dioecious, rarely incompletely so, with slender, often long-creeping rhizomes. Stems 1 to more commonly several, 10–40(–45) cm long (including inflorescence), erect or ascending, usually unbranched below the inflorescence, glabrous. Leaves alternate and basal (plants with a well-developed basal rosette at flowering). Ocreae membranous to papery, becoming irregularly dissected above the midpoint at maturity, light brown to reddish brown toward the base, grading to white or pale and translucent toward the tip. Basal leaves mostly long-petiolate, the blade similar to those of the lower stem leaves. Blades of stem leaves 2–6 cm long 0.3–2.0 cm wide, thin and herbaceous to slightly succulent, narrowly ovate to lanceolate-elliptic, lanceolate, or oblong-obovate, occasionally, narrowly lanceolate to nearly linear, with a pair of spreading, bluntly triangular basal lobes, occasionally unlobed, rarely with additional lobes, the margins otherwise entire, flat or nearly so, truncate to slightly concave at the base, angled or ta-

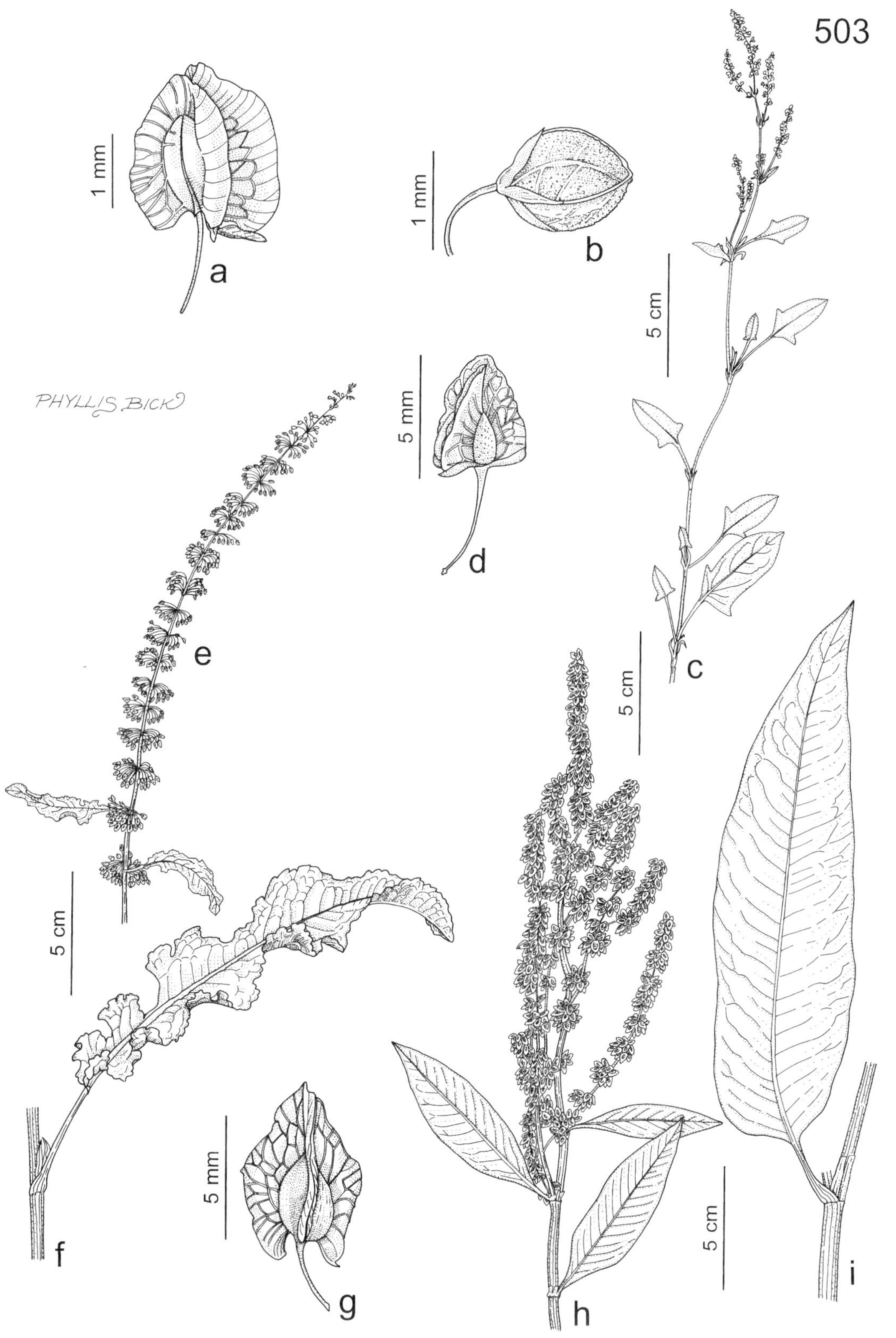

Plate 503. Polygonaceae. *Rumex hastatulus*, **a)** fruit in calyx. *Rumex acetosella*, **b)** fruit in calyx, **c)** habit. *Rumex crispus*, **d)** fruit in calyx, **e)** inflorescence, **f)** median node with leaf and remains of ocrea. *Rumex altissimus*, **g)** fruit in calyx, **h)** upper portion of stem with leaves and inflorescence, **i)** median node with leaf and ocrea.

pered to a bluntly or sharply pointed tip, rarely rounded at the tip, the surfaces glabrous. Inflorescences terminal, usually occupying the upper $^1/_2$–$^2/_3$ of the plant, broadly or narrowly paniculate, the main axis usually somewhat zigzag, occasionally reclining, interrupted. Flowers (3–)5–8(–10) per whorled fascicle, the stalks 1–3 mm, about as long as to about 2 times as long as the fruiting perianth, arched downward or drooping at maturity, not jointed. Pistillate flowers with the tepals not or only slightly enlarged at fruiting, 1.2–1.7(–2.0) mm long, 0.5–1.3 mm wide, not developing noticeable wings or nerves at fruiting, angled at the base, bluntly to more or less sharply pointed at the tip, the margins entire, the surfaces not strongly nerved; tubercles absent; outer whorl of tepals ascending at fruiting. Fruits 0.9–1.5 mm long, 0.6–0.9 mm wide, brown to dark brown. $2n$=14, 28, 42. May–September.

Introduced, scattered to common nearly throughout the state (native of Europe, Asia; introduced nearly worldwide). Margins of ponds and lakes, upland prairies, glades, and ledges and tops of bluffs; also pastures, fallow fields, old fields, railroads, roadsides, and open disturbed areas; usually on acidic substrates.

Steyermark (1963) noted that contact with this plant causes dermatitis in some individuals and that the large quantities of windborne pollen produced by the plants contributes to hay fever problems seasonally. The foliage has a sour flavor and has been eaten raw as a thirst quencher and an ingredient in salads, and also used as a flavoring in beverages and soups.

In the Old World, several infraspecific taxa have been recognized within the *R. acetosella* polyploid complex, but application of these names to North American material is presently uncertain. Steyermark (1963) assigned the Missouri plants to var. *acetosella*, which does not appear to be correct. Mitchell and Dean (1978) and den Nijs (1983) indicated that most of the North American specimens should be referred to as ssp. *angiocarpus* (Murb.) Murb. (now known as ssp. *pyrenaicus* (Pourr. ex Lapeyr.) Ackeroyd), but other infraspecific taxa probably also are present (Mosyakin, 2005).

2. **Rumex altissimus** Alph. Wood (tall dock, pale dock, peach-leaved dock)

Pl. 503 g–i; Map 2297

Plants perennial, monoecious (staminate flowers produced toward the inflorescence tips but generally not apparent at fruiting), with a stout, vertical rootstock. Stems 1 or few, 50–90(–120) cm long, erect or ascending, sometimes from a spreading base, few- to several-branched below the inflorescence, glabrous. Leaves alternate (plants lacking a well-developed basal rosette at flowering). Ocreae pale to white and membranous to papery, tearing and usually becoming deeply dissected at maturity. Leaf blades 10–15 cm long, 3.0–5.5 cm wide, thick and often somewhat leathery, lanceolate to elliptic-lanceolate, or narrowly ovate, unlobed, the margins entire, flat, broadly angled or rarely somewhat rounded at the base, angled or tapered to a sharply pointed tip, the surfaces glabrous. Inflorescences terminal and axillary, the terminal ones usually occupying the upper $^1/_5$–$^1/_3$ of the plant, normally broadly paniculate, dense and mostly continuous. Flowers 12–20 per whorled fascicle, the stalks (2–)3–7(–8) mm long, shorter than to about 2 times as long as the fruiting perianth, straight to somewhat curved, but not uniformly arched downward or drooping at maturity, jointed below the midpoint. Pistillate flowers with the inner whorl of tepals becoming enlarged to 4.5–6.0 mm long and 3.0–4.5(–5.0) mm wide at fruiting, moderately to broadly winged, broadly triangular to ovate-triangular, truncate or slightly cordate at the base, mostly bluntly pointed to narrowly rounded at the tip, the margins entire, the surfaces with a prominent network of nerves; tubercles 3, usually all similar in size, covering up to $^1/_2$ the width of the tepals (including the wings), glabrous, often minutely wrinkled; the outer whorl spreading at fruiting. Fruits 2.5–3.5 mm long, 1.8–2.3 mm wide, brown or dark reddish brown. $2n$=20. April–June.

Scattered nearly throughout the state (eastern U.S. west to Wyoming and Arizona; Canada, Mexico; introduced in Europe). Banks of streams and rivers, margins of ponds and lakes, oxbows, sloughs, swamps, bottomland prairies, bottomland forests, and marshes; also ditches, pastures, fencerows, railroads, roadsides, and open disturbed areas; sometimes emergent aquatics.

Steyermark (1953) noted that the young leaves can be eaten as a cooked vegetable. Vegetatively, this species is essentially indistinguishable from *R. triangulivalvis*. Both are distinctive in their flat, smooth-margined leaf blades that are light green to yellowish green.

3. **Rumex conglomeratus** Murray (clustered dock)

Pl. 504 a–c; Map 2298

Plants perennial, with perfect flowers, with a sometimes branched, tuberous-thickened, vertical rootstock. Stems few to several, 30–80(–120) cm long, erect or ascending, few- to several-branched below the inflorescence, glabrous. Leaves basal and

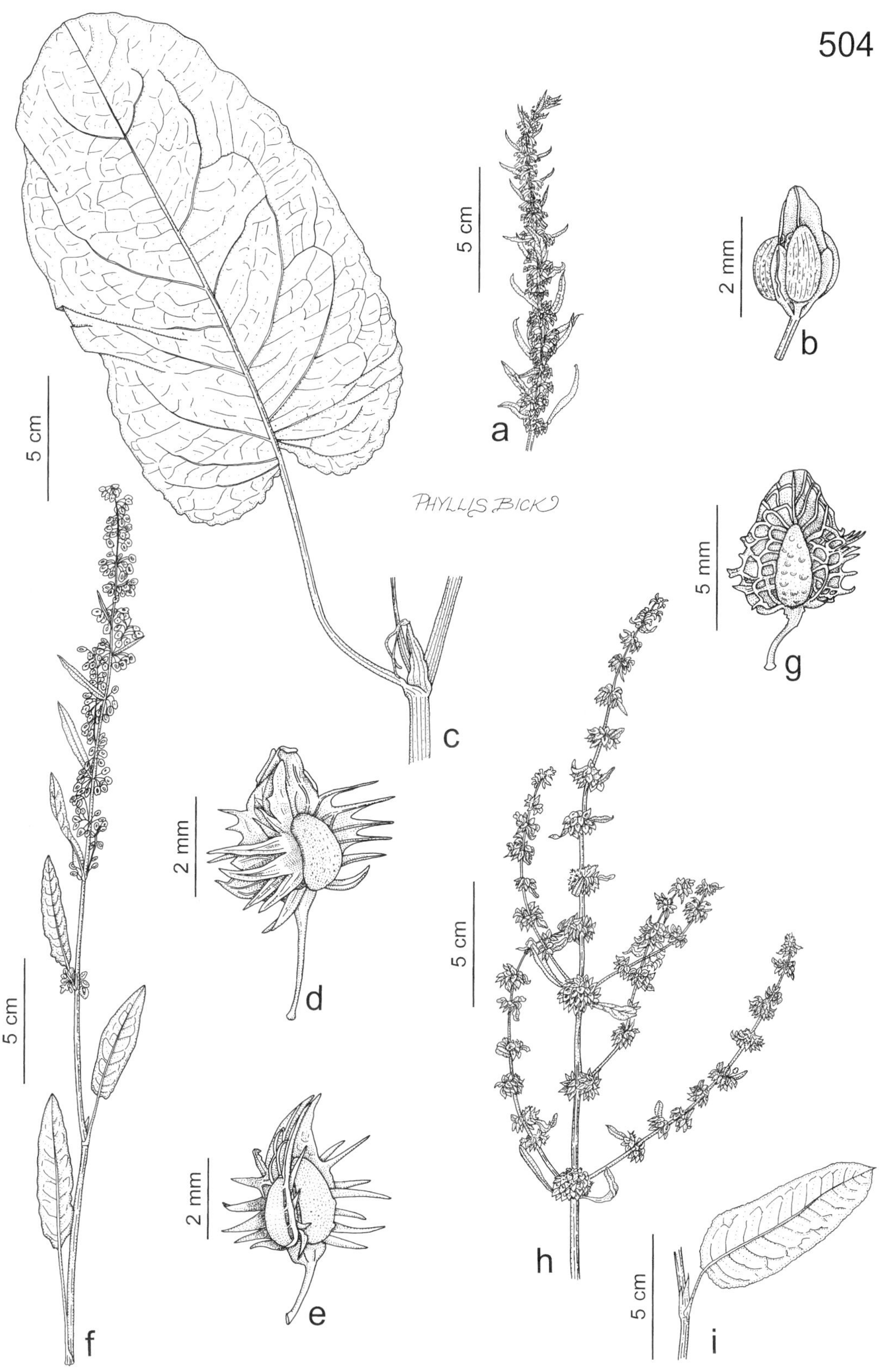

Plate 504. Polygonaceae. *Rumex conglomeratus*, **a)** portion of inflorescence, **b)** fruit in calyx, **c)** lower node with leaf and ocrea. *Rumex obtusifolius*, **d)** fruit in calyx. *Rumex dentatus*, **e)** fruit in calyx, **f)** fertile stem. *Rumex pulcher*, **g)** fruit in calyx, **h)** inflorescence, **i)** median node with leaf and ocrea.

2299. Rumex crispus 2300. Rumex cristatus 2301. Rumex dentatus

alternate (plants often with a well-developed basal rosette at flowering). Ocreae tan to white and papery, tearing and fragmenting, often partially or mostly shed at maturity. Basal leaves similar to the lowermost stem leaves, but often slightly larger and longer-petiolate. Blades of stem leaves 5–30 cm long (gradually reduced upward), 2.5–6.0 cm wide, thick and somewhat leathery, those of the lower leaves oblong to oblong-lanceolate or narrowly obovate, grading to lanceolate or narrowly elliptic-lanceolate toward the stem tip, unlobed, the margins entire, flat or very slightly undulate, those of the lower leaves truncate to cordate, grading to rounded or broadly angled and narrowly angled upward, broadly and bluntly pointed to more narrowly angled or tapered and more or less sharply pointed, the surfaces glabrous. Inflorescences terminal, usually occupying the upper 2/3 of the plant, consisting of leafy racemes, these usually grouped into relatively broad panicles, relatively open and with mostly well-spaced nodes. Flowers 10–20 per densely whorled fascicle, the stalks 1–4(–5) mm long, about as long as to slightly longer than the fruiting perianth, slightly to strongly downward-curved or -arched, jointed at or more commonly below the midpoint. Inner whorl of tepals becoming somewhat enlarged to 2–3 mm long and 1–2 mm wide at fruiting, very narrowly winged, oblong to oblong-ovate, truncate or broadly angled at the base, broadly and bluntly pointed at the tip, the margins entire, the surfaces with a prominent network of nerves; tubercles 3, all similar in size or nearly so, covering nearly the entire tepals (including the wings), glabrous, smooth or slightly wrinkled; the outer whorl loosely ascending at fruiting. Fruits 1.5–1.8 mm long, 1.0–1.4 mm wide, dark reddish brown, shiny. $2n$=20. April–June.

Introduced, known thus far only form a historical collection from Franklin County (native of Europe, Asia, Africa; introduced in the eastern U.S. west to Illinois and Texas, also the western U.S. east to New Mexico; Canada). Disturbed areas.

Steyermark's (1963) report of this species from St. Louis County could not be verified during the present study.

4. Rumex crispus L. (curly dock, sour dock, yellow dock)

Pl. 503 d–f; Map 2299

Plants perennial, with perfect flowers, with a thickened, somewhat turnip-shaped, vertical rootstock. Stems mostly 1 or few, 40–100(–150) cm long, erect or strongly ascending, unbranched below the inflorescence, glabrous or nearly so. Leaves alternate and basal (the rosette leaves occasionally withered by fruiting), the basal leaves larger and longer-petiolate than the lower stem leaves. Ocreae tan to reddish brown and membranous to papery, tearing and usually becoming deeply dissected or more commonly absent at maturity. Blades of main leaves 15–30(–35) cm long, 2–6 cm wide (more than 4 times longer than wide), herbaceous to somewhat leathery, oblong-lanceolate to lanceolate or sometimes nearly linear (upper leaves), those of the uppermost leaves progressively shorter and narrower, unlobed, the margins entire or nearly so, strongly undulate and crisped, angled to truncate, rounded, or shallowly cordate at the base, sharply pointed at the tip, the surfaces glabrous or the undersurface occasionally sparsely and minutely hairy along the main veins. Inflorescences terminal, usually occupying the upper 1/2 of the plant, narrowly paniculate with strongly ascending branches, dense and more or less continuous (at least at fruiting) except sometimes near the base. Flowers 10–25 per whorled fascicle, the stalks (3–)4–10 mm long, slightly longer than to about 2 times as long as the fruiting perianth, arched downward or nodding, jointed toward the base. Inner whorl of tepals becoming enlarged to 3.5–6.0 mm long and 3–5 mm wide at fruiting, with relatively broad wings, broadly ovate to ovate-triangular, truncate or very slightly cordate at the base, rounded to bluntly or rarely sharply pointed at the tip, the margins entire or slightly irregular but lacking teeth, the surfaces with a prominent network of nerves; tubercles 3, usually with 1 slightly larger than the other 2, the largest covering up to 1/2 of the width of the tepal (including the wings), glabrous, finely pitted; the outer whorl spreading at fruiting. Fruits 2–3 mm long, 1.5–2.0 mm wide, reddish brown. $2n=60$. April–June.

Common throughout the state (native of Europe, Asia; introduced nearly worldwide). Banks of streams and rivers, margins of ponds, lakes, and sinkhole ponds, sloughs, marshes, bottomland forests, bottomland prairies, upland prairies; also quarries, crop fields, fallow fields, pastures, gardens, ditches, levees, railroads, roadsides, and open disturbed areas; occasionally emergent aquatics.

Steyermark (1963) noted that the immature leaves of this species have been used as a wild green for salads, but that contact with the plant can cause dermatitis in some persons. The roots have been sold for their medicinal value under the name yellow dock.

Steyermark (1963) discussed the existence of rare hybrids between *R. crispus* and *R. obtusifolius* (*R.* ×*pratensis* Mert. & W.D.J. Koch), which occur sporadically where the two species grow in proximity.

5. Rumex cristatus DC. (crested dock, Greek dock)

Pl. 505 e; Map 2300

Plants perennial, with perfect flowers, with a thickened, somewhat turnip-shaped, vertical rootstock. Stems 1 to several, 70–200 cm long, erect or strongly ascending, unbranched below the inflorescence, glabrous. Leaves alternate and basal, the basal leaves larger and longer-petiolate than the lower stem leaves. Ocreae tan to reddish brown and more or less papery, tearing and usually becoming deeply dissected or absent at maturity. Blades of main leaves 15–35 cm long, 5–12 cm wide (mostly 3–4 times longer than wide), herbaceous to somewhat leathery, broadly lanceolate to oblong-lanceolate or sometimes nearly linear (upper leaves), those of the basal leaves sometimes narrowly oblong-ovate, those of the uppermost leaves progressively shorter and narrower, unlobed, the margins entire or nearly so, flat or more commonly slightly undulate and slightly crisped, angled (upper leaves) to broadly angled, truncate, or shallowly cordate at the base, sharply pointed at the tip, the surfaces glabrous. Inflorescences terminal, occupying the upper 1/2–2/3 of the plant, usually broadly paniculate, the main branches sometimes branching again, dense and more or less continuous (at least at fruiting) or interrupted toward the base. Flowers 15–20 per whorled fascicle, the stalks 6–14 mm long, as long as to about 1.5 times as long as the fruiting perianth, arched downward or nodding, jointed toward the midpoint. Inner whorl of tepals becoming enlarged to 6–9 mm long and 6–8 mm wide at fruiting, with broad wings, broadly ovate to nearly circular, distinctly cordate at the base, rounded to bluntly pointed or occasionally sharply pointed at the tip, the margins entire to somewhat irregular toward the tip, with irregular, triangular to narrowly triangular teeth 0.5–1.0 mm long below the midpoint, the surfaces with a prominent network of nerves; tubercles 3, unequal in size (1 tubercle distinctly larger and plumper than the other 2), the largest tubercle relatively small, covering much less than 1/2 of the width of the tepal (including the wings), glabrous, finely pitted; the outer whorl spreading to reflexed at fruiting. Fruits 2.8–3.5 mm long, 2.0–2.5 mm wide, brown to dark reddish brown. $2n=80$. April–June.

Introduced, uncommon, known thus far from St. Louis County and City (native of Europe; introduced sporadically in Illinois, Missouri, and Kansas). Railroads and open disturbed areas.

This plant was first reported for Missouri by Shildneck et al. (1981), who noted that it is easily confused with *R. patientia*. Both species are very robust, thick-stemmed, and tall, with large stem leaves. According to Mosyakin (2005), the two species are capable of hybridization where they grow in proximity; such hybrids have been called *R.* ×*xenogenus* Rech. f. and eventually may be discovered in Missouri. In addition to the production of tubercles on all 3 fruiting inner tepals (vs. only on 1 tepal in *R. patientia*), the larger leaves of *R. cristatus* have the secondary veins branching from the midvein at 45–60° (vs. 60–90° in *R. patientia*) and its fruiting inner tepals are noticeably reddish-tinged and toothed (vs. merely brown and entire or slightly irregular in *R. patientia*).

Steyermark (1963) reported *R. orbiculatus* A. Gray from St. Louis County, but Mühlenbach (1983) redetermined the Missouri material as *R. cristatus*. True *R. orbiculatus*, which is now known under the older name, *R. brittanica* L., is native to the northeastern United States and Canada, and disjunctly in California and Louisiana (Mosyakin, 2005). It differs from *R. cristatus* in its inner tepals entire or nearly so and also usually is a somewhat less robust plant.

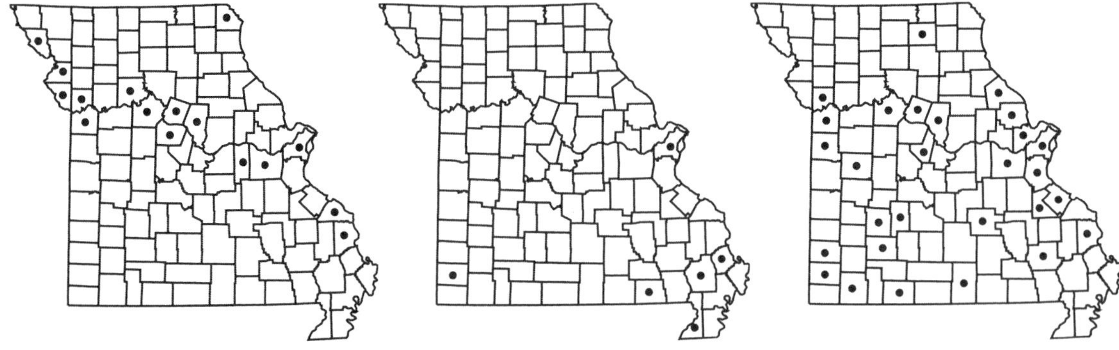
2302. Rumex fueginus 2303. Rumex hastatulus 2304. Rumex obtusifolius

6. Rumex dentatus L. (toothed dock)
Pl. 504 e, f; Map 2301

Plants annual or rarely biennial, with perfect flowers, with an often somewhat thickened taproot. Stems 1 or few, 20–70(–80) cm long, erect or strongly ascending, sometimes slightly zigzag toward the tip, unbranched or branched above the midpoint, glabrous. Leaves alternate and basal, the basal leaves longer-petiolate than the lower stem leaves. Ocreae pale to light brown and membranous to papery, tearing and usually becoming deeply dissected or absent at maturity. Blades of main leaves 3–8(–12) cm long, 2–5 cm wide (mostly 2–4 times longer than wide), relatively thin and herbaceous, oblong, ovate-elliptic, elliptic-lanceolate, or sometimes narrowly elliptic-lanceolate (upper leaves), those of the uppermost leaves progressively or relatively abruptly shorter and narrower, unlobed, the margins entire or nearly so, flat to somewhat undulate and sometimes slightly crisped, angled (upper leaves) to broadly angled, truncate, or shallowly cordate at the base, bluntly to sharply pointed at the tip, the surfaces glabrous. Inflorescences terminal, occupying the upper 1/2 of the plant, usually broadly paniculate (the branches spreading to loosely ascending), relatively open and interrupted. Flowers 10–20 per whorled fascicle, the stalks 2–5 mm long, as long as or slightly longer than the fruiting perianth, arched downward or nodding, jointed toward the base. Inner whorl of tepals becoming enlarged to 3.0–5.5(–6.0) mm long and 2–3 mm wide (excluding the teeth) at fruiting, with relatively broad wings, ovate-triangular, mostly truncate at the base, sharply pointed at the tip, each tepal with 4–8(–10) conspicuous narrowly triangular to spine-like teeth along the wing margins, these 1–3(–5) mm long, mostly 1–3 times as long as the width of the undivided portion, the surfaces with a prominent network of nerves; tubercles 3, all similar in size, covering about 1/2 of the width of the tepal (excluding the teeth), glabrous, smooth to finely pitted; the outer whorl spreading. Fruits 2.0–2.8 mm long, 1.4–1.8 mm wide, dark reddish brown. $2n$=40. May–July.

Introduced, known thus far only from the city of St. Louis (native of Europe, Asia, Africa; introduced sporadically in the U.S.; Canada). Railroads.

This species was first discovered in Missouri during Viktor Mühlenbach's (1979) botanical surveys of the St. Louis railyards. In the Old World several subspecies have been recognized. The scanty North American materials have not been determined satisfactorily yet at the infraspecific level, although Mosyakin (2005) suggested that most of the materials in the Flora of North America region corresponded to ssp. *halacsyi* (Rech. f.) Rech. f., which is native to the eastern portion of the Mediterranean region.

7. Rumex fueginus Phil. (golden dock, American golden dock)
R. maritimus L. ssp. *fueginus* (Phil.) Hultén
R. maritimus var. *fueginus* (Phil.) Dusén
Pl. 505 a, b; Map 2302

Plants annual, with perfect flowers, with a tapered taproot, sometimes rooting at the lower nodes. Stems 1 to several, 15–65 cm long, erect to loosely ascending, sometimes from a spreading base, few- to several-branched toward the tip, usually short-hairy, at least toward the tip. Leaves alternate (plants lacking a well-developed basal rosette at flowering). Ocreae pale to light brown and papery, tearing and usually becoming deeply dissected or absent at maturity. Leaf blades (3–) 5–25(–30) cm long, 1.5–3.0(–4.0) cm wide (4 or more times longer than wide), thick and often somewhat leathery, linear-lanceolate to lanceolate or occasionally oblong-lanceolate, those of the uppermost leaves sometimes nearly linear, unlobed, the margins entire or finely and unevenly scalloped, usually undulate and crisped, angled (up-

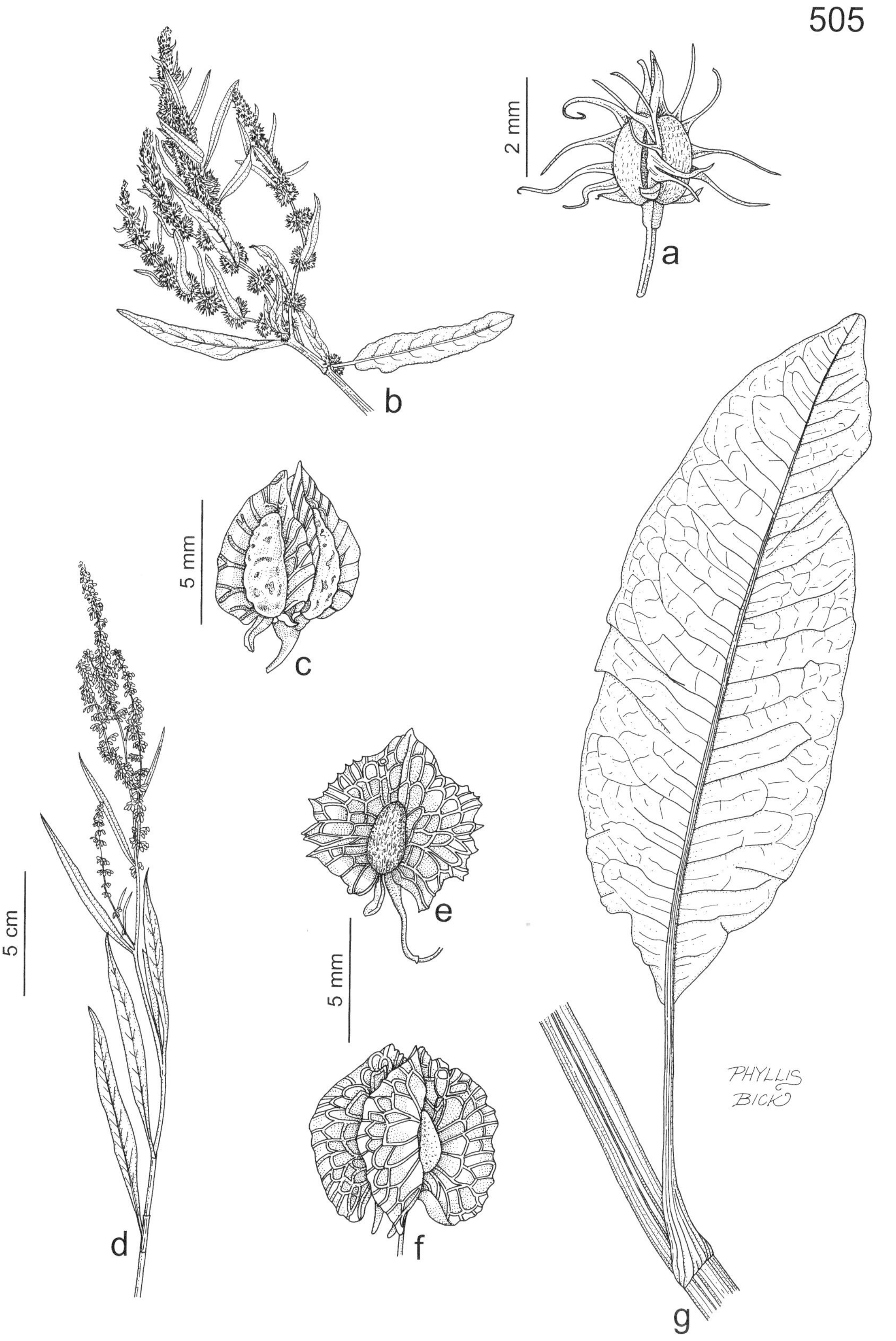

Plate 505. Polygonaceae. *Rumex fueginus*, **a)** fruit in calyx, **b)** inflorescence. *Rumex triangulivalvis*, **c)**, fruit in calyx, **d)** fertile stem. *Rumex cristatus*, **e)** fruit in calyx. *Rumex patientia*, **f)** fruit in calyx, **g)** lower node with leaf and ocrea.

per leaves) to truncate to shallowly cordate somewhat rounded at the base, angled or tapered to a usually sharply pointed tip, the surfaces glabrous or more commonly the upper surface sparsely to moderately short-hairy along the main veins. Inflorescences terminal and axillary, usually occupying the upper 1/2 of the plant, broadly paniculate, dense and mostly continuous. Flowers 15–30 per whorled fascicle, the stalks 3–7(–9) mm long, 1–2 times as long as the fruiting perianth, arched downward or nodding, jointed toward the base. Inner whorl of tepals becoming enlarged to 1.5–2.5 mm long and 0.7–0.9(–1.2) mm wide (excluding the teeth) at fruiting, with slender wings, narrowly triangular to narrowly rhombic-triangular, truncate or broadly angled at the base, sharply pointed at the tip, each tepal with 4–6 conspicuous long, slender, bristlelike or spinelike teeth along the wing margins, these 1–3 mm long, 1.5–2.5(–4.0) times as long as the width of the undivided portion, the surfaces with a prominent network of nerves; tubercles 3, usually more or less similar in size, covering up to 1/2 the width of the tepals (including the wings but not the teeth), glabrous, usually finely pitted; the outer whorl spreading to somewhat reflexed at fruiting. Fruits 1.1–1.4 mm long, 0.6–0.8 mm wide, light brown to yellowish brown. $2n$=40. May–July.

Scattered in counties bordering the Missouri and Mississippi Rivers (nearly throughout the U.S. [including Alaska] except for some southeastern States; Canada, South America, Europe). Banks of streams and rivers, saline seeps, oxbows, sloughs, and margins of lakes; also open disturbed areas; sometimes emergent aquatics.

This taxon has been treated by some botanists as a variety of *R. maritimus*. True *R. maritimus* is a Eurasian species that has become introduced rarely and sporadically in North America mostly in disturbed coastal areas. According to Mosyakin (2005) *R. fueginus* is more closely related to *R. persicarioides* L., another segregate of the complex that is native to northeastern and northwestern portions of the United States and adjacent Canada. *Rumex maritimus* in the strict sense has nearly glabrous inflorescence branches (vs. distinctly minutely hairy), narrowly tapered leaf bases (vs. cordate, rounded, or angled), and usually tubercles with the surface smooth (vs. with a network of shallow ridges and pits). *Rumex persicarioides* has the tubercles straw-colored at maturity, rounded to bluntly pointed at the tips, and nearly as wide as the associated tepals. *Rumex fueginus* has the tubercles greenish brown to reddish brown at maturity, more or less sharply pointed at the tips, and distinctly narrower than the associated tepals (Mosyakin, 2005).

8. Rumex hastatulus Baldwin (wild sorrel, sour dock, heartwing sorrel)

Acetosa hastatula (Baldwin) Á. Löve

Pl. 503 a; Map 2303

Plants annual (sometimes flowering the first year), dioecious, rarely incompletely so, with a slender, sometimes somewhat woody, vertical rootstock, often appearing taprooted. Stems 1 to several, 10–40(–45) cm long (including inflorescence), erect or ascending, usually unbranched below the inflorescence, glabrous. Leaves alternate and basal (plants with a well-developed basal rosette at flowering). Ocreae membranous to papery, becoming irregularly dissected to above or below the midpoint at maturity, light brown to reddish brown at the base, grading abruptly to white or pale and translucent toward the tip. Basal leaves mostly long-petiolate, the blades similar to those of the lower stem leaves. Blades of stem leaves 2–6(–10) cm long 0.5–2.0 cm wide, thin and herbaceous to slightly succulent, narrowly ovate to oblong-lanceolate, lanceolate, or oblong-obovate, with a pair of spreading, bluntly triangular basal lobes, occasionally unlobed, the margins otherwise entire, flat or nearly so, truncate to slightly concave at the base, angled or tapered to a bluntly or sharply pointed tip, rarely rounded at the tip, the surfaces glabrous. Inflorescences terminal, usually occupying the upper 1/2–2/3 of the plant, narrowly paniculate, usually somewhat zigzag, occasionally reclining, interrupted. Flowers 3–6(–8) per whorled fascicle, the stalks 1.5–3.0 mm, about as long as to slightly longer than the fruiting perianth, arched downward or nodding at maturity, jointed below the midpoint. Pistillate flowers with the inner whorl of tepals becoming enlarged to 2.5–3.2 mm long and 2.7–3.2 mm wide at fruiting, winged, broadly ovate to nearly circular, broadly cordate to rounded at the base, bluntly to more or less sharply pointed at the tip, the margins entire, the surfaces with a network of prominent nerves; tubercles absent or the midnerves slightly swollen toward the base; outer whorl of tepals reflexed at fruiting. Fruits 0.9–1.2 mm long, 0.6–0.8 mm wide, brown to dark brown. $2n$=8–10. April–June.

Uncommon in the southeastern portion of the state and locally in Newton County (eastern U.S. west to Illinois and Texas). Glades and upland prairies; also pastures, railroads, and open disturbed areas.

Steyermark (1963) and Mosyakin (2005) both noted that although the enlarged sepals at fruiting readily distinguish *R. hastatulus* from *R. acetosella*, the two are very difficult to distinguish during other times of the year.

2305. Rumex patientia 2306. Rumex pulcher 2307. Rumex stenophyllus

9. Rumex obtusifolius L. (bitter dock, blunt-leaved dock, broad-leaved dock, red-veined dock)

Pl. 504 d; Map 2304

Plants perennial, with perfect flowers, with a thickened, somewhat turnip-shaped, vertical rootstock. Stems 1 to several, 60–120(–150) cm long, erect or strongly ascending, sometimes arched upward, usually well-branched above the midpoint, glabrous or nearly so. Leaves alternate and basal, the basal leaves larger and usually longer-petiolate than the lower stem leaves. Ocreae pale to light brown and papery, tearing and usually becoming deeply dissected or absent at maturity. Blades of main leaves 20–40 cm long, 10–15 cm wide (mostly 2–4 times longer than wide), thick and often somewhat leathery, oblong to oblong-ovate or occasionally broadly ovate, those of the uppermost leaves progressively shorter and narrower, unlobed, the margins entire or nearly so, flat or more commonly somewhat undulate and sometimes slightly crisped, angled (upper leaves) to truncate, rounded, or cordate at the base, rounded or bluntly pointed at the tip, the surfaces glabrous or the undersurface minutely hairy along the main veins. Inflorescences terminal and axillary, usually occupying the upper 1/2–2/3 of the plant, narrowly to more broadly paniculate with ascending branches (diverging from the main axis at an angle of 30–45°), relatively dense but interrupted toward the base. Flowers 10–25 per whorled fascicle, the stalks 5–10 mm long, mostly 2–3 times as long as the fruiting perianth, arched downward or nodding, jointed toward the base or rarely near the midpoint. Inner whorl of tepals becoming enlarged to 3–6 mm long and 2.0–3.5 mm wide (excluding the teeth) at fruiting, with relatively broad wings, ovate to ovate-triangular, truncate or very broadly angled at the base, sharply or bluntly pointed at the tip, each tepal with 4–8 conspicuous, sharp, sometimes spinelike teeth along the wing margins, these 0.5–1.8 mm long, mostly shorter than the width of the undivided portion, the surfaces with a prominent network of nerves; tubercle usually 1 (2 of the inner tepals lacking tubercles), rarely with 1 larger and 2 smaller tubercles, the largest covering 1/2 or more of the width of the tepal (including the wings but not the teeth), glabrous, smooth; the outer whorl spreading at fruiting. Fruits 2.0–2.7 mm long, 1.2–1.7 mm wide, brown to reddish brown. $2n=40$. April–June.

Introduced, scattered nearly throughout the state (native of Europe, Asia; introduced nearly throughout the U.S.; Canada). Banks of streams, rivers, and springs, bottomland forests, mesic upland forests in ravines, and marshes; also pastures, quarries, gardens, railroads, roadsides, and disturbed areas; rarely emergent aquatics.

In the Old World *R. obtusifolius* has been divided into several subspecies (Mosyakin, 2005). However, application of these names to North American material is not well understood.

10. Rumex patientia L. (patience dock, monk's rhubarb, passion dock)

Pl. 505 f, g; Map 2305

Plants perennial, with perfect flowers, with a thickened, somewhat turnip-shaped, vertical rootstock. Stems 1 to several, 80–200 cm long, erect or strongly ascending, unbranched below the inflorescence, glabrous. Leaves alternate and basal, the basal leaves larger and longer-petiolate than the lower stem leaves. Ocreae tan to reddish brown and papery, tearing and usually becoming deeply dissected or absent at maturity. Blades of main leaves 30–45(–50) cm long, 10–15 cm wide (mostly 3–4 times longer than wide), herbaceous to somewhat leathery, broadly lanceolate to oblong-lanceolate or sometimes nearly linear (upper leaves), those of the basal leaves often oblong-ovate, those of the uppermost leaves progressively shorter and narrower, unlobed, the margins entire or nearly so, flat or slightly undulate and not or only slightly crisped, angled (upper leaves) to broadly angled, truncate, or shallowly cordate at the base, mostly sharply pointed at the tip, the surfaces glabrous. Inflorescences terminal, usually occupying the upper 1/2 of the plant, narrowly to broadly paniculate with loosely to strongly ascending or upward-arched branches, dense and more or less continuous (at least at fruiting). Flowers (1–)10–20(–25) per whorled fascicle, the stalks 5–11(–17) mm long, mostly about as long as the fruiting perianth, arched downward or nodding, jointed toward the base. Inner whorl of tepals becoming enlarged to 5.5–10.0 mm long and 5–9(–10) mm wide at fruiting, with broad wings, broadly ovate to nearly circular, usually distinctly cordate at the base, rounded to bluntly pointed at the tip, the margins entire or slightly irregular but lacking teeth, the surfaces with a prominent network of nerves; tubercle 1 (2 of the inner tepals not developing tubercles), this relatively small, covering much less than 1/2 of the width of the tepal (including the wings), glabrous, finely pitted; the outer whorl spreading to somewhat reflexed at fruiting. Fruits 3.0–3.5 mm long, 1.5–2.5 mm wide, brown to dark reddish brown. $2n$=60. April–June.

Introduced, uncommon, mostly in counties bordering the Missouri River, especially in the Kansas City and St. Louis metropolitan areas (native of Europe, Asia; introduced in the northern U.S. south to South Carolina, Oklahoma, and Utah; Canada). Banks of rivers; also pastures, railroads, roadsides, and open disturbed areas.

In the Old World, some botanists divide *R. patientia* into three or more subspecies (Mosyakin, 2005). Application of these names to North American plants is not well understood. Among the Missouri species, *R. patientia* is most easily confused with *R. cristatus*. For further discussion on the morphological differences between these two species, see the treatment of *R. cristatus*.

11. Rumex pulcher L. (fiddle dock)

Pl. 504 g–i; Map 2306

Plants perennial, with perfect flowers, with a thickened, somewhat turnip-shaped, vertical rootstock. Stems 1 to several, 20–60(–70) cm long, erect or strongly ascending, sometimes somewhat curved or zigzag toward the tip, usually well-branched nearly throughout, glabrous or sparsely to moderately and minutely hairy. Leaves alternate and basal, the basal leaves larger and usually longer-petiolate than the lower stem leaves. Ocreae pale to light brown and membranous to papery, tearing and usually becoming deeply dissected or absent at maturity. Blades of main leaves 4–10(–15) cm long, 2–3(–5) cm wide (mostly 2–4 times longer than wide), herbaceous to thick and somewhat leathery, oblong to oblong-ovate or occasionally broadly lanceolate, those of the basal leaves often somewhat fiddle-shaped (with a contracted portion below the midpoint), those of the uppermost leaves progressively shorter and narrower, lacking basal lobes, the margins entire or nearly so, flat or more commonly somewhat undulate but usually not or only slightly crisped, angled (upper leaves) to truncate, rounded, or shallowly cordate at the base, rounded or bluntly to sharply (upper leaves) pointed at the tip, the surfaces glabrous or the undersurface minutely hairy along the main veins. Inflorescences terminal and axillary, usually occupying the upper 2/3–3/4 of the plant, broadly paniculate with spreading to loosely ascending branches (diverging from the main axis at an angle of 60–90°, then arched upward), relatively open and discontinuous. Flowers 10–20 per whorled fascicle, the stalks 2–5 mm long, shorter than to about as long as the fruiting perianth, arched downward or nodding, jointed toward the base. Inner whorl of tepals becoming enlarged to 3–6 mm long and 2–3 mm wide (excluding the teeth) at fruiting, with relatively broad wings, oblong-ovate to ovate-triangular, truncate or very broadly rounded at the base, mostly bluntly pointed at the tip, each tepal with 2–5(–9) conspicuous, sharp, sometimes spinelike teeth along the wing margins, these 0.5–2.5 mm long, mostly shorter than to about as wide as the undivided portion, the surfaces with a prominent network of nerves; tubercles (1–)3, similar in size or 1 larger than the other 2, the largest covering about 1/2 of the width of the tepal (including the wings but not the teeth), glabrous, finely warty; the outer whorl spreading at fruiting. Fruits 2.0–2.8 mm long, 1.3–2.0 mm wide, dark reddish brown to nearly black. $2n$=20. April–July.

Introduced, uncommon and sporadic (native of Europe, Asia, Africa; introduced in the eastern U.S. west to Missouri and Texas, also the western U.S. east to Nevada and New Mexico; Canada). Bottomland forests; also railroads and disturbed areas.

In the Old World *R. pulcher* has been divided into several subspecies (Mosyakin, 2005). However, application of these names to North American material is not well understood.

12. Rumex stenophyllus Ledeb. (narrow-leaved dock)

Pl. 506 g, h; Map 2307

Plants perennial, with perfect flowers, with a somewhat thickened vertical rootstock. Stems 1

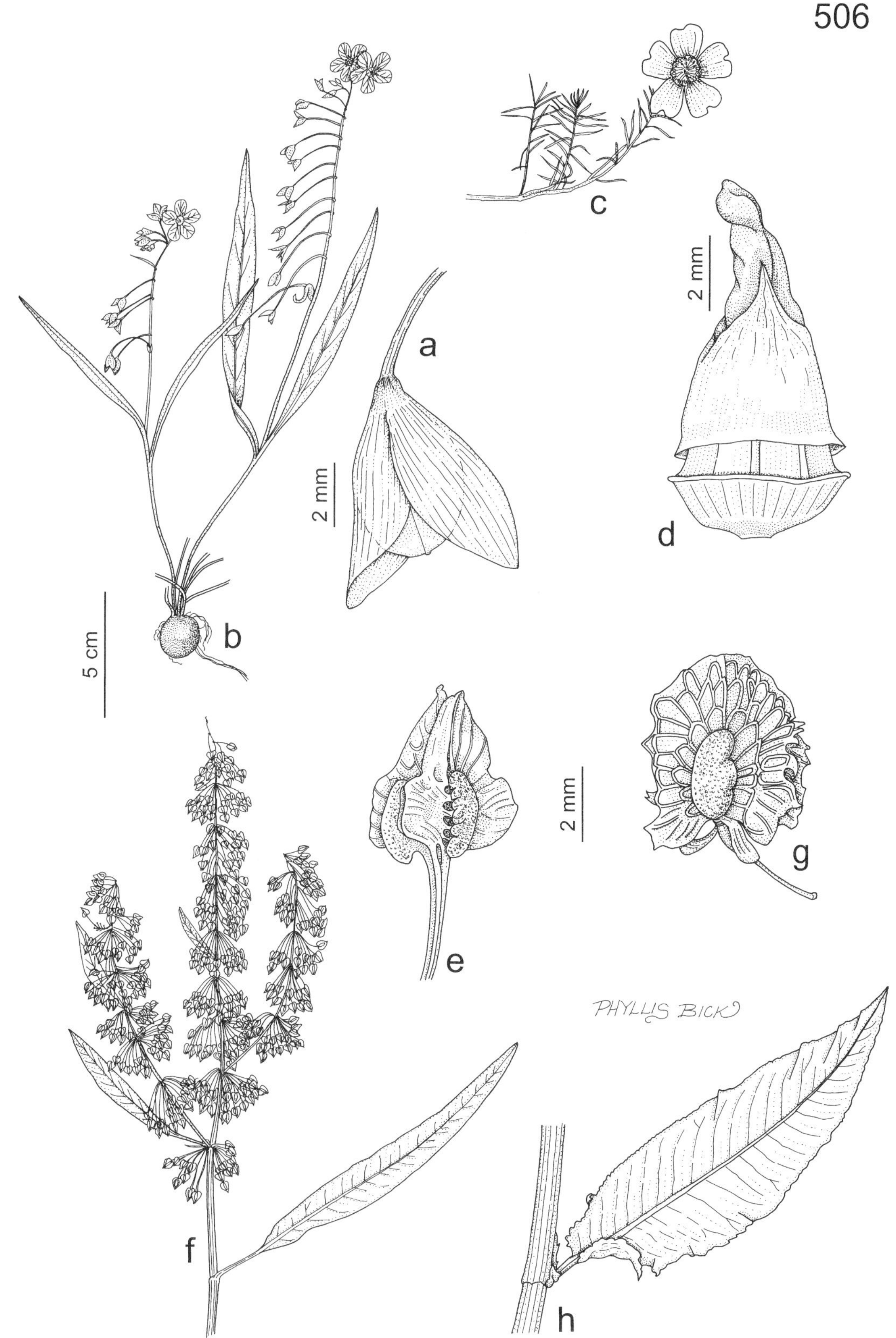

Plate 506. Polygonaceae. Portulacaceae. *Claytonia virginica*, **a)** habit, **b)** fruit in calyx. *Portulaca grandiflora*, **c)** habit, **d)** fruit starting to dehisce. *Rumex verticillatus*, **e)** fruit in calyx, **f)** upper portion of stem with leaf and inflorescence. *Rumex stenophyllus*, **g)** fruit in calyx, **h)** node with leaf and remains of ocrea.

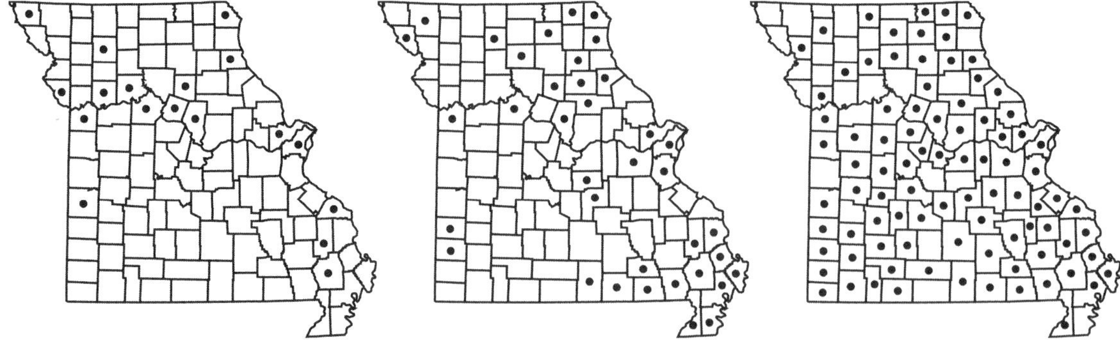

2308. Rumex triangulivalvis 2309. Rumex verticillatus 2310. Claytonia virginica

or few, 40–80(–130) cm long, erect or strongly ascending, usually well-branched above the midpoint, glabrous. Leaves alternate and basal, the basal leaves longer-petiolate than the lower stem leaves. Ocreae pale to light brown and membranous to papery, tearing and usually absent at maturity. Blades of main leaves 15–25(–30) cm long, 2–7 cm wide (mostly more than 4 times longer than wide), herbaceous to slightly leathery, oblong-lanceolate, lanceolate, or narrowly lanceolate, those of the uppermost leaves progressively or relatively abruptly shorter and narrower, unlobed, the margins entire or nearly so, usually undulate and somewhat crisped, rarely flat or nearly so, angled to truncate at the base, sharply pointed at the tip, the surfaces glabrous. Inflorescences terminal, occupying the upper 1/2 or more of the plant, relatively narrowly paniculate (the branches ascending or occasionally curved upward), relatively dense at fruiting continuous or interrupted toward the base. Flowers 20–25 per whorled fascicle, the stalks 3–8 mm long, 1.0–1.5 times as long as the fruiting perianth, arched downward or nodding, jointed toward the base. Inner whorl of tepals becoming enlarged to 3.5–5.0 mm long and 3–5 mm wide (excluding the teeth) at fruiting, with relatively broad wings, broadly ovate to ovate-triangular, truncate to slightly cordate at the base, bluntly to sharply pointed at the tip, each tepal with 8–20 conspicuous triangular to narrowly triangular teeth along the wing margins, these 0.2–1.5 mm long, much shorter than the width of the undivided portion, the surfaces with a prominent network of nerves; tubercles 3, all similar in size, covering less than 1/2 of the width of the tepal (excluding the teeth), glabrous, smooth to faintly and finely wrinkled; the outer whorl spreading. Fruits 2–3 mm long, 1.0–1.5 mm wide, light reddish brown to reddish brown or brown. $2n$=60. May–June.

Introduced, uncommon, known thus far from the St. Louis metropolitan region (native of Europe, Asia; introduced in the western U.S. east to Minnesota and Missouri, also South Carolina; Canada). Banks of streams and rivers and bottomland forests; also railroads, roadsides, and open disturbed areas.

Steyermark's reports of this species from Buchanan and Clay Counties could not be verified during the current study.

13. Rumex triangulivalvis (Danser) Rech. f. (willow-leaved dock)

R. salicifolius Weinm. var. *triangulivalvis* (Danser) J.C. Hickman

Pl. 505 c, d; Map 2308

Plants perennial, monoecious (staminate flowers produced toward the inflorescence tips but generally not apparent at fruiting), with a stout, vertical rootstock, occasionally with short rhizomes. Stems 1 or few, (30–)40–100 cm long, erect or ascending, sometimes from a spreading base, few- to several-branched below the inflorescence, glabrous. Leaves alternate (plants lacking a well-developed basal rosette at flowering). Ocreae tan to white and membranous to papery, tearing and usually shed early. Leaf blades 6–17 cm long, 1–4(–5) cm wide, sometimes relatively thick but usually not leathery, lanceolate to elliptic-lanceolate, those of the uppermost leaves sometimes nearly linear, unlobed, the margins entire, flat or only slightly undulate, angled at the base, angled or tapered to a sharply pointed tip, the surfaces glabrous. Inflorescences terminal and axillary, the terminal ones usually occupying the upper 1/5–1/3 of the plant, narrowly to broadly paniculate, dense and mostly continuous, sometimes somewhat interrupted near the base. Flowers 10–25 per whorled fascicle, the stalks 4–8 mm long, slightly longer than (to about 1.5 times) the fruiting perianth, straight to somewhat curved, but not uniformly arched downward or drooping at maturity, jointed below the midpoint. Pistillate flowers with the inner whorl of tepals becoming enlarged to 2.5–4.0 mm long and 2–3 mm wide at fruiting, moderately to broadly winged, broadly triangular, truncate or broadly rounded at the base, mostly sharply pointed at the tip, the margins entire, the surfaces with a prominent network of nerves; tubercles usually 1, rarely 2 or 3, but then with only 1 fully formed tubercle, the other(s) much smaller, the well-developed tubercle covering up to 1/2 of the width of the tepals (including the wings), glabrous, sometimes minutely pitted; the outer whorl spreading at fruiting. Fruits 1.7–2.2 mm long, 1.0–1.5 mm wide, brown or dark reddish brown. $2n$=20. April–June.

Scattered, mostly in counties bordering the Missouri and Mississippi Rivers (western U.S. east to Texas, Kentucky, and North Carolina; Canada; introduced in Europe). Banks of streams and rivers, margins of ponds and lakes, sloughs, swamps, and marshes; also ditches, railroads, roadsides, and open disturbed areas; sometimes emergent aquatics.

There continues to be controversy as to the distinctness of taxa in the *R. salicifolius* polyploid complex, with various authors accepting 1–16 species. Because most of the segregates can be distinguished only based on fruiting perianth characters, many botanists have not been able to determine significant proportions of their collections with confidence. The taxon in Missouri has at times been referred to as *R. salicifolius* (in the very broad sense), *R. mexicanus* Meisn. (in an intermediate classification), or *R. triangulivalvis* (when the complex is fully split). The present treatment follows that of the last monographer of the group (Rechinger, 1937) and Mosyakin (2005) in the Flora of North America series in recognizing several species in the United States, one of which occurs in the eastern half of the country. For a different viewpoint, see Hickman (1984). In the restricted sense, *R. salicifolius* occurs in the southwestern United States and adjacent Mexico, and is distinguished in having at least one of the tubercles nearly as wide as the tepal on which it is produced. *Rumex mexicanus* is a Mexican taxon entering the United States only in New Mexico; it has larger tepals and fruits than does *R. triangulivalvis*. Both Sarkar (1958), who studied the cytology of the group, and Mosyakin (2005) cautioned that more detailed studies are needed before a stable classification can be achieved.

Steyermark (1963), who called this species *R. mexicanus*, noted that the young foliage can be eaten as a cooked vegetable. Vegetatively, it is essentially indistinguishable from *R. altissimus*. Both are distinctive in their flat, smooth-margined leaf blades that are light green to yellowish green.

14. Rumex verticillatus L. (swamp dock, water dock)

Pl. 506 e, f; Map 2309

Plants perennial, monoecious (staminate flowers produced toward the inflorescence tips but generally not apparent at fruiting), with a stout, vertical rootstock, sometimes rooting at the lower nodes. Stems mostly solitary, 40–100 (–150) cm long, erect or ascending, sometimes from a spreading base, usually few- to several-branched below the inflorescence, glabrous or nearly so. Leaves alternate (plants lacking a well-developed basal rosette at flowering). Ocreae white or translucent and membranous to more or less papery, becoming shredded into fibers with age. Leaf blades 5–30(–40) cm long, 1–5 cm wide, usually relatively thin, not or only slightly leathery, narrowly lanceolate to lanceolate, those of the uppermost leaves sometimes nearly linear, unlobed, the margins entire, flat or somewhat crisped or undulate, narrowly angled at the base, angled or tapered to a usually sharply pointed tip, the surfaces glabrous. Inflorescences terminal and axillary, the terminal ones usually occupying the upper 1/3–1/2 of the plant, narrowly paniculate, the main axis often somewhat zigzag, mostly relatively open and interrupted, at least below the midpoint (the nodes often relatively well-separated along the axis). Flowers 10–15(–25) per whorled fascicle, the stalks 10–17 mm long, 2–5 times as long as the fruiting perianth, mostly strongly arched downward or nodding at maturity, jointed toward the base. Pistillate flowers with the inner whorl of tepals becoming enlarged to 3.5–5.0 mm long and 2.5–4.0 mm wide at fruiting, moderately to broadly winged, ovate to ovate-triangular, truncate or broadly rounded at the base, narrowly rounded or sharply pointed at the tip, the margins entire, the surfaces with a prominent network of nerves; tubercles 3, all of similar size or 1 of them slightly larger than the others, covering up to 1/2 of the width of the tepals (including the wings), glabrous, sometimes minutely pitted, pebbled, or slightly cross-wrinkled; the outer whorl spreading to somewhat reflexed at fruiting. Fruits 2.3–3.1 mm long, 1.6–2.2 mm wide, brown or dark brown. $2n$=60. April–June.

Scattered nearly throughout Missouri, but uncommon in the western portion of the Ozark Division (eastern U.S. west to South Dakota and Texas; Canada). Swamps, sloughs, oxbows, and bottomland forests; often emergent aquatics.

This species is usually easily recognized at fruiting because of the relatively long, nodding fruit stalks.

PORTULACACEAE (Purslane Family)

Plants annual or perennial herbs (woody elsewhere), often somewhat succulent, sometimes with tubers or with thickened or tuberous roots. Aerial stems simple or branched, sometimes reduced and inconspicuous (absent elsewhere), erect to spreading. Leaves basal and/or alternate or opposite, the leaves at each node equal in size or nearly so, sessile or petiolate. Stipules absent or uncommonly represented by hairs at the nodes (in *Portulaca*). Leaf blades simple, variously shaped, the margins usually entire. Inflorescences terminal and/or axillary clusters, racemes, or panicles, sometimes appearing umbellate, occasionally reduced to a solitary flower, the flowers and/or branch points sometimes associated with leaflike or scalelike bracts. Flowers actinomorphic, perfect, hypogynous or (in *Portulaca*) perigynous to epigynous. Calyces of 2 free sepals (3–9 elsewhere), these sometimes overlapping basally, persistent at fruiting or shed as the flower opens. Corollas of 4–6 free petals (rarely numerous in doubled horticultural forms, occasionally fused basally elsewhere, occasionally shed as the flower opens and thus appearing apetalous elsewhere), these variously shaped, sometimes rounded or shallowly notched at the tip. Stamens 4 to numerous, often appearing opposite the petals, the filaments distinct or fused basally into clusters, sometimes attached to the petal bases, the anthers small, various colors, dehiscing longitudinally. Staminodes absent. Pistil 1 per flower, the ovary superior or (in *Portulaca*) partially to completely inferior, of usually 3 fused carpels, with 1 locule, the placentation free central or basal. Style 1, 3–9-branched nearly to the base or nearly unbranched, the stigmas 1 per style branch (sometuimes appearing as a single, lobed unit when the styles are fused to near the tip), capitate, club-shaped, or linear. Fruits capsules, 1 to many-seeded, the dehiscence longitudinal or circumscissile, sometimes only with age. Seeds kidney-shaped to globose, sometimes somewhat flattened, the embryo appearing curved or coiled. Twenty to 30 genera, about 450 species, worldwide.

The family is treated here in the traditional sense, as the familial limits are still controversial. Several phylogenetic studies based on both morphological and molecular markers have provided evidence that, on the one hand, such families as Basellaceae, Cactaceae, and Didiereaceae are specialized derivatives from within the Portulacaceae as traditionally circumscribed, and on the other hand that there exist three or more major lineages within the traditional Portulacaceae that could be classified as separate families (Rodman, 1990, 1994; Hershkovitz and Zimmer, 1997; Applequist and Wallace, 2001; Edwards et al., 2005; Nyffeler, 2007; Nyffeler and Eggli, 2010). By this latter view, among the genera in Missouri only *Portulaca* would be retained in Portulacaceae in the strict sense; *Claytonia* and *Phemeranthus* would be classified in the Montiaceae and *Talinum* would be segregated into the Talinaceae.

A number of members of the Portulacaceae are cultivated as garden ornamentals, including various species of *Lewisia* Pursh, *Phemeranthus, Portulaca*, and *Talinum*. A few of the larger succulents are grown as specimen plants in homes, greenhouses, and conservatories.

1. Ovary about half to fully inferior; fruits with circumscissile dehiscence; inflorescences of solitary flowers or small, dense clusters, the flowers sessile or nearly so .... 3. PORTULACA
1. Ovary superior; fruits with longitudinal dehiscence (sometimes only with age); inflorescences racemes or panicles (sometimes appearing umbellate), the flowers noticeably stalked
    2. Sepals persistent and ascending at flowering and fruiting; inflorescences racemes; aerial stems with a single pair of opposite leaves (1 or more basal leaves also present); plants producing globose to ovoid, brown tubers usually well below the substrate surface .... 1. CLAYTONIA

2. Sepals shed as the flower opens, sometimes (in *Talinum*) briefly persistent during flowering, then strongly reflexed and shed before fruiting; inflorescences panicles, sometimes appearing umbellate; aerial stems with numerous leaves, in plants with very short stems these sometimes appearing basal or nearly so; plants sometimes with fleshy roots or short, succulent, reddish brown stems, but not producing tubers (the roots appearing tuberous-thickened in *Talinum*)
   3. Aerial stems (below the inflorescence) short to very short, sometimes apparently absent; leaves crowded, sometimes appearing as a basal rosette, the blade more or less tubular, strongly succulent, and linear; inflorescences terminal, flat-topped to shallowly dome-shaped panicles, sometimes appearing umbellate, long-stalked, occasionally 1- or few-branched toward the stalk base, the branch(es) producing a second inflorescence; petals (4–)5–16 mm long . . . . 2. PHEMERANTHUS
   3. Aerial stems elongate; leaves well-spaced, the blade flattened, only slightly succulent, elliptic, somewhat rhombic, or obovate; inflorescences axillary and terminal, elongate panicles (not appearing flat-topped, dome-shaped, or umbellate), short- to long-stalked; petals 2.5–4.0 mm long . . . . 4. TALINUM

1. **Claytonia** L. (spring beauty)
(J. M. Miller and Chambers, 2006)

Twenty-six species, North America, Central America, South America, Asia.

The tuber-forming species of *Claytonia* have a long history of use for food by both Native Americans and European colonists. The small tubers sometimes have been called by fanciful names, such as fairy spuds or fairy potatoes, and have a somewhat nutty flavor. Elsewhere, some of the annual species, such as *C. perfoliata* (see below), have been harvested for salads.

For a number of years, a small population of *C. perfoliata* Donn ex Willd. ssp. *perfoliata* persisted as weeds in a gravel waste area between two greenhouses at the Missouri Botanical Garden in St. Louis. The origins of this small population are unclear, but perhaps seeds were accidentally introduced as soil contaminants in plants that were cultivated in a greenhouse. This unusual occurrence was discovered by John MacDougal (then manager of conservatories at the institution) and is vouchered by his specimen from 2000 in the Garden's herbarium. However, the plants did not spread to other areas and apparently have since become extirpated from the site. Thus, this species is not fully treated in the present account. *Claytonia perfoliata* (miner's lettuce) is a many-stemmed spring annual with a wide native range from western Canada and the western United States discontinuously through Mexico to Central America and South America. It differs from the other Missouri species in its smaller flowers with usually white petals 2–6 mm long, as well as in its characteristic pair of stem leaves that are circular to somewhat quadrangular in outline, disc- to somewhat cup-shaped, and completely perfoliate around the stems.

1. Claytonia virginica L. (Virginia spring beauty, spring beauty

C. virginica f. *robusta* (Somes) E.J. Palmer & Steyerm.

C. ozarkensis John M. Mill. & K.L. Chambers

Pl. 506 a, b; Map 2310

Plants perennial herbs (annual elsewhere), with a globose to ovoid, brown, tuberous rootstock (with taproots or rhizomes elsewhere), this 8–20 mm wide, usually positioned well below the substrate surface. Aerial stems 1 to several, 5–20 cm long, usually well-developed, erect to loosely ascending, sometimes sprawling, not succulent or thickened, glabrous. Leaves basal and a single opposite pair along the stems, glabrous. Basal leaves 1 or few (rarely more), 6–20 cm long, the blade relatively thick, linear to narrowly oblanceolate or narrowly elliptic (rarely broader), angled or tapered to a sharply pointed tip, long-tapered to an indistinct, short to long petiole, green to dark green, some-

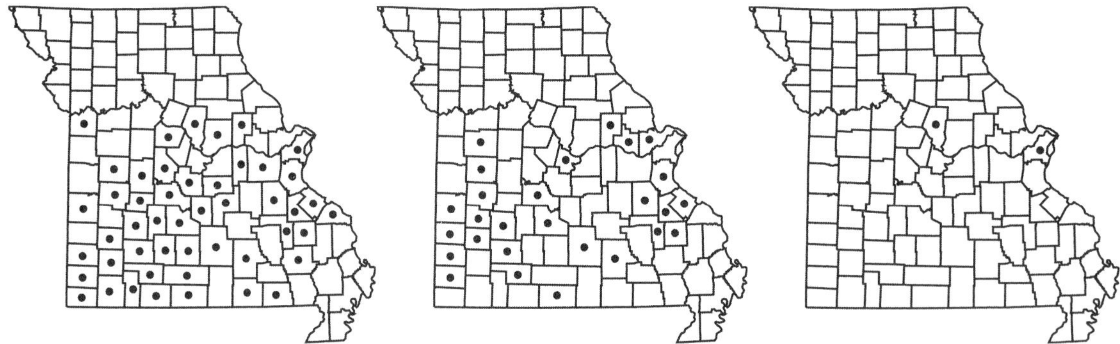

2311. Phemeranthus calycinus 2312. Phemeranthus parviflorus 2313. Portulaca grandiflora

times reddish-tinged. Stem leaves 4–15 cm long, sessile or with a short, indistinct petiole, the blade relatively thick, linear to narrowly oblanceolate or narrowly elliptic, occasionally more broadly elliptic, angled or tapered to a sharply pointed tip, long-tapered at the base a single pair, variously sessile to long-petiolate. Stipules absent. Inflorescences terminal, solitary racemes 4–18 cm long, short- to long-stalked (measured from above the pair of stem leaves), with 5–18 flowers, all tending to be oriented toward 1 side of the axis, only the basal 1(2) flower(s) subtended by a membranous to herbaceous bract, this 2–9 mm long, narrowly lanceolate to ovate, more or less sheathing the axis. Flowers mostly relatively long-stalked, the stalk continuing to elongate after flowering, hypogynous; cleistogamous flowers absent. Calyces 5–7 mm long, the sepals overlapping, persistent at fruiting, remaining ascending after flowering. Petals 5, 7–14 mm long, white or pinkish-tinged, usually with pink venation, withering after the flowering. Stamens 5, the anthers pink (white elsewhere). Ovary superior, the style 3-branched above the midpoint. Fruits 2.5–5.0 mm long, ellipsoid to ovoid, with longitudinal dehiscence, the valves remaining attached, the 1–6 seeds forcibly ejected. Seeds 1–3 mm long, more or less circular (often minutely notched at the base) in outline, somewhat flattened, the surface smooth, shiny, black (dull and/or pebbled elsewhere). $2n=12$ to ca. 190. February–May.

Common throughout the state (eastern U.S. west to Wisconsin and Texas; Canada). Bottomland forests, mesic upland forests, banks of streams and rivers, upland prairies, sand prairies, bases and ledges of bluffs, and occasionally margins of sinkhole ponds; also pastures, cemeteries, lawns, railroads, and roadsides.

Claytonia virginica holds the distinction of having the longest recorded series of aneuploid chromosome counts (W. H. Lewis and Semple, 1977; Doyle, 1981). In the St. Louis area alone, W. H. Lewis et al. (1967) recorded plants with $2n=22$–37 and suggested that broader leaves are correlated with higher chromosome numbers. Rare plants at the broadest-leaved extreme have been called f. *robusta*.

The rare *Claytonia ozarkensis* was first described in late 2006 (J. M. Miller and Chambers, 2006). It was documented from a dozen herbarium specimens in the Ozark portions of Arkansas and Missouri, as well as east-central Oklahoma. The Missouri distribution included the type specimen from Ozark County and single historical specimens from Jefferson and Stone Counties. These specimens were determined by Steyermark (1963) as *C. virginica* f. *robusta*. Recent study of the Missouri and Oklahoma specimens has determined that, in fact, Steyermark's determinations were correct and the plants in question are all merely broad-leaved examples of *C. virginica* (Yatskievych et al., 2013). Additionally, detailed searches by several botanists at the type locality in Ozark County and the site in Stone county failed to disclose plants of *C. ozarkensis* or suitable acidic rock outcrops for it to grow on. However, at the type locality, broad-leaved individuals of *C. virginica* similar to the *C. ozarkensis* type specimen were found intermixed with narrower-leaved plants of that species. For these reasons, *C. ozarkensis* must be excluded from the Missouri flora for the present and treated as endemic to Arkansas. Because the type specimen of the species was misdetermined, the taxon was of necessity renamed, *C. arkansana* Yatsk., R. Evans & Witsell, and the name *C. ozarkensis* technically becomes a synonym of *C. virginica*.

Claytonia arkansana is superficially less similar to *C. virginica* than to *C. caroliniana* Michx., which is widespread in the eastern

United States and adjacent Canada, and to *C. ogilviensis* McNeill, a rare endemic in northwestern Canada. It was distinguished morphologically from these taxa in the monograph of the genus (J. M. Miller and Chambers, 2006, as *C. ozarkensis*). Subsequently, Croft et al. (2011) provided genetic evidence to support the distinctness of the Arkansas populations. It is restricted to horizontal ledges and tops of sandstone bluffs where plants usually are rooted in relatively deeply incised rock seams. Seed dispersal in the plants is unusual in that the inflorescences become recurved back into the crevice as the fruits mature, depositing the seeds back into suitable habitat adjacent to the maternal parent (Matthew Albrecht, Missouri Botanical Garden, personal communication). When pressed and dried, specimens of *C. arkansana* are easily differentiated from broad-leaved plants of *C. virginica* because the leaves and stems become very flaccid, thin, and translucent (whereas, plants of the latter species generally remain firm and opaque). The key below, which is adapted from J. M. Miller and Chambers (2006), provides additional characters to distinguish the two. Because sandstone bluffs occur in several Missouri counties, it is possible that this unusual rare species eventually may be discovered growing in the state.

1. Basal leaves several to many, the petioles long, well-differentiated from the elliptic to ovate blades; stem leaves with the blade mostly elliptic to ovate; inflorescences with all or at least the basal 3 or more flowers each subtended by a small, membranous bract .... 1. C. ARKANSANA
1. Basal leaves usually 1 to few, the petioles short to long, but poorly differentiated from the linear to narrowly oblanceolate or narrowly elliptic blades; stem leaves with the blade linear to narrowly oblanceolate or narrowly elliptic (rarely more broadly elliptic); inflorescences with only the basal 1(2) flowers subtended by a small, membranous or herbaceous bract .... 2. C. VIRGINICA

2. **Phemeranthus** Raf. (rock pink)

Plants perennial herbs, with thickened, more or less fleshy roots, the rootstock and stems often brownish red. Aerial stems (below the inflorescence) short to very short, sometimes apparently absent, erect or ascending, occasionally sprawling, succulent-thickened, glabrous. Leaves alternate, but appearing densely crowded along the stems, sometimes appearing as a basal rosette or nearly so, sessile. Leaf blades more or less tubular and relatively strongly succulent, linear (broader elsewhere), truncate and often slightly expanded at the base, angled or tapered to a sharply pointed tip, glabrous, sometimes appearing slightly glaucous. Stipules absent. Inflorescences terminal, flat-topped to shallowly dome-shaped panicles, sometimes appearing umbellate, long-stalked, occasionally 1- or few-branched toward the stalk base, the branch(es) producing a second inflorescence. Flowers short- to long-stalked, occasionally appearing sessile, hypogynous, occasionally some functionally cleistogamous flowers present. Sepals not overlapping, shed as the flower opens (persistent elsewhere). Petals 5(6), (4–)5–16 mm long, pink to purplish red (other colors elsewhere), shed quickly after the flower opens. Stamens 4 to numerous, the anthers yellow to orangish yellow. Ovary superior, the style variously fused nearly to the tip or branched nearly to the base, mostly appearing 3-branched. Fruits ellipsoid to broadly ovoid, with longitudinal dehiscence, the valves becoming detached. About 30 species, North America, Central America, South America.

This genus has a variety of common names. Flower-of-an-hour is an allusion to the phenology of most of the species, whose flowers tend to open mid- to late afternoon and to last for only two to four hours. Yatskievych (1988) reviewed the application of the common names fame flower and flame flower for members of the genus (as *Talinum*), concluding that both have a long history of independent usage. For a discussion of the segregation of this genus from the related genus *Talinum*, see the treatment of that genus.

1. Petals 10–16 mm long; stamens 25–45; fruits 6–8 mm long . . . . 1. P. CALYCINUS
1. Petals (4–)5–7 mm long; stamens 4–8; fruits 3–5 mm long . . . . 2. P. PARVIFLORUS

1. Phemeranthus calycinus (Engelm.) Kiger (fame flower, flame flower, rock pink, large flower-of-an-hour)

Talinum calycinum Engelm.

Pl. 507 a–c; Map 2311

Rootstocks unbranched or more commonly with few to several branches, each developing a stem. Aerial stems 2–7(–10) cm long (below the inflorescence). Leaves 1.5–7.0 cm long. Inflorescences with the stalk 5–25 cm long. Flowers short- to long-stalked. Sepals 4–6(–8) mm long. Petals 10–16 mm long, 5–9 mm wide, bright reddish pink to reddish purple. Stamens 25–45. Style fused nearly to the tip, the stigma appearing solitary, irregularly capitate, 3-lobed. Fruits 6–8 mm long, ovoid to broadly ovoid. Seeds 0.8–1.2 mm long, the surface smooth, shiny, black. $2n$=24, 48. May–October.

Scattered in the Ozark, Ozark Border, and Unglaciated Plains Divisions, north locally to Jackson County (Illinois to Louisiana west to Nebraska, Colorado, and New Mexico). Glades and ledges and tops of bluffs, usually on acidic substrates but occasionally on calcareous substrates.

Reinhard and Ware (1989), who studied *Phemeranthus* (as *Talinum*) populations in the Ozarks and Ouachitas and also grew plants from seeds in different soils, concluded that rare occurrences of *P. calycinus* on calcareous substrates represent localized adaptations toward a slightly increased tolerance to limestone, rather than a specieswide phenomenon. These authors also failed to locate analogous limestone tolerant populations of *P. parviflorus*.

David Ferguson (personal communication) of the Rio Grande Botanic Garden, in New Mexico, has grown a number of samples of this taxon in common culture. He believes that the Ozarkian populations are distinct from those farther west in the Great Plains, based on a tendency to have a more branched habit that results in more inflorescences and flowers, as well as having slightly smaller, darker corollas. The differences in growth form may in part be a result of differing habitats. Plants from Nebraska westward tend to grow in sandy soils and to have a more elongate rootstock, whereas those found in glades and other rocky habitats in the Ozarks are constrained by the thin soil layer in which they grow. However, cultivated individuals of *P. calycinus* from Missouri sources, which increasingly are becoming popular in the wildflower nursery trade, tend to retain their growth form even when grown in richer, deeper soils in a garden situation. The situation requires further study.

2. Phemeranthus parviflorus (Nutt.) Kiger (prairie fame flower, rock pink, small flower-of-an-hour)

Talinum parviflorum Nutt.

Pl. 507 d, e; Map 2312

Rootstocks unbranched or occasionally branched, the plants thus appearing with 1 or less commonly 2 stems. Aerial stems 0.5–2.0 (–3.0) cm long (below the inflorescence). Leaves 1.5–5.0 cm long. Inflorescences with the stalk 3–15 cm long. Flowers mostly short-stalked, less commonly long-stalked or occasionally appearing sessile. Sepals 2.5–4.0 mm long. Petals (4–) 5–7 mm long, 2–3 mm wide, pink to more commonly reddish pink or reddish purple. Stamens 4–8. Style fused about 2/3 of its length, the 3 stigmas unlobed, linear. Fruits 3–5 mm long, ovoid to ellipsoid. Seeds 0.8–1.0 mm long, the surface smooth, shiny, black. $2n$=24, 48. May–September.

Scattered in the Ozark, Ozark Border, and Unglaciated Plains Divisions (Wyoming to Arizona east to Illinois and Louisiana, disjunct in Alabama; Mexico). Glades, ledges and tops of bluffs, on acidic substrates.

Missouri botanists should be on the lookout for the closely related *P. rugospermus* (Holz.) Kiger (rough-seeded fameflower). Cochrane (1993) discussed its distribution (as *Talinum rugospermum* Holz.), which is unusual in consisting of a number of sets of disjunct populations in the upper Midwest (Wisconsin to Indiana), the Great Plains (Nebraska and Kansas), and eastern Texas. MacRoberts and MacRoberts (1997) also discovered the species (as *Talinum*) in western Louisiana. In his treament in the Flora of North America series, Kiger (2003) noted that it would not be surprising for *P. rugospermus* to be discovered in Missouri, Arkansas, and/or Oklahoma in the future. Most recently, Theo Witsell (personal communication) of the Arkansas Natural Heritage Commission has indicated that some populations in Arkansas may indeed represent this species. *Phemeranthus rugospermus*, as the name denotes, differs from *T. parviflorum* and other related taxa in temperate North America in its wrinkled (vs. relatively smooth) seed coat. It tends to grow in deeper sands throughout its range, rather than in the thin sandy pockets and crevices of sandstone glades where *P. parviflorus* occurs. The corollas also tend to be a lighter pink than is the norm for those in *P. parviflorus*.

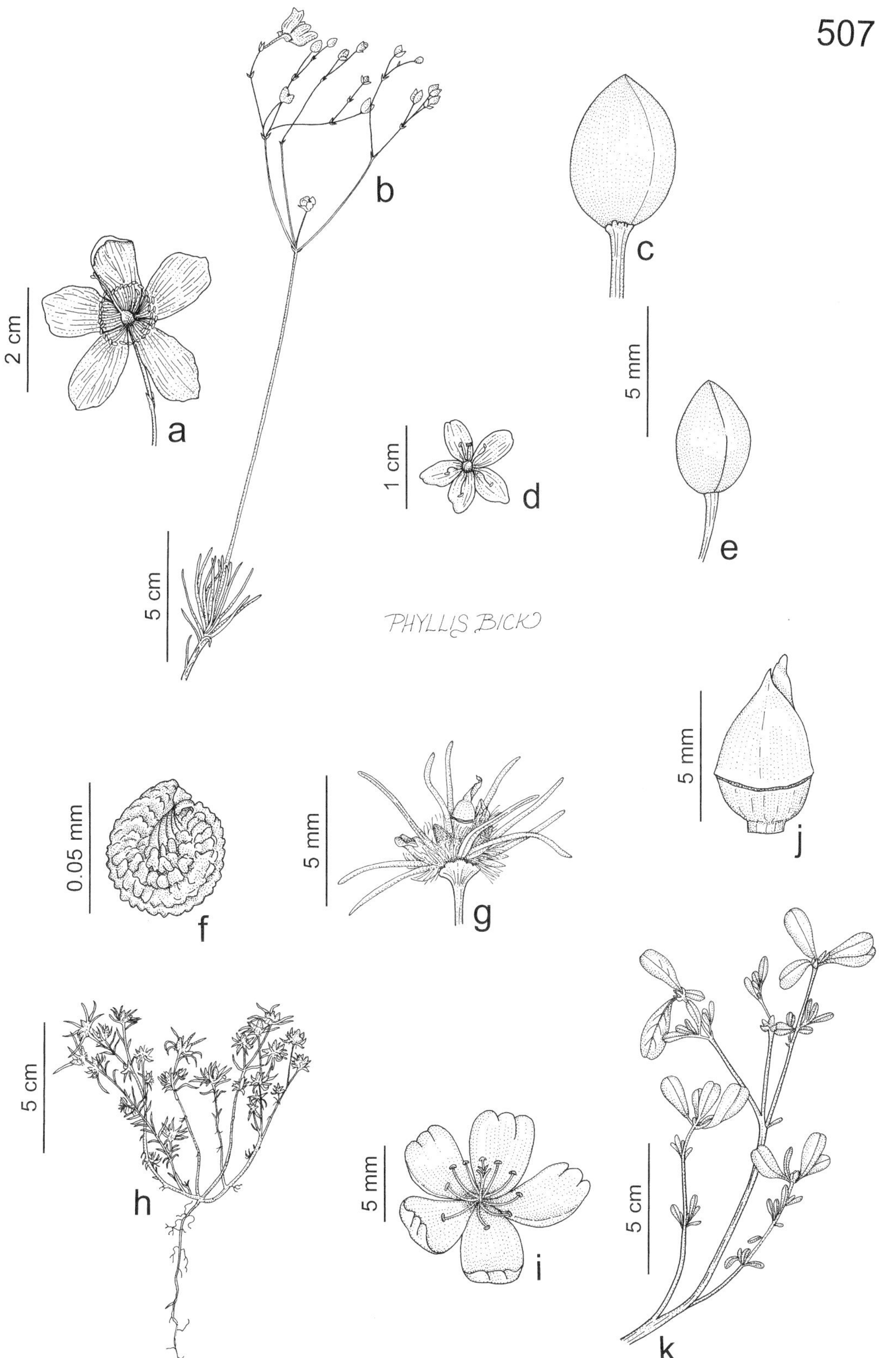

Plate 507. Portulacaceae. *Phemeranthus calycinus*, **a)** flower, **b)** habit, **c)** fruit. *Phemeranthus parviflorus*, **d)** flower, **e)** fruit. *Portulaca pilosa*, **f)**, seed, **g)** node with cluster of fruits, **h)** habit. *Portulaca oleracea*, **i)**, flower, **j)** fruit, **k)** habit.

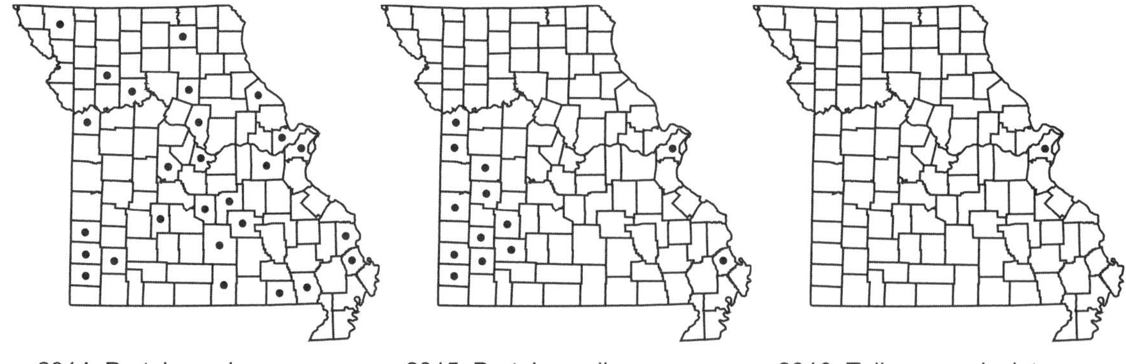

2314. Portulaca oleracea   2315. Portulaca pilosa   2316. Talinum paniculatum

3. **Portulaca** L. (purslane)

Plants annual (sometimes apparently perennial herbs in *P. pilosa*), taprooted, the lateral roots fibrous or somewhat fleshy-thickened, not producing tubers. Stems prostrate to ascending, often somewhat fleshy, glabrous or hairy, especially at the nodes and toward the tip. Leaves alternate and well-spaced along the stems (sometimes appearing congested and nearly whorled at the stem tips), sessile to short-petiolate. Leaf blades somewhat to strongly succulent, flattened or more or less tubular, variously shaped, glabrous. Stipules absent or represented by tufts of hairs in the leaf axils. Inflorescences axillary, of solitary flowers or small, dense clusters. Flowers sessile or nearly so, perigynous or epigynous. Sepals strongly overlapping, persistent after flowering, but usually shed as the fruit reaches maturity. Petals 4–6 (numerous in some doubled forms of *P. grandiflora*), 3–25 mm long, pink, red, purple, yellow, orange, or bronze-colored, occasionally white. Stamens 5 to numerous, the anthers usually yellow. Ovary about half to fully inferior (the perianth fused partially or completely to the ovary wall), the style 3–9-branched. Fruits ovoid, with circumscissile dehiscence. About 125 species, nearly worldwide.

1. Leaves flattened, spatulate to obovate, oblanceolate, or oblong-oblanceolate; stems glabrous at the nodes . . . . 2. P. OLERACEA
1. Leaves not flattened (circular to semicircular in cross-section), linear to narrowly triangular-lanceolate; stems hairy (sometimes only toward the tips, with pronounced tufts of hairs in the leaf axils
    2. Petals 15–25 mm long; seeds 0.7–1.0 mm long. . . . . 1. P. GRANDIFLORA
    2. Petals 4–8 mm long; seeds 0.4–0.6 mm long. . . . . 3. P. PILOSA

1. Portulaca grandiflora Hook. (portulaca, moss rose, rose moss)

Pl. 506 c, d; Map 2313

Plants with the lateral roots fibrous. Stems 5–30 cm long, prostrate to strongly ascending, sometimes hairy toward the tip. Leaf blades 5–30 mm long, 1–4 mm wide, not flattened (circular to semicircular in cross-section), linear to narrowly triangular-lanceolate. Stipules represented by dense, conspicuous tufts of long, somewhat contorted hairs in the leaf axils. Inflorescences with dense, conspicuous tufts of long, somewhat contorted hairs. Sepals 7–11 mm long (measured to the base of the ovary). Petals 15–25 mm long, pink, red, purple, yellow, orange, bronze-colored, or white. Stamens numerous (except rarely in some horticulatural forms with doubled flowers). Fruits 5–8 mm long, 3–6 mm wide. Seeds 0.7–1.0 mm long, variously orbicular to kidney-shaped, the surface smooth or with minute tubercles or spines, gray to dark gray, often somewhat iridescent. $2n$=18. July–September.

Introduced, known thus far only from Boone County and the city of St. Louis (native of South America; introduced widely but sporadically in the U.S. [except for the northwestern states], Canada,

Europe). Banks of rivers; also railroads and roadsides.

This attractive species produces brightly colored flowers in a variety of different colors and is grown commonly in North America as a bedding plant.

2. Portulaca oleracea L. (common purslane, pusley, pursley)

P. oleracea ssp. *granulato-stellulata* (Poelln.) Danin & H.G. Baker

P. oleracea ssp. *nicaraguensis* Danin & H.G. Baker

P. oleracea ssp. *nitida* Danin & H.G. Baker

P. neglecta Mack. & Bush

P. retusa Engelm.

Pl. 507 i–k; Map 2314

Plants with the lateral roots fibrous. Stems 5–50 cm long, prostrate or occasionally ascending, glabrous. Leaf blades 4–30(–50) mm long, 2–15 mm wide, flattened, spatulate to obovate, oblanceolate, or oblong-oblanceolate. Stipules absent (stem nodes glabrous). Inflorescences glabrous or rarely with sparse, inconspicuous, short hairs. Sepals 2.5–4.0 mm long (measured to the base of the ovary). Petals 3–5 mm long, light yellow to yellow. Stamens 6–12(–18). Fruits 5–9 mm long, 4–8 mm wide. Seeds 0.6–1.0 mm long, variously orbicular to more or less kidney-shaped, the surface smooth or more commonly appearing roughened or with minute, blunt tubercles, dark brown to black, not iridescent. $2n$=18, 36, 54. June–November.

Scattered to common nearly throughout the state (nearly worldwide). Banks of streams and rivers, margins of ponds, lakes, and marshes, seeps, bases, ledges and tops of bluffs, upland prairies, and glades; also crop fields, fallow fields, ditches, gardens, lawns, sidewalks, railroads, roadsides, and open disturbed areas.

Some North American authors have considered this species as possibly introduced from the Old World, but Matthews et al. (1993) summarized several sources of data that suggest a native presence in the Americas, including observations of plants growing in some states by the earliest plant explorers and the presence of seeds at archaeological sites and in sediment cores that predate the colonization of North America by Europeans. The species is considered among the worst agricultural weeds in the world and can become an aggressive pest in gardens. Moerman (1998) noted that various tribes of Native Americans ate plants raw and cooked as a vegetable and that the species also was used medicinally to treat worms, ear aches, bruises, and burns. In the Old World, some races of *P. oleracea* have been cultivated traditionally as fodder and vegetable crops in the Middle-East.

Opinions have varied greatly on the taxonomy of the *P. oleracea* complex, which is treated here in a broad sense. *Portulaca neglecta* and *P. retusa* are among the segregates that were reported from Missouri (Steyermark, 1963); the former was characterized as having ascending stems, relatively long leaves, and relatively numerous stamens compared to typical *P. oleracea*, whereas the latter was said to differ in its more rounded petals, slightly notched leaf tips, and more sharply pointed tubercles on the seed coat. In his taxonomic account of the species of *Portulaca* in the Americas, Legrand (1962) placed both of these names in synonymy under *P. oleracea* based on his observations that the descriptions and specimens he had examined fell within the overall range of variation in the species. Matthews and Levins (1985b) and Matthews et al. (1993) studied the *P. oleracea* complex in temperate North America and reached a similar conclusion, noting that this nearly cosmopolitan species reproduces mostly through self-pollination, which favors the maintenance of local races differing slightly but consistently in morphology. Danin et al. (1978) attempted to correlate ploidy and seed morphology and recognized nine subspecies of *P. oleracea*, of which ssp. *granulato-stellulata*, ssp. *nicaraguensis*, ssp. *nitida*, and ssp. *oleracea* were mapped as occurring in Missouri. None of the other named species-level variants was synonymized with these subspecies and in practice the minor differences in seed-coat morphology said to characterize these taxa do not serve to separate Missouri plants adequately (Matthews et al., 1993).

3. Portulaca pilosa L.

P. mundula I.M. Johnst.

P. parvula A. Gray

Pl. 507 f–h; Map 2315

Plants with the taproot and main lateral roots often somewhat thickened and fleshy. Stems 8–35 cm long, prostrate to strongly ascending, often sparsely pubescent, more densely so near the tip. Leaf blades 5–20 mm long, 1–3 mm wide, not flattened (circular to semicircular in cross-section), linear to narrowly oblong-lanceolate. Stipules represented by dense, conspicuous tufts of long, somewhat contorted hairs in the leaf axils. Inflorescences with dense, conspicuous tufts of long, somewhat contorted hairs. Sepals 3–7 mm long (measured to the base of the ovary). Petals 4–8 mm long, red to reddish purple. Stamens 10–15 (to numerous). Fruits 3.5–7.0 mm long, 2.0–4.5 mm wide. Seeds 0.4–0.6 mm long, variously orbicular

2317. Anagallis arvensis 2318. Anagallis minima 2319. Androsace occidentalis

to kidney-shaped, the surface pebbled or with minute, blunt, irregular tubercles, dark gray to black, not iridescent. $2n$=8, 16. May–October.

Scattered in the southwestern portion of the state north to Jackson County; uncommon and sporadic farther east (southern U.S. north to Arizona, Kansas, and North Carolina; Mexico, Central America, South America, Caribbean Islands). Glades, sand prairies, and ledges and tops of bluffs; also railroads, roadsdies, and open disturbed areas.

Many early botanists recognized three or more species within the *P. pilosa* polyploid complex (Steyermark, 1963). *Portulaca mundula* was described to account for plants with relatively large flowers and reddish purple corollas (this includes the plants growing in Missouri). True *P. pilosa* was said to have slightly smaller flowers with reddish purple corollas. Plants with relatively small flowers and yellow to orangish or bronze-colored corollas were described as *P. parvula*. Matthews and Levins (1985a, b) and Matthews et al. (1992) showed that there is almost complete overlap in flower sizes as well as for most other quantitative characters of these plants across their relatively large geographic range and that yellow- and red-flowered variants can grow in the same population. Matthews et al. (1992) also suggested that *P. pilosa* is a tetraploid ($2n$=16) in Mexico and portions of the southeastern United States and a diploid ($2n$=8) in much of the remainder of the North American portion of its range, with no measurable correlation between ploidy and any morphological features.

4. **Talinum** Adans.

About 15 species, southern U.S. to South America, Africa.

In recent decades, phylogenetic studies have supported the segregation of the New World *Phemeranthus* from the mainly African genus *Talinum* (Carolin, 1987; Hershkovitz, 1993; Hershkovitz and Zimmer, 1997; Applequist and Wallace, 2001; Edwards et al., 2005; Nyffeler, 2007). In fact, the two groups are not closely related within the family. Most species of *Phemeranthus* have elongate, tubular leaves, as opposed to the relatively broad, flattened leaf blades in *Talinum* in the strict sense. There are also differences in the morphology of the capsule wall, seed coat, and pollen. Species of *Phemeranthus* also have the seeds covered with a thin membrane at maturity, which is lacking in *Talinum*.

1. Talinum paniculatum (Jacq.) Gaertn. (pink baby's breath, jewels of Opar)

Map 2316

Plants perennial herbs (subshrubs elsewhere), with tuberous-thickened roots. Stems 25–90 cm long, loosely to strongly ascending, not or only slightly fleshy, glabrous. Leaves alternate and well-spaced along the stems (basal leaves sometimes also present at flowering), mostly short-petiolate. Leaf blades 2–12 cm long (those near the inflorescences abruptly smaller than those lower on the stem), strongly flattened and only slightly succulent, elliptic, somewhat rhombic, or obovate, rounded or tapered at the base, broadly angled or tapered abruptly to a bluntly or sharply pointed tip, glabrous. Stipules absent. Inflorescences terminal and sometimes also appearing axillary, diffuse, often elongate panicles. Flowers long-stalked, hypogynous. Sepals 2–4 mm long, not overlapping, occasionally persistent during flowering (then strongly reflexed and shed as the fruit matures), but more commonly shed as the flower opens. Petals 5(6), 3–6 mm long, bright pink or less commonly red, shed quickly after the flower opens. Stamens 15–20, the anthers yellow to orangish yellow. Ovary superior, the style often very short, 3-branched. Fruits 3–5 mm long globose or nearly so, with often tardy longitudinal dehiscence, the valves usually becoming detached. Seeds 0.9–1.2 mm long, broadly kidney-shaped, the surface smooth or occasionally with faint lines or tubercles, black, shiny. $2n=24$. June–October.

Introduced, known thus far only from the city of St. Louis (native of the southwestern U.S. south to South America; introduced sporadically in Missouri and the southeastern U.S.; also Asia, Africa). Open, disturbed areas.

The St. Louis record is based on a specimen collected by the author in October 2002 on a large pile of fill-dirt at a vacant lot. Plants flowered and fruited for several years before the area was graded and converted to a lawn. Because the species is sometimes cultivated in the state as an annual bedding plant and thus might reappear in Missouri in the future, it is given a full treatment in the present account.

PRIMULACEAE (Primula Family)

Contributed by David J. Bogler and George Yatskievych

Plants annual or perennial herbs. Aerial stems unbranched or branched, sometimes absent. Leaves basal and/or alternate, opposite, or whorled, sessile or relatively short-petiolate, the petiole sometimes poorly differentiated from the blade. Stipules absent. Leaf blades simple, with entire or less commonly shallowly toothed margins (pinnately dissected in *Hottonia*), glabrous or inconspicuously hairy, the hairs sometimes stellate. Inflorescences terminal and/or axillary, racemes, panicles, or umbels, or of solitary flowers, the flowers usually subtended by bracts. Flowers actinomorphic, hypogynous (perigynous by fusion of the ovary to the calyx tube in *Samolus*), perfect; cleistogamous flowers absent. Calyces shallowly to deeply (4)5(–7)-lobed, usually persistent at fruiting. Corollas shallowly to deeply (4)5(–7)-lobed (absent elsewhere), variously shaped and colored. Stamens (4)5(–7), the filaments opposite the petals, attached in the corolla tube, sometimes near the base, free (short and fused into a tube in some *Primula* species), the anthers exserted or not, appearing 2-locular, usually attached at or near the base, variously colored; small staminodes sometimes present, alternating with the stamens. Pistil 1 per flower, of 5 fused carpels. Ovary unlobed, 1-locular, with usually numerous ovules, the placentation free-central. Style 1, situated at the tip of the ovary, usually unbranched, the stigma usually capitate. Fruits capsules, dehiscent longitudinally by 5 valves (circumscissile in *Anagallis*), with few to numerous seeds. Twenty to 30 genera, about 1,000 species, nearly worldwide, most diverse in north-temperate regions.

The family Primulaceae is easily recognized by the combination of herbaceous habit, united petals, single whorl of stamens opposite the petals, valvate capsules, and free-central placentation. Primulaceae are similar to Theophrastaceae and Myrsinaceae, and these three families form a discrete lineage in phylogenetic analyses. Recent molecular studies, however, suggest that Primulaceae as traditionally circumscribed do not form a natural group, with *Anagallis*, *Lysimachia*, and several other genera allied with the Myrsinaceae, and *Samolus* closer to Theophrastaceae (Källersjö et al., 2000). For the present, the family is here circumscribed in the traditional sense, in anticipation of more detailed studies in the future.

The family is of relatively little economic importance, though numerous members are commonly grown as ornamentals and in rock gardens.

1. Leaves all basal; inflorescences umbels terminal on an elongate scape
   2. Plants small annuals with taproots and multiple scapes; corollas 2–4 mm long, inconspicuous, shorter than to about as long as the calyces . . . . 2. ANDROSACE
   2. Plants perennials with rhizomes and solitary scapes; corollas 12–25 mm long, showy, longer than the calyces . . . . 5. PRIMULA
1. At least some of the leaves alternate, opposite, or whorled on aerial stems (these often submerged in water in *Hottonia*); inflorescences axillary and/or terminal, racemes (sometimes spikelike), panicles, or of solitary flowers
   3. Leaves finely dissected; inflorescences terminal, spikelike racemes with the axis strongly inflated, the flowers in whorls . . . . 3. HOTTONIA
   3. Leaves entire or with the margins finely toothed; inflorescences axillary and/ or terminal, of solitary flowers or racemes or panicles, the axis and branches then not inflated
      4. Leaves all or mostly alternate
         5. Inflorescences axillary, of solitary flowers; ovary superior . . . 1. ANAGALLIS
         5. Inflorescences terminal racemes or panicles; ovary partly inferior (by fusion to the calyx tube) . . . . 6. SAMOLUS
      4. Leaves all or mostly opposite or whorled
         6. Plants annuals, lacking rhizomes; leaves 2 cm long or shorter; corollas red, reddish orange, or rarely blue or white; fruits with circum scissile dehiscence . . . . 1. ANAGALLIS
         6. Plants perennials, rhizomatous; leaves all or mostly longer than 2 cm, corollas yellow; fruits with longitudinal dehiscence by valves . . . . 4. LYSIMACHIA

1. **Anagallis** L.

Plants annual. Leaves alternate or opposite, sessile or nearly so. Leaf blades unlobed, the margins entire, the surfaces glabrous. Inflorescences axillary, of solitary flowers. Sepals distinct. Corollas, deeply (4)5(6)-lobed, saucer-shaped, the tube short, the limb spreading, shallowly lobed, variously colored. Stamens 5, the filaments attached in the corolla tube, pubescent or glabrous. Ovary ovoid, the style linear, the stigma minute, capitate. Fruits capsules, globose, membranous, with circumscissile dehiscence. Seeds numerous, angular. About 28 species, South America, Europe, Africa; introduced nearly worldwide.

1. Leaves opposite or whorled; corolla conspicuous . . . . 1. A. ARVENSIS
1. Leaves alternate; corolla inconspicuous . . . . 2. A. MINIMA

1. Anagallis arvensis L. (scarlet pimpernel, poor man's weatherglass)

A. arvensis var. *phoenicea* Gouan

Pl. 508 c, d; Map 2317

Stems 5–28 cm long, much-branched at the base, spreading, 4-sided in cross-section and slightly winged, glabrous. Leaves opposite. Leaf blades 1.0–1.6 cm long, 0.5–1.0 cm wide, ovate. Flower stalks 1–2 cm long, slender, recurved in fruit. Calyces of 5 free sepals, 3–4 mm long, membranous along the margins. Corollas deeply 5-lobed, 3–5 mm long, saucer-shaped, the lobes obovate, usually red or salmon-colored, less commonly white or blue, the margins with inconspicuous stalked glands or minutely toothed and lacking glands. Stamens 5, the filaments attached at the corolla base, hairy, purple toward the tip, the anthers yellow. Ovary globose, with numerous ovules, the style 1.5–2.0 mm long, slender. Fruits 3–4 mm long, globose, golden brown. Seeds numerous, angular, the surface tuberculate, dark brown. $2n=40$. May–October.

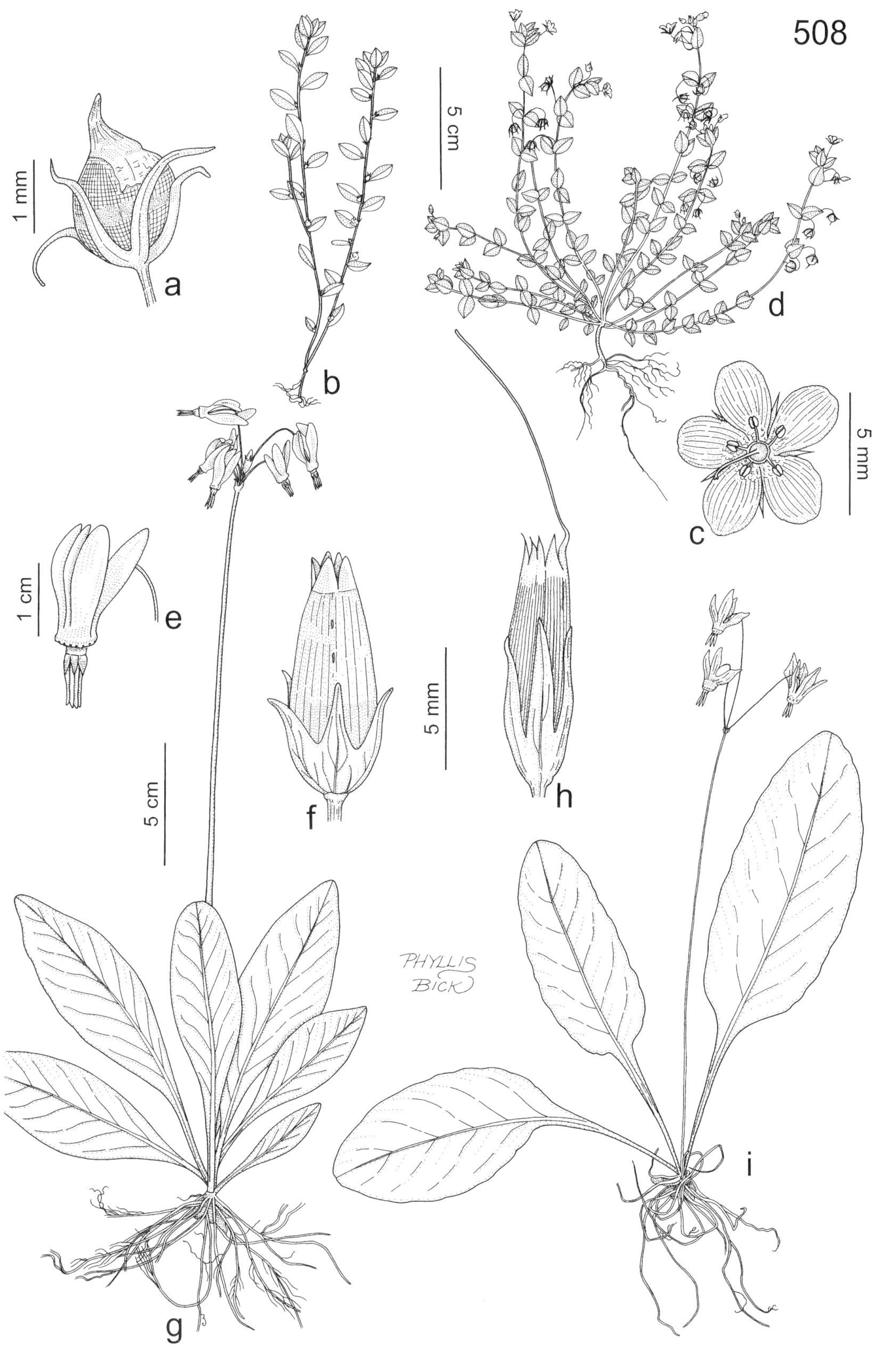

Plate 508. Primulaceae. *Anagallis minima*, **a)** fruit, **b)** habit. *Anagallis arvensis*, **c)** flower, **d)** habit. *Primula meadia*, **e)** flower, **f)** fruit, **g)** habit. *Primula fassettii*, **h)** fruit. *Primula frenchii*, **i)** habit.

Introduced, scattered in the state, mostly south of the Missouri River (native of Europe, Asia; widely introduced in North America and elsewhere in the world). Glades, ledges and tops of bluffs, banks of streams and rivers, and margins of ponds and lakes; also pastures, crop fields, fallow fields, railroads, roadsides, and rocky, open, disturbed areas.

Anagallis arvensis produces a variety of toxins that are poisonous to livestock, and an irritant in the plant hairs can cause dermatitis in humans. The common name poor man's weatherglass comes from the flowers, which close in cloudy weather and open again when the sun comes out. It is a frequent crop weed in warmer parts of the world.

This species has been treated as a complex of two or more infraspecific taxa (Channell and Wood, 1959), two of which have been reported from Missouri (Steyermark, 1963). These are treated as subspecies in the present account, following Marsden-Jones and Weiss (1938).

1. Leaf blades with the undersurface having prominent, scattered purple dots; flower stalks 12–28 mm long, greatly exceeding the subtending leaves; corollas usually red to salmon-colored, rarely white or blue, the margins fringed with minute gland-tipped hairs .... 1A. SSP. ARVENSIS
1. Leaf blades with the undersurface lacking dots or these present only near the tip; flower stalks 6–14 mm long, scarcely exceeding the subtending leaves; corollas blue, the margins uneven or finely toothed, lacking a fringe of gland-tipped hairs ... 1B. SSP. FOEMINA

1a. ssp. arvensis

Leaf blades with the undersurface having prominent, scattered purple dots throughout. Flowers relatively long-stalked, the stalks 12–28 mm long, greatly exceeding the subtending leaves. Corollas usually red or salmon-colored, rarely white or blue, the margins fringed with minute gland-tipped hairs. $2n$=40. May–September.

Introduced, scattered in the state, mostly south of the Missouri River (native of Europe, Asia; widely introduced in North America and elsewhere in the world). Glades, ledges and tops of bluffs, banks of streams and rivers, and margins of ponds and lakes; also pastures, crop fields, fallow fields, railroads, roadsides, and rocky, open, disturbed areas.

The ssp. *arvensis* is by far the more common of these in Missouri. Blue-flowered specimens of this subspecies are reported to be common in Europe and abundant in the tropics (P. Taylor, 1955; Blamey and Grey-Wilson, 1989), but are not known from Missouri. Note, however, that on herbarium specimens the reddish flowers frequently dry dark blue.

1b. ssp. foemina (Mill.) Schinz & Thell.

A. arvensis f. *caerulea* (L.) Lüdi

A. arvensis var. *caerulea* Gouan

Leaf blades with the undersurface lacking dots or these present only near the tip. Flowers relatively short-stalked, the stalks 6–14 mm long, scarcely exceeding the subtending leaves. Corollas blue, the margins uneven or finely toothed, lacking a fringe of gland-tipped hairs.

Introduced, uncommon, sporadic (native of Europe, Asia; widely introduced in North America and elsewhere in the world). Margins of ponds; also railroads. $2n$=40. May–October.

Crosses between ssp. *arvensis* and ssp. *foemina* are sterile except when the ssp. *arvensis* parent has salmon-colored flowers (Marsden-Jones and Weiss, 1938). The ssp. *foemina* is relatively rare in Missouri. It apparently has not been collected in ythe state since 1971. Most of the records are from along railroad tracks in counties around the St. Louis metropolitan area.

2. Anagallis minima (L.) E.H.L. Krause (chaffweed)

Centunculus minimus L.

Pl. 508 a, b; Map 2318

Stems 5–14 cm long, spreading with ascending tips, rooting at the nodes, often forming mats, more or less circular in cross-section, not winged, glabrous. Leaves mostly alternate. Leaf blades 4–10 mm long, 2–4 mm wide, oblong to obovate. Flowers sessile or the stalks 0.5–1.0 mm long. Calyces deeply 4(5)-lobed, the lobes 2–3 mm long. Corollas deeply 4(5)-lobed, 1–2 mm long, more or less saucer-shaped, the lobes broadly lanceolate, light pink, becoming erect after flowering, the corolla withering but persistent as a cap at fruiting. Stamens 4(5), the filaments fused to the corolla for about half their length, glabrous, the anthers yellow. Ovary globose, with numerous seeds, the style 0.4–0.6 mm long. Fruits 1.5–2.5 mm long, more or less globose, yellowish brown to straw-colored. Seeds numerous, angular, the surface pitted, dark brown. $2n$=22. May–August.

Scattered, mostly south of the Missouri River (nearly worldwide). Glades, openings of dry upland forests, savannas, rocky portions of upland prairies, banks of streams and rivers, bottomland prairies, margins of ponds, and fens; also fallow fields, old fields, roadsides, and open disturbed areas.

This species is often treated in the segregate

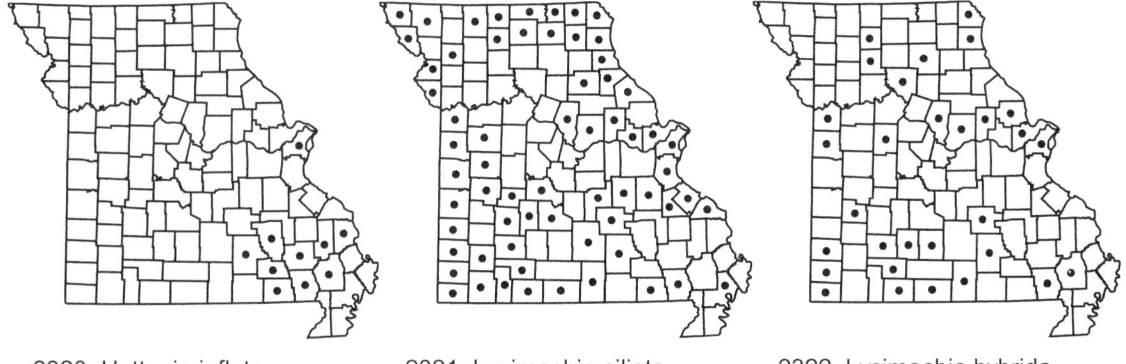

2320. Hottonia inflata 2321. Lysimachia ciliata 2322. Lysimachia hybrida

genus *Centunculus* L., which is separated from *Anagallis* by the alternate leaves and stamens fused to the minute corollas. Molecular studies have not entirely resolved resolved which placement is more appropriate (Martins et al., 2003; Manns and Anderberg, 2005).

2. **Androsace** L.
(Robbins, 1944)

About 100 species, North America, Europe, Asia.

Some of the showier, mainly Asian species of *Androsace* are grown as ornamentals in rock gardens.

1. Androsace occidentalis Pursh (western rock jasmine)

A. simplex Rydb.

Pl. 508 g, h; Map 2319

Plants annual. Aerial stems absent. Scapes 1 to more commonly several per plant, 3–12 cm long, erect or arched upward, finely pubescent with minute, branched hairs. Leaves all basal, sessile. Leaf blades 5–20 mm long, 2–4 mm wide, lanceolate to ovate, the margins inconspicuously toothed, the upper surface mintuely hairy, the undersurface glabrous. Inflorescences simple umbels with 2–10 flowers, with an involucre of bracts, these 3–4 mm long, 1.5–2.0 mm wide, the flower stalks 10–30 mm long, unequal, ascending, finely hairy. Calyces 3–6 mm long, the tube 5-ridged, glabrous or nearly so, the lobes shorter than the tube, lanceolate, ascending, minutely hairy. Corollas shorter than to about as long as the calyces, 2–4 mm long, white, the (4)5 lobes loosely ascending to spreading, withering but tending to persist at fruiting. Stamens 5, the short filaments attached near the middle of the corolla tube, the tiny anthers oval. Ovary 1 mm in diameter, hemispherical, the style about 0.2 mm long, the stigma minute, capitate. Fruits capsules, 1–2 mm long, globose, membranous toward the base, hardened toward the tip, dehiscing incompletely from the tip by 5 valves. Seeds 0.8–1.2 mm long, 0.8 mm wide, triangular in outline, the surface with a network of ridges and pits, dark brown. $2n=20$. March–June.

Scattered in the state, but uncommon or apparently absent from the Mississippi Lowlands Division, as well as the eastern portion of the Ozarks and eastern portion of the Glaciated Plains (Idaho to Ohio south to California, Texas, and Arkansas; Canada; introduced in Massachusetts). Glades, ledges and tops of bluffs, and openings of mesic to dry upland forests; also old fields, crop fields, pastures, lawns, cemeteries, railroads, roadsides, and open, disturbed areas.

This inconspicuous species is easily overlooked in the field.

3. **Hottonia** L.

Two species, North America, Europe, Asia.

1. Hottonia inflata Elliott (water violet, featherfoil)

Pl. 509 i, j; Map 2320

Plants biennial, emergent aquatics. Stems 25–90 cm long, spongy internally, all or mostly submerged. Leaves alternate, scattered along the submerged portion of the stem and clustered in a more or less floating whorl beneath the emergent inflorescence, 4–10 cm long, sessile or short-petiolate, finely pinnately dissected with numerous slender segments. Infloresences spikelike racemes grouped into a sessile umbel, the flowers borne in whorls of 2–10 on inflated, emergent axes, these 12–30 cm long, 0.5–1.5 cm thick, ascending at flowering, becoming more or less horizontal (floating) at fruiting, conspicuously segmented at the nodes, inconspicuously pubescent with minute, gland-tipped and nonglandular hairs. Bracts 3–8 mm long, linear, Flower stalks 3–10(–15) mm long, often with glandular dots or short hairs. Calyces deeply lobed, lobes 4–9 mm long, 1 mm wide, linear-oblong, with glandular dots or hairs. Corollas 4–5 mm long, white, 5-lobed to below the midpoint, the lobes rounded to bluntly pointed at the tips. Stamens 5, short, the filaments attached in the corolla tube. Ovary broadly ovoid to nearly globose, the style short, slender, the stigma club-shaped. Fruits capsules, 2.0–2.5 mm long more or less pear-shaped to nearly globose, green, membranous, dehiscing longitudinally by 5 valves. Seeds minute, 0.2–0.3 mm long, oblong in outline, the surface with faint lines. $2n=22$. April–July.

Uncommon in the northern portion of the Mississippi Lowlands Division and adjacent counties of the eastern Ozarks (eastern U.S. west to Missouri and Texas). Swamps, sloughs, oxbows, sinkhole ponds, and lakes; also ditches; emergent aquatics, sometimes stranded on mud.

In this unusual species, the seeds are dispersed into the water during the summer, germinating and producing stems with several usually submerged leaves by first frost. Plants survive under the ice or water's surface and grow to reach the surface late the following spring. After flowering, the inflated stems fall over and the leaves are shed as the seeds mature (Channel and Wood, 1959). The abundance of this plant appears to vary tremendously from year to year. It is considered uncommon in most of its range.

4. **Lysimachia** L. (loosestrife)

Plants perennial herbs, often with rhizomes, stolons, or offsets. Stems erect or ascending, or prostrate, branched or unbranched. Leaves opposite or whorled (the lowermost leaves sometimes merely subopposite in *L. vulgaris*), sessile or petiolate, the petiole sometimes flattened and fringed along the margins. Leaf blades unlobed, variously linear to circular, the margins entire, the surfaces sometimes gland-dotted (the dots sometimes slightly sunken into the tissue, the structure then termed punctate). Inflorescences of axillary, solitary flowers or of terminal and/or axillary racemes, these occasionally appearing paniculate. Flowers long-stalked. Calyces (4)5–7-lobed nearly to the base, the lobes spreading or arched outward. Corollas saucer-shaped to bell-shaped, the tube very short, the lobes spreading or arched outward, yellow (white elsewhere), sometimes with reddish purple markings on the upper surface near the base and/or along the margins, often densely glandular on the upper surface near the base, sometimes with small reddish purple to dark red or nearly black dots or lines on 1 or both surfaces. Stamens 5, the filaments attached at the corolla base, sometimes also fused into a short basal tube, the anthers yellow. Staminodes sometimes present, 5, alternating with the stamens, membranous or scalelike, narrowly lanceolate to nearly triangular. Ovary superior, with few to many ovules, the style slender, the stigma capitate. Fruits capsules, ovoid to globose, dehiscent longitudinally by 5 valves, the style usually persistent on 1 of the valves. Seeds few to many, often angular. About 200 species, nearly worldwide.

The flowers of *Lysimachia* have an interesting pollination syndrome. The flowers of several species are visited by bees in the genus *Macropis*. The bees collect the pollen, and the females harvest the secretions produced by the glandular trichomes on the inner surface of the petals and on the filaments (Simpson et al., 1983). Coffey and Jones (1980) reported that species in the subgenus *Seleucia* Bigelow are self-incompatible, and that interspecific crosses within subgenus *Seleucia* are fertile. However, Simpson et al. (1983) observed that at least

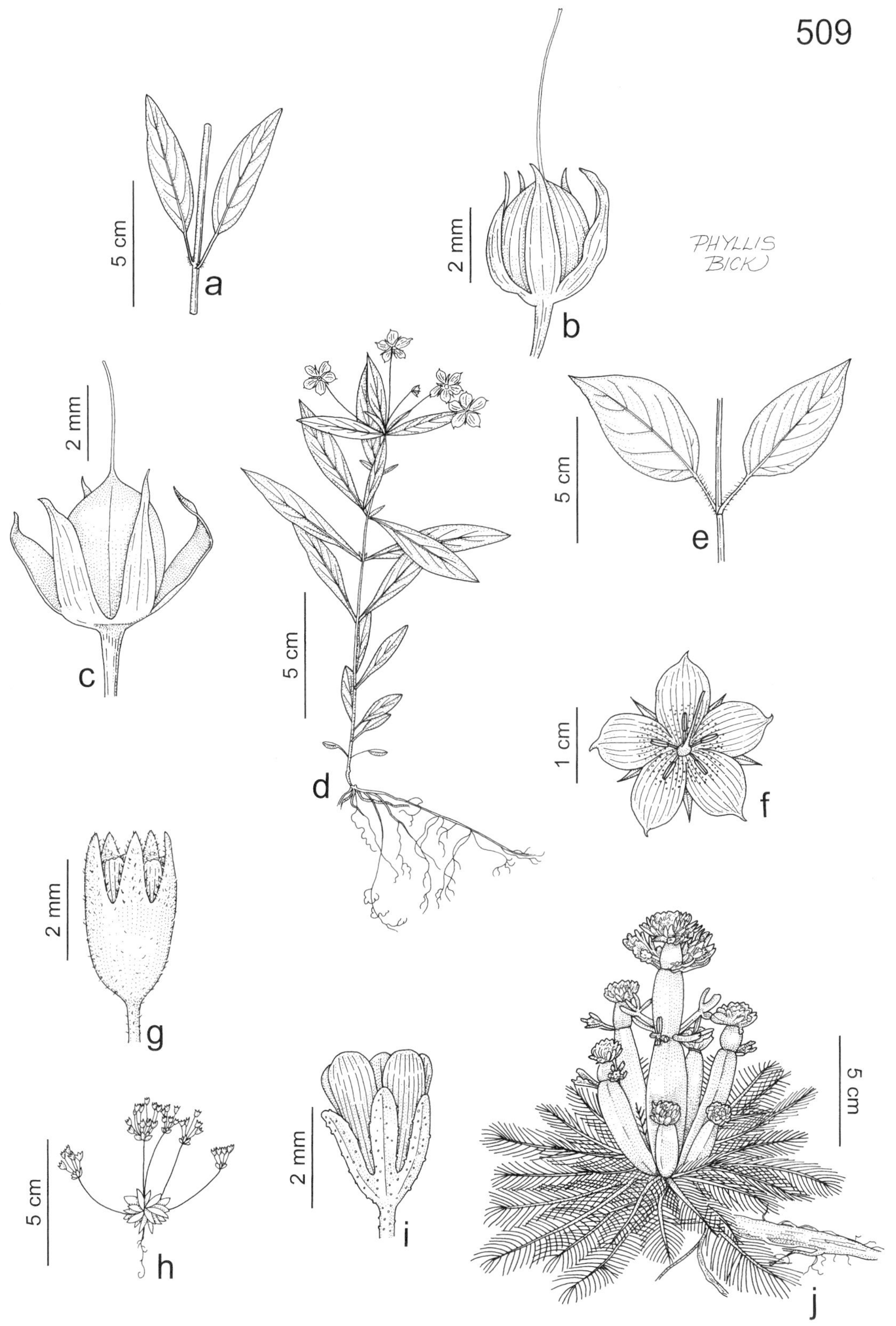

Plate 509. Primulaceae. *Lysimachia hybrida*, **a)** node with pair of leaves, **b)** fruit. *Lysimachia lanceolata*, **c)** fruit, **d)** habit. *Lysimachia ciliata*, **e)** node with pair of leaves, **f)** fruit. *Androsace occidentalis*, **g)** fruit, **h)** habit. *Hottonia inflata*, **i)** flower **j)** habit.

some individuals of *L. ciliata* were found to be self-fertile, and suggested that the incompatibility system might vary geographically or with ploidy level.

Three of the five subgenera of *Lysimachia* are recognized by J. D. Ray (1956) are represented in Missouri. In the past, each has, at times been treated as a separate genus. The largest of these is subgenus *Seleucia*, a group of species with staminodes alternating with the stamens, petals that each enclose a stamen in the bud, and leaves that are not gland-dotted. Included in this subgenus are *L. ciliata, L. lanceolata, L. hybrida, L. quadriflora,* and *L. radicans*. Subgenus *Lysimachia*, represented in Missouri by *L. nummularia, L. terrestris*, and *L. vulgaris*, is a group that lacks staminodes and has leaves that are gland-dotted. Subgenus *Naumbergia* (Moench) Hand.-Mazz., containing only *L. thyrsiflora*, has unique, long-stalked, short, dense racemes produced from the median leaf axils.

In addition to the species treated below, two other species of *Lysimachia* bear mention in the present account. *Lysimachia clethroides* Duby (gooseneck loosestrife; a member of the Old World subgenus *Pallida* (Moench) Hand.-Mazz.) is native to Asia, but is cultivated in the United States, both as an ornamental and as a bee plant. This taxon is strongly colonial from long-creeping, branched rhizomes and differs from other species of *Lysimachia* in Missouri in its alternate leaves, strongly arched, terminal racemes, and white corollas. It has not yet been recorded as an escape in Missouri, but has become naturalized widely in the northeastern United States, including four states adjoining Missouri: Iowa, Illinois, Kentucky, and Tennessee. Many gardeners learn to their dismay that this species is very aggressive in gardens and in some states it is considered an invasive exotic.

On the other hand, *L. quadrifolia* L. (whorled yellow loosestrife; subgenus *Naumbergia*) is a native wildflower that is widespread in the eastern United States and adjacent Canada, west to Minnesota, Illinois, and Alabama. It has been collected several times in Missouri, all from a single woodland site at the Shaw Nature Reserve (Franklin County) and in each case the collector thought that it was a native taxon. However, data in the Missouri Botanical Garden's archives indicates that former Garden director, Edgar Anderson, planted a small colony at this site sometime during the 1940s. Plants have continued to persist, but there is no evidence that the population is growing or spreading. In Steyermark's (1963) treatment, *L. quadrifolia* keys imperfectly to either *L. hybrida* or *L. ciliata* (both subgenus *Seleucia*), but in fact, the species is more closely related to *L. nummularia* and *L. vulgaris* (both subgenus *Lysimachia*) by virtue of its gland-dotted leaves and flowers lacking staminodes. It is a rhizomatous perennial with ascending stems, leaves mostly in whorls of 4, and flowers that are solitary from some of the leaf axils.

1. Stems prostrate, creeping; leaves circular .... 4. L. NUMMULARIA
1. Stems erect or ascending; leaves linear to elliptic or ovate
    2. Leaves gland-dotted or punctate; staminodes absent
        3. Stems glabrous (ignoring the glandular dots or lines); inflorescences from the median leaf axils, consisting of short, dense, ovoid, spikelike racemes on long stalks; calyx lobes lacking reddish margins; corolla lobes linear; stamens slightly longer than the corollas, the filaments not fused into a tube .... 8. L. THYRSIFLORA
        3. Stems glabrous or hairy (ignoring the glandular dots or lines); inflorescences terminal and sometimes also from the upper leaf axils, consisting of racemes or small panicles of loose clusters; corolla lobes narrowly elliptic to obovate; stamens shorter than the corollas, the filaments fused into a short tube at their bases
            4. Stems glabrous; calyx lobes lacking reddish margins; corolla lobes narrowly elliptic, with reddish to black glandular dots and/or lines, but with minute glandular hairs only at the base...... 7. L. TERRESTRIS

4. Stems hairy, usually somewhat sticky; calyx lobes with reddish margins; corolla lobes obovate, lacking reddish to black glandular dots or lines, but the upper surface with dense, minute, yellow, more or less spherical glands (usually also reddish-tinged or with reddish markings toward the base) . . . . 9. L. VULGARIS

2. Leaves not gland-dotted or punctate; staminodes present
   5. Corolla lobes 3–5 mm long; stems arched, trailing, or ascending from a spreading base (when young), often rooting at the nodes . . . . 6. L. RADICANS
   5. Corolla lobes 5–12 mm long; stems erect or ascending (rarely ascending from a more or less spreading base), not rooting at the nodes; petioles with marginal hairs throughout
      6. Leaf blades 2–5 cm wide, ovate or broadly lanceolate, angled or rounded at the base to a well-defined petiole
         7. Petioles with the margins evenly hairy throughout; plants with elongate, slender rhizomes . . . . 1. L. CILIATA
         7. Petioles with the margins hairy only below the midpoint or much more sparsely hairy above the midpoint; plants with short, stout rhizomes (these rarely to 5 cm long). 2. L. HYBRIDA
      6. Leaf blades 0.1–1.5(–2.0) cm wide, linear to lanceolate or oblong-lanceolate, narrowly angled or tapered at the base to an often poorly differentiated petiole (sessile or nearly so in *L. quadriflora*); petioles with the margins hairy only or mostly below the midpoint (glabrous or with a few hairs at the very base of the leaf in *L. quadriflora*)
         8. Leaves sessile or nearly so, glabrous or with a few hairs at the very base; leaf blades linear to very narrowly lanceolate, the margins curled-under; lateral veins not evident . . . . 5. L. QUADRIFLORA
         8. Leaves more or less petiolate (except sometimes in *L. lanceolata*), the petiole with a fringe of spreading hairs along the margins below the midpoint (or toward the leaf base in *L. lanceolata*), the hairs occasionally reduced to slender, stiff teeth; leaf blades linear to lanceolate or oblong-lanceolate, the margins flat; lateral veins evident
            9. Plants with short, stout rhizomes (these rarely to 5 cm long); stems relatively stout; leaves evenly green on both surfaces . . . . 2. L. HYBRIDA
            9. Plants with long, slender, stoloniferous rhizomes; stems relatively slender; leaves pale green on the undersurface . . . . . 3. L. LANCEOLATA

1. Lysimachia ciliata L. (fringed loosestrife)
Steironema ciliatum (L.) Baudo
S. pumilum Greene
Pl. 509 e, f; Map 2321

Plants with long, slender rhizomes. Stems 40–120 cm long, erect or strongly ascending, relatively stout, unbranched or branched toward the tip, glabrous or minutely glandular-hairy toward the tip, especially near the nodes. Basal leaves not persistent at flowering. Stem leaves opposite or occasionally in whorls of 3 at the uppermost nodes, with a well-defined petiole (0.5–)1.0–5.0 cm long, this somewhat flattened, narrowly winged, the margins evenly pubescent with long, spreading hairs the entire length. Leaf blades 3–13 cm long, 2–5 cm wide, ovate to broadly lanceolate, rounded, angled, or occasionally shallowly cordate at the base, angled or somewhat tapered to a sharply pointed tip, the margins short-hairy, the surfaces lacking gland dots, not punctate, the upper surface dark green to green, glabrous or nearly so, the undersurface light green, sparsely short hairy or glabrous; secondary veins evident. Inflorescences axillary from the uppermost nodes, of solitary flowers, the flower stalks 2–5(–8) cm long, minutely glandular-hairy. Calyces mostly 5-lobed, the lobes 4–9 mm long, narrowly triangular to lanceolate, not gland-dotted or punctate, sometimes reddish-veined. Corollas mostly 5-lobed, the lobes 5–12 mm long, obovate, bluntly pointed at the tip

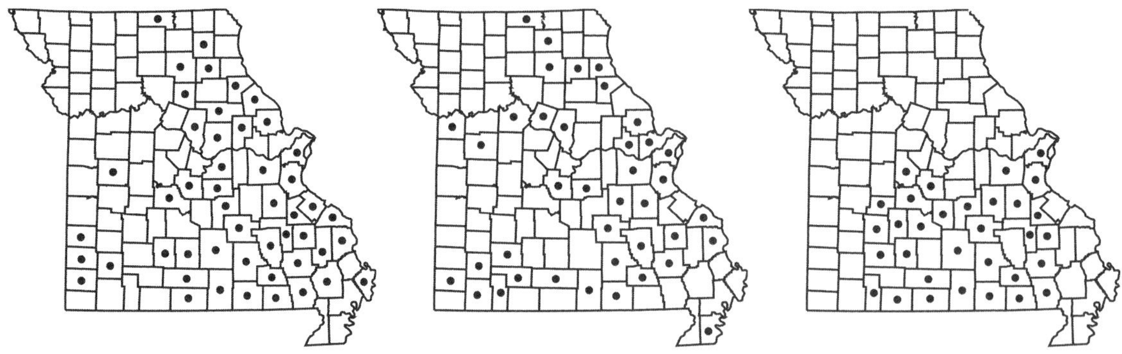
2323. Lysimachia lanceolata 2324. Lysimachia nummularia 2325. Lysimachia quadriflora

or short-tapered to a minute, sharp point, the margins sometimes somewhat uneven or toothed, yellow, densely glandular and with reddish markings on the upper surface toward the base, lacking purple spots or lines. Stamens shorter than the corollas, the filaments 2–3 mm long, not fused into a basal tube, glandular-hairy. Staminodes alternating with the stamens, slender. Styles 3–4 mm long. Fruits 3.5–6.5 mm long, broadly ovoid to globose. Seeds 1.9–2.2 mm long, irregularly elliptic, oblong, or rhombic in outline, triangular in cross-section, dark brown to black. 2*n*=34, 92, 96, 100, 108, 112. May–September.

Scattered nearly throughout the state (U.S. [including Alaska, excluding a few southwestern states], Canada). Banks of streams, rivers, and spring branches, swamps, margins of ponds and lakes, sloughs, bottomland forests, bottomland prairies, moist swales of upland prairies, and rarely wet deprseeions of glades; also ditches, levees, railroads, and roadsides.

2. Lysimachia hybrida Michx.

L. lanceolata Walter ssp. *hybrida* (Michx.) J.D. Ray

Steironema hybridum (Michx.) Raf. ex Small

S. lanceolatum (Walter) A. Gray var. *hybridum* (Michx.) A. Gray

Pl. 509 a, b; Map 2322

Plants with short, stout rhizomes (these rarely to 5 cm long). Stems 30–80 cm long, relatively stout (2–6 mm in diameter at the base), erect or strongly ascending (occasionally reclining), usually branched above the midpoint, glabrous or minutely glandular-hairy toward the tip, especially near the nodes. Basal leaves rarely persistent at flowering. Stem leaves opposite or occasionally in whorls of 3 at the uppermost nodes, the lowermost ones usually with a relatively well-defined petiole 0.5–1.5 cm long, the petioles progressively shorter and less well-defined toward the stem tip, the uppermost leaves often appearing essentially sessile, when present, the petiole somewhat flattened, narrowly winged, the margins pubescent with long, spreading hairs below the midpoint, the pubescence sparse or absent toward the tip. Leaf blades 3–12 cm long, 1.0–2.5 cm wide, those of the lowermost leaves ovate to lanceolate, becoming progressively narrower toward the stem tip, the uppermost leaf blades often linear to narrowly lanceolate, the bases accordingly rounded to angled or tapered, angled or somewhat tapered to a sharply pointed tip, the margins entire or more commonly roughened with minute papillae, the surfaces lacking gland dots, not punctate, green, glabrous; secondary veins evident. Inflorescences axillary from the uppermost nodes, of solitary flowers, the flower stalks 1–4 cm long, glabrous. Calyces mostly 5-lobed, the lobes 4–6 mm long, lanceolate, not gland-dotted or punctate, usually with 1–3(–5) evident veins. Corollas mostly 5-lobed, the lobes 5–10 mm long, obovate to broadly obovate, broadly rounded but usually with a minute sharp point at the tip, the margins sometimes somewhat uneven or irregularly toothed, yellow, densely glandular and with reddish markings on the upper surface toward the base, lacking purple spots or lines. Stamens shorter than the corollas, the filaments 2–3 mm long, not fused into a basal tube, glandular-hairy. Staminodes alternating with the stamens, slender. Styles 3–4 mm long. Fruits 3.5–5.0 mm long, broadly ovoid to globose. Seeds 1.2–1.8 mm long, irregularly elliptic, oblong, or rhombic in outline, triangular in cross-section, dark brown to black. 2*n*=34. May–September.

Scattered nearly throughout the state (eastern U.S. west to North Dakota and Oklahoma, sporadic farther west; Canada). Margins of ponds, lakes, and sinkhole ponds, swamps, fens, banks of streams, bottomland forests, and bottomland prairies; also pastures, railroads, roadsides, and moist disturbed areas.

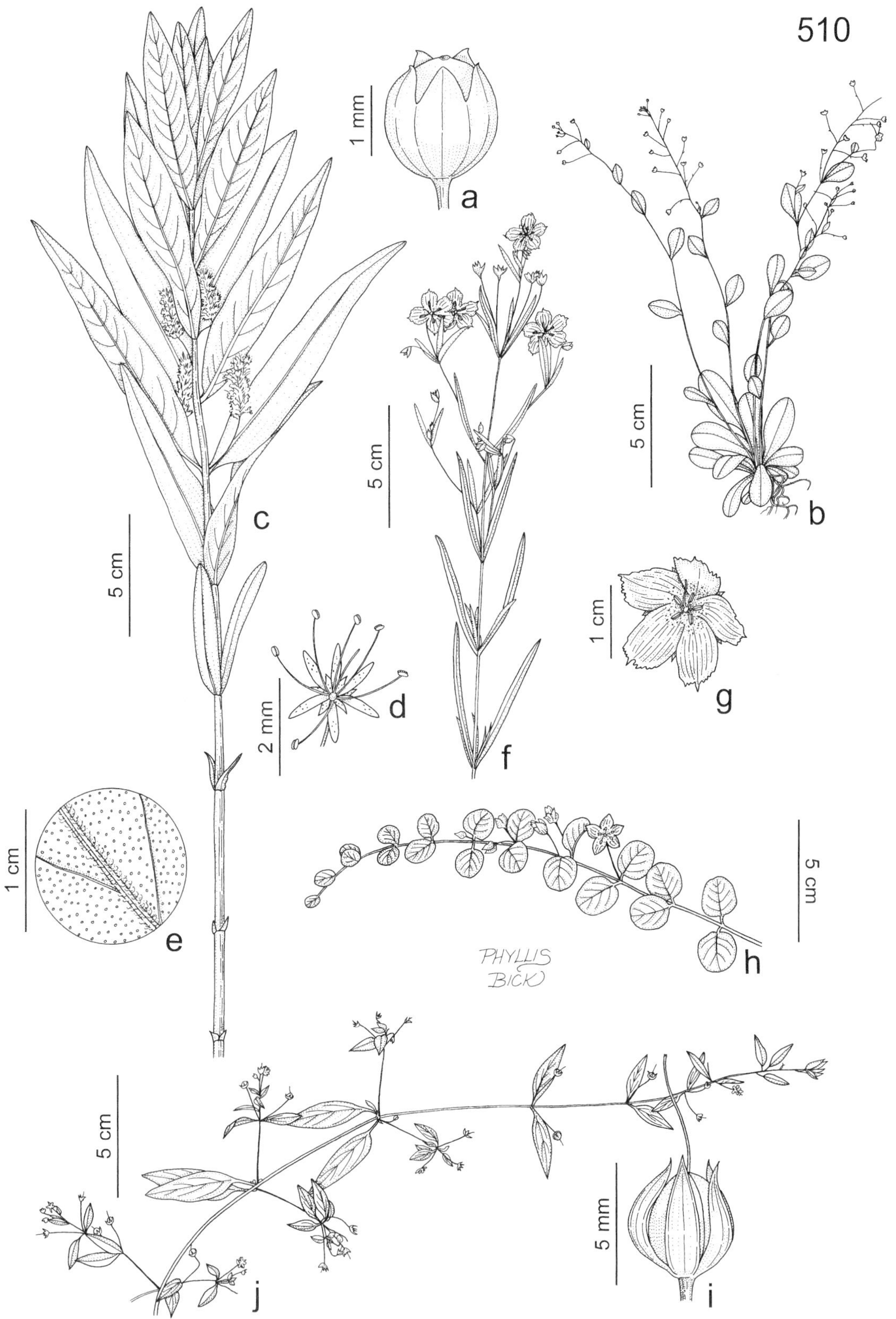

Plate 510. Primulaceae. *Samolus parviflorus*, **a)** fruit, **b)** habit. *Lysimachia thyrsiflora*, **c)** habit, **d)** flower, **e)** leaf detail. *Lysimachia quadriflora*, **f)** habit, **g)** flower. *Lysimachia nummularia*, **h)** habit. *Lysimachia radicans*, **i)** fruit, **j)** habit.

Lysimachia hybrida is closely related to *L. lanceolata*, and the two were once considered a single species by some botanists. However, *L. hybrida* is a larger plant with a thicker stem, lacks the slender stoloniferous rhizomes of *L. lanceolata*, has mostly narrower leaves, and longer stem internodes. *Lysimachia hybrida* also begins flowering slightly later than *L. lanceolata*.

3. Lysimachia lanceolata Walter

L. lanceolata var. *angustifolia* (Lam.) A. Gray
L. angustifolia Lam.
L. heterophylla Michx.
Steironema lanceolatum (Walter) A. Gray

Pl. 509 c, d; Map 2323

Plants with elongate, slender rhizomes. Stems 20–80 cm long, relatively slender (1–2 mm in diameter at the base), erect or strongly ascending, usually short-branched above the midpoint, glabrous or minutely glandular-hairy around the upper nodes. Basal and lower stem leaves often at least partially persistent at flowering, smaller than the median stem leaves. Stem leaves opposite, the lowermost ones usually with a relatively well-defined petiole 0.5–1.5 cm long, the petioles progressively shorter and less well-defined toward the stem tip, the uppermost leaves often appearing essentially sessile, when present, the petiole somewhat flattened, narrowly winged, the margins pubescent with long, spreading hairs below the midpoint, the pubescence sparse or absent toward the tip. Leaf blades 3–14 cm long, 0.7–2.0 cm wide, sometimes more or less folded lengthwise along the midvein, those of the lowermost leaves ovate to narrowly obovate or elliptic-lanceolate, becoming progressively narrower toward the stem tip, the uppermost leaf blades often linear to narrowly lanceolate, the bases accordingly rounded to angled or tapered, angled or somewhat tapered to a sharply pointed tip, the margins entire or more commonly roughened with minute papillae, the surfaces lacking gland dots, not punctate, glabrous, the upper surface green to dark green, the undersurface light green; secondary veins evident. Inflorescences axillary from the uppermost nodes, of solitary flowers, the flower stalks 2.0–4.5 cm long, glabrous. Calyces mostly 5-lobed, the lobes 4–7 mm long, lanceolate, not gland-dotted or punctate, usually with 1–3 faint veins. Corollas mostly 5-lobed, the lobes 5–10 mm long, obovate to oblong-obovate, broadly rounded or truncate but usually with a minute to somewhat elongate (tail-like) sharp point at the tip, the margins sometimes somewhat uneven or irregularly toothed, yellow, densely glandular and with reddish markings on the upper surface toward the base, lacking purple spots or lines. Stamens shorter than the corollas, the filaments 2–3 mm long, not fused into a basal tube, glandular-hairy. Staminodes alternating with the stamens, slender or somewhat broadened toward the base. Styles 3–4 mm long. Fruits 3.0–4.5 mm long, broadly ovoid to globose. Seeds 1.2–1.8 mm long, irregularly elliptic, oblong, or rhombic in outline, triangular in cross-section, dark brown to black. $2n=34$. May–August.

Scattered nearly throughout the state, but apparently absent from the western half of the Glaciated Plains Division (eastern U.S. west to Iowa and Texas; Canada). Banks of streams, rivers, and spring branches, margins of ponds, lakes, and sinkhole ponds, acid seeps, bottomland forests, mesic upland forests, bottomland prairies, upland prairies, sand prairies, and glades; also pastures, railroads, and roadsides.

Lysimachia lanceolata is recognized by its short stature, thin stems, stolons, dimorphic leaves with the lower leaves persistent at flowering, and pale lower leaf surfaces. It typically occurs in drier habitats than does *L. hybrida*, but can grow in wetter habitats.

4. Lysimachia nummularia L. (moneywort)

Pl. 510 h; Map 2324

Plants lacking rhizomes. Stems 10–40 cm long or longer, relatively slender, prostrate, creeping, sometimes mat-forming, rooting at the nodes, glabrous, but usually developing scattered, minute, gland-dots with age. Leaves opposite, with a mostly well-differentiated, but short petiole, this 0.2–0.5 cm long, somewhat flattened, narrowly winged, glabrous except for scattered, minute, gland-dots or punctations (these sometimes faint). Leaf blades (0.4–)1.0–2.5 cm long, (0.4–)1.0–2.5 cm wide, circular or nearly so, the bases rounded, rounded or very broadly angled to a bluntly pointed tip, the margins entire, the surfaces with scattered, minute, glandular dots, these orangish brown to reddish brown, sometimes relatively faint, otherwise glabrous, green to yellowish green; secondary veins faint but often evident. Inflorescences axillary from the nodes, of solitary flowers, the flower stalks 1–3 cm long, glabrous. Calyces (4)5-lobed, the lobes 6–9 mm long, narwowly ovate to ovate with a cordate base, gland-dotted, with a slightly thickened midvein. Corollas (4)5-lobed, the lobes 12–15 mm long, obovate, rounded or broadly angled to a bluntly pointed tip, the margins usually slightly uneven or toothed toward the tip, yellow, moderately to densely glandular but lacking reddish markings on the upper surface toward the base, both surfaces with scattered, reddish purple to nearly black glandular spots and short lines.

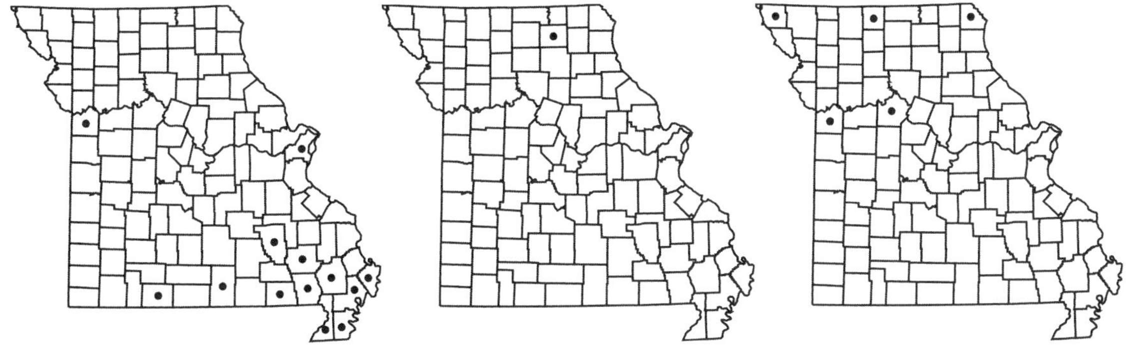

2326. Lysimachia radicans 2327. Lysimachia terrestris 2328. Lysimachia thyrsiflora

Stamens shorter than the corollas, the filaments 3–5 mm long, slightly unequal, fused basally into a short tube, glandular-hairy. Staminodes absent. Styles 4–5 mm long. Fruits and seeds not produced. $2n$=30, 32, 34, 36, 43, 45. May–August.

Introduced, scattered nearly throughout the state, but apparently absent from the Unglaciated Plains and the western half of the Glaciated Plains Divisions (native of Europe, Asia; introduced nearly throughout the U.S. and adjacent Canada, except so far for portions of the Great Plains and Rocky Mountains). Banks of streams, rivers, and spring branches; margins of ponds and lakes, sloughs, acid seeps, and bottomland forests; also lawns, railroads, roadsides, and moist disturbed areas.

This Eurasian species is easily distinguished by its creeping habit and round, punctate leaves. It is sometimes cultivated as a ground cover and several cultivars differing in leaf color exist, however, it can become very aggressive in gardens. It is considered an invasive exotic in some eastern states. In Missouri specimens, fruits apparently are not produced and the species appears to be spreading mainly vegetatively in the state, from stem pieces transported by water, and human-mediated means.

5. Lysimachia quadriflora Sims (narrow-leaved loosestrife)
L. longifolia Pursh
L. revoluta Nutt.
Steironema quadriflorum (Sims) Hitchc.
Pl. 510 f, g; Map 2325

Plants with elongate, slender rhizomes. Stems 20–70 cm long, relatively slender (2–3 mm in diameter at the base), erect or strongly ascending, unbranched or short-branched above the midpoint (but usually with very short branches represented by dense clusters of leaves at many of the nodes), glabrous. Basal leaves rarely present at flowering, shorter than the stem leaves, long- petiolate, the blade elliptic to obovate. Stem leaves opposite (sometimes appearing whorled because of the dense clusters of leaves in the axils), all sessile or nearly so. Leaf blades 3–9 cm long, 0.2–0.6 cm wide, linear to very narrowly lanceolate, the bases angled or tapered, angled or somewhat tapered to a sharply pointed tip, the margins entire, usually curled-under, glabrous or with a few, long, spreading hairs at the base, the surfaces lacking gland dots, not punctate, glabrous, the upper surface green to dark green, shiny, the undersurface slightly lighter green, not shiny; secondary veins not evident. Inflorescences axillary from the uppermost nodes, of solitary flowers, the flower stalks 0.5–3.5 cm long, glabrous. Calyces mostly 5-lobed, the lobes 4–6 mm long, lanceolate, not gland-dotted or punctate, usually with 3–5 evident veins. Corollas mostly 5-lobed, the lobes 7–10 mm long, obovate to rhombic-elliptic, angled or slightly tapered to a sharp, sometimes extended point at the tip, the margins otherwise entire or slightly uneven, yellow, densely glandular and with reddish markings on the upper surface toward the base, lacking purple spots or lines. Stamens shorter than the corollas, the filaments 2–3 mm long, not fused into a basal tube, glandular-hairy. Staminodes alternating with the stamens, slender or somewhat broadened toward the base. Styles 4–5 mm long. Fruits 3.5–5.0 mm long, broadly ovoid to globose. Seeds 1.1–1.4 mm long, irregularly elliptic, oblong, or rhombic in outline, triangular in cross-section with 1 side somewhat concave, dark brown to reddish brown, shiny. $2n$=34. June–August.

Scattered in the Ozark and Ozark Border Divisions (eastern U.S. west to Minnesota and Alabama; Canada). Fens, bases and ledges of bluffs, seepy banks of streams, rivers, and spring branches, and occasionally bottomland forests and swamps; usually on calcareous substrates.

The clusters of leaves that develop on very short branches in the main leaf axils, sometimes make the leaf distribution appear whorled. Plants with white or cream-colored flowers have been collected rarely.

6. Lysimachia radicans Hook. (creeping loosestrife, trailing loosestrife)
Steironema radicans (Hook.) A. Gray
Pl. 510 i, j; Map 2326

Plants with relatively short, slender rhizomes. Stems 40–100 cm long, relatively slender to more commonly relatively stout (2–5 mm in diameter at the base), arched, trailing, or ascending from a spreading base (when young), often rooting at some of the nodes, glabrous except for a fringe of spreading hairs at each node. Basal leaves rarely present at flowering, shorter than the stem leaves, long-petiolate, the blade elliptic to lanceolate. Stem leaves opposite or in whorls of 3 at the upper nodes, with a mostly well-differentiated petiole, this 0.5–1.5 cm long, somewhat flattened, narrowly winged, the margins pubescent with long, spreading hairs below the midpoint, the pubescence sparse or absent toward the tip. Leaf blades 1–9 cm long, 0.3–3.0 cm wide, the uppemost blades sometimes linear, those of the main leaves lanceolate to oblong-elliptic or ovate, the bases mostly rounded to broadly angled (more narrowly angled on the uppermost blades), angled or somewhat tapered to a sharply pointed tip, the margins entire or more commonly roughened with minute papillae, the surfaces lacking gland dots, not punctate, glabrous, the upper surface green to dark green, the undersurface lighter green; secondary veins evident. Inflorescences axillary from the uppermost nodes, of solitary flowers, the flower stalks 0.7–3.0 cm long, glabrous. Calyces (4)5-lobed, the lobes 3–4 mm long, lanceolate, not gland-dotted or punctate, usually with 3–5 relatively faint veins. Corollas (4)5-lobed, the lobes 3–5 mm long, obovate to broadly obovate, broadly rounded to nearly truncate at the tip, sometimes with a minute, extended point at the tip, the margins otherwise slightly uneven or toothed toward the tip, yellow, densely glandular and with reddish markings on the upper surface toward the base, lacking purple spots or lines. Stamens shorter than the corollas, the filaments 1.5–2.5 mm long, not fused into a basal tube, glandular-hairy. Staminodes alternating with the stamens, slender above a somewhat broadened base. Styles 3–4 mm long. Fruits 3–4 mm long, broadly ovoid to globose. Seeds 1.1–1.4 mm long, irregularly elliptic, oblong, or rhombic in outline, triangular in cross-section, dark brown to reddish brown, shiny. $2n=34$. June–September.

Scattered in the Mississippi Lowlands Division and uncommon in the southernmost portion of the Ozarks (southeastern U.S. west to Oklahoma and Texas). Swamps, bottomland forests, and margins of sinkhole ponds; also ditches and wet roadsides.

The two historical specimens originally thought to represent disjunct occurrences in Jackson County and the city of St. Louis lack detailed locality data. John Kellogg's 1910 collection from St. Louis may represent an introduced population or a cultivated plant. Steyermark (1963) noted that although Henry Eggert's 1892 collection from Blue Springs was attributed to Jackson County in J. D. Ray's (1956) monograph of the genus, Eggert never botanized in the Kansas City area and it is more likely to have originated from one of the Blue Springs that are in several counties in the southeastern portion of the state. Examination of specimen data entered into the Flora of Missouri Database discloses that Eggert collected only in Butler County on 19 August 1892, and a specimen of *Helenium amarum* collected that day from "near Blue Spring" is clearly labeled as having originated from Butler County.

7. Lysimachia terrestris (L.) Britton, Sterns & Poggenb. (swamp candles, bulbil loosestrife)
Map 2327

Plants with short to elongate, relatively stout, fleshy rhizomes. Stems 30–100 cm long, relatively relatively stout (2–5 mm in diameter at the base), erect or strongly ascending, not rooting at the nodes, unbranched or few-branched toward the tip, developing small, narrowly ellipsoid bulbils in the main leaf axils toward the end of the growing season, glabrous, but with scattered, glandular dots and lines. Lower stem leaves reduced to small scales, these sessile, ovate, grading into the main leaves in the lower 1/3 of the stem. Main stem leaves opposite, sessile or nearly so. Leaf blades (above the basal 1/3 of the stem) 3–10 cm long, 0.7–1.5 (–2.0) cm wide, narrowly lanceolate to narrowly oblanceolate, long-tapered at the base, angled or somewhat tapered to a bluntly or sharply pointed tip, the margins entire, often minutely curled-under, the surfaces with orangish red to reddish purple gland-dots or punctations, otherwise glabrous, the upper surface green to dark green, the undersurface lighter green, usually slightly glaucous; secondary veins usually evident but often faint. Inflorescences terminal and usually solitary, of racemes with numerous flowers, the flower stalks 0.3–0.8 cm long (elongating to 0.8–1.5 mm as the fruits mature), glabrous, but gland-dotted.

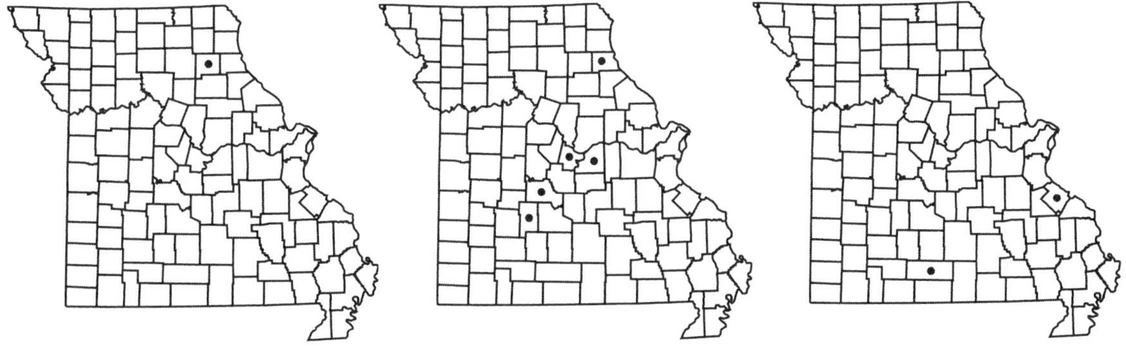

2329. Lysimachia vulgaris

2330. Primula fassettii

2331. Primula frenchii

Calyces 5(–7)-lobed, the lobes 3–4 mm long, narrowly lanceolate-triangular to lanceolate, gland-dotted, usually with 3 relatively faint veins. Corollas 5(–7)-lobed, the lobes 5–7 mm long, narrowly oblong-elliptic, angled to a bluntly pointed tip, the margins entire or slightly uneven, yellow, glabrous and occasionally with reddish or orangish markings on the upper surface toward the base, however both surfaces with reddish purple to nearly black lines. Stamens slightly shorter than the corollas, the filaments 3–6 mm long, fused into a short basal tube, this glandular-hairy. Staminodes absent. Styles 4–5 mm long. Fruits 2.0–2.5 mm long, globose, the surface gland-dotted. Seeds few, 1.0–1.2 mm long, irregularly elliptic, oblong, or rhombic in outline, triangular in cross-section, sometimes oblong-elliptic and rounded, black, with patches of lighter glaucous covering, shiny. $2n=84$. May–June.

Uncommon, known thus far from a single specimen from Adair County (eastern [mostly northeastern] U.S. west to Minnesota, Missouri, and Oklahoma, also Idaho, Oregon, and Washington; Canada). Acid seeps, on sandy substrate.

Steyermark (1963) included this species on a list of plants occurring in Illinois that he predicted would be discovered in Missouri in the future. It was added to the state's flora by T. E. Smith and Gremaud (2006), who documented a small population growing in an open seepage community on a bench of the Chariton River.

8. Lysimachia thyrsiflora L. (tufted loosestrife, swamp loosestrife)

Naumbergia thyrsiflora (L.) Rchb.

Pl. 510 c–e; Map 2328

Plants with usually elongate, relatively stout, stoloniferous rhizomes. Stems 30–80 cm long, relatively stout (3–7 mm in diameter at the base), erect or strongly ascending, not rooting at the nodes, unbranched or occasionally with 1 or few branches toward the tip, not developing bulbils in the leaf axils, glabrous, but with scattered, glandular dots and lines. Lower stem leaves reduced to small scales, these sessile, ovate, grading into the main leaves near the midpoint of the stem. Main stem leaves opposite or rarely in whorls of 3, sessile or nearly so. Leaf blades (above the stem midpoint) 4–12 cm long, 0.7–2.0 cm wide, narrowly lanceolate to narrowly oblong-elliptic or narrowly oblanceolate, variously angled or tapered at the base, sometimes more or less truncate and slightly clasping the stem, angled or somewhat tapered to a sharply pointed tip, the margins entire or slightly and finely wavy, usually also with scattered, crinkly, multicellular hairs, often minutely curled-under, the surfaces with dark purple to black gland-dots or punctations, the upper surface green to dark green, glabrous, the undersurface lighter green, sometimes slightly glaucous, with sparse to moderate, crinkly, multicellular hairs along the main veins; secondary veins usually evident but often faint. Inflorescences axillary from the median leaf axils, 1 to several, of short, dense, ovoid, spikelike racemes with numerous flowers, these long-stalked (1.5–5.5 cm), the individual flower stalks 0.1–0.4 cm long, glabrous, but gland-dotted. Calyces 5–7-lobed, the lobes 2–3 mm long, narrowly lanceolate-triangular to lanceolate, with glandular dots and short lines, with 1–3 faint veins (the midvein sometimes slightly thickened). Corollas 5–7-lobed, the lobes 4–6 mm long, linear, angled to a bluntly or sharply pointed tip, the margins entire, yellow, glabrous and lacking reddish markings on the upper surface toward the base, however both surfaces with reddish purple to nearly black, short lines. Stamens slightly longer than the corollas, appearing exserted at flowering, the filaments 4.5–6.0 mm long, not fused into a tube basally. Staminodes absent. Styles 5–7 mm long. Fruits 2–4 mm long, more or less globose, the surface gland-dotted. Seeds few, 1.2–1.5 mm long, irregularly elliptic, oblong, or rhombic in outline, triangular in cross-section, dark brown. $2n=54$. May–July.

Uncommon in the northern half of the state, mostly in counties adjoining the big rivers (northern U.S. [including Alaska] south to California and West Virginia; Europe, Asia). Marshes, fens, and bottomland prairies.

This distinctive species is easily recognized by the dense, ovoid inflorescences borne on long stalks from the axils of the median leaves, the exserted stamens, and the black-dotted stems, leaves, flowers, and fruits.

9. Lysimachia vulgaris L. (garden loosestrife) Map 2329

Plants with usually elongate, relatively stout, stoloniferous rhizomes. Stems 30–90 cm long, relatively stout (3–6 mm in diameter at the base), erect or strongly ascending, not rooting at the nodes, unbranched or occasionally with 1 or few branches toward the tip, not developing bulbils in the leaf axils, moderately to densely pubescent with short, spreading to somewhat crinkly, multicellular hairs, especially around the nodes, usually somewhat sticky to the touch, often also with faint, scattered, glandular dots and lines. Lower stem leaves usually shed by flowering, sometimes reduced to small, sessile, ovate, scales grading into the main leaves in the lower 1/3 of the stem. Main stem leaves opposite, or some of them in whorls of 3, sessile to very short-petiolate (petiole to 4 mm). Main leaf blades 5–12 cm long, 1–4 cm wide, lanceolate to narrowly ovate or elliptic-ovate, angled or occasionally more or less rounded at the base, angled or tapered to a sharply pointed tip, the margins entire or slightly wavy, flat, the surfaces with faint orangish red gland dots or punctations, the upper surface green to dark green, sparsely to moderately pubescent with minute, glandular hairs, the undersurface lighter green, with moderate, short, crinkly, multicellular hairs, especially along the veins, also with minute, glandular hairs; secondary veins evident. Inflorescences terminal and from the upper leaf axils, racemes and the axillary ones usually small panicles of loose clusters, the inflorescences relatively short-stalked (1–4 cm), the individual flower stalks 0.3–1.5 cm long, hairy (some of the hairs minute and glandular). Calyces (4)5(6)-lobed, the lobes 3–5 mm long, lanceolate, the margins with a reddish purple to nearly black line or band, usually with a very narrow thin, pale area just outside of this and finely glandular-hairy, the surface usually lacking evident gland-dots or lines, with a somewhat thickened and keeled midvein, this usually hairy. Corollas (4)5(6)-lobed, the lobes 8–12 mm long, obovate to broadly elliptic, rounded to bluntly pointed at the tip, the margins entire, yellow, densely pubescent with minute, yellow, more or less spherical glands on the upper surface and usually reddish-tinged or with reddish markings on the upper surface toward the base, lacking reddish gland dots, lines, or punctations. Stamens shorter than the corollas, the filaments 3–4 mm long, fused into a tube in the basal third, densely glandular. Staminodes absent. Styles 4–5 mm long. Fruits 3–4 mm long, broadly ovoid to nearly globose, the surface glabrous but with scattered gland-dots, sometimes in lines. Seeds 1.0–1.2 mm long, irregularly elliptic, oblong, or rhombic in outline, triangular in cross-section, dark brown. $2n=56$, 84. June–August.

Introduced, uncommon, known thus far only from Camden and Shelby Counties (native of Europe, Asia; introduced sporadically in the northern half of the United States and adjacent Canada). Margins of ponds and bottomland prairies; also moist, open, disturbed areas.

Lysimachia vulgaris was first brought to the authors' attention in 1994 by native plant enthusiast, Sue Vanderbilt (although an earlier specimen was later discovered during herbarium research). It was first reported for the state by T. E. Smith and Gremaud (2006), based on the Shelby County population. This species is no longer as commonly cultivated as an ornamental in the Midwest as it once apparently was. It is distinguished by its hairy stems, leaves dotted with orange or black glands, and flowers borne in terminal and axillary racemes and/or panicles.

Beckett (1995) studied this species in his garden in England and was surprised to find that the stoloniform, but well-branched rhizomes of *L. vulgaris* can spread horizontally at or just below the soil surface for distances as long as 2.5 m before tip-rooting and giving rise to a new plant.

5. Primula L.

Plants perennial herbs, with rhizomes. Aerial stems absent. Leaves all basal, in a dense rosette, with a winged, sometimes indistinct petiole. Leaf blades simple, lanceolate to ovate, oblong-ovate, or narrowly to broadly oblanceolate, the margins entire or shallowly toothed or scalloped. Inflorescences solitary, a simple umbel terminal on a scape, this erect or ascending, unbranched, glabrous or hairy. Calyces 5-lobed, the lobes lanceolate to triangular-lanceolate, angled to sharply pointed tips. Corollas shallowly or very deeply 5-lobed, the lobes spreading or strongly reflexed, variously colored. Stamens 5, inserted on the corolla tube, the filaments short, separate or united into a tube by a membrane, the anthers exserted and united around the style or not exserted and not united in *P. veris*. Ovary ovoid, the style slender, extended past the anthers (except in some flowers of *P. veris*), the stigma capitate. Capsule ellipsoid to ovoid, 1-celled, walls thin or thick, dehiscent incompletely longitudinally from the tip into 5 valves. Seeds numerous, in our species oblong-ellipsoid to cuboid or more or less wedge-shaped, the surface with a fine to coarse network of ridges and pits. About 430 species, nearly worldwide, most diverse in temperate and montane-tropical regions

In recent studies in molecular systematics (Mast et al., 2001, 2004), *Dodecatheon* L. (about 17 species from North America and Siberia) is placed in a lineage that lies entirely within the much larger genus *Primula* (Mast and Reveal, 2007). *Dodecatheon* differs from *Primula* in its reflexed petals, united filaments, thickened connectives, and distinctive anthers. The production of heterostylous flowers is common in *Primula*, but apparently has been lost in *Dodecatheon*.

The flowers of *Primula* sect. *Dodecatheon* (L.) Mast & Reveal are fragrant, but produce no nectar. The flowers are pollinated by bumblebees (*Bombus* spp.). The bees hang from the anther cone and buzz their wings to release the pollen, which they collect and transfer to other flowers. Exclusion studies indicate that self-fertilization does not occur, and that the plants are dependent on bees for pollen transport (Macior, 1970b; Mast et al., 2004).

1. Leaf blades with the margins somewhat curled-under and/or somewhat corrugated, also with shallow, broad teeth or scallops; corollas with the lobes spreading or loosely ascending; stamens with the anthers free, not exserted . . . . 4. P. VERIS
1. Leaf blades with the margins relatively flat, entire or (in *P. frenchii*) inconspicuously scalloped; corollas with the lobes strongly reflexed; stamens with the anthers fused into a tube around the style, exserted
    2. Leaves bluish green; corolla deep pink to purple; capsule cylindrical, more than 3 times as long as broad, thin-walled, light brown to golden yellow . . . . 1. P. FASSETTII
    2. Leaves green; corolla lilac to white; capsule stout, thick, less than 3 times as long as wide, thick-walled, dark reddish brown
        3. Leaves cordate at base, abruptly tapered to the petiole; flowers white; filaments separate nearly half their length; seeds oblong-ellipsoid, not angular . . . . 2. P. FRENCHII
        3. Leaves gradually tapered to the petiole; flowers white, pink, or purple; filaments united nearly to apex; seeds more or less cuboid, angular . . . . 3. P. MEADIA

1. Primula fassettii A.R. Mast & Reveal (jeweled shooting star)

Dodecatheon amethystinum (Fassett) Fassett

D. meadia L. var. *amethystinum* Fassett

D. radicatum Greene var. *radicatum*

Pl. 508 h; Map 2330

Basal leaves 10–25 cm long, the blades narrowly to broadly oblanceolate or elliptic-oblanceolate, gradually tapered at the base, lacking a distinct petiole, rounded or broadly angled to a blunt point at the tip, the margins entire to inconspicuously scalloped, the surfaces glabrous, lacking reddish coloration at the base, especially along the midvein. Scapes 15–25 cm tall, glabrous. Umbels with 2–7 flowers, the involucral bracts 3–6 mm long, lanceolate, the flower stalks 3–5 cm long, glabrous. Calyces 6–9 mm long, deeply lobed. Corollas with the tube 2–3 mm long, purple to maroon at the throat, the lobes 10–12 mm long, 3–4 mm wide, strongly reflexed, pink or lavender to rose-purple, rarely white, usually yellow at the base. Stamens exserted, the filaments fused into a membranous tube or nearly free, the anthers 6–7 mm long, fused into a tube around the style, the bases of their connectives expanded and purple. Style 7–9 mm long, slender, the stigma minute, capitate. Fruits 9–11 mm long, 3–4 mm wide, relatively thin and papery-walled, light brown to yellowish brown. Seeds 0.7–1.0 mm long, more or less cuboid, angular, brown. $2n$=88. May–June.

Uncommon in central Missouri, also in Marion County (Minnesota to Missouri east discontinuously to Pennsylvania and Virginia). Moist ledges and tops of limestone and dolomite bluffs.

Primula fassettii is recognized by its thin-walled, more or less cylindrical (vs. somewhat tapered) capsules, uniform rose-purple flower color, lack of reddish coloration on the leaf bases, and distinctive bluff-ledge habitat mostly along major rivers. The capsules are probably the most diagnostic character (Iltis and Shaughnessy, 1960; Schwegman, 1984). Those of *P fassettii* have walls approximately 35–120 microns thick, yielding easily to light pressure from a finger or dissecting probe, are more or less cylindrical in outline, and have a light brown, almost golden color. In contrast, the capsules of *P. meadia* are 130–325 microns thick, unyielding to light pressure, typically narrowly ovoid in shape, and reddish brown. The leaves of *P. fassettii* are said to have a bluish green cast (Steyermark, 1940). Although these characters are often distinctive on living plants, they tend to become obscured on herbarium specimens.

The taxonomy of *P. fassettii* remains controversial. Through most of its range, it is surrounded by populations of *P. meadia* and was once included as a variety of that species. Without fruits, the two species are very difficult to separate. Schwegman (1984) reported populations that were morphologically intermediate between the two species, and which he thought might represent hybrids between them. The distinctive capsule of *P. fassettii* is very similar in shape, color, and wall thickness to capsules of taxa found farther to the west. Thompson (1953) included *P. fassettii* (as *Dodecatheon*) with *P. pauciflora* (Greene) A.R. Mast & Reveal (*D. radicatum* Greene, *D. pulchellum* (Raf.) Merr.), a polyploid complex widespread in the western U.S. from southern Alaska to Arizona. The eastern limit of the *P. pauciflora* complex is said to be somewhere near the western edge of the Great Plains. Most recently, Oberle (2009) and Oberle and Schaal (2011) conducted studies that measured genetic variation in the complex (as *Dodecatheon*) using molecular markers. They concluded that although *P. fassettii* appears to be more or less immersed within the genetic variation recorded for western populations of the *P. pauciflora* complex in the portion of its range where it is isolated from related *Primula* species (and thus might best be considered an eastern outlier of a single, widespread species), the populations sampled in the Midwest show evidence of long-term hybridization with *P. meadia*. More research is needed to determine the taxonomic relationship between *P. fassettii* and its relatives.

The isolated presence of *P. fassettii* in unglaciated areas of Wisconsin has given rise to much speculation about the distribution and migration of this species in Pleistocene times (Fassett, 1931, 1944; Ugent et al. 1982; Schwegman, 1984). Fassett suggested that the species was widespread in the northeastern states in pre-glacial times and survived the Wisconsin glaciations in situ in the driftless areas, where it is still found today. Its presence on bluffs in northeastern Missouri along the Mississippi River was taken to indicate that this area too had escaped glaciation, the plants persisting today in habitats approximating those of its pre-glacial distribution, along with other plants having generally boreal distributions. At the Missouri site near Hannibal, *P. fassettii* is found with *Sambucus pubens* Michx. (Caprifoliaceae [Steyermark, 1963]; now Adoxaceae [Angiosperm Phylogeny Group, 2009]), *Aralia nudicaulis* L. (Araliaceae), and other taxa considered to be northern relicts (Steyermark, 1940). The resemblance of *P. fassettii* to *P. pauciflora* suggests an alternative hypothesis, that *P. fassettii* migrated eastward from the Rocky Mountains during more favorable glacial periods, and has since become restricted to specialized habitats (Iltis and Shaughnessy, 1960). More recently, the range of *P. fassettii* has been considerably enlarged by additional collections along bluffs in Illinois, at both unglaciated and glaciated sites (Ugent et al., 1982; Schwegman, 1984). The extended distribution to the south may support the hypothesis that *P. fassettii* existed well to the south of the glaciated areas and migrated into the southern Wisconsin area after the retreat of the ice (Ugent et al., 1982). Schwegman (1984) hypothesized that *P. fassettii* (as *D. amethystinum*) was much more widespread during periods of glaciation, and that it migrated eastward into Illinois along the Missouri River during Pleistocene times. As noted above, Oberle's (2009) genetic studies could not distinguish isolated eastern populations of this taxon from western populations of *P. pauciflora*.

2. Primula frenchii (Vasey) A.R. Mast & Reveal (French's shooting star)
Dodecatheon frenchii (Vasey) Rydb.
D. meadia L. var. *frenchii* Vasey
D. meadia ssp. *membranaceum* R. Knuth
Pl. 508 i; Map 2331

Basal leaves 10–28 cm long, the blades 8–14 cm long, narrowly to broadly ovate or elliptic,

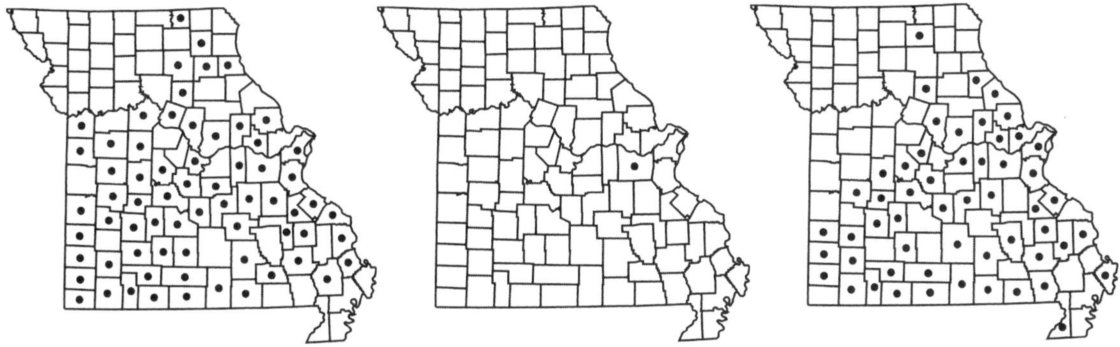

2332. Primula meadia 2333. Primula veris 2334. Samolus parviflorus

abruptly tapered at the base to a more or less well-defined, winged petiole 5–12 cm long, rounded or broadly angled to a blunt point at the tip, the margins usually inconspicuously scalloped, sometimes slightly wavy, the surfaces glabrous, lacking red coloration at the petiole base. Scapes 10–30 cm tall, glabrous. Umbels with 3–5 flowers, the involucral bracts 2–4 mm long, lanceolate, the flower stalks 2–5 cm long, glabrous. Calyces 6–9 mm long, deeply lobed. Corollas with the tube 1.5–2.0 mm long, purple at the throat, the lobes 13–16 mm long, 3–4 mm wide, strongly reflexed, white, sometimes pale pinkish-tinged. Stamens exserted, the filaments fused into a membranous tube or free to about the midpoint, the anthers 6–7 mm long, fused into a tube around the style, the bases of their connectives expanded and purple. Style 7–8 mm long, slender, the stigma minute, capitate. Fruits 6–9 mm long, 2.5–3.5 mm wide, thick- and hard-walled, dark brown. Seeds 0.6–1.0 mm long, oblong-ellipsoid, not angular, brown. $2n$=44. May–June.

Uncommon, known thus far only from Ste. Genevieve and possibly Douglas Counties (Arkansas, Missouri, Illinois, Indiana, and Alabama). Ledges of moist, shaded, usually overhung sandstone bluffs.

Much has been written about this species, which was considered to be an unusual variant of the widespread *P. meadia* (Fassett, 1944; Thompson, 1953) until studies in Illinois indicated that it is a morphologically and cytologically distinct, but relatively cryptic species (Voigt and Swayne, 1955; Olah and DeFilipps, 1968). More recently, Oberle and Esselman (2011) documented consistent differences in seed morphology between *P. frenchii* and *P. meadia*. They also detailed a population in southern Illinois in which plants with intergading morphologies co-occur with plants of typical *P. frenchii* leaves and fertile parts. Oberle and Esselman then noted that some other populations in the range of *P. frenchii* show intergrading morphotypes, including the one in Ste. Genevieve County. they mentioned in passing that the other Missouri population (in Douglas County) contained plants with intermediate but relatively stable (nonintergrading) morphology. Oberle et al. (2012) subsequently documented cytological variation within the intensively studied intergrading population in Illinois. In their ecophysiological and genetic research on the complex, Oberle (2009) and Oberle and Schaal (2011) were unable to clearly distinguish two taxa and suggested that *P. frenchii* might represent a stable ecotype of *P. meadia* rather than a taxonomically distinct species.

In Missouri, *P. frenchii* was not fully treated by Steyermark (1963), who merely noted its potential presence in Missouri. The species was first reported by P. W. Nelson (1979b). It is still known definitely only from the Hickory Canyons drainage system in Ste. Genevieve County, which, according to Oberle and Esselman (2011) contains individuals with both typical *P. frenchii* morphology and morphologies grading toward *P. meadia*. A potential second locality was discovered in 2008 in Douglas County by Brad Oberle during his doctoral research at Washington University of St. Louis. However, at this second locality the plants are anomalous in having uniformly morphologically intermediate leaves with less well-defined petioles. Further research is needed.

3. Primula meadia (L.) A.R. Mast & Reveal (common shooting star, bird's bill, prairie rooster bill, American cowslip)

Dodecatheon meadia L.
D. meadia var. *brachycarpum* (Small) Fassett
D. meadia var. *stanfieldii* (Small) Fassett
D. pauciflorum Greene
D. pulchellum (Raf.) Merr. ssp. *pauciflorum* (Greene) Hultén
D. stanfieldii Small

Pl. 508 e–g; Map 2332

Basal leaves 8–26 cm long, the blades narrowly to broadly oblanceolate or elliptic-oblanceolate, gradually tapered at the base, lacking a distinct petiole, rounded or broadly angled to a blunt point at the tip, the margins entire to inconspicuously scalloped, sometimes slightly wavy, the surfaces glabrous, sometimes with reddish coloration toward the base, especially along the midvein. Scapes 20–50 cm tall, glabrous. Umbels with 5–20 flowers, the involucral bracts 2–5 mm long, lanceolate, the flower stalks 3–8 cm long, glabrous. Calyces 5–9 mm long, deeply lobed. Corollas with the tube 2–4 mm long, purple at the throat, the lobes 13–18 mm long, 3–4 mm wide, strongly reflexed, white to pink or lavender, usually yellow at the base. Stamens exserted, the filaments free or fused into a membranous tube, the anthers 6–9 mm long, fused into a tube around the style, the bases of their connectives expanded and purple. Style 7–10 mm long, slender, the stigma minute, capitate. Fruits 8–14 mm long, 3–5 mm wide, thick- and hard-walled, dark brown. Seeds 0.6–0.8 mm long, more or less cuboid, angular, brown. $2n$=88. April–June.

Scattered nearly throughout the state, but uncommon or apparently absent from the western half of the Glaciated Plains Division and the southern portion of the Mississippi Lowlands (eastern U.S. west to Minnesota and Texas). Mesic to dry upland forests, upland prairies, glades, savannas, ledges and tops of bluffs, and banks of streams; also pastures, old fields, and roadsides.

Primula meadia exhibits a wide range of variation and consequently there have been attempts to divide it into subspecies and varieties (Fassett, 1944; Thompson, 1953; Gleason and Cronquist, 1991). Much of the variation involves quantitative differences in the length of the calyx lobes, anthers, and capsules. Two variants are said to occur in Missouri. Fassett (1944) described *D. meadia* "var. *genuinum*" (= var. *meadia*) as relatively large plants with 4–125 flowers per umbel, calyx lobes 3–7 mm long, anthers 6.5–10.0 mm long, capsules 10.5–18.0 mm long, and occurring primarily on previously glaciated areas. *Dodecatheon meadia* var. *brachycarpum* was described as relatively small plants with 1–14 flowers per umbel, calyx lobes 2.5–5.0 mm long, anthers 4–7 mm, capsules 7.5–10.0 mm long, and occurring mainly on unglaciated southern uplands. These two varieties were said to overlap through much of their ranges, occupy the same kinds of habitats, and to intergrade where they come in contact (Fassett, 1944). In practice, there is so much overlap in the measurements and lack of correlation between characters that it is difficult or impossible to assign most specimens to either variety with any confidence. Thus, infraspecific taxa are not recognized in the present treatment.

Primula meadia is gaining in popularity as an ornamental in native wildflower gardens and rock gardens. Steyermark (1963) noted that the flowers have an aroma reminiscent of grape juice, whereas the rootstock has an odor suggestive of canned corned beef. White-flowered plants have been called *D. meadia*. f. *album* J.F. Macbr., whereas plants lacking reddish coloration on their leaf bases have been called *D. meadia* f. *pallidum* Fassett and f. *sedens* Fassett.

4. Primula veris L. **ssp. veris** (English cowslip)

Map 2333

Basal leaves 3–20 cm long, the blades ovate, oblong-ovate, or oblong-obovate, abruptly tapered at the base to a more or less well-defined, winged petiole 2–9 cm long, rounded or broadly angled to a blunt point at the tip, the margins somewhat curled-under and/or somewhat corrugated, also with shallow, broad teeth or scallops, the upper surface sparsely to moderately pubescent with a mixture of short, stiff, unicellular hairs and more flexible multicellular hairs, slightly roughened to the touch, often slightly shiny, the undersurface moderately to densely pubescent with short, grayish white, multicellular hairs, especially along the main veins, pale green, sometimes with reddish coloration toward the base, especially along the broad midvein. Scapes 12–35 cm tall, densely pubescent with short, grayish white hairs. Umbels with 3–15 flowers, the involucral bracts 4–9 mm long, narrowly lanceolate, the flower stalks 0.5–2.0 cm long, densely and minutely grayish white-hairy. Calyces 8–15 mm long, lobed about 1/3 of the length. Corollas with the tube 9–15 mm long, yellow or orange at the throat, the lobes 5–14 mm long, 4–7 mm wide, spreading or loosely ascending, conspicuously notched at the tip, yellow, often with an orange to reddish spot on the upper surface at the base. Stamens not exserted, the filaments individually fused nearly their entire length to the inner surface of the corolla tube, the anthers 1.5–2.5 mm long, free, positioned either near the corolla throat (in short-styled plants) or toward the midpoint of the corolla tube (in long-styled plants). Style variously 4–15 mm long (in short- or long-styled flowers), slender, the stigma relatively conspicuous, capitate. Fruits 4–8 mm long, 3–5 mm wide, thick- and hard-walled, dark brown. Seeds 0.6–1.0 mm long, oblong-ellipsoid to more or less cuboid, mostly relatively angular, reddish brown. $2n$=22. April–June.

Introduced, known thus far only from Franklin County (native of Europe, introduced sproadically in the northeastern U.S. and adjacent Canada). Edges of mesic upland forests; also roadsides.

2335. Aconitum uncinatum 2336. Actaea pachypoda 2337. Actaea racemosa

6. Samolus L.

About 10 species, nearly worldwide.

1. Samolus parviflorus Raf. (water pimpernel, brookweed)
S. floribundus Kunth
Pl. 510 a, b; Map 2334

Plants perennial herbs, with fibrous roots, slightly succulent. Stems 5–35 cm tall, glabrous, unbranched or more commonly branched. Leaves basal and alternate, shortly petiolate. Leaf blades 2–9 cm long, oblanceolate to obovate, tapered at the base, rounded or broadly angled to a bluntly pointed tip, the margins entire, the surfaces glabrous. Inflorescences terminal and/or axillary racemes, these sessile or short-stalked, the axis slender, flexuous, the flowers long-stalked, with a small bract near the midpoint. Flowers perigynous by fusion of the ovary to the calyx tube. Calyces 0.7–1.5 mm long, the tube 0.5–1.0 mm, fused to ovary, the lobes 0.2–0.5 mm long, broadly triangular, sharply pointed at the tips. Corollas 1.0–1.5 mm long, white, deeply 5-lobed, the lobes oblong rounded or notched at the tips. Stamens 5, attached in the corolla tube. Staminodes minute, slender, scalelike, positioned in the sinuses between the corolla lobes. Ovary $\frac{1}{2}$–$\frac{2}{3}$ of the way inferior, the style minute, the stigma, minute, capitate. Fruits capsules 2–3 mm long, globose, green, the walls translucent, dehiscent partway from the tip by 5 valves. Seeds minute, more or less triangular in outline, brown. $2n=26$. April–September.

Scattered in the Ozark, Ozark Border, Unglaciated Plains, and Mississippi Lowlands Divisions, north locally in the Unglaciated Plains to Adair County (U.S., Canada, Mexico, Central America, South America, Caribbean Islands). Banks of streams, rivers, and spring branches, margins of ponds and lakes, sloughs, bottomland forests, and bases and ledges of bluffs, sometimes emergent aquatics in shallow water.

This species can grow underwater and sometimes is sold as an aquarium plant.

RANUNCULACEAE (Buttercup Family)

Contributed by Alan Whittemore

Plants annual or perennial herbs, occasionally subshrubs or lianas, sometimes monoecious, dioecious, or with a mixture of perfect and imperfect flowers. Leaves alternate, opposite, whorled, or basal, simple or compound. Stipules absent, but the petioles sometimes expanded or with small appendages. Flowers actinomorphic or zygomorphic, perfect or sometimes imperfect, hypogynous, occasionally with 3 bractlets subtending the calyx. Calyces of 3–6(–20) free sepals, sometimes shed as the flowers open. Corollas absent or of 3 to numerous, free petals, these showy or inconspicuous. Stamens 5 to numerous, free and distinct, the anthers attached at their bases, dehiscing by longitudinal slits. Pistils 1 to numerous per flower, each of 1 carpel (except in *Nigella*). Ovary superior, 1-locular, the placentation marginal, apical or basal, the ovules 1 to numerous. Style absent or 1, short or long, the stigma 1, often lateral. Fruits berries, follicles, capsules, or achenes, often aggregated on the dome-shaped or elongate receptacle, the seeds 1 to many per fruit. Fifty-six to 60 genera, 2,100–2,500 species, worldwide.

Many species of Ranunculaceae are poisonous due to the presence of benzylisoquiniline alkaloids, and some are a significant danger to livestock.

Nigella damascena L. (love-in-a-mist), an annual species sometimes grown as an ornamental, occasionally escapes in the eastern United States. It was mapped for Missouri by Ford (1997), but the specimen upon which this report was based was a weed in a garden. The species is not yet known to escape into uncultivated ground in Missouri, but it should be watched for. It has several alternate stem leaves and a whorl of leaflike involucral bracts that are 3 times pinnately dissected into narrowly linear segments; large, solitary flowers with showy blue, white, or pink sepals; and a single compound ovary with axile placentation and 5 or 6(–10) linear styles.

1. Flowers zygomorphic; upper sepal deeply concave or spurred, petals smaller than the sepals
  2. Upper sepal deeply concave (hoodlike) but not spurred; petals completely hidden by the sepals . . . . 1. ACONITUM
  2. Upper sepal spurred; petals projecting from the center of the calyx . . . . 9. DELPHINIUM
1. Flowers actinomorphic; sepals and petals various
  3. Stem leaves and bracts opposite or whorled
    4. Basal leaves absent at flowering; stems woody, at least at the base . . . . 8. CLEMATIS
    4. Basal leaves present at flowering; stems never woody
      5. Sepals yellow; fruits follicles . . . . 10. ERANTHIS
      5. Sepals white, pink, purple, blue, or green (sometimes pale cream-colored in *Anemone quinquefolia*); fruits achenes
        6. Style absent; achenes not beaked, achene wall with conspicuous veins . . . . 15. THALICTRUM
        6. Style present; achenes beaked, achene wall not veined . . . 4. ANEMONE
  3. Stem leaves and bracts alternate or sometimes absent
    7. Petals with long spurs; a ring of papery staminodes present between the stamens and pistils . . . . 5. AQUILEGIA
    7. Petals without spurs (note that the sepals are spurred in *Myosurus*); staminodes absent between the stamens and pistils
      8. Leaves all basal, the blade linear or dissected into linear segments; flowers solitary, without bracts; receptacle elongated at fruiting
        9. Leaf blades dissected into linear segments; sepals not spurred, persistent at fruiting; achenes ellipsoidal, the beak 3.5–4.5 mm long . . . . 7. CERATOCEPHALUS
        9. Leaf blades linear to very narrowly oblanceolate, entire; sepals spurred, shed before fruit is ripe; achenes prismatic, the beak 0.05–0.40 mm long . . . . 13. MYOSURUS
      8. Leaves or bracts alternate on the stems (inconspicuous and sepaloid in some *Anemone* species, then with broadly 3-lobed basal leaves); flowers solitary or not; receptacle not or only slightly elongated at fruiting
        10. Fruits follicles, berries, or capsules; each ovary with 2 to many ovules

11. Leaves compound
    12. Flowers many in more or less dense leafless racemes or panicles; leaflets jaggedly and irregularly toothed and lobed; petals present . . . . 2. ACTAEA
    12. Flowers solitary or 2–4 in open leafy racemes; leaflets deeply 2- or 3-lobed or -parted, the lobes sometimes again shallowly notched; petals absent . . . . 12. ISOPYRUM
11. Leaves simple, sometimes shallowly lobed
    13. Leaf blades palmately lobed; fruit a cluster of 5–15 red berries . . . . 11. HYDRASTIS
    13. Leaf blade unlobed; fruit a cluster of 3–12 brown to dark brown follicles . . . . 6. CALTHA

10. Fruits achenes or utricles; each ovary with 1 ovule
    14. Petals absent; flowers up to 200 per inflorescence, in panicles
        15. Leaves compound; fruits achenes . . . . 15. THALICTRUM
        15. Leaves simple, palmately lobed; fruits utricles . . . . 16. TRAUTVETTERIA
    14. Petals present (rarely absent in some flowers of *Ranunculus parviflorus*); flowers solitary or up to 20 per inflorescence, in open clusters
        16. Anthers purplish black (fading to olive green in old herbarium specimens); petals without nectaries; leaves dissected into linear segments . . . . 3. ADONIS
        16. Anthers yellow; each petal with a basal nectary; leaves variously lobed, dissected or compound (if dissected into linear segments then plants are submerged or emergent aquatics) . . . . 14. RANUNCULUS

1. Aconitum L. (monkshood)

About 100 species, North America, Europe, Asia.

Aconitum is closely related to *Delphinium*. A recent molecular phylogenetic study of the group even suggested that *Aconitum* might represent a basal group within a broadly circumscribed *Delphinium* (Jabbour and Renner, 2011), but more detailed research involving a broader sampling of species is needed to address this issue.

1. Aconitum uncinatum L. (wild monkshood) Map 2335

Plants perennial herbs, with rhizomes having tuberous roots. Stems 30–200 cm long, weak and arching, glabrous. Leaves alternate, glabrous, the basal leaves absent at flowering time. Well-developed lower leaves with the petiole 7–20 mm long, blade 5.3–6.2 cm long, 8.6–9.9 cm wide, 3-parted (sinuses ending 4–9 mm from leaf base), the base cordate, the three main segments rhombic or oblanceolate, narrowed to a sharp point at the tip, each with 4–10 coarse teeth or shallow lobes. Inflorescences terminal, narrow panicles, their main axes glabrous, the flower stalks with long, spreading hairs toward the tip. Flowers zygomorphic, perfect. Sepals 5, deep blue, 15–27 mm long, the upper sepal deeply hooded, lateral and lower sepals plane, not persistent at fruiting. Petals 2, 8–13 mm long, completely hidden by the sepals, spurred. Stamens hidden by the sepals, the anthers almost black. Staminodes absent. Pistils 3, each with 10–20 ovules, style present. Fruits cylindrical follicles, the body 12–14 mm long, the outer wall thick, prominently cross-veined, the beak about 3 mm long. Receptacle not much enlarged at fruiting. September.

Known thus far only from Shannon County (east-central U.S. from Maryland and Georgia west to Missouri). Tops of low dolomite bluffs and mesic upland forests.

This species is unusual in that its stems are strongly arched toward their tips resulting in flowers that appear inverted. Nonflowering material of *A. uncinatum* is superficially very similar to *Delphinium exaltatum*, and the two are easily confused. Differences are discussed under *D. exaltatum*. The few Missouri localities of *A. uncinatum* are disjunct from a similarly isolated station in southern Indiana. The species was first reported for the state by Summers (1997).

2. **Actaea** L. (baneberry, snakeroot)

Plants perennial herbs, with rhizomes or persistent, somewhat woody bases. Stems erect. Leaves alternate and sometimes also in a basal rosette, 2 or 3 times pinnately compound, the leaflets ovate to broadly lanceolate. Inflorescences terminal and axillary, many-flowered racemes or panicles with racemelike branches. Flowers actinomorphic, perfect. Sepals 3–5, greenish white, 2–6 mm long, plane or somewhat concave, not persistent at fruiting. Petals 4–10, white or cream-colored, 2–4 mm long, plane. Stamens showy, the anthers white or yellow. Staminodes absent. Pistils 1–8, each with 4 to numerous ovules. Style present or absent. Fruits follicles or berries, ovoid, ellipsoid, or spherical, the tip with a tubercle or beak 0.5–2.5 mm long, the outer wall thick, not prominently veined. Receptacle not much enlarged at fruiting. About 20 species, North America, Europe, Asia.

Species with follicles are often split off as a separate genus, *Cimicifuga* L. ex Wernisch., but the two groups are very closely related (Compton et al., 1998). Ramsey (1988) presented characters for separating vegetative specimens of Actaea from morphologically similar genera of Ranunculaceae.

1. Inflorescences racemes, 1–11 cm long; flower stalks 4–19 mm long, 1–2 mm thick, sparsely roughened with minute stiff forward-pointed hairs; fruits berries . . . . 1. A. PACHYPODA
1. Inflorescences panicles with several racemelike branches, 10–60 cm long; flower stalks 4-6 mm long, 0.2–0.5 mm thick, densely pubescent with short soft contorted hairs; fruits follicles . . . . 2. A. RACEMOSA

1. Actaea pachypoda Elliott (white baneberry, doll's eyes)

Pl. 511 c, d; Map 2336

Stems 38–50 cm long. Leaves 2 or 3 times pinnately compound, the leaflets ovate to broadly lanceolate, toothed and also sharply and irregularly lobed, the largest leaflets 5–11 cm long, 3.5–7.0 cm wide. Inflorescences racemes, 1–11 cm long. Flower stalks 4–19 mm long, 1–2 mm wide, sparsely roughened with minute stiff forward-pointing hairs. Fruits berries, 1 per flower stalk, 5–10 mm long, ellipsoid or spherical, white. $2n$=16. March–May.

Glaciated Plains, Ozark Border, Ozark and Crowley's Ridge Divisions (eastern U.S. west to Nebraska and Oklahoma; Canada). Mesic to dry upland forests, often in rocky places, bases and ledges of bluffs, sinkholes, and occasionally banks of streams.

A closely related species, *A. rubra* (Aiton) Willd. (red baneberry), is found near the Missouri border in southern Iowa and might be found in the future in northern Missouri. It differs from *A. pachypoda* primarily in having sharply or bluntly pointed (not truncate) petals and more slender fruit stalks (only about 0.5 mm thick) that are unpigmented, vs. bright red at fruiting in *A. pachypoda*.

2. Actaea racemosa L. (black cohosh, black snakeroot)

Cimicifuga racemosa (L.) Nutt.

Pl. 513 e, f; Map 2337

Stems 75–250 cm long. Leaves 2 times pinnately compound, the leaflets ovate to broadly lanceolate, toothed and irregularly sharply lobed, the largest leaflets 5–10 cm long, 3–7 cm wide. Inflorescences panicles with 2 or 3 racemelike branches, 15–43 cm long. Flower stalks 4–6 mm long, 0.2–0.5 mm wide, densely pubescent with soft contorted hairs. Stigma 0.5–1.0 mm wide. Fruits follicles, 1 per flower stalk, 5–10 mm long, ovoid or ellipsoid. Seeds smooth or ridged. $2n$=16. June–July.

Scattered in the Ozark Division and the eastern portion of the Ozark Border (eastern U.S. west to Iowa and Arkansas; Canada). Mesic upland forests, often in rocky places, bases of bluffs, mostly on calcareous and cherty substrates.

The roots and rhizomes of black cohosh have a long history of use among Native Americans for a variety of ailments, including malaria, rheumatism, tuberculosis, colds, kidney problems, and menstrual irregularity, as well as a general tonic and to induce abortion (Moerman, 1998). The potential pharmaceutical use of the species in modern medicine to treat problems associated with the onset of menopause has received considerable

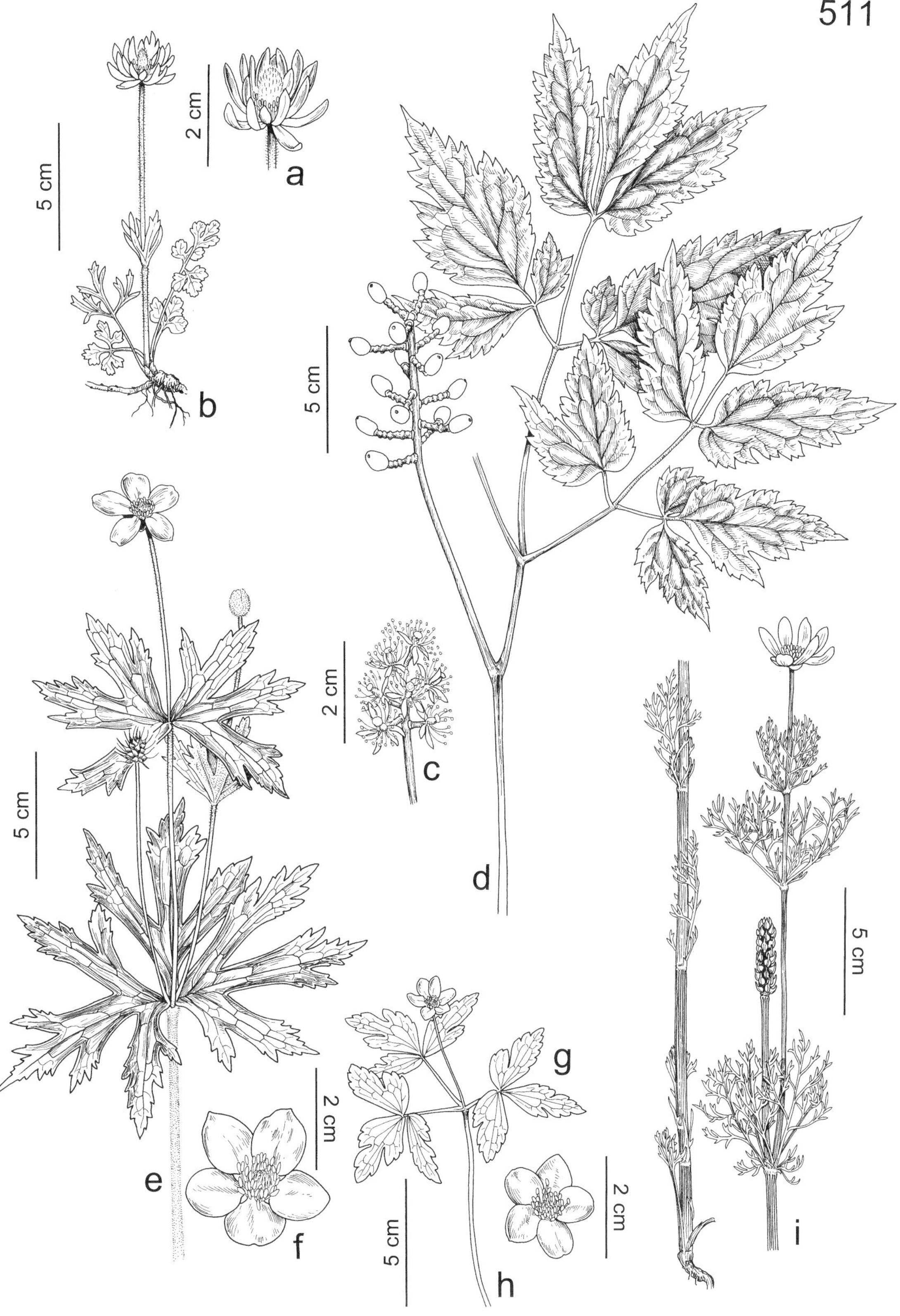

Plate 511. Ranunculaceae. *Anemone caroliniana*, **a)** flower, **b)** habit. *Actaea pachypoda*, **c)** inflorescence, **d)** habit at fruiting. *Anemone canadensis*, **e)** habit, **f)** flower. *Anemone quinquefolia*, **g)** habit, **h)** flower. *Adonis annua*, **i)** lower and upper portions of habit with flower and fruits.

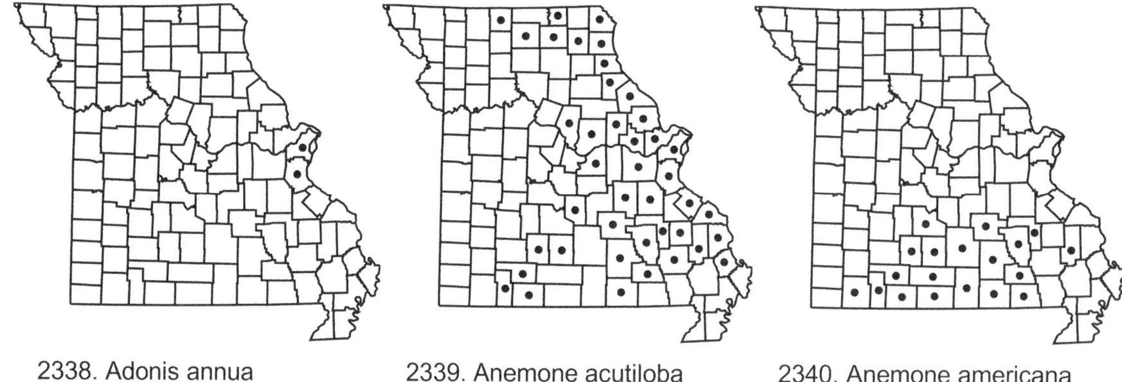

2338. Adonis annua 2339. Anemone acutiloba 2340. Anemone americana

study, mostly in Europe (Brinckmann and Wollschlager, 2003), and it is available as over the counter pills at some health food stores. As with nearly all members of the family, *A. racemosa* is considered toxic, thus readers should be wary of self-medicating using wild-collected plants.

3. **Adonis** L. (adonis, pheasant's eye) (Heyn and Pazy, 1989)

About 35 species, native to Eurasia and North Africa, introduced in North America.

1. Adonis annua L. (summer adonis)
A. autumnalis L.

Pl. 511 i; Map 2338

Plants annual. Stems 15–40 cm long, erect. Leaves alternate and in a basal rosette (this often absent at flowering), the blades 2–4 times pinnately dissected, the segments 0.5–1.0 mm wide, narrowly linear. Inflorescences of solitary flowers at the branch tips. Flowers actinomorphic, perfect. Sepals 5(–8), 4–9 mm long, oblong-elliptic, plane, green, purple, or nearly colorless, not persistent at fruiting. Petals 5–10, 7–15 mm long, 3–4 mm wide, plane, dark red, usually with a dark basal blotch. Stamens prominent but scarcely showy, the anthers dark purple (fading to olive green with age). Staminodes absent. Pistils 25–40, each with 1 ovule. Style present. Fruits achenes, in cylindrical heads 12–18 mm long and 6–9 mm wide, the body of each fruit about 3 mm long, broadly top-shaped, the outer wall thick, wrinkled or faintly veined, the tip tapered abruptly to a straight lanceolate beak about 1 mm long. Receptacle becoming elongated at fruiting, glabrous. $2n$=16. April–May.

Introduced, Jefferson County (native of Eurasia, escaped at scattered localities in the southeastern U. S. from Alabama to Texas and Missouri). Crop fields and open disturbed areas.

This species was once a well-known weed of grain fields all over the world, but it has largely disappeared from this habitat as farming techniques have changed. The few Missouri collections and reports were made in the 1930s and 1940s, thus it may no longer escape in the state. Steyermark (1963) reported this plant as *Adonis aestivalis* L. but our specimens are clearly *A. annua*. He also stated that the petals are yellow, rather than red. This may have been caused by the red petals sometimes fading to yellow in dried herbarium specimens.

4. **Anemone** L. (windflower, anemone)

Plants perennial herbs, with rhizomes or tubers, usually with 1–12 basal leaves and 3 or more opposite or whorled bracts along the stem (sometimes close to the flower and calyxlike). Stems erect or less commonly arching. Leaves deeply lobed or compound, the blades or leaflets kidney-shaped to obtriangular or lanceolate. Inflorescences of solitary flowers or open clusters or groups of 2–7 flowers at the stem tips. Flowers actinomorphic, perfect. Sepals 5–12, 7–

24 mm long, petaloid, white, pink, purple, blue, green, or pale cream-colored, plane, not persistent at fruiting. Petals absent. Stamens prominent but scarcely showy, the anthers yellow or white. Staminodes absent. Pistils many, each with 1 ovule. Style present. Fruits achenes, usually somewhat flattened, with a beak 0.5–6.0 mm long, the fruit wall thick, not veined. Receptacle sometimes elongated at fruiting. About 150 species, nearly worldwide.

1. Basal leaves usually absent, dying back, or darkened to maroon or dark purple at flowering, 3-lobed for 30–80% of their length, the lobes entire; involucre calyxlike, the bracts entire and unlobed, closely subtending the flower
   2. Leaf lobes bluntly or sharply pointed, the sinuses extending 60–80% of the leaf length . . . . 1. A. ACUTILOBA
   2. Leaf lobes rounded, the sinuses extending 30–50% of the leaf length . . . . 2. A. AMERICANA
1. Basal leaves usually present and still green at flowering, deeply parted (more than 90% of their length) or compound, the leaflets or segments toothed and/or cleft; involucre not at all calyxlike, the bracts deeply parted or compound, positioned at least 1.5 cm below the flower
   3. Sepals 10–20(–30) per flower; stems arising from an underground tuber; ultimate segments of leaf 1–7 mm wide . . . . 4. A. CAROLINIANA
   3. Sepals 5(6) per flower; stems arising from an underground rhizome; ultimate segments of leaf 3–26 mm wide
      4. Involucral bracts sessile; fruits with the beak 3–6 mm long; basal leaves deeply divided (to 97% of their length) but not compound . . . . 3. A. CANADENSIS
      4. Involucral bracts stalked; fruits with the beak (0.3–)0.5–2.0 mm long; basal leaves compound
         5. Basal leaf solitary or absent; stems 6–27 cm long; involucre one per inflorescence; fruits with sparse to moderate straight hairs not concealing the surface . . . . 6. A. QUINQUEFOLIA
         5. Basal leaves 2 or more; stems (20–)30–110 cm long; involucre usually double, an upper involucre on each branch above the main one; fruits with dense woolly hairs concealing the surface or nearly so
            6. Heads of fruits cylindrical, more than twice as long as wide; fruits with the beak 0.5–1.0 mm long; bracts 3–7 per involucre . . . . 5. A. CYLINDRICA
            6. Heads of fruits spherical or short-cylindrical, less than twice as long as wide; fruits with the beak 1.0–1.5 mm long; bracts 3(–5) per involucre . . . . 7. A. VIRGINIANA

1. Anemone acutiloba (DC.) G. Lawson (sharp-lobed hepatica, sharp-lobed liverleaf)
Hepatica acutiloba DC.
H. nobilis Mill. var. *acuta* (Pursh) Steyerm.
Pl. 515 e; Map 2339

Stems 8–17 cm long, unbranched, from rhizomes. Basal leaves 2–12, usually absent, dying back, or darkened to maroon or dark purple at flowering, long-petiolate. Leaf blades 3-lobed for 60–80% of their length (sinuses extending within 8–23 mm of the leaf base), the lobes 27–40 mm wide, entire, bluntly to sharply pointed at the tip. Each stem with 1 whorl of 3 involucral bracts closely subtending the flower, the bracts sessile, unlobed, ovate to lanceolate, bluntly pointed to rounded at the tip, the margins entire. Flowers solitary. Sepals 5–8(–12), 7–16 mm long, white, pink, or pale blue. Head of fruits 5–6 mm long, 9–11 mm in diameter, hemispherical. Fruits elliptic in outline, sparsely to moderately pubescent with straight hairs not concealing the surface, the beak 0.5–1.0 mm long, brittle and often broken off. Receptacle not enlarged at fruiting. $2n$=14. February–April.

Scattered, mostly in the eastern half of the state (eastern U.S. west to Minnesota and Arkansas; Canada). Mesic upland forests, mostly on north-facing slopes, rock outcrops, and shaded ledges of bluffs, on calcareous substrates, less commonly on sandstone.

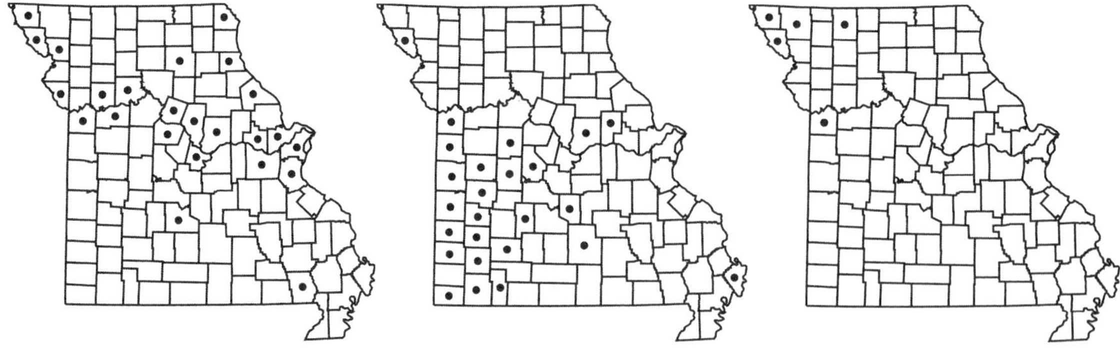

2341. Anemone canadensis 2342. Anemone caroliniana 2343. Anemone cylindrica

The two liverleaf species, *A. acutiloba* and *A. americana*, are very similar to one another, and very different in appearance from other North American species of *Anemone*. They have often been treated in a segregate genus, *Hepatica* Mill, by many authors (Steyermark, 1963). Although the superficial differences are striking, the basic structure of the flowers and inflorescences is identical and all of the characters of *Hepatica* (shallowly lobed leaves, unlobed involucral bracts, flowers close to the involucre) also are found individually in Asian species of *Anemone*. Recent genetic research also supports the inclusion of *Hepatica* within *Anemone* (Hoot and Reznicek, 1994).

The two liverleaf species seldom grow together, because *A. americana* prefers somewhat drier sites and more acidic soils than *A. acutiloba*, and its range in Missouri is more southern. Julian and Cora Steyermark (1960) reported finding populations intermediate between *A. acutiloba* and *A. americana*, probably of hybrid origin, in Carter and Reynolds Counties, and also in northern Illinois. *Anemone hepatica* L. (=*Hepatica nobilis* Mill.) of Europe also is very similar to *A. americana*. On average, *A. hepatica* has slightly larger flowers and its leaves have deeper lobes that are often obtuse, but many European plants closely resemble American plants, either *A. americana* or intermediates between *A. americana* and *A. acutiloba*. Because of the similarity between these three taxa and the presence of occasional intermediates between *A. acutiloba* and *A. americana*, Steyermark and Steyermark (1960) treated the two American liverleaf taxa as varieties of *A. hepatica*. This treatment does not accurately reflect the biology of these plants in North America, where *A. americana* and *A. acutiloba* behave as good species—intermediates between them are rare and local, and the two taxa seem amply distinct morphologically and ecologically despite the very large overlap in their distributions (over a large part of eastern North America).

Plants with white sepals have been called *Hepatica nobilis* f. *albiflora* (Ralph Hoffm.) Steyerm. and those with pink sepals have been called *H. nobilis* f. *rosea* (Ralph Hoffm.) Steyerm.

2. Anemone americana (DC.) H. Hara (round-lobed hepatica, round-lobed liverleaf)
Hepatica americana (DC.) Ker Gawl.
H. nobilis var. *obtusa* (Pursh) Steyerm.
Pl. 515 c, d; Map 2340

Stems 5–15 cm long, unbranched, from rhizomes. Basal leaves 2–13, usually absent, dying back, or darkened to maroon or dark purple at flowering, the blades 3-lobed for 30–50% of their length (sinuses extending within 6–22 mm of the leaf base), the lobes 18–31 mm wide, rounded at the tip, the margins entire. Each stem with 1 whorl of 3 involucral bracts closely subtending the flower, the bracts sessile, elliptic to ovate, rounded or broadly and bluntly pointed at the tip, the margins entire. Flowers solitary. Sepals 5 or 6(–12), 7–13 mm long, white, pink, or pale blue. Head of fruits 4–6 mm long, 7–10 mm in diameter, hemispherical. Fruits elliptic in outline, sparsely to moderately pubescent with straight hairs not concealing the surface, the beak 0.5–1.0 mm long, brittle and often broken off. Receptacle not enlarged at fruiting. $2n=14$. February–April.

Scattered in the eastern and southern parts of the Ozark Division (eastern U.S. west to Minnesota and Arkansas; Canada). Mesic upland forests, often at the bases and on north-facing slopes, rock outcrops and shaded ledges of bluffs, on acidic and less commonly calcareous substrates.

Plants with white sepals have been called *Hepatica nobilis* f. *candida* (Fernald) Steyerm. Alternate taxonomic placements of *A. americana* are discussed under the treatment of *A. acutiloba*.

3. Anemone canadensis L. (white anemone, meadow anemone)
Pl. 511 e, f; Map 2341

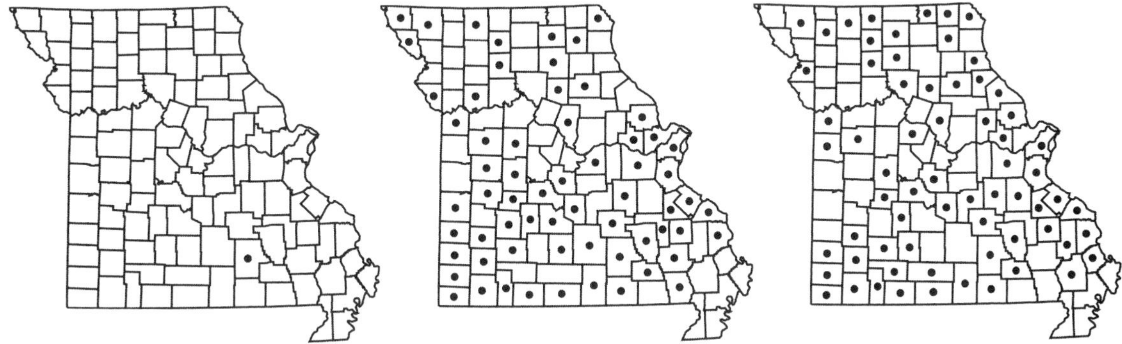

2344. Anemone quinquefolia 2345. Anemone virginiana 2346. Aquilegia canadensis

Stems 20–55 cm tall, usually branched, from rhizomes. Basal leaves 1–7, usually present and green at flowering, deeply 3-parted (to 97% of their length), the segments deeply parted again (the sinuses extending within (1–)2–5 mm of the leaf base, occasionally merely lobed), the ultimate leaf segments 5–16 mm wide, narrowed or tapered to a sharp or rarely blunt point at the tip, the margins toothed. Each stem with 1–3 involucres of opposite pairs or whorls of 3 bracts, the uppermost involucre positioned 3–9 cm below the flower, the bracts sessile, broadly wedge-shaped, 3-lobed or -parted, the segments toothed and sometimes lobed, narrowed or tapered to a sharp point at the tip. Flowers 1–5 per stem. Sepals 5 or 6, 9–24 mm long, white. Head of fruits 5–10 mm long, 8–11 mm in diameter, nearly spherical. Fruits obovate in outline, sparsely to moderately pubescent with straight hairs not concealing the surface, the beak 3–6 mm long. Receptacle not enlarged at fruiting. $2n$=14. May–July.

Scattered, mostly in counties adjacent to the Missouri and Mississippi Rivers (northern U.S. south to Virginia, Tennessee, and New Mexico; Canada). Bottomland forests, bottomland prairies, moist bases of limestone bluffs, also levees, moist roadsides, depressions along railroads, and moist open disturbed floodplain areas.

4. Anemone caroliniana Walter (prairie anemone, Carolina anemone)

Pl. 511 a, b; Map 2342

Stems 6–30 cm tall, unbranched, from tubers. Basal leaves 2–5, usually present and green at flowering, once-compound with 3 leaflets, the leaflets usually deeply parted (occasionally merely lobed), the ultimate segments 1–7 mm wide, rounded with a small sharp point to narrowed or tapered to a sharp point at the tip, the margins toothed or entire. Each stem technically leafless, but with 1 whorl of 3 involucral bracts positioned 3–16 cm below the flower, the bracts sessile, wedge-shaped, deeply cleft, narrowed or tapered to a sharp point at the tip. Flowers solitary. Sepals 13–21, 10–23 mm long, white or less commonly pale blue, pink, or purple. Head of fruits 14–18 mm long, 5–7 mm wide, cylindrical. Fruits elliptic in outline, densely pubescent with woolly hairs more or less concealing the surface, the beak 1.2–1.8 mm long. Receptacle elongated at fruiting. $2n$=16. March–April.

Scattered, mostly in the Unglaciated Plains Division and the western portion of the Ozarks, with a few disjunct populations farther north and east; also introduced in portions of the western and central Ozarks (southeastern and central U.S. from South Carolina to Texas and Indiana to South Dakota). Glades and upland prairies; also pastures, roadsides, vacant lots, cemeteries, and rocky or sandy open disturbed places, usually on acidic substrates.

5. Anemone cylindrica A. Gray (thimbleweed, candle anemone)

Pl. 512 d, e; Map 2343

Stems 35–70 cm long, usually branched, from rhizomes. Basal leaves (2–)6–13 (most or all may be withered at flowering time), once-compound with 3 leaflets, the leaflets deeply parted and lobed, the ultimate segments 3–11 mm wide, rounded with a small sharp point to narrowed or tapered to a sharp point at the tip, the margins toothed. Each stem with 2 or 3 whorls of 3 involucral bracts (but 2 whorls often crowded together and appearing as a single whorl of 6 bracts), the uppermost involucre positioned 16–29 cm below the flower, the bracts with stalks 10–50 mm long, once-compound with 3 leaflets, the leaflets lobed or deeply parted again, rounded with a small sharp point or narrowed to a sharp or blunt point at the tip. Flowers 1–3 per stem. Sepals 5 or 6, 8–10 mm long, green or white. Head of fruits 21–28 mm long, 8–10 mm wide, more than twice as long as wide, cylindrical. Fruits elliptic in outline, densely pubescent with woolly hairs more or less concealing the surface, the beak ca. 1 mm long Receptacle elongated at fruiting. 2*n*=16. June.

Uncommon, confined to the western portion of the Glaciated Plains Division (northern U.S. south to Arizona and New Jersey). Loess hill prairies and adjacent mesic upland forests, less commonly in other upland prairies.

6. Anemone quinquefolia L. **var. quinquefolia** (wood anemone, windflower)

Pl. 511 g, h; Map 2344

Stems 6–30 cm long, unbranched, from rhizomes. Basal leaf 1 or absent at flowering, if present, then once-compound with 3 leaflets, the leaflets often lobed, the ultimate segments 4–24 mm wide, rounded with a small sharp point to narrowed or tapered to a sharp point at the tip, the margins toothed. Each stem with 1 whorl of 3 involucral bracts positioned (1.5–)2.0–5.0 cm below the flower, the bracts with stalks 10–20 mm long, once-compound with 3 leaflets, the leaflets toothed and often lobed, rounded with a small sharp point or narrowed to a sharp or blunt point at the tip. Flowers solitary. Sepals (4)5(–7), 7–13 mm long, white or pale cream-colored, the outer (under) side sometimes pinkish-tinged. Head of fruits 4–5 mm long, 8–10 mm wide, hemispherical. Fruits elliptic in outline, sparsely to moderately pubescent with straight hairs not concealing the surface, the beak 1.0–1.5 mm long. Receptacle not enlarged at fruiting. 2*n*=32. April–May.

Uncommon, known thus far only from a single population in Shannon County (eastern U.S. west to North Dakota and Arkansas; Canada). Mesic upland forests on steep slopes.

The presence of this species in Missouri was first reported by Christ (1984), but no additional populations have been found to date. Because it also occurs in adjacent Iowa, Illinois, Kentucky, Tennessee, and Arkansas, it is expected to grow elsewhere in Missouri as well.

7. Anemone virginiana L. **var. virginiana** (thimbleweed)

Pl. 512 c; Map 2345

Stems 30–100 cm tall, usually branched, from rhizomes. Basal leaves 1–8, usually present and green at flowering, once-compound with 3 leaflets, the leaflets deeply parted or lobed, the ultimate segments 12–26 mm wide, rounded with a small sharp point to narrowed to a sharp or blunt point at the tip, the margins toothed. Each stem with 2 or 3 opposite pairs or whorls of 3 or 4 involucral bracts, the uppermost involucre positioned 11–42 cm below the flower, the bracts once-compound with 3 leaflets, the leaflets with stalks 9–64 mm long, lobed or parted and toothed, narrowed or tapered to a sharp point at the tip. Flowers 1–7 per stem. Sepals 5, 7–21 mm long, white or greenish-tinged. Head of fruits 12–28 mm long, 10–13 mm wide, less than twice as long as wide, cylindrical. Fruits obovate in outline, densely pubescent with woolly hairs more or less concealing only the basal half, the beak 1–2 mm long. Receptacle elongated at fruiting. 2*n*=16. June–August.

Scattered to common nearly throughout the state (eastern U.S. west to North Dakota, Wyoming, and Arkansas; Canada). Mesic to dry upland forests, upland prairies, savannas, glades, ledges and tops of bluffs, and rocky banks of streams; also roadsides and open disturbed areas.

Plants with white sepals have been called f. *leucosepala* Fernald. A plant collected once at Rockwoods Reservation, St. Louis County, with all of the stamens transformed into petaloid structures, has been called f. *plena* E.J. Palmer & Steyerm.

5. Aquilegia L. (columbine)

About 70 species, throughout the northern hemisphere.

1. Aquilegia canadensis L. (columbine, red columbine, wild honeysuckle)

Pl. 512 f, g; Map 2346

Plants perennial herbs. Stems 15–90 cm long, erect or arching, usually few-branched, glabrous or sparsely to moderately pubescent with fine spreading hairs. Leaves alternate and usually also in a basal rosette, short- to long-petiolate. Leaf blades twice ternately compound, the leaflets 17–52 mm long, ternately 1 or 2 times lobed or parted, glabrous or hairy, especially on the undersurface. Inflorescences of solitary flowers or open clusters or groups of up to 10 flowers at the branch tips. Flowers actinomorphic, pendant, perfect. Sepals 5, 8–18 mm long, 3–8 mm wide, spreading, plane, ovate to broadly ovate, red, sometimes green at the tip, not persistent at fruiting. Petals 5, consisting of a flat to somewhat cupped blade narrowed abruptly into a basal spur; the blade 5–9 mm long, oblong to oblong-circular, pale yellow or

Plate 512. Ranunculaceae. *Caltha palustris*, **a)** flower, **b)** habit. *Anemone virginiana*, **c)** habit with flower and fruits. *Anemone cylindrica*, **d)** flower, **e)** habit with flower and fruits. *Aquilegia canadensis*, **f)** flower, **g)** habit. *Thalictrum thalictroides*, **h)** fruits, **i)** habit.

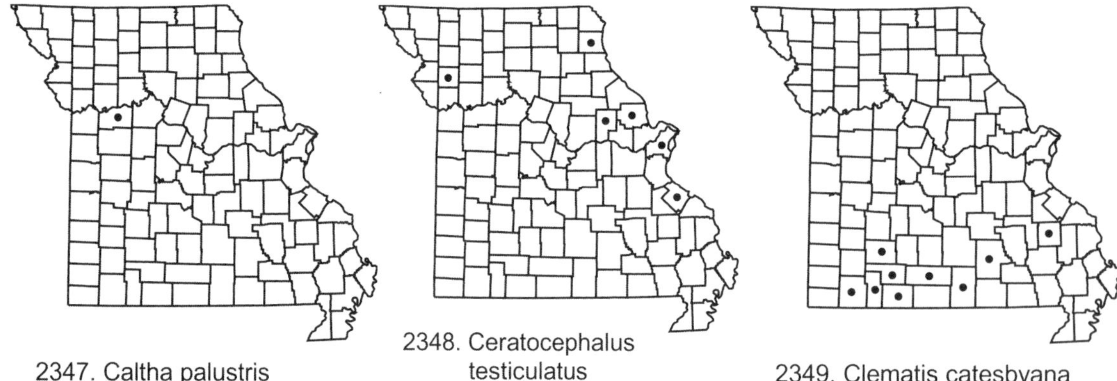

2347. Caltha palustris

2348. Ceratocephalus testiculatus

2349. Clematis catesbyana

yellowish green; the spur 13–25 mm long, straight to slightly hooked inward or somewhat spreading, relatively stout toward the base, abruptly narrowed near the midpoint, and with a somewhat club-shaped tip, red. Stamens not showy, the anthers yellow, the innermost usually replaced by a ring of papery staminodes. Pistils 5–10, each with many ovules. Style present. Fruits follicles, the body 15–31 mm long, cylindrical, the wall thick and prominently veined, the tip spreading or curved outward at maturity, the beak 10–18 mm long. Receptacle not much enlarged at fruiting, the fruits in a ring. $2n$=14. April–July.

Scattered to common nearly throughout the state (eastern U.S. west to North Dakota and Texas; Canada). Mesic upland forests, usually on moist rocky slopes, shaded ledges of bluffs, and rock outcrops, rarely bottomland forests; also occasionally roadsides; often on calcareous substrates, rarely on sandstone.

Aquilegia canadensis is becoming increasingly popular as an ornamental in gardens. Rare plants with entirely yellow flowers have been called f. *flaviflora* (Tenney) Britton, and color forms with white or salmon-colored flowers exist in other states.

6. **Caltha** L. (marsh marigold)

Ten species, temperate regions of both hemispheres.

1. Caltha palustris L. (marsh marigold, cowslip)

C. palustris var. *palustris*

Pl. 512 a, b; Map 2347

Plants perennial herbs, with thick, fleshy roots. Stems 20–60 cm long, erect or ascending, relatively thick and hollow. Leaves in a basal rosette of 3–6 and several alternate stem leaves, long-petiolate (the stem leaves with shorter petioles than the basal ones). Leaf blades 3.0–5.5 cm long, 5–10 cm wide, simple and unlobed, circular to somewhat kidney-shaped, the base deeply cordate, the tip broadly rounded, the margins finely and usually bluntly toothed. Inflorescences of solitary flowers or open clusters of up to 7 flowers, at the stem tip. Flowers actinomorphic, perfect. Sepals 5(–9), 11–22 mm long, 5–14 mm wide, ovate to obovate, rounded at the tip, plane, yellow, not persistent at fruiting. Petals absent. Stamens prominent but scarcely showy, the anthers yellow. Staminodes absent. Pistils 3–12, each with about 15 ovules, the style rather short. Fruits cylindrical follicles, the main body 7–10 mm long, the fruit wall thick, prominently veined or not, the outer layer brown to dark brown, the beak 1–2 mm long, straight or curved. Seeds 10–15 per fruit. Receptacle not much enlarged at fruiting, the fruits in a ring. $2n$=24–80. April.

Known thus far only from Lafayette County (northern U.S. [including Alaska] south to California, Missouri and North Carolina; Canada, Europe, Asia). Fens, in saturated soil, sometimes emergent aquatics.

Caltha palustris was first reported for Missouri by Gremaud (1988), based on his investigations of tiny, remnant fen communities entirely surrounded by crop fields in a heavily agriculturalized part of Lafayette County. These endangered habitats have yielded a number of other regionally rare plant species, including *Berula erecta* (Huds.) Coville (Apiaceae), *Doellingeria umbellata* (Mill.) Nees (Asteraceae, reported as *Aster pubentior* Cronquist), and *Eupatorium maculatum* L. (Asteraceae, reported by Ladd [1990]), that were state records.

7. **Ceratocephalus** Pers.

Ten species, Europe, Asia, and Africa; introduced in Australia and North America.

1. Ceratocephalus testiculatus (Crantz) Roth (bur buttercup, hornseed buttercup)
Ranunculus testiculatus Crantz
Pl. 520 f, g; Map 2348

Plants annual, stemless. Leaves in a basal rosette, 0.9–3.8 cm long, 0.5–1.5 cm wide, 1–2 times dissected, broadly spatulate in outline, the ultimate segments linear, entire, narrowed or tapered to blunt points at the tip, the surfaces sparsely to densely and finely woolly. Inflorescences of solitary flowers, long-stalked. Flowers actinomorphic, perfect, the stalk 3–7(–10) cm long, sparsely to densely and finely woolly. Sepals 5, 3–6 mm long, 1–2 mm wide, plane, green, sometimes almost hidden in the dense white woolly hairs, persistent at fruiting. Petals 5, 3–5 mm long, 1–3 mm wide, plane, yellow. Stamens not showy, the anthers yellow. Staminodes absent. Pistils 20–50, each with 1 ovule. Style present, persistent. Fruits ellipsoid achenes clustered in cylindrical heads, these 9–16 mm long and 8–10 mm wide; the main body of each fruit 1.6–2.0 mm long, 1.8–2.0 mm wide, the fruit wall thick, not veined, woolly-hairy, the beak 3.5–4.5 mm long, lanceolate. Receptacle becoming elongated and cylindrical in fruit, glabrous. March–May.

Introduced, uncommon and widely scattered in northeastern Missouri (native of Europe, Asia, widely naturalized in portions of the western U.S. and sporadically eastward to Ohio). Compacted gravel of parking lots and picnic grounds in campgrounds and urban parks.

Ladd and Schuette (1990) discussed the ecology of *C. testiculatus* in Missouri. The species is gradually becoming more widespread in the state, always forming relatively dense, localized populations at highly disturbed sites. The genus has often been treated as a subgenus of *Ranunculus*, but it may be more closely related to *Myosurus*. The structure of the achene in *Ceratocephalus* is unique, with two empty lateral cavities next to the seed.

8. **Clematis** L. (virgin's bower)
(Erickson, 1943; Essig, 1990)

Plants twining or erect lianas (sometimes woody only at the base), climbing by means of tendril-like petioles and leaf rachises, or erect perennial herbs from elongate rhizomes, dioecious or with perfect flowers. Leaves opposite, simple or 1–2 times pinnately or ternately compound. Leaf blades ovate or elliptic to lanceolate in outline. Inflorescences of solitary flowers, panicles, or dense clusters of many flowers, in the leaf axils and/or at stem tips. Flowers actinomorphic. Sepals usually 4, petaloid, white, blue, purple, or greenish, plane or somewhat thickened and incurved, not persistent at fruiting. Petals absent. Stamens sometimes prominent but scarcely showy, the anthers white or yellow. Staminodes absent. Pistils 5–150, each with 1 ovule. Style present, persistent and becoming elongated at fruiting. Fruits achenes, lenticular or flattened-ellipsoidal, with a long beak, the outer wall thick, not prominently veined. Receptacle not much enlarged at fruiting. About 300 species, worldwide.

1. Perianth saucer-shaped, the sepals thin, spreading; flowers in dense clusters
    2. Leaflets entire; flowers all or mostly perfect; sepals 9–15 mm long, their inner surfaces glabrous . . . . 5. C. TERNIFLORA
    2. Leaflets toothed and often shallowly lobed; flowers mostly or all staminate or pistillate; sepals 6–10 mm long, their inner surfaces with a few scattered hairs
        3. Leaflets 3 . . . . 8. C. VIRGINIANA
        3. Leaflets 5 to many . . . . 1. C. CATESBYANA
1. Perianth bell-shaped or urn-shaped, the sepals thick, incurved; flowers solitary or in groups of 2 or 3.

4. Leaves simple and unlobed; stems erect or nearly so, not twining or climbing . . . . 3. C. FREMONTII
4. Leaves compound (the uppermost leaves sometimes simple); stems twining or climbing
   5. Sepals 32–41 mm long, the membranous, crisped, apical margin (2–)3–5 mm wide . . . . 2. C. CRISPA
   5. Sepals 15–34 mm long, the membranous, crisped, apical margin lacking or to 1.5 mm broad
      6. Leaflets glaucous; outer surface of sepals glabrous (except along margin); main veins on underside of leaf glabrous or nearly so . . . . . . 6. C. VERSICOLOR
      6. Leaflets not glaucous; outer surface of sepals pubescent; main veins on underside of leaf hairy
         7. Achenes with the beak glabrous or with a few short appressed hairs; leaflets usually leathery, with the minor veins usually forming a raised network . . . . 4. C. PITCHERI
         7. Achenes with the beak plumose with long, spreading hairs; leaflets herbaceous, the minor veins not raised . . . . 7. C. VIORNA

1. Clematis catesbyana Pursh (Catesby's leather flower, satin curls)

Map 2349

Plants mostly dioecious, sometimes incompletely so, the stems woody (at least toward the base), twining, 2–6 m long. Well-developed leaves pinnately 5-foliate, herbaceous in texture, the minor veins not raised, the leaflets toothed and usually shallowly 3-lobed, the upper surface green, the undersurface hairy along the main veins, paler but not glaucous. Flowers in dense clusters (cymes), only occasionally perfect. Perianth saucer-shaped, the sepals 6–9(–14) mm long, spreading horizontally, white or cream-colored, not thickened or leathery, the margins relatively smooth, the outer surface hairy, the inner surface with a few scattered hairs. Fruits with the beak 2.5–3.5 cm long, plumose with long spreading hairs. June–August.

Scattered in the eastern and southern portions of the Ozark Division (southeastern U.S. west to Kansas and Louisiana). Openings and margins of mesic upland forests and banks of streams and rivers; also roadsides, mostly on dolomite and chert substrates.

Plants of this taxon were originally mistaken for *C. virginiana* (Steyermark, 1963). The species was first reported from Missouri by Essig (1990).

2. Clematis crispa L. (swamp leather flower, blue jasmine)

Pl. 513 a; Map 2350

Plants with perfect flowers, the stems woody (at least toward the base), twining, 1–3 m long. Well-developed leaves pinnately 5–7-foliate or twice pinnately compound, herbaceous in texture, the minor veins not raised, the leaflets usually undivided, sometimes 2- or 3-lobed, the margins entire, the upper surface green, the undersurface glabrous, paler but not glaucous. Flowers solitary. Perianth urn-shaped, the sepals 32–41 mm long, erect to somewhat incurved, reflexed toward the tip, purplish blue, thickened and leathery, with membranous crisped margins (2–)3–5 mm wide, at least toward the tip, the outer surface hairy (at least near the margins), the inner surface glabrous. Fruits with the beak 2.0–3.5 cm long, with short, often appressed hairs. $2n$=16. April–August.

Uncommon in the Mississippi Lowlands Division and the adjacent portion of the Ozarks (southeastern U.S. west to Missouri and Texas). Bottomland forests, mesic upland forests, swamps, and banks of spring branches; also fallow fields, sometimes in sandy soil.

3. Clematis fremontii S. Watson (Fremont's leather flower)

C. fremontii var. *riehlii* R.O. Erickson

Pl. 513 g, h; Map 2351

Plants with perfect flowers, with elongate rhizomes, the stems herbaceous, erect or nearly so, not twining or climbing, 0.2–0.5(–0.7) m long. Leaves all simple, leathery in texture, the minor veins forming a raised network, the margins entire, the upper surface green, the undersurface pubescent with long white hairs along the main veins, paler but not glaucous. Flowers solitary. Perianth urn-shaped, the sepals 18–32 mm long, erect to somewhat incurved, reflexed toward the tip, white to greenish, purplish or pink, thickened and leathery, with membranous, crisped margins 0.5–2.0 mm wide, the outer surface hairy, the inner surface glabrous. Fruits with the beak 1.5–3.5 cm long, glabrous. $2n$=16. April–May.

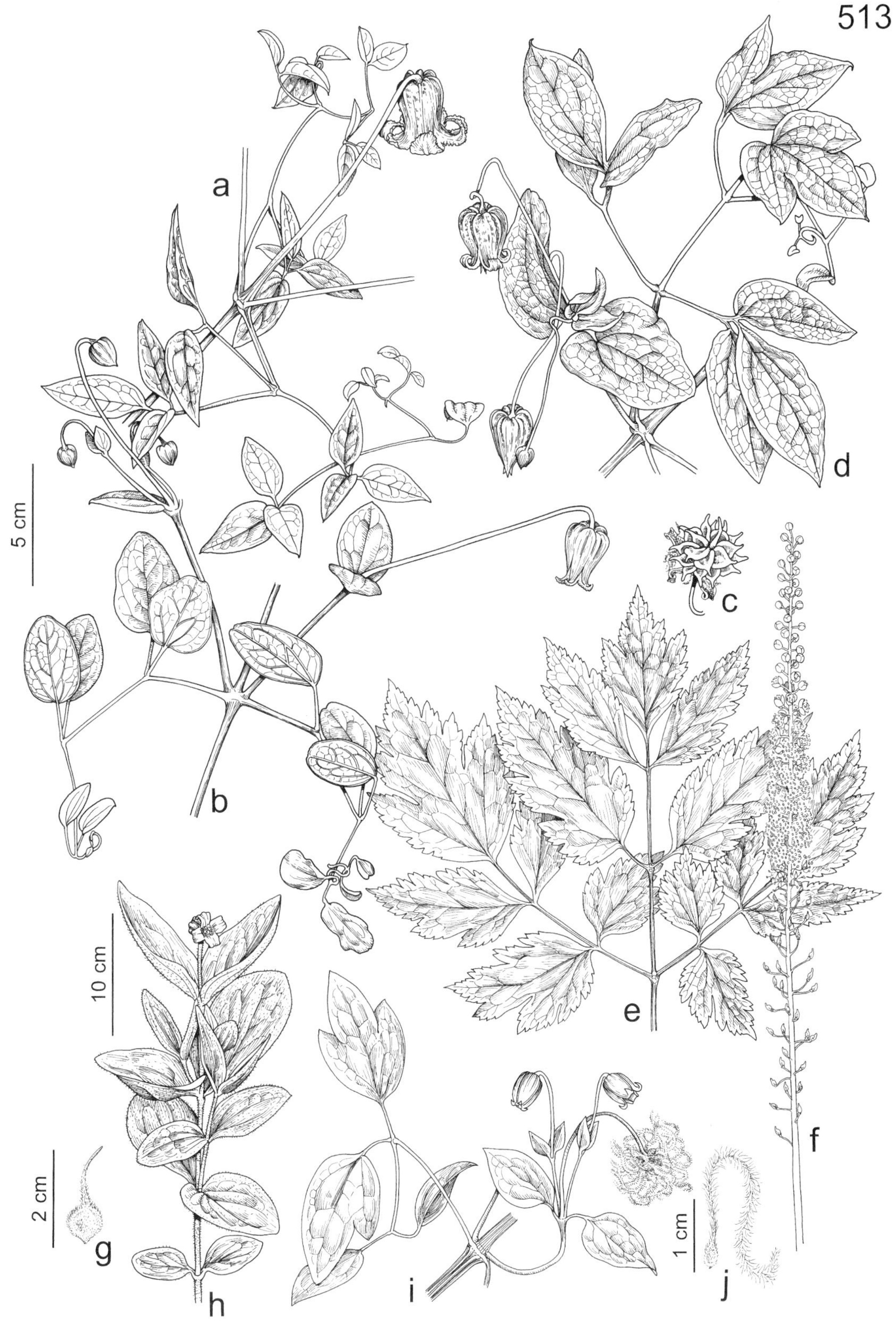

Plate 513. Ranunculaceae. *Clematis crispa*, **a)** habit. *Clematis versicolor*, **b)** habit. *Clematis pitcheri*, **c)** fruits, **d)** habit. *Actaea racemosa*, **e)** lower stem leaf, **f)** inflorescence. *Clematis fremontii*, **g)** fruit, **h)** habit. *Clematis viorna* **i)** habit, **j)** fruit.

2350. Clematis crispa

2351. Clematis fremontii

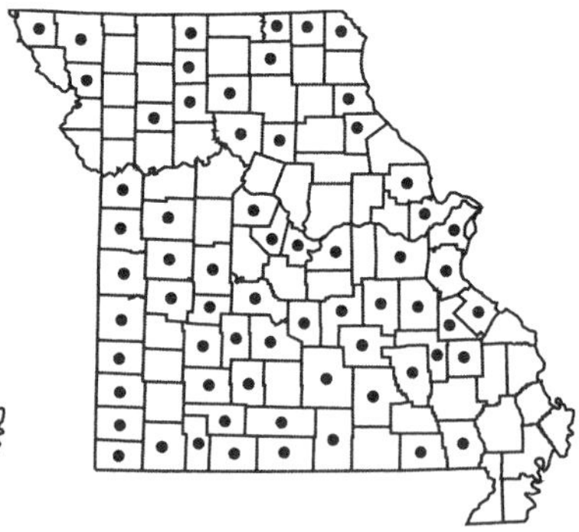

2352. Clematis pitcheri

Uncommon in the eastern portion of the Ozark Border Division and disjunctly in Ozark County (Kansas, Nebraska, Missouri, and Georgia). Dolomite glades and occasionally tops of bluffs; also occasionally roadsides.

In glades where it occurs, Fremont's leatherflower is often a conspicuous element. It is the only nontwining, herbaceous species of *Clematis* in the state. The stems persist through the winter, sometimes breaking off toward the base, and the leaves become "skeletonized," turning brown and papery, with the tissue drying and breaking away, leaving the lacelike network of veins intact. Ralph Erickson (unpublished observations), who studied Missouri populations over about a 50-year period, found that plants of this species are very slow-growing but long-lived perennials. Plants normally require 4 or more years to reach flowering size, and individuals transplanted by him as young plants in about 1940 along a transect up a gladey slope in southern Franklin County were nearly all still growing in place when revisited in 1988, but had not spread significantly.

Missouri populations of *C. fremontii* have sometimes been treated as var. *riehlii* (Steyermark, 1963), based upon supposed differences in stem height and leaf shape. Keener (1967) noted that these characters are too variable across the species' distribution to permit recognition of infraspecific taxa.

4. Clematis pitcheri Torr. & A. Gray **var. pitcheri** (leather flower, bluebell)

Pl. 513 c, d; Map 2352

Plants with perfect flowers, the stems woody (at least toward the base), twining, 1–4 m long. Well-developed leaves pinnately 4–8-foliate, green above, paler beneath but not glaucous, usually leathery in texture, the minor veins usually forming a raised network, the leaflets usually undivided, sometimes 3-lobed, the margins entire, the upper surface green, the undersurface hairy, at least along the main veins near the base, paler but not glaucous on the undersurface. Flowers solitary, rarely in groups of 3. Perianth cylindrical to urn-shaped, the sepals 18–34 mm long, erect to somewhat incurved, reflexed toward the tip, purple or blue, thickened and leathery, with thin, crisped margins 0.5–1.5 mm wide, at least toward the tip, the outer surface hairy, the inner surface glabrous. Fruits with the beak 1–3 cm long, glabrous. $2n=16$. May–August.

Scattered nearly throughout the state, but apparently absent from portions of the Mississippi Lowlands Division (Indiana to Tennessee west to Nebraska, New Mexico, and Texas; Mexico). Bottomland forests, mesic upland forests, ledges and tops of bluffs, banks of streams and rivers, and rarely bottomland prairies; also roadsides, fencerows, railroads, and fallow fields, on various substrates.

Clematis pitcheri usually is treated as consisting of two varieties. The var. *dictyota* (Greene) W.M. Dennis, is endemic to the mountains from southern New Mexico and Texas to adjacent northern Mexico. It differs from var. *pitcheri* in having smaller, thicker leaflets and less pubescent stamens.

5. Clematis terniflora DC. (virgin's bower)

C. dioscoreifolia H. Lév. & Vaniot

C. maximowicziana Franch. & Sav.

C. paniculata Thunb., an illegitimate name

Pl. 514 f–h; Map 2353

Plants with mostly perfect flowers (a few staminate flowers sometimes present), the stems woody (at least toward the base), twining, 3–6 m long. Well-developed leaves pinnately 5–7-foliate, herbaceous in texture, the minor veins not raised, the leaflets entire, the upper surface green, the undersurface glabrous or very sparsely hairy along the main veins, green or pale but not glaucous. Flowers in dense clusters (cymes). Perianth saucer-shaped, the sepals 9–15 mm long, spreading

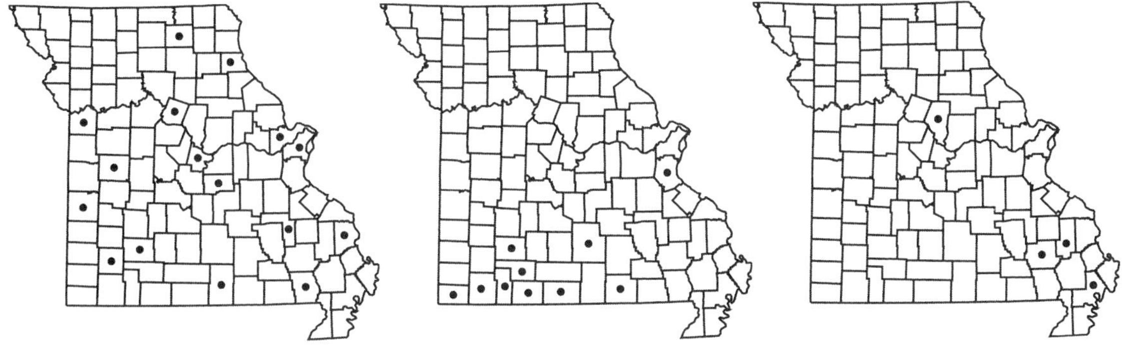

2353. Clematis terniflora 2354. Clematis versicolor 2355. Clematis viorna

horizontally, white, not thickened or leathery, the margins relatively smooth, the outer surface hairy, the inner surface glabrous. Fruits with the beak 2–6 cm long, plumose with long, spreading hairs. $2n$=16. August–September.

Introduced, scattered nearly throughout Missouri, especially in urban areas (native of eastern Asia, widely introduced in the eastern U.S. and adjacent Canada west to Minnesota, Nebraska, and Texas). Banks of streams; also fencerows, roadsides, railroads, and open disturbed areas, sometimes a weed in hedges and other woody plantings.

This species has increased in abundance in the state since Steyermark (1963) reported it from four counties in southwestern and eastern Missouri. It is cultivated for its dense attractive flowers, which have a sweet fragrance, and its "fuzzy" masses of fruits. However, it spreads aggressively by seeds in gardens and is a vigorous colonizer of disturbed habitats.

6. Clematis versicolor Small ex Rydb. (pale leather flower, leather flower)

Pl. 513 b; Map 2354

Plants with perfect flowers, the stems woody (at least toward the base), twining, 1–5 m long. Well-developed leaves pinnately 5–9-foliate, usually leathery in texture, the minor veins usually forming a raised network, the leaflets usually undivided, seldom 2- or 3-lobed, the margins entire, the upper surface green to grayish green, the undersurface pale and glaucous, glabrous. Flowers solitary or in groups of 2 or 3. Perianth more or less cylindrical, the sepals 15–18 mm long, erect or with only the apical 1–2 mm reflexed, purple, thickened and leathery, with membranous, crisped margins not evident, the inner and outer surfaces glabrous, the margins densely hairy. Fruits with the beak (3–)5–6 cm long, plumose with long, spreading hairs. $2n$=16. May–June.

Scattered in the southern portion of the Ozark Division (Kentucky to Alabama west to Oklahoma and Texas). Mesic to dry upland forests, ledges and tops of bluffs, glades, and banks of streams and rivers, most commonly on calcareous substrates, occasionally also on sandstone or chert.

7. Clematis viorna L. (leather flower, vase vine)

Pl. 513 i, j; Map 2355

Plants with perfect flowers, the stems woody (at least toward the base), twining, 1–4 m long. Well-developed leaves pinnately 5–9-foliate or twice pinnately compound, usually herbaceous in texture, the minor veins not raised, the leaflets undivided or 2- or 3-lobed, the margins entire, the upper surface green, the undersurface green or pale, hairy (sometimes very sparsely so), at least along the main veins. Flowers solitary or in groups of 2 or 3. Perianth more or less cylindrical, the sepals 17–25 mm long, erect or with only the apical 3–5 mm reflexed, purple, thickened and leathery, with thin, crisped margins absent or less than 1 mm wide, the inner surface glabrous, the outer surface hairy, the margins densely hairy. Fruits with the beak 2–6 cm long, plumose with long, spreading hairs. $2n$=16. May–June.

Scattered sites in the Ozark and Ozark Border Divisions (eastern United States, west to Missouri). Open or wooded ridgetops, bluffs, rocky slopes, around rock outcrops and edges of glades, on limestone, dolomite, and sandstone.

8. Clematis virginiana L. (virgin's bower)

Pl. 514 a–c; Map 2356

Plants dioecious, sometimes incompletely so, the stems woody (at least toward the base), twining, 2–6 m long. Well-developed leaves 3-foliate, herbaceous in texture, the minor veins not raised, the leaflets toothed and usually shallowly 3-lobed, the upper surface green, the undersurface glabrous or appressed-hairy along the main veins, paler but not glaucous. Flowers in dense clusters (cymes), imperfect. Perianth saucer-shaped, the sepals 6–10 mm long, spreading horizontally, white or cream-colored, not thickened or leathery, the margins relatively smooth, the outer surface hairy, the inner surface with a few scattered hairs. Fruits with the beak 1.5–4.0 cm long, plumose with long spreading hairs. July–August.

Scattered nearly throughout the state (eastern U.S., west to the Dakotas and Texas). Bottomland forests, mesic upland forests, swamps, banks of streams and rivers, and bases of bluffs; also fencerows, railroads, roadsides, and open, disturbed areas.

Plants in which the undersides of the leaflets are particularly densely pubescent have been called f. *missouriensis* (Rydb.) Fernald.

9. **Delphinium** L. (larkspur)

Plants taprooted annuals or perennial herbs with rhizomes or fascicles of fibrous to sometimes somewhat tuberous roots, occasionally dioecious but more commonly with perfect flowers. Stems erect or ascending, unbranched or with ascending branches. Leaves alternate and in a basal rosette (this sometimes absent at flowering time), when present the basal leaves usually larger than the stem leaves and with longer petioles. Leaf blades circular to broadly ovate, pentagonal, or kidney-shaped in outline, deeply palmately and/or ternately 2 or more times divided, the ultimate segments linear to lanceolate, oblanceolate, or narrowly oblong. Inflorescences terminal and sometimes also axillary racemes, these sometimes clustered into panicles, with few to many flowers. Flowers zygomorphic, perfect or occasionally imperfect. Sepals 5 (rarely more), petaloid, usually blue or purple to bluish- or purplish-tinged, less commonly pink or white, the upper sepal spurred, the lateral and lower sepals plane, not persistent at fruiting. Petals technically 2 (but then fused into a single 3-lobed structure) or 4 and not fused (the body then tapered abruptly to a stalklike base), white, to blue, purple, or pink, the fused petals or upper free petals spurred, the spur enclosed in the spur of the upper sepal. Stamens not showy, the anthers yellow to brown or almost black. Staminodes absent. Pistils 1 or 3(–5), each with 8–20 ovules. Style present, persistent at fruiting. Fruits follicles, more or less cylindric and somewhat curved, with a beak 1–4 mm long, the outer wall thick, prominently veined or not. Receptacle not enlarged at fruiting. Seeds semicircular to more or less trapezoidal in outline, sometimes with a longitudinal wing along 1 side, irregularly 3- or 4-angled in cross-section (sometimes appearing somewhat flattened), the surface sometimes scaly. About 340 species, North America, Europe, Asia, Africa.

Delphinium is here treated broadly to include the approxiametely 40 Old World species formerly segregated by some authors into the genus *Consolida* (DC.) Gray. These annuals differ from the remaining species of *Delphinium* in their flowers with only two petals that are fused into a 3-lobed structure and with only a single pistil. Recently, the molecular phylogenetic study of *Delphinium* and its relatives by Jabbour and Renner (2011) presented strong evidence that the *Consolida* group represents a specialized lineage within *Delphinium* that experienced a rapid diversification of species in the Mediterranean and adjacent regions.

Several larkspurs are cultivated as garden ornamentals and a number of cultivars exist, including some with doubled perianths. However, the exotic larkspurs generally do better in climates with the summers cooler than in Missouri, and only members of the *D. ajacis* group (rocket larkspurs) are grown very much in Missouri gardens. Aside from these, the principal species available horticulturally is the Eurasian *D. elatum* L. (tall larkspur), but *D. grandiflorum* L. (Chinese delphinium) and a variety of hybrids also are sold. In recent years, the native species of *Delphinium*, which are better adapted than the Old World taxa to Missouri's hot summers and cold winters, also have been brought into cultivation and are sold at some wildflower nurseries.

Members of the genus are rich in alkaloids, especially delphinine, which renders them toxic to livestock and humans. For the same reason, larkspurs are considered relatively free from browsing by deer.

Plate 514. Ranunculaceae. *Clematis virginiana*, **a)** flower, **b)** fruit, **c)** habit. *Delphinium consolida*, **d)** flower, **e)** habit. *Clematis terniflora*, **f)** flower, **g)** fruit, **h)** habit. *Delphinium exaltatum*, **i)** node with bract and fruits, **j)** upper and lower portions of habit.

1. Pistil 1; petals technically 2, but fused into a single lobed structure; plants annual
   2. Inflorescences narrow racemes; bract subtending the lowermost flower stalk on each inflorescence divided into several to many segments (occasionally only 3); inflorescence axis with some hairs having bulbous yellow bases, the nonbulbous hairs spreading or curved upward (toward the stem tip) to appressed . . . . 1. D. AJACIS
   2. Inflorescences open panicles; bracts subtending the flower stalks all undivided or occasionally the lowermost 3-parted; inflorescence axis without bulbous-based hairs, all of the hairs slender to the base, appressed, bent downward (toward the stem base) . . . . 3. D. CONSOLIDA
1. Pistils 3(–5); petals 4, free; plants perennial
   3. Stem moderately to densely short-hairy, sometimes also glandular; flower stalks erect or nearly so, appearing more or less appressed to the inflorescence axis, at least in the lower half of the inflorescence; seeds with the outer surface covered with dense, tall, thin ridges . . . . 2. D. CAROLINIANUM
   3. Stem glabrous or nearly so, at least below the inflorescence; flower stalks ascending, spreading at an angle from inflorescence axis; seeds with the outer surface finely wrinkled, but not appearing ridged
      4. Stems with 10–20 leaves above the basal rosette; most leaves with the blade divided to within 4–12 mm of the base (sometimes a minority of leaves cleft to the base or nearly so); ultimate segments of the leaf blades 7–17 mm wide; basal leaves absent at flowering . . . . 4. D. EXALTATUM
      4. Stems with 1–4 leaves above the basal rosette; most or all of the leaves with the blade divided to the base (sometimes a few leaves divided only to within 1–2(–3) mm of the base); ultimate segments of the leaf blades 1–11 mm broad; basal rosette of leaves present at flowering (but sometimes withered)
         5. Fruits erect, those developing from a given flower more or less parallel; inflorescence glabrous (sometimes tips of flower stalks minutely hairy); stalks of the lower flowers 3.0–9.5 cm long . . . . 5. D. TRELEASEI
         5. Fruits spreading, those developing from a given flower divergent; inflorescence axis short-hairy (except sometimes at the base); stalks of the lower flowers 0.5–2.5(–5.8) cm long . . . . 6. D. TRICORNE

1. Delphinium ajacis L. (garden larkspur, rocket larkspur)

Consolida ajacis (L.) Schur

C. ambigua (L.) P.W. Ball & Heyw., misapplied

Pl. 515 a, b; Map 2357

Plants annual, with taproots. Stems 30–80 (–100) cm long, pubescent (often very sparsely so) with slender, upward-curved or -appressed hairs, sometimes also glandular. Basal leaves absent at flowering; stem leaves 4 to several. Leaf blades 2–6 cm long, 1.5–6.0 cm wide, dissected 2 or 3 times, the ultimate segments 0.4–1.0 mm wide, linear, angled or tapered to a bluntly or sharply pointed tip. Inflorescences narrow racemes, the axis moderately to densely pubescent with short, spreading to upward-curved or -appressed slender hairs and scattered, longer spreading hairs with orangish bulbous bases, the longest flower stalks 10–32 mm, bract subtending the lowermost flower stalk on each inflorescence divided into several to many segments (occasionally only 3), progressively less divided toward the tip. Sepals 10–15 mm long, usually purplish blue, less commonly pink or white, the upper sepal spur 15–20 mm long. Corollas technically of 2 petals, but these fused into a single structure with 2 rounded lateral lobes and a strongly 2-cleft central lobe, 4–8 mm long, blue to purple, pink, or white, lacking a beard. Pistil 1. Fruits with the body 12–18 mm long, hairy, tapered abruptly into the beak, the beak 1.8–2.1 mm long, densely hairy, at least some of the hairs bulbous-based. May–June.

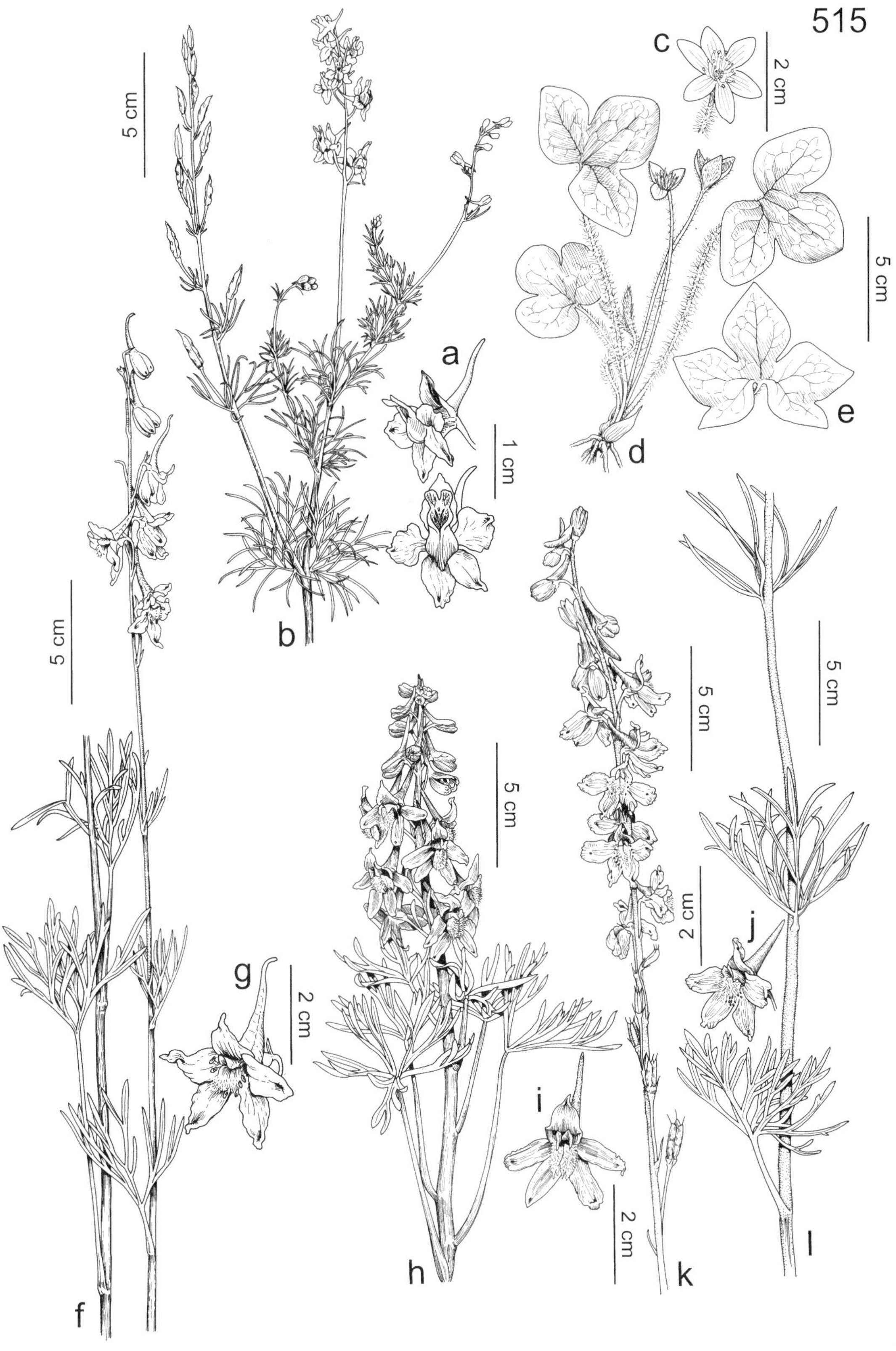

Plate 515. Ranunculaceae. *Delphinium ajacis*, **a)** flowers, **b)** habit. *Anemone americana*, **c)** flower, **d)** habit. *Anemone acutiloba*, **e)** leaf. *Delphinium treleasei*, **f)** upper and lower portions of habit, **g)** flower. *Delphinium tricorne*, **h)** habit, **i)** flower. *Delphinium carolinianum*, **j)** flower, **k)** inflorescence, **l)** lower portion of stem with leaves.

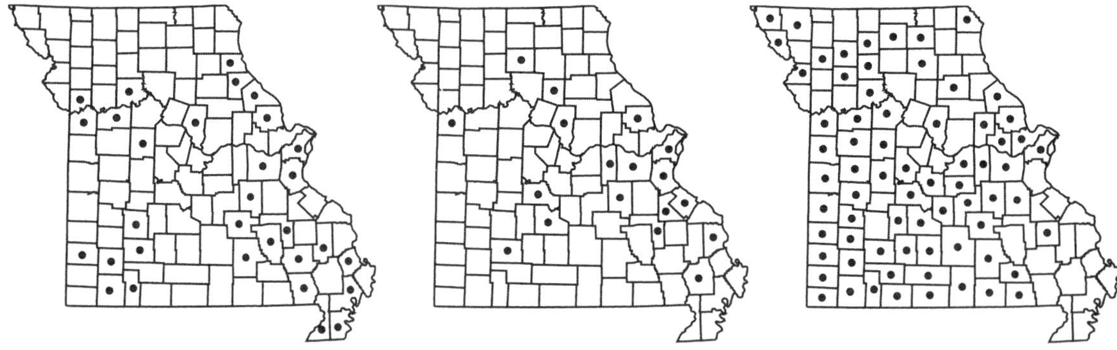

2356. Clematis virginiana 2357. Delphinium ajacis 2358. Delphinium carolinianum

Scattered, mostly in the eastern half of the state (native of Europe, Africa, introduced in temperate areas throughout the world). Banks of streams and rivers and ledges of bluffs; also pastures, farmyards, railroads, roadsides, and open, disturbed areas.

Delphinium orientale J. Gay ex Des Moul. (*Consolida orientalis* (J. Gay ex Des Moul.) Schrödinger), which was reported from Missouri by Warnock (1997a), resembles *D. ajacis* in most of its features. It is a smaller plant than *D. ajacis*; its stems are more pubescent; its leaves are only 2–4 cm long (vs. 2–6 cm); the slender-based hairs of its inflorescences are normally spreading or downward-curved; the uppermost bractlet of each flower stalk is elongate and strongly overlaps the base of the flower; the sepals are purple, with spurs only 7–10 mm long; and the tip of the follicles are truncate below the beak. Missouri specimens that have been called *D. orientale* have the long uppermost bractlet that overlaps the base of the flower, but are large plants that resemble *D. ajacis* in most other characters. They probably represent ornamental strains of *D. ajacis* that have been crossbred with *D. orientale* to get a more purplish perianth color (wild *D. ajacis* has dark blue sepals, but white, pink, and purple flowers can be seen in cultivated ornamentals and escaped populations of garden origin). Morphologically typical *D. orientale* has not been seen from Missouri.

2. Delphinium carolinianum Walter (Carolina larkspur, wild blue larkspur, prairie larkspur)

Pl. 515 j–l; Map 2358

Plants perennial, the roots fibrous, sometimes relatively stout, not tuberous. Stems (20–)35–90 cm long, moderately to densely short-hairy, less commonly with longer hairs, some of the hairs usually gland-tipped. Rosette of basal leaves often absent or withered at flowering; stem leaves 5–12 above the rosette. Leaf blades 2.2–7.5 cm long, 4.5–11.0 cm wide, the deepest divisions reaching the blade base (rarely within 2 mm of the base), the ultimate segments 1–3(–5) mm wide, linear or narrowly oblanceolate, angled to a bluntly or sharply pointed tip or rounded with an abrupt, minute, sharply pointed tip. Inflorescences narrow racemes, the axis moderately to densely pubescent with minute curled hairs, the flower stalks erect or nearly so, appearing more or less appressed to the inflorescence axis, at least in the lower half of the inflorescence, the lowermost stalks 1.2–3.3 cm long, the bracts subtending the flower stalks all undivided or occasionally the lowermost 3-parted. Sepals pale to deep blue, purple to white, or greenish, the lateral sepals 7–15 mm long, the spur 12–17 mm long, slightly curved upward. Corollas of 4 free petals, these with the body 4.5–7.5 mm long, white, but often purplish- or bluish-tinged (occasionally yellowish-tinged in flowers with whiter sepals), the lower pair 2-lobed to about the midpoint, bearded on the inner surface. Pistils 3(–5). Fruits 10–19(–28) mm long, erect, those developing from a given flower more or less parallel, glabrous or finely short-hairy. Seeds 1.4–2.0 mm long, the outer surface covered with dense, tall, thin, undulating, multicellular ridges (appearing scaly), yellowish brown to brown. $2n$=16, 32. May–June.

Scattered nearly throughout the state, but absent from the Mississippi Lowlands Division and the adjacent southeastern portion of the Ozark Division (North Dakota to Colorado and Texas east to South Carolina and Florida; Canada, Mexico). Upland prairies, glades, tops of bluffs, savannas, and opening of mesic to dry upland forests, less commonly banks of streams and rivers, margins of lakes, and bottomland prairies; also railroads and roadsides.

Steyermark (1963) treated this as a complex of two species (*D. carolinianum* and *D. virescens*

2359. Delphinium consolida 2360. Delphinium exaltatum 2361. Delphinium treleasei

Nutt.), each divided into two or more varieties. The present treatment follows the biosystematic work of Warnock (1981), who treated the taxonomy of members of subsect. *Virescens* M.J. Warnock more conservatively, with only a single species and two subspecies present in Missouri (an additional two subspecies grow elsewhere to the south and east of Missouri). Even these two subspecies are not always clearly distinguishable in Missouri and some populations combine the characteristics of the two in various ways. Other characters that have been used to separate infraspecific taxa in the past, such as width of the leaf lobes and development of a wing on the seed, and pubescence details, are very variable and poorly correlated with one another, and with upper petiole length, sepal color, and habitat.

1. Petioles of the uppermost stem leaves 1–5 mm long; sepals pale to deep blue or purple, or blue and white .... 2A. SSP. CAROLINIANUM
1. Petioles of the uppermost stem leaves 6–13 mm long; sepals white or greenish white to pale blue or occasionally bright blue .... 2B. SSP. VIRESCENS

2a. ssp. carolinianum (Carolina larkspur, wild blue larkspur)

D. carolinianum var. *crispum* L.M. Perry

D. carolinianum var. *nortonianum* (Mack. & Bush) L.M. Perry

Petioles of the uppermost stem leaves 1–5 mm long. Sepals pale to deep blue or purple, or blue and white. $2n$=16, 32. May–June.

Scattered nearly throughout the state, less abundantly north of the Missouri River, but absent from the Mississippi Lowlands Division and the adjacent southeastern portion of the Ozark Division (North Dakota to Colorado and Texas east to South Carolina and Florida; Canada, Mexico). Upland prairies, glades, tops of bluffs, savannas, and opening of mesic to dry upland forests, less commonly banks of streams and rivers, margins of lakes, and bottomland prairies; also railroads and roadsides.

Warnock (1997b) suggested that this subspecies occasionally hybridizes with *D. treleasei* where the two taxa grow in proximity. However, no specimens of putative hybrids from Missouri were located during the present study.

2b. ssp. virescens (Nutt.) R.E. Brooks (prairie larkspur)

D. carolinianum ssp. *penardii* (Huth) M.J. Warnock

D. virescens Nutt.

D. virescens var. *penardii* (Huth) L.M. Perry

Petioles of the uppermost stem leaves 6–13 mm long. Sepals white or greenish white to pale blue or occasionally bright blue. $2n$=16. May–June.

Scattered in the Unglaciated Plains Division and the western half of the Glaciated Plains; uncommon in the western half of the Ozark Division east to Texas County (North Dakota to Colorado and Texas east to Wisconsin and Missouri; Canada). Upland prairies, glades, and tops of bluffs; also railroads and roadsides.

3. Delphinium consolida L. (garden rocket)

Consolida regalis Gray

Pl. 514 d, e; Map 2359

Plants annual, with taproots. Stems 20–60 (–100) cm long, pubescent (often very sparsely so) with scattered downward-appressed, slender, white hairs. Basal leaves absent at flowering; stem leaves 3 to several. Leaf blades 1–4 cm long, 0.8–4.0 cm wide, dissected 2 or 3 times, the ultimate segments 0.4–1.0 mm wide, linear, angled or tapered to a bluntly or sharply pointed tip. Inflorescences with the axis pubescent with scattered, downward-appressed slender hairs, the longest flower stalks 25–50 mm, the bracts subtending the flower stalks all undivided or occasionally the lowermost 3-parted. Sepals 8–15 mm long, purplish blue, the upper sepal spur 16–21 mm long. Corollas technically of 2 petals, but these fused into a single structure with 2 rounded lateral lobes and a strongly 2-cleft central lobe, 3–5 mm long, blue to purple, pink, or white, lacking a beard. Pistil 1. Fruits with the body 10–13 mm long, glabrous, tapered abruptly into the beak, this 1–2 mm long. June.

Introduced, uncommon, known thus far only from the city of St. Louis (native of Europe, Asia; introduced sporadically in the eastern United States). Railroads.

Delphinium consolida is a grainfield weed, its seeds being transported with seed grain. The species was first reported for Missouri by Mühlenbach (1979), but plants apparently have not persisted at the only verified location. The species is probably no longer present in the state. However, specimens intermediate between D. *ajacis* and *D. consolida* have been found at several sites in eastern Missouri. This variable group of plants usually has rather narrow panicles with upright branches; the inflorescence branches have the pubescence typical of *D. ajacis* (a mixture of spreading hairs with bulbous orange bases and curved or appressed slender hairs). Such specimens apparently represent an escaped garden ornamental of hybrid origin, derived from crosses between *D. ajacis* and *D. consolida* or closely related species. They have sometimes been identified as *D. pubescens* DC. (*Consolida pubescens* (DC.) Soó), but true *D. pubescens* is very similar to *D. consolida*, having panicles whose branches have only slender hairs that are bent backwards (toward the base of the stem) and appressed. *Delphinium pubescens* differs from *D. consolida* primarily in being more densely pubescent and in having the middle lobe of the corolla smaller and undivided or only shallowly notched at the tip.

Delphinium tenuissimum Sibth. & Sm. (*Consolida tenuissima* (Sibth. & Sm.) Soó) has also been reported from Missouri (Warnock, 1997a), but no specimens to confirm this could be located during the present study. Both *D. tenuissimum* and *D. pubescens* are rarely cultivated, and are unlikely to be introduced in Missouri.

4. Delphinium exaltatum Aiton (tall larkspur)

Pl. 514 i, j; Map 2360

Plants perennial, the roots fibrous, usually relatively stout, not tuberous. Stems 60–150(–200) cm long, glabrous or nearly so. Rosette of basal leaves absent at flowering; stem leaves 10–20 above the rosette. Leaf blades 4.5–9.5 cm long, 7.5–16.5 cm wide, the deepest divisions ending within 4–12 mm of the blade base on most leaves (sometimes a minority of leaves cleft to the base or nearly so); ultimate leaf segments 7–17 mm wide, lanceolate to oblanceolate, angled or tapered to a sharply pointed tip. Inflorescences stout racemes, occasionally grouped into panicles, the axis sparsely to densely short-hairy; the flower stalks ascending, but generally angled away from the inflorescence axis, the lowermost stalks 1.0–2.5 cm long, the bracts subtending the flower stalks usually all undivided. Sepals pale to dark blue or purple to almost white, the lateral sepals 7–10 mm long, the spur 9–13 mm long, straight, but often somewhat upward- or downward-angled, occasionally bent downward near the tip. Corollas of 4 free petals, these with the body 3–5 mm long, white, the lower pair 2-lobed to above the midpoint, bearded on the inner surface near the notch. Pistils 3(–5). Fruits 7–12 mm long, erect, those developing from a given flower more or less parallel, finely short-hairy. Seeds 1.5–2.5 mm long, the outer surface finely wrinkled, brown to dark brown. July–August.

Uncommon, known thus far only from Howell and Shannon Counties (Pennsylvania and North Carolina west to Missouri). Rocky slopes in mesic upland forests, usually on dolomite substrate.

Steyermark (1963) noted the existence of a putative Greene County record in the herbarium of Drury University, but excluded this voucher as likely to have been mislabeled. That interpretation is supported in the present work.

Nonflowering material of *D. exaltatum* is very similar to *Aconitum uncinatum*. Both species have glabrous stems and leaves divided into a few deep divisions, with teeth and small lobes on their margins. These marginal teeth and small lobes are few and relatively narrow in *D. exaltatum*, and broader and usually more numerous in *A. uncinatum*. In addition, *A. uncinatum* is taller, with an arching, not erect, stem, and the flowers are resupinate (with the stalk twisted so that the flowers are right-side up along the arched stem, thus appearing inverted in pressed specimens).

5. Delphinium treleasei Bush ex K.C. Davis (Trelease's larkspur)

Pl. 515 f, g; Map 2361

Plants perennial, the roots fibrous,usually relatively stout, not tuberous. Stems 40–80(–120) cm long, glabrous. Rosette of basal leaves present, but sometimes withered at flowering; stem leaves 1–3 above the rosette. Leaf blades 3.5–8.5 cm long, 6.5–10.5(–14.5) cm wide, the deepest divisions reaching the blade base (sometimes a few leaves divided

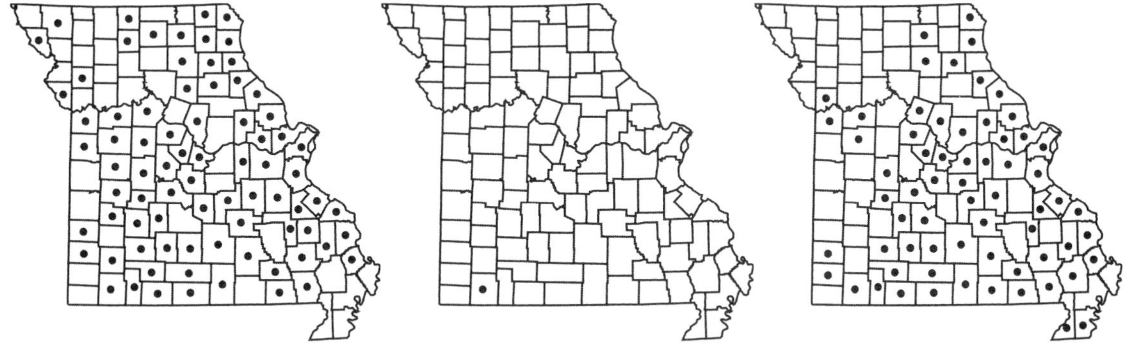

2362. Delphinium tricorne 2363. Eranthis hyemalis 2364. Hydrastis canadensis

only to within 1–2(–3) mm of the base); the ultimate leaf segments 1–4(–7) mm wide, linear, angled or tapered to a sharply or less commonly bluntly pointed tip. Inflorescences narrow racemes, occasionally grouped into panicles, the axis glabrous (the tips of the flower stalks sometimes minutely hairy); the flower stalks ascending, but generally angled away from the inflorescence axis, the lowermost stalks 3.0–9.5 cm long, the bracts subtending the flower stalks all undivided or occasionally the lowermost 3-parted, the bracts subtending the flower stalks undivided or the lowermost 3-parted. Sepals bright blue to dark blue, the lateral sepals 13–17 mm long, the spur 13–20 mm long, straight, but often slightly upward-angled. Corollas of 4 free petals, these with the body 5–8 mm long, white, sometimes tinged with purple or blue, the lower pair 2-lobed to at or below the midpoint, bearded with yellow hairs on the inner surface. Pistils 3(–5). Fruits 12–18(–22) mm long, more or less erect, those developing from a given flower more or less parallel, but arched outward near their tips, glabrous. Seeds 1.5–2.5 mm long, the outer surface finely wrinkled, brown to dark brown. $2n=16$. May–June.

Uncommon in the western half of the Ozark Division (Missouri and Arkansas). Limestone and dolomite glades, ledges and tops of bluffs, and occasionally openings and edges of dry upland forests.

Populations of *Delphinium treleasei* are most commonly associated with outcrops of the Ordovician Cotter dolomite (Warnock, 1987). This attractive species is endemic to the Ozarks.

6. Delphinium tricorne Michx. (dwarf larkspur)

Pl. 515 h, i; Map 2362

Plants perennial, the roots thickened and often somewhat tuberous. Stems 20–60 cm long, glabrous. Rosette of basal leaves present at flowering; stem leaves 2–4 above the rosette. Leaf blades 2.8–6.5 cm long, 4.5–13.0 cm wide, the deepest divisions reaching the blade base (sometimes a few leaves divided only to within 1–2(–3) mm of the base); the ultimate leaf segments 2–11 mm wide, oblong to oblanceolate or occasionally linear, rounded or angled to a bluntly pointed tip. Inflorescences narrow racemes, the axis sparsely to moderately short-hairy, occasionally glabrous; the flower stalks ascending, but generally angled away from the inflorescence axis, the lowermost stalks 0.5–2.5(–5.8) cm long. Sepals pale to deep purple or blue, often tinged with white or occasionally all white, the lateral sepals 10–17 mm long, the spur 12–18 mm long, straight, but sometimes slightly upward-angled. Corollas of 4 free petals, these with the body 6–10 mm long, blue or occasionally white, the lower pair shallowly notched or lobed near the tip, sparsely white-hairy on the inner surface toward the base. Pistils 3(–5). Fruits 14–22 mm long, more or less spreading, those developing from a given flower divergent from their bases, glabrous or nearly so. Seeds 1.5–2.5 mm long, the outer surface finely wrinkled, brown to dark brown. $2n=16$. April–May.

Scattered nearly throughout the state (eastern U.S. west to Nebraska and Oklahoma). Bottomland forests, mesic upland forests, banks of streams and rivers, bases and ledges of bluffs, and rarely margins of glades; also railroads and roadsides.

This species is an attractive, conspicuous, and common spring wildflower. Steyermark (1963) noted that none of the specimens he had observed had the sepals pure white and thus all of the Missouri materials were referable to f. *tricorne*. However, such plants with pure white sepals, which Steyermark called f. *albiflora* Millsp., have since been encountered as solitary or few individuals within populations of plants with otherwise colored sepals at scattered sites around the state.

10. **Eranthis** Salisb.

About 8 species, Europe, Asia.

1. Eranthis hyemalis (L.) Salisb. (winter aconite)

Map 2363

Plants perennial herbs, with short, stout, segmented, tuberous rhizomes. Basal leaves 1 or 2(3), usually present and green at or slightly after flowering, long petiolate, the blade 2–8 cm long, palmately compound or deeply cleft into 3 or 5 leaflets or lobes, these broadly spatulate to broadly fan-shaped or irregularly semicircular, often with 2 or 3 main lobes, the margins otherwise with few to several shorter, narrow lobes or teeth. Stems 1–8(–12) cm long at flowering, becoming somewhat elongated at fruiting, solitary from the rhizome joints, erect, unbranched, glabrous, technically leafless, but with a whorl of 3 leaflike or calyxlike bracts positioned immediately below the flower, these sessile, 1–3 cm long, otherwise similar to the leaflets of the basal leaves, persistent at fruiting. Flowers solitary, actinomorphic, perfect. Sepals 5–8, 15–22 mm long, petaloid, yellow, curved or somewhat cupped upward, usually not persistent at fruiting. Petals 5–8, inconspicuous, short, 2-lipped at the tip, yellow, nectar-secreting. Stamens prominent, relatively showy, the anthers yellow. Staminodes absent. Pistils 3–9, each with 6–9 ovules. Style present. Fruits 3–9 follicles in a ring at the tip of a short stalk, the body 8–44 mm long, oblong-lanceolate to narrowly oblong in outline, somewhat flattened, abruptly tapered to a beak 1–3 mm long, the fruit wall papery to somewhat hardened, with a conspicuous network of veins at maturity. Receptacle not elongated at fruiting. Seeds 1.6–2.2 mm long, ovoid to ellipsoid, somewhat flattened, the surface smooth, olive green. $2n=16$. January–March.

Introduced, uncommon, known thus far only from Barry County (native of Europe, Asia, introduced sporadically in the northeastern U.S.). Disturbed, rocky slopes.

Winter aconite is a beautiful garden perennial, prized for its very early flowering. The plants die back soon after the fruits mature. The species seldom escapes from cultivation in North America. In Missouri, it was first documented as an escape in 2010 by Susan Neahring and land owner Marvin Pottorff on a partially wooded slope below a residence overlooking Table Rock Lake, where plants had been growing and slowly naturalizing for at least a decade.

11. **Hydrastis** L.

One species, eastern North America.

Hydrastis appears to be a relatively early divergent, extant member of the Ranunculaceae lineage, a hypothesis supported by the available molecular data (Johansson and Jensen, 1993; Hoot, 1995). The genus has retained a number of anatomical features considered primitive within the family, as well as having flowers with 3 perianth parts. Some botanists have therefore considered it to represent a separate family intermediate between the Ranunculaceae and Berberidaceae (Tobe and Keating, 1985). Keener (1993) has argued convincingly for its continued inclusion in the Ranunculaceae. As this appears to be a case largely of lumpers vs. splitters, the traditional treatment of the genus in the Ranunculaceae is maintained in the present work.

1. Hydrastis canadensis L. (goldenseal)

Pl. 516 a–c; Map 2364

Plants perennial herbs, with thick, yellow, creeping, usually branched rhizomes. Basal leaf one, often small and scalelike or withering early when subtending fertile stems, well-developed when no fertile stem is produced, long-petiolate. Stems 15–35(–50) cm long, erect or ascending, moderately pubescent with short, fine, spreading hairs. Stem leaves 2, alternate near the stem tip, relatively short-petiolate. Leaf blades 2–6 cm long and 3–10 cm wide at flowering, developing to 12–25 cm wide at fruiting, palmately moderately to deeply (3–)5–9-lobed, broadly heart-shaped or kidney-shaped to nearly circular, the base deeply cordate, the lobes variously oblong-elliptic to elliptic, ovate, or somewhat rhombic, tapered to sharply pointed tips, the margins otherwise sharply and irregularly jagged-toothed or -lobed, the upper surface glabrous or nearly so, the undersurface

Plate 516. Ranunculaceae. *Hydrastis canadensis*, **a)** flower, **b)** habit, **c)** leaf with fruits. *Thalictrum dasycarpum*, **d)** leaf. *Thalictrum revolutum*, **e)** staminate flower, **f)** pistillate flower, **g)** leaf, **h)** habit. *Thalictrum dioicum*, **i)** node with leaf. *Trautvetteria caroliniensis*, **j)** flower, **k)** upper and lower portions of habit. *Isopyrum biternatum*, **l)** fruits, **m)** habit. *Myosurus minimus*, **n)** flower, **o)** habit.

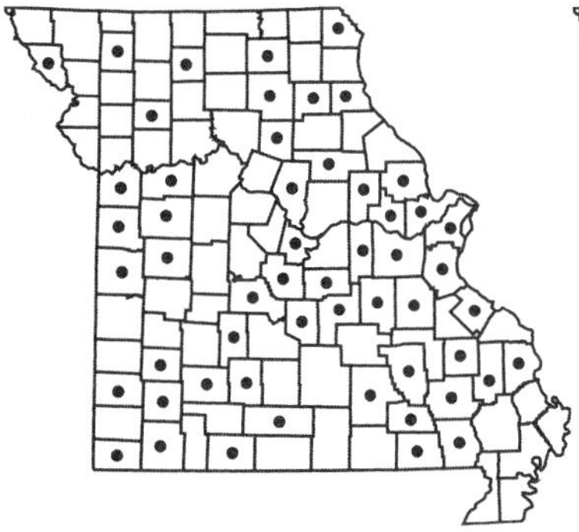

2365. Isopyrum biternatum

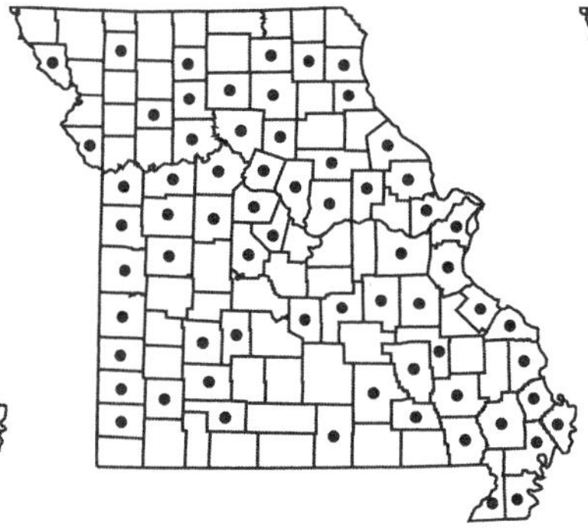

2366. Myosurus minimus

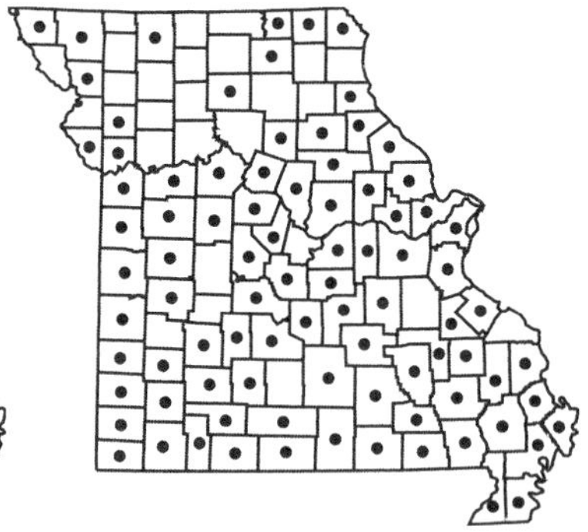

2367. Ranunculus abortivus

finely hairy along the veins. Inflorescences of solitary flowers at the stem tips, the stalk 5–35 mm long, produced as the leaves develop. Flowers actinomorphic, perfect. Sepals 3, 3.5–7.0 mm long, ovate to oval or elliptic, rounded or bluntly pointed at the tip, plane, green to creamy white, shed before the stamens and pistils become functional. Petals absent. Stamens showy, the anthers yellow. Staminodes absent. Pistils 5–15, each with 2 ovules, the style rather short. Fruits dense, globose aggregates of 5–15 berries, the clusters 15–26 mm in diameter, the individual berries 5–8 mm long, more or less globose to more commonly distended and/or somewhat flattened by adjacent berries, juicy, the outer layer thin, red, smooth or finely pebbled, not veiny, tipped with a peglike beak, this 1–2 mm long, straight, but usually somewhat angled. Seeds 1 or 2 per berry, 2.5–4.5 mm long, ellipsoid, the surface smooth, black, shiny. Receptacle not enlarged at fruiting. $2n=26$. April–May.

Scattered nearly throughout the state, but uncommon in the Unglaciated Plains Division and the western half of the Glaciated Plains (eastern U.S. west to Minnesota and Arkansas; Canada). Bottomland forests, mesic upland forests, banks of streams and rivers, and bases of bluffs.

Hydrastis canadensis makes an attractive, but relatively slow-growing, groundcover in shade gardens and is available at many wildflower nurseries. Goldenseal preparations (made from the rhizomes) are widely sold as herbal medicines, and the sale of wild-collected goldenseal generates more money than any other North American herbal remedy except ginseng. Goldenseal is becoming uncommon in many areas, probably due to commercial overcollecting by root diggers (Concannon and DeMeo, 1997). In recognition of conservation concerns about the species, international trade in the underground parts of *H. canadensis* is regulated under Appendix II of the Convention on International Trade in Endangered Species of Wild Fauna and Flora (CITES).

12. Isopyrum L.

Sixteen species, North America, Europe, Asia.

Some authors have treated the genus *Isopyrum* in a relatively restricted sense, with several species groups segregated as separate genera (Tamura and Lauener, 1968; Tamura, 1984, 1993). Among these, the eight species of *Isopyrum* in the strict sense have basally attached, tiny, nectar-secreting petals. Species that lack petals are then treated as the genus *Enemion* Raf., which includes *E. biternatum* of eastern temperate North America, four species that are each narrowly endemic in portions of the western United States and Canada, and one eastern Asian species.

1. Isopyrum biternatum (Raf.) Torr. & A. Gray (false rue-anemone)
Enemion biternatum Raf.
Pl. 516 l, m; Map 2365

Plants perennial herbs, with slender, woody rhizomes. Stems 15–35 cm long, erect or ascending. Leaves alternate and basal. Basal leaves 2 or 3, long-petiolate. Stem leaves 1–4, progressively shorter-petiolate, the uppermost sometimes sessile or nearly so. Leaf blades mostly twice ternately compound, those of the uppermost leaves sometimes only once ternate, the leaflets mostly 10–25 mm long, ovate to broadly lanceolate, deeply 2- or 3-lobed or -parted, the lobes sometimes again shallowly notched. Inflorescences of solitary flowers or loose clusters, sometimes appearing as short, leafy racemes of 2–4 flowers. Flowers long-stalked, actinomorphic, perfect. Sepals 5 or 6, 6–12 mm long, white, occasionally pinkish-tinged, plane, not persistent at fruiting. Petals absent. Stamens prominent but scarcely showy, the anthers yellow. Staminodes absent. Pistils 2–5 in a whorl, each with 2–6 ovules. Style present. Fruits follicles, obovate in outline, flattened; the body 4–5 mm long, tapered abruptly to a slender beak 1.5–2.0 mm long, the outer wall thick, prominently veined. Receptacle not much enlarged at fruiting. $2n$=14. March–May, rarely also October.

Scattered to common nearly throughout the state, but apparently absent from most of the Mississippi Lowlands Division (eastern U.S. west to South Dakota and Oklahoma; Canada). Bottomland forests, mesic upland forests, bases and ledges of bluffs, and banks of streams and rivers; also rarely roadsides.

Isopyrum biternatum is superficially very similar to *Thalictrum thalictroides*, and the two are sometimes confused by collectors. The easiest way to distinguish them is by the arrangement of stem leaves: *I. biternatum* has 1–4 alternate cauline leaves, whereas in *T. thalictroides* there is only a single, opposite pair of bracts (rarely a whorl of 3) above the basal leaves (these bracts have three leaflets each, and if they are sessile, they may resemble a whorl of six simple leaves). In addition, *T. thalictroides* tends to grow in somewhat drier habitats than does *Isopyrum biternatum*, has tuberous roots, and produces achenes rather than follicles.

13. **Myosurus** L.

Fifteen species, North America, Europe, Asia, Africa.

1. Myosurus minimus L. (mousetail)
Pl. 516 n, o; Map 2366

Plants annuals, taprooted. Stems absent. Leaves several to numerous in a basal rosette, lacking well-defined petioles, 2.0–11.5 cm long. Leaf blades simple, unlobed, linear to very narrowly oblanceolate, somewhat flattened but relatively thick, the margins entire, the surfaces glabrous. Inflorescences of solitary flowers at the tips of naked stalks, these erect or ascending, as long as or longer than the leaves. Flowers actinomorphic, perfect. Sepals (3–)5(–8), 1.5–4.0 mm long, green or pinkish- to reddish-tinged, plane, not persistent at fruiting, each with a slender spur 0.7–3.0 mm long, this appressed-descending along the tip of the flower stalk. Petals rarely absent, usually (1–)5, 1.0–2.5 mm long, slender, tapered to a stalklike base, more or less plane, white, not persistent at fruiting. Stamens not showy, the anthers yellow. Staminodes absent. Pistils numerous (to 400) in a dense spiral, each with 1 ovule. Style present. Fruits dense cylindric-tapered aggregates (1–3 mm wide) of achenes, these narrowly rhombic to elliptic or oblong in profile, 3-angled in cross-section; the body 4–5 mm long, the inner angle extended as a minute, appressed-ascending beak, the outer wall thick, not noticeably veined. Receptacle becoming elongated to 15–50 mm at fruiting. $2n$=16. March–May.

Scattered nearly throughout the state (nearly throughout the U.S.; Canada, Mexico, Europe, Asia, Africa). Banks of streams and rivers, margins of ponds, lakes, and sinkhole ponds, marshes, swamps, fens, and openings of bottomland forests; also crop fields, fallow fields, old fields, pastures, ditches, gardens, railroads, roadsides, and moist disturbed areas.

Steyermark (1963) noted that the petite plants are winter annuals with seeds that sprout in the fall and that persist through the winter as basal rosettes.

14. **Ranunculus** L. (buttercup, crowfoot)
(L. Benson, 1948, 1954; Whittemore, 1997)

Plants annual or perennial herbs, sometimes with tuberous roots or stolons. Stems variously erect to prostrate. Leaves alternate and sometimes also in a basal rosette, the blades variously simple to 2–4 times pinnately to palmately or dichotomously lobed, dissected, or compound, the segments usually more than 1 mm wide (if dissected into threadlike or linear segments then plants are submerged or emergent aquatics), the lobes, segments, or leaflets variously shaped. Inflorescences open clusters or panicles of up to 20 flowers at the branch tips or of solitary flowers, these then axillary or terminal. Flowers actinomorphic, perfect. Sepals 3–5, 1–15 mm long, variously shaped, plane or less commonly with a small, saclike appendage toward the base, green, yellowish or occasionally purple, not persistent at fruiting. Petals 3–22(–150) or in a few species sometimes absent, when present 1–27 mm long, plane or slightly cupped, yellow or less commonly white, each with a basal nectary. Stamens prominent but scarcely showy, the anthers yellow. Staminodes absent. Pistils 5 to numerous (to 300 or more), each with 1 ovule. Style absent or present. Fruits achenes, in globose to cylindrical heads or less commonly in small, dense clusters, the body of each fruit, discoid, lenticular, globose, obovoid, or cylindrical, the outer wall thick or papery, smooth to wrinkled, pebbled, or with fine, sharply pointed tubercles, the dorsal surface sometimes winged or keeled, the tip often obliquely beaked (the beak to 4.5 mm long, straight or hooked). Receptacle slightly to moderately elongated at fruiting, glabrous or hairy. About 300 species, nearly worldwide (except in the lowland tropics).

There is a great deal of morphological variation within *Ranunculus*, and several of the more divergent species groups are sometimes treated as separate genera (Tamura, 1993, 1995; Johansson, 1998). Two of the species in Missouri, *R. cymbalaria* and *R. ficaria*, might better be separated from the genus if future research confirms their phylogenetic distinctness (see further discussion under these species).

Collectors should seek to gather specimens that have both flowers and fruits. These are frequently present on the same plant. Vegetative specimens or those just starting to flower should be avoided.

1. All leaves simple and unlobed
    2. Petals 7–10, 10–15 mm long; sepals 3; roots tuberous; small, white bulbils formed in the leaf axils after flowering . . . . 8. R. FICARIA
    2. Petals 1–6, 1.5–7.0 mm long; sepals 4 or 5; roots never tuberous; bulbils absent
        3. Plants with erect flowering stems and prostrate stolons; leaf blades oblong to heart-shaped or circular; achenes with the walls thin, with strong raised veins . . . . 6. R. CYMBALARIA
        3. All stems erect or ascending, stolons absent; leaf blades ovate to lanceolate or narrowly elliptic; achenes with the walls thick, smooth
            4. Petals 1–3, 1.5–2.0 mm long; achenes 1.0–1.2 mm long . . . . . 15. R. PUSILLUS
            4. Petals 4–6, 2–6 mm long; achenes 0.8–1.0 mm long . . . 12. R. LAXICAULIS
1. Stem leaves (and often also basal leaves) deeply lobed, divided, or compound
    5. Some or all of the basal leaves simple and unlobed.
        6. Petals 4.0–7.5 mm long . . . . 10. R. HARVEYI
        6. Petals 1.5–3.5 mm long
            7. Stems glabrous; receptacle hairy, sometimes sparsely so; roots fibrous, not swollen or tuberous; leaf blades with the base more or less cordate . . . . 1. R. ABORTIVUS
            7. Stems hairy, sometimes sparsely so; receptacle glabrous; bases of some roots swollen and tuberous; leaf blades with the base usually broadly angled or truncate . . . . 13. R. MICRANTHUS
    5. Leaves all lobed, divided, or compound
        8. Styles absent; achenes 0.8–1.2 mm long, not beaked; erect plants of shallow water or very wet soil . . . . 19. R. SCELERATUS
        8. Styles present; achenes 1.6–6.5 mm long, beaked; terrestrial plants, erect or prostrate
            9. Achenes 4.0–6.5 mm long, covered with long, stiff, sharply pointed, spinescent tubercles . . . . 4. R. ARVENSIS

9. Achenes either 1.6–5.2 mm long and smooth or 1.7–3.0 mm long and pebbled or covered with short tubercles or hooked hairs
  10. Stems submerged or sometimes stranded on exposed mud (then weak and matted); leaves deeply lobed, parted, or dissected, but never clearly compound
    11. Petals white, occasionally yellow toward the base; heads of achenes 2–4 mm long . . . . 3. R. AQUATILIS
    11. Petals uniformly yellow; heads of achenes 8–10 mm long . . . 9. R. FLABELLARIS
  10. Stems erect or, if prostrate, then the plants terrestrial and the leaves compound
    12. Basal leaves simple, but usually moderately to deeply lobed or divided
      13. Petals 7–12 mm long; sepals spreading or weakly reflexed
        14. Achene beak 1.0–1.8 mm long, lanceolate; leaf blades semicircular to kidney-shaped in outline, their segments entire or scalloped; plants of wet floodplains, submerged in water or occasionally stranded on mud . . . . 9. R. FLABELLARIS
        14. Achene beak 0.2–1.0 mm long, broadly triangular and usually with a slender, tapered tip; leaves pentagonal in outline, their segments toothed or finely lobed; plants of dry, disturbed habitats . . . . 2. R. ACRIS
      13. Petals 1–5 mm long (rarely absent in *R. parviflorus*); sepals reflexed; basal leaves heart-shaped, kidney-shaped, or semicircular in outline
        15. Blades of basal leaves 2.0–7.5 cm long, 3.0–11.6 cm wide; petals 3–5 mm long; achenes with the lateral faces appearing smooth (minute pits visible under strong magnification), glabrous . . . . 16. R. RECURVATUS
        15. Blades of basal leaves 1.5–3.2 cm long, 1.0–2.4 cm wide; petals 1.1–1.8 mm long; achenes with the lateral faces covered with hooked hairs . . . . 14. R. PARVIFLORUS
    12. Basal leaves compound
      16. Sepals reflexed along a well-defined transverse fold 1–3 mm above the base
        17. Achenes with the beak slender, long-tapered, straight, 0.8–3.0 mm long; stems ascending from a spreading base, often rooting at the lower nodes . . . . 11. R. HISPIDUS
        17. Achenes with the beak oblong or triangular, curved, 0.2–0.8 mm long; stems erect, not rooting at the nodes
          18. Stems bulbous-thickened at the base; petals 9–13 mm long, 8–11 mm wide; achenes with the surfaces smooth . . . 5. R. BULBOSUS
          18. Stems not bulbous or noticeably thickened at the base; petals 7–10 mm long, 4–8 mm wide; achenes with the lateral surfaces usually pebbled or with minute papillae . . . . 18. R. SARDOUS
      16. Sepals spreading, lacking a transverse joint, sometimes reflexed from the base with age
        19. Most of the roots tuberous-thickened; leaf blades with the ultimate lobes or divisions much longer than wide, linear to narrowly oblong, bluntly pointed at their tips, the margins otherwise entire or occasionally with 1 or 2 blunt teeth . . . . 7. R. FASCICULARIS
        19. Roots fibrous, not tuberous or thickened; leaf blades with the ultimate lobes or divisions about as wide as long, ovate to rhombic or wedge-shaped, sharply pointed at their tips, the margins otherwise irregularly several-toothed

20. Achenes with the beak 0.8–1.4 mm long, curved, the dorsal margin bluntly angled to narrowly ribbed .... 17. R. REPENS
20. Achenes with the beak 0.8–3.0 mm long, straight or nearly so, the dorsal margin keeled (sometimes appearing 3-ridged) or somewhat winged .... 11. R. HISPIDUS

1. Ranunculus abortivus L. (small-flowered crowfoot, early wood buttercup)
R. abortivus var. *indivisus* Fernald
Pl. 519 h–j; Map 2367

Plants biennials or short-lived perennials. Roots not tuberous (but sometimes enlarged basally). Stems 10–70 cm long, erect or strongly ascending, not rooting at the lower nodes, glabrous, without bulbils, the base not bulbous-thickened. Basal leaves present at flowering, moderately to long-petiolate, the blade 1–4 cm long, 2–5 cm wide, kidney-shaped to more or less circular, simple, unlobed or sometimes the innermost 3-lobed, the base shallowly to deeply cordate, the margins otherwise finely to coarsely scalloped. Stem leaves sessile to short-petiolate, the blade deeply 3- or 5-lobed or -compound, the segments narrowly lanceolate to obovate, the broader ones toothed or narrowly lobed along the margins. Sepals 5, 2.0–2.5 mm long, spreading or reflexed from the base with age (lacking a transverse fold or joint), plane. Petals 5, 1.5–3.5 mm long, 1–2 mm wide, elliptic, slightly shorter than the sepals, yellow (sometimes fading to tan or white). Style present but minute. Head of achenes 3–6 mm long at maturity, ovoid-globose, the receptacle sparsely hairy. Achenes 1.4–1.6 mm long, turgid, the dorsal margin broadly and bluntly angled, the wall thick, smooth, glabrous, the beak 0.1–0.2 mm long, slender, curved. $2n=16$. March–June.

Scattered to common nearly throughout the state (nearly throughout the U.S. [including Alaska], but excluding some western states; Canada). Banks of streams and rivers, margins of ponds, lakes, and sinkhole ponds, bottomland forests, mesic to dry upland forests, bottomland prairies, loess hill prairies, and ledges of bluffs; also pastures, fallow fields, old fields, orchards, lawns, ditches, roadsides, and open, disturbed areas.

Ranunculus abortivus and *R. micranthus* are superficially very similar in appearance, but the latter species can be distinguished by its hairy lower stems and petioles, tuberous-thickened roots, glabrous receptacles, and fruits with slightly longer beaks. It also tends to occupy the drier portion of the range of habitats in which *R. abortivus* can be found. Attempts to divide *R. abortivus* into varieties are unwarranted (Whittemore, 1997).

2. Ranunculus acris L. (tall buttercup)
Pl. 518 d; Map 2368

Plants perennials. Roots not tuberous. Stems 40–90 cm long, erect or strongly ascending, not rooting at the lower nodes, glabrous or pubescent with spreading and/or appressed hairs, without bulbils, the base not bulbous-thickened. Basal leaves present at flowering, long-petiolate, the blade 1.8–6.0 cm long, 2.7–9.0 cm wide, pentagonal in outline, simple, but deeply 3- or 5-lobed or -parted, the base V-shaped, the margins otherwise irregularly lobed and/or toothed. Stem leaves grading from moderately petiolate toward the base to sessile near the tip, progressively reduced toward the stem tip, the lower leaves with the blade similar to those of the basal ones, the primary divisions wedge-shaped or rhombic, those of the upper leaves deeply 3(5)-lobed with slender, entire or few-toothed segments. Sepals 5, 4–7 mm long, spreading, plane. Petals 5, 8–11 mm long, 7–13 mm wide, broadly obovate, noticeably longer than the sepals, yellow. Style present. Head of achenes 5–7 mm long at maturity, more or less globose, the receptacle glabrous. Achenes 2–3 mm long, the dorsal margin keeled and usually narrowly winged, the wall thick, smooth, glabrous, the beak 0.3–0.7 mm long, flattened-triangular, tapered to a slender or threadlike tip, this 0.1–0.2 mm long, straight or curved. $2n=14$. May–August.

Introduced, uncommon, sporadic (native of Europe, Asia, introduced nearly throughout temperate North America). Pastures, railroads, and open, disturbed areas.

Steyermark (1963) noted that this is a problem pasture weed in the northern United States and can poison livestock. Also, milk from cattle that have ingested plants becomes tainted with a bitter flavor.

3. Ranunculus aquatilis L. **var. diffusus** With. (white water crowfoot)
R. longirostris Godr.
Pl. 517 f–h; Map 2369

Plants perennials. Roots not tuberous. Stems mostly 30–80 cm long, weak (usually submerged in water, when stranded on mud then appearing prostrate and matted), usually rooting at the lower nodes, glabrous or nearly so, without bulbils, the base not bulbous-thickened. Basal leaves absent at flowering. Stem leaves short- to moderately

Plate 517. Ranunculaceae. *Ranunculus sardous*, **a)** fruits, **b)** flower, **c)** habit. *Ranunculus hispidus*, **d)** flowers, **e)** habit. *Ranunculus aquatilis*, **f)** flower, **g)** fruits, **h)** habit. *Ranunculus fascicularis*, **i)** habit, **j)** fruits, **k)** flower.

2368. Ranunculus acris 2369. Ranunculus aquatilis 2370. Ranunculus arvensis

petiolate, the petioles usually appearing somewhat thickened or inflated, the blade 0.6–4.0 cm long, 1.5–5.0 cm wide, broadly fan-shaped to semicircular or kidney-shaped in outline, 2–4 times ternately then dichotomously dissected into linear or threadlike, sharply pointed segments, the base broadly angled to truncate or cordate, the margins otherwise entire. Sepals 5(6), 2–4 mm long, spreading or reflexed from the base with age (lacking a transverse fold or joint), plane. Petals 5, 4–10 mm long, 4–7 mm wide, obovate, noticeably longer than the sepals, white, occasionally yellow toward the base. Style present. Head of achenes 2–4 mm long at maturity, hemispheric to more or less globose, the receptacle hairy or rarely glabrous. Achenes 1.0–1.8 mm long, the dorsal margin keeled but usually unwinged, the wall thick, with coarse transverse ridges, glabrous or hairy, the beak 0.2–1.2 mm long, slender or threadlike, straight or curved. $2n$=16, 32, 48. May–July.

Scattered mainly in the Ozark Division (North America, Europe, Asia, Australia). Submerged aquatics in streams, rivers, spring branches, ponds, lakes, sloughs, swamps, and oxbows; uncommonly terrestrial when stranded in mud along receding shorelines.

Many floristic manuals have treated *R. aquatilis* as several closely related species on the basis of supposed differences in texture and leaf form, but these characters are not genetically based and instead are determined by environmental factors (Cook, 1966). Missouri material has been treated as *R. longirostris* (Steyermark, 1963), supposedly differing from *R. aquatilis* in the length of the achene beak, but beak length varies continuously from long to very short, and there is no biological basis for separating the long-beaked plants. In his treatment for the Flora of North America Project, Whittemore (1997) treated *R. aquatilis* in a broad sense, but separated some western North American populations as var. *aquatilis* based on their dimorphic leaves (submerged, highly dissected leaves and floating, less divided leaves).

Whittemore (1997) noted that, unlike most other species of *Ranunculus*, members of the aquatic subgenus *Batrachium* (DC.) A. Gray apparently are not poisonous.

4. Ranunculus arvensis L. (hungerweed, corn crowfoot)

Pl. 518 b, c; Map 2370

Plants annuals. Roots not tuberous. Stems 10–50 cm long, erect or ascending, not rooting at the lower nodes, sparsely pubescent with upward-curved hairs, without bulbils, the base not bulbous-thickened. Basal leaves present at flowering, long-petiolate, the blade 1.8–5.2 cm long, 1.5–4.2 cm wide, obovate to nearly circular, rhombic, or somewhat 5-angled in outline, simple but deeply lobed or compound, commonly 3- or 5-parted, the base cordate or V-shaped, the primary divisions or leaflets usually deeply 3-parted, the ultimate segments linear to oblanceolate or narrowly wedge-shaped, the margins otherwise entire or sharply few-toothed toward the tip. Stem leaves grading from moderately petiolate toward the base to short-petiolate or sessile near the tip, progressively reduced toward the stem tip, the lower leaves with the blade similar to those of the basal ones, those of the upper leaves often only 1 time deeply 3(5)-lobed. Sepals 5, 3.5–6.0 mm long, spreading, plane. Petals 5, 5–8 mm long, 2–4 mm wide, obovate, longer than the sepals, yellow. Style present. Head of achenes 5–8 mm long at maturity, hemispheric to more or less globose, the receptacle sparsely hairy. Achenes 4.0–6.5 mm long, the dorsal margin narrowly keeled and winged, the wall thick, covered with long, stiff, sharply pointed, spinescent tubercles, glabrous, the beak 1.5–3.8 mm long, slender, tapered, straight. $2n$=32. April–June.

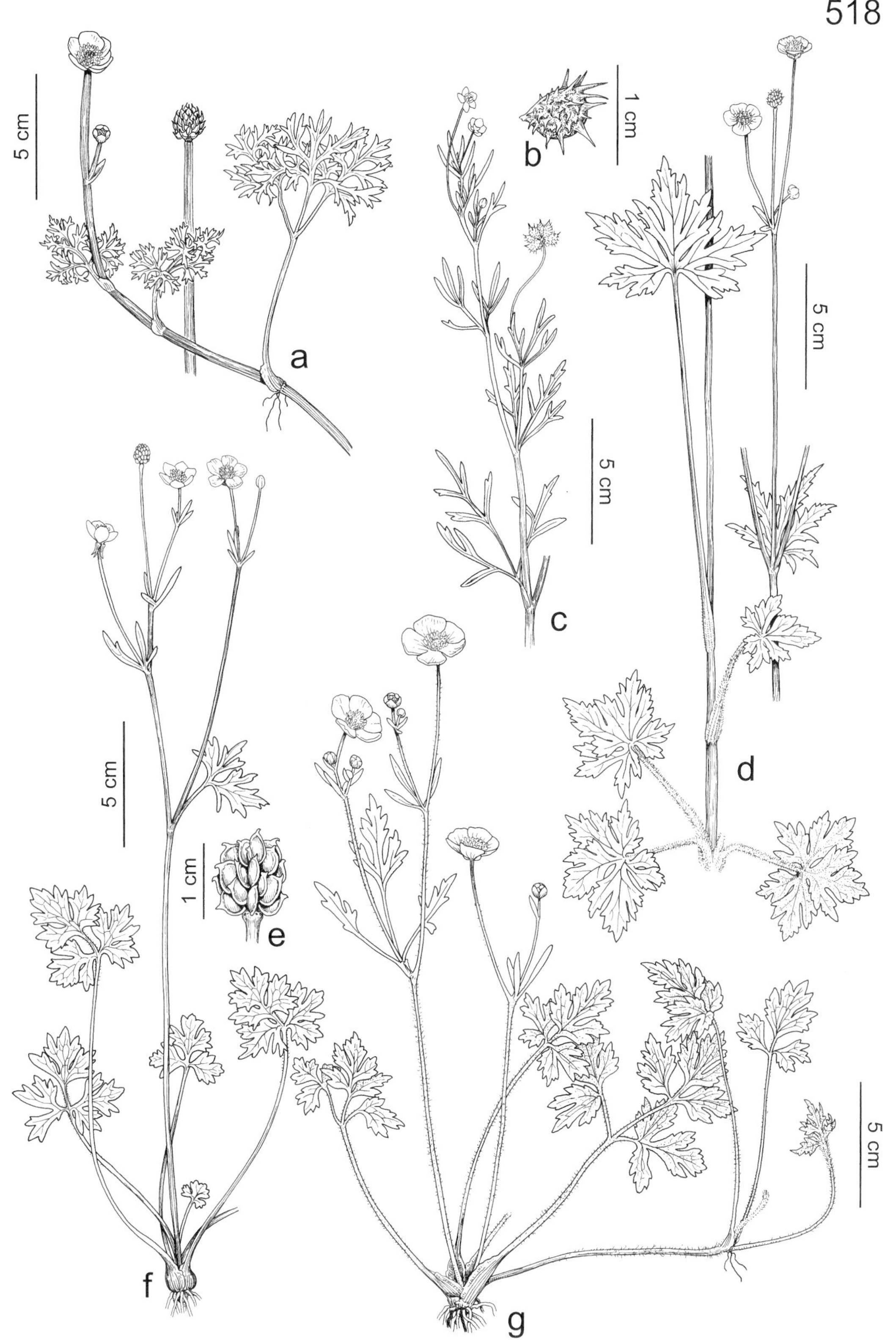

Plate 518. Ranunculaceae. *Ranunculus flabellaris*, **a)** habit with flower and detached fruits. *Ranunculus arvensis*, **b)** fruit, **c)** habit. *Ranunculus acris*, **d)** upper and lower portions of habit. *Ranunculus bulbosus*, **e)** fruits, **f)** habit. *Ranunculus repens*, **g)** habit.

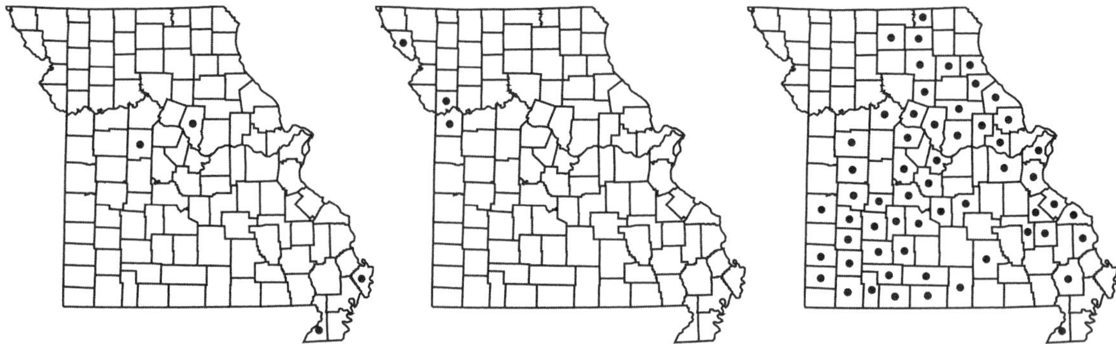
2371. Ranunculus bulbosus 2372. Ranunculus cymbalaria 2373. Ranunculus fascicularis

Introduced, uncommon, known thus far only from Dunklin County and the city of St. Louis (native of Europe, Asia; introduced in scattered portions of the U.S. [most abundantly in the western states], Canada, South America, Pacific Islands, and Australia). Railroads and open, disturbed areas.

5. Ranunculus bulbosus L. (bulbous buttercup)

Pl. 518 e, f; Map 2371

Plants perennials. Roots not tuberous. Stems 10–50(–70) cm long, erect or ascending, not rooting at the lower nodes, moderately pubescent with more or less spreading to appressed hairs, without bulbils, the base noticeably bulbous-thickened, sometimes appearing cormlike. Basal leaves present at flowering, long-petiolate, the blade 2.0–5.3 cm long, 2.5–4.5 cm wide, ovate to more or less heart-shaped in outline, usually compound (rarely only deeply divided), with 3 primary leaflets (or lobes), the base cordate, the terminal leaflet usually stalked, the primary leaflets (or lobes) deeply lobed, the segments oblong to obovate, the margins otherwise coarsely toothed, the teeth mostly rounded to bluntly pointed at their tips. Stem leaves mostly short-petiolate to sessile, progressively reduced toward the stem tip, the blade progressively deeply 3-parted into narrower, less divided and toothed, more sharply pointed segments. Sepals 5, 5–7 mm long, abruptly reflexed along a well-defined transverse fold 2–3 mm above the base, more or less plane. Petals 5, 9–13 mm long, 8–11 mm wide, broadly obovate, longer than the sepals, yellow. Style present. Head of achenes 6–9 mm long at maturity, globose to more or less ovoid, the receptacle hairy. Achenes 2.2–3.2 mm long, the dorsal margin keeled and sometimes narrowly winged, the wall thick, smooth, glabrous, the beak 0.3–0.8 mm long, flattened, lanceolate to broadly triangular, usually tapered to a slender, minute, hooked tip. $2n=16$. April–June.

Introduced, uncommon, sporadic (native of Europe, Asia, introduced widely in the U.S. [most abundantly in eastern states], Canada, South America, Pacific Islands, Australia). Pastures, fallow fields, orchards, lawns, gardens, and roadsides.

Plants of *R. bulbosus* are superficially similar to those of *R. sardous*, and can be separated most reliably by their smooth (vs. pebbled) achenes (Keener and Hoot, 1987). Steyermark (1963) noted a specimen that he collected from near Campbell (Dunklin County) in 1939 that was somewhat intermediate between the two species vegetatively, but that he determined as *R. bulbosus* based on its achenes.

6. Ranunculus cymbalaria Pursh (seaside crowfoot, shore buttercup)

R. cymbalaria f. *hebecaulis* Fernald

Halerpestes cymbalaria (Pursh) Greene

Pl. 520 l, m; Map 2372

Plants perennials. Roots not tuberous. Stems of 2 types, the main stems strongly stoloniferous, prostrate, rooting and forming rosettes of leaves at the nodes, forming mats to 70 cm in diameter, glabrous or nearly so; the flowering stems 3–15 (–25) cm long, erect or strongly ascending, leafless or nearly so above the base, not rooting at the lower nodes, glabrous or sparsely to moderately pubescent with short, loosely appressed hairs, without bulbils, the base not bulbous-thickened. Basal leaves long-petiolate, the blade 1–3 cm long, 0.8–2.6 cm wide, broadly ovate to heart-shaped, kidney-shaped, or nearly circular, simple, unlobed or shallowly lobed, the base broadly rounded to truncate or cordate, the margins otherwise scalloped. Stem leaves mostly near the base and reduced, appearing bractlike, narrowly lanceolate to lanceolate, simple, entire. Sepals 5, 3–5 mm long, spreading, more or less plane. Petals 5, 2–8 mm long, 1–3 mm wide, obovate, longer than the sepals, yellow (sometimes fading to tan or white).

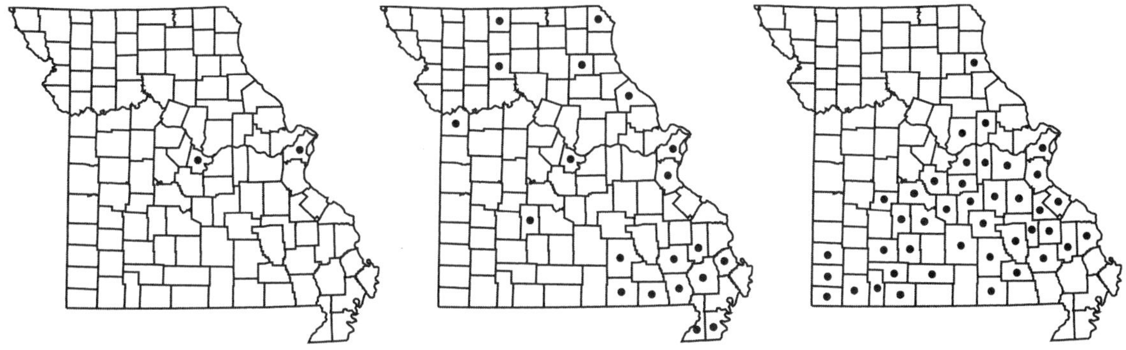
2374. Ranunculus ficaria 2375. Ranunculus flabellaris 2376. Ranunculus harveyi

Style present. Head of achenes 6–8 mm long at maturity, ovoid to cylindric, the receptacle glabrous or hairy. Achenes 1.2–1.8 mm long, the dorsal margin bluntly angled, unwinged, the wall papery, with 2–4 longitudinal ribs or prominent nerves on each side, glabrous, the beak 0.2–0.3 mm long, conic, straight. $2n=16$. July–September.

Uncommon, known only from historical collections from the northwestern portion of the state in the Big Rivers Division (western U.S. east to Michigan, Missouri, and Texas, also several northeastern states; Canada, Mexico, South America, Europe, Asia). Banks of rivers and margins of lakes, often in sandy soil.

Ranunculus cymbalaria is very different from other *Ranunculus* species in the United States, especially in the thin, papery walls of its achenes. It has sometimes been treated in a separate, small genus, *Halerpestes* Greene. Recent research suggests that it is more closely related to *Trautvetteria* (Ro and McPheron, 1997; Johansson, 1998), which also has a papery, strongly veined fruit wall. Further studies are needed before the genera can be redefined in a more natural way.

Ranunculus cymbalaria has not been collected in the state since 1934. It should be searched for on sandy washouts along the Missouri River and its major tributaries. Future floods may wash achenes or pieces of plants down from upstream that could form new populations in Missouri in the future. Native Americans used the species ceremonially, as well as medicinally to induce vomiting and to treat venereal diseases (Moerman, 1998).

7. Ranunculus fascicularis Muhl. ex Bigelow (early buttercup, prairie buttercup)

R. fascicularis var. *apricus* (Greene) Fernald

Pl. 517 i–k; Map 2373

Plants perennials. Roots a mixture of thin and fibrous and tuberous-thickened. Stems 5–30 cm long, erect or ascending, not rooting at the lower nodes, moderately to densely pubescent with spreading and/or appressed hairs (especially toward the base), without bulbils, the base not bulbous-thickened. Basal leaves present at flowering, long-petiolate, the blade 1.5–5.5 cm long, 1–4 cm wide, ovate to broadly ovate in outline, pinnately compound (the earliest leaves sometimes merely deeply lobed) with 3 or 5 primary leaflets, these variously unlobed or 2–5-lobed or -parted, the ultimate lobes or segments linear to oblong, oblanceolate, or obovate, the base rounded to more commonly angled or tapered, the margins otherwise entire or occasionally few-toothed toward the tip. Stem leaves sessile to short-petiolate, the blade entire or shallowly to deeply 3-lobed or -divided, the segments linear to narrowly oblong or narrowly oblong-oblanceolate, the margins otherwise mostly entire. Sepals 5, 5–7 mm long, spreading or reflexed from the base with age (lacking a transverse fold or joint), plane. Petals 5(–8), 8–14 mm long, 3–6 mm wide, oblong, oblong-elliptic or narrowly oblong-ovate, longer than the sepals, yellow. Style present. Head of achenes 5–9 mm long at maturity, globose to more or less ovoid, the receptacle glabrous or hairy. Achenes 2.0–2.8 mm long, turgid, the dorsal margin sharply angled, the wall thick, smooth, glabrous, the beak 1.2–2.8 mm long, slender, straight. $2n=32$. February–May.

Scattered to common in most of Missouri, but absent or uncommon in most of the western half of the Glaciated Plains Division and uncommon in the Mississippi Lowlands (eastern U.S. west to Minnesota and Texas; Canada). Mesic to dry upland forests, savannas, glades, upland prairies, dry sinkholes, and less commonly banks of streams and bottomland prairies; also pastures, lawns, farmyards, railroads, and roadsides; more abundant on acidic substrates.

Plants of *R. fascicularis* can be difficult to differentiate from those of *R. hispidus*, especially var.

hispidus. In general, *R. fascicularis* begins blooming earlier than does *R. hispidus* and tends to occur in drier habitats. Morphologically, the presence of tuberous roots is the most reliable character to distinguish *R. fascicularis* when fruits are not present, although its relatively blunt leaf divisions with few to no teeth are also more or less characteristic. In *R. fascicularis*, the petals also tend to be widest at or below the midpoint, whereas in *R. hispidus* they tend to be broadest above the midpoint.

8. Ranunculus ficaria L. **ssp. bulbifer** Lambinon (lesser celandine, pilewort)
R. ficaria var. *bulbifera* Albert
Ficaria verna Huds. ssp. *bulbifera* Á. Löve & D. Löve

Pl. 519 c, d; Map 2374

Plants perennials. Roots a mixture of slender, fibrous, and tuberous-thickened. Stems 5–30 cm long, loosely ascending, often from a spreading base, not rooting at the lower nodes, glabrous, the base not bulbous-thickened, but small, white, ellipsoid to globose bulbils forming in the leaf axils after flowering has begun. Basal leaves long-petiolate, the blade 1.8–4.0 cm long, 2–4 cm wide, broadly heart-shaped to kidney-shaped or semicircular, simple, unlobed, the base truncate or cordate, the margins otherwise scalloped to nearly entire. Stem leaves similar to the basal leaves, but smaller, short-petiolate, and with the margins more often entire. Sepals 3 or 4, 4–9 mm long, spreading, pouched or saclike at the base. Petals 7–10 (more in doubled forms), 10–17 mm long, 3–7 mm wide, broadly oblanceolate to obovate, longer than the sepals, yellow. Style absent. Fruits rarely produced (the flowers then withering soon after flowering), when present then the head of achenes up to 4 or 5 mm long at maturity, more or less hemispheric, the receptacle glabrous or short-hairy. Achenes (when produced) 2.6–2.8 mm long, turgid, not angular, the wall thick, smooth, minutely hairy, the beak absent. $2n=32$. March–May.

Introduced, presently known only from Cole and St. Louis Counties (native of Europe, introduced in the northeastern U.S. west to Wisconsin and Missouri, also Texas, Oregon, and Washington). Banks of streams and bottomland forests; also lawns, roadsides, and disturbed, usually shaded areas.

Ranunculus ficaria is not closely related to our other species of *Ranunculus* (Johansson, 1998). It and a few close relatives in Eurasia are sometimes treated as a separate genus, *Ficaria* Guett., differing chiefly in the presence of bulbils, the pouched bases of the sepals, and the absence of stylar beaks on the achenes.

Ranunculus ficaria has been divided into a complex series of subspecies in its native range (Sell, 1994). Recently, Post et al. (2009) produced a quantitative analysis of morphological variation among introduced populations in North America based on herbarium specimens and attempted to show that all five of the subspecies are present in the New World (four of these in Missouri). However, their study did not take into account seasonal and environmentally based differences in morphology, differences in ploidy among the Old World taxa, or the effects of selective breeding to produce horticultural forms. Some authors have maintained that there is too much intergradation between the subspecies to warrant formal taxonomic recognition of infraspecific taxa (Whittemore, 1997; Nesom, 2008). At a minimum, it seems possible to segregate a diploid taxon that does not produce axillary bulbils, but produces abundant viable seeds (ssp. *ficaria*) and a tetraploid taxon that produces axillary bulbils after flowering, but has mostly nonviable seeds (ssp. *bulbifer*). Notwithstanding the report of Post et al. (2009), field work in the state indicates that Missouri plants all correspond to the mostly sterile, tetraploid subspecies that produces bulbils. Further taxonomic splitting based on petal size, stem length, fruit pubescence, or other morphological characters would require corroboration from biosystematic or population-genetic studies.

Lesser celandine was introduced into the United States from Europe prior to 1867 as a garden ornamental (Post et al., 2009). It is an attractive groundcover and shade-tolerant flowering plant, and a number of cultivars are available, including some with doubled corollas. However, it can become overly aggressive in gardens, spreading by bulbils and pieces of the tuberous roots. These vegetative propagules also spread into natural plant communities through the action of wind and water, and the dumping of waste soil. The species then becomes an aggressive invader in disturbed floodplains. In Missouri, by the time that Yatskievych and Figg (1989) first reported the species as naturalized in the state, it was already well-established in several drainage systems in St. Louis County.

9. Ranunculus flabellaris Raf. (yellow water crowfoot)
R. flabellaris f. *riparius* Fernald

Pl. 518 a; Map 2375

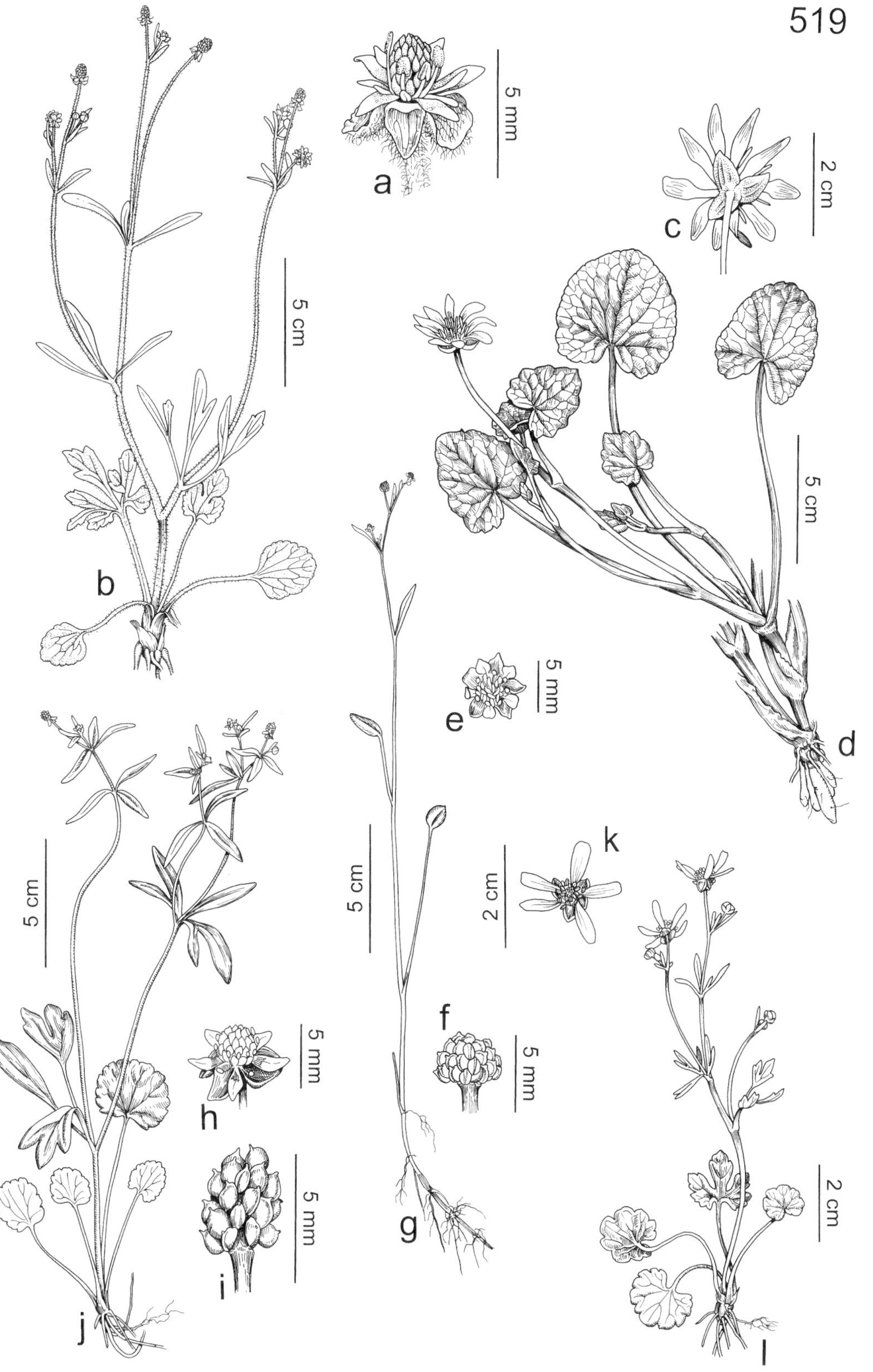

Plate 519. Ranunculaceae. *Ranunculus micranthus*, **a)** flower, **b)** habit. *Ranunculus ficaria*, **c)** flower, **d)** habit. *Ranunculus pusillus*, **e)** flower, **f)** fruits, **g)** habit. *Ranunculus abortivus*, **h)** flower, **i)** fruits, **j)** habit. *Ranunculus harveyi*, **k)** flower, **l)** habit.

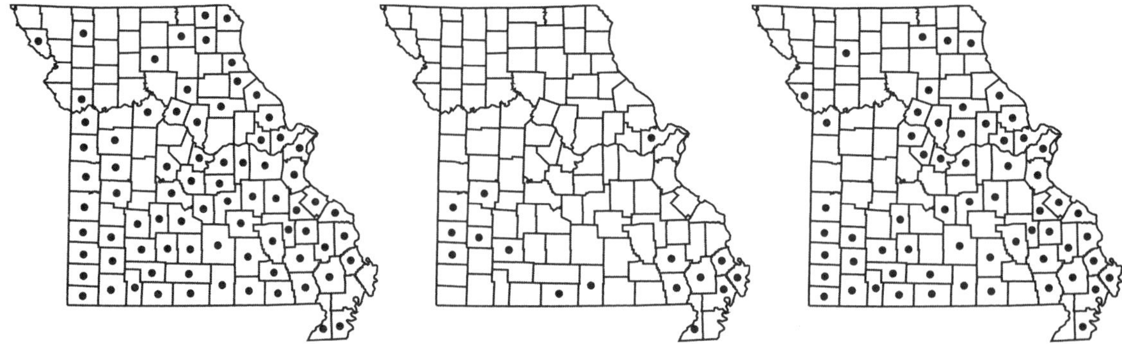

2377. Ranunculus hispidus   2378. Ranunculus laxicaulis   2379. Ranunculus micranthus

Plants perennials. Roots not tuberous. Stems mostly 30–70 cm long, shorter when terrestrial, weak (usually submerged in water, when stranded on mud then appearing prostrate), usually rooting at the lower nodes, glabrous or somewhat hairy (when terrestrial), without bulbils, the base not bulbous-thickened, but the stem sometimes appearing somewhat inflated above the base (when aquatic). Basal leaves absent at flowering when aquatic or sometimes present when terrestrial. Leaves mostly long-petiolate toward the stem base, short-petiolate to sessile near the tip, the blade 1.2–7.5 cm long, 1.5–11.0 cm wide, broadly fan-shaped to semicircular or kidney-shaped in outline; those of the submerged leaves 2–5 times pinnately, ternately, and occasionally partially dichotomously dissected into threadlike or linear flattened, mostly sharply pointed segments 0.2–2.0 mm wide; grading into those of emergent or terrestrial leaves, which are less divided, mostly ternately compound or deeply divided (the smaller leaves occasionally merely 3-lobed), truncate to deeply cordate at the base, the leaflets or segments variously undivided or 1 or 2 times shallowly to deeply lobed, variously rhombic to nearly circular, the ultimate segments rounded to sharply pointed at the tip, the margins otherwise entire or sharply few-toothed. Sepals 5, 5–8 mm long, spreading or slightly recurved, plane. Petals 5(6–14), 7–12 mm long, 5–9 mm wide, obovate, noticeably longer than the sepals, yellow. Style present. Head of achenes 7–12 mm long at maturity, ovoid to globose, the receptacle sparsely hairy. Achenes 1.8–2.2 mm long, the margins corky-thickened, especially toward the achene base, the wall thick, smooth, glabrous, the beak 1.0–1.8 mm long, flattened, lanceolate, straight. $2n$=32. April–July.

Scattered in the southeastern quarter of the state and uncommon to scattered in the northern half of the state, mostly in counties along the Missouri and Mississippi Rivers (nearly throughout the U.S. [except for some southern states]; Canada). Streams, rivers, spring branches, oxbows, swamps, ponds, lakes, sinkhole ponds, sloughs, and marshes; submerged aquatics, emergent in shallow water, or terrestrial in wet soils.

Terrestrial plants of *R. flabellaris* are similar in appearance to those of *R. sceleratus*, a taxon that is usually terrestrial but occasionally occurs in shallow water. Specimens of that species differ in their erect stems, shorter petals (2–5 mm) and nearly beakless achenes.

Ranunculus flabellaris was used medicinally by Native Americans to treat colds and respiratory ailments (Moerman, 1998).

10. Ranunculus harveyi (A. Gray) Britton **var. harveyi** (Harvey's buttercup)

R. harveyi f. *pilosus* (Benke) E.J. Palmer & Steyerm.

Pl. 519 k, l; Map 2376

Plants perennials. Roots a mixture of thin and fibrous and tuberous-thickened. Stems 10–45 cm long, erect or strongly ascending, not rooting at the lower nodes, glabrous toward the tip, usually with sparse to moderate, more or less spreading hairs toward the base, without bulbils, the base not bulbous-thickened. Basal leaves present at flowering, moderately to long-petiolate, the blade 1–3 cm long, 1.5–3.8 cm wide, more or less kidney-shaped to nearly circular, unlobed or occasionally the innermost 3-lobed, the base shallowly to deeply cordate, the margins otherwise finely to coarsely scalloped. Stem leaves with the petiole progressively shorter from base to tip, the lowermost leaves moderately to long-petiolate, those toward the stem tip usually sessile, the blade deeply 3-lobed or -compound, the segments narrowly lanceolate to obovate, the broader ones toothed or narrowly lobed along the margins. Sepals 5, 2–4 mm long, spreading or reflexed from the base (lacking a transverse fold or joint), plane or nearly so. Petals 5–7, 4.0–7.5 mm long, 1.5–3.0 mm wide, narrowly oblong-obovate, longer than the sepals, yellow (sometimes fading to tan or white). Style present. Head of achenes 3–6 mm long at maturity, hemispheric to more or less globose, the receptacle glabrous or very sparsely hairy. Achenes 1.4–1.7 mm long, turgid, the dorsal margin broadly and bluntly angled, the wall thick, smooth, glabrous, the beak 0.2–0.6 mm long, slender, straight. March–May.

Scattered in the Ozark and Ozark Border Divisions north locally to Marion County (Oklahoma to Indiana south to Louisiana and Alabama). Mesic to dry upland forests, savannas, ledges and tops of bluffs, and banks of streams; also roadsides; most abundant on acidic substrates.

The American plants are all var. *harveyi*. A second variety is endemic to Baja California, Mexico, and has been called var. *australis* (Brandegee) L.D. Benson. It differs in its uniformly stout but nontuberous roots, uniformly glabrous stems, somewhat more divided leaf blades, and achenes with slightly longer beaks (L. Benson, 1948). Our knowledge of the taxonomy of this disjunct taxon would benefit from more detailed taxonomic study.

Steyermark (1963) noted the existence of a historical specimen of *R. rhomboideus* Goldie (prairie buttercup) in the herbarium of the Academy of Natural Sciences of Philadelphia labeled as having been collected in Missouri. The collector name on the sheet's label is *Thenor* and the locality is listed as *Ft. Gratiot*. Steyermark excluded this species from the Missouri flora, based on difficulties that he had in interpreting the label data in a Missouri context. Fort Gratiot is a nineteenth-century army base located in present-day St. Clair County, Michigan, which supports Steyermark's decision to exclude the species. however, *R. rhomboideus*, which has a broad range in the north-central United States and Canada, occurs as far south as central Iowa and Nebraska; thus might plausibly be discovered growing in northern Missouri. Specimens of *R. rhomboideus* would key to *R. harveyi* in the key to species above, but differ from that species in their achenes, which are flattened (vs. turgid) toward the base and in their basal leaves, in which the blades are mostly somewhat longer than wide and variously angled to rounded or truncate at the base (vs. mostly wider than long to about as long as wide and mostly shallowly to deeply cordate).

11. Ranunculus hispidus Michx. (hispid buttercup, bristly buttercup, swamp buttercup)

Pl. 517 d, e; Map 2377

Plants perennials. Roots not tuberous. Stems 15–50(–70) cm long, variously strongly ascending to erect or lax to loosely ascending, often from a spreading base (at least with age), sometimes rooting at the lower nodes (then appearing stoloniferous, especially late in the growing season), sparsely to densely pubescent with spreading and/or appressed hairs (especially toward the base), without bulbils, the base not bulbous-thickened. Basal leaves present at flowering, moderately to long-petiolate, the blade 2–14 cm long, 2.5–16.0 cm wide, ovate to broadly ovate or nearly heart-shaped in outline, ternately compound (the earliest leaves sometimes merely deeply lobed), the 3 primary leaflets variously unlobed or 2–5-lobed or -parted, the ultimate lobes or segments narrowly elliptic or oblanceolate to circular, the base rounded to more commonly angled or tapered, the margins otherwise few- to several-toothed toward the tip. Stem leaves with progressively shorter petioles toward the stem tip, the blade moderately to deeply 3-lobed or -divided, the segments narrowly elliptic or oblanceolate to obovate, the margins otherwise usually few- to several-toothed above the midpoint. Sepals 5, 5–8 mm long, spreading or reflexed from at (then lacking a transverse fold or joint) or along a well-defined transverse fold about 1 mm above the base, plane. Petals 5, 8–16 mm long, 3–9 mm wide, obovate to oblong-obovate, longer than the sepals, yellow. Style present. Head of achenes 6–12 mm long at maturity, hemispheric to ovoid or globose, the receptacle hairy. Achenes 2.2–5.2 mm long, flattened, the dorsal margin sharply angled or narrowly winged (often the entire margin angled or winged), the wall thick, smooth, glabrous, the beak 0.8–2.6 mm long, narrowly triangular, tapered, straight or somewhat curved. $2n=32$, 64. March–June.

Scattered to common south of the Missouri River, progressively less abundant farther north (eastern U.S. west to North Dakota and Texas; Canada). Bottomland forests, mesic to dry upland forests, banks of streams and rivers, bases of bluffs, bottomland prairies, moist swales of upland prairies, margins of sinkhole ponds and sloughs, fens, acid seeps, marshes, and occasionally savannas and edges of glades; also pastures, ditches, railroads, and roadsides.

There has been a great deal of disagreement between botanists about how to classify these plants. *Ranunculus hispidus*, as defined here, has often been split up into two or three species, but different authors often disagree about how the segregates should be delimited, as well as whether they should be treated as species or varieties (L. Benson, 1948, 1954; Steyermark, 1963; T. Duncan, 1980; Nesom, 1993). The present treatment follows that of T. Duncan (1980), who recognized three varieties of *R. hispidus*, all of which have been recorded from Missouri.

1. Stems erect or nearly erect, never rooting at the nodes . . . 11B. VAR. HISPIDUS
1. Stems lax or loosely ascending, often from a spreading base, often rooting at the lower nodes
    2. Sepals spreading or reflexed from the base; achenes with the winged portion of the dorsal margin narrow, 0.1–0.2 mm wide . . . 11A. VAR. CARICETORUM
    2. Sepals reflexed from about 1 mm above the base; achenes with the winged portion of the dorsal margin relatively broad, 0.4–1.2 mm wide . . . . 11C. VAR. NITIDUS

11a. var. caricetorum (Greene) T. Duncan
R. septentrionalis Poir. var. *caricetorum* (Greene) Fernald

Stems lax or loosely ascending, often from a spreading base, often rooting at the lower nodes (then appearing stoloniferous, especially late in the growing season). Leaves with the petiole base having a stipular wing 2–4 mm wide, this rarely narrower or absent. Basal leaves with the lateral leaflets having stalklike bases 0–17 mm long. Sepals spreading or reflexed from the base, lacking a well-developed transverse fold above the base. Achenes with the winged portion of the dorsal margin narrow, 0.1–0.2 mm wide. $2n$=64. April–May.

Uncommon, widely scattered in the southern half of the state (northeastern U.S. west to North Dakota and Missouri; Canada). Fens, sloughs, and bottomland prairies, also wet, disturbed areas.

11b. var. hispidus
R. hispidus var. *marilandicus* (Poir.) L.D. Benson

Stems erect or strongly ascending, not rooting at the lower nodes. Leaves with the petiole base having a stipular wing 1–2 mm wide, this occasionally narrower or absent. Basal leaves with the lateral leaflets having stalklike bases 0–10 mm long. Sepals spreading or reflexed from the base, lacking a well-developed transverse fold above the base. Achenes with the winged portion of the dorsal margin narrow, 0.1–0.2 mm wide. $2n$=32. April–May.

Scattered in the Ozark and Ozark Border Divisions, uncommon to absent elsewhere in the state (eastern U.S. west to Illinois, Kansas, and Oklahoma; Canada). Bottomland forests, mesic to dry upland forests, bases of bluffs, banks of streams, margins of sinkhole ponds and sloughs, fens, acid seeps, marshes, and occasionally savannas and edges of glades; also pastures, railroads, and roadsides.

11c. var. nitidus (Chapm.) T. Duncan
R. carolinianus DC.
R. septentrionalis Poir.
R. septentrionalis var. *pterocarpus* L.D. Benson

Stems lax or loosely ascending, often from a spreading base, often rooting at the lower nodes (then appearing stoloniferous, especially late in the growing season). Leaves with the petiole base having a stipular wing 2–4 mm wide, this rarely narrower. Basal leaves with the lateral leaflets having stalklike bases 1–16(–32) mm long. Sepals reflexed abruptly along a transverse fold about 1 mm above the base. Achenes with the winged portion of the dorsal margin relatively broad, 0.4–1.2 mm wide. $2n$=32. March–June.

Scattered to common south of the Missouri River, progressively less abundant farther north (eastern U.S. west to South Dakota and Texas; Canada). Bottomland forests, mesic to dry upland forests, banks of streams and rivers, bases of bluffs, bottomland prairies, moist swales of upland prairies, margins of sinkhole ponds and sloughs, fens, acid seeps, marshes, and occasionally savannas and edges of glades; also pastures, ditches, railroads, and roadsides.

This is the most commonly encountered of the three varieties.

12. Ranunculus laxicaulis (Torr. & A. Gray) Darby (water-plantain, spearwort)
Pl. 520 a–c; Map 2378

Plants annuals. Roots not tuberous. Stems 15–80 cm long, erect or ascending, but often weak or

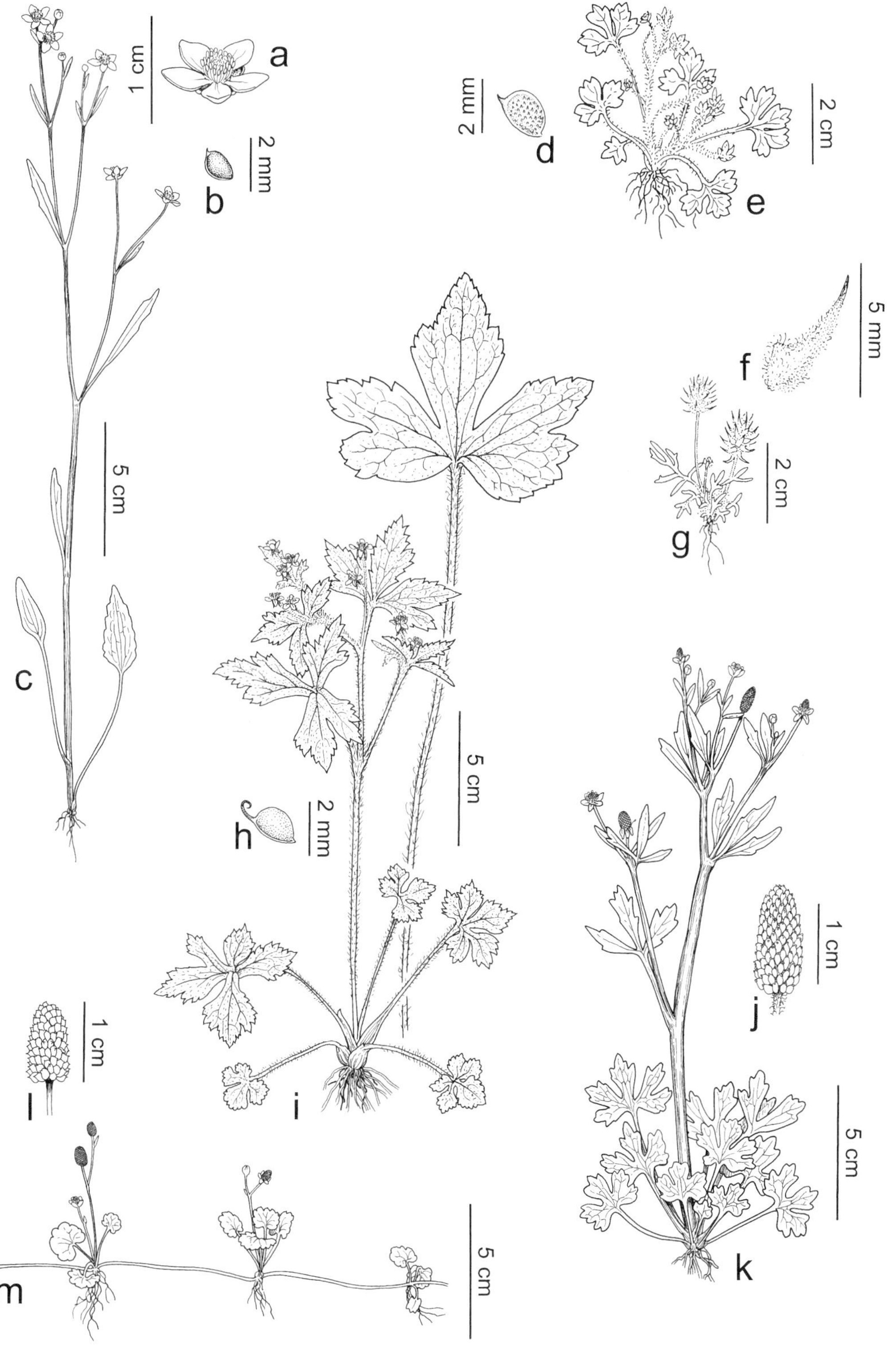

Plate 520. Ranunculaceae. *Ranunculus laxicaulis*, **a)** flower, **b)** fruit, **c)** habit. *Ranunculus parviflorus*, **d)** fruit, **e)** habit. *Ceratocephalus testiculatus*, **f)** fruit, **g)** habit. *Ranunculus recurvatus*, **h)** fruit, **i)** habit. *Ranunculus sceleratus*, **j)** fruits, **k)** habit. *Ranunculus cymbalaria*, **l)** fruits, **m)** habit.

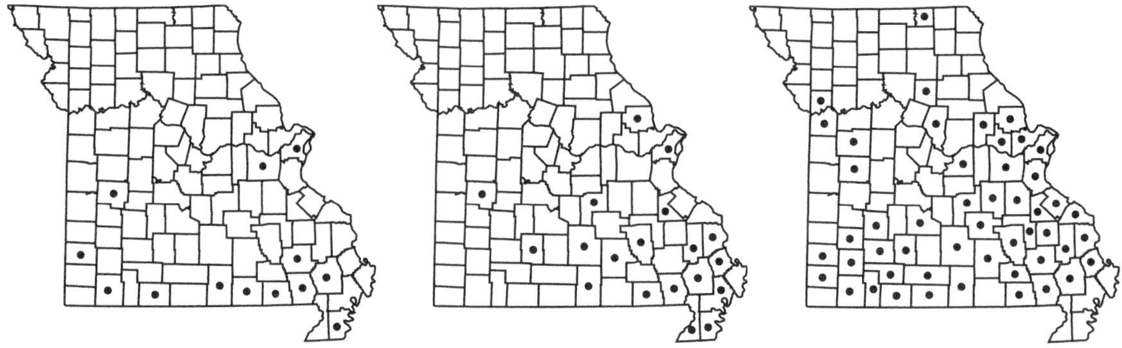

2380. Ranunculus parviflorus 2381. Ranunculus pusillus 2382. Ranunculus recurvatus

reclining in aquatic habitats, often rooting at the lower nodes, glabrous or uncommonly sparsely hairy, without bulbils, the base not bulbous. Basal leaves usually absent at flowering, when present long-petiolate, the blade 1.0–4.5 cm long, 0.6–1.8 cm wide, oblong to ovate, simple, the base rounded to truncate, truncate or angled to a bluntly pointed tip, the margins entire or shallowly toothed, the teeth mostly bluntly pointed at their tips. Stem leaves mostly sessile, the blade 1.5–7.0 cm long, 0.4–2.0 cm wide, simple, ovate to lanceolate or narrowly elliptic, the uppermost leaf blades mostly linear to narrowly oblong-elliptic, the margins entire or few-toothed toward the tip. Sepals 4 or 5, 1.5–3.0 mm long, spreading or reflexed from the base (lacking a transverse fold), more or less plane. Petals 4–6(–10), 2–6 mm long, 1–2 mm wide, oblong, about as long as to somewhat longer than the sepals, yellow. Style present, but the apical portion usually shed after flowering. Head of achenes 2–4 mm long at maturity, hemispheric to globose or ovoid, the receptacle glabrous. Achenes 0.8–1.0 mm long, the dorsal margin bluntly to sharply keeled and sometimes very narrowly winged, the wall thick, smooth, glabrous, the beak 0.1–0.2 mm long, flattened, triangular, straight or bent to the side. April–September.

Scattered in the southern third of the state, also disjunct in St. Charles County (eastern [mostly southeastern] U.S. west to Kansas and Texas). Ponds, sinkhole ponds, streams, banks of rivers, sloughs, bottomland prairies, and wet swales and depressions of upland prairies and sand prairies; also ditches and wet roadsides; terrestrial or more commonly emergent aquatics, occasionally appearing floating-leaved.

13. Ranunculus micranthus Nutt. (rock buttercup)

R. micranthus var. *delitescens* (Greene) Fernald

Pl. 519 a, b; Map 2379

Plants biennials or short-lived perennials. Roots mostly tuberous-thickened. Stems 10–50 cm long, erect or strongly ascending, not rooting at the lower nodes, sparsely to moderately pubescent with spreading hairs (more densely so toward the base), without bulbils, the base not bulbous-thickened. Basal leaves present at flowering, moderately to long-petiolate, the blade 1.0–3.3 cm long, 1–3 cm wide, broadly elliptic to broadly oblong, ovate or more or less circular, simple or compound, unlobed or sometimes the innermost 3(5)-lobed or with 3(5) leaflets, the base truncate to broadly angled or sometimes shallowly cordate, the margins otherwise finely to coarsely scalloped. Stem leaves sessile to short-petiolate, the blade deeply 3-lobed or compound with 3 leaflets, the segments narrowly lanceolate to obovate, the broader ones toothed or narrowly lobed along the margins. Sepals 5, 2–4 mm long, spreading or reflexed from the base with age (lacking a transverse fold or joint), plane. Petals 5, 1.5–3.5 mm long, 0.5–1.5 mm wide, elliptic, slightly shorter than the sepals, yellow (sometimes fading to tan or white). Style present but minute. Head of achenes 3–7 mm long at maturity, ovoid to more or less globose, the receptacle glabrous. Achenes 1.1–1.5 mm long, turgid, the dorsal margin broadly and bluntly angled, the wall thick, smooth, glabrous, the beak 0.2–0.3 mm long, slender, straight or curved. $2n=16$. March–May.

Scattered to common south of the Missouri River, uncommon farther north and apparently absent from most of the northwestern portion of the state (eastern U.S. west to Kansas and Oklahoma). Mesic to dry upland forests, savannas, glades, banks of streams, rivers, and spring branches, ledges and tops of bluffs, margins of sinkhole ponds, and less commonly bottomland forests; also fallow fields, cemeteries, ditches, roadsides, and open, disturbed areas.

Plants of *R. micranthus* are superficially very similar to those of *R. abortivus*. For further discussion, see the treatment of that species.

14. Ranunculus parviflorus L. (stickseed crowfoot, small-flowered crowfoot)
Pl. 520 d, e; Map 2380

Plants annuals. Roots not tuberous. Stems 10–30 cm long, erect to loosely ascending, not rooting at the lower nodes, moderately to densely pubescent with fine, spreading hairs, without bulbils, the base not bulbous. Basal leaves present at flowering, long-petiolate, the blade 1.5–3.2 cm long, 1.0–2.4 cm wide, heart-shaped, kidney-shaped, or semicircular, simple, the base shallowly to deeply cordate or less commonly broadly rounded, the lobes sometimes again 3-lobed, the lobes more or less rounded at the tip, the margins otherwise entire or sharply toothed. Stem leaves progressively and less divided from the stem base to tip, mostly short-petiolate or sessile. Sepals 5, 1.5–2.0 mm long, reflexed from the base (lacking a transverse fold), more or less plane. Petals 1–5 or rarely absent, when present 1.1–1.8 mm long, 0.2–0.7 mm wide, narrowly obovate to narrowly oblong-obovate, shorter than the sepals, yellow. Style present. Head of achenes 3–5 mm long at maturity, more or less globose, the receptacle glabrous. Achenes 1.7–2.0 mm long, the dorsal margin bluntly thickened, the wall thick, the lateral faces covered with hooked hairs, the beak flattened, triangular, strongly curved or hooked. $2n=32$. March–June.

Introduced, scattered in the southern half of the state, mostly abundantly along the southern tier of counties (native of Europe, Asia; introduced in the eastern U.S. west to Missouri and Texas, also the western U.S. from Washington to California; Canada, Australia, Pacific Islands). Upland prairies; also pastures, old fields, ditches, lawns, and open, disturbed areas.

15. Ranunculus pusillus Poir. (low spearwort)
Pl. 519 e–g; Map 2381

Plants annuals. Roots not tuberous. Stems 10–50 cm long, erect or ascending, but often weak or reclining in aquatic habitats, often rooting at the lower nodes, glabrous, without bulbils, the base not bulbous. Basal leaves usually absent at flowering, when present long-petiolate, the blade 1.0–4.5 cm long, 0.5–1.2 cm wide, ovate to lanceolate, simple, the base angled or tapered, occasionally truncate, rounded or more commonly angled or tapered to a bluntly or sharply pointed tip, the margins entire or shallowly toothed, the teeth bluntly or sharply pointed at their tips. Stem leaves progressively shorter and narrower from the stem base to tip, all but the lowermost sessile, the blade simple, linear to narrowly lanceolate, the margins entire or few-toothed toward the tip. Sepals 4 or 5, 1.5–3.0 mm long, spreading or reflexed from the base (lacking a transverse fold), more or less plane. Petals 1–3(–5), 1.5–2.0 mm long, 0.5–1.0 mm wide, obovate to oblong, about as long as to somewhat shorter than the sepals, yellow. Style present, but short and often shed after flowering. Head of achenes 2–8 mm long at maturity, hemispheric to ovoid or rarely cylindric, the receptacle glabrous. Achenes 1.0–1.2 mm long, the dorsal margin bluntly to sharply keeled and sometimes very narrowly winged, the wall thick, with at least a few minute tubercles or sometimes appearing somewhat pebbled, glabrous, the beak not apparent or minute (to 0.1 mm) and somewhat peglike or nipplelike, usually bent to the side. April–June.

Uncommon in the southeastern quarter of the state, rarely farther north and west (eastern United States west to Missouri and Texas, also California). Bottomland forests, swamps, sloughs, marshes, fens, banks of streams and rivers, and margins of sinkhole ponds; also ditches, fallow fields margins of crop fields, and borrow pits.

16. Ranunculus recurvatus Poir. **var. recurvatus** (hooked crowfoot)
Pl. 520 h, i; Map 2382

Plants perennials. Roots not tuberous. Stems 15–40(–60) cm long, erect or ascending, not rooting at the lower nodes, moderately to densely pubescent with fine, spreading hairs, without bulbils, the base bulbous-thickened, cormlike. Basal leaves present at flowering, long-petiolate, the blade 2.0–7.5 cm long, 3.0–11.6 cm wide, ovate to kidney-shaped, simple, moderately to deeply 3-lobed, the base shallowly to deeply cordate, the lobes oblong to rhombic, sometimes again 2- or 3-lobed, the lobes bluntly to sharply pointed at the tip, the margins otherwise scalloped or finely and mostly bluntly toothed. Stem leaves relatively few, similar to the basal leaves, but with shorter petioles (the uppermost often sessile or nearly so) and less divided blades. Sepals 5, 3–6 mm long, reflexed from at or near the base (lacking a transverse fold), more or less plane. Petals 5, 3–5 mm long, 1–2 mm wide, lanceolate to broadly oblong-oblanceolate, about as long as or slightly shorter than the sepals, yellow. Style present. Head of achenes 5–7 mm long at maturity, ovoid to more or less

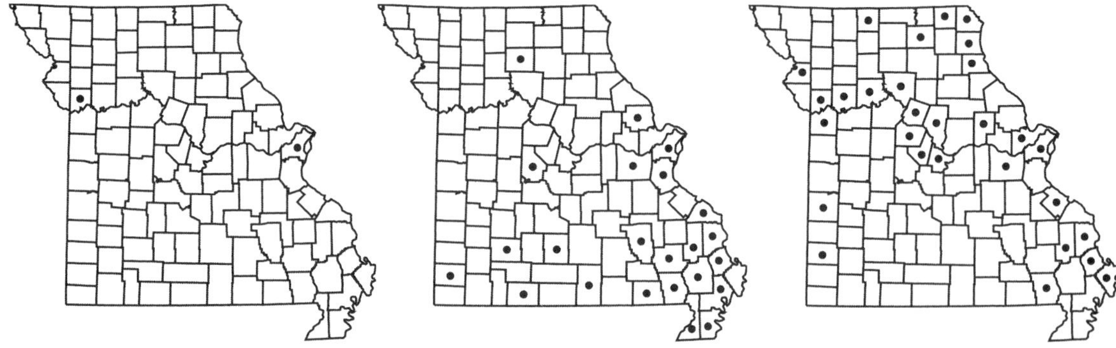

2383. Ranunculus repens 2384. Ranunculus sardous 2385. Ranunculus sceleratus

globose, the receptacle hairy. Achenes 1.6–2.2 mm long, flattened, the dorsal margin strongly keeled but usually unwinged, the wall thick, the lateral faces minutely pitted (appearing smooth except under strong magnification), glabrous, the beak flattened, lanceolate, hooked. $2n=32$. April–July.

Scattered to common south of the Missouri River, uncommon or absent farther north (eastern U.S. west to North Dakota and Texas; Canada). Bottomland forests, mesic to less commonly dry upland forests, banks of streams, rivers, and spring branches, margins of ponds and sinkhole ponds, swamps, fens, and ledges of bluffs; also moist roadsides.

This is a common spring wildflower in the Ozarks, although not a very showy one. The flowers are not elevated very far above the leaf bases and have relatively inconspicuous petals.

The other variety, var. *tropicus* (Griseb.) Fawc. & Rendle, is endemic to various Caribbean Islands and differs in its less bulbous stem bases, often somewhat less divided leaf blades with blunter teeth, slightly longer petals, and better-developed nectaries (L. Benson, 1948).

Native Americans used this species medicinally as a sedative and laxative, as well as for treating skin disorders, toothaches, and venereal discomfort (Moerman, 1998).

17. Ranunculus repens L. (creeping buttercup)

R. repens var. *pleniflorus* Fernald

Pl. 518 g; Map 2383

Plants perennials. Roots not tuberous. Stems 10–60 cm long, prostrate and mat-forming, occasionally some of the stems loosely ascending from a spreading base, rooting at the nodes, sparsely to densely pubescent with loosely appressed to spreading hairs, without bulbils, the base not bulbous. Basal leaves present at flowering, long-petiolate, the blade 1.0–8.5 cm long, 1.5–10.0 cm wide, ovate to kidney-shaped, usually compound, with 3 primary leaflets, the central leaflet usually stalked, the base mostly deeply cordate, the leaflets obovate to bluntly rhombic, 1 or 2 times again 3-lobed or -parted, the ultimate segments obovate to elliptic or occasionally narrowly oblong, bluntly to sharply pointed at the tip, the margins otherwise irregularly toothed. Stem leaves similar to basal leaves but with shorter petioles and somewhat less divided blades having narrower segments. Sepals 5, 4–7 mm long, spreading or reflexed from the base (lacking a transverse fold), more or less plane. Petals 5 (more in doubled forms), 6–18 mm long, 5–12 mm wide, obovate to broadly obovate, longer than the sepals, yellow. Style present. Head of achenes 5–10 mm long at maturity, ovoid to more or less globose, the receptacle hairy or rarely glabrous. Achenes 2.6–3.2 mm long, flattened, the dorsal margin bluntly angled or narrowly ribbed, unwinged, the wall thick, smooth, glabrous, the beak 0.8–1.4 mm long, flattened, triangular to narrowly triangular, usually curved. $2n=14, 32$. May–September.

Introduced, uncommon, sporadic, known only from historical specimens (native of Europe, Asia, introduced widely in North America south to South America, also Australia, Pacific Islands; in the U.S., widespread [including Alaska] except in the Great Plains region). Mesic upland forests; also moist, disturbed areas.

This species is grown as an ornamental and groundcover in gardens. A horticultural form with doubled corollas has been called var. *pleniflorus* and was collected once as an escape in Clay County (Steyermark, 1963).

18. Ranunculus sardous Crantz (hairy buttercup)

Pl. 517 a–c; Map 2384

Plants annuals. Roots not tuberous. Stems 5–50 cm long, erect or ascending, not rooting at the lower nodes, moderately to densely (rarely sparsely) pubescent with more or less spreading hairs, without bulbils, the base not bulbous. Basal leaves usually present at flowering (sometimes withered at fruiting), long-petiolate, the blade 1–7 cm long, 1–6 cm wide, ovate to more or less heart-shaped in outline, usually compound (rarely only deeply divided), with 3 leaflets (or lobes), the base rounded to cordate, the terminal leaflet often short-stalked, the primary leaflets (or lobes) usually again 2- or 3-lobed, the segments oblong to obovate or rhombic, the margins otherwise coarsely toothed, the teeth mostly rounded to bluntly pointed at their tips. Stem leaves mostly short-petiolate to sessile, progressively reduced toward the stem tip, the blade progressively deeply 3-parted into narrower, less divided and toothed, more sharply pointed segments, those of the uppermost leaves linear to narrowly oblong-oblanceolate. Sepals 5, 3–6 mm long, reflexed from ca. 1–3 mm above the base along a well-defined transverse fold, more or less plane (the apical portion sometimes appearing somewhat folded longitudinally). Petals 5, 7–10 mm long, 4–8 mm wide, obovate to broadly oblong-obovate, longer than the sepals, yellow. Style present. Head of achenes 5–8 mm long at maturity, globose to ovoid, the receptacle hairy. Achenes 2–3 mm long, the dorsal margin keeled and narrowly winged, the wall thick, usually pebbled or with minute papillae, occasionally smooth, glabrous, the beak 0.4–0.7 mm long, flattened, oblong to triangular, curved. $2n=16$. April–June, rarely July–October.

Introduced, scattered in the Mississippi Lowlands, Ozark, and Ozark Border Divisions and sporadic in the Glaciated Plains (native of Europe, Asia; introduced in the eastern U.S. west to Oklahoma and Texas, also in the western U.S. from Washington to California; Canada, Australia, Pacific Islands). Banks of streams and rivers, sloughs, acid seeps, margins of sinkhole ponds, and openings of bottomland forests; also ditches, levees, pastures, fallow fields, edges of crop fields, farm yards, railroads, roadsides, and open, disturbed areas.

Steyermark (1963) knew *R. sardous* from Missouri based only on a single historical specimen from St. Louis County, but commented that it was then already relatively common in adjacent portions of southern Illinois. It has spread dramatically in southern Missouri since the 1960s. This species is avoided by cattle and can become quite dense and showy in heavily grazed pastures.

19. Ranunculus sceleratus L. **var. sceleratus** (cursed crowfoot)

Pl. 520 j, k; Map 2385

Plants annuals. Roots not tuberous, relatively slender but often somewhat fleshy. Stems 10–60 cm long (longer in rare aquatic plants), erect or strongly ascending, rarely rooting at the lowermost nodes, glabrous, without bulbils, often appearing somewhat inflated, the base not bulbous. Basal leaves present at flowering, long-petiolate, the blade 1–7 cm long, 3–9 cm wide, kidney-shaped to more or less semicircular, deeply 3-lobed, the base broadly cordate or less commonly truncate, the lobes obovate to bluntly rhombic, at least the central lobe usually again 3-lobed, rounded or bluntly pointed at the tip, the margins otherwise scalloped. Stem leaves mostly short-petiolate to sessile, deeply 3-lobed or -parted nearly to the base, the lobes linear to oblanceolate or narrowly rhombic-elliptic, the margins entire or nearly so or those of the lower leaves shallowly lobed or bluntly toothed or scalloped. Sepals 3–5, 2–4 mm long, reflexed from at or just above the base (lacking a transverse fold), more or less plane. Petals 3–5, 2–5 mm long, 1–3 mm wide, obovate, as long as or slightly longer than the sepals, yellow. Style absent. Head of achenes 3–12 mm long at maturity, at first globose to ovoid but usually becoming cylindric, the receptacle sparsely hairy or glabrous. Achenes 0.8–1.2 mm long, flattened, the margins corky-thickened, unwinged, the wall thick, with a network of very fine cross-wrinkles, glabrous, the beak essentially absent (reduced to a minute, off-center, peglike structure). $2n=32$. April–August.

Scattered in counties bordering the Mississippi and Missouri Rivers, sporadically elsewhere in the state (eastern U.S. west to Minnesota, Nebraska, and Texas; Canada, Mexico, Europe, Asia; introduced from Washington to California). Banks of streams, rivers, and spring branches; margins of ponds and lakes, oxbows, sloughs, swamps, marshes; also ditches, pastures, margins of crop fields, roadsides, and moist, disturbed areas; terrestrial or emergent aquatics.

Immature or depauperate individuals of this species sometimes are morphologically similar to emergent forms of *R. flabellaris*. Both are members of sect. *Hecatonia* DC. The stems in *R. sceleratus* are more erect than those of *R. flabellaris* and do not commonly form roots above the base. Fertile individuals are more easily distinguished, in that *R. sceleratus* has shorter petals and smaller, beakless achenes.

This is a widespread species and the limits of its natural distribution are not fully understood.

2386. Thalictrum dasycarpum 2387. Thalictrum dioicum 2388. Thalictrum revolutum

There is some disagreement among botanists as to whether the species is native in the eastern United States (Whittemore, 1997), where it often is very weedy in wetlands. Steyermark (1963) accepted it as a native member of the Missouri flora and there presently is no strong evidence to contradict his interpretation. A second variety, var. *multifidus* Nutt., occurs in the western United States (including Alaska) and Canada, differing from var. *sceleratus* in its more deeply and regularly divided leaf blades and its achenes with the lateral faces smooth except for a ring of minute pits.

Steyermark (1963) noted that this species has been documented to poison cattle and also causes their milk to develop a bitter flavor. He also noted that exposure to the plant may cause dermatitis in susceptible humans.

15. **Thalictrum** L. (meadow-rue)

Plants perennial herbs, sometimes with rhizomes (these often somewhat woody in Missouri species) or tuberous roots, often dioecious (monoecious elsewhere). Stems erect or ascending. Leaves alternate (opposite or rarely whorled in *T. thalictroides*) and in a basal rosette, 2–4 times ternately or ternately-pinnately compound, the leaflets obovate to ovate, lanceolate, or kidney-shaped, sometimes with 3 or more lobes. Inflorescences terminal and sometimes also axillary, many-flowered (to 200) racemes or panicles with loose to dense clusters of flowers (reduced to few-flowered umbels or solitary flowers in *T. thalictroides*). Flowers actinomorphic, perfect or imperfect. Sepals 4–10, white to green, pink, or purple, 2–13 mm long, plane, not persistent at fruiting. Petals absent. Stamens sometimes relatively prominent but not showy, the anthers yellow. Staminodes absent. Pistils (3–)6–21, each with 1 ovule. Style present or absent. Fruits achenes, ellipsoid to narrowly ovoid, sometimes appearing short-stalked, the tip with or without a short beak (when present, this to about 0.5 mm long), the outer wall thick, prominently veined or ribbed. Receptacle not much enlarged at fruiting. About 150 species, nearly worldwide.

1. Inflorescences umbels or of solitary flowers, these always perfect; sepals 5–13 mm long; stem leaves a single opposite pair or rarely a whorl of 3; rhizomes absent, the roots tuberous . . . . 4. T. THALICTROIDES
1. Inflorescences panicles of numerous flowers, these all or mostly staminate or pistillate; sepals 2–5 mm long; stem leaves alternate at few to several nodes; rhizomes present, the roots never tuberous
    2. Leaflets all 3–7-lobed, the lobes broadly rounded or truncate at their tips, their margins often scalloped; all leaves petiolate; filaments of the stamens yellow to greenish yellow . . . . 2. T. DIOICUM

2. Leaflets unlobed or 3(–7)-lobed, the lobes rounded to more commonly angled or tapered to bluntly or sharply pointed tips, their margins entire or occasionally scalloped; the upper leaves sessile or nearly so, the lower leaves petiolate; filaments white or purplish-tinged
    3. Undersurface of leaflets, leaflet stalks, inflorescence branches, and/or fruits usually with gland-tipped hairs, never with nonglandular hairs; leaf blades papery to leathery in texture, the revolute margins 0.1–0.3 mm wide . . . . 3. T. REVOLUTUM
    3. Undersurface of leaflets, leaflet stalks, inflorescence branches, and/or fruits usually with nonglandular hairs, never with gland-tipped hairs (but rarely with small sessile glands); leaf blade papery or rather stiff in texture, the revolute margins 0.07–0.16 mm wide . . . . 1. T. DASYCARPUM

1. Thalictrum dasycarpum Fisch. & Avé-Lall. (purple meadow-rue)

T. dasycarpum var. *hypoglaucum* (Rydb.) B. Boivin

Pl. 516 d; Map 2386

Plants dioecious (rarely with a few perfect flowers), with rhizomes and slender, nontuberous roots. Stems 60–150(–200) cm long. Stem leaves alternate at few to several nodes; the lower leaves petiolate, the upper leaves sessile or nearly so; well-developed lower leaves 3 times ternately or ternately-pinnately compound, the largest leaflets 1.5–5.5 cm long, mostly longer than wide, oblong to obovate or oblanceolate, unlobed or 3(–7)-lobed, the lobes rounded to more commonly angled or tapered to bluntly or sharply pointed tips, papery or rather stiff in texture, the margins entire or occasionally scalloped, narrowly revolute (the curled-under portion 0.07–0.16 mm wide), the undersurface and stalk glabrous or pubescent with nonglandular hairs, sometimes also with sessile glands, the minor veins somewhat raised from the surface. Inflorescences panicles, their branches glabrous or pubescent with nonglandular hairs, sometimes also with sessile glands. Flowers all or nearly all imperfect. Sepals 4, 2–3 mm long, white or sometimes pale pinkish- to purplish-tinged. Filaments white or purplish-tinged. Fruits 3–5 mm long, flattened-ellipsoid, not appearing stalked, the beak 3–4 mm long, becoming brittle with age and often shed, the surface glabrous or pubescent with nonglandular hairs, sometimes also with inconspicuous sessile glands. May–July.

Scattered in the northern and western halves of the state, but uncommon or absent from most of the southeastern quarter (nearly throughout the U.S. except for some far-eastern and far-western states; Canada). Banks of streams, rivers, and oxbows, bases of bluffs, bottomland forests, and openings and edges of mesic upland forests; also old fields, fencerows, ditches, railroads, and roadsides.

Glabrous forms of this species have been called var. *hypoglaucum*. They do not differ from typical plants in any other way and both forms are sometimes found in the same population, so it seems best not to assign a formal name to this variant. These glabrous plants are very difficult to distinguish from glabrous forms of *T. revolutum* (see further discussion under that species).

2. Thalictrum dioicum L. (early meadow-rue, quicksilver weed)

Pl. 516 i; Map 2387

Plants dioecious, with short rhizomes and more or less slender, nontuberous roots. Stems 30–70 cm long. Stem leaves alternate at few to several nodes; all noticeably petiolate; well-developed lower leaves 3 or 4 times ternately or ternately-pinnately compound, the largest leaflets 1.5–2.8 cm long, shorter than to slightly longer than wide, broadly elliptic to obovate, kidney-shaped, broadly heart-shaped, or nearly circular, 3–7-lobed, the lobes rounded to truncate, relatively thin and membranous in texture, the margins entire or more commonly scalloped, sometimes slightly thickened or inconspicuously revolute (the curled-under portion 0.05–0.10 mm wide), the undersurface and stalk glabrous or with inconspicuous sessile glands, the minor veins not raised from the surface. Inflorescences panicles, their branches glabrous or with inconspicuous sessile glands. Flowers imperfect. Sepals 4, 3–5 mm long, green or purple. Filaments yellow to greenish yellow. Fruits 3–4 mm long, narrowly ellipsoid, sometimes appearing short-stalked, the beak 1.5–2.0 mm long, often shed except for a minute, peglike base, the surface glabrous or with inconspicuous sessile glands. April–May.

Scattered, mostly south of the Missouri River (eastern U.S. west to North Dakota and Oklahoma; Canada). Bases, ledges, and tops of bluffs, mesic upland forests, and banks of streams and rivers.

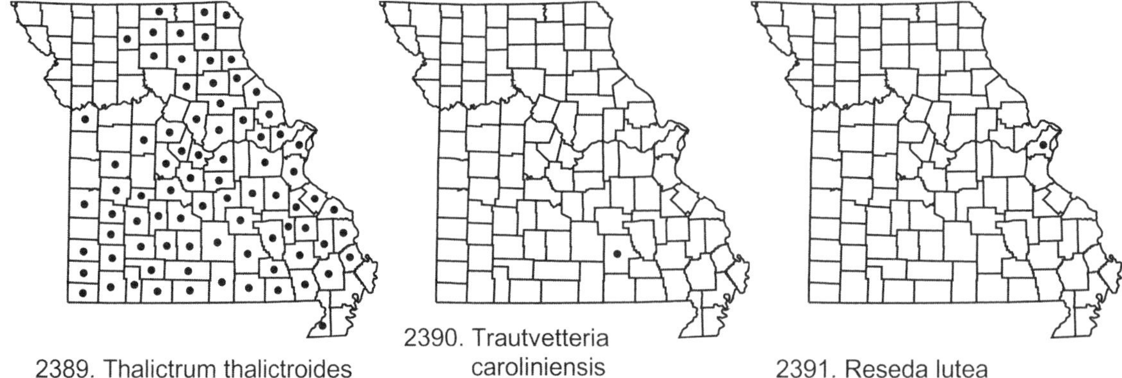

2389. Thalictrum thalictroides

2390. Trautvetteria caroliniensis

2391. Reseda lutea

3. Thalictrum revolutum DC. (wax-leaved meadow-rue, waxy meadow-rue)

Pl. 516 e–h; Map 2388

Plants dioecious (rarely with a few perfect flowers), with rhizomes and slender, nontuberous roots. Stems 50–150(–200) cm long. Stem leaves alternate at few to several nodes; the lower leaves petiolate, the upper leaves sessile or nearly so; well-developed lower leaves 3 times ternately or ternately-pinnately compound, the largest leaflets 1.5–5.5 cm long, mostly longer than wide, oblong to obovate, ovate, or lanceolate, unlobed or 3(–7)-lobed, the lobes rounded to angled or tapered to bluntly or sharply pointed tips, papery or rather stiff in texture, the margins entire or occasionally scalloped, narrowly revolute (the curled-under portion 0.1–0.3 mm wide), the undersurface and stalk usually pubescent with gland-tipped hairs, less commonly glabrous or with only inconspicuous sessile glands, the minor veins often relatively strongly raised from the surface. Inflorescences panicles, their branches glabrous or pubescent with gland-tipped hairs, sometimes only with sessile glands. Flowers all or nearly all imperfect. Sepals 4, 3–4 mm long, white or sometimes pale pinkish- to purplish-tinged. Filaments white or purplish-tinged. Fruits 3–5 mm long, flattened-ellipsoid, not appearing stalked, the beak 2–3 mm long, becoming brittle with age and often shed, the surface glabrous or pubescent with gland-tipped hairs, sometimes only with sessile glands. $2n=140$. May–August.

Uncommon in the Mississippi Lowlands division and the western half of the Glaciated Plains, scattered elsewhere in the state (eastern U.S. west to Minnesota and Texas, west sporadically to South Dakota, Nevada, and Arizona; Canada). Banks of streams and rivers, margins of ponds, bases of bluffs, fens, bottomland prairies, bottomland forests, openings and edges of mesic upland forests, and glades; also old fields, fencerows, ditches, railroads, and roadsides.

Glabrous forms of this species have been called *T. revolutum* f. *glabrum* Pennell. They do not differ from the typical plant in any other way, and both forms are sometimes found in the same population, so it seems best not to assign a formal name to this variant.

Thalictrum revolutum can normally be distinguished from *T. dasycarpum* by the form of the hairs found on many parts of the plant: tipped with knob-shaped glands in *T. revolutum*, nonglandular and tapered at the tips in *T. dasycarpum*. In most specimens of both species, careful search of the leaflets (especially the bases of the main veins), rachises, inflorescence branches, and achenes will turn up at least a few hairs, but completely glabrous specimens are sometimes found, and they are very difficult to name. M. M. Park (1992) showed that glabrous forms of the two species can be distinguished reliably by multivariate statistical analysis of multiple measurements of the leaflets, filaments, anthers, stigmas, and stalklike bases of the achenes, but these characters are so variable, and the ranges of variation overlap so broadly, that they are useless for routine identification. Steyermark (1963) and Mohlenbrock (1981) tried to distinguish glabrous forms of *T. dasycarpum* and *T. revolutum* based on the thickness of the lamina, the revolution of the margin, and the degree to which the minor veins project on the underside of the leaflet, but these characters are also very variable. On average, *T. revolutum* does have thicker, more revolute leaves, but many specimens of both species have moderately thick leaflets with veins that project weakly or moderately.

4. Thalictrum thalictroides (L.) A.J. Eames & B. Boivín (rue-anemone)

Anemonella thalictroides (L.) Spach
Pl. 512 h, i; Map 2389

Plants lacking rhizomes, the roots swollen and tuberous. Stems 8–25 cm long. Stem leaves (technically bracts) opposite or rarely in a whorl of 3 at a single node immediately subtending the inflorescence (note that in some cases the leaflets may resemble a whorl of six simple leaves), sessile, once ternately compound (the basal leaves twice ternately compound), the largest leaflets 0.8–2.5 cm long, wider than long to about as long as wide, broadly ovate to obovate, broadly oblong, or nearly circular, 3(5)-lobed, the lobes rounded or broadly and bluntly pointed at the tips, relatively thin and membranous in texture, the margins entire or occasionally scalloped, sometimes slightly thickened or inconspicuously revolute (the curled-under portion 0.05–0.10 mm wide), the undersurface and stalk glabrous and lacking sessile glands, the minor veins not noticeably raised from the surface. Inflorescences umbels of 3–6 flowers or solitary flowers, the stalks glabrous. Flowers all perfect. Sepals 5–7, 5–13 mm long, white or sometimes pale pink to purple, rarely green. Filaments yellow to greenish yellow. Fruits 4–5 mm long, narrowly ellipsoid to narrowly ovoid, not appearing stalked, beakless. March–June.

Scattered to common nearly throughout the state, but uncommon or apparently absent from most of the western half of the Glaciated Plains Division (eastern U.S. west to Minnesota and Texas; Canada). Mesic to dry upland forests, bases, ledges, and tops of bluffs, margins of glades, and occasionally banks of streams, also roadsides.

Plants from Polk County with green, leaflike sepals have been called *Anemonella thalictroides* f. *chlorantha* Fassett and plants from Putnam County with all of the stamens and pistils transformed into petaloid structures have been called *A. thalictroides* f. *favilliana* Bergseng. *Thalictrum thalictroides* is superficially very similar to *Isopyrum biternatum*, and the two are often confused. Differences are discussed under that species.

16. **Trautvetteria** Fisch. & C.A. Mey.

One species, North America, Asia.

Recent studies (Ro and McPheron, 1997; Johansson, 1998) suggest that *Trautvetteria* is very closely related to a group of species that have traditionally been placed in *Ranunculus*. Additional research is necessary to reevaluate generic limits within this group. For further discussion, see the treatment of *R. cymbalaria*.

1. Trautvetteria caroliniensis (Walter) Vail
(false bugbane)
Pl. 516 j, k; Map 2390

Plants perennial herbs, with short, slender rhizomes. Basal leaves 2–5, well-developed and persistent at flowering, long-petiolate. Stems 50–150 cm long, erect or ascending, glabrous or nearly so. Stem leaves 1–3, alternate, sessile or short-petiolate. Blades of the basal leaves 7–14 cm long and 7–20 cm wide at flowering, becoming enlarged to 20–30 cm at fruiting, palmately deeply 5–9 (–11)-lobed, kidney-shaped to nearly circular, wider than long to about as long as wide, the base truncate to deeply cordate but with a small tapered area along the tip of the petiole, the lobes oblong to oblong-elliptic or oblong-oblanceolate, angled or tapered to bluntly or sharply pointed tips, the margins otherwise sharply toothed (occasionally scalloped) and sometimes irregularly lobed, the surfaces glabrous or nearly so; blades of the stem leaves similar to those of the basal leaves but much smaller and usually somewhat less divided. Inflorescences dense, flat-topped panicles or large clusters of 10–20 (or more) flowers at the stem tips, the stalk 10–60 mm long, produced more or less after the leaves develop. Flowers actinomorphic, perfect, the stalks with dense, minute, hooked hairs. Sepals 3–5(–7), 3–6 mm long, broadly ovate to obovate, rounded at the tip, deeply cupped (concave), pale green to greenish white, shed as the flowers open. Petals absent. Stamens showy (to 7 mm long), forming a globose mass, the filaments flattened and broadened toward the tips (wider than the anthers), white, the anthers yellow. Staminodes absent. Pistils 10–15, each with 1 ovule, the style present, hooked. Fruits utricles in dense clusters, each 3–4 mm long, more or less semicircular to asymmetrically obovate in outline, 4-angled in cross-section, the outer layer thin, papery, brown, prominently veined on at least 2 of the faces, tipped with a beak, this 0.4–0.8 mm long, strongly curved or hooked. Seeds 1 per utricle, 1.5–2.5 mm long, narrowly ellipsoid, the surface smooth, olive green. Receptacle not enlarged at fruiting. $2n=16$. June–August.

Uncommon, known thus far only from Shannon County (eastern and western U.S. [absent from the Great Plains region]; Canada, Asia). Bases and ledges of dolomite bluffs.

In Missouri, this species is restricted to a few north-facing bluffs along stretches of the Current and Jack's Fork Rivers where cold-air drainage creates a microhabitat that preserves populations of several plant species with main ranges to the north of Missouri. Steyermark (1963) considered *Trautvetteria* and other species at these sites to represent glacial relicts. For further discussion, see the discussion on *Affinities of the Flora* in Volume 1 of the present series (Yatskievych 1999: 86–87).

Some botanists have treated the populations west of the Great Plains as var. *borealis* (H. Hara) T. Shimizu (var. *occidentalis* (A. Gray) C.L. Hitchc.) and the Asian plants as var. *japonica* (Siebold & Zucc.) T. Shimizu (Shimizu, 1981). The differences between plants from these geographically isolated areas do not appear to warrant formal taxonomic recognition (Parfitt, 1997).

Native Americans in the western United States applied a poultice derived from the rootstock of this species in the treatment of boils (Moerman, 1989).

RESEDACEAE (Mignonette Family)

Six genera, about 70 species, North America, Mexico, Europe, Asia, Africa.

1. **Reseda** L. (mignonette)
(Abdallah and de Wit, 1978)

Plants annual, biennial, or perennial herbs, usually with taproots. Stems often angled or longitudinally ridged, often branched, glabrous. Leaves alternate and basal, short-petiolate, glabrous. Stipules absent. Leaf blades mostly oblanceolate in outline, simple, entire or pinnately lobed, the lobes usually above the blade midpoint. Inflorescences terminal, slender, elongate, sometimes spikelike racemes, these sometimes branched basally and thus appearing paniculate, the axis and stalk sometimes roughened with minute, prickly outgrowths, otherwise glabrous. Flowers perfect. Calyces actinomorphic or nearly so, of 4–7 sepals, these free or fused at the bases. Corollas of 4–6 free petals, zygomorphic or asymmetrically irregular, the petals of different sizes and shapes, yellow to greenish yellow, usually narrowed to stalklike bases, the body of at least the larger petals irregularly appendaged on the dorsal surface (sometimes more or less at the tip), the appendage variously lobed or dissected. Stamens and pistil on a short stalk above the perianth. Stamens 10–25, the filaments inserted on an asymmetric, fleshy disc around the ovary base, the anthers attached near their midpoints, yellow. Pistil 1 per flower, superior, of (2)3(4) fused carpels, the ovary with 1 locule, bluntly angled, usually noticeably open apically, the angles extending into 3 thickened, triangular teeth, the stigmatic regions along the irregular terminal portions of the teeth. Ovules numerous, the placentation parietal. Fruits capsules that shed seeds apically through the openings. Seeds numerous, broadly ovoid to globose, sometimes slightly kidney-shaped, with curved embryos, the surface smooth, dark brown to black, shiny. Fifty-five species, Europe, Asia, Africa, Atlantic islands.

Several species of *Reseda* are cultivated as garden ornamentals. In addition to the use of *R. luteola* for dyes, the European *R. odorata* L. (fragrant mignonette) produces a fragrant volatile oil that is used in some perfumes.

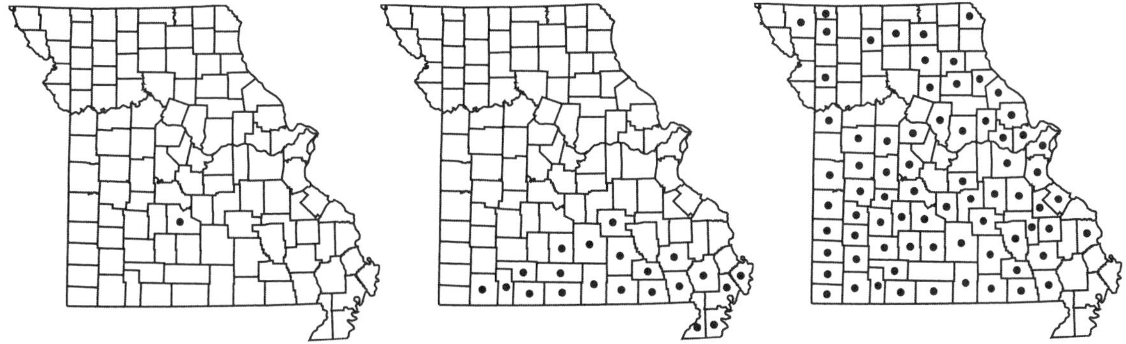

2392. Reseda luteola 2393. Berchemia scandens 2394. Ceanothus americanus

1. Sepals 6; petals 6; leaf blades frequently with 1–3 lobes per side; bracts not persistent at fruiting; seeds 1.4–2.0 mm long . . . . 1. R. LUTEA
1. Sepals 4; petals 4; leaf blades unlobed; bracts persistent at fruiting; seeds 0.6–0.9 mm long; . . . . 2. R. LUTEOLA

1. Reseda lutea L. **ssp. lutea** (wild mignonette, reseda)

Pl. 521 c–e; Map 2391

Plants annual (perennial herbs elsewhere). Stems 30–70 cm long, erect or strongly ascending. Leaf blades 5–10 cm long, entire or with 1–3 lobes on each side, the lobes linear, sometimes deep, occasionally lobed again, the margins otherwise often somewhat wavy and occasionally hairy. Inflorescences 10–20 cm, the flowers subtended by linear to narrowly oblong bracts 1.5–2.5 mm long, these shed before the fruits mature. Flower stalks 3–6 mm long. Sepals 6, 2.5–3.5 mm long, spreading to reflexed, linear to narrowly oblong. Petals 6, 2–4 mm long, the lower pair the smallest, the appendage usually entire, the lateral pair with the appendage 2- or 3-parted, the upper pair the largest, the appendage 3- or more-parted. Stamens (12–)15–20. Fruits 7–8(–10) mm long. Seeds 1.4–2.0 mm long. $2n=12$, 24, 48. May–September.

Introduced, uncommon, known thus far only from the city of St.Louis (native of Europe, Asia; introduced widely but sporadically in North America). Railroads.

This species was first reported from Missouri by Mühlenbach (1983) and is still known in the state only from his original collection from the St. Louis railyards. Abdallah and de Wit (1978) divided the species into two subspecies, a widespread ssp. *lutea* and a northern African endemic, ssp. *neglecta* (differing in capsules with more seeds that are individually smaller than in ssp. *lutea*). Within ssp. *lutea*, they accepted a complex series of five varieties, differing in various minor details of leaf division, fruit orientation, and degree of petal lobing. Most of the North American introduced plants appear to correspond to var. *lutea*.

2. Reseda luteola L. **ssp. luteola** (Dyer's rocket, yellow weed)

Pl. 521 f, g; Map 2392

Plants biennial. Stems (20–)40–80(–100) cm long, erect or strongly ascending. Leaf blades 5–10 cm long, unlobed, the margins otherwise often slightly wavy and occasionally hairy. Inflorescences 10–20 cm, the flowers subtended by linear to narrowly lanceolate-triangular bracts 2.0–3.5 mm long, these persistent at fruiting. Flower stalks 1.0–1.5 mm long. Sepals 4, 1.5–2.5 mm long, loosely ascending to ascending, oblong-elliptic to oblong-ovate. Petals 4, 2–4 mm long, the lower pair the smallest, the appendage often entire, the lateral pair with the appendage 2- or more commonly 3-parted, the upper pair the largest, the appendage 3- or more-parted. Stamens 20–25. Fruits 7–8(–10) mm long. Seeds 0.6–0.9 mm long. $2n=24$, 26, 28. May–October.

Introduced, uncommon, known thus far only from a single historical collection from Laclede County (native of Europe introduced widely but sporadically in North America). Open, disturbed areas.

This species has a long history of use as a dye plant in Europe, yielding a bright yellow dye.

It is also occasionally cultivated as an ornamental. The other subspecies accepted by Abdallah and de Wit (1978), the rare ssp. *dimerocarpa* (Müll. Arg.) Abdallah & de Wit, differs in its 2-carpellate pistils and apparently is endemic to Sardinia.

RHAMNACEAE (Buckthorn Family)

Plants shrubs, trees, or lianas, sometimes dioecious (sometimes incompletely so), sometimes the branchlets thorny at the tip. Leaves alternate or opposite. Stipules absent or more commonly present, but minute, scalelike, and shed early (spinescent elsewhere). Leaf blades simple, unlobed, the venation with 3 main veins from the base or with 1 midvein and pinnate secondary veins, these sometimes strongly arched toward the blade tip. Inflorescences terminal or axillary clusters, umbels, or panicles, sometimes reduced to solitary flowers, the branch points or flowers sometimes subtended by small bracts, but these usually shed early. Flowers relatively small, actinomorphic, perfect or imperfect, shallowly perigynous, the hypanthium variously saucer-shaped to somewhat cup-shaped, usually persistent and becoming disc-shaped, even after the fruits have been shed. Calyces of 4 or 5 apparently free sepals, these attached to the hypanthium rim (and thus can be interpreted as fused into the hypanthium basally), usually not persistent above the hypanthium at fruiting. Corollas of 4 or 5 free petals, these often tapered to a stalklike base, white or whitish green. Stamens 4 or 5, positioned opposite the petals, the filaments usually distinct, attached at the petal bases and often partially enclosed by the concave petals, reduced or absent in pistillate flowers, the anthers attached near the midpoint or at the base, usually yellow. Nectar disc present between the stamens and pistil, more or less cushion-shaped. Pistil of 2 or 3(4) fused carpels, reduced or absent in staminate flowers. Ovary superior, 2- or 3(4)-locular, the locules sometimes incomplete toward the ovary tip, the placentation more or less axile (often appearing nearly basal), the ovules 1 per locule. Style 1 and entire or 2–4-branched, the stigmas variously 2- or 3-lobed or entire, more or less capitate. Fruits drupes or appearing capsular (schizocarps or samaras elsewhere), indehiscent or more or less dehiscent, the stones 1–4, bony. About 55 genera, about 950 species, nearly worldwide, most diverse in tropical and warm-temperate regions.

The main use of members of the Rhamnaceae in the United States is as ornamental plants in gardens. However, some of the genera contain medicinally important species (particularly as laxatives). Some of the tropical genera provide lumber for buildings and furniture, as well as handcrafts and musical instruments. Species of jujube are members of the genus *Ziziphus* Mill., some of which are cultivated for their edible fruits

1. Leaf blades with 3 main veins from at or just above the base; fruits appearing capsular, breaking open longitudinally and explosively at maturity . . . . . 2. CEANOTHUS
1. Leaf blades with a single midvein and pinnately arranged secondary veins; fruits drupes, not breaking apart at maturity
    2. Plants lianas, the stems twining, sometimes forming loose thickets in open areas, but usually climbing in trees . . . . 1. BERCHEMIA
    2. Plants shrubs or small trees, the stems not twining, usually erect or ascending. . . . . 3. RHAMNUS

1. Berchemia Neck. ex DC.

About 12 species, U.S., Mexico, Central America, Asia, Africa.

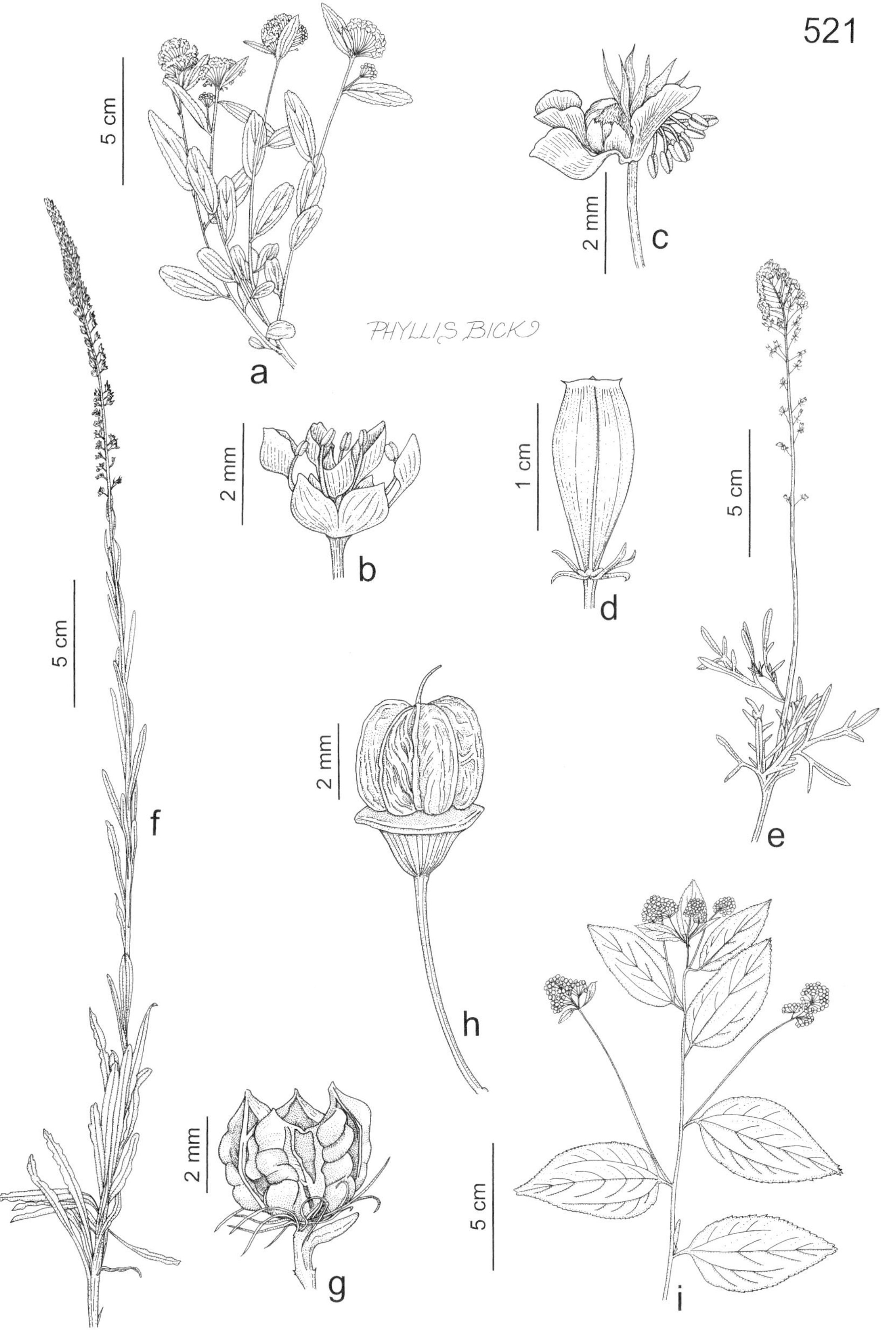

Plate 521. Resedaceae. Rhamnaceae. *Ceanothus herbaceus*, **a)** fertile stem, **b)** flower. *Reseda lutea*, **c)** flower, **d)** fruit, **e)** fertile stem. *Reseda luteola*, **f)** fertile stem, **g)** fruit following dehiscence. *Ceanothus americanus*, **h)** fruit, **i)** fertile stem.

1. Berchemia scandens (Hill) K. Koch (supple-jack, rattan vine)

Pl. 522 h, i; Map 2393

Plants lianas, lacking tendrils. Stems to 30 m or more long, branched, twining, tough and flexible, the bark at first green and smooth, becoming gray or brown with age and developing fine, pale, longitudinal streaks as well as scattered, raised lenticels and branch scars in cross lines, the branches not spine-tipped. Twigs green to reddish brown, glabrous, the leaf scars mostly peglike, the winter buds small, ovate, flattened, with a few overlapping scales. Leaves alternate, short-petiolate. Leaf blades 2–6 cm long, usually somewhat leathery, ovate, oval, or oblong to narrowly oblong, angled or rounded at the base, slightly tapered to a sharply pointed tip, the margins entire or slightly wavy, sometimes with scattered, minute, blunt teeth, the upper surface glabrous, green, shiny, the undersurface glabrous, pale green, not shiny, the venation pinnate with a single midvein and 8–12 (–16) pairs of lateral veins. Inflorescences terminal and often also axillary, of loose clusters or small panicles. Flowers apparently perfect (but apparently functionally staminate or pistillate). Hypanthium minute, 1–2 mm in diameter at fruiting. Sepals 5, 1–2 mm long, triangular. Petals 5, 1.0–1.5 mm long, greenish yellow. Stamens 5. Ovary 2-locular, the style short, unbranched. Fruits drupes, 5–8 mm long, oblong-ellipsoid, with 2 stones, the outer surface thin, leathery, bluish black, glaucous. May–June.

Scattered in the southern portion of the Ozark Division and the Mississippi Lowlands (southeastern U.S. west to Missouri and Texas; Mexico, Central America). Glades, openings of dry upland forests,bottomland forests, swamps, and bases of bluffs; also roadsides.

As noted by Steyermark (1963), Kurz (1997), and others, this species is remarkable in its ability to survive in habitats with drastically different moisture levels, from the driest Ozark glades to the wettest Bootheel swamps. The most robust plants occur in bottomland forests and swamps in southeastern Missouri, where stems can reach diameters of 18 cm (Foote and Jones, 1989). Kurz (1997) noted that the twining stems can eventually girdle and kill trees upon which they grow.

2. **Ceanothus** L.

Plants small shrubs (larger and rarely treelike elsewhere). Stems branched, not twining, brittle or flexible, green to brown, sometimes developing longitudinal fissures with age, the branches not spine-tipped. Twigs green to brown or light yellowish brown, glabrous to densely and minutely hairy. Leaves alternate, mostly short-petiolate. Leaf blades variously shaped, the margins sharply and finely toothed, the teeth gland-tipped when young, the surfaces glabrous to densely short-hairy, not shiny, the venation with 3 main veins that diverge at or just above the blade base and are connected by numerous, pinnate finer veins. Inflorescences terminal on present year's growth, either on the main branchlets or at the tips of short axillary branches, of relatively dense, small panicles, the branches often somewhat umbellate, occasionally reduced to umbels. Flowers perfect, with relatively long, slender stalks. Hypanthium small, 3–4 mm in diameter at fruiting. Sepals 5, oblong, somewhat concave and the tips incurved at flowering. Petals 5, tapered to long stalked bases, white. Stamens 5, exserted. Ovary 3-locular, shallowly 3-lobed at the tip, the style 3-branched toward the tip. Fruits capsulelike, modified drupes, 4–6 mm long, depressed-obovoid, with usually 3 stones, the outer surface thin, leathery, black, not glaucous, the stones dehiscing explosively at maturity, tearing open the outer fruit layers and expelling the seeds. Seeds 1.5–2.0 mm long, more or less obovate in outline, somewhat anglar, the surface smooth, reddish brown to brown, shiny. About 60 species, North America, Central America, most diverse in the western U.S.

Nearly 45 species of *Ceanothus* occur in California, where some of them are conspicuous elements of chaparral and woodland plant communities. These shrubs flower profusely in the spring with showy displays of fragrant, white to purple flowers. They are appreciated in their natural environments and also prized in cultivation as ornamental shrubs. Further, it has been known for some time that some species in the genus (including *C. americanus*) produce root nodules (Furman, 1959) similar to those found in many Fabaceae. These nodules contain symbiotic, nitrogen-fixing bacteria capable of converting atmospheric nitrogen into ntirates

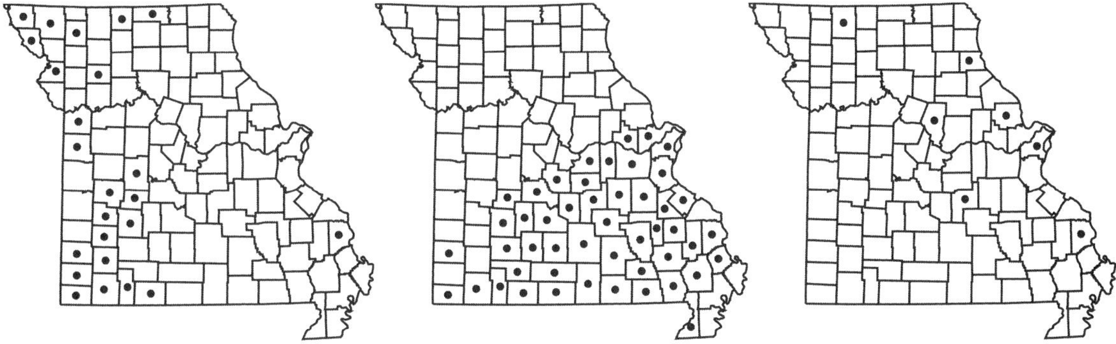

2395. Ceanothus herbaceus 2396. Rhamnus caroliniana 2397. Rhamnus cathartica

to improve soil fertility (Bond, 1967). The two Missouri species are western-occurring components of a small group of only four or five species that occur natively east of the Mississippi River. They were among the first members of the genus to be brought into cultivation, with many cultivars produced, but seem to have fallen from favor during the first half of the twentieth century. In recent years, they have had a revival of sorts and have become increasingly available again as ornamentals, especially through wildflower nurseries. The leaves of the Missouri species were used by Native Americans in a tea and the often extensive, woody rootstocks were used for fuelwood in areas where aboveground biomass of other woody species was low (Kurz, 1997; Moerman, 1998). Medicinally they have been used as a coagulant and historically were an ineffective treatment for syphilis.

The fruits of *Ceanothus* have sometimes been described as capsules, but technically they are modified drupes (Brizicky, 1964a). The somewhat fleshy fruits contain usually three stones. At maturity, these split open explosively, tearing open the outer fruit layers and expelling the seeds. The remainder of the fruit is shed at that time or soon thereafter, leaving the persistent hypanthia.

1. Leaf blades mostly oblong or oblong-ovate; inflorescences terminal on short branchlets that are axillary on the main branchlets of the current-year's growth, relatively long-stalked, the inflorescence stalk generally longer than the subtending leaf . . . . 1. C. AMERICANUS
1. Leaf blades mostly elliptic to elliptic-lanceolate; inflorescences terminal on the main branchlets of the present year's growth, relatively short-stalked, the inflorescence stalk generally shorter than the subtending leaf. . . . . 2. C. HERBACEUS

1. Ceanothus americanus L. (New Jersey tea, wild snowball)

C. americanus var. *intermedius* (Pursh) Torr. & A. Gray

C. americanus var. *pitcheri* Torr. & A. Gray

Pl. 521 h, i; Map 2394

Plants shrubs, 40–100 cm tall, the branches ascending. Leaf blades 1–8(–10) cm long, 0.5–5.0 cm wide, oblong or oblong-ovate, sometimes broadly so, broadly angled to rounded or shallowly cordate at the base, angled or slightly tapered to a bluntly or sharply pointed tip, the upper surface sparsely to moderately pubescent with short, somewhat cobwebby hairs, rarely glabrous, the undersurface moderately to densely short-hairy, rarely glabrous or nearly so. Inflorescences terminal on short branchlets that are axillary on the main branchlets of the current-year's growth, relatively long-stalked, the inflorescence stalk generally longer than the subtending leaf. Sepals 0.5–1.0 mm long. Petals 1.5–2.5 mm long. $2n=24$. May–November.

Scattered nearly throughout the state, but absent from most of the Mississippi Lowlands Division (eastern U.S. west to Minnesota, Nebraska, and Texas; Canada).

Some authors separate this species into three varieties. The var. *americanus*, with the leaves glabrous or sparsely hairy, mostly along the veins, occurs to the east of Missouri. Plants with the leaves persistently and densely short-hairy have been called var. *pitcheri* and occur nearly thoughout the species range, but tend to predominate in the western portion, including Missouri. An uncommon variant that has a more densely branched habit than that of the other varieties and relatively small leaves (mostly 1–4 vs. 2.5–6.0 cm) is known as var. *intermedius*. It occurs mostly on the Coastal Plain in the southeastern United States, but has been found sporadically in the eastern half of Missouri. Brizicky (1964a) noted that various authors have circumscribed this variant in somewhat different ways, resulting in different distributional ranges for the taxon. He suggested that until further taxonomic research could be completed, the complex should probably be regarded as a single polymorphic taxon.

2. Ceanothus herbaceus Raf. (redroot, inland New Jersey tea)
C. herbaceus var. *pubescens* (Torr. & A. Gray ex S. Watson) Shinners
C. ovatus Desf.
C. ovatus var. *pubescens* Torr. & A. Gray ex S. Watson

Pl. 521 a, b; Map 2395

Plants shrubs, 30–100 cm tall, the branches ascending. Leaf blades 1–6 cm long, 0.5–2.0 cm wide, elliptic to elliptic-lanceolate, sometimes narrowly so, occasionally oblong-lanceolate or oblanceolate, angled to rounded at the base, angled or slightly tapered to a usually bluntly pointed tip, the upper surface sparsely to moderately pubescent with short, somewhat cobwebby hairs, rarely glabrous, the undersurface moderately to densely short-hairy, rarely glabrous or nearly so. Inflorescences terminal on the main branchlets of the present year's growth, relatively short-stalked, the inflorescence stalk generally shorter than the subtending leaf. Sepals 0.5–1.0 mm long. Petals 1.5–2.5 mm long. $2n=24$. April–June.

Uncommon in the western half of the state, sporadic farther east (Vermont to Montana south to Virginia, Louisiana, and New Mexico; Canada). Upland prairies, loess hill prairies, glades, and openings of dry upland forests.

Brizicky (1964a, b) outlined the reasons for changing the long-accepted name, *C. ovatus*, to the more obscure epithet, *C. herbaceus*. When it was described, *C. herbaceus* was mistakenly thought to be a herbaceous perennial rather than a small shrub, which contributed to earlier confusion about the taxonomy of the group. Some authors have segregated plants with the leaf blades relatively densely hairy on the undersurface as var. *pubescens* (Steyermark, 1963), but most recent authors have accepted Soper's (1941) conclusion that this represents merely pubescent forms of the species that occur throughout its range.

3. **Rhamnus** L. (buckthorn)

Plants shrubs or small trees, sometimes dioecious (or incompletely so). Trunks or main stems erect or ascending, branched, not twining, usually relatively tough, the bark and twigs various, the winter buds naked or with several overlapping scales. Leaves alternate or opposite, short-petiolate. Leaf blades variously shaped, the margins sharply and finely toothed, the teeth gland-tipped when young, the surfaces glabrous to densely short-hairy, shiny or not, the venation pinnate with a single midvein and 3–11 or rarely more pairs of lateral veins, these relatively straight to strongly arched toward the tip. Inflorescences axillary on present year's growth, either along the main branchlets or along short axillary branches, of small clusters or umbels, sometimes reduced to a solitary flower. Flowers perfect or imperfect, with very short to relatively long stalks. Hypanthium minute to small, variously 1.5–3.0 mm in diameter at fruiting. Sepals 4 or 5, triangular. Petals 5, longer in staminate flowers than in pistillate ones, greenish yellow. Stamens 4 or 5 (abortive and shed early in pistillate flowers), not exserted. Ovary 2–4-locular (reduced and rudimentary in staminate flowers), unlobed, the style unbranched or 2–4-branched toward the tip. Fruits drupes, globose to broadly oblong-ovoid, with 2–4 stones, the outer surface thin, sometimes somewhat leathery, red or black, not glaucous, the stones indehiscent. Stones wedge-shaped in cross-section, asymmetric in outline with a convex to broadly dorsal (outer) side and a more or less straight ventral angle. About 200 species, North America, Central America, Europe, Asia, Africa.

There has been growing momentum in recent years to split the genus *Rhamnus* into two or more genera. Among the potential segregates, the genus *Frangula* Mill. has received the most support, which would affect just one of the Missouri species, *R. caroliniana*. The approximately 50 species of *Rhamnus* sect. *Frangula* (Mill.) DC. are distributed in North America, Europe, Asia, and Africa. They differ fairly consistently from the remainder of *Rhamnus* in their thornless stems, naked winter buds, leaf blades with pinnate, relatively straight secondary veins, perfect flowers, and seeds possessing a terminal, beaklike projection but lacking a longitudinal furrow. In spite of the morphological differences, the two groups have been considered close relatives, even by those botanists who have chosen to treat them as separate genera. Molecular studies (Richardson et al., 2000; Bolmgren and Oxelman, 2004) have confirmed that each group represents a natural, discrete lineage, but that the two groups are each other's closest relatives. Thus the taxonomic level at which to recognize the distinctions remains largely a matter of taxonomic opinion (lumping vs. splitting). A survey of the recent floristic literature shows that most North American authors have accepted a broadly circumscribed *Rhamnus*, as has the Flora of China Project (Y. Chen and Schirarend, 2007). Also, the last comprehensive taxonomic survey of the genera of Rhamnaceae did not accept the genus *Frangula* (Medan and Schirarend, 2004). A few authors have segregated two genera, notably Kartesz and Meacham (1999) in their annotated checklist of temperate North American plants, as well as the forthcoming treatment in the Flora of North America series (Guy Nesom, personal communication). For the present, it seems most prudent to continue to recognize a single genus.

A number of species of *Rhamnus* native to Asia are the overwintering hosts of the soybean aphid, *Aphis glycines* Matsum. (Homoptera: Aphididae) (Venette and Ragsdale, 2004). This Asian insect was first found to have invaded the United States in 2000 in Wisconsin fields, and by 2005 had spread to portions of 22 states (including Missouri). During the warmer portions of the year, *A. glycines* feeds and reproduces on soybeans and related legumes, reducing agricultural yields. These crop pests also are a vector for the spread and transmission of viral diseases of these economically important legumes.

1. Leaves mostly alternate; plants unarmed
    2. Winter buds naked, hairy; leaves with mostly 6–11 pairs of secondary veins; flowers perfect; sepals and petals 5; styles unbranched ..... 1. R. CAROLINIANA
    2. Winter buds with several overlapping scales, the scales glabrous or sparsely hairy only along the margins toward the tip; leaves with mostly 4–7 pairs of lateral veins; flowers imperfect; sepals and petals 4; styles noticeably 2-branched toward the tip .... 5. R. LANCEOLATA
1. Leaves mostly opposite; some of the branches thorn-tipped
    3. Leaf blades with the upper surface minutely hairy, at least toward the tip and/or along the main veins; fruits with 2 stones .... 4. R. JAPONICA
    3. Leaf blades with the upper surface glabrous; fruits with 2–4 stones
        4. Leaf blades ovate to broadly elliptic or nearly circular, less commonly oblong-elliptic, mostly 1–2 times as long as wide and widest at or below the midpoint, with 2–4 pairs of secondary veins; styles (3)4-branched near the tip; fruits with 3 or 4 stones .... 2. R. CATHARTICA
        4. Leaf blades oblong-lanceolate to more commonly oblong-elliptic, elliptic, oblong-oblanceolate, or oblanceolate, mostly 2–3 times as long as wide and widest at or above the midpoint, with 3–6 pairs of secondary veins; styles 2-branched toward the tip; fruits with 2 stones .... 3. R. DAVURICA

1. Rhamnus caroliniana Walter (Carolina buckthorn, Indian cherry)

R. caroliniana var. *mollis* Fernald

Frangula caroliniana (Walter) A. Gray

Pl. 522 f, g; Map 2396

Plants shrubs or rarely small trees, 2–5(–12) m tall. Main stems usually several, the branches all ascending and elongate, none of them thorn-tipped. Bark gray to brown, sometimes with lighter blotches, shallowly furrowed on larger stems, rela-

2398. Rhamnus davurica 2399. Rhamnus japonica 2400. Rhamnus lanceolata

tively smooth. Twigs slender, green to reddish brown, becoming gray with age, glabrous to densely short-hairy, the winter buds slender, naked, reddish brown, densely and minutely hairy. Leaves alternate (occasionally a few appearing subopposite), the petioles 6–20 mm long. Leaf blades 3–12 cm long, 2–3 times as long as wide, rounded or broadly angled at the base, angled or slightly tapered to a bluntly or sharply pointed tip, the upper surface green to dark green, glabrous or occasionally minutely hairy along the midvein, shiny, the undersurface light green, glabrous to densely and minutely hairy, especially along the veins, the lateral veins mostly 6–11 pairs, these straight or slightly curved, mainly toward their tips. Inflorescences axillary, small clusters of 2–8 flowers or occasionally reduced to solitary flowers, the clusters with a stalk 3–10 mm long, the individual flower stalks 3–6 mm long. Flowers perfect. Sepals 5, 1.3–2.0 mm long, often white to greenish white on the upper surface. Petals 5, 1.0–1.5 mm long, broadly obovate, notched at the tips. Style unbranched. Fruits 7–10 mm long, globose, with 3 stones, red at maturity, sometimes becoming black with age. May–June.

Scattered in the Ozark and Ozark Border Divisions and in the Crowley's Ridge and Sikeston Ridge portions of the Mississippi Lowlands (southeastern U.S. west to Missouri and Texas). Mesic to dry upland forests, banks of streams and rivers, glades, and ledges and tops of bluffs; also roadsides.

Plants in the western portion of the species range with relatively densely and persistently hairy leaves have been segregated as var. *mollis*. Steyermark (1963) accepted this taxon with great hesitation, noting that the two variants are not clearly distinct in Missouri, with all gradations of pubescence present in the state. The species is treated as a single variable taxon in the present account.

The superficially similar glossy false buckthorn (European alder-buckthorn), *R. frangula* L. (*Frangula alnus* Mill.) is widely naturalized in the United States, including several states adjoining Missouri (Nebraska, Iowa, Illinois, Kentuicky, and Tennessee). This aggressive, exotic invader of wetlands is native to Europe, but has escaped from cultivation in North America, Asia, and Africa. It differs from *R. caroliniana* in its somewhat shorter (4–8 cm) leaves with the blades mostly less than twice (vs. two to three times) as long as wide and with the margins entire or at most with a few, minute, glandular teeth near the tip, as well as its flower stalks and hypanthia glabrous or nearly so (vs. hairy). Botanists should watch for this species in the future, particularly in northern Missouri wetlands.

2. Rhamnus cathartica L. (common buckthorn, European buckthorn)

Pl. 522 d, e; Map 2397

Plants shrubs or small trees, 2–8 m tall, dioecious, often incompletely so. Main stems 1 to several, the main branches ascending to loosely ascending and elongate, also with shorter, spreading to loosely ascending, relatively straight branches, these mostly thorn-tipped. Bark gray to reddish or purplish brown, sometimes with lighter blotches, relatively smooth, but with relatively prominent, raised lenticels in cross lines and branch scars, somewhat peeling on older, larger stems. Twigs relatively stout, slightly flattened, gray to yellowish brown, with small, dark lenticels, glabrous, the winter buds narrowly ovoid, with several overlapping scales, these reddish or purplish brown, glabrous except for marginal hairs. Leaves mostly opposite (occasionally subopposite on new growth, the petioles 9–28 mm long. Leaf blades 2–6(–9) cm long, mostly 1–2 times as long as wide, ovate to broadly elliptic or nearly circular, less commonly oblong-elliptic, mostly widest

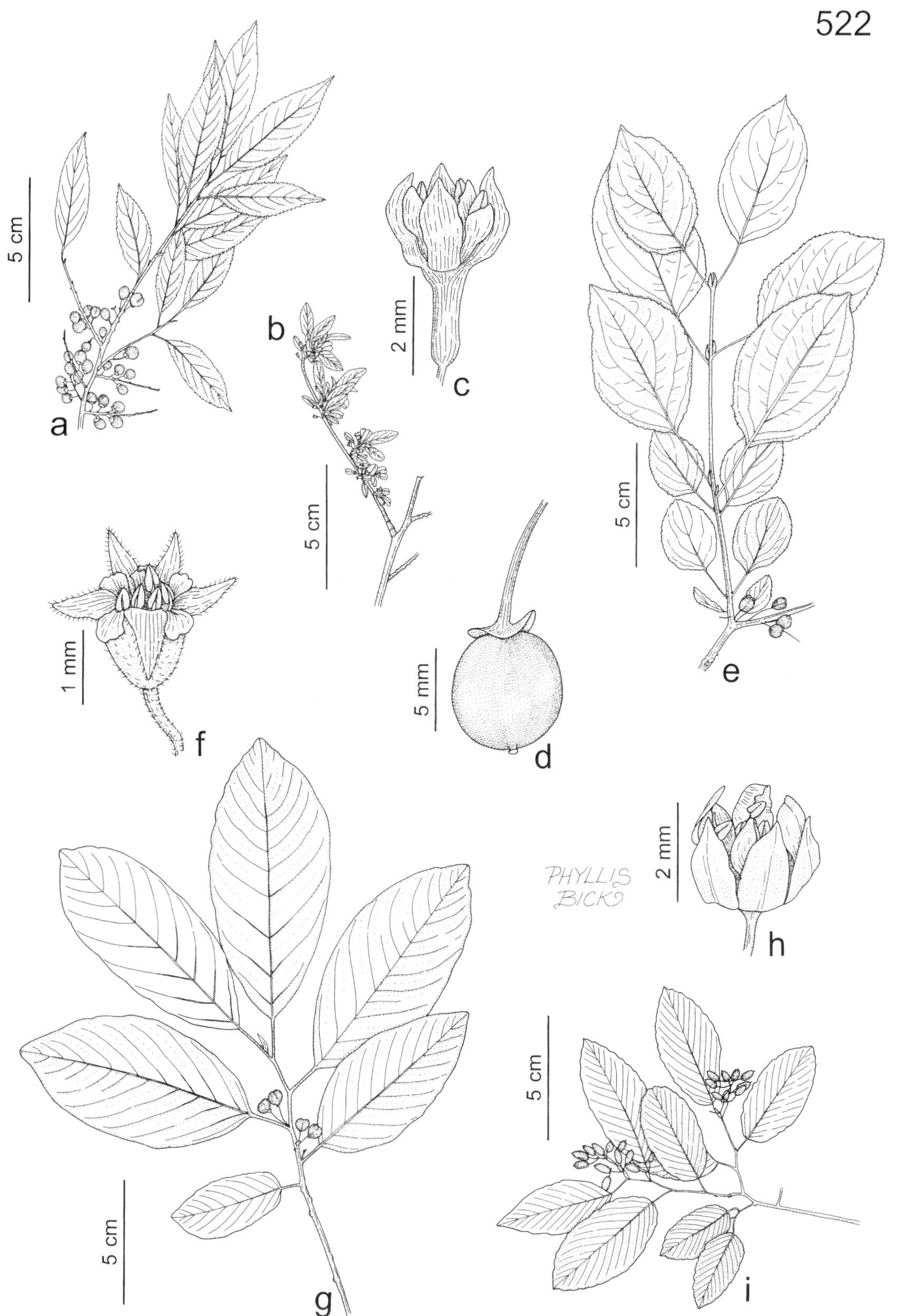

Plate 522. Rhamnaceae. *Rhamnus lanceolata*, **a)** fruiting branch, **b)** flowering branch, **c)** flower. *Rhamnus cathartica*, **d)** fruit, **e)** fruiting branch. *Rhamnus caroliniana*, **f)** flower, **g)** fruiting branch. *Berchemia scandens*, **h)** flower, **i)** fruiting branch.

at or below the midpoint, angled at the base, tapered to a sharply pointed tip or the smaller leaves sometimes merely angled or rounded, the upper surface green to dark green, glabrous, slightly shiny, the undersurface light green, glabrous, the lateral veins mostly 2–4 pairs, these strongly arched toward the blade tip. Inflorescences axillary, small clusters of 2–9 staminate or pistillate flowers or occasionally reduced to solitary flowers, the clusters sessile, the individual flower stalks 2–6 mm long. Flowers imperfect. Sepals 4, 1.3–2.0 mm long. Petals 4, 1.0–1.3 mm long in staminate flowers, 0.6–0.9 mm long in pistillate flowers, lanceolate, entire at the tips. Style noticeably (3)4-branched toward the tip. Fruits 5–8 mm long, more or less globose or broadly oblong-obovoid, with 3 or 4 stones, black at maturity. $2n=24$. April–June.

Introduced, scattered sporadically, mostly in the eastern half of the state (native of Europe, Asia; introduced widely in the northern U.S. south to Utah, Tennessee, and Virginia; Canada). Bottomland forests, mesic upland forests, banks of streams, and edges of upland prairies; also fencerows, cemeteries, and roadsides.

The *R. cathartica* complex comprises a group of closely related species of Eurasian buckthorns in *Rhamnus* subg. *Rhamnus*. The other two taxa reported as escapes in Missouri have commonly been misdetermined as *R. cathartica* by some collectors. The proliferation of these three morphologically similar species may be attributable to historical misdetermination of nursery stocks sold in this country. Although *R. cathartica* is still the most abundant member of the complex in Missouri and the one that has caused the most problems for land managers as an invasive exotic species, there is evidence that *R. davurica* is becoming more widespread in recent decades. Where members of this complex become established, they can spread aggressively by seed and eventually can form nearly impenetrable thickets. Seltzner and Eddy (2003) showed that the fruits and leaves of *R. cathartica* can have an allelopathic effect, that is, substances that leach from the fruits and leaves can inhibit the germination and growth of other plant species.

Steyermark (1963) noted that the dark green foliage of *R. cathartica* persists late into the autumn and that the bark, leaves, and fruits have a strong purgative effect when ingested. He further noted that a fruit extract was used historically in the preparation of a stain for maps and a pigment used by artists known as sap green. Although the species was not then considered to be a big problem in native midwestern ecosystems, Steyermark (1963) discussed that: "It seeds easily and sometimes becomes too plentiful in some areas."

3. Rhamnus davurica Pall. (Dahurian buckthorn)

R. citrifolia (Weston) W.J. Hess & Stearn, an illegitimate name

Map 2398

Plants shrubs or small trees, 2–10 m tall, dioecious, often incompletely so. Main stems 1 to several, the main branches ascending to loosely ascending and elongate, also with shorter, spreading to loosely ascending, relatively straight branches, these mostly thorn-tipped. Bark grayish brown to reddish brown, sometimes with lighter blotches, relatively smooth, but with relatively prominent, raised lenticels in cross lines and branch scars, somewhat peeling on older, larger stems. Twigs relatively stout, slightly flattened, gray to reddish brown, with small, dark lenticels, glabrous, the winter buds narrowly ovoid, with several overlapping scales, these light brown, glabrous except for marginal hairs. Leaves mostly opposite (occasionally subopposite on new growth, the petioles 5–25 mm long. Leaf blades 4–13 cm long, mostly 2–3 times as long as wide, oblong-lanceolate to more commonly oblong-elliptic, elliptic, oblong-oblanceolate, or oblanceolate, mostly widest at or above the midpoint, angled or tapered at the base, tapered to a sharply pointed tip, the upper surface green to dark green, glabrous, slightly shiny, the undersurface light green, glabrous, the lateral veins 3–6 pairs, these strongly arched toward the blade tip. Inflorescences axillary, small clusters of 2–20 staminate or pistillate flowers or occasionally reduced to solitary flowers, the clusters sessile, the individual flower stalks 6–18 mm long. Flowers imperfect. Sepals 4, 1.3–2.0 mm long. Petals 4, 1.0–1.3 mm long in staminate flowers, 0.6–0.9 mm long in pistillate flowers, lanceolate, entire at the tips. Style noticeably 2-branched toward the tip. Fruits 5–7 mm long, more or less globose, with 2 stones, black at maturity. $2n=24$. April–June.

Introduced, uncommon, sporadic (native of Asia; introduced in the northeastern U.S. west to North Dakota, Nebraska, and Missouri). Disturbed, mesic upland forests and banks of streams; also pastures and old fields.

This species was first reported for Missouri by Basinger (2002), although older specimens had already been collected that had been misdetermined as *R. cathartica*. See the treatment of that species for further discussion. Naturalized populations of *R. davurica* in North America are not clearly assignable to the varieties that are rec-

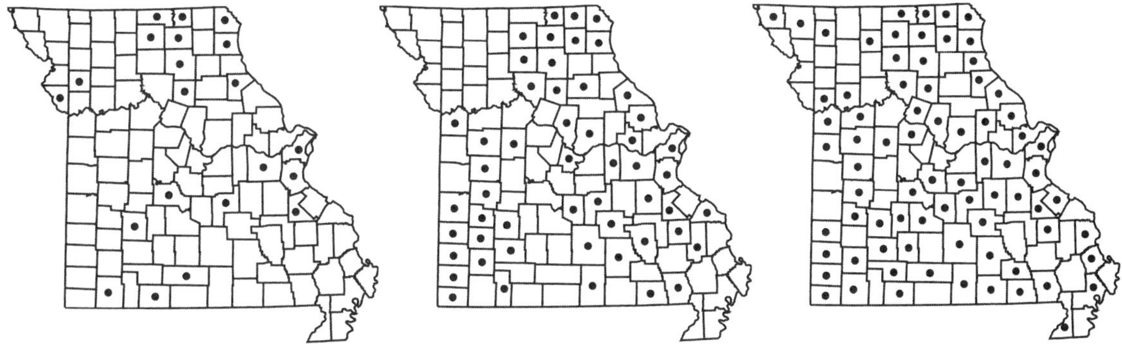
2401. Agrimonia gryposepala 2402. Agrimonia parviflora 2403. Agrimonia pubescens

ognized by some Old World botanists within the native range of the species.

4. Rhamnus japonica Maxim. (Japanese buckthorn)

Map 2399

Plants shrubs or occasionally small trees, 1–5 m tall, dioecious, often incompletely so. Main stems 1 to several, the main branches ascending to loosely ascending and elongate, also with shorter, spreading to loosely ascending, relatively straight branches, these mostly thorn-tipped. Bark gray to reddish brown, sometimes with lighter blotches, relatively smooth, but with relatively prominent, raised lenticels in cross lines and branch scars, somewhat peeling on older, larger stems. Twigs relatively stout, slightly flattened, gray to yellowish brown, with small, dark lenticels, initially often minutely hairy, but soon glabrous or nearly so, the winter buds narrowly ovoid, with several overlapping scales, these reddish brown, glabrous except for marginal hairs. Leaves mostly opposite (occasionally subopposite on new growth, the petioles 5–18 mm long. Leaf blades 2–8 cm long, 1.5–3.0 times as long as wide, obovate to elliptic-oblanceolate or broadly oblanceolate, mostly widest at or below the midpoint, angled at the base, tapered to a sharply pointed tip or the smaller leaves sometimes merely angled or rounded, the upper surface green to dark green, minutely hairy, at least toward the tip and/or along the main veins, slightly shiny, the undersurface light green, glabrous, the lateral veins mostly 4 or 5 pairs, these strongly arched toward the blade tip. Inflorescences axillary, small clusters of 2–6 staminate or pistillate flowers or reduced to solitary flowers, the clusters sessile, the individual flower stalks 3–9 mm long. Flowers imperfect. Sepals 4, 1.3–2.0 mm long. Petals 4, 1.0–1.3 mm long in staminate flowers, 0.6–0.9 mm long in pistillate flowers, lanceolate, entire at the tips. Style noticeably 2-branched toward the tip. Fruits 5–7 mm long, more or less globose, with 2 stones, black at maturity. May–June.

Introduced, uncommon, known thus far only from single specimens from Boone and Jackson Counties (native of Japan; introduced sporadically in Illinois, Missouri). Disturbed, mesic upland forests.

The first collection of this species to date (*Smith & McKenzie 3569* on 19 May 2000), which is in the Missouri Botanical Garden Herbarium, originally was misdetermined as the superficially very similar *R. cathartica*. It was redetermined during studies by Guy Nesom toward a forthcoming treatment of the genus in the *Flora of North America* series. Elsewhere in the United States, this taxon is still known thus far only from the Chicago metropolitan area (Swink and Wilhelm, 1994), where it has become increasingly frequent during the past few decades.

5. Rhamnus lanceolata Pursh

R. lanceolata ssp. *glabrata* (Gleason) Kartesz & Gandhi

R. lanceolata var. *glabrata* Gleason

Pl. 522 a–c; Map 2400

Plants shrubs, 1–3 m tall, dioecious. Main stems usually several, the branches all ascending to loosely ascending and elongate, none of them thorn-tipped. Bark gray, sometimes with lighter blotches, relatively smooth, but with relatively prominent lenticels in raised cross lines, somewhat peeling on older, larger stems. Twigs slender, reddish brown, becoming gray with age, initially often minutely hairy, but soon glabrous or nearly so, the winter buds ovate in outline, somewhat flattened, with several overlapping scales, these reddish brown, glabrous except for sparse, marginal hairs near the tip. Leaves alternate (occasionally a few appearing subopposite), the petioles 3–9 mm long. Leaf blades 2–8 cm long, 2–4 times as long as wide, rounded or angled at the base, angled or tapered to a bluntly or more commonly sharply pointed tip, the upper surface green to dark green, glabrous or minutely hairy, mostly along the veins, shiny, the undersurface light green, glabrous to densely and minutely hairy, especially along the veins, the lateral veins mostly 4–7 pairs, these more or less arched toward the blade tip. Inflorescences axillary, small clusters of 2 or 3 staminate flowers or solitary (rarely paired) pistillate flowers, the clusters sessile, the individual flower stalks 2–3 mm long. Flowers imperfect. Sepals 4, 1.3–2.0 mm long. Petals 4, 1.0–1.2 mm long in staminate flowers, 0.5–0.9 mm long in pistillate flowers, broadly obovate, notched at the tips. Style noticeably 2-branched toward the tip. Fruits 4–7 mm long, globose, with 2 stones, black at maturity. April–June.

Scattered nearly throughout the state, but apparently absent from the Mississippi Lowlands Division (eastern U.S. west to South Dakota and Texas). Glades, upland prairies, ledges and tops of bluffs, banks of streams and rivers, savannas, and openings of dry upland forests.

This species can easily be overlooked in its natural habitats. In the field, plants sometimes are mistaken for *Ilex decidua* Walter (deciduous holly, Aquifoliaceae). Plants with the leaf blades glabrous or nearly so on the undersurface, which are the common phase in Missouri, have been called var. *glabrata*. Plants with relatively densely and minutely hairy leaves (var. *lanceolata*) are far less common and occur mainly in the eastern half of the state. In her taxonomic revision of the *R. serrata* complex, Johnston (1975) considered these to represent merely trivial forms of a variable species. Intermediate plants are relatively widespread in the midwestern states.

ROSACEAE (Rose Family)

Plants annual or more commonly perennial, herbs, shrubs, or trees, sometimes armed with thorns or prickles. Leaves alternate (opposite in *Rhodotypos* and sometimes in *Geum*) or all basal. Stipules present, herbaceous, scalelike, or spinescent. Leaf blades simple or compound, the margins usually toothed or lobed, the teeth often with small glandular points. Inflorescences axillary and/or terminal clusters, racemes, spikes, panicles, or umbels, rarely solitary flowers. Flowers usually perfect (except in the mostly dioecious *Aruncus*), actinomorphic (slightly zygomorphic in *Gillenia*), perigynous or epigynous, often with a noticeable hypanthium (floral cup or disc), but sometimes appearing nearly hypogynous. Sepals 4 or 5, free, sometimes appearing fused basally in species with a well-developed hypanthium, often persistent at fruiting. Petals 4 or 5 (to numerous in some horticultural variants), rarely absent, free. Stamens (1–)10 to numerous, often in multiples of 5, usually free, the anthers with 2 locules, dehiscing by longitudinal slits. Pistils 1 to many per flower, each apparently with 1–5 carpels. Ovary superior or inferior, with 1 or less commonly 2 locules, each with 1 to few ovules. Styles 1–5, free or fused toward the base, each with 1 stigma, this minute to disc-shaped and terminal or linear and lateral toward the style tip. Fruits diverse. About 100 genera, about 3,000 species, worldwide.

The Rosaceae are an extremely diverse group morphologically, and the characters that hold members of the family together may not be immediately obvious to botany students. A complex classification exists involving 3 or more subfamilies with up to 13 tribes, but many details of generic relationships remain controversial. The most recognizable traditionally accepted subfamilies are: Spiraeoideae Arn. (including *Aruncus, Gillenia, Physocarpus*, and *Spiraea*), with flowers containing 2 or more pistils that ripen into follicles; Amygdaloideae Arn. (in Missouri, including only *Prunus*), with the flowers having only 1 pistil ripening into a drupe; and Maloideae C. Weber (including *Amelanchier, Aronia, Chaenomeles, Crataegus, Malus, Pyracantha*, and *Pyrus*), with inferior ovaries ripening into a specialized fruit type known as a pome, which is a fleshy fruit with a core containing seeds surrounded by the hardened or papery, modified inner ovary walls. Most recently, the Amygdaloideae and Maloideae have been suggested to represent merely specialized tribes within the subfamily

Spiraeoideae (Potter et al., 2007). The other subfamily, Rosoideae, with indehiscent fruits of various kinds, consists of genera with a diversity of habits, flower types, and other features.

Various members of the family have great economic value as fruit plants, ornamentals, ingredients in cosmetics and soaps, and minor timber trees. A number of species are used medicinally and as sources of vitamins.

1. Plants annual or perennial herbs, sometimes woody at the very base
    2. Stems armed with prickles, sometimes also with long, spreading bristles . . . . 25. RUBUS
    2. Stems not armed, lacking thorns and prickles
        3. Leaves less than 1 cm long, the blades deeply 3-lobed, the lobes each 3-lobed or 3-toothed; flowers in small clusters opposite the leaves, mostly hidden by the leaflike stipules; petals absent; stamen 1 per flower . . . . 3. APHANES
        3. Leaves more than 1 cm long, the blades of at least the larger or lower leaves variously compound; flowers solitary or in various inflorescence types, but not hidden by the stipules; petals present; stamens 10 to numerous
            4. Leaf blades ternately or ternately then pinnately 2–3 times compound; flowers mostly unisexual (plants incompletely dioecious) . . . . 5. ARUNCUS
            4. Leaves pinnately or palmately once-compound or with only 3 leaflets; flowers perfect
                5. At least some of the leaves with the blades pinnately once-compound
                    6. Pistils 2 per flower (usually 1 aborting before fruiting); fruits contained within and hidden by the hardened hypanthium; hypanthium covered with hooked bristles . . . . 1. AGRIMONIA
                    6. Pistils 5 to numerous; fruits free from and not hidden by the hypanthium; hypanthium glabrous or hairy, but without hooked bristles (do not confuse the hooked styles of *Geum* with the hooked hypanthium bristles of *Agrimonia*)
                        7. Styles noticeably jointed at or above the middle, the two portions dissimilar, the apical portion shed as the fruit matures, the basal portion hooked at the tip and becoming elongated, persistent at fruiting . . . . 14. GEUM
                        7. Styles not jointed, withering after flowering or, if persistent, then not conspicuous or hooked at the tip
                            8. Terminal leaflet palmately 5–9-lobed; pistils 5–15, ripening into a single whorl of fruits . . . 12. FILIPENDULA
                            8. Terminal leaflet unlobed, 3-lobed, or pinnately lobed; pistils numerous, ripening into a headlike cluster of fruits
                                9. Styles attached laterally near the ovary base; fruits lacking a lateral appendage . . . 9. DRYMOCALLIS
                                9. Styles attached at the ovary tip or nearly so; fruits with a corky, winglike appendage along the inner side . . . . 19. POTENTILLA
                5. Leaf blades palmately once-compound or with only 3 leaflets (the uppermost leaves sometimes simple)

9. Styles noticeably jointed at or above the middle, the two portions dissimilar, the apical portion shed as the fruit matures, the basal portion hooked at the tip and becoming elongated, persistent at fruiting . . . . 14. GEUM
9. Styles not jointed, withering after flowering or, if persistent, then not conspicuous or hooked at the tip
    10. Flowers not subtended by bractlets at maturity; pistils 2–6; hypanthium cup-shaped or obconic
        11. Plants with erect or ascending, leafy stems; leaves sessile or with petioles less than 1 cm long; petals white or pinkish-tinged; fruits follicles, not hidden by the cup-shaped hypanthium . . . . 15. GILLENIA
        11. Plants with rhizomes, the leaves in a basal rosette, with petioles 4–15 cm long; petals yellow; fruits achenes, more or less hidden in the obconic hypanthium . . . 14. GEUM
    10. Flowers closely subtended by 5 bractlets, the calyx thus appearing more or less composed of 10 sepals; pistils numerous, ripening into a headlike cluster of fruits or the fruits covering the surface of an expanded receptacle; the hypanthium saucer-shaped
        12. Bractlets noticeably wider than the sepals, prominently 3-toothed or -lobed at the tip . . . . 10. DUCHESNEA
        12. Bractlets mostly similar in width to the sepals, entire or occasionally 2-lobed at the tip, rarely with a few inconspicuous teeth along the sides
            13. Leaves all basal; petals white, rarely tinged with pink; receptacle becoming greatly enlarged, fleshy, and red at fruiting, the tiny brown to black achenes scattered over the surface . . . . 13. FRAGARIA
            13. Leaves basal and alternate on the stems (sometimes appearing all basal in *Potentilla canadensis*); petals cream-colored or pale to bright yellow, sometimes becoming nearly white when dried; receptacle only slightly enlarged at fruiting, not visible between the densely overlapping green to straw-colored or brown achenes . . . . 19. POTENTILLA

1. Plants trees or shrubs, strongly woody
    14. At least some of the leaves compound; stems armed with spines and/or prickles, sometimes also with long, spreading bristles

15. Leaves pinnately compound (some of the leaves reduced to 3 leaflets or rarely simple); hypanthium urn-shaped, concealing the ovaries; fruits hips (the hypanthium becoming fleshy and usually reddish-colored, enclosing several to many achenes) . . . . 24. ROSA
15. Leaves palmately compound or with only 3 leaflets; hypanthium flat (at flowering) or elongate, the ovaries exposed on the surface; fruits headlike masses of drupelets . . . . 25. RUBUS

14. Leaves all simple, sometimes lobed; stems unarmed or sometimes armed with thorns
16. Leaves opposite; flowers with 4 petals . . . . 23. RHODOTYPOS
16. Leaves alternate; flowers mostly with 5 petals (4-petaled flowers occasionally found among the 5-petaled flowers)
17. Ovaries 1–5, superior, separate or fused only toward the base (fused their entire length in *Exochorda*); fruits solitary drupes or capsules, or 2 or more follicles or achenes
18. Ovary 1 per flower; fruits solitary drupes or capsules
19. Fruits 5-lobed capsules, brown to gray at maturity; seeds mostly 1 per locule, winged; stipules absent . . . . 11. EXOCHORDA
19. Fruits drupes, red to black or bluish black at maturity; seed 1, embedded in a hard stone, both the seed and the stone unwinged; stipules present, but sometimes small, shed early . . . . 20. PRUNUS
18. Ovaries 2–5 per flower; fruits follicles or achenes
20. Petals absent; sepals 3.5–10 mm long, deeply toothed above the middle; stamens numerous (100 or more) . . . . 17. NEVIUSIA
20. Petals present, white or pink; sepals 0.8–4.5 mm long, the margins entire; stamens 15 to numerous (50 or fewer)
21. Stipules present, herbaceous (on older branches, check for small stipular scars); at least some of the leaves lobed, the margins also toothed. . . . . 18. PHYSOCARPUS
21. Stipules absent; leaves not lobed, the margins only toothed . . . . 26. SPIRAEA
17. Ovaries 2–5, all or mostly inferior, fused together in the basal half or more at flowering (check the number of free styles if unsure) and also to the hypanthium; fruits solitary pomes (1 per flower)
22. Flowers in elongate racemes, some inflorescences rarely reduced to 1 or 2 flowers; petals narrowly oblanceolate to narrowly strap-shaped . . . . 2. AMELANCHIER
22. Flowers solitary or more commonly in rounded to flat-topped clusters, these sometimes branched toward the base or appearing umbellate (sometimes appearing short-racemose at flowering in *Pyrus calleryana*); petals obovate or broadly elliptic to nearly circular
23. Petals usually bright red, rarely pink; seeds numerous in each locule . . . . 6. CHAENOMELES
23. Petals white to pink; seeds 1 or 2 per locule, the whole fruit thus with 10 or fewer seeds or stones
24. At least some of the leaf blades lobed
25. Carpel walls turning hard and bony, the pome thus with 2–5 seedlike stones, each containing 1 seed; thorns usually present, determinate, never elongating into leafy branches . . . . 8. CRATAEGUS
25. Carpel walls turning papery or leathery, the pome thus with 2–10 easily exposed seeds surrounded by the “core”

of leathery or papery remains of the carpel walls; thorns, if present, indeterminate, consisting of short branches that eventually become elongated and leafy . . . . 16. MALUS

24. Leaf blades all unlobed
  26. Leaf blades with a line of small peglike reddish purple glands on the upper surface along the midrib; fruits purplish black to black at maturity; petals 4–7 mm long . . . . 4. ARONIA
  26. Leaf blades glabrous or hairy, but lacking a row of glands along the midrib; fruits red to brown at maturity, sometimes darkening upon drying; petals 7–18 mm long
    27. Carpel walls turning papery or leathery, the pome thus with 2–10 easily exposed seeds surrounded by the "core" of leathery or papery remains of the carpel walls; thorns, if present, indeterminate, consisting of short branches that eventually become elongated and leafy
      28. Styles fused toward the base, the fused portion with dense white hairs; petals pink or pinkish white, often fading to white with age; pulpy portion of the fruit lacking evident stone cells . . . . 16. MALUS
      28. Styles free to the base, glabrous; petals white at flowering, sometimes somewhat pinkish-tinged in bud; pulpy portion of the fruit with small clusters of gritty stone cells . . . . 22. PYRUS
    27. Carpel walls turning hard and bony, the pome thus with 2–5 seedlike stones, each containing 1 seed; thorns, if present, determinate, never elongating into leafy branches
      29. Leaf blades not toothed or lobed; thorns absent; sepals at fruiting appearing as small, fleshy, incurved teeth . . . . 7. COTONEASTER
      29. Leaf blades toothed and/or lobed; thorns usually present, but sometimes sparse; sepals shed as the fruits mature or, if persistent at fruiting, then remaining herbaceous and spreading to downward-curved
        30. Leaves deciduous; thorns usually determinate, rarely become elongated or leafy; styles (1–)2–5, free or less commonly fused at the base, the stigmas club-shaped; ovules usually 2 per locule but only 1 of these fertile (stones 1-seeded) . . . . 8. CRATAEGUS
        30. Leaves more or less evergreen or partially overwintering; thorns mostly indeterminate (eventually growing into branches) and often with 1 or a few leaves toward the base, even when short; styles 5, fused at the base, the stigmas more or less capitate; ovules 2 per locule (stones 2-seeded) . . . . 21. PYRACANTHA

1. **Agrimonia** L. (agrimony)

Plants perennial herbs, lacking spines and thorns, with short usually knotty rhizomes. Stems erect or ascending. Leaves alternate and often also basal, pinnately compound with small leaflets interspersed among the larger primary ones, short-petiolate, the leaf blade elliptic to broadly obovate in overall outline, the leaflets with the margins coarsely toothed. Stipules leaflike, the pair at each node fused to the petiole toward the base (those of basal leaves fused most of their length with only the slender tips free), lanceolate to broadly and asymmetrically ovate, rounded or somewhat cordate at the base, the outer margin with several lobes or coarse teeth (sometimes more finely toothed in *A. rostellata*). Inflorescences terminal, spikelike racemes, the flower stalks very short, each subtended by a small deeply lobed bract (the hypanthium also subtended by a pair of inconspicuous 3-lobed bractlets). Flowers ascending, becoming spreading to reflexed as the fruits mature, perigynous, the hypanthium appearing obconic, deeply cup-shaped with a nectar disc nearly closing the opening, armed with dense hooked bristles toward the rim, developing 10 longitudinal ridges and grooves as the fruits mature. Sepals 5, short, oblong-elliptic, spreading at flowering, becoming erect or somewhat incurved with age and developing into a small beak on the fruit. Petals 5, 2–6 mm long, broadly elliptic to nearly circular, yellow. Stamens 5–15. Pistils 2 per flower. Ovaries superior, hidden in the hypanthium, with 1 ovule. Style 1 per ovary, terminal, the stigma somewhat 2-lobed. Fruits consisting of the hardened obconic hypanthium containing 1(2) small globose to somewhat ovoid achene, the hooked bristles becoming somewhat elongated and hardened. About 18 species, North America, South America, Europe, Asia.

The fruits of *Agrimonia* species are dispersed when the hooked bristles along the hypanthium margin become entangled in the fur (or clothing) of passing mammals. The glands referred to in the key and descriptions are somewhat sticky and resinous, sessile, yellow to yellowish brown (sometimes darkening upon drying), and globose to depressed-globose on inflorescence axes. When present on leaflets, they appear often flattened or even impressed, especially when dried.

Species of agrimony were used medicinally by various tribes of Native Americans, mostly for intestinal problems, particularly diarrhea. Bush (1916) and Palmer and Steyermark (1935) reported specimens of *A. microcarpa* Wallr. (low agrimony) from various parts of Missouri, many of these under the name *A. platycarpa* Wallr., which is now considered a synonym of the former species. Steyermark (1963) excluded this species from the flora, stating that the specimens actually represented a mixture of *A. pubescens* and *A. rostellata*. *Agrimonia microcarpa* grows to the east and south of Missouri and differs from other Missouri agrimonies most noticeably in its leaves with usually only 3 primary leaflets.

1. Median and lower leaves with 11–23 primary (larger) leaflets . . . . . 2. A. PARVIFLORA
1. Leaves all with 3–9 primary (larger) leaflets
    2. Axis of the inflorescence not glandular, but densely pubescent with short ascending hairs and usually also sparse, long, spreading hairs; leaflets with the undersurface not glandular, but densely pubescent with appressed velvety hairs; fruiting hypanthium minutely hairy in the grooves, often also sparsely hairy along the ridges, not glandular . . . . 3. A. PUBESCENS
    2. Axis of the inflorescence sparsely to densely glandular, often also hairy; leaflets with the undersurface usually glandular and often also sparsely short-hairy, especially along the veins; fruiting hypanthium not hairy but often glandular in the grooves, also glandular and occasionally hairy along the ridges
        3. Fruiting hypanthium at maturity 3–5 mm long (fruits 6–8 mm long, including the beak); axis of the inflorescence glandular and also pubescent with sparse to moderate, long, spreading hairs, these mostly more than 1 mm long, longer than the width of the axis at the attachment point . . . . 1. A. GRYPOSEPALA
        3. Fruiting hypanthium at maturity 2.0–2.5 mm long (fruits 3.5–4.5 mm long, including the beak); axis of the inflorescence glandular (sometimes sparsely so) and usually also with sparse to moderate short ascending hairs toward the base (rarely only glandular), occasionally also with sparse, longer, spreading hairs, but these 1 mm or shorter, mostly shorter than the width of the axis at the attachment point . . . . 4. A. ROSTELLATA

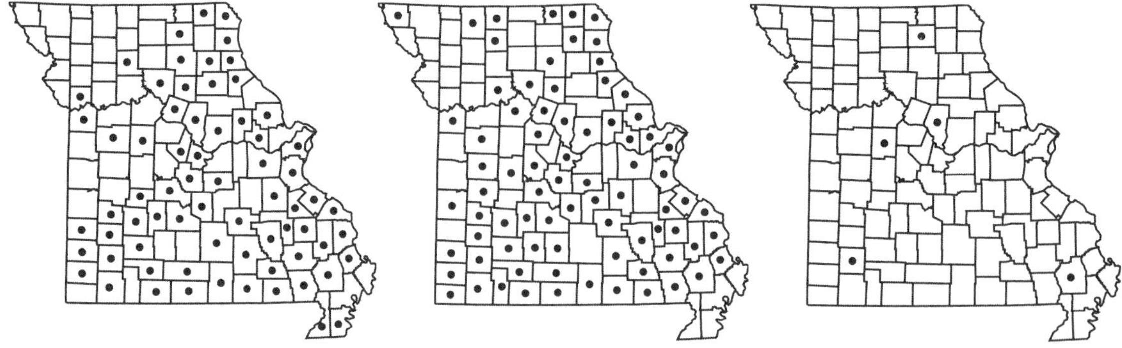

2404. Agrimonia rostellata 2405. Amelanchier arborea 2406. Amelanchier humilis

1. Agrimonia gryposepala Wallr. (tall agrimony, hooked agrimony)

Pl. 523 d, e; Map 2401

Roots all fibrous. Stems 30–150 cm long, glandular and sparsely to densely pubescent with long, spreading hairs. Leaves 6–30 cm long, the primary (larger) leaflets (3–)5–9, these 3.0–11.5 cm long, elliptic or obovate to broadly elliptic or ovate-lanceolate, mostly tapered to a sharply pointed tip, the margins with mostly relatively blunt teeth, the upper surface glabrous or with sparse, long, spreading hairs, the undersurface glandular and with sparse to moderate, short, appressed hairs, especially along the veins, sometimes also with a few longer, spreading hairs. Inflorescence axis glandular and also pubescent with sparse to moderately long, spreading hairs, these mostly more than 1 mm long, longer than the width of the axis at the attachment point. Stamens mostly 15. Hypanthium at fruiting 3–5 mm long (fruits 6–8 mm long, including the beak), deeply grooved, glandular, sometimes also sparsely hairy toward the base along the ridges. $2n=28$. July–September.

Uncommon in the Glaciated Plains Division with a disjunct occurrence in Douglas County (U.S., except some western and southern states; Canada). Mesic to dry upland forests, upland prairies, savannas, and banks of streams.

Immature specimens of *A. gryposepala* can be difficult to separate from *A. rostellata*. In general, plants of *A. rostellata* tend to have more slender stems and somewhat smaller, narrower, more delicate leaves than does *A. gryposepala*, but there is too much overlap in these quantitative features to allow their incorporation into a key.

2. Agrimonia parviflora Aiton (swamp agrimony, many-flowered agrimony, harvest lice)

Pl. 523 f, g; Map 2402

Roots usually all fibrous (rarely tuberous-thickened). Stems 30–120(–200) cm long, glandular and densely pubescent with short, ascending and long, spreading hairs, the hairs becoming sparser toward the tip. Leaves 3–30 cm long, the median and lower leaves with the primary (larger) leaflets 11–23, these 1.5–7.0 cm long, mostly narrowly elliptic or lanceolate, tapered to a sharply pointed tip, the margins with mostly relatively sharply pointed teeth, the upper surface glabrous or nearly so, the undersurface glandular and sparsely to moderately short-hairy, especially along the veins, sometimes also with sparse, longer, spreading hairs. Inflorescence axis glandular and pubescent with short ascending and often also sparse, long, spreading hairs. Stamens 5–10. Hypanthium at fruiting 2–3 mm long (4–5 mm long, including the beak), deeply grooved, glandular, sometimes also sparsely hairy along the ridges. $2n=56$. July–August.

Scattered nearly throughout the state, but apparently absent from portions of northwestern and southeastern Missouri (eastern U.S. west to South Dakota and Texas; Canada, Mexico, Caribbean Islands). Bottomland forests, mesic upland forests in ravines, banks of streams and spring branches, swamps, fens, bottomland prairies, upland prairies; also roadsides, ditches, margins of crop fields, and moist disturbed areas.

In addition to having leaves with more primary leaflets, the inflorescences of this species tend to have more and denser flowers than in other Missouri agrimonies.

3. Agrimonia pubescens Wallr. (downy agrimony, soft agrimony)

Pl. 523 l, m; Map 2403

Roots fibrous mixed with tuberous-thickened. Stems 30–100(–120) cm long, moderately to densely pubescent with short, ascending to incurved, and long, spreading to reflexed hairs. Leaves 1.5–30.0 cm long, the primary (larger) leaf-

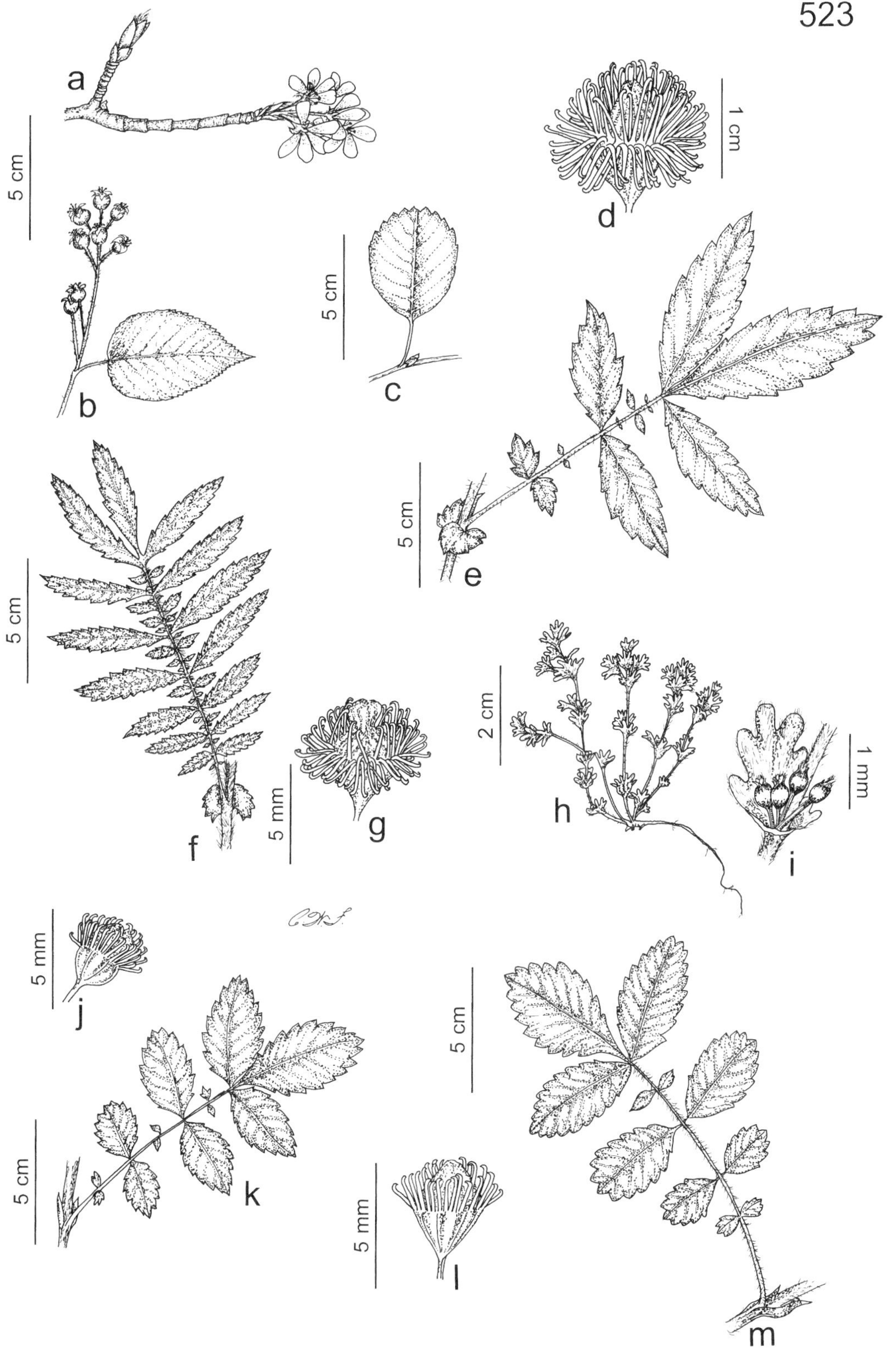

Plate 523. Rosaceae. *Amelanchier arborea*, **a)** flowering branch, **b)** fruiting branch. *Amelanchier humilis*, **c)** leaf. *Agrimonia gryposepala*, **d)** fruit, **e)** node with stipules and leaf. *Agrimonia parviflora*, **f)** node with stipules and leaf, **g)** fruit. *Aphanes australis*, **h)** habit, **i)** node with fused stipules and fruits. *Agrimonia rostellata*, **j)** fruit, **k)** node with stipules and leaf. *Agrimonia pubescens*, **l)** fruit, **m)** node with stipules and leaf.

lets 3–9, these 1–10 cm long, mostly lanceolate to elliptic or narrowly obovate, narrowed or tapered to a bluntly or sharply pointed tip, the margins with mostly relatively blunt teeth, the upper surface glabrous or pubescent with sparse to moderate, short, appressed hairs along the midrib and sometimes also sparse, long, spreading hairs, the undersurface not glandular, but densely velvety-pubescent with hairs of various lengths and orientations. Inflorescence axis not glandular, but densely pubescent with short ascending hairs and usually also sparse, long, spreading hairs. Stamens mostly 10. Hypanthium at fruiting 2.5–5.0 mm long (5–7 mm long, including the beak), deeply grooved, not glandular, but minutely hairy in the grooves, usually with sparse longer hairs along the ridges. July–September.

Scattered to common nearly throughout the state, although apparently absent from portions of southeastern Missouri (eastern U.S. west to South Dakota and Oklahoma; Canada). Bottomland forests, mesic to dry upland forests, savannas, glades, upland prairies, banks of streams and rivers, and tops of bluffs; also old fields.

4. Agrimonia rostellata Wallr. (woodland agrimony)

Pl. 523 j, k; Map 2404

Roots fibrous mixed with tuberous-thickened. Stems 25–70(–100) cm long, glandular and sparsely pubescent with short, ascending and long, spreading hairs, especially toward the base. Leaves 3–20 cm long, the primary (larger) leaflets 3–9, these 1.5–10.0 cm long, mostly elliptic to broadly elliptic or obovate, narrowed or tapered to a bluntly or sharply pointed tip, the margins with mostly relatively blunt teeth, the upper surface glabrous or nearly so, the undersurface glandular (sometimes sparsely so) and often also sparsely pubescent with short, appressed and long, spreading hairs along the veins. Inflorescence axis glandular (sometimes sparsely so) and usually also with sparse to moderate, short, ascending hairs toward the base (rarely only glandular), occasionally also with sparse, longer, spreading hairs, but these 1 mm or shorter, mostly shorter than the width of the axis at the attachment point. Stamens 10–15. Hypanthium at fruiting 2.0–2.5 mm long (fruits 3.5–4.5 mm long, including the beak), often only faintly grooved, glandular but otherwise glabrous. July–September.

Scattered nearly throughout the state, but apparently absent from portions of the Glaciated Plains Division (eastern U.S. west to Minnesota and Texas). Bottomland forests, mesic to dry upland forests, savannas, margins of glades, ledges and tops of bluffs, and banks of streams.

See the treatment of *A. gryposepala* for comments on difficulties in distinguishing *A. rostellata* from that species.

2. **Amelanchier** Medik. (shadbush, serviceberry)

Plants trees or shrubs, sometimes forming colonies, lacking spines and thorns. Bark gray, smooth, becoming shallowly furrowed and with scaly ridges with age. Twigs orangish brown to purplish brown, with small white lenticels, glabrous, usually glaucous, the leaf buds narrowly elliptic lanceolate, sharply pointed at the tip, with several overlapping scales, these greenish yellow to light brown, at maturity often strongly tinged with red or purple. Leaves alternate, folded lengthwise during development, relatively long-petiolate, the petioles lacking glands, densely pubescent with soft matted hairs when young, glabrous or nearly so at maturity. Stipules 9–15 mm long, linear, pinkish to reddish purple, shed during leaf development. Leaf blades simple, unlobed, the margins sharply toothed, especially above the middle. Inflorescences short racemes at the tips of normal branches, produced before the leaves or as the leaves begin to unfold, the axis densely hairy at flowering but becoming nearly glabrous by fruiting, the flowers subtended by slender, reddish purple, hairy bracts that are shed as the flowers open. Flowers epigynous, the hypanthium fused to the ovaries, hairy at flowering, the rim with a nectar disc. Sepals 5, spreading to recurved, hairy on the outer surface, persistent at fruiting. Petals 5, narrowly oblanceolate to narrowly strap-shaped, often somewhat lax or drooping, white or rarely pink. Stamens 15–20, the anthers yellow. Pistils 5. Ovaries inferior, fused along the lateral surfaces and dorsally to the hypanthium at flowering, but with the inner sides free until the fruit develops, each with 2 ovules. Styles 5, fused toward the base, the stigmas more or less disc-shaped. Fruits pomes, globose to somewhat oblong-elliptic in outline, glabrous, with 4–10 easily exposed seeds embedded in the "core" of leathery or papery carpel wall remains and the fleshy portion. About 20 species, North America, Europe, Asia, Africa.

Amelanchier species are browsed by deer and livestock. The fruits are relished by wildlife; in fact, mature fruits are often relatively uncommonly encountered, birds having consumed them as they ripened. Some additional species are cultivated as ornamentals, in particular *A. spicata* (Lam.) K. Koch.

1. Plants shrubs or more commonly small to medium-sized trees, mostly single-trunked; leaf blades narrowed or more commonly tapered to a sharp point at the tip; petals 10–17 mm long; fruits red, becoming reddish purple at maturity, not glaucous . . . . 1. A. ARBOREA
1. Plants shrubs, usually suckering to form colonies; leaf blades rounded, bluntly pointed, or less commonly abruptly short-pointed at the tip; petals 7–10 mm long; fruits reddish purple, becoming purplish black at maturity, glaucous . . . . 2. A. HUMILIS

1. Amelanchier arborea (F. Michx.) Fernald **var. arborea** (downy serviceberry, downy shadbush, sarviss berry, sarviss tree, June berry, shadblow, sugar plum)

Pl. 523 a, b; Map 2405

Plants shrubs or more commonly trees, 2–16 m tall, mostly single-trunked. Leaf blades 4–10 cm long, ovate to obovate, narrowed or more commonly tapered to a sharp point at the tip, rounded or shallowly cordate at the base. Inflorescences loosely ascending or somewhat nodding. Sepals 2.5–3.5 mm long, oblong-triangular, rounded to abruptly short-pointed at the tip. Petals 10–17 mm long. Ovaries usually glabrous at the tip. Fruits 6–10 mm long, turning red, then reddish purple at maturity, not glaucous. Seeds 3.5–5.0 mm long, obliquely elliptic to lanceolate in outline, somewhat flattened, the surface smooth to slightly roughened, dark purplish brown. $2n$=mostly 34. March–May.

Scattered to common nearly throughout the state, but absent from portions of the Mississippi Lowlands (eastern U.S. west to Minnesota and Oklahoma; Canada). Mesic to dry upland forests, often on steep slopes, tops and ledges of bluffs, and margins of glades.

The showy flowers of this species are conspicuous elements of Missouri upland forests in the early spring, and in the autumn the foliage turns pleasing shades of yellow to pale orange often mottled or streaked with various shades of green and red. Trees with pink or pinkish-tinged petals occur rarely and sporadically within populations of white-flowered individuals. Downy shadbush is becoming popular as an ornamental tree in gardens, but, like most members of the subfamily Maloideae, plants can be susceptible to cedar apple rust (see the treatment of *Crataegus*). The wood is sometimes used for handcrafts and ax handles, and is among the heaviest of North American woods. The fruits are relatively bland and mealy when eaten raw, but the flavor intensifies when they are used in baked goods. Native Americans prepared a paste from the fruits, which was dried and mixed with flour in bread making.

2. Amelanchier humilis Wiegand (low shadbush, low serviceberry)

Pl. 523 c; Map 2406

Plants shrubs, 0.3–2.5 m tall, usually suckering to form colonies. Leaf blades 2.5–5.0 cm long, oblong-elliptic, to oblong-obovate, rounded, bluntly pointed, or less commonly abruptly short-pointed at the tip, rounded at the base. Inflorescences ascending to erect. Sepals 2–3 mm long, triangular, narrowed to a sharply pointed tip. Petals 7–10 mm long. Ovaries densely hairy at the tip. Fruits 6–10 mm long, turning reddish purple, then purplish black at maturity, glaucous. Seeds 3.0–4.5 mm long, obliquely elliptic to lanceolate in outline, somewhat flattened, the surface smooth or with faint lines, brown. $2n$=62, 64, 68. April–May.

Introduced, uncommon and widely scattered in Missouri (northeastern U.S. west to North Dakota, Iowa, Indiana, and South Carolina). Margins of prairies; also fence rows and roadsides.

Low shadbush occasionally is cultivated as an ornamental. The fruits have sweet flesh and can be eaten raw, prepared into jellies, or used in baked goods. The nomenclature and taxonomy of this species remain controversial. Some authors (Jones, 1946; Gleason and Cronquist, 1991) place *A. humilis* in synonymy under *A. spicata* (Lam.) K. Koch, but relationships in the complex are clouded by hybridization, polyploidy, and apomixis. Until future studies with new data can clarify species limits in the group, it seems safest to continue using the name *A. humilis* for our plants.

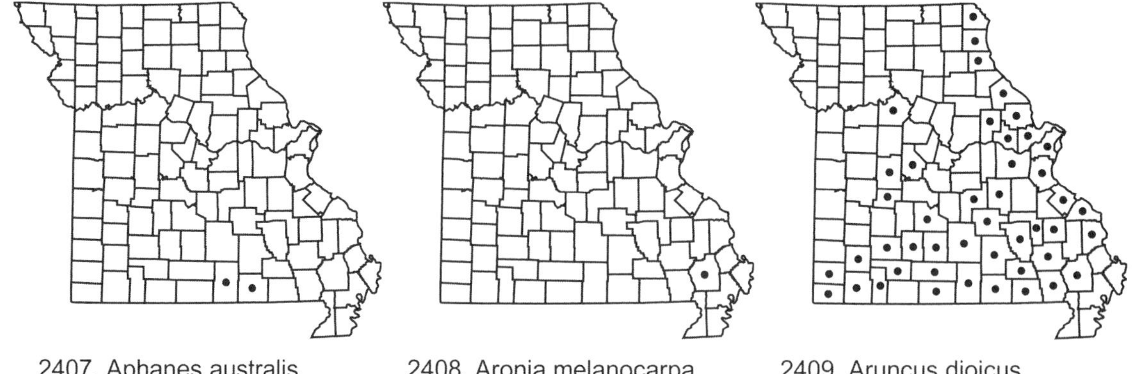
2407. Aphanes australis 2408. Aronia melanocarpa 2409. Aruncus dioicus

3. **Aphanes** L.

About 20 species, North America, South America, Europe, Asia, Africa; introduced elsewhere.

The genus is sometimes treated as a subgenus of annual plants with reduced stamen number within the larger more cosmopolitan *Alchemilla* L. K. R. Robertson (1974) and some other authors have suggested that future studies may reveal that the number of species in *Aphanes* may be fewer than generally has been accepted, with most of the American and other taxa possibly representing mere variants of a couple of European species that have become disseminated widely during the last few centuries.

1. Aphanes australis Rydb. (parsley piert)
A. inexspectata W. Lippert
A. microcarpa (Boiss. & Reut.) Rothm., misapplied
Alchemilla microcarpa Boiss. & Reut., misapplied

Pl. 523 h, i; Map 2407

Plants annual, lacking spines and thorns, the stems and foliage moderately to densely pubescent with silky ascending and/or appressed hairs. Stems 3–10 cm long, spreading to ascending, usually somewhat 4-angled, often reddish-tinged. Leaves alternate, short-petiolate, sometimes appearing sessile. Stipules leaflike, the pair at each node fused in the basal half into a cup and also usually fused to the base of the leaf, palmately several-lobed to about the middle. Leaf blades 3–6 mm long, broadly obovate to nearly circular in outline, deeply 3-lobed, the lobes each 3-lobed or 3-toothed. Inflorescences small clusters opposite the leaves, mostly hidden by the fused stipules. Flowers perigynous, the hypanthium urn-shaped, 0.5–0.7 mm long at flowering, elongating to about 1.5 mm at fruiting, hairy. Sepals 4, less than 0.5 mm long, triangular, hairy, especially along the margins, persistent at fruiting. Petals absent. Stamen 1(2). Pistil 1 per flower. Ovary superior, nearly filling the hypanthium. Style 1, attached toward the base of the ovary, the stigma disc-shaped. Fruits achenes, 1.0–1.2 mm long, elliptic in outline, yellowish brown, glabrous. $2n=16$. April–May.

Introduced, uncommon, known thus far only from Howell County (native of Europe, widely introduced in the southeastern U.S.). Disturbed open areas at edges of upland prairies, roads, and pastures.

Aphanes australis was long known as *A. microcarpa* until Lippert (1984) concluded that true *A. microcarpa* is a different species that is endemic to the western Mediterranean region and has not become introduced into the New World. The next-oldest name applied to our weedy, introduced taxon is *A. australis*. Lippert overlooked this in redescribing the widespread species as *A. inexspectata*.

This species was first reported from Missouri by Yatskievych and Summers (1991, as *A. microcarpa*). It is so inconspicuous that it is likely present at other sites in southern Missouri, but has been overlooked by collectors. In some southeastern states, parsley piert is a lawn weed.

4. **Aronia** Medik. (chokeberry)

Two species, eastern U.S., Canada.

1. Aronia melanocarpa (Michx.) Elliott (black chokeberry)
Photinia melanocarpa (Michx.) K.R. Robertson & J.B. Phipps
Pyrus melanocarpa (Michx.) Willd.
Pl. 524 a–d; Map 2408

Plants shrubs, 0.3–3.0 m tall. Stems lacking spines and thorns. Bark reddish brown to grayish brown, with inconspicuous lighter lenticels, smooth or somewhat roughened. Leaves alternate, rolled during development, mostly relatively short-petiolate, the petioles with scattered small peglike glands in the groove on the upper surface. Stipules 2.5–3.5 mm long, herbaceous, linear to narrowly oblong, the margins with reddish glandular teeth, shed soon after the leaves develop. Leaf blades 1.5–9.0 cm long, simple, unlobed, elliptic to broadly elliptic, mostly long-tapered at the tip and base, the margins finely and sharply toothed, the upper surface glabrous but with a line of small peglike reddish purple glands along the midrib (and sometimes also the main lateral veins), the undersurface glabrous or finely hairy when young. Inflorescences branched clusters at the tips of normal branches, produced as the leaves uncurl, the axis glabrous, the flowers with a small linear bract toward the middle of the stalk, this shed before the flower opens. Flowers epigynous, the hypanthium fused to the ovaries, glabrous. Sepals 5, 1.3–1.8 mm long, spreading, triangular, the margins thin and somewhat irregular or glandular, the inner surface densely woolly, persistent at fruiting. Petals 5, 4–7 mm long, broadly obovate to nearly circular, white, occasionally tinged with pink. Stamens 15–20, the anthers yellow. Pistil 1 per flower. Ovary inferior, the tip densely woolly, with 5 locules, each with 1 ovule. Styles 5, fused toward the base, the stigmas more or less capitate. Fruits berrylike pomes, 7–10 mm long, globose, glabrous, purplish black to black at maturity, with (3–)5 easily exposed seeds embedded in the "core" of inconspicuous papery carpel wall remains and the fleshy portion. $2n=34$. April–May.

Uncommon, known only from the Crowley's Ridge Section of the Mississippi Lowlands, in Stoddard County (eastern U.S. west to Minnesota and Arkansas; Canada). Margins of acid seeps.

The fruits of this species are relatively tart when fresh, but have been used in jellies. This species was first discovered growing in the state by Steyermark (1958a), who catalogued an interesting series of species restricted in Missouri to the sandy ravines along the base of Crowley's Ridge. Steyermark (1963) and some other authors have included *Aronia* in an expanded concept of the genus *Pyrus*, but morphological studies (Phipps et al., 1991; K. R. Robertson et al., 1991) have suggested that this small genus may be more closely allied to the otherwise Asian genus, *Photinia* Lindl.

The other member of the genus, *A. arbutifolia* (L.) Pers. is an eastern and southern species whose range in Arkansas approaches southern Missouri (Hardin, 1973). It differs from *A. melanocarpa* in its more or less hairy leaves, red fruits (that persist on the plants for far longer than do those of *A. melanocarpa*), and leaves that turn bright red (rather than brown) in the autumn.

5. **Aruncus** L. (goat's beard)

One species, North America, Europe, Asia.

1. Aruncus dioicus (Walter) Fernald **var. pubescens** (Rydb.) Fernald (goat's beard)
Pl. 524 e, f; Map 2409

Plants perennial herbs, dioecious, often incompletely so, with short thick woody rhizomes. Stems (0.7–)1.0–2.0 m long, ascending or arching, lacking spines and thorns, glabrous or sparsely hairy toward the tip. Leaves alternate and basal, mostly relatively long-petiolate, at least the lowermost petioles somewhat expanded at the base and partially sheathing the stem, the sheathes extended downward along the stem as fine longitudinal ridges. Stipules absent. Leaf blades 5–50 cm long, ovate to ovate-triangular in outline, 2–3 times irregularly ternately or pinnately compound, the leaflets 2–15 cm long, narrowly to broadly ovate, elliptic, or oblong-elliptic, occasionally with a single deep lobe, long-tapered at the tip, rounded to shal-

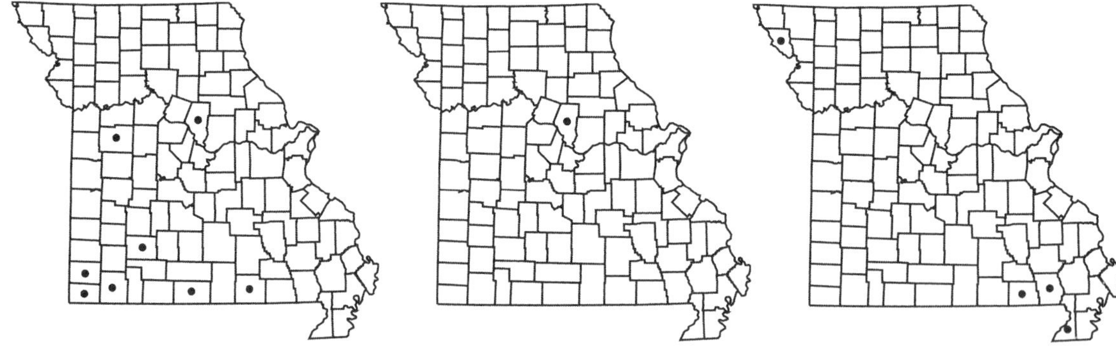

2410. Chaenomeles speciosa 2411. Cotoneaster acutifolius 2412. Crataegus marshallii

lowly cordate at the sessile to long-stalked, often asymmetrical base (this often with minute, reddish purple, stalked glands in addition to hairs at the attachment point to the rachis), the margins sharply and doubly toothed, the upper surface glabrous, sometimes with sparse, minute, stalked, reddish purple glands along the midrib and main lateral veins, the undersurface sparsely to moderately pubescent with appressed, fine, curved hairs, especially along the veins. Inflorescences terminal panicles, the ultimate branches densely racemose, the axes sparsely to densely pubescent with curved hairs, the flowers short-stalked, each subtended by a minute linear to narrowly triangular bract. Flowers mostly unisexual (sometimes a few flowers at the tips of the inflorescence branches perfect on otherwise staminate plants), perigynous, the hypanthium saucer-shaped to shallowly cup-shaped, with a nectar disc around the margin (this poorly developed in pistillate flowers), glabrous or sparsely hairy. Sepals 5, 3–6 mm long, ascending, triangular, glabrous, white to cream-colored in staminate flowers, green and often with thinner pale margins in pistillate flowers, persistent at fruiting. Petals 5, 0.7–1.0 mm long, obovate to elliptic-obovate, white. Stamens (15–)20(–30), highly reduced in pistillate flowers, the anthers yellow. Pistils 2 or 3(–5), highly reduced in staminate flowers. Ovary superior, glabrous, with 1 locule, with 2–4 ovules. Style 1 per pistil, persistent and forming a slender beak at fruiting, the stigma capitate. Fruits follicles dehiscing longitudinally along the inner surface (and often also toward the tip along the outer surface), pendant at maturity, 1.7–2.2 mm long, narrowly oblong to narrowly bean-shaped in outline, glabrous, shiny, brown at maturity, with 2–4 seeds. Seeds 1.5–2.0 mm long, linear to narrowly oblong in outline, 2- or 3-angled and usually narrowly ribbed, the ribs extended as small wings at each end, brown. $2n=18$. May–July.

Scattered in southern and eastern Missouri (Virginia to Iowa and Oklahoma, Washington to California, Alaska; Canada). Bases and ledges of bluffs, mesic upland forests, mostly on rocky lower slopes or along the forest margin, and banks of streams.

Overall, the range of *A. dioicus* extends across most of the eastern U.S. and adjacent Canada westward to the Great Plains, disjunctly from Alaska southward to California, and circumboreally in parts of Europe and Asia. Some botanists accept as many as four varieties in North America (Kartesz and Meacham, 1999) and as many as five species were once recognized for North America (K. R. Robertson, 1974). Most botanists today maintain only two varieties, but these are not sharply distinct, with many intermediate plants and a broad region of overlap. The var. *pubescens* is the predominant phase of the species in the western portion of the range, and differs from the eastern var. *dioicus* in its relatively dull, hairy foliage and more slender follicles.

Goat's beard is an attractive ornamental in gardens, especially staminate-flowered plants, which have somewhat showier inflorescences. Medicinally, the species has been used to treat infections, fevers, and stomach aches, but readers should note that the plants are suspected to contain cyanogenic glucosides (K. R. Robertson, 1974) and thus may be poisonous.

6. **Chaenomeles** Lindl. (flowering quince) (C. Weber, 1964)

Four species, Asia.

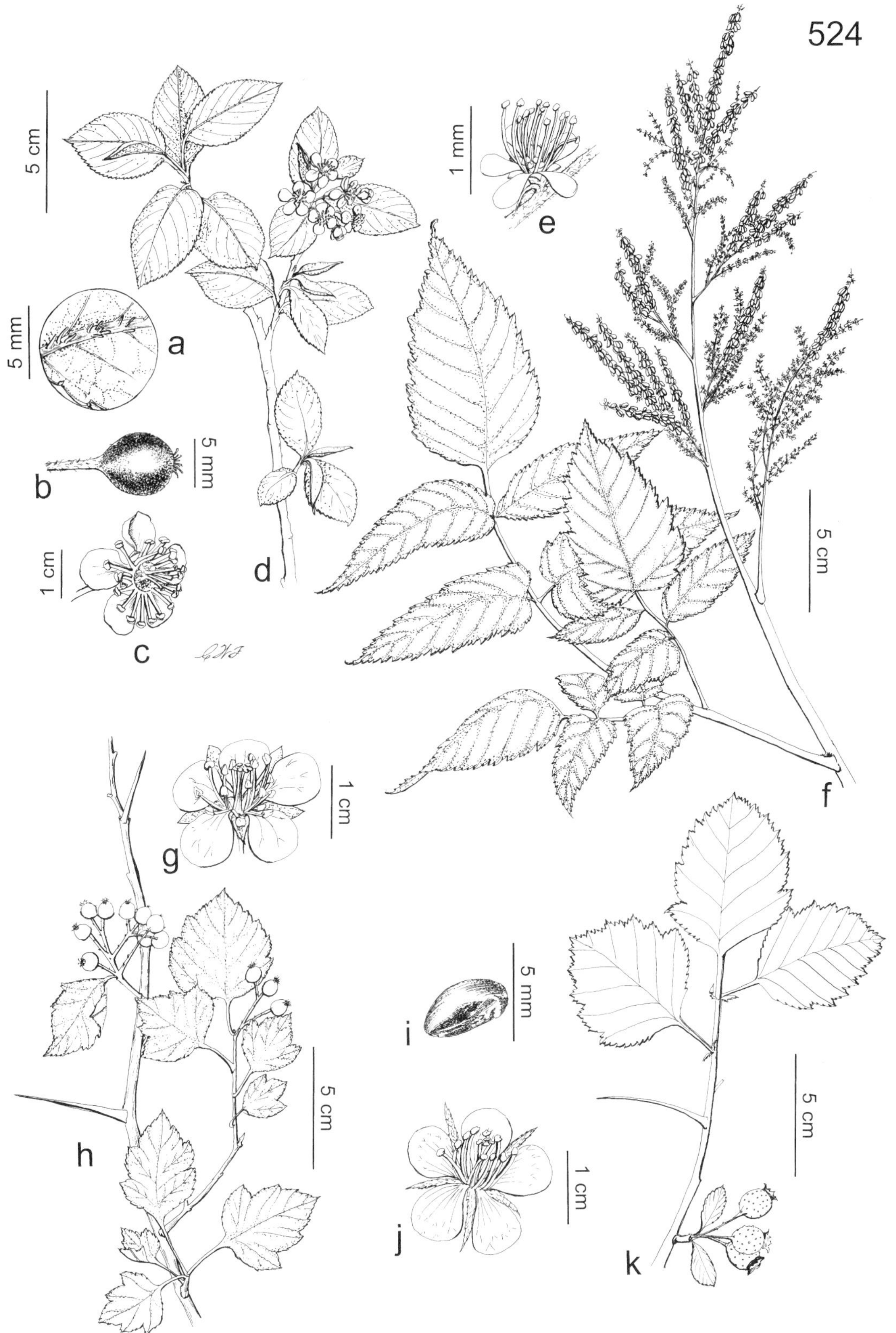

Plate 524. Rosaceae. *Aronia melanocarpa*, **a)** leaf detail, **b)** fruit, **c)** flower, **d)** flowering branch. *Aruncus dioicus* var. *pubescens*, **e)** flower, **f)** fruiting branch. *Crataegus phaenopyrum*, **g)** flower, **h)** fruiting branch. *Crataegus fecunda*, **i)** seed, **j)** flower, **k)** fruiting branch.

1. Chaenomeles speciosa (Sweet) Nakai (common flowering quince, Japanese quince)

C. lagenaria (Loisel.) Koidz.

Map 2410

Plants shrubs, 1.0–2.5 m tall, sometimes suckering to form colonies. Stems usually armed with stout straight thorns, these indeterminate, consisting of short branches that eventually become elongated and leafy. Bark reddish brown to grayish brown, smooth or somewhat roughened. Leaves alternate or appearing fascicled at the tips of short-shoots arising laterally from older branches, folded lengthwise during development, short-petiolate to nearly sessile, the petioles glabrous. Stipules present only on vigorously elongating young branches, 5–10 mm long, herbaceous, kidney-shaped to heart-shaped, the margins finely toothed, shed as the leaves develop. Leaf blades 1.5–10.0 cm long, simple, unlobed, elliptic to elliptic-obovate, tapered at the base, rounded or more commonly narrowed or tapered to a sharply pointed tip, the margins finely and sharply toothed, the upper surface glabrous, the undersurface glabrous or sparsely hairy along the main veins when young. Inflorescences small clusters or single flowers along second-year or older branches, produced before or as the leaves unfold (often a few flowers also produced later in the year), the stalks very short, glabrous, with a small linear bract, this usually shed before the flower opens. Flowers epigynous, the hypanthium fused to the ovaries, glabrous. Sepals 5, 3–4 mm long, cupped upward or inward, oblong with broadly rounded tips, the margins usually reddish purple and hairy, the inner surface moderately to densely woolly, usually persistent and beaklike at fruiting. Petals 5, 13–22 mm long, broadly obovate, orangish red to red, less commonly pink or white. Stamens numerous, the anthers yellow. Pistil 1 per flower. Ovary inferior, the tip usually moderately hairy, with 5 locules, each with numerous ovules. Styles 5, fused toward the base, the stigmas 2-lobed, the lobes often somewhat elongate along opposite sides toward the tip of the style. Fruits rarely produced in Missouri, pomes, 30–70 mm long, globose or somewhat ovoid, glabrous, yellow to greenish or brownish yellow, with small lighter-colored spots at maturity, with numerous easily exposed seeds embedded in the "core" of leathery carpel wall remains and the fleshy portion, this with small clusters of gritty stone cells. $2n=34$. March–May.

Introduced, uncommon and widely scattered (native of Asia, widely cultivated and escaped sporadically in the eastern U.S. west to Wisconsin and Louisiana). Roadsides and disturbed areas, especially around old homesites.

This species is commonly cultivated as an ornamental shrub and is sometimes trimmed into hedges. It was first reported for Missouri by T. E. Smith (2001) without citation of specimens. Most of the few collections from Missouri represent plants merely persisting from old plantings. However, at one site along a dirt roadway on a narrow ridgetop in Oregon County, shrubs had become naturalized to form large thickets.

7. Cotoneaster Medik. (cotoneaster)

Ninety to 260 species, Europe, Asia, Africa.

The taxonomy of *Cotoneaster* is in urgent need of more detailed study. As in the related genera *Amelanchier* and *Crataegus*, *Cotoneaster* contains a number of complexes characterized by apomixis and polyploidy.

Numerous species of *Cotoneaster* are cultivated as ornamental shrubs in the United States, often because of their unusual growth forms involving intricate branching patterns. The plants also are treasured by bee-keepers and the sometimes abundant fruits provide food for birds (the main dispersal agents) and small mammals. Within their native ranges, many species are used medicinally and for handcrafts.

1. Cotoneaster acutifolius Turcz. **var. lucidus** (Schltdl.) L.T. Lu (hedge cotoneaster, shiny cotoneaster)

C. lucidus Schltdl.

Map 2411

Plants shrubs, 1.5–2.5 m tall, densely branched from at or near the base, the relatively stout main stems ascending (more spreading in some cultivars), more or less straight, with relatively short secondary branches; unarmed. Bark brown, thin, smooth when young, developing fine longitudinal fissures with age. Twigs brown to reddish brown, short-hairy, soon becoming glabrous. Leaves alternate, the petioles 4–8 mm long. Stipules 2–5 mm long, linear to narrowly lanceolate, hairy, the margins entire, usually shed early. Leaf blades 2.0–4.5 cm long (to 6.5 cm on some rapidly elongating vegetative shoots), broadly elliptic to ovate, broadly angled or rounded at the base, broadly angled or occasionally short-tapered to a bluntly or less commonly sharply pointed tip, somewhat leathery, the margins flat or slightly curled under, the margins entire, the upper surface dark green, sparsely hairy when young, but glabrous at maturity, shiny, the undersurface lighter green, sparsely to moderately but inconspicuously hairy. Inflorescences terminating short lateral branches (these few-leaved), small, flat-topped to dome-shaped, relatively small panicles (appearing as loose, compound clusters) with 2–5(–12) flowers, the stalks finely hairy, at least when young; the flowers subtended by short, linear to lanceolate bracts, these shed early. Flowers epigynous, the hypanthium glabrous or sparsely hairy when young, red to pinkish or purplish red. Sepals 5, 1–2 mm long, triangular, red to pinkish or purplish red, becoming fleshy, toothlike, and incurved at fruiting. Petals 5, 3.5–4.5 mm long, broadly obovate to oblong-obovate, white, sometimes pinkish- or reddish-tinged or greenish-tinged. Stamens 15–20, the filaments 2.0–3.5 mm long, pink, the anthers white. Pistil 1 per flower. Ovary inferior, with 2 or 3 locules, each with 2 ovules. Styles 2 or 3, more or less free, the stigmas more or less club-shaped. Fruits berrylike pomes, 7–10 mm in diameter, broadly ellipsoid or globose, glabrous (short-hairy when young), red during development, but soon turning black, more or less glaucous, shiny, with 2 or 3 stones, these 3–5 mm long, narrowly and asymmetrically elliptic in outline, wedge-shaped in cross-section, the surface hard (bony), with a pair of longitudinal, dorsal grooves, light brown. $2n$=68. April–May.

Introduced, known thus far only from a single collection from Boone county (native of Asia, introduced sporadically in the northern U.S. and adjacent Canada south locally to Colorado and Indiana). Strip mine reclamation sites.

The sole Missouri specimens were collected in 2002 by Baadi Tadych of the University of Missouri Dunn-Palmer herbarium. Within its native range, several other varieties are recognized within *C. acutifolius*, which differ in pubescence characters of the leaves and hypanthium. None of these is commonly cultivated in the United States.

This species is commonly cultivated, particularly in the northern states, as an accent shrub or pruned into a hedge. A number of cultivars are sold, including some with variegated leaves or more spreading stems. In the autumn, the leaves turn various shades of dark red, orange, or yellow.

8. **Crataegus** L. (hawthorn)
Contributed by James B. Phipps

Plants shrubs or small trees. Trunks sometimes armed with simple or branched thorns. Branches often producing straight to slightly curved thorns, these usually determinate, persistent for a few years, rarely becoming elongated or leafy. Bark various shades of tan to orangish brown and brown, appearing mottled or checkered, usually peeling in irregular flakes or thin patches, in a few species not flaking, but instead tight and furrowed. Twigs brown to dark brown, becoming gray or grayish brown with age, glabrous or hairy. Winter buds terminal and lateral, small, broadly ovoid to nearly globose, with few to several overlapping scales, these dark red, somewhat fleshy, glabrous. Leaves deciduous, alternate but sometimes appearing clustered at the tips of short branchlets, folded during development, mostly short- to long-petiolate, the petioles grooved on the upper surface, with or without glands along the margins. Stipules relatively large and persistent on rapidly elongating shoots, smaller and shed early on flowering branches, membranous or more commonly leafy, the margins with gland-tipped teeth. Leaf blades simple, those of rapidly elongating branches often larger and more deeply lobed, those of flowering branches smaller, unlobed or lobed, variously shaped, the margins bluntly or sharply and finely to coarsely toothed, the teeth sometimes gland-tipped, the surfaces glabrous or variously hairy at maturity, the upper surface lacking glands. Inflorescences appearing terminal on lateral, short branchlets, dome-shaped to more or less flat-topped, small panicles of several to numerous short- to long-stalked flowers, occasionally reduced to umbellate clusters or solitary flowers, produced as the leaves develop or later, the axis and stalks glabrous, hairy, or stalked-glandular, the stalks each with a small bract at the base, this linear to narrowly oblong-elliptic, membranous to subherbaceous, usually gland-toothed along the margins, often shed early. Flowers epigynous, usually malodorous, the hypanthium fused to the ovaries, globose or slightly urn-shaped, with a greenish yellow nectar disc at the mouth (this usually turning red after pollination has occurred), glabrous or hairy. Sepals 5, spreading to somewhat reflexed at flowering, mostly triangular to narrowly triangular, angled or tapered to a sharply pointed tip, the margins entire to toothed or shallowly lobed, the teeth or lobes gland-tipped, the outer surface glabrous or hairy, persistent until fruiting or shed after flowering. Petals 5, broadly obovate-elliptic to nearly circular, usually slightly concave (cupped) at flowering, the margins entire or sometimes slightly uneven, white or less commonly pinkish-tinged at flowering, usually pinkish-tinged in bud, fading to white with age, overlapping in bud. Stamens (5–)10–20(–25), the filaments often somewhat unequal in length, attached to the margin of the nectar disc, the anthers (0.3–)0.5–1.0(–1.8) mm, ivory to cream-colored or pink to red, rarely dark pinkish purple. Pistil 1 per flower, of (1–)2–5 fused carpels. Ovary inferior, sometimes protruding slightly from the hypanthium at flowering, with (1–)2–5 locules, each with usually 2 ovules (but 1 of these usually abortive). Styles (1–)2–5, free or fused at the base, exserted from the hypanthium, the stigmas club-shaped. Fruits pomes, globose or broadly ellipsoid, glabrous or hairy at maturity, dark to bright red to less commonly orange or yellow, the surface sometimes dotted, with (1–)2–5 stones (usually called nutlets in *Crataegus*), these hard or bony, wedge-shaped when more than 2, dorsally usually slightly concave or shallowly grooved, indehiscent, 1(2)-seeded, embedded in the fleshy to mealy middle layer of the fruit. About 200 species, most diverse in temperate regions of the northern hemisphere, a few in the montane tropics and temperate southern hemisphere.

Where abundant, hawthorns are important wildlife plants, providing shelter and protection for numerous small mammals, excellent nesting sites for birds, and autumnal food for medium-sized passerine birds, ground-birds, rodents, deer, cattle, and even bears. Cattle and deer browse young hawthorn shoots before the thorns have hardened, but can actually promote the successful growth of hawthorn plants by browsing back many competitors more severely. Bees and other Hymenoptera, as well as syrphid flies and numerous other Diptera, find the mass-flowering of hawthorns a major seasonal source of pollen and/or nectar. The fruits of most species are considered unpalatable in modern times, but those of a few species are used in jams and jellies. Some species of hawthorns are cultivated as ornamentals in Missouri, although there are relatively few disease-resistant cultivars sold at nurseries in the state. The species that is the most popular is *C. phaenopyrum*, but a few mainly thornless cultivars of *C. crus-galli* and *C. viridis* also are grown, and additional species are occasionally cultivated as specimen plants. The very hard wood is sometimes used in handicrafts. A few species have medicinal value, principally in treating hypertension (Kurz, 2003).

In 1923, the "white hawthorn blossom" was designated as Missouri's official state floral emblem by the state legislature. This was not without controversy, as some botanists complained that establishing a state flower based on such a taxonomically cantankerous genus and without designation of a particular species was inappropriate (Bush, 1927). However, intense lobbying, mainly by the Daughters of the American Revolution, swayed the politicians. Because the precise wording of the bill mentioned red haw, most of the few botanists and other natural history enthusiasts who care about such details have interpreted *C. mollis* to be the species involved, but this interpretation stands on shaky legal ground.

In Missouri, hawthorns are much less numerous in nature than they were 100 years ago, probably due to several interacting factors, including urbanization, conversion of land to crop fields, fire suppression, fencing of open range, and other changes in land-management practices, as well as the increasing abundance of eastern red cedar (see below) and possibly even increasing populations of nonnative songbirds (which can serve as seed dispersal agents for junipers). In general, hawthorns grow, flower, and become established less well under increasing canopy density in woodlands. The decline in hawthorn abundance in the state is reflected in the fact that the vast majority of the more than 4,000 herbarium specimens of *Crataegus* accessioned in various herbaria were collected before 1940. However, in spite of the major ecological changes in the state since colonization by settlers of European descent, the genus continues to be moderately common in Missouri.

Hawthorns are susceptible to cedar apple rust (caused by *Gymnosporangium juniperi-virginianae* Schwein. and related species), a fungus with a complex life cycle involving species of *Juniperus* L. (in Missouri, mainly *J. virginiana* L., eastern red cedar) as the alternate host. Symptoms of infection by cedar apple rust in hawthorns include deformation of branches and leaves, the presence of powdery orange patches on the plant surfaces, a reduction in the amount of flowering (correlated with malformation of branchlets), and abnormal development of fruits (resulting in amorphously shaped fruits with small, orange, columnar outgrowths on the surface and usually aborted seeds). Although small hawthorns can become so weakened that the infection proves fatal, a main effect of this fungal disease is the reduction in seed production, which is detrimental to recruitment of new individuals over time. It is worthy to note, that of the nearly 1,600 specimens of Missouri *Crataegus* accessioned at the Missouri Botanical Garden herbarium up to and including 1937, very few show evidence of infection by rust, but many of the more recently collected specimens show signs of such damage. For more information, see the treatment of *Juniperus* (Cupressaceae) in Yatskievych (1999).

There is not strong consensus on the species-level taxonomy in *Crataegus*. Missouri is especially problematic in this regard, because some of the premier collectors of hawthorn specimens in the country during the period of 1880 to 1930 resided in the state, notably B. F. Bush, John Davis, Henry Eggert, John H. Kellogg, Kenneth Mackenzie, and Ernest J. Palmer (see biographical notes on these botanists in Yatskievych [1999]). The specimens collected by these botanists fueled the studies of hawthorn specialists of the era, notably Charles S. Sargent (1908, 1912), who eventually named about 130 species based on type specimens from Missouri (Phipps et al., 2007). Palmer (1963) regarded some of these as hybrids and reduced many of the older species to varietal status or synonymy, and accepted 50 species in the state's hawthorn flora. At the opposite extreme, treatments such as those of E. L. Little (1979) or Gleason and Cronquist (1991) include only about a dozen species for the state. Many of the taxonomic ambiguities in species recognition have been attributed to the high incidence of hybridization, polyploidy, and apomixis in the genus. However, in studying specimens mainly from Missouri, Phipps (2005) concluded that putative hybrids are restricted to relatively few, very local, sometimes apparently now extinct entities. Instead, Phipps attributed much of the morphological variation to the rather vigorous formation of local races within mainly the series *Crus-galli, Punctatae, Pruinosae*, and *Virides*. On the other hand, in a survey of rangewide cellular DNA content in hawthorns (using flow cytometry), Talent and Dickinson (2005) found that tetraploids far outnumbered diploids, with triploid taxa a relatively small, third group of taxa. Talent and Dickinson further noted that diploids are self-sterile (requiring cross-pollination), whereas most tetraploids are self-fertile and some are apomictic. Thus polyploidy and apomixis have a role in the maintenance of local races and the fixation of minor morphological differences within the reduced gene pools of such potentially inbreeding or asexually (apomictically) reproducing plants.

A number of putative hybrids are suspected among the Missouri *Crataegus* taxa treated in the earlier literature. They were mostly discovered in the Ozarks during the great period of exploration for Missouri *Crataegus*, roughly 1890–1935, and most were described with binomials as good species rather than as interspecific hybrids. Hawthorns were more abundant then and therefore the opportunities for interserial hybridization were greater. It is notable that relatively few of these persist to this date. The following is an alphabetical list of these putative hybrids, with notes on their presumptive parentage and abundance historically and presently in the Missouri flora. *Crataegus sicca*, whose hybrid status is not fully understood, but which is extant at scattered sites in the Ozarks and has been divided into two varieties by some authors, is treated as a species in ser. *Rotundifoliae* and merely mentioned in the list below. *Crataegus rupicola*, which may prove to be of hybrid origin in the future, is discussed briefly under *C. calpodendron*.

C. ×*atrorubens* Ashe (ser. *Molles* × ser. *Virides*). Described from the St. Louis area, last collected in 1897, and apparently extinct in the wild; however still extant in cultivation as specimen trees at a number of botanical gardens.

C. coccinioides Ashe × *C. mollis* (Torr. & A. Gray) Scheele (ser. *Dilatatae* × ser. *Molles*). Known from a few historical collections from Jefferson, Ripley, and Shannon Counties; it was last collected in 1910.

C. collina Chapm. × *C. margarettae* Ashe (ser. *Punctatae* × ser. *Rotundifoliae*). Known from a few historical collections from Lincoln and Marion Counties; it has not been collected since 1911.

C. collina Chapm. × *C. padifolia* Sarg. (ser. *Intricatae* × ser. *Punctatae*). Known from a single historical specimen collected in 1924 in Benton County.

C. ×*danielsii* E.J. Palmer (ser. *Crus-galli* × ser. *Virides* [possibly *C. engelmannii*]). Rickett (1937) regarded this as being quite common near Columbia (Boone County), but it has not been collected since 1952.

C. ×*dispessa* Ashe (*C. treleasei* Sarg., *C. pyriformis* Britton) (ser. *Molles* × ser. *Punctatae*) (Pl. 528 f). Scattered historically in southern Missouri north locally to Boone County; it was last collected in 1964 in Dallas County.

C. ×*incaedua* Sarg. (*C. pudens* Sarg.) (ser. *Crus-galli* × ser. *Punctatae* [*C. collina*]). Scattered historically in southern Missouri; it was last-collected in 1987 in Camden County.

C. ×*kelloggii* Sarg. (ser. *Molles* × ser. *Rotundifoliae* [*C. margarettae*]). Originally collected in Marion County and the city of St. Louis; it was last collected in Greene County in 1926. This striking taxon produces yellow fruits.

C. ×*latebrosa* Sarg. (*C. noelensis* Sarg.) (ser. *Molles* × ser. *Punctatae*) (Pl. 528 h). Collected historically more than 50 times in the southern half of the state; it was last-collected in 1953 in Dade County.

C. ×*lawrencensis* Sarg. (probably ser. *Crus-galli* × ser. *Virides*). Not seen since the few original collections in 1903 and 1908, all from Jasper County.

C. neobushii Sarg. × *C. pruinosa* (H.L. Wendl.) K. Koch (ser. *Intricatae* × ser. *Pruinosae*). Known from a single historical specimen collected in 1907 in Taney County.

C. ×*nuda* Sarg. (ser. *Crus-galli* × ser. *Macracanthae* [perhaps *C. succulenta*]). Described from a collection made in Taney County; it was last-collected in 1952 in Lawrence County.

C. ×*permixta* E.J. Palmer (*C. intermixta* Sarg., an illegitimate name) (ser. *Crus-galli* × ser. *Virides*) (Pl. 525 p). Described from the Hannibal area (Marion, Pike Counties), it has not been collected since 1913.

C. ×*persimilis* Sarg. (*C. swanensis* Sarg.) (ser. *Crus-galli* × ser. *Macracanthae*). 2n=68. Known from a few historical collections from Taney County (the original material upon which C. swanensis was described), but also a Dent County specimen collected in 2004 by Alan Brant.

C. sicca Sarg. (possibly ser. *Pruinosae* × ser. *Rotundifoliae*). See the treatment in ser. Rotundifoliae.

C. ×*simulata* Sarg. (ser. *Crus-galli* × ser. *Macracanthae* [possibly *C. macracantha*]) (Pl. 526 c). Known from a number of historical collections from Jasper and Lawrence Counties; it was last-collected in 1956 in Lawrence County.

C. ×vailiae Britton (ser. *Macracanthae* [probably *C. calpodendron*] × ser. *Parvifoliae* [*C. uniflora*]). Known from a number of historical collections in the southern half of the state and still extant at several sites; it was last collected in 1997 in Howell County. See the treatments of *C. uniflora* and ser. *Macracanthae* for further discussion.

C. ×verruculosa Sarg. (ser. *Crus-galli* × ser. *Punctatae* [probably *C. collina*]) (Pl. 530 i–k). Scattered historically in the southern half of the state and apparently still extant at a few sites; it was last-collected in 2002 in Iron County.

The series *Crus-galli, Punctatae, Pruinosae*, and *Virides* mainly appear recalcitrant to any straightforward taxonomy, and the account given here gives priority to the ability to unequivocally recognize taxa. The descriptions provided below for those Missouri series with more than one species apply only to Missouri taxa and mostly also exclude putative interserial hybrids. In using the keys, the leaf shapes referred to are those near the center of variation on the flowering branches, except where extension shoots are specifically indicated. The values given for leaf lobing are characteristic for the deepest lobes on a leaf. Because most hawthorns are considered difficult to identify, botanists should give priority to collecting fresh flowering material or ripe fruiting material with accurate color notes on the anthers or fruits (these can change color upon dessication). When fruits are present, it is also helpful to remove stones (nutlets) from a few representatives, as this is easier to accomplish while the pomes are fresh rather than waiting until they are dried. Serious students of the genus generally attempt to collect matching flowering and fruiting specimens from marked individuals. However, material with spent flowers and immature fruits usually can be determined to species.

As has been traditional, the Missouri species of *Crataegus* are here classified into series. A key to the determination of series follows.

1. Leaf blades lobed, with veins running both to the tips of the lobes and to the major sinuses (check particularly carefully when the sinuses are very deep)
    2. Fruits mostly ellipsoid; leaf blades nearly as wide as long, mostly deeply and irregularly lobed, hairy, at least when young, not leathery, the upper surface dull; styles and nutlets 1 or 2(3); petals elliptic . . . . 1. APIIFOLIAE
    2. Fruits more or less globose; leaf blades longer than wide, or, if not, then with relatively shallow sinuses, somewhat leathery, glabrous, the upper surface more or less shiny; styles and nutlets 3–5; petals broadly obovate to circular
        3. Leaves 3/4 as wide as long or wider, broadly rounded to shallowly cordate at the base, angled or tapered to a sharply pointed tip; bark rough, exfoliating in small, checkered flakes; fruits 5–8 mm wide, red . . . . 2. CORDATAE
        3. Leaves 1/2 as wide as long or narrower, tapered at the base, rounded or bluntly to infrequently sharply pointed at the tip; bark smooth, exfoliating in large, irregular flakes or patches; fruits 3–5 mm wide, red to orange . . . . 7. MICROCARPAE
1. Leaf blades unlobed, or, if lobed, then with veins running to only the tips of the lobes, not to the major sinuses (note that leaves on rapidly growing vegetative shoots may have unusual morphology and sometimes cannot be keyed using lobing or venation patterns)
    4. Leaf blades unlobed or with extremely shallow lobes restricted to the terminal portion; angle at the base of the leaf blade less, often much less, than 90°
        5. Plants relatively small shrubs, 0.4–1.5 m tall; leaf blades 1.5–4.0(–5.0) cm long, but mostly less than 3 cm long; inflorescences of solitary flowers or less commonly small clusters of 2–4(5) flowers; sepals slightly longer than the petals, more or less leaflike, the margins strongly toothed and/or with slender lobes, the teeth and lobes gland-tipped . . . . 9. PARVIFOLIAE

5. Plants usually medium-sized to large shrubs or small trees, (2–)3–15 m tall; leaf blades 1.5–7.0 cm long, but mostly more than 2 cm long; sepals much shorter than the petals, the margins entire or sparsely glandular-toothed
  6. Leaf blades dull, hairy, sometimes densely so on the upper surface, at least when young, not leathery; bractlets linear to oblong or oblong-obovate, with numerous marginal glands or gland-tipped teeth
    7. Leaf blades seldom less than 2 cm wide at maturity, usually widest at or below the midpoint; bractlets more than 2 mm wide; flowers often more than 20 mm in diameter; sepals irregularly incised with slender, gland-tipped lobes; petioles mostly $^{1}/_{3}$–$^{1}/_{2}$ as long as the blades . . . . 8. MOLLES
    7. Leaf blades seldom more than 2 cm wide at maturity, usually widest towards the tip; bractlets 2 mm wide or narrower; flowers often less than 20 mm in diameter; sepals more or less entire to glandular-toothed; petioles about $^{1}/_{8}$–$^{1}/_{4}$ as long as the blades . . . . 11. PUNCTATAE
  6. Leaf blades often more or less shiny and commonly glabrous on the upper surface (sometimes sparsely and inconspicuously appressed-hairy along the midvein), often more or less leathery; bractlets linear to filiform, nonglandular or with only sparse glandular teeth
    8. Fruits mostly 3–5 mm wide; leaves herbaceous or slightly leathery; leaf blades usually at least slightly asymmetric in outline . . . . 14. VIRIDES
    8. Fruits 6–14 mm wide (rarely to 25 mm in ser. *Crus-galli*); leaves leathery or papery; leaf blades usually symmetric in outline
      9. Sepals entire or slightly irregular, occasionally with a few inconspicuous teeth toward the tip, only the teeth gland-tipped; petioles less than $^{1}/_{5}$ as long as the blades; fruits round in cross-section, lacking a raised collar at the tip (but the withered sepals sometimes obscuring the tip), not glaucous . . . . 3. CRUS-GALLI
      9. Sepals glandular-toothed; petioles at least $^{1}/_{3}$ as long as the blades; fruits frequently bluntly angled in cross-section, with a distinct raised collar at the tip (but the withered sepals sometimes obscuring the tip), usually glaucous . . . 10. PRUINOSAE

4. Leaf blades distinctly lobed, although lobing may be quite shallow, with the deeper sinuses only 10–15% of the way toward the midvein; angle at the base of the leaf blade usually more than 90°
  10. Lateral faces of nutlets pitted (note that in the putative interserial hybrid, *C. ×vailiae* [ser. Macracanthae × ser. Parvifoliae] the nutlets are only slightly pitted), ... 6. MACRACANTHAE
  10. Lateral faces of nutlets flat to slightly convex
    11. Bractlets linear-filiform, the margins nonglandular or nearly so; undersurface of leaf blades often with inconspicuous tufts of hairs in the axils of the lower secondary veins . . . . 14. VIRIDES
    11. Bractlets oblong or spatulate to linear, the margins noticeably glandular; undersurface of leaf blades variously glabrous or hairy, but lacking tufts of hairs in the axils of the secondary veins
      12. Leaf blades generally large (5–9 cm long), the upper surface densely hairy, at least when young; fruits hairy

13. Fruits with a small, but distinct, raised collar at the tip (but the withered sepals sometimes obscuring the tip); fruits orange to brownish red or dark red; margins of leaf blades and petioles conspicuously stalked-glandular; inflorescences 2–8-flowered . . . . 5. INTRICATAE
13. Fruits lacking a distinct raised collar at the tip (but the withered sepals sometimes obscuring the tip), bright red, rarely yellow; margins of leaf blades and petioles nonglandular; inflorescences mostly 7–15-flowered. . . . . 8. MOLLES
12. Leaf blades small to large (3–9 cm long), glabrous or the upper surface less commonly roughened or hairy when young (glabrous or nearly so at maturity except in uncommon variants of *C. pruinosa*); fruits glabrous
14. Leaf blades 2–6 cm long, ovate, elliptic, or rhombic-elliptic to obovate, broadly obovate, or nearly circular, the secondary veins often 3 or 4 on each side of the midvein . . . . 12. ROTUNDIFOLIAE
14. Leaf blades 4–8 cm long, mostly ovate to ovate-triangular, occasionally broadly elliptic, the secondary veins usually 4 or more on each side of the midvein
15. Leaf blades with the upper surface roughened-hairy when young, becoming glabrous or nearly so at maturity (except in uncommon variants of *C. pruinosa*)
16. Fruits glaucous, with a distinct raised collar at the tip (but the withered sepals sometimes obscuring the tip); leaf texture papery to somewhat leathery. . . . . 10. PRUINOSAE
16. Fruits not glaucous, lacking a distinct raised collar at the tip (but the withered sepals sometimes obscuring the tip); leaf texture herbaceous to somewhat papery. . . . . 13. TENUIFOLIAE
15. Leaf blades with the upper surface glabrous or soft-hairy when young, then becoming glabrous or nearly so at maturity (except in uncommon variants of *C. pruinosa*)
17. Stamens (5–)10; fruits yellow to red or dark red. . . . . 5. INTRICATAE
17. Stamens (10–)20; fruits green tinged with pink or purple to bright pink or red
18. Sepals glandular-lobed; leaves, flowers more than 20 mm in diameter; fruits circular in cross-section, lacking a distinct raised collar at the tip (but the withered sepals sometimes obscuring the tip) . . . . 4. DILATATAE
18. Sepals glandular-toothed; flowers less than 20 mm in diameter; sepals at fruits frequently bluntly angled in cross-section, with a distinct raised collar at the tip (but the withered sepals some times obscuring the tip). . . . . 10. PRUINOSAE

1. Series Apiifoliae (Loudon) Rehder (Phipps, 1998)

One species, southeastern U.S.

Series *Apiifoliae*, with its deeply divided leaves and secondary veins extending to the major sinuses between the lobes, is much like the speciose western Eurasian ser. *Crataegus* but differs in having only determinate thorns. Other interesting features are the elliptic petals; 20 reddish anthers; small, usually ellipsoid, shiny red fruits, and low number of styles and

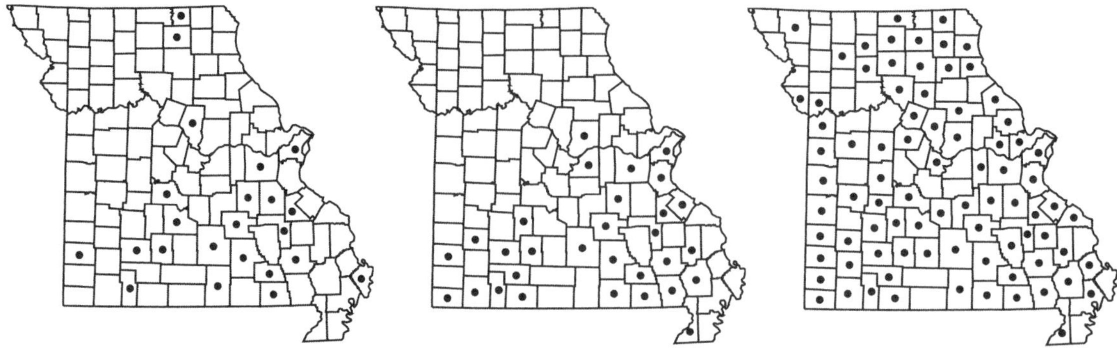

2413. Crataegus phaenopyrum 2414. Crataegus berberifolia 2415. Crataegus crus-galli

nutlets. It also has beautifully patterned, smooth, exfoliating bark. Members of this section deserve more attention for their horticultural potential in the Southeast and Lower Midwest.

1. Crataegus marshallii Eggl. (parsley haw)
C. apiifolia (Marshall) Michx., an illegitimate name

Pl. 528 j, k; Map 2412

Plants shrubs or more commonly small trees, 2–8 m tall, occasionally bigger, usually single-trunked, the trunks lacking branched thorns, the bark gray, smooth, mottled, peeling in large, irregular, thin patches exposing an orangish brown to grayish green underlayer. Branchlets unarmed or variably thorny, the thorns mostly 2–3 cm long, slender, straight or slightly curved, dark and shiny at second year. Twigs mostly densely appressed-hairy in first year, becoming glabrous with age, dark purplish brown the first year, grayish brown with age, sometimes slightly glaucous. Petioles 10–20 mm long, hairy, nonglandular. Leaf blades 1.5–3.0 cm long, nearly as wide as long, broadly ovate to ovate-triangular in outline, sometimes slightly asymmetric in outline, broadly rounded to truncate or shallowly cordate at the base, the tip and lobe tips mostly sharply but sometimes broadly pointed, with (2)3–4(5) mostly deep, jagged lobes per side and narrow sinuses, the margins otherwise sharply and irregularly toothed, nonglandular, herbaceous in texture, the upper surface dull, moderately hairy when young, becoming glabrous at maturity, the undersurface densely hairy when young, becoming glabrous at maturity except along the main veins, at least some of the secondary veins extending to the sinuses between the larger lobes, the other veins with the ultimate branches extending to the tips of the lobes and teeth. Inflorescences 3–8-flowered, the branches densely hairy, the bractlets membranous, linear, entire to obscurely toothed, with glandular margins, shed early. Flowers 12–17 mm in diameter, the hypanthium hairy. Sepals 3–4 mm long, narrowly triangular, the margins finely toothed, the teeth gland-tipped, glabrous on the inner surface, hairy on the outer surface. Petals 6–8 mm long, elliptic, white. Stamens 20, the anthers deep pink to red or rarely yellow. Styles 1 or 2(3). Fruits 4–6(–8) mm long, 3–4 mm wide, ellipsoid to occasionally subglobose, lacking a raised collar at the tip, the surface bright red at maturity, glabrous, shiny. Nutlets 1 or 2, the lateral faces not pitted. Diploid or triploid (by flow cytometry; see Talent and Dickinson [2005]). Late March–early May.

Uncommon, known only from the western portion of the Mississippi Lowlands Division (southeastern U.S. west to Missouri, Oklahoma, and Texas). Swamps and bottomland forests; also rarely old fields in bottoms.

Yellow-anthered forms of this attractive species have been called *C. apiifolia* var. *flavanthera* Sarg.

2. Series Cordatae (Beadle) Rehder (Phipps, 1998)

One species, southeastern U.S., introduced farther north.

Series *Cordatae*, with its deeply angular-lobed, leathery leaves and secondary veins extending to the sinuses between the major lobes is allied to Eurasian ser. *Crataegus*. It differs from the similar ser. *Apiifoliae* in leaf shape, its numerous small flowers with cream-colored anthers, later average start of flowering, small globose fruits, and rougher bark.

2. Crataegus phaenopyrum (L.f.) Medik. (Washington thorn)
C. cordata (Mill.) Aiton
Mespilus phaenopyrum L.f.
Pl. 524 g, h; Map 2413

Plants small trees, 4–10 m tall, the trunks armed with branched thorns, the older bark gray, rough, peeling in small, irregular, checkered flakes, often exposing an orangish brown underlayer. Branchlets with scattered thorns, these mostly 2–5 cm long, dark brown at second year, eventually becoming gray, slender, straight. Twigs glabrous or nearly so, purplish brown by the second year. Petioles 13–25 mm long, glabrous, nonglandular. Leaf blades 2–5(–7) cm long, wider than long to about 3/4 as wide as long, ovate-triangular to broadly ovate-triangular in outline, sometimes slightly asymmetric in outline, broadly rounded to truncate or rarely shallowly cordate at the base, the tip and large lobe tips more or less tapered and sharply pointed, with 2–5 lobes per side (these sometimes appearing palmate or nearly so), the basal pair shallow to more commonly about 50% of the way to the midvein, the remaining lobes much shallower, the sinuses narrow to relatively broad, the margins otherwise sharply and irregularly but often relatively sparsely toothed, nonglandular, the texture somewhat leathery, the surfaces glabrous or nearly so, the upper surface usually at least slightly shiny, the secondary veins extending to the sinuses between the larger lobes and to the tips of the lobes and teeth. Inflorescences 20–30-flowered, the branches glabrous, the bractlets membranous to herbaceous, linear to narrowly oblong-lanceolate, entire to obscurely toothed, with glandular margins, shed early. Flowers 10–12 mm in diameter, the hypanthium glabrous. Sepals 1.5–2.0 mm long, broadly triangular, the margins entire, usually finely and densely hairy, nonglandular, glabrous on the inner and outer surfaces. Petals 4–5 mm long, broadly obovate to circular, white. Stamens 20, the anthers ivory-colored. Styles 3(–5). Fruits 5–8 mm long, 5–8 mm wide, globose, lacking a raised collar at the tip, the surface bright orangish red to red at maturity, glabrous, shiny. Nutlets 3(–5), the lateral faces not pitted. $2n=72$; triploid, tetraploid, or higher ploidy (by flow cytometry; see Talent and Dickinson [2005]). Late April–late May.

Scattered in the southern half of the state (except for the Glaciated Plains Division) north locally to Boone and Putnam Counties (eastern U.S. west to Missouri and Louisiana; introduced north to Canada). Bottomland forests, mesic upland forests, banks of streams, margins of sinkhole ponds, marshes, and seeps, ledges of bluffs, and rarely glades; also railroads and roadsides.

Washington thorn is one of the finest ornamentals in the genus. It is often among the latest-flowering members of the genus. In Missouri it is perhaps the most commonly planted species of hawthorn, especially around businesses and along highways in urban areas. The trees are striking and possess beautiful foliage that is remarkably disease-resistant. The bright red fruits are particularly attractive against the glossy bronzed fall foliage.

3. Series Crus-galli (Loudon) Rehder

Plants shrubs or small trees, (2–)4–8 m tall, the trunks often armed with branched thorns, the bark gray to brownish gray, finely furrowed and often with small plates, peeling in small, irregular, checkered flakes or longitudinal strips, usually exposing an orangish brown underlayer. Branchlets with scattered to relatively dense thorns (sometimes thornless in cultivated plants), these mostly 3–8 cm long, slender to more commonly relatively stout, mostly slightly curved. Twigs glabrous or densely short-hairy, gray to reddish brown, orangish brown, or green tinged with red, shiny or dull, occasionally somewhat glaucous. Petioles less than 20% as long as the blade, glabrous or sparsely short-hairy, nonglandular. Leaf blades 1.5–6.5 cm long, varying from less than 30% as wide as long to nearly as wide as long, narrowly oblanceolate to broadly oblong-elliptic or broadly obovate (those of extension shoots sometimes nearly circular in *C. reverchonii*), usually symmetric in outline, tapered at the base, the tip rounded or broadly and often bluntly pointed, unlobed (except sometimes those of rapidly elongating shoots), the margins sharply and finely or less commonly relatively coarsely and then irregularly toothed, mostly above the midpoint, nonglandular, the texture somewhat to strongly leathery, the upper surface glabrous (sometimes with scattered appressed hairs along the main veins) or moderately to densely roughened with short, stiff hairs, at least when young, dull or more commonly shiny, the undersurface glabrous or soft-hairy along the main veins, the secondary veins mostly 4–9 per side, the ultimate branches extending to the tips of the teeth. Inflorescences 8–32-flowered, the branches glabrous or short-hairy, the bractlets membranous, linear to filiform, entire, the margins then nonglandular or with sparse stalked glands, or sparsely glandular-toothed, shed early. Flowers 10–20 mm in diameter, the hypanthium glabrous or hairy. Sepals 3–6 mm long, shorter than the petals, narrowly triangular to lanceolate-triangular, the margins entire or slightly irregular, occasionally with a few inconspicuous teeth toward the tip, only the teeth gland-tipped, glabrous or the inner surface inconspicuously hairy. Petals 6–8 mm long, broadly obovate to circular, white. Stamens 10–20, the anthers cream-colored to pink, reddish pink, or rarely dark pinkish purple. Styles (1)2–4(5). Fruits 6–14 mm long, mostly 7–14 mm wide, globose or subglobose or rarely slightly obovoid, lacking a raised collar at the tip (but the withered calyces usually more or less persistent), the surface red to orange at maturity, glabrous or minutely hairy, shiny or dull, occasionally somewhat glaucous. Nutlets 1–4, the lateral faces not pitted. About 7 species, eastern U.S., Canada; introduced in Europe.

Members of ser. *Crus-galli* are normally thorny, large shrubs to small trees with glossy, more or less leathery, unlobed (very rarely shallowly lobed), usually narrow leaves. They have very narrow bractlets that normally are only sparsely glandular along the margins and red fruits. These are among the most common of Missouri hawthorns, especially at upland sites.

1. Leaf blades with the upper surface roughened-hairy, at least when young, inflorescence branches hairy
   2. Margins of the leaf blades simply toothed, the teeth less than 1 mm long, even; blades 2.0–4.5 cm long, the upper surface relatively densely and persistently roughened-hairy; inflorescences 8–12-flowered, the branches usually densely and persistently hairy . . . . 3. C. BERBERIFOLIA
   2. Margins of the leaf blades doubly toothed, the larger teeth mostly 2–3 mm long, appearing somewhat uneven; blades mostly 5.0–6.5 cm long, the upper surface sparsely to moderately roughened-hairy when young, becoming nearly glabrous (except along the midvein) at maturity; inflorescences mostly 10–25-flowered, the branches sparsely to moderately hairy . . . . 5. C. FECUNDA
1. Leaf blades with the upper surface glabrous except sometimes for scattered appressed hairs along the main veins; inflorescence branches usually glabrous
   3. Extension shoot leaves at most broadly elliptic to broadly obovate-elliptic; styles and nutlets 1–3(4); leaf blades 2–7 cm long . . . . 4. C. CRUS-GALLI
   3. Extension shoot leaves frequently circular or nearly so; styles and nutlets 3 or 4; leaf blades 2–5 cm long . . . . 6. C. REVERCHONII

3. Crataegus berberifolia Torr. & A. Gray (barberry-leaved hawthorn, Engelmann hawthorn)

C. barbata Sarg.
C. engelmannii Sarg.
C. hirtella Sarg.
C. munita Sarg.
C. pilifera Sarg.
C. setosa Sarg.
C. tenuis Sarg.
C. vallicola Sarg.
C. villiflora Sarg.

Pl. 525 q, r; Map 2414

Plants shrubs or small trees to 6 m tall. Branchlets with the thorns widely to more densely scattered, mostly 4.0–6.5 cm long, slender or relatively stout, mostly chestnut brown at second year. Twigs orangish brown or green tinged with red when young, becoming brown or grayish tinged by the second year, eventually turning gray. Petioles 3–6 mm long, nearly glabrous at maturity, nonglandular. Leaf blades 2–4 cm long, oblanceolate to narrowly obovate, the margins finely and singly toothed except toward the base, or toothed only

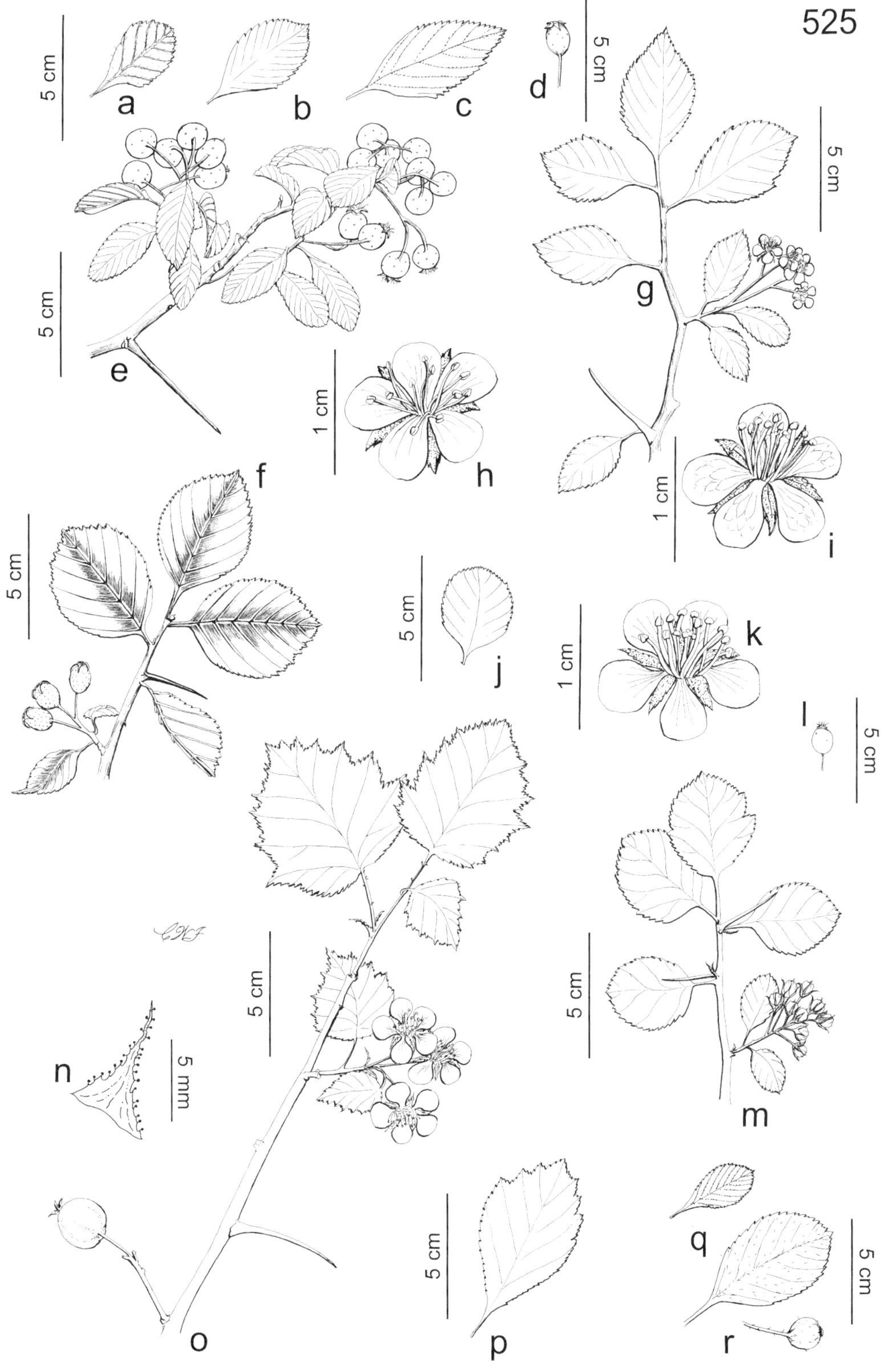

Plate 525. Rosaceae. *Crataegus crus-galli*, **a, b, c)** representative leaf shapes, **d)** fruit, **e)** fruiting branch, variant with typical leaves, **f)** fruiting branch, variant with broad leaves, **g)** flowering branch, **h)** 10-stamened flower, **i)** 20-stamened flower. *Crataegus reverchonii* var. *reverchonii*, **j)** leaf. *Crataegus reverchonii* var. *palmeri*, **k)** flower, **l)** fruit, **m)** branch with buds. *Crataegus coccinioides*, **n)** detached sepal, **o)** flowering branch. *Crataegus* ×*permixta* (ser. *Crus-galli* × ser. *Virides*), **p)** leaf. *Crataegus berberifolia*, **q)** representative leaf shapes, **r)** fruit.

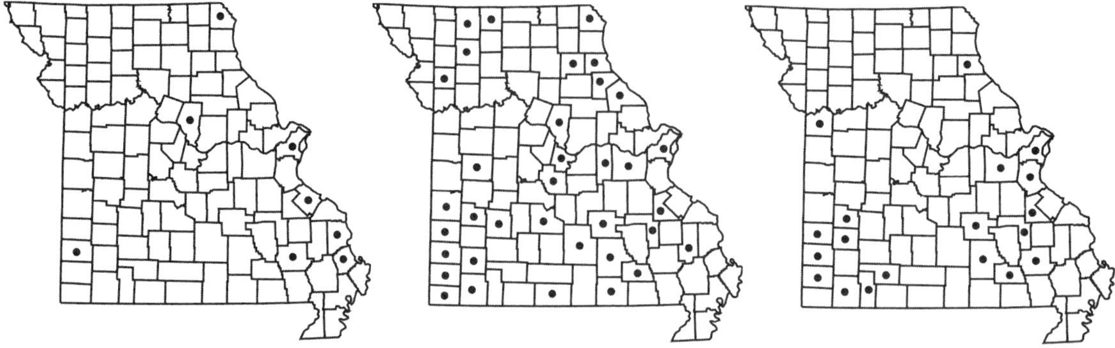

2416. Crataegus fecunda 2417. Crataegus reverchonii 2418. Crataegus coccinioides

beyond the broadest portion, the teeth less than 1 mm long, relatively even, leathery, the upper surface strongly roughened-hairy, shiny, the undersurface soft-hairy on the veins, the secondary veins (4)5 or 6 per side. Inflorescences 8–12-flowered, the branches hairy, the bractlets sparsely glandular-toothed or entire but with scattered stalked glands. Flowers 10–20 mm in diameter, the hypanthium hairy or occasionally nearly glabrous. Sepals 5–6 mm long; the margins entire. Stamens 10 and with pink (rarely light yellow) anthers or 20 with cream-colored anthers. Styles 2 or 3. Fruits 8–10 mm in diameter, more or less globose, the surface red to orangish red, more or less glabrous, usually not glaucous. Nutlets 2 or 3. Late April–early June.

Scattered in the southern half of the state, except for the Glaciated Plains Division (southeastern U.S. west to Missouri and Texas). Mesic to dry upland forests, glades, bottomland and upland prairies, ledges and tops of bluffs, banks of streams and rivers, and margins of fens; also old fields, pastures, fence rows, and roadsides.

Barberry-leaved hawthorn is little-differentiated from some forms of *C. crus-galli* except in its roughened-hairy upper leaf-surface, but it tends to have somewhat smaller leaves and is more restricted in its range. Most of the Missouri plants have 10 stamens with pink (or rarely light yellow) anthers and somewhat smaller flowers (10–15 mm in diameter), which is characteristic of plants attributed to var. *engelmannii* (Sarg.) Eggl. (Phipps, 2006). Plants attributable to var. *berberifolia* produce somewhat larger flowers (16–20 mm in diameter) with 20 stamens and cream-colored anthers. They occur uncommonly in southernmost Missouri.

4. Crataegus crus-galli L. (cockspur hawthorn)

C. crus-galli f. *spinulosa* H.W. Rickett
C. crus-galli f. *staminea* H.W. Rickett
C. crus-galli f. *vulgaris* H.W. Rickett
C. acutifolia Sarg.
C. albanthera Sarg.
C. arborea Beadle
C. barrettiana Sarg.
C. calophylla Sarg.
C. candens Sarg.
C. consueta Sarg.
C. efferta Sarg.
C. effulgens Sarg.
C. erecta Sarg.
C. ferox Sarg.
C. hamata Sarg.
C. hannibalensis E.J. Palmer
C. infera Sarg.
C. infesta Sarg.
C. leptophylla Sarg.
C. ludoviciensis Sarg.
C. monosperma Sarg.
C. pachyphylla Sarg.
C. paradoxa Sarg.
C. parkiae Sarg.
C. permera Sarg.
C. phaneroneura Sarg.
C. polyclada Sarg.
C. pyracanthifolia Wender.
C. pyracanthoides Beadle
C. regalis Beadle
C. rubrifolia Sarg.
C. rudis Sarg.
C. severa Sarg.
C. strongylophylla Sarg.
C. tantula Sarg.
C. tardiflora Sarg.
C. tenuispina Sarg.
C. truncata Sarg.

Pl. 525 a–i; Map 2415

Plants shrubs or small trees to 6 m tall. Branchlets with the thorns often relatively numerous, mostly 3–6 cm long, slender or relatively stout, dark gray to nearly black at second year. Twigs pale brown by the second year, becoming gray with age. Petioles absent or to 5 mm long, glabrous, nonglandular. Leaf blades 2–7 cm long, narrowly oblanceolate to elliptic or narrowly oblanceolate (those on extension shoot leaves sometimes broadly elliptic to broadly obovate), the margins finely and singly toothed or scalloped sometimes only above the midpoint or beyond the broadest portion, the teeth less than 1 mm long, relatively even, leathery, the upper surface glabrous or inconspicuously pubescent with short, scattered, appressed hairs along the midvein, often shiny, the undersurface glabrous, the secondary veins 4–7 per side. Inflorescences mostly 12–17-flowered, the branches glabrous or rarely finely hairy, the bractlets sparsely glandular-toothed or entire but with scattered stalked glands. Flowers 10–17 mm in diameter, the hypanthium glabrous. Sepals 5–6 mm long, the margins entire or slightly irregular. Stamens 10–20, the anthers ivory or pink to reddish pink. Styles 1–3. Fruits 8–14 mm in diameter, globose or subglobose or rarely slightly obovoid, the surface red to dark red, glabrous, often glaucous. Nutlets 1–3. $2n$=24, 51, 64, 68; triploid (by anatomical sections; see Longley [1924]); diploid to tetraploid (by flow cytometry; see Talent and Dickinson [2005]). Late April–mid-June.

Scattered nearly throughout the state (eastern U.S. west to Iowa and Texas; southeastern Canada; introduced in Europe). Mesic to dry upland forests, glades, savannas, upland prairies, sand prairies, bottomland forests, banks of streams, rivers, and spring branches, and margins of ponds, lakes, and sinkhole ponds; also quarries, mine spoils, old fields, fallow fields, pastures, fence rows, and roadsides.

This is one of the more commonly cultivated of the hawthorns, and some disease-resistant, as well as thornless, cultivars exist. Well-grown specimens with their neat rows of branches bearing glossy leaves and dense clusters of fruit are extremely attractive. It is sometimes planted along highways by transportation departments.

Crataegus crus-galli is not easily confused with other hawthorns, other than those in its series and hybrids involving the species. It is one of the commonest hawthorns in the eastern half of the United States and correspondingly variable. Very many segregates have been described based on minor differences in fruit shape and color, nutlet number, anther color, typical leaf shape, flower size, etc. Twenty-seven of these occur in Missouri. One of the more distinct is *C. ludoviciensis*, with relatively large, broad leaves that are not particularly glossy but have good autumn color. Several varieties may be recognized, three of which have been documented from Missouri. An overwhelming proportion of the specimens in the herbaria key to var. *crus-galli*, which remains morphologically variable. Phipps (2006) reported var. *regalis* (Beadle) J.B. Phipps from the state, but no Missouri specimens of this taxon could be located during the present study. Plants of var. *regalis* differ from those of var. *crus-galli* in their very dark green (often somewhat bluish green) mature leaves with relatively coarse marginal teeth. The var. *capillata* is a trivial variant with somewhat pubescent inflorescences, but the narrow-leaved var. *pyracanthifolia* is a more distinct variant that can be locally abundant in some portions of its range. The key below serves to separate the major variants that occur in Missouri.

1. Flower stalks hairy ..... var. *capillata* Sarg.
1. Flower stalks glabrous
   2. Leaves relatively broad, 1.5–2.0 times as long as wide ..... var. *crus-galli*
   2. Most of the leaves very narrow, 2–3 times as long as wide .... var. *pyracanthifolia* Aiton

5. Crataegus fecunda Sarg. (St. Clair hawthorn, fecund hawthorn)

Pl. 524 i–k; Map 2416

Plants small trees to 8 m tall. Branchlets with the thorns often relatively numerous, mostly 5–7 cm long, slender, chestnut brown to nearly black at second year. Twigs green when young, becoming orangish brown by the end of the first year and eventually turning light gray. Petioles (3–)10–32 mm long, finely hairy when young, becoming glabrous or nearly so at maturity, usually narrowly winged, at least toward the tip, often with scattered stalked glands. Leaf blades mostly 5.0–6.5 cm long, oblong-obovate to oval or broadly ovate, the margins relatively coarsely and doubly toothed for most of their length, the larger teeth mostly 2–3 mm long, relatively jagged and somewhat uneven, leathery, the upper surface sparsely to moderately roughened-hairy when young, becoming nearly glabrous at maturity (usually persistently hairy along the midvein), somewhat shiny, the undersurface glabrous or sparsely hairy along the main veins, the pubescence sometimes restricted to the axils of the main veins toward the base, the secondary veins 7–9 per side. Inflorescences mostly 10–25-flowered, the branches sparsely to moderately hairy, the bractlets sparsely glandular-toothed. Flowers 17–20 mm in diameter, the hypanthium sparsely hairy. Sepals 3.0–4.5 mm long; the margins sparsely and irregularly glandular-toothed to nearly entire. Stamens 10(–15), the anthers dark pinkish purple. Styles 2 or 3. Fruits 11–14(–25) mm in diameter, globose or subglobose or rarely slightly oblong-ellipsoid, the surface orangish red to red, with scattered, long, fine hairs during development, becoming nearly glabrous at maturity, not shiny (but dotted with numerous small, pale lenticels). Nutlets 2 or 3. Late April–late June.

Uncommon, widely scattered in the state (Missouri, Arkansas, Illinois, and Kentucky). Bottomland forests, mesic upland forests, ledges of bluffs, banks of streams, and rarely glades; also open disturbed areas and railroads.

Crataegus fecunda is known from a number of specimens collected mainly near the Mississippi and Ohio Rivers, but apparently was never common. Except for two 1983 specimens from Clark County, this species is known from Missouri only from historical collections. The several early collections of *C. fecunda* are all quite distinctive, so this taxon is maintained as a separate species in the present work. According to Sargent (1902), this is an extremely handsome species with large fruits and lustrous green summer leaves that in the autumn assume brilliant shades of orange and scarlet or deep bronze. *Crataegus fecunda* should be highly prized as an ornamental, yet it has not been brought into commercial cultivation in the United States.

6. Crataegus reverchonii Sarg. (Reverchon hawthorn)

C. reverchonii var. *palmeri* (Sarg.) J.B. Phipps
C. reverchonii var. *stevensiana* (Sarg.) E.J. Palmer
C. discolor Sarg.
C. jasperensis Sarg.
C. palmeri Sarg.
C. parciflora Sarg.
C. rotunda Sarg.
C. rubrisepala Sarg.
C. stevensiana Sarg.

Pl. 525 j–m; Map 2417

Plants shrubs to 4 m tall. Branchlets with the thorns often relatively numerous, mostly 2.5–6.0 cm long, relatively stout, dark brown to nearly black at second year. Twigs reddish-tinged when young, becoming chestnut brown during first year and dark gray at second year. Petioles 2–5 mm long, glabrous, nonglandular. Leaf blades mostly 2–6 cm long, broadly oblong-elliptic to broadly elliptic or broadly obovate-elliptic (those on extension shoot leaves frequently circular or nearly so), the margins finely and simply toothed, mostly above the midpoint, the teeth less than 1 mm long, even, leathery, the upper surface sparsely hairy when young, becoming glabrous at maturity or inconspicuously pubescent with short, scattered, appressed hairs along the midvein, shiny, the undersurface glabrous, the secondary veins mostly 5 or 6 per side. Inflorescences mostly 8–15-flowered, the branches glabrous or occasionally sparsely short-hairy, the bractlets sparsely glandular-toothed or entire but with scattered stalked glands. Flowers 8–12 mm in diameter, the hypanthium glabrous or rarely sparsely hairy. Sepals 3–4 mm long, the margins entire or slightly irregular. Stamens 10–15(–20), the anthers cream-colored or pink to reddish pink. Styles (2)3(4). Fruits 6–10 mm in diameter, globose or nearly so, the surface red to orangish red, glabrous, not glaucous. Nutlets (2)3(4). Triploid (as *C. palmeri*, by anatomical sections; see Longley [1924]). Mid-April–early June.

Scattered nearly throughout the state, but apparently absent from the Mississippi Lowlands Division (southeastern U.S. west to Kansas and Texas). Mesic to dry upland forests, glades, upland prairies, banks of streams, rivers, and spring branches, margins of lakes and sinkhole ponds, and bottomland forests; also mine spoils, pastures, and old fields.

As with many wide-ranging hawthorn species, *C. reverchonii* has been treated variously as a series of segregate species or a single species with multiple varieties. It is distinguished from other members of series *Crus-galli* by the circular to broadly elliptic extension shoot leaves. Two varieties occur in Missouri. The var. *reverchonii* (Pl. 525 j) has a southwestern range and is uncommon in Missouri, known from only a few specimens collected in the southwestern portion of the state. Nearly all of the Missouri materials are var. *palmeri* (Sarg.) J.B. Phipps (Pl. 525 k–m). Plants of typical var. *palmeri* differ from those of var. *reverchonii* mainly in their larger leaves (mostly 4–6 vs. 2–4 cm long), slightly larger flowers (10–12 mm vs. 8–10 mm in diameter), and generally fewer styles and nutlets (2 or more, commonly 3, vs. 3–5). However, in practice, these varieties tend to be more or less confluent morphologicallyin our portion of the species range and there is not good correlation between the characters said to separate them.

2419. Crataegus biltmoreana 2420. Crataegus neobushii 2421. Crataegus padifolia

4. Series Dilatatae (Sarg.) Rehder

One species, eastern U.S. and adjacent Canada

Series *Dilatatae* comprises large hawthorns that are in some ways intermediate between some members of ser. *Molles* and ser. *Pruinosae*. The series is recognized by the combination of its large, somewhat shallowly lobed leaves, stout thorns, large flowers and fruits, large, conspicuous, glandular-margined bractlets, and sepal margins with conspicuocus teeth, lobes, and glands.

7. Crataegus coccinioides Ashe (Kansas hawthorn)

C. callicarpa Sarg.

C. declivitatis Sarg.

C. dilatata Sarg.

C. locuples Sarg.

C. speciosa Sarg.

Pl. 525 n, o; Map 2418

Plants shrubs or more commonly small trees, 4–7 m tall, the trunks often armed with branched and unbranched thorns, the bark dark gray, furrowed and with small plates, peeling tardily in small, irregular, checkered flakes, usually exposing a brown underlayer. Branchlets with scattered to relatively dense thorns, these mostly 3–6 cm long, slender, mostly slightly curved, blackish brown at second year. Twigs often short-hairy when young, becoming glabrous, tan to yellowish brown or dark brown and shiny at second year. Petioles 6–30(–50) mm long, glabrous or with a few widely scattered hairs, nonglandular or sparsely glandular. Leaf blades 4–9(–11) cm long, about 2/3 as wide as long or wider, ovate to broadly ovate, more or less symmetric in outline, broadly rounded to truncate or rarely shallowly cordate at the base, the tip and large lobe tips angled to slightly tapered and mostly sharply but sometimes broadly pointed, with 3–5 lobes per side, these usually less than 20% of the way to the midvein, the sinuses relatively broad, the margins otherwise sharply and irregularly toothed, nonglandular, the texture herbaceous, the upper surface glabrous or sparsely to moderately hairy along the midvein (sometimes more hairy when young), dull or only slightly shiny, the undersurface glabrous or nearly so, sometimes with small tufts of hairs in the axils of the main veins, the secondary veins mostly 5–7 per side, the ultimate branches extending to the tips of the lobes and teeth. Inflorescences (3–)5–12-flowered, the branches glabrous or less commonly finely hairy, the bractlets membranous (then colorless to tan or reddish-tinged) to herbaceous and opaque green, linear to narrowly oblong, mostly entire to obscurely toothed, the lowermost sometimes with a pair of irregular basal lobes, with glandular margins, mostly shed by fruiting. Flowers 20–25 mm in diameter, the hypanthium glabrous or sparsely and finely hairy. Sepals 4–5 mm long, lanceolate-triangular, the margins strongly toothed and with slender lobes, the teeth and lobes gland-tipped, moderately hairy in a patch toward the tip on the inner surface, glabrous or sparsely and finely hairy on the outer surface. Petals 10–12 mm long, broadly obovate to circular, white. Stamens 20, the anthers ivory or more commonly pink to red. Styles (4)5. Fruits 10–17(–25) mm long, 10–16(–23) mm wide, globose to subglobose, lacking a raised collar at the tip or this only slightly developed and indistinct, the surface red or somewhat pinkish-tinged at maturity, glabrous, often somewhat glaucous, not or only slightly shiny. Nutlets (4)5, the lateral faces not pitted. Triploid (by anatomical sections; see Longley [1924]). Late April–late May.

Scattered mostly in the southern half of the state, north locally to Marion County (Massachusetts to Wisconsin and Iowa, south to Oklahoma, Arkansas, Illinois, and Indiana; southeastern Canada). Mesic to dry upland forests, glades, ledges and tops of bluffs, banks of streams, and bottomland forests; also fence rows and roadsides.

Superficially, the leaves of *C. coccinioides* resemble larger leaves of *C. pruinosa*. However, the flowers and fruits tend to be larger than in that species and the sepals are very different. Hairy forms of *C. coccinioides* have been called *C. locuples*, *C. declivitatis*, and, to a lesser extent, *C. dilatata*. White-anthered forms have been called *C. conspecta*. The typical form is glabrous and rose-anthered. All can be found in Missouri in various mixes of these different characters.

Kansas hawthorn is one of the most strikingly handsome of all hawthorns when well-grown.

5. Series Intricatae (Sarg.) Rehder

Plants shrubs (rarely small trees), mostly 2–6 m tall, the main stems or trunks with sparse to dense, unbranched thorns, the bark relatively thin, gray to dark gray, peeling in small, thin, irregular, sometimes longitudinal flakes to expose light brown to grayish brown underlayers. Branchlets essentially unarmed to more commonly moderately to densely thorny, the thorns 2–6 cm long, slender or the shorter ones relatively stout, straight to slightly curved. Twigs glabrous or densely pubescent with somewhat tangled hairs, reddish brown to blackish gray, dull, becoming gray with age. Petioles mostly 30–50% as long as the blade, glabrous or densely and more or less persistently pubescent with short, spreading hairs, also with few to several short, stalked (peglike), dark glands. Leaf blades mostly 3–7 cm long (occasionally to 9 cm long on extension shoots), varying from about 40% as wide as long to nearly as wide as long (seldom less than 2 cm wide), narrowly ovate to elliptic, rhombic, broadly ovate, or occasionally nearly circular, widest at or below the midpoint, usually symmetric in outline, angled or rounded to less commonly more or less truncate at the base, the tip sharply pointed, unlobed or more commonly with 1–5 shallow (at most 25% of the way to the midvein, rarely deeper on extension shoot leaves) lobes per side, the margins finely to coarsely and simply toothed above the base, the teeth sharp but gland-tipped, the basal portion (below the teeth) with short, stalked (peglike), dark glands, the texture herbaceous to somewhat papery, the upper surface glabrous or sparsely to moderately pubescent, with fine, straight hairs, sometimes only when young or only along the basal portion of the midvein, dark green to green, dull or less commonly somewhat shiny, the undersurface glabrous or sparsely to relatively densely hairy, sometimes mostly along the veins, the secondary veins mostly 4–7 per side, the ultimate branches extending to the tips of the teeth. Inflorescences 2–8-flowered, the branches glabrous or woolly, the bractlets membranous, conspicuous, linear, the margins irregular or minutely toothed, with numerous stalked glands, shed early. Flowers (13–)15–25 mm in diameter, the hypanthium glabrous or densely woolly. Sepals 4–7 mm long, shorter than the petals, triangular to lanceolate-triangular, the margins glandular-toothed or irregularly incised with slender, gland-tipped lobes, the inner surface glabrous or woolly. Petals 7–12 mm long, broadly obovate to circular, white. Stamens (5–)10, the anthers cream-colored or pink to pinkish purple. Styles 2–5. Fruits 9–16 mm long, 9–15 mm wide, globose to subglobose or occasionally somewhat oblong-obovoid, often with a short, raised collar at the tip (but the withered sepals mostly persistent and often obscuring the actual tip), the surface orange to red, brownish red, or dark red, occasionally yellow, glabrous or short-hairy, shiny or dull, not glaucous. Nutlets 2–5, the lateral faces not pitted. About 10 species, eastern U.S., Canada.

The series consists of mostly shrubs and small trees, and the most distinctive features include the petioles usually bearing small dark glands and the calyx in most species borne on a small raised collar (most easily observed at fruiting). The species-level taxonomy of ser.

Intricatae remains controversial. Kruschke (1965) and Lance (2011) have advocated different schemes for reducing most of the taxa in the series to varieties of a broadly circumscribed *C. intricata* Lange.

1. Inflorescence branches at flowering densely woolly; fruits finely but sometimes sparsely hairy at maturity . . . . 8. C. BILTMOREANA
1. Inflorescence branches at flowering glabrous; fruits glabrous at maturity
    2. Fruiting sepals more or less sessile; fruits orangish red to red . . . 10. C. PADIFOLIA
    2. Fruiting sepals clearly elevated on a small collar; fruits yellow or brownish red to dark red
        3. Leaf blades mostly narrowly ovate to oblong-ovate (occasionally somewhat rhombic), sharply but sometimes shallowly lobed; fruits globose to subglobose . . . . 9. C. NEOBUSHII
        3. Leaf blades mostly elliptic to narrowly ovate; unlobed or very shallowly lobed; fruits subglobose to somewhat oblong-obovoid . . . . 11. C. RUBELLA

8. Crataegus biltmoreana Beadle (Biltmore hawthorn)

C. intricata Lange var. *biltmoreana* (Beadle) R.W. Lance

C. villicarpa Sarg.

Pl. 526 a, b; Map 2419

Plants shrubs, usually only 2–3 m tall. Branchlets essentially unarmed to moderately thorny, the thorns mostly 3–5 cm long, dark brown to black and usually shiny at second year, sometimes becoming gray with age. Twigs reddish brown when young, becoming dark gray at second year, densely and persistently pubescent with somewhat tangled hairs. Petioles 15–30 mm long, densely and more or less persistently pubescent with short, spreading hairs, also with several short, stalked (peglike), dark glands, sometimes winged toward the tip. Leaf blades 4–7 cm long, narrowly ovate to broadly ovate or elliptic, the lobes 3–5 per side, mostly 10–25% of the way to the midvein, broadly angled to very short-tapered at the base, the lobes and tip sharply but sometimes broadly pointed, the upper surface moderately and persistently pubescent with fine, straight hairs, the undersurface similarly hairy, most densely along the veins, the secondary veins mostly 4 or 5 pairs per side. Inflorescences 2–5-flowered, the branches woolly, the bractlets relatively persistent. Flowers (13–)15–20 mm in diameter, the hypanthium densely woolly. Sepals 5–7 mm long, lanceolate-triangular, the inner surface woolly. Stamens 10, the anthers cream-colored. Styles 3–5. Fruits 9–14 mm in diameter, globose to subglobose, with a short, raised collar at the tip, finely but sometimes sparsely hairy at maturity, orange to brownish-reddish, occasionally yellow (but then often mottled with red). Nutlets 3–5. Tetraploid (by anatomical sections; see Longley [1924]). May.

Scattered, mostly in the southern half of the Ozark Division; disjunct in Caldwell and Callaway Counties (eastern U.S. west to Missouri and Arkansas). Mesic to dry upland forests, savannas, glades, banks of streams, and margins of sinkhole ponds.

This is a distinctive species with its very hairy growing parts, fairly large flowers, small stature, fairly long thorns, and orange to brownish red or occasionally yellow fruits. The leaf shape is much more variable than in *C. neobushii*, ranging from broadly ovate to more or less elliptic and shallowly lobed.

9. Crataegus neobushii Sarg. (thicket hawthorn)

C. intricata Lange var. *neobushii* (Sarg.) Kruschke

C. leioclada Sarg.

Pl. 526 d, e; Map 2420

Plants shrubs, mostly 2–4 m tall. Branchlets essentially unarmed to moderately thorny, the thorns mostly 3–6 cm long, dark purplish brown to black, and more or less shiny at second year. Twigs reddish brown when young, becoming brownish purple by end of the first year, dark gray with age, glabrous. Petioles 10–20 mm long, glabrous or nearly so, but with several, short, stalked (peglike), dark glands, usually unwinged. Leaf blades 3.5–6.0 cm long, elliptic to narrowly ovate or oblong-ovate, occasionally somewhat rhombic, the lobes 3 or 4 per side, 10–15(–20)% of the way to the midvein, narrowly to broadly angled at the base, the lobes and tip sharply but sometimes broadly pointed, the upper surface glabrous or sparsely hairy along the midvein toward the base, the undersurface glabrous or nearly so, the secondary veins mostly 4–6 pairs per side. Inflorescences 3–7-flowered, the branches glabrous, the

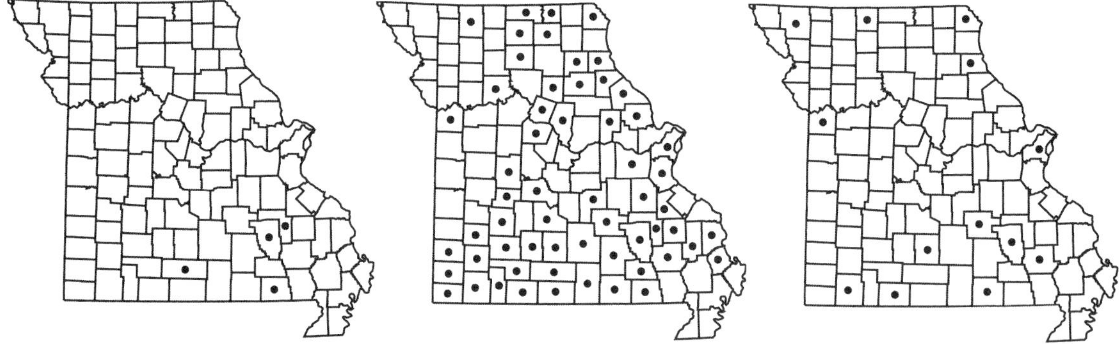

bractlets mostly shed early. Flowers 18–22(–25) mm in diameter, the hypanthium glabrous. Sepals 5–7 mm long, triangular, the inner surface glabrous. Stamens 10, the anthers pink. Styles 3–5. Fruits 10–12 mm in diameter, globose to subglobose, with a short, raised collar at the tip, glabrous at maturity, yellow or brownish red to dark red. Nutlets 3–5. May.

Scattered in the Ozark Division, uncommon farther north (southeastern U.S. west to Missouri and Arkansas). Mesic to dry upland forests, savannas, glades, and banks of streams; also pastures, old fields, and fence rows.

This species has sometimes been combined with *C. intricata*. The leaf shape of Missouri *C. neobushii* is generally quite different from that of typical *C. intricata*, which is more like *C. padifolia* var. *incarnata*. Moreover, the fresh anthers are pink, as opposed to ivory in *C. intricata*. Several historical specimens, provisionally included in this species and collected by E.J. Palmer from near Ironton (Iron County), have the leaves similar to those in *C intricata*. However, these still produced flowers with pink anthers.

10. Crataegus padifolia Sarg. (bird-cherry hawthorn)

C. intricata Lange var. *padifolia* (Sarg.) R.W. Lance

Pl. 526 f, g; Map 2421

Plants shrubs or rarely small trees, mostly 4–6 m tall. Branchlets moderately to densely thorny, the thorns mostly 2–4 cm long, dark purplish brown to black at second year, dull or shiny. Twigs reddish green to reddish brown when young, deep reddish brown by end of the first year, becoming gray to brownish gray with age, glabrous. Petioles 10–20 mm long, glabrous, but with sparse to moderate, short, stalked (peglike), dark glands, sometimes winged toward then tip. Leaf blades 3–5 cm long, elliptic to broadly elliptic or ovate, occasionally nearly circular, unlobed or the lobes 1–4 per side, to 20% of the way to the midvein, short-tapered or broadly angled to rounded or nearly truncate at the base, the lobes and tip sharply but often broadly pointed, the upper surface glabrous or sparsely hairy along the midvein toward the base, the undersurface glabrous, the secondary veins mostly 4–7 pairs per side. Inflorescences 3–6-flowered, the branches glabrous, the bractlets relatively persistent. Flowers 15–18 mm in diameter, the hypanthium glabrous. Sepals 4–5 mm long, triangular, the inner surface glabrous. Stamens (7–)10, the anthers cream-colored, sometimes pinkish-tinged. Styles 2 or 3. Fruits 12–15 mm in diameter, globose to subglobose, lacking a noticeable raised collar at the tip, glabrous at maturity, brownish orange to reddish orange or red. Nutlets 2 or 3. Triploid (by anatomical sections; see Longley [1924]). Mid-April–late May.

Scattered in the southern half of the Ozark Division; disjunct historically in Marion County (Missouri, Arkansas, and Oklahoma). Mesic to dry upland forests, savannas, glades, and banks of streams.

Crataegus padifolia is nearly endemic to the Ozark and Ouachita Mountains. It has been separated into two varieties that occupy similar ranges in Missouri. The variety *incarnata* is similar to *C. intricata* except for the sessile fruiting calyx. The varieties may be distinguished as follows.

1. Leaf blades 1.0–1.4 times as long as wide, broad angled to rounded or nearly truncate at the base, with 3 or 4 evident but often shallow and blunt lobes per side; secondary veins 4–6 per side . . . . var. *incarnata* Sarg.
1. Leaf blades 1.4–2.0 times as long as wide, tapered at the base, unlobed or with 1 or 2 pairs of shallow, broad lobes per side; secondary veins mostly 5–7 per side . . . . var. *padifolia* (Pl. 526 f, g)

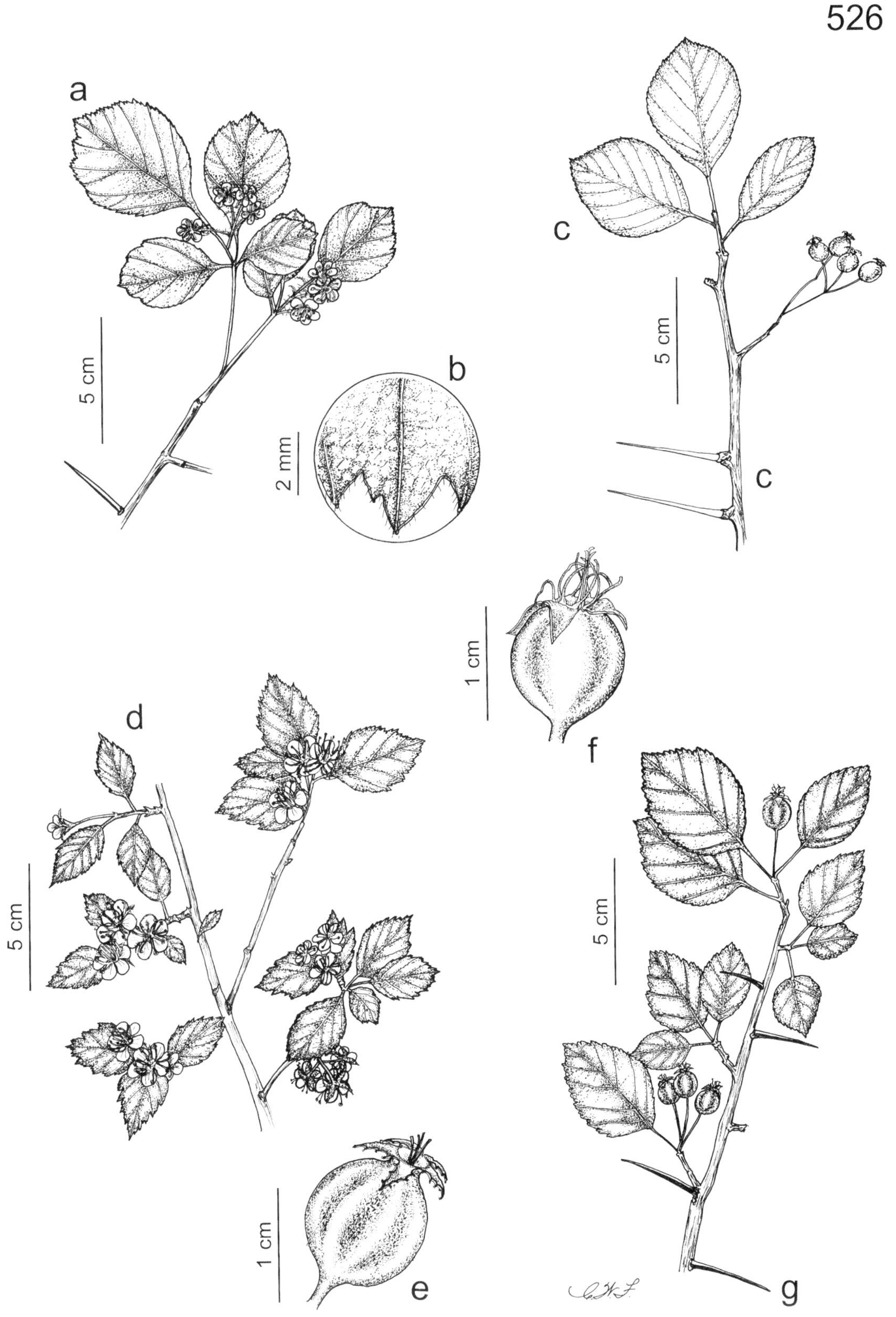

Plate 526. Rosaceae. *Crataegus biltmoreana*, **a)** flowering branch, **b)** leaf detail. *Crataegus ×simulata* (ser. *Crus-galli* × ser. *Macracanthae*), **c)** fruiting branch. *Crataegus neobushii*, **d)** fruit, **e)** flowering branch. *Crataegus padifolia* var. *padifolia*, **f)** fruit, **g)** fruiting branch.

11. Crataegus rubella Beadle
C. apposita Sarg.
C. intricata Lange var. *rubella* (Beadle) Kruschke

Map 2422

Plants shrubs, mostly 2–4 m tall. Branchlets essentially unarmed to moderately thorny, the thorns mostly 2.0–3.5 cm long, dark purplish brown to black, and more or less shiny at second year. Twigs reddish green to reddish brown when young, dull reddish brown by end of first year, usually becoming gray at second year and dark gray with age, glabrous. Petioles 15–30 mm long, glabrous, but with sparse to moderate, short, stalked (peglike), dark glands, sometimes winged toward the tip. Leaf blades (2.5–)3.0–5.0 cm long (occasionally to 9.0 cm on extension shoot leaves), elliptic to narrowly ovate or occasionally broadly elliptic to rhombic-elliptic, unlobed or the shallow lobes 1 or 2 per side, 10–15% (occasionally deeper on extension shoot leaves) of the way to the midvein, narrowly to broadly angled or short-tapered at the base, the lobes and tip sharply but sometimes broadly pointed, the upper surface glabrous or sparsely hairy along the midvein toward the base, the undersurface glabrous (sometimes sparsely hairy along the main veins toward the base when young), the secondary veins mostly 4 or 5 pairs per side. Inflorescences 3–8-flowered, the branches glabrous, the bractlets shed early. Flowers 15–18 mm in diameter, the hypanthium glabrous. Sepals 4–5 mm long, triangular, the inner surface glabrous. Stamens 10, the anthers pinkish purple. Styles 2–4. Fruits 10–12 mm in diameter, subglobose to somewhat oblong-obovoid with a short, raised collar at the tip, glabrous at maturity, red to orangish red. Nutlets 2–4. May.

Uncommon, widely scattered in the Ozark Division (southeastern U.S. west to Missouri and Louisiana). Banks of streams, openings and edges of mesic upland forests, and glades.

6. Series Macracanthae (Loudon) Rehder

Plants shrubs or small trees, 3–8 m tall, the trunks unarmed or more commonalty with sparse to dense unbranched and sometimes also branched thorns, the bark relatively thin, brownish gray to gray, with slender longitudinal furrows separating flat plates, these peeling in small, thin, irregular flakes to expose orangish brown to light brown underlayers. Branchlets essentially unarmed to relatively densely thorny, the thorns 2.5–7.0(–11.0) cm long, slender (sometimes stouter in *C. succulenta*), the shorter ones straight, the longer ones usually slightly curved, brown to dark gray at second year, often becoming lighter gray with age. Twigs glabrous to densely pubescent with soft, fine, often somewhat tangled hairs, green to grayish brown or reddish brown, often shiny, becoming brown to dark gray with age. Petioles mostly 10–50% as long as the blade, often narrowly winged for most of the length, glabrous to finely hairy when young, sometimes only along the upper side, nonglandular. Leaf blades 3–12 cm long, varying from about 50% as wide as long to nearly as wide as long (seldom less than 2 cm wide), elliptic to oval, rhombic, ovate, oblong-ovate, obovate, or occasionally nearly circular, widest at or below the midpoint, usually symmetric in outline, broadly angled or short-tapered at the base, the tip sharply but sometimes broadly pointed, unlobed or more commonly with 2–8 shallow (at most 10–15% of the way to the midvein) lobes per side, the margins usually coarsely and often doubly toothed above the base, the teeth sharp and often appearing somewhat jagged, sometimes inconspicuously gland-tipped when young, the texture herbaceous to somewhat papery, the upper surface sparsely to moderately appressed- or roughened-hairy, dull or less commonly somewhat shiny, the undersurface moderately to densely pubescent with soft, sometimes velvety or shaggy hairs when young, especially along the veins, in some species becoming glabrous or nearly so at maturity, the secondary veins mostly 5–8 per side, the ultimate branches extending to the tips of the teeth. Inflorescences 10–30(–45)-flowered, the branches densely appressed-hairy, the bractlets membranous, inconspicuous, linear, shed early, more or less entire, the margins with numerous stalked glands. Flowers 12–19 (–22) mm in diameter, the hypanthium soft-hairy. Sepals 3–5 mm long, shorter than the petals, triangular to narrowly lanceolate-triangular, the margins glandular-toothed or irregularly incised with slender, gland-tipped lobes, the inner surface hairy. Petals 6–9 mm long, broadly obovate to circular, white. Stamens 10–20, the anthers ivory or pink, red, or purple. Styles 2 or 3. Fruits 8–12(–14) mm long, mostly 7–12 mm wide, globose to subglobose or oblong-ellipsoid, lacking a distinct raised collar at the tip (but the withered sepals mostly persistent and often obscuring the actual tip), the surface orangish red to red, short-hairy during development, usually becoming less densely hairy at maturity, mostly toward the tip and base, dull or somewhat shiny, not glaucous. Nutlets 2 or 3(4), the lateral faces pitted. Three or 4 species, nearly throughout the U.S. (absent from the Gulf Coast and southwestern states); southern Canada.

Crataegus ×*vailiae* Britton (*C. missouriensis* Ashe) is a putative hybrid involving *C. calpodendron* and *C. uniflora* (ser. *Parvifoliae*). It forms sparsely thorny shrubs 2–4(–5) m tall and is thus much larger than the relatively dwarf *C. uniflora* but mostly smaller than *C. calpodendron*. It also has leaves somewhat different in size and shape than in either parent (to 6 cm long, mostly elliptic to rhombic-elliptic, lacking marginal glands, and sharply toothed) and inflorescences with 2–8(–20) flowers. In many respects it is more or less intermediate between the parents. The nutlets are slightly pitted. Individuals of this hybrid will key imperfectly to ser. *Macracanthae*, but differ in their relatively large sepals with the margins divided into gland-tipped lobes (characters inherited from the *C. uniflora* parent). Unlike most other putative *Crataegus* hybrids, which are known only historically from Missouri, *C.* ×*vailiae* is known from several extant populations in Douglas, Howell, Oregon, and Shannon Counties, sometimes in the absence of one or both parental taxa.

1. Twigs densely hairy, at least along rapidly growing shoots; branchlets often unarmed or nearly so at maturity; mature leaf blades with the upper surface green to yellowish green, the venation not conspicuously impressed, usually persistently short-hairy on the undersurface; stamens 20 . . . 12. C. CALPODENDRON
1. Twigs usually glabrous; branchlets usually strongly thorny; mature leaf blades with the upper surface usually dark green to bluish green, the venation impressed, glabrous or densely hairy on the lower surface; stamens 10 or 20
    2. Stamens 10; leaf blades with the undersurface persistently velvety- to shaggy-hairy . . . . 13. C. MACRACANTHA
    2. Stamens 20; leaf blades with the undersurface glabrous or nearly so at maturity, but soft-hairy along the veins when young . . . . 14. C. SUCCULENTA

12. Crataegus calpodendron (Ehrh.) Medik. (late hawthorn)

C. calpodendron var. *globosa* (Sarg.) E.J. Palmer
C. calpodendron var. *hispidula* (Sarg.) E.J. Palmer
C. calpodendron var. *microcarpa* (Chapm.) E.J. Palmer
C. calpodendron var. *mollicula* (Sarg.) E.J. Palmer
C. globosa Sarg.
C hispidula Sarg.
C. insperata Sarg.
C. mollicula Sarg.
C. mollita Sarg.
C. obesa Ashe
C. obscura Sarg.
C. scabera Sarg.
C. spinulosa Sarg.
C. tomentosa L. var. *microcarpa* Chapm.
Mespilus calpodendron Ehrh.

Pl. 527 a–f; Map 2423

Plants shrubs or more commonly small trees, 4–7 m tall, the trunks variously unarmed to relatively densely thorny. Branchlets often unarmed or nearly so, occasionally sparsely thorny, the thorns mostly 2.5–6.0 cm long. Twigs dark brown and shiny at second year, densely pubescent (at least along rapidly growing shoots) with soft, fine, often somewhat tangled, yellowish brown hairs when young, becoming less hairy at second year, eventually glabrous with age. Petioles 10–30 mm long, narrowly winged for most of the length, finely hairy, sometimes only on the upper side. Leaf blades 3–12 cm long, ovate to oblong-elliptic or rhombic, rarely nearly circular, with (2–)5–8 shallow lobes per side, rarely unlobed, the upper surface green to yellowish green, roughened-hairy when young, becoming glabrous or nearly so at maturity, the undersurface usually persistently pubescent with soft, more or less appressed hairs, especially on the veins, the veins not conspicuously

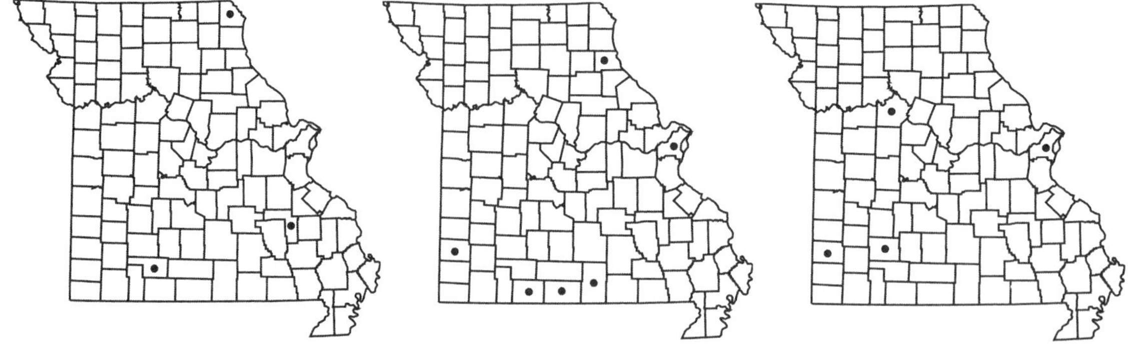

2425. Crataegus succulenta 2426. Crataegus spathulata 2427. Crataegus lanuginosa

impressed on the upper surface. Inflorescences 12–25(–45)-flowered, the branches densely hairy. Flowers 13–15(–22) mm in diameter. Stamens (10–)20, the anthers pink, red, or pinkish purple, occasionally ivory. Fruits (5–)7–9(–10) mm in diameter, oblong-ellipsoid to less commonly globose, orangish red to red. $2n$=34, 56, 68; diploid or triploid (by flow cytometry; see Talent and Dickinson [2005]). Late April–mid-June.

Scattered nearly throughout the state, but apparently uncommon or absent from many of the westernmost counties (eastern U.S. west to Wisconsin and Texas; southeastern Canada). Mesic to dry upland forests, savannas, glades, bases, ledges, and tops of bluffs, banks of streams and rivers, edges of marshes, and bottomland forests; also fence rows and roadsides.

This is a widespread and usually distinctive species. It is one of the latest of the native hawthorns to begin flowering (only *C. phaenopyrum* is sometimes later) in all parts of its range. Several varieties have been recognized in *C. calpodendron*, most of which, if accepted, can be found in Missouri. These have been said to differ in degree of leaf lobing, density of pubescence, prominence of venation, and slight differences in fruit shape. In general those forms with narrower, more elliptic, less deeply lobed leaves have shorter and somewhat blunter leaf-teeth, whereas forms with more nearly isodiametric leaves have the deepest and sharpest lobes as well as the biggest and sharpest marginal teeth. The number of lateral veins increases with narrowness of the leaf blade. Both extremes, as well as intermediates, are well-represented in Missouri herbaria. Examination of a large number of specimens suggests that the different leaf-shapes do not correlate well with the varieties historically recognized primarily on fruit characteristics. *Crataegus hispidula, C. insperata, C. obscura*, and *C. spinulosa* represent the broader-leaved, more sharply lobed forms, and *C. calpodendron* var. *calpodendron*, var. *globosa*, and var. *mollicula*, as well as *C. mollita* include the more elliptic-leaved forms. *Crataegus scabera* and *C. spinulosa* are intermediates. In practice, there is far too much variability among the Missouri plants to allow the formal recognition of any infraspecific taxa.

Crataegus rupicola Sarg. (*C. harveyana* Sarg.) represents a rare form with a few glands at the base of the leaf and on the petiole, which is an indication that it is perhaps of hybrid origin, involving a cross between *C. calpodendron* and a species in some other series (such glands occur most frequently in ser. *Intricatae*, but are produced by some taxa in a few other series as well).

13. Crataegus macracantha Lodd. ex Loudon (largethorn hawthorn)

C. divida Sarg

C. pertomentosa Ashe

C. succulenta Schrad. ex Link var. *macracantha* (Lodd. ex Loudon) Eggl.

Pl. 527 i, j; Map 2424

Plants shrubs or more occasionally small trees, 3–6(–8) m tall, the trunks usually with branched thorns. Branchlets sparsely to more commonly relatively densely thorny, the thorns mostly 3–7 (–11) cm long, shiny and dark blackish brown in second year, becoming gray with age. Twigs usually blackish gray at second year, becoming dark gray to gray with age, glabrous. Petioles 10–20 mm long, narrowly winged toward the tip, glabrous. Leaf blades 4–7(–10) cm long, elliptic to broadly elliptic or somewhat rhombic, occasionally elliptic-obovate, with (2–)3–5 shallow to very shallow lobes per side, the upper surface dark green to bluish green, roughened-hairy when young, becoming glabrous or nearly so at maturity, the undersurface densely and persistently pubescent with soft, sometimes velvety or shaggy hairs, especially along the veins (glabrous at maturity else-

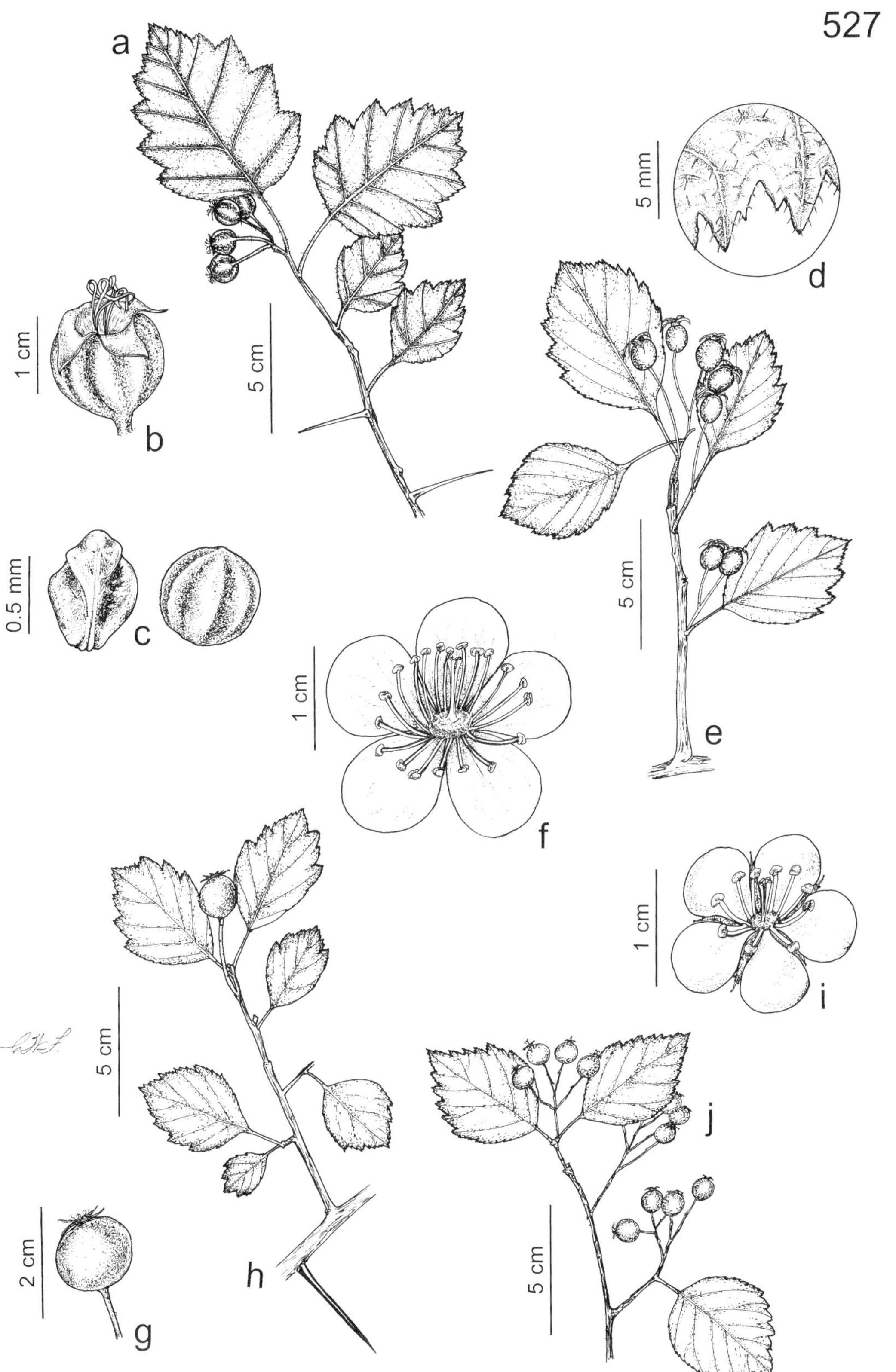

Plate 527. Rosaceae. *Crataegus calpodendron*, **a)** fruiting branch, **b)** fruit, **c)** ventral and dorsal views of seeds, **d)** leaf detail, **e)** fruiting branch, **f)** flower. *Crataegus sicca*, **g)** fruit, **h)** fruiting branch. *Crataegus macracantha*, **i)** flower, **j)** fruiting branch.

where), the veins relatively conspicuously impressed on the upper surface. Inflorescences (10–)15–30-flowered, the branches moderately to densely hairy (glabrous elsewhere). Flowers 13–19 mm in diameter. Stamens 10, the anthers ivory or pink. Fruit 7–12 mm in diameter, globose to subglobose, red. $2n$=68; tetraploid (by flow cytometry; see Talent and Dickinson [2005]). Late April–late May.

Uncommon, widely scattered in the state (nearly throughout the U.S.; southern Canada). Mesic to dry upland forests, glades, ledges of bluffs, banks of streams, and margins of lakes.

Crataegus macracantha is the most wide-ranging of all North American hawthorns. Many authors have included it in *C. succulenta*, but substantial range differences, anther size, and stamen number seem adequate for specific distinction.

This species is sometimes divided into two varieties. Missouri materials all belong to var. *pertomentosa* (Ashe) Kruschke, which is distinguished from var. *macracantha* in its leaves more densely shaggy-hairy on the undersurface and its more densely hairy hypanthia, flower stalks, and fruits. This variety occurs from Montana and South Dakota through eastern Kansas, Iowa, and Missouri to Illinois, most commonly in the northern Great Plains.

14. Crataegus succulenta Schrad. ex Link (succulent hawthorn)

C. ensifera Sarg.

Map 2425

Plants shrubs or small trees, 4–6(–8) m tall, the trunks usually with branched thorns. Branchlets sparsely to more commonly relatively densely thorny, the thorns mostly 3–6(–8) cm long, shiny and dark blackish brown at second year, becoming dark gray with age. Twigs reddish green when young, usually blackish brown at second year, eventually becoming dark gray and sometimes lighter gray with age, glabrous. Petioles 10–20 mm long, narrowly winged toward the tip, glabrous. Leaf blades 4–8 cm long, rhombic-elliptic to broadly elliptic, broadly oval, or rhombic-ovate, unlobed or with (2–)3–5 shallow to very shallow lobes per side, the upper surface often dark green to bluish green, roughened-hairy when young, becoming glabrous or nearly so at maturity, the undersurface moderately pubescent with soft, sometimes somewhat shaggy hairs when young, becoming glabrous or nearly so at maturity, the veins relatively conspicuously impressed on the upper surface. Inflorescences (10–)15–30-flowered, the branches glabrous to moderately hairy. Flowers 12–17 mm in diameter. Stamens 20, the anthers pink to red or rarely ivory. Fruit 7–12 mm in diameter, globose to subglobose, red. $2n$=34, 51, 72; diploid or triploid (by flow cytometry; see Talent and Dickinson [2005]; triploid by chromosome count [Muniyamma and Phipps, 1979]). May.

Uncommon, sporadic and known only from historical collections in the state (northeastern U.S. west to Minnesota and Missouri, south along the Appalachian Mountains to North Carolina; Canada). Mesic upland forests and banks of streams.

This is a widespread species that is very rare in the state. Missouri specimens are referable to var. *succulenta*. Additional varieties are known, mostly from states to the northeast of Missouri. Specimens referable to var. *neofluvialis* (Ashe) E.J. Palmer, which Palmer (1963) recorded from Christian and Randolph Counties, could not be verified during the present study.

7. Series Microcarpae (Loudon) Rehder (Phipps, 1998)

One species, southeastern U.S.

Series *Microcarpae* is characterized by smooth, thinly exfoliating, often honey-colored, bark, as well as small flowers, leaves and fruit. As in ser. *Apiifoliae* and ser. *Cordatae*, this series also has leaves with veins extending to the sinuses, although this may not be obvious in the most deeply cut leaves.

15. Crataegus spathulata Michx. (littlehip hawthorn)

C. microcarpa Lindl.

Pl. 528 i, j; Map 2426

Plants small trees, 3–7 m tall, the trunks unarmed or with unbranched or occsionally branched thorns, the bark greenish gray, appearing smooth, peeling in large irregular, checkered flakes or patches, exposing a gray to cream-colored and/or orangish brown underlayer. Branchlets with scattered (sometimes sparse) thorns, these 1–4 cm long, slender to less commonly stout, straight or nearly so, black or nearly so at second year. Twigs dark reddish brown at end of first year, turning deep purplish brown at second year, eventually becoming dark gray, often finely and somewhat woolly-hairy when young, glabrous with age. Petioles relatively indistinct, winged all or most of the length, essentially absent or the slender portion to 6(–11) mm long, then glabrous, nonglandular. Leaf blades 1–3(–5) cm long, ½ as wide as long or narrower, oblanceolate to spatulate or occasionally obovate in outline, sometimes slightly asymmetric in outline, tapered to long-tapered at the base, the tip and (when present) large lobe tips rounded to bluntly or infrequently sharply pointed, unlobed or more commonly with 1 or 2 lobes per side toward the tip, these mostly relatively shallow (deeper on extension shoot leaves), commonly less than 30% of the way to the midvein, the sinuses narrow to occasionally relatively broad, the margins otherwise bluntly toothed to entire or nearly so, nonglandular, the texture somewhat to strongly leathery, the upper surface usually sparsely and finely hairy at maturity (often more densely hairy when young; often glabrous or nearly so elsewhere), somewhat shiny, the undersurface glabrous or nearly so, sometimes somewhat glaucous, at least some of the secondary veins extending to the sinuses between the lobes, the other veins with the ultimate branches extending to the tips of the lobes or teeth. Inflorescences 20–30-flowered, the branches glabrous or sparsely to moderately and finely hairy, the bractlets membranous to herbaceous, linear to narrowly oblong-elliptic, toothed and sometimes with a few short, linear lobes, these gland-tipped, mostly withered or shed by flowering. Flowers 8–12 mm in diameter, the hypanthium glabrous. Sepals 1.5–2.0 mm long, broadly triangular, the margins entire, glabrous or finely hairy, sometimes in patches, nonglandular, glabrous or sparsely and finely hairy on the inner surface, glabrous on the outer surface. Petals 3.5–5.5 mm long, more or less circular, white. Stamens 20, the anthers pale yellow. Styles 3–5. Fruits 3–5 mm long, 3–5 mm wide, globose, lacking a raised collar at the tip, the surface bright red to orange at maturity, glabrous, slightly shiny. Nutlets 3–5, the lateral faces not pitted. Diploid or triploid (by flow cytometry; see Talent and Dickinson [2005]). Early April–mid-May.

Uncommon, known thus far only from the southern portion of the Ozark Division; disjunct historically in Marion County (southeastern U.S. west to Missouri, Oklahoma, and Texas). Mesic to dry upland forests, and banks of streams and rivers.

8. Series Molles (Sarg.) Rehder (Phipps, 2012)

Plants shrubs or small trees, 3–10 m tall, the trunks often unarmed, especially on smaller plants, occasionally with sparse unbranched and/or branched thorns, the bark relatively thick, brownish gray to dark gray, with slender longitudinal furrows separating flat plates, these peeling in small, thin, irregular flakes to expose orangish brown to brown underlayers. Branchlets essentially unarmed to densely thorny, the thorns (2–)3–7(–9) cm long, stout, straight to slightly curved. Twigs densely pubescent with soft, fine, often somewhat tangled or woolly hairs, green to grayish tan and dull, becoming gray and often somewhat shiny with age. Petioles mostly 20–50% as long as the blade, densely woolly when young, becoming less hairy or sometimes nearly glabrous at maturity, nonglandular or with sparse small glands. Leaf blades 2–9 cm long (sometimes to 12 cm in extension shoot leaves), varying from about 40% as wide as long to nearly as wide as long (seldom less than 2 cm wide), narrowly elliptic or narrowly rhombic to oblong-elliptic, rhombic, rhombic-ovate, broadly ovate, or nearly circular, widest at or below the midpoint, usually symmetric in outline, broadly angled or rounded to less commonly more or less truncate or shallowly cordate at the base, the tip rounded to sharply pointed, unlobed or with (1–)3–5 shallow to moderately deep (at most 60% of the way to the midvein) lobes per side (sometimes more deeply lobed on extension shoots), the margins usually coarsely and sometimes doubly toothed above the base, the teeth sharp and often appearing somewhat jagged, sometimes gland-tipped, at least when young, the texture herbaceous to less commonly somewhat leathery, the upper surface densely appressed- or roughened-hairy when young, sometimes becoming nearly glabrous with age, dull or less commonly somewhat shiny, the undersurface densely white-woolly when young (less so in *C. texana*), especially along the veins, becoming only a little less densely hairy at maturity, the secondary veins mostly 3–7 per side, the ultimate branches extending to the tips of the teeth. Inflorescences 5–17-flowered, the branches woolly, the bractlets membranous to herbaceous, either inconspicuous (but mostly more than 2 mm wide), linear to narrowly lanceolate, and shed early or relatively conspicuous, oblong to oblong-obovate and persistent, more or less entire and the margins then with numerous stalked glands, or finely glandular-toothed. Flowers (14–)18–25 mm in diameter, the hypanthium woolly. Sepals 4–6 mm long, shorter than the petals, often persistent and spreading at fruiting, triangular to narrowly triangular or lanceolate-triangular, the margins glandular-toothed or irregularly incised with slender, gland-tipped lobes, the inner surface hairy. Petals (7–)9–12 mm long, broadly obovate to circular, white. Stamens (10–)20, the anthers ivory to cream-colored or pink to red, rarely yellow. Styles 4 or 5. Fruits 10–18(–25) mm long, mostly 10–16 mm wide, globose to subglobose or occasionally oblong-obovoid, lacking a distinct raised collar at the tip (but the withered sepals mostly persistent and often obscuring the actual tip), the surface red to reddish orange, rarely yellow, densely short-hairy during development, usually becoming more or less glabrous at maturity, sometimes with small persistent patches of hairs, especially toward the tip and base, shiny or dull, not glaucous. Nutlets (3–)5, the lateral faces not pitted. About 5 species, eastern U.S., southeastern Canada.

Members of ser. *Molles* commonly are large hawthorns with medium to very large leaves, stout thorns, and relatively large flowers and fruits. Characteristic in most specimens are the large, semi-persistent, and more or less herbaceous bractlets, which are relatively broad, and the often-persistent and spreading fruiting calyces. The series name refers to the dense woolly hairs covering all developing parts of the plants that may be whitish at first and are variably persistent until fruiting. Most species have 20 stamens. Some members of ser. *Molles* are relatively common in Missouri.

1. Mature leaf blades 2–6(–8) cm long, broadest at or near the midpoint, the lobes if any, very short
   2. Mature leaf blades broadly ovate-elliptic to nearly circular, bluish-tinged and often somewhat shiny at maturity; bractlets somewhat herbaceous, conspicuous; anthers pink to red . . . . 16. C. LANUGINOSA
   2. Mature leaf blades narrowly elliptic or narrowly rhombic to oblong-elliptic or nearly circular, green to dark green (not bluish-tinged) and usually dull at maturity; bractlets membranous, inconspicuous, shed early; anthers ivory to cream-colored, rarely yellow . . . . 17. C. MOLLIS
1. Mature leaf blades 5–9 mm long, usually broadest well below the midpoint, commonly lobed
   3. Anthers ivory to cream-colored, rarely yellow; leaf blades most frequently noticeably lobed and broadest in the lower third . . . . 17. C. MOLLIS
   3. Anthers pink to red, leaf blades frequently very shallowly lobed to unlobed or broadest toward the midpoint . . . . 18. C. TEXANA

16. Crataegus lanuginosa Sarg.

C. mollis (Torr. & A. Gray) Scheele var. *lanuginosa* (Sarg.) R.W. Lance

Pl. 528 g; Map 2427

Plants shrubs, 2–4 m tall, or rarely small trees to 9 m tall, Trunks and sometimes also larger branches usually armed with conspicuous branched thorns. Branchlets somewhat zigzag, grayish brown at second year, densely thorny, the thorns 3–7(–9) cm long, purplish black and shiny at second year, becoming gray and dull with age. Petioles 10–15(–20) mm long, densely woolly when young, becoming somewhat less densely hairy with age. Leaf blades (2–)4–5(–8) cm long, (2–)3–4(–6) cm wide, broadly ovate-elliptic to nearly circular, broadest at or near the midpoint, broadly angled to rounded at the base, rounded or more commonly angled or tapered to a sharply (but sometimes broadly) pointed tip, unlobed or more commonly with 1–3 very short, somewhat irregular lobes

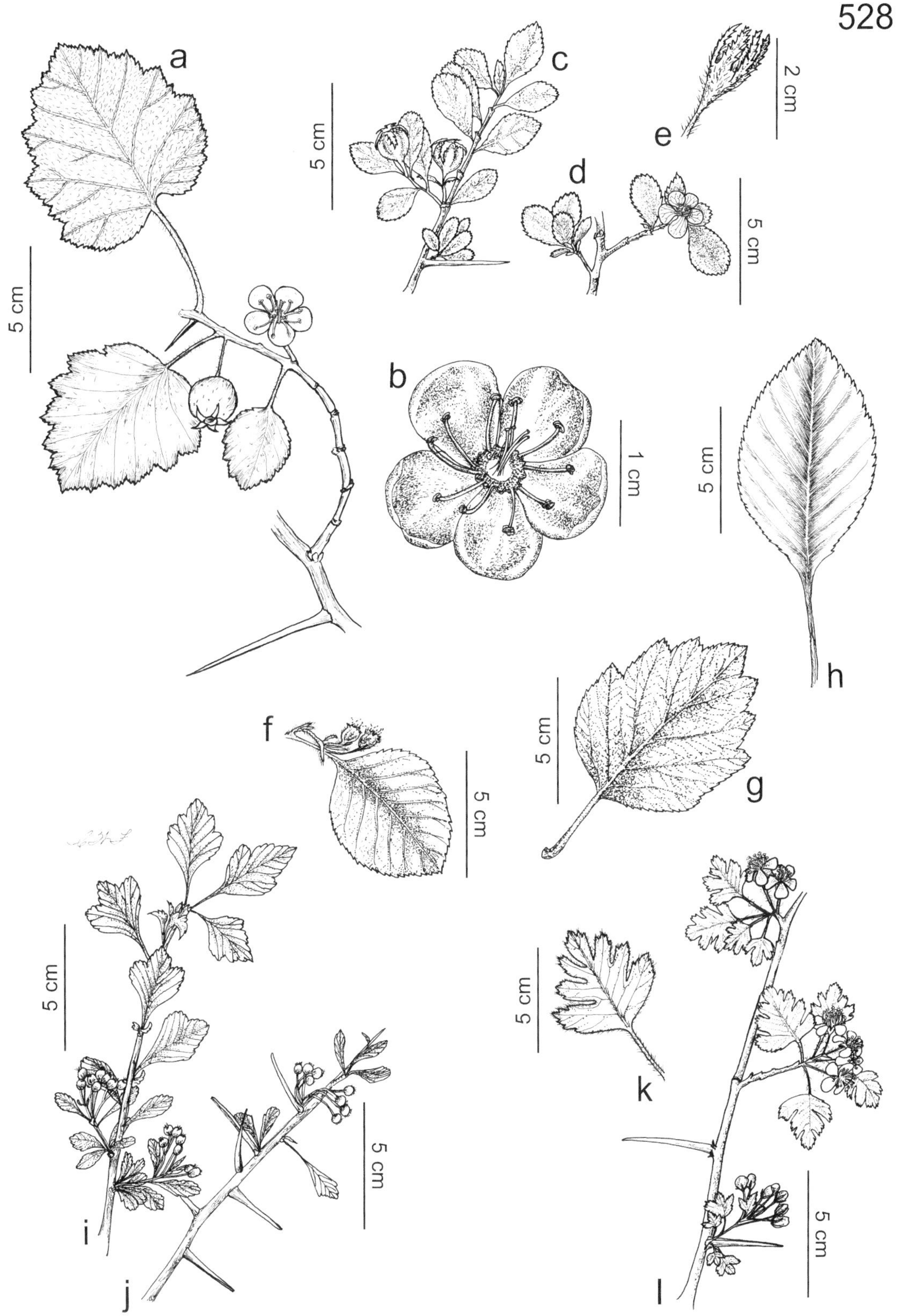

Plate 528. Rosaceae. *Crataegus mollis*, **a)** fruiting and flowering branch, **b)** flower. *Crataegus uniflora*, **c)** fruiting branch, **d)** flowering branch, **e)** bud showing calyx. *Crataegus ×dispessa* (ser. *Molles* × ser. *Punctatae*), **f)** node with leaf and young fruits. *Crataegus lanuginosa*, **g)** leaf. *Crataegus ×latebrosa* (ser. *Molles* × ser. *Punctatae*), **h)** leaf. *Crataegus spathulata*, **i, j)** fruiting branches. *Crataegus marshallii*, **j)** leaf, **k)** flowering branch.

2428. Crataegus mollis 2429. Crataegus texana 2430. Crataegus uniflora

above the midpoint (sometimes more deeply lobed on extension shoot leaves), the margins otherwise sharply and often doubly toothed, the teeth often gland-tipped, at least when young, the secondary veins mostly 3–5 pairs, somewhat leathery, the upper surface moderately to densely short-hairy, especially along the main veins, bluish-tinged and usually somewhat shiny at maturity, the undersurface densely and persistently woolly (velvety to the touch), yellowish green at maturity. Inflorescences relatively dense, the branches densely woolly, the bractlets somewhat herbaceous, relatively conspicuous, oblong to oblong-obovate, rounded to sharply pointed at the tip, finely glandular-toothed, persistent. Hypanthium densely woolly. Sepals 4–5 mm long. Petals 8–10 mm long. Stamens 20, the anthers pink to red. Styles usually 5. Fruits 10–15 mm in diameter, subglobose to slightly oblong-ovoid, usually with small persistent patches of hairs at the ends, dark red, occasionally with scattered, large, pale dots. Nutlets usually 5. Mid-April–mid-May.

Uncommon, sporadic in the state (Kansas, Oklahoma, Missouri, and Arkansas). Mesic to dry upland forests and banks of streams; also pastures and roadsides.

This species was recorded first from near Webb City, and the overwhelming majority of the Missouri specimens collected thus far have originated from Jasper County. This species was last recorded in the state by Palmer in 1957 from a site in Greene County.

Crataegus lanuginosa is readily distinguished from the other species of the ser. *Molles* by the distinctly bluish-tinged color of the relatively small, often shiny leaves. The dark red, hard fruits, extremely dense pubescence, and the remarkable development of the thorns are also unusual in ser. *Molles*. The species usually flowers before the leaves are half-grown.

17. Crataegus mollis (Torr. & A. Gray) Scheele (red haw, downy hawthorn, summer haw, turkey apple)

C. arkansana Sarg.
C. berlandieri Sarg.
C. coccinea L. var. *mollis* Torr. & A. Gray
C. dumetosa Sarg.
C. invisa Sarg.
C. lasiantha Sarg.
C. limaria Sarg
C. macrophylla Sarg.
C. meridionalis Sarg.
C. sera Sarg.
C. transmississippiensis Sarg.
C. umbrosa Sarg.

Pl. 528 a, b; Map 2428

Plants large shrubs or small trees to 10 m tall, Trunks (and sometimes also larger branches) variously unarmed or with sparse to occasionally relatively dense branched and/or unbranched thorns. Branchlets straight to slightly zigzag, tan at second year, becoming pale gray with age, essentially unarmed to relatively densely thorny, the thorns (2–)3–6 cm long, blackish brown to nearly black and shiny or dull at second year, becoming dark gray and dull with age. Petioles 10–40 mm long, densely woolly when young, becoming less hairy or sometimes nearly glabrous with age. Leaf blades (2–)4–9 cm long (sometimes to 12 cm on extension shoot leaves), (2–)3–8 cm wide, narrowly elliptic or narrowly rhombic to oblong-elliptic, rhombic, rhombic-ovate, broadly ovate, or nearly circular, broadest variously at or near the midpoint or toward the base, broadly angled or rounded to less commonly more or less truncate or shallowly cordate at the base, rounded or angled to short-tapered to a sharply (but sometimes broadly) pointed tip, unlobed or with 3–5 shallow to occasionally moderately deep (at most 60% of the way to the midvein) lobes per side (sometimes more deeply lobed on extension shoot leaves), the margins otherwise sharply and occasionally doubly toothed, the teeth usually lacking glandular tips, the secondary veins mostly 4–7(–9) pairs, herbaceous to somewhat papery or only slightly leathery, the upper surface moderately to densely short-hairy when young, the hairs more or less appressed and often stiff (roughened to the touch), often becoming sparsely hairy to glabrous at maturity, green to dark green and usually dull at maturity, the undersurface densely woolly when young (often somewhat velvety to the touch), especially along the veins, usually becoming sparsely pubescent mostly along the main veins with age, light green at maturity. Inflorescences relatively dense to more open, the branches woolly, the bractlets membranous, inconspicuous, linear to narrowly lanceolate, sharply pointed at the tip, more or less entire and with numerous stalked glands along the margins to finely glandular-toothed, shed early. Hypanthium densely woolly. Sepals 4–6 mm long. Petals (7–)9–12 mm long. Stamens (10–)20, the anthers ivory to cream-colored, rarely yellow. Styles (4)5. Fruits 10–15 mm in diameter, globose to subglobose or occasionally oblong-obovoid, glabrous or with small persistent patches of hairs at the ends, bright red to dark red or rarely yellow, sometimes with scattered, inconspicuous, pale dots or slightly glaucous (most visible when dried). Nutlets (4)5. $2n$=68; diploid ($2n$=16, by anatomical sections; see Longley [1924]); diploid or triploid (by flow cytometry; see Talent and Dickinson [2005]). Late March–mid-May.

Scattered in the northern and western halves, but uncommon in the southeastern quarter of the state (eastern U.S. from west of the Appalachians to North Dakota and Texas; southeastern Canada). Mesic to dry upland forests, savannas, edges of loess hill prairies, glades, ledges of bluffs, banks of streams, rivers, and spring branches, margins of lakes, and occasionally bottomland forests; also old fields, pastures, fence rows, quarries, and cemeteries.

This is a common and widespread hawthorn. Red haw can be a handsome tree when well-grown in gardens.

Crataegus mollis is very variable, particularly in leaf shape and size but other characters are more constant. The leaves are generally shorter, and broader in Texas specimens than in those collected from more northern sites. The most superficially distinct form is var. *incisifolia*, with sharply pointed lobes and deeper sinuses (blades lobed to at least 40% of the way to the midvein). The lobing varies from this varietal extreme to the nearly unlobed var. *dumetosa*, which is a rare form from Missouri with suborbiculate, more or less unlobed leaves. *Crataegus meridionalis* is somewhat similar, but has narrower leaves. *Crataegus mollis* usually has about 20 ivory to cream-colored anthers that distinguish it from pink-anthered species, such as *C. texana*. Addressing whether *C. mollis* is truly distinct from the ppink-anthered *C. texana* will require future detailed populational studies and biosystematic work. *Crataegus invisa*, with slightly lobed, ovate to broadly elliptic leaf blades, is a small-anthered variant of *C. mollis* var. *dumetosa* from Arkansas and southwestern Missouri.

Rare historical specimens (last collected in 1926) from Barry, McDonald, and Stone Counties have 10 stamens with yellow anthers 1.2–1.8 mm long (vs. 0.6–1.2 mm in other Missouri members of ser. *Molles*) and nonglandular petioles. These have been called *C. transmississippiensis*. Sargent (1921) likened this taxon to *C. submollis* Sarg., a more northern 10-stamened member of the complex said to differ in its larger, more coarsely and deeply toothed leaves, larger, more densely flowered but less hairy inflorescences, and somewhat pear-shaped fruits. Palmer (1946) went so far as to make it a synonym of *C. submollis*. Phipps (2012) noted its close relationship to *C. mollis*. Some authors have even placed the taxon into synonymy under *C. mollis*, but more likely it should be recognized at least at varietal level. Unfortunately a valid infraspecific combination appears to be lacking. The relationships among the named variants within the *C. mollis* complex await more detailed study. The following key serves to distinguish the major variants in Missouri, of which var. *mollis* is by far the most common.

1. Stamens 10 . . *C. transmississippiensis* Sarg.
1. Stamens 20 (*C. mollis*)
    2. Leaves sharply lobed, the deepest lobing 20–40(–60)% of the way to the midvein . . . . var. *incisifolia* . . . . Kruschke ex J.B. Phipps
    2. Leaves unlobed or shallowly lobed, the deepest lobing up to 20% of the way to the midvein
        3. Leaves nearly circular to more or less elliptic, broadest well above the base, essentially unlobed . . . . . var. *dumetosa* (Sarg.) Kruschke
        3. Leaves more or less ovate, broadest toward the base, mostly shallowly lobed . . . . var. *mollis*

18. Crataegus texana Buckley (Texas red hawthorn)

C. mollis (Torr. & A. Gray) Scheele var. *texana* (Buckley) R.W. Lance
C. dasyphylla Sarg.
C. induta Sarg.
C. dallasiana Sarg.
C. quercina Ashe

Map 2429

Plants large shrubs or small trees to 10 m tall, Trunks and larger branches variously unarmed or with sparse to occasionally relatively dense branched and/or unbranched thorns. Branchlets straight to slightly zigzag, pale grayish tan at first year (but this often obscured by densely woolly hairs), becoming gray with age, essentially unarmed to relatively densely thorny, the thorns (2–)3–5 cm long, blackish brown to nearly black and shiny at second year, sometimes becoming dark gray and dull with age. Petioles 15–30 mm long, densely woolly when young, becoming less hairy or sometimes nearly glabrous with age. Leaf blades 4–9 cm long (sometimes to 12 cm on extension shoot leaves), 3–8 cm wide, elliptic to rhombic, rhombic-ovate, or broadly ovate, broadest below but sometimes near the midpoint, broadly angled or rounded to less commonly shallowly cordate at the base, rounded or more commonly angled to short-tapered to a sharply (but sometimes broadly) pointed tip, unlobed or with 2–4 shallow (at most 25% of the way to the midvein) lobes per side (sometimes more deeply lobed on extension shoot leaves), the margins otherwise sharply, coarsely, and irregularly doubly toothed, the teeth usually gland-tipped, the secondary veins mostly 4 or 5 pairs, herbaceous to somewhat papery, the upper surface moderately to densely short-hairy when young, the hairs more or less appressed and often stiff (roughened to the touch), often becoming sparsely hairy at maturity, green to dark green and shiny at maturity, the undersurface sparsely to moderately woolly when young, especially along the veins, becoming somewhat less densely hairy with age, light green at maturity. Inflorescences relatively dense to more open, 7–12-flowered, the branches woolly, the bractlets membranous to somewhat herbaceous, inconspicuous, linear, sharply pointed at the tip, more or less entire and with numerous stalked glands along the margins, often shed early. Hypanthium woolly. Sepals 4–6 mm long. Petals 7–10 mm long. Stamens 20, the anthers pink to red. Styles 4 or 5. Fruits 10–14(–25) mm in diameter, globose to subglobose, with small persistent patches of hairs at the ends, red to reddish orange, rarely yellow, occasionally with scattered, inconspicuous, pale dots. Nutlets 4 or 5. Late April–mid-May.

Uncommon, sporadic in the state (Texas, Oklahoma, Arkansas, and Missouri). Banks of streams and bottomland forests.

Crataegus texana is most abundant in the southern two-thirds of Texas, mainly east of the Edwards Plateau. In Missouri it reaches its extreme northeastern limit. However, all of the Missouri specimens are historical; the taxon was last collected in the state in 1930.

Missouri specimens of *C. texana* plants have leaves with relatively few, shallow, blunt lobes and have been called var. *dasyphylla* (Sarg.) J.B. Phipps. They are similar in leaf morphology to some Texas forms of the wide-ranging *C. mollis* but have ivory or cream-colored anthers. Plants of var. *texana*, which are more abundant but restricted to Texas, have sharply and quite deeply lobed leaves. *Crataegus induta* is a particularly large-fruited variant, sometimes known as turkey haw.

9. Series Parvifoliae (Loudon) Rehder (Phipps and Dvorsky, 2006)

Two to 4 species, eastern (mostly southeastern) U.S.; Mexico.

Members of ser. *Parvifoliae* are distinctive hawthorns of unusually small stature, with small, usually unlobed, leathery leaves, very fine thorns, few-flowered inflorescences and usually yellow to ruddy fruits. However, their most distinctive feature is their long sepals, equaling or somewhat exceeding the petals in length.

19. Crataegus uniflora Münchh. (one-flowered hawthorn)

C. parvifolia Aiton
C. trianthophora Sarg.

Pl. 528 c–e; Map 2430

Plants relatively small shrubs, 0.4–1.5 m tall, the larger stems unarmed or with unbranched thorns, the bark gray, relatively smooth to finely roughened, occasionally splitting longitudinally for short stretches, exposing a yellowish brown

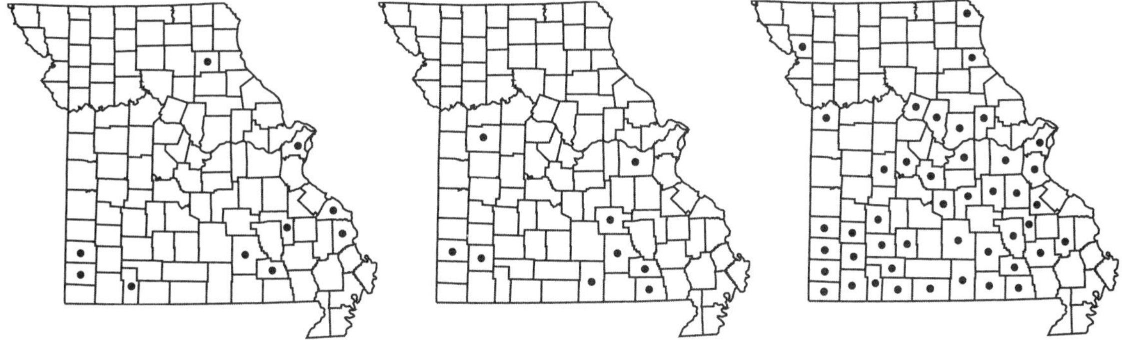

2431. Crataegus gattingeri 2432. Crataegus pruinosa 2433. Crataegus collina

underlayer. Branchlets with scattered (rarely sparse or absent) thorns, these 2–5(–8) cm long, slender, straight or nearly so at maturity, black or nearly so at first year. Twigs brown at second year, densely and finely hairy when young, becoming glabrous and gray with age. Petioles absent or to 2(–6) mm long, then densely hairy, nonglandular or with a few long, slender, stalked glands on the upper surface. Leaf blades 1.5–4.0(–5.0) cm long, 1/2 as wide as long or usually wider, oblanceolate to spatulate, elliptic, or obovate in outline, usually symmetric in outline, angled or slightly tapered at the base, the tip rounded to bluntly or broadly (but sharply) pointed, unlobed, the margins finely to relatively coarsely bluntly toothed or scalloped, mostly at or above the midpoint, the teeth often gland-tipped when young, the glands often shed with age, the texture somewhat leathery, the upper surface moderately to densely pubescent with short stiff hairs when young, becoming sparsely hairy to glabrous with age, usually somewhat shiny, the undersurface densely pubescent with softer hairs along the main veins, more sparsely hairy between the veins, the secondary veins 2–4(–5) per side, the ultimate branches extending to the teeth. Inflorescences of solitary flowers or less commonly small clusters of 2 or 3(4) flowers, the stalks densely hairy and occasionally with 1 or 2 slender, stalked glands, the bractlets mostly membranous, small, linear to narrowly elliptic, with small gland-tipped teeth, shed early. Flowers 10–15 mm in diameter, the hypanthium densely woolly. Sepals 5–8 mm long, persistent and conspicuous at fruiting, lanceolate-triangular (2–4 mm wide), the margins strongly toothed and/or with slender lobes, the teeth and lobes gland-tipped, sparsely to moderately hairy on the inner surface, moderately to densely hairy on the outer surface. Petals 5–7 mm long, more or less circular, white. Stamens 20, the anthers pale yellow. Styles 5. Fruits 8–10 mm long, 8–10 mm wide, globose or subglobose, sometimes with a low, poorly developed, raised collar at the tip, the surface yellowish green, yellow, yellowish orange, or occasionally reddish orange at maturity, woolly during development, but usually becoming more or less glabrous with age, not shiny. Nutlets 4 or 5, the lateral faces not pitted. $2n=51$; triploid or tetraploid (by flow cytometry; see Talent and Dickinson [2005]). Early May–late June.

Scattered in the Ozark Division north locally to Callaway and St. Louis Counties; also a single historical collection from the Crowley's Ridge Section of the Mississippi Lowlands Division in Dunklin County (eastern [mostly southeastern] U.S. west to Missouri, Oklahoma, and Texas; Mexico). Mesic to dry upland forests, glades, banks of streams, and ledges and tops of bluffs; occasionally also pastures, fence rows, and roadsides.

Crataegus uniflora is somewhat variable in plant size, leaf shape, number of flowers per inflorescence (usually one), and anther color. *Crataegus trianthophora* (described from Carter County, Missouri), for instance, nearly always has inflorescences with more than one flower. Lance (2011) suggested the recognition of a second variety, var. *brittonii* (Eggl.) R.W. Lance, based on somewhat taller plants in the southeastern United States with larger, more noticeably lobed leaves. However, Phipps and Dvorsky (2006) presented evidence in support of maintaining this taxon as a separate species, *C. brittonii* Eggl.

Crataegus uniflora is distinctive among Missouri hawthorns, particularly in its small habit, slender spines, and large sepals. However, a putative hybrid with *C. calpodendron* known as *C.* ×*vailiae* shares some of these features. It is a larger shrub with up to 8-flowered inflorescences. It is discussed further in the treatment of ser. *Macracanthae*.

10. Series Pruinosae (Sarg.) Rehder

Plants shrubs or small trees, mostly 4–8 m tall, the trunks often armed with branched thorns, the bark relatively thin, grayish brown to gray, divided into small, irregular plates, these peeling tardily in small, thin, irregular flakes to expose orangish brown to light brown underlayers. Branchlets moderately to densely thorny, the thorns mostly 2.5–5.0 cm long, slender, straight to slightly curved. Twigs glabrous but sometimes somewhat glaucous, reddish brown to purplish brown, becoming grayish brown to gray, dull. Petioles mostly 50–90% as long as the blade, glabrous, nonglandular or with a few, inconspicuous, sessile, dark glands. Leaf blades 2.5–6.0 cm long (rarely slightly longer on extension shoots), mostly 60–90% as wide as long (seldom less than 2 cm wide), ovate to oblong-ovate or triangular-ovate, widest below or occasionally near the midpoint, more or less symmetric in outline, broadly angled to rounded, truncate, or shallowly cordate at the base, the tip sharply pointed or occasionally nearly rounded, with 3 or 4 relatively shallow (at most 30% of the way to the midvein; occasionally more deeply lobed on extension shoots), sharp lobes per side (except in *C. pruinosa* var. *magnifolia*, with some or most of the leaves unlobed or merely with short to obscure and often blunt or rounded lobes), the margins finely and usually simply toothed above the base, the teeth sharp and often appearing somewhat jagged, nonglandular, the texture thick-herbaceous to papery or slightly leathery, the upper surface glabrous (except sometimes with short hairs along the midvein toward the base) or less commonly moderately to densely short-hairy or roughened-hairy, dark green to bluish green and dull, the undersurface sometimes sparsely to moderately hairy when young, usually glabrous or nearly so at maturity, the secondary veins mostly 4–6 per side, the ultimate branches extending to the tips of the teeth. Inflorescences 4–10-flowered, the branches glabrous or rarely inconspicuously hairy, the bractlets membranous, inconspicuous, linear, entire, the margins with numerous stalked glands, shed early. Flowers 12–22 mm in diameter, the hypanthium glabrous or rarely short-hairy. Sepals 3–5(–7) mm long, shorter than the petals, triangular to narrowly triangular or linear above a triangular base, the margins with a few glandular teeth toward the tip or entire, then nonglandular or rarely with a few short, stalked, dark glands, the inner surface glabrous. Petals 7–11 mm long, broadly obovate to circular, white. Stamens (10–)20, the anthers ivory to cream-colored or pink to red. Styles 3–5. Fruits 6–16 mm long, 6–15 mm wide, globose to subglobose, with a distinct raised collar at the tip (but the withered sepals sometimes obscuring the actual tip), the surface green with reddish tinge to reddish pink, dull red, brownish red, or dark red, glabrous, dull and slightly to strongly glaucous. Nutlets 2–5, the lateral faces not pitted. About 7 species, eastern U.S., Canada.

Members of ser. *Pruinosae* are most commonly glabrous, rather thorny hawthorns, characterized by their calyces persistent on a short, elevated collar at fruiting and often somewhat angular fruit. Most of the species also are notable for their strongly glaucous fruits. If the glaucous covering wears away, the fruit color tends towards crimson unless, in some cases, it remains greenish until late in development. In other respects the series is quite average for American hawthorns in overall dimensions, numbers of parts, leaf lobing, etc.

Species delimitations within ser. *Pruinosae* remain controversial, with some botanists treating the series as comprising but a single variable species.

1. Leaf blades mostly 2.5–4.0 cm long, often rather narrow (usually 1.5 or more times as long as wide), the base very broadly angled to truncate or shallowly cordate; flowers 12–17 mm in diameter; anthers 0.3–0.5 mm long; fruits mostly 6–10 mm in diameter . . . . 20. C. GATTINGERI
1. Leaf blades mostly 5–8 cm long, mostly ovate to broadly elliptic, relatively broad or, if relatively narrow, then rounded to angled at the base; flowers 15–22 mm in diameter; anthers 0.7–1.0 mm long; fruits 9–16 mm in diameter . . . . 21. C. PRUINOSA

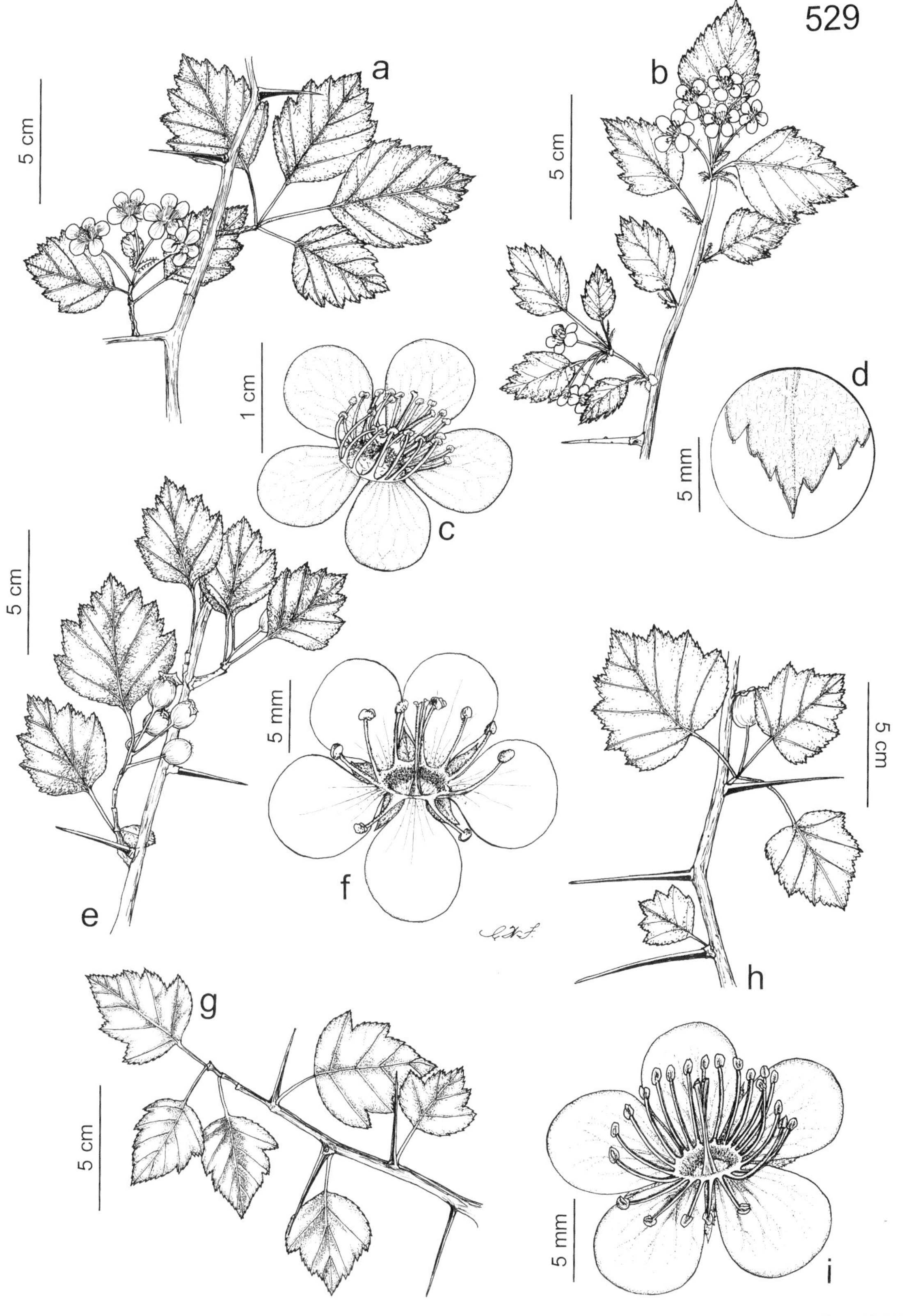

Plate 529. Rosaceae. *Crataegus pruinosa* var. *pruinosa*, **a, b)** flowering branches, **c)** flower, **d)** leaf detail. *Crataegus pruinosa* var. *dissona*, **e)** fruiting branch, **f)** flower. *Crataegus gattingeri*, **g)** vegetative branch. *Crataegus pruinosa* var. *rugosa*, **h)** vegetative branch, **i)** flower.

20. Crataegus gattingeri Ashe (Gattinger's hawthorn)
C. pruinosa (H.L. Wendl.) K. Koch var. *gattingeri* (Ashe) R.W. Lance
Pl. 529 g; Map 2431

Plants shrubs or small trees, 3–8 m tall. Thorns mostly 2.5–4.0 cm long, very dark brown to purplish black or black at second year. Twigs reddish tinged when young, deep reddish brown at end of first year, becoming dark grayish brown at second year, deep gray with age, glabrous. Petioles 12–25 mm long, glabrous, nonglandular. Leaf blades 2.5–5.0 cm long, often relatively narrow (more than 1.5 times as long as wide), ovate to broadly ovate or triangular-ovate, very broadly angled to truncate or shallowly cordate at the base, with 3 or 4 shallow (at most 15% of the way to the midvein), sharp lobes per side, mostly with 5 pairs of secondary veins per side, the surfaces glabrous. Inflorescences mostly 4–8-flowered; the branches glabrous. Flowers 12–17 mm in diameter. Sepals 3–4 mm long, the lobes triangular. Stamens 20, the anthers 0.3–0.5 mm long, pink to red. Styles mostly 4. Fruits 6–10(–12) mm in diameter, green with reddish tinge to brownish red or dark red, slightly to strongly glaucous. Nutlets mostly 4. Mid-April–late May.

Scattered in the southern half of the state; disjunct in Shelby County (eastern U.S. west to Missouri and Louisiana). Mesic to dry upland forest, banks of streams, glades, edges of upland prairies, and tops of bluffs; also pastures and roadsides.

21. Crataegus pruinosa (H.L. Wendl.) K. Koch (frosted hawthorn)
C. aperta Sarg.
C. aspera Sarg.
C. brachypoda Sarg.
C. bracteata Sarg.
C. calliantha Sarg.
C. decorata Sarg.
C. disjuncta Sarg.
C. dissona Sarg.
C. eggertii Britton
C. leiophylla Sarg.
C. mackenzii Sarg.
C. mackenzii var. *aspera* (Sarg.) E.J. Palmer
C. magnifolia Sarg.
C. patrum Sarg.
C. platycarpa Sarg.
C. rigida Sarg.
C. rubicundula Sarg.
C. rugosa Ashe
C. seclusa Sarg
C. seducta Sarg.
C. tumida Sarg.
C. virella Ashe
Mespilus pruinosa H.L. Wendl.
Pl. 529 a–f, h, i; Map 2432

Plants shrubs, 2–4 m tall. Twigs reddish-brown when young, dull purple-brown at end of first year, becoming dull gray at second year, paler gray with age, glabrous .Thorns on branchlets mostly 3–5 cm long, purplish brown to black at second year. Petioles 15–30 mm long, very slender, glabrous, nonglandular or with a few, inconspicious, sessile, dark glands. Leaf blades 2.5–6.0 cm long, often relatively broad (to 90% as wide as long) and broadly angled to rounded, truncate, or shallowly cordate at the base or, if relatively narrow, then rounded to angled at the base, ovate to broadly ovate or triangular-ovate, usually with 3 or 4 shallow to moderate (at most 30% of the way to the midvein), sharp lobes per side (except in var. *magnifolia*, with has some or most of the leaves unlobed or merely with short to obscure and often blunt or rounded lobes), mostly with 4–6 pairs of secondary veins per side, the upper surface glabrous or less commonly sparsely to densely roughened-hairy when young (then often becoming glabrous or nearly so at maturity), the undersurface glabrous or occasionally sparsely to moderately soft-hairy along the veins when young (then often becoming glabrous or nearly so at maturity). Inflorescences mostly 5–10-flowered; the branches glabrous or rarely sparsely hairy. Flowers 15–22 mm in diameter. Sepals 4–5(–7) mm long, the lobes triangular to narrowly triangular or linear above a triangular base. Stamens (10–)20, the anthers 0.7–1.0 mm long, ivory to cream-colored or pale to bright pink, rarely dull purple. Styles 3–5. Fruits 9–16 mm in diameter, reddish pink to red or green with pink to purple areas, often strongly glaucous. Nutlets 2–5. $2n$=51, 68, 72; triploid (by anatomical sections; see Longley [1924]); triploid to tetraploid (by flow cytometry; see Talent and Dickinson [2005]). Mid-April–late May.

Scattered nearly throughout the state (eastern U.S. west to Iowa, Kansas, and Oklahoma; southeastern Canada). Mesic to dry upland forests, banks of streams and rivers, glades, bases and ledges of bluffs, and occasionally bottomland forests; also pastures, old fields, fence rows, and roadsides.

Frosted hawthorn is a widespread species and very variable in leaf morphology. It is characterized (except for uncommon forms) in being glabrous and having strongly glaucous (pruinose) fruits, at least the immature ones, hence the vernacular name. There are several more or less well-marked varieties of *C. pruinosa* in Missouri, as well as a large array of less distinctive forms. The major variants may be distinguished using the following key. Plants with 10 stamens (var. *dissona*) usually have pink anthers, but occasionally they produce pale, cream-colored anthers. Distinctive plants with relatively small, moderately to densely hairy leaves may be referred to as *C. pruinosa* var. *virella*. However, sparsely hairy plants are occasionally encountered and have been called *C. mackenzii* and *C. decorata* in the older literature. These differ somewhat in leaf shape from var. *virella* and thus are better considered unusual components of var. *pruinosa*. *Crataegus aspera* is a poorly understood taxon based on a few historical specimens from Jasper County. It has strongly roughened-hairy foliage and inflorescences, as well as somewhat larger leaves than is typical in var. *virella*. Plausibly it may reflect past hybridization between *C. pruinosa* and *C. biltmoreana* (ser. *Intricatae*). If future studies find that this taxon should be combined with var. *virella*, then the epithet *aspera* has priority at the varietal rank and a new combination will have to be published under *C. pruinosa*.

1. Stamens about 10 . . . . var. *dissona* . . . . (Sarg.) Eggl. (Pl. 529 e, f)
1. Stamens 20
   2. Anthers ivory or pale cream-colored . . . . var. *cognata* (Sarg.) J.B. Phipps
   2. Anthers pale to bright pink
      3. Some or most well-developed leaf blades unlobed or merely with short to obscure and often obtuse lobes . . . . var. *magnifolia* . . . . (Sarg.) J.B. Phipps
      3. Leaf blades all with 3 or 4 distinct, sharp lobes per side
         4. Leaf blades widest near the base, 1.0–1.3 times as long as wide . . . . var. *rugosa* (Ashe) . . . . Kruschke (Pl. 529 h, i)
         4. Leaf blades widest near the midpoint, or, if widest closer to the base, then more than 1.3 times as long as wide
            5. Leaf blades glabrous or nearly so on the upper surface; inflorescence branches glabrous . . . . var. *pruinosa* (Pl. 529 a–d)
            5. Leaf blades with the upper surface roughened-hairy when young, usually becoming glabrous or nearly so with age; inflorescence branches often sparsely hairy . . . . var. *virella* . . . . (Ashe) Kruschke

11. **Series Punctatae** (Loudon) Rehder (Phipps, 2006)

Plants shrubs or small trees, 3–8 m tall, the trunks often armed with branched thorns, the bark relatively thin, grayish brown to gray, divided into small, irregular plates, these peeling in small, thin, irregular flakes to expose orangish brown to light brown underlayers. Branchlets sparsely to moderately thorny, the thorns 2–6 cm long, slender, straight to slightly curved. Twigs glabrous or nearly so or densely pubescent with soft, fine, often somewhat tangled hairs, at least when young, grayish brown or reddish brown and dull to somewhat shiny at second year. Petioles mostly 10–25% as long as the blade (but usually winged toward the tip and thus sometimes appearing shorter), sparsely to densely hairy when young, sometimes only on the upper side, often becoming less hairy or sometimes nearly glabrous at maturity, usually nonglandular (rarely with a few sessile glands). Leaf blades 2.0–6.0(–7.5) cm long (rarely slightly longer on extension shoots), mostly 50–80% as wide as long (seldom more than 2 cm wide), oblanceolate to more commonly somewhat rhombic, elliptic-obovate or obovate, mostly widest above the midpoint, more or less symmetric in outline, mostly short-tapered at the base to the winged apical portion of the petiole, occasionally appearing angled or rounded, the tip bluntly to sharply pointed or occasionally rounded, unlobed or with 1–4 shallow, tooth-like lobes per side (occasionally more deeply lobed on extension shoots), the margins finely to coarsely and usually simply toothed above the base, the teeth sharp and often appearing somewhat jagged, sometimes inconspicuously gland-tipped, at least when young, the texture thick-herbaceous to slightly leathery, the upper surface sparsely to moderately pubescent when young, sometimes mostly along the main veins, often becoming glabrous or nearly so at maturity, sometimes roughened-hairy, the hairs either short, stiff, and appressed or longer, soft, and curved, dull, the undersurface sparsely to densely short-hairy, mostly along the veins, the secondary veins 4–10 per side, the ultimate branches extending to the tips of the teeth. Inflorescences 5–12-flowered, the branches glabrous to more commonly sparsely to densely hairy, the bractlets more or less membranous, inconspicuous, linear (2 mm wide or narrower), entire, the margins with numerous stalked glands, shed early. Flowers 12–22 mm in diameter, the hypanthium woolly. Sepals 4–6 mm long, shorter than the petals, narrowly triangular or lanceolate-triangular, the margins more or less entire to glandular-toothed, the inner surface hairy. Petals (7–)9–12 mm long, broadly obovate to circular, white. Stamens 5–20, the anthers ivory to cream-colored, yellow, or pink to red. Styles (3–)5. Fruits 8–14 mm long, mostly 8–14 mm wide, more or less globose, lacking a distinct raised collar at the tip (but the withered sepals mostly persistent and often obscuring the actual tip), the surface red or dark red to orange, densely pubescent with short, spreading hairs during development, usually becoming more or less glabrous at maturity, sometimes with small persistent patches of hairs, especially toward the tip and base, dull or somewhat shiny, not glaucous. Nutlets (2–)3–5, the lateral faces not pitted. Four or 5 species, eastern U.S., Canada, Mexico.

Members of ser. *Punctatae* are characterized by dense short hairs on young growth , as well as leaves that nearly always are relatively narrow, and unlobed or only shallowly lobed. The young twigs soon become pale gray to nearly white in most species, a character especially noticeable in the winter.

1. Largest leaf blades 3.0–4.5 cm long (to 7.5 cm in rare broad-leaved plants), narrowly obovate to broadly elliptic, mostly rounded or broadly angled at the tip, unlobed or with 1–4 very shallow lobes per side; twigs grayish brown at second year, glabrous or sparsely short-hairy when young, soon becoming glabrous; flowers mostly 13–18 mm in diameter .... 22. C. COLLINA
1. Largest leaf blades mosly 4–6 cm long, broad-elliptic to rhombic-elliptic, tapered at the tip, with 2 or 3(4) short, sharp lobes per side; twigs reddish brown at second year, densely hairy when young, becoming sparsely hairy to nearly glabrous at second year; flowers mostly 20–24 mm in diameter .... 23. C. SPES-AESTATUM

22. Crataegus collina Chapm. (hillside hawthorn)

C. angustata Sarg.
C. hirtiflora Sarg.
C. lettermanii Sarg.
C. macropoda Sarg.
C. secta Sarg.
C. sordida Sarg.
C. succincta Sarg.
C. sucida Sarg.
C. vicina Sarg.

Pl. 530 e–h; Map 2433

Plants usually shrubs, 3–8 m tall. Twigs grayish brown at second year, paler gray with age, glabrous or sparsely short-hairy when young, soon becoming glabrous. Petioles 7–10 mm long (sometimes winged most of the length and thus appearing shorter), sparsely to densely hairy when young, becoming sparsely hairy to nearly glabrous with age. Leaf blades 2.0–4.5(–7.5) cm long, narrowly obovate to broadly elliptic, widest either just above the midpoint or toward the tip, rounded or broadly angled to a bluntly or sharply pointed tip, unlobed or with 1–4 very short, sharp lobes per side, the margins otherwise usually relatively finely and sharply toothed, the upper surface sparsely to moderately short-hairy or roughened-hairy when young, sometimes mostly along the veins, often becoming glabrous or nearly so at maturity, the undersurface sparsely to densely hairy, mostly along the veins, the secondary veins 4–10 per side. Inflorescences 5–10-flowered, the branches usually appressed-hairy. Flowers mostly 12–18 mm in diameter. Sepals more or less entire to glandular-toothed. Stamens 5–20, the anthers ivory to cream-colored, yellow, or pink to red. Styles (2–)5. Fruits 8–14 mm wide, the surface red or dark red to orange, dull or somewhat shiny. Nutlets (2–)5. Triploid ($2n$=24, as *C. vicina*, by anatomical sections; see Longley [1924]); diploid to tetraploid (by flow cytometry; see Talent and Dickinson [2005]). Late March–mid-May.

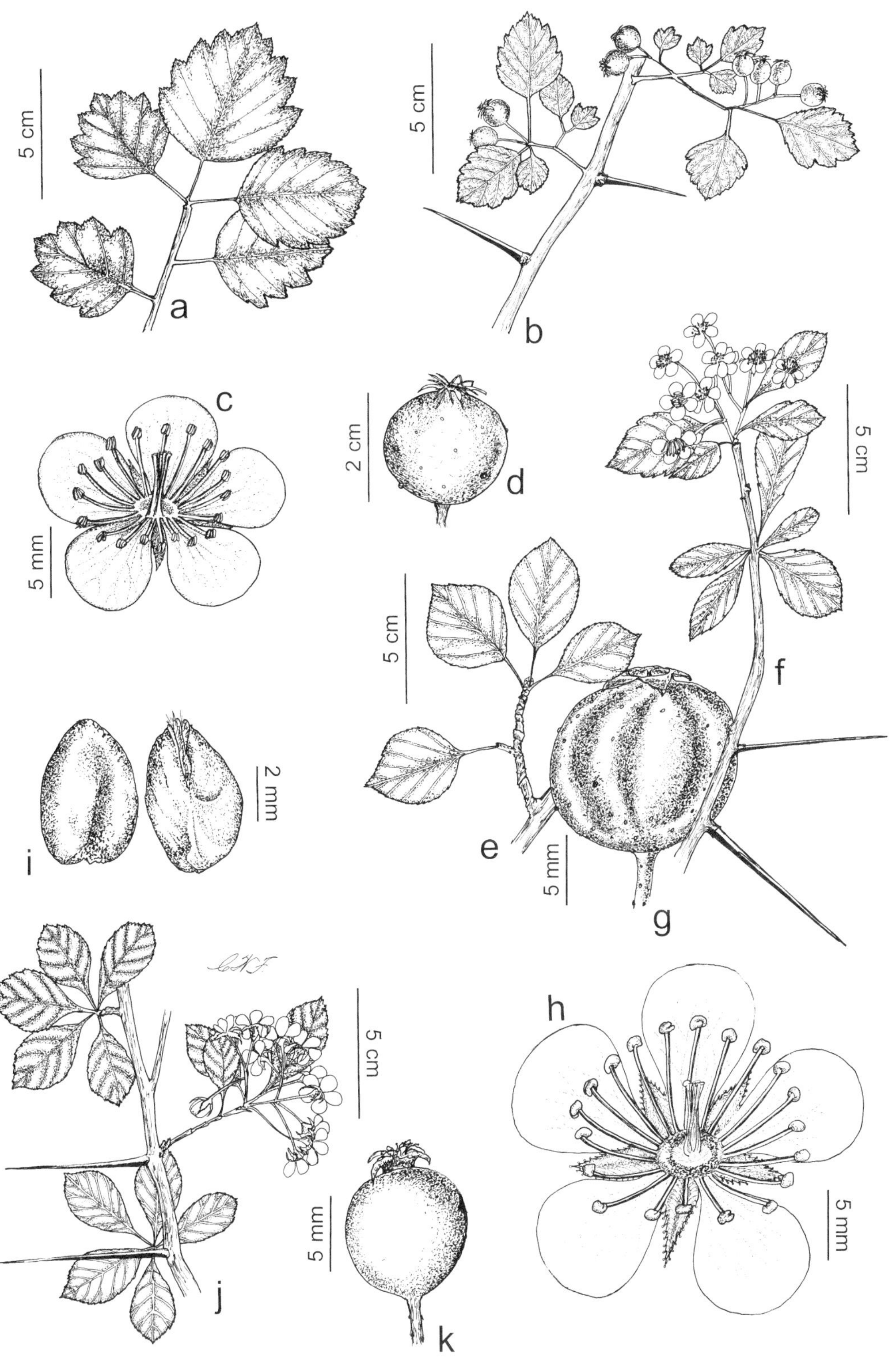

Plate 530. Rosaceae. *Crataegus margarettae*, **a)** vegetative branch, **b)** fruiting branch, **c)** flower, **d)** fruit. *Crataegus collina*, **e)** vegetative branch, **f)** flowering branch, **g)** fruit, **h)** flower. *Crataegus* ×*verruculosa* (ser. *Crus-galli* × ser. *Punctatae*), **i)** seeds, lateral and ventral views, **j)** flowering branch, **k)** fruit.

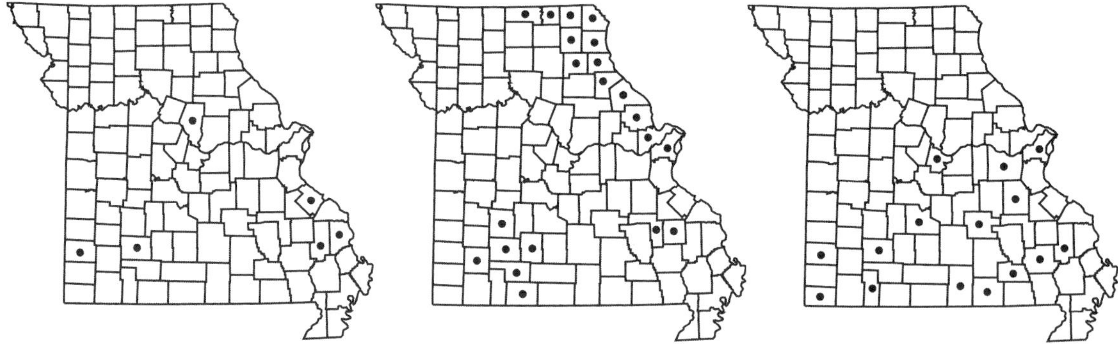
2434. Crataegus spes-aestatum 2435. Crataegus margarettae 2436. Crataegus sicca

Scattered, mostly in the southern half of the state, northward locally to Clark County; apparently absent from the Mississippi Lowlands Division (southeastern U.S. west to Kansas and Oklahoma). Mesic to dry upland forests, savannas, glades, bases, ledges, and tops of bluffs, banks of streams, margins of sinkhole ponds, and rarely bottomland forests; also pastures, old fields, fence rows, and roadsides.

Hillside hawthorn is one of the first hawthorns to begin flowering in its area of distribution. Only the mayhaws (*C. aestivalis* (Walter) Torr. & A. Gray and other species in the ser. *Aestivales* (C. K. Schneid.) Rehder), all of which occur to the south of Missouri, are substantially earlier-flowering.

Crataegus collina is a rather variable species and has been separated into several varieties, five of which have been reported from Missouri. Of these, only var. *collina* and var. *hirtiflora* are known in the state from modern collections. The following key serves to separate the major named variants in the state.

1. Leaves 2/3 to fully developed at flowering, the blades mostly 1.2–1.3 times as long as wide; lobes, if present, mostly broadly or bluntly pointed
   2. Stamens 15–20; leaf blades mostly 3.0–4.5 cm long, unlobed or obscurely lobed . . . . var. *hirtiflora* . . . . (Sarg.) J.B. Phipps
   2. Stamens 5–15(–20); leaf blades mostly 5.0–7.5 cm long, evidently lobed . . . . var. *lettermanii* (Sarg.) Eggl.
1. Leaves usually only 1/3–1/2 developed at flowering, the blades mostly 1.3–2.0 times as long as wide; lobes, if present, angled or short-tapered to sharply pointed tips
   3. Anthers white or cream-colored
      4. Leaf blades at flowering mostly 1.3–1.6 times as long as wide; flowers 14–22 mm in diameter . . . . var. *collina*
      4. Leaf blades at flowering mostly 1.7–2.0 times as long as wide; flowers 12–15 mm in diameter . . . var. *succincta* (Sarg.) E.J. Palmer
   3. Anthers pink to red
      5. Leaf blades at flowering mostly 1.3–1.5 times as long as wide . . . . var. *sordida* (Sarg.) Eggl.
      5. Leaf blades at flowering mostly 1.7–2.0 times as long as wide . . . . var. *succincta* (*'angustata'* form)

23. Crataegus spes-aestatum J.B. Phipps (Summers hawthorn)

Map 2434

Plants shrubs or small trees, 3–6 m tall. Twigs reddish brown at second year, densely pubescent with soft, fine, often somewhat tangled hairs when young, often becoming glabrous or sparsely hairy at second year, Petioles 7–15 mm long, densely hairy when young, sometimes only on the upper side, becoming sparsely hairy to nearly glabrous with age. Leaf blades 3–5 cm long, broad-elliptic to rhombic-elliptic, widest mostly just above the midpoint, angled or short-tapered to a sharply pointed tip, with 2 or 3 short, sharp lobes per side, the margins otherwise relatively coarsely and sharply toothed, the upper surface moderately roughened-hairy when young, becoming less densely hairy at maturity, the undersurface densely hairy when young, mostly along the veins, becoming somewhat less densely hairy at maturity, the secondary veins 4–6 per side. Inflorescences 6–12-flowered, the branches densely pubescent with short, spreading to loosely appressed hairs. Flowers mostly 20–24 mm in diameter. Sepals mostly glandular-toothed. Stamens 20, the anthers yellow or cream-colored. Styles 5. Fruits 8–12 mm wide, the surface red, dull. Nutlets 3–5. Mid-April–late May.

Uncommon in the southern half of the state (Illinois, Missouri). Mesic to less commonly dry upland forests, banks of streams, and bases of bluffs; also pastures.

Crataegus spes-aestatum was first described in 2006 to honor one of Missouri's best field botanists, Bill Summers. The specific epithet translates as *hope of Summers*, in the hope that Bill Summers would locate new populations of this interesting Missouri hawthorn. The status of this species in Missouri remains poorly understood. It was once widely distributed in the state, but was last collected in 1957 in Boone County. *Crataegus spes-aestatum* is morphologically quite similar to *C. collina*. However, the two are fairly easily distinguished and it is relatively constant in its morphology. In the future, this taxon may be better treated as a variety of *C. collina*.

12. Series Rotundifoliae Rehder

Plants shrubs or rarely small trees, 2–7 m tall, the trunks unarmed or less commonly with unbranched thorns, the bark relatively thin, grayish brown to gray, divided into small, irregular plates, these peeling tardily in small, thin, irregular flakes to expose orangish brown to light brown underlayers. Branchlets unarmed or sparsely to occasionally moderately thorny, the thorns mostly 2.5–7.0 cm long, slender, straight or slightly curved. Twigs brownish tan to purplish brown and shiny at end of first year, becoming brownish gray to dark grayish brown at second year, dull gray with age, glabrous (even when young). Petioles mostly 30–60% as long as the blade, glabrous or sparsely hairy, nonglandular or with a few stalked (peglike), dark glands, these often shed early. Leaf blades 2–6 cm long, 2 times as long as wide (narrower elsewhere) to slightly wider than long, ovate, elliptic, or rhombic-elliptic to obovate, broadly obovate, or nearly circular, widest below, at, or less commonly above the midpoint, occasionally slightly asymmetric in outline, angled to broadly angled, rounded, or short-tapered at the base, the tip bluntly to sharply pointed or occasionally more or less rounded, with 1–4 shallow or very shallow (at most 15% of the way to the midvein; occasionally more deeply lobed on extension shoots), rounded to sharp lobes per side, the margins finely and simply or occasionally doubly toothed above the base, the teeth blunt to sharp, not or inconspicuously glandular, the texture herbaceous to slightly leathery, the upper surface sparsely to moderately pubescent with fine, straight, appressed hairs when young, becoming more sparsely hairy or nearly glabrous at maturity, green to dark green or occasionally bluish green and dull to somewhat shiny, the undersurface sparsely to moderately hairy when young, especially along the main veins toward the base, sometimes becoming glabrous or nearly so at maturity, the secondary veins 3–6 per side, the ultimate branches extending to the tips of the teeth. Inflorescences (4–)6–15-flowered, the branches glabrous or occasionally sparsely hairy, the bractlets membranous to more or less herbaceous, inconspicuous, linear, entire or somewhat irregular, the margins with numerous stalked glands, shed early or less commonly somewhat more persistent. Flowers (8–)12–20 mm in diameter, the hypanthium glabrous. Sepals 4–5 mm long, shorter than the petals, narrowly triangular, the margins more or less entire, with several stalked, dark glands, the inner surface glabrous. Petals 6–9 mm long, broadly obovate to circular, white. Stamens 10 or 20, the anthers ivory to cream-colored, rarely pale pink. Styles 2–5. Fruits 7–12 mm long, mostly 7–12 mm wide, globose to subglobose or rarely ellipsoid, often with a distinct but short, raised collar at the tip (but the withered sepals sometimes obscuring the actual tip), the surface pale orange to red or dark red, occasionally yellow, glabrous, dull or somewhat shiny. Nutlets 2–5, the lateral faces not pitted. About 9 species, northern U.S. south to Utah, New Mexico, Missouri, and North Carolina; Canada.

Series *Rotundifoliae* is not particularly easy to characterize owing to of its morphological diversity. The species group to which *C. margaretta* belongs is rather small-leaved, with lobeless or only shallowly lobed leaves, usually glabrous except for the undersurfaces of the young leaves, and is one of the few hawthorn groups in which yellow fruit is common (although the color ranges to red or even burgundy). The often densely branched habit can impart a characteristic appearance.

1. Stamens 20; petioles nonglandular; leaf blades 2–5 cm long, usually with 3 or 4 lateral veins, green to more commonly dark green or bluish green at maturity . . . . 24. C. MARGARETTAE
1. Stamens 10; petioles usually more or less glandular; leaf blades 3–6 cm long, usually with 4–6 lateral veins, green or occasionally dark green at maturity . . . . 25. C. SICCA

24. Crataegus margarettae Ashe (Margaret's hawthorn)
C. brownii Britton
C. meiophylla Sarg.

Pl. 530 a–d; Map 2435

Plants shrubs or rarely small trees, mostly 4–7 m tall. Branchlets essentially unarmed or the sparse (denser elsewhere) thorns 2.5–5.0 cm long, straight or slightly curved, black or nearly so and usually somewhat shiny at second year. Petioles 7–20 mm long, sometimes narrowly winged toward the tip, nonglandular or the glands small, inconspicuous, and shed early. Leaf blades 2–5 cm long, ovate, elliptic, or rhombic-elliptic to obovate, broadly obovate, or nearly circular (lanceolate-ecliptic elsewhere), often widest at or above the midpoint, rounded to angled or occasionally very short-tapered at the base, rounded to bluntly or occasionally more or less sharply pointed at the tip, with 1–3 shallow lobes per side, these rounded to bluntly or broadly pointed (occasionally more sharply pointed) at their tips, the margins bluntly to sharply toothed, the secondary veins usually 3 or 4 per side, the upper surface sparsely to moderately appressed-hairy when young, becoming less densely hairy to nearly glabrous at maturity, green to dark green or bluish green, the undersurface sparsely to moderately appressed-hairy when young, especially along the main veins, becoming less densely hairy to glabrous at maturity. Inflorescences 6–15-flowered; the branches glabrous or rarely sparsely and finely hairy, the bractlets sparse to relatively numerous, often at least some of them more or less persistent at flowering. Flowers (8–)12–17 mm in diameter. Stamens 20, the anthers ivory. Styles and nutlets 2 or 3(4). $2n$=32, 34; *diploid* ($2n$=16, by anatomical sections; see Longley [1924]). Early April–mid-June.

Scattered in the northeastern quarter of the state, mostly in counties along the Mississippi River; also scattered in the western portion of the Ozark Division and disjunctly in Iron and Madison Counties (northeastern U.S. west to Wisconsin, Iowa, and Missouri; southeastern Canada). Bottomland forests, mesic upland forests, banks of streams and rivers, and occasionally glades; also mine spoils, pastures, fence rows, and roadsides.

The leaves of *C. margarettae* are very variable in shape, but typically are relatively small and have a pronouncedly blunt or rounded appearance to the lobes. Several uncommon regional variants differing in details of leaf shape and fruit color have been named as varieties. The var. *margarettae*, which is by far the most common of these variants and occurs nearly throughout the range of the species, is also relatively variable. The var. *brownii* (Britton) Sarg. has smaller flowers (8–12 mm in diameter) than var. *margarettae* and ellipsoid fruits, and occurs sporadically from Missouri (where known from a single historical collection from Clark County) to Virginia. It may simply be a product of dry, hot summers. The var. *angustifolia* E.J. Palmer has relatively narrow leaves (lanceolate-elliptic or oblong-lanceolate) that are pointed at the tip, angled at the base, and only slightly lobed. It is known from Indiana, southern Michigan and extreme southwestern Ontario. The var. *meiophylla* (Sarg.) E.J. Palmer has small leaves not over 2.5 cm long and has been found only in Ohio. Both of these variants may be found in Missouri in the future but require further research to substantiate their distinctness. The foliage of *C. margarettae* is often somewhat bluish green at maturity.

25. Crataegus sicca Sarg.
C. glabrifolia Sarg.

Pl. 527 g, h; Map 2436

Plants shrubs, 2–4(–5) m tall. Branchlets essentially unarmed or the sparse to occasionally moderate thorns 3–7 cm long, straight, black at second year. Petioles 10–15 mm long, often narrowly winged toward the tip, usually with a few stalked (peglike), dark glands. Leaf blades 3–6 cm long, rhombic-ellip-

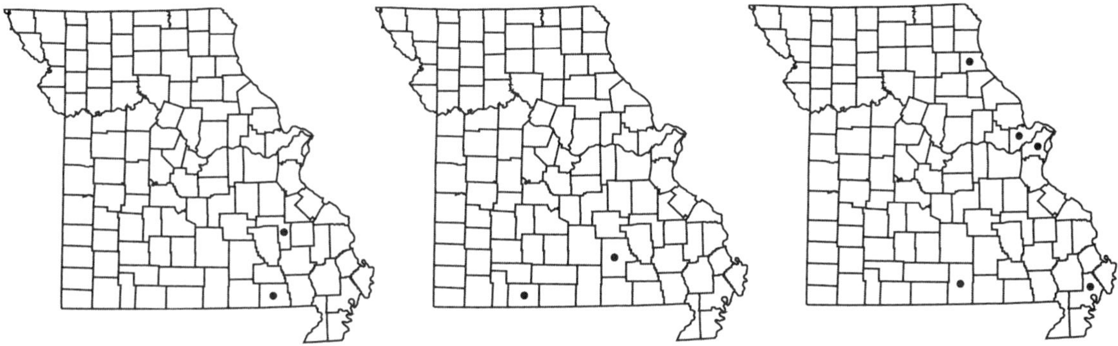

2437. Crataegus macrosperma 2438. Crataegus schuettei 2439. Crataegus nitida

tic to rhombic-ovate, usually widest at or below the midpoint, angled or occasionally very short-tapered at the base, sharply pointed at the tip, with 2–4 shallow lobes per side, these bluntly to sharply pointed at their tips, the margins sharply toothed, the secondary veins usually 4–6 per side, the upper sparsely appressed-hairy when young, usually becoming glabrous or nearly so at maturity, green to occasionally dark green, the undersurface sparsely to moderately appressed-hairy when young, especially along the main veins toward the base, often becoming somewhat less densely hairy at maturity. Inflorescences 4–6-flowered; the branches glabrous or relatively sparsely (occasionally densely) and finely hairy, the bractlets relatively numerous, but shed early. Flowers 10–20 mm in diameter. Stamens 10, the anthers cream-colored, rarely pale pink. Styles and nutlets 3–5. Late April–mid-May.

Scattered in the Ozark and Ozark Border Divisions (Missouri, Ohio). Mesic to dry upland forests, savannas, banks of streams, and bases of bluffs; also old fields, pastures and fencerows.

Crataegus sicca is a poorly understood species that has been suggested as possibly having originated through hybridization. The small size of the plant and leaves, the latter with a certain amount of pubescence, suggest *C. margarettae* as one parent, whereas the thorns and flowers, as well as fruit and leaf-shape suggest *C. pruinosa* as a possible other parent. Two strongly marked varieties have been recognized, whose relationships also require more intensive study. Of these, plants of var. *sicca* are more widely distributed in the state. The few occurrences of var. *glabrifolia* have been recorded from the western half of the Ozark Division. The following key summarizes the major morphological differences between them.

1. Plants medium-sized shrubs, usually 2–3 m tall; flowers relatively large, 18–20 mm in diameter; fruits orange; inflorescences and leaves hairy, at least when young . . . . var. *sicca*
1. Plants mostly tall shrubs, usually 3–5 m tall; flowers relatively small, 10–14 mm in diameter; fruits deep red; inflorescences and leaves glabrous . . . . var. *glabrifolia* (Sarg.) E.J. Palmer

13. Series Tenuifoliae (Sarg.) Rehder

Plants shrubs or small trees, 2–7 m tall, the trunks unarmed or with usually unbranched thorns, the bark relatively thin, brownish gray to gray, divided into irregular, longitudinal plates, these peeling in small, thin, irregular flakes to expose orangish brown to light brown underlayers. Branchlets moderately to densely thorny, the thorns mostly 3–5 cm long, usually relatively stout, slightly curved, gray with age. Twigs yellowish green or reddish-tinged when young, dark reddish brown and shiny at first year, becoming gray and dull with age. Petioles 30–60% as long as the blade, glabrous, nonglandular or more commonly with few to several short, more or less sessile, dark glands. Leaf blades 3–6(–8) cm long, varying from about 50% as wide as long to nearly as wide as long (seldom less than 2 cm wide), elliptic to ovate, broadly ovate, or triangular-ovate, widest at or below the midpoint, usually symmetric in outline, angled to broadly angled, rounded, more or less truncate, or occasionally shallowly cordate at the base, the tip sharply pointed, with 3–6 shallow to moderately deep (at most 30% of the way to the midvein) lobes per side (sometimes more deeply lobed on extension shoots), the margins finely and simply or doubly toothed nearly to the base, the teeth sharp and often appearing somewhat jagged, gland-tipped (less conspicuously so above the lower teeth), the texture herbaceous to somewhat papery, the upper surface sparsely to densely roughened-hairy when young, usually becoming glabrous or nearly so with age, dark green, dull or only slightly shiny, the undersurface glabrous or nearly so, the secondary veins mostly 4–7 per side, the ultimate branches extending to the tips of the teeth. Inflorescences 5–10(–12)-flowered, the branches glabrous or shaggy-hairy, the bractlets absent or few, then membranous, inconspicuous, linear, finely glandular-toothed or more or less entire and with more or less sessile glands, shed early. Flowers 13–17(–20) mm in diameter, the hypanthium glabrous. Sepals 3–5 mm long, shorter than the petals, narrowly triangular, the margins unlobed or finely glandular-toothed, the inner surface hairy. Petals 6–8 mm long, broadly obovate to circular, white. Stamens 5–10 or (16–)20, the anthers pink to red or reddish purple. Styles 3–5. Fruits 8–17 mm long, mostly 6–15 mm wide, globose to subglobose or occasionally oblong-ellipsoid, lacking a distinct raised collar at the tip (but the withered sepals sometimes persistent and obscuring the actual tip), the surface red, glabrous, dull, not glaucous. Nutlets 3–5, the lateral faces not pitted. About 6 species, eastern U.S., southeastern Canada.

The core of ser. *Tenuifoliae* contains the typical American hawthorns: thorny bushes with conspicuously lobed leaves, red fruit and many other features unremarkable in the genus. The series was named for its rather thin leaves and delicate petioles, which often enable the leaves to flutter in the wind, especially when young. Series *Tenuifoliae* is widespread in the Great Lakes, northeastern, and Appalachian regions, and also has a sporadic, but apparently extensive, range farther to the southwest. Although its status in Missouri requires further study, recent finds in Arkansas of both of the Missouri taxa lend support to its presence in the state. Vegetatively, plants of ser. *Tenuifoliae* might be mistaken for members of ser. *Pruinosae*, but the older thorns are a different color and the young leaves are always appressed-hairy on the upper surface.

1. Stamens 5–10 . . . . 26. C. MACROSPERMA
1. Stamens (16–)20 . . . . 27. C. SCHUETTEI

26. Crataegus macrosperma Ashe (eastern hawthorn)

C. acutiloba Sarg.
C. brumalis Ashe
C. demissa Sarg.
C. matura Sarg.
C. pastorum Sarg.
C. pentandra Sarg.
C. roanensis Ashe
C. tenuifolia Britton

Map 2437

Plants shrubs, mostly 3–4 m tall, or small trees to 7 m tall. Branchlets with the thorns 2.5–4.0 cm long and relatively stout (4–6 cm and more slender elsewhere), dark brown to purplish black at second year, becoming gray with age. Petioles 10–30 mm long, Leaf blades 3–6(–8) cm long, ovate to triangular-ovate or broadly ovate, the upper surface usually relatively densely roughened with stiff appressed hairs when young, the lateral veins 4–7 pairs. Inflorescences 5–10(–12)-flowered, the branches glabrous or shaggy-hairy, the bracts few and shed early or frequently not produced. Flowers 13–17 mm in diameter. Sepals 4–5 mm long, the margins more or less entire. Stamens 5–10, the anthers pink to reddish purple. Fruits 8–15 mm in diameter, bright to deep red. $2n=68$; tetraploid (by flow cytometry; see Talent and Dickinson [2005]). April–May.

Uncommon, known thus far only from historical collections from Iron and Ripley Counties (eastern [mostly northeastern] U.S. west to Minnesota and Arkansas; southeastern Canada). Habitat unknown.

The status of this species as a member of the state's flora is uncertain. All four of the specimens said to have originated from Missouri were collected from plants cultivated at the Arnold Arboretum said to have been grown from seeds collected in Missouri by Bush and Palmer.

27. Crataegus schuettei Ashe (royal hawthorn, Schuette hawthorn)

C. basilica Beadle

C. ferrissii Ashe

Map 2438

Plants shrubs, mostly 4–5 m tall, or more commonly small trees to 7 m tall. Branchlets with the thorns 3–6 cm long and often relatively stout, dark brown to black at second year, becoming gray with age. Petioles 20–30 mm long, Leaf blades 3–6 cm long, elliptic to ovate, broadly ovate, or triangular-ovate, the upper surface sparsely to moderately roughened-hairy when young, the lateral veins 4–6 pairs. Inflorescences 5–10-flowered, the branches glabrous, the bracts few and shed early or sometimes not produced. Flowers 15–17(–20) mm in diameter. Sepals 3–4 mm long, the margins nearly entire to finely toothed. Stamens (16–)20, the anthers red. Fruits 10–17 mm in diameter, red. $2n$=51;triploid (by flow cytometry; see Talent and Dickinson [2005]). April–May.

Uncommon, known thus far only from historical specimens from Shannon and Taney Counties (eastern [mostly northeastern] U.S. west to Minnesota and Arkansas; southeastern Canada). Banks of streams and rivers.

In addition to the above, specimens of *C. ×atrorubens* Ashe will key out to *C. schuettei*. *Crataegus ×atrorubens* is an interserial hybrid with parents in ser. *Molles* and ser. *Virides*. Nevertheless, it has frequently been mentioned in the floristic and horticultural literature, and has often been misnamed in botanical gardens. It is a small tree with densely hairy petioles, leaf blades with the undersurface having small tufts of hairs in the vein axils, and fruits that are 9–11 mm in diameter. *Crataegus ×atrorubens* was described from material originally collected in 1888 by Henry Eggert in St. Louis and now appears to be extinct in the wild. Its parentage is still subject to discussion.

14. Series Virides (Gordon) Rehder

Plants shrubs or small trees, (2–)3–15 m tall, the trunks sometimes armed with branched thorns, the bark either thin, gray to dark gray, relatively smooth, mottled, and peeling in large, irregular patches to expose lighter gray and orangish brown underlayers, or thicker, dark gray, divided into small irregular plates, these peeling tardily to expose lighter gray and yellowish brown underlayers. Branchlets nearly unarmed or with scattered to relatively dense thorns, these mostly (1–)2–4(–7) cm long, slender or less commonly relatively stout, mostly straight. Twigs glabrous, green or reddish tinged and shiny or dull when young, gray to chestnut brown or reddish brown at first year, becoming gray and dull with age. Petioles mostly 30–50% as long as the blade, glabrous, nonglandular (rarely with sparse stalked glands in *C. nitida*). Leaf blades 1.5–8.0 cm long (rarely to 11 cm in extension shoot leaves), varying from less than 30% as wide as long to nearly as wide as long, narrowly elliptic to oblong elliptic, rhombic, or broadly ovate, usually at least slightly asymmetric in outline, tapered or less commonly narrowly angled at the base (those of extension shoot leaves sometimes rounded), the tip rounded to sharply pointed, unlobed or with 1–3 shallow lobes per side toward the tip (sometimes more deeply lobed throughout on rapidly elongating shoots), the margins variously finely scalloped to coarsely and sharply toothed, sometimes mostly above the midpoint, nonglandular, the texture herbaceous to slightly leathery, the upper surface glabrous (sometimes with scattered appressed hairs along the main veins), dull or occasionally somewhat shiny, the undersurface glabrous or with small tufts of hairs in the axils of the main veins, the secondary veins mostly 3–6 per side, the ultimate branches extending to the tips of the teeth. Inflorescences 5–18(–30)-flowered, the branches glabrous, the bractlets membranous, linear to filiform, entire, the margins then nonglandular or with sparse stalked glands, or sparsely glandular-toothed, shed early. Flowers 10–18 mm in diameter, the hypanthium glabrous or rarely sparsely and finely hairy. Sepals 1.5–3.0 mm long, shorter than the petals, triangular to broadly triangular, the margins entire or slightly irregular, occasionally sparsely glandular-toothed, glabrous or the inner surface rarely sparsely and inconspicuously hairy, the tip sometimes glandular. Petals 5–9 mm long, broadly obovate to circular, white. Stamens usu-

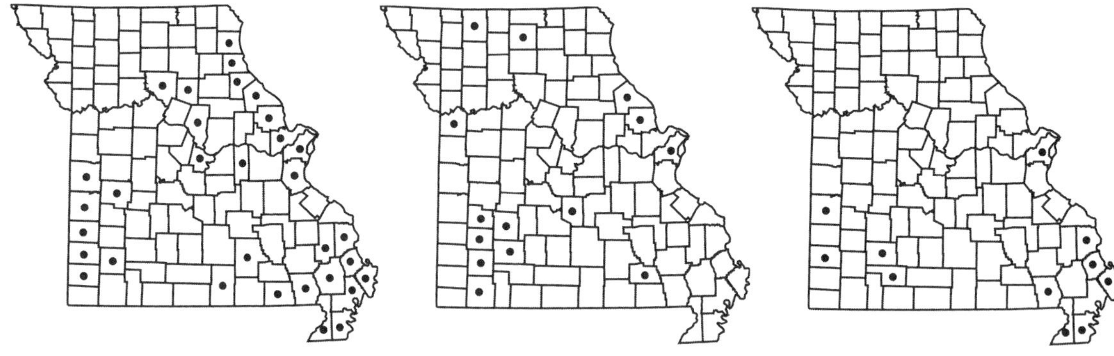

2440. Crataegus viridis   2441. Drymocallis arguta   2442. Duchesnea indica

ally 20, the anthers ivory to cream-colored, occasionally pale pink. Styles 3–5. Fruits 5–12 mm long, mostly 4–12 mm wide, subglobose, lacking a raised collar at the tip, the surface orange to deep red at maturity, glabrous, shiny or dull, sometimes glaucous. Nutlets 3–5, the lateral faces not pitted. About 2 species, eastern U.S.

Despite its relatively great variation in leaf shape, the ser. *Virides* is among the more easily identified groups of American hawthorns. Characteristically, the species are large plants, often open-branched small trees, and are notable for very narrow bractlets with the margins lacking glands or nearly so, as well as small fruits. The series is relatively common and widespread in Missouri.

1. Leaf blades 4–8 cm long (to 12 cm on extension shoots); fruits (5–)8–12 mm in diameter, glaucous . . . . 28. C. NITIDA
1. Leaf blades mostly 2–6 cm long (rarely to 11 cm on extension shoots); fruits 5–8 mm in diameter, not glaucous . . . . 29. C. VIRIDIS

28. Crataegus nitida (Engelm. ex Britton & A. Br.) Sarg.

C. viridis var. *nitida* Engelm. ex Britton & A. Br.

Map 2439

Plants trees to 10 m tall. Bark moderately thick, dark gray, divided into small irregular plates, these peeling tardily to expose lighter gray and yellowish brown underlayers. Branchlets with scattered thorns, these mostly 2.5–5.0 cm long, slender or more commonly relatively stout, chestnut-brown to nearly black at second year, eventually becoming dark gray with age. Petioles 1.5–2.5 cm long, finely hairy on the upper side when young, becoming glabrous or nearly so at maturity, occasionally with very sparse stalked glands. Leaf blades 4–8 cm long (to 12 cm on extension shoots), broadly lanceolate to narrowly rhombic-elliptic, tapered at the base, angled or tapered to a sharply pointed tip, unlobed or with 1 or 2(3) sharply pointed, but shallow lobes per side, the margins relatively coarsely, sharply, and usually doubly toothed, the larger teeth mostly 2–3 mm deep, leathery, the upper surface somewhat shiny, the secondary veins usually 5 or 6 per side. Inflorescences mostly 10–18-flowered, the branches glabrous, bractlets not produced. Flowers 15–18 mm in diameter. Sepals glabrous. Petals 7–9 mm long. Stamens 20, the anthers cream-colored or rarely pale pink. Fruits (6–)8–12 mm in diameter, brick red to red, somewhat glaucous (appearing very glaucous on drying). Diploid (by anatomical sections; see Longley [1924]; also by flow cytometry; see Talent and Dickinson [2005]). Late April to late May.

Uncommon, widely scattered, mostly in counties along the Mississippi River (Arkansas, Illinois, and Missouri). Bottomland forests, sloughs, and banks of streams and rivers.

This taxon differs from *C. viridis* only slightly, principally in the thicker bark divided into small plates, somewhat larger and more glaucous fruits, and leaf blades somewhat more leathery at maturity and with more pronounced marginal teeth. In fact, originally it was described as a variety of *C. viridis*. However, Sargent (1901) dismissed this idea because historically the plants were: "...too numerous and too generally distributed over the Mississippi bottoms near St. Louis to make such a supposition probable..." In the present work

Plate 531. Rosaceae. *Crataegus viridis*, **a)** fruiting branch, **b)** flower. *Gillenia stipulata*, **c)** leaf, **d)** flower, **e)** fruiting branch. *Gillenia trifoliata*, **f)** fruiting branch, **g)** flower. *Filipendula rubra*, **h)** stem tip with inflorescence and detached lower leaf, **i)** fruits, **j)** flower.

C. nitida is recognized as a separate species because the morphological distinctions all lie outside of the normal range of *C. viridis*. Its current rarity has also given rise to the suggestion that *C. nitida* might be of hybrid origin between *C. crus-galli* and *C. viridis* (Gleason and Cronquist, 1991). Horticultural plants seen by the author that are labeled as *C. nitida* usually appear to be a genuine *C. crus-galli* × *C. viridis* cross and are not morphologically identical to specimens of *C. nitida* collected from the wild. Thus the idea that *C. nitida* is of hybrid origin has little support.

29. Crataegus viridis L. (green hawthorn)
C. davisii Sarg.
C. furcata Sarg.
C. lanceolata Sarg.
C. glabriuscula Sarg.
C. lutensis Sarg.
C. nitens Sarg.
C. ovata Sarg.
C. pechiana Sarg.

Pl. 531 a, b; Map 2440

Plants shrubs or trees, (2–)3–15 m tall. Bark thin, gray to dark gray, relatively smooth, mottled, peeling in large, irregular patches exposing lighter gray and orangish brown underlayers. Branchlets nearly unarmed or with scattered to relatively dense thorns, these(1–)2–4(–7) cm long, relatively slender, very dark brown to dark purplish brown or nearly black at second year, occasionally dark gray or with age. Petioles 0.7–2.5 cm long, glabrous or nearly so, nonglandular. Leaf blades 1.5–6.0 (–8.0) cm long (rarely to 11 cm on extension shoots), narrowly elliptic to oblong elliptic, rhombic, or broadly ovate, tapered or less commonly narrowly angled at the base, (those of extension shoot leaves sometimes rounded), rounded to sharply pointed at the tip, unlobed or with 1–3 sharply (but sometimes broadly) pointed, but shallow lobes per side, the margins variously finely scalloped to coarsely and sharply but simply toothed, the largest teeth mostly 1–2 mm deep, papery to slightly leathery, the upper surface dull to less commonly somewhat shiny, the secondary veins usually 3–5 per side. Inflorescences 5–18(–30)-flowered, the branches usually glabrous, the bractlets usually relatively numerous, but shed early. Flowers 10–18 mm in diameter. Sepals glabrous. Petals 5–9 mm long. Stamens usually 20, the anthers ivory to cream-colored. Fruits 5–8 mm in diameter, orange to deep red, not glaucous. Diploid to triploid or higher (by flow cytometry; see Talent and Dickinson [2005]). Early April–late May.

Scattered in the state; uncommon to absent from most of the Glaciated Plains Division and the western and northern portions of the Ozarks (eastern U.S. west to Kansas and Texas). Bottomland forests, swamps, fens, marshes, margins of ponds, lakes, and sinkhole ponds, mesic upland forests, savannas, glades, edges of sand prairies, banks of streams and rivers, ledges, bases, and tops of bluffs; edges of pastures, margins of crop fields, fencerows, and roadsides.

Crataegus viridis is the most commonly encountered Missouri hawthorn at bottomland sites, however, the species also can occur in drier upland situations. Selected rust-resistant cultivars such as cv. 'Winter King' make excellent ornamentals. The best forms color well in the autumn and have attractive fruits. The fresh green foliage is attractive in the spring, against which the white flowers are well set off.

This is an extremely variable species in Missouri as to leaf size and shape, and to some extent fruit size. The bractlets of *C. viridis* are mostly nonglandular and are exceptionally narrow and shed very early. Forms with red anthers or hairy flower stalks, petioles, and leaves are very rare and may represent past hybridization with other hawthorn species. The major variants of *C. viridis* have been separated as varieties by some authors. In its main range, the most distinctive of these is var. *glabriuscula*, which is southwestern in its distribution and enters southern Missouri, where it becomes somewhat less distinct, intergrading somewhat with var. *viridis*. The other varieties all grade into one another and show no ecological segregation. Missouri specimens attributable to var. *nitens* were all collected in Marion County by the Rev. John Davis in the early 1900s. These plants possibly represent a hybrid, perhaps with *C. pruinosa*; their leaves are more leathery and fruits a little larger than those of the other varieties.The remaining varieties have been documented from both historical and modern collections. The following key is provided for those who wish to determine specimens below the species level.

1. Leaf blades broadly lanceolate to narrowly rhombic, often with 2 or 3 sharp lobes per side; lateral veins 4 or 5 per side . . . . var. *viridis*
1. Leaf blades without the above combination of characteristics; lateral veins 3–7 per side
   2. Leaf blades mostly 3–4 cm long, broadly elliptic to ovate, relatively thick and leathery at maturity, distinctly lobed, the teeth much smaller than the lobes; second-year twigs reddish brown . . . . var. *nitens* . . . . (Sarg.) J.B. Phipps

2. Leaf blades 2.0–6.5 cm long, broadly lanceolate to rhombic-ovate, unlobed or with 1 shallow lobe per side, relatively thin and herbaceous to papery at maturity; second-year twigs dark gray to reddish brown
   3. Leaf blades relatively narrow, at least 1.75 times as long as wide; secondary veins mostly 4–7 per side . . . . var. *lanceolata* . . . . (Sarg.) E.J. Palmer
   3. Leaf blades relatively broad, at most 1.5 times as long as wide; lateral veins 3–5 per side
      4. Leaf blades mostly 2–4 cm long, narrowly ovate to nearly circular, oblong, rhombic-elliptic or narrowly obovate; marginal teeth to 2 mm at maturity; secondary veins 3(4) per side . . . . var. *glabriuscula* . . . . (Sarg.) J.B. Phipps
      4. Leaf blades mostly 3.5–6.0 cm long, narrowly ovate to oblong; marginal teeth less than 1 mm long at maturity (the margins sometimes nearly entire); lateral veins mostly 4–5 per side . . . . var. *ovata* . . . . (Sarg.) E.J. Palmer

9. **Drymocallis** Fourr. ex Rydb.

About 30 species, North America, Europe, Asia.

As noted in the discussion under the genus *Potentilla*, in recent years, phylogenetic analysis of the fragarioid Rosaceae has tended to support a relatively restricted circumscription of that genus, with various species groups now treated as segregate genera (Kurtto and Eriksson, 2003; Ertter, 2007). For the Missouri flora, this affects only *Drymocallis arguta*, which is part of a north-temperate assemblage that traditionally was included within *Potentilla* sect. *Rupestres* Poeverl. Members of the group differ consistently from species of true *Potentilla* in several subtle characters, notably their horseshoe-shaped anthers with a single locule and slender, basally attached achenes (Ertter, 2007). Eriksson et al. (1998), who studied the molecular phylogenetics of the fragarioid Rosaceae using molecular data, also found that this group forms a lineage distinct from true *Potentilla*. Thus, support for this group as a segregate genus is relatively strong. Within, *Drymocallis*, *D. arguta* is the only one of the fifteen species in temperate North America whose distribution extends significantly east of the Great Plains.

1. Drymocallis arguta (Pursh) Rydb. (tall cinquefoil, prairie cinquefoil)
Potentilla arguta Pursh
Pl. 534 f–h; Map 2441

Plants perennial herbs, with usually branched, short, stout, woody rootstocks. Stems 30–100 cm long, erect or ascending, unarmed, usually unbranched below the inflorescence, moderately to densely pubescent with short, brown, multicellular, mostly gland-tipped hairs, sticky to the touch. Leaves basal and 2 or 3 on the stems below the inflorescences, 4–40 cm long, the basal leaves long-petiolate, the stem leaves short-petiolate to nearly sessile, the petioles glandular-hairy. Stipules 10–30 mm long, lanceolate to ovate, entire or toothed, the surfaces hairy. Blades of the basal leaves pinnately compound with 7–11 main leaflets (sometimes alternating with additional, much smaller leaflets, those of the stem leaves mostly with 5–9 leaflets, the leaflets 1–8 cm long (the terminal leaflet longer than the progressively shorter lateral ones), sessile or short-stalked (to 7 mm on terminal leaflets), ovate to elliptic-obovate, or oblong-obovate, bluntly to sharply pointed at the tips, the margins doubly and often coarsely toothed, the surfaces green, moderately to densely pubescent with short, glandular and nonglandular hairs, occasionally only sparsely hairy to nearly glabrous. Inflorescences terminal, more or less flat-topped to dome-shaped panicles of relatively dense clusters of flowers (sometimes more elongate), the axes and stalks densely glandular- and nonglandular-hairy. Flowers short- to long-stalked, perigynous. Bractlets 5, alternating with the sepals (the calyx thus appearing 10-parted), each 4–7 mm long, elliptic to lanceolate, hairy. Hypanthia 2–3 mm in diameter, shallowly cup-shaped to saucer-shaped, glandular-hairy. Sepals 6–9 mm long (becoming elongated to 12–16 mm at fruiting), oblong-lanceolate, bluntly to sharply pointed and with a short extension of the midvein at the tips. Petals 5, 5–9 mm long, oblong-obovate to obovate, broadly rounded at the tip, white, cream-colored, or pale yellow. Stamens 25(–30), the anthers yellow. Ovary

2443. Exochorda racemosa 2444. Filipendula rubra 2445. Fragaria vesca

superior, glabrous, with 1 locule, with 1 ovule. Style 1 per pistil, attached laterally toward the ovary base, usually jointed at the base and shed as the fruits mature, the stigma slender. Fruits achenes, densely aggregated on the surface of the hemispheric receptacle, 0.9–1.3 mm long, asymmetrically ovate in outline, glabrous or nearly so, the surface faintly ribbed or wrinkled, light brown, unwinged, 1-seeded. $2n=14$. May–August.

Uncommon, widely scattered in the state, most abundantly in the western portion of the Ozark Division (northeastern U.S. west to Montana and Colorado; Canada). Glades, upland prairies, tops of bluffs; also open, disturbed areas.

A group of about 30 species from North America and Eurasia that are members of *Potentilla* sect. *Rupestres* differs consistently from the rest of the genus in several subtle characters, notably its horseshoe-shaped anthers and slender, basally attached achenes (Ertter, 2007). Eriksson et al. (1998), who studied the molecular phylogenetics of the fragarioid Rosaceae using molecular data, also found that this group forms a lineage distinct from true *Potentilla*. Thus, many authors have begun to segregate this group as the genus *Drymocallis*. Within, *Drymocallis*, *D. arguta* is the only one of the fifteen species in temperate North America to occur east of the Rocky Mountains.

10. **Duchesnea** Sm.

Two or more species, Asia south to Java.

1. Duchesnea indica (Andrews) Focke (Indian strawberry)

Pl. 532 n–p; Map 2442

Plants perennial herbs with short rhizomes. Stems 8–80 cm or more long, prostrate and stolonlike, rooting at some nodes, lacking spines and thorns, hairy. Leaves in rosettes at the rhizome tip and where the aerial stems root, otherwise alternate, long-petiolate. Stipules 4–9 mm long, herbaceous, narrowly oblong-elliptic, hairy, those of the rosette leaves fused toward the base. Leaf blades 1–7 cm long, broadly ovate-triangular in outline, divided into 3 similar leaflets, these 1.0–6.5 cm long, elliptic or ovate, rounded to more commonly pointed at the tip, rounded or narrowed to a short-stalked base, the margins bluntly toothed or scalloped, the surfaces sparsely to moderately appressed-hairy, especially along the veins. Flowers solitary in the leaf axils (and among the rosette leaves), long-stalked, perigynous, the hypanthium saucer-shaped, with a poorly developed nectar disc around the margin, sparsely to moderately hairy, each flower with 5 bractlets alternating with the sepals (the calyx thus appearing more or less 10-parted), these noticeably wider than the sepals, broadly obovate, prominently 3-toothed or -lobed at the tip, becoming somewhat enlarged and reflexed at fruiting. Sepals 5, 4–7 mm long, usually arched upward, ovate-triangular, sparsely to moderately hairy, somewhat enlarged at fruiting. Petals 5, 5–9 mm long, narrowly obovate, yellow. Stamens 15–25, the anthers yellow. Pistils numerous, densely covering the surface of the obconic expanded receptacle. Ovary superior, glabrous, with 1 locule, with 1 ovule. Style 1 per pistil, attached laterally, shed before the fruit matures, the stigma disc-shaped. Fruits achenes, in a headlike cluster densely covering the surface of the enlarged expanded receptacle, but shed

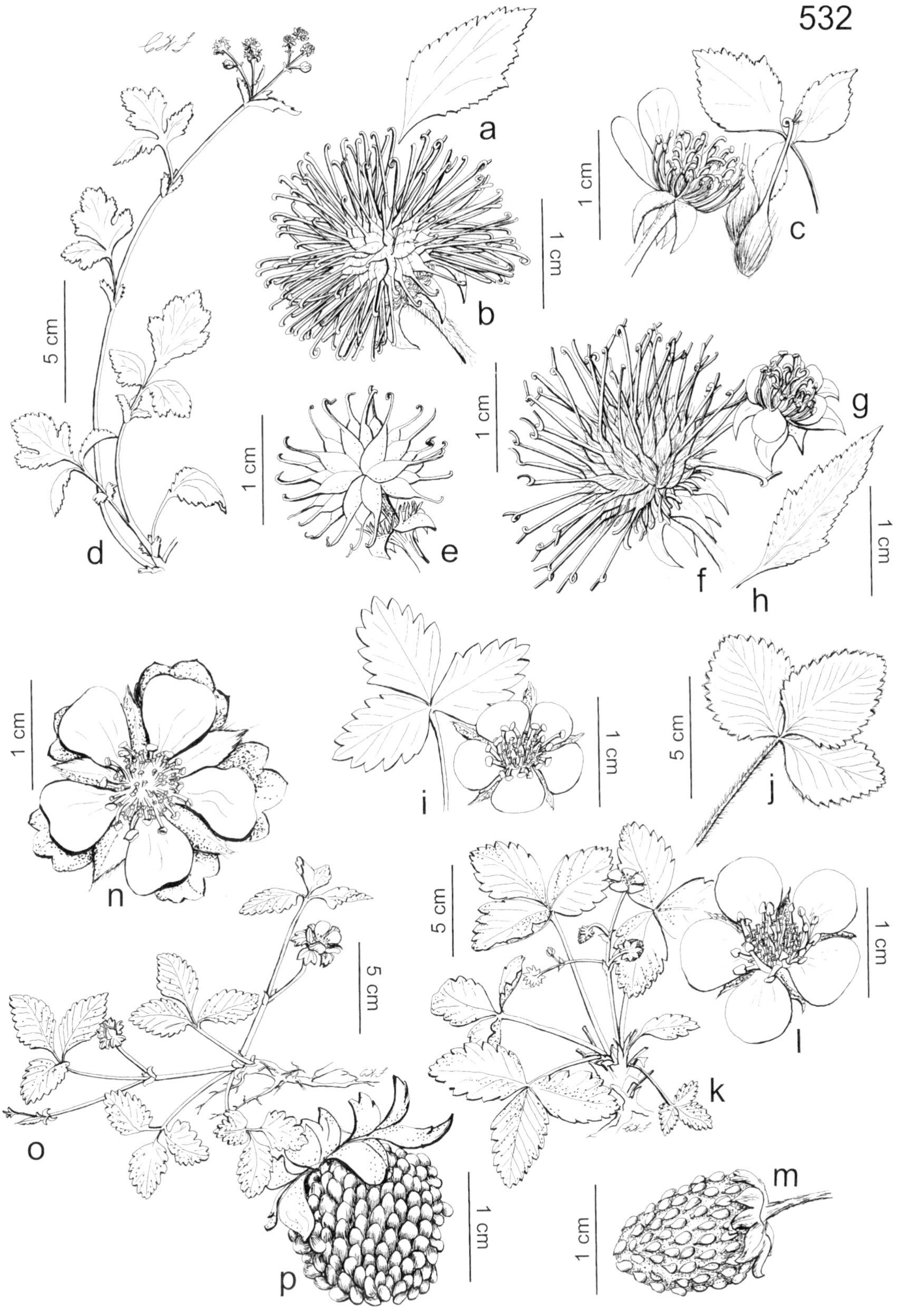

Plate 532. Rosaceae. *Geum laciniatum*, **a)** leaf, **b)** fruits. *Geum canadense*, **c)** flower, fruits, and leaf. *Geum vernum*, **d)** habit, **e)** fruits. *Geum virginianum*, **f)** fruit, **g)** flower, **h)** upper leaf. *Fragaria vesca*, **i)** flower and leaf. *Fragaria virginiana*, **j)** leaf, **k)** habit, **l)** flower, **m)** fruit. *Duchesnea indica*, **n)** flower, **o)** habit, **p)** fruit.

eventually, 0.9–1.5 mm long, asymmetrically ovate in outline, glabrous, shiny, red, with 1 seed. $2n=42$, 84. April–June, sometimes also sporadically until September.

Introduced, widely scattered in the southern half of the state (native to Asia south to the Philippines, introduced widely in Europe and the U.S.). Lawns and open disturbed areas.

The bright red fruits of *Duchnesnea* resemble those of *Fragaria* (strawberry), but are mealy and not sweet, hence the common name, barren strawberry.

11. **Exochorda** Lindl. (pearlbush)

One or 5 species, Asia; cultivated widely in temperate regions.

Exochorda usually is treated as comprising four or five closely related species, but in her unpublished dissertation, Gao (1998) presented anatomical, morphological, cytological, and genetic evidence for an alternative classification involving a single species (*E. racemosa*) consisting of three subspecies. As her subspecific combinations remain unpublished, this treatment has not been adopted elsewhere.

Exochorda species are sometimes cultivated as ornamentals. although they are extremely showy when in flower, their flowering period tends to be rather short. Plants that are not pruned also tend to become leggy, with arching stems. The most commonly cultivated taxon, *E. racemosa*, escapes sporadically, but does not appear to be sufficiently aggressive in nature to pose an invasive threat.

1. Exochorda racemosa (Lindl.) Rehder (common pearlbush)

Map 2443

Plants shrubs, 1–3 m tall, with several slender trunks. Branches sometimes developing lateral short shoots, but unarmed. Bark orangish brown, brown, and gray, peeling in variously colored plates on older trunks, usually reddish brown to dark gray and with scattered small lenticels on branches. Twigs tan to reddish brown, becoming dark gray, angled, glabrous. Winter buds lateral and terminal, narrowly ovoid to narrowly conic, bluntly to occasionally sharply pointed at the tip, with several overlapping scales, glabrous. Leaves alternate, occasionally appearing more or less whorled at the tips of short branches, rolled lengthwise during development, sessile or more commonly short-petiolate, the petioles glabrous. Stipules absent. Leaf blades simple, unlobed, elliptic to oblong-elliptic, oblong-obovate or narrowly obovate, rounded or bluntly and broadly pointed at the tip, often with a minute, sharply pointed extension of the midvein at the very tip, tapered at the base, the margins entire or shallowly wavy, scalloped, or toothed toward the tip, the teeth not gland-tipped, the surfaces glabrous, the undersurface pale, the venation pinnate. Inflorescences terminal, racemes of 4–10 flowers, these produced after the leaves develop, the axis and stalks glabrous at maturity, sometimes sparsely hairy when young, the flower stalks 3–12 mm long, each with a small bract at the base, the lowest of these leaflike, grading upward to small, narrowly oblong-elliptic, and scalelike, mostly persistent, each flower also with a pair of small bractlets near the tip of the stalk, these linear to narrowly oblong, white, shed early. Flowers perfect or usually functionally staminate or pistillate (mixed in the same inflorescence), shallowly perigynous, not fragrant, the hypanthium not fused to the ovary above its base, saucer-shaped to shallowly bell-shaped, persistent as a small disc, glabrous. Sepals 5, 2–3 mm long, loosely ascending to spreading at flowering, triangular, mostly sharply pointed at the tip, the margins entire, the surfaces glabrous, shed after flowering. Petals 5, 12–20 mm long, obovate to nearly circular, short-tapered to a short, stalklike base, white. Stamens 15–20, exserted, the filaments attached on the disc, the anthers yellow. Pistil 1 per flower, of 5 loosely fused carpels. Ovary superior, with 5 locules containing 2 ovules (1 of these usually abortive). Styles 5, erect and appressed but not fused, the stigmas small, capitate. Fruits capsules, 6–10 mm long, broadly obovate to broadly oblong or depressed-circular in outline, deeply 5-lobed, the lobes appearing more or less flattened and semicircular in outline, glabrous, brown to gray at maturity, dehiscing from the tip, persistent after the seeds are shed, the lobes sometimes breaking apart (the joined surfaces usually pale and appearing densely short-hairy), with usually 1 seed per locule, this 5–9 mm long, irregularly oval to oblong-elliptic or semicircular, strongly flattened, with a complete or incomplete wing around the margin, the surface brown to dark brown, glabrous, smooth. $2n$=16, 18. April–May.

Introduced, uncommon, known thus far only from Franklin County (native of Asia, introduced sporadically in the eastern U.S. west to Missouri and Louisiana). Disturbed mesic upland forests.

This species was first collected in 2006 by Nels Holmberg at the Shaw Nature Reserve, where plants escaped from an old planting into an adjacent forest.

12. **Filipendula** Mill.
(Schanzer, 1994)

Fifteen species, U.S., Europe, Asia.

1. Filipendula rubra (Hill) B.L. Rob. (queen of the prairie)

Pl. 531 h–j; Map 2444

Plants perennial herbs, with stout rhizomes. Stems 0.8–2.5 m long, erect or ascending, lacking spines and thorns, glabrous, hollow, with fine longitudinal ridges. Leaves alternate and basal, short- to long-petiolate. Stipules large, herbaceous, asymmetrically ovate to kidney-shaped, the rounded to cordate bases clasping the stem, those of the basal leaves fused to the petioles. Leaf blades (2–)8–50 cm long, broadly obovate to oblong-ovate in outline, once pinnately compound (the uppermost leaves sometimes merely palmately lobed) with small leaflets at the base and irregularly interspersed among the larger ones, the margins sharply toothed and shallowly pinnately lobed, the upper surface glabrous, the undersurface finely hairy, mostly along the veins, the primary lateral leaflets mostly 2–10, 2–15 cm long, mostly ovate-circular and deeply palmately 3–7-lobed (sometimes without palmate lobes and narrowly oblong-elliptic), the terminal leaflets larger than the lateral ones, nearly circular to broadly fan-shaped or kidney-shaped, palmately 5–9-lobed. Inflorescences dense terminal panicles of numerous flowers, without bracts or bractlets. Flowers perigynous, the hypanthium shallowly cup-shaped, glabrous, somewhat fleshy. Sepals 5(–7), 1.0–1.6 mm long, reflexed, oblong a with rounded tip, glabrous, white- to pink-tinged and usually noticeably veiny, persistent at fruiting. Petals 5(–7), 2–4 mm long, obovate, pink, rarely white. Stamens 15 to numerous, the anthers pink. Pistils 5–15 in a single whorl. Ovary superior, glabrous, with 1 locule, with 2 ovules. Style 1 per pistil, curved, persistent at fruiting, the stigma capitate. Fruits achenes (but appearing similar to follicles), ascending, 5–8 mm long, linear to narrowly oblong, straight or somewhat curved (banana-shaped), often tapered to a short stalk at maturity, glabrous, with 1 seed. June–August.

Uncommon, widely scattered in the Ozark Division (northeastern U.S. west to Wisconsin and Missouri; escaped elsewhere in the U.S. and adjacent Canada). Fens.

This attractive large perennial is quite uncommon and sporadic in the wild, but is widely cultivated as an ornamental in gardens. Steyermark (1963) treated this species as a relict from past times when Missouri had a cooler climate, now with a generally more northern distribution and restricted to a cool microhabitat in southern Missouri. He also noted that the foliage is fragrant. *Filipendula rubra*, which is uncommon throughout its range, is an obligate outcrosser that often produces relatively small quantities of seed, apparently because many adjacent individuals are clonal, resulting from spread by rhizomes rather than seeds (Aspinwall and Christian, 1992a, b). The main pollinators, at least in Missouri, are bees.

13. **Fragaria** L. (strawberry)

Plants perennial herbs with short, stout rhizomes. Stems absent, but slender, prostrate to arched stolons present, these rooting at the tip, lacking spines and thorns, hairy, usually with 1 scalelike leaf toward the midpoint, this 3–10 mm long, simple, linear to narrowly lanceolate, entire, often wrapped around the stolon at the base. Leaves in rosettes at the rhizome tip and where the stolons root, long-petiolate, the petioles usually sparsely to densely hairy. Stipules

10–20 mm long, herbaceous, oblong-lanceolate to ovate, narrowed or tapered to a sharply pointed tip, fused to the petiole above the middle, glabrous or more commonly hairy, especially along the midvein and margins, green to orangish brown and often purple-tinged, persistent after the leaves are shed. Leaf blades broadly triangular in outline, divided into 3 more or less similar leaflets (lateral leaflets sometimes somewhat symmetric at the base; seedling leaves simple), these elliptic to obovate, rounded or bluntly pointed at the tip, narrowed at the base, the margins toothed or scalloped, sparsely to densely pubescent with appressed to spreading, silky hairs, at least on the undersurface. Inflorescences open few-flowered panicles, rarely reduced to a solitary flower, sometimes appearing umbellate or racemose, produced on long usually hairy stalks from the leaf axils at the rhizome tip, the branch points with stipule-like bracts, the basal branch point with an additional 1 or 2 reduced leaf-like bracts. Flowers short- to long-stalked, perigynous, the hypanthium saucer-shaped, lacking a nectar disc, hairy, each flower with 5 bractlets alternating with the sepals (the calyx thus appearing 10-parted), these similar to but somewhat narrower than the sepals, linear-lanceolate to lanceolate, tapered to the sharply pointed tip, not becoming noticeably enlarged at fruiting. Sepals 5, sometimes slightly unequal in length, ascending to reflexed, lanceolate to broadly lanceolate, tapered to a sharply pointed tip, the margins often with a pair of small teeth or lobes near the tip, sparsely to moderately hairy, not noticeably enlarged at fruiting. Petals 5, obovate, white. Stamens 20–35, some of the outer ones sometimes reduced to staminodes, the anthers yellow. Pistils numerous, densely covering the surface of the obconic or columnar receptacle. Ovary superior, glabrous, with 1 locule, with 1 ovule. Style 1 per pistil, attached laterally near the ovary base, persistent at fruiting, the stigma disc-shaped. Fruits achenes, 0.9–1.5 mm long, moderately spaced (not overlapping) in shallow pits or on the surface of the enlarged, top-shaped to ovoid or ellipsoid, bright red receptacle, asymmetrically ovate in outline, glabrous, shiny, yellowish brown to dark brown, with 1 seed. About 20 species, North America, Hawaii, South America, Europe, Asia.

The garden strawberry, *F.* ×*ananassa* Duchesne ex Rozier, is a commercially important crop plant that was initially developed in Europe during the first half of the eighteenth century from spontaneous hybrids of the American species *F. chiloensis* (L.) Mill. and *F. virginiana* Mill. that arose from plants cultivated there (K. R. Robertson, 1974; Staudt, 1999). Extensive continuing breeding programs have given rise to numerous strawberry cultivars differing in yield, ripening times, and flavor, but also in such morphological characters as calyx and bractlet size and the size and shape of the "berry" (which is actually a greatly expanded receptacle covered with tiny achenes). *Fragaria* ×*ananassa* rarely escapes from cultivation in Missouri (Steyermark, 1963; Mühlenbach, 1979). It differs from the two native species in the generally larger size of the plants; its larger flowers (petals 10–15 mm long), sometimes with extra sepals and petals, and more foliacous sepals; and its larger fruits (1.5–3.0 cm long); as well as in several subtle characters of leaf texture and pubescence.

1. Leaflets with the tip of the terminal tooth usually extended past those of the immediately adjacent lateral ones; fruits positioned along the surface of the receptacle, not in pits; petals mostly 5–7 mm long . . . . 1. F. VESCA
1. Leaflets with the tip of the terminal tooth not extended beyond those of the immediately adjacent lateral ones; fruits positioned in small shallow pits along the surface of the receptacle; petals mostly 7–10 mm long . . . . 2. F. VIRGINIANA

1. Fragaria vesca L. (woodland strawberry, wild strawberry)

F. vesca var. *americana* Porter

F. vesca ssp. *americana* (Porter) Staudt

F. americana (Porter) Britton

Pl. 532 i; Map 2445

Leaflets 1.5–5.5 cm long, sessile or nearly so, thin-textured, the margins relatively sharply toothed, the terminal tooth more than 1/2 as wide as and with the tip usually extended past those of

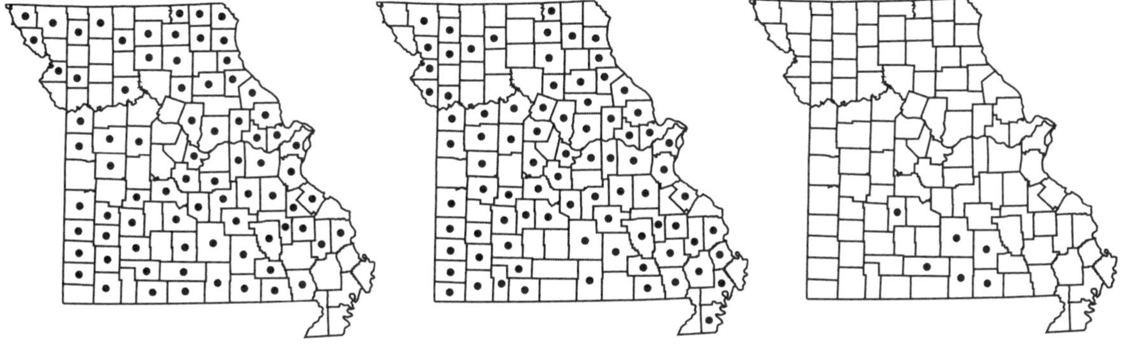
2446. Fragaria virginiana
2447. Geum canadense
2448. Geum fragarioides

the immediately adjacent lateral teeth, the upper surface sparsely to moderately hairy, bright green to yellowish green, the undersurface moderately to densely silky-hairy, grayish-tinged. Inflorescences usually shorter than the leaves at flowering, elongating with age, sometimes appearing racemelike (but the lower flowers on longer stalks than the upper ones), the main stalk with mostly appressed or ascending, fine long hairs (similar to those of the petioles). Sepals 3–5 mm long, usually reflexed at fruiting. Petals 5–7(–9) mm long. Fruits positioned along the surface of the receptacle, not in pits. $2n=14$. April–May.

Uncommon in the northeastern quarter of the state (northern U.S. west to Virginia, Missouri, and Nebraska; Canada). Mesic upland forests, usually on steep, north-facing slopes, and shaded ledges of bluffs.

The above key and description covers only plants assigned by some botanists to var. or ssp. *americana*. The group's most recent monographer, Staudt (1999), recognized four weakly defined subspecies (and several additional forms) within *F. vesca*, of which only this one has been found to occur in Missouri thus far. The Eurasian ssp. *vesca*, which has escaped from cultivation in the northeastern U.S. and Canada (also Hawaii), and which eventually may be found in Missouri, differs from ssp. *americana* in its generally spreading pubescence and more globose fruits. The other two North American subspecies, ssp. *bracteata* (A. Heller) Staudt and ssp. *californica* (Cham. & Schltdl.) Staudt, occur in the western United States (the former also in Canada and Mexico), and differ in subtle details of petal length and fruit morphology. These designations may be of practical utility for plant breeders or others seeking to compartmentalize the complex and subtle morphological variation in the species, but are of limited use to field botanists for determining most wild populations.

2. Fragaria virginiana Mill. (wild strawberry)
F. virginiana ssp. *grayana* (E. Vilm. ex J. Gay) Staudt
F. virginiana var. *illinoensis* A. Gray
Pl. 532 j–m; Map 2446

Leaflets 1.5–11.0 cm long, sessile or more commonly short-stalked, firm-textured, the margins scalloped or toothed, the terminal tooth mostly less than 1/2 as wide as and with the tip not extended beyond those of the immediately adjacent lateral ones, the upper surface sparsely to moderately hairy, green to dark green or bluish green, the undersurface nearly glabrous to densely silky-hairy, sometimes yellowish- or grayish-tinged. Inflorescences usually shorter than the leaves at both flowering and fruiting, sometimes appearing umbellate, not appearing racemelike (the flower stalks all about the same length), the main stalk with appressed-ascending or more commonly spreading, fine long hairs (similar to those of the petioles). Sepals 5–10 mm long, ascending, spreading, or reflexed at fruiting. Petals (6–)7–10(–11) mm long. Fruits positioned in shallow pits in the surface of the receptacle. $2n=56$. April–May.

Scattered nearly throughout Missouri (nearly throughout the U.S., including Alaska; Canada). Margins and openings of mesic to dry upland forests, savannas, upland prairies, ledges and tops of bluffs, and calcareous glades; also old fields, railroads, roadsides, and open disturbed areas.

As with *F. vesca*, Staudt (1999) subdivided *F. virginiana* into a series of weakly defined subspecies (and forms), which are not accepted here. Of these, the widespread eastern and northern ssp. *virginiana* and the less common ssp. *grayana*, mostly of the midwestern and south-central United States, were treated as occurring in Missouri. The ssp. *grayana* differs from ssp. *virginiana* in its petioles, as well as inflorescence and flower stalks, having only spreading pubescence, and in its leaflet margins with the teeth usually more sharply pointed. Overall, Missouri populations are quite variable morphologically, although plants within a given population usually are relatively uniform.

14. **Geum** L. (avens)

Plants perennial herbs, lacking spines and thorns, with short erect rootstocks or rhizomes. Stems erect or ascending, sometimes somewhat arched. Leaves alternate and basal (all basal in *G. fragarioides*, opposite in *G. triflorum*), sometimes entire, ternately, and/or pinnately compound, in some species the most divided leaves often with smaller leaflets interspersed with the larger primary ones, short- to long-petiolate, the leaf blade broadly ovate, triangular, or kidney-shaped to lanceolate or oblanceolate in overall outline, the leaflets with the margins finely to coarsely toothed and often also lobed. Stipules herbaceous, those of the basal leaves fused to the petiole, those of the stem leaves free and more or less leaflike, elliptic-lanceolate to broadly and asymmetrically ovate, rounded or narrowed at the base, the outer margin often with few to several lobes and also toothed. Inflorescences terminal, solitary or more commonly few-flowered, open, branched clusters or panicles, the flowers stalks mostly relatively long, the branch points usually with a stipulelike bract, the stalks often also with a pair of reduced stipulelike bracts near the midpoint. Flowers ascending or drooping, perigynous, the hypanthium deeply cup-shaped to bell-shaped, sometimes appearing obconic, with a more or less well-developed nectar disc along the rim, each flower (except in *G. fragarioides* and *G. vernum*) with 5 bractlets alternating with the sepals (the calyx thus appearing more or less 10-parted), these 1–3 mm long, inconspicuous, linear to narrowly oblong, not becoming enlarged at fruiting. Sepals 5, loosely ascending to spreading or reflexed at flowering, usually reflexed at fruiting. Petals 5, narrowly to broadly elliptic, white, cream-colored, or yellow (pink to pinkish purple in *G. triflorum*). Stamens (10–)20 to numerous, the anthers yellow. Pistils numerous, densely covering the surface of the more or less spherical to short-columnar receptacle, this glabrous or hairy. Ovaries superior, not hidden in the hypanthium, with 1 ovule. Style 1 per ovary, terminal, noticeably jointed at the base (in *G. fragarioides*) or above the midpoint (not noticeably jointed in *G. triflorum*), in the jointed species the 2 portions dissimilar, the apical segment shed as the fruit matures, the basal segment (absent in *G. fragarioides*) hooked at the tip and becoming elongated, persistent at fruiting. Stigma minute. Fruits achenes, free from and not hidden by the hypanthium (mostly hidden in *G. fragarioides*), densely covering the receptacle (sometimes few in *G. fragarioides*) and forming a globose to obovoid mass, the main body asymmetrically elliptic to ovate or obovate in outline, sometimes narrowly so, somewhat flattened, tapered abruptly to the hardened style at the tip (except in *G. fragarioides*). About 50 species, North America, South America, Europe, Asia, Africa, New Zealand, Tasmania, Auckland Islands.

The persistent, hooked basal portions of the styles in most species of *Geum* are an adaption to dispersal of the fruits in the fur (or clothing) of passing mammals. In the spring, the first leaves of the basal rosette produced in all species (except *G. fragarioides*) usually are simple and entire, with subsequent basal leaves progressively more divided.

Rydberg (1908–1918) reported two additional species, *G. aleppicum* Jacq. (yellow avens, as *G. strictum* Aiton, an illegitimate name) and *G. rivale* L. (water avens, purple avens), as occurring in Missouri, but because his treatment did not include specimen citations it has not been possible to account for these conclusively. As far as can be determined from examination of older specimens, his reports were based on misdetermined specimens. Both of these widespread species otherwise occur only to the north of Missouri in the midwestern portion of their ranges.

1. Stem leaves opposite; styles not noticeably jointed, persistent and becoming elongated to 3–5 cm at fruiting . . . . 4. G. TRIFLORUM
1. Leaves all basal or the stem leaves alternate; styles conspicuously jointed above the midpoint and persistent at fruiting or, if inconspicuously jointed at the base, then withering and shed after flowering
   2. Styles jointed inconspicuously at the base, withering and shed after flowering, not conspicuous or hooked at the tip; leaves all basal; corollas yellow . . . . 2. G. FRAGARIOIDES
   2. Styles noticeably jointed at or above the middle, the two portions dissimilar, the apical portion shed as the fruit matures, the basal portion hooked at the tip and becoming elongated, persistent at fruiting; leaves basal and along the stems; corollas white (except in *G. vernum* and *G. virginianum*)
      3. Cluster of fruits raised above the calyx on a noticeable stalk; sepals 2–3 mm long, not alternating with bractlets (bractlets absent); petals 1–2 mm long, yellow; apical segment of style glabrous . . . . 5. G. VERNUM
      3. Cluster of fruits sessile or nearly so, not evidently raised above the calyx; sepals 3–10 mm long, alternating with shorter bractlets; petals 2–9 mm long, white (except in *G. virginianum*); apical segment of style hairy
         4. Receptacle glabrous or sparsely and inconspicuously hairy between the fruits (this character difficult to see at flowering because the carpels may be hairy); flower stalks densely pubescent with conspicuous (visible without magnification), spreading to somewhat downward-pointing hairs . . . . 3. G. LACINIATUM
         4. Receptacle densely bristly-hairy between the fruits (remove a fruit to observe this); flower stalks densely pubescent with minute velvety hairs, sometimes also with sparse, longer, spreading hairs
            5. Petals (3–)4–9 mm long, as long as or longer than the sepals, white, sometimes fading to cream-colored with age; lower portion of stem glabrous or sparsely pubescent with appressed to spreading hairs . . . . 1. G. CANADENSE
            5. Petals 2–4 mm long, noticeably shorter than the sepals, cream-colored to lemon yellow; lower portion of stem moderately to densely pubescent with spreading to somewhat downward-pointing hairs . . . . 6. G. VIRGINIANUM

1. Geum canadense Jacq. (white avens, redroot)

G. canadense var. *camporum* (Rydb.) Fernald & Weath.

Pl. 532 c; Map 2447

Stems 30–100 cm long, the lower portion glabrous or sparsely pubescent with appressed to spreading hairs, grading upward into dense pubescence of minute velvety hairs and sometimes also sparse, longer, spreading hairs. Basal leaves simple or pinnately compound with 3–5 primary leaflets, long-petiolate; those of the stems few to several, alternate, mostly ternately compound or deeply lobed with progressively shorter petioles, the uppermost usually simple and nearly sessile. Stipules 8–24 mm long, oblong-ovate, often with few jagged lobes. Primary leaflets 2.5–15.0 cm long, broadly ovate to rhombic or lanceolate-elliptic, sometimes shallowly to deeply 3-lobed, the surfaces glabrous or more commonly sparsely to moderately short-hairy. Inflorescence branches relatively straight, the tips not drooping. Flower stalks densely pubescent with minute velvety hairs, sometimes also with sparse, longer, spreading hairs. Receptacle densely bristly-hairy. Sepals 4–8 mm long, lanceolate to oblong-ovate, alternating with shorter narrower bractlets. Petals (3–)4–9 mm long, as long as or longer than the sepals, white, sometimes fading to cream-colored with age. Apical segment of style 1–2 mm long, sparsely to moderately pubescent with short bristly hairs toward the base. Cluster of fruits 10–16 mm in diameter, sessile or nearly so, not evidently raised above the calyx. Fruits with the main body 2.5–

3.5 mm long, flattened but without thickened angles, sparsely to moderately bristly-hairy, the persistent stylar beak 4–7 mm long, glabrous or sparsely hairy toward the base. $2n=42$. May–October.

Scattered to common nearly throughout the state (eastern U.S. west to Montana, Wyoming, and Texas; Canada). Bottomland forests, mesic upland forests, banks of streams and rivers, moist depressions of upland prairies, and rarely margins of glades; also pastures, fallow fields, roadsides, and railroads.

This species is quite variable morphologically. Occasional Missouri specimens have scattered glandular trichomes in addition to the other types of hairs. Sometimes the upper leaf surface is lighter adjacent to the main veins. Fernald and Weatherby (1922) divided *G. canadense* into a complicated series of varieties and forms, which they themselves considered merely "recognizable trends," based on minor differences in pubescence patterns and floral structures. Their var. *camporum* was based upon robust plants at the larger end of the species' range of variation for achene size and number per flower, but there is nearly continuous variation in the species for these characters.

2. Geum fragarioides (Michx.) Smedmark
(barren strawberry)
Waldsteinia fragarioides (Michx.) Tratt.
Pl. 543 g–i; Map 2448

Aerial stems absent above the slender, branched rhizomes. Leaves in basal rosettes at the rhizome tip (stem leaves absent), long-petiolate, the winged petiole bases persistent after the leaves are shed. Stipules 4–9 mm long, fused with the petiole and appearing as a triangular wing at the petiolar base, purplish red, sparsely to moderately pubescent with short to long yellow hairs along the margins and sometimes also the surface. Leaf blades 2–7 cm long, kidney-shaped to broadly ovate-triangular in outline, divided into 3 similar leaflets (rarely only deeply lobed), these 1.5–6.0 cm long, broadly and sometimes unevenly spatulate (lateral leaflets) to fan-shaped (middle leaflet), rounded, narrowed to a sessile or short-stalked base, the margins broadly (but with a sharp point) toothed, often also shallowly lobed, sparsely to moderately pubescent with short to long yellow hairs, the surfaces glabrous or sparsely to moderately hairy. Inflorescence branches relatively straight, the tips not drooping. Flower stalks sparsely to moderately pubescent with longer nonglandular hairs and shorter glandular hairs. Receptacle densely bristly-hairy. Sepals 3–6 mm long, spreading, triangular to narrowly ovate-triangular, moderately to densely pubescent, often felty-hairy, at least on the upper surface, persistent at fruiting, not subtended by bractlets. Petals 5–11 mm long, noticeably longer than the sepals, yellow. Style 2–3 mm long, attached terminally, shed from a basal joint soon after flowering. Cluster of fruits 4–9 mm in diameter, sessile or nearly so, not evidently raised above the calyx. Fruits 1.5–2.2 mm long, flattened but without thickened angles, densely hairy. $2n=14, 21$. April–May.

Uncommon in the Ozark Division (eastern U.S. west to Minnesota and Arkansas; Canada). Mesic upland forests, usually on north-facing slopes, and ledges of bluffs, on acidic substrates.

Geum fragarioides and the four or five related species that form a circumboreal complex have been treated in the segregate genus *Waldsteinia* Willd. in most of the previous botanical literature (Steyermark, 1963), based on their nonpersistent styles. However, molecular phylogenetic studies by Smedmark and Eriksson (2002), Smedmark et al. (2003), and Smedmark (2006) have provided strong evidence that *Waldsteinia* should be considered one of several groups embedded within a more broadly circumscribed *Geum*.

Steyermark (1963) considered the Missouri occurrences of this species to be relictual from times when the climate was cooler, by which he explained the sporadic distribution of the taxon in the state. Some authors have segregated a second subspecies, *W. fragarioides* ssp. *doniana* (Tratt.) Teppner (an invalid name) for plants with shorter, narrower, more pointed petals that are restricted mostly to the southern portion of the Appalachians. However, other botanists consider this a separate species, for which the name *G. donianum* (Tratt.) Weakley & Gandhi is available.

3. Geum laciniatum Murray (rough avens)
G. laciniatum var. *trichocarpum* Fernald
Pl. 532 a, b; Map 2449

Stems 40–100 cm long, the entire length moderately to densely pubescent with spreading to somewhat downward-pointing hairs. Basal leaves simple or more commonly pinnately compound with 3–7 primary leaflets, long-petiolate, those of the stems few to several, alternate, mostly ternately compound or deeply lobed with progressively shorter petioles, the uppermost simple and nearly sessile. Stipules 7–15 mm long, oblong-ovate to broadly ovate, often with few to several jagged lobes. Primary leaflets 2.5–10.0 cm long, broadly ovate or nearly circular to rhombic or lanceolate-elliptic, often shallowly to deeply 3- or more-lobed, the surfaces sparsely to moderately hairy. Inflorescence branches relatively straight, the tips not drooping. Flower stalks densely pubescent with conspicuous, spreading to somewhat downward-pointing hairs. Receptacle glabrous or sparsely and inconspicuously hairy between the fruits (this character difficult to see at flowering because the carpels may be hairy). Sepals 4–10 mm long, ovate-triangular, alternating with shorter narrower bractlets. Petals 3–5 mm long, usually shorter than the sepals, white, sometimes fading to cream-colored with age. Apical segment of style 1.0–1.5 mm long, sparsely to moderately pubescent with short bristly hairs toward the base. Cluster of fruits 12–20 mm in diameter, sessile or nearly so, not evidently raised above the calyx. Fruits with the main body 3–5 mm long, flattened, sometimes slightly thickened along the angles, glabrous or more commonly sparsely to moderately pubescent with minute and/or bristly hairs, the persistent stylar beak 4–6 mm long, usually sparsely hairy toward the base. $2n=42$. May–July.

Scattered in the Glaciated Plains Division (eastern [mostly northeastern] U.S. west to Nebraska and Kansas; Canada). Bottomland forests, bottomland prairies, moist depressions of upland prairies, banks of streams and rivers, margins of sloughs, and bases of bluffs; also ditches, roadsides, and railroads.

The name *G. virginianum* was applied by some earlier authors (Rydberg, 1908–1918) to plants properly called *G. laciniatum*, and plants known to modern authors as *G. virginianum* were treated under the name *G. hirsutum*. This resulted in some confusion in the literature and herbaria as to the status and distribution of the two species. Plants with sparsely bristly-hairy fruits occur mainly in the southern and western parts of the range (including Missouri) and have been called var. *trichophorum*, but this minor variation seems unworthy of taxonomic recognition.

4. Geum triflorum Pursh **var. triflorum**
(prairie smoke, old man's whiskers)
Map 2450

Stems 20–40 cm long, the entire length moderately to more commonly densely pubescent with soft, spreading to somewhat downward-angled, short hairs and sparse to moderate, longer, spreading hairs. Basal leaves pinnately deeply lobed and/or compound with 7–17 primary leaflets, short-petiolate, the petiole frequently winged; the primary leaflets 1–5 cm long, gradually smaller from the blade tip to base, irregularly and often jaggedly, sharply lobed or toothed, the surfaces and margins sparsely to moderately and finely hairy (more densely so when young). Stem leaves 2(4), opposite, much smaller than the basal leaves, pinnately or nearly palmately deeply and jaggedly divided into 3–7 main segments, these often jaggedly toothed toward the tips. Stipules 4–12 mm long, broadly ovate to ovate, often with few to several jagged lobes. Inflorescence branches nodding. Flower stalks densely pubescent with conspicuous, fine, woolly to spreading, shorter and longer hairs, some of the hairs sometimes gland-tipped. Receptacle short-hairy between the fruits. Sepals 7–13 mm long, ovate-triangular to broadly lanceolate, pink to reddish pink, alternating with shorter to longer, much narrower bractlets (these usually arched outward at flowering). Petals 10–14 mm long, slightly longer than the sepals (but often shorter than the bractlets), white, sometimes pinkish- or purplish-tinged, especially with age. Styles not noticeably jointed, persistent, mostly straight or only slightly curved toward the tip, becoming elongated to 3–5 cm at fruiting, the lower portion densely pubescent (plumose) with long, soft, spreading hairs, the apical 2–4(–6) mm hairy only toward its base. Cluster of fruits 8–12 mm in diameter (not including the styles), sessile or nearly so, not evidently raised above the calyx. Fruits with the main body 2.5–3.5 mm long, somewhat flat-

2452. Geum virginianum 2453. Gillenia stipulata 2454. Gillenia trifoliata

tened, densely pubescent with soft hairs. 2n=42. April–May.

Possibly introduced, uncommon, known thus far only from a single specimen from Andrew County (northern U.S. south to Colorado and Illinois; Canada). Fencerows and pastures.

The status of this species in Missouri is uncertain. The Andrew County record was collected in 2007 by Nicholas Cole, then an undergraduate student in a plant taxonomy class at the University of Missouri. It originated from the edge of a pasture on a farm. However, attempts to secure more information from the collector were in vain and a search in 2008 of nearby public lands by members of the Missouri Native Plant Society was unsuccessful. Northwestern Missouri is not too far outside the overall range of this northern Great Plains species, thus Andrew County plausibly might represent a range extension for a native species. However, plants are also cultivated as ornamentals and the specimen in question might have escaped from a nearby garden. Further attempts to locate this species in Missouri should be undertaken in the future.

Three other varieties are often recognized, all inhabiting portions of the western United States and western Canada. They differ in details of floral morphology, style length at fruiting, and the size and degree of division of the basal leaves.

5. Geum vernum (Raf.) Torr. & A. Gray (spring avens, early water avens)

Pl. 532 d, e; Map 2451

Stems 25–60 cm long, glabrous or the lower portion moderately to densely pubescent with short, soft, spreading to ascending hairs, becoming sparser toward the tip. Basal leaves simple (these larger and long-petiolate) and also pinnately compound with 3–9 primary leaflets (these smaller and shorter-petiolate); those of the stems few to several, alternate, those above the lower portion of the stems progressively less divided, short-petiolate to sessile, the uppermost often merely simple and 3-lobed. Stipules 6–18 mm long, lanceolate to broadly ovate or kidney-shaped, usually with few to several jagged lobes. Primary leaflets 0.5–6.5 (–10) cm long, broadly ovate or nearly circular to rhombic or lanceolate-elliptic, often shallowly to deeply 3- or more-lobed, the surfaces sparsely to moderately hairy. Inflorescence branches relatively often drooping, curled, or hooked toward the tips. Flower stalks glabrous or sparsely and inconspicuously pubescent with short soft hairs. Receptacle glabrous. Sepals 2–3 mm long, triangular, not alternating with bractlets. Petals 1–2 mm long, slightly shorter than the sepals, yellow. Apical segment of style 0.7–1.0 mm long, glabrous. Cluster of fruits 8–14 mm in diameter, raised above the calyx on a noticeable stalk 1–4 mm long. Fruits with the main body 2–3 mm long, flattened, somewhat thickened along the angles, minutely (visible only with strong magnification) and inconspicuously hairy, the persistent stylar beak 1.5–3.0 mm long, glabrous. 2n=42. April–June.

Scattered to common in the Ozark, Ozark Border, and Unglaciated Plains Divisions, but uncommon in the Glaciated Plains and Mississippi Lowlands (eastern U.S. west to Nebraska and Texas; Canada). Bottomland forests, mesic upland forests, banks of streams and rivers, and bases of bluffs.

6. Geum virginianum L. (pale avens)

G. hirsutum Muhl. ex Link

Pl. 532 f–h; Map 2452

Stems 30–90(–120) cm long, the lower portion moderately to densely pubescent with spreading to somewhat downward-pointing hairs, grading upward into dense pubescence of minute velvety hairs and sometimes also sparse, longer, spreading hairs. Basal leaves simple or less commonly pinnately compound with 3–5 primary leaflets, long-petiolate, those of the stems mostly ternately compound or deeply lobed with progressively shorter petioles, the uppermost simple and nearly sessile. Stipules 7–28 mm long, oblong-ovate, often with few to several jagged lobes. Primary leaflets 2.5–12.0 cm long, broadly ovate to rhombic or lanceolate-elliptic, often shallowly to deeply 3-lobed, the surfaces sparsely to moderately hairy. Inflorescence branches relatively straight, the tips not drooping. Flower stalks densely pubescent with minute velvety hairs, usually also with sparse, longer, spreading hairs. Receptacle densely bristly-hairy. Sepals 3.5–5.0 mm long, lanceolate-triangular, alternating with shorter narrower bractlets. Petals 2–4 mm long, noticeably shorter than the sepals, cream-colored to lemon yellow. Apical segment of style 1–2 mm long, sparsely to moderately pubescent with short bristly hairs toward the base. Cluster of fruits 10–15 mm in diameter, sessile or nearly so, not evidently raised above the calyx. Fruits with the main body 2–3 mm long, flattened but without thickened angles, sparsely to moderately bristly-hairy, the persistent stylar beak 4–7 mm long, glabrous or sparsely hairy toward the base. $2n=42$. May–August.

Uncommon and widely scattered in the southern half of the state (eastern U.S. and adjacent Canada west to Minnesota, Missouri, and Alabama). Mesic to dry upland forests and banks of streams, often in disturbed areas.

Specimens of *G. canadense* have sometimes been misdetermined as *G. virginianum*, probably because of confusion surrounding petal colors: those of the former are white when the flower opens, but tend to fade to cream-colored with age; those of the latter are cream-colored to lemon yellow from the start. Other characters that help to separate the two species include the petal length relative to the sepal length and the dense long, spreading hairs on the lower stems and petioles of *G. virginianum* (vs. glabrous to sparsely hairy in *G. canadense*). *Geum virginianum* also tends to have stipules and leaves with more lobes and the terminal leaflets more noticeably larger than the lateral ones, but these characters are fairly variable in both species.

See the treatment of *G. laciniatum* for mention of the earlier nomenclatural confusion between this species and *G. virginianum*. *Geum virginianum* was first reported (with a question mark) for Missouri by Rydberg (1908–1918, as *G. hirsutum*), based on a 1905 specimen from Barry County accessioned at the Missouri Botanical Garden, but, in spite of this, the species was not mentioned by Steyermark (1963). Solecki (1983) first confirmed its presence in the Missouri flora.

15. **Gillenia** Moench (Indian physic)

Plants perennial herbs, with woody rhizomes. Stems erect or ascending, lacking spines and thorns, glabrous or hairy. Leaves alternate, sessile or with petioles less than 1 cm long. Stipules herbaceous, conspicuous or inconspicuous. Leaf blades broadly ovate to triangular in outline, mostly trifoliate, the uppermost sometimes simple, the leaflets sessile or nearly so, variously finely and sharply toothed to deeply lobed (toward the stem base), the upper surface glabrous or sparsely hairy, the undersurface sparsely to densely hairy. Inflorescences loose few-flowered panicles, terminal or sometimes also from the upper leaf axils, occasionally reduced to a single flower, with reduced leaflike bracts at the branch points. Flowers perigynous, slightly zygomorphic, the hypanthium deeply cup-shaped at flowering, persistent but tearing apart by fruiting, 10-nerved, usually dark red or purple, not subtended by bractlets. Sepals 5, 1–2 mm long, erect, triangular, with small peglike reddish purple glands and sometimes also short white hairs along the margins, persistent at fruiting. Petals 5, spreading to slightly drooping, linear to narrowly oblanceolate, white or pinkish-tinged. Stamens (10–)20, the anthers pink. Pistils 5 in a single whorl, appearing fused toward the base at flowering, but becoming separate by fruiting. Ovary superior, with 1 locule, with 2–4 ovules. Style 1 per pistil, erect, persistent at fruiting, the stigma minute. Fruits follicles, not hidden by the hypanthium, ascending, ovoid, somewhat curved toward the tip, dehiscing along the inner suture and also partially along the outer (dorsal) suture, with 2–4 seeds. Seeds 2.5–3.5 mm long, asymmetrically ovoid and somewhat angular, the surface usually finely wrinkled, reddish brown to dark brown. Two species, eastern U.S., Canada.

In many floristic manuals, the name *Porteranthus* Britton is used for this genus. *Gillenia* is the older of the two names, but is very similar in spelling to the unrelated *Gillena* Adans., a synonym of the non-Missouri genus *Clethra* (Clethraceae). Because the two spellings are so similar and *Gillena* was the earliest validly published name, many authors have considered *Gillenia* to represent a mere spelling variant and later homonym of *Gillena* (K. R. Robertson, 1974). Hunt (1982) officially proposed to conserve *Gillenia* (Rosaceae) against the little-used *Gillena* (Clethraceae), but his proposal was rejected by the Committee for Spermatophyta of the International Botanical Congress on the grounds that the two names do have different spellings, are not truly homonyms, and are not likely to be confused (Voss, 1986; Brummitt, 1988). Because of the language of this ruling, *Gillenia* must continue to be used as the oldest valid name for this genus.

In recent years, species of *Gillenia* have increased in popularity as garden ornamentals. The common names, Indian physic and American ipecac, refer to the medicinal use by Native Americans of a decoction of the rhizomes of both species for emetic purposes. Among other uses, they were applied externally to relieve symptoms of rheumatism (Moerman, 1998).

1. Stipules 10–30 mm long, conspicuous and persistent, broadly ovate to nearly circular; petals 10–13 mm long . . . . 1. G. STIPULATA
1. Stipules 5–10 mm long, inconspicuous and usually shed soon after the leaves develop, linear; petals 12–22 mm long . . . . 2. G. TRIFOLIATA

1. Gillenia stipulata (Muhl. ex Willd.) Nutt. (midwestern Indian physic, American ipecac)

Porteranthus stipulatus (Muhl. ex Willd.) Britton

Pl. 531 c–e; Map 2453

Stems 40–120 cm long, glabrous or sparsely to densely hairy. Stipules 10–30 mm long, conspicuous and persistent, broadly ovate to nearly circular, the margins with numerous sharp teeth, the upper surface glabrous or with sparse nonglandular hairs, the undersurface with sessile glands, usually also with sparse to dense nonglandular hairs. Leaflets 2–10 cm long, those of the lowermost leaves usually deeply parted into narrow lobed and/or toothed lobes, those of the median and upper leaves lanceolate to narrowly elliptic, the margins sharply toothed, the upper surface usually glabrous, the undersurface with sessile glands and usually also nonglandular hairs. Hypanthium 4.5–5.5 mm long, glabrous. Petals 10–13 mm long. Fruits 6–8 mm long, glabrous or very sparsely hairy. $2n=18$. May–July.

Scattered nearly throughout the state, but absent from western and northern portions of the Glaciated Plains Division (eastern U.S. west to Kansas and Texas). Mesic to dry upland forests; also railroads and roadsides.

2. Gillenia trifoliata (L.) Moench (mountain Indian physic, Bowman's root)

Porteranthus trifoliatus (L.) Britton

Pl. 531 f, g; Map 2454

Stems 50–100 cm long, glabrous or sparsely hairy. Stipules 5–10 mm long, inconspicuous and usually shed soon after the leaves develop, linear, the margins smooth or more commonly with few glandular teeth, the upper surface glabrous or with sparse to moderate nonglandular hairs, usually also with scattered minute, stalked, dark-tipped glands, the undersurface with sparse to moderate nonglandular hairs. Leaflets 1.7–10.0 cm long, none of them deeply lobed, lanceolate or elliptic to oblong-oblanceolate, the margins sharply toothed, the upper surface glabrous or with sparse to moderate nonglandular hairs, usually also with scattered minute, stalked, dark-tipped glands, the undersurface with sparse to moderate nonglandular hairs. Hypanthium 5.5–8.0 mm long, glabrous or sparsely hairy. Petals 12–22 mm long. Fruits 5–8 mm long, sparsely to moderately hairy. $2n=18$. May–June.

Known only from a single historical collection from Lawrence County (eastern U.S. west to Missouri and Arkansas; Canada [extirpated]). Margin of mesic upland forest on limestone substrate.

K. R. Robertson (1974) suggested that the Missouri specimen may represent an introduced population, but the mere proximity of a railroad to the site is insufficient evidence that the occurrence was nonnative, particularly as the species grows natively in northwestern Arkansas.

16. **Malus** Mill. (apple)

Plants shrubs or small to medium trees. Branches sometimes producing short, stout branchlets with thorny tips, these indeterminate, mostly eventually become elongated and leafy. Bark dark brown to gray, on younger trunks relatively smooth but with prominent, raised branch scars and lenticels, on older trunks sometimes developing a network of ridges, these breaking up into more or less rectangular, fine, scaly plates, sometimes peeling. Winter buds ovoid to more or less conic, with several overlapping scales. Leaves alternate but often appearing clustered at the tips of short branchlets, rolled or folded longitudinally during development, mostly long-petiolate, the petioles lacking glands. Stipules small, membranous to papery, shed early (sometimes larger, more herbaceous, and more persistent on rapidly growing vegetative shoots). Leaf blades simple, unlobed or shallowly, pinnately few-lobed, variously shaped, the margins bluntly to sharply toothed, the surfaces glabrous or hairy at maturity, the upper surface lacking glands. Inflorescences terminal on lateral short branchlets, dome-shaped, umbellate clusters (lacking a noticeable central axis) of 2–6 long-stalked flowers, produced as the leaves uncurl or later, the axis and stalks glabrous or hairy, the stalks each with a small bract at the base, this linear to narrowly oblong-elliptic, brown to reddish brown, shed early. Flowers epigynous, often fragrant, the hypanthium fused to the ovaries, cup-shaped to ellipsoid or more or less urn-shaped, relatively open at the tip, glabrous or hairy. Sepals 5, spreading to somewhat reflexed at flowering, lanceolate, angled or tapered to a sharply pointed tip, the margins and/or upper surface hairy, sometimes persistent at fruiting. Petals 5 (except in rare doubled forms), broadly obovate to nearly circular, pink or pinkish white, in some species white at flowering but pinkish-tinged in bud, often fading to white with age. Stamens 15 to numerous, the anthers yellow or pink. Pistil 1 per flower, of 2–5 fused carpels. Ovary inferior, but sometimes protruding slightly from the hypanthium, with 2–5 locules, each with usually 2 ovules. Styles 2–5, fused toward the base, the fused portion densely hairy, protruding from the hypanthium the stigmas capitate or club-shaped. Fruits pomes, globose or nearly so, glabrous at maturity, variously colored, the surface not dotted, with 2–10 more or less easily exposed seeds embedded in the "core" of leathery to papery carpel wall remains and the fleshy portion, this lacking stone cells. Seeds obovoid to narrowly obovoid, the outer surface smooth, light brown to black. Twenty-five to 47 species, North America, Europe, Asia, introduced widely.

Steyermark (1963) and some other authors have classified *Malus* as a subgenus of a broadly defined *Pyrus*. For further discussion, see the treatment of that genus. Species limits among the apples and crab apples remain poorly understood.

1. Leaf blades unlobed, the margins finely and evenly toothed; anthers yellow; leaves curled inward during development; branchlets thornless
   2. Leaf blades broadly elliptic to broadly ovate-elliptic, sometimes nearly circular, 1–2 times as long as wide; fruits 2–6 cm long (bigger in cultivated plants); calyces withered but more or less persistent at fruiting; corollas pink to pinkish white at flowering (sometimes fading to white with age) . . . . 4. M. DOMESTICA
   2. Leaf blades variously oblong-elliptic to ovate-elliptic or ovate, 1.5–4.0 times as long as wide; fruits 0.8–1.4 cm long; calyces not persistent at fruiting; corollas sometimes white or nearly so at flowering (pinkish-tinged in bud)
      3. Leaf blades oblong-elliptic to ovate-elliptic, mostly 3–4 times as long as wide, the margins usually bluntly toothed; calyces not persistent at fruiting . . . . 2. M. BACCATA
      3. Leaf blades ovate to oblong-ovate, mostly 1.5–2.5 times as long as wide, the margins sharply toothed; calyces not persistent at fruiting . . . . 5. M. FLORIBUNDA
1. Blades of at least the larger leaves shallowly lobed, the margins somewhat irregularly and sharply toothed; anthers pink to red; leaves folded lengthwise during development; at least some of the branchlets thorn-tipped
   4. Hypanthia and calyces glabrous or very sparsely hairy externally (note that the calyces are densely hairy on the inner [upper] surface)
      5. Leaf blades oblong-lanceolate to narrowly elliptic, 3 or more times as long as wide, rounded to bluntly or broadly pointed at the tip . . . . 1. M. ANGUSTIFOLIA
      5. Leaf blades broadly lanceolate to broadly ovate, 1.5–3.0 times as long as wide, angled or tapered to a sharply pointed tip . . . . 3. M. CORONARIA
   4. Hypanthia and calyces densely hairy externally
      6. Calyces persistent at fruiting; fruits 2.0–3.5 cm long; styles 5 . . . 6. M. IOENSIS
      6. Calyces not persistent at fruiting; fruits 0.6–0.9 cm long; styles 3 or 4 . . . . 7. M. SIEBOLDII

1. Malus angustifolia (Aiton) Michx. (wild crab, narrow-leaved crab apple)
Pyrus angustifolia Aiton
P. angustifolia var. *spinosa* (Rehder) L.H. Bailey

Pl. 533 a, b; Map 2455

Plants shrubs or small trees to 5(–10) m tall, often colonial from root suckers. Branchlets mostly thorn-tipped. Twigs sparsely to densely short-hairy. Leaf blades folded lengthwise during development, 3–7 cm long, 3 or more times as long as wide, oblong-lanceolate to narrowly elliptic, angled or tapered at the base, rounded to bluntly or broadly pointed at the tip, the margins somewhat irregularly, sharply toothed, those of at least the larger leaves usually shallowly lobed, the surfaces glabrous at maturity, the undersurface sparsely hairy when young. Flower stalks and hypanthia glabrous or sparsely hairy. Calyces more or less persistent at fruiting, the sepals 3–5 mm long, triangular to narrowly triangular, the outer surface glabrous or very sparsely hairy, the inner surface densely woolly. Petals 1.5–2.5 cm long, the body ovate to oblong-ovate, short-tapered to a noticeable stalklike base, pink or pinkish-tinged at flowering, often fading to white. Anthers pink to orangish red. Styles 5, the stigmas narrowly club-shaped. Fruits 2.5–3.5 cm long, green to yellowish green, often somewhat glaucous. $2n=34$, 68. April–May.

Uncommon in the Mississippi Lowlands Division north to Cape Girardeau County (eastern [mostly southeastern] U.S. west to Mississippi and Texas). Bottomland forests, banks of spring branches, edges of mesic upland forests, savannas, and sand prairies; also railroads and roadsides.

Malus angustifolia is closely related to and perhaps not distinct from *M. coronaria* and *M. ioensis*. For discussion of problems with the taxonomy of these species, see the discussion of *M. coronaria*.

2. Malus baccata (L.) Borkh. (Siberian crab)

Pl. 533 g; Map 2456

Plants shrubs or small trees to 5 m tall, not colonial. Branchlets thornless. Twigs initially densely

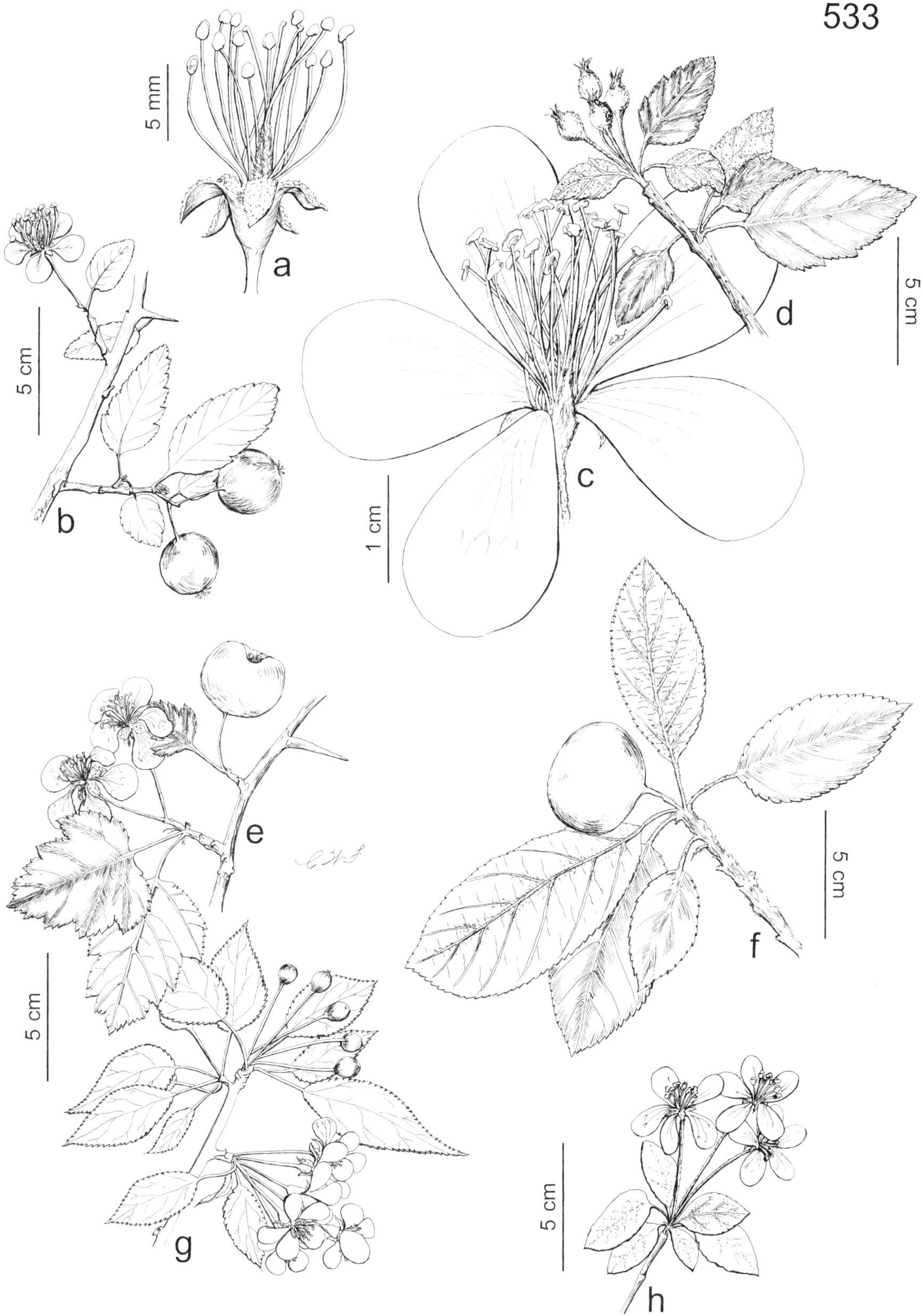

Plate 533. Rosaceae. *Malus angustifolia*, **a)** flower with petals removed, **b)** flowering and fruiting branch. *Malus ioensis*, **c)** flower, **d)** branch with young fruits. *Malus coronaria*, **e)** flowering and fruiting branch. *Malus ×soulardii*, **f)** fruiting branch. *Malus baccata*, **g)** flowering and fruiting branch. *Malus domestica*, **h)** flowering branch.

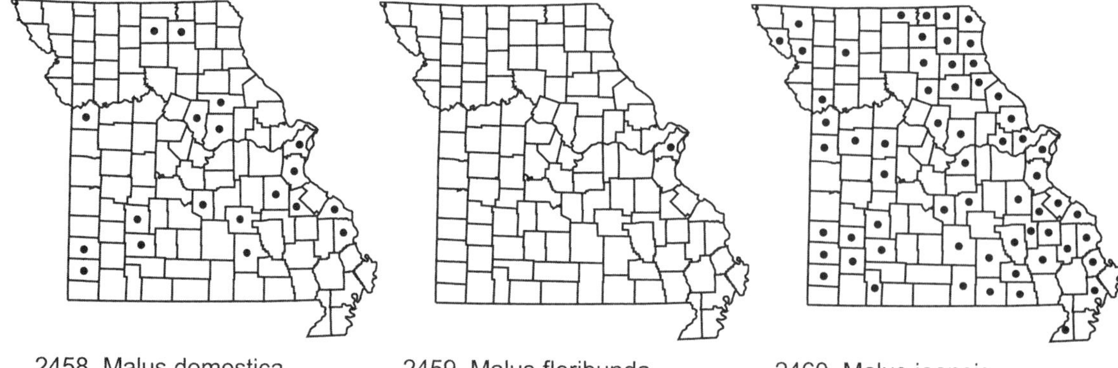
2458. Malus domestica 2459. Malus floribunda 2460. Malus ioensis

hairy but soon glabrous. Leaf blades rolled inward during development, 3–8 cm long, mostly 3–4 times as long as wide, oblong-elliptic to ovate-elliptic, mostly angled or short-tapered at the base, angled or tapered to a sharply pointed tip, the margins finely, evenly, and bluntly toothed, unlobed, the surfaces glabrous at maturity, the undersurface often minutely hairy when young. Flower stalks and hypanthia glabrous. Calyces not persistent at fruiting, the sepals 5–7 mm long, triangular to narrowly triangular, the outer surface glabrous or very sparsely hairy, the inner surface densely woolly. Petals 1.5–2.5 cm long, the body obovate to broadly obovate, short-tapered to a short, stalklike base, usually white at flowering (pinkish-tinged in bud). Anthers yellow. Styles (4)5, the stigmas broadly club-shaped to more or less capitate. Fruits 0.8–1.4 cm long, red to yellow or yellowish green, not glaucous. $2n=34$. April–May.

Introduced, known thus far only from St. Louis City and County (native of Europe, Asia; introduced sporadically in the northeastern U.S. west to Minnesota and Missouri; Canada). Gardens and railroads.

This species was first reported for Missouri by Mühlenbach (1979) during his inventories of the St. Louis railyards. It is a commonly cultivated ornamental tree, with a variety of cultivars.

3. Malus coronaria (L.) Mill. (wild crab, sweet crab apple)

Pyrus coronaria L.

P. coronaria var. *lancifolia* (Rehder) Fernald

Pl. 533 e; Map 2457

Plants shrubs or small trees to 5(–10) m tall, often colonial from root suckers. Branchlets mostly thorn-tipped. Twigs glabrous or nearly so. Leaf blades folded lengthwise during development, 3–10 cm long, 1.5–3.0 times as long as wide, broadly lanceolate to broadly ovate, broadly angled to rounded or shallowly cordate at the base, angled or tapered to a sharply pointed tip, the margins somewhat irregularly, sharply toothed, those of at least the larger leaves usually shallowly lobed, the surfaces glabrous at maturity, the undersurface sparsely hairy when young. Flower stalks and hypanthia glabrous or sparsely hairy. Calyces more or less persistent at fruiting, the sepals 3–5 mm long, triangular to narrowly triangular, the outer surface glabrous or very sparsely hairy, the inner surface densely woolly. Petals 1.5–2.5 cm long, the body ovate to oblong-ovate, short-tapered to a noticeable stalklike base, pink or pinkish-tinged at flowering, often fading to white. Anthers pink to orangish red. Styles 5, the stigmas narrowly club-shaped. Fruits 2–3 cm long, green to yellowish green, often somewhat glaucous. $2n=68$. April–May.

Scattered in southeastern and central Missouri (eastern U.S. west to Wisconsin and Arkansas). Bottomland forests, mesic upland forests, banks of streams, margins of sinkhole ponds, and upland prairies; also roadsides.

Steyermark (1963) noted that the fruits of this species sometimes are used in jams and jellies, and that early settlers fermented the fruits into a cider.

The key characters will serve to adequately distinguish most but not all specimens of *M. angustifolia, M. coronaria*, and *M. ioensis*. These taxa are part of a polyploid complex within *Malus* sect. *Chloromeles* (Decne.) Rehder. Some authors have suggested that distinctions among the taxa are complicated not only by the variable morphology within them, but also because of past hybridization (K. R. Robertson, 1974). In a study of genetic variation within and between populations of American *Malus* species using allozyme markers, Dickson et al. (1991) found little variation within sect. *Chloromeles* and no genetic markers to distinguish any of the species. K. R. Robertson (1974) suggested that the three species present in Missouri should be maintained provisionally until more detailed taxonomic studies could be completed.

4. Malus domestica Borkh. (apple)
M. pumila Mill.
M. pumila var. *domestica* (Borkh.) C.K. Schneid.
Pyrus malus L.
P. malus var. *paradisiaca* L.
Pl. 533 h; Map 2458

Plants small trees to 10(–15) m tall, not colonial. Branchlets thornless. Twigs densely short-hairy to woolly, often becoming glabrous with age. Leaf blades rolled inward during development, 4–10 cm long, 1–2 times as long as wide, broadly elliptic to broadly ovate-elliptic, sometimes nearly circular, broadly angled or rounded to occasionally shallowly cordate at the base, angled or tapered to a sharply pointed tip, the margins finely, evenly, and usually bluntly toothed, unlobed, the surfaces densely short-hairy during development, the upper surface often becoming glabrous or nearly so at maturity. Flower stalks and hypanthia densely short-hairy. Calyces withered but more or less persistent at fruiting, the sepals 7–9 mm long, narrowly triangular to triangular, both surfaces densely woolly. Petals 2–3 cm long, the body obovate to broadly obovate, short-tapered to a short, stalklike base, pink to pinkish white at flowering (sometimes fading to white with age). Anthers yellow. Styles 5, the stigmas broadly club-shaped to more or less capitate. Fruits 2–6 cm long (bigger in cultivated plants), indented at each end, red to yellow or green, sometimes mottled, not glaucous. $2n$=34. April–May.

Introduced, sporadic (cultigen of Eurasian origin, introduced widely but sporadically in North America). Edges of mesic upland forests, banks of streams and rivers, and margins of ponds and sinkhole ponds; also fencerows, pastures, old fields, old orchards, old mines, old homesites, railroads, and roadsides.

Apples are an extremely important fruit crop and have a long history of cultivation. Current evidence suggests that the cultivated apple originated through hybridization between two or more closely related Eurasian species in the ser. *Malus* (Robinson et al., 2001), including the European wild apple, *M. sylvestris* Mill. (Coart et al., 2006). However, the precise parentage and sequence of events leading to the development of the cultigen are still not well understood. The correct name for the cultivated apple also remains somewhat controversial. The oldest validly published name is *Pyrus malus*, but that epithet cannot be transferred to the genus *Malus* as the species epithet for a plant name cannot be identical to the genus name. Mabberley et al. (2001) noted that in the genus *Malus*, various authors have applied the names *M. domestica, M. pumila*, and *M. sylvestris* to this taxon. Currently, the name *M. sylvestris* is applied to the European wild apple, which is not the same taxon as the cultivated species. The names *M. domestica* (considered to be illegitimate by some botanists [Qian et al., 2010]) and *M. pumila* were both based on cultivated material, and part of the problem with picking the correct name revolves around whether the orchard apple (first described as *M. domestica*) is taxonomically separable from the paradise apple (a dwarf fruit tree first described as *P. malus* var. *paradisiaca* and *M. pumila*). The conclusion of Mabberley et al. (2001) that these two cultigens should be referred to by the same species name is accepted here. Qian et al. (2010) recently published a proposal to officially conserve the name *M. domestica* against the names *M. pumila, M. communis, M. frutescens*, and *Pyrus dioica* Moench, should these names be considered to represent a single species. Although formal approval of this proposal is still pending, it is provisionally followed here.

For a discussion of hybridization between *M. domestica* and *M. ioensis*, see the treatment of that species.

5. Malus floribunda Siebold ex Van Houtte (Japanese flowering crab apple, purple chokeberry)
Pyrus floribunda G. Kirchn., an illegitimate name
Map 2459

Plants shrubs or small trees to 5(–8) m tall, not colonial. Branchlets thornless. Twigs densely short-hairy when young, soon becoming glabrous. Leaf blades rolled inward during development, 3–8 cm long, mostly 1.5–2.5 times as long as wide, ovate to oblong-ovate, broadly angled or rounded at the base, tapered to a sharply pointed tip, the margins finely, evenly, and sharply toothed, unlobed, the upper surface glabrous at maturity, the undersurface usually minutely hairy. Flower stalks and hypanthia glabrous or the stalks sparsely hairy. Calyces withered but more or less persistent at fruiting, the sepals 3–6 mm long, narrowly triangular to triangular, the outer surface glabrous or very sparsely hairy, the inner surface densely woolly. Petals 2–3 cm long, the body oblong-ovate to broadly ovate, short-tapered to a short or noticeable, stalklike base, pink to nearly white at flowering (usually fading to white with age). Anthers yellow. Styles usually 4, the stigmas capitate. Fruits 0.8–1.0 cm long, red to yellow, shiny, not glaucous. April–May.

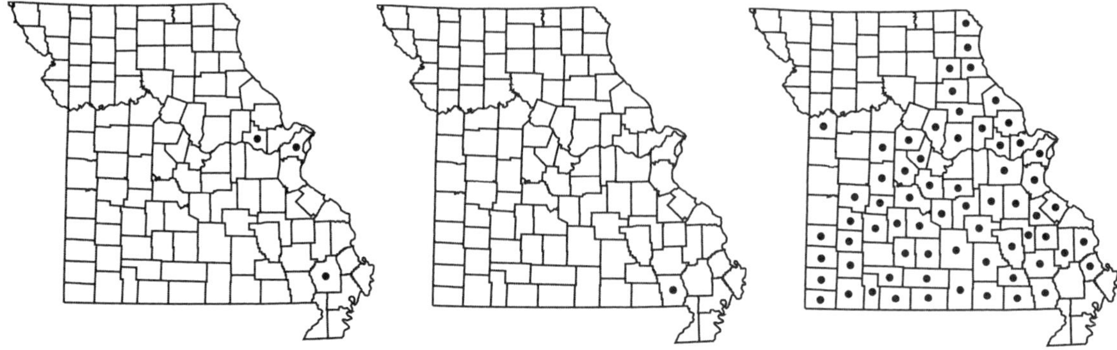

Introduced, uncommon, known thus far only from the city of St. Louis (native of Asia; introduced sporadically in the United States). Railroads.

Malus floribunda is a commonly cultivated crab apple with showy reddish to dark purple branches, large flowers, and brightly colored fruits. The branches often are spreading and weeping cultivars exist. The species was first collected in Missouri in 1977 by Viktor Mühlenbach during his inventories of the St. Louis railyards, but was not included in his publications on this project.

6. Malus ioensis (Alph. Wood) Britton (prairie crab apple, wild crab)
M. ioensis var. *bushii* Rehder
M. ioensis var. *palmeri* Rehder
Pyrus ioensis (Alph. Wood) L.H. Bailey
Pl. 533 c, d; Map 2460

Plants shrubs or small trees to 8(–10) m tall, sometimes colonial from root suckers. Branchlets mostly thorn-tipped. Twigs short-hairy. Leaf blades folded lengthwise during development, 4–10 cm long, 2–3 times as long as wide, oblong-ovate to broadly elliptic, angled to broadly angled at the base, angled or tapered to a sharply pointed tip, the margins somewhat irregularly, sharply toothed, those of at least the larger leaves usually shallowly lobed, densely woolly on both surfaces when young, the upper surface sometimes becoming nearly glabrous at maturity. Flower stalks and hypanthia finely woolly. Calyces persistent at fruiting, the sepals 6–9 mm long, narrowly triangular, the outer surface glabrous or very sparsely hairy, the inner surface densely woolly. Petals 1.2–2.0 cm long, the body obovate to ovate, short-tapered to a short stalklike base, pink or pinkish-tinged to nearly white at flowering, pinkish-tinged in bud, often fading to white with age. Anthers pink to orangish red. Styles 5, the stigmas narrowly club-shaped. Fruits 2–3 cm long, green to yellowish green, often somewhat glaucous. $2n=34$. April–May.

Scattered nearly throughout the state (South Dakota to Texas east to West Virginia and Mississippi). Upland prairies, sand prairies, glades, savannas, edges of mesic upland forests, banks of streams and rivers, and edges of bottomland forests; also old fields, fencerows, old mines, railroads, roadsides, and open, disturbed areas.

Malus ioensis is closely related to and perhaps not distinct from *M. angustifolia* and *M. coronaria*. For discussion of problems with the taxonomy of these species, see the discussion of *M. coronaria*.

Prairie crab is a conspicuous and attractive native species. Steyermark (1963) noted that the foliage turns a dull rose color blended with yellow and dull green at the end of the growing season. He also noted that plants have deep-seated rootstocks, making them difficult to transplant, but that they are propagated relatively easily from seeds. A horticultural variant that is slightly later-blooming and has doubled flowers (f. *plena* Rehder) is sold under the name Bechtel's crab. The fruits are sometimes used in jams and jellies, and early settlers fermented the fruits into a cider.

The Soulard crab, *M.* ×*soulardii* (L.H. Bailey) Britton (Pl. 533 f) arose as a spontaneous hybrid between the cultivated apple (*M. domestica*) and the native wild crab apple, *M. ioensis*. According to L. H. Bailey (1911), it was first discovered in Illinois near the city of St. Louis, Missouri, but several discordant accounts existed as to the credit for its discovery. The hybrid has leaves similar to those of *M. ioensis*, but has a larger fruit. The fruit is too tart to be eaten raw, but can be used in baking and preserves.

7. Malus sieboldii (Regel) Rehder (toringo crab)
M. toringo (K. Koch) Carrière
Pyrus sieboldii Regel
Map 2461

Plants small trees to 8(–10) m tall, usually not

colonial. Branchlets mostly thorn-tipped. Twigs short-hairy. Leaf blades folded lengthwise during development, 2–6 cm long, 2–4 times as long as wide, ovate to elliptic or oblong-elliptic, angled at the base, angled or tapered to a sharply pointed tip, the margins somewhat irregularly, sharply toothed, those of at least the larger leaves usually shallowly lobed, moderately to densely short-hairy on both surfaces when young, the upper surface sometimes becoming nearly glabrous at maturity. Flower stalks and hypanthia densely short-hairy to woolly. Calyces not persistent at fruiting, the sepals 3–5 mm long, narrowly triangular, the outer surface densely short-hairy to woolly, the inner surface densely woolly. Petals 1.0–1.6 cm long, the body oblong-obovate, tapered to a short stalklike base, white or pinkish-tinged at flowering, pinkish-tinged in bud. Anthers pink to orangish red. Styles 3 or 4, the stigmas club-shaped. Fruits 0.6–0.8 cm long, red to brownish yellow, not glaucous. $2n$=34, 58, 85. April–May.

Introduced, uncommon, known thus far only from St. Louis and Warren Counties (native of Asia; introduced sporadically in the northeastern U.S.). Ditches, fencerows, old homesites, and disturbed areas.

The toringo crab is commonly cultivated as an ornamental tree and has been used as grafting stock in apple breeding.

17. **Neviusia** A. Gray (snow wreath)

2 species, southeastern U.S., California.

1. Neviusia alabamensis A. Gray (Alabama snow wreath)

Pl. 534 a, b; Map 2462

Plants shrubs, 1–2 m tall, forming dense colonies from root suckers. Stems ascending or arching, lacking spines and thorns. Bark light brown to purplish brown, smooth or finely ridged. Leaves alternate, mostly relatively short-petiolate, the petioles pubescent with appressed hairs and often with scattered small reddish purple peglike glands on the upper surface. Stipules 2–5 mm long, herbaceous, linear to narrowly lanceolate, pubescent with appressed hairs, the margins often with a few reddish purple glandular teeth, all but a small persistent scalelike basal portion shed soon after the leaves develop. Leaf blades 1–6(–9) cm long, simple, unlobed or less commonly with a few very shallow irregularly rounded lobes, ovate, mostly long-tapered at the tip, rounded or less commonly shallowly cordate at the base, the margins finely and sharply toothed, the upper surface pubescent with sparse to moderate appressed hairs, the undersurface sparsely to moderately pubescent with appressed hairs, especially along the veins, often also with sparse to moderate, small, reddish purple, peglike or hemispherical glands. Inflorescences small clusters at the tips of normal branches, sometimes appearing umbellate, produced as the leaves uncurl, the stalks appressed-hairy, the flowers with 1(2) small linear to linear-elliptic bract above the middle of the stalk, this usually shed by flowering. Flowers perigynous, the hypanthium shallowly cup-shaped, appressed-hairy on the outer surface. Sepals 5, 5–12 mm long, reflexed at flowering, narrowly elliptic to obovate, the margins sharply toothed above the middle, the surfaces usually sparsely appressed-hairy, persistent (and spreading) at fruiting. Petals absent. Stamens numerous, the filaments relatively long, showy, white, the anthers yellow. Pistils 2–5. Ovary superior, densely woolly, with 1 locule and 1 ovule. Styles 1 per pistil, with a short, linear, stigmatic area along the inner side toward the tip. Fruits seldom produced, often only 1 per flower, drupelike achenes (the outer layer slightly fleshy), 4–6 mm long, asymmetrically ovoid, appearing attached laterally toward the base, tapered to a short beak at the tip, usually sparsely appressed-hairy, olive green to brownish green at maturity, 1-seeded. $2n$=14, 18. April–May.

Known only from a single historical report from Butler County (Georgia, Alabama, Mississippi, Tennessee, Arkansas, Missouri). Habitat: "...growing in a southeastern exposure on a somewhat sandy-loam slope of a small hill near the bank of a little creek..." (Uphof, 1922).

Although *N. alabamensis* presently is not listed under the federal Endangered Species Act, it is rare throughout its range and under a number of threats, including grazing, erosion, and habitat destruction for quarrying. The Missouri station was the first to be reported outside of Alabama, where the species was initially discovered. A single individual was located by Uphof (1921, 1922) during his research on the vegetational ecology of southeastern Missouri. Unfortunately, Uphof never deposited a specimen in an herbarium, but his descriptions and drawing

2464. Potentilla argentea
2465. Potentilla canadensis
2466. Potentilla norvegica

leave no doubt as to the correctness of his determination that the plant was *Neviusia*. Steyermark and others have searched repeatedly in the vicinity of the original site, but the species has yet to be relocated in Missouri. Alabama snow wreath is slowly gaining popularity as an ornamental shrub, but isolated clones of this self-sterile species develop no fruits, so horticultural propagation has been mostly from root divisions and softwood cuttings.

18. **Physocarpus** (Cambess.) Raf.

Six species, U.S. (including Alaska), Canada, eastern Asia.

1. Physocarpus opulifolius (L.) Maxim. **var. intermedius** (Rydb.) B.L. Rob. (ninebark) Pl. 534 c–e; Map 2463

Plants shrubs, 0.4–2.0(–3.0) m tall, sometimes suckering from the roots. Stems mostly ascending, lacking spines and thorns. Bark yellowish brown to brown, peeling in narrow strips in several thin layers. Leaves alternate, short- to long-petiolate, the petioles glabrous. Stipules 2–6 mm long, scalelike, linear to narrowly oblong-lanceolate, pubescent with stellate hairs, mostly shed soon after the leaves develop. Leaf blades 1–6(–9) cm long, simple, unlobed or more commonly (1)3(5)-lobed, ovate to kidney-shaped, rounded or narrowed to a point at the tip, narrowed, rounded, or shallowly cordate at the base, the margins doubly (finely and more coarsely) toothed, the upper surface glabrous, the undersurface glabrous or sparsely stellate-hairy, especially along the veins. Inflorescences clusters or more commonly short racemes at the tips of branches, often appearing umbellate, produced after the leaves mature, the flower stalks stellate-hairy, the flowers with a small stipulelike bract at the stalk base. Flowers perigynous, the hypanthium shallowly cup-shaped, with a prominent yellow nectar ring, stellate-hairy on the outer surface. Sepals 5, 2.0–3.5 mm long, reflexed to loosely ascending at flowering, triangular, the surfaces stellate-hairy, persistent (and loosely ascending) at fruiting. Petals 5, 4–8 mm long, broadly obovate to broadly elliptic, white, occasionally faintly pinkish-tinged. Stamens 20 to numerous, the anthers pink. Pistils (1–)3–5, sometimes fused toward the base. Ovary superior, usually densely stellate-hairy, with 1 locule and 2–4 ovules. Styles 1 per pistil, the basal portion persistent, the stigma more or less capitate. Fruits follicles, ascending, 4–10 mm long, elliptic-ovate in outline, narrowed abruptly to a slender spreading beak at the tip, sparsely to moderately stellate-hairy, tan to brown and leathery to papery at maturity, dehiscing along the inner suture and also partially along the outer (dorsal) suture, 1–4-seeded. Seeds 1.8–2.5 mm long, asymmetrically ovoid, the surface hard, smooth, yellow, shiny. $2n$=18. May–June.

Scattered nearly throughout the state, but absent from much of the western portion of the Glaciated Plains Division, as well as most of the Unglaciated Plains and Mississippi Lowlands (New York to North Dakota south to Colorado and Arkansas; Canada). Banks of streams, rivers, and temporary water courses, bases and lower ledges of bluffs, and margins of fens; also roadsides.

This species has become popular in native plant horticulture for its attractive flowers, foliage, and shredding bark. The var. *opulifolius* is widespread in the eastern United States (to the north and east of Missouri) and Canada, and differs mainly in its glabrous fruits.

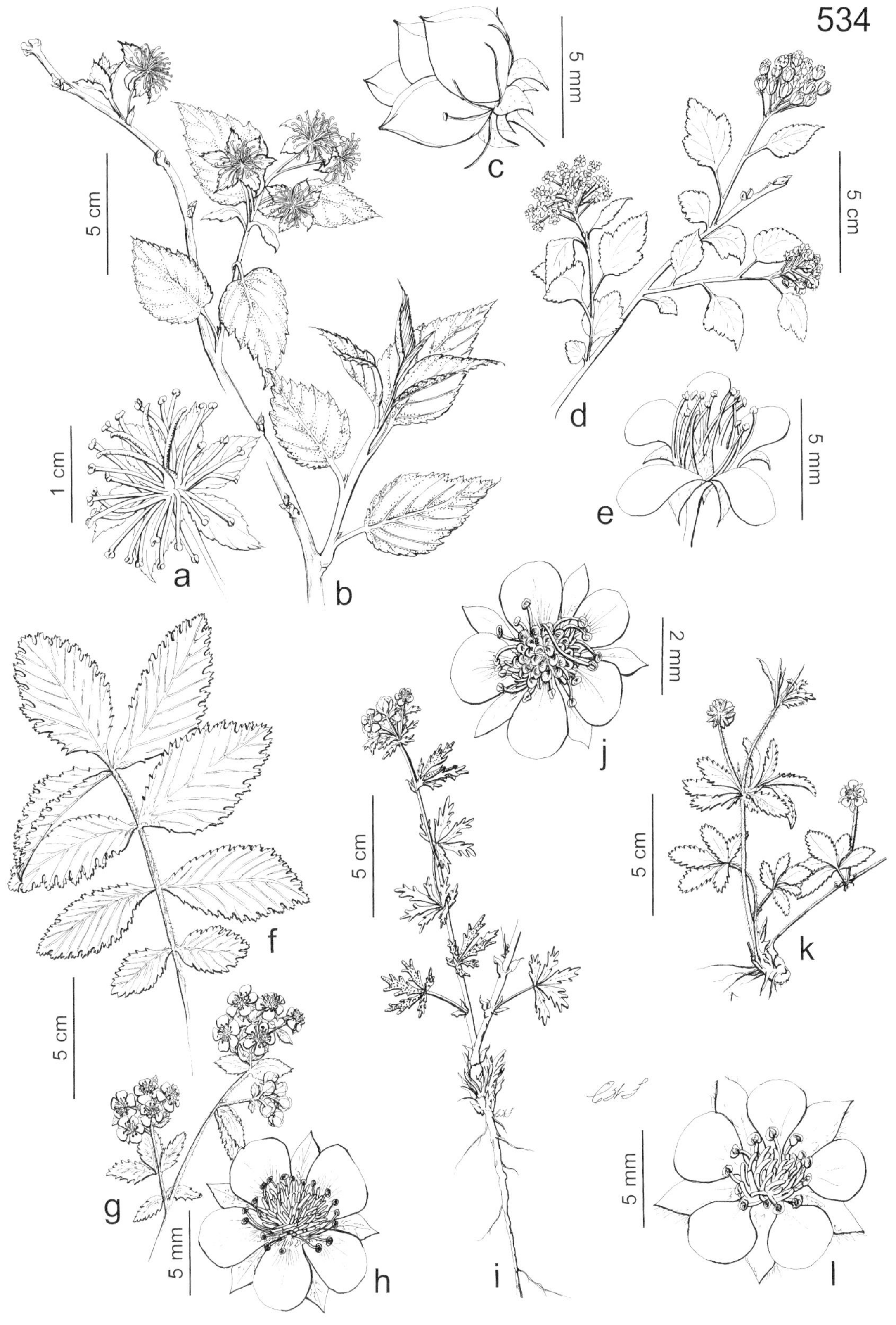

Plate 534. Rosaceae. *Neviusia alabamensis*, **a)** flower, **b)** flowering branch. *Physocarpus opulifolius*, **c)** fruits, **d)** flower, **e)** flowering and fruiting branch. *Drymocallis arguta*, **f)** basal leaf, **g)** flowering branch, **h)** flower. *Potentilla argentea*, **i)** habit, **j)** flower. *Potentilla canadensis*, **k)** habit, **l)** flower.

19. **Potentilla** L. (cinquefoil)

Plants annual, biennial, or perennial herbs (shrubs elsewhere) with taproots, woody rootstocks, or rhizomes. Stems variously erect or ascending to arched or spreading, in a few species stoloniferous and rooting at the tips and some of the nodes, unarmed, glabrous or hairy. Leaves alternate and sometimes also basal, sometimes appearing in clusters on short shoots, long-petiolate to sessile, the petioles glabrous to densely hairy. Stipules of various size, herbaceous, fused to the petiole but free in the terminal half, sometimes lobed or toothed, glabrous or hairy, green, in perennials those of the basal leaves usually persisting as withered scalelike remains. Leaf blades pinnately or palmately compound with 5–11 leaflets or trifoliate, the leaflets variously shaped, sessile or short-stalked, the margins usually toothed or lobed, the surfaces variously glabrous or hairy. Inflorescences terminal or axillary, clusters, panicles, or solitary flowers, the branch points often with small bracts, these usually shed early. Flowers short- to long-stalked, perigynous, the hypanthium saucer-shaped to cup-shaped or rarely disc-shaped, with a nectar disc, usually hairy, each flower with 5 bractlets alternating with the sepals (the calyx thus appearing 10-parted), these occasionally becoming enlarged at fruiting. Sepals 5, similar in size and shape, ascending to spreading, variously shaped, usually hairy, occasionally becoming enlarged at fruiting. Petals 5, more or less obovate, yellow or less commonly pale yellow, cream-colored, or white (red or purple elsewhere). Stamens 5 to numerous, the anthers yellow. Pistils 10 to numerous, densely covering the surface of the obconic or columnar receptacle. Ovary superior, glabrous, with 1 locule, with 1 ovule. Style 1 per pistil, attached terminally or nearly so on the ovary, usually jointed at the base and shed as the fruits mature, the stigma slender or somewhat club-shaped, sometimes somewhat curved. Fruits achenes, densely aggregated on the surface of the hemispheric to somewhat elongate receptacle, asymmetrically ovate in outline, glabrous or nearly so (hairy elsewhere), with 1 seed. About 400 species, North America, South America, Asia south to Australia.

Potentilla is a taxonomically difficult genus, with relatively widespread polyploidy, hybridization, and apomixis. In recent years, phylogenetic analysis of the fragarioid Rosaceae has tended to support a relatively restricted circumscription of *Potentilla*, with various species groups segregated generically, those with representatives in North America as *Comarum* L., *Dasiphora* Raf., *Drymocallis*, and *Sibbaldiopsis* Rydb. (Kurtto and Eriksson, 2003; Ertter 2007). For the Missouri flora, this affects only *Drymocallis arguta*, which is treated under that genus.

Potentilla contains a number of species that are cultivated as ornamentals. Some species also have minor, historical, medicinal uses.

1. Inflorescences of solitary, axillary flowers (these long-stalked); stem leaves often produced in clusters on short axillary branches
    2. Lowest flower produced from the axil of the lowest stem leaf; leaf blades with the central leaflet usually 2 or more times as long as wide . . . . . 2. P. CANADENSIS
    2. Lowest flower produced from the axil of the second-lowest stem leaf; leaf blades with the central leaflet usually 2 times as long as wide or wider . . . . 6. P. SIMPLEX
1. Inflorescences of terminal and sometimes also axillary clusters of flowers, these sometimes grouped into panicles (the individual flowers short- to long-stalked); stem leaves alternate, not clustered at the nodes
    3. Leaf blades with the undersurface strongly whitened with a dense mat of white to silvery, felty to woolly hairs . . . . 1. P. ARGENTEA
    3. Leaf blades with the undersurface glabrous or hairy, if hairy then the hairs relatively straight, not felty or woolly, the undersurface at most slightly whitened

4. Stem leaf blades all or nearly all trifoliate
   5. Petals 3–5 mm long, 3/4 as long to about as long as the sepals; stamens (15–) 20; fruits with the surface usually coarsely wrinkled or longitudinally ribbed . . . . 3. P. NORVEGICA
   5. Petals 1.5–3.0 mm long, about 1/2 as long as the sepals; stamens (5–)10(–15); fruits with the surface not noticeably wrinkled or ribbed . . . . 5. P. RIVALIS
4. Leaf blades palmately or pinnately compound with 5 or more leaflets, the uppermost leaves sometimes merely trifoliate
   6. Leaf blades pinnately compound; fruits with a corky, winglike appendage along the inner side . . . . 7. P. SUPINA
   6. Leaf blades palmately compound; fruits lacking a corky, winglike appendage along the inner side
      7. Petals 4–13 mm long, as long as or longer than the sepals; stamens 25–30; fruits 1.2–1.8 mm long, the surface usually coarsely wrinkled . . . . 4. P. RECTA
      7. Petals 1.5–3.0, about 1/2 as long as the sepals; stamens (5–)10(–15); fruits 0.7–0.9 mm long, the surface smooth or occasionally slightly wrinkled . . . . 5. P. RIVALIS

1. Potentilla argentea L. (silvery cinquefoil)
Pl. 534 i, j; Map 2464

Plants perennial herbs, with a compact, slender to stout rootstock and usually with a woody taproot. Stems 10–50 cm long, loosely ascending to spreading, not rooting at the tip or nodes, densely pubescent with woolly hairs or moderately pubescent with straight, appressed, silky hairs, sometimes becoming glabrous in patches with age. Leaves alternate (basal leaves usually withered at flowering), the petioles 0.5–7.0 cm long or the uppermost leaves often sessile, densely woolly. Stipules 4–9 mm long, lanceolate, entire, with pubescence as in the leaf blades. Leaf blades 1–3(–5) cm long, palmately compound with usually 5 leaflets (the smaller leaves occasionally trifoliate), these 0.5–3.0(–5.0) cm long (the basal or outermost leaflets the smallest), sessile, narrowly oblanceolate to narrowly obovate, divided above the midpoint into 2–4(–6), ascending, deep lobes and/or coarse teeth, these linear to narrowly oblong-triangular, bluntly to sharply pointed at the tips, the margins otherwise slightly curled under, the upper surface green to dark green, glabrous or sparsely to moderately pubescent with long, silky hairs along the veins, the undersurface strongly whitened with a dense mat of white to silvery, felty to woolly hairs. Inflorescences terminal and often also axillary, of dense clusters, occasionally reduced to a solitary flower, the terminal ones usually grouped into small, open, leafy panicles. Bractlets 1.5–3.0 mm long, elliptic-lanceolate to narrowly ovate, entire, densely silky-hairy to woolly. Hypanthia 2.5–5.0 mm in diameter, cup-shaped. Sepals 2–4 mm long, broadly elliptic to narrowly ovate, sharply pointed at the tips, densely silky-hairy to woolly. Petals 2.0–4.5 mm long, obovate, rounded to shallowly and broadly notched at the tip, yellow. Stamens 20. Ovaries with the styles attached near the tips. Fruits 0.5–0.9 mm long, the surface faintly to relatively strongly ribbed, brown, unwinged. $2n=14$. May–August.

Introduced, uncommon, known thus far only from Lewis County and the city of St. Louis (native of Europe, Asia; introduced widely in the U.S. [except in some southern states]; Canada). Lawns and railroads.

This species was first reported from Missouri by Mühlenbach (1979) during his floristic inventories of the St. Louis railyards.

2. Potentilla canadensis L. (Canada cinquefoil, five-finger)
P. canadensis var. *villosissima* Fernald
Pl. 534 k, l; Map 2465

Plants perennial herbs, with very short rhizomes. Stems 5–120 cm long, erect or ascending when young (sometimes flowering at this stage), but soon becoming spreading and stoloniferous, often rooting at the tip and occasionally also at some of the moderately spaced nodes (but not producing tubers at the rooted stem tips), sparsely to moderately pubescent with appressed hairs or glabrous. Leaves often produced in small clusters on short, axillary branches (also usually in a prominent basal rosette), the petioles of basal leaves 1–10 cm long, those of stem leaves absent or to 5 cm, when present then glabrous or finely appressed- or spreading-hairy. Stipules of basal leaves 2–12

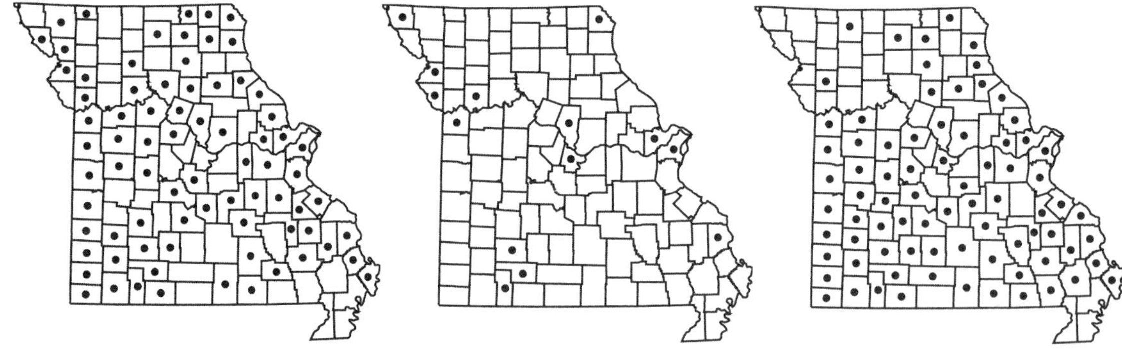
2467. Potentilla recta 2468. Potentilla rivalis 2469. Potentilla simplex

mm long, narrowly oblong, entire or with a pair of lateral lobes, those of the stem leaves variously 5–15 mm long, narrowly elliptic to narrowly oblong, with 1 or 2 pairs of slender, ascending, lateral lobes, with pubescence as in the leaf blades. Leaf blades 1–4 cm long, palmately compound with usually 5 leaflets (the smaller leaves occasionally trifoliate), these 0.5–4.0 cm long (the basal or outermost leaflets the smallest), sessile, oblanceolate to elliptic or obovate, the central leaflet usually 2 times as long as wide or wider, bluntly pointed at the tips, the margins moderately toothed usually above the midpoint, the upper surface green, glabrous or sparsely to moderately hairy, the undersurface lighter green, sparsely to moderately hairy (the hairs sometimes silky and appressed, the surface then appearing somewhat whitened). Inflorescences axillary, of solitary, long-stalked flowers, the lowermost flower produced from the axil of the lowermost stem leaf. Bractlets 2–6 mm long, linear to narrowly lanceolate, sometimes pinnately few-lobed or coarsely few-toothed, moderately hairy. Hypanthia 2.5–5.0 mm in diameter, cup-shaped. Sepals 2–5 mm long, lanceolate, sharply pointed and sometimes with a minute extension of the midvein at the tips. Petals 5–10 mm long, obovate, rounded to very shallowly and broadly notched at the tip, yellow or occasionally light yellow. Stamens 20. Ovaries with the styles attached near the tips. Fruits 1.2–1.4 mm long, the surface smooth, brown, unwinged. $2n=28$. April–May.

Scattered in the Ozark and Ozark Border Divisions (northeastern U.S. west to Michigan and Tennessee, disjunct in Missouri; Canada). Glades, savannas, dry upland forests, tops of bluffs, and banks of springs and spring branches; also pastures and open, dry, disturbed areas; often in acidic soils.

This species is closely related to and sometimes difficult to distinguish from *P. simplex*. In addition to the key characters, it tends to be a more slender plant with more finely toothed leaflets.

Steyermark (1963) noted that *P. canadensis* makes a good ground cover, forming mats in rock gardens and in open, dry sites.

3. Potentilla norvegica L. (rough cinquefoil, Norwegian cinquefoil)

P. norvegica var. *hirsuta* (Michx.) Lehm.

P. norvegica var. *labradorica* (Lehm.) Fernald

Pl. 535 h–j; Map 2466

Plants annual or biennial, rarely short-lived perennial herbs. Stems 5–60 cm long, erect or ascending, moderately to densely pubescent with straight to somewhat woolly hairs, the hairs often with minute, pustular bases, rarely glabrous or nearly so. Leaves with the petioles 1–6 cm long, sometimes nearly sessile toward the stem tips, these hairy. Stipules 5–20 mm long, ovate, sharply pointed at the tip, the margins entire or toothed and often hairy, the surfaces hairy. Leaf blades 2–6(–8) cm long, palmately or pinnately trifoliate, a few of the leaves rarely with 5 leaflets, these 1–6(–8) cm long (the central leaflet the largest, sessile or short-stalked), broadly oblanceolate or elliptic to obovate or broadly obovate, rounded to bluntly pointed at the tips, the margins coarsely toothed, the upper surface green, glabrous or sparsely to moderately hairy, especially along the veins, the undersurface lighter green, moderately hairy, especially along the veins. Inflorescences terminal, relatively small, flat-topped to shallowly dome-shaped clusters of several flowers. Bractlets 4–8 mm long, oblanceolate to narrowly elliptic, moderately hairy. Hypanthia 4–5 mm in diameter, shallowly cup-shaped to nearly disc-shaped. Sepals 3.5–6.0 mm long, broadly ovate, sharply pointed at the tips. Petals 3–5 mm long, obovate, shallowly and broadly notched at the tip, yellow. Stamens (15–)20. Ovaries with the styles attached near the tips. Fruits 0.8–1.2 mm long, the surface

Plate 535. Rosaceae. *Potentilla supina*, **a)** fertile stem, **b)** flower, **c)** inflorescence. *Potentilla simplex*, **d)** seed, **e)** flower, **f)** habit. *Potentilla rivalis*, **g)** fertile stem. *Potentilla norvegica*, **h)** flower, **i)** seed, **j)** fertile stem. *Potentilla recta*, **k)** flower, **l)** seed, **m)** habit.

usually coarsely wrinkled or longitudinally ribbed, brown to greenish brown, unwinged. $2n$=42, 56, 63, 70. May–October.

Scattered, mostly in the western and northern halves of the state (U.S. [including Alaska]; Canada, Mexico, Greenland, Europe, Asia). Banks of streams and rivers, margins of ponds, lakes, and oxbows, bases of bluffs, bottomland prairies, moist swales of upland prairies, sand prairies, and bottomland forests; also pastures, fallow fields, crop fields, barnyards, railroad, roadsides, and open, disturbed areas.

4. Potentilla recta L. (rough-fruited cinquefoil, sulphur cinquefoil)

Pl. 535 k–m; Map 2467

Plants perennial herbs, with rhizomes. Stems 15–80 cm long, erect or ascending, moderately to densely pubescent (more densely so toward the tip) with longer and shorter, appressed and spreading hairs, the shorter hairs usually gland-tipped, the longer hairs often with minute, pustular bases. Leaves with the petioles 3–8 cm long, these densely hairy with glandular and nonglandular hairs. Stipules 8–30(–40) mm long, lanceolate to ovate, with several deep, slender lobes, these sharply pointed at the tip, hairy. Leaf blades 3–10 cm long, palmately compound with 5 or 7 leaflets, these 1.5–10 cm long, sessile, narrowly oblanceolate to oblanceolate (rarely wider), rounded to bluntly pointed at the tips, the margins coarsely toothed, the upper surface green, glabrous or sparsely to moderately hairy, the undersurface lighter green, moderately hairy, especially along the veins. Inflorescences terminal, relatively open, flat-topped to shallowly dome-shaped panicles of several flowers. Bractlets 5–12 mm long, narrowly lanceolate to lanceolate, moderately hairy. Hypanthia 5–9 mm in diameter, cup-shaped. Sepals 4–10(–12) mm long, oblanceolate to elliptic, sharply pointed at the tips. Petals 4–13 mm long, more or less heart-shaped to broadly obovate with a shallowly and broadly notched tip, pale yellow to lemon yellow. Stamens 25–30. Ovaries with the styles attached near the tips. Fruits 1.2–1.8 mm long, the surface usually coarsely wrinkled, brown, unwinged. $2n$=28, 42. May–August.

Introduced, scattered nearly throughout the state (native of Europe; introduced widely in the U.S., Canada). Upland prairies, loess hill prairies, bases and tops of bluffs, glades, and banks of streams; also pastures, fallow fields, old fields, railroads, roadsides, and open, disturbed areas.

The New World plants have not usually been treated at the subspecies level. Within the native range in Europe, a complex classification of up to six subspecies is often accepted (Kurtto et al., 2004). The midwestern plants tend to key closest to ssp. *recta*. *Potentilla recta* is considered a problem weed throughout most of its introduced range. In some states it is considered an invasive exotic of native grasslands.

A specimen from Shell Knob (Stone County) in the Missouri Botanical Garden herbarium that was collected by B. F. Bush in 1933 was annotated as representing a putative hybrid by Rudolf Kamelin of St. Petersburg University. Although Dr. Kamelin did not determine the other parent, it is possible that this represents a cross between *P. recta* and *P. norvegica*. Such hybrids are formed sporadically, but generally do not proliferate.

5. Potentilla rivalis Nutt. (brook cinquefoil)

P. rivalis var. *millegrana* (Engelm. ex Lehm.) S. Watson
P. rivalis var. *pentandra* (Engelm.) S. Watson
P. millegrana Engelm. ex Lehm.
P. pentandra Engelm.

Pl. 535 g; Map 2468

Plants annual or biennial. Stems 5–60 cm long, ascending to loosely ascending from a spreading base, moderately to densely pubescent with longer, straight, spreading to loosely appressed hairs and shorter, crinkly hairs. Leaves with the petioles mostly 1–6 cm long (those toward the stem tips usually sessile), when present these hairy. Stipules 5–15 mm long, narrowly ovate to ovate, sharply pointed at the tip, the margins entire or toothed and often hairy, the surfaces hairy. Leaf blades 2–6 cm long, palmately compound or the largest leaves sometimes with a short rachis, with 3 or 5 leaflets, these 0.5–6.0 cm long (the central or terminal leaflet the largest), oblanceolate to obovate or oval, rounded to bluntly pointed at the tips, the margins coarsely toothed or scalloped, the upper surface green, glabrous or sparsely to moderately hairy, the undersurface lighter green, moderately hairy. Inflorescences terminal panicles, sometimes reduced to racemes, mostly with numerous flowers. Bractlets 3–6 mm long, narrowly elliptic, moderately hairy. Hypanthia 4–5 mm in diameter, shallowly cup-shaped. Sepals 3–7 mm long, broadly ovate, sharply pointed at the tips. Petals 1.5–3.0 mm long, oblanceolate to obovate, yellow to pale yellow. Stamens (5–)10(–15). Ovaries with the styles attached near the tips. Fruits 0.7–0.9 mm long, the surface smooth or occasionally slightly wrinkled, brown, unwinged. $2n$=14, 70. May–October.

Scattered in counties adjacent to the Mississippi, Missouri, and Des Moines Rivers (western U.S. east to Wisconsin and Arkansas; sporadically

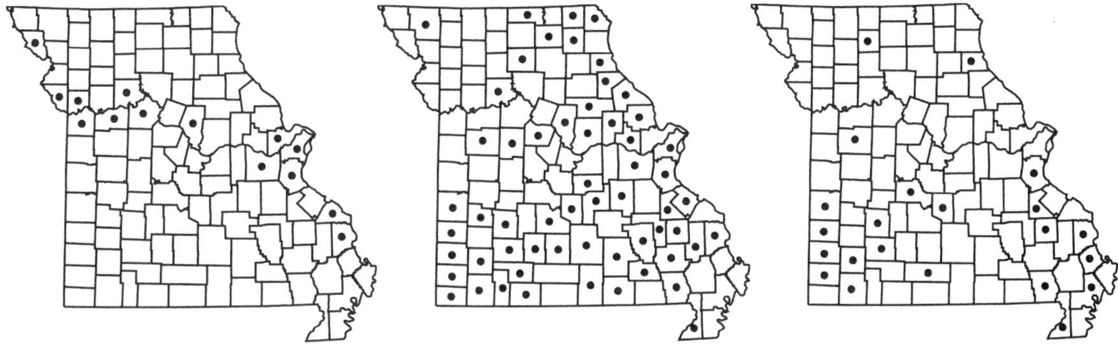

2470. Potentilla supina 2471. Prunus americana 2472. Prunus angustifolia

east to Maine and Virginia; Canada, Mexico). Banks of streams and rivers, openings of bottomland forests and swamps, and rarely ledges of bluffs; also pastures and open, disturbed areas.

Opinions have varied on whether to segregate three varieties or species from the *P. rivalis* complex, based on whether some of the leaves are palmately rather than pinnately compound and whether all of the leaves have three leaflets or the larger ones have five. They seem best-accounted for informally.

6. Potentilla simplex Michx. (common cinquefoil, old-field cinquefoil, five-finger)
P. simplex var. *argyrisma* Fernald
P. simplex var. *calvescens* Fernald
Pl. 535 d–f; Map 2469

Plants perennial herbs, with short rhizomes. Stems 20–150 cm long, erect or ascending when young (sometimes flowering at this stage), but soon becoming arched to spreading and stoloniferous, often rooting at the tip (then producing small tubers that are infrequently collected, these giving rise to new plants) and occasionally also rooting at some of the widely spaced nodes, sparsely to moderately pubescent with spreading to appressed hairs. Leaves sometimes produced in small clusters on short, axillary branches (also usually in a prominent basal rosette), the petioles of basal leaves 1–15 cm long, those of stem leaves absent or to 1.5 cm, when present usually hairy. Stipules of basal leaves 10–18 mm long, narrowly lanceolate, entire or with a pair of lateral lobes, those of the stem leaves variously 5–30 mm long, lanceolate to broadly elliptic, pinnately several-lobed, glabrous or hairy, the surfaces hairy. Leaf blades 3–8 cm long, palmately compound with usually 5 leaflets (the smaller leaves occasionally trifoliate), these 2–8 cm long (the basal or outermost leaflets the smallest), sessile, oblanceolate to elliptic or narrowly obovate, the central leaflet usually 2 or more times as long as wide, bluntly pointed at the tips, the margins coarsely toothed, the upper surface green, glabrous or sparsely to moderately hairy, the undersurface lighter green, sparsely to moderately hairy (the hairs sometimes silky and appressed, the surface then appearing somewhat whitened). Inflorescences axillary, of solitary, long-stalked flowers, the lowermost flower produced from the axil of the second-lowest stem leaf. Bractlets 2–5 mm long, linear to narrowly lanceolate, sometimes pinnately few-lobed or coarsely few-toothed, moderately hairy. Hypanthia 3–5 mm in diameter, cup-shaped. Sepals 4–6 mm long, lanceolate to elliptic-lanceolate, sharply pointed and sometimes with a minute extension of the midvein at the tips. Petals 4–7 mm long, obovate, rounded to very shallowly and broadly notched at the tip, yellow. Stamens 20. Ovaries with the styles attached near the tips. Fruits 0.9–1.2 mm long, the surface smooth or faintly ribbed or wrinkled, yellowish brown, unwinged. April–June.

Scattered nearly throughout the state, but uncommon or absent from the western portion of the Glaciated Plains Division (northern U.S. south to Oregon, New Mexico, Louisiana, and Pennsylvania; Canada, Mexico, Europe, Asia). Upland prairies, sand prairies, openings of mesic to dry upland forests, glades, savannas, tops of bluffs, banks of streams and rivers, and margins of sinkhole ponds; also pastures, old fields, fallow fields, crop fields, fencerows, old mines, old quarries, railroads, roadsides, and open, disturbed areas.

K. R. Robertson (1974) considered the varieties of this species that were accepted by Steyermark (1963) to represent trivial variants within a morphologically variable species.

7. Potentilla supina L. **ssp. paradoxa** (Nutt.) Soják (bushy cinquefoil)
P. supina var. *nicolletii* S. Watson
P. supina var. *paradoxa* (Nutt.) Th. Wolf

P. nicolletii (S. Watson) E. Sheld.
P. paradoxa Nutt.

Pl. 535 a–c; Map 2470

Plants annual or biennial, rarely short-lived perennial herbs. Stems 10–50 cm long, spreading to ascending, moderately to densely pubescent with relatively long, fine, spreading hairs, occasionally glabrous toward the base. Leaves with the petioles 1–6(–10) cm long, hairy. Stipules 5–15 mm long, broadly lanceolate to broadly ovate, bluntly pointed at the tip, the margins entire or occasionally toothed and often hairy, the surfaces hairy, especially along the veins. Leaf blades 3–15 cm long, pinnately compound with (3)5 or 7(9) leaflets, these 1–3 cm long, oblanceolate to elliptic or obovate, rounded to bluntly pointed at the tips, the margins coarsely toothed or scalloped, the upper surface green, sparsely to moderately hairy, the undersurface lighter green, moderately hairy. Inflorescences terminal panicles, sometimes reduced to racemes, mostly with numerous flowers. Bractlets 2–5 mm long, narrowly elliptic, sparsely hairy, more densely so along the margins. Hypanthia 3–4 mm in diameter, saucer-shaped to shallowly cup-shaped. Sepals 3–6 mm long, broadly ovate, sharply pointed at the tips. Petals 2.5–4.0 mm long, obovate, yellow. Stamens 20(–25). Ovaries with the styles attached near the tips. Fruits 0.7–1.3 mm long, the surface smooth or somewhat wrinkled, brown, with a large, tan, corky, winglike appendage along the inner margin. $2n=28$. May–September.

Scattered in counties adjacent to the Mississippi and Missouri Rivers (northern U.S. south to Oregon, New Mexico, Louisiana, Ohio, and Pennsylvania; Canada, Europe, Asia). Banks of rivers and margins of lakes; also railroads and roadsides.

The present treatment follows that of Soják (1987) and Kurtto et al. (2004), in recognizing a broadly circumscribed, circumboreal *P. supina* with four subspecies, the others all restricted to the Old World. Some authors have gone even further, considering New World plants part of the variation in ssp. *supina* (C.-l. Li et al., 2003). Steyermark (1963) also segregated *P. nicolletii* from the complex, but the minor differences in leaflet number and shape said to separate the two taxa are not taxonomically significant and can vary within populations.

20. **Prunus** L. (cherry, plum, peach, apricot)

Plants shrubs or small to medium trees, sometimes colonial from root suckers. Branches sometimes producing short, stout branchlets with thorny tips. Bark dark reddish brown to brown or dark gray, on younger trunks relatively smooth and sometimes somewhat shiny, but with prominent, raised branch scars and light-colored lenticels in transverse lines, on older trunks usually developing more or less rectangular, fine, scaly plates, these sometimes peeling. Winter buds lateral and terminal or pseudoterminal (then produced in groups of 2 or 3 at the twig tips), narrowly ovoid to narrowly conic, usually sharply pointed at the tip, with several overlapping scales. Leaves alternate but sometimes appearing clustered at the tips of short branchlets, rolled or folded lengthwise during development, usually short-petiolate, the petioles sometimes with 1 to several variously shaped glands near the tip. Stipules linear to lanceolate, membranous to papery, the margins toothed or lobed and frequently glandular, shed early. Leaf blades simple, unlobed, variously shaped, the margins entire or more commonly bluntly to sharply toothed, occasionally appearing finely scalloped, the teeth sometimes gland-tipped, the surfaces glabrous or hairy, the upper surface lacking glands, sometimes shiny, the venation with a midvein and pinnate secondary veins, these sometimes faint. Inflorescences terminal and/or axillary, on main branchlets or more commonly on lateral, short branchlets, variously solitary flowers, umbellate clusters, or racemes of few to many, short- to long-stalked flowers, produced before or as the leaves uncurl, the axis and stalks glabrous or hairy, the stalks each with a small bract at the base, this linear to narrowly oblong-elliptic, shed early. Flowers deeply perigynous, sometimes sweetly fragrant or with an unpleasant odor, the hypanthium not fused to the ovary above its base, cup-shaped to bell-shaped or more or less urn-shaped, shed after flowering or less commonly persistent as a small disc, glabrous or hairy. Sepals 5, ascending to reflexed at flowering, triangular to oblong-ovate or oblong, sharply pointed to rounded at the tip, the margins entire or toothed (the teeth often gland-tipped), the inner surface usually woolly, the outer surface glabrous or hairy, shed with the hypanthium after flowering (persistent in *P. serotina*). Petals 5 (except in doubled forms), elliptic to obovate or nearly circular, rounded or short-tapered to a short, stalklike base, white or less commonly pink. Stamens usually 10–20(–30) (usually fewer in doubled flowers), exserted, the anthers yellow or pink to red. Pistil 1 per flower, of 1 carpel. Ovary superior, with 1 locule containing 2 ovules (1 of these usually abortive). Style 1, the stigma capitate to more or less disc-shaped. Fruits drupes, globose to oblong-ovoid or broadly ellipsoid, in some species with a longitudinal groove along 1 side, glabrous or less commonly hairy at maturity, variously colored, the surface usually not dotted (white-dotted in *P. hortulana*), with 1 seed embedded in a hard stone (sometimes called a pit), this mostly thick-walled, indehiscent, globose to ovoid, sometimes somewhat flattened, the surface smooth or with a network of ridges, furrows, and pits. About 200 species, nearly worldwide, most diverse in temperate regions of the northern hemisphere.

The genus *Prunus* contains a number of important fruit crops, including peaches, nectarines, apricots, pluots, plums, and cherries. Secondarily, the fleshy-fruited species are used in baked goods, juices, jams and preserves, and as flavorants. The almond (*P. dulcis* (Mill.) D.A. Webb) is anomalous among the fruit crops in that its fruit has a leathery rather than fleshy middle layer of the fruit wall, which is removed, and the thin-walled stone is marketed as a nut. A large number of species, hybrids, and cultivars are important as flowering ornamentals and the larger species also are planted as shade trees. The wood of various species has a wide variety of uses, from firewood to pipe stems, and has been used in furniture, tool handles, rifle stocks, handcrafts, veneers, paneling, flooring, scientific and musical instruments, and caskets. Some of the species have been used in a variety of ways medicinally.

Some botanists have split *Prunus* into a number of smaller genera, including *Amygdalus* L., *Armeniaca* Scop., *Cerasus* Mill., *Lauro-cerasus* Duhamel, *Padus* Mill., and/or *Persica* Mill. These correspond more or less to groups recognized as subgenera and sections by other botanists. However, there is no agreement on how many units should be recognized (K. R. Robertson, 1974), and preliminary molecular studies (Bortiri et al., 2001, 2002; Lee and Wen, 2001) have cast doubt on whether most of these groups are natural. A taxonomic reappraisal of the genus must await more detailed studies. The present account follows closely the forthcoming treatment of the genus *Prunus* in the Flora of North America series by Joseph Rohrer.

Key emphasizing flowering characters

1. Flowers sessile or on stalks to 3 mm long
    2. Petals white at flowering, often slightly pinkish-tinged in bud . . . . . 3. P. ARMENIACA
    2. Petals pink or strongly pinkish-tinged both in flower and bud . . . . . . 8. P. PERSICA
1. Flowers on noticeable stalks mostly more than 4 mm long (often relatively short in *P. spinosa*)
    3. Inflorescences elongate racemes with 20 or more flowers, the axis longer than the flower stalks
        4. Sepals persistent after flowering, sharply pointed at the tip, the margins lacking glands or sparsely glandular-toothed . . . . 9. P. SEROTINA
        4. Sepals shed after flowering, bluntly pointed at the tip, the margins noticeably glandular-toothed . . . . 11. P. VIRGINIANA
    3. Inflorescences of solitary flowers, umbellate clusters of 2–6 flowers per bud (note that 2 or more buds may be clustered at some nodes), or short racemes of 4–12 flowers (then with the axis shorter than the flower stalks)
        5. Flowers produced when the leaves are half- or fully grown; inflorescences short, dome-shaped racemes of 4–12 flowers, ... 6. P. MAHALEB
        5. Flowers produced before the leaves or before the leaves are half-grown

6. Sepals with the margins noticeably glandular-toothed for most of their length
   7. Inflorescences of 1 or 2 flowers (sometimes 2 or more inflorescences produced on the same short branchlet) . . . . 10. P. SPINOSA
   7. Inflorescences umbellate clusters of (2)3–6 flowers
      8. Sepals glabrous, petals 10–14 mm long . . . . 4. P. CERASUS
      8. Sepals densely hairy on the inner surface toward the base, petals 4–9 mm long . . . . 5. P. HORTULANA
6. Sepals with the margins lacking glands or sparsely glandular-toothed only toward the tip
   9. Sepals 1–2 mm long, ascending to spreading at flowering; petals 3–6 mm long; developing leaves with large, conic, reddish glands on the marginal teeth. . . . . 2. P. ANGUSTIFOLIA
   9. Sepals 1.5–4.0(–5.0) mm long, spreading to reflexed at flowering; petals 6–15 mm long; developing leaves with the marginal teeth glandless or with small, spherical, black glands (species difficult to distinguish)
      10. Hypanthia usually glabrous; twigs usually hairy; petioles usually lacking glands; plants shrubs or small trees, usually suckering and strongly colonial . . . . 1. P. AMERICANA
      10. Hypanthia usually moderately hairy; twigs usually glabrous; petioles usually with 1 or 2(–4) glands near the tip; plants small trees, usually not suckering and not colonial. . . . . 7. P. MEXICANA

Key emphasizing vegetative and fruiting characters

1. Leaf blade with the margin sharply toothed, the teeth relatively straight and ascending to spreading, not associated with glands (except sometimes near the blade base)
   2. Leaf blades with the undersurface glabrous or hairy only along the main veins
      3. Winter buds pseudoterminal, produced in groups of 2 or 3 at the twig tips; infructescences small umbellate clusters of 2–5; fruits 15–30 mm long, subglobose to ellipsoid, the surface red, orange, or yellowish orange, glaucous; stones strongly flattened . . . . 1. P. AMERICANA
      3. Winter buds truly terminal, solitary at the twig tips; infructescences elongate racemes of 18 to numerous fruits; fruits 6–14 mm long, globose, the surface red, purple, dark purple, or black, not glaucous; stones not flattened . . . . 11. P. VIRGINIANA
   2. Leaf blades with the undersurface evenly hairy (species difficult to distinguish)
      4. Petioles usually hairy only on the upper side, occasionally glabrous, rarely hairy on all sides, usually lacking glands, occasionally with 1 or 2 disclike glands near the tip; leaf blades mostly angled to broadly angled at the base; plants shrubs or small trees, usually suckering and strongly colonial . . . . 1. P. AMERICANA
      4. Petioles usually evenly hairy on all sides, rarely only on the upper side, usually with 1 or 2(–4) club-shaped glands near the tip; leaf blades broadly angled to more commonly rounded or shallowly cordate at the base; plants small trees, usually not suckering and not colonial. . . . . 7. P. MEXICANA
1. Leaf blades with the margins bluntly toothed (occasionally only slightly so in *P. armeniaca*) or the teeth incurved, with glands at the tips or along the side of the teeth (except in *P. serotina*), the glands sometimes shed, but then leaving a minute scar at the attachment points

5. Leaf blades broadly elliptic, broadly ovate, oblong-ovate, nearly circular, less than 2 times as long as wide
    6. Infructescences short, dome-shaped racemes of 4–12 fruits . . . . 6. P. MAHALEB
    6. Infructescences of small clusters of 2–4 fruits or the fruits solitary
        7. Fruit stalks 1.5–3.0 mm long
            8. Petioles (12–)20–45 mm long; leaf blades (3–)5–9 cm long, angled or short-tapered to a sharply pointed tip; fruits 25–60 mm long, yellow to orange, sometimes tinged with red, densely and minutely hairy, but often becoming nearly glabrous at maturity, somewhat flattened; plants not thorny . . . . 3. P. ARMENIACA
            8. Petioles 4–6 mm long; leaf blades 1.5–4.0 cm long, angled to a broadly or bluntly pointed tip; fruits 10–15 mm long, bluish black, glabrous, not flattened; plants producing thorn-tipped branchlets . . . 10. P. SPINOSA
        7. Fruit stalks 6–40 mm long
            9. Petioles 16–20 mm long; leaf blades (4.5–)6.0–8.0 cm long, angled or short-tapered to a sharply pointed tip; fruits 13–20 mm long, bright red; plants not thorny . . . . 4. P. CERASUS
            9. Petioles 4–6 mm long; leaf blades 1.5–4.0 cm long, angled to a broadly or bluntly pointed tip; fruits 10–15 mm long, bluish black; plants strongly thorny. . . . . 10. P. SPINOSA
5. Leaf blades lanceolate to narrowly elliptic, oblanceolate, or oblong-obovate, more than 2 times as long as wide
    10. Leaf blades somewhat folded lengthwise at maturity, forming a keel along the midvein on the undersurface; fruit stalks 1–3 mm long; fruits 25–80 mm long, the surface densely short-hairy; plants not thorny . . . . 8. P. PERSICA
    10. Leaf blades flat or slightly concave at maturity, the midvein sometimes raised, but not angled or keeled on the undersurface (sometimes somewhat folded and keeled in *P. angustifolia*); fruit stalks 6–40 mm long; fruits 10–20 mm long, the surface glabrous; plants producing thorn-tipped branchlets or not thorny
        11. Plants not thorny; twigs producing a true, terminal, winter bud; fruits with the stone not flattened
            12. Infructescences small umbellate clusters of 2–4 fruits or the fruits solitary; fruits 13–20 mm, bright red; petioles glandless or occasionally with 1 or 2 glands near the tip; leaf blades broadly elliptic to ovate or obovate, the margins doubly toothed, the teeth relatively straight, ascending, gland-tipped or the glands appearing nearly between the teeth . . . . 4. P. CERASUS
            12. Infructescences elongate racemes with 18 or more fruits; fruits 5–10 mm long, dark purple to nearly black; petioles with 2(–6) glands near the tip; leaf blades narrowly elliptic to elliptic or lanceolate, rarely obovate, the margins simply toothed, the teeth incurved or appressed, glandless . . . . 9. P. SEROTINA
        11. Plants producing thorn-tipped branchlets; twigs producing pseudoterminal winter buds (these usually in a cluster of 2 or 3 at the tip); fruits with the stones somewhat flattened
            13. Leaf blades (5–)7–11(–13) cm long, (2.0–)3.0–5.5 cm wide, long-tapered at the tip, the margins sometimes doubly toothed; fruits 20–30(–40) mm long, the surface red to yellowish orange with conspicuous white dots . . . . 5. P. HORTULANA
            13. Leaf blades 1.5–6.0 cm long, 0.8–2.2 cm wide; angled or short-tapered at the tip, the margins simply toothed; fruits 10–20 mm long, the surface bluish black or red to yellow, but not noticeably dotted

14. Twigs glabrous; leaf blades lanceolate to narrowly elliptic, angled or short-tapered to a sharply pointed tip, usually somewhat folded lengthwise, the midvein keeled on the undersurface; fruits 15–20 mm long, the surface red to yellow, the stone ovoid . . . . 2. P. ANGUSTIFOLIA
14. Twigs minutely hairy, at least when young; leaf blades elliptic to obovate, angled to a broadly or bluntly pointed tip, flat or nearly so, the midvein not keeled on the undersurface; fruits 10–15 mm long, the surface bluish black, the stone subglobose . . . . 10. P. SPINOSA

1. Prunus americana Marshall (wild plum)
P. americana var. *lanata* Sudw.
Pl. 536 c; Map 2471

Plants shrubs, 2–5 m tall, or less commonly small trees to 8 m tall, usually strongly suckering to form dense thickets. Branches moderately thorny. Twigs usually short-hairy, occasionally glabrous, producing pseudoterminal winter buds (these usually in a cluster of 2 or 3 at the tip). Petioles 4–19 mm, usually hairy only on the upper side, occasionally glabrous, rarely hairy on all sides, usually lacking glands, but occasionally with 1 or 2 disclike glands near the tip. Leaf blades 5–11 cm long, 2.0–5.5 cm wide, mostly 2 or more times as long as wide, elliptic to broadly elliptic or obovate, rarely ovate, flat to slightly concave at maturity, not keeled along the midvein, angled to broadly angled or infrequently rounded at the base, abruptly short-tapered to tapered at the sharply pointed tip, rarely, merely angled, the margins coarsely and doubly toothed, the relatively straight teeth sharply pointed, lacking glandular tips, the upper surface glabrous to appressed-hairy, the undersurface glabrous except along the midvein or occasionally glabrous or more evenly short-hairy. Inflorescences produced before or as the leaves develop, umbellate clusters of 2–5 flowers per bud, the flower stalks (4–)8–20 mm long, glabrous or less commonly hairy. Flowers with the hypanthium 2.5–5.0 mm long, conic, usually glabrous but occasionally sparsely to densely hairy. Sepals 2–4(–5) mm long, spreading to reflexed at flowering, lanceolate to ovate, the margins hairy, entire or sparsely glandular-toothed toward the tip, the inner surface densely woolly, the outer surface glabrous to moderately short-hairy. Petals 7–15 mm long, oval to oblong-obovate, white. Fruits 15–30 mm long, subglobose to ellipsoid, shallowly longitudinally grooved on 1 side, the surface red, orange, or yellowish orange, glabrous, somewhat glaucous, the fleshy layer well-developed, the stone ovoid, strongly flattened, the surface veiny or slightly wrinkled. $2n=16$. April–May.

Scattered to common nearly throughout the state, but relatively uncommon in the northwestern portion (nearly throughout the U.S. [except Texas and some southwestern states]; Canada). Banks of streams, rivers, and spring branches, mesic upland forests, margins of sinkhole ponds, bases and tops of bluffs, and edges of glades; also pastures, old fields, fencerows, railroads, and roadsides.

This widespread, native species is sometimes planted as an ornamental or for wildlife. Steyermark (1963) noted that it is one of the earliest-flowering woody species in native woodlands. A number of cultivars exist, but the usually strongly colonial habit allows plantings to spread too aggressively for small gardens.

The key characters serve to separate the majority of specimens in the state, but occasional plants appear intermediate between *P. americana* and other plum species. Shaw and Small (2005), who studied the geography of molecular markers in this group, suggested that periodic hybridization contributes to these problems. In Missouri, there are particular problems with the separation of *P. americana* from the closely related *P. mexicana* (Rohrer et al., 2004). These are reasonably distinct in most other parts of their strongly overlapping ranges. Steyermark (1963) noted that although *P. americana* is usually found as a colonial shrub or small tree, sometimes isolated, single-trunked small trees are encountered that key to this species. Conversely, although *P. mexicana* tends not to produce abundant root suckers, occasionally it forms thickets. Both species can produce individuals with hairy leaves and twigs.

2. Prunus angustifolia Marshall (Chickasaw plum)
P. angustifolia var. *variens* W. Wight & Hedrick
P. angustifolia var. *watsonii* (Sarg.) Waugh
Pl. 536 d, e; Map 2472

Plants shrubs or small trees, 3–5 m tall, often suckering to produce thickets. Branches thorny. Twigs glabrous, producing pseudoterminal winter buds (these usually in a cluster of 2 or 3 at the tip). Petioles 2–14 mm, usually sparsely short-hairy on the upper side, rarely entirely glabrous or hairy all around, usually glandless, less com-

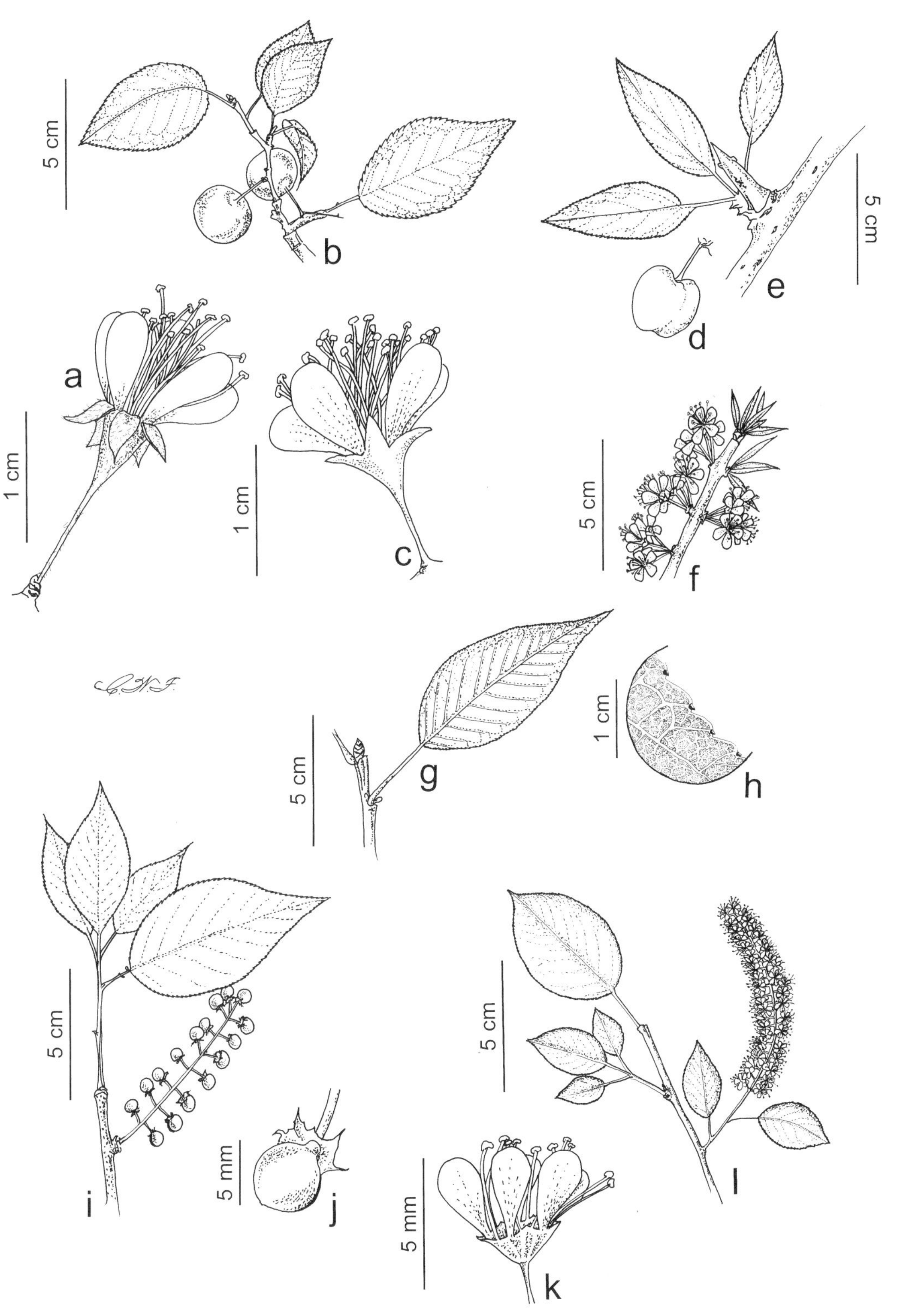

Plate 536. Rosaceae. *Prunus mexicana*, **a)** flower, **b)** fruiting branch. *Prunus americana*, **c)** flower. *Prunus angustifolia*, **d)** fruit, **e)** node with thorn and cluster of leaves. *Prunus munsoniana*, **f)** flowering branch, **g)** twig with leaf and winter buds, **h)** leaf detail. *Prunus serotina*, **i)** fruiting branch, **j)** fruit. *Prunus virginiana*, **k)** flower, **l)** flowering branch.

2473. Prunus armeniaca 2474. Prunus cerasus 2475. Prunus hortulana

monly with 1 or 2 glands near the tip. Leaf blades 1.5–6.0 cm long, 0.8–2.0 cm wide, more than 2 times as long as wide, lanceolate to oblong-lanceolate or narrowly elliptic, often somewhat arched, sometimes somewhat folded longitudinally, then keeled along the midvein, angled to broadly angled or occasionally rounded at the base, angled or short-tapered to a sharply pointed tip, the margins finely and simply toothed, the blunt, incurved teeth gland-tipped, the upper surface glabrous, the undersurface nearly glabrous, with only a few hairs along the midvein. Inflorescences produced before or as the leaves develop, umbellate clusters of 2–4 flowers per bud, the flower stalks 3–10 mm, glabrous. Flowers with the hypanthium 1–3 mm long, more or less bell-shaped, glabrous. Sepals 1–2 mm long, erect to spreading at flowering, ovate, the margins sparsely hairy, entire, nonglandular, the inner surface hairy, especially toward the base. Petals 3–6 mm, obovate to nearly circular, white. Fruits 15–20 mm long, globose to ellipsoid, shallowly longitudinally grooved on 1 side, the surface red to yellow, glabrous, slightly glaucous, the fleshy layer well-developed, the stone ovoid, somewhat flattened, the surface roughened or somewhat pitted. $2n=16$. March–April.

Scattered, mostly south of the Missouri River (eastern [mostly southeastern] U.S. west to Colorado, New Mexico, and California). Swales of upland prairies, sand prairies, savannas, and openings of mesic upland forests; also pastures, ditches, old mines, railroads, roadsides, and open, disturbed areas.

Prunus angustifolia is a relatively distinctive plum species because of its relatively small, glandular-margined leaves that are folded lengthwise during development. For a discussion of hybridization with *P. hortulana*, see the treatment of that species.

3. Prunus armeniaca L. (apricot)
Armeniaca vulgaris Lam.
Pl. 537 k, l; Map 2473

Plants trees, to 10 m tall, not suckering. Branches unarmed. Twigs glabrous, producing pseudoterminal winter buds (these usually in a cluster of 2 or 3 at the tip). Petioles (12–)20–45 mm long, glabrous or somewhat hairy toward the tip, with 1–5 stout, stalked glands near the tip and/or along the basal portions of the blade margins. Leaf blades (3–)5–9 cm long, (2–)4–8 cm wide, less than 2 times as long as wide, ovate to broadly ovate or slightly heart-shaped, rounded to broadly angled to occasionally truncate or shallowly cordate at the base, angled or short-tapered to a sharply pointed tip, the margins simply or doubly toothed, the more or less blunt and usually at least slightly incurved teeth gland-tipped, the upper surface glabrous or occasionally with scattered, short, stiff hairs, the undersurface with main veins hairy, mostly in the basal half of the leaf or more commonly with only small tufts of hairs in the axils of the main veins. Inflorescences produced before the leaves develop, of solitary flowers per bud (2 or more buds may be clustered at some nodes), the flower stalks absent or to 3 mm long, hairy. Flowers with the hypanthium 4–6 mm long, narrowly bell-shaped, glabrous or sparsely hairy. Sepals 4–6 mm long, reflexed at flowering, oblong-ovate, the margins nearly entire, with at most a few glandular teeth, the inner surface sparsely hairy. Petals 8–12 mm long, broadly elliptic to nearly circular, pink or pinkish-tinged while in bud, but becoming white at flowering. Fruits 25–60 mm long, broadly ellipsoid to more or less globose, somewhat flattened, shallowly to deeply longitudinally grooved on 1 side, the surface yellow to orange, sometimes tinged with red, densely and minutely hairy, but often becoming nearly glabrous at maturity, the fleshy layer well-developed, the stone ellipsoid to more or less globose, strongly flattened, the surface more or less smooth. $2n=16$. March–April.

Introduced, uncommon, known thus far only from Franklin and Stoddard Counties and the city

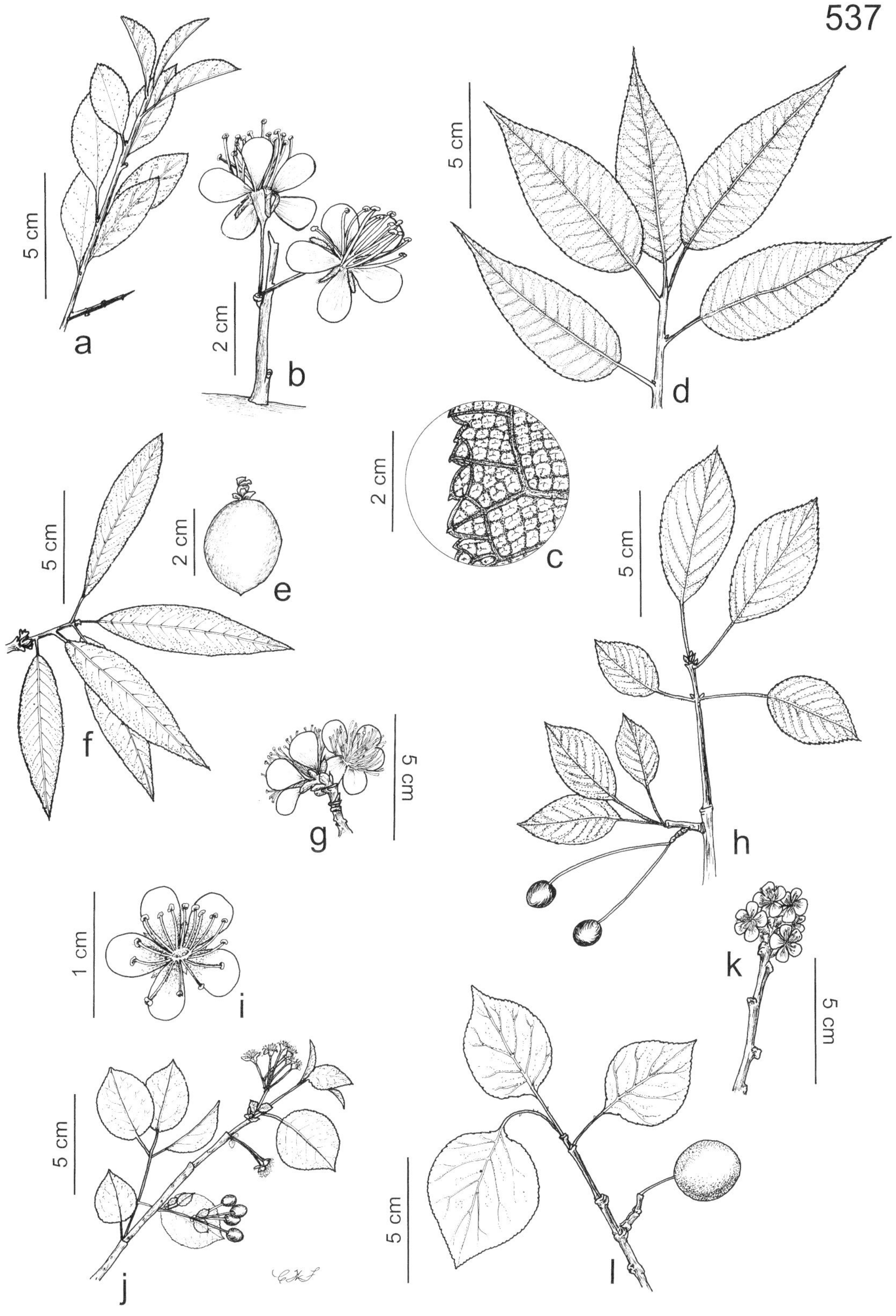

Plate 537. Rosaceae. *Prunus spinosa*, **a)** twig with leaves and thorn, **b)** node with flowers. *Prunus hortulana*, **c)** leaf detail, **d)** twig with leaves. *Prunus persica*, **e)** fruit, **f)** twig with leaves, **g)** inflorescence. *Prunus cerasus*, **h)** fruiting branch. *Prunus mahaleb*, **i)** flower, **j)** flowering and fruiting branch. *Prunus armeniaca*, **k)** inflorescence, **l)** fruiting branch.

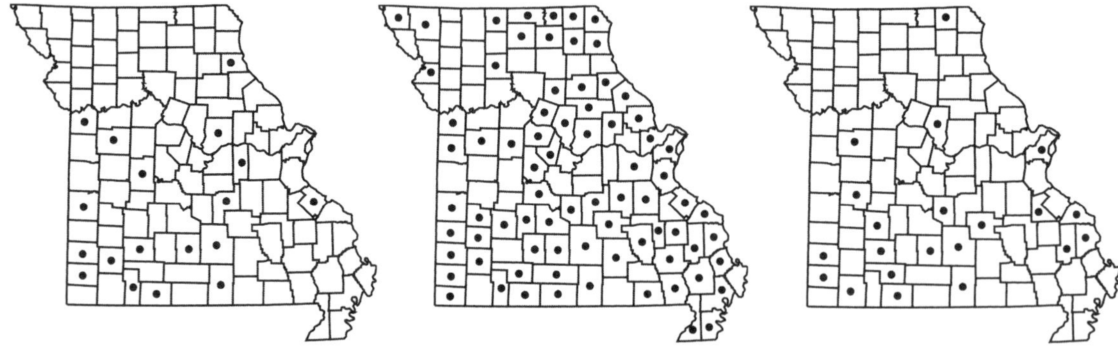
2476. Prunus mahaleb 2477. Prunus mexicana 2478. Prunus persica

of St. Louis (native of Asia; introduced sporadically in the northeastern and western U.S.; Canada). Railroads and open, disturbed areas.

Typically, the leaf blades of *P. armeniaca* are glabrous, except for tufts of short, stiff hairs in the axils of the main veins toward the base of the undersurface. However, some of the North American material has the hairs extending up the midvein or other main veins and in a few specimens the upper surface has scattered, minute, stiff, sharply pointed hairs. These differences undoubtedly are an indirect result of plant breeding for various fruit variants in different cultivars.

Apricots are popular fruits when dried, canned, or used in jams and preserves. They also are sold as fresh fruits. This species was first reported for Missouri by Mühlenbach (1979) from the St. Louis railyards.

4. Prunus cerasus L. (sour cherry)

Pl. 537 h; Map 2474

Plants shrubs or small trees, 3–5(–10) m tall, suckering, sometimes producing thickets. Branches unarmed. Twigs glabrous, producing a terminal winter bud. Petioles 16–20 mm long, glabrous, lacking glands or sometimes with 1 or 2 large discoid glands along the margins of the blade base. Leaf blades 4.5–6.0(–8.0) cm long, 2.8–4.0 (–6.0) cm wide, less than 2 times as long as wide, broadly elliptic to ovate or obovate, angled to broadly angled or rounded at the base, angled or short-tapered to a sharply pointed tip, the margins finely and doubly toothed or scalloped, the more or less rounded teeth gland-tipped, the upper surface glabrous, the undersurface glabrous or sparsely hairy along the main veins. Inflorescences mostly produced as the leaves develop (rarely before the leaves, more commonly continuing to develop after the leaves are more than half-grown), of solitary flowers per bud or umbellate clusters of 2–4 flowers per bud, the flower stalks 8–37 mm long, glabrous. Flowers with the hypanthium 4–6 mm long, narrowly bell-shaped, glabrous. Sepals 4–7 mm long, reflexed at flowering, oblong, the margins glandular-toothed, the inner surface glabrous. Petals 10–14 mm long, nearly circular, white. Fruits 13–20 mm long, globose, not grooved, the surface bright red, glabrous, not glaucous, the fleshy layer well-developed, the stone subglobose, not flattened, the surface relatively smooth. $2n=32$. April–May.

Introduced, uncommon, widely scattered in the state (cultigen of Eurasian origin; introduced widely but sporadically in the U.S., Canada). Edges of mesic upland forests; also old homesites and open, disturbed areas.

Sour cherries are eaten fresh, but more often are used in pies, cobblers, other baked goods, jams, and jellies, or dried. In addition to its use as a commercial fruit tree, several cultivars of *P. cerasus* exist that are grown as ornamentals, including some with doubled flowers. Steyermark (1963) noted that this taxon is more shade-tolerant than most other fruit tree species.

5. Prunus hortulana L.H. Bailey (wild goose plum, hortulan plum)

P. hortulana var. *mineri* L.H. Bailey

Pl. 537 c, d; Map 2475

Plants shrubs, 3–5 m tall, or more commonly trees to 6(–10) m tall, sometimes suckering to form thickets. Branches moderately thorny. Twigs glabrous, producing pseudoterminal winter buds (these usually in a cluster of 2 or 3 at the tip). Petioles 6–20 mm long, hairy on the upper side, often with 1 to several glands near the tip. Leaf blades (5–)7–11(–13) cm long, (2.0–)3.0–5.5 cm wide, more than 2 times as long as wide, lanceolate to oblong-lanceolate, oblanceolate, or narrowly elliptic, broadly angled to rounded at the base, long-tapered at the tip, the margins finely and simply to doubly toothed, the blunt incurved teeth gland-tipped, the upper surface glabrous or with a few hairs along the midvein, the undersurface sparsely to moderately hairy along the main veins. Inflorescences produced when the leaves are about half-grown, umbellate clusters of 2–4(5) flowers per bud, the flower stalks 8–20 mm long, glabrous. Flowers with the hypanthium 2–3 mm long, bell-shaped, glabrous. Sepals 1.5–3.0 mm long, ascending to reflexed at flowering, ovate, the margins glandular-toothed, the inner surface densely short-hairy below the midpoint. Petals 4–9 mm long, obovate, white. Fruits 20–30(–40) mm long, globose, shallowly longitudinally grooved on 1 side, the surface red to yellowish, with conspicuous white dots, glabrous, not or only slightly glaucous, the fleshy layer well-developed, the stone ovoid-ellipsoid, somewhat flattened, the surface shallowly pitted. $2n=16$. March–May.

Scattered to common nearly throughout the state, but relatively uncommon in the northwestern portion (eastern U.S. west to Nebraska and Texas). Banks of streams and rivers, margins of ponds, lakes, and sinkhole ponds, upland prairies, edges of bottomland prairies, and swamps; also pastures, fencerows, old homesites, railroads, and roadsides.

This species is sometimes cultivated as an ornamental, and a number of cultivars exist. Steyermark (1963) noted the existence in southwestern Missouri of occasional putative hybrids between *P. hortulana* and *P. mexicana*, to which he applied the name *P. palmeri* Sarg.

Most recent floristic works for the eastern half of the United States have included the wild goose plum (*P. munsoniana* W. Wight & Hedrick; Pl. 536 f–h) as a distinct species, and it is the source of a number of ornamental cultivars. Steyermark (1963) recorded it from scattered populations nearly throughout the state. However, in practice, the distinctions between *P. hortulana* and *P. munsoniana* (earlier-flowering relative to leaf development and inflorescences on long branches [vs. short spur shoots] in *P. munsoniana*, as well as its lateral [vs. terminal] gland-tips on the leaf teeth and a usually strongly suckering habit) have been difficult to apply to specimens and trees in nature. The molecular phylogenetic studies of Rohrer (2006) have provided evidence that plants attributed to *P. munsoniana* constitute a series of fertile hybrids between *P. angustifolia* and *P. hortulana*.

6. Prunus mahaleb L. (perfumed cherry)

Pl. 537 i, j; Map 2476

Plants shrubs or trees to 15 m tall, not suckering. Branches unarmed. Twigs densely pubescent with short hairs (at least some of these glandular) when young, becoming glabrous and glaucous with age, producing a terminal winter bud. Petioles 4–20 mm, glabrous or less commonly short-hairy on the upper side, glandless or with 1 or 2 large discoid glands at or near the tip. Leaf blades 1.5–4.5 cm long, 1.2–3.4 cm wide, less than 2 times as long as wide, oblong-ovate to more commonly broadly ovate or nearly circular, rounded to truncate or occasionally shallowly cordate at the base, abruptly short-tapered to a bluntly pointed tip, the margins finely and simply toothed to more or less scalloped, the blunt to rounded teeth with lateral glands (appearing positioned more or less in the sinus between adjacent teeth), the upper surfaces glabrous, the undersurface glabrous or finely hairy along the main veins. Inflorescences produced when the leaves are half- or more grown, short, dome-shaped racemes (the axis shorter than the flower stalks) of 4–12 flowers, the flower stalks 6–18 mm, glabrous. Flowers with the hypanthium 2–3 mm long, conic to somewhat bell-shaped, glabrous. Sepals 1.3–2.0 mm long, reflexed at flowering, oblong, the margins entire, nonglandular, the inner surface glabrous. Petals 6–7 mm long, elliptic to obovate, white. Fruits 6–10 mm long, ovoid, not grooved, the surface dark red to black, glabrous, not glaucous, the fleshy layer poorly developed, thin and dry, the stone ellipsoid to subglobose, somewhat flattened, the surface smooth. $2n=16$. April–May.

Introduced, scattered in the southern half of the Ozark Division, uncommon and sporadic elsewhere in the state (native of Europe, Asia; introduced widely in the eastern and western U.S.; Canada). Glades, mesic to dry upland forests, banks of streams and rivers, ledges of bluffs; also pastures, old fields, fencerows, cemeteries, roadsides and open, disturbed areas.

Prunus mahaleb is cultivated as an ornamental and has also been used as grafting stock for other cherries. In the past, its wood also was used in the manufacture of cherrywood pipestems. Steyermark (1963) noted that the taxon is often encountered without flowers or fruits in Missouri and is then sometimes confused for *Pyrus communis* (pear). In such cases, the presence of small glands apparently between the teeth serve to distinguish *P. mahaleb* from *P. communis*.

7. Prunus mexicana S. Watson (big tree plum, wild plum)

Pl. 536 a, b; Map 2477

Plants trees to 12 m tall, usually not suckering or colonial. Branches sparsely thorny. Twigs usually glabrous, but occasionally hairy, producing

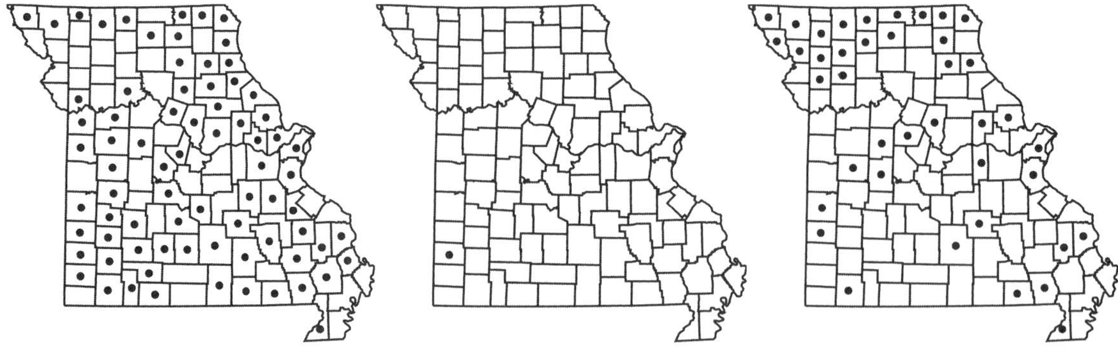

2479. Prunus serotina 2480. Prunus spinosa 2481. Prunus virginiana

pseudoterminal winter buds (these usually in a cluster of 2 or 3 at the tip). Petioles 4–18 mm long, usually evenly hairy on all sides, usually with 1 or 2(–4) large, dark, club-shaped glands near the tip. Leaf blades 6–12 cm long, 3–7 cm wide, 2 times as long as wide or wider, obovate to oblong-ovate, less commonly elliptic, broadly elliptic, or ovate, broadly angled to more commonly rounded or shallowly cordate at the base, short-tapered or tapered to rarely angled to a sharply pointed tip, the margins coarsely and doubly toothed, the sharp teeth relatively straight, glandless, the upper surface glabrous to moderately short-hairy and appearing somewhat wrinkled, the undersurface densely short-hairy. Inflorescences produced before the leaves, umbellate clusters of 2–5 flowers per bud, the flower stalks 4–20 mm long, glabrous or sparsely hairy. Flowers with the hypanthium 2.0–4.5 mm long, conic, usually glabrous, rarely finely hairy. Sepals 1.5–4.0 mm long, reflexed at flowering, lanceolate to ovate, rounded or sometimes minutely notched or toothed at the tip, the margins hairy, entire or obscurely glandular, the inner surface moderately to densely hairy. Petals 6–12 mm long, elliptic to obovate, white, sometimes fading to pink. Fruits 15–30 mm long, ellipsoid to nearly globose, shallowly longitudinally grooved on 1 side, the surface purplish red to dark blue, glabrous, glaucous, the fleshy layer well-developed, the stone ovoid-ellipsoid, somewhat flattened, the surface usually smooth. $2n=16$. April–May.

Scattered to common nearly throughout the state (South Dakota to Texas east to Indiana and Georgia; Mexico). Banks of streams and rivers, mesic upland forests, margins of ponds, bases and tops of bluffs, edges of glades, and edges of sand prairies; also pastures, old fields, fencerows, old mines, railroads, and roadsides.

Prunus mexicana is sometimes planted as an ornamental or as a wildlife food plant. Specimens attributed to this species sometimes are morphologically intermediate with *P. americana* and *P. hortulana*. For a discussion of putative hybridization in this complex see the treatments of those species.

8. Prunus persica (L.) Batsch (peach)
Amygdalus persica L.
Persica vulgaris Mill.

Pl. 537 e–g; Map 2478

Plants trees to 6(–10) m tall, not suckering. Branches unarmed. Twigs glabrous, producing a terminal winter bud. Petioles 5–10(–15) mm long, glabrous, sometimes with 1–4 sessile discoid glands near the tip. Leaf blades (5–)7–15 cm long, 2–4 cm wide, more than 2 times as long as wide oblong to oblong-lanceolate, somewhat arched and folded longitudinally, forming a keel along the midvein on the undersurface, angled to broadly angled at the base, more or less tapered to a sharply pointed tip, the margins simply toothed, the blunt teeth gland-tipped, the surfaces glabrous. Inflorescences produced before the leaves, of solitary or rarely paired flowers per bud (2 or more buds may be clustered at some nodes), sessile or the flower stalks to 3 mm long. Flowers with the hypanthium 4–5 mm long, more or less hemispheric, glabrous. Sepals 3.5–5.0 mm long, spreading at flowering, oblong-ovate, the margins entire, hairy, glandless, the inner surface glabrous. Petals 10–17 mm long, obovate to nearly circular, pink or strongly pinkish-tinged. Fruits 40–80 mm long, globose, usually deeply longitudinally grooved on 1 side, the surface yellow to orange tinged with red, finely felty-hairy (glabrous in nectarines), the fleshy layer well-developed, the stone ellipsoid, strongly flattened, the surface deeply pitted and furrowed. $2n=16$. March–April.

Introduced, uncommon and sporadic in the state (native of Asia, introduced widely but sporadically in the eastern and western U.S.; Canada). Mesic upland forests, banks of streams, and bases of bluffs; also old homesites, railroads, roadsides, and open, disturbed areas.

Prunus persica includes both peaches and nectarines, which are very important fruit crops in the United States. Some cultivars also are grown as ornamentals, as are some hybrids with related species, such as almonds (*P. dulcis* (Mill.) D.A. Webb).

9. Prunus serotina Ehrh. **ssp. serotina** (black cherry, wild cherry, rum cherry)
Pl. 536 i, j; Map 2479

Plants shrubs or more commonly trees, mostly 10–15 m tall, not suckering. Branches unarmed. Twigs glabrous or sparsely pubescent with short, reddish brown, glandular hairs, producing a terminal winter bud. Petioles (7–)10–18 mm long (to 30 mm on rapidly elongating shoots), glabrous, with 2(–6) glands at or near the tip. Leaf blades 4.0–13.5 cm long, 1.7–5.8 cm wide, more than 2 times as long as wide, narrowly elliptic to elliptic or lanceolate, rarely obovate, angled to rounded at the base, angled or more commonly tapered to a sharply pointed tip, the margins simply toothed, the blunt to more or less sharp teeth mostly incurved to appressed, lacking glands, the upper surface glabrous, usually somewhat shiny, the undersurface pale, glabrous or short-hairy along the midvein toward the base. Inflorescences produced after the leaves develop, terminal on current year's growth, racemes of 18–55 or rarely more flowers, (5–)8–16 cm long, the flower stalks 2–6(–10) mm long, usually glabrous. Flowers with the hypanthium 1.5–3.0 mm long, hemispheric to more or less bell-shaped, glabrous. Sepals 0.5–1.5 mm long, ascending to spreading or reflexed at flowering, oblong-triangular (sharply pointed at the tip and persistent after flowering), the margins lacking glands or sparsely glandular-toothed, the inner surface glabrous. Petals 2–4 mm long, obovate to nearly circular, white. Fruits 5–10 mm long globose, not grooved, the surface dark purple to nearly black, glabrous, not glaucous, the fleshy layer well-developed, the stone subglobose, not flattened, the surface smooth to finely wrinkled. $2n=32$. April–May.

Scattered to common nearly throughout the state (eastern U.S. west to Minnesota and Texas; Canada). Bottomland forests, mesic upland forests, upland prairies, banks of streams and rivers, margins of lakes and sinkhole ponds, fens; also pastures, fencerows, old mines, railroads, roadsides, and open to shaded, disturbed areas.

Steyermark (1963) noted that the reddish brown wood of *P. serotina* has been used in furniture, veneers, flooring, paneling, tool handles, scientific instruments, musical instruments, and caskets. The bark has astringent properties and has been used medicinally for sore throats, coughs, and as an expectorant. The leaves produce prussic acid, which can lead to cyanide poisoning when ingested by livestock or humans. The fruits, however, are used in beverages, baked goods, jams, jellies, and as a flavorant. Plants are sometimes cultivated as shade trees, but mature individuals tend to lose their tops and branches in storms.

This species has a broad distribution from southern Canada through the United States, the mountains of Mexico and Central America, and into Andean South America. McVaugh (1951) divided *P. serotina* into a complex series of five subspecies and one additional variety. However, all of the plants in the eastern half of the country except for some populations with relatively dense, rusty pubescence in Alabama and Georgia (ssp. *hirsuta* (Elliott) McVaugh) were treated as ssp. *serotina*, which McVaugh held to be morphologically relatively homogeneous.

10. Prunus spinosa L. (blackthorn, sloe)
Pl. 537 a, b; Map 2480

Plants shrubs to 4 m, usually strongly suckering to form dense thickets. Branches strongly thorny. Twigs minutely hairy when young, usually glabrous at maturity, producing pseudoterminal winter buds (these usually in a cluster of 2 or 3 at the tip). Petioles 4–6 mm long, minutely hairy, occasionally only on the upper side, glandless. Leaf blades 1.5–4.0 cm long, 1.0–2.2 cm wide, less than 2 times as long as wide, elliptic to obovate, broadly angled to rounded at the base, angled to a broadly or bluntly pointed tip, the margins simply toothed or scalloped, the short, blunt to rounded teeth mostly gland-tipped, the upper surface glabrous or sparsely short-hairy along the midvein near the base, the undersurface short-hairy along the main veins, at least when young. Inflorescences produced before the leaves, of solitary or less commonly paired flowers per bud, the flower stalks 1.5–6.0(–8.0) mm long, glabrous or rarely short-hairy. Flowers with the hypanthium 1.5–2.5 mm long, hemispheric, glabrous. Sepals 1.5–2.5 mm long, reflexed at flowering, ovate to triangular-ovate, the margins glandular-toothed, the inner surface glabrous. Petals 4–8 mm, elliptic to broadly elliptic or broadly oblong-elliptic, white. Fruits 10–15 mm long, globose, shallowly longitudinally grooved on 1 side, the surface bluish black, glabrous, glaucous, the fleshy layer well-developed but sometimes relatively thin, the stone subglobose, somewhat flattened, the surface smooth or nearly so. $2n=32$. April–May.

2482. Pyracantha coccinea 2483. Pyrus calleryana 2484. Pyrus communis

Introduced, uncommon, known only from two historical specimens from Jasper County (native of Europe, Asia, Africa; introduced in the northeastern and northwestern U.S., also Missouri, Tennessee). Open, disturbed areas.

This shrubby species produces abundant thorn-tipped branches and often forms dense thickets. It flowers abundantly and the small, bluish black fruits are attractive but inedible. Apparently it is no longer as commonly cultivated in the Midwest as it once was.

Trees of the common European plum and Damson plum (*Prunus domestica* L. var *domestica* and var. *insititia* (L.) Fiori & Paol.) occasionally persist at old home sites, but thus far have not escaped into the wild in Missouri. Members of the *P. domestica* complex are superficially similar to *P. spinosa*, differing among other characters in their mostly paired flowers and glabrous twigs, and are somewhat larger trees with larger fruits. These taxa are no longer as commonly cultivated in the Midwest as they once were. Currently, most of the plums available as fresh fruits in markets in the United States are derived from the Japanese plum, *P. salicina* Lindl. However, cultivars of *P. domestica* are still the source of most prunes. Fresh damson plums occasionally are offered for sale in midwestern markets, and are recognizable as relatively small, bluish black, glaucous plums. Damson plums also are the basis of plum brandy (slivovitz).

11. Prunus virginiana L. **var. virginiana**
(choke cherry, eastern choke cherry)
Padus virginiana (L.) Mill.

Pl. 536 k, l; Map 2481

Plants shrubs or small trees, 2–6(–10) m tall, often suckering. Branches unarmed. Twigs glabrous or occasionally short-hairy, producing a terminal winter bud. Petioles (4–)9–28 mm long, usually glabrous, rarely short-hairy, often with 2 large discoid glands at or near the tip. Leaf blades 2.5–12.0 cm long, 1.2–6.5 cm wide, those of at least larger leaves less than 2 times as long as wide, ovate to broadly elliptic or obovate, broadly angled to rounded or occasionally shallowly cordate at the base, angled or tapered to a sharply or occasionally bluntly pointed tip, the margins finely to moderately toothed, the relatively straight teeth sharply pointed, lacking glands, the upper surface glabrous or less commonly sparsely short-hairy along the main veins, sometimes somewhat shiny, the undersurface pale, glabrous or sparsely to densely hairy along the main veins, occasionally the pubescence restricted to small tufts in the vein axils. Inflorescences produced after the leaves develop, terminal on current year's growth, racemes of 18–64 flowers, (2–)4–12 cm long, the flower stalks 2–8(–12) mm long, usually glabrous. Flowers with the hypanthium 1.5–2.0 mm long, hemispheric to more or less bell-shaped, glabrous. Sepals 0.7–1.0 mm long, ascending to reflexed at flowering, semicircular to broadly oblong-triangular (bluntly pointed at the tip and shed after flowering), the margins noticeably glandular-toothed, the inner surface glabrous. Petals 2.5–4.0 mm long, obovate to nearly circular, white. Fruits 6–14 mm long, globose, not grooved, the surface red, purple, dark purple, or black, glabrous, not glaucous, the fleshy layer well-developed, the stone subglobose to ellipsoid, not flattened, the surface smooth to slightly and finely wrinkled. $2n$=16, 32. April–May.

Scattered to common north of the Missouri River, uncommon farther south, mostly in the Ozark Border Division (eastern U.S. west to North Dakota and Texas; Canada). Banks of streams and rivers, upland prairies, loess hill prairies, bottomland forests, mesic upland forests, margins of sinkhole ponds, and bases of bluffs; also pastures, ditches, fencerows, railroads, roadsides, and open, disturbed areas.

Plants with the undersurface of the leaves glabrous or with the pubescence merely in small tufts in the vein axils have been called f. *deamii* G.N. Jones. Plants in the western United States with somewhat narrower leaves, longer inflorescences, and larger flowers are known as var. *demissa* (Nutt) Torr.

21. **Pyracantha** M. Roem. (firethorn)

Ten species, Europe, Asia.

1. Pyracantha coccinea M. Roem. (pyracantha, scarlet firethorn)

Map 2482

Plants shrubs (small trees elsewhere), 2–4 m tall, densely branched, the short shoots usually armed with thorns, these mostly indeterminate (growing into branches) and often with 1 or a few leaves toward the base, even when short. Bark gray to light grayish brown, thin, smooth when young, developing fine longitudinal fissures with age. Twigs gray, short-hairy, soon becoming glabrous. Leaves often more or less evergreen or partially overwintering, alternate or sometimes appearing in fascicles on short shoots, the petioles 2–5 mm long. Stipules 4–8 mm long, lanceolate, the margins few-toothed, shed early. Leaf blades 1.5–4.0 cm long, lanceolate to oblanceolate or obovate, narrowly angled at the base, angled to a bluntly or more commonly sharply pointed tip, somewhat leathery, the margins flat or slightly curled under, finely scalloped or toothed, the upper surface bright green, glabrous, shiny, the undersurface lighter green, sparsely and inconspicuously hairy, at least when young. Inflorescences terminal, small, flat-topped to dome-shaped, relatively dense panicles with up to 40 flowers, the stalks finely pubescent with short, spreading hairs; the flowers subtended by short, lanceolate bracts, these shed early. Flowers epigynous, the hypanthia finely hairy. Sepals 5, 1.0–1.5 mm long, triangular. Petals 5, 4–5 mm long, broadly rhombic-obovate to nearly circular, white. Stamens 15–20, the filaments 2–4 mm long, white, the anthers yellow. Pistil 1 per flower. Ovary inferior, with 5 locules, each with 2 fertile ovules. Styles 5, fused toward the base, the stigmas more or less capitate. Fruits berrylike pomes, 5–7 mm in diameter, globose, finely hairy when young, becoming glabrous or nearly so at maturity, orangish red to red at maturity (rarely yellow elsewhere), shiny, with 5 stones, these 3–5 mm long, narrowly and asymmetrically elliptic in outline, wedge-shaped in cross-section, indehiscent, 2-seeded, embedded in the fleshy middle layer of the fruit, the surface hard, more or less smooth, light brown. May–June.

Introduced, known thus far only from a single specimen from Taney County (native of Europe, Asia, introduced widely in the U.S. from Washington to California, east to Texas and Florida, and north to Indiana and New York; Canada). Ledges of bluffs.

This species is commonly cultivated as an ornamental and in hedges. The bright-colored fruits are long-persistent and provide food for songbirds, which sometimes become intoxicated when they ingest older fruits in which the fleshy layer has become fermented. The shrubs are relatively hardy, but are susceptible to fire blight, a disease caused by the bacterium, *Erwinia amylovora* (Burrill) Winslow, which causes a rapid wilt of infected foliage and flowers and the formation of small cankers on the affected branches, often followed by death of the plant during the same growing season. This disease also affects pears and some other members of the maloid Rosaceae.

22. **Pyrus** L. (pear)

Plants small to medium trees. Branches sometimes producing short, stout branchlets with thorny tips. Bark dark brown to gray, on younger trunks relatively smooth but with prominent, raised branch scars and lenticels, on older trunks developing a network of flat-topped ridges, these breaking up into more or less rectangular, fine, scaly plates. Winter buds ovoid, with several overlapping scales. Leaves alternate but often appearing clustered at the tips of short branchlets, rolled during development, long-petiolate, the petioles lacking glands. Stipules small, membranous to papery, shed early. Leaf blades simple, unlobed, broadly ovate, broadly elliptic, or broadly oblong-obovate to nearly circular, mostly rounded at the base, tapered to a sharply pointed tip, the margins finely, evenly, and bluntly toothed, the surfaces cottony-to cobwebby-hairy during development, glabrous at maturity, the upper surface lacking glands, often shiny. Inflorescences terminal on main branchlets or more commonly on lateral, short branchlets, dome-shaped clusters (often with a short central axis and thus technically racemose) of 4–12, long-stalked flowers, produced before or as the leaves uncurl, the axis and stalks glabrous or hairy, the stalks each with a small bract at the base, this linear to narrowly oblong-elliptic, brown to reddish brown, hairy, mostly shed early. Flowers epigynous, usually strongly malodorous, the hypanthium fused to the ovaries, cup-shaped to more or less urn-shaped, with a free, thickened rim nearly closing the tip, glabrous or hairy. Sepals 5, spreading to somewhat reflexed at flowering, lanceolate-triangular to lanceolate, long-tapered to a sharply pointed tip, the margins and/or upper surface hairy, sometimes persistent at fruiting. Petals 5 (except in rare doubled forms), broadly obovate to nearly circular, white, occasionally tinged with pink. Stamens usually (15–)20, the anthers pink to red. Pistil 1 per flower, of 2–5 fused carpels. Ovary inferior, with 2–5 locules, each with 2 ovules. Styles 2–5, free, protruding from the rim of the hypanthium the stigmas more or less capitate. Fruits pomes, globose to obovoid or pear-shaped, glabrous at maturity, variously colored, the surface often with small, light or dark dots, with 4–10 more or less easily exposed seeds embedded in the "core" of leathery to papery carpel wall remains and the fleshy portion, this with abundant stone cells (creating a gritty texture). Seeds more or less obovoid, the outer surface smooth, dark brown to black. About 25 species, Europe, Asia, Africa, introduced widely.

The present treatment follows a relatively restricted concept of *Pyrus*, in contrast to the much broader circumscription favored by Steyermark (1963) and K. R. Robertson (1974), which includes species here segregated into such genera as *Aronia* and *Malus*. Morphologically, *Pyrus* in the strict sense is distinctive mainly in the relatively cryptic anatomical feature of stone cells (sclereids) in its fruits (which give the flesh a gritty texture in fruits of edible species), which are shared with the genus of quince, *Cydonia* Mill. (Aldasoro et al., 1998). Evidence from details of floral morphology (Rohrer et al., 1994) and phytochemical data (A. H. Williams, 1982) support the splitting of the two genera. Molecular studies (C. S. Campbell et al., 2007; Potter et al., 2007) also have tended to reinforce *Pyrus* as a natural genus and to suggest that some of the other segregates of *Pyrus* are more closely related to other genera. It should be noted that even with analysis of massive molecular data sets, the phylogenetic relationships among *Pyrus* and related genera still are not fully understood (C. S. Campbell et al., 2007).

1. Styles 2 or 3; petals 5–7(–10) mm long; fruits 0.8–1.5 cm long, globose, the surface green to blackish brown, with pale dots (to 2.0 cm, brownish yellow, and with dark dots in rare hybrids; see discussion under the species treatment) . . . . 1. P. CALLERYANA
1. Styles usually 5; petals 10–15 mm long; fruits (2–)3–8(–16) cm long, ovoid, obovoid, or pear-shaped, the surface yellowish green, brown, or red, with or without pale dots . . . . 2. P. COMMUNIS

1. Pyrus calleryana Decne. (Bradford pear)
Map 2483

Plants trees to 15(–20) m tall, usually thornless, rarely thorny. Twigs glabrous or rarely short-hairy, the winter buds relatively large, the scales densely cobwebby-hairy and fringed. Petioles 20–45 mm, moderately hairy when young, glabrous or sparsely hairy at maturity. Leaf blades 4–9 cm long, elliptic to ovate or more commonly broadly ovate, occasionally nearly circular, mostly short-tapered to a sharply pointed tip, rounded to shallowly cordate at the base, the margins finely scalloped or finely and bluntly toothed, the surfaces somewhat cobwebby-hairy during development, glabrous and shiny at maturity, green to dark green. Inflorescences variously umbellate clusters

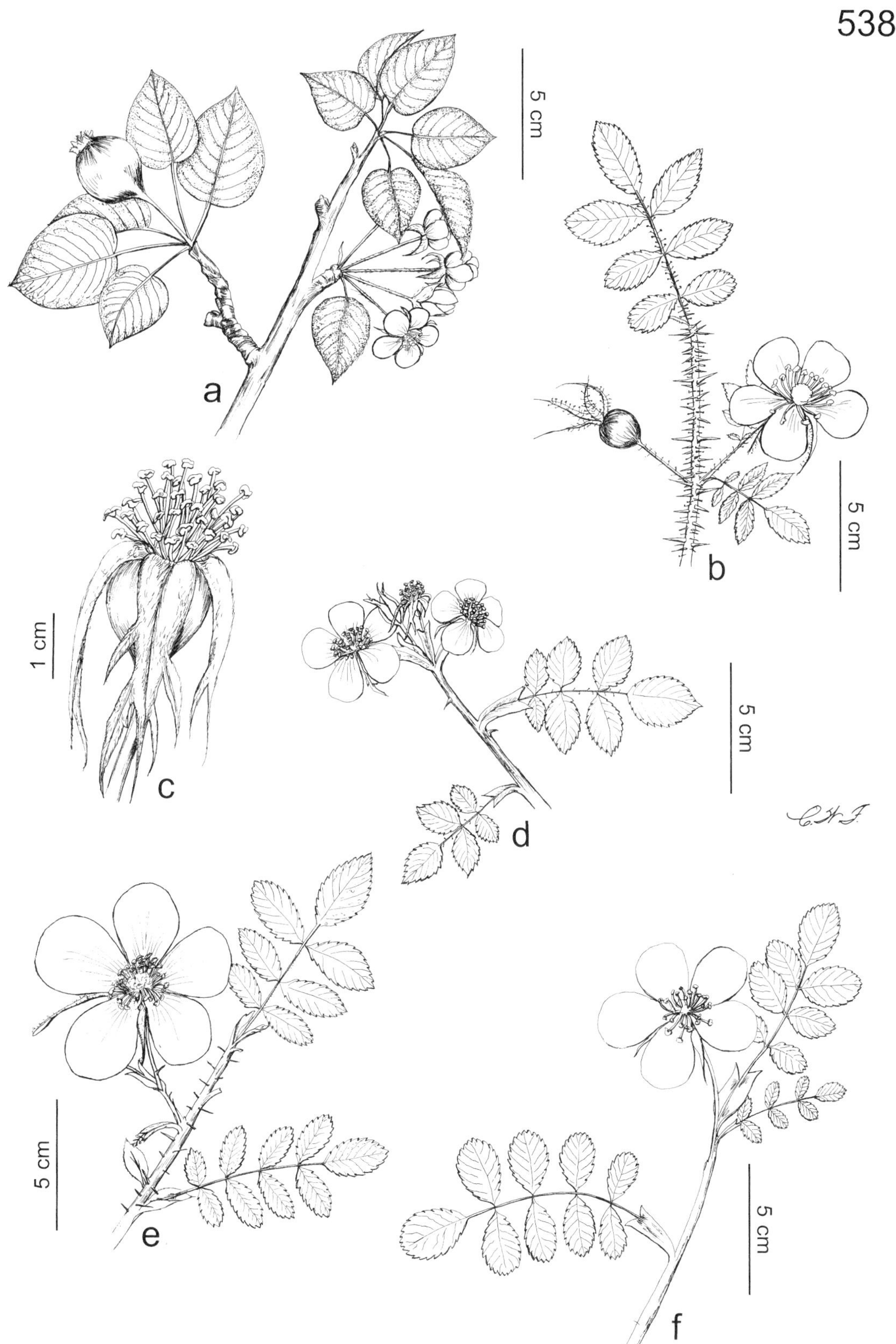

Plate 538. Rosaceae. *Pyrus communis*, **a)** flowering and fruiting branch. *Rosa carolina* ssp. *subserrulata*, **b)** flowering and fruiting branch. *Rosa canina*, **c)** flower with petals removed, **d)** flowering branch. *Rosa arkansana*, **e)** flowering branch. *Rosa blanda*, **f)** flowering branch.

or short, broad racemes with 5–12 flowers. Sepals 2–4 mm long, triangular, the upper surface densely hairy, the undersurface glabrous, not persistent at fruiting. Petals 5–7(–10) mm long, white. Styles 2 or 3. Fruits 0.8–1.5 cm long, globose to slightly oblong-globose, the surface green to yellowish brown or blackish brown, with pale dots. Seeds 1 or 2(–4). $2n=34$. March–May, rarely also September–October.

Introduced, scattered in the southern half of the state and continuing to spread (native of Asia, introduced widely in the eastern U.S. west to Illinois, Kansas, and Texas, also California). Banks of streams and mesic upland forests; also pastures, old fields, ditches, roadsides, and disturbed areas.

The history of the introduction of this species was reviewed in detail by Vincent (2005) and Culley and Hardimann (2007). *Pyrus calleryana* was imported into the United States in 1917 by the U.S. Department of Agriculture, which was tasked with locating *Pyrus* germplasm that was resistant to the fungal disease, fire blight (see also the treatment of *Pyracantha* in the present volume). It was used in breeding programs with the cultivated pear and also was used as root stock onto which other pears could be grafted. It was not until the 1950s that breeding of *P. calleryana* as an ornamental was begun, and the popular cv. 'Bradford' was released to the nursery trade in 1960. Bradford pear was touted as having low fruit set and was rated highly as a hardy street and yard tree tolerant of a wide variety of environmental conditions. It soon became one of the most popular ornamental trees in horticulture in many parts of the country, with an estimated 300,000 trees planted by about 1980 (Dirr, 1981). Concerns about the overuse and mass-planting of the trees initially was aimed at the potential impacts of new diseases on them, as well as the observation that the brittle wood tended to cause excessive breakage and trunk-splitting following ice- and wind-storms. By the late 1990s, flags were being raised about the invasive potential of the species into natural plant communities. In addition to the documented escape of plants into natural habitats, cases were noted in which trees reverted to the so-called wild-type for some characters, including production of more fruits and thorn-tipped branches.

Although the earliest specimen documenting the escape of *P. calleryana* in Missouri dates to 1952 in Barton County, virtually all of the other specimens in the state's herbaria were collected on or after 1990. In portions of southern Missouri, the species has developed extensive populations during the past two decades, particularly in area that are recovering from heavy disturbance, such as highway improvement projects. Recent observations suggest that abundant fruits are produced in some populations, which are spread by birds and mammals. Culley and Hardimann (2007) proposed a broad potential range for the species nearly throughout the United States and considered it to represent an invasive exotic that was in the early phases of its epidemiology.

An unusual variant with brownish yellow, globose fruits to 2 cm in diameter has been collected sporadically in fencerows and pastures, thus far in Howell, Jefferson, Phelps, and Washington Counties. This was originally noted by Doug Stevens of the Missouri Botanical Garden, who says that the fermented pomes can be used to make a superior fruit wine. Flowering and fruiting specimens appeared initially to key to *P. serrulata* Rehder. However, following his studies of the Bradford pear complex in North America, Michael Vincent of Miami University (personal communication) redetermined these as probable hybrids of horticultural origin developed from crosses of *P. calleryana* and some other species, perhaps *P. serrulata*.

2. Pyrus communis L. (pear)

Pl. 538 a; Map 2484

Plants trees to 15 m tall, usually at least somewhat thorny. Twigs glabrous or nearly so, the winter buds relatively small, the scales glabrous but hairy along the margins. Petioles 15–40 mm, moderately to densely hairy when young, becoming glabrous at maturity. Leaf blades 4–8 cm long, elliptic to ovate or broadly ovate, mostly short-tapered to a sharply pointed tip, broadly angled to rounded or shallowly cordate at the base, the margins finely and bluntly toothed, the surfaces somewhat cobwebby-hairy during development, glabrous and shiny at maturity, green to dark green. Inflorescences short, broad, dome-shaped racemes with 4–9 flowers. Sepals 6–9 mm long, triangular, the margins usually with a narrow, pale band and slightly uneven, the upper surface moderately to densely hairy, at least toward the base, the undersurface glabrous or with patches of hair basally where adjacent sepals adjoin, persistent at fruiting. Petals 10–15 mm long, white. Styles usually 5. Fruits (2–)3–8(–16) cm long, ovoid, obovoid, or pear-shaped, the surface yellowish green, brown, or red, with or without pale dots. Seeds 5–10. $2n=34$. April–May.

Introduced, uncommon, mostly in the eastern half of the state (cultigen of European or Asian origin, introduced widely in the U.S. [except for some of the northern Plains states]; Canada). Banks of streams, bottomland forests, and mesic

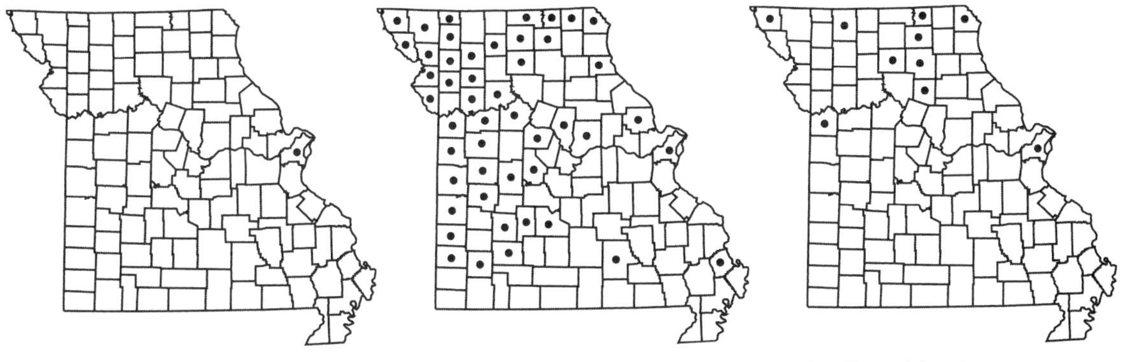

2485. Rhodotypos scandens 2486. Rosa arkansana 2487. Rosa blanda

upland forests; also pastures, old fields, old homesites, railroads, roadsides, and disturbed areas.

Pears are a very old fruit crop. Escaped plants usually do not breed true to the parental tree, producing bitter, relatively inedible fruits.

23. **Rhodotypos** Siebold & Zucc.

One species, Asia.

This is one of the few genera of Rosaceae with opposite leaves. The flowers superficially resemble those of *Philadelphus* (Hydrangeaceae).

1. Rhodotypos scandens (Thunb.) Makino (jet bead)

Map 2485

Plants shrubs, 0.5–2.0 m tall. Stems arched, slender, branched, unarmed. Twigs green, becoming reddish brown, glabrous, the winter buds small, with several overlapping scales. Leaves opposite, the petioles 2–10 mm long, short-hairy. Stipules 2–4 mm long, linear, the margins hairy, shed early. Leaf blades 4–8(–10) cm long, narrowly to broadly ovate, rounded to broadly angled at the base, tapered to a sharply pointed tip, the margins sharply, irregularly, and doubly toothed, the upper surface green, sparsely pubescent with short, straight hairs, sometimes somewhat shiny, with the main veins impressed, the undersurface slightly lighter green, sparsely to moderately hairy, with raised veins. Inflorescences terminal, of solitary flowers, the stalks 6–13 mm long at flowering, becoming elongated to 12–30 mm at fruiting; lacking bracts. Flowers perigynous, the hypanthia saucer-shaped, with a prominent, ridgelike, irregularly 4-lobed disc just inside its rim, closely subtended by 4 bractlets, these alternating with the sepals, 2.5–5.0 mm long, linear to narrowly lanceolate, usually shed as the fruits mature. Sepals 4, 8–16 mm long, ovate to oblong-ovate or ovate-elliptic, thickened toward the base, tapered to sharply pointed tips, the margins with numerous, irregular, sharp, slender teeth, sparsely to moderately short-hairy, at least toward the tips, persistent at fruiting. Petals 4, 17–25 mm long, broadly obovate to nearly circular, white, the margins often slightly uneven, the upper surface sometimes appearing slightly corrugated or finely wrinkled. Stamens numerous, the filaments often grouped into more or less 4 clusters, 4–6 mm long, white, the anthers yellow. Pistils usually 4 per flower, each with 1 carpel, more or less covered by the hypanthial disc at flowering. Ovaries superior, each with 1 locule, each with 2 ovules. Styles 1 per pistil, slender, the stigma capitate. Fruits 1–4, aggregated on the slightly expanded receptacle, drupelike (the inner layer thin and hard, the middle layer dry, papery, the outer layer thin, hard), 7–9 mm in the longest dimension, asymmetrically broadly obovoid to broadly ellipsoid, the outer surface glabrous, smooth, black or brownish black, shiny. April–June.

Introduced, uncommon, known thus far only from Cape Girardeau County and the city of St. Louis (native of Asia; introduced sporadically in the northeastern U.S. west to Wisconsin and Missouri). Mesic upland forests.

This species was first reported for Missouri by T. E. Smith (1998b) without citation of specimens. In some states, jet bead has become widely enough established in native forests and woodlands to be considered an invasive exotic.

24. Rosa L. (rose)

Plants shrubs, sometimes lianas or the stems leaning or climbing over other vegetation, sometimes colonial from rhizomes or root suckers. Stems prostrate to erect or ascending, sometimes climbing or loosely twining, all those found in Missouri armed with prickles, at least toward the stem base, these variously slender and straight or broad-based, flattened, and curved. Leaves petiolate, the petioles sometimes armed with prickles. Stipules conspicuous, usually persistent, leaflike, fused to the petiole laterally, each with an earlike basal auricle, the margins entire, toothed, or less commonly dissected into slender lobes. Leaf blades trifoliate or pinnately compound with 5–11 leaflets, the rachis with the underside sometimes armed with prickles, the leaflets sessile or short-stalked (the terminal leaflet almost always short-stalked), variously shaped, the margins bluntly to sharply toothed, the surfaces glabrous or hairy, sometimes with gland-tipped hairs. Inflorescences terminal on long branches or on short, lateral branches, loose clusters or panicles, sometimes reduced to a solitary flower, the stalks with bracts, these of various size, shape, and number. Flowers perfect (but occasionally functionally imperfect), deeply perigynous (but appearing epigynous unless dissected), the hypanthium globose to urn-shaped with a circular opening at the tip, the enclosed ovaries not fused to the inner wall above their bases. Sepals 5, sometimes persistent at fruiting, erect, horizontally spreading, or reflexed, mostly lanceolate to narrowly ovate, angled to a long-tapered tip, the margins entire or toothed (in a few species dilated and with lateral appendages). Petals usually 5 (except in doubled flowers), white to pink or red (yellow or purple in some cultivated forms), obovate to broadly obovate, broadly rounded and usually shallowly and broadly notched at the tip. Stamens numerous (fewer in doubled flowers) in several dense whorls, the filaments relatively short, free, attached inside the hypanthium rim, the anthers mostly yellow to orangish yellow. Pistils mostly numerous, attached along the inner surface of the hollow hypanthium, each with 1 carpel. Ovaries free, each with 1 ovule. Style 1 per ovary, slender, the styles within a flower either free or more or less fused into a column toward the tip, protruding slightly from the mouth of the hypanthium, the stigma disc-shaped, the stigmas in a flower often tightly clustered into a flattened mass. Fruits hips, the hypanthium becoming fleshy and usually reddish-colored, enclosing the several to numerous, small, hard-walled achenes. Achenes asymmetrically obovoid to ellipsoid, usually more or less rounded but sometimes longitudinally grooved dorsally and angled to wedge-shaped on the other side, the surface appearing smooth, light yellow to more commonly tan or light brown, densely pubescent with long hairs (glabrous in *R. rugosa*), often mostly along 1 side and/or only toward the tip, sometimes becoming glabrous or nearly so with age. About 150 species, North America, Europe, Asia, Africa.

Rosa is a taxonomically difficult genus, with abundant hybridization, polyploidy, and apomixis. A complex classification exists and numerous cultivars have been developed. Interested readers should consult the literature and electronic resources of the American Rose Society or the Royal Horticultural Society as a gateway to the massive amounts of information available on all aspects of the genus. Numerous species are cultivated as ornamentals and the genus also provides an extremely important source of cut flowers. The fleshy hips can be consumed raw or made into jellies, jams, or preserves. They are an important food source for wildlife and also are an important commercial source of vitamin C. Some species are cultivated for their fragrances and used in rose water and perfumes.

The present account follows closely the forthcoming treatment of the genus *Rosa* in the *Flora of North America* series by Walter Lewis, Barbara Ertter, and Anne Bruneau.

1. Styles fused into a column toward the tip, this protruding from the mouth of the hypanthium, the mass of stigmas thus elevated 3–6 mm above the hypanthium

2. Leaves with 3 or 5 leaflets, these 3–10 cm long; stipules with the margins entire, but pubescent with stalked glands; petals 1.8–3.0 cm long; inflorescences more or less flat-topped clusters of 3–6(–15) flowers; stems climbing, sometimes loosely twining . . . . 11. R. SETIGERA
2. Leaves with (5)7 or 9 leaflets, these 0.9–3.0(–4.5) cm long; stipules with the margins toothed or deeply divided into slender lobes, often also pubescent with stalked glands; petals 0.7–1.5 cm long; inflorescences panicles of (5–) 10–30 or more flowers; stems spreading to scrambling or strongly arching from an ascending base
3. Stipules with the margins deeply toothed; stems spreading or scrambling; flowers (at least in Missouri plants) usually doubled; styles hairy . . . . 6. R. LUCIEAE
3. Stipules with the margins deeply divided into slender lobes; stems strongly arching from an ascending base; flowers usually not doubled; styles glabrous . . . . 7. R. MULTIFLORA
1. Styles free, not protruding noticeably from the mouth of the hypanthium, the mass of stigmas appearing sessile or protruding up to 2.5 mm from the hypanthium, generally appearing as a plug closing the mouth of the hypanthium
4. Leaflets densely dotted on at least the undersurface with small, sessile, rust-colored glands, these resinous, apple-scented when disturbed . . . . 9. R. RUBIGINOSA
4. Leaflets variously glabrous or hairy, sometimes with stalked glands, but lacking sessile, resinous glands
5. Young stems and branch tips densely and finely woolly; prickles hairy, at least toward their bases; flowers often doubled . . . . 10. R. RUGOSA
5. Stems and branch tips glabrous (some of the prickles sometimes slender and bristly), but stalked glands present in some species; prickles glabrous; flowers not doubled (except in *R. gallica* and sometimes *R. spinosissima*)
6. Flowers solitary at the branch tips (inflorescences rarely small clusters of 2 or 3 flowers), the stalk bractless; flowers often doubled
7. Petals 27–35 mm long; leaflets 3 or 5(7) per leaf, 2–6 cm long; hips dark red at maturity . . . . 5. R. GALLICA
7. Petals 15–25 mm long; leaflets 5–11 per leaf, 0.5–2.2 cm long; hips black or dark brown at maturity. . . . . 12. R. SPINOSISSIMA
6. Flowers 2 to several in loose clusters or, if solitary, then the stalk with 1 or more bracts at or toward the base; flowers not doubled
8. Sepals strongly unequal, some of them pinnately deeply divided with 2 or 3 pairs of lanceolate lobes; mouth of the hypanthium about 1 mm in diameter, styles exserted 1–2 mm beyond the mouth of the hypanthium, the mass of stigmas thus slightly elevated and not closing the hypanthium mouth . . . . 3. R. CANINA
8. Sepals equal or unequal, some of them sometimes with elongate, expanded, somewhat leaflike tips, sometimes also pinnately shallowly to moderately toothed or divided with 1 or 2 pairs of linear lobes; mouth of the hypanthium about 2–3 mm in diameter, styles not or very slightly extending beyond the mouth of the hypanthium, the mass of stigmas thus not elevated and closing the hypanthium mouth

9. Sepals mostly persistent at fruit maturity, usually arched inward, erect, or ascending after flowering (occasionally spreading to reflexed in *R. woodsii*); flower stalks, hypanthia, sepals, and hips usually without stalked glands, but these occasionally present
   10. Leaflets (5)7 or 9(11); inflorescences at the tips of current year's stems and often also on side branches developing from second year's stems . . . . 1. R. ARKANSANA
   10. Leaflets 5 or 7(9); inflorescences only at the tips of side branches developing from second year's stems
       11. Stems armed only toward the base with dense, slender prickles, essentially unarmed in the apical 1/3–1/2 . . . . 2. R. BLANDA
       11. Stems armed more or less evenly from base to tip with a mixture of slender and stouter prickles . . . . 13. R. WOODSII
9. Sepals eventually shed at fruit maturity, usually spreading or reflexed after flowering; flower stalks, hypanthia, and hips usually with prominent stalked glands, these sometimes sparse, occasionally absent
   12. Stems 30–100(–130) cm long, spreading to strongly ascending, usually few-branched; leaflets relatively coarsely toothed, the teeth mostly 8–14 per side . . . . 4. R. CAROLINA
   12. Stems 100–250(–300) cm long, erect, usually with several arching to spreading branches; leaflets finely toothed, the teeth mostly 20–30 per side . . . . 8. R. PALUSTRIS

1. Rosa arkansana Porter (prairie wild rose)
R. arkansana var. *suffulta* (Greene) Cockerell
R. suffulta Greene

Pl. 538 e; Map 2486

Stems 60–100(–150) cm long, low, spreading to loosely ascending, slender or stout, unbranched or few-branched, dull red to purplish red, relatively densely and evenly covered with small prickles and bristlelike prickles, occasionally unarmed toward the stem tips, the prickles 1–3(–4) mm long, 0.5–2.0 mm wide, the bases rarely broadened to 3 mm, mostly not flattened, straight, the smallest often gland-tipped, not paired at the nodes. Leaves 5–10(–16) cm long, the petiole and rachis glabrous or more commonly hairy, rarely with stalked glands, occasionally with small prickles and scattered bristles. Stipules 18–24 mm long, the margins undulating, coarsely or shallowly glandular-toothed, the auricle flared, (2.5–)4.0–7.0 mm long. Leaflets (5)7 or 9(11), 15–40 mm long, 8–20 mm wide, the terminal leaflet with a stalk 4–12 mm long, the blades obovate or occasionally elliptic, angled at the base, sharply pointed at the tip, the margins simply or rarely doubly toothed with 8–16 teeth per side, these rarely gland-tipped, the upper surface dull, green, often somewhat glaucous, the undersurface pale green, usually hairy, at least along the veins (glabrous elsewhere). Inflorescences at the tips of current year's stems and/or on lateral branches from second year's stems, of solitary flowers or more commonly small clusters of 2–16 flowers, the flower stalks 10–20 mm long, glabrous, with 1–3 bracts. Flowers rarely doubled, the hypanthium 5.0–6.5 mm long, glabrous, the mouth about 2–3 mm in diameter. Sepals 11–20(–30) mm long, (1.5–)3.0–4.0 mm wide, often somewhat unequal, some of them extending into dilated, somewhat leaflike tips and with the margins toothed or with 1 or 2 pairs of linear lobes, the undersurface of at least the outer sepals with stalked glands, erect, ascending or somewhat incurved after flowering, mostly persistent at fruit maturity. Petals 22–26 mm long, light to bright pink, rarely white. Pistils 26–43, the styles free, hairy, not or very slightly extending beyond the mouth of the hypanthium, the mass of stigmas thus not elevated, closing the hypanthium mouth. Hips 10–11 mm long, 7.5–13.0 mm wide, globose, subglobose, or oblong-ellipsoid, fleshy, the surface dull orangish red to red, lacking stalked glands or these occasionally present. Achenes 12–15, 4.5–5.0 mm long. $2n=28$. May–July.

Scattered in the Glaciated Plains and Unglaciated Plains Divisions and the western and northern portions of the Ozark and Ozark Border Divisions (Montana to New Mexico east to Arkansas, Indiana, and Maine; Canada). Bottomland prairies, upland prairies, loess hill prairies, sand prairies, savannas, glades, tops of bluffs, edges and openings of mesic to dry upland forests, margins of lakes, and swamps; also pastures, railroads, and roadsides.

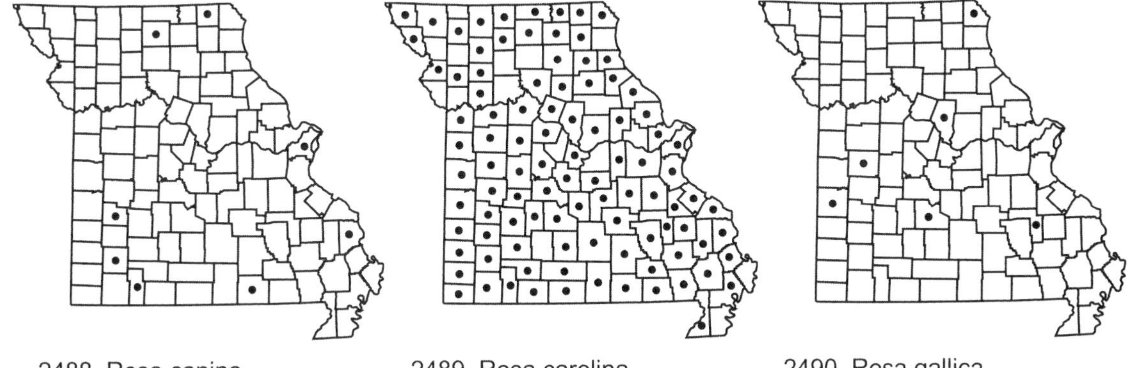

2488. Rosa canina

2489. Rosa carolina

2490. Rosa gallica

Rosa arkansana is a widespread species that occurs in a number of habitats. According to the molecular studies of Jolie et al. (2006), this tetraploid taxon originated from diploids in the *R. blanda* complex and had an independent origin from that of *R. carolina*, which is also a tetraploid, but apparently originating from past hybridization between diploid members of the *R. blanda* and *R. palustris* complexes. Many midwestern botanists have had difficulties in distinguishing between *R. arkansana* and *R. carolina*, particularly in western Missouri, which is the region of greatest overlap in their distributions. Part of the problems stem from the shared evolutionary history of these two tetraploids. The study of Jolie et al. (2006) also hinted at hybridization between *R. arkansana* and *R. carolina*. Steyermark (1963) had already noted this phenomenon, also noting that Erlanson (1931) had performed experimental crosses between the two species to produce plants with morphology similar to some of the Missouri intermediates. Erlanson (1934) had suggested the name *R. rudiuscula* for such hybrids, but Steyermark (1954, 1963) rejected that name as being of uncertain application. More recently, W. H. Lewis (2008) reaffirmed introgression between the two tetraploids in nature, but noted that the type specimen of *R. rudiuscula* is assignable to *R. carolina* var. *carolina*. He coined the new epithet, *R.* ×*medioccidentalis* W.H. Lewis for hybrids between *R. arkansana* and *R. carolina*.

Rare plants of *R. arkansana* with doubled flowers have been called f. *plena* W.H. Lewis and a white-flowered variant is known as f. *alba* (Rehder) W.H. Lewis. The var. *suffulta* refers to plants that are prevalent in the eastern half of the species range (including nearly all Missouri materials) with the leaflets hairy on the undersurface. Rangewide, there is total intergradation between glabrous and hairy plants, thus this distinction is not worthy of formal taxonomic recognition.

Native Americans used *R. arkansana* to treat convulsions, bleeding wounds, eye sores, and as a stimulant and tonic, as well as a fragrant additive to hair oils (Moerman 1998).

2. Rosa blanda Aiton (smooth wild rose, meadow rose)

Pl. 538 f; Map 2487

Stems (50–)100–200 cm long, erect or ascending, slender, unbranched or few-branched, green to red or orangish red, somewhat glaucous, densely covered with bristlelike prickles and bristles toward the base, grading fairly abruptly to essentially unarmed toward the tip, the prickles 1–5 mm long, 0.5–1.5 mm wide, the bases not broadened, mostly not flattened, straight, not gland-tipped, not paired at the nodes. Leaves 8.5–11.0 cm long, the petiole and rachis glabrous or hairy, sometimes with stalked glands, occasionally with small prickles and/or scattered bristles. Stipules (9–)15–25 mm long, the margins entire or shallowly glandular-toothed, the auricle mostly flared, (2.5–)4.0–6.0 mm long. Leaflets 5 or 7(9), (15–)25–40(–55) mm long, (8–)12–20(–30) mm wide, the terminal leaflet with a stalk 5–9(–13) mm long, the blades ovate to obovate, angled at the base, the margins simply toothed with 12–16(–20) teeth per side, these mostly not gland-tipped, often darkened at the sharply or occasionally bluntly pointed tips, the upper surface dull, green, the undersurface pale green, glabrous or minutely hairy mainly along the veins. Inflorescences at the tips of lateral branches from second year's stems, of solitary flowers or more commonly small clusters of 2–4 (–10) flowers, the flower stalks 14–25 mm long, glabrous or rarely with a few stalked glands, with usually 1 bract. Flowers not doubled, the hypanthium 4–5 mm long, glabrous or occasionally with a few stalked glands, the mouth about 2–3 mm in diameter. Sepals (12–)20–30 mm long, 2.5–3.5 mm wide, sometimes unequal, then some of them extended into dilated, somewhat leaflike tips and with the margins toothed or with 1 or 2 pairs of linear lobes, the undersurface glabrous or more commonly hairy and with stalked glands, erect, ascending or somewhat incurved after flowering, mostly persistent at fruit maturity. Petals 13–26 mm long, light to bright pink. Pistils 32–55, the styles free, hairy, not or very slightly extending beyond the mouth of the hypanthium, the mass of stigmas thus not elevated, closing the hypanthium mouth. Hips 8–11 mm long, 8–11 mm, globose or subglobose, fleshy, the surface red, lacking stalked glands or occasionally a few present. Achenes 16–28, 3.5–4.5 mm long. $2n$=14. May–July.

Uncommon in the Glaciated Plains Division south locally to Jackson and St. Louis Counties (northeastern U.S. west to Minnesota and Missouri; Canada; introduced in Europe). Bottomland prairies, upland prairies, sand prairies, and edges and openings of mesic upland forests; also roadsides.

The name "smooth wild rose" is somewhat misleading, for although the stems mostly lack prickles, frequently there is a relatively abrupt transition to dense, bristlelike prickles in the basal third. *Rosa blanda* is closely related to *R. woodsii*. For further discussion on the taxonomic controversy surrounding these species, see the treatment of *R. woodsii*.

3. Rosa canina L. (dog rose, dog briar)

Pl. 538 c, d; Map 2488

Stems 50–250(–500) cm long, arching or sprawling, less commonly ascending, slender, few- to several-branched, green, with prickles mostly confined to and paired at the nodes, these 5–7 mm long, 4–9 mm wide, the bases broadened, flattened, curved or hooked, not gland-tipped. Leaves 6–11 cm long, the petiole and rachis glabrous, occasionally with a few small prickles. Stipules 10–22 mm long, the margins entire or shallowly glandular-toothed, the auricle flared, 3–5 mm long. Leaflets 5 or 7, 15–40 mm long, 12–20 mm wide, the terminal leaflet with a stalk 5–11 mm long, the blades elliptic to obovate, mostly rounded to broadly angled at the base, the margins simply or doubly toothed with 20–30 teeth per side, these sharply or bluntly pointed, usually gland-tipped, the upper surface dull or shiny, dark green, sometimes glaucous, the undersurface green to light green, glabrous or minutely hairy along the veins. Inflorescences at the tips of current year's stems, sometimes also on lateral branches from second year's stems, of solitary flowers or more commonly small clusters of 2–6 flowers, the flower stalks 8–20 mm long, glabrous or with stalked glands, with 1 or more bracts. Flowers not doubled, the hypanthium 7–9 mm long, glabrous, the mouth about 1.0–1.5 mm in diameter. Sepals 10–17 mm long, 3–5 mm wide, strongly unequal, some of them with prolonged dilated, somewhat leaflike tips and the margins with 2 or 3 pairs of deep, lanceolate lobes, the undersurface glabrous, spreading to recurved after flowering, usually shed by fruit maturity. Petals 18–25 mm long, light to bright pink or white. Pistils 25–36, the styles free, hairy, extending 1–2 mm beyond the mouth of the hypanthium, the mass of stigmas thus slightly elevated above the hypanthium mouth. Hips 10–16(–24) mm long, 6–16 mm wide, globose or ellipsoid-ovoid, fleshy, the surface red, lacking stalked glands. Achenes 14–23, 5–6 mm long. $2n$=35. May–July.

Introduced, uncommon and sporadic in the state (native of Europe, Asia, Africa; introduced widely but sporadically in the eastern and western U.S.; Canada, Mexico). Edges of mesic upland forests; also fencerows and roadsides.

Steyermark (1963) and some other authors recognized two varieties within *R. canina*, with the name var. *canina* applied to plants with the leaflet undersurface hairy with glandular and often also nonglandular hairs, whereas var. *dumetorum* (Thuill.) Poir. (*R. corymbifera* Borkh.; *R.* ×*dumetorum* Thuill.) was applied to plants with glabrous leaflets. In their molecular study of then dog roses, Ritz et al. (2005) found no significant differences between the ITS sequences of plants assignable to the two types. The exact disposition of these varieties will have to await more intensive research, but for now there is no evidence to support their distinctness.

The dog roses (sect. *Caninae* DC. ex Ser.), with about 20 species native to Europe and adjacent portions of Asia and Africa) are all stable pentaploids and have a unique mechanism of meiosis that results in the formation of tetraploid eggs and haploid pollen, which fertilize to result in a pentaploid zygote. The evolution of the group appears to have involved multiple hybridization events between some extinct or as-yet undiscovered diploid dog rose with members of one or more other sections of *Rosa* (Ritz et al., 2005).

4. Rosa carolina L. (pasture rose)

R. carolina var. *grandiflora* (Baker) Rehder

Pl. 538 b, 540 e, f; Map 2489

Stems 30–100(–130) cm long, low or weak and spreading to strongly ascending, mostly slender, dull reddish brown, the relatively few branches with mostly paired prickles at the nodes, these (2–) 3–9 mm long, 1.5–3.0 mm wide, sometimes slightly to moderately broadened at the base, straight, occasionally downward-angled or rarely curved, flattened or not, sometimes also with sparse, scattered bristles between the nodes, rarely unarmed or densely covered throughout with larger prickles at the nodes and shorter prickles, bristlelike prickles, and stalked glands between the nodes. Leaves 5–10(–16) cm long, the petiole and rachis glabrous or less commonly hairy, occasionally with sparse stalked glands, occasionally with small prickles and/or scattered bristles. Stipules 10–18(–23) mm long, the margins entire or finely glandular-toothed toward the tip, the auricle flared, 2–4 mm long. Leaflets (3)5 or 7(9), 18–50 mm long, 9–28 mm wide, the terminal leaflet with a stalk 4–11 mm long, the blades ovate, elliptic, or lanceolate, angled at the base, angled or tapered at the sharply or occasionally bluntly pointed tip, the margins relatively coarsely simply or rarely doubly (or more) toothed with 8–14 teeth per side, these sometimes gland-tipped, the upper surface dull or slightly shiny, green, the undersurface glabrous or rarely hairy or glandular on the midvein. Inflorescences on lateral branches from second year's stems, of solitary flowers or less commonly clusters of 2 or 3(–6) flowers, the flower stalks 5–19 mm long, with few to many stalked glands, rarely nonglandular, with usually several bracts. Flowers not doubled, the hypanthium 4–6(–8) mm long, with sparse to dense stalked glands, rarely nonglandular, the mouth usually 1.5–2.0 mm in diameter. Sepals 16–22 mm long, 3–4 mm wide, often unequal, some of them usually extending into dilated, somewhat leaflike tips and with the margins toothed or with 1 or few pairs of linear lobes, the undersurface nonglandular or occasionally sparsely glandular, becoming reflexed after flowering, usually shed by fruit maturity. Petals 15–24 mm long, pink to less commonly reddish pink, rarely white. Pistils 32–46, the styles free, hairy, not extending or extending up to 1 mm beyond the mouth of the hypanthium, the mass of stigmas thus not or only slightly elevated, more or less closing the hypanthium mouth. Hips 7–14 mm long, 6–15 mm wide, globose to slightly depressed-globose or rarely ellipsoid, fleshy, the surface dull to more commonly shiny, red to orangish red, lacking glands or with sparse to dense stalked glands. Achenes mostly 2–6 (note additional presence of up to 16 undeveloped ovaries), 4–5 mm long. $2n=28$. May–July.

Scattered to common nearly throughout the state (eastern U.S. west to Minnesota and Texas; Canada; Mexico). Upland prairies, loess hill prairies, sand prairies, edges and openings of mesic to dry upland forests, savannas, glades, ledges and tops of bluffs, banks of streams, and rarely bottomland prairies, levees, and swamps; also pastures, old fields, fencerows, cemeteries, railroads, and roadsides.

Rosa foliolosa Nutt. is a distinctive species related to *R. carolina* that has a fairly restricted range in portions of eastern Texas, Oklahoma, and western Arkansas. It differs from *R. carolina* in its narrower leaflets, shorter (6–14 mm) stipules, and white corollas. W. H. Lewis (1958a) reported *R. foliolosa* from northeasternmost Oklahoma, close to the Missouri border, and suggested that it might be discovered in southwestern Missouri in the future. This is the only native American rose with routinely white corollas, which is a feature that collectors should focus on in searching for Missouri populations. Other normally pink-flowered roses do have rare forms with white corollas. However, these normally exist as white-flowered individuals within pink-flowered populations.

Rosa carolina is a morphologically variable and broadly distributed tetraploid taxon that can be difficult to distinguish from *R. arkansana*. For further discussion, see the treatment of that species. W. H. Lewis (2008) treated *R. carolina* as comprising three subspecies. The third subspecies, ssp. *mexicoensis* W.H. Lewis, occurs in Mexico and differs in its smaller, more densely branched habit, smaller leaves, and more finely and doubly toothed leaflet margins with the teeth gland-tipped.

1. Stems relatively weak, often spreading or strongly arched, with long, thin prickles paired at the nodes, at least toward the tips, but lacking prickles or hairs between the nodes or nearly so, rarely unarmed; flower stalks, hypanthia, and hips usually with stalked glands . . . . 4A. SSP. CAROLINA
1. Stems robust, straight, with prickles paired at the nodes and also relatively densely with unpaired prickles and/or bristles, and stalked glands between the nodes; hypanthia, hips, and flower stalks often lacking stalked glands . . . . 4B. SSP. SUBSERRULATA

4a. ssp. carolina

R. carolina var. *villosa* (Best) Rehder

Pl. 540 e, f

Stems relatively weak, often spreading or strongly arched, armed with long, thin prickles paired at the nodes, at least toward the stem tips, but lacking prickles or hairs between the nodes or nearly so, rarely totally unarmed. Flower stalks usually with stalked glands. Hypanthia and hips usually with stalked glands. $2n=28$. May–July.

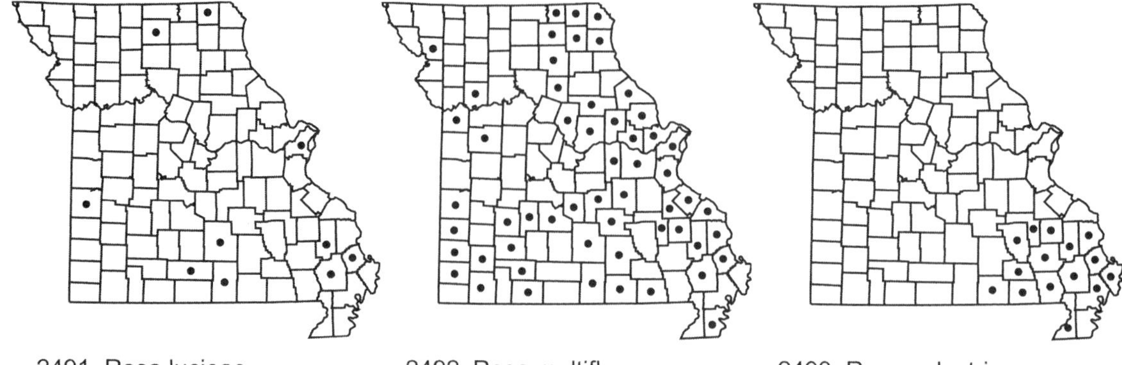

Scattered to common nearly throughout the state (eastern U.S. west to Minnesota and Texas; Canada). Upland prairies, sand prairies, edges and openings of mesic upland forests, savannas, glades, ledges and tops of bluffs, banks of streams, and rarely bottomland prairies, levees, and swamps; also pastures, old fields, fencerows, cemeteries, railroads, and roadsides.

Plants with stalked glands on the leaves have been called f. *glandulosa* (Crép.) Fernald, and plants with moderately to densely hairy leaves have been called var. *villosa*.

4b. ssp. subserrulata (Rydb.) W.H. Lewis
R. palmeri Rydb.
R. subserrulata Rydb.

Pl. 538 b

Stems robust, straight, loosely ascending to spreading, with long, thin prickles paired at the nodes, also with relatively dense, but somewhat shorter, unpaired prickles between the nodes, bristles and stalked glands also present between the nodes. Flower stalks with stalked glands or more commonly nonglandular. Hypanthia and hips with stalked glands or more commonly nonglandular. $2n=28$. May–July.

Uncommon and widely scattered (eastern U.S. west to Kansas and Texas; Canada). Glades, upland prairies, loess hill prairies, openings of dry upland forests, ledges and tops of bluffs.

Steyermark (1963) referred to these plants as *R. virginiana* Mill., which is a distinct species occurring only to the east of Missouri. True *R. virginiana* is an erect shrub to 2 m tall that is more similar in general appearance to *R. palustris* than to *R. carolina*. It differs from *R. palustris* in its stouter, more densely branched stems and its shiny, more coarsely toothed leaflets.

5. Rosa gallica L. (French rose)

Pl. 539 d–g; Map 2490

Stems 40–150 cm long, often forming colonies, ascending to erect, slender, green to reddish-tinged or reddish brown, the prickles paired at the nodes and individual between the nodes, 3–6 mm long, 1–4 mm wide, broadened at the base, at least the larger ones curved, relatively stout and somewhat flattened, often also with bristlelike prickles, these often gland-tipped. Leaves 10.5–12.0 cm long, the petiole and rachis minutely hairy, also with dense stalked glands and scattered short, small prickles. Stipules 14–24 mm long, the margins glandular-toothed, the auricle flared or erect, 4–10 mm long. Leaflets 3 or 5(7), 20–60 mm long, 18–30 mm wide, the terminal leaflet with a stalk 6–8 mm long, the blades ovate to narrowly elliptic or nearly circular, broadly angled to shallowly cordate at the base, broadly to narrowly angled or tapered at the tip, the margins mostly simply toothed with 14–23 teeth per side, leathery and often appearing somewhat wrinkled, the upper surface often somewhat shiny, bluish green, glabrous, the undersurface paler, short-hairy, and glandular. Inflorescences terminal at tips of current year's growth, usually of solitary flowers, the flower stalks (10–)20–60 mm long, stout, with stalked glands, appearing bractless. Flowers usually doubled, the hypanthium 5–7 mm long, with stalked glands, the mouth usually 1.5–2.0 mm in diameter. Sepals 16–23 mm long, 2.5–3.5 mm wide, unequal, some of them extending into dilated, somewhat leaflike tips and the margins with few pairs of deep lobes, the undersurface glandular, becoming reflexed after flowering, usually shed by fruit maturity. Petals 27–35 mm long, deep pink to dark red. Pistils 30–47, the styles free, usually hairy, not extending or extending up to 0.5 mm beyond the mouth of the hypanthium, the mass of stigmas thus not or only slightly elevated, more or less closing the hypanthium mouth. Hips uncommonly maturing, 8–9 mm long, 7–10 mm wide, broadly ellipsoid to subglobose, leathery, the surface dark

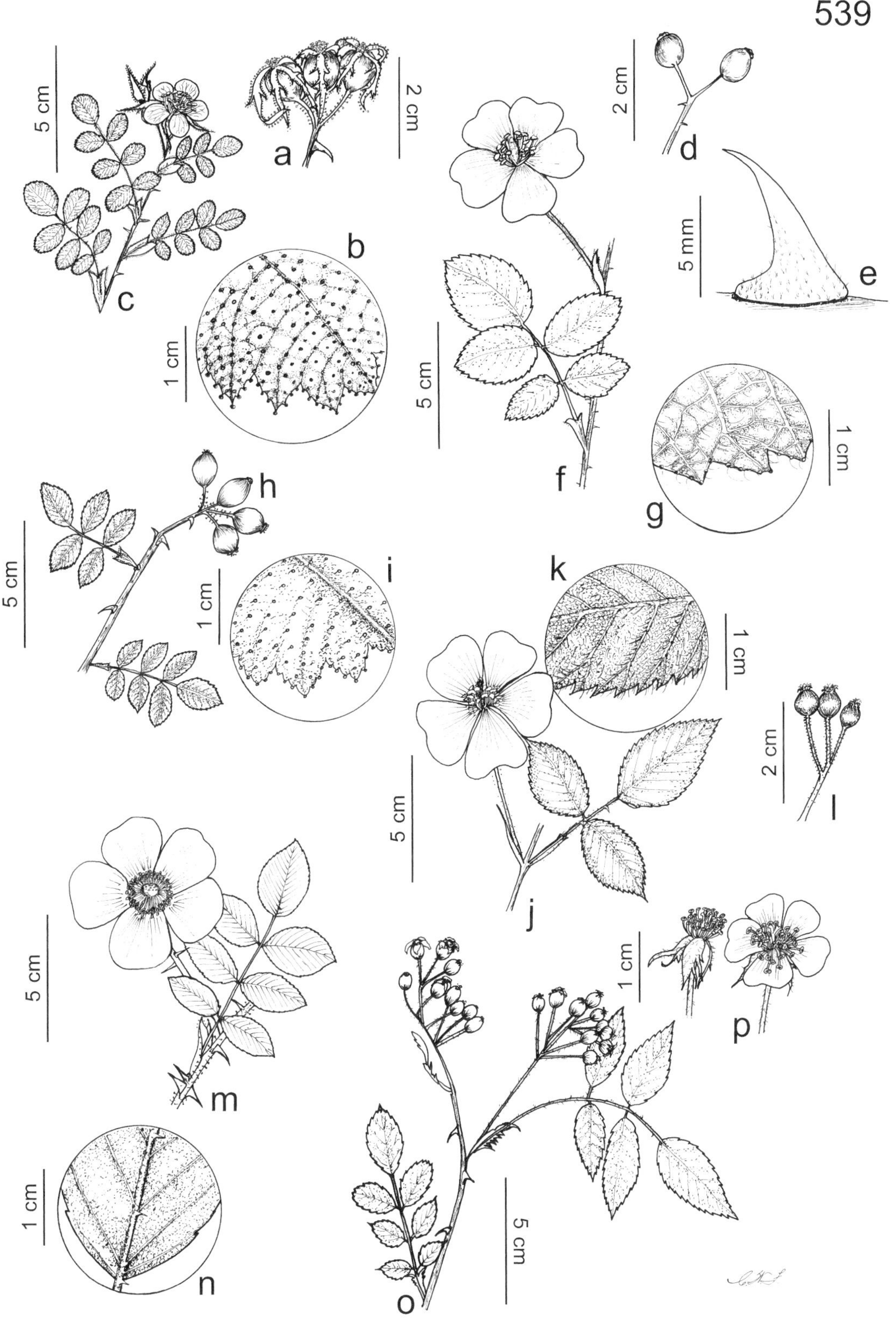

Plate 539. Rosaceae. *Rosa rubiginosa*, **a)** fruits, **b)** leaf detail, **c)** flowering and fruiting branch. *Rosa gallica*, **d)** fruits, **e)** prickle, **f)** flowering branch, **g)** leaf detail. *Rosa micrantha*, **h)** leaf detail, **i)** fruiting branch. *Rosa rugosa*, **j)** flowering branch, **k)** leaf detail, **l)** fruits. *Rosa palustris*, **m)** flowering branch, **n)** leaf detail. *Rosa multiflora*, **o)** fruiting branch, **p)** flower with petals removed and flower with petals intact.

red, with dense stalked glands. Achenes usually 3 (note additional presence of up to 21 undeveloped ovaries), 4–5 mm long. $2n$=28. May–July.

Introduced, uncommon, sporadic (native of Europe; introduced widely but sporadically in the New World, in U.S. in the eastern states west to Missouri and Texas). Banks of streams and margins of sinkhole ponds; also ditches, old homesites, old farmyards, railroads, and roadsides.

Steyermark (1963) reported *R.* ×*centifolia* L. (cabbage rose) from Missouri, based on a historical collection from Clark County. A second historical specimen from McDonald County also exists. This taxon represents an ancient, horticulturally derived hybrid involving *R. gallica* and possibly *R. moschata* Mill. (musk rose) (Heath, 1992; Bruneau et al., 2007). It is sterile and thus cannot spread by seed. A second cultigen that possibly shares the same parentage is *R.* ×*damascena* Mill. (Damask rose), which Steyermark (1963) reported from Clark, Dunklin, and Grundy Counties. These derivatives are said to differ from *R. gallica* in their more robust growth form, stouter, more curved prickles, leaflets with more sharply toothed margins, and longer, more glandular sepals. In practice, the cultivars of *R. gallica* are so variable that these distinctions are not readily apparent.

This species has numerous cultivars. It is commonly cultivated as an ornamental, but seldom escapes. Steyermark (1963) called plants with doubled flowers "var. *officinalis* Thory" (apothecary's rose, red rose of Lancaster), but this variant is actually a cultivar (Krüssman, 1981). French rose is a staple of the cut flower industry and also a popular rose bush in gardens. The petals have a strong, sweet fragrance and are often used in potpourri. *Rosa* ×*centifolia* is a commercial source of rose water.

6. Rosa lucieae Franch. & Rochebr. ex Crép. (Lucie rose, memorial rose)

R. wichurana Crép. (often misspelled *R. wichuraiana* or *R. wichuriana*)

Map 2491

Stems 100–300 cm long, often rooting at the nodes, low and trailing to scrambling or climbing on fences or other vegetation, moderately stout, green with prickles paired at the nodes and usually also unpaired between the nodes, these 4–5 mm long, 2–4 mm wide broadened at the base, curved or downward-angled, flattened. Leaves often somewhat evergreen, 7–10 cm long, the petiole and rachis glabrous but with stalked glands and usually sparse sessile glands, also with small prickles. Stipules 10–12 mm long, the margins deeply toothed, the auricle flared, 2–4 mm long. Leaflets 7 or 9, 9–30 mm long, 7–20 mm wide, the terminal leaflet with a stalk 5–9 mm long, the blades broadly ovate to obovate, angled at the base, angled or tapered at the tip, the margins simply toothed (occasionally a few second-order teeth present) with 12–16 teeth per side, these gland-tipped, the upper surface glossy, green, the undersurface glabrous except for a few glands along the midvein. Inflorescences on lateral branches from second year's stems and/or at tips of current year's growth, panicles of (5–)12–20 flowers, the flower stalks 18–25 mm long, glabrous, lacking glands or prickles, with 1–3 bracts (these similar to stipules but broader and longer). Flowers usually doubled (at least in Missouri materials), the hypanthium 4.0–6.5 mm long, glabrous, lacking glands, the mouth 1.5–2.0 mm in diameter. Sepals 6–8 mm long, 1.0–1.5 mm wide, usually more or less similar, the margins divided into slender lobes, the undersurface glabrous, becoming reflexed after flowering, usually shed by fruit maturity. Petals 13–15 mm long, pink, rarely white or red. Pistils 12–21, the styles fused but often becoming separated with age, hairy, extending 3.5–5.0 mm beyond the mouth of the hypanthium, thus protruding noticeably. Hips rarely maturing, 5–10 mm long, 5–9 mm wide, more or less globose, fleshy, red, glabrous, lacking glands. Achenes 1–11, 4.0–4.5 mm long. $2n$=14, 28. May–July.

Introduced, widely scattered, but most abundant in the southeastern portion of the state (native of Asia, introduced widely in the eastern U.S. west to Illinois, Missouri, and Louisiana). Edges of mesic upland forests; also fencerows, roadsides, and open, disturbed areas.

This species has not previously been reported from Missouri, and early collections were misdetermined as *R. multiflora*. The Missouri plants, which all possess doubled flowers, are mostly comparable with cv. 'Dorothy Perkins', a cultivar that was released in the early 1900s and was a popular climbing rose. The petals have a fragrance reminiscent variously of apples or cloves.

Many authors have referred to this species by the name *R. wichurana* (or variants of that spelling; Gleason and Cronquist, 1991). Ohba (2000) was the first to note that *R. lucieae* and *R. wichurana* represent the same taxon, with the former having nomenclatural priority.

7. Rosa multiflora Thunb. (multiflora rose, Japanese rose)

Pl. 539 o, p; Map 2492

Stems 150–250(–500) cm long, sometimes forming colonies, erect and strongly arching or occasionally sprawling, mostly relatively slender, green to reddish brown, the prickles paired at the nodes, 4–6 mm long, 2–3 mm wide, often broadened at the base, curved or rarely straight, flattened. Leaves 5–12 cm long, the petiole and rachis finely short-hairy and/or pubescent with longer more woolly hairs, usually with stalked glands and small prickles. Stipules 8–13 mm long, the margins deeply divided into numerous slender lobes, the auricle free, similar to the adjacent lobes, 4–8 mm long. Leaflets (5)7 or 9, 10–45 mm long, 8–25 mm wide, the terminal leaflet with a stalk 7–13 mm long, the blades obovate to elliptic, angled at the base, angled or tapered at the tip, the margins simply or occasionally somewhat doubly toothed with 12–20 teeth per side, these rarely gland-tipped, the upper surface shiny or dull, glabrous or nearly so, green, the undersurface glabrous to more commonly finely hairy, especially along the midvein. Inflorescences on lateral branches from second year's stems and at tips of current year's growth, panicles of (5–)10–30 flowers, the flower stalks 5–12 mm long, woolly, especially toward the tip, with stalked glands but lacking prickles, with several to numerous bracts (these similar to the stipules, but shed before fruiting). Flowers rarely doubled (more frequently so in cultivated plants), the hypanthium 2–3 mm long, glabrous, sometimes with stalked glands, the mouth 0.5–1.0 mm in diameter. Sepals 6–10 mm long, 1.5–2.0 mm wide, all similar, the margins with slender lobes, the undersurface glabrous, but often with stalked glands, becoming reflexed and shed soon after flowering. Petals 7–13 mm long, white or less commonly pink. Pistils 6–11, the styles fused but sometimes becoming separated toward the tips, glabrous, extending 3–4 mm beyond the mouth of the hypanthium, thus protruding noticeably. Hips (4–)5–7(–10) mm long, 5–7 mm wide, ovoid to globose, orangish red to red, glabrous, lacking glands or with a few stalked glands. Achenes 1–11, tan, 3.8–5.0 mm long. $2n$=14. May–June.

Introduced, scattered nearly throughout the state, but uncommon in the western half of the Glaciated Plains Division (native of Asia; introduced widely in the U.S.). Openings and edges of mesic upland forests, banks of streams, rivers, and spring branches, edges of marshes and fens; margins of ponds, lakes, and sinkhole ponds, upland prairies, and glades; also pastures, old fields, fallow fields, fencerows, railroads, roadsides, and open, disturbed areas.

Multiflora rose was first imported into the United State during the nineteenth century as grafting stock for other roses. Improved cultivars were bred by the U.S. Soil Conservation Service in the 1930s. The species was promoted by many state and federal wildlife agencies in the eastern half of the country as an ornamental, wildlife food plant, soil binder, and living fence between pastures. Sadly, much of the pioneering work in this regard was done by the Missouri Department of Conservation, which embraced the plant without proper data on its potential effects on the environment (Klimstra, 1956). This aggressive species soon began to spread and quickly became a problem in crop lands, pastures, and native plant communities. It was officially designated as a noxious weed by the Missouri legislature in 1983 and has been similarly designated in a number of other states. Many plants of multiflora rose in Missouri currently show evidence of a plant disease that stunts their growth and eliminates flowering, eventually killing the plant. This phenomenon is known as rose rosette disease (also called witch's brooms of rose) and is caused by an as-yet unknown agent, probably a virus or virus-like pathogen. It originated in the western United States and has been migrating eastward naturally. The disease is spread by a species of minute eriophyid mite (*Phyllocoptes fructiphilus* Keifer) and although multiflora rose is the main species infected in the Midwest, some other cultivated roses also are susceptible (Doudrick et al., 1986). This fatal disease causes disfiguring, abnormal, dense, highly branched growth of stems with anomalous, reduced leaves and bright reddish mottling.

8. Rosa palustris Marshall (swamp rose)
Pl. 539 m, n; Map 2493

Stems 100–250(–300) cm long, sometimes forming colonies or thickets, erect, relatively slender, reddish brown, the branches spreading to arched, with mostly paired prickles at the nodes, these 3.5–8.0 mm long, 2–5(–10) mm wide, stout, mostly broadened at the base, curved or occasionally straight, flattened, the uppermost branches or rapidly elongating shoots sometimes with prickles and/or bristles between the nodes, lacking glands, the stems rarely unarmed or nearly so. Leaves 8–11 cm long; the petiole and rachis glabrous or sparsely pubescent with short or longer and somewhat woolly hairs, lacking glands or with few stalked glands, sometimes with short, straight or curved prickles. Stipules 10–22(–35) mm long, 2.5–4.0 mm wide, the margins entire (but sometimes with glandular hairs) and usually somewhat curled under, the auricle straight or occasionally flared, 2.5–4.5(–8.0) mm long. Leaflets 5 or 7, 20–45 mm long 10–18 mm wide, the terminal leaflet with a stalk 5–10 mm long, the blades narrowly ovate to elliptic or elliptic-oblanceolate, angled to more or less rounded at the base, angled at the

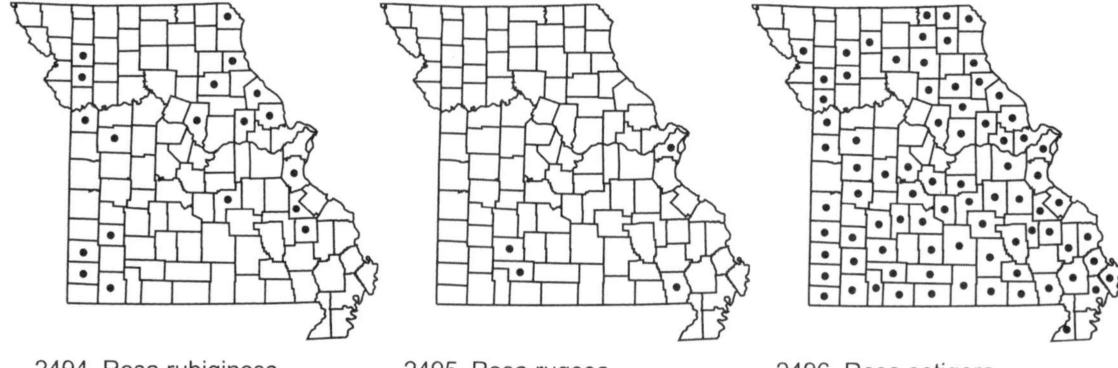

2494. Rosa rubiginosa

2495. Rosa rugosa

2496. Rosa setigera

sharply or bluntly pointed tip, the margins finely simply or occasionally doubly toothed with mostly 20–30 teeth per side, these sometimes gland-tipped, the upper surface dull or somewhat shiny, green to dark green, the undersurface pale green, glabrous or finely hairy along the main veins. Inflorescences on lateral branches from second year's stems and/or terminal on current year's growth, of solitary flowers or more commonly clusters of 2–10 flowers (rarely with panicles of up to 40 flowers elsewhere), the flower stalks 6–15 mm long, with many stalked glands, with 2 bracts. Flowers not doubled, the hypanthium 2–4 mm long, with sparse to dense stalked glands, the mouth 1.0–1.5 mm in diameter. Sepals 15–30(–40) mm long, 2.0–3.5 mm wide, usually slightly unequal, tapered to sharply pointed tips, the margins entire or with a few slender toothlike lobes, the undersurface with sparse to more commonly dense stalked glands, spreading to reflexed or erect and somewhat incurved after flowering, shed by fruit maturity or occasionally some of them persistent but withered. Petals 14–28 mm long, light pink to deep pink, frequently shriveling to form a small cap on the hip at fruiting. Pistils 24–50, the styles free, hairy, extending 0.5–1.0 mm beyond the mouth of the hypanthium, the mass of stigmas thus at most slightly elevated, more or less closing the hypanthium mouth. Hips 7–11 mm long, 7–11 mm wide, globose to subglobose or rarely slightly pear-shaped, fleshy, the surface deep red, with sparse to more commonly dense stalked glands. Achenes 12–28, 2.5–3.5 mm long. $2n$=14. May–July.

Scattered in the Mississippi Lowlands Division and the adjacent eastern portion of the Ozarks (eastern U.S. west to Iowa and Louisiana; Canada). Bottomland forests, swamps, fens, marshes, seeps, and margins of ponds and sinkhole ponds; also ditches and canals; often in shallow water.

This species is notable for its usually dense shrubby habit and slender stipules. Swamp rose is sometimes cultivated in gardens. It does best in moist, acidic soils and can be trimmed into hedges.

9. Rosa rubiginosa L. (sweetbrier, sweet briar, eglantine rose)

R. eglanteria L., an officially rejected name

Pl. 539 a–c; Map 2494

Stems 100–300 cm long, compact, erect to somewhat arching, slender, reddish brown, occasionally with paired prickles at the nodes, but mostly with scattered individual prickles at and between the nodes, these 6–10 mm long, 3–8 mm wide, broadened at the base, stout, curved, sometimes mixed with slender, bristlelike prickles and gland-tipped bristles, particularly on flowering branches. Leaves 4.0–6.5 cm long, the petiole and rachis sparsely hairy, also with stalked glands and small prickles. Stipules 6–10 mm long, the margins finely glandular-hairy, the auricle flared, 3–5 mm long. Leaflets 5 or 7; 10–22 mm long, 8–15 mm wide, the terminal leaflet with a stalk 5–10 mm long, the blades nearly circular to ovate or obovate, broadly angled to more or less rounded at the base, angled to a bluntly or sharply pointed tip, the margins irregularly doubly (or more) toothed with 10–18 teeth per side, these gland-tipped, the upper surface usually dull, green to dark green, glabrous or nearly so (rarely glandular-hairy), the undersurface with dense, sessile, rust-colored resin glands, often also finely hairy, occasionally also with scattered stalked glands along the midvein. Inflorescences on lateral branches from second year's stems or at the tips of current year's growth, of solitary flowers or less commonly clusters of 3–7 flowers, the flower stalks 6–9 mm long, with dense, stalked glands and/or fine gland-tipped bristles, with 1 or few bracts, these shed early. Flowers not doubled, the hypanthium 5–6 mm long (including the conspicuous neck), glabrous, but often with scattered to relatively dense stalked glands and/or finer gland-tipped hairs, the mouth

1.2–2.0 mm in diameter. Sepals 14–18 mm long, 2–3 mm wide, usually somewhat unequal, some of them sometimes extending into short, dilated, somewhat leaflike tips, the margins variously entire to having a few, long, slender lobes, the undersurface with dense stalked glands, usually becoming more or less erect (to rarely spreading) after flowering, more or less persistent at fruiting. Petals pink, 11–20 mm long. Pistils 25–45, the styles free, densely hairy, fused throughout, extending 1–2 mm beyond the mouth of the hypanthium. Hips 10–25 mm long, 10–22 mm wide, ellipsoid to somewhat pear-shaped, less commonly subglobose to broadly ovoid, fleshy to somewhat leathery, the surface red, glabrous or with short, stiff, often gland-tipped hairs, especially toward the base. Achenes 15–25, 3.5–4.0(–5.0) mm long. $2n$=35. May–July.

Introduced, scattered, mostly north of the Missouri River, known mostly from historical collections (native of Europe, Asia, Africa; introduced widely but sporadically in the U.S.). Banks of streams, and bases and ledges of bluffs; also pastures, railroads, and roadsides.

Sweetbrier is no longer as widely cultivated in Midwestern gardens as it was in the past. It is thus rarely encountered as an escape. When crushed, the resinous glands on the undersurface of the leaflets release an odor similar to that of fermenting apples. The hips of this species are crushed to extract rose hip oil, which is used in some skin care products. It is said to be useful in minimizing scars, wrinkles, and sun damage to the skin.

The name *Rosa eglanteria* predates the publication of the name *R. rubiginosa*, but because of ambiguities in its application to different species Turland (1996b) proposed that it be officially rejected and banned from further use. This was approved at the subsequent International Botanical Congress.

The small-flowered sweetbrier (*R. micrantha* Borrer, *R. rubiginosa* var. *nemoralis* (Léman) Thory, *R. rubiginosa* var. *micrantha* (Borrer) Lindl., *R. nitidula* Besser; Pl. 539 h, i) is a horticulturally derived taxon of uncertain origin. It may be simply part of the variation within *R. rubiginosa* or may have originated through hybridization between that species and *R. canina*. It is very similar to *R. rubiginosa* in general morphology, including the presence of resinous glands on the leaflets, but is usually distinguished by its ovate terminal leaflets that are angled or short-tapered to sharply pointed tips, slightly smaller flowers, paler petals, more or less glabrous styles, and less persistent sepals. Steyermark (1963) reported this taxon from historical collections from Jackson and McDonald Counties and it also is known from historical collections from Knox and Marion Counties.

10. Rosa rugosa Thunb. (Japanese rose, rugosa rose)

Pl. 539 j–l; Map 2495

Stems 100–250 cm long, potentially forming large colonies or thickets, erect to somewhat arched or reclining on surrounding vegetation, stout, initially pale green, turning purplish black, densely pubescent, at least when young, with fine woolly hairs, with paired prickles at the nodes and dense unpaired prickles between the nodes, these often intermixed with needlelike prickles and stalked glands, the largest prickles 6–10 mm long, 2–4 mm wide, gradually or abruptly broadened at the hairy base, straight. Leaves 7–11 cm, the petiole and rachis densely short-hairy, sometimes also with scattered small prickles. Stipules 20–30 mm long, 4–7 mm wide, the margins often with sessile glands, the auricle flared, 4–6 mm long. Leaflets 5–9, 15–55 mm long, 8–35 mm wide, the terminal leaflet with a stalk 8–18 mm long, the blades broadly elliptic to ovate or rarely obovate, leathery, angled to broadly angled or rounded at the base, angled to a bluntly or sharply pointed tip, the margins more or less curved downward, singly and mostly bluntly toothed with 11–17 teeth per side, the teeth usually gland-tipped, the upper surface glossy, dark green, appearing wrinkled with impressed veins, the undersurface grayish green, finely hairy, strongly veined. Inflorescences on short lateral branches from first or second year's stems, appearing axillary, of solitary flowers or less commonly clusters of 3–7 flowers, the flower stalks 10–15 mm long, finely short-hairy, the hairs often somewhat woolly, often also with stalked and/or sessile glands and short stiff bristles, with 1 or few bracts, these usually persistent. Flowers often doubled, the hypanthium 6–8 mm long (including the conspicuous neck), 5–6 mm wide, glabrous or with short bristly hairs toward the base and/or tip, the mouth 3–5 mm in diameter. Sepals 20–37 mm long, 4–6 mm wide, more or less similar to slightly unequal, usually extended into short, dilated tips, the margins entire but often glandular and/or hairy, the undersurface glandular and often with moderate to dense, short, bristly hairs, becoming erect after flowering, often persistent at fruiting. Petals pink to reddish pink or purplish pink (occasionally white in cultivated plants), 35–50 mm long. Pistils 48–60, the styles densely hairy, free, extending 1–2 mm beyond the mouth of the hypanthium. Hips 18–20 mm long, 20–25 mm

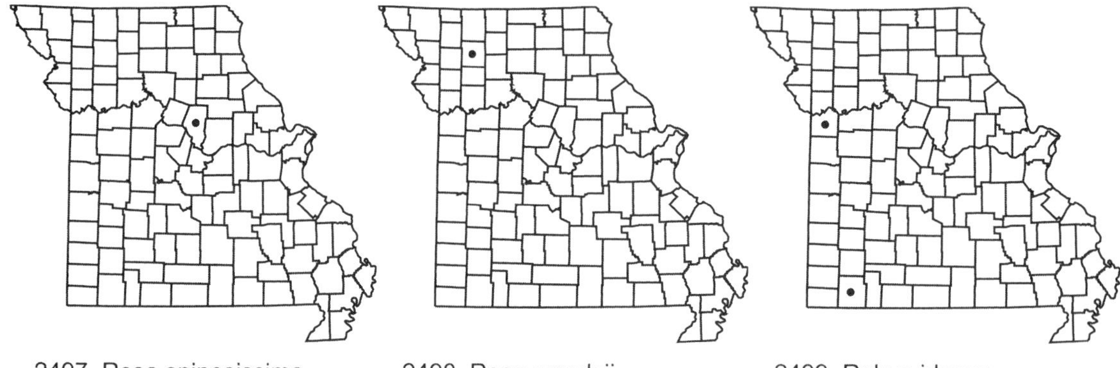

2497. Rosa spinosissima 2498. Rosa woodsii 2499. Rubus idaeus

wide, depressed-globose, leathery, the surface red, glabrous or occasionally with a few short, bristly hairs. Achenes 15–40, 4–6 mm long. $2n$=14. May–September.

Introduced, uncommon, known thus far only from Greene and Butler Counties (native of Asia, introduced in the northeastern U.S. west to Minnesota and Missouri, sporadically west to Oregon and Washington; Canada). Railroads and roadsides.

This is a common garden rose and numerous cultivars have been developed. The petals are sweetly scented and are sometimes used in potpourri. It seldom escapes from cultivation in Missouri (but is considered an invasive exotic in a few northern states and provinces). The doubled cultivars tend to produce relatively few achenes.

11. Rosa setigera Michx. (prairie rose, climbing rose)

R. setigera var. *tomentosa* Torr. & A. Gray

Pl. 540 a, b; Map 2496

Stems 100–200(–600) cm long, sometimes tip-rooting, trailing or more commonly climbing on and through other vegetation, occasionally loosely twining, the tips often arched downward, usually relatively slender, green to light brown, glabrous or occasionally finely hairy, the prickles paired at the nodes and individual between the nodes, 5–9 mm long, 3–5 mm wide, broadened at the base, curved or occasionally some of them straight, stout, flattened, rarely unarmed. Leaves 8–12 cm long, the petiole and rachis usually finely hairy and with stalked glands, occasionally glabrous, sometimes also with small prickles. Stipules 12–15 mm long, the margins entire, but sparsely stalked-glandular, sometimes somewhat curved under, the auricle flared, 3–4 mm long. Leaflets 3 or 5, 30–50 (–100) mm long, 15–40 mm wide, the terminal leaflet with a stalk 10–16 mm long, the blade ovate to elliptic-ovate, broadly angled or rounded at the base, angled or more commonly tapered to a sharply pointed tip, the margins simply or doubly toothed with 18–42 teeth per side, these relatively coarse, gland-tipped, the upper surface glossy, green to dark green, usually glabrous, the undersurface lighter green, glabrous or finely hairy (sometimes finely woolly), sometimes also glandular. Inflorescences on lateral branches from second year's stems and axillary toward tips of current year's growth, of solitary flowers or more commonly more or less flat-topped clusters or small panicles of 3–6(–15) flowers, the flower stalks 15–25 mm long, with stalked glands but usually lacking prickles, with 1 to several bracts, these slender, often shed early. Flowers not doubled, the hypanthium 4–6 mm long, with usually dense stalked glands, the mouth 0.5–1.0 mm in diameter. Sepals 10–18 mm long, 2–4 mm wide, all similar, the margins entire, the undersurface finely woolly and stalked-glandular, becoming reflexed and shed soon after flowering. Petals 18–25(–30) mm long, pink to reddish pink, rarely white. Pistils 20–25, the styles fused, glabrous, extending 5–6 mm beyond the mouth of the hypanthium, thus protruding noticeably. Hips 6–10 mm long, 6–9 mm wide, globose to subglobose, more or less fleshy, the surface red. glabrous or with scattered stalked glands. Achenes 17–22, 4–5 mm long. $2n$=14. May–July.

Scattered nearly throughout the state (eastern U.S. [except some far-eastern states] west to Nebraska and Texas; introduced along the Atlantic Seaboard; Europe). Bottomland forests, mesic upland forests, banks of streams and rivers, margins of ponds, lakes, and sinkhole ponds, upland prairies, glades, bases, ledges, and tops of bluffs, edges of fens, and rarely swamps; also pastures, old fields, fencerows, ditches, old quarries, old mines, railroads, and roadsides.

Although the flowers of *R. setigera* appear perfect, they are functionally either staminate or pis-

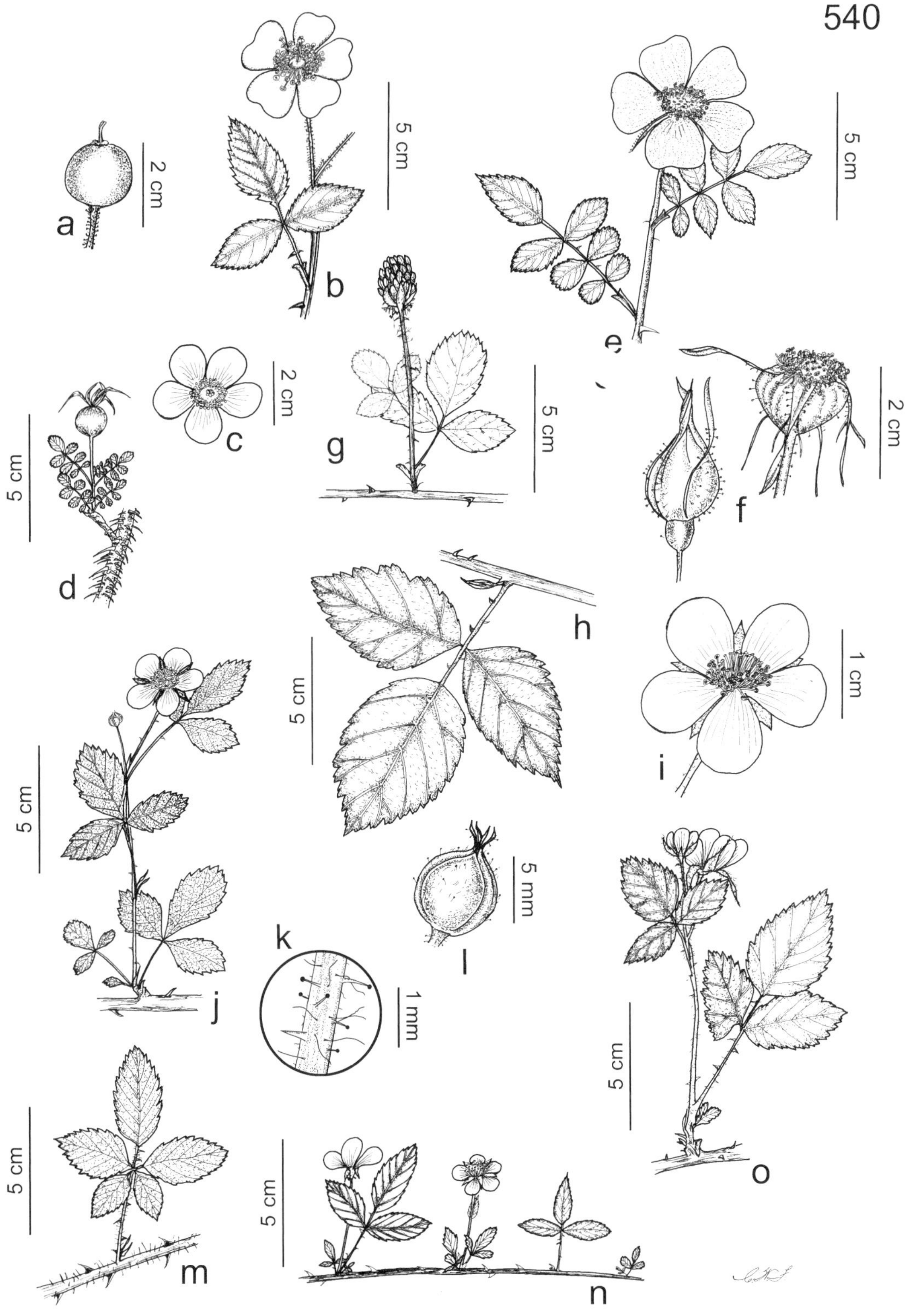

Plate 540. Rosaceae. *Rosa setigera*, **a)** fruit, **b)** flowering branch. *Rosa spinosissima*, **c)** flower, **d)** fruiting branch. *Rosa carolina* ssp. *carolina*, **e)** flowering branch, **f)** bud and fruit. *Rubus enslenii*, **g)** floricane node with leaves and fruits, **h)** primocane node with leaf, **i)** flower. *Rubus leviculus*, **j)** floricane branch with inflorescence, **k)** detail of flower stalk showing prickle, hairs and stalked glands, **l)** bud. *Rubus trivialis*, **m)** primocane node with leaf, **n)** flowering branch with flowers. *Rubus steelei*, **o)** floricane branch with inflorescence.

tillate (W. H. Lewis, 1958b). This results in some plants that do not set fruit and the necessity of having staminate and pistillate plants in proximity to ensure seed production.

Rosa setigera is sometimes cultivated and also has been used in horticultural crosses with European roses. Plants with prickly stems and hairy leaves have sometimes been called var. *tomentosa* Torr. & A. Gray, but this combination of characters does not circumscribe a discrete taxon (W. H. Lewis, 1958b). Rare plants with unarmed stems and hairy leaves have been called f. *serena* (E.J. Palmer & Steyerm.) Fernald, whereas those with unarmed stems and glabrous leaves have been called f. *inermis* E.J. Palmer & Steyerm. White-flowered individuals are f. *alba* Steyerm. All of these variants have been reported from Missouri (Steyermark, 1963).

12. Rosa spinosissima L. (Scotch rose, burnet)
Pl. 540 c, d; Map 2497

Stems 40–150 cm long, often forming colonies, ascending to erect, slender, green to reddish-tinged or reddish brown, the prickles dense and longer on main stems, but usually sparse and smaller on the flowering branches, paired at the nodes and individual between the nodes, 4–6 mm long, 1–3 mm wide, at least the larger prickles abruptly broadened at the base, straight, slender, unequal, scattered, mixed with bristlelike prickles, usually nonglandular. Leaves 2.5–8.0 cm long, the petiole and rachis glandular-hairy, also scattered small prickles. Stipules 9–14 mm long, the margins entire or glandular-toothed, the auricle flared, 2.5–5.0 mm long. Leaflets 5–11, 5–11(–22) mm long, 5–8 mm wide, the terminal leaflet with a stalk 2–4 mm long, the blades oblong-ovate to nearly circular, broadly angled to broadly rounded at the base, rounded or broadly angled to a bluntly or sharply pointed tip, the margins mostly simply toothed with 8–12 teeth per side, not appearing wrinkled, the upper surface dull, green to dark green, glabrous, the undersurface green, nonglandular. Inflorescences terminal at tips of present year's growth, usually of solitary flowers, the flower stalks 15–35 mm long, stout, usually nonglandular but often with sparse bristlelike prickles, appearing bractless. Flowers sometimes doubled, the hypanthium 3–5 mm long, glabrous, the mouth usually 2.0–2.5 mm in diameter. Sepals 10–12 mm long, 2.5–3.0 mm wide, equal or nearly so, all long-tapered to sharply pointed tips and with the margins entire, the undersurface glabrous, becoming erect or ascending and often somewhat incurved after flowering, persistent at fruiting. Petals 15–25 mm long, white or rarely pink (deep pink to red in some cultivated plants). Pistils 18–23, the styles free, densely hairy, extending 1.5–2.0 mm beyond the mouth of the hypanthium, the mass of stigmas thus protruding noticeably, not closing the hypanthium mouth. Hips 8–13 mm long, 8–16 mm wide, globose to somewhat depressed-globose, fleshy, the surface black or dark brown, glabrous, shiny. Achenes 8–12, 3–4 mm long. $2n=28$. May–July.

Introduced, uncommon, known thus far from a couple of historical specimens from Boone County (native of Asia; introduced sporadically in the northeastern U.S. west to Wisconsin and Kansas; Canada). Roadsides and disturbed areas.

13. Rosa woodsii Lindl. **ssp. woodsii** (Woods' rose)
Map 2498

Stems 40–120(–200) cm long, sometimes forming thickets, erect or ascending, slender, unbranched or few-branched, dark red, the oldest stems sometimes turning grayish and developing peeling bark, the prickles usually moderate to dense the length of the stem, paired at the nodes and individual between the nodes, 3–7 mm long, 1–5 mm wide, broadened at the base, curved or occasionally some of them straight, slender to stout, flattened or not, those between the nodes sometimes more slender than those at the nodes. Leaves 2.5–8.0(–12.0) cm long, the petiole and rachis glabrous or rarely hairy, sometimes with sparse stalked glands, unarmed. Stipules (6–)9–16(–25) mm long, the margins wavy or with a few shallow, glandular teeth, the auricle usually flared, 2–6 mm long. Leaflets 5 or 7(9), 8–25(–40) mm long, 6–20(–26) mm wide, the terminal leaflet with a stalk 3–12 mm long, the blades obovate to less commonly elliptic or ovate, angled to occasionally broadly rounded at the base, rounded to broadly angled at the tip, the margins simply or occasionally doubly toothed with (5–)7–14 teeth per side, these mostly not gland-tipped, the upper surface dull, green, rarely somewhat glaucous, the undersurface lighter green, glabrous or minutely hairy, mainly along the veins. Inflorescences at the tips of lateral branches from second year's stems, of solitary flowers or small clusters of 2 or 3 flowers, the flower stalks 10–20(–33) mm long, glabrous or rarely with a few stalked glands, with usually 1 bract. Flowers not doubled, the hypanthium 3–6 mm long, glabrous, the mouth 2.0–2.5 mm in diameter. Sepals 8–14(–21) mm long, 1.5–2.5 mm wide, sometimes unequal, then some of them extended into dilated, somewhat leaflike tips and with the margins toothed or with 1 or 2 pairs of linear lobes, the undersurface glabrous, erect or ascending to occasionally spreading or reflexed after flowering, mostly persistent at fruit maturity. Petals (15–)20–32 mm long, pink to deep pink. Pistils (16–)20–40(–50), the styles free, hairy, not or very slightly extending beyond the mouth of the hypanthium, the mass of stigmas thus not elevated, closing the hypanthium mouth. Hips (8–)10–20 mm long, (5–)10–20 mm wide, globose or subglobose to ovoid, fleshy, the surface orangish red to purplish red, glabrous, not glandular. Achenes 15–35, 3.0–4.5 mm long. $2n$=14. May–June.

Introduced, known thus far from a single historical collection from Daviess County (western U.S. east to Wisconsin and Texas, eastward locally to Illinois; Canada, Mexico; introduced sporadically farther east in the U.S.). Railroads.

This species was reported from Missouri by Fernald (1950) and Gleason and Cronquist (1963, 1991), but was excluded from the state's flora by Steyermark (1963), who indicated that the specimens attributed to this species were instead examples of *R. carolina*. However, during his research toward the *Rosa* treatment for the as-yet unpublished Rosaceae in Volume 9 of the *Flora of North America*, Walter Lewis (Washington University) discovered a historical Daviess County specimen (*Kellogg 26131*, collected on 28 May 1934 near Pattonburg) at the Missouri Botanical Garden herbarium.

Rosa woodsii has been regarded as a close relative of *R. blanda* and is widespread in the western and northern United States, as well as portions of Canada and Mexico. The species is said to differ from *R. blanda* mainly on the basis of its well-developed prickles between the nodes, even toward the stem tips. However, in their studies of molecular phylogeny of the genus *Rosa*, Joly and Bruneau (2006) concluded that *R. blanda* and *R. woodsii* represent an interbreeding set of populations whose taxonomy should be reassessed. They noted that hybrids between the *R. blanda* and *R. woodsii* morphotypes (*R.* ×*dulcissima* Lunell) are fertile. The controversy surrounding the *R. blanda* / *woodsii* complex requires further study; the present treatment provisionally accepts the traditional classification of the two taxa as separate species.

Rosa woodsii has been divided into a series three to five weak infraspecific taxa, based on such characters as overall plant size, density and distribution of prickles and presence or absence of stalked glands. The Missouri sheet is referable to ssp. *woodsii* by these criteria. It should be noted that plants of *R. woodsii* might key to *R. arkansana* in the key to species above, but in *R. arkansana* at least some of the inflorescences always develop at the tips of current year's stems, whereas in *R. blanda* and *R. woodsii* the inflorescences always develop at the tips of side branches from second year's stems.

25. **Rubus** L. (raspberry, blackberry, bramble)
Contributed by Mark P. Widrlechner

Plants shrubs (perennial herbs elsewhere), sometimes suckering or reproducing vegetatively from rooted stem tips. Stems (referred to as canes) biennial or (in sect. *Rubus*) occasionally persisting longer, prostrate, climbing, arched, or erect, all those found in Missouri armed with prickles, sometimes also with long, stiff bristles (except for horticultural selections); first year's stems generally vegetative (called primocanes), unbranched or few-branched (moderately branched in a few species); second year's stems fertile (called floricanes), usually dying back to the rootstock at the end of the second growing season. Leaves rarely evergreen. Stipules small to conspicuous, persistent, leaflike, fused to the petiole laterally or attached at the junction of the stem and petiole. Leaf blades palmately or pinnately compound, the margins toothed and sometimes lobed, the surfaces often hairy, sometimes with gland-tipped hairs or bristles, the main veins sometimes also with prickles. Inflorescences typically axillary on floricanes, of branched or simple clusters, racemes, panicles, or occasionally solitary flowers. Flowers perfect (imperfect elsewhere), with a noticeable hypanthium. Sepals 5, fused basally, persistent at fruiting, erect, horizontally spreading, or reflexed. Petals usually 5 (rarely more in horticultural selections), white or less commonly pale pink to rose pink, glabrous or hairy, the margins usually entire, rarely lobed (in *R. laciniatus*). Stamens numerous, the filaments attached at the mouth of the hypanthium. Pistils many, attached to an enlarged receptacle, this hemispheric, or becoming conic or cylindric as the fruits mature, each with 1 carpel and 2 ovules, 1 of these aborting as the fruit develops. Style 1, threadlike or occasionally narrowly club-shaped, glabrous or hairy. Fruits drupelets positioned on the enlarged receptacle to form an aggregate, which either separates freely from the receptacle when ripe (raspberries) or is shed with the receptacle attached (blackberries). About 800 species (although species concepts and the resulting estimates differ widely), worldwide, particularly abundant in temperate regions of the northern hemisphere.

Rubus is a large genus that is diverse in the midwestern United States. Like *Crataegus*, its reputation for hybridization, polyploidy, and apomixis has given it a well-deserved reputation as a taxonomically difficult genus. Additionally the prickly canes that change appearance from the first to second growing seasons have caused many collectors to avoid the plants. *Rubus* is thus underrepresented in herbaria and many of the specimens that are present are relatively incomplete. Plants often thrive in disturbed, successional habitats, sometimes in mixed-species thickets. A number of mixed collections in various herbaria involve the primocanes of one species and the floricanes of a second species growing at the same site.

The genus produces a number of important fruit crops, including raspberries, blackberries, and loganberries. In addition to being consumed as fresh fruits, the fruits are used extensively in baked goods, jams, jellies, preserves, and as juices and flavorants. The fruits also are an important food source for wildlife. A few species are cultivated as ornamentals.

Whenever possible, users of this treatment should examine primocane leaves for identification characters involving leaves. References to primocane leaves and leaflets in the descriptions that follow are based on fully expanded leaves. It should be noted that in leaves with five leaflets, the apical or middle leaflet is called the central leaflet, the adjoining pair of lateral leaflets are termed the middle leaflets, and the pair positioned closest to the petiole are known as basal leaflets. Floricane foliage is primarily found on inflorescence branches, but can also be found on sterile branches. Large, sterile branches that sometimes form near the base of floricanes are called parcifronds and can easily be confused with primocanes. Floricane leaves on sterile branches generally resemble those of primocanes, but often are smaller and have only 3 leaflets. Floricane foliage on inflorescence branches is denoted in the species descriptions below as bracts or leafy bracts.

In the present treatment, inflorescences are described and measured from well-developed examples on floricanes. Inflorescences can also emerge directly from the crown, generally when a primocane has been damaged or winter-killed. These atypical inflorescences are called novirames, can flower out of season, and are extremely difficult to identify to species.

North American herbaria hold thousands of fragmentary *Rubus* specimens that cannot be determined to species. When vouchering *Rubus*, it is important to collect pieces of both the primocane and floricane, with the primocane displaying both sides of fully-expanded leaves and the floricane displaying well-developed inflorescences, collected at any time between the start of flowering and the formation of ripe fruits. These pieces should be kept together, either by mounting on the same sheet or through appropriate labeling. On labels, it is also useful to note the overall plant habit and the presence or absence of rooting at the tips of the canes.

The genus *Rubus* generally is divided into five subgenera and numerous sections. The relationships among the sections are poorly understood, and it is not clear if the traditional classification circumscribes natural subgroups. The nomenclature of the sections is also still somewhat controversial, as Liberty Hyde Bailey (1858–1954; the leading specialist on the genus in the first half of the twentieth century) initially treated these as unranked groups and by the time that he validated the sectional names, some might already have been published by other authors. Thus the names applied to the sections should be treated as provisional, pending further review of the literature on *Rubus*. Grouping the species into morphologically based subgenera and sections, as is done in the present treatment, has the advantage of saving space by requiring that characters shared among species within a section be listed only once in the section description. Users are cautioned to read the subgeneric and sectional descriptions carefully in addition to the species descriptions for help in confirming the identity of plants being determined. The following key to the subgenera and sections of *Rubus* uses a combination of characters from primocanes and floricanes, but can be navigated with only one type of cane, with the exception of couplet 4, which requires inflorescences.

1. Leaves whitish beneath; petals inconspicuous, generally 4–7 mm long, spatulate to obovate, greenish white, white, or rose pink; ripe fruits readily falling intact from the receptacle . . . . 1. SUBG. IDAEOBATUS
1. Leaves green to gray beneath (except for *R. armeniacus* in Sect. *Rubus*, which is whitish beneath); petals showy, generally 7–25 mm long, obovate to nearly orbicular, white to light pink (primarily in sect. *Rubus*); ripe fruits not separating from the receptacle (2. SUBG. RUBUS)
    2. Canes arching to prostrate; primocanes with indeterminate growth resulting in whiplike tips that can take root when contacting soil; floricanes with inflorescences borne irregularly along the canes based on access to light, the most fully developed inflorescences often found in the middle third of the canes; prickles downward-angled to downward-curved, 1–5 mm long; leaves nearly glabrous to softly pubescent beneath
        3. Canes bearing prickles and sometimes glands with translucent or pale stalks; leaves nearly glabrous to softly pubescent beneath, deciduous . . . . 2C. SECT. FLAGELLARES
        3. Canes bearing prickles and stiff, dark red hairs, some tipped with glands; leaves nearly glabrous beneath, more or less persistent through winter . . . . 2F. SECT. VEROTRIVIALES
    2. Canes erect to arching; primocanes with determinate growth resulting in condensed nodes at tips that do not take root; floricanes with inflorescences typically borne in the terminal third of the cane, the most fully developed inflorescences often found a few nodes below the tips; prickles straight or downward-curved, if curved then 3–8 mm long or occasionally longer; leaves softly pubescent beneath
        4. Inflorescences racemose, paniculate, or flaring in a broomlike form, often branched
            5. Inflorescences paniculate; canes branched; primocane leaves either white-felted beneath or green, margins deeply toothed to irregularly divided into slender jagged lobes; linear or filiform stipules clearly lateral, diverging 3–6 mm or more from the base of the petiole . . . . 2D. SECT. RUBUS
            5. Inflorescences paniculate or racemose, varying on the same cane from large, branched, and broomlike to reduced and unbranched, both cylindric and more-or-less flat-topped; canes with no or few vegetative branches; primocane leaves green beneath, the margins sharply toothed; lanceolate stipules basal to somewhat lateral, diverging 0–3 mm from the base of the petiole . . . . 2E. SECT. SETOSI
        4. Inflorescences racemose, cylindric or flaring toward the apex, rarely branched
            6. Primocane tips, inflorescences, and often also the flower stalks with stalked glands (typically 0.5–1.5 mm long, obvious at 10× magnification) . . . . 2A. SECT. ALLEGHENIENSES
            6. Plants lacking stalked glands
                7. Primocane leaflets softly pubescent, but not typically gray-felted beneath, petals white . . . . 2B. SECT. ARGUTI
                7. Primocane leaflets gray-felted beneath, petals light pink (but fading to white) . . . . 2D. SECT. RUBUS

1. Subgenus Idaeobatus (Focke) Focke (raspberry)

Primocane leaves pinnately or palmately compound, felted-hairy beneath. Stipules persistent, fused to the petiole. Fruits freely falling intact from the receptacle. Sepals long-tapered to a filiform tip. About 135 species, throughout the northern hemisphere, also in New Guinea and Australia, most diverse in China.

1. Canes erect to arching, not rooting at the tips, ripe fruits purplish red (rarely yellow) . . . . 1. R. IDAEUS
1. Canes arching to ascending from a spreading base, often rooting at the tips, ripe fruits red, purple, or purplish black (rarely amber)
   2. Canes with slender, straight prickles and red to purple hairs, many gland-tipped . . . . 4. R. PHOENICOLASIUS
   2. Canes with slender, downward-angled to downward-curved prickles or broad-based prickles, lacking red to purple hairs or stalked glands
      3. Primocanes purple (or rarely yellowish green) in winter, often glaucous; primocane central leaflets ovate to elliptic, tapered to a sharply pointed tip, cordate to truncate at the base; ripe fruits purplish black (or rarely amber) . . . . 2. R. OCCIDENTALIS
      3. Primocanes reddish brown in winter, not glaucous; primocane central leaflets obovate to broadly subrhombic, angled or tapered to a bluntly pointed tip, angled at the base; ripe fruits red . . . . 3. R. PARVIFOLIUS

1. Rubus idaeus L. **var. strigosus** (Michx.) Maxim. (red raspberry)

R. idaeus ssp. *sachalinensis* (H. Lév.) Focke

R. strigosus Michx.

Pl. 543 a, b; Map 2499

Canes to 170 cm long, erect to high-arching, to 150 cm tall, not rooting at the tips, but forming colonies from rhizomes. Primocanes light green, tan, or light purple, 3–5 mm in diameter. Bristles and prickles dense, 15–50 bristles or prickles per cm of cane, the prickles fine and needlelike, straight to downward-angled, 1.0–3.5 mm long, some bristles and prickles gland-tipped. Petioles with gland-tipped hairs and bristles. Stipules 5–10 mm long, threadlike. Primocane leaves pinnately compound with 3–5 leaflets, the margins sharply toothed, sometimes doubly toothed, the upper surface glabrous to thinly hairy, the undersurface silvery gray to grayish green-felted. Central primocane leaflets 6–12 cm long, 5–8 cm wide, elliptic to ovate, occasionally 3-lobed, base rounded to cordate, tapered to a sharply pointed tip, the leaflet stalk generally 1/4–1/3 as long as the leaflet blade in examples with 3 leaflets; lateral leaflets ovate to elliptic, rounded to cordate at the base, tapered to a sharply pointed tip, sessile or nearly so. Inflorescences in simple to complex clusters, in extreme cases paniculate, 7–35 cm long, with 1–18 flowers often concentrated near the tip and 3–8 leafy bracts, these mostly with 3 leaflets; flower and inflorescence stalks thinly to densely covered with gland-tipped hairs and bristles, the flower stalks branched on the most vigorous inflorescences. Sepals 5–9 mm long, 2–3 mm wide, triangular-ovate, often with gland-tipped hairs. Petals 4–6 mm long, narrowly obovate, white. Fruits 13–17 mm long, 13–17 mm wide, conic, purplish red (or rarely yellow) when ripe, the drupelets often separating. $2n$=14. May–June.

Introduced, known thus far only from historical specimens from Barry and Jackson Counties (northeastern U.S. and adjacent Canada south to North Carolina, Tennessee, Illinois, and Iowa and west to Nebraska and North Dakota). Habitat unknown, but presumably brushy disturbed areas.

The only records supporting the naturalization of red raspberries in Missouri come from Courtney, in Jackson County, where B. F. Bush collected them in 1927 and 1932, and from a 1974 collection by J. T. Shelton of a young red raspberry seedling along a logging road in Barry County. Steyermark (1963) included these specimens as examples of *R. strigosus*, herein treated as a variety of *R. idaeus* native to eastern North America. This variety is not widely cultivated, but has contributed to the parentage of some domestic raspberry cultivars. No specimens have been located to support Steyermark's (1963) report that native red raspberries occur in Atchison or Holt Counties. Cultivars of the domesticated Eurasian red raspberry (var. *idaeus*) are widely cultivated in Missouri, but are not known to naturalize in the state.

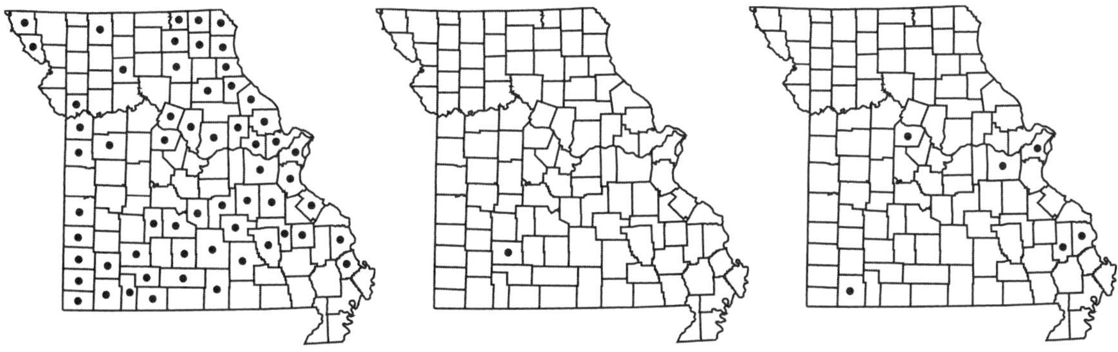
2500. Rubus occidentalis 2501. Rubus parvifolius 2502. Rubus phoenicolasius

2. Rubus occidentalis L. (black raspberry)
Pl. 542 j, k; Map 2500

Canes to 400 cm long, erect, to 200 cm tall when young, but ultimately arching and often rooting at the tips. Primocanes bluish green in summer, turning purple (or rarely yellowish green) in winter, often glaucous, 3–6 mm in diameter. Prickles sparse to moderate, 0.2–1.7 prickles per cm of cane, broad-based and downward-curved, 3–5 mm long. Petioles armed with broad-based, downward-curved prickles 1–2 mm long. Stipules 4–8 mm long, threadlike. Primocane leaves palmately compound with 3(5) leaflets, the margins coarsely and doubly toothed, the upper surface glabrous to thinly hairy, the undersurface white-felted. Central primocane leaflets 7–12 cm long, 4–9 cm wide, ovate to elliptic, cordate to rounded at the base, angled or tapered to a sharply pointed tip, the leaflet stalk about 1/4–1/2 as long as the leaflet blade; lateral leaflets ovate to ovate-elliptic, rounded to broadly angled at the base, tapered to a sharply pointed tip, the middle pair (when present) stalked, the basal pair nearly sessile, occasionally with asymmetric basal lobes. Inflorescences simple to complex clusters, in extreme cases paniculate, 8–40 cm long, with (3–)5–19(–27) flowers and (3–)5–13 leafy bracts, these nearly all with 3 leaflets; flower and inflorescence stalks with fine hairs, needlelike prickles (that at 10× magnification seem to be made of ivory), and occasional small, broad-based prickles, the flower stalks branched on the most vigorous inflorescences. Sepals 5–9 mm long, 2–3 mm wide, triangular-ovate to triangular-elliptic. Petals 2–5 mm long, narrowly obovate, white. Fruits 12–15 mm long, 12–15 mm wide, hemispheric, purplish black (rarely amber) when ripe. $2n$=14. April–June.

Scattered nearly throughout the state (northeastern U.S. and adjacent Canada south to South Carolina, Georgia, and Arkansas and west to Oklahoma, Nebraska, and North Dakota). Mesic upland forests, bottomland forests, bases and ledges of bluffs, banks of streams, upland prairies, loess hill prairies, and savannas; also pastures, fencerows, ditches, railroads, and roadsides.

3. Rubus parvifolius L. (Japanese raspberry)
Map 2501

Canes to 300 cm long, low-arching to prostrate, to 70 cm tall, often rooting at the tips and occasionally at other nodes. Primocanes green to purplish green in summer, turning reddish brown in winter, 2.0–3.5 mm in diameter. Prickles moderate to dense, 0.5–4.0 prickles per cm of cane, slender and downward-angled or downward-curved, 1–2 mm long. Petioles armed with slender, downward-curved prickles 1–2 mm long. Stipules 8–18 mm long, linear-lanceolate and sometimes cleft into 2 lobes. Primocane leaves with 3 leaflets or pinnately compound with 5 leaflets, margins coarsely and broadly toothed, the upper surface glabrous, the undersurface white-felted. Central primocane leaflets of leaves with only 3 leaflets 3.5–8.5 cm long, 3.0–7.5 cm wide, obovate to broadly subrhombic, often lobed, angled at the base, angled or tapered to a bluntly pointed tip, the leaflet stalk about 2/5–1/2 as long as the leaflet blade when only 3 leaflets are present; lateral leaflets rhombic (or elliptic when 5 leaflets are present), nearly sessile. Inflorescences simple to complex clusters, in extreme cases paniculate, 15–45 cm long, with (3–)10–19 flowers and 4–10 leafy bracts, these mostly with 3 leaflets; flower and inflorescence stalks with stiff hairs, and small needlelike and broad-based prickles, the flower stalks branched on vigorous inflorescences. Sepals 6–10 mm long and 3–4 mm wide at flowering, expanding to 8–12 mm long and 4–5 mm wide at fruiting, triangular-ovate. Outer surface of sepals prickly. Petals 4–6 mm long, spatulate, rose pink. Fruits 10–15 mm long, 10–17 mm wide, hemispheric to short-conic when well formed (some

2503. Rubus allegheniensis 2504. Rubus alumnus 2505. Rubus rosa

populations produce few drupelets, if any), bright, glossy red when ripe. $2n$=14, 21, 28. May.

Introduced, known thus far only from Greene County (native to Asia, Australia; naturalized in Illinois, Iowa, Massachusetts, Missouri, Virginia). Edges of mesic upland forests; also roadsides and open, disturbed areas.

This unusual member of subgen. *Idaeobatus* was first collected by Michael Currier in 1989, but was not correctly determined for another nine years. The species is first reported from Missouri in the present account. The population, which is near the Dickerson Park Zoo in Springfield, has continued to expand locally. In Iowa, populations of this species also have been noted as expanding, and may invade prairies and savannas (Drobney and Widrlechner, 2010).

4. Rubus phoenicolasius Maxim. (wineberry)
Map 2502

Canes to 400 cm long, erect, to 200 cm tall when young, but ultimately arching and often rooting at the tips. Primocanes dark purple, 4–7 mm in diameter, densely covered with stiff purple hairs, many gland-tipped, and occasional prickles. Prickles moderate, 1.0–2.3 prickles per cm of cane, slender and straight, 5–8 mm long. Petioles densely covered with stiff purple hairs, many gland-tipped, and downward-curved prickles to 3 mm long. Stipules 5–14 mm long, linear, glandular. Primocane leaves with 3 leaflets, the margins doubly toothed, the upper surface appearing wrinkled (having deeply impressed veins), grayish and hairy, the undersurface densely white-felted but with dark red veins bearing gland-tipped hairs. Central primocane leaflets 6.5–11.5 cm long, 5.5–12.0 cm wide, broadly ovate, often 3-lobed, cordate to nearly truncate at the base, abruptly tapered to a short, sharply pointed tip, the leaflet stalk about 1/4–2/5 as long as the leaflet blade; lateral leaflets much smaller, oblong or oval, nearly sessile. Inflorescences paniculate, 30–42 cm long, with 6–22 flowers and 6–8 leafy bracts, these mostly with 3 leaflets; flower and inflorescence stalks densely covered with stiff hairs, many gland-tipped, inflorescence stalks also bearing slender, straight prickles. Sepals 7–8 mm long, 2.5–3.5 mm wide, lanceolate, the outer surface with gland-tipped hairs. Petals 4–5 mm long, spatulate, minutely toothed near the tip, white to pale pink. Fruits 10–15 mm long, 10–15 mm wide, hemispheric, enclosed in the calyx when young, orangish red. $2n$=14. May–June.

Introduced, uncommon and sporadic south of the Missouri River (native of Asia; naturalized in the eastern U.S.). Bottomland forests, disturbed openings of mesic upland forests, and bases and ledges of bluffs; also railroads and roadsides.

Although it was first-collected in 1949 in Cape Girardeau County, *R. phoenicolasius* was not included in Steyermark's (1963) treatment of the genus in Missouri. The species is first-reported from Missouri in the present account. In some other states, it is considered an invasive exotic.

2. Subgenus Rubus (blackberry, dewberry)

Primocane leaves palmately compound. Stipules persistent, fused to the petiole laterally or basally. Fruits purplish black except in horticultural selections (see discussion of *R. louisianus* A. Berger in the treatment of *R. argutus*), remaining attached to the receptacle when ripe.

About 600 species, throughout the northern hemisphere, extending south to the higher elevations of South America, most diverse in Europe and eastern North America.

Subgenus *Rubus* has been divided into many sections, eight of which are native to eastern temperate North America and five (thus far) to Missouri. Sectional delimitations are somewhat artificial, in some cases relying only on differences in presence vs. absence of hairs or of gland-tipped vs. nonglandular hairs.

2a. Section Alleghenienses L.H. Bailey (highbush blackberry)

Canes erect to high-arching the first year, generally more clearly arched the second year; clonal by root-suckering, but rarely observed to tip-root. Prickles broad-based, straight or somewhat downward-curved. Primocane leaves with (3)5 leaflets. Sepals tapered to long-tapered at the tips, the outer surface often bearing stalked glands. Petals white. Some part of the plant always bearing stalked glands, most often the flower stalks. About 6 species, eastern U.S. and adjacent Canada.

Alice et al. (in press) treat members of this section under a broadly circumscribed *R. allegheniensis*. On the other hand, Steyermark (1963) divided Missouri's representatives of sect. *Alleghenienses* into two taxa: *R. allegheniensis*, with narrow, cylindric inflorescences, and *R. orarius* Blanch., with wider, upward flaring ones. Yet, W. H. Blanchard's (1906) description of *R. orarius* is so vague that this species could be placed in three different sections, and he designated no holotype. L. H. Bailey (1944) treated it in sect. *Alleghenienses*, but also stated that it might as easily be included in either sect. *Arguti* or sect. *Canadenses* L.H. Bailey. He did not designate a lectotype. Davis et al. (1969) discussed it in their revision of sect. *Arguti*. All of these authors, along with Fernald (1921), restricted *R. orarius* to Maine, Massachusetts, and Nova Scotia. Specimens that would key out as *R. orarius* in Steyermark (1963) are treated herein as *R. alumnus* and *R. rosa*, two polyploid taxa.

1. Inflorescences narrowly racemose, usually appearing almost leafless (the bracts much smaller than the leaves), at least twice as long as wide; sepals 5–7(–8) mm long and 2.0–3.5 mm wide . . . . 5. R. ALLEGHENIENSIS
1. Inflorescences racemose, flaring widely toward the tip, or leafy and less than twice as long as wide; calyx lobes at least 7 mm long and 3 mm wide
   2. Primocane central leaflets ovate to elliptic-oblong, usually about 2/3 as wide as long . . . . 6. R. ALUMNUS
   2. Primocane central leaflets broadly ovate, usually at least 3/4 as wide as long . . . . 7. R. ROSA

5. Rubus allegheniensis Porter (Allegheny blackberry)

R. allegheniensis var. *plausus* L.H. Bailey

R. villosus Thunb. var. *engelmannii* Focke

Pl. 541 c–e; Map 2503

Canes to 370 cm long and 100–250 cm tall, (3–)4–6(–9) mm in diameter; primocanes occasionally with scattered gland-tipped hairs toward the apex but lacking nonglandular hairs. Prickles sparse to dense, 0.1–3.0 per cm of cane, 4–6(–8) mm long. Petioles with sparse to dense nonglandular hairs and sometimes gland-tipped hairs, armed with broad-based, downward-curved prickles to 3 mm long. Stipules 6–12(–16) mm long, linear to linear-lanceolate. Primocane leaflet margins sharply toothed, the upper surface thinly hairy, the undersurface velvety hairy. Central primocane leaflets 8.5–13.5 cm long, 4–9 cm wide, ovate to elliptic-ovate or elliptic, cordate to subcordate at the base, tapered to long-tapered to a sharply pointed or filiform tip, the leaflet stalk about 1/4–1/3 as long as the leaflet blade; middle leaflets ovate to ovate-elliptic, rounded to subcordate at the base, tapered to a sharply pointed tip, stalked; basal leaflets ovate-elliptic to elliptic, angled to rounded at the base, tapered to a sharply pointed tip, nearly sessile to stalked, asymmetrically lobed when only 3 leaflets are present. Inflorescences narrowly racemose, 8–22 cm long, 3–7 cm wide (at least twice as long as wide), with 7–25 flowers and 2–6 bracts, these usually much smaller than the leaves and mostly simple; flower and inflorescence stalks with short, dense, nonglandular hairs, longer gland-tipped hairs, and sometimes slender or broad-based prickles. Sepals 5–7(–8) mm long, 2.0–3.5 mm wide, triangular-elliptic. Petals 10–16 mm long, obovate to broadly obovate. Fruits 8–18 mm long, 8–15 mm wide, more or less globose to short-cylindric. $2n$=14. April–June.

Scattered to common nearly throughout the state (northeastern U.S. and adjacent Canada south to South Carolina, Tennessee, and Arkansas, and west to Oklahoma, Nebraska, and Minnesota; introduced in Europe). Upland prairies, banks of streams and rivers, margins of ponds and lakes, sloughs, bottomland forests, mesic upland forests, bases, ledges, and tops of bluffs, savannas, and glades; also old fields, fallow fields, fencerows, old mines, levees, railroads, roadsides, and open, disturbed areas.

Rubus villosus var. *engelmannii*, a synonym of *R. allegheniensis* (Widrlechner, 1998), was described from collections first made in Missouri, although Focke (1914) designated no type. *Rubus allegheniensis* has been reported to naturalize in Europe (Edees and Newton, 1988; Zielinski, 2004).

6. Rubus alumnus L.H. Bailey

R. pubifolius L.H. Bailey

Pl. 541 a, b; Map 2504

Canes to 300 cm long and 150–250 cm tall, 5–7 mm in diameter; primocanes occasionally with scattered, gland-tipped hairs toward the tip but lacking nonglandular hairs. Prickles moderate, 0.7–1.5 per cm of cane, 4–8 mm long. Petioles with sparse to dense nonglandular hairs and sometimes gland-tipped hairs, armed with broad-based, downward-curved prickles to 4 mm long. Stipules 7–21 mm long, linear to linear-lanceolate. Primocane leaflet margins sharply toothed, the upper surface thinly hairy, the undersurface velvety hairy. Central primocane leaflets 8.5–15.5 cm long, 5–10 cm wide, elliptic to elliptic-oblong, cordate to subcordate at the base, tapered to a sharply pointed tip, the leaflet stalk about 1/4–1/3 as long as the leaflet blade; middle leaflets ovate to ovate-elliptic, angled to rounded at the base, tapered to a sharply pointed tip, stalked; basal leaflets ovate-elliptic, angled at the base, tapered to a sharply pointed tip, nearly sessile. Inflorescences racemose, flaring broadly toward the apex, 6–18 cm long, 6–10 cm wide, with 5–19 flowers, appearing leafy with 2–7 bracts, about evenly divided between simple bracts and those with 3 leaflets; flower and inflorescence stalks with dense nonglandular hairs, gland-tipped hairs, and scattered needlelike prickles. Sepals 6–10 mm long, 3–5 mm wide, triangular to narrowly triangular. Petals 10–22 mm long, broadly obovate. Fruits 12–21 mm long, 9–18 mm wide, short cylindric. $2n$=21, 28. May–June.

Scattered, mostly south of the Missouri River (northeastern U.S. west to Kansas, Iowa, and Minnesota). Bottomland forests, mesic upland forests, banks of streams and rivers, bottomland prairies, upland prairies, bases, ledges, and tops of bluffs; also pastures, old fields, fallow fields, roadsides, and open, disturbed areas.

Rubus alumnus was described by L. H. Bailey (1923) from Missouri, with the type material collected by B. F. Bush in Jackson County. *Rubus pubifolius*, a synonym of *R. alumnus* (Widrlechner, 1998), was also described from a Missouri type (L. H. Bailey 1945) collected by J. H. Kellogg near Eagle Rock, in Barry County.

7. Rubus rosa L.H. Bailey

Map 2505

Canes to 270 cm long and 150–250 cm tall, 5–8 mm in diameter, primocanes with gland-tipped hairs, at least toward the apex. Prickles moderate to dense, 0.3–3.0 per cm of cane, 6–8 mm long. Petioles with sparse to dense nonglandular hairs and often gland-tipped hairs, armed with broad-based, downward-curved prickles to 4 mm long. Stipules 12–20 mm long, linear to lanceolate. Primocane leaflets soft and clothlike, the margins sharply toothed, the upper surface thinly hairy, the undersurface velvety hairy. Central primocane leaflets 9.5–17.0 cm long, 8.5–13.0 cm wide, broadly ovate or suborbicular, cordate to rounded at the base, tapered to a sharply pointed tip, the leaflet stalk about 1/3 as long as the leaflet blade; middle leaflets ovate to oblong, rounded to subcordate at the base, tapered to a sharply pointed tip, stalked; basal leaflets elliptic, angled at the base, tapered to a sharply pointed tip, nearly sessile. Inflorescences mostly racemose, occasionally flaring and with some secondary branching, especially near the base, 5–36 cm long, 6–13 cm wide, with 5–24 flowers and 1–6 bracts, these often leafy and mostly with 3 leaflets; flower and inflorescence stalks with nonglandular hairs, gland-tipped hairs, and variously shaped prickles. Sepals 7–10(–16) mm long, 3–5 mm wide, triangular. Petals 13–18 mm long, obovate to broadly obovate. Fruits 12–20 mm long, 8–16 mm wide, short-cylindric to short-conic. $2n$=21, 28. May.

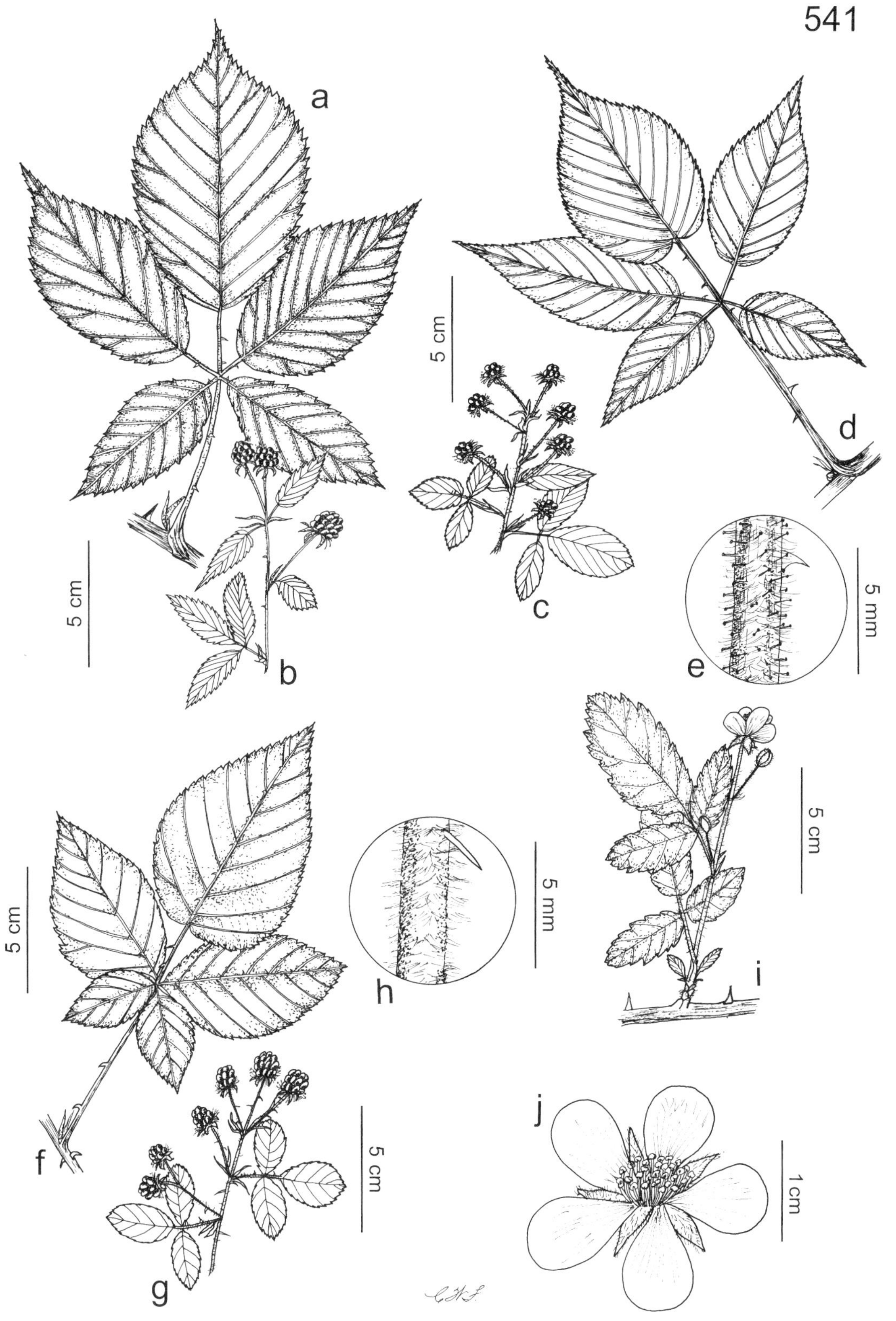

Plate 541. Rosaceae. *Rubus alumnus*, **a)** primocane node with leaf, **b)** apical portion of inflorescence with fruits. *Rubus allegheniensis*, **c)** apical portion of inflorescence with fruits, **d)** primocane node with leaf, **e)** inflorescence axis. *Rubus frondosus*, **f)** primocane node with leaf, **g)** apical portion of inflorescence with fruits, **h)** inflorescence axis. *Rubus curtipes*, **i)** floricane node with flowers and buds, **j)** flower.

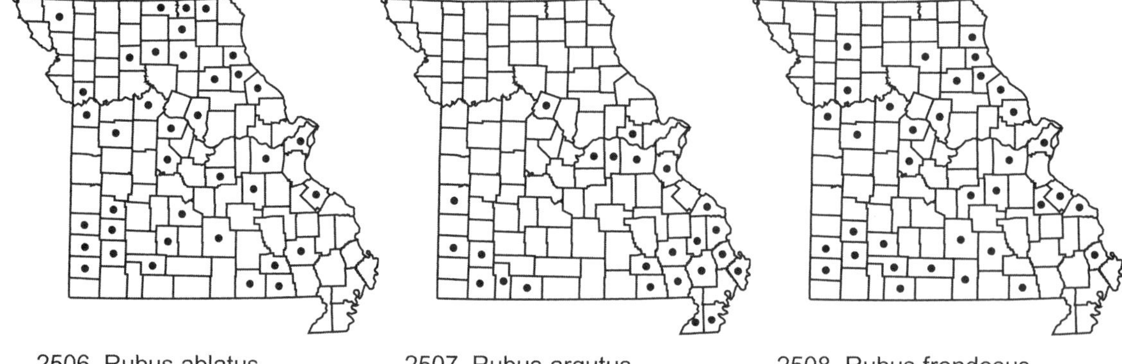

2506. Rubus ablatus

2507. Rubus argutus

2508. Rubus frondosus

Possibly introduced, uncommon, known thus far only from historical collections from Jackson County (northeastern U.S. and adjacent Canada west to Nebraska and Minnesota). Bottomland forests and bottomland prairies.

Large-fruited selections of *R. rosa* have been cultivated and may escape or persist from gardens. Few collections of *R. rosa* have been made in Missouri, and it is not clear whether any of these represent native populations that predate European settlement.

2b. Section Arguti Rydb. (highbush blackberry)

Canes typically glabrous, erect to arching the first year, generally more clearly arched the second year; clonal by root-suckering, but rarely observed to tip-root. Prickles broad-based, straight or somewhat downward-curved. Primocane leaves mostly with 5 leaflets, sometimes with only 3 leaflets. Sepals never bearing gland-tipped hairs. Petals white. About 50 species, primarily in the eastern U.S. and adjacent Canada.

Alice et al. (in press) take a very different approach to this geographically and morphologically diverse section, treating most members as synonyms of a single, highly variable taxon, *R. pensilvanicus* Poir. Steyermark (1963) recognized four taxa within sect. *Arguti*. The treatment and descriptions of *R. argutus* and *R. mollior* herein resemble those of Steyermark. Steyermark's other two taxa, *R. ostryifolius* Rydb. and *R. pensilvanicus*, are excluded. Davis et al. (1969) concluded that the combination of fragmentary type material along with Britton's (1901) vague description makes it impossible to understand *R. ostryifolius*, and indicated that L. H. Bailey (1945) probably misinterpreted it in his treatment of sect. *Arguti*. Most collections that had been treated as *R. ostryifolius* are included here under *R. ablatus* and *R. laudatus*. *Rubus pensilvanicus* is native to the eastern United States and resembles its widespread congener, *R. frondosus*, but lacks that species' broadly ovate central leaflets and its short, leafy inflorescences.

1. Primocane central leaflets ovate-elliptic to nearly orbicular; floricanes generally arching to low-arching, 50–130 cm tall; inflorescences leafy-racemose . . . . 10. R. FRONDOSUS
1. Primocane central leaflets narrowly elliptic, elliptic, or obovate; floricanes generally erect to arching, 70–400 cm tall; inflorescences various
    2. Primocane central leaflets narrowly elliptic, at least twice as long as wide; inflorescences condensed at the ends of weak, leafy racemes or small, flaring clusters, the bracteal leaflets with jaggedly incised margins . . . . . . 9. R. ARGUTUS
    2. Primocane central leaflets elliptic to obovate, less than twice as long as wide; inflorescences with substantial flower stalks and bracteal leaflets with regularly toothed margins

3. Primocane central leaflets generally widest above the midpoint, grayish green beneath; inflorescences, with 5–11 flowers often condensed at the ends of stout, usually well-armed racemes . . . . 12. R. MOLLIOR
3. Primocane central leaflets generally widest at or below the midpoint, softly pubescent beneath but not noticeably grayish green; inflorescences with 5–16 (–20) flowers, typically racemose (but some populations of *R. laudatus* in southwestern Missouri have short, flaring inflorescences)
    4. Primocane central leaflets tapered or long-tapered to a sharply pointed or filiform tip; inflorescences narrowly racemose . . . . 8. R. ABLATUS
    4. Primocane central leaflets angled or short-tapered to a pointed tip; inflorescences racemose to widened at apex . . . . 11. R. LAUDATUS

8. Rubus ablatus L.H. Bailey
Pl. 542 f, g; Map 2506

Canes to 250 cm long and 100–200 cm tall, 4–6 mm in diameter. Prickles sparse to moderate, 0.3–1.5 per cm of cane, 4–7 mm long. Petioles with sparse to dense nonglandular hairs, armed with broad-based, downward-curved to downward-angled prickles to 3 mm long. Stipules 10–15 mm long, linear-lanceolate to threadlike. Primocane leaflet margins sharply toothed, the upper surface thinly hairy, the undersurface velvety hairy. Primocane central leaflets 7–13 cm long, 4.5–8.5 cm wide, elliptic-ovate, elliptic, or elliptic-oblong, subcordate to truncate at the base, long-tapered to a sharply pointed or filiform tip, the leaflet stalk about 1/4–2/5 as long as the leaflet blade; middle leaflets elliptic, rounded at the base, long-tapered to a sharply pointed or filiform tip, stalked; basal leaflets elliptic, angled at the base, tapered to a sharply pointed tip, sessile to short-stalked, asymmetrically lobed when only 3 leaflets are present. Inflorescences racemose, 6–15 cm long, 4–7 cm wide, with 5–16(–20) flowers and (1)2–4 bracts, these about evenly divided between simple bracts and those with 3 leaflets; flower and inflorescence stalks with nonglandular hairs and occasional needlelike prickles. Sepals 5–7 mm long, 2.5–4.0 mm wide, triangular to oblong, tapered to a sharply pointed tip or abruptly tapered to a short, slender point. Petals 8–14 mm long, obovate. Fruits 8–15 mm long, 8–15 mm wide, short cylindric to globose. May.

Scattered nearly throughout the state but apparently absent from the Mississippi Lowlands Division (north-central U.S. from Kansas north to Minnesota and east to Ohio and West Virginia). Mesic upland forests, upland prairies, savannas, and ledges of bluffs; also pastures, fencerows, and railroads.

Rubus ablatus was described by L. H. Bailey (1945) from Missouri, with the type material collected by B. F. Bush from a prairie in Morgan County. It has many characteristics that suggest an *R. allegheniensis* lacking gland-tipped hairs, and specimens of *R. allegheniensis* with few (or overlooked) gland-tipped hairs could easily be misdetermined as *R. ablatus*.

9. Rubus argutus Link
Pl. 542 h, i; Map 2507

Canes to 450 cm long and 120–400 cm tall, 3–7 mm in diameter. Prickles sparse, 0.3–1.0 per cm of cane, 4–5 mm long, sometimes strongly downward-curved near the tip of the cane. Petioles with sparse to dense nonglandular hairs, armed with downward-curved prickles to 2 mm long. Stipules 6–13 mm long, linear to threadlike. Primocane leaflet margins finely to sharply toothed, the upper surface thinly hairy, the undersurface velvety hairy. Central primocane leaflets 5.0–13.5 cm long, 2.5–5.5 cm wide, narrowly elliptic, often less than 1/2 as wide as long, rounded to angled at the base, angled (or occasionally tapered) to a sharply pointed tip, the leaflet stalk about 1/10–1/4(–1/3) as long as the leaflet blade; lateral leaflets resembling but smaller than the central leaflet, the middle pair stalked, the basal pair nearly sessile. Inflorescences typically racemose or clustered, (4.5–)6.0–12.0 cm long, 3–5 cm wide, with 2–5 (–11) flowers and 3–7 bracts, these typically jaggedly toothed, about evenly divided between simple bracts and those with 3 leaflets; flower and inflorescence stalks rather delicate, with nonglandular hairs and occasional needlelike or small, downward-curved prickles. Sepals 4–5 mm long, 2.5–3.0 mm wide, ovate-triangular, tapered to a sharply pointed tip or abruptly tapered to a short, slender point. Petals 7–13 mm long, narrowly obovate. Fruits 9–14 mm long, 7–12 mm wide, short cylindric. $2n$=14, 21. May–June.

Scattered, mostly south of the Missouri River (southeastern U.S. west to Texas and Oklahoma). Bottomland forests, mesic upland forests, banks of streams and spring branches, swamps, ledges and tops of bluffs, glades, and upland prairies; also pastures, fencerows, margins of crop fields, and roadsides.

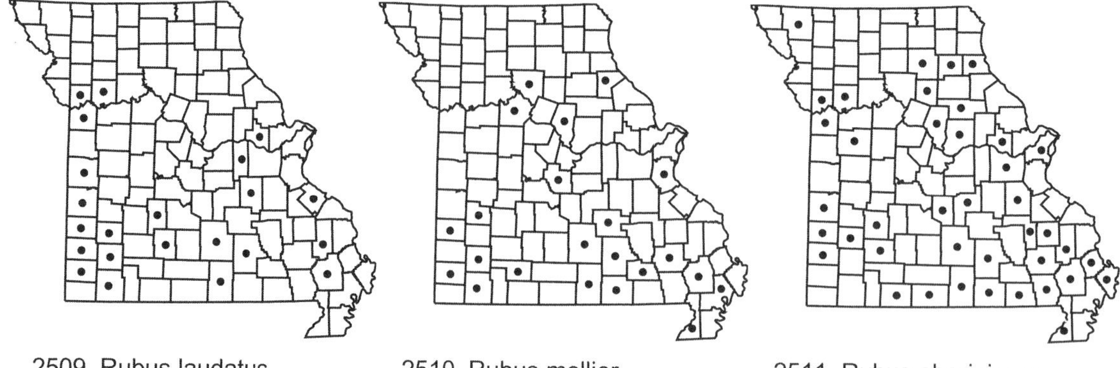
2509. Rubus laudatus 2510. Rubus mollior 2511. Rubus aboriginum

A white-fruited, horticultural variant, cv. 'Crystal White', closely resembles *R. argutus*, but has been recognized as a distinct species, *R. louisianus* A. Berger by some botanists. Specimens from Oregon and Texas Counties may represent escapes of cv. 'Crystal White' from cultivation, but such plants would key to *R. argutus* in the present treatment and can only be distinguished from the range of variation present in *R. argutus* when fruits are present.

10. Rubus frondosus Bigelow (Yankee blackberry)
R. pratensis L.H. Bailey

Pl. 541 f–h; Map 2508

Canes to 200 cm long and 70–130 cm tall, 3–7 mm in diameter. Prickles moderate, 0.5–1.2(–2.0) per cm of cane, 2.5–5.0 mm long. Petioles with sparse to dense nonglandular hairs, armed with fine, needlelike, downward-angled prickles to 3 mm long. Stipules 10–18 mm long, linear to linear-lanceolate. Primocane leaflet margins sharply toothed, the upper surface thinly hairy, the undersurface velvety hairy. Central primocane leaflets (7–)8–14 cm long, (5–)6–11 cm wide, nearly orbicular to ovate-elliptic, cordate to subcordate at the base, tapered to a sharply pointed tip, the leaflet stalk about 1/6–3/10 as long as the leaflet blade; middle leaflets rhombic to elliptic or elliptic-obovate, angled to rounded at the base, tapered to a sharply pointed tip, stalked; basal leaflets elliptic, angled at the base, tapered or short-tapered to a sharply pointed tip, sessile. Inflorescences racemose, sometimes flaring toward the apex, but typically compact, 3.0–17.5 cm long, 5–11 cm wide, with 1–11 flowers and 3–9 bracts, these leafy, about evenly divided between simple bracts and those with 3 leaflets; flower and inflorescence stalks with nonglandular hairs and rarely with needlelike prickles. Sepals 6–10 mm long, 3–4 mm wide, triangular-elliptic, tapered to a sharply pointed tip or abruptly tapered to a short, slender point. Petals 8–14 mm long, obovate. Fruits 10–15 mm long, 10–16 mm wide, globose. May–June.

Scattered nearly throughout the state (northeastern U.S. and adjacent Canada west to Kansas, Iowa, and Minnesota). Bottomland forests, mesic to dry upland forests, banks of streams, bases, ledges, and tops of bluffs, and upland prairies; also pastures, old fields, old quarries, cemeteries, railroads, and roadsides.

Rubus pratensis, a synonym of *R. frondosus* (Widrlechner, 1998), was described based on a Missouri type (L. H. Bailey, 1945) collected by B. F. Bush from a prairie in Cooper County. A white-fruited form from Howell County was described as *R. pensilvanicus* Poir. f. *albinus* E.J. Palmer & Steyerm.

11. Rubus laudatus A. Berger
R. bushii L.H. Bailey
R. sertatus L.H. Bailey
R. virilis L.H. Bailey

Map 2509

Canes to 270 cm long and 100–200 cm tall, 4–10 mm in diameter. Prickles sparse to moderate, 0.2–1.2 per cm of cane (except in southwestern Missouri and Kansas, where more densely armed populations are found with up to 3 prickles per cm), 4–6 mm long. Petioles with sparse to dense nonglandular hairs, armed with broad-based downward-curved to downward-angled prickles to 3 mm long. Stipules 12–16 mm long, linear. Primocane leaflet margins finely toothed, the upper surface thinly hairy, the undersurface velvety hairy, in developing leaves appearing grayish. Central primocane leaflets 7.5–13.0 cm long, 3–7 cm wide, elliptic-ovate, elliptic, elliptic-oblong, or narrowly-elliptic, subcordate to rounded at the base, angled or short-tapered to a sharply pointed tip, the leaflet stalk about 1/4–1/3 as long as the leaflet blade; lateral leaflets elliptic, angled at the base, angled to a sharply pointed tip, the middle pair stalked, the basal pair nearly sessile. Inflo-

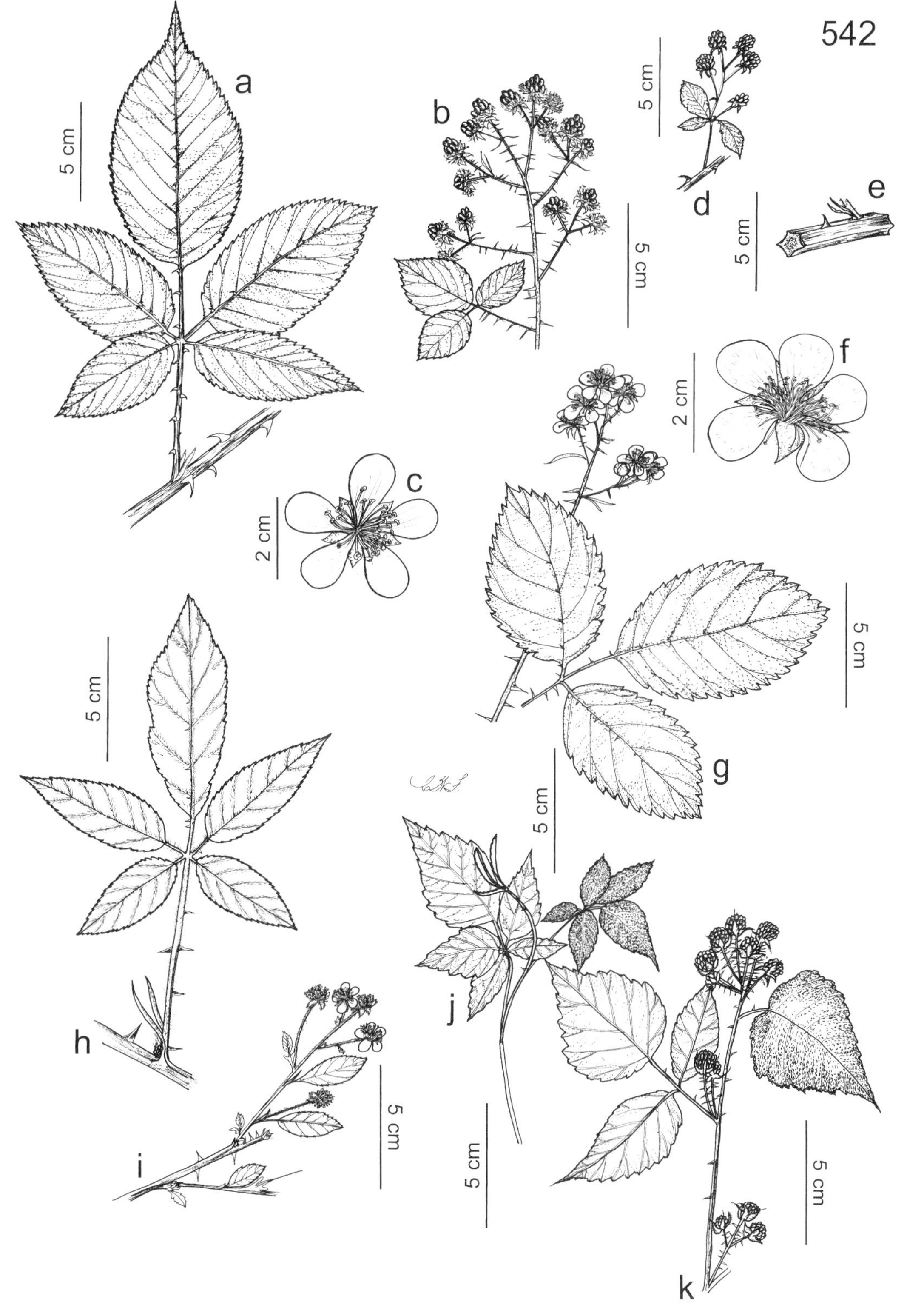

Plate 542. Rosaceae. *Rubus armeniacus*, **a)** primocane node with leaf, **b)** apical portion of inflorescence with fruits and bract, **c)** flower. *Rubus serissimus*, **d)** floricane node with fruits, **e)** section of primocane showing angled cross-section. *Rubus ablatus*, **f)** flower, **g)** apical portion of inflorescence and detached floricane leaf. *Rubus argutus*, **h)** primocane node with leaf, **i)** portion of floricane with inflorescence. *Rubus occidentalis*, **j)** tip of primocane, **k)** portion of floricane with fruits.

rescences typically racemose, 5.0–15.5 cm long, 4–6 cm wide, with 5–13 flowers (western populations can bear short, flaring inflorescences with as few as 4 flowers, especially near the floricane tips) and 1–5 bracts, these sometimes leafy, about evenly divided between simple bracts and those with 3 leaflets; flower and inflorescence stalks with nonglandular hairs, and occasional needlelike prickles (western populations sometimes bear downward-curved, broad-based prickles). Sepals 5–7 mm long, 2.5–4.0 mm wide, narrowly triangular, tapered to a sharply pointed tip. Petals 12–22 mm long, obovate. Fruits 12–20 mm long, 8–14 mm wide, ovoid to cylindric. May.

Scattered, mostly south of the Missouri River (central U.S. from Pennsylvania south to Virginia, Arkansas and Texas west to Oklahoma and Kansas; introduced in Australia). Bottomland forests, mesic upland forests, upland prairies, and banks of streams and rivers; also pastures, old fields, fencerows, ditches, railroads, and, roadsides.

Large-fruited selections of *R. laudatus* have been cultivated and may escape or persist from gardens. The species is known to have become naturalized in Australia (Evans et al., 2007). It was originally described by Berger (Hedrick et al., 1925) from the cv. 'Bundy', which was selected by T. B. Bundy from Wayne County, Missouri. *Rubus bushii*, *R. sertatus*, and *R. virilis*, three synonyms of *R. laudatus*, also were described from Missouri types. The type of *R. bushii* was collected by E. J. Palmer from Jasper County in 1929 (L. H. Bailey, 1932), that of *R. sertatus* by Bush from Jackson County in 1926 (L. H. Bailey, 1945), and that of *R. virilis* by J. H. Kellogg from Texas County in 1933 (L. H. Bailey, 1945).

12. Rubus mollior L.H. Bailey

Map 2510

Canes to 250 cm long and 100–180 cm tall, 4–9 mm in diameter. Prickles moderate, 0.7–2.2 per cm of cane, 5–7 mm long. Petioles with dense nonglandular hairs, armed with broad-based downward-curved prickles to 3 mm long. Stipules 7–10 mm long, linear. Primocane leaflet margins finely toothed, the upper surface thinly hairy, the undersurface velvety hairy, often appearing grayish with a rusty midvein. Central primocane leaflets 7.5–11.0 cm long, 4.0–6.5 cm wide, elliptic-obovate or oblong-obovate, rounded at the base, short-tapered to a sharply pointed tip, the leaflet stalk about 1/5–3/10 as long as the leaflet blade; lateral leaflets elliptic-obovate, rounded to angled at the base, abruptly short-tapered to a sharply pointed tip, the middle pair stalked, the basal pair nearly sessile. Inflorescences typically racemose, 9–12 cm long, 4–7 cm wide, with 5–9 flowers and 3–7 bracts, these fairly small at flowering, but sometimes becoming leafy and prominent as the fruits develop, hiding the fruits later in development; flower and inflorescence stalks with nonglandular hairs and occasional needlelike prickles. Sepals 5–6 mm long, 2–4 mm wide, triangular-ovate to elliptic, abruptly tapered to a short, slender point. Petals 10–12 mm long, obovate. Fruits 10–15 mm long, 7–12 mm wide, globose to cylindric. May.

Scattered, mostly south of the Missouri River (Missouri, Arkansas, Oklahoma, and Kansas). Bottomland forests, mesic upland forests, swamps, upland prairies, and margins of ponds and lakes; also fencerows.

Rubus mollior is nearly an Ozark endemic.

2c. Section Flagellares L.H. Bailey (dewberry)

Canes arching to prostrate the first year, generally low-arching to prostrate the second year; clonal by both root-suckering and tip-rooting. Prickles mostly broad-based, typically downward-curved (exceptions noted below). Primocane leaves with 3–5(–7) leaflets. Petals white. About 50 species, predominantly apomictic polyploids, primarily in the eastern U.S. and adjacent Canada.

Alice et al. (in press) take a very different approach to this geographically and morphologically diverse section, treating most of these species as synonyms of a single taxon, *Rubus flagellaris*. Steyermark (1963) recognized three species in Missouri within this section. His treatment of the members of this section designates all populations that bear gland-tipped hairs as *R. invisus* (L.H. Bailey) Britton. Good examples of glandular members of sect. *Flagellares* in Missouri are scarce but they clearly can be separated into two distinct taxa with small primocane leaves, typically with 3 leaflets: *R. deamii* and *R. leviculus* (distinguished in the key to species below). *Rubus invisus*, native to the east of Missouri, has larger primocane leaves, usually with 5 leaflets, and open inflorescences with up to 7 flowers on long, ascending stalks. Steyermark's (1963) nonglandular taxa, *R. enslenii* and *R. flagellaris*, were separated on the basis of the shape of their floricane leaflets, a problematic and often variable character.

Rubus occidualis (L.H. Bailey) L.H. Bailey (*R. flagellaris* var. *occidualis* L.H. Bailey) was described from Iowa and is best considered as a synonym of *R. roribaccus*. However, plants that keyed to *R. flagellaris* var. *occidualis* in Steyermark's (1963) treatment are herein divided among five species under the second lead of couplet 7. Similarly, the type specimen of *R. flagellaris* f. *roseus* Steyerm., which was said to differ from typical var. *occidualis* in its pinkish-tinged petals, cannot be determined to species within sect. *Flagellares* with confidence.

Rubus hancinianus L.H. Bailey, a large, mounding member of sect. *Flagellares* that is endemic to Kansas and Oklahoma, has been reported from Jackson County, Missouri (Davis et al., 1968). The collection upon which this report was based (*A.M. Fuller, H.A. Davis & T. Davis F-65-11, 13791*, accessioned in the herbaria at the Carnegie Museum of Natural History and the Milwaukee Public Museum) has inflorescences typical of *R. meracus* and *R. roribaccus*, but with primocane leaflets of a form intermediate between those two taxa. Although it is clearly not *R. hansonianus*, the collection is not sufficient to show the distinction between the other two species in question.

1. Leaflets thinly hairy beneath at maturity, not soft to the touch at maturity
    2. Primocane leaves with 3 leaflets; inflorescences generally 5–8 cm long, with 1 flower . . . . 17. R. ENSLENII
    2. Primocane leaves with 3 and/or 5 leaflets; inflorescences generally 5–20 cm long, with 1–12 flowers
        3. Canes 2–3 mm in diameter with prickles 1–2 mm long; primocane leaves mostly with 3 leaflets; inflorescences with 1–6 flowers . . . . 23. R. STEELEI
        3. Canes 3–5 mm or more in diameter with prickles 2–4 mm long; primocane leaves mostly with 5 leaflets; inflorescences with 4–12 flowers
            4. Primocanes with 1–3 prickles per cm; central primocane leaflets gradually tapered to relatively long-tapered at the tip . . . . 14. R. CELER
            4. Primocanes with 3–5 prickles per cm; central primocane leaflets shouldered, abruptly tapered at the tip . . . . 18. R. FLAGELLARIS
1. Leaflets densely hairy beneath at maturity, soft to the touch at maturity
    5. Inflorescences bearing stalked glands (obvious at 10× magnification)
        6. Primocane central leaflets ovate to elliptic, with a subcordate base; at least some inflorescences with 3–6 flowers . . . . 16. R. DEAMII
        6. Primocane central leaflets elliptic, with an angled to rounded base; inflorescences with 1 or 2 flowers . . . . 19. R. LEVICULUS
    5. Inflorescences lacking stalked glands
        7. Canes robust, forming a large, mounding tangle, cane diameter frequently exceeding 4 mm; inflorescences with 4–8 flowers, mostly racemose . . . . 22. R. SATIS
        7. Canes whiplike, low-arching to prostrate; cane diameter mostly 2.5–5.0 mm; inflorescences with 1–10 flowers, not racemose
            8. Inflorescences condensed in a terminal cluster with flower stalks at most 3.5 cm long . . . . 15. R. CURTIPES
            8. Inflorescences 1-flowered or in loose clusters with long, ascending flower stalks often exceeding 3.5 cm long
                9. Primocane central leaflets with jagged-incised margins; inflorescences usually with 1 flower; sepals often with a prominent spatulate tip 1–3 mm long . . . . 13. R. ABORIGINUM

9. Primocane central leaflets with regularly toothed margins; inflorescences usually with 2–8 flowers; sepals tapered to a sharply pointed tip, with a leafy tip or abruptly tapered to a short, slender point
   10. Primocane central leaflets elliptic, rounded at base . . . . 20. R. MERACUS
   10. Primocane central leaflets ovate, cordate at base . . . . 21. R. RORIBACCUS

13. Rubus aboriginum Rydb.
R. mundus L.H. Bailey

Map 2511

Canes to 350 cm long and 40 cm tall, 2.5–5.0 mm in diameter. Prickles usually moderate, rarely dense, 1–3(–6) per cm of cane, 1.5–3.0 mm long (but cultivated selections exist that lack cane prickles). Petioles with sparse to dense nonglandular hairs, armed with downward-curved to downward-angled prickles to 2 mm long. Stipules 12–21 mm long, linear to lanceolate. Primocane leaflets 3, 5, or rarely 7, the margins irregularly jagged, often doubly toothed, the upper surface thinly hairy, the undersurface velvety hairy. Central primocane leaflets (5.5–)6.0–9.5 cm long and (4.0–)4.5–7.0 (–7.5) cm wide, ovate to elliptic or elliptic-obovate, subcordate to rounded at the base, tapered to a sharply pointed tip, the leaflet stalk extremely variable in length relative to the leaflet blade; middle leaflets ovate to elliptic or rhombic, rounded to angled at the base, angled to tapered to a sharply pointed tip, stalked; basal leaflets ovate to elliptic, angled at the base, angled to a sharply pointed tip, nearly sessile, often asymmetrically lobed when only 3 leaflets are present. Inflorescences 3–15(–22) cm long, with 1–5 flowers on long, ascending stalks and 1–6 leafy bracts, these mostly with 3 leaflets; flower and inflorescence stalks with nonglandular hairs and sometimes with needle-like prickles. Sepals 6–8 mm long, 3–4 mm wide, triangular, often with a prominent spatulate tip 1–3 mm long. Petals 10–22 mm long, broadly obovate. Fruits 10–30 mm long, 12–20 mm wide, subglobose to cylindric. $2n$=56. April–June.

Scattered, mostly south of the Missouri River (southeastern U.S. west to Kansas, Oklahoma and Texas). Bottomland forests, mesic upland forests, swamps, banks of streams and rivers, bottomland prairies, upland prairies, and fens; also old fields, fallow fields, fencerows, railroads, and roadsides.

Large-fruited and unarmed selections of *R. aboriginum* have been cultivated and may escape or persist from gardens. In addition to rooting at the tips, the canes of this species sometimes also root at the nodes, an unusual character in the section. Another unusual character is the presence of extra petals. A collection of *R. aboriginum* with 10 petals was made by J. Pinkston in Macon County in 1997 and is accessioned at the herbarium of Missouri State University.

14. Rubus celer L.H. Bailey

Map 2512

Canes to 200 cm long and 50 cm tall, 3–5 mm in diameter. Prickles usually moderate, 1–3 per cm of cane, 2–3 mm long. Petioles with sparse nonglandular hairs, armed with downward-curved to downward-angled prickles to 2 mm long. Stipules 10–16 mm long, linear. Primocane leaflets mostly 5, sometimes 3, the margins sharply toothed, the upper surface glabrous, the undersurface thinly hairy, especially along the veins. Central primocane leaflets 6.0–8.5 cm long and 4–6 cm wide, ovate-elliptic to elliptic, rounded at the base, tapered to a sharply pointed tip, the leaflet stalk about 1/4 as long as the leaflet blade; middle leaflets elliptic-obovate, angled at the base, angled to a sharply pointed tip, stalked; basal leaflets elliptic-obovate, angled at the base, angled to a sharply pointed tip, and sessile when 5 leaflets are present, tapered to a sharply pointed tip, short-stalked, and often asymmetrically lobed when only 3 leaflets are present. Inflorescences (7–)10–20(–32) cm long, with 4–12 flowers on long, ascending stalks (often with a more-or-less flat-topped arrangement) and 4–7 leafy bracts, these mostly with 3 leaflets; flower and inflorescence stalks with spreading, nonglandular hairs and sometimes with small, downward-curved prickles. Sepals 6–7 mm long, 3–4 mm wide, triangular-ovate, abruptly tapered to a short, slender point. Petals 10–13 mm long, obovate. Fruits 13–20 mm long, 11–20 mm wide, globose to short-cylindric. May–June.

Uncommon, mostly in the eastern half of the state (northeastern U.S. and adjacent Canada west to Wisconsin, Iowa, and Missouri). Upland prairies; also pastures, fencerows, ditches, roadsides, and open, sandy, disturbed areas.

15. Rubus curtipes L.H. Bailey

Pl. 541 i, j; Map 2513

Canes to 200 cm long and 50 cm tall, 2.5–4.0 mm in diameter. Prickles moderate to dense, 1–4(–7) per cm of cane, 1.0–3.5 mm long. Petioles with nonglandular hairs, armed with downward-curved prickles to 2 mm long. Stipules 8–15 mm long, linear to narrowly elliptic, sometimes notched. Primocane leaves with 3 and/or 5 leaflets, margins irregularly to doubly serrate, up-

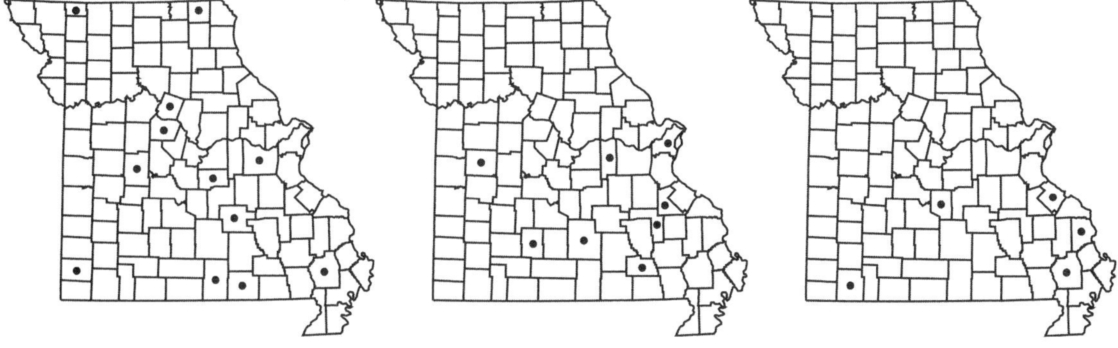

2512. Rubus celer 2513. Rubus curtipes 2514. Rubus deamii

per surface thinly hairy, the undersurface velvety hairy. Central primocane leaflets 4.5–8.0 cm long and (3.5–)4.0–5.5(–6.5) cm wide, ovate to elliptic, base subcordate to rounded, tapered to a sharply pointed tip, the leaflet stalk about 1/3 as long as the leaflet blade; middle leaflets rhombic or elliptic to obovate, base angled, angled to tapered to a sharply pointed tip, stalked; basal leaflets elliptic to elliptic-obovate and sessile, when 5 leaflets, ovate and stalked, often asymmetrically lobed, when 3 leaflets, base angled, angled to a sharply pointed tip. Inflorescences, 5–10 cm long, with 1–7 flowers condensed in clusters near the apex, with 1–6 leafy bracts, about evenly divided between simple bracts and those with 3 leaflets; flower and inflorescence stalks with dense nonglandular hairs, with the longest flower stalk at most 3.5 mm long. Sepals 6–8 mm long and 3–4 mm wide, triangular-ovate to oblong, abruptly tapered to a short, slender point. Petals 10–12 mm long, obovate. Fruits 10 mm long, 10 mm wide, globose. April–May.

Uncommon, south of the Missouri River (northeastern U.S. west to Wisconsin, Iowa, and Missouri). Mesic upland forests, glades, and tops of bluffs; also old fields and railroads.

Another member of sect. *Flagellares*, *R. plicatifolius* Blanch., resembles *R. curtipes* in having condensed inflorescences with small flowers held on short flower stalks, but typically *R. plicatifolius* has less hairy leaves that appear slightly corrugated or pleated at each major vein. *Rubus plicatifolius* is known from eastern and southern Iowa. A single, incomplete specimen is accessioned at the Iowa State University herbarium that was collected northwest of Chillicothe (Livingston County) (*S. Sparling 818* on 25 May 1951) may represent a native Missouri population. This species should be searched for in northern Missouri at well-drained, open sites, such as on railroad ballast.

16. Rubus deamii L.H. Bailey

Map 2514

Canes to 250 cm long and 50 cm tall, 3–5 mm in diameter. Primocanes rarely with gland-tipped hairs. Prickles moderate to dense, (2–)3–5(–8) per cm of cane, (1–)2–3 mm long. Petioles with nonglandular hairs, occasional gland-tipped hairs and downward-curved prickles to 2 mm long. Stipules 8–12 mm long, linear-lanceolate, glandular. Primocane leaves mostly with 3 leaflets, rarely with 5, margins irregularly, doubly serrate, upper surface thinly hairy, the undersurface velvety hairy. Central primocane leaflets (4.0–)5.0–6.5 (–7.0) cm long and (2.5–)3.0–4.0(–4.5) cm wide, ovate to elliptic, base rounded to subcordate, tapered to a sharply pointed tip, the leaflet stalk about 1/6–1/4 as long as the leaflet blade; middle leaflets ovate to elliptic; basal leaflets ovate, often asymmetrically lobed, base rounded, angled to a sharply pointed tip. Inflorescences (4–)6–11(–12) cm long, with 1–6 flowers on ascending stalks, with (1)2 or 3(–5) leafy bracts, about evenly divided between simple bracts and those with 3 leaflets; flower and inflorescence stalks with nonglandular and glandular hairs, and small downward-angled, needlelike prickles. Sepals 7–8 mm long and 3.0–3.5 mm wide, oblong, tapered to a sharply pointed tip or abruptly tapered to a short, slender point. Petals 8–14 mm long, obovate. Fruits 9–10 mm long, 9–10 mm wide, globose. April–May.

Uncommon, known thus far mostly from solitary specimens from five scattered counties (central U.S. from Virginia and West Virginia west to Indiana, Kentucky, and Tennessee, disjunct in Missouri). Bases of sandstone bluffs.

The Missouri record dates to 1933 to a specimen of primocane and floricane material gathered by Julian Steyermark near Chimney Rocks in a bluffy area with large sandstone boulders. The specimens originally were determined as *R. invisus* (see discussion under the sect. *Flagellares* treatment above).

2515. Rubus enslenii
2516. Rubus flagellaris
2517. Rubus leviculus

17. Rubus enslenii Tratt.
R. nefrens L.H. Bailey
Pl. 540 g–i; Map 2515

Canes to 180 cm long, typically prostrate, 2–3 mm in diameter, sometimes so delicate as to seem herbaceous. Prickles moderate, 1–2 per cm of cane, 1–2 mm long. Petioles armed with downward-curved prickles to 1.5 mm long. Stipules 8–10 mm long, linear. Primocane leaves almost always with 3 leaflets, rarely a few with 5 leaflets, the margins finely toothed to more coarsely and sharply toothed, the upper surface glabrous, the undersurface thinly hairy. Central primocane leaflets (4–)5–6(–7) cm long and (2.0–)2.5–4.5(5.0) cm wide, elliptic-rhombic, rounded at the base, tapered to a pointed or sharply pointed tip, the leaflet stalk about 1/8–1/5 as long as the leaflet blade; basal leaflets ovate-elliptic, often asymmetrically lobed, angled at the base, short-tapered to a sharply pointed tip, sessile. Inflorescences (4.5–)5.0–8.0 (–12.0) cm long, with 1 or 2(–4) flowers on long, ascending stalks, with (1)2 or 3(–5) narrow, leafy bracts, about evenly divided between simple bracts and those with 3 leaflets; flower and inflorescence stalks nearly glabrous or with relatively few nonglandular hairs, sometimes with small downward-angled or downward-curved prickles. Sepals 6–10 mm long, 2.5–4.0 mm wide, elliptic, with expanded, leafy tips or abruptly tapered to a short, slender point. Petals 8–12 mm long, obovate. Fruits 8–12 mm long, 8–12 mm wide, short-conic to short-cylindric. April–May.

Scattered nearly throughout the state, more abundantly south of the Missouri River (southeastern U.S. west to Kansas and Oklahoma) Bottomland forests, mesic upland forests, banks of streams and rivers, ledges of bluffs, margins of sinkhole ponds, and upland prairies; also pastures and old fields.

Rubus nefrens is here treated as a synonym of *R. enslenii* (Widrlechner, 1998). It was described from Missouri, with the type material collected by B. F. Bush in Jackson County in 1923 (L. H. Bailey, 1925).

18. Rubus flagellaris Willd. (northern dewberry)
Map 2516

Canes to 250(–450) cm long and 60 cm tall, 3–6 mm in diameter. Prickles moderate to dense, (2)3–5 per cm of cane, (2–)3–4 mm long. Petioles with nonglandular hairs, armed with downward-curved prickles to 3 mm long. Stipules 14–18 mm long, linear to linear-lanceolate or narrowly elliptic. Primocane leaves mostly with 5 leaflets, rarely with 3, margins coarsely, doubly, and sharply toothed, the upper surface glabrous, the undersurface thinly hairy, often with small prickles on the veins. Primocane central leaflets 6–9 cm long and 4–7 cm wide, broadly ovate to ovate-elliptic, occasionally obovate, especially in late-season growth, cordate to rounded at the base, shouldered to an abruptly tapered, sharply pointed tip, the leaflet stalk about 1/4 as long as the leaflet blade; middle leaflets elliptic-obovate, angled at the base, short-tapered to a sharply pointed tip, short-stalked; basal leaflets elliptic, angled at the base, angled to a sharply pointed tip, sessile. Inflorescences 9–22 cm long, with 1–6 flowers on long, ascending stalks, with 2–6 leafy bracts, these mostly simple or mostly with 3 leaflets; flower and inflorescence stalks nearly glabrous, with small downward-curved prickles. Sepals 6–8 mm long, 3–4 mm wide, narrowly elliptic to triangular-elliptic, abruptly tapered to a short, slender point. Petals 12–20 mm long, obovate. Fruits 10–20 mm long, 10–15 mm wide, globose to broadly oblong in outline. $2n$=28, 35, 49. May–June.

Scattered nearly throughout the state, more abundantly south of the Missouri River (northeastern U.S. west to Wisconsin and Oklahoma). Openings of mesic to dry upland forests, upland prai-

2518. Rubus meracus

2519. Rubus roribaccus

2520. Rubus satis

ries, banks of streams, tops of bluffs, and margins of sinkhole ponds; also pastures, old fields, old mines, cemeteries, railroads, roadsides, and open, disturbed areas.

Palmer and Steyermark (1958) published the name *R. flagellaris* f. *roseoplenus* E.J. Palmer & Steyermark based on a specimen collected by Palmer (*59655*) in 1955 along a railroad in Barton County. This form was stated to differ from the typical one in its doubled corollas with numerous pinkish-tinged petals. However, the type specimen instead appears to represent an unusual plant of sect. *Arguti* (although the specimen is incomplete and thus cannot be determined to species with confidence).

19. Rubus leviculus L.H. Bailey

Pl. 540 j–l; Map 2517

Canes to 200 cm long, typically prostrate, 2–4 mm in diameter, sometimes so delicate as to seem herbaceous. Primocanes sometimes with gland-tipped hairs. Prickles moderate, 1.0–3.5 per cm of cane, 0.5–2.0 mm long. Petioles with nonglandular hairs, occasional gland-tipped hairs, and downward-curved prickles to 1 mm long. Stipules 8–15 mm long, linear-lanceolate. Primocane leaves mostly with 3 leaflets, rarely with 5, margins sharply and sometimes coarsely toothed, the upper surface thinly hairy, the undersurface thinly to velvety hairy. Central primocane leaflets 5–8 cm long and 2.5–4.0 cm wide, elliptic, rounded to angled at the base, tapered to a sharply pointed tip, the leaflet stalk about 1/5–3/10 as long as the leaflet blade; basal leaflets asymmetrically ovate-elliptic to elliptic, angled at the base, angled to a sharply pointed tip, short-stalked to nearly sessile. Inflorescences 4.0–12.5 cm long, with 1–3 flowers on long, ascending stalks, with 1–4 narrow, leafy bracts, these mostly with 3 leaflets; flower and inflorescence stalks densely covered with nonglandular and glandular hairs and small downward-angled, needlelike prickles or downward-curved prickles. Sepals 5.5–7.0 mm long, 3–4 mm wide, triangular to elliptic-ovate, tapered to a sharply-pointed tip. Petals 10–14 mm long, obovate. Fruits 10–14 mm long, 10–15 mm wide, short-cylindric. April–May.

Uncommon, mostly south of the Missouri River (southeastern U.S. west to Missouri and Iowa). Mesic upland forests, upland prairies, sand prairies, and banks of streams and rivers; also pastures, old fields, railroads, roadsides, and open, disturbed areas.

20. Rubus meracus L.H. Bailey
R. frustratus L.H. Bailey
R. kelloggii L.H. Bailey

Map 2518

Canes to 250 cm long and 60 cm tall, 2.5–4.0 mm in diameter. Prickles usually moderate, 1–4 per cm of cane, 1.5–2.5 mm long. Petioles with dense, nonglandular hairs, armed with downward-curved prickles to 2 mm long. Stipules 12–15 mm long, linear to linear-lanceolate. Primocane leaves with 3 or 5 leaflets, margins coarsely and sharply toothed, the upper surface thinly hairy, the undersurface velvety hairy. Central primocane leaflets (6.5–)7.0–9.0 cm long and (4–)5–6 cm wide, elliptic, rounded at the base, tapered to a short, sharply pointed tip, the leaflet stalk about 1/5–3/10 as long as the leaflet blade; middle leaflets elliptic-obovate, angled at the base, short-tapered to a sharply pointed tip, stalked; basal leaflets, when 3 leaflets are present, ovate, often with asymmetric lobes, rounded at the base, angled to tapered to a sharply pointed tip, short-stalked, when 5 leaflets are present, elliptic to elliptic-obovate, angled at the base, angled to a sharply pointed tip, sessile. Inflorescences 5–20 cm long, with 1–6(–8) flowers on long, ascending stalks, with 2–6 leafy bracts, these about evenly divided between simple bracts and those with 3 leaflets; flower and inflorescence stalks with dense, nonglandular hairs and downward-angled to downward-curved prickles. Sepals 7–9 mm long, 3.0–4.5 mm wide, triangular-elliptic, tapered to a sharply pointed tip or abruptly tapered to a short, slender point. Petals (9–)12–20 mm long, obovate to broadly obovate. Fruits 12–17 mm long, 10–15 mm wide, globose to short-cylindric. 2*n*=49. April–May.

Uncommon, mostly south of the Missouri River (eastern U.S. west to Kansas and Oklahoma). Mesic to dry upland forests, upland prairies, sand prairies, savannas, glades, and margins of sinkhole ponds; also roadsides.

Two taxa here treated as synonyms of *R. meracus* (Widrlechner, 1998), *R. frustratus* and *R. kelloggii*, were described from Missouri, with the type material of *R. frustratus* collected by B. F. Bush in Cooper County in 1935 (L. H. Bailey, 1943a) and that of *R. kelloggii* by J. H. Kellogg in Stoddard County in 1933 (L. H. Bailey, 1945).

This dewberry can closely resemble *R. roribaccus*. The primary difference is expressed in the narrower leaflets of *R. meracus*. Some Missouri collections are difficult, if not impossible, to assign to one species or the other with certainty, and the county distribution map for *R. meracus* may contain a few records that will in the future be redetermined as *R. roribaccus*. This is especially true of fragmentary collections and of late-season primocanes, which can have relatively narrow leaflets in both taxa.

21. Rubus roribaccus (L.H. Bailey) Rydb. (Lucretia dewberry)

R. occidualis (L.H. Bailey) L.H. Bailey

Map 2519

Canes to 250 cm long and 50 cm tall, 2.5–5.0 mm in diameter. Prickles moderate to dense, 1.0–4.5 per cm of cane, 1.0–3.5 mm long. Petioles with dense, nonglandular hairs, armed with downward-curved prickles to 2 mm long. Stipules 12–18 mm long, linear to linear-lanceolate, sometimes notched. Primocane leaves with 3 or 5 leaflets, margins coarsely and sharply toothed, the upper surface thinly hairy, the undersurface velvety hairy. Central primocane leaflets (6.5–)7.5–9.5 (–11.0) cm long and (5.5–)6.0–8.5(–9.0) cm wide, nearly orbicular to ovate or ovate-elliptic, sometimes lobed, cordate to truncate at the base, tapered to a sharply pointed tip, the leaflet stalk about 1/4–2/5 as long as the leaflet blade; middle leaflets elliptic to elliptic-obovate, rounded to angled at the base, short-tapered to a sharply pointed tip; basal leaflets, when 3 leaflets are present, ovate, rounded at the base, short-tapered to a sharply pointed tip, sessile, when 5 leaflets are present, elliptic to elliptic-obovate, angled at the base, angled to a sharply pointed tip, short-stalked to sessile. Inflorescences (4.0–)6.5–25.0 (–35.0) cm long, with (1–)3–5(–8) flowers on long, ascending stalks, with 2–7 leafy bracts, these about evenly divided between simple bracts and those with 3 leaflets; flower and inflorescence stalks with dense, nonglandular hairs and downward-angled to downward-curved prickles. Sepals 7–16 mm long, 4–5 mm wide, triangular-ovate to elliptic, tapered to a sharply pointed tip or expanded into a lobed, leafy tip. Petals 12–20 mm long, obovate to broadly obovate. Fruits 13–25 mm long, 15–25 mm wide, globose to cylindric or long-conic. 2*n*=49. April–June.

Scattered nearly throughout the state but apparently absent from most of the western portion of the Glaciated Plains Division (northeastern U.S. west to Iowa, Kansas, and Oklahoma). Openings of mesic to dry upland forests, upland prairies, banks of streams and rivers, margins of ponds and sinkhole ponds, ledges of bluffs; also pastures, old fields, old mines, cemeteries, railroads, roadsides, and open, disturbed areas.

Large-fruited selections of *R. roribaccus* have been cultivated and may escape or persist from gardens. This species has escaped from cultivation in Australia (Evans et al., 2007).

22. Rubus satis L.H. Bailey

Map 2520

Canes to 300 cm long and 40–120 cm tall, 4–7 mm in diameter, forming a dense, mounding tangle. Prickles usually moderate, 0.5–3.0 per cm of cane, 1.5–5.0 mm long. Petioles with nonglandular hairs, armed with downward-curved prickles to 1.5 mm long. Stipules 10–14 mm long, linear to linear-lanceolate. Primocane leaves mostly with 5 leaflets, rarely with 3 leaflets, the margins sharply toothed, the upper surface thinly hairy, the undersurface velvety hairy. Central primocane leaflets 8–14 cm long and 6.0–10.5 cm wide, broadly ovate to ovate, cordate at the base, tapered or long-tapered to a sharply pointed or filiform tip, the leaflet stalk about 1/3 as long as the leaflet blade; middle leaflets elliptic, angled at the base, long-tapered to a sharply pointed tip, stalked; basal leaflets elliptic-obovate, angled at the base, tapered to a sharply pointed tip, sessile. Inflorescences mostly racemose, occasionally appearing flat-topped or flaring toward the apex, (4–)9–21 cm long, with 4–8 flowers and 1–6 leafy bracts, these mostly with 3 leaflets; flower and inflorescence stalks with dense, nonglandular hairs and downward-curved prickles. Sepals 6–7 mm long, 3–4 mm wide, triangular-ovate, tapered to a

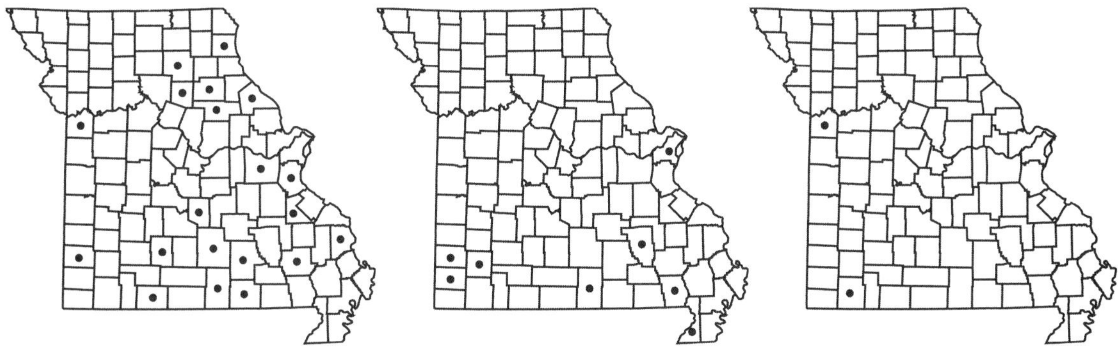

2521. Rubus steelei 2522. Rubus armeniacus 2523. Rubus laciniatus

sharply pointed tip. Petals 14–18 mm long, obovate to broadly obovate. Fruits 10–18 mm long, 9–17 mm wide, globose to short-cylindric. $2n$=63. May–June.

Uncommon in the eastern portion of the Ozark Division (northeastern U.S. and adjacent Canada west to Minnesota, Iowa, and Missouri). Mesic upland forests; also roadsides and open, disturbed areas.

The Missouri populations are somewhat disjunct from the main range of the species. The habit of *R. satis* fits Steyermark's (1963) erroneous description of *R. missouricus*, as expressed in his key. *Rubus missouricus* is not known to tip-root. The growth habit of *R. satis* also resembles that of *R. hancinianus*, which is found on dry, upland prairies in central Kansas and which could occur on similar sites in western Missouri (for further discussion on this species' current exclusion from the flora, see the treatment of sect. *Flagellares*). *Rubus hancinianus* differs from *R. satis* in having smaller, elliptical to obovate central primocane leaflets with rounded bases, as well as flowers that often include more than 5 petals.

23. Rubus steelei L.H. Bailey

Pl. 540 o; Map 2521

Canes to 180 cm long, typically prostrate, 2–3 mm in diameter, sometimes so delicate as to seem herbaceous. Prickles moderate, 0.5–2.0(–3.0) per cm of cane, 1–2(–3) mm long. Petioles with nonglandular hairs, armed with downward-angled, needlelike prickles to 1.5 mm long. Stipules 5–12 (–20) mm long, linear to lanceolate. Primocane leaves almost always with 3 leaflets, rarely a few with 5 leaflets, the margins bluntly to sharply toothed, the teeth varying widely in coarseness, the upper surface glabrous, the undersurface thinly hairy. Central primocane leaflets 5.0–8.5 (–9.0) cm long and 3–6(–7) cm wide, elliptic-ovate to elliptic-rhombic, subcordate to angled at the base, tapered to a pointed or sharply pointed tip, the leaflet stalk about 1/6–1/3 as long as the leaflet blade; basal leaflets asymmetrically ovate, sometimes lobed, rounded at the base, angled to a pointed tip, sessile or nearly so. Inflorescences 5.0–20.5(–28.0) cm long, with 1–6(–10) flowers on long, ascending stalks, if only a single flower is present, then generally the inflorescences are at least 8 cm long, with (2–)3–5(–7) bracts, these usually mostly simple; flower and inflorescence stalks nearly glabrous, sometimes with small downward-angled, needlelike prickles. Sepals 6–12 mm long, 3–4 mm wide, triangular-ovate to elliptic, with slender to filiform or lobed and leafy tips. Petals 8–12 mm long, obovate. Fruits 8–15 mm long, 10–15 mm wide, globose to short-cylindric. May–June.

Scattered, mostly in the Ozark and Ozark Border Divisions and the eastern portion of the Glaciated Plains (eastern U.S. and adjacent Canada west to Minnesota, Kansas, and northeastern Texas). Mesic to dry upland forests, savannas, glades, and tops of bluffs; also fallow fields, old fields, fencerows, railroads, and roadsides.

2d. Section Rubus

Canes stout with considerable branching, sometimes persisting more than two years. Stipules lateral. Perhaps as many as 300 species, native to the Old World, primarily in western and central Europe.

Treatments vary widely in the number of species recognized. Steyermark (1963) accepted three naturalized taxa in sect. *Rubus*: *R. bifrons* Vest, *R. laciniatus*, and *R. procerus*. *Rubus laciniatus* is a distinctive taxon that is clearly defined. *Rubus procerus* was the name that Steyermark applied to the Himalayan blackberry, an invasive species whose correct name is a matter of some controversy. The present treatment follows those of Evans and Weber (2003) and Zielinski (2004) in using the name *R. armeniacus*. No specimens confirming the presence of *R. bifrons* in the Missouri flora have been located.

1. Canes typically less than 150 cm tall, arching; leaves softly pubescent, but green, beneath; central primocane leaflets irregularly divided or lobed, deeply incised and tapered to a narrow, sharply pointed tip . . . . 25. R. LACINIATUS
1. Canes robust, often more than 150 cm tall, high-arching to erect; leaves white to gray-felted beneath; central primocane leaflets elliptic to broadly elliptic to oblong, sharply, but not deeply, serrate
   2. Primocane leaves with (3–)5 leaflets, white to light gray-felted beneath, central leaflet typically at least 6 cm wide; canes with sparse tufted or simple hairs, nearly glabrous at maturity; inflorescences paniculate, with 5–30+ flowers on strongly branching stalks, leafy bracts concentrated at the basal nodes, lower inflorescence stalks covered with tufted clusters of hairs (some times appearing stellate-hairy) . . . . 24. R. ARMENIACUS
   2. Primocane leaves with 3 or 4(5) leaflets, gray-felted beneath, central leaflet 3.5–4.5 cm wide; canes hairy, inflorescences racemose, typically with 3–8 flowers on unbranched stalks; leafy bracts found nearly the entire length of the inflorescence, lower inflorescence stalks covered with simple, spreading hairs . . . . 26. R. SERISSIMUS

24. Rubus armeniacus Focke (Himalayan blackberry)
R. discolor Weihe & Nees
R. procerus P.J. Muell., misapplied
Pl. 542 a–c; Map 2522

Canes to 350 cm long and 100–300 cm tall, 5–15 mm in diameter, often branched. Canes with sparse tufted or simple hairs, becoming nearly glabrous at maturity. Prickles moderate, 0.6–1.4 per cm of cane, 8–11 mm long, often with red at or near their bases. Petioles with nonglandular hairs, armed with many broad-based, downward-curved prickles to 5 mm long. Stipules 10–12 mm long, linear to filiform, clearly lateral, diverging 3–6+ mm from the base of the petiole. Primocane leaves with (3)5 leaflets, margins irregularly, coarsely, and sharply toothed, the upper surface nearly glabrous, the undersurface white- to light gray-felted, the midvein on the undersurface drying to a light yellow. Central primocane leaflets 5.0–12.5 cm long and 6.0–9.5 cm wide, suborbicular to broadly elliptic-obovate or oblong-obovate, cordate to rounded at the base, abruptly short-tapered to a sharply pointed tip, the leaflet stalk about $1/3$–$2/5$ as long as the leaflet blade; lateral primocane leaflets resembling the central leaflet but somewhat smaller, the basal pair with stalks 3–6 mm long. Inflorescences typically paniculate, 28–38 cm long and 8–18 cm wide, with 5–31 flowers and 3–7 leafy bracts, these concentrated at the basal nodes, a mixture of simple bracts, and those with 3 and 5 leaflets; flower and inflorescence stalks with dense nonglandular hairs, and strong straight to downward-angled prickles; the flower stalks sometimes with a few gland-tipped hairs and lanceolate, stipulelike bractlets to 15 mm long, the lower inflorescence stalks covered with tufted clusters of hairs (sometimes appearing stellate-hairy). Sepals 3.0–3.5 mm long, 6–8 mm wide, triangular, abruptly tapered to a sharp tip. Petals 13–20 mm long, broadly elliptic, pale pink. Fruits 10–20 mm long, 10–20 mm wide, oblong. $2n=28$. May–June.

Introduced, scattered in the southern portion of the Ozark Division (likely native to the Caucasus; naturalized widely in Europe, U.S., and adjacent Canada). Banks of streams, also pastures and roadsides.

This species is likely represented in Missouri only by escaped or persisting populations of cv. 'Himalaya' or similar horticultural selections. It is a serious, invasive pest in the Pacific Northwest. It was treated by Steyermark (1963) as *R. procerus*, a synonym of the Eurasian species, *R. praecox* Bertol. (Zielinski, 2004).

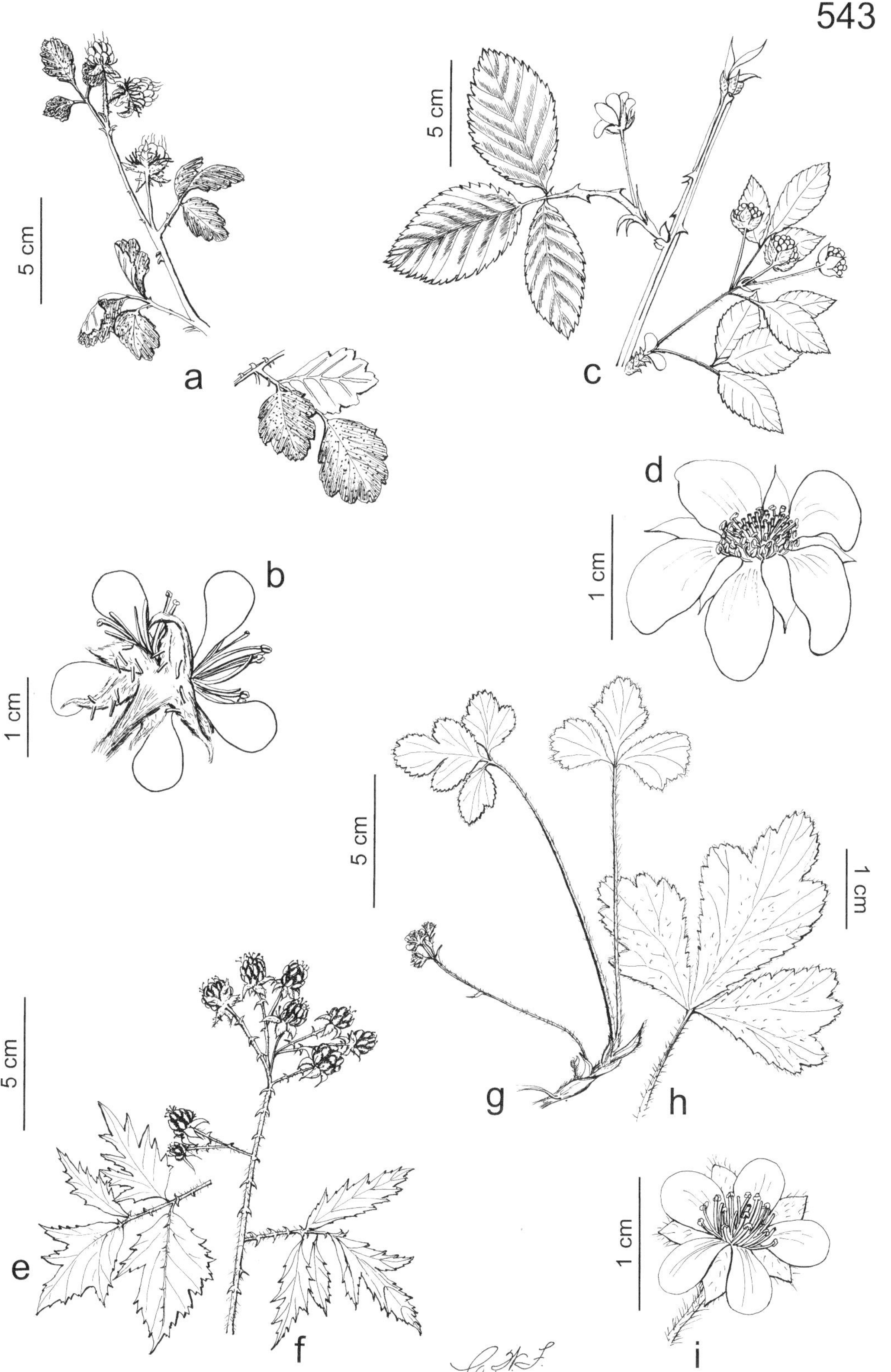

Plate 543. Rosaceae. *Rubus idaeus*, **a)** tip of floricane and larger floricane leaf, **b)** flower from below. *Rubus missouricus*, **c)** flower, **d)** floricane node with fruits. *Rubus laciniatus*, **e)** primocane leaf, **f)** apical portion of inflorescence with fruits. *Geum fragarioides*, **g)** habit, **h)** leaf, **i)** flower.

2524. Rubus serissimus 2525. Rubus missouricus 2526. Rubus trivialis

25. Rubus laciniatus Willd. (cutleaf blackberry)

Pl. 543 e, f; Map 2523

Canes to 150 cm long and 80–120 cm tall, 3–10 mm in diameter, arching and often branched. Canes at maturity with sparse, fine hairs. Prickles moderate, 1.0–2.4 per cm of cane, 5–7 mm long, broad-based and strongly downward-curved. Petioles with nonglandular hairs, armed with many broad-based, downward-curved prickles to 3 mm long. Stipules 6–13 mm long, linear to filiform, clearly lateral, diverging (2–)3–6+ mm from the base of the petiole. Primocane leaves with 5 or more leaflets, the margins irregularly and sharply toothed and deeply incised, sometimes strongly lobed, the upper surface thinly to moderately hairy, the undersurface densely and softly hairy. Central primocane leaflets 6.0–11.5 cm long and 4–11 cm wide, ovate-triangular in general outline, angled to truncate at the base, angled ultimately to a narrow, sharply pointed tip, the leaflet stalk about 2/5–1/2 as long as the leaflet blade; lateral primocane leaflets also strongly lobed or divided but typically also ovate-triangular in general outline, stalked. Inflorescences typically paniculate, 23–58 cm long and 10–27 cm wide, with 10–40 flowers, and 4–8 leafy bracts, these concentrated at the basal nodes, a mixture of simple bracts and those with 3 or 5 leaflets; flower stalks with spreading, nonglandular hairs, inflorescence stalks zigzag, with sparse to dense nonglandular hairs, both flower and inflorescence stalks with strong downward-curved prickles. Sepals 8–13(–25) mm long, 2–4 mm wide, lanceolate, armed with small, yellowish, needlelike prickles, sometimes with leafy, jagged tips. Petals 10–12 mm long, obovate, often with 2 or 3 lobes, pale pink. Fruits 15–25 mm long, 10–25 mm wide, short-cylindrical. $2n=28$. June.

Introduced, uncommon, sporadic (may be of European garden origin; widely naturalized in Europe, the U.S., and Australia). Edges of mesic upland forests; also pastures and open, disturbed areas.

This distinctive taxon appears to be represented in Missouri only by escaped or persisting populations of European horticultural selections. Steyermark (1963) also considered it nonnative in the United States.

26. Rubus serissimus L.H. Bailey (everbearing blackberry)

Pl. 542 d, e; Map 2524

Canes to 300 cm long and 125–200 cm tall, 3–8 mm in diameter, with the main stem fairly erect and with strong branches that often arch down nearly to the ground, but without tip-rooting. Canes covered with fine hairs, at least on new growth. Prickles moderate, 1.2–1.8 per cm of cane, 5–9 mm long, broad-based and nearly straight, often with red at or near the base. Petioles with dense, nonglandular hairs, armed with many broad-based, downward-curved prickles to 4 mm long. Stipules 10–14 mm long, filiform, clearly lateral, diverging 3 mm from the base of the petiole. Primocane leaves typically with 3 leaflets, a few with 4 or rarely 5 (when more than 3 leaflets are present, each basal leaflet is attached in a more or less pedate fashion to the lateral leaflet stalk), the margins finely, doubly, and sharply toothed, the upper surface thinly hairy, the undersurface gray-felted. Central primocane leaflets 5–7 cm long and 3.5–4.5 cm wide, elliptic, subcordate to rounded at the base, short-tapered to a sharply pointed tip, the leaflet stalk about 1/4–1/3 as long as the leaflet blade; lateral and basal primocane leaflets asymmetrically elliptic to elliptic-obovate, sometimes lobed, rounded at the base, angled to a pointed tip, basal leaflets (when present) with stalks 1–2 mm long. Inflorescences typically racemose, well-developed examples can be compactly paniculate, 8–28 cm long and 4–10 cm wide, with 3–8 flowers concentrated near the apex and 3–7 leafy bracts, these mostly with 3 leaflets, positioned nearly the entire length of the axis; flower and inflorescence stalks dense with simple, spreading hairs, strong straight to downward-angled prickles, and rarely lanceolate, stipulelike bractlets to 5 mm long. Sepals 6–8 mm long, 2–3 mm wide, triangular, tapered to a sharply pointed tip. Petals 7–11 mm long, obovate, pale pink, fading to white. Fruits 15–20 mm long, 14–20 mm wide, globose to short-cylindric. May–July.

Introduced, scattered south of the Missouri River (likely native to Europe, perhaps of garden origin, naturalized in Arkansas, Missouri, Oklahoma, and Texas). Upland prairies and sand prairies; also gardens, railroads, roadsides, and open, disturbed areas.

This unusual blackberry was described by L.H. Bailey (1943b) from cultivated material of unknown nativity supplied to him by a nursery in Denton, Texas. At that time, it was evidently also being grown in Oklahoma. Bailey classified it in sect. *Cuneifolii* L.H. Bailey, a group native to the southeastern United States, but he was unable to locate any native populations. Its combination of light pink petals, large prickles with reddish markings, and lateral stipules all suggest that it is actually of Old World origin and not a member of sect. *Cuneifolii*. It was first collected in Missouri in 1969 by Viktor Mühlenbach during his inventories of the St. Louis railyards. Its most common habitats in the state (near railroad rights-of-ways and in gardens) also suggest escape from cultivation, rather than being an overlooked member of the native flora. However, attempts to key it out to any of the numerous described species of Sect. *Rubus* native to Europe have so far proven unsuccessful. Perhaps it is of garden origin. Alice et al. (in press) also consider it to be of New World origin and include *R. serissimus* under a broadly circumscribed taxon, *R. longii* Fernald, consisting of various members of sect. *Cuneifolii*.

Until recently, specimens of *R. serissimus* in Missouri herbaria were misdetermined either as *R. bifrons* or *R. armeniacus* (for further discussion, see the sectional treatment above). In fact, *R. serissimus* is more widespread and abundant in the state than is *R. armeniacus* and now represents an important invasive element that is forming dense thickets in various natural habitats.

2e. Section Setosi L.H. Bailey

About 25 species, primarily in the northeastern U.S. and adjacent Canada, particularly on acidic or nutrient-poor soils.

27. Rubus missouricus L.H. Bailey

Pl. 543 c, d; Map 2525

Canes 80–170 cm long and 70–150 cm tall, 5–7 mm in diameter, erect to arching the first year, arching the second year; clonal by root-suckering but not tip-rooting. Prickles moderate to dense, 1–5(–8) per cm of cane, 2–5 mm long, needlelike, straight or somewhat downward angled. Petioles with soft, nonglandular hairs, armed with downward-angled or downward-curved, needlelike prickles to 3.5 mm long. Stipules 15–20 mm long, lanceolate, basal to somewhat lateral, diverging 0–3 mm from the base of the petiole. Primocane leaves with 3 or 5 leaflets, the margins sharply and somewhat irregularly toothed, the upper surface thinly hairy, the undersurface velvety hairy. Central primocane leaflets 7–10 cm long and 4.0–5.5 cm wide, elliptic to obovate-elliptic, rounded to angled at the base, tapered to a sharply pointed tip, the leaflet stalk about 1/5–1/3 as long as the leaflet blade; lateral primocane leaflets elliptic to obovate-elliptic, often asymmetrically lobed when only 3 leaflets are present, angled at the base, tapered to a sharply pointed tip, stalked (the stalks much shorter in the basal pair). Inflorescences varying on a single cane from large, compound, flaring inflorescences, almost broomlike in appearance, to reduced, simple racemes, 8–21(–35) cm long and 6.5–15.0 cm wide, with 7–20 flowers and 2–5 bracts, mostly with 3 leaflets; flower and inflorescence stalks with dense nonglandular hairs, rarely with a few hairlike prickles. Sepals 5–7 mm long, 2.5–4.0 mm wide, ovate-elliptic to triangular, tapered abruptly to a short, slender point. Petals 10–14 mm long, obovate, typically white, but sometimes drying to a light rose pink. Fruits 10–20 mm long and 12–17(–19) mm wide, globose to short cylindric or short-conic. $2n$=21, 28. May–June.

Uncommon, mostly north of the Missouri River (north-central U.S. from Missouri north to Minnesota and east to Michigan). Upland prairies, bottomland prairies, edges of bottomland forests, and banks of streams; also pastures and fencerows.

Rubus missouricus was described by L. H. Bailey (1932) from Missouri, with the type mate-

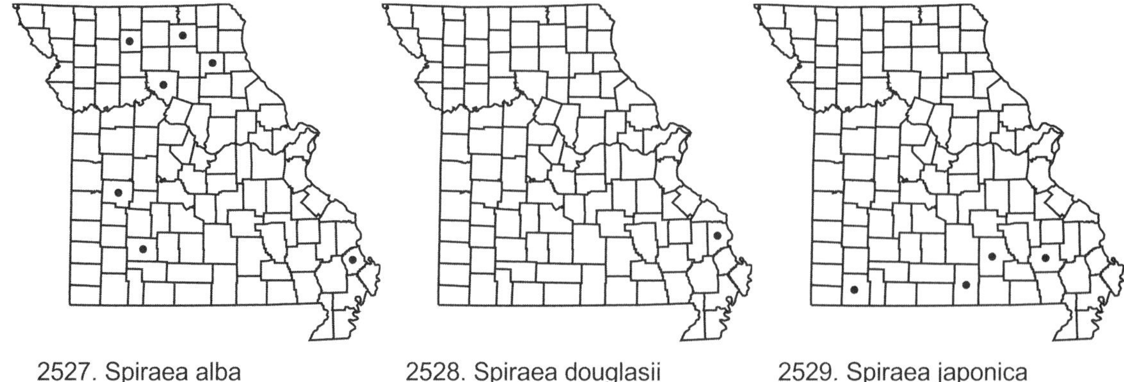
2527. Spiraea alba
2528. Spiraea douglasii
2529. Spiraea japonica

rial collected by B. F. Bush from Jackson County. It was considered endemic to Missouri, even in Steyermark's (1963) treatment. The species has not often been collected in the state and most of the specimens in herbaria are historical. The species typically begins to flower a week or two later than neighboring populations of other native members of subgenus *Rubus*.

2f. Section Verotriviales L.H. Bailey

About 10 species, primarily in the southeastern U.S. and adjacent northeastern Mexico; most taxa treated as one highly polymorphic taxon, *R. trivialis*, by Alice et al. (in press).

28. Rubus trivialis Michx. (southern dewberry)
Pl. 540 m, n; Map 2526

Canes to 250 cm long and to 70 cm tall, arching to prostrate, typically branched, 3–4 mm in diameter, reddish, often rooting at tips and sometimes at nodes. Primocanes with prickles and dark red hairs, some or all gland-tipped. Prickles moderate, 1–3 per cm of cane, 3–5 mm long, downward angled or downward curved. Petioles with stiff, dark red hairs, some or all gland-tipped, and downward-angled prickles to 2 mm long. Stipules 5–14 mm long, linear to filiform. Primocane leaves mostly with 5 leaflets, less often 3, the margins sharply or bluntly toothed, the upper surface nearly glabrous, the undersurface hairy only on the veins. Central primocane leaflets 5.0–7.5 cm long and 2.5–5.0 cm wide, narrowly elliptic to ovate-elliptic to obovate-elliptic, truncate, rounded, or angled at the base, usually tapered to a sharply pointed tip, but occasionally rounded at tip, the leaflet stalk about 1/6–1/3 as long as the leaflet blade; basal leaflets generally of same form as the central leaflet, but smaller, short-stalked. Primocane leaflets often persist until flowering the following spring, by winter typically appearing bronzed. Inflorescences 5–18 cm long, with 1–3(4) flowers on long, ascending stalks and 2–6 leafy bracts, these mostly with 3 leaflets; flower and inflorescence stalks thinly covered with light-colored, nonglandular hairs and sometimes with red, glandular hairs, and needlelike or downward-curved prickles. Sepals 4–6(–9) mm long, 2–5 mm wide, narrowly triangular, tapered to a pointed tip or abruptly tapered to a short, slender point. Petals 10–14(–18) mm long, obovate to broadly obovate, white or sometimes pinkish-tinged. Fruits 10–30 mm long, 9–12 mm wide, oblong. $2n=14$. April–June.

Scattered, mostly south of the Missouri River, most abundantly in the Mississippi Lowlands Division (southeastern U.S. west to Oklahoma and Texas; Mexico). Banks of streams and rivers, bases and ledges of bluffs, bottomland forests, swamps, and sloughs; also pastures, fallow fields, levees, and roadsides.

This attractive species is distinctive in its abundant, red, prickles and bristly hairs, some of which are gland-tipped. As in some other species of *Rubus*, it produces long, white roots that are fleshy and somewhat thickened, which can give rise to new canes at quite some distance from the original clump.

26. **Spiraea** L. (spiraea)

Plants shrubs, often colonial from rhizomes and/or root sprouts. Stems ascending or less commonly strongly arched, branched or unbranched, unarmed. Twigs glabrous or hairy, the winter buds small, with 2 to several overlapping scales. Leaves alternate, sessile or short-petiolate. Stipules absent. Leaf blades simple, unlobed, variously shaped, the margins simply or less commonly doubly toothed. Inflorescences axillary clusters or small umbels, or terminal panicles, the flowers relatively small, with a small linear bract along the stalk. Flowers perigynous, the hypanthium cup-shaped to somewhat conic, with a usually prominent nectar ring, glabrous or hairy on the outer surface. Sepals 5, spreading to ascending at flowering, triangular, persistent at fruiting or shed after flowering. Petals 5, broadly obovate to nearly circular, white or pink. Stamens 15 to numerous (except in doubled flowers), the anthers white or pink. Pistils 5 (except in doubled flowers), free. Ovary superior, glabrous or hairy, with 1 locule and 2 to several ovules. Style 1 per pistil, persistent, the stigma more or less capitate. Fruits follicles, ascending, elliptic-ovate in outline, tapered to an erect beak at the tip, glabrous or hairy, tan to brown and leathery to papery at maturity, dehiscing along the inner suture and also partially along the outer (dorsal) suture, 1–4-seeded. Seeds 1.5–2.5 mm long, narrowly ellipsoid, the surface with faint longitudinal lines or a fine network of slender ridges and quadrangular pits, yellowish to reddish brown. Seventy to 120 species, North America, Europe, Asia.

A number of species of *Spiraea* are cultivated as ornamental shrubs. Although the individual flowers are small, they are often produced in profusion.

1. Inflorescences axillary, umbellate clusters of 2–6 flowers; corollas appearing doubled, with numerous petals . . . . 4. S. PRUNIFOLIA
1. Inflorescences terminal panicles of numerous flowers; corollas not doubled, with 5 petals
    2. Leaves with the undersurface glabrous or nearly so; corollas white or pink
        3. Corollas white; inflorescences longer than wide, ovoid to more or less pyramid-shaped, sometimes elongate; leaf blades narrowly elliptic to oblanceolate . . . . 1. S. ALBA
        3. Corollas pink; inflorescences wider than long, more or less flat-topped to broadly dome-shaped; leaf blades lanceolate to narrowly ovate . . . . 3. S. JAPONICA
    2. Leaves with the undersurface moderately to densely hairy; corollas pink
        4. Leaf blades elliptic to oblong-elliptic or nearly oval, the undersurface densely woolly with grayish white hairs; ovaries and fruits glabrous . . . . 2. S. DOUGLASII
        4. Leaf blades ovate to lanceolate or oblong-elliptic, the undersurface densely woolly with tan to yellowish white hairs; ovaries and fruits hairy . . . . 5. S. TOMENTOSA

1. Spiraea alba Du Roi **var. alba** (meadowsweet)

Pl. 544 d, e; Map 2527

Plants shrubs, 0.5–2.0 m tall. Twigs reddish brown to grayish brown with prominent, small lenticels, somewhat angular, minutely hairy toward the tip when young, glabrous or nearly so at maturity. Leaves mostly short-petiolate. Leaf blades 3–6 cm long, narrowly elliptic to oblanceolate, angled or tapered at the base, angled or tapered to a sharply pointed tip, the margins finely and sharply toothed, the surfaces glabrous or nearly so. Inflorescences terminal panicles of numerous flowers, longer than wide, ovoid to more or less pyramid-shaped, sometimes elongate. Hypanthia 1.5–2.0 mm wide, cup-shaped, glabrous or minutely hairy. Sepals 1.0–1.5 mm long, triangular, bluntly pointed at the tip. Corollas not doubled,

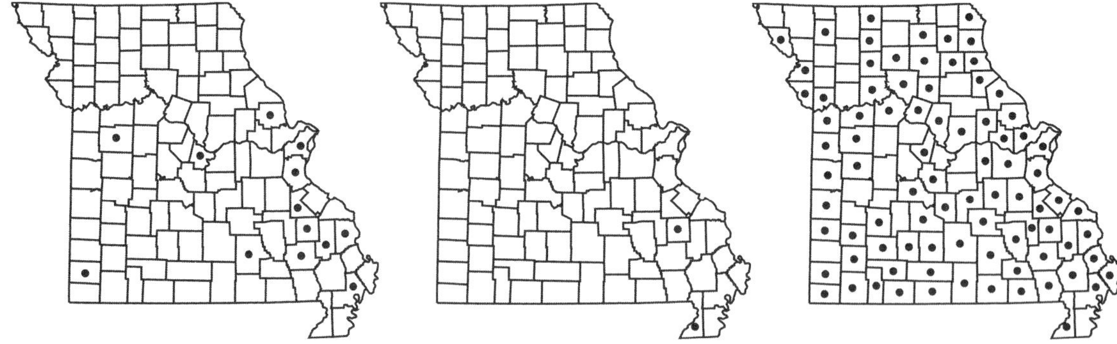

2530. Spiraea prunifolia 2531. Spiraea tomentosa 2532. Cephalanthus occidentalis

with 5 petals. Petals 2.7–3.5 mm long, white. Ovaries glabrous. Fruits 3.0–3.5 mm long, glabrous. $2n$=36. June–August.

Uncommon in the Glaciated Plains Division, with solitary, disjunct occurrences in Greene, St. Clair, and Scott Counties (northeastern U.S. west to North Dakota and Missouri; Canada). Bottomland prairies, marshes, margins of lakes, edges of bottomland forests, and edges of mesic upland forests (in sand); also ditches, fencerows, railroads, and roadsides.

The rather weakly distinguishable var. *latifolia* (Aiton) H.E. Ahles differs from var. *alba* in its purplish to reddish brown twigs, somewhat broader (mostly less than three times as long as wide) and more coarsely toothed leaves. and relatively glabrous inflorescences. It tends to occur in the eastern portion of the species range and often in somewhat drier habitats.

2. Spiraea douglasii Hook. **var. douglasii**

Pl. 544 a–c; Map 2528

Plants shrubs, 0.7–2.0 m tall. Twigs reddish brown, usually rounded but often with fine longitudinal lines or ridges, finely and persistently woolly. Leaves mostly short-petiolate. Leaf blades 3–9 cm long, elliptic to oblong-elliptic or nearly oval, angled or rounded at the base, angled to a sharply or bluntly pointed tip, the margins finely and sharply toothed, the upper surface sparsely and finely hairy at maturity, green, the undersurface persistently and densely pubescent with grayish white, finely woolly hairs. Inflorescences terminal panicles of numerous flowers, longer than wide, narrowly ovoid to narrowly pyramid-shaped, often elongate. Hypanthia 1.0–1.5 mm wide, cup-shaped to somewhat conic, finely woolly. Sepals 0.5–1.0 mm long, triangular, sharply pointed at the tip. Corollas not doubled, with 5 petals. Petals 1.3–1.7 mm long, pink. Ovaries glabrous. Fruits 3.5–4.5 mm long, glabrous. $2n$=36. June–August.

Introduced, uncommon, known thus far from a single, historical collection from Cape Girardeau County (native of the western U.S. [including Alaska] and adjacent Canada east to Montana and Colorado; introduced sporadically east of the Great Plains). Pastures.

3. Spiraea japonica L. f.

Map 2529

Plants shrubs, 0.5–1.5 m tall. Twigs reddish brown to grayish brown with scattered, small, dark lenticels, somewhat angular, minutely hairy toward the tip when young, sometimes glabrous or nearly so at maturity. Leaves sessile or short-petiolate. Leaf blades 6–15 cm long, lanceolate to narrowly ovate, angled at the base, tapered to a sharply pointed tip, the margins finely and sharply toothed, the surfaces glabrous or nearly so. Inflorescences terminal panicles of numerous flowers, wider than long, more or less flat-topped to broadly dome-shaped. Hypanthia 1.5–2.0 mm wide, cup-shaped, minutely hairy. Sepals 1.0–1.5 mm long, triangular, sharply pointed at the tip. Corollas not doubled, with 5 petals (doubled in some cultivated forms). Petals 1.8–2.5 mm long, pink. Ovaries glabrous. Fruits 2.5–3.0 mm long, glabrous. $2n$=18. June–August.

Introduced, uncommon, (native of Japan, introduced widely in the eastern U.S. west to Illinois, Missouri, and Alabama). Seepy banks of streams.

For many years, Stan Hudson observed a solitary, mature shrub persisting from an old planting near a former sawmill in Wayne County, but this individual never reproduced itself by seed or vegetatively. However, in 2001 Bill Summers discovered a large population of reproducing shrubs scattered on the seepy banks of a stream in Shannon County.

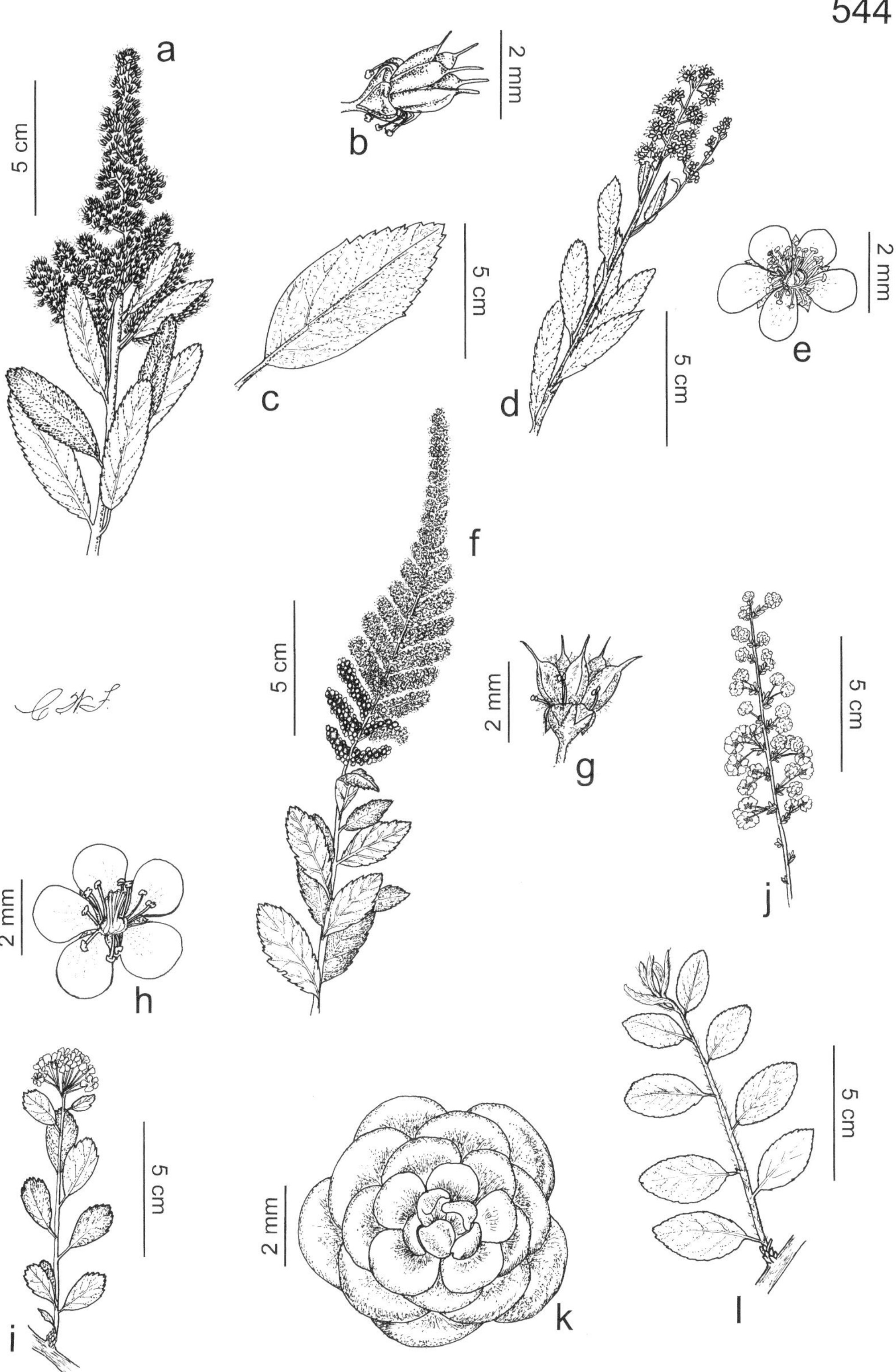

Plate 544. Rosaceae. *Spiraea douglasii*, **a)** stem tip with leaves and inflorescence, **b)** lower stem leaf, **c)** fruits. *Spiraea alba*, **d)** stem tip with leaves and inflorescence, **e)** flower. *Spiraea tomentosa*, **f)** stem tip with leaves and inflorescence, **g)** fruits. *Spiraea ×vanhouttei*, **h)** flower, **i)** branch tip with leaves and inflorescence. *Spiraea prunifolia*, **j)** inflorescence, **k)** branch with leaves, **l)** flower.

4. Spiraea prunifolia Siebold & Zucc. **var. prunifolia** (bridal wreath)

S. prunifolia var. *plena* C.K. Schneid.

Pl. 544 j–l; Map 2530

Plants shrubs, 0.8–1.5 m tall (but the stems often longer and strongly arching). Twigs reddish brown to dark grayish brown, somewhat angular, moderately to densely and minutely hairy, at least when young, often also with scattered, longer, more or less spreading hairs. Leaves short-petiolate. Leaf blades 1–4 cm long, elliptic to ovate, angled to short-tapered at the base, rounded or angled to a bluntly pointed tip, the margins finely and sharply toothed, the upper surface glabrous or sparsely hairy along the main veins, sometimes somewhat shiny, the undersurface sparsely to moderately hairy. Inflorescences axillary, umbellate clusters of 2–6 flowers. Hypanthia 1.5–2.0 mm wide, cup-shaped to somewhat conic, glabrous or sparsely hairy. Sepals 0.8–1.2 mm long, triangular to broadly triangular, sharply pointed at the tip. Corollas appearing doubled, with numerous petals (not doubled, with 5 petals elsewhere). Petals 2.0–2.5 mm long, white. Ovaries glabrous. Fruits infrequently produced, 1.5–2.5 mm long, glabrous. $2n$=18. April–May.

Introduced, uncommon, mostly in the southeastern quarter of the state (native of Asia; introduced widely in the eastern U.S. west to Missouri and Louisiana; Canada). Tops of bluffs and banks of streams; also old homesites, fencerows, pastures, gardens, cemeteries, railroads, and roadsides.

Missouri plants routinely have the corollas doubled, with numerous extra petals at the expense of all or most of the stamens and pistils. Such plants have been called var. *plena*, but are better treated as a cultivar. They reproduce primarily vegetatively by root sprouts.

Hornberger (1980) and Ohmart (1987) reported occurrences of *S.* ×*vanhouttei* (Briot) Carrière (bridal wreath; Pl. 544 h, i) as having escaped from cultivation in the state. This horticultural hybrid was developed through crosses, possibly between two Asian species, *S. cantoniensis* Lour. and *S. trilobata* L. It is similar to *S. prunifolia* in its small axillary umbels, but differs in having its inflorescences noticeably stalked, as well as in its rhombic-ovate to obovate leaves with the margins few-toothed toward the tip. The flowers also usually do not appear doubled.

5. Spiraea tomentosa L. (hardhack, steeple bush)

S. tomentosa var. *rosea* (Raf.) Fernald

Pl. 544 f, g; Map 2531

Plants shrubs, 0.5–1.2 m tall. Twigs reddish brown to orangish brown, usually rounded but often with fine longitudinal lines or ridges, finely and persistently woolly. Leaves sessile or short-petiolate. Leaf blades 3–7 cm long, ovate to lanceolate or oblong-elliptic, angled or tapered at the base, rounded or angled to a usually bluntly pointed tip, the margins finely and sharply toothed, the upper surface sparsely to moderately and finely hairy at maturity, green to dark green, the undersurface persistently and densely pubescent with tan to yellowish white, finely woolly hairs. Inflorescences terminal panicles of numerous flowers, longer than wide, narrowly ovoid to narrowly pyramid-shaped, often elongate. Hypanthia 1.0–1.5 mm wide, cup-shaped to somewhat conic, finely woolly. Sepals 0.5–1.0 mm long, triangular, bluntly to sharply pointed at the tip. Corollas not doubled, with 5 petals. Petals 1.5–2.0 mm long, pink (white elsewhere). Ovaries woolly-hairy. Fruits 2.5–4.0 mm long, woolly-hairy. $2n$=24. June–August.

Uncommon, known thus far only from Dunklin and Madison Counties (eastern U.S. west to Minnesota, Arkansas, and possibly Louisiana; Canada). Mesic upland forests and margins of lakes.

Steyermark (1963) knew this species only from a historical collection from Crowley's Ridge in Dunklin County. However, in 1998 an apparently native population was discovered in Madison County by David Lindsay, a naturalist at the S Bar F Boy Scout Ranch. Steyermark (1963) noted that although this species is widely cultivated elsewhere in the United States it is not commonly grown in Missouri as it requires moist, acidic soils.

Some authors have separated *S. tomentosa* into two varieties. Missouri plants correspond to the var. *rosea*, a widespread variant with slightly less densely flowered inflorescence branches. Plants from the northwestern United States correspond to var. *rosea*, with fewer flowers per cm of inflorescence branches. However, there is too much overlap between these variants in the eastern states to make their recognition feasible (K. R. Robertson, 1974).

RUBIACEAE (Madder Family)

Contributed by George Yatskievych and Charlotte M. Taylor

Plants annual or perennial herbs, shrubs, or small trees. Leaves opposite or whorled. Stipules present or absent (the leaves then whorled), when present often interpetiolar (see discussion below), persistent or shed early, of various shapes, most often small and triangular, but frequently bearing 2 to several marginal lobes or bristles. Leaf blades simple, entire. Inflorescences terminal or axillary, of solitary flowers or more commonly clusters or panicles of several to numerous flowers, dense, globose, and headlike in *Cephalanthus*. Flowers perfect (occasionally some of the flowers imperfect in *Galium*), epigynous or nearly so (occasionally with the tip of the ovary slightly protruding above the fused portion of the calyx), actinomorphic, sometimes subtended by bracts. Calyces 4- or 5-lobed or sometimes with an entire margin, occasionally reduced to 2 or 3 lobes or apparently absent in *Galium*. Corollas (3)4- or 5-lobed, saucer-shaped to bowl-shaped, or funnelform (nearly cylindric in *Cephalanthus*), white, blue, purple, or pink, the lobes overlapping or not in bud. Stamens (3)4 or 5, alternating with the corolla lobes, the filaments attached in the corolla tube, the anthers attached near the midpoint or the base, exserted or not, variously colored. Style 1 or 2, the stigmas 1–2, of various forms, included or exserted. Pistil 1 per flower, of 2 fused carpels. Ovaries inferior or nearly so (occasionally with the tip of the ovary slightly protruding above the fused portion of the calyx in *Houstonia* and *Oldenlandia*), 2-locular, the placentation axile or occasionally nearly apical, the ovules 1 to numerous in each locule. Fruits capsules, schizocarps (with 2 indehiscent mericarps), achenelike (2-lobed and 2-seeded, but not splitting into mericarps or dehiscing), or fleshy and drupelike (the ovaries of 2 closely adjacent flowers becoming fused into a single drupelike fruit in *Mitchella*). Seeds small, variously shaped. About 563 genera, 10,000–11,000 species, worldwide.

The Rubiaceae are a well-delimited family, and can be recognized by the combination of their simple, entire, opposite (or whorled) leaves, interpetiolar stipules (in most genera; see below), corollas of fused petals, and inferior ovaries. The vast majority of the genera and species are shrubs and small trees of wet tropical regions. All our genera range into tropical regions, and *Galium* additionally is found to the northern and southern extremes, respectively, of the Americas, Eurasia, and Africa.

In spite of its large number of species (currently fourth in size among flowering plants), the family includes relatively few plants of economic importance. The principal commercially useful genera include *Cinchona* L., the source of the antimalarial drug quinine, and *Coffea* L., the source of the popular beverage coffee. The alkaloid caffeine, which acts as a psychoactive stimulant in the nervous system of humans, was first isolated and characterized chemically from *Coffea* (although it was later determined that the earlier-described compound theanine from tea leaves represents the same compound). Various species in other genera are used for their wood, medicinally, as dye plants, and in tanning leathers. A number of species are cultivated as ornamentals, mostly in warmer regions.

The interpetiolar stipules found in most opposite-leaved species of Rubiaceae are formed by fusion of adjacent stipules between the two leaves at each node and are unusual among the flowering plants and therefore distinctive. Stipules normally are produced in pairs, one on each side of the base of the petiole of an individual leaf; thus each leaf normally bears two stipules and the two stipules of each leaf are not fused to those of any other leaves. However, in many Rubiaceae, each stipule is fused to the stipule next to it that belongs to the other leaf at that node. This fused structure looks like a single stipule and is connected to the petioles of two different leaves. Consequently, it stretches between them across the stem node, thus its technical name interpetiolar. Interpetiolar stipules are rarely found in other families of flow blades with toothed margins. In Missouri, aside from the Rubiaceae, interpetiolar stipules are found only in some members of the Urticaceae. In using this feature to diagnose the family, it should be noted that in several species of Rubiaceae the stipules are shed quickly, leaving only a scar in the form of a line between the petioles as a sign of their former presence. This becomes problematic, because several other plant families with opposite entire leaves lack interpetiolar stipules but do develop a low ridge or line that runs across the stem at each node and connects the leaves. Missouri plant families that include such species are Acanthaceae, Caprifoliaceae, Gentianaceae, and Loganiaceae. In such cases, botanists should take care to observe young growth where any stipules produced are still attached.

Another characteristic of many Rubiaceae, including some of the Missouri species of *Houstonia*, is the production of distylous (heterostylous) flowers. In such species, two different kinds of flowers are produced: a so-called pin flower that has a relatively long style and short stamens such that the stigma is positioned above the anthers; a so-called thrum flower that has a relatively short style and long stamens such that the stigma is positioned below the anthers. Each individual plant bears only one kind of flower. The different flower forms function to promote cross-pollination in the species, and additionally they restrict the plant to breeding with only about half of the other plants in the population. This is because even though they belong to the same species, flowers of the same form mostly cannot successfully cross-pollinate other flowers of the same form, for two reasons. First, because the pollen of a short-styled flower is deposited in a different position on the body of a visiting insect than that from a long-styled flower, so that when this insect visits another flower, the pollen will be transferred successfully to the stigma of only the other kind of flower. Additionally, Rubiaceae species with heterostylous flowers have been found to have a chemical incompatibility system that prevents the pollen of one type of flower from germinating on the stigmas of that same flower type. Thus, even accidental or artificial pollinations involving the same flower type are unsuccessful. Species that lack heterostyly (all flowers have stamens and styles of the same relative lengths) are called homostylous. At least half of the species in the Rubiaceae are distylous.

1. Plants woody, shrubs tor small trees 1.5–5.0 m tall; inflorescences long-stalked, dense, globose heads . . . . 1. CEPHALANTHUS
1. Plants annual or perennial herbs (sometimes woody at the base and/or with a woody rootstock) to 1 m tall; inflorescences not dense heads or, if dense clusters, then sessile or nearly so in the leaf axils
    2. Leaves mostly in whorls of 4–12 (rarely opposite at a few nodes); nodes lacking evident stipules
        3. Flowers arranged variously, but not in headlike clusters subtended by involucral bracts; calyces lacking lobes; corollas 0.5–2.0 mm long, the tube minute, shorter than the lobes . . . . 3. GALIUM
        3. Flowers subsessile in small headlike clusters that are closely subtended by whorls of (4)6 or 8, small, leaflike, involucral bracts; calyces with minute, triangular lobes; corollas 4.0–5.5 mm long, the tube well-developed, longer than the lobes . . . . 7. SHERARDIA
    2. Leaves all opposite or occasionally in whorls of 3 at a few nodes, with evident (but often small) interpetiolar stipules
        4. Stipules truncate to broadly rounded at the tip, the margins with 3–11 bristles, at least some of these 3 mm or more long
            5. Flowers 1 or 2 per leaf axil (2–4 per node); corollas 6–13 mm long; fruits 3–5 mm wide, schizocarps splitting into 2 indehiscent mericarps at maturity . . . . 2. DIODIA
            5. Flowers 3–20 per leaf axil (6–40 per node); corollas 3.0–3.5 mm long; fruits 2–3 mm wide, achenelike, not splitting or dehiscing at maturity . . . . 8. SPERMACOCE

4. Stipules truncate to triangular, the margins entire, irregular or with 2 or 3 narrow lobes up to 2 mm long, lacking conspicuous marginal bristles
   6. Stems creeping, rooting at the nodes; flowers in closely adjacent pairs, fused at the ovaries; fruits drupelike, orange to red, each with 2 calyx scars at the tip . . . . 5. MITCHELLA
   6. Stems erect or ascending, not rooting at the nodes; flowers sometimes adjacent, but all separate; fruits capsules, green becoming pale brown, each with only 1 calyx scar at the tip
      7. Corollas 3.5–12.0 mm long, longer than the calyces . . . . 4. HOUSTONIA
      7. Corollas 0.8–1.0 mm long, shorter than the calyces . . . . 6. OLDENLANDIA

1. Cephalanthus L.

About 7 species, North America, South America, Asia, Africa.

1. Cephalanthus occidentalis L. (buttonbush, globe flower, honeyball, pond dogwood)
C. occidentalis var. *pubescens* Raf.
Pl. 545 e, f; Map 2532

Plants shrubs or small trees, 1.5–5.0 m tall. Stems somewhat angled to rounded, glabrous, the twigs sometimes finely short-hairy. Leaves variously opposite or in whorls of 3 per node, the petioles 10–25 mm long. Stipules interpetiolar, 2–4 mm long, triangular, generally persisting with the leaves, angled to a sharply pointed tip, sometimes with minute fingerlike glands on the margins. Leaf blades 2.5–16.0(–23.0) cm long, 2–6(–10) cm wide, elliptic to elliptic-oblong or ovate, angled or slightly tapered to a sharply pointed tip, rounded or angled at the base, the margins flat, the upper surface glabrous, sometimes shiny, the undersurface glabrous or with small tufts of hair in the vein axils, occasionally more uniformly short-hairy, the midvein and 6–8 pairs of secondary veins visible. Inflorescences dense, globose heads 1–4 cm in diameter, in a loose terminal cluster and sometimes also solitary from the uppermost leaf axils, each with a stalk 1.5–6.0 cm long. Flowers homostylous, numerous, intermingled with short, linear to narrowly club-shaped bracts. Calyces 0.8–1.1 mm long, persistent at fruiting, 4-lobed, the lobes 0.4–0.5 mm long (becoming enlarged to 1.0–1.5 mm at fruiting), rounded at the tips. Corollas 9.0–12.5 mm long, narrowly funnelform to nearly cylindric, white, externally glabrous, internally pubescent with short wavy hairs in upper part of the tube and sometimes also the lobes, the tube 7–10 mm long, 4-lobed, the lobes 1.5–2.5 mm long, narrowly elliptic, bluntly pointed at the tips. Stamens 4, the anthers exserted. Ovary fully inferior, 2-locular, the ovules 1 per locule. Style 1, the stigma 1, cylindric, exserted. Fruits apparently multiple in a single globose mass but at maturity separating from each other, the individual fruits schizocarps, 5.0–5.5 mm long, 1.5–2.0 mm wide, dry, narrowly obconic, at maturity splitting from the base into 2 mericarps. $2n=44$. June–August.

Scattered to common throughout the state (eastern U.S. west to Minnesota and Texas, also Arizona, California; Mexico, Central America, Caribbean Islands). Swamps, sloughs, oxbows, bottomland forests, banks of streams, rivers, and spring branches, margins of ponds, lakes, and sinkhole ponds, and marshes; also ditches, pastures, railroads, and wet roadsides; sometimes emergent aquatics.

The leaves vary notably in number per node, size, shape, petiole length, and pubescence. *Cephalanthus occidentalis* can be reliably recognized even without flowers, however, as our only Rubiaceae shrub, that is, our only shrub with entire leaves and interpetiolar stipules. The stipules are sometimes shed toward the end of the growing season, but usually at least a few persist on the younger stems.

Buttonbush is occasionally cultivated as an ornamental and for pond-bank stabilization, and also is a good bee-plant. Steyermark (1963) noted that: "the leaves contain a bitter substance used in medicine. A tea made from the root of the bark has sometimes been employed in the treatment of diabetes. Animals may be poisoned by feeding on the leaves." Old shrubs, particularly in sinkhole pond communities, tend to develop hummocky bases, on which a characteristic herbaceous flora develops, notably *Carex decomposita* Muhl., an unusual species of Cyperaceae.

Some authors have separated two varieties within this species: var. *pubescens* has been distinguished by its leaves and young twigs having at least sparse pubescence (vs. glabrous in var. *occidentalis*). The var. *pubescens* is most often found in the southeastern U.S., including south-

2533. Diodia teres 2534. Diodia virginiana 2535. Galium anglicum

eastern Missouri (Steyermark, 1963). However as noted by various authors, pubescent plants are found sporadically more or less throughout the range of the species. And, as discusssed by Steyermark, there is a wide and continuous range of pubescence density among individual plants. Thus these varieties are not recognized here. The occurrence of tufts of hairs in the axils where the secondary veins meet the midvein on the leaf underside is a common feature in woody Rubiaceae species, and one whose occurrence is known to be independant of other pubescence. These tufts are known as acarodomatia, that is, structures that shelter very tiny mites. These mites pay for their shelter by cleaning the plant's leaves of attackers such as fungal spores, which they eat.

2. **Diodia** L. (buttonweed)

Plants annual or perennial herbs, sometimes woody at the base. Stems erect or ascending to prostrate or loosely ascending form a spreading base, sometimes mat-forming, usually 4-angled. Leaves opposite, subsessile, with interpetiolar stipules, these generally truncate to broadly rounded at the tips, persisting with the leaves, fused to the leaf bases on either side, membranaceous, bearing along the margins 3–11 bristles, at least some of which are more than 3 mm long. Inflorescences axillary, of solitary, sessile flowers or sessile, 2–4-flowered clusters. Flowers homostylous, often subtended by small bracts. Calyces deeply 2- or 4-lobed, usually persistent at fruiting. Corollas 4-lobed, funnelform to saucer-shaped, white to light pink. Stamens 4, inserted near the top of the corolla tube, the anthers exserted or not. Ovary fully inferior, 2-locular, the ovules 1 per locule. Style 1, slender, either unbranched and with a 2-lobed, capitate stigma or 2-branched and with long linear stigmas, exserted or not. Fruits schizocarps, dry, hairy, at maturity splitting from the base into 2 mericarps. Five to about 30 species, North America, Central America, South America, Caribbean Islands, and possibly Africa; introduced in Asia, Pacific islands.

In their taxonomic revision of *Diodia*, Bacigalupo and Cabral (1999) treated the American species traditionally ascribed to the genus as belonging to four different genera. They transferred *D. teres* and about fifteen other species to the genus *Diodella* Small, based mainly only differences in fruit morphology. However, the taxonomic relationships among these genera are still not well-understood. The as-yet not fully published doctoral studies of Dessein (2003) include a preliminary molecular analysis of the tribe Spermacoceae Bercht. & J. Presl, which supports the division of *Diodia* into several groups, but also does not resolve the relationships between these groups. *Diodia teres* is thus retained in the genus *Diodia* for the present, with the knowledge that it will likely become recognized as part of a segregate genus (likely *Diodella* Small) in the future. *Diodia* is related to *Spermacoce*. For further discussion, see the treatment of that genus.

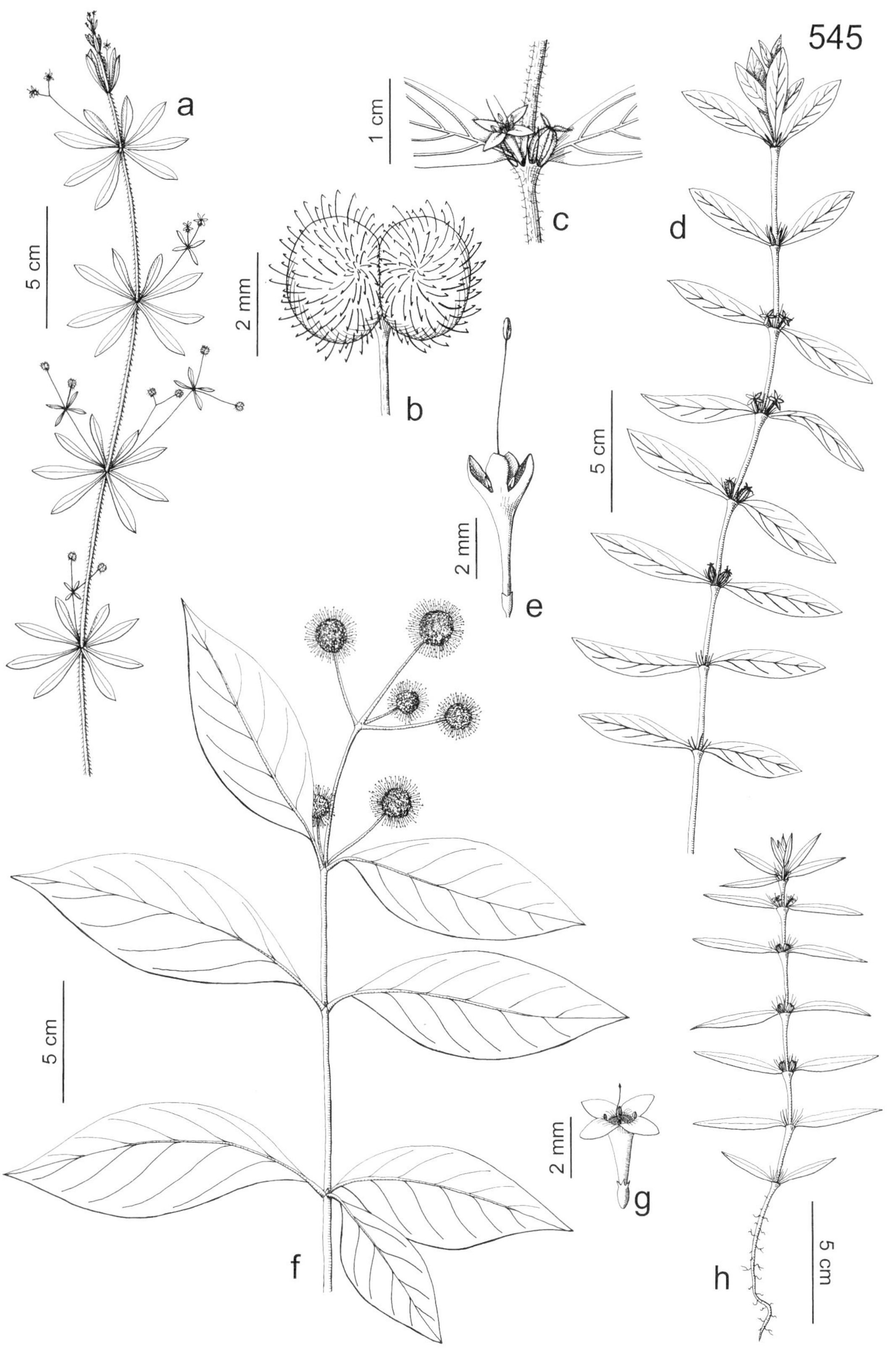

Plate 545. Rubiaceae. *Galium aparine*, **a)** stem with inflorescences, **b)** fruit. *Diodia virginiana*, **c)** node with leaf bases, flower, and fruit, **d)** stem with leaves and fruits. *Cephalanthus occidentalis*, **e)** flower, **f)** branch with leaves and inflorescence. *Diodia teres*, **g)** flower, **h)** habit.

1. Stipules with the marginal bristles 5–11 on each side; calyx lobes 4, 1–2 mm long; corollas light pink to less commonly mauve or pale purple, the tube 4–5 mm long; styles unbranched, with a 2-lobed, capitate stigma; fruits 3.0–3.5 mm long, 3.0–3.5 mm wide, rounded to broadly angled on the sides; stems usually erect or strongly ascending . . . . 1. D. TERES
1. Stipules with the marginal bristles 3–7 on each side; calyx lobes 2, 2–8 mm long; corollas white, the tube 6–7 mm long; style 2-branched, with long, linear stigmas; fruits 6–7 mm long, 4–5 mm wide, with well-developed longitudinal ridges on the sides; stems prostrate or loosely ascending from spreading bases . . . . 2. D. VIRGINIANA

1. Diodia teres Walter (rough buttonweed)
Diodella teres (Walter) Small
Pl. 545 g, h; Map 2533

Plants annual, sometimes becoming woody at the base. Stems 15–35(–80) cm long, usually erect or strongly ascending, minutely roughened and becoming more or less glabrous to sparsely or moderately pubescent with short, straight to wavy hairs of various lengths. Stipules 1–2 mm long, the bristles 5–11 on each side, variously 1–10 mm long. Leaf blades 1.0–3.5 cm long, 2–5 mm wide, narrowly lanceolate to narrowly elliptic or nearly linear, angled to somewhat rounded at the base, angled to a sharply pointed tip, the margins minutely pubescent with stiff, spinescent hairs, frequently becoming curled under (especially in plants from particularly dry microsites), the surfaces glabrous or nearly so, the venation with the midvein and usually 2 or 3 pairs of pinnate, secondary veins visible. Flowers usually solitary in the leaf axils, produced at most of the nodes. Calyces deeply 4-lobed, the lobes 1–2 mm long, triangular, often unequal. Corollas 6–7 mm long, pink to less commonly mauve or pale purple, externally glabrous or hairy, internally glabrous, the tube 4–5 mm long, the lobes 2.0–2.5 mm long, lanceolate. Styles unbranched, with a 2-lobed, capitate stigma. Fruits 3.0–3.5 mm long, 3.0–3.5 mm wide, broadly obovoid, smoothly rounded to broadly angled on the sides. $2n$=28. June–October.

Scattered nearly throughout the state, but apparently absent from most of the western half of the Glaciated Plains Division (eastern U.S. to Kansas and Texas, also New Mexico, Arizona, and California; Mexico, Central America, South America, Caribbean Islands). Glades, thin-soil areas of upland prairies, sand prairies, ledges and tops of bluffs, and banks of streams and rivers; also pastures, old fields, fallow fields, old quarries and mines, railroads, roadsides, and open, disturbed areas; often in sandy or acidic soils.

As is common in annual species of weedy habit and habitat, *D. teres* shows considerable variation in plant size, leaf size, and pubescence, among other characters. This variation is continuous and mostly found throughout the range of the species, often within populations. Some authors have attempted to classify this variation (Fernald and Griscom, 1937), but, as was noted by Steyermark (1963), these intergrade greatly. With more collections and information on genetics now available, the five North American varieties of *D. teres* are now considered to represent only minor variants within a variable species, and morphological expression also is thought to be strongly influenced by the plant's environment.

2. Diodia virginiana L. (large buttonweed)
D. virginiana var. *attenuata* Fernald
Pl. 545 c, d; Map 2534

Plants annual, perhaps sometimes short-lived perennial herbs. Stems 20–40(–80) cm long, prostrate or loosely ascending from a spreading base, sparsely to moderately pubescent with short or sometimes longer, straight to wavy hairs. Stipules 2.0–4.5 mm long, the bristles 3–7 on each side, variously 2–7 mm long. Leaf blades 1.2–7.0 cm long, 2–15 mm wide, narrowly elliptic to oblong-lanceolate, angled at the base, angled to a sharply pointed tip, the margins minutely pubescent with stiff, spinescent hairs, flat, the surfaces glabrous short-hairy, the undersurface finely glandular-hairy along the midvein, the venation with the midvein and usually 2–4 pairs of pinnate, secondary veins visible. Flowers solitary or paired in the leaf axils, produced at most of the nodes. Calyces deeply 2-lobed, the lobes 2–8 mm long, lanceolate to triangular, often unequal. Corollas 10–13 mm long, bright white, externally glabrous or nearly so, internally bearded in the throat with the pubescence usually extending onto the lobes, the tube 6–7 mm long, the lobes 5–6 mm long, lanceolate. Styles 2-branched, with long, linear stigmas. Fruits 6–7 mm long, 4–5 mm wide, ellipsoid to broadly ellipsoid, with well-developed longitudinal ridges on the sides, becoming corky, eventually splitting longitudinally into halves. $2n$=28. June–September.

Scattered in the Mississippi Lowlands Division and the southern portion of the Ozarks north locally to Lincoln and Hickory Counties (eastern [mostly southeastern] U.S. west to Kansas and Texas). Margins of ponds, lakes, and sinkhole ponds, bottomland prairies, mesic swales of sand prairies, fens, banks of streams and rivers; also levees, ditches, roadsides and moist, open, disturbed areas.

The corky fruits of this species apparently float and are probably dispersed at least in part by water. As in most weedy plants growing in seasonally wet sites, the plants are notably variable in overall size as well as the sizes of the leaves and stipules. And, as in most Rubiaceae, there is a range of variation in pubescence among different plants. Noticeably hairy plants have been called f. *hirsuta* (Pursh) Fernald.

3. **Galium** L. (bedstraw, cleavers)

Plants annual or perennial herbs, rarely woody at the base, the rootstock usually slender, sometimes reddish-tinged. Stems usually 4-sided, erect to loosely ascending, sometimes matted or clambering, sometimes notably hairy, roughened, and/or minutely prickly. Leaves in whorls of (3–)4–8(–12), rarely opposite at a few of the nodes, sessile or nearly so. Stipules absent. Inflorescences terminal and/or axillary, of solitary flowers or in clusters or panicles, not in headlike clusters subtended by involucral bracts. Flowers sessile or stalked, some of the flowers sometimes imperfect or sterile (mixed with perfect ones). Calyces essentially absent, occasionally represented by a minute rim at the flower base. Corollas 0.5–2.0 mm long, with a minute tube, deeply (3)4-lobed, the lobes spreading, not overlapping in bud, variously white, yellow, dark reddish purple, maroon, or yellowish green. Stamens (3)4, attached in the corolla tube, the anthers exserted. Style 2-lobed, the stigmas 2, capitate. Ovary inferior, 2-locular, the ovules 1 per locule. Fruits rather small (1–3 mm long), 2-lobed, dry (fleshy elsewhere), glabrous to tuberculate, sometimes densely pubescent with hooked hairs, separating into 2 indehiscent mericarps at maturity, these subglobose or less commonly broadly kidney-shaped. About 400 species, nearly worldwide, most diverse in temperate regions.

The Eurasian genus *Asperula* L. is very closely related to *Galium* (or perhaps not distinct), differing only in its trumpet-shaped corollas with well-developed tubes. The Eurasian species *A. odorata* L. (*G. odoratum* (L.) Scop.; sweet woodruff, waldmeister), is commonly cultivated as an ornamental groundcover, including in Missouri, and is similar in general aspect to *G. circaezans*. *Asperula odorata* has white flowers with a sweet, pleasant aroma, and in Europe is also used to flavor drinks, including May wine (Rombauer and Becker, 1975).

Readers should beware of confusing *Galium* with *Mollugo* (Molluginaceae). Although the flowers and inflorescences of the two genera are very different, both are weak-stemmed plants with whorled, relatively slender leaves. The development of the whorled, apparently nonstipulate leaves of *Galium* and a few related Rubiaceae genera (*Rubia* L., *Sherardia*) has been debated. The whorled leaves are distinctive for their lack of stipules (in a family otherwise characterized by the presence of stipules) and their arrangement commonly in even numbers. The leaves are sometimes considered actually to be produced in pairs, as in most other Rubiaceae, and the interpetiolar stipules in this case to be present and to have expanded to appear similar to the leaves, thus producing an apparent whorl of four leaves at each node. The leaves that are supposed to be derived from enlarged stipules cannot be separated from the true leaves, at least by an observer of the whole plant. In species with more than four leaves at a node, either more "true" leaves are assumed to have been produced, or the leaf-like stipules are thought to have become divided into several equal segments.

Galium generally has not been studied well in North America. Its taxonomy is complicated by the variability of the plant size, leaf size and shape, and development of the inflorescences within a single species, and also by the varied and perhaps variable breeding system of some species. Species of *Galium* range from occasionally with all the flowers perfect through polygamous to fully dioecious; so the flowers may be variously staminate, pistillate, perfect, or even completely sterile, all on one plant. The presence of hooked hairs on the fruits and ovaries has been considered informative taxonomically and is apparently consistent within at least most of the species that occur in Missouri. When these hairs are present, they are usually visible on the ovaries of the flowers as well as the fruits. The number of leaves at a node is also generally consistent for many species, such as *G. circaezans*, although it is quite variable in others, such as *G. aparine*. Leaf size may be widely variable, among plants but also along a single stem, which often shows a marked increase in size in the later-season leaves at the top of the stem. Most of the species have short, curved, sharply pointed hairs with stout bases; these have been described as prickly or rough; their orientation varies on different parts of the plant. Some of the species also possess embedded glands, which appear as small lines or streaks on the undersurface of the leaf blades. Soft or fine (nonprickly) hairs are also present in some species. The taxonomy of *Galium* is also complicated by the circumpolar distribution of some species (or groups of closely related species), so that plants from all three continents must be studied together to make sense of a given species.

Measurements of leaves given below include the smaller leaves at the stem bases; some other authors give leaf measurements only for the largest leaves on a stem, which should be kept in mind when comparing different treatments of *Galium*. Also, fruit size is given by some authors only for the individual mericarps of the fruit, at least for species with dry fruit; here the fruit size is that of the whole fruit, that is, of the paired mericarps when they are apparently mature. However, fruit measurements here are only for the fruit itself, not including any hairs. Occasionally only one of the two mericarps develops, and in this case the single mericarp is often a bit larger than those of a pair. Measurements given here of corollas report the length from the corolla base to the tips of the lobes; some other authors have given measurements of *Galium* flowers (but not those of other Rubiaceae) in terms of the diameter of the open flower.

Plants of some *Galium* species have a persistent sweet odor when dried, and in olden times were mixed with bedding to freshen it; *G. boreale* was one species used in this way, which is apparently the origin of the common name bedstraw. Other species, such as *G. verum*, were added to milk to curdle it. The hooked hairs of the fruits cause them to adhere to animal fur or clothing, apparently for dispersal. Even the vegetative portions of some species, for example *G. aparine* and *G. asprellum*, are sticky in this way, due to small spinules on their surfaces, and the whole plant can adhere to clothing or climb on other plants in this manner. This surface stickiness is apparently the origin of the common name cleavers (meaning cleaving to or sticking to). The flowers are varied in their pollinators and attractive mechanisms; they may be nearly odorless (to us), or sweetly fragrant, or in some fly-pollinated species may offer odors that only those insects can truly appreciate.

1. Leaves 5–10 at the majority of stem nodes
    2. Fruits and ovaries pubescent with hooked hairs
        3. Mature fruits (including both mericarps) 2–3 mm long, 3.5–5.5 mm (not including the hairs), when only 1 mericarp is developed then 1/2 this size or a little larger; leaves 1–5 mm wide; flowers sessile or with stalks to 1 mm long, the stalk elongating to as much as 8 mm long in fruit . . . . . 2. G. APARINE
        3. Mature fruits (including both mericarps) ca. 2 mm long, 2 mm wide (not including the hairs), when only 1 mericarp is developed then 1/2 this size or a little larger; leaves 2–15 mm wide; flowers on stalks 1–7 mm long, the stalk elongating to as much as 10 mm long in fruit . . . . 13. G. TRIFLORUM
    2. Fruits and ovaries glabrous
        4. Leaves broadly pointed to rounded at the tip; stems glabrous or pubescent, but usually with a distinct ring of short hairs around the nodes

5. Flowers mostly 4–12 per inflorescence; corollas with 4 lobes . . . . 7. G. CONCINNUM
5. Flowers 2–4 per inflorescence; corollas with 3 or less commonly 4 lobes (the number often variable on a single plant) . . . 12. G. TINCTORIUM
4. Leaves narrowly pointed at the tip; stems uniformly glabrous or pubescent
6. Plants perennial, sometimes fairly tall; stems glabrous or pubescent with short wavy hairs and also roughened with minute, tuberculate spinules; corollas 1.0–1.6 mm long
7. Stems roughened with minute, tuberculate spinules; corollas 1.0–1.2 mm long, white . . . . 4. G. ASPRELLUM
7. Stems densely pubescent with soft wavy hairs but not scabrous; corollas 1.4–1.6 mm long, light yellow . . . . 14. G. VERUM
6. Plants annual, low-growing (stems to 40 cm long, but often loosely ascending or clambering); stems minutely roughened but otherwise glabrous; corollas 0.4–0.6 mm long
8. Inflorescences with mostly 2 or 3 branch points between the base and each flower, the branches ascending to less commonly spreading . . . . 1. G. ANGLICUM
8. Inflorescences with mostly 3–6 branch points between the base and each flower, the branches loosely ascending to spreading . . . . 8. G. DIVARICATUM
1. Leaves (2–)4 at the majority of stem nodes
9. Fruits and ovaries pubescent with hooked hairs
10. Plants annual; leaves 2–7 mm long, 1–2 mm wide, with one vein (the midvein) visible; flowers 1–2 at each stem node, subsessile, these and the fruits borne below and largely covered by the leaves . . . . 15. G. VIRGATUM
10. Plants perennial; leaves 5–42 mm long, 2–22 mm wide, with 3 veins (the midvein and a pair of lateral veins) visible; flowers 4 to many at each stem node, sessile to noticeably stalked in cymes 1–4 cm long, these usually longer than and spreading above the leaves
11. Flowers and fruits sessile; corollas pale green . . . . 6. G. CIRCAEZANS
11. Flowers and fruits with stalks 1–10 mm long; corollas dark purple to maroon . . . . 11. G. PILOSUM
9. Fruits and ovaries of flowers glabrous or sparsely pubescent with straight, appressed hairs
12. Leaves with 3 veins (the midvein and a pair of lateral veins) visible, although the lateral veins may be short and weakly developed
13. Flower stalks 10–28 mm long; corollas dark red or maroon . . . . 3. G. ARKANSANUM
13. Flower stalks 1–2 mm long; corollas white . . . . 5. G. BOREALE
12. Leaves with only 1 vein (the midvein) visible
14. Plants perennial; leaves 5–23 mm long; corollas 0.8–2 mm long
15. Corollas 1.5–2.0 mm long, 4-lobed . . . . 9. G. OBTUSUM
15. Corollas 0.8–1.2 mm long, 3- or sometimes 4-lobed . . . . 12. G. TINCTORIUM
14. Plants annual; leaves 2–10 mm long; corollas ca. 0.5 mm long
16. Plants little or not branched; stems pubescent throughout with straight slender hairs; flowers borne in very small, pendant clusters that hang below and are largely covered by the leaves . . . . 10. G. PEDEMONTANUM

16. Plants usually branched; stems minutely roughened but otherwise glabrous; flowers borne in small panicles or clusters that spread or angle over the leaves
  17. Inflorescences with mostly 2 or 3 branch points between the base and each flower, the branches ascending to less commonly spreading . . . . . 1. G. ANGLICUM
  17. Inflorescences with mostly 3–6 branch points between the base and each flower, the branches loosely ascending to spreading . . . . 8. G. DIVARICATUM

1. Galium anglicum Huds.
G. parisiense L. var. *leiocarpum* Tausch
Map 2535

Plants annual, sometimes becoming slightly hardened at the base at maturity. Stems 10–30 cm long, usually weak, erect to loosely ascending or clambering, often branched and/or tufted, roughened with minute, prickly, downward-curved hairs on the angles. Leaves (2–)4–6(–8) per node, spreading or downward-angled in orientation. Leaf blades 1–6 mm long, 0.5–2.0 mm wide, narrowly elliptic to narrowly oblong or linear, angled or short-tapered to a sharply pointed tip, the midvein sometimes extended into a minute, sharp point, angled to truncate at the base, not glandular on the undersurface, with only the midvein visible, the margins with minute, stiff, prickly hairs and usually curled under. Inflorescences terminal and also axillary from the uppermost leaves, the axillary ones not pendant, positioned over the leaves, consisting of small clusters or fascicles, these usually grouped into small panicles with mostly 2 or 3(4) branch points and relatively short, ascending to less commonly spreading branches. Flowers relatively few, the stalks 0.5–2.0 mm long. Corollas 0.4–0.6 mm long, 4-lobed, white. Fruits about 1 mm long, 1.5 mm wide, the surface glabrous, smooth to granular. May–June.

Introduced, uncommon in the southern half of the state (native of Europe; introduced sporadically in eastern U.S. west to Missouri, Oklahoma, and Texas; also California, Oregon, Hawaii, Canada, New Zealand, and Australia). Glades; also ditches, cemeteries, lawns, and open, disturbed areas.

This species was first reported for Missouri by Lipscomb and Nesom (2007), who studied the introduced populations of the *G. parisiense* complex in the United States and determined that three species should be recognized. *Galium anglicum* was long-misdetermined as *G. divaricatum* or treated as a glabrous-fruited form or variety of *G. parisiense*. In the eastern half of the United tates, true *G. parisiense* appears to be established only in Alabama, Louisiana, and Mississippi, as well as sporadically in the southern 2/3 of Arkansas. Surprisingly, there are currently more specimens to document *G. anglicum* in Missouri than there are of *G. divaricatum*. The two may be separated by characters in the key to species above.

2. Galium aparine L. (cleavers, goose grass, annual bedstraw)
Pl. 545 a, b; Map 2536

Plants annual. Stems 10–100 cm long usually weak, spreading to loosely ascending or clambering, often few- to several-branched (small plants may be unbranched), minutely roughened with minute, prickly, downward-curved hairs on and sometimes also between the angles, otherwise glabrous or only sparsely pubescent with short hairs. Leaves 6–8(10) per node, generally spreading or somewhat ascending in orientation. Leaf blades 8–60 mm long, 1–5 mm wide, narrowly oblanceolate, short-tapered to a sharply pointed tip, the midvein usually extended into a short, sharp point, angled or tapered at the base, not glandular on the undersurface, with only the midvein visible, the margins with minute, stiff, prickly hairs and usually somewhat curved under. Inflorescences axillary, occurrring mostly at nodes above the stem midpoint, not pendant, positioned over the leaves, consisting of small clusters or fascicles (occasionally reduced to solitary flowers), these sometimes grouped into small panicles with mostly 1 or 2 branch points and relatively short, ascending branches. Flowers relatively few, the stalks absent or to 1 mm long at flowering, becoming elongated to as much as 8 mm at fruiting. Corollas 1.0–1.2 mm long, 4-lobed, white. Fruits 2–3 mm long, 3.5–5.5 mm wide, the surface densely pubescent with hooked hairs 0.5–0.8 mm long. $2n$=22, 42, 44, 63, 64, 65, 66, 86. April–July.

Scattered to common throughout the state (U.S. [including Alaska]; Canada, Greenland, Europe, Asia, Africa; introduced in the southern hemisphere). Bottomland forests, mesic upland forests, banks of streams and rivers, margins of ponds, lakes, and oxbows, bottomland prairies, marshes, sloughs, and occasionally glades; also pastures, ditches, old mines, gardens, lawns, railroads, roadsides, and disturbed areas.

Steyermark (1963) commented on use of the fruits of this plant as a coffee substitute. Native Americans used the plant medicinally for renal, urinary, and dermatological problems (including treatment for poison ivy rashes), and also as an antihemorrhagic, diuretic, and love potion (Moerman, 1998). *Galium aparine* is also a minor crop weed that impacts mainly crop species har-

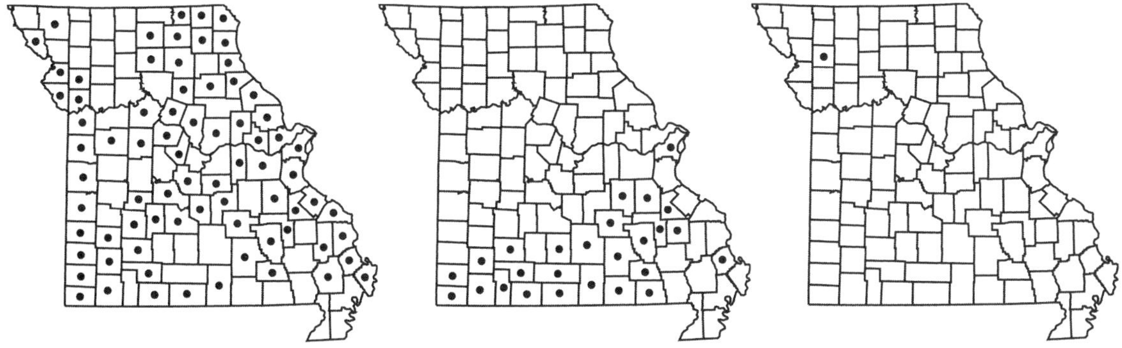

2536. Galium aparine 2537. Galium arkansanum 2538. Galium asprellum

vested relatively early in the growing season, such as winter wheat crops and some members of the mustard family. The status of the species in North America has been a source of controversy. R. J. Moore (1975) in his study of the taxon in Canada, suggested that although it apparently was native in mostly coastal portions of northwestern North America, most or all of the inland populations represented an introduction of the species from Europe by early settlers as a contaminant in crop seeds. In the Midwest, Deam (1940) considered it to be native in Indiana and Steyermark (1963) also treated it as native in Missouri. The oldest herbarium specimens collected in the state, which date from the 1860s, were collected from natural habitats. Thus, although it is not possible to confirm the status of the taxon with certainty, there is no compelling evidence to support that it is not native to the region.

One clue to the identification of *Galium aparine* is its roughened surfaces: not only the fruits but the entire plant will adhere to clothing or animal hair. The stickiness of the leaves and stems is due to tiny downward-curved prickles on their surfaces. These prickles are single-celled, hardened protuberances of the plant epidermis. They adhere to the stems of other plants and even to each other, so the plants often form extensive tangles, which disappear by late summer. Some other species (notably *G. asprellum*) are similarly sticky.

Smaller-leaved plants from throughout the range of this species were once treated as *G. vaillantii* DC. or *G. aparine* var. *vaillantii* (DC.) Koch, but this variety has not been recognized taxonomically since the early twentieth century. Even Steyermark (1963) declined to separate it, suggesting instead that it represented plants growing in poorer soils in drier, more exposed habitats.

Galium spurium L. of Eurasia and Africa is similar to *G. aparine*. Their alleged (Voss, 1996) distinctions, although clearly stated in keys can be rather subtle to diagnose on living plants. Flowers are 0.8–1.3 mm in diameter and greenish yellow in *G. spurium*, vs. 1.5–1.8 mm in diameter and white in *G. aparine*. Fruits are 2–3 mm in diameter in *G. spurium*, vs. 3–5 mm in diameter in *G. aparine*. Chromosome numbers also are different, $2n=20$, 40 in *G. spurium*, vs. $2n=$ mostly some multiple of 22 in *G. aparine*. *Galium spurium* has been reported from Canada by Moore (1975), but in the United States only a few reports from western states exist and these require verification. Although Moore suggested that it might be present in Missouri, the specimens so far collected from Missouri under the name *G. aparine* have been reviewed during the present study for clear signs of falsity, which has not been found. In the past, the name *G. spurium* has been applied in error to some North American plants of *G. aparine*, which has further confused the situation. For the technically minded, the separation of these two species is touched on by Verdcourt (1976) and was also discussed by Moore (1975) and Malik and Vanden Born (1988).

3. **Galium arkansanum** A. Gray (Arkansas bedstraw)

Pl. 546 a, b; Map 2537

Plants perennial. Stems 15–35(–50) cm long, erect or ascending, sometimes from a spreading base, unbranched or few-branched from the base, glabrous or sparsely to moderately pubescent with short, straight, soft, spreading to loosely upward-angled hairs, especially at the nodes. Leaves 4 per node, spreading in orientation or very slightly arched downward. Leaf blades 7–45 mm long, 2–9 mm wide, lanceolate to narrowly lanceolate, angled to a bluntly or more commonly sharply pointed tip, angled or slightly tapered at the base, the undersurface with impressed, linear glands (appearing as small streaks or lines), otherwise glabrous or roughened with short, stiff hairs along the midvein, the venation palmate with 3 veins (the midvein and 2 finer lateral veins) visible, but the lateral veins often weak, the margins with short, stiff, prickly hairs and flat or only slightly curved under. Inflorescences terminal and axillary from the upper leaves, not pendant, positioned over the leaves, consisting of small panicles (occasionally some of these reduced to simple clusters) with mostly 1–3 branch points and relatively long, ascending branches. Flowers relatively few to more numerous, the stalks 10–20(–28) long. Corollas 1.5–2.0 mm long, 4-lobed, dark red, purple, or maroon. Fruits 2–4 mm long, 3.5–5.0 mm wide, the surface minutely roughened (the tubercles less than 0.1 mm long), otherwise glabrous. May–June.

Scattered in the Ozark Division and also in Scott County (Arkansas, Missouri, and Oklahoma). Mesic to dry upland forests, edges of glades, and ledges of bluffs; also roadsides.

The corolla lobes often terminate in rather long filaments that are sometimes paler in color. Two varieties of *Galium arkansanum* have been recognized by some botanists: var. *arkansanum* and var. *pubiflorum* E.B. Sm. E. B. Smith (1979) separated var. *pubiflorum* based on its corollas that are densely pubescent on the exterior with long silky hairs (vs. glabrous in var. *arkansanum*) and its leaves 2.5–3.5 mm wide with generally only the midvein visible (vs. 5–8 mm wide and 1- or 3-veined in var. *arkansanum*). He noted that pure var. *pubiflorum* is known only from one county in southwestern Arkansas, and that continuous variation from fully glabrous to fully pubescent flowers is found among the rest of the populations in southern Arkansas, which frequently also contain plants spanning the full range of leaf widths for the species. In his discussion, E. B. Smith referred to his these varieties as phases, and documented the many intermediate forms. Thus, no infraspecific taxa are recognized in the present treatment.

Galium latifolium Michx. of the Appalachian Mountains is similar to *G. arkansanum*. It differs in its broader leaves (9–20 mm), its generally well-developed lateral leaf veins, and its shorter flower stalks (4–8 mm).

4. Galium asprellum Michx. (rough bedstraw)
Map 2538

Plants perennial. Stems 15–120 cm long, spreading to loosely ascending or clambering, usually much-branched, roughened with minute, prickly, downward-curved hairs on and usually also between the angles, sometimes also pubescent around the stem nodes with short straight hairs. Leaves 5 or 6 per node, mostly spreading to very slightly ascending in orientation. Leaf blades 4–15 mm long, 1–5 mm wide, lanceolate to narrowly lanceolate or linear, angled to a sharply pointed tip, the midvein usually extended into a minute, sharp point, angled or truncate at the base, the undersurface not glandular, glabrous, the venation with only the midvein visible, the margins with short, stiff, prickly hairs and often somewhat curved under. Inflorescences terminal and axillary from the upper leaves, not pendant, positioned over the leaves, consisting of small panicles (occasionally some of these reduced to simple clusters) with mostly 1–3 branch points and short to relatively long, ascending branches. Flowers mostly several to numerous, the stalks 1–5 mm long. Corollas 1.0–1.2 mm long, 4-lobed, white. Fruits 0.8–1.0 mm long, 1.5–2.0 mm wide, the surface glabrous, smooth. August.

Uncommon, known thus far only from a single, historical specimen from DeKalb County (northeastern U.S. west to Wisconsin and Missouri; Canada). Habitat unknown, but possibly bottomland forests (based on specimens from other states).

The presence of this species in Missouri was first suggested without documentation by Gleason (1952), but was not confirmed until the present research. *Galium asprellum* can be distinguished vegetatively from some other *Galium* species by its stem nodes that are surrounded with a ring of spreading pubescence. This feature is also found in *G. boreale*, *G. concinnum*, *G. obtusum*, and *G. tinctorium*. The Missouri collection represents an unusual, somewhat disjunct, southwestern locality for *G. asprellum*. If more data were available about the population from which it was harvested, a case could be made that the species represents a relict in Missouri, surviving from glaciated times when our state had a cooler climate.

5. Galium boreale L. (northern bedstraw)
Pl. 546 c–e; Map 2539

Plants perennial, often somewhat woody at the base and/or with a woody rootstock. Stems 30–50 cm long, erect or ascending, sometimes pendant, usually much-branched and sometimes rather bushy, pubescent with soft, sometimes wavy or curved, short hairs, sometimes glabrous or nearly so between the nodes, the minute, prickly, downward-angled hairs on the angles, when present, frequently restricted to the basal portions. Leaves 4 per node (sometimes apparently more because of short axillary stems with additional leaves), mostly spreading in orientation. Leaf blades 5–36 mm long, 1–5 mm wide, narrowly oblong-lanceolate to narrowly triangular, angled to a sharply pointed tip, the midvein usually extended into a minute, sharp point, angled at the base, the

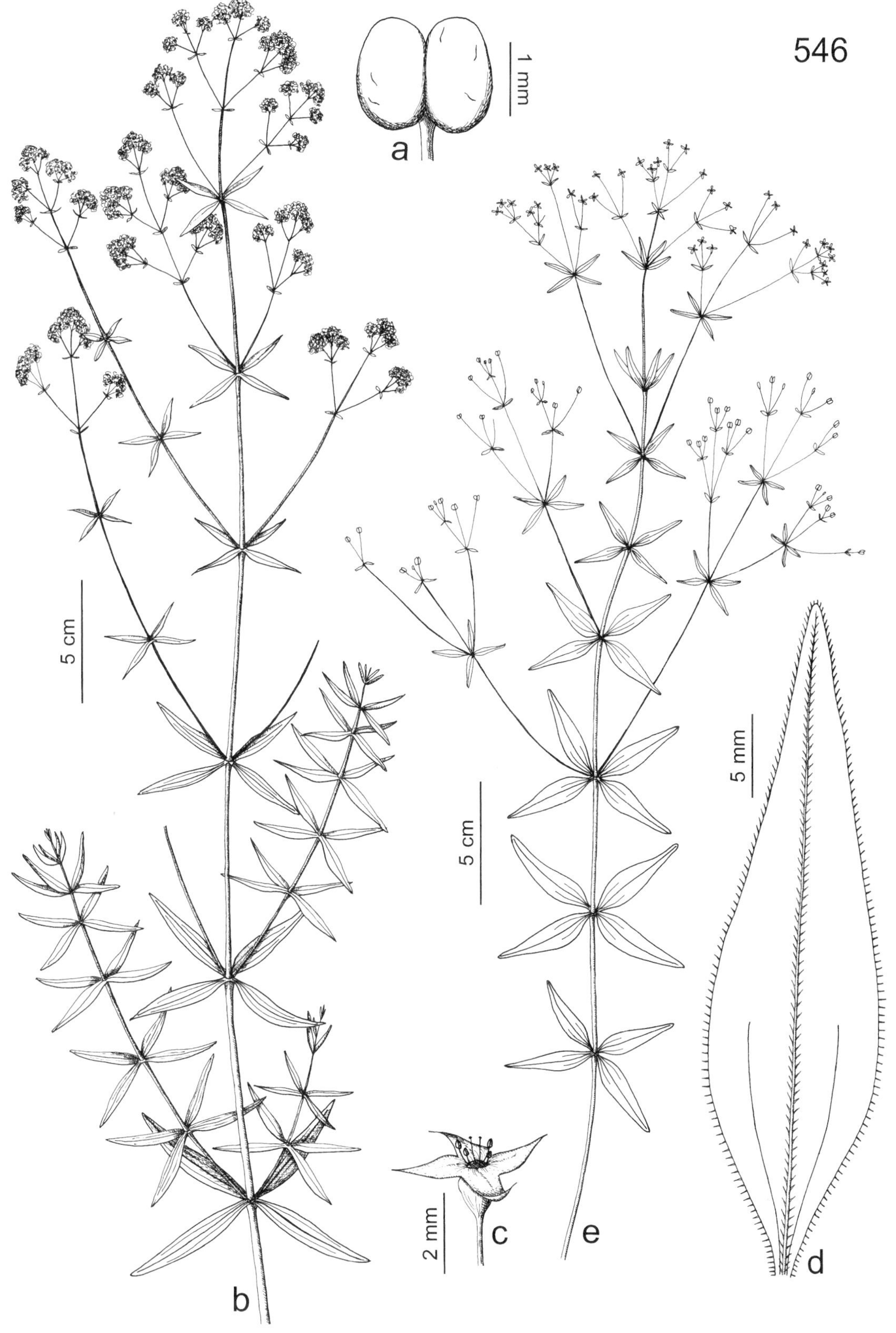

Plate 546. Rubiaceae. *Galium arkansanum*, **a)** fruit, **b)** stem with inflorescences. *Galium boreale*, **c)** flower, **d)** leaf, **e)** stem with inflorescences.

2539. Galium boreale 2540. Galium circaezans 2541. Galium concinnum

undersurface with impressed, fine, oval to circular glands (appearing as pale dots, but sometimes faint), otherwise glabrous or with short, fine, often stiff hairs along the main veins, the venation palmate with 3 veins (the midvein and 2 finer lateral veins) visible, especially on the upper surface, the margins with short, stiff, prickly hairs and usually curved under. Inflorescences terminal and axillary from leaves on the upper 1/3 of the stem, not pendant, positioned over the leaves, consisting of small panicles with mostly 2–4 branch points and relatively short, mostly spreading branches. Flowers mostly several to numerous, the stalks 1–2 mm long. Corollas 1.0–1.2 mm long, 4-lobed, white. Fruits 1.2–1.6 mm long, 1.7–2.2 mm wide, the surface smooth, glabrous or with sparse minute, stout, straight hairs. $2n$=22, 44, 66. May–July.

Uncommon, known only from Shannon County (northern U.S. [including Alaska] south to California, New Mexico, Missouri, Tennessee, and Virginia; Canada, Greenland, Europe, Asia). Bases and ledges of north-facing dolomite bluffs.

Steyermark (1963) considered the Missouri populations of the circumboreal *G. boreale* to represent relicts from the Pleistocene Epoch. He noted that a small set of species exists in the cool moist microclimate in the area where the species occurs that all have their present-day main ranges to the north of Missouri and were stranded when the surrounding climate became warmer as the glaciers receded. These include *Campanula rotundifolia, Galium boreale, Trautvetteria caroliniensis*, and *Zigadenus elegans*, among others. See the introductory section on origins of the Missouri flora in Yatskievych (1999) for further discussion.

Galium boreale can be distinguished vegetatively from some other *Galium* species by its stem nodes that are surrounded with a ring of spreading pubescence. This feature is also found in *G. asprellum*, *G. concinnum*, *G. obtusum*, and *G. tinctorium*. *Galium boreale* is wide-ranging and variable, and unsurprisingly several varieties have been recognized among the North American plants. Both locally and across its range, *G. boreale* varies markedly in the size and shape of the leaves, in chromosome number (from diploid to hexaploid, although the counts are all based on Old World populations), and in the pubescence of its fruits. Several varieties have been named based on this last character: var. *hyssopifolium* (Hoffm.) DC. includes plants with smooth, glabrous fruits; var. *boreale* includes plants with fruits that are densely pubescent with relatively long, straight hairs; and var. *intermedium* DC. includes plants with fruits that are sparsely to densely pubescent with relatively short, curved hairs. All of these fruit forms are found generally throughout the North American range of *G. boreale* without apparent correlation to range or habitat and have not been considered informative enough to be worth separating by most recent authors (Voss, 1996). Steyermark (1963) struggled mightily to classify the Missouri plants into these varieties. He noted that the Missouri plants belong to a small, isolated, local population and have fruits that vary from glabrous (var. *hyssopifolium*) to pubescent (var. *intermedium*). In spite of this variation, he chose to call all of the material in the state var. *hyssopifolium*. Steyermark's own analysis provides justification for not accepting infraspecific taxa in this species.

6. Galium circaezans Michx. (forest bedstraw, licorice bedstraw)

G. circaezans var. *hypomalacum* Fernald

Pl. 547 g, h; Map 2540

Plants perennial. Stems 20–50 cm long, erect or ascending, sometimes from a spreading base, unbranched or few- to several-branched from the base, glabrous or moderately to densely pubescent on and between the angles with fine, straight, soft,

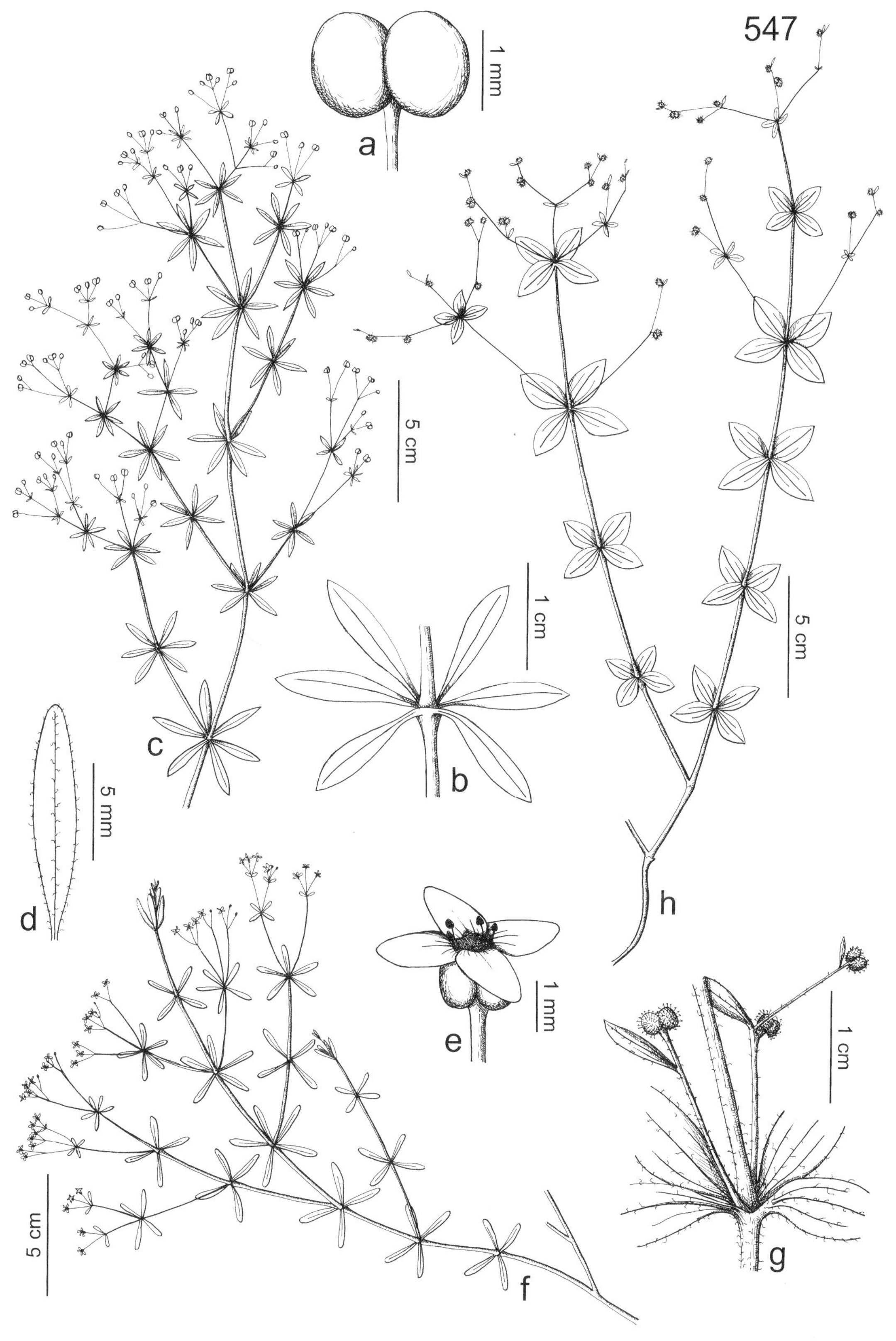

Plate 547. Rubiaceae. *Galium concinnum*, **a)** fruit, **b)** node with leaves, **c)** stem with leaves and inflorescences. *Galium obtusum*, **d)** leaf, **e)** flower, **f)** stem with leaves and inflorescences. *Galium circaezans*, **g)** node with leaf bases and fruits, **h)** habit.

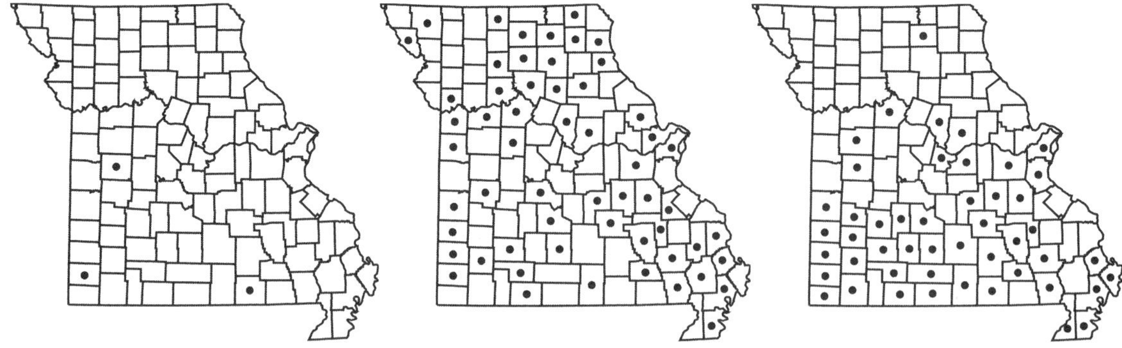

2542. Galium divaricatum 2543. Galium obtusum 2544. Galium pedemontanum

spreading to somewhat upward-curved hairs. Leaves 4 per node, spreading in orientation. Leaf blades 5–42 mm long, 2–22 mm wide, elliptic to ovate or lanceolate, angled to a bluntly (and sometimes broadly) or less commonly sharply pointed tip, angled to somewhat rounded at the base, the undersurface with impressed, linear glands (appearing as small streaks or lines), otherwise glabrous or with short, fine to stiff, straight hairs along and often also between the veins, the venation palmate with 3 veins (the midvein and 2 finer lateral veins) visible, the margins with short, stiff, prickly hairs and flat. Inflorescences terminal and sometimes also axillary from the upper leaves, not pendant, positioned over the leaves, consisting of small panicles (1–4 cm long) with mostly 1–3 branch points or reduced and appearing as unbranched spikes, the branches very short to relatively long, mostly spreading. Flowers relatively few to more numerous, sessile. Corollas 1.0–1.5 mm long, 4-lobed, pale green. Fruits 2–3 mm long, 3.5–4.0 mm wide, the surface densely pubescent with short (ca. 1 mm long), hooked hairs. $2n$=22. May–July.

Scattered to common nearly throughout the state (eastern U.S. west to Minnesota, Nebraska, and texas; Canada). Mesic to dry upland forests, ledges of bluffs, margins of sinkhole ponds, glades, upland prairies, savannas, banks of streams, and occasionally bottomland forests; also pastures and roadsides.

In many species of *Galium* that bear hooked hairs on the fruits, these hairs are developed and easily seen on the ovaries of the flowers. However, in *G. circaezans,* these hairs are hardly developed and not readily visible at flowering, so they can easily be overlooked. They enlarge rapidly as the fruits develop. Steyermark recognized two varieties of this species in Missouri, var. *circaezans* and var. *hypomalacum*, and noted that most of the Missouri plants fall into the latter group. These varieties were distinguished by patterns of pubescence and leaf size, and Steyermark himself discussed the extensive intergradation and difficulties in how to distinguish them, their lack of geographic or ecological separation, and their lack of recognition by some previous authors. They are not accepted in the present treatment.

7. Galium concinnum Torr. & A. Gray (shining bedstraw)

Pl. 547 a–c; Map 2541

Plants perennial. Stems 20–80 cm long, spreading to loosely ascending,weak and often trailing, clambering on other vegetation, and/or matted, usually well-branched, glabrous or moderately to densely pubescent with short, straight, spreading hairs, these frequently restricted to the nodes. Leaves (4–)6 per node, mostly spreading in orientation. Leaf blades 5–16 mm long, 1–3 mm wide, narrowly oblanceolate to linear, rounded or angled to a bluntly pointed tip, the midvein often extended into a minute, sharp point, angled or tapered at the base, the undersurface not glandular, glabrous, the venation with only the midvein visible, the margins glabrous (often appearing somewhat thickened) and flat or only slightly curled under. Inflorescences terminal and sometimes also axillary from the upper leaves, not pendant, positioned over the leaves, consisting of small panicles with mostly 1 or 2(3) branch points, the branches relatively short, spreading to more commonly loosely ascending. Flowers mostly 4–12, the stalks 1–7 mm long. Corollas 1.0–1.2 mm long, 4-lobed, white. Fruits 1.6–2.0 mm long, 3.0–3.5 mm wide, the surface glabrous, smooth. April–July.

Scattered nearly throughout the state (northeastern U.S. west to Nebraska and Oklahoma; Canada). Bottomland forests, mesic upland forests, banks of streams, fens, bases and ledges of bluffs, and rarely edges of glades; also roadsides.

Galium concinnum can be distinguished vegetatively from most other *Galium* species by its stem nodes that are surrounded with a ring of very short, spreading pubescence; this feature is also found in *G. asprellum*, *G. boreale*, *G. obtusum*, and *G. tinctorium*. For some reason Steyermark says "it is reported that tea prepared from the plant is a remedy for dropsy and kidney disorders."

8. Galium divaricatum Pourr. ex Lam.
(Lamarck's bedstraw)
Pl. 548 a, b; Map 2542

Plants annual, sometimes becoming slightly hardened at the base at maturity. Stems 10–40 cm long, usually weak,erect to loosely ascending or clambering, often branched or tufted, roughened with minute, prickly, downward-angled hairs on the angles, otherwise glabrous. Leaves (2)4–6(–8) per node, spreading or downward-angled in orientation. Leaf blades 2–6 mm long, 0.5–2.0 mm wide, narrowly elliptic to narrowly oblong or linear, angled or short-tapered to a sharply pointed tip, the midvein sometimes extended into a minute, sharp point, angled to truncate at the base, not glandular on the undersurface, glabrous, with only the midvein visible, the margins with minute, stiff, prickly hairs and usually curled under. Inflorescences terminal and also axillary from the uppermost leaves, the axillary ones not pendant, positioned over the leaves, consisting of small clusters or fascicles, these usually grouped into small panicles with mostly 3–6 branch points and relatively short, loosely ascending to spreading branches. Flowers relatively few, the stalks 0.5–2.0 mm long. Corollas 0.4–0.6 mm long, 4-lobed, white. Fruits about 1 mm long, 1.5 mm wide, the surface glabrous, smooth to granular. $2n$=22, 44. May–June.

Introduced, uncommon, known thus far only from Henry and Oregon Counties (native of Europe; introduced widely but sporadically in the U.S.). Cemeteries and railroads.

This species was first reported for Missouri by Castaner (1982a). Its taxonomy was clarified by Lipscomb and Nesom (2007), who redetermined a number of collections from the eastern United States as either *G. anglicum* Hudson or *G. parisiense* L. (the latter not yet known from Missouri). *Galium divaricatum* is similar to *Galium parisiense* L., which differs in its fruits that are densely covered with spreading, hooked hairs; also *G. parisiense* is usually a taller, rather leggier plant. The two species have mistakenly been combined in some herbaria and literature, with *G. divaricatum* at times mistakenly equated with a smooth-fruited European variant of *G. parisiense*, *G. parisiense* var. *leiocarpum* Tausch.

The mericarps of the fruits are relatively elongated and curved, that is rather sausage-shaped, and as they mature they grow apart until they are not even touching, with each of them connected separately to the stalk.

9. Galium obtusum Bigelow **ssp. obtusum**
(blunt-leaved bedstraw, wild madder)
Pl. 547 d–f; Map 2543

Plants perennial. Stems 15–50 cm long, spreading to loosely ascending, weak and often trailing, clambering on other vegetation, and/or matted, usually well-branched, glabrous or sparsely pubescent with minute to short, straight, spreading hairs, these mostly restricted to the nodes. Leaves (3)4(5) per node, more or less spreading in orientation. Leaf blades 5–23 mm long, 1–5 mm wide, narrowly elliptic to narrowly lanceolate or narrowly oblanceolate, rounded or angled to a bluntly pointed tip, the midvein not extended into a point, angled or tapered at the base, the undersurface not glandular, glabrous or sparsely pubescent with minute to short, stout, stiff hairs along the midvein toward the base, the venation with only the midvein visible, the margins glabrous or more commonly roughened with minute, ascending, spine-like outgrowths, usually somewhat curled under. Inflorescences terminal and sometimes also axillary from the upper leaves, not pendant, positioned over the leaves, consisting of small (0.5–1.5 cm long), stalked clusters or less commnonly panicles with 1(2) branchpoint, the branches, when present, relatively short to moderately long, loosely ascending. Flowers mostly few to several, the stalks 1–6 mm long. Corollas 1.5–2.5 mm long, 4-lobed, white. Fruits 1.5–2.0 mm long, 3–4 mm wide, the surface glabrous, smooth. $2n$=48. May–July.

Scattered nearly throughout the state, more abundantly outside the Ozark Division (eastern U.S. west to South Dakota and Texas; Canada). Bottomland forests, swamps, sloughs, margins of sinkhole ponds and oxbows, banks of streams, marshes, fens, bottomland prairies, moist swales in upland prairies, and rarely wet depressions of glades; also ditches, railroads, and roadsides.

Galium obtusum can be distinguished vegetatively from most other *Galium* species by its stem nodes that are surrounded with a ring of spreading pubescence; this feature is also found in *G. asprellum*, *G. boreale*, *G. concinnum*, and *G. tinctorium*. The taxonomy of the complex was investigated by Puff (1977), who recognized two subspecies. The other subspecies, ssp. *filifolium* (Wiegand) Puff (var. *filifolium* (Wiegand) Fernald), is distinguished by its narrower (0.8–2.0 mm) leaves and occurs mainly in the Coastal Plain and

2545. Galium pilosum 2546. Galium tinctorium 2547. Galium triflorum

Piedmont Regions of the southeastern United States. Historically, *G. obtusum* was sometimes confused with *G. tinctorium* by botanists.

10. Galium pedemontanum (Bellardi) All.
Cruciata pedemontana (Bellardi) Ehrend.
Pl. 548 h, i; Map 2544

Plants annual. Stems 5–40 cm long, relatively weak, variously erect to loosely ascending or clambering, unbranched or few-branched from near the base, evenly and sparsely to densely pubescent with relatively long, straight, spreading, slender hairs, usually also roughened with minute, prickly, downward-curved hairs on the angles. Leaves 4 per node, weakly ascending to spreading or downward-angled (with age) in orientation. Leaf blades 3–10 mm long, 1.5–5.0 mm wide, narrowly elliptic to elliptic or lanceolate, the smaller ones occasionally ovate, rounded or angled to a bluntly or sharply pointed tip, the midvein not extended into a point, rounded or angled at the base, not glandular, hairy along the midvein on the undersurface, with only the midvein visible, the margins with relatively long, spreading to ascending, slender hairs, flat or only slightly curled under. Inflorescences axillary from all but the lowermost leaves, pendant, hanging below and largely covered by the leaves, consisting of solitary flowers or more commonly small clusters or fascicles. Flowers (1)2 or 3, the stalks 1–2 mm long. Corollas 0.4–0.6 mm long, 4-lobed, yellowish green to pale yellow. Fruits 1.0–1.5 mm long, 1–2 mm wide, the surface glabrous, smooth to granular. $2n=18$. April–July.

Introduced, scattered, mostly south of the Missouri River (native of Europe, Asia; eastern U.S. west to Kansas and Texas, also northwestern U.S.). Upland prairies, glades, ledges and tops of bluffs, bottomland prairies, upland prairies, sand prairies, margins of ponds, and fens; also pastures, old fields, margins of crop fields, cemeteries, lawns, ditches, railroads, roadsides, and open disturbed areas.

This species is sometimes placed in the segregate genus *Cruciata* L., which includes about 16 species native mainly to the Mediterranean region. *Cruciata* is is said to differ from *Galium* in its combination of leaves 4 per node and mixed perfect and imperfect flowers that are yellow, pendulous, and generally held below the leaves. However *Cruciata* is here considered to represent merely a specialized group within *Galium*. Morphologically, *G. pedemontanum* is similar to the native *G. virgatum*; their distinctions are discussed under that species. It shows marked morphological variation locally, probably in response to microenvironment. Most notably it varies from short-stemmed plants with the leaves relatively large and the nodes about a leaf-length apart, to quite lanky plants with the leaves rather small and the nodes well-separated. Plants of *G. pedemontanum* also change markedly in general aspect near the end of their lives: as the last fruits mature, the leaves quickly dry and sometimes become oriented downward.

This species is a relative newcomer to the United States. It was first reported from Kentucky (Bartholomew, 1941). Its rapid spread was documented by Sanders (1976), who also reported it for the first time from Missouri. Since that time, it has become increasingly common in the southern half of the state and is now likely to be present in every county south of the Missouri River. It has also begun to make inroads in northern Missouri.

11. Galium pilosum Aiton (hairy bedstraw)
G. pilosum var. *puncticulosum* (Michx.) Torr. & A. Gray
Pl. 548 e–g; Map 2545

Plants perennial. Stems 20–80 cm long, variously erect to loosely ascending, sometimes weak and spreading or reclining, unbranched or few- to

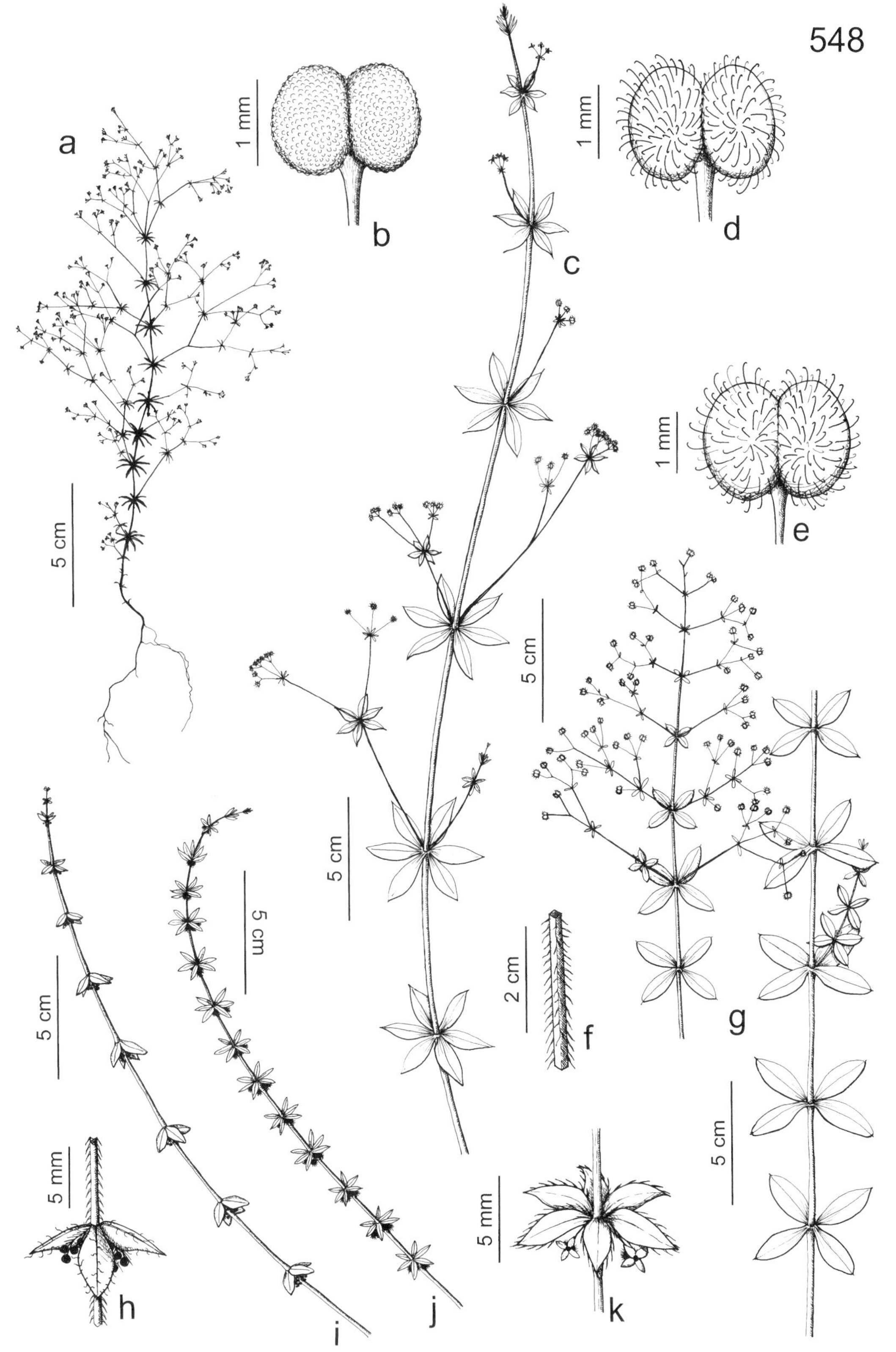

Plate 548. Rubiaceae. *Galium divaricatum*, **a)** habit, **b)** fruit. *Galium triflorum*, **c)** habit, **d)** fruit. *Galium pilosum*, **e)** fruit, **f)** stem, **g)** upper and lower portions of fertile stem. *Galium pedemontanum*, **h)** node with leaves and fruits, **i)** fertile stem. *Galium virgatum*, **j)** fertile stem, **k)** node with leaves and flowers.

several-branched from the base, rarely also with few branches toward the tip, moderately to densely pubescent with fine, straight or curved, soft, more or less spreading hairs (rarely those toward the stem base upwardly incurved), rarely glabrous or nearly so. Leaves 4 per node, spreading in orientation. Leaf blades 5–25 mm long, 2–14 mm wide, elliptic to oblong-elliptic or narrowly oval, rounded or more commonly angled to a usually bluntly pointed tip, the midvein occasionally extended into a minute, sharp point (especially on smaller leaves), angled at the base, the undersurface with minute, impressed, linear to oblong or circular glands (appearing as small dots, streaks, and/or lines), otherwise sparsely to moderately pubescent with short, fine, straight to curved hairs along and often also between the veins, the venation palmate with 3 veins (the midvein and 2 finer lateral veins, these sometimes obscure or produced only toward the blade base) visible, the margins with fine, soft hairs, flat or only slightly curled under. Inflorescences terminal and usually also axillary from the upper leaves, not pendant, positioned over the leaves, consisting of small panicles (1–4 cm long) with mostly 1–3 branch points or reduced and appearing as unbranched racemes, the branches short to relatively long, spreading to ascending. Flowers relatively few to more commonly several to numerous, the stalks 1–3 mm long (becoming elongated to 4–10 mm at fruiting). Corollas 1.0–1.5 mm long, 4-lobed, dark purple or maroon. Fruits 2.0–2.5 mm long, 2.5–3.5 mm wide, the surface densely pubescent with short (ca. 1 mm long), hooked hairs. $2n=22$. June–August.

Scattered, mostly south of the Missouri River (eastern U.S. west to Kansas and New Mexico; Canada). Mesic to dry upland forests, bases and ledges of bluffs, savannas, edges of glades, upland prairies, sand prairies, banks of streams and rivers, and margins of ponds and lakes; also old quarries, railroads, and roadsides.

Two varieties of *Galium pilosum* have sometimes been recognized, var. *pilosum* and var. *puncticulosum* (Steyermark, 1963). These are separated only by the density, length, and orientation of the stem hairs, and they intergrade widely. Steyermark noted that although both varieties grow in Missouri, plants assignable to var. *puncticulosum* are very rarely encountered nearly throughout the range of var. *pilosum* in the state. The taxonomic recognition of only the two extremes of a continuous range of morphological variation does not seem informative and is not done here.

Occasional specimens of *G. pilosum* from throughout its range (including Missouri) are malformed by an undetermined pathogen, such that their inflorescences are enlarged, with groups of small leaves or bracts and apparently numerous flowers. The identification of such plants may be problematic if their abnormal development is not recognized.

12. Galium tinctorium L. (southern three-lobed bedstraw)

G. tinctorium ssp. *floridanum* (Wiegand) Puff
G. tinctorium var. *floridanum* Wiegand
G. claytonii Michx.
G. trifidum L. ssp. *tinctorium* (L.) H. Hara

Pl. 549 a, b; Map 2546

Plants perennial. Stems 15–50 cm long, spreading to loosely ascending, weak and often trailing, clambering on other vegetation, and/or matted, usually well-branched, pubescent with short, spreading hairs at the nodes, otherwise glabrous or rarely sparsely roughened with minute, downward-angled hairs on the angles. Leaves 4–6 per node, more or less spreading in orientation. Leaf blades 5–20 mm long, 1–5 mm wide, narrowly elliptic to narrowly oblanceolate or linear, rounded or angled to a bluntly pointed tip, the midvein not extended into a point, angled at the base, the undersurface not glandular, glabrous or sparsely roughened with minute, stout, pricklelike hairs along the midvein, the venation with only the midvein visible, the margins glabrous or roughened with minute, ascending to spreading hairs, slightly to moderately curled under. Inflorescences terminal and usually also axillary from the upper leaves, not pendant, positioned over the leaves, usually consisting of small (0.5–1.5 cm long), stalked clusters. Flowers 2 or 3, rarely solitary, the stalks 1–10 mm long. Corollas 0.8–1.2 mm long, 3- or less commonly 4-lobed (the number often variable on a single plant), white. Fruits 1.5–2.0 mm long, 2.5–3.0 mm wide, the surface glabrous, smooth. $2n=24$. May–September.

Scattered, mostly south of the Missouri River, most abundantly in the eastern half of the state (eastern U.S. west to Minnesota, Nebraska, and Texas; Canada, Europe, Asia). Bottomland forests, swamps, margins of ponds, lakes, sinkhole ponds, and oxbows, marshes, fens, sloughs, bottomland prairies, mesic swales of upland prairies, and seepy ledges of bluffs; also ditches.

Galium tinctorium can be distinguished vegetatively from some other *Galium* species by its stem nodes that are surrounded with a ring of spreading pubescence. This feature is also found in *G. asprellum*, *G. boreale*, *G. concinnum*, and *G. obtusum*. Puff (1976) noted that historically *G. tinctorium* was sometimes confused with *G. obtusum*, both because of misdeterminations and

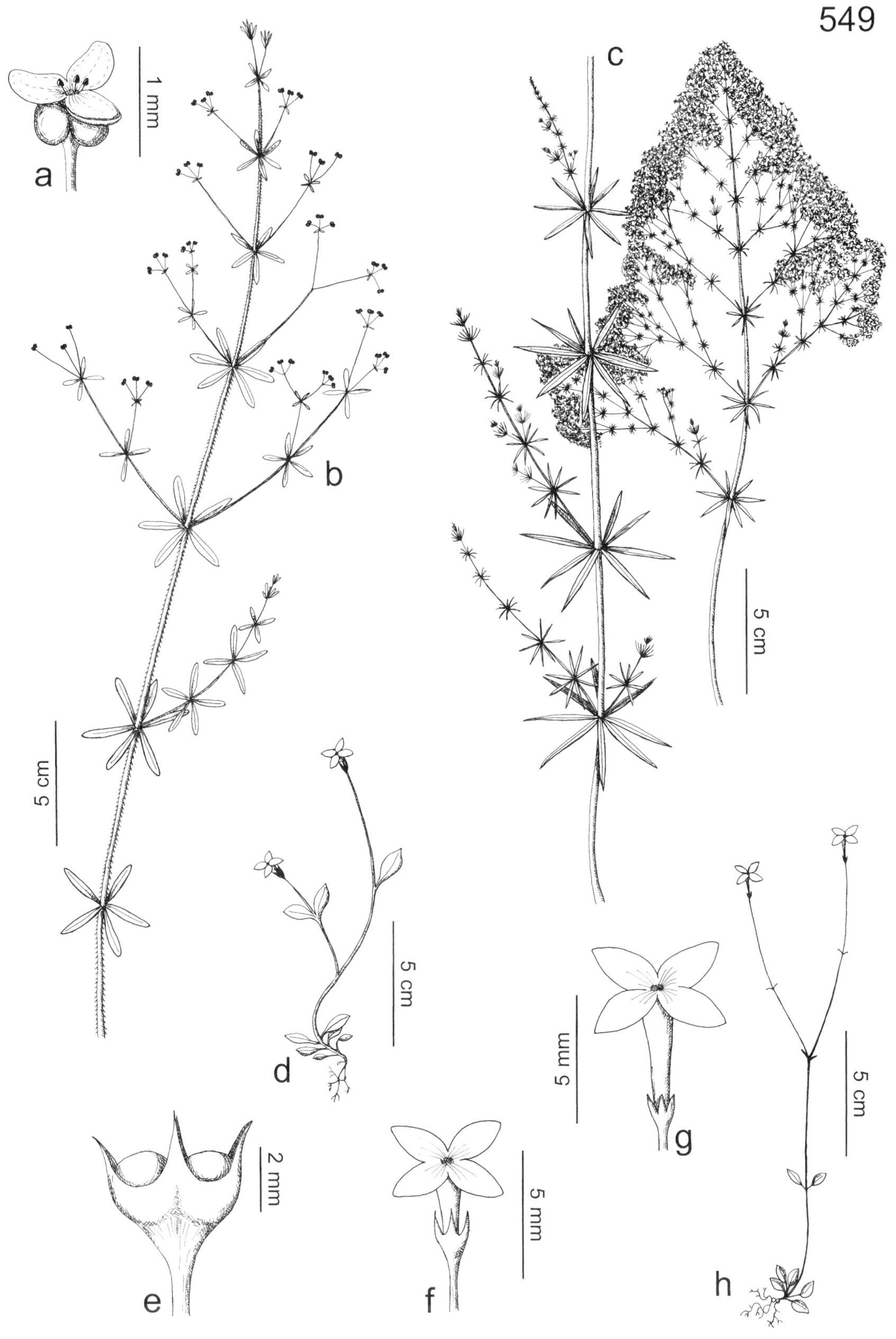

Plate 549. Rubiaceae. *Galium tinctorium*, **a)** flower, **b)** fertile stem. *Galium verum*, **c)** upper and lower portions of fertile stem. *Houstonia pusilla*, **d)** habit, **e)** fruit, **f)** flower. *Houstonia caerulea*, **g)** flower, **h)** habit.

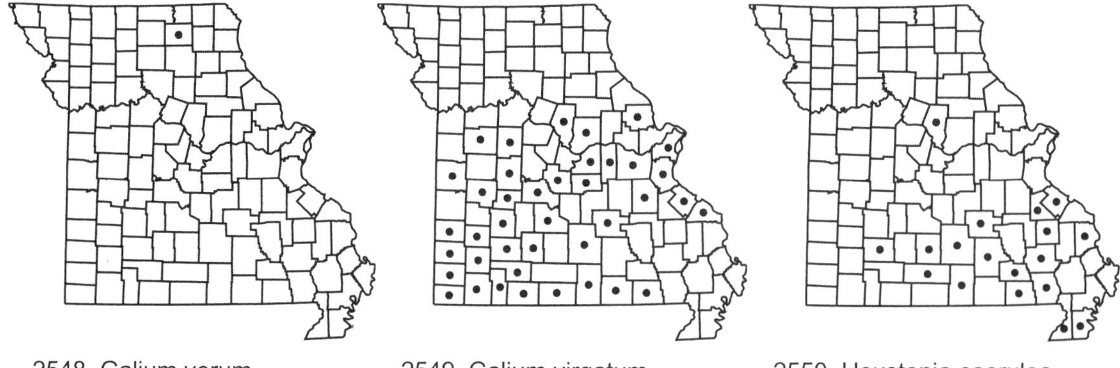
2548. Galium verum 2549. Galium virgatum 2550. Houstonia caerulea

because of misapplication of the names. The two species can be difficult to distinguish, especially without mature flowers. Steyermark (1963) noted the existence of at least one apparently intermediate specimen from Wayne County. *Galium tinctorium* also has sometimes been considered part of the closely related *G. trifidum* (northern three-lobed bedstraw), a circumboreal species that is widely distributed in the United States, but in the Midwest occurs mainly to the north of Missouri. *Galium trifidum* might plausibly be discovered in northern Missouri in the future; it differs in its mostly solitary flowers developing curved (vs. straight) and often roughened (vs. smooth) stalks at fruiting, as well as leaves mostly in whorls of 4.

Puff (1976) recognized two subspecies within *G. tinctorium*. The ssp. *floridanum* was said to differ in its longer fruit stalks, slightly larger mericarps, more densely branched inflorescences, and more robust habit. Plants with this morphology grow in the southeastern United States west to Missouri and Texas. However, Puff himself discussed the broad morphological overlap between the two taxa and the gradual replacement of plants with the ssp. *floridanum* morphology by plants with the ssp. *tinctorium* morphology. Maintenance of these infraspecific taxa does not seem warranted.

13. Galium triflorum Michx. (fragrant bedstraw)

G. triflorum var. *asprelliforme* Fernald

Pl. 548 c, d; Map 2547

Plants perennial. Stems 20–70 cm long, spreading to loosely ascending, weak and often trailing, clambering on other vegetation, and/or matted, usually well-branched, sometimes only near the base, glabrous or sparsely to moderately pubescent with short, spreading to downward-angled hairs on the angles. Leaves (4)5 or 6 per node, more or less spreading in orientation. Leaf blades 5–35(–45) mm long, 2–10(–15) mm wide, narrowly elliptic to elliptic or oblanceolate, the smallest occasionally linear or ovate, rounded or angled to a bluntly pointed tip, the midvein extended into a short, sharp point, angled or tapered at the base, the undersurface not glandular, glabrous or sparsely roughened with minute, downward-angled, pricklelike hairs along the midvein, the venation with only the midvein visible, the margins roughened with minute to short, stiff, ascending hairs, flat or slightly curled under. Inflorescences terminal and usually also axillary from the upper leaves, not pendant, positioned over the leaves, usually consisting of small (1–5 cm long), stalked clusters or small panicles, with mostly 1–3 branch points. Flowers 3 to several, the stalks 1–6 mm long (becoming elongated to 10 mm at fruiting). Corollas 1.0–1.2 mm long, 4-lobed, white. Fruits 1.5–2.5 mm long, 1.5–2.5 mm wide, the surface densely pubescent with hooked hairs ca. 0.8 mm long. $2n$=44, 66. May–September.

Scattered nearly throughout the state (U.S. [including Alaska]; Canada, Mexico, Greenland, Europe, Asia). Bottomland forests, mesic upland forests, banks of streams, rivers, and spring branches, margins of sinkhole ponds, fens, and ledges of bluffs; also pastures, old quarries, railroads, and roadsides.

Galium triflorum is sometimes used to flavor drinks, including wine, and the flowers have a sweet aroma reminiscent of vanilla. The epithet *triflorum* refers to the usual arrangement of the flowers in clusters of three. This arrangement is quite characteristic of *G. triflorum* plants to our north, and also very common but not so consistent in Missouri plants. As with most wide-ranging species, several varieties have been recognized within *G. triflorum*. Steyermark treated our plants as var. *triflorum*, commenting that the other North American variety, var. *asprelliforme* of the eastern United States and adjacent Canada, is distinguished by its more numerous inflorescences with more branching and consequently more flowers, with these accompanied by reduced rather than full-sized leaves. With more collections of *G. triflorum* now available for study, some plants from Missouri could be classified as var. *asprelliforme*, whereas others are intermediate in morphology. The degree of inflorescence development appears to be related to size and age of the plant, with most of the compounded clusters on larger, older plants. Therefore, these varieties are not recognized formally here.

Galium triflorum is superficially similar to and sometimes confused with *G. aparine*, but the latter has larger fruits, narrower leaves, and is strongly roughened, whereas *G. triflorum* is smooth or only slightly roughened. Voss (1996) noted that these species can be distinguished in Michigan by the tiny spinules or barbs on their leaf margins, which are retrorse in *G. aparine* but curved toward the leaf tip in *G. triflorum*. Although the marginal spinules of *G. aparine* are evident with careful observation under a good handlens, those of Missouri's *G. triflorum* at best are only minutely discernible with significant magnification and sometimes spreading rather than curved.

14. Galium verum L. (yellow bedstraw, Our Lady's bedstraw)

Pl. 549 c; Map 2548

Plants perennial, sometimes somewhat woody at the base and/or with a woody rootstock. Stems 25–100 cm long, erect to loosely ascending, often weak and clambering on other vegetation and/or matted, usually well-branched toward the tip, also usually with relatively well-developed but often short axillary branches below the flowering portion, evenly and moderately to densely pubescent with minute to short, wavy, soft, spreading or somewhat downward-curved hairs on and between the angles. Leaves 6–8(–12) per node, sometimes apparently more numerous because of short axillary branches, more or less spreading in orientation. Leaf blades 5–20 mm long (sometimes 0.2–0.4 mm along the flowering branches), 0.5–2.0 mm wide, very narrowly triangular to more commonly linear, angled or tapered to a sharply pointed tip, the midvein usually extended into a minute, sharp point, truncate or slightly tapered at the base, the undersurface (difficult to observe) not glandular, densely and evenly pubescent with minute curled hairs (sometimes appearing mealy), usually also with dense, slightly longer, straight, stiff hairs along the midvein, the venation with only the midvein visible, the margins and upper surface roughened with minute to short, stiff, ascending hairs (or the margins glabrous, but this usually impossible to observe), strongly curled under. Inflorescences terminal and also axillary from leaves in the apical 1/3 of the stem, not pendant, positioned over the leaves, consisting of small (1–2 cm long), stalked panicles, with mostly 3–6 branch points, the apical 1/3 of the stem with its reduced leaves and dense inflorescences often resembling a single, highly branched panicle. Flowers mostly numerous, the stalks 1–2 mm long. Corollas 1.2–1.5 mm long, 4-lobed, yellow. Fruits 0.8–1.2 mm long, 0.8–1.2 mm wide, the surface glabrous, smooth. $2n=22$, 44. May–July.

Introduced, uncommon, known thus far only from a single historical specimen from Adair County (native of Europe, Asia; introduced widely but sporadically in the northern U.S. south to California, Colorado, and North Carolina; Canada, Greenland). Habitat unknown.

This native of the Mediterranean region is cultivated as an ornamental, mostly in shade gardens. The flowers have a strong, sweet odor, which some persons find pleasant and others definitely do not. Steyermark also noted some uses of *G. verum*, as follows: "This is the species of bedstraw which is claimed to have been the one which filled the Christ Child's manger at Bethlehem, and later on was used for stuffing mattresses. The distilled flowering tops are made into a beverage, and a type of cheese, made from the milk of sheep and goats, is prepared from the plant when mixed with rennet from calves." *Galium verum* additionally has provided a medicinal styptic, a red dye, and, no doubt, other useful substances.

Also cultivated in Missouri is *Galium mollugo* L. (false baby's breath), which occasionally escapes and becomes naturalized in the eastern United States. This species is quite similar to *G. verum* but can be distinguished by its glabrous, smooth stems and white to pale green flowers.

15. Galium virgatum Nutt. ex Torr. & A. Gray (southwestern bedstraw)

Pl. 548 j, k; Map 2549

Plants annual. Stems 8–20(–40) cm long, erect or ascending, sometimes from a spreading base, with minute, prickly, more or less spreading hairs on the angles, otherwise usually glabrous. Leaves 4 per node (but sometimes apparently more because of short axillary stems with additional leaves), mostly ascending in orientation. Leaf blades 2–7 mm long, 1–2 mm wide, narrowly elliptic to lanceolate, occasionally the smaller ones elliptic, angled to a bluntly or sharply pointed tip, the midvein sometimes extended into a minute, sharp point, rounded or angled at the base, the undersurface with impressed, round to linear glands (appearing as faint, irregular dots, streaks, and/or lines), otherwise glabrous or with sparse, stiff hairs along the midvein, the margins flat, pubescent with relatively long, stiff, spreading to more commonly ascending hairs. Inflorescences axillary from all but the lowermost leaves, pendant, hanging below and largely covered by the leaves, consisting of solitary flowers. Flowers subsessile, the stalks 0.3–0.8 mm long. Corollas 0.4–0.6 mm long, 4-lobed, pale yellow or cream-colored. Fruits 1.5–2.0 mm long, 2–3 mm wide, the surface densely pubescent with hooked hairs to 1 mm long. April–June.

Scattered in the Ozark, Ozark Border, and Unglaciated Plains Divisions, north locally to Lincoln County (Illinois to Alabama west to Kansas and Texas). Glades, ledges and tops of bluffs, and rocky portions of upland prairies; rarely also gardens and roadsides.

Steyermark (1963) reported that the species dries up and disappears by midsummer. In Missouri, the native *G. virgatum* sometimes has been confused with the superficially similar, nonnative taxon, *G. pedemontanum*. That species differs in its clusters of mostly two or three flowers, as well as fruits and ovaries lacking hooked hairs. Also, plants of *G. virgatum* are mostly encountered as single-stemmed individuals, whereas those of *G. pedemontanum* are mostly several stemmed and strongly colonial.

4. **Houstonia** L. (bluets)
(Terrell, 1996)

Plants annual or perennial herbs, the perennials sometimes with rhizomes or the rootstock sometimes woody. Stems erect or ascending to somewhat spreading (creeping elsewhere), rounded to somewhat 4-angled or channeled on 2 opposing sides. Leaves all basal or the stem leaves opposite (rarely in whorls of 3 at some nodes in *H. longifolia*), short-petiolate to nearly sessile. Stipules interpetiolar, usually persistent but sometimes becoming shredded with age, fused to the leaf bases on either side, membranous to papery, more or less triangular, entire, toothed, or with 2(3) slender lobes toward the tip, the margins glabrous or glandular. Inflorescences terminal and sometimes also axillary, of solitary flowers or in open to dense clusters, these sometimes appearing as small, flat-topped to dome-shaped panicles. Flowers sessile to long-stalked, the stalks sometimes subtended by bracts, in some species distylous. Calyces deeply 4-lobed. Corollas funnelform or trumpet-shaped, white to pink, purple, or blue, often with a contrasting, yellow or reddish eye at the tip of the throat, 4-lobed, the lobes shorter than or equal to the tube, not overlapping in bud. Stamens 4, attached near the top or the base of the corolla tube, the anthers exserted or not. Ovary fully inferior (but at fruiting usually expanding above the calyx tube), 2-locular, the ovules several to numerous in each locule. Style 1, slender, relatively long or short, exserted or not, the stigmas 2, linear. Fruits capsules, 2-lobed and somewhat flattened, dehiscing longitudinally from the tip, sometimes beaked. Seeds 0.5–1.5 mm long, more or less globose to ellipsoid, but concave on 1 face, the cavity sometimes with a longitudinal ridge, the surface otherwise black, with minute tubercles. About 25 species, North America, Caribbean Islands.

Generic limits in this complex of genera remain controversial. *Hedyotis* L. is the oldest, most inclusive name for the worldwide group of about 300 species that are slender herbs and low shrubs with rather small leaves, interpetiolar stipules often with 2–5 lobes, small white to purple flowers, and small, stiffly papery capsules with numerous very small seeds. Some authors have favored this broad view of a single, polymorphic genus (Merrill and Metcalf, 1942; W. H. Lewis, 1961, 1962; G. K. Rogers, 1987; Gleason and Cronquist, 1991; B. L. Turner, 1995), but others have treated the group as an assemblage of three or more narrowly circumscribed genera, including (for the Missouri species) such segregates as *Houstonia*, *Oldenlandia* L., and *Stenaria* (Raf.) Terrell (Bremekamp, 1952; Steyermark, 1963; Terrell, 1990, 1991, 1996, 2001a, b; Terrell and Robinson, 2006). Much of the recent taxonomic work has been biased toward sampling of North American species, which makes it difficult to evaluate the classification on a worldwide basis. Possible relationships among species from several continents has not yet been evaluated adequately to give a clear picture of the species relationships on a world-wide scale. The current situation is regrettable, as two polarized camps have arisen among botanists in the United States, with both of the very different generic taxonomies reflected in the recent floristic, taxonomic, and conservation literature.

Initial molecular studies of the group have contributed to an understanding of species-level questions among North American taxa (Church, 2003; Church and Taylor, 2005), but did not include sufficient geographic and taxonomic sampling to address problems of generic limits, even among the North American species. The most recently published study (Groeninckx et al., 2009) did include a broader taxonomic sampling, but should still be viewed as somewhat preliminary. This molecular study of the phylogeny of tribe Spermacoceae Bercht. & J. Presl used data from three plastid DNA markers and concluded that: 1) Seed and fruit morphology, which had been viewed by some botanists as fundamental characters for delimiting genera within the tribe (Terrell and Wunderlin, 2002), are less useful than first thought; 2) The American species variously referred to as *Hedyotis* and *Houstonia* are not closely related to the Old World complex of *Hedyotis* species (which includes the type of that name) and thus should be treated as a separate genus, *Houstonia*; 3) *Houstonia nigricans* and its relatives, which some taxonomists have segregated into the genus *Stenaria* based mainly on seed morphology (Terrell, 2001a), is merely a specialized group within *Houstonia*; and 4) The species of *Oldenlandia* should continue to be recognized as a genus separate from both *Hedyotis* and *Houstonia*. For further discussion of *Oldenlandia*, see the treatment of that genus. Generic concepts in the present volume largely mirror those suggested by the molecular studies of Groeninckx et al. (2009).

Species circumscriptions within *Houstonia* in the present work follow those of the excellent species-level studies by Edward Terrell (1996, 2001a). Flower color is sometimes quite variable in both intensity and hue, within as well as between populations. For example, flowers of individual plants of *Houstonia pusilla* in a local population often range from deep lilac-blue to pale sky blue, and also fade with age of the flower.

1. Flowers borne in terminal clusters of 3–12 or more per stem; corollas trumpet-shaped to funnelform
    2. Blades of stem leaves (3–)5–35 mm wide, lanceolate to ovate, with 3 or 5 main veins observable (note that these may be less easily observed in specimens with narrower leaves) . . . . 5. H. PURPUREA
    2. Blades of stem 1–6 mm wide, leaves narrowly lanceolate to narrowly oblong, narrowly elliptic, or linear, with only the midvein observable (rarely with a faint pair of lateral veins in *H. nigricans*).
        3. Plants often with a rosette of basal leaves at flowering, herbaceous throughout, the roots slender and fibrous; flower stalks 2–8 mm long . . . . 2. H. LONGIFOLIA
        3. Plants without basal leaves at flowering, sometimes woody at the base, with a usually woody and often branched rootstock; flowers sessile or the stalks to 3 mm long . . . . 4. H. NIGRICANS
1. Flowers solitary from the axils of the uppermost pair of leaves, rarely also terminal (thus appearing terminal and often in pairs or rarely in clusters of 3), relatively long-stalked; corollas trumpet shaped
    4. Stems 1–4(–7) cm tall; corollas white to pink, lacking a ring of contrasting color in the throat . . . . 7. H. ROSEA
    4. Stems 2–15 cm tall; corollas blue to violet, less commonly pink or white, often with a ring of contrasting yellow or red at the top of the throat

5. Flowers distylous, the anthers and stigmas borne at different levels in the upper half of the corolla tube; corollas blue (rarely white) with a yellow eye at the tip of the throat, the tube 4–7 mm long, as long as to much longer than the calyx lobes; plants perennial (but sometimes flowering the first year), often with a rosette of basal leaves . . . . 1. H. CAERULEA
5. Flowers homostylous, the anthers and stigmas borne at the same level in the lower part of the corolla tube; corollas variously blue to purple or less commonly pink or white, if blue then usually with a reddish eye at the tip of the throat, the tube 0.8–5.0 mm long, much shorter than to slightly longer than the calyx lobes; plants annual, with at most a pair of basal leaves
    6. Corollas white, the tube 0.8–2.5 mm long, about as long as or slightly longer than the calyx lobes . . . . 3. H. MICRANTHA
    6. Corollas variously blue to purple or less commonly pink or white, the tube 2–5 mm long, much shorter than to slightly longer than the calyx lobes . . . . 6. H. PUSILLA

1. Houstonia caerulea L. (big bluets, Quaker ladies, innocence)
Hedyotis caerulea (L.) Hook.
Pl. 549 g, h; Map 2550

Plants perennial herbs (sometimes flowering the first year, thus potentially annual), often producing short, inconspicuous, threadlike rhizomes. Stems 1 to several, 3–15 cm long, erect or strongly ascending, glabrous or slightly roughened. Leaves opposite and frequently also in a basal rosette, sometimes slightly succulent. Stipules 0.5–1 mm long, oblong to broadly triangular, truncate to broadly pointed at the tip, entire or minutely glandular along the margins. Leaf blades 3–15 mm long, 1–3 mm wide, oblanceolate to narrowly oblong-spatulate, rounded or broadly pointed at the tip, angled or tapered at the base, the margins entire and generally flat, the upper surface glabrous or sparsely hairy, the undersurface glabrous, without visible venation or sometimes the midvein visible. Flowers distylous (the anthers and stigmas borne at different levels in the upper half of the corolla tube), solitary in the axils of the uppermost pair of leaves, rarely also terminal (thus appearing terminal and often in pairs or rarely in a cluster of 3), on stalks 15–50 mm long. Calyx lobes 1.5–2.5 mm long, sometimes slightly unequal, up to 1/2 as long as the adjacent corolla tube, triangular. Corollas trumpet-shaped, sky blue to light blue, rarely white, usually with a yellow contrasting ring at the top of the throat, externally glabrous, internally pubescent in the upper part of the tube, the tube 4–7 mm long, the lobes 5–8 mm long, lanceolate to triangular. Fruits 2–3 mm long, 3–5 mm wide, subglobose to depressed-obovoid, somewhat flattened laterally, partially inferior, sometimes slightly 2-lobed, glabrous or nearly so. Seeds 0.5–1.0 mm long. $2n$=16, 32, 48. April–May.

Scattered in southeastern part of our state; distributed from Nova Scotia through southern Wisconsin to Iowa, south to Georgia and northern Louisiana. Occasional and sometimes locally frequent in sunny sites in woodlands including post oak land, along streams and roads, on banks and ledges, and in meadows, frequently in seepy or ephemerally wet spots, often on poor soils and even growing mixed with mosses.

Terrell (1996) described this species as perennial, which may be true, but usually is not evident on herbarium collections. He noted that when the plants overwinter they form a basal rosette of leaves, but that in first-year plants, the leaves are all borne along the stem, and thus these plants either are or appear annual.

Steyermark (1963) mentioned the existence of white-flowered plants (f. *albiflora* Millsp.) in this species, but had no records of these from Missouri. Such individuals have since been discovered to be of rare occurrence within a few, scattered populations in the state. Steyermark (1963) also noted the existence of an uncommon northeastern variant with a relatively low habit and short white flowers that is sometimes recognized taxonomically as var. *faxonorum* Pease & A.H. Moore. However, Terrell (1996) did not recognize this variety, noting that these unusual plants represent only the endpoint of continuous variation from lower to higher elevations in that region and cannot be separated clearly. He also noted that the supposedly unique characters appear sporadically in populations elsewhere in the species range.

2. Houstonia longifolia Gaertn. (long-leaved bluets, slender-leaved bluets)
Ho. longifolia var. *ciliolata* (Torr.) Alph. Wood

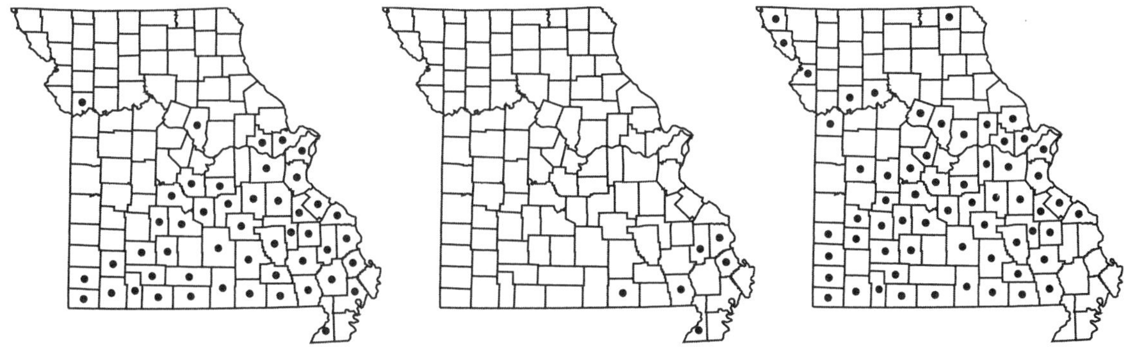

2551. Houstonia longifolia 2552. Houstonia micrantha 2553. Houstonia nigricans

Ho. longifolia var. *tenuifolia* (Nutt.) Alph. Wood
Hedyotis longifolia (Gaertn.) Hook., an illegitimate name

Pl. 550 e–h; Map 2551

Plants perennial herbs, sometimes with threadlike rhizomes. Stems usually few to several, 6–30 cm long, erect or ascending, glabrous or minutely hairy, especially at the nodes. Leaves opposite or occasionally in whorls of 3 at a few nodes, also basal in a rosette, not succulent. Stipules 1.0–3.5 mm long, triangular to ovate, bluntly or sharply pointed, entire along the margins. Leaf blades 5–40 mm long, 0.5–6.0 mm wide (to 10 mm on basal leaves), linear to narrowly oblong (elliptic to oblanceolate on basal leaves), bluntly pointed at the tip (sometimes rounded on basal leaves), narrowly angled or tapered at the base, the margins entire and often roughened or minutely hairy, flat or sometimes slightly rolled under, the upper surface glabrous or sparsely short-hairy, the undersurface glabrous, the venation with only the midvein observable. Flowers distylous (the anthers and stigmas borne at different levels in the upper half of the corolla tube), terminal and sometimes also from the upper leaf axils, in small clusters of 2–7, these sometimes grouped into small panicles, the flower stalks 2–8 mm long. Calyx lobes 1.0–2.5 mm long, sometimes slightly unequal, up to 1/2 as long as the adjacent corolla tube, narrowly triangular. Corollas funnelform, white to to pink, pale purple, or bluish-tinged, lacking a ring of contrasting color in the throat, externally glabrous, internally pubescent in most of the tube, sometimes sparsely so, the tube 4–5 mm long, the lobes 2–4 mm long, triangular. Fruits 2.0–2.5 mm long, 2.0–2.5 mm wide, subglobose to ovoid, not or only slightly flattened, partially inferior, sometimes slightly notched at the tip, glabrous or nearly so. Seeds 0.8–1.4 mm long. $2n=12, 24$. April–July, rarely also October.

Scattered in the Ozark, Ozark Border, and Mississippi Lowlands Divisions (eastern U.S. west to North Dakota and Oklahoma; Canada). Mesic to dry upland forests, savannas, upland prairies, glades, ledges and tops of bluffs, margins of sinkhole ponds, and banks of streams and rivers; also pastures, lawns, railroads, and roadsides.

Several authors have recognized varieties within this species, although they have not agreed on the number or delimitation of these. Steyermark (1963) recognized two varieties among the Missouri plants, var. *longifolia* and var. *tenuifolia*, the latter known only from a few localities in the southeastern part of our state. He separated var. *tenuifolia* by its shorter flower stalks and characteristically narrow leaves, but noted that it overlaps morphologically with plants of var. *longifolia* in leaf width. More recently, Terrell (1996) analyzed, characterized, and mapped the variation in this species across its entire range, and declined to formally recognize any infraspecific taxa. Plants that are just starting to flower often have the inflorescences rather compact and shorter than the leaves, whereas plants that have been flowering for a while usually have the flower stalks more expanded and the inflorescences spreading and longer than the leaves. Thus, the age of the plant affects expression of the character used to delineate the Missouri variants.

3. Houstonia micrantha (Shinners) Terrell (southern bluets)
Hedyotis australis W.H. Lewis & Dw. Moore
He. crassifolia Raf. var. *micrantha* Shinners

Map 2552

Plants annual, not producing rhizomes. Stems 2–11 cm long, glabrous. Leaves opposite (plants not producing a basal rosette, but the lower nodes sometimes relatively closely spaced), herbaceous or somewhat succulent. Stipules 0.8–2.5 mm long,

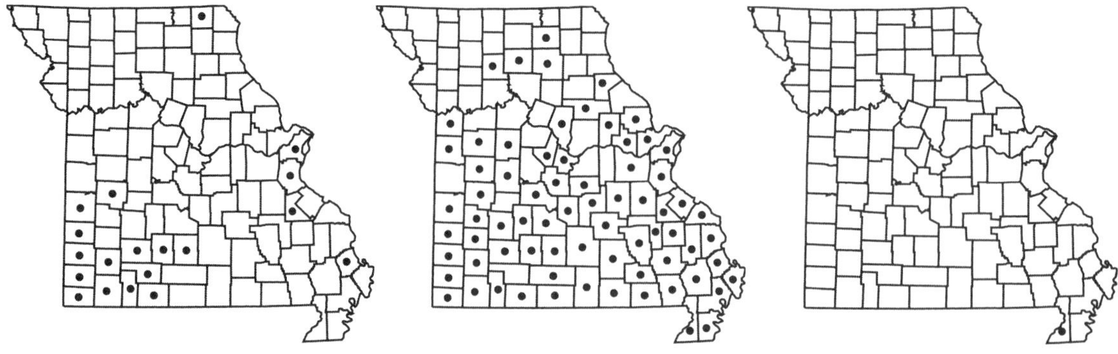

2554. Houstonia purpurea 2555. Houstonia pusilla 2556. Houstonia rosea

triangular, broadly to narrowly pointed at the tip, entire or inconspicuously toothed along the margins. Leaf blades 1.0–5.5 mm long, 1–4 mm wide, narrowly elliptic to elliptic, or ovate (those of the lowermost leaves sometimes spatulate, rounded to pointed at the tip, rounded to angled or tapered at the base, the margins entire and generally flat, sometimes minutely hairy, the surfaces glabrous, without visible venation. Flowers homostylous (the anthers and stigmas borne at the same level in the lower part of the corolla tube), solitary in the axils of the uppermost pair of leaves, rarely also terminal (thus appearing terminal and often in pairs or rarely in a cluster of 3), on stalks 2–25 mm long. Calyx lobes 1–3 mm long, slightly shorter than to about as long as the adjacent corolla tube, lanceolate to triangular-ovate. Corollas trumpet-shaped, white, occasionally with a contrasting yellow ring at the top of the throat, externally glabrous, internally glabrous or sparsely pubescent in the upper part of the tube, the tube 0.8–2.5 mm long, the lobes 0.8–2.5 mm long, ovate. Fruits 2.0–3.5 mm long, 3.0–5.5 mm wide, subglobose to depressed-obovoid, somewhat flattened laterally, partially inferior, usually slightly 2-lobed, glabrous. Seeds 0.5–1.0 mm long. $2n=32$. March–April.

Uncommon, known thus far only from Butler, Dunklin, and Oregon Counties (southeastern U.S. west to Missouri and Texas). Cemeteries, lawns, and open disturbed areas, usually on sandy substrates.

This species was first reported from Missouri by W. H. Lewis (2006), as *Hedyotis australis*.

4. Houstonia nigricans (Lam.) Fernald **var. nigricans** (narrow-leaved bluets, diamond flower)

Ho. angustifolia Michx.

Hedyotis nigricans (Lam.) Fosberg

Stenaria nigricans (Lam.) Terrell

Pl. 551 f, g; Map 2553

Plants perennial herbs, sometimes woody at the base, the rootstock usually woody and branched. Stems 10–40 cm long, glabrous or minutely roughened. Leaves opposite (plants lacking basal rosettes), usually also produced in dense fascicles from some of the main leaf axils, not succulent. Stipules 1.0–3.5 mm long, triangular, sharply pointed or with 2(3) slender lobes at the tip, somewhat uneven along the margins. Leaf blades 7–40 mm long, 0.5–4.5 mm wide, linear to very narrowly elliptic or very narrowly lanceolate, sharply pointed at the tip, narrowly angled or tapered at the base, the margins entire, but roughened with minute, stiff hairs, rolled under, the upper surface sparsely roughened or minutely hairy, the undersurface sparsely roughened or sparsely to moderately and minutely hairy, the venation with only the midvein observable or rarely with a faint pair of lateral veins. Flowers distylous (the anthers and stigmas borne at different levels in the upper half of the corolla tube), terminal and usually also from the upper leaf axils, in relatively dense clusters of 3–9, these often grouped into small panicles, the flowers sessile or on stalks to 3 mm long. Calyx lobes 1–2 mm long, sometimes slightly unequal, less than 1/2 as long as the adjacent corolla tube, narrowly triangular to ovate. Corollas funnelform, white to pink or pale purple, lacking a ring of contrasting color in the throat, externally glabrous, internally pubescent in most of the tube, the tube 3–5 mm long, the lobes 2.5–4.0 mm long, triangular. Fruits 3–4 mm long, 1–2 mm wide, subglobose to more commonly obovoid or oblong-ellipsoid, not or only slightly flattened, partially inferior, rounded to slightly notched at the tip, glabrous or sparsely roughened. Seeds 0.5–1.0 mm long. $2n=18$, 20. May–October.

Scattered to common in the Ozark and Ozark Border Divisions, uncommon in portions of the Glaciated Plains and Unglaciated Plains Divisions

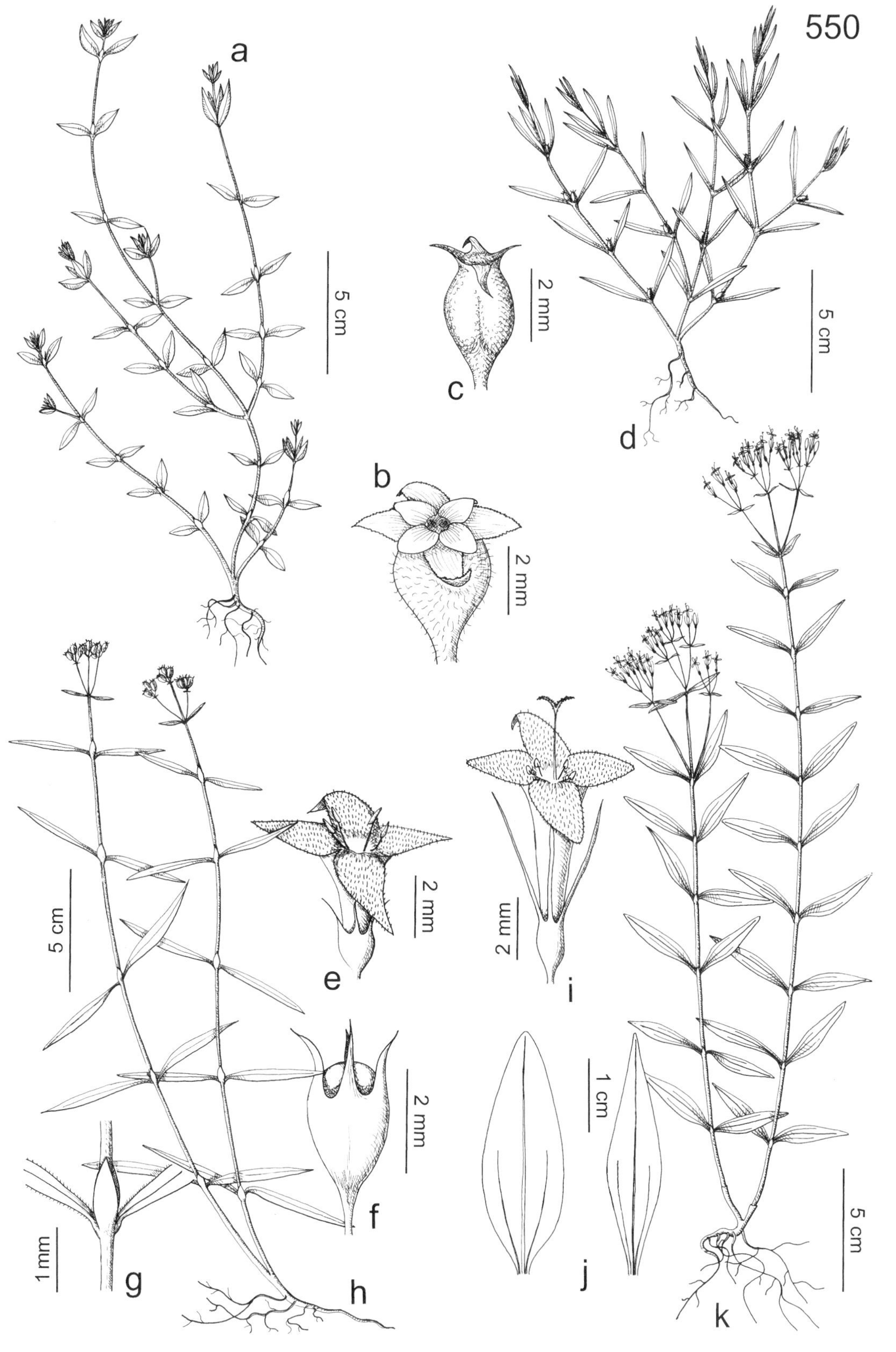

Plate 550. Rubiaceae. *Oldenlandia uniflora*, **a)** habit, **b)** flower. *Oldenlandia boscii*, **c)** flower, **d)** habit. *Houstonia longifolia*, **e)** flower, **f)** fruit, **g)** node with leaf bases and stipule, **h)** habit. *Houstonia purpurea*, **i)** flower, **j)** leaves, **k)** habit.

(eastern U.S. west to Nebraska and New Mexico; Mexico). Glades, ledges and tops of bluffs, upland prairies, loess hill prairies, openings of mesic to dry upland forests, and banks of streams and rivers; also pastures, old fields, fencerows, railroads, and roadsides.

The plants often turn dark or blacken when dried, particularly if they dry slowly, which is apparently the origin of the scientific name. Often, the buds in the axils of the leaves produce several well-developed leaves without an apparent branch, so the plants appear very bushy.

This species was transferred by Terrell (2001a) into the segregate genus *Stenaria*, into which he classified five species with a main center of diversity in Texas and northeastern Mexico. This genus is included in a broadly circumscribed *Houstonia* in the present work, as discussed at the head of the generic treatment, above.

This relatively widespread species is morphologically very variable. Several varieties have been described and variously recognized or synonymized by different authors (B. L. Turner, 1995). Most recently and with very thorough study, Terrell (2001a) recognized four varieties of *Stenaria nigricans*, but combinations for most of these have not yet been published under *Houstonia*. Most of the plants fall into var. *nigricans*, fortunately including all of those from Missouri. The relatively small-leaved and small-flowered *S. nigricans* var. *breviflora* Terrell and the relatively broad-leaved *S. nigricans* var. *gypsophila* (B.L. Turner) Terrell (*Hedyotis nigricans* var. *gypsophila* B.L. Turner) are restricted to northern Mexico, where their respective ranges overlap that of var. *nigricans*. *Stenaria nigricans* var. *floridana* (Standl.) Terrell (*Houstonia nigricans* var. *floridana* (Standl.) Terrell) replaces var. *nigricans* in southern Florida and the Bahamas, and has subglobose fruits.

5. Houstonia purpurea L. (mountain houstonia, broad-leaved bluets)

Ho. purpurea var. *calycosa* A. Gray
Hedyotis purpurea (L.) Torr. & A. Gray
He. purpurea var. *calycosa* (A. Gray) Fosberg

Pl. 550 i–k; Map 2554

Plants perennial herbs, the rootstock sometimes somewhat woody. Stems 1 to several, 8–40 cm long, erect or ascending, glabrous toward the tip, usually with short, stiff hairs toward the base. Leaves opposite, not succulent. Stipules 1–7 mm long, triangular to ovate, bluntly or sharply pointed or minutely notched at the tip, entire or minutely few-toothed along the margins. Leaf blades 8–60 mm long, (3–)5–35 mm wide, lanceolate to ovate, sharply pointed at the tip, broadly angled or rounded to very shallowly cordate at the base, the margins entire and generally flat, sometimes minutely hairy, the upper surface glabrous or short-hairy, the undersurface glabrous or short-hairy along the veins, the venation with 3 or 5 main veins observable (note that these may be less easily observed in specimens with narrower leaves), the lateral veins branching from the midvein at or near its base. Flowers distylous (the anthers and stigmas borne at different levels in the upper half of the corolla tube), terminal in small clusters of 3–15, these sometimes grouped into small panicles, the flower stalks 2–5 mm long. Calyx lobes 2–5 mm long, sometimes slightly unequal, more than 1/2 as long as the adjacent corolla tube, narrowly triangular. Corollas funnelform, white to light purple, lacking a ring of contrasting color in the throat, externally glabrous, internally pubescent in most of the tube, the tube 4–6 mm long, the lobes 3–4 mm long, triangular. Fruits 2–4 mm long, 2–4 mm wide, subglobose to ovoid, not flattened, partially inferior, sometimes slightly notched at the tip, glabrous or nearly so. Seeds 0.8–1.5 mm long. $2n$=12, 24. May–June.

Scattered in the southwestern quarter of the state, uncommon and sporadic farther north and east (eastern U.S. west to Nebraska and Texas). Bottomland forests, mesic to dry upland forests, ledges of bluffs, glades, and banks of streams and rivers; also pastures and roadsides.

Three varieties of *H. purpurea* have recently been recognized: var. *montana* (Small) Terrell, which is narrowly endemic in the Roan Mountain area of the Appalachians; var. *calycosa*, found nearly throughout the range of the species; and var. *purpurea*, found throughout the range of the species. Terrell (1996) commented that var. *montana* is local and is listed as endangered on the federal Endangered Species List; it is distinguished by its relatively broad leaves and relatively large, deep purple corollas, but these features appear to overlap somewhat with those of var. *purpurea*. The var. *calycosa* has been distinguished from var. *purpurea* by its relatively narrower leaves and relatively long calyx lobes, but, as Terrell noted, these varieties intergrade extensively and are separated only arbitrarily. After careful study of the problem, Terrell (1996) did recognize var. *calycosa* as separate, but with the comment, "In summary, var. *calycosa* is a weakly differentiated variety that is most distinct in the cedar glades, has a preference for more xeric habitats, and is centered west of the Appalachians." In light of the continuous morphological variation, which is also found in Missouri plants and links the extremes that equate with the two named

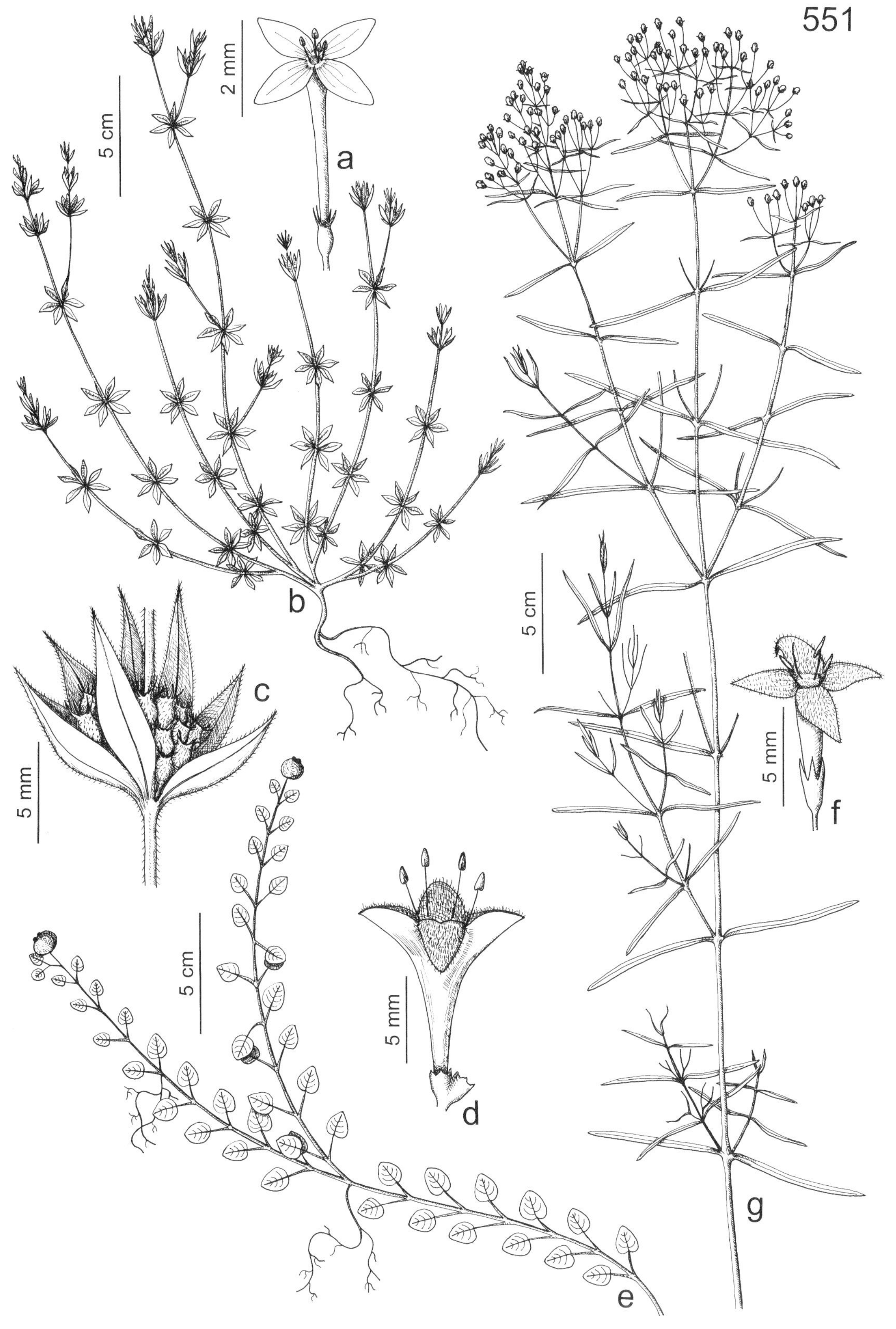

Plate 551. Rubiaceae. *Sherardia arvensis*, **a)** flower, **b)** habit, **c)** node with leaves and fruits. *Mitchella repens*, **d)** flower, **e)** habit. *Houstonia nigricans*, **f)** flower, **g)** fertile stem.

phases occurring in the state, these varieties are not accepted in the present treatment.

6. Houstonia pusilla Schöpf (small bluets, star violet)
Ho. minima L.C. Beck
Hedyotis crassifolia Raf.
He. minima (L.C. Beck) Torr. & A. Gray
He. pusilla (Schöpf) Mohlenbr., an invalid name

Pl. 549 d–f; Map 2555

Plants annual, not producing rhizomes. Stems 1 or few, 2–12 cm long, erect or ascending to less commonly somewhat sprawling, glabrous. Leaves opposite (plants not producing a basal rosette, but the lower nodes sometimes relatively closely spaced), somewhat succulent. Stipules 0.8–1.5 mm long, triangular, broadly to narrowly pointed at the tip, entire along the margins. Leaf blades 4–11 mm long, 1–4 mm wide, elliptic to oblanceolate, broadly to narrowly pointed at the tip, angled or tapered at the base, the margins entire and generally flat, sometimes minutely hairy, the surfaces glabrous or minutely hairy, without visible venation. Flowers homostylous (the anthers and stigmas borne at the same level in the lower part of the corolla tube), solitary in the axils of the uppermost pair of leaves, rarely also terminal (thus appearing terminal and often in pairs or rarely in a cluster of 3), on stalks 5–35 mm long. Calyx lobes 1–4 mm long, sometimes slightly unequal, ½ as long to slightly longer than the adjacent corolla tube, triangular. Corollas trumpet-shaped, sky blue to deep blue, lavender, or violet, occasionally pink or white, often with a contrasting reddish ring at the top of the throat, externally glabrous, internally pubescent in the upper part of the tube, the tube 2–5 mm long, the lobes 2.0–5.5 mm long, lanceolate. Fruits 1.5–3.0 mm long, 2.5–5.0 mm wide, subglobose to depressed-obovoid, somewhat flattened laterally, partially inferior, sometimes slightly 2-lobed, glabrous or nearly so. Seeds 0.5–1.0 mm long. $2n$=16. March–April.

Widespread in the central and southern part of our state; distributed from Maryland to Iowa and central Kansas, south to the Florida panhandle and Texas. Frequent in sunny, often seepy sites in woodlands, glades on various substrates, meadows, river or spring margins, mowed lawns, pastures, and roadsides, on poor to rich soils.

This species is often ephemeral, sometimes present in great numbers one year and then not seen again. It is often introduced into new sites in transported soil. The relatively large capsules sometimes look out of place on this small, slender plant. Especially in fruit, with its sometimes rather bilobed capsules, this species may bear a superficial resemblance to some of the annual species of *Veronica* (Plantaginaceae).

Steyermark (1963) treated this taxon as two species, *H. pusilla* and *H. minima*, based on minor differences in calyx morphology. The latter name is of interest in having been described based on material from the St. Louis area, but otherwise its relatively long and broad calyx lobes represent merely an endpoint of continuous variation in calyx shape and size. As Steyermark noted, this is the most commonly encountered phase in Missouri, but he also noted the continuous variation among plants in the state. Terrell (1996) did not recognize *H. minima* as a separate species, treating it a as mainly western and northwestern variant of *H. pusilla* that also occurs sporadically farther eastward and sometimes in mixed collections.

This species is quite variable in the size of plants and flowers. It is also notably variable in flower color (Terrell, 1996). Steyermark (1963) recognized several forms of this species based on these color variations, including white-flowered mutations (*H. pusilla* f. *albiflora* Standl., *He. minima* f. *albiflora* Lathrop) that can exist as entirely white-flowered populations (Terrell, 1996), and a pink-flowered variant (*H. pusilla* f. *rosea* Steyerm.).

7. Houstonia rosea (Raf.) Terrell (rose bluets, pygmy bluets)
Hedyotis pygmaea Roem. & Schult.
He. rosea Raf.

Map 2556

Plants annual, not producing rhizomes. Stems usually few to several, 1–4(–7) cm long, loosely to more strongly ascending, sometimes from a more or less spreading base, glabrous, somewhat roughened, or minutely hairy. Leaves opposite (plants not producing a basal rosette, but lower nodes sometimes relatively closely spaced), not succulent. Stipules 0.8–1.0 mm long, oblong, truncate at the tip, entire along the margins. Leaf blades 3–20 mm long, 0.5–3.0 mm wide, oblanceolate to narrowly spatulate, rounded or broadly pointed at the tip, angled or tapered at the base, the margins entire and generally flat, the surfaces glabrous or minutely hairy, without visible venation or sometimes with the midvein visible. Flowers homostylous (the anthers and stigmas borne at the same level in the lower part of the corolla tube), solitary in the axils of the uppermost pair of leaves (thus appearing terminal and sometimes in pairs), on stalks 6–12 mm long. Calyx lobes 1.0–2.5 mm long, usu-

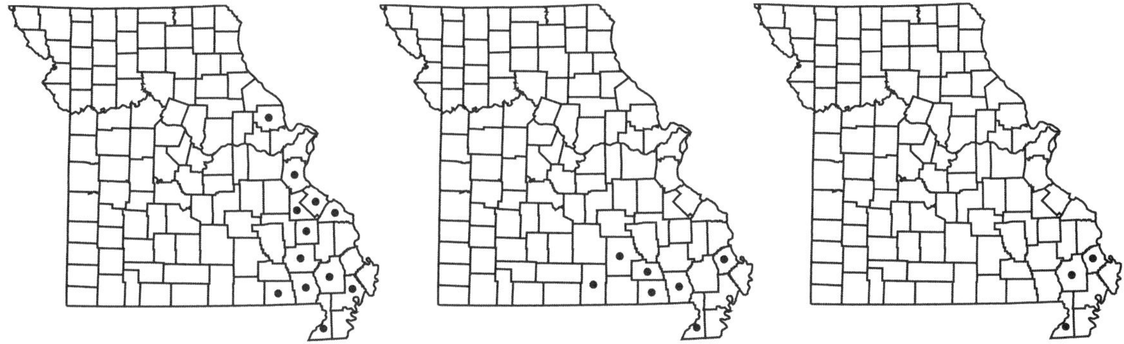

2557. Mitchella repens 2558. Oldenlandia boscii 2559. Oldenlandia uniflora

ally less than ½ as long as the adjacent corolla tube, triangular. Corollas trumpet-shaped, white to pink, lacking a ring of contrasting color in the throat, externally glabrous, internally pubescent in the upper part of the tube, the tube 3.5–8.0 mm long, the lobes 2–4 mm long, lanceolate. Fruits 1.5–3.5 mm long, 2.5–4.0 mm wide, subglobose to depressed-obovoid, somewhat flattened laterally, partially inferior, sometimes slightly 2-lobed, glabrous or nearly so. Seeds 0.5–1.0 mm long. $2n$=16. March–April.

Uncommon, known thus far only from historical specimens from Dunklin county (Alabama to Texas north to Missouri and Oklahoma). Openings of bottomland forests.

This species was first reported from Missouri by Terrell (1996). W. H. Lewis (2006, as *Hedyotis*) noted that the locality is about 175 miles to the north of the closest sites in central Arkansas. He viewed the Missouri specimens as outliers in the distribution and suggested that the species possibly had become extirpated from the state since its original collection in 1931.

The diminutive plants of *H. rosea* are dwarfed by their flowers. An unusual feature of this species compared with the closely related *H. micrantha* and *H. pusilla* is that the seeds are more or less ellipsoid, with an elliptic opening to the concave side and a ridge extending along the middle of the pit (vs. nearly globose, with a more or less circular opening to the concave side and lacking a ridge in the pit).

5. **Mitchella** L.

Two species, North America, Asia.

1. Mitchella repens L. (partridge berry, twinberry)

Pl. 551 d, e; Map 2557

Plants perennial herbs. Stems 10–50 cm long, prostrate, creeping (sometimes pendant from bluff ledges), usually rooting at the nodes, often forming mats, rounded or sometimes channeled on 2 opposing sides, minutely pubescent with short, straight to curved hairs or nearly glabrous. Leaves usually evergreen, opposite, with petioles 1–15 mm long, somewhat thickened, somewhat leathery, slightly succulent. Stipules interpetiolar, 0.5–1.0 mm long, triangular, generally persistent, sharply pointed at the tip or sometimes divided into 3 slender lobes, these often minutely gland-tipped. Leaf blades 5–20 mm long, 5–20 mm wide, broadly ovate to nearly circular, rounded to broadly pointed at the tip, sometimes with a small sharp point at the very tip, broadly rounded to truncate or slightly cordate at the base, the margins glabrous, flat or occasionally very slightly rolled under, the upper surface green to dark green, glabrous, usually glossy, the undersurface green, glabrous, the venation with the midvein and usually 2 or 3 pairs of pinnate, secondary veins visible (the midvein often somewhat whitened on the upper surface). Inflorescences terminal and occasionally also axillary, of 2(3) flowers at the tip of a common stalk 1–4 mm long. Flowers distylous (the anthers and stigmas borne at different levels in the upper half of the corolla tube), the hypanthia of each pair fused into a single unit. Calyces deeply (3)4(5, 6)-lobed, the lobes 0.5–1.0 mm long, broadly triangular. Corollas funnelform, white, externally glabrous, the tube 7–10 mm long, densely bearded in the throat with the pubescence usually extending onto the lobes, the lobes (3)4(5, 6), 3–4 mm long, ovate-triangular, spreading to slightly recurved, not overlapping in bud. Stamens (3)4(5, 6), attached in the corolla throat, in short-styled flowers the anthers well exerted, in long-styled flowers the anthers situated in the throat and not or only partially visible from the outside of the corolla. Ovaries of each flower pair(trio) fused, inferior, each 4-locular, the ovules 1 per locule. Style 1, slender, terminating in 4, short, linear stigmas, in short-styled flowers these positioned near the midpoint in the corolla tube, in long-styled flowers well-exserted. Fruits of the 2(3) flowers in each inflorescence fused into a single unit with a pair of minute, persistent calyces at the tip, berry-like drupes, 6–9 mm long, 7–10 mm wide, subglobose to depressed-globose, bright orange to red, with a total of 4–8(–12) small stones. $2n=22$. May–July.

Scattered in the Mississippi Lowlands and the eastern portion of the Ozark and Ozark Border Divisions, north locally to Lincoln County (eastern U.S. west to Minnesota and Texas; Canada, Mexico, Central America). Ledges and bases of bluffs, bottomland forests, and banks of rivers, often in sandy soils or on sandstone substrates.

The fused ovaries of the flower pairs are unusual and give rise to the common name twinberry. Very rarely, the fruits are even formed from three flowers. Other occasional, aberrant fruits have a pair of rudimentary leaves developed from the sides. The fruits of *Mitchella* are often called berries due to their size and juicy texture, but are actually drupes with stones (the seeds each enclosed in a hard covering that forms around it from the ovary wall). A rare form with white fruits has been called f. *leucocarpa* Bissell, but has not yet been found in Missouri. Another sporadic mutation, with the flowers completely fused throughout their corollas, remains unnamed and also has not yet been found in Missouri.

Gleason and Cronquist (1991) referred to the fruits as insipid. As noted by Steyermark (1963), this is an attractive plant, which he recommended be cultivated in a terrarium. It apparently is not easily grown in gardens.

6. **Oldenlandia** L.
(Terrell and Robinson, 2006)

Plants annual or perennial herbs, the rootstock sometimes somewhat woody. Stems erect or ascending to loosely ascending from a spreading base or prostrate and creeping, rounded or somewhat 4-angled, often channeled on 2 opposing sides. Leaves opposite, sessile or nearly so. Stipules interpetiolar, oblong to triangular, persistent but sometimes becoming degraded with age, membranous, fused to the leaf bases on either side, the margin with 2–5 slender lobes. Inflorescences axillary and/or terminal, of solitary flowers or small clusters of 2–4 flowers. Flowers sessile or short-stalked, the stalks sometimes with small bracts at the bases, homostylous or distylous. Calyces deeply 4-lobed. Corollas shorter than the calyces, saucer-shaped, white or occasionally pale pink, the 4 short lobes longer than the minute tube, not overlapping in bud. Stamens 4, attached in the corolla tube, the anthers more or less exserted. Ovary fully inferior, 2-locular, the ovules several to numerous in each locule. Style 1, slender, 2-branched, with long, linear stigmas, these more or less exserted. Fruits capsules, dehiscent longitudinally from the tip. Seeds several to numerous, minute (about 0.2 mm long and wide), irregularly angular (often appearing somewhat triangular in outline), sometimes slightly flattened, the surface with a network of very fine ridges and pits, purplish black to black. About 100 species, tropical and warm-temperate regions nearly worldwide.

This genus has sometimes been included within a broadly circumscribed *Hedyotis* (W. H. Lewis, 1961, 1962; Yatskievych and Turner, 1990; Gleason and Cronquist, 1991). For a discussion of problems of generic delimitations in *Hedyotis* and its relatives, see the treatment of that genus. The recent molecular phylogenetic analysis by Groeninckx et al. (2009) indicates that the genus *Oldenlandia* in the broad sense is highly unnatural (polyphyletic), with the one North American species sampled (*O. uniflora*) grouping with some morphologically similar African *Oldenlandia* species and the Old World species of *Hedyotis* in the strict sense (rather than with the New World members of *Houstonia* or with the group that includes

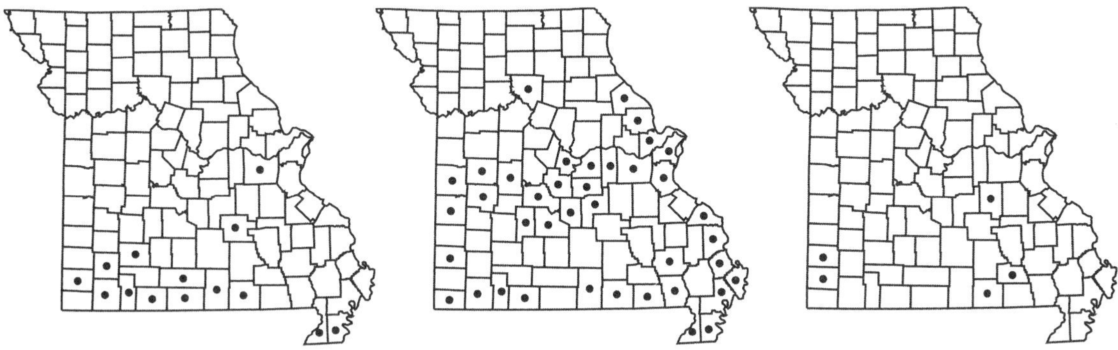

2560. Sherardia arvensis   2561. Spermacoce glabra   2562. Citrus trifoliata

O. corymbosa L., a neotropical species that is the type of the genus). However, more species will need to be sampled before the classification of *Oldenlandia* becomes stabilized. The two Missouri species are retained in the genus for now, but with the knowledge that in the future they will most likely be reclassified.

1. Stems and leaves glabrous; leaf blades 1–2 mm wide, linear to very narrowly oblong-elliptic; fruits roughened with minute tubercles or occasionally minutely hairy . . . . 1. O. BOSCII
1. Stems and leaves pubescent with short, soft hairs, at least on young growth; leaf blades 2–12 mm wide, elliptic to lanceolate; fruits pubescent with short, soft hairs . . . . 2. O. UNIFLORA

1. Oldenlandia boscii (DC.) Chapm.
Hedyotis boscii DC.

Pl. 550 c, d; Map 2558

Plants perennial herbs, the rootstock sometimes somewhat woody. Stems 5–20 cm long, loosely ascending from a spreading base or prostrate and creeping, glabrous. Stipules with the body 0.5–1.0 mm long, the lobes 2–5, 0.5–2.0 mm long. Leaf blades 7–22 mm long, 1–2 mm wide, linear to very narrowly oblong-elliptic, angled at the base, angled to a sharply pointed tip, the margins glabrous, flat or slightly curled under, the surfaces glabrous, the venation with only the midvein visible. Inflorescences axillary at most of the nodes and sometimes also terminal, of solitary or paired flowers. Flowers sessile or nearly so (the stalk becoming elongated to about 1 mm at fruiting). Calyx lobes 1.0–1.5 mm long, lanceolate to narrowly lanceolate-triangular, tapered at the tips. Corolla lobes 0.8–1.0 mm long, white or occasionally pale pink. Fruits 2.0–2.5 mm long, 2.0–2.5 mm wide, oblong-ellipsoid, somewhat flattened laterally, roughened with minute tubercles or occasionally minutely hairy. $2n=36$. June–October.

Uncommon in the Mississippi Lowlands Division and the southeastern portion of the Ozarks (New York to Florida west to Oklahoma and Texas). Moist swales of sand prairies, margins of ponds and sinkhole ponds; also borrow pits, railroads, roadsides, and open, wet, disturbed areas.

This taxon is easily confused with some genera of Lythraceae, such as *Rotala ramosior* (L.) Koehne (toothcup), because of its similar habit, axillary flowers, and occurrence in similar habitats.

2. Oldenlandia uniflora L.
Hedyotis uniflora (L.) Lam.

Pl. 550 a, b; Map 2559

Plants annual. Stems 5–25 cm long, erect or ascending, occasionally loosely ascending from a spreading base, moderately pubescent with short, soft hairs, at least on young growth. Stipules with the body 1.2–2.0 mm long, the lobes 2, 1.5–3.0 mm long. Leaf blades 5–20 mm long, 2–12 mm wide, elliptic to lanceolate, angled to broadly angled at the base, angled to a bluntly or sharply pointed tip, the margins minutely roughened or hairy, flat, the surfaces sparsely to moderately pubescent with short, soft hairs, the venation with the midvein and often 1 or 2 faint pairs of pinnate, lateral veins visible. Inflorescences clusters of 2–4 flowers at the upper nodes and usually also terminal, the terminal clusters appearing headlike and subtended by leafy bracts. Flowers short-stalked, the stalks 0.5–1.0 mm long at flowering (becoming elongated to 1–2 mm at fruiting). Calyx lobes 1.0–1.2 mm long (usually becoming enlarged to 2–3 mm at fruiting), lanceolate to narrowly lanceolate-triangular, tapered at the tips. Corolla lobes 0.8–1.0 mm long, white. Fruits 1.0–1.5 mm long, 2.0–2.5 mm wide, subglobose to somewhat depressed-globose, somewhat flattened laterally, densely pubescent with short, soft hairs. $2n$=36, 72. June–October.

Uncommon in the Mississippi Lowlands Division (New York to Florida west to Oklahoma and Texas; Caribbean Islands). Moist swales of sand prairies, banks of spring branches, and margins of ponds, often in sandy soil.

7. **Sherardia** L.

One species, Europe, Asia, introduced widely.

In its whorled leaves without apparent stipules *Sherardia* is similar to *Galium*. It differs in its headlike inflorescences, pink corollas with a well-developed tube, and schizocarpous fruits. For comments about this leaf arrangement, see the discussion under that genus.

1. Sherardia arvensis L. (field madder)
Pl. 551 a–c; Map 2560

Plants annual herbs, glabrous or more commonly minutely roughened or pubescent throughout with wavy, soft hairs, the stem nodes sometimes ringed with dense, short hairs. Stems 5–40 cm long, spreading to loosely ascending from a spreading base, rarely strongly ascending, sometimes forming loose mats, 4-angled. Leaves in whorls of (4)6, sessile or nearly so. Stipules absent. Leaf blades 4–20 mm long, 1.5–3.0 mm wide, narrowly elliptic to nearly linear, angled at the base, angled to a sharply pointed tip, the margins entire, flat to slightly rolled under, glabrous or minutely roughened or hairy, the upper surface glabrous to slightly roughened or sparsely soft-hairy, the undersurface soft-hairy, not glandular, the venation with only the midvein visible. Inflorescences terminal and axillary, of small headlike clusters that are closely subtended by whorls of (4)6 or 8, small, leaflike, involucral bracts. Flowers sessile or nearly so, 2–4 per cluster. Calyces 0.5–1 mm long, minutely 4-lobed, the lobes triangular. Corollas 4.0–5.5 mm long, trumpet-shaped, the slender tube well-developed, longer than the 4 spreading lobes, pink (blue elsewhere), the lobes not overlapping in bud. Stamens 4, attached in the corolla throat, the anthers exserted. Style unequally 2-branched, the stigmas capitate. Ovary fully inferior, 1-locular, the ovules 1 per locule. Fruits schizocarps, 1–2 mm long, 2–4 mm wide, 2-lobed, splitting from the base into 2 mericarps, these globose or nearly so, indehiscent. $2n$=22. April–June.

Introduced, scattered in the southern portion of the Ozark and Mississippi Lowlands Divisions (native of Europe, Asia; introduced in the eastern U.S. west to Iowa and Texas, also Washington to California, Nevada, and Arizona, Hawaii; Canada, Mexico). Glades, ledges of bluffs, and banks of streams and rivers; also pastures, cemeteries, lawns, gardens, sidewalks, roadsides, and open, disturbed areas.

Steyermark (1963) knew this weedy species only from a few collections from Barry and Newton Counties. It has become much more widely distributed in southern Missouri since that time.

8. **Spermacoce** L.

About 30–250 species (depending on generic circumscription), in the broad sense nearly worldwide.

The genus *Spermacoce* has been delimited in several very different ways. There is much variation in the fruits of the group, and this has been given differing weights by various authors. In general *Spermacoce* and its relatives are herbs and low shrubs with bristly stipules and 2-seeded, small, dry fruits. *Diodia* of our flora is related; this genus is distinguished by its fruits that split into two parts, with each part indehiscent, and is native to the New World. About 100 other species found in tropical and warm-temperate regions around the world are similar to *Diodia*, but have fruits that dehisce to some extent, or sometimes do not dehisce or

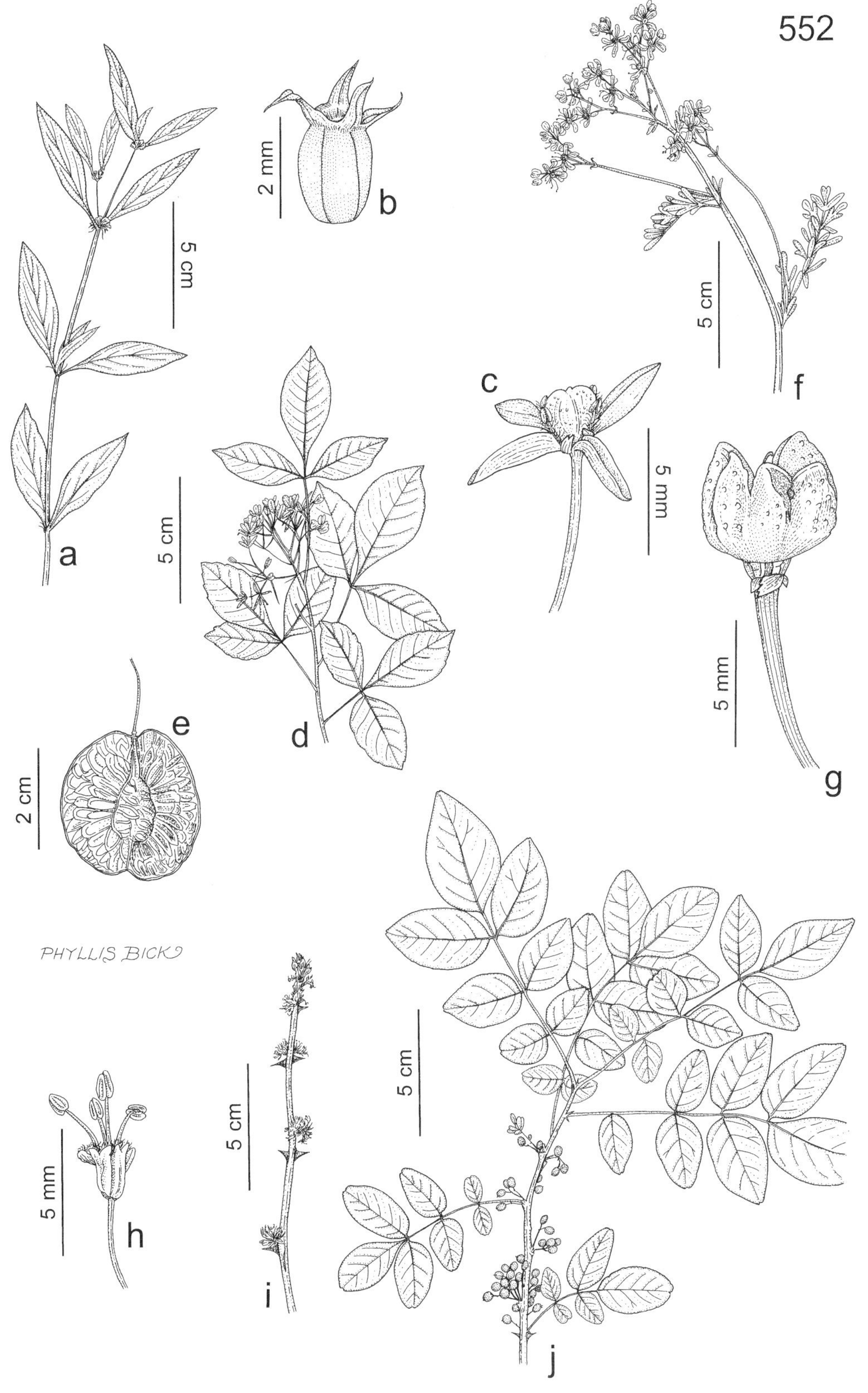

Plate 552. Rubiaceae. Rutaceae. *Spermacoce glabra*, **a)** fertile stem, **b)** fruit. *Ptelea trifoliata*, **c)** flower, **d)** fertile stem, **e)** fruit. *Ruta graveolens*, **f)** fertile stem, **g)** fruit. *Zanthoxylum americanum*, **h)** flower, **i)** flowering stem, **j)** fruiting branch.

split apart at all. Some authors, led by Verdcourt (1958, 1976) and Fosberg et al. (1993), have considered the fruit type to be relatively unimportant for classification in this group, and have included all these species in one cosmopolitan genus, *Spermacoce*. However, other authors, including Steyermark (1963) and Bacigalupo and Cabral (1999), have considered fruit type very important in this group, and have used it along with several other features, especially pollen morphology (which is quite variable among these plants), to separate various genera.

Bacigalupo and Cabral (1999) included in their version of the genus *Spermacoce* only about 30 New World tropical species with unusual fruits that are asymmetrical, with the larger half opening to release its seed and the smaller half remaining closed (dispersing with its seed still inside). In their classification, the Missouri species of *Spermacoce*, which does not have this fruit type, was segregated into its own genus, *Spermacoceodes* Kuntze (although the Caribbean species *Tobagoa maleolens* Urb. has similar fruits and may be closely related). Until this entire group is better understood through careful worldwide study, the generic classification will not be agreed upon and consequently will continue changing. Therefore, the genus *Spermacoce* is retained in the traditional sense in the present treatment to include the Missouri species.

1. Spermacoce glabra Michx. (buttonweed)
Spermacoceodes glabrum (Michx.) Kuntze
Pl. 552 a, b; Map 2561

Plants annual herbs, glabrous at maturity or sometimes somewhat roughened. Stems 10–60 cm long, erect to loosely ascending or trailing, 4-angled or often longitudinally 6-ridged or slightly channeled on opposing sides, glabrous at maturity or sometimes somewhat roughened along the angles. Leaves opposite, sessile or nearly so. Stipules interpetiolar, papery to scalelike, generally truncate to broadly rounded at the tips, usually persistent with the leaves (but sometimes becoming shredded with age), fused to the leaf bases on either side, the sheath 1.5–3.5 mm long, membranous, bearing along its margin 5–7 bristles, these 1.5–6.0 mm long, usually somewhat unequal with the middle ones longer. Leaf blades 2–8 cm long, 3–24 mm wide, narrowly elliptic, angled at the base, angled to a sharply pointed tip, the margins flat, entire, sometimes with minute spinules angled toward the leaf tip, the surfaces glabrous, not glandular, the venation with the midvein and 2 or 3 pairs of secondary veins visible. Inflorescences axillary at mostly the upper stem nodes, dense sessile clusters of 3–20 flowers. Calyces deeply 4-lobed, the lobes 1.5–2.0 mm long, narrowly triangular, glabrous. Corollas 2.5–4.0 mm long, funnelform, 4-lobed to about the midpoint, white, the outer surface glabrous, densely bearded in the throat with the pubescence often extending onto the upper surface of the lobes, these not overlapping in bud. Stamens 4, attached in the corolla throat, partially exserted. Stigmas 2, linear, not exserted. Ovary fully inferior, 2-locular, the ovules 1 per locule. Fruits achenelike, 3–4 mm long, 2–3 mm wide, obconic and somewhat flattened laterally, indehiscent, the surface smooth, leathery to stiffly papery. $2n=28$. June–October.

Scattered in the southern half of the state, north locally to Boone and Pike Counties (southeastern U.S. west to Kansas and Texas). Banks of streams and rivers, margins of ponds, lakes, sinkhole ponds, and oxbows, bottomland forests, bottomland prairies, sloughs, and moist depressions of sand prairies; also ditches and wet roadsides.

This species is superficially similar to members of *Lycopus* L. (bugleweed) in the Lamiaceae, which also has dense axillary clusters of small white flowers. *Spermacoce glabra* differs in its ovary position (inferior vs. superior); fruit type (unlobed and achenelike vs. nutlets), deeply lobed, unnerved calyces (vs. shallowly lobed or to about the midpoint and strongly nerved), and 4 (vs. 2) stamens, among other characters.

RUTACEAE (Citrus Family)

Plants mostly shrubs or trees (perennial herbs in *Ruta*), sometimes more or less dioecious, with internal glands containing aromatic, volatile oils (these visible externally as dots on twig, leaf, flower, and often also fruit tissue, imparting a strong odor when plant parts are bruised or crushed), sometimes armed with thorns or prickles. Stems or trunks 1 to several from the base, erect or ascending, usually branched. Leaves opposite or alternate, petiolate. Stipules absent (subopposite prickles appearing as pseudostipules in *Zanthoxylum*). Leaf blades trifoliate or pinnately compound (sometimes merely deeply lobed in *Ruta*), the leaflets (or lobes) usually sessile, variously entire, toothed, scalloped, or deeply lobed. Inflorescences terminal or axillary, many-branched panicles or small clusters, sometimes reduced to solitary flowers, the branch points often with small bracts, these sometimes reduced to glandular structures. Flowers perfect or imperfect, hypogynous, actinomorphic, sessile to long-stalked. Calyces of 4 or 5(–7) sepals (absent in some *Zanthoxylum* species), these free or fused at the base, usually overlapping in bud, ascending, not persistent at fruiting. Corollas of 4 or 5(–7) petals, these free, white, greenish white, greenish yellow, or cream-colored. Stamens 4–8 or numerous, the filaments distinct or fused at the base, the anthers attached toward the dorsal midpoint or near the base, yellow; staminodes absent, except sometimes in pistillate, imperfect flowers. Nectar disc present, more or less flat to cup-shaped. Pistil(s) 1 or 3–5 per flower, of 3–8 (more elsewhere) fused or separate carpels, sometimes slightly elevated above the perianth at flowering (the stalks sometimes continuing to elongate as the fruits mature). Ovaries superior, 1-locular (when more than 1 per flower) or 4–8-locular (when 1 per flower, then often indented at the tip), the placentation respectively axile or lateral. Style usually 1 per flower (even flowers with multiple ovaries usually have fused styles), sometimes with longitudinal lines or ridges equal to the carpel number, the stigmas capitate, sometimes lobed. Ovules (1)2 to several per carpel. Fruits follicles, capsules, samaras, or hesperidia (other types elsewhere), variously shaped. Seeds various in size and appearance. About 158 genera, about 1,900 species, nearly worldwide.

The family Rutaceae contains the most widely cultivated tropical fruit trees in the world (see the treatment of the genus *Citrus*), a few of which also contain minor timber trees. A number of genera have been used medicinally. Some members of the family also provide scented extracts used in perfumes, soaps, and creams, as well as in mosquito repellents, and stain and glue removers. A few genera contain species cultivated as ornamentals.

1. Plants perennial herbs; leaf blades pinnately compound or deeply lobed, the leaflets again deeply pinnately lobed . . . . 3. RUTA
1. Plants shrubs or trees; leaf blades ternately or pinnately compound, the leaflets unlobed (the margins entire, somewhat wavy or finely toothed to scalloped, not lobed
    2. Stems armed with thorns or prickles
        3. Stems (even the older branches ant trunks) green; branches armed with thorns, these solitary at the nodes, large and stout, strongly flattened; leaf blades with 3 leaflets; flowers perfect; fruits 1 hesperidium per flower, this globose, 30–45 cm in diameter . . . . 1. CITRUS
        3. Stems (except the young twigs) dark brown; branches armed with prickles, these solitary (and frequently sparse) between the nodes and more or less paired at the nodes, stout, strongly flattened; leaf blades with 5–13 leaflets; flowers imperfect (plants dioecious); fruits 2–5 follicles per flower, these ellipsoid, 4–6 mm long . . . . 5. ZANTHOXYLUM
    2. Stems unarmed
        4. Leaves alternate; leaf blades ternately compound; fruits samaras, indehiscent and encircled by a pair of longitudinal wings . . . . 2. PTELEA
        4. Leaves opposite (occasionally a few leaves subopposite); leaf blades pinnately compound with 5–11 leaflets; fruits follicles, dehiscent longitudinally and unwinged . . . . 4. TETRADIUM

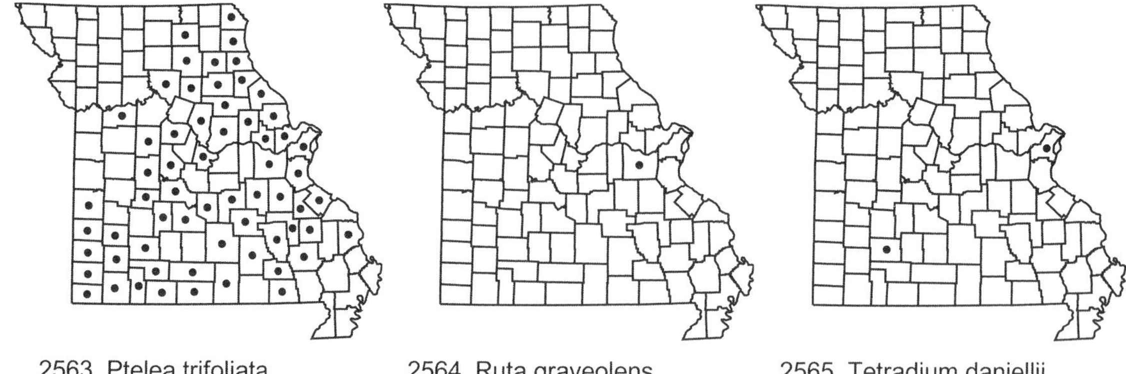

2563. Ptelea trifoliata 2564. Ruta graveolens 2565. Tetradium daniellii

1. Citrus L. (citrus)

Sixteen to 25 species, Asia, Australia, New Caledonia.

The citrus fruits, including orange, lemon, lime, grapefruit, tangelo, pomello, bergamot, and others, are extremely important agricultural crops, providing fruits, juices, and flavorants for human consumption. A number of species also are cultivated as ornamentals for their fragrant, attractive flowers and brightly colored fruits (in the Midwest, this principally comprises dwarf cultivars that are grown in pots). Other uses of various species in the genus include as a bitter additive in some tonic waters (in place of quinine), as raw material for the artificial sweetener naringin, medicinally to raise blood pressure, to counteract (and prevent) the symptoms of scurvy, and in some weight loss preparations.

1. Citrus trifoliata L. (trifoliate citrus, bitter orange)
Poncirus trifoliata (L.) Raf.
Map 2562

Plants dense shrubs (rarely becoming small trees elsewhere), 1–5 m tall. Stems much-branched, the bark relatively smooth, green, at least the younger branches somewhat flattened and/or angled, armed with large, stout, flattened thorns produced from primary leaf axils, glabrous. Leaves alternate or appearing whorled, the petiole shorter than to about as long as the blade, jointed at the tip, narrowly winged. Leaf blades trifoliate (simple in young plants), the leaflets 2–5 cm long, obovate-elliptic, the lateral pair usually somewhat oblique, angled or short-tapered to a sessile base, rounded or bluntly and broadly pointed at the tip, the margins entire or minutely scalloped, the surfaces glabrous or the undersurface minutely hairy along the midvein (especially when young). Inflorescences axillary on older portions of branches, of solitary or paired flowers, these nearly sessile. Flowers perfect. Sepals (4)5(–7), 3–6 mm long, oblong-ovate, glabrous, not persistent at fruiting. Petals (4)5(–7), 15–30 mm long, obovate, overlapping, white, glabrous. Stamens numerous, the filaments free, attached to a flat to cup-shaped, hairy nectar disc encircling the ovary base. Ovary 6–8-locular, densely and finely hairy, each locule with 4–8 ovules, short-tapered to a short, stout style, the stigma capitate. Fruits hesperidia, 3.0–4.5 cm in diameter, globose or somewhat pear-shaped, with 6–8 locules, the matrix relatively dry, containing a network of cobwebby hairlike structures, the rind relatively thin, finely wrinkled to nearly smooth, densely and finely hairy, yellow to orangish yellow at maturity. Seeds usually numerous, 8–12 mm long, more or less ovoid, plump, the surface leathery, tan to bone-colored, smooth or faintly ridged. $2n=18$. April–June.

Introduced, uncommon, known thus far from widely scattered sites in the Ozark Division (native of Asia; introduced widely from Pennsylvania to Florida west to Missouri, Oklahoma, and Texas). Mesic upland forests and spring branches; also fence rows and disturbed areas.

This species can form dense, impenetrable thickets of thorny branches and is sometimes planted under windows for security purposes. It is a relatively recent escape in Missouri, the oldest collection known dating to 1959. However, it is known from well over half the counties in Arkansas (E. B. Smith, 1988) and may be expected to spread in southern Missouri as well. *Citrus trifoliata* is facultatively apomictic and thus produces viable seeds in every fruit. The attractive fruits produce small quantities of a very bitter and acidic juice.

The trifoliate citrus has often been treated as a segregate genus, *Poncirus* Raf., based on its unusual morphology, including deciduous, trifoliate leaves and winter buds with scales (vs. naked). However, this taxon can be crossed successfully with the more typical *Citrus* species, *C. aurantium* L. (yielding citrange, which is used as a rootstock onto which other economically important citrus taxa and cultivars can be grafted). A recent, comprehensive, molecular phylogenetic study of the orange subfamily (Rutaceae subfam. Aurantioideae Horan.) has shown that *C. trifoliata* is well embedded within a broadly defined concept of the genus *Citrus* (R. J. Bayer et al., 2009).

2. **Ptelea** L. (hop tree)
(V. L. Bailey, 1962)

Three species, U.S., Mexico.

1. Ptelea trifoliata L. **ssp. trifoliata** (common hop tree, stinking ash)

P. trifoliata var. *mollis* Torr. & A. Gray

Pl. 552 c–e; Map 2563

Plants shrubs or rarely small trees, incompletely dioecious (producing at least a few apparently perfect flowers), 1–2(–8) m tall. Stems with ascending branches, the bark relatively smooth or somewhat roughened, brown with prominent lenticels (appearing as short or longer, incomplete cross-ridges), occasionally developing a network of shallow, longitudinal ridges on large old trunks, unarmed, the twigs green, becoming brown, glabrous or sparsely pubescent with short, inconspicuous hairs, rarely densely hairy. Leaves alternate, long-petiolate, the petiole not jointed at the tip, unwinged (sometimes slightly expanded at the tip). Leaf blades trifoliate, the leaflets 2–15 cm long, the terminal leaflet longer than the lateral pair, ovate-elliptic, elliptic, rhombic, or oblong-ovate, the lateral pair somewhat oblique, angled or short-tapered to a sessile base, angled or tapered to a bluntly or sharply pointed tip, the margins minutely scalloped or toothed and sometimes slightly wavy, the upper surface green, glabrous, the undersurface pale green to light green, glabrous or sparsely hairy, mostly along the main veins, rarely densely hairy. Inflorescences terminal (sometimes on short branches and thus appearing lateral), rounded panicles appearing as loose, branched clusters, the individual flowers long-stalked. Flowers imperfect or less commonly perfect. Sepals 4 or 5, 1–2 mm long, ovate to oblong-ovate, minutely hairy, at least along the margins, not persistent at fruiting. Petals 4 or 5, 4–6 mm long, narrowly elliptic-ovate to narrowly oblong, overlapping in bud, usually minutely hairy, white. Stamens in staminate and perfect flowers 4 or 5 (reduced to 4 or 5 small staminodes in pistillate flowers), alternating with the petals, the filaments free, flattened, and densely hairy toward the base, attached to a lobed, glabrous nectar disc below the slightly elevated ovary base. Ovary 2-locular (3-locular elsewhere), flattened (reduced to a more or less conic rudiment in staminate flowers), glabrous, each locule with 2 ovules, rounded or short-tapered to a short, slender style, the stigma more or less 2-lobed. Fruits samaras, 2.0–3.0 cm long, with a broad lateral wing, this circular to obovate-circular, papery, with a network of strong veins, short-tapered to cordate at the base, shallowly to deeply notched at the tip, the body 5–7 mm long, obovate to nearly circular in outline, flattened, with 2 locules, the outer layer hard, glabrous, strongly veined, pale green to straw-colored at maturity. $2n=42$. April–June.

Scattered nearly throughout the state, but apparently absent from the Mississippi Lowlands Division and most of the western half of the Glaciated Plains (eastern U.S. west to Minnesota, Utah, and Arizona; Canada, Mexico). Glades, savannas, upland prairies, bases, ledges, and tops of bluffs, mesic to dry upland forests, banks of streams and rivers, margins of ponds and lakes, and occasionally bottomland forests; also fencerows, pastures, railroads, and roadsides.

V. L. Bailey (1962) reduced the 59 species accepted by Greene (1906) to only 3, but she subdivided *P. trifoliata* into a complex array of five subspecies and eleven total varieties. Of these, she mapped only ssp. *trifoliata* var. *trifoliata* for Missouri. Steyermark (1963) included both var. *trifoliata* and var. *mollis* in Missouri, the latter from a single historical specimen in Madison County. The two were stated to differ in the densely hairy (vs. glabrous or nearly so) leaf undersurfaces and branchlets of var. *mollis*. However, across the ranges of the two varieties (including Missouri), there is a great deal of intergradation between the pubescence extremes. Subsequently, V. L. Bailey et al. (1970) examined the geography, morphology, and chemotaxonomy of populations of ssp. *trifoliata* and determined that plants attributed to the various varieties had a strong correlation with habitat types. Most subsequent botanists have either treated *P. trifoliata* as a single polymorphic species without recognizing infraspecific taxa or have recognized at most the five subspecies.

V. L. Bailey (1962) also noted an occurrence of ssp. *polyadenia* (Greene) V.L. Bailey from Cleburne County, Arkansas, near the Missouri border. This subspecies differs from ssp. *trifoliata* in its leaves having relatively large, conspicuous gland-dots that are visible without magnification, as well as fruits that often are 3-winged and 3-locular. It occurs mostly in the southwestern United States, with scattered occurrences eastward to Oklahoma, Texas, and Arkansas. It might be discovered in Missouri in the future.

3. **Ruta** L. (rue)

Eight species, Europe, Asia, Africa, Macaronesia.

1. Ruta graveolens L. (common rue)
Pl. 552 f, g; Map 2564

Plants perennial herbs (shrubs elsewhere), 0.3–1.0 m tall. Stems with ascending branches, unarmed, glabrous. Leaves alternate, the lowermost usually long-petiolate, the petioles progressively shorter to the stem tip, the uppermost leaves often sessile, the petiole not jointed at the tip, unwinged. Leaf blades 2–12 cm long, pinnately compound or deeply lobed, the leaflets or primary segments on all but the uppermost leaves again deeply pinnately lobed, the ultimate segments 0.3–1.5 cm long, oblanceolate to oblong-oblanceolate, mostly somewhat narrowed toward the base, rounded or broadly and bluntly pointed at the tip, the margins entire or minutely scalloped, the upper surface green to bluish green, glabrous, sometimes slightly glaucous, the undersurface similar but often slightly lighter green. Inflorescences terminal, more or less flat-topped panicles, the individual flowers short- to long-stalked. Flowers perfect. Sepals 4 or 5, 1–3 mm long and ovate to narrowly ovate at flowering, persistent, becoming elongate to 4–5 mm, and narrowly triangular to lanceolate-triangular at fruiting, glabrous. Petals 4 or 5, 4–7 mm long, narrowly spatulate to narrowly oblong-obovate, overlapping in bud, narrowed to a stalklike basal portion, incurved and somewhat hooded toward the rounded tip, glabrous, greenish yellow to yellow. Stamens 8 or 10, in 2 unequal, alternating series, the filaments free, slender, slightly expanded toward the base, attached to a cushionlike, glabrous nectar disc below the slightly elevated ovary base. Ovary 4- or 5-locular, deeply 4- or 5-lobed from the tip, glabrous, each locule with usually numerous ovules, the slender style attached at the concave ovary tip, the stigma minute, more or less 4- or 5-lobed. Fruits capsules, 5–7 mm long, 6–8 mm wide, 4- or 5-lobed from the tip to about the midpoint, each or the blunt, hornlike lobes dehiscent along the inner suture from the tip to the fusion point, the outer surface leathery, slightly wrinkled and/or pitted, yellowish green, glabrous. Seeds few to several per locule, 1.3–2.0 mm long, more or less semicircular (asymmetrically elliptic) in outline, wedge-shaped to triangular in cross-section, the outer surface with dense, small tubercles, yellowish brown to dark brown or nearly black, somewhat shiny. $2n=72$, 81. June–July.

Introduced, uncommon, sporadic (native of Europe; widely cultivated but introduced sporadically in the New World). Open disturbed areas.

The Missouri collections are historical and it is likely that the species does not persist long outside of cultivation. The common rue has a long history of use as a garden ornamental, spice, and medicinal plant. However, Brizicky (1962) noted that the volatile oil has unpleasant side effects (gastric inflammation) when ingested in sufficient quantity and that the species should be considered poisonous. He also noted that the foliage causes dermatitis in some individuals. Poutaraud et al. (2000) noted that among the array of secondary metabolites produced in this species are furanocoumarins, which induce photodermatitis. They investigated methods of cultivating and harvesting *R. graveolens* to maximize the yield of furanocoumarins, which are used in treating psoriasis and some types of skin cancers, as well as in the treatment of multiple sclerosis.

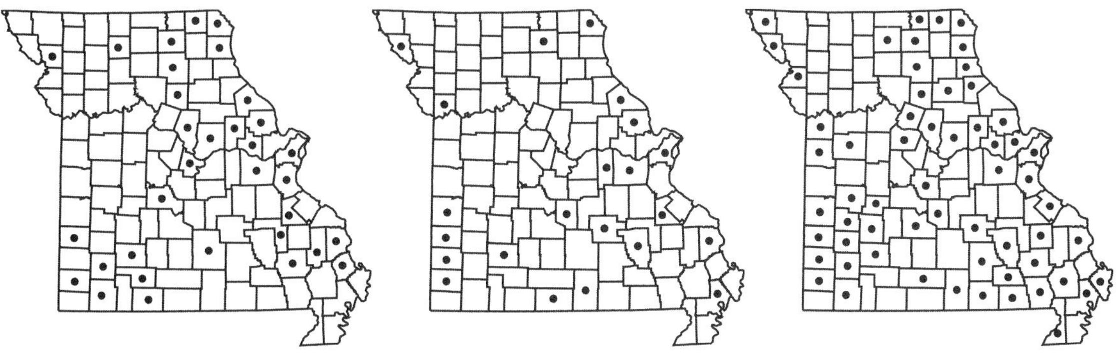
2566. Zanthoxylum americanum 2567. Populus alba 2568. Populus deltoides

4. **Tetradium** Lour.

Nine species, Asia.

1. Tetradium daniellii (Benn.) T.G. Hartley (Korean evodia)
Euodia daniellii (Benn.) Hemsl. (often spelled *Evodia*)

Map 2565

Plants small trees, dioecious, 5–12 m tall. Stems with ascending branches, the bark relatively smooth, gray to dark gray with scattered, lighter gray and brownish blotches, becoming somewhat roughened on old trunks, unarmed, the twigs olive brown, densely short-hairy when young, becoming gray and glabrous with age. Winter buds, naked, partially sunken into the petiole base (but mostly exposed). Leaves opposite or occasionally a few subopposite, mostly short-petiolate, the petiole not jointed at the tip, unwinged. Leaf blades 15–38 cm long, pinnately compound with 5–11 leaflets, these 5–13 cm long (progressively longer from the basal pair), oblong-elliptic to elliptic or ovate, the lateral pairs usually somewhat oblique, rounded to short-tapered to a sessile or short-stalked base, short-tapered or tapered to a sharply pointed tip, the margins minutely scalloped, the upper surface green, glabrous or sparsely and inconspicuously hairy toward the base of the midvein, the undersurface pale green to light green, sparsely and inconspicuously hairy when young, generally appearing glabrous at maturity. Inflorescences terminal, much-branched, dome-shaped panicles, the individual flowers short-stalked. Flowers imperfect. Sepals (4)5, 0.5–1.5 mm long, broadly triangular, fused toward the base, the margins hairy, usually persistent at fruiting. Petals (4)5, 3–5 mm long, narrowly oblong-ovate to narrowly oblong-elliptic, overlapping in bud, white to cream-colored, mostly with thin, minutely uneven margins, sparsely hairy on the upper surface. Stamens in staminate flowers (4)5 (reduced to short, strap-shaped staminodes in pistillate flowers), alternating with the petals, the filaments free, slender, sparsely hairy toward the base, attached to a more or less cup-shaped to short-cylindric nectar disc around the ovary base. Ovary (4)5-locular, deeply (4)5-lobed from the tip (reduced to a more or less lobed rudiment in staminate flowers), sparsely hairy toward the tip, each locule with 2 ovules, the short, stout style attached at the concave ovary tip, the stigma more or less disc-shaped. Fruits (4)5 follicles per flower, these fused at the base or nearly separate, 6–11 mm long, asymmetrically obovoid to ellipsoid, tapered or narrowed to the fused basal portion, abruptly tapered to a minute, oblique beak at the tip, the outer surface leathery, smooth to finely pitted, red to dark red or nearly black, sometimes brownish yellow with a pinkish tinge, glabrous or nearly so. Seeds 1 or 2 per fruit, 2.5–4.0 mm long, usually only 1 of the pair fertile, oblong-ellipsoid, the outer surface smooth, black, shiny. $2n=78$. June–August.

Introduced, uncommon, known thus far only from Greene County (native of Asia; introduced uncommonly thus far in Missouri, Ohio, and Pennsylvania). Disturbed mesic upland forests.

This species was first reported for Missouri by Bowe and Redfearn (2002, as *Evodia*). Vincent (2004b) discussed its recent spread in the United States. Likely, this species will become more widely naturalized in the future. Vincent (2004b) also noted that in its native range, the species has a long history of medicinal use for intestinal ailments, arthritis, and other problems. He further noted that the species is highly prized as a bee plant, especially as the trees flower heavily at a time of year when relatively few other resources are available for honeybees.

Another Asian taxon in the Rutaceae that likely will become a member of the Missouri flora in the near future is *Phellodendron amurense* Rupr. (Amur corkbark tree). In 2008, Glenn Beffa of the staff of the Shaw Nature Reserve (Franklin County) reported a large individual persisting from an old planting at a presently undeveloped site at the property. He noted further that seedlings were present under and around the mature tree. Shortly thereafter, Steve Buback of Forest Park (St. Louis City) reported a similar situation at that property. There has been an effort to eradicate individuals at Shaw Nature Reserve, which is managed as a natural landscape, and as of this writing Forest Park is considering a similar course of action. Amur corkbark tree is cultivated as a shade tree in temperate North America and has escaped in scattered eastern and Midwestern states, as well as in eastern Canada. In some eastern states, it is considered an invasive exotic. Like *Tetradium, P. amurense* is a small to medium-sized, dioecious tree with large, opposite, pinnately compound leaves and large panicles of numerous small flowers terminal on the branches. It differs from *Tetradium* in its relatively thick, corky bark, fruits that are fleshy, drupelike berries, leaves with 7–13 leaflets, and in having axillary buds that are concealed in the hollow, expanded bases of the petioles.

5. **Zanthoxylum** L. (prickly ash)
(D. M. Porter, 1976)

About 200 species, North America to South America, Caribbean Islands, Africa, Asia to Australia, and Pacific oceanic islands.

A number of *Zanthoxylum* species are used medicinally by various cultures within their native ranges, principally for antimicrobial and pain-relieving properties of the bark and fruit extracts, and as a tonic. A few species, such as *Z. piperitum* DC. (Japanese pepper) and *Z. simulans* Hance (Sichuan pepper), are cultivated for their spicy fruits, which are used in some Oriental cuisines, especially as a condiment.

1. Zanthoxylum americanum Mill. (common prickly ash, northern prickly ash, toothache tree)

Pl. 552 h–j; Map 2566

Plants shrubs or rarely small trees, colonial from rhizomes, dioecious, 0.5–3.0(–8.0) m tall. Stems with usually few, ascending branches, the bark relatively smooth, gray to dark brown with lighter blotches and scattered, small, light-colored, more or less circular lenticels, occasionally becoming longitudinally fissured on old trunks, armed with small, flattened, slightly downward-curved prickles, these mostly subopposite and stipulelike at the nodes, the twigs gray to dark brown, moderately to densely short-hairy when young, becoming glabrous with age. Leaves alternate, short-petiolate, the petiole not jointed at the tip, unwinged. Leaf blades pinnately compound with 5–11(–13) leaflets, these 1–6(–8) cm long (progressively longer from the basal pair), oblong to elliptic, or ovate, the lateral pairs sometimes very slightly oblique, angled or short-tapered to a sessile or minutely stalked base, rounded or more commonly angled to a bluntly or sharply pointed tip, the margins entire or minutely scalloped, the upper surface dark green, sparsely to moderately and minutely hairy along the veins, becoming glabrous with age, the undersurface pale green to light green, moderately pubescent with somewhat longer, fine, somewhat tangled hairs, mostly along the veins and near the margins, becoming nearly glabrous with age. Inflorescences produced before the leaves, axillary, small umbellate clusters, the individual flowers short-stalked (sometimes appearing nearly sessile at flowering, the stalks elongating as the fruits develop). Flowers imperfect. Sepals absent. Petals 4 or 5, 1.5–2.5 mm long, narrowly oblong-obovate, overlapping in bud, yellowish green, with a minutely uneven apical portion and usually a fringe of short, crinkly reddish brown hairs. Stamens in staminate flowers 4 or 5 (absent or reduced to 4 or 5 minute staminodes in pistillate flowers), alternating with the petals, the filaments free, slender, and glabrous, attached outside of a small, often lobed or incomplete nectar disc around the ovary base. Ovaries 2–5 per flower, each short-stalked (reduced to a more or less lobed rudiment in staminate flowers), glabrous, 1-locular with 2 ovules, rounded or short-tapered at the tip, the styles fused above their bases, the stigma minute, with 2–5 lobes. Fruits 2–5 separate follicles per flower, each 4.5–6.0 mm long, ellipsoid to ellipsoid-obovoid, tapered to a short stalk at the base, abruptly tapered to a minute, oblique beak at the tip, the outer surface leathery, pitted, red to dark red or reddish brown, sometimes brownish yellow with a reddish tinge, glabrous or minutely hairy. Seeds 1 or 2 per fruit, 4.0–4.5 mm long, obovoid to nearly globose, sometimes (in 2-seeded follicles) somewhat flattened, the outer surface with faint, coarse facets or pits and finely pebbled, black, shiny. $2n=34$. April–May.

Scattered in the state, most abundantly outside the Ozark Division (eastern U.S. west to North Dakota and Oklahoma; Canada). Openings of mesic to dry upland forests, tops of bluffs, savannas, glades, banks of streams and rivers, and margins of ponds, lakes, and sinkhole ponds; also pastures, old fields, old strip mines, and railroads.

Uphof (1922) reported the closely related *Z. clava-herculis* L. (Hercules club, southern prickly ash) from southeastern Missouri, but no herbarium specimens have been located to substantiate his observations. This species occurs from Virginia to Florida west to Oklahoma and Texas and differs in flowering after the leaves expand and in having leaves with mostly 9–13 leaflets, the lateral ones strongly oblique at their bases, more or less terminal, looser inflorescences, and calyces of 5 sepals.

SALICACEAE (Willow Family)

Plants shrubs or more commonly small to large trees, dioecious, sometimes colonial from root suckers or stem fragmentation. Leaves alternate (rarely opposite or whorled in *Salix purpurea*), short- to long-petiolate, sometimes with glandular dots or short, conical, toothlike glands along the petiole or at the blade base. Stipules usually present (sometimes minute), herbaceous, sometimes fused around the stem along 1 margin, often shed early. Leaf blades simple, unlobed (5-lobed in *Populus alba*), pinnately veined with a single, well-developed midvein or with 3(5) main veins, the margins finely to coarsely toothed, the teeth usually minutely gland-tipped. Staminate and pistillate inflorescences on separate plants, often flowering before the leaves develop, consisting of solitary, more or less dense catkins (inflorescence type and position variable elsewhere), these terminal on short branches or appearing axillary, each flower subtended by a small bract, this sometimes shed early. Flowers imperfect, technically hypogynous, but the perianth highly reduced to a small, flat to cup-shaped disc (in *Populus*) or 1 or 2, fringed and sometimes irregularly lobed or fused nectaries (in *Salix*). Corollas absent. Staminate flowers with the stamens (1)2 to numerous, the filaments free or variously fused, slender, the anther attached at its base, purple or yellow; lacking a vestigial pistil. Pistillate flowers with the ovary superior, sessile or short-stalked, composed of 2–4 carpels, 1-locular, with 2 to numerous ovules, the placentation parietal; staminodes absent. Style 1, usually with 2(–4) minute branches toward the tip, occasionally very short or apparently absent, the stigmas 2(–4), variously shaped. Fruits capsules, dehiscent most of the way from the tip by 2–4 persistent valves. Seeds 0.8–3.0 mm long, narrowly oblong in outline to ovoid, green to reddish brown, surrounded by a dense tuft of long, silky hairs attached at the seed base. Traditionally two genera, about 450 species (but see discussion below), nearly worldwide (but absent from Australia and Malaysia), most diverse in the northern hemisphere.

Traditionally, the family Salicaceae in North America was treated as comprising only 2 genera, *Populus* and *Salix*, with about 450 total species (Steyermark, 1963; Cronquist, 1981; Gleason and Cronquist, 1991). However, molecular studies have shown that the members of several tropical families, notably portions of the Flacourtiaceae, should be considered part of the Salicaceae in order to circumscribe a natural group (Savolainen et al., 2000; Soltis et al., 2000; Chase et al., 2002). This results in an enlarged family of nearly worldwide distribution that includes about 58 genera and more than 1,200 total species, although the exact limits are still under study (Chase et al., 2002; Alford, 2006). The familial description above focuses on the Missouri representatives (which are included in the tribe Saliceae within the broadly conceived family), in order to maximize the usefulness of the treatment for determination of Missouri specimens. Some of the discordant characters of the more tropical genera include: perfect flowers that occasionally are not hypogynous and are organized into in a variety of different inflorescence types; presence of 3–8 or more unmodified sepals, these distinct or fused basally; presence of 3–8 or more small petals; in some genera the presence of staminodes among the stamens; fruits variously berries, capsules, or drupes; and seeds commonly hairy or with an aril, but lacking the tuft of long hairs (the coma) that is characteristic in *Populus* and *Salix*.

The two Missouri genera of Salicaceae contain numerous species that are prized horticulturally as shade and specimen trees, as well as for wind breaks, erosion control, and revegetation of highly impacted sites, such as mines and quarries. Some of the species also are used commercially for lumber, fence posts (including living fences), pulp for paper, handcrafts, and medicinally.

1. Leaf blades triangular to broadly ovate or occasionally nearly circular, less than 2 times as long as wide, with 3 or 5 main veins; catkins arched and drooping to pendulous; winter buds with several overlapping bud scales; perianth appearing as a small disc-shaped to cup-shaped structure subtending the stamens or pistil; staminate flowers with 7 to numerous stamens . . . . 1. POPULUS
1. Leaf blades variously narrowly lanceolate to ovate, narrowly oblanceolate to obovate, or narrowly elliptic to elliptic, 2 or more times as long as wide, with a single midvein; catkins straight or slightly curved, erect to spreading; winter bud with 1 bud scale; perianth absent, replaced by 1 or 2, entire or sometimes irregularly lobed or fused nectaries; staminate flowers with (1)2–8 stamens . . . . 2. SALIX

1. **Populus** L. (poplar)
(Eckenwalder, 1977a, b, 2010)

Plants medium to often large trees, sometimes suckering and colonial. Twigs often with prominent lenticels, these mostly circular to oval, sometimes appearing slightly raised, usually lighter than the surrounding surface. Winter buds lateral and usually also terminal, ovoid to ellipsoid, with 3–10 more or less resinous-sticky scales. Leaves mostly long-petiolate, the petiole lacking glands. Stipules usually minute (sometimes prominent on sucker shoots), shed early. Leaf blades usually heterophyllous (with leaves developing in the spring from the winter buds usually differing in shape from those produced later in the year), triangular to broadly ovate or occasionally nearly circular (narrower elsewhere), less than 2 times as long as wide, with 3 or 5 main veins; sometimes with 1–6 glands at the base; the margins faintly to strongly and finely to coarsely toothed, the teeth usually blunt and sometimes uneven. Catkins arched and drooping to pendulous, appearing before the leaves on twigs of previous year's growth. Flowers each subtended by an irregularly and deeply lobed bract (merely toothed in *P. alba*), this glabrous or with a patch of hairs on the undersurface, in some species with dense long, silky hairs along the margins, shed early. Perianth appearing as a small disc-shaped to cup-shaped structure subtending the stamens or pistil; nectaries absent. Staminate flowers with 7 to numerous stamens, the filaments free. Pistillate flowers with the pistil composed of 2–4 carpels. Stigmas 2–4, variously linear to disc-shaped, often inrolled or convoluted, sometimes irregularly 2-lobed. Ovules 6 to numerous. Capsules narrowly to broadly ovoid or more or less globose, dehiscing by 2–4 valves. About 30 species, North America, Asia, Africa.

Most of the species of *Populus* produce leaves with two kinds of morphology. These are associated with differences between so-called early-season leaves, which are produced from meristems in the winter bud and are the first flush of leaves to develop in the spring, vs. so-called late-season leaves, which are produced from lateral meristems as the twigs continue to elongate after the first flush of growth in the spring. The two types of leaves sometimes differ somewhat in size and shape, but principally are distinguished in the number and relative coarseness of the marginal teeth. The early-season leaves usually persist until the end of the growing season, and collectors should attempt to gather both kinds when specimens are collected. The leaves of most *Populus* species turn yellow in the autumn.

Species of *Populus* often are ecologically important as primary colonizers of flooded, burned, or otherwise highly disturbed sites, where they are important in soil stabilization. Poplars and cottonwoods generally grow quickly and a number of species are cultivated as ornamental and specimen trees. Some species can develop massive trunks. The wood has been used for flooring, veneers, fence posts (including living fences), palettes, barrel staves, handcrafts, toys, musical instruments, popsickle sticks, and wood chips. Because it tends to burn slowly, it also has been used for match sticks. In some portions of Europe, the fluffy seed hairs were used as a kapok substitute in pillows and life vests, and poplar wood also was a popular material for stakes to strike through a vampire's heart (Mabberley, 2008). Poplars were of minor medicinal and ceremonial use by some Native American tribes (Moerman, 1998). The staminate catkins produce abundant, wind-dispersed pollen and thus are one cause of hay fever in the springtime.

In the early spring, in addition to honey, bees harvest the resinous exudate from the winter bud scales of several *Populus* species, which they use as a natural adhesive in hive construction and maintenance, and in chores such as encasement of hive invaders (Hausen et al., 1987a). This fragrant, yellowish brown to dark brown substance, known as propolis or bee glue (one kind of balm-of-Gilead in Europe), has a long history of use medicinally, as an ingredient in facial creams and ointments, and in high-quality polishes and varnishes used on violins. Ghisalberty (1979) documented hundreds of historical and present-day applications for this substance. Chemically, propolis is a complex mixture of lipophilic compounds, including flavonoid aglycones, substituted benzoic acids and esters, and substituted phenolic acids and esters (Hausen et al., 1987a, b; Wollenweber et al., 1987; Greenaway et al., 1988; Hashimoto et al., 1988). Some of these ingredients have been reported to cause contact dermatitis in susceptible individuals, thus propolis harvested from bee hives or purchased at health food and vitamin stores should be used with care.

Key emphasizing vegetative characters

1. Mature leaves with the undersurface densely pubescent with white, felty hairs; leaves produced after the first spring flush palmately 3- or 5-lobed . . . . 1. P. ALBA
1. Mature leaves with the undersurface glabrous or sparsely pubescent with nonfelty hairs, mostly along the veins (sometimes more densely pubescent during development); leaves all unlobed, sometimes with relatively coarse, broad, pinnately arranged teeth
    2. Petioles circular or slightly 5-angled in cross-section, not or only very slightly flattened (then mostly on the top and bottom sides) . . . . 4. P. HETEROPHYLLA
    2. Petioles noticeably flattened along the sides, at least near the tip
        3. Blades of leaves developing early in the season with the margins relatively coarsely toothed, the teeth mostly 8–14 per side (leaves developing in the summer sometimes with more and finer teeth)

4. Leaf blades more or less triangular to broadly ovate-triangular, broadest at or near the base, with a well-defined, narrow, pale yellow or translucent line or band along the margins (viewed under magnification), the marginal teeth with a minute, incurved, somewhat thickened point at the tip; bark brown, with a network of coarse, longitudinal ridges and deep furrows; winter buds shiny, glabrous or sparsely spreading-hairy, resinous-sticky .... 2. P. DELTOIDES
4. Leaf blades elliptic to ovate, broadest well above the base, lacking a pale yellow or translucent line or band along the margins(even when viewed under magnification), the marginal teeth bluntly to sharply pointed with a straight, unthickened point at the tip; bark light gray, relatively smooth, developing darker, roughened cross-ridges, becoming dark brown and longitudinally ridged and furrowed only near the base of older trees; winter buds dull, sparsely to moderately pubescent with gray, cobwebby to somewhat woolly hairs, not or only slightly resinous or sticky .... 3. P. GRANDIDENTATA

3. Leaf blades of early-season and late-season leaves similar, the margins finely to moderately toothed, the teeth mostly 15–50 teeth per side
5. Leaf blades more or less triangular to broadly rhombic-triangular or broadly ovate-triangular, broadest at or near the base, with a well-defined, narrow, pale yellow or translucent line or band along the margins (viewed under magnification), the marginal teeth moderate, with a minute, incurved, somewhat thickened point at the tip; bark of older trunks dark gray to dark grayish brown, with a network of longitudinal ridges and deep furrows; winter buds resinous-sticky .... 5. P. NIGRA
5. Leaf blades ovate to broadly ovate or occasionally nearly circular, lacking a pale yellow or translucent line or band along the margins (even when viewed under magnification), the marginal teeth fine, rounded or with a blunt, incurved, unthickened point at the tip; bark light gray or greenish gray to white, relatively smooth with darker, roughened spots, cross-ridges, and branch scars; winter buds not or only slightly resinous or sticky .... 6. P. TREMULOIDES

Key emphasizing reproductive characters

1. Bracts subtending the flowers with the margins toothed or with 3–7 narrow, deep lobes, the teeth or lobes with long, silky hairs along the margins, at least toward the tips; staminate flowers with 6–10(–12) stamens; terminal winter buds 3–10 mm long
2. Bracts with the margins toothed, young twigs and winter buds densely pubescent with white, felty hairs (becoming glabrous or nearly so with age) .... 1. P. ALBA
2. Bracts with the margins appearing cut into 3–7 narrow, deep lobes; twigs and winter buds glabrous or sparsely to moderately pubescent with gray, cobwebby to somewhat woolly hairs
3. Terminal winter buds dull, sparsely to moderately pubescent with gray, cobwebby to somewhat woolly hairs; inflorescences 5–10 cm long; blades of early-season and late-season leaves dissimilar, those of early-season leaves mostly with 5–14 relatively coarse teeth per side, those of l ate-season leaves with 15–60 finer scallops or teeth per side .... 3. P. GRANDIDENTATA

3. Terminal winter buds shiny, glabrous or nearly so; inflorescences 4–7 cm long; blades of early-season and late-season leaves more or less similar, all with the margins relatively finely toothed, the teeth mostly 20–80 per side . . . . 6. P. TREMULOIDES

1. Bracts subtending the flowers with the margins appearing cut or fringed, with 9 to more commonly numerous, narrow, deep lobes, otherwise with glabrous margins; staminate flowers with (10–)12 to numerous stamens; terminal winter buds 10–25 mm long

4. Styles branched to about the midpoint, the stigmas thus appearing distinctly elevated from the ovary on a stalk; stamens 12–20; bracts with the undersurface finely hairy; winter buds not or only slightly resinous or sticky; developing leaves pubescent with dense, fine, felty to woolly hairs; twigs with the pith orange . . . . 4. P. HETEROPHYLLA

4. Styles branched to the base, the stigmas thus appearing sessile on the ovary; stamens (15–)20 to numerous; bracts with the undersurface glabrous; winter buds resinous-sticky; developing leaves glabrous or sparsely to moderately pubescent, the hairs then not felty or woolly; twigs with the pith white

5. Stamens numerous (mostly 40 or more); relatively open, widely branched trees with mostly spreading to loosely ascending branches; winter buds with yellowish resin . . . . 2. P. DELTOIDES

5. Stamens (15–)20–30; plants relatively slender, columnar trees with strongly ascending branches; winter buds with reddish resin . . . . . 5. P. NIGRA

1. Populus alba L. (silver poplar, white poplar)
Pl. 553 g; Map 2567

Plants trees 5–25 m tall, open, the main trunk relatively short, widely branched with spreading to ascending branches, strongly colonial from root suckers. Bark greenish white, becoming gray to grayish brown, relatively smooth with prominent, darker, raised, corky branch scars and lenticels, developing a network of coarse, longitudinal ridges and furrows on older trunks. Twigs slender, greenish gray, densely pubescent with white, felty hairs (becoming glabrous or nearly so with age), the pith white. Winter buds 3–10 mm long, reddish brown, not resinous-sticky, densely felty when young, becoming thinly woolly with age. Leaves strongly heterophyllous (the early-season leaves unlobed, those produced after the first spring flush palmately 3- or 5-lobed), the petiole shorter than the blade, not flattened (more or less circular in cross-section), densely woolly. Leaf blades 1–7 cm long, those produced in early spring ovate, longer than wide, angled to a bluntly pointed tip, rounded to broadly cordate at the base, the margins minutely hairy, lacking a yellow or translucent line or band, each margin with 3–8 coarse, broad, blunt teeth or scallops; those produced later in the season palmately 3- or 5-lobed, the margins otherwise irregularly and relatively finely toothed or scalloped; all of the leaf blades sometimes with a pair of cup-shaped glands at or near the base, the upper surface green to dark green, shiny, thinly hairy along the main veins (more densely so when young), the undersurface densely pubescent with white, felty hairs. Inflorescences 2–8 cm long (the pistillate ones elongating slightly to 4–10 cm at fruiting), the bracts unlobed, shallowly toothed along the silky-hairy margins, the undersurface glabrous. Staminate flowers with 6–10(–12) stamens. Pistillate flowers with mostly 2 carpels, the style branched nearly to the base, the stigmas slender. Fruits (2–)3–5 mm long, narrowly ovoid, glabrous, with usually 2 valves. $2n=38$. March–May.

Introduced, widely scattered in the state (native of Europe, Asia; introduced widely in temperate North America). Mesic upland forests and banks of streams; also pastures, old fields, fencerows, old homesites, old mines, roadsides, and open disturbed areas.

This species was formerly more popular in cultivation than it is today (Steyermark, 1963). It tends to persist as large clonal colonies of root suckers around old plantings. Because most colonies are a single clone and thus exclusively staminate or pistillate (mostly the latter), seed production is rare in the state.

Steyermark (1963) reported the European aspen, *P. tremula* L., from Missouri based on a single specimen that he collected in 1936 in Washington County. However, this specimen was redetermined as *P. ×canescens* (Aiton) Sm. (Pl. 554 l, m), the

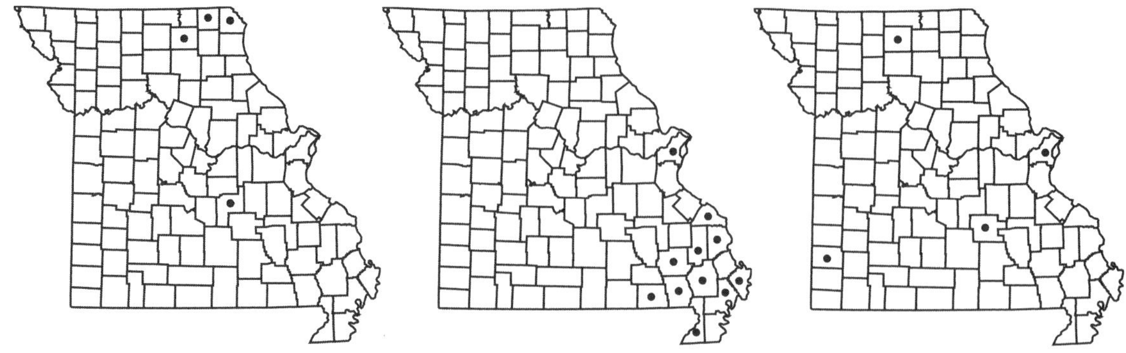

2569. Populus grandidentata 2570. Populus heterophylla 2571. Populus nigra

hybrid between *P. alba* and *P. tremula*, in 1990 by James Eckenwalder during his doctoral research on the genus. According to Eckenwalder (2010), *P.* ×*canescens*, the gray poplar, is a hybrid that occurs naturally and in cultivation in Europe, and also is cultivated in temperate North America. It apparently is represented in the New World primarily as a pistillate clone, which persists from old plantings and forms colonies from vigorous root sprouts. *Populus* ×*canescens* is morphologically similar to *P. alba*, but differs in its gray- to tan-tinged (vs. white) hairs (the undersides of the leaves appear generally less hairy), its late-season leaves more irregularly and somewhat more pinnately lobed, and 10–18 (vs. 2–6) ovules per ovary or seeds per capsule.

2. Populus deltoides W. Bartram ex Marshall (cottonwood, eastern cottonwood, southern cottonwood)

Pl. 553 h–k; Map 2568

Plants trees 15–35 m tall, relatively open, widely branched with mostly spreading to loosely ascending branches. Bark light to dark brown, with a network of coarse, longitudinal ridges and deep furrows. Twigs slender to more commonly stout, yellowish brown to reddish brown, becoming tan or gray with age, glabrous or sparsely pubescent with short, spreading to somewhat curved, white hairs, the pith white. Winter buds 6–18 mm long, greenish yellow, strongly sticky with yellowish resin, glabrous or sparsely spreading-hairy. Leaves somewhat heterophyllous (the early-season leaves with fewer, deeper teeth than those produced later), the petiole about as long as the blade, noticeably flattened on the sides, at least near the tip, glabrous or hairy. Leaf blades 3–12(–14) cm long, more or less triangular to broadly ovate-triangular, noticeably wider than long to slightly longer than wide, broadest near the base, abruptly tapered to a sharply pointed tip, truncate to broadly cordate at the base, the margins minutely pubescent with moderate to dense, curved hairs, with a well-defined, narrow, pale yellow or translucent line or band (viewed under magnification), finely and unevenly scalloped or bluntly toothed, the teeth mostly 5–20 and relatively coarse (early-season) or 20–50 finer (late-season) per side, these 0.5–7.0 mm deep, incurved and minutely thickened at their tips, sometimes also with 1–6 club-shaped gland(s) at or near the base, the upper and undersurface grayish green to bright green, glabrous. Inflorescences 5–18 cm long (the pistillate ones elongating to 10–24 cm at fruiting), the bracts deeply cut into 9 to numerous irregular lobes, the margins glabrous, the undersurface glabrous. Staminate flowers with mostly numerous (40 or more) stamens. Pistillate flowers with 3 or 4 carpels, the style variously unbranched nearly to the tip or branched nearly to the base, the stigmas flattened. Fruits 6–13 mm long, ovoid, glabrous, with 3 or 4 valves. $2n=38$. March–May.

Scattered nearly throughout the state, most abundant in counties bordering large rivers (nearly throughout the U.S. except for some far-western states; Canada, Mexico). Bottomland forests, banks of streams and rivers, margins of ponds and lakes, sloughs, swamps, marshes, and bottomland prairies; also ditches, strip mines, quarries, railroads, roadsides, and moist disturbed areas.

Cottonwood is a fast-growing, large, stately tree that can attain massive trunk diameters. It is cultivated as a shade tree and for erosion control and mine reclamation, but is considered a relatively short-lived species.

Two hybrids that are sometimes also planted in Missouri have been recorded as uncommon escapes (Steyermark, 1963; Eckenwalder, 1977a). Carolina poplar, *P.* ×*canadensis* Moench is a hybrid between *P. deltoides* and the nonnative *P. nigra*. It is known thus far from historical collections from Iron and Washington Counties. This hybrid

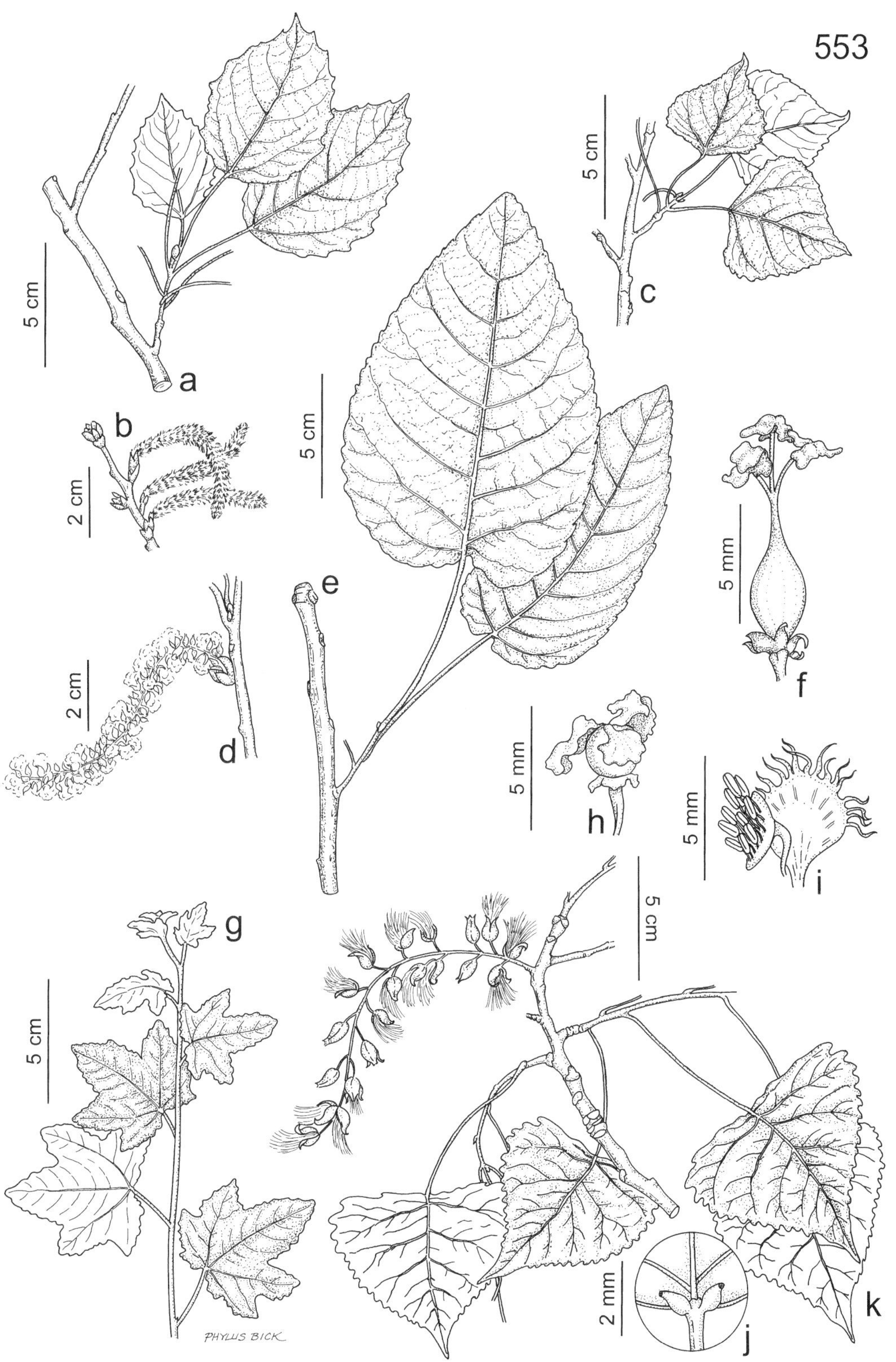

Plate 553. Salicaceae. *Populus grandidentata*, **a)** portion of branch with leaves, **b)** staminate catkins. *Populus nigra*, **c)** portion of branch with leaves. *Populus heterophylla*, **d)** portion of branch with pistillate catkin at fruiting, **e)** portion of branch with leaves, **f)** pistillate flower. *Populus alba*, **g)** branch tip with leaves. *Populus deltoides*, **h)** pistillate flower, **i)** staminate flower with bract, **j)** leaf detail, **k)** portion of branch with leaves and pistillate catkin at fruiting.

has the more columnar growth form of its Lombardy poplar parent and also has early-season leaves that are somewhat longer than wide, with more numerous, finer teeth along the margins than in *P. deltoides*. It is usually a staminate clone that reproduces vegetatively from root sprouts. Balm-of-Gilead, *P. ×jackii* Sarg. (*P. ×gileadensis* Rouleau) is a mostly sterile hybrid between the cultivated *P. balsamifera* L. and *P. deltoides*. In Europe it is cultivated for its winter bud resins. It usually occurs as a pistillate clone that reproduces vegetatively from root sprouts. In Missouri, this hybrid is known thus far from historical collections from Reynolds and St. Louis Counties. It differs from *P. deltoides* in its petioles that are not or only slightly flattened laterally and in its more narrowly ovate leaf blades with somewhat rounded bases.

Eckenwalder (1977a, b) treated the *P. deltoides* complex as consisting of three intergrading subspecies, two of which are found in Missouri. The third subspecies, ssp. *wislizeni* (S. Watson) Eckenwalder (Rio Grande cottonwood) occurs in the western United States and adjacent Mexico. It is most similar to ssp. *deltoides*, differing from that subspecies in its leaf blades mostly lacking basal glands and its sparsely hairy winter buds.

1. Early season leaves with the blades relatively short-tapered at the tip, usually with 3–6 basal glands; inflorescences with relatively variable, short to long flower stalks; twigs mostly reddish brown, becoming tan or gray with age . . . . 2A. SSP. DELTOIDES
1. Early season leaves with the blades relatively long-tapered at the tip, usually with 2 basal glands; inflorescences with relatively uniform, short flower stalks; twigs mostly yellowish brown, becoming tan with age . . . . 2B. SSP. MONILIFERA

2a. ssp. deltoides (southern cottonwood)

P. deltoides var. *missouriensis* (A. Henry) A. Henry

P. deltoides var. *pilosa* (Sarg.) Sudw.

Twigs mostly reddish brown, becoming tan or gray with age, the winter buds usually glabrous. Later-season leaves often slightly longer than wide. Blades of early-season leaves relatively short-tapered at the tip, usually with 3–6 basal glands. Inflorescences with the flower stalks 1–14 mm long, progressively longer from the catkin tip to base. $2n=38$. March–May.

Uncommon, widely scattered nearly throughout the state (eastern U.S. west to Iowa and Texas; Canada). Bottomland forests, banks of streams and rivers, margins of ponds and lakes, sloughs, swamps, marshes, and bottomland prairies; also ditches, strip mines, quarries, railroads, roadsides, and moist disturbed areas.

This subspecies occupies the eastern portion of the species range, overlapping the distribution of ssp. *monilifera* in states bordering the eastern edge of the Great Plains and eastward along the Great Lakes through a line of states from Illinois to Pennsylvania. In Missouri, ssp. *deltoides* is approaching the western edge of its range and is much less common than ssp. *monilifera*. It is most abundant along the Mississippi River from Ste. Genevieve County northward, but has also been collected sporadically to the western edge of the state.

2b. ssp. monilifera (Aiton) Eckenw. (Plains cottonwood)

P. sargentii Dode

Twigs mostly yellowish brown, becoming tan with age, the winter buds usually sparsely pubescent with spreading or somewhat curved hairs. Later-season leaves about as long as wide or slightly wider than long. Blades of early-season leaves relatively long-tapered at the tip, usually with 3–6 basal glands. Inflorescences with the flower stalks 2–6 mm long, relatively uniform from the catkin tip to base. $2n=38$. March–May.

Scattered nearly throughout the state, most abundant in counties bordering large rivers (Montana to New Mexico east to Pennsylvania, Missouri, and Texas; Canada). Bottomland forests, banks of streams and rivers, margins of ponds and lakes, sloughs, swamps, marshes, and bottomland prairies; also ditches, strip mines, quarries, railroads, roadsides, and moist disturbed areas.

Steyermark excluded this taxon (as *P. sargentii*) from the state's flora based on a faulty understanding of morphological variation within the complex. This is the more abundant of the two subspecies in Missouri and occupies a range mainly in the center of the country, from the eastern portion of the Rocky Mountains through the Great Plains, eastward along the Great Lakes. For further discussion, see the treatment of ssp. *deltoides*.

3. Populus grandidentata Michx. (large-toothed aspen)

Pl. 553 a, b; Map 2569

Plants trees 8–20 m tall, relatively open (but sometimes slender in aspect), widely branched with mostly spreading to loosely ascending branches. Bark light gray, relatively smooth, developing darker, roughened cross-ridges, becoming dark brown and longitudinally ridged and furrowed only near the base of older trees. Twigs slender to moderately stout, reddish brown, becoming gray with age, sparsely to moderately pubescent with gray, cobwebby to somewhat woolly hairs when young, becoming glabrous or nearly so with age, the pith white. Winter buds 3–8 mm long, reddish brown, dull, not or only slightly resinous, cobwebby-hairy when young, becoming glabrous or nearly so with age. Leaves heterophyllous (the blades of early-season leaves with fewer, coarser teeth than those developing later in the season), the petiole half as long as to about as long as the blade, noticeably flattened on the sides, at least toward the tip, glabrous or sparsely hairy at the tip. Leaf blades (2–)4–10(–20) cm long, ovate, longer than wide, tapered to a sharply pointed tip, broadly angled or broadly rounded to more commonly truncate at the base, the margins lacking a pale yellow or translucent line or band (viewed under magnification), glabrous or sparsely to moderately pubescent with silky hairs, finely to coarsely scalloped or toothed, mostly with 5–14 relatively coarse teeth 0.5–5.0 mm deep (early-season) or 15–60 finer scallops or teeth 0.5–2.0 mm deep (late-season) per side, the teeth spreading or incurved at their tips but not strongly thickened, often also with 2(–4) shallowly cup-shaped glands at or near the base, the upper surface dark green, glabrous and usually shiny, the undersurface pale green and usually somewhat glaucous, glabrous or sparsely pubescent with white, silky hairs. Inflorescences 5–10 cm long (the pistillate ones elongating to 7–14 cm at fruiting), the bracts deeply cut into 3–7 narrow, deep lobes, the margins silky-hairy, the undersurface glabrous. Staminate flowers with 6–12 stamens. Pistillate flowers with usually 2 carpels, the style branched above the midpoint, the stigmas linear and often deeply 2-lobed (then sometimes appearing as 4 stigmas). Fruits 3–6 mm long, ovoid, glabrous, with usually 2 valves. $2n=38$. March–May.

Uncommon, known thus far from northeasternmost Missouri and a single disjunct population in Phelps County (northeastern U.S. west to North Dakota and Missouri; Canada). Mesic upland forests (mostly in openings and along edges), mostly near heads of ravines, and tops of low bluffs.

There has been debate on whether the Missouri populations represent escapes from cultivation or native occurrences that are relictual from past times when the climate was cooler (Steyermark, 1963; Kurz, 2003). However, the species is not commonly planted in Missouri and even the Phelps County population appears to be part of a native plant community without evidence of nearby old homesites. The Missouri Natural Heritage Program tracks the taxon as being of conservation concern in the state. Thus large-toothed aspen is treated as a native member of the flora in the present work.

4. Populus heterophylla L. (swamp cottonwood)

Pl. 553 d–f; Map 2570

Plants trees 10–28 m tall, relatively open, widely branched with mostly spreading to loosely ascending branches. Bark dull brown to reddish brown, with a network of coarse, longitudinal ridges and deep furrows, sometimes flaking off in long, slender plates. Twigs stout, reddish brown, becoming gray with age, sparsely to moderately pubescent with gray, felty or woolly hairs when young, becoming glabrous or nearly so with age, the pith orange. Winter buds 4–7 mm long, reddish brown, not or only slightly resinous, relatively dull, cobwebby-hairy when young, becoming glabrous or nearly so with age. Leaves not noticeably heterophyllous (the early-season and late-season leaves similar in shape and marginal toothing), the petiole about half as long as the blade, slightly 5-angled in cross-section, not or only very slightly flattened (then mostly on the top and bottom sides near the tip), glabrous or hairy. Leaf blades (4–)8–20(–25) cm long, ovate to broadly ovate, mostly longer than wide, rounded or angled to a bluntly pointed tip, sometimes short-tapered to a minute, sharp point at the very tip, cordate to deeply cordate at the base, the margins lacking a pale yellow or translucent line or band (viewed under magnification), glabrous, finely and unevenly scalloped or bluntly toothed, the teeth mostly 30–60 per side, 0.3–1.0 mm deep, incurved and minutely thickened at their tips, often also with 2 shallowly cup-shaped glands at or near the base, the upper surface dark green, glabrous or sparsely pubescent with white cobwebby hairs at maturity (more densely woolly when young), the undersurface pale green, glabrous or sparsely pubescent with white, woolly to cobwebby hairs at maturity (densely felty or woolly when young). Inflorescences 1.5–8.0 cm long (the pistillate ones elongating to 6–18 cm at fruiting), the bracts deeply cut into numerous irregular lobes, the margins glabrous, the undersurface somewhat hairy. Staminate flowers with 15–35 stamens. Pistillate flowers with 2 or more, commonly 3, carpels, the style unbranched nearly to the tip, the stigmas convoluted, flattened. Fruits 8–14 mm long, ovoid, glabrous, with 2 or 3 valves. March–April.

Uncommon in the Mississippi Lowlands Division (eastern U.S. west to Missouri and Louisiana).

2572. Populus tremuloides 2573. Salix alba 2574. Salix amygdaloides

Swamps, bottomland forests, sloughs, and margins of sinkhole ponds.

This species has been used for wood pulp in the paper industry. It also has been used for making boxes and crates, as well as for interior panels of some furniture.

5. Populus nigra L. (Lombardy poplar, black poplar)

Pl. 553 c; Map 2571

Plants trees 15–35 m tall, relatively slender, columnar, with strongly ascending branches. Bark gray to grayish brown, relatively smooth with prominent, darker, raised, corky branch scars and lenticels, becoming dark gray to dark grayish brown and developing a network of coarse, longitudinal ridges and deep furrows on older trunks. Twigs mostly relatively slender, orangish brown to yellowish brown, becoming reddish gray or gray with age, glabrous, the pith white. Winter buds 7–19 mm long, reddish brown, strongly sticky with reddish resin, glabrous. Leaves strongly heterophyllous (the early-season leaves differently shaped and with fewer, deeper teeth than those produced later), the petiole about as long as the blade, noticeably flattened on the sides, at least near the tip, glabrous. Leaf blades 2–10 cm long, those of early-season leaves rhombic-ovate to broadly ovate, mostly slightly longer than wide, broadest below the midpoint, angled or slightly tapered to a sharply pointed tip, and angled to broadly angled at the base; those produced later in the season broadly triangular to rhombic-triangular or broadly ovate-triangular, mostly wider than long, broadest at or near the base, abruptly tapered to a sharply pointed tip, truncate to very broadly angled at the base; both types with the margins glabrous, with a well-defined, narrow, pale yellow or translucent line or band (viewed under magnification), finely and unevenly toothed, the teeth mostly 10–22 per side (early-season leaves) or 20–50 per side (later-season leaves), these 0.5–7.0 mm deep, with a minute, incurved, somewhat thickened point at the tip, usually lacking glands at the base, the upper surface green to bright green, glabrous, shiny, the undersurface light green, glabrous (finely hairy when young), somewhat shiny at maturity. Inflorescences 2–6 cm long (the pistillate ones elongating to 5–10 cm at fruiting), the bracts deeply cut into 9 to numerous irregular lobes, the margins glabrous, the undersurface glabrous. Staminate flowers with (15–)20–30 stamens. Pistillate flowers with 2 carpels, the style variously mostly branched nearly to the base, the stigmas flattened. Fruits 5–9 mm long, ovoid, glabrous, with 2 valves. $2n=38$. March–May.

Introduced, uncommon, sporadic (species native to Europe, Asia; widely planted and persisting or escaping in temperate North America). Edges of mesic upland forests; also fencerows, old homesites, lawns, and roadsides; persisting after cultivation, but sometimes appearing escaped.

The description above refers to cv. 'Italica' (var. *italica* Münchh.), the Lombardy poplar, a horticultural variant of black poplar with a columnar growth form. This cultivar exists mainly as a staminate clone in the United States (Eckenwalder, 2010), although pistillate clones with a slightly broader growth form are known in Europe. It thus does not reproduce by seed in Missouri and merely persists from old plantings. However, groups of individuals persisting in a natural habitat can be mistaken for escaped plants. Other forms of black poplar also are planted in the Midwest. These exist as both staminate and pistillate clones and are characterized by a slightly shorter, more openly branched growth form. These have not yet escaped in Missouri.

Lombardy poplar is popular in horticulture, although trees tend to be disease-prone and relatively short-lived.

6. Populus tremuloides Michx. (quaking aspen, trembling aspen)

Pl. 554 j, k; Map 2572

Plants trees 8–20 m tall, relatively open (but sometimes slender in aspect), widely branched with mostly spreading to loosely ascending branches. Bark light gray or greenish gray to white, relatively smooth, developing darker, roughened or corky spots, cross-ridges, and branch scars, occasionally becoming dark brown and longitudinally ridged and furrowed only near the base of older trees. Twigs slender to moderately stout, reddish brown, becoming grayish yellow with age, glabrous, the pith white. Winter buds 3–6(–8) mm long, reddish brown, shiny, not or only slightly resinous, glabrous. Leaves slightly heterophyllous (the blades of early-season leaves with somewhat fewer teeth than those developing later in the season), the petiole about as long as the blade, noticeably flattened on the sides, at least toward the tip, glabrous. Leaf blades 1.5–8.0(–12.0) cm long, ovate to broadly ovate or occasionally nearly circular, longer than to about as long as wide, broadly angled or abruptly tapered to a sharply pointed tip, broadly angled or broadly rounded to more commonly truncate at the base, rarely shallowly cordate, the margins with an inconspicuous, very thin, pale yellow or translucent line or band (viewed under magnification), this sometimes lacking, glabrous or sparsely pubescent with silky hairs, finely toothed, mostly with 20–80 rounded to bluntly pointed teeth 0.1–1.5 mm deep per side, the teeth incurved at their tips but not strongly thickened, sometimes also with 1 or 2 shallowly cup-shaped glands at or near the base, the upper surface dark green, glabrous, the undersurface pale green and often slightly glaucous, glabrous. Inflorescences 4–7 cm long (the pistillate ones elongating to 7–12 cm at fruiting), the bracts deeply cut into 3–7 narrow, deep lobes, the margins silky-hairy, the undersurface glabrous. Staminate flowers with 6–12 stamens. Pistillate flowers with usually 2 carpels, the style usually branched nearly to the base, the stigmas linear and often deeply 2-lobed. Fruits 2.5–5.5 mm long, narrowly ovoid, glabrous, with usually 2 valves. $2n$=38, 57, 76. March–April.

Uncommon, known thus far from northeasternmost Missouri and a single historical, population in Dent County (western and northeastern U.S. (including Alaska); Canada, Mexico). Mesic upland forests (mostly in openings and along edges), edges of upland prairies, and banks of streams; also ditches and roadsides.

Quaking aspen tends to be an early-successional in montane habitats, where it colonizes areas opened by fires or other natural disturbances and gradually is replaced by other tree species. The quaking portion of the common name refers to the movement of the leaves in the wind, which (in this species and many other taxa of *Populus*) is enhanced by the flattened petioles. *Populus tremuloides* is quite uncommon in Missouri. Steyermark (1963) discussed the disjunct, historical station in Dent County, which he interpreted to represent a native occurrence.

Steyermark (1963) reported the European aspen, *P. tremula* L., from Missouri based on a single specimen that he collected in 1936 in Washington County. However, this specimen was redetermined as *P.* ×*canescens* Sm., the hybrid between *P. alba* and *P. tremula*, in 1990 by James Eckenwalder during his doctoral research on the genus. Eckenwader (2010) noted that *P. tremula* is sometimes cultivated in temperate North America, particularly a columnar cultivar that exists only as a staminate clone. The species is closely related to *P. tremuloides*, but differs in the following characters: terminal winter buds with cobwebby hairs toward the base (vs. glabrous or nearly so) and blades of later-season leaves with the margins relatively coarsely toothed (vs. finely toothed), the teeth mostly 8–14 (vs. 20–80) per side. *Populus tremula* can persist from old plantings, but has not become established outside of cultivation in Missouri thus far.

2. **Salix** L. (willow)
Contributed by George W. Argus

Plants shrubs or small to trees, sometimes colonial from stem or root suckers, stem fragmentation, or stem layering. Branchlets often with prominent lenticels, these variously shaped, lighter or darker than the surrounding surface. Winter buds lateral and sometimes also appearing terminal, variously shaped, with a single scale, this not sticky, consisting of 2 opposite scales that are fused totally into a conical cap or with free, overlapping margins on the side facing the stem. Leaves short-petiolate, the petiole sometimes with glandular dots or short, flat glands. Stipules usually present (in some species minute or apparently absent, especially on early-season shoots), herbaceous, variously shaped, often shed early. Leaf blades not or only slightly heterophyllous (leaves developing in the spring from the winter buds sometimes differing slightly in size and/or shape from those produced later in the year), variously linear, narrowly lanceolate to ovate, narrowly oblanceolate to obovate, or narrowly elliptic to elliptic, 2 or more times as long as wide, with a single midvein, the margins entire or more commonly faintly to strongly toothed, the teeth blunt (sometimes appearing scalloped) or sharply pointed, sometimes gland-tipped (the glands sometimes also present along the margins on the untoothed portions). Catkins straight or slightly curved, erect to spreading, appearing before or as the leaf buds open, terminal on short, axillary branches or lateral on branchlets of usually the second year's growth, sessile or nearly so. Flowers each subtended by an entire or somewhat uneven-margined to toothed bract, this usually hairy on the margins and/or surfaces, and persistent or shed early. Perianth absent, apparently replaced by 1 or rarely 2 nectaries, these entire or sometimes irregularly lobed, distinct or fused into a cup. Staminate flowers with (1)2–8 stamens, the filaments distinct or fused toward the base. Pistillate flowers with the pistil composed of 2 carpels. Stigmas 2, usually linear, occasionally 2-lobed. Ovules 2 to numerous. Capsules ovoid to obovoid, ellipsoid, or somewhat club-shaped, dehiscing by 2 valves. About 450 species, nearly worldwide, but absent (except for introductions) from Oceania and Australia.

Some species of *Salix* are ecologically important as primary colonizers of seasonally flooded or otherwise highly disturbed, wet sites, where they are important in soil stabilization. Steyermark (1963) noted that they are used by beavers in lodge construction, and they are also an important food source for these mammals. Willows also are hosts for a bewildering array of parasitic insects whose larvae form galls on the stems and leaves (Redfern and Askew, 1992). These include mainly sawflies (Hymenoptera: Tenthredinidae) and gall midges (Diptera: Cecidomyiidae). Nyman et al. (2000) noted that more than 200 species of sawflies alone form galls on various *Salix* species.

A number of species of willows are cultivated commonly as ornamentals and specimen plants, as well as for wind breaks, erosion control, bioremediation, and reclamation of severely impacted sites, such as mines and quarries. The leaves of most species turn yellow in the autumn. Many cultivars have been developed for some of the more commonly grown species and a number of hybrids exist as well. To the chagrin of many home owners, the roots of willows are adept at invading sewer lines, thus planting of this genus near underground pipes is to be avoided. The wood is relatively soft, but tends to resist warping and splitting, and has been used (at least historically) for packing crates, palettes, furniture, cricket bats, and paper pulp. Many children of a bygone era learned the penalty for misbehavior, which involved application of a willow switch.

The bitter inner bark of *Salix* species is one of the original sources of salicylic acid (2-hydrobenzoic acid), and willow bark has an extremely long history of medicinal use for fevers, aches, skin conditions, and head aches, and also is an effective anti-clotting agent (Jeffries, 2005). Edmund Stone (1763) noted its astringent properties and experimented with the use of dried, powdered bark of *S. alba* in reducing fevers; he brought the plant to the attention of the European scientific community. Today, salicylic acid is a common ingredient in some acne, psoriasis, wart, and callus treatments, as well as in some dandruff shampoos, and also is used as an antiseptic additive in some toothpastes. Aspirin (acetylsalicylic acid) is a biochemical derivative of salicylic acid. It was first synthesized by the French chemist, Charles Gerhardt (1853). A team of German chemists subsequently purified the compound and determined its biochemical structure (Schröder et al., 1869). During the period of 1897–1899, the German pharmaceutical firm of Bayer developed the commercial drug, for which it coined the name aspirin (Jeffries, 2005). Aspirin apparently was the first mass-market drug developed for commercial profit and continues to be one of the most-used medicines in the world.

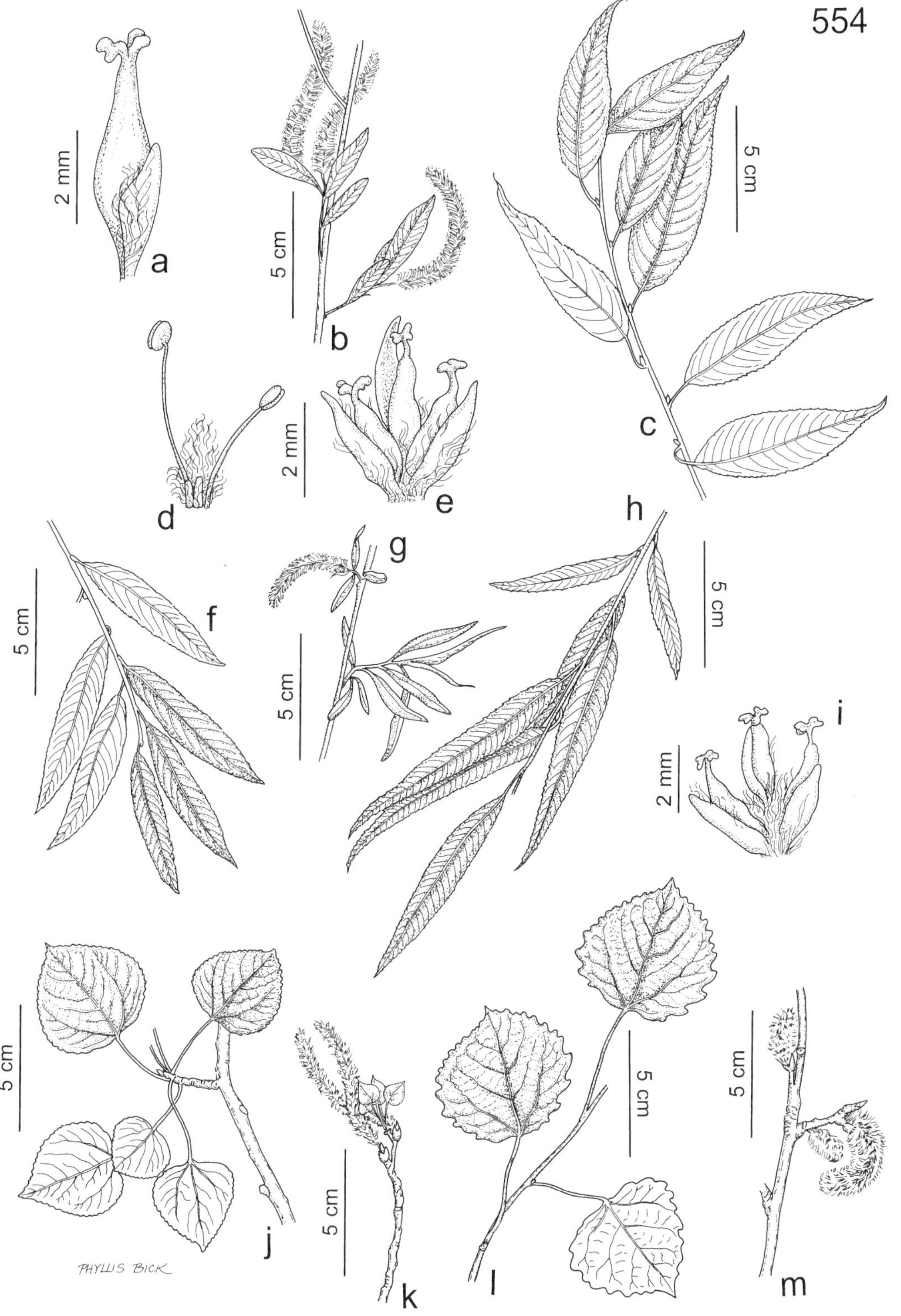

Plate 554. Salicaceae. *Salix amygdaloides*, **a)** pistillate flower with bract, **b)** portion of branch with staminate catkins and leaves, **c)** branch tip with leaves. *Salix alba*, **d)** staminate flower, **e)** cluster of pistillate flowers with bracts, **f)** branch tip with leaves, **g)** portion of branch with staminate catkins and developing leaves. *Salix ×sepulcralis*, **h)** branch tip with leaves, **i)** pistillate flower. *Populus tremuloides*, **j)** portion of branch with leaves, **k)** branch tip with pistillate catkins and developing leaves. *Populus ×canescens*, **l)** portion of branch with leaves, **m)** portion of branch with staminate catkins.

Species of *Salix* have a reputation for hybridization. Elsewhere in the United States this may be true, but in Missouri hybrids appear to be uncommon. Although several different hybrids have been reported, these mainly are documented by historical specimens. This may be correlated with the dramatic decline of wetlands in the state during the first half of the twentieth century.

Note that in the present treatment the term *branchlets* to denote first-season growth at the tips of branches is used in place of the more frequently encountered term *twig*, as this terminology has been used in other recent literature on the genus.

Key emphasizing vegetative characters

1. At least some of the leaves opposite or subopposite . . . . 11. S. PURPUREA
1. Leaves all alternate
   2. Plants shrubs, rarely small trees
      3. Leaf margins entire, scalloped, or somewhat wavy
         4. Leaf blades with the margins flat, the undersurface glabrous or sparsely hairy, at least when young . . . . 4. S. DISCOLOR
         4. Leaf blades with the margins rolled under, the undersurface finely woolly-hairy (occasionally becoming more or less glabrous with age) . . . . 7. S. HUMILIS
      3. Leaf margins sharply toothed
         5. Leaf blades linear to narrowly lanceolate or narrowly oblong-elliptic, mostly 10–20 times as long as wide, the margins with slender, widely spaced teeth, the blade narrowly angled at the base; petioles 1–5(–9) mm long, quite short in relation to leaf length . . . . 8. S. INTERIOR
         5. Leaf blades narrowly lanceolate or narrowly elliptic to oblanceolate, oblong-oblanceolate, or occasionally obovate, 2.5–10.0 times as long as wide, the margins with broad, closely spaced teeth, the blade angled to rounded or shallowly cordate at the base; petioles 3–27 mm long, relatively long in relation to leaf length
            6. Leaves glabrous or sparsely hairy on the undersurface; stipules well-developed (but occasionally shed early)
               7. Branches reddish brown . . . . 5. S. ERIOCEPHALA
               7. Branches yellow to yellowish or grayish brown (branchlets sometimes reddish brown, but turning yellowish in subsequent years) . . . . 6. S. FAMELICA
            6. Leaves densely hairy on the undersurface; stipules absent or highly reduced
               8. Juvenile leaves pubescent with long, silky hairs; largest mature leaf blades sparsely to densely pubescent with relatively long, silky hairs; branches flexible at base . . . . 10. S. PETIOLARIS
               8. Juvenile leaves pubescent with short, silky hairs; largest mature leaf blades sparsely to densely pubescent with relatively short, silky hairs on the undersurface; branches brittle at base . . . . 12. S. SERICEA
   2. Plants medium to large trees
      9. Buds sharply pointed at the tip; bud scale margins free and overlapping on the side facing the stem
         10. Leaves not glaucous on the undersurface . . . . 9. S. NIGRA
         10. Leaves glaucous on the undersurface

11. Stipules absent or highly reduced on early-season shoots, but often well-developed on late-season and vigorous shoots; branches yellowish to gray-brown, glabrous; leaf blades mostly 3–6 times as long as wide, the upper surface dull . . . . 2. S. AMYGDALOIDES
11. Stipules well-developed; branches red-brown to gray-brown, usually villous or tomentose to glabrescent; leaf blades 5–10 times as long as wide, the upper surface glossy . . . . 3. S. CAROLINIANA

9. Buds bluntly pointed at the tip; bud scale margins fused (forming a cap)
12. Branches and branchlets pendulous (weeping) (introduced hybrids; see discussion under *S. alba*) . . . . *S. ×pendulina, S. ×sepulcralis*
12. Branchlets and branchlets erect or ascending
13. Plants medium to large trees, (3–)10–25 m tall; petioles with glandular dots or stalked glands toward the tip; branches brittle at the base
14. Leaf blades persistently long-silky on the undersurface . . . . 1. S. ALBA
14. Leaf blades sparsely to moderately short silky-hairy when young, but becoming glabrous at maturity, (introduced hybrid; see discussion under *S. alba*). . . . . *S. ×fragilis*
13. Plants shrubs or small trees, 1–7 m tall; petioles not glandular at the tip; branches flexible at base
15. Leaves linear to very narrowly lanceolate or narrowly elliptic, mostly 9–20 times as long as wide, the margins with slender, widely spaced teeth; plants strongly colonial. . . . . 8. S. INTERIOR
15. Leaves narrowly oblong, narrowly elliptic, or lanceolate, 2.5–9.0 times as long as wide, the margins with broad, closely spaced teeth, sometimes nearly entire; plants not colonial
16. Stipules on at least shoots produced late in the season well-developed; branches yellow to yellowish or grayish brown . . . . 6. S. FAMELICA
16. Stipules absent or rudimentary; branches reddish brown to purple . . . . 10. S. PETIOLARIS

Key emphasizing staminate flowers

1. Stamens 3 or more
2. Leaf blades not glaucous on the undersurface . . . . 9. S. NIGRA
2. Leaf blades glaucous on the undersurface
3. Leaf blades mostly 3–6 times as long as wide; branchlets tan or grayish to yellowish brown, sometimes darkening and becoming reddish-tinged with age, glabrous . . . . 2. S. AMYGDALOIDES
3. Leaf blades mostly 5–10 times as long as wide; branchlets light brown to dark reddish brown, sometimes hairy . . . . 3. S. CAROLINIANA
1. Stamens 2 (the filaments fused but the 2 anthers distinct in *S. purpurea*)
4. Filaments fused (but the 2 anthers distinct); at least some of the leaves (and catkins) opposite or subopposite . . . . 11. S. PURPUREA
4. Filaments and anthers distinct or the filaments fused only near the base; leaves (and catkins) all alternate
5. Plants flowering before the leaves emerge; catkins sessile or on very short flowering branchlets
6. Branches highly brittle at the base, grayish brown to purple . . . . 12. S. SERICEA
6. Branches flexible at the base, reddish to yellowish brown

7. Plants shrubs or trees, 2–8 m tall; 3; anthers 0.5–1.0 mm long; catkin axis sometimes twisted; flowering branchlets 0–5 mm long . . . . 4. S. DISCOLOR
7. Plants shrubs, 0.3–3.0 m tall; anthers 0.4–0.6 mm long, catkin axis straight; flowering branchlets 0–1 mm long . . . . 7. S. HUMILIS

5. Plants flowering as or just before leaves emerge, catkins on distinct, leafy flowering branchlets (absent or to 2 mm long in *S. sericea*)
   8. Plants medium to large trees, (3–)10–25 m tall
      9. Branchlets and leaves densely pubescent with long, silky hairs; branches erect or strongly ascending; petioles long-silky . . . . 1. S. ALBA
      9. Branchlets and leaves hairy at first but soon becoming glabrous; branches pendant or drooping, sometimes erect in *S. ×fragilis*; petioles short-hairy (introduced hybrids; see discussion under *S. alba*) . . . . *S. ×fragilis, S. ×pendulina, S.×sepulcralis*
   8. Plants shrubs or small trees, 0.5–8.0 m tall, rarely taller
      10. Leaf blades mostly 10–20 times as long as wide, linear to very narrowly lanceolate or narrowly elliptic, the margins with widely spaced, sharply pointed teeth . . . . 8. S. INTERIOR
      10. Leaf blades mostly 2.5–9.0 times as long as wide, narrowly elliptic to narrowly oblong-elliptic, lanceolate, oblanceolate, or occasionally obovate, the margins with closely spaced teeth (sometimes scalloped in *S. famelica*)
         11. Filaments hairy toward the base
            12. Leaf blades with the undersurface densely pubescent with long, silky hairs, sometimes becoming nearly glabrous with age; branches flexible at the base . . . . 10. S. PETIOLARIS
            12. Leaf blades with the undersurface densely and persistently pubescent with short, silky hairs; branches brittle at the base . . . . 12. S. SERICEA
         11 Filaments glabrous
            13. Leaf blades with the undersurface densely hairy . . . . . . 12. S. SERICEA
            13. Leaf blades with the undersurface sparsely hairy
               14. Branches reddish brown . . . . 5. S. ERIOCEPHALA
               14. Branches yellowish to grayish brown . . . . 6. S. FAMELICA

Key emphasizing pistillate flowers

1. Ovaries hairy
   2. Pistillate catkins with the bracts shed after flowering; plants flowering as the leaves emerge . . . . 8. S. INTERIOR
   2. Pistillate catkins with the bracts persistent at fruiting; plants flowering before the leaves emerge (occasionally just as the leaves begin to appear in *S. sericea*)
      3. Fruits sessile; at least some of the leaves (and catkins) opposite or subopposite . . . . 11. S. PURPUREA
      3. Fruits on stalks 0.6–4.0 mm long; leaves (and catkins) alternate
         4. Stipules well-developed (occasionally highly reduced in *S. discolor*)
            5. Branchlets sparsely to densely pubescent with short, straight hairs, becoming glabrous or nearly so with age; leaf blades sparsely to moderately pubescent with short, curved hairs on the undersurface, the margins flat; plants tall shrubs, 2–6 m tall . . . . 4. S. DISCOLOR

5. Branches moderately to densely pubescent with velvety or woolly hairs, sometimes glabrous in patches with age; leaf blades sparsely to densely woolly on the undersurface, the margins rolled under; plants short to mid-sized shrubs, 0.3–3.0 m tall . . . . 7. S. HUMILIS
4. Stipules absent or highly reduced
6. Juvenile leaves sparsely to densely woolly; stems ascending from more or less spreading basal portion; margins of the median leaves strongly revolute . . . . 7. S. HUMILIS
6. Juvenile leaves silky-hairy; stems erect or strongly ascending; margins of the median leaves flat
7. Juvenile leaves pubescent with relatively long, silky hairs; branches flexible at the base; ovaries pear-shaped . . . . 10. S. PETIOLARIS
7. Juvenile leaves pubescent with relatively long, silky hairs; branches highly brittle at the base; ovaries ovoid . . . . 12. S. SERICEA
1. Ovaries glabrous
8. Plants shrubs or rarely small trees
9. Branches reddish brown . . . . 5. S. ERIOCEPHALA
9. Branches yellowish to grayish brown . . . . 6. S. FAMELICA
8. Plants medium to large trees
10. Buds bluntly pointed at the tip; bud scale margins fused (forming a cap)
11. Brances and branchlets erect or strongly ascending; leaf blades with the undersurface persistently long-silky . . . . 1. S. ALBA
11. Branches and branchlets weakly to strongly pendant (weeping); leaf blades with the undersurface becoming glabrous or nearly so at maturity (introduced hybrids; see discussion under *S. alba*) . . . . *S. ×fragilis, S. ×pendulina, S. ×sepulcralis*
10. Buds sharply pointed at the tip; bud scale margins free and overlapping on the side facing the stem
12. Leaves not glaucous on the undersurface . . . . 9. S. NIGRA
12. Leaves glaucous on the undersurface
13. Stipules absent or highly reduced (except on vigorous shoots); branchlets glabrous . . . . 2. S. AMYGDALOIDES
13. Stipules prominent; branchlets sparsely to densely hairy, becoming glabrous or nearly so with age . . . . 3. S. CAROLINIANA

1. Salix alba L. (white willow)
S. alba var. *caerulea* (Sm.) Sm.
S. alba var. *calva* G. Mey.
S. alba var. *vitellina* (L.) Stokes
Pl. 554 d–g; Map 2573

Plants medium to large trees, 10–25 m tall, usually not reproducing clonally. Trunks with the bark becoming deeply ridged and furrowed, brown to grayish brown. Branches flexible to somewhat brittle at the base, yellowish, grayish, or reddish brown. Branchlets yellowish or grayish to reddish brown, not glaucous, sparsely to occasionally densely pubescent with short to long, curved or spreading, sometimes silky hairs. Winter buds blunt at the tip, the scale margins fused. Leaves alternate. Petioles 3–13 mm long, with paired or clustered glandular dots, or stalked glands toward the tip, the upper side hairy. Stipules minute to well-developed, sharply pointed at the tip. Leaf blades 6–12 cm long, 4–8 times as long as wide, narrowly oblong to elliptic or lanceolate, usually tapered to a sharply pointed tip, angled or tapered at the base, the margins flat, sharply toothed, the upper surface dull, sparsely hairy to nearly glabrous, the undersurface glaucous, sparsely to densely pubescent with long, white, silky, appressed hairs. Catkins flowering as the leaves appear, on distinct, leafy, flowering branchlets; the bracts 1.5–2.8 mm long, entire, rounded at the tip, tawny, glabrous or sparsely and evenly hairy, those of the pistillate catkins not persistent at fruiting; the staminate catkins 2.5–6.0 cm long; the pistillate catkins 3–5 cm long. Staminate flowers with 2 stamens, the filaments free, hairy toward the base; nectaries 2, free. Pistillate flowers with the styles fused nearly to the tip, the stigmas flat or broadly cylindric; nectary 1. Fruits 3.5–5.0 mm long, on stalks 0.2–0.8 mm long. $2n$=76. April–May

Introduced, scattered widely in the state, most commonly in the western half (native of Europe, Asia; introduced widely in the U.S. [including Alaska and Hawaii], except for most of the Great Plains; also Canada). Banks of streams, margins of ponds, and edges of bottomland prairies; also moist disturbed areas.

White willow is an attractive, often large cultivated tree that escapes sporadically. Traditionally, several infraspecific taxa were recognized within *S. alba* (Steyermark, 1963). However, several authors, including Argus (1986), have treated all of these as cultivars (horticulturally developed variants). Of these, plants referred to as var. *vitellina*, which are characterized by erect to slightly pendant, yellowish brown, sparsely hairy branches, often are planted and occasionally may escape

Weeping willow, *S. babylonica* L. was treated by Steyermark (1963) as having become established outside of cultivation in Dent and Jackson Counties. He noted, however, that true *S. babylonica* is cold-hardy only in the southern third of the United States. Argus (1986, 2007) similarly mapped its range outside cultivation as occurring to the south of Missouri, and the species thus has been excluded from the Missouri flora. It should be noted that a few historical specimens document its cultivation in the St. Louis region in the 1830s, but there is no information on how successful such trees were during that era. Argus (1986) noted that plants growing farther north in the United States mostly represent a horticultural hybrid between *S. alba* and *S. babylonica* known as *S.* ×*sepulcralis* Simonk. (Pl. 554 h, i). This hybrid differs from *S. babylonica* in its usually hairy branches, slender pistillate catkins (2.8–4.8 times longer than wide) that are borne on noticeably elongate (3–14 mm) lateral branches, ovaries pear-shaped, and capsules 1–2 mm long. In contrast, true *S. babylonica* has glabrous branches, relatively stout pistillate catkins (1.3–2.2 times as long as wide) that are borne on at most short (to 4 mm) branches, ovaries egg- or turnip-shaped, and capsules 2.0–2.7 mm. *Salix* ×*sepulcralis* has been collected in the wild sporadically, primarily in the eastern portion of the state.

A second weeping willow that is less commonly cultivated in Missouri, *S.* ×*pendulina* Wender., arose as a horticultural hybrid between *S. babylonica* and *S. euxina* J. Belyaeva (see the discussion of *S.* ×*fragilis* below). It is morphologically very similar to *S.* ×*sepulcralis*, but differs in its more coarsely toothed leaves, its relatively loosely flowered staminate catkins, and in its fused, more or less cup-shaped staminate nectaries (vs. 2 free staminate nectaries in *S.* ×*sepulcralis*). In Missouri, *S.* ×*pendulina* is known as an apparent escape thus far only from historical specimens from Jasper and Marion Counties.

Another horticulturally derived hybrid with ascending (nonweeping) branches that has escaped sporadically in Missouri is *S.* ×*fragilis* L. (*S. alba* × *S. euxina*; Pl. 555 a–c) This hybrid was long known by the name *S.* ×*rubens* Shrank, but a recently approved proposal to conserve the type of the name *S. fragilis* based upon a specimen of the hybrid necessitates the name change (Belyaeva, 2009; Argus, 2010). True crack willow, which recently was renamed *S. euxina* (Belyaeva, 2009), is a western Asian species that is widely cultivated in Europe and occasionally also in North America, where it rarely if ever escapes (Argus, 2007). The Russian willow specialist, A. K. Skvortsov (1973), stated that all of the crack willow specimens collected in North America that he had examined turned out to represent the hybrid, rather than the parent. This hybrid was treated under the name *S. fragilis* by Steyermark (1963). It was first reported from Missouri as a hybrid by Mühlenbach (1983, as *S.* ×*rubens*). It differs from *S. alba* in having mature leaves glabrous (vs. densely long-silky-hairy), petioles sparsely hairy (vs. densely long-silky-hairy), branches highly brittle at the base (vs. flexible or only somewhat brittle), styles 0.4–1.0 mm (vs. 0.2–0.4 mm) long, and ovaries tapering to the styles (vs. slightly bulged below the styles).

2. Salix amygdaloides Andersson (peach-leaved willow)

Pl. 554 a–c; Map 2574

Plants small to large trees, 4–20 m tall, usually not clonal by suckering but sometimes clonal by stem fragmentation. Trunks with the bark becoming deeply ridged and furrowed, reddish brown to more commonly dark grayish brown. Branches flexible or less commonly somewhat brittle at the base, reddish brown to dark grayish brown. Branchlets tan or grayish to yellowish brown, sometimes darkening and becoming reddish-tinged with age, not glaucous, glabrous. Winter buds sharply pointed at the tip, the scale margins free

2575. Salix caroliniana 2576. Salix discolor 2577. Salix eriocephala

and overlapping along the side facing the stem. Petioles 7–21 mm long, lacking glands or with paired glandular dots at the tip, the upper side sometimes hairy. Leaves alternate. Stipules minute to well-developed in age, sometimes persistent through the growing season, rounded at the tip. Leaf blades 6–13 cm long, mostly 3–6 times as long as wide, narrowly elliptic to elliptic or lanceolate to narrowly oblanceolate or oblanceolate, tapered to a sharply pointed tip, angled, rounded, or shallowly cordate at the base, the margins flat, sharply toothed, the upper surface dull, glabrous to sparsely and inconspicuously hairy along the midvein, the undersurface glaucous, glabrous. Catkins flowering slightly before or as the leaves appear, on distinct, leafy, flowering branchlets; the bracts 1.5–2.8 mm long, entire or toothed, rounded to pointed at the tip, tawny, glabrous or sparsely to moderately hairy toward the base, those of the pistillate catkins not persistent at fruiting; the staminate catkins 2.3–8.0 cm long; the pistillate catkins 4–11 cm long. Staminate flowers with 3–7 stamens, the filaments free, hairy toward the base; nectaries 2, free. Pistillate flowers with the styles fused nearly to the tip, the stigmas short; nectary 1. Fruits 3–7 mm long, on stalks 1–3 mm long. $2n=38$. April–June.

Scattered in the state, most commonly in counties adjoining the Mississippi and Missouri Rivers (Washington, New York, south to Arizona, Texas, and West Virginia; Canada). Banks of streams and rivers, margins of ponds and lakes, bottomland forests, swamps, edges of fens and marshes, and bases of bluffs; occasionally also mine spoils.

Salix amygdaloides sometimes can be difficult to distinguish from the more abundant *S. nigra*. Further complicating the situation, where the species grow in proximity occasional hybrids have been reported. These have been called *S.* ×*glatfelteri* C.K. Schneid., based on plants studied by Noah Glatfelter in the St. Louis area. Glatfelter (1894), who investigated this hybrid complex in the field, estimated that more than forty percent of the populations in the St. Louis area contained hybrids. However, there are no recent collections, perhaps because of the widespread destruction of wetlands during subsequent urban expansion and the alteration of hydrological processes following damming and channelization of rivers. The hybrids have leaves with the narrower shape characteristic of *S. nigra*, but with at least some glaucousness on the undersurface, as in *S. amygdaloides*.

Glatfelter (1898) also reported the hybrid between *S. amygdaloides* and *S. caroliniana*, but this could not be confirmed (Argus, 1986, 2010).

3. Salix caroliniana Michx. (Carolina willow, Ward's willow)
S. longipes Shuttlew. ex Andersson
S. wardii Bebb

Pl. 555 f, g; Map 2575

Plants shrubs or small to medium trees, 3–10 m tall, clonal by suckers or stem fragmentation. Trunks with the bark variable, shallowly to deeply ridged and furrowed, reddish brown or gray. Branches brittle or occasionally somewhat flexible at the base, reddish brown to grayish brown. Branchlets light brown to dark reddish brown, not glaucous to strongly glaucous, sparsely to densely hairy, becoming glabrous or nearly so with age. Winter buds sharply pointed at the tip, the scale margins free and overlapping along the side facing the stem. Leaves alternate. Petioles 3–22 mm long, with paired glandular dots, clusters of dots, or rarely stalked glands at the tip, the upper side hairy. Stipules usually well-developed, rounded to pointed at the tip. Leaf blades 5–22 cm long, mostly 5–10 times as long as wide, narrowly oblong-elliptic or narrowly lanceolate to lanceolate, tapered to a sharply pointed tip, variously angled to rounded or rarely shallowly cordate at the base, the margins flat, sharply and finely toothed, the upper surface glossy, sparsely and inconspicuously hairy to glabrous, the undersurface glaucous, glabrous or more commonly sparsely hairy along the midvein. Catkins flowering as the leaves appear, on distinct, leafy, flowering branchlets; the bracts 1–3 mm long, entire or slightly uneven, rounded to pointed at the tip, tawny, sparsely and evenly hairy, those of the pistillate catkins not persistent at fruiting; the staminate catkins 3–10 cm long; the pistillate catkins 3–10 cm long. Staminate flowers with 4–7 stamens, the filaments free or fused toward the base, hairy at the base; nectaries 2, free. Pistillate flowers with the styles fused completely or nearly to the tip, the stigmas 2, flattened; nectary 1. Fruits 4–6 mm long, on stalks 1.3–5.3 mm long. April–May.

Scattered nearly throughout the state, except for the Mississippi Lowlands Division and the western portion of the Glaciated Plains (eastern [mostly southeastern] U.S. west to Kansas and Texas; Mexico, Central America, Caribbean Islands). Banks of streams, rivers, and spring branches, fens, and occasionally margins of ponds and lakes.

This relatively common willow mostly occurs in wetlands with some water flow. Glatfelter (1898) reported the uncommon hybrid between *S. caroliniana* and *S. nigra* from trees that he studied in the St. Louis area. Some of his suspected hybrids have leaves sparsely glaucous below but show no other signs of potential hybridization. Because the specimens could have lost some of their glaucousness in drying, all were redetermined as *S. caroliniana* (Argus 1986, 2010). Glatfelter also reported the hybrid between *S. amygdaloides* and *S. caroliniana*; but this could not be confirmed (Argus, 1986). Steyermark (1963) discussed hybrids between *S. caroliniana* and *S. eriocephala* (as *S. rigida*), but subsequent research also has not confirmed this report (Argus, 2010).

4. Salix discolor Muhl. (pussy willow)
Pl. 555 l, m; Map 2576

Plants shrubs or occasionally small trees, 2–4 m tall, sometimes clonal from suckers or stem fragmentation. Trunks with the bark finely ridged, gray to grayish brown. Branches flexible at the base, dark reddish brown or yellowish brown. Branchlets yellowish brown or more commonly reddish brown to dark purplish brown, variously glaucous or not, sparsely to densely pubescent with short, straight hairs, becoming glabrous or nearly so with age. Winter buds blunt at the tip, the scale margins fused. Leaves alternate. Petioles 6–17 mm long, lacking glands, the upper side hairy. Stipules minute to well-developed, sometimes persistent through the growing season, pointed at the tip. Leaf blades 3–9(–12) cm long, mostly 2–6 times as long as wide, narrowly elliptic to elliptic, oblanceolate, or obovate, angled or short-tapered to a bluntly or sharply pointed tip, broadly to less commonly narrowly angled at the base, the margins flat or somewhat wavy, bluntly and finely toothed or scalloped (rarely nearly entire), the upper surface shiny or dull, sparsely and inconspicuously hairy to glabrous, the undersurface glaucous, sparsely hairy or glabrous. Catkins flowering before the leaves appear, sessile or on very short flowering branchlets; the bracts 1.5–2.5 mm long, entire, pointed (sometimes very bluntly so) at the tip, black with brown marginal bands and/or a pale base, moderately to densely and evenly hairy, persistent at fruiting; the staminate catkins 2–5 cm long; the pistillate catkins 3–11 cm long. Staminate flowers with 2 stamens, the filaments free, sometimes hairy at the base; nectary 1. Pistillate flowers with the style fused to the tip, unbranched, the stigmas 2, linear; nectary 1. Fruits 6–11 mm long, on stalks 1–3 mm long. $2n$=76, 114. February–April.

Uncommon, northeastern Missouri (northeastern U.S. west to North Dakota, Missouri, and disjunctly Idaho, Montana, Wyoming, and South Dakota; also Canada). Fens; also railroads and moist disturbed areas.

The fertile branches of the pussy willow often are cut and gathered in late-winter and early spring for display indoors in vases. The attractive budding catkins with their bracts bearing soft, silky hairs have a fanciful resemblance to the pads of cats' feet. Two Old World species, *S. caprea* L. (goat willow) and *S. cinerea* L. (gray willow) are also cultivated in the Midwest as pussy willows. These have not escaped from cultivation in Missouri, but may persist from plantings for many years. A specimen of *S. cinerea* accessioned at the Missouri Botanical Garden herbarium was collected in 2004 in Webster County (*Yatskievych et al., 04-210*) along the wooded margin of a sinkhole pond. There was only a solitary old tree (which thus could not reproduce by seed), obviously long-persistent from a planting at a former homesite. *Salix cinerea* resembles the native *S. discolor*, but differs in its much more densely woolly young leaves with the fine tertiary veins more prominently raised and often more or less parallel (vs. not raised and irregular). Also, if the bark is removed from three- to five-year old branches, the wood of *S. cinerea* has prominent longitudinal ridges and lines called striae (these absent or short and indistinct in *S. discolor*).

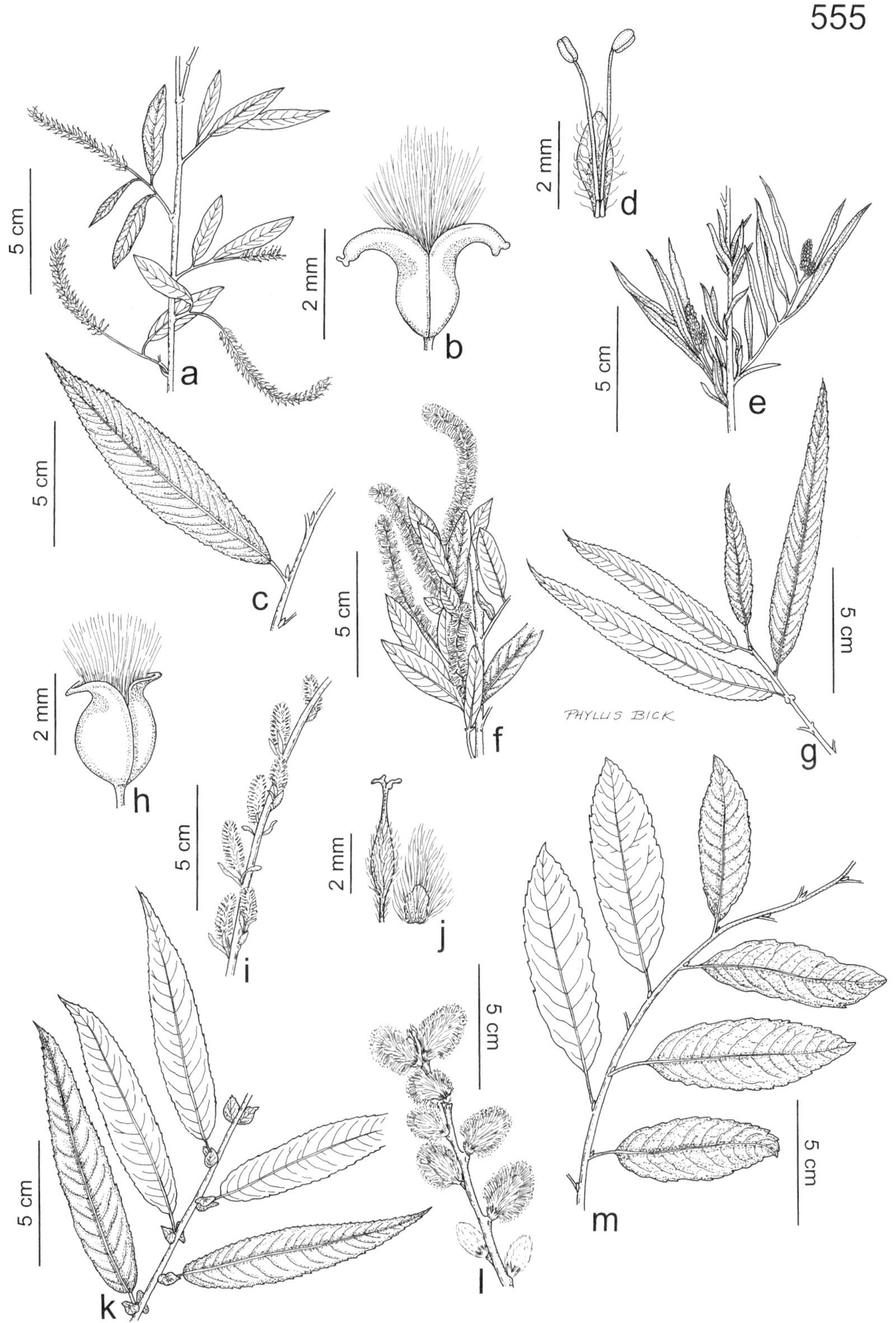

Plate 555. Salicaceae. *Salix ×fragilis*, **a)** portion of branch with pistillate catkins and developing leaves, **b)** fruit, **c)** portion of branch with leaf. *Salix interior*, **d)** staminate flower, **e)** portion of branch with pistillate catkins and leaves. *Salix caroliniana*, **f)** portion of branch with staminate catkins and developing leaves, **g)** branch tip with leaves. *Salix eriocephala*, **h)** fruit, **i)** portion of branch with staminate catkins and developing leaves, **j)** pistillate flower with bract, **k)** portion of branch with leaves and stipules. *Salix discolor*, **l)** portion of branch with staminate catkins, **m)** portion of branch with leaves.

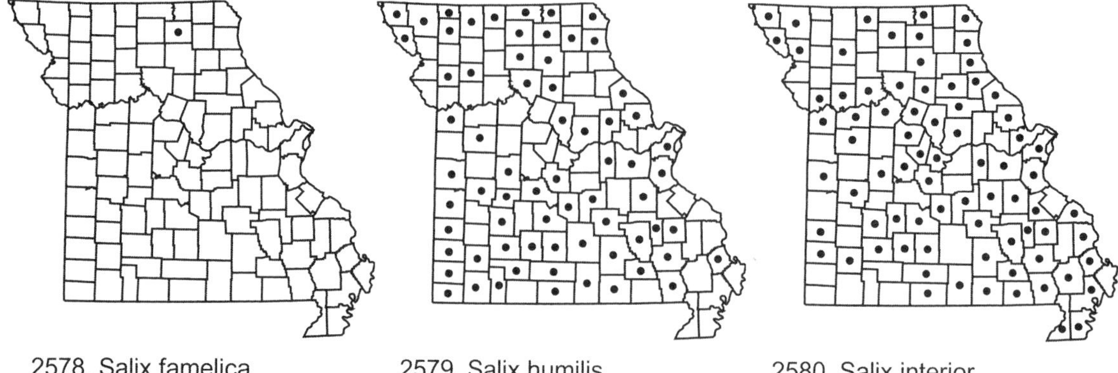

2578. Salix famelica 2579. Salix humilis 2580. Salix interior

Currently, *Salix discolor* has been documented as a native component of the flora only from a historical collection from Clark County and a more recent one from Schuyler County. It should be searched for in other wetlands in the northeastern portion of the state. Isolated occurrences farther south represent escapes from cultivation. A hybrid between *S. discolor* and *S. humilis* (*Salix* ×*conifera* Wangenh.) has been documented from the mixed population in northern Schuyler County. It has the woolly leaf pubescence of *S. humilis* and the longer catkins and styles of *S. discolor*.

5. Salix eriocephala Michx. (diamond willow, Missouri willow)

S. missouriensis Bebb

S. rigida Muhl.

S. rigida f. *mollis* (E.J. Palmer & Steyerm.) Fernald

S. rigida f. *subintegra* (E.J. Palmer & Steyerm.) Steyerm.

S. rigida var. *angustata* (Pursh) Fernald

S. rigida var. *vestita* (Andersson) C.R. Ball

Pl. 555 h–k; Map 2577

Plants shrubs or small to rarely medium trees, 2–6(–15) m tall, clonal from suckers or stem fragmentation. Trunks with the bark relatively smooth or on older trees sometimes finely furrowed, grayish brown to dark brown. Branches usually brittle at the base, reddish brown. Branchlets yellowish brown to reddish brown, not glaucous, sparsely to densely hairy, occasionally becoming glabrous or nearly so with age. Winter buds blunt or rounded at the tip, the scale margins fused toward the base. Leaves alternate. Petioles 3–18 mm long, lacking glands, the upper side hairy. Stipules well-developed, often shed early, rounded to pointed at the tip. Leaf blades 5–10(–14) cm long, mostly 2–5 (–8) times as long as wide, narrowly oblong to narrowly elliptic or occasionally obovate, angled or short-tapered to a sharply pointed tip, broadly angled to rounded or shallowly cordate at the base, the margins flat, finely toothed, the upper surface shiny or dull, sparsely hairy, the undersurface glaucous, sparsely hairy, often glabrous with age. Catkins flowering as or slightly before the leaves appear, on distinct, leafy, flowering branchlets; the bracts 0.8–1.6 mm long, entire, rounded at the tip, dark brown, sometimes with lighter margins and/or a pale base, moderately to densely and evenly hairy, persistent at fruiting; the staminate catkins 2.0–4.5 cm long; the pistillate catkins 2–7 cm long. Staminate flowers with 2 stamens, the filaments free or fused toward the base, glabrous; nectary 1. Pistillate flowers with the styles fused to the tip, unbranched, the stigmas 2, relatively broad or plump; nectary 1. Fruits 3.5–7.0 mm long, on stalks 0.6–1.6(–2.8) mm long. $2n$=38. April–May.

Scattered nearly throughout the state, but uncommon or absent from most of the Mississippi Lowlands Division (eastern [mostly northeastern] U.S. west to North Dakota, Colorado, and Kansas; Canada). Banks of streams, rivers, and spring branches, sloughs, bases of bluffs, bottomland prairies, wet swales in upland prairies, and fens; also old mine sites, ditches, and moist disturbed areas.

Hybrids between *S. eriocephala* and *S. sericea* are produced where the ranges of the species overlap (Argus, 2010). However, the only examples known thus far from Missouri were collected in the St. Louis area by Noah Glatfelter in the 1890s. Generally, such hybrids tend to resemble *S. eriocephala*, in their prominent stipules, but have leaves that are sparsely to moderately densely short-silky on the undersurface and ovaries that are hairy, as in *S. sericea*. Hardig et al. (2000) studied morphological variation in a mixed population in New York, concluding that the variation is complex and that hybrid plants are not always intermediate between the parents. Some authors have referred to this hybrid under the names *S.* ×*myricoides* Muhl. and *S.* ×*bebbii* Gand.

(Steyermark, 1963; Yatskievych and Turner, 1990), but the former refers to a different species that does not grow in Missouri and the latter name is invalid. Steyermark (1963) also discussed hybrids between *S. caroliniana* and *S. eriocephala* (as *S. rigida*), but subsequent research has not confirmed this report.

6. Salix famelica (C.R. Ball) Argus (hungry willow)

S. eriocephala Michx. var. *famelica* (C.R. Ball) Dorn

Map 2578

Plants shrubs or small trees, 1.5–7 m tall, sometimes clonal by suckers. Trunks with the bark relatively smooth or on older trees sometimes finely ridged, usually grayish brown. Branches flexible at the base, yellowish brown to grayish brown. Branchlets yellow to yellowish or grayish brown, not or only slightly glaucous (then appearing somewhat sparkly), glabrous or sparsely to moderately hairy). Winter buds blunt at the tip, the scale margins fused. Petioles 3–27 mm long, lacking glands, the upper side glabrous or hairy. Leaves alternate. Stipules well-developed, rounded to pointed at the tip. Leaf blades 3–12 cm long, mostly 2.5–7.0 times as long as wide, narrowly oblong-lanceolate to narrowly elliptic or lanceolate, angled or tapered to a sharply pointed tip, broadly angled to rounded or rarely shallowly cordate at the base, the margins flat, finely toothed or scalloped, the upper surface shiny or dull, sparsely hairy, especially along the midvein, the undersurface glaucous, sparsely hairy, especially along the midvein. Catkins flowering slightly before or with the leaves, on distinct, leafy, flowering branchlets; the bracts 0.8–1.6 mm long, entire, rounded at the tip, tawny to dark brown, moderately to densely hairy, the hairs sometimes mostly toward the base, persistent at fruiting; the staminate catkins 1.5–4.5 cm long; the pistillate catkins 1.6–7.5 cm long. Staminate flowers with 2 stamens, the filaments free or fused toward the base, glabrous; nectary 1. Pistillate flowers with the styles fused to the tip, unbranched, the stigmas 2, linear but flattened; nectary 1. Fruits 4–6 mm long, on stalks 0.7–2.8 mm long. $2n$=38. April–May.

Uncommon, known thus far from a single population in Adair County (Montana to Minnesota south to Wyoming and Missouri; Canada). Banks of streams.

Salix famelica was first collected in Missouri by Craig Anderson (then a biologist with the Missouri Department of Conservation) from near a bridge crossing a creek along one of the hiking trails in Thousand Hills State Park. It is still known definitely only from this population, which includes both staminate and pistillate plants. However, *S. famelica* should be searched for in other northeastern Missouri wetlands, because a specimen from a fen in northern Schuyler County apparently represents the hybrid between *Salix famelica* and *S. petiolaris*. For further discussion of this hybrid, see the treatment of *S. petiolaris*.

7. Salix humilis Marshall (prairie willow)

Pl. 556 g, h; Map 2579

Plants shrubs, 0.5–3.0 m tall, strongly clonal by layering. Stems with the bark roughened, on older stems usually with shallow, irregular fissures, gray to grayish brown. Branches flexible at the base, reddish brown. Branchlets yellowish to reddish or greenish brown, not or only slightly glaucous, moderately to densely velvety- or woolly-hairy, sometimes glabrous in patches with age. Winter buds blunt at the tip, the scale margins fused. Petioles 0.5–7.0(–12.0) mm long, lacking glands, the upper side hairy. Leaves alternate. Stipules minute or well-developed, pointed at the tip. Leaf blades (1–)2–9(–13) cm long, mostly 2–9 times as long as wide, narrowly oblong, narrowly elliptic, elliptic, oblanceolate, obovate or broadly obovate, angled or short-tapered to a bluntly or sharply pointed tip, angled at the base, the margins strongly rolled under (to flat), the edge finely scalloped or wavy to nearly entire, the upper surface glossy, glabrous to sparsely or moderately hairy, especially along the midvein, the undersurface glaucous, moderately to densely pubescent with usually fine, woolly hairs. Catkins flowering before the leaves appear, sessile or on very short flowering branchlets; the bracts 0.8–2.0 mm long, entire, rounded or pointed at the tip, dark brown to black, sparsely to moderately and evenly hairy, persistent at fruiting; the staminate catkins 0.6–3.5 cm long; the pistillate catkins 1–5 cm long. Staminate flowers with 2 stamens, the filaments free, glabrous or hairy; nectary 1. Pistillate flowers with the styles fused to the tip, unbranched, the stigmas 2, linear; nectary 1. Fruits 5–12 mm long, on stalks 1.2–2.0 mm long. $2n$=38, 76. March–May.

Scattered nearly throughout the state (eastern U.S. west to North Dakota and Texas; Canada). Upland prairies, bottomland prairies, loess hill prairies, savannas, edges of mesic to dry upland forests, glades, bases, ledges, and tops of bluffs, and less commonly fens and banks of streams and spring branches; also pastures, old fields, roadsides, and open disturbed areas.

This distinctive willow tends to form small thickets in upland habitats. The present treatment

2581. Salix nigra   2582. Salix petiolaris   2583. Salix purpurea

follows that of Argus (1986, 2010), who treated *S. humilis* as comprising two varieties.

1. Stems erect or ascending; stipules well-developed, leaflike; leaf blades (2–)5–9(–13) cm long; staminate catkins 1.5–3.5 cm long; pistillate catkins (1.5–)2.0–4.5 cm long . . . . 7A. VAR. HUMILIS
1. Stems spreading to loosely ascending with more strongly ascending tips; stipules absent or minute, sometimes absent; leaf blades (1.5–)2.0–5.0(–7.0) cm long; staminate catkins 0.6–1.4 cm long; pistillate catkins 1.0–1.8 cm long . . . . 7A. VAR. TRISTIS

7a. var. humilis (prairie willow, gray willow)
S. humilis var. *hyporhysa* Fernald

Plants 0.3–3.0 m tall, the stems erect or ascending. Branchlets mostly reddish to greenish brown or brown. Petioles (1.5–)3.0–7.0(–12.0) mm long. Stipules well-developed, leaflike, pointed at the tip. Leaf blades (2–)5–9(–13) cm long. Staminate catkins 1.5–3.5 cm long. Pistillate catkins (1.5–)2.0–4.5 cm long. Fruits 7–12 mm long. $2n$=38, 76. March–May.

Scattered nearly throughout the state (eastern U.S. west to North Dakota and Texas; Canada). Upland prairies, bottomland prairies, loess hill prairies, savannas, edges of mesic to dry upland forests, glades, bases, ledges, and tops of bluffs, and less commonly fens and banks of streams and spring branches; also pastures, old fields, roadsides, and open disturbed areas.

This is the more commonly encountered variety in the state. The hybrid between *S. discolor* and *S. humilis* (*S.* ×*conifera* Wangenh.) has been documented from a mixed population at a fen in northern Schuyler County. It has the woolly leaf pubescence of *S. humilis* and the longer catkins and styles of *S. discolor*.

7b. var. tristis (Aiton) Griggs (dwarf gray willow, sage willow)
S. humilis var. *microphylla* (Andersson) Fernald

Plants 0.3–1.0 m tall, the stems spreading to loosely ascending with more strongly ascending tips. Branchlets mostly yellowish brown. Petioles 0.5–3.0(–6.0) mm long. Stipules absent or minute. Leaf blades (1.5–)2.0–5.0(–7.0) cm long. Staminate catkins 0.6–1.4 cm long. Pistillate catkins 1.0–1.8 cm long. Fruits 5–9 mm long. March–May.

Scattered to uncommon nearly throughout the state, but apparently absent from most of the Mississippi Lowlands Division (eastern U.S. west to Minnesota and Oklahoma). Upland prairies, loess hill prairies, savannas, glades, edges of mesic to dry upland forests, and less commonly fens and banks of streams; also pastures, old fields, roadsides, and open disturbed areas.

The dwarf variety of *S. humilis* occurs at scattered sites within the range of var. *humilis*. Argus (1986) called this taxon var. *microphylla*, but the epithet var. *tristis* is the oldest validly published varietal epithet. Some herbarium specimens of *S. humilis* are difficult to determine at the varietal level. This may stem from incomplete herbarium specimens comprising only a small portion of a plant rather than of actual intermediacy for vegetative characters.

8. Salix interior Rowlee (sandbar willow)
S. interior f. *wheeleri* (Rowlee) Rouleau
S. exigua Nutt. ssp. *interior* (Rowlee) Cronquist
S. exigua var. *sericans* (Nees) Dorn
Pl. 555 d, e; Map 2580

Plants shrubs or small trees, 4–6(–10) m tall, strongly clonal from root suckers. Trunks with the bark relatively smooth or on older trees sometimes finely plated or ridged, gray or brown. Branches flexible at the base, gray or yellowish brown.

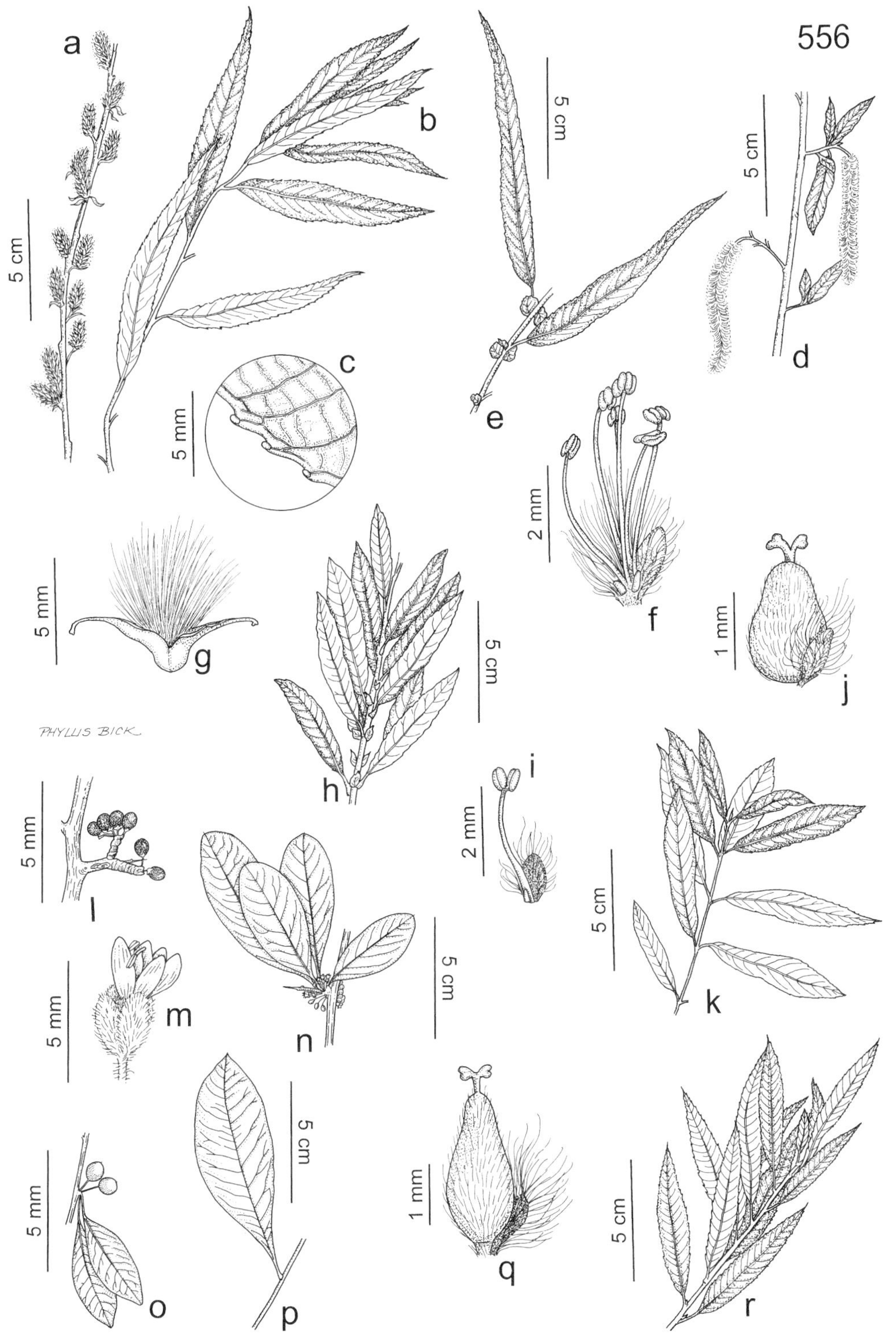

Plate 556. Salicaceae. Sapotaceae. *Salix petiolaris*, **a)** portion of branch with staminate catkins and developing leaves, **b)** branch tip with leaves, **c)** leaf detail. *Salix nigra*, **d)** portion of branch with staminate catkins and developing leaves, **e)** portion of branch with leaves and stipules, **f)** staminate flower with bract. *Salix humilis*, **g)** fruit, **h)** portion of branch with leaves and stipules. *Salix purpurea*, **i)** staminate flower with bract, **j)** pistillate flower with bract, **k)** branch tip with leaves. *Sideroxylon lanuginosum*, **l)** portion of branch with leaves and fruits, **m)** flower, **n)** node with leaves, thorn, and inflorescence. *Sideroxylon lycioides*, **o)** portion of branch with leaves and fruits, **p)** node with leaf. *Salix sericea*, **q)** pistillate flower with bract, **r)** branch tip with leaves.

Branchlets yellowish or reddish brown, not glaucous, sparsely to densely hairy, sometimes becoming glabrous or nearly so with age. Winter buds blunt at the tip, the scale margins fused. Leaves alternate. Petioles 1–5(–9) mm long, lacking glands, the upper side glabrous or hairy. Stipules absent or minute, those of leaves produced later in the season occasionally well-developed, then shed early, rounded at the tip. Leaf blades 6–16 cm long, mostly 10–20 times as long as wide, linear to narrowly lanceolate or narrowly elliptic, tapered to a sharply pointed tip, angled at the base, the margins flat, with widely spaced, elongate, sharply pointed teeth, the upper surface shiny, sparsely to moderately hairy, mostly along the midvein, the undersurface glaucous, sparsely to densely hairy. Catkins flowering as the leaves appear (plants elsewhere sometimes flowering from secondary shoots throughout the growing season), on distinct, leafy, flowering branchlets; the bracts 1.5–3.5 mm long, entire or somewhat irregular to toothed, rounded or pointed at the tip, tawny, glabrous or sparsely hairy at the base and tip, those of the pistillate catkins not persistent at fruiting; the staminate catkins 2–6 cm long; the pistillate catkins 2–7 cm long. Staminate flowers with 2 stamens, the filaments free, hairy toward the base; nectaries 2, free. Pistillate flowers with the styles fused to the tip, unbranched, the stigmas 2, flattened to broadly cylindric or slightly club-shaped; nectary 1. Fruits 4–10 mm long, on stalks 0.5–0.8 mm long. $2n=38$. May–June.

Scattered throughout the state (Washington to Maine south to Texas, Mississippi, and Tennessee; Canada, Mexico).

This species is relatively easy to recognize because of the often yellowish green, nonglaucous, narrow leaves with widely spaced, slender teeth along the margins. It belongs to the *S. exigua* complex. The taxonomic level at which *S. interior* should be recognized continues to be controversial. Argus (1986) treated it informally within a broadly circumscribed *S. exigua*. Dorn (1998, 2001) accepted it as a subspecies and then a variety of *S. exigua*. Both of these authors discussed morphological overlap in the region where the ranges of the two taxa overlap. However, Brunsfeld et al. (1991, 1992) presented evidence from analyses of allozyme and chloroplast DNA restriction site variation that within the complex *S. exigua* and *S. interior* are closely related, but reasonably distinct genetically. The present treatment follows that of Argus (2007, 2010) in tentatively recognizing *S. interior* as a species separate from *S. exigua*, which is widespread in the western United States and adjacent Canada. The ranges of the two species overlap in the western Great Plains and portions of the Pacific Northwest.

9. Salix nigra Marshall (black willow)

Pl. 556 d–f; Map 2581

Plants medium to large trees, 5–20(–30) m tall, not producing suckers but occasionally clonal by stem fragmentation. Trunks with the bark becoming deeply furrowed, variously light grayish brown to black. Branches brittle at the base, reddish brown to yellowish brown. Branchlets grayish to reddish brown, not glaucous, glabrous or sparsely hairy, rarely moderately to densely hairy. Winter buds sharply pointed at the tip, the scale margins free and overlapping along the side facing the stem. Leaves alternate. Petioles (2–)3–10(–15) mm long, with a pair of glandular dots at the tip, the upper side sometimes hairy. Stipules usually well-developed, at least on shoots produced later in the season, rounded or more commonly pointed at the tip. Leaf blades (5–)7–12(–19) cm long, mostly 6–13 times as long as wide, narrowly elliptic, narrowly lanceolate, or linear, occasionally lanceolate or elliptic, tapered to a sharply pointed tip, angled at the base, the margins flat, finely toothed, the upper surface shiny, glabrous to sparsely hairy, the undersurface not glaucous, sparsely hairy. Catkins flowering as the leaves appear, on distinct, leafy, flowering branchlets; the bracts 1–3 mm long, entire, rounded or pointed at the tip, tawny, sparsely and evenly hairy, those of the pistillate catkins not persistent at fruiting; the staminate catkins 3.5–8.0 cm long; the pistillate catkins 2.5–7.0 cm long. Staminate flowers with 4–6 stamens, the filaments free or occasionally fused at the base, hairy near the base; nectaries 2, free or fused into a cup-shaped structure. Pistillate flowers with the style fused to the tip, unbranched, the stigmas 2, short; nectary 1. Fruits 3–5 mm long, on stalks 0.5–1.5 mm long. $2n=38$. April–May.

Scattered to common throughout the state (eastern U.S. west to Minnesota, Kansas, and Texas; Canada, Mexico). Banks of streams, rivers, and spring branches, margins of ponds, lakes, oxbows, and sinkhole ponds, sloughs, bottomland forests, swamps, marshes, and bases of bluffs; also ditches, bases of levees, railroads, roadsides, and moist disturbed areas.

This is the commonest willow in Missouri. Although it grows in a variety of different wetlands, it is most commonly found in muddy soils associated with still waters. *Salix nigra* hybridizes with *S. amygdaloides* where the two species grow in proximity. Such hybrids have been called *S. ×glatfelteri* C.K. Schneid. (not a validly published name). Glatfelter (1898) also reported the uncommon hybrid between *S. caroliniana* and *S. nigra* from trees that he studied in the St. Louis area, but these specimens have been redetermined as *S. caroliniana*.

10. Salix petiolaris Sm. (meadow willow)
S. gracilis Andersson
S. gracilis var. *textoris* Fernald
Pl. 556 a–c; Map 2582

Plants shrubs, 1–3(–6) m tall, sometimes clonal from suckers. Trunks with the bark relatively smooth or somewhat roughened, on older trees becoming flaky, grayish green to reddish brown, on older trunks becoming brown. Branches flexible at the base, reddish brown to dark purple. Branchlets yellowish to reddish brown, not or only slightly glaucous, sparsely to densely velvety-hairy. Winter buds blunt at the tip, the scale margins fused. Leaves alternate. Petioles 3–12 mm long, lacking glands, the upper side usually hairy. Stipules minute or absent. Leaf blades 3–15 cm long, mostly 5–9 times as long as wide, very narrowly elliptic to narrowly oblong-elliptic or narrowly oblong-oblanceolate, angled or tapered to a bluntly or sharply pointed tip, angled at the base, the margins flat or slightly rolled under, finely toothed (sometimes entire elsewhere), the upper surface dull or shiny, glabrous or sparsely silky-hairy, the undersurface glaucous, densely pubescent with long, silky hairs, sometimes becoming nearly glabrous with age. Catkins flowering as the leaves appear, on distinct, leafy, flowering branchlets; the bracts 1–2 mm long, entire, rounded at the tip, tawny, often pinkish-tinged or paler toward the margins, sparsely and evenly hairy, not persistent at fruiting; the staminate catkins 1–3 cm long; the pistillate catkins 1–4 cm long. Staminate flowers with 2 stamens, the filaments free, hairy at the base; nectary 1. Pistillate flowers with the styles fused to the tip or nearly so, often minutely branched at the tip, the stigmas 2, short, linear; nectary 1. Fruits 5–9 mm long, on stalks 1.5–4.0 mm long. $2n=38$. April–June.

Uncommon, known thus far from a single historical collection from Dent County (northeastern U.S. west to South Dakota and Nebraska, also disjunct in Colorado and Missouri; Canada). Banks of streams and rivers.

Steyermark's (1963) report of the isolated occurrence of this species in Dent County is still the only verified specimen of *S. petiolaris* from Missouri. However, the species should be searched for in the northeastern portion of the state. A specimen at the Missouri Botanical Garden Herbarium collected in 1991 in northern Schuyler County along the Iowa border (*Ladd & Heumann 15127B*) appears to represent the hybrid *Salix famelica* × *S. petiolaris*. The locality, known as State Line Fen, harbors a diversity of willows, but neither of the parents has yet been documented from the site. More intensive inventory of *Salix* species should be conducted at this fen. This interspecific hybrid resembles *S. famelica* in its well-developed stipules on later leaves and its yellowish brown branchlets. It resembles *S. petiolaris* in producing rust-colored hairs on the juvenile leaves and is intermediate in leaf shape and in ovary pubescence.

11. Salix purpurea L. (basket willow, purple osier)
Pl. 556 i–k; Map 2583

Plants shrubs or small trees, 1.5–5.0 m tall, usually not suckering but sometimes clonal by stem fragmentation. Trunks with the bark relatively smooth or somewhat roughened, on older trees becoming finely and irregularly furrowed, grayish brown to gray. Branches flexible to somewhat brittle at the base, yellowish brown to olive brown. Branchlets yellowish brown to greenish brown, sometimes reddish- or purplish-tinged, not glabrous. Winter buds blunt at the tip, the scale margins fused. Leaves mostly alternate, but at least some of them opposite to subopposite. Petioles 2–7 mm long, lacking glands, the upper side glabrous. Stipules minute, often appearing absent. Leaf blades 3–8 cm long, mostly 3–9 times as long as wide, narrowly oblong-elliptic to narrowly oblanceolate or oblanceolate, angled or tapered to a sharply pointed tip, rounded or angled at the base, the margins rolled under, finely toothed or occasionally entire or nearly so, the upper surface more or less dull, glaucous, glabrous or sparsely hairy. Catkins flowering before the leaves appear, sessile or on very short flowering branchlets; the bracts 0.8–1.5 mm long, entire, rounded at the tip, dark brown to black, sometimes paler toward the margins and/or base, sparsely and evenly hairy, persistent at fruiting; the staminate catkins 2.5–3.5 cm long; the pistillate catkins 1.5–3.5 cm long. Staminate flowers with 2 stamens, the filaments fused, often to the tip (then only 1 apparent stamen, but with 2 anthers), hairy at the base; nectary 1. Pistillate flowers with the styles fused to the tip, very short, unbranched, the stigmas 2, short and flattened; nectary 1. Fruits 2.5–5.0 mm long, sessile. $2n=38$. April–May.

Introduced, uncommon, known thus far only from historical collections in Ralls and St. Louis Counties (native of Europe; introduced widely in the northeastern U.S. west to Minnesota and Missouri, disjunctly in California, Oregon, and Utah; also Canada). Margins of ponds and lakes; also moist disturbed areas.

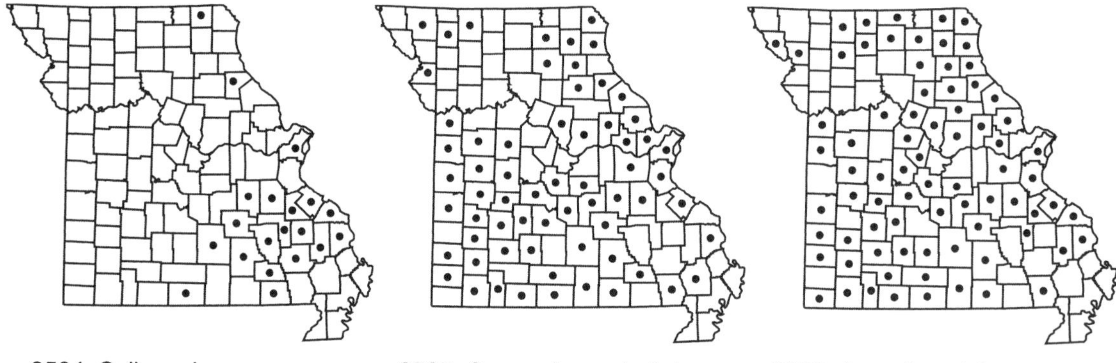

2584. Salix sericea 2585. Comandra umbellata 2586. Aesculus glabra

Steyermark (1963) noted that *Salix purpurea* was brought to the United States in early colonial times and that its branches have been used in basketry.

12. Salix sericea Marshall (silky willow)
S. sericea f. *glabra* E.J. Palmer & Steyerm.
Pl. 556 q, r; Map 2584

Plants shrubs or less commonly small trees, 0.5–4.0 m tall, usually not suckering but sometimes clonal by stem fragmentation. Trunks with the bark somewhat roughened or on older trees becoming finely furrowed, greenish brown to grayish brown or gray. Branches brittle at the base, grayish brown to dark purple. Branchlets dark reddish brown to dark purple, sometimes mottled with yellowish brown, not glaucous, sparsely to densely velvety-hairy. Winter buds blunt at the tip, the scale margins fused. Leaves alternate. Petioles 3.5–12.0(–21.0) mm long, sometimes with a pair of glandular dots at the tip, the upper side hairy. Stipules minute or absent. Leaf blades (5–)7–10(–13) cm long, mostly 5–11 times as long as wide, narrowly oblong-lanceolate to narrowly elliptic or narrowly oblong-elliptic, angled or tapered to a sharply pointed tip, angled at the base, the margins flat, finely toothed, the upper surface dull, sparsely hairy, often becoming glabrous or nearly so at maturity, glaucous (this sometimes obscured by pubescence), densely pubescent with short, silky hairs (especially when young). Catkins flowering before the leaves appear or occasionally just as the leaves begin to appear, sessile or on very short flowering branchlets; the bracts 0.8–1.5 mm long, entire, rounded at the tip, dark brown to black, sparsely and evenly hairy, persistent at fruiting; the staminate catkins 1.5–4.0 cm long; the pistillate catkins 1.8–4.5 cm long. Staminate flowers with 2 stamens, the filaments free or fused toward the base, glabrous or hairy toward the base; nectary 1. Pistillate flowers with the styles fused to about the midpoint or to the tip, unbranched or 2-branched, the stigmas 2, cylindric; nectary 1. Fruits 2.5–4.0 mm long, on stalks 0.6–1.5 mm long. March–May.

Scattered in the southeastern quarter of the state, north locally to Ralls and Scotland Counties (eastern [mostly northeastern] U.S. west to Iowa and Arkansas; Canada). Banks of streams, rivers, and spring branches, margins of ponds and lakes, bottomland forests, swamps, and fens; also roadsides.

Steyermark (1963) noted that in Missouri this species tends to form densely rounded shrubs and that the grayish silvery leaf undersurfaces contrast nicely with the dark reddish brown to dark purple branchlets. He considered the species to be a member of the relictual flora persisting somewhat disjunctly in cool microclimates in Missouri from a generally broader distribution during Pleistocene glaciation. Argus (1986) remarked that specimens of *S. sericea* with mature foliage sometimes are misdetermined as *S. eriocephala*. For a discussion of hybridization between *S. sericea* and *S. eriocephala*, see the treatment of that species.

SANTALACEAE (Sandalwood Family)

About 35 genera, 400–540 species.

Although it is not apparent from examination of the above-ground portions of the plants, members of the Santalaceae are all obligate parasites that derive water and part of their nutrition from the roots of host plants, in spite of being green and carrying out photosynthesis. Many members of the family are trees or shrubs, and it was only when their root systems were studied by botanists that the haustorial connections to host species were discovered. The genus *Santalum* L., sandalwood, produces a commercially important, resinous wood used in cabinet making. The ground wood and the extracted oils have a long history of use in incense, perfumes, and other cosmetics.

Recent molecular studies have suggested that the traditional Santalaceae might best be broken up into several smaller families within the order Santalales (Der and Nickrent, 2008). In the restricted sense, the family Comandraceae would comprise only two genera, *Comandra* and the North American endemic, *Geocaulon* Fernald (one species). A more conservative interpretation of these data results in an enlarged concept of Santalaceae to include the aforementioned genera, as well as the genera traditionally included in the dwarf mistletoe family, Viscaceae (Angiosperm Phylogeny Group, 2009).

1. **Comandra** Nutt. (bastard toadflax) (Piehl, 1965)

One species, North America, Europe.

1. Comandra umbellata (L.) Nutt. **ssp. umbellata**

C. richardsiana Fernald

Pl. 557 e–g; Map 2585

Plants parasitic on the roots of other plants, perennial herbs with an often extensive system of deepset rhizomes. Stems 10–25(–40) cm long, ascending, glabrous. Leaves alternate, sessile or short-petiolate, glabrous. Stipules absent. Leaf blades 0.7–5.0 cm long (short and scalelike near the stem base), simple, elliptic to oblanceolate, lanceolate, or ovate, the margins entire, narrowed to a bluntly or sharply pointed tip, narrowed or tapered at the base, sometimes somewhat lighter-colored on the undersurface. Inflorescences dense somewhat umbellate clusters of flowers axillary in the uppermost leaves, these short-stalked, the whole inflorescence thus appearing as a leafy panicle. Flowers perfect, epigynous, actinomorphic, the hypanthium 2–3 mm long, narrowly bell-shaped to obconical, with a prominent shallowly lobed nectary on the inner surface. Sepals (4)5, 2–3 mm long, oblong-lanceolate to narrowly ovate, white, loosely ascending at flowering, persistent at fruiting, usually minutely hairy on the upper surface and with a small tuft of longer hairs at the base, these hairs more or less fused with the adjacent anther. Petals absent. Stamens (4)5, attached opposite the sepals along the hypanthium rim, the filaments short, the anthers yellow. Pistil 1 per flower, usually of 3 fused carpels. Ovary inferior, 1-locular, the placentation free-central, with (2)3 ovules. Style 1, the stigma capitate. Fruits 1-seeded drupes, 4–6 mm in diameter, spherical or nearly so, dark brown at maturity. $2n=28, 52$. May–July.

Scattered nearly throughout the state (eastern U.S. west to North Dakota and Oklahoma; Canada). Dry upland forests, ledges and tops of bluffs, glades, savannas, and upland prairies.

Piehl (1965) reduced *Comandra* to a single species composed of four subspecies differing in details of sepal shape and size, rhizome anatomy, fruit shape, and degree of glaucousness of the plants. The Missouri subspecies is the only one in the eastern United States; ssp. *elegans* (Rochel ex Rchb.) Piehl is the Old World phase of the species and ssp. *pallida* (A. DC.) Piehl and ssp. *californica* (Eastw. ex Rydb.) Piehl occur in western North America.

Native Americans used this species variously for cuts and sores, and also for colds and other respiratory ailments (Moerman, 1998). *Comandra umbellata* is a nonspecialized parasite in that it appears to parasitize any and all species with which it is able to come into contact. Hedgcock (1915) recorded it on more than 50 host species in a number of different families, including herbs and woody plants, monocots and dicots. Later botanists have since added other species to this list.

SAPINDACEAE (Soapberry Family)

Plants mostly monoecious or dioecious, often incompletely so, variously herbs, shrubs, or trees, occasionally vining. Leaves alternate or opposite, petiolate, the petiole bases usually noticeably swollen or expanded. Stipules mostly lacking (often present in *Cardiospermum*, occasionally partly free from the petiole base in *Acer*), but small tufts of purplish cobwebby hairs or minute dark scales often present at the leaflet bases in *Aesculus*. Leaf blades variously palmately, pinnately, or ternately compound or lobed, the leaflets of various shapes, the margins usually toothed or lobed. Inflorescences terminal or appearing lateral, panicles or clusters, usually lacking bracts and bractlets. Flowers actinomorphic or zygomorphic, hypogynous (the staminate ones often perigynous in *Acer*), mostly imperfect (a small number of perfect flowers often present in some inflorescences. Calyces of 4 or 5(6) sepals, these sometimes fused, often colored, variously shaped, sometimes persistent at fruiting. Corollas absent or of 4 or 5(6) free petals. Stamens 3–8 (reduced to tiny rudiments in pistillate flowers or appearing short but fully formed in *Koelreuteria*), sometimes exserted, the filaments free, usually attached to the inside of a sometimes inconspicuous nectar-ring, the anthers usually more or less exserted, attached at the base. Pistil 1 per flower (usually reduced to a peglike rudiment in staminate flowers), superior, of 2 or 3 fused carpels. Ovary usually with 2 or 3 locules, sometimes strongly lobed or flattened, the placentation axile. Styles 1 (then sometimes deeply 2-lobed) or 2 per flower, each (or each branch) with 1 stigma, this entire (capitate to club-shaped) or deeply 3-lobed. Ovules 1 or 2 per locule. Fruits samaras, drupelike berries, or capsules, if capsules then dehiscent longitudinally, 1–3-seeded. Seeds various, often with arils. About 130 genera, 1,450–1,900 species, nearly worldwide, most diverse in tropical regions.

Recent morphological and molecular studies have suggested that genera traditionally classified in the families Aceraceae (2 genera, about 115 species) and Hippocastanaceae (2 or 3 genera, 16–20 species) are better treated within an expanded circumscription of the Sapindaceae (Judd et al., 1994; Harrington et al., 2005; Buerki et al., 2009; Harris et al., 2009). Although the three families have traditionally been thought to be closely related (Cronquist, 1981, 1991), their combination adds further variation to an already morphologically variable family. Molecular phylogenetic research (Harrington et al., 2005; Buerki et al., 2009) has supported the close relationship between Aceraceae and Hippocastanaceae near the base of the lineage that includes the remaining genera of Sapindaceae. However, these studies have agreed that at the base of the entire lineage (below Aceraceae and Hippocastanaceae) is the odd, monospecific, Chinese genus *Xanthoceras* Bunge, which traditionally has always been classified in the Sapindaceae. The alternative approach of segregating *Xanthoceras* into its own family and resplitting the Aceraceae and Hippocastanaceae has also been proposed (Buerki et al., 2010), but has not yet met with widespread acceptance by botanists. The family Aceraceae was treated in Volume 2 of the present work (Yatskievych, 2006). The Missouri treatment of the maples was completed before most of the compelling molecular work on the group was published. Because of this, users of the present work will have to consult Volume 2 for the treatment of the genus *Acer*, which most botanists now agree should be included in a broadly circumscribed Sapindaceae. For convenience, the family description above includes those characters unique to *Acer* and the genus is included in the key to genera of Sapindaceae below. *Aesculus*, which was treated in the Hippocastanaceae by Steyermark (1963), is here included in the Sapindaceae.

The Sapindaceae include a number of tropical timber trees. Many of the species produce saponins and nonprotein amino acids that render them toxic to mammals, but surprisingly a number of the tropical genera produce edible fruits, several of which are commercially cultivated, including *Blighia* K.D. Koenig (ackee, akee), *Dimocarpus* Lour. (longan), *Litchi* Sonn. (litchi, lychee), *Nephelium* L. (rambutan), and *Paullinia* L. (guaraná). Often the edible portion is a fleshy aril surrounding the seed. Some of the edible species also are sold as juices. The hard seeds of some species are used as beads in handcrafts. The saponins of some members of the family have been extracted to make soap. However, in the United States, the main economic benefit from the family is in the variety of horticulturally important species, including maples, buckeyes (horse chestnuts), golden rain tree, and soapberry.

Steyermark (1963) included another member of the order Sapindales, *Melia azedarach* L. (Meliaceae; chinaberry tree, china tree, pride of India), in the flora without any definite records, stating that because the species was commonly cultivated in southeastern Missouri it likely would be discovered as an escape along a railroad or in a waste area. Since that time, no specimens have been collected in the state and the species is no longer very commonly grown as an ornamental shade tree in the region. It has been recorded from adjacent counties in northeastern Arkansas, but for the present must be excluded from formal treatment in the Missouri flora. *Melia azederach* is a tree to 15 m tall with large leaves that are 2-times pinnately compound and similar in appearance to those of *Koelreuteria paniculata*. The fruits (which contain poisonous tetranorterpenes known as meliatoxins) are superficially similar in size and appearance to those of *Sapindus saponaria*, but are drupes with a 5-locular stone rather than 1-seeded berries. The flowers of *Melia* differ from all members of the Sapindaceae in their unusual staminal structure: the filaments form a slender, dark purple tube terminating in 20–24 sharply pointed teeth, and the yellow anthers are attached inside the tube below the apex. The flowers of *Melia* also differ from those of Missouri members of the Sapindaceae in their 5(6) petals, which are 8–11 mm long, slender, stiffly spreading, and lilac to pale pinkish purple.

1. Leaves opposite, palmately lobed or compound (pinnately compound in 1 species of *Acer*)
    2. Leaves palmately lobed (pinnately compound in 1 species); corollas absent or actinomorphic, relatively small and usually not showy; fruits paired samaras, each with an elongate, oblique wing . . . . 1. ACER
    2. Leaves palmately compound; corollas zygomorphic (sometimes not strongly so), relatively large and showy; fruits capsules, unwinged . . . . 2. AESCULUS
1. Leaves alternate, pinnately or ternately compound
    3. Plants annual vines, the stems and often also inflorescence stalks bearing tendrils; leaves 1 or 2 times ternately compound . . . . 3. CARDIOSPERMUM
    3. Plants shrubs or trees, lacking tendrils; leaves 1-time pinnately compound, (occasionally some of the larger leaves 2-times pinnately compound in *Koelreuteria*)
        4. Leaflets with the margins coarsely toothed and/or lobed, occasionally some of the leaflets more or less compound; corollas bright yellow, asymmetrical (zygomorphic with the petals all oriented more or less upward); fruits capsular, with the outer wall inflated and papery, containing 3–6 seeds . . . . 4. KOELREUTERIA
        4. Leaflets with the margins entire; corollas white, actinomorphic; fruits drupelike berries with a leathery outer layer and a waxy middle layer surrounding a single seed . . . . 5. SAPINDUS

1. **Acer** L. (maple)

Plants shrubs or more commonly small or large trees, lacking tendrils. Bark and twigs various. Winter buds ovate to elliptic-ovate, bluntly to sharply pointed at the tip, mostly with 4–12 overlapping scales. Leaves opposite, petiolate. Leaf blades variously shaped, usually palmately lobed, less commonly pinnately compound. Inflorescences terminal or lateral toward the branch tips, sometimes appearing axillary, ranging from small clusters to racemes or small panicles. Flowers actinomorphic, hypogynous, the staminate ones often perigynous. Calyces of 4 or 5(6) sepals, these sometimes fused, often colored. Corollas absent or of 4 or 5(6) free petals. Stamens 3–8 (except in pistillate flowers), usually strongly exserted, the anthers red, purple, yellow, or yellowish green. Pistil of 2 fused carpels (except in staminate flowers), usually with 2 locules, flattened at right angles to the septum. Styles 2 per flower or sometimes 1 and deeply 2-lobed, the stigmas club-shaped. Ovules usually 2 per locule. Fruits consisting of 2 samaras that are initially fused at the base but break apart at maturity and are dispersed independantly, each with a single basal seed and an obliquely terminal wing. About 115 species, widespread in temperate portions of the northern hemisphere and in mountains in the tropics.

For the treatment of the 5 species of *Acer* present in Missouri, please refer to Volume 2 (pp. 7–16) of the present work (Yatskievych, 2006).

2. **Aesculus** L. (buckeye)

Plants shrubs or trees, lacking tendrils. Bark variously light to dark-colored, smooth to scaly. Twigs gray to reddish brown, with light yellowish lenticels and relatively large leaf scars, the winter buds ovoid to more or less conical, sharply pointed (blunt in *A. pavia*), with several pairs of overlapping scales, these often somewhat keeled, rounded at the tip and often with a minute sharp point. Leaves opposite, long-petiolate. Stipules lacking, but small tufts of purplish cobwebby hairs or minute dark scales often present at the leaflet bases. Leaf blades palmately compound, pentagonal to kidney-shaped or nearly circular in overall outline, with 5–11 leaflets, these often short-stalked, oblanceolate to obovate or elliptic, angled or tapered at the base, angled or more commonly tapered to a sharply pointed tip, the margins finely and sharply toothed. Inflorescences terminal on the branches, cylindrical to ovoid panicles, the main axis usually densely pubescent with short, curly hairs, sometimes also with tufts of longer, cobwebby, reddish purple hairs at the branch points, the flowers relatively densely racemose along the pinnate main branches. Flowers zygomorphic, hypogynous. Calyces 5-lobed, usually not green, variously shaped but usually only slightly zygomorphic, with relatively shallow, rounded to bluntly pointed lobes, the outer surface densely pubescent with minute, curved or curled hairs, often also with sparse to moderate, gland-tipped hairs, usually not persistent at fruiting. Corollas of 4 free petals (5 elsewhere), the lowermost petal usually abortive, slightly to relatively strongly zygomorphic, the upper pair usually somewhat longer and narrower than the lower pair, all tapered to flattened or winged, stalklike bases, variously colored, the margins and outer surface variously hairy. Stamens 6–8 (except in pistillate flowers), the anthers yellow, orange, or red. Pistil of (2)3 fused carpels (except in staminate flowers), usually 3-locular. Style 1, elongate, unbranched, the stigma entire or only slightly lobed. Fruits capsules, dehiscent longitudinally, the surface sometimes with spinelike tubercles, 1–3-seeded. Seeds 2–4 cm long, variously depressed-globose, often with a somewhat flattened area, the surface often faintly wrinkled, smooth, tan to dark reddish brown with a large, pale, attachment scar.

The horse chestnut, which is native to Europe and Asia, is widely cultivated in temperate North America as an ornamental shade tree and for its seeds, which are roasted and eaten. It

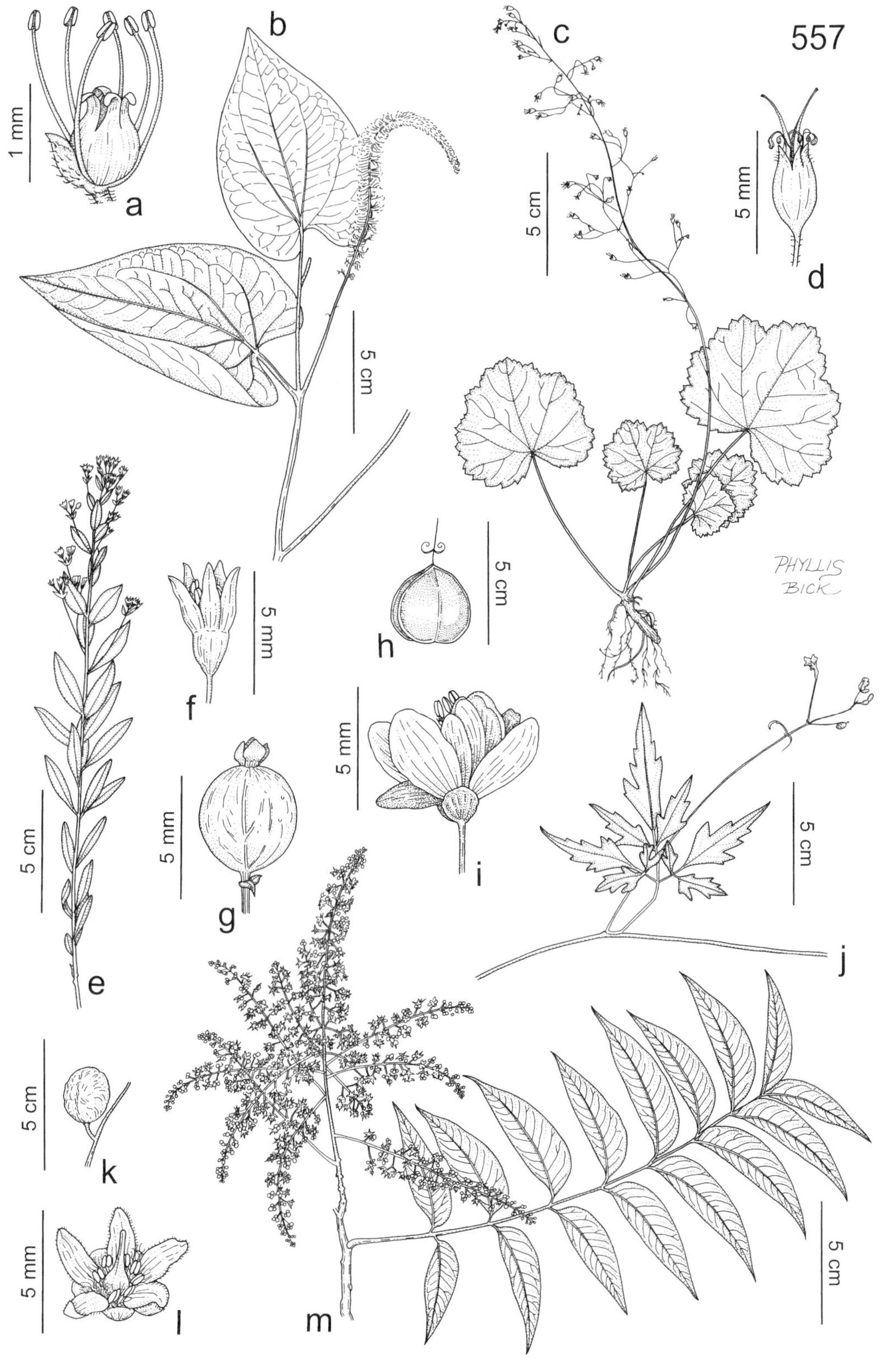

Plate 557. Santalaceae. Sapindaceae. Saururaceae. Saxifragaceae. *Saururus cernuus*, **a)** flower, **b)** stem tip with leaves and inflorescence. *Heuchera parviflora*, **c)** habit, **d)** flower. *Comandra umbellata*, **e)** fertile stem, **f)** flower, **g)** fruit. *Cardiospermum halicacabum*, **h)** fruit, **i)** flower, **j)** node with leaf and inflorescence. *Sapindus saponaria*, **k)** fruit, **l)** flower, **m)** branch tip with leaf and inflorescence.

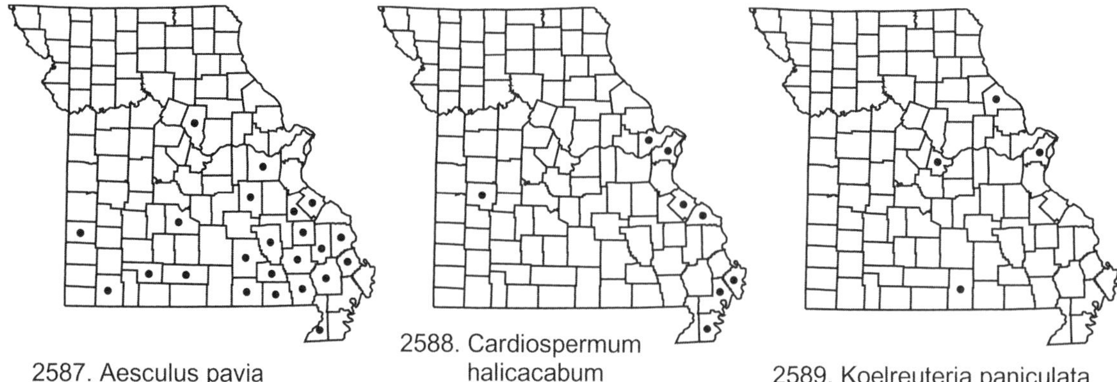
2587. Aesculus pavia
2588. Cardiospermum halicacabum
2589. Koelreuteria paniculata

may be recognized by its large (often 10–25 cm long) leaflets, 5-petaled flowers (the lower petal well-developed), and very large (often more than 5 cm in diameter), spiny fruits. It has escaped from cultivation only rarely in North America has not escaped thus far in Missouri (Steyermark, 1963).

The two species native to Missouri also have been roasted for food, but they have smaller seeds sometimes with inferior flavor. It should be noted that raw seeds (or other portions of the plant) are considered mildly toxic to livestock and humans, containing a complex array of chemical compounds that can cause symptoms of intoxication, including an unbalanced stance, staggering or exaggerated movements, and rarely muscle spasms and collapse (Burrows and Tyrl, 2001). The crushed seeds have been scattered into ponds as a poison to stun fish (Steyermark, 1963).

Buckeyes are of only minor value for their lightweight, relatively brittle timber, which occasionally has been used in furniture, musical instruments, or implements. The roots occasionally were used by pioneers to produce a soap. The large seeds of buckeyes have been carried in the pocket for their pleasant, smooth texture, as a good luck charm, and as a folk remedy for rheumatism. The principal use of both of the Missouri species is for cultivation as ornamentals.

Buckeyes produce new foliage relatively early in the spring and are among the earliest woody species to drop their leaves in the autumn. The seeds tend to germinate very quickly after dispersal, but enter a deep physiological dormancy that is very difficult to break if allowed to dry (Baskin and Baskin, 1998). Thus, they cannot be stored for future germination using standard seed storage protocols.

1. Calyces yellow to greenish yellow; corollas only slightly zygomorphic, greenish yellow; stamens long-exserted; fruits usually appearing spiny . . . . 1. A. GLABRA
1. Calyces dark red to dark reddish brown; corollas moderately to strongly zygomorphic, red (rarely yellow); stamens not or only slightly exserted; fruits spineless . . . . 2. A. PAVIA

1. Aesculus glabra Willd. (Ohio buckeye)
Pl. 425 k–m; Map 2586

Plants small to more commonly medium to large trees to 20 m tall, occasionally flowering as shrubs 3–5 m tall (these sometimes caused by flood damage followed by resprouting). Bark relatively smooth on young trees, developing shallow fissures and small scaly plates with age, dark grayish brown, often becoming lighter gray on older trees. Leaf blades palmately compound with 5–11 leaflets, these 5–16 cm long, the upper surface glabrous or sparsely to moderately short-hairy along the veins, yellowish green to bright green, the undersurface sparsely to densely pubescent with short, curly, sometimes tangled or woolly hairs, sometimes only along the veins or as small tufts in the vein axils, green to dark green or noticeably pale. Inflorescences 10–15 cm long. Calyces 3–8 mm long, more or less bell-shaped, relatively symmetrical at the base, yellow to greenish yellow, the lobes more or less similar. Corollas only slightly zygomorphic, greenish yellow, sometimes the upper pair of petals marked with orangish or reddish spots and/or central region, the petals 10–19 mm long, the upper pair slightly longer and more or less oblanceolate, gradually tapered to the stalklike base, the lower pair slightly shorter and with the blade broadly oblong to oblong-ovate or nearly circular, more abruptly tapered to the stalklike base. Stamens 7, the filaments 15–23 mm long, strongly exserted, hairy below the midpoint, the anthers orange. Fruits 2–4(–5) cm long, ovoid to obovoid or nearly globose, the outer wall leathery, light brown to brown, usually with abundant, irregular, spinelike tubercles, these sometimes shed with age, otherwise slightly roughened or warty. $2n$=40. April–May.

Scattered to common nearly throughout the state, but apparently absent from most of the Mississippi Lowlands Division (eastern U.S. west to Minnesota, Nebraska, and Texas; Canada). Banks of streams and rivers, bottomland forests, mesic upland forests, and bases of bluffs; also margins of pastures and old fields.

Steyermark (1963) noted that during his monographic research on the genus, Hardin (1957a, b) annotated several specimens from Missouri as representing putative hybrids of intermediate morphology between *A. glabra* and *A. flava* Sol. (the latter parent under the name *A. octandra* Marshall). Some of these have been called *A.* ×*marylandica* Booth ex Dippel. The yellow buckeye, *A. flava*, occurs from western Pennsylvania southwest along the Ohio River to southern Illinois. It differs from *A. glabra* in its more strongly zygomorphic flowers, stamens that are not or only slightly exserted, and nonspiny fruits. The hybrids are characterized by somewhat exserted stamens, scattered gland-tipped hairs mixed in with the nonglandular hairs on the calyx and flower stalk, somewhat more zygomorphic corollas, and fruits with irregular clusters of spines. For a discussion of putative hybrids between *A. glabra* and *A. pavia*, see the treatment of that species.

Steyermark (1963) treated *A. glabra* in Missouri as comprising three varieties and one additional form. He did not agree with the earlier treatment of Hardin (1957a, b), who accepted only two varieties. However, Hardin argued persuasively that plants attributed to var. *leucodermis*, characterized by paler bark and pale undersurface of the leaflets, appeared sporadically throughout the species range and that these two characters did not correlate in many specimens. He studied plants at the type locality of var. *leucodermis* in Arkansas, where he observed the existence of a large putatively interbreeding population of *A. glabra* and *A. pavia*. Hardin also noted that pubescence density, which was said to characterize f. *pallida*, varied too much within populations to be useful taxonomically. These views are accepted in the present treatment, with acknowledgment that occasional specimens are difficult to determine to variety.

1. Leaflets mostly 7–11, 1–3(–5) cm wide .... 1A. VAR. ARGUTA
1. Leaflets mostly 5–7, (2–)3–6(–8) cm wide .... 1B. VAR. GLABRA

1a. var. arguta (Buckley) B.L. Rob.

Plants usually small trees to 6 m tall. Leaflets (5–)7–11, 5–16 cm long, 1–3(–5) cm wide, mostly oblanceolate to narrowly elliptic. $2n$=40. April–May.

Scattered in the Unglaciated Plains Division and sporadically farther east, mostly south of the Missouri River (Iowa to Arkansas west to Nebraska and Texas). Banks of streams and rivers, bottomland forests, mesic upland forests, and bases of bluffs.

1b. var. glabra

A. glabra var. *leucodermis* Sarg.

A. glabra f. *pallida* (Willd.) Fernald

Plants usually medium to large trees to 20 m tall, rarely shorter trees or shrubs. Leaflets 5–7, 6–16 cm long, (2–)3–6(–8) cm wide, mostly obovate to elliptic. $2n$=40. April–May.

Scattered to common nearly throughout the state, but apparently absent from most of the Mississippi Lowlands Division (eastern U.S. west to Minnesota, Kansas, and Texas; Canada). Banks of streams and rivers, bottomland forests, mesic upland forests, and bases of bluffs; also margins of pastures and old fields.

2. Aesculus pavia L. (red buckeye)

A. discolor Pursh

A. discolor var. *mollis* Sarg.

Pl. 425 h–j; Map 2587

Plants shrubs or small trees to 5(–10) m tall. Bark relatively smooth on young trees, developing shallow fissures and small scaly plates with age, gray to brown. Leaf blades palmately compound with 5(–7) leaflets, these 5–17 cm long, the upper surface glabrous or sparsely short-hairy along the midvein, dark green, the undersurface sparsely to densely pubescent with short, curly, sometimes tangled or woolly hairs, sometimes mainly along the veins or as small tufts in the vein axils, green to pale green. Inflorescences 10–25 cm long. Calyces 10–18 mm long, more or less cylindrical, often somewhat asymmetrically pouched at the base, dark red to dark reddish brown, the upper 3 lobes slightly longer than the lower 2. Corollas moderately to strongly zygomorphic, red (rarely yellow), the petals 20–40 mm long, the upper pair longer and more or less strap-shaped with a long, broad stalklike base and a short, oblong-obovate to hemispheric or nearly circular expanded tip, the lower pair shorter and with the better-developed blade broadly oblong to oblong-obovate. Stamens 6–8, the filaments 20–35 mm long, not or only slightly exserted, hairy below the midpoint, the anthers dark red to dark reddish purple. Fruits 3–6 cm long, obovoid or nearly globose, the outer wall leathery, light brown to yellowish brown, lacking spinelike tubercles, the surface roughened or finely and shallowly pitted. $2n=40$. April–May.

Scattered in the Mississippi Lowlands Division and sporadically in the Ozark Division north to Crawford and St. Francois Counties and west to Barry, Texas, and Laclede Counties; introduced uncommonly farther west and north (southeastern U.S. west to Missouri, Oklahoma, and Texas). Banks of streams and rivers, bottomland forests, mesic upland forests, and bases of bluffs; also margins of pastures and old fields.

Hardin (1957a, b) and Steyermark (1963) discussed the existence of scattered specimens from southern Missouri representing putative hybrids between *A. pavia* and *A. glabra*. Some of these plants have been called *A.* ×*bushii* C.K. Schneid. They may be recognized by a yellowish tinge in the corollas, somewhat exserted stamens, at least some glandular hairs on the petals, and irregularly spiny fruits (Hardin, 1957a, b).

3. **Cardiospermum** L. (balloon vine, heartseed)

About 12 species, North America to South America, Caribbean Islands, Asia, Africa.

1. Cardiospermum halicacabum L. (common balloon vine, love-in-a-puff)

Pl. 557 h–j; Map 2588

Plants annuals (perennial herbs farther south), vines. Stems 30–400 cm long, climbing, much-branched, with branched, axillary tendrils (these with tiny bracts similar to those of the inflorescences at the branch points). Leaves alternate, short- to often long-petiolate. Stipules lacking. Leaf blades pentagonal to ovate-triangular in overall outline, 1 or 2 times ternately compound, the primary leaflets 1–8 cm long, usually stalked (the stalks sometimes winged to the base), narrowly to broadly lanceolate to ovate or ovate-rhombic, not noticeably curved, the ultimate leaflets variously broadly angled to narrowly tapered at the base, angled or tapered to a usually sharply pointed tip, often with a minute, sharp point at the very tip, the margins coarsely toothed and/or pinnately few-lobed, also minutely hairy, the surfaces sparsely pubescent with short, curved hairs along the main veins. Inflorescences axillary clusters or small panicles, these long-stalked, the stalk with a pair of tendril-branches at or more commonly above its midpoint, the branches and stalks glabrous or sparsely and minutely hairy, the branch points (and ultimate branches) with inconspicuous bracts, these in groups of 2 or 3, linear to narrowly lanceolate or narrowly triangular 0.5–1.3 mm long. Flowers somewhat zygomorphic, hypogynous, short- to relatively long-stalked. Calyces deeply 4-lobed, relatively strongly zygomorphic with 2 shorter opposite lobes 0.8–1.5 mm long and 2 longer opposite lobes 2.5–3.5 mm long, the lobes light green to yellowish green, oblong to oblong-obovate, rounded to broadly pointed at their tips, especially the larger pair concave (cupped), glabrous, usually persistent but inconspicuous and withered at fruiting. Corollas of 4 free petals, only slightly zygomorphic with opposite slightly longer and shorter pairs, 2.5–3.5 mm long, white, the blade obovate, relatively flat and spreading, glabrous, narrowed to a short, slender, stalklike base, with a petaloid appendage on the upper surface at the base (these about ⅔ as long as the petal and narrower, erect and surrounding the stamens and ovary) on the upper surface near the base. Stamens 8 (appearing short but fully formed in pistillate flowers), the filaments glabrous, erect, somewhat exserted beyond the petal appendages, somewhat unequal, the anthers yellow. Pistil of 3 fused carpels (except in staminate flowers), 3-locular, with 1 ovule per locule. Style 1, slightly exserted at flowering, relatively short, unbranched, the stigma deeply 3-lobed. Fruits capsular, 30–45 mm long, more or less circular in profile, broadly 3-lobed in cross-section (ridged on the angles and with a prominent longitudinal vein along the sinus), short-tapered at the sometimes short-stalked base, rounded to short-tapered at the tip, the outer wall inflated and papery, appearing veiny, light green during development, turning reddish brown

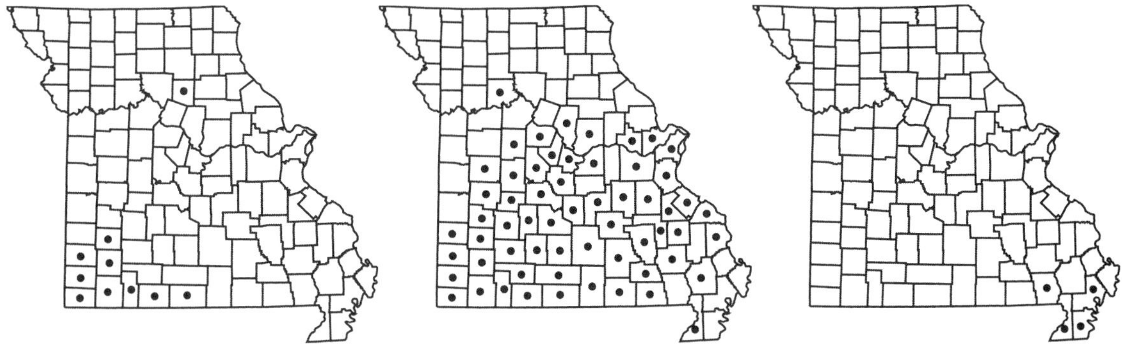

2590. Sapindus saponaria 2591. Sideroxylon lanuginosum 2592. Sideroxylon lycioides

to straw-colored at maturity, but then sometimes pinkish-tinged, usually finely short-hairy, 3-locular, mostly 3-seeded (1 per locule). Seeds 4.5–5.5 mm long, globose or nearly so, glabrous at maturity, the surface appearing smooth to faintly wrinkled, slightly shiny, black, with a well-developed aril toward the base, this sometimes extending nearly to the midpoint of the fruit, broadly heart-shaped to deeply kidney-shaped, white, finely granular in texture. $2n=22$. July–September.

Introduced, scattered in easternmost Missouri, thus far only in counties bordering the Mississippi River (native to both the New and Old World tropics, introduced in the eastern U.S. west to Kansas and Texas, also Hawaii). Banks of streams and rivers, marshes, and edges of bottomland forests; also levees, ditches, fallow fields, roadsides, and open, moist disturbed areas.

This species is most successful in full-sun sites and most often is an early colonizer of highly disturbed areas. On a very local scale, plants of *C. halicacabum* can form relatively dense mats over other low vegetation. The species is considered a noxious weed by Agriculture Departments in several states to the south and east of Missouri. It is sometimes grown as an ornamental in gardens, mostly on trellises and for its attractive fruits. The seeds sometimes are used as beads in jewelry. The plants also have been used medicinally, mainly in the Old World, as an emetic and laxative, and for treatment of a variety of ailments, including rheumatoid arthritis (Subramanyam et al., 2007).

4. **Koelreuteria** Laxm. (Meyer, 1976)

Three or 4 species, China, Taiwan, Fiji.

1. Koelreuteria paniculata Laxm. (golden rain tree, star of India)

Map 2589

Plants trees to 12 m tall, lacking tendrils. Bark gray to grayish brown, with relatively thick ridges and plates and often reddish brown, longitudinal furrows. Twigs light reddish brown, somewhat angular or with irregular lines of low corky ridges, glabrous or sometimes minutely hairy near the tips of fertile branchlets, with inconspicuous pale lenticels and relatively prominent leaf scars, the winter buds oblong-obovoid, with usually 1 pair of exposed, overlapping scales, these usually glabrous. Leaves alternate, short-petiolate. Stipules lacking. Leaf blades elliptic to narrowly elliptic in overall outline, pinnately compound with 7–17 leaflets (terminal leaflet present), these mostly subopposite along the rachis, 1–10 cm long, sessile or short-stalked, ovate to broadly ovate, elliptic, or obovate, usually not noticeably curved, rounded or abruptly tapered at the often slightly asymmetric base, tapered to a sharply pointed tip, the margins coarsely toothed and/or pinnately lobed, sometimes some of the largest leaflets more or less fully compound, also usually minutely hairy, the upper surface glabrous or more commonly minutely hairy along the midvein, the undersurface usually moderately pubescent with fine, curved hairs along the relatively prominent, main veins, sometimes also somewhat glandular. Inflorescences terminal on the branches, broadly ovoid panicles, well-branched and with numerous flowers, the branches and stalks densely and minutely hairy. Flowers zygomorphic (asymmetric), hypogynous, short-stalked. Calyces deeply 5-lobed, more or less zygomorphic (2 of the lobes slightly smaller than the other 3), yellowish green, the lobes 1.5–2.5 mm long, ovate, finely irregular and glandular-hairy along the margins, glabrous or sparsely glandular-hairy on the outer surface, usually persistent but inconspicuous and withered at fruiting. Corollas asymmetric (the 4 free petals all oriented more or less toward the top of the flower), each 5–7 mm long, bright yellow, the blade narrowly oblong, flat, usually bent backward, glabrous, orange basally, narrowed abruptly to a slender, hairy, stalklike base, with a small, lobed, orange appendage on the upper surface where the blade meets the stalk. Stamens 8(–10) (appearing short but fully formed in pistillate flowers), the filaments long-hairy, angled downward, the anthers brownish purple. Pistil of 3 fused carpels (except in staminate flowers), 3-locular, with 2 ovules per locule. Style 1, exserted at flowering, elongate, unbranched, the stigma deeply 3-lobed. Fruits capsular, 45–70 mm long, oblong-ovate in profile, triangular in cross-section, truncate to slightly concave at the base, tapered to a sharply pointed tip, the outer wall inflated and papery, appearing veiny, dark brown to straw-colored at maturity, but then sometimes pinkish-tinged, incompletely 3-locular (only toward the base), 3–6-seeded (seeds 1 or 2 per locule, attached at the tip of the 3-locular portion). Seeds 6–8 mm long, globose or nearly so, glabrous, the surface appearing smooth to finely granular, often slightly shiny, black, sometimes with a white coating around the base. May–July, occasionally reblooming August–September.

Introduced, uncommon and widely scattered, mostly in and around urban areas (native of Asia, introduced widely but sporadically in the eastern U.S. west to Kansas and Texas, rarely farther west). Disturbed mesic upland forests and ledges and tops of bluffs; also fencerows, railroads, roadsides, and open, disturbed areas.

Long cultivated as a shade tree and ornamental in Asia, the species was brought into European horticulture in the late 1700s and was introduced in the United States as early as 1809, when Thomas Jefferson received seeds from France (Meyer, 1976). The plants are fast-growing, disease resistant, and tolerate a variety of growing conditions. The attractive, broad-crowned trees produce wind-dispersed fruits prolifically and anywhere that the species is successful numerous seedlings begin to grow in the vicinity. Thus, it is not surprising that *K. paniculata* has escaped to disturbed areas adjacent to plantings. In recent years, conservationists in the Midwest have begun to notice populations developing in more natural forest and bluff habitats, which has led to predictions that the species will become a problem invasive exotic in the future.

5. **Sapindus** L. (soapberry)

Ten species, nearly worldwide, most diverse in the Old World tropics.

1. Sapindus saponaria L. **var. drummondii** (Hook. & Arn.) L.D. Benson (western soapberry, chinaberry)
S. drummondii Hook. & Arn.
Pl. 557 k–m; Map 2590

Plants (in Missouri) shrubs or small trees to 6 m tall (to 15 m tall elsewhere), sometimes colonial from root sprouts, lacking tendrils. Bark grayish brown to occasionally reddish brown, on larger trunks breaking into narrow elongate plates, on smaller trunks merely shallowly ridged and grooved. Twigs yellowish green to grayish tan or gray, somewhat angular, finely hairy, with inconspicuous pale lenticels and relatively prominent leaf scars, the winter buds small, globose, with 1 or 2 pairs of overlapping scales, these finely and densely hairy. Leaves alternate, short-petiolate. Stipules lacking, but small tufts of purplish cobwebby hairs or minute dark scales often present at the leaflet bases. Leaf blades oblong-elliptic to oblong-oblanceolate in overall outline, pinnately compound with (8–)12–20 leaflets (terminal leaflet absent), these alternate along the rachis, 2–9 cm long, mostly short-stalked, asymmetrically lanceolate to elliptic-lanceolate, usually curved (sickle-shaped), angled or tapered at the base, tapered to a sharply pointed tip, the margins entire, the surfaces glabrous or the undersurface sometimes finely short-hairy, the midvein often appearing off-center. Inflorescences terminal on the branches, ovoid to pyramid-shaped panicles, well-branched and with numerous flowers, the branches and stalks finely hairy. Flowers actinomorphic, hypogynous, short-stalked. Calyces deeply (4)5-lobed, yellowish green to pale yellowish green, the lobes 1–2 mm long, elliptic to more or less ovate, somewhat concave (cupped), glabrous or nearly so except along the margins, usually not persistent at fruiting. Corollas of (4)5 free petals, these 2.5–4.0 mm long, abruptly tapered to a hairy, stalklike base, slightly concave (cupped), white, the margins minutely hairy. Stamens (7)8–10 (except in pistillate flowers), the filaments long-hairy, the anthers yellow. Pistil of 3 fused carpels (except in staminate flowers), usually 3-locular (2 of the carpels aborting during fruit development). Style 1, somewhat exserted before the bud opens, elongate, unbranched, the stigma usually shallowly 3-lobed often asymmetrically so. Fruits drupelike berries, 12–15 mm long, globose (often slightly asymmetrical), long-persistent on the trees, the outer surface translucent, greenish yellow to amber, darkening to yellowish brown and becoming wrinkled with age, 1-seeded, the middle layer waxy. Seed 8–9 mm long, obovoid, with a tuft of minute, pale hairs at the base, the surface appearing smooth and somewhat shiny (but appearing minutely pitted with magnification), black. May–July.

Uncommon in the southwestern portion of the Ozark Division (Missouri to Louisiana west to Colorado and Arizona; Mexico. Glades, banks of streams, and tops and ledges of bluffs; also roadsides; usually on limestone and dolomite.

The taxonomic relationship of this taxon to var. *saponaria* (wing-leaved soapberry) requires further study; L. Benson's (1943) transfer of the North American plants from specific to varietal status was not adequately justified and many authors continue to treat the two taxa as distinct species. The overall distribution of the species is quite large, from North America to South America, various Caribbean Islands, on a number of Pacific Islands west to New Caledonia, and in parts of Africa. The var. *drummondii*, which occupies the northern portion of the range in the New World, differs mainly in the lack of a winged leaf rachis and slight differences in leaflet shape and size.

Both the scientific name *Sapindus* and the vernacular name soapberry refer to the practice of extracting the saponins from crushed fruits for use in soaps. Steyermark (1963) noted that in Mexico the fruits also are used as a fish poison and the seeds are used as beads in jewelry. However, handling of the fruits can cause contact-dermatitis in some individuals. In Missouri, the species is at the northern edge of its range and there is evidence of die-back from winter freezes. Most of the plants are shrubs or small trees less than 5 m tall that do not produce flowers or fruits during many years. Interestingly, there are cold-hardy genotypes that can grow into attractive shade trees to the north of the native distribution of the species. Steyermark (1963) remarked on trees in cultivation on the campus of the University of Missouri (Boone County) and a large specimen tree likewise has been grown for many years at the Missouri Botanical Garden (St. Louis City).

SAPOTACEAE (Sapodilla Family)

(Pennington, 1990, 1991)

Fifty-three genera, about 1,100 species, nearly worldwide, mostly in tropical and subtropical regions.

1. **Sideroxylon** L. (buckthorn)

(Pennington, 1990)

Plants shrubs or small trees with more or less milky sap, often with short spine-tipped branches along the main stems. Leaves alternate or appearing fascicled at the tips of short condensed branches, lacking stipules. Petioles absent or short. Leaf blades simple, with entire margins, the smaller veins forming an intricate network. Flowers axillary, solitary or in umbellate clusters, actinomorphic, perfect, hypogynous. Calyx of 5 free or nearly free sepals, these overlapping, often persisting in fruit. Corolla 5-lobed, white or sometimes yellowish-tinged, the tube shorter than to about as long as the lobes, the lobes ascending to spreading at maturity, each with a pair of short, petal-like lobes or appendages at the base. Stamens 5, fused to the corolla tube, positioned opposite the corolla lobes. Staminodes 5, similar in size, shape, and color to the corolla lobes but usually somewhat more pointed at the tip, fused to the corolla tube to the inside of the fertile stamens, positioned alternate with the corolla lobes. Pistil 1 per flower, of (2–)5(–8) fused carpels. Ovary superior, with (2–)5(–8) locules, usually hairy, at least toward the base, tapered to the single style. Ovules usually 1 per locule, attached toward the locule base. Stigma not conspicuous, a small obliquely positioned receptive area at the tip of the style. Fruits drupelike berries, ovoid to ellipsoid, green turning purplish black or black at maturity, the outer layer leathery, usually shiny when ripe, the pulp somewhat milky, usually with 1 seed. Seeds obovoid to ellipsoid or nearly globose, the seedcoat relatively thick, hard, shiny, light to dark brown, sometimes mottled on one side, the base with a noticeable attachment scar. About 75 species, North America to South America, Asia, Africa, and Madagascar.

Cronquist (1945, 1946) recognized three American genera as distinct from *Sideroxylon*, which he restricted to the Old World. The 23 species of *Bumelia* Sw. were said to differ from the closely related *Mastichodendron* (Engl.) H.J. Lam and *Dipholis* A. DC. primarily in their divided corolla lobes and seeds lacking endosperm. Pennington (1991), who reviewed the generic-level taxonomy of the family worldwide, noted however that no single character served to adequately separate these segregates and that some recently discovered neotropical species blurred the perceived boundaries between the groups. Accordingly, he chose to accept *Sideroxylon* in a broad sense, to include a number of both New and Old World genera treated as distinct by earlier authors (Pennington, 1990, 1991). His circumscription is adopted here.

The small flowers of *Sideroxylon* species have an odd appearance. The stamens appear to be sandwiched between two whorls of corolla lobes. The inner whorl, however, is composed of petaloid staminodes.

1. Undersurface of young and mature leaves usually moderately pubescent with matted, somewhat spreading hairs, these with 2 straight or wavy arms attached to a short central stalk . . . . 1. S. LANUGINOSUM
1. Undersurface of mature leaves glabrous or sparsely hairy, when young with the hairs appressed, straight, attached in the middle, the 2 arms not attached to a central stalk . . . . 2. S. LYCIOIDES

1. Sideroxylon lanuginosum Michx. **ssp. oblongifolium** (Nutt.) T.D. Penn. (gum bumelia, woolly buckthorn, chittim wood, false buckthorn, gum-elastic)

Bumelia lanuginosa (Michx.) Pers. ssp. *oblongifolia* (Nutt.) Cronquist

B. lanuginosa var. *albicans* Sarg.

B. lanuginosa var. *oblongifolia* (Nutt.) R.B. Clark

Pl. 556 l–n; Map 2591

Plants shrubs or trees to 15 m tall. Bark dark gray, becoming furrowed into narrow scaly ridges. Twigs gray to dark gray, with rust-colored to reddish gray hairs, at least when young, the buds lateral, ovoid, often partially embedded in the twig, the several overlapping scales hairy. Leaf blades 2–10 cm long, narrowly oblanceolate to broadly elliptic, tapered at the base, rounded to bluntly pointed at the tip, the undersurface usually moderately pubescent with matted, somewhat spreading, silvery hairs, these with 2 straight or wavy arms attached to a short central stalk, becoming glabrous with age, the upper surface glabrous except along the midvein, often shiny. Calyx 1.5–2.0 mm long, the lobes broadly elliptic, rounded at the tip, with rust-colored to reddish gray hairs, the lateral margins thinner, glabrous. Corollas 3–4 mm long. Fruits 9–15 mm long. Seeds 6–9 mm long, the attachment scar irregularly triangular to circular, appearing flat to somewhat notched, usually bissected by a low, narrow ridge. $2n$=24. June–August.

Scattered, mostly in the Ozark, Ozark Border, and Mississippi Lowlands Divisions (Missouri to Louisiana west to Kansas, Oklahoma, and Texas). Dry upland forests, glades, and tops and ledges of exposed bluffs, often but not always on calcareous substrates, less commonly mesic upland forests, bottomland forests, and streambanks; also railroads and fencerows.

Sideroxylon lanuginosum has been divided into several infraspecific taxa, three of which have been accepted by the last two monographers of the group (Cronquist, 1945; Pennington, 1990). The ssp. *oblongifolium* is endemic to the south-central states and is characterized by grayish

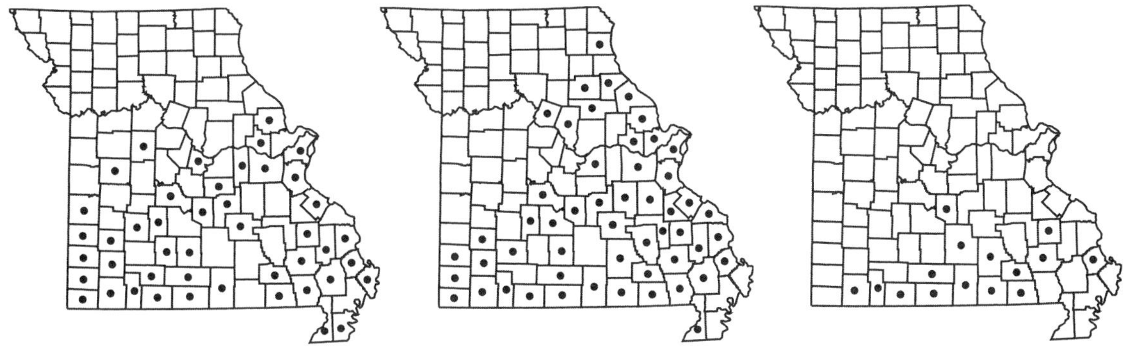

2593. Saururus cernuus   2594. Heuchera americana   2595. Heuchera parviflora

silvery pubescence. The very similar ssp. *lanuginosum*, which is widely distributed in the southeastern states east of the Mississippi River, is characterized by brown to bronze- or rusty-colored hairs. The ssp. *rigidum* (A. Gray) T.D. Penn. has white to gray pubescence and all of the leaves less than 5 cm long. It occurs from Oklahoma to Arizona and in northern Mexico at elevations above 600 m (higher than the other subspecies).

This species is mostly encountered as a shrub or small tree in dry areas, but can grow to 15 m tall. The wood of *S. lanuginosum* has been used occasionally for cabinets and tool handles. Native Americans sometimes used the latex released from the inner bark (chicle) as chewing gum, and the milky latex exuded from cut or bruised parts of the plants gave rise to the common name, gum bumelia. The fruits are eaten by wildlife, but apparently not very palatable to humans.

2. Sideroxylon lycioides L. (buckthorn bumelia, southern buckthorn, Carolina buckthorn, ironwood, smooth bumelia)
Bumelia lycioides (L.) Pers.
Pl. 556 o, p; Map 2592

Plants shrubs or trees to 12 m tall. Bark thin, smooth or with irregular peeling plates, gray to orangish gray. Twigs light to dark reddish brown, glabrous or sparsely hairy, the buds terminal and lateral, the terminal bud flattened and narrowly ovate in outline, the lateral buds depressed-globose, often partially embedded in the twig, the several overlapping scales glabrous or sparsely hairy. Leaf blades 2–12 cm long, narrowly oblanceolate to broadly elliptic, tapered at the base, rounded to bluntly pointed at the tip, the undersurface of mature leaves glabrous or sparsely hairy, when young with the hairs appressed, straight, attached in the middle, the 2 arms not attached to a central stalk, the upper surface glabrous except sometimes along the midvein, often shiny. Calyx 1.5–2.5 mm long, the lobes broadly elliptic, rounded to bluntly pointed at the tip, glabrous, usually appearing wrinkled or warty when dry, the lateral margins thinner and white. Corollas 2.0–3.5 mm long. Fruits 7–13 mm long. Seeds 0.7–10 mm long, the attachment scar oval to circular, flat. June–July.

Uncommon, restricted to the Mississippi Lowlands Division (southeastern U.S. west to Missouri and Louisiana). Swamps, bottomland forests, and banks of streams and rivers.

SAURURACEAE (Lizard's Tail Family)

Four genera, about six species, North America, Asia south to Indonesia.

In addition to the species native to Missouri, *Houttuynia cordata* Thunb. is sometimes cultivated in the state, especially its variegated cultivar 'Chameleon.' It differs from *Saururus* in producing relatively short, dense, spikes, each subtended by an involucre of large, white, petaloid bracts (the entire inflorescence has the appearance of a single large flower or a radiate head of an Asteraceae). This species can become aggressive in gardens and has become naturalized outside of cultivation at scattered sites in a few states to the east of Missouri.

1. **Saururus** (lizard's tail)

Two species, North America, Asia.

1. Saururus cernuus L.

Pl. 557 a, b; Map 2593

Plants perennial herbs, strongly colonial from stout, fleshy, jointed, often branched rhizomes. Aerial stems 40–120 cm long, erect or ascending, unbranched or few-branched above the midpoint, appearing jointed at the nodes, sparsely to densely pubescent with slender, multicellular hairs, at least toward the tip. Leaves alternate, mostly short-petiolate, the petioles channeled on the upper surface, somewhat winged at the base and partially sheathing the stem. Stipules appearing absent. Leaf blades 6–15 cm long, ovate-heart-shaped, oblong-heart-shaped, or narrowly heart-shaped, deeply cordate with a broadly rounded sinus at the base, tapered to a sharply pointed tip, the margins entire, usually with a very slender, thickened, yellowish marginal band, the upper surface glabrous or sparsely hairy along the midvein, the undersurface sparsely to moderately and finely hairy, especially along the main veins, sometimes becoming glabrous at maturity, the venation more or less palmate (the main lateral veins departing the midvein above the blade base). Inflorescences moderately dense, spikelike racemes of numerous flowers, these solitary or less commonly paired, appearing positioned opposite the leaves, 6–15 cm long, ascending with strongly drooping tips, usually with a few widely spaced, small, oblong bracts along the stalk well below the flowers, the stalk and central axis pale green to white, moderately to densely pubescent with slender, multicellular hairs. Flowers appearing actinomorphic, perfect, hypogynous, each closely subtended by a minute bractlet, the short, slender, hairy stalks expanded at the base. Calyces and corollas absent. Stamens (4–)6–8, free, the slender filaments longer than the pistils, white, 1.5–4.5 mm long, often somewhat unequal in length, the anthers attached basally at the slightly expanded filament tips, opening by longitudinal slits, pale yellow. Pistils 3–5, fused only basally. Ovaries superior, 1-locular, the placentation lateral, the ovules (1)2(–4). Style 1 per pistil, stout, slightly flattened, and outward-curved or -curled, the stigmatic region along the upper side. Fruits 2–3 mm long, globose, an aggregation of 3–5, loosely fused achenelike carpels, these somewhat fleshy, indehiscent, the outer surface wrinkled. Seeds 1 per carpel. $2n=22$. May–September.

Scattered south of the Missouri River (progressively more common southeastward), north locally to Lincoln County (eastern U.S. west to Kansas and Texas; Canada). Swamps, bottomland forests, margins of oxbows and sloughs, and banks of streams, rivers, and spring branches; also ditches; often emergent aquatics.

Lizard's tail is sometimes grown in gardens, often in low, wetter areas where a limited array of shade-tolerant species do not grow well. More frequently it is grown as a container plant in pools or as an aquatic in water gardens. Variegated and red-stemmed cultivars have been developed.

SAXIFRAGACEAE (Saxifrage Family)

Plants perennial herbs with rhizomes. Aerial stems sometimes absent, the inflorescence then directly from the rhizome tip. Leaves all or mostly basal, these mostly long-petiolate, those of the apparent flowering stems (actually the inflorescence stalks) either 1 to few, alternate and reduced to scalelike bracts, or a single pair and then conspicuous; these sessile or short-petiolate. Stipules absent or scalelike. Leaf blades simple, the margins entire or more commonly scalloped or toothed. Inflorescences panicles or racemes. Flowers actinomorphic to somewhat zygomorphic, perfect, perigynous. Hypanthium urn-shaped to bell-shaped, partially fused to the ovary. Calyces of 4 or 5 free sepals, these attached along the hypanthium rim. Corollas of 5 free petals, these attached between the sepals along the hypanthium rim. Stamens 5 or 10, the filaments attached just inside the hypanthium rim, the anthers attached at their bases or toward their midpoints. Staminodes absent. Pistil 1 per flower, divided into 2 (–4) separate carpels toward the tip, the fused portion 1- or 2-locular. Ovary partially inferior, the hypanthium fused half or more the length of the ovary. Style 1 per free carpel tip, persistent at fruiting. Ovules numerous. Fruits capsules (sometimes appearing as a pair of follicles), the tip of each carpel tapered to a conspicuous stylar beak, dehiscing variously longitudinally. Seeds tiny, often numerous. About 30 genera, about 550 species, North America to South America, Europe, Africa, Asia to New Guinea.

Taxonomic circumscription of the Saxifragaceae is still somewhat controversial. Traditionally (Steyermark, 1963), the family was treated in a relatively broad sense to include about 80 genera and 1,200 total species grouped into 15–17 tribes, but a number of generic groups are now thought to have different affinities. Recent taxonomic and phylogenetic studies (Morgan and Soltis, 1993; Soltis and Soltis, 1997; Bohm et al., 1999) have shown that the woody genera formerly classified in the Saxifragaceae should be segregated into the Grossulariaceae, Hydrangeaceae, and Iteaceae (and several other families not present in Missouri). Two morphologically unusual genera, *Parnassia* and *Penthorum*, have been elevated to Parnassiaceae and Penthoraceae, respectively. This dismemberment results in a smaller but morphologically more cohesive Saxifragaceae and more accurately reflects modern hypotheses of phylogeny in the Saxifragales. See the treatments of the various segregate families for further discussion.

A number of other genera of Saxifragaceae are cultivated as garden ornamentals, but have not escaped into the wild in Missouri, including *Astilbe* and *Tiarella*.

1. Apparent flowering stem (actually the main stalk of the inflorescence) with a pair of conspicuous opposite bracteal leaves; petals pinnately dissected . . . 3. MITELLA
1. Apparent flowering stem (actually the main stalk of the inflorescence) leafless or with 1 or a few small bracteal leaves, these alternate; petals with the margins entire or finely toothed
    2. Blades of at least the larger leaves more than 3 times as long as wide, narrowed or tapered at the base; stamens 10. 2. MICRANTHES
    2. Leaf blades mostly about as long as wide, cordate at the base; stamens 5
        3. Ovary 1-locular, with parietal placentation; anthers attached toward their midpoints, white, pink, or orange . . . . 1. HEUCHERA
        3. Ovary 2-locular, with axile placentation; anthers attached at their bases, yellow . . . . 4. SULLIVANTIA

1. **Heuchera** L. (alum root)
(Wells, 1984; Kallhoff and Yatskievych, 2001)

Plants with short, stout rhizomes. Leaves all basal (rarely 1 or a few highly reduced bracteal leaves alternate on the inflorescence stalks), short- to more commonly long-petiolate. Stipules scalelike, inconspicuous, fused to the petiole base to above the midpoint, the margins fringed, persistent on the rhizome after the leaves die back. Leaf blades mostly about as long as wide, circular to broadly ovate or kidney-shaped, the base cordate, the tip rounded or narrowed to a blunt or sharp point, the margins with 3–7 shallow lobes, also finely to coarsely scalloped or toothed and with short spreading hairs, palmately veined with usually 7 primary veins, the upper surface glabrous to hairy and/or glandular, green, sometimes with lighter mottling, the undersurface glabrous to hairy and/or glandular, usually grayish green or reddish- or purplish-tinged. Inflorescences panicles with usually numerous flowers, usually long-stalked, glabrous or more commonly hairy and/or glandular, with small scalelike or leaflike bracts at the branch points. Flowers somewhat zygomorphic or less commonly appearing actinomorphic (often appearing more zygomorphic upon pressing), each subtended by a small, linear bract at the base of the flower stalk. Hypanthium obconic to bell-shaped, fused to half or more the length of the ovary. Sepals oblong to triangular-oblong, rounded at the tip. Petals glabrous to minutely hairy, the margins entire or finely toothed, green, white, or pink. Stamens 5, barely included to long-exserted from the calyx, the anthers small, attached toward their midpoints, white, pink, or orange. Ovary 1-locular, the placentation parietal. Styles tapered, persistent and often arched or spreading at fruiting, the stigmas more or less capitate. Fruits ovoid, 2-beaked, dehiscing longitudinally from between the beaks. Seeds numerous, variously shaped, usually somewhat asymmetrical in outline, smooth or with fine tubercles or spines, dark brown to nearly black. About 35 species, North America.

Several species of *Heuchera* are cultivated to some extent as ornamentals in wildflower gardens and rock gardens, including all of the species occurring in Missouri. However, the plants most common in horticulture are the coralbells, with red flowers, representing *H. sanguinea* Engelm. (native to the southwestern United States and adjacent Mexico) and its assortment of cultivars and hybrids. *Heuchera* species also have a long history of medicinal use, taken both internally for various antidiarrheal, astringent, and analgesic properties, and externally as a dermatological and antirheumatic aid and a styptic (Spongberg, 1972; Moerman, 1998).

1. Hypanthium and calyx 3–12 mm long at flowering, 2.5–7.5 mm in diameter, the free portion above the ovary 0.6–7.0 mm long, the outer surface with minute glandular hairs, green to yellowish green; petals elliptic-spatulate, shorter than to slightly longer than the sepals
    2. Hypanthium at flowering with the free portion 0.6–2.0 mm long, weakly asymmetrical . . . . 1. H. AMERICANA
    2. Hypanthium at flowering with the free portion 2–7 mm long, strongly asymmetrical . . . . 3. H. RICHARDSONII
1. Hypanthium and calyx 1.3–3.3 mm long at flowering, 1.1–2.9 mm in diameter, the free portion above the ovary 0.1–0.4 mm long, the outer surface with shaggy glandular hairs, white to pink, the sepals often green-tipped; petals linear or oblanceolate, 2–3 times as long as the sepals
    3. Leaf blades with the lobes rounded; petals oblanceolate; pubescence of the petioles and main inflorescence axis of longer, straight, spreading hairs and/or minute glandular hairs; seeds smooth . . . . 2. H. PARVIFLORA
    3. Leaf blades with the lobes widely to narrowly triangular, narrowed to pointed tips; petals linear; pubescence of the petioles and main inflorescence axis of woolly hairs; seeds spiny . . . . 4. H. VILLOSA

1. Heuchera americana L. (common alum root)

Pl. 558 i, j; Map 2594

Petioles glabrous to densely pubescent with relatively long spreading hairs with minute glandular tips, often also with moderate to dense minute glandular hairs. Leaf blades (0.5–)2.0–11.0 cm long, circular to broadly ovate, the upper surface glabrous or sparsely hairy, the undersurface glabrous or sparsely to moderately hairy, the lobes more or less rounded, the margins scalloped or toothed. Inflorescences 25–75 cm long, erect or ascending, the axes glabrous or sparsely to densely hairy with mostly spreading hairs to 3 mm long having minute glandular tips, usually also with scattered minute glandular hairs, especially on the branches and toward the inflorescence tip. Hypanthium at flowering 2–6 mm long, 2–4 mm in diameter, the free portion 0.6–2.0 mm long, weakly zygomorphic, urn-shaped to bell-shaped when fresh, glabrous or minutely glandular on the outer surface, green to yellowish green, sometimes reddish-tinged. Sepals 1–3 mm long, glabrous or minutely glandular on the outer surface and margins, the sinuses between the sepals relatively narrow. Petals 1–4 mm long, oblanceolate to narrowly spatulate, glabrous or minutely glandular on the outer surface and margins, green or white, sometimes pinkish-tinged. Fruits with the body 4–10 mm long, tapered into the 3.0–4.5 mm long styles. Seeds 0.6–0.9 mm long, ovoid, the surface with fine tubercles or spines. $2n$=14. April–June.

Scattered in the Ozark, Ozark Border, and Mississippi Lowlands divisions, northward locally

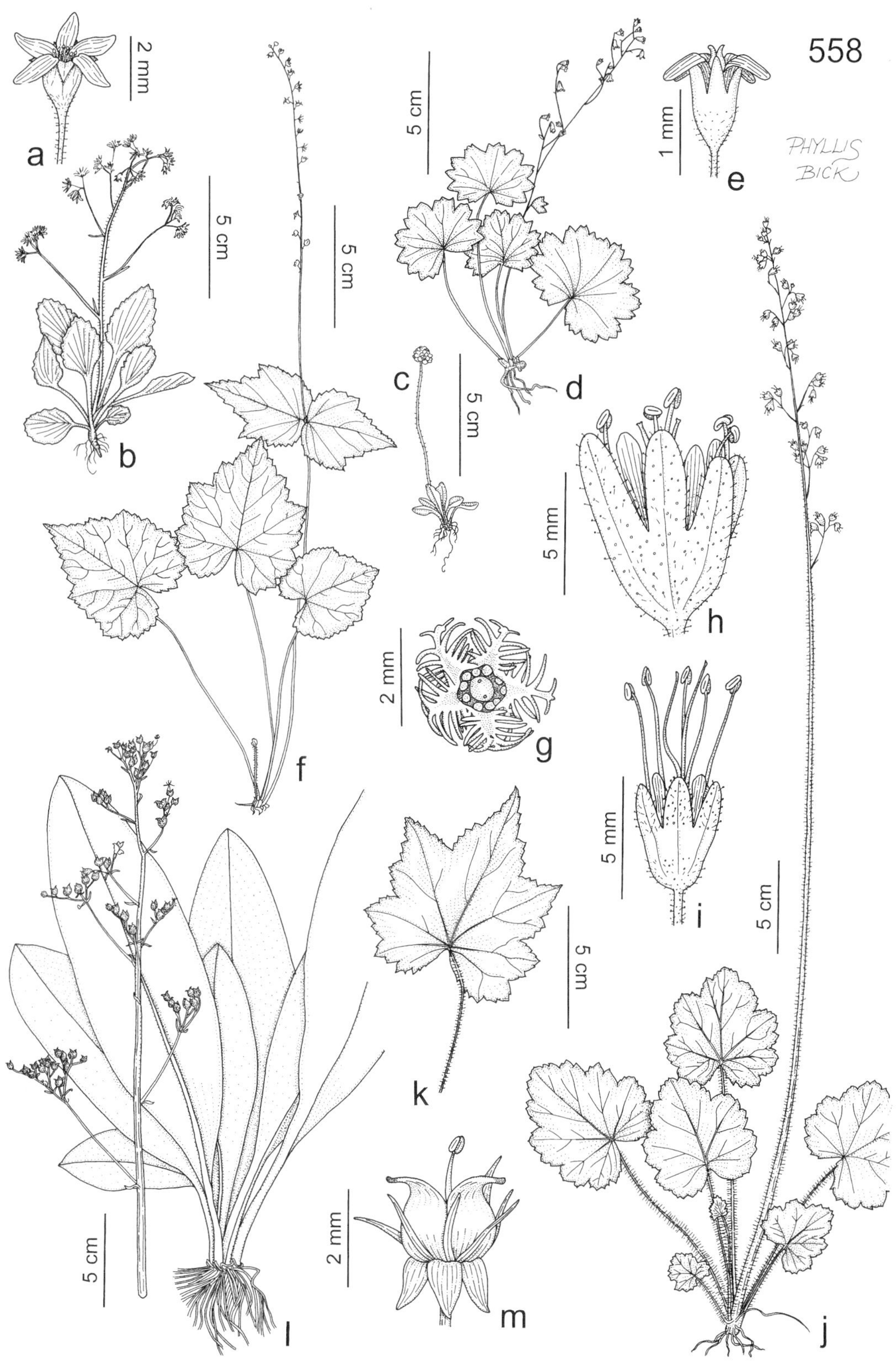

Plate 558. Saxifragaceae. *Micranthes virginiensis*, **a)** flower, **b)** habit. *Micranthes texana*, **c)** habit. *Sullivantia sullivantii*, **d)** habit, **e)** flower. *Mitella diphylla*, **f)** habit, **g)** flower. *Heuchera richardsonii*, **h)** flower. *Heuchera americana*, **i)** flower, **j)** habit. *Heuchera villosa*, **k)** leaf. *Micranthes pensylvanica*, **l)** upper and lower portions of habit, **m)** fruit with remains of corolla and stamen.

to Lewis County in counties along the Mississippi River (eastern U.S. and adjacent Canada west to Nebraska and Oklahoma). Bases and ledges of bluffs and rock outcrops, rocky banks of streams, mesic to dry upland forests, and edges of glades; also roadsides, roadcuts, and railroads.

Wells (1984) treated *H. americana* as comprising a complex of three varieties, only two of which have been found in Missouri. Her var. *hispida* (Pursh) E.F. Wells is restricted to mountains of Virginia and adjacent states and is characterized by more or less glabrous petioles, hypanthia with the free portion 1.5–2.0 mm long, and petals wider than the sepals.

1. Petioles glabrous or sparsely hairy, sometimes also glandular; hypanthium with the free portion 0.6–1.5 mm long .... 1A. VAR. AMERICANA
1. Petioles moderately to densely hairy, also glandular; hypanthium with the free portion 1.5–2.0 mm long .... 1B. VAR. HIRSUTICAULIS

1a. var. americana

Petioles glabrous or sparsely hairy, the hairs mostly less than 1 mm long, sometimes also glandular. Hypanthium urn-shaped at flowering, the free portion 0.6–1.5 mm long. Petals narrower than the sepals. $2n$=14. April–June.

Possibly introduced, known only from McDonald, Newton, and St. Louis Counties (eastern U.S. and adjacent Canada west to Illinois, Oklahoma, and Louisiana). Bluffs; also railroads.

Wells (1984) reported the distribution of var. *americana* as including a number of counties in northwestern Arkansas, but cited only a single McDonald County specimen from Missouri (the first report for the state). Of the three specimens representing var. *americana* located thus far (Kallhoff and Yatskievych, 2001), the historical collection from a railroad embankment in McDonald county (cited by Wells [1984]) presumably is a nonnative occurrence, but the more recent specimen from a bluff along a creek in Newton County may in fact represent a native population. The St. Louis County specimen collected by Sherff contains no further locality data and may represent a gathering from plants cultivated in a garden (Steyermark, 1963). discussed the problems of interpreting Sherff's collections, many of which apparently were made from plants cultivated in gardens, and the inclusion of this record in the distribution of the variety is equivocal.

1b. var. hirsuticaulis (Wheelock) Rosend., Butters & Lakela

H. americana var. *interior* Rosend., Butters & Lakela

H. hirsuticaulis (Wheelock) Rydb.

Petioles moderately to densely hairy, the hairs to 3 mm long, also glandular. Hypanthium urn-shaped to bell-shaped at flowering, the free portion 1.5–2.0 mm long. Petals narrower than the sepals. April–June.

Scattered in the Ozark, Ozark Border, and Mississippi Lowlands Divisions, northward locally to Lewis County in counties along the Mississippi River (Ohio to Nebraska south to Arkansas and Oklahoma). Bases and ledges of bluffs and rock outcrops, rocky banks of streams, mesic to dry upland forests, and edges of glades; also roadsides, roadcuts, and railroads.

The classification of this taxon, which was originally described from plants collected by Engelmann south of St. Louis, remains controversial. Some authors have treated it as *H. ×hirsuticaulis*, a presumed hybrid between the more eastern *H. americana* and the more western *H. richardsonii* (Gleason and Cronquist, 1991). Wells (1984) included it within her concept of *H. americana*, but discussed the possibility that some of its features might have resulted from past hybridization between *H. americana* (var. *americana*) and *H. richardsonii*. She demonstrated that these species can hybridize easily and that crosses between them as well as backcrosses to either parent give rise to fertile progeny (Wells, 1979), but suggested that differences in habitats and cold tolerance keep the taxa spacially and reproductively isolated throughout most of their ranges. The zone of geographic overlap between *H. americana* var. *americana* and *H. richardsonii* coincides almost entirely with that stated for *H. americana* var. *hirsuticaulis* above. Nevertheless, Wells (1984) was able to show a relatively sharp discontinuity between populations of *H. americana* and *H. richardsonii* in this region for every character separating the two species, with only occasional intermediates, and the populations of var. *hirsuticaulis* fell within the morphological variation that she ascribed to *H. americana*. As Missouri specimens are relatively uniform morphologically and nearly all are separated easily from *H. richardsonii* by the characters in the key, the present treatment follows that of Wells (1984) in accepting var. *hirsuticaulis* as a component of *H. americana*.

2. Heuchera parviflora Bartl. (small-flowered alum root)

Pl. 557 c, d; Map 2595

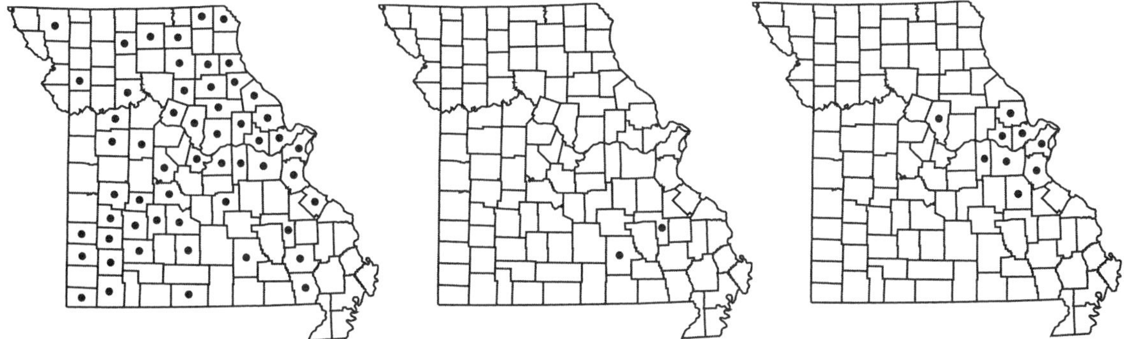

2596. Heuchera richardsonii 2597. Heuchera villosa 2598. Micranthes pensylvanica

Petioles sparsely to densely pubescent with minute glandular hairs, sometimes also with sparse to dense longer spreading hairs having minute glandular tips. Leaf blades 2–13 cm long, circular to kidney-shaped, the surfaces glabrous or sparsely to moderately hairy and sometimes also minutely glandular-hairy, the lobes more or less rounded, the margins scalloped or toothed. Inflorescences 9–45 cm long, ascending to pendant or spreading, the axes sparsely to densely pubescent with minute glandular hairs, sometimes also with sparse to dense longer spreading hairs having minute glandular tips. Hypanthium at flowering 1.3–3.5 mm long, 1–2 mm in diameter, the free portion 0.1–0.3 mm long, appearing actinomorphic or very slightly zygomorphic, obconic to somewhat bell-shaped when fresh, glabrous or minutely glandular on the outer surface, green to yellowish green, rarely reddish-tinged. Sepals 0.5–1.5 mm long, glabrous or minutely glandular on the outer surface and margins, the sinuses between the sepals relatively narrow. Petals 1.5–3.5 mm long, oblanceolate, glabrous, white, rarely pinkish-tinged. Fruits with the body 2.8–5.5 mm long, tapered into the 1.5–4.0 mm long styles. Seeds 0.4–0.6 mm long, ellipsoid-ovoid, the surface smooth. $2n=14$. July–November.

Scattered in the Ozark Division (southeastern U.S. west to Missouri and Mississippi). Bluffs and rock outcrops, mostly on limestone or dolomite substrates.

Wells (1984) recognized two varieties of *H. parviflora*, both present in Missouri and differing in trichome characteristics. Occasional nearly glabrous plants, which Steyermark (1963) called *H. puberula* f. *glabrata*, are essentially examples of var. *puberula* with the minute glandular trichomes mostly absent except on the inflorescence branches and flowers. These unusual individuals are not easily accommodated by the key to varieties.

1. Petioles and inflorescence axis sparsely to moderately pubescent with longer hairs, the minute glandular hairs sparse or absent except on the inflorescence branches .... 2A. VAR. PARVIFLORA
1. Petioles and inflorescence axis densely pubescent only with minute glandular hairs............... 2B. VAR. PUBERULA

2a. var. parviflora

H. parviflora var. *rugelii* (Shuttlew. ex Kunze) Rosend., Butters & Lakela

H. missouriensis Rosend.

Petioles sparsely to moderately pubescent with gland-tipped hairs 0.7–2.5 mm long. Leaf blades with the upper surface glabrous or sparsely pubescent with shorter nonglandular hairs, the undersurface sparsely to moderately pubescent with longer gland-tipped hairs. Inflorescences with the main axis sparsely to moderately pubescent with gland-tipped hairs 0.7–2.5 mm long, these grading into minute glandular hairs on the ultimate branches, the small bracts at the branch points scalelike or leaflike, with toothed margins. $2n=14$. July–November.

Scattered in the Ozark Division (southeastern U.S. west to Missouri and Arkansas). Bluffs and rock outcrops, mostly on limestone or dolomite substrates, less commonly on sandstone.

Based on materials sent to him by Julian Steyermark, Rosendahl (1951) described *H. missouriensis* as a rare species endemic to limestone bluffs in Madison and Wayne Counties, Missouri. He contrasted this from his *H. parviflora* var. *rugelii* (also now considered a synonym of var. *parviflora*) by its denser hairiness, slightly narrower inflorescences, more numerous flowers with shorter stalks, and less round-based, slightly shorter fruits. Steyermark (1963) accepted *H. missouriensis* and separated it from *H. puberula* (now treated as *H. parviflora* var. *puberula*) on the basis of whether the hairs of the hypanthium and sepals were gland-tipped or not. Since that time, many more specimens have accumulated in herbaria. It is apparent that Steyermark's observations on hypanthium pubescence were spurious, as all members of the *H. parviflora* complex have similar, short, gland-tipped hairs. Also, the characters used by Rosendahl to characterize *H. missouriensis* are all expressed singly or together in plants of other parts of the range of the complex and are part of the continuum of morphological variation in *H. parviflora*.

2b. var. puberula (Mack. & Bush) E.F. Wells
H. puberula Mack. & Bush
H. puberula f. *glabrata* Steyerm.

Petioles densely pubescent with minute (less than 0.6 mm long) glandular hairs, rarely glabrous or nearly so. Leaf blades with the surfaces moderately to densely pubescent with minute glandular hairs, rarely glabrous or nearly so. Inflorescences with the main axis moderately to densely pubescent with minute (less than 0.6 mm long) glandular hairs, the small bracts at the branch points scalelike, with entire margins. $2n=14$. July–November.

Scattered in the Ozark Division (Indiana to Tennessee west to Missouri and Arkansas). Bluffs, apparently restricted to limestone or dolomite substrates.

3. Heuchera richardsonii R. Br. (prairie alum root)
H. richardsonii var. *affinis* Rosend., Butters & Lakela
H. richardsonii var. *grayana* Rosend., Butters & Lakela

Pl. 558 h; Map 2596

Petioles usually densely pubescent with relatively long spreading hairs having minute glandular tips, usually also with moderate to dense minute glandular hairs. Leaf blades 1.5–9.5 cm long, circular to broadly ovate, the upper surface glabrous or sparsely hairy, the undersurface sparsely to densely hairy, the lobes more or less rounded, the margins scalloped or toothed. Inflorescences 25–100 cm long, erect or ascending, the axes glabrous or sparsely to densely hairy with mostly spreading hairs to 5 mm long having minute glandular tips, also with scattered minute glandular hairs, especially on the branches and toward the inflorescence tip. Hypanthium at flowering 5–11 mm long, 4–8 mm in diameter, the free portion 2–7 mm long, conspicuously zygomorphic, broadly bell-shaped when fresh, minutely glandular on the outer surface, green to yellowish green, sometimes reddish-tinged. Sepals 1.5–4.0 mm long, minutely glandular on the outer surface and margins, the sinuses between the sepals relatively broad. Petals 1.5–4.0 mm long, oblanceolate to narrowly spatulate, usually minutely glandular on the outer surface and margins, green or white, sometimes pinkish-tinged. Fruits with the body 7.0–14.5 mm long, tapered into the 3.0–4.5 mm long styles. Seeds 0.6–0.9 mm long, ovoid, the surface with fine tubercles or spines. $2n=14, 28$. April–June.

Scattered nearly throughout the state, most abundant in the Glaciated Plains and Unglaciated Plains Divisions, apparently absent from the Mississippi Lowlands (Michigan to Montana south to Indiana, Oklahoma, and Colorado; Canada). Upland prairies, glades, ledges and tops of bluffs, and rock outcrops in mesic to dry upland forests; also roadsides and railroads.

4. Heuchera villosa Michx. **var. villosa** (maple-leaved alum root)
H. villosa var. *macrorhiza* (Small) Rosend., Butters & Lakela

Pl. 558 k; Map 2597

Petioles sparsely to densely pubescent with relatively long, usually woolly hairs, some of which have minute glandular tips, lacking minute glandular hairs. Leaf blades 2–19 cm long, circular to broadly ovate-triangular, the upper surface glabrous to moderately pubescent with straight hairs, the undersurface glabrous to moderately pubescent with usually woolly hairs, sometimes somewhat glandular-sticky, the lobes widely to narrowly triangular, the margins usually sharply toothed. Inflorescences 20–75 cm long, erect or ascending, the axes sparsely to densely hairy with woolly hairs to 4 mm long, the ultimate branches with shorter straighter gland-tipped hairs. Hypanthium at flowering 1.5–3.5 mm long, 2–3 mm in diameter, the free portion 0.2–0.5 mm long, appearing actinomorphic or very slightly zygomorphic, obconic to bell-shaped when fresh, with minute to short straight gland-tipped hairs on the outer surface, green to yellowish green or whitish green, sometimes reddish-tinged. Sepals 0.5–1.7 mm long, glabrous or with minute gland-tipped hairs on the outer surface and margins, the sinuses between the sepals relatively narrow. Petals 1–3 mm long, linear, often curled or coiled, usually glabrous, green or white, sometimes pinkish-tinged. Fruits with the body 3.0–6.5 mm long, tapered into the 1.5–3.5 mm long styles. Seeds 0.6–0.9 mm long, ellipsoid, the surface with fine spines. $2n=14$. June–August, occasionally also November.

Uncommon, known only from historical specimens from Iron County and a more recent specimen from Shannon County (eastern U.S. west to Illinois and Mississippi; Missouri). Limestone bluffs.

Steyermark (1963) discussed the problems of accepting this species as a member of the flora. The voucher specimens he examined were from cultivated material supposedly gathered from two sites in Iron County and grown in St. Louis in the garden of Albert Chandler. However the taxon was rediscovered in 1992 at a site in the Ozark National Scenic Riverways (Shannon County) by Mike Skinner, which confirms its inclusion in the Missouri flora. Wells (1984) accepted the name var. *arkansana* (Rydb.) E.B. Sm. for dwarf plants with very short flower stalks and less hairy, narrower calyces. Interestingly, this variety is considered endemic to just eight counties in the northwestern quarter of Arkansas. Thus, it may eventually be found to occur in adjacent counties in Missouri.

2. **Micranthes** Haw. (saxifrage)

Plants fibrous rooted, sometimes with short rhizomes. Leaves all basal (rarely 1 or a few highly reduced bracteal leaves alternate on the inflorescence stalks), sessile or petiolate. Stipules absent. Leaf blades of at least the larger leaves more than 3 times as long as wide, narrowly lanceolate to oblanceolate, ovate, or obovate, the base narrowed or tapered, the tip rounded or narrowed to a blunt or less commonly sharp point, the margins unlobed, but usually finely to coarsely scalloped or toothed, the teeth often with a small reddish purple (sometimes white-centered) spot on the upper surface near the tip, also usually with short crinkly multicellular hairs, mostly pinnately veined with a single prominent midvein but often with several finer arched veins from at or near the blade base, often somewhat thickened and/or slightly succulent, the surfaces glabrous or hairy, the undersurface sometimes reddish- or purplish-tinged. Inflorescences panicles with usually numerous flowers, erect or ascending, condensed and appearing as 1 or a few dense capitate clusters of flowers early in the season and later often becoming more open, lax, and elongated, usually long-stalked, glabrous or hairy, often with small linear bracts at the branch points. Flowers actinomorphic. Hypanthium saucer-shaped to more or less obconic, fused for less than half the length of the ovary. Sepals oblong to lanceolate, ovate, or elliptic, rounded or bluntly to sharply pointed at the tip, usually glabrous, sometimes reddish- or purplish-tinged or tipped. Petals glabrous, the margins entire, white or cream-colored to greenish yellow. Stamens 10, usually exserted from the calyx, the anthers small, attached at their often deeply cordate bases (often appearing attached toward their midpoints after dehiscing), yellow or orange. Ovary 2(–4)-locular, the placentation axial. Styles tapered, persistent and arched to spreading at fruiting, the stigmas capitate or arc-shaped. Fruits capsules but sometimes appearing more like follicles, broadly ovoid to narrowly ovoid, 2-beaked, dehiscing longitudinally along the inner suture of each beak, the free tips divergent. Seeds numerous, 0.7–1.1 mm long, asymmetrically ovoid to narrowly ovoid in outline, with several fine longitudinal ribs, these usually with minute tubercles or papillae, reddish brown or dark brown. Three hundred and fifty to 440 species, North America, Europe, Asia, most diverse in cool-temperate and arctic regions; a few species in South America, Africa.

Until recently, most authors of floras have treated *Saxifraga* in an inclusive sense to include up to 440 species. However, molecular research by Soltis et al. (2001) has shown clearly that there are two major groups within the Saxifragaceae (in the strict sense), with a restricted version of *Saxifraga* well separated from a large group of genera related to *Heuchera*. Among these, *Micranthes* and some smaller North American relatives form a well-resolved lineage. Nomenclatural combinations for some of these taxa have only recently become available.

A number of species of *Saxifraga* and a smaller number of species of *Micranthes* are cultivated as ornamental, principally in rock gardens.

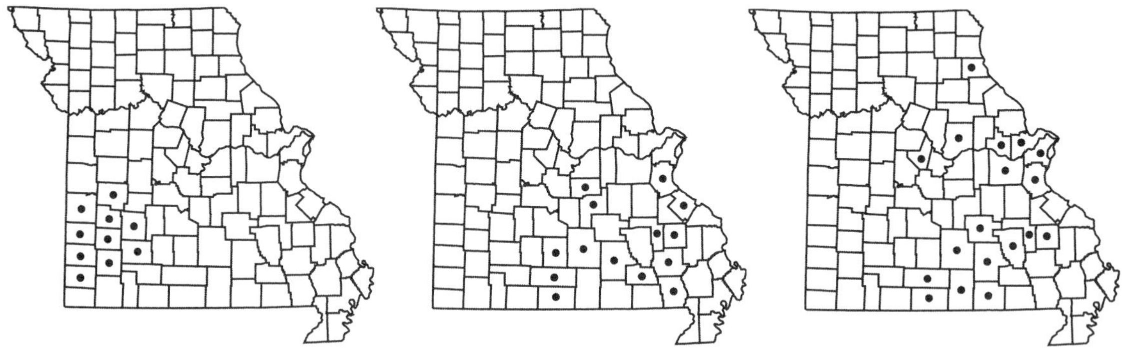

2599. Micranthes texana 2600. Micranthes virginiensis 2601. Mitella diphylla

1. Largest leaves (8–)10–30 cm long; inflorescences 35–95 cm long; sepals becoming reflexed soon after the flower opens; petals cream-colored or light yellow to greenish yellow . . . . 1. M. PENSYLVANICA
1. Largest leaves 1–5(–8) cm long; inflorescences 3–30 cm long; sepals ascending at fruiting; petals white
    2. Inflorescences remaining more or less condensed, not becoming more open and lax later in the flowering season, the branches glabrous or sparsely hairy; sepals often prominently reddish-tinged; petals 2.0–3.5 mm long; carpels mostly 3 or 4 . . . . 2. M. TEXANA
    2. Inflorescences more or less condensed early in the growing season, becoming more open and lax later, the branches moderately to densely hairy; sepals not or only slightly reddish-tinged; petals 3.5–6.0 mm long; carpels usually 2 . . . . 3. M. VIRGINIENSIS

1. Micranthes pensylvanica (L.) Haw. **var. forbesii** (Vasey) Bush (Forbes's saxifrage)

Saxifraga forbesii Vasey

S. pensylvanica L. var. *forbesii* (Vasey) Engl. & Irmsch.

Pl. 558 l, m; Map 2598

Leaf blades (3–)6–30 cm long, the longest ones at flowering (8–)10–30 cm, tapered or narrowed to a sessile base or to a flattened mostly winged petiole to 40 mm long, the surfaces sparsely to moderately pubescent with usually crinkly multicellular hairs, the margins nearly entire or relatively finely scalloped or toothed. Inflorescences 35–95 cm long, condensed at the start of flowering but eventually becoming more open and lax, the main axis moderately pubescent with somewhat spreading, crinkly, multicellular, gland-tipped hairs to 2 mm long, these becoming progressively shorter and denser toward the branch tips. Sepals 0.6–1.2 mm long at flowering, becoming elongated to 1.4–1.8 mm at fruiting, becoming reflexed soon after the flower opens, ovate to oblong-ovate, mostly sharply pointed at the tip, usually lacking reddish coloration. Petals 2–4 mm long, linear to narrowly oblong or narrowly oblanceolate, cream-colored or light yellow to greenish yellow. Pistils with usually 2 carpels, mostly narrowly ovoid. Fruits (including the stylar beaks) 3.5–5.5 mm long. $2n=56$–ca. 280. April–June.

Uncommon, east-central Missouri, with a disjunct occurrence in Boone County (Indiana to Missouri north to Iowa, Minnesota, and Wisconsin). In and around moist, usually north-facing, somewhat protected sandstone bluffs.

The taxonomy of the *M. pensylvanica* complex is very confusing and requires further study. G. W. Burns (1942) presented evidence that var. *forbesii* (as *Saxifraga*) should be regarded as a distinct tetraploid ($2n=56$) species, based upon morphological, cytological, and phytogeographic studies. More recently, M. Phillips and Kowal (1983, as *Saxifraga*) questioned the validity of Burns's data and whether it should be recognized taxonomically at all. They pointed out that the cytological and morphological situation in this group is considerably more complex than thought by Burns. M. Phillips and Kowal found extreme cytological variation within the complex, including counts from the type locality of var. *forbesii* of plants up to the dodecaploid ($2n=280$) level. Throughout the range of the *M. pensylvanica* complex, they documented a complex pattern of polyploid variation ranging from 4x ($2n$=56) to 14x ($2n$=392), and they suggested that these ploidy differences did not correlate with morphological variation in the group. However, M. Phillips and Kowal's paper (available only as an abstract of a presentation at a conference) has not been published fully, and it apparently does not account for all of the morphological differences between taxa suggested by G. W. Burns (1942), nor does it address differences in habitat requirements. Thus, var. *forbesii* is here retained at varietal status, until more detailed studies can account for the confusing patterns of variation in the group.

2. Micranthes texana (Buckley) Small
Saxifraga texana Buckley
Pl. 558 c; Map 2599

Leaf blades 1–5(–8) cm long, tapered or narrowed to a sessile base or to a flattened mostly winged petiole to 15 mm long, the surfaces glabrous or nearly so, the margins somewhat wavy to relatively coarsely and irregularly scalloped or toothed. Inflorescences 3–30 cm long, remaining more or less condensed, not becoming more open and lax later in the flowering season, the main axis sparsely moderately pubescent with somewhat spreading, crinkly, multicellular, gland-tipped hairs, these minute or up to about 1 mm long, becoming sparsely hairy or glabrous toward the short branch tips. Sepals 1.2–2.0 mm long at flowering, not becoming elongated at fruiting, remaining ascending after the flower opens, oblong-ovate, rounded to bluntly pointed at the tip, often prominently reddish-tinged, especially along the margins. Petals 2.0–3.5 mm long, oblong-elliptic to obovate, white. Pistils with usually 3 or 4 carpels, mostly broadly ovoid. Fruits (including the stylar beaks) 3–4 mm long. April–May.

Scattered, restricted to the Unglaciated Plains Division (Missouri and Oklahoma south to Texas; disjunct in Georgia). Glades and rock outcrops in upland prairies, on sandstone and chert substrates.

3. Micranthes virginiensis (Michx.) Small (early saxifrage)
Saxifraga virginiensis Michx.
Pl. 558 a, b; Map 2600

Leaf blades 1–5(–8) cm long, tapered or narrowed to a petiole 5–25 mm long, the surfaces glabrous or sparsely to moderately pubescent with usually crinkly multicellular hairs, the margins finely to relatively coarsely scalloped or toothed. Inflorescences 3–30 cm long, condensed at the start of flowering but eventually becoming more open and lax, the main axis moderately pubescent with somewhat spreading, crinkly, multicellular, gland-tipped hairs, these minute or up to about 1 mm long, becoming densely hairy toward the branch tips. Sepals 1.2–2.0 mm long at flowering, not becoming elongated at fruiting, remaining ascending after the flower opens, ovate to triangular-ovate or oblong-ovate, mostly bluntly to sharply pointed at the tip, lacking reddish coloration or only slightly reddish-tinged at the tip. Petals 3.5–6.0 mm long, narrowly oblanceolate to spatulate or obovate, white. Pistils with usually 2 carpels, mostly broadly ovoid. Fruits (including the stylar beaks) 3–4 mm long. $2n$=20 ($2n$=38 and possibly 28 elsewhere). February–June.

Scattered in portions of the Ozark and Ozark Border Divisions (eastern U.S. west to Minnesota and Oklahoma; Canada). Glades, ledges and tops of bluffs, and rock outcrops in mesic to dry upland forests, often on acidic substrates.

The closely related *M. palmeri* Bush (*Saxifraga palmeri* Bush) is endemic to approximately the western half of Arkansas and adjacent portions of eastern Oklahoma (Steyermark, 1959; E. B. Smith, 1988), with occurrences in a few counties adjacent to southwestern Missouri. It differs from *M. virginiensis* in its leaf blades with entire or nearly entire margins, as well as inflorescence axes with somewhat longer, less crinkly, less glandular hairs, the branches becoming sparsely hairy or glabrous toward the tips. This species should be searched for in southwestern counties, in acidic glade habitats similar to those of *M. texana*.

3. Mitella L. (mitrewort, bishop's cap)

Twelve to 20 species, North America, Asia.

1. Mitella diphylla L.
Pl. 558 f, g; Map 2601

Plants with short stout rhizomes. Leaves basal and mostly long-petiolate, the apparent flowering stems (actually the main inflorescence stalks) with a single pair of opposite bracteal leaves toward the midpoint, these sessile or short-petiolate, the petioles pubescent with moderate to dense, longer, downward-pointing hairs having minute dark glandular tips and sparse to dense, minute stalked

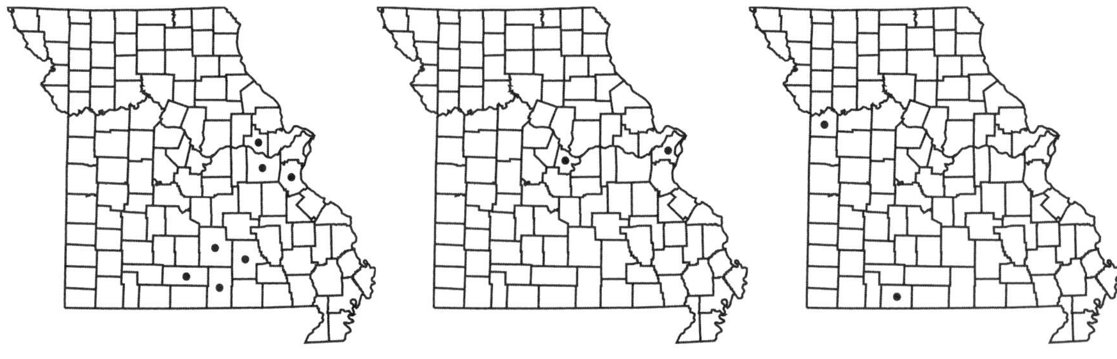
2602. Sullivantia sullivantii 2603. Buddleja davidii 2604. Scrophularia lanceolata

glands. Stipules 1–5 mm long, scalelike, ovate to oblong-ovate, the margins entire to somewhat irregular. Leaf blades 1–9 cm long, slightly longer than to about as long as wide (the flowering stem leaves to about 2 times as long as wide), ovate to nearly circular, the base cordate (often broadly rounded to truncate on flowering stem leaves), the tip rounded or pointed, the margins usually with 3 or less commonly 5 shallow lobes, also finely to coarsely scalloped or toothed, palmately veined with mostly 5 primary veins, the upper surface glabrous or sparsely to moderately hairy, sometimes sparsely glandular toward the base, green, the undersurface sparsely to moderately hairy and glandular, especially along the veins, pale green to grayish green. Inflorescences 10–40 cm long (measured from the rhizome tip), spikelike racemes with 5–20 flowers, erect, long-stalked, moderately to densely pubescent with minute stalked glands below the pair of bracteal leaves, also with longer, spreading to downward-pointing hairs having minute glandular tips. Flowers actinomorphic, subtended by a minute oblong-ovate bract. Hypanthium shallowly bell-shaped, fused to the basal portion of the ovary. Sepals 1.0–1.5 mm long, oblong-triangular, glabrous, greenish white. Petals 2.0–2.5 mm long, deeply pinnately dissected into linear segments, glabrous, white. Stamens 10, much shorter than the calyx, the anthers small, attached at their cordate bases, pale yellow. Ovary 1-locular in the fused portion, the placentation parietal. Styles short, the stigmas crescent-shaped or lozenge-shaped. Fruits 2.5–3.0 mm long, broadly ovoid, 2-beaked, dehiscing longitudinally from between the beaks. Seeds 5 to numerous, 0.8–1.2 mm long, narrowly obovoid, somewhat angular, the surface appearing smooth (microscopically pebbled), black, shiny. $2n=14$. April–June.

Uncommon in the eastern half of the state, mostly in the Ozark and Ozark Border Divisions (eastern U.S. west to Minnesota and Arkansas; Canada). Shaded ledges of north-facing dolomite or sandstone bluffs and rocky slopes in mesic upland forests.

Both the generic name *Mitella* and the common name mitrewort refer to the dehiscing fruits, which take on the appearance of a bishop's cap (mitre). Although the flowers are often oriented toward the side in the inflorescence, by fruiting they have moved to a position facing upward. The seeds are held in the cuplike basal portion following dehiscence and apparently are dispersed by water droplets striking the fruit during rainfall (Spongberg, 1972). The stiff lacy petals with obliquely ascending linear lobes of this beautiful little wildflower are unique among Missouri plants in their pattern of dissection.

4. **Sullivantia** Torr. & A. Gray (Soltis, 1991)

Three species, United States.

1. Sullivantia sullivantii (Torr. & A. Gray) Britton

S. renifolia Rosend.

Pl. 558 d, e; Map 2602

Plants with short rhizomes. Leaves all basal (sometimes a few small leaves alternate on the inflorescence stalk), mostly long-petiolate, the petiole glabrous or with sparse stalked glands. Stipules

1–3 mm long, scalelike, ovate to oblong-ovate, the margins entire to irregularly toothed or lobed. Leaf blades (of basal leaves) 1–10 cm long, shorter than to about as long as wide, kidney-shaped to nearly circular, the base cordate, the tip rounded, the margins usually with 7–13 shallow lobes, also finely to coarsely scalloped or toothed (the teeth with glandular tips) and often with sparse stalked glands, palmately veined with 7–13 primary veins, the surfaces glabrous, often somewhat shiny. Inflorescences 5–30 cm long, panicles with 12 to numerous flowers, erect to spreading or pendant, usually long-stalked, moderately to densely pubescent with stalked glands, with small linear to oblong bracts at the branch points. Flowers actinomorphic, sometimes subtended by a small linear bract. Hypanthium more or less bell-shaped, fused to about half the length of the ovary. Sepals 0.8–1.5 mm long, triangular-ovate to oblong-lanceolate, glabrous or with sparse stalked glands. Petals 1.5–3.5 mm long, glabrous, white, the margins entire or slightly irregular. Stamens 5, slightly shorter than the calyx, the anthers small, attached at their cordate bases, yellow. Ovary 2-locular in the fused portion, the placentation axile. Styles tapered, the stigmas capitate. Fruits 3–7 mm long, narrowly ovoid to narrowly ellipsoid, 2-beaked, dehiscing longitudinally from each of the beaks. Seeds numerous, 0.7–1.2 mm long, narrowly oblong-elliptic in outline, somewhat flattened and narrowly winged, the surface smooth, tan to light brown. $2n=14$. May–August.

Uncommon in the central portion of the Ozark Division and the northeastern Ozark Border (Virginia to Missouri north to Ohio and Minnesota). Shaded moist north-facing bluffs on both dolomite and sandstone substrates.

Earlier authors (Rosendahl, 1927; Steyermark, 1963) attempted to separate western populations of this species under the name *S. renifolia*, based on subtle differences in leaf, sepal, petal, and capsule characters. However, Soltis (1991), who had many additional collections to examine and who studied the biosystematics of the genus in the field, greenhouse, and herbarium, concluded that all of these characters exhibited more or less continuous variation. Steyermark (1963) discussed the hypothesis that this taxon represents a relict that had a wider distributional range before the last advance of Pleistocene glaciers and whose range became highly fragmented as a result of changes in climate during and after the Ice Age. *Sullivantia sullivantii* is not particularly common anywhere in its present distribution, and populations maintain themselves in localized microhabitats where cool moist conditions are more or less constant throughout the year.

SCROPHULARIACEAE (Figwort Family)

Plants biennial or perennial herbs or shrubs (annuals elsewhere). Leaves basal and/or opposite or alternate along the stems, sessile to long-petiolate. Stipules absent or inconspicuous, then herbaceous, attached along a line or slender ridge, the adjacent stipules of the leaves at each node fused into a single triangular to depressed-ovate structure. Leaf blades unlobed or less commonly pinnately lobed, variously shaped, the margins usually scalloped or toothed. Inflorescences racemes (sometimes spikelike) or panicles of usually numerous flowers, at least the lower nodes subtended by bracts, inconspicuous bractlets sometimes also present below each flower. Flowers perfect, hypogynous. Calyces actinomorphic or nearly so (zygomorphic elsewhere), 4- or 5-lobed, persistent at fruiting. Corollas actinomorphic or zygomorphic and sometimes bilabiate, variously colored, 4- or 5-lobed, the tube variously short to long. Stamens 4 or 5, alternating with the corolla lobes, the filaments attached in the corolla tube (sometimes near its base), sometimes unequal, the anthers exserted or not, attached near their midpoints or sometimes fused to the filament tips for 1/2 or more of the length, often appearing U-shaped or 1-locular, yellow or orange. Staminodes absent or (sometimes in flowers with 4 fertile stamens) present, then well differentiated from the fertile stamens. Pistil 1 per flower, of 2 fused carpels. Ovary 2-locular (sometimes appearing 4-locular elsewhere), with numerous ovules, the placentation axile. Style 1, often persistent at fruiting (but sometimes shriveled or shed toward the tip), the stigma 1, variously shaped, entire or 2-lobed. Fruits capsules (drupes or schizocarps elsewhere), dehiscent longitudinally from the tip. Seeds numerous, minute. About 52 genera, about 1,680 species, worldwide.

Perhaps no other family in the traditional Englerian classification has undergone as profound a reevaluation as the Scrophulariaceae. In the traditional broad sense, the family once included nearly 200 genera and about 4,000 total species. However, a number of botanists had already noted the extreme morphological diversity within the group and the difficulties involved in using morphological characters to separate the Scrophulariaceae from some related families, such as the Acanthaceae, Bignoniaceae, Globulariaceae, and Orobanchaceae (Cronquist, 1981). The application of molecular data from selected DNA marker sequences has had a profound effect on the classification of families and orders across the angiosperms and has greatly altered the classification and circumscription of the groups associated with the Scrophulariaceae, Lamiaceae, and related families. As presently circumscribed, a very large order Lamiales includes the families formerly included in that order, as well as those previously segregated into Bignoniales and Scrophulariales by some botanists (see Judd et al. [2008] for a brief review). As researchers began to sample a larger suite of species within the various groups contained in the Lamiales, it soon became apparent that the Scrophulariaceae, as traditionally treated, was an unnatural group (as reviewed by Tank et al. [2006]). This has resulted in the break-up of the Scrophulariaceae into seven to eleven families, but the number of families and the generic limits of each have not yet become fully stabilized (Angiosperm Phylogeny Group, 1998, 2003, 2009). Additionally, a few groups formerly classified as distinct families have been included within some of these recently recircumscribed families. The introduced genus *Buddleja* was not known to be a member of the state's flora when Steyermark (1963) completed his volume. Many botanists have treated it in the segregate family Buddlejaceae (Leeuwenberg and Leenhouts, 1980; Cronquist 1981, 1991; G. K. Rogers, 1986; Norman, 2000), but it currently is considered by most botanists to belong in the Scrophulariaceae.

A summary of the current familial placement of the various genera included by Steyermark in the Scrophulariaceae is presented below, along with other genera that are now classified into this group of families. Members of the Acanthaceae, Bignoniaceae, Martyniaceae, and Oleaceae, families in the Lamiales whose circumscriptions for the genera present in Missouri have not changed since Steyermark's (1963) time, are not included in the list.

Linderniaceae .... *Lindernia* (Albach et al., 2005; Oxelman et al., 2005; Rahmanzadeh et al., 2005; Tank et al., 2006)

Orobanchaceae .... *Agalinis* (*Gerardia*), *Aureolaria* (*Gerardia*), *Buchnera, Castilleja, Dasistoma, Epifagus* (formerly Orobanchaceae), *Orobanche* (formerly Orobanchaceae), *Pedicularis*

Paulowniaceae .... *Paulownia* (Olmstead and Reeves, 1995)

Phrymaceae .... *Mazus, Mimulus, Phryma* (formerly Phrymaceae) (Oxelman et al., 1999; Beardsley and Olmstead, 2002; Tank et al., 2006; Albach et al., 2009)

Plantaginaceae .... *Callitriche* (formerly Callitrichaceae), *Plantago* (formerly Plantaginaceae), *Antirrhinum, Bacopa, Chaenorhinum, Chelone, Collinsia, Cymbalaria, Gratiola, Kickxia, Leucospora* (formerly *Conobea*), *Linaria, Mecardonia, Nuttallanthus* (formerly *Linaria*), *Penstemon, Veronica, Veronicastrum* (Albach and Chase, 2004; Albach et al., 2005; Oxelman et al., 2005)

Scrophulariaceae ...... *Buddleja* (see note above), *Scrophularia, Verbascum* (Olmstead et al., 2001; Oxelman et al., 2005)

Tetrachondraceae ..... *Polypremum* (formerly Loganiaceae) (Oxelman et al., 2005)

In its diminished current form, the family Scrophulariaceae has relatively limited economic importance. Members of several genera are cultivated as ornamentals, including *Buddleja* and *Verbascum*. Conversely, in some parts of the United States, species in these genera are considered troublesome weeds. Many species in the family are considered at least mildly toxic to livestock and a few genera contain plants that have been used medicinally.

In addition to the taxa treated below, Steyermark (1963) also included a report of *Limosella aquatica* L. (common mudwort; Pl. 559 a, b) from Missouri. He noted that the inclusion of this species was based on a single historical voucher specimen collected by B. F. Bush in Jackson County, but did not cite a collection number or herbarium where the specimen was accessioned. Palmer and Steyermark (1935) had stated that the plant occurred on sand bars of the Missouri River (as *L. subulata* E. Ives), but gave no other clues to the location of a voucher. Pennell (1935), in his study of the eastern temperate North American Scrophulariaceae, did not cite Missouri in the range of any species of *Limosella*, and Missouri has not been cited as part of the species range in other publications. During the present research, despite diligent searches, no specimens were discovered to support the inclusion of *Limosella* in the Missouri flora, and it is thus excluded for the present. *Limosella aquatica* is widespread but of sporadic occurrence in Europe and Asia, and in the New World from Canada south to South America. In the United States it is known from most western states eastward to Minnesota and Nebraska. It is a submerged or emergent aquatic that likely is spread by mud on the feet and feathers of migratory waterfowl. The species is inconspicuous; plants are rarely 15 cm tall, although the stems are stoloniferous and creeping. The leaves occur in dense tufts at the rooted stem nodes and consist of a long, slender petiole terminating in a short, lanceolate to elliptic blade. The inconspicuous flowers are solitary at the tip of a stalk that is shorter than the leaves. Corollas are 3–4 mm long, more or less actinomorphic, with a slender tube and 5 short lobes, white or pale pinkish- to purplish-tinged. Botanists who do field work in wetlands that are stopover points for ducks and other migratory waterfowl should be alert for new records of this diminutive member of the Scrophulariaceae.

1. Plants shrubs; corollas actinomorphic, 4-lobed . . . . 1. BUDDLEJA
1. Plants herbs; corollas slightly to strongly zygomorphic, sometimes bilabiate, 5-lobed
    2. Stem leaves opposite; corollas reddish brown or partially yellowish green; fertile stamens 4, the staminode well differentiated, consisting of a filament with a flattened, expanded tip . . . . 2. SCROPHULARIA
    2. Stem leaves alternate; corollas yellow or white, sometimes with purplish markings at the base of the lobes; fertile stamens 5, staminodes absent . . . . 3. VERBASCUM

1. **Buddleja** L. (butterfly bush)

About 100 species, North America to South America, Caribbean Islands, Asia, Africa, Madagascar, Pacific Islands.

Several species of *Buddleja* are cultivated as ornamentals and as a long-blooming nectar source for insects, particularly butterflies. Of these, only *B. davidii* has been recorded as an escape in Missouri, but two other species are present outside of cultivation with increasing frequency in the southeastern United States (G. K. Rogers, 1986). *Buddleja madagascariensis* Lam. (native to Madagascar) tends to have a climbing habit and is also characterized by tan pubescence and yellow to yellowish orange corollas. *Buddleja lindleyana* Fortune (native to China) differs from *B. davidii* in its curved, externally hairy and/or glandular corolla tubes, generally more glandular pubescence, somewhat less branched and often more or less 1-sided inflorescences, and less sharply toothed leaf margins.

1. Buddleja davidii Franch. (orange eye, summer lilac, common butterfly bush)

Map 2603

Plants shrubs with spreading or arching branches, 0.5–4.0 m tall, pubescent with minute, white, somewhat shiny, stellate hairs and small, globose, light yellow glands. Bark smooth or shallowly ridged or furrowed, gray to tan. Twigs rounded to slightly angled, purplish brown to green but often appearing grayish-tinged, with moderate to dense hairs and scattered glands. Leaves opposite, sessile or very short-petiolate. Stipules herbaceous, appearing as small, bluntly triangular to depressed-ovate flaps of tissue between the leaf bases. Leaf blades 2–15 cm long, narrowly elliptic to ovate-elliptic, tapered to a sharply pointed tip, narrowed or tapered at the base, the margins finely and sharply toothed, the upper surface sparsely hairy and sparsely glandular, appearing green, the undersurface with dense matted hairs and sparse to moderate glands, appearing silvery gray or white. Inflorescences terminal on the branches, narrow panicles to 20 cm long, sometimes appearing racemose, the flowers in clusters along short branches from the main axis. Flowers often subtended by 2 or 3 short linear bracts. Calyces 2.5–4.0 mm long, actinomorphic, shallowly 4-lobed, tubular to narrowly bell-shaped, sparsely to moderately hairy and glandular, the lobes triangular to narrowly triangular. Corollas 9–12 mm long, actinomorphic, 4-lobed, trumpet-shaped, the tube moderately hairy internally and usually more than twice as long as the lobes, the lobes with rounded, irregular margins, reddish purple (rarely pink or white in cultivation) with the inner side of the tube and throat orange. Stamens 4, not exserted, the free portions of the filaments very short, the anthers appearing U-shaped, more or less 2-locular, yellow or orange. Staminodes absent. Ovary ovoid, 2-locular. Style about as long as the ovary, not exserted, the stigma 1, sometimes 2-lobed, club-shaped. Fruits 6–8 mm long, narrowly ellipsoid to cylindric-ovoid. Seeds 2–4 mm long, the body more or less ellipsoid, slightly flattened, tapered to a slender wing at each end, the surface smooth, brown. $2n=76$. July–October.

Introduced, uncommon, widely scattered, usually in urban areas (native of Asia, introduced sporadically nearly throughout the New World, including the eastern and western U.S.). Roadsides, gardens, and open disturbed areas.

This species was first reported from the city of St. Louis by Mühlenbach (1979), based on a single plant growing immediately adjacent to a cultivated planting of the species. As the species continues to become more common in cultivation in the state, populations are sure to become established in other areas. It has become a problem invasive exotic in portions of the Pacific Northwest.

2. **Scrophularia** L. (figwort)

Plants perennial herbs, fibrous-rooted. Stems 1 to several from the base, erect or strongly ascending, unbranched or few-branched, strongly 4-angled, glabrous or moderately pubescent with minute, mostly glandular hairs, at least toward the tip. Leaves opposite, short- to long-petiolate. Stipules absent (note that short, leafy axillary branches are frequently produced). Leaf blades lanceolate to ovate, variously tapered or angled to rounded, truncate, or shallowly cordate at the base, angled or tapered to a sharply pointed tip, the margins sharply toothed, the upper surface glabrous or sparsely and minutely hairy along the main veins or toward the base, the undersurface sparsely to moderately and minutely hairy, the venation pinnate, but the lowermost pair of lateral veins sometimes slightly more prominent than the others. Inflorescences terminal, elongate panicles of usually numerous flowers, the lowermost branches usually with pairs of bracts resembling reduced leaves, the flowers mostly long-stalked, the stalk with a pair of minute, linear bractlets at the base, the axes, flower stalks, and bractlets usually with dark, stalked glands. Calyces actinomorphic, 5-lobed to about the midpoint, the lobes ovate to broadly ovate, with irregular margins. Corollas bilabiate, reddish brown or partially yellowish green, the tube urn-shaped to broadly urn-shaped, the upper lip more or less straight, about as long as to slightly longer than the lower lip, notched at the tip, the lower lip strongly 3-lobed, the lateral lobes broadly ovate to nearly circular, more or less straight, the central lobe, triangular-ovate, strongly reflexed. Stamens 4, not exserted, the free portions of the filaments short, the anthers appearing 1-celled, yellow. Staminode 1, well differentiated, consisting of a filament with a flattened, expanded tip, oriented under the upper lip. Ovary broadly ovoid to nearly globose, 2-locular. Style usually longer than the ovary, usually slightly exserted, the stigma 1, unlobed, capitate. Fruits with the body ovoid to broadly ovoid

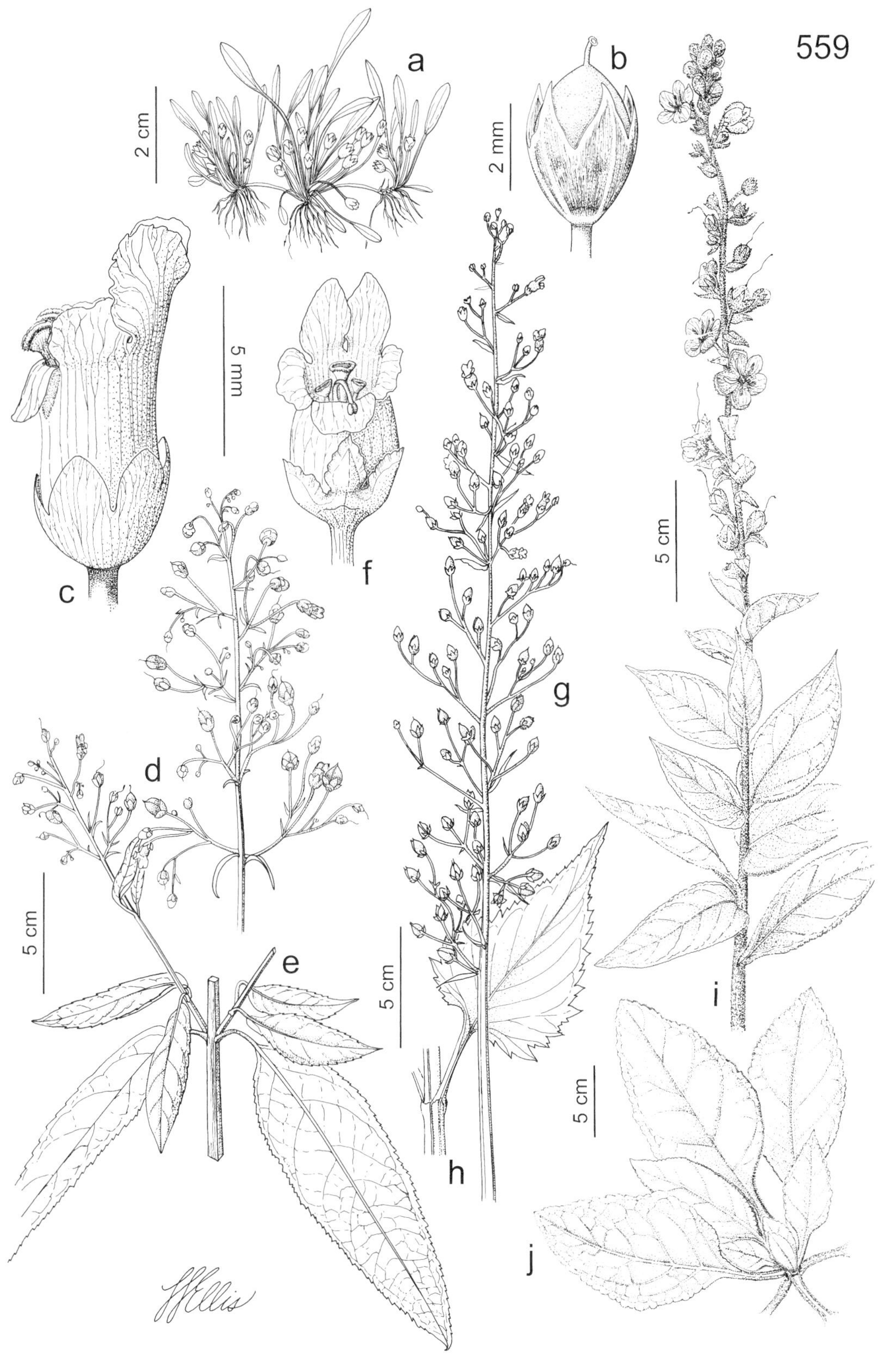

Plate 559. Scrophulariaceae. *Limosella aquatica*, **a)** habit, **b)** fruit. *Scrophularia marilandica*, **c)** flower, **d)** inflorescence, **e)** node with leaves and axillary inflorescence. *Scrophularia lanceolata*, **f)** flower, **g)** inflorescence, **h)** node with leaf and leaf base. *Verbascum phlomoides*, **i)** stem tip with leaves and inflorescence, **j)** basal leaves.

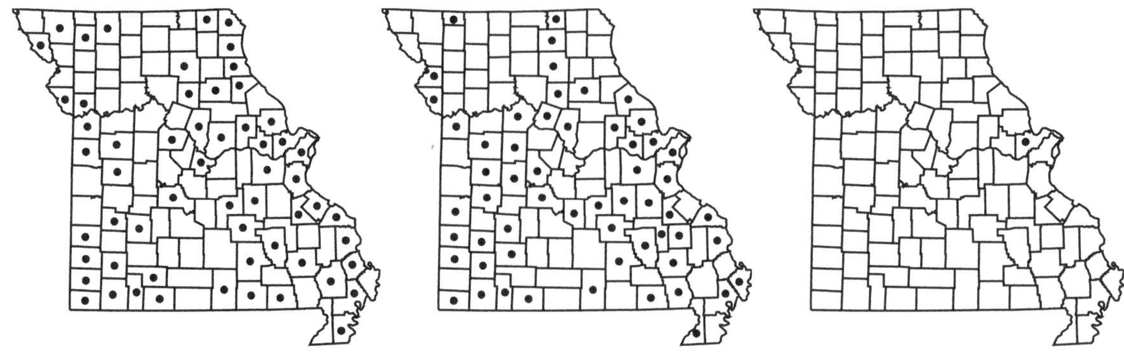

2605. Scrophularia marilandica 2606. Verbascum blattaria 2607. Verbascum lychnitis

or nearly globose, broadly angled to tapered to a beaked tip. Seeds 0.6–0.9 mm long, oblong-ellipsoid, turgid, unwinged, the surface with a network of ridges, dark brown. About 200 species, North America, South America, Caribbean Islands, Europe, Asia, Africa.

1. Staminode yellowish green (often drying darker), the tip more or less fan-shaped, wider than long to slightly longer than wide; fruits 6–10 mm long, the body ovoid, tapered at the tip, the surface dull at maturity; leaf blades sharply, irregularly, and often coarsely toothed along the margins . . . . 1. S. LANCEOLATA
1. Staminode dark purple to purplish brown, elliptic to ovate, longer than wide; fruits 4–7 mm long, the body broadly ovoid to nearly globose, broadly angled to broadly tapered at the tip, the surface often somewhat shiny at maturity; leaf blades bluntly to sharply and often finely toothed along the margins . . . . 2. S. MARILANDICA

1. Scrophularia lanceolata Pursh (American figwort, early figwort)

Pl. 559 f–h; Map 2604

Stems 70–150(–200) cm long, with 4 blunt angles and flat to slightly concave sides, moderately pubescent with minute, glandular hairs, at least toward the tip. Petioles of the larger leaves 1–4 cm long, winged toward the tip, grooved on the upper side, the margins of the groove with pale, slightly thickened ridges. Leaf blades 4–20 cm long, lanceolate to narrowly ovate, variously tapered or angled to rounded, truncate, or shallowly cordate at the base, sharply, irregularly, often coarsely, and sometimes doubly toothed along the margins, the upper surface glabrous, the undersurface glabrous or sparsely and minutely glandular-hairy along the main veins. Inflorescences tending to be cylindric in shape, mostly 4–8 cm wide, the main branches ascending to strongly ascending. Calyces 2–4 mm long. Corollas 7–12 mm long, yellowish green, usually with pale reddish-brown to brown mottling, sometimes appearing pale reddish brown except for the inner surface of the lowermost lobe. Staminode with the strongly expanded tip yellowish green (often drying darker), the tip more or less fan-shaped, wider than long to slightly longer than wide. Fruits 6–10 mm long, the body ovoid, tapered at the tip, the surface dull at maturity. $2n$=92–96. May–June.

Uncommon, known only from a historical collection from Jackson county (northern U.S. south to Virginia, Oklahoma, and California; Canada). Habitat unknown, but presumably the edge of an upland forest.

This species is known in Missouri only from the 1912 collection by B. F. Bush. The specimen was collected in Atherton (now part of the Kansas City metropolitan area) and the label listed only “dry ground” as the habitat.

2. Scrophularia marilandica L. (eastern figwort, carpenter’s square, late figwort)

Pl. 559 c–e; Map 2605

Stems 60–170(–260) cm long, with 4 blunt angles and usually somewhat concave sides, moderately pubescent with minute, glandular hairs, at least toward the tip. Petioles of the larger leaves 1.5–8.0 cm long, uniformly slender above the slightly expanded base, grooved and often paler on the upper side. Leaf blades 3–15(–25) cm long, lanceolate to ovate, all but the uppermost leaves rounded to shallowly cordate at the base, bluntly to sharply and often relatively finely toothed along the margins, the upper surface glabrous or sparsely and minutely glandular-hairy along the midvein, the undersurface glabrous or sparsely to moderately and minutely glandular-hairy. Inflorescences tending to be pyramid-shaped, mostly 5–18 cm wide, the main branches mostly spreading to loosely ascending. Calyces 2.0–3.5 mm long. Corollas 5–9 mm long, yellowish green, usually with reddish brown to brownish red mottling at least on the inner surface and along the margins, sometimes appearing nearly entirely reddish brown to brownish red. Staminode with the somewhat expanded tip dark purple to purplish brown, elliptic to ovate, longer than wide. Fruits 4–7 mm long, the body broadly ovoid to nearly globose, broadly angled to broadly tapered at the tip, the surface often somewhat shiny at maturity. July–October.

Scattered nearly throughout the state (eastern U.S. west to South Dakota and Texas). Bottomland forests, edges of mesic upland forests, banks of streams and rivers, and bottomland prairies.

Plants with the leaf undersurface relatively densely hairy have been called f. *neglecta* (Rydb. ex Small) Pennell, but grade fully into the less hairy, typical form. This species can be very difficult to distinguish from *S. lanceolata*.

3. **Verbascum** L. (mullein)

Plants biennial herbs (perennial elsewhere), taprooted. Stems usually solitary, unbranched or branched, circular in cross-section or 4-angled, variously glandular- or nonglandular-pubescent, the hairs unbranched or branched, sometimes glabrous or nearly so toward the base. Leaves in a dense basal rosette and alternate on the stems, the stem leaves sessile, often clasping the stems or decurrent along the stems, the basal leaves sessile or less commonly tapered to a short petiole. Stipules absent. Leaf blades variously shaped, unlobed or pinnately lobed, the margins otherwise toothed or scalloped, sometimes finely so, the upper surface variously sparsely to dense hairy (the hairs glandular or not, branched or unbranched), the undersurface variously glabrous to densely hairy (the hairs branched or unbranched), the venation pinnate. Inflorescences terminal, elongate spikes or racemes, these sometimes grouped into panicles, of usually numerous flowers, each flower (or flower stalk) subtended by a small leaflike or scalelike bract; bractlets absent. Calyces actinomorphic, deeply 5-lobed, the lobes oblanceolate to oblong-lanceolate or lanceolate, with entire margins. Corollas slightly to moderately zygomorphic, not markedly bilabiate, more or less saucer-shaped, 5-lobed, the lobes spreading, broadly obovate to nearly circular, the 2 upper lobes somewhat shorter than the other 3, yellow or white, sometimes with purplish markings at the base of the lobes. Stamens 5, usually appearing exserted, the free portions of the filaments elongate, all or some of them densely hairy, the anthers appearing 1-celled, sometimes unequal in size. Staminodes absent. Ovary ovoid to globose, 2-locular. Style usually longer than the ovary, exserted, angled downward, but sometimes curved upward slightly toward the tip, the stigma 1, unlobed, capitate. Fruits ovoid to globose, the style base usually persistent as a slender, short beak. Seeds 0.6–1.0 mm long, broadly conic-quadrate in outline (truncate at each end, slightly tapered from tip to base), turgid, unwinged, the surface with thick wavy longitudinal ridges, these sometimes with cross-ridges to form a loose network, brown to dark brown. About 350 species, Europe, Asia, Africa, introduced elsewhere.

A number of mulleins are cultivated as ornamentals and numerous hybrids and cultivars exist. Several species have been used medicinally, mainly for respiratory ailments. Native Americans began using *V. thapsus* soon after its introduction by early European colonists, developing a number of medicinal uses and incorporating the leaves (as a tobacco substitute) into ceremonies (Moerman, 1998). The seeds are very long-lived; samples of *V. blattaria* and *V. thapsus* stored in moist sand still exhibited significant viability after 120 years (Telewski and Zeevaart, 2002).

1. Hairs unbranched, glandular . . . . 1. V. BLATTARIA
1. Hairs branched, not glandular
   2. Corollas white; inflorescences elongate panicles, the branches of numerous spikelike racemes or sometimes slender panicles . . . . 2. V. LYCHNITIS
   2. Corollas yellow; inflorescences spikelike racemes, unbranched or occasionally few-branched from near the base
      3. Leaf blades less densely woolly on the upper surface than on the undersurface, the bases rounded to clasping, but not decurrent . . . . 3. V. PHLOMOIDES
      3. Leaf blades densely woolly on both surfaces, the bases decurrent as wings along the stems . . . . 4. V. THAPSUS

1. Verbascum blattaria L. (moth mullein)
Pl. 560 a–c; Map 2606

Stems 40–150 cm long, erect, 4-angled, sometimes branched, moderately pubescent toward the tip with unbranched, glandular hairs, often glabrous or nearly so toward the base. Leaves dark green, those of the basal rosettes 4–20 cm long, sessile or with a very short, winged petiole, the blade oblanceolate, irregularly coarsely toothed and/or scalloped to pinnately lobed; stem leaves progressively shorter and more finely toothed or scalloped (sometimes entire or nearly so) toward the stem tip, sessile, mostly lanceolate, sometimes somewhat clasping the stems, but the bases not decurrent, grading into the inflorescence bracts; leaf blades with the upper surface moderately pubescent with unbranched, glandular-hairs, the undersurface glabrous. Inflorescences open racemes (rarely appearing paniculate in branched plants), the flowers solitary at the nodes, the flower stalks 8–15(–25) mm long, glandular-hairy. Calyces 5–8 mm long, the lobes narrowly elliptic-lanceolate to narrowly lanceolate, glandular-hairy. Corollas 9–15 mm long, white or yellow, with a reddish purple ring at the lobe bases, glandular-hairy. Stamens unequal, the upper 3 with the filaments shorter, straight, bearded with purplish red and usually also white hairs; the lower 2 with the filaments longer, angled downward, bearded with purplish red hairs, the anthers orange, those of the lower pair fused to the filaments from the base to about the midpoint. Fruits 5–8 mm long, more or less globose, minutely glandular-hairy. $2n$=18, 30, 32. May–September.

Introduced, scattered nearly throughout the state (native of Europe, Asia; introduced widely, in North America). Banks of streams and rivers; also old fields, pastures, fallow fields, ditches, railroads, roadsides, and open, disturbed areas.

Two forms of the species are widespread in Missouri. In many cases entire populations comprise only one form or the other, but mixed populations are known. Typical plants have yellow corollas, whereas white-flowered plants have been called f. *albiflora* (G. Don) House (f. *erubescens* Brug.). Aside from the different corolla colors, there appear to be no morphological characters to distinguish the two morphs. Although it is common for species with corollas ranging from pink to red, purple, or blue to include mutants with white flowers, it is very rare for species with yellow corollas to develop white-flowered mutations.

2. Verbascum lychnitis L. (white mullein)
Pl. 560 d, e; Map 2607

Stems 40–150 cm long, erect, circular in cross-section or 4-angled to bluntly polygonal toward the tip, unbranched or sometimes few- to several-branched from near the base, densely hairy when young with minute stellate, nonglandular hairs (appearing scurfy), becoming glabrous in patches at maturity. Leaves appearing green to light green or grayish green, those of the basal rosettes 8–45 cm long, short-petiolate, the blade oblanceolate to obovate or elliptic-obovate, the margins unlobed and shallowly scalloped to bluntly toothed, grading into the stem leaves, these progressively shorter above the stem midpoint, entire to finely scalloped or toothed, those toward the stem base usually short-petiolate, the others sessile, oblanceolate to elliptic or ovate, the bases variously tapered to rounded or shallowly cordate (then somewhat clasping the stem), but not decurrent, grading into the inflorescence bracts; leaf blades with the upper surface glabrous or minutely stellate-hairy along the midvein, the undersurface densely felty with minute, stellate, nonglandular hairs. Inflorescences elongate panicles, the branches of numerous spikelike racemes or sometimes slender panicles, the flowers in clusters of 2–7 at the nodes, the flower stalks 1–6 mm long, densely stellate-hairy. Calyces 2–4 mm long, the lobes lanceolate to narrowly triangular, densely stellate-hairy. Corollas 5–9 mm long, white (yellow elsewhere), lacking reddish markings, the margins and usually also the outer surface minutely stellate-

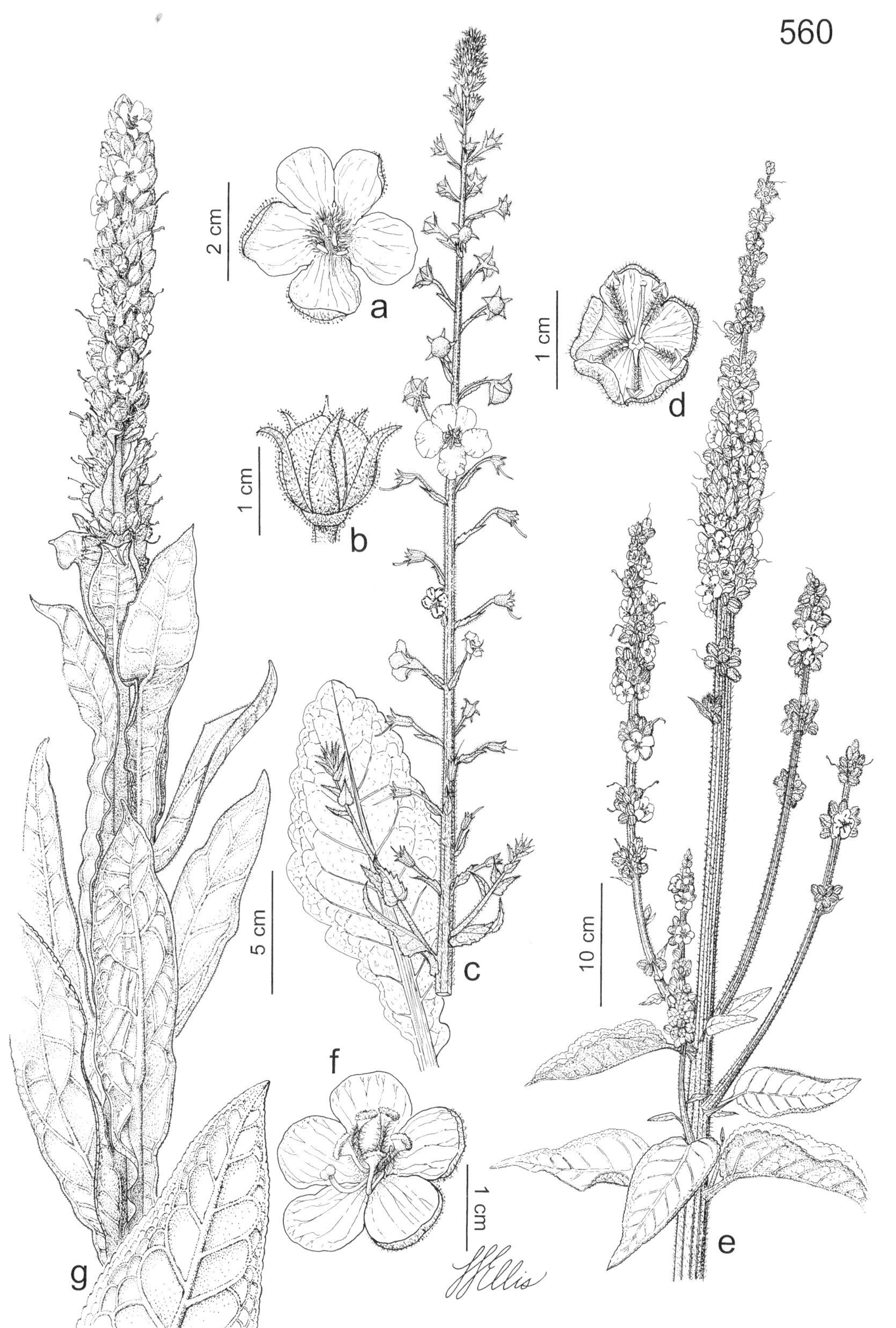

Plate 560. Scrophulariaceae. *Verbascum blattaria*, **a)** flower, **b)** fruit, **c)** inflorescence and leaf. *Verbascum lychnitis*, **d)** flower, **e)** stem tip with leaves and inflorescence. *Verbascum thapsus*, **f)** flower, **g)** stem tip with leaves, also inflorescence and tip of lower leaf.

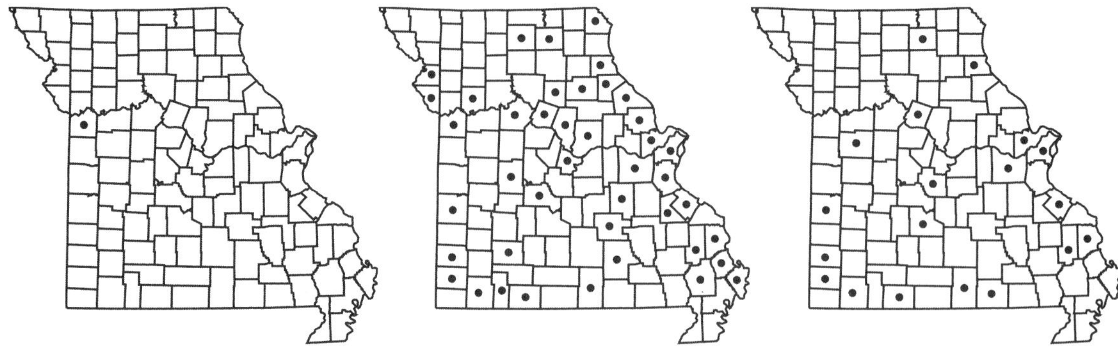

2608. Verbascum phlomoides 2609. Verbascum thapsus 2610. Ailanthus altissima

hairy, the inner surface often faintly gland-dotted. Stamens slightly unequal, the upper 3 with the filaments and anthers slightly shorter, straight, the filaments densely bearded with white to whitish yellow or greenish white hairs; the lower 2 similar, but with the filaments slightly longer, the anthers orange, all kidney-shaped. Fruits 4–7 mm long, broadly ovoid to nearly globose, densely and minutely stellate-hairy. $2n$=32, 34. June–August.

Introduced, uncommon, known thus far from a single specimen from St. Charles County (native of Europe, Asia; introduced sporadically, mostly in the northeastern U.S., Canada). Roadsides.

This species was first reported for Missouri by Denison (1975), based on a collection made by Art Christ near a truck stop along Interstate 70 near Wentzville. the plants do not appear to have persisted at this site, but the species may be reintroduced in the state in the future.

3. Verbascum phlomoides L. (clasping mullein)

Pl. 559 i, j; Map 2608

Stems 25–130 cm long, erect or strongly ascending, circular in cross-section or bluntly polygonal, unbranched or occasionally few-branched near the tip, densely woolly or scurfy with minute, stellate, and slightly larger, branched (having an axis), nonglandular hairs, sometimes becoming glabrous in patches at maturity. Leaves appearing green to light green or grayish green, those of the basal rosettes 8–35 cm long, short-petiolate, the blade oblong-elliptic to elliptic, oblong-oblanceolate, or oblong-obovate, the margins unlobed and shallowly scalloped to bluntly toothed, grading into the stem leaves, these progressively shorter toward the stem tip, mostly finely scalloped or toothed, those toward the stem base usually short-petiolate, the others sessile, oblanceolate to elliptic or ovate, the bases variously tapered to rounded or shallowly cordate (then somewhat clasping the stem), but not decurrent, grading into the inflorescence bracts; leaf blades with the upper surface moderately stellate-hairy (often also with at least a few branched hairs), the undersurface densely woolly with stellate and branched (having an axis), nonglandular hairs, especially along the main veins. Inflorescences relatively dense spikelike racemes (occasionally appearing paniculate in branched plants), the flowers solitary or more commonly in small clusters of 2–6 at the nodes, the flower stalks 3–9 mm long, densely woolly. Calyces 5–7 mm long, the lobes lanceolate to narrowly triangular-ovate, densely stellate-hairy. Corollas 12–19 mm long, yellow, lacking reddish markings, the margins and the outer surface minutely stellate-hairy. Stamens unequal, the upper 3 with the filaments shorter, straight, densely bearded with pale yellow to nearly white hairs; the lower 2 with the filaments longer, angled downward, glabrous, the anthers orange, those of the lower pair fused laterally to the filaments for most of their length. Fruits 5–8 mm long, broadly elliptic-ovoid to broadly oblong-ovoid, densely stellate-hairy, sometimes becoming glabrous in patches. $2n$=32, 34. August–September.

Introduced, uncommon, known thus far from a single historical specimen from Jackson County (native of Europe; introduced widely but sporadically in North America). Railroads.

This plant was first reported for Missouri by Yatskievych and Summers (1993). It has not been collected in the state since the initial discovery, in 1915.

4. Verbascum thapsus L. (mullein, flannel plant)

Pl. 560 f, g; Map 2609

Stems 30–230 cm long, erect, circular in cross-section or slightly polygonal, unbranched or occasionally few-branched toward the tip, densely woolly with branched (having an axis) and stellate, nonglandular hairs. Leaves appearing grayish green or light yellowish green (the green surfaces obscured by pubescence), those of the basal rosettes 8–55 cm long, sessile or with a short, winged petiole, the blade oblanceolate to obovate, the margins unlobed and entire or shallowly scalloped to bluntly toothed; stem leaves progressively shorter toward the stem tip, entire to finely scalloped or toothed, sessile, oblong-oblanceolate to oblanceolate, the bases decurrent down the stems as a pair of wings, grading fairly abruptly into the inflorescence bracts; leaf blades with the surfaces densely woolly with branched (having an axis) and stellate, nonglandular hairs. Inflorescences dense spikelike racemes (occasionally appearing paniculate in branched plants), the flowers solitary or more commonly in small, irregular clusters at the nodes, the flower stalks absent or to 4 mm long, densely woolly. Calyces 5–12 mm long, the lobes lanceolate to triangular-lanceolate, densely woolly. Corollas 8–18 mm long, yellow, lacking reddish markings, the margins minutely stellate-hairy. Stamens unequal, the upper 3 with the filaments and anthers shorter, straight, the filaments densely bearded with yellow hairs; the lower 2 with the filaments and anthers longer, glabrous or sparsely hairy, the anthers orange, those of the lower pair fused laterally to the filaments for most of their length. Fruits 7–10 mm long, broadly ovoid, densely stellate-hairy. $2n$=32, 36. May–September.

Introduced, scattered to common nearly throughout the state (native of Europe, Asia; introduced widely nearly throughout temperate North America and sporadically farther south). Banks of streams and rivers, margins of ponds, lakes, marshes, and oxbows, and disturbed portions of glades and upland prairies; also old fields, pastures, fallow fields, farm yards, ditches, railroads, roadsides, and open, disturbed areas.

Verbascum thapsus plants are avoided by grazing mammals and thus can become problem weeds in pastures. The fuzzy, first-year rosettes are distinctive, as are the tall, stout, second-year flowering stems. A rare, white-flowered mutant has been called f. *candicans* House, but has not yet been recorded from Missouri.

SIMAROUBACEAE (Quassia Family)

Plants shrubs or trees, functionally dioecious (but sometimes incompletely so or the flowers sometimes appearing perfect; monoecious elsewhere), often colonial from root sprouts, usually with internal secretory canals, but these not visible externally, the tissues sometimes with a strong odor when plant parts are bruised or crushed. Stems or trunks 1 to several from the base, erect or ascending, branched or less commonly unbranched, unarmed. Leaves alternate, petiolate. Stipules absent (occasionally represented by a pair of minute glands in *Ailanthus*). Leaf blades simple and unlobed or pinnately compound, the leaflets then opposite, sessile, unlobed. Inflorescences terminal, many-branched panicles or axillary, dense catkins. Flowers usually imperfect (but sometimes appearing perfect by well-developed appearance of staminodes and/or nonfunctional pistils), hypogynous, actinomorphic, sessile or stalked. Calyces absent or of (3)4 or 5(–8) sepals, these sometimes minute, free or fused at their bases, usually overlapping in bud, ascending or spreading, often persistent at fruiting. Corollas absent or of 5 petals, these free, not overlapping in bud, spreading to curved or curled downward. Staminate flowers with (3)8–12(–15) stamens (individual flowers difficult to distinguish in *Leitneria*), the filaments distinct, sometimes hairy or with small appendages near the base, the anthers attached at or above the base, yellow; pistillate flowers with staminodes absent (*Leitneria*) or 3–5 and stamenlike in appearance (*Ailanthus*). Nectar disc sometimes present, often lobed. Pistil(s) 1–5 per flower, of 2–5 carpels, these fused, at least below the midpoint; absent or rudimentary in staminate flowers (well-developed externally elsewhere). Ovaries superior, 1–5-locular, the placentation lateral (when 1-locular) or axile (when 2–5-locular). Style 1 per flower, sometimes with twisted, longitudinal lines equal to the carpel number, the stigma either a receptive region near the grooved style tip or capitate and 2–5-lobed. Ovule 1 per carpel. Fruits drupes or technically schizocarps, but then appearing as small clusters of samaras. Seeds various in size and appearance. Thirteen genera, about 95 species, North America to South America, Asia to Australia.

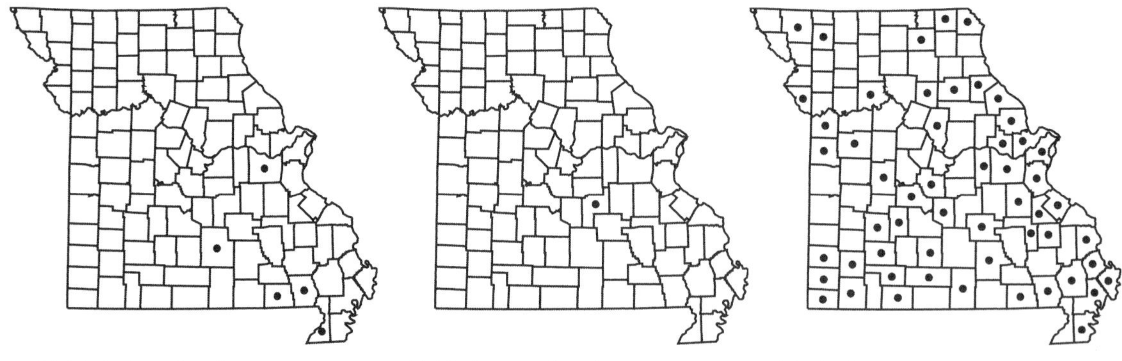

2611. Leitneria pilosa 2612. Datura inoxia 2613. Datura stramonium

The familial circumscription of Simaroubaceae has changed greatly since the application of molecular techniques to the study of phylogeny in the family and its relatives. Traditionally, the family was thought to comprise about 30 genera and 200 species that were most diverse in the tropics of both the Old and New Worlds, but affinities of the family within the order Sapindales were poorly understood (Channell and Wood, 1962). Sequence analyses of the *rbcL* marker (Fernando and Quinn, 1995; Fernando et al., 1995; Gadek et al., 1996) disclosed that several genera should be transferred to the closely related family Rutaceae and that others were even more distantly related and should be classified into segregate families, some of them in other orders. Perhaps most surprisingly, these studies indicated strongly that the genus *Leitneria*, previously treated in its own taxonomically isolated family (Cronquist, 1981, 1991), should be included in the Simaroubaceae in the strict sense. This reclassification results in a family that is morphologically very diverse, but whose genera share a suite of anatomical and phytochemical features.

Members of the family have relatively minor economic importance. Some species are harvested for timber in Asia and a number of species have been used medicinally, particularly for their inner bark and seeds, which are rich in triterpenoid lactones (thus bitter, astringent). A few species are cultivated shade trees or specimen plants.

1. Plants medium to large trees; leaves pinnately compound; inflorescences large panicles; fruits samaras . . . . 1. AILANTHUS
1. Plants shrubs or occasionally small trees; leaves simple, unlobed; inflorescences dense catkins; fruits drupes . . . . 2. LEITNERIA

1. **Ailanthus** Desf. (tree-of-heaven)

Five species, Asia to Australia, introduced widely.

1. Ailanthus altissima (Mill.) Swingle (tree-of-heaven, cancer tree, stinkweed)

Pl. 561 a–c; Map 2610

Plants small mostly medium-sized trees to 20 m tall (but frequently flowering when much smaller), incompletely dioecious, colonial from root suckers with age. Trunks 1 or few, stout, the bark light grayish brown, initially relatively smooth, eventually developing thin, pale grooves and thin, darker ridges. Twigs stout, tan to reddish brown with numerous, slender, pale lenticels and large, prominent, more or less kidney-shaped leaf scars, minutely hairy (difficult to see without magnification) the winter buds axillary (sometimes appearing asymmetrically terminal), depressed-globose, with several broadly rounded scales, densely short-hairy. Leaves (20–)40–100(–130) cm long, relatively short-petiolate. Leaf blades pinnately compound with (7–)11–41 leaflets (the terminal leaflet sometimes reduced or absent), these oppo-

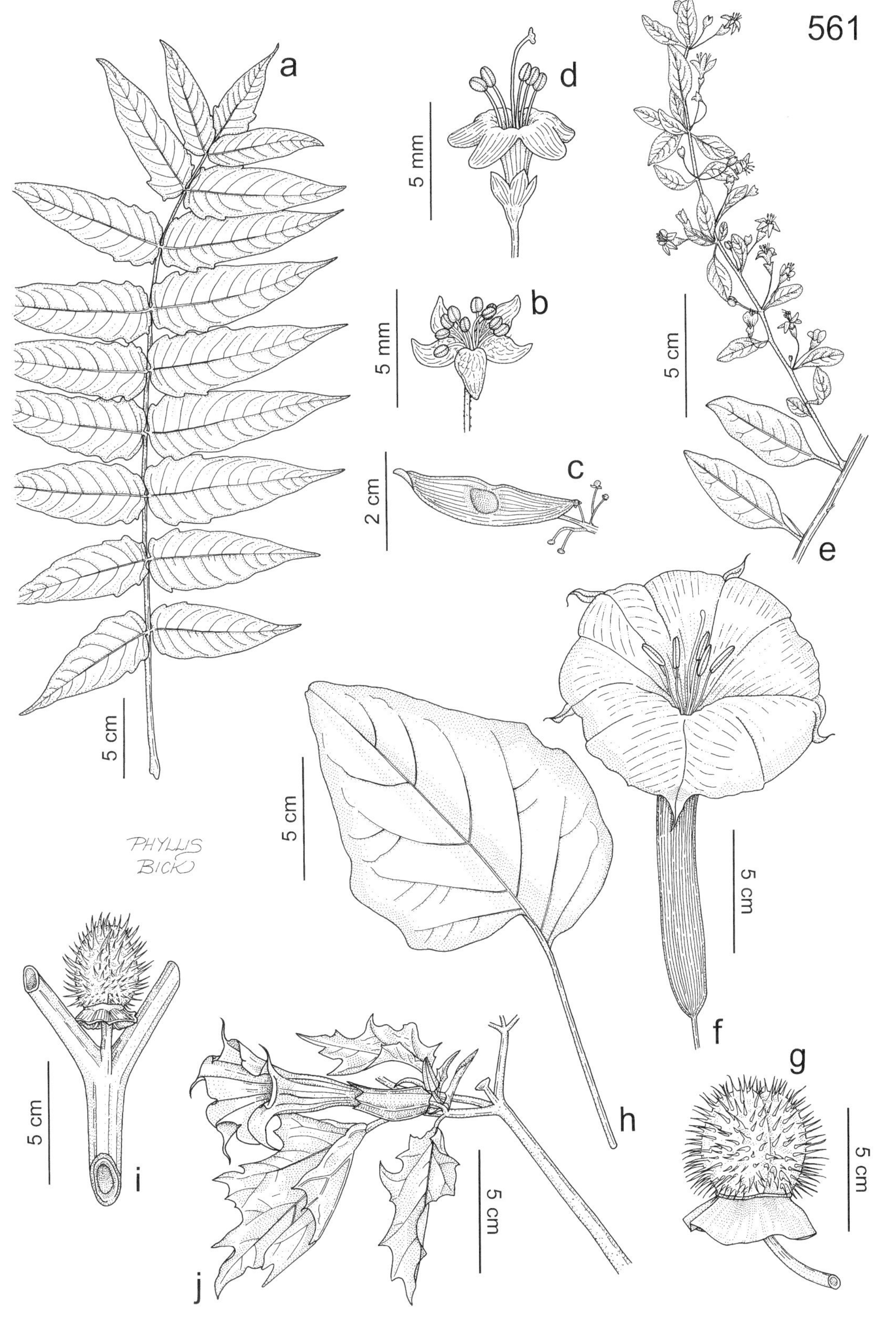

Plate 561. Simbaroubaceae. Solanaceae. *Ailanthus altissima*, **a)** leaf, **b)** flower, **c)** fruit. *Lycium barbarum*, **d)** flower, **e)** branch with leaves and flowers. *Datura wrightii*, **f)** flower, **g)** fruit, **h)** leaf. *Datura stramonium*, **i)** fruit, **j)** portion of branch with leaves and flower.

site, lanceolate to oblong-lanceolate or narrowly oblong-ovate, the lateral ones often somewhat curved toward the leaf tip, rounded to shallowly cordate and (at least the lateral ones) usually somewhat asymmetrical at the base, tapered to a sharply pointed tip, the margins entire or with 1–3(–5) pairs of blunt teeth toward the base, sometimes also hairy, the surfaces sparsely to moderately short-hairy and also with scattered, minute glands, especially near the midvein, sometimes becoming nearly glabrous with age, the upper surface dark green, the undersurface pale or lighter green. Inflorescences terminal, large, diffusely branched panicles, lacking bracts, the numerous, individually stalked flowers solitary or in small clusters along the ultimate branches, each cluster subtended by a pair of minute, linear, reddish bracts that are shed before the flowers open. Flowers mostly functionally imperfect, but often the pistillate ones appearing perfect by production of stamenlike staminodes. Calyces of (4)5 sepals, these 0.8–1.2 mm long, free or fused at the base, triangular, spreading with ascending tips, often persistent at fruiting. Corollas of (4)5 petals, these 2–4 mm long, narrowly oblong-elliptic, somewhat concave (scoop-shaped), green to greenish yellow, the upper surface (and basal portion of the margins) densely and minutely woolly. Staminate flowers with (8)10 stamens; pistillate flowers with 3–5 stamenlike staminodes, these rarely absent. Nectar disc present, irregularly lobed, dark purple to purplish brown. Pistillate flowers with the ovary of 2–5 carpels, 2–5-locular and lobed at flowering, the placentation axile; absent or rudimentary in staminate flowers. Style 1 per flower, with twisted, longitudinal lines equal to the carpel number, the stigma capitate and deeply 2–5-lobed. Fruits technically schizocarps, but through separation of the carpels appearing as small clusters of 2–5 samaras (sometimes only a solitary samara developing), these sessile, 3.0–5.5 cm long, 7–13 mm wide, narrowly oblong-elliptic (usually with a shallow, broad, median notch along 1 side), often somewhat spirally twisted toward the tip, usually strongly dark reddish-tinged at maturity, but becoming straw-colored to tan before dispersal. Seed 1 per samara, positioned near its midpoint, 5–8 mm long, ovate, strongly flattened, the seed coat fused to the samara wall. $2n$=80. May–June.

Introduced, scattered, mostly south of the Missouri River, most abundantly in and around urban areas (native of Asia, introduced nearly throughout the U.S., also Canada, Mexico). Bottomland forests, mesic upland forests, banks of streams and rivers, and bases of bluffs; also old mines and quarries, fencerows, railroads, roadsides, and disturbed areas.

Tree-of-heaven is fast-growing and has been planted widely as a street tree, as a shade tree in yards, and as a soil binder in eroded areas or recovering strip mines. It is disease-resistant and also survives well in areas with high levels of air pollution. It was an inspiration for the popular novel *A Tree Grows in Brooklyn* by Betty Smith (1943), in which the tree is a metaphor for the heroine, a young woman who manages to flourish under difficult circumstances.

Ailanthus altissima mainly becomes naturalized in highly disturbed areas, but in Missouri and some other states it occasionally becomes strongly invasive in forests. Once established, the species is very difficult to control or eradicate, resprouting readily from remaining roots after the aboveground portions have been killed by application of herbicides or mechanical removal. The relatively high growth rate and ability to sucker from roots allows the species to efficiently colonize forest margins and areas disturbed by tree falls. Established trees also produce allelopathic compounds, principally the quassinoid triterpenoid, ailanthone, that inhibit the germination and growth of neighboring plants of other species (Heisey, 1990a, b, 1996; J. G. Lawrence et al., 1991). The leaves and twigs emit a disagreeable aroma when bruised or crushed, and the staminate flowers also are malodorous. Contact with the plants can cause dermatitis in some individuals and the plants reputedly are toxic when ingested (Burrows and Tyrl, 2001), but the strong bitter flavor makes it unlikely that the plant would be eaten in quantity. Bisognano et al. (2005) described a case in which absorption of the sap through an open wound resulted in myocarditis (inflammation of the heart muscle) in a person working as a tree surgeon.

2. Leitneria Chapm. (corkwood)
Contributed by Alan Whittemore

One species (but see the discussion below), southeastern United States.

As noted above in the discussion of the family, *Leitneria* was treated traditionally as a monospecific family of uncertain taxonomic affinity (Cronquist, 1981, 1991), the only angiosperm family entirely endemic to the United States. This was in large part because the highly reduced flowers of *L. floridana* offer few clues as to relationships with other families. It was initially surprising when molecular data suggested that *Leitneria* was, in fact, a highly modified member of the Simaroubaceae in which floral reduction and condensation of inflorescences correlated with a shift from insect pollination to wind pollination. Also, most Simaroubaceae are aromatic, at least when tissue is crushed or bruised, but not *Leitneria*. Features supporting the genetic data turn out to be mostly nonmorphological, for example that *Leitneria* produces secretory canals. Other evidence supporting a relationship between *Leitneria* and the Simaroubaceae comes from serological studies of pollen proteins (this involves injecting protein samples into laboratory animals to produce antisera, which are then isolated and tested for their relative reactivity to protein samples from other plant species) by Petersen and Fairbrothers (1983), as well as comparative embryological research (Tobe, 2011).

1. Leitneria floridana Chapm. (corkwood)
L. pilosa J.A. Schrad. & W.R. Graves
L. pilosa ssp. *ozarkana* J.A. Schrad. & W.R. Graves
Pl. 445 g, h; dust jacket; Map 2611

Plants shrubs or occasionally small trees, usually dioecious, suckering from creeping rootstocks and forming dense thickets. Trunks 1.0–4.0(–7.5) m tall, the bark smooth, gray to dark reddish brown, with prominent lenticels. Twigs 3–4 mm thick, densely grayish-hairy, the winter buds more or less triangular, with several scales, densely white-hairy. Leaves with the petioles 2.0–5.5 cm long. Leaf blades 6–23 cm long, 2–10 cm wide, simple and unlobed, lanceolate to narrowly elliptic, occasionally narrowly ovate or oblanceolate, somewhat leathery, angled or tapered to less commonly rounded at the base, narrowed or tapered to a bluntly or sharply pointed (rarely rounded) tip, the margins entire, the upper surface green, sparsely to densely hairy, the undersurface pale green or light yellow, densely felty-hairy, the veins strongly raised. Inflorescences dense, ascending catkins from second-year twigs, 1–4 cm long at flowering, the pistillate catkins shorter and narrower than the staminate ones. Flowers apparently actinomorphic. Staminate flowers difficult to differentiate in the catkins, with small groups of flowers in dense clusters, each cluster subtended by a small hairy ovate-triangular bract, the calyces rudimentary or absent, the corollas absent. Stamens 3 to apparently 10–12(–15), the filaments short. Pistillate flowers each subtended by a small, ovate-triangular, hairy bract and 2 minute scale-like bractlets, the calyces of (3)4(–8) minute sepals, the corollas and staminodes absent. Pistils 1-locular, finely hairy, the placentation lateral. Style 1, linear, often somewhat twisted, dark red, not persistent, the stigmatic area lateral in a minute groove toward the tip. Fruits drupes, 15–20 mm long, 6–7 mm wide, narrowly ellipsoid to narrowly obovoid, slightly flattened, green becoming olive green to greenish brown with age. $2n=32$. February–April.

Uncommon in the Mississippi Lowlands Division (southeastern United States, from Missouri south to Florida and Texas). Bottomland forests, swamps, and rarely margins of ponds and lakes; also ditches and wet roadsides.

Schrader and Graves (2011) recently divided *L. floridana* into three taxa with disjunct geographic ranges. According to their taxonomy, only plants native to northern Florida and southern Georgia are *L. floridana*, whereas material from Missouri and Arkansas is *L. pilosa* ssp. *ozarkana* (var. *pilosa* is found in eastern Texas). Plants from the western portion of the range were indicated by these authors to differ from the populations in Florida in their smaller leaves and at least the undersurface of the leaf blades relatively densely hairy. However, These conclusions were based on the study of only seven populations, and the authors cited no specimens except for their own collections. Examination of herbarium material from a wider range of sites indicates that these plants are much more variable morphologically than the keys and descriptions provided by Schrader and Graves (2011) indicate. Plants from the three geographical areas cannot be distinguished consistently using the characters given in their paper. Separation of Missouri material from *L. floridana* cannot be justified unless reliable characters can be discovered and confirmed through study of a wide spectrum of specimens from throughout the range of the genus.

Corkwood formerly formed dense thickets in bottomlands of the Mississippi Lowlands Division, although in gardens it flourishes much farther to the north in well-drained soils. Much of its former habitat has been eliminated by clearing and draining of the Bootheel region and today the species survives mainly along ditches, which are prone to periodic dredging and spraying of herbicides. Because of habitat destruction throughout its range, in recent years there have been conservation concerns about the decline of this unusual plant.

The wood of corkwood is soft and is the lightest wood of any species native to North America. It has been used locally as a cork substitute, especially for fishing floats. The species is sometimes cultivated as a specimen plant. It can form dense colonies and does well in gardens, even far to the north of its native range (Channell and Wood, 1962). In mixed plantings, over time there is a tendency for staminate clones to outcompete pistillate ones.

SOLANACEAE (Nightshade Family)

Plants annual or perennial herbs or shrubs (small trees elsewhere), sometimes with rhizomes or tubers, occasionally climbing or twining, but lacking tendrils, sometimes armed with short thorns or prickles. Stems unbranched or more commonly branched, glabrous or hairy and/or glandular, the hairs lacking persistent pustular bases, sometimes stellate or appearing in fascicles. Leaves alternate (sometimes a pair of leaves appearing on the same side of the stem from closely adjacent nodes or [in *Datura*] the uppermost leaves sometimes appearing opposite) or appearing fasciculate at the tips of very short branches (in *Lycium*), well-developed, mostly petiolate. Stipules absent. Leaf blades simple or less commonly pinnately lobed to deeply divided, the surfaces glabrous or hairy and/or glandular, the hairs lacking persistent pustular bases and not roughened to the touch, sometimes stellate or appearing in fascicles. Inflorescences variously lateral (opposite the leaves), axillary, or terminal, of solitary flowers or clusters, racemes or small panicles, not appearing paired or coiled, the individual flowers not subtended by bracts. Flowers actinomorphic (slightly zygomorphic in some *Solanum* species), hypogynous, perfect; cleistogamous flowers absent. Calyces usually 5-lobed (sometimes appearing 2–4-lobed in *Lycium* by fusion of lobes, sometimes 6-lobed in *Solanum dimidiatum*), sometimes nearly to the base, persistent at fruiting (only the basal portion in *Datura*), sometimes becoming enlarged, sometimes becoming inflated and enclosing the fruit. Corollas usually 5-lobed, the lobes sometimes reduced to small teeth or absent, variously arranged in bud, often with longitudinal pleats, funnel-shaped, bell-shaped, or saucer-shaped. Stamens usually 5, the filaments attached in the corolla tube, not subtended by scales, the anthers often exserted, sometimes fused into a ring, appearing 2-locular (sometimes 1 of the sacs slightly reduced), usually attached on the dorsal side below the midpoint, white, yellow, purplish blue, or blue. Pistil 1 per flower, of 2 fused carpels. Ovary unlobed to bluntly 3–5-angled, 2–5-locular, with usually numerous ovules per locule, the placentation axile. Style 1, situated at the tip of the ovary, the stigma 2-lobed. Fruits berries or capsules, if the latter, then dehiscent longitudinally from the tip, sometimes irregularly so, with several to numerous seeds. Ninety to 102 genera, 3,000–4,000 species, nearly worldwide, most diverse in the New World tropics.

The Solanaceae are an important family economically. Members that are major sources of foods and/or spices includes potatoes (*Solanum*), tomatoes (*Solanum*), tomatillos (*Physalis*), eggplants (*Solanum*), and chili peppers (*Capsicum* L.). Most genera in the family produce poisonous alkaloids, some of which also have narcotic properties. One such highly addictive compound is nicotine, and *Nicotiana* is the basis for the tobacco industry. Some of the species produce alkaloids that are also important for pharmaceuticals. Well-known alkaloid-producing genera include *Atropa* L. (belladonna), *Datura* (jimsonweed), *Hyoscyamus* L. (henbane), *Mandragora* L. (mandrake), and *Nicotiana* (tobacco). The family is the source of a large number of horticulturally important plants, including species of *Browallia* L. (amethyst flower), *Brugmansia* Pers. (angel trumpets), *Brunfelsia* L. (lady of the night; yesterday, today, and tomorrow), *Cestrum* L. (night-blooming jessamine), *Datura*, *Lycium*, *Nicotiana*, *Petunia*, *Physalis*, *Salpiglossis* Ruiz & Pav. (painted tongue, velvet trumpet flower), *Schizanthus* Ruiz & Pav. (butterfly flower, fringeflower, poor-man's-orchid), and *Solanum*. Some species are noxious weeds of crop fields and rangelands.

1. Stems woody; plants shrubs or lianas
   2. Stems usually with short thorns at some of the nodes; leaf blades lanceolate to elliptic, oblanceolate, ovate, or rhombic-ovate, angled or tapered at the base, unlobed . . . . 2. LYCIUM
   2. Stems unarmed; leaf blades ovate to halberd-shaped, rounded or cordate at the base, sometimes with 1 or 2 spreading lobes at the base . . . . 7. SOLANUM
1. Stems not woody or woody only at the base; plants annual or perennial herbs
   3. Plants flowering
      4. Corollas more or less saucer-shaped
         5. Stamens with the anthers free and separate . . . . 6. PHYSALIS
         5. Stamens with the anthers fused into a ring . . . . 7. SOLANUM
      4. Corollas funnelform to bell-shaped or trumpet-shaped
         6. Inflorescences terminal panicles or racemes . . . . 4. NICOTIANA
         6. Inflorescences axillary, of solitary flowers or occasionally small, few-flowered clusters
            7. Calyces 3.5–10.0 cm long, narrowly tubular, shallowly lobed . . . . 1. DATURA
            7. Calyces 0.3–1.8 cm long, broadly tubular or narrowly to broadly bell-shaped, shallowly to deeply lobed (sometimes nearly to the base)
               8. Calyces deeply concave at the base, with 5 angular, basal auricles (1 per lobe); ovary 3–5-locular . . . . 3. NICANDRA
               8. Calyces rounded or truncate at the base, or, if concave, then lacking basal auricles; ovary 2-locular
                  9. Flowers ascending; corollas 3–6 cm long (sometimes longer in cultivated plants) . . . . 5. PETUNIA
                  9. Flowers spreading to more commonly nodding; corollas 0.6–2.0 cm long . . . . 6. PHYSALIS
   3. Plants fruiting
      10. Fruits capsules, dry and dehiscent at maturity
         11. Fruits prickly on the surface; calyces with a circumscissile zone of abscission toward the base, the apical portion shed after flowering, leaving a small, persistent, circular disc at the base of the fruit . . . . 1. DATURA
         11. Fruits unarmed; calyces remaining intact at fruiting, the apical portion not shed
            12. Fruits in terminal panicles or racemes . . . . 4. NICOTIANA
            12. Fruits axillary or lateral (the uppermost sometimes appearing terminal), solitary . . . . 5. PETUNIA
      10. Fruits berries (burlike in *S. rostratum*), fleshy (mealy in *Nicandra*) and indehiscent at maturity
         13. Calyces not or only slightly enlarged, or, if enlarged and closely surrounding all or part of the fruit, then not balloonlike or papery, not angled . . . . 7. SOLANUM
         13. Calyces becoming enlarged and balloonlike, entirely and loosely enclosing the fruit, 5- or 10-angled
            14. Fruits mealy; 3–5-locular; calyces lobed to at or below the midpoint (excluding the 5 prominent, angular, basal auricles), the fruit usually easily visible when viewed from the tip . . . . 3. NICANDRA
            14. Fruits more or less juicy, 2-locular; calyces shallowly lobed at the tip (lacking basal auricles), the fruit visible only when the calyx is torn apart . . . . 6. PHYSALIS

1. **Datura** L. (jimsonweed, thorn-apple)

Plants annual or perennial herbs, sometimes with a somewhat thickened or woody rootstock, unarmed (except for the prickly fruits). Stems erect to loosely ascending from a spreading base, relatively stout, with few to several ascending to spreading branches (sometimes appearing equally forked or with whorls of 3 branches), sometimes mostly toward the tip, glabrous or more commonly short-hairy. Leaves alternate, but those near the stem tip sometimes appearing opposite, moderately to long-petiolate. Leaf blades simple, unlobed or irregularly pinnately lobed, the margins otherwise wavy or coarsely toothed, the surfaces usually hairy. Inflorescences initially terminal, of solitary flowers, but later appearing produced in the axils of forked or 3-branched nodes. Flowers erect or ascending, the fruits sometimes nodding or pendant. Calyces 3.5–10.0 cm long, shallowly and unequally 5-lobed at the tip, narrowly tubular at flowering, rounded to more or less truncate at the base, lacking basal auricles, the sides rounded or 5-angled to narrowly winged, with a circumscissile zone of abscission toward the base, the apical portion shed after flowering, leaving a small, persistent, circular disc at the base of the fruit, this strongly reflexed in our species (more or less flat elsewhere). Corollas 6–20 cm long, funnelform at full flowering (appearing tubular at other times), very shallowly 5- or 10-lobed, the lobes toothlike, abruptly tapered to often slender, sharply pointed tips, usually appearing spirally twisted and pleated in bud, white, sometimes purplish-tinged or pale to light purple. Stamens with relatively long filaments, the anthers free, erect, positioned in a loose ring, not exserted, dehiscent longitudinally, light yellow to nearly white, usually hairy. Ovary 2-carpellate but appearing 4-locular, the style elongate, positioned at about the level of the anthers, green. Fruits capsules, dry, globose to broadly oblong-ovoid, 4-locular, pale green at maturity, becoming tan with age, dehiscent longitudinally from the tip (sometimes irregularly so), with numerous seeds, the surface armed (unarmed elsewhere) with dense prickles, these slender above the expanded base, straight. Seeds 3–6 mm in longest dimension, more or less kidney-shaped to asymmetrically ovate, flattened, the surface smooth, minutely pitted or finely wrinkled, sometimes only faintly so, more or less shiny, variously colored, lacking wings, the dorsal margin sometimes bordered by a pair of parallel grooves. Eight to 11 species, North America, introduced widely.

Plants of *Datura* are highly toxic to humans and other mammals, and contain principal alkaloids similar to those of *Atropa* (belladonna) and related genera (Burrows and Tyrl, 2001). The species have long been used medicinally (in small dosage) to treat asthma and for their narcotic properties, among other maladies. The genus is important as a ceremonial hallucinogen for a number of North American Indian tribes (Moerman, 1998). However, euphoria-seeking amateurs who have ingested *Datura* intending to reproduce the drug effects ascribed to the plants have fairly frequently become poisoned instead, leading to seizures, coma, and sometimes death.

The flowers of *Datura* species, which are pale-colored, night-flowering, fragrant, and long-tubed, are adapted to pollination by long-tongued nocturnal hawk moths (H. G. Baker, 1961; V. Grant and K. Grant, 1983). However, although the flowers of *D. stramonium* are visited by both hawk moths and bees, apparently they are mostly self-pollinated (Motten and Antonovics, 1992).

Some authors have accepted an expanded circumscription of *Datura* to include five to seven additional arborescent, South American species (Hammer et al., 1983), but molecular studies have confirmed that this lineage should be treated in the segregate genus *Brugmansia* Pers. (Mace et al., 1999). Members of this genus also produce showy, often fragrant flowers with large, funnelform corollas, but the flowers are pendant and in some species the corollas are yellow, pink, or orange toward the tip. Some species of *Brugmansia* and their hybrids also are cultivated as ornamentals, but none of these can overwinter outdoors in Missouri.

1. Stems glabrous or sparsely and minutely hairy toward the tip; leaf blades with coarsely and sharply lobed and/or toothed margins; corollas 6–10 cm long; calyces 3–5 cm long, strongly 5-angled to narrowly winged; fruits erect . . . . 2. D. STRAMONIUM
1. Stems moderately to densely pubescent with minute (0.1–0.3 mm), stout, curled or appressed hairs or with longer (0.5–2.0 mm), spreading, multicellular hairs; leaves variously entire or with slightly wavy to shallowly few-toothed margins; corollas (12–)15–20 cm long; calyces 6–12(–15) cm long, rounded along the sides (not angled or winged); fruits nodding (species difficult to distinguish)
    2. Leaf blades with the surfaces (especially the undersurface) relatively densely pubescent along the main veins with spreading, multicellular hairs 0.5–2.0 mm long, the surfaces between the veins usually only sparsely hairy; corollas glabrous on the outer surface . . . . 1. D. INOXIA
    2. Leaf blades with the surfaces (especially the undersurface) moderately to densely pubescent along and between the veins with minute, stout, curled or appressed hairs 0.1–0.3 mm long, occasionally also with a few longer, spreading multicellular hairs along the main veins; corollas inconspicuously glandular-hairy on the outer surface, mostly along the veins . . . . 3. D. WRIGHTII

1. Datura inoxia Mill.
D. meteloides Dunal

Map 2612

Plants perennial herbs (but often flowering the first year), often with a somewhat thickened or woody rootstock, often forming mounds about as tall as wide. Stems 80–150 cm long, erect to loosely ascending from a spreading base, usually several-branched, moderately to densely pubescent with spreading multicellular hairs (0.5–2.0 mm long), some of these usually gland-tipped. Leaf blades 6–18(–20) cm long, ovate to broadly ovate, angled or tapered to a sharply pointed tip, rounded to truncate at the often oblique base, the margins variously entire or slightly wavy to shallowly few-toothed, the surfaces (especially the undersurface) relatively densely pubescent along the main veins with spreading, multicellular hairs 0.5–2.0 mm long, the surfaces between the veins usually only sparsely hairy, sparsely to densely pubescent at maturity with minute (0.1–0.3 mm), stout, curled or appressed hairs, more densely so along the veins of the undersurface and when young, sometimes also with sparse, longer, spreading multicellular hairs (0.5–2.0 mm long), some of these usually gland-tipped. Flower stalks 1.2–3.0 cm long. Calyces 7–10 cm long, tubular, rounded along the sides (not angled or winged), finely hairy, the persistent base 5–9 mm long, spreading to more commonly reflexed at fruiting. Corollas (12–)15–20 cm long, white or partially pale to dark purplish-tinged. Fruits 2.5–3.5 cm long, nodding, usually more or less globose, the surface minutely hairy and densely prickly, the prickles 3–8 mm long. Seeds 4.5–6.0 mm in longest dimension, the surface not wrinkled, with a pair of fine dorsal grooves parallel on opposite sides of the blunt margin, more or less shiny, tan to yellowish brown or grayish brown. $2n=24$. August–October.

Introduced, uncommon, known thus far only from a single historical collection from Phelps county (apparently native to eastern Mexico and possibly the adjacent United States). Railroads.

The taxonomy of the group that includes this taxon is in need of more thorough study. Several species are cultivated as ornamentals in the eastern United States and a number of cultivars exist. Horticultural sources have not always been able to apply scientific names accurately to the materials for sale. In Missouri, four names have been applied. *Datura metel* L. is relatively distinct and is discussed separately in the next paragraph. The other three names have often been confused. The name *D. meteloides* has generally been treated as a synonym of either *D. inoxia* or *D. wrightii*, but current opinion seems to favor the former interpretation (Haegi, 1976; Hammer et al., 1983). Steyermark (1963) and other botanists mostly have called the plants collected in Missouri and surrounding states by the name *D. inoxia*, a mainly eastern Mexican species that originally was described from material cultivated in Europe. Gleason and Cronquist (1991) were the first to suggest that the large-flowered species that has escaped in the northeastern and upper midwestern states has long been misdetermined and is actually the closely related *D. wrightii*. Comparison of the available specimens collected in Missouri with

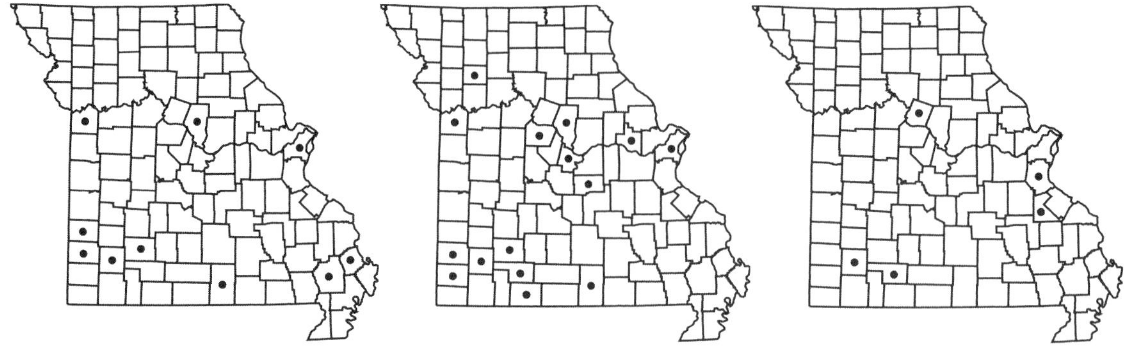

2614. Datura wrightii

2615. Lycium barbarum

2616. Lycium chinense

authentic material of both species from outside the state supports the notion that nearly all of these represent *D. wrightii*. The exception is a single specimen collected in 1929 by John Kellogg (housed at the herbarium of the Missouri Botanical Garden). Because both species are cultivated in Missouri gardens, it is possible that *D. inoxia* will be recorded from the state as an escape at more sites in the future. The differences between the two are subtle. In *D. inoxia*, the tertiary veinlets of the leaf blade are faint; the pubescence of the leaf undersurface, which is concentrated along the main veins, consists of relatively long (0.5–2.0 mm), spreading multicellular hairs; the corolla is glabrous externally; and the corolla margin between the tips of the five pleats appears relatively strongly angled or toothed (thus the corolla often appears shallowly 10-lobed). In *D. wrightii*, the tertiary veinlets of the leaf blade connecting the secondary veins are somewhat more prominent; the pubescence of the leaf undersurface is often relatively dense on and between the veins and consists entirely or mainly of minute (0.1–0.3 mm), stout, curled or appressed hairs; the corolla is inconspicuously glandular-hairy externally, mostly along the veins; and the corolla margin between the tips of the five pleats is relatively weakly toothed or appears merely curved (thus the corolla appears shallowly 5-lobed). Other characters of fruit size and shape, the length of the prickles on the fruit, seed size, and the diameter of the expanded corolla are too subtle to allow determination with much certainty. Note that the epithet has sometimes been erroneously spelled *D. innoxia* in some of the older literature.

A single specimen vouchers the occurrence of *D. metel* as an escape in Missouri. The specimen in the herbarium of the Missouri Botanical Garden was collected by Moses Craig in August 1912 and originated from the Missouri Botanical Garden grounds (the label reads “weed patch”). Although the label suggests that the specimen represents a spontaneous occurrence, the plant apparently was not observed in subsequent years. The species is thus excluded from the Missouri flora, at least for now. *Datura metel* is easily distinguished from *D. inoxia* and *D. wrightii* by its glabrous or very sparsely hairy stems and leaves, as well as its pendant fruits, which at maturity are covered with soft conic tubercles rather than sharp prickles.

2. Datura stramonium L. (common jimsonweed)

D. stramonium var. *tatula* (L.) Torr.

Pl. 561 i, j; Map 2613

Plants annuals, taprooted. Stems 50–120(–150) cm long, erect or strongly ascending, few- to several-branched toward the tip, sparsely to moderately pubescent toward the tip with curled or loosely appressed, nonglandular hairs 0.1–0.4 mm long. Leaf blades 5–20(–25) cm long, ovate to narrowly ovate, angled or tapered to a sharply pointed tip, broadly angled to truncate at the sometimes oblique base, the margins coarsely and sharply lobed and/or toothed, the surfaces (especially the undersurface) sparsely pubescent along and between the veins at maturity with minute hairs 0.1–0.4 mm long, more densely so when young. Flower stalks 0.5–1.5 cm long. Calyces 3–5 cm long, tubular, but strongly 5-angled to narrowly winged along the sides, finely hairy, the persistent base 4–7 mm long, usually reflexed at fruiting. Corollas 6–10 cm long, white to partially pale to dark purplish-tinged or more uniformly light purple. Fruits 3–5 cm long, erect, more or less globose to more commonly broadly oblong-ovoid, the surface minutely hairy and densely prickly (more or less smooth elsewhere), the prickles 3–5(–9) mm long. Seeds 3–4 mm in longest dimension, the surface finely pitted and often finely wrinkled, sometimes only faintly so, lacking a pair of dorsal grooves, more or less shiny, dark gray to black. $2n=24$. May–October.

Introduced, scattered nearly throughout the state (native range uncertain, presumably North America [Hammer et al. (1983) suggested the southwestern U.S.], now widely distributed in tropical and warm-temperate regions of both hemispheres). Banks of streams and rivers; also farmyards, pastures, fallow fields, margins of crop fields, railroads, roadsides, and open, disturbed areas.

Several varieties and forms have been named within *D. stramonium* (Hammer et al., 1983), but these appear to be minor variants unworthy of formal taxonomic recognition. Plants with more strongly purplish-tinged corollas and stems, as well as mostly equal prickles on the fruits, have been called var. *tatula*. Plants with this morphology appear sporadically throughout the range of the species. Other variants include those with more or less unarmed capsules and plants with a mixture of prickly and unarmed capsules, but these have not yet been found in Missouri.

2. Datura wrightii Regel (hairy jimsonweed, sacred thorn-apple)

Pl. 561 f–h; Map 2614

Plants perennial herbs (but often flowering the first year), often with a somewhat thickened or woody rootstock, often forming mounds about as tall as wide. Stems 80–150 cm long, erect to loosely ascending from a spreading base, usually several-branched, moderately to densely pubescent with a mixture of minute (0.1–0.3 mm), curled or appressed hairs and longer, spreading multicellular hairs (0.5–2.0 mm long), some of these usually gland-tipped. Leaf blades 6–20 cm long, ovate to broadly ovate, angled or tapered to a sharply pointed tip, rounded to truncate at the often oblique base, the margins variously entire or slightly wavy to shallowly few-toothed, the surfaces moderately to densely pubescent at maturity with minute (0.1–0.3 mm), stout, curled or appressed hairs, more densely so along the veins of the undersurface and when young, sometimes also with sparse, longer, spreading multicellular hairs (0.5–2.0 mm long), some of these usually gland-tipped. Flower stalks 0.5–3.0 cm long. Calyces 7–10 cm long, tubular, rounded along the sides (not angled or winged), finely hairy, the persistent base 5–9 mm long, spreading to more commonly reflexed at fruiting. Corollas (12–)15–20 cm long, white or partially pale to dark purplish-tinged, the outer surface inconspicuously glandular-hairy, mostly along the veins. Fruits 2.5–3.5 cm long, nodding, usually more or less globose, the surface minutely hairy and densely prickly, the prickles 3–8 mm long. Seeds 4.5–6.0 mm in longest dimension, the surface not wrinkled, with a pair of fine dorsal grooves parallel on opposite sides of the blunt margin, more or less shiny, tan to yellowish brown or light brown. $2n=24$. August–October.

Introduced, uncommon, mostly in the southern half of the state (native of the southwestern U.S. east to Oklahoma and Texas; Mexico; introduced sporadically farther north and east). Railroads, roadsides, and open, disturbed areas.

For a discussion of the past confusion between this species and the closely related *D. inoxia*, see the treatment of that species.

2. **Lycium** L. (wolfberry)

Plants shrubs, usually armed with short thorns 5–20 mm long at some nodes. Stems 1–2 (–4) m long, arched or spreading, sometimes climbing, many-branched, glabrous. Twigs gray to yellowish tan, the winter buds ovoid to nearly globose, with 3 or more appressed, glabrous scales. Leaves alternate or appearing fasciculate at the tips of very short branches, sessile or short-petiolate, the petiole often mostly winged. Leaf blades, simple, unlobed, lanceolate to elliptic, oblanceolate, ovate, or rhombic-ovate, rounded or more commonly angled or slightly tapered to a bluntly or sharply pointed tip, angled or tapered at the base, the margins otherwise entire or rarely minutely scalloped or slightly wavy, the surfaces green to grayish green, glabrous. Inflorescences axillary, of solitary flowers or small clusters of 2 or 3 flowers. Flowers mostly spreading, the stalks often slightly curved upward, thickened toward the tips. Calyces 2–5-lobed, fused to the midpoint or slightly less, when fewer than 5 lobes present then the lobes frequently few-toothed, bell-shaped to broadly tubular, rounded at the base, lacking basal auricles, persistent intact at fruiting, not or only slightly enlarged, not balloonlike or papery, closely cupped around (and often ruptured by) the base of the fruit, not angled. Corollas narrowly bell-shaped to more or less trumpet-shaped, the 5(6) lobes spreading, not appearing pleated in bud, the tube white to pale green, sometimes pinkish- or pale purplish-tinged, the lobes light pink to pale purple or lilac, the throat usually with a network of darker lines. Stamens with the relatively long filaments attached at or near the tip of the corolla tube, usually densely hairy basally, the anthers free, dehiscent longitudinally, exserted, light yellow. Ovary 2-locular, the style exserted, about as long as to slightly longer then the stamens, the stigma green. Fruits berries, 8–20 mm long, juicy, ellipsoid to slightly ovoid or obovoid, 2-locular, red to reddish orange, with several to numerous seeds, lacking stony granules among the seeds. Seeds 2–3 mm in longest dimension, somewhat flattened, circular to oval in outline, notched at the attachment point, lacking wings, the surface finely pitted, yellowish brown. About 80 species, North America, Central America, South America, Europe, Africa, Asia south to Australia.

Most species of *Lycium* are native to the New World, but a minority of species are widespread in the Old World. The species in Missouri are introductions from Eurasia that formerly were popular in cultivation in the Midwest, but are no longer as widely sold by plant nurseries. The fruits of *L. barbarum* and *L. chinense* are marketed by the health food industry under the name goji berry. They are purported to be rich in antioxidants and are sold as a general immune system booster and tonic. Both have a long history of medicinal use in China.

The two species reported as escapes from cultivation in Missouri require further study. Some American authors have gone so far as to treat them as a single species (Gleason and Cronquist, 1991). None of the specimens is a perfect match for the morphology displayed by *L. chinense* in its native range, and there appears to be less than perfect correlation between leaf shape, calyx lobing, and pubescence characters ascribed to *L. chinense* in the introduced populations. Because our plants are derived from horticultural materials, it is likely that they are atypical as a result of breeding and selection of cultivars.

Capsicum annuum L. (bird pepper) is a species that is frequently cultivated for its brightly colored fruits. The genus *Capsicum* is also the source of the chili peppers used for food and spice. For a lucid, nontechnical discussion of chili peppers, see Heiser (1969). Occasionally when plants or fruits are discarded or added to compost piles, seeds germinate. The spontaneous forms of the species, whether wild within the native range or escaped elsewhere, have been referred to the ssp. *annuum*. However, although two specimens have been collected in noncultivated situations in the state, there is no evidence that the species ever reproduces successfully in the wild in Missouri. A specimen at the University of Missouri Herbarium collected by David Dunn in October, 1966, documents a solitary plant of *C. annuum* growing as a volunteer in a student housing area in Columbia (Boone County). The plant did not reappear during subsequent years. Similarly, a plant was collected in 1993 by James Miller (then of the Missouri Botanical Garden) on a gravel bar of the Jacks Fork River in Shannon County. Such rare individuals can grow from seeds accidentally scattered into a highly disturbed habitat, but show no evidence of persistence in subsequent years. The genus *Capsicum* is superficially similar to *Lycium*, but lacks thorns and has saucer-shaped corollas. Although within its native range *C. annuum* is a shrub to 3 m or more tall, plants sold in the horticultural industry in the midwestern United States generally are grown as annuals.

1. Leaf blades mostly lanceolate to oblanceolate; flower stalks 10–23 mm long; corolla tube longer than the lobes . . . . 1. L. BARBARUM
1. Leaf blades elliptic, ovate, or rhombic-ovate, occasionally lanceolate; flower stalks 3–15 mm long; corolla tube shorter than to about as long as the lobes . . . . 2. L. CHINENSE

1. Lycium barbarum L. (matrimony vine)
L. halimifolium Mill.

Pl. 561 d, e; Map 2615

Leaf blades 0.8–2.0 cm long (to 5[–7] cm on rapidly elongating young shoots) mostly lanceolate to oblanceolate, rarely elliptic or ovate-elliptic. Flower stalks 10–23 mm long. Calyces 3–6 mm long, the lobes rounded or angled to a

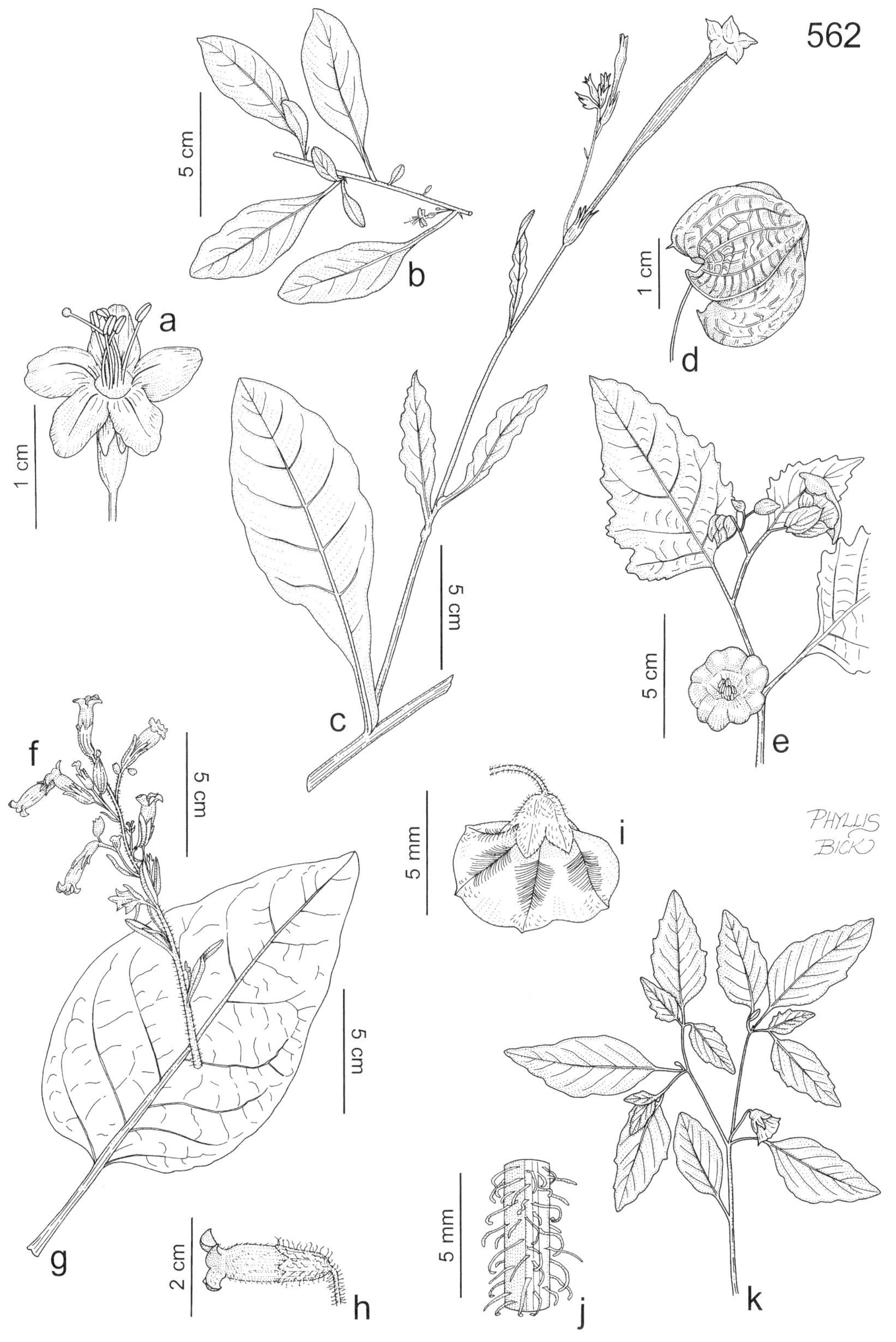

Plate 562. Solanaceae. *Lycium chinense*, **a)** flower, **b)** branch with leaves and flower. *Nicotiana longiflora*, **c)** portion of branch with leaves, flower, and bud. *Nicandra physalodes*, **d)** fruit, **e)** tip of branch with leaves and flowers. *Nicotiana rustica*, **f)** inflorescence, **g)** leaf, **h)** flower. *Physalis virginiana*, **i)** flower, **j)** stem detail, **k)** tip of fertile branch.

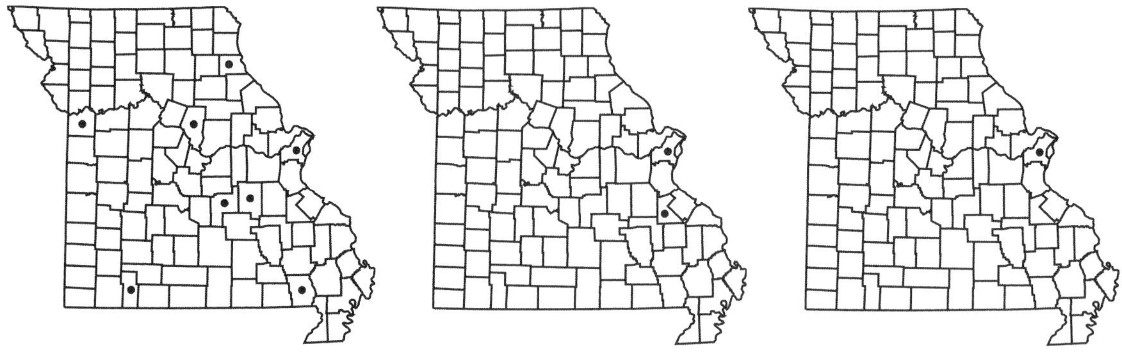

2617. Nicandra physalodes 2618. Nicotiana longiflora 2619. Nicotiana rustica

bluntly or sharply pointed tip. Corollas 9–14 mm long, purple, narrowly bell-shaped; the tube longer than the lobes, more or less funnel-shaped, the lobes spreading, mostly glabrous to sparsely and minutely hairy along the margins. $2n=24$. May–September.

Introduced, widely scattered, mostly around urban areas and in southwestern Missouri (native of Europe, Asia, introduced sporadically nearly throughout North America). Banks of streams and rivers; also old homesites, farmyards, pastures, fencerows, levees, railroads, roadsides, and open, disturbed areas.

Feinbrun and Stearn (1963) studied type specimens in the *L. barbarum* complex and determined that the epithets *L. barbarum* and *L. halimifolium* apply to the same species, with *L. barbarum* the oldest name.

2. Lycium chinense Mill. (Chinese matrimony vine)

Pl. 562 a, b; Map 2616

Leaf blades 1.2–4.0 cm long (to 5[–7] cm on rapidly elongating young shoots) elliptic, ovate, or rhombic-ovate, occasionally lanceolate. Flower stalks 3–15 mm long. Calyces 3–4 mm long, the lobes angled to a sharply pointed tip. Corollas 9–12 mm long, purple, narrowly bell-shaped; the tube shorter than to about as long as the lobes, tubular to more or less funnel-shaped, the lobes spreading, moderately to densely and minutely hairy along the margins. $2n=24$, 36, 48. June–October.

Introduced, uncommon, sporadic (native of Europe, Asia, introduced sporadically in the eastern U.S. and California; Canada). Banks of streams and rivers; also old homesites, farmyards, roadsides, and open, disturbed areas.

3. Nicandra Adans.

One species, South America.

1. Nicandra physalodes (L.) Gaertn. (apple of Peru, shoo-fly plant)

Pl. 562 d, e; Map 2617

Plants annuals, taprooted, unarmed. Stems 20–120 cm long, erect or ascending, unbranched or with few to several, ascending to spreading branches toward the tip, glabrous. Leaves moderately to long-petiolate, the petioles winged toward the tip or occasionally most of their length. Leaf blades 5–25 cm long, ovate to broadly ovate, simple, with shallow, broad, sometimes irregular, pinnate lobes, tapered to a sharply pointed tip, tapered at the base, the margins usually also with broad, sharp teeth, glabrous or nearly so at maturity, sometimes sparsely and minutely hairy when young. Inflorescences axillary or lateral, of solitary flowers. Flowers nodding, the flower stalk 0.8–1.5 cm, becoming only slightly elongated (to 1.2–2.0 cm) at fruiting. Calyces 1–2 cm long (in total length) at flowering, lobed to at or below the midpoint (excluding the 5 prominent, angular, basal auricles), sometimes nearly to the base, broadly tubular but strongly 5-angled at flowering, deeply concave at the base and with 5 angular to arrowhead-shaped, basal auricles (1 per lobe), persistent at fruiting, the tube becoming enlarged, papery, balloonlike, and with a strong network of raised veins, loosely enclosing the fruit (this usually easily visible when viewed from the tip), strongly 5-angled, green to tan. Corollas 1.5–3.0 cm long, broadly bell-shaped to occasionally nearly saucer-shaped, shallowly 5-lobed (the margin sometimes appearing merely scalloped), appearing pleated in bud, blue to light blue, usually with a lighter throat, this sometimes with 5 bluish markings (rarely entirely white elsewhere). Stamens with the filaments mostly somewhat longer than the anthers, hairy toward the base, the anthers erect, not fused laterally, dehiscent longitudinally, white to pale yellow, not exserted. Ovary 3–5-locular, the stigma usually positioned at about the same height as the anthers. Fruits berries, 1–2 cm long, mealy (not juicy), globose, 3–5-locular, green to greenish brown, with numerous seeds, lacking stony granules (these present elsewhere). Seeds 1.5–1.8 mm in longest dimension, more or less kidney-shaped to asymmetrically ovate or nearly circular, flattened, the surface with a fine network of ridges and pits, more or less shiny, orangish brown, lacking wings. $2n=20$. July–September.

Introduced, scattered, sporadic (native of South America; introduced sporadically nearly throughout North America). Crop fields, old homesites, farmyards, railroads, and open, disturbed areas.

Steyermark (1963) noted that historically this species was grown to eradicate house flies. Stems were pulped and moistened with a small amount of milk to attract flies, which purportedly succumbed to lethal intoxication within a half hour after ingesting the mixture.

4. **Nicotiana** L. (tobacco) (Goodspeed, 1954)

Plants annual or perennial herbs (shrubs or small trees elsewhere), unarmed. Stems erect or strongly ascending, often relatively stout, unbranched or with few to several ascending branches, mostly toward the tip, sparsely to densely glandular-hairy (the plants somewhat sticky to the touch). Leaves alternate and sometimes also basal, sessile to long-petiolate. Leaf blades simple, variously shaped, unlobed or with a pair of basal auricles, the margins entire or at most wavy, the surfaces glandular-hairy and sometimes also with sessile glands. Inflorescences terminal, panicles or racemes. Flowers ascending and/or spreading, the fruits usually more or less ascending. Calyces 8–20 cm long, unequally 5-lobed, broadly tubular to bell-shaped at flowering, rounded at the base, lacking basal auricles, the sides rounded or more commonly more or less 5- or 10-ridged, persistent intact but often slightly enlarged and distended at fruiting. Corollas 1.5–11.0 cm long, trumpet-shaped to somewhat funnelform, relatively shallowly 5-lobed, the lobes variously shaped, often appearing spirally twisted and pleated in bud, variously colored. Stamens with filaments longer than the anthers, unequal in length and/or attachment point in the corolla tube (usually 1 filament longer than and/or attached below the other 4), the anthers free, erect or incurved, not exserted, dehiscent longitudinally, light yellow to nearly white. Ovary 2-carpellate the style elongate, positioned at or slightly above the level of the anthers, often green. Fruits capsules, dry, globose to ovoid or ellipsoid, 2-locular, tan or light brown, dehiscent longitudinally from the tip, with numerous seeds, unarmed. Seeds 0.4–1.1 mm in longest dimension, more or less circular ro broadly elliptic, oval, or angular, flattened, the surface with a fine network of ridges or appearing finely wrinkled, light to dark brown, lacking wings. About 76 species, North America, Central America, South America, Africa, Australia.

1. Corollas 1.5–2.0 cm long, less than 2 times as long as the calyces; stem leaves short- to moderately petiolate . . . . 2. N. RUSTICA
1. Corollas 5–11 cm long; 3 or more times as long as the calyces; stem leaves sessile, expanded at the base, either auriculate or decurrent along the stem
    2. Stem leaves relatively few, noticeably smaller than the basal ones, with a pair of small auricles at the base, the stems appearing relatively naked; inflorescences short panicles or racemes, relatively few-flowered; corollas 7–11 cm long . . . . 1. N. LONGIFLORA

2. Lower stem leaves only slightly smaller than the basal ones, decurrent at the base, the stems appearing leafy to the inflorescence; inflorescences elongate to somewhat pyramid-shaped panicles, mostly with numerous flowers; corollas 5–6 cm long . . . . 3. N. TABACUM

1. Nicotiana longiflora Cav. (long-flowered tobacco)

Pl. 562 c; Map 2618

Plants annual (short-lived perennials elsewhere), more or less fleshy-rooted. Stems 30–60 (–100) cm long, relatively slender, sparsely short-hairy toward the tip. Leaves few and much-reduced above the persistent basal rosettes (the stems usually appearing relatively naked), the margins entire or somewhat wavy, sometimes somewhat corrugated, the surfaces moderately to densely and minutely hairy. Basal rosette leaves 5–30 cm long, the blades oblanceolate to elliptic-obovate, rounded or more commonly angled to a bluntly or sharply pointed tip, tapered to a short, winged petiole at the base, some of the hairs gland-tipped. Stem leaves 1–9 cm long, linear to narrowly lanceolate or narrowly oblong-lanceolate, tapered to a sharply pointed tip, sessile, slightly expanded or with a pair of small, clasping auricles at the base. Inflorescences short panicles or racemes, relatively few-flowered. Calyces 13–17 mm long, 10-ribbed, glandular-hairy, the tube about as long as to slightly shorter than the 5 narrowly triangular lobes, these long-tapered to sharply pointed tips. Corollas 7–11 cm long, trumpet-shaped, glandular-hairy, the tube slender, pale yellow to cream-colored, sometimes purplish-tinged, the lobes 7–12 mm long, narrowly ovate, white, pale yellow or light purple. Fruits 11–16 mm long, ovoid. Seeds 0.4–0.6 mm in longest dimension, the surface with a fine network of ridges or appearing finely wrinkled, light brown. $2n=20$. May–August.

Introduced, uncommon, sporadic (native of South America, introduced sporadically in the eastern half of the U.S.; Canada). Railroads, roadsides, and open, disturbed areas.

For many years a population of this species existed on ballast ground along the Mississippi River in St. Louis, but it now appears to have become extirpated.

Nicotiana ×sanderae W. Watson (red-flowered garden nicotiana) is a horticulturally derived hybrid bred from crosses between *N. alata* Link & Otto and *N. forgetiana* Hemsl. It has been recorded from a single historical collection from St. Louis County. It is superficially similar to *N. longiflora*, but differs in its broader stem leaves, corollas variously white, pink, red, or purple (occasionally even yellow) with a somewhat stouter tube and larger lobes, and its somewhat longer calyces.

2. Nicotiana rustica L. (wild tobacco)

Pl. 562 f–h; Map 2619

Plants annual, taprooted. Stems 40–150 cm long, relatively stout, densely pubescent with short, often woolly or matted, multicellular, mostly gland-tipped hairs, scattered, longer spreading hairs sometimes also present. Leaves relatively numerous above the usually nonpersistent basal rosettes and only gradually reduced toward the stem tip (the stems appearing relatively densely leafy), the lower leaves moderately petiolate, the upper leaves short-petiolate. Leaf blades 6–15 (–30) cm long, those of the lower leaves broadly ovate to nearly heart-shaped, rounded to bluntly pointed at the tip, rounded to truncate or shallowly cordate at the often oblique base; those of the upper leaves ovate to elliptic, angled or slightly tapered to a bluntly or sharply pointed tip; the margins entire to slightly wavy, relatively flat, the surfaces sparsely to moderately pubescent with short, mostly gland-tipped hairs, more densely so along the veins, often also with sessile glands. Inflorescences elongate to somewhat pyramidal panicles, mostly with numerous flowers. Calyces 8–15 mm long, faintly 5- or 10-ribbed, glandular-hairy, the tube longer than the 5, ovate to broadly ovate-triangular lobes, these angled or curved to mostly bluntly pointed tips. Corollas 1.5–2.0 cm long, glandular-hairy, lemon yellow to greenish yellow, the tube relatively broad, narrowly funnelform to nearly tubular but usually slightly constricted at the tip, the lobes 2.5–4.0 mm long, more or less semicircular or the margin appearing merely scalloped. Fruits 7–16 mm long, broadly oblong-ovoid to nearly globose. Seeds 0.7–1.1 mm in longest dimension, the surface with a network of ridges, brown to dark brown. $2n=48$. July–October.

Introduced, uncommon, known thus far only from a single historical collection from the city of St. Louis (native of South America, introduced sporadically in mostly the northeastern U.S.; Canada). Habitat unknown, but presumably open, disturbed areas.

Even in pre-Columbian times, Native Americans cultivated *N. rustica* for medicinal use and to smoke as a tobacco (Steyermark, 1963; Epstein, 1981; Moerman, 1998). In more recent years, with increased reliance on commercially available *N. tabacum*, cultivation of the species has declined sharply and it does not appear as an escape any longer. Heiser (1969) noted that although this spe-

2620. Nicotiana tabacum 2621. Petunia axillaris 2622. Petunia integrifolia

cies had a South American origin similar to that of *N. tabacum*, it apparently was not smoked in South America. Its mode of arrival in North America is not well understood.

3. Nicotiana tabacum L. (tobacco)

Map 2620

Plants annual (biennial or short-lived perennial elsewhere), sometimes taprooted. Stems 100–250 cm long, stout, densely pubescent with a mixture of shorter and longer, woolly or matted to mostly spreading, multicellular, gland-tipped hairs. Leaves numerous above the often persistent basal rosettes and only gradually reduced toward the stem tip (the stems appearing densely leafy), the lowermost leaves sometimes with short, winged petioles, the other stem leaves sessile, the bases expanded and decurrent along the stem for a short distance below the attachment point. Leaf blades (8–)12–50 cm long, lanceolate or oblanceolate to elliptic or ovate; the margins entire to slightly wavy, relatively flat, the surfaces sparsely to densely pubescent with short, mostly gland-tipped hairs, more densely so along the veins, usually also with sessile glands. Inflorescences short to more commonly elongate or somewhat pyramidal panicles, mostly with numerous flowers. Calyces 12–20 mm long, faintly 5-ribbed, glandular-hairy, the tube longer than the 5, triangular lobes, these tapered to sharply pointed tips. Corollas 5–6 cm long, glandular-hairy, the tube funnelform (the tubular base expanded abruptly above the midpoint), white, often greenish-tinged toward the base and pinkish- or reddish-tinged toward the tip, the lobes 5–8 mm long, triangular to broadly triangular, pink or red, rarely white. Fruits 15–20 mm long, ellipsoid to ovoid, oblong-obovoid, broadly ovoid, or nearly globose. Seeds 0.4–0.6 mm in longest dimension, the surface with a network of ridges, brown to dark brown. $2n$=48. July–October.

Introduced, uncommon, known thus far only from a single historical collection from St. Louis County (native of South America, introduced sporadically in the U.S.; Canada). Habitat unknown, but presumably open disturbed areas.

Tobacco is among the most controversial of crop plants in the United States. Originally cultivated in Andean South America, the species is an example of the ancient exchange of goods between South American and North American cultures, having become widely cultivated in the North America during pre-Columbian times (Heiser, 1969). Early Spanish and English explorers in North America were quick to notice this plant, which was used medicinally for a number of ailments and also ceremonially by the peoples they encountered. During the 1600s, tobacco became an important export item from the Colonies to Europe, and over time its cultivation proliferated as European expansion pushed westward on the North American continent. Because tobacco cultivation is a labor-intensive practice, the industry became closely associated with slavery prior to the American Civil War. During the first half of the nineteenth century, tobacco was a major crop plant in Missouri, but the industry began a long decline following the Civil War. Today, a few farms continue to produce slightly more than 1,700 tons of dried leaves annually, down from more than 3.3 million tons in 1902 (W. Williams, 1904). The history of Missouri's tobacco industry is interpreted at Weston Bend State Park (Platte County).

Tobacco is grown primarily for its leaves, which are harvested and then cured through drying. The leaves contain thousands of biochemically active compounds (National Cancer Institute, 1998), but the mild sedative and euphoric effect is caused primarily by nicotine, a highly addictive alkaloid. The fermented and dried, ground leaves most frequently are burned and the smoke inhaled through smoking in cigars, cigarettes, and pipes. They are also taken orally by allowing the juice to diffuse from a small amount inserted into the mouth (between cheek and gum). To a lesser extent, the finely ground leaves also have been inhaled directly as snuff. Tobacco is a controlled substance in the United States, a legal drug whose production and distribution are heavily regulated and taxed by federal and state governments. Ingestion of tobacco is considered very unhealthy (with direct causal links established to various cancers [especially lung cancer] and leukemia). Since the early 1960s, the federal government has invested heavily in attempts to educate Americans about the evils of tobacco products. Although this has led to a decline in the sales of cigarettes, it has not made the desired impact on the public demand for tobacco products. Smokers also should note that nicotine and its biosynthetic derivatives are an important ingredient in some poisons used as insecticides.

5. **Petunia** Juss. (petunia)

Plants annual (perennial herbs elsewhere), unarmed. Stems erect to loosely ascending or ascending from a spreading base, with few to several ascending branches, moderately to densely pubescent with short, spreading, multicellular, glandular-hairs (the plants somewhat sticky to the touch). Leaves alternate, often appearing opposite near the stem tip, sessile or with short, mostly winged petioles. Leaf blades simple, oblanceolate to elliptic or ovate, unlobed, the margins entire, the surfaces moderately to densely pubescent with short, spreading, multicellular, glandular-hairs. Inflorescences axillary or lateral (the uppermost flowers sometimes appearing terminal), of solitary flowers. Flowers ascending and/or spreading, the fruits either ascending or nodding. Calyces 1.2–1.8 cm long, deeply 5-lobed (often nearly to the base), bell-shaped at flowering, rounded at the base, lacking basal auricles, the lobes usually slightly unequal, linear to narrowly oblong, rounded or bluntly pointed at the tips, glandular-hairy, persistent more or less intact at fruiting. Corollas 2.5–6.0 cm long (sometimes longer in cultivated plants), funnelform to trumpet-shaped, very shallowly 5-lobed, the lobes broadly rounded to broadly and bluntly pointed, appearing pleated toward the tip in bud (but not spirally twisted), variously colored. Stamens with filaments longer than the anthers, unequal in length, attached at or below the midpoint of the corolla tube, the anthers free, erect or incurved, not exserted, dehiscent longitudinally, light yellow. Ovary 2-carpellate the style elongate, positioned above the level of the uppermost anthers, green. Fruits capsules, dry, (7–)9–15 mm long, ovoid to narrowly ovoid, 2-locular, tan or light brown, dehiscent longitudinally from the tip, with numerous seeds, unarmed. Seeds 0.5–0.8 mm in longest dimension, more or less circular ro broadly elliptic, oval, or angular, flattened, the surface with a fine network of ridges, light to dark brown, lacking wings. About 40 species, South America.

Some botanists accept an expanded generic concept of *Petunia* that increases the number of included species to about 45 and expands the native range northward into the United States. The Missouri species are among those that are important in horticulture, generally as bedding plants grown for their bright flowers and long flowering period.

1. Corolla tube slender, expanded abruptly toward the tip; stamens with the filaments attached near the midpoint of the corolla tube, the anthers of the longest stamens positioned in the throat . . . . 1. P. AXILLARIS
1. Corolla tube funnelform, expanded gradually from the base to the tip; stamens with the filaments attached near the base of the corolla tube, the anthers of the longest stamens positioned near the midpoint of the corolla tube . . . 2. P. INTEGRIFOLIA

1. Petunia axillaris (Lam.) Britton, Sterns & Poggenb. (white petunia)

Pl. 563 f; Map 2621

Stems 20–50 cm long, usually erect or ascending. Leaf blades 1–7 cm long, lanceolate to ovate or narrowly to broadly elliptic, those of the lowermost leaves sometimes oblanceolate. Flower stalks (15–)25–55 mm long. Corollas 4.5–6.5 cm long, the

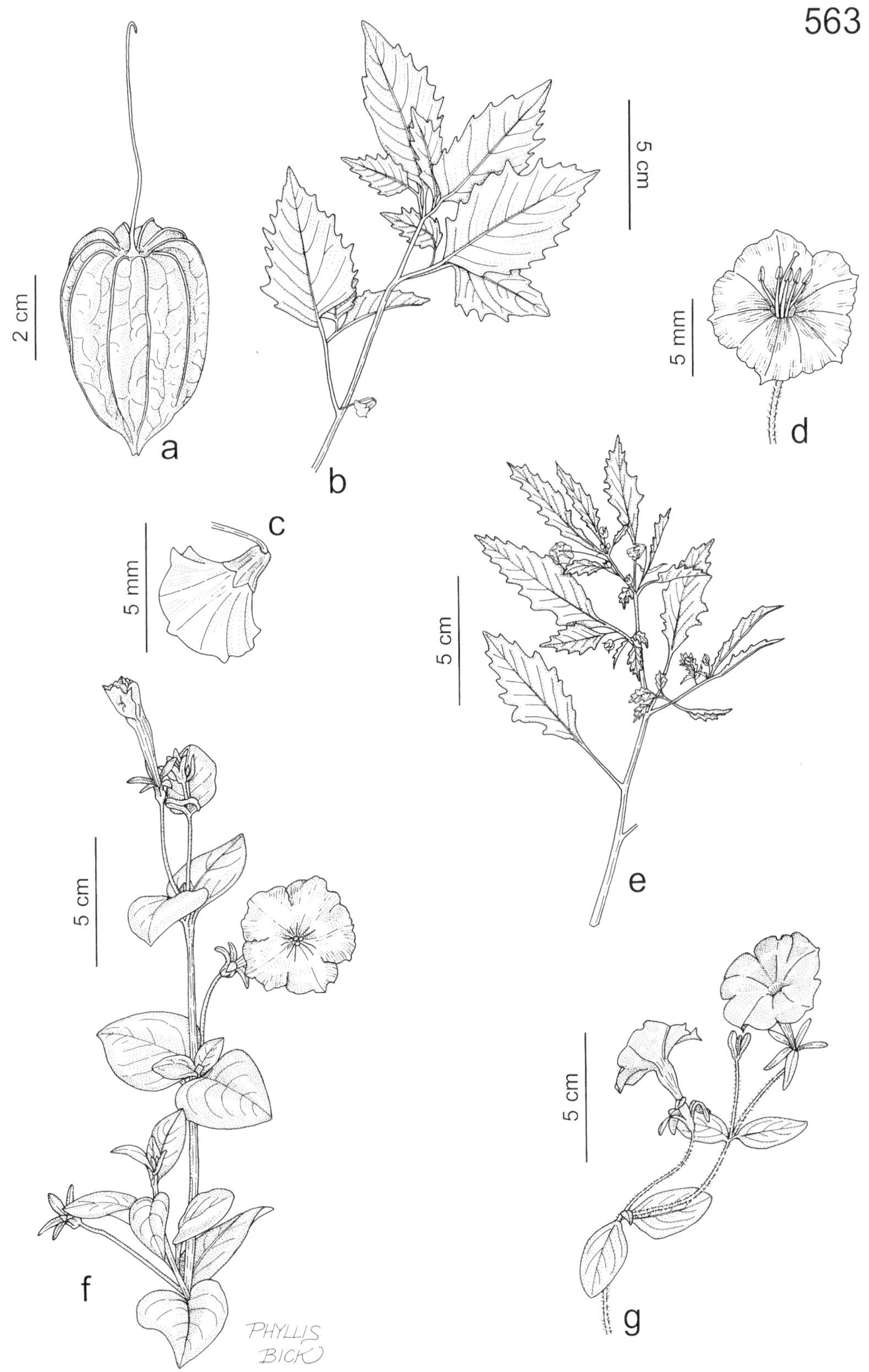

Plate 563. Solanaceae. *Physalis angulata*, **a)** fruit, **b)** tip of branch with leaves and flower, **c)** flower. *Physalis acutifolia*, **d)** flower, **e)** tip of branch with leaves and flowers. *Petunia axillaris*, **f)** tip of branch with leaves and flowers. *Petunia integrifolia*, **g)** tip of branch with leaves and flowers.

2623. Physalis acutifolia
2624. Physalis alkekengi
2625. Physalis angulata

tube slender, expanded abruptly toward the tip, white, the rim mostly 4–6 cm in diameter, the expanded portion white (occasionally pink to red or purple-striped, -mottled, or -colored in cultivated plants). Stamens with the filaments attached near the midpoint of the corolla tube, the anthers of the 2 longest stamens positioned in the throat, those of the other 3 positioned in the upper portion of the tube. $2n=14$. July–September.

Introduced, uncommon, known thus far only from the southwestern and eastern portions of the state (native of South America; introduced sporadically in mostly the eastern half of the U.S.; Canada). Open, disturbed areas.

2. Petunia integrifolia (Hook.) Schinz & Thell. (violet petunia)
P. violacea Lindl.

Pl. 563 g; Map 2622

Stems 15–45 cm long, loosely ascending or spreading with ascending tips, uncommonly erect. Leaf blades 1–5 cm long, lanceolate to ovate-elliptic, elliptic, or oblanceolate. Flower stalks 12–30 mm long. Corollas 2.5–4.5 cm long, the tube funnelform, expanded gradually from the base to the tip, white to strongly purplish-tinged, rarely pinkish-tinged, the rim mostly 3.0–4.5 cm in diameter, the expanded portion usually reddish purple to bluish purple, rarely pink to red (striped or mottled in cultivated plants). Stamens with the filaments attached near the base of the corolla tube, the anthers of the 2 longest stamens positioned near the midpoint of the corolla tube, those of the other 3 positioned slightly lower in the tube. $2n=14$. July–September.

Introduced, uncommon, sporadic (native of South America; introduced sporadically in mostly the eastern half of the U.S.; Canada). Railroads and open, disturbed areas.

The present work follows Wijsman (1982) in treating the name *P. violacea*, which is still widely used in horticulture, as a synonym of *P. integrifolia*. However, Wijsman's infraspecific classification within *P. integrifolia* is not followed, as Ando et al. (2008) have shown that the complex comprises two species, rather than one species with three subspecies.

The garden petunia is a horticulturally derived, fertile hybrid thought to have been developed from crosses between *P. axillaris* and *P. integrifolia* (Wijsman, 1982; Sink, 1984). There has been controversy over the correct scientific name for this hybrid taxon, with two names and several authorships cited in the botanical literature. Of these, the earliest correct name appears to be *P. ×atkinsiana* (Sweet) D. Don ex W.H. Baxter (*P. ×hybrida* E. Vilm.). The taxon has a long history of cultivation as a bedding plant in gardens, as well as in planters and pots, and has escaped from cultivation sporadically in the southern half of the state. Plants of *P. ×atkinsiana* resemble those of *P. violacea*, but differ in their usually larger (mostly 5–8 cm long), more broadly flaring corollas of varying colors (white, red, and/or purple). Numerous cultivars exist and the taxon is quite variable morphologically. In some cases, it can be very difficult to discriminate the hybrids from *P. integrifolia*.

6. Physalis L. (ground cherry, husk tomato)

Plants in our species annual or perennial herbs, sometimes with rhizomes, unarmed. Stems erect or ascending, less commonly loosely ascending from a spreading base, with several to numerous branches, mostly toward the tip, glabrous or hairy. Leaf blades simple, unlobed or with shallow, broad, pinnate lobes, glabrous or hairy. Inflorescences axillary, of solitary flowers (few-flowered clusters elsewhere). Flowers spreading to more commonly nodding. Calyces 0.3–1.6 cm long at flowering, shallowly lobed at the tip, broadly tubular or narrowly to broadly bell-shaped at flowering, rounded to truncate or occasionally slightly concave at the base, lacking basal auricles, persistent at fruiting, the tube becoming enlarged and balloonlike, entirely and loosely enclosing the fruit (this visible only when the calyx is torn apart), 5- or 10-angled to 10-ribbed (then more or less rounded). Corollas 0.6–2.0 cm long, broadly bell-shaped to saucer-shaped, shallowly 5-lobed to nearly entire, usually appearing pleated in bud, light yellow to lemon yellow, often with 5 large, prominent dark purplish brown to dark reddish brown spots on the inner surface toward the base (these sometimes merged into a ring or appearing smudged), rarely white. Stamens with relatively short filaments, the anthers erect, positioned in a loose ring but not fused laterally, dehiscent longitudinally, light yellow to yellow or less commonly purplish blue to dark blue (sometimes only bluish-tinged or -lined). Ovary 2-locular, the style usually protruding through the center of the loose ring of anthers. Fruits berries, more or less juicy, globose, 2-locular, green, yellow, or orange, sometimes purplish-streaked or -mottled, with numerous seeds, lacking stony granules (these present elsewhere). Seeds 1.5–2.5 mm in longest dimension, more or less kidney-shaped to asymmetrically ovate, flattened, the surface minutely pitted, sometimes only faintly so, more or less shiny, light yellow to orangish yellow or light yellowish brown, lacking wings. About 75 species, nearly worldwide.

The species of *Physalis* are variously known as husk tomato, ground cherry, and tomatillo. The cultivated tomatillo, which is a staple in Mexican cooking and salsas, is *P. philadelphica*, an annual species native to Mexico and the adjacent southwestern United States. Tomatillos have successfully bridged the gap from an ingredient in a regional cuisine to a more general food source, and usually are available in the produce section of grocery stores year-round. Steyermark (1963) noted that although plants (including immature fruits) of *Physalis* should be considered toxic to livestock and humans, the mature berries of several of the juicier-fruited Missouri species can be prepared into jams and preserves or eaten raw.

When collecting specimens of *Physalis*, it is important to note whether the plant is annual or perennial. In annuals, the rootstock generally is easily pulled up with the rest of the plant. The perennials mostly have deepset rhizomes that are collected infrequently. In both cases, the stems generally do not appear colonial. Also, a notation of whether the corollas have dark spots facilitates determination, as the flowers often press in a closed position. Finally, if fruits are present, whether the inflated calyces are sharply 5-angled or bluntly 10-angled is important to note, as this character can be difficult to diagnose in pressed materials. Fertile material, preferably with both flowers and fruits, is necessary to determine some of the taxa with confidence.

The taxonomy and nomenclature of *Physalis* are complex and there have been numerous changes since the last comprehensive taxonomic treatment of the temperate North American species by Waterfall (1958, 1968). The present work closely follows that of J. R. Sullivan (2004), whose excellent treatment of the species present in the southeastern United States includes most of the Missouri taxa.

Steyermark (1963) discussed the existence of a historical specimen from the Allenton area (St. Louis County) of what he called *Physalis lobata* Torr. This species is now segregated by most botanists into a separate genus, *Quincula* Raf. (Barboza, 2000), comprising only the species *Q. lobata* (Torr.) Raf. Purple ground cherry is native to the southwestern United States (west to Kansas and Texas), as well as northern Mexico. It is similar in general morphology to *Physalis*, but differs from at least the Missouri species in its small clusters of ascending flowers with purple corollas, its dull seeds with a network of ridges on the surface, and details of its calyx venation and trichome structure. Steyermark (1963) chose to exclude the sole Missouri specimen from the state's flora in the belief that it was mislabeled and actually collected in another state. Steyermark was personally acquainted with the collector, John Kellogg, and presumably was able to confirm that Kellogg had not collected the species in eastern Missouri. Unfortunately, the specimen in question could not be located during the present study. Thus, Steyermark's exclusion of the taxon from the Missouri treatment is continued in the present work.

1. Plants perennial, with deep-set, long-creeping rhizomes; corollas 10–18 cm long
    2. Corollas white, sometimes with 5 green spots on the inner surface toward the base; fruiting calyces orangish red to red . . . . 2. P. ALKEKENGI
    2. Corollas light yellow to yellow, with 5 prominent dark purplish brown to dark reddish brown spots on the inner surface toward the base, these sometimes merged into a ring or appearing smudged (in *P. pumila* only faintly greenish- or brownish-tinged and lacking distinct discolorations); fruiting calyces green, drying pale brown to tan
        3. Leaf blades pubescent with at least some of the hairs branched (at least on the undersurface)
            4. Corollas with 5 large, prominent dark purplish brown to dark reddish brown spots on the inner surface toward the base; calyces at flowering densely pubescent with minute, relatively fine, matted or woolly hairs 0.1–0.3 mm long having 3 or more branches (check the calyx lobes if only fruits are present) . . . . 4. P. CINERASCENS
            4. Corollas greenish- or pale brownish-tinged on the inner surface toward the base, lacking distinct spots or discolorations; calyces at flowering densely pubescent with relatively stout, woolly hairs of various lengths, the longest of these 1–2 mm long, with 2 or 3 branches (check the calyx lobes if only fruits are present) . . . . . 12. P. PUMILA
        3. Leaf blades with the undersurface glabrous or pubescent with all of the hairs unbranched
            5. Leaf blades broadly ovate to nearly circular; stems and leaves with at least some of the hairs gland-tipped. . . . . 7. P. HETEROPHYLLA
            5. Leaf blades lanceolate to oblanceolate or ovate; stems and leaves with all of the hairs nonglandular
                6. Stems and calyces glabrous or sparsely pubescent with minute, appressed-ascending hairs 0.1–0.5 mm long . . . . 8. P. LONGIFOLIA
                6. Stems and calyces moderately to densely hairy with a mixture of minute loosely appressed hairs 0.1–0.5 mm long and scattered, spreading, longer hairs to 2.0 mm long
                    7. Corollas greenish- or pale brownish-tinged on the inner surface toward the base, lacking distinct spots or discolorations; stems, flower stalks, and calyces with a mixture of longer, spreading hairs and shorter upward-angled or -curved hairs . . . . 12. P. PUMILA
                    7. Corollas with 5 large, prominent dark purplish brown to dark reddish brown spots on the inner surface toward the base (these sometimes merged into a ring or appearing smudged); stems, flower stalks, and calyces with a mixture of longer, spreading hairs and shorter downward-angled or -curved hairs . . . . 13. P. VIRGINIANA
1. Plants annual, more or less taprooted; corollas 6–11 mm long (occasionally to 15 mm in *P. philadelphica*)

8. Stamens with the anthers blue, strongly coiled after dehiscence; fruits at maturity 2.0–2.8 cm in diameter, pale green, often purplish-tinged or with purple streaks . . . . 10. P. PHILADELPHICA
8. Stamens with the anthers yellow, blue, or purplish blue, arched but not coiled after dehiscence; fruits at maturity 1.0–1.8 cm in diameter (occasionally to 2.5 cm in *P. pubescens*), orange, yellow, or green, not streaked
    9. Leaf blades with the surfaces glabrous or sparsely pubescent with short, appressed, nonglandular hairs
        10. Corollas with the inner surface uniformly pale yellow to nearly white or at most pale purplish-tinged toward the base (greenish-tinged or with a greenish yellow region toward the base in *P. acutifolia*); calyces shallowly 10-angled or 10-ribbed at fruiting (sometimes appearing shallowly 5-angled in *P. missouriensis*); stamens with slender filaments (half as wide as the anther or less)
            11. Stems moderately to densely pubescent with gland-tipped hairs (nonglandular hairs often also present); leaf blades broadly ovate to nearly circular, rounded to cordate at the base; flower stalks 4–7 mm long, elongating only slightly to 5–10 mm at fruiting; anthers yellow . . . . 9. P. MISSOURIENSIS
            11. Stems glabrous or sparsely pubescent with nonglandular hairs toward the tip; leaf blades narrowly lanceolate to elliptic or ovate, rounded to angled or tapered at the base; flower stalks 7–40 mm long, elongating to 15–60 mm at fruiting; anthers yellow or blue to purplish blue or less commonly merely bluish- or greenish-tinged
                12. Corollas pale yellow to nearly white, the inner surface greenish-tinged or with a greenish yellow region toward the base, this densely hairy; anthers 2.5–3.0 mm long, yellow or less commonly bluish- or greenish-tinged . . . . 1. P. ACUTIFOLIA
                12. Corollas light yellow to lemon yellow, the inner surface not tinged or pale purplish-tinged toward the base; anthers 1.5–2.5 mm long, blue to purplish blue or merely bluish-tinged . . . . 3. P. ANGULATA
        10. Corollas with 5 large, prominent dark purplish brown to dark reddish brown spots on the inner surface toward the base (these sometimes merged into a ring or appearing smudged, or interrupted by lighter-colored venation); calyces sharply 5-angled at fruiting; stamens with broad filaments (about as wide as the anther)
            13. Stems glabrous or sparsely pubescent toward the tip with minute, appressed, nonglandular hairs 0.1–0.3 mm long; leaf blades moderately toothed (the larger ones with 8 or more teeth along each side); fruit stalk (10–)15–35 mm long, calyces at fruiting glabrous (except sometimes hairy on the lobes). . . . . 5. P. CORDATA
            13. Stems sparsely to moderately pubescent (at least toward the tip) with minute and slightly longer, spreading, mostly gland-tipped hairs 0.1–0.5 mm long, often also with moderate to dense, longer (1–3 mm), spreading nonglandular hairs; leaf blades entire or relatively sparsely toothed (with 1–3(–7) teeth along each side); fruit stalk 6–15 mm long; calyces at fruiting hairy . . . . 11. P. PUBESCENS
    9. Leaf blades with the surfaces moderately to densely pubescent with various kinds of glandular and/or nonglandular hairs

14. Corollas with the inner surface uniformly pale yellow or at most lightly purplish-tinged toward the base; calyces shallowly 10-angled or 10-ribbed at fruiting (sometimes appearing shallowly 5-angled in *P. missouriensis*)
    15. Stems glabrous or sparsely pubescent with nonglandular hairs toward the tip; leaf blades narrowly lanceolate to elliptic or ovate, rounded to angled or tapered at the base; flower stalks 7–17 mm long, elongating to 15–30 mm at fruiting; anthers blue to purplish blue or less commonly merely bluish-tinged . . . . 3. P. ANGULATA
    15. Stems moderately to densely pubescent with gland-tipped hairs (nonglandular hairs often also present); leaf blades broadly ovate to nearly circular, rounded to cordate at the base; flower stalks 4–7 mm long, elongating only slightly to 5–10 mm at fruiting; anthers yellow. . . . 9. P. MISSOURIENSIS
14. Corollas with 5 large, prominent dark purplish brown to dark reddish brown spots on the inner surface toward the base (these sometimes merged into a ring or appearing smudged, or interrupted by lighter-colored venation); calyces sharply 5-angled at fruiting
    16. Stems glabrous or sparsely pubescent toward the tip with minute, appressed, nonglandular hairs 0.1–0.3 mm long; fruit stalk (10–)15–35 mm long, calyces at fruiting glabrous (except sometimes hairy on the lobes) . . . . 5. P. CORDATA
    16. Stems sparsely to more commonly densely pubescent (at least toward the tip) with minute and longer, spreading, gland-tipped and/or nonglandular hairs 0.1–3.0 mm long (sessile glands may also be present in *P. grisea*); fruit stalk 5–15 mm long; calyces at fruiting hairy
        17. Leaf blades grayish green when fresh, drying orange, orangish-tinged, or with orangish patches, the margins coarsely and unevenly toothed (the larger ones with 8 or more teeth along each side); stamens with slender filaments (half as wide as the anther or less) . . . . 6. P. GRISEA
        17. Leaf blades green when fresh, drying uniformly green (lacking orangish tinging or patches), the margins entire or relatively sparsely toothed (with 1–3[–7] main teeth along each side); stamens with broad filaments (about as wide as the anther) . . . . 11. P. PUBESCENS

1. Physalis acutifolia (Miers) Sandwith
Pl. 563 d, e; Map 2623

Plants annual, taprooted. Stems 15–80 cm long, erect to loosely ascending, with several to many, loosely ascending branches, glabrous or sparsely pubescent toward the tip with short, upward-appressed, unicellular and few-celled, nonglandular hairs 0.1–0.5 mm long. Leaves mostly long-petiolate. Leaf blades 2–12 cm long, narrowly lanceolate to lanceolate or narrowly elliptic to oblong-elliptic or elliptic, angled or tapered to a sharply pointed tip, narrowly angled or tapered at the base, the margins entire or more commonly few- to several-toothed, minutely nonglandular-hairy, the teeth mostly sharply pointed, shallow to deep, relatively narrow to broad, the surfaces green to more commonly dark green when fresh, drying uniformly green (lacking orangish tinging or patches), glabrous or sparsely pubescent with minute, appressed, nonglandular, unicellular or few-celled hairs. Flower stalks 5–40 mm long, becoming elongated to 25–60 mm at fruiting. Calyces 3–7 mm long at flowering, the lobes 1–3 mm long, the outer surface sparsely to moderately pubescent with minute nonglandular hairs at flowering, glabrous or sparsely hairy along the main veins and lobes at fruiting, at fruiting becoming elongated to 20–25 mm long, shallowly 10-angled or 10-ribbed, rounded to very shallowly concave at the base, mostly remaining green, occasionally pale brown to tan with age. Corollas 6–12 mm long, pale yellow to nearly white, the inner surface greenish-tinged or with a greenish yellow region toward the base, this densely hairy. Stamens with slender filaments half as wide as the anthers or narrower, the anthers 2.5–3.0 mm long, yellow or less commonly bluish- or greenish-tinged, arched but not coiled after dehiscence. Fruits 1.0–1.5 cm long, green or yellow to orangish yellow. June–October.

Introduced, uncommon, known thus far only from the city of St. Louis (native of the southwest-

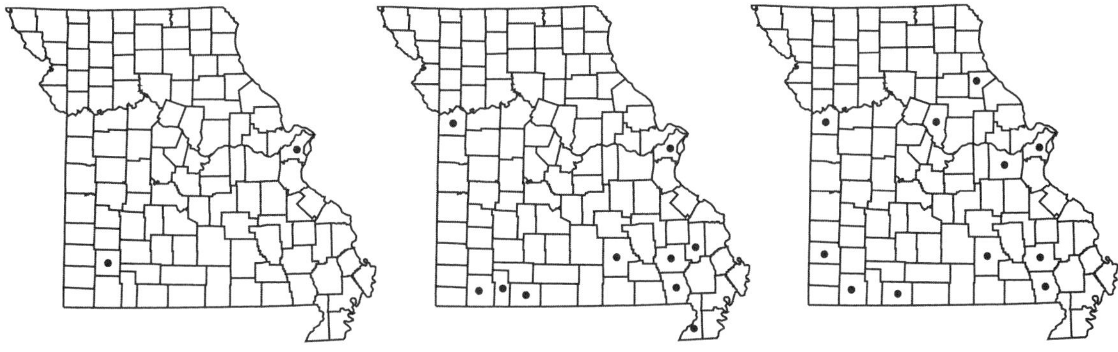

2626. Physalis cinerascens 2627. Physalis cordata 2628. Physalis grisea

ern U.S. east to Louisiana, Mexico; introduced sporadically farther north and east). Railroads.

This species was first reported from Missouri by Mühlenbach (1979). For a discussion of difficulties in distinguishing this species from the very similar *P. angulata*, see the treatment of that species.

2. Physalis alkekengi L. (Chinese lantern plant)

Pl. 564 g; Map 2624

Plants perennial, with deep-set, long-creeping rhizomes. Stems 30–90 cm long, erect or ascending, unbranched or with few, ascending branches, glabrous or sparsely pubescent with spreading, multicellular, nonglandular hairs 0.5–1.5 mm long. Leaves short- to long-petiolate. Leaf blades 3–12 cm long, ovate to broadly ovate, angled or tapered to a sharply pointed tip, rounded to truncate or occasionally shallowly cordate at the base, the margins entire or with several, widely spaced, irregular, shallow to moderately deep, blunt to more commonly sharply pointed teeth per side, minutely nonglandular-hairy, the surfaces green when fresh, drying uniformly green or rarely yellowish- or orangish-tinged, glabrous or sparsely pubescent with short, nonglandular hairs. Flower stalks 9–13 mm long, becoming elongated to 20–40 mm at fruiting. Calyces 4–7 mm long at flowering, the lobes 2.0–3.5 mm long, the outer surface sparsely to moderately pubescent with short to moderately long (to 1 mm), nonglandular hairs at flowering, persistently hairy at fruiting, at fruiting becoming elongated to 30–55 mm long, shallowly 10-angled or 10-ribbed, concave at the base, bright orangish red to red. Corollas 10–15 mm long, white, the inner surface unmarked or sometimes with 5 green spots toward the base (these often merged into a ring). Stamens with slender filaments half as wide as the anthers or narrower, the anthers 2.5–3.0 mm long, yellow, arched but not coiled after dehiscence. Fruits 1.0–1.8 cm long, orangish red to red. $2n=24$. June–September.

Introduced, uncommon, sporadic (Europe, Asia; introduced sporadically in the northern U.S.; Canada). Gardens, fencerows, and open, disturbed areas.

This species was first reported for Missouri by Dunn (1982). It is cultivated for its attractive reddish fruits and occasionally grows as a weed or escapes into adjacent, uncultivated areas. In their molecular study of the physaloid genera, Whitson and Manos (2005) concluded that *P. alkekengi* and its few relatives are not particularly closely related to the remainder of the genus *Physalis*. However, they considered their study too preliminary to permit them to propose a revised generic classification. Additionally, the species-level taxonomy of the *P. alkekengi* complex requires further research and the species is here circumscribed in a relatively inclusive sense to include more or less glabrous plants with somewhat larger fruiting calyces that have sometimes been segregated as var. *franchetii* (Mast.) Makino (*P. franchetii* Mast.).

3. Physalis angulata L. (cutleaf ground cherry)
P. angulata var. *pendula* (Rydb.) Waterf.
P. pendula Rydb.

Pl. 563 a–c; Map 2625

Plants annual, taprooted. Stems 15–80(–150) cm long, erect or ascending, with several to many, ascending branches, glabrous or sparsely pubescent toward the tip with short, upward-appressed, unicellular and few-celled, nonglandular hairs 0.1–0.5 mm long. Leaves moderately to long-petiolate. Leaf blades 2–12 cm long, narrowly to broadly lanceolate, elliptic, oblong-elliptic, or ovate, tapered to a sharply pointed tip, rounded to angled or tapered at the base, the margins relatively sparsely toothed (with 2–9 teeth along each side), minutely nonglandular-hairy, the teeth mostly sharply pointed, irregularly shallow and broad, the surfaces green when fresh, drying uniformly green (lacking orangish tinging or patches), glabrous or sparsely to occasionally moderately pubescent (more densely on the undersurface) with minute, appressed, nonglandular, unicellular or few-celled hairs. Flower stalks 7–17(–22) mm long, becoming elongated to 15–30 mm at fruiting. Calyces 3–5 mm long at flowering, the lobes 1–3 mm long, the outer surface glabrous or sparsely pubescent with minute nonglandular hairs (these denser along the lobe margins) at flowering, glabrous or sparsely hairy along the main veins and lobes at fruiting, at fruiting becoming elongated to 20–40 mm long, shallowly 10-angled or 10-ribbed, rounded to very shallowly concave at the base, mostly remaining green, occasionally pale brown to tan with age. Corollas 6–10 mm long, uniformly light yellow to lemon yellow, the inner surface occasionally slightly purplish-tinged toward the base. Stamens with slender filaments half as wide as the anthers or narrower, the anthers 1–3 mm long, blue or bluish-tinged, arched but not coiled after dehiscence. Fruits 1.0–1.5 cm long, green or yellow to orangish yellow. $2n$=24, 48. May–September.

Scattered, mostly south of the Missouri River (southern U.S. north to California, Kansas, Illinois, and Virginia, north locally to Massachusetts; Mexico, Central America, South America, Caribbean Islands). Banks of streams and rivers, sloughs, margins of ponds, lakes, and sinkhole ponds, and moist depressions and disturbed portions of upland prairies; also ditches, levees, margins of crop fields, fallow fields, railroads, and moist, disturbed areas.

Steyermark (1963) followed Waterfall (1958) in recognizing two varieties of *P. angulata* in Missouri. Plants of western Missouri with narrower leaves, longer flower and fruit stalks, and slightly larger flowering and fruiting calyces having slightly longer teeth were segregated as var. *pendula*. J. R. Sullivan (2004) concluded that there was too much intergradation between such plants and those ascribed to the typical variety to allow recognition of infraspecific taxa in the species. She also discussed the difficulties in separating plants ascribed to var. *pendula* with those of *P. acutifolia*, noting that vegetative material could not be determined to species with confidence. Aside from the differences in the key above, the following characters are useful in separating *P. acutifolia* from *P. angulata*: a widely flaring, very flat corolla (vs. broadly bell-shaped to more or less saucer-shaped); fruit stalks averaging somewhat longer (25–60 vs. 15–40 mm); and a slightly smaller fruiting calyx that often is nearly filled by the enclosed berry.

4. Physalis cinerascens (Dunal) Hitchc. var. **cinerascens**

P. viscosa L. ssp. *mollis* (Nutt.) Waterf. var. *cinerascens* (Dunal) Waterf.

Pl. 565 a–c; Map 2626

Plants perennial, with deep-set, long-creeping rhizomes. Stems (5–)15–50 cm long, erect or ascending to more commonly ascending from a spreading base, with few to several, spreading to ascending branches, sparsely to densely pubescent with minute, nonglandular, multicellular hairs, these 0.1–0.3(–0.6) mm long, occasionally also with widely scattered, spreading, longer, unbranched hairs 0.5–1.0 mm long. Leaves short- to long-petiolate. Leaf blades 1.5–8.0 cm long, ovate to broadly ovate or nearly circular, angled to a usually bluntly pointed tip, broadly angled or broadly rounded to truncate or shallowly cordate at the base, the margins entire or wavy to sparsely, bluntly and irregularly few- to moderately toothed, minutely branched-hairy, the surfaces grayish green to dark green when fresh, drying uniformly green (lacking orangish tinging or patches), moderately to densely pubescent with nonglandular, multicellular, branched hairs, these 0.1–0.4 mm long, with 3 or more branches, sometimes appearing somewhat woolly or matted. Flower stalks 10–32 mm long, becoming elongated to 15–60 mm at fruiting. Calyces (4–)5–9 mm long at flowering, the lobes 1.5–4.0 mm long, the outer surface densely pubescent at flowering with minute, nonglandular, multicellular hairs, these relatively fine, matted or woolly, 0.1–0.3 mm long having 3 or more branches, sparsely to moderately hairy at fruiting (more densely so on the lobes), at fruiting, becoming elongated to (15–)20–35 mm long, shallowly 10-angled or 10-ribbed, concave at the base, mostly remaining green, occasionally pale brown to tan with age. Corollas 8–16 mm long, pale yellow to lemon yellow or yellow, the inner surface with 5 large, prominent dark purplish brown to dark reddish brown spots toward the base (these sometimes merged into a ring or appearing smudged, occasionally disrupted by yellowish venation). Stamens with slender filaments half as wide as the anthers or less, the anthers 2–4 mm long, yellow or rarely purplish-tinged, arched but not coiled after dehiscence. Fruits 0.9–1.5 cm long, green, yellow, or orangish yellow. May–October.

Introduced, uncommon, known thus far only from historical collections from Lawrence County and St. Louis City (New Mexico to Kansas and Louisiana; introduced sporadically farther east and in California). Railroads.

The taxonomy adopted here follows that of J. R. Sullivan (1985). The other variety, var.

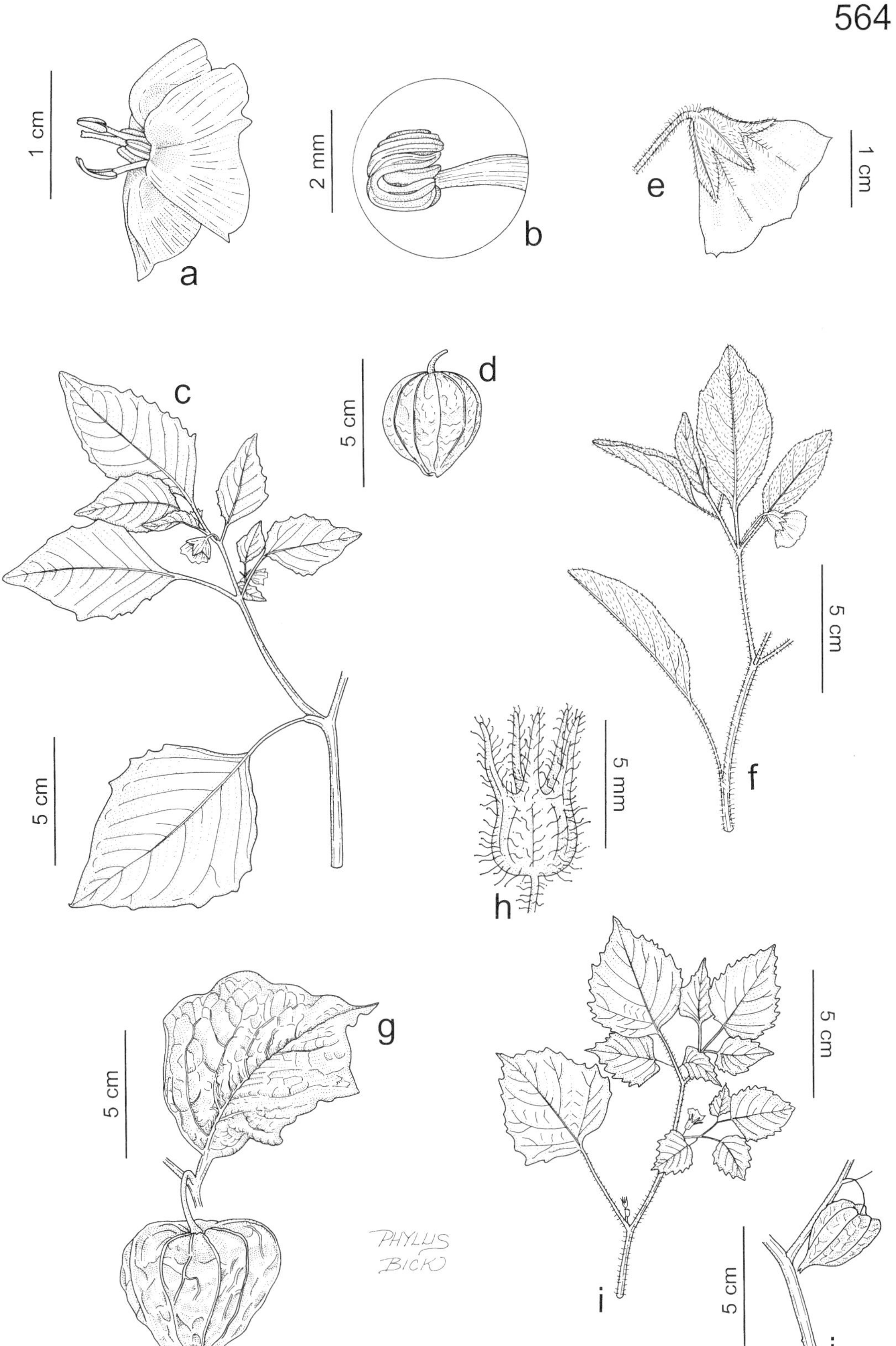

Plate 564. Solanaceae. *Physalis philadelphica*, **a)** flower, **b)** stamens, **c)** portion of branch with leaves and flowers, **d)** fruit. *Physalis pumila*, **e)** flower, **f)** portion of branch with leaves and flower. *Physalis alkekengi*, **g)** node with leaf and fruit. *Physalis cordata*, **h)** calyx, **i)** portion of branch with leaves and flowers, **j)** node with fruit.

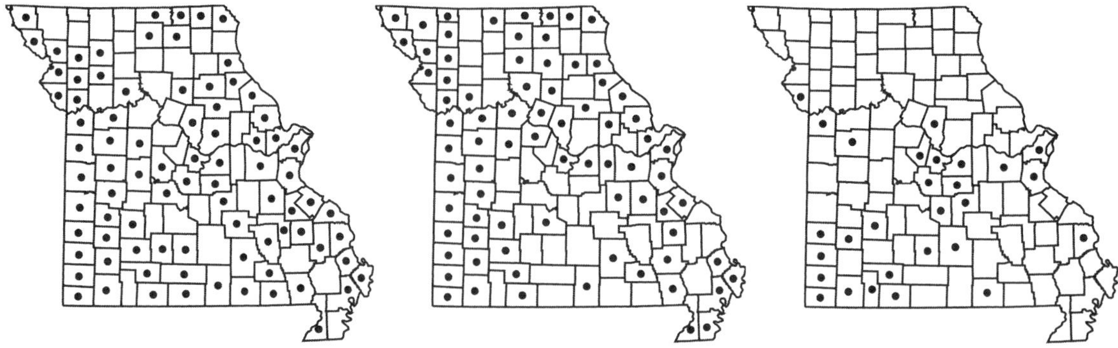

2629. Physalis heterophylla 2630. Physalis longifolia 2631. Physalis missouriensis

spathulifolia (Torr.) J.R. Sullivan, is endemic to portions of Texas, Louisiana, and adjacent northeastern Mexico. It differs from var. *cinerascens* in its spatulate leaf blades and strongly spreading to somewhat reflexed corollas at full flowering.

5. Physalis cordata Mill. (ground cherry)
Pl. 564 h–j; Map 2627

Plants annual, more or less taprooted. Stems 15–50 cm long, erect or ascending, with several to many ascending branches, glabrous or sparsely pubescent toward the tip with minute, upward-appressed, nonglandular hairs 0.1–0.3 mm long. Leaves mostly long-petiolate. Leaf blades 1.5–8.0 cm long, ovate to broadly ovate or occasionally nearly circular, angled or short-tapered to a sharply pointed tip, rounded to truncate or shallowly cordate at the base, occasionally angled or broadly short-tapered, the margins moderately toothed (the larger ones with 8 or more teeth per side), minutely hairy, the teeth bluntly or sharply pointed, shallow and broad to deeper and narrower, the surfaces green to dark green when fresh, not drying orangish-tinged, sparsely pubescent with minute, nonglandular hairs only along the veins. Flower stalks (4.5–)6–11 mm long, becoming elongated to (10–)15–35 mm at fruiting. Calyces 3.5–6.5 mm long at flowering, the lobes 2.0–4.5 mm long, the outer surface glabrous or sparsely and minutely appressed-hairy at flowering, glabrous (except sometimes for the lobes) at fruiting, at fruiting becoming elongated to (25–)30–40 mm long, sharply 5-angled, concave at the base, green or pale brown to tan. Corollas 6.5–9.5 mm long, pale yellow to lemon yellow, the inner surface with 5 large, prominent dark purplish brown to dark reddish brown spots toward the base (these sometimes merged into a ring or appearing smudged). Stamens with broad filaments about as wide as the anthers, the anthers 1.5–2.5 mm long, blue or bluish-tinged, arched but not coiled after dehiscence. Fruits 1.2–1.5 cm long, green, yellow, or orangish yellow. July–October.

Uncommon, widely scattered in the southern half of the state (southeastern U.S. west to Missouri and Texas; Mexico, Central America, South America, Caribbean Islands, Pacific Islands; introduced in Asia). Banks of streams and rivers; also pastures, quarries, railroads, and open, often sandy, disturbed areas.

Steyermark (1963) did not mention *P. cordata*, but specimens of this species were included in his concept of *P. pubescens* var. *glabra*. In her taxonomic study of the *P. pubescens* complex, Martínez (1998) discussed the distinctness of *P. cordata* and its distribution, including Missouri.

6. Physalis grisea (Waterf.) M. Martínez
P. pruinosa L., misapplied
P. pubescens var. *grisea* Waterf.
Map 2628

Plants annual, more or less taprooted. Stems 20–50 cm long, erect or ascending, with several to many loosely ascending to spreading branches, sparsely to more commonly densely pubescent (at least toward the tip) with minute to longer, spreading, multicellular, gland-tipped and/or nonglandular hairs 0.1–1.0 mm long (sessile glands often also be present). Leaves moderately to long-petiolate. Leaf blades 2–12 cm long, ovate to broadly ovate, angled or short-tapered to a bluntly or sharply pointed tip, broadly rounded to shallowly cordate at the often oblique base, the margins coarsely, unevenly, and bluntly toothed (the larger ones with 8 or more teeth per side), minutely hairy, the teeth bluntly or sharply pointed, shallow and broad to deeper and narrower, the surfaces grayish green when fresh, drying orange, orangish-tinged, or with orangish patches, moderate to densely pubescent (more densely on the undersurface, especially along the veins) with short, nonglandular and gland-tipped hairs (sessile

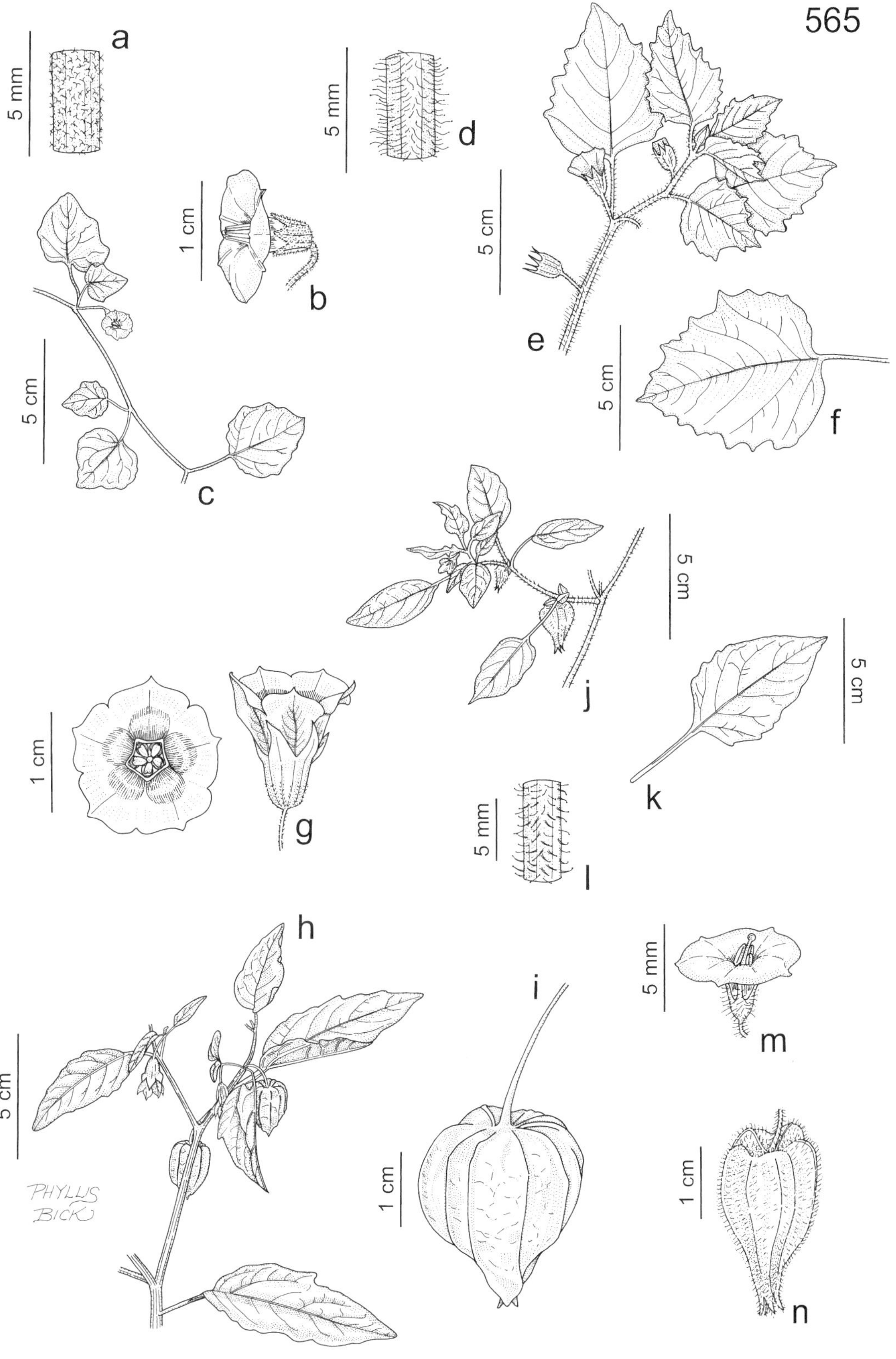

Plate 565. Solanaceae. *Physalis cinerascens*, **a)** stem detail, **b)** flower, **c)** portion of branch with leaves and flower. *Physalis heterophylla*, **d)** stem detail, **e)** portion of fertile branch, **f)** median leaf. *Physalis longifolia* var. *longifolia*, **g)** two views of flower, **h)** portion of fertile branch, **i)** fruit. *Physalis pubescens*, **j)** portion of fertile branch, **k)** median leaf, **l)** stem detail, **m)** flower, **n)** fruit.

glands usually also present). Flower stalks 4–6 mm long, becoming slightly elongated to 5–12 mm at fruiting. Calyces 3–5 mm long at flowering, the lobes 1.5–2.5 mm long, the outer surface with relatively dense, short, multicellular, nonglandular hairs at flowering, persistently hairy at fruiting, at fruiting becoming elongated to 20–35 mm long, sharply 5-angled, concave at the base, green or pale brown to tan. Corollas 5–8 mm long, pale yellow to lemon yellow, the inner surface with 5 large, prominent dark purplish brown to dark reddish brown spots toward the base (these sometimes merged into a ring or appearing smudged). Stamens with slender filaments half as wide as the anthers or narrower, the anthers 1–2 mm long, blue or bluish-tinged, arched but not coiled after dehiscence. Fruits 1.0–1.5 cm long, green, yellow, or orange. $2n=24$. May–November.

Uncommon, widely scattered, mostly in the southern half of the state (eastern U.S. west to Nebraska and Louisiana, also Washington to California; Canada). Banks of streams and rivers, upland prairies, and glades; also farmyards, pastures, railroads, roadsides, and open, disturbed areas.

7. Physalis heterophylla Nees (clammy ground cherry)

Pl. 565 d–f; Map 2629

Plants perennial, with deep-set, long-creeping rhizomes. Stems 15–100 cm long, erect or ascending, unbranched or with few to many, loosely ascending to spreading branches, densely pubescent with varying proportions of shorter, gland-tipped and longer, nonglandular hairs, variously 0.1–2.0 (–3.0) mm long, spreading, multicellular. Leaves moderately to long-petiolate. Leaf blades 3–12 (–15) cm long, broadly ovate or nearly circular, angled or short-tapered to a sharply pointed tip, broadly rounded to more or less truncate or shallowly cordate at the base, the margins entire or sparsely to moderately and irregularly toothed, glandular- and nonglandular-hairy, the teeth bluntly to sharply pointed (occasionally rounded), coarse, relatively shallow, and broad, the surfaces green to dark green when fresh, drying uniformly green (lacking orangish tinging or patches), moderately to densely pubescent with short, gland-tipped, and longer, nonglandular, multicellular hairs. Flower stalks 9–15 mm long, becoming elongated to 20–30 mm at fruiting. Calyces 6–12 mm long at flowering, the lobes 3–6 mm long, the outer surface with dense, multicellular, long, nonglandular hairs and often also short, gland-tipped hairs at flowering, persistently hairy at fruiting, at fruiting, becoming elongated to 25–40 mm long, shallowly 10-angled or 10-ribbed, shallowly concave at the base, mostly remaining green, occasionally pale brown to tan with age. Corollas 10–17 mm long, pale yellow to lemon yellow or yellow, the inner surface with 5 prominent purplish brown to bluish purple spots toward the base (these often merged into a ring or appearing smudged). Stamens with broad, club-shaped filaments about as wide as the anthers, the anthers 2.5–4.5 mm long, yellow, rarely bluish-tinged or each anther sac with a bluish longitudinal line along the zone of dehiscence, arched but not coiled after dehiscence. Fruits 0.8–1.2 cm long, green or yellow. $2n=24$. May–September.

Scattered nearly throughout the state (nearly throughout the U.S.; Canada). Banks of streams and rivers, upland prairies, loess hill prairies, sand prairies, glades, bottomland forests, openings of mesic upland forests, and edges of marshes and fens; also old fields, pastures, strip mines, quarries, railroads, roadsides, and open, disturbed areas.

Waterfall (1958) separated this species into three varieties, based on stem stoutness and density of longer hairs, with the Missouri plants corresponding to var. *heterophylla* (with stems not strongly thickened toward the base and less frequent longer hairs). J. R. Sullivan (2004) concluded that there was too much overlap in these characters to permit recognition of infraspecific taxa.

8. Physalis longifolia Nutt. (common ground cherry)

Pl. 565 g–i; Map 2630

Plants perennial, with deep-set, long-creeping rhizomes. Stems 20–60(–80) cm long, erect or ascending, unbranched or with few to several, ascending to loosely ascending branches, often purplish-tinged, glabrous or sparsely pubescent with short, upward-appressed, unicellular and few-celled, nonglandular hairs 0.1–0.5 mm long. Leaves short- to long-petiolate. Leaf blades 2.5–10.0(–13.0) cm long, lanceolate to ovate or elliptic-ovate, angled or tapered to a sharply pointed tip, broadly angled to rounded or more or less truncate at the base, the margins entire or sparsely and irregularly toothed, minutely nonglandular-hairy, the teeth mostly bluntly pointed, relatively shallow and broad, the surfaces green to dark green when fresh, drying uniformly green (lacking orangish tinging or patches), glabrous or sparsely pubescent with minute, appressed, nonglandular, mostly unicellular hairs. Flower stalks 5–20 mm long, becoming elongated to 15–35 mm at fruiting. Calyces (5–)7–12(–15) mm long at flowering, the lobes 3–6 mm long, the outer surface glabrous or sparsely pubescent with minute, appressed-ascending hairs at flowering, glabrous or very sparsely hairy at fruiting, at fruiting, becoming elongated to 20–40 mm long, shallowly 10-angled or 10-ribbed, rounded to very shallowly concave at the base, mostly remaining green, occasionally pale brown to tan with age. Corollas 10–20 mm long, pale yellow to lemon yellow or yellow, the inner surface with 5 prominent purplish brown to bluish purple spots toward the base (these often merged into a ring or appearing smudged). Stamens with broad filaments about as wide as (or occasionally wider than) the anthers, the anthers 2–4 mm long, yellow, occasionally bluish-tinged or each anther sac with a bluish longitudinal line along the zone of dehiscence, arched but not coiled after dehiscence. Fruits 0.8–1.0 (–1.5) cm long, green or yellow. $2n=24$, 48. May–September.

Scattered nearly throughout the state (nearly throughout the U.S.; Canada, Mexico). Bottomland forests, mesic upland forests, banks of streams, rivers, and spring branches, bottomland prairies, swales in upland prairies, saline marshes, fens, margins of ponds and lakes, swamps, sloughs, savannas, bases and ledges of bluffs, and rarely glades; also ditches, pastures, fallow fields, crop fields, railroads, roadsides, and open to shaded, disturbed areas.

The two varieties accepted by most botanists exhibit considerable morphological overlap in Missouri.

1. Leaf blades lanceolate to narrowly elliptic . . . . 8A. VAR. LONGIFOLIA
1. Leaf blades ovate to elliptic-ovate . . . . 8B. VAR. SUBGLABRATA

8a. var. longifolia

Pl. 565 g–i

Leaf blades 2.5–8.0 cm long, lanceolate to narrowly elliptic, often angled at the base. Flower stalks 5–15 mm long. Corollas 10–15 mm long. $2n=24$, 48. May–September.

Scattered nearly throughout the state (nearly throughout the U.S.; Canada, Mexico). Bottomland forests, mesic upland forests, banks of streams, rivers, and spring branches, bottomland prairies, swales in upland prairies, saline marshes, fens, margins of ponds and lakes, swamps, sloughs, savannas, bases and ledges of bluffs, and rarely glades; also ditches, pastures, fallow fields, crop fields, railroads, roadsides, and open to shaded, disturbed areas.

Although the ranges of the two varieties are similar, var. *longifolia* tends to tolerate slightly more sunny, dry conditions than does var. *subglabrata*.

8b. var. subglabrata (Mack. & Bush) Cronquist

P. subglabrata Mack. & Bush

P. virginiana Mill. var. *subglabrata* (Mack. & Bush) Waterf.

Leaf blades 3–10(–13) cm long, ovate to elliptic-ovate, often rounded or more or less truncate at the base. Flower stalks 8–20 mm long. Corollas 12–20 mm long. $2n=24$. May–September.

Scattered in the northern and western halves of the state, uncommon farther south and east (nearly throughout the U.S.; Canada). Bottomland forests, mesic upland forests, banks of streams, rivers, and spring branches, bottomland prairies, swales in upland prairies, saline marshes, fens, margins of ponds and lakes, swamps, sloughs, savannas, bases and ledges of bluffs, and rarely glades; also ditches, pastures, fallow fields, crop fields, railroads, roadsides, and open to shaded, disturbed areas.

Plants of var. *subglabrata* with relatively large fruiting calyces have been called f. *macrophysa* (Rydb.) Steyerm. (*P. macrophysa* Rydb.). This is the more abundant of the two varieties in Missouri and the one more likely to be encountered in bottomland situations.

9. Physalis missouriensis Mack. & Bush

P. pubescens var. *missouriensis* (Mack. & Bush) Waterf.

Map 2631

Plants annual, more or less taprooted. Stems 10–55 cm long, erect or ascending, with several to many, loosely ascending to spreading branches, moderately to densely pubescent with short, spreading, multicellular, gland-tipped and sometimes also nonglandular hairs 0.1–0.5(–0.8) mm long. Leaves mostly long-petiolate. Leaf blades 2–10 cm long, broadly ovate or nearly circular, angled or short-tapered to a usually bluntly pointed tip, broadly rounded to more or less truncate or occasionally shallowly cordate at the base, the margins relatively sparsely toothed (with 1–5[–7] teeth along each side), glandular-hairy, the teeth bluntly pointed, shallow and broad, the surfaces green to dark green when fresh, drying uniformly green (lacking orangish tinging or patches), sparsely to moderately pubescent (more densely on the undersurface) along the veins with short, gland-tipped, multicellular hairs. Flower stalks 4–7 mm long, becoming slightly elongated to 5–10 mm at fruiting. Calyces 2.5–4.0 mm long at flowering, the lobes 1–2 mm long, the outer surface with dense, short, multicellular, gland-tipped hairs at flowering, persistently hairy at fruiting, at fruiting becoming elongated to 13–25 mm long, shallowly 10-

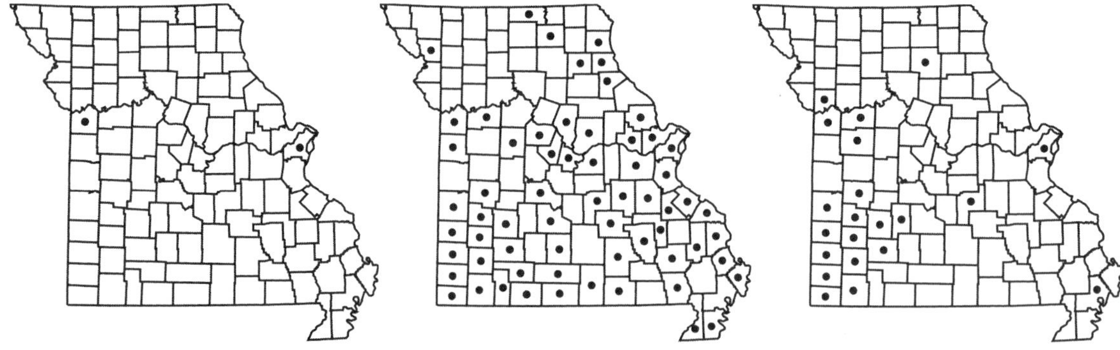

2632. Physalis philadelphica 2633. Physalis pubescens 2634. Physalis pumila

angled or 10-ribbed (sometimes appearing merely shallowly 5-angled), rounded to very shallowly concave at the base, mostly remaining green, occasionally pale brown to tan with age. Corollas 5–7 mm long, uniformly pale yellow to lemon yellow, the inner surface occasionally slightly purplish-tinged toward the base. Stamens with slender filaments half as wide as the anthers or narrower, the anthers 1.0–1.5 mm long, yellow, arched but not coiled after dehiscence. Fruits 1.0–1.5 cm long, green or yellow to orangish yellow. June–October.

Scattered, mostly south of the Missouri River (Arkansas, Oklahoma, Kansas, Nebraska, and Missouri). Banks of streams and rivers, ledges and tops of bluffs, glades, and openings of mesic to dry upland forests; also railroads, roadsides, and open, disturbed areas.

10. Physalis philadelphica Lam. (strawberry tomato, tomatillo)

P. ixocarpa Brot. ex Hornem., misapplied

Pl. 564 a–d; Map 2632

Plants annual, more or less taprooted. Stems 15–100 cm long, erect or ascending, with several to many, ascending to spreading branches, glabrous or sparsely pubescent toward the tip with short, mostly downward-angled, unicellular and few-celled, nonglandular hairs 0.1–0.3 mm long. Leaves moderately to long-petiolate. Leaf blades 2–7 cm long, broadly lanceolate to ovate, tapered to a sharply pointed tip, rounded to angled or tapered at the base, the margins entire or with several, widely spaced, shallow, blunt teeth per side, minutely nonglandular-hairy, the surfaces green when fresh, drying uniformly green (lacking orangish tinging or patches), glabrous or sparsely pubescent with minute, appressed, nonglandular, mostly unicellular hairs. Flower stalks 3–6 mm long, becoming slightly elongated to 3–8 mm at fruiting. Calyces 5–7(–10) mm long at flowering, the lobes 2–4 mm long, the outer surface sparsely to moderately pubescent with minute nonglandular hairs (these denser along the lobe margins) at flowering, glabrous except for the lobes at fruiting, at fruiting becoming elongated to 25–35 mm long, shallowly 10-angled or 10-ribbed, rounded to shallowly concave at the base, mostly remaining green, occasionally brownish- or purplish-tinged with age. Corollas 7–15 mm long, light yellow to lemon yellow, the inner surface with 5 prominent purplish brown to bluish purple spots toward the base (these often merged into a ring or appearing smudged). Stamens with slender filaments half as wide as the anthers or narrower, the anthers 1–3 mm long, blue, becoming strongly coiled after dehiscence. Fruits 2.0–2.8 cm long, sticky, pale green, often purplish-tinged or with purple streaks. $2n=24$. July–September.

Introduced, uncommon, known thus far only from Jackson County and the city of St. Louis (native of Mexico, introduced sporadically in the U.S.; Canada). Railroads and open, disturbed areas.

Steyermark (1963) and most earlier authors called this species *P. ixocarpa*. In her study of the ground cherries naturalized in Portugal, Fernandes (1970) studied the typification and application of these names and concluded that they represent two distinct species, with the cultivated tomatillo associated with the epithet *P. philadelphica*.

Tomatillo has been cultivated since pre-Columbian times (W. D. Hudson, 1986). It is an important ingredient in Mexican cuisine, both in salsas and in prepared dishes. It is sufficiently popular that large-fruited cultivars are available in most supermarkets.

11. Physalis pubescens L. (downy ground cherry, annual ground cherry)

P. pubescens var. *glabra* (Michx.) Waterf.

P. pubescens var. *integrifolia* (Dunal) Waterf.

2635. Physalis virginiana

2636. Solanum americanum

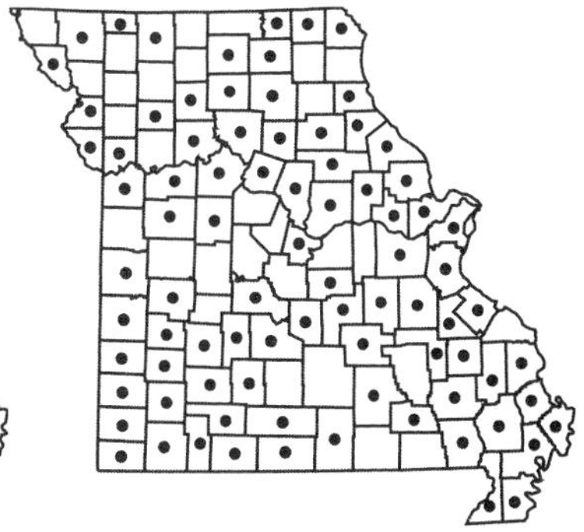

2637. Solanum carolinense

P. barbadensis Jacq.
P. barbadensis var. *glabra* (Michx.) Fernald
Pl. 565 j–n; Map 2633

Plants annual, more or less taprooted. Stems (5–)10–80 cm long, erect or ascending, unbranched or with relatively few, loosely ascending to spreading branches, glabrous or sparsely to moderately pubescent (at least toward the tip) with short, spreading, multicellular, mostly gland-tipped hairs 0.1–0.5 mm long, often also with moderate to dense, longer (1–3 mm), spreading, nonglandular hairs. Leaves short- to long-petiolate. Leaf blades 2–8 cm long, ovate to broadly ovate or nearly circular, angled or short-tapered to a sharply pointed tip, broadly rounded to more or less truncate at the base, the margins entire or relatively sparsely toothed (with 1–3[–7] main teeth along each side), minutely hairy, the teeth bluntly or sharply pointed, shallow and broad, the surfaces green when fresh, drying uniformly green (lacking orangish tinging or patches), sparsely to moderately pubescent (more densely on the undersurface) with short, mostly gland-tipped, multicellular hairs. Flower stalks 3.5–9.0 mm long, becoming elongated to 6–15 mm at fruiting. Calyces 3–6 mm long at flowering, the lobes 1.0–3.5 mm long, the outer surface with sparse to dense, short, multicellular, nonglandular hairs at flowering, persistently hairy at fruiting, at fruiting becoming elongated to 20–30 mm long, sharply 5-angled, concave at the base, green or pale brown to tan. Corollas 6–11 mm long, pale yellow to lemon yellow, the inner surface with 5 large, prominent dark purplish brown to dark reddish brown spots toward the base (these sometimes merged into a ring or appearing smudged). Stamens with broad filaments about as wide as the anthers, the anthers 1–2 mm long, blue or bluish-tinged, arched but not coiled after dehiscence. Fruits 1.0–1.2 cm long, green or yellow, occasionally purplish-tinged. $2n=24$. May–November.

Scattered nearly throughout the state, but uncommon in or absent from many counties in the northwestern quarter (nearly throughout the U.S. [except for some of the northern Plains and Rocky Mountains states]; Canada, Mexico, Central America, South America, Caribbean Islands, Pacific Islands).

Steyermark (1963) followed Waterfall (1958) in dividing this species into a number of varieties. Subsequent taxonomic studies have resulted in the segregation of *P. cordata* and *P. grisea* as distinct species, and the recognition of the remaining plants as a morphologically variable *P. pubescens* without the separation of further infraspecific taxa (Martínez, 1998; J. R. Sullivan, 1984, 2004).

12. Physalis pumila Nutt. (prairie ground cherry)

Pl. 564 e, f; Map 2634

Plants perennial, with deep-set, long-creeping rhizomes. Stems 15–45 cm long, erect or ascending, unbranched or with few, ascending branches toward the tip, moderately to densely pubescent with nonglandular, multicellular hairs, variously of minute loosely appressed, unbranched hairs 0.1–0.5 mm long and scattered to abundant, spreading, longer hairs 0.5–2.0 mm long, these either all or nearly all unbranched or many to most 2- or 3-branched. Leaves short- to moderately petiolate. Leaf blades 2–8(–10) cm long, lanceolate to elliptic or ovate, angled to a sharply pointed tip, tapered at the base, the margins entire or wavy to sparsely, shallowly and bluntly few-toothed, hairy, the surfaces grayish green to dark green when fresh, drying uniformly green (lacking orangish tinging or patches), moderately to densely pubescent with nonglandular, multicellular hairs, variously of minute loosely appressed, unbranched hairs 0.1–0.5 mm long and scattered to abundant, spreading, longer hairs 0.5–2.0 mm long, these either all or nearly all unbranched or many to most

2- or 3-branched. Flower stalks (7–)14–30 mm long, becoming elongated to 25–55 mm at fruiting. Calyces 6–12 mm long at flowering, the lobes 2.5–5.0 mm long, the outer surface densely pubescent at flowering with nonglandular, multicellular hairs, variously of minute loosely appressed, unbranched hairs 0.1–0.5 mm long and scattered to abundant, spreading, longer hairs 0.5–2.0 mm long, these either all or nearly all unbranched or many to most 2- or 3-branched, sparsely to moderately hairy at fruiting (more densely so on the lobes), at fruiting, becoming elongated to 25–40 mm long, shallowly 10-angled or 10-ribbed, concave at the base, mostly remaining green, occasionally pale brown to tan with age. Corollas 11–17 mm long, pale yellow to lemon yellow, the inner surface greenish- or pale brownish-tinged toward the base, lacking distinct spots or discolorations. Stamens with broad filaments about as wide as the anthers, the anthers 2–3 mm long, yellow, arched but not coiled after dehiscence. Fruits 1.0–1.5 cm long, green or yellow. $2n=24$. May–September.

Scattered in the southwestern quarter of the state, uncommon farther north and east (Illinois to Louisiana west to Wyoming and New Mexico; introduced sporadically farther east and west). Glades and upland prairies; also strip mines, railroads and open, disturbed areas.

The present treatment follows that of W. F. Hinton (1976) in recognizing two subspecies within *P. pumila*.

1. Stems and leaf blades pubescent with the hairs all or nearly all unbranched . . . . 1A. SSP. HISPIDA
1. Stems and leaf blades pubescent with many or most of the hairs branched . . . . SSP. PUMILA

12a. ssp. hispida (Waterf.) W.F. Hinton

P. longifolia Nutt. var. *hispida* (Waterf.) Steyerm.

P. hispida (Waterf.) Cronquist

P. virginiana Mill. var. *hispida* Waterf.

Stems moderately to densely pubescent with the hairs all or nearly all unbranched. Leaf blades 2–7 cm long, pubescent on the undersurface with the hairs all or nearly all unbranched. $2n=24$. May–September.

Uncommon in western Missouri (South Dakota to Missouri and Texas west to Wyoming and Colorado; introduced sporadically farther east and west). Railroads and open, disturbed areas.

The taxonomic status of this entity requires further study. Steyermark (1963) included it as a variety under *P. longifolia*, but noted that the plants were otherwise morphologically similar to *P. pumila*. W. F. Hinton (1976) studied morphology and crossing relationships in the group, and concluded that it was best treated as a subspecies of *P. pumila*, separable from ssp. *pumila* mainly in the proportion of branched vs. unbranched hairs. Cronquist (1991) elevated the taxon to species status, based on his conclusion that it was distinct enough morphologically and occupied a fairly discrete range and habitat. However, in their study of pubescence types in the genus, Seithe and Sullivan (1990) found that there is a developmental series from unbranched to branched trichomes and discussed the difficulties of applying hair-type characters to the taxonomic classification within the genus. Because ssp. *hispida* and ssp. *pumila* produce fertile offspring following artificial hybridization and because there is some limited overlap in the density of hair types in specimens of the complex, the two taxa are here retained as subspecies, as proposed by W. F. Hinton (1976).

The status of this taxon in Missouri is unclear. It is known only from historical collections, most of which indicate that the collector thought the plants to be introduced. The specimen cited by Steyermark (1963, as *P. longifolia* var. *hispida*) from Johnson County could not be located during the present study. Because at least a couple of the Jackson County specimens do not include the word introduced on their labels and the ambiguity of the putative Johnson County specimen, as well as the fact that the Missouri distribution is adjacent to occurrences in Kansas, the taxon remains tentatively accepted as a native element in the flora.

12b. ssp. pumila

Stems densely pubescent with some or most of the hairs 2- or 3-branched. Leaf blades 3–8(–10) cm long, pubescent with many or most of the hairs 2- or 3-branched. $2n=24$. May–August.

Scattered in the southwestern quarter of the state, uncommon farther north and east (Nebraska to Illinois south to Texas and Louisiana). Glades and upland prairies; also strip mines, railroads and open, disturbed areas.

13. Physalis virginiana Mill. (Virginia ground cherry)

Pl. 562 i–k; Map 2635

Plants perennial, with deep-set, long-creeping rhizomes. Stems 15–60 cm long, erect or ascending, with few to several, ascending branches, moderately pubescent with short, downward-angled, unicellular and few-celled, nonglandular hairs 0.1–0.5 mm long, usually also with scattered to commonly, longer, spreading, multicellular, nonglandular hairs 0.5–1.0 mm long. Leaves short- to moderately petiolate. Leaf blades 2–7 cm long, narrowly to broadly lanceolate, elliptic or ovate, angled to a sharply pointed tip, angled to tapered at the base, the margins entire, wavy, or sparsely to moderately toothed, minutely nonglandular-hairy, the teeth bluntly pointed to rounded, relatively shallow and broad to coarse, the surfaces green to dark green when fresh, drying uniformly green (lacking orangish tinging or patches), moderately pubescent with a mixture of shorter, more or less appressed, and longer, spreading, nonglandular hairs. Flower stalks (6–)9–20(–27) mm long, becoming elongated to 15–30 mm at fruiting. Calyces 6–12(–14) mm long at flowering, the lobes 3–6 mm long, the outer surface moderately to densely pubescent (especially along the nerves) with a mixture of longer, spreading hairs and shorter downward-angled or -curved hairs at flowering, sparsely to moderately hairy at fruiting, at fruiting, becoming elongated to 20–40 mm long, shallowly 10-angled or 10-ribbed, concave at the base, mostly remaining green, occasionally pale brown to tan with age. Corollas 10–17(–20) mm long, pale yellow to lemon yellow or yellow, the inner surface with 5 prominent purplish brown to bluish purple spots toward the base (these often merged into a ring or appearing smudged, occasionally only the venation strongly darkened). Stamens with broad filaments about as wide as (or occasionally wider than) the anthers, the anthers 2–3 mm long, yellow, occasionally bluish-tinged or each anther sac with a bluish longitudinal line along the zone of dehiscence, arched but not coiled after dehiscence. Fruits 1.0–1.5(–1.8) cm long, green or yellow to orangish yellow. $2n=24$. April–October.

Scattered nearly throughout the state, but absent or uncommon in many northwestern counties (eastern U.S. west to North Dakota and New Mexico; Canada, Mexico). Mesic to dry upland forests, upland prairies, loess hill prairies, savannas, glades, ledges and tops of bluffs, and occasionally banks of streams and rivers; also pastures, old fields, railroads, roadsides, and open, disturbed areas.

7. **Solanum** L. (nightshade)

Plants in our species annual or perennial herbs (the stems climbing, twining, and often woody toward the base in *S. dulcamara*), sometimes with rhizomes or tubers, sometimes armed with prickles. Stems variously erect to loosely ascending or spreading (climbing in *S. dulcamara*), unbranched or more commonly branched, glabrous or hairy. Leaf blades simple or less commonly 1 or 2 times pinnately deeply lobed to compound, glabrous or variously hairy. Inflorescences mostly axillary or lateral (positioned opposite the leaves), clusters, racemes, or small panicles, rarely of solitary flowers. Flowers variously ascending, spreading, or nodding. Calyces fused to the midpoint or less, bell-shaped to saucer-shaped, rounded to truncate at the base, lacking basal auricles, persistent intact at fruiting (the tips of the lobes sometimes breaking off with age), not or only slightly enlarged, or, if enlarged and closely surrounding all or part of the fruit, then not balloonlike or papery, not angled. Corollas saucer-shaped to broadly bell-shaped, sometimes with the lobes more or less reflexed, usually appearing pleated in bud, white to purple or blue, less commonly cream-colored or yellow. Stamens with short filaments, the anthers fused or merely positioned into a ring, dehiscent longitudinally (in *S. lycopersicum*) or by terminal pores (these sometimes narrow and slitted), light yellow to orangish yellow, occasionally reddish- or purplish-tinged. Ovary 2(–4)-locular, the style often protruding through the center of the anther ring. Fruits berries (burlike in *S. rostratum*), more or less juicy, variously shaped, 2(–4)-locular, variously green, yellow, orange, or red, occasionally with lighter mottling, with several to numerous seeds, sometimes with irregularly globose groups of hard granules (composed of stony cells) among the seeds. Seeds various, lacking wings. About 1,500 species, nearly worldwide.

Solanum is the largest genus in the family and is particularly diverse in neotropical regions, where many of the species are shrubs or small trees. It is economically very important, including such food plants as tomatoes, eggplants, and potatoes. Some of the species in the United States are important weeds of crop fields and rangelands. As with most other members of the Solanaceae, plants of *Solanum* species (other than the edible portions of particular species) contain alkaloids that are poisonous to humans and livestock. The species with prickly stems and leaves also can sometimes cause injuries to the mouthparts and feet of livestock. For a wealth of information on the genus, users of the internet are directed to the excellent *Solanaceae Source* website (http://www.nhm.ac.uk/research-curation/research/projects/solanaceaesource/).

The unusual morphology of the stamens is an adaptation to buzz pollination (also known as sonication) by certain bees (Buchmann, 1983). This adaptation has evolved independantly in some members of the Solanaceae and in several other plant families, for example, in the shooting star group of *Primula* (Primulaceae) and in some members of the Ericaceae. A rapid contraction of the indirect flight muscles of the bee causes the anther to vibrate (sonicate), which causes the pollen to be expelled through the apical openings where it can be collected easily by the bee.

1. Plants armed with prickles on the stems and/or leaves, also pubescent with stellate hairs
    2. Leaf blades 1 or 2 times deeply lobed or parted; calyces armed with dense prickles, totally enclosing the mature fruits; corollas bright yellow ..... 10. S. ROSTRATUM
    2. Leaf blades entire to shallowly lobed; calyces unarmed or with only scattered prickles toward the base, spreading or loosely cupped around only the basal portion of the fruits; corollas white to purple or bluish purple
        3. Stems and leaves densely pubescent, the green tissue totally obscured by the densely matted stellate hairs, appearing silvery-white ..... 5. S. ELAEAGNIFOLIUM
        3. Stems and leaves moderately to densely pubescent, but the green tissue not obscured by the stellate hairs
            4. Stems with the larger hairs appearing spreading (stellate hairs with the central branch erect); leaf blades with the stellate hairs sessile, the central branch vertically spreading, the other branches appressed; calyces 5–7 mm long; inflorescences simple racemes (these short and often appearing as a cluster at flowering) ..... 2. S. CAROLINENSE
            4. Stems with the hairs appearing appressed and matted (stellate hairs with all of the branches horizontal); leaf blades with the hairs short-stalked (visible only with magnification), all of the branches appressed; calyces 8–13 mm long; inflorescences often paniculate, consisting of 2 or 3 racemes (usually short and appearing as compound clusters at flowering) branching from the tip of a common stalk ..... 3. S. DIMIDIATUM
1. Plants unarmed, glabrous or mostly pubescent with unbranched hairs (with sparse, apically 2-branched or pinnately branched, stalked hairs in *S. pseudocapsicum*)
    5. Stems climbing, twining, and often woody toward the base; at least some of the leaf blades with 1 or 2(–4) shallow to deep lobes at the base ..... 4. S. DULCAMARA
    5. Stems not climbing or twining, herbaceous above the rootstock; leaves either entire or shallowly to deeply pinnately lobed to compound (the lobes or divisions not restricted to the base)
        6. Leaf blades 1 or 2 times deeply lobed to compound (some of the leaves only moderately lobed in *S. triflorum*)
            7. Corollas yellow; anthers tapered to a sterile appendage, dehiscing longitudinally ..... 6. S. LYCOPERSICUM

7. Corollas white to purple; anthers lacking a sterile apical portion, dehiscing by terminal pores
    8. Leaves moderately to deeply pinnately lobed but not fully compound, the lobes broadly attached; plants annual, tap-rooted, lacking tubers . . . . 12. S. TRIFLORUM
    8. At least the larger leaves pinnately compound, the leaflets narrowly attached, with at least the larger leaflets short-stalked; plants perennial, fibrous-rooted, producing tubers . . . . 13. S. TUBEROSUM
6. Leaf blades entire, shallowly wavy, scalloped, or toothed, occasionally shallowly lobed
    9. Stems and leaves sparsely to moderately pubescent with apically 2-branched or pinnately branched, stalked hairs, more densely so on young growth; leaf blades narrowly oblong-elliptic to narrowly oblanceolate . . . . . . 8. S. PSEUDOCAPSICUM
    9. Stems and leaves glabrous or sparsely to densely pubescent with unbranched hairs; leaf blades ovate, triangular-ovate, oval, broadly lanceolate, or elliptic-lanceolate
        10. Calyces somewhat enlarged at fruiting, the tube closely cupping the basal half of the fruit; fruits brownish green at maturity; corollas white or pale cream-colored; stems and leaves moderately to densely pubescent with gland-tipped hairs . . . . 11. S. SARRACHOIDES
        10. Calyces not or only slightly enlarged at fruiting, spreading or only slightly cupping the very base of the fruit; fruits green or more commonly dark purplish black to black at maturity; corollas white; stems and leaves glabrous or sparsely to moderately hairy with nonglandular hairs (more densely so in young growth); species very difficult to distinguish
            11. Inflorescences very short racemes, not strictly umbellate (a short axis present); fruit stalks thickened toward the tips; corollas 6–8 mm long; anthers 1.8–2.5 mm long; fruits lacking stony granules; seeds 1.8–2.1 mm long . . . . 7. S. NIGRUM
            11. Inflorescences umbels; fruit stalks slender throughout; corollas 4–7 mm long; anthers 1.2–1.8 mm long; fruits with stony granules; seeds 1.4–1.8 mm long
                12. Anthers 1.2–1.5 mm long; corollas 4–6 mm long; seeds 1.4–1.6 mm long; green fruits usually dull, with several to many small, white flecks; granules usually 2–5 per fruit . . . . 1. S. AMERICANUM
                12. Anthers 1.4–1.8 mm long; corollas 5–7 mm long; seeds 1.5–1.8 mm long; green fruits shiny, with few or no white flecks; granules usually 6–15 per fruit . . . . 9. S. PTYCHANTHUM

1. Solanum americanum Mill. (black nightshade, American black nightshade)

S. americanum var. *nodiflorum* (Jacq.) Edmonds

S. nigrum L. var. *americanum* (Mill.) O.E. Schulz

S. nodiflorum Jacq.

Map 2636

Plants annuals or short-lived, herbaceous perennials, more or less taprooted. Stems 10–100 cm long, strongly to loosely ascending or occasionally spreading, glabrous or sparsely pubescent with short, curved to appressed, 1- or few-celled, nonglandular hairs; unarmed. Leaves unarmed, the lower long-petiolate, the petioles progressively shorter toward the stem tips, the petiole often very narrowly winged, at least above the midpoint. Leaf blades 2–10 cm long, simple, variable on the same plant, lanceolate to oval, elliptic, ovate, or triangular-ovate, mostly angled or tapered to a sharply pointed tip, less commonly bluntly pointed or rounded, short-tapered to rounded, truncate, or cordate at the base, unlobed, the margins entire, somewhat wavy, or irregularly scalloped, the surfaces glabrous or sparsely pubescent along the main veins with short, curved to appressed, 1- or

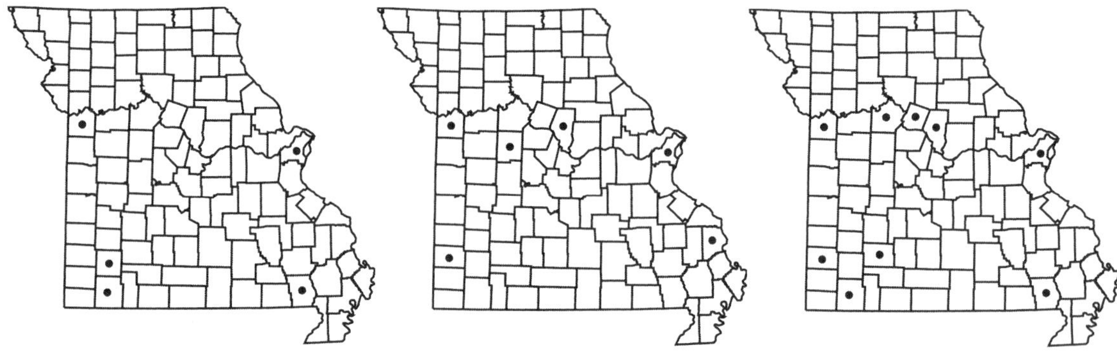
2638. Solanum dimidiatum 2639. Solanum dulcamara 2640. Solanum elaeagnifolium

few-celled, nonglandular hairs (more densely so on young growth). Inflorescences axillary, umbels of (5–)7–15 flowers, the flower stalks often unequal, not jointed (but the fruits usually dispersed with the stalks attached), slender throughout. Flowers spreading to more or less pendant. Calyces 1.1–2.0 mm long, 5-lobed to about the midpoint, the lobes sometimes somewhat unequal, oblong-lanceolate to broadly triangular, the outer surface glabrous or sparsely short-hairy, the hairs nonglandular. Corollas 4–6 mm long, white, sometimes yellow in the throat, lobed to at or below the midpoint, the lobes broadly lanceolate to oblong-ovate, recurved at full flowering, the inner surface glabrous, the outer surface minutely nonglandular-hairy, especially toward the tip. Anthers 1.2–1.5 mm long, oblong, lacking a sterile tip, dehiscing by terminal pores. Ovary 2-locular, the surface glabrous, the style exserted from the anther ring. Fruits 0.6–0.9 cm long, globose, with usually 2–5 stony granules, the surface glabrous, purplish black at full maturity but often remaining green for a relatively long time, then usually dull, with several to many small, white flecks. Seeds 1.4–1.6 mm in longest dimension, broadly obovate to nearly circular in outline, often minutely notched at the attachment point, strongly flattened, unwinged, the surface minutely pitted or with a very fine network of ridges, yellow to tan. $2n=24$. May–November.

Uncommon, widely scattered, most abundantly in the Mississippi Lowlands Division (Washington to California east to Florida, the northern limits uncertain; Mexico, Central America, South America, Europe, Asia south to Australia). Bottomland forests, mesic upland forests, and banks of streams and rivers; also ditches, railroads, roadsides, and shaded to open, disturbed areas.

Solanum americanum is a member of the taxonomically difficult *S. nigrum* complex. For further discussion of this complex in Missouri, see the treatment of *S. ptychanthum*.

2. Solanum carolinense L. **var. carolinense** (horse nettle, bull nettle)

Pl. 566 b–d; Map 2637

Plants perennial herbs, with relatively deep rhizomes. Stems 20–100 cm long, erect or strongly ascending, unbranched or with few to several, ascending branches, moderately to densely pubescent with minute, stellate, nonglandular hairs, also with moderate to dense, larger, stellate hairs having the central branch erect and elongate, the green tissue not obscured by hairs; armed with scattered to relatively common, flattened, straight, straw-colored to greenish yellow prickles (1–)2–4(–7) mm long. Leaves armed with scattered prickles, mostly along the petiole and main veins, mostly short-petiolate, the petiole often winged toward the tip, moderately to densely stellate-hairy, the hairs mostly with the central branch erect and elongate. Leaf blades 2–8(–12) cm long, simple, ovate to elliptic-ovate or oblong-elliptic, rounded or more commonly angled or tapered to a bluntly or sharply pointed tip, variously tapered, angled, truncate, or occasionally rounded at the base, often somewhat oblique, the margins with 2–5 coarse teeth or shallow lobes per side or sometimes merely wavy, the surfaces moderately to densely pubescent with stellate hairs, these sessile, with 4–9 branches, the central branch vertically spreading, the green tissue not obscured by the hairs. Inflorescences axillary near the stem tips, sometimes appearing terminal, simple (unbranched) racemes (short, often appearing clustered at flowering, the axis with sparse to moderate, short prickles, pubescent similar to the stems, elongating as the fruits develop) of 5–20 flowers, the flower stalks not jointed, slender to moderately stout, thickened toward the tip. Flowers spreading to ascending, the stalks becoming nodding as the fruits develop. Calyces 5–7 mm long at flowering, becoming slightly enlarged to 8–9 mm at fruiting (but the tips of the lobes often breaking off), the tube loosely

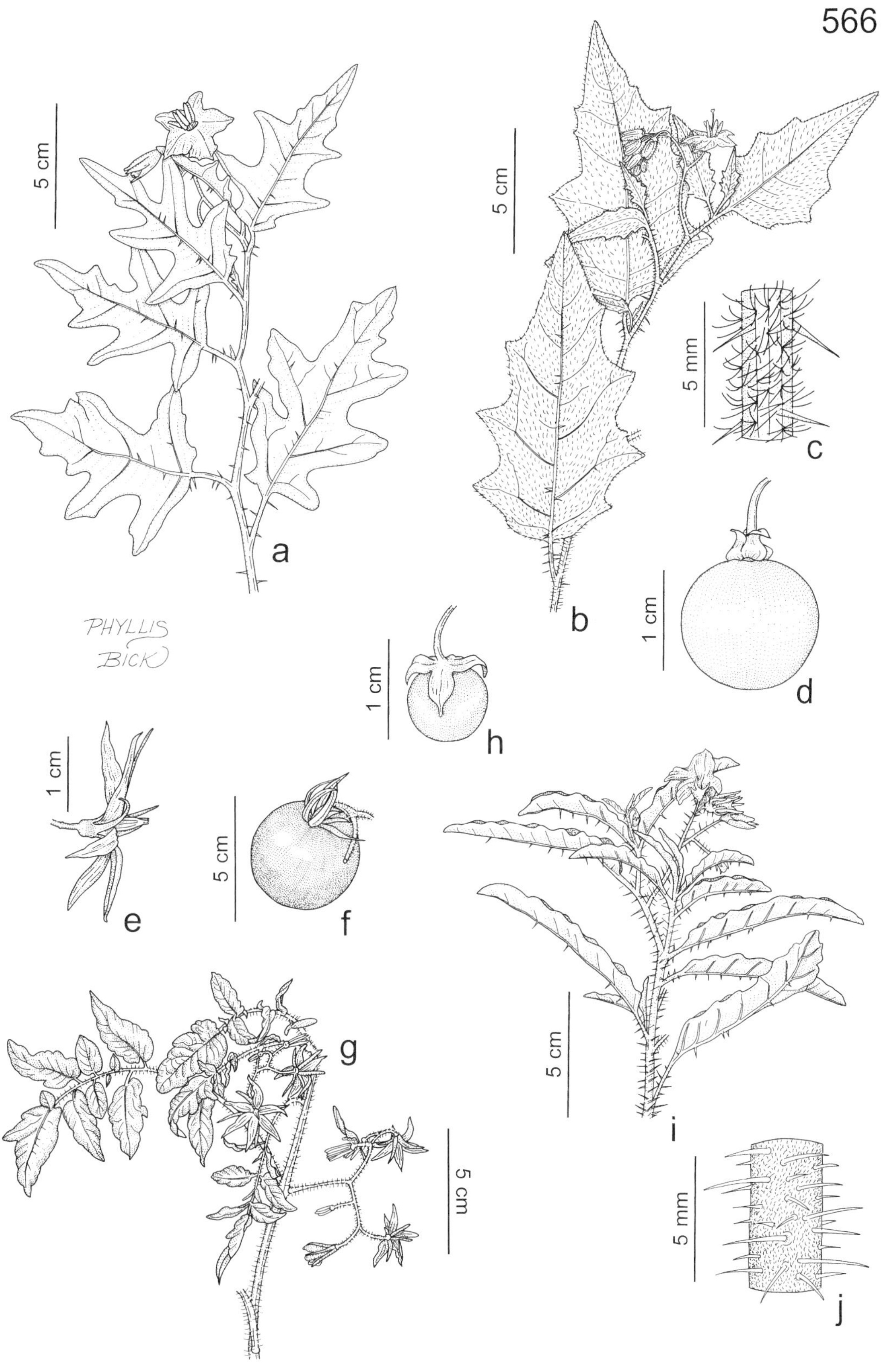

Plate 566. Solanaceae. *Solanum dimidiatum*, **a)** fertile stem. *Solanum carolinense*, **b)** fertile stem, **c)** stem detail, **d)** fruit. *Solanum lycopersicum*, **e)** flower, **f)** fruit, **g)** fertile stem. *Solanum elaeagnifolium*, **h)** fruit, **i)** fertile stem, **j)** stem detail.

cupping the base of the fruit, deeply 5-lobed at flowering, the lobes much longer than the tube, more or less equal, lanceolate to narrowly lanceolate, the outer surface unarmed, moderately to densely stellate-hairy. Corollas 10–15 mm long, white or purplish-tinged, sometimes more uniformly pale purple to pale bluish purple, lobed to at or more commonly above the midpoint, the lobes broadly triangular, usually appearing entire along the margins, tapered abruptly to bluntly or sharply pointed tips, spreading to slightly reflexed at flowering, the inner surface glabrous, the outer surface stellate-hairy. Anthers equal, 6–9 mm long, narrowly oblong (tapered toward the tip), yellow, dehiscing by terminal pores. Ovary 2-locular, the surface sparsely to moderately stellate-hairy, the style exserted from the anther ring. Fruits 1–2 cm long, globose, lacking stony granules, green mottled with darker green, becoming bright yellow at maturity, shiny, unarmed, glabrous or very sparsely stellate-hairy, often with a few, small, white flecks, usually becoming somewhat wrinkled with age (frequently persisting until the next growing season). Seeds 1.5–3.0 mm in longest dimension, broadly ovate to broadly oblong in outline, sometimes bluntly angular, moderately flattened, unwinged, the surface smooth or very slightly few-wrinkled, brown to dark brown or grayish brown, dull or somewhat shiny. $2n=24$. May–October.

Scattered nearly throughout the state (eastern U.S. west to South Dakota and Texas; Canada, Mexico; introduced farther west). Banks of streams, rivers, and spring branches, margins of ponds, sloughs, marshes, and oxbows, edges and openings of bottomland forests, upland prairies, and glades; also old fields, pastures, fallow fields, crop fields, strip mines, railroads, roadsides, and open, disturbed areas.

The bright yellow berries of this species are potentially attractive to children, but are very poisonous. They often persist through the winter into the next growing season and are mainly a problem in pastures and hayfields. Plants with all-white corollas have been called f. *albiflorum* (Kuntze) Benke. A second variety, var. *floridanum* (Shuttlew. ex Dunal) Chapm. is endemic to northern florida and adjacent Georgia and differs in its more deeply divided leaves (D'Arcy, 1974).

3. Solanum dimidiatum Raf. (western horse nettle)

S. torreyi A. Gray

Pl. 566 a; Map 2638

Plants perennial herbs, with relatively deep rhizomes. Stems 30–100 cm long, erect or ascending, unbranched or more commonly with few to several, ascending branches, moderately to densely pubescent with stellate, nonglandular hairs, these appearing more or less appressed and matted (with 9–13 horizontal branches), the green tissue not fully obscured by hairs; armed with scattered to relatively common, flattened, straight, straw-colored to greenish yellow prickles 2–7 mm long. Leaves armed with scattered prickles, mostly along the petiole and main veins, mostly short-petiolate, the petiole sometimes winged toward the tip, moderately to densely stellate-hairy, the hairs short-stalked (visible only with magnification), all of the branches appressed. Leaf blades 2–8(–12) cm long, simple, ovate to elliptic-ovate or oblong-elliptic, or elliptic-lanceolate, rounded or more commonly angled or tapered to a bluntly or sharply pointed tip, angled or tapered to occasionally truncate at the base, the margins with 2–5 coarse teeth or shallow, sometimes irregular lobes per side or sometimes merely wavy, the surfaces moderately to densely pubescent with stellate hairs, these sessile, with 4–9 branches, all appressed, the green tissue not obscured by the hairs. Inflorescences terminal or sometimes appearing axillary from the upper nodes, often paniculate, consisting of a group of 2 or 3 racemes from a common stalk (these short, often appearing clustered at flowering, the axis unarmed or with sparse, short prickles, pubescent similar to the stems, elongating as the fruits develop), each raceme of 3–7 flowers, the flower stalks not jointed, moderately stout, thickened toward the tip. Flowers spreading to ascending, the stalks becoming nodding as the fruits develop. Calyces 8–13 mm long at flowering, not or only slightly enlarged at fruiting (the tips of the lobes often breaking off), the tube loosely cupping the base of the fruit, deeply 5- or 6-lobed at flowering, the lobes much longer than the tube, more or less equal, linear above a short, ovate, basal portion, the outer surface unarmed, moderately to densely stellate-hairy. Corollas 14–25 mm long, lavender to light purple or light bluish purple (white elsewhere), often with a green, star-shaped area in the throat, lobed to at or slightly below the midpoint, the lobes broadly ovate, appearing nearly entire to wavy or ruffled along the margins, tapered abruptly to bluntly or sharply pointed tips, spreading at flowering, the inner surface glabrous, the outer surface stellate-hairy. Anthers equal, 8–10 (–12) mm long, narrowly oblong (tapered toward the tip), yellow, dehiscing by terminal pores. Ovary 2-locular, the surface sparsely to moderately stellate-hairy, the style exserted from the anther ring. Fruits 2.2–3.0 cm long, globose, lacking stony granules, green mottled with darker green, becoming pale yellow at maturity, more or less shiny, un-

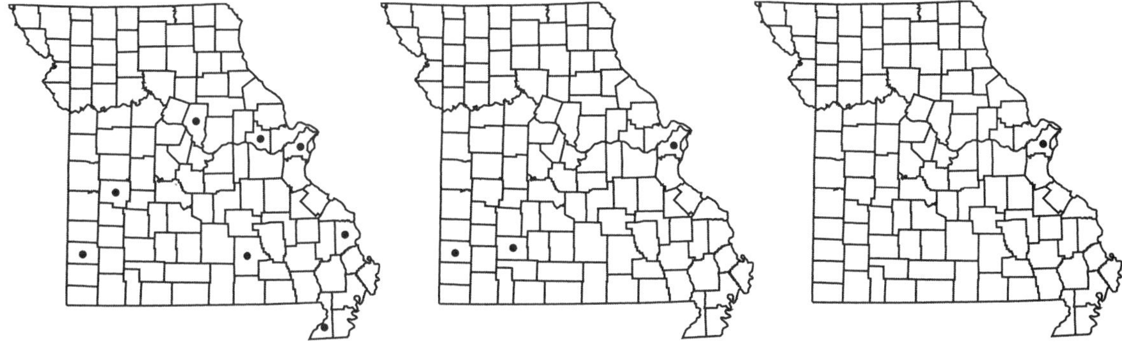

2641. Solanum lycopersicum 2642. Solanum nigrum 2643. Solanum pseudocapsicum

armed, glabrous or very sparsely stellate-hairy, occasionally with a few, small, white flecks, usually becoming somewhat wrinkled with age (frequently persisting until the next growing season). Seeds 3.8–4.3 mm in longest dimension, broadly ovate to broadly oval or nearly circular in outline, flattened, unwinged, the surface finely pitted or appearing pebbled, light brown to yellowish brown, dull or somewhat shiny. $2n=72$. May–September.

Introduced, uncommon, widely scattered in the southern half of the state (Florida to New Mexico north to Arkansas and Kansas; introduced farther north and in California). Gardens, railroads, roadsides, and open, disturbed areas.

4. Solanum dulcamara L. (climbing nightshade, bittersweet)

Pl. 567 a–c; Map 2639

Plants herbaceous perennials or more commonly more or less lianas, usually with rhizomes. Stems 100–300 cm long, climbing, twining, and often woody toward the base, glabrous or occasionally sparsely pubescent with short, curved to appressed, 1- or few-celled, nonglandular hairs; unarmed. Leaves unarmed, mostly long-petiolate, the petiole often very narrowly winged. Leaf blades 2–6(–10) cm long, simple, ovate to broadly ovate, more or less heart-shaped, or halberd-shaped (occasionally some of the upper leaves oblong-lanceolate), tapered to a sharply pointed tip, truncate to cordate at the base, those of at least some of the leaves with 1 or 2(–4) shallow to deep, spreading to somewhat downward-angled lobes at the base, the margins otherwise entire, the surfaces glabrous or sparsely pubescent along the main veins with short, curved to appressed, 1- or few-celled, nonglandular hairs, some of these sometimes branched. Inflorescences terminal and sometimes also appearing lateral opposite the upper leaves, clusters or small panicles of 5–14(–30) flowers, the flower stalks jointed at the base. Flowers more or less pendant. Calyces 2.0–2.5 mm long, 5-lobed (sometimes irregularly so) to above the midpoint, the lobes broadly triangular, the outer surface sparsely short-hairy, the hairs nonglandular. Corollas 6–9 mm long, purple to bluish purple with a pair of yellow to yellow and green spots at the base of each lobe, deeply lobed, the lobes lanceolate, recurved at full flowering, the inner surface glabrous, the outer surface minutely nonglandular-hairy toward the tip. Anthers 3.5–4.5 mm long, oblong, lacking a sterile tip, dehiscing by terminal pores. Ovary 2-locular, the surface glabrous, the style exserted from the anther ring. Fruits 0.8–1.1 cm long, broadly ellipsoid to elliptic-ovoid, lacking granules, the surface glabrous, red, shiny, lacking white flecks. Seeds 2.5–3.0 mm long, broadly ovate to nearly circular in outline, often minutely notched at the attachment point, strongly flattened, unwinged, the surface with a minute network of ridges, often appearing pebbled, orangish brown to brown. $2n=24, 48, 72$. May–November.

Introduced, uncommon, mostly in counties bordering the Missouri and Mississippi Rivers (native of Europe, Asia; introduced nearly throughout the U.S.; Canada). Bottomland forests and banks of streams and rivers; also farmyards, fencerows, roadsides, and open, disturbed areas.

Steyermark (1963) stated that *S. dulcamara* might be present nearly throughout Missouri, although it had been collected in only a few counties. Few collections have been made since then, which suggests that although the species may be undercollected it is neither abundant nor widespread in the state.

5. Solanum elaeagnifolium Cav. (white horse nettle, silverleaf nightshade)

Pl. 566 h–j; Map 2640

Plants perennial herbs (sometimes somewhat woody at the base), with deep, long-creeping rhizomes. Stems 15–70(–100) cm long, erect or strongly ascending, unbranched or with few to several, ascending branches, densely pubescent with stellate, nonglandular hairs (these a mixture of sessile and short-stalked, with 9–13 branches), the green tissue totally obscured by the densely matted, silvery-white hairs; unarmed or more commonly armed with scattered, slender, straight, tan to orangish brown prickles 1–4 mm long. Leaves unarmed or with scattered, short prickles, mostly along the petiole and midvein, mostly short-petiolate, the petiole unwinged, densely stellate-hairy, the hairs sessile, with 9–13 branches). Leaf blades 3–10(–15) cm long, simple, narrowly lanceolate to oblong-lanceolate or narrowly oblong, angled to a bluntly or sharply pointed tip, obliquely rounded to more commonly tapered at the base, the margins entire or shallowly to moderately wavy, the surfaces densely pubescent with stellate hairs, the green tissue totally obscured by the densely matted, silvery-white hairs. Inflorescences axillary near the stem tips, sometimes appearing terminal, of solitary flowers or more commonly racemes (short, often appearing clustered at flowering, the axis with sparse to moderate, short prickles, pubescent similar to the stems, elongating as the fruits develop) of 3–7 flowers, the flower stalks not jointed, slender to moderately stout, thickened toward the tip. Flowers spreading to ascending, the stalks often becoming more or less nodding as the fruits develop. Calyces 9–14 mm long at flowering, the tube somewhat 5-angled at flowering, not or only slightly enlarged at fruiting, the tube spreading or only loosely cupping the base of the fruit, deeply 5-lobed at flowering, the lobes as long as to longer than the tube, slightly unequal, linear above a triangular base, the outer surface unarmed or with only scattered small prickles toward the base, also densely stellate-hairy. Corollas 10–16 mm long, purple to bluish purple or lavender (rarely white elsewhere), lobed to at or above the midpoint, the lobes broadly ovate-triangular, sometimes appearing minutely wavy or irregular along the margins, tapered abruptly to bluntly or sharply pointed tips, spreading at flowering, the inner surface glabrous, the outer surface stellate-hairy. Anthers equal, 6–9 mm long, narrowly oblong (slightly tapered toward the tip), yellow, dehiscing by terminal pores. Ovary 2-locular, the surface stellate-hairy, the style slightly exserted from the anther ring. Fruits 1.0–1.5 cm long, globose, lacking stony granules, yellow (eventually turning black with age), dull to somewhat shiny, unarmed, glabrous or sparsely stellate-hairy. Seeds 3–5 mm in longest dimension, broadly ovate to broadly oblong in outline, sometimes bluntly angular, moderately flattened, unwinged, the surface smooth or very slightly few-wrinkled, brown to dark brown or grayish brown, dull or somewhat shiny. $2n=24$, 72. July–September.

Uncommon, mostly in the western half of the state (Missouri and Kansas south to Texas and Arizona; Mexico; introduced farther east and west). Glades, upland prairies, tops of bluffs, and margins of salt springs; also pastures, railroads, roadsides, and open, disturbed areas.

6. Solanum lycopersicum L. (tomato, garden tomato)

Lycopersicon esculentum Mill.

L. esculentum var. *leptophyllum* (Dunal) D'Arcy

Pl. 566 e–g; Map 2641

Plants annual (biennial or perennial elsewhere), taprooted. Stems 15–50 cm long (–400 cm in cultivation), initially erect or ascending, but weak and eventually reclining or scrambling, pubescent with a mixture of dense, short, unicellular hairs and sparse to moderate, long, spreading, multicellular hairs, many of the hairs gland-tipped; unarmed. Leaves unarmed, mostly short-petiolate. Leaf blades 10–35 cm long, irregularly pinnately compound, the leaflets strongly unequal, the main leaflets 5–11, 2–7 cm long, interspersed along the rachis with much smaller leaflets, lanceolate, ovate, or elliptic, angled or tapered to a bluntly or sharply pointed tip, oblique at the base, the margins irregularly scalloped or toothed, sometimes shallowly to deeply lobed or fully compound again, the surfaces pubescent with a mixture of dense, short, unicellular hairs and sparse to moderate, long, spreading, multicellular hairs, many of the hairs gland-tipped. Inflorescences usually racemes (unbranched or in some smaller-fruited cultivars 2-branched from the top of the stalk), with (3–)5–15 well-spaced flowers, the flower stalks inconspicuously jointed above the midpoint. Flowers often nodding. Calyces 8–10 mm long, deeply 5-lobed, the lobes linear, the outer surface densely hairy, many of the hairs gland-tipped. Corollas (7–)9–15(–24) mm long, bright yellow, lobed half way or more deeply to the midpoint, the lobes narrowly lanceolate to narrowly triangular, recurved at full flowering (loosely ascending before and after), the inner surface glabrous, the outer surface usually minutely hairy and glandular, at least toward the tip. Anthers 6–8 mm long, the body tapered to a sterile, often slightly outward-curved tip, dehiscent longitudinally by slits. Ovary 2–4-locular, the surface glandular, the style usually not exserted from the anther ring. Fruits 1.0–2.5 cm long (to 10 cm in some cultivars), variously depressed-globose to globose, oblong-ellip-

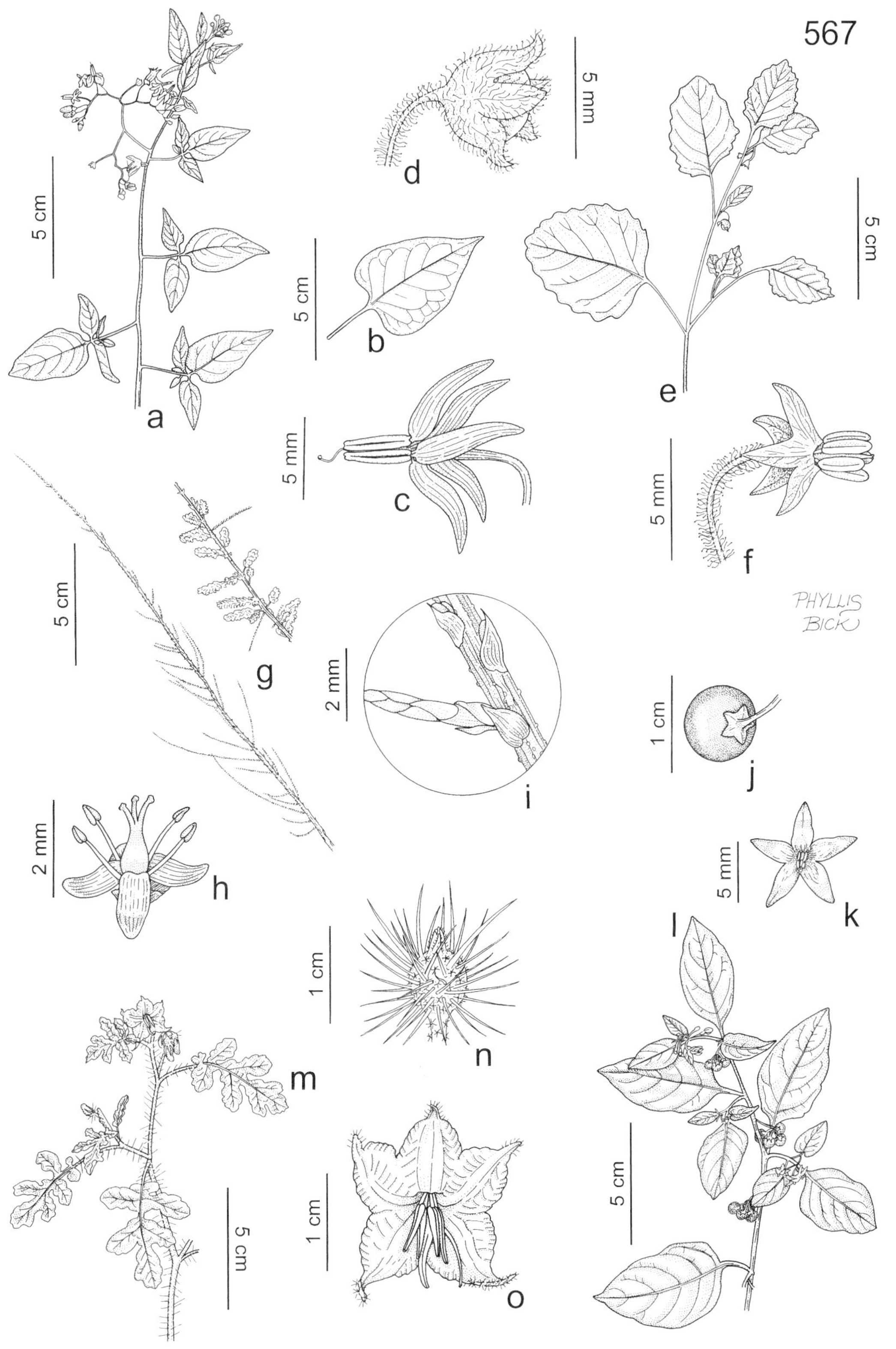

Plate 567. Solanaceae. Tamaricaceae. *Solanum dulcamara*, **a)** fertile stem, **b)** simple leaf, **c)** flower. *Solanum sarrachoides*, **d)** fruit, **e)** fertile stem, **f)** flower. *Tamarix ramosissima*, **g)** vegetative and fertile branches, **h)** flower, **i)** branchlet detail with leaves. *Solanum ptychanthum*, **j)** fruit, **k)** flower, **l)** fertile stem. *Solanum rostratum*, **m)** fertile stem, **n)** fruit, **o)** flower.

soid, oblong-ovoid, or somewhat pear-shaped, lacking granules, the surface glabrous or nearly so, yellow to orange or red, usually somewhat shiny, lacking white flecks. Seeds 2.5–3.3 mm long, obovate in outline, strongly flattened, narrowly winged, at least toward the tip, the surface pebbled or appearing minutely bumpy to hairy, pale brown to cream-colored. $2n=24$. May–September.

Introduced, uncommon, sporadic (cultigen of south American origin; introduced nearly worldwide). Banks of streams and rivers; also railroads and open disturbed areas.

The tomato and related species originally were described by Linnaeus (1753) as members of the genus *Solanum*. However, during most of the 19th and 20th centuries, botanists treated them in a segregate genus, *Lycopersicon* Mill., based mostly on differences in anther morphology and dehiscence. The taxonomic level at which the group should be recognized was controversial even during these centuries. Molecular studies (Spooner et al., 1993; Rodriguez et al., 2009) provided firm support for the inclusion of the tomatoes as a section within *Solanum* subgenus *Potatoe* (Dumort.) D'Arcy (which includes the potatoes and their relatives). In their monograph of the tomato and its wild relatives, Peralta et al. (2008) summarized the morphological, cytological, biochemical, and molecular evidence for the unity of tomatoes and potatoes, and provided the most recent taxonomic classification for the group.

Tomatoes are a major fruit crop in most countries. In addition to the raw fruits, processed tomatoes are sold commercially canned as whole fruits, chunks, paste, and sauces. Tomatoes also are an important component in many cuisines, including everything from Italian and Mexican dishes to hamburgers and fried green tomatoes. Tomatoes also are the basis for catsup (ketchup), a nearly universal condiment that can improve the flavor of almost any food. The breeding, propagation, and sale of tomatoes as plants and seeds is also a commercially important part of the horticulture industry. The tomato is a cultigen, that is, the species exists only in cultivation and as an escape. A large number of cultivars exist and there is great variation in fruit size, shape, color, texture, and flavor. It presumably was first developed in western South America, but whether it arose from some now-extinct wild populations that would be classified as the same species or was bred from populations of the closely related *S. pimpinellifolium* L. (*Lycopersicon pimpinellifolium* (L.) Mill.) is still not known with any degree of certainty (Peralta et al., 2008).

7. Solanum nigrum L. (black nightshade)
Map 2642

Plants annuals or short-lived, herbaceous perennials, more or less taprooted. Stems 15–60 cm long, strongly to loosely ascending or occasionally spreading, moderately to densely pubescent with short, curved to spreading, multicellular, hairs, some of these usually gland-tipped; unarmed. Leaves unarmed, the lower long-petiolate, the petioles progressively shorter toward the stem tips, the petiole often very narrowly winged, at least above the midpoint. Leaf blades 2–10 cm long, simple, elliptic to ovate, those of the uppermost leaves sometimes lanceolate, angled or tapered to a sharply pointed tip, tapered to occasionally rounded or truncate at the base, the margins wavy to coarsely toothed or shallowly few-lobed, the surfaces moderately to densely pubescent with short, curved to appressed, 1- or few-celled, nonglandular hairs, often also with gland-tipped hairs toward the blade base. Inflorescences axillary, very short racemes (not strictly umbellate, a short axis present), of 5–7 flowers, the flower stalks more or less equal, not jointed (but the fruits usually dispersed with the stalks attached), thickened toward the tips. Flowers spreading to more or less pendant. Calyces 1.5–2.0 mm long, 5-lobed to at or more commonly below the midpoint, the lobes equal, oblong to oblong-obovate, the outer surface moderately to densely short-hairy, the hairs nonglandular. Corollas 6–8 mm long, white, sometimes yellow to greenish yellow in the throat, lobed to about the midpoint, the lobes ovate to triangular-ovate, spreading at full flowering, the inner surface glabrous, the outer surface minutely nonglandular-hairy, especially toward the tip. Anthers 1.8–2.5 mm long, oblong, lacking a sterile tip, dehiscing by terminal pores. Ovary 2-locular, the surface glabrous, the style slightly exserted from the anther ring. Fruits 0.7–1.0 cm long, globose, lacking stony granules, the surface glabrous, purplish black at full maturity but often remaining green for a relatively long time, then dull, lacking small white flecks. Seeds 1.8–2.1 mm in longest dimension, broadly obovate to nearly circular in outline, often minutely notched at the attachment point, strongly flattened, unwinged, the surface minutely pitted or with a very fine network of ridges, yellow to tan. $2n=72$. August–November.

Introduced, uncommon, known thus far only from the city of St. Louis (native of Europe, Asia; introduced sporadically in the U.S., mostly in easternmost and westernmost states; Canada). Railroads and open, disturbed areas.

There are three morphologically similar members of the *S. nigrum* complex in Missouri. For

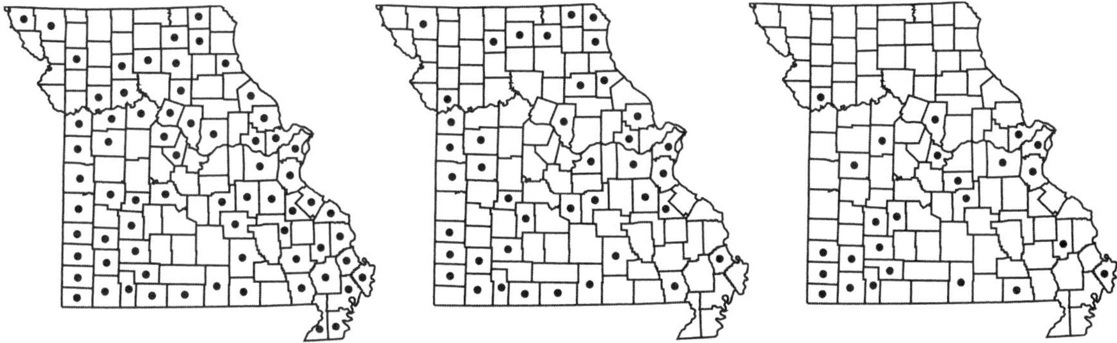
2644. Solanum ptychanthum 2645. Solanum rostratum 2646. Solanum sarrachoides

further discussion of this complex, see the treatment of *S. ptychanthum*.

8. Solanum pseudocapsicum L. (Jerusalem cherry)

Map 2643

Plants perennials (shrubs elsewhere), with woody rootstocks. Stems 45–80(–200) cm long, sometimes woody at the very base (more strongly woody elsewhere), sparsely to moderately pubescent with apically 2-branched or pinnately branched, stalked hairs, more densely so on young growth (glabrous elsewhere); unarmed. Leaves unarmed, sessile to short-petiolate, the petiole usually winged toward the tip. Leaf blades 2–8 (–12) cm long, simple, narrowly oblong-elliptic to narrowly oblanceolate, rounded or angled to a bluntly pointed tip (sharply pointed elsewhere), tapered at the base, the margins entire to finely wavy or irregularly scalloped, the surfaces glabrous or sparsely pubescent with apically 2-branched or pinnately branched, stalked hairs when young. Inflorescences usually lateral (occasionally axillary), of solitary flowers or small clusters of 2 or 3 flowers (to 8 elsewhere), the flower stalks inconspicuously jointed at the base. Flowers often nodding. Calyces 4–6 mm long, deeply 5-lobed, the lobes narrowly oblong to oblong-lanceolate, the outer surface sparsely branched-hairy, especially toward the tip. Corollas 5–7 mm long, white, deeply lobed, the lobes narrowly lanceolate to narrowly oblong-lanceolate, spreading to more commonly recurved, the inner surface glabrous, the outer surface usually minutely hairy toward the tip. Anthers 3–4 mm long, oblong to oblong-elliptic, lacking a sterile tip, dehiscing by terminal pores. Ovary 2-locular, the surface glabrous at maturity, the style exserted from the anther ring. Fruits 1.2–1.8 cm long, globose, lacking granules, the surface glabrous, yellow to more commonly orangish red or red, usually somewhat shiny, lacking white flecks. Seeds 2.5–3.5 mm in longest dimension, broadly kidney-shaped to nearly circular in outline, strongly flattened, unwinged, the surface minutely pitted or appearing minutely pebbled, yellow to yellowish tan. $2n=24$. July–October.

Introduced, uncommon, known thus far only from the city of St. Louis (native of South America, introduced in tropical and warm-temperate regions nearly worldwide, in the U.S. mostly in southeastern states). Open, disturbed areas.

This species is cultivated for its attractive fruits, which mature late in the year. The description above applies mainly to the scanty Missouri materials and observation of plants cultivated in the state. Plants elsewhere in the distribution of the species may appear different in height, leaf size and shape, and pubescence density. The flowers are somewhat similar to those of species in the *S. nigrum* complex, but the plants at flowering are much more densely leafy.

9. Solanum ptychanthum Dunal (eastern black nightshade)

S. americanum Mill., misapplied

S. nodiflorum Jacq., misapplied

Pl. 567 j–l; Map 2644

Plants annuals or short-lived, herbaceous perennials, more or less taprooted. Stems 7–120 cm long, strongly to loosely ascending or occasionally spreading, glabrous or sparsely pubescent with short, curved to appressed, 1- or few-celled, nonglandular hairs; unarmed. Leaves unarmed, the lower long-petiolate, the petioles progressively shorter toward the stem tips, the petiole often very narrowly winged, at least above the midpoint. Leaf blades 2–12 cm long, simple, variable on the same plant, lanceolate to oval, elliptic, ovate, or triangular-ovate, mostly angled or tapered to a sharply pointed tip, less commonly bluntly pointed or rounded, short-tapered to rounded, truncate, or cordate at the base, unlobed, the margins entire, somewhat wavy, or irregularly scalloped, the surfaces glabrous or sparsely to moderately pubescent with short, curved to appressed, 1- or few-celled, nonglandular hairs (sometimes mostly along the veins). Inflorescences axillary, umbels of 3–6 flowers, the flower stalks often unequal, not jointed (but the fruits usually dispersed with the stalks attached), slender throughout. Flowers spreading to more or less pendant. Calyces 1.5–3.0 mm long, 5-lobed to about the midpoint, the lobes sometimes somewhat unequal, oblong-lanceolate to broadly triangular, the outer surface glabrous or sparsely short-hairy, the hairs nonglandular. Corollas 5–7 mm long, white, sometimes yellow in the throat, lobed to at or below the midpoint, the lobes broadly lanceolate to oblong-ovate, spreading to recurved at full flowering, the inner surface glabrous, the outer surface minutely nonglandular-hairy, especially toward the tip. Anthers 1.4–1.8 mm long, oblong, lacking a sterile tip, dehiscing by terminal pores. Ovary 2-locular, the surface glabrous, the style usually exserted from the anther ring. Fruits 0.7–1.0 cm long, globose, with usually 6–15 stony granules, the surface glabrous, purplish black at full maturity but often remaining green for a relatively long time, then shiny, with few or no small white flecks. Seeds 1.5–1.8 mm in longest dimension, broadly obovate to nearly circular in outline, often minutely notched at the attachment point, strongly flattened, unwinged, the surface minutely pitted or with a very fine network of ridges, yellow to tan. $2n$=24. May–November.

Scattered to common nearly throughout the state (eastern U.S. west to Montana and New Mexico). Banks of streams and rivers, bases and ledges of bluffs, bottomland forests, margins of ponds, lakes, and sinkhole ponds, savannas, and openings of mesic upland forests; also ditches, levees, pastures, fallow fields, crop fields, gardens, railroads, roadsides, and open, disturbed areas.

Solanum ptychanthum is by far the most common member of the *S. nigrum* complex in Missouri. The *S. nigrum* complex includes four morphologically similar species that are present in Missouri (Schilling, 1981). Of these, *S. sarrachoides* is relatively easily distinguished by pubescence and floral characters. However, the other three taxa continue to be problematic to identify and have a long history of confused taxon circumscriptions and ranges, as well as resultant misapplication of names. The introduced hexaploid ($2n$=72), *S. nigrum* is the most distinctive member of the three by virtue of its slightly nonumbellate inflorescences with a short central axis and thickened flower stalks. The main problems of identification have involved the two diploid, native members. The present work follows that of Schilling (1981), who circumscribed *S. americanum* as a widespread taxon with small flowers and dull, moderately to densely white-flecked developing fruits (and also green seedling leaves), vs. the eastern North American *S. ptychanthum* with slightly larger flowers, shiny developing fruits with white specks absent or sparse (and also seedling leaves strongly reddish-tinged on the undersurface). These two species are so similar that flowering specimens of the complex can be very difficult to determine to species. In Missouri, the number of stony granules per fruit appears to be more variable and thus somewhat less diagnostic than treated by Schilling (1981). To add to the confusion, plants now called *S. ptychanthum* were known as *S. americanum* in some of the earlier literature (Steyermark, 1963; Schilling, 1978). Still, it is telling that even weed scientists (who need to distinguish these species for practical, economically driven reasons) appear to have adopted Schilling's species circumscriptions and to follow his nomenclature (Ogg and Rogers, 1989). A recent molecular study of mainly the African members of the complex (but with worldwide sampling) using AFLP markers (Manoko et al., 2007) reinforced the taxonomic separation between *S. americanum* and *S. nigrum*, but appears to have confused the issue of the separation of *S. americanum* from *S. ptychanthum*.

Members of the *S. nigrum* complex are important agricultural weeds, both because they potentially lower crop yields and because the toxic seeds and other plant parts can contaminate the harvest of particularly soybean and other nongraminoid seed crops (Ogg and Rogers, 1989).

10. Solanum rostratum Dunal (buffalo bur, Kansas thistle)
S. cornutum Lam.

Pl. 567 m–o; Map 2645

Plants annuals, taprooted. Stems 20–70 cm long, erect to loosely ascending, the branches spreading, densely pubescent with stellate, nonglandular hairs, these sessile or stalked; armed with numerous, straight, straw-colored to light yellow prickles 3–8 mm long (the stem sometimes purplish-tinged at the prickle attachment points). Leaves armed with scattered to relatively common, straight, straw-colored to light yellow prickles (these along the main veins of both surfaces), short- to moderately petiolate, the petiole unwinged. Leaf blades 3–15 cm long, simple but deeply 1- or 2-times pinnately lobed or parted, elliptic or oblong to broadly ovate, the lobes narrowly to broadly oblong or obovate, rounded at the tips, broadly attached at their bases, the margins otherwise entire, the surfaces moderately to densely pubescent with sessile, stellate, nonglandular hairs, these mostly with the central branch erect and elongate. Inflorescences axillary, racemes (short and often appearing clustered at flowering, the axis with scattered prickles, elongating as the fruits develop) of 3–9(–15) flowers, the flower stalks not jointed, uniformly slender to moderately stout. Flowers spreading, the stalks becoming more or less ascending as the fruits develop. Calyces 7–12 mm long at flowering, the tube 1.5–2.5 mm long at flowering, becoming enlarged to 7–12 mm at fruiting, enclosing the fruit, deeply 5-lobed at flowering, the lobes more or less equal, linear to very narrowly oblong-triangular, the outer surface densely prickly and stellate-hairy (the hairs sometimes breaking off after flowering, leaving the short, erect stalk). Corollas 8–14 mm long, slightly zygomorphic, bright yellow, shallowly lobed, the lobes broadly triangular, often appearing somewhat ruffled or irregular along the margins, tapered abruptly, the lower 2 lobes more strongly extended into slender, tips, spreading at flowering, the inner surface glabrous, the outer surface stellate-hairy. Anthers unequal, 4 of them 7–9 mm long, more or less oblong (slightly tapered toward the tip), yellow, the anther of the lowermost stamen 11–14 mm long, long-tapered and curved upward toward its tip, usually reddish- or purplish-tinged, all dehiscing by terminal pores. Ovary 2-locular, fused to the inner surface of the calyx tube, the style angled downward and to the side from the anther ring, somewhat upward-curved. Fruits 0.7–1.1 cm long, more or less globose, lacking stony granules, hidden within the burlike, densely prickly calyx, green (eventually tan to light brown), with straw-colored to light yellow, straight prickles. Seeds 2.3–2.7 mm in longest dimension, broadly ovate to nearly circular in outline, moderately flattened, unwinged, the surface minutely pitted or with a fine network of ridges, usually also coarsely wrinkled, dark brown to nearly black, the ridges sometimes grayish brown. $2n=24$. May–October.

Introduced, scattered nearly throughout the state (native of the central U.S., Mexico; introduced farther to the west, north, and east; also sporadically in the Old World and southern hemisphere). Banks of streams and rivers and disturbed portions of upland prairies; also pastures, farmyards, roadsides, and open, disturbed areas.

Although *S. rostratum* is considered introduced in Missouri, it is interesting that a specimen exists (duplicates in the herbaria of the Natural History Museum in England and the Missouri Botanical Garden) that apparently was collected in Missouri by one of the two earliest botanists in the state. Thomas Nuttall collected this species, presumably in 1810 or 1811, but the label lacks more detailed collection information. His collecting activities in what is now the state of Missouri were mainly in the area around St. Louis and along the banks of the Missouri River westward and northward to the state's border with Nebraska and Iowa. For more information, see the chapter on the history of floristic botany in Volume 1 of the present work (Yatskievych, 1999).

Whalen (1979) included *S. rostratum* in *Solanum* sect. *Androceras* (Nutt.) Marzell, a group of twelve mostly annual species of open habitats that are native to Mexico and adjacent countries. The pollination syndrome and associated floral morphology in members of this section are relatively specialized (Bowers, 1975; Whelan, 1979). The downward-angled, upward-curved, enlarged lowermost anther is angled to the left or right on alternating flowers of an inflorescence, with the style positioned to the side of the flower opposite the lower anther. Bees that land on top of the four, shorter anthers to extract pollen from them are positioned so that the longer lower anther and the stigma contact opposite sides of their bodies. A bee that has had pollen deposited on the lefthand side of its body effects pollen transfer when it subsequently visits a flower with the stigma on that side (and/or vice versa). This mechanism enforces cross-pollination.

Plants of *S. rostratum* (and its relatives) produce berries enclosed by the burlike, prickly calyces and thus are animal-dispersed. Steyermark (1963) noted that the prickles on the plants can cause mechanical injury to livestock (and barefoot humans). Also the fruits can become entangled in the fleece of sheep, causing damage as well as contaminating and thus reducing the value of the wool.

11. Solanum sarrachoides Sendtn. (hairy nightshade, viscid nightshade)

Pl. 567 d–f; Map 2646

Plants annuals or short-lived, herbaceous perennials, more or less taprooted. Stems 10–60 (–80) cm long, loosely ascending, sometimes from a spreading base, densely pubescent with short, spreading, multicellular, gland-tipped hairs and scattered, longer, spreading, often nonglandular hairs; unarmed, somewhat sticky. Leaves unarmed, all or mostly long-petiolate, the petiole often narrowly winged, at least above the midpoint. Leaf blades 1–8(–12) cm long, simple, ovate to nar-

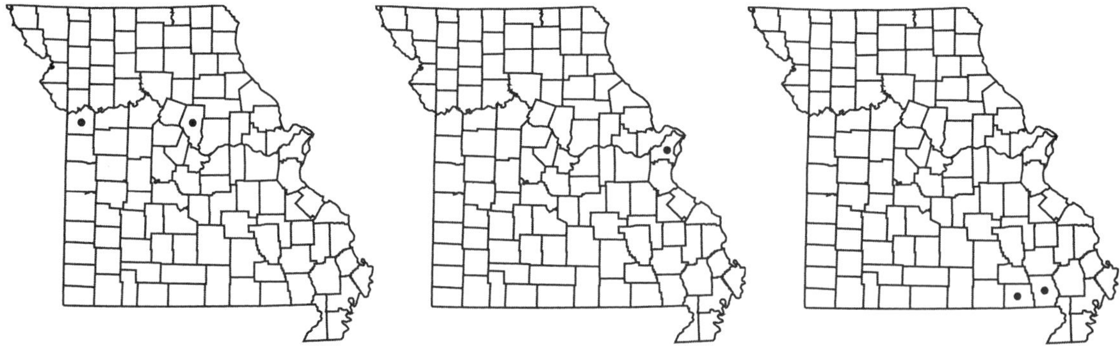

2647. Solanum triflorum 2648. Solanum tuberosum 2649. Sphenoclea zeylanica

rowly ovate, oblong-ovate, or triangular-ovate, angled or tapered to a usually sharply pointed tip, angled to rounded or occasionally truncate to slightly cordate at the base, the margins entire or more commonly wavy to bluntly or occasionally sharply toothed, the surfaces moderately to densely pubescent with short, spreading, multicellular, gland-tipped hairs, somewhat sticky. Inflorescences axillary, umbels of 3–5 flowers, the flower stalks more or less equal, not noticeably jointed (but the fruits usually dispersed with the stalks attached), thickened toward the tips. Flowers spreading to more or less pendant. Calyces 3–6 mm long at flowering, becoming enlarged to 6–11 mm at fruiting, the tube closely cupping the basal half of the fruit, 5-lobed to below the midpoint, the lobes equal, oblong-lanceolate (broadening as the fruits mature), the outer surface densely glandular-hairy. Corollas 5–8 mm long, white to pale cream-colored, usually yellow to yellowish green in the throat, lobed to at or above the midpoint, the lobes broadly triangular, spreading at full flowering, the inner surface glabrous, the outer surface minutely nonglandular-hairy, especially toward the tip. Anthers 1.6–2.0 mm long, oblong, lacking a sterile tip, dehiscing by terminal pores. Ovary 2-locular, the surface glabrous, the style not or only slightly exserted from the anther ring. Fruits 0.6–0.8(–1.0) cm long, globose, with usually 4–6 stony granules, the surface glabrous, olive green to brownish green at full maturity, shiny when young, but dull at maturity, when immature often with scattered, small white flecks or lighter mottling. Seeds 1.3–1.5 mm in longest dimension, broadly obovate to nearly circular in outline, often minutely notched at the attachment point, moderately to strongly flattened, unwinged, the surface faintly and minutely pitted, often appearing nearly smooth, yellow. $2n=24$. July–October.

Introduced, scattered, mostly south of the Missouri River (native of South America, introduced widely in North America and Europe). Banks of streams and rivers, bottomland forests, and glades; also pastures, fencerows, farmyards, gardens, railroads, roadsides, and open, disturbed areas.

Steyermark (1963) treated the limited number of Missouri specimens known to him under the name *S. villosum* Mill. He knew the taxon only from specimens collected by Viktor Mühlenbach in the St. Louis railyards. Mühlenbach (1979) redetermined these as *S. sarrachoides*. Since that time, the range of *S. sarrachoides* in Missouri has expanded greatly, especially as a pasture weed in the Ozarks. *Solanum villosum* is a tetraploid ($2n=48$) Eurasian species that has become introduced rarely and sporadically in the northeasternmost United States and also Florida and California. According to Schilling (1981), it occurs mainly around shipyards and along railroads, and does not persist long at any site. The species differs from *S. sarrachoides* in its yellow to orangish red berries, and smaller fruiting calyces.

Solanum sarrachoides has also been confused with the superficially similar *S. physalifolium* Rusby (Gleason and Cronquist, 1991; Kartesz and Meacham, 1999), another weedy species of South American origin. According to Edmonds (1986), who studied the distinctions between the two species, *S. physalifolium* occurs sporadically throughout the United States and adjacent Canada, but most commonly in the Plains and western States (although she cited no specimens). Thus, it is to be expected to be found eventually in Missouri. The North American populations have been referred to var. *nitibaccatum* (Bitter) Edmonds, which differs from *S. sarrachoides* in its more prostrate growth form, less sticky herbage, short-racemose inflorescences, smaller calyces (2–3 mm long at flowering) that cup around only the bases of the fruits, shorter anthers (0.4–0.5 mm), and shiny berries, with fewer, larger seeds (1.8–2.4 mm) (Edmonds, 1986).

12. Solanum triflorum Nutt. (cutleaf nightshade)

Pl. 568 i, j; Map 2647

Plants annuals, taprooted (not producing tubers). Stems 10–40(–60) cm long, prostrate or spreading with loosely ascending tips, glabrous or sparsely pubescent with short, curved to appressed or less commonly spreading, few- to several-celled, nonglandular hairs (more densely so toward the tips); unarmed. Leaves unarmed, short- to long-petiolate, the petiole often narrowly winged, at least above the midpoint. Leaf blades 1–5 cm long, simple but deeply pinnately several-lobed with rounded sinuses, oblong to ovate in outline, angled or tapered to a sharply pointed tip, short-tapered to broadly angled at the base, the lobes broadly attached, lanceolate to nearly linear, the margins otherwise entire or with a few coarse, blunt teeth, usually rolled under, the surfaces glabrous or more commonly sparsely to moderately pubescent with short, appressed, 1- or few-celled, nonglandular hairs (sometimes mostly along the veins). Inflorescences axillary, of solitary flowers or umbels of 2 or 3 flowers, the flower stalks not jointed (but the fruits usually dispersed with the stalks attached), slender to somewhat stout, sometimes slightly thickened toward the tips. Flowers spreading to more or less pendant. Calyces 1.5–2.5 mm long at flowering, becoming enlarged to 4–6 mm at fruit, more or less cupped around the basal 1/3 of the fruit, 5-lobed (sometimes slightly unequally so) to below the midpoint, the lobes oblong-lanceolate to narrowly triangular, the outer surface sparsely to moderately pubescent (mostly toward the base) with short, relatively stout, nonglandular hairs. Corollas 4–7 mm long, white, sometimes pale purplish-tinged, yellow to yellowish green in the throat, lobed to at or below the midpoint, the lobes broadly lanceolate to bluntly triangular, spreading to somewhat recurved at full flowering, the inner surface glabrous, the outer surface minutely nonglandular-hairy, especially toward the tip. Anthers 2.0–2.5 mm long, oblong, lacking a sterile tip, dehiscing by terminal pores. Ovary 2-locular, the surface glabrous, the style usually exserted from the anther ring. Fruits 0.9–1.2 cm long, globose, with usually 4–15 stony granules, the surface glabrous, purplish black at full maturity but often remaining green for a relatively long time, then dull or shiny, often with lighter mottling or streaks. Seeds 2.0–2.5 mm in longest dimension, broadly obovate to nearly circular in outline, often minutely notched at the attachment point, strongly flattened, unwinged, the surface minutely pitted or with a very fine network of ridges, yellow. $2n=24$. June–August.

Introduced, uncommon, sporadic (native of the western U.S. east to Minnesota and Oklahoma; Canada; introduced farther east). Railroads and open to semi-shaded, disturbed areas.

This species is generally described as having a foetid odor (Gleason and Cronquist, 1991).

13. Solanum tuberosum L. (potato)

Pl. 568 k, l; Map 2648

Plants perennial (but usually blooming the first year), fibrous-rooted, with slender, fleshy offsets bearing narrowly oblong to ovoid, oblong-ellipsoid, or depressed-globose tubers. Stems 15–70(–100) cm long, erect or ascending, sometimes reclining at maturity, sometimes angled or with narrow, wavy wings of green tissue decurrent from the leaf bases, sparsely to moderately pubescent with short, spreading, multicellular, nonglandular hairs, sometimes also sticky; unarmed. Leaves unarmed, short-petiolate. Leaf blades 7–18 cm long, those of the larger leaves irregularly pinnately compound (the smaller leaves merely deeply pinnately divided), the leaflets strongly unequal, the main leaflets 5–9, 1–8 cm long, the terminal leaflet usually larger than the lateral ones, narrowly attached and mostly short-stalked, interspersed along the rachis with much smaller leaflets, ovate to broadly ovate, oblong-ovate, or elliptic, rounded or more commonly tapered to a bluntly or sharply pointed tip, tapered to rounded at the sometimes somewhat oblique base, the margins entire, the surfaces sparsely to moderately pubescent with short, nonglandular hairs, especially along the main veins on the undersurface. Inflorescences terminal or appearing lateral from the uppermost nodes, rounded or flat-topped panicles, with (3–)7–25 well-spaced flowers, the flower stalks inconspicuously jointed at or above the midpoint. Flowers ascending to spreading. Calyces 4–9 mm long, deeply 5-lobed, spreading to slightly reflexed at fruiting, the lobes variously ovate-triangular to linear above a short, ovate-triangular base, the outer surface nonglandular-hairy, sometimes also with scattered, sessile to short-stalked glands. Corollas 12–28 mm long, white to purple, shallowly lobed to above the midpoint, the lobes broadly ovate to broadly triangular, spreading at full flowering, the margins usually somewhat corrugated or slightly ruffled, the inner surface glabrous, the outer surface minutely hairy toward the tip. Anthers 5–7 mm long, narrowly oblong, lacking a sterile tip, dehiscent by terminal pores. Ovary 2-locular, the surface glabrous, the style exserted from the anther ring. Fruits 1.0–2.5 cm long, globose, lacking granules, the surface glabrous, green mottled with dark green, becoming yellow or more

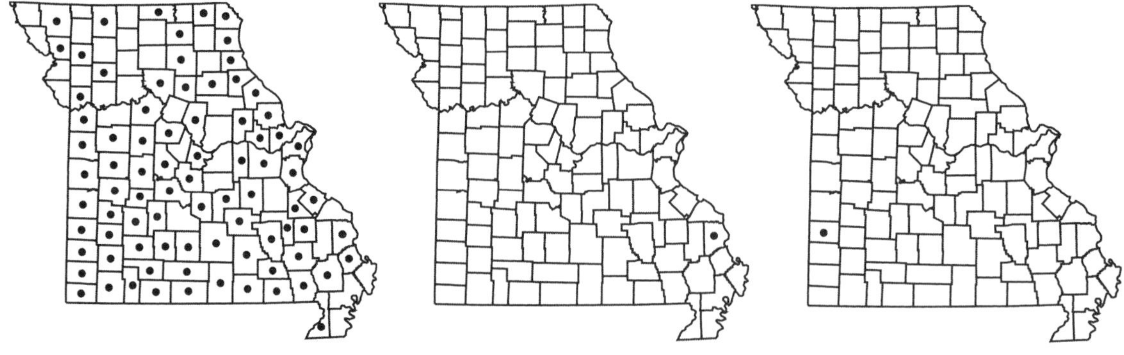
2650. Staphylea trifolia 2651. Halesia carolina 2652. Halesia diptera

commonly strongly purplish-mottled with age, somewhat shiny, sometimes with widely scattered, small, white flecks. Seeds 1.8–2.2 mm long, ovate to broadly oblong-ovate in outline, flattened, not winged, the surface appearing fuzzy (covered with an outer layer of disrupted cell walls that render the surface mucilaginous when wetted), whitish green to light brown. $2n$=24, 36, 48. June–August.

Introduced, uncommon, known thus far only from a historical collection from the city of St. Louis (native of South America, escaped sporadically in North America). Railroads and open disturbed areas.

Solanum tuberosum is an extremely important food crop and a principal source of carbohydrates for humans around the world. Potato starch also has a variety of uses, ranging from cooking to the basis for a kind of electrophoretic gel in laboratory analyses. In recent years, potato cultivation has become more widespread in the sandy soils of the Sikeston Ridge in the northern portion of the Mississippi Lowlands Division. Potato domestication occurred more than 7,000 years ago in Andean South America (Hawkes, 1990), and it is possible that independant domestication also occurred in Chile. A large number of cultivars of the potato are grown, differing most notably in a suite of tuber characteristics involving size, shape, texture, color, etc. The cultivated potatoes have variously been classified taxonomically as seven or more species with additional infrataxa (Hawkes, 1990) or as a single species with varying numbers of cultivar groups (Huamán and Spooner, 2002).

SPHENOCLEACEAE (Sphenoclea Family)

One genus, 1 or 2 species, nearly worldwide.
Some authors include this family within the Campanulaceae.

1. **Sphenoclea** Gaertn.

One or 2 species, nearly worldwide.

1. Sphenoclea zeylanica Gaertn. (chicken spike, gooseweed)

Pl. 568 m, n; Map 2649

Plants annual, glabrous, usually somewhat glaucous. Stems 25–150 cm long, erect or ascending, sometimes rooting from prostrate lower nodes, hollow, usually with numerous branches. Leaves alternate, petiolate, lacking stipules. Leaf blades 1–12 cm long, ovate to elliptic, rounded to pointed at the tip, narrowed at the base, the margins entire. Inflorescences dense spikes at stem and branch tips, with numerous flowers. Flowers zygomorphic, perfect, epigynous, subtended by 3 unequal, inconspicuous, linear to narrowly spatulate, green-tipped bracts. Calyx 0.4–0.7 mm long in flower, deeply 5-lobed, enlarging to 1.5 mm long

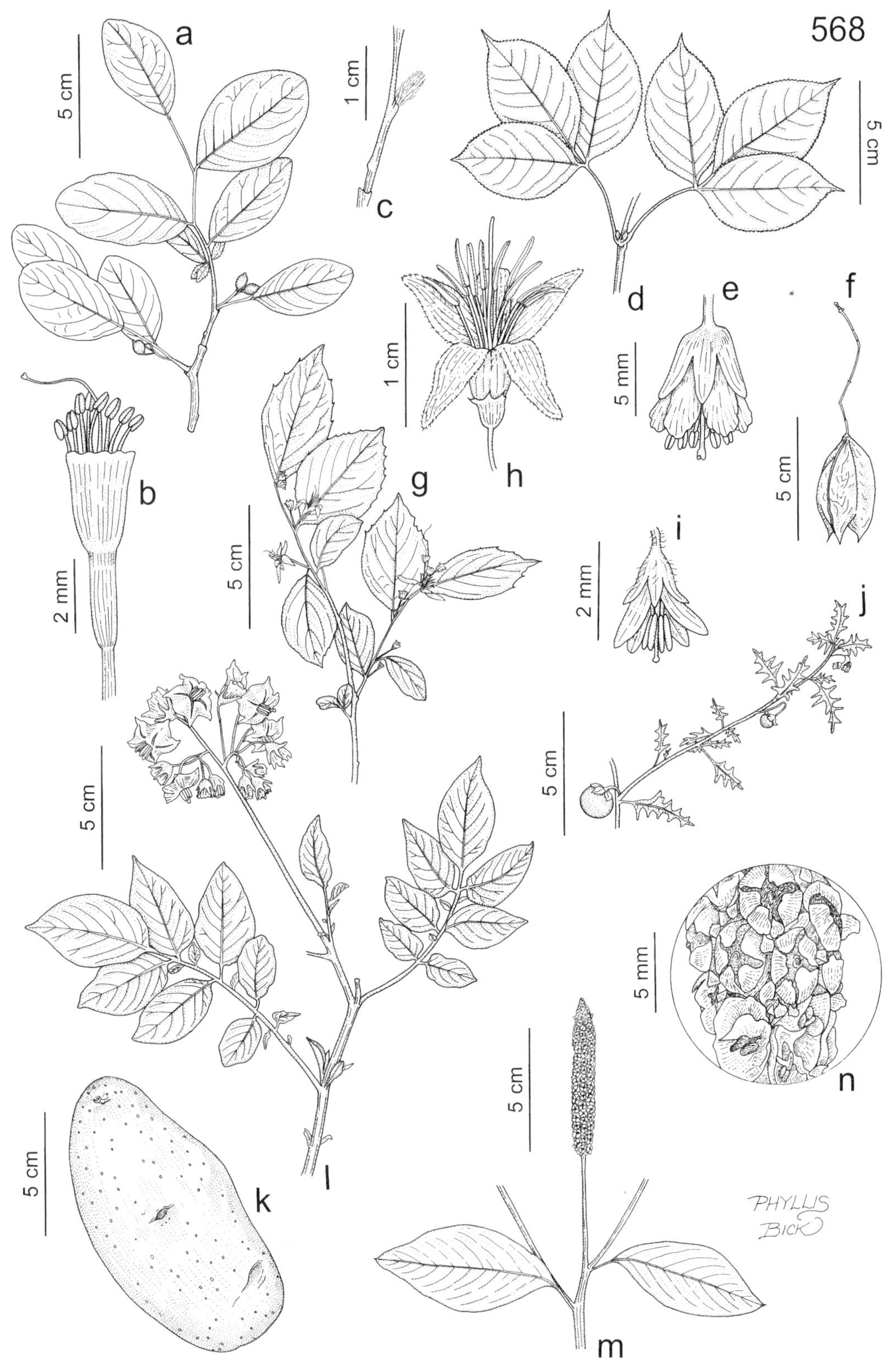

Plate 568. Solanaceae. Sphenocleaceae. Staphyleaceae. Styracaceae. Thymelaeaceae. *Dirca palustris*, **a)** fertile branch, **b)** flower, **c)** twig with winter bud. *Staphylea trifolia*, **d)** node with leaves, **e)** flower, **f)** fruit. *Styrax americanus*, **g)** fertile branch, **h)** flower. *Solanum triflorum*, **i)** flower, **j)** fertile stem. *Solanum tuberosum*, **k)** tuber, **l)** fertile stem. *Sphenoclea zeylanica*, **m)** portion of stem with leaves and inflorescence, **n)** inflorescence detail.

in fruit, the lobes triangular to broadly triangular, irregularly pointed to rounded or notched at the tip, the margins sometimes also shallowly and irregularly toothed, persistent on the fruit. Corollas 5-lobed, 2.0–2.5 mm long, white, the lobes oblong-triangular, rounded to bluntly pointed at the tip. Stamens 5, the filaments fused with the basal portion of the corolla tube. Pistil 1 per flower, of 2 fused carpels. Ovary inferior, with 2 locules, the placentation axile. Style minute, the stigma capitate, slightly 2-lobed. Fruits capsules, 3–4 mm long, broadly obconic, slightly 4- or 5-angled, the dehiscence circumscissile by a flattened lid with the attached calyx. Seeds numerous, 0.3–0.5 mm long, narrowly oblong, with 4–6 irregular longitudinal ridges, yellowish tan, shiny. $2n$=24. July–November

Introduced, uncommon, restricted to the Mississippi Lowlands Division (nearly worldwide weed in rice fields). Rice fields and ditches, usually emergent aquatics.

Sphenoclea zeylanica was first reported as occurring in Missouri by S. Hudson (1992). Chicken spike is nearly cosmopolitan in areas where rice is cultivated. In addition to its use as a potherb in portions of the Old World tropics, it is tolerated as a weed because an exudate from its roots apparently provides a control for rice-root nematodes, *Hirschmanniella* spp. (Rosatti, 1986), The natural range of *S. zeylanica* is poorly understood because it became widely distributed through human agricultural activities before it could be studied botanically. It is native to some portion of the Old World tropics. The second species, *S. pongatia* A. DC., which is often treated as a synonym of *S. zeylanica*, is endemic to western Africa (Rosatti, 1986) and thus may provide a clue to the origins of the genus.

STAPHYLEACEAE (Bladdernut Family)

Five genera, 27–50 species, North America, Europe, Asia, Malesia.

1. **Staphylea** L. (bladdernut)

About 10 species, North America, Europe, Asia.

1. Staphylea trifolia L. (American bladdernut) Pl. 568 d–f; Map 2650

Plants shrubs or rarely small trees to 3 m tall, colonial by root suckers. Bark smooth or slightly furrowed with age, often appearing somewhat striped on branches, sometimes peeling in fine flakes with age. Twigs not angled, glabrous or minutely and inconspicuously hairy, greenish brown to reddish brown, the buds lateral, with several scales, ovate in outline, reddish brown, glabrous. Leaves opposite, long-petiolate. Stipules inconspicuous, membranous, linear, shed as the leaves develop. Leaf blades trifoliate, the lateral leaflets short-stalked to nearly sessile, the middle leaflet longer-stalked. Leaflets 4–10 cm long, ovate to elliptic or oblong-obovate, short-tapered to a sharply pointed tip, narrowed or tapered at the base (the lateral ones sometimes asymmetric at the base), the margins finely toothed, the upper surface glabrous at maturity or sparsely pubescent with shorter hairs, the undersurface moderately to densely pubescent with longer, crinkled, mostly multicellular hairs. Inflorescences axillary or terminal on second year's branch growth, loose racemelike clusters 4–10 cm long, pendant, produced as the leaves develop or afterward. Flowers perfect, hypogynous, actinomorphic. Calyx of 5 sepals, these fused at the base, 6–8 mm long, oblong-ovate to oblong-triangular, erect, green, glabrous. Corolla of 5 petals, these 8–10 mm long, obovate, erect with a usually spreading tip, white to greenish white. Stamens 5, small, free, the filaments attached at the base of a nectar disc, hairy below the midpoint, the anthers attached at the notch of the cordate base, exserted, yellow. Pistil 1 per flower, of 3 fused carpels. Ovary superior, 3 lobed, 3-locular, the placentation axile or appearing basal, with 4–12 ovules per locule. Styles 3, fused toward the base and also at the very tip, the stigmas forming a more or less disc-shaped mass. Fruits strongly inflated pendant capsules, 3–5 cm long, obovoid, 3-lobed at maturity toward the tip, the lobes narrowed to sharply pointed tips, the walls thin and papery to somewhat leathery, brown, indehiscent or eventually dehiscent from the tip along the inner sutures of the lobes, usually persistent through the winter. Seeds 1–4 per locule, 5.5–6.0 cm long, ellipsoid to nearly globose, slightly flattened, the surface smooth, brown, somewhat shiny. $2n=78$. April–May.

Scattered to common throughout the state (eastern U.S. west to Minnesota, Nebraska, and Oklahoma; Canada). Banks of streams and rivers, bases and sheltered ledges of bluffs, and mesic upland forests in ravines, often on north- to east-facing slopes on calcareous substrates.

Bladdernut is sometimes cultivated as an ornamental shrub for its unusual fruits and attractive flowers. The leaves remain green until late in autumn, eventually turning yellowish green. It is easily distinguished vegetatively from the superficially similar *Ptelea trifoliata* L. (Rutaceae, hop tree) by its opposite leaves with a relatively long-stalked central leaflet.

STYRACACEAE (Styrax Family)

Plants shrubs or small trees, usually pubescent with minute, stellate hairs, the wood often resinous, the bark, twigs, and winter buds various. Leaves alternate, short- to moderately petiolate, the fall foliage turning yellow. Stipules absent. Leaf blades simple, narrowly to broadly elliptic, oblong-elliptic, ovate, or oblong-obovate, tapered to a sharply pointed tip, angled or rounded at the base, the margins entire or finely toothed, the surfaces usually stellate-hairy when young, often becoming glabrous or nearly so at maturity. Inflorescences axillary (sometimes appearing terminal on short branches in *Styrax*), of solitary flowers or small clusters or racemes of 2–6 flowers. Flowers perfect, actinomorphic, perigynous to epigynous (hypogynous elsewhere), relatively long-stalked, not subtended by bractlets. Calyces 4- or 5-lobed, obconic to narrowly obconic, the lobes mostly small, triangular, toothlike, glabrous or sparsely stellate-hairy, persistent but often inconspicuous at fruiting. Corollas deeply 4- or 5-lobed, bell-shaped or more or less saucer-shaped, white, the lobes rounded to truncate and sometimes slightly irregular, the outer surface moderately to densely stellate-hairy. Stamens 8 or 10, the filaments usually attached in 1 row at the base of the corolla tube, sometimes fused into a minute crown basally, glabrous, the anthers attached at their bases, linear, dehiscing longitudinally. Pistil 1 per flower, of 2–4 fused carpels. Ovary partially to entirely inferior (superior elsewhere), with 2–4, often incomplete locules, the placentation axile. Style 1, unbranched, the stigmas minute or capitate, entire or shallowly 3- or 4-lobed. Ovules 4–6 per locule. Fruits various, mostly modified capsules (often indehiscent or irregularly and tardily dehiscent), drupelike or samaralike in Missouri genera, variously shaped, the surface glabrous or sparsely stellate-hairy. Seeds 1–3(4) per fruit, at most 1 per locule, narrowly ellipsoid or oblong-ellipsoid, the surface sometimes shiny, sometimes longitudinally several-grooved, brown. Eleven genera, about 160 species, North America to South America, Caribbean Islands, Europe, Asia south to Indonesia.

The family Styracaceae has an extensive fossil history and some of the genera were once more widespread than they are in the present day (Manchester et al., 2009). Today, the majority of the species are in the widely distributed genus *Styrax*. The main economic use of the family is for horticulture; several genera include commonly cultivated species (Spongberg, 1976). A few of the larger tropical species furnish wood for handcrafts. Some of the species also have medicinal uses.

1. Petioles 10–35 mm long; twigs with chambered pith; flowers epigynous; calyces with the tube fused to the ovary for their entire length, shallowly 4-lobed (or the lobes sometimes absent); corollas 4-lobed; fruits samaralike, appearing somewhat flattened, winged . . . . 1. HALESIA

1. Petioles 3–10 mm long; twigs with solid pith; flowers perigynous; calyces fused to the ovary for about 1/2 their length, shallowly 5-lobed; corollas 5-lobed; fruits drupelike, subglobose to somewhat obovoid, unwinged . . . . 2. STYRAX

1. **Halesia** J. Ellis ex L. (silverbell)

Plants shrubs or more commonly small trees. Bark gray and brown, initially with a network of shallow, longitudinal ridges (often appearing striped), on older trunks eventually flaking off in small plates. Twigs green, darkening to grayish brown or more commonly reddish brown, often with darker lenticels, shiny, pubescent with small stellate hairs when very young, becoming sparsely hairy or glabrous at maturity, the outer layer usually peeling or shredding in thin strips with age, the winter buds ovoid to ellipsoid, with several overlapping scales, the pith chambered. Petioles 10–35 mm long. Leaf blades narrowly to broadly ovate to obovate, rarely oblanceolate, rounded to broadly angled at the base, bluntly and broadly pointed to noticeably but abruptly tapered at the tip, the margins minutely toothed, the surfaces felty with minute stellate hairs when young, these shed in patches, essentially glabrous at maturity, the upper surface green, the undersurface usually pale green to grayish green. Inflorescences axillary, small clusters or short racemes of 2–6 flowers, rarely reduced to solitary flowers. Flowers epigynous, the calyces fused to the full length or the ovary, the outer surface stellate-hairy, the free portion absent or extending above the ovary as 4 small, more or less triangular lobes to 2.5 mm long, the stalk jointed at the calyx base. Corollas shallowly or deeply 4-lobed, bell-shaped, the lobes more or less ascending at maturity. Stamens 8–16. Fruits samaralike, 2.5–6.0 cm long green, turning tan to reddish brown at maturity, variously shaped, beaked at the tip, appearing somewhat flattened, 2- or 4-winged, the surface glabrous or nearly so at maturity, the middle layer more or less mealy, usually developing a hollow chamber above the seeds, indehiscent. Seeds relatively thin-walled, narrowly ellipsoid, pointed at each end. Three species, U.S., Asia.

As many as five New World species (plus one Asian taxon) have been recognized by some authors, but Fritsch and Lucas (2000) determined that only two morphologically separable species exist in North America.

1. Fruits strongly 4-winged, the wings about equally developed; corollas shallowly lobed (the lobes much shorter than the tube), glabrous . . . . 1. H. CAROLINA
1. Fruits strongly 2-winged, but the middle of each face strongly ridged or occasionally narrowly winged; corollas deeply lobed (the lobes noticeably longer than the tube), hairy on the outer surface . . . . 2. H. DIPTERA

1. Halesia carolina L. (Carolina snowbells)
H. tetraptera J. Ellis

Map 2651

Plants shrubs or more commonly small trees to 4 m tall (to 20 m or more elsewhere). Leaf blades 6–22 cm long, oblong-ovate to elliptic or obovate, those of the smaller leaves rarely oblanceolate, bluntly and broadly pointed to tapered at the tip, the margins with mostly closely spaced (usually 12–40 per side) fine teeth, the upper surface stellate-hairy, usually dull at maturity, the undersurface moderately stellate-hairy when young, especially along the main veins, sometimes glabrous or nearly so at maturity. Calyces (including the hypanthium) 3–5 mm long. Corollas 10–25 mm long, shallowly 4-lobed (the lobes much shorter than the tube), glabrous. Stamens 12–16, the filaments usually hairy only along the inner side. Style glabrous. Fruits 2.5–6.0 cm long (not counting the persistent, hardened, stylar beak), elliptic to oblong-obovoid in outline, strongly 4-winged, the wings about equally developed. $2n=24$. May–June.

Introduced (but see discussion below), known thus far only from Cape Girardeau and possibly Barton Counties (southeastern U.S. west to Arkan-

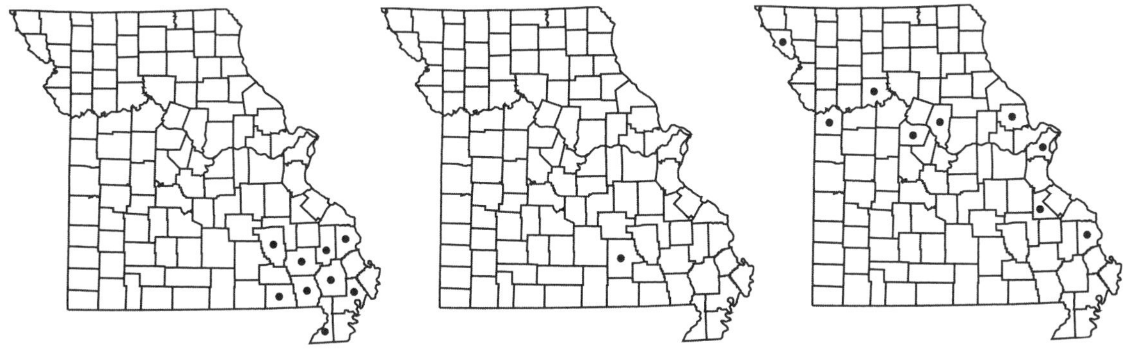

2653. Styrax americanus 2654. Styrax grandifolius 2655. Tamarix ramosissima

sas and Texas; introduced farther north). Disturbed mesic upland forests.

Steyermark (1963) excluded this species from the Missouri flora, stating that no specimens could be located to confirm an earlier mention of Missouri in the range by Fernald (1950). However, specimens collected in 1980 by Otto Ohmart are accessioned at the herbarium of Southeast Missouri State University. These originated from the university campus at a former plant nursery site and were indicated as likely to be spreading on the specimen labels. The Barton County record is based on anecdotal evidence. A photo of an obviously four-winged, detached fruit was taken in 1994 by Mike Skinner at the Lester R. Davis Memorial Conservation Area, where the previous owner of the property planted numerous woody species to test their hardiness in the region and allowed these to naturalize if able. However, shrubs to document the species were not located during subsequent visits to the property by several botanists and no voucher specimen exists for this county.

Halesia carolina may be discovered as an escape at other sites in the future, as it is hardy in at least the southern half of the state and relatively commonly grown in woodland gardens. Additionally, it has been documented from native populations in several counties in northwestern Arkansas, as well as in southernmost Illinois and adjacent western portions of Kentucky and Tennessee (Fritsch and Lucas, 2000). Thus it is plausible that a native population might exist in some remote spot in southern Missouri.

2. Halesia diptera J. Ellis (two-winged silverbell, American snowdrop tree)
H. diptera var. *magniflora* R.K. Godfrey
Map 2652

Plants shrubs or small trees to 4 m tall (to 11 m elsewhere). Leaf blades 6–18 cm long, ovate to elliptic or obovate, usually noticeably tapered at the tip, the margins with mostly well-spaced (usually 6–14 per side) fine teeth, at least above the midpoint, the upper surface stellate-hairy along the main veins, often shiny at maturity, the undersurface moderately stellate-hairy when young, especially along the main veins, usually glabrous or nearly so at maturity. Calyces (including the hypanthium) 4–6 mm long, the outer surface stellate-hairy, the tube (hypanthium) narrowly obconic at flowering. Corollas 10–18 mm long (to 30 mm elsewhere), deeply 4-lobed (the lobes noticeably longer than the tube), densely pubescent on the outer surface with short, curled hairs. Stamens 8(–10), the filaments usually evenly hairy on both sides. Style hairy. Fruits 2.5–5.5 cm long (not counting the persistent, hardened, stylar beak), oblong-elliptic to elliptic, oblanceolate or narrowly oblong-obovate in outline, with a pair of broad, corky, longitudinal wings on opposing sides, the middle of each face strongly ridged or occasionally narrowly winged. $2n=24$. May–June.

Introduced, uncommon, known thus far only from a single specimen from Barton County (southeastern U.S. west to Arkansas and Texas). Disturbed mesic upland forests.

Larger-flowered plants (corollas to 30 mm long) endemic to portions of southern Georgia and adjacent Florida have been called var. *magniflora*, but do not appear sharply distinct.

This species has been documented from the Lester R. Davis Memorial Conservation Area, where the previous owner of the property planted numerous woody species to test their hardiness in the region and allowed these to naturalize if able. It was first collected as an escape in 2008 by Mike Skinner of the Missouri Department of Conservation.

2. **Styrax** L. (styrax, storax)

Plants shrubs (small trees elsewhere). Bark gray, roughened. Twigs green to gray or reddish brown, with lighter lenticels, shiny, pubescent with small stellate hairs when very young, becoming glabrous with age, the winter buds ovoid to ellipsoid, lacking scales, the pith solid. Petioles 3–10 mm long. Leaf blades narrowly to broadly ovate or obovate, rarely oblanceolate, rounded to broadly angled at the base, bluntly and broadly pointed to noticeably but abruptly tapered at the tip, the margins minutely toothed to entire or nearly so, the upper surface pale to dark green, glabrous or nearly so at maturity (sometimes minutely stellate-hairy along the main veins when young), the undersurface pale green or sometimes appearing grayish green to somewhat silvery, glabrous or sparsely and minutely stellate-hairy at maturity. Inflorescences terminal and/or axillary, clusters or short racemes of 2–20 flowers, sometimes of solitary flowers. Flowers perigynous, the calyces fused to the ovary for about half of their length, the stalk not jointed at the calyx base. Corollas deeply 5-lobed, bell-shaped to broadly bell-shaped or more or less saucer-shaped, the lobes often curled outward at full flowering. Stamens 10. Fruits drupelike, 5–11 mm long, gray to grayish green, subglobose to somewhat obovoid, not or only minutely beaked at the tip, not flattened, lacking wings, the surface densely pubescent with a covering of minute stellate hairs, the middle layer waxy or mealy, solid, indehiscent. Seeds with a relatively thick, hard coat, subglobose to ovoid or ellipsoid, rounded or blunt at each end. About 130 species, North America to South America, Caribbean Islands, Asia south to Indonesia.

Some species of *Styrax* are cultivated as ornamentals. A resin collected from the injured trunks of some of the Old World species is known as benzoin and gum benjamin. It has been used in incense, perfumes, and flavored cigarettes. It also was used medicinally in mixtures as a disinfectant and for the treatment of bronchitis and asthma.

It should be noted that there has been considerable controversy over the gender of the generic name *Styrax* (Wood and Channell, 1960; Howard, 1974). Current practice favors neuter terminations (with epithets mostly ending in *-us*), as Linnaeus (1753) intended in his original description of the genus.

1. Leaf blades mostly elliptic to oblong or ovate, rarely broadly elliptic to obovate, the margins entire or with small teeth scattered on each side, the undersurface glabrous; inflorescences clusters or less commonly short racemes of 2–5 flowers, sometimes of solitary flowers; calyces (including the hypanthium) 2.5–4.0 mm long; corollas 10–16 mm long . . . . 1. S. AMERICANUS
1. Leaf blades mostly broadly obovate to nearly circular, less commonly broadly elliptic to ovate, the margins entire or with a few blunt teeth on each side toward the tip, the undersurface usually sparsely to densely stellate-hairy (rarely glabrous); inflorescences racemes of 5–20 flowers, rarely reduced to small clusters or solitary flowers; calyces 4–6 mm long; corollas 15–22 mm long . . . . 2. S. GRANDIFOLIUS

1. Styrax americanus Lam. (snowbell, American snowbell, mock orange)

S. americanus var. *pulverulentus* (Michx.) Rehder

Pl. 568 g, h; Map 2653

Plants shrubs to 4 m tall. Petioles 3–5 mm long. Leaf blades 1.5–10.0 cm long, mostly elliptic to oblong or ovate, rarely broadly elliptic to obovate, the margins entire or with small teeth scattered on each side, the upper surface glabrous, dark green to green, the undersurface glabrous or sparsely and minutely stellate-hairy at maturity (more densely hairy elsewhere), pale green. Inflorescences axillary, sometimes appearing terminal at the ends of short branches, clusters or less commonly short racemes of 2–5 flowers, sometimes of solitary flowers. Calyces (including the hypanthium) 2.5–4.0 mm long, the tube (hypanthium) obconic at flowering, sparsely stellate-hairy to nearly glabrous, shallowly 5-lobed, the lobes 0.5–1.0 mm long, triangular, sharply pointed. Corollas 10–16 mm long. Fruits (5–)8–10 mm long, 6–8 mm wide, subglobose to somewhat obovoid. 2*n*=16. April–May.

Uncommon in the Mississippi Lowlands Division (southeastern U.S. west to Missouri and Texas). Swamps and bottomland forests; also levees, ditches, and wet roadsides.

Steyermark (1963) noted that *S. americanus* is a very showy shrub when in flower and has been underutilized in Missouri landscaping, perhaps in part because the species requires a moist environment to reach its potential.

Gonsoulin (1974) separated plants with more persistently hairy foliage and slightly broader leaves as var. *pulverulentus* and cited a specimen collected by Julian Steyermark in Butler County as this variety. However, Steyermark (1963) himself had studied the situation earlier and concluded that the hairier variety did not occur in Missouri. Part of the disagreement stems from the fact that the differences between the varieties are not sufficient to permit formal taxonomic recognition and numerous intermediates with varying degrees of leaf pubescence exist.

2. Styrax grandifolius Aiton (bigleaf snowbell) Map 2654

Plants shrubs to 4 m tall (to 9 m elsewhere). Petioles (3–)5–10 mm long. Leaf blades 4–10(–20) cm long, mostly broadly obovate to nearly circular, less commonly broadly elliptic to ovate, the margins entire or with a few blunt teeth on each side toward the tip, the upper surface glabrous or nearly so at maturity (sometimes minutely stellate-hairy along the main veins when young), dark green to pale green, the undersurface sparsely to densely stellate-hairy, rarely glabrous, pale green or grayish green to silvery. Inflorescences terminal and axillary (terminal at the ends of short branches), racemes of 5–20 flowers, rarely reduced to small clusters or solitary flowers. Calyces (including the hypanthium) 4–6 mm long, the tube (hypanthium) obconic at flowering, densely stellate-hairy, shallowly 5-lobed, the lobes 0.5–1.0 mm long, triangular to broadly triangular, sharply pointed. Corollas 15–22 mm long. Fruits 7–11 mm long, 7–10 mm wide, subglobose. 2*n*=32. April–May.

Uncommon, known thus far only from a single collection from Shannon County (southeastern U.S. west to Missouri and Texas). Mesic upland forests.

This species was first reported for Missouri by Gonsoulin (1974), based on a specimen collected by Julian Steyermark in New Madrid County (*Steyermark 69683*, in the herbarium of the Field Museum). Unfortunately, the report was erroneous; that specimen had actually determined correctly as *S. americanus* by Gonsoulin during his studies. Kurz (1997) and others suggested that the species might eventually be discovered in southeastern Missouri. The first verified population of *S. grandifolius* was not discovered until 2012 by Justin Thomas and Jacob Hadle.

In most of its range, *S. grandifolius* can occur in habitats similar to those in which *S. americanus* grows. However, in several states it also thrives at more upland sites, where it often produces leaves that are somewhat smaller than those of bottomland plants. Plants at these upland sites frequently produce very few fruits. However, the species can reproduce vegetatively through often extensive root suckering; at the Shannon County site, plants formed an almost continuous expanse covering about half an acre (Justin Thomas, personal communication).

W. H. Duncan and M. B. Duncan (1988) noted that vegetative material of *S. grandifolius* from smaller-leaved plants can be difficult to distinguish from similar material of *Halesia diptera*. They suggested that the winter buds are slightly different in the two taxa. In both genera the axillary winter buds are *superposed*, that is, each smaller axillary bud has an additional bud produced above it. In *Styrax*, the upper of the two buds tends to be slightly separated from the lower one and is more rounded at the tip (thumb-shaped), whereas in *Halesia*, the upper bud is always contiguous with the lower one and is more pointed apically (conic).

2656. Polypremum procumbens 2657. Dirca decipiens 2658. Dirca palustris

TAMARICACEAE (Tamarisk Family)

Five genera, about 100 species, Europe, Asia, Africa

1. Tamarix L. (tamarisk) (Baum, 1967, 1978)

About 54 species, Europe, Asia, Africa, mostly in the Middle East and Mediterranean regions, widely introduced elsewhere, especially in arid regions.

1. Tamarix ramosissima Ledeb. (salt cedar)
Pl. 567 g–i; Map 2655

Plants shrubs or small trees to 5 m tall, highly branched, deciduous and often shedding branchlets in autumn. Bark reddish brown to dark purplish gray, smooth or becoming ridged with age. Twigs slender, light green, glabrous. Leaves alternate, sessile. Stipules absent. Leaf blades 0.8–1.5 mm long, scalelike, glabrous, light green, lanceolate to ovate, the base clasping the stem, the tip sharply pointed, the margins entire. Inflorescences panicles with racemose branches (rarely reduced to simple racemes) of numerous tiny flowers. Flowers perfect, actinomorphic, hypogynous, short-stalked, each subtended by a minute scalelike bract, the perianth persisting in fruit. Sepals 5, 0.5–1.0 mm long, lanceolate to elliptic or ovate, bluntly to sharply pointed at the tip (rarely rounded), light green, the margins irregularly and minutely toothed, often thinner and white. Petals 5, 1.0–1.8 mm long, obovate to oblong-obovate, light pink to pink, fading to grayish white. Stamens 5, exserted. the filaments attached between the shallow lobes of a nectar disc. Pistil 1 per flower, of 3(–5) fused carpels. Ovary superior, flask-shaped, with 1 locule, the placentation appearing more or less basal. Styles 3(–5), usually appearing twisted, each with 1 club-shaped stigma. Ovules numerous. Fruits capsules, 3–4 mm long, tapered to a beaklike tip, dehiscing longitudinally into 3(–5) valves. Seeds 0.5–0.7 mm long, the body minute, obovoid, yellowish tan, with a slender tail-like appendage at the tip, this with a coma of dense long fine hairs that spreading at right angles to the axis when dry. $2n=24$. April–September.

Introduced, uncommon, mostly in the Big Rivers Division, but occasionally escaped from cultivation elsewhere in the state (native of Europe and Asia, widely naturalized in the southeastern U.S., more sporadically in the central and southeastern U.S.). Sand and gravel bars along rivers; also mine spoils, roadsides, alleys, and railroads.

Of the 54 species of *Tamarix*, 10 have been introduced in the United States (Crins, 1989). They were originally cultivated as ornamentals, for wind screens, and to a lesser extent for mine spoil reclamation, and most subsequently escaped. Especially in the arid southwestern states, a few of these species, including *T. chinensis* Lour., *T. ramosissima*, and their naturally occurring hybrid have become extremely bad invasive weeds in floodplain habitats (Gaskin and Schaal, 2002), where they crowd out native riparian plant communities and, by virtue of their extensive root systems and high water use, drastically lower the water table. Nearly all of the species thrive in salty soils and the genus is characterized by glands that secrete salts taken up by the roots and translocated. The plants usually have minute white beadlike salt secretions on the surfaces of the branches, leaves, and calyces.

Steyermark (1963) appears to have used a very preliminary account of the genus in the southwestern United States by Shinners (1957), which included only five species, as the basis for his acceptance of *T. gallica* L. (French tamarisk) in Missouri. Baum (1967) presented the first comprehensive account of the naturalized tamarisks, but curiously included only *T. parviflora* DC. (small-flowered tamarisk) for Missouri (later erroneously accepted by Yatskievych and Turner [1990]). *Tamarix parviflora* has 4-parted flowers, whereas all of the fertile specimens from Missouri examined thus far are clearly 5-merous and key to *T. ramosissima* in Baum's (1967, 1978) keys. Crins (1989) and others have questioned whether *T. ramosissima* is distinct from *T. chinensis*, which differs only in subtle characters of the sepal shape and attachment points of the stamens on the nectar disc. However, Gaskin and Schaal (2002, 2003) presented molecular evidence that the two should be treated as closely related but separate species. Baum himself suggested that further studies were desirable in regions where these two taxa grow together. It should be noted that most tamarisk species are indistinguishable if no flowers or fruits are present.

TETRACHONDRACEAE (Juniper-Leaf Family)

Two genera, 3 species, North America to South America, Caribbean Islands, New Zealand.

The family Tetrachondraceae is relatively isolated within the order Lamiales, with distant affinities to either the Gesneriaceae or Oleaceae. *Polypremum* traditionally was classified in the Loganiaceae (Steyermark, 1963; G. K. Rogers, 1986) and more recently in the Buddlejaceae (Cronquist, 1981, 1991). The other genus, *Tetrachondra* Petrie ex Oliv. (2 species of South America and New Zealand), formerly was included in the Lamiaceae. Data from molecular phylogenetic studies (Oxelman et al., 1999; Backlund et al., 2000) provided evidence that the two genera are closely related and that they should be segregated into their own small family. However, some botanists have been uncomfortable placing both of these genera in the same family, and the name Polypremaceae Takht. ex Reveal (Reveal 2011) was recently validated to accommodate those taxonomists who feel that *Polypremum* is sufficiently distinct to merit its own family.

1. **Polypremum** L.

One species, North America to South America, Caribbean Islands.

1. Polypremum procumbens L. (juniper-leaf, rustweed)

Pl. 569 a–c; Map 2656

Plants perennial herbs, often forming dense tufts of short, densely leafy, overwintering stems toward the end of the growing season, pubescent with minute unbranched appressed hairs. Flowering stems 6–25 cm long, spreading to loosely ascending, often branched dichotomously, angled or ridged, at least toward the base, slightly roughened with sparse hairs along the angles. Leaves basal and opposite along the stems, sessile, the bases of each leaf in a pair connected by a small stipular membrane. Leaf blades 0.4–2.5 cm long, linear or nearly so, often relatively stiff, tapered to a sharp point at the tip, the margins slightly roughened with sparse hairs. Inflorescences dense, small, sometimes branched clusters, terminal on the branches, rarely reduced to solitary axillary flowers, with short leaflike bracts. Flowers perfect, perigynous, subtended by a pair of short leaflike bracts. Calyces 2–3 mm long, persistent and sometimes becoming slightly enlarged at fruiting, deeply 4(5)-lobed, the lobes lanceolate, relatively thick, sharply pointed at the tips. Corollas 2–3 mm long, 4-lobed to about the midpoint, white, the lobes broadly elliptic, broadly spreading at maturity, the throat with a dense ring of short hairs.

Stamens 4, alternating with the corolla lobes, the filaments attached in the corolla tube, the anthers not exserted, attached at their midpoints, yellow. Pistil 1 per flower, of 2 fused carpels. Ovary about 1/2-inferior, 2-locular, with numerous ovules, the placentation axile. Style 1, very short, the stigma 1, capitate. Fruits capsules, 1.8–2.5 mm long, obovate to elliptic in outline, somewhat flattened, shallowly notched to rounded or very bluntly pointed at the tip, each face with a slightly concave suture line, dehiscent from the tip along the sutures, glabrous. Seeds 0.2–0.4 mm long, irregularly cube-shaped, the surface somewhat translucent, yellow. $2n$=20, 22, 33. June–October.

Uncommon in southeastern Missouri, nearly confined to the Mississippi Lowlands Division, with a disjunct historical occurrence in Clark County (eastern U.S. west to Missouri and Texas; Mexico, Central America, Caribbean Islands, Paraguay; introduced in Hawaii, Micronesia). Margins of ponds, lakes and swales of sand prairies; also roadsides, wheel ruts in unpaved roads, margins of crop fields, fallow fields, and open disturbed areas, usually in sandy soil.

THYMELAEACEAE (Mezereum Family)

About 45 genera, 850 species, nearly worldwide.

Most species of Thymelaeaceae produce a variety of potentially toxic compounds including coumarins (and their glucosides and glycosides) and diterpenoid esters (phorbol esters), and are variously considered mildly to strongly poisonous. The family has limited economic importance, including some minor tropical timber trees and incense plants. The bark of species in several genera has been used in the production of paper and cordage in various parts of the world. The best-known genus is *Daphne* L., which comprises about 95 species native to Europe and Asia. Several *Daphne* species are cultivated as ornamental shrubs, a few (such as *D. papyracea* Wall. ex G. Don) are used in the production of thin, pliable papers, and others have been used medicinally as purgatives, emetics, and abortifacients. Additionally, bitter-tasting resins from some *Daphne* species have purported insecticidal properties and have been added to pills to encourage their swallowing without chewing.

Thymelaea passserina (L.) Coss. & Germ. (spurge flax) is a native of Europe that has become established sporadically in several states, including portions of Alabama, Illinois, Iowa, Kansas, Mississippi, Nebraska, Ohio, Texas, Washington, and Wisconsin (Holmes et al., 2000). Unlike most members of the family, it is a tap-rooted annual. Plants have ascending, slender stems to 60 cm tall, linear-lanceolate leaves to 15 mm long and small axillary clusters of flowers. The flowers are produced in the summer and are 2–3 mm long, perigynous with an urn-shaped, hairy hypanthium tipped with 4, minute, greenish yellow calyx lobes, but lacking a corolla. The 8 stamens are included in the hypanthium. Fruits are 1-seeded and indehiscent, with an ovoid body tapered into a short beak, and remain enclosed in the persistent hypanthium. Because of its sporadic occurrence to the east and west of Missouri, this nondescript, weedy species should be watched for in the state.

1. **Dirca** L.(leatherwood)
(Floden et al., 2009)

Plants shrubs, occasionally somewhat colonial from rhizomes. Stems ascending, branched, thickened and jointed at the nodes. Twigs yellowish brown or occasionally yellowish green, glabrous or hairy, with scattered small lenticels, the developing buds hidden in the hollow petiole bases of existing leaves, eventually exposed, densely hairy. Leaves alternate, simple, short-petiolate to nearly sessile. Stipules absent. Leaf blades simple and unlobed, oblong-

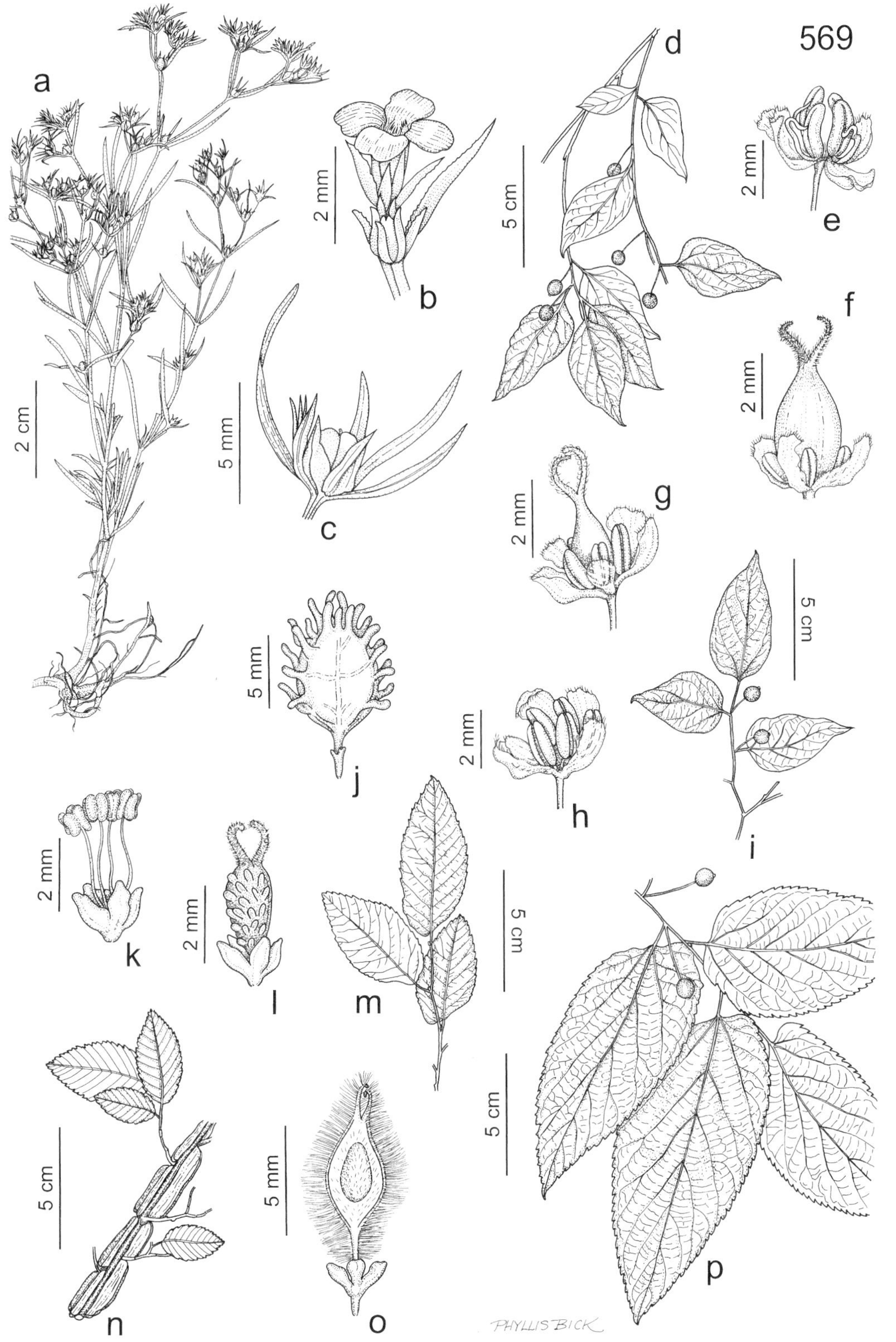

Plate 569. Tetrachondraceae. Ulmaceae. *Polypremum procumbens*, **a)** habit, **b)** flower, **c)** branch tip with leaves and fruit. *Celtis laevigata*, **d)** fruiting branch, **e)** staminate flower, **f)** young fruit. *Celtis pumila*, **g)** perfect flower, **h)** staminate flower, **i)** fruiting branch. *Planera aquatica*, **j)** mature fruit, **k)** staminate flower, **l)** young fruit, **m)** twig with leaves. *Ulmus alata*, **n)** portion of branch with winged bark and leafy twigs, **o)** fruit. *Celtis occidentalis*, **p)** fruiting branch.

ovate to broadly elliptic, ovate, or broadly oblong-ovate, the surfaces variously glabrous or finely hairy (especially the undersurface), the margins entire but often with fine, more or less appressed to spreading hairs. Inflorescences produced before the leaves or as they develop, axillary, sessile or stalked clusters of 2–6 nodding flowers, the bud ellipsoid to oblong-obovoid, with 4 bracts, the outer 2 shed as the flowers develop, the inner 2 persistent and becoming leaflike, sparsely hairy on the upper surface, densely woolly on the undersurface. Flowers perfect, strongly perigynous (the hypanthium appearing as a calyx tube), actinomorphic, not subtended by bractlets. Calyces appearing narrowly funnelform to nearly tubular, consisting of a calyxlike, green to yellowish green basal portion (the hypanthium) and a slightly expanded, petaloid (yellow to lemon yellow), apical portion (the limb) formed from 4 fused sepals, the tip truncate or with 4 shallow, uneven, rounded to broadly triangular lobes, otherwise appearing somewhat irregularly scalloped and/or toothed, not persistent at fruiting. Corollas absent, but sometimes minute petaloid scales are present on the inner surface of the calyx between the attachment points of the stamens. Stamens 8, exserted, the filaments of 2 lengths, attached in a ring to the inner surface of the calyx above the hypanthium portion, the anthers small, attached basally, yellow. Pistil 1 per flower, appearing composed of 1 carpel, the superior ovary sometimes with a minute, irregularly lobed nectar disc at the base. Style 1, slender, exserted and extending slightly beyond the anthers, not persistent at fruiting, the stigma 1, minute, capitate. Ovules 1, the placentation more or less basal. Fruits drupes, shed quickly, the seed enclosed in a thin, hard (stony) inner layer, covered with a thin fleshy layer and a thin, leathery outer layer, this pale yellowish green to nearly white, turning orange to reddish purple with age. Seed 1, broadly ovoid, the bony outer coating with a shallow longitudinal groove along 1 side, smooth, brown to dark brown, sometimes finely and faintly mottled. Four species, North America.

The genus *Dirca* is currently known to consist of one relatively widespread species and three uncommon species with limited distributions. In addition to the two species treated below, *D. occidentalis* A. Gray is endemic to coastal northern California and *D. mexicana* G.L. Nesom & Mayfield is known only from a small area in northeastern Mexico.

The stems of *Dirca* species are relatively slightly lignified, with very light wood (Nevling, 1962). They are surprisingly flexible and have been used in cordage and baskets. Like most members of the Thymelaeaceae, *Dirca* is considered to be toxic. The plants contain daphnane ester diterpenoids that can cause strong contact dermatitis (Burrows and Tyrl, 2001), but the tissues have a bitter flavor; thus are not likely to be consumed in quantity by humans or livestock. The fruits can also cause a strong allergic reaction in the mouth and throat, and apparently also have narcotic effects (Burrows and Tyrl, 2001), although reports of dizziness and stupor are largely anecdotal to date.

1. Twigs hairy; flower clusters sessile or developing a stalk to 1 mm at flowering, this not elongating noticeably as the fruits mature, the individual flowers with hairy stalks to 2 mm long (even at fruiting); bracts pubescent on the undersurface with white to gray or light brown hairs; calyces with 4 irregular lobes 1–3 mm long, the margins also irregular or minutely scalloped; ovaries and fruits with an inconspicuous tuft of short hairs at the tip .... 1. D. DECIPIENS
1. Twigs glabrous; flower clusters stalked, the stalk often short at flowering but eventually elongating to 5–13 mm as the fruits mature, the individual flowers with glabrous stalks 2–10 mm long; bracts pubescent on the undersurface with dark (rarely light or grayish) brown hairs; calyces unlobed, merely irregular and/or minutely scalloped along the margin; ovaries and fruits glabrous at the tip .... 2. D. PALUSTRIS

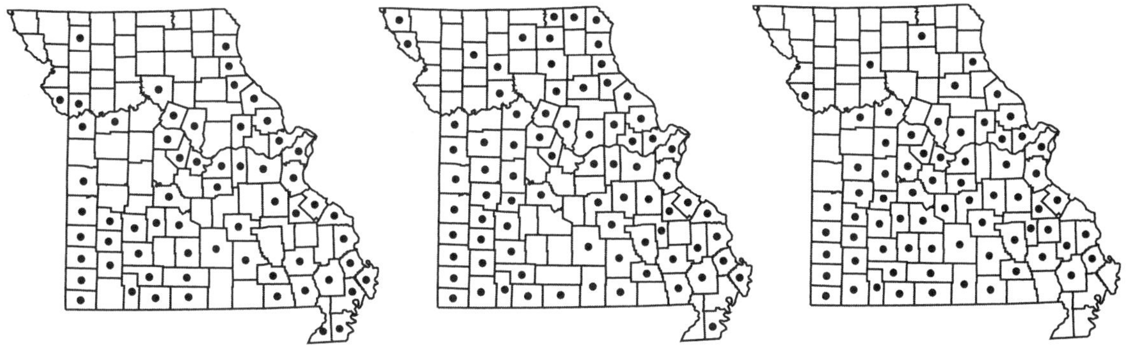

2659. Celtis laevigata

2660. Celtis occidentalis

2661. Celtis pumila

1. Dirca decipiens Floden

Map 2657

Plants shrubs, 1–2(–3) m tall, several stemmed, the bark tan to light grayish brown. Twigs 3–5 mm wide, yellowish brown, sparsely to moderately pubescent with short, appressed, white hairs. Leaves with the petioles 1–4 mm long, densely appressed-hairy. Leaf blades 2–9 cm long (those of the persistent pair of leaflike bracts at the base of each year's growth shorter), rounded or broadly angled to a bluntly pointed tip, rounded or angled at the base, the margins usually relatively densely and finely hairy (mostly 11–18 hairs per mm). Flower clusters sessile or developing a stalk to 1 mm at flowering, this not elongating noticeably as the fruits mature, the individual flowers with hairy stalks to 2 mm long (even at fruiting). Persistent bracts 6–15 mm at flowering, continuing to enlarge to 40 mm or more as the fruits develop, oblong-obovate, the upper surface usually sparsely hairy along the veins, the undersurface woolly with dense, white to gray or light brown hairs. Calyces 7–11 mm long (including the hypanthium), with 4 distinct but irregular lobes 1–3 mm, the margin also irregular and/or minutely scalloped. Fruits 7–9 mm long, ovoid to ellipsoid-ovoid or slightly pear-shaped, with an inconspicuous tuft of short hairs at the tip. March–April.

Uncommon in the southern half of the state (Arkansas, Kansas, Missouri). Bases and ledges of dolomite bluffs and steep slopes in mesic upland forests.

This species, whose fragmented range is at the western edge of that of *D. palustris*, was described very recently (Floden et al., 2009). Because its distribution and abundance are still not well understood, additional populations should be searched for in Missouri.

2. Dirca palustris L. (leatherwood)

Pl. 568 a–c; Map 2658

Plants shrubs, 0.7–2.0(–3.3) m tall, several stemmed, the bark usually light grayish brown. Twigs 2–4 mm wide, yellowish brown or occasionally reddish brown, glabrous (sometimes hairy when very young elsewhere). Leaves with the petioles 1–5 mm long, densely appressed-hairy. Leaf blades 2–10 cm long (those of the persistent pair of leaflike bracts at the base of each year's growth shorter), rounded or broadly angled to a bluntly or less commonly sharply pointed tip, rounded or angled at the base, the margins glabrous or sparsely to moderately hairy (0–8 hairs per mm). Flower clusters stalked, the stalks 2–7 mm long at flowering, elongating to 5–13 mm at fruiting, the individual flowers with glabrous stalks 1–10 mm long (sometimes 2 or more of the stalks in a cluster fused nearly to the tips), elongating slightly as the fruits mature. Persistent bracts 5–10 mm at flowering, continuing to enlarge to 35 mm or more as the fruits develop, oblong-obovate to narrowly oblong-obovate, the upper surface glabrous or sparsely hairy along the veins, the undersurface woolly with dense, dark (rarely light or grayish) brown hairs. Calyces 5–8 mm long (including the hypanthium), truncate, unlobed, the margin merely irregular and/or minutely scalloped. Fruits 8–12 mm long, ovoid to ellipsoid-ovoid, glabrous at the tip. March–April.

Scattered in the Ozark and Ozark Border Divisions north locally to Audrain County (eastern U.S. west to North Dakota and Oklahoma; Canada). Bottomland forests, mesic upland forests (along mostly north- and east-facing slopes of valleys), banks of streams and rivers, and bases and ledges of bluffs.

Specimens mapped from Mercer and Grundy Counties by Steyermark (1963) were not located during the present study. Leatherwood occasionally is cultivated as a specimen plant in gardens, but is slow-growing. Steyermark (1963) speculated that a shrub with stems 5 cm in diameter may be more than a century in age. *Dirca palustris* is one of the earliest-flowering shrubs in Missouri. According to Floden et al. (2009), on average the species even begins to flower about a week earlier than does *D. decipiens*. Steyermark also noted that the fruits mature quickly (often by the end of April) and perpetuated rumors that these have narcotic properties. Fernald (1945) noted that although the basic fruit color tends to be pale green or yellowish at maturity, plants in the northern portion of the species range routinely produce fruits that are strongly reddish- or purplish-tinged. In Missouri, both color morphs can be found, but yellowish fruits appear to predominate and even these sometimes become darker and redder with age.

ULMACEAE (Elm Family)

Contributed by Alan Whittemore

Plants trees, sometimes monoecious, unarmed, not producing milky sap. Branches unarmed (occasionally thorny elsewhere). Leaves alternate, petiolate. Stipules small, linear or lanceolate, shed early. Leaf blades simple, pinnately veined or 3-veined from the base, the margins entire or toothed. Inflorescences axillary, small clusters or of solitary flowers, sometimes from the axils of bud scales, thus clustered at the base of new growth, in monoecious species the staminate flowers below the perfect flowers on the same shoot. Flowers perfect or imperfect, actinomorphic (sometimes appearing somewhat zygomorphic at fruiting in *Ulmus*), hypogynous, stalked or less commonly sessile. Calyces of 4–9 separate or fused sepals, in perfect and pistillate flowers persistent at fruiting. Corollas absent. Stamens 4–9 (absent or highly reduced in pistillate flowers), the filaments more or less straight in the bud, free or fused to the basal portion of the calyx tube, the anthers exserted, attached toward their bases, variously colored, dehiscent by longitudinal slits. Pistil 1 per flower (absent in staminate flowers), of 2 fused carpels, the ovary superior, sometimes short-stalked, 1-locular, with 1 ovule, the placentation apical. Styles 2 or 1 deeply 2-branched, stigmatic along the inner side. Fruits drupes, samaras, or ellispoidal and with a leathery outer layer, 1-seeded. About 16 genera, 165 species, widespread in temperate and tropical regions, mainly in the northern hemisphere.

Recent research involving anatomical features and molecular genetics of the group has established that *Ulmus* and *Planera* are not the closest relatives of *Celtis*. The traditional genera of Ulmaceae are now typically treated in two families. Ulmaceae in the strict sense comprises 7 genera (including *Ulmus* and *Planera*) and about 60 total species. However, *Celtis* and its relatives are now considered members of an expanded Cannabaceae, with 11 genera (including *Cannabis* and *Humulus* as well as *Celtis*) and about 110 total species (Sytsma et al., 2002; Judd et al., 2008). In Missouri, the families are most easily distinguished by the morphology of the leaves (3-veined or palmately compound in Cannabaceae, simple and pinnately veined in Ulmaceae), flowers (sepals free and anthers oriented toward the center of the flower in Cannabaceae, sepals united below and anthers oriented toward the outer margin of the flower in Ulmaceae), and seeds (embryo curved or coiled in Cannabaceae, straight in Ulmaceae). Because the treatment of Cannabaceae that was published in Volume 2 of the present work (Yatskievych, 2006) still treated that family in the traditional sense to exclude *Celtis*, the account of Ulmaceae below includes the genus *Celtis* for purely practical reasons.

1. Leaves spirally alternate in several planes, not 2-ranked; secondary veins forming a network of loops before reaching the entire or toothed margins, the basal pair longer and more prominent than the others, the blade thus appearing more or less 3-veined from base; bark smooth or with irregular corky warts and/or ridges; fruits drupes with soft flesh, a large, stony endocarp, and a smooth outer surface . . . . 1. CELTIS

1. Leaves 2-ranked; secondary veins not forming a network, extending straight to the margins, each ending in a tooth, the basal pair shorter than the median veins, thus the venation more or less regularly pinnate; bark scaly or longitudinally fissured into parallel ridges, without corky protuberances; fruits dry, either leathery and covered with soft blunt protuberances or papery, 2-winged samaras
   2. Leaf blades lanceolate to narrowly ovate, widest well below the midpoint; secondary veins curved, often not parallel, irregularly spaced, variable in strength; bark scaly, not deeply fissured; fruits wingless, the outer surface leathery, covered with soft blunt cylindric or ridged protuberances; flowers staminate and perfect . . . . 2. PLANERA
   2. Leaf blades elliptic, oblong, or ovate to obovate or broadly oblanceolate, widest slightly below to slightly above the midpoint; secondary veins straight, parallel, evenly spaced, uniform in strength; bark (in our species) longitudinally fissured into parallel ridges (forming thin, elongate plates on younger trunks of *U. crassifolia*); fruits 2-winged samaras, papery, glabrous or finely hairy but lacking outgrowths; flowers all perfect . . . . 3. ULMUS

1. **Celtis** L. (hackberry)

Plants trees or shrubs to 35 m tall. Bark smooth or with irregular corky warts and/or ridges, pale gray or almost white. Twigs sometimes slightly zigzag, the winter buds axillary, usually with several overlapping scales (naked elsewhere). Leaves spirally alternate in several planes, not 2-ranked. Leaf blades lanceolate, ovate, or triangular, slightly to strongly asymmetric at the base, widest 1/5–1/3 from the base, the secondary veins turning aside and anastomosing before reaching the entire or toothed margins, the basal pair longer and more prominent than the others, thus the blade appearing more or less palmately 3-veined from the base. Flowers staminate and apparently perfect, appearing before the leaves, the staminate flowers in small, dense clusters at the base of current-year's growth, the perfect flowers solitary in the axils of the lowest expanding leaves of the current season. Calyces 2–3 mm long, shallowly to deeply 4(5)-lobed, but the lobes usually shed early, green to greenish yellow, turning brown after flowering. Stamens with the anthers yellow to greenish yellow. Fruits drupes, the outer layer thin, glabrous, smooth, sometimes somewhat glaucous; the middle layer thin, fleshy, sweet-tasting, the seeds covered by a large, stony endocarp. About 70 species, worldwide.

The genus *Celtis* is taxonomically very difficult. All characters are variable and identifications are best made using several characters in combination. In the past, this variability was often explained by postulating high levels of hybridization between species. Recent studies have shown that this is false. It is difficult to cross the species in the garden (Whittemore and Townsend, 2007) and hybrids among sexually reproducing taxa are rare or absent in nature (Whittemore, 2005). Instead, *C. pumila* is apomictic, the seeds developing without pollination, and as in many apomicts there is a great deal of variation from population to population. Interestingly, it is possible to obtain hybrids by placing pollen from sexual species on stigmas of apomictic plants (A. Whittemore, unpublished), and some puzzling forms found in the wild may have originated by pollination of apomictic *C. pumila* by pollen from *C. laevigata* or *C. occidentalis*.

The bark of native *Celtis* species is characterized by the formation of distinctive corky warts and/or ridges. The species differ in the amount and distribution of corky outgrowths, but the degree of cork development varies from tree to tree and varies with the age of the tree, so this character is not always reliable for species identification. Cork also forms in response to injury, so trunks that have been damaged may form corky outgrowths in unusual places.

Leaves from different parts of the plant or at different stages in its life may be very different in form. In particular, leaves from vigorous leading shoots are much larger than leaves from lateral branches and may differ strikingly in shape and marginal toothing. The shape, texture, and toothing of juvenile leaves may be very different from leaves of adult plants. Unless otherwise stated, descriptions apply to a leaf that is subtending a flower or fruit. Vegetative plants or vegetative, leafy twigs in the herbarium are not always identifiable.

The fruit of hackberries is eaten by many species of mammals and birds. The flesh of the fruit is sweet, with a pleasant flavor, but the stone is large and there is relatively little flesh in the drupe. Some scholars have suggested that the nettle tree of Europe and the Near East, *C. australis* L., was the lotus of Homer's *Odyssey*, the fruit eaten by the Lotus-Eaters of North Africa. Others have felt that Homer's lotus was more likely a species of *Ziziphus* Mill. (Rhamnaceae). Fruit of our native *Celtis* often remain on the tree long into the winter, and sometimes a few fruits even remain on the tree until the following spring.

Hackberry wood is soft and it is not widely used. It is sometimes used for furniture and handcrafts, where its light color is valued.

Some species, especially *C. occidentalis*, are used in horticulture, as they are fast-growing trees with great tolerance to heat, cold, drought, periodic flooding, rapid changes in temperature and moisture, and poor soils. They are widely grown in dry areas with extreme temperature fluctuations (especially the western Great Plains and the intermountain western states), but their use in Missouri is limited by their susceptibility to several pathogens (Elias, 1970), especially witch's brooms (caused by mites [*Eriophyes* sp.] and powdery mildew [*Sphaerotheca phytoptophila* Kellerm. & Swingle]) and leaf galls (caused by hemipteran insects, *Pachypsylla* spp.). These pathogens do not harm the plant, but they cause cosmetic damage and reduce its popularity as a shade tree. Hackberries also tend to become weedy in garden settings.

1. Leaf blades (applies only to fully formed leaves subtending fruits) 7–12 cm long, the longer side with 23–40 teeth; secondary veins 5–8 on each side of midvein, the second areole 5–9 times as long as wide; fruits 8–10 mm long . . . . 2. C. OCCIDENTALIS
1. Leaf blades (applies only to fully formed leaves subtending fruits) 3.5–8.5 cm long, the long side entire or with no more than 13 teeth; secondary veins 3–5 on each side of midvein, the second areole 2.5–5.0(–7.0) times as long as wide; fruits 5–8 mm long
    2. Plants usually trees to 8–30 m tall (rarely shrublike and shorter), typically with one strong, vertical trunk (occasionally 2 or 3 parallel, strongly ascending trunks), if small then with erect leaders; bark with corky warts or ridges over the whole trunk and major branches; leaf blades with the undersurface usually bright green, only rarely somewhat paler than the upper surface; flower stalks glabrous (rarely with a few scattered, spreading hairs); twigs glabrous or sparsely hairy . . . . 1. C. LAEVIGATA
    2. Plants shrubs or small trees, 3–7 m tall, with ascending trunks and horizontal or arching leaders, or (with age) multiple, spreading trunks; bark with large corky warts at base of trunk and around old wounds, otherwise smooth; leaf blades with the undersurface pale green or glaucous, usually much lighter than the upper surface; flower stalks usually hairy, at least toward the base (but occasionally glabrous throughout); twigs usually moderately to densely hairy, occasionally glabrous . . . . 3. C. PUMILA

1. Celtis laevigata Willd. **var. laevigata**
(sugarberry, southern hackberry)
C. laevigata var. *texana* (Scheele) Sarg.
Pl. 569 d–f; Map 2659

Plants usually trees 8–30 m tall (rarely shrublike and shorter at dry sites), typically with one strong, vertical trunk (occasionally with 2 or 3 parallel, strongly ascending trunks), if small then with erect leaders. Bark of trunk and large branches with prominent corky warts and sometimes also ridges. Twigs glabrous or sparsely pubescent with stiff, spreading hairs, the winter buds 2–3 mm, ovoid or flattened, bluntly pointed at the tip, orange-brown or purple, glabrous or minutely hairy. Leaves thin and flexible when dry (rarely thicker and somewhat leathery in plants from dry sites), both surfaces the similarly bright green to yellowish green (rarely the undersurface somewhat paler), the upper surface plane or weakly grooved over the midvein and secondary veins, the undersurface usually with sparse hairs along the main veins and/or with dense tufts of hairs in the vein axils, sometimes also hairy between the veins, or sometimes the blade totally glabrous, the margins plane or narrowly curved under. Leaves subtending fruits with the petiole 6–12 mm long, glabrous or the upper surface sparsely hairy; the blade 4.5–8.5 cm long, 2–4 cm wide, lanceolate or narrowly ovate, asymmetrically rounded or obliquely truncate at the base; gradually tapered to a slender, sharply pointed tip, the margins entire or occasionally with 1 or 2 small teeth on one side; the upper surface smooth, rarely very slightly roughened, the secondary veins 3–5 on each side, the basal secondary veins extending to $^1/_3$–$^1/_2$ of blade length (occasionally slightly longer on one side). Leaves of vigorous leading shoots to 12 × 5 cm, lanceolate, usually entire but sometimes toothed (occasionally sharply toothed for their whole length); leaves of juvenile plants ovate or oblong-ovate, the base symmetric, the margins strongly toothed, the tip broadly angled or broadly short-tapered, the upper surface roughened. Flower stalks 7–16 mm long, 1–2 times as long as the subtending petiole, glabrous (rarely with a few scattered hairs). Stamens with the anthers always dehiscent, the pollen copious and well-formed (95–100% stainable with acetocarmine). Fruits spherical, 7–8 mm long, orangish brown or red when fully mature. $2n$=20. April–May.

Scattered south of the Missouri River and in the Big Rivers Division, uncommon to absent elsewhere northward (southeastern U.S. west to Missouri and Texas; Mexico). Bottomland forests, mesic upland forests, banks of streams and rivers, margins of lakes and sinkhole ponds, bases of bluffs, and occasionally dry upland forests, glades, and tops of bluffs; rarely also fencerows.

Celtis laevigata specimens often have a very distinctive appearance, with thin, entire, yellowish green leaves having a rounded or obtuse base. However, these characters can vary, and some specimens of *C. laevigata* and *C. pumila* have very similar foliage. The drupes of *C. laevigata* are brownish orange or red at maturity, without any purplish pigments. Drupes of our other species, *C. occidentalis* and *C. pumila*, mature to a dark brownish purple or purplish red color (but nearly mature drupes of these species may be yellow, orange, or red).

Populations growing in dry habitats sometimes have somewhat thicker leaves. Such plants have been called *C. laevigata* var. *texana*, but they differ from typical *C. laevigata* in no other characters and seem to be the result of convergent evolution due to local selection in dry sites (Whittemore, 2005). The name *C. laevigata* var. *texana* is technically a synonym of var. *laevigata*, but it has often been applied to forms of *C. pumila* as well as specimens of *C. laevigata*.

In *C. laevigata* var. *laevigata*, the leaves that subtend the fruits are evenly tapered or gradually narrowed to a slender, sharply pointed tip, and their margins are usually entire, seldom with 1 or 2 small (0.5–1.0 mm long) teeth on one side. The other variety, var. *smallii* (Beadle) Sarg., occurs along the Piedmont from Virginia south to Georgia, and on the Cumberland Plateau from Kentucky south to northern Alabama. In var. *smallii*, the leaves that subtend the fruit are more abruptly tapered to the long, slender, curved, sharply pointed tip, and each side has 2–15 sharp, forward-pointed, irregular teeth 1–2 mm long. Morphologically, *C. laevigata* var. *smallii* appears intermediate between *C. laevigata* var. *laevigata* and *C. occidentalis*, but in its native range the var. *smallii* forms uniform populations over extensive areas where *C. laevigata* var. *laevigata* and *C. occidentalis* are absent, so they cannot be incidental hybrids. This variety has been reported from Missouri, but all of the Missouri specimens determined as var. *smallii* either were misidentified or are vegetative and not determinable to variety. Vegetative specimens of *C. laevigata* with toothed leaves are often annotated in herbaria as var. *smallii*, but in fact such specimens cannot be named to variety. Leaves on juvenile plants of both varieties are normally toothed and leaves of vigorous leaders also may sometimes be strongly toothed in var. *laevigata*.

2. Celtis occidentalis L. (northern hackberry, nettle tree)

C. occidentalis var. *canina* (Raf.) Sarg.

Pl. 569 p; Map 2660

Plants medium to large trees 10–35 m tall (rarely shrublike and shorter at dry sites), typically with one strong, vertical trunk (occasionally with 2 or 3 parallel, strongly ascending trunks). Bark of trunk and large branches with prominent

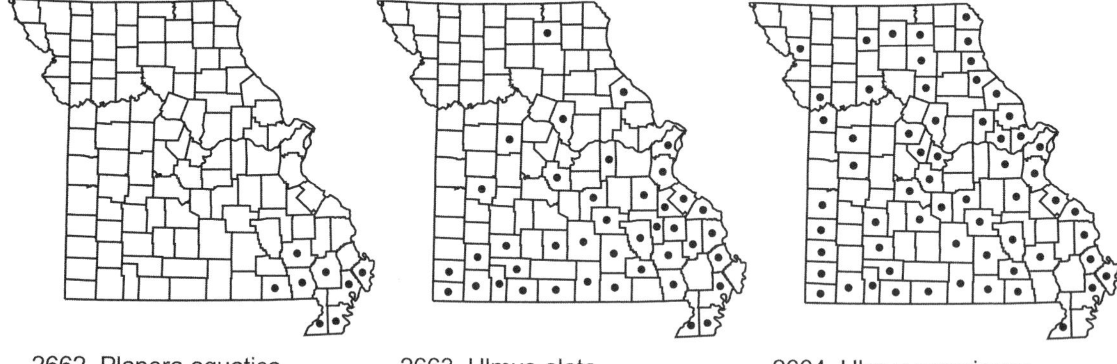
2662. Planera aquatica 2663. Ulmus alata 2664. Ulmus americana

corky ridges and/or warts. Twigs glabrous (rarely sparsely pubescent with stiff, spreading hairs, especially on juvenile plants), the winter buds 1.5–5.0 mm, ovoid or flattened, bluntly pointed at the tip, orange-brown or purple, glabrous or minutely hairy. Leaves thin and flexible when dry, both surfaces the similarly bright green or the undersurface somewhat paler, the upper surface plane or weakly grooved over the midvein and secondary veins (sometimes also over some of the minor veins), the undersurface with sparse hairs along the main veins and often also with dense tufts of hairs in the vein axils, or sometimes the blade totally glabrous, the margins plane or narrowly curved under. Leaves subtending fruits with the petiole 9–15 mm long, glabrous or the upper surface sparsely hairy; the blade 7–12 cm long, 4.0–7.7 cm wide, obliquely triangular-ovate, very obliquely truncate or sometimes slightly asymmetrically rounded at the base; gradually tapered to a slender, sharply pointed tip, the margins relatively evenly toothed with 12–27 teeth on the shorter side and 23–40 teeth on the longer side; the upper surface smooth or slightly roughened, the secondary veins 5–8 on each side, the basal secondary veins extending to 1/3–2/5 of blade length (occasionally to 1/2 the blade length on one side). Leaves of vigorous leading shoots 8.5–18.0 × 3–11 cm, narrowly triangular to triangular-ovate, evenly toothed for their whole length; leaves of juvenile plants triangular-ovate, the base sometimes symmetric, the margins sharply toothed, the tip gradually long-tapered, the upper surface smooth or roughened. Flower stalks 12–24 mm long, 1.2–2.2 times as long as the subtending petiole, glabrous. Stamens with the anthers always dehiscent, the pollen copious and well-formed (95–100% stainable with acetocarmine). Fruits more or less spherical, 8–10 mm long, dark brownish purple when fully mature. $2n$=20. March–April.

Scattered nearly throughout the state, but uncommon to absent from the western portion of the Glaciated Plains Division (eastern U.S. west to Montana and New Mexico; Canada). Bottomland forests, mesic upland forests, banks of streams, rivers, and spring branches, margins of sinkhole ponds, bases of bluffs, and occasionally edges of upland prairies and glades, and tops of bluffs; also old fields, cemeteries, and roadsides.

Celtis occidentalis is distinctive in having larger leaf blades that are evenly toothed for their whole length, with more numerous secondary veins that are more nearly parallel so the areoles are long and narrow, whereas our other species have areoles that are wider relative to their length; the areoles often are widest near the midvein and gradually narrowed for their whole length, but venation is variable. Leaves that subtend fruit in *C. occidentalis* are gradually tapered for at least half of their length. The bark of *C. occidentalis* often shows much heavier development of corky warts and ridges than in other Missouri species, but this character is very variable.

For further discussion on the differences between this species and *C. pumila*, see the treatment of that species.

3. Celtis pumila Pursh (dwarf hackberry)
C. pumila var. *georgiana* (Small) Sarg.
C. georgiana Small
C. occidentalis L. var. *pumila* (Pursh) A. Gray
C. tenuifolia Nutt.
C. tenuifolia var. *georgiana* (Small) Fernald & B.G. Schub.

Pl. 569 g–i; Map 2661

Plants shrubs or small, spreading trees, 3–7 m tall, with ascending trunks and loosely ascending to spreading or arched leaders, or (with age) developing multiple, loosely ascending trunks. Bark with corky warts only at base of trunk and around old wounds, otherwise smooth. Twigs usually moderately to densely pubescent with stiff, spreading hairs (sometimes only sparsely hairy or becoming nearly glabrous with age), the winter buds 1–3 mm, ovoid or flattened, bluntly pointed at the tip, light brown to dark purplish brown, minutely hairy. Leaves thin and flexible or more commonly stiff and somewhat leathery when dry, the undersurface pale green or glaucous, usually much lighter than the upper surface, the upper surface plane or weakly grooved over the midvein and secondary veins (occasionally also over some of the minor veins), the undersurface hairy along the main veins and often also the minor veins, the margins plane or narrowly curved under. Leaves subtending fruits with the petiole 3–8 mm long, hairy, sometimes only on the upper surface; the blade 3.5–8.5 cm long, 1.8–5.5 cm wide, lanceolate-triangular to triangular-ovate, obliquely truncate or asymmetrically shallowly cordate at the base; broadly angled to a bluntly or sharply pointed tip or sometimes more narrowly angled or short-tapered to a sharply pointed tip, the margins entire or with 1–13, sometimes irregularly spaced small teeth on each side; the upper surface smooth to strongly roughened, the secondary veins 3 or 4 on each side, the basal secondary veins extending to $^1/_2$–$^2/_3$ of blade length (occasionally slightly shorter on one side). Leaves of vigorous leading shoots to 10.5 × 4.5 cm, ovate to triangular-ovate to lanceolate, entire or with up to 14 teeth on each side; leaves of juvenile plants ovate or oblong-ovate, the base symmetric, the margins strongly toothed, the tip broadly angled or broadly short-tapered, the upper surface roughened. Flower stalks 4–13 mm long, 0.7–2.3 times as long as the subtending petiole, usually hairy, at least toward the base (but occasionally glabrous throughout). Stamens with the anthers often small and indehiscent, the pollen mostly or entirely aborted (seldom more than 10% of the grains stainable with acetocarmine, these very variable in size and number of pores). Fruits spherical, 5–8 mm long, dark brownish purple when fully mature. $2n$=30. April–May.

Absent from the southern portion of the Mississippi Lowlands Division and uncommon to absent from much of the Glaciated and Unglaciated Plains; scattered elsewhere in the state (eastern U.S. west to Kansas and Texas; Canada). Glades, ledges and tops of bluffs, edges of dry upland forests, savannas, and occasionally banks of streams and margins of sinkhole ponds; also old fields and fencerows.

Celtis pumila is often a successional species, colonizing old fields, fence lines, and open woodland margins. It is not tolerant of deep shade, and trees tend to lose vigor and die out as they are shaded by the growth of taller trees. Plants growing in shade also tend to flower and fruit infrequently.

Unlike other *Celtis* species in Missouri, *C. pumila* is apomictic. The seeds develop without pollination (Whittemore and Tobe, unpublished). When flowers are present, the species can be recognized because the anthers either do not dehisce or are filled with scanty, malformed pollen. As in other apomictic groups, *C. pumila* is very variable morphologically. It has confused taxonomists and authors of regional accounts in the past and it has been necessary to recheck determinations on all of the older materials in order to accurately portray the range of the species. Plants with strongly toothed leaves have often been confused with *C. occidentalis*, and entire-leaved plants have been misdetermined as *C. laevigata* (often as var. *texana*) or gone under the name *C. tenuifolia*.

When anthers are not available, the growth form is the most distinctive feature of the species. Vigorous leading stems are loosely ascending to spreading or arched, and they bear very large leaves that are much larger than leaves on the lateral shoots and also different in shape. With age, *C. pumila* can develop into a small tree with a leaning trunk or multiple spreading trunks. Our other native species (*C. laevigata* and *C. occidentalis*) develop into tall canopy trees with one strong vertical trunk (occasionally 2 or 3 parallel upright trunks). All of our *Celtis* species sometimes grow in shallow soil over rock, and then they may mature as shrubs or small trees, but even dwarfed trees of *C. laevigata* and *C. occidentalis* will have erect leaders and one or few upright trunks.

Shrubs of *C. pumila* that are growing vigorously have quite a distinctive appearance, with their large-leaved horizontal or arching leaders, but the species often grows in harsh sites, where growth is slow, vigorous leaders seldom form, and shrubs may flower during alternate years or even less frequently. Leaves of vegetative stem sectors may be very variable: they may be obtuse or rounded, subsymmetrical, and truncate or shallowly cordate, whereas leaves of vegetative depauperate plants may even be lanceolate or lanceolate-oblong and up to 2.5 times as long as wide. *Celtis pumila* can be difficult to distinguish from other *Celtis* species when only isolated twigs and foliage are available. Then, *C. pumila* must be recognized by a combination of characters: the shorter petioles, paler leaf undersurfaces, and usually hairy flower stalks. The pubescence of the twigs and petioles is usually denser than in the other two species, but sparsely pubescent or even glabrous forms are sometimes encountered. The illustrations and discussion in W. H. Wagner (1974, as *C. tenuifolia*) are useful in understanding *C. pumila*, but the species is more variable in Missouri than in Michigan. In particular, leaves from Missouri plants often have much longer, more gradually tapered apices and more oblique bases than the leaves that Wagner illustrated. However's Wagner's comments on the differences in venation between *C. occidentalis* and *C. pumila* (as *C. tenuifolia*) are quite valid for Missouri materials.

2. **Planera** J.F. Gmel.

One species, southeastern United States.

Although there is only one extant species in the genus, several additional species are known from the fossil record in various parts of North America (Elias, 1970), dating back to the Cretaceous Period.

1. Planera aquatica J.F. Gmel. (water elm, planer tree)

Pl. 569 j–m; Map 2662

Plants trees 5–15 m tall. Bark shed in thin narrow vertical strips, orange-brown where freshly exposed, weathering to pale gray. Twigs often somewhat zigzag, dark purplish brown, minutely hairy, sometimes becoming nearly glabrous late in the growing season, never developing corky wings or outgrowths, the winter buds 1.5–3.0 mm long, ovoid or ellipsoid, with several, closely overlapping scales, dark purplish brown, glabrous or minutely hairy. Leaves 2-ranked, the petioles 3–4 mm, minutely hairy. Leaf blades 5–12 cm long, 2–7 cm wide, lanceolate to narrowly ovate, widest 1/4–1/3 from the base, asymmetrically rounded at the base, angled from the widest point to a bluntly or sharply pointed tip, the margins doubly toothed; the undersurface much paler than the upper surface, glabrous or the undersurface minutely hairy along the main veins, the venation pinnate, with secondary veins 8–11 on each side of the midvein, not forming loops or a network, extending to the margins, each ending in a tooth, unequal in development, commonly forked, somewhat curved and not strictly parallel, the basal pair shorter than the median veins. Flowers staminate and bisexual, appearing after the leaves develop, the staminate flowers in clusters from buds on previous year's growth or clustered at the base of current-year's growth, the perfect flowers solitary or paired in the axils of leaves on current-year's growth. Calyces 1–2 mm long, 4- or 5-lobed to about the midpoint, greenish yellow, turning brown after flowering. Stamens with the anthers red to purplish red. Fruits 5–7 mm long above a stalk 1–3 mm long, ellipsoid, green, the outer layer leathery, covered with soft, blunt, cylindric or flattened protuberances, each 1–2 mm long. April.

Uncommon, restricted to the Mississippi Lowlands Division (southeastern U.S. west to Missouri and Texas). Swamps, bottomland forests, and sloughs.

The bark of *Planera* is very different from any of our native or naturalized elms. It bears some resemblance to the bark of *Ulmus parvifolia* Jacq., but the bark exfoliates as long strips in *Planera* and as irregular, woody scales 2–8 cm across in *U. parvifolia* (see further discussion under *Ulmus*).

The fruits of *P. aquatica* have been described in the literature variously as dry drupes, soft nutlets, or soft burs. None of these descriptions is accurate, and in fact *Planera* fruits do not fit into any of the standard fruit types (the author thanks Steven Manchester [Florida State Museum] and Walter Judd [University of Florida] for supplemental information on the fresh fruit). *Planera aquatica* grows most commonly in forests that are seasonally flooded, and the fruits float on water. They are eaten by ducks and rodents.

3. **Ulmus** L. (elm)

Plants shrubs or more commonly trees, 2–35 m tall. Bark longitudinally fissured into more or less parallel ridges (in our species; see the note below on *U. parvifolia*), not breaking into loose plates or strips (forming thin, elongate plates on younger trunks of *U. crassifolia*, but these remaining relatively tight), lacking warty protuberances, medium to dark gray. Twigs often somewhat zigzag, sometimes developing corky wings or outgrowths, the winter buds of various sizes and shapes, with several overlapping scales. Leaves 2-ranked. Leaf blades elliptic, oblong, or ovate to obovate or broadly oblanceolate, widest slightly below to slightly above the midpoint, rounded to more or less bluntly angled at the base, variously symmetrically to strongly asymmetrically so, sometimes with a minute notch at the attachment point, tapered to a sharply pointed tip (except in *U. crassifolia* and sometimes *U. pumila*), the margins bluntly or sharply and doubly toothed (often more or less singly toothed in *U. pumila*), the upper surface smooth to strongly roughened, the undersurface much paler than the upper surface, glabrous or minutely hairy along the veins, the hairs occasionally occurring as dense tufts in the vein axils, the venation pinnate, with secondary veins not forming loops or a network, extending to the margins, each ending in a tooth, equal in development and spacing, unbranched or occasionally forked, straight and more or less parallel, the basal pair shorter than median veins. Flowers all perfect (but see the note below), appearing in clusters (these sometimes umbellate or short-racemose) either before the leaves develop in the spring from buds on previous year's growth or in autumn in the axils of leaves on current-year's growth. Calyces 1.5–3.0 mm long, shallowly to deeply 4–9-lobed, green to greenish yellow or reddish brown, turning tan to brown after flowering. Stamens with the anthers red to brownish red or purple to nearly black. Fruits samaras, appearing flattened, with 2 wings spreading in the same plane on either side of the seed and an apical notch, papery, the surface glabrous or finely hairy, lacking outgrowths. About 40 species, North America, Central America, Europe, Asia.

Vegetative specimens of *Ulmus* cannot be identified with any high degree of confidence, and should be avoided by collectors. Critical identification requires flowers or fruits. Leaves of juvenile plants, or of fast-growing shoots from plants that have been cut back or damaged, tend to be large, strongly toothed, strongly roughened on the upper surface, and uniformly pubescent on the underside. They are similar in all species of elm, and are very difficult to identify. The keys and descriptions below all refer to leaves of fertile adult trees. Some species may develop corky wings or outgrowths on their twigs, vs. others that never do so. However, trees normally develop cork only on some twigs, and some trees may not show cork formation on any twigs even if it is usually characteristic of the species. The presence of corky twigs on a specimen can be very useful for identification, but it is important not to read too much into the absence of cork on twigs of a particular branch.

Flowers of *Ulmus* have both staminate and pistillate organs well-developed and are generally described as having perfect flowers. It has recently been shown that many individuals of *U. minor* are functionally staminate and set no fruit (López-Almansa et al., 2003), and the reproductive biology of other elm species should be checked.

The seeds and buds of elm are eaten by deer, rodents, and birds. Elm wood is hard and strong, and it is used for furniture, fenceposts, flooring, and general construction, and as a veneer for other woods. In the past, it was widely used for manufacturing wheel hubs, saddle trees, ship's hulls, and agricultural and kitchen implements, uses where its hardness and flexibility are especially valuable. It has interlocked grain and is difficult to split, making it especially suitable for items that require bending, such as hockey sticks or curved pieces in furniture. Elm bark is used for tanning leather. Elms were once widely used for construction of bark canoes, and fiber from elm bark has been used in the past to make ropes, cords, and roofing felt. Elm foliage was long used as fodder for livestock in Eurasia, and is still important for this purpose in northern India. Many bird, insect, and mammal species feed on the buds, twigs, and seeds of various elm species.

Elms are very widely grown as street trees and shade trees, but their use has been limited by an exotic disease, known as Dutch elm disease, which was first noted in Europe in 1919 and subsequently spread across North America in the mid-twentieth century (Strohel and Lanier, 1981; Wolkomir, 1998). Dutch elm disease is caused by exotic wilt fungi (*Ophiostoma* spp.); the native range of these fungi is still unknown. All of our native elm species are impacted by the fungus. In the best-studied species, *U. americana*, there is great variation in disease resistance from tree to tree, and several disease-tolerant trees of *U. americana* have been identified (Wolkomir, 1998) and propagated, and are now available commercially (Townsend et al., 2005).

Ulmus serotina Sarg. (September elm or red elm) occurs in northern Arkansas and southern Illinois. It resembles *U. thomasii* in most of its characters, but it flowers in the fall rather than the spring, and also has glabrous twigs and buds and usually smaller leaves. Old reports of *U. serotina* from Missouri are based on a mislabeled specimen that was actually collected in Arkansas (Steyermark, 1963). Another fall-flowering (but non-native) elm that is commonly cultivated in Missouri is *U. parvifolia* Jacq, (Chinese elm). For further discussion of this species, see the treatment of *U. pumila*.

1. Leaf blades angled to a broadly or bluntly pointed tip; flowering and fruiting in the fall; calyx lobes linear . . . . 3. U. CRASSIFOLIA
1. Leaf blades tapered to a sharply pointed tip (sometimes merely angled in *U. pumila*); flowering and fruiting before the leaves develop in the spring; calyx lobes broadly rounded
    2. Fruits with the margins of the wings with dense, fine, spreading hairs, the surfaces of the wings and the body variously glabrous or hairy; flowers noticeably stalked, in drooping umbellate clusters or short racemes; leaves lacking forked veins or uncommonly with more than 1 secondary vein on each side of the midvein forked toward their tips
        3. Leaf blades 1.8–4.0 cm wide, the base more or less symmetric, the marginal teeth 1(–2) mm deep; fruits 7–8 × 2–3 mm; winter buds 2.5–3.0 mm long, dark brown or purplish brown to nearly black; at least some of the twigs almost always developing 1 or 2, opposite, flat, irregular wings . . . . 1. U. ALATA
        3. Leaf blades 3.3–9.5 cm wide, the base strongly asymmetric, the marginal teeth 2–4 mm deep; fruits 9–23 × 6–13 mm; winter buds 3–8 mm long, reddish brown; twigs either unwinged or developing corky outgrowths that form several irregular, more or less parallel wings
            4. Inflorescences umbellate clusters; fruits not appearing inflated, the surfaces glabrous, only the wing margins hairy; leaf blades with the undersurface sparsely hairy on and between the veins, also with small dense tufts of hairs in the vein axils; branches never developing wings or woody outgrowths . . . . 2. U. AMERICANA
            4. Inflorescences short, pendant racemes; fruits appearing inflated, the surfaces and margins evenly hairy; leaf blades with the undersurface moderately hairy, the hairs denser on the veins, sometimes also tufted in the vein axils; usually at least some older twigs developing irregular, corky wings anastomosing in several ranks . . . . 7. U. THOMASII
    2. Fruits with the surfaces and margins of the wings glabrous (except for hairs on the stigmatic surface in the apical notch; the body densely hairy on the surfaces in *U. rubra*); flowers sessile or nearly so, in dense clusters; many leaves with 1–3 secondary veins on each side of the midvein prominently forked toward their tips
        5. Leaves 1.2–3.5 cm wide, elliptic-ovate to elliptic-lanceolate, symmetric or only weakly asymmetric at the base, angled or gradually tapered at the tip, the margins singly or doubly toothed, the teeth 0.5–1.0 mm deep . . . . 5. U. PUMILA
        5. Leaves 3.5–10.0 cm wide, broadly oblanceolate to obovate, oblong, or elliptic, strongly asymmetric at the base, tapered to short-tapered at the tip, the margins doubly toothed, the major teeth 1.5–3.0 mm deep

6. Winter buds 3.0–4.5 mm long, broadly ovoid to subglobose, bluntly pointed to more or less rounded, reddish brown to nearly black, the scales moderately to densely pubescent with minute, sometimes appressed, white or occasionally red hairs; fruits with the body glabrous (except for hairs on the stigmatic surface in the apical notch) . . . . 4. U. MINOR
6. Winter buds 4–8 mm long, ovoid, bluntly pointed, dark red to orangish brown, at least the inner scales densely pubescent with spreading and shaggy, red to brownish red hairs (these sometimes becoming bleached in late winter); fruits with the body densely hairy . . . . 6. U. RUBRA

1. Ulmus alata Michx. (winged elm, wahoo)
Pl. 569 n, o; Map 2663

Plants shrubs or trees, 2–15(–30) m tall. Twigs dark brown to purplish brown, usually hairy; almost always at least some of them with 1 or 2, opposite, flat, thin, corky wings. Winter buds 2.5–3.0 mm long, narrowly ovoid to more or less conic, sharply pointed, dark brown to purplish brown or nearly black, minutely pubescent with white or off-white hairs. Petioles 1.5–4.0 mm long. Leaf blades 2–10 cm long, 1–4 cm wide, narrowly elliptic or lanceolate-elliptic, the base nearly asymmetric, gradually tapered to a sharply pointed tip, the major marginal teeth 0.8–1.2(–2.0) mm deep, blunt or more commonly sharp, all or most with 1–3 smaller secondary teeth, the upper surface smooth or slightly roughened, the undersurface short-hairy along and often also between the main veins, not tufted in the vein axils, the secondary veins 10–15 on each side of the midvein, the lateral veins seldom forked toward their tips; juvenile leaves never lobed. Inflorescences short racemes, appearing in the spring before the leaves develop on second-year twigs. Flowers with the stalks 2–7 mm long, the calyces deeply 5-lobed, the tube glabrous, the lobes broadly rounded, glabrous or with a few marginal hairs near their tips. Fruits 0.7–0.8 cm long, 0.2–0.4 cm wide, elliptic to oblong-elliptic or ovate-elliptic, tan, the body and wings densely and finely hairy, the wing margins also densely hairy. $2n = 28$. March.

Scattered in the Ozark, Ozark Border, and Mississippi Lowlands Divisions, uncommon or absent elsewhere in the state (eastern [mostly southeastern] U.S. west to Kansas and Texas). Glades, savannas, tops of bluffs, banks of streams and rivers, margins of sinkhole ponds, bottomland forests, mesic to dry upland forests, and sand prairies; also roadsides.

Ulmus alata is most frequently encountered in Missouri as a shrub or small tree in dry upland areas. However, the species can also occur in bottomlands, where it sometimes grows into a mid-story or canopy tree. Trees in wet areas tend to develop fewer corky branches than do those at drier sites.

2. Ulmus americana L. (American elm, white elm)
Pl. 570 k, l; Map 2664

Plants trees to 35 m tall. Twigs tan to reddish brown, usually hairy when young, often becoming nearly glabrous with age, rarely completely glabrous, never with corky outgrowths or wings. Winter buds 3–8 mm long, narrowly ovoid to conic, sharply pointed, brown to reddish or purplish brown, minutely pubescent with white to red hairs. Petioles (3–)4–9 mm long. Leaf blades 6.5–12.5 cm long, 3.3–7.5 cm wide, elliptic to oblong-elliptic or occasionally oblong, the base strongly asymmetric, short-tapered to a sharply pointed tip, the major marginal teeth 2–4 mm deep, sharp, all or most with 1–3 smaller secondary teeth, the upper surface smooth or somewhat roughened, the undersurface sparsely hairy along and between the main veins, usually also tufted in the vein axils, the secondary veins 11–21 on each side of the midvein, the lateral veins seldom (occasionally 1 or 2 per side) forked toward their tips; juvenile leaves never lobed. Inflorescences drooping umbellate clusters appearing before the leaves develop in the spring on second-year twigs. Flowers with the stalks 10–20 mm long, the calyces shallowly (5–)7–9-lobed, the tube glabrous, the lobes broadly rounded, glabrous or the margins sparsely hairy. Fruits 0.9–1.3 cm long, 0.6–0.8 cm wide, elliptic, tan, not appearing inflated, the body and wings glabrous on the surfaces, but the wing margins densely hairy. $2n=28, 56$. March–April.

Scattered nearly throughout the state, but apparently absent from the northwestern corner (eastern U.S. west to Montana and Texas; Canada). Bottomland forests, mesic upland forests, banks of streams and rivers, margins of ponds, sinkhole ponds, and oxbows, and tops and ledges of bluffs; also roadsides and disturbed areas.

Plants vary in the relative roughness of the leaves and the pubescence of the twigs, and these have been named as follows: Plants with roughened leaves and pubescent twigs have been called f. *alba* (Aiton) Fernald; plants with roughened leaves and glabrous twigs have been called

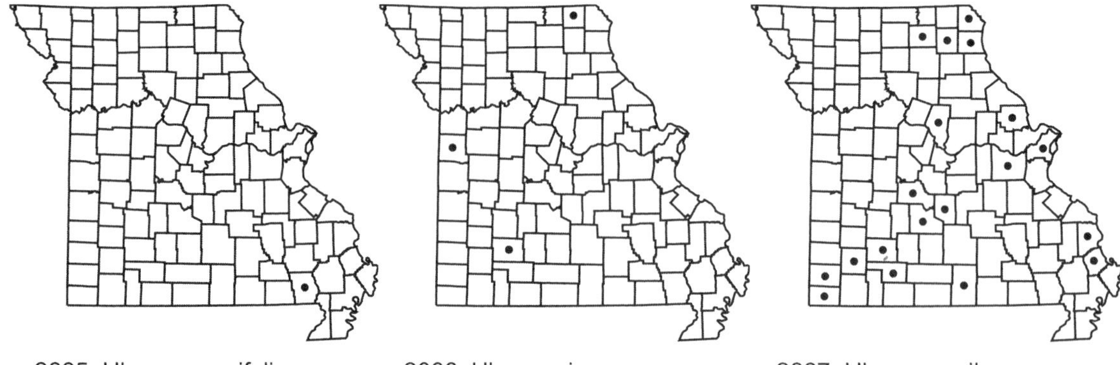
2665. Ulmus crassifolia 2666. Ulmus minor 2667. Ulmus pumila

f. *intercedens* Fernald; plants with smooth leaves and pubescent twigs have been called f. *pendula* (Aiton) Fernald; and plants with smooth leaves and glabrous twigs have been called f. *laevior* Fernald. Elias (1970) noted that smooth and roughened leaves may be found on the same tree; thus these minor variants are unworthy of formal taxonomic recognition.

Most references list *U. americana* as a tetraploid, with $2n$=56, but actually both diploid and tetraploid races are found in Missouri, sometimes coexisting in the same stand; triploids are known from elsewhere (Whittemore and Olsen, 2011). Further study is needed to determine the relationship between these genetic forms.

Ulmus americana formerly was grown widely as a street tree and shade tree, but the use of the species has been limited by Dutch elm disease, which has killed millions of elms since its introduction into North America (for further discussion, see the paragraph under the generic treatment of *Ulmus*).

3. Ulmus crassifolia Nutt. (cedar elm)

Pl. 570 a, b; Map 2665

Plants trees to 15(–30) m tall. Twigs purplish brown, hairy; usually at least some of them with 1 or 2 opposite, irregular, corky wings. Winter buds 1.5–2.5 mm long, ovoid, bluntly pointed, dark red, minutely hairy. Petioles 2–3 mm long. Leaf blades (1.6–)3.2–6.0 cm long, (0.9–)1.6–3.0 cm wide, oblong or narrowly elliptic, the base weakly to moderately asymmetric, angled to a broadly or bluntly pointed tip, the major marginal teeth 0.7–1.2 mm deep, blunt, most with 1 or 2 smaller secondary teeth, the upper surface roughened, the undersurface short-hairy, at least along the main veins, not tufted in the vein axils, the secondary veins 8–12 on each side of the midvein, many leaves with 2 or more lateral veins on each side forked toward their tips; juvenile leaves never lobed. Inflorescences short umbellate clusters appearing in the fall in the leaf axils from the older parts of the current-year's growth. Flowers with the stalks 7–10 mm long, the calyces deeply 6–9-lobed, the tube pubescent with crisped white hairs, the lobes linear, glabrous except for red hairs along the margins near their tips or sometimes also with scattered hairs on the outer surface. Fruits 0.9–1.1 cm long, 0.5–0.7 cm wide, elliptic, tan, the body and wings finely hairy, the wing margins also densely hairy. $2n$ = 28. September.

Uncommon, known thus far only from Butler County (Florida to Texas north to Tennessee, Missouri, and Oklahoma; Mexico). Bottomland forests; also levees and roadsides.

This species was first reported for Missouri from Dunklin County by Heineke (1987), but his voucher specimen could not be located during the present work.

4. Ulmus minor Mill. (English elm)

U. carpinifolia Gled.

U. procera Salisb.

Map 2666

Plants trees to 30 m tall. Twigs tan to reddish brown, glabrous or hairy, often at least a few of them with 1 to several corky outgrowths that form more or less parallel wings. Winter buds 3.0–4.5 mm long, short-ovoid to subglobose, rounded to bluntly pointed, reddish brown to nearly black, minutely pubescent with appressed or spreading, white or occasionally red hairs. Petioles 4–13 mm long. Leaf blades 5.0–11.5 cm long, 3.5–8.5 cm wide, ovate to elliptic or oblong-elliptic, the base strongly asymmetric, short-tapered to an often broadly but usually sharply pointed tip, the major marginal teeth 2–3 mm deep, sharp, all or most with 1–3 smaller secondary teeth, the upper surface smooth or roughened, the undersurface sparsely to moderately hairy along and between the main veins, or glabrous except for tufts in the

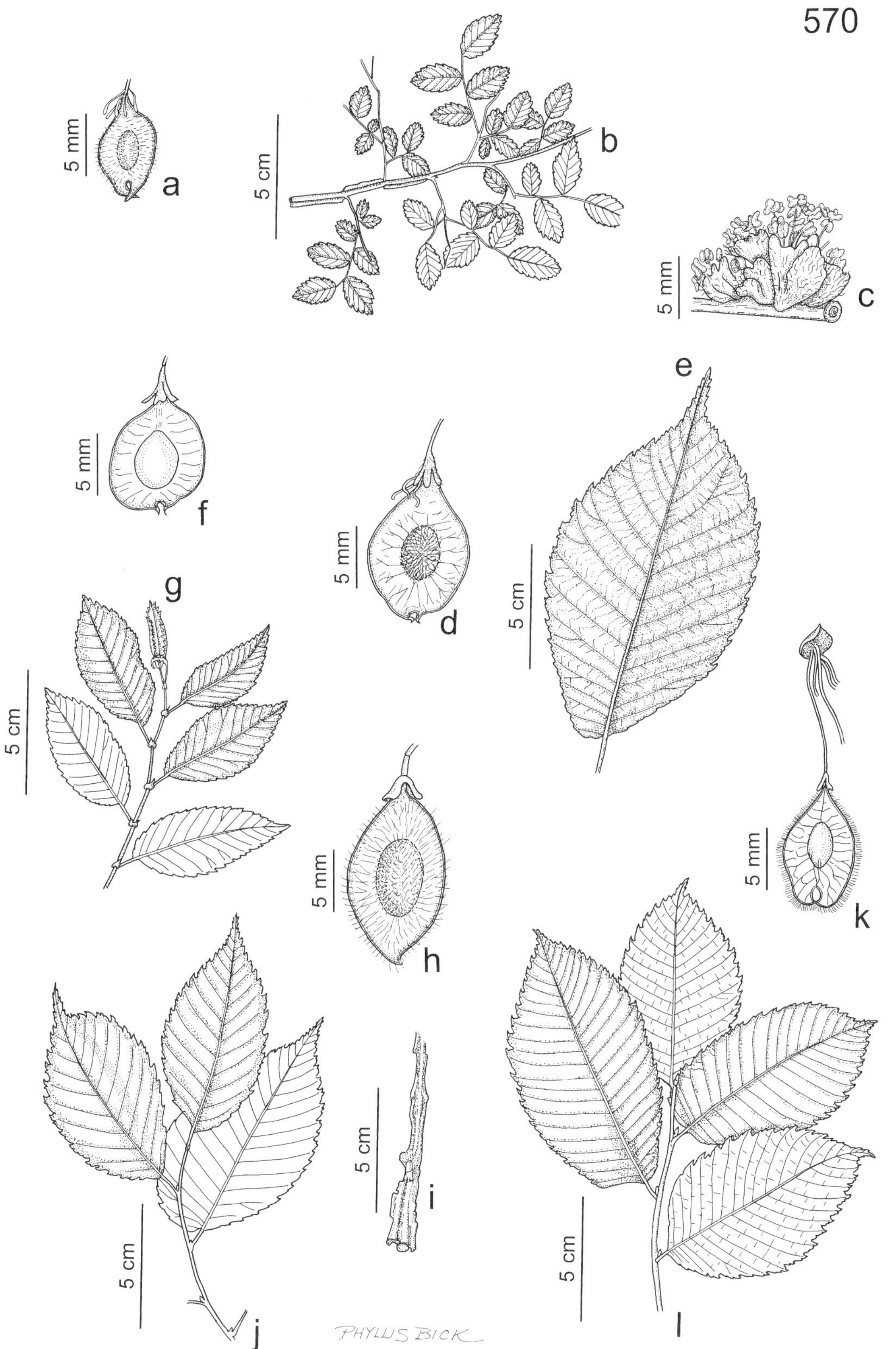

Plate 570. Ulmaceae. *Ulmus crassifolia*, **a)** fruit, **b)** vegetative branch with winged bark and leaves. *Ulmus rubra*, **c)** inflorescence, **d)** fruit, **e)** leaf. *Ulmus pumila*, **f)** fruit, **g)** branch tip with leaves and stipules. *Ulmus thomasii*, **h)** fruit, **i)** twig with winged bark, **j)** leaf. *Ulmus americana*, **k)** fruit, **l)** branch tip with leaves.

vein axils, the secondary veins 10–17 on each side of the midvein, those of many leaves with 1–3 secondary veins on each side of the midvein prominently forked toward their tips; juvenile leaves never lobed. Inflorescences dense clusters appearing before the leaves develop in the spring on second-year twigs. Flowers sessile or nearly so, the calyces shallowly 5–8-lobed, the tube glabrous or sparsely short-hairy, the lobes broadly rounded, glabrous or sparsely hairy on the outer surface, the margins sparsely to more commonly densely hairy. Fruits 1.6–2.0 cm long, 1.1–2.0 cm wide, nearly circular to broadly elliptic, pale tan (usually darker over the seed), the body glabrous, the wings glabrous on the surfaces and margins (except for hairs on the stigmatic surface in the apical notch). $2n=28$. April.

Introduced, uncommon, sporadic (native of Europe; introduced sporadically in the northeastern U.S. west to Kansas, also California and Nevada). Edges of glades and upland prairies; also open, disturbed areas.

Ulmus minor suckers heavily, and is easily propagated by transplanting suckers. Partly because of this, *U. minor* has been economically very important in Europe for thousands of years (Fuentes-Utrilla et al., 2004), and especially in Roman times, when the planting of elms was widely recommended by agricultural writers. Elm foliage was widely used as feed for cattle and sheep, and the wood was used for fuel and construction. Trees were pollarded to produce ample foliage low enough for easy harvesting, and were often used as supports for grape vines, which were planted near young elms and trained to climb the trees. The taxonomy of *U. minor* was considered very controversial until recently. Plants from different areas of western Europe can often be distinguished by minor morphological differences, and these forms were treated as distinct species by some authors (Richens, 1983). Recent research has made it clear that many of these recognizable minor variants are not natural entities at all, but are clones that were selected and widely propagated in past millennia, which have survived and reproduced by suckering through the intervening two thousand years. The commonest of these clones is the English elm, previously known as *U. procera*. Recent research has demonstrated that this clone originated in the hills north of Rome, and is probably the clone called Atinian elm by Roman authors such as Columella (Gil et al., 2004). The shapely crowns of this elm formed a characteristic feature of landscapes in western Europe for centuries. Unfortunately, most of these trees have been infected by Dutch elm disease during the twentieth century (see further discussion under the generic treatment). The fungus girdles the trunk and kills the crown of the tree, but the root system survives and continues to produce suckers, which grow until they, too, are infected and girdled by the fungus. The plants thus survive in many areas, but the graceful trees of the past have been replaced by low thickets of suckers.

5. Ulmus pumila L. (Siberian elm, dwarf elm)
Pl. 570 f, g; Map 2667

Plants trees to 25(–35) m tall but often much shorter. Twigs tan to yellowish gray, glabrous or hairy, never with corky outgrowths or wings. Winter buds 3–4 mm long, ovoid to subglobose, rounded to bluntly pointed, dark brown to reddish brown, glabrous or minutely pubescent with white hairs. Petioles 4–10 mm long. Leaf blades 2–8 cm long, 1.2–3.5 cm wide, elliptic-ovate to elliptic-lanceolate, the base symmetric or only weakly asymmetric, angled or gradually tapered to a usually sharply pointed tip, the major marginal teeth 0.5–1.0 mm deep, sharp, simple or often many of them with 1 smaller secondary tooth, the upper surface smooth, the undersurface glabrous or sparsely to moderately hairy along the main veins, sometimes also tufted in the vein axils, the secondary veins 9–16 on each side of the midvein, those of many leaves with 1–3 secondary veins on each side of the midvein forked toward their tips; juvenile leaves never lobed. Inflorescences dense clusters appearing before the leaves develop in the spring on second-year twigs. Flowers sessile or nearly so, the calyces shallowly 4- or 5-lobed, the tube glandular, the lobes broadly rounded, glabrous or the margin sparsely hairy. Fruits 1–2 cm long, 1.0–1.5 cm wide, nearly circular to occasionally broadly obovate to broadly elliptic, pale tan, the body glabrous, the wings glabrous on the surfaces and margins (except for hairs on the stigmatic surface in the apical notch). $2n=28$. March.

Introduced, sporadic in the state, but relatively common in southwesternmost and northeasternmost Missouri and the St. Louis metropolitan region (native of Europe, Asia; introduced nearly throughout the U.S., Canada). Mesic upland forests, upland prairies, sand prairies, and banks of streams; also old fields, fencerows, cemeteries, gardens, railroads, roadsides, and disturbed areas.

Ulmus pumila has a remarkable ability to tolerate extreme heat, cold, and drought, and it thrives in difficult sites. It was introduced to the United States to provide windbreaks and wood for fuel and construction on farms in the Great Plains and arid interior of the western United States, and

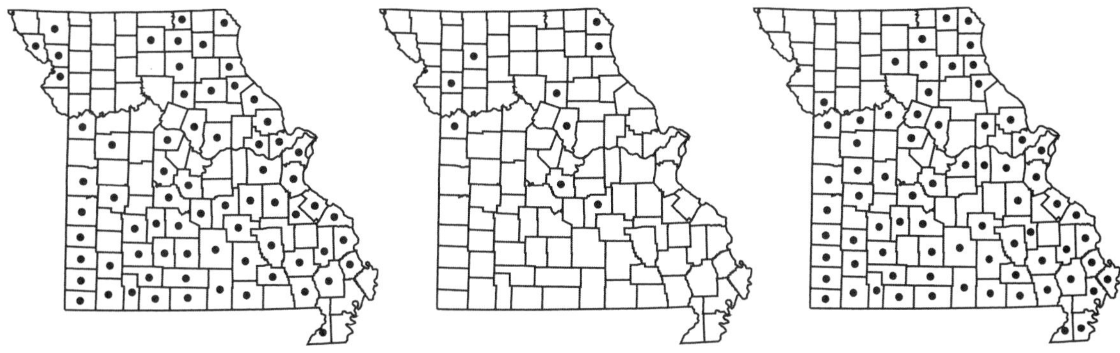

2668. Ulmus rubra 2669. Ulmus thomasii 2670. Boehmeria cylindrica

it was widely planted in the early twentieth century. It has become an aggressive invader in many areas in the western part of the U. S., but in Missouri it is primarily found in areas of moderate to severe human disturbance. It is known to hybridize with *U. rubra* where they occur together (for further discussion, see the treatment of that species).

The Chinese elm, *U. parvifolia* Jacq., is widely cultivated in Missouri. It is rather weedy in cultivated situations, and it is known to be invasive in disturbed habitats in several states, including Arkansas and Kentucky (Medley and Thieret, 1991, Sherman-Broyles, 1997), so it should be watched for in Missouri. *Ulmus parvifolia* has foliage very similar to that of *U. pumila*; in fact, they often cannot be distinguished confidently based on vegetative twigs. *Ulmus parvifolia* can be recognized by its distinctive bark, which is not furrowed, but exfoliates as irregular woody scales 2–8 cm across, and is orange-brown where freshly exposed and soon weathering ashy gray, and because it flowers and fruits in autumn rather than spring.

6. Ulmus rubra Muhl. (slippery elm)
U. fulva Michx.

Pl. 570 c–e; Map 2668

Plants trees to 35 m tall. Twigs tan to reddish brown or purplish brown, hairy, never with corky outgrowths or wings. Winter buds 4–8 mm long, ovoid to narrowly ovoid, bluntly pointed, dark red to orangish brown, at least the inner scales densely pubescent with short, spreading and shaggy, red to brownish red hairs (these sometimes becoming bleached in late winter). Petioles 4–9 mm long. Leaf blades 10–19 cm long, 5–10 cm wide, broadly oblanceolate to obovate or elliptic, the base strongly asymmetric, tapered to short-tapered to a sharply pointed tip, the major marginal teeth 1.5–3.0 mm deep, blunt to more or less sharp, most with 1–5 smaller secondary teeth, the upper surface strongly roughened, the undersurface moderately to densely hairy along and between the main veins, also noticeably tufted in the vein axils, the secondary veins 10–20 on each side of the midvein, those of many leaves with 3 or more secondary veins on each side of the midvein forked toward their tips; juvenile leaves unlobed or with 1 or 2 tapered lobes per side toward the tip. Inflorescences dense clusters appearing before the leaves develop in the spring on second-year twigs. Flowers sessile or nearly so, the calyces shallowly 5–9-lobed, the tube pubescent with straight white hairs, the lobes broadly rounded, densely pubescent with red to reddish brown hairs on the outer surface and margins. Fruits 1.4–1.9 cm long, 1.2–1.8 cm wide, nearly circular to broadly elliptic, pale tan (usually darker over the seed), the body densely short-hairy, the wings glabrous or minutely hairy on the surfaces, but the margins glabrous. $2n=28$. March–April.

Common nearly throughout the state (eastern U.S. west to North Dakota and Texas; Canada). Bottomland forests, mesic upland forests, bases, ledges, and tops of bluffs, banks of streams and rivers, and margins of ponds and sinkhole ponds; also old fields, fencerows, strip mines, roadsides, and disturbed areas.

This is the most abundant species of *Ulmus* in the state. The inner bark is very mucilaginous, unlike that of our other elms. Native Americans had multiple uses for *U. rubra*, ranging from a variety of medicinal applications to a kind of tea, a food preservative, cordage, and building material (Moerman, 1998). Slippery elm inner bark is still used medicinally for soothing sore throats, treating colds, and sometimes for lower digestive problems like diverticulitis. It is among the natural products that is wild-harvested, particularly in the Ozarks and sold by individuals to wholesale distributors.

Ulmus rubra hybridizes with *U. pumila* where the two species grow together, and the hybrids are fertile and backcross to the parental species (Zalapa et al., 2009). Such hybrids occur in St. Louis County, but specimens observed thus far are all vegetative and display juvenile leaf morphology. The hybrid is called *U. ×intermedia* Elowsky, but some specimens in herbaria have been annotated with the unpublished name "*U. ×notha.*"

7. Ulmus thomasii Sarg. (rock elm, cork elm)
Pl. 570 h–j; Map 2669

Plants trees to 30 m tall. Twigs tan to reddish brown, hairy when young, sometimes sparsely so, often becoming nearly glabrous with age, usually at least some older twigs developing irregular, corky wings anastomosing in several ranks. Winter buds 4–8 mm long, narrowly ovoid to narrowly ellipsoid, sharply pointed, brown to more commonly reddish brown, minutely pubescent with white to red hairs. Petioles 3–5(–7) mm long. Leaf blades 8.5–15.0 cm long, 4.5–9.5 cm wide, elliptic to more or less rhombic-elliptic, the base strongly asymmetric, short-tapered to a sharply pointed tip, the major marginal teeth 2–4 mm deep, sharp, all or most with 1–3 smaller secondary teeth, the upper surface smooth, the undersurface moderately hairy, more densely so along the main veins, sometimes also tufted in the vein axils, the secondary veins 15–29 on each side of the midvein, the lateral veins rarely forked toward their tips; juvenile leaves never lobed. Inflorescences short pendant racemes appearing before the leaves develop in the spring on second-year twigs. Flowers with the stalks 5–10 mm long, the calyces (5–)7- or 8-lobed to about the midpoint, the tube glabrous or sparsely pubescent with short straight hairs, the lobes broadly rounded, glabrous or with a few marginal hairs. Fruits 1.7–2.3 cm long, 1.1–1.3 cm wide, elliptic, tan, somewhat inflated, the body and wings evenly and finely hairy on the surfaces, the wing margins also densely hairy. $2n=28$. April.

Uncommon in the northern and central portions of the state (Eastern North America, from southern Ontario and Minnesota south to Virginia and Arkansas). Bottomland forests, mesic upland forests, tops and ledges of bluffs, and rarely savannas.

Ulmus thomasii resembles *U. americana* in the form and serration of its leaves and in its long, narrowly pointed winter buds. When vegetative, the two species are easily confused. *Ulmus thomasii* is considered to have the best-quality wood of any of our elms, and it was extensively logged in the past. Likely it was never common in Missouri and many of the herbarium records documenting its distribution are historical.

URTICACEAE (Nettle Family)

Plants annual or perennial herbs (shrubs elsewhere), usually monoecious or dioecious, sometimes incompletely so, sometimes armed with stinging hairs, not producing milky sap. Leaves alternate or opposite, petiolate. Stipules absent or present, when present usually small and herbaceous to scalelike, sometimes (in species with opposite leaves) fused between the adjacent leaf bases, often shed early. Leaf blades simple, pinnately veined or with 3 main veins, the margins entire, scalloped, or toothed, the surfaces usually appearing minutely dotted, the dots due to small structures (known as cystoliths) containing calcium carbonate crystals present in enlarged epidermal cells. Inflorescences axillary or terminal, small clusters, spikes, or spikelike racemes, these sometimes arranged into panicles, the flowers sometimes subtended by short bractlets. Flowers minute, actinomorphic (except sometimes in *Pilea*), hypogynous, mostly imperfect. Staminate flowers usually short-stalked, the calyces of 4 or 5, small, free sepals (usually absent in *Parietaria*). Pistillate flowers sessile or short-stalked, the calyces of (2)3–5, small, free or fused sepals, persistent at fruiting. Corollas absent. Stamens 4 or 5 (absent in pistillate flowers, except in *Pilea*, where reduced to 3 strongly inward-curved, scalelike staminodes), free, opposite the sepals, the filaments bent inward in the bud, reflexing suddenly as the bud opens and ejecting the pollen explosively, the anthers attached basally, yellow, dehiscing by longitudinal slits. Pistil 1 per flower (reduced to a small peglike structure in staminate flowers), of 1 carpel, the ovary superior, 1-locular, with 1 ovule, the placentation basal. Style absent or 1, the stigma 1, linear or capitate. Fruits achenes, not fused into groups. About 55 genera, about 1,650 species, nearly worldwide.

Members of the Urticaceae mostly are of relatively little economic importance (N. G. Miller, 1971). The principal crop plant is *Boehmeria nivea* (L.) Gaudich. (ramie), which originated in eastern Asia. After chemical treatment to remove pectins and other impurities, the inner bark of this herbaceous perennial yields a strong fiber that for many centuries has been woven into fine fabrics and yarns. Some species of *Urtica* were used similarly historically until ramie production became more widespread. Otherwise, members of a few genera, such as *Pilea*, are cultivated as ornamentals, outdoors in warmer climates and as house plants farther north. Some of the species produce sufficient windborne pollen to be of significance as allergens, particularly in portions of Europe. The species with stinging hairs have sometimes been used for culinary purposes and medicinally.

The stinging hairs of some genera act as syringes. They consist of a bulbous reservoir at the base and a slender shaft, which breaks off along a zone of weakness below a slightly bulbous tip, leaving a sharp beveled tip to pierce the skin. The hairs contain a cocktail of alkaloids and other toxic substances, including some similar in action to acetylcholine, and cause a strong histamine reaction that involves contact dermatitis and a lingering burning sensation (Burrows and Tyrl, 2001). The hairs of some *Urtica* species have been touted as a treatment for various ailments, including rheumatism. However, some members of the Australian genus of stinging trees, *Dendrocnide* Miq., have been documented to cause fatalities in humans and other mammals.

1. Plants armed with stinging hairs (with magnification, note the well-differentiated, expanded base below the stiff, straight, slender shaft), sometimes sparsely so; pistillate flowers with 4 free sepals, the calyx consisting of smaller and larger opposite pairs (occasionally the smaller pair totally absent)
    2. Leaves alternate; ovary with an elongate style (this persistent at fruiting and sometimes hooked), the stigmatic region linear . . . . 2. LAPORTEA
    2. Leaves opposite; ovary and fruit lacking a style, the stigma capitate and papillose (appearing bushy) . . . . 5. URTICA
1. Plants unarmed (the hairs, if present, soft, frequently curved, and lacking an expanded base); pistillate flowers either with 3 free sepals or with 4 sepals fused nearly to the tip (the calyx then tubular to flask-shaped)
    3. Leaves alternate, the margins entire . . . . 3. PARIETARIA
    3. Leaves all opposite or occasionally a few pairs subopposite, the margins scalloped or toothed
        4. Plants hairy; pistillate flowers with 4 sepals fused nearly to the tip (the calyx appearing more or less flask-shaped, but flattened); inflorescences of small dense clusters arranged into interrupted axillary spikes . . . . 1. BOEHMERIA
        4. Plants glabrous; pistillate flowers with 3 free sepals, these equal or occasionally 1 somewhat enlarged and hoodlike; inflorescences of small axillary clusters, these often arranged into small panicles . . . . 4. PILEA

1. **Boehmeria** Jacq.
(Wilmot-Dear and Friis, 1995)

About 50 species, nearly worldwide.

1. Boehmeria cylindrica (L.) Sw. (false nettle)
B. cylindrica var. *drummondiana* (Wedd.) Wedd.

Pl. 571 a, b; Map 2670

Plants perennial, unarmed, but often with sparse to dense, short, fine, soft or stiff (then roughened to the touch), nonstinging hairs, with rhizomes, the roots fibrous. Stems 40–150 cm long,

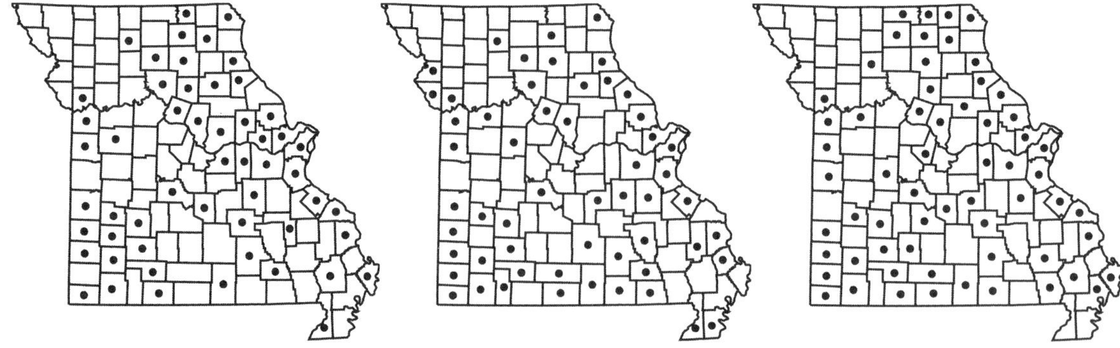

2671. Laportea canadensis   2672. Parietaria pensylvanica   2673. Pilea pumila

erect or strongly ascending, unbranched or less commonly few- to several-branched. Leaves opposite or occasionally a few pairs subopposite, short- to long-petiolate, stipulate. Leaf blades 3–15 cm long, sometimes the pair at a node slightly unequal in size, lanceolate to elliptic, broadly elliptic, ovate, or broadly ovate, somewhat asymmetrically angled to rounded at the base, tapered at the tip, the margins sharply toothed, the venation with the 2 basalmost lateral veins more developed than the others; cystoliths rounded. Inflorescences axillary, small dense clusters arranged into short or elongate and interrupted spikes, these sometimes leafy toward the tips, the staminate flowers scattered among the pistillate flowers or in separate spikes (rarely the plants totally dioecious). Bractlets not forming an involucre. Staminate flowers with 4 sepals, these 0.7–1.1 mm long, cupped around the stamens. Stamens 4. Pistillate flowers with 4 sepals fused nearly to the minutely 2- or 4-toothed tip (the calyx appearing more or less flask-shaped, but flattened), 0.6–1.0 mm long, enclosing and fused to the ovary and fruit. Style elongate (exserted from the calyx), the stigmatic region linear. Fruits 0.6–0.8 mm long, remaining enclosed in the persistent calyx, which appears flattened, with a pair of thick wings, the surface with sparse to dense, ascending, sometimes hooked hairs. $2n=28$. June–October.

Scattered to common nearly throughout the state, but uncommon or absent from northwesternmost Missouri (eastern U.S. west to Minnesota, Utah, and Arizona; Canada, Mexico, Central America, South America, Caribbean Islands; introduced in California). Bottomland forests, banks of streams, rivers, and spring branches, and fens; also roadsides.

Steyermark (1963) and some other authors have separated this species into two varieties. The typical variety, which is by far the most abundant and widespread of the two in Missouri, has glabrous or sparsely hairy stems; ascending to spreading, relatively long-petiolate leaves; leaf blades that are relatively thin, flat, long-tapered at the tip, only slightly roughened on the upper surface, and glabrous or sparsely hairy on the undersurface; and fruiting calyces that are glabrous or sparsely hairy with straight hairs and lack purple mottling. The var. *drummondiana*, which in Missouri tends to grow around the margins of some fens and spring branches in the Ozark Division, has been characterized as having moderately to densely hairy stems; drooping, relatively short-petiolate leaves; leaf blades that tend to be relatively thick, folded longitudinally, short-tapered at the tip, strongly roughened on the upper surface, and moderately hairy on the undersurface; and fruiting calyces that are densely hairy with straight and hooked hairs and with purple mottling. Although the extreme forms of the species can appear strikingly different, Wilmot-Dear and Friis (1995) and Boufford (1997b) noted that across the broad distributional range of the species, the characters listed tend to vary independantly and numerous intermediates are known. N. G. Miller (1971) suggested that the morphological differences may be reflections of habitat differences, rather than genetically based variation. The unpublished thesis studies of Stuyvesant (1984) tended to corroborate this.

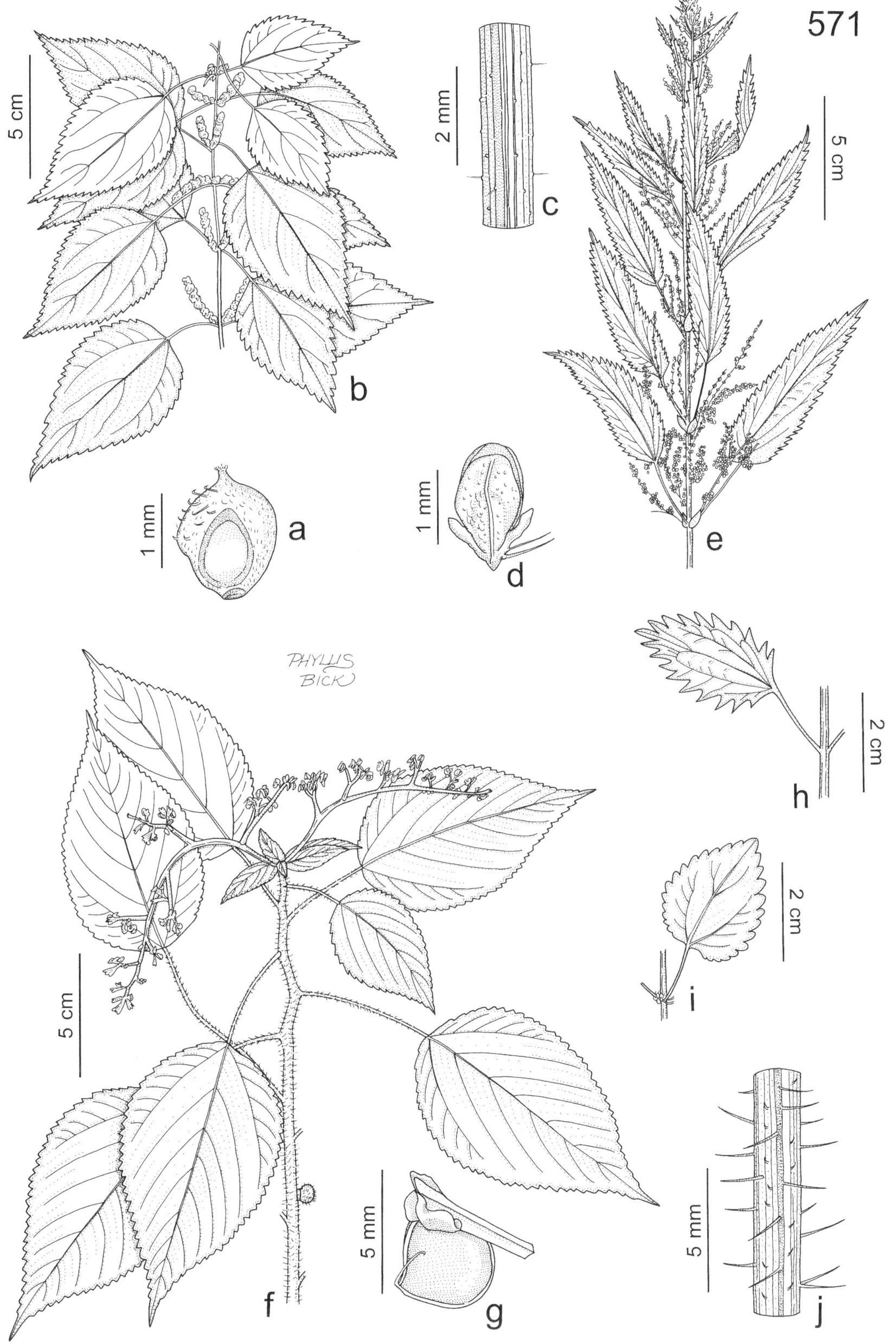

Plate 571. Urticaceae. *Boehmeria cylindrica*, **a)** fertile stem, **b)** fruit. *Urtica dioica* ssp. *gracilis*, **c)** stem detail, **d)** pistillate flower, **e)** fertile stem. *Laportea canadensis*, **f)** fertile stem, **g)** fruit. *Urtica urens*, **h)** node with leaf. *Urtica chamaedryoides*, **i)** node with leaf, **j)** stem detail.

2. **Laportea** Gaudich.
(Chew, 1969)

About 22 species, North America to South America, Caribbean Islands, Europe, Asia, Africa, Pacific Islands, most diverse in Africa and Madagascar.

1. Laportea canadensis (L.) Gaudich. (wood nettle)

Pl. 571 f, g; Map 2671

Plants perennial (but often flowering the first year), armed with sparse to dense, long, stinging hairs, also with sparse to dense, shorter, finer, nonstinging hairs, with rhizomes, the roots fibrous, sometimes somewhat tuberous. Stems 35–150 cm long, erect or strongly ascending, usually unbranched, often slightly zigzag. Leaves alternate, long-petiolate, stipulate. Leaf blades 6–20(–30) cm long, ovate to broadly ovate, elliptic, or broadly elliptic, broadly angled to rounded or shallowly cordate at the base, short-tapered at the tip, the margins coarsely toothed, the venation pinnate or the 2 basalmost lateral veins slightly more developed than the others; cystoliths rounded. Inflorescences axillary (but sometimes also appearing terminal), small clusters arranged in panicles, the staminate panicles usually shorter-stalked then the subtending petiole and positioned at nodes below the pistillate ones, which usually are longer than the subtending petiole. Bractlets absent. Staminate flowers with 5 sepals, these 0.8–1.1 mm long, cupped around the stamens. Stamens 5. Pistillate flowers with 4 free sepals, the calyx consisting of 2 smaller sepals, these 0.2–0.3 mm long (occasionally totally absent), spreading, alternating with 2 larger sepals, these 0.8–1.1 mm long, loosely cupped around but not fused to the ovary and fruit. Style elongate (persistent at fruiting and sometimes hooked), the stigmatic region linear. Fruits 2–3 mm long, obliquely attached at the tip of a short, winged stalk, strongly flattened, the body more or less circular in outline, greenish brown to dark brown, glabrous. $2n=26$. May–August.

Scattered to common throughout the state (eastern U.S. west to North Dakota and Oklahoma; Canada). Bottomland forests, swamps, margins of sloughs and oxbows, and banks of streams and rivers; also roadsides.

Steyermark (1963) noted that this species is one of the greatest nuisances to those hiking in bottomlands in Missouri, especially because it tends to occur in dense, large stands. In deference to his adventures in the neotropics, he was quick to point out that the burning sensation imparted by *Laportea* cannot compare to the severity of the stinging hairs of some tropical genera.

3. **Parietaria** L. (pellitory)

Twenty to 30 species, nearly worldwide.

1. Parietaria pensylvanica Muhl. ex Willd. (Pennsylvania pellitory, pellitory)

P. pensylvanica var. *obtusa* (Rydb. ex Small) Shinners

Pl. 572 d, e; Map 2672

Plants annual, unarmed, finely and often densely pubescent with minute, nonstinging hairs (the leaves sometimes, except for the margins, glabrous or nearly so), the hairs at least in part curved and/or hooked, usually with a short taproot, rarely sparse, longer, spreading hairs also present. Stems 5–40(–60) cm long, variously erect to loosely ascending, often weak and reclining on other vegetation or the ground, unbranched or branched. Leaves alternate, short-petiolate, lacking stipules. Leaf blades 1–7 cm long, narrowly to broadly lanceolate, oblong-lanceolate, ovate, oblong-ovate, or somewhat rhombic, mostly narrowly angled at the base, narrowly angled to tapered at the tip, the margins entire, with 3 main veins (the lateral pair branching above the blade base); cystoliths rounded. Inflorescences axillary, small, sessile clusters, the perfect and pistillate flowers usually more or less mixed in the same cluster. Bractlets 3.5–5.0 mm long, longer than the calyces, linear to narrowly oblong, 2–6 bractlets forming an involucre around each cluster. Staminate flowers usually absent, the perfect flowers with 4 sepals, these 1–2 mm long, ascending to slightly incurved at the tips. Stamens 4. Pistillate flowers with 4 sepals fused to about the midpoint, these 1.5–2.0 mm long, equal. Style absent or minute, the stigma capitate (papillose and appearing bushy, not persistent at fruiting). Fruits 0.9–1.2 mm long,

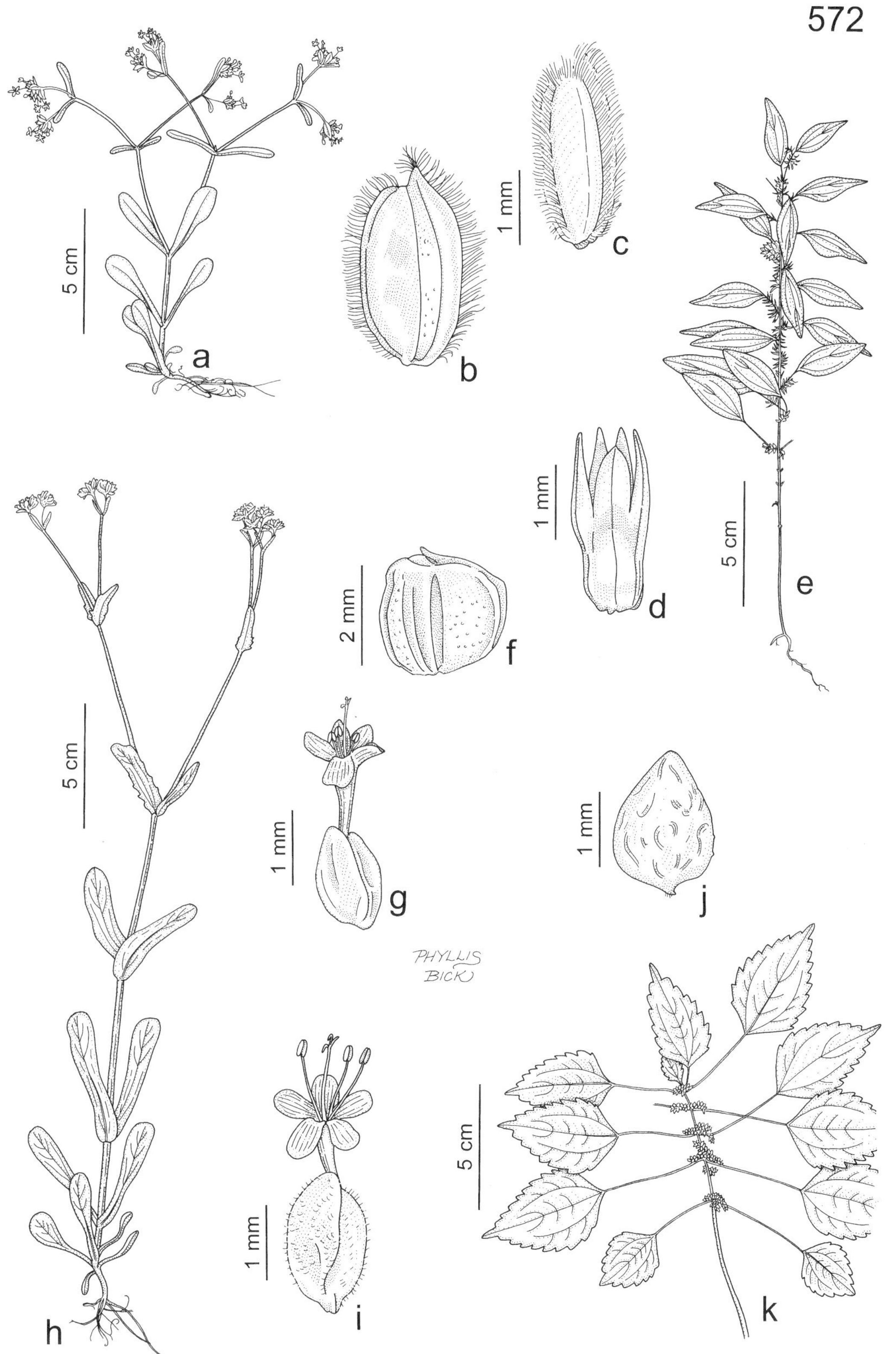

Plate 572. Urticaceae. Valerianaceae. *Valerianella ozarkana*, **a)** habit, **b)** fruit, lateral view, **c)** fruit, ventral view. *Parietaria pensylvanica*, **d)** flower (calyx), **e)** habit. *Valerianella locusta*, **f)** fruit, **g)** flower. *Valerianella radiata*, **h)** flower, **i)** habit. *Pilea pumila*, **j)** fruit, **k)** fruiting stem.

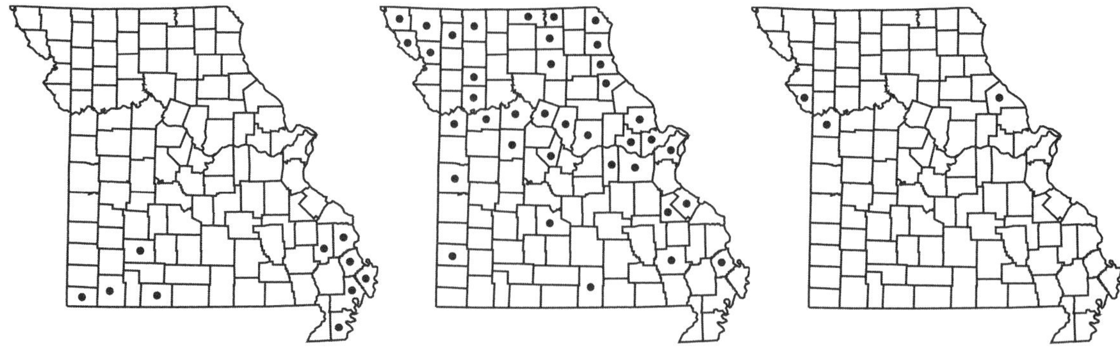

2674. Urtica chamaedryoides 2675. Urtica dioica 2676. Urtica urens

symmetrically attached at the tip of a short, basally expanded stalk, flattened, the body ovate in outline, the surface smooth, tan to yellowish brown. $2n=14$, 16. May–October.

Scattered nearly throughout the state (nearly throughout the U.S.; Canada, Mexico). Glades, bases, ledges, and tops of bluffs, savannas, mesic to dry upland forests, and banks of streams and rivers; also fencerows, gardens, railroads, roadsides, and open disturbed areas.

Uncommon plants with slightly denser pubescence and slightly shorter involucres, as well as sometimes also slightly blunter leaf tips, have been called var. *obtusa*. This variant intergrades freely with the more typical form (Boufford, 1997b).

During his studies of the genus *Parietaria* in the United States and Canada, B. D. Hinton (1968a) discovered an unusual specimen collected by John Kellogg in 1913 near Jerome (Phelps County) that he determined to be *P. praetermissa* B.D. Hinton. This species, which formerly was known under the misapplied name *P. floridana* Nutt. (B. D. Hinton, 1968b), is otherwise endemic to the Atlantic and Gulf Coastal Plains from North Carolina south to Florida and west to Louisiana (B. D. Hinton, 1968a; Boufford, 1997b). During his studies for the flora of North America Project, Boufford (1997b) confirmed this determination, but doubted the occurrence, suggesting that the specimen must be mislabeled. *Parietaria praetermissa* occurs mainly in coastal areas near sea level and it is very unlikely that a chance introduction into Missouri would survive for very long. Thus, the species is excluded from the Missouri flora for the present. However, botanists working in disturbed, sandy areas of central Missouri should keep watch for unusual specimens of pellitory. *Parietaria praetermissa* differs from *P. pensylvanica* very subtly; its fruits have the apical point and basal attachment point positioned slightly off-center (vs. symmetrically in *P. pensylvanica*) and its perianth is slightly longer (1.7–2.3 mm).

4. **Pilea** Lindl.

Six hundred to 715 species, nearly worldwide, except Australia and New Zealand.

1. Pilea pumila (L.) A. Gray (clearweed)
P. pumila var. *deamii* (Lunell) Fernald
Pl. 572 j, k; Map 2673

Plants annual, unarmed, glabrous, the stems and leaves relatively translucent, usually with a short taproot. Stems 7–50(–70) cm long, erect or strongly ascending, occasionally from a spreading base, usually unbranched, stout and slightly succulent. Leaves opposite, long-petiolate, stipulate. Leaf blades 2–12(–15) cm long, elliptic to broadly elliptic or ovate, broadly angled to rounded at the base, occasionally more narrowly angled, tapered at the tip, the margins bluntly toothed to nearly scalloped or occasionally sharply toothed, with 3 main veins; cystoliths linear. Inflorescences axillary, small clusters, these often arranged into small panicles, the staminate and pistillate flowers usually on different branches of the same panicle. Bractlets not forming an involucre. Staminate flowers with 4 sepals, these 0.7–1.1 mm long, loosely cupped around the stamens. Stamens 4. Pistillate flowers with 3 free sepals, these 0.8–1.2 mm long, equal or occasionally 1 somewhat enlarged and hoodlike. Style absent, the stigma capitate (papillose and appearing bushy, not persistent at fruiting). Fruits 1.3–1.8 mm long, flattened, ovate in outline, the surface smooth or nearly so, green to straw-colored, often with fine purple streaks. $2n=24$, 26. July–October.

Scattered to common throughout the state (eastern U.S. west to North Dakota and Texas; Canada; Asia). Bottomland forests, banks of streams and rivers, margins of ponds and lakes; also roadsides and moist, disturbed areas.

Steyermark (1963) noted that some plants in Missouri with more narrowly angled leaf bases and margins with denser, sharper teeth were referable to var. *deamii*, a variant that occurs mostly in the northeastern portion of the species range. However, he discussed the numerous intermediates and chose not to formally recognize this variety, a practice followed by most subsequent botanists (Boufford, 1997b).

Steyermark (1963) also noted that clearweed has sometimes been used as a teaching tool in biology classes to demonstrate the passage of substances through a plant from its roots into the aboveground portions. Because the stems are relatively translucent, water that has been dyed with a stain such as eosin can be seen to be uptaken by roots or cut stems and transported into the upper stems and leaves.

Botanists in northern Missouri should watch for the closely related *P. fontana* (Lunell) Rydb., which has a broad range in the eastern and northern United States that includes the southern half of Iowa and adjacent portions of Illinois and Nebraska. The two species can be very difficult to distinguish in the field. Compared with *P. pumila*, *P. fontana* tends to have slightly smaller, more opaque, less shiny leaves. The main technical differences are in the achenes. Those of *P. fontana* are dark purple or dark olive to black, except for narrow, pale, marginal bands, with a noticeably pebbled surface (Boufford, 1997b).

5. **Urtica** L. (nettle)

Plants annual or perennial (but then sometimes flowering the first year), armed with sparse to dense, long, stinging hairs, often also with sparse to dense, shorter, finer, nonstinging hairs, more or less taprooted or with rhizomes. Stems erect or ascending to more or less spreading with ascending tips, unbranched or branched. Leaves opposite, short- to more commonly long-petiolate, stipulate. Leaf blades lanceolate to ovate, elliptic, or somewhat heart-shaped, variously angled to rounded, truncate, or shallowly cordate at the base, short-tapered to tapered at the tip, the margins variously toothed, more or less with 3 main veins, sometimes appearing weakly palmately veined; cystoliths more or less rounded or less commonly linear. Inflorescences axillary, small clusters, these often arranged in panicles or occasionally dense spikelike racemes, the staminate flowers variously in the same inflorescences as the pistillate ones or in different inflorescences on the same or different plants. Staminate flowers with 4 sepals, these 0.8–1.5 mm long, cupped around the stamens. Stamens 4. Pistillate flowers with 4 free sepals, the calyx consisting of 2 smaller, spreading sepals, alternating with 2 larger sepals, these loosely cupped around but not fused to the ovary and fruit. Style absent, the stigma capitate (papillose and appearing bushy, not persistent at fruiting), the stigmatic region linear. Fruits 1–2 mm long, flattened, ovate or ovate-elliptic in outline, tan to brown, glabrous, the surface often appearing finely pebbled. About 45 species, nearly worldwide.

Stinging hairs are relatively sparse in the taxa of *Urtica* native to Missouri and care must be taken not to miss these during identification of the genus. Steyermark (1963) noted that the young herbage of nettle species is sometimes boiled and eaten as a vegetable.

1. Stipules 5–15 mm long, narrowly lanceolate; inflorescences mostly longer than the subtending petioles; plants perennials with rhizomes . . . . 1. U. DIOICA
1. Stipules 1–4 mm long, oblong to narrowly oblong or linear; inflorescences mostly shorter than the subtending petioles; plants annuals, usually with taproots
    2. Leaves noticeably smaller toward the stem tip; leaf blades narrowly ovate to ovate or somewhat heart-shaped, rounded to truncate or shallowly cordate at the base (those of the reduced upper leaves sometimes narrower and more or less angled at the base), the margins bluntly and relatively finely toothed . . . . 1. U. CHAMAEDRYOIDES

2. Leaves more or less the same size along the stem (the upper leaves only slightly smaller than the others at maturity); leaf blades elliptic to broadly elliptic or less commonly broadly ovate, angled or occasionally rounded at the base, the margins mostly sharply and relatively coarsely toothed . . . . 3. U. URENS

1. Urtica chamaedryoides Pursh (nettle, weak nettle)

Pl. 571 i, j; Map 2674

Plants annual, with short taproots. Stems 15–80(–100) cm long, erect or ascending, but often weak and reclining on surrounding vegetation at maturity, unbranched or more commonly branched from the base, sparsely pubescent with stinging hairs, otherwise glabrous. Stipules 1–4 mm long, oblong to narrowly oblong or linear. Leaf blades 1–8 cm long, noticeably smaller toward the stem tip, narrowly ovate to ovate or somewhat heart-shaped, rounded to truncate or shallowly cordate at the base (those of the reduced upper leaves sometimes narrower and more or less angled at the base), the margins bluntly and relatively finely toothed, the surfaces sparsely to moderately short-hairy, the upper surface (rarely also the undersurface) also with scattered stinging hairs along the main veins, the undersurface sometimes purplish-tinged; cystoliths rounded or short-linear. Inflorescences shorter than the subtending petioles, small, globose clusters, these sometimes appearing as short, dense, spikelike racemes, the staminate and pistillate flowers mixed in the same inflorescence. Pistillate flowers with the 2 smaller sepals 0.4–0.8 mm long, linear, the 2 larger sepals 1.4–2.0 mm long, ovate. Fruits 1.0–1.5 mm long. $2n=26$. April–September.

Uncommon in southwestern and southeastern Missouri (eastern [mostly southeastern] U.S. west to Kansas and Texas; Mexico). Banks of streams and rivers, bottomland forests, bases of bluffs, and less commonly sand savannas; also railroads and moist disturbed areas.

This species routinely occurs in shaded bottomland sites (Woodland et al., 1976), but in the sandy soils of the Sikeston Ridge in the northern portion of the Mississippi Lowlands Division it rarely may be found in association with a more upland flora.

2. Urtica dioica L. (tall nettle)

Pl. 571 c–e; Map 2675

Plants perennial, usually densely colonial from rhizomes. Stems 50–200(–250) cm long, erect or ascending, unbranched or less commonly branched from the base, sparsely to moderately pubescent with stinging hairs, otherwise glabrous or sparsely to densely pubescent with short, nonstinging hairs. Stipules 5–15 mm long, narrowly lanceolate. Leaf blades 4–15(–18) cm long, more or less the same size along the stem (the upper leaves only slightly smaller than the others at maturity), elliptic to lanceolate or narrowly to less commonly broadly ovate, rounded to truncate or shallowly cordate at the base, the margins sharply and relatively coarsely toothed (sometimes appearing doubly toothed, the main teeth having smaller teeth along their margins), the surfaces glabrous or the undersurface sparsely to moderately short-hairy, one or both surfaces often also with scattered stinging hairs along the main veins, the undersurface sometimes lighter green but not purplish-tinged; cystoliths rounded. Inflorescences mostly longer than the subtending petioles, small globose clusters, grouped into panicles with the branches of spikelike racemes, the staminate and pistillate flowers in different inflorescences either on the same or on different plants. Pistillate flowers with the 2 smaller sepals 0.8–1.2 mm long, linear to narrowly lanceolate or oblanceolate, the 2 larger sepals 1.4–1.8 mm long, ovate to broadly ovate. Fruits 1.0–1.5 mm long. $2n=26$, 52. May–October.

Scattered north of the Missouri River, sporadic farther south (nearly throughout the United States but less abundant in the southeastern states; Canada, Mexico, Europe, Asia). Bottomland forests, banks of streams and rivers, sloughs, and bases of bluffs; also levees, ditches, railroads, roadsides, and moist disturbed areas.

Dennis Woodland and his colleagues studied the biosystematics and taxonomy of the *U. dioica* complex (Woodland, 1982a, 1982b; Woodland et al., 1982). They concluded that it was best treated as a series of three subspecies with mostly nonoverlapping ranges. Two of these occur in Missouri. The third, ssp. *holosericea* (Nutt.) Thorne, occupies the western portion of the North American range of the species, overlapping with ssp. *gracilis* in the northwestern and Intermountain states. The ssp. *holosericea* differs from ssp. *gracilis* in its more densely soft-hairy stems that also have more abundant stinging hairs and in the more densely hairy undersurface of the leaves. Whereas, both diploid and tetraploid cytotypes occur in ssp. *gracilis*, thus far ssp. *holosericea* has only been documented as a diploid ($2n=26$).

1. Plants dioecious; stems with relatively dense stinging hairs; leaf blades with both surfaces having stinging hairs . . . . 2A. SSP. DIOICA

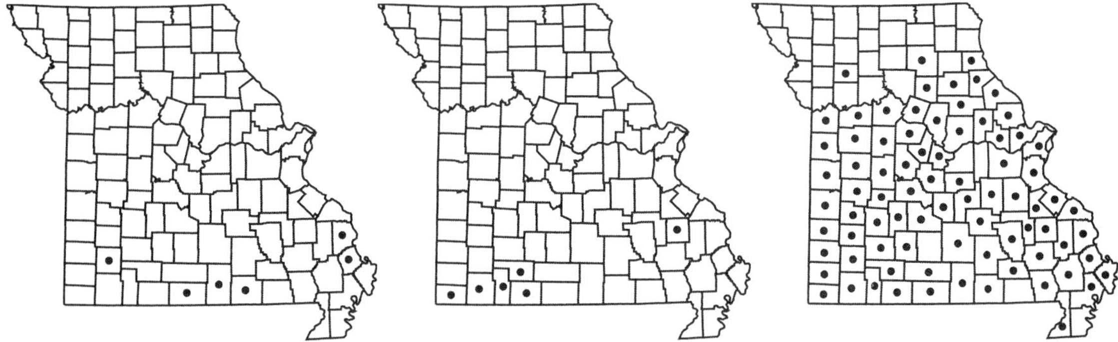

2677. Valerianella locusta 2678. Valerianella ozarkana 2679. Valerianella radiata

1. Plants monoecious; stems with relatively sparse stinging hairs; leaf blades with stinging hairs mostly only on the undersurface, the upper surface rarely with a few stinging hairs .... 2B. SSP. GRACILIS

2a. ssp. dioica

Plants dioecious, the staminate and pistillate flowers on separate plants. Stems moderately pubescent with stinging hairs, otherwise moderately to densely pubescent with short, nonstinging hairs. Leaf blades with both surfaces having stinging hairs, the upper surface otherwise glabrous or nearly so, the undersurface moderately pubescent with short, nonstinging hairs. $2n=52$. May–October.

Introduced, known thus far from a single historical collection from St. Louis City (native of Europe, Asia; introduced in the northeastern U.S. west to Ohio and Georgia, sporadically farther west and in Canada). Habitat unknown, but presumably a moist, disturbed area.

This subspecies was first reported for Missouri by Woodland (1982a).

2b. ssp. gracilis (Aiton) Selander

U. dioica var. *procera* (Muhl. ex Willd.) Wedd.

Pl. 571 c–e

Plants monoecious, the staminate and pistillate flowers in different inflorescences on the same plants. Stems mostly sparsely pubescent with stinging hairs, otherwise glabrous or sparsely to moderately pubescent with short, nonstinging hairs, at least when young. Leaf blades with stinging hairs mostly only on the undersurface, the upper surface only rarely with a few stinging hairs, otherwise glabrous or nearly so, the undersurface glabrous or sparsely to moderately pubescent with short, nonstinging hairs, mostly along the veins. $2n=26$, 52. May–October.

Scattered north of the Missouri River, sporadic farther south (nearly throughout the U.S., but sporadic in the southeastern states; Canada). Bottomland forests, banks of streams and rivers, sloughs, and bases of bluffs; also levees, ditches, railroads, roadsides, and moist disturbed areas.

3. Urtica urens L. (dwarf nettle, burning nettle)

Pl. 571 h; Map 2676

Plants annual, with taproots. Stems 15–50 (–80) cm long, erect or ascending, unbranched or branched from the base, moderately to densely pubescent with stinging hairs, otherwise moderately pubescent with short, nonstinging hairs. Stipules 1–4 mm long, oblong to narrowly oblong or linear. Leaf blades 2–9 cm long, more or less the same size along the stem (the upper leaves only slightly smaller than the others at maturity), elliptic to broadly elliptic or occasionally ovate, narrowly to broadly angled or occasionally rounded at the base, the margins mostly sharply and relatively coarsely toothed (sometimes appearing doubly toothed, the main teeth having smaller teeth along their margins), the surfaces sparsely to moderately short-hairy, the undersurface or both surfaces also with scattered stinging hairs along the main veins, sometimes lighter green but not purplish-tinged; cystoliths rounded. Inflorescences mostly shorter than the subtending petioles, small, globose clusters, these more commonly appearing as short, dense, spikelike racemes, rarely appearing as small panicles, the staminate and pistillate flowers mixed in the same inflorescence. Pistillate flowers with the 2 smaller sepals 0.5–0.7 mm long, ovate, the 2 larger sepals 0.6–0.9 mm long, broadly ovate. Fruits 1.5–1.8 mm long. $2n=24$, 26. May–September.

Introduced, (native of Europe, Asia; introduced widely, in temperate North America most abundantly in the western U.S., sporadically farther east and in Canada). Bottomland forests; also moist disturbed areas.

Steyermark (1963) mentioned an additional occurrence of this species in Clark County, but this could not be verified during the present study.

VALERIANACEAE (Valerian Family)

Ten to 13 genera, about 300 species, nearly worldwide.

The circumscription of the Valerianaceae in the present treatment is the traditional one, as in Steyermark (1963), Cronquist (1981, 1991), among other publications. In recent decades, phylogenetic studies (Judd et al., 1994; Bell et al., 2001; Donoghue et al., 2001) have suggested the need for a renovation of familial limits in the order Dipsacales (which includes such Missouri families as the Caprifoliaceae, Dipsacaceae, and Valerianaceae). However, the level at which the component lineages should be recognized taxonomically is still somewhat controversial. Some specialists continue to recognize the Dipsacaceae and Valerianaceae as separate, closely related families (Bell, 2004, 2007; Bell and Donoghue, 2005; Bell et al., 2012). Others have gone so far as to combine both families within an enlarged and recircumscribed Caprifoliaceae (Judd et al., 1994, 2008; Angiosperm Phylogeny Group, 2009), treating them as subfamilies or informally designated lineages within that family. For practical reasons, it is desirable to continue to treat the Valerianaceae as a separate family, as the traditional version of Caprifoliaceae was already published in Volume 2 of the present work (Yatskievych, 2006).

Species of Valerianaceae possess specialized ethereal oil cells in their tissues that contain a diversity of terpenoids. These are the chemical basis of a variety of medicinal uses of several members of the family and also tend to render the plants malodorous, especially when dried. The Valerianaceae section of most herbaria can be located easily by the nearly overpowering characteristic aroma persisting on the specimens. The oil of the European *Valeriana officinalis* L. (garden heliotrope, all-heal) is sometimes prescribed by physicians for problems of nervousness and insomnia, and as an anticonvulsant. It and some of the of the other showier species also have been cultivated as garden ornamentals.

Steyermark (1963) noted earlier reports of *Valeriana pauciflora* Michx. (large-flowered valerian) from Missouri, but excluded the species because the only specimen he could verify (said possibly to have originated from plants transplanted either from Lincoln or Warren Counties) was of a plant cultivated in a wildflower garden. He also noted, however, that this widespread eastern species is common in southern Illinois adjacent to the Missouri border, and it thus might eventually be discovered somewhere in the eastern portion of the state. *Valeriana pauciflora* is a tall rhizomatous perennial with the leaves pinnately divided into 3–7 lobes, the several calyx lobes apparent and becoming elongated and more or less plumose-hairy (resembling a pappus) at fruiting, and corollas 14–20 mm long.

1. **Valerianella** Mill. (corn salad, lamb's lettuce)

Plants annual or less commonly biennial. Stems usually dichotomously forked several times, angled, glabrous or sparsely and inconspicuously hairy along the angles. Basal leaves sessile or short-petiolate, the blades simple, spatulate to obovate, mostly rounded at the tip, the margins mostly entire, grading into the stem leaves; stem leaves mostly sessile, often with the bases fused around the stem, oblong-spatulate to lanceolate, rounded to bluntly pointed at the tip, the margins entire or few-toothed near the base, glabrous or more commonly inconspicuously hairy along the margins and the midvein of the undersurface. Stipules absent. Inflorescences dense small headlike clusters, these solitary or more commonly paired at the branch tips, with small lanceolate to narrowly elliptic or narrowly spatulate bracts surrounding each cluster. Flowers mostly perfect, epigynous, each subtended by a small bractlet. Calyces absent or reduced to a minute crown. Corollas actinomorphic or slightly zygomorphic, 5-lobed, trumpet-shaped to narrowly bell-shaped, the tube often inconspicuously pouched at the base. Stamens 3, the filaments attached near the top of the corolla tube, the anthers attached at their midpoint, white. Pistils 1 per flower, apparently of 3 carpels. Ovary inferior, with 3 locules, but only 1 of these fertile, with 1 ovule, the others empty. Style 1, the stigma usually slightly 3-lobed. Fruits achenelike, usually straw-colored at maturity, the fertile locule becoming leathery and indehiscent, the sterile locules persistent and differentiated in shape and size from the fertile one. About 60 species, North America, Europe, Asia.

Species of *Valerianella* are classified principally on the basis of fruit morphology, thus some specimens lacking fruits can be difficult to determine with confidence. However, crossing experiments have shown that distinctive fruit morphologies within some species complexes are actually the result of minor genetic variations (see below for more discussion). The vernacular names "corn salad" and "lamb's lettuce" refer to the use of young plants of several species in the genus as potherbs and salad greens.

1. Corollas light pink to rose-pink or less commonly lilac, 10–12 mm long, the tube 3–4 times as long as the expanded upper portion (limb and lobes); bracts with minute, gland-tipped teeth . . . . 2. V. OZARKANA
1. Corollas pale blue or white, rarely pinkish-tinged, 1.5–5.0 mm long, the tube shorter than to about as long as the expanded upper portion (limb and lobes); bract glabrous or with short, nonglandular, bristly teeth along the margin
    2. Corollas pale blue; stamens included in the corolla or only slightly exserted; fertile locule of the fruit with a corky mass on the back, opposite the sterile locules . . . . 1. V. LOCUSTA
    2. Corollas white, rarely pinkish-tinged; stamens noticeably exserted from the corolla; fertile locule of the fruit lacking a corky mass on the back, smooth or at most with a fine nerve or slight ridge . . . . 3. V. RADIATA

1. Valerianella locusta (L.) Laterr. (corn salad, lamb's lettuce)

V. olitoria (L.) Pollich

Pl. 572 f, g; Map 2677

Stems 8–35 cm long. Leaf blades 1–8 cm long, the margins of stem leaves entire or less commonly few-toothed near the base. Bracts and bractlets with short, nonglandular, bristly hairs along the margin. Corollas 1.5–4.0 mm long, the tube shorter than to about as long as the expanded upper portion (limb and lobes), pale blue. Stamens included in the corolla or only slightly exserted. Fruits 1.9–2.3(–3.5) mm long, oblong-elliptic to ovate in dorsal view, glabrous, minutely hairy, or sparsely bristly-hairy, the fertile locule much wider than the sterile ones, with a dome-shaped corky mass on the back, smooth or at most with a fine nerve or slight ridge, the sterile locules more or less parallel, with a shallow, narrow, relatively inconspicuous groove between them. $2n$=14, 16, 18. April–May.

Introduced, uncommon in southern Missouri (native of Europe, widely but sporadically naturalized in the eastern and western U.S. and Canada). Bottomland forests; also margins of crop fields, fallow fields, and roadsides.

2. Valerianella ozarkana Dyal (Ozark corn salad)

V. bushii Dyal

Pl. 572 a–c; Map 2678

Stems 6–45 cm long. Leaf blades 1–6 cm long, the margins of stem leaves usually entire. Bracts and bractlets glabrous or with minute, gland-tipped teeth. Corollas 10–12 mm long, the tube 3–4 times as long as the expanded upper portion (limb and lobes), light pink to rose-pink or less commonly lilac. Stamens noticeably exserted from the corolla. Fruits 1.9–4.0 mm long, elliptic-ovate to oblong or narrowly elliptic in dorsal view, glabrous or with dense lines of short bristly hairs along the angles, the fertile locule narrower than to wider than the

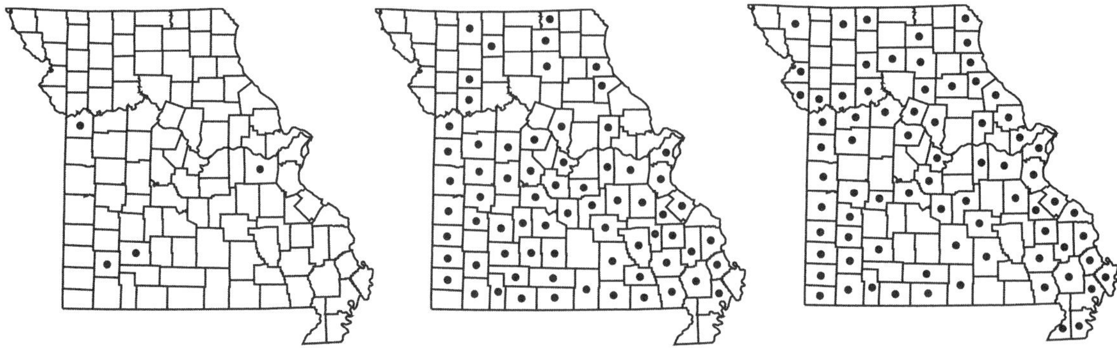

2680. Glandularia bipinnatifida 2681. Glandularia canadensis 2682. Lippia lanceolata

sterile ones, lacking a corky mass on the back, smooth or with a strongly developed, angled or keeled midrib, the sterile locules parallel, with a shallow or deep longitudinal groove between them. $2n$=32. April–May.

Uncommon in the southwestern portion of the Ozark Division and disjunct in Madison County (Missouri, Arkansas, Oklahoma, Texas). Glades and openings of dry upland forests; also roadsides, usually on calcareous substrates.

Steyermark (1963) accepted two taxa in the *V. ozarkana* complex, but commented that it was highly questionable whether they should be maintained as distinct species as their fruit morphologies were the only distinguishing characters and occurred in mixed populations. Ware (1983; see also Eggers, 1969) performed controlled crosses between plants of *V. ozarkana* and *V. bushii*, and determined that the difference in fruit morphologies could be attributed to a single gene difference. She accorded these minor genetic variants the status of forms. Typical *V. ozarkana* has the fertile and sterile locules expanded and longitudinally compressed, giving the fruits a larger, strongly 3-angled appearance, and has lines of hairs along the angles. In f. *bushii* (Dyal) Egg. Ware, the locules are not flattened or expanded, giving the fruits a smaller rounded appearance, without lines of hairs.

3. Valerianella radiata (L.) Dufr.
V. radiata var. *fernaldii* Dyal
V. radiata var. *missouriensis* Dyal
V. stenocarpa Krok var. *parviflora* Dyal
V. woodsiana (Torr. & A. Gray) Walp.
Pl. 572 h, i; Map 2679

Stems 8–40(–60) cm long. Leaf blades 1–10 cm long, the margins of stem leaves often few-toothed near the base. Bracts and bractlets glabrous or with short, nonglandular, bristly hairs along the margin. Corollas 1.5–5.0 mm long, the tube shorter than to about as long as the expanded upper portion (limb and lobes), white, rarely pinkish-tinged. Stamens noticeably exserted from the corolla. Fruits 1.7–2.5 mm long, narrowly oblong-elliptic to more commonly ovate in dorsal view, glabrous, minutely hairy, or sparsely bristly-hairy, the fertile locule usually about as wide as the sterile ones, lacking a corky mass on the back, smooth or at most with a fine nerve or slight ridge, the sterile locules usually more or less parallel, with a shallow or more commonly relatively deep longitudinal groove between them. $2n$=64, 90. April–May.

Scattered to common in the southern 2/3 of the state (eastern U.S. west to Kansas and Texas). Upland prairies, glades, savannas, fens, openings and edges of bottomland and mesic upland forests, and banks of streams and rivers; also pastures, fallow fields, roadsides, railroads, and open disturbed areas.

Much as she did in the *V. ozarkana* complex, Ware (1983; see also Eggers, 1969) suggested that the differences in fruit morphology of taxa in the *V. radiata* complex were due to minor genetic polymorphisms, and she demoted these variants from species to forms. Plants of f. *fernaldii* (Dyal) Egg. Ware have the fruits shaped like those of f. *radiata* in dorsal view, but have the two sterile locules noticeably narrower than the fertile locule and with a shallow longitudinal groove between them. Plants of f. *parviflora* (Dyal) Egg. Ware have fruits narrowly oblong-elliptic in dorsal view, and 2.5–3.0 times as long as wide, but the sterile locules shaped much like those of f. *radiata*. Plants of typical *V. radiata* have the fruits ovate in dorsal view, 1.5–2.5 times as long as wide, with the sterile locules about as wide as the fertile locule, more or less parallel, and with a pronounced groove between them. Ware (1983) noted that the variants occur sporadically within populations of typical *V. radiata*.

Most authors have treated *V. woodsiana* as a distinct species, but it differs from *V. radiata* only in the larger and more divergent sterile locules of the fruits. Eggers (1969) and Ware (1983) accepted it as distinct, but noted that further study might result in its being treated as simply another fruit variant of *V. radiata*, a position that is tentatively accepted here. *Valerianella woodsiana* occurs in Texas, Oklahoma, and Kansas, with disjunct localities in Louisiana, Georgia, and South Carolina. Eggers (1969) first reported this taxon for Missouri, based on a single historical specimen from St. Louis that apparently represents an introduced population. She also listed and mapped a potentially native occurrence in Lawrence County, based on a single specimen accessioned at the herbarium of Southwest Missouri State University. However, this appears to represent a recording error, as the same collection was also cited in her dissertation under *V. radiata* and the actual specimen contains only plants of that species.

VERBENACEAE (Vervain Family)

Plants annual or perennial herbs (shrubs or trees elsewhere). Stems usually branched, usually strongly 4-angled (square) in cross-section (occasionally 6-angled), often hairy, sometimes bristly or roughened, or some or all of the hairs gland-tipped. Leaves opposite (occasionally some of them whorled), well-developed, petiolate or more commonly sessile, the petiole often winged. Stipules absent. Leaf blades simple, but sometimes deeply lobed, the margins usually toothed, the surfaces usually hairy, the hairs sometimes with pustular bases, sometimes stiff and calcified, some or all of the hairs sometimes gland-tipped. Inflorescences terminal and/or axillary spikes, these occasionally short and headlike, occasionally grouped into terminal panicles, the flowers subtended by persistent bracts. Flowers zygomorphic, often weakly so, hypogynous, perfect; cleistogamous flowers absent. Calyces usually deeply 5-lobed (shallowly lobed elsewhere), the lobes then slightly unequal, in *Lippia* with only 2(4) incurved, usually shallow, keeled lobes, persistent at fruiting. Corollas 4- or more commonly 5-lobed, funnelform or more or less trumpet-shaped, often with a slender tube that is slightly expanded near the throat, less commonly 2-lipped, the inside of the throat sometimes with a band of hairs, but not appendaged. Stamens 4, the filaments attached in the corolla tube, short, the anthers not exserted, attached at their base, usually yellow, sometimes with a minute, glandular appendage; staminodes absent (staminode 1 elsewhere). Pistil 1 per flower, of 2 fused carpels (but often appearing 4-carpellate), sometimes 1 carpel aborting during development. Ovary often with a basal nectar disc or a very short, stout stalk (gynophore), entire to more commonly deeply 4-lobed, 2- or more commonly 4-locular, with 1 ovule per locule, the placentation basal. Style 1, usually not persistent at fruiting, situated at the tip of the ovary, not exserted, appearing broadly notched or with a pair of very short, unequal, triangular branches at the slightly expanded and often somewhat flattened tip, the stigma consisting of a swollen glandular area, usually along only 1 of the style branches (in *Lippia*, the branches very short and inconspicuous, the stigmatic region appearing terminal and more or less capitate). Fruits usually schizocarps splitting (sometimes tardily so) into (1)2 or 4 nutlets, these 1-seeded, indehiscent, with a hardened sometimes bony outer wall (drupes with 1, 2, or 4 stones elsewhere). About 35 genera, about 1,000 species, nearly worldwide, most diverse in tropical and warm-temperate regions.

In recent decades phylogenetic analyses based on both morphological (Cantino 1992a, 1992b; Judd et al., 1994, 2008) and molecular (Wagstaff and Olmstead, 1997; Marx et al., 2010) data sets has clarified the familial limits of the closely related families, Verbenaceae and Lamiaceae. Interestingly, the revised classification is similar to one first suggested more than 75 years ago (Junell, 1934). Traditionally, the two families were distinguished morphologically mainly based on the position of the style relative to the ovary: toward the ovary base in a deep central depression in Lamiaceae vs. at the tip of the ovary in Verbenaceae. This character has since been reinterpreted as a specialization within a more broadly circumscribed Lamiaceae. Separation of the two families now rests on variation of a series of subtle characters, including indeterminate vs. determinate inflorescence axes, differences in the attachment patterns of the ovules, ultrastructure of the pollen grains, and production of more or less discrete (often more or less globose), well-developed, stigmatic regions on 1 or both of the 2 stylar lobes in Verbenaceae vs. indistinct receptive areas near the tips of the style branches in Lamiaceae.

The result of this reclassification has been a considerable reduction in the size of the Verbenaceae, from the traditional view of about 100 genera and more than 2,600 species to only about 35 remaining genera with about 1,000 species. For Missouri, the effects of the changes have been relatively modest: the two woody genera, *Callicarpa* and *Vitex*, have been transferred to the Lamiaceae in the present treatment.

Several genera of Verbenaceae contain species that are cultivated as garden ornamentals or house plants. Some of the subtropical and tropical groups, such as *Lantana* L., are frost-tender perennials that are grown as annuals in Missouri. The wood of some tropical tree species of *Citharexylum* L. (fiddlewood) is prized for use in cabinetry, inlays, and musical instruments. *Aloysia citriodora* Paláu (lemon verbena) is used in herbal teas, as a flavoring, and as an ingredient in sachets and potpourri. Some South American species of *Lippia* have been used as spices as a substitute for oregano. Various species have purported medicinal properties.

1. Flowers in dense, spherical to more commonly oblong-ellipsoidal (knob-shaped), axillary heads, these solitary at the tips of elongate, erect, naked stalks; corollas 4-lobed; calyces 2-lobed; fruits consisting of 2 nutlets .... 2. LIPPIA
1. Flowers in dense to open, terminal spikes, sometimes appearing as dense, flat-topped clusters when young, these not associated with specialized branches; corollas 5-lobed; calyces 5-lobed; fruits consisting of (2–)4 nutlets
   2. Corollas 12–30 mm long, the limb 8–20 mm in diameter; calyces 7–13 mm long, the lobes narrowly triangular to nearly linear, becoming contorted as the fruits mature; inflorescences dense, flat-topped clusters at initial flowering, becoming elongated into spikes with age .... 1. GLANDULARIA
   2. Corollas 4–10 mm long, the limb 2–9 mm in diameter; calyces 1.5–5.0 mm long, the lobes remaining erect or becoming slightly incurved as the fruits mature; inflorescences not appearing as flat-topped clusters, but appearing as short to elongate spikes throughout flowering, sometimes appearing as more or less globose (not flat-topped) heads .... 3. VERBENA

1. **Glandularia** J.F. Gmel. (verbena) (Umber, 1979)

Plants perennial herbs (annual elsewhere). Stems few to several from the rootstock, loosely ascending or creeping with ascending tips, rarely erect, sometimes rooting at the lower nodes, weakly 4-angled, hairy. Leaves usually with a partially winged petiole, variously 3-lobed to ternately or pinnately 1 or 2 times deeply divided, the segments narrowly oblong to linear, pointed at the tip, the margins sometimes coarsely few-toothed, sparsely to moderately hairy. Inflorescences terminal on the branches, not associated with slender, elongate stalks, initially dense, flat-topped clusters, becoming elongated into spikes with age. Calyces narrowly tubular, 7–13 mm long, 5-lobed, the lobes somewhat unequal in length, narrowly triangular to nearly linear, erect at flowering, becoming contorted as the fruits mature, hairy on the outer surface and along the margins. Corollas 12–30 mm long (shorter elsewhere), trumpet-shaped, slightly zygomorphic, 5-lobed, pink to lavender or light bluish purple, rarely white (red elsewhere), usually fading to dark blue, the limb 8–20 mm in diameter, the lobes often shallowly and broadly notched at the tip. Stamens inserted at 2 levels toward the tip of the corolla tube, usually with a small glandular appendage positioned laterally between the anther sacs. Ovary 4-locular, appearing 4-lobed, slightly concave at the tip. Style 15–22 mm long (shorter elsewhere), the sterile lobe extending noticeably beyond the fertile lobe (this obscured by the globose stigmatic area), flattened and narrowly triangular. Fruits consisting of (2–)4 nutlets, these more or less cylindric, usually rounded at the tip, somewhat asymmetrically concave at the base, the surface variously wrinkled, lined, and/or with a network of blunt ridges, sometimes also with small papillae, especially along the inner surface, grayish black to black (tan to brown elsewhere). About 50 species, North America to South America, Caribbean Islands; introduced in the Old World.

Traditionally, *Glandularia* was regarded as a section within the genus *Verbena* (Steyermark, 1963). Umber (1979) and Sanders (2001) have summarized morphological, cytological, and phytochemical data supporting the separation of the two groups as genera, which is how most South American botanists have been treating them for many years (Schnack and Covas, 1944).

Several species of *Glandularia* are cultivated as ornamentals in the Midwest, including both of the taxa growing wild in Missouri. *Glandularia canadensis* is popular in native plant gardening because it is easy to grow and flowers over a long period, but it can be aggressive under some conditions. Most of the other species that are cultivated in the region are grown as annual groundcovers in sunny sites. Among these, *G. ×hybrida* (Groenl. & Rumpler) G.L. Nesom & Pruski (garden verbena) is the most striking taxon, a complex hybrid of unconfirmed parentage (Pruski and Nesom, 1992) with relatively large flowers (calyces 10–15 mm long, corollas 25–40 mm long), with the corollas peach-colored to pink, blue, or white, but most commonly deep red. This hybrid has escaped sporadically in a number of southeastern states and eventually may be recorded as an escape in Missouri, although it does not appear to be very commonly grown in the Midwest in recent decades.

Cronquist (1991) cited the existence of moss verbena, *G. pulchella* (Sweet) Tronc. (as *Verbena tenuisecta* Briq.), as an escape in St. Louis, but the specimens upon which this report apparently was based were gathered from plants cultivated at the Missouri Botanical Garden. Thus, this widely cultivated South American species has been excluded for the present from the Missouri flora. It is somewhat similar to *G. bipinnatifida*, but has 2 times pinnately deeply lobed leaves with linear, tapered segments, the floral bracts shorter than the calyx (thus the spikes appearing more slender), and brown nutlets that are beaked at the tip.

1. Corollas 12–15 mm long, the tube slightly longer than to about 1.5 times as long as the calyx, the limb 8–11 mm in diameter; bracts mostly longer than the calyx; leaf blades deeply lobed nearly to the midvein . . . . 1. G. BIPINNATIFIDA
1. Corollas 18–30 mm long, the tube about 2 times as long as the calyx, the limb 10–20 mm in diameter; bracts as long as or shorter than the calyx; leaf blades irregularly incised to lobed, most of the lobes extending half way or less to the midvein . . . . 2. G. CANADENSIS

1. Glandularia bipinnatifida (Nutt.) Nutt. (Dakota vervain, cutleaf vervain)

Verbena bipinnatifida Nutt.

Pl. 573 e, f; Map 2680

Stems 5–60 cm long, moderately to densely pubescent with stiff, tapered, spreading hairs. Leaves short- to moderately petiolate, the petiole winged above the midpoint. Leaf blades 1–6 cm long, variously spatulate, elliptic, ovate, or nearly circular in outline, tapered into the petiole basally, 1- or 2- times pinnately or ternately then pinnately deeply lobed nearly to the midvein, the segments narrowly oblong to less commonly narrowly oblanceolate or nearly linear, the margins entire or few toothed, sometimes somewhat curled under, the surfaces moderately pubescent with stiff, tapered, more or less appressed hairs. Inflorescences 1–4 cm long and compact at flowering, becoming elongated to 5–20 cm at fruiting, the bract subtending each flower mostly slightly longer than the calyx, narrowly lanceolate. Calyces 7–10 mm long, the lobes 1–4 mm long, hairy along the margins and sometimes also the outer surface. Corollas 12–15 mm long, the tube slightly longer than to about 1.5 times as long as the calyx, the limb 8–11 mm in diameter, pink to lavender or purple. Nutlets 2–3 mm long. $2n$=20, 30. May–September.

Uncommon in western Missouri and introduced in Franklin County (Wyoming to Texas east to South Dakota, Missouri, and Louisiana; Mexico; introduced farther east to Maryland, Canada, Mexico). Openings of mesic to dry upland forests; also pastures, railroads, and open disturbed areas.

Steyermark (1963) knew this species only from a small set of historical collections from Jackson County, from which he recorded it as introduced. However, the two occurrences in southwestern Missouri appear to represent native populations along the edge of the species' native distribution. The disjunct Franklin County specimen almost certainly came from a nonnative source.

Umber (1979) treated this species as consisting of two varieties. Within the southern portion of the range of var. *bipinnatifida*, he segregated more upright plants with shorter fruiting spikes, glandular inflorescences, relatively short calyx teeth, and relatively long corolla tubes as var. *brevispicata* Umber. In his treatment of Texas *Glandularia*, B. L. Turner (1998a) noted the existence of an earlier epithet for this glandular variant and renamed the taxon as var. *ciliata* (Benth.) B.L. Turner. More recently, Nesom (2010c) treated the species in a more restricted sense, raising to species level several taxa treated either in synonymy or as infraspecific taxa by earlier authors. His treatment, which recognizes no infraspecific taxa, is followed here.

Putative hybrids between *G. bipinnatifida* and *G. canadensis* have been called *Verbena* ×*oklahomensis* Moldenke. However, B. L. Turner (1998a) expressed doubts about the hybrid nature of this variant, noting that the parental species rarely, if ever, grow together, and suggested that this taxon might represent aberrant individuals of *G. bipinnatifida*.

2. Glandularia canadensis (L.) Small (rose verbena, rose vervain, eastern verbena)
Verbena canadensis (L.) Britton

Pl. 574 h, i; Map 2681

Stems 25–60 cm long, sparsely to moderately pubescent with stiff, tapered, more or less spreading hairs. Leaves short- to long-petiolate, the petiole winged toward the tip. Leaf blades 2–9 cm long, variously ovate to oblong-ovate, triangular-ovate, or lanceolate in outline, tapered (sometimes abruptly so) into the petiole basally, 1-time pinnately or occasionally ternately, irregularly incised to lobed, most of the lobes extending half way to the midvein or less, the lobes variously triangular to oblong, the margins usually prominently few-toothed, flat, the surfaces sparsely to moderately pubescent with stiff, tapered, more or less appressed hairs, occasionally nearly glabrous. Inflorescences 1–5 cm long and compact at flowering, becoming elongated to 5–15 cm at fruiting, the bract subtending each flower mostly shorter than to about as long as the calyx, linear. Calyces 10–14 mm long, the lobes 4–6 mm long, hairy along the margins and pubescent with gland-tipped hairs on the outer surface. Corollas 18–30 mm long, the tube about 2 times as long as the calyx, the limb 10–20 mm in diameter, pink to lavender or purple, rarely white. Nutlets 3.0–3.5 mm long. $2n$=30. March–November.

Scattered to common nearly throughout the state (eastern [mostly southeastern] U.S. west to Nebraska and New Mexico; introduced in some northeastern states). Glades, openings of mesic to dry upland forests, savannas, banks of streams and rivers, and ledges and tops of bluffs; also old fields, pastures, railroads, and roadsides.

Rose verbena is a conspicuous spring wildflower with a blooming period that can last into the autumn. Rare white-flowered plants have been called "f. *candidissima* (J.N. Haage & Schmidt) Umber" or "*V. canadensis* f. *candidissima* (Haage & Schmidt) E.J. Palmer & Steyerm.," but this epithet has not been validly published.

2. **Lippia** L. (fog fruit, frog fruit)

Plants perennial herbs (shrubs elsewhere), sometimes with rhizomes. Stems few to several from the rootstock, prostrate, sometimes loosely ascending at the tip (ascending elsewhere), irregularly branched, usually rooting at the nodes, moderately 4-angled in young growth but often nearly circular in cross-section at maturity, pubescent with appressed hairs, these with 2 opposite branches, thus appearing attached medially along a straight line (hairs

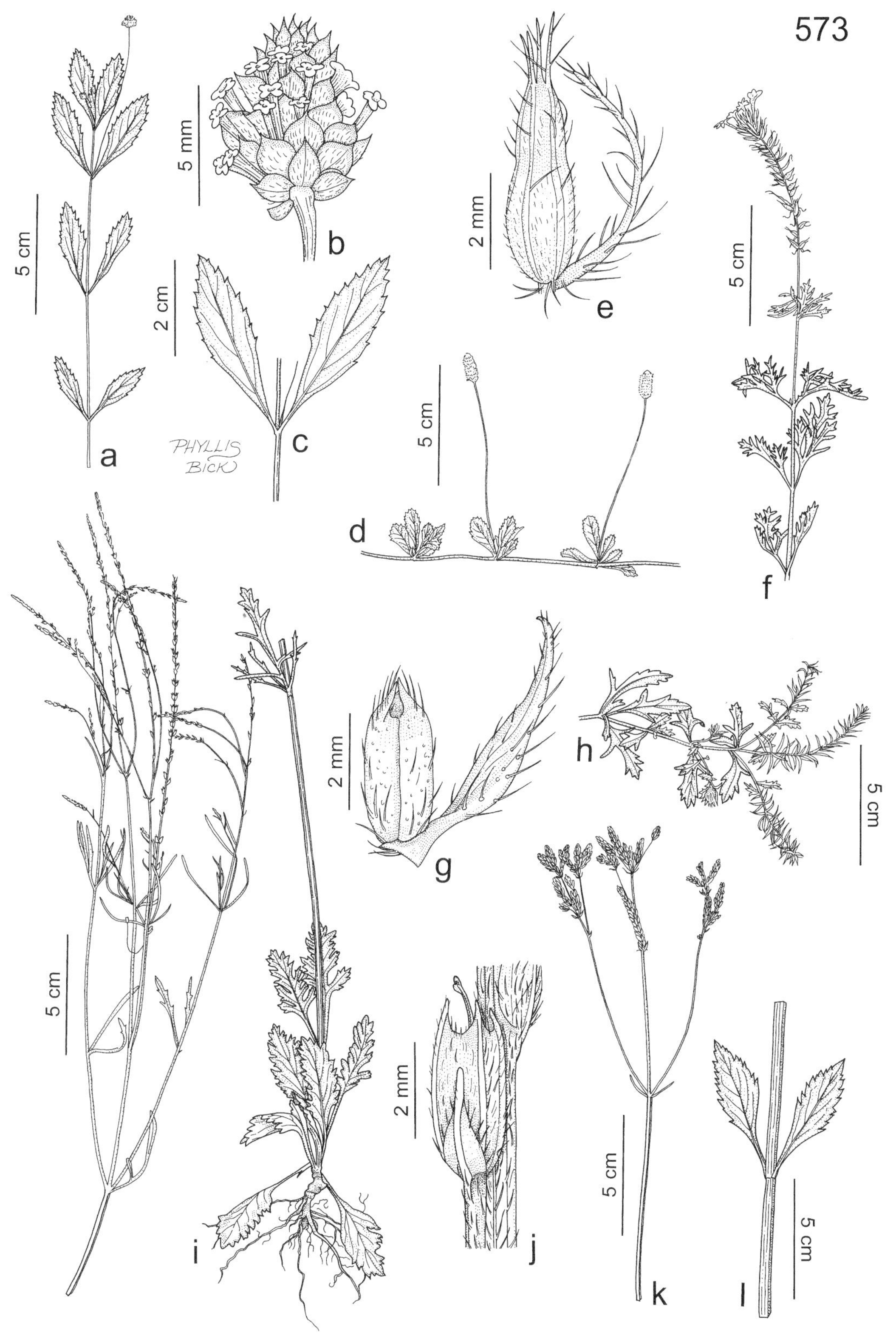

Plate 573. Verbenaceae. *Lippia lanceolata*, **a)** fertile stem, **b)** inflorescence, **c)** node with pair of leaves. *Lippia nodiflora*, **d)** habit. *Glandularia bipinnatifida*, **e)** fruiting calyx, **f)** fertile stem. *Verbena bracteata*, **g)** fruiting calyx with bract, **h)** fertile stem. *Verbena halei*, **i)** upper and lower portions of habit, **j)** node with fruit in calyx and bract. *Verbena brasiliensis*, **k)** inflorescence, **l)** node with pair of leaves.

unbranched elsewhere). Leaves sessile or with an indistinct, short (to 5 mm), winged petiole. Leaf blades unlobed, variously shaped, rounded to sharply pointed at the tip, the margins toothed, mostly above the midpoint, pubescent with hairs similar to those of the stems. Inflorescences axillary, dense, spherical to more commonly oblong-ellipsoidal (knob-shaped), axillary heads, these solitary at the tips of slender, elongate, erect, naked stalks. Calyces flattened, ovate to nearly circular in outline, 1.5–2.0 mm long, 2-lobed, the lobes triangular, somewhat incurved, and keeled, glabrous or sparsely hairy, sometimes with a line of hairs along each angle. Corollas 3.5–4.5 mm long, moderately to strongly zygomorphic, somewhat 2-lipped, 4-lobed, white to pinkish- or lavender-tinged, usually with a yellow central spot at the base of the limb, the limb 2–4 mm in diameter, the lobes slightly irregular or shallowly and broadly notched. Stamens inserted at 2 levels toward the tip of the corolla tube, lacking glandular appendages. Ovary 2-locular (1 of the carpels aborting early in development), not appearing lobed, rounded at the tip. Style 0.2–0.4 mm long, the minute lobes spreading, but obscured by the confluent stigmatic regions, these appearing as a single capitate stigma. Fruits 1.0–1.2 mm long (larger elsewhere), appearing somewhat flattened longitudinally, circular to broadly obovate in outline, consisting of 2 nutlets, these more or less hemispheric, usually rounded at the tip and base, the surface finely pebbled to minutely pitted, olive green to yellowish brown. About 50 species, North America to South America, Caribbean Islands; introduced in the Old World.

The five to nine herbaceous species of *Lippia* with trailing stems that range into temperate North America sometimes have been segregated into the genus *Phyla* Lour. More recently, Sanders (2001) concluded that, in the context of the overall generic classification within the tribe Lantaneae Endl., the species sometimes segregated as *Phyla* are better treated merely as a specialized subgroup within *Lippia*. He noted, however, that the taxonomy of the tribe is in need of more detailed study. A recent, comprehensive, molecular study of phylogenetic relationships in the family similarly concluded that generic limits within the Lataneae require reevaluation following more detailed taxon sampling (Marx et al., 2010). The most recent taxonomic revision of the group accepted the segregation of *Phyla* from *Lippia* based on differences in habit and pubescence types. However, the situation requires more detailed study and the taxonomic limits of *Lippia*, in the broad sense accepted here, remain controversial.

The vernacular names fog fruit and frog fruit have been used relatively interchangeably for the North American members of this genus.

1. Leaf blades mostly broadest at or below the midpoint, tapered to a sharply pointed tip, the margins toothed from at or below the midpoint to the tip .... 1. L. LANCEOLATA
1. Leaf blades mostly broadest above the midpoint, rounded or angled to a bluntly pointed tip, the margins toothed only above the midpoint .... 2. L. NODIFLORA

1. Lippia lanceolata Michx. (northern fog fruit, northern frog fruit)

Phyla lanceolata (Michx.) Greene

Pl. 573 a–c; Map 2682

Plants with a sometimes rhizomatous, often branched, somewhat woody rootstock. Stems 15–60(–100) cm long, rooting at some of the nodes, sparsely to moderately hairy. Leaf blades 1–6(–8) cm long, lanceolate to narrowly elliptic, narrowly rhombic, oblong-lanceolate, or occasionally ovate, mostly broadest at or below the midpoint, tapered to a sharply pointed tip, the margins with 5–11, broad, sharp teeth on each side from at or below the midpoint to the tip, the surfaces moderately to densely hairy. Inflorescences 5–7 mm in diameter, at first globose to ovoid, elongating to 3.5 cm with age, the stalk 4–9 cm long, longer than the subtending leaves. Bracts 2.5–3.0 mm long, obovate, often with purplish-tinged margins. Calyces 1.5–1.8 mm long, about as long as the corolla tube, glabrous except for a line of hairs along the keels. Corollas 2.5–3.5 mm long. Nutlets 0.9–1.2 mm long. $2n=32$. May–September.

Scattered to common nearly throughout the state (nearly throughout the U.S. except some northwestern and a few northeastern states; Canada; introduced uncommonly in Europe). Banks of streams and rivers, margins of ponds,

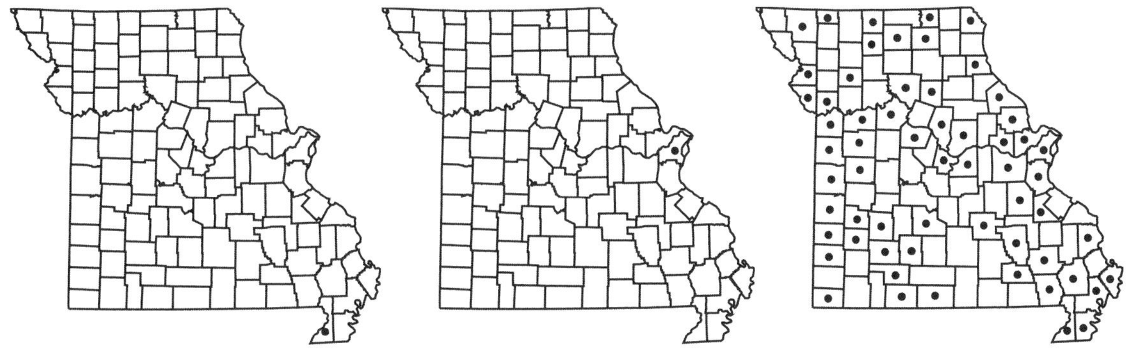
2683. Lippia nodiflora 2684. Verbena bonariensis 2685. Verbena bracteata

sloughs, swamps, marshes, openings of bottomland forests; also ditches, lawns, edges of crop fields, railroads, roadsides, and moist, open, disturbed areas.

2. Lippia nodiflora (L.) Michx. (common fog fruit, common frog fruit, turkey tangle)
Phyla nodiflora (L.) Greene
P. nodiflora var. *incisa* (Small) Moldenke
Pl. 573 d; Map 2683

Plants with a nonrhizomatous, usually unbranched, not or only slightly woody rootstock. Stems 15–60(–100) cm long, rooting at many of the nodes, sparsely to moderately hairy, the older stems often nearly glabrous. Leaf blades 1–5 cm long, narrowly oblanceolate to narrowly elliptic-oblanceolate, less commonly spatulate or broadly oblanceolate, mostly broadest above the midpoint, rounded or angled to a bluntly pointed tip, the margins with 3–7, broad, sharp teeth on each side from above the midpoint to the tip, the surfaces moderately to densely hairy. Inflorescences 5–8 (–10) mm in diameter, at first globose to ovoid, elongating to 2.5 cm with age, the stalk 3–10 cm long, mostly longer than the subtending leaves. Bracts 2–3 mm long, obovate, sometimes with purplish-tinged margins. Calyces 1.5–2.0 mm long, about as long as the corolla tube, the surface glabrous or nearly so, but with a line of hairs along the keels. Corollas 3–4 mm long. Nutlets 1.0–1.3 mm long. $2n$=18, 32, 36. May–September.

Uncommon, known thus far only from a couple of historical collections from Dunklin County (southern U.S. from California to Kansas and Maryland, sporadically farther north to Oregon and Pennsylvania; Mexico, Central America, South America, Caribbean Islands; introduced widely in the Old World). Habitat not known with certainty, but listed as sandy ground.

During the course of his lengthy career, Harold Moldenke described ten varieties of *Phyla nodiflora*, based primarily on subtle variations in leaf morphology. Most of these names have never been formally transferred to *Lippia*, and most botanists working with regional floras have not dealt with these variants. In her taxonomic revision of *Phyla*, Kennedy (1992) did not accept varieties within the species. O'Leary and Múlgara (2012) took an intermediate approach to the *P. nodiflora* complex and recognized two varieties: the mainly South American var. *minor* (Gillies & Hook.) N. O'Leary & Múlgara (*L. nodiflora* var. *minor* Gillies & Hook. and the var. *reptans* Kunth) Moldenke (*L. nodiflora* var. *reptans* (Kunth) Kuntze), which is widespread in Latin America. If their varieties are accepted, then the Missouri plants are all referable to var. *nodiflora*. More detailed studies are needed.

Kennedy (1992) reported a specimen of *L. cuneifolia* (Torr.) Steud. (wedgeleaf fog fruit; reported as *Phyla*) collected in 1896 by Henry Eggert in East St. Louis (St. Clair County), Illinois. An even older collection exists, made by George Engelmann in 1845 on sandy banks of the Mississippi River across from St. Louis, also in the Missouri Botanical Garden herbarium, providing evidence that the species was once native to the Mississippi River Valley. Kennedy also mapped an occurrence apparently in Jackson County, Missouri, but did not include Missouri in her summary of the species range and did not cite any voucher specimens. Steyermark (1963) had earlier excluded the species from the Missouri flora based on a small fragment of *P. cuneifolia* mounted on a specimen otherwise consisting of *P. lanceolata* and collected by Eggert in 1876 in the city of St. Louis. Aside from the outlying populations in Illinois, *P. cuneifolia* otherwise has a range extending from South Dakota to Texas west to Wyoming, Utah, and California, as well as in adjacent northern Mexico. It is similar to *P. nodiflora*, but differs in its longer bracts (4–5 mm), consequently stouter spikes (8–12 vs. 6–9 mm in diameter), more slender leaves (2–8 mm) that are somewhat more pointed at the tip, larger fruits (1.7–2.0 mm), and stems that only infrequently root at the nodes (Kennedy, 1992). Because this species has been documented to grow in eastern Nebraska and Kansas, it is possible that it may be discovered in northwestern Missouri in the future, or that a relict population might be located eventually in the St. Louis region.

3. **Verbena** L. (vervain)

Plants annual or perennial herbs (shrubs elsewhere). Stems 1 to several from the rootstock, strongly ascending to erect, in *V. bracteata* usually loosely ascending or creeping with ascending tips, not rooting at the nodes, weakly to more commonly strongly 4-angled, glabrous or more commonly hairy. Leaves sessile or with a partially to entirely winged petiole, variously unlobed to 3-lobed or pinnately deeply divided, the segments then variously shaped, the margins usually sharply toothed, sparsely to densely hairy. Inflorescences terminal on the branches, not associated with slender, elongate stalks, short to elongate, dense to open spikes, sometimes appearing as more or less globose (not flat-topped) heads, sometimes grouped into panicles. Calyces tubular to narrowly bell-shaped, 1.5–5.0 mm long, 5-lobed, the lobes somewhat unequal in length, narrowly to broadly triangular, erect at flowering and fruiting or sometimes becoming slightly incurved as the fruits mature, sparsely to densely hairy on the outer surface and usually also along the margins. Corollas 2–10 mm long, funnelform to trumpet-shaped (sometimes narrowly so), slightly zygomorphic, 5-lobed, lavender to purple, purplish blue, or white, rarely pink, often fading to dark blue, the limb 2–9 mm in diameter, the lobes sometimes shallowly and broadly notched at the tip. Stamens inserted at 2 levels below the tip of the corolla tube, lacking a glandular appendage between the anther sacs. Ovary 4-locular, appearing 4-lobed, slightly concave at the tip. Style 2–3 mm long, the sterile lobe not extending beyond the fertile lobe, flattened and triangular. Fruits consisting of (2–)4 nutlets, these more or less oblong in outline, 3-angled in cross-section, mostly rounded to truncate at the tip, not concave at the base, the surface variously wrinkled, lined, and/or with a network of blunt ridges, sometimes also with small papillae, especially along the inner surface, grayish brown to dark brown or less commonly black. About 150 species, nearly worldwide, but most diverse in warm-temperate to tropical regions of the New World.

Mühlenbach (1979) reported a nonnative occurrence of the Texas vervain, *V. halei* Small (*V. officinalis* L. ssp. *halei* (Small) S.C. Barber), in Missouri, based on plants that he encountered during his botanical inventories of the St. Louis railyards. However, no voucher specimens could be located during the present study to support this claim. For the present, the species is thus excluded from the Missouri flora. *Verbena halei* differs from the species in the genus documented to grow in Missouri in having at least the lower leaves one or two times pinnately divided nearly to the midvein (Pl. 573 i, j). It is widespread in the southern United States and adjacent Mexico.

Several species of vervains are cultivated as ornamentals. Some species also have a long history of medicinal use for various ailments. The native species of *Verbena* are notorious for their frequent interspecific hybridization. Users of the present work should be aware that such hybrids, which are more or less intermediate in morphology between their avowed parental species, are not accommodated in the key to species below, but are to be expected to occur sporadically throughout the state. The following putative hybrids have been recorded from Missouri or are to be expected in the state:

V. ×*blanchardii* Moldenke (*V. hastata* × *V. simplex*) (to be expected)
V. ×*deamii* Moldenke (*V. bracteata* × *V. stricta*)
V. ×*engelmannii* Moldenke (*V. hastata* × *V. urticifolia*)
V. ×*illicita* Moldenke (*V. stricta* × *V. urticifolia*)

V. ×*moechina* Moldenke (*V. simplex* × *V. stricta*)
V. ×*perriana* Moldenke (*V. bracteata* × *V. urticifolia*)
V. ×*rydbergii* Moldenke (*V. hastata* × *V. stricta*)
V. ×*stuprosa* Moldenke (*V. simplex* × *V. urticifolia*) (to be expected)

1. Bracts 8–15 mm long, 2–4 times as long as the calyces; stems usually spreading to loosely ascending or spreading with ascending tips . . . . 2. V. BRACTEATA
1. Bracts 2–6 mm long, shorter than to slightly longer than the calyces; stems erect or strongly ascending
    2. Spikes dense, short and relatively stout, not elongating much with age; corollas narrowly trumpet-shaped, sometimes nearly tubular, the limb 1.5–3.0 mm in diameter
        3. Spikes grouped into dense, globose or headlike clusters; leaves narrowly lanceolate to narrowly oblong, narrowly oblong-lanceolate, or narrowly oblong-elliptic, often slightly narrowed toward the base, but with small auricles and noticeably clasping the stem . . . . 1. V. BONARIENSIS
        3. Spikes more open, noticeable as individual units, not appearing as headlike clusters; leaves elliptic to elliptic-obovate, tapered to a slender, nonclasping base . . . . 3. V. BRASILIENSIS
    2. Spikes dense to open, slender, not appearing headlike, becoming greatly elongated with age; corollas funnelform to more or less trumpet-shaped, the limb 3–9 mm in diameter (smaller in *V. urticifolia*)
        4. Leaf blades all 2–10(–15) mm wide, narrowly lanceolate to oblanceolate or linear . . . . 5. V. SIMPLEX
        4. Largest leaves with the blade 15–70(–120) mm wide, variously ovate, broadly elliptic, oblong, or lanceolate
            5. Spikes open, the flowers not overlapping; corollas 2–4 mm long, the limb 1–2(–3) mm in diameter, white . . . . 7. V. URTICIFOLIA
            5. Spikes moderately dense, the flowers strongly overlapping, the lowermost few flowers sometimes more widely spaced; corollas 6–9 mm long, the limb 3–9 mm in diameter, purple to purplish blue (rarely white or pink)
                6. Largest leaves noticeably petiolate (mostly 10–25 mm long); leaves glabrous or sparsely to moderately, but inconspicuously hairy (not appearing grayish); spikes mostly 5 or more at the stem tip; corolla limb 3.0–4.5 mm in diameter . . . . 4. V. HASTATA
                6. Largest leaves sessile or short-petiolate (to 5 mm); leaves densely hairy (appearing grayish); spikes mostly solitary or in groups of 3 at the stem tip; corolla limb 7–9 mm in diameter . . . . 6. V. STRICTA

1. Verbena bonariensis L. (South American vervain)

Map 2684

Plants annual (perennial farther south). Stems 60–150(–250) cm long, stiffly erect with ascending branches, strongly 4-angled, moderately to densely pubescent with a mixture of gland-tipped and nonglandular, spreading to ascending hairs, usually strongly roughened to the touch. Leaves sessile, the blades 2–10(–15) cm long, narrowly lanceolate to narrowly oblong, narrowly oblong-lanceolate, or narrowly oblong-elliptic, often slightly narrowed toward the base, but with small auricles and noticeably clasping the stem, rounded or more commonly angled to a bluntly or sharply pointed tip, the margins finely and often irregularly toothed, sparsely to moderately pubescent on the upper surface with stiff, loosely ascending, sometimes pustular-based hairs, the undersurface densely pubescent with a mixture of finer gland-tipped and nonglandular hairs. Inflorescences dense clusters of spikes, 0.5–2.0(–4.0) cm long, short and broad, usually appearing as globose or headlike clusters, not elongating much with age.

2686. Verbena brasiliensis 2687. Verbena hastata 2688. Verbena simplex

Bracts 2.5–4.0 mm long, slightly shorter than to slightly longer than the calyx, narrowly lanceolate to narrowly triangular. Calyces 2.0–3.5 mm long. Corollas 5–8 mm long, the outer surface densely hairy, narrowly trumpet-shaped, sometimes nearly tubular, purple to purplish blue, the tube slender, the limb 1.5–3.0 mm in diameter. Nutlets 1.3–2.0 mm long, narrowly oblong in outline, the inner surface usually pale and with dense, minute papillae, the outer surface brown, with several longitudinal ridges, these with a few cross-ridges toward the tip. $2n=28$. April–October.

Introduced, known thus far only from the city of St. Louis (native of South America, introduced sporadically in the U.S. mostly in southern states and Hawaii; also Caribbean Islands, Australia, New Zealand). Open disturbed areas.

This common garden species was first collected as an escape in Missouri by the author in 2002.

2. Verbena bracteata Lag. & Rodr. (creeping vervain, prostrate vervain)

Pl. 573 g, h; Map 2685

Plants annual or perennial, sometimes forming mats. Stems 10–50 cm long, usually spreading to loosely ascending or spreading with ascending tips, rarely more strongly ascending, slightly to moderately 4-angled, moderately to densely pubescent with nonglandular, straight to slightly curved, spreading, often pustular-based hairs. Leaves sessile or with a winged petiole, the blades 1–5(–7) cm long, lanceolate to ovate-lanceolate or oblanceolate, tapered to a slender, nonclasping base, rounded or broadly angled to a bluntly pointed tip, variously unlobed to pinnately or ternately deeply lobed, the margins also finely to coarsely toothed, occasionally only above the midpoint, both surfaces moderately to densely pubescent with loosely ascending, nonglandular, sometimes pustular-based hairs. Inflorescences usually solitary spikes, 2–20 cm long, dense (the flowers strongly overlapping), relatively stout, elongating greatly with age. Bracts 8–15 mm long, 2–4 times as long as the calyx, narrowly lanceolate to narrowly elliptic. Calyces 3–4 mm long. Corollas 4–6 mm long, the outer surface glabrous, narrowly funnelform to somewhat trumpet-shaped, purplish blue, the tube relatively slender, the limb 2–3 mm in diameter. Nutlets 2.0–2.5 mm long, oblong to narrowly oblong in outline, the inner surface usually pale and with dense, minute papillae, the outer surface yellowish brown to reddish brown, with several longitudinal ridges, these with several cross-ridges above the midpoint. $2n=14, 28$. April–October.

Scattered nearly throughout the state (throughout the U.S.; Canada, Mexico). Banks of streams and rivers, glades, and disturbed portions of upland prairies; also old fields, pastures, railroads, roadsides, and open disturbed areas.

This species is mostly observed in highly disturbed areas, such as along sidewalks, alleys, and roadsides, sometimes forming mats, especially in sandy soils. It is distinctive in its long bracts, which are loosely ascending to loosely reflexed (toward the spike bases). The inflorescences, including the bracts, are mostly 1–2 cm in diameter. In addition to the putative hybrids listed at the beginning of the generic treatment, Steyermark (1963) also reported a specimen from Grundy County deposited at the University of Missouri herbarium (*Crookshanks 142*) that he suspected to represent *V. bracteata* × *V. hastata*. However, the late Harold Moldenke, who was the specialist on hybridization in *Verbena*, later redetermined this specimen as *V. ×perriana* (*V. bracteata* × *V. urticifolia*).

3. Verbena brasiliensis Vell. (Brazilian vervain)

Pl. 573 k, l; Map 2686

Plants annual (perennial farther south). Stems 60–150(–250) cm long, stiffly erect with ascend-

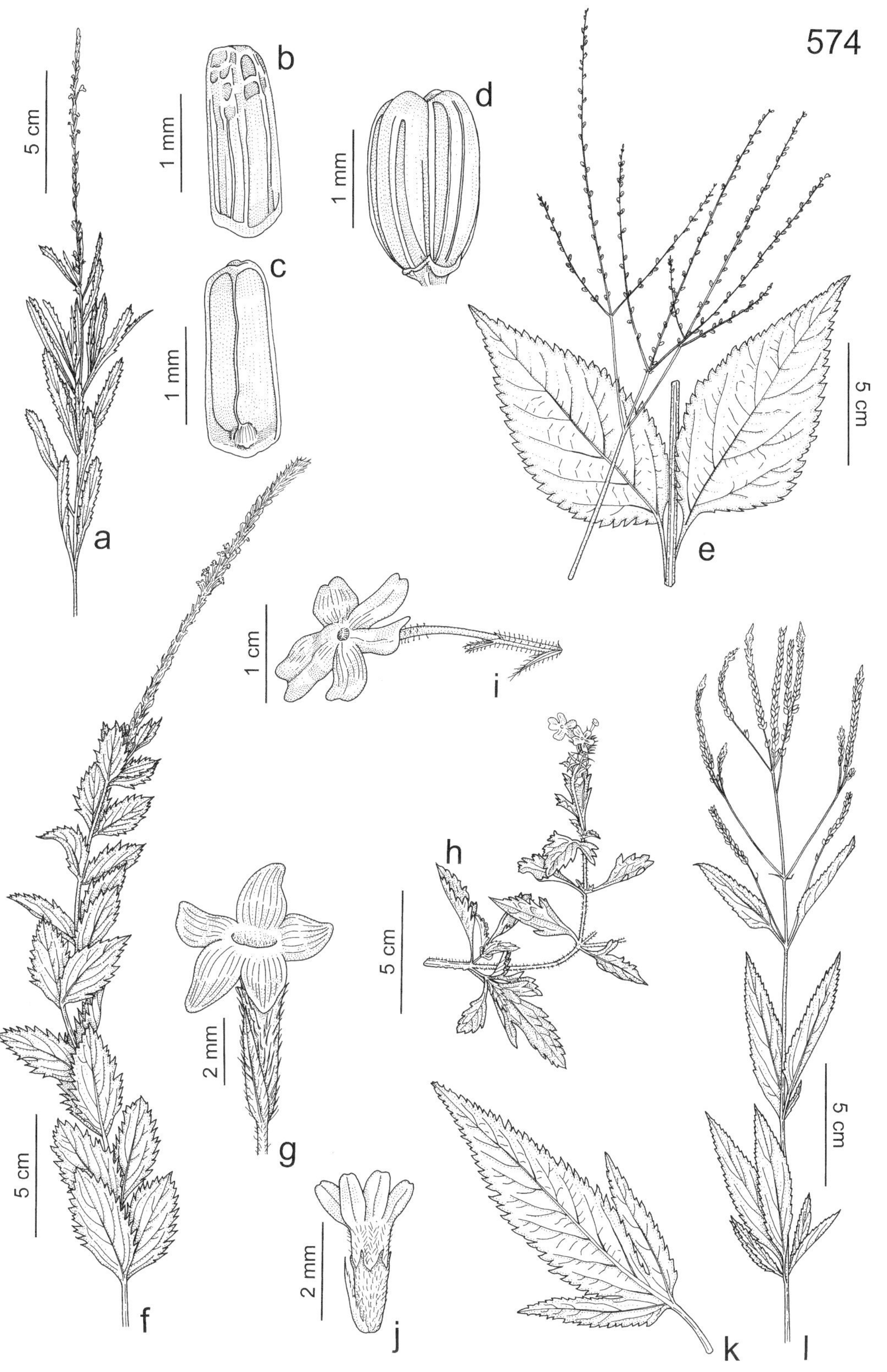

Plate 574. Verbenaceae. *Verbena simplex*, **a)** fertile stem, **b)** fruit, dorsal view, **c)** fruit, ventral view. *Verbena urticifolia*, **d)** fruit, **e)** inflorescence and median node with pair of leaves. *Verbena stricta*, **f)** fertile stem, **g)** flower. *Glandularia canadensis*, **h)** fertile stem, **i)** flower. *Verbena hastata*, **j)** flower, **k)** lower leaf, **l)** fertile stem.

ing branches, strongly 4-angled, the angles and sometimes also the surfaces sparsely to moderately pubescent with nonglandular, strongly ascending hairs, usually strongly roughened to the touch. Leaves sessile or nearly so, the blades 2–10(–15) cm long, elliptic to elliptic-obovate, tapered to a slender, nonclasping base, rounded or angled to a bluntly or sharply pointed tip, the margins finely to coarsely toothed, often only above the midpoint, both surfaces moderately to densely pubescent with stiff, loosely ascending to more or less appressed, nonglandular, sometimes pustular-based hairs. Inflorescences solitary or in clusters of 3(5) spikes, noticeable as individual units, dense (the flowers strongly overlapping) but not appearing as headlike clusters, slender to moderately stout, 1–4 cm long, not elongating much with age. Bracts 2.5–4.0 mm long, usually slightly longer than the calyx, narrowly lanceolate to narrowly triangular. Calyces 2.0–3.5 mm long. Corollas 3–6 mm long, the outer surface densely hairy, narrowly trumpet-shaped, sometimes nearly tubular, purple to purplish blue, the tube slender, the limb 2–3 mm in diameter. Nutlets 1.2–1.8 mm long, oblong to elliptic-oblong in outline, the inner surface usually pale and with dense, minute papillae, the outer surface brown, with several longitudinal ridges, these with several cross-ridges toward the tip. $2n=28$. May–October.

Introduced, known thus far only from the city of St. Louis (native of South America, introduced sporadically in the U.S., mostly in southeastern states; also Mexico, Caribbean Islands and in the Old World). Railroads.

This species was first reported for Missouri by Mühlenbach (1979) from his inventories of the St. Louis railyards. Moldenke (1982) reported another occurrence in Washington County without supplemental information or citation of a voucher specimen. No specimens could be located during the present study to support this claim.

4. Verbena hastata L. (blue vervain)
Pl. 574 j–l; Map 2687

Plants perennial. Stems 40–150(–220) cm long, erect or strongly ascending, moderately to strongly 4-angled, moderately to densely pubescent with nonglandular, somewhat curved, loosely ascending to appressed, often pustular-based hairs. Leaves moderately petiolate, the petioles mostly 10–25 mm long, usually winged toward the tip, the blades 4–20 cm long, at least those of the largest leaves 15–45 mm wide, lanceolate to oblong-lanceolate or narrowly ovate, rounded, angled, or short-tapered to a nonclasping base, tapered to a sharply pointed tip, unlobed or the larger leaves with a pair of spreading basal lobes, the margins relatively coarsely and doubly toothed, both surfaces glabrous or sparsely to moderately, but inconspicuously pubescent with short, loosely appressed, nonglandular, occasionally pustular-based hairs (not appearing grayish), sometimes roughened to the touch. Inflorescences usually panicles of 5 to numerous spikes, these 2–20 cm long, moderately dense (the flowers strongly overlapping except sometimes the lowermost ones), appearing stout when young but relatively slender at maturity, elongating greatly with age. Bracts 2.0–2.5 mm long, slightly shorter than the calyx, narrowly lanceolate. Calyces 2.3–3.0 mm long. Corollas 6–10 mm long, the outer surface sparsely to moderately hairy near the tip of the tube, funnelform, purple to purplish blue (rarely white or pink), the tube slightly broadened toward the tip, the limb 3.0–4.5 mm in diameter. Nutlets 1.5–2.0 mm long, narrowly oblong to narrowly oblong-elliptic in outline, the inner surface usually slightly pale and with sparse to moderate, minute, appressed hairs, the outer surface reddish brown, smooth or with several faint longitudinal ridges, these sometimes with a few, faint cross-ridges toward the tip. $2n=14$. June–October.

Scattered nearly throughout the state, but less common in the Ozark Division than elsewhere (throughout the U.S.; Canada). Banks of streams and rivers, margins of ponds, sloughs, and lakes, bottomland prairies, fens, bottomland forests, mesic upland forests, and ledges of bluffs; also fencerows, margins of crop fields, fallow fields, ditches, railroads, roadsides, and open, disturbed areas.

Plants with the pubescence tending to be stiffer and the stems and leaves thus more roughened to the touch have been called var. *scabra* Moldenke, but these intergrade fully with plants having somewhat softer pubescence. Rare plants with white corollas have been called f. *albiflora* Moldenke, and plants with pink corollas have been called f. *rosea* C.I. Cheney.

5. Verbena simplex Lehm. (narrow-leaved vervain)
V. angustifolia Michx., an illegitimate name
Pl. 574 a–c; Map 2688

Plants perennial. Stems 10–70 cm long, erect or strongly ascending, moderately to strongly 4-angled, sparsely to moderately pubescent with nonglandular, straight, strongly ascending, occasionally pustular-based hairs. Leaves sessile or with a winged petiole, the blades 2–8(–10) cm long, 1–10(–15) mm wide, narrowly lanceolate to oblanceolate or linear, tapered to a slender, nonclasping base, mostly angled to a bluntly or sharply pointed

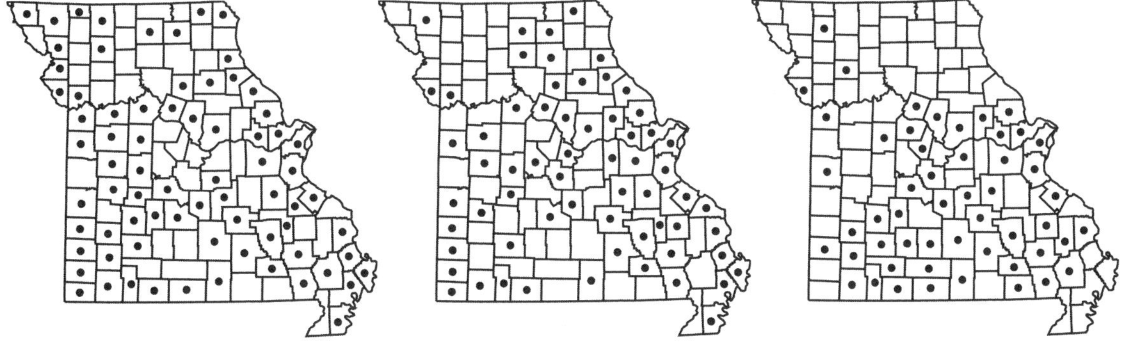

2689. Verbena stricta 2690. Verbena urticifolia 2691. Cubelium concolor

tip, unlobed, the margins relatively finely toothed, both surfaces sparsely pubescent with appressed, nonglandular, occasionally pustular-based hairs, especially along the veins, the upper surface sometimes nearly glabrous. Inflorescences usually solitary spikes, 4–25 cm long, moderately dense (the flowers strongly overlapping except sometimes the lowermost ones), slender, elongating greatly with age. Bracts 3–5 mm long, slightly shorter than to slightly longer than the calyx, lanceolate. Calyces (2–)3–4 mm long. Corollas 4–6 mm long, the outer surface sparsely hairy toward the tip of the tube, funnelform to somewhat trumpet-shaped, dark lavender or purple to white or bluish-tinged, the tube relatively slender, the limb 4–6 mm in diameter. Nutlets 2–3 mm long, narrowly oblong to narrowly oblong-elliptic in outline, the inner surface usually slightly pale and smooth or with sparse to moderate, minute papillae, the outer surface greenish brown to reddish brown, with several longitudinal ridges, these with several cross-ridges above the midpoint. $2n=14$. May–September.

Scattered nearly throughout the state, but apparently absent from the Mississippi Lowlands Division and uncommon in the northernmost counties (eastern U.S. west to Minnesota and Texas; Canada). Glades, tops of bluffs, upland prairies, and banks of streams and rivers; also pastures, old fields, railroads, roadsides, and open disturbed areas.

The name *V. angustifolia* Michx., which was applied to this taxon in some of the older botanical literature, is a later homonym of *V. angustifolia* Mill., a different species that is widespread in the Neotropics and is now called *Stachytarpheta angustifolia* (Mill.) Vahl by most botanists. Some authors (O'Leary et al., 2010) segregate populations of *Verbena simplex* from Baja California, Mexico with more spreading hairs on the foliage and somewhat glandular calyces as var. *orcuttiana* (L.M. Perry) N. O'Leary.

6. Verbena stricta Vent. (hoary vervain)
Pl. 574 f, g; Map 2689

Plants perennial. Stems 20–120(–150) cm long, erect or strongly ascending, moderately to strongly 4-angled, moderately to densely pubescent with nonglandular, somewhat curved, more or less spreading, often pustular-based hairs, usually also with moderate to dense, shorter, more appressed hairs. Leaves sessile or short-petiolate, the petioles (when present) to 5 mm long, winged, the blades 1–9 cm long, at least those of the largest leaves 15–50 mm wide, ovate to elliptic or nearly circular, those of the uppermost leaves sometimes only lanceolate to narrowly elliptic, rounded to more commonly angled or short-tapered to a nonclasping base, rounded to more commonly angled or short-tapered to a bluntly or sharply pointed tip, unlobed, the margins relatively coarsely and sometimes doubly toothed, both surfaces densely pubescent with longer and shorter, appressed, nonglandular, sometimes pustular-based hairs (appearing grayish), felty or roughened to the touch. Inflorescences mostly solitary spikes, these sometimes grouped into small panicles of 3(5) spikes, each 4–20 cm long, moderately dense (the flowers strongly overlapping except sometimes the lowermost ones), appearing stout when young but often relatively slender at maturity, elongating greatly with age. Bracts 2.0–2.5 mm long, slightly shorter than to slightly longer than the calyx, narrowly lanceolate. Calyces 3.5–5.0 mm long. Corollas 7–10 mm long, the outer surface sparsely to moderately hairy, especially near the tip of the tube, funnelform, purple to purplish blue (rarely white or pink), the tube slightly broadened toward the tip, the limb 7–9 mm in diameter. Nutlets 2–3 mm long, narrowly oblong in outline, the inner surface usually pale and with dense, minute papillae, the outer surface grayish brown, with several longitudinal ridges, these with several cross-ridges above the midpoint. $2n=14$. June–October.

Scattered nearly throughout the state (nearly throughout the U.S.; Canada, but introduced east of Ohio). Glades, upland prairies, loess hill prairies, sand prairies, savannas, tops of bluffs, banks of streams and rivers, margins of lakes, and openings of mesic to dry upland forests; also pastures, old fields, fallow fields, railroads, roadsides, and open, disturbed areas.

Rare plants with white corollas have been called f. *albiflora* Wadmond, and rare plants with pink corollas have been called f. *roseiflora* Benke.

Moldenke (1980) reported *V. xutha* Lehm. from the city of St. Louis, but without supplemental information or citation of a voucher specimen. Intensive herbarium searches, including at Moldenke's personal herbarium (now at the University of Texas), failed to disclose any support for the inclusion of this species in the flora. Additionally, O'Leary et al. (2010) did not cite any Missouri vouchers in their recent taxonomic revision of *Verbena* series *Verbena*. Thus, the species has been excluded from the state's flora. *Verbena xutha* occurs in a broad band across the southern United States from Arizona east to Alabama. It differs from *V. stricta* in its deeply divided leaf blades and its somewhat less dense mature inflorescences.

7. Verbena urticifolia L. (white vervain, nettle-leaved vervain)

Pl. 574 d, e; Map 2690

Plants perennial. Stems 50–150(–250) cm long, erect or strongly ascending, moderately to strongly 4-angled, moderately pubescent with nonglandular, straight or somewhat curved, spreading, sometimes pustular-based hairs, often with shorter, ascending hairs toward the tip. Leaves short- to moderately petiolate, the petiole usually winged above the midpoint, the blades (2–)5–12 (–20) cm long, at least those of the largest leaves (10–)20–70(–120) mm wide, broadly lanceolate to oblong-lanceolate or ovate, rounded or short-tapered to a nonclasping base, tapered to a sharply pointed tip, unlobed, the margins relatively coarsely and sometimes doubly toothed, both surfaces glabrous to moderately (rarely densely) pubescent with spreading to loosely appressed, nonglandular, sometimes pustular-based hairs. Inflorescences usually panicles of several to numerous spikes, these 8–50 cm long, relatively open (the flowers not overlapping), slender, elongating greatly with age. Bracts 0.5–1.5 mm long, shorter than the calyx, ovate to narrowly ovate. Calyces 1.5–2.5 mm long. Corollas 2–4 mm long, the outer surface sparsely to moderately hairy, especially toward the tip of the tube, funnelform, white, the tube relatively slender, the limb 1–2(–3) mm in diameter. Nutlets 1.5–2.0 mm long, oblong to narrowly oblong or oblong-elliptic in outline, the inner surface usually slightly pale and smooth or with sparse to moderate, minute papillae, the outer surface yellowish brown to reddish brown, smooth or with several longitudinal ridges, these sometimes with a few cross-ridges toward the tip. $2n=14$. May–October.

Scattered nearly throughout the state (eastern U.S. west to North Dakota and Texas; Canada, Mexico). Banks of streams and rivers, margins of ponds, and lakes, openings of bottomland and mesic upland forests, and occasionally savannas; also pastures, old fields, fencerows, margins of crop fields, railroads, roadsides, and open disturbed areas.

Plants with the leaf undersurface densely short-hairy and slightly shorter calyces and nutlets have been called var. *leiocarpa* L.M. Perry & Fernald. This variety, which occurs nearly throughout the species range, seems to intergrade freely with the less hairy var. *urticifolia* and thus should not be provided with formal taxonomic recognition. Steyermark (1963) noted the existence of such plants elsewhere, but excluded the taxon from the Missouri flora. Moldenke (1980) reported the variety from Greene and Shannon Counties, but without supplemental information or citation of a voucher specimen. Thus far, no specimens have been discovered with this extreme morphology, although a number of seemingly intermediate specimens from mostly eastern Missouri have been collected.

VIOLACEAE (Violet Family)

Contributed by Harvey E. Ballard Jr.

Plants annual or perennial herbs (shrubs, small trees, or rarely lianas elsewhere), sometimes with rhizomes or stolons. Aerial stems absent or, if present, then erect or ascending (sometimes becoming lax with age), unbranched or few-branched. Leaves all basal or (in taxa with aerial stems) also alternate (the lowermost leaves sometimes opposite or subopposite), short- to long-petiolate, the blades simple, but sometimes deeply lobed, the margins otherwise entire to more commonly toothed or scalloped. Stipules often conspicuous, usually herbaceous, entire or more commonly toothed or lobed, sometimes irregularly so. Inflorescences of solitary or rarely clusters of 2 or 3 axillary flowers (these appearing basal in taxa not producing aerial stems; elsewhere grouped into heads, racemes, or panicles), the flowers short- to long-stalked, bractless or (in *Viola*) with a pair of small, herbaceous bracts positioned variously on the stalk, in some species the late-season flowers usually cleistogamous. Flowers mostly strongly zygomorphic (less so in *Cubelium*), perfect, hypogynous. Calyces of 5 free sepals, these somewhat overlapping, some or all of these sometimes with a small pouchlike auricle at the base, usually persistent at fruiting. Corollas of 5 free petals (these undeveloped in cleistogamous flowers), the lowermost petal slightly longer than (in *Cubelium*) to somewhat shorter than (in *Viola*) the upper pair, often (in *Viola*) with a basal spur. Stamens 5 (reduced to 1 or 2 in cleistogamous flowers), the filaments sometimes somewhat fused, appearing very short but sometimes extending beyond the anthers into tapered, terminal appendages (in *Viola*), the relatively stout anthers appressed to and usually fused into a tube around the pistil, the lowermost pair (or sometimes all of the anthers elsewhere) often with a basal spur. Pistil 1 per flower, of 3 fused carpels, the ovary superior, 1-locular, the placentation parietal. Ovules 3 to many Style 1, often asymmetrically expanded toward the tip, the stigma often positioned obliquely or appearing obliquely lobed. Fruits capsules, with 3 relatively thick valves, dehiscing longitudinally between the placentae (sometimes explosively), the edges of the valves usually curling inward along the sides as they dry. Seeds sometimes with arils. Twenty-three genera, about 830 species, nearly worldwide.

The family's main economic importance is in horticulture. A number of species of *Viola* are cultivated as ornamentals in gardens as are a few species in other genera. However, African violets belong to the genus *Saintpaulia* H. Wendl., which is in the tropical family Gesneriaceae. The flowers of some *Viola* species are occasionally used fresh as garnishes in salads, cooked as potherbs, or candied as decorations for other foods. Members of the Midwestern genera have a limited use in folk medicine, mainly as an infusion or syrup for respiratory ailments or digestive troubles. In folklore, violets are a symbol of fertility and have thus been used in love potions.

Although Midwestern botanists are familiar with members of the Violaceae as relatively small herbs, most of the family's taxonomic diversity in tropical regions is in the form of shrubs and small trees (the largest tropical genus, *Rinorea* Aubl., contains about 300 species).

1. Plants commonly 40 cm or more tall at flowering, with 8 or more nodes per stem; corollas 2–3 mm long, uniformly greenish white, the lowermost petal usually slightly pouched, but not spurred; sepals lacking auricles; fruits 17–26 mm long . . . . 1. CUBELIUM
1. Plants mostly less than 35 cm tall at flowering, lacking aerial stems or with up to 6 nodes per stem; corollas 4–30 mm long, white, cream-colored, yellow, bluish purple or multi-colored, the lowermost petal prominently spurred; sepals with basal auricles (these sometimes small and inconspicuous); fruits 4–15 mm long . . . . 2. VIOLA

1. **Cubelium** Raf. ex Britton & A. Br.

One species, eastern U.S., Canada.

Traditionally, most botanists classified this species in the genus *Hybanthus* Jacq., which was treated in a broad sense to include about 70 species distributed from North America to South America and also portions of Asia, Africa, and Australia (but most diverse in tropical

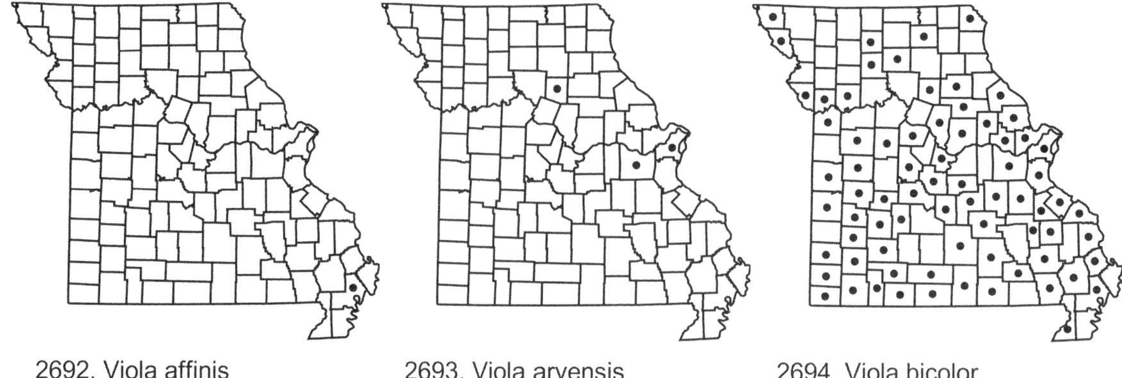

2692. Viola affinis

2693. Viola arvensis

2694. Viola bicolor

portions of South America). Despite relatively strong uniformity in floral morphology across much of the genus, molecular data (supported by chromosome numbers, inflorescence and vegetative morphology, and biogeography) have revealed that *Hybanthus* in the traditional sense is an unnatural assemblage consisting of 5–8 evolutionarily distinct lineages (Tokuoka, 2008). *Cubelium* is one of these segregates, but was not sampled in earlier molecular phylogenetic studies of the family. Like some other segregates of *Hybanthus*, *Cubelium* is herbaceous. Unlike most of these groups, the flowers are in reduced, 1–3-flowered clusters (other groups having solitary flowers or long-stalked, many-flowered, paniculate or racemose inflorescences), and the flowers are small and weakly zygomorphic with the bottom petal only a little longer than the others. All of the stamens are fused, lacking distinct filaments (shared only with *Hybanthus* in the strict sense), vs. stamens with the filaments partially or entirely separate in other segregates. The bottom pair of stamens lack glands but share a thin ridge at the base (other groups bear a gland on each lower stamen or share a large, shield-like gland). The capsule in *Cubelium* is relatively large, with each valve bearing (3)4(5) seeds, which are large, ivory-colored, and smooth (other groups have smaller capsules, with 1–3(4) seeds per valve, these generally small to medium-sized and pale yellow to dark brown or black, and typically pitted, wrinkled, prismatic or longitudinally ribbed). The lowermost petal in *Cubelium*, as in flowers of other New World groups, has a somewhat pouched base (vs. almost always conspicuously spurred in Old World taxa). All eight segregate molecular lineages are distinct morphologically and biogeographically. They also differ in basal chromosome numbers (x=6, 8) and in the shapes and sizes of calcium oxalate crystals in their tissues.

1. Cubelium concolor (T.F. Forst.) Raf. ex Britton & A. Br. (green violet)
Hybanthus concolor (T.F. Forst.) Spreng.
Pl. 575 h–j; Map 2691

Plants perennial herbs, with prostrate to ascending rhizomes 1–10 cm long and 4–10 mm thick, these often obscured by dense fibrous roots. Aerial stems 40–100 cm long at flowering, erect or strongly ascending, with 8 to numerous nodes, usually unbranched, glabrous or more commonly moderately pubescent with fine, mostly spreading hairs. Leaves alternate or the basalmost sometimes opposite or subopposite, subsessile to short-petiolate, the petiole narrowly winged. Stipules relatively inconspicuous, 4–15 mm long, linear to narrowly lanceolate, the margins entire. Leaf blades 4–12 cm long, unlobed, oblanceolate, tapered to a usually sharply pointed tip, angled at the base, the margins entire or with a few, small, irregularly spaced teeth toward the tip, the surfaces glabrous or sparsely to moderately hairy, sometimes only along the veins of the undersurface. Inflorescences of solitary flowers or less commonly some of the nodes with small clusters of 2 or 3 flowers, the stalks, short, drooping at flowering (usually becoming straighter and spreading at fruiting), bractless, but jointed near the midpoint. Cleistogamous flowers produced, but rarely observed in Missouri plants. Sepals 1–3 mm long, narrowly lanceolate to nearly linear, S-shaped or inward-curved, sharply pointed at the tip, the margins entire but sometimes hairy, lacking auricles. Corollas (except in cleistogamous flowers) 2–3 mm long, uniformly greenish white, the pet-

Plate 575. Violaceae. *Viola arvensis*, **a)** habit. *Viola tricolor*, **b)** habit, **c)** stipule from lower node. *Viola bicolor*, **d)** fruit, **e)** flower, **f)** habit, **g)** stipule. *Cubelium concolor*, **h)** flower, **i)** fruit, **j)** fertile stem.

als oriented forward or spreading in life, the lowermost petal often slightly pouched but not spurred. Stamens with the filaments free or partially fused, the dorsal surface of the 2 lowermost (or less commonly all) often bearing a small gland, the anther appendages small or relatively large, oblong to ovate, entire, membranous, orange to white. Style slender, somewhat S-shaped, slightly thickened at the tip, the stigma directed downward, slender, lobed or tapered. Fruits 17–26 mm long, oblong-ellipsoid, green when fresh, drying to tan, glabrous, dehiscing explosively. Seeds 1 to several per valve, 3–4 mm long, mostly subglobose, the surface smooth, white to cream-colored, lacking an aril or rarely with a small aril. $2n=48$. April–June.

Scattered to common in the Ozark and Ozark Border Divisions, uncommon elsewhere in the state, apparently absent from most of the eastern half of the Glaciated Plains (eastern U.S. west to Minnesota, Kansas, and Oklahoma; Canada). Bottomland forests, mesic upland forests, banks of streams and rivers, bases and ledges of bluffs, and rarely tops of bluffs and margins of glades, often on calcareous substrates.

Green violets are easily overlooked in the field, owing to their nondescript morphology, "unviolетlike" appearance and bottomland forest habitat. Comparatively large plant size, small greenish white flowers, and numerous leaves make it easy to distinguish from members of *Viola*, once its membership in the Violaceae has been established. Most Missouri plants have hairy stems and leaves. Glabrous plants are encountered occasionally in the state and are called f. *subglabrum* Eames.

2. **Viola** L. (violet)

Plants annual or perennial herbs, with taproots or rhizomes, respectively. Aerial stems absent or, if present, then mostly less than 35 cm tall at flowering, erect, ascending, or spreading with ascending tips, unbranched or more or less dichotomously few-branched, with up to 6 nodes, glabrous or hairy. Leaves all basal or (in species with aerial stems) also alternate, the basal leaves long-petiolate, the stem leaves short- to more commonly long-petiolate. Stipules various, often relatively conspicuous, entire, toothed, or dissected into several, deep, jagged lobes. Inflorescences of solitary flowers, these axillary in species with aerial stems, appearing basal in taxa not producing aerial stems, the flowers mostly long-stalked, with a pair of small, herbaceous bracts positioned variously on the jointless stalk. Calyces of 5 free sepals, each of these with a sometimes inconspicuous saclike auricle at the base. Corollas of 5 free petals, 4–30 mm long, white or pale cream-colored to yellow, orange, bluish purple, or multicolored, the throat yellow or greenish white, the lowermost petal short- to long-spurred. Stamens 5, the filaments adjacent but free, the lowermost pair bearing nectaries inserted into the petal spur, the anther appendages relatively large, oblong-ovate, entire, membranous, orange. Style linear or strongly thickened toward the tip (club-shaped), often appearing slightly curved above the midpoint, the stigma variously shaped, often lobed or positioned obliquely. Fruits 4–15 mm long, oblong-ovoid to oblong-ellipsoid, usually green at maturity, sometimes darker in cleistogamous flowers, in some species dehiscing explosively. Seeds 9 to many per carpel, narrowly ovoid to obovoid or globose, the surface usually appearing dull, finely pebbled or minutely roughened, variously colored, usually with an aril. About 525 species, nearly worldwide.

Viola is the largest genus in the family and has been classified historically as comprising several sections and subsections, some of which have a well-deserved reputation as taxonomically difficult (especially in the so-called stemless groups that do not produce aerial stems). Much of the controversy involves differing opinions as to whether perceived morphological differences are representative of relatively cryptic species or whether these differences are the result of minor differentiation between populations that is maintained through the production of cleistogamous flowers. Cleistogamous flowers are produced later in the growing season by many species and are obligately self-pollinated. Their stalks often are a different length or positioned differently, and at maturity such flowers resemble young buds (the corollas are poorly developed and the calyces remain closed over the stamens and ovary at maturity).

For many years, most botanists followed the taxonomic summary of North American violets by Russell (1965), although this relatively preliminary treatment contained insufficient comparative taxonomic data to address difficulties encountered in determining specimens of *Viola*. More recently, three researchers have focused on the taxonomy of the stemless blue violets of eastern temperate North America, with the result that botanists have a choice between three somewhat discordant treatments published over a five-year period (McKinney, 1992; Ballard, 1994; Gil-Ad, 1997; and other papers by these authors). Ultimately, resolution of taxonomic controversies in some groups must await further, more intensive studies involving a broad sampling of populations and analysis of molecular markers. Further notes are included below in the discussions of selected taxa. The present treatment is adapted and updated from the author's earlier study of the Michigan violets (Ballard, 1994).

The attractive flowers of the stemless blue-flowered species are used in a delicious and beautiful jelly, candied for unique confections, and eaten in salads. The leaves also harbor significant amounts of vitamin C and are edible as a nutritious potherb in spring. Steyermark (1963) noted that the young foliage of most of the stemless blue violets is mucilaginous and can be eaten raw or cooked, with a texture similar to that of okra.

1. Plants producing aerial stems, these erect, ascending, or spreading with ascending tips, the leaves alternate (but sometimes also basal)
    2. Plants perennial, with a stout, prostrate or ascending rhizome; corollas with the petals oriented forward or spreading in life; stipules entire or irregularly toothed or lobed
        3. Corollas yellow; spur not or only slightly exserted beyond the sepals, the lower petal often appearing merely pouched; stipules entire or sparsely, shallowly, and irregularly toothed; leaf blades ovate to heart-shaped or somewhat kidney-shaped, angled (often broadly so) or broadly rounded at the tip . . . . 12. V. PUBESCENS
        3. Corollas white to pale cream-colored; spur well-exserted beyond the sepals; stipules with a fringe of relatively slender, deep teeth or lobes; leaf blades ovate to heart-shaped, narrowly angled or tapered at the tip . . . . 15. V. STRIATA
    2. Plants annual, with a slender vertical taproot; corollas appearing frontally flattened in life; stipules deeply lobed with a fringe of several, long, linear or oblong-lanceolate segments on each side
        4. Petals shorter than the sepals (rarely surpassing them by up to 2 mm); corollas white or pale cream-colored (with a yellow throat) . . . . 2. V. ARVENSIS
        4. Petals much longer than sepals; corollas pale blue (to occasionally light blue or nearly white) or yellowish orange to purple (with a yellow throat)
            5. Corollas pale blue (to occasionally light blue or nearly white), the upper petals lacking dark blue tips; leaf blades and terminal lobes of the stipules entire or with 1(2) pair(s) of blunt or rounded teeth along the margins . . . . 3. V. BICOLOR
            5. Corollas yellowish orange to purple, the upper petals commonly tinged or tipped with dark blue; leaf blades and terminal lobes of the stipules with 3 or more pairs of blunt or rounded teeth along the margin . . . . 16. V. TRICOLOR
1. Plants not producing aerial stems, the leaves all basal from a rhizome (note that a few species produce node-rooting stolons at fruiting)

6. Blades of at least the mid- and late-season leaves shallowly to deeply lobed or dissected into linear or lanceolate segments (in some species, the first leaves produced each season may be unlobed)
   7. Leaf blade outline much longer than wide, elliptic to narrowly triangular, tapered at apex, the lobes produced only near the blade base and less than 1/3 the length of the midvein . . . . 13. V. SAGITTATA
   7. Leaf blade outline slightly longer than wide to wider than long, ovate to kidney-shaped, broadly and bluntly pointed or broadly rounded at the tip, the lobes not confined to the blade base or at least 1/3 the length of the midvein
      8. Leaf blades deeply and ternately dissected into uniformly slender, linear segments 1–3 mm wide, at least the lateral segments ternately lobed again; leaves glabrous or more commonly minutely hairy, the margins also commonly short-hairy
         9. Corollas appearing frontally flattened in life, the petals all glabrous on the upper surface; stamens exserted beyond the corolla throat, the orange tips noticeable; stipules fused to the petiole for part of their length; central segment of the leaf blade typically undivided or with a few irregular teeth . . . . 10. V. PEDATA
         9. Corollas with the petals oriented more or less forward in life but curved or arched outward or backward toward their tips, the lateral and lower (spurred) petals bearded with slender hairs on the upper surface, stamens not exserted, the tips therefore not visible without dissection of the flower; stipules free from the petioles; central main segment of the leaf blade ternately divided similar to the lateral lobes . . . . 11. V. PEDATIFIDA
      8. Leaf blades shallowly to deeply lobed, but the segments not lobed again, at least the central one lanceolate to triangular-ovate and mostly more than 3 mm wide; leaves glabrous or moderately to densely pubescent with spreading or shaggy hairs
         10. Leaf blades hairy; cleistogamous fruits on short, arched or spreading stalks, the surface dark green mottled with purple . . . . 9. V. PALMATA
         10. Leaf blades glabrous; cleistogamous fruits on tall erect stalks, the surface uniformly green . . . . 17. V. VIARUM
6. Leaf blades unlobed, the margins finely to coarsely scalloped or toothed to occasionally entire or nearly so
   11. Most or all of the leaf blades distinctly longer than wide
      12. Leaf blades lanceolate, angled or tapered at the base; corollas 4–10 mm long, always white; rhizomes slender, less than 2 mm thick; threadlike stolons produced at fruiting; seeds 1.0–1.5 mm long . . . . 5. V. LANCEOLATA
      12. Leaf blades narrowly ovate to ovate, heart-shaped, or triangular; corollas (6–)10–18 mm long, blue to purplish blue or purple (rarely white or mottled); rhizomes stout, more than 3 mm thick; stolons not produced; seeds 1.2–2.2 mm long
         13. Leaf blades narrowly triangular to ovate-triangular, truncate to shallowly cordate at the base, the margins subentire in the apical third, otherwise finely to coarsely toothed; sepals oblong-lanceolate to broadly lanceolate, angled to a rounded or bluntly pointed tip, the auricles 0.5–1.0 mm long at flowering, not becoming noticeably enlarged at fruiting; lateral petals bearded on the upper surface with uniformly slender hairs (a few of these occasionally slightly thickened toward their tips) . . . . 6. V. MISSOURIENSIS

13. Leaf blades narrowly to broadly ovate or heart-shaped, cordate at the base, the margins finely toothed their entire length; sepals narrowly triangular-lanceolate, tapered to a sharply pointed tip, the auricles 2–6 mm long at flowering, elongating to 3–10 mm at fruiting; lateral and sometimes lower petals bearded on the upper surface with knob-shaped to club-shaped hairs
    14. Lateral petals bearded on the upper surface with club-shaped to narrowly club-shaped hairs, these also present but often sparse toward the base of the upper surface on the lower (spurred) petal; sepal spurs relatively inconspicuous, 0.6–1.5 mm long at flowering, elongating only slightly at fruiting; seeds orangish yellow to yellowish brown . . . . 1. V. AFFINIS
    14. Lateral petals bearded on the upper surface with knob-shaped to club-shaped hairs, these absent on the lower (spurred) petal; sepal spurs relatively conspicuous, 2–6 mm long at flowering, elongating to 3–10 mm at fruiting; seeds olive black . . . 4. V. CUCULLATA

11. Most or all of the leaf blades distinctly wider than to about as wide as long
    15. Corollas always white, the lateral petals not noticeably bearded, the upper surface glabrous or nearly so; rhizomes slender, less than 2 mm thick; thread like stolons produced at fruiting; seeds 1.0–1.4 mm long . . . . 8. V. PALLENS
    15. Corollas blue to purplish blue or purple (rarely white or mottled), the lateral petals bearded on the upper surface; rhizomes stout, more than 3 mm thick; stolons not produced; seeds 1.2–2.5 mm long
        16. Sepals narrowly triangular-lanceolate, tapered to a sharply pointed tip, the auricles 2–6 mm long at flowering, elongating to 3–10 mm at fruiting; lateral petals bearded on the upper surface with knob-shaped to club-shaped hairs . . . . 4. V. CUCULLATA
        16. Sepals oblong to broadly lanceolate, rounded to more commonly bluntly or broadly pointed at the tip, the auricles 0.5–1.0 mm long at flowering, not becoming noticeably enlarged at fruiting; lateral petals bearded on the upper surface with uniformly slender hairs
            17. Leaves glabrous except for tiny, scattered hairs on upper surface of the blade; lower (spurred) petal densely bearded on the upper surface toward the base; cleistogamous fruits on ascending to erect stalks, uniformly green; seeds olive black . . . . 7. V. NEPHROPHYLLA
            17. Leaves glabrous or more commonly moderately to densely, but finely hairy, especially on underside of the leaf blade and petiole (if nearly glabrous, then the hairs not confined to the upper blade surface); lower (spurred) petal glabrous or bearing a few hairs on the upper surface toward the base; cleistogamous fruits on prostrate or abruptly arched stalks; green to dark green, usually mottled with purple; seeds brown . . . . 14. V. SORORIA

1. Viola affinis Leconte (sand violet)
Pl. 577 j; Map 2692

Plants perennial, to 20 cm tall, with a prostrate to loosely ascending rhizome 4–6 mm thick. Aerial stems not produced. Leaves in a basal rosette, long-petiolate, the petiole glabrous. Stipules conspicuous, membranous, free from the petiole, narrowly lanceolate, the margins entire but glandular-hairy to glandular-toothed. Leaf blades 1–6(–8) cm long, about as wide as long (early-season leaves) to distinctly longer than wide, narrowly to broadly heart-shaped, all unlobed, angled or slightly tapered to a sharply pointed tip, those of early-season leaves sometimes broadly angled to bluntly angled or rarely rounded, cordate at the base, the margins finely, evenly, and bluntly toothed to somewhat scalloped their entire length, the upper surface glabrous or sparsely pubescent toward the base with short, relatively stiff hairs, the undersurface glabrous. Cleistogamous flowers produced. Flower stalks mostly about as long as the leaves (those of the cleistogamous flowers loosely ascending to arched or prostrate). Sepals 6–10 mm long, narrowly triangular-lanceolate to narrowly oblong-lanceolate, tapered to a bluntly or sharply pointed tip, the margins glabrous but sometimes sparsely toothed, the basal auricles relatively inconspicuous (0.6–1.5 mm long at flowering, elongating only slightly at fruiting). Corollas (6–)11–20 mm long (except in cleistogamous flowers), the petals oriented more or less forward in life but curved or arched outward or backward toward their tips, longer than the sepals, bluish purple (with a greenish white throat) with darker veins, the lateral petals bearded on the upper surface with club-shaped to narrowly club-shaped hairs, the lower petal at least sparsely bearded toward the base, the hairs distributed into the mouth of the spur, the spur conspicuous, usually well-exserted beyond the sepal auricles, stout and often somewhat hemispheric in shape. Stamens not exserted, typically not visible without dissection of the flower. Style relatively stout, slightly expanded into a more or less scoop-shaped, hollow, broadly rounded tip. Fruits 6–11 mm long, ellipsoid to narrowly ellipsoid, green to olive green, sometimes mottled or streaked with dark purple, usually drying to tan or olive brown, the surface glabrous. Seeds 1.5–1.9 mm long, orangish yellow to yellowish brown. $2n$=54. April–June.

Uncommon, known thus far definitely only from New Madrid Counties (northeastern U.S. west to Minnesota and Texas; Canada). Swamps, bottomland forests, and shaded levees.

This species was mapped from Stoddard County by Russell (1965), but without citation of a voucher specimen. The first documented occurrence is based on a specimen collected in April 2009 at Big Oak Tree State Park (New Madrid County) by George Yatskievych. *Viola affinis* differs from *V. cucullata* in its longer petal beard hairs that are present on the lateral and lower (spurred) petals, its shorter sepal spurs, its open flowers positioned at about the same height as the leaves, and its cleistogamous flowers on loosely ascending to arched or spreading stalks. Superficially, it resembles *V. missouriensis* in its relatively long-tapered leaf tips, but the blades are more deeply cordate at the base and more finely and regularly toothed along the margins. See the treatment of *V. sororia* for a discussion of the controversy surrounding application of the name *V. papilionacea*, which may be the oldest validly published epithet for the taxon currently called *V. affinis*.

2. Viola arvensis Murray (wild pansy, European field pansy)

Pl. 575 a; Map 2693

Plants annual, with a slender, vertical taproot 1–2 mm thick. Stems 8–35 cm long, erect or ascending. Leaves alternate and basal, subsessile to long-petiolate, the petiole glabrous or minutely and inconspicuously hairy. Stipules relatively large and leaflike, free from the petiole, deeply lobed with a fringe of several, long, linear or oblong-lanceolate segments on each side, the terminal segment much longer and broader than lateral ones, usually entire. Leaf blades 0.5–2.5 cm long, unlobed, obspatulate to obovate, broadly to narrowly angled to a bluntly or sharply pointed tip, broadly angled to rounded at the base, the margins otherwise bluntly toothed, the surfaces glabrous or minutely and inconspicuously hairy. Cleistogamous flowers not produced. Flower stalks not or only slightly overtopping the leaves. Sepals 7–15 mm long, lanceolate, angled to a sharply pointed tip, the margins glabrous, the basal auricles well-developed. Corollas 4–12 mm long, appearing strongly frontally flattened in life, the petals shorter than the sepals, white or pale cream-colored with a yellow throat (this usually best developed on the lower petal and there usually also with a few, dark purple lines), the lateral petals bearded on the upper surface with mostly knob-shaped hairs, the lowermost petal glabrous on the upper surface, the spur 1.0–1.5 mm long, well-exserted beyond the sepal auricles, relatively stout. Stamens not exserted, typically not visible without dissection of the flower. Style enlarged into a globose, hollow stigmatic tip. Fruits 5–10 mm long, broadly ellipsoid, green, drying to tan, the surface glabrous. Seeds 1.5–1.7 mm long, tan. $2n$=34. April–June.

Introduced, uncommon in the eastern half of the state (native of Europe, introduced nearly throughout the U.S. [except in some portions of the Great Plains], Canada, Greenland). Edges of forests; also old fields and railroads.

Steyermark (1963) excluded this species from the Missouri flora, having concluded that a historical specimen from Johnson County that was reported as this species by Palmer and Steyermark (1935) had been misdetermined. The specimens currently accepted as escapes of *V. arvensis* in Missouri were all collected after 1970.

3. Viola bicolor Pursh (Johnny-jump-up, field pansy, wild pansy)

V. kitaibeliana Roem. & Schult. var. *rafinesquei* (Greene) Fernald

V. rafinesquei Greene

Pl. 575 d–g; Map 2694

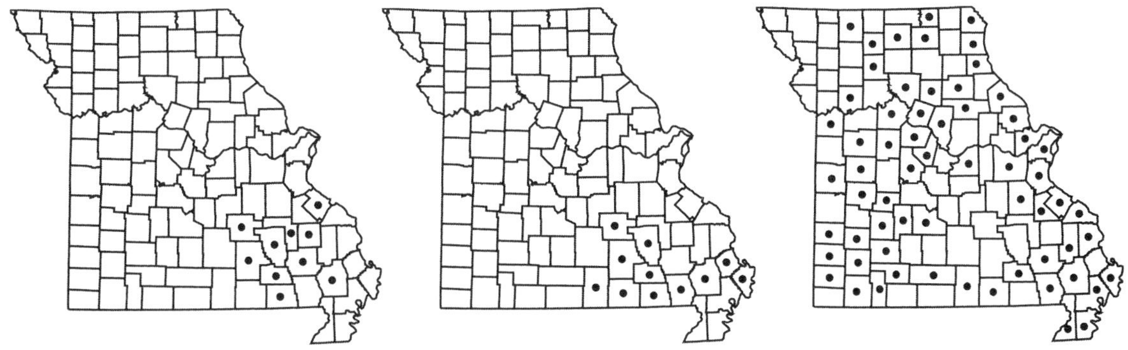

2695. Viola cucullata 2696. Viola lanceolata 2697. Viola missouriensis

Plants annual, with a slender, vertical taproot 1–2 mm thick. Stems 4–25 cm long, erect or ascending. Leaves alternate and basal, subsessile to long-petiolate, the petiole glabrous. Stipules relatively large and leaflike, free from the petiole, deeply lobed with a fringe of several, long, linear or oblong-lanceolate segments on each side, the terminal segment similar to the lateral ones, entire or with 1(2) pair(s) of blunt or rounded teeth along the margins. Leaf blades 0.7–3.0 cm long, unlobed, obspatulate to nearly circular, rounded or angled to a bluntly or sharply pointed tip, truncate to rounded at the base, the margins otherwise entire or with 1(2) pair(s) of blunt or rounded teeth, the surfaces glabrous. Cleistogamous flowers rarely produced (not observed in Missouri specimens). Flower stalks not or only slightly overtopping the leaves (those of the cleistogamous flowers erect or ascending). Sepals 3–5 mm long, lanceolate, angled to a sharply pointed tip, the margins glabrous, the basal auricles well-developed. Corollas 4–12 mm long, appearing strongly frontally flattened in life, the petals longer than the sepals, pale blue to occasionally light blue or nearly white (the upper petals lacking dark blue tips), often with darker veins or dark purple lines, with a yellow throat (this usually best-developed on the lower petal), the lateral petals bearded on the upper surface with mostly knob-shaped hairs, the lowermost petal glabrous on the upper surface, the spur 1.0–1.5 mm long, well-exserted beyond the sepal auricles, relatively stout. Stamens not exserted, typically not visible without dissection of the flower. Style enlarged into a globose, hollow stigmatic tip. Fruits 4–6 mm long, broadly ellipsoid, green, drying to tan, the surface glabrous. Seeds 1.3–1.5 mm long, tan. $2n=34$. March–May.

Common south of the Missouri River, scattered farther north (eastern U.S. west to South Dakota, Arizona, and Idaho, Canada). Glades, disturbed margins of loess hill prairies, ledges and tops of bluffs, savannas, openings of bottomland and upland forest, and banks of streams, rivers, and spring branches; also old fields, pastures, fallow fields, margins of crop fields, gardens, railroads, roadsides, and open disturbed areas.

Steyermark (1963) reviewed the long-standing controversy on whether this taxon is native in North America, but did not express an opinion. Russell (1965) also discussed the topic, concluding that Shinners (1961) had provided a convincing case for its nativity (albeit with weedy tendencies) in eastern temperate North America. It is also accepted as a member of the native flora in the present treatment. Although some botanists have treated the taxon as a variety of the European *V. kitaibeliana*, its morphology and chromosome number differ from that species.

As in some other members of the Violaceae, *V. bicolor* contains quantities of methyl salicylate (oil of wintergreen) in its taproot. It is a well-known winter annual with a complex seed dormancy. Unlike other pansies in Europe (or introduced as weeds here in North America), it apparently rarely produces cleistogamous flowers.

4. Viola cucullata Aiton (marsh violet)
Pl. 577 c, d; Map 2695

Plants perennial, to 40 cm tall, with a prostrate to loosely ascending rhizome 4–6 mm thick. Aerial stems not produced. Leaves in a basal rosette, long-petiolate, the petiole glabrous. Stipules conspicuous, membranous to somewhat herbaceous, free from the petiole, narrowly lanceolate, the margins entire, glandular-hairy. Leaf blades 1–5 cm long, about as wide as long (especially early-season leaves) to distinctly longer than wide, narrowly to broadly ovate, heart-shaped, triangular, or nearly circular (occasionally in early-season leaves), all unlobed, angled or tapered to a sharply pointed tip, cordate at the base, the margins finely toothed or scalloped their entire length, the upper surface sparsely pubescent with short, relatively stiff hairs, the undersurface glabrous. Cleistogamous flowers produced. Flower stalks mostly strongly overtopping the leaves (those of the cleistogamous flowers erect or strongly ascending). Sepals 7–10 mm long, narrowly triangular-lanceolate, tapered to a sharply pointed tip, the margins glabrous, the basal auricles conspicuous (2–6 mm long at flowering, elongating to 3–10 mm at fruiting). Corollas (6–)10–14 mm long (except in cleistogamous flowers), the petals oriented more or less forward in life but curved or arched outward or backward toward their tips, longer than the sepals, bluish purple to light purple (with a greenish white throat usually with a dark border) with darker veins, the lateral petals bearded on the upper surface with knob-shaped to more commonly club-shaped hairs, the lower petal glabrous, the spur conspicuous, usually well-exserted beyond the sepal auricles, stout and often somewhat hemispheric in shape. Stamens not exserted, typically not visible without dissection of the flower. Style slender, slightly expanded into a narrow, more or less scoop-shaped, hollow, truncate tip. Fruits 8–15 mm long, narrowly ellipsoid, green, usually drying to tan or olive brown, the surface glabrous. Seeds 1.2–1.7 mm long, olive black. $2n$=54. April–June.

Uncommon in the eastern half of the Ozark Division and adjacent portions of the Ozark Border (eastern U.S. west to Minnesota and Arkansas; Canada). Fens, acid seeps, and bases of bluffs.

This species is distinctive in its flowers that tend to strongly overtop the leaves, and in the club-shaped hairs of the petal beards. Other useful field features for flowering plants include the corollas with a distinctly darkened eyespot surrounding the throat; sepals often noticeably 3-veined and uniformly tapering to very sharply pointed tips; and the leaf blades with scattered short stiff hairs on the upper surfaces. It produces perhaps the longest capsules and longest sepal auricles of any stemless violet. Steyermark (1963) considered *V. cucullata* to represent a relict from times when Missouri's climate was cooler.

Voss (1985), McKinney (1992), and Gil-Ad (1994) discussed the potential use of the name *V. obliqua* Hill for this species. Although publication of the name predates that of *V. cucullata*, most botanists who have studied the problem agree that *V. obliqua* is of uncertain application, with neither the vague text nor the ambiguous plate that form the original description clearly referable to any particular species. Although the epithet *V. obliqua* has not been formally rejected nomenclaturally, its use for the marsh violet should be discouraged.

5. Viola lanceolata L. **ssp. lanceolata** (lance-leaved violet)

Pl. 578 a, b; Map 2696

Plants perennial, to 35 cm tall, with a prostrate to erect rhizome 1–2 mm thick, also producing threadlike stolons at fruiting. Aerial stems not produced. Leaves in a basal rosette, short- to moderately petiolate, the petiole glabrous. Stipules inconspicuous, membranous, free from the petiole, narrowly lanceolate, the margins entire, glandular-hairy. Leaf blades 1–12 cm long, mostly more than 3 times as long as wide, lanceolate, all unlobed, angled to a sharply pointed tip, angled or tapered at the base, the margins finely toothed or scalloped, the surfaces glabrous. Cleistogamous flowers produced. Flower stalks not or only slightly overtopping the leaves (those of the cleistogamous flowers relatively long, erect or strongly ascending). Sepals 3–10 mm long, lanceolate, angled to a sharply pointed tip, the margins glabrous, the basal auricles short and inconspicuous. Corollas 4–10 mm long (except in cleistogamous flowers), the petals oriented more or less forward in life but curved or arched outward or backward toward their tips, longer than the sepals, white (with a greenish white throat), the lateral and lower petals often with dark purple or brownish purple veins toward the bases, the lateral petals sometimes bearded on the upper surface with slightly club-shaped hairs, the lower petal glabrous, the spur conspicuous, well-exserted beyond the sepal auricles, stout and often somewhat hemispheric in shape. Stamens not exserted, typically not visible without dissection of the flower. Style slender, slightly expanded into a narrow, more or less scoop-shaped, hollow, truncate tip. Fruits 5–8 mm long, narrowly ellipsoid, green, the surface glabrous. Seeds 1.0–1.5 mm long, black. $2n$=24. April–June.

Scattered in the eastern half of the Ozark Division and adjacent northern portions of the Mississippi Lowlands (eastern U.S. west to Minnesota and Texas; Canada; South America). Fens, acid seeps, marshes, swamps, margins of sinkhole ponds, wet swales in sand prairies, and occasionally banks of streams and rivers; also moist roadsides; occasionally epiphytic on rotting logs.

This species is distinctive among eastern temperate North American violets in its slender leaf blades. The widespread morphotype with glabrous, lanceolate leaves represents var. *lanceolata*. Populations in the Atlantic and Gulf Coastal Plains tend to have hairy, nearly linear leaves and have been segregated as ssp. *vittata* (Greene) N. Russell (var. *vittata* (Greene) Weath. & Griscom). Russell (1965) also treated plants growing from northern California to Washington as ssp. *occidentalis* (A. Gray)

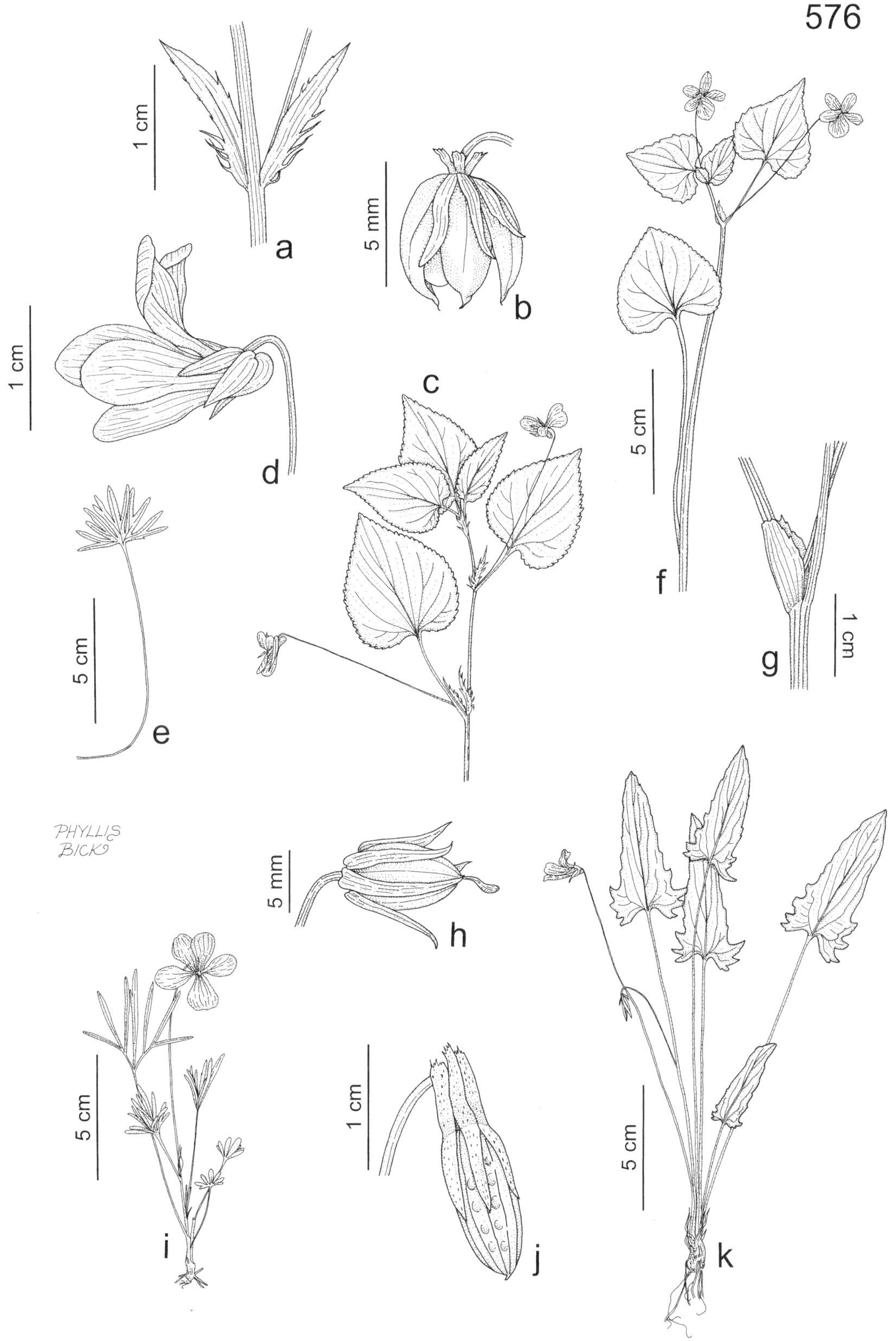

Plate 576. Violaceae. *Viola striata*, **a)** node with stipules, **b)** dehiscing fruit, **c)** fertile stem. *Viola pedatifida*, **d)** flower, **e)** leaf. *Viola pubescens*, **f)** habit, **g)** node with stipules. *Viola pedata*, **h)** fruit, **i)** habit. *Viola sagittata*, **j)** fruit, **k)** habit.

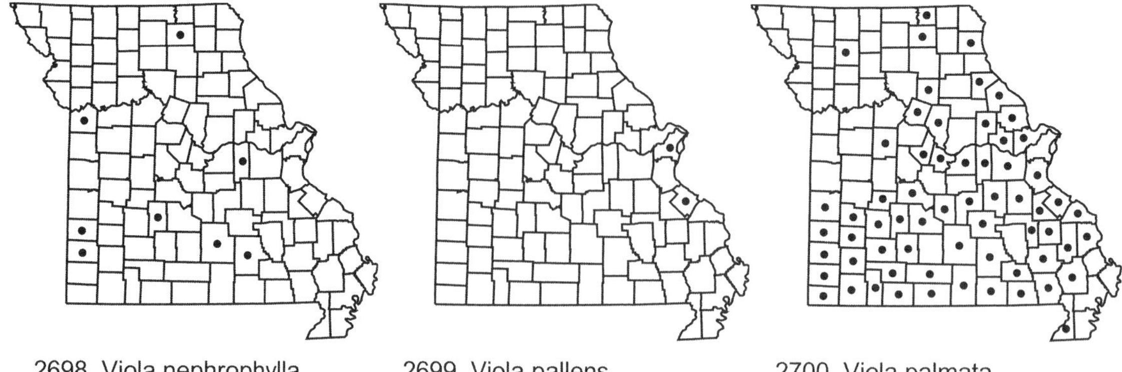

2698. Viola nephrophylla 2699. Viola pallens 2700. Viola palmata

N. Russell. More often, recent authors have treated the western plants as either a separate species, *V. occidentalis* (A. Gray) Howell or as a western subspecies of the related, otherwise eastern species, *V. primulifolia* L. ssp. *occidentalis* (Gray) L.E. McKinney & R.J. Little.

Aside from its strikingly different leaf-blade shape, the species is very similar to *V. pallens*, with which it frequently comes into contact and sometimes hybridizes elsewhere in the northeastern portion of its range. Putative hybrids are difficult to detect morphologically (except in their lack of pollen and seed viability) from the Appalachian and Coastal Plain species *V. primulifolia* L. (primrose-leaved violet), whose origin may indeed have been derived from an ancient interspecific hybridization event (Ballard, 1994).

6. Viola missouriensis Greene (Missouri violet)

V. sororia Willd. var. *missouriensis* (Greene) L.E. McKinney

Pl. 577 e, f; Map 2697

Plants perennial, to 40 cm tall, with a prostrate to ascending rhizome 4–6 mm thick. Aerial stems not produced. Leaves in a basal rosette, long-petiolate, the petiole glabrous. Stipules conspicuous, membranous to somewhat herbaceous, free from the petiole, narrowly lanceolate, the margins entire, glandular-hairy. Leaf blades 2–8 cm long, distinctly longer than wide, narrowly triangular to ovate-triangular, all unlobed, angled or tapered to a usually sharply pointed tip, truncate to shallowly cordate at the base, the margins finely to coarsely toothed, mostly in the basal 2/3 (subentire in the apical 1/3), the surfaces glabrous. Cleistogamous flowers produced. Flower stalks not or only slightly overtopping the leaves (those of the cleistogamous flowers arched or spreading). Sepals 6–8 mm long, oblong-lanceolate to broadly lanceolate, angled to a rounded or bluntly pointed tip, the margins glabrous, the basal auricles inconspicuous (0.5–1.0 mm long at flowering, not becoming noticeably enlarged at fruiting). Corollas (8–)10–18 mm long (except in cleistogamous flowers), the petals oriented more or less forward in life but curved or arched outward or backward toward their tips, longer than the sepals, bluish purple to light purple (with a greenish white throat sometimes with a dark border) often with somewhat darker veins, the lateral petals bearded on the upper surface with uniformly slender hairs (a few of these occasionally slightly thickened toward their tips)), the lower petal glabrous or nearly so, the spur conspicuous, usually well-exserted beyond the sepal auricles, stout and often somewhat hemispheric in shape. Stamens not exserted, typically not visible without dissection of the flower. Style slender, slightly expanded into a narrow, more or less scoop-shaped, hollow, truncate tip. Fruits 7–9 mm long, broadly ellipsoid to broadly ovoid, green to olive green, often mottled with dark purple, the surface glabrous. Seeds 1.5–2.2 mm long, brown. $2n=54$. March–May.

Scattered to common nearly throughout the state (Ohio to South Carolina west to North Dakota and New Mexico). Bottomland forests, mesic upland forests, banks of streams and rivers, sloughs, margins of ponds, lakes, and sinkhole ponds, bottomland prairies, and marshes; also ditches, fallow fields, lawns, railroads, roadsides and disturbed areas.

The relationship of *V. missouriensis* with the widespread *V. sororia* remains controversial. Some authors have merged the two under the latter name (Gleason and Cronquist, 1991) or treated the former as a variety of *V. sororia* (McKinney, 1992). Although they occasionally grow together and some plants, particularly those collected at fruiting, can be difficult to determine, the two taxa tend to differ consistently in leaf morphology and habitat preference. Thus, they are retained as sepa-

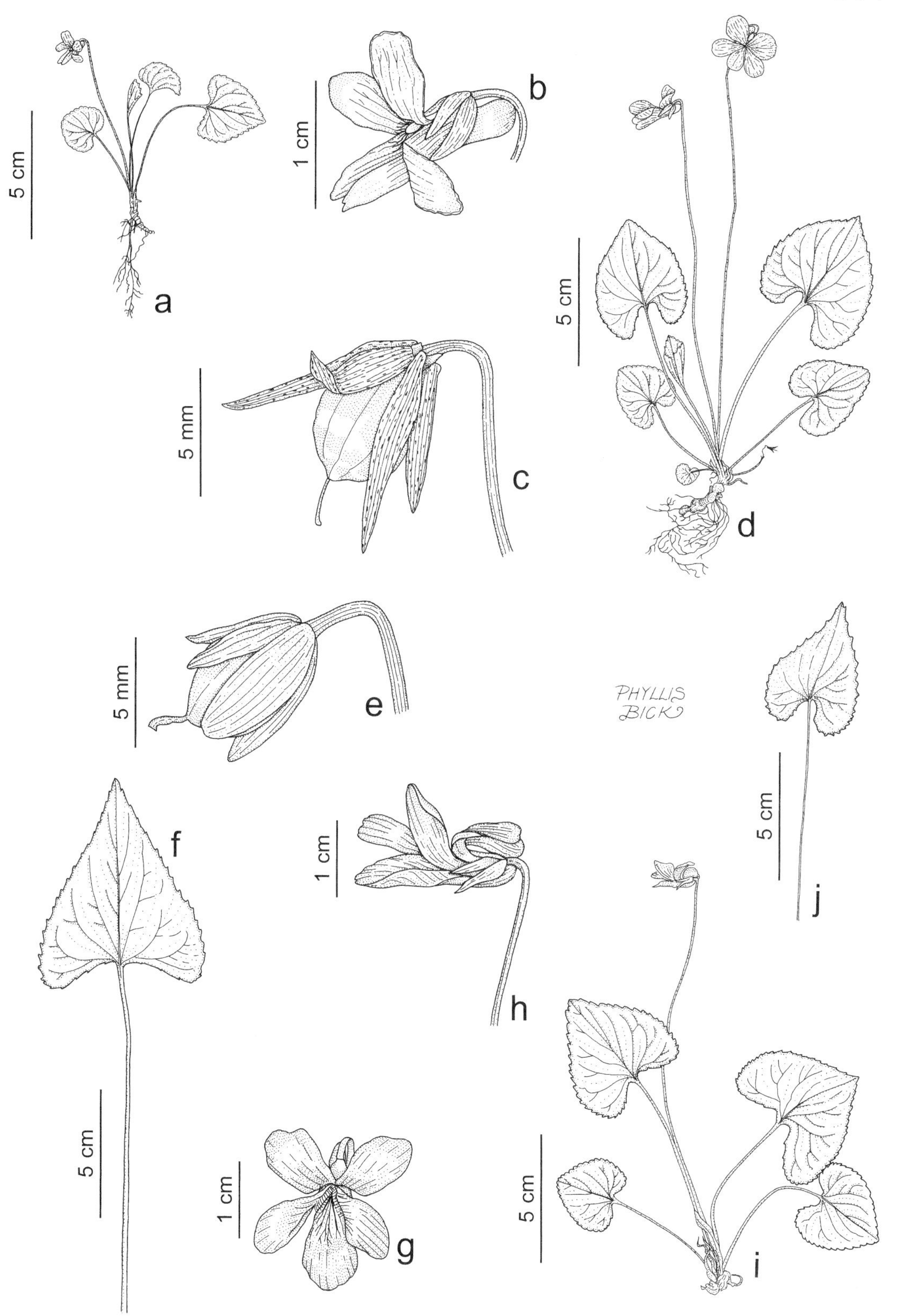

Plate 577. Violaceae. *Viola nephrophylla*, **a)** habit, **b)** flower. *Viola cucullata*, **c)** fruit, **d)** habit. *Viola missouriensis*, **e)** fruit, **f)** leaf. *Viola sororia*, **g)** flower, lateral view, **h)** flower, face view, **i)** habit. *Viola affinis*, **j)** leaf.

rate species in the present treatment. However, further intensive comparative studies are needed on the morphological and genetic variation within and between populations rangewide.

Steyermark (1963) reported the presence of rare, putative hybrids between *V. missouriensis* and *V. sororia* (including *V. papilionacea*), but these have been redetermined as *V. sororia*. In addition, rare putative hybrids between *V. missouriensis* and both *V. pedatifida* and *V. viarum* have been collected in the state.

Viola missouriensis reportedly hybridizes with most other stemless blue violets. Hybrids represented in Missouri consist of a few collections of apparent crosses with *V. pedatifida* and (possibly) *V. viarum*.

7. Viola nephrophylla Greene (blue prairie violet)

V. pratincola Greene

Pl. 577 a, b; Map 2698

Plants perennial, to 20 cm tall, with a prostrate to loosely ascending rhizome 4–6 mm thick. Aerial stems not produced. Leaves in a basal rosette, long-petiolate, the petiole glabrous. Stipules conspicuous, membranous to somewhat herbaceous, free from the petiole, narrowly lanceolate, the margins entire, glandular-hairy. Leaf blades 1.5–5.0 cm long, slightly longer than wide to slightly wider than long, ovate to broadly ovate, heart-shaped, or somewhat kidney-shaped, all unlobed, rounded or broadly angled to a bluntly or less commonly sharply pointed tip, cordate at the base, the margins finely toothed or scalloped their entire length, the upper surface sparsely and minutely hairy, the undersurface glabrous. Cleistogamous flowers produced. Flower stalks mostly not or only slightly overtopping the leaves, less commonly the stalks extending noticeably past the leaves (those of the cleistogamous flowers erect or strongly ascending). Sepals 4–7 mm long, oblong-lanceolate, rounded or more commonly angled to a bluntly or broadly pointed tip, the margins glabrous, the basal auricles small and inconspicuous (0.5–1.0 mm long at flowering). Corollas 7–12 mm long (except in cleistogamous flowers), the petals oriented more or less forward in life but curved or arched outward or backward toward their tips, longer than the sepals, bluish purple to purple (with a greenish white throat usually lacking a dark border) with darker veins, the lateral and lower petals bearded on the upper surface with uniformly slender hairs (the lower petal sometimes glabrous or nearly so elsewhere), the spur conspicuous, usually well-exserted beyond the sepal auricles, stout and often somewhat hemispheric in shape. Stamens not exserted, typically not visible without dissection of the flower. Style slender, slightly expanded into a narrow, more or less scoop-shaped, hollow, truncate tip. Fruits 7–10 mm long, narrowly ellipsoid, green, usually drying to tan, the surface glabrous. Seeds 1.3–1.8 mm long, olive black to black. $2n=54$. April–May, rarely also October–November.

Uncommon, widely scattered, mostly south of the Missouri River (western U.S. east to Arkansas, Illinois, and Maine; Canada). Banks of streams and rivers, fens, margins of ponds, and other seepy areas; also mine tailings, lawns, railroads, and roadsides.

This widely distributed taxon is the preeminent member of the *V. sororia* complex in the western United States, where it occurs outside the ranges of other closely related taxa. It has not been universally accepted as a distinct species by botanists in the central and eastern states. Steyermark (1963) did not even mention the species in his treatment of the genus for the Missouri flora, but it was mapped from several counties (mostly in the southwestern portion of the state) by Russell (1965) under the name *V. pratincola*. Russell followed earlier authors in segregating a series of taxa from *V. sororia*. He treated *V. pratincola* as a species mainly of prairie borders in the Great Plains and adjoining Midwestern states having glabrous leaf blades with relatively sharply pointed tips. Russell segregated *V. nephrophylla* as a species of moist sites with a much larger range, leaves usually short-hairy on the upper surface, and producing relatively firm, more or less oval early-season leaves with the undersurface frequently purplish-tinged. Subsequent students of the group have treated the complex in various ways. McKinney (1992) treated both *V. nephrophylla* and *V. pratincola* as synonyms under his concept of *V. sororia* var. *sororia*, but has since concluded that *V. nephrophylla* may represent a distinct species (Landon McKinney, personal communication, 2009). Ballard (1994), after examining the type specimens for each of the names, concluded that *V. pratincola* and *V. nephrophylla* represented the same taxon, which he considered to be potentially more closely related to *V. cucullata* than to *V. sororia*. Gil-Ad (1997) considered *V. nephrophylla* to be a distinct species, based primarily on certain cellular features of its seed coat, but treated the name *V. pratincola* as representing a series of putative hybrids involving *V. nephrophylla* and other stemless blue violet taxa. Thus, although there appears to be a developing consensus among students of the genus that *V. nephrophylla* should be recognized as a distinct species, its ecological and morphological circumscription remains controversial. Further studies are needed.

In overall habit, *V. nephrophylla* appears most similar to some of the less hairy forms of *V. sororia*, although the leaves tend to occupy the smaller end of the size range of that species. The purplish pigmentation on the undersurface of at least the early-season leaves is relatively frequent in *V. nephrophylla*, but is rarely observed in *V. sororia*. The small, oval, early-season leaves frequently produced in *V. nephrophylla* are also relatively uncommon in *V. sororia*. The leaf blades of *V. nephrophylla* also tend to be relatively firm and finely and evenly toothed along their entire margins, whereas in *V. sororia* the blades often are relatively thin, with the margins often more coarsely toothed and with the teeth larger and more widely spaced toward the tip. However, *V. sororia* is so variable in its leaf morphology that each of these leaf characters may occur in some plants of that species. More diagnostic are the cleistogamous fruits, which in *V. nephrophylla* are uniformly green and positioned on erect or ascending stalks, whereas those of *V. sororia* are frequently mottled with purple and have the stalks strongly arched and/or spreading.

Although Russell (1965) treated the name *V. pratincola* (here reduced to synonymy under *V. nephrophylla*) as a taxon of upland prairies and prairie borders, in Missouri it does not appear to occur in this type of habitat. Instead, botanists seeking this species should search for new records in various wetlands, particularly in fens and other seepage areas.

8. Viola pallens (Banks ex DC.) Brainerd (smooth white violet, northern white violet)

V. macloskeyi F.E. Lloyd ssp. *pallens* (Banks ex DC.) M.S. Baker

V. macloskeyi var. *pallens* (Banks ex DC.) C.L. Hitchc.

Pl. 578 c, d; Map 2699

Plants perennial, to 10 cm tall, with a prostrate to ascending rhizome 1–2 mm thick, also producing threadlike stolons at fruiting. Aerial stems not produced. Leaves in a basal rosette, long-petiolate, the petiole glabrous or hairy. Stipules inconspicuous, membranous, free from the petiole, narrowly lanceolate, the margins entire, glandular-hairy. Leaf blades 0.8–3.7 cm long, mostly wider than to about as wide as long, broadly ovate to nearly circular, all unlobed, rounded or broadly and bluntly pointed at the tip, cordate at the base, the margins finely toothed or scalloped to occasionally nearly entire, the surfaces glabrous. Cleistogamous flowers produced. Flower stalks not or only slightly overtopping the leaves (those of the cleistogamous flowers often relatively short, erect or strongly ascending). Sepals 3–4 mm long, lanceolate, angled to a sharply pointed tip, the margins glabrous, the basal auricles short and inconspicuous. Corollas 6–10 mm long (except in cleistogamous flowers), the petals oriented more or less forward in life but curved or arched outward or backward toward their tips, longer than the sepals, white (with a greenish white throat), the lower petal usually with dark purple or brownish purple veins toward the base, the lateral and lower petals not noticeably bearded, the upper surface glabrous or nearly so (somewhat bearded with slightly club-shaped hairs elsewhere), the spur conspicuous, well-exserted beyond the sepal auricles, stout and often somewhat hemispheric in shape. Stamens not or only slightly exserted, often not visible without dissection of the flower. Style slender, slightly expanded into a narrow, more or less scoop-shaped, hollow, truncate tip. Fruits 4–6 mm long, narrowly ellipsoid, green, drying to tan, the surface glabrous. Seeds 1.0–1.4 mm long, black. $2n=24$. April–May.

Uncommon, known thus far only from Ste. Genevieve County (widespread in the U.S. [except for most of the Great Plains]; Canada). Bases and ledges of bluffs and banks of streams; on sandstone substrate.

Steyermark (1963) discussed the biogeography of this uncommon, dainty little member of the Missouri flora. He concluded that it represents a relict taxon more widely distributed during Pleistocene times when Missouri's climate was cooler and wetter, and currently is limited to cool, moist microhabitats in the state. Its main distribution in the eastern half of the United States is in the Great Lakes region and the Appalachian Mountains, with isolated occurrences farther south and west. Russell (1956, 1965), who treated the taxon as a subspecies of *V. macloskeyi*, discussed the interesting western and eastern division in the North American distribution of the species, suggesting that its historical range had become fragmented at least twice in the history of the complex. Typical *V. macloskeyi* is restricted to the northwestern United States, where its range overlaps that of *V. pallens*. The relationship between the two taxa requires further study, as does their relationship to the Caribbean endemic, *V. domingensis* Urb.

2701. Viola pedata 2702. Viola pedatifida 2703. Viola pubescens

9. Viola palmata L. (cleft violet, three-lobed violet)

V. palmata var. *dilatata* Elliott

V. falcata Greene

V. triloba Schwein.

V. triloba f. *dilatata* (Elliott) E.J. Palmer & Steyerm.

V. triloba var. *dilatata* (Elliott) Brainerd

Pl. 578 e, f; Map 2700

Plants perennial, to 30 cm tall, with a short, prostrate to sometimes ascending rhizome 4–6 mm thick. Aerial stems not produced. Leaves in a basal rosette, long-petiolate, the petiole finely hairy. Stipules conspicuous, membranous to somewhat herbaceous, free from the petiole, narrowly lanceolate, the margins usually entire, glandular-hairy. Leaf blades 1.5–9.5 cm long, about as long to about 2 times as long as wide, variously ovate to kidney-shaped, heart-shaped, or triangular-ovate, those produced early and late in the season unlobed, the main leaves shallowly to deeply lobed, often somewhat irregularly so, the segments not lobed again, at least the central one lanceolate to triangular-ovate and mostly more than 3 mm wide, the margins otherwise entire or toothed toward the segment tips (sometimes also minutely hairy), the surfaces moderately pubescent with fine, spreading or shaggy hairs. Cleistogamous flowers produced (frequently dark green mottled with purple). Flower stalks not or only slightly overtopping the leaves (those of the cleistogamous flowers short and arched or spreading). Sepals 3–9 mm long, oblong-lanceolate, rounded or broadly angled to a bluntly pointed tip, the margins minutely hairy, the basal auricles short and inconspicuous. Corollas 9–20(–26) mm long (except in cleistogamous flowers), the petals oriented more or less forward in life but curved or arched outward or backward toward their tips, longer than the sepals, bluish purple (with a greenish white throat and usually darker veins), rarely all white, the lateral petals bearded on the upper surface with slender hairs, the lower petal glabrous, the spur conspicuous, well-exserted beyond the sepal auricles, stout and often somewhat hemispheric in shape. Stamens not exserted, typically not visible without dissection of the flower. Style slender, slightly expanded into a narrow, more or less scoop-shaped, hollow, truncate tip. Fruits 8–10 mm long, broadly ellipsoid to broadly ovoid, green to olive green, often tinged or mottled with purple, the surface glabrous. Seeds 1.7–2.0 mm long, brown to tan or orangish yellow, sometimes mottled. $2n=54$. April–May.

Scattered nearly throughout the state, but less abundant in the Glaciated Plains and Unglaciated Plains Divisions (eastern U.S. west to Illinois, Kansas, and Texas; Canada). Mesic upland forests, bases, ledges, and tops of bluffs, and banks of streams and rivers; also railroads and roadsides.

Viola palmata has a heterophyllous pattern of leaf production in that, in contrast to the leaves mostly present at flowering time, the earliest spring and latest summer and fall leaves are unlobed, a feature not always observable on herbarium specimens. As circumscribed in the present treatment, it consists of two apparently intergradient extremes: var. *palmata* with most of the mid-season leaves shallowly lobed with broadly triangular or ovate segments, primarily northeastern and Appalachian in distribution, often thriving in moist to wet forests on heavier loamy to clayey soils; and var. *dilatata*, with most mid-season leaves deeply lobed with linear to lanceolate segments, primarily southeastern and Lower Midwest in distribution, thriving principally in open upland forests on light sandy or gravelly soils. Most of the Missouri materials appear assignable to var. *dilatata*, however a substantial number of intermediates cannot reliably be assigned to either variety, discouraging recognition of taxonomic subdivisions in this species without

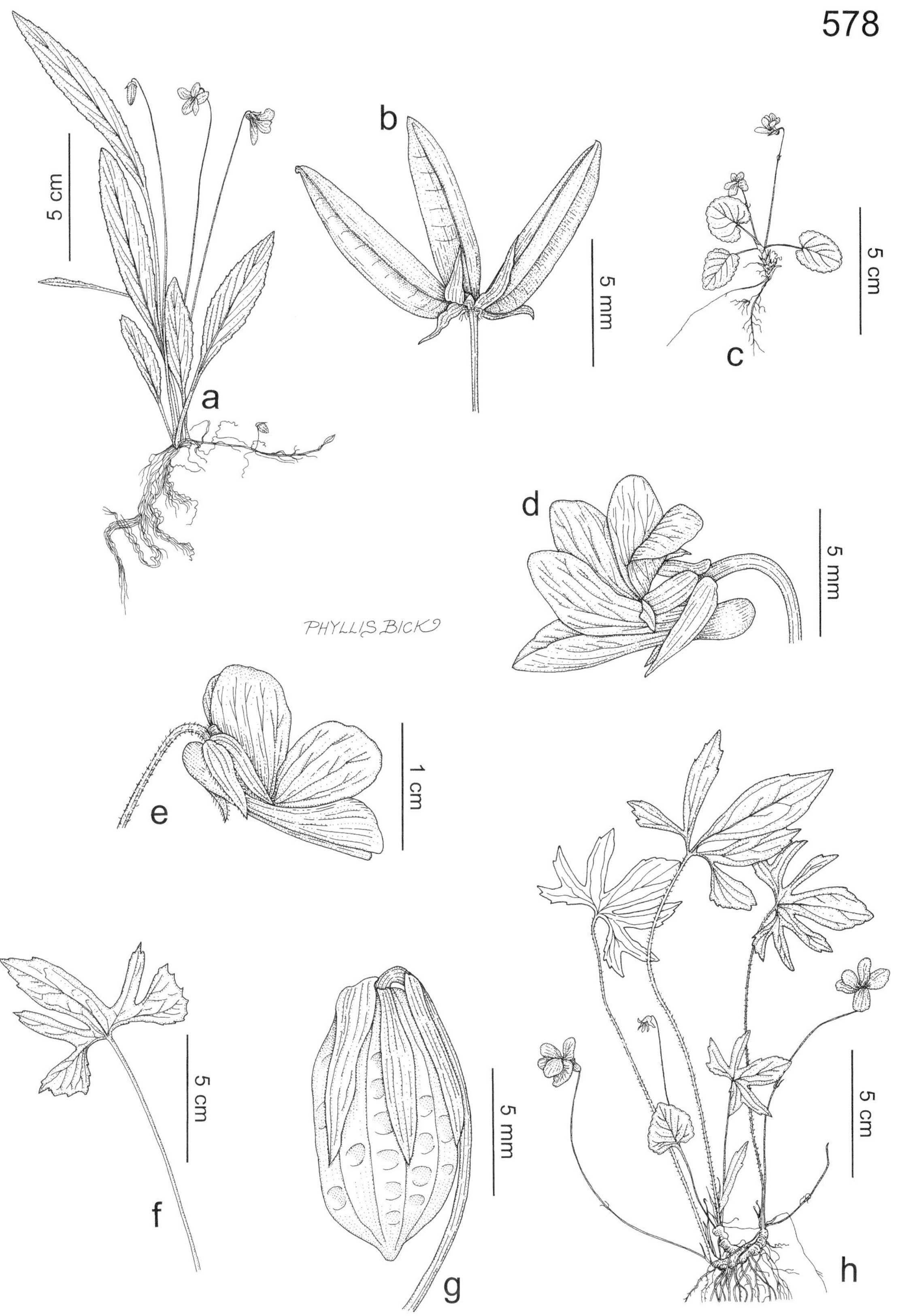

Plate 578. Violaceae. *Viola lanceolata*, **a)** habit, **b)** dehisced fruit. *Viola pallens*, **c)** habit, **d)** flower. *Viola palmata*, **e)** flower, **f)** leaf. *Viola viarum*, **g)** fruit, **h)** habit.

further illumination on the complex. Rare white-flowered plants have been called *V. triloba* f. *albida* Steyerm.

This species was long known as *V. triloba* (Steyermark, 1963; Russell, 1965). However, most botanists currently believe that the type specimen of *V. palmata* represents this heterophyllous taxon (McKinney, 1992). Previously, the name *V. palmata* had been applied to an eastern species (now called *V. subsinuata* (Greene) Greene)) in which all of the leaves produced are lobed. However, this opinion has not been held universally. Gleason and Cronquist (1991) combined the homophyllous and heterophyllous populations into a single species (along with *V. pedatifida*) under the name *V. palmata*, apparently as a matter of convenience. Conversely, Gil-Ad (1997) suggested that the name *V. palmata* probably should be restricted to hybrids involving *V. pedatifida* and some other stemless blue violet taxon with unlobed leaves, while recognizing separate homophyllous and heterophyllous species under the names *V. subsinuata* and *V. triloba*.

Rare putative hybrids between *V. palmata* and *V. sororia* have been called *V.* ×*populifolia* Greene. These have been encountered sporadically in southern Missouri, but most of the specimens are historical.

10. Viola pedata L. (bird's foot violet, pansy violet, hens and roosters)

Pl. 576 h, i; dust jacket; Map 2701

Plants perennial, to 20 cm tall, with a short, erect rhizome 4–8 mm thick. Aerial stems not produced. Leaves in a basal rosette, long-petiolate, the petiole glabrous or minutely and inconspicuously hairy. Stipules conspicuous, membranous, fused to the petiole for about $^{2}/_{3}$ of their length, lanceolate, the margins narrowly lobed or toothed, glandular-hairy. Leaf blades 0.5–5.0 cm long, slightly longer than wide to wider than long, ovate to kidney-shaped, deeply and ternately dissected into uniformly slender, linear to oblanceolate segments 1–3 mm wide, the central segment typically undivided or with a few irregular teeth, the lateral segments ternately lobed again, the ultimate segments mostly rounded to broadly or bluntly pointed at the tips, entire or irregularly few-toothed (usually also minutely hairy) along the margins, the surfaces usually minutely hairy. Cleistogamous flowers not produced. Flower stalks not or only slightly overtopping the leaves. Sepals 9–13 mm long, lanceolate, narrowly angled to a sharply pointed tip, the margins sometimes minutely hairy, the basal auricles well-developed. Corollas 12–22(–30) mm long, appearing strongly frontally flattened in life, the petals longer than the sepals, all lavender, pale blue, or bluish purple (with a white throat and sometimes greenish veins) or the upper pair conspicuously darker purple to purplish black, rarely all entirely white, all glabrous (beardless) on the upper surface, the spur conspicuous, well-exserted beyond the sepal auricles, stout and often somewhat hemispheric in shape, often somewhat purplish-tinged. Stamens exserted beyond the corolla throat, the orange tips noticeable without dissection of the flower. Style club-shaped with an oblique concave area near the tip. Fruits 8–13 mm long, narrowly ellipsoid to ellipsoid, green, drying to tan or olive-colored, the surface glabrous. Seeds 1.5–1.7 mm long, tan. $2n$=54. April–June, frequently also September–December.

Scattered to common nearly throughout the state, but uncommon to absent in the western portion of the Glaciated Plains Division and the southern portion of the Mississippi Lowlands (eastern U.S. west to Minnesota and Texas; Canada). Glades, upland prairies, savannas, openings and edges of mesic to dry upland forests, tops of bluffs, banks of streams, and margins of sinkhole ponds; also pastures, old fields, fencerows, and roadsides.

Viola pedata is widely regarded by taxonomists and naturalists as the queen of eastern North American violets and sometimes forms large displays on glades and dry, poor-soil, cherty road banks in the Ozarks. The species was originally described based on material of the striking bicolorous form with the three lower petals light blue-violet and the upper pair velvety purple-black. The more concolorous form with uniformly lighter-colored corollas often occurs near the bicolorous one and has been called f. *rosea* A.L. Sanders or var. *lineariloba* DC. The two color morphs are represented about equally in Missouri. Additionally, rare plants with all-white corollas occur within populations of the other two taxa and have been called f. *alba* (Thurb.) Britton.

The species is unusual in the genus in never producing cleistogamous flowers in the wild and in being functionally outcrossing. It is very distinctive in its frontally flattened, pansy-like corollas and strongly exserted orange stamens. *Viola pedata* is nevertheless sometimes confused with *V. pedatifida,* with which it grows occasionally. The two differ substantially in rhizome morphology and strikingly in stipule morphology, with stipules of *V. pedata* fused to the petiole for most of their length (vs. free in *V. pedatifida*). In addition, the central segment of the dissected leaf blades is undivided or bears one or two irregular short teeth in *V. pedata* but is consistently ternately divided in *V. pedatifida*. Remarkable leaf morphs of *V. pedata* are very rarely encountered in herbarium collections, specifically on plants subjected to developmental trauma, such as from fires; these peculiarly lobed plants—reproductively mature specimens of which always bear typical *V. pedata* flowers and good pollen—have been the basis for erroneous reports of hybridization with other violet species.

Bird's foot violet also is gaining popularity in horticulture. J. K. Small (1935) was an early advocate of its cultivation as an ornamental. Although in nature the species thrives in seemingly sterile, droughty, and inhospitable environments and would seem to be a natural for pampered rock gardens, transplantation of individuals once in the ground is reportedly difficult. However, potted plants grown from seed have increasingly become available at wildflower nurseries in recent years.

11. Viola pedatifida G. Don (prairie violet, larkspur violet)

V. delphiniifolia Nutt.

V. palmata L. var. *pedatifida* (G. Don) Cronquist

Pl. 576 d, e; Map 2702

Plants perennial, to 30 cm tall, with a usually short, prostrate to more commonly ascending or erect rhizome 4–6 mm thick. Aerial stems not produced. Leaves in a basal rosette, long-petiolate, the petiole glabrous or finely hairy. Stipules conspicuous, membranous (sometimes purple-spotted or mottled), free from the petiole, narrowly lanceolate, the margins usually entire, glandular-hairy. Leaf blades 0.5–6.0 cm long, slightly longer than wide to wider than long, ovate to kidney-shaped, deeply and ternately dissected into uniformly slender, linear to oblanceolate segments 1–3 mm wide, the central and lateral primary segments all ternately lobed again, the ultimate segments mostly rounded to broadly or bluntly pointed at the tips, entire or irregularly few-toothed (usually also minutely hairy) along the margins, the surfaces glabrous or finely hairy. Cleistogamous flowers produced. Flower stalks not or only slightly overtopping the leaves (those of the cleistogamous flowers erect or ascending). Sepals 5–10 mm long, lanceolate, narrowly angled to a sharply pointed tip, the margins glabrous, the basal auricles well-developed, elongating as the fruits develop. Corollas 10–16(–20) mm long (except in cleistogamous flowers), the petals oriented more or less forward in life but curved or arched outward or backward toward their tips, longer than the sepals, lavender, pale blue, or bluish purple (with a greenish white throat and usually darker veins), the lateral and lower petals bearded on the upper surface with slender hairs, the spur conspicuous, well-exserted beyond the sepal auricles, stout and often somewhat hemispheric in shape. Stamens not exserted, typically not visible without dissection of the flower. Style slender, slightly expanded into a narrow, more or less scoop-shaped, hollow, truncate tip. Fruits 7–13 mm long, narrowly ellipsoid to ellipsoid, green, drying to tan or olive-colored, the surface glabrous. Seeds 1.7–2.2 mm long, brown to tan. $2n=54$. April–May.

Scattered in the western half of the state, uncommon historically farther east (Ohio to North Dakota south to Arkansas and Arizona; Canada). Upland prairies, loess hill prairies, tops of bluffs, glades, and occasionally bottomland prairies; also railroads and roadsides.

For a discussion of the separation of this species from the superficially similar *V. pedata*, see the treatment of that species. Although the two species are similar morphologically in having ternately dissected leaves, they do not appear to be closely related taxonomically within the genus. *Viola pedatifida* hybridizes with *V. sororia* in Missouri (such hybrids have been called *V. ×bernardii* Greene), and rare putative hybrids with *V. sagittata* have also been collected in Jackson, Jasper, and Newton Counties.

12. Viola pubescens Aiton (yellow violet, smooth yellow violet)

V. pubescens var. *eriocarpa* (Schwein.) N. Russell, an illegitimate name

V. pubescens var. *scabriuscula* Torr. & A. Gray

V. eriocarpa Schwein.

V. eriocarpa var. *leiocarpa* Fernald & Wiegand

V. pensylvanica Michx.

V. pensylvanica var. *leiocarpa* (Fernald & Wiegand) Fernald

Pl. 576 f, g; Map 2703

Plants perennial, with a stout, prostrate to ascending rhizome 3–5 mm thick. Stems 10–35 cm long, erect or ascending, sometimes from a spreading base. Leaves alternate (some nodes sometimes appearing subopposite) and sometimes also basal at flowering, short- to long-petiolate, the petiole glabrous or hairy. Stipules relatively small and inconspicuous, free from the petiole, unlobed, the margins entire or shallowly and irregularly toothed. Leaf blades 1.5–7.0 cm long, unlobed, ovate to heart-shaped or somewhat kidney-shaped, angled (often broadly so) or broadly rounded at the tip, rounded to cordate at the base, the margins evenly and bluntly toothed, the surfaces glabrous

2704. Viola sagittata 2705. Viola sororia 2706. Viola striata

or hairy. Cleistogamous flowers produced. Flower stalks not or only slightly overtopping the leaves (those of the cleistogamous flowers erect or ascending). Sepals 3–6 mm long, lanceolate, angled to a sharply pointed tip, the margins sometimes minutely hairy, the basal auricles short and inconspicuous. Corollas 8–12 mm long (except in cleistogamous flowers), the petals oriented forward with arched or outward-curved apical portions, longer than the sepals, yellow, the upper and lateral petals sometimes brownish-tinged on the undersurface, the lateral and especially the lower petals usually with dark purple to brownish purple veins, the lateral petals bearded on the upper surface with mostly knob-shaped hairs, the lowermost petal glabrous on the upper surface, the spur minute (the lower petal often appearing merely pouched), not or only slightly exserted beyond the sepal auricles, relatively stout. Stamens not exserted, typically not visible without dissection of the flower. Style club-shaped, with a capitate, bearded tip. Fruits 9–12 mm long, narrowly ellipsoid to ellipsoid, green, drying to tan to olive-colored, the surface glabrous or densely woolly. Seeds 2.1–2.4 mm long, brown. $2n=12$. March–May.

Scattered to common throughout the state (eastern U.S. west to North Dakota, Wyoming, and Texas; Canada). Bottomland forests, mesic upland forests, bases and ledges of bluffs, banks of streams and rivers, and margins of sinkhole ponds; also edges of pastures and shaded roadsides.

This species is represented over much of eastern North America by two morphological and ecological extremes treated by many botanists as varieties (Steyermark, 1963; Ballard, 1994). Plants of var. *pubescens* tend to produce a single, erect or strongly ascending aerial stem, 1 or no basal leaves at flowering, relatively densely hairy foliage, ovate to somewhat kidney-shaped leaf blades with broadly angled to truncate bases, and herbaceous, ovate upper stipules, whereas plants of var. *scabriuscula* usually produce several aerial stems that are ascending from spreading bases, several basal leaves at flowering, glabrous to sparsely hairy leaf blades, heart-shaped to broadly ovate leaf blades with usually cordate bases, and semi-herbaceous lanceolate upper stipules. Unfortunately, a substantial proportion of collections across the range of the species defy confident assignment to either variety; indeed, the vast majority of Missouri specimens are at least partially referable to var. *scabriuscula*, but few specimens match the extreme morphologies of either variety. Neither variety is accepted here, given the difficulty of interpreting Missouri material.

13. Viola sagittata Aiton (arrow-leaved violet, arrowhead violet)

V. sagittata var. *subsagittata* (Greene) Pollard

V. arkansana Greene

V. emarginata (Nutt.) Leconte

Pl. 576 j, k; Map 2704

Plants perennial, to 30 cm tall, with a prostrate to ascending rhizome 4–6 mm thick. Aerial stems not produced. Leaves in a basal rosette, long-petiolate (except sometimes in early-season leaves), the petiole glabrous or finely hairy. Stipules conspicuous, free from the petiole, membranous to somewhat herbaceous, narrowly lanceolate, the margins toothed and glandular-hairy. Leaf blades 1.5–8.5 cm long, all except sometimes those of the first leaves of the growing season lobed (occasionally appearing only coarsely toothed), the lobes produced only near the blade base and less than 1/3 the length of the midvein, the blade much longer than wide, elliptic to narrowly triangular, narrowly angled to a rounded to sharply pointed tip, broadly rounded to truncate or cordate at the base, the margins otherwise bluntly toothed below an often more or less entire apical portion, the surfaces glabrous or finely hairy. Cleistogamous flowers produced. Flower stalks not or only slightly overtopping the leaves (those of the cleistogamous flowers erect or ascending). Sepals 4–8 mm long, lanceolate, narrowly angled to a sharply pointed tip, the margins sometimes minutely hairy, the basal auricles well-developed, elongating as the fruits develop. Corollas 8–14 mm long (except in cleistogamous flowers), the petals oriented forward with arched or outward-curved apical portions, longer than the sepals, bluish purple to purple (with a greenish white throat) with darker veins (rarely all white elsewhere), the lateral and lower petals bearded on the upper surface with slender hairs, the spur conspicuous, well-exserted beyond the sepal auricles, stout and often somewhat hemispheric in shape. Stamens not exserted, typically not visible without dissection of the flower. Style slender, slightly expanded into a narrow, more or less scoop-shaped, hollow, truncate tip. Fruits 11–18 mm long, narrowly ellipsoid to ellipsoid, green, drying to tan, the surface glabrous. Seeds 1.2–1.5 mm long, tan to brown. $2n$=54. April–June.

Scattered nearly throughout the state, but uncommon or absent from the Mississippi Lowlands Division and the western half of the Glaciated Plains (eastern U.S. west to Minnesota and Texas; Canada). Dry prairies, sandstone glades and edges or clearings of open oak forests, on dry to seasonally somewhat moist sand or gravelly sand.

This species is relatively widespread in Missouri, but apparently is nowhere very abundant in the state. Steyermark (1963) noted the extreme seasonal variation in leaf morphology within *V. sagittata*, with early-season leaves small and tending to have relatively short petioles and relatively entire blades. Blades of late-season leaves tend to be much shorter relative to their width than those produced earlier in the growing season. Most botanists recognize two varieties within the species (McKinney, 1992; Ballard, 1994). Missouri plants generally correspond to the widespread var. *sagittata*, with leaves becoming erect and long-petiolate at fruiting and with the blades of all but early-season leaves lobed. The var. *ovata* (Nutt.) Torr. and A. Gray (*V. fimbriatula* Sm.) comprises populations in the Great Lakes region and portions of the Appalachians that have the leaves spreading and relatively short-petiolate at fruiting, as well as blades that are at most coarsely toothed, rather than lobed. Yost (1987) studied plants of this complex that were clonally produced or grown from seed planted into shaded and sunny plots in a forest and adjacent field in New York. She concluded that leaf size and petiole length were strongly influenced by environmental factors and were not good characters for taxonomic distinctions in the group, but did not address whether other potential distinguishing features might be under genetic control. Gil-Ad (1997), who treated the two taxa as separate species, tabulated a number of additional (often subtle or overlapping) morphological characters said to separate them and also noted apparent differences in habitat preferences.

Viola sagittata produces putative hybrids with several other species of stemless blue violets. Russell and Risser (1960) studied plants that had been called *V. emarginata* and concluded that these represented a series of such interspecific hybrids, but McKinney (1992) noted that the type specimen of this name appears to be referable to *V. sagittata*. Occasional hybrids with *V. sororia* were studied by Russell and Cooperrider (1955) and have been called *V.* ×*greenei* House. Steyermark (1963) also reported rare putative hybrids between *V. sagittata* and *V. pedatifida*, but these have not been confirmed by later specialists.

14. Viola sororia Willd. (common violet, meadow violet, butterfly violet)
V. domestica E.P. Bicknell
V. papilionacea Pursh, apparently misapplied
V. septentrionalis Greene
Pl. 577 g–i; Map 2705

Plants perennial, to 30(–45) cm tall, with a prostrate to ascending rhizome 4–7 mm thick. Aerial stems not produced. Leaves in a basal rosette, long-petiolate, the petiole glabrous or sparsely to moderately pubescent with fine, spreading hairs. Stipules conspicuous, membranous to somewhat herbaceous, free from the petiole, narrowly lanceolate, the margins entire, glandular-hairy. Leaf blades 1.5–9.0 cm long, slightly longer than wide to noticeably wider than long, ovate to broadly ovate, heart-shaped, or kidney-shaped, all unlobed, rounded or broadly angled to a bluntly or less commonly sharply pointed tip (occasionally abruptly tapered at the very tip), cordate at the base, the margins finely to relatively coarsely toothed or scalloped their entire length or with the teeth somewhat larger and more widely spaced in the apical portions, the surfaces glabrous to more commonly sparsely to densely, but finely hairy, when pubescent then both surfaces usually hairy. Cleistogamous flowers produced. Flower stalks mostly not or only slightly overtopping the leaves (those of the cleistogamous flowers prostrate or abruptly arched). Sepals 4–7 mm long, oblong-lanceolate to broadly lanceolate, rounded or more commonly angled to a bluntly or broadly pointed tip, the margins glabrous or minutely hairy, the basal auricles small and inconspicuous (0.5–1.0

mm long at flowering). Corollas (6–)10–18 mm long (except in cleistogamous flowers), the petals oriented more or less forward in life but curved or arched outward or backward toward their tips, longer than the sepals, bluish purple to purple (with a greenish white throat usually lacking a dark border) with darker veins, rarely all white or white with purple spots or mottling, the lateral petals bearded on the upper surface with uniformly slender hairs (rarely a few of the hairs may be somewhat club-shaped), the lower petal glabrous or with a few slender hairs on the upper surface toward the base, the spur conspicuous, usually well-exserted beyond the sepal auricles, stout and often somewhat hemispheric in shape. Stamens not exserted, typically not visible without dissection of the flower. Style slender, slightly expanded into a narrow, more or less scoop-shaped, hollow, truncate tip. Fruits 10–12 mm long, broadly ellipsoid to broadly ovoid or nearly globose, green to dark green or olive green, usually mottled with purple, drying to tan or olive brown, the surface glabrous. Seeds 1.8–2.5 mm long, brown to dark brown, sometimes mottled with purple. $2n$=54. March–June, sometimes also October–December

Common throughout the state (eastern U.S. west to North Dakota, Utah, and New Mexico; Canada; Mexico). Bottomland forests, mesic to dry upland forests, banks of streams, rivers, and spring branches, margins of ponds, lakes, oxbows, and sinkhole ponds, bases, ledges, and tops of bluffs, upland prairies, glades, and savannas; also ditches, fallow fields, orchards, pastures, lawns, gardens, cemeteries, railroads, roadsides, and disturbed areas.

As noted by McKinney (1992) and other authors, among the stemless blue violets *V. sororia* has the most tortuous nomenclatural history (and longest synonymy). In recent decades, this taxon has been interpreted broadly to include both typical hairy plants and glabrous plants once segregated by Steyermark (1963) and other early botanists as *V. papilionacea*. Such glabrous or nearly glabrous plants grow spontaneously among otherwise typical, hairy individuals and especially become common in disturbed areas, such as along trails and heavily timbered woodlots. Numerous intermediates exist, some of which were treated as hybrids between the glabrous and hairy species under the name *V.* ×*napae* House. Further complicating the issue, widely distributed cultivars developed from *V. sororia*, also with glabrous foliage, may be found naturalized in cemeteries, waste places, gardens, and city woodlots. However, the name *V. papilionacea*, as originally applied by Pursh (1814), apparently does not refer to glabrous plants of *V. sororia* (as the name has commonly been interpreted by American taxonomists). Instead, the type specimen at the herbarium of the Royal Botanic Gardens in Kew, England, appears identical with hairy-leaved plants currently referred to as *V. affinis* LeConte.

Plants in undisturbed or minimally disturbed forest sites appear to reproduce predominately or exclusively by seed from fruits developing from self-pollinating, cleistogamous flowers. It is perhaps this preponderance of inbreeding that is responsible for promoting the wealth of minor populational variants involving foliage pubescence, corolla color, and seed color. Co-occurrence with other species over a broad geographic range, and potential opportunities for hybridization, likely also contributes to the morphological variation in the species. Densely hairy individuals are distinct from other species of stemless blue violets with unlobed leaves. However, glabrous or nearly glabrous specimens may possibly be confused with *V. affinis, V. missouriensis*, or *V. nephrophylla* (for further discussion, see the treatments of those species).

Steyermark (1963) reported the presence of rare, putative hybrids between *V. missouriensis* and *V. sororia* (including *V. papilionacea*), but these have been redetermined as *V. sororia*. However, in Missouri *V. sororia* sometimes can hybridize with *V. pedatifida* in prairies of northern and western Missouri (such hybrids have been called *V.* ×*bernardii* Greene), and also rarely with *V. sagittata* in southern and eastern counties (such hybrids have been called *V.* ×*greenei* House). Steyermark (1963) also reported a rare putative hybrid between *V. sororia* and *V. palmata* (as *V. triloba*) under the name *V.* ×*populifolia* Greene.

The most common phase of this species produces deep bluish purple corollas. However, numerous other color variants have been named. In Missouri, plants with lavender or more reddish purple petals are encountered occasionally within populations. Steyermark treated glabrous and hairy plants with all-white corollas as *Viola papilionacea* f. *alba* (Torr. & A. Gray) Farw. and *V. sororia* f. *beckwithae* House, respectively. He also mentioned a form with nearly white corollas and only a small amount of purple coloration on the veins of the lower petal, which was described from Ohio plants as *V. papilionacea* f. *albiflora* Grover. The Confederate violet (*V. sororia* f. *priceana* (Pollard) Cooperr.), a form with grayish white corollas marked with violet or blue veins and sometimes also the lower petal spotted or mottled with purple, has been found at a few sites in Missouri in recent years, but probably represents plants that have escaped from cultivation rather than truly native

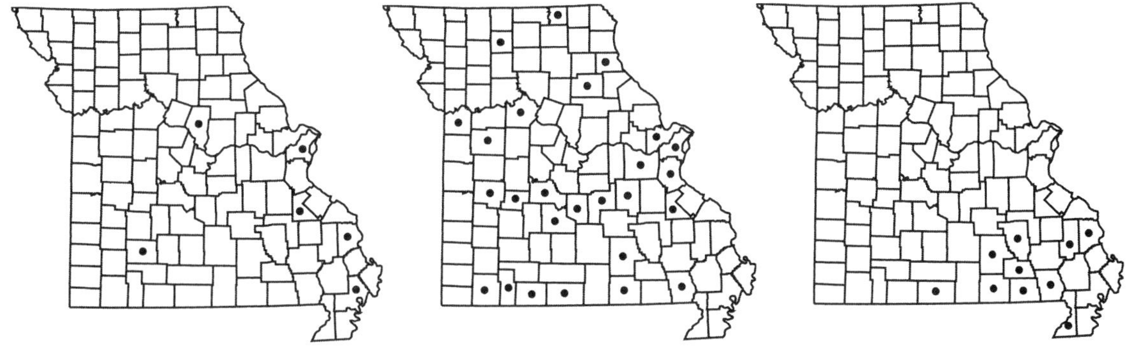

2707. Viola tricolor 2708. Viola viarum 2709. Phoradendron leucarpum

occurrences. Where such plants grow within natural populations of plants with bluish purple petals, individuals with intermediate corolla color patterns may also occur.

15. Viola striata Aiton (pale violet, cream violet)

Pl. 576 a–c; Map 2706

Plants perennial, with a slender to stout, prostrate to ascending rhizome 2–4 mm thick. Stems 10–40 cm long, erect or ascending, sometimes from a spreading base. Leaves alternate (some nodes sometimes appearing subopposite) and also basal at flowering, mostly long-petiolate, the petiole glabrous. Stipules relatively large and conspicuous, herbaceous, free from the petiole, the margins with a fringe of relatively slender, deep teeth or lobes. Leaf blades 1.0–5.5 cm long, unlobed, ovate to heart-shaped, narrowly angled or tapered at the tip, truncate to cordate at the base, the margins evenly scalloped or bluntly toothed, the surfaces glabrous or sparsely hairy on the upper surface. Cleistogamous flowers produced. Flower stalks not or only slightly overtopping the leaves (those of the cleistogamous flowers erect or ascending). Sepals 7–12 mm long, lanceolate, narrowly angled to a sharply pointed tip, the margins sometimes minutely hairy (especially on cleistogamous flowers), the basal auricles well-developed. Corollas 8–13 mm long (except in cleistogamous flowers), the petals oriented forward with arched or outward-curved apical portions, longer than the sepals, white to pale cream-colored, the lower petal usually with dark purple to brownish purple veins, the lateral petals bearded on the upper surface with slender hairs, the lowermost petal glabrous on the upper surface, the spur well-exserted beyond the sepal auricles, relatively stout. Stamens not exserted, typically not visible without dissection of the flower. Style slender, bent at the tip. Fruits 3–9 mm long, broadly ellipsoid to ellipsoid, green, drying to tan to olive-colored, the surface glabrous. Seeds 1.8–2.0 mm long, tan. $2n=20$. April–June.

Scattered to common in the Ozark, Ozark Border, and Mississippi Lowlands Divisions, uncommon or absent elsewhere in the state (eastern U.S. west to Iowa and Oklahoma; Canada). Bottomland forests, mesic upland forests, banks of streams and rivers, fens, bases and ledges of bluffs, and margins of upland prairies, glades, and savannas; also fencerows and roadsides.

This species is analogous to *V. bicolor*, which frequents similar sites, in being particularly abundant in somewhat disturbed sites, such as along woodland paths and frequently flooded riparian areas. It is a decided calciphile, often abundant in moist to wet silty loam soils. It is distinctive among stemmed Missouri violets in its white corolla, and, even when vegetative, can be distinguished by its lacerate stipules. Steyermark (1963) noted that it grows well in woodland gardens.

Recent genetic studies have shown that although *V. striata* has a mixed mating system (outcrossing open flowers followed by obligately selfing cleistogamous flowers), most of the mature fruits produced (in an Ohio population) during a given year originated from self-pollination (Cortés-Palomec et al., 2006). This type of reproductive system has been speculated to result in the maintenance of high genetic variation between populations and the retention of morphological variants within the populations in which they have developed.

Steyermark (1963) noted the existence of a 1936 specimen in the herbarium of William Jewell College that represents a potential Missouri record of *V. canadensis* L. (tall white violet). This taxon would key imperfectly above to either *V. pubescens* or *V. striata*. It differs from both in its petals, which are white on the upper surface, but are strongly purplish-tinged on the undersurface. Its stipules are similar to those of *V. pubescens*, and its flowers tend to have relatively long, slender stalks and relatively small corollas. Steyermark doubted that the specimen was correctly labeled as having been collected in Clinton County and noted that repeated searches for the species in the vicinity of the locality indicated on the Missouri label had not resulted in the rediscovery of the species in the state. The specimen in question, which appears to represent a student collection, is correctly determined, but probably mislabeled, and the species thus continues to be excluded from the Missouri flora. However, *V. canadensis* has been documented from portions of several states surrounding Missouri and may eventually be located growing in the state.

16. Viola tricolor L. (miniature pansy, Johnny-jump-up, heart's ease)

Pl. 575 b, c; Map 2707

Plants annual, with a slender, vertical taproot 1–3 mm thick. Stems 10–35 cm long, erect or ascending, often from a spreading base. Leaves alternate and basal, subsessile to long-petiolate, the petiole glabrous. Stipules relatively large and leaflike, free from the petiole, deeply lobed with a fringe of several, long, linear or oblong-lanceolate segments on each side, the terminal segment much longer and broader than lateral ones, with 3 or more pairs of shallow, blunt or rounded teeth along the margins. Leaf blades 0.5–2.5 cm long, unlobed, obspatulate to nearly circular, rounded or angled to a bluntly or sharply pointed tip, narrowly angled to truncate at the base, the margins otherwise with 3 or more pair(s) of blunt or rounded teeth, the surfaces glabrous. Cleistogamous flowers not produced. Flower stalks only slightly overtopping the leaves. Sepals 5–14 mm long, lanceolate, angled to a sharply pointed tip, the margins glabrous, the basal auricles well-developed. Corollas 10–25 mm long, appearing strongly frontally flattened in life, the petals longer than the sepals, yellowish orange to purple on the same flower, the upper petals commonly tinged or tipped with dark blue, the lateral and lower petals often with darker veins or dark purple lines, with a yellow throat (this usually best developed on the lower petal), the lateral petals bearded on the upper surface with mostly knob-shaped hairs, the lowermost petal glabrous on the upper surface, the spur 1–2 mm long, well-exserted beyond the sepal auricles, relatively stout. Stamens not exserted, typically not visible without dissection of the flower. Style enlarged into a globose, hollow stigmatic tip. Fruits 6–10 mm long, broadly ellipsoid, green, drying to tan, the surface glabrous. Seeds 1.5–1.7 mm long, tan. $2n$=26. March–May.

Introduced, uncommon and sporadic (native of Europe; escaped sporadically in the northeastern U.S. west to Utah). Lawns and open, disturbed areas.

Miniature pansies are one of the parents (crossed with a series of wild Old World pansies, including *V. altaica* Ker Gawl. and *V. lutea* Huds.) of the garden pansies, which are collectively referred to as *V. ×wittrockiana* Gams and were developed in the early 1800s. The species itself has a long history of cultivation and breeding in Europe, and also was important in folklore and for various medicinal uses. It apparently does not persist long outside of cultivation in North America.

17. Viola viarum Pollard (Plains violet)

Pl. 578 g, h; Map 2708

Plants perennial, to 30 cm tall, with a short, prostrate to sometimes ascending rhizome 4–6 mm thick. Aerial stems not produced. Leaves in a basal rosette, long-petiolate, the petiole glabrous. Stipules conspicuous, membranous to somewhat herbaceous, free from the petiole, narrowly lanceolate, the margins usually entire, glandular-hairy. Leaf blades 2–9 cm long, about as long to about 2 times as long as wide, variously ovate to kidney-shaped, heart-shaped, or triangular ovate, those produced early and late in the season unlobed, the main leaves shallowly to deeply lobed, often somewhat irregularly so, the segments not lobed again, at least the central one lanceolate to triangular-ovate and mostly more than 3 mm wide, the margins otherwise entire or toothed toward the segment tips, the surfaces glabrous. Cleistogamous flowers produced (frequently dark green mottled with purple). Flower stalks not or only slightly overtopping the leaves (those of the cleistogamous flowers relatively long, erect or strongly ascending). Sepals 4–8 mm long, lanceolate, angled to a sharply pointed tip, the margins glabrous, the basal auricles short and inconspicuous. Corollas 8–22 mm long (except in cleistogamous flowers), the petals oriented more or less forward in life but curved or arched outward or backward toward their tips, longer than the sepals, bluish purple (with a greenish white throat and usually darker veins), the lateral and lower petals bearded on the upper surface with slender hairs, the spur conspicuous, well-exserted beyond the sepal auricles, stout and often somewhat hemispheric in shape. Stamens not exserted, typically not visible without dissection of the flower. Style slender, slightly expanded into a narrow, more or less scoop-shaped, hollow, truncate tip. Fruits 7–13 mm long, ellipsoid to ovoid, uniformly green, the surface glabrous. Seeds 1.5–1.9 mm long, brown to olive brown. $2n$=54. April–May.

Uncommon, widely scattered in the state (Illinois to South Dakota south to Arkansas and Oklahoma). Banks of streams and rivers, bases and crevices of bluffs, and rarely bottomland prairies; also railroads and roadsides.

This unusual taxon, which was described based on plants collected along railroads in the St. Louis area, appears to be relatively uncommon throughout its range. Russell (1965) concluded that *V. viarum* is closely related to *V. missouriensis*. However, its acceptance as a species has not been universal. Some authors, including Gleason and Cronquist (1991), McKinney (1992), and Ballard (1994), have treated it merely as part of the morphological variation within *V. palmata*. Other botanists, including Gil-Ad (1997) have considered it to represent a series of putative hybrids between *V. pedatifida* and some other species of stemless blue violets with unlobed leaves, such as *V. missouriensis* or *V. nephrophylla*. Although the species sometimes has been described as homophyllous, plants of all sizes with early-season unlobed leaves have been observed in otherwise typical populations at a considerable distance from other violet species. The heterophyllous putative hybrids described from Missouri by Gil-Ad (1997) represent part of the developmentally variable morphology referable to this species. Additional studies of this and related lobed-leaved violets are warranted.

Viola viarum f. *piliferum* E.J. Palmer & Steyerm. was based on rare plants from Gentry County that are morphologically similar to typical *V. viarum*, but differ in having the petioles sparsely to moderately pubescent with short, spreading hairs. Whether this name should be placed in synonymy under *V. viarum* or *V. palmata*, or represents putative hybrids between *V. viarum* and some hairier violet species, requires further study.

VISCACEAE (Mistletoe Family)

Seven genera, about 400 species, nearly worldwide, most diverse in tropical and subtropical regions.

Although all mistletoes were traditionally classified in a single family, Loranthaceae, most botanists (Cronquist, 1981, 1991) currently accept two families, the Loranthaceae (with showy mostly perfect flowers) and Viscaceae (with minute, unisexual flowers), the latter occurring in temperate North America. Note, however, that the Angiosperm Phylogeny Group (2009) places the Viscaceae into a broadly circumscribed version of Santalaceae, based on a conservative interpretation of molecular phylogenetic data (Der and Nickrent, 2008).

The European mistletoe (*Viscum album* L.) has a long tradition of medicinal applications, as well as great significance in folklore and religion. Decorative use of this species at Christmas time dates back to the Middle Ages or even earlier. In North America, superficially similar native species of *Phoradendron* are usually substituted for *Viscum*, and the practice of harvesting fruiting plants, from the wild for commercial sale has caused the extirpation of some populations, particularly of *P. leucarpum*.

1. **Phoradendron** Nutt. (American mistletoe) (Wiens, 1964; Kuijt, 2003)

About 235 species, U.S., Mexico, Central America, South America, Caribbean Islands.

1. **Phoradendron leucarpum** (Raf.) Reveal & M.C. Johnst. **ssp. leucarpum** (eastern mistletoe, American Christmas mistletoe)
   P. flavescens Engelm.
   P. serotinum (Raf.) M.C. Johnst.

Pl. 579 g–i; Map 2709

Plants dioecious, parasitic on the above-ground portions of trees, the external growth forming somewhat woody clumps, spreading internally in the host branches by rootlike filaments. Stems

2710. Ampelopsis arborea

2711. Ampelopsis brevipedunculata

2712. Ampelopsis cordata

15–40 cm long, glabrous or minutely stellate-hairy, stout, yellowish green to olive green. Leaves opposite, simple mostly short-petiolate, thick and leathery, the blades 1.2–4.0(–6.0) cm long, oblong to oblanceolate or obovate, rounded at the tip, tapered at the base, the margins entire, the surfaces usually minutely stellate-hairy, especially toward the base. Stipules absent. Inflorescences short axillary spikes, these mostly paired on opposite sides of the nodes, with the flowers somewhat sunken in whorls at the nodes of the axis. Sepals (2)3(4), 5–7 mm long, incurved to erect, scalelike, broadly ovate-triangular, persistent at fruiting. Petals absent. Staminate flowers with 3(4) stamens, the anthers sessile and attached toward the sepal bases. Pistillate flowers epigynous, the pistil 1, the ovary inferior, with 1 locule and 1(2) functional ovule, the style absent, the stigma more or less globose but poorly differentiated. Fruits berries 4–6 mm long, depressed-globose, white, glabrous, dull. Seed 1(2), 2.5–3.0 mm long, flattened, elliptic in outline, narrowed to an abrupt point at the tip, white, embedded in a sticky pulp. $2n=28$. October–November.

Uncommon in the southern portion of the state (southeastern U.S. west to Kansas and Texas). Bottomland forests and banks of streams and rivers. Parasitic on a number of deciduous trees, in Missouri mostly on species of *Betula, Nyssa, Platanus*, and *Ulmus*.

Plants of eastern mistletoe, like other species of *Phoradendron*, are moderately poisonous, containing phoratoxin and related amines that cause severe irritation to the stomach and intestinal walls in both humans and livestock. The species parasitizes a large variety of dicot tree hosts throughout its range, including (in addition to those listed above) species of *Acer, Broussonetia, Carya, Celtis, Diospyros, Fraxinus, Gleditsia, Juglans, Maclura, Morus, Pyrus, Quercus, Robinia*, and several others not found in Missouri, but appears to favor particular hosts in different portions of its range (Baldwin and Speese, 1957; Overlease and Overlease, 2011). Seeds are dispersed by birds, and the principal deterrents to potential establishment of the parasite appear to be the degree of bark peeling and the thickness of the host bark. The parasite causes abnormal, swollen and distorted growth of host tissues that become infected, and new aerial stems may rupture through the host bark at some distance from the initial infection site. In addition to gradually killing the branches of the host, the mistletoe also weakens the host tree to invasion by wood-boring insects, fungi, and other disease-causing organisms.

There has been confusion in the literature as to the flowering period of *P. leucarpum*. For example, Steyermark (1963) indicated that flowering occurred in Missouri from March to August, but Gleason and Cronquist (1991) stated October to December as the months of flowering for plants throughout the species range. Part of this problem arises from difficulty in ascertaining, particularly for pistillate plants, exactly when the plants are in flower. Allard (1943) studied phenology of the species and concluded that flowering in the eastern states occurs from late October to December. Baldwin and Speese's (1957) study of meiosis in developing flowers supported the autumnal flowering of the species. Apparently, staminate inflorescences develop during the summer months, flower in autumn, and wither by the following spring. Pistillate inflorescences also develop during the summer and flower in autumn, but these flowers persist relatively unchanged through the following year, with fruits not maturing until about a year after flowering, when the next set of inflorescences to be produced is also mature.

The complicated nomenclatural history of this species was discussed by Reveal and Johnston (1989). Some authors (Kuijt, 2003) have preferred

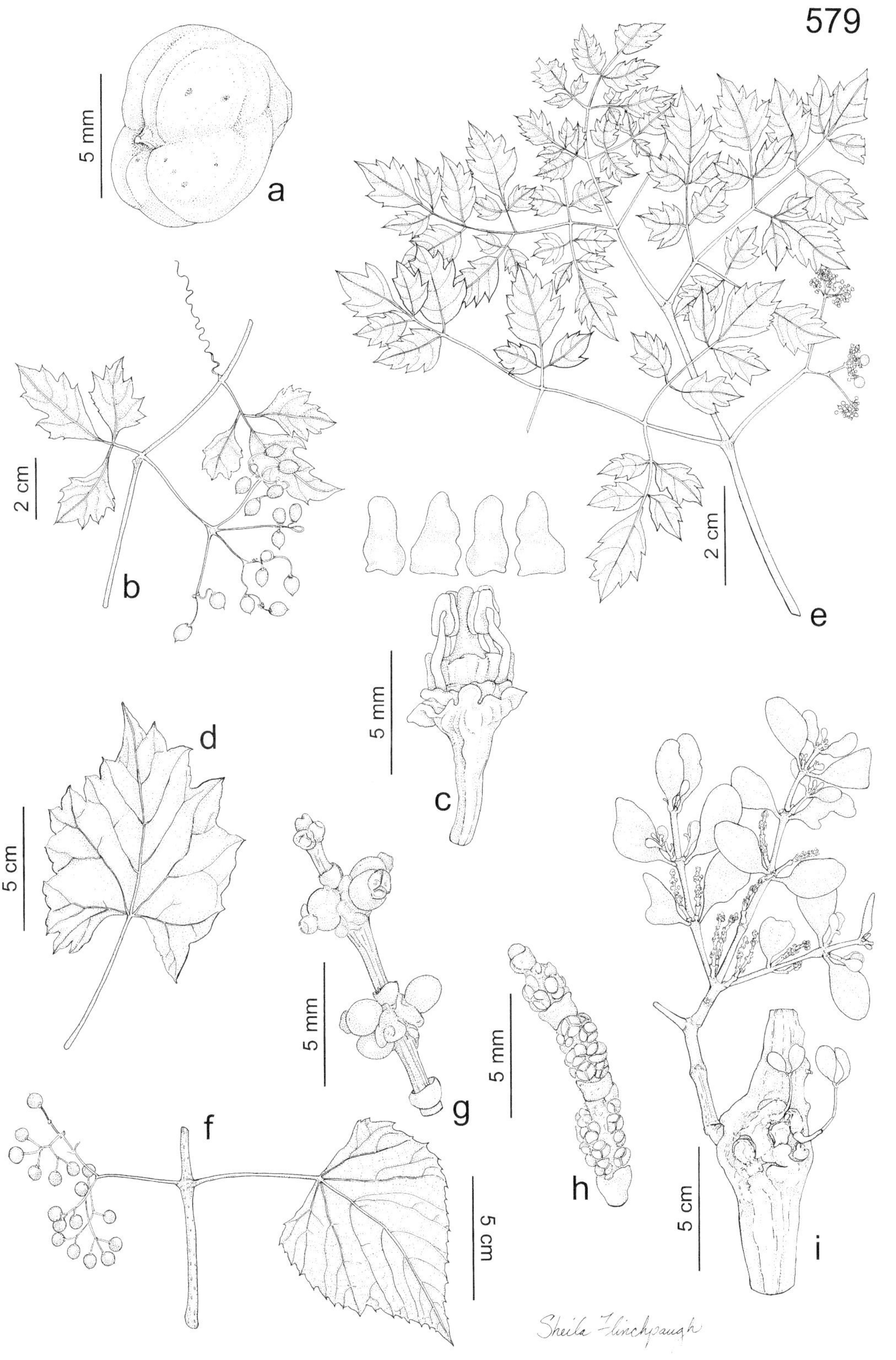

Plate 579. Viscaceae. Vitaceae. *Cissus trifoliata*, **a)** fruit, **b)** portion of fruiting stem, **c)** flower (4 free petals artificially detached and drawn above remainder of flower), **d)** simple leaf. *Ampelopsis arborea*, **e)** portion of fertile branch. *Ampelopsis cordata*, **f)** node with leaf and inflorescence. *Phoradendron leucarpum*, **g)** portion of pistillate inflorescence, **h)** portion of staminate inflorescence, **i)** habit.

to call the species *P. serotinum* based on the belief that *P. leucarpum* is a later homonym of a different species, *P. leucocarpum* Pacz., but the Committee for Spermatophyta of the International Botanical Congress (Brummitt, 1988) ruled that this is not the case, and Reveal and Johnston (1989) also presented a strong argument to the contrary. Recently, the epithet *P. serotinum* was proposed for nomenclatural conservation over *P. leucarpum* (Nickrent et al., 2010). However, this proposal was rejected by the International Botanical Congress in July, 2011, and the name *P. leucarpum* therefore must be used for the taxon as currently circumscribed.

In his monograph of the genus, Kuijt (2003) treated this species in a broad sense under the name *P. serotinum*, making new subspecific combinations within the species for three additional taxa growing in the southwestern United States and Mexico that previous authors had accepted as separate species. These other subspecies all have thicker, more brittle leaves that tend to be elliptic to ovate (vs. mostly oblanceolate to obovate). All plants in the eastern United States were included in the nominate subspecies by Kuijt. Recently, Abbott and Thompson (2011) published a new set of combinations for these three taxa under *P. leucarpum*.

VITACEAE (Grape Family)

Plants vines, usually lianas, with tendrils (except occasionally in *Vitis rupestris*) positioned opposite leaves and sometimes also in the inflorescence, these often branched (except in *Cissus* and *Vitis rotundifolia*), lacking thorns or spines, sometimes incompletely or nearly completely monoecious or dioecious. Stems often swollen at the nodes. Leaves alternate, simple or compound, short to more commonly long-petiolate. Stipules small, scalelike, shed before the leaves mature. Inflorescences compound umbels or panicles of small flowers. Flowers perfect or functionally staminate or pistillate, hypogynous, actinomorphic. Calyces fused into a low spreading collar, this sometimes shallowly 4- or 5-lobed. Corollas of mostly 4 or 5 petals, these free or (in *Vitis*) fused at the tips. Stamens mostly 4 or 5 (produced but somewhat reduced and nonfunctional in pistillate flowers), opposite the petals, the anthers minute, attached near their midpoint. Nectar disc present, unlobed or the lobes alternating with the stamens (reduced and fused to the ovary in *Parthenocissus*). Pistil 1 per flower (reduced and nonfunctional in staminate flowers), of usually 2 fused carpels. Ovary superior (but sometimes appearing somewhat sunken into the nectar disc), 2-locular, each locule with 2 ovules, the placentation more or less basal. Style 1, the stigma minute, unlobed or slightly 2-lobed. Fruits 1–4-seeded berries, globose to obovoid. Eleven to 14 genera, 700–850 species, nearly worldwide, most diverse in tropical regions.

Seeds of the subfamily Vitoideae, which includes all of the Missouri genera, are distinctive. They vary in shape depending upon the number in the particular berry, but all have a broadly convex dorsal surface and a longitudinally angled inner surface. The angle of the inner surface has a pair of slender longtiudinal grooves along its spine that wrap around to the dorsal surface and encircle in a well-marked circular to teardrop-shaped plate known as a "chalazal knot." This distinctive feature allows ready familial determination of seeds, even from archaeological sites or the fossil record.

1. Leaf blades simple, although sometimes 3- or 5-lobed (occasionally a few of the largest leaves completely trifoliate in *Parthenocissus*)
    2. Petals fused at the tip, shed as a caplike unit as the flower opens; inflorescences longer than wide, more or less pinnately branched panicles, the ultimate branches sometimes umbellate; pith (most easily seen on branches less than 1 cm in diameter) brown . . . . 4. VITIS

2. Petals free, persistent and spreading at flowering; inflorescences flat-topped or dome-shaped, wider than long or slightly longer than wide, repeatedly dichotomously forked or with 3–5 umbellate branches, the flowers in small umbellate clusters at the tips of the ultimate branches; pith white
   3. Leaf blades herbaceous to somewhat papery, not thickened; nectar disc noticeable under magnification, cup-shaped, the basal portion fused to the ovary, the rim free . . . . 1. AMPELOPSIS
   3. Leaf blades somewhat leathery, thickened; nectar disc indistinct, entirely fused to the ovary . . . . 3. PARTHENOCISSUS

1. Leaf blades compound (rarely a few simple leaves present in *Cissus*)
   4. Leaf blades with 3 leaflets; stems herbaceous or woody only toward the base; petals and stamens 4 per flower . . . . 2. CISSUS
   4. Leaf blades all or mostly with 5 or more leaflets
      5. Leaf blades twice pinnately or pinnately then ternately compound, with 9–35 leaflets . . . . 1. AMPELOPSIS
      5. Leaf blades once palmately (rarely ternately) compound, with (3)5(7) leaflets . . . . 3. PARTHENOCISSUS

1. **Ampelopsis** Michx.

Plants lianas, often more or less completely monoecious. Young stems often slightly angled or ridged, green or reddish-tinged, glabrous or hairy. Older stems to 20 m or more long, gray to dark brown, often appearing somewhat warty with oval lenticels, eventually developing deeply fissured nonshredding bark, the pith white, not or only rarely chambered. Tendrils at scattered nodes, sometimes few, occasionally also in the inflorescence, branched, the tips slender. Leaf blades simple or twice pinnately or pinnately then ternately compound with 9–35 leaflets. Inflorescences usually opposite the leaves, panicles, flat-topped or dome-shaped, wider than long or slightly longer than wide, repeatedly dichotomously forked or with 3–5 umbellate branches, the flowers in small umbellate clusters at the tips of the ultimate branches. Petals 5, free, 2.0–2.8 mm long, persistent and spreading at flowering, greenish yellow. Stamens 5. Nectar disc noticeable under magnification, about half as long as the ovary, cup-shaped, the basal portion fused to the ovary, the rim free, entire or irregularly scalloped. Style often very short, sometimes persistent at fruiting. Fruits globose, often with small warty dots at maturity. Seeds 1–3 per fruit, 3.5–4.5 mm long, asymmetrically obovoid to broadly obovoid, somewhat longitudinally angled along the inner side, reddish brown to yellowish brown. About 25 species, most diverse in Asia, but also in North America, Central America.

The berries of *Ampelopsis* species have thin flesh and are not palatable to humans, but are a food source for birds and small mammals. The stems are sometimes used in basketry and other handcrafts.

1. Leaf blades twice pinnately or pinnately then ternately compound, with 9–35 leaflets . . . . 1. A. ARBOREA
1. Leaf blades unlobed or palmately 3- or 5-lobed
   2. Leaf blades unlobed or less commonly shallowly 3-lobed; twigs glabrous . . . . 3. A. CORDATA
   2. At least some of the leaves with the blades moderately to deeply 3- or 5-lobed; twigs hairy when young (becoming glabrous with age) . . . . 2. A. BREVIPEDUNCULATA

1. Ampelopsis arborea (L.) Koehne (peppervine)

Pl. 579 e; Map 2710

Twigs glabrous or sparsely and inconspicuously hairy when young, becoming glabrous with age. Leaf blades 10–25 cm long, twice pinnately or pinnately then ternately compound, broadly triangular in outline, with 9–35 leaflets. Leaflets 2–6 cm long, mostly ovate, sessile or more commonly short-stalked, mostly truncate at the base, narrowed or tapered to a sharply pointed tip, the margins with a few broad coarse teeth, dark green, the upper surface shiny, glabrous or very sparsely hairy, the undersurface glabrous or sparsely hairy along the veins. Inflorescences much shorter than the leaves. Fruits 7–10 mm long, shiny at maturity, not glaucous, changing from green to pink or bluish gray and eventually to dark purple or black. $2n=40$. June–August.

Scattered in southern and eastern Missouri; introduced in Boone and Jackson Counties (southeastern U.S. west to Illinois, Oklahoma, and New Mexico). Bottomland forests, swamps, and banks of streams and rivers; also wooded roadsides.

2. Ampelopsis brevipedunculata (Maxim.) Trautv. (porcelain berry, amur peppervine, turquoise berry)

A. glandulosa (Wall.) Momiy. var. *brevipedunculata* (Maxim.) Momiy.

Map 2711

Twigs moderately pubescent with upward-curled hairs when young, becoming glabrous with age. Leaf blades 3–12 cm long, simple, the uppermost entire to more commonly shallowly 3-lobed, the larger leaves moderately to deeply palmately 3- or 5-lobed, broadly ovate to ovate-triangular in outline, mostly shallowly or broadly cordate at the base, tapered to a sharply pointed tip, the margins finely to coarsely and bluntly to sharply toothed sometimes nearly entire on deeply lobed leaves), often also minutely hairy, green to olive green, the upper surface shiny, glabrous or minutely hairy along the veins, the undersurface with relatively stout, curved hairs along the veins. Inflorescences shorter than to slightly longer than the leaves. Fruits 7–10 mm long, shiny, not or only slightly glaucous at maturity, changing from green or white to pale yellow to pink, to purple, and eventually turquoise blue, sometimes speckled or mottled at maturity. $2n=20$. July–September.

Introduced, uncommon, known thus far only from the city of St. Louis (native of Asia, introduced in the eastern [mostly northeastern] U.S. west to Iowa and Missouri). Railroads and fencerows.

This species was once grown widely as an ornamental, but is no longer commonly sold because of its aggressive tendencies in gardens. It is considered an invasive exotic in several northeastern states. The first specimen documenting its escape in Missouri was collected by the author in 2003.

Ampelopsis brevipedunculata is superficially similar to some *Vitis* species, but differs in several characters, including the stout, curved hairs on the leaf undersurfaces, the flat-topped to dome-shaped inflorescences, details of the perianth and nectar disc, fruit color, and a later flowering time.

3. Ampelopsis cordata Michx. (raccoon grape, false grape, heartleaf ampelopsis)

Pl. 579 f; Map 2712

Twigs glabrous. Leaf blades 5–12 cm long, simple, unlobed or less commonly shallowly 3-lobed, broadly ovate to ovate-triangular in outline, truncate or less commonly shallowly cordate at the base, tapered to a sharply pointed tip, the margins coarsely and sharply toothed, green to olive green, the upper surface not or only slightly shiny, glabrous or very sparsely hairy, the undersurface glabrous or sparsely hairy along the veins. Inflorescences shorter than to slightly longer than the leaves. Fruits 7–10 mm long, not shiny or glaucous at maturity, changing from green to pink or orange, to purple, and eventually turquoise blue. $2n=40$. May–July.

Scattered in the southern $^{2}/_{3}$ of the state and locally northward along the Missouri and Mississippi Rivers (southeastern U.S. west to Nebraska and Texas and adjacent Mexico; introduced sporadically in New England).

Vegetatively, *A. cordata* might be confused with some *Vitis* species, but differs in its white pith, generally fewer tendrils, and tight bark.

2. Cissus L.
(Lombardi, 2000)

About 200 species, widespread in tropical and warm-temperate regions of both the Old World and New World.

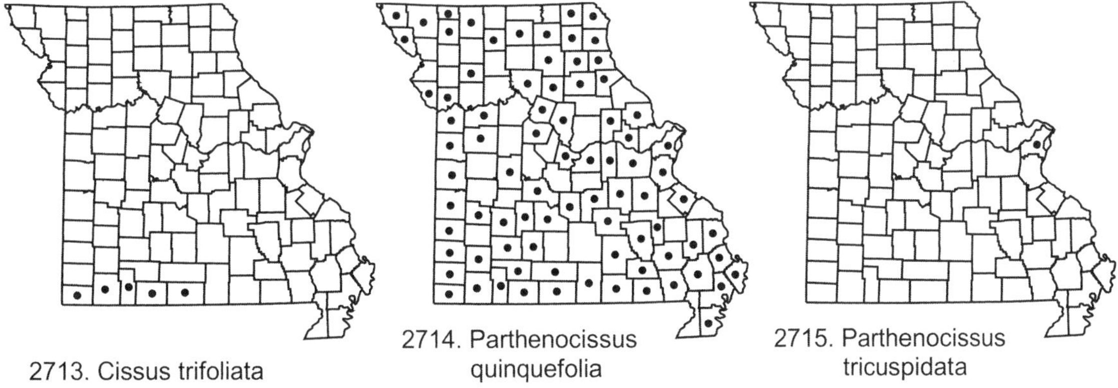
2713. Cissus trifoliata
2714. Parthenocissus quinquefolia
2715. Parthenocissus tricuspidata

Some exotic species of *Cissus* are cultivated as house plants and in greenhouses, and have large tuberous rootstocks and/or succulent sometimes winged stems. The rootstocks of various species reputedly are poisonous.

1. Cissus trifoliata (L.) L. (marine vine, marine ivy)

C. incisa (Nutt. ex Torr. & A. Gray) Des Moul.

Pl. 579 a–d; Map 2713

Plants lianas (in Missouri, woody only toward the base), but sometimes dying back to the ground during cold winters and thus appearing herbaceous, with tuberous roots, often mostly with perfect flowers. Young stems somewhat angled and often somewhat succulent, pale green to green or reddish-tinged, glabrous. Older stems 2–6(–10 m) long, gray to less commonly brown, warty with orange to reddish orange lenticels, the pith white, not chambered. Tendrils at scattered nodes, sometimes few, occasionally also in the inflorescence, mostly unbranched, the tip slender. Leaves fleshy, the blades 1–7 cm long, ternately compound (rarely a few simple leaves present), ovate to triangular in general outline. Leaflets 1.0–6.5 cm long, ovate to obovate, spatulate, or narrowly fan-shaped, narrowed or tapered at the base (sometimes cordate in simple leaves), narrowed to sharply pointed tips, the margins coarsely and irregularly toothed, sometimes also with 1 or 2 lobes toward the base, glabrous or both surfaces very sparsely hairy. Inflorescences opposite the leaves or appearing terminal on branches, appearing as compound umbels, mostly longer than the leaves, flat-topped or somewhat dome-shaped, wider than long or slightly longer than wide, usually with 3–5 umbellate branches, the flowers in small umbellate clusters at the branch tips. Petals 4, free, 1.5–2.5 mm long, persistent and spreading at flowering, greenish yellow. Stamens 4. Nectar disc noticeable under magnification, cup shaped, fused to the ovary most of its length, the rim free, more or less entire. Style short, persistent at fruiting. Fruits 5–8 mm long, globose or obovoid, becoming shiny and black at maturity, sometimes with sparse small warty dots. Seeds usually 1 per fruit, 4.5–5.0 mm long, asymmetrically broadly obovoid, somewhat longitudinally angled along the inner side, brown. June–July.

Uncommon in southwestern Missouri (Florida to Arizona north to Missouri and Oklahoma; Mexico, Central America, South America, Caribbean Islands). Tops and ledges of dolomite bluffs.

Plants from the northern portion of the overall range of *C. trifoliata* were once separated as *C. incisa* by some authors (Steyermark, 1963; Brizicky, 1965), based on the presence of some simple leaves, more irregularly toothed leaflet margins, and slightly larger fruits, but most authors now consider them to represent merely a part of this variable species. In Missouri, the leaves of marine vine usually are deciduous and the stems tend to die back during cold winters, but farther south the leaves tend to be evergreen and the stems become much longer and woodier. Although it has been found in very different locations and habitats in the state, *C. trifoliata* possibly might be confused with the introduced *Parthenocissus tricuspidata*. The latter species differs in its branched tendrils with preformed adhesive discs at the tips, abundant aerial roots where stems climb up or over substrates, usually simple, 3-lobed leaves, and flowers with 5 petals and stamens.

3. **Parthenocissus** Planch. (woodbine)

Plants lianas, sometimes only scrambling on the ground, with all perfect or occasionally a few functionally staminate flowers. Young stems often with a shallow longitudinal groove, grayish brown to reddish brown, glabrous or minutely hairy. Older stems gray to dark brown, sometimes appearing somewhat warty with small lenticels, eventually developing dark brown, deeply fissured, nonshredding (but sometimes breaking into irregular plates) bark, the pith white, not or only rarely chambered. Tendrils at scattered nodes, absent from the inflorescence, few- to many-branched, the tips slender or with small circular adhesive discs. Leaf blades once palmately (rarely ternately) compound, with (3)5(7) leaflets (usually simple in *P. tricuspidata*). Inflorescences opposite the leaves (sometimes appearing terminal on short branches), panicles, flat-topped to dome-shaped or somewhat pyramid-shaped, wider than long to longer than wide. Petals 5, free, 2–3 mm long, persistent and spreading at flowering, yellowish green. Stamens 5. Nectar disc reduced, indistinct, entirely fused to the ovary. Style short, not persistent at fruiting. Fruits globose, not warty, dark purple or dark blue to nearly black, sometimes slightly glaucous. Seeds 1–4 per fruit, obovoid to broadly obovoid, somewhat longitudinally angled along the inner side, light brown to dark brown. About 15 species, most diverse in Asia, but also in North America.

The berries of *Parthenocissus* species have thin flesh and are not palatable to humans (reputedly, at least those of *P. quinquefolia* are poisonous), but are a food source for birds and small mammals. Turkey and deer sometimes eat the young shoots and leaves,. The flowers are visited by a variety of bees, flies, wasps, and beetles (Brizicky, 1965). The bark, which is sometimes eaten by animals in the winter, has been used medicinally in an infusion as a tonic and expectorant. The stems are sometimes used in basketry and other handcrafts.

1. Leaves simple and deeply 3-lobed or rarely ternately compound, thickened and somewhat leathery . . . . 2. P. TRICUSPIDATA
1. Leaves palmately compound with 5(7) leaflets (except in seedlings), herbaceous or somewhat papery, not thickened
    2. Tendrils with (3–)5–8(–12) branches, most of the tips with small circular adhesive discs; inflorescences usually with a well-defined central axis . . . . 1. P. QUINQUEFOLIA
    2. Tendrils with 3–5 branches, none of the tips with adhesive discs; inflorescences lacking a well-defined central axis . . . . 3. P. VITACEA

1. Parthenocissus quinquefolia (L.) Planch. (Virginia creeper, woodbine, five-leaved ivy)

P. quinquefolia f. *hirsuta* (Pursh) Fernald
P. inserta (A. Kern.) Fritsch

Pl. 580 f–h; Map 2714

Stems to 30 m or more long, the older stems producing coarse aerial roots for anchors when climbing. Young stems glabrous or minutely hairy. Tendrils with (3–)5–8(–12) branches, most of the tips with small circular adhesive discs. Leaf blades once palmately compound, with (3)5(7) leaflets. Leaflets 5–15 cm long, elliptic to obovate, tapered or narrowed at the base, tapered to a sharply pointed tip, the margins sharply toothed mostly above the midpoint, herbaceous or somewhat papery, not thickened, dull, glabrous, and dark green on the upper surface, paler, glabrous or hairy, and sometimes slightly glaucous on the undersurface. Inflorescences mostly longer than wide, usually with a zigzag central axis and the main branches repeatedly dichotomously forked, the 25–200 flowers in small umbellate clusters at the branch tips. Fruits 5–8 mm in diameter. Seeds 3.5–4.0 mm long, the surface smooth to finely wrinkled, shiny, brown to dark brown. $2n$=40. May–August.

Common throughout the state (eastern U.S. west to South Dakota, Utah, and Texas). Bottomland forests, mesic upland forests, bases, ledges, and tops of bluffs, and banks of streams and rivers; also fencerows, pastures, roadsides, railroads, and shaded disturbed areas.

The application of species epithets in North American *Parthenocissus* was confused in much

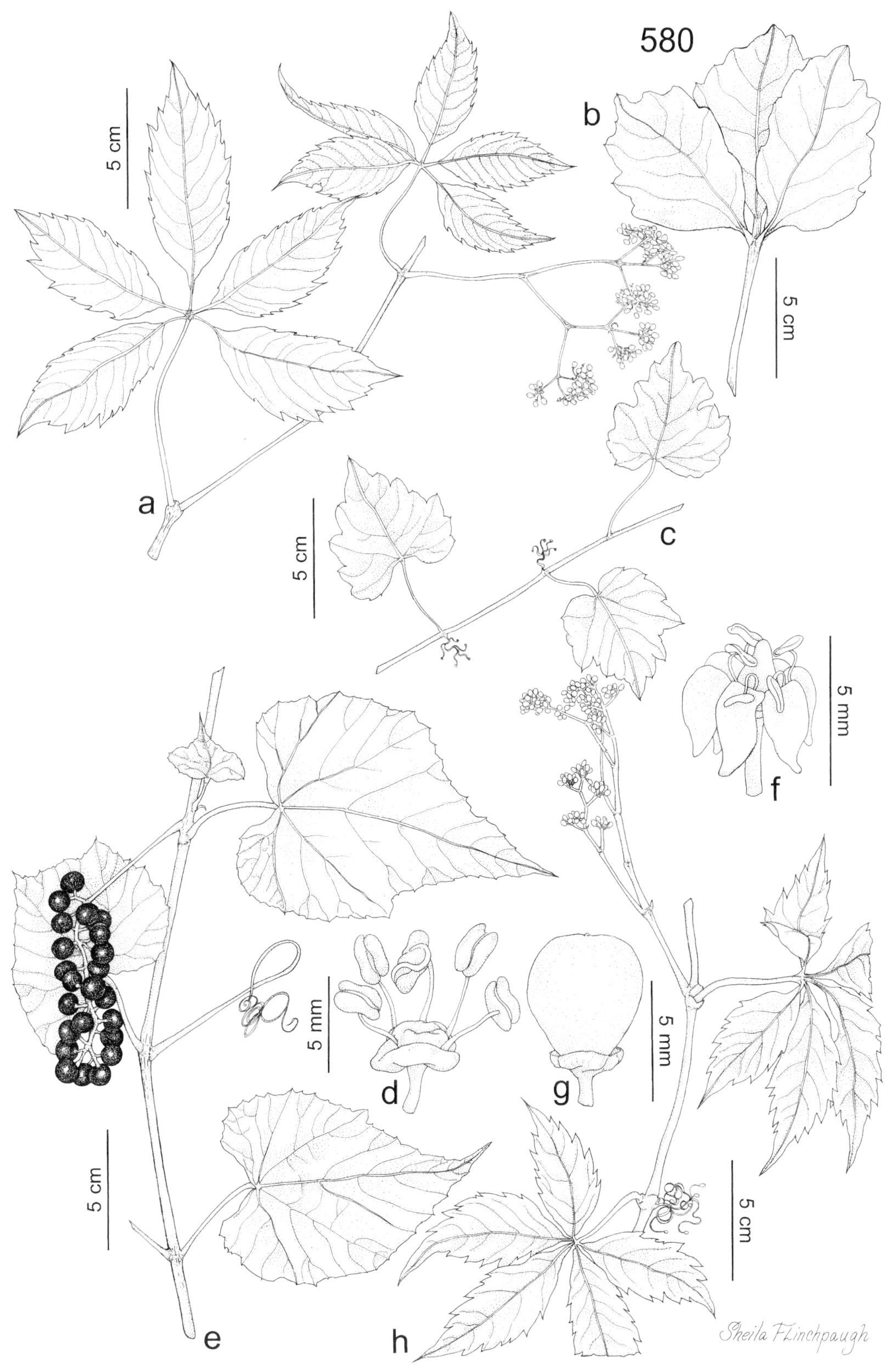

Plate 580. Vitaceae. *Parthenocissus vitacea*, **a)** fertile branch. *Parthenocissus tricuspidata*, **b)** more-divided leaf, **c)** fruiting branch with less-divided leaves. *Vitis aestivalis*, **d)** staminate flower, **e)** fruiting branch. *Parthenocissus quinquefolia*, **f)** flower, **g)** fruit, **h)** fertile branch.

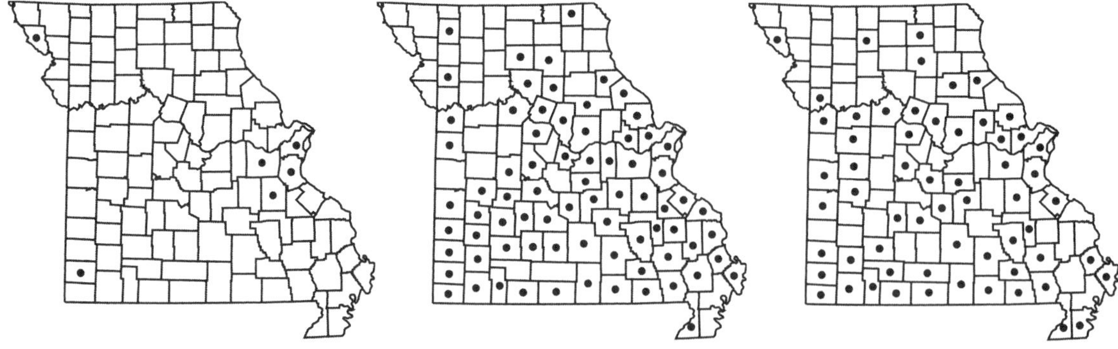
2716. Parthenocissus vitacea 2717. Vitis aestivalis 2718. Vitis cinerea

of the older botanical literature because of disagreement over whether the name *P. quinquefolia* should be applied to the more northern, glossy-leaved species lacking tendril discs or the more southern taxon with dull leaves and well-developed tendril discs. Brizicky (1965) and D. A. Webb (1967) independently established that the name corresponded to the latter taxon, and this interpretation has been followed by most subsequent authors (Gleason and Cronquist, 1991) and in the present treatment.

Flowering in *P. quinquefolia* is often restricted to portions of the plant exposed to fairly direct sunlight, such as the forest canopy, and fertile specimens are often located most easily along open forest margins or when plants are climbing an isolated substrate, like a telephone pole. The foliage of Virginia creeper turns various shades of bright red and purple in the autumn, and the species has a long history of horticultural use, but it can become too large and heavy for cultivation in some situations. Seedlings or occasional sprouts of older plants having leaves with only three leaflets are sometimes confused with poison ivy (*Toxicodendron radicans*; Anacardiaceae), with which Virginia creeper often grows. The presence of tendrils and the middle leaflet with the stalk lacking or about as long as those of the other leaflets in Virginia creeper are characters to help distinguish between the two species.

2. Parthenocissus tricuspidata (Siebold & Zucc.) Planch. (Boston ivy)

Pl. 580 b, c; Map 2715

Stems to 10 m or more long, the older stems often producing aerial roots. Young stems hairy. Tendrils short, with 3–7 branches, most of the tips with small circular adhesive discs. Leaf blades mostly simple and shallowly to deeply 3-lobed, usually toward the tip (rarely unlobed), occasionally a few of the largest leaves completely trifoliate, 3–15(–20) cm long, broadly ovate to obovate or obhemispheric in outline, broadly rounded or more commonly cordate at the base, the lobes tapered to sharply pointed tips, the margins sharply and finely to coarsely toothed, thickened and somewhat leathery, the upper surface shiny, glabrous, and dark green, the undersurface paler and often sparsely hairy along the main veins. Inflorescences wider than long to longer than wide, lacking a well-defined central axis, with 2(3) main branches at the tip of the stalk, each of these 1–3 times dichotomously (less commonly trichotomously) forked, the 10–90 flowers single or in small clusters at the branch tips. Fruits 6–8 mm in diameter. Seeds 3.5–4.0 mm long, the surface somewhat roughened or wrinkled, light brown to brown. $2n=40$. May–July.

Introduced, known thus far only from the city of St. Louis (native of eastern Asia, introduced sporadically in the northeastern U.S. and adjacent Canada). Railroads and disturbed open areas.

Boston ivy has long been valued horticulturally as a cover for arbors, walls, and rocky areas. It was first reported as an escape in Missouri by Mühlenbach (1979).

3. Parthenocissus vitacea (Knerr) Hitchc. (woodbine, thicket creeper)

P. inserta (A. Kern.) Fritsch, misapplied

P. inserta f. *dubia* Rehder

Pl. 580 a; Map 2716

Stems to about 10 m long, the older stems often not producing aerial roots. Young stems glabrous. Tendrils with 3–5 branches, the tips slender, very rarely a few with circular adhesive discs. Leaf blades once palmately compound, with (3)5(7) leaflets. Leaflets 3–12 cm long, elliptic to obovate, tapered or narrowed at the base, tapered to a sharply pointed tip, the margins sharply and usually coarsely toothed mostly above the midpoint, herbaceous or somewhat papery, not thickened, shiny, glabrous, and dark green on the upper surface, paler and glabrous or hairy on the undersurface. Inflorescences mostly wider than long, lacking a well-defined central axis, with 2(3) main branches at the tip of the stalk, each of these 1–3 times dichotomously (less commonly trichotomously) forked, the 10–60 flowers single or in small clusters at the branch tips. Fruits 10–12 mm in diameter. Seeds 4.5–5.0 mm long, the surface somewhat roughened or wrinkled, light brown to brown. $2n=40$. May–July.

Uncommon and widely scattered in the state (western and northern U.S. east and south to Texas, Illinois, and New Jersey; Canada). Banks of streams and rivers, bases and ledges of bluffs, less commonly bottomland forests and mesic upland forests; also fencerows.

The foliage of woodbine turns yellow or more commonly various shades of red in the autumn, making it an attractive ornamental for arbors or rocky areas. In nature, plants of *P. vitacea* are most often found scrambling across gravel bars and bluff ledges or through low thickets, whereas *P. quinquefolia*, which can become a much more robust liana, often climbs up trees and telephone poles. Steyermark (1963) and many other earlier authors used the name *P. inserta* for this species, but Gleason (1947) determined that this name is merely a synonym of *P. quinquefolia*. See the treatment of that species for further discussion of nomenclatural confusion in North American *Parthenocissus*.

4. **Vitis** L. (grape)
(Moore, 1991)

Plants lianas, sometimes only scrambling on the ground, usually incompletely or functionally dioecious. Twigs and young stems circular in cross-section or somewhat angled, green or reddish-tinged, glabrous to densely hairy. Older stems to 20 m or more long, gray to dark brown, often appearing somewhat warty with oval lenticels, eventually developing shredding bark (nonshredding in *V. rotundifolia*), the pith brown, usually broken into chambers at maturity (at least on older branches), interrupted by a diaphragm of lighter unchambered tissue at the nodes (except in *V. rotundifolia*). Tendrils at pairs of nodes (at several adjacent nodes in *V. labruscana* L.H. Bailey), sometimes few, occasionally also in the inflorescence, 2- or 3-branched (unbranched in *V. rotundifolia*) toward the tip, the branch tips slender. Leaf blades simple, often shallowly to deeply palmately 3- or more-lobed, cordate to broadly cordate at the base, the margins with relatively broad coarse teeth abruptly tapered or narrowed to sharply pointed tips. Inflorescences usually opposite the leaves, more or less pinnately-branched panicles, the ultimate branches sometimes in small umbellate clusters at the tips of the ultimate branches. Petals mostly 5, fused at the tip, 1–3 mm long, shed as a caplike unit as the flower opens, greenish yellow. Stamens mostly 5. Nectar disc inconspicuous, short, deeply 5-lobed or divided into 5 separate glands alternating with the stamens. Style often very short, usually not persistent at fruiting. Fruits globose or less commmonly ellipsoid, usually lacking warty dots at maturity. Seeds 1–4 per fruit, asymmetrically obovoid with a small nipplelike base, thus appearing somewhat pear-shaped, somewhat longitudinally angled along the inner side. About 65 species, North America, Central America, South America, Caribbean Islands, Europe, Asia.

Some specimens of *Vitis* bear a superficial resemblance to *Ampelopsis cordata*, but the brown, usually chambered (vs. white, unchambered) pith serves to distinguish them easily. The flowers of grape species are too similar to provide diagnostic characters for species determination. However, collectors should note the abundance, positions, and degree of branching of any tendrils and inflorescences present, as well as the appearance of the bark of larger stems, as these figure prominently in the key below. Also, a branch should be shaved or sliced longitudinally through one or more nodes to reveal the size and configuration of the pith.

Morphological and molecular studies have provided strong evidence that *Vitis* comprises two well-differentiated subgenera (summarized in Aradhya et al., 2013). Breeders have also noted the existence of genetic barriers that restrict (but do not totally prevent) hybridization between members of subg. *Vitis* and subg. *Muscadinia* (Planch.) Rehder. Some botanists have interpreted these data as justification for segregation of *V. rotundifolia* into the segregate genus *Muscadinia* (Planch.) Small (Weakley et al., 2011). However, whether interpreted as two subgenera of *Vitis* or two closely related genera, there is consensus among botanists that the grapes form a natural group. Treatment as separate genera seems counterproductive to any discussion of the economic botany of grapes.

Fruits of *Vitis* provide an important food source for birds, mammals, and even box turtles. Deer and other mammals also browse the young foliage. The stems are used in the construction of baskets and other handcrafts. Grapes also are an important crop of worldwide importance dating back to the dawn of agriculture. The berries are eaten fresh as table grapes, dried as raisins, and processed into jellies, jams, and preserves. An extract from the skin of the fruit is used as a colorant for other foods and drinks. The berries also produce a juice that is used as a beverage either fresh or fermented as wine (or further-processed into beverages like champagne, sherry, and cognac), and which is also used as a flavoring and base for other blended fruit drinks. In addition to its beverage use, wine has been associated with religious ceremonies since antiquity. According to the online FAOSTAT databases of the Food and Agriculture Organization of the United Nations (FAO), in the year 2000 the United States produced about 6.8 million metric tons of grapes and processed about 2.5 million metric tons of wine. The principal species involved in table grapes are the eastern North American *V. labrusca* L. (fox grape), from which Concord, Catawba, and other dark and rosy grape cultivars were developed, and *V. vinifera* L. (a native of the Middle East), from which popular lighter selections such as Thompson seedless were developed. Most wines (except Concord wines) are made by fermenting the fruits of various cultivars and hybrids involving *V. vinifera*.

In Missouri, the grape-growing industry began with the influx of European colonists into the Missouri River Valley and portions of the Ozarks during the early nineteenth century. The first commercial vineyards were established by German settlers in the 1830s around the town of Hermann (Montgomery County). By the 1870s, when growers in the state were producing nearly two million gallons annually, the state's wine industry ranked second only to that of California in the United States. The wine industry collapsed in the 1920s with the onset of the Prohibition Era, and did not begin to rebound until the 1960s, when a small number of enterprising families rebuilt the Missouri wine industry from scratch by restoring some of the original wineries. Some growers were able to remain in business following the Volstead Act's ban on alcoholic beverages by converting their businesses to juice grape production, and Missouri has been an important source of such grapes for companies like Welch's. During the 1990s, with the increasing availability of less expensive grapes from Asia and Latin America, the juice grape industry in Missouri has endured economic hardships.

In about 1860, enologists (wine scientists) experimenting with North American samples in an effort to breed new and better European grape cultivars accidentally introduced into France the grape phylloxera, *Daktulosphaira vitifoliae* (Fitch), a small aphidlike insect native to North America that parasitizes *Vitis* species (Downie et al., 2000; Sorensen et al., 2008). Within a relatively short time, this insect wiped out nearly a third of France's vineyards and spread to all grape-growing regions of Europe. A French commission sent to the United States determined that the most effective way to save their continent's wine industry was to graft their precious vines onto rootstocks of hardy North American species, which were naturally resistant to phylloxera infestation (C. Campbell, 2004). Missouri and Texas furnished the majority of the rootstocks that eventually saved Europe's wine industry. Secondarily, these grafted grapes were imported into California to improve the wine industry there as well. Missouri botanists like Benjamin F. Bush and Henry Eggert (see introductory chapter on the History of Floristic Botany in Missouri) amassed small fortunes through the sale of rootstocks of various *Vitis* species wild-collected from Missouri and surrounding states.

1. Bark remaining tight on older stems (not shredding), appearing irregularly warty; pith continuous through the nodes; tendrils unbranched . . . . 6. V. ROTUNDIFOLIA

1. Bark shredding on older branches, not appearing warty; pith interrupted by a diaphragm of lighter unchambered tissue at the nodes; tendrils mostly branched toward the tip (occasionally absent in *V. rupestris*)
    2. Leaf blades when young with the undersurface densely pubescent with long cobwebby hairs more or less appressed to the surface, these persistent (but sometimes becoming sparser) at maturity (except in some specimens of *V. aestivalis* that appear glabrous at maturity); shorter, straight, more or or less spreading hairs sometimes also present along the veins
        3. Tendrils and/or inflorescences present in at least some groups of 3 or more adjacent nodes . . . . 3. V. LABRUSCA
        3. Tendrils and/or inflorescences present at no more than 2 adjacent nodes
            4. Petioles and first-year twigs glabrous or sparsely hairy at flowering time; leaf blades at maturity with the undersurface hairy and also somewhat to strongly glaucous; nodes of younger branches often glaucous; fruits 7–15 mm in diameter, usually glaucous . . . 1. V. AESTIVALIS
            4. Petioles and first-year twigs moderately to densely hairy at flowering time; leaf blades at maturity with the undersurface hairy but not glaucous; nodes of younger branches not glaucous; fruits 4–8 mm in diameter, not or only slightly glaucous . . . . 2. V. CINEREA
    2. Leaf blades when young with the undersurface pubescent with relatively short, straight, more or less spreading hairs; at maturity often becoming glabrous except in the axils of the larger veins
        5. Tendrils absent or restricted to the uppermost nodes of the main stem and branches; leaves mostly wider than long, kidney-shaped to depressed-ovate, rounded to very short-tapered at the tip, often somewhat folded longitudinally at maturity . . . . 7. V. RUPESTRIS
        5. Tendrils regularly present at most pairs of nodes; leaves as long as or often somewhat longer than wide, ovate to broadly ovate or nearly circular, noticeably narrowed or tapered at the tip, flat or slightly convex at maturity
            6. Diaphragms interrupting the pith at the nodes 0.5–1.0 mm wide on new growth, eventually thickening to 1–2 mm wide on older branches . . . . 5. V. RIPARIA
            6. Diaphragms interrupting the pith at the nodes 1–4 mm wide on new growth, eventually thickening to 2.5–6.0 mm wide on older branches
                7. Diaphragms interrupting the pith at the nodes 2.5–4.0 mm wide on new growth, eventually thickening to 4–6 mm wide on older branches; new twigs of season uniformly bright red to purplish red; fruits not or only slightly glaucous . . . . 4. V. PALMATA
                7. Diaphragms interrupting the pith at the nodes 1.0–2.5 mm wide on new growth, eventually thickening to 2–5 mm wide on older branches; new twigs of season green, gray, or brown, occasionally with some purplish red coloration along 1 side; fruits strongly glaucous . . . . 8. V. VULPINA

1. Vitis aestivalis Michx. **var. aestivalis** (summer grape, pigeon grape)

V. aestivalis var. *argentifolia* (Munson ex L.H. Bailey) Fernald

V. aestivalis var. *bicolor* A. Gray

Pl. 580 d, e; Map 2717

Young stems circular in cross-section, sparsely cobwebby-hairy or glabrous at flowering time, green, gray, or brown, the nodes sometimes glaucous, not reddish-tinged. Pith interrupted at the nodes, the diaphragms 1–3 mm wide on new growth, eventually thickening to 2–4 mm wide on

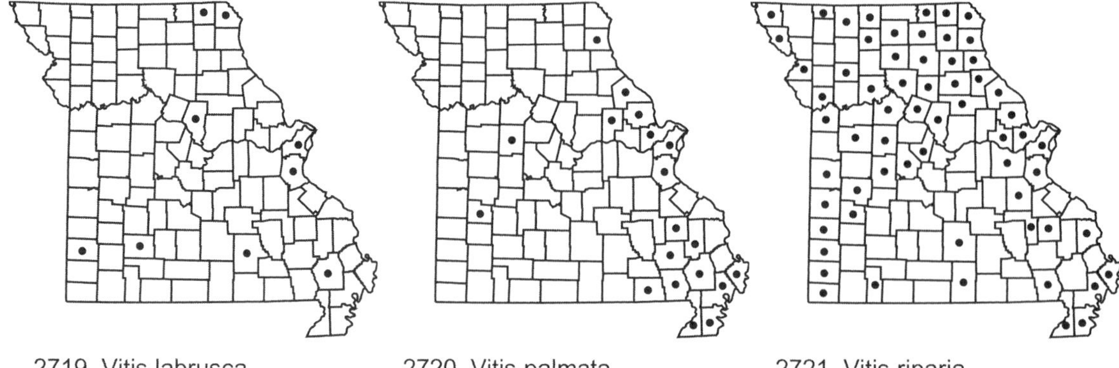

2719. Vitis labrusca

2720. Vitis palmata

2721. Vitis riparia

older branches. Older stems with the bark shredding, not appearing warty. Tendrils common, present at no more than 2 adjacent nodes (every third node lacking both a tendril and an inflorescence), 2- or 3-branched. Leaves with the petiole 2/3 as long to about as long as the blades, sparsely hairy to glabrous at flowering time. Leaf blades mostly 5–22 cm long, mostly longer than wide, ovate to broadly ovate or nearly circular in outline, flat or slightly convex at maturity, mostly 3- or 5-lobed, less commonly unlobed or nearly so, the sinuses U-shaped or V-shaped, the lobes usually tapered or narrowed to a sharply pointed tip, the upper surface glabrous or minutely hairy, usually not shiny. Undersurface of young leaves moderately to densely pubescent with grayish to rusty brown cobwebby hairs that are more or less appressed to the surface, these persistent but usually becoming sparser (or sometimes appearing glabrous) at maturity; shorter, straight, more or or less spreading hairs sometimes also present along the veins and the undersurface also somewhat to strongly glaucous, this sometimes obscured by the pubescence. Inflorescences at no more than 2 adjacent nodes, 7–20 cm long, mostly narrowly pyramid-shaped. Fruits mostly more than 25 per infructescence, 7–15 mm in diameter, the surface with lenticels few and inconspicuous or absent, black, usually glaucous. Seeds 3–7 mm long, tan to brown. $2n$=38. May–July.

Common in the southern 2/3 of Missouri, but absent from portions of the Glaciated Plains Division (eastern U.S. west to Minnesota, Nebraska, and Texas; Canada). Mesic to dry upland forests, ledges and tops of bluffs, glades, margins of upland prairies, and less commonly banks of streams and bottomland forests; also fencerows, old fields, roadsides, and railroads.

Moore (1991) recognized three varieties within this species. The var. *linsecomii* (Buckl.) Munson is endemic to eastern Texas and adjacent Louisiana, Arkansas, and Oklahoma, and differs from var. *aestivalis* in a suite of minor branchlet, leaf, and fruit characters. Moore also noted that plants with glabrous or nearly glabrous leaf undersurfaces tend to be more glaucous and have slightly smaller fruits. Steyermark (1963) and others referred to such plants as var. *argentifolia*, but Moore argued that Deam (1924) had validly published the name var. *bicolor* earlier for this taxon. However, although Deam did attribute the name var. *bicolor* to such plants, he noted that all degrees of intermediates existed between these plants and the more typical phase and believed that such plants should be regarded merely as a northern form of the species. Moore and others also did not realize that the the name var. *bicolor* had already been validated in 1890 by Asa Gray. Steyermark (1963) also discussed the lack of correlation between and continuous variation in the characters said to separate the varieties of *V. aestivalis*. Although the morphological extremes can be striking, field observations in Missouri tend to confirm the view that plants of *V. aestivalis* form a continuum from the more densely hairy, less glaucous phase to glabrescent, more strongly glaucous individuals. Also, although plants of the latter phase appear to be fairly uncommon, there do not appear to be geographic or ecological correlations with the morphological extremes. Thus, the two Missouri varieties are not accepted in the present treatment.

2. Vitis cinerea (Engelm.) Engelm. ex Millardet **var. cinerea** (graybark grape, sweet winter grape, pigeon grape)

Pl. 581 a, b; Map 2718

Young stems slightly to moderately angled, moderately to densely pubescent with short, straight, spreading hairs at flowering and also with sparse to dense cobwebby hairs, green, gray, or brown, the nodes not glaucous, but sometimes red-

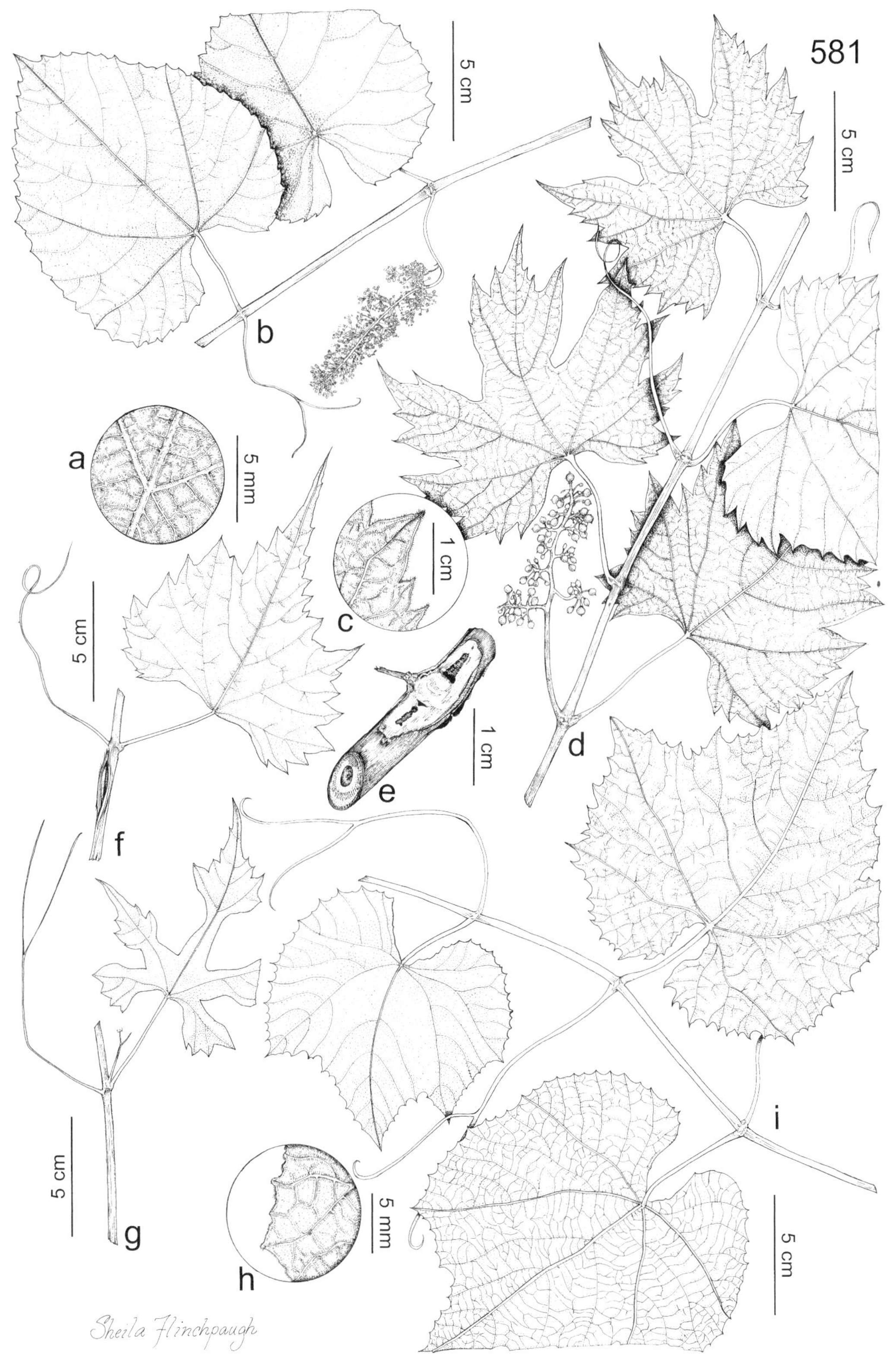

Plate 581. Vitaceae. *Vitis cinerea*, **a)** leaf undersurface detail, **b)** portion of fertile branch. *Vitis riparia*, **c)** leaf margin detail, **d)** fertile branch. *Vitis palmata*, **e)** longitudinally sectioned node showing interrupted pith, **f)** node with less-divided leaf and tendril, **g)** node with more-divided leaf and tendril. *Vitis labrusca*, **h)** leaf margin detail, **i)** portion of stem with leaves and tendrils (3 consecutive nodes).

dish-tinged. Pith interrupted at the nodes, the diaphragms 1–3 mm wide on new growth, eventually thickening to 2–4 mm wide on older branches. Older stems with the bark shredding, not appearing warty. Tendrils common, present at no more than 2 adjacent nodes (every third node lacking both a tendril and an inflorescence), 2- or 3-branched. Leaves with the petiole ⅔ as long to about as long as the blades, moderately to densely hairy at flowering time. Leaf blades mostly 5–22 cm long, mostly longer than wide, ovate to broadly ovate or nearly circular in outline, flat or slightly convex at maturity, unlobed or shallowly to less commonly deeply 3-lobed, the sinuses mostly U-shaped, the lobes usually tapered or narrowed to a sharply pointed tip, the upper surface glabrous or minutely hairy, usually not shiny. Undersurface of young leaves moderately to densely pubescent with light gray to gray (rarely brownish-tinged) cobwebby hairs that are more or less appressed to the surface, these persistent but often becoming sparser (or occasionally appearing nearly glabrous) at maturity; shorter, straight, more or or less spreading hairs usually also present along the veins but the undersurface not glaucous. Inflorescences at no more than 2 adjacent nodes, 10–25 cm long, narrowly to broadly pyramid-shaped. Fruits mostly more than 25 per infructescence, 4–8 mm in diameter, the surface with lenticels few and inconspicuous or absent, black, not or only slightly glaucous. Seeds 2–4 mm long, brown. $2n$=38. May–July.

Common in the southern ⅔ of Missouri, but absent from portions of the Glaciated Plains Division (Indiana to Florida west to Nebraska and Texas). Bottomland forests, mesic upland forests, banks of streams and rivers, and margins of ponds and lakes; also fencerows, old fields, roadsides, and railroads.

Moore (1991) accepted a complex series of four varieites within *V. cinerea*. Of these, the least widely distributed is the Texas endemic, var. *helleri* (L.H. Bailey) M.O. Moore, with its strongly glaucous berries and relatively small nearly glabrous leaves. The var. *floridana* Munson is restricted to the Atlantic and Gulf Coastal Plains (Maryland to Louisiana), and var. *baileyana* (Munson) Comeaux is widespread in the eastern United States (west to Indiana and Mississippi); both of these weakly separable taxa differ from var. *cinerea* in the absence of near absence of short straight spreading hairs on the branchlets and leaves, differing from each other in the density of cobwebby pubescence.

3. Vitis labrusca L. (fox grape, Labruscan vineyard grape)

Pl. 581 h, i; Map 2719

Young stems circular in cross-section or slightly angled, densely pubescent with cobwebby hairs at flowering and sometimes also with sparse stiff, spreading, gland-tipped hairs, green, gray, or brown, the nodes not glaucous, not reddish-tinged. Pith interrupted at the nodes, the diaphragms 0.5–1.5 mm wide on new growth, eventually thickening to 1.0–2.5 mm wide on older branches. Older stems with the bark shredding, not appearing warty. Tendrils common, present at most nodes, not regularly grouped at pairs of nodes, 2- or 3-branched. Leaves with the petiole ⅔ as long to about as long as the blades, glabrous or sparsely to moderately hairy at flowering time. Leaf blades mostly 5–25 cm long, slightly shorter than to slightly longer than wide, broadly ovate or nearly circular in outline, flat or slightly convex at maturity, unlobed or shallowly 3-lobed, less commonly deeply 3- or 5-lobed, the sinuses mostly U-shaped, the lobes usually tapered or narrowed to a sharply pointed tip, the upper surface glabrous or sparsely and minutely hairy along the main veins, not shiny. Undersurface of young leaves densely pubescent with light gray to more commonly tan cobwebby hairs that are more or less appressed to the surface, these persistent at maturity and obscuring the surface, the undersurface not glaucous. Inflorescences often at 3 or more adjacent nodes, 6–14 cm long, narrowly to broadly pyramid-shaped or less commonly cylindrical. Fruits mostly 10–25 per infructescence, 12–20 mm in diameter, the surface with lenticels absent, dark reddish purple to black, not or only slightly glaucous. Seeds 5–8 mm long, brown. $2n$=38. May–June.

Introduced, uncommon and widely scattered in the state (northeastern U.S. west to Michigan and Tennessee; Canada; escaped from cultivation farther south and west). Fencerows, old fields, roadsides, railroads, and open disturbed areas.

Some botanists prefer to restrict the use of the name *V. labrusca* to the wild type of this taxon and apply the name *V. labruscana* collectively to the confusing series of cultivars and hybrids that have been developed from this species, and which have escaped from cultivation. Missouri plants all belong to this latter group. However, this taxonomic interpretation seems untenable, both because of the heterogeneous origins of the cultivated plants and the fact that they all clearly can be traced back to wild *V. labrusca* plants, from which they were modified during historical times. The cultivated grapes developed from the wild fox grape include Concord, Catawba, Niagara, Chautauqua, Worden, and many other dark and rosy types.

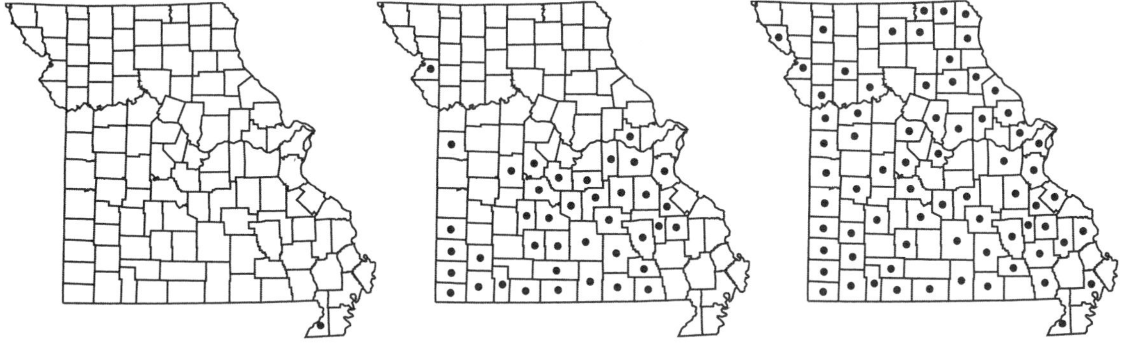

2722. Vitis rotundifolia

2723. Vitis rupestris

2724. Vitis vulpina

4. Vitis palmata Vahl (red grape, cat grape, catbird grape, Missouri grape)

Pl. 581 e–g; Map 2720

Young stems circular in cross-section or slightly angled, glabrous or sparsely and inconspicuously cobwebby-hairy at flowering time, uniformly bright red to purplish red, the nodes not glaucous. Pith interrupted at the nodes, the diaphragms 2.5–4.0 mm wide on new growth, eventually thickening to 4–6 mm wide on older branches. Older stems with the bark shredding, not appearing warty. Tendrils common, present at no more than 2 adjacent nodes (every third node lacking both a tendril and an inflorescence), 2- or 3-branched. Leaves with the petiole 2/3 as long to as long as the blades, sparsely and minutely hairy to glabrous at flowering time. Leaf blades mostly 4–15(–20) cm long, slightly longer than to about as long as wide, ovate to broadly ovate or nearly circular in outline, flat at maturity, shallowly or more commonly deeply 3-lobed, sometimes 5-lobed, the sinuses U-shaped or V-shaped, the lobes mostly tapered or long-tapered to a sharply pointed tip, the upper surface glabrous, occasionally somewhat glaucous, usually not shiny. Undersurface of young leaves glabrous or sparsely to moderately pubescent with minute, straight, more or or less spreading hairs along and in the axils of the main veins, not glaucous. Inflorescences at no more than 2 adjacent nodes, 4–18 cm long, mostly narrowly pyramid-shaped. Fruits mostly more than 25 per infructescence, 5–10 mm in diameter, the surface with lenticels absent, bluish black to black, not or only slightly glaucous. Seeds 4–6 mm long, dark brown. May–July.

Scattered in eastern Missouri, mostly in counties adjacent to the Mississippi and Missouri Rivers, with sporadic occurrences farther west (Indiana to Florida west to Oklahoma and Texas). Bottomland forests, swamps, sloughs, banks of streams and rivers, and edges of ponds and lakes.

5. Vitis riparia Michx. (riverbank grape, frost grape)

V. riparia var. *praecox* Engelm. ex L.H. Bailey

V. riparia var. *syrticola* (Fernald & Wiegand) Fernald

Pl. 581 c, d; Map 2721

Young stems circular in cross-section, glabrous or sparsely and minutely hairy at flowering time, green, gray, or brown, the nodes sometimes glaucous, not reddish-tinged. Pith interrupted at the nodes, the diaphragms 0.5–1.0 mm wide on new growth, eventually thickening to 1–2 mm wide on older branches. Older stems with the bark shredding, not appearing warty. Tendrils common, present at no more than 2 adjacent nodes (every third node lacking both a tendril and an inflorescence), 2- or 3-branched. Leaves with the petiole 1/2–2/3 as long as the blades, sparsely hairy to glabrous at flowering time. Leaf blades mostly 3–20 cm long, slightly longer than to about as long as wide, ovate to broadly ovate or nearly circular in outline, flat at maturity, mostly shallowly 3-lobed, less commonly unlobed or more deeply 3-lobed, the sinuses U-shaped or more commonly V-shaped, the lobes tapered or narrowed to a sharply pointed tip, the upper surface glabrous (occasionally sparsely and minutely hairy when young), usually not shiny. Undersurface of young leaves sparsely to moderately pubescent with minute, straight, more or or less spreading hairs along the main veins, not glaucous. Inflorescences at no more than 2 adjacent nodes, 4–12 cm long, mostly narrowly pyramid-shaped. Fruits mostly more than 25 per infructescence, 6–12 mm in diameter, the surface with lenticels absent, black, glaucous. Seeds 5–6 mm long, dark brown. $2n$=38. April–June.

Scattered in eastern Missouri, becoming uncommon farther south, and absent from most of the Ozark Division (northern U.S. south to

Virginia, Louisiana, and Oregon; Canada). Bottomland forests, mesic upland forests, bases and ledges of bluffs, edges of bottomland prairies, and banks of streams and rivers; also fencerows, and roadsides.

Steyermark (1963) split *V. riparia* into three varieties differing only in pubescence density and quantitative details of inflorescences, fruits, and seeds. Moore (1991) did not accept these varieties, which appear to represent no more than extremes in a continuum of independant character variations.

6. Vitis rotundifolia Michx. **var. rotundifolia** (muscadine, scuppernong, southern fox grape)

Muscadinia rotundifolia (Michx.) Small

Pl. 582 h; Map 2722

Young stems circular in cross-section or slightly angled, minutely hairy or glabrous at flowering time, green, gray, or brown, the nodes not glaucous, but sometimes reddish-tinged. Pith continuous through the nodes. Older stems with the bark becoming fissured and remaining tight or occasionally peeling irregularly in small plates, not shredding, appearing somewhat warty with oval lenticels. Tendrils common, present at no more than 2 adjacent nodes (every third node lacking both a tendril and an inflorescence), unbranched. Leaves with the petiole about as long as or slightly longer than the blades, minutely hairy or glabrous at flowering time. Leaf blades mostly 4–12 cm long, mostly about as long as wide or slightly wider than long, broadly ovate or nearly circular in outline, relatively flat at maturity, mostly unlobed, less commonly shallowly 3-lobed, the sinuses more or less U-shaped, rounded to broadly and abruptly narrowed or short-tapered to a sharply pointed tip, the upper surface glabrous, shiny. Undersurface of young leaves glabrous or sparsely to moderately pubescent with minute appressed hairs along the main veins; small patches of cobwebby hairs sometimes also present in the axils of main veins, the undersurface also somewhat shiny, not glaucous. Inflorescences at no more than 2 adjacent nodes, 3–8 cm long, mostly broadly pyramid-shaped. Fruits mostly less than 12 per infructescence, 8–25 mm in diameter, the surface with relatively prominent lenticels, black to purplish black or bronze-colored, usually not glaucous. Seeds 5–8 mm long, brown. $2n$=40. May–June.

Uncommon, known only from Dunklin County (eastern [mostly southeastern] U.S. west to Oklahoma and Texas). Bottomland forests, swamps, margins of sloughs, and sand savannas.

Vitis rotundifolia is the ancestor of the cultivated muscadine grapes, which have been used for wines, table grapes, and raisins. A second variety, var. *munsoniana* (J.H. Simpson ex Planch.) M.O. Moore, is restricted to portions of Alabama, Georgia, and Florida, and differs in its somewhat smaller leaves and infructescences with a somewhat larger number of fruits that are individually smaller. Some botanists recognize a third variety, var. *pygmaea* McFarlin ex D.B. Ward, which is a small-leaved taxon that generally does not climb and is endemic to the central Florda scrub community. The report by Steyermark (1963) of var. *rotundifolia* in Madison and Pemiscot Counties could not be verified during the present study, although it is to be expected in the latter county.

Some botanists prefer to transfer the muscadine to its own genus, *Muscadinia*. For further discussion, see the paragraph under the *Vitis* generic description.

7. Vitis rupestris Scheele (sand grape)

V. rupestris f. *dissecta* (Eggert ex L.H. Bailey) Fernald

Pl. 582 a; Map 2723

Young stems circular in cross-section or somewhat angled, glabrous or sparsely and minutely hairy at flowering time, green, gray, or brown, the nodes not glaucous, not reddish-tinged. Pith interrupted at the nodes, the diaphragms 0.3–0.5 mm wide on new growth, eventually thickening to 0.5–1.0 mm wide on older branches. Older stems with the bark shredding, not appearing warty. Tendrils absent or restricted to the uppermost nodes of the main stem and branches, present at no more than 2 adjacent nodes (every third node lacking both a tendril and an inflorescence), 2- or 3-branched. Leaves with the petiole 1/2–2/3 as long as the blades, sparsely hairy to more commonly glabrous at flowering time. Leaf blades mostly 3–10 cm long, mostly wider than long, kidney-shaped to depressed-ovate in outline, often somewhat folded longitudinally at maturity, unlobed or shallowly 3-lobed, the sinuses mostly broadly U-shaped, the lobes rounded or abruptly tapered or narrowed to a sharp but broadly pointed tip, the upper surface glabrous (occasionally sparsely and minutely hairy when young), often somewhat shiny. Undersurface of young leaves glabrous or sparsely pubescent with minute, straight, more or or less spreading hairs along and/or in the axils of the main veins. Inflorescences at no more than 2 adjacent nodes, 2–7 cm long, broadly pyramid-shaped to nearly globose. Fruits mostly less than 15 per infructescence, 6–12 mm in diameter, the

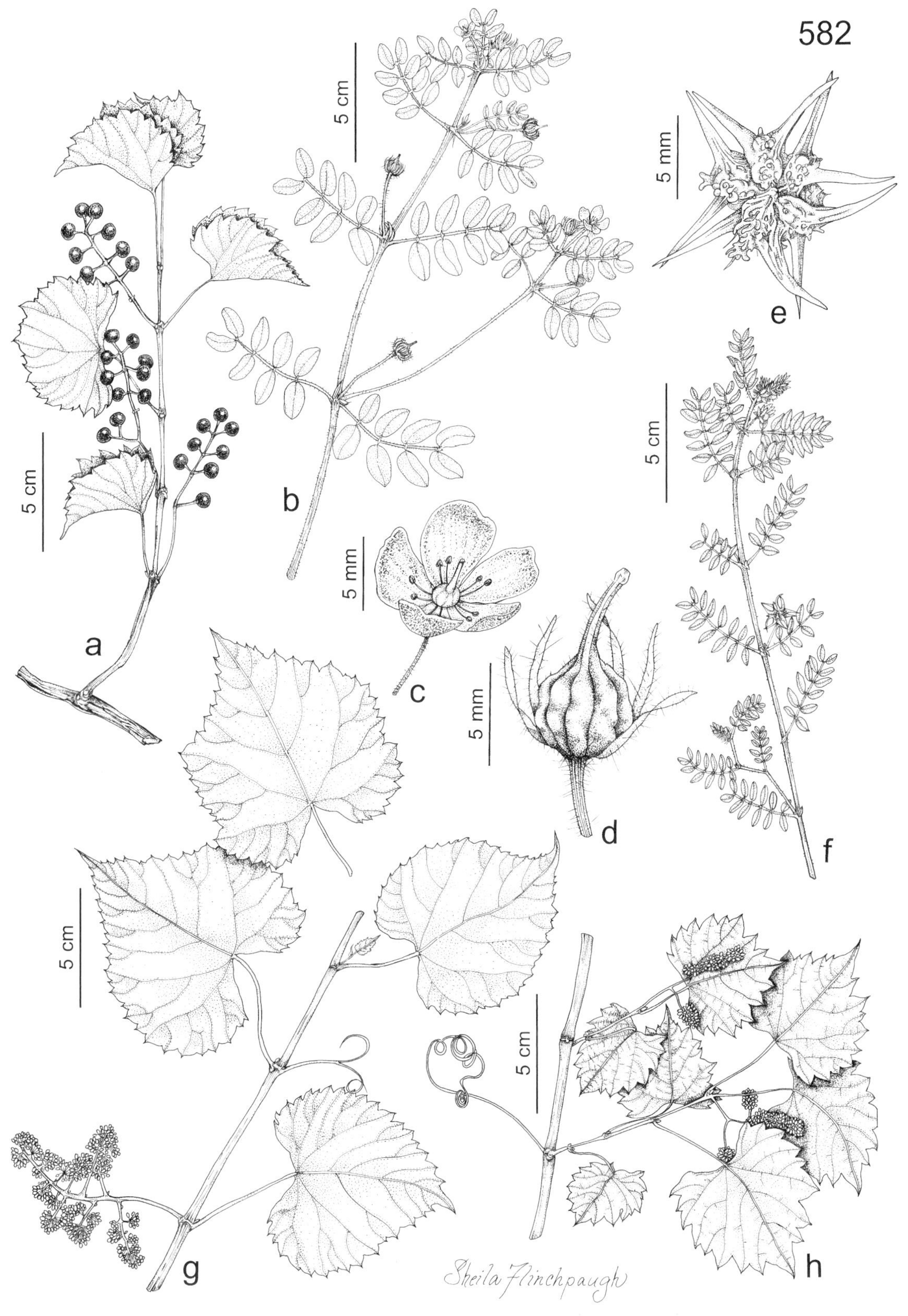

Plate 582. Vitaceae. Zygophyllaceae. *Vitis rupestris*, **a)** fruiting branch. *Kallstroemia parviflora*, **b)** fertile stem, **c)** flower, **d)** fruit. *Tribulus terrestris*, **e)** fruit, **f)** fertile branch. *Vitis vulpina*, **g)** fertile branch and detached leaf. *Vitis rotundifolia*, **h)** fertile branch.

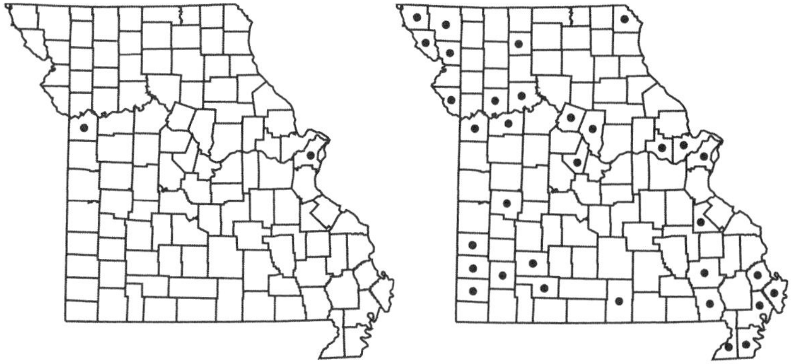
2725. Kallstroemia parviflora 2726. Tribulus terrestris

surface with lenticels absent, black, sometimes glaucous. Seeds 5–6 mm long, light brown. $2n$=38. May–June.

Scattered in southern and central Missouri, mostly in the Ozark and Ozark Border Divisions (Pennsylvania to Virginia west to Louisiana, Texas, and Missouri). Banks of streams and rivers and bases of bluffs.

Unlike other native Missouri species of grapes, which generally climb over other vegetation and eventually up into the forest canopy, plants of *V. rupestris* appear to be specialized to a scrambling habit, forming low colonies on open gravel bars. Most commonly, plants are encountered without flowers and fruits, and tendrils are not very abundant on the stems, occurring infrequently toward the tips of some branches. Moore (1991) indicated that although this species is quite important in the grape-growing industry as a source of rootstocks for grafting, it has become extirpated from much of its range outside Missouri. Steyermark (1963) indicated that this attractive species should be investigated for possible horticultural use as an ornamental groundcover.

8. Vitis vulpina L. (winter grape, frost grape, chicken grape)
V. cordifolia Michx.
Pl. 582 g; Map 2724

Young stems circular in cross-section or somewhat angled, glabrous at flowering time (sometimes sparsely and inconspicuously cobwebby-hairy when very young), green, gray, or brown, occasionally with some purplish red coloration along 1 side, the nodes not glaucous, not reddish-tinged. Pith interrupted at the nodes, the diaphragms 1.0–2.5 mm wide on new growth, eventually thickening to 2–5 mm wide on older branches. Older stems with the bark shredding, not appearing warty. Tendrils common, present at no more than 2 adjacent nodes (every third node lacking both a tendril and an inflorescence), 2- or 3-branched. Leaves with the petiole 2/3 as long to as long as the blades, sparsely to moderately and minutely hairy or glabrous at flowering time. Leaf blades mostly 4–15(–20) cm long, slightly longer than to about as long as wide, ovate to broadly ovate or nearly circular in outline, flat at maturity, unlobed or shallowly (rarely deeply) 3-lobed, the sinuses mostly broadly U-shaped, the lobes narrowed or tapered to short-tapered to a sharply pointed tip, the upper surface glabrous or sparsely and minutely hairy along the main veins, sometimes somewhat shiny. Undersurface of young leaves glabrous or more commonly sparsely to fairly densely pubescent with minute, straight, more or or less spreading hairs along and in the axils of the main veins, not glaucous. Inflorescences at no more than 2 adjacent nodes, 6–19 cm long, mostly narrowly pyramid-shaped. Fruits mostly more than 25 per infructescence, 5–12 mm in diameter, the surface with lenticels absent, black, not or only slightly glaucous. Seeds 3–5 mm long, dark brown. May–June.

Scattered nearly throughout the state (eastern U.S. west to Nebraska and Texas; Canada). Bottomland forests, mesic upland forests, bases and ledges of bluffs, banks of streams and rivers, and edges of ponds and lakes; also roadsides and railroads.

ZYGOPHYLLACEAE (Caltrop Family)

(D. M. Porter, 1972)

Plants annual, the stems often with somewhat swollen nodes. Leaves opposite, the pair at each node often somewhat unequal in size, short-petiolate, pinnately compound, usually with an even number of leaflets (lacking a terminal leaflet), the leaflets sessile or nearly so. Stipules herbaceous. Inflorescences of solitary flowers in the leaf axils, usually only 1 per node. Flowers actinomorphic, perfect, hypogynous. Calyces of 5 free sepals. Corollas of 5 free petals, these spreading to slightly cupped. Stamens 10(–12), in 2 whorls opposite the sepals and petals, the anthers usually attached toward the base, yellow. Ovary 1 per flower, superior, of 5 fused carpels, with 5 or 10 locules. Style 1 per flower, the stigma 1, club-shaped or oblong-capitate, usually somewhat 10-lobed. Ovules 1 or few per locule. Fruits schizocarps, 5–10-lobed, splitting at maturity into 5–10 mericarps, each with 1 or few seeds. About 30 genera, about 285 species, nearly worldwide, but most diverse in regions with warm arid climates.

The description above applies only to the Missouri species. Elsewhere, many members of the family are perennial herbs, shrubs, or trees with simple or bifoliate to twice compound leaves, more complex inflorescences, variously colored corollas, and/or various fruit types, among other differences.

Although some species of *Kallstroemia* have been noted to provide good forage for livestock (D. M. Porter, 1969a), there are abundant reports of livestock poisoning attributed to members of the family (Burrows and Tyrl, 2001), including both of the genera growing in Missouri. It is fortunate that apparently large quantities must be ingested to produce the symptoms of poisoning. Species of *Kallstroemia* can cause neurological problems and death in cattle, sheep, and goats. The chemical cause of the symptoms is as yet poorly understood, but indole alkaloids have been implicated as one possible group of toxins (Burrows and Tyrl, 2001). In *Tribulus*, in addition to neurological effects, ingestion by livestock can lead to photosensitization and liver damage, a syndrome known in veterinary medicine as “bighead” (D. M. Porter, 1969a). Again, the causes of the symptoms are not fully understood, but suggestions as potential toxins include saponins in conjunction with indole alkaloids (Burrows and Tyrl, 2001), possibly combined with the effects of selenium or nitrate accumulated in the plants from the soil (D. M. Porter, 1969a).

1. Fruits with a beak 3–9 mm long, breaking into 10 mericarps, leaving a persistent central axis, the mericarps wrinkled and/or warty, each with 1 seed; petals orange, sometimes fading to yellow or white with age; leaflets 3–5 pairs per leaf . . . . 1. KALLSTROEMIA
1. Fruits not or minutely beaked, breaking into 5 mericarps, without a persistent central axis, the mericarps spiny, each with 2–5 seeds; petals yellow, sometimes fading to white with age; leaflets (4–)6–8 pairs per leaf . . . . 2. TRIBULUS

1. Kallstroemia Scop.

(D. M. Porter, 1969a, b)

Seventeen species, North America to South America, Caribbean Islands.

1. Kallstroemia parviflora Norton (small-flowered caltrop, warty caltrop)
K. intermedia Rydb.
Pl. 582 b–d; Map 2725

Stems to 120 cm long, prostrate or loosely ascending, moderately to densely pubescent with minute upwardly curled hairs and sparser longer stiff hairs. Leaf blades 1–6 cm long, elliptic in outline, the leaflets in (3)4(5) pairs, 7–18 mm long, elliptic to oblong-obovate, asymmetrically rounded at the base, bluntly to sharply pointed at the tip, the upper surface more or less glabrous, the undersurface and margins moderately pubescent with stiff mostly appressed hairs. Stipules 5–7 mm long, somewhat curved, narrowly lanceolate. Flower stalks 10–40 mm long, hairy. Sepals 4–7 mm long, narrowly lanceolate, densely pubescent with stiff, mostly pustular-based hairs, withered but persistent at fruiting. Petals 5–10 mm long, orange, sometimes fading to yellow or white with age. Stamens attached to an inconspicuous nectar disc. Fruits with the body 3–4 mm long, 4–6 mm in diameter, broadly ovoid, tapered abruptly to a relatively slender beak 3–9 mm long, more or less 10-lobed, at maturity breaking into 10 mericarps, leaving the persistent central axis and beak, the mericarps 1-seeded, the surface wrinkled and/or warty, somewhat obscured by dense minute hairs. June–September.

Introduced, uncommon, known only from historical collections from Jackson County and St. Louis City and County (native from Kansas to Nevada south to Texas and California; introduced farther east and in South America). Railroads and open disturbed areas.

Steyermark (1963) treated *K. parviflora* (under the name *K. intermedia*) as a synonym of the related *K. hirsutissima* Vail. However, D. M. Porter (1969a, b) showed that these are two distinct species and that Missouri plants are referable to *K. parviflora*. *Kallstroemia hirsutissima* grows mostly in the Chihuahuan Desert from northern Mexico northward to southern Arizona, New Mexico and Texas, and differs in its fruits with short nipplelike beaks and obovate leaves. Steyermark's (1963) report of this genus from Jasper and Lawrence Counties could not be confirmed during the present study.

2. **Tribulus** L.

About 30 species, Europe, Asia, Africa, most diverse in the Middle East.

1. Tribulus terrestris L. (puncture vine, caltrop)

Pl. 582 e, f; Map 2726

Stems to 100 cm long, prostrate, sparsely to densely pubescent with minute upwardly curled hairs and sparse to moderate longer stiff hairs. Leaf blades 1.5–5.0 cm long, elliptic in outline, the leaflets (4–)6–8 pairs, 3–14 mm long, elliptic to oblong or narrowly ovate, asymmetrically rounded at the base, bluntly to more commonly sharply pointed at the tip, the upper surface moderately to densely pubescent with stiff to silky, appressed hairs, sometimes only along the midvein, the undersurface and margins usually densely pubescent with stiff to silky mostly appressed hairs. Stipules 3–6 mm long, somewhat curved, narrowly lanceolate. Flower stalks 5–16 mm long, hairy. Sepals 2–4 mm long, narrowly lanceolate to narrowly ovate, densely pubescent with stiff appressed hairs shed soon after flowering. Petals 3–6 mm long, yellow, sometimes fading to white with age. Stamens not attached to a nectar disc, the outer 5 attached to the petal bases, the inner 5 attached at the ovary base but each flanked by an inconspicuous pair of nectar glands. Fruits with the body 4–6 mm long, 7–12 mm in diameter, depressed-globose, not or minutely beaked, strongly 5-lobed, at maturity breaking into 5 mericarps, without a persistent central axis, the mericarps each obliquely partitioned into 2–5 locules, each locule 1-seeded, the mericarp surface with a pair of stout dorsal spines situated on either side of the keel and also additional smaller spines, or tubercles, and/or wrinkles, usually also with sparse to moderate minute hairs. $2n$=12, 24, 36, 48. June–September.

Introduced, scattered nearly throughout the state (apparently originally native to the Mediterranean Region, now known from every continent except Antarctica; nearly throughout the U.S.; also adjacent Canada). Banks of streams and rivers; also margins of crop fields, roadsides, railroads, and open disturbed areas, often in sandy soil.

In a number of western states, *T. terrestris* is considered a noxious weed. In California, biological controls in the form of two Indian weevil species were started during the 1960s and appear to have been successful in reducing the abundance of the species (D. M. Porter, 1972). In addition to the poisonous properties mentioned in the discussion at the head of the family treatment, the spiny fruits can cause mechanical injury to unwary livestock and humans, as well as potentially puncturing bicycle and automobile tires. The misfortune of kneeling in a sandy area to examine some wildflower only to grind a fruit of puncture vine into the knee with one's full body weight is a crippling experience generally avoided enthusiastically after the first incident. The common name "caltrop" is an allusion to the apparent morphological similarities between the fruits of some Zygophyllaceae and the ancient military device consisting of a metal ball with sharply pointed spikes mounted so that one is always facing upright, which was used by armies to impede the advance of enemy soldiers and horses.

GLOSSARY

a-. Prefix indicating lack of, as in asymmetrical, indicating lack of symmetry.

achene. Indehiscent, dry, single-seeded fruit with the seed attached to the fruit wall in only one place. Achenelike refers to a fruit similar to an achene, but lacking one or more of the technical characters that define this fruit type.

actinomorphic. Having radial symmetry, such as flowers with parts arranged so that opposite sides of a line drawn through the middle in any plane are mirror images.

acuminate. Gradually tapered to a sharp point with the sides somewhat concave near the tip.

acute. Narrowed to a pointed tip with the sides relatively straight.

adventitious. Developing in an irregular pattern, as in a root system lacking a single taproot or with roots produced along a stem.

allele. Variant of a gene locus differing in its expression or its attributes (molecular weight and/or charge) as resolved in an electrophoretic study.

allelopathy. The process whereby exudate chemical compounds in the soil from the roots or shed parts of a given plant species inhibit the germination and/or growth of other plant species.

allotetraploid. Plant or taxon formed through hybridization between two diploid parental species that regains fertility through a doubling of its genome (thus containing two sets of chromosomes from each parent).

allozyme. Genetically based variant of an enzyme produced by an organism that differs in size and/or charge of the molecule and can be separated from other forms of the same kind of enzyme (produced by the same gene) using the technique of gel electrophoresis. Allozymes are studied to document levels of genetic variation within and between populations and between species. See also isozyme.

alternate. With parts or structures attached singly at each point or node, as in stems with a single leaf at each node. Contrasted with opposite and whorled.

anastomosing. Forming a network. In plants, usually applied to veins or nerves.

anatomy. Study of the internal organization and structures of plants. See also morphology.

androgynophore. In some plants (for example, members of the genus *Passiflora*) a structure in the flower formed by an extension of the receptacle into a stalk with the stamens and pistil(s) attached at its tip.

aneuploid. Cells or plants in which the chromosome number differs from the usual number for that taxon by other than a strict multiple of the basic number. For example, in *Claytonia virginica* (Portulacaceae), the base number is x=6, but plants have been recorded with a lengthy series of different chromosome numbers, 2n=12 to ca. 190.

annual. Plant that germinates, grows, flowers, produces mature fruits, and dies during a single year. Also used generally for processes that occur during a single year. A variant known as biennial refers to plants that germinate during the autumn, overwinter in a rosette stage, and flower and fruit the following spring.

anther. Saclike, pollen-containing portion of the stamen, usually two- or four-lobed, the pollen-containing lobes known as anther sacs.

apical. Located at the tip of a structure.

apogamy. Type of asexual reproduction in which seeds are formed directly from maternal cells rather than following fertilization of an egg by a pollen nucleus. Also referred to as apomixis (technically apogamy means without gametes and apomixis means without fertilization). Apomictic plants frequently are difficult or impossible to distinguish morphologically from sexually reproducing ones and apomixis must be confirmed through cytological and/or anatomical studies. One clue is that often in apogamous plants the proportion of flowers developing into fruits is very high.

apomixis. Synonymous with apogamy.
areole. In leaves with a network of anastomosing veins, an area of the surface surrounded by veins.
aril. Fleshy, often brightly colored appendage of the seed in some plants (for example, members of the genus *Oxalis*), sometimes appearing restricted at or near the hilum, sometimes covering all or nearly all of the seed. See also caruncle.
asexual reproduction. Any form of reproduction by which offspring are formed directly from maternal tissue without fertilization. Includes apogamy, gemmae, bulblets, and stem and root divisions.
asymmetrical. Uneven, often applied to leaf bases in which the tissue is narrowed differently on opposite sides of the attachment point and to flowers in which the corolla is contorted or twisted such that the flower has neither bilateral nor radial symmetry.
-ate. Suffix indicating a two-dimensional shape, such as ovate, meaning a structure that is egg-shaped in outline.
auricle. Small, earlike lobe of tissue. Usually applied to the base of a leaf blade or leaflet.
awn. Slender, bristle-like structure, usually slightly broadened at the base, often attached to the tip of a leaf, fruit, or flower part.
axile. Kind of placentation in an ovary or fruit with two or more locules in which the ovules and seeds are attached centrally where the walls of adjacent locules intersect.
axillary. Located in the angle (axil) between a structure and its axis, such as the angle between a leaf and the stem to which it is attached.
axis. Stemlike longitudinal support structure, such as the rachis of a compound leaf or the stemlike central portion of an inflorescence.
backcross. Offspring of a mating between a hybrid and one of its parental taxa.
banner. Uppermost, usually showy petal in many species of Fabaceae.
basionym. In botanical nomenclature, the initial publication of an epithet for a species or other level in a classification, which may be changed later as the classification becomes modified by transfer of the epithet into a new combination. For example, the water pepper, *Persicaria hydropiper* (L.) Delarbre, originally was described under the basionym *Polygonum hydropiper* by Linnaeus before subsequent taxonomic studies determined that the traditional, broadly circumscribed *Polygonum* should be broken into several segregate genera.
beak. Hard or firm, narrow point or projection, usually applied to appendages of fruits or seeds, less commonly to petals or other flower parts.
beard. Tuft or line of dense, long hairs.
berry. Fleshy fruit developing from a single pistil, with several to many seeds (and lacking a stone or hard core).
bicolorous. Having two colors. Usually applied to petals, leaves, or scales with marginal or central bands different in color than the rest of the tissue. See also concolorous.
biconvex. Applied to structures, such as seeds and fruits, that are elliptic or oval in cross-section and thus appear convex on both sides. Structures that are unequally convex are rounded on one side and nearly flat on the other.
biennial. Plant that germinates and forms a rosette during the autumn, overwinters in this state, and produces flowers and fruits the following spring, eventually dying during the second growing season. See also annual.
bilabiate. Two-lipped. Usually applied to zygomorphic corollas or calyces of some plants that are lobed into upper and lower main components (lips), one or both or which may be further toothed or more finely lobed.
bilocular. Structure that is hollow and divided into two chambers (locules); often used in reference to an ovary or fruit.

biosystematics. The science involving the study of living plants and the use of comparative data from cytology, reproductive behavior, biogeography, and other sources to evaluate the evolutionary relationships within a taxonomic group.

bract. A leaf, often reduced in size or otherwise modified, that subtends a flower or part of an inflorescence. In a complex inflorescence with more than one order of such structures, those subtending individual flowers or that are attached at the base of a flower stalk are referred to as bractlets.

bristle. Hairlike structure, often firm or stiff.

bulb. Underground modified leaf bud, usually covered with fleshy scales. Small structure similar in appearance to bulbs, produced anywhere on a plant, are called bulblets; in many cases these serve as asexual propagules.

calcareous. Limey, referring to substrates derived from limestone or dolomite bedrock.

calyx. Outermost perianth whorl of a flower, consisting of sepals. When only one perianth whorl is present, it is called a calyx by default.

capitate. Headlike, as in a dense cluster of flowers, or a stigma that is entire, more or less globose, and expanded abruptly from the tip of the style.

capsule. Dry, dehiscent fruit composed of more than one carpel and usually containing two or more seeds.

carpel. In angiosperms, the modified fertile leaf (sporophyll) bearing the ovules. The pistil is composed of one or more carpels.

caruncle. Outgrowth or appendage near the hilum of a seed, usually hard and not fleshy. See also aril.

catkin. Type of inflorescence that is a dense spike or slender raceme with numerous small flowers, these all either staminate or pistillate and lacking corollas; catkins are often associated with wind-pollination.

caudex. Short perennial stem at or just below the ground surface producing new aerial stems each year that die back at the end of the growing season.

chasmogamous. Open-flowering, referring to flowers whose buds open at flowering time to allow for dispersal of pollen and potential cross-pollination. Contrasted with cleistogamous (closed-flowering).

chlorophyll. Green pigment in plants associated with photosynthesis.

chromosome. DNA-containing element of a cell nucleus, becoming condensed and visible prior to and during nuclear division.

circumboreal. With a more or less continuous northern distribution in both North America and Eurasia.

circumscissile. Pattern of dehiscence in a transverse line, so that the top of a structure separates from the base like a lid; often applied to the dehiscence of some capsular fruits, for example, members of the genus *Plantago*.

clade. Natural (monophyletic) group in a phylogenetic study; that is, a group perceived to have a single genealogical origin, represented as a discrete branch of an evolutionary tree.

clasping. Enlarged at the base and partially surrounding the structure to which an object is attached, as in the bases of some leaves clasping the stem.

class. Rank in a taxonomic classification consisting of a collection of orders (groups of families) with a shared genealogy. For example, the dicots are class Magnoliopsida of the division Magnoliophyta (angiosperms).

cleistogamous. Closed-flowering, referring to modified flowers of some species that remain closed and set seed without exposing the stamens and stigmas, thus are obligately self-pollinated. Contrasted with chasmogamous (open-flowering).

colonial. Growth pattern in which plants form patches or colonies through vegetative spread by rhizomes, root sprouts, or similar structures.

column. A hollow tube formed from the fused filaments of stamens in some taxa; for example many genera of Fabaceae and Malvaceae. In the flowers and fruits (schizocarps) of the Geraniaceae and a few other families, the slender central structure to which the styles and mericarps are attached.

compound. Divided into two or more discrete parts, as in a compound leaf divided into two or more distinct leaflets. Also applied to a pistil composed of more than one carpel.

concave. Referring to a surface that is hollowed out or sunken.

concolorous. Having a single color. See also bicolorous.

conic. Three-dimensional shape, cone-shaped, with the broad end at the base. Cone-shaped structures attached at the narrow end are referred to as obconic.

contorted. Asymmetrically and abruptly bent, curved, or looped, as in the corollas of some species.

conserve, conservation. Preservation of species and their habitats. Also, in botanical nomenclature, the official retention of a particular name for a taxon (in order to stabilize the nomenclature) that is not the oldest validly published name for that taxon. This requires a formal printed proposal, a review of the situation by an official nomenclatural committee under the auspices of the International Association for Plant Taxonomy, and a vote on the proposal at an International Botanical Congress.

convex. Referring to a surface that is rounded, bulged outward, or dome-shaped.

cordate. Heart-shaped, with the point of attachment at the notched end. Usually applied to the base of a leaf (note that in some other texts this term refers to a heart-shaped leaf or structure, but in the present volume it is restricted to the base of a leaf or other structure).

corm. Short, solid, erect, underground stem with short internodes, usually thickened and with leaves closely overlapping, these sometimes modified and papery.

corolla. Innermost perianth whorl of a flower, consisting of petals.

corona. In some flowers, an extra whorl of petal-like structures located between the corolla and stamens.

corrugated. Folded or wrinkled into a regular series of ridges and valleys, as in some leaves that are zigzag in cross-section or some fruits with the surface having a regular series of parallel ridges and furrows.

cotyledon. Embryonic leaf in the seed, often developing into the first leaf of a seedling.

cultigen. Type of plant that arose from a cultivated progenitor. Such plants, often have specialized morphological features that enhance their yield or harvestability, but inhibit their ability to survive in the wild.

cupule. In the fruits of the Fagaceae, a hardened or woody structure with a spiny or scaly outer surface that encloses all or the basal portion of the nut.

cuticle. Outermost waxy layer of a leaf or stem.

cyme. Branched determinate inflorescence, that is, one in which the terminal flowers bloom first. In the present volume, cymes are included in the general definition of panicles.

cystoliths. Calcium carbonate inclusions in the cells of some plants. Frequently associated with the hardened, swollen bases of hairs in some families (Boraginaceae, Cannabaceae, etc.). Sometimes also used to refer to the swollen or pustular basal portion of such hairs.

cytogenetics. Study of the cytological aspects of inheritance and other genetic phenomena.

cytology. Study of cell function and structure. As applied to taxonomy (cytotaxonomy), mostly referring to the comparative study of chromosome number, morphology, and behavior at meiosis.

cytotype. Variant of a taxon differing in chromosome number from other members of the taxon.

deciduous. Referring to a structure that is shed at the end of the same growing season in which it is produced. Usually refers to leaves that are shed in autumn.

decurrent. Extended downward from the point of attachment, as in a leaf base that extends downward as two wings of tissue fused to the stem below the main attachment point of the leaf.

dehiscence. Method or pattern of breaking open, as in a fruit releasing its seeds.

depauperate. Stunted or poorly developed, as in a plant that is noticeably smaller than normal for the species because of adverse growing conditions.

dichotomous. Divided into two equal parts, as in a vein that forks evenly into two veins of equal size.

dicot. Member of the division Magnoliophyta class Magnoliopsida; a group delineated by a syndrome of features, among them the production by the embryo of a pair of cotyledons. Note that molecular phylogenetic studies have shown this to be an unnatural group because the monocots evolved from within it.

dimorphic. With two morphologically different forms, as when leaves on various parts of a plant appear differently shaped.

dioecious. Plants that produce exclusively staminate or pistillate flowers, these located only on different individuals, not on the same plant.

diploid. Containing two sets of chromosomes in each cell; the basic level of ploidy in the sporophyte generation of plants. Usually contrasted with polyploid.

disc. Refers to the general shape of an object that is relatively flat and more or less circular in outline, as in the stigmas of some flowers. Discoid refers to a structure with a disclike shape, such as a broad, flat stigma.

disjunct. Widely separated geographically from the rest of a taxon's distribution.

distyly. In some species, the production of flowers with two kinds of reproductive morphology: some flowers with relatively elongate styles and short stamens such that the stigmas are positioned above the anthers; other flowers with shorter styles and longer stamens such that the stigmas are positioned below the level of the anthers. Distyly is an adaptation to promote cross-pollination between flowers by limiting the ability of stamens to shed pollen directly onto the stigma(s) within the same flower.

division. Highest level in the taxonomic hierarchy of plant classification. Equivalent to the term phylum in animals.

dorsal. Referring to the back or outer (under) surface of a structure in relation to the axis. Also applied to the side of a leaf sheath where the leaf blade is attached to the sheath.

drupe. Usually fleshy, indehiscent fruit containing a stone. In drupes, the fruit wall is differentiated into three layers: the outer skin; a middle, usually fleshy layer; and a hardened inner layer surrounding the usually solitary seed.

elaiosome. Oily, usually white, appendage on some seeds that provides food for ants that act as dispersal agents for the seed.

elastically dehiscent. The springlike dehiscence pattern of some fruits in which the fruit wall creates physical stress on portions of the fruit as it matures, resulting in sudden breaking open and often violent dispersal of the seeds.

ellipsoid. Three-dimensional shape about two to three times as long as wide, broadest at the middle, and tapering equally to each end.

elliptic. Two-dimensional shape about two to three times as long as wide, broadest at the middle, and tapering equally to each end.

embryology. Study of the morphology and development of embryos as well as ovules and the cells and nuclei of germinating pollen grains.

endemic. Restricted to a particular geographic area, as in Ozark endemics, which occur only on the Ozark Plateau of Arkansas, Missouri, and/or Oklahoma.

endosperm. Nutritive tissue surrounding the embryo in some seeds.

ephemeral. Lasting only a short time, such as the aboveground growth of some plants.

epicalyx. Whorl of bracts closely subtending a flower and appearing similar to the calyx.

epigynous. Flowers having the perianth and stamens appearing attached at the tip of the ovary, the ovary appearing entirely inferior to the other floral parts.
epiphyte. Plant that is rooted on the surface of another plant, rather than in soil, but derives no water or nutrient from its host's tissues.
evergreen. Referring to a structure that persists for more than one growing season. Usually applied to leaves that remain green and function for two or more years.
exserted. Extending beyond the surrounding parts. Usually applied to stamens or styles that extend beyond the surrounding perianth. Contrasted with included.
extinct. No longer occurring anywhere; with no remaining living individuals.
extirpated. No longer occurring in a portion of its former range; eliminated from an area.
extrafloral nectary. Nectar-producing gland located on some portion of the plant other than the flower.
family. Rank in a taxonomic classification consisting of a collection of genera with a shared genealogy and one or more cytological, morphological, or other shared distinguishing features.
fascicle. Tight bundle or cluster, often applied to clusters of leaves or stamens.
fertile. Capable of bearing spores, ovules, seeds, or stamens. Contrasted with sterile.
filament. Stalklike or threadlike portion of a stamen to which the anther is attached.
filiform. Very slender and threadlike.
floricane. In the genus *Rubus,* the second year's growth of the biennial stem, which bears leaves, flowers and fruits. Contrast with primocane.
flow cytometry. Laser-based laboratory technology for the counting of cells or sorting of cells or nuclei for various parameters; in plant systematics it is often used to distinguish members of polyploid complexes.
flower. Reproductive structure of angiosperms, consisting of one or more pistils and/or stamens usually surrounded by one or two whorls of perianth (corolla and/or calyx).
foliaceous. Leaflike in appearance and texture, often in reference to bracts or sepals.
follicle. Dry, dehiscent fruit composed of one carpel and usually containing two or more seeds that dehisce longitudinally along only one side.
form. Trivial variant within a species, often produced by a minor mutation, such as white-flowered individuals in an otherwise red-flowered species. Technically called forma.
fragarioid. In the Rosaceae, an informal designation for the group of genera related to *Fragaria*, the genus of strawberries.
fruit. Ripened ovary and associated structures; organ containing the seed(s) of angiosperms.
funnelform. A three-dimensional shape, funnel-shaped, a tubular structure that is narrowest at the base and expanded gradually toward the tip; often applied to corollas.
fusiform. A three-dimensional shape, spindle-shaped, broadest at the middle, and tapering equally to each end.
gall. Abnormal growth of plant tissue caused by insects, microogansims, or physical injury, often appearing as a swollen area or an anomalous structure (sometimes superficially resembling a fruit).
genealogy. The study of the descent over time of an individual or taxon from its ancestors. In the context of plant evolution and phylogeny, the pattern of presumed ancestral relationships that gave rise to various taxa in a natural group.
genus. Rank in a taxonomic classification consisting of a collection of species with a shared genealogy.
glabrescent. Initially hairy, but becoming glabrous at maturity or with age.
glabrous. Lacking vestiture such as hairs or glands. Sometimes ambiguously referred to as smooth.
gland. Secretory structure.
glaucous. With a white, gray, or light blue waxy coating on the surface.

globose. A three-dimensional shape, globe-shaped or spherical.

grain. Generally applied to small rounded structures, such as a unit of pollen or the seedlike structure on the surface of fruits in *Rumex*. Sometimes applied generally to plants producing seeds or fruits used in cereals and/or flour.

gynoecium. Collective term for the carpels or pistils in a flower.

gynophore. In some flowers, a stalklike extension of the receptacle that elevates the pistil(s) above the attachment points of the perianth and stamens.

habit. Growth form of a plant.

hastate. Halberd-shaped, as in leaf blades that are triangular with the basal lobes spreading rather than pointed downward.

haustorium. In parasitic plants, the organ that forms the physical connection between the parasite and the tissue of the host plant.

head. Dense cluster of flowers that are sessile or nearly so, sometimes surrounded by bracts.

herbaceous. Referring to plants that are not woody and/or die back to the ground every year. Also sometimes applied to tissues that are leaflike or relatively soft in texture (but not papery or membranous).

heterostyly. In some species, the production of flowers with more than one kind of reproductive morphology: some flowers with relatively elongate styles and short stamens such that the stigmas are positioned above the anthers; other flowers with shorter styles and longer stamens such that the stigmas are positioned below the level of the anthers. Rarely a third floral morphology is produced involving stamens of varying lengths in the same flower and short to intermediate length styles. Heterostyly is an adaptation to promote cross-pollination between flowers by limiting the ability of stamens to shed pollen directly onto the stigma(s) within the same flower.

hexaploid. Containing six sets of chromosomes in each cell. Usually contrasted with diploid.

hip. In the genus *Rosa*, a compound fruit in which the urn-shaped hypanthium surrounding the ovaries becomes enlarged, fleshy, and colored as the contained achenes develop.

historical. Event that occurred long ago; for purposes of the present volume, historical refers to events or collections having occurred more than 50 years ago.

hood. A concave, hollow or arched covering, as in a modified petal in the corollas of some flowers.

horn. In some flowers or fruits, an often arched or curved, tapered extension or appendage resembling the horn of a cow.

hybrid. Offspring of a mating between two parents belonging to different taxa. Interspecific hybrids often are sterile or have reduced fertility.

hypanthium. Ring or cup of tissue completely or partially surrounding (but not fused to) the ovary, formed either by the marginal expansion of the receptacle or the basal fusion of stamens and perianth.

hypogynous. Flowers having the perianth and stamens appearing attached at the base of the ovary, the ovary appearing entirely superior to the other floral parts.

imperfect. Applied to an individual flower containing functional stamens or pistils, but not both (often termed unisexual). Plants with imperfect flowers are either monoecious or dioecious.

included. Not extending beyond the surrounding parts. Usually applied to stamens or styles that do not extend beyond the surrounding perianth and therefore can only be observed by dissecting the flower. Contrasted with exserted.

incurved. Referring to a structure with the tip curved inward, toward the middle, or toward the tip of the axis.

indehiscent. Not breaking open or apart at maturity to release its contents, as in some kinds of fruits.

indicator. In the present volume, referring to a taxon with strong fidelity to a particular habitat, such that the finding of the taxon at a given site is indicative of the presence of a particular community or substrate there.

inferior ovary. Ovary with the other flower parts (stamens, petals, and/or sepals) attached at its tip. Flowers with inferior ovaries are referred to as epigynous.

inflorescence. Flowering portion of a plant. The arrangement pattern of the flowers on a plant determine the inflorescence type.

infraspecific. In botanical classification the taxonomic categories within a given species, including subspecies, varieties, and forms.

infructescence. Fruiting portion of a plant; the inflorescence at fruiting time. The arrangement pattern of the fruits on a plant determine the infructescence type.

internode. Portion of a stem, root, or inflorescence axis between the attachment points of structures such as a leaves and branches.

interpetiolar. In species with opposite leaves, structures positioned between the opposing petioles or attachment points in each leaf pair. Usually applied to the stipules in most opposite-leaved species of Rubiaceae, which are formed by fusion of adjacent stipules between the two leaves at each node, such that a single stipular structure is present on each side of the node.

introgression. Also called introgressive hybridization. Widespread hybridization over several generations between two at least partially interfertile taxa resulting in a mixed population of individuals with a range of morphologies variously intermediate between the parents.

invasive. Species that aggressively colonizes plant communities in which it did not occur naturally prior to human-mediated disturbances in recent times. Invasive exotics are non-native plant species that spread aggressively into native plant communities in a region.

involucre. Two or more bracts closely subtending an inflorescence.

iridescent. Shiny with many colors; reflecting a rainbow pattern of colors.

keel. Prominent longitudinal ridge, like the keel of a boat. Also applied to the two, usually fused, lowermost petals of flowers in many species of Fabaceae.

lacerate. Irregularly cut or divided, appearing torn.

lacticifer. Latex canal; latex-conducting tissue found in the tissues of latex-producing species; for example, in Nymphaeaceae.

lanceolate. Two-dimensional shape, lance-shaped, about three to six times as long as wide, broadest below the middle, and tapering to the tip.

lanceoloid. Three-dimensional shape, lance-shaped, about three to six times as long as wide, broadest below the middle, and tapering to the tip.

lateral. Attached to the side of a structure, as opposed to the tip. Often applied to inflorescences originating from the axils of leaves along the stem.

leaf axil. Angle between a leaf and the stem to which it is attached.

leaf blade. The expanded usually flattened portion of a leaf.

leaf sheath. Basal portion of some leaves that is wrapped around the stem.

legume. Dry fruit derived from a single carpel and usually dehiscent longitudinally along a pair of sutures along opposite margins of the fruit. The main fruit type in the Fabaceae. Also, a vernacular name applied generally to members of the Fabaceae.

lenticel. Small, often slightly raised, corky pore or line on the surface of a stem that facilitates gas exchange between the atmosphere and the interior tissues.

liana. Woody climbing plant, for example a grape vine.

linear. Two-dimensional shape that is long and narrow (more than ten times as long as wide) with more or less parallel sides.

lip. Variously shaped projecting segment of a zygomorphic corolla or calyx, sometimes positioned lowermost in the flower. Corollas having both upper and lower lips are termed bilabiate.

lobe. Often rounded segment of a structure that is partially divided (not all the way to the base or center). Usually applied to a part of a leaf blade that is noticeably divided but not compound.

locule. Chamber or compartment of an ovary, fruit, anther, or other structure.

loess. Type of substrate composed of loosely compacted silty material thought to have been created by the grinding action of Pleistocene glaciers and later dispersed by wind across the landscape, creating deposits of uneven thickness.

loment. Dry fruit derived from a single carpel that breaks apart transversely at maturity into one-seeded segments enclosed in the fruit walls. Most commonly found in some genera of Fabaceae.

meiosis. Form of nuclear division normally resulting in four daughter cells, each with half as many sets of chromosomes as the parental cell. One of the two essential stages of sexual reproduction (along with fertilization).

membranous. Thin, soft, and more or less transparent.

mericarp. In the fruit type known as a schizocarp, the seed-containing structures that break apart from one another into separate units and eventually are dispersed intact as the fruit reaches maturity.

meristem. Tissue consisting of undifferentiated cells that divide actively to produce cells that become differentiated into various structures or tissues. Often located at the tip or margins of structures such as stems or leaves.

midnerve. Central nerve or vein of a structure other than a leaf, such as a sepal, petal, or fruit.

midrib. Central vein of a leaf, bract, or scale that is raised above the surface.

midvein. Central vein of a leaf with pinnate venation, often heavier or more conspicuous than the other veins.

mitosis. Form of nuclear division normally resulting in two daughter cells, each with the same number of sets of chromosomes as the parental cell; the cellular process that results in plant growth.

molecular. Dealing with molecules. In botany, usually referring to nucleic acids.

mono-. Prefix meaning one or single, as in a monotypic genus containing only a single species.

monocarpic. Type of life-history strategy in which a perennial plant produces only vegetative growth (usually a basal rosette of leaves) for multiple years before flowering, then dies after flowering and fruiting only once.

monoculture. Habitat or site containing only a single species.

monoecious. Plant that produces exclusively staminate and pistillate flowers, these both produced on the same plant, but sometimes in different positions.

monomorphic. Having only one type of shape or form. May refer to various structures, including leaves, fruits, and flower parts. Contrasted with dimorphic.

monophyletic. Having a shared genealogy; part of the same clade (branch in a phylogenetic tree).

montane. Of or pertaining to mountains.

morphology. Study of the external organization and structures of plants. See also anatomy.

morphometrics. Comparative study of taxa involving the statistical analysis of quantitative (and to some extent qualitative) morphological data.

morphotype. Morphological variant within a taxon.

mucilaginous. Slimy and moist; snotty.

mycorrhizal. Symbiotic relationship between plant roots and soilborne fungi. The plant receives nutrients and water from the fungus and the fungus presumably receives an enclosed, protected environment in which to grow.

mycotrophic. Organism that receives nutrients and water from soilborne fungi through a mycorrhizal association.

naturalized. Non-native taxon that has become established at a site and has formed a viable self-reproducing population.

nectary. Nectar-producing secretory structure, often but not always associated with the flower. Sometimes applied to an organ containing a nectar-producing gland.

nerve. Usually unbranched vein or rib of a structure other than a leaf, such as a corolla segment or fruit.

node. Place on a stem, root, or inflorescence axis at which structures such as leaves and branches are attached.

nodulation. In many members of the Fabaceae (and some other plant families), a symbiotic relationship between the plant and bacteria of the genus of *Rhizobium* resulting in the development of specialized root nodules in which the bacteria fix atmospheric nitrogen into nitrates, the nitrogen-rich compounds used by plants as fertilizers.

nomenclature. System of names for a process or classification. As applied to botany, the system of names of taxa at all levels in the taxonomic classification.

nutlet(s). Appearing similar to small nuts. In some plant families, such as Boraginaceae and Lamiaceae, the one-seeded, hard-walled segments or lobes (mericarps) of the indehiscent dry fruits (schizocarps) that break apart at maturity and are dispersed intact.

ob-. Prefix indicating inversion, such as oblanceolate, meaning lanceolate, but with the attachment point at the narrow end and the widest point above the middle.

oblong. Two-dimensional shape about two to four times as long as wide, broadest at the middle, and rounded or tapering at each end, with the sides more or less parallel.

ocrea. In some genera of Polygonaceae, the modified stipules, which are fused into a papery tubular sheath around the stem.

ocreola. In some genera of Polygonaceae, the modified bracts, which are fused into a papery tubular sheath around the inflorescence axis.

-oid. Suffix indicating similarity, such as petaloid sepals that are colored or otherwise resemble petals. Frequently applied to three-dimensional shapes, such as ovoid, meaning a structure that is egg-shaped.

opposite. With a pair of parts or structures attached on opposite sides of a single point or node, as in stems with two leaves at each node. Contrasted with alternate and whorled.

orbicular. Two-dimensional shape more or less circular in outline.

ovary. Expanded basal portion of the pistil containing the ovule(s).

ovate. Two-dimensional shape, egg-shaped, about two to three times as long as wide, broadest below the middle, and tapering to the tip.

ovoid. Three-dimensional shape, egg-shaped, about two to three times as long as wide, broadest below the middle, and tapering to the tip.

ovule. In seed plants, the structure containing the egg nucleus, associated tissue, and surrounding layer(s) of integument; after fertilization developing into the seed.

palate. In the fused corollas of some species, a raised or swollen portion at the base of the lower lip that partially or completely closes the throat.

palmate. With all of the parts or divisions originating from a single point. Referring to leaves and other lobed or compound structures with the parts all attached to the tip of the subtending petiole or stem, or to a venation pattern with the main veins all arising from a single point at or near the base of the leaf blade.

panicle. Branched inflorescence, technically restricted to inflorescences with the ultimate branches having flowers that mature from the base of each branch toward the tip, but in the present volume referring generally to any inflorescence with the flowers attached to branches rather than to a single central axis.

papilionaceous. Referring to the flowers of species in the Fabaceae subfamily Faboideae, whose zygomorphic corollas are differentiated into a banner, two wings, and a keel.

papillose. Referring to a structure with tiny nipple-like projections (papillae) on the surface.

pappus. In the Asteraceae, the apparent outermost whorl of perianth in some or all flowers of most species, consisting variously of awns, bristles, and/or scales, sometimes reduced to a low crown or rim.

parasitic. Organism that derives all or part of its food and/or water directly from the tissues of another organism.

parietal. Referring to the inner wall of a hollow structure. In botany, a kind of placentation in which ovules or seeds are attached to placentae along the outer wall of a locule (vs. axile placentation, in which the placentae are located centrally where the walls of adjacent locules intersect).

pebbled. With an uneven, sometimes roughened surface having a series of minute, low bumps or papillae.

peltate. Umbrella-shaped, with a relatively flat structure having a stalk attached on the undersurface inward from the base.

pentaploid. Containing five sets of chromosomes in each cell. Because they contain an odd number of chromosome sets, pentaploid plants usually are either sterile or reproduce apomictically.

perennial. Plant that lives for three or more years. Also sometimes applied to stems that do not die back in the winter.

perfect. Applied to an individual flower containing one or more functional stamens and pistils (often termed bisexual).

perfoliate. Referring to a leaf or leaves with the margins entirely surrounding the stem, such that the stem appears to pass through the leaf tissue.

perianth. General term referring collectively to any and all calyx (sepals) and corolla (petals) present in a flower.

perigynous. Flowers having the perianth and stamens appearing attached at the tip of a hypanthium, this saucer-shaped to deeply urn-shaped, sometimes more or less enclosing but not actually fused to the superior ovary.

persistent. Not shed with age, as in the calyx of some plant species that remains attached to the mature fruit.

petal. Structure or unit of the innermost perianth whorl of a flower, often white or brightly colored.

petaloid. Similar in appearance to a petal; for example the colored sepals of flowers in some species.

petiole. Stalklike base of a leaf to which the blade is attached. Leaves bearing petioles are termed petiolate.

photosynthesis. Physiological process by which plants convert water and carbon dioxide into sugars using light for energy.

phylogenetic. Referring to the genealogical history of a group of organisms as represented by ancestor/descendant relationships. Phylogenetics is the branch of systematics concerned with reconstructing phylogenies by hypothesizing and analyzing the primitive and derived states in a data set of various characters for a study group and its presumed closest relatives.

phylum. Highest level in the taxonomic hierarchy of animal classification; equivalent to the term division in plants.

physiology. Study of the processes and mechanisms by which organisms live and function.

phytochemistry. Study of chemical compounds produced by plants.

pinna. Primary division of a compound leaf, which may be further subdivided into pinnules.

pinnate. With the divisions originating at various points along opposite sides of an axis. Usually referring to compound leaves with the leaflets attached singly or in pairs along the central axis (rachis), or to a venation pattern with the main veins arising at different points along a longitudinal midvein.

pinnule. Ultimate discrete division (leaflet) of a compound leaf. Usually applied to leaves that are two or more times compound.

pistil. Ovule-producing organ of a flower, consisting of an ovary (which contains one or more ovules) and usually a style and stigma.

pistillate. Referring to a flower bearing only one or more pistils and no functional stamens, or to a portion of a plant or inflorescence where the pistil-bearing flowers are situated. Sometimes also applied to scales or other structures associated with pistil-bearing flowers.

pith. Soft or spongy tissue at the center of some stems and roots.

placenta. Portion of the ovary to which the ovules are attached. The organization and position of this tissue in the ovary is called placentation.

pleurogram. In the seeds of some species of Fabaceae, a region of the seed coat demarcated by a fine line or groove and differing in color and/or texture.

pneumatophore. Specialized roots in some aquatic plants that grow up to or out along the water surface and often appear inflated, white, or spongy; they facilitate gas exchange between the submerged portions of the plant and the atmosphere.

pollen. In seed plants, the minute structure containing and dispersing the usually immature, highly reduced male gametophyte; formed in an anther. Often referred to as a pollen grain.

polyembryony. In some plant species, the production of more than one embryo within a single seed. In flowering plants, this unusual phenomenon often is associated with apomixis.

polymorphic. With several different forms, referring to plants or structures of plants that have different morphologies in different individuals or populations.

polyploid. Containing more than two sets of chromosomes in each cell. Usually contrasted with diploid.

prickle. Sharply pointed outgrowth of the epidermis or bark. Because prickles are surface structures, they often are relatively easily dislodged from the plant. See also spine, thorn.

primocane. In the genus *Rubus,* the first year's growth of the biennial stem, which bears only leaves and not flowers or fruits. Contrast with floricane.

propagule. A structure that gives rise to a new plant; may refer to a seed, fruit, or asexual structure.

prostrate. Positioned flat on the ground.

protogynous. Condition in some flowers in which the style matures before the stamens and the stigma becomes receptive before the pollen is dispersed. This mechanism promotes cross-pollination. Contrast with protandrous.

pubescent. General term for a surface covered with any type of hair.

pulvinus. Thickened or swollen portion at the base of a petiole or leaflet stalk. Common in the Fabaceae.

punctate. Referring to a structure dotted with colored spots of pigment or with small glands (punctations) that are sessile or often sunken into shallow pits.

pustular-based. Referring to a hair with the base expanded into a small blisterlike mass (pustule), this usually persistent after the threadlike portion of the hair has worn away.

raceme. Unbranched inflorescence, technically restricted to inflorescences with stalked flowers maturing from the base toward the tip of an unbranched axis, but in the present volume referring generally to any inflorescence with stalked flowers attached to an unbranched central axis.

rachis. In a compound leaf, the stemlike axis to which the pinnae are attached. Sometimes also applied to the central axis of an inflorescence.

rank. Level in a taxonomic hierarchy, such as the rank of species. Also applied to the pattern of attachment of structures to an axis, such as a stem with 2-ranked leaves in which the leaves are attached in alternating fashion on opposite sides of the stem, but all in the same plane (as in some species of Ulmaceae).

receptacle. Expanded tip of a stem (pedicel) to which the flower parts are attached. Also applied to the expanded tip of a stem (peduncle) to which the flowers in a head are attached.

recurved. Referring to a structure with the tip curved outward, downward, or away from the tip of the axis.

reflexed. Referring to a structure abruptly bent backward or downward.

relictual. Historical distributional pattern in which populations that occur disjunctly at some significant distance from the main range of a species are thought to have been trapped in specialized microhabitats as climates have changed. Often used in reference to glacial relicts trapped in cool, moist habitats when glaciers retreated and warming climate became inhospitable in intervening areas.

resin-dotted. Referring to a structure dotted with small glands that are sessile or often sunken into shallow pits and produce a resinous exudate.

resupinate. Referring to flowers that are twisted at the base during development so that the top of the flower is oriented toward the bottom at maturity.

reticulate. Forming a network, as in anastomosing veins of a leaf. Also applied to a type of phylogeny in which new species evolve as the result of interspecific hybridization followed by a stabilization of the hybrid and regaining of fertility, such that the offspring becomes a self-reproducing taxon reproductively isolated from both parental taxa.

rhizome. Modified stem occurring all or mostly underground, often creeping horizontally.

rhombic. Two-dimensional shape, more or less diamond-shaped.

rib. Prominent longitudinal vein or line raised above the surface of a structure such as a leaf, bract, scale, or fruit.

rootstock. General term for the underground structures of a plant including rhizomes and other rootlike stems and usually also the roots.

rosette. Group or cluster of leaves positioned in a radial pattern at or near gound level.

samara. Dry, indehiscent, usually single-seeded, winged fruit; for example, the fruits of *Acer* and *Fraxinus*.

scale. Small, flattened, usually thin structure resembling the scale of a fish or reptile. Applied to structures covering some rhizomes, stems, leaves, or other plant parts that may be related to hairs or highly modified leaves. In the Asteraceae, a type of pappus segment.

schizocarp. A dry fruit formed from a compound ovary that breaks apart at maturity into two or more (depending on the number of carpels) indehiscent structures known as mericarps, which eventually are dispersed intact.

scorpioid. Coiled, like the tail of a scorpion. Usually applied to inflorescences with a central axis that uncurls from the tip as it elongates and develops, as in many species of Boraginaceae. In the present volume, this category includes both the truly coiled inflorescences more precisely called helicoid and those derived from a more or less spirally zigzag axis to which the term scorpioid is restricted in some other works.

seed. Ripened ovule, consisting of an embryo and surrounding structures and/or layers.

self. In botany, usually refering to the ability of a flower to self pollinate, that is for pollen from the stamens to be capable of successfully fertilizing the ovules in an ovary of the same flower.

sepal. Structure or unit of the outermost or only perianth whorl of a flower, usually green, but less commonly white or brightly colored.

sepaloid. Similar in appearance to a sepal, as in petals that are green rather than colored, or a bract that closely subtends a calyx.

septum. Partition between the chambers (locules) in an ovary or fruit. Also applied to one of the cross-partitions occurring in the stems or inflated leaves of some plant species.

sessile. Referring to structures that are not stalked, as in leaves lacking petioles.

sexual reproduction. Type of reproduction involving fusion of gametes, such as the fertilization of the egg and polar nuclei by sperm nuclei in angiosperms.

shrub. Woody plant with usually several main stems that are shorter and thinner (generally less than four inches in dbh [diameter at breast height]) at maturity than those of trees.

simple. In the present volume, opposite of compound. Usually applied to leaves with the blade entire or lobed but not fully divided into leaflets.

sinus. In plants, usually applied to the concave area between two adjacent lobes of a leaf or other divided structure.

spatulate. Spatula-shaped, widest near the broad rounded tip and tapering gradually to the base. Sometimes spelled spathulate in the older literature.

species. Basic unit of taxonomic classification; variously defined, but generally consisting of a group of usually interfertile populations of individual organisms with shared distinctive morphological or other features and a shared genealogy that maintain their discrete identity over time.

spherical. Three-dimensional shape, globe-shaped or ball-shaped.

spike. Unbranched inflorescence, technically restricted to inflorescences with sessile flowers maturing from the base toward the tip of an unbranched axis, but in the present volume referring generally to any inflorescence with sessile flowers attached to an unbranched central axis.

spine. Sharply pointed structure derived from a modified leaf or stipule and arising from beneath the epidermis or bark. Often imprecisely applied to any sharply pointed structure on a plant. See also prickle, thorn.

spinescent. Modified into a sharply pointed, stiff or hard, spinelike tip, for example, the involucral bracts of many thistles and some other species of Asteraceae. Note that the term spinescent usually is not applied to true spines, but rather spinelike structures.

spinule. Minute spine or spinelike structure.

spur. In some flowers, a hollow saclike or tubular appendage of a calyx (or individual sepal) or corolla (or individual petal), often containing nectar.

stamen. Pollen-producing organ of a flower, consisting of an anther (containing the pollen in usually two or four sacs) and often a stalklike or threadlike filament.

staminate. Referring to a flower bearing only one or more stamens and no functional pistils, or to a portion of a plant or inflorescence where the stamen-bearing flowers are situated. Sometimes also applied to scales or other structures associated with stamen-bearing flowers.

staminode. Sterile stamen (not producing pollen), usually modified in appearance from that of the fertile stamen(s).

stellate. Star-shaped, usually applied to hairs with several branches in a radial pattern from the base.

sterile. Unable to reproduce sexually, usually because of chromosomal problems during meiosis and/or failure of spores or fruits to mature.

stigma. Portion of the pistil that is receptive to pollen grains.

stipel. Small stipule-like structures at the base of leaflets in a compund leaf. Common in the Fabaceae.

stipule. Appendage at a leaf base. May be leaflike, glandular, spiny, or shaped otherwise. Occurring in a pair on opposite sides of the leaf base and free, united with each other, or partially fused to the petiole or stem.

stolon. Modified stem elongated and arching or creeping above the ground, rooting at nodes or at the tip and giving rise to new plants asexually.

style(s). In some pistils, one or more structures connecting the ovary and stigma(s).

sub-. Prefix indicating below or less than. See subfamily or subtending in this glossary for examples.

subfamily. Infrafamilial rank in a taxonomic classification; a group of genera in a particular family with a shared genealogy and shared cytological, morphological, or other distinguishing features differing from other genera of the same family; sometimes divided into tribes.

subspecies. Infraspecific rank in a taxonomic classification; a group of plants or populations with characteristic cytology, morphology, or other distinguishing features differing from other members of the same species; often used interchangeably with variety, but sometimes used to denote an infraspecific taxon occupying a discrete portion of the range of a species; occasionally divided into varieties.

subtending. Attached or positioned immediately below some structure, such as a bract subtending a flower.

succulent. Referring to plants or structures that are thickened, fleshy, and juicy, as in the stems of most species of Cactaceae.

superior ovary. Ovary with the other flower parts (stamens, petals, and/or sepals) attached at its base or along the rim of a hypanthium that is not fused to the sides of the ovary. Flowers with superior ovaries are referred to as hypogynous (flower parts attached at the ovary base) or perigynous (flower parts attached to a hypanthium).

suture. Line or seam between adjacent fused structures. Often applied to the lines that separate the fused carpels in a compound ovary or the seams along which some fruits dehisce.

synonym. In taxonomic nomenclature, a plant name that refers to the same taxon as another name, but for which the other name has priority.

systematics. Study of the kinds and diversity of organisms and their interrelationships, including identification, classification, nomenclature, and taxonomy.

tapered. Gradually narrowed or drawn out to a usually sharp point, usually with noticeably concave sides. Contrasted with narrowed (wedge-shaped, with more or less straight sides) and rounded (not pointed, but with strongly convex sides).

taproot. Solitary vertical primary root, often with fine spreading branch roots.

taxon. General term for an entity of any rank in a system of biological classification, such as a variety, species, genus, or family.

taxonomy. Technically, the study of the classification of objects into systems, including sets of principles, procedures, and rules, but in practice used interchangeably with the more general term systematics.

tendril. Slender twining organ derived from a modified stem or leaf and used to attach some climbing plants to supporting vegetation or other structures.

tepal. Segment of a perianth that is not differentiated into a calyx and corolla. May be applied to a perianth with numerous parts that grade gradually from green sepal-like segments into white or colored petal-like segments (as in many species of Cactaceae) or to a perianth in which the sepals and petals are identical in morphology and coloration.

terminal. Attached at the tip of a structure, as opposed to the sides, as in an inflorescence attached at the tip of a stem, a flower at the tip of an inflorescence or inflorescence branch, or a leaflet at the tip of a compound leaf.

ternate. Divided into three parts, as in some lobed compound leaves that are divided into lobes or leaflets in groups of threes.

tetraploid. Containing four sets of chromosomes in each cell. Usually contrasted with diploid.

thorn. Modified branch sharply pointed at the tip. Determinate thorns tend to cease growth after reaching a particular length, whereas indeterminate thorns potentially elongate with age into a normal branch with leaves and axillary buds. See also prickle, spine.

tree. Woody plant with usually one or few main stems that are taller and thicker (generally more than four inches in dbh [diameter at breast height]) at maturity than those of shrubs.

tribe. Infrafamilial rank in a taxonomic classification; a group of genera with a shared genealogy in a particular family that share cytological, morphological, or other distinguishing features differing from other genera of the same family.

trichome. General term for any kind of hair or hairlike structure on a plant surface.

trifoliate. Technically, a leaf pattern in which three leaves are attached in a whorl at a node of a stem. However, in many treatments, including the present volume, applied inaccurately to a compound leaf with three leaflets (which technically should be called trifoliolate). Trifoliate leaves may be pinnately or palmately derived, which usually can be determined by whether or not the terminal leaflet has a stalk that is different from the condition of the lateral leaflets.

trigonous. Referring to structures appearing more or less triangular in cross-section. Usually applied to stems or fruits.

triploid. Containing three sets of chromosomes in each cell. Because they contain an odd number of chromosome sets, triploid plants are either sterile or reproduce apomictically.

truncate. Appearing squared off or cut straight-across at the tip or base.

tuber. Underground stem, thickened and with several to numerous buds on the surface, modified for food storage. Also applied to a thickened, modified portion of a rhizome.

tuberous. In reference to a structure that is thickened and fleshy, thus similar in appearance to a tuber; often applied to roots or a modified base of an aerial stem.

tubercle. Small rounded protuberance on a surface or structure.

turion. Small modified budlike branch that acts as an overwintering structure, germinating to regenerate a plant asexually the following spring, as in some species of Lemnaceae.

umbel. Unbranched, flat-topped or convex inflorescence with the stalked flowers attached in a more or less radial pattern at the tip of the main stalk. Also applied to branched inflorescences with a radial pattern of branching and the individual flowers in umbels at the tips of the ultimate branches. In compound umbels, the ultimate clusters of stalked flowers are referred to as umbellets. Also, a vernacular name applied to a member of the plant family Apiaceae.

undulate. Referring to structures with a regular wavy pattern, as in the margins of some leaves..

unisexual. Referring to imperfect flowers or plants producing imperfect flowers. Commonly used but technically incorrect, as angiosperm plants are sporophytes and sexuality applies to gametophytes.

utricle. Indehiscent, dry, thin-walled, more or less inflated or bladder-like, single-seeded fruit.

valve. One of the segments of a dehiscent fruit, usually separated from other segments prior to dehiscence by sutures.

variety. Basic infraspecific rank in a taxonomic classification; a group of plants or populations with characteristic cytology, morphology, or other distinguishing features differing from other members of the same species.

vascular. The tissue (differentiated into xylem and phloem) that conducts the movement of water, nutrients, and sugars between the various organs in plants.

vascular plant. Plant possessing vascular tissue (xylem and phloem for conducting water and sugars, respectively). Although a few mosses and algae have small amounts of vascular tissue, the term is usually reserved for pteridophytes, gymnosperms, and angiosperms.

vegetative. Referring to structures or growth not associated with flowers or fruits.

ventral. Referring to the inner (upper) surface of a structure in relation to the axis. Also applied to the side of a leaf sheath opposite to the attachment point of the leaf blade.

vestiture. General term for any hairs, glands, scales, or other coverings produced on the surface of plant structures.

whorl(ed). With three or more parts or structures attached at a single point or node, as in stems with a whorl of four leaves at each node. Contrasted with alternate and opposite.

wing. Thin or membranous, flattened projection or flap of tissue along a surface or the margin of a structure, such as some stems and fruits. Also applied to a lateral petal in the strongly zygomorphic flowers of many species of Fabaceae.

woody. Producing wood (quantities of hardened secondary vascular tissue). As applied in the present volume, indicating a tree, shrub, or perennial with a woody caudex.

xeric. In reference to a place having a very dry climate.

zygomorphic. Having bilateral symmetry, such as flowers with parts arranged so that opposite sides of a line drawn through the middle are mirror images in only one plane.

LITERATURE CITED IN VOLUME 3

Abbott, J. R. 2011. Notes on the disintegration of *Polygala* (Polygalaceae), with four new genera for the Flora of North America. J. Bot. Res. Inst. Texas 5: 125–137.

Abbott, J. R., and R. L. Thompson. 2011. New combinations in *Phoradendron leucarpum* (Viscaceae). J. Bot. Res. Inst. Texas 5: 139–141.

Abdallah, M. S., and H. C. D. de Wit. 1978. The Resedaceae: A taxonomical revision of the family (final installment). Meded. Landbouwhogeschool 78(14): i–iv, 99–416, fig. 19–91.

Adolphi, H. 1982. Proposal to conserve *Glechoma* L. against *Glecoma* L. Taxon 31: 118.

Aedo, C. 2000. The genus *Geranium* in North America. I. Annual species. Anales Jard. Bot. Madrid 58: 38–82.

———. 2001. The genus *Geranium* in North America. II. Perennial species. Anales Jard. Bot. Madrid 59: 3–65.

Aiken, S. G. 1981. A conspectus of *Myriophyllum* (Haloragaceae) in North America. Brittonia 33: 57–69.

Akiyama, S. 1988. A revision of the genus *Lespedeza* section *Macrolespedeza* (Leguminosae). Univ. Mus., Univ. Tokyo, Bull. 33: i–vii, 1–170, pl. 1–27.

Akiyama, S., and H. Ohba, 1985. The branching of the inflorescence and vegetative shoot and taxonomy of the genus *Kummerowia* (Leguminosae). Bot. Mag. (Tokyo) 98: 137–150.

Albach, D. C., and M. W. Chase. 2004. Incongruence in Veroniceae (Plantaginaceae): Evidence from two plastid and a nuclear ribosomal DNA region. Molec. Phylogen. Evol. 32: 183–197.

Albach D. C., M. M. Martínez-Ortega, M. A. Fischer, and M. W. Chase. 2004. A new classification of the tribe Veroniceae—Problems and a possible solution. Taxon 53: 429–452.

Albach, D. C., H. M. Meudt, and B. Oxelman. 2005. Piecing together the "new" Plantaginaceae. Amer. J. Bot. 92: 297–315.

Albach D. C., M. M. Martínez-Ortega, L. Delgado, H. Weiss-Schneeweis, F. Özgökce, and M. A. Fischer. 2008. Chromosome numbers in Veroniceae (Plantaginaceae): Review and several new counts. Ann. Missouri Bot. Gard. 95: 543–566.

Albach, D. C., K. Yan, S. R. Jensen, and H.-q. Li. 2009. Phylogenetic placement of *Triaenophora* (formerly Scrophulariaceae) with some implications for the phylogeny of the Lamiales. Taxon 58: 749–756.

Aldasoro, J. J., C. Aedo, and C. Navarro. 1998. Pome anatomy of Rosaceae subfam. Maloideae, with special reference to *Pyrus*. Ann. Missouri Bot. Gard. 85: 518–527.

Alford, M. H. 2006. Gerrardinaceae: A new family of African flowering plants unresolved among Brassicales, Huerteales, Malvales, and Sapindales. Taxon 55: 959–964.

Alice, L. A., D. H. Goldman, G. Moore, and J. A. Macklin. In press. *Rubus*. *in* Flora of North America Editorial Committee, eds. Flora of North America North of Mexico. Volume 9. Picramniaceae to Rosaceae. Oxford University Press, New York.

Allan, G. P., and J. M. Porter. 2000. Tribal delimitation and phylogenetic relationships of Loteae and Coronilleae (Faboideae: Fabaceae) with special reference to *Lotus*: Evidence from nuclear ribosomal ITS sequences. Amer. J. Bot. 87: 1871–1881.

Allard, H. A. 1943. The eastern false mistletoe (*Phoradendron flavescens*); when does it flower? Castanea 8: 72–78.

Allen, O. N., and E. K. Allen. 1981. The Leguminosae: A Source Book of Characteristics, Uses, and Nodulation. University of Wisconsin Press, Madison.

Al-Shehbaz, I. A. 1991. The genera of Boraginaceae in the southeastern United States. J. Arnold Arbor., suppl. ser. 1: 1–169.

Alston, R. E., and B. L. Turner. 1963. Natural hybridization among four species of *Baptisia* (Leguminosae). Amer. J. Bot. 50: 159–173.

Alverson, W. S., B. A. Whitlock, R. Nyffeler, C. Bayer, and D. A. Baum. 1999. Phylogeny of the core Malvales: evidence from *ndhF* sequence data. Amer. J. Bot. 86: 1474–1486.

Anderson, C. L., K. Bremer, and E. M. Friis. 2005. Dating phylogenetically basal eudicots using *rbcL* sequences and multiple fossil reference points. Amer. J. Bot. 92: 1737–1748.

Anderson, G. J. 1976. The pollination of *Tilia*. Amer. J. Bot. 63: 1203–1212.

Anderson, G. J., and J. D. Hill. 2002. Many to flower, few to fruit: The reproductive biology of *Hamamelis virginiana* (Hamamelidaceae). Amer. J. Bot. 89: 67–78.

Ando, T., N. Ishikawa, H. Watanabe, H. Kokubun, Y. Yanagisawa, G. Hashimoto, E. Marchesi, and E. Suárez. 2005. A morphological study of the *Petunia integrifolia* complex (Solanaceae). Ann. Bot. 96: 887–900.

Angiosperm Phylogeny Group. 1998. An ordinal classification for the families of flowering plants. Ann. Missouri Bot. Gard. 85: 531–553.

———. 2003. An update of the Angiosperm Phylogeny Group classification for the orders and families of flowering plants: APG II. Bot. J. Linn. Soc. 141: 399–436.

———. 2009. An update of the Angiosperm Phylogeny Group classification for the orders and families of flowering plants: APG III. Bot. J. Linn. Soc. 161: 105–121.

Applequist, W. L. 2006. Proposal to reject *Plantago psyllium* L. (Plantaginaceae). Taxon 55: 235–236.

Applequist, W. L., and R. S. Wallace. 2001. Phylogeny of the portulacaceous cohort based on *ndhF* sequences. Syst. Bot. 26: 406–419.

Aradhya, M., Y. Wang, M. A. Walker, B. H. Prins, A. M. Koehmstedt, D. Velasco, J. M. Gerrath, G. S. Dangl, and J. E. Preece. 2013. Genetic diversity, structure, and patterns of differentiation in the genus *Vitis*. Pl. Syst. Evol. 299: 317–330.

Arambarri, A. M., S. A. Stenglen, M. N. Colares, and M. C. Novoa. 2005. Taxonomy of the New World species of *Lotus* (Leguminosae: Loteae). Austral. J. Bot. 53: 797–812.

Argus, G. W. 1986. The genus *Salix* (Salicaceae) in the southeastern United States. Syst. Bot. Monogr. 9: 1–170.

———. 2007. *Salix* (Salicaceae) distribution maps and a synopsis of their classification in North America, north of Mexico. Harvard Pap. Bot. 12: 333–368.

———. 2010. *Salix*. Pp. 23–162 *in* Flora of North America Editorial Committee, eds., Flora of North America North of Mexico. Volume 7. Magnoliophyta: Salicaceae to Brassicaceae. Oxford University Press, New York.

Arkansas Vascular Flora Committee. 2006. Checklist of the Vascular Plants of Arkansas. University of Arkansas, Fayetteville.

Armstrong, J. E. 1985. The delimitation of Bignoniaceae and Scrophulariaceae based on floral anatomy, and the placement of problem genera. Amer. J. Bot. 72: 755–766.

Aspinwall, N., and T. Christian. 1992a. Clonal structure, genotypic diversity, and seed production in populations of *Filipendula rubra* (Rosaceae) from the Northcentral United States. Amer. J. Bot. 79: 294–299.

———. 1992b. Pollination biology, seed production, and population structure in queen-of-the-prairie, *Filipendula rubra* (Rosaceae) at Botkin Fen, Missouri. Amer. J. Bot. 79: 488–494.

Bacigalupo, N. M., and E. L. Cabral. 1999. Revisión de las especies Americanas del género *Diodia* (Rubiaceae, Spermacoceae). Darwiniana 37: 153–165.

Backlund, M., B. Oxelman, and B. Bremer. 2000. Phylogenetic relationships within the Gentianales based on *ndhF* and *rbcL* sequences, with particular reference to the Loganiaceae. Amer. J. Bot. 87: 1029–1043.

Bailey, J. P., and C. A. Stace. 1992. Chromosome number, morphology, pairing, and DNA values of species and hybrids in the genus *Fallopia* (Polygonaceae). Pl. Syst. Evol. 180: 29–52.

Bailey, L. H. 1911. Sketch of the Evolution of our Native Fruits. MacMillan Company, New York.

———. 1923. V. Certain cultivated rubi. Gentes Herb. 1: 139–200.

———. 1925. VI. Rubus: Enumeration of the *Eubati* (dewberries and blackberries) native in North America. Gentes Herb. 1: 203–297.

———. 1932. Addenda in *Eubatus*. Gentes Herb. 2: 442–471.

———. 1943a. Species batorum. The genus *Rubus* in North America. V. *Flagellares*. Gentes Herb. 5: 229–422.

———. 1943b. Species batorum. The genus *Rubus* in North America. V. *Cuneifolii*. Gentes Herb. 5: 423–461.

———. 1944. Species batorum. The genus Rubus in North America. VIII. *Alleghenienses*. Gentes Herb. 5: 504–588.

———. 1945. Species batorum. The genus *Rubus* in North America. IX. *Arguti*. Gentes Herb. 5: 589–835.

Bailey, V. L. 1962. Revision of the genus *Ptelea* (Rutaceae). Brittonia 14: 1–45.

Bailey, V. L., S. B. Herlin, and H. E. Bailey. 1970. *Ptelea trifoliata* ssp. *trifoliata* (Rutaceae) in deciduous forest regions of eastern North America. Brittonia 22: 346–358.

Baker, H. G. 1961. The adaptation of flowering plants to nocturnal and crepuscular pollinators. Quart. Rev. Biol. 36: 64–73.

Baker, M., and W. M. Williams. 1987. White Clover. CAB International, Wallingford, Oxon, Great Britain.

Baldwin, J. T., Jr., and B. M. Speese. 1957. *Phoradendron flavescens*: Chromosomes, seedlings, and hosts. Amer. J. Bot. 44: 136–140.

Ballard, H. E., Jr. 1994. Violets of Michigan. Michigan Bot. 33: 131–199.

Ballard, H. E., [Jr.], G. A. Wahlert, J. de Paula-Souza, and M. Feng. 2009. Molecular phylogenetic relationships, infrafamilial groups, and a proposed classification for the violet family (Violaceae). P. 188, in Botany and Mycology 2009. Snowbird, Utah, July 25–29, Abstract Book. [Abstract, see also http://2009.botanyconference.org/engine/search/index.php?func=detail&aid=660].

Baranski, M. J. 1975. An analysis of variation within white oak (*Quercus alba* L.) Techn. Publ. North Carolina Agric. Exp. Sta. 236: i–iv, 1–176.

Barboza, G. E. 2000. Rehabilitación del genero *Quincula* (Solanaceae, Solaneae). Kurtziana 28: 69–79.

Barker, W. R., G. L. Nesom, P. M. Beardsley, and N. S. Fraga. 2012. A taxonomic conspectus of Phrymaceae: A narrowed circumscription for *Mimulus*, new and resurrected genera, and new names and combinations. Phytoneuron 2012–39: 1–60.

Barneby, R. C. 1952. A revision of North American species of *Oxytropis* DC. Proc. Calif. Acad. Sci., ser. IV 27: 177–312.

———. 1964. Atlas of North American *Astragalus*. Mem. New York Bot. Gard. 13: 1–1188 (2 parts).

———. 1977. Daleae imagines: an illustrated revision of *Errazurizia* Philippi, *Psorothamnus* Rydberg, *Marina* Liebmann, and *Dalea* Lucanus emend. Barneby, including all species of Leguminosae tribe Amorpheae Borissova ever referred to *Dalea*. Mem. New York Bot. Gard. 27: i–viii, 1–891.

Barnes, D. K., B. P. Goplen, and J. E. Baylor. 1988. Highlights in the USA and Canada. Pp. 1–22 *in* A. A. Hanson, D. K. Barnes, and R. R. Hill Jr., eds., Alfalfa and Alfalfa Improvement. Agronomy Monograph 29. American Society of Agronomy Press, Madison, WI.

Bartgis, R. L. 1985. Rediscovery of *Trifolium stoloniferum* Muhl. ex A. Eaton. Rhodora 87: 425–429.

Bartholomew, E. A. 1941. *Galium pedemontanum* in North America. Castanea 6: 141–142.

Basinger, M. A. 2002. Notable plant records from Missouri. Trans. Illinois State Acad. Sci. 95: 247–249.

Baskin, C. C., and J. M. Baskin. 1998. Seeds: Ecology, Biogeography, and Evolution of Dormancy and Germination. Academic Press, San Diego.

Baskin, J. M., and C. C.Baskin. 1978. Leaf temperatures in *Heliotropium tenellum* and their ecological implications. Amer. Midl. Naturalist 100: 488–492.

Bassett, I. J. 1966. Taxonomy of North American *Plantago* L. section *Micropsyllium* Decne. Canad. J. Bot. 44: 467–479.

Bates, D. M. 1967. A reconsideration of *Sidopsis* Rydberg and notes on *Malvastrum* A. Gray (Malvaceae). Rhodora 69: 9–28.

Baudoin, J. P. 1988. Genetic resources, domestication, and evolution of Lima bean, *Phaseolus lunatus*. Pp. 125–142 *in* P. Gepts, ed., Genetic Resources of *Phaseolus* Beans. Kluwer Academic Publishers, Dordrecht, The Netherlands.

Baum, B. R. 1967. Introduced and naturalized tamarisks in the United States and Canada (Tamaricaceae). Baileya 15: 19–25.

———. 1978. The Genus *Tamarix*. Israel Academy of Science and Humanities, Jerusalem.

Bayer, C., M. F. Fay, A. Y. De Bruijn, V. Savolainen, C. M. Morton, K. Kubitzki, W. S. Alverson, and M. W. Chase. 1999. Support for an expanded family concept of Malvaceae within a recircumscribed order Malvales: A combined analysis of plastid *atp*B and *rbc*L DNA sequences. Bot. J. Linn. Soc. 129: 267–303.

Bayer, R. J., D. J. Mabberley, C. Morton, C. H. Miller, I. K. Sharma, B. E. Pfeil, S. Rich, R. Hitchcock, and S. Sykes. 2009. A molecular phylogeny of the orange subfamily (Rutaceae: Aurantioideae) using nine cpDNA sequences. Amer. J. Bot. 96: 668–685.

Beal, E. O. 1956. Taxonomic revision of the genus *Nuphar* Sm. of North America and Europe. J. Elisha Mitchell Sci. Soc. 72: 317–346.

Beardsley, P. M., and R. G. Olmstead. 2002. Redefining Phrymaceae: the placement of *Mimulus*, tribe Mimuleae, and *Phryma*. Amer. J. Bot. 89: 1093–1102.

Beardsley, P. M., S. E. Schoenig, J. B. Whittall, and R. G. Olmstead. 2004. Patterns of evolution in western North American *Mimulus* (Phrymaceae). Amer. J. Bot. 91: 474–489.

Beckett, K. A. 1995. Miscellaneous note 5: Yellow loosestrife (*Lysimachia vulgaris*): An observation. New Plantsman 2: 140–141.

Bell, C. D. 2004. Preliminary phylogeny of Valerianaceae (Dipsacales) inferred from nuclear and chloroplast DNA sequence data. Molec. Phylogenet. Evol. 31: 340–350.

———. 2007. Phylogenetic placement and biogeography of the North American species of *Valerianella* (Valerianaceae: Dipsacales) based on chloroplast and nuclear DNA. Molec. Phylogenet. Evol. 44: 929–941.

Bell, C. D., and M. J. Donoghue, 2005. Phylogeny and biogeography of Valerianaceae (Dipsacales) with special reference to the South American valerians. Organisms Divers. Evol. 5: 147–159.

Bell, C. D., E. J. Edwards, S.-t. Kim, and M. J. Donoghue. 2001. Dipsacales phylogeny based on chloroplast DNA sequences. Harvard Pap. Bot. 6: 481–499.

Bell, C. D., A. Kutschker, and M. K. T. Arroyo. 2012. Phylogeny and diversification of Valerianaceae (Dipsacales) in the southern Andes. Molec. Phylogenet. Evol. 63: 724–737.

Bell, C. R., and L. J. Musselman. 1982. Unilateral hybridization in *Aureolaria* Raf. (Scrophulariaceae). Amer. J. Bot. 69: 647–649.

Belyaeva, I. 2009. Nomenclature of *Salix fragilis* L. and a new species, *S. euxina* (Salicaceae). Taxon 58: 1344–1348.

Bendiksby, M., A. K. Brysting, L. Thorbek, G. Gussarova, and O. Ryding. 2011. Molecular phylogeny and taxonomy of the genus *Lamium* L. (Lamiaceae): Disentangling origins of presumed allotetraploids. Taxon 60: 986–1000.

Bennett, J. R., and S. Mathews. 2006. Phylogeny of the parasitic plant family Orobanchaceae inferred from phytochrome A. Amer. J. Bot. 93: 1039–1051.

Benson, L. 1943. Revisions of status of southwestern desert trees and shrubs. II. Amer. J. Bot. 30: 630–632.

———. 1948. A treatise on the North American *Ranunculi*. Amer. Midl. Naturalist 40: 1–261.

———. 1954. Supplement to a treatise on the North American *Ranunculi*. Amer. Midl. Naturalist 52: 328–369.

Benson, W. W., K. S. Brown Jr., and L. E. Gilbert. 1975. Coevolution of plants and herbivores: passion flower butterflies. Evolution 29: 659–680.

Bentz, G. D., and T. S. Cooperrider. 1978. The Scrophulariaceae subfamily Rhinanthoideae of Ohio. Castanea 43: 145–154.

Bernström, P. 1955. Cytogenetic studies on relationships between annual species of *Lamium*. Hereditas 41: 1–122.

Bisognano, J. D., K. S. McGrody, and A. M. Spence. 2005. Myocarditis from the Chinese sumac tree. Ann. Intern. Med. 143: 159–160.

Blackmon, W. J., and B. D. Reynolds. 1986. The crop potential of *Apios americana*—preliminary evaluations. HortScience 21: 1334–1336.

Black-Schaefer, C. L., and R. L. Beckmann. 1989. Foliar flavonoids and the determination of ploidy and gender in *Fraxinus americana* and *F. pennsylvanica* (Oleaceae). Castanea 54: 115–118.

Blamey, M., and C. Grey-Wilson. 1989. The Illustrated Flora of Britain and Northern Europe. Hodder & Stoughton, London.

Blanchard, O. J. 1976. A revision of species segregated from *Hibiscus* sect. *Trionum* (Medicus) de Candolle *sensu lato*. Ph.D. dissertation, Cornell University, Ithaca, NY.

Blanchard, W. H. 1906. Some Maine rubi. The blackberries of the Kennebunks and Wells, II. Rhodora 8: 169–180.

Boeshore, I. 1920. The morphological continuity of Scrophulariaceae and Orobanchaceae. Contr. Bot. Lab. Univ. Pennsylvania 5: 139–177.

Bogle, A. L. 1970. The genera of Molluginaceae and Aizoaceae in the southeastern United States. J. Arnold Arbor. 51: 431–462.

———. The genera of Nyctaginaceae in the southeastern United States. J. Arnold Arbor. 55: 1–37.

Bohm, B. A., J. Y. Yang, J. E. Page, and D. S. Soltis. 1999. Flavonoids, DNA, and relationships of *Itea* and *Pterostemon*. Biochem. Syst. Ecol. 27: 79–83.

Böllmann, J., and M. Scholler. 2004. *Glechoma hederacea* (Lamiaceae) in North America: Invasion history and current distribution. Feddes Repert. 115: 178–188.

Bolmgren, K., and B. Oxelman. 2004. Generic limits in *Rhamnus* s. l. L. (Rhamnaceae) inferred from nuclear and chloroplast DNA sequence phylogenies. Taxon 53: 383–390.

Bond, G. 1967. Nitrogen fixation in some non-legume root nodules. Phyton (Buenos Aires) 24: 57–66.

Borsch, T., and W. Barthlott. 1994. Classification and distribution of the genus *Nelumbo* Adans. (Nelumbonaceae). Beitr. Biol. Pflanzen 68: 421–450.

Bortiri, E., S.-h. Oh, J. Jiang, S. Baggett, A. Granger, C. Weeks, M. Buckingham, D. Potter, and D. E. Parfitt. 2001. Phylogeny and systematics of *Prunus* (Rosaceae) as determined by sequence analysis of ITS and the chloroplast *trnL-trnF* spacer DNA. Syst. Bot. 26: 797–807.

Bortiri, E., S.-h. Oh, F.-y. Gao, and D. Potter. 2002. The phylogenetic utility of nucleotide sequences of sorbitol 6-phosphate dehydrogenase in *Prunus* (Rosaceae). Amer. J. Bot. 89: 1697–1708.

Boufford, D. E. 1982. The systematics and evolution of *Circaea* (Onagraceae). Ann. Missouri Bot. Gard. 69: 804–994.

———. 1997a. *Fumaria*. Pp. 356–357 *in* Flora of North America Editorial Committee, eds., Flora of North America North of Mexico. Volume 3. Magnoliophyta: Magnoliidae and Hamamelidae. Oxford University Press, New York.

———. 1997b. Urticaceae. Pp. 400–413 *in* Flora of North America Editorial Committee, eds., Flora of North America North of Mexico. Volume 3. Magnoliophyta: Magnoliidae and Hamamelidae. Oxford University Press, New York.

———. 2005. *Circaea lutetiana* sensu lato (Onagraceae) reconsidered. Harvard Pap. Bot. 9: 255–256.

Bowe, L. M., and P. L. Redfearn Jr. 2002. A new escaped species in Missouri: *Evodia daniellii* (Rutaceae). Missouriensis 23: 28–31.

Bowers, K. A. W. 1975. The pollination ecology of *Solanum rostratum*. Amer. J. Bot. 62: 635–638.

Braham, R. R. 1977. Crown position and heterophylly in white oak. Michigan Bot. 16: 141–147.

Bremekamp, C. E. B. 1952. The African species of *Oldenlandia* L. *sensu* Hiern & K. Schumann. Verh. Kon. Ned. Akad. Wetensch., Afd. Natuurk., Tweede Reeks 48: 1–297.

Bretting, P. K. 1981. A systematic and ethnobotanical survey of *Proboscidea* and allied genera of Martyniaceae. Ph.D. dissertation, Indiana University, Bloomington.

———. 1983. The taxonomic relationship between *Proboscidea louisianica* and *Proboscidea fragrans* (Martyniaceae). SouthW. Naturalist 28: 445–449.

Brinckmann, J., and B. Wollschlager. 2003. American Botanical Council Black Cohosh Education Module. American Botanical Council, Austin, TX.

Britton, N. L. 1901. Manual of the Flora of the Northern States and Canada. Henry Holt & Co., New York.

Brizicky, G. K. 1961. The genera of Turneraceae and Passifloraceae in the southeastern United States. J. Arnold Arbor. 42: 204–218.

———. 1962. The genera of Rutaceae in the southeastern United States. J. Arnold Arbor. 43: 1–22.

———. 1964a. The genera of Rhamnaceae in the southeastern United States. J. Arnold Arbor. 45: 439–463.

———. 1964b. A further note on *Ceanothus herbaceus* versus *C. ovatus*. J. Arnold Arbor. 45: 471–473.

———. 1965. The genera of Vitaceae in the southeastern United States. J. Arnold Arbor. 46: 48–67.

Brooks, C. J. 1977. A revision of the genus *Forestiera* (Oleaceae). Ph.D. dissertation, University of Alabama, Tuscaloosa, AL.

Brooks, R. E. 1976. A new *Veronica* (Scrophulariaceae) hybrid from Nebraska. Rhodora 78: 773–775.

———. 1983. *Trifolium stoloniferum*, running buffalo clover: Description, distribution, and current status. Rhodora 85: 343–354.

———. 1986. Lamiaceae Lindl. Pp. 708–740 *in* Great Plains Flora Association, eds., Flora of the Great Plains. University Press of Kansas, Lawrence.

Brouillet, L. 2008. The taxonomy of North American Loti (Fabaceae: Loteae): new names in *Acmispon* and *Hosackia*. J. Bot. Res. Inst. Texas 21: 387–394.

Brown, D. K., and R. B. Kaul. Floral structure and mechanism in Loasaceae. Amer. J. Bot. 68: 361–372.

Brown, G. K., and G. S. Varadarajan. 1985. Studies in Caryophyllales I: Reevaluation of classification of Phytolaccaceae s. l. Syst. Bot. 10: 49–63.

Brummitt, R. K. 1988. Report of the Committee for Spermatophyta: 34. Synopsis of decisions Sydney 1981–Berlin 1987. Taxon 37: 139–140.

Brummitt, R. K. 2009. Report of the Committee for Vascular Plants: 60. Taxon 58: 280–292.

Bruneau, A., and G. J. Anderson. 1988. Reproductive biology of diploid and triploid *Apios americana* (Leguminosae). Amer. J. Bot. 75: 1876–1883.

———. 1994. To bee or not to bee?: The pollination biology of *Apios americana* (Leguminosae). Pl. Syst. Evol. 192: 147–149.

Bruneau, A., J. R. Starr, and S. Joly. 2007. Phylogenetic relationships in the genus *Rosa*: New evidence from chloroplast DNA sequences and an appraisal of current knowledge. Syst. Bot. 32: 366–378.

Brunsfeld, S. J., D. E. Soltis, and P. S. Soltis. 1991. Patterns of genetic variation in *Salix* section *Longifoliae* Salicaceae). Amer. J. Bot. 78: 855–869.

———. 1992. Evolutionary patterns and processes in *Salix* section *Longifoliae*: Evidence from chloroplast DNA. Syst. Bot. 17: 239–256.

Buchanan, R. A., I. M. Cull, F. H. Otey, and C. R. Russell. 1978. Hydrocarbon- and rubber-producing crops; evaluation of 100 U.S. plant species. Econ. Bot. 32: 146–153.

Buchmann, S. L. 1983. Buzz pollination in angiosperms. Pp. 73–113 *in* C. E. Jones and R. J. Little, eds., Handbook of Experimental Pollination Biology. Van Nostrand Reinhold, New York.

Buerki, S., F. Forest, P. Acevedo-Rodríguez, M. W. Callmander, J. A. A. Nylander, M. Harrington, I. Sanmartín, F. Küpfer, and N. Alvarez. 2009. Plastid and nuclear DNA markers reveal intricate relationships at subfamilial and tribal levels in the soapberry family (Sapindaceae). Molec. Phylogen. Evol. 51: 238–258.

Buerki, S., P. P. Lowry II, N. Alvarez, S. G. Razafimandimbison, P. Küpfer, and M. W. Callmander. 2010. Phylogeny and circumscription of Sapindaceae revisited: Molecular sequence data, morphology and biogeography support recognition of a new family, Xanthoceraceae. Pl. Ecol. Evol. 143: 148–159.

Burnett, J. H. 1950. The correct name for *Veronica aquatica* Bernhardi. Watsonia 1: 349–353.

Burnham, C. R. 1988. The restoration of the American chestnut. Amer. Sci. 76: 478–487.

Burns, G. W. 1942. The taxonomy and cytology of *Saxifraga pensylvanica* L. and related forms. Amer. Midl. Naturalist 28: 127–160.

Burns, J. F. 1986. The Polygalaceae of Ohio. Castanea 51: 137–144.

Burrows, G. E., and R. J. Tyrl. 2001. Toxic Plants of North America. Iowa State University Press, Ames.

Bush, B. F. 1916. The Missouri agrimonies. Ann. Missouri Bot. Gard. 3: 309–318.

———. 1927. A great blunder. Amer. Midl. Naturalist 10: 401–403.

Callahan, H. S. 1997. Infraspecific differentiation in the *Amphicarpaea bracteata* (Fabaceae) species complex: varieties and ecotypes. Rhodora 99: 64–82.

Callen, E. O. 1959. Studies in the genus *Lotus* (Leguminosae) I. Limits and subdivision of the genus. Canad. J. Bot. 37: 157–165.

Cameron, D. D., and J. F. Bolin. 2010. Isotopic evidence of partial mycoheterotrophy in the Gentianaceae: *Bartonia virginica* and *Obolaria virginica* as case studies. Amer. J. Bot. 97: 1272–1277.

Campbell, C. 2004. *Phylloxera*. How Wine Was Saved for the World. HarperCollins Publishers, London.

Campbell, C. S., R. C. Evans, D. R. Morgan, T. A. Dickinson, and M. P. Arsenault. 2007. Phylogeny of subtribe Pyrinae (formerly the Maloideae, Rosaceae): limited resolution of a complex evolutionary history. Pl. Syst. Evol. 266: 119–145.

Campbell, J. J. N., M. Evans, M. E. Medley, and N. L. Taylor. 1988. Buffalo clovers in Kentucky (*Trifolium stoloniferum* and *T. reflexum*): historical records, presettlement environment, rediscovery, endangered status, cultivation, and chromosome number. Rhodora 90: 399–418.

Canne, J. M. 1981. Chromosome counts in *Agalinis* and related taxa (Scrophulariaceae). Canad. J. Bot. 59: 1111–1116.

Canne-Hilliker, J. M. 1987. Status Report on Skinner's Purple False Foxglove, *Agalinis skinneriana* (Wood) Britton, an Endangered Species in Canada. Unpublished report to the Plants Subcommittee, Committee on the Status of Endangered Wildlife of Canada, National Museum of Natural Sciences, Ottawa.

———. 1988. *Agalinis* (Scrophulariaceae) in Peru and Bolivia. Brittonia 40: 433–440.

Canne-Hilliker, J. M., and C. M. Kampny. 1991. Taxonomic significance of leaf and stem anatomy of *Agalinis* (Scrophulariaceae) from the U.S.A. and Canada. Canad. J. Bot. 69: 1935–1950.

Cantino, P. D. 1982. A monograph of the genus *Physostegia* (Labiatae). Contr. Gray Herb. 211: 1–105.

———. 1992a. Evidence for a polyphyletic origin of the Labiatae. Ann. Missouri Bot. Gard. 79: 361–379.

———. 1992b. Toward a phylogenetic classification of the Labiatae. Pp. 27–37 *in* R. M. Harley and T. Reynolds, eds., Advances in Labiat Science. Royal Botanic Gardens, Kew, Great Britain.

Cantino, P. D., and S. J. Wagstaff. 1998. A reexamination of North American *Satureja* s.l. (Lamiaceae) in light of molecular evidence. Brittonia 50: 63–70.

Carolin, R. C. 1987. A review of the family Portulacaceae. Austral. J. Bot. 35: 383–412.

Carr, B. L., D. P. Gregory, P. H. Raven, and W. Tai. 1986. Experimental hybridization and chromosomal diversity within *Gaura* sect. *Gaura* (onagraceae). Syst. Bot. 11: 98–111.

Carulli, J. P., and D. G. Fairbrothers. 1988. Allozyme variation in three eastern United States species of *Aeschynomene* (Fabaceae), including the rare *A. virginica*. Syst. Bot. 13: 559–566.

Carulli, J. P., A. O. Tucker, and N. H. Dill. 1988. *Aeschynomene rudis* Benth. (Fabaceae) in the United States. Bartonia 54: 18–20.

Carver, G. W. 1916. How to grow the peanut and 105 uses for preparing it for human consumption. Tuskegee Institute Experimental Station Bulletin 31. Tuskegee Institute Press, Tuskegee, AL [7th ed. (1940) online at: http://aggie-horticulture.tamu.edu/publications/guides/carver_peanut.html].

Castaner, D. 1982a. *Galium divaricatum* (Rubiaceae), new to Missouri. Sida 9: 368–369.

———. 1982b. Two speedwells new to Missouri. Missouriensis 4: 49.

Castaner, D., and D. LaPlante. 1992. Plants of the Little Bean Marsh Wildlife Area, Platte County, Missouri. Missouriensis 13: 27–49.

Castaner, D., and T. Priesendorf. 1989. *Leonurus marrubiastrum* L. (Lamiaceae), new to Missouri. Sida 13: 383–384.

Catling, P. M. 1998. A synopsis of the genus *Proserpinaca* in the southeastern United States. Castanea 63: 408–414.

Caulkins, D. B., and R. Wyatt. Variation and taxonomy of *Phytolacca americana* and *P. rigida* in the southeastern United States. Bull. Torrey Bot. Club 117: 357–367.

Chacón S., M. I., B. Pickersgill, and D. G. Debouck. 2005. Domestication patterns in common bean (*Phaseolus vulgaris* L.) and the origin of the Mesoamerican and Andean cultivated races. Theor. Appl. Genet. 110: 432–444.

Chang, M.-c., L.-q. Qiu, and P. S. Green. 1996. Oleaceae. Pp. 272–319 *in* Wu Zheng-yi and Peter H. Raven, eds., Flora of China, Volume 15, Myrsinaceae through Loganiaceae. Science Press, Beijing.

Channell, R. B., and C. E. Wood Jr. 1959. The genera of the Primulales of the southeastern United States. J. Arnold Arbor. 40: 268–288.

———. 1962. The Leitneriaceae in the southeastern United States. J. Arnold Arbor. 43: 435–438.

Chase, M. W, S. Zmarzty, M. D. Lledó, K. J. Wurdack, S. M. Swensen, and M. F. Fay. 2002. When in doubt, put it in Flacourtiaceae: A molecular phylogenetic analysis based on plastid *rbcL* DNA sequences. Kew Bull. 57: 141–181.

Chassot, P., S. Nemomissa, Y.-m. Yuan, and P. Küpfer. 2001. High paraphyly of *Swertia* L. (Gentianaceae) in the *Gentianella*-lineage as revealed by nuclear and chloroplast DNA sequence variation. Pl. Syst. Evol. 229: 1–21.

Chen, S.-l., and M. G. Gilbert. 1994. Verbenaceae. Pp. 1–49 *in* Wu Zheng-yi and Peter H. Raven, eds., Flora of China. Volume 17. Verbenaceae through Solanaceae. Science Press, Beijing.

Chen, Y., and C. Schirarend. 2007. Rhamnaceae. Pp. 115–168 *in* Wu Zheng-yi and Peter H. Raven, eds., Flora of China. Volume 12. Hippocastanaceae through Theaceae. Science Press, Beijing.

Chew, W.-l. 1969. A monograph of *Laportea* (Urticaceae). Gard. Bull. Straits Settlem. 25: 111–178.

Christ, A. 1984. Missouri's interesting flora: Wood anemone. Missouriensis 5: 129.

Chrtková-Zertová, A. 1973. A monographic study of *Lotus corniculatus* L. I. Central and northern Europe. Rozpr. Ceskoslov. Akad. Ved. Rada Mat. Prír. Ved 83(4): 1–94, tab. 1.

Chuang, T. I., and L. Constance. 1977. Cytogeography of *Phacelia ranunculacea* (Hydrophyllaceae). Rhodora 79: 115–122.

Church, S. A. 2003. Molecular phylogenetics of *Houstonia* (Rubiaceae): descending aneuploidy and breeding system evolution in the radiation of the lineage across North America. Molec. Phylogen. Evol. 27: 223–238.

Church, S. A., and D. R. Taylor. 2005. Speciation and hybridization among *Houstonia* (Rubiaceae) species: the influence of polyploidy on reticulate evolution. Amer. J. Bot. 92: 1732–1780.

Claßen-Bockhoff, R., P. Wester, and E. Tweraser. 2003. The staminal lever mechanism in *Salvia* L. (Lamiaceae)—a review. Pl. Biol. 5: 33–41.
Clewell, A. F. 1964. The biology of the common native lespedezas in southern Indiana. Brittonia 16: 208–220.
———. 1966a. Natural history, cytology, and isolating mechanisms of the native American lespedezas. Bull Tall Timbers Res. Sta. 6: 37, 1 pl.
———. 1966b. Native North American species of *Lespedeza* (Leguminosae). Rhodora 68: 359–405.
———. 1966c. Identification of the lespedezas in North America. Bull. Tall Timbers Res. Sta. 7: 1–25.
Clinebell, R. R. II, and P. Bernhardt. 1998. The pollination ecology of five species of *Penstemon* (Scrophulariaceae) in the tallgrass prairie. Ann. Missouri Bot. Gard. 85: 126–136.
Coart, E., S. van Glabeke, M. De Loose, A. S. Larsen, and I. Roldán-Ruiz. 2006. Chloroplast diversity in the genus *Malus*: new insights into the relationship between the European wild apple (*Malus sylvestris* (L.) Mill.) and the domesticated apple (*Malus domestica* Borkh.). Molec. Ecol. 15: 2171–2182.
Cochrane, T. S. 1993. Status and distribution of *Talinum rugospermum* Holz. (Portulacaceae). Nat. Areas J. 33: 13: 33–41.
Coffey, V. J., and S. B. Jones. 1980. Biosystematics of *Lysimachia* section *Seleucia* (Primulaceae). Brittonia 32: 309–322.
Colgan, N. 1896. The shamrock in literature: a critical chronology. J. Roy. Soc. Antiquaries Ireland 26: 211–226, 349–361.
Collins, J. L. 1976. A revision of the annulate *Scutellaria* (Labiatae). Ph.D. dissertation, Vanderbilt University, Nashville, TN.
Collins, L. T. 1973. Systematics of *Orobanche* section *Myzorrhiza* (Orobanchaceae). Ph.D. dissertation, University of Wisconsin, Milwaukee.
Collins, L. T., A. E. L. Colwell, and G. Yatskievych. 2009. *Orobanche riparia* (Orobanchaceae), a new species from the American Midwest. J. Bot. Res. Inst. Texas 3: 3–11.
Compton, J. [A.] 1994. Mexican salvias in cultivation. The Plantsman 15: 193–215.
Compton, J. A., A. Culham, and S. L. Jury. 1998. Reclassification of *Actaea* to include *Cimicifuga* and *Souliea* (Ranunculaceae): Phylogeny inferred from morphology, nrDNA ITS, and cpDNA *trn*L-F sequence variation. Taxon 47: 593–634.
Concannon, J. A., and T. E. DeMeo. 1997. Goldenseal: facing a hidden crisis. Endang. Sp. Techn. Bull. 22: 10–12.
Conrad, M. 1984. [Miscellaneous county records]. Pp. 68–89 *in* W. R. Weber and J. Raveill, eds., Missouri botanical record. Missouriensis 5: 68–89.
Consaul, L. L., S. I. Warwick, and J. McNeill. 1991. Allozyme variation in the *Polygonum lapathifolium* complex. Canad. J. Bot. 69: 2261–2270.
Constance, L. 1949a. The genus *Ellisia*. Rhodora 42: 33–39.
———. 1949b. A revision of *Phacelia* subgenus *Cosmanthus* (Hydrophyllaceae). Contr. Gray Herb. 168: 1–48.
Cook, C. D. K. 1966. A monographic study of *Ranunculus* subgenus *Batrachium* (DC.) A. Gray. Mitt. Bot. Staatssamml. München 6: 47–237.
Coon, N. 1980. Using Wild and Wayside Plants. Dover Publications, New York.
Cooper, A. W., and E. P. Mercer. 1977. Morphological variation in *Fagus grandifolia* Ehrh. in North Carolina. J. Elisha Mitchell Sci. Soc. 93: 136–149.
Cooperrider, T. S. 1976. Notes on Ohio Scrophulariaceae. Castanea 41: 223–226.
Cooperrider, T. S., and G. A. McCready. 1975. On separating Ohio specimens of *Lindernia dubia* and *L. anagallidea* (Scrophulariaceae). Castanea 40: 191–197.
Cortés-Palomec, A. C., R. A. McCauley, and H. E. Ballard Jr. 2006. Population genetic structure in temperate and tropical species of *Viola* (Violaceae) with a mixed breeding system. Int. J. Pl. Res. 167: 503–512.
Costea, M., and F. J. Tardif. 2003a. Nomenclatural changes in the genus *Polygonum* section *Polygonum* (Polygonaceae). Sida 20: 987–997.
———. 2003b. *Polygonum aviculare* subsp. *rurivagum* (Polygonaceae) in North America. Sida 20: 1709–1711.
Costea, M., F. J. Tardif, and H. R. Hinds. 2005. *Polygonum*. Pp. 547–571 *in* Flora of North America Editorial Committee, eds., Flora of North America North of Mexico. Volume 5. Magnoliophyta: Caryophyllidae, Part 2. Oxford University Press, New York.

Couch, R., and E. Nelson. 1986. *Myriophyllum spicatum* in North America. Pp. 8–18 *in* L. W. J. Anderson, ed., Proceedings, 1st International Symposium on Watermilfoil (*Myriophyllum spicatum*) and Related Haloragaceae Species. Aquatic Plant Management Society, Vicksburg, Mississippi.

Crank, E. 1990. *Castilleja* growth with and without host species. Wildflower (Austin): 3(1)6–9, 16.

Cranmer, M. F., and T. J. Mabry. 1966. The lupine alkaloids of the genus *Baptisia* (Leguminosae). Phytochemistry 5: 1133–1138.

Crawford, D. J., E. J. Esselman, J. L. Windus, and C. S. Pabin. 1998. Genetic variation in Running buffalo clover (*Trifolium stoloniferum:* Fabaceae) using random amplified polymorphic DNA markers (RAPDs). Ann. Missouri Bot. Gard. 85: 81–89.

Crins, W. J. 1989. The Tamaricaceae in the southeastern United States. J. Arnold Arbor. 70: 403–425.

Croft, G., G. Yatskievych, and B. Schaal. 2011. *Claytonia ozarkensis* (Montiaceae): Genetic variation in a rare spring beauty endemic to the Ozarks and its relatives. Botany 2011 Conference Abstracts: 247 [Abstract].

Cronquist, A. 1945. Studies in the Sapotaceae. III. *Dipholis* and *Bumelia*. J. Arnold Arbor. 26: 435–471.

———. 1946. Studies in the Sapotaceae–2. Survey of the North American genera. Lloydia 9: 241–292.

———. 1981. An Integrated System of Classification of Flowering Plants. Columbia University Press, New York.

———. 1991. Synoptical arrangement of the subclasses, orders, and families of Liliopsida, as represented in our flora. Pp. lxxiii–lxxv *in* H. A. Gleason and A. Cronquist. Manual of Vascular Plants of Northeastern United States and Adjacent Canada, 2d ed. New York Botanical Garden, Bronx, NY.

Crosswhite, F. S. 1965. Variation in *Chelone glabra* in Wisconsin (Scrophulariaceae). Michigan Bot. 4: 62–66.

Crow, G. E., and C.B. Helquist. 2000. Aquatic and Wetland Plants of Northeastern North America. Volume 1. University of Wisconsin Press, Madison.

Culley, T. M., and N. E. Hardimann. 2007. The beginning of a new invasive plant: a history of the ornamental Callery pear in the United States. Bioscience 57: 956–964.

Cunningham, M., and P. D. Parr. 1990. Successful culture of the rare annual hemiparasitic *Tomanthera auriculata* (Michx.) Raf. (Scrophulariaceae). Castanea 55: 266–271.

Cusick, A. W. 1989. *Trifolium stoloniferum* (Fabaceae) in Ohio: History, habitats, decline and rediscovery. Sida 13: 467–480.

Danin, A., I. Baker, and H. G. Baker. 1978. Cytogeography and taxonomy of the *Portulaca oleracea* L. polyploid complex. Israel J. Bot. 27: 177–211.

D'Arcy, W. G. 1974. *Solanum* and its close relatives in Florida. Ann. Missouri Bot. Gard. 61: 819–867.

———. 1978. Names in *Agalinis* for some plants that were called *Gerardia* and *Virgularia* (Scrophulariaceae). Ann. Missouri Bot. Gard. 65: 769–771.

———. Proposal to conserve the name *Agalinis* Raf. (1837) against *Virgularia* Ruiz & Pavon (1794) (Scrophulariaceae). Taxon 28: 419–422.

Darlington, J. A. 1934. A monograph of the genus *Mentzelia*. Ann. Missouri Bot. Gard. 21: 103–226.

Darwin, C. 1877. The Different Forms of Flowers on Plants of the Same Species. Appleton and Company, New York.

Davenport, L. J. 1988. A monograph of *Hydrolea* (Hydrophyllaceae). Rhodora 90: 169–208.

Davis, H. A., A. M. Fuller, and T. Davis. 1968. Contributions toward the revision of the Eubati of Eastern North America. III. *Flagellares*. Castanea 33: 206–241.

———. 1969. Contributions toward the revision of the Eubati of eastern North America. V. *Arguti*. Castanea 34: 235–266.

De Steven, D. 1983. Floral ecology of witch-hazel (*Hamamelis virginiana*). Michigan Bot. 22: 163–171.

Deam, C. C. 1924. Shrubs of Indiana. Indiana Department of Conservation, Indianapolis.

———. 1940. Flora of Indiana. Indiana Department of Conservation, Indianapolis.

Degtjareva, G. V., C. M. Valiejo Roman, T. E. Kramina, E. M. Mironov, T. H. Samigullin, and D. D. Sokoloff. 2003 Taxonomic and phylogenetic relationships between Old World and New World members of the tribe Loteae (Leguminosae): new insights from molecular and morphological data, with special emphasis on *Ornithopus*. Wulfenia 10: 15–50.

Delgado-Salinas, A. 1985. Systematics of the genus *Phaseolus* (Leguminosae) in North and Central America. Ph.D. dissertation, The University of Texas, Austin.

Delgado-Salinas, A., A. Bruneau, and J. J. Doyle. 1993. Chloroplast DNA phylogenetic studies in New World Phaseolineae (Leguminosae: Papilionoideae: Phaseoleae). Syst. Bot. 18: 6–17.

den Nijs, J. C. M. 1983. Biosystematic studies of the *Rumex acetosella* complex (Polygonaceae). Ph.D. dissertation, University of Amsterdam [published as Contr. Bot. Inst. Univ. Amsterdam 1983–1984(31): 1–157 + 6 separately paginated appendices].

Denison, E. 1975. *Verbascum lichnytis* [sic]—A new plant for Missouri. Nature Notes (Webster Groves) 47: 72–73.

dePamphilis, C. W., N. D. Young, and A. D. Wolfe. 1997. Evolution of plastid gene *rps2* in a lineage of hemiparasitic and holoparasitic plants: Many losses of photosynthesis and complex patterns of rate variation. Proc. Natl. Acad. Sci. U.S.A. 94: 7367–7372.

Der, J. P., and D. L. Nickrent. 2008. A molecular phylogeny of Santalaceae (Santalales). Syst. Bot. 33: 107–116.

Dessein, A. 2003. Systematic studies in the Spermacoceae (Rubiaceae). Ph.D. dissertation, Katholieke Universiteit Leuven, Leuven, Belgium.

DeWolf, G. P. 1954. Notes on cultivated labiates 4. *Satureja* and some related genera. Baileya 2: 143–150.

Diane, N., H. Förther, and H. H. Hilger. 2002. A systematic analysis of *Heliotropium, Tournefortia*, and allied taxa of the Heliotropiaceae (Boraginales) based on ITS1 sequences and morphological data. Amer. J. Bot. 89: 287–295.

Dickson, E. E., S. Kresovich, and N. F. Weeden. 1991. Isozymes in North American *Malus* (Rosaceae) species: Hybridization and species differentiation. Syst. Bot. 16: 363–375.

Dieringer, G., and L. Cabrera R. 2002. The interaction between pollinator size and the bristle staminode of *Penstemon digitalis* (Scrophulariaceae). Amer. J. Bot. 89: 991–997.

Dierker, W. 1989. [Miscellaneous county records]. Pp. 55–77 *in* W. R. Weber and W. Corcoran, eds., Missouri botanical record 11. Missouriensis 10: 54–77.

Dietrich, W., and W. L. Wagner. 1988. Systematics of *Oenothera* section *Oenothera* subsection *Raimannia* and subsection *Nutantigemma* (Onagraceae). Syst. Bot. Monogr. 24: 1–91.

Dietrich, W., W. L. Wagner, and P. H. Raven. 1997. Systematics of *Oenothera* sect. *Oenothera* subsect. *Oenothera* (Onagraceae). Syst. Bot. Monogr. 50: 1–234.

Diggs, G. M., Jr., B. L. Lipscomb, and R. O'Kennon. 1999. Shinners and Mahler's Illustrated Flora of North Central Texas. Sida Bot. Misc. 16: i–xii, 1–1626.

Dirr, M. A. 1981. What do we know about cultivars? Amer. Nurseryman 154: 16–17, 88.

Donoghue, M. J., T. Eriksson, P. A. Reeves, and R. G. Olmstead. 2001. Phylogeny and phylogenetic taxonomy of Dipsacales, with special reference to *Sinadoxa* and *Tatradoxa* (Adoxaceae). Harvard. Pap. Bot. 6: 459–479.

Dorn, R. D. 1988. Vascular Plants of Wyoming, 1st ed. Mountain West Publishing, Cheyenne, WY.

———. 1998. A taxonomic study of *Salix* section *Longifoliae* (Salicaceae). Brittonia 50: 193–210.

———. 2001. Vascular Plants of Wyoming, edition 3. Mountain West Publishing, Cheyenne, WY.

Doroszenko, A. 1985. Taxonomic studies on the *Satureja complex* (Labiatae). Ph.D. dissertation, Edinburgh University, Edinburgh, Great Britain.

Dorr, L. J. 1986. Jean-Baptiste Duerinck (1809–1857) and his collections from the middle western United States. Bull. Jard. Bot. Belg. 56: 397–416.

———. 1990. A revision of the North American Genus *Callirhoe* (Malvaceae). Mem. New York Bot. Garden 56: 1–74.

Doudrick, R. L., W. R. Enns, M. F. Brown, and D. F. Millikan. 1986. Characteristics and role of the mite *Phyllocoptes fructiphilus* (Acari, Eriophyidae) in the etiology of rose rosette. Entomol. News 97: 163–168.

Downie, D. A., J. Granett, and J. R. Fisher. 2000. Distribution and abundance of leaf galling and foliar sexual morphs of grape phylloxera (Hemiptera: Phylloxeridae) and *Vitis* species in the central and eastern United States. Environm. Entomol. 29: 979–986.

Doyle, J. J. 1981. Biosystematic studies on the *Claytonia virginica* aneuploid complex. Ph.D. dissertation, Indiana University, Bloomington.

Doyle, J. J., J. L. Doyle, J. A. Ballenger, E. A. Dickson, T. Kajita, and H. Ohashi. 1997. A phylogeny of the chloroplast gene *rbcL* in the Leguminosae: Taxonomic correlations and insights into the evolution of nodulation. Amer. J. Bot. 84: 541–554.

Drobney, P. M., and M. P. Widrlechner. 2010. Japanese raspberry (*Rubus parvifolius* L.): An invasive species threat in savanna and prairie. Pp. 148–152 *in* D. Williams, B. Butler, and D. Smith, eds., Restoring a National Treasure. Proceedings of the 22nd North American Prairie Conference, Held August 1–5, 2010, Tallgrass Prairie Center, University of Northern Iowa, Cedar Falls.

Duke, J. A. 1985. CRC Handbook of Medicinal Herbs. CRC Press, Boca Raton, FL.

———. 1993. White teeth and white flowers (*Sanguinaria canadensis*). Wildflower (Austin) 6(1): 10–15.

Duley, M. L., and M. A. Vincent. 2003 A synopsis of the genus *Cladrastis* (Leguminosae). Rhodora 105: 205–239.

Duncan, T. 1980. A taxonomic study of the *Ranunculus hispidus* Michaux complex in the western hemisphere. Univ. Calif. Publ. Bot. 77: 1–125.

Duncan, W. H. 1979. Changes in *Galactia* (Fabaceae) of the Southeastern United States. Sida 8: 170–180.

Duncan, W. H., and M. B. Duncan. 1988. Trees of the Southeastern United States. University of Georgia Press, Athens.

Dunn, D. B. 1982. Problems in "keeping-up" with the flora of Missouri. Trans. Missouri Acad. Sci. 16: 95–98.

Dunn, D. [B.], and F. Knauer. 1975. Plant introductions by waterfowl to Mingo National Wildlife Refuge, Missouri. Trans. Missouri Acad. Sci. 9: 27–28.

Durkee, L. T. 1983. The extra-floral nectaries of Passiflora. II. The extra-floral nectary. Amer. J. Bot. 69: 1420–1428.

Earle, A. S., and J. L. Reveal. 2003. Lewis and Clark's Green World: The Expedition and its Plants. Farcountry Press, Helena, MT.

Eckenwalder, J. E. 1977a. Systematics of *Populus* L. (Salicaceae) in southwestern North America with special reference to sect. *Aigeiros* Duby. Ph.D. dissertation, University of California, Berkeley.

———. 1977b. North American cottonwoods (*Populus*, Salicaceae) of sections *Abaso* and *Aigeiros*. J. Arnold Arbor. 58: 193–208.

———. 2010. *Populus*. Pp. 5–22 *in* Flora of North America Editorial Committee, eds., Flora of North America North of Mexico. Volume 7. Magnoliophyta: Salicaceae to Brassicaceae. Oxford University Press, New York.

Edees, E. S., and A. Newton. 1988. Brambles of the British Isles. The Ray Society, London.

Edmonds, J. M. 1986. Biosystematics of *Solanum sarrachoides* Sendtner and *S. physalifolium* Rusby (*S. nitidibaccatum* Bitter). Bot. J. Linn. Soc. 92: 1–38.

Edwards, E. J., R. Nyffeler, and M. J. Donoghue. 2005. Basal cactus phylogeny: Implications of *Pereskia* (Cactaceae) paraphyly for the transition to the cactus life form. Amer. J. Bot. 92: 1177–1178).

Egan, A. N., and K. A. Crandall. 2008. Incorporating gaps as phylogenetic characters across eight DNA regions: ramifications for North American Psoraleeae (Leguminosae). Molec. Phylogenet. Evol. 46: 532–546.

Egan, A. N., and J. L. Reveal. 2009. A new combination in *Pediomelum* and a new genus, *Ladeania*, from western North America (Fabaceae, Psoraleeae). Novon 19: 310–314.

Eggers, D. M. 1969. A revision of *Valerianella* in North America. Ph.D. dissertation, Vanderbilt University, Nashville, TN.

Eiten, G. 1963. Taxonomy and regional variation of *Oxalis* section *Corniculatae*. I. Introduction, keys, and synopsis of the species. Amer. Midl. Naturalist 69: 257–309.

Elias, T. S. 1970. The genera of Ulmaceae in the southeastern United States. J. Arnold Arbor. 51: 18–40.

Epling, C. 1938–1939. A revison of *Salvia*, subgenus *Calosphace*. Repert. Spec. Nov. Regni Veg. Beih. 110(1): 1–380, maps 1–33, pl. 1–33 [reprinted 1940 in: Publ. Univ.Calif. Los Angeles Biol. Sci. 2: 1–383, , maps 1–33, pl. 1–33].

———. 1942. The American species of *Scutellaria*. Univ. Calif. Publ. Bot. 20: 1–145.

Epling, C., and C. Játiva. 1966. A descriptive key to the species of *Satureja* indigenous to North America. Brittonia 18: 244–248.

Epstein, D. B. 1981. Plants used in pipe smoking by the Indians of the United States of America east of the Rocky Mountains. B.A. thesis, Washington University, St. Louis, MO.

Erickson, R. O. 1943. Taxonomy of *Clematis* section *Viorna*. Ann. Missouri Bot. Gard. 30: 1–62.

Eriksson, T., M. J. Donoghue, and M. S. Hibbs. 1998. Phylogenetic analysis of *Potentilla* using DNA sequences of ribosomal internal transcribed spacers (ITS), and implications for the classification of Rosoideae (Rosaceae). Pl. Syst. Evol. 211: 155–179.

Erlanson, E. W. 1931. Sterility in wild rose and in some species hybrids. Genetics 16: 75–96.

———. 1934. Experimental data for a revision of the North American wild roses. Bot. Gaz. (Crawfordsville) 96: 197–259.

Ernst, W. R. 1962. The genera of Papaveraceae and Fumariaceae in the southeastern United States. J. Arnold Arbor. 43: 315–343.

———. 1964. The genera of Berberidaceae, Lardizabalaceae, and Menispermaceae in the southeastern United States. J. Arnold Arbor. 45: 1–35.

Ernst, W. R., and H. J. Thompson. 1963. The Loasaceae in the southeastern United States. J. Arnold Arbor. 154: 138–142.

Ertter, B. 2007. Generic realignments in tribe Potentilleae and revision of *Drymocallis* (Rosoideae: Rosaceae) in North America. J. Bot. Res. Inst. Texas 1: 31–46.

Essig, F. 1990. The *Clematis virginiana* (Ranunculaceae) complex in the southeastern United States. Sida 14: 49–68.

Estes, J. R., and P. M. Hall. 1975. Pollination of *Ipomopsis rubra* by ruby-throated hummingbirds. Bull. Torrey Bot. Club 102: 413–415.

Evans, K. J., and H. E. Weber. 2003. *Rubus anglocandicans* (Rosaceae) is the most widespread taxon of European blackberry in Australia. Austral. Syst. Bot. 16:527–537.

Evans, K. J., D. E. Simon, M. A. Whalen, J. R. Hosking, R. M. Barker, and J. A. Oliver. 2007. Systematics of the *Rubus fruticosus* aggregate (Rosaceae) and other exotic *Rubus* taxa in Australia. Austral. Syst. Bot. 20:187–251.

Everett, T. H. 1971. Some facts and fallacies about the shamrock. Gard. J. 21: 24–26.

Fantz, P. R. 1977. A monograph of the genus *Clitoria* (Leguminosae: Glycineae). Ph.D dissertation, University of Florida, Gainesville.

———. 1991. Ethnobotany of *Clitoria* (Leguminosae). Econ. Bot. 45: 511–520.

———. 1995. Taxonomic notes on new varieties of species of *Clitoria* (Leguminosae–Phaseoleae–Clitoriinae). Sida 16: 721–730.

———. 2000. Nomenclatural notes on the genus *Clitoria* for the Flora North American Project. Castanea 65: 89–92.

———. 2002a. Distribution of *Centrosema* (Leguminosae: Phaseoleae: Clitoriinae) for the Flora of North America Project. Vulpia 1: 41–81.

———. 2002b. Distribution of *Clitoria* (Leguminosae: Phaseoleae: Clitoriinae) for the Flora of North America Project. Vulpia 1: 82–132.

Fantz, P. R., and S. V. Predeep. 1992. Comments on four legumes (*Clitoria, Centrosema*) reported as occurring in India. Sida 15: 1–7.

Fassett, N. C. 1931. Notes from the herbarium of the University of Wisconsin—VII. Rhodora 33: 224–228.

———. 1944. *Dodecatheon* in eastern North America. Amer. Midl. Naturalist 13: 455–486.

Fassett, N. C. 1949. The variations of *Polygonum punctatum*. Brittonia 6: 369–393.

Feinbrun, N., and W. T. Stearn. 1963. Typification of *Lycium barbarum* L., *L. afrum* L., and *L. europaeum* L. Israel J. Bot. 12: 114–123.

Ferguson, C. J. 1998. Molecular systematics of eastern *Phlox* L. (Polemoniaceae). Ph.D. dissertation, University of Texas, Austin.

Ferguson, C. J., and R. K. Jansen. 2002. A chloroplast DNA phylogeny of eastern *Phlox* (Polemoniaceae): Implications of congruence and incongruence with the ITS phylogeny. Amer. J. Bot. 89: 1324–1335.

Ferguson, D. M. 1999. Phylogenetic analysis and relationships in Hydrophyllaceae based on *ndhF* sequence data. Syst. Bot. 23: 253–268.

Fernald, M. L. 1911. The variations of *Lathyrus palustris* in eastern America. Rhodora 13: 47–52.

———. 1921. The Gray Herbarium expedition to Nova Scotia, 1920. Rhodora 23: 85–111, 130–171, 184–195, 223–245, 257–278, 284–300.

———. 1944. The confused publication of *Monarda russeliana*. Rhodora 46: 491–493.

———. 1945. The fruit of *Dirca palustris*. Rhodora 43: 117–119.

———. 1950. Gray's Manual of Botany, edition 8. American Book Co., New York.

Fernald, M. L., and L. Griscom. 1935. Three days of botanizing in southeastern Virginia. Rhodora 37: 167–189.

———. 1937. Notes on *Diodia*. Rhodora 39: 306–308.

Fernald, M. L., and C. A. Weatherby. 1922. Varieties of *Geum canadense*. Rhodora 24: 47–50.

Fernandes, R. B. 1970. Sur l'identification d'une espèce de *Physalis* souspontanèe au Portugal. Bol. Soc. Brot., ser. 2, 44: 343–367, pl. 1–10.

Fernando, E. S., and C. J. Quinn. 1995. Picramniaceae, a new family, and a recircumscription of Simaroubaceae. Taxon 44: 177–181.

Fernando, E. S., P. A. Gadek, and C. J. Quinn. 1995. Simaroubaceae, an artificial construct: evidence from *rbc*L sequence variation. Amer. J. Bot. 82: 92–103.

Figlar, R. B. 1981. A last stand in Arkansas. J. Magnolia Soc. 17(1): 17–20.

Fischer, E. 2004. Scrophulariaceae. Pp. 333–432 *in* K. Kubitzki, ed., The Families and Genera of Vascular Plants. Vol. 7. Flowering Plants—Dicotyledons. Lamiales (except for Acanthaceae including Avicenniaceae) (J. W. Kadereit, vol. ed.). Springer-Verlag, Berlin.

Fischer, M. A. 1987. On the origin of *Veronica persica* (Scrophulariaceae)—a contribution to the history of a neophytic weed. Pl. Syst. Evol. 155: 105–132.

Fisher, C. 1995. Horse care: Perilous pasture plants. Rural Heritage 20: 44–45.

Floden, A. J., M. H. Mayfield, and C. J. Ferguson. 2009. A new narrowly endemic species of *Dirca* (Thymelaeaceae) from Kansas and Arkansas, with a phylogenetic overview and taxonomic synopsis of the genus. J. Bot. Res. Inst. Texas 3: 485–499.

Focke, W. O. 1914. Species ruborum. Monographiae generis rubi prodromus. Pars 3. Biblioth. Bot. 83:1–274.

Foote, L. E., and S. B. Jones Jr. 1989. Native Shrubs and Woody Vines of the Southeast. Timber Press, Portland, OR.

Ford, B. A. 1997. *Nigella*. P. 184 *in* Flora of North America Editorial Committee, eds., 1996. Flora of North America North of Mexico. Volume 3. Magnoliophyta: Magnoliidae and Hamamelidae. Oxford University Press, New York.

Förther, H. 1998. Die infragenerische Gliederung der Gattung *Heliotropium* L. und ihre Stellung innerhalb der subfam. Heliotropioideae (Schrad.) Arn. (Boraginaceae). Sendtnera 5: 35–241.

Fosberg, F. R., M.-h. Sachet, and R. L. Oliver. 1993. Flora of Micronesia: 5. Bignoniaceae–Rubiaceae. Smithsonian Contrib. Bot. 81: i–iii, 1–135.

Fraedrich, S. W., T. C. Harrington, R. J. Rabaglia, M. D. Ulyshen, A. E. Mayfield III, J. L. Hanula, J. M. Eickwort, and D. R. Miller. 2008. A fungal symbiont of the redbay ambrosia beetle causes a lethal wilt in redbay and other Lauraceae in the southeastern United States. Pl. Dis. 92: 215–224.

Freeman, C. C. 2005. Polygonaceae subfam. Polygonoideae. Pp. 479–601 *in* Flora of North America Editorial Committee, eds., Flora of North America North of Mexico. Volume 5. Magnoliophyta: Caryophyllidae, Part 2. Oxford University Press, New York.

Freeman, C. C., and H. R. Hinds. 2005a. *Fallopia*. Pp. 541–546 *in* Flora of North America Editorial Committee, eds., Flora of North America North of Mexico. Volume 5. Magnoliophyta: Caryophyllidae, Part 2. Oxford University Press, New York.

———. 2005b. *Persicaria*. Pp. 574–594 *in* Flora of North America Editorial Committee, eds., Flora of North America North of Mexico. Volume 5. Magnoliophyta: Caryophyllidae, Part 2. Oxford University Press, New York.

Freeman, C. C., and J. L. Reveal. 2005. Polygonaceae. Pp. 216–601 *in* Flora of North America Editorial Committee, eds., Flora of North America North of Mexico. Volume 5. Magnoliophyta: Caryophyllidae, Part 2. Oxford University Press, New York.

Freytag, G. F., and D. G. DeBouck. 2002. Taxonomy, distribution, and ecology of the genus *Phaseolus* (Leguminosae–Papilionoideae) in North America, Mexico, and Central America. Sida Bot. Misc. 23: i–xviii, 1–300.

Friesner, R. C. 1930. Chromosome numbers in ten species of *Quercus*, with some remarks on the contribution of cytology to taxonomy. Butler Univ. Bot. Stud. 1: 77–103.

Fritsch, P. W., and S. D. Lucas. 2000. Clinal variation in the *Halesia carolina* complex (Styracaceae). Syst. Bot. 25: 197–210.

Fryxell, P. A. 1979. Natural History of the Cotton Tribe (Malvaceae: Tribe Gossipieae). Texas A & M University Press, College Station, TX.

———. 1985. Sidus sidarum—V. The North and Central American species of *Sida*. Sida 11: 62–91.
———. 1987. Revision of the genus *Anoda* (Malvaceae). Aliso 11: 485–522.
———. 1988. Malvaceae of Mexico. Syst. Bot. Monogr. 25: 1–522.
———. 1997. The American genera of Malvaceae—II. Brittonia 49: 204–269.
Fuckel, [K. W. G.] L. 1846. Ueber die Honigabsonderung der Nebenblättchen (Stipulae) bei *Vicia sativa* L. Flora 29: 417–418.
Fuentes-Utrilla, P., R. A. López-Rodríguez, and L. Gil. 2004. The historical relationship of elms and vines. Invest. Agrar., Sist. Recurs. Forest. 13: 37–45.
Furman, T. E. 1959. The structure of the root nodules of *Ceanothus sanguineus* and *Ceanothus velutinus*, with special reference to the endophyte. Amer. J. Bot. 46: 698–703.
Gadek, P. A., E. S. Fernando, C. J. Quinn, S. B. Hoot, T. Terrazas, M. C. Sheahan, and M. W. Chase. 1996. Sapindales: molecular delimitation and infraordinal groups. Amer. J. Bot. 83: 802–811.
Galla, S. J. B. L. Viers, P. E. Gradie, and D. E. Saar. 2009. *Morus murrayana* (Moraceae): A new mulberry from easteern North America. Phytologia 91: 105–116.
Gammon, M. A., J. L. Grimsby, D. Tsirelson, and R. Kesseli. 2007. Molecular and morphological evidence reveals introgression in swarms of the invasive taxa *Fallopia japonica, F. sachalinensis*, and *F.* ×*bohemica*. Amer. J. Bot. 94: 948–956.
Gandhi, K. N., and R. D. Thomas, 1983. Stem pubescence in the *Salvia azurea* var. *azurea* and var. *grandiflora* complex. Phytologia 4: 283–284.
Gao, F.-y. 1998. *Exochorda*: five species or one? A biosystematic study of the rosaceous genus *Exochorda*. Ph.D. dissertation, Wangeningen Agricultural University, Wangeningen, The Netherlands.
Gard, M. 2010. The toxicity of extracts of *Tephrosia virginiana* (Fabaceae) in Oklahoma. Oklahoma Native Pl. Rec. 10: 54–64.
Gaskin, J. F., and B. A. Schaal. 2002. Hybrid *Tamarix* widespread in U.S. invasion and undetected in native Asian range. Proc. Natl. Acad. Sci. U.S.A. 99: 11256–11259.
———. 2003. Molecular phylogenetic investigation of U.S. invasive *Tamarix*. Syst. Bot. 28: 86–95.
Geesink, R. L., W. L. Wagner, and D. R. Herbst. 1999. 47. Fabaceae, pea family. Pp. 629–721 *in* W. L. Wagner, D. R. Herbst, and S. H. Sohmer. Manual of the Flowering Plants of Hawai'i., edition 2 (2 vols.). University of Hawai'i Press, Honolulu.
Gentry, H. S. 1957. Gum tragacanth in Iran. Econ. Bot 11: 40–63.
Gerhardt, C. 1853. Untersuchungen über die wasserfreien organischen Säuren. Annal. Chem. Pharmac. 87: 57–84.
Ghebrehiwet, M. 2001. Taxonomy, phylogeny, and biogeography of *Kickxia* and *Nanorrhinum* (Scrophulariaceae). Nordic J. Bot. 20: 655–690.
Ghisalberty, E. L. 1979. Propolis. Bee World 60: 59–84.
Gibbons, E. 1962. Stalking the Wild Asparagus. McKay, New York.
Gil, L., P. Fuentes-Utrilla, Á. Soto, M. T. Cervera, and C. Collada. 2004. Phylogeography: English elm is a 2,000-year-old Roman clone. Nature 451: 1053.
Gil-Ad, N. L. 1997. Systematics of *Viola* subsection *Boreali-Americanae*. Boissiera 53: 1–130.
Gilbert, L. E. 1982. The coevolution of a butterfly and a vine. Sci. Amer. 247: 110–121.
Gill, L. S. 1977. A cytosystematics study of the genus *Monarda* L. (Labiatae) in Canada. Caryologia 30: 381–394.
Gillett, G. W. 1964. Genetic barriers in the *Cosmanthus* phacelias. Rhodora 66: 359–370.
———. 1965. Genotypic variation in the *Phacelia hirsuta* complex. Rhodora 67: 42–48.
Gillett, J. M. 1957. A revision of the North American species of *Gentianella* Moench. Ann. Missouri Bot. Gard. 44: 195–269.
———. 1959. A revision of *Bartonia* and *Obolaria*. Rhodora 61: 43–62.
———. 1985. Taxonomy and morphology. Pages 7–69 *in* N.L. Taylor, ed., Clover Science and Technology. Agronomy 25. American Society of Agronomy, Madison, WI.
Gillett, J. M., and T. S. Cochrane. 1973. Preliminary reports on the flora of Wisconsin. No. 63. The genus *Trifolium*—the clovers. Trans. Wisconsin Acad. Sci. 61: 59–74.
Gillett, J. M., and N. L. Taylor. 2001. The World of Clovers. Iowa State University Press, Ames.
Gillis, W. T. 1980. *Wisteria* in the Great Lakes region. Michigan Bot. 19: 79–83.
Glad, J. B., and R. R. Halse. 1993. Invasion of *Amorpha fruticosa* L. (Leguminosae) along the Columbia and Snake Rivers in Oregon and Washington. Madroño 40: 62–63.

Glatfelter, N. M. 1894. Study of the relations of *Salix nigra* and *Salix amygdaloides*, together with the hybrids arising from them as these species exhibit themselves in the vicinity of St. Louis. Trans. St. Louis Acad. Sci. 6: 427–431.

———. 1898. Notes on *Salix longipes*, Shuttlw. and its relations to *S. nigra*, Marsh. Annual Rep. Missouri Bot. Gard. 9: 43–51, 3 pl.

Gleason, H. A. 1947. The preservation of well known binomials. Phytologia 2: 201–212.

Gleason, H. A. 1952. The New Britton and Brown Illustrated Flora of the Northeastern United States and Adjacent Canada, 3 vols. New York Botanical Garden, Bronx.

Gleason, H. A., and A. Cronquist. 1963. Manual of Vascular Plants of Northeastern United States and Adjacent Canada. D. Van Nostrand Co., Princeton, NJ.

———. 1991. Manual of Vascular Plants of Northeastern United States and Adjacent Canada, 2d ed. New York Botanical Garden, Bronx.

Gobert, V., S. Moja, M. Colson, and P. Taberlet. 2002. Hybridization in the section *Mentha* (Lamiaceae) inferred from AFLP markers. Amer. J. Bot. 89: 2017–2023.

Goelz, J. C. G., and D. W. Carlson. 1997. Growth and seed production of sawtooth oak (*Quercus acutissima*) 22 years after direct seeding. Res. Note S. O., U.S. Forest Serv. SO-386: 1–7.

Goldberg, A. 1967. The genus *Melochia* L. (Sterculiaceae). Contr. U.S. Natl. Herb. 34: 191–363, pl. 1–9.

Gonsoulin, G. J. 1974. A revision of *Styrax* (Styracaceae) in North America, Central America, and the Caribbean. Sida 5: 191–258.

Goodman, G. J., and C. A. Lawson. 1992. Two new combinations and a name change from the Long Expedition of 1820. Rhodora 94: 381–382.

———. 1995. Retracing Major Stephen H. Long's 1820 Expedition: The Itinerary and Botany. University of Oklahoma Press, Norman.

Goodspeed, T. H. 1954. The Genus *Nicotiana*; Origins, Relationships, and Evolution of its Species in the Light of their Distribution, Morphology, and Cytogenetics. Chronica Botanica Company, Waltham, MA.

Gottschling, M., H. H. Hilger, M. Wolf, and N. Diane. 2001. Secondary structure of the ITS1 transcript and its application in a reconstruction of the phylogeny of Boraginales. Pl. Biol. (Stuttgart) 3: 629–636.

Graham, S. A. 1964. The genera of Lythraceae in the southeastern United States. J. Arnold Arbor. 45: 235–250.

———. 1975. Taxonomy of the Lythraceae in the southeastern United States. Sida 6: 80–103.

———. 1979. The origin of *Ammannia ×coccinea*. Taxon 28: 169–178.

———. 1985. A revision of *Ammannia* (Lythraceae) in the western hemisphere. J. Arnold Arbor. 66: 395–420.

Graham, S. A., and R. Kleiman. 1985. Fatty acid composition in *Cuphea* seed oils from Brazil and Nicaragua. J. Amer. Oil Chem. Soc. 62: 81–82.

Graham, S. A., and C. E. Wood Jr. 1965. The genera of Polygonaceae in the southeastern United States. J. Arnold Arbor. 46: 91–121.

Graham, S. A., F. Hirsinger, and G. Röbbelen. 1981. Fatty acids of *Cuphea* (Lythraceae) seed lipids and their systematic significance. Amer. J. Bot. 68: 908–917.

Graham, S. A., J. Hall, K. Sytsma, and S.-h. Shi. 2005. Phylogenetic analysis of the Lythraceae based on four gene regions and morphology. Int. J. Pl. Sci. 166: 995–1017.

Grant, A. L. 1924 [1925]. A monograph of the genus *Mimulus*. Ann. Missouri Bot. Gard. 11: 99–389, pl. 1–10.

Grant, E., and C. Epling. 1943. A study of *Pycnanthemum* (Labiatae). Univ. Calif. Publ. Bot. 20: 195–240.

Grant, V. 1959. Natural History of the Phlox Family: Systematic Botany. Martinus Nijhoff, The Hague, Netherlands.

Grant, V., and K. Grant. 1965. Flower Pollination in the Phlox Family. Columbia University Press, New York.

———. 1983. Behavior of hawkmoths on flowers of *Datura meteloides*. Bot. Gaz. (Crawfordsville) 144: 280–284.

Grear, J. W. 1978. A revision of the New World species of *Rhynchosia* (Leguminosae—Faboideae). Mem. New York Bot. Gard. 31: 1–168.

Greenaway, W., E. Wollenweber, T. Scaysbrook, and F. R. Whatley. 1988. Novel isoferulate esters identified by gas chromatography–mass spectrophotometry in bud exudates of *Populus nigra*. J. Chromatogr. 448: 284–290.

Greene, E. L. 1906. The genus *Ptelea* in the western and southwestern United States and Mexico. Contr. U.S. Natl. Herb. 10: 49–79.

Gregory, W. C., A. Krapovickas, and M. P. Gregory. 1980. Structure, variation, evolution, and classification in *Arachis*. Pp. 469–481 *in* R. J. Summerfield and A. H. Bunting, eds., Advances in Legume Science. Royal Botanic Gardens, Kew, Great Britain.

Gremaud, G. 1988. Three species new to Missouri. Missouriensis 9: 15–17.

Grimes, J. W. 1990. A revision of the New World species of Psoraleeae (Leguminosae: Papilionoideae). Mem. New York Bot. Gard. 61: 1–113.

Groeninckx, I., S. Dessein, H. Ochoterena, C. Persson, T. J. Motley, J. Kårehed, B. Bremer, S. Huysmans, and E. Smets. 2009. Phylogeny of the herbaceous tribe Spermacoceae (Rubiaceae) based on plastid DNA data. Ann. Missouri Bot. Gard. 96: 109–132.

Gunn, C. R. 1968. The *Vicia americana* complex (Leguminosae). Iowa State J. Sci. 43: 171–214.

———. 1979. Genus *Vicia* with notes about tribe Vicieae (Fabaceae) in México and Central America. Techn. Bull. U.S.D.A. 1601: 1–41.

Gustafsson, Å. 1979. Linnaeus' *Peloria*: The history of a monster. Theoret. Appl. Genet. 54: 241–248.

Gustafsson, M. H. G., and K. Bremer. 1995. Morphology and phylogenetic interrelationships of the Asteraceae, Calyceraceae, Campanulaceae, Goodeniaceae, and related families (Asterales). Amer. J. Bot. 82: 250–265.

Gustafsson, M. H. G., A. Backlund, and K. Bremer. 1996. Phylogeny of the Asterales sensu lato based on *rbcL* sequences with particular reference to the Goodeniaceae. Pl. Syst. Evol. 199: 217–242.

Gutierrez, R. 2008. Preliminary chloroplast DNA studies in the flowering plant family Martyniaceae (order Lamiales). J. Arizona-Nevada Acad. Sci. 40: 105–110.

Gutiérrez Salgado, A., P. Gepts, and D. G. Debouck. 1995. Evidence for two gene pools of the Lima bean, *Phaseolus lunatus* L., in the Americas. Genet. Res. Crop Evol. 42: 15–28.

Ha, S. 1993. Genetical studies in interspecific differentiation in the genus *Melilotus*. Mem. Fac. Agric. Hokkaido Univ. 18: 67–107.

Haddock, R. C., and S. J. Chaplin. 1982. Pollination and seed production in two phenologically divergent prairie legumes (*Baptisia leucophaea* and *B. leucantha*). Amer. Midl. Naturalist 108: 175–186.

Haegi, L. 1976. Taxonomic account of *Datura* L. (Solanaceae) in Australia with a note on *Brugmansia*. Austral. J. Bot. 24: 415–435.

Hager, H. A., and R. D. Vinebrooke. 2004. Positive relationships between purple loosestrife (*Lythrum salicaria*) and plant species diversity in Minnesota wetlands. Canad. J. Bot. 82: 763–773.

Haining, Q., S. Graham, and M. G. Gilbert. 2007. Lythraceae. Pp. 274–289 *in* Wu Zheng-yi and Peter H. Raven, eds., Flora of China. Volume 13. Clusiaceae through Araliaceae. Science Press, Beijing.

Hall, T. F., and W. T. Penfound. 1944. The biology of the American lotus, *Nelumbo lutea* (Willd.) Pers. Amer. Midl. Naturalist 31: 744–758.

Hammer, K., A. Romeike, and C. Tittel. 1983. Vorarbeiten zur mongraphischen Darstellung von Wildpflanzensortrimenten: *Datura* L., sectiones *Dutra* Bernh., *Ceratocaulis* Bernh., et *Datura*. Kulturpflanze 31: 13–75.

Hara, H. 1953. *Ludwigia* vs. *Jussiaea*. J. Jap. Bot. 28: 289–294.

Haraldson, K. 1978. Anatomy and taxonomy in Polygonaceae subfam. Polygonoideae Meisn. emend Jaretzky. Symb. Bot. Upsal. 22(1): 1–95.

Hardig, T. M., S. J. Brunsfeld, R. S. Fritz, M. Morgan, and C. M. Orians. 2000. Morphological and molecular evidence for hybridization and introgression in a willow (*Salix*) hybrid zone. Molec. Ecol. 9: 9–24.

Hardin, J. W. 1957a. A monographic study of the American Hippocastanaceae. Ph.D. dissertation, University of Michigan, Ann Arbor.

———. 1957b. A revision of the American Hippocastanaceae—II. Systematic treatment. Brittonia 9: 173–195.

———. 1973. The enigmatic chokeberries (*Aronia*, Rosaceae). Bull. Torrey Bot. Club 100: 178–184.

———. 1974. Studies of the southeastern United States flora. IV. Oleaceae. Sida 5: 274–285.
———. 1975. Hybridization and introgression in *Quercus alba*. J. Arnold Arbor. 56: 336–363.
———. 1979. *Quercus prinus* L.—nomen ambiguum. Taxon 28: 355–357.
———. 1990. Variation patterns and recognition of varieties of *Tilia americana* s. l. Syst. Bot. 15: 33–48.
Hardin, J. W., and J. M. Arena. 1974. Human Poisoning from Native and Cultivated Plants. 2nd ed. Duke University Press, Durham, NC.
Hardin, J. W., and R. L. Beckmann. 1982. Atlas of foliar surface features in woody plant, V. *Fraxinus* (Oleaceae) of eastern North America. Brittonia 34: 129–140.
Harley, R. M. 1965. The spicate mints. Bot. Soc. Brit. Isles Proc. 6: 369–373.
———. 1972. Notes on the genus *Mentha* (Labiatae). Pp. 250–253 *in* V. H. Heywood, ed., Flora Europaea notulae systematicae ad floram Europaeam spectantes, no. 12. Bot. J. Linn. Soc. 65: 223–269.
Harley, R. M., and C. A. Brighton. 1977. Chromosome numbers in the genus *Mentha* L. Bot. J. Linn. Soc. 74: 71–96.
Harley, R. [M.], and A. Paton. 2001. *Leonurus japonicus* Houtt. (Labiatae): the correct name for a common tropical weed. Kew Bull. 56: 243–244.
Harper, B. L. 1986. A Minnesota counterattack on purple loosestrife (*Lythrum salicaria*). Pp. 262–264 *in* R. L. Stuckey and K. J. Reese, eds., The prairie peninsula—in the "shadow" of Transeau: proceedings of the sixth North American prairie conference. Biol. Notes Ohio Biol. Surv. 15: i–x, 1–279.
Harriman, N. A. 1969. *Magnolia tripetala* L. and *Aralia spinosa* L. in St. Louis County, Missouri. Rhodora 71: 478–479.
Harrington, M. G., K. J. Edwards, S. A. Johnson, M. W. Chase, and P. A. Gadek. 2005. Phylogenetic inference in Sapindaceae sensu lato using plastid *matK* and *rbcL* DNA sequences. Systematic Botany 30: 366–382.
Harris, A. J., Q.-y. Xiang, and D. T. Thomas. 2009. Phylogeny, origin, aand biogeographic history of *Aesculus* L. (Sapindales)—An update from combined analysis of DNA sequences, morphology, and fossils. Taxon 58: 108–126.
Harvey, A. W. 2009. Extrafloral nectaries in kudzu, *Pueraria montana* (Lour.) Merr., and groundnut, *Apios americana* Medicus (Fabaceae). Castanea 74: 360–371.
Hashimoto, T., M. Tori, Y. Asakawa, and E. Wollenweber. 1988. Synthesis of two allergenic constituents of propolis and poplar bud excretion. Z. Naturforsch. 43c: 470–472.
Hausen, B. M., E. Wollenweber, H. Senff, and B. Post. 1987a. Propolis allergy (I). Origin, properties, usage and literature review. Contact Dermatitis 17: 163–170.
———. 1987b. Propolis allergy (II). The sensitizing properties of 1,1-dimethyl caffeic acid ester. Contact Dermatitis 17: 171–177.
Havananda, T., E. C. Brummer, I. J. Maureira-Butler, and J. J. Doyle. 2010. Relationships among diploid members of the *Medicago sativa* (Fabaceae) species complex based on chloroplast and mitochondrial DNA sequences. Syst. Bot. 35: 140–150.
Hawkes, J. G. 1990. The Potato: Evolution, Biodiversity, and Genetic Resources. Belhaven Press, London.
Hays, J. F. 1998. *Agalinis* (Scrophulariaceae) in the Ozark Highlands. Sida 18: 555–577.
Heath, P. V. 1992. Preliminary notes on British roses. Calyx 1: 141–196.
Hedgcock, G. G. 1915. Parasitism of *Comandra umbellata*. J. Agric. Res. 5: 133–135.
Hedrick, V. P., G. H. Howe, O. M. Taylor, A. Berger, G. L. Slate, and O. Einset. 1925. The Small Fruits of New York. New York Department of Farms & Markets, 33rd Ann. Report. J.B. Lyon Co., Albany, NY.
Heineke, T. E. 1987. Cedar elm in Missouri. Castanea 52: 229.
Heiser, C. B., Jr. 1962. Some observations on pollination and compatibility in *Magnolia*. Proc. Indiana Acad. Sci. 72: 259–266.
———. 1969. Nightshades, the Paradoxical Plants. W. H. Freeman and Company, San Francisco, CA.
Heisey, R. M. 1990a. Allelopathic and herbicidal effects of extracts from tree-of-heaven (*Ailanthus altissima*). Amer. J. Bot. 77: 662–670.
——. 1990b. Evidence for allelopathy by tree-of-heaven (*Ailanthus altissima*). J. Chem. Ecol. 16: 2039–2055.

——. 1996. Identification of an allelopathic compound from *Ailanthus altissima* (Simaroubaceae) and characterization of its herbicidal activity. Amer. J. Bot. 83: 192–200.

Heitzman, J. R., and J. E. Heitzman. 1987. Butterflies and Moths of Missouri. Missouri Department of Conservation, Jefferson City.

Hempel, A. C., P. A. Reeves, R. G. Olmstead, and R. K. Jansen. 1995. Implications of rbcL sequence data for higher order relationships of the Loasaceae and the anomalous aquatic plant *Hydrostachys* (Hydrostachyaceae). Pl. Syst. Evol. 194: 25–37.

Henderson, N. C. 1962. A taxonomic revision of the genus *Lycopus* (Labiatae). Amer. Midl. Naturalist 68: 95–138.

———. 1980. Additions to the Flora of Missouri. Natural History Note. Missouri Department of Conservation, Jefferson City.

Henson, P. R., and E. A. Hollowell. 1960. Winter annual legumes for the south. USDA Farm. Bull. 2146: 1–24.

Hermann, F. J. 1960. *Vicia*: Vetches in the United States—native, naturalized and cultivated. Agric. Handb. 168: 1–84.

———. 1962. A revision of the genus *Glycine* and its immediate allies. Techn. Bull. U.S.D.A. 1268: 1–82.

Hershkovitz, M. A. 1993. Revised circumscriptions and subgeneric taxonomies of *Calandrinia* and *Montiopsis* with notes on the phylogeny of the portulacaceous alliance. Ann. Missouri Bot. Gard. 80: 366–396.

Hershkovitz, M. A., and E. A. Zimmer. 1997). On the evolutionary origins of the cacti. Taxon 46: 217–232.

Hess, W. D., and N. A. Stoynoff. 1988. Taxonomic status of *Quercus acerifolia* (Fagaceae) and a morphological comparison of four members of the *Quercus shumardii* complex. Syst. Bot. 23: 89–100.

Heyn, C. C., and B. Pazy. 1989. The annual species of *Adonis* (Ranunculaceae)—a polyploid complex. Pl. Syst. Evol. 168: 181–193.

Hickey, R. J., M. A. Vincent, and S. I. Guttman. 1991. Genetic variation in Running buffalo clover (*Trifolium stoloniferum*, Fabaceae). Conservation Biol. 5: 309–316.

Hickman, J. C. 1984. Nomenclatural changes in *Persicaria*, *Polygonum*, and *Rumex* (Polygonaceae). Madroño 31: 249–252.

Hilger, H. H., and N. Diane. 2003. A systematic analysis of Heliotropiaceae (Boraginales) based on *trnL* and ITS1 sequence data. Bot. Jahrb. Syst. 125: 19–51.

Hill, R. J. 1976. Taxonomic and phylogenetic significance of seed coat microscupturing in *Mentzelia*. Brittonia 28: 86–112.

———. 1985. Kudzu-vine, *Pueraria lobata* (Willd.) Ohwi. Pennsylvania Department of Agriculture, Weed Circular 11(1: 9): 23–30.

Hill, S. R. 1982. A monograph of the genus *Malvastrum* A. Gray (Malvaceae: Malveae). Rhodora 84: 1–83, 159–264, 317–409.

———. 2002. *Malvastrum hispidum* (Pursh) Hochr. (Malvaceae) in Illinois: Status, distribution, and nomenclature. Botany 2002 Conference Abstracts: 127.

Hinds, H. R., and C. C. Freeman. 2005 *Persicaria*. Pp. 574–594 *in* Flora of North America Editorial Committee, eds., Flora of North America North of Mexico. Volume 5. Magnoliophyta: Caryophyllidae, Part 2. Oxford University Press, New York.

Hinton, B. D. 1968a. The native annual *Parietaria* (Urticaceae) of the conterminous United States and Canada. M.S. thesis, University of Southwestern Louisiana, Monroe.

———. 1968b. *Parietaria praetermissa* (Urticaceae), a new species from the southeastern United States. Sida 3: 191–194.

Hinton, W. F. 1976. The systematics of *Physalis pumila* ssp. *hispida* (Solanaceae). Syst. Bot. 1: 188–193.

Hipp, A. L., and J. A. Weber. 2008. Taxonomy of Hill's oak (*Quercus ellipsoidalis*: Fagaceae): evidence from AFLP data. Syst. Bot. 33: 148–158.

Hitchcock, C. L. 1952. A revision of the North American species of *Lathyrus*. Univ. Wash. Publ. Biol. 15: 1–104.

Hoggard, R. K. 1998. A taxonomic treatment of the Oklahoma species of *Plantago*. M.S. thesis, University of Central Oklahoma, Edmond.

Holmes, W. C., J. F. Pruski, and J. R. Singhurst. 2000. *Thymelaea passerina* (Thymelaeaceae) new to Texas. Sida 19: 403–406.

Holmgren, N. H. 1986. Scrophulariaceae. Pp. 751–797 *in* Great Plains Flora Association, eds., Flora of the Great Plains. University Press of Kansas, Lawrence.

Holub, J. 1993. *Leonurus intermedius*, species nova: with additional notes on some other *Leonurus* taxa. Preslia 65: 97–115.

Hong, S.-p., L. P. Ronse De Craene, and E. Smets. 1998. Systematic significance of tepal surface morphology in tribes Persicarieae and Polygoneae (Polygonaceae). Bot. J. Linn. Soc. 127: 91–116.

Hoot, S. B. 1995. Phylogeny of the Ranunculaceae based on preliminary *atpB, rbcL* and 18S nuclear ribosomal DNA sequence data. Pl. Syst. Evol., Suppl. 9: 241–251.

Hoot, S. B., and A. A. Reznicek. 1994. Phylogenetic relationships in *Anemone* (Ranunculaceae) based on morphology and chloroplast DNA. Syst. Bot. 19: 169–200.

Hopp, R., and B. M. Lawrence. 2007. Natural and synthetic menthol. Pp. 371–397 *in* B. M. Lawrence, ed., Mint: The Genus *Mentha*. CRC Press, Boca Raton, FL.

Hornberger, K. L. 1980. The vascular flora of Roaring River State Park, Barry County, Missouri. M.S. thesis, Southwest Missouri State University, Springfield.

Horton, J. H. 1963. A taxonomic revision of *Polygonella* (Polygonaceae). Brittonia 15: 177–203.

———. 1972. Studies of the southeastern United States flora. IV. Polygonaceae. J. Elisha Mitchell Sci. Soc. 88: 92–102.

Howard, R. A. 1974. Further comments on *Styrax* L. Sida 5: 334–337.

Hsiao, J.-y. 1973. A numerical taxonomic study of the genus *Platanus* based on morphological and phenolic characters. Amer. J. Bot. 60: 678–684.

Hu, J.-m., M. Lavin, M. Wojciechowski, and M. J. Sanderson. 2002. Phylogenetic analysis of nuclear ribosomal ITS/5.8S sequences in the tribe Millettieae (Fabaceae): *Poecilanthe-Cyclolobium*, the core Millettieae, and the *Callerya* Group. Syst. Bot. 27: 722–733.

Hu, S. Y. 1954–1956. A monograph of the genus *Philadelphus*. J. Arnold Arbor. 35: 275–333; 36: 52–109, 325–368; 37: 15–90.

Huamán, Z., and D. M. Spooner. 2002. Reclassification of landrace populations of cultivated potatoes (*Solanum* sect. *Petota*). Amer. J. Bot. 89: 947–965.

Huang, M., D. J. Crawford, J. V. Freudenstein, and P. D. Cantino. 2008. Systematics of *Trichostema* (Lamiaceae): evidence from ITS, *ndh*F, and morphology. Syst. Bot. 33: 437–446.

Hudson, S. 1992. Chicken spike (*Sphenoclea zeylanica*, Campanulaceae), new to Missouri. Missouriensis 13: 50–52.

———. 1994. Three new plants for southeastern Missouri. Missouriensis 15(2): 13–18.

Hudson, W. D., Jr. 1986. Relationships of domesticated and wild *Physalis philadelphica*. Pp. 416–432 *in* W. G. D'Arcy, ed., Solanaceae, Biology and Systematics. Columbia University Press, New York.

Hufford, L. 1992. 1997. A phylogenetic analysis of Hydrangeaceae based on morphological data. Int. J. Plant Sci. 158: 652–672.

Hunt, D. R. 1982. Proposal to conserve 3325 *Gillenia* Moench (1802) [Rosaceae] against *Gillena* Adans. (1763) [Clethraceae]. Taxon 31: 568.

———. 1990. A systematic review of *Quercus* ser. *Laurifoliae, Marilandicae* and *Nigrae*. Ph. D. dissertation, University of Georgia, Athens.

Hussain, F., F. Mobeen, B.-s. Kil, and S. O. Yoo. 1997. Allelopathic suppression of wheat and mustard by *Rumex dentatus* ssp. *klotzschianus*. J. Pl. Biol. 40: 120–124.

Hymowitz, T. 1970. On the domestication of the soybean. Econ. Bot. 26: 49–60.

Ihlenfeldt, H.-d. 2004. Trapellaceae. Pp. 445–448 *in* K. Kubitzki, ed., The Families and Genera of Vascular Plants. Vol. 7. Flowering Plants. Dicotyledons. Lamiales (except Acanthaceae including Avicenniaceae) (J. W. Kadereit, vol. ed.). Springer-Verlag, Berlin.

Iltis, H. H., and W. M. Shaughnessy. 1960. Preliminary notes on the flora of Wisconsin no. 43. Primulaceae—primrose family. Trans. Wisconsin Acad. Sci. 49: 113–135.

Irving, R. S. 1980. The systematics of *Hedeoma* (Labiatae). Sida 8: 218–295.

Isely, D. 1948. *Lespedeza striata* and *L. stipulacea*. Rhodora 50: 21–27.

———. 1978. New varieties and combinations in *Lotus, Baptisia, Thermopsis*, and *Sophora* (Leguminosae). Brittonia 30: 466–472.

———. 1981. Leguminosae of the United States. III. Subfamily Papilionoideae: tribes Sophoreae, Podalyrieae, Loteae. Mem. New York Bot. Gard. 25(3): 1–264.

———. 1983a. *Astragalus* L. (Leguminosae: Papilionoideae) I: keys to United States species. Iowa State J. Res. 58: 1–172.

———. 1983b. The *Desmodium paniculatum* (L.) DC. (Fabaceae) complex revisited. Sida 10: 142–158.

———. 1984. *Astragalus* L. (Leguminosae: Papilionoideae) II: species summary A–E. Iowa State J. Res. 59: 99–209.

———. 1985. *Astragalus* L. (Leguminosae: Papilionoideae) III: species summary F–M. Iowa State J. Res. 60: 183–320.

———. 1986a. *Astragalus* L. (Leguminosae: Papilionoideae) IV: species summary N–Z. Iowa State J. Res. 61: 157–289.

———. 1986b. Notes about *Psoralea* sensu auct., *Amorpha*, *Baptisia*, *Sesbania*, and *Chamaecrista* (Leguminosae) in the southeastern United States. Sida 11: 429–440.

———. 1990. Vascular Flora of the Southeastern United States, Vol. 3, Part 2, Leguminosae (Fabaceae). University of North Carolina Press, Chapel Hill.

———. 1998. Native and Naturalized Leguminosae (Fabaceae) of the United States (Exclusive of Alaska and Hawaii). Monte L. Bean Life Science Museum, Brigham Young University, Provo, UT.

Isely, D., and F. J. Peabody. 1984. *Robinia* (Leguminosae: Papilionoideae). Castanea 49: 187–202.

Ishikawa, N., J. Yokoyama, and H. Tsukaya. 2009. Molecular evidence of reticulate evolution in the subgenus *Plantago* (Plantaginaceae). Amer. J. Bot. 96: 1627–1635.

Iwatsubo, Y., Y. Souma, N. Miura, and N. Naruhashi. 2004. Polyploidy of *Glechoma hederacea* subsp. *grandis* (Labiatae). J. Phytogeogr. Taxon. 52: 67–71.

Jabbour, F., and S. S. Renner. 2011. *Consolida* and *Aconitella* are an annual clade of *Delphinium* (Ranunculaceae) that diversified in the Mediterranean basin and the Irano-Turanian region. Taxon 60: 1029–1040.

James, E., ed. 1823. Account of an Expedition from Pittsburgh to the Rocky Mountains, Performed in the Years 1819 and '20, by Order of the Hon. J. C. Calhoun, Sec'y of War: Under the Command of Major Stephen H. Long. Two volumes. H. C. Carey and I. Lea, Philadelphia [also in 3 volumes, Longman, Hurst, Rees, Orme, and Brown, London].

Jeffreys, D. 2005. Aspirin: The Remarkable Story of a Wonder Drug. Bloomsbury, London.

Jensen, A. R., H.-q. Li, D. C. Albach, and C. H. Godfredsen. 2008. Phytochemistry and molecular systematics of *Triaenophora rupestris* and *Oreosolon wattii* (Scrophulariaceae). Phytochem. 69: 2162–2166.

Jensen, L. P. 1932. New plant immigrants in the Gray Summit extension of the Garden. Bull. Missouri Bot. Gard. 20: 41–43, pl. 13–14.

Jensen, R. J. 1977. A preliminary numerical analysis of the red oak complex in Michigan and Wisconsin. Taxon 26: 399–407.

———. 1986. Geographic spatial autocorrelation in *Quercus ellipsoidalis*. Bull. Torrey Bot. Club 113: 431–439.

———. 1989. The *Quercus falcata* Michaux complex in the Land Between the Lakes, Kentucky and Tennessee: A study of morphological variation. Amer. Midl. Naturalist 121: 245–255.

———. 1997. *Quercus* sect. *Lobatae*. Pp. 447–468 *in* Flora of North America Editorial Committee, eds., Flora of North America North of Mexico. Volume 3. Magnoliophyta: Magnoliidae and Hamamelidae. Oxford University Press, New York.

Jobson, R. W., J. Playford, K. M. Cameron, and V. A. Albert. 2003. Molecular phylogenetics of Lentibulariaceae inferred from plastid *rps16* intron and *trnL-F* DNA sequences: Implications for character evolution and biogeography. Syst. Bot. 28: 157–171.

Johansson, J. T. 1998. Chloroplast DNA restriction site mapping and the phylogeny of *Ranunculus* (Ranunculaceae). Pl. Syst. Evol. 213: 1–19.

Johansson, J. T., and R. K. Jansen. 1993. Chloroplast DNA variation and phylogeny of the Ranunculaceae. Pl. Syst. Evol. 187: 29–49.

Johnson, G. P. 1988. Revision of *Castanea* sect. *Balanocaston* (Fagaceae). J. Arnold Arbor. 69: 25–49.

Johnson, L. A., L. M. Chan, T. L. Weese, L. D. Busby, and S. McMurry. 2008. Nuclear and cpDNA sequences combined provide strong inference of higher phylogenetic relationships in the phlox family (Polemoniaceae). Molec. Phylogenet. Evol. 48: 997–1012.

Johnsson, A., P. I. Johnsen, T. Rinnan, and D. Skrove. 2006. Basic properties of the circadian leaf movements of *Oxalis regnellii*, and period change due to lithium ions. Physiol. Pl. 53: 361–367.
Johnston, L. A. 1975. Revision of the *Rhamnus serrata* complex. Sida 6: 67–79.
Joly, S., J. R. Starr, W. H. Lewis, and A. Bruneau. 2006. Polyploid and hybrid evolution in roses east of the Rocky Mountains. Amer. J. Bot. 93: 412–425.
Jones, G. N. 1946. American species of *Amelanchier*. Illinois Biol. Monogr. 20(2): 1–126.
———. 1968. Taxonomy of American species of linden (*Tilia*). Illinois Biol. Monogr. 39: 1–156.
Judd, W. S., Sanders, R. W., and M. J. Donoghue. 1994. Angiosperm family pairs: Preliminary cladistic analyses. Harvard Pap. Bot. No. 5: 1–51.
Judd, W. S., C. S. Campbell, E. A. Kellogg, P. F. Stevens, and M. J. Donoghue. 2002. Plant Systematics, a Phylogenetic Approach. 2d ed. Sinauer Associates, Sunderland, MA.
———. Judd, W. S., C. S. Campbell, E. A. Kellogg, P. F. Stevens, and M. J. Donoghue. 2008. Plant Systematics, a Phylogenetic Approach. 3rd ed. Sinauer Associates, Sunderland, MA.
Junell, S. 1934. Zur Gynäceummorphologie und Systematik der Verbenaceen und Labiaten. Symb. Bot. Upsal. 4: 1–219.
Kajita, T., H. Ohashi, Y. Tateishi, C. D. Bailey, and J. J. Doyle. 2001. *rbcL* and legume phylogeny with particular reference to Phaseoleae, Millettieae, and allies. Syst. Bot. 26: 515–536.
Källersjö, M., G. Bergqvist, and A. A. Anderberg. 2000. Generic realignment in primuloid families of the Ericales s.l.: a phylogenetic analysis based on DNA sequences from three chloroplast genes and morphology. Amer. J. Bot. 87: 1325–1341.
Kallhoff, V., and G. Yatskievych. 2001. Revision of the genus *Heuchera* in Missouri. Missouriensis 22: 22–38.
Kaplan, L., and L. N. Kaplan. 1988. *Phaseolus* in archaeology. Pp. 125–142 *in* P. Gepts, ed., Genetic Resources of *Phaseolus* Beans. Kluwer Academic Publishers, Dordrecht, The Netherlands.
Kartesz, J. T., and K. N. Gandhi. 1989. Nomenclatural notes for the North American flora. I. Phytologia 67: 461–467.
———. 1991a. Nomenclatural notes for the North American flora. VII. Phytologia 71: 87–100.
———. 1991b. Nomenclatural notes for the North American flora. VIII. Phytologia 71: 269–280.
Kartesz, J. T., and C. A. Meacham. 1999. Synthesis of the North American Flora, ver. 1.0. North Carolina Botanical Garden, Chapel Hill [CD-ROM].
Kaul, R. B. 1986. Polygonaceae. Pp. 214–235 *in* Great Plains Flora Association, eds., Flora of the Great Plains. University Press of Kansas, Lawrence.
Kearney, T. H. 1935. The North American species of *Sphaeralcea* subgenus *Eusphaeralcea*. Univ. California Publ. Bot. 19: 1–128, pl. 1–12.
Keeling, K. H. 1981. Function of *Mentzelia nuda* (Loasaceae) postfloral nectaries in seed defense. Amer. J. Bot. 68: 295–299.
Keener, C. S. 1967. A biosystematic study of *Clematis* subsection *Integrifoliae* (Ranunculaceae). J. Elisha Mitchell Sci. Soc. 83: 1–41.
———. 1993. A review of the classification of the genus *Hydrastis* (Ranunculaceae). Aliso 13: 551–558.
Keener, C. S., and S. B. Hoot. 1987. *Ranunculus* section *Echinella* (Ranunculaceae) in the southeastern United States. Sida 12: 57–68.
Kennedy, K. 1992. A systematic study of the genus *Phyla* Lour, (Verbenaceae: Verbenoideae, Lantanae). Unpublished Ph.D. dissertation, University of Texas, Austin.
Key, J. S. 1980. Ivy-leaved speedwell in Missouri. Missouriensis 2(2): 14.
Kiger, R. W. 1975. *Papaver* in North America North of Mexico. Rhodora 77: 410–422.
———. 1997. *Papaver*. Pp. 323–333 *in* Flora of North America Editorial Committee, eds., Flora of North America North of Mexico. Volume 3. Magnoliophyta: Magnoliidae and Hamamelidae. Oxford University Press, New York.
———. 2003. *Phemeranthus*. Pp. 488–495 *in* Flora of North America Editorial Committee, eds., Flora of North America North of Mexico. Volume 4. Magnoliophyta: Caryophyllidae, Part 1. Oxford University Press, New York.
Kim, K.-j. 1998. A new species of *Fontanesia* (Oleaceae) from China and taxonomic revision of the genus. J. Pl. Biol. 41: 142–145.
Kim, S.-t., and M. J. Donoghue. 2008a. Molecular phylogeny of *Persicaria* (Persicarieae, Polygonaceae). Syst. Bot. 33: 77–86.
———. 2008b. Incongruence between cpDNA and nrITS tree indicates extensive hybridization within *Eupersicaria* (Polygonaceae). Amer. J. Bot. 95: 1122–1135.

Kim, S.-t., M.-h. Kim, and C.-w. Park. 2000. A systematic study on *Fallopia* section *Fallopia* (Polygonaceae). Korean J. Pl. Taxon. 30: 35–54 [in Korean].

Klimstra, W. D. 1956. Problems in the use of multiflora rose. Trans. Illinois Acad. Sci. 48: 66–72.

Knight, W. E. 1985. Crimson clover. Pp. 491–502 *in* N. L. Taylor, ed., Clover Science and Technology. Agronomy 25. American Society of Agronomy, Madison, WI.

Koelling, A. C. 1964. Taxonomic studies in *Penstemon deamii* and its allies. Ph.D. dissertation, University of Illinois, Urbana.

Koenig, R., and P. Gepts. 1989. Allozyme diversity in wild *Phaseolus vulgaris*: Further evidence for two major centers of genetic diversity. Theor. Appl. Genet. 78: 809–817.

Kooiman, P. 1974. Iridoid substances in the Loasaceae and the taxonomic position of the family. Acta Bot. Neerl. 23: 677–679.

Koptur, S. 1979. Facultative mutualism between weedy vetches bearing extrafloral nectaries and weedy ants in California. Amer. J. Bot. 66: 1016–1020.

Kosnik, M. A., G. M. Diggs, P. A. Redshaw, and B. L. Lipscomb. 1996. Natural hybridization among three sympatric *Baptisia* (Fabaceae) species in north central Texas. Sida 17: 479–500.

Krakos, K. 2011. The evolution and reproductive ecology of *Oenothera* (Onagraceae). Ph.D. dissertation, Washington University, St. Louis, MO.

Kral, R., and P. E. Bostick. 1969. The genus *Rhexia* (Melastomataceae). Sida 3: 387–440.

Krapovickas, A., and W. C. Gregory. 2007. Taxonomy of the genus *Arachis* (Leguminosae). English translation by D. E. Williams and C. E. Simpson. Bonplandia 16 (Suppl.): 1–205 [originally published in Spanish in 1994 as Taxonomia del genero *Arachis* (Leguminosae). Bonplandia 8: 1–186].

Krestovskaja, T. V. 1987. Chtoe takoe *Leonurus heterophyllus* Sweet (Lamiaceae)? [What is *Leonurus heterophyllus* Sweet (Lamiaceae)?]. Novosti Sist. Vyssh. Rast. 24. 156–158. [in Russian].

———. 1992. Systematics and phytogeography of *Leonurus* L. Pp. 139–148 *in* R. M. Harley and T. Reynolds, eds., Advances in Labiat Science. Royal Botanic Gardens, Kew, Great Britain.

Kruschke, E. P. 1965. Contributions to the taxonomy of *Crataegus*. Milwaukee Public Mus. Publ. Bot. 3: 1–273.

Krüssman, G. 1981. The Complete Book of Roses. Timber Press, Portland, OR.

Kuijt, J. 1969. The Biology of Parasitic Flowering Plants. University of California Press, Berkeley.

———. 2003. Monograph of *Phoradendron* (Viscaceae). Syst. Bot. Monogr. 66: 643.

Kurtto, A., and T. Eriksson. 2003. Atlas Florae Europaeae notes. 15. Generic delimitations and nomenclatural adjustments in Potentilleae. Ann. Bot. Fennici 40: 135–141.

Kurtto, A., R. Lampinen, and L. Junikka, eds., 2004. Atlas Florae Europaeae. Distribution of vascular plants in Europe. Volume 13. Rosaceae (*Spiraea* to *Fragaria*, ex *Rubus*). Committee for Mapping the Flora of Europe, Helsinki.

Kurz, D. 1997. Shrubs and Woody Vines of Missouri. Missouri Department of Conservation, Jefferson City.

———. 1999. Ozark Wildflowers. Falcon Publishing, Helena, MT.

———. 2003. Trees of Missouri. Missouri Department of Conservation, Jefferson City.

Lackey, J. A. 1981. Tribe Phaseoleae. Pp. 301–327 *in* R. M. Polhill and P. H. Raven, eds., Advances in Legume Systematics, Part 1. Royal Botanic Gardens, Kew, Great Britain.

———. 1983 [1984]. A review of generic concepts in American Phaseolinae (Fabaceae; Faboideae). Iselya 2: 21–64.

Lackney, V. K. 1981. The parasitism of *Pedicularis lanceolata* Michx., a root parasite. Bull. Torrey Bot. Club 108: 422–429.

Ladd, D. 1990. Noteworthy collections. Missouri. Castanea 55: 293–294.

———. 1994. Three plants new to Missouri. Missouriensis 15(2): 28–30.

Ladd, D., and B. Schuette. 1990. *Ranunculus testicularis* [sic], a new, weedy buttercup for Missouri. Missouriensis 11: 36–39.

Ladd, D., G. Gremaud, and B. Heumann. 1991. New and noteworthy Missouri vascular plants. Missouriensis 12: 36–44.

Lamb Frye, A. S., and K. A. Kron. 2003. *RbcL* phylogeny and character evolution in Polygonaceae. Syst. Bot. 28: 326–322.

Lance, R. W. 2011. New alignments in North American *Crataegus* (Rosaceae). Phytoneuron 2011-3: 1–8.

Lane, T. M. 1986. *Scutellaria*. Pp. 733–737 *in* Great Plains Flora Association, eds., Flora of the Great Plains. University Press of Kansas, Lawrence.
Lapin, B. 1995. Control of false indigo (*Amorpha fruticosa*), a non-native plant, in riparian areas in Connecticut. Nat. Areas J. 15: 279.
Larisey, M. M. 1940. A monograph of the genus *Baptisia*. Ann. Mo. Bot. Gard. 27: 119–244.
Lassen, P. 1989. A new delimitation of the genera *Coronilla, Hippocrepis*, and *Securigera* (Fabaceae). Willdenowia 19: 49–62.
Lassetter, J. S. 1978. Seed characters in some native American vetches. Sida 7: 255–263.
———. 1984. Taxonomy of the *Vicia ludoviciana* complex (Leguminosae). Rhodora 86: 475–505.
Laughlin, K. 1969. *Quercus shumardii* var. *stenocarpa* Laughlin, stenocarp Shumard oak. Phytologia 19: 57–64.
Lavin, M., and J. J. Doyle. 1991. Tribal relationships of *Sphinctospermum* (Leguminosae): integration of traditional and chloroplast DNA characters. Syst. Bot. 16: 162–172.
Lavin, M., and M. Sousa S. 1995. Phylogenetic systematics and biogeography of the tribe Robineae (Leguminosae). Syst. Bot. Monogr. 45: 1–165.
Lavin, M., J. J. Doyle, and J. D. Palmer. 1990. Evolutionary significance of the loss of the chloroplast-DNA inverted repeat in the Leguminosae subfamily Papilionoideae. Evolution 44: 390–402.
Lawrence, G. H. M. 1954. Studies in the genus *Ceratostigma*. Gentes Herb. 8: 410–420.
Lawrence, J. G., A. Colwell, and O. J. Sexton. 1991. The ecological impact of allelopathy in *Ailanthus altissima* (Simaroubaceae). Amer. J. Bot. 78: 948–958.
Lawrence, R. 2005. Emerald ash borer: A serious threat to North American ash trees. Missouri Nat. Areas Newslet. 6(1): 5–6.
Lee, S., and J. Wen. 2001. A phylogenetic analysis of *Prunus* and the Amygdaloideae (Rosaceae) using ITS sequences of nuclear ribosomal DNA. Amer. J. Bot. 88: 150–160.
Leeuwenberg, A. J. M., and P. W. Leenhouts. 1980. Taxonomy. Chapter 2, pp. 8–96 *in* A. Engler, H. Harms, J. Mattfield, E. Werdermann, and T. Eckardt, eds., Die Natürlichen Pflanzenfamilien, Band 28BI, Angiospermae: Ordnung Gentianales Fam. Loganiaceae. 2d ed. Duncker and Humblot, Berlin.
Legrand, C. D. 1962. Las especies Americanas de *Portulaca*. Anales Mus. Nac. Montevideo, ser. 2, 7: 1–147, 29 pl.
Les, D. H., E. L. Schneider, D. J. Padgett, P S. Soltis, D. E. Soltis, and M. Zanis. 1999. Phylogeny, classification and floral evolution of water lilies (Nymphaeaceae; Nymphaeales): A synthesis of non-molecular, *rbcL, matK*, and 18S rDNA data. Syst. Bot. 24: 28–46.
Lesins, K. A., and I. Lesins. 1979. Genus *Medicago* (Leguminosae). W. Junk, The Hague, The Netherlands.
Levy, M., and D. A. Levin. 1974. Novel flavonoids and reticulate evolution in the *Phlox pilosa-P. drummondii* complex. Amer. J. Bot. 61: 156–167.
Lewis, D. Q. 2000. A revision of the New World species of *Lindernia* (Scrophulariaceae). Castanea 65: 93–122.
Lewis, G., B. Schrire, B. Mackinder, and M. Lock, eds. 2005. Legumes of the World. Royal Botanic Gardens, Kew, Great Britain.
Lewis, H. 1945. A revision of the genus *Trichostema*. Brittonia 5: 276–303.
Lewis, W. H. 1958a. A monograph of the genus *Rosa* in North America. II. *R. foliolosa*. SouthW. Naturalist 3: 145–153.
———. 1958b. A monograph of the genus *Rosa* in North America. III. *R. setigera*. SouthW. Naturalist 3: 154–174.
———. 1961. Merger of the North American *Houstonia* and *Oldenlandia* under *Hedyotis*. Rhodora 63: 216–223.
———. 1962. Phylogenetic study of *Hedyotis* (Rubiaceae) in North America. Amer. J. Bot. 49: 855–865.
———. 2006. *Hedyotis australis* (Rubiaceae) new to Missouri and Florida and related species in the south-central United States. Sida 22: 831–836.
———. 2008. *Rosa carolina* (Rosaceae) subspecies and hybrids in eastern and midwestern United States, Canada, and Mexico. Novon 18: 192–198.
Lewis, W.H., and J. C. Semple. 1977. Geography of *Claytonia virginica* cytotypes. Amer. J. Bot. 64: 1078–1082.
Lewis, W. H., Y. Suda, and B. MacBryde. 1967. Chromosome numbers of *Claytonia virginica* in the St. Louis, Missouri area. Ann. Missouri Bot. Gard. 54: 147–152.

Li, A.-j., B.-j. Bao, A. E. Grabovskaya-Borodina, S.-p. Hong, J. McNeill, S. L. Mosyakin, H. Ohba, and C.-w. Park. 2003. Polygonaceae. Pp. 277–350 *in* Wu Zheng-yi and Peter H. Raven, eds., Flora of China. Volume 5. Ulmaceae through Basellaceae. Science Press, Beijing.

Li, C.-l., H. Ikeda, and H. Ohba. 2003. *Potentilla*. Pp. 291–328 *in* Wu Zheng-yi and Peter H. Raven, eds., Flora of China. Volume 9. Pittosporaceae to Connaraceae. Science Press, Beijing.

Li, H.-l. 1954. The genus *Mazus* (Scrophulariaceae). Brittonia 8: 29–38.

Li, H.-w., and I. C. Hedge. 1994. Lamiaceae. Pp. 50–299 *in* Wu Zheng-yi and Peter H. Raven, eds., Flora of China. Volume 17. Verbenaceae through Solanaceae. Science Press, Beijing.

Lidén, M. 1986. Synopsis of Fumarioideae (Papaveraceae) with a monograph of the tribe Fumarieae. Opera Bot. 88: 1–133.

Linnaeus, C. 1753. Species Plantarum: Exhibentes Plantas Rite Cognitas Ad Genera Relatas, cum Differentiis Specificis, Nominibus Trivialibus, Synonymis Selectis, Locis Natalibus, Secundum Systema Sexuale Digestas, 2 vols. Impensis Laurentii Salvii, Holmiae [Stockholm].

Lippert, W. 1984. Zur Kentniss des *Aphanes microcarpa*-Komplexes. Mitt. Bot. München 20: 451–464.

Lippok, B., A. A. Gardine, P. S. Williamson, and S. S. Renner. 2000. Pollination by flies, bees, and beetles of *Nuphar ozarkana* and *N. advena* (Nymphaeaceae). Amer. J. Bot. 87: 898–902.

Lipscomb, B. L., and G. L. Nesom. 2007. *Galium anglicum* (Rubiaceae) new for Texas and notes on the taxonomy of the *G. parisiense/divaricatum* complex. J. Bot. Res. Inst. Texas 1: 1269–1276.

Little, E. L. 1979. Checklist of United States Trees (Native and Naturalized) U.S.D.A. Agriculture Handbook 541. U.S. Department of Agriculture, Washington, DC.

Little, V. A. 1931. Devil's shoestring as an insecticide. Science 73: 315–316.

Locklear, J. H. 2011. Phlox: a Natural History and Gardener's Guide. Timber Press, Portland, OR.

Lombardi, J. A. 2000. Vitaceae—Gêneros *Ampelocissus, Ampelopsis* e *Cissus*. Fl. Neotrop. Monogr. 80: 1–250.

Longley, A. E. 1924. Cytological studies in the genus *Crataegus*. Amer. J. Bot. 11: 103–115.

López-Almansa, J. C., J. R. Pannell, and L. Gil. 2003. Female sterility in *Ulmus minor* (Ulmaceae): A hypothesis invoking the cost of sex in a clonal plant. Amer. J. Bot. 90: 603–609.

Lourteig, A. 1979. Oxalidaceae extra-austroamericanae. II. *Oxalis* L. sectio *Corniculatae* DC. Phytologia 42: 57–198.

Löve, Á., and D. Löve. 1956. Chromosomes and taxonomy of eastern North American *Polygonum*. Canad. J. Bot. 34: 501–521.

Lovett Doust, L., J. Lovett Doust, and P. B. Cavers. 1981. Fertility relationships in closely related taxa of *Oxalis*, section *Corniculatae*. Canad. J. Bot. 59: 2603–2609.

Lutz, M. V. R. 1986. Systematics and evolution of *Ludwigia* sect. *Ludwigia* (Onagraceae). M.S. thesis, Saint Louis University, St. Louis, MO.

Mabberley, D. J. 1997. The Plant Book: A Portable Dictionary of the Higher Plants. 2d ed. Cambridge University Press, Cambridge.

———. 2008. The Plant Book: A Portable Dictionary of the Higher Plants. 3rd ed. Cambridge University Press, Cambridge.

Mabberley, D. J., C. E. Jarvis, and B. E. Juniper. 2001. The name of the apple. Telopea 9: 421–430.

MacDougal, J. M. 1994. Revision of *Passiflora* subgenus *Decaloba* section *Pseudodysosmia* (Passifloraceae). Syst. Bot. Monogr. 41: 1–146.

Mace, E. S., C. G. Gebhardt, and R. N. Lester. 1999. AFLP analysis of genetic relationships in the tribe Datureae (Solanaceae). Theor. Appl. Genet. 99: 634–641.

Macior, L. W. 1970a. The pollination ecology of *Dicentra cucullaria*. Amer. J. Bot. 57: 6–11.

———. 1970b. Pollination ecology of *Dodecatheon amethystinum* (Primulaceae). Bull. Torrey Bot. Club 97: 150–153.

———. 1978a. Pollination interactions in sympatric *Dicentra* species. Amer. J. Bot. 65: 57–62.

———. 1978b. Pollination ecology of vernal angiosperms. Oikos 30: 452–460.

Mackenzie, K. K., and B. F. Bush 1902. The *Lespedeza* of Missouri. Trans. Acad. Sci. St. Louis 12: 11–19.

MacRoberts, M. H., and B. R. MacRoberts. 1997. *Talinum rugospermum* Holz., new to Louisiana with notes on terete-leaved *Talinum* in Louisiana. Phytologia 82: 86–93.

Makasheva, R. K. 1984. The Pea. A. A. Balkema, Rotterdam, The Netherlands.

Malcolm, W. M. 1966. Root parasitism of *Castilleja coccinea*. Ecology 47: 179–186.

Malecki, R. A., B. Blossey, S. D. Hight, D. Schroeder, L. T. Kok, and J. R. Coulson. 1993. Biological control of purple loosestrife. Biosci. 43: 680–686.

Malik, N., and W. H. Vanden Born. 1988. The biology of Canadian weeds: 86. *Galium aparine* L. and *Galium spurium* L. Canad. J. Pl. Sci. 68: 481–499.

Manchester, S. R., Z.-d. Chen A.-m. Lu, and K. Uemura. 2009. Eastern Asian endemic seed plant genera and their paleogeographic history throughout the northern hemisphere. J. Syst. Evol. 47: 1–41.

Manhart, J. R., and J. H. Rettig. 1994. Gene sequence data. Pp. 235–246 *in* H.-d. Behnke and T.J. Mabry, eds., Caryophyllales. Evolution and Systematics. Springer-Verlag, Berlin.

Mann, W. F., Jr., and L. J. Musselman, 1981. Autotrophic growth of southern root parasites. Amer. Midl. Naturalist 106: 203–205.

Manning, S. D. 1991. The genera of Pedaliaceae in the southeastern United States. J. Arnold Arbor., suppl. ser. 1: 313–347.

Manns, U., and A. A. Anderberg. 2005. Molecular phylogeny of *Anagallis* (Myrsinaceae) based on ITS, *trn*-F, and *ndh*F sequence data. Int. J. Pl. Sci. 166: 1019–1028.

Manoko, M. L. K., R. G. van den Berg, R. M. C. Feron, G. M. van der Weerden, and C. Mariani. 2007. AFLP markers support separation of *Solanum nodiflorum* from *Solanum americanum* sensu stricto (Solanaceae). Pl. Syst. Evol. 267: 1–11.

Marsden-Jones, E. M., and F. E. Weiss. 1938. The essential differences between *Anagallis arvensis* Linn. and *Anagallis foemina* Mill. Proc. Linn. Soc. London 150 (Sess.): 146–155.

Martin, A. C., H. S. Zim, and A. L. Nelson. 1951. American Wildlife & Plants: A Guide to Wildlife Food Habits; the Use of Trees, Shrubs, Weeds, and Herbs by Birds and Mammals of the United States. McGraw-Hill, New York.

Martínez, M. 1998. Revision of *Physalis* section *Epeteiorhiza* (Solanaceae). Anales Inst. Biol. Univ. Nac. Autón. México, Bot. 69: 71–117.

Martínez-Ortega, M. M., and E. Rico. 2001. Taxonomy of *Veronica* subsect. *Serpyllifoliae* (Scrophulariaceae). Bot. J. Linn. Soc. 135: 179–194.

Martínez-Ortega, M. M., A. Herrero, and L. M. Muñoz-Centeno. 2006. Proposal to reject the name *Veronica latifolia* (Scrophulariaceae). Taxon 55: 538–539.

Martins, L., C. Oberprieler, and F. H. Hellwig. 2003. A phylogenetic analysis of Primulaceae s.l. based on internal transcribed spacer (ITS) DNA sequence data. Pl. Syst. Evol. 237: 75–85.

Marx, H. E., N. O'Leary, Y.-w. Yuan, P. Lu-Irving. D. C. Tank, M. E. Múlgura, and R. G. Olmstead. 2010. A molecular phylogeny and classification of Verbenaceae. Amer. J. Bot. 97: 1647–1663.

Mast, A. R., and J. L. Reveal. 2007. Transfer of *Dodecatheon* to *Primula*. Brittonia 59: 79–82.

Mast, A. R., S. Kelso, A. J. Richards, D. J. Lang., D. M. S. Feller, and E. Conti. 2001. Phylogenetic relationships in *Primula* L. and related genera *(Primulaceae)* based on noncoding chloroplast DNA. Int. J. Pl. Sci. 162: 1381–1400.

Mast, A. R., D. M. S. Feller, S. Kelso, and E. Conti. 2004. Buzz-pollinated *Dodecatheon* originated from within the heterostylous *Primula* subgenus *Auriculastrum* (Primulaceae): A seven-region cpDNA phylogeny and its implications for floral evolution. Amer. J. Bot. 91: 926–942.

Mathews, K. G., N. Dunne, E. York, and L. Struwe. 2009. A phylogenetic analysis and taxonomic revision of *Bartonia* (Gentianaceae: Gentianeae), based on molecular and morphological evidence. Syst. Bot. 34: 162–172.

Matthews, J. F., and P. A. Levins. 1985a. *Portulaca pilosa* L., *P. mundula* I. M. Johnst. and *P. parvula* Gray in the Southwest. Sida 11: 45–61.

———. 1985b. The genus *Portulaca* in the southeastern United States. Castanea 50: 96–104.

Matthews, J. F., D. V. Ketron, and S. F. Zane. 1992. The reevaluation of *Portulaca pilosa* and *P. mundula* (Portulacaceae). Sida 15: 71–89.

———. 1993. The biology and taxonomy of the *Portulaca oleracea* L. (Portulacaceae) complex in North America. Rhodora 95: 166–183. [erratum by J.F. Matthews (1994): Rhodora 96: 109].

McClintock, E. 1957. A monograph of the genus *Hydrangea*. Proc. Calif. Acad. Sci. 29: 147–256.

McClintock, E., and C. Epling. 1942. A review of the genus *Monarda* (Labiatae). Univ. Calif. Publ. Bot. 20: 147–194.

———. 1946. A revision of *Teucrium* in the New World, with observations on its variation, geographical distribution and history. Brittonia 5: 491–510.

McCrae, J. 1919. In Flanders Fields and Other Poems, With an Essay in Character by Sir Andrew Macphail. G. P. Putnam's Sons, New York. [In Flanders Fields originally published anonymously in Punch Magazine 149(3883), 1915].
McDonald, C. B. 1980. A biosystematic study of the *Polygonum hydropiperoides* (Polygonaceae) complex. Amer. J. Bot. 67: 664–670.
McGregor, R. L. 1986. Fabaceae. Pp. 416–490 *in* Great Plains Flora Association, eds., Flora of the Great Plains. University Press of Kansas, Lawrence.
McGregor, R. L., and R. E. Brooks. 1986. Plantaginaceae. Pp. 742–747 *in* Great Plains Flora Association, eds., Flora of the Great Plains. University Press of Kansas, Lawrence.
McKenzie, P. M., T. Smith, and N. Holmberg. 2006. Two new localities for *Persicaria glabra* (Polygonaceae) in Missouri and comments on its identification and habitat requirements. Missouriensis 27: 11–16.
McKinney, L. E. 1992. A taxonomic revision of the acaulescent blue violets (*Viola*) of North America. Sida, Bot. Misc. 7: i–vi, 1–60.
McLain, D. K. 1983. Ants, extrafloral nectaries and herbivory on the passion vine. Amer. Midl. Naturalist 110: 433–439.
McNeill, J. 1981. The taxonomy and distribution in eastern Canada of *Polygonum arenastrum* (4*x*=40) and *P. monspeliense* (6*x*=60), introduced members of the *P. aviculare* complex. Canad. J. Bot. 59: 2744–2751.
McNeill, J., F. R. Barrie, H. M. Burdet, V. Demoulin, D. L. Hawksworth, K. Marhold, D. H. Nicolson, J. Prado, P. C. Silva, J. E. Skog, J. H. Wiersema, and N. J. Turland, eds. 2006. International code of botanical nomenclature (Vienna code). Regnum Veg. 146: i–xviii, 1–568.
McVaugh, R. 1951. A revision of the North American black cherries (*Prunus serotina* Ehrh., and relatives). Brittonia 7: 279–315.
———. 1987. Flora Novo-Galiciana, Volume 5, Leguminosae. University of Michigan Press, Ann Arbor.
McWilliam, A. L. 1967. A redefinition of *Penstemon arkansanus* Pennell (Scrophulariaceae). Phytologia 15: 233–235.
Medan, D., and C. Schirarend. 2004. Rhamnaceae. Pp. 320–338 *in* K. Kubitzki, ed., The Families and Genera of Vascular Plants. Vol. 6. Flowering Plants. Dicotyledons. Celastrales, Oxalidales, Rosales, Cornales, Ericales (K. Kubitzki, vol. ed.). Springer-Verlag, Berlin.
Medley, M. E., and J. W. Thieret. 1991. *Ulmus parvifolia* (Ulmaceae) naturalized in Kentucky. Sida 14: 610–612.
Mendel, J. G. 1865 [1866]. Versuche über Pflanzenhybriden. Verh. Naturf. Vereins Brünn 4: 3–47.
Mennema, J. 1989. A taxonomic reivsion of *Lamium* (Lamiaceae). Leiden Bot. Ser. 11: i–vi, 1–198.
Merrill, E. D., and S.-y. Hu. 1949. Work and publications of Henry Muhlenberg, with special attention to unrecorded or incorrectly recorded binomials. Bartonia 25: 1–66, 1 pl.
Merrill, E. D., and F. P. Metcalf. 1942. *Hedyotis* Linnaeus versus *Oldenlandia* Linnaeus and the status of *Hedyoptis lancea* Thunberg in relation to *H. sanguinea* Hance. J. Arnold Arbor. 23: 226–230, pl. 1.
Mertens, T. R., and P. H. Raven. 1965. Taxonomy of *Polyognum*, section *Polygonum* (*Avicularia*) in North America. Madroño 18: 85–92.
Meyer, F. G. 1976. A revision of the genus *Koelreuteria* (Sapindaceae). J. Arnold Arbor. 57: 129–166.
Meyerowitz, E. M., D. R. Smyth, and J. L. Bowman 1989. Abnormal flowers and pattern formation in floral development. Development 106: 209–217.
Miller, G. N. 1955. The genus *Fraxinus*, the ashes, in North America, north of Mexico. Mem. Cornell Univ. Agric. Exp. Sta. 33: 1–64.
Miller, J. M., and K. L. Chambers. 2006. Systematics of *Claytonia* (Portulacaceae). Syst. Bot. Monogr. 78: 1–236, 1 pl.
Miller, N. G. 1971. The genera of Urticaceae in the southeastern United States. J. Arnold Arbor. 52: 40–68.
———. 2001 The Callitrichaceae in the southeastern United States. Harvard Pap. Bot. 5: 277–301.
Millspaugh, C. F. 1974. American Medicinal Plants: An Illustrated and Descriptive Guide to Plants Indigenous to and Naturalized in the United States Which Are Used in Medicine. Dover Publications, New York [reprint of the 1892 edition with a new table of revised classification and nomenclature by E. S. Harrar].

Mitchell, R. S. 1968. Variation in the *Polygonum amphibium* complex and its taxonomic significance. Univ. Calif. Publ. Bot. 45: i–vi, 1–65.

———. 1976. Submergence experiments on nine species of semi-aquatic *Polygonum*. Amer. J. Bot. 63: 1158–1165.

Mitchell, R.S., and E. O. Beal. 1979. Magnoliaceae through Ceratophyllaceae of New York. New York State Mus. Bull 435: i–vi, 1–62.

Mitchell, R. S., and J. K. Dean. 1978. Polygonaceae (buckwheat family) of New York State. Bull. New York State Mus. 431: i–vi, 1–79, index.

Moerman, D. E. 1998. Native American Ethnobotany. Timber Press, Portland, OR.

Mohlenbrock, R. H. 1957. A revision of the genus *Stylosanthes*. Ann. Missouri Bot. Gard. 45: 299–355.

———. 1963. Further considerations in *Stylosanthes* (Leguminosae). Rhodora 65: 245–258.

———. 1981. The Illustrated Flora of Illinois. Flowering Plants: Magnolias to Pitcher Plants. Southern Illinois University Press, Carbondale.

———. 2002. Vascular Flora of Illinois. Southern Illinois University Press, Carbondale.

Moldenke, H. N. 1980. A sixth summary of the Verbenaceae, Avicenniaceae, Stilbaceae, Chloanthaceae, Symphoremaceae, Nyctanthaceae, and Eriocaulaceae of the world as to valid taxa, geographic distribution and synonymy. Phytologia Mem. 2: 1–629.

———. 1982. A sixth summary of the Verbenaceae, Avicenniaceae, Stilbaceae, Chloanthaceae, Symphoremaceae, Nyctanthaceae, and Eriocaulaceae of the world as to valid taxa, geographic distribution and synonymy. Supplement 2. Phytologia 52: 110–129.

Moody, M. L., and D. H. Les. 2002. Evidence of hybridity in invasive watermilfoil (*Myriophyllum*) populations. Proc. Natl. Acad. Sci. U.S.A. 99: 14867–14781.

———. 2010. Systematics of the aquatic angiosperm genus *Myriophyllum* (Haloragaceae). Syst. Bot. 35: 121–139.

Moore, M. O. 1991. Classification and systematics of eastern North American *Vitis* L. (Vitaceae) north of Mexico. Sida 14: 339–367.

Moore, R. J. 1975. The *Galium aparine* complex in Canada. Canad. J. Bot. 53: 877–893.

Morgan, D. R., and D. E. Soltis. 1993. Phylogenetic relationships among members of Saxifragaceae sensu lato based on *rbcL* sequence data. Ann. Missouri Bot. Gard. 80: 631–660.

Morris, J. A. 2007. A molecular phylogeny of the Lythraceae and inference of the evolution of heterostyly. Ph.D. Dissertation, Kent State University, Kent, OH.

Morris, M., and G. Yatskievych. 2000. The genus *Callirhoe* in Missouri. Missouriensis 21: 1–20.

Mosyakin, S. I. 2005. *Rumex*. Pp. 489–533 *in* Flora of North America Editorial Committee, eds., Flora of North America North of Mexico. Volume 5. Magnoliophyta: Caryophyllidae, Part 2. Oxford University Press, New York.

Motten, A. F., and J. Antonovics. 1992. Determinants of outcrossing rate in a predominantly self-fertilizing weed, *Datura stramonium* (Solanaceae). Amer. J. Bot. 79: 419–427.

Mühlenbach, V. 1979. Contributions to the synanthropic (adventive) flora of the railroads in St. Louis, Missouri, U.S.A. Ann. Missouri Bot. Gard. 66: 1–108.

———. 1983. Supplement to the contributions to the synanthropic (adventive) flora of the railroads in St. Louis, Missouri, U.S.A. Ann. Missouri Bot. Gard. 70: 170–178.

Muirhead, J. R., B. Leung, C. van Overdijk, D. W. Kelly, K. Nandakumar, K. R. Marchant, and D. J. MacIsaac. 2006. Modeling local and long-distance dispersal of invasive emerald ash borer *Agrilus planipennis* (Coleoptera) in North America. Diversity & Distrib. 12: 71–79.

Mulcahy, D. L., and D. Caporello. 1970. Pollen flow within a tristylous species: *Lythrum salicaria*. Amer. J. Bot. 57: 1027–1030.

Muller, C. H. 1952. Ecological control of hybridization in *Quercus*: A factor in the mechanism of evolution. Evolution 6: 147–161.

Mulligan, G. A., and D. B. Munro. 1983. The status of *Stachys palustris* (Labiatae) in North America. Canad. J. Bot. 61: 679–682.

———. 1989. Taxonomy of species of North American *Stachys* (Labiatae) found north of Mexico. Naturaliste Canad. 116: 35–51.

Muniyamma, M., and J. B. Phipps. 1979. Meiosis and polyploidy in Ontario species of *Crataegus* in relation to their systematics. Canad. J. Bot. 21: 231–241.

Müntzing, A. 1930. Outlines to a genetic monograph of the genus *Galeopsis* with special reference to the nature and inheritance of partial sterility. Hereditas 13: 185–341.

———. 1932. Cytogenetic investigations on synthetic *Galeopsis tetrahit*. Hereditas 16: 105–154.

Munz, P. A. 1932. Studies in Onagraceae VIII. The subgenera *Hartmannia* and *Gauropsis* of the genus *Oenothera*. The genus *Gayophytum*. Amer. J. Bot. 19: 755–778.

Murray, M. J., and D. E. Lincoln. 1970. The genetic basis of acyclic oil constituents in *Mentha citrata* Ehrh. Genetics 65: 457–471.

Murray, M. J., D. Lincoln, and P. M. Marble. 1972. Oil composition of *Mentha aquatica* × *M. spicata* F_1 hybrids in relation to the origin of *M* ×*piperata*. Canad. J. Genet. Cytol. 14: 13–29.

Murrell, C., E. Gerber, C. Krebs, M. Parepa, U. Schaffner, and O. Bossdorf. 2011 Invasive knotweed affects native plants through allelopathy. Amer. J. Bot. 98: 38–43.

Musselman, L. J., and W. F. Mann Jr. 1978. Root parasites of southern forests. Gen. Techn. Rep. S. O., U.S. Forest Service 20: 1–76.

Musselman, L. J., C. S. Harris, and W. F. Mann Jr. 1978. *Agalinis purpurea*: a parasitic weed on sycamore, sweetgum, and loblolly pine. Tree Planters' Notes 29(4): 24–25.

Mymudes, M. S., and D. H. Les. 1993. Morphological and genetic variability in *Plantago cordata* (Plantaginaceae), a threatened aquatic plant. Amer. J. Bot. 80: 351–359.

Nabhan, G., A. Whiting, H. Dobyns, R. Hevly, and R. Euler. 1981. Devil's claw domestication: evidence from southwestern Indian fields. J. Ethnobiol. 1: 135–164.

Namestnik, S. A., J. R. Thomas, and B. S. Slaughter. 2012. Two recent plant discoveries in Missouri: *Cladium mariscus* subsp. *jamaicense* (Cyperaceae) and *Utricularia minor* (Lentibulariaceae). Phytoneuron 2012-92: 1–6.

Nandini, A. V., B. G. Murray, I. E. W. O'Brien, and K. R. W. Hammett. 1997. Intra- and interspecific variation in genome size in *Lathyrus* (Leguminosae). Bot. J. Linn. Soc. 125: 359–366.

Narayan, R. K. J. 1982. Discontinuous DNA variation in the evolution of plant species: The genus *Lathyrus*. Evolution 36: 877–891.

Nation, P. N. 1989. Alsike clover poisoning: a review. Canad. Vet. J. 30: 410–415.

National Cancer Institute. 1998. Cigars: Health Effects and Trends. Smoking and Tobacco Control Monograph No. 9. Bethesda (MD): National Institutes of Health, National Cancer Institute, Bethesda, MD [http://cancercontrol.cancer.gov/tcrb/monographs/9/m9_complete.pdf].

Neel, M. C., and M. P. Cummings. 2004. Section-level relationships of North American *Agalinis* (Orobanchaceae) based on DNA sequence analysis of three chloroplast gene regions. BMC Evol. Biol. 4: 15. doi: 10.1186/1471-2148-4-15 [online journal].

Nelson, A. D., and W. J. Elisens. 1999. Polyploid evolution and biogeography in *Chelone* (Scrophulariaceae): morphological and isozyme evidence. Amer. J. Bot. 86: 1487–1501.

Nelson, A. D., W. J. Elisens, and D. Benesh. 1998. Notes on chromosome numbers in *Chelone* (Scrophulariaceae). Castanea 63: 183–187.

Nelson, A. P. 1964. Relationships between two subspecies in a population of *Prunella vulgaris*. Evolution 18: 43–51.

Nelson, E. C. 1991. Shamrock: Botany and History of an Irish Myth. Borthius Press, Aberystwyth, Wales.

Nelson, E. N., and R. W. Couch. 1986. History of the introduction and distribution of *Myriophyllum aquaticum* in North America. Pp. 19–26 *in* L. W. J. Anderson, eds., Proceedings, 1st International Symposium on Watermilfoil (*Myriophyllum spicatum*) and Related Haloragaceae Species. Aquatic Plant Management Society, Vicksburg, Mississippi.

Nelson, J. B. 1980. *Mitreola* vs. *Cynoctonum*, and a new combination for the southeastern United States. Phytologia 46: 338–340.

Nelson, P. W. 1979a. A new halophyte for Missouri. Castanea 44: 246–247.

———. 1979b. Frenches [sic] shooting star (*Dodecatheon frenchii*) revealed in Missouri. Missouriensis 1(1): 7–8.

Nemoto, T., and H. Ohashi, 1993. The inflorescence structure of *Kummerowia* (Leguminosae). J. Linn. Soc. Bot. 111: 281–294.

Nepal. M. P., M. H. Mayfield, and C. J. Ferguson. 2012. Identification of eastern North American *Morus* (Moraceae): taxonomic status of *M. murrayana*. Phytoneuron 2012-26: 1–6.

Nesom, G. L. 1993. *Ranunculus* (Ranunculaceae) in Nuevo Leon, with comments on the *R. petiolaris* group. Phytologia 75: 391–398.

———. 2008. *Ranunculus ficaria* (Ranunculaceae) naturalized in Texas. J. Bot. Res. Inst. Texas 2: 741–742.

———. 2009. Again: taxonomy of yellow-flowered caulescent *Oxalis* (Oxalidaceae) in eastern North America. J. Bot. Res. Inst. Texas 3: 727–738.

———. 2010a. *Fraxinus biltmoreana* and *Fraxinus smallii* (Oleaceae), forest trees of the eastern United States. Phytoneuron 2010-51: 1–30.

———. 2010b. Notes on *Fraxinus profunda* (Oleaceae). Phytoneuron 2010-532: 1–6.

———. 2010c. Taxonomy of the *Glandularia bipinnatifida* group (Verbenaceae). Phytoneuron 2010-46: 1–20.

Nevling, L. I. Jr. 1962. The Thymelaeaceae in the southeastern United States. J. Arnold Arbor. 43: 428–434.

Nickrent, D. L., R. J. Duff, A. E. Colwell, A. D. Wolfe, N. D. Young, K. E. Steiner, and C. W. dePamphilis. 1998. Molecular phylogenetic and evolutionary studies of parasitic plants. Pp. 211–241 *in* D. Soltis, P. Soltis, and J. Doyle, eds., Molecular Systematics of Plants II. DNA Sequencing. Kluwer Academic Publishers, Boston, MA.

Nickrent, D. L., D. E. Boufford, and J. Kuijt. 2010. Proposal to conserve the name *Viscum serotinum* (*Phoradendron serotinum*) against *V. leucarpum* (Viscaceae). Taxon 59: 1903–1904.

Nicolson, D. H., and J. H. Wiersema. 2004. Proposal to conserve *Sesamum indicum* against *Sesamum orientale* (Pedaliaceae). Taxon 53: 210–211.

Nie, Z-l., H. Sun, P. M. Beardsley, R. G. Olmstead, and J. Wen. 2006. Evolution of biogeographic disjunction between eastern Asia and eastern North America in *Phryma* (Phrymaceae). Amer. J. Bot. 93: 1343–1356.

Nigh, T. [A.], and D. Ladd. 1987. *Cynoctonum mitreola* rediscovered in Missouri. Missouriensis 8: 1–3.

Nightingale, A., and K. C. Olson. 1984. *Veronica anagallis-aquatica* L. in Taney County. Missouriensis 5: 136–137.

Nitta, M., and O. Ohnishi. 1999. Genetic relationships among two *Perilla* crops, shiso and egoma, and the weedy type revealed by RAPD markers. Genes Genet. Syst. 74: 43–48.

Nitta, M., J. K. Lee, C. W. Kang, M. Katsuta, S. Yasumoto, D. Liu, T. Nagamine, and O. Ohnishi. 2005. The distribution of *Perilla* species. Genet. Resources Crop Evol. 52: 797–804.

Nixon, K. C., and C. H. Muller. 1997. *Quercus* sect. *Quercus*. Pp. 471–506 *in* Flora of North America Editorial Committee, eds., Flora of North America North of Mexico. Volume 3. Magnoliophyta: Magnoliidae and Hamamelidae. Oxford University Press, New York.

Nold, R. 1999. Penstemons. Timber Press, Portland, OR.

Norman, E. 2000. Buddlejaceae. Fl. Neotrop. Monogr. 81: 1–225.

Nowicke, J. W. 1969. Palynotaxonomic study of the Phytolaccaceae. Ann. Missouri Bot. Gard. 55: 294–363.

Nyffeler, R. 2007. The closest relatives of cacti: Insights from phylogenetic analyses of chloroplast and mitochondrial sequences with special emphasis on relationships in the tribe Anacampseroteae. Amer. J. Bot. 94: 89–101.

Nyffeler, R., and U. Eggli. 2010. Disintegrating Portulacaceae: A new familial classification of the suborder Portulacineae (Caryophyllales) based on molecular and morphological data. Taxon 59: 227–240.

Nyman, T., A. Widmer, and H. Roininen. 2000. Evolution of gall morphology and host-plant relationships in willow-feeding sawflies (Hymenoptera: Tenthredinidae). Evolution 54: 526–533.

Oakley, R. A., and H. L. Westover. 1922. How to grow alfalfa. Farmers' Bull. U.S.D.A. 1283: i–ii, 1–35.

Oberle, B. J. 2009. Historical and geographic context for the evolution of climate niche breadth in temperate plants. Ph.D. dissertation, Washington University, St. Louis, MO.

Oberle, B. [J.], and E. J. Esselman. 2011. Fruit and seed characters help distinguish southern Illinois *Dodecatheon* (Primulaceae) species and highlight unusual intergrading populations Rhodora 113: 280–295.

Oberle, B. [J.], and B. A. Schaal. 2011. Responses to historical climate change identify contemporary threats to diversity in *Dodecatheon*. Proc. Amer. Acad. Sci. 108: 5655–5660.

Oberle, B. [J.], R. A. Montgomery, J. E. Beck, and E. J. Esselman. 2012. A morphologically intergrading population facilitates plastid introgression from diploid to tetraploid *Dodecatheon* (Primulaceae). Bot. J. Linn. Soc. 168: 91–100.

Ogg, A. G., Jr., and B. S. Rogers. 1989. Taxonomy, distribution, biology, and control of black nightshade (*Solanum nigrum*) and related species in the United States and Canada. Rev. Weed Sci. 4: 25–58.

Oh, S.-h., and P. S. Manos. 2008. Molecular phylogenetics and cupule evolution in Fagaceae as inferred from nuclear CRABS CLAW sequences. Taxon 57: 434–451.

Ohashi, H. 1981. Notes on *Lespedeza thunbergii* (DC) Nakai. J. Jap. Bot. 56: 244 [in Japanese with English summary].

———. 2005. Tribe Desmodieae. Pp. 433–445 *in* G. Lewis, B. Shrire, B. Mackinder, and M. Lock, eds., Legumes of the World. Royal Botanic Gardens, Kew, Great Britain.

Ohashi, H., and R. R. Mill. 2000. *Hylodesmum*, a new name for *Podocarpium* (Leguminosae). Edinburgh J. Bot. 57: 171–188.

Ohba, H. 2000. The type and identity of *Rosa luciae* Rochebr. & Franch. ex Crép. and the varieties described by Franchet and Savatier. J. Jap. Bot. 75: 148–163.

Ohmart, O. 1987. [Miscellaneous county records]. Pp. 33–47 *in* W. R. Weber, J. P. Rebman, ad W. Corcoran, eds., Missouri botanical record. Missouriensis 8: 32–47.

Ohnishi, O. 1998. Search for the wild ancestor of buckwheat III. The wild ancestor of common cultivated buckwheat and of Tartary buckwheat. Econ. Bot. 52: 123–133.

Ohnishi, O., and Y. Matsuoka. 1996. Search for the wild ancestor of buckwheat II. Taxonomy of *Fagopyrum* (Polygonaceae) species based on morphology, isozymes and cpDNA variability. Genes Genet. Systems 71: 383–390.

Ohsako, T., and O. Ohnishi. 2000. Intra- and interspecific phylogeny of wild *Fagopyrum* (Polygonaceae) species based on nucleotide of noncoding regions of chloroplast DNA. Amer. J. Bot. 87: 573–582.

Olah, L. V., and R. A. DeFillips. 1968. A cytotaxonomic study of French's shooting star. Bull. Torrey Bot. Club 95: 186–198.

O'Leary, N., and M. E. Múlgara. 2012. A taxonomic revision of the genus *Phyla* (Verbenaceae). Ann. Missouri Bot. Gard. 89: 578–596.

O'Leary, N., M. E. Múlgura, and O. Morrone. 2010. Revisión taxonómic de las especies del género Verbena (Verbenaceae). II. Serie Verbena. Ann. Missouri Bot. Garden 97: 365–424.

Olmstead, R. G., and P. A. Reeves. 1995. Evidence for the polyphyly of the Scrophulariaceae based on chloroplast *rbc*L and *ndh*F sequences. Ann. Missouri Bot. Gard. 82: 176–193.

Olmstead, R. G., C. W. dePamphilis, A. D. Wolfe, N. D. Young, W. J. Elisens, and P. A. Reeves. 2001. Disintegration of the Scrophulariaceae. Amer. J. Bot. 88: 348–361.

Ornduff, R. 1972. The breakdown of trimorphic incompatibility in *Oxalis* section *Corniculatae*. Evolution 20: 52–65.

Orzell, S. L. 1982. *Menyanthes trifoliata* in Missouri? Missouriensis 4: 29–30.

Ostry, M. E., M. E. Mielke, and D. D. Skilling. 1994. Butternut—Strategies for managing a threatened tree. Gen. Techn. Rep. N. C., U.S. Forest Serv. NC-165: 1–7.

Ottley, A. M. 1944. The American loti with special consideration of a proposed new section, *Simpeteria*. Brittonia 5: 81–123.

Overlease, W. R. 1975. A study of variation in black oak (*Quercus velutina* Lam.) populations from unglaciated southern Indiana to the range limits in northern Michigan. Proc. Pennsylvania Acad. Sci. 49: 141–144.

———. 1977. A study of the relationship between scarlet oak (*Quercus coccinea* Muenchh.) and Hill Oak (*Quercus ellipsoidalis* E.J. Hill) in Michigan and nearby states. Proc. Pennsylvania Acad. Sci. 51: 47–50.

Overlease, W. R., and E. Overlease. 2011. A note on the host species of mistletoe (*Phoradendron leucarpum*) in the eastern United States. Bartonia 65: 105–111.

Ownbey, G. B. 1947. Monograph of the North American species of *Corydalis*. Ann. Missouri Bot. Gard. 34: 187–259.

———. 1958. Monograph of the genus *Argemone* for North America and the West Indies. Mem. Torrey Bot. Club 21(1): 1–159.

Oxelman, B., M. Backlund, and B. Bremer. 1999. Relationships of the Buddlejaceae s. l. investigated using parsimony jackknife and branch support analysis of chloroplast *ndhF* and *rbcL* sequence data. Syst. Bot. 24: 164–182.

Oxelman, B., P. Kornhall, R. G. Olmstead, and B. Bremer. 2005. Further disintegration of Scrophulariaceae. Taxon 54: 411–425.

Padgett, D. J. 2001. Noteworthy collections and spread of exotic aquatics in Missouri. Castanea 66: 303–306.

———. 2007. A monograph of *Nuphar* (Nymphaeaceae). Rhodora 109: 1–95.
Padgett, D. J., and H. Parker. 1998. *Castanea mollissima* in southwest Missouri. Missouriensis 19: 37–39.
Padgett, D. J., D. H. Les, and G. E. Crow. 1999. Phylogenetic relationships in *Nuphar* (Nymphaeaceae): evidence from morphology, chloroplast DNA, and nuclear ribosomal DNA. Amer. J. Bot. 86: 1316–1324.
Paillet, F. L. 1993. Growth form and life histories of American chestnut and Allegheny and Ozark chinquapin at various North American sites. Bull. Torrey Bot. Club 120: 257–268.
Palmer, E. J. 1931. Conspectus of the genus *Amorpha*. J. Arnold Arbor. 12: 157–197.
———. 1946. *Crataegus* in the northeastern and central United States and adjacent Canada. Brittonia 5: 471–490.
———. 1948. Hybrid oaks of North America. J. Arnold Arbor. 29: 1–48.
———. 1953. A hybrid *Amorpha* and new forms and records from Missouri. Rhodora 55: 157–160.
———. 1961. *Mentzelia albescens* and *Lonicera xylosteum* in Missouri. Rhodora 63: 118–119.
———. 1963. *Crataegus* L. Hawthorn, red haw. Pp. 802–822 *in* J. A. Steyermark. Flora of Missouri. Iowa State University Press, Ames.
Palmer, E. J., and J. A. Steyermark. 1935. An annotated catalogue of the flowering plants of Missouri. Ann. Missouri Bot. Gard. 22: 375–758, 21 pl.
———. 1958. Plants new to Missouri. Brittonia 10: 109–120.
Parfitt, B. D. 1997. *Trautvetteria*. Pp. 138–139 *in* Flora of North America Editorial Committee, eds., 1996. Flora of North America North of Mexico. Volume 3. Magnoliophyta: Magnoliidae and Hamamelidae. Oxford University Press, New York.
Park, C.-w. 1988. Taxonomy of *Polygonum* section *Echinocaulon* (Polygonaceae). Mem. New York Bot. Gard. 47: 1–82.
Park, J.-m., J.-f. Manen, A. E. Colwell, and G. M. Schneeweis. 2008. A plastid phylogeny of the non-photosynthetic parasitic *Orobanche* (Orobanchaceae) and related genera. J. Pl. Res. 121: 365–376.
Park, M. M. 1992. A biosystematic study of *Thalictrum* section *Leucocoma*. Ph.D. dissertation, Pennsylvania State University, University Park.
Parker, M. A. 1996. Cryptic species within *Amphicarpaea bracteata* (Leguminosae): Evidence from isozymes, morphology, and pathogen specificity. Canad. J. Bot. 74: 1640–1650.
Paterson, A. K. 2000. Range expansion of *Polygonum caespitosum* var. *longisetum* in the United States. Bartonia 60: 57–69.
Peirson, J. A., P. D. Cantino, and H. E. Ballard Jr. 2006. A taxonomic revision of *Collinsonia* (Lamiaceae) based on phenetic analysis of morphological variation. Syst. Bot. 31: 398–409.
Pelotto J. P., and M. A. Del Pero Martínez. 1998 Flavonoids in *Strophostyles* species and the related genus *Dolichopsis* (Phaseolinae, Fabaceae): Distribution and phylogenetic significance. Sida 18: 213–222.
Peng, C.-i. 1989. The systematics and evolution of *Ludwigia* sect. *Microcarpium* (Onagraceae). Ann. Missouri Bot. Gard. 76: 221–302.
Peng, C.-i., C. L. Schmidt, P. C. Hoch, and P. H. Raven. 2005. Systematics and evolution of *Ludwigia* section *Dantia* (Onagraceae). Ann. Missouri Bot. Gard. 92: 307–359.
Pennell, F. W. 1921. *Veronica* in North and South America. Rhodora 23: 1–22, 29–41.
———. 1928. *Agalinis* and allies in North America I. Proc. Acad. Nat. Sci. Philadelphia 80: 339–449.
———. 1929. *Agalinis* and allies in North America II. Proc. Acad. Nat. Sci. Philadelphia 81: 111–249.
———. 1935. The Scrophulariaceae of eastern temperate North America. Acad. Nat. Sci. Philadelphia Monogr. 1: i–xv, 1–650, 1 map.
———. 1946. Reconsideration of the *Bacopa-Herpestis* problem of the Scrophulariaceae. Proc. Acad. Nat. Sci. Philadelphia 98: 83–98.
Pennington, T. D. 1990. Sapotaceae. Fl. Neotrop. Monogr. 52: 1–770.
———. 1991. The genera of Sapotaceae. Royal Botanic Gardens, Kew, Great Britain.
Peralta, I. E., S. Knapp, and D. M. Spooner. 2008. Taxonomy of wild tomatoes and their relatives (*Solanum* sect. *Lycopersicoides*, sect. *Juglandifolia*, sect. *Lycopersicon*; Solanaceae). Syst. Bot. Monogr. 84: 1–186, fronticepiece.
Perry, J. E., D. M. E. Ware, and A. McKenney-Mueller. 1998. *Aeschynomene indica* L. (Fabaceae) in Virginia. Castanea 63: 191–194.
Persson, C. 2001. Phylogenetic relationships in Polygalaceae based on plastid DNA sequences from the *trnL-F* region. Taxon 50: 763–779.

Petersen, F. P., and D. E. Fairbrothers. 1983. A serotaxonomic appraisal of *Amphipterygium* and *Leitneria*, two amentiferous taxa of Rutiflorae (Rosidae). Syst. Bot. 8: 134–148.
Peterson, C. J., and J. Ems-Wilson. 2003. Catnip essential oil as a barrier to subterranean termites (Isoptera: Rhinotermitidae) in the laboratory. J. Econ. Entomol. 96: 1275–1282.
Peterson, C. J., L. T. Nemetz, L. M. Jones, and J. R. Coats. 2002. Behavioral activity of catnip (Lamiaceae) essential oil components to the German cockroach (Blattodea: Blattellidae). J. Econ. Entomol. 95: 377–380.
Peterson, K. M. 1978. Systematic studies of *Salvia* L. subgenus *Calospace* (Benth.) in Benth. & Hook. section *Farinaceae* (Epling) Epling (Lamiaceae). Ph.D. dissertation, University of Maryland, College Park.
Pettengill, J. B., and M. C. Neel. 2008. Phylogenetic patterns and conservation among North American members of the genus *Agalinis* (Orobanchaceae). BMC Evol. Biol. 8: 264. doi: 10.1186/1471-2148-8-264 [online journal].
Philbrick, C. T. 1984. Pollen tube growth within vegetative tissues of *Callitriche* (Callitrichaceae). Amer. J. Bot. 71: 882–886.
———. 1989. Systematic studies of North American Callitriche (Callitrichaceae). Ph.D. dissertation, University of Connecticut, Storrs.
Philcox, D. 1965. Revision of the New World species of *Buchnera* L. (Scrophulariaceae. Kew Bull. 18: 275–316.
Phillips, J. 1998. Wild Edibles of Missouri. 2d ed. Missouri Department of Conservation, Jefferson City.
Phillips, M., and R. A. Kowal. 1983. Systematics of *Saxifraga pensylvanica* and a putative segregate species, *S. forbesii* (Saxifragaceae). Amer. J. Bot. 70(5, pt. 2): 125–126 [abstract].
Phillips, W. W. 2003. Plants of the Lewis and Clark Expedition. Mountain Press, Missoula, MT.
Phipps, J. B. 1998. Synopsis of *Crataegus* series *Apiifoliae, Cordatae, Microcarpae*, and *Brevispinae* (Rosaceae subfam. Maloideae). Ann. Missouri Bot. Gard. 85: 475–491.
———. 2005. Review of hybridization in *Crataegus*—another look at "the *Crataegus* problem." Ann. Missouri Bot. Gard. 92: 113–127.
———. 2006. *Crataegus spes-aestivum*, a new species in series *Punctatae* (Rosaceae), and six new varietal names from the Missouri *Crataegus* flora. Novon 16: 381–387.
———. 2012. Critical taxa in *Crataegus* series *Molles* (Rosaceae): Typifications, new combinations, and taxonomic review. Phytoneuron 2012-78: 1–23.
Phipps, J. B., and K. Dvorsky. 2006. *Crataegus* series *Parvifoliae* and its putative hybrids in the southeastern United States. Sida 22: 423–445.
Phipps, J. B., K. R. Robertson, J. R. Rohrer, and P. G. Smith. 1991. Origins and evolution of subfam. Maloideae (Rosaceae). Syst. Bot. 16: 303–332.
Phipps, J. B., G. Yatskievych, and K. Wood. 2007. Typification of *Crataegus* (Rosaceae) names from the Missouri Flora. Harvard Pap. Bot. 11: 179–197.
Piehl, M. A. 1962. The parasitic behavior of *Dasistoma macrophylla*. Rhodora 64: 331–336.
———. 1963. Mode of attachment, haustorium structure, and hosts of *Pedicularis canadensis*. Amer. J. Bot. 50: 978–985.
———. 1965. The natural history and taxonomy of *Comandra* (Santalaceae). Mem. Torrey Bot. Club 22(1): 1–97.
Piehl, M. A., and P. N. Piehl. 1973. *Orobanche* in Tennessee. J. Tennessee Acad. Sci. 48: 39.
Pieters, A. J. 1920. The Hop Clovers. USDA Office of Forage Crop Investigations, Washington DC. [mimeographed circular].
Pilatowski, R. E. 1982. A taxonomic study of the *Hydrangea arborescens* complex. Castanea 47: 84–98.
Piper, C. V. 1924. Forage Plants and Their Culture. Revised ed. Macmillan, New York.
Pittman, A. B. 1988. Systematic studies in *Scutellaria* section *Mixtae* (Labiatae). Ph.D. dissertation, Vanderbilt University, Nashville, TN.
Poland, T. M., and D. G. McCullough. 2006. Emerald ash borer: invasion of the urban forest and the threat to North America's ash resource. J. Forest. (Washington, DC) 104: 118–124.
Polhill, R. M. 1981. Papilionoideae. Pp. 191–208 *in* R. M. Polhill and P. H. Raven, eds., Advances in Legume Systematics, Part 1. Royal Botanic Gardens, Kew, Great Britain.
Polhill, R. M., P. H. Raven, and C. H. Stirton. 1981. Evolution and systematics of the Leguminosae. Pp. 1–26 *in* R. M. Polhill and P. H. Raven, eds., Advances in Legume Systematics, Part 1. Royal Botanic Gardens, Kew, Great Britain.

Porter, D. M. 1969a. The genus *Kallstroemia* (Zygophyllaceae). Contr. Gray Herb. 198: 41–153.

———. 1969b. *Kallstroemia* (Zygophyllaceae) in Missouri. Ann. Missouri Bot. Gard. 56: 290.

———. 1972. The genera of Zygophyllaceae in the southeastern United States. J. Arnold Arbor. 53: 531–552.

———. 1976. *Zanthoxylum* (Rutaceae) in North America north of Mexico. Brittonia 28: 443–447.

Porter, J. M., and L. A. Johnson. 2000. A phylogenetic classification of the Polemoniaceae. Aliso 19: 55–91.

Post, A. R., A. Krings, W. A. Wall, and J. C. Neal. 2009. Introduced lesser celandine (*Ranunculus ficaria*, Ranunculaceae) and its putative subspecies in the United States: A morphometric analysis J. Bot. Res. Inst. Texas 3: 193–209.

Potter, D., T. Eriksson, R. C. Evans, S. Oh, J. E. E. Smedmark, D. R. Morgan, M. Kerr, K. R. Robertson, M. Arsenault, T. A. Dickinson, and C. S. Campbell. 2007. Phylogeny and classification of Rosaceae. Pl. Syst. Evol. 266: 5–43.

Potts, L. K. 1996. Rattlesnake weed. Missouri Conservationist 57(8): 20–21.

Poutaraud, A., F. Bourgaud, P. Girardin, and E. Gontier. 2000. Cultivation of rue (*Ruta graveolens* L., Rutaceae) for the production of furanocoumarins of therapeutic use. Canad. J. Bot. 78: 1326–1335.

Pringle, J. S. 1967. Taxonomy of *Gentiana*, section *Pneumonanthe*, in eastern North America. Brittonia 19: 1–32.

———. 1990. Taxonomic notes on western American Gentianaceae. Sida 14: 179–187.

Pruski, J. F., and G. L. Nesom. 1992. *Glandularia ×hybrida* (Verbenaceae), a new combination for a common horticultural plant. Brittonia 44: 494–496.

Puff, C. 1976. The *Galium trifidum* group (*Galium* sect. *Aparinoides*, Rubiaceae). Canad. J. Bot. 54: 1911–1925.

———. 1977. The *Galium obtusum* group (*Galium* sect. *Aparinoides*, Rubiaceae). Bull. Torrey Bot. Club 104: 202–208.

Pursh, F. 1814 [1813]. Flora Americae Septentrionalis or, a Systematic Arrangement and Description of the Plants of North America: Containing, Besides What Have Been Described by Preceding Authors, Many New and Rare Species, Collected During Twelve Years Travels and Residence in That Country. White, Cochrance, and Company, London.

Qian, G., L.-f. Liu, and G. G. Tang. 2010. (1933) Proposal to conserve the name *Malus domestica* against *M. pumila, M. communis, M. frutescens*, and *Pyrus dioica* (Rosaceae). Taxon 59: 650–652.

Qin, H.-n. 1997. A taxonomic revision of the Lardizabalaceae. Cathaya 8–9: 1–214.

Rabeler, R. K. 1984. An unnecessary infraspecific combination in *Medicago* (Fabaceae). Taxon 33: 321–322.

Rahmanzadeh, R., K. Müller, E. Fischer, D. Bartels, and T. Borsch. 2005. The Linderniaceae and Gratiolaceae are further lineages distinct from the Scrophulariaceae (Lamiales). Pl. Biol. 7: 67–76.

Rahn, K. 1978. *Plantago* ser. *Gnaphalodes* Rahn a taxonomic revision. Bot. Tidsskr. 73: 137–154.

Ramamoorthy, T. P., and E. M. Zardini. 1987. The systematics and evolution of *Ludwigia* sect. *Myrtocarpus* sensu lato (Onagraceae). Monogr. Syst. Bot. Missouri Bot. Gard. 19: i–v, 1–120.

Ramsey, G. W. 1988. A comparison of vegetative characteristics of several genera with those of the genus *Cimicifuga* (Ranunculaceae). Sida 13: 57–63.

Raveill, J. A. 1995. Hybridization and population structure in the Desmodium paniculatum species complex. Ph.D. dissertation, Vanderbilt University, Nashville, TN.

———. 2002. Allozyme evidence for the hybrid origin of *Desmodium humifusum* (Fabaceae). Rhodora 104: 253–270.

Raveill, J. A., and G. Yatskievych. 2008. *Heliotropium europaeum* (Heliotropiaceae) new to Missouri. Missouriensis 28/29: 10–17.

Raven, P. H. 1963. The Old World species of *Ludwigia* (including *Jussiaea*), with a synopsis of the genus (Onagraceae). Reinwardtia 6: 327–427.

———. 1964. The generic subdivision of Onagraceae, tribe Onagreae. Brittonia 16: 276–288.

———. 1965. An earlier name for *Ludwigia natans* (Onagraceae). Rhodora 67: 83–84.

Raven, P. H., and D. P. Gregory. 1972. A revision of the genus *Gaura* (Onagraceae). Mem. Torrey Bot. Club 23: 1–96.

Raven, P. H., W. Dietrich, and W. Stubbe. 1979. An outline of the systematics of *Oenothera* subsect. *Euoenothera* (Onagraceae). Syst. Bot. 4: 242–252.

Ray, J. D., Jr. 1956. The genus *Lysimachia* in the New World. Illinois Biol. Monogr. 24(3–4): 1–160.

Ray, M. F. 1998. New combinations in Malva (Malvaceae: Malveae). Novon 8: 288–295.

Reader's Digest Association. 1984. Magic and Medicine of Plants. Reader's Digest Association, Pleasantville, NY.

Rebman, J. P., and W. R. Weber. 1988. Two woody taxa of noteworthy occurrence in southwest Missouri. Missouriensis 9: 58–61.

Rechinger, K. H. 1937. The North American species of *Rumex*. Publ. Field Mus. Nat. Hist., Bot. Ser. 17(1): 1–151.

Reddy, K. N. 2005. Deep tillage and glyphosate-reduced redvine (*Brunnichia ovata*) and trumpetcreeper (*Campsis radicans*) populations in glyphosate-resistant soybean. Weed Technol. 19: 713–718.

Redfern, M., and R. R. Askew. 1992. Plant Galls (Naturalists' Handbooks, vol. 17). Richmond Publishing Company, Slough, Great Britain.

Reed, C. F. 1977. History and distribution of Eurasian watermilfoil in United States and Canada. Phytologia 36: 417–436.

Reed, H. S., and I. Smoot. 1906. The mechanism of seed-disperal in *Polygonum virginianum*. Bull. Torrey Bot. Club 33: 377–386.

Rehder, A. 1921. *Philadelphus verrucosus* Schrader spontaneous in Illinois. J. Arnold Arbor. 2: 153–156.

———. 1940. Manual of Cultivated Trees and Shrubs Hardy in North America, Exclusive of the Subtropical and Warmer Temperate Regions. Revised ed. MacMillan Company, New York.

———. 1945. *Carya alba* proposed as *nomen ambiguum*. J. Arnold Arbor. 26: 482–483.

Reinhard, R. T., and S. Ware. 1989. Adaptation to substrate in rock outcrop plants: Interior Highlands *Talinum* (Portulacaceae). Bot. Gaz. (Crawfordsville) 150: 449–453.

Reveal, J. L. 2005. Polygonaceae subfam. Eriogonoideae. Pp. 218–478 *in* Flora of North America Editorial Committee, eds., Flora of North America North of Mexico. Volume 5. Magnoliophyta: Caryophyllidae, Part 2. Oxford University Press, New York.

———. 2011. Summary of recent systems of angiosperm classification. Kew Bull. 66: 5–48.

Reveal, J. L., and F. R. Barrie. 1992. On the identity of *Hedysarum violaceum* Linnaeus (*Fabaceae*). Phytologia 71: 456–461.

Reveal, J. L., and M. C. Johnston. 1989. A new combination in *Phoradendron* (Viscaceae). Taxon 38: 107–108.

Rice, E. L. 1984. Allelopathy. 2d ed. Academic Press, Orlando, FL.

Richardson, J. E., M. F. Fay, Q. C. B. Cronk, D. Bowman, and M. W. Chase. 2000. A phylogenetic analysis of Rhamnaceae using *rbcL* and *trnL-F* plastid DNA sequences. Amer. J. Bot. 87: 1309–1324.

Richens, R. H. 1983. Elm. Cambridge University Press, Cambridge, Great Britain.

Rickett, H. W. 1937. Forms of *Crataegus crus-galli*. Bot. Gaz. (Crawfordsville) 98: 609–616.

Riley-Hulting, E. T., A. Delgado-Salinas, and M. Lavin. 2004. Phylogenetic systematics of *Strophostyles* (Fabaceae): a North American temperate genus within a neotropical diversification. Syst. Bot. 29: 627–653.

Ritchason, J. 1995. The Little Herb Encyclopedia. 3rd ed. Woodland Health Books, Pleasant Grove, UT.

Ritz, C. M., H. Schmuths, and V. Wissemann. 2005. Evolution by reticulation: European dogroses originated by multiple hybridization across the genus *Rosa*. J. Hered. 96: 4–14.

Ro, K.-e., and B. A. McPheron. 1997. Molecular phylogeny of the *Aquilegia* group (Ranunculaceae) based on internal transcribed spacers and 5.8S nuclear ribosomal DNA. Biochem. Syst. Ecol. 25: 445–461.

Robbins, G. T. 1944. North American species of *Androsace*. Amer. Midl. Naturalist 32: 137–163.

Robertson, C. [R.] 1889a. Flowers and insects. I. Bot. Gaz. (Crawfordsville) 14: 120–126.

———. 1889b. Flowers and insects. III. Bot. Gaz. (Crawfordsville) 14: 297–404.

———. 1891. Flowers and insects, Asclepiadaceae to Scrophulariaceae. Trans. Acad. Sci. St. Louis 5: 569–598.

Robertson, K. R. 1972. The genera of Geraniaceae in the southeastern United States. J. Arnold Arbor. 53: 182–201.

———. 1974. The genera of Rosaceae in the southeastern United States. J. Arnold Arbor. 55: 303–401, 611–662.

———. 1975. The Oxalidaceae in the southeastern United States. J. Arnold Arbor. 56: 223–239.

———. 1977. *Cladrastis*: the yellow-woods. Arnoldia 37: 137–150.

Robertson, K. R., J. B. Phipps, J. R. Rohrer, and P. G. Smith. 1991. A synopsis of genera in Maloideae (Rosaceae). Syst. Bot. 16: 376–394.

Robinson, J. P., S. A. Harris, and B. E. Juniper. 2001. Taxonomy of the genus *Malus* Mill. (Rosaceae) with emphasis on the cultivated apple, *Malus* ×*domestica* Borkh. Pl. Syst. Evol. 226: 35–58.

Rodman, J. E. 1990. Centrospermae revisited, part 1. Taxon 39: 383–393.

———. 1994. Cladistic and phenetic studies. Pp. 279–301 *in* H.-d. Behnke and T. J. Mabry, eds., Caryophyllales: Evolution and Systematics. Springer-Verlag, Berlin.

Rodriguez, F., F. Wu, C. Ané, S. Tanksley, and D. M Spooner. 2009. Do potatoes and tomatoes have a single evolutionary history, and what proportion of the genome supports this history. BMC Evol. Biol. 9: 191. doi: 10.1186/1471-2148-9-191 [online journal].

Rogers, C. M. 1963. Yellow flowered species of *Linum* in eastern North America. Brittonia 15: 97–122.

———. 1968. Yellow flowered species of *Linum* in Central America and western North America. Brittonia 20: 107–135.

———. 1984. Linaceae. N. Amer. Fl., ser. 2 12: 1–48.

Rogers, G. K. 1985. The genera of Phytolaccaceae in the southeastern United States. J. Arnold Arbor. 66: 1–37.

———. 1986. The genera of Loganiaceae in the southeastern United States. J. Arnold Arbor. 67: 143–185.

———. 1987. The genera of Cinchonoideae (Rubiaceae) in the southeastern United States. J. Arnold Arbor. 68: 137–183.

Rohrer, J. R. 2006. Phylogenetic relationships among species of North American plums inferred from nuclear DNA sequences. Botany 2006 Abstracts 296 [abstract].

Rohrer J. R., K. R. Robertson, and J. B. Phipps. 1994. Floral morphology of Maloideae (Rosaceae) and its systematic relevance. Amer. J. Bot. 81: 574–581.

Rohrer, J. R., R. Ahmad, S. M. Southwick, and D. Potter. 2004. Microsatellite analysis of relationships among North American plums (*Prunus* sect. *Prunocerasus*, Rosaceae). Pl. Syst. Evol. 244: 69–75.

Rombauer, I. S., and M. R. Becker. 1975. Joy of Cooking. 1975 ed. Bobbs-Merrill Company, Indianapolis, IN.

Ronse De Craene, L. P., S.-p. Hong, and E. F. Smets. 2000. Systematic significance of fruit morphology and anatomy in tribes Persicarieae and Polygoneae (Polygonaceae). Bot. J. Linn. Soc. 134: 301–337.

———. 2004. What is the taxonomic status of *Polygonella*? Evidence of floral morphology. Ann. Missouri Bot. Gard. 91: 320–345.

Ronse Decraene, L.-p., and J. R. Akeroyd. 1988. Generic limits in *Polygonum* and related genera (Polygonaceae) on the basis of floral characters. Bot. J. Linn. Soc. 98: 321–371.

Rosatti, T. J. 1984. The Plantaginaceae in the southeastern United States. J. Arnold Arbor. 65: 533–562.

———. 1986. The genera of Sphenocleaceae and Campanulaceae in the southeastern United States. J. Arnold Arbor. 67: 1–64.

Rosendahl, C. O. 1927. A revision of the genus *Sullivantia*. Minnesota Stud. Pl. Sci. 1: 401–427.

———. 1951. A new *Heuchera* from Missouri together with some notes on the *Heuchera parviflora* complex. Rhodora 53: 105–109.

Rowan, D. 1994. Rediscovery of *Trifolium stoloniferum* in Missouri. Missouriensis 15(2): 1–5.

———. 1995. Final Report on the Missouri Natural Features Inventory of Bollinger, Butler, Madison, and Wayne Counties. Missouri Department of Conservation, Jefferson City.

Rudd, V. E. 1955. The American species of *Aeschynomene*. Contr. U.S. Natl. Herb. 32: 1–172.

———. 1971. Studies in the Sophoreae (Leguminosae): 1. Phytologia 21: 327.

Russell, N. H. 1956. Regional variation patterns in the stemless white violets. Amer. Midl. Naturalist 56: 491–503.

———. 1965. Violets (*Viola*) of central and eastern United States: an introductory survey. Sida 2: 1–113.

Russell, N. H., and M. Cooperrider. 1955. Predictions of introgression in *Viola*. Amer. Midl. Naturalist 54: 42–51.

Russell, N. H., and A. C. Risser. 1960. The hybrid nature of *Viola emarginata* (Nuttall) LeConte. Brittonia 12: 298–305.

Ryan, J. 1994. *Menyanthes trifoliata* rediscovered in Missouri. Missouriensis 15(1): 17–21.

Rydberg, P. A. 1908–1918. Rosaceae. Fl. N. Amer. 22(3–6): 239–533.

———. 1929. Astragalinae. N. Amer. Fl., ser. 1 24: 251–462.

Saar, D. E., N. C. Bundy, L. J. Potts, and M. O. Saar. 2012. Status of *Morus murrayana* (Moraceae). Phytologia 94: 245–252.

Salah, S. 2006. Genetic and morphological characterization of red mulberry (*Morus rubra* L.), white mulberry (*Morus alba* L.) and their hybrids. M.S. thesis, University of Central Missouri, Warrensburg.

Sanchez, A, and K. A. Kron. 2008. Phylogenetics of Polygonaceae with an emphasis on the evolution of Eriogonoideae. Syst. Bot. 33: 87–96.

Sand, S. 1992. The American yellowwood. Amer. Horticulturist 71: 35–38.

Sanders, R. W. 1976. Distributional history and probable ultimate range of *Galium pedemontanum* (Rubiaceae) in North America. Castanea 41: 73–80.

———. 2001. The genera of Verbenaceae in the southeastern United States. Harvard Pap. Bot. 5: 303–358.

Sanderson, M. J., and M. F. Wojciechowski. 1996. Diversification rates in a temperate legume clade: are there "so many species" of *Astragalus* (Fabaceae)? Amer. J. Bot. 83: 1488–1502.

Sano, Y., and F. Kita. 1975. Cytological studies of several interspecific F_1 hybrids in the subgenus *Eumelilotus*. J. Fac .Agr. Hokkaido Univ. 58: 225–246.

Santamour , F. S., Jr. 1962. The relation between polyploidy and morphology in white and biltmore ashes. Bull. Torrey Bot. Club 89: 228–232.

Sargent, C. S. 1901. New or little-known North American trees. III. Bot. Gaz. (Crawfordsville) 31: 217–240.

———. 1902. Silva of North America: A Description of the Trees Which Grow Naturally in North America Exclusive of Mexico. Vol. XIII. Supplement: Rhamnaceæ–Rosaceae. Houghton, Mifflin and Company, Boston.

———. 1908. *Crataegus* in Missouri. Ann. Rep. Missouri Bot. Gard. 19: 35–126.

———. 1912. *Crataegus* in Missouri, II. Ann. Rep. Missouri Bot. Gard. 22: 67–83.

———. 1921 [1922]. Notes on North American trees. IX. J. Arnold Arbor. 1–11.

———. 1922. Manual of the Trees of North America (Exclusive of Mexico). 2d ed. Houghton, Mifflin and Company, Boston.

Sarkar, N. M. 1958. Cytotaxonomic studies on *Rumex* section *Axillares*. Canad. J. Bot. 36: 947–996.

Sauer, J. D. 1950. Pokeweed, an old American herb. Missouri Bot. Gard. Bull. 38: 82–88.

———. 1952. A geography of pokeweed. Ann. Missouri Bot. Gard. 39: 113–125.

———. 1964. Revision of *Canavalia*. Brittonia 16: 106–181.

———. 1993. Historical Geography of Crop Plants—A Select Roster. CRC Press, Boca Raton, FL.

Savage, A. D., and T. R. Mertens. 1968. A taxonomic study of genus *Polyognum*, section *Polygonum* (*Avicularia*) in Indiana and Wisconsin. Proc. Indiana Acad. Sci. 77: 357–369.

Savolainen, V, M. F. Fay, D. C. Albach, A. Backlund, M. van der Bank, K. M. Cameron, S. A. Johnson, M. D. Lledó, J. C. Pintaud, M. Powell, M. C. Sheahan, D. E. Soltis, P. S. Soltis, P. Weston, W. M. Whitten, K. J. Wurdack, and M. W. Chase. 2000. Phylogeny of the eudicots: A nearly complete familial analysis based on *rbcL* gene sequences. Kew Bull. 55: 257–309.

Schanzer, I. A. 1994. Taxonomic revision of the genus *Filipendula* Mill. (Rosaceae). J. Jap. Bot. 69: 290–319.

Schilling, E. E. 1978. A systematic study of the *Solanum nigrum* in North America. Ph.D. dissertation, Indiana University, Bloomington.

———. 1981. Systematics of *Solanum* sect. *Solanum* (Solanaceae) in North America. Syst. Bot. 6: 172–185.

Schnack, B., and G. Covas. 1944. Nota sobre la validez del género *Glandularia* (Verbenáceas). Darwiniana 6: 469–476.

Schnee, B. K., and D. M. Waller. 1986. Reproductive behavior of *Amphicarpaea bracteata* (Leguminosae), an amphicarpic annual. Amer. J. Bot. 73: 376–386.

Schneider, E. L., and J. D. Buchanan. 1980. Morphological studies of the Nymphaeaceae. XI. The floral biology of *Nelumbo pentapetala*. Amer. J. Bot. 67: 189–193.

Schneider, E. L., and T. Chaney. 1981. The floral biology of *Nymphaea odorata* (Nymphaeaceae). SouthW. Naturalist 26: 159–165.

Schneider, E. L., and L. A. Moore. 1977. Morphological studies of the Nymphaeaceae. VII. The floral biology of *Nuphar lutea* subsp. *macrophylla*. Brittonia 29: 88–99.

Schrader, J. A., and W. R. Graves. 2011. Taxonomy of *Leitneria* (Simaroubaceae) resolved by ISSR, ITS, and morphometric characterization. Castanea 76: 313–338.

Schrire, B. D. 2005. Tribe Phaseoleae. Pp. 392–431 *in* G. Lewis, B. Schrire, B. Mackinder, and M. Lock, eds., Legumes of the World. Royal Botanic Gardens, Kew, Great Britain.

Schröder, A., A. Prinzhorn, and K. Kraut. 1869. Über Salicylverbindungen. Annal. Chem. Pharmac. 150: 1–20.

Schuster, T. M., J. L. Reveal, and K. A. Kron. 2011. Phylogeny of Polygoneae (Polygonaceae: Polygonoideae). Taxon 60: 1653–1666.

Schwartz, S. 2005. Psychoactive Herbs in Veterinary Behavior Medicine. Blackwell Publishing Professional, Ames, IA.

Schwartz, T. 1995. *Lespedeza bicolor*. Amer. Nurseryman 181: 130.

Schwarzwalder, R. N. 1986. Systematics and early evolution of the Platanaceae. Ph.D. dissertation, Indiana University, Bloomington.

Schwegman, J. E. 1982. A new species of *Oxalis*. Phytologia 50: 463–467.

———. 1984 The jeweled shooting star (*Dodecatheon amethystinum*) in Illinois. Castanea, 49: 74–82.

Scora, R. W. 1965. The confused nomenclature of *Monarda bradburiana* Beck. Bull. Torrey Bot. Club 92: 492–493.

———. 1967. Interspecific relationships in the genus *Monarda* (Labiatae). Univ. Calif. Publ. Bot. 41: 1–71.

Seabrook, J. A., and L. A. Dionne. 1976. Studies on the genus *Apios* I. Chromosome number and distribution of *Apios americana* and *A. priceana*. Canad. J. Bot. 54: 2567–2572.

Seithe, A., and J. R. Sullivan. 1990. Hair morphology and systematics of *Physalis* (Solanaceae). Pl. Syst. Evol. 170: 193–204.

Sell, P. D. 1994. *Ranunculus ficaria* L. sensu lato. Watsonia 20: 41–50.

Seltzner, S., and T. L. Eddy. 2003. Allelopathy in *Rhamnus cathartica*, European buckthorn. Michigan Bot. 42: 51–61.

Senn, H. 1939. The North American species of *Crotalaria*. Rhodora 41: 317–366.

Settergren, C., and R. E. McDermott. 1962. Trees of Missouri. University of Missouri—Columbia Agricultural Experiment Station, Columbia [reprinted several times without changes].

Sewell, M., and M. A. Vincent. 2006. Biosystematics of the *Phacelia ranunculacea* complex (Hydrophyllaceae). Castanea 71: 192–209.

Shaw, J., and R. L. Small. 2005. Chloroplast DNA phylogeny and phylogeography of the North American plums (*Prunus* subgenus *Prunus* section *Prunocerasus*, Rosaceae). Amer. J. Bot. 92: 2011–2030.

Shen-Miller, J., J. W. Schopf, J. Harbottle, R.-i. Cao, S. Ouyang, K.-s. Zhou, J. R. Southon, and G.-h. Liu. 2002. Long-lived lotus: Germination and soil-irradiation of centuries-old fruits, and cultivation, growth, and phenotypic abnormalities of offspring. Amer. J. Bot. 89: 236–247.

Sherman-Broyles, S. L. 1997. *Ulmus*. Pp. 369–375 *in* Flora of North America Editorial Committee, eds., Flora of North America North of Mexico. Volume 3. Magnoliophyta: Magnoliidae and Hamamelidae. Oxford University Press, New York.

Shildneck, P., A. G. Jones, and V. Mühlenbach. 1981. Additions to the vouchered records of Illinois plants and a note on the occurrence of *Rumex cristatus* in North America. Phytologia 47: 265–290.

Shimizu, T. 1981. Miscellaneous notes from my Appalachian trip. Hikobia, Suppl. 1: 445–453.

Shinners, L. H. 1951. *Phlox drummondii* Hook. var. *mcallisteri* (Whitehouse) Shinners, comb. nov. Field & Lab. 19: 127.

———. 1956. Authorship and nomenclature of bur clovers (*Medicago*) found wild in the United States. Rhodora 58: 1–13.

———. 1957. Salt cedars (*Tamarix*, Tamaricaceae) of the Soviet Union, by S. G. Gorschkova, translation and commentary. SouthW. Naturalist 2: 48–73 [appendix: notes on *Tamarix* in the southwestern United States, pp. 71–73].

———. 1958. Spring Flora of the Dallas-Fort Worth Area, Texas. Published by the author, Dallas, TX.

———. 1961. *Viola rafinesquii*: nomenclature and native status. Rhodora 63: 327–335.

———. 1963. The varieties of *Teucrium canadense* (Labiatae). Sida 1: 182–183.

———. 1967. Stray notes on Texas *Plantago* (Plantaginaceae). Sida 3: 120–122.

Shurtleff, W., and A. Aoyagi. 1985. The Book of Kudzu: A Culinary and Healing Guide. Autumn Press, Brookline, MA.

Siedo, S. J. A taxonomic treatment of *Sida* sect. *Ellipticifoliae* (Malvaceae). Lundellia 2: 100–127.

Sievers, A. F., G. A. Russell, M. S. Lowman, E. D. Fowler, C. O. Erlanson, and V. A. Little. 1938. Studies on the possibilities of devil's shoestring (*Tephrosia virginiana*) and other native species of *Tephrosia* as commercial sources of insecticides. Techn. Bull. U.S.D.A. 595: 1–40.

Sievers, A. F., M. S. Lowman, G. A. Russell, and W. N. Sullivan. 1940. Changes in the insecticidal value of the roots of cultivated devil's shoestring, *Tephrosia virginiana*, at four seasonal growth periods. Amer. J. Bot. 27: 284–289.

Simmers, R. W., and R. Kral. 1992. A new species of *Blephilia* (Lamiaceae) from northern Alabama. Rhodora 94: 1–14.

Simpson, B. B., Neff, J. L., and D. S. Seigler. 1983. Floral biology and floral rewards of *Lysimachia* (Primulaceae). Amer. Midl. Naturalist 110: 249–256.

Sink, K. C. 1984. Taxonomy. Pp. 3–9 *in* K. C. Sink, ed., Petunia. Monogr. Theor. Appl. Genet. 9: 1–256.

Sinnot, Q. P. 1985. A revision of *Ribes* L. subg. *Grossularia* (Mill.) Pers. sect. *Grossularia* (Mill.) Nutt. (Grossulariaceae) in North America. Rhodora 87: 189–286.

Skvortzov, A. K. 1973. Present distribution and probable primary range of brittle willow (*Salix fragilis* L.). Pp. 263–280 *in* A. K. Skvortsov. Problemy Biogeotsenologii, Geobotaniki I Botanicheskoy Geografii. Nauka, Leningrad [in Russian; online English translation by I. Kadis available at: http://salicicola.com/translations/Skv1973SF.html].

Small, E., and B. S. Brookes. 1984. Taxonomic circumscription and identification in the *Medicago sativa-falcata* (alfalfa) continuum. Econ. Bot. 38: 83–96.

Small, E., and M. Jomphe. 1989. A synopsis of the genus *Medicago* (Leguminosae). Canad. J. Bot. 67: 3260–3294, 1 foldout poster.

Small, E., P. Lassen, and B. S. Brookes. 1987. An expanded circumscription of *Medicago* (Leguminosae, Trifolieae) based on explosive flower tripping. Willdenowia 16: 415–437.

Small, J. K. 1935. An ever-flowering violet. J. New York Bot. Gard. 36: 240–243.

Smartt, J. 1990. Grain Legumes, Evolution and Genetic Resources. Cambridge University Press, Cambridge, Great Britain.

Smedmark, J. E. E. 2006. Recircumscription of *Geum* (Colurieae: Rosaceae). Bot. Jahrb. Syst. 126: 409–417.

Smedmark, J. E. E., and T. Eriksson. 2002. Phylogenetic relationships of *Geum* (Rosaceae) and relatives inferred from the nrITS and *trnL–trnF* regions. Syst. Bot. 27: 302–317.

Smedmark, J. E. E., T. Eriksson, R. C. Evans, and C. C. Campbell. 2003. Ancient allopolyploid speciation in Geinae (Rosaceae): Evidence from nuclear granule-bound starch synthesis (GBSSI) gene sequences. Syst. Biol. 52: 374–385.

Smith, B. 1943. A Tree Grows in Brooklyn. Harper & Brothers, New York.

Smith, E. B. 1979. A new variety of *Galium arkansanum* (Rubiaceae) endemic to the Ouachita Mountains of Arkansas. Brittonia 31: 279–283.

———. 1988. An Atlas and Annotated List of Vascular Plants of Arkansas, edition 2. Published by the author, Fayetteville, AR.

Smith, D. M., and D. A. Levin. 1967. Karyotypes of eastern North American *Phlox*. Amer. J. Bot. 54: 324–334.

Smith, M., and N. Parker. 2005. *Quercus montana* (Fagaceae), new to Missouri. Sida 21: 1921–1922.

Smith, R. R., N. L. Taylor, and S. R. Bowley. 1985. Red clover. Pp. 457–470 *in* N.L. Taylor, ed., Clover Science and Technology. Agronomy 25. American Society of Agronomy, Madison, WI.

Smith, T. E. 1990. Two noteworthy collections of *Orobanche* L. (Orobanchaceae) from Missouri. Missouriensis 11: 5–6.

———. 1994. *Carya pallida*, a new hickory for Missouri. Missouriensis 15(1): 5–7.

———. 1998a. The reintroduction of running buffalo clover (*Trifolium stoloniferum*) in Missouri, 1989–1998. Missouriensis 19: 12–22.

———. 1998b. Highlights of Missouri field botany (1997–1998). Missouriensis 19: 54–58.

———. 2001. Highlights of Missouri field botany (1999–2000). Missouriensis 22: 16–21.
Smith, T. E., and G. K. Gremaud. 2006 [2007]. *Lysimachia terrestris* (Primulaceae) new to Missouri. Missouriensis 27: 4–8.
Sohmer, S. H., and D. F. Sefton. 1978. The reproductive biology of *Nelumbo pentapetala* (Nelumbonaceae) on the upper Mississippi River II: The insects associated with the transfer of pollen. Brittonia 30: 355–364.
Soják, J. 1987. Notes on *Potentilla paradoxa* and *P. supina*. Preslia 59: 271–272.
Sokoloff, D. 2000. New combinations in *Acmispon* (Leguminosae, Loteae). Ann. Bot. Fennici 37: 125–131.
Sokolov, R. 1981. Roots. Nat. Hist. 90(9): 98–100.
———. 1984. Broad bean universe. Nat. History 93(12): 84–87.
Solecki, M. K. 1983. Vascular plant communities and noteworthy taxa of Hawn State Park, Ste. Genevieve County, Missouri. Castanea 48: 50–55.
———. 1984. [Miscellaneous county records]. Pp. 143–156 *in* W. R. Weber and J. Raveill, eds., Missouri botanical record. Missouriensis 5: 143–156.
Soltis, D. E. 1991. A revision of *Sullivantia* (Saxifragaceae). Brittonia 43: 27–53.
Soltis, D. E., and P. S. Soltis. 1997. Phylogenetic relationships in Saxifragaceae sensu lato: a comparison of topologies based on 18S rDNA and *rbcL* seuqences. Amer. J. Bot. 84: 504–522.
Soltis, D. E., Q. Y. Xiang, and L. Hufford. 1995. Relationships and evolution of Hydrangeaceae based on *rbcL* sequence data. Amer. J. Bot. 82: 504–514.
Soltis, D. E., P. S. Soltis, M. W. Chase, M. E. Mort, D. C. Albach, M. Zanis, V. Savolainen, W. H. Hahn, S. B. Hoot, M. F. Fay, M. Axtell, S. M. Swensen, L. M. Prince, W. J. Kress, K. C. Nixon, and J. S. Farris. 2000. Angiosperm phylogeny inferred from 18S rDNA, *rbcL*, and *atpB* sequences. Bot. J. Linn. Soc. 133: 381–461.
Soltis, D. E., R. K. Kuzoff, M. E. Mort, M. Zanis, M. Fishbein, L. Hufford, J. Koontz, and M. K. Arroyo. 2001. Elucidating deep-level phylogenetic relationships in Saxifragaceae using sequences for six chloroplast and nuclear regions. Ann. Missouri Bot. Gard. 88: 669–693.
Soper, J. H. 1941. Pubescent form of *Ceanothus ovatus*. Rhodora 43: 82–83.
Sorensen, W. C., E. H. Smith, J. Smith, and Y. Carton. 2008. Charles V. Riley, France, and *Phylloxera*. Amer. Entomol. 54: 134–149.
Spellenberg, R. 2002. Varietal status for a hirsute phase of *Mirabilis linearis* (Nyctaginaceae). Novon 12: 270–271.
———. 2003. *Mirabilis*. Pp. 40–57 *in* Flora of North America Editorial Committee, eds., Flora of North America North of Mexico. Volume 4. Magnoliophyta: Caryophyllidae, Part 1. Oxford University Press, New York.
Spongberg, S. A. 1972. The genera of Saxifragaceae in the southeastern United States. J. Arnold Arbor. 53: 409–498.
———. 1976. Styracaceae hardy in temperate North America. J. Arnold Arbor. 57: 54–73.
Spongberg, S. A., and J. Ma. 1997. *Cladrastis* (Leguminosae subfamily Faboideae tribe Sophoreae): a historic and taxonomic overview. Int. Dendrol. Soc. Year Book 1996: 27–35.
Spooner, D. M. 1984. Infraspecific variation in *Gratiola viscidula* Pennell (Scrophulariaceae). Rhodora 86: 79–87.
Spooner, D. M., G. J. Anderson, and R. K. Jansen. 1993. Chloroplast DNA evidence for the interrelationships of tomatoes, potatoes, and pepinos (Solanaceae). Amer. J. Bot. 80: 676–688.
St. John, H. 1941. Revision of the genus *Swertia* (Gentianaceae) of the Americas and the reduction of *Frasera*. Amer. Midl. Naturalist 26: 1–29.
Starkey, D. A., F. Oliveria, A. Mangini, and M. Mielke. 2004. Oak decline and red oak borer in the Interior Highlands of Arkansas and Missouri: Natural phenomena, severe occurrences. Pp. 217–222 *in* M. A. Spetich, ed., Upland oak ecology symposium: History, current conditions, and sustainability. Gen. Techn. Rep. S. R., U.S. Forest Serv. SRS-73: 1–311.
Staudt, G. 1999. Systematics and geographic distribution of the American strawberry species. Univ. Calif. Publ. Bot. 81: i–xii, 1–162.
Steele, K. P., S. M. Ickert-Bond, S. Zarre, and M. F. Wojciechowski. 2010. Phylogeny and character evolution in *Medicago* (Leguminosae): Evidence from analyses of plastid *trnK/matK* and nuclear *GA3ox1* sequences. Amer. J. Bot. 97: 1142–1155.

Stern, K. R. 1961. Revision of *Dicentra* (Fumariaceae). Brittonia 13: 1–57.
———. 1997a. *Corydalis*. Pp. 348–355 *in* Flora of North America Editorial Committee, eds., Flora of North America North of Mexico. Volume 3. Magnoliophyta: Magnoliidae and Hamamelidae. Oxford University Press, New York.
———. 1997b. *Dicentra*. Pp. 341–347 *in* Flora of North America Editorial Committee, eds., Flora of North America North of Mexico. Volume 3. Magnoliophyta: Magnoliidae and Hamamelidae. Oxford University Press, New York.
Steyermark, J. A. 1940. *Dodecatheon amethystinum* Fassett. Rhodora 42: 102.
———. 1949a. *Lindera melissaefolia*. Rhodora 51: 153–162.
———. 1949b. New Missouri plant records (1946–1948). Rhodora 51: 115–119.
———. 1952. An example of how dams destroy valuable scientific records. Sci. Monthly 74: 231–233.
———. 1954. Notes on some roses in the Gray's Manual range. Rhodora 56: 70–79.
———. 1958a. An unusual botanical area in Missouri. Rhodora 60: 205–208.
———. 1958b. *Floerkea proserpinacoides* in Missouri. Brittonia 10: 150–152.
———. 1959. The taxonomic status of *Saxifraga palmeri*. Brittonia 11: 71–77.
———. 1963. Flora of Missouri. Iowa State University Press, Ames, IA [with errata, printings 2–7, 1969–1996].
Steyermark, J. A., and C. S. Steyermark. 1960. *Hepatica* in North America. Rhodora 62: 223–232.
Stone, E. 1763. An account of the success of the bark of the willow in the cure of agues. Philos. Trans. Biol. Sci. 53: 195–200.
Straley, G. B. 1977. Systematics of *Oenothera* sect. *Kneiffia* (Onagraceae). Ann. Missouri Bot. Gard. 64: 381–424.
Straw, R. M. 1966. A redefinition of *Penstemon*. Brittonia 18: 80–95.
Strickland, W. 1801. Observations on the Agriculture of the United States of America. W. Bulmer, London.
Stritch, L. R. 1984. Nomenclatural contributions to a revision of the genus *Wisteria*. Phytologia 56: 183–184.
———. 1985. A revision of the leguminous genus *Wisteria* Nuttall. Ph.D dissertation, Southern Illinois University, Carbondale.
Strohel, G. A., and G. N. Lanier. 1981. Dutch elm disease. Sci. Amer. 245(2): 56–66.
Stuckey, R. L. 1974. The introduction and spread of *Nymphoides peltatum* (Menyanthaceae) in North America. Bartonia 42: 14–23.
———. 1980. Distributional Hisotry of *Lythrum salicaria* in North America. Bartonia 47: 3–20.
Stuyvesant, R. L. 1984. A biosystematic study of the *Boehmeria cylindrica* (Urticaceae) complex. MS. thesis, Andrews University, Berrien Springs, MI.
Styles, B. T. 1962. The taxonomy of *Polygonum aviculare* and its allies in Britain. Watsonia 5: 177–214, 2 pl.
Subramanyam, R., S. G. Newmaster, G. Paliyath, and C. B. Newmaster. 2007. Exploring ethnobiological classifications for novel alternative medicine: a case study of *Cardiospermum halicacabum* L. (modakathon, balloon vine) as a traditional herb for treating rheumatoid arthritis. Ethnobotany 19: 1–18.
Sullivan, J. M. 1995. *Utricularia subulata* in Missouri. Missouriensis 16: 39–41.
Sullivan, J. R. 1984. Systematic studies in *Physalis* (Solanaceae). Ph.D. dissertation, University of Oklahoma, Norman.
———. 1985. Systematics of the *Physalis viscosa* complex (Solanaceae). Syst. Bot. 10: 426–444.
———. 2004. The genus *Physalis* (Solanaceae) in the southeastern United States. Rhodora 106: 305–326.
Summers, B. 1997. Two new native plants in Missouri. Missouriensis 18: 15–18.
Summers, B., M. Skinner, and G. Yatskievych. 1995. *Dalea gattingeri*, a cedar glade endemic new to Missouri. Missouriensis 16: 5–9.
Sutton, D. A. 1988. A Revision of the Tribe Antirrhineae. British Museum (Natural History), London. vi, 575 pp., 4 microfiche.
Svenson, H. K. 1940. Plants of southern United States. Rhodora 42: 1–19, pl. 586–587.
Swink, F., and G. Wilhelm. 1994. Plants of the Chicago Region. 4th ed. Indiana Academy of Science, Indianapolis.

Sytsma, K. J., J. Morawetz, J. C. Pires, M. Nepokroeff, E. Conti, M. Zihra, J. C. Hall, and M. W. Chase. 2002. Urticalean rosids: Circumscription, rosid ancestry, and phylogenetics based on *rbcL, trnL-F*, and *ndhF* sequences. Amer. J. Bot. 89: 1531–1546.

Talent, N., and T. A. Dickinson. 2005. Polyploidy in *Crataegus* and *Mespilus* (Rosaceae, Maloideae): evolutionary inferences from flow cytometry of nuclear DNA amounts. Canad. J. Bot. 83: 1268–1304.

Tamura, M. 1984. Phylogenetical considerations on the Ranunculaceae. J. Korean Pl. Taxon. 14: 33–42.

———. 1993. Ranunculaceae. Pp. 563–583 *in* K. Kubitzki, ed., The Families and Genera of Vascular Plants. Volume 2. Flowering Plants—Dicotyledons. Magnoliid, Hamamelid, and Caryophyllid Families (K. Kubitzki, J. G. Rohwer, and V. Bittrich, vol. eds.). Springer-Verlag, Berlin.

———. 1995. Ranunculaceae. Systematic part. Pp. 223–554 *in* P. Hiepko, ed., Die Natürlichen Pflanzenfamilien, Volume 17a(IV). Duncker and Humblot, Berlin.

Tamura, M., and L. A. Lauener. 1968. A revision of *Isopyrum, Dichocarpum* and their relatives. Notes Roy. Bot. Gard. Edinburgh 28: 267–273.

Tank, D. C., P. M. Beardsley, S. A. Kelchner, and R. G. Olmstead. 2006. Review of the systematics of Scrophulariaceae s.l. and their current disposition. Austral. Syst. Bot. 19: 289–307.

Taylor, N. L. 1975. Red clover and Alsike clover. Pp. 148–158 *in* M. E. Heath, D. S. Metcalfe, and R. E. Barnes, eds., Forages. Iowa State University Press, Ames.

Taylor, N. L., and K. H. Quesenberry. 1996. Red Clover Science. Kluwer, Boston, MA.

Taylor, N. L., J. M. Gillett, J. J. N. Campbell, and S. Berger. 1994. Crossing and morphological relationships among native clovers of eastern North America. Crop Sci. 34: 1097–1100.

Taylor, P. 1955. The genus *Anagallis* in tropical and South Africa. Kew Bull. 1955: 321–350.

———. 1989. The genus *Utricularia*%a taxonomic monograph. Kew Bull., additional ser. 14: i–xi, 1–724.

Taylor, R. J. 1990. Relationships of *Lamium amplexicaule, L. purpureum*, and *L. hybridum*, the last (?) word. Fourth International Congress of Systematic and Evolutionary Biology, Abstracts. University of Maryland, College Park [abstract].

Taylor-Lehman, J. W. 1987. Variation in *Polygonum pensylvanicum* L. Polygonaceae) with an emphasis on variety *eglandulosum* Myers. M.S. thesis, Ohio State University, Columbus.

Telewski, F. W., and J. A. D. Zeevaart. 2002. The 120-yr period for Dr. Beal's seed viability experiment. J. Bot. 89: 1285–1288.

Terrell, E. E. 1990. Synopsis of *Oldenlandia* (Rubiaceae) in the United States. Phytologia 68: 125–133.

———. 1991. Overview and annotated list of North American species of *Hedyotis*, *Houstonia*, *Oldenlandia* (Rubiaceae) and related genera. Phytologia 71: 212–243.

———. 1996. Revision of *Houstonia* (Rubiaceae-Hedyotideae). Syst. Bot. Monogr. 48: 1–118.

———. 2001a. Taxonomy of *Stenaria* (Rubiaceae: Hedyotideae), a new genus including *Hedyotis nigricans*. Sida 19: 591–614.

———. 2001b. Taxonomic review of *Houstonia acerosa* and *H. palmeri*, with notes on *Hedyotis* and Oldenlandia (Rubiaceae). Sida 19: 913–922.

Terrell, E. E., and H. Robinson. 2006. Taxonomy of North American species of *Oldenlandia* (Rubiaceae). Sida 22: 305–329.

Terrell, E. E., and R. P. Wunderlin. 2002. Seed and fruit characters in selected Spermacoceae and comparison with Hedyotideae (Rubiaceae). Sida 20: 549–557.

Tessene, M. F. 1968. Preliminary notes on the flora of Wisconsin no. 59. Plantaginaceae%plantain family. Trans. Wisconsin Acad. Sci. 56: 281–313.

Thieret, J. W. 1958. Proposal for the conservation of the generic name *Agalinis* Rafinesque, New Fl. N. Amer. 2: 61 (1837) versus *Chytra* Gaertn. f., Fruct. et Sem. Plant. 3: 184, t. 214 (1805) (both of them Scrophulariaceae). Taxon 7: 142–143.

———. 1964. *Fatoua villosa* (Moraceae) in Louisiana: New to North America. Sida 1: 248.

———. 1969a. *Baptisia lactea* (Rafinesque) Thieret, comb. nov. (Leguminosae). Sida 3: 446.

———. 1969b. Notes on *Epifagus*. Castanea 34: 397–402.

———. 1971. The genera of Orobanchaceae in the southeastern United States. J. Arnold Arbor. 52: 404–434.

———. 1972. The Phrymaceae in the southeastern United States. J. Arnold Arbor. 53: 226–233.

———. 1977. The Martyniaceae in the southeastern United States. J. Arnold Arbor. 58: 25–39.

Thomas, J. R. 2005. Vegetative key to *Polygonum* in Missouri. Missouriensis 26: 22–35.

Thompson, H. J. 1953. The biosystematics of *Dodecatheon*. Contr. Dudley Herb. 4: 73–154.

Thomson, P. M. 1977. *Quercus* ×*introgressa*, a new hybrid oak. Rhodora 79: 453–464.

Threadgill, P. F., J. M. Baskin, and C. C. Baskin. 1981. The ecological life cycle of *Frasera caroliniensis*, a long-lived monocarpic perennial. Amer. Midl. Naturalist 105: 277–289.

Timme, S. L. 1997. Two new and interesting records for Missouri plants. Missouriensis 18: 21–23.

Tindall, H. D. 1983. Vegetables in the Tropics. MacMillan Press, London.

Tobe, H. 2011. Embryological evidence supports the transfer of *Leitneria floridana* to the family Simaroubaceae. Ann. Missouri Bot. Gard. 98: 277–293.

Tobe, H., and R. C. Keating. 1985. The morphology and anatomy of *Hydrastis* (Ranunculales): systematic reevaluation of the genus. Bot. Mag. (Tokyo) 98: 291–316.

Tokuoka, T. 2008. Molecular phylogenetic analysis of Violaceae (Malpighiales) based on plastid and nuclear DNA sequences. J. Pl. Res. 121: 253–260.

Towner, H. F. 1977. The biosystematics of *Calylophus* (Onagraceae). Ann. Missouri Bot. Gard. 64: 48–120.

Townsend, A. M., S. E. Bentz, and L. W. Douglass. 2005. Evaluation of 19 American elm clones for tolerance to Dutch elm disease. J. Environm. Hort. 23: 21–24.

Trapp, E. J. 1988. Dispersal of heteromorphic seeds in *Amphicarpaea bracteata* (Fabaceae). Amer. J. Bot. 75: 1535–1539.

Trauth-Nare, A. E., and R. F. C. Naczi. 1997. Taxonomic status of the varieties of Seneca snakeroot, *Polygala senega* (Polygalaceae). Trans. Kentucky Acad. Sci. 58: 39 [abstract].

———. 1998. Taxonomic status of the varieties of Seneca snakeroot, *Polygala senega* (Polygalaceae). Amer. J. Bot. 85(6, suppl.): 163 [abstract].

Tucker, A. O., and H. L. Chambers. 2002. *Mentha canadensis* L. (Lamiaceae): a relict amphidiploid from the Lower Tertiary. Taxon 51: 703–718.

Tucker, A. O., and D. E. Fairbrothers. 1990. The origin of *Mentha* ×*gracilis* (Lamiaceae). I. Chromosome numbers, fertility, and three morphological characters. Econ. Bot. 44: 183–213.

Tucker, A. O., and R. F. C. Naczi. 2007. *Mentha*: An overview of its classification and relationships. Pp. 1–39 *in* B. M. Lawrence, ed., Mint: The Genus *Mentha*. CRC Press, Boca Raton, FL.

Tucker, A. O., and S. S. Tucker. 1988. Catnip and the catnip response. Econ. Bot. 42: 214–231.

Tucker, A. O., H. Hendriks, R. Bos, and D. E. Fairbrothers. 1991. The origin of *Mentha* ×*gracilis* (Lamiaceae). II. Essential oils. Econ. Bot. 45: 200–215.

Turkington, R. A., P. B. Cavers, and E. Rempel. 1978. The biology of Canadian weeds: 29. *Melilotus alba* Desr. and *M. officinalis* (L.) Lam. Canad. J. Pl. Sci. 58: 523–537.

Turland, N. 1996a. Proposal to reject the name *Malva rotundifolia* L. Taxon 45: 707–708.

———. 1996b. Proposal to reject the name *Rosa eglanteria* (Rosaceae). Taxon 45: 565–566.

Turner, B. H., and E. Quarterman. 1975. Allelochemic effects of *Petalostemon gattingeri* on the distribution of *Arenaria patula* in cedar glades. Ecology 56: 924–932.

Turner, B. L. 1981. Thermopsideae. Pp. 403–407 *in* R. M. Polhill and P. H. Raven, eds., Advances in Legume Systematics, Part 1. Royal Botanic Gardens, Kew, Great Britain.

———. 1982. Review. Leguminosae of the United States. III. Subfamily Papilionoideae: Tribes Sophoreae, Podalyrieae, Lotae, by Duane Isely. Syst. Bot. 7: 350–352.

———. 1993. Texas species of *Mirabilis* (Nyctaginaceae). Phytologia 75: 432–451.

———. 1994. Taxonomic treatment of *Monarda* (Lamiaceae) for Texas and Mexico. Phytologia 77: 56–79.

———. 1995. Taxonomic overview of *Hedyotis nigricans* (Rubiaceae) and closely allied taxa. Phytologia 79: 12–21.

———. 1998a. Texas species of *Glandularia* (Verbenaceae). Lundellia 1: 3–16.

———. 1998b. *Phlox drummondii* (Polemoniaceae) revisited. Phytologia 85: 280–287.

———. 2006. Overview of the genus *Baptisia* (Leguminosae). Phytologia 88: 253–268.

———. 2008. Revision of the genus *Orbexilum* (Fabaceae: Psoraleeae). Lundellia 11: 1–7.

Turner, B. L., and C. C. Cowan. 1993. Taxonomic overview of *Stemodia* (Scrophulariaceae) for North America and the West Indies. Phytologia 74: 61–103.

Turner, B. L., and O. S. Fearing. 1964. A taxonomic study of the genus *Amphicarpaea* (Leguminosae). SouthW. Naturalist 9: 207–218.

Tyson, W. S., and J. E. Ebinger. 1999. Distribution and ecology of *Corydalis curvisilqua* Engelm. var. *grandibracteata* Fedde (Papaveraceae) in Illinois. Trans. Illinois State Acad. Sci 92: 53–58.

Ugent, D., M. Verdun, and M. Mibb. 1982 The jeweled shooting star (*Dodecatheon amethystinum*): a post glacial migrant in the Mississippi Valley. Phytologia 51: 323–329.

Umber, R. E. 1979. The genus *Glandularia* (Verbenaceae) in North America. Syst. Bot. 4: 72–102.

Uphof, J. C. T. 1921. Das Vorkommen von *Neviusia alabamensis* Gray im Süden von Missouri. Mitt. Deutsche Dendrol. Ges. 1921: 282–283.

———. 1922. Ecological relations of plants in southeastern Missouri. Amer. J. Bot. 9: 1–17, 2 pl.

Valder, P. 1995. Wisterias: a Comprehensive Guide. Timber Press, Portland, OR.

Valls, J. F. M., and C. E. Simpson. 2005. New species of *Arachis* (Leguminosae) from Brazil, Paraguay and Bolivia. Bonplandia 14: 35–63.

Van Bruggen, T. 1986. Crassulaceae. Pp. 556–558 *in* Flora of the Great Plains Editorial Committee, eds. Flora of the Great Plains. University Press of Kansas, Lawrence.

Van Brunt, G. 2011. November botany report. Nature Notes (Webster Groves) 83(2): 4–11.

Van der Maesen, L. J. G. 1985. Revision of the genus *Pueraria* DC with some notes on *Teyleria* Backer. Agric. Univ. Wageningen Pap. 85(1): 1–132.

———. 1988. Two corrections to the nomenclature in the revision of *Pueraria* DC. J. Bombay Nat. Hist. Soc. 85: 233–234.

van Steenis, C. G. G. J. 1958. Miscellaneous notes on New Guinea plants V. Nova Guinea, new ser., 9: 31.

Vanderplank, J. 2000. Passion Flowers. 2d ed. MIT Press, Cambridge, MA.

Vegter, I. H. 1986. Index herbariorum, part 2(6), collectors: S. Regnum Veg. 114: 805–895.

Venette, R. C., and D. W. Ragsdale. 2004. Assessing the invasion by soybean aphid (Homoptera: Aphididae): where will it end? Ann. Entomol. Soc. Amer. 97: 219–226.

Verdcourt, B. 1958. Remarks on the classification of the Rubiaceae. Bull. Jard. Bot. Bruxelles 28: 209–290.

———. 1970a. Studies in the Leguminosae-Papilionoideae for the 'Flora of Tropical East Africa': III. Kew Bull. 24: 379–447.

———. 1970b. Studies in the Leguminosae-Papilionoïdeae for the 'Flora of Tropical East Africa': IV. Kew Bull. 24: 507–569.

———. 1976. Rubiaceae (Part 1). Pp. 1–415 *in* R. M. Polhill, ed., Flora of Tropical East Africa. Crown Agents for Overseas Governments and Administrations, London.

———. 1997. Proposals to conserve the name *Phaseolus helvolus* with a conserved type and spelling, and *Strophostyles* against *Phacellus* (Leguminosae). Taxon 46: 357–359.

Vincent, M. A. 2004a. Spread of *Fatoua villosa* (mulberry weed; Moraceae) in North America. J. Kentucky Acad. Sci. 65: 67–75.

———. 2004b. *Tetradium daniellii* (Korean evodia; Rutaceae) as an escape in North America. Michigan Bot. 43: 21–24.

———. 2005. On the spread and current distribution of *Pyrus calleryana* in the United States. Castanea 70: 20–31.

Vogel, S., and Hadacek, F. 2004. Contributions to the functional anatomy and biology of *Nelumbo nucifera* (Nelumbonaceae) III. An ecological reappraisal of floral organs. Pl. Syst. Evol. 249: 173–189.

Vogelmann, J. E. 1983. A biosystematic study of *Agastache* section *Agastache* (Labiatae). Ph.D. dissertation, Indiana University, Bloomington.

Voigt, J. W., and J. R. Swayne. 1955. French's shooting star in southern Illinois. Rhodora 57: 325–332.

von Hagen, K. B., and J. K. Kadereit. 2002. Phylogeny and flower evolution of the Swertiinae (Gentianaceae—Gentianeae): Homoplasy and the principle of variable proportions. Syst. Bot. 27: 548–572.

Voss, E. G. 1985. Michigan Flora. Part II. Dicots (Saururaceae–Cornaceae). Bull. Cranbrook Inst. Sci. 59. Cranbrook Institute of Science, Bloomfield Hills, MI.

———. 1986. General committee report 1982–1985. Taxon 35: 551–552.

———. 1996. Michigan Flora. Part III. Dicots (Pyrolaceae–Compositae). Bull. Cranbrook Inst. Sci. 61. Cranbrook Institute of Science, Bloomfield Hills, MI.

Wagner, W. H., Jr. 1974. Dwarf hackberry (Ulmaceae: *Celtis tenuifolia*) in the Great Lakes region. Michigan Bot. 13: 73–79.

Wagner, W. L. 1981. Our 'Missouri' primrose now properly *Oenothera macrocarpa*! Missouriensis 2(3): 14–17.

———. 1983. New species and combinations in the genus *Oenothera* (Onagraceae). Ann. Missouri Bot,. Gard. 70: 194–196.

Wagner, W. L., and P. Hoch. 2000. Proposal to reject the name *Gaura mollis* (Onagraceae). Taxon 49: 101–102.

Wagner, W. L., P. C. Hoch, and P. H. Raven. 2007. Revised classification of the Onagraceae. Syst. Bot. Monogr. 83: 1–240.

Wagstaff, S. J., and R. G. Olmstead. 1997. Phylogeny of the Labiatae and Verbenaceae inferred from *rbcL* sequences. Syst. Bot. 22: 165–179.

Wagstaff, S. J., R. G. Olmstead, and P. D. Cantino. 1995. Parsimony analysis of cpDNA restriction site variation in subfamily Nepetoideae (Lamiaceae). Amer. J. Bot. 82: 886–892.

Walker, J. B., K. J. Sytsma, J. Treutlein, and M. Wink. 2004. *Salvia* (Lamiaceae) is not monophyletic: implications for the systematics, radiation, and ecological specializations of *Salvia* and tribe Mentheae. Amer. J. Bot. 91: 1115–1123.

Wallander, E. 2008. Systematics of *Fraxinus* (Oleaceae) and evolution of dioecy. Pl. Syst. Evol. 273: 25–49.

Walraven, W. C. 1967. Taxonomy of the genus *Rhynchosia* (Leguminosae) of the United States. Ph.D. Dissertation, University of Georgia, Athens.

Walters, S. M., and D. A. Webb. 1972. *Veronica*. Pp. 242–251 *in* T. G. Tutin, V. H. Heywood, N. A. Burges, D. M. Moore, D. H. Valentine, S. M. Walters, and D. A. Webb, eds., Flora Europaea, volume 3, Diapensaceae to Myoporaceae. Cambridge University Press, Cambridge, Great Britain.

Wann, J. D., and J. R. Akeroyd. 2000. *Vitex* Linnaeus. P. 150 *in* J. Cullen, J. C. M. Alexander, C. D. Brickell, J. R. Edmondson, P. S. Green, V. H. Heywood, P.-m. Jørgensen, S. L. Jury, S. G. Knees, H. S. Maxwell, D. M. Miller, N. K. B. Robson, S. M. Walters, and P. F. Yeo, eds., The European Garden Flora. Volume VI. Dicotyledons (Part IV). Cambridge University Press, Cambridge Great Britain.

Ward, D. B. 1977. *Nelumbo lutea*, the correct name for the American lotus. Taxon 26: 227–234.

———. 1998. *Pueraria montana*: The correct scientific name of the kudzu. Castanea 63: 76–77.

———. 2004. Keys to the flora of Florida—9, *Oxalis* (Oxalidaceae). Phytologia 86: 32–41.

Ward, D. B., and J. Wiersema. 2008. Proposal to reject the name *Juglans alba* (*Carya alba*) (Juglandaceae). Taxon 57: 308–310.

Ward, L. F. 1893. Frost freaks of the dittany. Bot. Gaz. (Crawfordsville) 18: 183–186, pl. 19.

Ware, D. M. E. 1983. Genetic fruit polymorphism in North American *Valerianella* (Valerianaceae) and its taxonomic implications. Syst. Bot. 8: 33–44.

Warnock, M. J. 1981. Biosystematics of the *Delphinium carolinianum* complex (Ranunculaceae). Syst. Bot. 6: 38–54.

———. 1987. Vicariant distribution of two *Delphinium* species in the southeastern United States. Bot. Gaz. (Crawfordsville) 148: 90–95.

———. 1997a. *Consolida*. Pp. 240–242 *in* Flora of North America Editorial Committee, eds., 1996. Flora of North America North of Mexico. Volume 3. Magnoliophyta: Magnoliidae and Hamamelidae. Oxford University Press, New York.

———. 1997b. *Delphinium*. Pp. 196–240 *in* Flora of North America Editorial Committee, eds., 1996. Flora of North America North of Mexico. Volume 3. Magnoliophyta: Magnoliidae and Hamamelidae. Oxford University Press, New York.

Waterfall, U. T. 1958. A taxonomic study of the genus *Physalis* in North America north of Mexico. Rhodora 60: 107–114, 128–142, 152–173.

———. 1968. Emendations in United States *Physalis*. Rhodora 70: 574–576.

Waterfall, U. T. 1970. *Monarda stipitatoglandulosa*, a new species from Oklahoma. Rhodora 72: 502–504.

Watling, J. R., S. A. Robinson, and R. S. Seymour. 2006. Contribution of the alternative pathway to respiration during thermogenesis in flowers of the Sacred Lotus. Pl. Physiol. 140: 1367–1373.

Weakley, A. S., R. J. LeBlond, B. A. Sorrie, C. T. Witsell, L. D. Estes, K. Gandhi, K. G. Mathews, and A. Ebihara. 2011. New combinations, rank changes, and nomenclatural and taxonomic comments in the vascular flora of the southeastern United States. J. Bot. Res. Inst. Texas 5: 437–455.

Webb, D. A. 1967. What is *Parthenocissus quinquefolia* (L.) Planchon? Pp. 6–10 *in* V. H. Heywood, ed., Flora Europaea, notulae systematicae ad floram Europaeam spectantes, no. 6. Feddes Repert. Spec. Nov. Regni. Veg 74: 1–38.

———. 1968. Malvaceae. Pp. 248–255 *in* T. G. Tutin, V. H. Heywood, N. A. Burges, D. M. Moore, D. H. Valentine, S. M. Walters, and D. A. Webb, eds., Flora Europaea. Volume 2, Rosaceae to Umbelliferae. Cambridge University Press, Cambridge, Great Britain.

———. 1972. Taxonomic and nomenclatural notes on *Veronica* L. Pp. 266–269 *in* V. H. Heywood, ed., Flora Europaea notulae systematicae ad floram europaeum spectantes. Bot. J. Linn. Soc. 65: 233–269.

Webb, W. P. 1931. The Great Plains. Ginn and Company, Boston, MA.

Weber, C. 1964. The genus *Chaenomeles* (Rosaceae. J. Arnold Arbor. 45: 161–205, 302–345.

Weber, W. R., W. T. Corcoran, M. S. Brunell, and P. L. Redfearn Jr. 2000. Atlas of Missouri Vascular Plants. Ozarks Regional Herbarium, Southwest Missouri State University, Springfield.

Wells, E. F. 1979. Interspecific hybridization in eastern North American *Heuchera* (Saxifragaceae). Syst. Bot. 4: 319–338.

———. 1984. A revision of the genus *Heuchera* (Saxifragaceae) in eastern North America. Syst. Bot. Monogr. 3: 45–121.

Welsh, S. L. 2001. Revision of North American Species of *Oxytropis* de Candolle. E. P. S., Inc., Orem, UT.

———. 2007. North American Species of *Astragalus* Linnaeus (Leguminosae): A Taxonomic Revision. Monte L. Bean Life Science Museum, Brigham Young University, Provo, UT.

Wemple, D. K. 1970. Revision of the genus *Petalostemon* (Leguminosae). Iowa State J. Sci. 45: 1–102.

Wemple, D. K., and N. R. Lersten. 1966. An interpretation of the flower of *Petalostemon* (Leguminosae). Brittonia 18: 117–126.

Werth, C. R., and J. L. Riopel. 1979. A study of the host range of *Aureolaria pedicularia* (L.) Raf. (Scrophulariaceae). Amer. Midl. Naturalist 102: 300–306.

Westerkamp, C., and H. Paul. 1993. *Apios americana*, a fly-pollinated papilionaceous flower? Pl. Syst. Evol. 187: 135–144.

Whalen, M. D. 1979. Taxonomy of *Solanum* section *Androceras*. Gentes Herb. 11: 359–426.

Wherry, E. T. 1931. The eastern long-styled phloxes, part 1. Bartonia 13: 18–37.

———. 1932. The eastern long-styled phloxes, part 2. Bartonia 14: 14–26.

———. 1945. The *Phlox carolina* complex. Bartonia 23: 1–9.

———. 1955. The genus *Phlox* (Morris Arboretum Monographs 3). Morris Arboretum, Philadephia.

Whitlock, B. A., C. Bayer, and D. A. Baum. 2001. Phylogenetic relationships and floral evolution of the Byttnerioideae ("Sterculiaceae" or Malvaceae s.l.) based on sequences of the chloroplast gene, *ndhF*. Syst. Bot. 26: 420–437.

Whitson, M., and P. S. Manos. 2005. Untangling *Physalis* (Solanaceae) from the physaloids: a two-gene phylogeny of the Physalinae. Syst. Bot. 30: 216–230.

Whittemore, A. T. 1997. *Ranunculus*. Pp. 88–135 *in* Flora of North America Editorial Committee, eds., Flora of North America North of Mexico. Volume 3. Magnoliophyta: Magnoliidae and Hamamelidae. Oxford University Press, New York.

———. 2004. Sawtooth oak (*Quercus acutissima*, Fagaceae) in North America. Sida 21: 447–454.

———. 2005. Genetic structure, lack of introgression, and taxonomic status in the *Celtis laevigata–C. reticulata* complex (Cannabaceae). Syst. Bot. 30: 809–817.

Whittemore, A. T., and K. C. Nixon. 2005. Proposal to reject the name *Quercus prinus* (Fagaceae). Taxon 54: 213–214.

Whittemore, A. T., and R. T. Olsen. 2011. *Ulmus americana* (Ulmaceae) is a polyploid complex. Amer. J. Bot. 98: 754–760.

Whittemore, A. [T.], and B. A. Schaal. 2001. Interspecific gene flow in sympatric oaks. Proc. Natl. Acad. Sci. U.S.A. 88: 2540–2544.

Whittemore, A. T., and A. M. Townsend. 2007. Hybridization and self-compatibility in *Celtis*: AFLP analysis of controlled crosses. J. Amer. Soc. Hort. Sci. 132: 368–373.

Widrlechner, M. 1983. Historical and phenological observations on the spread of *Chaenorhinum minus* across North America. Canad. J. Bot. 61: 179–187.

———. 1998. The genus *Rubus* L. in Iowa. Castanea 63: 415–465.

Wiens, D. 1964. Revision of the acataphyllous species of *Phoradendron*. Brittonia 16: 11–54.

Wiersema, J. H. 1997. Nelumbonaceae. Pp. 64–65 *in* Flora of North America Editorial Committee, eds., Flora of North America North of Mexico. Volume 3. Magnoliophyta: Magnoliidae and Hamamelidae. Oxford University Press, New York.

Wiersema, J. H., and C. B. Hellquist. 1994. Nomenclatural notes in Nymphaeaceae for the North American flora. Rhodora 96: 170–178.

———. 1997. Nymphaeaceae. Pp. 66–77 *in* Flora of North America Editorial Committee, eds., Flora of North America North of Mexico. Volume 3. Magnoliophyta: Magnoliidae and Hamamelidae. Oxford University Press, New York.

Wight, M. F. 1907. The history of the cowpea and its introduction into America. Bull. Bur. Pl. Industr. U.S.D.A. 102(6): 1–21, pl. 1–3.

Wijsman, H. J. W. 1982. On the inter-relationship of certain species of *Petunia*. I. Taxonomic notes on the parental species of *Petunia hybrida*. Acta Bot. Neerlandica 31: 477–490.

Wilbur, R. L. 1955. A revision of the North American genus *Sabatia* (Gentianaceae). Rhodora 57: 1–33, 43–71, 78–104.

———. 1964. A revision of the dwarf species of *Amorpha* (Leguminosae). J. Elisha Mitchell Sci. Soc. 80: 51–65.

———. 1975. A revision of the North American genus *Amorpha* (Leguminosae-Psoraleae). Rhodora 77: 337–409.

———. 1988. The correct name of the pale, yellow or white gentian of the eastern United States. Sida 13: 161–165.

———. 1989. The status of the binomial *Sesbania macrocarpa* Muhl. (Leguminosae). Castanea 54: 121–127.

Wilken, D. H. 1986. Polemoniaceae. Pp. 666–677 *in* Flora of the Great Plains Editorial Committee, eds., Flora of the Great Plains. University Press of Kansas, Lawrence.

Williams, A. H. 1982. Chemical evidence from flavonoids relevant to the classification of *Malus* species. Bot. J. Linn. Soc. 84: 31–39.

Williams, G. R. 1970. Investigations in the white waterlilies (*Nymphaea*) of Michigan. Michigan Bot. 9: 72–86.

Williams, W., ed. 1904. The State of Missouri, an Autobiography. Press of E.W. Stephens, Columbia, MO.

Wilmot-Dear, C. M., and I. Friis. 1995. The New World species of *Boehmeria* and *Pouzolzia* (Urticaceae, tribus Boehmerieae). A taxonomic revision. Opera Bot. 129: 1–103.

Wilson, K. A., and C. E. Wood Jr. 1959. The genera of Oleaceae in the southeastern United States. J. Arnold Arbor. 40: 369–384.

Wilson, K. L. 1990. Some widespread species of *Persicaria* (Polygonaceae) and their allies. Kew Bull. 45: 621–636.

Windler, D. R. 1973. Field and garden studies in *Crotalaria sagittalis* L. and related species. Phytologia 26: 289–354.

———. 1974. A systematic treatment of the native unifoliolate crotalarias of North America (Leguminosae). Rhodora 76: 151–204.

Wojciechowski, M. F., M. J. Sanderson, and J. M. Hu. 1999. Evidence on the monophyly of *Astragalus* (Fabaceae) and its major subgroups based on nuclear ribosomal DNA ITS and chloroplast DNA *trnL* intron data. Syst. Bot. 24: 409–437.

Wojciechowski, M. F., M. Lavin, and M. J. Sanderson. 2004. A phylogeny of legumes (Leguminosae) based on analysis of the plastid *matK* gene resolves many well-supported subclades within the family. Amer. J. Bot. 91: 1846–1862.

Wolf, S. J., and J. McNeill. 1986. Synopsis and achene morphology of *Polygonum* section *Polygonum* (Polygonaceae) in Canada. Rhodora 88: 457–479.

———. 1987. Cytotaxonomic studies on *Polygonum* section *Polygonum* in eastern Canada and the adjacent United States. Canad. J. Bot. 65: 647–652.

Wolfe, A. D., and C. W. dePamphilis. 1998. The effect of relaxed functional constraints on the photosynthetic gene *rbcL* in photosynthetic and nonphotosynthetic parasitic plants. Molec. Biol. Evol. 15: 1243–1258.

Wolfe, A. D., C. P. Randle, L. Liu, and K. E. Steiner. 2005. Phylogeny and biogeography of Orobanchaceae. Folia Geobot. 40: 115–134.

Wolfe, A. D., C. P. Randle, S. L. Detwyler, J. J. Morawetz, N. Arguedas, and J. Diaz. 2006. Phylogeny, taxonomic affinities, and biogeography of *Penstemon* (Plantaginaceae) based on ITS and cpDNA sequence data. Amer. J. Bot. 93: 1699–1713.

Wolkomir, R. 1998. Racing to revive our embattled elms. Smithsonian 29(3): 40–49.

Wollenweber, E., Y. Asakawa, D. Schillo, U. Lehmann, and H. Weigel. 1987. A novel caffeic acid derivative and other constituents of *Populus* bud excretion and propolis (bee-glue). Z. Naturforsch. 42c: 1030–1034.

Wood, C. E., Jr. 1949. The American barbistyled species of *Tephrosia* (Leguminosae). Rhodora 51: 193–231, 233–302, 305–364, 369–384.

———. 1983. The genera of Menyanthaceae in the southeastern United States. J. Arnold Arbor. 64: 431–445.

Wood, C. E., Jr., and R. C. Channell. 1960. The genera of Ebenales in the southeastern United States. J. Arnold Arbor. 41: 1–35.

Woodland, D. W. 1982a. Biosystematics of the perennial North American species of *Urtica*. Ph.D. dissertation, Iowa State University, Ames.

———. 1982b. Biosystematics of the perennial North American taxa of *Urtica*. II. Taxonomy. Syst. Bot. 7: 282–290.

Woodland, D. W., I. J. Bassett, and C. W. Crompton. 1976. The annual species of stinging nettle (*Hesperocnide* and *Urtica*) in North America. Canad. J. Bot. 54: 374–383.

Woodland, D. W., I. J. Bassett, C. Crompton, and S. Forget. 1982. Biosystematics of the perennial North American taxa of *Urtica*. I. Chromosome number, hybridization, and palynology. Syst. Bot. 7: 269–281.

Woods, K., K. W. Hilu, J. H. Wiersema, and T. Borsch. 2005a. Pattern of variation and systematics of *Nymphaea odorata*: I. Evidence from morphology and inter-simple sequence repeats (ISSRs). Syst. Bot. 20: 471–480.

Woods, K., K. W. Hilu, T. Borsch, and J. H. Wiersema. 2005b. Pattern of variation and systematics of *Nymphaea odorata*: II. Sequence information from ITS and *trnL-trnF*. Syst. Botany 20: 481–493.

Woods, M. 2005. A revision of the North American species of *Apios* (Fabaceae). Castanea 70: 85–100.

Wright, J. W. 1944. Epidermal characters in *Fraxinus*. Proc. Indiana Acad. Sci. 54: 84–90.

———. 1965. Green ash (*Fraxinus pennsylvanica* Marsh.). P. 185–190 *in* Fowells, H. A., ed., Silvics of forest trees of the United States. Agric. Handb. No. 271: i–vi, 1–762.

Wunderlin, R. P., B. F. Hansen, and D. W. Hall. 1985. The vascular flora of central Florida: taxonomic and nomenclatural changes, additional taxa. Sida 11: 232–244.

Xia, Z., Y.-z. Wang, and J. F. Smith. 2009. Familial placement and relations of *Rehmannia* and *Triaenophora* (Scrophulariaceae s.l.) inferred from five gene regions. Amer. J. Bot. 96: 519–530.

Yamazaki, T. 1957. Taxonomical and phylogenetic studies of Scrophulariaceae–Veroniceae with special reference to *Veronica* and *Veronicastrum* in eastern Asia. J. Fac. Sci. Univ. Tokyo, Sect. 3, Bot. 7: 92–162.

Yasui, Y., and O. Ohnishi. 1998. Phylogenetic relationships among *Fagopyrum* species revealed by the nucleotide sequences of the ITS region of the nuclear rRNA gene. Genes Genet. Systems 73: 201–210.

Yatskievych, G. 1988. The confusing case of the flame fame flower. Missouriensis 9: 10–12.

———. 1990. Studies in the flora of Missouri, II. Missouriensis 11: 2–6.

———. 1999. Steyermark's Flora of Missouri. Volume 1. Revised ed. Missouri Department of Conservation, Jefferson City.

———. 2006. Steyermark's Flora of Missouri. Volume 2. Revised ed. Missouri Botanical Garden Press, St. Louis.

Yatskievych, G., and A. Brant. 1994. The *Polygonum aviculare* complex in Missouri. Missouriensis 15(1): 22–44.

Yatskievych, G., and D. Figg. 1989. Studies in the flora of Missouri, I. New records for introduced taxa. Missouriensis 10: 16–19.

Yatskievych, G., and J. A. Raveill. 2001. Notes on the increasing proportion of non-native angiosperms in the Missouri flora, with reports of three new genera for the state. Sida 19: 701–709.

Yatskievych, G., and J. Sullivan. 1992. Yet another speedwell invades Missouri. Nature Notes (Webster Groves) 64: 75–76.

Yatskievych, G., and B. Summers. 1991. Studies in the flora of Missouri, III. Missouriensis 12: 4–11.

———. 1993. Studies in the flora of Missouri, IV. Missouriensis 14: 27–42.

Yatskievych, G., and J. Turner. 1990. Catalogue of the Flora of Missouri. Monogr. Syst. Bot. Missouri Bot. Gard. 37: i–xii, 1–345.

Yatskievych, G., R. J. Evans, and C. T. Whitsell. 2013. A reevaluation of the Ozark endemic, *Claytonia ozarkensis* (Montiaceae). Phytoneuron 2013-XX: 1–XXX.

Yinger, B. R. 1992. Hardscrabble elegance: the bush clovers. Fl. Gard., Mid-Amer. Ed. 36(5): 36–37.

Yost, S. E. 1987. The effects of shade on petiole length in the *Viola fimbriatula-sagittata* complex (Violaceae). Brittonia 39: 180–187.

Young, N. D., K. E. Steiner, and C. W. dePamphilis. 1999. The evolution of parasitism in Scrophulariaceae/Orobanchaceae: plastid gene sequences refute an evolutionary transition series. Ann. Missouri Bot. Gard. 86: 876–893.

Young, V. 1990. State nut: senators approve choice, after some snickers, sneers. St. Louis Post Dispatch 112(54): A1.

Yu, H.-c., K. Kosuna, and M. Haga, eds. 1997. Perilla: The Genus *Perilla*. Harwood Academic Publishers, Amsterdam, Netherlands.

Zalapa, J. E., J. Brunet, and R. P. Guries. 2009. Patterns of hybridization and introgression between invasive *Ulmus pumila* (Ulmaceae) and native *U. rubra*. Amer. J. Bot. 96: 1116–1128.

Zardini, E. M., and P. H. Raven, 1992. Erratum: On the separation of two species within the *Ludwigia uruguayensis* complex (Onagraceae). Syst. Bot. 17: 692.

Zardini, E. M., H. Gu, and P. H. Raven. 1991. On the separation of two species within the *Ludwigia uruguayensis* complex (Onagraceae). Syst. Bot. 16: 242–244.

Zhang L.-b., and M. P. Simmons. 2006. Phylogeny and delimitation of the Celastrales inferred from nuclear and plastid genes. Syst. Bot. 31: 122–137.

Zielinski, J. 2004. The genus *Rubus* (Rosaceae) in Poland. Polish Bot. Studies 16: 1–300.

Zika, P. F., and A. L. Jacobson. 2003. An overlooked hybrid Japanese knotweed (*Polygonum cuspidatum* × *sachalinense*; Polygonaceae) in North America.

Zohary, M. 1963a. Taxonomical studies in *Alcea* of south-western Asia. Part I. Bull. Res. Council Israel 11D: 210–229.

———. 1963b. Taxonomical studies in *Alcea* of southwestern Asia. Part II. Israel J. Bot. 12: 1–26.

Zohary, M., and D. Heller. 1984. The genus *Trifolium*. Israel Academy of Sciences and Humanities, Jerusalem.

INDEX TO VOLUME 3

Principal entries for accepted scientific names have the page numbers in **boldface**. For scientific names treated as synonyms in the principal entries, both the names and page numbers are in *italics*. Vernacular names and all other occurrences of scientific names are in plain type.

A

B

C

D

E

G

H

I

J

N

P

Q

R

S

T

X

Y

Z

INDEX TO FAMILIES TREATED IN VOLUME 3